Algebra Foundations

Prealgebra, Introductory Algebra, and Intermediate Algebra

Elayn Martin-Gay

University of New Orleans

PEARSON

Boston Columbus Hoboken Indianapolis New York San Francisco
Amsterdam Cape Town Dubai London Madrid Milan Munich Paris Montréal Toronto
Delhi Mexico City São Paulo Sydney Hong Kong Seoul Singapore Taipei Tokyo

Editorial Director, Mathematics: *Christine Hoag*
Editor-in-Chief: *Michael Hirsch*
Acquisitions Editor: *Mary Beckwith*
Project Team Lead: *Christina Lepre*
Project Manager: *Lauren Morse*
Assistant Editor: *Matthew Summers*
Editorial Assistant: *Megan Tripp*
Development Editor: *Dawn Nuttall*
Program Team Lead: *Karen Wernholm*
Program Manager: *Patty Bergin*
Cover and Illustration Design: *Tamara Newnam*
Program Design Lead: *Heather Scott*
Director, Course Production: *Ruth Berry*
Executive Content Manager, MathXL: *Rebecca Williams*
Senior Content Developer, TestGen: *John Flanagan*
Media Producer: *Audra Walsh*
Director of Marketing, Mathematics: *Michelle Renda*
Senior Marketing Manager: *Rachel Ross*
Marketing Assistant: *Kelly Cross*
Senior Author Support/Technology Specialist: *Joe Vetere*
Senior Procurement Specialist: *Carol Melville*
Interior Design, Production Management, Answer Art, and Composition: *Integra Software Services Pvt. Ltd.*
Text Art: *Scientific Illustrators*

Acknowledgments of third party content appear on page P1, which constitutes an extension of this copyright page.

Library of Congress Cataloging-in-Publication Data
Martin-Gay, K. Elayn
Algebra foundations: prealgebra, introductory algebra, and intermediate algebra/ Elayn Martin-Gay, University of New Orleans. – 1st edition.
pages cm
ISBN 0-321-97865-X (alk. paper)
1. Algebra–Textbooks. I. Title.
QA152.3.M3525 2015
512.9–dc23

2014030692

www.pearsonhighered.com

ISBN-10: 0-321-97865-X (Student Edition)
ISBN-13: 978-0-321-97865-3

This book is dedicated to students everywhere—
and we should all be students. After all, is there anyone among
us who really knows too much? Take that hint and continue
to learn something new every day of your life.

Best of wishes from a fellow student:
Elayn Martin-Gay

Contents

13 Factoring Polynomials 925

14 Rational Expressions 993

15 More on Functions and Graphs 1062

16 Inequalities and Absolute Value 1103

17 Rational Exponents, Radicals, and Complex Numbers 1139

18 Quadratic Equations and Functions 1204

19 Exponential and Logarithmic Functions 1265

20 Conic Sections 1331

21 Sequences, Series, and the Binomial Theorem 1365

Appendices

Preface

Algebra Foundations: Prealgebra, Introductory Algebra, and Intermediate Algebra was written to provide a solid foundation in arithmetic and algebra as well as to develop problem-solving skills. It is intended for a course sequence covering prealgebra, introductory algebra, and intermediate algebra. Specific care was taken to make sure students have the most up-to-date, relevant text preparation for their next mathematics course or for nonmathematical courses that require an understanding of algebraic fundamentals. I have tried to achieve this by writing a user-friendly text that is keyed to objectives and contains many worked-out examples. As suggested by AMATYC and the NCTM Standards (plus Addenda), real-life and real-data applications, data interpretation, conceptual understanding, problem solving, writing, cooperative learning, appropriate use of technology, number sense, estimation, critical thinking, and geometric concepts are emphasized and integrated throughout the book.

Highlights of the Program

- **The Martin-Gay Program** provides a design in the text and MyMathLab® to actively encourage students to use the text, video program, and Video Organizer as an integrated learning system.
- **The Video Organizer** is designed to help students take notes and work practice exercises while watching the Interactive Lecture Series videos (available in MyMathLab and on DVD). All content in the Video Organizer is presented in the same order as it is presented in the videos, making it easy for students to create a course notebook and build good study habits.
  - Covers all of the video examples in order.
  - Provides ample space for students to write down key definitions and properties.
  - Includes "Play" and "Pause" button icons to prompt students to follow along with the author for some exercises while they try others on their own.

  The Video Organizer is available in a loose-leaf, notebook-ready format. It is also available for download in MyMathLab.
- **Vocabulary, Readiness & Video Check** questions appear prior to every section exercise set. These exercises quickly check a student's understanding of new vocabulary words. The **readiness** exercises center on a student's understanding of a concept that is necessary in order to continue to the exercise set. **Video check questions for the Martin-Gay Interactive Lecture videos** are included in every section for each learning objective. **These exercises are all available for assignment in MyMathLab** and are a great way to assess whether students have viewed and understand the key concepts presented in the videos.
- **Student Success Tips Videos** are 3- to 5-minute video segments designed to be daily reminders to students to continue practicing and maintaining good organizational and study habits. They are organized in three categories and are available in MyMathLab and the Interactive Lecture Series. The categories are:
  1. Success Tips that apply to any course in college in general, such as Time Management.
  2. Success Tips that apply to any mathematics course. One example is based on understanding that mathematics is a course that requires homework to be completed in a timely fashion.

3. Section- or Content-specific Success Tips to help students avoid common mistakes or to better understand concepts that often prove challenging. One example of this type of tip is how to apply the order of operations to simplify an expression.

- **Interactive DVD Lecture Series,** featuring your text author (Elayn Martin-Gay), provides students with active learning at their own pace. The videos offer the following resources and more:

  A complete lecture for each section of the text highlights key examples and exercises from the text. "Pop-ups" reinforce key terms, definitions, and concepts.

  An interface with menu navigation features allows students to quickly find and focus on the examples and exercises they need to review.

  Interactive Concept Check exercises measure students' understanding of key concepts and common trouble spots.

  Student Success Tips Videos.

- **The Interactive DVD Lecture Series** also includes the following resources for test prep:

  The Chapter Test Prep Videos help students during their most teachable moment—when they are preparing for a test. This innovation provides step-by-step solutions for the exercises found in each Chapter Test. The chapter test prep videos are also available on YouTube™. The videos are captioned in English and Spanish.

  The Practice Final Exam Videos help students prepare for an end-of-course final. Students can watch full video solutions to each exercise in the Practice Final Exam at the end of this text.

- **The Martin-Gay MyMathLab** course provides exercise coverage of odd-numbered exercises in the text, including assignable video check questions and robust video program. There are section lecture videos for every section, which students can also access at the specific objective level; Student Success Tips videos; and an extensive number of watch clips at the exercise level to help students while doing homework in MathXL. Suggested homework assignments have been premade for assignment at the instructor's discretion.

- **MyMathLab Ready to Go Courses** (access code required) provide students with all the same great MyMathLab features that you're used to, but make it easier for instructors to get started. Each course includes preassigned homework and quizzes to make creating your course even simpler. Ask your Pearson representative about the details for this particular course or to see a copy of this course.

Key Pedagogical Features

The following key features are found throughout the text.

Problem-Solving Process This is formally introduced in Chapter 3 with a four-step process that is integrated throughout the text. The four steps are **Understand, Translate, Solve,** and **Interpret.** The repeated use of these steps in a variety of examples shows their wide applicability. Reinforcing the steps can increase students' comfort level and confidence in tackling problems.

Exercise Sets The exercise sets have been carefully designed. Special focus was placed on making sure that even- and odd-numbered exercises are paired and that real-life applications are current and relevant.

Examples Detailed, step-by-step examples are included. Many examples reflect real life. Additional instructional support is provided in the annotated examples.

Practice Exercises Throughout the text, each worked-out example has a parallel Practice exercise. These invite students to be actively involved in the learning process. Students should try each Practice exercise after finishing the corresponding example. Learning by doing will help students grasp ideas before moving on to other concepts.

Helpful Hints Helpful Hints contain practical advice on applying mathematical concepts. Strategically placed where students are most likely to need immediate reinforcement, Helpful Hints help students avoid common trouble areas and mistakes.

Concept Checks This feature allows students to gauge their grasp of an idea as it is being presented in the text. Concept Checks stress conceptual understanding at the point-of-use and help suppress misconceived notions before they start. Answers appear at the bottom of the page. Exercises related to Concept Checks are included in the exercise sets.

Mixed Practice Exercises In the section exercise sets, these exercises require students to determine each problem type and strategy needed to solve it just as they would need to do on a test.

Integrated Reviews This unique, mid-chapter exercise set helps students assimilate new skills and concepts that they have learned separately over several sections. These reviews provide yet another opportunity for students to work with "mixed" exercises as they master the topics.

Vocabulary Check This feature provides an opportunity for students to become more familiar with the use of mathematical terms as they strengthen their verbal skills. These appear at the end of each chapter before the Chapter Highlights. Vocabulary, Readiness & Video exercises provide practice at the section level.

Chapter Highlights Found at the end of every chapter, these contain key definitions and concepts with examples to help students understand and retain what they have learned and help them organize their notes and study for tests.

Chapter Review The end of every chapter contains a comprehensive review of topics introduced in the chapter. The Chapter Review offers exercises keyed to every section in the chapter, as well as Mixed Review exercises that are not keyed to sections.

Chapter Test and Chapter Test Prep Videos The Chapter Test is structured to include those problems that involve common student errors. The **Chapter Test Prep Videos** give students instant access to a step-by-step video solution of each exercise in the Chapter Test.

Cumulative Review This review follows every chapter in the text (except Chapter 1). Each odd-numbered exercise contained in the Cumulative Review is an earlier worked example in the text that is referenced in the back of the book along with the answer.

Writing Exercises These exercises appear in almost every exercise set and require students to provide a written response to explain concepts or justify their thinking.

Applications Real-world and real-data applications appear throughout the text. These exercises occur in almost every exercise set and show the relevance of mathematics and help students gradually and continuously develop their problem-solving skills.

Review Exercises These exercises occur in each exercise set (except in Chapter 1) and are keyed to earlier sections. They review concepts learned earlier in the text that will be needed in the next section or chapter.

Exercise Set Resource Icons Located at the opening of each exercise set, these icons remind students of the resources available for extra practice and support:

See Student Resources descriptions on page xix for details on the individual resources available.

Exercise Icons These icons facilitate the assignment of specialized exercises and let students know what resources can support them.

- DVD Video icon: exercise worked on the Interactive DVD Lecture Series.
- △ Triangle icon: identifies exercises involving geometric concepts.
- Pencil icon: indicates a written response is needed.
- Calculator icon: optional exercises intended to be solved using a scientific or graphing calculator.

Optional: Calculator Exploration Boxes and Calculator Exercises The optional Calculator Explorations provide keystrokes and exercises at appropriate points to give students an opportunity to become familiar with these tools. Section exercises that are best completed by using a calculator are identified by for ease of assignment.

Student and Instructor Resources

STUDENT RESOURCES

<table>
<tr>
<td>Interactive DVD Lecture Series Videos
Provides students with active learning at their pace. The videos offer:
• A complete lecture for each text section. The interface allows easy navigation to examples and exercises students need to review.
• Interactive Concept Check exercises
• Student Success Tips Videos
• Practice Final Exam
• Chapter Test Prep Videos</td>
<td>Video Organizer
Designed to help students take notes and work practice exercises while watching the Interactive Lecture Series videos.
• Covers all of the video examples in order.
• Provides ample space for students to write down key definitions and rules.
• Includes “Play” and “Pause” button icons to prompt students to follow along with the author for some exercises while they try others on their own.
• Includes Student Success Tips Outline and Questions
Available in loose-leaf, notebook-ready format and in MyMathLab.</td>
</tr>
<tr>
<td></td>
<td>Student Solutions Manual
Provides completely worked-out solutions to the odd-numbered section exercises and to all exercises in the Integrated Reviews, Chapter Reviews, Chapter Tests, and Cumulative Reviews.
Available to print on demand.</td>
</tr>
</table>

INSTRUCTOR RESOURCES

<table>
<tr>
<td>Annotated Instructor’s Edition
Contains all the content found in the student edition, plus the following:
• Answers to exercises on the same text page
• Teaching Tips throughout the text placed at key points
• Available to instructors in MyMathLab and to adopters of the program.</td>
<td>Instructor’s Resource Manual with Tests and Mini-Lectures—available in MyMathLab.
• Mini-lectures for each text section
• Additional practice worksheets for each section
• Several forms of test per chapter—free response and multiple choice
• Answers to all items
Instructor’s Solutions Manual—available in MyMathLab.
TestGen® (Available for download from the IRC)</td>
</tr>
<tr>
<td>Instructor-to-Instructor Videos—available in the Instructor Resources section of the MyMathLab course.</td>
<td>Online Resources
MyMathLab® (access code required)

MathXL® (access code required)</td>
</tr>
</table>

Acknowledgments

There are many people who helped me develop this text, and I will attempt to thank some of them here. Emily Keaton, Courtney Slade, and Cindy Trimble were *invaluable* for contributing to the overall accuracy of the text. Dawn Nuttall was *invaluable* for her many suggestions and contributions during the development and writing of this text. Allison Campbell and Lauren Morse provided guidance throughout the production process.

A very special thank you goes to my editor, Mary Beckwith, for being there 24/7/365, as my students say. And, my thanks to the staff at Pearson for all their support: Heather Scott, Patty Bergin, Matt Summers, Michelle Renda, Michael Hirsch, Chris Hoag, and Greg Tobin.

I would like to thank the following reviewers for their input and suggestions:

Lisa Angelo, *Bucks Community College*
Sandi Athanassiou, *University of Missouri–Columbia*
Victoria Baker, *Nicholls State College*
Teri Barnes, *McLennan Community College*
Michelle Beerman, *Pasco-Hernandez Community College*
Monika Bender, *Central Texas College*
Laurel Berry, *Bryant & Stratton College*
Thomas Blackburn, *Northeastern Illinois University*
Gail Burkett, *Palm Beach Community College*
Marie Caruso and students, *Middlesex Community College*
Anita Collins, *Mesa Community College*
Lois Colpo, *Harrisburg Area Community College*
Fay Dang, *Joliet Junior College*
Robert Diaz, *Fullerton College*
Tamie Dickson, *Reading Area Community College*
Latonya Ellis, *Gulf Coast Community College*
Sonia Ford, *Midland College*
Cheryl Gibby, *Cypress College*
Kathryn Gunderson, *Three Rivers Community College*
Elizabeth Hamman, *Cypress College*
Craig Hardesty, *Hillsborough Community College*
Lloyd Harris, *Gulf Coast Community College*
Teresa Hasenauer, *Indian River College*
Julia Hassett, *Oakton Community College*
Bob Hervey, *Hillsborough Community College*
Jeff Koleno, *Lorain County Community College*
Judy Langer, *Westchester Community College*
Sandy Lofstock, *St. Petersburg College*
Michael Maltenfort, *Truman College*
Stan Mattoon, *Merced College*
Dr. Kris Mudunuri, *Long Beach City College*
Carol Murphy, *San Diego Miramar College*
Greg Nguyen, *Fullerton College*
Jean Olsen, *Pikes Peak Community College*
Darlene Ornelas, *Fullerton College*
Warren Powell, *Tyler Junior College*
Jorge Romero, *Hillsborough Community College*
Jeanette Shea, *Central Texas College*
Katerina Vishnyakova, *Collin College*
Corey Wadlington, *West Kentucky Community and Technical College*
Edward Wagner, *Central Texas College*
Joseph Wakim, *Brevard Community College*
Flo Wilson, *Central Texas College*
Jenny Wilson, *Tyler Junior College*

I would also like to thank the following dedicated group of instructors who participated in our focus groups, Martin-Gay Summits, and our design review for the series. Their feedback and insights have helped to strengthen this text. These instructors include:

Billie Anderson, *Tyler Junior College*
Cedric Atkins, *Mott Community College*
Lois Beardon, *Schoolcraft College*
Laurel Berry, *Bryant & Stratton College*
John Beyers, *University of Maryland*
Bob Brown, *Community College of Baltimore County–Essex*
Lisa Brown, *Community College of Baltimore County–Essex*
NeKeith Brown, *Richland College*

Gail Burkett, *State College of Florida*
Cheryl Cantwell, *Seminole Community College*
Ivette Chuca, *El Paso Community College*
Jackie Cohen, *Augusta State College*
Julie Dewan, *Mohawk Valley Community College*
Monette Elizalde, *Palo Alto College*
Kiel Ellis, *Delgado Community College*
Janice Ervin, *Central Piedmont Community College*
Richard Fielding, *Southwestern College*
Dena Frickey, *Delgado Community College*
Cindy Gaddis, *Tyler Junior College*
Gary Garland, *Tarrant County Community College*
Kim Ghiselin, *State College of Florida*
Nita Graham, *St. Louis Community College*
Kim Granger, *St. Louis Community College*
Pauline Hall, *Iowa State College*
Pat Hussey, *Triton College*
Dorothy Johnson, *Lorain County Community College*
Sonya Johnson, *Central Piedmont Community College*
Ann Jones, *Spartanburg Community College*
Irene Jones, *Fullerton College*
Paul Jones, *University of Cincinnati*
Mike Kirby, *Tidewater Community College*
Kathy Kopelousous, *Lewis and Clark Community College*
Tara LaFrance, *Delgado Community College*
John LaMaster, *Indiana Purdue University Fort Wayne*
Nancy Lange, *Inver Hills Community College*
Judy Langer, *Westchester Community College*
Lisa Lindloff, *McLinnan Community College*
Sandy Lofstock, *St. Petersburg College*
Kathy Lovelle, *Westchester Community College*
Nicole Mabine, *North Lake College*
Jean McArthur, *Joliet Junior College*
Kevin McCandless, *Evergreen Valley College*
Ena Michael, *State College of Florida*
Daniel Miller, *Niagara County Community College*
Marica Molle, *Metropolitan Community College*
Carol Murphy, *San Diego Miramar College*
Charlotte Newsom, *Tidewater Community College*
Cao Nguyen, *Central Piedmont Community College*
Greg Nguyen, *Fullerton College*
Eric Oilila, *Jackson Community College*
Linda Padilla, *Joliet Junior College*
Armando Perez, *Laredo Community College*
Davidson Pierre, *State College of Florida*
Marilyn Platt, *Gaston College*
Chris Riola, *Moraine Valley College*
Carole Shapero, *Oakton Community College*
Janet Sibol, *Hillsborough Community College*
Anne Smallen, *Mohawk Valley Community College*
Barbara Stoner, *Reading Area Community College*
Jennifer Strehler, *Oakton Community College*
Ellen Stutes, *Louisiana State University Eunice*
Tanomo Taguchi, *Fullerton College*
Robyn Toman, *Anne Arundel Community College*
MaryAnn Tuerk, *Elsin Community College*
Walter Wang, *Baruch College*
Leigh Ann Wheeler, *Greenville Technical Community College*
Darlene Williams, *Delgado Community College*
Valerie Wright, *Central Piedmont Community College*

A special thank you to those students who participated in our design review: Katherine Browne, Mike Bulfin, Nancy Canipe, Ashley Carpenter, Jeff Chojnachi, Roxanne Davis, Mike Dieter, Amy Dombrowski, Kay Herring, Todd Jaycox, Kaleena Levan, Matt Montgomery, Tony Plese, Abigail Polkinghorn, Harley Price, Eli Robinson, Avery Rosen, Robyn Schott, Cynthia Thomas, and Sherry Ward.

Elayn Martin-Gay

About the Author

Elayn Martin-Gay has taught mathematics at the University of New Orleans for more than 25 years. Her numerous teaching awards include the local University Alumni Association's Award for Excellence in Teaching, and Outstanding Developmental Educator at University of New Orleans, presented by the Louisiana Association of Developmental Educators.

Prior to writing textbooks, Elayn Martin-Gay developed an acclaimed series of lecture videos to support developmental mathematics students in their quest for success. These highly successful videos originally served as the foundation material for her texts. Today, the videos are specific to each book in the Martin-Gay series. The author has also created Chapter Test Prep Videos to help students during their most "teachable moment"–as they prepare for a test–along with Instructor-to-Instructor videos, which provide teaching tips, hints, and suggestions for each developmental mathematics course, including basic mathematics, prealgebra, beginning algebra, and intermediate algebra.

Elayn is the author of 12 published textbooks as well as multimedia and interactive mathematics, all specializing in developmental mathematics courses. She has also published series in Algebra 1, Algebra 2, and Geometry. She has participated as an author across the broadest range of educational materials: textbooks, videos, tutorial software, and courseware. This provides the opportunity of various combinations for an integrated teaching and learning package that offers great consistency for the student.

Applications Index

Automotive

Aviation

Business

Chemistry and physics

Entertainment and recreation

Finance

Food and nutrition

Geography and geology

Time and distance

1

The Whole Numbers

A Selection of Resources for Success in this Mathematics Course

Textbook

Instructor

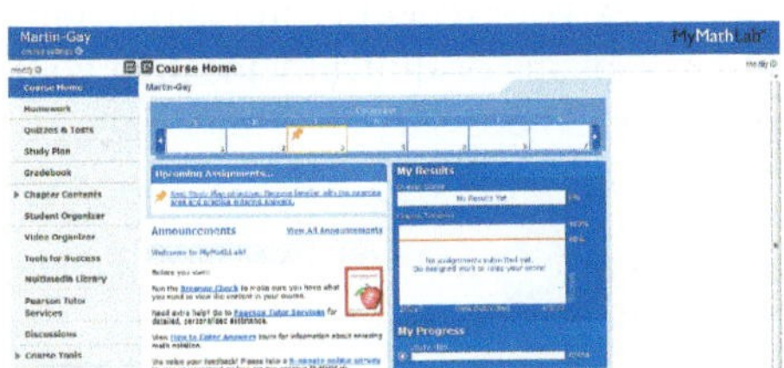

MyMathLab and MathXL

Video Organizer

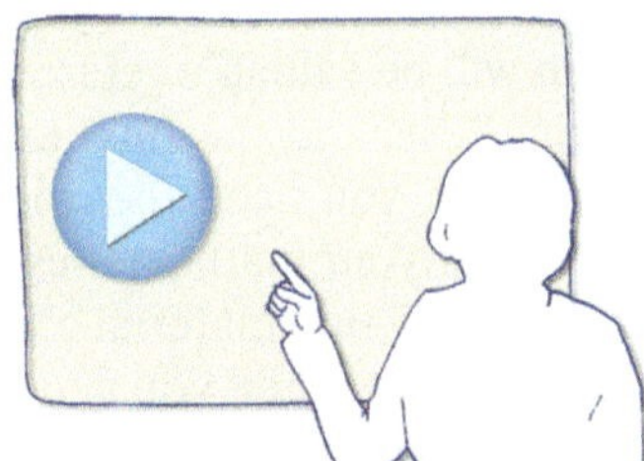

Interactive Lecture Series

For more information about the resources illustrated above, read Section 1.1.

Sections

Check Your Progress

Whole numbers are the basic building blocks of mathematics. The whole numbers answer the question "How many?"

This chapter covers basic operations on whole numbers. Knowledge of these operations provides a good foundation on which to build further mathematical skills.

1.1 Study Skill Tips for Success in Mathematics 

Objectives

A Get Ready for This Course.

B Understand Some General Tips for Success.

C Know How to Use This Text.

D Know How to Use Text Resources.

E Get Help as Soon as You Need It.

F Learn How to Prepare for and Take an Exam.

G Develop Good Time Management.

Before reading Section 1.1, you might want to ask yourself a few questions.

1. When you took your last math course, were you organized? Were your notes and materials from that course easy to find, or were they disorganized and hard to find—if you saved them at all?
2. Were you satisfied—really satisfied—with your performance in that course? In other words, do you feel that your outcome represented your best effort?

If the answer is "no" to these questions, then it is time to make a change. Changing to or resuming good study skill habits is not a process you can start and stop as you please. It is something that you must remember and practice each and every day. To begin, continue reading this section.

Objective A Getting Ready for This Course

Now that you have decided to take this course, remember that a *positive attitude* will make all the difference in the world. Your belief that you can succeed is just as important as your commitment to this course. Make sure you are ready for this course by having the time and positive attitude that it takes to succeed.

Make sure that you are familiar with the way that this course is being taught. Is it a traditional course, in which you have a printed textbook and meet with an instructor? Is it taught totally online, and your textbook is electronic and you e-mail your instructor? Or is your course structured somewhere in between these two methods? (Not all of the tips that follow will apply to all forms of instruction.)

Also make sure that you have scheduled your math course for a time that will give you the best chance for success. For example, if you are also working, you may want to check with your employer to make sure that your work hours will not conflict with your course schedule.

On the day of your first class period, double-check your schedule and allow yourself extra time to arrive on time in case of traffic problems or difficulty locating your classroom. Make sure that you are aware of and bring all necessary class materials.

Objective B General Tips for Success 

Below are some general tips that will increase your chance for success in a mathematics class. Many of these tips will also help you in other courses you may be taking.

Most important! Organize your class materials. In the next couple pages, many ideas will be presented to help you organize your class materials—notes, any handouts, completed homework, previous tests, etc. In general, you MUST have these materials organized. All of them will be valuable references throughout your course and when studying for upcoming tests and the final exam. One way to make sure you can locate these materials when you need them is to use a three-ring binder. This binder should be used solely for your mathematics class and should be brought to each and every class or lab. This way, any material can be immediately inserted in a section of this binder and will be there when you need it.

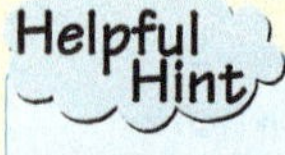

MyMathLab® and MathXL®
When assignments are turned in online, keep a hard copy of your complete written work. You will need to refer to your written work to be able to ask questions and to study for tests later.

Form study groups and/or exchange names and e-mail addresses. Depending on how your course is taught, you may want to keep in contact with your fellow students. Some ways of doing this are to form a study group—whether in person or through the Internet. Also, you may want to ask if anyone is interested in exchanging e-mail addresses or any other form of contact.

Choose to attend all class periods. If possible, sit near the front of the classroom. This way, you will see and hear the presentation better. It may also be easier for you to participate in classroom activities.

Do your homework. You've probably heard the phrase "practice makes perfect" in relation to music and sports. It also applies to mathematics. You will find that the more time you spend solving mathematics exercises, the easier the process becomes. Be sure to schedule enough time to complete your assignments before the due date assigned by your instructor.

MyMathLab® and MathXL®
If you are doing your homework online, you can work and re-work those exercises that you struggle with until you master them. Try working through all the assigned exercises twice before the due date.

Check your work. Review the steps you took while working a problem. Learn to check your answers in the original exercises. You may also compare your answers with the "Answers to Selected Exercises" section in the back of the book. If you have made a mistake, try to figure out what went wrong. Then correct your mistake. If you can't find what went wrong, **don't** erase your work or throw it away. Show your work to your instructor, a tutor in a math lab, or a classmate. It is easier for someone to find where you had trouble if he or she looks at your original work.

MyMathLab® and MathXL®
If you are completing your homework online, it's important to work each exercise on paper before submitting the answer. That way, you can check your work and follow your steps to find and correct any mistakes.

Learn from your mistakes and be patient with yourself. Everyone, even your instructor, makes mistakes. (That definitely includes me—Elayn Martin-Gay.) Use your errors to learn and to become a better math student. The key is finding and understanding your errors.

Was your mistake a careless one, or did you make it because you can't read your own math writing? If so, try to work more slowly or write more neatly and make a conscious effort to carefully check your work.

Did you make a mistake because you don't understand a concept? Take the time to review the concept or ask questions to better understand it.

Did you skip too many steps? Skipping steps or trying to do too many steps mentally may lead to preventable mistakes.

Know how to get help if you need it. It's all right to ask for help. In fact, it's a good idea to ask for help whenever there is something that you don't understand. Make sure you know when your instructor has office hours and how to find his or her office. Find out whether math tutoring services are available on your campus. Check on the hours, location, and requirements of the tutoring service.

Don't be afraid to ask questions. You are not the only person in class with questions. Other students are normally grateful that someone has spoken up.

Turn in assignments on time. This way, you can be sure that you will not lose points for being late. Show every step of a problem and be neat and organized. Also be sure that you understand which problems are assigned for homework. If allowed, you can always double-check the assignment with another student in your class.

MyMathLab® and MathXL®
Be aware of assignments and due dates set by your instructor. Don't wait until the last minute to submit work online.

Objective C Knowing and Using Your Text

Flip through the pages of this text or view the e-text pages on a computer screen. Start noticing examples, exercise sets, end-of-chapter material, and so on. Every text is organized in some manner. Learn the way this text is organized by reading about and then finding an example in your text of each type of resource listed below. Finding and using these resources throughout your course will increase your chance of success.

- *Practice Exercises.* Each example in every section has a parallel Practice exercise. As you read a section, try each Practice exercise after you've finished the corresponding example. Answers are at the bottom of the page. This "learn-by-doing" approach will help you grasp ideas before you move on to other concepts.
- *Symbols at the Beginning of an Exercise Set.* If you need help with a particular section, the symbols listed at the beginning of each exercise set will remind you of the resources available.

- *Objectives.* The main section of exercises in each exercise set is referenced by an objective, such as **A** or **B**, and also an example(s). There is also often a section of exercises entitled "Mixed Practice," which is referenced by two or more objectives or sections. These are mixed exercises written to prepare you for your next exam. Use all of this referencing if you have trouble completing an assignment from the exercise set.
- *Icons (Symbols).* Make sure that you understand the meaning of the icons that are beside many exercises. ▶ tells you that the corresponding exercise may be viewed on the video Lecture Series that corresponds to that section. ✎ tells you that this exercise is a writing exercise in which you should answer in complete sentences. △ tells you that the exercise involves geometry.
- *Integrated Reviews.* Found in the middle of each chapter, these reviews offer you a chance to practice—in one place—the many concepts that you have learned separately over several sections.
- *End-of-Chapter Opportunities.* There are many opportunities at the end of each chapter to help you understand the concepts of the chapter.

  Vocabulary Checks contain key vocabulary terms introduced in the chapter.

  Chapter Highlights contain chapter summaries and examples.

  Chapter Reviews contain review problems. The first part is organized section by section and the second part contains a set of mixed exercises.

  Chapter Tests are sample tests to help you prepare for an exam. The Chapter Test Prep Videos found in the Interactive Lecture Series, MyMathLab, and YouTube provide the video solution to each question on each Chapter Test.

  Cumulative Reviews start at Chapter 2 and are reviews consisting of material from the beginning of the book to the end of that particular chapter.
- *Student Resources in Your Textbook.* You will find a **Student Resources** section at the back of this textbook. It contains the following to help you study and prepare for tests:

  Study Skill Builders contain study skills advice. To increase your chance for success in the course, read these study tips, and answer the questions.

  Bigger Picture—Study Guide Outline provides you with a study guide outline of the course, with examples.

  Practice Final provides you with a Practice Final Exam to help you prepare for a final.
- *Resources to Check Your Work.* The **Answers to Selected Exercises** section provides answers to all odd-numbered section, chapter review, and cumulative review exercises and to all integrated review and chapter test exercises.

MyMathLab®

In MyMathLab, you have access to the following video resources:

- Lecture Videos for each section
- Chapter Test Prep Videos

Use these videos provided by the author to prepare for class, review, and study for tests.

Objective D Knowing and Using Video and Notebook Organizer Resources ▶

Video Resources

Below is a list of video resources that are all made by me—the author of your text, Elayn Martin-Gay. By making these videos, I can be sure that the methods presented are consistent with those in the text.

- *Interactive DVD Lecture Series.* Exercises marked with a ▶ are fully worked out by the author on the DVDs and within MyMathLab. The lecture series provides approximately 20 minutes of instruction per section and is organized by Objective.

- *Chapter Test Prep Videos.* These videos provide solutions to all of the Chapter Test exercises worked out by the author. They can be found in MyMathLab, the Interactive Lecture series, and You Tube. This supplement is very helpful before a test or exam.
- *Student Success Tips.* These video segments are about 3 minutes long and are daily reminders to help you continue practicing and maintaining good organizational and study habits.
- *Final Exam Videos.* These video segments provide solutions to each question. These videos can be found within MyMathLab and the Interactive Lecture Series.

Video Organizer

This resource is in three-ring notebook ready form. It is to be inserted in a three-ring binder and completed. The notebook is numbered according to the sections in your text to which it refers. It is closely tied to the Interactive Lecture (Video) Series. Each section should be completed while watching the lecture video on the same section. Once completed, you will have a set of notes to accompany the Lecture (Video) Series section by section.

Objective E Getting Help 

If you have trouble completing assignments or understanding the mathematics, get help as soon as you need it! This tip is presented as an objective on its own because it is so important. In mathematics, usually the material presented in one section builds on your understanding of the previous section. This means that if you don't understand the concepts covered during a class period, there is a good chance that you will not understand the concepts covered during the next class period. If this happens to you, get help as soon as you can.

Where can you get help? Many suggestions have been made in this section on where to get help, and now it is up to you to get it. Try your instructor, a tutoring center, or a math lab, or you may want to form a study group with fellow classmates. If you do decide to see your instructor or go to a tutoring center, make sure that you have a neat notebook and are ready with your questions.

Helpful Hint

MyMathLab® and MathXL®

- Use the **Help Me Solve This** button to get step-by-step help for the exercise you are working. You will need to work an additional exercise of the same type before you can get credit for having worked it correctly.
- Use the **Video** button to view a video clip of the author working a similar exercise.

Objective F Preparing for and Taking an Exam 

Make sure that you allow yourself plenty of time to prepare for a test. If you think that you are a little "math anxious," it may be that you are not preparing for a test in a way that will ensure success. The way that you prepare for a test in mathematics is important. To prepare for a test:

1. Review your previous homework assignments.
2. Review any notes from class and section-level quizzes you have taken. (If this is a final exam, also review chapter tests you have taken.)
3. Review concepts and definitions by reading the Chapter Highlights at the end of each chapter.
4. Practice working out exercises by completing the Chapter Review found at the end of each chapter. (If this is a final exam, go through a Cumulative Review. There is one found at the end of each chapter except Chapter 1. Choose the review found at the end of the latest chapter that you have covered in your course.) *Don't stop here!*

Helpful Hint

MyMathLab® and MathXL®

Review your written work for previous assignments. Then, go back and re-work previous assignments. Open a previous assignment, and click **Similar Exercise** to generate new exercises. Re-work the exercises until you fully understand them and can work them without help features.

5. It is important that you place yourself in conditions similar to test conditions to find out how you will perform. In other words, as soon as you feel that you know the material, get a few blank sheets of paper and take a sample test. There is a Chapter Test available at the end of each chapter, or you can work selected problems from the Chapter Review. Your instructor may also provide you with a review sheet. During this sample test, do not use your notes or your textbook. Then check your sample test. If your sample test is the Chapter Test in the text, don't forget that the video solutions are in MyMathLab, the Interactive Lecture Series, and YouTube. If you are not satisfied with the results, study the areas that you are weak in and try again.
6. On the day of the test, allow yourself plenty of time to arrive at where you will be taking your exam.

When taking your test:

1. Read the directions on the test carefully.
2. Read each problem carefully as you take the test. Make sure that you answer the question asked.
3. Watch your time and pace yourself so that you can attempt each problem on your test.
4. If you have time, check your work and answers.
5. Do not turn your test in early. If you have extra time, spend it double-checking your work.

Objective G Managing Your Time 

As a college student, you know the demands that classes, homework, work, and family place on your time. Some days you probably wonder how you'll ever get everything done. One key to managing your time is developing a schedule. Here are some hints for making a schedule:

1. Make a list of all of your weekly commitments for the term. Include classes, work, regular meetings, extracurricular activities, etc. You may also find it helpful to list such things as laundry, regular workouts, grocery shopping, etc.
2. Next, estimate the time needed for each item on the list. Also make a note of how often you will need to do each item. Don't forget to include time estimates for the reading, studying, and homework you do outside of your classes. You may want to ask your instructor for help estimating the time needed.
3. In the exercise set that follows, you are asked to block out a typical week on the schedule grid given. Start with items with fixed time slots like classes and work.
4. Next, include the items on your list with flexible time slots. Think carefully about how best to schedule items such as study time.
5. Don't fill up every time slot on the schedule. Remember that you need to allow time for eating, sleeping, and relaxing! You should also allow a little extra time in case some items take longer than planned.
6. If you find that your weekly schedule is too full for you to handle, you may need to make some changes in your workload, classload, or other areas of your life. You may want to talk to your advisor, manager or supervisor at work, or someone in your college's academic counseling center for help with such decisions.

1.1 Exercise Set MyMathLab®

1. What is your instructor's name?

2. What are your instructor's office location and office hours?

3. What is the best way to contact your instructor?

4. Do you have the name and contact information of at least one other student in class?

5. Will your instructor allow you to use a calculator in this class?

6. Why is it important that you write step-by-step solutions to homework exercises and keep a hard copy of all work submitted?

7. Is there a tutoring service available on campus? If so, what are its hours? What services are available?

8. Have you attempted this course before? If so, write down ways that you might improve your chances of success during this next attempt.

9. List some steps that you can take if you begin having trouble understanding the material or completing an assignment. If you are completing your homework in MyMathLab® and MathXL®, list the resources you can use for help.

10. How many hours of studying does your instructor advise for each hour of instruction?

11. What does the icon in this text mean?

12. What does the △ icon in this text mean?

13. What does the icon in this text mean?

14. Search the minor columns in your text. What are Practice exercises?

15. When might be the best time to work a Practice exercise?

16. Where are the answers to Practice exercises?

17. What answers are contained in this text and where are they?

18. What are Study Skill Tips of the Day and where are they?

19. What and where are Integrated Reviews?

20. How many times is it suggested that you work through the homework exercises in MathXL® before the submission deadline?

21. How far in advance of the assigned due date is it suggested that homework be submitted online? Why?

22. Chapter Highlights are found at the end of each chapter. Find the Chapter 1 Highlights and explain how you might use it and how it might be helpful.

23. Chapter Reviews are found at the end of each chapter. Find the Chapter 1 Review and explain how you might use it and how it might be helpful.

24. Chapter Tests are found at the end of each chapter. Find the Chapter 1 Test and explain how you might use it and how it might be helpful when preparing for an exam on Chapter 1. Include how the Chapter Test Prep Videos may help. If you are working in MyMathLab® and MathXL®, how can you use previous homework assignments to study?

25. What is the Video Organizer? Explain the contents and how it might be used.

26. Read or reread objective **G** and fill out the schedule grid on the next page.

	Monday	Tuesday	Wednesday	Thursday	Friday	Saturday	Sunday
4:00 a.m.							
5:00 a.m.							
6:00 a.m.							
7:00 a.m.							
8:00 a.m.							
9:00 a.m.							
10:00 a.m.							
11:00 a.m.							
12:00 p.m.							
1:00 p.m.							
2:00 p.m.							
3:00 p.m.							
4:00 p.m.							
5:00 p.m.							
6:00 p.m.							
7:00 p.m.							
8:00 p.m.							
9:00 p.m.							
10:00 p.m.							
11:00 p.m.							
Midnight							
1:00 a.m.							
2:00 a.m.							
3:00 a.m.							

1.2 Place Value, Names for Numbers, and Reading Tables

Objectives

A Find the Place Value of a Digit in a Whole Number.

B Write a Whole Number in Words and in Standard Form.

C Write a Whole Number in Expanded Form.

D Read Tables.

The **digits** 0, 1, 2, 3, 4, 5, 6, 7, 8, and 9 can be used to write numbers. For example, the **whole numbers** are

0, 1, 2, 3, 4, 5, 6, 7, 8, 9, 10, 11, . . .

and the **natural numbers** are 1, 2, 3, 4, 5, 6, 7, 8, 9, 10, 11, . . .

The three dots (. . .) after the 11 mean that this list continues indefinitely. That is, there is no largest whole number. The smallest whole number is 0.

Objective A Finding the Place Value of a Digit in a Whole Number

The position of each digit in a number determines its **place value.** For example, the average distance (in miles) between the planet Mercury and the planet Earth can be represented by the whole number 48,337,000. On the next page, is a place-value chart for this whole number.

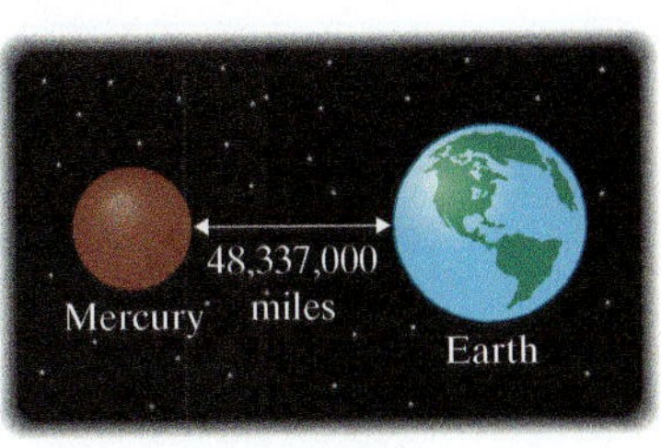

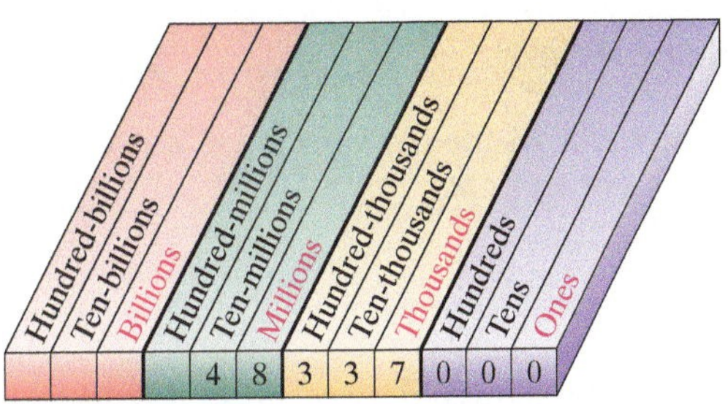

The two 3s in 48,337,000 represent different amounts because of their different placements. The place value of the 3 on the left is hundred-thousands. The place value of the 3 on the right is ten-thousands.

Examples Find the place value of the digit 3 in each whole number.

1. 396,418
↑
hundred-thousands

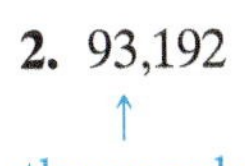

2. 93,192
↑
thousands

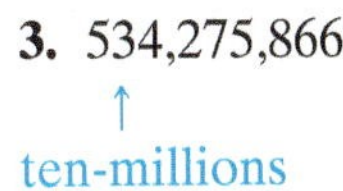

3. 534,275,866
↑
ten-millions

Work Practice 1–3

Practice 1–3

Find the place value of the digit 8 in each whole number.

1. 38,760,005

2. 67,890

3. 481,922

Objective B Writing a Whole Number in Words and in Standard Form

A whole number such as 1,083,664,500 is written in **standard form.** Notice that commas separate the digits into groups of three, starting from the right. Each group of three digits is called a **period.** The names of the first four periods are shown in red.

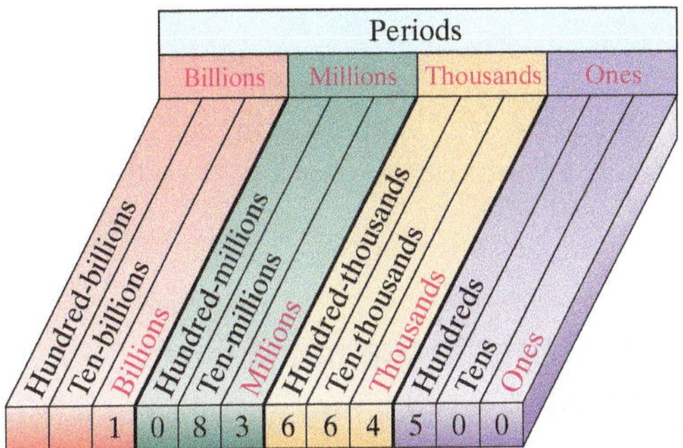

Writing a Whole Number in Words

To write a whole number in words, write the number in each period followed by the name of the period. (The ones period name is usually not written.) This same procedure can be used to read a whole number.

Fox example, we write 1,083,664,500 as

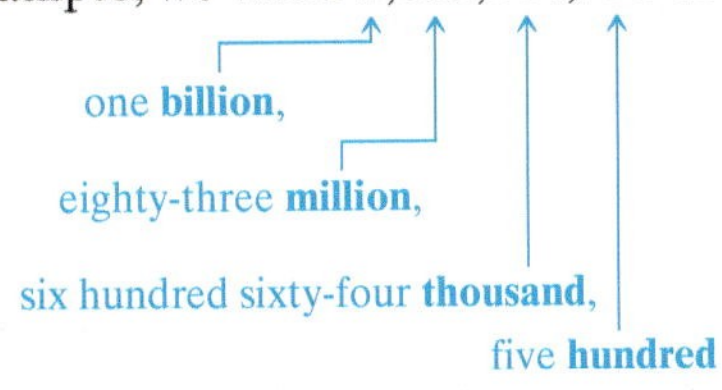

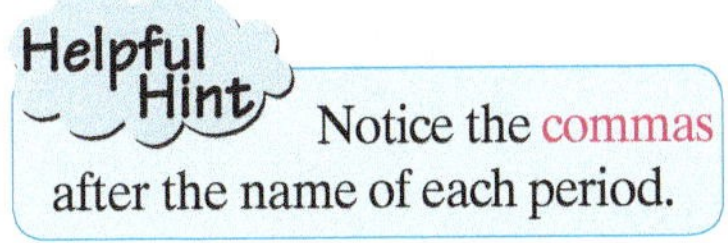

Answers

1. millions **2.** hundreds **3.** ten-thousands

Helpful Hint

The name of the ones period is not used when reading and writing whole numbers. For example,

9,265

is read as

"nine **thousand**, two hundred sixty-five."

Practice 4-6

Write each whole number in words.

4. 54

5. 678

6. 93,205

Examples Write each whole number in words.

4. 72 seventy-two

5. 546 five hundred forty-six

6. 27,034 twenty-seven thousand, thirty-four

Work Practice 4–6

Helpful Hint The word "and" is *not* used when reading and writing whole numbers. It is used only when reading and writing mixed numbers and some decimal values, as shown later in this text.

Practice 7

Write 679,430,105 in words.

Example 7 Write 308,063,557 in words.

Solution: 308,063,557 is written as

three hundred eight **million**, sixty-three **thousand**, five hundred fifty-seven

Work Practice 7

✓Concept Check True or false? When writing a check for $2600, the word name we write for the dollar amount of the check is "two thousand sixty." Explain your answer.

Practice 8-11

Write each whole number in standard form.

8. thirty-seven

9. two hundred twelve

10. eight thousand, two hundred seventy-four

11. five million, fifty-seven thousand, twenty-six

Answers

4. fifty-four

5. six hundred seventy-eight

6. ninety-three thousand, two hundred five

7. six hundred seventy-nine million, four hundred thirty thousand, one hundred five

8. 37 **9.** 212

10. 8,274 or 8274 **11.** 5,057,026

✓Concept Check Answer

false

Writing a Whole Number in Standard Form

To write a whole number in standard form, write the number in each period, followed by a comma.

Examples Write each whole number in standard form.

8. forty-one 41 **9.** seven hundred eight 708

10. six thousand, four hundred ninety-three

6,493 or 6493

11. three million, seven hundred forty-six thousand, five hundred twenty-two

3,746,522

Work Practice 8–11

Helpful Hint

A comma may or may not be inserted in a four-digit number. For example, both

6,493 and 6493

are acceptable ways of writing six thousand, four hundred ninety-three.

Objective C Writing a Whole Number in Expanded Form

The place value of a digit can be used to write a number in expanded form. The **expanded form** of a number shows each digit of the number with its place value. For example, 5672 is written in expanded form as

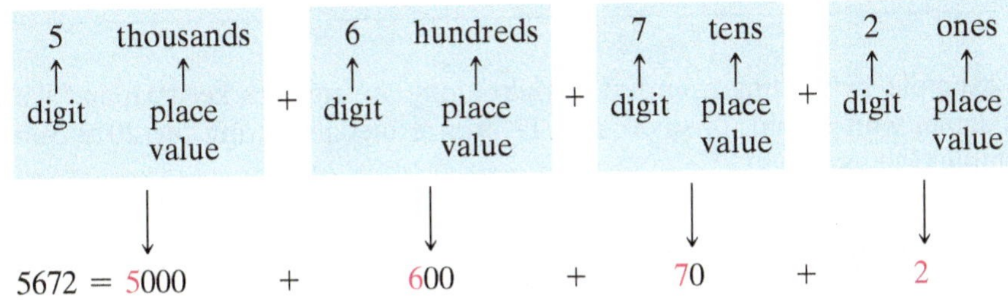

$5672 = 5000 + 600 + 70 + 2$

Practice 12

Write 4,026,301 in expanded form.

Example 12 Write 5,207,034 in expanded form.

Solution: $5{,}000{,}000 + 200{,}000 + 7000 + 30 + 4$

Work Practice 12

We can visualize whole numbers by points on a line. The line below is called a **number line.** This number line has equally spaced marks for each whole number. The arrow to the right simply means that the whole numbers continue indefinitely. In other words, there is no largest whole number.

Number Line

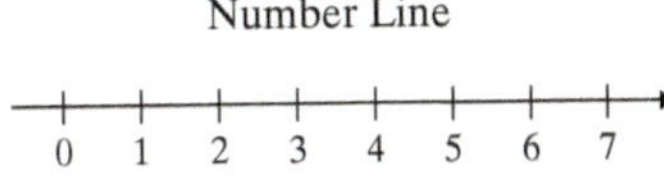

We will study number lines further in Section 1.4.

Objective D Reading Tables

Now that we know about place value and names for whole numbers, we introduce one way that whole number data may be presented. **Tables** are often used to organize and display facts that contain numbers. The following table shows the countries that won the most medals during the Summer Olympic Games in London in 2012. (Although the medals are truly won by athletes from the various countries, for simplicity we will state that countries have won the medals.)

Answer

12. $4{,}000{,}000 + 20{,}000 + 6000 + 300 + 1$

2012 London Summer Olympics Medal Count

Country	Gold	Silver	Bronze	Total
United States	46	29	29	104
China	38	27	23	88
Russian Federation	24	26	32	82
Great Britain	29	17	19	65
Germany	11	19	14	44
Japan	7	14	17	38
Australia	7	16	12	35
France	11	11	12	34
South Korea	13	8	7	28
Italy	8	9	11	28
Netherlands	6	6	8	20
Ukraine	6	5	9	20

(*Source:* International Olympic Committee)

For example, by reading from left to right along the row marked "Japan," we find that Japan won 7 gold, 14 silver, and 17 bronze medals during the 2012 Summer Olympic Games.

Example 13 Use the 2012 Summer Games table to answer each question.

a. How many silver medals did the Russian Federation win during the 2012 Summer Olympic Games?

b. Which countries shown won fewer gold medals than Italy?

Solution:

a. Find "Russian Federation" in the left-hand column. Then read from left to right until the "silver" column is reached. We find that the Russian Federation won 26 silver medals.

b. Italy won 8 gold medals. Of the countries shown, Australia, the Netherlands, Ukraine, and Japan each won fewer than 8 gold medals.

Work Practice 13

Practice 13

Use the 2012 Summer Games table to answer each question.

a. How many bronze medals did Great Britain win during the 2012 Summer Olympic Games?

b. Which countries won more than 80 medals?

Answers

13. a. 19 **b.** United States, China, Russia

Vocabulary, Readiness & Video Check

Use the choices below to fill in each blank.

standard form	period	whole
expanded form	place value	words

1. The numbers 0, 1, 2, 3, 4, 5, 6, 7, 8, 9, 10, 11, 12, … are called ____________ numbers.
2. The number 1286 is written in ____________.
3. The number "twenty-one" is written in ____________.
4. The number 900 + 60 + 5 is written in ____________.
5. In a whole number, each group of three digits is called a(n) ____________.
6. The ____________ of the digit 4 in the whole number 264 is ones.

Martin-Gay Interactive Videos *Watch the section lecture video and answer the following questions.*

See Video 1.2

Objective A 7. In Example 1, what is the place value of the digit 6?

Objective B 8. Complete this statement based on Example 3. To read (or write) a number, read from ________ to ________.

Objective C 9. In Example 5, what is the expanded form value of the digit 8?

Objective D 10. Use the table given in Example 6 to determine which breed shown has the fewest American Kennel Club registrations.

1.2 Exercise Set MyMathLab®

Objective A *Determine the place value of the digit 5 in each whole number. See Examples 1 through 3.*

1. 657 **2.** 905 **3.** 5423 **4.** 6527

5. 43,526,000 **6.** 79,050,000 **7.** 5,408,092 **8.** 51,682,700

Objective B *Write each whole number in words. See Examples 4 through 7.*

9. 354 **10.** 316 **11.** 8279 **12.** 5445

13. 26,990 **14.** 42,009 **15.** 2,388,000 **16.** 3,204,000

17. 24,350,185 **18.** 47,033,107

Write each number in the sentence in words. (Do not write the years in words.) See Examples 4 through 7.

19. In 2013, the population of Iceland was 321,800. (*Source: Statistics: Iceland*)

20. Between 1990 and 2010, Brazil lost 553,170 square kilometers of forest area.

21. The Burj Khalifa, in Dubai, United Arab Emirates, a hotel and office building, is currently the tallest in the world at a height of 2720 feet. (*Source:* Council on Tall Buildings and Urban Habitat)

22. In 2012, there were 124,681 patients in the United States waiting for an organ transplant. (*Source:* United Network for Organ Sharing)

23. Each day, UPS delivers an average of 16,300,000 packages and documents worldwide. (*Source:* UPS)

24. Each day, FedEx delivers an average of 3,300,000 packages and documents worldwide. (*Source:* FedEx)

25. The highest point in Colorado is Mount Elbert, at an elevation of 14,433 feet. (*Source:* U.S. Geological Survey)

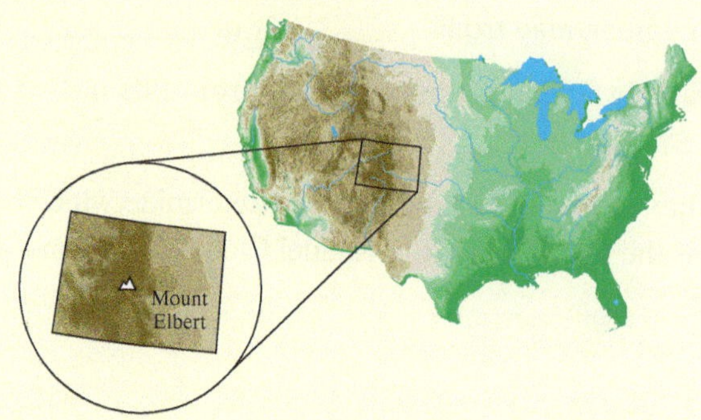

26. The highest point in Oregon is Mount Hood, at an elevation of 11,239 feet. (*Source:* U.S. Geological Survey)

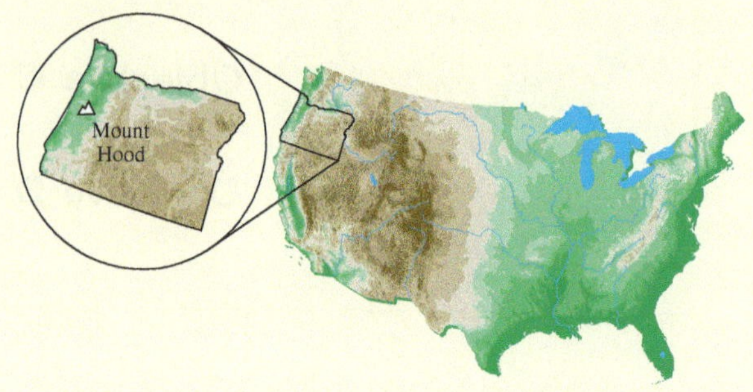

27. The average low price for a 2013 Honda Civic was \$18,027. (*Source: U.S. News*)

28. The Goodyear blimp *Eagle* holds 202,700 cubic feet of helium. (*Source:* The Goodyear Tire & Rubber Company)

Write each whole number in standard form. See Examples 8 through 11.

29. Six thousand, five hundred eighty-seven

30. Four thousand, four hundred sixty-eight

31. Fifty-nine thousand, eight hundred

32. Seventy-three thousand, two

33. Thirteen million, six hundred one thousand, eleven

34. Sixteen million, four hundred five thousand, sixteen

35. Seven million, seventeen

36. Two million, twelve

37. Two hundred sixty thousand, nine hundred ninety-seven

38. Six hundred forty thousand, eight hundred eighty-one

Write the whole number in each sentence in standard form. See Examples 8 through 11.

39. The Mir Space Station orbits above Earth at an average altitude of three hundred ninety-five kilometers. (*Source:* Heavens Above)

40. The average distance between the surfaces of the Earth and the Moon is about two hundred thirty-four thousand miles.

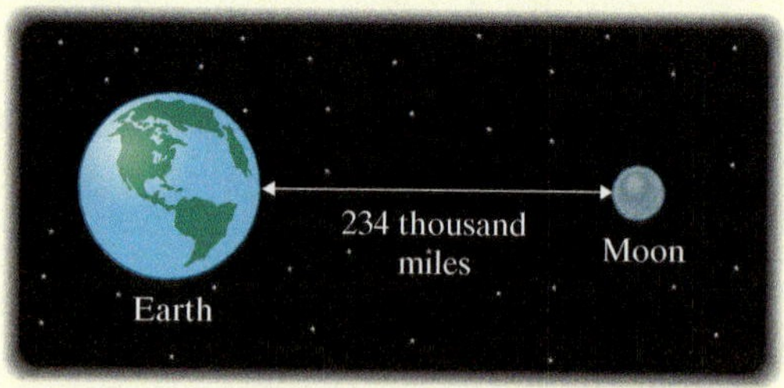

41. The world's tallest free-standing tower is the Tokyo Sky Tree in Tokyo, Japan. Its height is two thousand eighty feet. (*Source:* Council on Tall Buildings and Urban Habitat)

42. The world's second-tallest free-standing tower is the Canton Tower in Guanzhou, China. Its height is one thousand nine hundred sixty-nine feet. (*Source:* Council on Tall Buildings and Urban Habitat)

43. The film *The Avengers* grabbed hold of the world record for opening weekend income when it took in two hundred million, three hundred thousand dollars in 2012. (*Source: Guinness World Records*)

44. The film *Harry Potter and the Deathly Hallows—Part II* holds the record for second-highest opening weekend income; it took in one hundred sixty-nine million, two hundred thousand dollars in 2011. (*Source: Guinness World Records*)

45. Morten Anderson, who played in the National Football League in 1982–2007, holds the record for most career field goals at five hundred sixty-five. (*Source:* NFL)

46. Morten Anderson also holds the record for the most field goals attempted in a career at seven hundred nine. (*Source:* NFL)

Objective C *Write each whole number in expanded form. See Example 12.*

47. 209 **48.** 789 **49.** 3470 **50.** 6040

51. 80,774 **52.** 20,215 **53.** 66,049 **54.** 99,032

55. 39,680,000 **56.** 47,703,029

Objective D *The table shows the beginning year of recent eruptions of major volcanoes in the Cascade Mountains. Use this table to answer Exercises 57 through 62. See Example 13.*

Recent Eruptions of Major Cascade Mountain Volcanoes (1750–present) *

		Year of Eruption (beginning year)					
Volcano	**State Location**	**1750–1799**	**1800–1849**	**1850–1899**	**1900–1949**	**1950–1999**	**2000–present**
Mt. Baker	Washington	1792	1843	1870, 1880			
Glacier Peak	Washington	1750(?)					
Mt. Rainier	Washington		1841, 1843	1854			
Mt. St. Helens	Washington		1800			1980	
Mt. Hood	Oregon			1854, 1859, 1865			
Three Sisters	Oregon			1853(?)			
Medicine Lake	California				1910		
Mt. Shasta	California	1786		1855			
Cinder Cone*	California			1850			
Lassen Peak	California				1914		
Chaos Crags*	California			1854			

Other major volcanoes in the Cascades have had no eruptions from 1750 to present; *Source:* Harris
*Cinder Cone and Chaos Crags are located by Lassen Peak.

57. Mount Shasta erupted in the 1700s. Locate, then write this eruption year in standard form.

58. Mount Baker erupted in the 1700s. Locate, then write this eruption year in standard form.

59. Which volcano in the table has had the most eruptions?

60. Which volcano(es) in the table has had two eruptions?

61. Which volcano in the table had the earliest eruption?

62. Which volcano in the table had the most recent eruption?

The table shows the top ten popular breeds of dogs in 2012 according to the American Kennel Club. The breeds are listed in the order of number of registrations. Use this table to answer Exercises 63 through 68. See Example 13.

Top Ten American Kennel Club Registrations in 2012

Breed	Average Dog Maximum Height (in inches)	Average Dog Maximum Weight (in pounds)
Labrador retriever	25	75
German shepherd	26	95
Golden retriever	24	80
Beagle	15	30
Bulldog	26	90
Yorkshire terrier	9	7
Boxer	25	70
Poodle (standard, miniature, and toy)	standard: 26	standard: 70
Rottweiler	26	none given
Dachshund	9	25

(*Source:* American Kennel Club)

63. Which breed has a greater average weight, the Bulldog or the German shepherd?

64. Which breed has more dogs registered, Golden retriever or German shepherd?

65. Which breed is the most popular dog? Write the maximum weight for this breed in words.

66. Which of the listed breeds has the fewest registrations? Write the average weight for this breed in words.

67. What is the maximum weight of an average-size Boxer?

68. What is the maximum height of an average-size standard poodle?

Concept Extensions

69. Write the largest four-digit number that can be made from the digits 1, 9, 8, and 6 if each digit must be used once. ____ ____ ____ ____

70. Write the largest five-digit number that can be made using the digits 5, 3, and 7 if each digit must be used at least once. ____ ____ ____ ____ ____

Check to see whether each number written in standard form matches the number written in words. If not, correct the number in words. See the Concept Check in this section.

71.

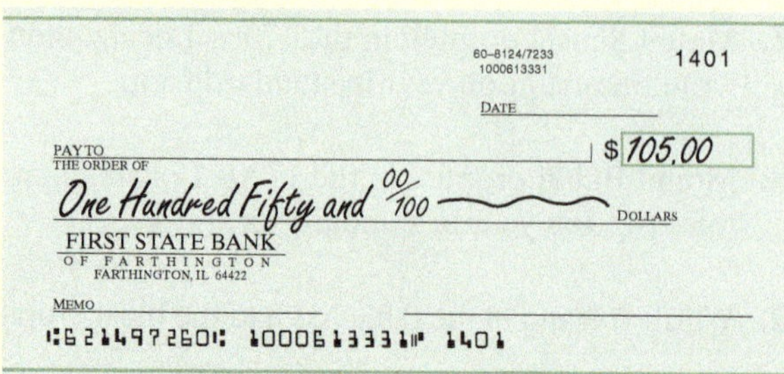
60-8124/7233
1000613331
1401
DATE
PAY TO THE ORDER OF $ 105.00
One Hundred Fifty and 00/100 DOLLARS
FIRST STATE BANK OF FARTHINGTON
FARTHINGTON, IL 64422
MEMO
⑆621497260⑆ 1000613331⑈ 1401

72.

60-8124/7233
1000613331
1402
DATE
PAY TO THE ORDER OF $ 7030.00
Seven Thousand Thirty and 00/100 DOLLARS
FIRST STATE BANK OF FARTHINGTON
FARTHINGTON, IL 64422
MEMO
⑆621497260⑆ 1000613331⑈ 1402

73. If a number is given in words, describe the process used to write this number in standard form.

74. If a number is written in standard form, describe the process used to write this number in expanded form.

75. How large is a trillion? To get an idea, one trillion seconds is over 31 thousand years. Look up "trillion" in a dictionary and use the definition to write this number in standard form.

76. How large is a quadrillion? To get an idea, if one quadrillion dollars were divided among the population of the United States, we would each receive over 9 million dollars. Look up "quadrillion" (in the American system) and write this number in standard form.

1.3 Adding and Subtracting Whole Numbers, and Perimeter

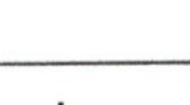

Objective A Adding Whole Numbers

Objectives

A Add Whole Numbers. 

B Subtract Whole Numbers. 

C Find the Perimeter of a Polygon.

D Solve Problems by Adding or Subtracting Whole Numbers.

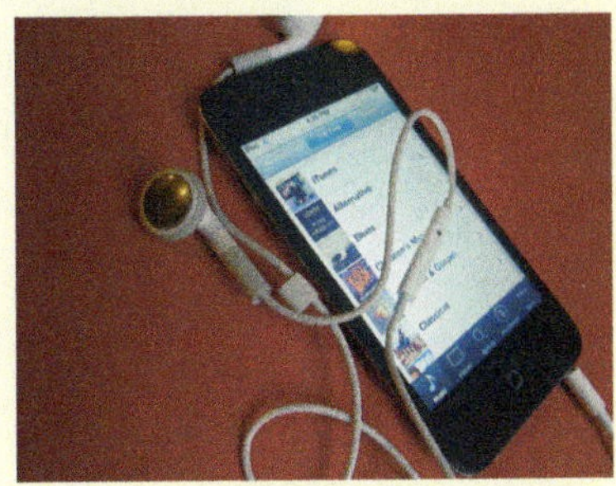

An iPod is a hard drive–based portable audio player. In 2013, it was still the most popular digital music player in the United States.

Suppose that an electronics store received a shipment of two boxes of iPods one day and an additional four boxes of iPods the next day. The **total** shipment in the two days can be found by adding 2 and 4.

2 boxes of iPods + 4 boxes of iPods = 6 boxes of iPods

The **sum** (or total) is 6 boxes of iPods. Each of the numbers 2 and 4 is called an **addend,** and the process of finding the sum is called **addition.**

$$\underset{\text{addend}}{2} + \underset{\text{addend}}{4} = \underset{\text{sum}}{6}$$

To add whole numbers, we add the digits in the ones place, then the tens place, then the hundreds place, and so on. For example, let's add 2236 + 160.

$$\begin{array}{r} 2236 \\ +\ 160 \\ \hline 2396 \end{array}$$

Line up numbers vertically so that the place values correspond. Then add digits in corresponding place values, starting with the ones place.

2396: 6 — sum of ones; 9 — sum of tens; 3 — sum of hundreds; 2 — sum of thousands

Example 1 Add: 46 + 713

Solution:

$$\begin{array}{r} 46 \\ +713 \\ \hline 759 \end{array}$$

Work Practice 1

Practice 1

Add: 4135 + 252

Adding by Carrying

When the sum of digits in corresponding place values is more than 9, **carrying** is necessary. For example, to add 365 + 89, add the ones-place digits first.

Carrying

$$\begin{array}{r} \scriptstyle 1 \\ 365 \\ +\ \ 89 \\ \hline 4 \end{array}$$

5 ones + 9 ones = **14 ones** or **1 ten** + **4 ones**

Write the 4 ones in the ones place and carry the 1 ten to the tens place.

Next, add the tens-place digits.

$$\begin{array}{r} \scriptstyle 11 \\ 365 \\ +\ \ 89 \\ \hline 54 \end{array}$$

1 ten + 6 tens + 8 tens = **15 tens** or **1 hundred** + **5 tens**

Write the 5 tens in the tens place and carry the 1 hundred to the hundreds place.

Answer

1. 4387

Next, add the hundreds-place digits.

$$\begin{array}{r} \scriptstyle 1\,1 \\ 365 \\ +\ \ 89 \\ \hline 454 \end{array}$$

1 hundred + 3 hundreds = 4 hundreds
Write the 4 hundreds in the hundreds place.

Practice 2

Add: 47,364 + 135,898

Example 2 Add: 46,278 + 124,931

Solution:

$$\begin{array}{r} \scriptstyle 1\,1\,1 \\ 46{,}278 \\ +\ 124{,}931 \\ \hline 171{,}209 \end{array}$$

Work Practice 2

✓**Concept Check** What is wrong with the following computation?

$$\begin{array}{r} 394 \\ +\ 283 \\ \hline 577 \end{array}$$

Before we continue adding whole numbers, let's review some properties of addition that you may have already discovered. The first property that we will review is the **addition property of 0.** This property reminds us that the sum of 0 and any number is that same number.

Addition Property of 0

The sum of 0 and any number is that number. For example,

$7 + 0 = 7$

$0 + 7 = 7$

Next, notice that we can add any two whole numbers in any order and the sum is the same. For example,

$4 + 5 = 9$ and $5 + 4 = 9$

We call this special property of addition the **commutative property of addition.**

Commutative Property of Addition

Changing the **order** of two addends does not change their sum. For example,

$2 + 3 = 5$ and $3 + 2 = 5$

Another property that can help us when adding numbers is the **associative property of addition.** This property states that when adding numbers, the grouping of the numbers can be changed without changing the sum. We use parentheses to group numbers. They indicate which numbers to add first. For example, let's use two different groupings to find the sum of 2 + 1 + 5.

$$(2 + 1) + 5 = 3 + 5 = 8$$

Also,

$$2 + (1 + 5) = 2 + 6 = 8$$

Both groupings give a sum of 8.

Answer

2. 183,262

✓**Concept Check Answer**

forgot to carry 1 hundred to the hundreds place

Associative Property of Addition

Changing the **grouping** of addends does not change their sum. For example,

$$3 + \underbrace{(5 + 7)} = 3 + 12 = 15 \quad \text{and} \quad \underbrace{(3 + 5)} + 7 = 8 + 7 = 15$$

The commutative and associative properties tell us that we can add whole numbers using any order and grouping that we want.

When adding several numbers, it is often helpful to look for two or three numbers whose sum is 10, 20, and so on. Why? Adding multiples of 10 such as 10 and 20 is easier.

Example 3 Add: $13 + 2 + 7 + 8 + 9$

Solution: $13 + 2 + 7 + 8 + 9 = 39$

$20 + 10 + 9$

39

Work Practice 3

Practice 3

Add: $12 + 4 + 8 + 6 + 5$

Feel free to use the process of Example 3 anytime when adding.

Example 4 Add: $1647 + 246 + 32 + 85$

Solution:

```
 122
 1647
  246
   32
+  85
 2010
```

Work Practice 4

Practice 4

Add: $6432 + 789 + 54 + 28$

Objective B Subtracting Whole Numbers 

If you have \$5 and someone gives you \$3, you have a total of \$8, since $5 + 3 = 8$. Similarly, if you have \$8 and then someone borrows \$3, you have \$5 left. **Subtraction** is finding the **difference** of two numbers.

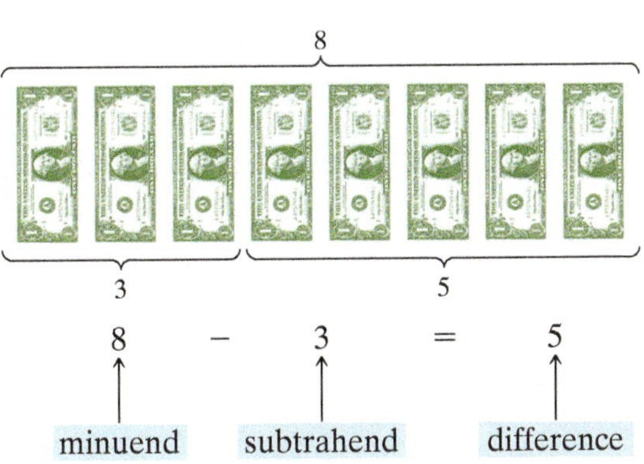

$$8 - 3 = 5$$

minuend subtrahend difference

Answers

3. 35 **4.** 7303

In this example, 8 is the **minuend,** and 3 is the **subtrahend.** The **difference** between these two numbers, 8 and 3, is 5.

Notice that addition and subtraction are very closely related. In fact, subtraction is defined in terms of addition.

$8 - 3 = 5$ because $5 + 3 = 8$

This means that subtraction can be *checked* by addition, and we say that addition and subtraction are reverse operations.

Practice 5

Subtract. Check each answer by adding.

a. $14 - 6$

b. $20 - 8$

c. $93 - 93$

d. $42 - 0$

Example 5

Subtract. Check each answer by adding.

a. $12 - 9$ **b.** $22 - 7$ **c.** $35 - 35$ **d.** $70 - 0$

Solution:

a. $12 - 9 = 3$ because $3 + 9 = 12$

b. $22 - 7 = 15$ because $15 + 7 = 22$

c. $35 - 35 = 0$ because $0 + 35 = 35$

d. $70 - 0 = 70$ because $70 + 0 = 70$

Work Practice 5

Look again at Examples 5(c) and 5(d).

5(c) $35 - 35 = 0$ (35 and 35: same number; 0: difference is 0)

5(d) $70 - 0 = 70$ ($70 - 0$: a number minus 0; 70: difference is the same number)

These two examples illustrate the subtraction properties of 0.

Subtraction Properties of 0

The difference of any number and that same number is 0. For example,

$11 - 11 = 0$

The difference of any number and 0 is that same number. For example,

$45 - 0 = 45$

To subtract whole numbers we subtract the digits in the ones place, then the tens place, then the hundreds place, and so on. When subtraction involves numbers of two or more digits, it is more convenient to subtract vertically. For example, to subtract $893 - 52$,

$$\begin{array}{r} 893 \\ -\ 52 \\ \hline 841 \end{array}$$

893 ⟵ minuend
52 ⟵ subtrahend
841 ⟵ difference

Line up the numbers vertically so that the minuend is on top and the place values correspond. Subtract in corresponding place values, starting with the ones place.

$3 - 2$
$9 - 5$
$8 - 0$

To check, add.

$$\begin{array}{r} \text{difference} \\ +\ \text{subtrahend} \\ \hline \text{minuend} \end{array} \quad \text{or} \quad \begin{array}{r} 841 \\ +\ 52 \\ \hline 893 \end{array}$$

⟵ Since this is the original minuend, the problem checks.

Answers

5. a. 8 **b.** 12 **c.** 0 **d.** 42

Example 6 Subtract: 7826 − 505. Check by adding.

Solution:

$$\begin{array}{r} 7826 \\ -\ 505 \\ \hline 7321 \end{array} \qquad \text{Check: } \begin{array}{r} 7321 \\ +\ 505 \\ \hline 7826 \end{array}$$

Work Practice 6

Practice 6

Subtract. Check by adding.

a. 9143 − 122

b. 978 − 851

Subtracting by Borrowing

When subtracting vertically, if a digit in the second number (subtrahend) is larger than the corresponding digit in the first number (minuend), **borrowing** is necessary. For example, consider

$$\begin{array}{r} 81 \\ -63 \\ \hline \end{array}$$

Since the 3 in the ones place of 63 is larger than the 1 in the ones place of 81, borrowing is necessary. We borrow 1 ten from the tens place and add it to the ones place.

Borrowing

8 tens − 1 ten = 7 tens → $\begin{array}{r} \overset{7}{\cancel{8}}\overset{11}{\cancel{1}} \\ -6\ 3 \\ \hline \end{array}$ ← 1 ten + 1 one = 11 ones

Now we subtract the ones-place digits and then the tens-place digits.

$$\begin{array}{r} \overset{7}{\cancel{8}}\overset{11}{\cancel{1}} \\ -6\ 3 \\ \hline 1\ 8 \end{array} \qquad \text{Check: } \begin{array}{r} 18 \\ +63 \\ \hline 81 \end{array} \text{ The original minuend.}$$

8 ← 11 − 3 = 8

1 ← 7 − 6 = 1

Example 7 Subtract: 543 − 29. Check by adding.

Solution:

$$\begin{array}{r} 5\overset{3}{\cancel{4}}\overset{13}{\cancel{3}} \\ -\ 2\ 9 \\ \hline 5\ 1\ 4 \end{array} \qquad \text{Check: } \begin{array}{r} 514 \\ +\ 29 \\ \hline 543 \end{array}$$

Work Practice 7

Practice 7

Subtract. Check by adding.

a. $\begin{array}{r} 697 \\ -\ 49 \\ \hline \end{array}$

b. $\begin{array}{r} 326 \\ -245 \\ \hline \end{array}$

c. $\begin{array}{r} 1234 \\ -\ 822 \\ \hline \end{array}$

Sometimes we may have to borrow from more than one place. For example, to subtract 7631 − 152, we first borrow from the tens place.

$$\begin{array}{r} 76\overset{2}{\cancel{3}}\overset{11}{\cancel{1}} \\ -\ 15\ 2 \\ \hline 9 \end{array} \leftarrow 11 - 2 = 9$$

In the tens place, 5 is greater than 2, so we borrow again. This time we borrow from the hundreds place.

6 hundreds − **1 hundred** = 5 hundreds

$$\begin{array}{r} 7\overset{5}{\cancel{6}}\overset{12}{\cancel{2}}\overset{11}{\cancel{1}} \\ -\ 1\ 5\ 2 \\ \hline 7\ 4\ 7\ 9 \end{array} \qquad \text{Check: } \begin{array}{r} 7479 \\ +\ 152 \\ \hline 7631 \end{array} \text{ The original minuend.}$$

1 hundred + 2 tens

or

10 tens + 2 tens = 12 tens

Answers

6. a. 9021 **b.** 127

7. a. 648 **b.** 81 **c.** 412

Practice 8

Subtract. Check by adding.

a. $\begin{array}{r} 400 \\ -164 \\ \hline \end{array}$

b. $\begin{array}{r} 1000 \\ -\ \ 762 \\ \hline \end{array}$

Example 8 Subtract: $900 - 174$. Check by adding.

Solution: In the ones place, 4 is larger than 0, so we borrow from the tens place. But the tens place of 900 is 0, so to borrow from the tens place, we must first borrow from the hundreds place.

$$\begin{array}{r} \overset{8}{\cancel{9}}\ \overset{10}{\cancel{0}}\ 0 \\ -1\ 7\ 4 \\ \hline \end{array}$$

Now borrow from the tens place.

$$\begin{array}{r} \overset{8}{\cancel{9}}\ \overset{\overset{9}{\cancel{10}}}{\cancel{0}}\ \overset{10}{\cancel{0}} \\ -1\ 7\ 4 \\ \hline 7\ 2\ 6 \end{array} \qquad \text{Check: } \begin{array}{r} 726 \\ +174 \\ \hline 900 \end{array}$$

Work Practice 8

Objective C Finding the Perimeter of a Polygon

In geometry, addition is used to find the perimeter of a polygon. A **polygon** can be described as a flat figure formed by line segments connected at their ends. (For more review, see Appendix A.1.) Geometric figures such as triangles, squares, and rectangles are called polygons.

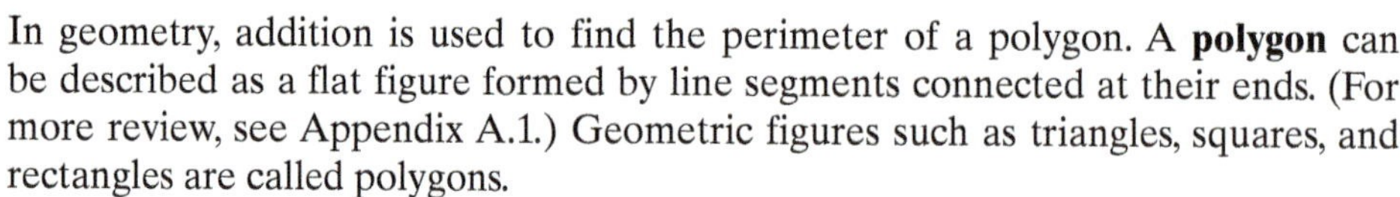
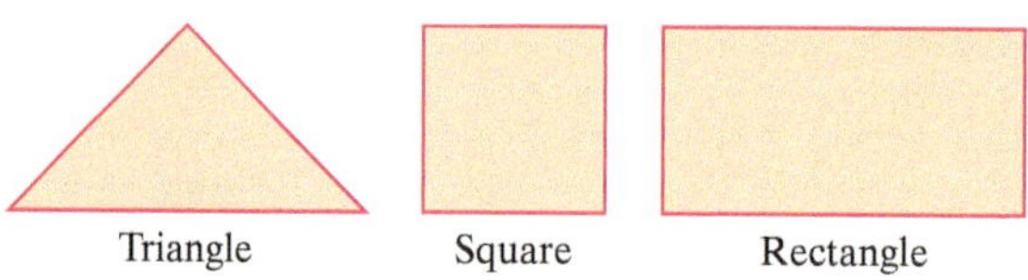

The **perimeter** of a polygon is the *distance around* the polygon, shown in red above. This means that the perimeter of a polygon is the sum of the lengths of its sides.

Practice 9

Find the perimeter of the polygon shown. (A centimeter is a unit of length in the metric system.)

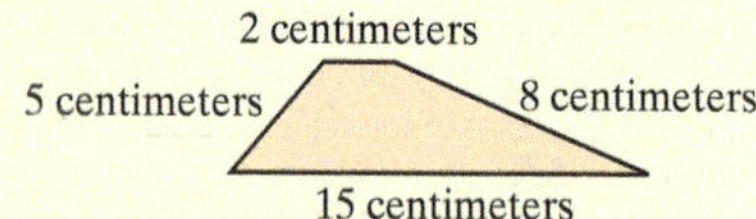

Example 9 Find the perimeter of the polygon shown.

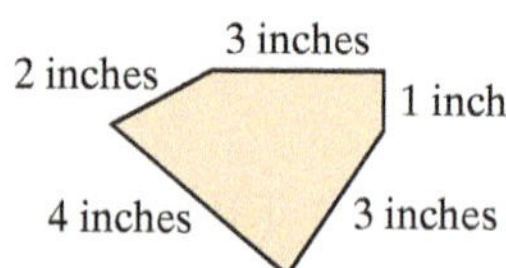

Solution: To find the perimeter (distance around), we add the lengths of the sides.

$$2 \text{ in.} + 3 \text{ in.} + 1 \text{ in.} + 3 \text{ in.} + 4 \text{ in.} = 13 \text{ in.}$$

The perimeter is 13 inches.

Work Practice 9

To make the addition appear simpler, we will often not include units with the addends. If you do this, make sure units are included in the final answer.

Answers
8. a. 236 **b.** 238
9. 30 cm

Example 10 Calculating the Perimeter of a Building

The largest commercial building in the world under one roof is the flower auction building of the cooperative VBA in Aalsmeer, Netherlands. The floor plan is a rectangle that measures 776 meters by 639 meters. Find the perimeter of this building. (A meter is a unit of length in the metric system.) (*Source: The Handy Science Answer Book*, Visible Ink Press)

Solution: Recall that opposite sides of a rectangle have the same length. To find the perimeter of this building, we add the lengths of the sides. The sum of the lengths of its sides is

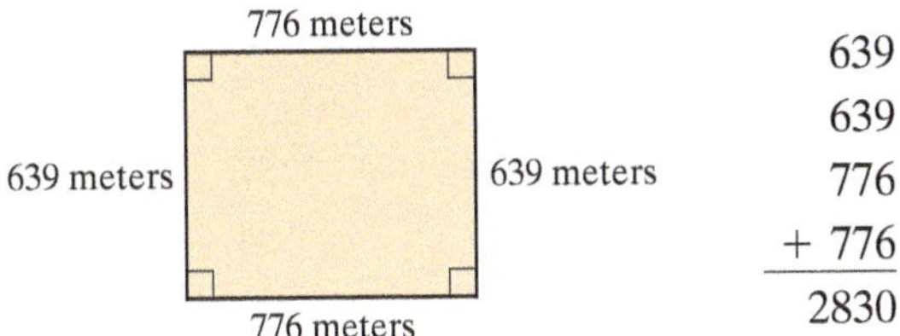

The perimeter of the building is 2830 meters.

Work Practice 10

Practice 10

A park is in the shape of a triangle. Each of the park's three sides is 647 feet. Find the perimeter of the park.

Objective D Solving Problems by Adding or Subtracting

Often, real-life problems occur that can be solved by adding or subtracting. The first step in solving any word problem is to *understand* the problem by reading it carefully.

Descriptions in problems solved through addition or subtraction *may* include any of these key words or phrases:

Helpful Hint Be careful when solving applications that suggest subtraction. Although order *does not* matter when adding, order *does* matter when subtracting. For example, 20 − 15 and 15 − 20 do not simplify to the same number.

Addition

Key Words or Phrases	Examples	Symbols
added to	5 added to 7	7 + 5
plus	0 plus 78	0 + 78
increased by	12 increased by 6	12 + 6
more than	11 more than 25	25 + 11
total	the total of 8 and 1	8 + 1
sum	the sum of 4 and 133	4 + 133

Subtraction

Key Words or Phrases	Examples	Symbols
subtract	subtract 5 from 8	8 − 5
difference	the difference of 10 and 2	10 − 2
less	17 less 3	17 − 3
less than	2 less than 20	20 − 2
take away	14 take away 9	14 − 9
decreased by	7 decreased by 5	7 − 5
subtracted from	9 subtracted from 12	12 − 9

✓**Concept Check** In each of the following problems, identify which number is the minuend and which number is the subtrahend.

a. What is the result when 6 is subtracted from 40?

b. What is the difference of 15 and 8?

c. Find a number that is 15 fewer than 23.

To solve a word problem that involves addition or subtraction, we first use the facts given to write an addition or subtraction statement. Then we write the corresponding solution of the real-life problem. It is sometimes helpful to write the statement in words (brief phrases) and then translate to numbers.

Answer

10. 1941 ft

✓**Concept Check Answers**

a. minuend: 40; subtrahend: 6

b. minuend: 15; subtrahend: 8

c. minuend: 23; subtrahend: 15

Practice 11

The radius of Uranus is 15,759 miles. The radius of Neptune is 458 miles less than the radius of Uranus. What is the radius of Neptune? (*Source:* National Space Science Data Center)

Example 11 Finding the Radius of a Planet

The radius of Jupiter is 43,441 miles. The radius of Saturn is 7257 miles less than the radius of Jupiter. Find the radius of Saturn. (*Source:* National Space Science Data Center)

Solution:

In Words		Translate to Numbers
radius of Jupiter	→	43,441
− 7257	→	− 7257
radius of Saturn	→	36,184

The radius of Saturn is 36,184 miles.

Helpful Hint Since subtraction and addition are reverse operations, don't forget that a subtraction problem can be checked by adding.

Work Practice 11

Graphs can be used to visualize data. The graph shown next is called a **bar graph.** For this bar graph, the height of each bar is labeled above the bar. To check this height, follow the top of each bar to the vertical line to the left. For example, the first bar is labeled 214. Follow the top of that bar to the left until the vertical line is reached, a bit more than halfway between 200 and 225, or 214.

Practice 12

Use the graph in Example 12 to answer the following:

a. Which country shown has the fewest threatened amphibians?

b. Find the total number of threatened amphibians for Brazil, Peru, and Mexico.

Example 12 Reading a Bar Graph

As years pass, the number of endangered species per country increases. In the following graph, each bar represents a country and the height of each bar represents the number of endangered species identified in that country.

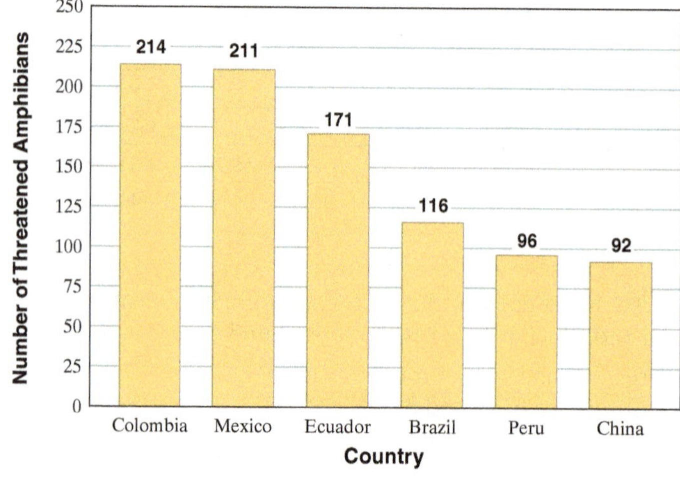

(*Source: The Top 10 of Everything*, 2013)

a. Which country shown has the greatest number of threatened amphibians?

b. Find the total number of threatened amphibians for Ecuador, China, and Colombia.

Answers

11. 15,301 miles

12. a. China **b.** 423

Solution:

a. The country with the greatest number of threatened amphibians corresponds to the tallest bar, which is Colombia.

b. The key word here is "total." To find the total number of threatened amphibians for Ecuador, China, and Colombia, we add.

In Words		Translate to Numbers
Ecuador	→	171
China	→	92
Colombia	→	+ 214
		Total 477

The total number of threatened amphibians for Ecuador, China, and Colombia is 477.

Work Practice 12

Calculator Explorations Adding and Subtracting Numbers

Adding Numbers

To add numbers on a calculator, find the keys marked [+] and [=] or [ENTER].

For example, to add 5 and 7 on a calculator, press the keys [5] [+] [7] then [=] or [ENTER].

The display will read [12].

Thus, 5 + 7 = 12.

To add 687, 981, and 49 on a calculator, press the keys [687] [+] [981] [+] [49] then [=] or [ENTER].

The display will read [1717].

Thus, 687 + 981 + 49 = 1717. (Although entering 687, for example, requires pressing more than one key, here numbers are grouped together for easier reading.)

Use a calculator to add.

1. 89 + 45

2. 76 + 97

3. 285 + 55

4. 8773 + 652

5.
$$\begin{array}{r} 985 \\ 1210 \\ 562 \\ +\quad 77 \\ \hline \end{array}$$

6.
$$\begin{array}{r} 465 \\ 9888 \\ 620 \\ +1550 \\ \hline \end{array}$$

Subtracting Numbers

To subtract numbers on a calculator, find the keys marked [−] and [=] or [ENTER].

For example, to find 83 − 49 on a calculator, press the keys [83] [−] [49] then [=] or [ENTER].

The display will read [34]. Thus, 83 − 49 = 34.

Use a calculator to subtract.

7. 865 − 95

8. 76 − 27

9. 147 − 38

10. 366 − 87

11. 9625 − 647

12. 10,711 − 8925

Vocabulary, Readiness & Video Check

Use the choices below to fill in each blank. Some choices may be used more than once and some may not be used at all.

0	order	addend	associative
sum	number	grouping	commutative
perimeter	minuend	subtrahend	difference

1. The sum of 0 and any number is the same ________ .
2. In $35 + 20 = 55$, the number 55 is called the ________ and 35 and 20 are each called a(n) ________ .
3. The difference of any number and that same number is ________ .
4. The difference of any number and 0 is the same ________ .
5. In $37 - 19 = 18$, the number 37 is the ________ , the 19 is the ________ , and the 18 is the ________ .
6. The distance around a polygon is called its ________ .
7. Since $7 + 10 = 10 + 7$, we say that changing the ________ in addition does not change the sum. This property is called the ________ property of addition.
8. Since $(3 + 1) + 20 = 3 + (1 + 20)$, we say that changing the ________ in addition does not change the sum. This property is called the ________ property of addition.

Martin-Gay Interactive Videos

See Video 1.3

Watch the section lecture video and answer the following questions.

Objective A 9. Complete this statement based on the lecture before Example 1. To add whole numbers, we line up ________ values and add from ________ to ________.

Objective B 10. In Example 5, explain how we end up subtracting 7 from 12 in the ones place.

Objective C 11. In Example 7, the perimeter of what type of polygon is found? How many addends are in the resulting addition problem?

Objective D 12. Complete this statement based on Example 8. To find the sale price, subtract the ________ from the ________ price.

1.3 Exercise Set MyMathLab®

Objective A *Add. See Examples 1 through 4.*

1. $14 + 22$

2. $27 + 31$

3. $62 + 230$

4. $37 + 542$

5. $12 + 13 + 24$

6. $23 + 45 + 30$

7. $5267 + 132$

8. $236 + 6243$

9. $22{,}781 + 186{,}297$

10. $17{,}427 + 821{,}059$

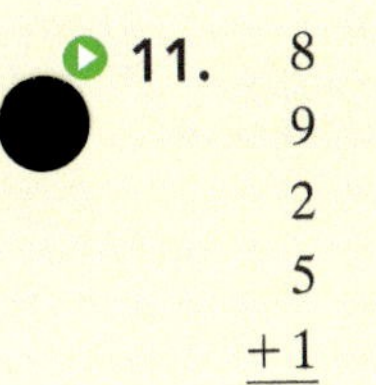

11. 8
9
2
5
+1

12. 3
5
8
5
+7

13. 81
17
23
79
+12

14. 64
28
56
25
+32

15. 24 + 9006 + 489 + 2407

16. 16 + 1056 + 748 + 7770

17. 6820
4271
+5626

18. 6789
4321
+5555

19. 49
628
5 762
+29,462

20. 26
582
4 763
+62,511

21. 121,742
57,279
26,586
+426,782

22. 504,218
321,920
38,507
+594,687

Objective B *Subtract. Check by adding. See Examples 5 through 8.*

23. 749
−149

24. 957
−257

25. 62
−37

26. 55
−29

27. 922
−634

28. 674
−299

29. 600
−432

30. 300
−149

31. 6283
− 560

32. 5349
− 720

33. 533
− 29

34. 724
− 16

35. 1983 − 1904

36. 1983 − 1914

37. 50,000 − 17,289

38. 40,000 − 23,582

39. 7020 − 1979

40. 6050 − 1878

41. 51,111 − 19,898

42. 62,222 − 39,898

Objectives A B Mixed Practice *Add or subtract as indicated. See Examples 1 through 8.*

43. 986
+ 48

44. 986
− 48

45. 76 − 67

46. 80 + 93 + 17 + 9 + 2

47. 9000
− 482

48. 10,000
− 1786

49. 10,962
4851
+ 7063

50. 12,468
3211
+ 1988

Objective C *Find the perimeter of each figure. See Examples 9 and 10.*

51.

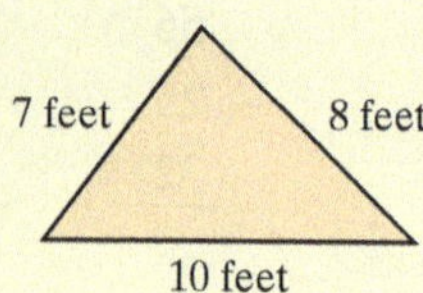

△ 52.

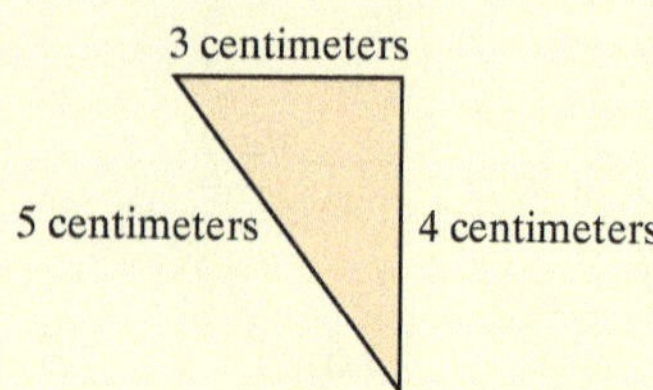

△ 53.

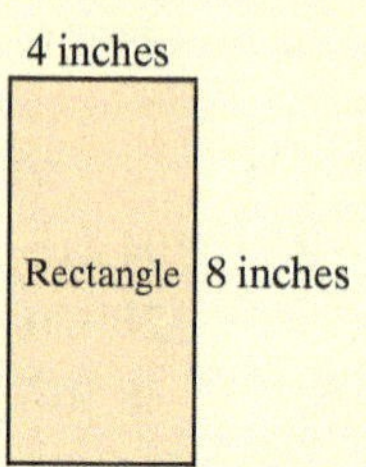

△ 54.

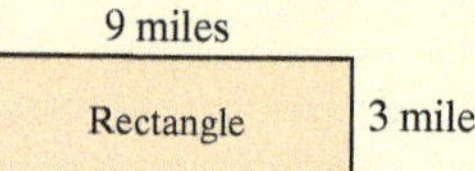

△ 55.

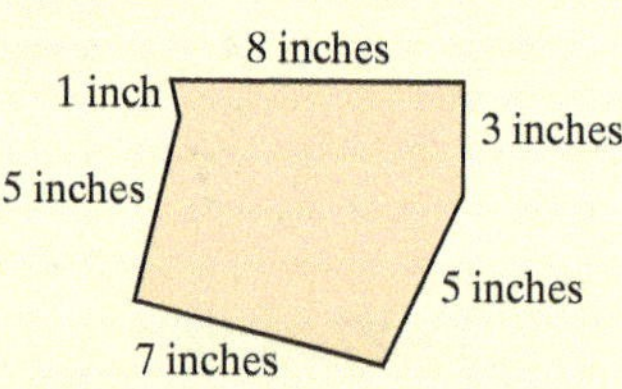

△ 56.

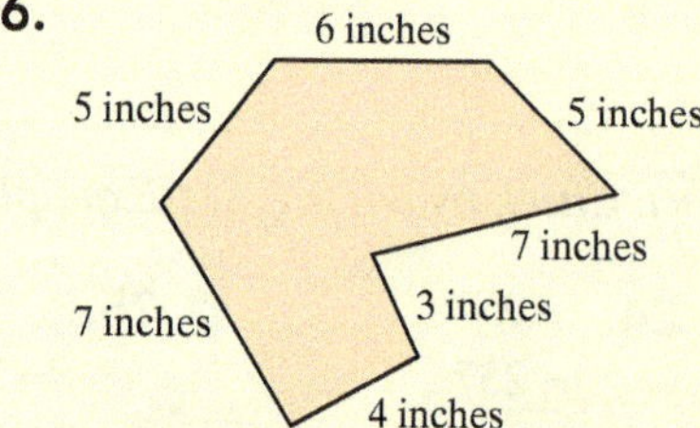

△ 57.

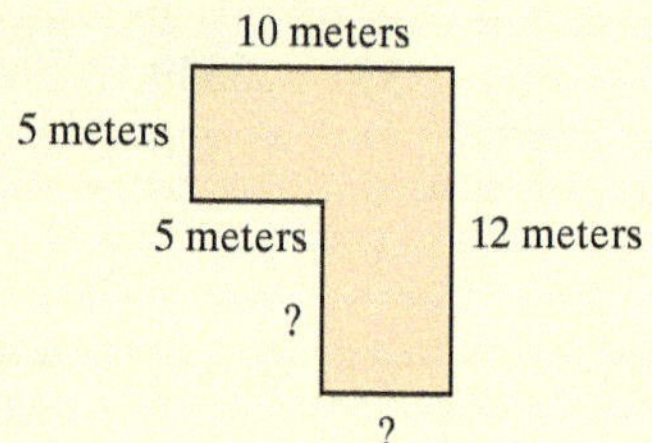

△ 58.

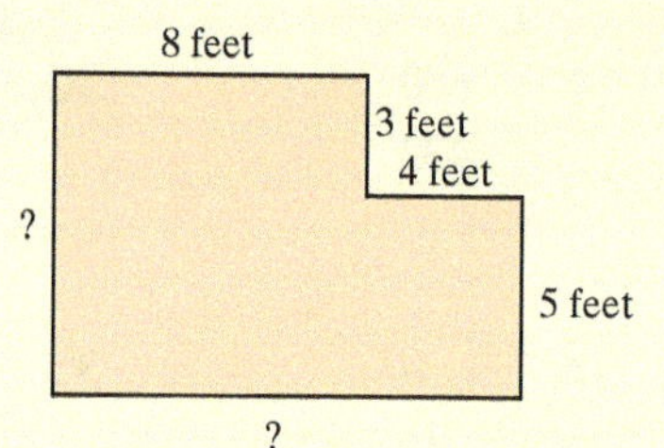

Objectives A B D Mixed Practice–Translating *Solve. See Examples 9 through 12.*

59. Find the sum of 297 and 1796.

60. Find the sum of 802 and 6487.

61. Find the total of 76, 39, 8, 17, and 126.

62. Find the total of 89, 45, 2, 19, and 341.

63. Find the difference of 41 and 21.

64. Find the difference of 16 and 5.

65. What is 452 increased by 92?

66. What is 712 increased by 38?

67. Find 108 less 36.

68. Find 25 less 12.

69. Find 12 subtracted from 100.

70. Find 86 subtracted from 90.

Solve.

71. The population of Florida is projected to grow from 19,308 thousand in 2010 to 22,478 thousand in 2020. What is Florida's projected population increase over this time period?

72. The population of California is projected to grow from 39,136 thousand in 2010 to 44,126 thousand in 2020. What is California's projected population increase over this time period?

73. A new DVD player with remote control costs $295. A college student has $914 in her savings account. How much will she have left in her savings account after she buys the DVD player?

74. A stereo that regularly sells for $547 is discounted by $99 in a sale. What is the sale price?

A river basin is the geographic area drained by a river and its tributaries. The Mississippi River Basin is the third largest in the world and is divided into six sub-basins, whose areas are shown in the following bar graph. Use this graph for Exercises 75 through 78.

75. Find the total U.S. land area drained by the Upper Mississippi and Lower Mississippi sub-basins.

76. Find the total U.S. land area drained by the Ohio and Tennessee sub-basins.

77. How many more square miles of land are drained by the Missouri sub-basin than the Arkansas Red-White sub-basin?

78. How many more square miles of land are drained by the Upper Mississippi sub-basin than the Lower Mississippi sub-basin?

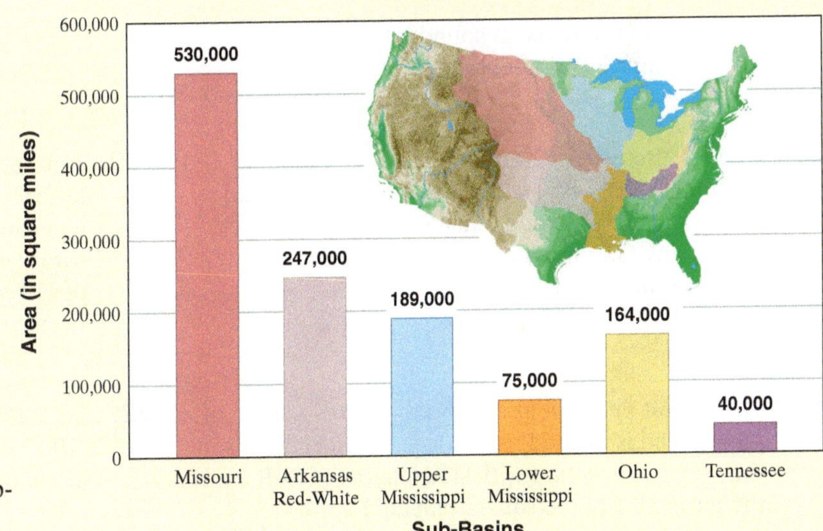

△ **79.** A homeowner is installing a fence in his backyard. How many feet of fencing are needed to enclose the yard below?

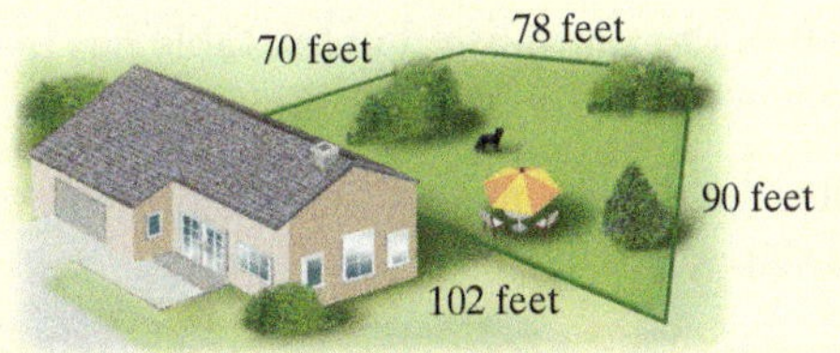

△ **80.** A homeowner is considering installing gutters around her home. Find the perimeter of her rectangular home.

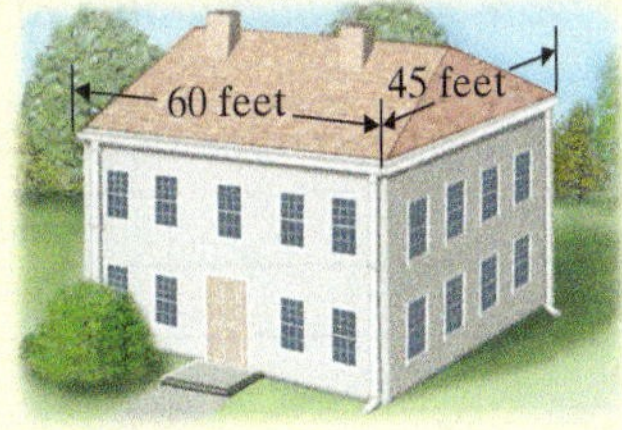

81. Professor Graham is reading a 503-page book. If she has just finished reading page 239, how many more pages must she read to finish the book?

82. When a couple began a trip, the odometer read 55,492. When the trip was over, the odometer read 59,320. How many miles did they drive on their trip?

83. In 2013, the country of New Zealand had 26,820,424 more sheep than people. If the human population of New Zealand in 2013 was 4,479,576, what was the sheep population? (*Source: Statistics: New Zealand*)

84. During one month in 2013, the two top-selling vehicles in the United States were the Ford F-Series and the Chevrolet Silverado, both trucks. There were 60,449 F-Series trucks and 42,080 Silverados sold that month. What was the total number of these trucks sold in that month? (*Source:* cars.com)

The decibel (dB) is a unit of measurement for sound. Every increase of 10 dB is a tenfold increase in sound intensity. The bar graph shows the decibel levels for some common sounds. Use this graph for Exercises 85 through 88.

85. What is the dB level for live rock music?

86. Which is the quietest of all the sounds shown in the graph?

87. How much louder is the sound of snoring than normal conversation?

88. What is the difference in sound intensity between live rock music and loud television?

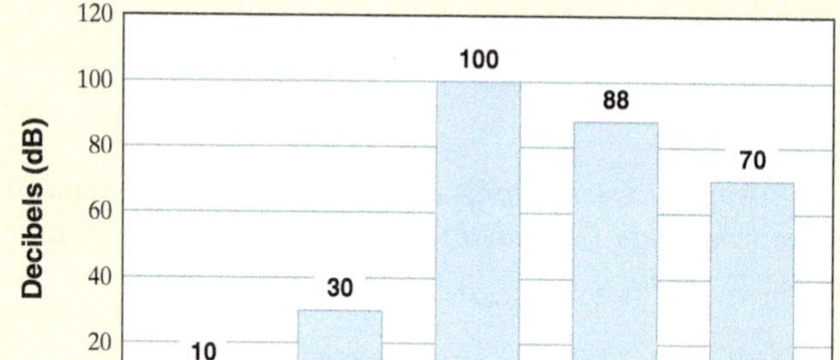

89. In 2013, there were 2410 Gap Inc. (Gap, Banana Republic, Old Navy) stores located in the United States and 1034 located outside the United States. How many Gap Inc. stores were located worldwide? (*Source:* Gap, Inc.)

90. Automobile classes are defined by the amount of interior room. A subcompact car is defined as a car with a maximum interior space of 99 cubic feet. A midsize car is defined as a car with a maximum interior space of 119 cubic feet. What is the difference in volume between a midsize and a subcompact car?

91. The largest permanent Monopoly board is made of granite and is located in San Jose, California. It is in the shape of a square with side lengths of 31 ft. Find the perimeter of the square playing board.

92. The smallest commercially available jigsaw puzzle is a 1000-piece puzzle manufactured in Spain. It is in the shape of a rectangle with length of 18 inches and width of 12 inches. Find the perimeter of this rectangular-shaped puzzle.

The table shows the number of Target stores in ten states. Use this table to answer Exercises 93 through 98.

The Top States for Target Stores in 2012

State	Number of Stores
Pennsylvania	63
California	257
Florida	123
Virginia	57
Illinois	89
New York	67
Michigan	59
Minnesota	75
Ohio	64
Texas	149

(*Source:* Target Corporation)

93. Which state has the most Target stores?

94. Which of the states listed in the table has the fewest Target stores?

95. What is the total number of Target stores located in the three states with the most Target stores?

96. How many Target stores are located in the ten states listed in the table?

97. Which pair of neighboring states have more Target stores combined, Pennsylvania and New York or Michigan and Ohio?

98. There are 775 Target stores located in the states not listed in the table. How many Target stores are in the United States?

99. The state of Delaware has 2029 miles of urban highways and 3865 miles of rural highways. Find the total highway mileage in Delaware. (*Source:* U.S. Federal Highway Administration)

100. The state of Rhode Island has 5193 miles of urban highways and 1222 miles of rural highways. Find the total highway mileage in Rhode Island. (*Source:* U.S. Federal Highway Administration)

Concept Extensions

For Exercises 101–104, identify which number is the minuend and which number is the subtrahend. See the second Concept Check in this section.

101. $\begin{array}{r} 48 \\ -\ 1 \\ \hline \end{array}$

102. $\begin{array}{r} 2863 \\ -1904 \\ \hline \end{array}$

103. Subtract 7 from 70.

104. Find 86 decreased by 25.

105. In your own words, explain the commutative property of addition.

106. In your own words, explain the associative property of addition.

Check each addition below. If it is incorrect, find the correct answer. See the first Concept Check in this section.

107. $\begin{array}{r} 566 \\ 932 \\ +\ 871 \\ \hline 2369 \end{array}$

108. $\begin{array}{r} 773 \\ 659 \\ +\ 481 \\ \hline 1913 \end{array}$

109. $\begin{array}{r} 14 \\ 173 \\ 86 \\ +257 \\ \hline 520 \end{array}$

110. $\begin{array}{r} 19 \\ 214 \\ 49 \\ +651 \\ \hline 923 \end{array}$

Identify each answer as correct or incorrect. Use addition to check. If the answer is incorrect, write the correct answer.

111. $\begin{array}{r} 741 \\ -\ 56 \\ \hline 675 \end{array}$

112. $\begin{array}{r} 478 \\ -\ 89 \\ \hline 389 \end{array}$

113. $\begin{array}{r} 1029 \\ -\ 888 \\ \hline 141 \end{array}$

114. $\begin{array}{r} 7615 \\ -\ 547 \\ \hline 7168 \end{array}$

Fill in the missing digits in each problem.

115. $\begin{array}{r} 526\_ \\ -2\_85 \\ \hline 28\_4 \end{array}$

116. $\begin{array}{r} 10,\_4\_ \\ -\ 85\_4 \\ \hline \_710 \end{array}$

117. Is there a commutative property of subtraction? In other words, does order matter when subtracting? Why or why not?

118. Explain why the phrase "Subtract 7 from 10" translates to "10 − 7."

119. The local college library is having a Million Pages of Reading promotion. The freshmen have read a total of 289,462 pages; the sophomores have read a total of 369,477 pages; the juniors have read a total of 218,287 pages; and the seniors have read a total of 121,685 pages. Have they reached a goal of one million pages? If not, how many more pages need to be read?

1.4 Rounding and Estimating 

Objectives

A Round Whole Numbers.

B Use Rounding to Estimate Sums and Differences.

C Solve Problems by Estimating.

Objective A Rounding Whole Numbers

Rounding a whole number means approximating it. A rounded whole number is often easier to use, understand, and remember than the precise whole number. For example, instead of trying to remember the Colorado state population as 5,188,000, it is much easier to remember it rounded to the nearest million: 5,000,000, or 5 million people. (*Source:* U.S. census)

Recall from Section 1.2 that the line below is called a number line. To **graph** a whole number on this number line, we darken the point representing the location of the whole number. For example, the number 4 is graphed below.

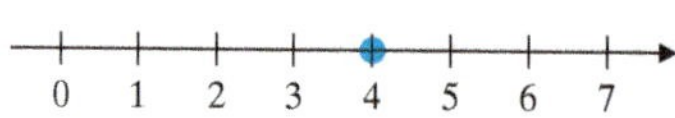

On the number line, the whole number 36 is closer to 40 than 30, so 36 rounded to the nearest ten is 40.

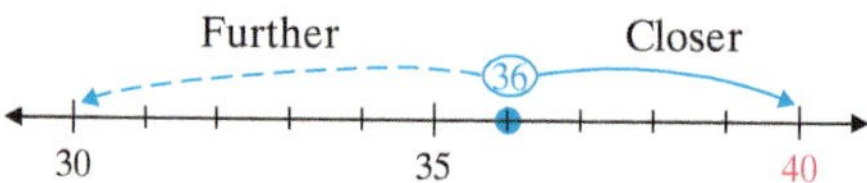

The whole number 52 is closer to 50 than 60, so 52 rounded to the nearest ten is 50.

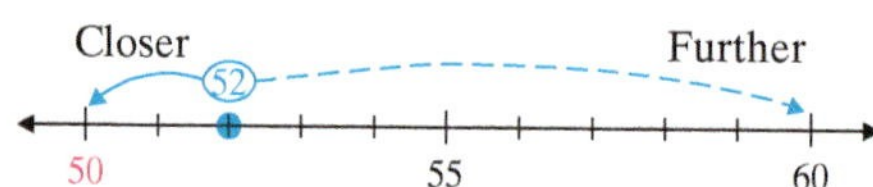

In trying to round 25 to the nearest ten, we see that 25 is halfway between 20 and 30. It is not closer to either number. In such a case, we round to the larger ten, that is, to 30.

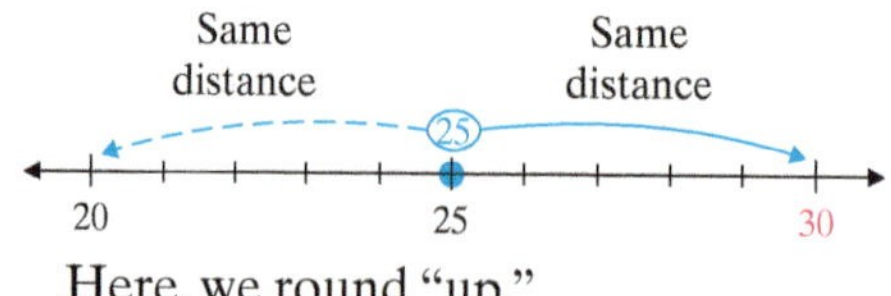

Here, we round "up."

To round a whole number without using a number line, follow these steps:

Rounding a Whole Number to a Given Place Value

Step 1: Locate the digit to the right of the given place value.

Step 2: If this digit is 5 or greater, add 1 to the digit in the given place value and replace each digit to its right by 0.

Step 3: If this digit is less than 5, replace it and each digit to its right by 0.

Example 1 Round 568 to the nearest ten.

Solution: 5 6 ⑧ The digit to the right of the tens place is the ones place, which is circled.

↑ tens place

5 6 ⑧ Since the circled digit is 5 or greater, add 1 to the 6 in the tens place and replace the digit to the right by 0.

↑ Add 1. ↖ Replace with 0.

We find that 568 rounded to the nearest ten is 570.

Work Practice 1

Practice 1

Round to the nearest ten.

a. 57

b. 641

c. 325

Example 2 Round 278,362 to the nearest thousand.

Solution:

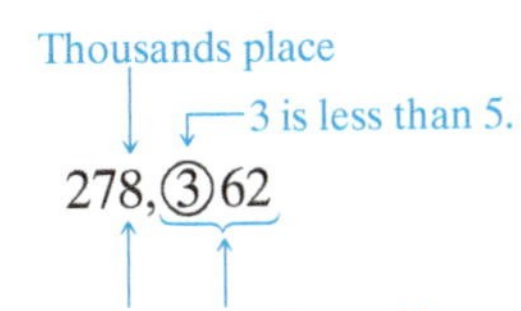

The number 278,362 rounded to the nearest thousand is 278,000.

Work Practice 2

Practice 2

Round to the nearest thousand.

a. 72,304

b. 9222

c. 671,800

Example 3 Round 248,982 to the nearest hundred.

Solution:

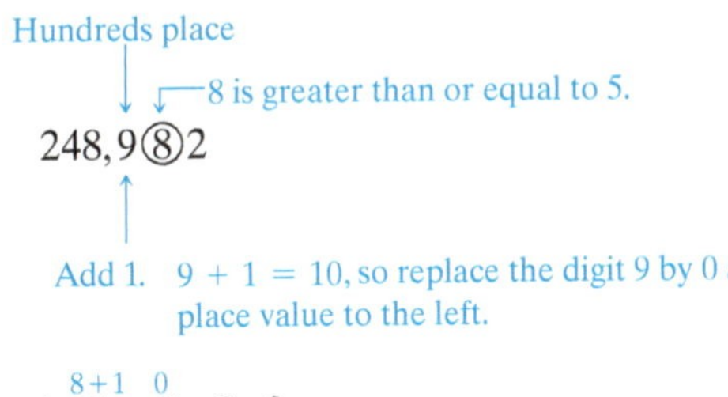

2 4 8, 9 8 2 (8 replaced by 8+1, 9 replaced by 0)

↑ Add 1. ↑ Replace with zeros.

The number 248,982 rounded to the nearest hundred is 249,000.

Work Practice 3

Practice 3

Round to the nearest hundred.

a. 3474

b. 76,243

c. 978,965

✓**Concept Check** Round each of the following numbers to the nearest *hundred*. Explain your reasoning.

a. 59 **b.** 29

Objective B Estimating Sums and Differences 

By rounding addends, minuends, and subtrahends, we can **estimate** sums and differences. An estimated sum or difference is appropriate when the exact number is not necessary. Also, an estimated sum or difference can help us determine if we made

Answers

1. a. 60 **b.** 640 **c.** 330
2. a. 72,000 **b.** 9000 **c.** 672,000
3. a. 3500 **b.** 76,200 **c.** 979,000

✓**Concept Check Answers**

a. 100 **b.** 0

a mistake in calculating an exact amount. To estimate the sum below, round each number to the nearest hundred and then add.

768	rounds to	800
1952	rounds to	2000
225	rounds to	200
+ 149	rounds to	+ 100
		3100

The estimated sum is 3100, which is close to the **exact** sum of 3094.

Practice 4

Round each number to the nearest ten to find an estimated sum.

49
25
32
51
+98

Example 4 Round each number to the nearest hundred to find an estimated sum.

294
625
1071
+ 349

Solution:

Exact:		**Estimate:**
294	rounds to	300
625	rounds to	600
1071	rounds to	1100
+ 349	rounds to	+ 300
		2300

The estimated sum is 2300. (The exact sum is 2339.)

Work Practice 4

Practice 5

Round each number to the nearest thousand to find an estimated difference.

3785
− 2479

Example 5 Round each number to the nearest hundred to find an estimated difference.

4725
− 2879

Solution:

Exact:		**Estimate:**
4725	rounds to	4700
−2879	rounds to	−2900
		1800

The estimated difference is 1800. (The exact difference is 1846.)

Work Practice 5

Objective C Solving Problems by Estimating

Making estimates is often the quickest way to solve real-life problems when solutions do not need to be exact.

Answers
4. 260
5. 2000

Example 6 Estimating Distances

A driver is trying to quickly estimate the distance from Temple, Texas, to Brenham, Texas. Round each distance given on the map to the nearest ten to estimate the total distance.

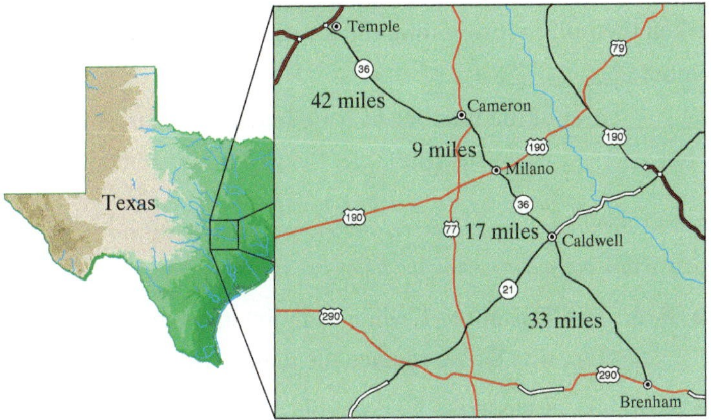

Solution:

Exact Distance:		Estimate:
42	rounds to	40
9	rounds to	10
17	rounds to	20
+33	rounds to	+30
		100

It is approximately 100 miles from Temple to Brenham. (The exact distance is 101 miles.)

Work Practice 6

Practice 6

Tasha Kilbey is trying to estimate how far it is from Gove, Kansas, to Hays, Kansas. Round each given distance on the map to the nearest ten to estimate the total distance.

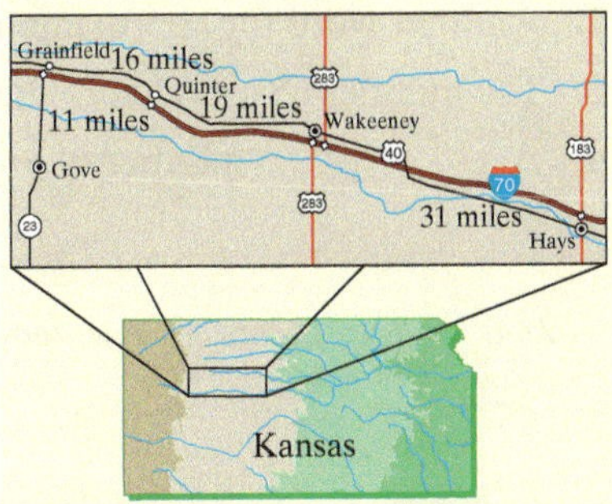

Example 7 Estimating Data

In three recent years the numbers of tons of air cargo and mail that went through Hartsfield-Jackson Atlanta International Airport were 629,700, 685,550, and 737,655. Round each number to the nearest thousand to estimate the tons of mail that passed through this airport.

Solution:

Exact Tons of Cargo Mail:		Estimate:
629,700	rounds to	630,000
685,550	rounds to	686,000
+737,655	rounds to	+738,000
		2,054,000

The approximate tonnage of mail that moved through Atlanta's airport over this 3-year period was 2,054,000 tons. (The exact tonnage was 2,052,905 tons.)

Work Practice 7

Practice 7

In 2010, there were 15,427 reported cases of chicken pox, 2612 reported cases of mumps, and 27,550 reported cases of pertussis (whooping cough). Round each number to the nearest thousand to estimate the total number of cases reported for these preventable diseases. (*Source:* Centers for Disease Control and Prevention)

Answers

6. 80 mi

7. 46,000 total cases

Vocabulary, Readiness & Video Check

Use the choices below to fill in each blank.

60	rounding	exact
70	estimate	graph

1. To ________ a number on a number line, darken the point representing the location of the number.
2. Another word for approximating a whole number is ________.
3. The number 65 rounded to the nearest ten is ________, but the number 61 rounded to the nearest ten is ________.
4. A(n) ________ number of products is 1265, but a(n) ________ is 1000.

Martin-Gay Interactive Videos *Watch the section lecture video and answer the following questions.*

See Video 1.4 

Objective A 5. In Example 1, when rounding the number to the nearest ten, why do we replace the digit 3 with a 4?

Objective B 6. As discussed in Example 3, explain how a number line can help us understand how to round 22 to the nearest ten.

Objective C 7. What is the significance of the circled digit in each height value in Example 5?

1.4 Exercise Set MyMathLab®

Objective A *Round each whole number to the given place. See Examples 1 through 3.*

1. 423 to the nearest ten
2. 273 to the nearest ten
3. 635 to the nearest ten
4. 846 to the nearest ten
5. 2791 to the nearest hundred
6. 8494 to the nearest hundred
7. 495 to the nearest ten
8. 898 to the nearest ten
9. 21,094 to the nearest thousand
10. 82,198 to the nearest thousand
11. 33,762 to the nearest thousand
12. 42,682 to the nearest ten-thousand
13. 328,495 to the nearest hundred
14. 179,406 to the nearest hundred
15. 36,499 to the nearest thousand
16. 96,501 to the nearest thousand
17. 39,994 to the nearest ten
18. 99,995 to the nearest ten
19. 29,834,235 to the nearest ten-million
20. 39,523,698 to the nearest million

Complete the table by estimating the given number to the given place value.

		Tens	Hundreds	Thousands
21.	5281			
22.	7619			
23.	9444			
24.	7777			
25.	14,876			
26.	85,049			

Round each number to the indicated place.

27. The state of Texas contains 310,850 miles of urban and rural highways. Round this number to the nearest thousand. (*Source:* U.S. Federal Highway Administration)

28. The state of California contains 171,874 miles of urban and rural highways. Round this number to the nearest thousand. (*Source:* U.S. Federal Highway Administration)

29. It takes 60,149 days for Neptune to make a complete orbit around the Sun. Round this number to the nearest hundred. (*Source:* National Space Science Data Center)

30. Kareem Abdul-Jabbar holds the NBA record for points scored, a total of 38,387 over his NBA career. Round this number to the nearest thousand. (*Source:* National Basketball Association)

31. In 2013, the most valuable brand in the world was Apple, Inc. The estimated brand value at this time of Apple was $185,000,000,000. Round this to the nearest ten billion. (*Source:* Millward Brown)

32. According to the U.S. Population Clock, the population of the United States was 316,539,415 in August 2013. Round this population figure to the nearest million. (*Source:* U.S. Census population clock)

33. The average salary for a baseball player in 2012 was $3,213,479. Round this average salary to the nearest hundred-thousand. (*Source:* Major League Baseball Players Association)

34. The average salary for a football player in 2012 was $1,900,000. Round this average salary to the nearest million. (*Source:* Businessweek.com)

35. The United States currently has 331,600,000 cellular phone users, while India has 893,862,500 users. Round each of the user numbers to the nearest million. (*Source: World Almanac*, 2013)

36. U.S. farms produced 3,112,500,000 bushels of soybeans in 2011. Round the soybean production figure to the nearest ten-million. (*Source:* U.S. Department of Agriculture)

U.S.
India

Objective B *Estimate the sum or difference by rounding each number to the nearest ten. See Examples 4 and 5.*

37. 39 + 45 + 22 + 17

38. 52 + 33 + 15 + 29

39. 449 − 373

40. 555 − 235

Estimate the sum or difference by rounding each number to the nearest hundred. See Examples 4 and 5.

41. 1913 + 1886 + 1925

42. 4050 + 3133 + 1220

43. 1774 − 1492

44. 1989 − 1870

45. 3995 + 2549 + 4944

46. 799 + 1655 + 271

Three of the given calculator answers below are incorrect. Find them by estimating each sum.

47. 463 + 219   600

48. 522 + 785   1307

49. 229 + 443 + 606   1278

50. 542 + 789 + 198   2139

51. 7806 + 5150   12,956

52. 5233 + 4988   9011

Helpful Hint Estimation is useful to check for incorrect answers when using a calculator. For example, pressing a key too hard may result in a double digit, while pressing a key too softly may result in the digit not appearing in the display.

Objective C *Solve each problem by estimating. See Examples 6 and 7.*

53. An appliance store advertises three refrigerators on sale at $899, $1499, and $999. Round each cost to the nearest hundred to estimate the total cost.

54. Suppose you scored 89, 97, 100, 79, 75, and 82 on your biology tests. Round each score to the nearest ten to estimate your total score.

55. The distance from Kansas City to Boston is 1429 miles and from Kansas City to Chicago is 530 miles. Round each distance to the nearest hundred to estimate how much farther Boston is from Kansas City than Chicago is.

56. The Gonzales family took a trip and traveled 588, 689, 277, 143, 59, and 802 miles on six consecutive days. Round each distance to the nearest hundred to estimate the distance they traveled.

57. The peak of Mt. McKinley, in Alaska, is 20,320 feet above sea level. The top of Mt. Rainier, in Washington, is 14,410 feet above sea level. Round each height to the nearest thousand to estimate the difference in elevation of these two peaks. (*Source:* U.S. Geological Survey)

58. A student is pricing new car stereo systems. One system sells for $1895 and another system sells for $1524. Round each price to the nearest hundred dollars to estimate the difference in price of these systems.

59. In 2012, the population of Springfield, Illinois, was 117,126, and the population of Champaign, Illinois, was 82,517. Round each population to the nearest ten-thousand to estimate how much larger Springfield was than Champaign. (*Source:* U.S. Census Bureau)

60. Round each distance given on the map to the nearest ten to estimate the total distance from North Platte, Nebraska, to Lincoln, Nebraska.

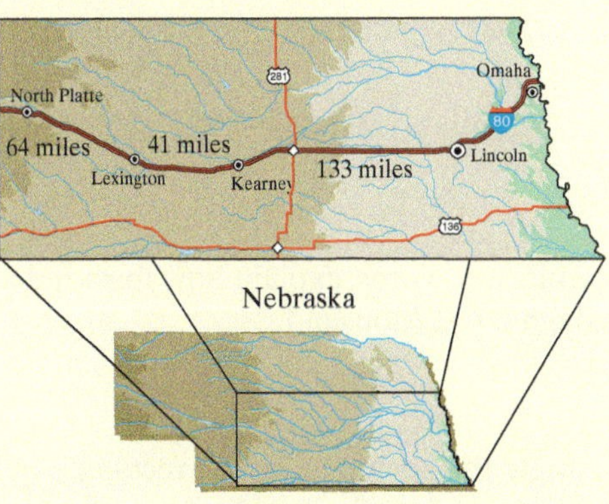

61. Head Start is a national program that provides developmental and social services for America's low-income preschool children ages three to five. Enrollment figures in Head Start programs showed an increase from 1,073,440 in 2010 to 1,128,030 in 2012. Round each number of children to the nearest thousand to estimate this increase. (*Source:* U.S. Department of Health and Human Services)

62. Enrollment figures at a local community college showed an increase from 49,713 credit hours in 2005 to 51,746 credit hours in 2006. Round each number to the nearest thousand to estimate the increase.

Mixed Practice (*Sections 1.2 and 1.4*) *The following table shows a few of the airports in the United States with the largest volumes of passengers. Complete this table. The first line is completed for you. (Source:* 2011 World Annual Trafic Report*)*

	City Location of Airport	Total Passengers in 2011 (in hundred-thousands of passengers)	Amount Written in Standard Form	Standard Form Rounded to the Nearest Million	Standard Form Rounded to the Nearest Ten-Million
	Atlanta, GA	924	92,400,000	92,000,000	90,000,000
63.	Chicago, IL	667			
64.	Los Angeles, CA	619			
65.	Dallas/Fort Worth, TX	578			
66.	Denver, CO	528			

Concept Extensions

67. Find one number that when rounded to the nearest hundred is 5700.

68. Find one number that when rounded to the nearest ten is 5700.

69. A number rounded to the nearest hundred is 8600.

a. Determine the smallest possible number.

b. Determine the largest possible number.

70. On August 23, 1989, it was estimated that 1,500,000 people joined hands in a human chain stretching 370 miles to protest the fiftieth anniversary of the pact that allowed what was then the Soviet Union to annex the Baltic nations in 1939. If the estimate of the number of people is to the nearest hundred-thousand, determine the largest possible number of people in the chain.

71. In your own words, explain how to round a number to the nearest thousand.

72. In your own words, explain how to round 9660 to the nearest thousand.

△ **73.** Estimate the perimeter of the rectangle by first rounding the length of each side to the nearest ten.

△ **74.** Estimate the perimeter of the triangle by first rounding the length of each side to the nearest hundred.

1.5 Multiplying Whole Numbers and Area 

Objectives

A Use the Properties of Multiplication.

B Multiply Whole Numbers.

C Find the Area of a Rectangle.

D Solve Problems by Multiplying Whole Numbers.

Multiplication Shown as Repeated Addition Suppose that we wish to count the number of laptops provided in a computer class. The laptops are arranged in 5 rows, and each row has 6 laptops.

Adding 5 sixes gives the total number of laptops. We can write this as $6 + 6 + 6 + 6 + 6 = 30$ laptops. When each addend is the same, we refer to this as **repeated addition.**

Multiplication is repeated addition but with different notation.

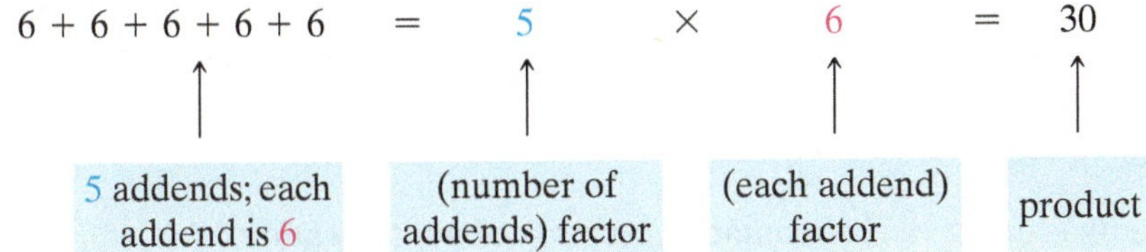

The × is called a **multiplication sign.** The numbers 5 and 6 are called **factors.** The number 30 is called the **product.** The notation 5 × 6 is read as "five times six." The symbols · and ( ) can also be used to indicate multiplication.

$$5 \times 6 = 30, \quad 5 \cdot 6 = 30, \quad (5)(6) = 30, \quad \text{and} \quad 5(6) = 30$$

✓Concept Check

a. Rewrite 5 + 5 + 5 + 5 + 5 + 5 + 5 using multiplication.

b. Rewrite 3 × 16 as repeated addition. Is there more than one way to do this? If so, show all ways.

Objective A Using the Properties of Multiplication

As with addition, we memorize products of one-digit whole numbers and then use certain properties of multiplication to multiply larger numbers. (If necessary, review the multiplication of one-digit numbers.)

Notice that when any number is multiplied by 0, the result is always 0. This is called the **multiplication property of 0.**

Multiplication Property of 0

The product of 0 and any number is 0. For example,

$$5 \cdot 0 = 0 \quad \text{and} \quad 0 \cdot 8 = 0$$

Also notice that when any number is multiplied by 1, the result is always the original number. We call this result the **multiplication property of 1.**

Multiplication Property of 1

The product of 1 and any number is that same number. For example,

$$1 \cdot 9 = 9 \quad \text{and} \quad 6 \cdot 1 = 6$$

Example 1 Multiply.

a. 4×1 **b.** $0(3)$ **c.** $1 \cdot 64$ **d.** $(48)(0)$

Solution:

a. $4 \times 1 = 4$ **b.** $0(3) = 0$
c. $1 \cdot 64 = 64$ **d.** $(48)(0) = 0$

Work Practice 1

Practice 1

Multiply.

a. 6×0

b. $(1)8$

c. $(50)(0)$

d. $75 \cdot 1$

✓Concept Check Answers

a. 7×5

b. 16 + 16 + 16; yes, 3 + 3 + 3 + 3 + 3 + 3 + 3 + 3 + 3 + 3 + 3 + 3 + 3 + 3 + 3 + 3

Answers

1. a. 0 **b.** 8 **c.** 0 **d.** 75

Like addition, multiplication is commutative and associative. Notice that when multiplying two numbers, the order of these numbers can be changed without changing the product. For example,

$$3 \cdot 5 = 15 \quad \text{and} \quad 5 \cdot 3 = 15$$

This property is called the **commutative property of multiplication.**

Commutative Property of Multiplication

Changing the **order** of two factors does not change their product. For example,

$$9 \cdot 2 = 18 \quad \text{and} \quad 2 \cdot 9 = 18$$

Another property that can help us when multiplying is the **associative property of multiplication.** This property states that when multiplying numbers, the grouping of the numbers can be changed without changing the product. For example,

$$\underbrace{(2 \cdot 3)} \cdot 4 = 6 \cdot 4 = 24$$

Also,

$$2 \cdot \underbrace{(3 \cdot 4)} = 2 \cdot 12 = 24$$

Both groupings give a product of 24.

Associative Property of Multiplication

Changing the **grouping** of factors does not change their product. From the previous work, we know that, for example,

$$(2 \cdot 3) \cdot 4 = 2 \cdot (3 \cdot 4)$$

With these properties, along with the **distributive property,** we can find the product of any whole numbers. The distributive property says that multiplication **distributes** over addition. For example, notice that $3(2 + 5)$ simplifies to the same number as $3 \cdot 2 + 3 \cdot 5$.

$$3\underbrace{(2 + 5)} = 3(7) = 21$$

$$\underbrace{3 \cdot 2} + \overbrace{3 \cdot 5} = 6 + 15 = 21$$

Since $3(2 + 5)$ and $3 \cdot 2 + 3 \cdot 5$ both simplify to 21, then

$$3(2 + 5) = 3 \cdot 2 + 3 \cdot 5$$

Notice in $3(2 + 5) = 3 \cdot 2 + 3 \cdot 5$ that each number inside the parentheses is multiplied by 3.

Distributive Property

Multiplication distributes over addition. For example,

$$2(3 + 4) = 2 \cdot 3 + 2 \cdot 4$$

Example 2 Rewrite each using the distributive property.

a. $5(6 + 5)$ b. $20(4 + 7)$ c. $2(7 + 9)$

Solution: Using the distributive property, we have

a. $5(6 + 5) = 5 \cdot 6 + 5 \cdot 5$
b. $20(4 + 7) = 20 \cdot 4 + 20 \cdot 7$
c. $2(7 + 9) = 2 \cdot 7 + 2 \cdot 9$

Work Practice 2

Practice 2

Rewrite each using the distributive property.

a. $6(4 + 5)$
b. $30(2 + 3)$
c. $7(2 + 8)$

Objective B Multiplying Whole Numbers 

Let's use the distributive property to multiply 7(48). To do so, we begin by writing the expanded form of 48 (see Section 1.2) and then applying the distributive property.

$$
\begin{aligned}
7(48) &= 7(40 + 8) && \text{Write 48 in expanded form.}\\
&= 7 \cdot 40 + 7 \cdot 8 && \text{Apply the distributive property.}\\
&= 280 + 56 && \text{Multiply.}\\
&= 336 && \text{Add.}
\end{aligned}
$$

This is how we multiply whole numbers. When multiplying whole numbers, we will use the following notation.

First:

$$\begin{array}{r} {}^{5}\\ 48 \\ \times\ 7 \\ \hline 336 \end{array} \longleftarrow 7 \cdot 8 = 56$$

Write 6 in the ones place and carry 5 to the tens place.

Next:

$$\begin{array}{r} {}^{5}\\ 48 \\ \times\ 7 \\ \hline 336 \end{array} \quad 7 \cdot 4 + 5 = 28 + 5 = 33$$

The product of 48 and 7 is 336.

Example 3 Multiply:

a. $\begin{array}{r} 25 \\ \times\ 8 \\ \hline \end{array}$ b. $\begin{array}{r} 246 \\ \times\ 5 \\ \hline \end{array}$

Solution:

a. $\begin{array}{r} {}^{4}\\ 25 \\ \times\ 8 \\ \hline 200 \end{array}$ b. $\begin{array}{r} {}^{23}\\ 246 \\ \times\ 5 \\ \hline 1230 \end{array}$

Work Practice 3

Practice 3

Multiply.

a. $\begin{array}{r} 29 \\ \times\ 6 \\ \hline \end{array}$ b. $\begin{array}{r} 648 \\ \times\ 5 \\ \hline \end{array}$

Answers

2. a. $6(4 + 5) = 6 \cdot 4 + 6 \cdot 5$ b. $30(2 + 3) = 30 \cdot 2 + 30 \cdot 3$ c. $7(2 + 8) = 7 \cdot 2 + 7 \cdot 8$
3. a. 174 b. 3240

To multiply larger whole numbers, use the following similar notation. Multiply 89×52.

Step 1

$$\begin{array}{r} \scriptstyle 1 \\ 89 \\ \times\ 52 \\ \hline 178 \end{array} \leftarrow \text{Multiply } 89 \times 2.$$

Step 2

$$\begin{array}{r} \scriptstyle 4 \\ 89 \\ \times\ 52 \\ \hline 178 \\ \times 4450 \\ \hline \end{array} \leftarrow \text{Multiply } 89 \times 50.$$

Step 3

$$\begin{array}{r} 89 \\ \times\ 52 \\ \hline 178 \\ 4450 \\ \hline 4628 \end{array} \quad \text{Add.}$$

The numbers 178 and 4450 are called **partial products.** The sum of the partial products, 4628, is the product of 89 and 52.

Practice 4

Multiply.

$$\begin{array}{r} 306 \\ \times\ 81 \\ \hline \end{array}$$

Example 4 Multiply: 236×86

Solution:

$$\begin{array}{rl} 236 & \\ \times\ \ 86 & \\ \hline 1416 & \leftarrow 6(236) \\ 18880 & \leftarrow 80(236) \\ \hline 20{,}296 & \text{Add.} \end{array}$$

Work Practice 4

Practice 5

Multiply.

$$\begin{array}{r} 726 \\ \times 142 \\ \hline \end{array}$$

Example 5 Multiply: 631×125

Solution:

$$\begin{array}{rl} 631 & \\ \times\ \ 125 & \\ \hline 3155 & \leftarrow 5(631) \\ 12620 & \leftarrow 20(631) \\ 63100 & \leftarrow 100(631) \\ \hline 78{,}875 & \text{Add.} \end{array}$$

Work Practice 5

✓**Concept Check** Find and explain the error in the following multiplication problem.

$$\begin{array}{r} 102 \\ \times\ 33 \\ \hline 306 \\ 306 \\ \hline 612 \end{array}$$

Answers

4. 24,786 **5.** 103,092

✓**Concept Check Answer**

$$\begin{array}{r} 102 \\ \times\ 33 \\ \hline 306 \\ 3060 \\ \hline 3366 \end{array}$$

Objective C Finding the Area of a Rectangle

A special application of multiplication is finding the **area** of a region. Area measures the amount of surface of a region. For example, we measure a plot of land or the living space of a home by its area. The figures on the next page show two examples of units of area measure. (A centimeter is a unit of length in the metric system.)

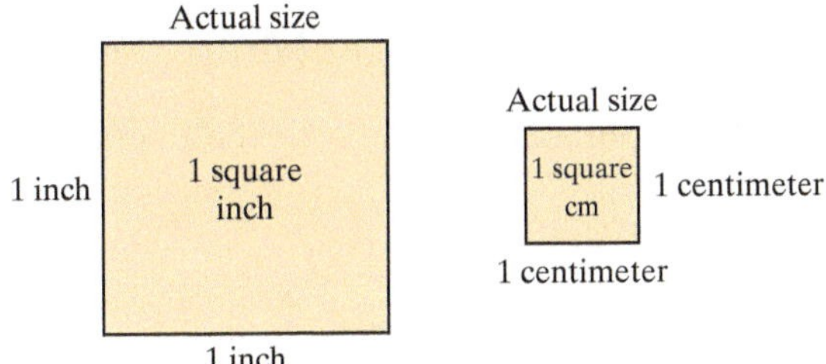

For example, to measure the area of a geometric figure such as the rectangle below, count the number of square units that cover the region.

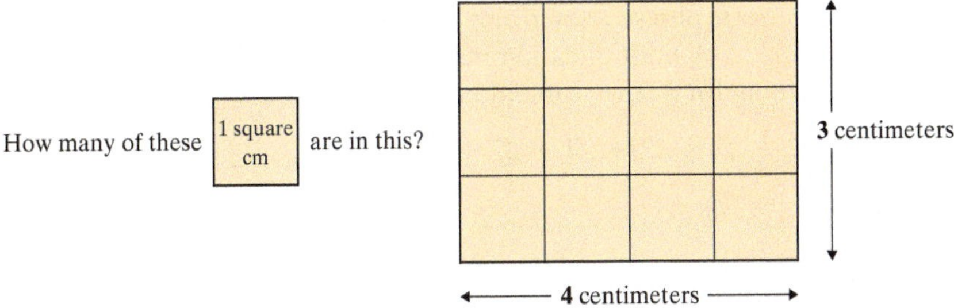

This rectangular region contains 12 square units, each 1 square centimeter. Thus, the area is 12 square centimeters. This total number of squares can be found by counting or by multiplying **4 · 3** (length · width).

$$\begin{aligned}\text{Area of a rectangle} &= \text{length} \cdot \text{width}\\ &= (4 \text{ centimeters})(3 \text{ centimeters})\\ &= 12 \text{ square centimeters}\end{aligned}$$

In this section, we find the areas of rectangles only. In later sections, we will find the areas of other geometric regions.

Helpful Hint

Notice that area is measured in **square** units while perimeter is measured in units.

Example 6 Finding the Area of a State

The state of Colorado is in the shape of a rectangle whose length is 380 miles and whose width is 280 miles. Find its area.

Solution: The area of a rectangle is the product of its length and its width.

$$\begin{aligned}\text{Area} &= \text{length} \cdot \text{width}\\ &= (380 \text{ miles})(280 \text{ miles})\\ &= 106{,}400 \text{ square miles}\end{aligned}$$

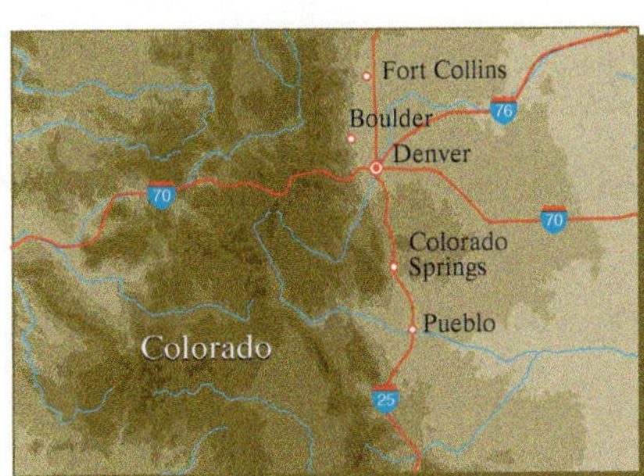

The area of Colorado is 106,400 square miles.

Work Practice 6

Practice 6

The state of Wyoming is in the shape of a rectangle whose length is 360 miles and whose width is 280 miles. Find its area.

Answer

6. 100,800 sq mi

Objective D Solving Problems by Multiplying

There are several words or phrases that indicate the operation of multiplication. Some of these are as follows:

Multiplication		
Key Words or Phrases	**Examples**	**Symbols**
multiply	multiply 5 by 7	$5 \cdot 7$
product	the product of 3 and 2	$3 \cdot 2$
times	10 times 13	$10 \cdot 13$

Many key words or phrases describing real-life problems that suggest addition might be better solved by multiplication instead. For example, to find the **total** cost of 8 shirts, each selling for $27, we can either add

$$27 + 27 + 27 + 27 + 27 + 27 + 27 + 27$$

or we can multiply 8(27).

Practice 7

A particular computer printer can print 16 pages per minute in color. How many pages can it print in 45 minutes?

Example 7 Finding DVD Space

A digital video disc (DVD) can hold about 4800 megabytes (MB) of information. How many megabytes can 12 DVDs hold?

Solution: Twelve DVDs will hold 12×4800 megabytes.

In Words		Translate to Numbers
megabytes per disc	→	4800
× DVDs	→	× 12
		9600
		48000
total megabytes		57,600

Twelve DVDs will hold 57,600 megabytes.

Work Practice 7

Practice 8

A professor of history purchased DVDs and CDs through a club. Each DVD was priced at $11 and each CD cost $9. He bought eight DVDs and five CDs. Find the total cost of the order.

Example 8 Budgeting Money

A woman and her friend plan to take their children to the Georgia Aquarium in Atlanta, the world's largest aquarium. The ticket price for each child is $22 and for each adult, $26. If five children and two adults plan to go, how much money is needed for admission? (*Source:* GeorgiaAquarium.org)

Solution: If the price of one child's ticket is $22, the price for 5 children is $5 \times 22 = \$110$. The price of one adult ticket is $26, so the price for two adults is $2 \times 26 = \$52$. The total cost is:

In Words		Translate to Numbers
cost for 5 children	→	110
+ cost for 2 adults	→	+ 52
total cost		162

The total cost is $162.

Work Practice 8

Answers
7. 720 pages 8. $133

Example 9 Estimating Word Count

The average page of a book contains 259 words. Estimate, rounding each number to the nearest hundred, the total number of words contained on 212 pages.

Solution: The exact number of words is 259×212. Estimate this product by rounding each factor to the nearest hundred.

$$\begin{array}{rcr} 259 & \text{rounds to} & 300 \\ \times\, 212 & \text{rounds to} & \times\, 200, \end{array} \quad 300 \times 200 = 60{,}000$$

$$3 \cdot 2 = 6$$

There are approximately 60,000 words contained on 212 pages.

Work Practice 9

Practice 9

If an average page in a book contains 163 words, estimate, rounding each number to the nearest hundred, the total number of words contained on 391 pages.

Answer

9. 80,000 words

Calculator Explorations Multiplying Numbers

To multiply numbers on a calculator, find the keys marked [×] and [=] or [ENTER]. For example, to find $31 \cdot 66$ on a calculator, press the keys [31] [×] [66] then [=] or [ENTER]. The display will read [2046]. Thus, $31 \cdot 66 = 2046$.

Use a calculator to multiply.

1. 72×48

2. 81×92

3. $163 \cdot 94$

4. $285 \cdot 144$

5. 983(277)

6. 1562(843)

Vocabulary, Readiness & Video Check

Use the choices below to fill in each blank.

area	grouping	commutative	1	product	length
factor	order	associative	0	distributive	number

1. The product of 0 and any number is ___________.

2. The product of 1 and any number is the ___________.

3. In $8 \cdot 12 = 96$, the 96 is called the ___________ and 8 and 12 are each called a(n) ___________.

4. Since $9 \cdot 10 = 10 \cdot 9$, we say that changing the ___________ in multiplication does not change the product. This property is called the ___________ property of multiplication.

5. Since $(3 \cdot 4) \cdot 6 = 3 \cdot (4 \cdot 6)$, we say that changing the ___________ in multiplication does not change the product. This property is called the ___________ property of multiplication.

6. ___________ measures the amount of surface of a region.

7. Area of a rectangle = ___________ · width.

8. We know $9(10 + 8) = 9 \cdot 10 + 9 \cdot 8$ by the ___________ property.

Martin-Gay Interactive Videos *Watch the section lecture video and answer the following questions.*

See Video 1.5

Objective A 9. The expression in Example 3 is rewritten using what property?

Objective B 10. During the multiplication process for Example 5, why is a single zero placed at the end of the second partial product?

Objective C 11. Why are the units to the answer to Example 6 not just meters? What are the correct units?

Objective D 12. In Example 7, why can "total" imply multiplication as well as addition?

1.5 Exercise Set MyMathLab®

Objective A *Multiply. See Example 1.*

1. $1 \cdot 24$
2. $55 \cdot 1$
3. $0 \cdot 19$
4. $27 \cdot 0$
5. $8 \cdot 0 \cdot 9$
6. $7 \cdot 6 \cdot 0$
7. $87 \cdot 1$
8. $1 \cdot 41$

Use the distributive property to rewrite each expression. See Example 2.

9. $6(3 + 8)$
10. $5(8 + 2)$
11. $4(3 + 9)$
12. $6(1 + 4)$
13. $20(14 + 6)$
14. $12(12 + 3)$

Objective B *Multiply. See Example 3.*

15. $\begin{array}{r} 64 \\ \times\ 8 \\ \hline \end{array}$
16. $\begin{array}{r} 79 \\ \times\ 3 \\ \hline \end{array}$
17. $\begin{array}{r} 613 \\ \times\ \ 6 \\ \hline \end{array}$
18. $\begin{array}{r} 638 \\ \times\ \ 5 \\ \hline \end{array}$
19. 277×6
20. 882×2
21. 1074×6
22. 9021×3

Objectives A B Mixed Practice *Multiply. See Examples 1 through 5.*

23. $\begin{array}{r} 89 \\ \times 13 \\ \hline \end{array}$
24. $\begin{array}{r} 91 \\ \times 72 \\ \hline \end{array}$
25. $\begin{array}{r} 421 \\ \times\ 58 \\ \hline \end{array}$
26. $\begin{array}{r} 526 \\ \times\ 23 \\ \hline \end{array}$
27. $\begin{array}{r} 306 \\ \times\ 81 \\ \hline \end{array}$
28. $\begin{array}{r} 708 \\ \times\ 21 \\ \hline \end{array}$
29. $(780)(20)$
30. $(720)(80)$
31. $(495)(13)(0)$
32. $(593)(47)(0)$
33. $(640)(1)(10)$

34. (240)(1)(20) **35.** 1234×39 **36.** 1357×79 **37.** 609×234 **38.** 807×127

39. $\begin{array}{r} 8649 \\ \times\ 274 \\ \hline \end{array}$ **40.** $\begin{array}{r} 1234 \\ \times\ 567 \\ \hline \end{array}$ **41.** $\begin{array}{r} 589 \\ \times 110 \\ \hline \end{array}$ **42.** $\begin{array}{r} 426 \\ \times 110 \\ \hline \end{array}$ **43.** $\begin{array}{r} 1941 \\ \times 2035 \\ \hline \end{array}$ **44.** $\begin{array}{r} 1876 \\ \times 1407 \\ \hline \end{array}$

Objective C Mixed Practice (*Section 1.3*) *Find the area and the perimeter of each rectangle. See Example 6.*

45.

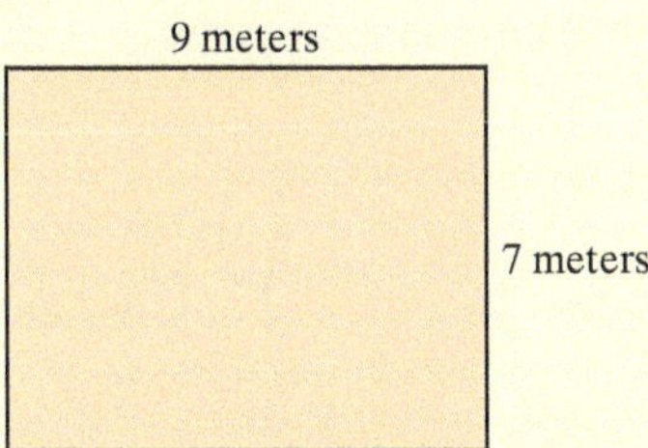

△ **46.**

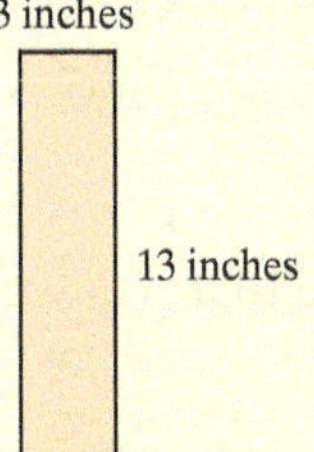

△ **47.**

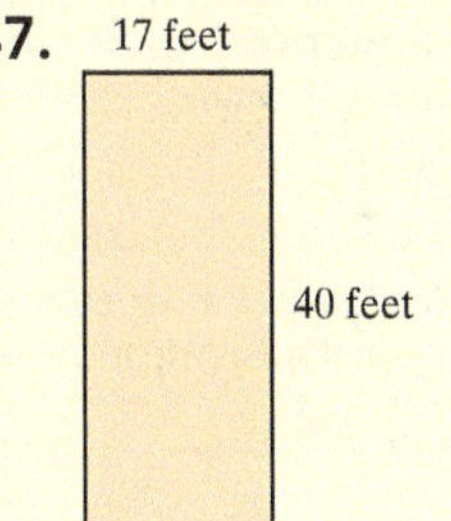

△ **48.**

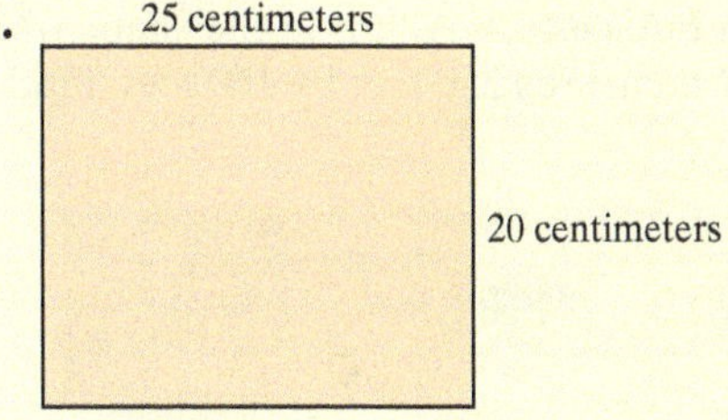

Objective D *Estimate the products by rounding each factor to the nearest hundred. See Example 9.*

49. 576×354 **50.** 982×650 **51.** 604×451 **52.** 111×999

Without actually calculating, mentally round, multiply, and choose the best estimate.

53. $38 \times 42 =$
a. 16
b. 160
c. 1600
d. 16,000

54. $2872 \times 12 =$
a. 2872
b. 28,720
c. 287,200
d. 2,872,000

55. $612 \times 29 =$
a. 180
b. 1800
c. 18,000
d. 180,000

56. $706 \times 409 =$
a. 280
b. 2800
c. 28,000
d. 280,000

Objectives C D Mixed Practice–Translating *Solve. See Examples 6 through 9.*

57. Multiply 80 by 11.

58. Multiply 70 by 12.

59. Find the product of 6 and 700.

60. Find the product of 9 and 900.

61. Find 2 times 2240.

62. Find 3 times 3310.

63. One tablespoon of olive oil contains 125 calories. How many calories are in 3 tablespoons of olive oil? (*Source: Home and Garden Bulletin No. 72,* U.S. Department of Agriculture).

64. One ounce of hulled sunflower seeds contains 14 grams of fat. How many grams of fat are in 8 ounces of hulled sunflower seeds? (*Source: Home and Garden Bulletin No. 72,* U.S. Department of Agriculture).

65. The textbook for a course in biology costs \$94. There are 35 students in the class. Find the total cost of the biology books for the class.

66. The seats in a large lecture hall are arranged in 14 rows with 34 seats in each row. Find how many seats are in this room.

67. Cabot Creamery is packing a pallet of 20-lb boxes of cheddar cheese to send to a local restaurant. There are five layers of boxes on the pallet, and each layer is four boxes wide by five boxes deep.

a. How many boxes are in one layer?

b. How many boxes are on the pallet?

c. What is the weight of the cheese on the pallet?

68. An apartment building has *three floors*. Each floor has five rows of apartments with four apartments in each row.

a. How many apartments are on 1 floor?

b. How many apartments are in the building?

△ **69.** A plot of land measures 80 feet by 110 feet. Find its area.

△ **70.** A house measures 45 feet by 60 feet. Find the floor area of the house.

△ **71.** The largest hotel lobby can be found at the Hyatt Regency in San Francisco, CA. It is in the shape of a rectangle that measures 350 feet by 160 feet. Find its area.

△ **72.** Recall from an earlier section that the largest commercial building in the world under one roof is the flower auction building of the cooperative VBA in Aalsmeer, Netherlands. The floor plan is a rectangle that measures 776 meters by 639 meters. Find the area of this building. (A meter is a unit of length in the metric system.) (*Source: The Handy Science Answer Book,* Visible Ink Press)

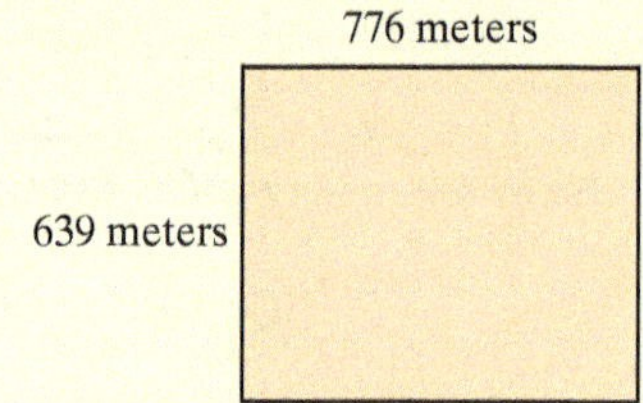

73. A pixel is a rectangular dot on a graphing calculator screen. If a graphing calculator screen contains 62 pixels in a row and 94 pixels in a column, find the total number of pixels on a screen.

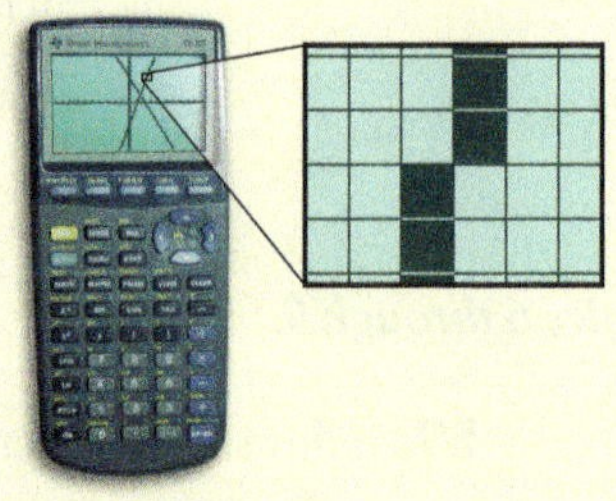

74. A certain compact disc (CD) can hold 700 megabytes (MB) of information. How many MBs can 17 discs hold?

75. A line of print on a computer contains 60 characters (letters, spaces, punctuation marks). Find how many characters there are in 35 lines.

76. An average cow eats 3 pounds of grain per day. Find how much grain a cow eats in a year. (Assume 365 days in 1 year.)

77. One ounce of Planters® Dry Roasted Peanuts has 160 calories. How many calories are in 8 ounces? (*Source:* RJR Nabisco, Inc.)

78. One ounce of Planters® Dry Roasted Peanuts has 13 grams of fat. How many grams of fat are in 16 ounces? (*Source:* RJR Nabisco, Inc.)

79. The Thespian club at a local community college is ordering T-shirts. T-shirts size S, M, or L cost $10 each and T-shirts size XL or XXL cost $12 each. Complete the table below and use it to find the total cost. (The first row is filled in for you.)

T-Shirt Size	Number of Shirts Ordered	Cost per Shirt	Cost per Size Ordered
S	4	$10	$40
M	6		
L	20		
XL	3		
XXL	3		

80. The student activities group at North Shore Community College is planning a trip to see the local minor league baseball team. Tickets cost $5 for students, $7 for nonstudents, and $2 for children under 12. Complete the following table and use it to find the total cost.

Person	Number of Persons	Cost per Person	Cost per Category
Student	24	$5	$120
Nonstudent	4		
Children under 12	5		

81. Celestial Seasonings of Boulder, Colorado, is a tea company that specializes in herbal teas. Their plant in Boulder has bagging machines capable of bagging over 1000 bags of tea per minute. If the plant runs 24 hours day, how many tea bags are produced in one day? (*Source:* Celestial Seasonings)

82. There were about 3 million "older" Americans (ages 65 and older) in 1900. By 2020, this number is projected to increase eighteen times. Find this projected number of "older" Americans in 2020. (*Source:* Administration on Aging, U.S. Census Bureau)

Mixed Practice (*Sections 1.3, 1.5*) *Perform each indicated operation.*

83. $\begin{array}{r} 128 \\ +\ 7 \\ \hline \end{array}$

84. $\begin{array}{r} 126 \\ -\ 8 \\ \hline \end{array}$

85. $\begin{array}{r} 134 \\ \times\ 16 \\ \hline \end{array}$

86. $47 + 26 + 10 + 231 + 50$

87. Find the sum of 19 and 4.

88. Find the product of 19 and 4.

89. Find the difference of 19 and 4.

90. Find the total of 14 and 9.

Concept Extensions

Solve. See the first Concept Check in this section.

91. Rewrite $6 + 6 + 6 + 6 + 6$ using multiplication.

92. Rewrite $11 + 11 + 11 + 11 + 11 + 11$ using multiplication.

93. **a.** Rewrite $3 \cdot 5$ as repeated addition.
b. Explain why there is more than one way to do this.

94. **a.** Rewrite $4 \cdot 5$ as repeated addition.
b. Explain why there is more than one way to do this.

Find and explain the error in each multiplication problem. See the second Concept Check in this section.

95.
```
   203
 × 14
   812
   203
  1015
```

96.
```
   31
 × 50
  155
```

Fill in the missing digits in each problem.

97.
```
    4_
 × _3
   126
  3780
  3906
```

98.
```
   _7
 × 6_
   171
  3420
  3591
```

99. Explain how to multiply two 2-digit numbers using partial products.

100. In your own words, explain the meaning of the area of a rectangle and how this area is measured.

101. A window washer in New York City is bidding for a contract to wash the windows of a 23-story building. To write a bid, the number of windows in the building is needed. If there are 7 windows in each row of windows on 2 sides of the building and 4 windows per row on the other 2 sides of the building, find the total number of windows.

102. During the NBA's 2012–2013 regular season, Carmelo Anthony of the New York Knicks scored 157 three-point field goals, 512 two-point field goals, and 425 free throws (worth one point each). How many points did Carmelo Anthony score during the 2012–2013 regular season? (*Source:* NBA)

1.6 Dividing Whole Numbers 

Objectives

A Divide Whole Numbers.

B Perform Long Division.

C Solve Problems That Require Dividing by Whole Numbers.

D Find the Average of a List of Numbers.

Suppose three people pooled their money and bought a raffle ticket at a local fund-raiser. Their ticket was the winner and they won a $75 cash prize. They then divided the prize into three equal parts so that each person received $25.

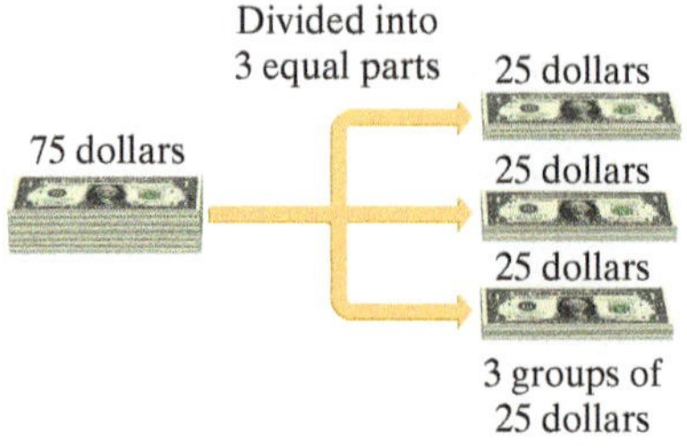

Objective A Dividing Whole Numbers 

The process of separating a quantity into equal parts is called **division.** The division above can be symbolized by several notations.

quotient → 25
$3\overline{)75}$ ← dividend
↑ divisor

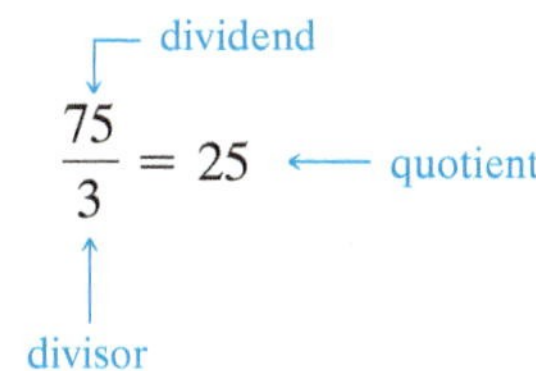

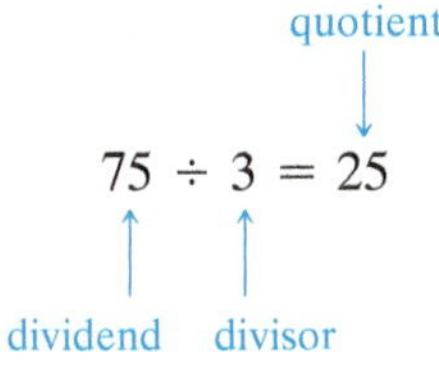

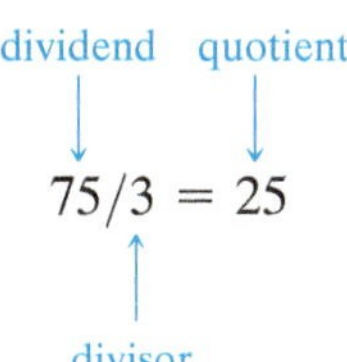

(In the notation $\frac{75}{3}$, the bar separating 75 and 3 is called a **fraction bar.**) Just as subtraction is the reverse of addition, division is the reverse of multiplication. This means that division can be checked by multiplication.

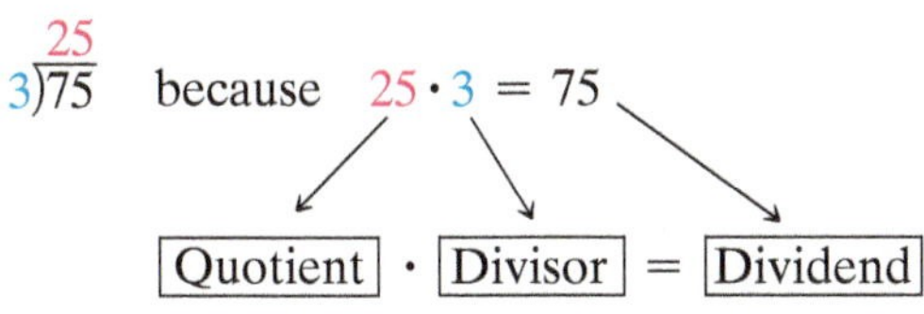

Since multiplication and division are related in this way, you can use your knowledge of multiplication facts to review quotients of one-digit divisors if necessary.

Example 1 Find each quotient. Check by multiplying.

a. $42 \div 7$ **b.** $\frac{64}{8}$ **c.** $3\overline{)21}$

Solution:

a. $42 \div 7 = 6$ because $6 \cdot 7 = 42$

b. $\frac{64}{8} = 8$ because $8 \cdot 8 = 64$

c. $3\overline{)21}$ = 7 because $7 \cdot 3 = 21$

Work Practice 1

Practice 1

Find each quotient. Check by multiplying.

a. $9\overline{)72}$

b. $40 \div 5$

c. $\frac{24}{6}$

Example 2 Find each quotient. Check by multiplying.

a. $1\overline{)7}$ **b.** $12 \div 1$ **c.** $\frac{6}{6}$ **d.** $9 \div 9$ **e.** $\frac{20}{1}$ **f.** $18\overline{)18}$

Solution:

a. $1\overline{)7}$ = 7 because $7 \cdot 1 = 7$

b. $12 \div 1 = 12$ because $12 \cdot 1 = 12$

c. $\frac{6}{6} = 1$ because $1 \cdot 6 = 6$

d. $9 \div 9 = 1$ because $1 \cdot 9 = 9$

e. $\frac{20}{1} = 20$ because $20 \cdot 1 = 20$

f. $18\overline{)18}$ = 1 because $1 \cdot 18 = 18$

Work Practice 2

Practice 2

Find each quotient. Check by multiplying.

a. $\frac{7}{7}$ **b.** $5 \div 1$

c. $1\overline{)11}$ **d.** $4 \div 1$

e. $\frac{10}{1}$ **f.** $21 \div 21$

Answers

1. a. 8 **b.** 8 **c.** 4 **2. a.** 1 **b.** 5 **c.** 11 **d.** 4 **e.** 10 **f.** 1

Example 2 illustrates the important properties of division described next:

Division Properties of 1

The quotient of any number (except 0) and that same number is 1. For example,

$$8 \div 8 = 1 \qquad \frac{5}{5} = 1 \qquad 4\overline{)4}^{\,1}$$

The quotient of any number and 1 is that same number. For example,

$$9 \div 1 = 9 \qquad \frac{6}{1} = 6 \qquad 1\overline{)3}^{\,3} \qquad \frac{0}{1} = 0$$

Practice 3

Find each quotient. Check by multiplying.

a. $\frac{0}{7}$ **b.** $8\overline{)0}$

c. $7 \div 0$ **d.** $0 \div 14$

Example 3 Find each quotient. Check by multiplying.

a. $9\overline{)0}$ **b.** $0 \div 12$ **c.** $\frac{0}{5}$ **d.** $\frac{3}{0}$

Solution:

a. $9\overline{)0}^{\,0}$ because $0 \cdot 9 = 0$

b. $0 \div 12 = 0$ because $0 \cdot 12 = 0$

c. $\frac{0}{5} = 0$ because $0 \cdot 5 = 0$

d. If $\frac{3}{0}$ = a *number*, then the *number* times $0 = 3$. Recall from Section 1.5 that any number multiplied by 0 is 0 and not 3. We say, then, that $\frac{3}{0}$ is **undefined.**

Work Practice 3

Example 3 illustrates important division properties of 0.

Division Properties of 0

The quotient of 0 and any number (except 0) is 0. For example,

$$0 \div 9 = 0 \quad \frac{0}{5} = 0 \quad 14\overline{)0}^{\,0}$$

The quotient of any number and 0 is not a number. We say that

$$\frac{3}{0}, \quad 0\overline{)3}, \quad \text{and} \quad 3 \div 0$$

are **undefined.**

Objective B Performing Long Division 

When dividends are larger, the quotient can be found by a process called **long division.** For example, let's divide 2541 by 3.

divisor → $3\overline{)2541}$ ← dividend

We can't divide 3 into 2, so we try dividing 3 into the first two digits.

$3\overline{)2541}$ with 8 over the 5: $25 \div 3 = 8$ with 1 left, so our best estimate is 8. We place 8 over the 5 in 25.

Answers

3. a. 0 **b.** 0 **c.** undefined **d.** 0

Next, multiply 8 and 3 and subtract this product from 25. Make sure that this difference is less than the divisor.

```
    8
 3)2541
  -24       8(3) = 24
   1        25 - 24 = 1, and 1 is less than the divisor 3.
```

Bring down the next digit and go through the process again.

```
    84      14 ÷ 3 = 4 with 2 left
 3)2541
  -24↓
    14
   -12      4(3) = 12
     2      14 - 12 = 2
```

Once more, bring down the next digit and go through the process.

```
    847     21 ÷ 3 = 7
 3)2541
  -24
    14
   -12↓
     21
    -21     7(3) = 21
      0     21 - 21 = 0
```

The quotient is 847. To check, see that $847 \times 3 = 2541$.

Example 4 Divide: $3705 \div 5$. Check by multiplying.

Solution:

```
    7       37 ÷ 5 = 7 with 2 left. Place this estimate, 7, over the 7 in 37.
 5)3705
  -35↓      7(5) = 35
    20      37 - 35 = 2, and 2 is less than the divisor 5.
            Bring down the 0.

    74      20 ÷ 5 = 4
 5)3705
  -35
    20
   -20↓     4(5) = 20
    05      20 - 20 = 0, and 0 is less than the divisor 5.
            Bring down the 5.

    741     5 ÷ 5 = 1
 5)3705
  -35
    20
   -20↓
      5
     -5     1(5) = 5
      0     5 - 5 = 0
```

(Continued on next page)

Practice 4

Divide. Check by multiplying.

a. $4908 \div 6$

b. $2212 \div 4$

c. $753 \div 3$

Answers

4. a. 818 **b.** 553 **c.** 251

Since division and multiplication are reverse operations, don't forget that a division problem can be checked by multiplying.

Practice 5

Divide and check by multiplying.

a. $7\overline{)2128}$

b. $9\overline{)45{,}900}$

Check:

$$\begin{array}{r} 741 \\ \times\ \ 5 \\ \hline 3705 \end{array}$$

Work Practice 4

Example 5 Divide and check: $1872 \div 9$

Solution:

$$\begin{array}{rl} 208 & \\ 9\overline{)1872} & \\ -18\downarrow & 2(9) = 18 \\ \hline 07 & 18 - 18 = 0\text{; bring down the 7.} \\ -0\downarrow & 0(9) = 0 \\ \hline 72 & 7 - 0 = 7\text{; bring down the 2.} \\ -72 & 8(9) = 72 \\ \hline 0 & 72 - 72 = 0 \end{array}$$

Check: $208 \cdot 9 = 1872$

Work Practice 5

Naturally, quotients don't always "come out even." Making 4 rows out of 26 chairs, for example, isn't possible if each row is supposed to have exactly the same number of chairs. Each of 4 rows can have 6 chairs, but 2 chairs are still left over.

We signify "leftovers" or **remainders** in this way:

$$\begin{array}{r} 6\ \text{R}\,2 \\ 4\overline{)26} \end{array}$$

The **whole number part of the quotient** is 6; the **remainder part of the quotient** is 2. Checking by multiplying,

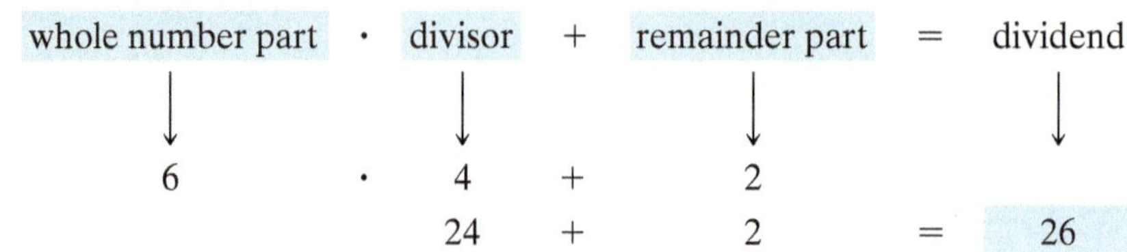

whole number part	·	divisor	+	remainder part	=	dividend
↓		↓		↓		↓
6	·	4	+	2		
		24	+	2	=	26

Answers

5. a. 304 **b.** 5100

Example 6 Divide and check: $2557 \div 7$

Solution:

$$\begin{array}{r} 365 \text{ R } 2 \\ 7\overline{)2557} \\ -21 \\ \hline 45 \\ -42 \\ \hline 37 \\ -35 \\ \hline 2 \end{array}$$

3(7) = 21
25 − 21 = 4; bring down the 5.
6(7) = 42
45 − 42 = 3; bring down the 7.
5(7) = 35
37 − 35 = 2; the remainder is 2.

Check: $365 \cdot 7 + 2 = 2557$

whole number part · divisor + remainder part = dividend

Work Practice 6

Practice 6

Divide and check.

a. $4\overline{)939}$

b. $5\overline{)3287}$

Example 7 Divide and check: $56{,}717 \div 8$

Solution:

$$\begin{array}{r} 7089 \text{ R } 5 \\ 8\overline{)56717} \\ -56 \\ \hline 07 \\ -0 \\ \hline 71 \\ -64 \\ \hline 77 \\ -72 \\ \hline 5 \end{array}$$

7(8) = 56
Subtract and bring down the 7.
0(8) = 0
Subtract and bring down the 1.
8(8) = 64
Subtract and bring down the 7.
9(8) = 72
Subtract. The remainder is 5.

Check: $7089 \cdot 8 + 5 = 56{,}717$

whole number part · divisor + remainder part = dividend

Work Practice 7

Practice 7

Divide and check.

a. $9\overline{)81{,}605}$

b. $4\overline{)23{,}310}$

When the divisor has more than one digit, the same pattern applies. For example, let's find $1358 \div 23$.

$$\begin{array}{r} 5 \\ 23\overline{)1358} \\ -115 \\ \hline 208 \end{array}$$

135 ÷ 23 = 5 with 20 left over. Our estimate is 5.
5(23) = 115
135 − 115 = 20. Bring down the 8.

(Continued on next page)

Answers

6. a. 234 R 3 **b.** 657 R 2
7. a. 9067 R 2 **b.** 5827 R 2

Now we continue estimating.

```
     59  R 1     208 ÷ 23 = 9 with 1 left over.
23)1358
  −115
    208
   −207          9(23) = 207
      1          208 − 207 = 1. The remainder is 1.
```

To check, see that $59 \cdot 23 + 1 = 1358$.

Practice 8

Divide: $8920 \div 17$

Example 8 Divide: $6819 \div 17$

Solution:

```
     401  R 2
17)6819
  −68↓|          4(17) = 68
    01|          Subtract and bring down the 1.
    −0↓          0(17) = 0
     19          Subtract and bring down the 9.
    −17          1(17) = 17
      2          Subtract. The remainder is 2.
```

To check, see that $401 \cdot 17 + 2 = 6819$.

Work Practice 8

Practice 9

Divide: $33{,}282 \div 678$

Example 9 Divide: $51{,}600 \div 403$

Solution:

```
      128  R 16
403)51600
   −403↓|        1(403) = 403
    1130|        Subtract and bring down the 0.
    −806↓        2(403) = 806
     3240        Subtract and bring down the 0.
    −3224        8(403) = 3224
       16        Subtract. The remainder is 16.
```

To check, see that $128 \cdot 403 + 16 = 51{,}600$.

Work Practice 9

Division Shown as Repeated Subtraction To further understand division, recall from Section 1.5 that addition and multiplication are related in the following manner:

$$\underbrace{3 + 3 + 3 + 3}_{\text{4 addends; each addend is 3}} = 4 \times 3 = 12$$

In other words, multiplication is repeated addition. Likewise, division is repeated subtraction.

For example, let's find

$35 \div 8$

Answers

8. 524 R 12 **9.** 49 R 60

by repeated subtraction. Keep track of the number of times 8 is subtracted from 35. We are through when we can subtract no more because the difference is less than 8.

35 ÷ 8: Repeated subtraction

35 −8	1 time
27 −8	2 times
19 −8	3 times
11 −8	4 times
3	⟵ Remainder (We cannot subtract 8 again.)

35 dollars → 8 dollars 1 time; 8 dollars 2 times; 8 dollars 3 times; 8 dollars 4 times; 3 dollars left over

Thus, 35 ÷ 8 = 4 R 3.

To check, perform the same multiplication as usual, but finish by adding in the remainder.

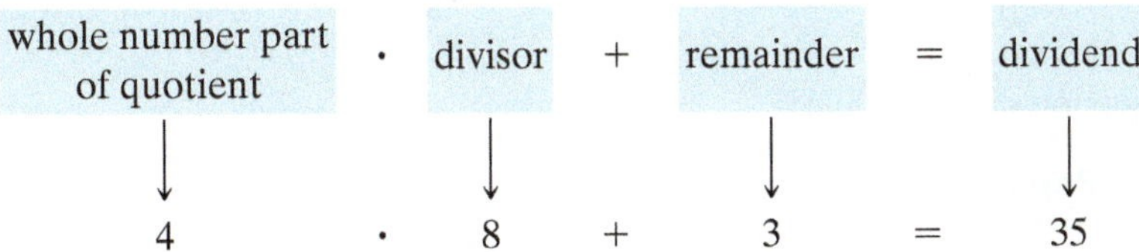

Objective C Solving Problems by Dividing

Below are some key words and phrases that may indicate the operation of division:

Division

Key Words or Phrases	Examples	Symbols
divide	divide 10 by 5	$10 \div 5$ or $\frac{10}{5}$
quotient	the quotient of 64 and 4	$64 \div 4$ or $\frac{64}{4}$
divided by	9 divided by 3	$9 \div 3$ or $\frac{9}{3}$
divided or shared equally among	$100 divided equally among five people	$100 \div 5$ or $\frac{100}{5}$
per	100 miles per 2 hours	$\frac{100 \text{ miles}}{2 \text{ hours}}$

✓**Concept Check** Which of the following is the correct way to represent "the quotient of 60 and 12"? Or are both correct? Explain your answer.

a. 12 ÷ 60

b. 60 ÷ 12

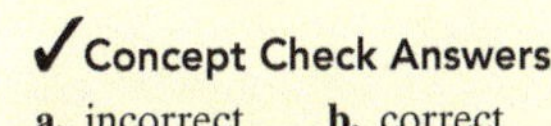
✓Concept Check Answers
a. incorrect **b.** correct

Practice 10

Three students bought 171 blank CDs to share equally. How many CDs did each person get?

Example 10 Finding Shared Earnings

Three college freshmen share a paper route to earn money for expenses. The total in their fund after expenses is \$2895. How much is each person's equal share?

Solution:

In words:	Each person's share	=	total money	÷	number of persons
	↓		↓		↓
Translate:	Each person's share	=	2895	÷	3

Then

$$\begin{array}{r} 965 \\ 3\overline{)2895} \\ -27 \\ \hline 19 \\ -18 \\ \hline 15 \\ -15 \\ \hline 0 \end{array}$$

Each person's share is \$965.

Work Practice 10

Practice 11

Printers can be packed 12 to a box. If 532 printers are to be packed but only full boxes are shipped, how many full boxes will be shipped? How many printers are left over and not shipped?

Example 11 Dividing Number of Downloads

As part of a promotion, an executive receives 238 cards, each good for one free song download. If she wants to share them evenly with 19 friends, how many download cards will each friend receive? How many will be left over?

Solution:

In words:	Number of cards for each person	=	number of cards	÷	number of friends
	↓		↓		↓
Translate:	Number of cards for each person	=	238	÷	19

$$\begin{array}{r} 12 \text{ R } 10 \\ 19\overline{)238} \\ -19 \\ \hline 48 \\ -38 \\ \hline 10 \end{array}$$

Each friend will receive 12 download cards. The cards cannot be divided equally among her friends since there is a nonzero remainder. There will be 10 download cards left over.

Work Practice 11

Answers
10. 57 CDs
11. 44 full boxes; 4 printers left over

Objective D Finding Averages

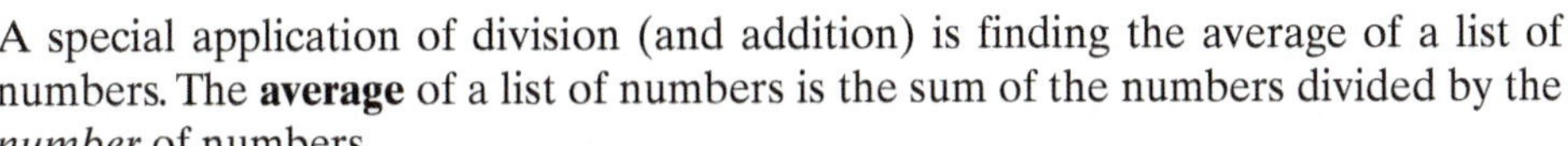

A special application of division (and addition) is finding the average of a list of numbers. The **average** of a list of numbers is the sum of the numbers divided by the *number* of numbers.

$$\text{average} = \frac{\text{sum of numbers}}{\textit{number}\text{ of numbers}}$$

Example 12 Averaging Scores

A mathematics instructor is checking a simple program she wrote for averaging the scores of her students. To do so, she averages a student's scores of 75, 96, 81, and 88 by hand. Find this average score.

Solution: To find the average score, we find the sum of the student's scores and divide by 4, the number of scores.

$$\begin{array}{r} 75 \\ 96 \\ 81 \\ +88 \\ \hline 340 \end{array} \text{ sum} \qquad \text{average} = \frac{340}{4} = 85 \qquad \begin{array}{r} 85 \\ 4\overline{)340} \\ -32 \\ \hline 20 \\ -20 \\ \hline 0 \end{array}$$

The average score is 85.

Work Practice 12

Practice 12

To compute a safe time to wait for reactions to occur after allergy shots are administered, a lab technician is given a list of elapsed times between administered shots and reactions. Find the average of the times 4 minutes, 7 minutes, 35 minutes, 16 minutes, 9 minutes, 3 minutes, and 52 minutes.

Answer

12. 18 min

Calculator Explorations Dividing Numbers

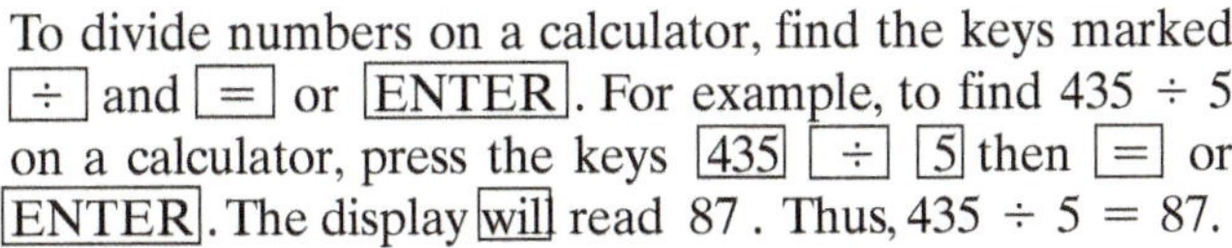

To divide numbers on a calculator, find the keys marked [÷] and [=] or [ENTER]. For example, to find 435 ÷ 5 on a calculator, press the keys [435] [÷] [5] then [=] or [ENTER]. The display will read 87 . Thus, 435 ÷ 5 = 87.

Use a calculator to divide.

1. 848 ÷ 16

2. 564 ÷ 12

3. $95\overline{)5890}$

4. $27\overline{)1053}$

5. $\frac{32,886}{126}$

6. $\frac{143,088}{264}$

7. 0 ÷ 315

8. 315 ÷ 0

Vocabulary, Readiness & Video Check

Use the choices below to fill in each blank. Some choices may be used more than once.

1	number	divisor	dividend
0	undefined	average	quotient

1. In 90 ÷ 2 = 45, the answer 45 is called the ________, 90 is called the ________, and 2 is called the ________.

2. The quotient of any number and 1 is the same ________.

3. The quotient of any number (except 0) and the same number is ________.

4. The quotient of 0 and any number (except 0) is ________.

5. The quotient of any number and 0 is ________.

6. The ________ of a list of numbers is the sum of the numbers divided by the ________ of numbers.

Martin-Gay Interactive Videos

See Video 1.6

Watch the section lecture video and answer the following questions.

Objective A 7. Look at Examples 6–8. What number can never be the divisor in division?

Objective B 8. In Example 10, how many 102s are in 21? How does this result affect the quotient?

9. What calculation would you use to check the answer in Example 10?

Objective C 10. In Example 11, what is the importance of knowing that the distance to each hole is the same?

Objective D 11. As shown in Example 12, what two operations are used when finding an average?

1.6 Exercise Set MyMathLab®

Objective A *Find each quotient. See Examples 1 through 3.*

1. $54 \div 9$ **2.** $72 \div 9$ **3.** $36 \div 3$ **4.** $24 \div 3$ **5.** $0 \div 8$

6. $0 \div 4$ **7.** $31 \div 1$ **8.** $38 \div 1$ **9.** $\frac{18}{18}$ **10.** $\frac{49}{49}$

11. $\frac{24}{3}$ **12.** $\frac{45}{9}$ **13.** $26 \div 0$ **14.** $\frac{12}{0}$ **15.** $26 \div 26$

16. $6 \div 6$ **17.** $0 \div 14$ **18.** $7 \div 0$ **19.** $18 \div 2$ **20.** $18 \div 3$

Objectives A B Mixed Practice *Divide and then check by multiplying. See Examples 1 through 5.*

21. $3\overline{)87}$ **22.** $5\overline{)85}$ **23.** $3\overline{)222}$ **24.** $8\overline{)640}$ **25.** $3\overline{)1014}$ **26.** $4\overline{)2104}$

27. $\frac{30}{0}$ **28.** $\frac{0}{30}$ **29.** $63 \div 7$ **30.** $56 \div 8$ **31.** $150 \div 6$ **32.** $121 \div 11$

Divide and then check by multiplying. See Examples 6 and 7.

33. $7\overline{)479}$ **34.** $7\overline{)426}$ **35.** $6\overline{)1421}$ **36.** $3\overline{)1240}$

37. $305 \div 8$ **38.** $167 \div 3$ **39.** $2286 \div 7$ **40.** $3333 \div 4$

Divide and then check by multiplying. See Examples 8 and 9.

41. $55\overline{)715}$ **42.** $23\overline{)736}$ **43.** $23\overline{)1127}$ **44.** $42\overline{)2016}$ **45.** $97\overline{)9417}$

46. $44\overline{)1938}$ **47.** $3146 \div 15$ **48.** $7354 \div 12$ **49.** $6578 \div 13$ **50.** $5670 \div 14$

51. $9299 \div 46$ **52.** $2505 \div 64$ **53.** $\frac{12,744}{236}$ **54.** $\frac{5781}{123}$ **55.** $\frac{10,297}{103}$

56. $\frac{23,092}{240}$ **57.** $20,619 \div 102$ **58.** $40,853 \div 203$ **59.** $244,989 \div 423$ **60.** $164,592 \div 543$

Divide. See Examples 1 through 9.

61. $7\overline{)119}$ **62.** $8\overline{)104}$ **63.** $7\overline{)3580}$ **64.** $5\overline{)3017}$

65. $40\overline{)85,312}$ **66.** $50\overline{)85,747}$ **67.** $142\overline{)863,360}$ **68.** $214\overline{)650,560}$

Objective C Translating *Solve. See Examples 10 and 11.*

69. Find the quotient of 117 and 5.

70. Find the quotient of 94 and 7.

71. Find 200 divided by 35.

72. Find 116 divided by 32.

73. Find the quotient of 62 and 3.

74. Find the quotient of 78 and 5.

Solve.

75. Martin Thieme teaches American Sign Language classes for $65 per student for a 7-week session. He collects $2145 from the group of students. Find how many students are in the group.

76. Kathy Gomez teaches Spanish lessons for $85 per student for a 5-week session. From one group of students, she collects $4930. Find how many students are in the group.

77. The gravity of Jupiter is 318 times as strong as the gravity of Earth, so objects on Jupiter weigh 318 times as much as they weigh on Earth. If a person would weigh 52,470 pounds on Jupiter, find how much the person weighs on Earth.

78. Twenty-one people pooled their money and bought lottery tickets. One ticket won a prize of $5,292,000. Find how many dollars each person received.

79. An 18-hole golf course is 5580 yards long. If the distance to each hole is the same, find the distance between holes.

80. A truck hauls wheat to a storage granary. It carries a total of 5768 bushels of wheat in 14 trips. How much does the truck haul each trip if each trip it hauls the same amount?

81. There is a bridge over highway I-35 every three miles. The first bridge is at the beginning of a 265-mile stretch of highway. Find how many bridges there are over 265 miles of I-35.

82. The white stripes dividing the lanes on a highway are 25 feet long, and the spaces between them are 25 feet long. Let's call a "lane divider" a stripe followed by a space. Find how many whole "lane dividers" there are in 1 mile of highway. (A mile is 5280 feet.)

83. Ari Trainor is in the requisitions department of Central Electric Lighting Company. Light poles along a highway are placed 492 feet apart. The first light pole is at the beginning of a 1-mile strip. Find how many poles he should order for the 1-mile strip of highway. (A mile is 5280 feet.)

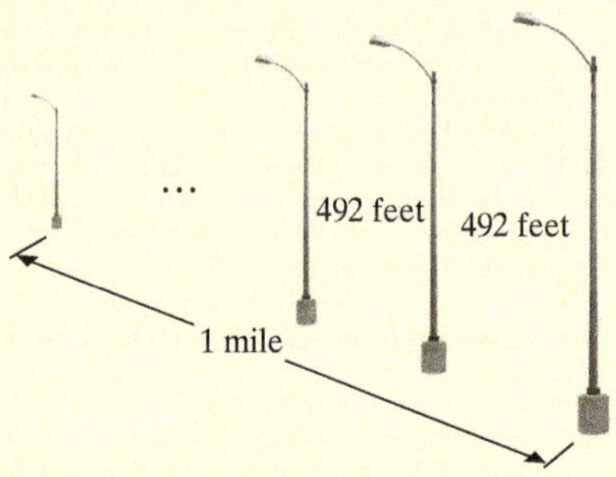

84. Professor Lopez has a piece of rope 185 feet long that she wants to cut into pieces for an experiment in her physics class. Each piece of rope is to be 8 feet long. Determine whether she has enough rope for her 22-student class. Determine the amount extra or the amount short.

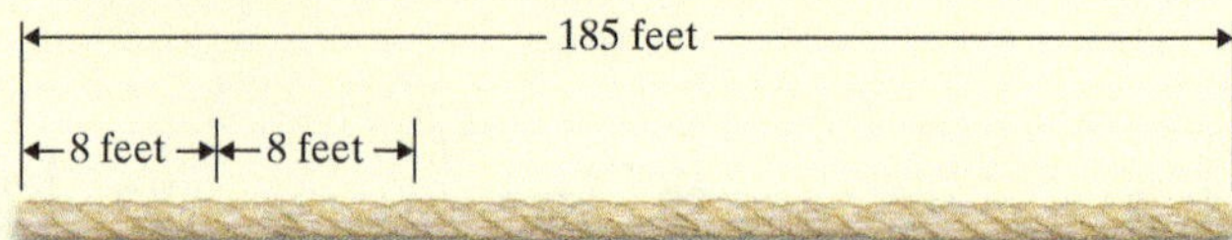

85. Broad Peak in Pakistan is the twelfth-tallest mountain in the world. Its elevation is 26,400 feet. A mile is 5280 feet. How many miles tall is Broad Peak? (*Source:* National Geographic Society)

86. Randy Moss of the New England Patriots led the NFL in touchdowns during the 2007 regular football season, scoring a total of 138 points from touchdowns. If a touchdown is worth 6 points, how many touchdowns did Moss make during the 2007 season? (*Source:* NFL)

87. Find how many yards are in 1 mile. (A mile is 5280 feet; a yard is 3 feet.)

1 foot | 1 foot | 1 foot | 1 foot | 1 foot | 1 foot | ...

1 yard | 1 yard | ...

5280 feet

88. Find how many whole feet are in 1 rod. (A mile is 5280 feet; 1 mile is 320 rods.)

Objective D *Find the average of each list of numbers. See Example 12.*

89. 10, 24, 35, 22, 17, 12

90. 37, 26, 15, 29, 51, 22

91. 205, 972, 210, 161

92. 121, 200, 185, 176, 163

93. 86, 79, 81, 69, 80

94. 92, 96, 90, 85, 92, 79

The normal monthly temperatures in degrees Fahrenheit for Salt Lake City, Utah, are given in the graph. Use this graph to answer Exercises 95 and 96. (Source: National Climatic Data Center*)*

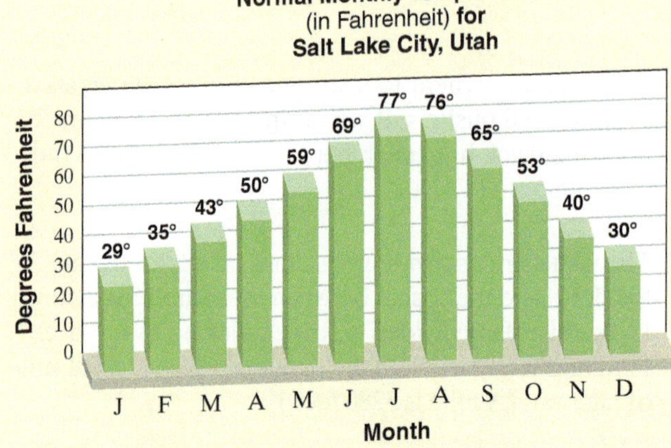

95. Find the average temperature for June, July, and August.

96. Find the average temperature for October, November, and December.

Mixed Practice (Sections 1.3, 1.5, 1.6) *Perform each indicated operation. Watch the operation symbol.*

97. $82 + 463 + 29 + 8704$

98. $23 + 407 + 92 + 7011$

99. 546×28

100. 712×54

101. $722 - 43$

102. $712 - 54$

103. $\frac{45}{0}$

104. $\frac{0}{23}$

105. $228 \div 24$

106. $304 \div 31$

Concept Extensions

Match each word phrase to the correct translation. (Not all letter choices will be used.) See the Concept Check in this section.

107. The quotient of 40 and 8

108. The quotient of 200 and 20

109. 200 divided by 20

110. 40 divided by 8

a. $20 \div 200$ **b.** $200 \div 20$
c. $40 \div 8$ **d.** $8 \div 40$

The following table shows the top five leading U.S. television advertisers during the first half of 2013 and the amount of money spent that half-year on advertising. Use this table to answer Exercises 111 and 112. (Source: Local Media Marketing Solutions*)*

Advertiser	Amount Spent on Television Advertising in 1st half of 2013
Ford Motor Company, dealers and corporate	$191,055,900
AT&T Inc.	$170,634,500
Comcast Corp	$157,780,500
Toyota, Dealers and corporate	$153,946,200
Chrysler-Cerberus	$122,784,200

111. Find the average amount of money spent on television ads for the half-year by the top two advertisers.

112. Find the average amount of money spent on television advertising by the top four advertisers.

In Example 12 in this section, we found that the average of 75, 96, 81, and 88 is 85. Use this information to answer Exercises 113 and 114.

113. If the number 75 is removed from the list of numbers, does the average increase or decrease? Explain why.

114. If the number 96 is removed from the list of numbers, does the average increase or decrease? Explain why.

115. Without computing it, tell whether the average of 126, 135, 198, and 113 is 86. Explain why it is possible or why it is not.

116. Without computing it, tell whether the average of 38, 27, 58, and 43 is 17. Explain why it is possible or why it is not.

117. If the area of a rectangle is 60 square feet and its width is 5 feet, what is its length?

118. If the area of a rectangle is 84 square inches and its length is 21 inches, what is its width?

119. Write down any two numbers whose quotient is 25.

120. Write down any two numbers whose quotient is 1.

121. Find $26 \div 5$ using the process of repeated subtraction.

122. Find $86 \div 10$ using the process of repeated subtraction.

Integrated Review Sections 1.1–1.6

Operations on Whole Numbers

Answers

1. ____
2. ____
3. ____
4. ____
5. ____
6. ____
7. ____
8. ____
9. ____
10. ____
11. ____
12. ____
13. ____
14. ____
15. ____
16. ____
17. ____
18. ____
19. ____
20. ____
21. ____
22. ____
23. ____
24. ____
25. ____
26. ____
27. ____
28. ____
29. ____
30. ____

Perform each indicated operation.

1. $\begin{array}{r} 42 \\ 63 \\ +\ 89 \\ \hline \end{array}$

2. $\begin{array}{r} 7006 \\ -\ 451 \\ \hline \end{array}$

3. $\begin{array}{r} 87 \\ \times 52 \\ \hline \end{array}$

4. $8\overline{)4496}$

5. $1 \cdot 67$

6. $\frac{36}{0}$

7. $16 \div 16$

8. $5 \div 1$

9. $0 \cdot 21$

10. $7 \cdot 0 \cdot 8$

11. $0 \div 7$

12. $12 \div 4$

13. $9 \cdot 7$

14. $45 \div 5$

15. $\begin{array}{r} 207 \\ -\ 69 \\ \hline \end{array}$

16. $\begin{array}{r} 207 \\ +\ 69 \\ \hline \end{array}$

17. $3718 - 2549$

18. $1861 + 7965$

19. $7\overline{)1278}$

20. $\begin{array}{r} 1259 \\ \times\ \ 63 \\ \hline \end{array}$

21. $7\overline{)7695}$

22. $9\overline{)1000}$

23. $32\overline{)21,240}$

24. $65\overline{)70,000}$

25. $4000 - 2963$

26. $10,000 - 101$

27. $\begin{array}{r} 303 \\ \times 101 \\ \hline \end{array}$

28. $(475)(100)$

29. Find the total of 62 and 9.

30. Find the product of 62 and 9.

31. Find the quotient of 62 and 9.

32. Find the difference of 62 and 9.

33. Subtract 17 from 200.

34. Find the difference of 432 and 201.

Complete the table by rounding the given number to the given place value.

		Tens	Hundrds	Thousands
35.	9735			
36.	1429			
37.	20,801			
38.	432,198			

Find the perimeter and area of each figure.

△ **39.** Square 6 feet

△ **40.** 14 inches Rectangle 7 inches

Find the perimeter of each figure.

△ **41.**

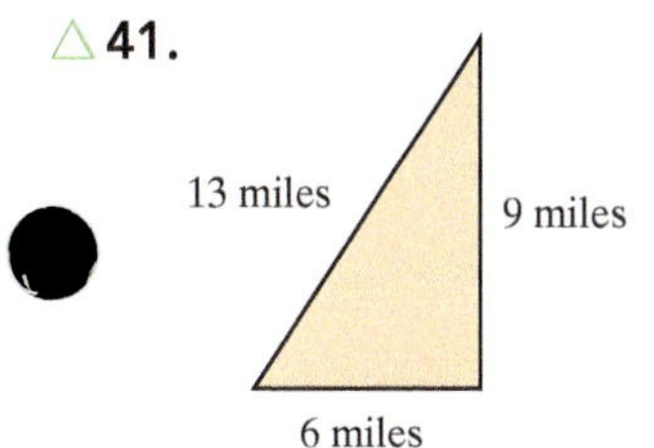

△ **42.** 3 meters 4 meters 3 meters 3 meters

Find the average of each list of numbers.

43. 19, 15, 25, 37, 24

44. 108, 131, 98, 159

45. The Mackinac Bridge is a suspension bridge that connects the lower and upper peninsulas of Michigan across the Straits of Mackinac. Its total length is 26,372 feet. The Lake Pontchartrain Bridge is a twin concrete trestle bridge in Slidell, Louisiana. Its total length is 28,547 feet. Which bridge is longer and by how much? (*Sources:* Mackinac Bridge Authority and Federal Highway Administration, Bridge Division)

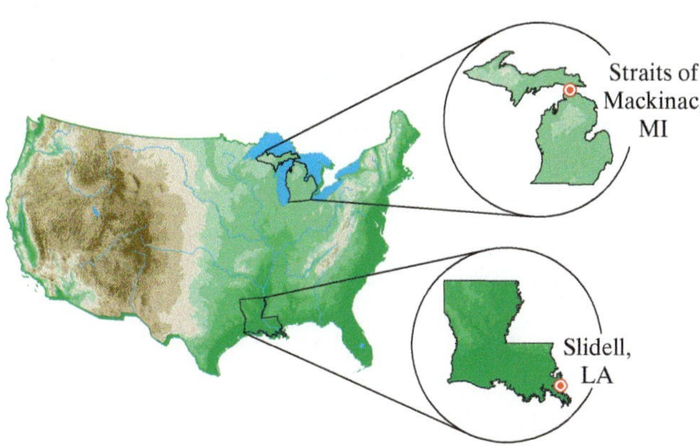

46. In the United States, the average toy expenditure per child is $309 per year. On average, how much is spent on toys for a child by the time he or she reaches age 18? (*Source:* statista)

31. ______
32. ______
33. ______
34. ______
35. ______
36. ______
37. ______
38. ______
39. ______
40. ______
41. ______
42. ______
43. ______
44. ______
45. ______
46. ______

1.7 Exponents and Order of Operations 

Objectives

A Write Repeated Factors Using Exponential Notation.

B Evaluate Expressions Containing Exponents.

C Use the Order of Operations.

D Find the Area of a Square.

Objective A Using Exponential Notation

In the product $3 \cdot 3 \cdot 3 \cdot 3 \cdot 3$, notice that 3 is a factor several times. When this happens, we can use a shorthand notation, called an **exponent,** to write the repeated multiplication.

$\underbrace{3 \cdot 3 \cdot 3 \cdot 3 \cdot 3}_{\text{3 is a factor 5 times}}$ can be written as

3^5 (exponent: 5; base: 3) Read as "three to the fifth power."

This is called **exponential notation.** The **exponent,** 5, indicates how many times the **base,** 3, is a factor.

The table below shows examples of reading exponential notation in words.

Expression	In Words
5^2	"five to the second power" or "five squared"
5^3	"five to the third power" or "five cubed"
5^4	"five to the fourth power"

Usually, an exponent of 1 is not written, so when no exponent appears, we assume that the exponent is 1. For example, $2 = 2^1$ and $7 = 7^1$.

Practice 1–4

Write using exponential notation.

1. $8 \cdot 8 \cdot 8 \cdot 8$
2. $3 \cdot 3 \cdot 3$
3. $10 \cdot 10 \cdot 10 \cdot 10 \cdot 10$
4. $5 \cdot 5 \cdot 4 \cdot 4 \cdot 4 \cdot 4 \cdot 4 \cdot 4$

Examples Write using exponential notation.

1. $7 \cdot 7 \cdot 7 = 7^3$
2. $3 \cdot 3 = 3^2$
3. $6 \cdot 6 \cdot 6 \cdot 6 \cdot 6 = 6^5$
4. $3 \cdot 3 \cdot 3 \cdot 3 \cdot 9 \cdot 9 \cdot 9 = 3^4 \cdot 9^3$

Work Practice 1–4

Objective B Evaluating Exponential Expressions

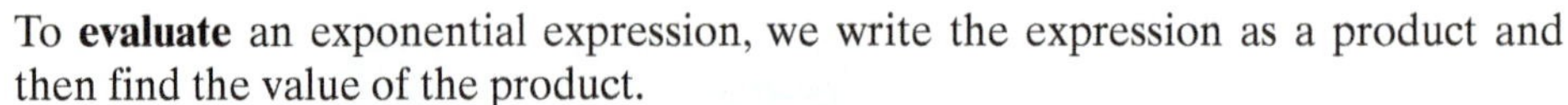

To **evaluate** an exponential expression, we write the expression as a product and then find the value of the product.

Practice 5–8

Evaluate.

5. 4^2
6. 7^3
7. 11^1
8. $2 \cdot 3^2$

Answers

1. 8^4 2. 3^3 3. 10^5 4. $5^2 \cdot 4^6$
5. 16 6. 343 7. 11 8. 18

Examples Evaluate.

5. $9^2 = 9 \cdot 9 = 81$
6. $6^1 = 6$
7. $3^4 = 3 \cdot 3 \cdot 3 \cdot 3 = 81$
8. $5 \cdot 6^2 = 5 \cdot 6 \cdot 6 = 180$

Work Practice 5–8

Example 8 illustrates an important property: An exponent applies only to its base. The exponent 2, in $5 \cdot 6^2$, applies only to its base, 6.

Helpful Hint

An exponent applies only to its base. For example, $4 \cdot 2^3$ means $4 \cdot 2 \cdot 2 \cdot 2$.

Helpful Hint

Don't forget that 2^4, for example, is *not* $2 \cdot 4$. The expression 2^4 means repeated multiplication of the same factor.

$$2^4 = 2 \cdot 2 \cdot 2 \cdot 2 = 16, \quad \text{whereas } 2 \cdot 4 = 8$$

✓Concept Check Which of the following statements is correct?

a. 3^5 is the same as $5 \cdot 5 \cdot 5$.
b. "Ten cubed" is the same as 10^2.
c. "Six to the fourth power" is the same as 6^4.
d. 12^2 is the same as $12 \cdot 2$.

Objective C Using the Order of Operations

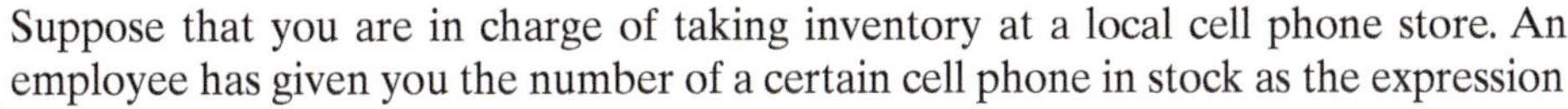

Suppose that you are in charge of taking inventory at a local cell phone store. An employee has given you the number of a certain cell phone in stock as the expression

$$6 + 2 \cdot 30$$

To calculate the value of this expression, do you add first or multiply first? If you add first, the answer is 240. If you multiply first, the answer is 66.

Mathematical symbols wouldn't be very useful if two values were possible for one expression. Thus, mathematicians have agreed that, given a choice, we multiply first.

$$\begin{aligned} 6 + 2 \cdot 30 &= 6 + 60 \quad \text{Multiply.} \\ &= 66 \quad \text{Add.} \end{aligned}$$

This agreement is one of several **order of operations** agreements.

Order of Operations

1. Perform all operations within parentheses ( ), brackets [ ], or other grouping symbols such as fraction bars, starting with the innermost set.
2. Evaluate any expressions with exponents.
3. Multiply or divide in order from left to right.
4. Add or subtract in order from left to right.

✓Concept Check Answer
c

Below we practice using order of operations to simplify expressions.

Practice 9

Simplify: $9 \cdot 3 - 8 \div 4$

Example 9 Simplify: $2 \cdot 4 - 3 \div 3$

Solution: There are no parentheses and no exponents, so we start by multiplying and dividing, from left to right.

$$\begin{aligned} 2 \cdot 4 - 3 \div 3 &= 8 - 3 \div 3 && \text{Multiply.} \\ &= 8 - 1 && \text{Divide.} \\ &= 7 && \text{Subtract.} \end{aligned}$$

Work Practice 9

Practice 10

Simplify: $48 \div 3 \cdot 2^2$

Example 10 Simplify: $4^2 \div 2 \cdot 4$

Solution: We start by evaluating 4^2.

$$4^2 \div 2 \cdot 4 = 16 \div 2 \cdot 4 \quad \text{Write } 4^2 \text{ as 16.}$$

Next we multiply or divide *in order* from left to right. Since division appears before multiplication from left to right, we divide first, then multiply.

$$\begin{aligned} 16 \div 2 \cdot 4 &= 8 \cdot 4 && \text{Divide.} \\ &= 32 && \text{Multiply.} \end{aligned}$$

Work Practice 10

Practice 11

Simplify: $(10 - 7)^4 + 2 \cdot 3^2$

Example 11 Simplify: $(8 - 6)^2 + 2^3 \cdot 3$

Solution:

$$\begin{aligned} (8 - 6)^2 + 2^3 \cdot 3 &= 2^2 + 2^3 \cdot 3 && \text{Simplify inside parentheses.} \\ &= 4 + 8 \cdot 3 && \text{Write } 2^2 \text{ as 4 and } 2^3 \text{ as 8.} \\ &= 4 + 24 && \text{Multiply.} \\ &= 28 && \text{Add.} \end{aligned}$$

Work Practice 11

Practice 12

Simplify:

$36 \div [20 - (4 \cdot 2)] + 4^3 - 6$

Example 12 Simplify: $4^3 + [3^2 - (10 \div 2)] - 7 \cdot 3$

Solution: Here we begin with the innermost set of parentheses.

$$\begin{aligned} 4^3 + [3^2 - (10 \div 2)] - 7 \cdot 3 &= 4^3 + [3^2 - 5] - 7 \cdot 3 && \text{Simplify inside parentheses.} \\ &= 4^3 + [9 - 5] - 7 \cdot 3 && \text{Write } 3^2 \text{ as 9.} \\ &= 4^3 + 4 - 7 \cdot 3 && \text{Simplify inside brackets.} \\ &= 64 + 4 - 7 \cdot 3 && \text{Write } 4^3 \text{ as 64.} \\ &= 64 + 4 - 21 && \text{Multiply.} \\ &= 47 && \text{Add and subtract from left to right.} \end{aligned}$$

Work Practice 12

Answers

9. 25 **10.** 64 **11.** 99 **12.** 61

Example 13 Simplify: $\frac{7 - 2\cdot3 + 3^2}{5(2 - 1)}$

Solution: Here, the fraction bar is a grouping symbol. We simplify above and below the fraction bar separately.

$$\frac{7 - 2\cdot3 + 3^2}{5(2 - 1)} = \frac{7 - 2\cdot3 + 9}{5(1)}$$ Evaluate 3^2 and $(2 - 1)$.

$$= \frac{7 - 6 + 9}{5}$$ Multiply $2\cdot3$ in the numerator and multiply 5 and 1 in the denominator.

$$= \frac{10}{5}$$ Add and subtract from left to right.

$$= 2$$ Divide.

Work Practice 13

Practice 13

Simplify: $\frac{25 + 8\cdot2 - 3^3}{2(3 - 2)}$

Example 14 Simplify: $64 \div 8\cdot2 + 4$

Solution: $64 \div 8\cdot2 + 4 = 8\cdot2 + 4$ Divide.
$= 16 + 4$ Multiply.
$= 20$ Add.

Work Practice 14

Practice 14

Simplify: $36 \div 6\cdot3 + 5$

Objective D Finding the Area of a Square

Since a square is a special rectangle, we can find its area by finding the product of its length and its width.

Area of a rectangle = length · width

By recalling that each side of a square has the same measurement, we can use the following procedure to find its area:

Area of a square = length · width
= side · side
= $(\text{side})^2$

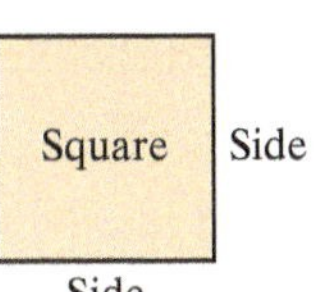

Helpful Hint

Recall from Section 1.5 that area is measured in **square** units while perimeter is measured in units.

Example 15 Find the area of a square whose side measures 4 inches.

Solution: Area of a square = $(\text{side})^2$
= $(4 \text{ inches})^2$
= 16 square inches

4 inches

The area of the square is 16 square inches.

Work Practice 15

Practice 15

Find the area of a square whose side measures 12 centimeters.

Answers
13. 7 **14.** 23 **15.** 144 sq cm

Calculator Explorations Exponents

To evaluate an exponential expression such as 4^7 on a calculator, find the keys marked $\boxed{y^x}$ or $\boxed{\wedge}$ and $\boxed{=}$ or $\boxed{\text{ENTER}}$. To evaluate 4^7, press the keys $\boxed{4}$ $\boxed{y^x}$ (or $\boxed{\wedge}$) $\boxed{7}$ then $\boxed{=}$ or $\boxed{\text{ENTER}}$. The display will read $\boxed{\quad 16384}$. Thus, $4^7 = 16{,}384$.

Use a calculator to evaluate.

1. 4^6 **2.** 5^6 **3.** 5^5

4. 7^6 **5.** 2^{11} **6.** 6^8

Order of Operations

To see whether your calculator has the order of operations built in, evaluate $5 + 2 \cdot 3$ by pressing the keys $\boxed{5}$ $\boxed{+}$ $\boxed{2}$ $\boxed{\times}$ $\boxed{3}$ then $\boxed{=}$ or $\boxed{\text{ENTER}}$. If the display reads $\boxed{11}$, your calculator does have the order of operations built in. This means that most of the time, you can key in a problem exactly as it is written, and the calculator will perform operations in the proper order. When evaluating an expression containing parentheses, key in the parentheses. (If an expression contains brackets, key in parentheses.) For example, to evaluate $2[25 - (8 + 4)] - 11$, press the keys $\boxed{2}$ $\boxed{\times}$ $\boxed{(}$ $\boxed{25}$ $\boxed{-}$ $\boxed{(}$ $\boxed{8}$ $\boxed{+}$ $\boxed{4}$ $\boxed{)}$ $\boxed{)}$ $\boxed{-}$ $\boxed{11}$ then $\boxed{=}$ or $\boxed{\text{ENTER}}$.

The display will read $\boxed{\quad 15}$.

Use a calculator to evaluate.

7. $7^4 + 5^3$

8. $12^4 - 8^4$

9. $63 \cdot 75 - 43 \cdot 10$

10. $8 \cdot 22 + 7 \cdot 16$

11. $4(15 \div 3 + 2) - 10 \cdot 2$

12. $155 - 2(17 + 3) + 185$

Vocabulary, Readiness & Video Check

Use the choices below to fill in each blank.

addition	multiplication	exponent
subtraction	division	base

1. In $2^5 = 32$, the 2 is called the __________ and the 5 is called the __________.

2. To simplify $8 + 2 \cdot 6$, which operation should be performed first? __________

3. To simplify $(8 + 2) \cdot 6$, which operation should be performed first? __________

4. To simplify $9(3 - 2) \div 3 + 6$, which operation should be performed first? __________

5. To simplify $8 \div 2 \cdot 6$, which operation should be performed first? __________

Martin-Gay Interactive Videos

See Video 1.7

Watch the section lecture video and answer the following questions.

Objective A 6. In the Example 1 expression, what is the 3 called and what is the 12 called?

Objective B 7. As mentioned in Example 4, what "understood exponent" does any number we've worked with before have?

Objective C 8. List the three operations needed to evaluate Example 7 in the order they should be performed.

Objective D 9. As explained in the lecture before Example 10, why does the area of a square involve an exponent whereas the area of a rectangle usually does not?

1.7 Exercise Set MyMathLab® 

Objective A *Write using exponential notation. See Examples 1 through 4.*

1. $4 \cdot 4 \cdot 4$
2. $5 \cdot 5 \cdot 5 \cdot 5$
3. $7 \cdot 7 \cdot 7 \cdot 7 \cdot 7 \cdot 7$
4. $6 \cdot 6 \cdot 6 \cdot 6 \cdot 6 \cdot 6 \cdot 6$
5. $12 \cdot 12 \cdot 12$
6. $10 \cdot 10 \cdot 10$
7. $6 \cdot 6 \cdot 5 \cdot 5 \cdot 5$
8. $4 \cdot 4 \cdot 3 \cdot 3 \cdot 3$
9. $9 \cdot 8 \cdot 8$
10. $7 \cdot 4 \cdot 4 \cdot 4$
11. $3 \cdot 2 \cdot 2 \cdot 2 \cdot 2$
12. $4 \cdot 6 \cdot 6 \cdot 6 \cdot 6$
13. $3 \cdot 2 \cdot 2 \cdot 2 \cdot 2 \cdot 5 \cdot 5 \cdot 5 \cdot 5 \cdot 5$
14. $6 \cdot 6 \cdot 2 \cdot 9 \cdot 9 \cdot 9 \cdot 9$

Objective B *Evaluate. See Examples 5 through 8.*

15. 8^2
16. 6^2
17. 5^3
18. 6^3
19. 2^5
20. 3^5
21. 1^{10}
22. 1^{12}
23. 7^1
24. 8^1
25. 2^7
26. 5^4
27. 2^8
28. 3^3
29. 4^4
30. 4^3
31. 9^3
32. 8^3
33. 12^2
34. 11^2
35. 10^2
36. 10^3
37. 20^1
38. 14^1
39. 3^6
40. 4^5
41. $3 \cdot 2^6$
42. $5 \cdot 3^2$
43. $2 \cdot 3^4$
44. $2 \cdot 7^2$

Objective C *Simplify. See Examples 9 through 14.*

45. $15 + 3 \cdot 2$
46. $24 + 6 \cdot 3$
47. $14 \div 7 \cdot 2 + 3$
48. $100 \div 10 \cdot 5 + 4$
49. $32 \div 4 - 3$
50. $42 \div 7 - 6$
51. $13 + \frac{24}{8}$
52. $32 + \frac{8}{2}$
53. $6 \cdot 5 + 8 \cdot 2$
54. $3 \cdot 4 + 9 \cdot 1$
55. $\frac{5 + 12 \div 4}{1^7}$
56. $\frac{6 + 9 \div 3}{3^2}$
57. $(7 + 5^2) \div 4 \cdot 2^3$
58. $6^2 \cdot (10 - 8)$
59. $5^2 \cdot (10 - 8) + 2^3 + 5^2$
60. $5^3 \div (10 + 15) + 9^2 + 3^3$
61. $\frac{18 + 6}{2^4 - 2^2}$
62. $\frac{40 + 8}{5^2 - 3^2}$
63. $(3 + 5) \cdot (9 - 3)$
64. $(9 - 7) \cdot (12 + 18)$
65. $\frac{7(9 - 6) + 3}{3^2 - 3}$

66. $\frac{5(12-7)-4}{5^2-18}$

67. $8 \div 0 + 37$

68. $18 - 7 \div 0$

69. $2^4 \cdot 4 - (25 \div 5)$

70. $2^3 \cdot 3 - (100 \div 10)$

71. $3^4 - [35 - (12 - 6)]$

72. $[40 - (8 - 2)] - 2^5$

73. $(7 \cdot 5) + [9 \div (3 \div 3)]$

74. $(18 \div 6) + [(3 + 5) \cdot 2]$

75. $8 \cdot [2^2 + (6 - 1) \cdot 2] - 50 \cdot 2$

76. $35 \div [3^2 + (9 - 7) - 2^2] + 10 \cdot 3$

77. $\frac{9^2 + 2^2 - 1^2}{8 \div 2 \cdot 3 \cdot 1 \div 3}$

78. $\frac{5^2 - 2^3 + 1^4}{10 \div 5 \cdot 4 \cdot 1 \div 4}$

79. $\frac{2 + 4^2}{5(20 - 16) - 3^2 - 5}$

80. $\frac{3 + 9^2}{3(10 - 6) - 2^2 - 1}$

81. $9 \div 3 + 5^2 \cdot 2 - 10$

82. $10 \div 2 + 3^3 \cdot 2 - 20$

83. $[13 \div (20 - 7) + 2^5] - (2 + 3)^2$

84. $[15 \div (11 - 6) + 2^2] + (5 - 1)^2$

85. $7^2 - \{18 - [40 \div (5 \cdot 1) + 2] + 5^2\}$

86. $29 - \{5 + 3[8 \cdot (10 - 8)] - 50\}$

△ **Objective D Mixed Practice (*Section 1.3*)** *Find the area and perimeter of each square. See Example 15.*

87.

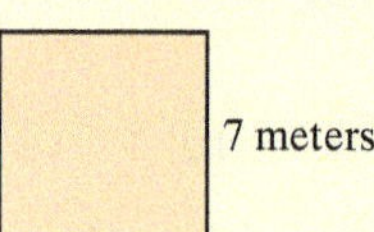

△ **88.**

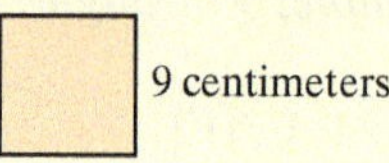

△ **89.**

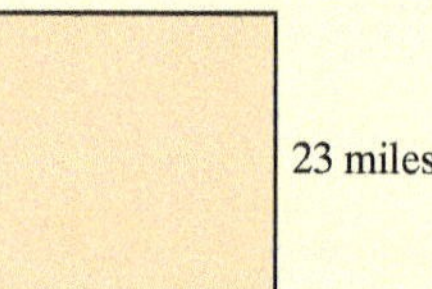

△ **90.**

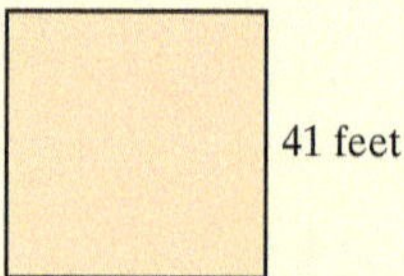

Concept Extensions

Answer the following true or false. See the Concept Check in this section.

91. "Six to the fifth power" is the same as 6^5.

92. "Seven squared" is the same as 7^2.

93. 2^5 is the same as $5 \cdot 5$.

94. 4^9 is the same as $4 \cdot 9$.

Insert grouping symbols (parentheses) so that each given expression evaluates to the given number.

95. $2 + 3 \cdot 6 - 2$; evaluates to 28

96. $2 + 3 \cdot 6 - 2$; evaluates to 20

97. $24 \div 3 \cdot 2 + 2 \cdot 5$; evaluates to 14

98. $24 \div 3 \cdot 2 + 2 \cdot 5$; evaluates to 15

△**99.** A building contractor is bidding on a contract to install gutters on seven homes in a retirement community, all in the shape shown. To estimate the cost of materials, she needs to know the total perimeter of all seven homes. Find the total perimeter.

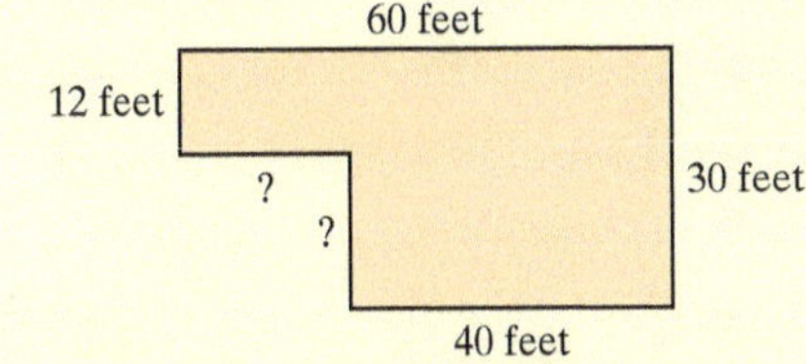

100. The building contractor from Exercise 99 plans to charge $4 per foot for installing vinyl gutters. Find the total charge for the seven homes given the total perimeter answer to Exercise 99.

Simplify.

101. $(7 + 2^4)^5 - (3^5 - 2^4)^2$

102. $25^3 \cdot (45 - 7 \cdot 5) \cdot 5$

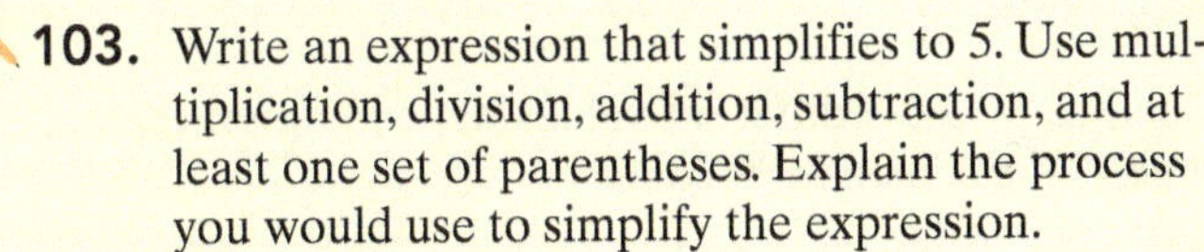

103. Write an expression that simplifies to 5. Use multiplication, division, addition, subtraction, and at least one set of parentheses. Explain the process you would use to simplify the expression.

104. Explain why $2 \cdot 3^2$ is not the same as $(2 \cdot 3)^2$.

1.8 Introduction to Variables, Algebraic Expressions, and Equations

Objective A Evaluating Algebraic Expressions 

Perhaps the most important quality of mathematics is that it is a science of patterns. Communicating about patterns is often made easier by using a letter to represent all the numbers fitting a pattern. We call such a letter a **variable.** For example, in Section 1.3 we presented the addition property of 0, which states that the sum of 0 and any number is that number. We might write

$0 + 1 = 1$
$0 + 2 = 2$
$0 + 3 = 3$
$0 + 4 = 4$
$0 + 5 = 5$
$0 + 6 = 6$
⋮

Objectives

A Evaluate Algebraic Expressions Given Replacement Values.

B Identify Solutions of Equations.

C Translate Phrases into Variable Expressions.

continuing indefinitely. This is a pattern, and all whole numbers fit the pattern. We can communicate this pattern for all whole numbers by letting a letter, such as a, represent all whole numbers. We can then write

$$0 + a = a$$

Using variable notation is a primary goal of learning **algebra.** We now take some important first steps in beginning to use variable notation.

A combination of operations on letters (variables) and numbers is called an **algebraic expression** or simply an **expression.**

Algebraic Expressions

$$3 + x \qquad 5 \cdot y \qquad 2 \cdot z - 1 + x$$

If two variables or a number and a variable are next to each other, with no operation sign between them, the operation is multiplication. For example,

$$2x \quad \text{means} \quad 2 \cdot x$$

and

$$xy \text{ or } x(y) \quad \text{means} \quad x \cdot y$$

Also, the meaning of an exponent remains the same when the base is a variable. For example,

$$x^2 = \underbrace{x \cdot x}_{\text{2 factors of } x} \quad \text{and} \quad y^5 = \underbrace{y \cdot y \cdot y \cdot y \cdot y}_{\text{5 factors of } y}$$

Algebraic expressions such as $3x$ have different values depending on replacement values for x. For example, if x is 2, then $3x$ becomes

$$3x = 3 \cdot 2$$
$$= 6$$

If x is 7, then $3x$ becomes

$$3x = 3 \cdot 7$$
$$= 21$$

Replacing a variable in an expression by a number and then finding the value of the expression is called **evaluating the expression** for the variable. When finding the value of an expression, remember to follow the order of operations given in Section 1.7.

Practice 1

Evaluate $x - 2$ if x is 7.

Example 1 Evaluate $x + 6$ if x is 8.

Solution: Replace x with 8 in the expression $x + 6$.

$$x + 6 = 8 + 6 \quad \text{Replace } x \text{ with 8.}$$
$$= 14 \quad \text{Add.}$$

Work Practice 1

When we write a statement such as "x is 5," we can use an equal sign ($=$) to represent "is" so that

x is 5 can be written as $x = 5$.

Answer

1. 5

Example 2 Evaluate $2(x - y)$ for $x = 6$ and $y = 3$.

Solution:

$$2(x - y) = 2(6 - 3) \quad \text{Replace } x \text{ with 6 and } y \text{ with 3.}$$
$$= 2(3) \quad \text{Subtract.}$$
$$= 6 \quad \text{Multiply.}$$

Work Practice 2

Practice 2

Evaluate $y(x - 3)$ for $x = 8$ and $y = 4$.

Example 3 Evaluate $\frac{x - 5y}{y}$ for $x = 35$ and $y = 5$.

Solution:

$$\frac{x - 5y}{y} = \frac{35 - 5(5)}{5} \quad \text{Replace } x \text{ with 35 and } y \text{ with 5.}$$
$$= \frac{35 - 25}{5} \quad \text{Multiply.}$$
$$= \frac{10}{5} \quad \text{Subtract.}$$
$$= 2 \quad \text{Divide.}$$

Work Practice 3

Practice 3

Evaluate $\frac{y + 6}{x}$ for $x = 6$ and $y = 18$.

Example 4 Evaluate $x^2 + z - 3$ for $x = 5$ and $z = 4$.

Solution:

$$x^2 + z - 3 = 5^2 + 4 - 3 \quad \text{Replace } x \text{ with 5 and } z \text{ with 4.}$$
$$= 25 + 4 - 3 \quad \text{Evaluate } 5^2.$$
$$= 26 \quad \text{Add and subtract from left to right.}$$

Work Practice 4

Practice 4

Evaluate $25 - z^3 + x$ for $z = 2$ and $x = 1$.

Helpful Hint

If you are having difficulty replacing variables with numbers, first replace each variable with a set of parentheses, then insert the replacement number within the parentheses.

Example:

$$x^2 + z - 3 = (\ )^2 + (\ ) - 3$$
$$= (5)^2 + (4) - 3$$
$$= 25 + 4 - 3$$
$$= 26$$

✓Concept Check What's wrong with the solution to the following problem?

Evaluate $3x + 2y$ for $x = 2$ and $y = 3$.

Solution:

$$3x + 2y = 3(3) + 2(2)$$
$$= 9 + 4$$
$$= 13$$

Answers

2. 20 **3.** 4 **4.** 18

✓Concept Check Answer

$$3x + 2y = 3(2) + 2(3)$$
$$= 6 + 6$$
$$= 12$$

Practice 5

Evaluate $\frac{5(F-32)}{9}$ for $F = 41$.

Example 5 The expression $\frac{5(F-32)}{9}$ can be used to write degrees Fahrenheit F as degrees Celsius C. Find the value of this expression for $F = 86$.

Solution:

$$\frac{5(F-32)}{9} = \frac{5(86-32)}{9} = \frac{5(54)}{9} = \frac{270}{9} = 30$$

Thus 86°F is the same temperature as 30°C.

■ Work Practice 5

Objective B Identifying Solutions of Equations 

In Objective **A**, we learned that a combination of operations on variables and numbers is called an algebraic expression or simply an expression. Frequently in this book, we have written statements like $7 + 4 = 11$ or area = length · width. Each of these statements is called an **equation.** An equation is of the form

expression = expression

Helpful Hint An equation contains "=," while an expression does not.

An equation can be labeled as

$$\underbrace{x + 7}_{\text{left side}} \overset{\text{equal sign}}{=} \underbrace{10}_{\text{right side}}$$

When an equation contains a variable, deciding which values of the variable make an equation a true statement is called **solving** an equation for the variable. A **solution** of an equation is a value for the variable that makes an equation a true statement. For example, 2 is a solution of the equation $x + 5 = 7$, since replacing x with 2 results in the *true* statement $2 + 5 = 7$. Similarly, 3 is not a solution of $x + 5 = 7$, since replacing x with 3 results in the *false* statement $3 + 5 = 7$.

Practice 6

Determine whether 8 is a solution of the equation $3(y - 6) = 6$.

Example 6 Determine whether 6 is a solution of the equation $4(x - 3) = 12$.

Solution: We replace x with 6 in the equation.

$4(x - 3) = 12$

$4(6 - 3) \stackrel{?}{=} 12$ Replace x with 6.

$4(3) \stackrel{?}{=} 12$

$12 = 12$ True

Since $12 = 12$ is a true statement, 6 *is* a solution of the equation.

■ Work Practice 6

Answers
5. 5 **6.** yes

A collection of numbers enclosed by braces is called a set. For example,

$$\{0, 1, 2, 3, \ldots\}$$

is the set of whole numbers that we are studying about in this chapter. The three dots after the number 3 in the set mean that this list of numbers continues in the same manner indefinitely.

The next example contains set notation.

Example 7 Determine which numbers in the set {26, 40, 20} are solutions of the equation $2n - 30 = 10$.

Solution: Replace n with each number from the set to see if a true statement results.

Let n be 26.	Let n be 40.	Let n be 20.
$2n - 30 = 10$	$2n - 30 = 10$	$2n - 30 = 10$
$2 \cdot 26 - 30 \stackrel{?}{=} 10$	$2 \cdot 40 - 30 \stackrel{?}{=} 10$	$2 \cdot 20 - 30 \stackrel{?}{=} 10$
$52 - 30 \stackrel{?}{=} 10$	$80 - 30 \stackrel{?}{=} 10$	$40 - 30 \stackrel{?}{=} 10$
$22 = 10$ False	$50 = 10$ False	$10 = 10$ True ✓

Thus, 20 is a solution while 26 and 40 are not solutions.

Work Practice 7

Practice 7

Determine which numbers in the set {10, 6, 8} are solutions of the equation $5n + 4 = 34$.

Objective C Translating Phrases into Variable Expressions

To aid us in solving problems later, we practice translating verbal phrases into algebraic expressions. Certain key words and phrases suggesting addition, subtraction, multiplication, or division are reviewed next.

Addition (+)	Subtraction (−)	Multiplication (·)	Division (÷)
sum	difference	product	quotient
plus	minus	times	divide
added to	subtract	multiply	shared equally among
more than	less than	multiply by	per
increased by	decreased by	of	divided by
total	less	double/triple	divided into

Example 8 Write as an algebraic expression. Use x to represent "a number."

a. 7 increased by a number
b. 15 decreased by a number
c. The product of 2 and a number
d. The quotient of a number and 5
e. 2 subtracted from a number

Solution:

a. In words: 7 | increased by | a number
Translate: 7 | + | x

(Continued on next page)

Practice 8

Write as an algebraic expression. Use x to represent "a number."

a. Twice a number
b. 8 increased by a number
c. 10 minus a number
d. 10 subtracted from a number
e. The quotient of 6 and a number

Answers

7. 6 is a solution.
8. a. $2x$ **b.** $8 + x$ **c.** $10 - x$ **d.** $x - 10$ **e.** $6 \div x$ or $\frac{6}{x}$

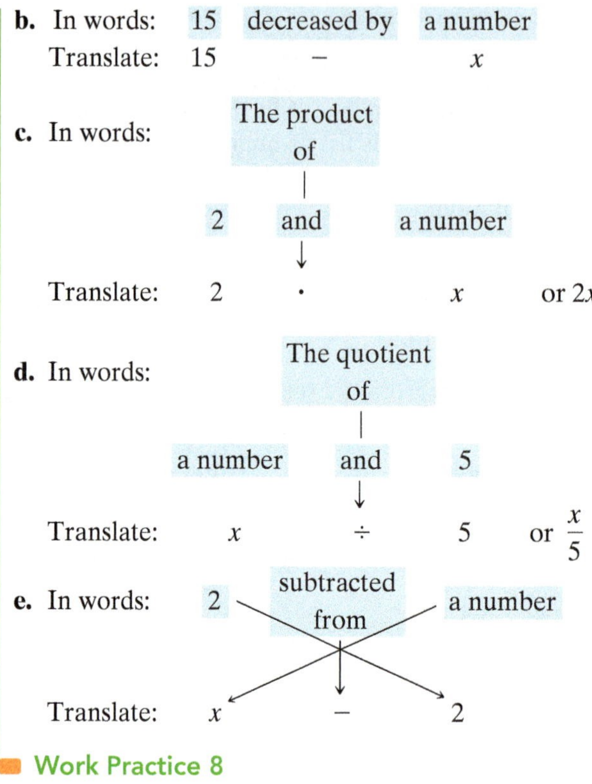

Work Practice 8

Helpful Hint

Remember that order is important when subtracting. Study the order of numbers and variables below.

Phrase	Translation
a number *decreased by* 5	$x - 5$
a number *subtracted from* 5	$5 - x$

Vocabulary, Readiness & Video Check

Use the choices below to fill in each blank. You may use each choice more than once.

evaluating the expression | variable(s) | expression | equation | solution

1. A combination of operations on letters (variables) and numbers is a(n) ________________.
2. A letter that represents a number is a(n) ________________.
3. $3x - 2y$ is called a(n) ________________ and the letters x and y are ________________.
4. Replacing a variable in an expression by a number and then finding the value of the expression is called ________________.
5. A statement of the form "expression = expression" is called a(n) ________________.
6. A value for the variable that makes an equation a true statement is called a(n) ________________.

Martin-Gay Interactive Videos

See Video 1.8

Watch the section lecture video and answer the following questions.

Objective A 7. Complete this statement based on the lecture before Example 1: When a letter and a variable are next to each other, the operation is an understood ____________.

Objective B 8. In Example 5, why is a question mark written over the equal sign?

Objective C 9. In Example 6, what phrase translates to subtraction?

1.8 Exercise Set MyMathLab®

Objective A *Complete the table. The first row has been done for you. See Examples 1 through 5.*

	a	b	$a + b$	$a - b$	$a \cdot b$	$a \div b$
	45	9	54	36	405	5
1.	21	7				
2.	24	6				
3.	152	0				
4.	298	0				
5.	56	1				
6.	82	1				

Evaluate each following expression for $x = 2$, $y = 5$, and $z = 3$. See Examples 1 through 5.

7. $3 + 2z$

8. $7 + 3z$

9. $3xz - 5x$

10. $4yz + 2x$

11. $z - x + y$

12. $x + 5y - z$

13. $4x - z$

14. $2y + 5z$

15. $y^3 - 4x$

16. $y^3 - z$

17. $2xy^2 - 6$

18. $3yz^2 + 1$

19. $8 - (y - x)$

20. $3 + (2y - 4)$

21. $x^5 + (y - z)$

22. $x^4 - (y - z)$

23. $\frac{6xy}{z}$

24. $\frac{8yz}{15}$

25. $\frac{2y - 2}{x}$

26. $\frac{6 + 3x}{z}$

27. $\frac{x + 2y}{z}$

28. $\frac{2z + 6}{3}$

29. $\frac{5x}{y} - \frac{10}{y}$

30. $\frac{70}{2y} - \frac{15}{z}$

31. $2y^2 - 4y + 3$

32. $3x^2 + 2x - 5$

33. $(4y - 5z)^3$

34. $(4y + 3z)^2$

35. $(xy + 1)^2$

36. $(xz - 5)^4$

37. $2y(4z - x)$

38. $3x(y + z)$

39. $xy(5 + z - x)$

40. $xz(2y + x - z)$

41. $\frac{7x + 2y}{3x}$

42. $\frac{6z + 2y}{4}$

43. The expression $16t^2$ gives the distance in feet that an object falls after t seconds. Complete the table by evaluating $16t^2$ for each given value of t.

t	1	2	3	4
$16t^2$				

44. The expression $\frac{5(F - 32)}{9}$ gives the equivalent degrees Celsius for F degrees Fahrenheit. Complete the table by evaluating this expression for each given value of F.

F	50	59	68	77
$\frac{5(F - 32)}{9}$				

Objective B *Decide whether the given number is a solution of the given equation. See Example 6.*

45. Is 10 a solution of $n - 8 = 2$?

46. Is 9 a solution of $n - 2 = 7$?

47. Is 3 a solution of $24 = 80n$?

48. Is 50 a solution of $250 = 5n$?

49. Is 7 a solution of $3n - 5 = 10$?

50. Is 8 a solution of $11n + 3 = 91$?

51. Is 20 a solution of $2(n - 17) = 6$?

52. Is 0 a solution of $5(n + 9) = 40$?

53. Is 0 a solution of $5x + 3 = 4x + 13$?

54. Is 2 a solution of $3x - 6 = 5x - 10$?

55. Is 8 a solution of $7f = 64 - f$?

56. Is 5 a solution of $8x - 30 = 2x$?

Determine which numbers in each set are solutions to the corresponding equations. See Example 7.

57. $n - 2 = 10$; $\{10, 12, 14\}$

58. $n + 3 = 16$; $\{9, 11, 13\}$

59. $5n = 30$; $\{6, 25, 30\}$

60. $3n = 45$; $\{15, 30, 45\}$

61. $6n + 2 = 26$; $\{0, 2, 4\}$

62. $4n - 14 = 6$; $\{0, 5, 10\}$

63. $3(n - 4) = 10$; $\{5, 7, 10\}$

64. $6(n + 2) = 23$; $\{1, 3, 5\}$

65. $7x - 9 = 5x + 13$; $\{3, 7, 11\}$

66. $9x - 15 = 5x + 1$; $\{2, 4, 11\}$

Objective C Translating *Write each phrase as a variable expression. Use x to represent "a number." See Example 8.*

67. Eight more than a number

68. The sum of three and a number

69. The total of a number and eight

70. The difference of a number and five hundred

71. Twenty decreased by a number

72. A number less thirty

73. The product of 512 and a number

74. A number times twenty

75. The quotient of eight and a number

76. A number divided by 11

77. The sum of seventeen and a number added to the product of five and the number

78. The quotient of twenty and a number, decreased by three

79. The product of five and a number

80. The difference of twice a number, and four

81. A number subtracted from 11

82. Twelve subtracted from a number

83. A number less 5

84. The sum of a number and 7

85. 6 divided by a number

86. The product of a number and 7

87. Fifty decreased by eight times a number

88. Twenty decreased by twice a number

Concept Extensions

For Exercises 89 through 92, use a calculator to evaluate each expression for $x = 23$ and $y = 72$.

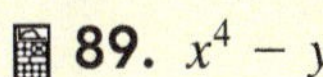

89. $x^4 - y^2$

90. $2(x + y)^2$

91. $x^2 + 5y - 112$

92. $16y - 20x + x^3$

93. If x is a whole number, which expression is the largest: $2x$, $5x$, or $\frac{x}{3}$? Explain your answer.

94. If x is a whole number, which expression is the smallest: $2x$, $5x$, or $\frac{x}{3}$? Explain your answer.

95. In Exercise 43, what do you notice about the value of $16t^2$ as t gets larger?

96. In Exercise 44, what do you notice about the value of $\frac{5(F - 32)}{9}$ as F gets larger?

Chapter 1 Group Activity

Investigating Endangered and Threatened Species

An **endangered** species is one that is thought to be in danger of becoming extinct throughout all or a major part of its habitat. A **threatened** species is one that may become endangered. The Division of Endangered Species at the U.S. Fish and Wildlife Service keeps close tabs on the state of threatened and endangered wildlife in the United States and around the world. The table below was compiled from 2013 data in the Division of Endangered Species' box score. The "Total Species" column gives the total number of endangered and threatened species for each group.

1. Round each number of *endangered animal species* to the nearest ten to estimate the Animal Total.
2. Round each number of *endangered plant species* to the nearest ten to estimate the Plant Total.
3. Add the exact numbers of endangered animal species to find the exact Animal Total and record it in the table in the Endangered Species column. Add the exact numbers of endangered plant species to find the Plant Total and record it in the table in the Endangered Species column. Then find the total number of endangered species (animals and plants combined) and record this number in the table as the Grand Total in the Endangered Species column.
4. Find the Animal Total, Plant Total, and Grand Total for the Total Species column. Record these values in the table.
5. Use the data in the table to complete the Threatened Species column.
6. Write a paragraph discussing the conclusions that can be drawn from the table.

Endangered and Threatened Species Worldwide

	Group	**Endangered Species**	**Threatened Species**	**Total Species**
Animals	Mammals	325		361
	Birds	286		317
	Reptiles	84		126
	Amphibians	25		37
	Fishes	95		166
	Snails	34		47
	Clams	74		86
	Crustaceans	20		23
	Insects	61		71
	Arachnids	12		12
	Corals	0		2
	Animal Total			
Plants	Flowering Plants	571		820
	Conifers	2		5
	Ferns and others	26		30
	Lichens	2		2
	Plant Total			
	Grand Total			

Chapter 1 Vocabulary Check

Fill in each blank with one of the words or phrases listed below.

difference	factor	perimeter	dividend	minuend	
place value	whole numbers	equation	divisor	variable	
sum	set	addend	exponent	expression	
solution	quotient	subtrahend	product	digits	area

1. The ____________ are 0, 1, 2, 3, . . .
2. The ____________ of a polygon is its distance around or the sum of the lengths of its sides.
3. The position of each digit in a number determines its ____________.
4. A(n) ____________ is a shorthand notation for repeated multiplication of the same factor.
5. To find the ____________ of a rectangle, multiply length times width.
6. The ____________ used to write numbers are 0, 1, 2, 3, 4, 5, 6, 7, 8, and 9.
7. A letter used to represent a number is called a(n) ____________.
8. A(n) ____________ can be written in the form "expression = expression."
9. A combination of operations on variables and numbers is called a(n) ____________.
10. A(n) ____________ of an equation is a value of the variable that makes the equation a true statement.
11. A collection of numbers (or objects) enclosed by braces is called a(n) ____________.

Use the facts below for Exercises 12 through 21.

$2 \cdot 3 = 6 \quad 4 + 17 = 21 \quad 20 - 9 = 11 \quad 5\overline{)35}$ with quotient 7

12. The 21 above is called the ____________.
13. The 5 above is called the ____________.
14. The 35 above is called the ____________.
15. The 7 above is called the ____________.
16. The 3 above is called a(n) ____________.
17. The 6 above is called the ____________.
18. The 20 above is called the ____________.
19. The 9 above is called the ____________.
20. The 11 above is called the ____________.
21. The 4 above is called a(n) ____________.

Are you preparing for your test? Don't forget to take the Chapter 1 Test on page 95. Then check your answers at the back of the text and use the Chapter Test Prep Videos to see the fully worked-out solutions to any of the exercises you want to review.

1 Chapter Highlights

Definitions and Concepts	Examples
Section 1.2 Place Value, Names for Numbers, and Reading Tables	
The **whole numbers** are 0, 1, 2, 3, 4, 5, ...	Examples of whole numbers: 0, 14, 968, 5,268,619
The position of each digit in a number determines its **place value.** A place-value chart is shown next with the names of the periods given.	

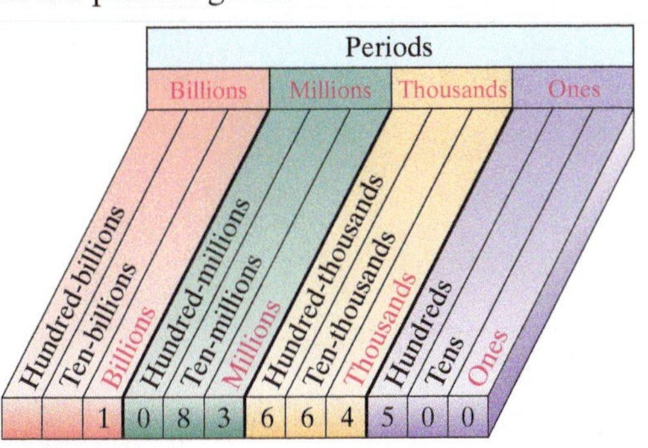

(continued)

Definitions and Concepts	Examples
Section 1.2 Place Value, Names for Numbers, and Reading Tables (*continued*)	
To write a whole number in words, write the number in each period followed by the name of the period. (The name of the ones period is not included.)	9,078,651,002 is written as nine billion, seventy-eight million, six hundred fifty-one thousand, two.
To write a whole number in standard form, write the number in each period, followed by a comma.	Four million, seven hundred six thousand, twenty-eight is written as 4,706,028.
Section 1.3 Adding and Subtracting Whole Numbers, and Perimeter	
To add whole numbers, add the digits in the ones place, then the tens place, then the hundreds place, and so on, carrying when necessary.	Find the sum: 211 (carries) 2689 ← addend 1735 ← addend + 662 ← addend 5086 ← sum
To subtract whole numbers, subtract the digits in the ones place, then the tens place, then the hundreds place, and so on, borrowing when necessary.	Subtract: 8 15 (borrows) 7954 ← minuend −5673 ← subtrahend 2281 ← difference
The **perimeter** of a polygon is its distance around or the sum of the lengths of its sides.	△ Find the perimeter of the polygon shown. 5 feet, 2 feet, 3 feet, 9 feet The perimeter is 5 feet + 3 feet + 9 feet + 2 feet = 19 feet.
Section 1.4 Rounding and Estimating	
Rounding Whole Numbers to a Given Place Value **Step 1:** Locate the digit to the right of the given place value. **Step 2:** If this digit is 5 or greater, add 1 to the digit in the given place value and replace each digit to its right with 0. **Step 3:** If this digit is less than 5, replace it and each digit to its right with 0.	Round 15,721 to the nearest thousand. 15, ⑦ 21 Add 1 — Replace with zeros. Since the circled digit is 5 or greater, add 1 to the given place value and replace digits to its right with zeros. 15,721 rounded to the nearest thousand is 16,000.

Definitions and Concepts	Examples
Section 1.5 Multiplying Whole Numbers and Area	
To multiply 73 and 58, for example, multiply 73 and 8, then 73 and 50. The sum of these partial products is the product of 73 and 58. Use the notation to the right.	73 ← factor × 58 ← factor 584 ← 73 × 8 3650 ← 73 × 50 4234 ← product
To find the **area** of a rectangle, multiply length times width.	△ Find the area of the rectangle shown. 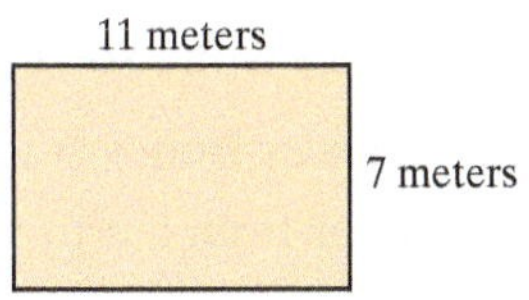 area of rectangle = length · width = (11 meters)(7 meters) = 77 square meters
Section 1.6 Dividing Whole Numbers	
Division Properties of 0 The quotient of 0 and any number (except 0) is 0.	$\frac{0}{5} = 0$
The quotient of any number and 0 is not a number. We say that this quotient is undefined.	$\frac{7}{0}$ is undefined
To divide larger whole numbers, use the process called **long division** as shown to the right.	507 R 2 ← quotient divisor → 14)7100 ← dividend −70 5(14) = 70 10 Subtract and bring down the 0. −0 0(14) = 0 100 Subtract and bring down the 0. −98 7(14) = 98 2 Subtract. The remainder is 2. To check, see that 507 · 14 + 2 = 7100.
The **average** of a list of numbers is $\text{average} = \frac{\text{sum of numbers}}{\textit{number} \text{ of numbers}}$	Find the average of 23, 35, and 38. $\text{average} = \frac{23 + 35 + 38}{3} = \frac{96}{3} = 32$

Definitions and Concepts	Examples
Section 1.7 Exponents and Order of Operations	
An **exponent** is a shorthand notation for repeated multiplication of the same factor.	$3^4 = \underbrace{3\cdot3\cdot3\cdot3}_{\text{4 factors of 3}} = 81$ (3 is the base; 4 is the exponent)
Order of Operations **1.** Perform all operations within parentheses ( ), brackets [ ], or other grouping symbols such as fraction bars, starting with the innermost set. **2.** Evaluate any expressions with exponents. **3.** Multiply or divide in order from left to right. **4.** Add or subtract in order from left to right.	Simplify: $\frac{5+3^2}{2(7-6)}$ Simplify above and below the fraction bar separately. $\frac{5+3^2}{2(7-6)} = \frac{5+9}{2(1)}$ Evaluate 3^2 above the fraction bar. Subtract $7-6$ below the fraction bar. $= \frac{14}{2}$ Add. Multiply. $= 7$ Divide.
The **area of a square** is $(\text{side})^2$.	Find the area of a square with side length 9 inches. Area of the square $= (\text{side})^2$ $= (9 \text{ inches})^2$ $= 81$ square inches
Section 1.8 Introduction to Variables, Algebraic Expressions, and Equations	
A letter used to represent a number is called a **variable.**	Variables: $x,\ y,\ z,\ a,\ b$
A combination of operations on variables and numbers is called an **algebraic expression.**	Algebraic expressions: $3+x,\ 7y,\ x^3+y-10$
Replacing a variable in an expression by a number, and then finding the value of the expression, is called **evaluating the expression** for the variable.	Evaluate $2x+y$ for $x=22$ and $y=4$. $2x+y = 2\cdot22+4$ Replace x with 22 and y with 4. $= 44+4$ Multiply. $= 48$ Add.
A statement written in the form "expression = expression" is an **equation.**	Equations: $n-8=12$ $2(20-7n)=32$ Area = length · width
A **solution** of an equation is a value for the variable that makes the equation a true statement.	Determine whether 2 is a solution of the equation $4(x-1)=7$. $4(2-1) \stackrel{?}{=} 7$ Replace x with 2. $4(1) \stackrel{?}{=} 7$ Subtract. $4 = 7$ False No, 2 is not a solution.

Chapter 1 Review

(1.2) *Determine the place value of the digit 4 in each whole number.*

1. 7640

2. 46,200,120

Write each whole number in words.

3. 7640

4. 46,200,120

Write each whole number in expanded form.

5. 3158

6. 403,225,000

Write each whole number in standard form.

7. Eighty-one thousand, nine hundred

8. Six billion, three hundred four million

The following table shows the Internet use by world regions. Use this table to answer Exercises 9 through 12 and other exercises throughout this review. (*Source:* International Telecommunications Union and Internet World Stats)

Internet Use by World Regions (in millions)

World Region	2004	2008	2013
Africa	21	51	140
Asia	296	579	1268
Europe	241	385	467
Middle East	29	42	141
North America	218	248	302
Latin America/ Caribbean	51	139	280
Oceania/ Australia	12	20	145

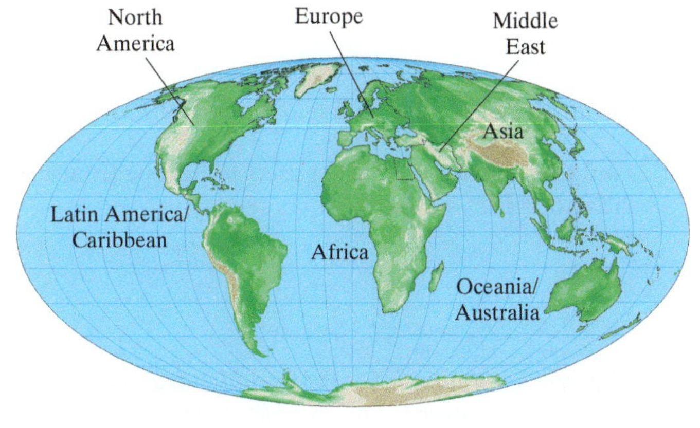

9. Find the number of Internet users in 2013 in Europe. Write your answer in standard form.

10. Find the number of Internet users in Oceania/ Australia in 2013. Write your answer in standard form.

11. Which world region had the smallest number of Internet users in 2008?

12. Which world region had the greatest number of Internet users in 2008?

(1.3) *Add or subtract as indicated.*

13. 18 + 49

14. 28 + 39

15. 462 − 397

16. 583 − 279

17. 428 + 21

18. 819 + 21

19. 4000 − 86

20. 8000 − 92

21. 91 + 3623 + 497

22. 82 + 1647 + 238

Translating *Solve.*

23. Find the sum of 74, 342, and 918.

24. Find the sum of 49, 529, and 308.

25. Subtract 7965 from 25,862.

26. Subtract 4349 from 39,007.

27. The distance from Washington, DC, to New York City is 205 miles. The distance from New York City to New Delhi, India, is 7318 miles. Find the total distance from Washington, DC, to New Delhi if traveling by air through New York City.

28. Susan Summerline earned salaries of $62,589, $65,340, and $69,770 during the years 2004, 2005, and 2006, respectively. Find her total earnings during those three years.

Find the perimeter of each figure.

△ **29.**

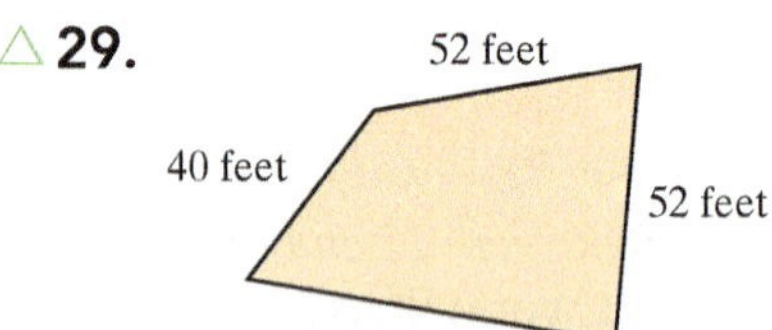

△ **30.**

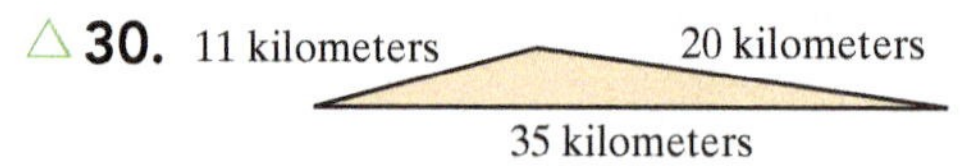

Use the Internet Use by World Regions table for Exercises 31 and 32.

31. Find the increase in Internet users in Europe from 2008 to 2013.

32. Find the difference in the number of Internet users in 2013 between Oceania/Australia and the Middle East.

The following bar graph shows the monthly savings account balances for a freshman attending a local community college. Use this graph to answer Exercises 33 through 36.

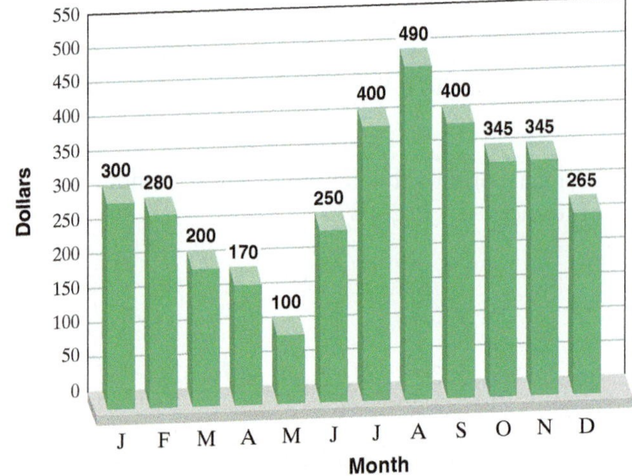

33. During what month was the balance the least?

34. During what month was the balance the greatest?

35. By how much did the balance decrease from February to April?

36. By how much did the balance increase from June to August?

(1.4) *Round to the given place.*

37. 43 to the nearest ten

38. 45 to the nearest ten

39. 876 to the nearest ten

40. 493 to the nearest hundred

41. 3829 to the nearest hundred

42. 57,534 to the nearest thousand

43. 39,583,819 to the nearest million

44. 768,542 to the nearest hundred-thousand

Estimate the sum or difference by rounding each number to the nearest hundred.

45. $3785 + 648 + 2866$

46. $5925 - 1787$

47. A group of students took a week-long driving trip and traveled 630, 192, 271, 56, 703, 454, and 329 miles on seven consecutive days. Round each distance to the nearest hundred to estimate the distance they traveled.

48. In 2012, the population of Europe was 820,918,446 while the population of Latin America/Caribbean was 593,688,638. Round each number to the nearest million, and estimate their difference in population in 2012. (Source: Internet World Stats)

(1.5) *Multiply.*

49. $\begin{array}{r} 276 \\ \times \ 8 \\ \hline \end{array}$

50. $\begin{array}{r} 349 \\ \times \ 4 \\ \hline \end{array}$

51. $\begin{array}{r} 57 \\ \times 40 \\ \hline \end{array}$

52. $\begin{array}{r} 69 \\ \times 42 \\ \hline \end{array}$

53. $20(7)(4)$

54. $25(9)(4)$

55. $26 \cdot 34 \cdot 0$

56. $62 \cdot 88 \cdot 0$

57. $\begin{array}{r} 586 \\ \times \ 29 \\ \hline \end{array}$

58. $\begin{array}{r} 242 \\ \times \ 37 \\ \hline \end{array}$

59. $\begin{array}{r} 642 \\ \times 177 \\ \hline \end{array}$

60. $\begin{array}{r} 347 \\ \times 129 \\ \hline \end{array}$

61. $\begin{array}{r} 1026 \\ \times 401 \\ \hline \end{array}$

62. $\begin{array}{r} 2107 \\ \times \ 302 \\ \hline \end{array}$

Translating *Solve.*

63. Find the product of 6 and 250.

64. Find the product of 6 and 820.

65. A golf pro orders shirts for the company sponsoring a local charity golfing event. Shirts size large cost \$32 while shirts size extra large cost \$38. If 15 large shirts and 11 extra-large shirts are ordered, find the cost.

66. The cost for a South Dakota resident to attend Black Hills State University full-time is \$6112 per semester. Determine the cost for 20 students to attend full-time. (*Source:* Black Hills State University)

Find the area of each rectangle.

△ **67.**

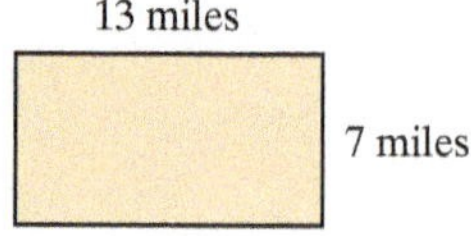

△ **68.**

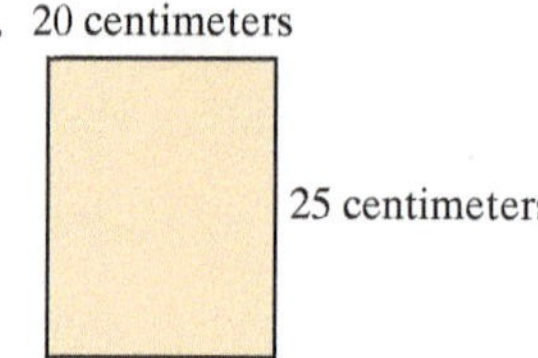

(1.6) *Divide and then check.*

69. $\frac{49}{7}$

70. $\frac{36}{9}$

71. $27 \div 5$

72. $18 \div 4$

73. $918 \div 0$

74. $0 \div 668$

75. $5\overline{)167}$

76. $8\overline{)159}$

77. $26\overline{)626}$

78. $19\overline{)680}$

79. $47\overline{)23{,}792}$

80. $53\overline{)48{,}111}$

81. $207\overline{)578{,}291}$

82. $306\overline{)615{,}732}$

Translating *Solve.*

83. Find the quotient of 92 and 5.

84. Find the quotient of 86 and 4.

85. A box can hold 24 cans of corn. How many boxes can be filled with 648 cans of corn?

86. One mile is 1760 yards. Find how many miles there are in 22,880 yards.

87. Find the average of the numbers 76, 49, 32, and 47.

88. Find the average of the numbers 23, 85, 62, and 66.

(1.7) *Simplify.*

89. 8^2

90. 5^3

91. $5 \cdot 9^2$

92. $4 \cdot 10^2$

93. $18 \div 2 + 7$

94. $12 - 8 \div 4$

95. $\frac{5(6^2 - 3)}{3^2 + 2}$

96. $\frac{7(16 - 8)}{2^3}$

97. $48 \div 8 \cdot 2$

98. $27 \div 9 \cdot 3$

99. $2 + 3[1^5 + (20 - 17) \cdot 3] + 5 \cdot 2$

100. $21 - [2^4 - (7 - 5) - 10] + 8 \cdot 2$

101. $19 - 2(3^2 - 2^2)$

102. $16 - 2(4^2 - 3^2)$

103. $4 \cdot 5 - 2 \cdot 7$

104. $8 \cdot 7 - 3 \cdot 9$

105. $(6 - 4)^3 \cdot [10^2 \div (3 + 17)]$

106. $(7 - 5)^3 \cdot [9^2 \div (2 + 7)]$

107. $\dfrac{5 \cdot 7 - 3 \cdot 5}{2(11 - 3^2)}$

108. $\dfrac{4 \cdot 8 - 1 \cdot 11}{3(9 - 2^3)}$

Find the area of each square.

△ **109.** A square with side length of 7 meters.

△ **110.** 3 inches

(1.8) *Evaluate each expression for $x = 5$, $y = 0$, and $z = 2$.*

111. $\dfrac{2x}{z}$

112. $4x - 3$

113. $\dfrac{x + 7}{y}$

114. $\dfrac{y}{5x}$

115. $x^3 - 2z$

116. $\dfrac{7 + x}{3z}$

117. $(y + z)^2$

118. $\dfrac{100}{x} + \dfrac{y}{3}$

Translating *Translate each phrase into a variable expression. Use x to represent a number.*

119. Five subtracted from a number

120. Seven more than a number

121. Ten divided by a number

122. The product of 5 and a number

Decide whether the given number is a solution of the given equation.

123. Is 5 a solution of $n + 12 = 20 - 3$?

124. Is 23 a solution of $n - 8 = 10 + 6$?

125. Is 14 a solution of $30 = 3(n - 3)$?

126. Is 20 a solution of $5(n - 7) = 65$?

Determine which numbers in each set are solutions to the corresponding equations.

127. $7n = 77$; $\{6, 11, 20\}$

128. $n - 25 = 150$; $\{125, 145, 175\}$

129. $5(n + 4) = 90$; $\{14, 16, 26\}$

130. $3n - 8 = 28$; $\{3, 7, 15\}$

Mixed Review

Perform the indicated operations.

131. $485 - 68$

132. $729 - 47$

133. 732×3

134. 629×4

135. $374 + 29 + 698$

136. $593 + 52 + 766$

137. $13\overline{)5962}$

138. $18\overline{)4267}$

139. 1968×36

140. 5324×18

141. $2000 - 356$

142. $9000 - 519$

Round to the given place.

143. 842 to the nearest ten

144. 258,371 to the nearest hundred-thousand

Simplify.

145. $24 \div 4 \cdot 2$

146. $\dfrac{(15 + 3) \cdot (8 - 5)}{2^3 + 1}$

Solve.

147. Is 9 a solution of $5n - 6 = 40$?

148. Is 3 a solution of $2n - 6 = 5n - 15$?

149. A manufacturer of drinking glasses ships his delicate stock in special boxes that can hold 32 glasses. If 1714 glasses are manufactured, how many full boxes are filled? Are there any glasses left over?

150. A teacher orders 2 small whiteboards for $27 each and 8 boxes of dry erase pens for $4 each. What is her total bill before taxes?

Step-by-step test solutions are found on the Chapter Test Prep Videos. Where available: MyMathLab® or YouTube

Chapter 1 Test

Simplify.

1. Write 82,426 in words.

2. Write "four hundred two thousand, five hundred fifty" in standard form.

3. $59 + 82$

4. $600 - 487$

5. $\begin{array}{r} 496 \\ \times \ \ 30 \\ \hline \end{array}$

6. $52{,}896 \div 69$

7. $2^3 \cdot 5^2$

8. $98 \div 1$

9. $0 \div 49$

10. $62 \div 0$

11. $(2^4 - 5) \cdot 3$

12. $16 + 9 \div 3 \cdot 4 - 7$

13. $6^1 \cdot 2^3$

14. $2[(6 - 4)^2 + (22 - 19)^2] + 10$

15. $5698 \cdot 1000$

16. Find the average of 62, 79, 84, 90, and 95.

17. Round 52,369 to the nearest thousand.

Estimate each sum or difference by rounding each number to the nearest hundred.

18. $6289 + 5403 + 1957$

19. $4267 - 2738$

Solve.

20. Subtract 15 from 107.

21. Find the sum of 15 and 107.

22. Find the product of 15 and 107.

23. Find the quotient of 107 and 15.

Answers

1. ____________
2. ____________
3. ____________
4. ____________
5. ____________
6. ____________
7. ____________
8. ____________
9. ____________
10. ____________
11. ____________
12. ____________
13. ____________
14. ____________
15. ____________
16. ____________
17. ____________
18. ____________
19. ____________
20. ____________
21. ____________
22. ____________
23. ____________

24. ______
25. ______
26. ______
27. ______
28. ______
29. ______
30. ______
31. ______
32. a. ______
b. ______
33. ______
34. ______

24. Twenty-nine cans of Sherwin-Williams paint cost \$493. How much was each can?

25. Jo McElory is looking at two new refrigerators for her apartment. One costs \$599 and the other costs \$725. How much more expensive is the higher-priced one?

26. One tablespoon of white granulated sugar contains 45 calories. How many calories are in 8 tablespoons of white granulated sugar? (*Source: Home and Garden Bulletin No. 72,* U.S. Department of Agriculture)

27. A small business owner recently ordered 16 digital cameras that cost \$430 each and 5 printers that cost \$205 each. Find the total cost for these items.

Find the perimeter and the area of each figure.

△**28.** △**29.**

30. Evaluate $5(x^3 - 2)$ for $x = 2$.

31. Evaluate $\frac{3x - 5}{2y}$ for $x = 7$ and $y = 8$.

32. Translate the following phrases into mathematical expressions. Use x to represent "a number."

a. The quotient of a number and 17

b. Twice a number, decreased by 20

33. Is 6 a solution of the equation $5n - 11 = 19$?

34. Determine which number in the set is a solution to the given equation.
$n + 20 = 4n - 10$; $\{0, 10, 20\}$

Integers and Introduction to Solving Equations

2

Director James Cameron made the deepest solo descent so far into the Mariana Trench in the Pacific Ocean. He reached a depth of 35,756 feet in the Deepsea Challenger, shown above. Next, may be Richard Branson?

The Krubera Cave now holds the title of deepest. In this cave, many new depth records have been set—each one deeper than the last. The latest record is 7188 feet but who knows how deeply this cave will be explored next?

Where do we explore next? Throughout this chapter, we present many applications having to do with water depths below sea level and land depths below the surface of Earth by way of mines and caves. Recently, there has been a surge of interest in exploring these depths. Although we have already reached the deepest-known part of our oceans—the Mariana Trench in the Pacific Ocean—cave exploration is a little more tricky. For example, with so many "branches" of a cave, we are never certain that it has been totally explored. New caves are being discovered and explored even as this is written. The deepest-known cave in the world, the Krubera, was not discovered until 2001 by Ukrainian cave explorers. See exercises throughout this chapter.

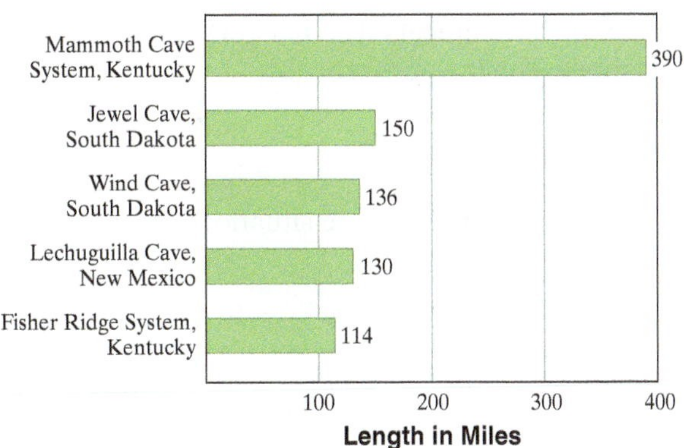

Source: Top 10 of Everything, 2013

Sections

Check Your Progress

Thus far, we have studied whole numbers, but these numbers are not sufficient for representing many situations in real life. For example, to express 5 degrees below zero or $100 in debt, numbers less than 0 are needed. This chapter is devoted to integers, which include numbers less than 0, and operations on these numbers.

2.1 Introduction to Integers 

Objectives

A Represent Real-Life Situations with Integers.

B Graph Integers on a Number Line.

C Compare Integers.

D Find the Absolute Value of a Number.

E Find the Opposite of a Number.

F Read Bar Graphs Containing Integers.

Objective A Representing Real-Life Situations

Thus far in this text, all numbers have been 0 or greater than 0. Numbers greater than 0 are called **positive numbers.** However, sometimes situations exist that cannot be represented by a number greater than 0. For example,

To represent these situations, we need numbers less than 0.

Extending the number line to the left of 0 allows us to picture **negative numbers,** which are numbers that are less than 0.

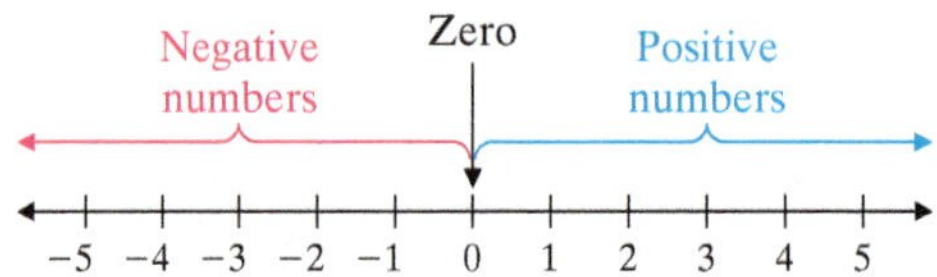

When a single + sign or no sign is in front of a number, the number is a positive number. When a single − sign is in front of a number, the number is a negative number. Together, we call positive numbers, negative numbers, and zero the **signed numbers.**

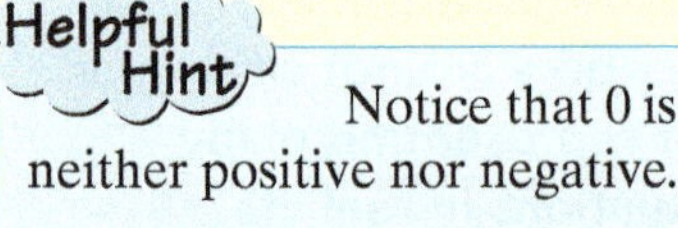

Notice that 0 is neither positive nor negative.

-5 indicates "negative five."

5 and $+5$ both indicate "positive five."

The number 0 is neither positive nor negative.

Some signed numbers are integers. The **integers** consist of the numbers labeled on the number line above. The integers are

$\ldots, -3, -2, -1, 0, 1, 2, 3, \ldots$

Now we have numbers to represent the situations previously mentioned.

5 degrees below 0 $-5°$

20 feet below sea level -20 feet

Helpful Hint

A − sign, such as the one in -1, tells us that the number is to the left of 0 on the number line. -1 is read "negative one."

A + sign or no sign tells us that a number lies to the right of 0 on the number line. For example, 3 and $+3$ both mean "positive three."

Example 1 Representing Depth with an Integer

The world's deepest cave is Krubera (or Voronja), in the country of Georgia, located by the Black Sea in Asia. It has been explored to a depth of 7188 feet below the surface of Earth. Represent this position using an integer. (*Source:* MessagetoEagle.com and Wikipedia)

Solution: If 0 represents the surface of Earth, then 7188 feet below the surface can be represented by -7188.

Work Practice 1

Objective B Graphing Integers 

Example 2 Graph 0, -3, 5, and -5 on the number line.

Solution:

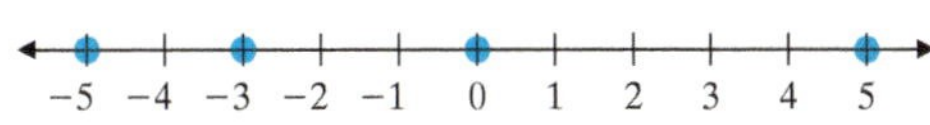

Work Practice 2

Objective C Comparing Integers 

We can compare integers by using a number line. For any two numbers graphed on a number line, the number to the **right** is the **greater number** and the number to the **left** is the **smaller number.** Also, the symbols $<$ and $>$ are called **inequality symbols.**

The inequality symbol $>$ means **"is greater than"** and
the inequality symbol $<$ means **"is less than."**

For example, both -5 and -7 are graphed on the number line below.

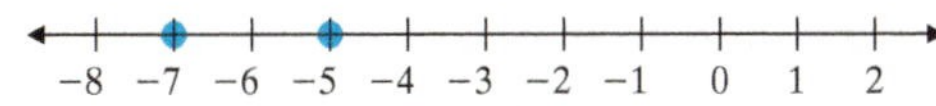

On the graph, -7 is **to the left of** -5, so -7 **is less than** -5, written as

$$-7 < -5$$

We can also write

$$-5 > -7$$

since -5 is **to the right of** -7, so -5 **is greater than** -7.

✓**Concept Check** Is there a largest positive number? Is there a smallest negative number? Explain.

Practice 1

a. The world's deepest bat colony spends each winter in a New York zinc mine at a depth of 3805 feet. Represent this position with an integer. (*Source: Guinness Book of World Records*)

b. The tamarack tree, a type of conifer, commonly grows at the edge of the arctic tundra and survives winter temperatures of 85 degrees below zero, Fahrenheit. Represent this temperature with an integer in degrees Fahrenheit.

Practice 2

Graph -4, -1, 2, and -2 on the number line.

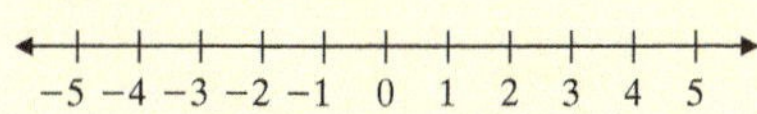

Answers

1. a. -3805 **b.** $-85°F$

2. −5 −4 −3 −2 −1 0 1 2 3 4 5

✓**Concept Check Answer**

no

Practice 3

Insert $<$ or $>$ between each pair of numbers to make a true statement.

a. $0 \quad -5$ **b.** $-3 \quad 3$

c. $-7 \quad -12$

Example 3

Insert $<$ or $>$ between each pair of numbers to make a true statement.

a. $-7 \quad 7$ **b.** $0 \quad -4$ **c.** $-9 \quad -11$

Solution:

a. -7 is to the left of 7 on a number line, so $-7 < 7$.

b. 0 is to the right of -4 on a number line, so $0 > -4$.

c. -9 is to the right of -11 on a number line, so $-9 > -11$.

Work Practice 3

Helpful Hint

If you think of $<$ and $>$ as arrowheads, notice that in a true statement the arrow always points to the smaller number.

$5 > -4$ (smaller number: -4) $\qquad -3 < -1$ (smaller number: -3)

Objective D Finding the Absolute Value of a Number 

The **absolute value** of a number is the number's distance from 0 on the number line. The symbol for absolute value is $| \;|$. For example, $|3|$ is read as "the absolute value of 3."

$|3| = 3$ because 3 is 3 units from 0.

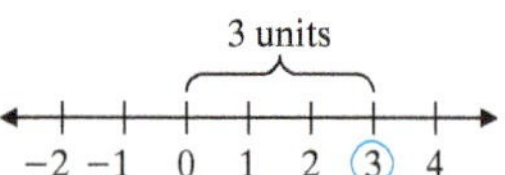

$|-3| = 3$ because -3 is 3 units from 0.

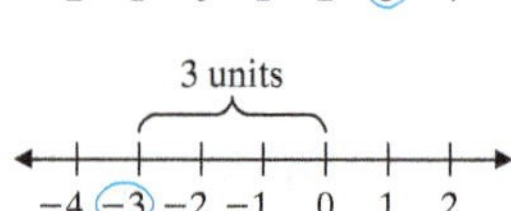

Practice 4

Simplify.

a. $|-6|$

b. $|4|$

c. $|-12|$

Example 4

Simplify.

a. $|-9|$ **b.** $|8|$ **c.** $|0|$

Solution:

a. $|-9| = 9$ because -9 is 9 units from 0.

b. $|8| = 8$ because 8 is 8 units from 0.

c. $|0| = 0$ because 0 is 0 units from 0.

Work Practice 4

Helpful Hint

Since the absolute value of a number is that number's *distance* from 0, the absolute value of a number is always 0 or positive. It is never negative.

$|0| = 0$ (zero) $\qquad |-6| = 6$ (a positive number)

Answers

3. a. $>$ **b.** $<$ **c.** $>$

4. a. 6 **b.** 4 **c.** 12

Objective E Finding Opposites 

Two numbers that are the same distance from 0 on the number line but are on opposite sides of 0 are called **opposites.**

4 and −4 are opposites.

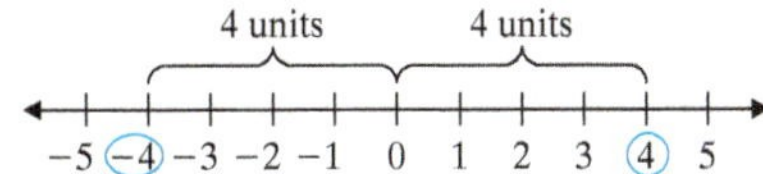

When two numbers are opposites, we say that each is the opposite of the other. Thus **4 is the opposite of −4** and **−4 is the opposite of 4.**

The phrase "the opposite of" is written in symbols as "−". For example,

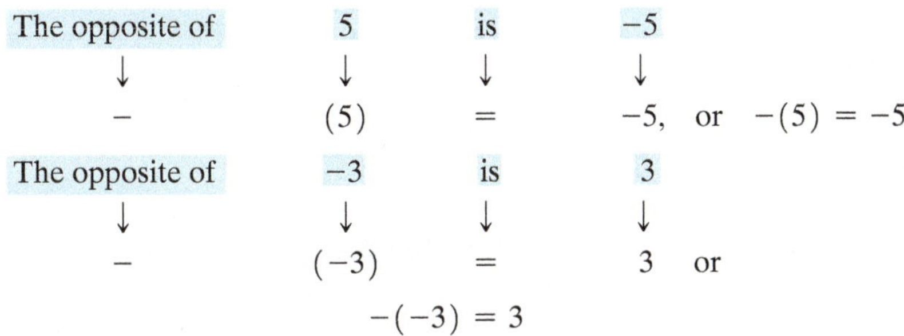

In general, we have the following:

Opposites

If a is a number, then $-(-a) = a$.

Notice that because "the opposite of" is written as "−", to find the opposite of a number we place a "−" sign in front of the number.

Example 5 Find the opposite of each number.

a. 13 **b.** −2 **c.** 0

Solution:

a. The opposite of 13 is −13.

b. The opposite of −2 is $-(-2)$ or 2.

c. The opposite of 0 is 0.

Helpful Hint Remember that 0 is neither positive nor negative.

Work Practice 5

Practice 5

Find the opposite of each number.

a. 14 **b.** −9

✓**Concept Check** True or false? The number 0 is the only number that is its own opposite.

Example 6 Simplify.

a. $-(-4)$ **b.** $-|-5|$ **c.** $-|6|$

Solution:

a. $-(-4) = 4$ The opposite of negative 4 is 4.

b. $-|-5| = -5$ The opposite of the absolute value of −5 is the opposite of 5, or −5.

c. $-|6| = -6$ The opposite of the absolute value of 6 is the opposite of 6, or −6.

Work Practice 6

Practice 6

Simplify.

a. $-|-7|$

b. $-|4|$

c. $-(-12)$

Answers

5. a. −14 **b.** 9

6. a. −7 **b.** −4 **c.** 12

✓**Concept Check Answer**

true

Practice 7

Evaluate $-|x|$ if $x = -6$.

Example 7 Evaluate $-|-x|$ if $x = -2$.

Solution: Carefully replace x with -2; then simplify.

$-|-x| = -|-(-2)|$ Replace x with -2.

Then $-|-(-2)| = -|2| = -2$.

Work Practice 7

Objective F Reading Bar Graphs Containing Integers

The bar graph below shows the average daytime surface temperatures (in degrees Fahrenheit) of the eight planets, excluding the newly classified "dwarf planet," Pluto. Notice that a negative temperature is illustrated by a bar below the horizontal line representing 0°F, and a positive temperature is illustrated by a bar above the horizontal line representing 0°F.

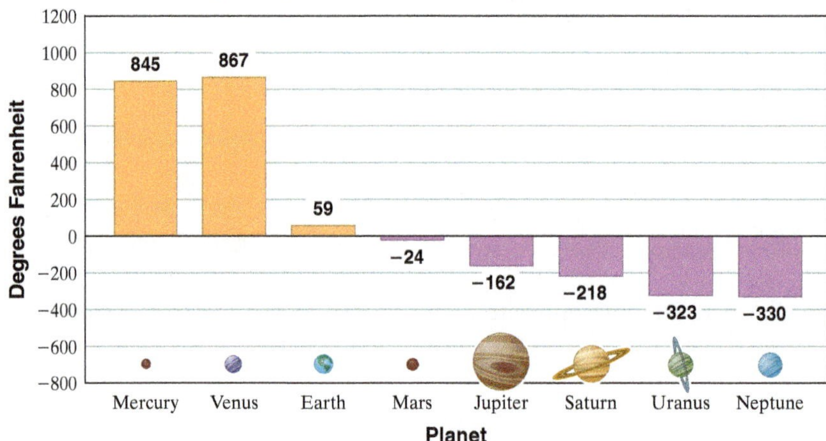

Source: The World Almanac, 2013

* For some planets, the temperature given is the temperature where the atmospheric pressure equals 1 Earth atmosphere.

Practice 8

Which planet has the highest average daytime surface temperature?

Answers

7. −6 8. Venus

Example 8 Which planet has the lowest average daytime surface temperature?

Solution: The planet with the lowest average daytime surface temperature is the one that corresponds to the bar that extends the farthest in the negative direction (downward). Neptune has the lowest average daytime surface temperature, −330°F.

Work Practice 8

Vocabulary, Readiness & Video Check

Use the choices below to fill in each blank. Not all choices will be used.

opposites	absolute value	right	is less than
inequality symbols	negative	positive	left
signed	integers	is greater than	

1. The numbers ... −3, −2, −1, 0, 1, 2, 3, ... are called ______________.
2. Positive numbers, negative numbers, and zero together are called ______________ numbers.
3. The symbols "<" and ">" are called ______________.

4. Numbers greater than 0 are called ______________ numbers while numbers less than 0 are called ______________ numbers.
5. The sign "<" means ______________ and ">" means ______________.
6. On a number line, the greater number is to the ______________ of the lesser number.
7. A number's distance from 0 on a number line is the number's ______________.
8. The numbers -5 and 5 are called ______________.

Martin-Gay Interactive Videos *Watch the section lecture video and answer the following questions.*

See Video 2.1

Objective A 9. In Example 1, what application is used to represent a negative number?

Objective B 10. In Example 2, the tick marks are labelled with what numbers on the number line?

Objective C 11. From Example 3 and your knowledge of a number line, complete this statement: 0 will always be greater than any of the ______________ integers.

Objective D 12. What is the answer to Example 5? The absolute value of what other integer has this same answer?

Objective E 13. Complete this statement based on Example 10: A negative sign can be translated to the phrase "______________."

Objective F 14. In Examples 13 and 14, what other lake has a negative integer elevation?

2.1 Exercise Set MyMathLab®

Objective A *Represent each quantity by an integer. See Example 1.*

1. A worker in a silver mine in Nevada works 1235 feet underground.

2. A scuba diver is swimming 25 feet below the surface of the water in the Gulf of Mexico.

3. The peak of Mount Elbert in Colorado is 14,433 feet above sea level. (*Source:* U.S. Geological Survey)

4. The lowest elevation in the United States is found at Death Valley, California, at an elevation of 282 feet below sea level. (*Source:* U.S. Geological Survey)

5. The record high temperature in Arkansas is 120 degrees above zero Fahrenheit. (*Source:* National Climatic Data Center)

6. The record high temperature in California is 134 degrees above zero Fahrenheit. (*Source*: National Climatic Data Center.)

7. The average depth of the Atlantic Ocean is 11,810 feet below its surface. (*Source: The World Almanac*, 2013)

8. The average depth of the Pacific Ocean is 14,040 feet below its surface. (*Source*: *The World Almanac*, 2013)

9. Sears had a loss of \$3140 million for the fiscal year 2011. (*Source:* CNN Money)

10. Rite Aid had a loss of \$555 million for the fiscal year 2011. (*Source:* CNN Money)

11. Two divers are exploring the wreck of the *Andrea Doria,* south of Nantucket Island, Massachusetts. Guillermo is 160 feet below the surface of the ocean and Luigi is 147 feet below the surface. Represent each quantity by an integer and determine who is deeper.

12. The temperature on one January day in Minneapolis was 10° below 0° Celsius. Represent this quantity by an integer and tell whether this temperature is cooler or warmer than 5° below 0° Celsius.

13. For the first half of 2013, digital track sales declined 2 percent when compared to the first half of 2012. (*Source:* Nielsen Sound Scan)

14. In a recent year, the number of CDs shipped to music retailers reflected a 23 percent decrease from the previous year. Write an integer to represent the percent decrease in CDs shipped. (*Source:* Recording Industry Association of America)

Objective B *Graph each integer in the list on the same number line. See Example 2.*

15. 0, 3, 4, 6

16. 7, 5, 2, 0

17. 1, −1, 2, −2, −4

18. 3, −3, 5, −5, 6

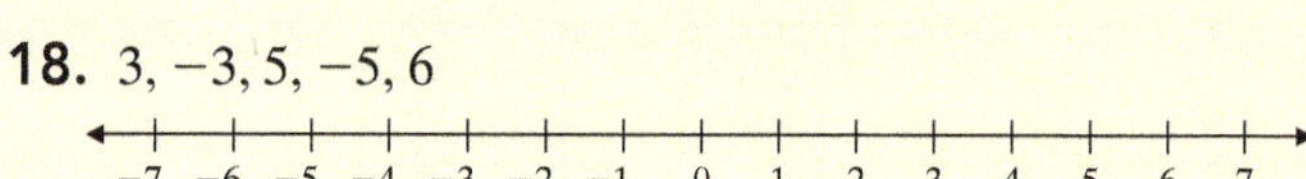

19. 0, 1, 9, 14

20. 0, 3, 10, 11

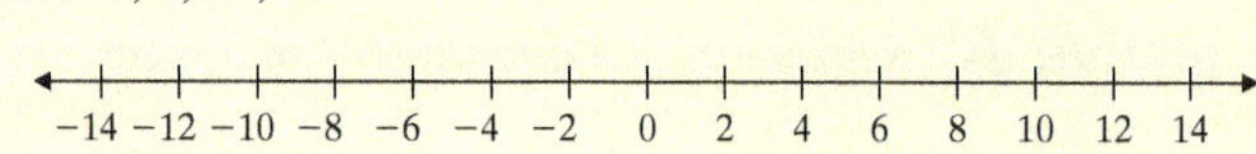

21. 0, −2, −7, −5

22. 0, −7, 3, −6

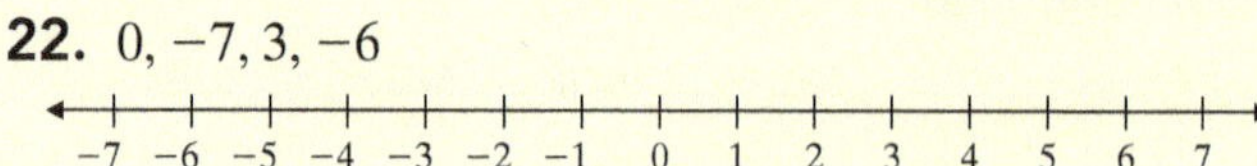

Objective C *Insert $<$ or $>$ between each pair of integers to make a true statement. See Example 3.*

23. 0 −7 **24.** −8 0 **25.** −7 −5 **26.** −12 −10

27. −30 −35 **28.** −27 −29 **29.** −26 26 **30.** 13 −13

Objective D *Simplify. See Example 4.*

31. $|5|$ **32.** $|7|$ **33.** $|-8|$ **34.** $|-19|$

35. $|0|$ **36.** $|100|$ **37.** $|-55|$ **38.** $|-10|$

Objective E *Find the opposite of each integer. See Example 5.*

39. 5
40. 8
41. −4
42. −6
43. 23
44. 123
45. −85
46. −13

Objectives D E Mixed Practice *Simplify. See Example 6.*

47. $|-7|$
48. $|-11|$
49. $-|20|$
50. $-|43|$
51. $-|-3|$
52. $-|-18|$
53. $-(-43)$
54. $-(-27)$
55. $|-15|$
56. $-(-14)$
57. $-(-33)$
58. $-|-29|$

Evaluate. See Example 7.

59. $|-x|$ if $x = -6$
60. $-|x|$ if $x = -8$
61. $-|-x|$ if $x = 2$
62. $-|-x|$ if $x = 10$
63. $|x|$ if $x = -32$
64. $|x|$ if $x = 32$
65. $-|x|$ if $x = 7$
66. $|-x|$ if $x = 1$

Insert $<$, $>$, or $=$ between each pair of numbers to make a true statement. See Examples 3 through 6.

67. $-12 \quad -6$
68. $-4 \quad -17$
69. $|-8| \quad |-11|$
70. $|-8| \quad |-4|$
71. $|-47| \quad -(-47)$
72. $-|17| \quad -(-17)$
73. $-|-12| \quad -(-12)$
74. $|-24| \quad -(-24)$
75. $0 \quad -9$
76. $-45 \quad 0$
77. $|0| \quad |-9|$
78. $|-45| \quad |0|$
79. $-|-2| \quad -|-10|$
80. $-|-8| \quad -|-4|$
81. $-(-12) \quad -(-18)$
82. $-22 \quad -(-38)$

Objectives D E Mixed Practice *Fill in the chart. See Examples 4 through 7.*

	Number	Absolute Value of Number	Opposite of Number
83.	31		
85.			−28

	Number	Absolute Value of Number	Opposite of Number
84.	−13		
86.			90

Objective F *The bar graph shows the elevations of selected lakes. Use this graph For Exercises 87 through 90. (Source: U.S. Geological Survey) See Example 8.*

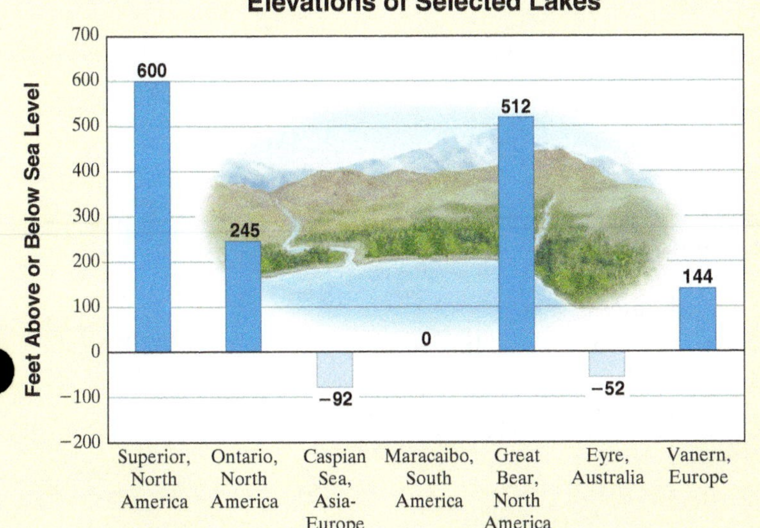

87. Which lake shown has the lowest elevation?

88. Which lake has an elevation at sea level?

89. Which lake shown has the highest elevation?

90. Which lake shown has the second-lowest elevation?

The following bar graph represents the boiling temperature, the temperature at which a substance changes from liquid to gas at standard atmospheric pressure. Use this graph to answer Exercises 91 through 94.

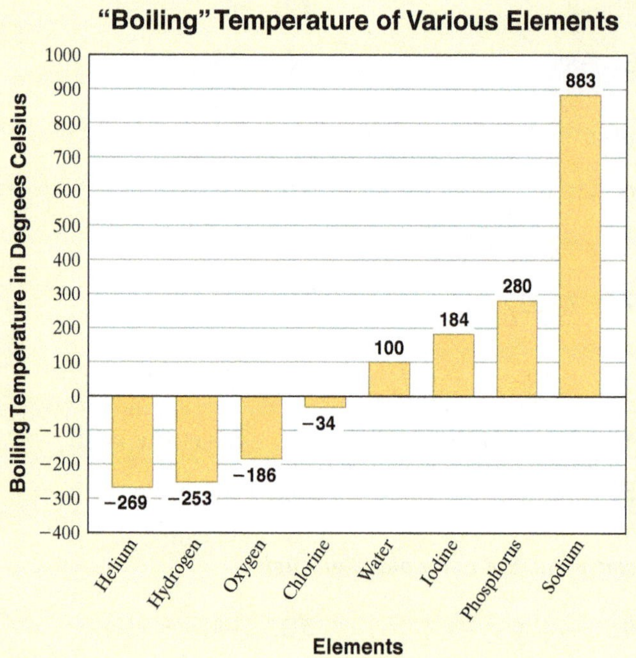

91. Which element has a positive boiling temperature closest to that of water?

92. Which element has the lowest boiling temperature?

93. Which element has a boiling temperature closest to −200°C?

94. Which element has an average boiling temperature closest to +300°C?

Review

Add. See Section 1.3.

95. $0 + 13$

96. $9 + 0$

97. $15 + 20$

98. $20 + 15$

99. $47 + 236 + 77$

100. $362 + 37 + 90$

Concept Extensions

Write the given numbers in order from least to greatest.

101. $2^2, -|3|, -(-5), -|-8|$

102. $|10|, 2^3, -|-5|, -(-4)$

103. $|-1|, -|-6|, -(-6), -|1|$

104. $1^4, -(-3), -|7|, |-20|$

105. $-(-2), 5^2, -10, -|-9|, |-12|$

106. $3^3, -|-11|, -(-10), -4, -|2|$

Choose all numbers for x from each given list that make each statement true.

107. $|x| > 8$

a. -9 **b.** -5 **c.** 8 **d.** -12

108. $|x| > 4$

a. 0 **b.** -4 **c.** 5 **d.** -100

109. Evaluate: $-(-|-8|)$

110. Evaluate: $(-|-(-7)|)$

Answer true or false for Exercises 111 through 115.

111. If $a > b$, then a must be a positive number.

112. The absolute value of a number is *always* a positive number.

113. A positive number is always greater than a negative number.

114. Zero is always less than a positive number.

115. The number $-a$ is always a negative number. (*Hint:* Read "−" as "the opposite of.")

116. Given the number line is it true that $b < a$?

117. Write in your own words how to find the absolute value of a signed number.

118. Explain how to determine which of two signed numbers is larger.

For Exercises 119 and 120, see the first Concept Check in this section.

119. Is there a largest negative number? If so, what is it?

120. Is there a smallest positive number? If so, what is it?

2.2 Adding Integers

Objectives

A Add Integers.

B Evaluate an Algebraic Expression by Adding.

C Solve Problems by Adding Integers.

Objective A Adding Integers

Adding integers can be visualized using a number line. A positive number can be represented on the number line by an arrow of appropriate length pointing to the right, and a negative number by an arrow of appropriate length pointing to the left.

Both arrows represent 2 or +2. They both point to the right and they are both 2 units long.

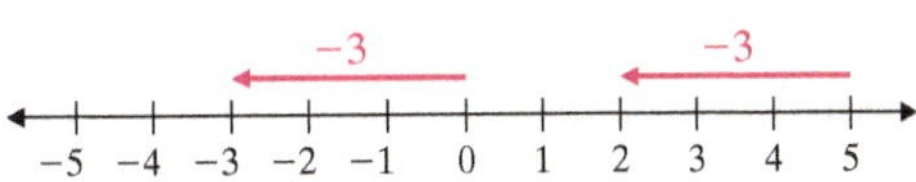

Both arrows represent −3. They both point to the left and they are both 3 units long.

Example 1 Add using a number line: $5 + (-2)$

Solution: To add integers on a number line, such as $5 + (-2)$, we start at 0 on the number line and draw an arrow representing 5. From the tip of this arrow, we draw another arrow representing −2. The tip of the second arrow ends at their sum, 3.

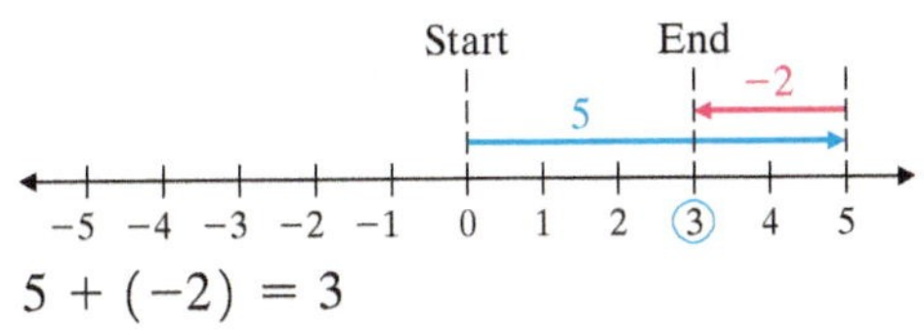

$$5 + (-2) = 3$$

Work Practice 1

Practice 1

Add using a number line:
$5 + (-1)$

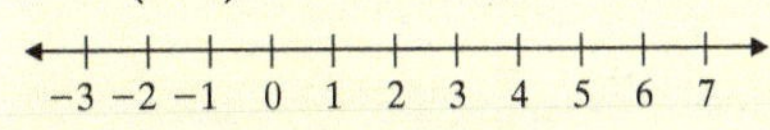

Answer

1.

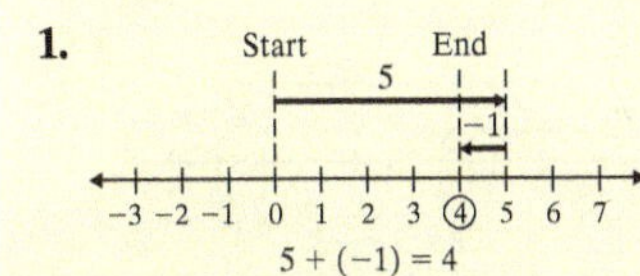

Practice 2

Add using a number line:
$-6 + (-2)$

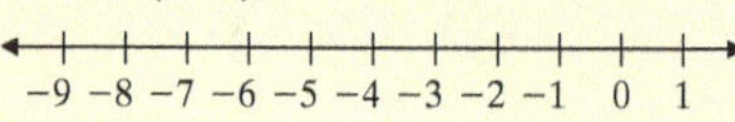

Example 2

Add using a number line: $-1 + (-4)$

Start at 0 and draw an arrow representing -1. From the tip of this arrow, we draw another arrow representing -4. The tip of the second arrow ends at their sum, -5.

Solution:

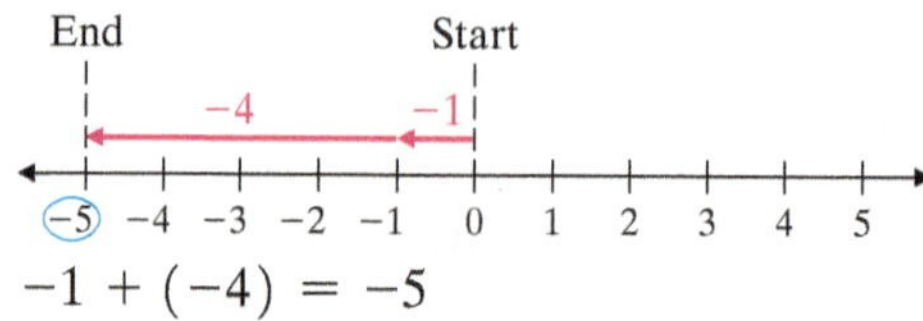

Work Practice 2

Practice 3

Add using a number line:
$-8 + 3$

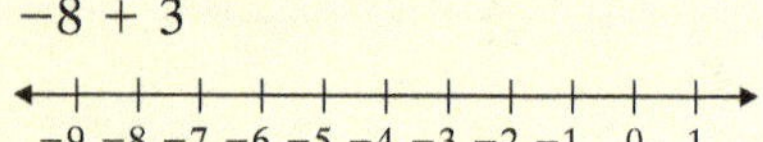

Example 3

Add using a number line: $-7 + 3$

Solution:

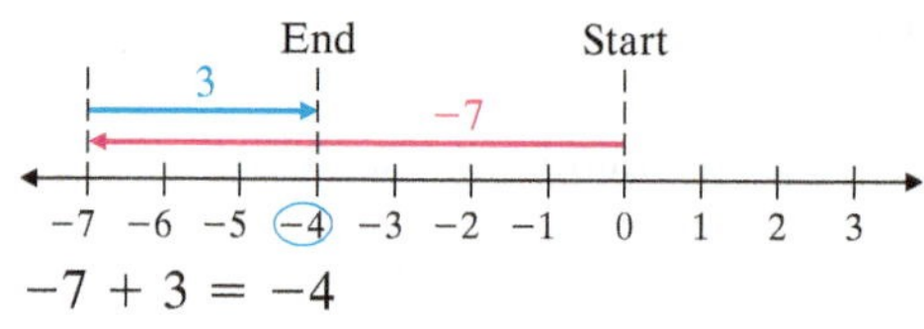

$-7 + 3 = -4$

Work Practice 3

Using a number line each time we add two numbers can be time consuming. Instead, we can notice patterns in the previous examples and write rules for adding signed numbers.

Rules for adding signed numbers depend on whether we are adding numbers with the same sign or different signs. When adding two numbers with the same sign, as in Example 2, notice that the sign of the sum is the same as the sign of the addends.

Adding Two Numbers with the Same Sign

Step 1: Add their absolute values.

Step 2: Use their common sign as the sign of the sum.

Practice 4

Add: $(-3) + (-19)$

Practice 5–6

Add.

5. $-12 + (-30)$ **6.** $9 + 4$

Example 4

Add: $-2 + (-21)$

Solution:

Step 1: $|-2| = 2$, $|-21| = 21$, and $2 + 21 = 23$.

Step 2: Their common sign is negative, so the sum is negative:

$$-2 + (-21) = -23$$

Work Practice 4

Examples

Add.

5. $-15 + (-10) = -25$

6. $2 + 6 = 8$

Work Practice 5–6

When adding two numbers with different signs, as in Examples 1 and 3, the sign of the result may be positive or negative, or the result may be 0.

Answers

2.

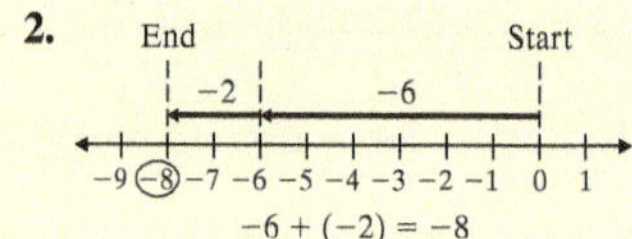

3.

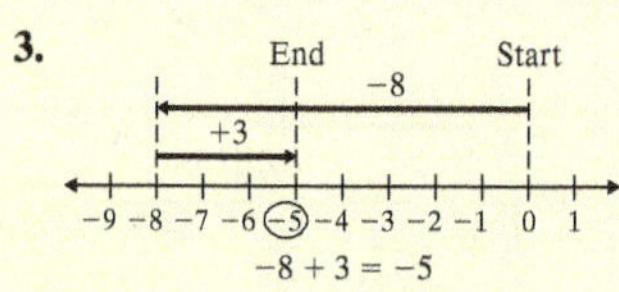

4. -22 **5.** -42 **6.** 13

Adding Two Numbers with Different Signs

Step 1: Find the larger absolute value minus the smaller absolute value.

Step 2: Use the sign of the number with the larger absolute value as the sign of the sum.

Example 7 Add: $-2 + 25$

Solution:

Step 1: $|-2| = 2$, $|25| = 25$, and $25 - 2 = 23$.

Step 2: 25 has the larger absolute value and its sign is an understood +:
$-2 + 25 = +23$ or 23

Work Practice 7

Practice 7

Add: $-1 + 26$

Example 8 Add: $3 + (-17)$

Solution:

Step 1: $|3| = 3$, $|-17| = 17$, and $17 - 3 = 14$.

Step 2: -17 has the larger absolute value and its sign is $-$:
$3 + (-17) = -14$

Work Practice 8

Practice 8

Add: $2 + (-18)$

Examples Add.

9. $-18 + 10 = -8$

10. $12 + (-8) = 4$

11. $0 + (-5) = -5$ The sum of 0 and any number is the number.

Work Practice 9–11

Practice 9–11

Add.

9. $-54 + 20$

10. $7 + (-2)$

11. $-3 + 0$

Recall that numbers such as 7 and -7 are called opposites. In general, the sum of a number and its opposite is always 0.

$7 + (-7) = 0$ (opposites) $\quad -26 + 26 = 0$ (opposites) $\quad 1008 + (-1008) = 0$ (opposites)

If a is a number, then

$-a$ is its opposite. Also,

$$\left.\begin{aligned} a + (-a) &= 0 \\ -a + a &= 0 \end{aligned}\right\}$$ The sum of a number and its opposite is 0.

Examples Add.

12. $-21 + 21 = 0$

13. $36 + (-36) = 0$

Work Practice 12–13

Practice 12–13

Add.

12. $18 + (-18)$

23. $-64 + 64$

✓**Concept Check** What is wrong with the following calculation?

$5 + (-22) = 17$

Answers

7. 25 **8.** -16 **9.** -34 **10.** 5 **11.** -3 **12.** 0 **13.** 0

✓ **Concept Check Answer**

$5 + (-22) = -17$

In the following examples, we add three or more integers. Remember that by the associative and commutative properties for addition, we may add numbers in any order that we wish. In Examples 14 and 15, let's add the numbers from left to right.

Practice 14

Add: $6 + (-2) + (-15)$

Example 14 Add: $(-3) + 4 + (-11)$

Solution: $(-3) + 4 + (-11) = 1 + (-11)$

$= -10$

Work Practice 14

Practice 15

Add: $5 + (-3) + 12 + (-14)$

Example 15 Add: $1 + (-10) + (-8) + 9$

Solution: $1 + (-10) + (-8) + 9 = -9 + (-8) + 9$

$= -17 + 9$

$= -8$

Work Practice 15

Helpful Hint Don't forget that addition is commutative and associative. In other words, numbers may be added in any order.

A sum is the same if we add the numbers in any order. To see this, let's add the numbers in Example 15 by first adding the positive numbers together and the negative numbers together.

$1 + (-10) + (-8) + 9 = 10 + (-18)$ Add the positive numbers: $1 + 9 = 10$. Add the negative numbers: $(-10) + (-8) = -18$.

$= -8$ Add these results.

The sum is -8.

Objective B Evaluating Algebraic Expressions 

We can continue our work with algebraic expressions by evaluating expressions given integer replacement values.

Practice 16

Evaluate $x + 3y$ for $x = -6$ and $y = 2$.

Example 16 Evaluate $2x + y$ for $x = 3$ and $y = -5$.

Solution: Replace x with 3 and y with -5 in $2x + y$.

$2x + y = 2 \cdot 3 + (-5)$

$= 6 + (-5)$

$= 1$

Work Practice 16

Practice 17

Evaluate $x + y$ for $x = -13$ and $y = -9$.

Example 17 Evaluate $x + y$ for $x = -2$ and $y = -10$.

Solution: $x + y = (-2) + (-10)$ Replace x with -2 and y with -10.

$= -12$

Work Practice 17

Objective C Solving Problems by Adding Integers

Next, we practice solving problems that require adding integers.

Practice 18

If the temperature was $-7°$ Fahrenheit at 6 a.m., and it rose 4 degrees by 7 a.m. and then rose another 7 degrees in the hour from 7 a.m. to 8 a.m., what was the temperature at 8 a.m.?

Example 18 Calculating Temperature

In Philadelphia, Pennsylvania, the record extreme high temperature is 104°F. Decrease this temperature by 111 degrees, and the result is the record extreme low temperature. Find this temperature. (*Source:* National Climatic Data Center)

Answers

14. -11 **15.** 0 **16.** 0 **17.** -22 **18.** 4°F

Solution:

In words:	extreme low temperature	=	extreme high temperature	+	decrease of of 111°
	↓		↓		↓
Translate:	extreme low temperature	=	104	+	(−111)
		= −7			

The record extreme low temperature in Philadelphia, Pennsylvania, is −7°F.

Work Practice 18

Calculator Explorations Entering Negative Numbers

To enter a negative number on a calculator, find the key marked [+/−]. (Some calculators have a key marked [CHS] and some calculators have a special key [(−)] for entering a negative sign.) To enter the number −2, for example, press the keys [2] [+/−]. The display will read [−2].

To find −32 + (−131), press the keys
[32] [+/−] [+] [131] [+/−] [=] or
[(−)] [32] [+] [(−)] [131] [ENTER]

The display will read [−163].

Thus −32 + (−131) = −163.

Use a calculator to perform each indicated operation.

1. −256 + 97

2. 811 + (−1058)

3. 6(15) + (−46)

4. −129 + 10(48)

5. −108,650 + (−786,205)

6. −196,662 + (−129,856)

Vocabulary, Readiness & Video Check

Use the choices below to fill in each blank. Not all choices will be used.

$-a$ a 0   commutative   associative

1. If n is a number, then $-n + n =$ ______.

2. Since $x + n = n + x$, we say that addition is ______.

3. If a is a number, then $-(-a) =$ ______.

4. Since $n + (x + a) = (n + x) + a$, we say that addition is ______.

Martin-Gay Interactive Videos

See Video 2.2

Watch the section lecture video and answer the following questions.

Objective A **5.** What is the sign of the sum in Example 6 and why?

Objective B **6.** What is the sign of the sum in Example 8 and why?

Objective C **7.** What does the answer to Example 10, −231, mean in the context of the application?

2.2 Exercise Set MyMathLab® 

Objective A *Add using a number line. See Examples 1 through 3.*

1. $-1 + (-6)$

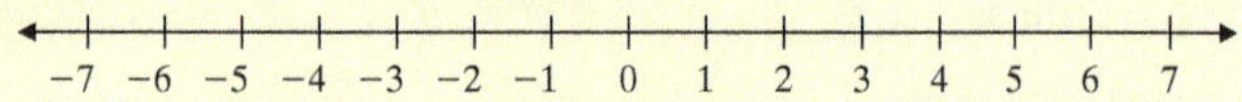

2. $-6 + (-5)$

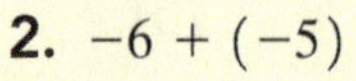

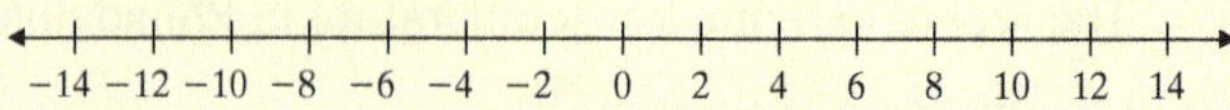

3. $-4 + 7$

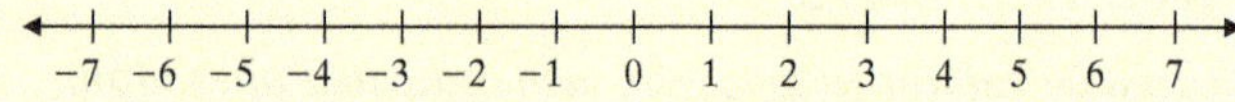

4. $10 + (-3)$

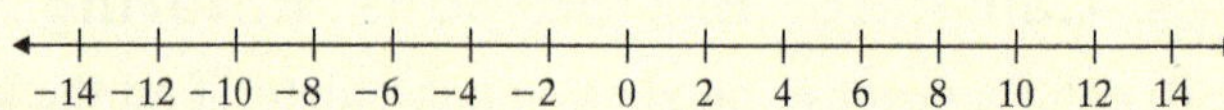

5. $-13 + 7$

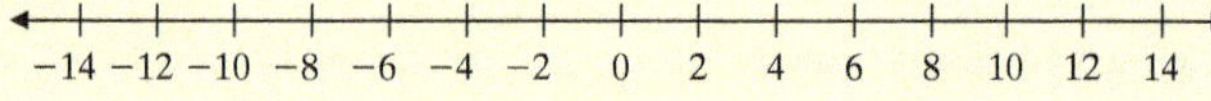

6. $9 + (-4)$

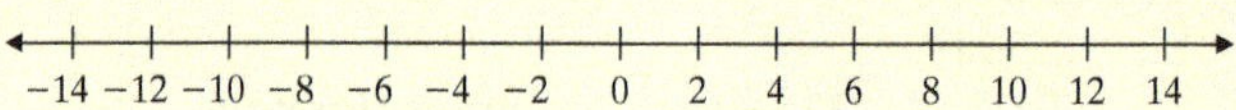

Add. See Examples 4 through 13.

7. $46 + 21$
8. $15 + 42$
9. $-8 + (-2)$
10. $-5 + (-4)$
11. $-43 + 43$
12. $-62 + 62$
13. $6 + (-2)$
14. $8 + (-3)$
15. $-6 + 0$
16. $-8 + 0$
17. $3 + (-5)$
18. $5 + (-9)$
19. $-2 + (-7)$
20. $-6 + (-1)$
21. $-12 + (-12)$
22. $-23 + (-23)$
23. $-640 + (-200)$
24. $-400 + (-256)$
25. $12 + (-5)$
26. $24 + (-10)$
27. $-6 + 3$
28. $-8 + 4$
29. $-56 + 26$
30. $-89 + 37$
31. $-45 + 85$
32. $-32 + 62$
33. $124 + (-144)$
34. $325 + (-375)$
35. $-82 + (-43)$
36. $-56 + (-33)$

Add. See Examples 14 and 15.

37. $-4 + 2 + (-5)$
38. $-1 + 5 + (-8)$
39. $-52 + (-77) + (-117)$
40. $-103 + (-32) + (-27)$
41. $12 + (-4) + (-4) + 12$
42. $18 + (-9) + 5 + (-2)$
43. $(-10) + 14 + 25 + (-16)$
44. $34 + (-12) + (-11) + 213$

Objective A Mixed Practice *Add. See Examples 1 through 15.*

45. $-6 + (-15) + (-7)$

46. $-12 + (-3) + (-5)$

47. $-26 + 15$

48. $-35 + (-12)$

49. $5 + (-2) + 17$

50. $3 + (-23) + 6$

51. $-13 + (-21)$

52. $-100 + 70$

53. $3 + 14 + (-18)$

54. $(-45) + 22 + 20$

55. $-92 + 92$

56. $-87 + 0$

57. $-13 + 8 + (-10) + (-27)$

58. $-16 + 6 + (-14) + (-20)$

Objective B *Evaluate $x + y$ for the given replacement values. See Examples 16 and 17.*

59. $x = -20$ and $y = -50$

60. $x = -1$ and $y = -29$

Evaluate $3x + y$ for the given replacement values. See Examples 16 and 17.

61. $x = 2$ and $y = -3$

62. $x = 7$ and $y = -11$

63. $x = 3$ and $y = -30$

64. $x = 13$ and $y = -17$

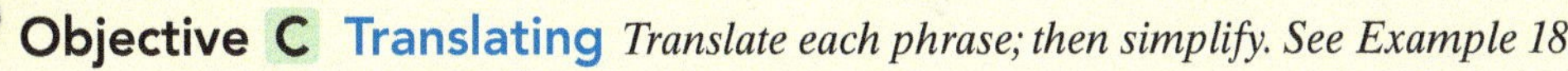

Objective C Translating *Translate each phrase; then simplify. See Example 18.*

65. Find the sum of −6 and 25.

66. Find the sum of −30 and 15.

67. Find the sum of −31, −9, and 30.

68. Find the sum of −49, −2, and 40.

Solve. See Example 18.

69. Suppose a deep-sea diver dives from the surface to 215 feet below the surface. He then dives down 16 more feet. Use positive and negative numbers to represent this situation. Then find the diver's present depth.

70. Suppose a diver dives from the surface to 248 meters below the surface and then swims up 8 meters, down 16 meters, down another 28 meters, and then up 32 meters. Use positive and negative numbers to represent this situation. Then find the diver's depth after these movements.

In golf, it is possible to have positive and negative scores. The following table shows the results of the eighteen-hole Round 2 for Jim Furyk and Jason Dufner at the 2013 PGA Championship in Rochester, New York. Use the table to answer Exercises 71 and 72.

Player/Hole	1	2	3	4	5	6	7	8	9	10	11	12	13	14	15	16	17	18
Furyk	−1	0	0	0	0	0	0	0	0	−1	0	0	0	0	0	−1	1	0
Dufner	0	−2	0	−1	−1	0	0	0	0	0	−1	0	−1	0	0	−1	0	0

(*Source:* Professional Golfers' Association)

71. Find the total score for each of the athletes in the round.

72. In golf, the lower score is the winner. Use the result of Exercise 71 to determine who won Round 2.

The following bar graph shows the yearly net income for Apple, Inc. Net income is one indication of a company's health. It measures revenue (money taken in) minus cost (money spent). Use this graph to answer Exercises 73 through 76. (Source: Apple, Inc.*)*

73. What was the net income (in dollars) for Apple, Inc. in 2012?

74. What was the net income (in dollars) for Apple, Inc. in 2001?

75. Find the total net income for the years 2011 and 2012.

76. Find the total net income for the years 2007, 2009, and 2011.

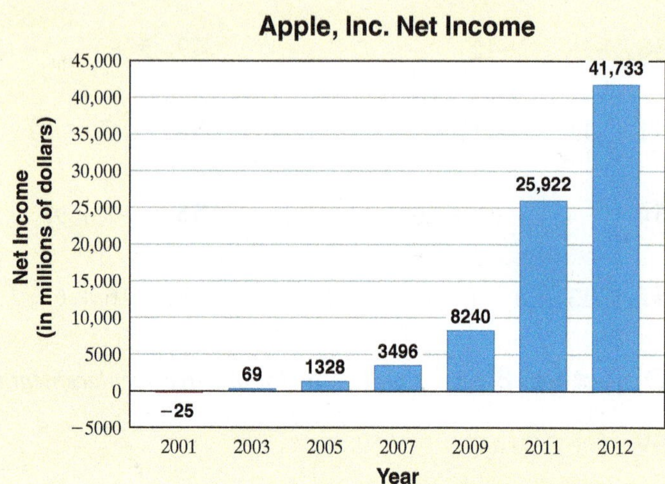

77. The temperature at 4 p.m. on February 2 was −10° Celsius. By 11 p.m. the temperature had risen 12 degrees. Find the temperature at 11 p.m.

78. In some card games, it is possible to have both positive and negative scores. After four rounds of play, Michelle had scores of 14, −5, −8, and 7. What was her total score for the game?

A small business reports the following net incomes. Use this table to answer Exercises 79 and 80.

Year	Net Income (in dollars)
2009	−$10,412
2010	−$1786
2011	$15,395
2012	$31,418

79. Find the sum of the net incomes for 2010 and 2011.

80. Find the sum of the net incomes for all four years shown.

81. The all-time record low temperature for Texas is −23°F. Florida's all-time record low temperature is 21°F higher than Texas' record low. What is Florida's record low temperature? (*Source:* National Climatic Data Center)

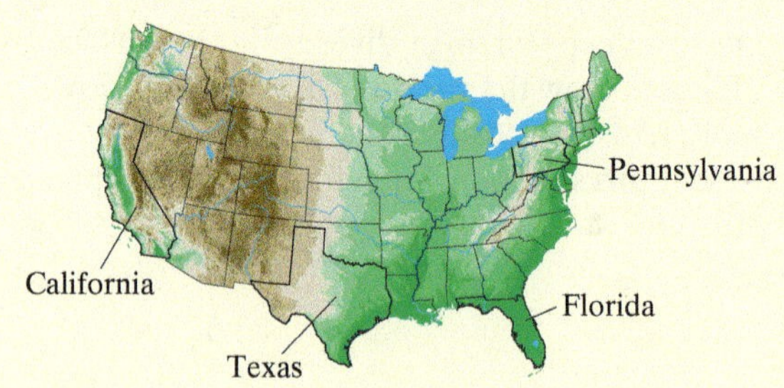

82. The all-time record low temperature for California is −45°F. In Pennsylvania, the lowest temperature ever recorded is 3°F higher than California's all-time low temperature. What is the all-time record low temperature for Pennsylvania? (*Source:* National Climatic Data Center)

83. The deepest spot in the Atlantic Ocean is the Puerto Rico Trench, which has an elevation of 8605 meters below sea level. The bottom of the Atlantic's Cayman Trench has an elevation 1070 meters above the level of the Puerto Rico Trench. Use a negative number to represent the depth of the Cayman Trench. (*Source:* Defense Mapping Agency)

84. The deepest spot in the Pacific Ocean is the Mariana Trench, which has an elevation of 10,924 meters below sea level. The bottom of the Pacific's Aleutian Trench has an elevation 3245 meters higher than that of the Mariana Trench. Use a negative number to represent the depth of the Aleutian Trench. (*Source:* Defense Mapping Agency)

Review

Subtract. See Section 1.3.

85. $44 - 0$

86. $91 - 0$

87. $200 - 59$

88. $400 - 18$

Concept Extensions

89. Name 2 numbers whose sum is -17.

90. Name 2 numbers whose sum is -30.

Each calculation below is incorrect. Find the error and correct it. See the Concept Check in this section.

91. $7 + (-10) \stackrel{?}{=} 17$

92. $-4 + 14 \stackrel{?}{=} -18$

93. $-10 + (-12) \stackrel{?}{=} -120$

94. $-15 + (-17) \stackrel{?}{=} 32$

For Exercises 95 through 98, determine whether each statement is true or false.

95. The sum of two negative numbers is always a negative number.

96. The sum of two positive numbers is always a positive number.

97. The sum of a positive number and a negative number is always a negative number.

98. The sum of zero and a negative number is always a negative number.

99. In your own words, explain how to add two negative numbers.

100. In your own words, explain how to add a positive number and a negative number.

2.3 Subtracting Integers 

Objectives

A Subtract Integers.

B Add and Subtract Integers.

C Evaluate an Algebraic Expression by Subtracting.

D Solve Problems by Subtracting Integers.

In Section 2.1, we discussed the opposite of an integer.

The opposite of 3 is -3.
The opposite of -6 is 6.

In this section, we use opposites to subtract integers.

Objective A Subtracting Integers 

To subtract integers, we will write the subtraction problem as an addition problem. To see how to do this, study the examples below.

$$10 - 4 = 6$$
$$10 + (-4) = 6$$

Since both expressions simplify to 6, this means that

$$10 - 4 = 10 + (-4) = 6$$

Also,

$$3 - 2 = 3 + (-2) = 1$$
$$15 - 1 = 15 + (-1) = 14$$

Thus, to subtract two numbers, we add the first number to the opposite of the second number. (The opposite of a number is also known as its **additive inverse.**)

Subtracting Two Numbers

If a and b are numbers, then $a - b = a + (-b)$.

Practice 1–4

Subtract.

1. $13 - 4$
2. $-8 - 2$
3. $11 - (-15)$
4. $-9 - (-1)$

Examples Subtract.

	subtraction	=	first number	+	opposite of the second number		
1.	$8 - 5$	=	8	+	(-5)	=	3
2.	$-4 - 10$	=	-4	+	(-10)	=	-14
3.	$6 - (-5)$	=	6	+	5	=	11
4.	$-11 - (-7)$	=	-11	+	7	=	-4

Work Practice 1–4

Practice 5–7

Subtract.

5. $6 - 9$
6. $-14 - 5$
7. $-3 - (-4)$

Examples Subtract.

5. $-10 - 5 = -10 + (-5) = -15$

6. $8 - 15 = 8 + (-15) = -7$

7. $-4 - (-5) = -4 + 5 = 1$

Work Practice 5–7

Helpful Hint

To visualize subtraction, try the following:

The difference between 5°F and −2°F can be found by subtracting. That is,

$$5 - (-2) = 5 + 2 = 7$$

Can you visually see from the thermometer on the right that there are actually 7 degrees between 5°F and −2°F?

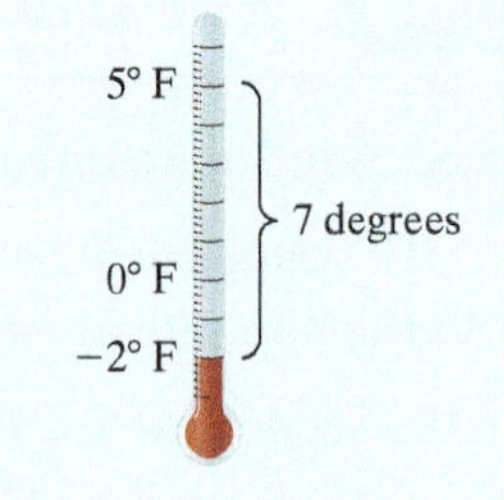

✓Concept Check What is wrong with the following calculation?

$$-9 - (-5) = -14$$

Practice 8

Subtract 6 from −15.

Example 8 Subtract 7 from −3.

Solution: To subtract 7 *from* −3, we find

$$-3 - 7 = -3 + (-7) = -10$$

Work Practice 8

Answers

1. 9 2. −10 3. 26 4. −8
5. −3 6. −19 7. 1 8. −21

✓Concept Check Answer

$-9 - (-5) = -9 + 5 = -4$

Objective B Adding and Subtracting Integers 

If a problem involves adding or subtracting more than two integers, we rewrite differences as sums and add. Recall that by associative and commutative properties, we may add numbers in any order. In Examples 9 and 10, we will add from left to right.

Example 9 Simplify: $7 - 8 - (-5) - 1$

Solution:

$$\begin{aligned} 7 - 8 - (-5) - 1 &= \underbrace{7 + (-8)} + 5 + (-1) \\ &= \underbrace{-1 \quad + 5} + (-1) \\ &= \underbrace{4 \quad + (-1)} \\ &= 3 \end{aligned}$$

Work Practice 9

Practice 9

Simplify: $-6 - 5 - 2 - (-3)$

Example 10 Simplify: $7 + (-12) - 3 - (-8)$

Solution:

$$\begin{aligned} 7 + (-12) - 3 - (-8) &= \underbrace{7 + (-12)} + (-3) + 8 \\ &= \underbrace{-5 \quad + (-3)} + 8 \\ &= -8 \quad + 8 \\ &= 0 \end{aligned}$$

Work Practice 10

Practice 10

Simplify: $8 + (-2) - 9 - (-7)$

Objective C Evaluating Expressions

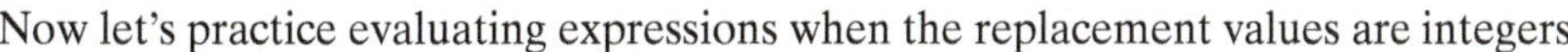

Now let's practice evaluating expressions when the replacement values are integers.

Example 11 Evaluate $x - y$ for $x = -3$ and $y = 9$.

Solution: Replace x with -3 and y with 9 in $x - y$.

$$\begin{aligned} & x \quad - \quad y \\ & \downarrow \quad \downarrow \quad \downarrow \\ &= (-3) - 9 \\ &= (-3) + (-9) \\ &= -12 \end{aligned}$$

Work Practice 11

Practice 11

Evaluate $x - y$ for $x = -5$ and $y = 13$.

Example 12 Evaluate $2a - b$ for $a = 8$ and $b = -6$.

Solution: Watch your signs carefully!

$$\begin{aligned} & 2a \quad - \quad b \\ & \downarrow \quad \downarrow \quad \downarrow \\ &= 2 \cdot 8 - (-6) && \text{Replace } a \text{ with 8 and } b \text{ with } -6. \\ &= 16 + 6 && \text{Multiply.} \\ &= 22 && \text{Add.} \end{aligned}$$

Helpful Hint Watch carefully when replacing variables in the expression $2a - b$. Make sure that all symbols are inserted and accounted for.

Work Practice 12

Practice 12

Evaluate $3y - z$ for $y = 9$ and $z = -4$.

Answers

9. −10 10. 4 11. −18 12. 31

Objective D Solving Problems by Subtracting Integers

Solving problems often requires subtraction of integers.

Practice 13

The highest point in Asia is the top of Mount Everest, at a height of 29,028 feet above sea level. The lowest point is the Dead Sea, which is 1312 feet below sea level. How much higher is Mount Everest than the Dead Sea? (*Source:* National Geographic Society)

Example 13 Finding a Change in Elevation

The highest point in the United States is the top of Mount McKinley, at a height of 20,320 feet above sea level. The lowest point is Death Valley, California, which is 282 feet below sea level. How much higher is Mount McKinley than Death Valley? (*Source:* U.S. Geological Survey)

Solution:

1. UNDERSTAND. Read and reread the problem. To find "how much higher," we subtract. Don't forget that since Death Valley is 282 feet *below* sea level, we represent its height by −282. Draw a diagram to help visualize the problem.

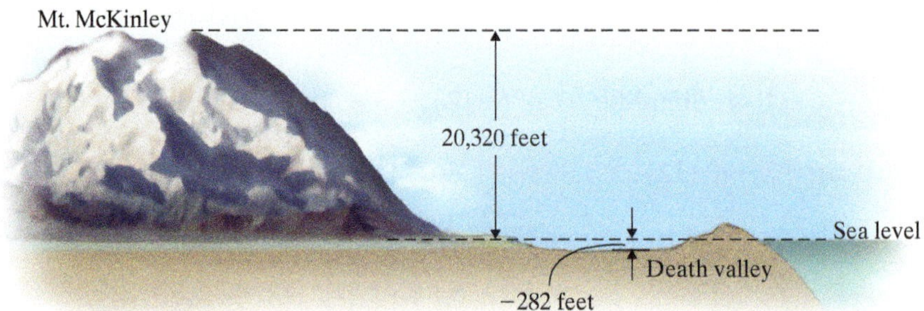

2. TRANSLATE.

In words:	how much higher is Mt. McKinley	=	height of Mt. McKinley	minus	height of Death Valley
	↓	↓	↓	↓	↓
Translate:	how much higher is Mt. McKinley	=	20,320	−	(−282)

3. SOLVE:

$20{,}320 - (-282) = 20{,}320 + 282 = 20{,}602$

4. INTERPRET. Check and state your conclusion: Mount McKinley is 20,602 feet higher than Death Valley.

Work Practice 13

Answer
13. 30,340 ft

Vocabulary, Readiness & Video Check

Multiple choice: Select the correct lettered response following each exercise.

1. It is true that $a - b =$ ________.
a. $b - a$ **b.** $a + (-b)$ **c.** $a + b$

2. The opposite of n is ________.
a. $-n$ **b.** $-(-n)$ **c.** n

3. To evaluate $x - y$ for $x = -10$ and $y = -14$, we replace x with -10 and y with -14 and evaluate ________.
a. $10 - 14$ **b.** $-10 - 14$ **c.** $-14 - 10$ **d.** $-10 - (-14)$

4. The expression $-5 - 10$ equals ________.
a. $5 - 10$ **b.** $5 + 10$ **c.** $-5 + (-10)$ **d.** $10 - 5$

Watch the section lecture video and answer the following questions.

Objective A 5. In the lecture before Example 1, what can the "opposite" of a number also be called?

Objective B 6. In Example 7, how is the example rewritten in the first step of simplifying and why?

Objective C 7. In Example 8, why do we multiply first?

Objective D 8. What does the answer to Example 9, 265, mean in the context of the application?

2.3 Exercise Set MyMathLab®

Objective A *Subtract. See Examples 1 through 7.*

1. $-8-(-8)$ **2.** $-6-(-6)$ **3.** $19-16$ **4.** $15-12$

5. $3-8$ **6.** $2-5$ **7.** $11-(-11)$ **8.** $12-(-12)$

9. $-4-(-7)$ **10.** $-25-(-25)$ **11.** $-16-4$ **12.** $-2-42$

13. $3-15$ **14.** $8-9$ **15.** $42-55$ **16.** $17-63$

17. $478-(-30)$ **18.** $844-(-20)$ **19.** $-4-10$ **20.** $-5-8$

21. $-7-(-3)$ **22.** $-12-(-5)$ **23.** $17-29$ **24.** $16-45$

Translating *Translate each phrase; then simplify. See Example 8.*

25. Subtract 17 from −25.

26. Subtract 10 from −22.

27. Find the difference of −22 and −3.

28. Find the difference of −8 and −13.

29. Subtract −12 from 2.

30. Subtract −50 from −50.

Mixed Practice (Sections 2.2, 2.3) *Add or subtract as indicated.*

31. $-37+(-19)$ **32.** $-35+(-11)$ **33.** $8-13$ **34.** $4-21$

35. $-56-89$ **36.** $-105-68$ **37.** $30-67$ **38.** $86-98$

Objective B *Simplify. See Examples 9 and 10.*

39. $8-3-2$ **40.** $8-4-1$ **41.** $13-5-7$

42. $30-18-12$ **43.** $-5-8-(-12)$ **44.** $-10-6-(-9)$

45. $-11+(-6)-14$ **46.** $-15+(-8)-4$ **47.** $18-(-32)+(-6)$

48. $23 - (-17) + (-9)$

49. $-(-5) - 21 + (-16)$

50. $-(-9) - 14 + (-23)$

51. $-10 - (-12) + (-7) - 4$

52. $-6 - (-8) + (-12) - 7$

53. $-3 + 4 - (-23) - 10$

54. $5 + (-18) - (-21) - 2$

Objective C *Evaluate* $x - y$ *for the given replacement values. See Examples 11 and 12.*

55. $x = -4$ and $y = 7$

56. $x = -7$ and $y = 1$

57. $x = 8$ and $y = -23$

58. $x = 9$ and $y = -2$

Evaluate $2x - y$ *for the given replacement values. See Examples 11 and 12.*

59. $x = 4$ and $y = -4$

60. $x = 8$ and $y = -10$

61. $x = 1$ and $y = -18$

62. $x = 14$ and $y = -12$

Objective **D** *Solve. See Example 13.*

The bar graph shows the monthly average temperatures in Fairbanks, Alaska. Notice that a negative temperatures is illustrated by a bar below the horizontal line representing 0°F. Use this graph to answer Exercises 63 through 66.

63. Find the difference in temperature between the months of March and February.

64. Find the difference in temperature between the months of November and December.

65. Find the difference in temperature between the two months with the lowest temperatures.

66. Find the difference in temperature between the month with the warmest temperature and the month with the coldest temperature.

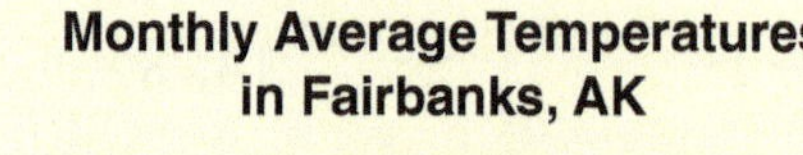

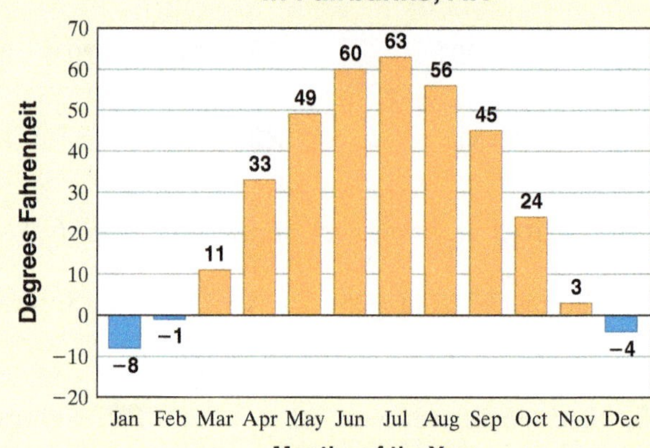

Source: National Climatic Data Center

Solve.

67. The coldest temperature ever recorded on Earth was −129°F in Antarctica. The warmest temperature ever recorded was 134°F in Death Valley, California. How many degrees warmer is 134°F than −129°F? (*Source: The World Almanac,* 2013)

68. The coldest temperature ever recorded in the United States was −80°F in Alaska. The warmest temperature ever recorded was 134°F in California. How many degrees warmer is 134°F than −80°F? (*Source: The World Almanac,* 2013)

69. Adam Scott from Australia finished first in the 2013 PGA Master's Tournament at Augusta National Golf Club with a score of −9, or nine strokes under par. Tied for 46th place was Thomas Bjorn from Denmark, with a score of +5, or 5 strokes over par. What was the difference in scores between Scott and Bjorn?

70. A woman received a statement of her charge account at Old Navy. She spent \$93 on purchases last month. She returned an \$18 blouse because she didn't like the color. She also returned a \$26 pajama set because it was damaged. What does she actually owe on her account?

71. The temperature on a February morning was $-4°$ Celsius at 6 a.m. If the temperature drops 3 degrees by 7 a.m., rises 4 degrees between 7 a.m. and 8 a.m., and then drops 7 degrees between 8 a.m. and 9 a.m., find the temperature at 9 a.m.

72. Mauna Kea in Hawaii has an elevation of 13,796 feet above sea level. The Mid-America Trench in the Pacific Ocean has an elevation of 21,857 feet below sea level. Find the difference in elevation between those two points. (*Source:* National Geographic Society and Defense Mapping Agency)

Some places on Earth lie below sea level, which is the average level of the surface of the oceans. Use this diagram to answer Exercises 73 through 76. (Source: Fantastic Book of Comparisons, Russell Ash*)*

73. Find the difference in elevation between Death Valley and Qattâra Depression.

74. Find the difference in elevation between the Danakil and Turfan Depressions.

75. Find the difference in elevation between the two lowest elevations shown.

76. Find the difference in elevation between the highest elevation shown and the lowest elevation shown.

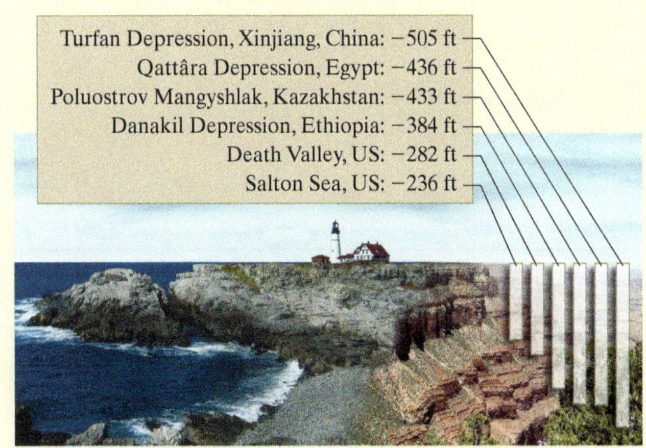

The bar graph from Section 2.1 shows heights of selected lakes. For Exercises 77 through 80, find the difference in elevation for the lakes listed. (Source: U.S. Geological Survey*)*

77. Lake Superior and Lake Eyre

78. Great Bear Lake and Caspian Sea

79. Lake Maracaibo and Lake Vanern

80. Lake Eyre and Caspian Sea

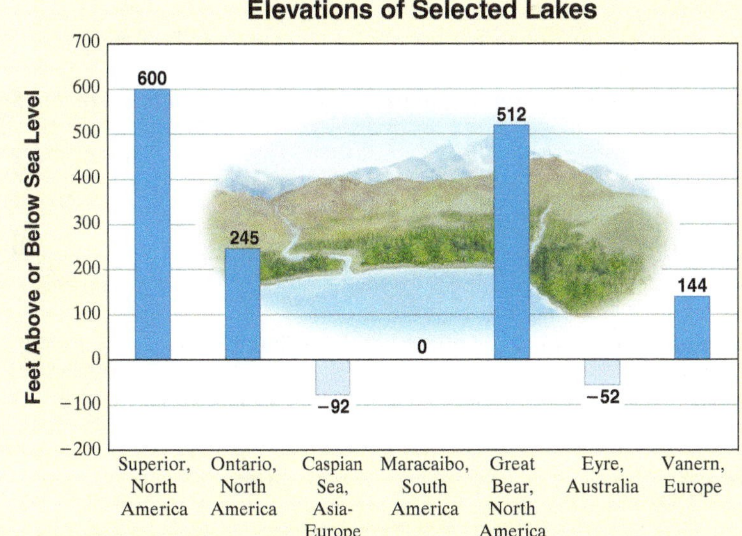

Solve.

81. The average daytime surface temperature of the hottest planet, Venus, is 867°F, while the average daytime surface temperature of the coldest planet, Neptune, is -330°F. Find the difference in temperatures.

82. The average daytime surface temperature of Mercury is 845°F, while the average daytime surface temperature of Jupiter is -162°F. Find the difference in temperatures.

83. The difference between a country's exports and imports is called the country's *trade balance*. In June 2013, the United States had \$191 billion in exports and \$225 billion in imports. What was the U.S. trade balance in June 2013? (*Source:* U.S. Department of Commerce)

84. In 2012, the United States exported 1165 million barrels of petroleum products and imported 3878 million barrels of petroleum products. What was the U.S. trade balance for petroleum products in 2012? (*Source:* U.S. Energy Information Administration)

Mixed Practice–Translating (*Sections 2.2, 2.3*)

Translate each phrase to an algebraic expression. Use "x" to represent "a number."

85. The sum of -5 and a number.

86. The difference of -3 and a number.

87. Subtract a number from -20.

88. Add a number and -36.

Review

Multiply or divide as indicated. See Sections 1.5 and 1.6.

89. $\frac{100}{20}$

90. $\frac{96}{3}$

91. $\begin{array}{r} 23 \\ \times\ 46 \\ \hline \end{array}$

92. $\begin{array}{r} 51 \\ \times\ 89 \\ \hline \end{array}$

Concept Extensions

93. Name two numbers whose difference is -3.

94. Name two numbers whose difference is -10.

Each calculation below is ***incorrect.*** *Find the error and correct it. See the Concept Check in this section.*

95. $9 - (-7) \stackrel{?}{=} 2$

96. $-4 - 8 \stackrel{?}{=} 4$

97. $10 - 30 \stackrel{?}{=} 20$

98. $-3 - (-10) \stackrel{?}{=} -13$

Simplify. (Hint: Find the absolute values first.)

99. $|-3| - |-7|$

100. $|-12| - |-5|$

101. $|-5| - |5|$

102. $|-8| - |8|$

103. $|-15| - |-29|$

104. $|-23| - |-42|$

For Exercises 105 and 106, determine whether each statement is true or false.

105. $|-8 - 3| = 8 - 3$

106. $|-2 - (-6)| = |-2| - |-6|$

107. In your own words, explain how to subtract one signed number from another.

108. A student explains to you that the first step to simplify $8 + 12 \cdot 5 - 100$ is to add 8 and 12. Is the student correct? Explain why or why not.

2.4 Multiplying and Dividing Integers 

Multiplying and dividing integers is similar to multiplying and dividing whole numbers. One difference is that we need to determine whether the result is a positive number or a negative number.

Objectives

A Multiply Integers.

B Divide Integers.

C Evaluate an Algebraic Expression by Multiplying or Dividing.

D Solve Problems by Multiplying or Dividing Integers.

Objective A Multiplying Integers

Consider the following pattern of products.

First factor decreases by 1 each time.

$$3 \cdot 2 = 6$$
$$2 \cdot 2 = 4$$
$$1 \cdot 2 = 2$$
$$0 \cdot 2 = 0$$

Product decreases by 2 each time.

This pattern can be continued, as follows.

$$-1 \cdot 2 = -2$$
$$-2 \cdot 2 = -4$$
$$-3 \cdot 2 = -6$$

This suggests that the product of a negative number and a positive number is a negative number.

What is the sign of the product of two negative numbers? To find out, we form another pattern of products. Again, we decrease the first factor by 1 each time, but this time the second factor is negative.

$$2 \cdot (-3) = -6$$
$$1 \cdot (-3) = -3$$
$$0 \cdot (-3) = 0$$

Product increases by 3 each time.

This pattern continues as:

$$-1 \cdot (-3) = 3$$
$$-2 \cdot (-3) = 6$$
$$-3 \cdot (-3) = 9$$

This suggests that the product of two negative numbers is a positive number. Thus we can determine the sign of a product when we know the signs of the factors.

Multiplying Numbers

The product of two numbers having the same sign is a positive number.

Product of Like Signs

$(+)(+) = +$

$(-)(-) = +$

The product of two numbers having different signs is a negative number.

Product of Different Signs

$(-)(+) = -$

$(+)(-) = -$

Examples Multiply.

1. $-7 \cdot 3 = -21$
2. $-3(-5) = 15$
3. $0 \cdot (-4) = 0$
4. $10(-8) = -80$

Work Practice 1–4

Practice 1–4

Multiply.

1. $-3 \cdot 8$
2. $-5(-2)$
3. $0 \cdot (-20)$
4. $10(-5)$

Answers

1. -24 **2.** 10 **3.** 0 **4.** -50

Recall that by the associative and commutative properties for multiplication, we may multiply numbers in any order that we wish. In Example 5, we multiply from left to right.

Practice 5–7

Multiply.

5. $8(-6)(-2)$
6. $(-9)(-2)(-1)$
7. $(-3)(-4)(-5)(-1)$

Examples Multiply.

5. $7(-6)(-2) = -42(-2)$
$= 84$

6. $(-2)(-3)(-4) = 6(-4)$
$= -24$

7. $(-1)(-2)(-3)(-4) = -1(-24)$ We have -24 from Example 6.
$= 24$

Work Practice 5–7

✓Concept Check What is the sign of the product of five negative numbers? Explain.

Recall from our study of exponents that $2^3 = 2 \cdot 2 \cdot 2 = 8$. We can now work with bases that are negative numbers. For example,

$$(-2)^3 = (-2)(-2)(-2) = -8$$

Practice 8

Evaluate $(-2)^4$.

Example 8 Evaluate: $(-5)^2$

Solution: Remember that $(-5)^2$ means 2 factors of -5.

$$(-5)^2 = (-5)(-5) = 25$$

Work Practice 8

Helpful Hint

Have you noticed a pattern when multiplying signed numbers?

If we let $(-)$ represent a negative number and $(+)$ represent a positive number, then

$(-)(-) = (+)$

$(-)(-)(-) = (-)$

$(-)(-)(-)(-) = (+)$

$(-)(-)(-)(-)(-) = (-)$

The product of an even number of negative numbers is a positive result.

The product of an odd number of negative numbers is a negative result.

Notice in Example 8 the parentheses around -5 in $(-5)^2$. With these parentheses, -5 is the base that is squared. Without parentheses, such as -5^2, only the 5 is squared. In other words, $-5^2 = -(5 \cdot 5) = -25$.

Practice 9

Evaluate: -8^2

Example 9 Evaluate: -7^2

Solution: Remember that without parentheses, only the 7 is squared.

$$-7^2 = -(7 \cdot 7) = -49$$

Work Practice 9

Answers

5. 96 6. −18 7. 60 8. 16 9. −64

✓Concept Check Answer

negative; answers may vary

Helpful Hint

Make sure you understand the difference between Examples 8 and 9.

parentheses, so -5 is squared

$(-5)^2 = (-5)(-5) = 25$

no parentheses, so only the 7 is squared

$-7^2 = -(7 \cdot 7) = -49$

Objective B Dividing Integers 

Division of integers is related to multiplication of integers. The sign rules for division can be discovered by writing a related multiplication problem. For example,

$\frac{6}{2} = 3$ because $3 \cdot 2 = 6$

$\frac{-6}{2} = -3$ because $-3 \cdot 2 = -6$

$\frac{6}{-2} = -3$ because $-3 \cdot (-2) = 6$

$\frac{-6}{-2} = 3$ because $3 \cdot (-2) = -6$

Helpful Hint Just as for whole numbers, division can be checked by multiplication.

Dividing Numbers

The quotient of two numbers having the same sign is a positive number.

The quotient of two numbers having different signs is a negative number.

Quotient of Like Signs

$\frac{(+)}{(+)} = + \qquad \frac{(-)}{(-)} = +$

Quotient of Different Signs

$\frac{(+)}{(-)} = - \qquad \frac{(-)}{(+)} = -$

Examples Divide.

10. $\frac{-12}{6} = -2$

11. $-20 \div (-4) = 5$

12. $\frac{48}{-3} = -16$

Work Practice 10–12

Concept Check What is wrong with the following calculation?

$\frac{-36}{-9} = -4$

Practice 10–12

Divide.

10. $\frac{42}{-7}$

11. $-16 \div (-2)$

12. $\frac{-80}{10}$

Answers

10. -6 **11.** 8 **12.** -8

Concept Check Answer

$\frac{-36}{-9} = 4$

Practice 13–14

Divide, if possible.

13. $\frac{-6}{0}$ **14.** $\frac{0}{-7}$

Examples Divide, if possible.

13. $\frac{0}{-5} = 0$ because $0 \cdot -5 = 0$

14. $\frac{-7}{0}$ is undefined because there is no number that gives a product of -7 when multiplied by 0.

Work Practice 13–14

Objective C Evaluating Expressions

Next, we practice evaluating expressions given integer replacement values.

Practice 15

Evaluate xy for $x = 5$ and $y = -8$.

Example 15 Evaluate xy for $x = -2$ and $y = 7$.

Solution: Recall that xy means $x \cdot y$.

Replace x with -2 and y with 7.

$$xy = -2 \cdot 7$$
$$= -14$$

Work Practice 15

Practice 16

Evaluate $\frac{x}{y}$ for $x = -12$ and $y = -3$.

Example 16 Evaluate $\frac{x}{y}$ for $x = -24$ and $y = 6$.

Solution: $\frac{x}{y} = \frac{-24}{6}$ Replace x with -24 and y with 6.

$$= -4$$

Work Practice 16

Objective D Solving Problems by Multiplying and Dividing Integers

Many real-life problems involve multiplication and division of signed numbers.

Practice 17

A card player had a score of -13 for each of four games. Find the total score.

Example 17 Calculating a Total Golf Score

A professional golfer finished seven strokes under par (-7) for each of three days of a tournament. What was his total score for the tournament?

Solution:

1. UNDERSTAND. Read and reread the problem. Although the key word is "total," since this is repeated addition of the same number, we multiply.
2. TRANSLATE.

In words:	golfer's total score	=	number of days	·	score each day
	↓	↓	↓	↓	↓
Translate:	golfer's total	=	3	·	(−7)

3. SOLVE: $3 \cdot (-7) = -21$
4. INTERPRET. Check and state your conclusion: The golfer's total score was -21, or 21 strokes under par.

Work Practice 17

Answers
13. undefined **14.** 0 **15.** −40 **16.** 4 **17.** −52

Vocabulary, Readiness & Video Check

Use the choices below to fill in each blank. Each choice may be used more than once.

negative	0
positive	undefined

1. The product of a negative number and a positive number is a(n) ________ number.
2. The product of two negative numbers is a(n) ________ number.
3. The quotient of two negative numbers is a(n) ________ number.
4. The quotient of a negative number and a positive number is a(n) ________ number.
5. The product of a negative number and zero is ________.
6. The quotient of 0 and a negative number is ________.
7. The quotient of a negative number and 0 is ________.

Martin-Gay Interactive Videos

See Video 2.4

Watch the section lecture video and answer the following questions.

Objective A 8. Explain the role of parentheses when comparing Examples 3 and 4.

Objective B 9. Complete this statement based on the lecture before Example 6: We can find out about sign rules for division because we know sign rules for ____________.

Objective C 10. In Example 10, what are you asked to remember about the algebraic expression *ab*?

Objective D 11. In Example 12, how do we know the example will involve a negative number?

2.4 Exercise Set MyMathLab®

Objective A *Multiply. See Examples 1 through 4.*

1. $-6(-2)$
2. $5(-3)$
3. $-4(9)$
4. $-7(-2)$
5. $9(-9)$
6. $-9(7)$
7. $0(-11)$
8. $-6(0)$

Multiply. See Examples 5 through 7.

9. $6(-2)(-4)$
10. $-2(3)(-7)$
11. $-1(-3)(-4)$
12. $-8(-3)(-3)$
13. $-4(4)(-5)$
14. $2(-5)(-4)$
15. $10(-5)(0)(-7)$
16. $3(0)(-4)(-8)$
17. $-5(3)(-1)(-1)$
18. $-2(-1)(3)(-2)$

Evaluate. See Examples 8 and 9.

19. -3^2 **20.** -2^4 **21.** $(-3)^3$ **22.** $(-1)^4$

23. -6^2 **24.** -4^3 **25.** $(-4)^3$ **26.** $(-3)^2$

Objective **B** *Find each quotient. See Examples 10 through 14.*

27. $-24 \div 3$ **28.** $90 \div (-9)$ **29.** $\frac{-30}{6}$ **30.** $\frac{56}{-8}$

31. $\frac{-77}{-11}$ **32.** $\frac{-32}{4}$ **33.** $\frac{0}{-21}$ **34.** $\frac{-13}{0}$

35. $\frac{-10}{0}$ **36.** $\frac{0}{-15}$ **37.** $\frac{56}{-4}$ **38.** $\frac{-24}{-12}$

Objectives A B Mixed Practice *Multiply or divide as indicated. See Examples 1 through 14.*

39. $-14(0)$ **40.** $0(-100)$ **41.** $-5(3)$ **42.** $-6 \cdot 2$

43. $-9 \cdot 7$ **44.** $-12(13)$ **45.** $-7(-6)$ **46.** $-9(-5)$

47. $-3(-4)(-2)$ **48.** $-7(-5)(-3)$ **49.** $(-7)^2$ **50.** $(-5)^2$

51. $-\frac{25}{5}$ **52.** $-\frac{30}{5}$ **53.** $-\frac{72}{8}$ **54.** $-\frac{49}{7}$

55. $-18 \div 3$ **56.** $-15 \div 3$ **57.** $4(-10)(-3)$ **58.** $6(-5)(-2)$

59. $-30(6)(-2)(-3)$ **60.** $-20 \cdot 5 \cdot (-5) \cdot (-3)$ **61.** $\frac{-25}{0}$ **62.** $\frac{0}{-14}$

63. $\frac{120}{-20}$ **64.** $\frac{63}{-9}$ **65.** $280 \div (-40)$ **66.** $480 \div (-8)$

67. $\frac{-12}{-4}$ **68.** $\frac{-36}{-3}$ **69.** -1^4 **70.** -2^3

71. $(-2)^5$ **72.** $(-11)^2$ **73.** $-2(3)(5)(-6)$ **74.** $-1(2)(7)(-3)$

75. $(-1)^{32}$ **76.** $(-1)^{33}$ **77.** $-2(-3)(-5)$ **78.** $-2(-2)(-3)(-2)$

79. $-48 \cdot 23$ **80.** $-56 \cdot 43$ **81.** $35 \cdot (-82)$ **82.** $70 \cdot (-23)$

Objective C *Evaluate ab for the given replacement values. See Example 15.*

83. $a = -8$ and $b = 7$

84. $a = 5$ and $b = -1$

85. $a = 9$ and $b = -2$

86. $a = -8$ and $b = 8$

87. $a = -7$ and $b = -5$

88. $a = -9$ and $b = -6$

Evaluate $\frac{x}{y}$ for the given replacement values. See Example 16.

89. $x = 5$ and $y = -5$

90. $x = 9$ and $y = -3$

91. $x = -15$ and $y = 0$

92. $x = 0$ and $y = -5$

93. $x = -36$ and $y = -6$

94. $x = -10$ and $y = -10$

Evaluate xy and also $\frac{x}{y}$ for the given replacement values. See Examples 15 and 16.

95. $x = -8$ and $y = -2$

96. $x = 20$ and $y = -5$

97. $x = 0$ and $y = -8$

98. $x = -3$ and $y = 0$

Objective D Translating *Translate each phrase; then simplify. See Example 17.*

99. Find the quotient of -54 and 9.

100. Find the quotient of -63 and -3.

101. Find the product of -42 and -6.

102. Find the product of -49 and 5.

Translating *Translate each phrase to an expression. Use x to represent "a number." See Example 17.*

103. The product of -71 and a number

104. The quotient of -8 and a number

105. Subtract a number from -16.

106. The sum of a number and -12

107. -29 increased by a number

108. The difference of a number and -10

109. Divide a number by -33.

110. Multiply a number by -17.

Solve. See Example 17.

111. A football team lost four yards on each of three consecutive plays. Represent the total loss as a product of signed numbers and find the total loss.

112. An investor lost $400 on each of seven consecutive days in the stock market. Represent his total loss as a product of signed numbers and find his total loss.

113. A deep-sea diver must move up or down in the water in short steps in order to keep from getting a physical condition called the "bends." Suppose a diver moves down from the surface in five steps of 20 feet each. Represent his total movement as a product of signed numbers and find the product.

114. A weather forecaster predicts that the temperature will drop five degrees each hour for the next six hours. Represent this drop as a product of signed numbers and find the total drop in temperature.

The graph shows melting points in degrees Celsius of selected elements. Use this graph to answer Exercises 115 through 118.

115. The melting point of nitrogen is 3 times the melting point of radon. Find the melting point of nitrogen.

116. The melting point of rubidium is -1 times the melting point of mercury. Find the melting point of rubidium.

117. The melting point of argon is -3 times the melting point of potassium. Find the melting point of argon.

118. The melting point of strontium is -11 times the melting point of radon. Find the melting point of strontium.

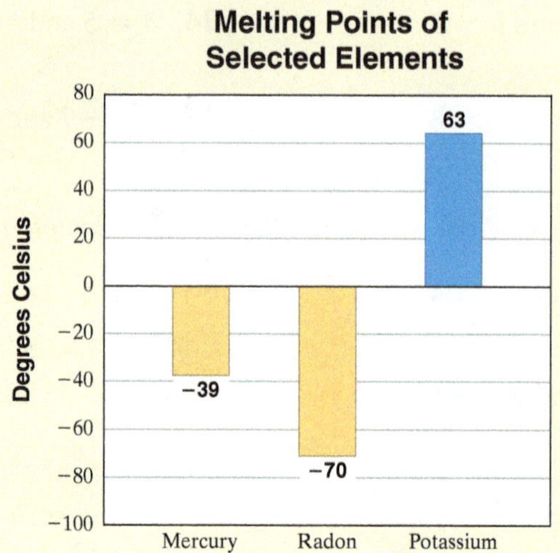

Solve. See Example 17.

119. For the first quarter of 2013, Wal-Mart, Inc. posted a loss of $33 million in membership and other income. If this trend was consistent for each month of the quarter, how much would you expect this loss to have been for each month? (*Source*: Wal-Mart Stores, Inc.)

120. For the first quarter of 2013, Chrysler Group LLC, maker of Jeep vehicles, posted a loss of about 30,000 Jeep Liberty shipments because they had stopped producing the vehicle in 2012. If this trend was consistent for each month of the quarter, how much would you expect this loss to have been for each month? (*Source*: Chrysler Group, LLC)

121. In 2008, there were 33,319 analog (nondigital) U.S. movie screens. In 2012, this number of screens dropped to 6387. (*Source:* Motion Picture Association: Worldwide Market Research)

a. Find the change in the number of U.S. analog movie screens from 2008 to 2012.

b. Find the average change per year in the number of analog movie screens over this period.

122. In 1987, the California Condor was all but extinct in the wild, with about 30 condors in the world. The condors in the wild were captured by the U.S. Fish and Wildlife Service in an aggressive move to rebuild the population by breeding them in captivity and releasing the chicks in to the wild. The condor population increased to approximately 405 birds in 2012. (*Source:* Arizona Game and Fish Department)

a. Find the change in the number of California Condors from 1987 to 2012.

b. Find the average change per year in the California Condor population over the period in part **a**.

California Condor: Average life span, 60 years

Review

Perform each indicated operation. See Section 1.7.

123. $90 + 12^2 - 5^3$

124. $3 \cdot (7 - 4) + 2 \cdot 5^2$

125. $12 \div 4 - 2 + 7$

126. $12 \div (4 - 2) + 7$

Concept Extensions

Mixed Practice (*Sections 2.2, 2.3, 2.4*) *Perform the indicated operations.*

127. $-57 \div 3$

128. $-9(-11)$

129. $-8 - 20$

130. $-4 + (-3) + 21$

131. $-4 - 15 - (-11)$

132. $-16 - (-2)$

Solve. For Exercises 133 and 134, see the first Concept Check in this section.

133. What is the sign of the product of seven negative numbers?

134. What is the sign of the product of ten negative numbers?

Without actually finding the product, write the list of numbers in Exercises 135 and 136 in order from least to greatest. For help, see a helpful hint box in this section.

135. $(-2)^{12}, (-2)^{17}, (-5)^{12}, (-5)^{17}$

136. $(-1)^{50}, (-1)^{55}, 0^{15}, (-7)^{20}, (-7)^{23}$

137. In your own words, explain how to divide two integers.

138. In your own words, explain how to multiply two integers.

Integrated Review Sections 2.1–2.4

Answers

1. ____
2. ____
3. ____
4. ____
5. ____
6. ____
7. ____
8. ____
9. ____
10. ____
11. ____
12. ____
13. ____
14. ____
15. ____
16. ____
17. ____
18. ____
19. ____
20. ____
21. ____
22. ____

Integers

1. The record low temperature in New Mexico is 50 degrees Fahrenheit below zero. The highest temperature in that state is 122 degrees above zero. Represent each quantity by an integer.

2. Graph the signed numbers on the given number line. $-4, 0, -1, 3$

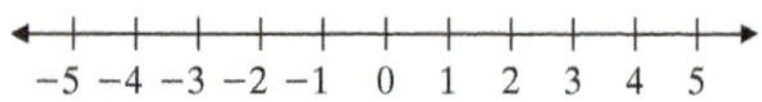

Insert $<$ or $>$ *between each pair of numbers to make a true statement.*

3. $0 \quad -10$ **4.** $-4 \quad 4$ **5.** $-15 \quad -5$ **6.** $-2 \quad -7$

Simplify.

7. $|-3|$ **8.** $|-9|$ **9.** $-|-4|$ **10.** $-(-5)$

Find the opposite of each number.

11. 11 **12.** -3 **13.** 64 **14.** 0

Perform the indicated operation.

15. $-3 + 15$ **16.** $-9 + (-11)$ **17.** $-8(-6)(-1)$ **18.** $-18 \div 2$

19. $65 + (-55)$ **20.** $1000 - 1002$ **21.** $53 - (-53)$ **22.** $-2 - 1$

23. $\frac{0}{-47}$

24. $\frac{-36}{-9}$

25. $-17 - (-59)$

26. $-8 + (-6) + 20$

27. $\frac{-95}{-5}$

28. $-9(100)$

29. $-12 - 6 - (-6)$

30. $-4 + (-8) - 16 - (-9)$

31. $\frac{-105}{0}$

32. $7(-16)(0)(-3)$

Translating *Translate each phrase; then simplify.*

33. Subtract -8 from -12.

34. Find the sum of -17 and -27.

35. Find the product of -5 and -25.

36. Find the quotient of -100 and -5.

Translating *Translate each phrase to an expression. Use x to represent "a number."*

37. Divide a number by -17

38. The sum of -3 and a number

39. A number decreased by -18

40. The product of -7 and a number

Evaluate the expressions below for $x = -3$ and $y = 12$.

41. $x + y$

42. $x - y$

43. $2y - x$

44. $3y + x$

45. $5x$

46. $\frac{y}{x}$

23. ______
24. ______
25. ______
26. ______
27. ______
28. ______
29. ______
30. ______
31. ______
32. ______
33. ______
34. ______
35. ______
36. ______
37. ______
38. ______
39. ______
40. ______
41. ______
42. ______
43. ______
44. ______
45. ______
46. ______

2.5 Order of Operations 

Objectives

A Simplify Expressions by Using the Order of Operations.

B Evaluate an Algebraic Expression.

C Find the Average of a List of Numbers.

Objective A Simplifying Expressions

We first discussed the order of operations in Chapter 1. In this section, you are given an opportunity to practice using the order of operations when expressions contain signed numbers. The rules for the order of operations from Section 1.7 are repeated here.

Order of Operations

1. Perform all operations within parentheses ( ), brackets [ ], or other grouping symbols such as fraction bars, starting with the innermost set.
2. Evaluate any expressions with exponents.
3. Multiply or divide in order from left to right.
4. Add or subtract in order from left to right.

Before simplifying other expressions, make sure you are confident simplifying Examples 1 through 3.

Practice 1-3

Find the value of each expression.

1. $(-2)^4$
2. -2^4
3. $3 \cdot 6^2$

Examples Find the value of each expression.

1. $(-3)^2 = (-3)(-3) = 9$ The base of the exponent is −3.

2. $-3^2 = -(3)(3) = -9$ The base of the exponent is 3.

3. $2 \cdot 5^2 = 2 \cdot (5 \cdot 5) = 2 \cdot 25 = 50$ The base of the exponent is 5.

Work Practice 1–3

Helpful Hint

When simplifying expressions with exponents, remember that parentheses make an important difference.

$(-3)^2$ and -3^2 **do not** mean the same thing.

$(-3)^2$ means $(-3)(-3) = 9$.

-3^2 means the opposite of $3 \cdot 3$, or -9.

Only with parentheses around it is the −3 squared.

Practice 4

Simplify: $\dfrac{-25}{5(-1)}$

Example 4 Simplify: $\dfrac{-6(2)}{-3}$

Solution: First we multiply −6 and 2. Then we divide.

$$\frac{-6(2)}{-3} = \frac{-12}{-3} = 4$$

Work Practice 4

Answers

1. 16 **2.** −16 **3.** 108 **4.** 5

Example 5 Simplify: $\frac{12 - 16}{-1 + 3}$

Solution: We simplify above and below the fraction bar separately. Then we divide.

$$\frac{12 - 16}{-1 + 3} = \frac{-4}{2}$$
$$= -2$$

Work Practice 5

Practice 5

Simplify: $\frac{-18 + 6}{-3 - 1}$

Example 6 Simplify: $60 + 30 + (-2)^3$

Solution:

$$60 + 30 + (-2)^3 = 60 + 30 + (-8) \quad \text{Write } (-2)^3 \text{ as } -8.$$
$$= 90 + (-8) \quad \text{Add from left to right.}$$
$$= 82$$

Work Practice 6

Practice 6

Simplify: $30 + 50 + (-4)^3$

Example 7 Simplify: $-4^2 + (-3)^2 - 1^3$

Solution:

$$-4^2 + (-3)^2 - 1^3 = -16 + 9 - 1 \quad \text{Simplify expressions with exponents.}$$
$$= -7 - 1 \quad \text{Add or subtract from left to right.}$$
$$= -8$$

Work Practice 7

Practice 7

Simplify: $-2^3 + (-4)^2 + 1^5$

Example 8 Simplify: $3(4 - 7) + (-2) - 5$

Solution:

$$3(4 - 7) + (-2) - 5 = 3(-3) + (-2) - 5 \quad \text{Simplify inside parentheses.}$$
$$= -9 + (-2) - 5 \quad \text{Multiply.}$$
$$= -11 - 5 \quad \text{Add or subtract from left to right.}$$
$$= -16$$

Work Practice 8

Practice 8

Simplify:
$2(2 - 9) + (-12) - 3$

Example 9 Simplify: $(-3) \cdot |-5| - (-2) + 4^2$

Solution:

$$(-3) \cdot |-5| - (-2) + 4^2 = (-3) \cdot 5 - (-2) + 4^2 \quad \text{Write } |-5| \text{ as } 5.$$
$$= (-3) \cdot 5 - (-2) + 16 \quad \text{Write } 4^2 \text{ as } 16.$$
$$= -15 - (-2) + 16 \quad \text{Multiply.}$$
$$= -13 + 16 \quad \text{Add or subtract from left to right.}$$
$$= 3$$

Work Practice 9

Practice 9

Simplify:
$(-5) \cdot |-8| + (-3) + 2^3$

Answers

5. 3 **6.** 16 **7.** 9 **8.** −29 **9.** −35

Practice 10

Simplify:
$-4[-6 + 5(-3 + 5)] - 7$

Example 10 Simplify: $-2[-3 + 2(-1 + 6)] - 5$

Solution: Here we begin with the innermost set of parentheses.

$$\begin{aligned} -2[-3 + 2(-1 + 6)] - 5 &= -2[-3 + 2(5)] - 5 && \text{Write } -1 + 6 \text{ as } 5. \\ &= -2[-3 + 10] - 5 && \text{Multiply.} \\ &= -2(7) - 5 && \text{Add.} \\ &= -14 - 5 && \text{Multiply.} \\ &= -19 && \text{Subtract.} \end{aligned}$$

Work Practice 10

✓**Concept Check** True or false? Explain your answer. The result of

$-4(3 - 7) - 8(9 - 6)$

is positive because there are four negative signs.

Objective B Evaluating Expressions

Now we practice evaluating expressions.

Practice 11

Evaluate x^2 and $-x^2$ for $x = -15$.

Example 11 Evaluate x^2 and $-x^2$ for $x = -11$.

Solution: $x^2 = (-11)^2 = (-11)(-11) = 121$

$-x^2 = -(-11)^2 = -(-11)(-11) = -121$

Work Practice 11

Practice 12

Evaluate $5y^2$ for $y = 4$ and $y = -4$.

Example 12 Evaluate $6z^2$ for $z = 2$ and $z = -2$.

Solution: $6z^2 = 6(2)^2 = 6(4) = 24$

$6z^2 = 6(-2)^2 = 6(4) = 24$

Work Practice 12

Practice 13

Evaluate $x^2 + y$ for $x = -6$ and $y = -3$.

Example 13 Evaluate $x + 2y - z$ for $x = 3$, $y = -5$, and $z = -4$.

Solution: Replace x with 3, y with -5, z with -4, and simplify.

Helpful Hint

Remember to rewrite the subtraction sign.

$$\begin{aligned} x + 2y - z &= 3 + 2(-5) - (-4) && \text{Let } x = 3, y = -5, \text{ and } z = -4. \\ &= 3 + (-10) + 4 && \text{Replace } 2(-5) \text{ with its product, } -10. \\ &= -3 && \text{Add.} \end{aligned}$$

Work Practice 13

Answers

10. -23 **11.** 225; -225 **12.** 80; 80 **13.** 33

✓**Concept Check Answer**

false; $-4(3 - 7) - 8(9 - 6) = -8$

Example 14 Evaluate $7 - x^2$ for $x = -4$.

Solution: Replace x with -4 and simplify carefully!

$$7 - x^2 = 7 - (-4)^2$$

$= 7 - 16$ $\quad (-4)^2 = (-4)(-4) = 16$

$= -9$ $\quad$ Subtract.

Work Practice 14

Practice 14

Evaluate $4 - x^2$ for $x = -8$.

Objective C Finding Averages 

Recall from Chapter 1 that the average of a list of numbers is

$$\text{average} = \frac{\text{sum of numbers}}{\textit{number} \text{ of numbers}}$$

Example 15 The graph shows the monthly normal temperatures for Barrow, Alaska. Use this graph to find the average of the temperatures for the months January through April.

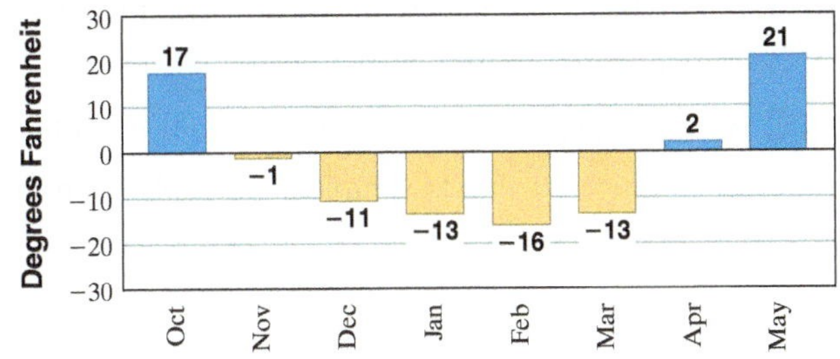

Solution: By reading the graph, we have

$$\text{average} = \frac{-13 + (-16) + (-13) + 2}{4} \quad \text{There are 4 months from January through April.}$$

$$= \frac{-40}{4}$$

$$= -10$$

The average of the temperatures is $-10°$F.

Work Practice 15

Practice 15

Find the average of the temperatures for the months October through April.

Answers

14. -60 **15.** $-5°$F

Calculator Explorations Simplifying an Expression Containing a Fraction Bar

Recall that even though most calculators follow the order of operations, parentheses must sometimes be inserted. For example, to simplify $\frac{-8 + 6}{-2}$ on a calculator, enter parentheses around the expression above the fraction bar so that it is simplified separately.

To simplify $\frac{-8 + 6}{-2}$, press the keys

[( ] [8] [+/−] [+] [6] [ )] [÷] [2] [+/−] [=] or

[( ] [(−)] [8] [+] [6] [ )] [÷] [(−)] [2] [ENTER]

The display will read [1].

Thus, $\frac{-8 + 6}{-2} = 1$.

Use a calculator to simplify.

1. $\frac{-120 - 360}{-10}$

2. $\frac{4750}{-2 + (-17)}$

3. $\frac{-316 + (-458)}{28 + (-25)}$

4. $\frac{-234 + 86}{-18 + 16}$

Vocabulary, Readiness & Video Check

Use the choices below to fill in each blank. Not all choices will be used.

average	subtraction	division	$-7 - 3(1)$
addition	multiplication	$-7 - 3(-1)$	

1. To simplify $-2 \div 2 \cdot (3)$, which operation should be performed first? __________
2. To simplify $-9 - 3 \cdot 4$, which operation should be performed first? __________
3. The __________ of a list of numbers is $\dfrac{\text{sum of numbers}}{\textit{number}\text{ of numbers}}$.
4. To simplify $5[-9 + (-3)] \div 4$, which operation should be performed first? __________
5. To simplify $-2 + 3(10 - 12) \cdot (-8)$, which operation should be performed first? __________
6. To evaluate $x - 3y$ for $x = -7$ and $y = -1$, replace x with -7 and y with -1 and evaluate __________.

Martin-Gay Interactive Videos *Watch the section lecture video and answer the following questions.*

See Video 2.5

Objective A 7. In Example 1, what two things about the fraction bar are we reminded of?

Objective B 8. In Example 5, why is it important to place the replacement value for x within parentheses?

Objective C 9. From the lecture before Example 6, explain why finding the average is a good example of an application for this section.

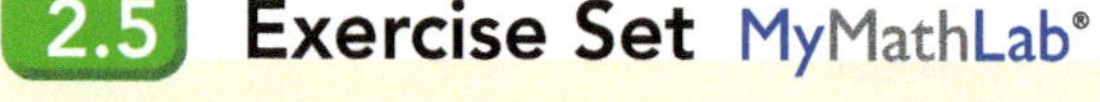

2.5 Exercise Set MyMathLab®

Objective A *Simplify. See Examples 1 through 10.*

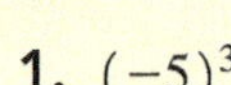

1. $(-5)^3$	**2.** -2^4	**3.** -4^3	**4.** $(-2)^4$
5. $8 \cdot 2^2$	**6.** $5 \cdot 2^3$	**7.** $8 - 12 - 4$	**8.** $10 - 23 - 12$
9. $7 + 3(-6)$	**10.** $-8 + 4(3)$	**11.** $5(-9) + 2$	**12.** $7(-6) + 3$
13. $-10 + 4 \div 2$	**14.** $-12 + 6 \div 3$	**15.** $6 + 7 \cdot 3 - 10$	**16.** $5 + 9 \cdot 4 - 20$
17. $\dfrac{16 - 13}{-3}$	**18.** $\dfrac{20 - 15}{-1}$	**19.** $\dfrac{24}{10 + (-4)}$	**20.** $\dfrac{88}{-8 - 3}$
21. $5(-3) - (-12)$	**22.** $7(-4) - (-6)$	**23.** $[8 + (-4)]^2$	**24.** $[9 + (-2)]^3$

25. $8 \cdot 6 - 3 \cdot 5 + (-20)$

26. $7 \cdot 6 - 6 \cdot 5 + (-10)$

27. $4 - (-3)^4$

28. $7 - (-5)^2$

29. $|7 + 3| \cdot 2^3$

30. $|-3 + 7| \cdot 7^2$

31. $7 \cdot 6^2 + 4$

32. $10 \cdot 5^3 + 7$

33. $7^2 - (4 - 2^3)$

34. $8^2 - (5 - 2)^4$

35. $|3 - 15| \div 3$

36. $|12 - 19| \div 7$

37. $-(-2)^6$

38. $-(-2)^3$

39. $(5 - 9)^2 \div (4 - 2)^2$

40. $(2 - 7)^2 \div (4 - 3)^4$

41. $|8 - 24| \cdot (-2) \div (-2)$

42. $|3 - 15| \cdot (-4) \div (-16)$

43. $(-12 - 20) \div 16 - 25$

44. $(-20 - 5) \div 5 - 15$

45. $5(5 - 2) + (-5)^2 - 6$

46. $3 \cdot (8 - 3) + (-4) - 10$

47. $(2 - 7) \cdot (6 - 19)$

48. $(4 - 12) \cdot (8 - 17)$

49. $(-36 \div 6) - (4 \div 4)$

50. $(-4 \div 4) - (8 \div 8)$

51. $(10 - 4^2)^2$

52. $(11 - 3^2)^3$

53. $2(8 - 10)^2 - 5(1 - 6)^2$

54. $-3(4 - 8)^2 + 5(14 - 16)^3$

55. $3(-10) \div [5(-3) - 7(-2)]$

56. $12 - [7 - (3 - 6)] + (2 - 3)^3$

57. $\dfrac{(-7)(-3) - (4)(3)}{3[7 \div (3 - 10)]}$

58. $\dfrac{10(-1) - (-2)(-3)}{2[-8 \div (-2 - 2)]}$

59. $-3[5 + 2(-4 + 9)] + 15$

60. $-2[6 + 4(2 - 8)] - 25$

Objective B *Evaluate each expression for* $x = -2$, $y = 4$, *and* $z = -1$. *See Examples 11 through 14.*

61. $x + y + z$

62. $x - y - z$

63. $2x - 3y - 4z$

64. $5x - y + 4z$

65. $x^2 - y$

66. $x^2 + z$

67. $\dfrac{5y}{z}$

68. $\dfrac{4x}{y}$

Evaluate each expression for $x = -3$ *and* $z = -4$. *See Examples 11 through 14.*

69. x^2

70. z^2

71. $-z^2$

72. $-x^2$

73. $2z^3$

74. $3x^2$

75. $10 - x^2$

76. $3 - z^2$

77. $2x^3 - z$

78. $3z^2 - x$

Objective C *Find the average of each list of numbers. See Example 15.*

79. $-10, 8, -4, 2, 7, -5, -12$

80. $-18, -8, -1, -1, 0, 4$

81. $-17, -26, -20, -13$

82. $-40, -20, -10, -15, -5$

Scores in golf can be 0 (also called par), a positive integer (also called above par), or a negative integer (also called below par). The bar graph shows final scores of selected golfers from a 2013 tournament. Use this graph for Exercises 83 through 88. (Source: LPGA*)*

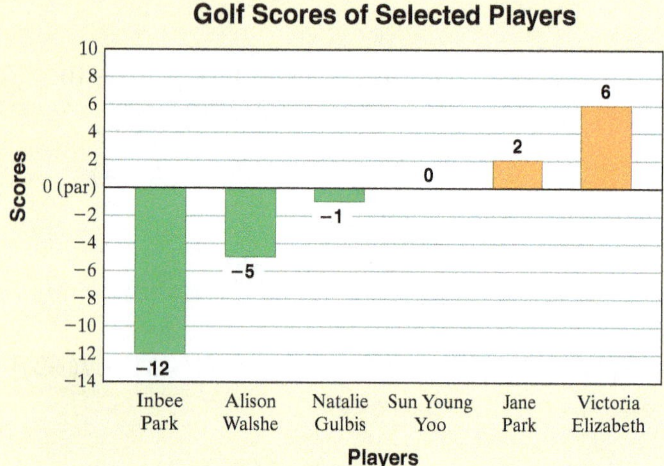

83. Find the difference between the lowest score shown and the highest score shown.

84. Find the difference between the two lowest scores.

85. Find the average of the scores for Walshe, Gulbis, Yoo, and Jane Park. (*Hint:* Here, the average is the sum of the scores divided by the number of players.)

86. Find the average of the scores for Inbee Park, Walshe, Gulbis, and Elizabeth.

87. Can the average for the scores in Exercise **86** be greater than the highest score, 6? Explain why or why not.

88. Can the average of the scores in Exercise **86** be less than the lowest score, −12? Explain why or why not.

Review

Perform each indicated operation. See Sections 1.3, 1.5, and 1.6.

89. $45 \cdot 90$

90. $90 \div 45$

91. $90 - 45$

92. $45 + 90$

Find the perimeter of each figure. See Section 1.3.

93. Square

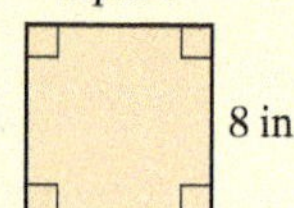

94. Parallelogram

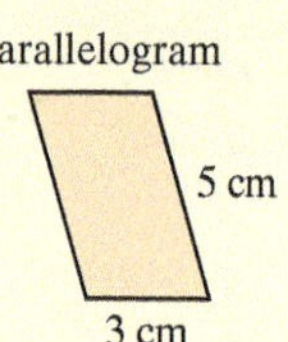

95. Rectangle

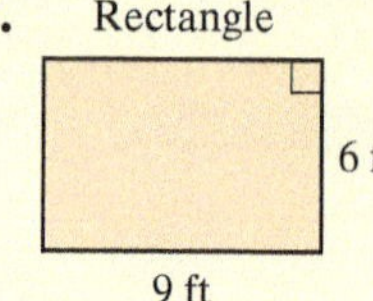

96. Triangle

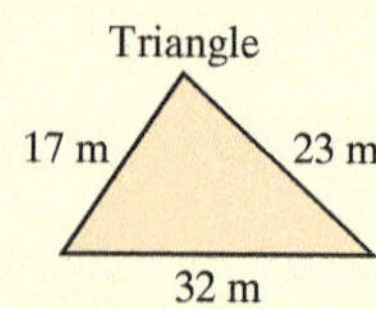

Concept Extensions

Insert parentheses where needed so that each expression evaluates to the given number.

97. $2 \cdot 7 - 5 \cdot 3$; evaluates to 12

98. $7 \cdot 3 - 4 \cdot 2$; evaluates to 34

99. $-6 \cdot 10 - 4$; evaluates to -36

100. $2 \cdot 8 \div 4 - 20$; evaluates to -36

101. Are parentheses necessary in the expression $3 + (4 \cdot 5)$? Explain your answer.

102. Are parentheses necessary in the expression $(3 + 4) \cdot 5$? Explain your answer.

103. Discuss the effect parentheses have in an exponential expression. For example, what is the difference between $(-6)^2$ and -6^2?

104. Discuss the effect parentheses have in an exponential expression. For example, what is the difference between $(2 \cdot 4)^2$ and $2 \cdot 4^2$?

Evaluate.

105. $(-12)^4$

106. $(-17)^6$

107. $x^3 - y^2$ for $x = 21$ and $y = -19$

108. $3x^2 + 2x - y$ for $x = -18$ and $y = 2868$

109. $(xy + z)^x$ for $x = 2$, $y = -5$, and $z = 7$

110. $5(ab + 3)^b$ for $a = -2$, $b = 3$

2.6 Solving Equations: The Addition and Multiplication Properties

In this section, we introduce properties of equations and we use these properties to begin solving equations. Now that we know how to perform operations on integers, this is an excellent way to practice these operations.

First, let's recall the difference between an equation and an expression. From Section 1.8, a combination of operations on variables and numbers is an expression, and an equation is of the form "expression = expression."

Equations	Expressions
$3x - 1 = -17$	$3x - 1$
area = length · width	$5(20 - 3) + 10$
$8 + 16 = 16 + 8$	y^3
$-9a + 11b = 14b + 3$	$-x^2 + y - 2$

Objectives

A Identify Solutions of Equations.

B Use the Addition Property of Equality to Solve Equations.

C Use the Multiplication Property of Equality to Solve Equations.

Helpful Hint

Simply stated, an equation contains "=" while an expression does not. Also, we *simplify* expressions and *solve* equations.

Objective A Identifying Solutions of Equations 

Let's practice identifying solutions of equations. Recall from Section 1.8 that a solution of an equation is a number that when substituted for a variable makes the equation a true statement.

For example,

-8 is a solution of

$\frac{x}{2} = -4$, because

$\frac{-8}{2} = -4$, or

$-4 = -4$ is true.

Also,

-8 is not a solution of

$x + 6 = 2$, because

$-8 + 6 = 2$ is false.

Let's practice determining whether a number is a solution of an equation. In this section, we will be performing operations on integers.

Practice 1

Determine whether -2 is a solution of the equation $-4x - 3 = 5$.

Example 1 Determine whether -1 is a solution of the equation $3y + 1 = 3$.

Solution:

$$\begin{aligned} 3y + 1 &= 3 \\ 3(-1) + 1 &\stackrel{?}{=} 3 \\ -3 + 1 &\stackrel{?}{=} 3 \\ -2 &= 3 \quad \text{False} \end{aligned}$$

Since $-2 = 3$ is false, -1 is *not* a solution of the equation.

Work Practice 1

Now we know how to check whether a number is a solution. But, given an equation, how do we find its **solution**? In other words, how do we find a number that makes the equation true? How do we solve an equation?

Objective B Using the Addition Property to Solve Equations

To solve an equation, we use properties of equality to write simpler equations, all equivalent to the original equation, until the final equation has the form

$x =$ **number** or **number** $= x$

Equivalent equations have the same solution, so the word "number" above represents the solution of the original equation. The first property of equality to help us write simpler, equivalent equations is the **addition property of equality.**

Answer

1. yes

Addition Property of Equality

Let a, b, and c represent numbers. Then

$a = b$ and $a + c = b + c$ are equivalent equations.	Also, $a = b$ and $a - c = b - c$ are equivalent equations.

In other words, the same number may be added to or subtracted from both sides of an equation without changing the solution of the equation. (Recall from Section 2.3 that we defined subtraction as addition of the first number and the opposite of the second number. Because of this, the addition property of equality also allows us to subtract the same number from both sides.)

A good way to visualize a true equation is to picture a balanced scale. Since it is balanced, each side of the scale weighs the same amount. Similarly, in a true equation the expressions on each side have the same value. Picturing our balanced scale, if we add the same weight to each side, the scale remains balanced.

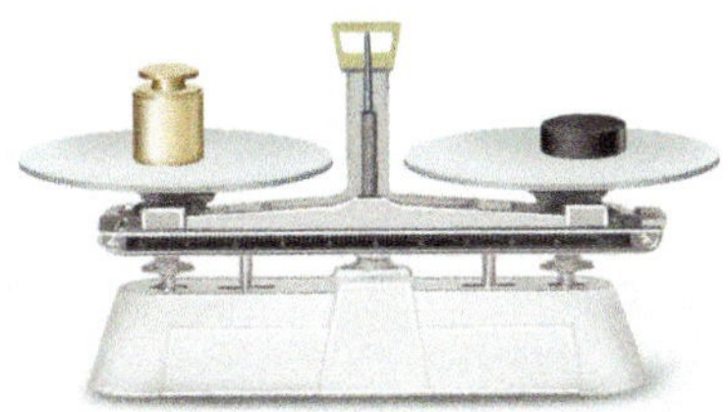

Example 2 Solve: $x - 2 = -1$ for x.

Solution: To solve the equation for x, we need to rewrite the equation in the form $x =$ number. In other words, our goal is to get x alone on one side of the equation. To do so, we add 2 to both sides of the equation.

$x - 2 = -1$

$x - 2 + 2 = -1 + 2$ Add 2 to both sides of the equation.

$x + 0 = 1$ Replace $-2 + 2$ with 0.

$x = 1$ Simplify by replacing $x + 0$ with x.

Check: To check, we replace x with 1 in the *original* equation.

$x - 2 = -1$ Original equation

$1 - 2 \stackrel{?}{=} -1$ Replace x with 1.

$-1 = -1$ True

Since $-1 = -1$ is a true statement, 1 is the solution of the equation.

Work Practice 2

Practice 2

Solve the equation for y: $y - 6 = -2$

Helpful Hint

Note that it is always a good idea to check the solution in the *original* equation to see that it makes the equation a true statement.

Let's visualize how we used the addition property of equality to solve an equation. Picture the equation $x - 2 = 1$ as a balanced scale. The left side of the equation has the same value as the right side.

Answer

2. 4

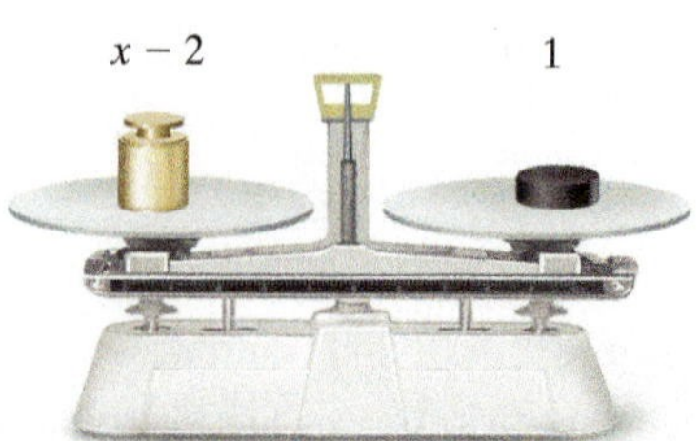

If the same weight is added to each side of a scale, the scale remains balanced. Likewise, if the same number is added to each side of an equation, the left side continues to have the same value as the right side.

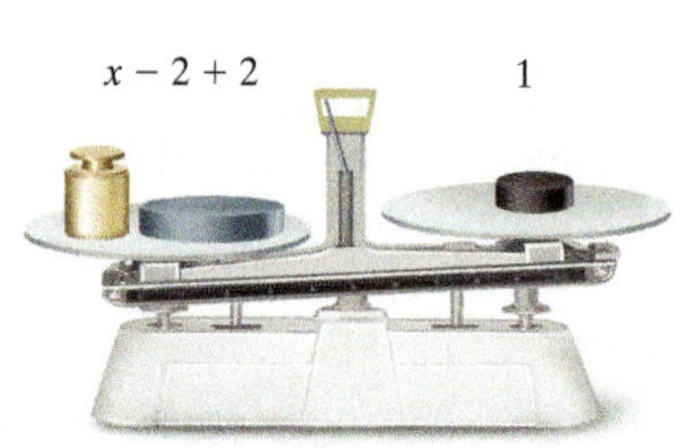

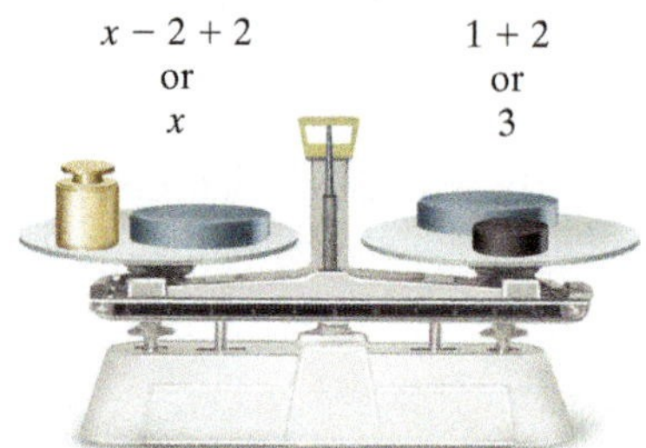

Practice 3

Solve: $-2 = z + 8$

Example 3 Solve: $-8 = n + 1$

Solution: To get n alone on one side of the equation, we subtract 1 from both sides of the equation.

$$-8 = n + 1$$
$$-8 - 1 = n + 1 - 1 \quad \text{Subtract 1 from both sides.}$$
$$-9 = n + 0 \quad \text{Replace } 1 - 1 \text{ with } 0.$$
$$-9 = n \quad \text{Simplify.}$$

Check:

$$-8 = n + 1$$
$$-8 \stackrel{?}{=} -9 + 1 \quad \text{Replace } n \text{ with } -9.$$
$$-8 = -8 \quad \text{True}$$

The solution is -9.

Work Practice 3

Helpful Hint Remember that we can get the variable alone on either side of the equation. For example, the equations $-9 = n$ and $n = -9$ both have the solution of -9.

✓**Concept Check** What number should be added to or subtracted from both sides of the equation in order to solve the equation $-3 = y + 2$?

Practice 4

Solve: $x = -2 + 90 + (-100)$

Example 4 Solve: $x = -60 + 4 + 10$

Solution: Study this equation for a moment. Notice that our variable x is alone on the left side. Thus, we only need to add on the right side to find the value of x.

$$x = -60 + 4 + 10$$
$$x = -56 + 10 \quad \text{Add } -60 \text{ and } 4.$$
$$x = -46 \quad \text{Add } -56 \text{ and } 10.$$

Check to see that -46 is the solution.

Work Practice 4

Answers
3. -10 4. -12

✓**Concept Check Answer**
Subtract 2 from both sides.

Objective C Using the Multiplication Property to Solve Equations

Although the addition property of equality is a powerful tool for helping us solve equations, it cannot help us solve all types of equations. For example, it cannot help us solve an equation such as $2x = 6$. To solve this equation, we use a second property of equality called the **multiplication property of equality.**

Multiplication Property of Equality

Let a, b, and c represent numbers and let $c \neq 0$. Then

$a = b$ and $a \cdot c = b \cdot c$ are equivalent equations.	Also, $a = b$ and $\frac{a}{c} = \frac{b}{c}$ are equivalent equations.

In other words, both sides of an equation may be multiplied or divided by the same nonzero number without changing the solution of the equation. (We will see in Chapter 4 how the multiplication property allows us to divide both sides of an equation by the same nonzero number.)

To solve an equation like $2x = 6$ for x, notice that 2 is *multiplied* by x. To get x alone, we use the multiplication property of equality to *divide* both sides of the equation by 2, and simplify as follows:

$$2x = 6$$

$$\frac{2 \cdot x}{2} = \frac{6}{2} \quad \text{Divide both sides by 2.}$$

Then it can be shown that an expression such as $\frac{2 \cdot x}{2}$ is equivalent to $\frac{2}{2} \cdot x$, so

$$\frac{2 \cdot x}{2} = \frac{6}{2} \quad \text{can be written as} \quad \frac{2}{2} \cdot x = \frac{6}{2}$$

$$1 \cdot x = 3 \quad \text{or} \quad x = 3$$

Picturing again our balanced scale, if we multiply or divide the weight on each side by the same nonzero number, the scale (or equation) remains balanced.

Example 5 Solve: $-5x = 15$

Solution: To get x alone, divide both sides by -5.

$$-5x = 15 \quad \text{Original equation}$$

$$\frac{-5x}{-5} = \frac{15}{-5} \quad \text{Divide both sides by } -5.$$

$$\frac{-5}{-5} \cdot x = \frac{15}{-5}$$

$$1x = -3 \quad \text{or} \quad x = -3 \quad \text{Simplify.}$$

(Continued on next page)

Practice 5

Solve: $3y = -18$

Answer

5. -6

Check: To check, replace x with -3 in the original equation.

$$-5x = 15 \quad \text{Original equation}$$
$$-5(-3) \stackrel{?}{=} 15 \quad \text{Let } x = -3.$$
$$15 = 15 \quad \text{True}$$

The solution is -3.

Work Practice 5

Practice 6

Solve: $-32 = 8x$

Example 6 Solve: $-27 = 3y$

Solution: To get y alone, divide both sides of the equation by 3.

$$-27 = 3y$$
$$\frac{-27}{3} = \frac{3y}{3} \quad \text{Divide both sides by 3.}$$
$$\frac{-27}{3} = \frac{3}{3} \cdot y$$
$$-9 = 1y \quad \text{or} \quad y = -9$$

Check to see that -9 is the solution.

Work Practice 6

Practice 7

Solve: $-3y = -27$

Example 7 Solve: $-12x = -36$

Solution: To get x alone, divide both sides of the equation by -12.

$$-12x = -36$$
$$\frac{-12x}{-12} = \frac{-36}{-12}$$
$$\frac{-12}{-12} \cdot x = \frac{-36}{-12}$$
$$x = 3$$

Check: To check, replace x with 3 in the original equation.

$$-12x = -36$$
$$-12(3) \stackrel{?}{=} -36 \quad \text{Let } x = 3.$$
$$-36 = -36 \quad \text{True}$$

Since $-36 = -36$ is a true statement, the solution is 3.

Work Practice 7

✓**Concept Check** Which operation is appropriate for solving each of the following equations, addition or division?

a. $12 = x - 3$

b. $12 = 3x$

The multiplication property also allows us to solve equations like

$$\frac{x}{5} = 2$$

Answers

6. -4 **7.** 9

✓**Concept Check Answers**

a. addition

b. division

Here, x is *divided* by 5. To get x alone, we use the multiplication property to *multiply* both sides by 5.

$5 \cdot \frac{x}{5} = 5 \cdot 2$ Multiply both sides by 5.

Then it can be shown that

$5 \cdot \frac{x}{5} = 5 \cdot 2$ can be written as $\frac{5}{5} \cdot x = 5 \cdot 2$

$1 \cdot x = 10$ or $x = 10$

Example 8 Solve: $\frac{x}{3} = -2$

Solution: To get x alone, multiply both sides by 3.

$$\frac{x}{3} = -2$$

$$3 \cdot \frac{x}{3} = 3 \cdot (-2) \quad \text{Multiply both sides by 3.}$$

$$\frac{3}{3} \cdot x = 3 \cdot (-2)$$

$$1x = -6 \quad \text{or} \quad x = -6 \quad \text{Simplify.}$$

Check: Replace x with -6 in the original equation.

$$\frac{x}{3} = -2 \quad \text{Original equation}$$

$$\frac{-6}{3} \stackrel{?}{=} -2 \quad \text{Let } x = -6.$$

$$-2 = -2 \quad \text{True}$$

The solution is -6.

Work Practice 8

Practice 8

Solve: $\frac{x}{-4} = 7$

Answer

8. -28

Vocabulary, Readiness & Video Check

Use the choices below to fill in each blank. Some choices may be used more than once.

equation	multiplication	addition
expression	solution	equivalent

1. A combination of operations on variables and numbers is called a(n) ______________.

2. A statement of the form "expression = expression" is called a(n) ______________.

3. A(n) ______________ contains an equals sign (=) while a(n) ______________ does not.

4. A(n) ______________ may be simplified and evaluated while a(n) ______________ may be solved.

5. A(n) ______________ of an equation is a number that when substituted for a variable makes the equation a true statement.

6. ______________ equations have the same solution.

7. By the ____________ property of equality, the same number may be added to or subtracted from both sides of an equation without changing the solution of the equation.
8. By the ____________ property of equality, both sides of an equation may be multiplied or divided by the same non-zero number without changing the solution of the equation.

Martin-Gay Interactive Videos *Watch the section lecture video and answer the following questions.*

See Video 2.6

Objective A 9. From the lecture before Example 1, what does an equation have that an expression does not?

Objective B 10. In the lecture before Example 2, what does the addition property of equality mean in words?

Objective C 11. Complete this statement based on Example 7: To check a solution, we go back to the ____________ equation, replace the variable with the proposed solution, and see if we get a ____________ statement.

2.6 Exercise Set MyMathLab®

Objective A *Determine whether the given number is a solution of the given equation. See Example 1.*

1. Is 6 a solution of $x - 8 = -2$?
2. Is 9 a solution of $y - 16 = -7$?
3. Is −5 a solution of $x + 12 = 17$?
4. Is −7 a solution of $a + 23 = -16$?
5. Is −8 a solution of $-9f = 64 - f$?
6. Is −6 a solution of $-3k = 12 - k$?
7. Is 3 a solution of $5(c - 5) = -10$?
8. Is 1 a solution of $2(b - 3) = 10$?

Objective B *Solve. Check each solution. See Examples 2 through 4.*

9. $a + 5 = 23$
10. $f + 4 = -6$
11. $d - 9 = -21$
12. $s - 7 = -15$
13. $7 = y - 2$
14. $1 = y + 7$
15. $-7 + 10 - 20 = x$
16. $-50 + 40 - 5 = z$

Objective C *Solve. Check each solution. See Examples 5 through 8.*

17. $5x = 20$
18. $6y = 48$
19. $-3z = 12$
20. $-2x = 26$

21. $\frac{n}{7} = -2$

22. $\frac{n}{11} = -5$

23. $2z = -34$

24. $7y = -21$

25. $-4y = 0$

26. $-9x = 0$

27. $-10x = -10$

28. $-31x = -31$

Objectives B C Mixed Practice *Solve. See Examples 2 through 8.*

29. $5x = -35$

30. $3y = -27$

31. $n - 5 = -55$

32. $n - 4 = -48$

33. $-15 = y + 10$

34. $-36 = y + 12$

35. $\frac{x}{-6} = -6$

36. $\frac{x}{-9} = -9$

37. $n = -10 + 31$

38. $z = -28 + 36$

39. $-12y = -144$

40. $-11x = -121$

41. $\frac{n}{4} = -20$

42. $\frac{n}{5} = -20$

43. $-64 = 32y$

44. $-81 = 27x$

Review

Translate each phrase to an algebraic expression. Use x to represent "a number." See Section 1.8.

45. A number decreased by −2

46. A number increased by −5

47. The product of −6 and a number

48. The quotient of a number and −20

49. The sum of −15 and a number

50. −32 multiplied by a number

51. −8 divided by a number

52. Subtract a number from −18.

Concept Extensions

Solve.

53. $n - 42{,}860 = -1286$

54. $n + 961 = 120$

55. $-38x = 15{,}542$

56. $\frac{y}{-18} = 1098$

57. Explain the differences between an equation and an expression.

58. Explain the differences between the addition property of equality and the multiplication property of equality.

59. Write an equation that can be solved using the addition property of equality.

60. Write an equation that can be solved using the multiplication property of equality.

Chapter 2 Group Activity

Magic Squares

Sections 2.1–2.3

A magic square is a set of numbers arranged in a square table so that the sum of the numbers in each column, row, and diagonal is the same. For instance, in the magic square below, the sum of each column, row, and diagonal is 15. Notice that no number is used more than once in the magic square.

2	9	4
7	5	3
6	1	8

The properties of magic squares have been known for a very long time and once were thought to be good luck charms. The ancient Egyptians and Greeks understood their patterns. A magic square even made it into a famous work of art. The engraving titled *Melencolia I,* created by German artist Albrecht Dürer in 1514, features the following four-by-four magic square on the building behind the central figure.

16	3	2	13
5	10	11	8
9	6	7	12
4	15	14	1

Exercises

1. Verify that what is shown in the Dürer engraving is, in fact, a magic square. What is the common sum of the columns, rows, and diagonals?

2. Negative numbers can also be used in magic squares. Complete the following magic square:

	-1	
0		-4

3. Use the numbers $-16, -12, -8, -4, 0, 4, 8, 12,$ and 16 to form a magic square:

Chapter 2 Vocabulary Check

Fill in each blank with one of the words or phrases listed below.

inequality symbols	addition	solution	is less than	integers
expression	average	negative	absolute value	equation
positive	opposites	is greater than	multiplication	

1. Two numbers that are the same distance from 0 on the number line but are on opposite sides of 0 are called ______.

2. The ______ of a number is that number's distance from 0 on a number line.

3. The ______ are $\ldots, -3, -2, -1, 0, 1, 2, 3, \ldots$.

4. The ______ numbers are numbers less than zero.

5. The ______ numbers are numbers greater than zero.

6. The symbols "<" and ">" are called ______.

7. A(n) ______ of an equation is a number that when substituted for a variable makes the equation a true statement.

8. The ______ of a list of numbers is $\dfrac{\text{sum of numbers}}{\textit{number} \text{ of numbers}}$.

9. A combination of operations on variables and numbers is called a(n) ______.

10. A statement of the form "expression = expression" is called a(n) ______.

11. The sign "<" means ______ and ">" means ______.

12. By the ______ property of equality, the same number may be added to or subtracted from both sides of an equation without changing the solution of the equation.

13. By the ______ property of equality, both sides of an equation may be multiplied or divided by the same nonzero number without changing the solution of the equation.

Helpful Hint

Are you preparing for your test? Don't forget to take the Chapter 2 Test on page 158. Then check your answers at the back of the text and use the Chapter Test Prep Videos to see the fully worked-out solutions to any of the exercises you want to review.

2 Chapter Highlights

Definitions and Concepts	Examples
Section 2.1 Introduction to Integers	
Together, positive numbers, negative numbers, and 0 are called **signed numbers.**	$-432, -10, 0, 15$
The **integers** are $\ldots, -3, -2, -1, 0, 1, 2, 3, \ldots$.	
The **absolute value** of a number is that number's distance from 0 on a number line. The symbol for absolute value is $\lvert\ \ \rvert$.	$\lvert -2 \rvert = 2$ 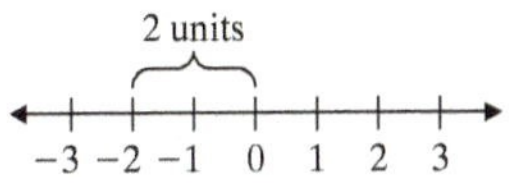 $\lvert 2 \rvert = 2$ 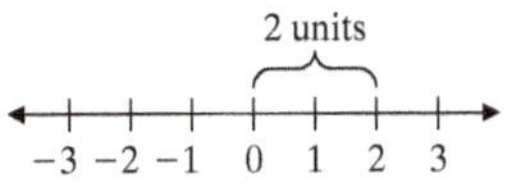

(continued)

Definitions and Concepts	Examples
Section 2.1 Introduction to Integers (*continued*)	
Two numbers that are the same distance from 0 on the number line but are on opposite sides of 0 are called **opposites**.	5 and −5 are opposites. 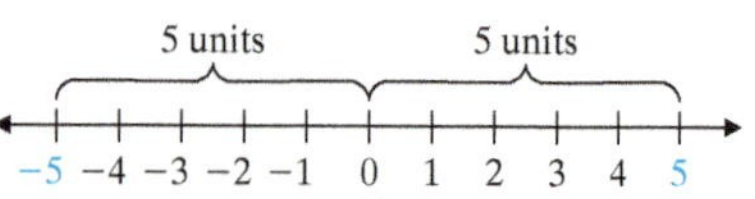
If a is a number, then $-(-a) = a$.	$-(-11) = 11$. Do not confuse with $-\lvert -3 \rvert = -3$
Section 2.2 Adding Integers	
Adding Two Numbers with the Same Sign **Step 1:** Add their absolute values. **Step 2:** Use their common sign as the sign of the sum.	Add: $-3 + (-2) = -5$ $-7 + (-15) = -22$
Adding Two Numbers with Different Signs **Step 1:** Find the larger absolute value minus the smaller absolute value. **Step 2:** Use the sign of the number with the larger absolute value as the sign of the sum.	$-6 + 4 = -2$ $17 + (-12) = 5$ $-32 + (-2) + 14 = -34 + 14$ $= -20$
Section 2.3 Subtracting Integers	
Subtracting Two Numbers If a and b are numbers, then $a - b = a + (-b)$.	Subtract: $-35 - 4 = -35 + (-4) = -39$ $3 - 8 = 3 + (-8) = -5$ $-10 - (-12) = -10 + 12 = 2$ $7 - 20 - 18 - (-3) = 7 + (-20) + (-18) + (+3)$ $= -13 + (-18) + 3$ $= -31 + 3$ $= -28$
Section 2.4 Multiplying and Dividing Integers	
Multiplying Numbers The product of two numbers having the same sign is a positive number. The product of two numbers having different signs is a negative number.	Multiply: $(-7)(-6) = 42$ $9(-4) = -36$ Evaluate: $(-3)^2 = (-3)(-3) = 9$
Dividing Numbers The quotient of two numbers having the same sign is a positive number. The quotient of two numbers having different signs is a negative number.	Divide: $-100 \div (-10) = 10$ $\frac{14}{-2} = -7$, $\frac{0}{-3} = 0$, $\frac{22}{0}$ is undefined.

Definitions and Concepts	Examples
Section 2.5 Order of Operations	
Order of Operations 1. Perform all operations within parentheses ( ), brackets [ ], or other grouping symbols such as fraction bars, starting with the innermost set. 2. Evaluate any expressions with exponents. 3. Multiply or divide in order from left to right. 4. Add or subtract in order from left to right.	Simplify: $3 + 2\cdot(-5) = 3 + (-10)$ $= -7$ $\frac{-2(5-7)}{-7+\lvert -3\rvert} = \frac{-2(-2)}{-7+3}$ $= \frac{4}{-4}$ $= -1$
Section 2.6 Solving Equations: The Addition and Multiplication Properties	
Addition Property of Equality Let a, b, and c represent numbers. If $a = b$, then $a + c = b + c$ and $a - c = b - c$ In other words, the same number may be added to or subtracted from both sides of an equation without changing the solution of the equation.	Solve: $x + 8 = 1$ $x + 8 - 8 = 1 - 8$ Subtract 8 from both sides. $x = -7$ Simplify. The solution is -7.
Multiplication Property of Equality Let a, b, and c represent numbers and let $c \neq 0$. If $a = b$, then $a\cdot c = b\cdot c$ and $\frac{a}{c} = \frac{b}{c}$ In other words, both sides of an equation may be multiplied or divided by the same nonzero number without changing the solution of the equation.	Solve: $-6y = 30$ $\frac{-6y}{-6} = \frac{30}{-6}$ Divide both sides by -6. $\frac{-6}{-6}\cdot y = \frac{30}{-6}$ $y = -5$ Simplify. The solution is -5.

Chapter 2 Review

(2.1) *Represent each quantity by an integer.*

1. A gold miner is working 1572 feet down in a mine.

2. Mount Hood, in Oregon, has an elevation of 11,239 feet.

Graph each integer in the list on the same number line.

3. $-3, -5, 0, 7$

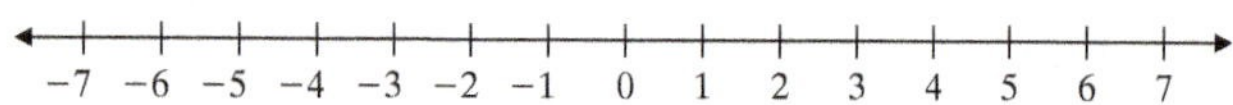

4. $-6, -1, 0, 5$

−7 −6 −5 −4 −3 −2 −1 0 1 2 3 4 5 6 7

Simplify.

5. $|-11|$

6. $|0|$

7. $-|8|$

8. $-(-9)$

9. $-|-16|$

10. $-(-2)$

Insert $<$ or $>$ between each pair of integers to make a true statement.

11. $-18 \quad -20$

12. $-5 \quad 5$

13. $|-123| \quad -|-198|$

14. $|-12| \quad -|-16|$

Find the opposite of each integer.

15. -18

16. 42

Answer true or false for each statement.

17. If $a < b$, then a must be a negative number.

18. The absolute value of an integer is always 0 or a positive number.

19. A negative number is always less than a positive number.

20. If a is a negative number, then $-a$ is a positive number.

Evaluate.

21. $|y|$ if $y = -2$

22. $|-x|$ if $x = -3$

23. $-|-z|$ if $z = -5$

24. $-|-n|$ if $n = -10$

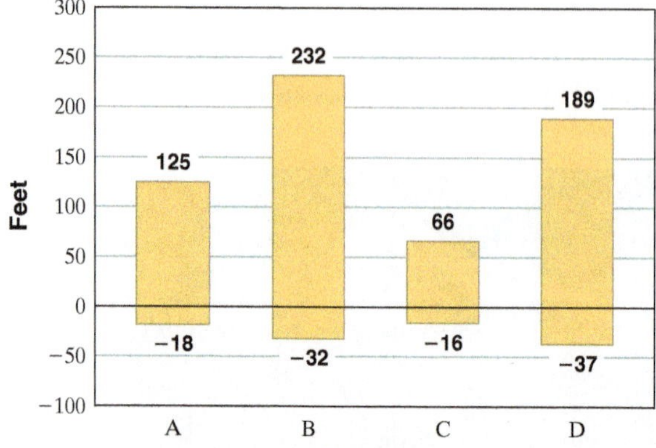

Elevator shafts in some buildings extend not only above ground, but in many cases below ground to accommodate basements, underground parking, etc. The bar graph shows four such elevators and their shaft distance above and below ground. Use the bar graph to answer Exercises 25 and 26.

25. Which elevator shaft extends the farthest below ground?

26. Which elevator shaft extends the highest above ground?

(2.2) *Add.*

27. $5 + (-3)$

28. $18 + (-4)$

29. $-12 + 16$

30. $-23 + 40$

31. $-8 + (-15)$

32. $-5 + (-17)$

33. $-24 + 3$

34. $-89 + 19$

35. $15 + (-15)$

36. $-24 + 24$

37. $-43 + (-108)$

38. $-100 + (-506)$

Solve.

39. The temperature at 5 a.m. on a day in January was $-15°$ Celsius. By 6 a.m. the temperature had fallen 5 degrees. Use a signed number to represent the temperature at 6 a.m.

40. A diver starts out at 127 feet below the surface and then swims downward another 23 feet. Use a signed number to represent the diver's current depth.

41. During the PGA 2008 Wyndham Championship tournament, the winner, Carl Pettersson, had scores of -6, -9, -4, and -2. What was his total score for the tournament? (*Source:* Professional Golfers' Association)

42. The Solheim Cup, a biannual pro women golfers' tournament between an American team and a European team, scores holes won. During the 2007 Solheim Cup, the winners, the American team, had a score of 16. The losing team, the Europeans, had a score 4 less than the Americans' score. What was the European team's score? (*Source:* Professional Golfers' Association)

(2.3) *Subtract.*

43. $12 - 4$

44. $-12 - 4$

45. $-7 - 17$

46. $7 - 17$

47. $7 - (-13)$

48. $-6 - (-14)$

49. $16 - 16$

50. $-16 - 16$

51. $-12 - (-12)$

52. $-5 - (-12)$

53. $-(-5) - 12 + (-3)$

54. $-8 + (-12) - 10 - (-3)$

Solve.

55. If the elevation of Lake Superior is 600 feet above sea level and the elevation of the Caspian Sea is 92 feet below sea level, find the difference of the elevations.

56. Josh Weidner has \$142 in his checking account. He writes a check for \$125, makes a deposit of \$43, and then writes another check for \$85. Represent the balance in his account by an integer.

57. Some roller coasters travel above and below ground. One such roller coaster is Tremors, located in Silverwood Theme Park, Athol, Idaho. If this coaster rises to a height of 85 feet above ground, then drops 99 feet, how many feet below ground are you at the end of the drop? (*Source:* ultimaterollercoaster.com)

58. Go to the bar graph for Review Exercises **25** and **26** and find the total length of the elevator shaft for Elevator C.

Answer true or false for each statement.

59. $|-5| - |-6| = 5 - 6$

60. $|-5 - (-6)| = 5 + 6$

(2.4) *Multiply.*

61. $-3(-7)$

62. $-6(3)$

63. $-4(16)$

64. $-5(-12)$

65. $(-5)^2$

66. $(-1)^5$

67. $12(-3)(0)$

68. $-1(6)(2)(-2)$

Divide.

69. $-15 \div 3$

70. $\frac{-24}{-8}$

71. $\frac{0}{-3}$

72. $\frac{-46}{0}$

73. $\frac{100}{-5}$

74. $\frac{-72}{8}$

75. $\frac{-38}{-1}$

76. $\frac{45}{-9}$

Solve.

77. A football team lost 5 yards on each of two consecutive plays. Represent the total loss by a product of integers, and find the product.

78. A horse race bettor lost $50 on each of four consecutive races. Represent the total loss by a product of integers, and find the product.

79. A person has a debt of $1024 and is ordered to pay it back in four equal payments. Represent the amount of each payment by a quotient of integers, and find the quotient.

80. Overnight, the temperature dropped 45 degrees Fahrenheit. If this took place over a time period of nine hours, represent the average temperature drop each hour by a quotient of integers. Then find the quotient.

(2.5) *Simplify.*

81. $(-7)^2$

82. -7^2

83. $5 - 8 + 3$

84. $-3 + 12 + (-7) - 10$

85. $-10 + 3 \cdot (-2)$

86. $5 - 10 \cdot (-3)$

87. $16 \div (-2) \cdot 4$

88. $-20 \div 5 \cdot 2$

89. $16 + (-3) \cdot 12 \div 4$

90. $-12 + 10 \div (-5)$

91. $4^3 - (8 - 3)^2$

92. $(-3)^3 - 90$

93. $\frac{(-4)(-3) - (-2)(-1)}{-10 + 5}$

94. $\frac{4(12 - 18)}{-10 \div (-2 - 3)}$

Find the average of each list of numbers.

95. $-18, 25, -30, 7, 0, -2$

96. $-45, -40, -30, -25$

Evaluate each expression for $x = -2$ and $y = 1$.

97. $2x - y$

98. $y^2 + x^2$

99. $\frac{3x}{6}$

100. $\frac{5y - x}{-y}$

(2.6) *For Exercises 101 and 102, answer "yes" or "no."*

101. Is -5 a solution of $2n - 6 = 16$?

102. Is -2 a solution of $2(c - 8) = -20$?

Solve.

103. $n - 7 = -20$

104. $-5 = n + 15$

105. $10x = -30$

106. $-8x = 72$

107. $-20 + 7 = y$

108. $x - 31 = -62$

109. $\frac{n}{-4} = -11$

110. $\frac{x}{-2} = 13$

111. $n + 12 = -7$

112. $n - 40 = -2$

113. $-36 = -6x$

114. $-40 = 8y$

Mixed Review

Perform the indicated operations.

115. $-6 + (-9)$

116. $-16 - 3$

117. $-4(-12)$

118. $\frac{84}{-4}$

119. $-76 - (-97)$

120. $-9 + 4$

Solve.

121. Wednesday's lowest temperature was -18°C. The cold weather continued and by Friday, it had dropped another 9°C. What was the temperature on Friday?

122. The temperature at noon on a Monday in December was -11°C. By noon on Tuesday, it had warmed by 17°C. What was the temperature at noon on Tuesday?

123. The top of a mountain has an altitude of 12,923 feet. The bottom of a valley is 195 feet below sea level. Find the difference between these two elevations.

124. Joe owed his mother \$32. He gave her \$23. Write his financial situation as a signed number.

Simplify.

125. $(3 - 7)^2 \div (6 - 4)^3$

126. $3(4 + 2) + (-6) - 3^2$

127. $2 - 4 \cdot 3 + 5$

128. $4 - 6 \cdot 5 + 1$

129. $\frac{-|-14| - 6}{7 + 2(-3)}$

130. $5(7 - 6)^3 - 4(2 - 3)^2 + 2^4$

Solve.

131. $n - 9 = -30$

132. $n + 18 = 1$

133. $-4x = -48$

134. $9x = -81$

135. $\frac{n}{-2} = 100$

136. $\frac{y}{-1} = -3$

Chapter 2 Test

Step-by-step test solutions are found on the Chapter Test Prep Videos. Where available: MyMathLab® or YouTube

Answers

1. ______
2. ______
3. ______
4. ______
5. ______
6. ______
7. ______
8. ______
9. ______
10. ______
11. ______
12. ______
13. ______
14. ______
15. ______
16. ______
17. ______
18. ______

Simplify each expression.

1. $-5 + 8$

2. $18 - 24$

3. $5 \cdot (-20)$

4. $-16 \div (-4)$

5. $-18 + (-12)$

6. $-7 - (-19)$

7. $-5 \cdot (-13)$

8. $\frac{-25}{-5}$

9. $|-25| + (-13)$

10. $14 - |-20|$

11. $|5| \cdot |-10|$

12. $\frac{|-10|}{-|-5|}$

13. $-8 + 9 \div (-3)$

14. $-7 + (-32) - 12 + 5$

15. $(-5)^3 - 24 \div (-3)$

16. $(5 - 9)^2 \cdot (8 - 2)^3$

17. $-(-7)^2 \div 7 \cdot (-4)$

18. $3 - (8 - 2)^3$

19. $\frac{4}{2} - \frac{8^2}{16}$

20. $\frac{-3(-2) + 12}{-1(-4 - 5)}$

21. $\frac{|25 - 30|^2}{2(-6) + 7}$

22. $5(-8) - [6 - (2 - 4)] + (12 - 16)^2$

Evaluate each expression for $x = 0, y = -3,$ *and* $z = 2$.

23. $7x + 3y - 4z$

24. $10 - y^2$

25. $\frac{3z}{2y}$

26. Mary Dunstan, a diver, starts at sea level and then makes 4 successive descents of 22 feet. After the descents, what is her elevation?

27. Aaron Hawn has \$129 in his checking account. He writes a check for \$79, withdraws \$40 from an ATM, and then deposits \$35. Represent the new balance in his account by an integer.

28. Mt. Washington in New Hampshire has an elevation of 6288 feet above sea level. The Romanche Gap in the Atlantic Ocean has an elevation of 25,354 feet below sea level. Represent the difference in elevation between these two points by an integer. (*Source:* National Geographic Society and Defense Mapping Agency)

29. Lake Baykal in Siberian Russia is the deepest lake in the world, with a maximum depth of 5315 feet. The elevation of the lake's surface is 1495 feet above sea level. What is the elevation (with respect to sea level) of the deepest point in the lake? (*Source:* U.S. Geological Survey)

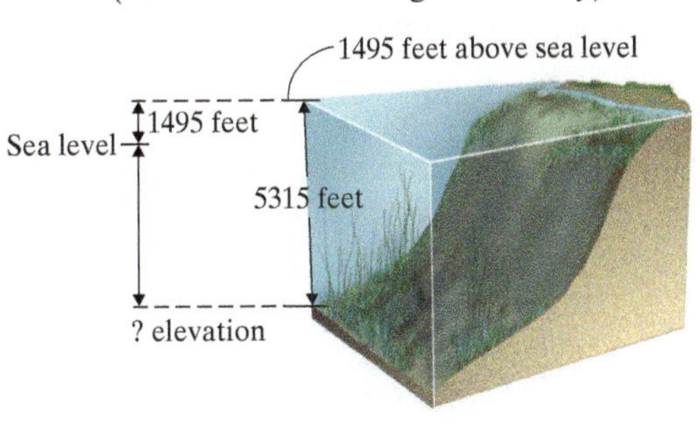

30. Find the average of −12, −13, 0, 9.

31. Translate the following phrases into mathematical expressions. Use x to represent "a number."

a. The product of a number and 17

b. A number subtracted from 20

Solve.

32. $-9n = -45$

33. $\frac{n}{-7} = 4$

34. $x - 16 = -36$

35. $-20 + 8 + 8 = x$

19. ______
20. ______
21. ______
22. ______
23. ______
24. ______
25. ______
26. ______
27. ______
28. ______
29. ______
30. ______
31. a. ______
b. ______
32. ______
33. ______
34. ______
35. ______

Chapters 1–2 Cumulative Review

Answers

1. ______
2. ______
3. ______
4. ______
5. ______
6. ______
7. a. ______
 b. ______
 c. ______
8. a. ______
 b. ______
 c. ______
9. ______
10. ______
11. ______
12. ______
13. ______
14. ______
15. ______
16. ______
17. ______
18. ______
19. a. ______
 b. ______
 c. ______
20. a. ______
 b. ______
 c. ______
21. ______
22. ______
23. a. ______
 b. ______
 c. ______
24. a. ______
 b. ______
 c. ______

Find the place value of the digit 3 in each whole number.

1. 396,418

2. 4308

3. 93,192

4. 693,298

5. 534,275,866

6. 267,301,818

7. Insert $<$ or $>$ to make a true statement.

a. $-7 \quad 7$

b. $0 \quad -4$

c. $-9 \quad -11$

8. Insert $<$ or $>$ to make a true statement

a. $12 \quad -4$

b. $-13 \quad -31$

c. $-82 \quad 79$

9. Add: $13 + 2 + 7 + 8 + 9$

10. Add: $11 + 3 + 9 + 16$

11. Subtract: $7826 - 505$ Check by adding.

12. Subtract: $3285 - 272$ Check by adding.

13. The radius of Jupiter is 43,441 miles. The radius of Saturn is 7257 miles less than the radius of Jupiter. Find the radius of Saturn. (*Source:* National Space Science Data Center)

14. C. J. Dufour wants to buy a digital camera. She has \$762 in her savings account. If the camera costs \$237, how much money will she have in her account after buying the camera?

15. Round 568 to the nearest ten.

16. Round 568 to the nearest hundred.

17. Round each number to the nearest hundred to find an estimated difference.

$$\begin{array}{r} 4725 \\ -2879 \\ \hline \end{array}$$

18. Round each number to the nearest thousand to find an estimated difference.

$$\begin{array}{r} 8394 \\ -2913 \\ \hline \end{array}$$

19. Rewrite each using the distributive property.

a. $5(6 + 5)$

b. $20(4 + 7)$

c. $2(7 + 9)$

20. Rewrite each using the distributive property.

a. $5(2 + 12)$

b. $9(3 + 6)$

c. $4(8 + 1)$

21. Multiply: 631×125

22. Multiply: 299×104

23. Find each quotient. Check by multiplying.

a. $42 \div 7$

b. $\dfrac{64}{8}$

c. $3\overline{)21}$

24. Find each quotient. Check by multiplying.

a. $\dfrac{35}{5}$

b. $64 \div 8$

c. $4\overline{)48}$

25. Divide: $3705 \div 5$. Check by multiplying.

26. Divide: $3648 \div 8$. Check by multiplying.

27. As part of a promotion, an executive receives 238 cards, each good for one free song download. If she wants to share them evenly with 19 friends, how many download cards will each friend receive? How many will be left over?

28. Mrs. Mallory's first-grade class is going to the zoo. She pays a total of $324 for 36 admission tickets. How much does each ticket cost?

Evaluate.

29. 9^2

30. 5^3

31. 6^1

32. 4^1

33. $5 \cdot 6^2$

34. $2^3 \cdot 7$

35. Simplify: $\dfrac{7 - 2 \cdot 3 + 3^2}{5(2 - 1)}$

36. Simplify: $\dfrac{6^2 + 4 \cdot 4 + 2^3}{37 - 5^2}$

37. Evaluate $x + 6$ if x is 8.

38. Evaluate $5 + x$ if x is 9.

39. Simplify:
   a. $|-9|$
   b. $|8|$
   c. $|0|$

40. Simplify:
   a. $|4|$
   b. $|-7|$

41. Add: $-2 + 25$

42. Add: $8 + (-3)$

43. Evaluate $2a - b$ for $a = 8$ and $b = -6$.

44. Evaluate $x - y$ for $x = -2$ and $y = -7$.

45. Multiply: $-7 \cdot 3$

46. Multiply: $5(-2)$

47. Multiply: $0 \cdot (-4)$

48. Multiply: $-6 \cdot 9$

49. Simplify: $3(4 - 7) + (-2) - 5$

50. Simplify: $4 - 8(7 - 3) - (-1)$

25. ______
26. ______
27. ______
28. ______
29. ______
30. ______
31. ______
32. ______
33. ______
34. ______
35. ______
36. ______
37. ______
38. ______
39. a. ______
b. ______
c. ______
40. a. ______
b. ______
41. ______
42. ______
43. ______
44. ______
45. ______
46. ______
47. ______
48. ______
49. ______
50. ______

3

Solving Equations and Problem Solving

Sections

3.1 Simplifying Algebraic Expressions

3.2 Solving Equations: Review of the Addition and Multiplication Properties

Integrated Review—Expressions and Equations

3.3 Solving Linear Equations in One Variable

3.4 Linear Equations in One Variable and Problem Solving

Check Your Progress

Vocabulary Check

Chapter Highlights

Chapter Review

Chapter Test

Cumulative Review

In this chapter, we continue making the transition from arithmetic to algebra. Recall that in algebra, letters (called variables) represent unknown quantities. Using variables is a very powerful method for solving problems that cannot be solved with arithmetic alone. This chapter introduces operations on algebraic expressions, and we continue solving variable equations.

Coca-Cola recently celebrated the 125th anniversary of its namesake brand. The illumination in the photo was on display for a month in Atlanta, Georgia. At this writing, the illumination holds the world record for largest single-building illumination. Also, Coca-Cola was recently named the Most Valuable Global Brand, as shown in the graph below. What is a global brand? There are many definitions, but a brand is the identity of a product or a service. The identity can be projected through its name, a symbol, a slogan, or something similar. A global brand, then, is a product whose brand is easily recognized throughout the world. Read the list of company names in the graph below and see how many of these brand names you know.

In Section 3.4, Exercises 31 and 32, we will solve applications that give us the unknown bar heights in the bar graph.

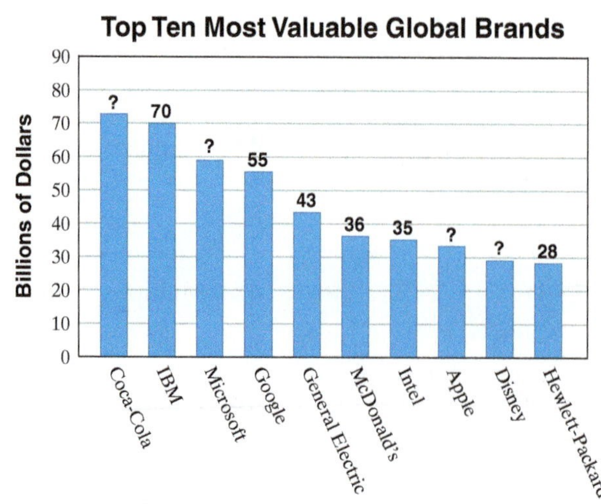

Source: Interbrand, *Top 10 of Everything*, 2013

3.1 Simplifying Algebraic Expressions 

Recall from Section 1.8 that a combination of numbers, letters (variables), and operation symbols is called an **algebraic expression** or simply an **expression.** Examples of expressions are below.

Algebraic Expressions

$4 \cdot x$, $n + 7$, and $3y - 5 - x$

Recall that if two variables or a number and a variable are next to each other, with no operation sign between them, the indicated operation is multiplication. For example,

$3y$ means $3 \cdot y$

and

xy or $x(y)$ means $x \cdot y$

Also, the meaning of an exponent remains the same when the base is a variable. For example,

$$y^2 = \underbrace{y \cdot y}_{\text{2 factors of } y} \quad \text{and} \quad x^4 = \underbrace{x \cdot x \cdot x \cdot x}_{\text{4 factors of } x}$$

Just as we can add, subtract, multiply, and divide numbers, we can add, subtract, multiply, and divide algebraic expressions. In previous sections we evaluated algebraic expressions like $x + 3$, $4x$, and $x + 2y$ for particular values of the variables. In this section, we explore working with variable expressions without evaluating them. We begin with a definition of a term.

Objectives

A Use Properties of Numbers to Combine Like Terms.

B Use Properties of Numbers to Multiply Expressions.

C Simplify Expressions by Multiplying and Then Combining Like Terms.

D Find the Perimeter and Area of Figures.

Objective A Combining Like Terms 

The addends of an algebraic expression are called the **terms** of the expression.

$x + 3$ — 2 terms

$3y^2 + (-6y) + 4$ — 3 terms

A term that is only a number has a special name. It is called a **constant term,** or simply a **constant.** A term that contains a variable is called a **variable term.**

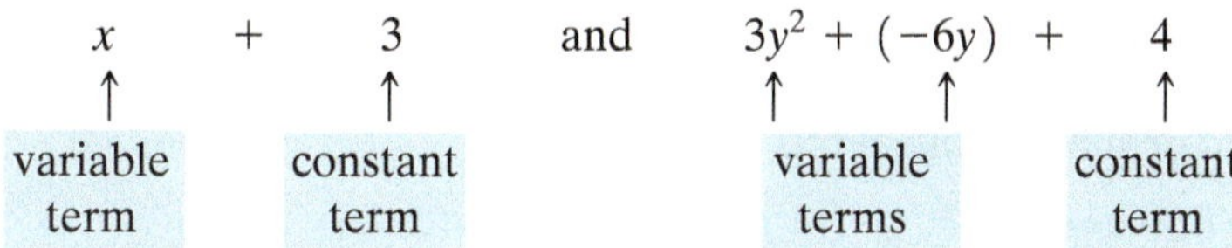

The number factor of a variable term is called the **numerical coefficient.** A numerical coefficient of 1 is usually not written.

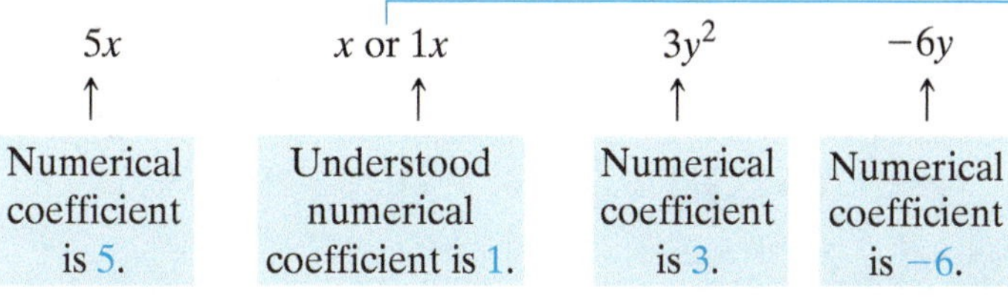

Helpful Hint

Recall that $1 \cdot$ any number = that number. This means that

$1 \cdot x = x$ or that $1x = x$

Thus x can always be replaced by $1x$ or $1 \cdot x$.

Terms with the same variable factors, except that they may have different numerical coefficients, are called **like terms.**

Like Terms	Unlike Terms
$3x, -4x$	$5x, x^2$
$-6y, 2y, y$	$7x, 7y$

✓Concept Check True or false? The terms $-7xz^2$ and $3z^2x$ are like terms. Explain.

A sum or difference of like terms can be simplified using the **distributive property.** Recall from Section 1.5 that the distributive property says that multiplication distributes over addition (and subtraction). Using variables, we can write the distributive property as follows:

$$(a + b)c = ac + bc.$$

If we write the right side of the equation first, then the left side, we have the following:

Distributive Property

If a, b, and c are numbers, then

$$ac + bc = (a + b)c$$

Also,

$$ac - bc = (a - b)c$$

The distributive property guarantees that, no matter what number x is, $7x + 2x$ (for example) has the same value as $(7 + 2)x$, or $9x$. We then have that

$$7x + 2x = (7 + 2)x = 9x$$

This is an example of **combining like terms.** An algebraic expression is **simplified** when all like terms have been combined.

Practice 1

Simplify each expression by combining like terms.

a. $8m - 14m$

b. $6a + a$

c. $-y^2 + 3y^2 + 7$

Example 1 Simplify each expression by combining like terms.

a. $4x + 6x$ **b.** $y - 5y$ **c.** $3x^2 + 5x^2 - 2$

Solution: Add or subtract like terms.

a. $4x + 6x = (4 + 6)x$
$= 10x$

Understood 1

b. $y - 5y = 1y - 5y$
$= (1 - 5)y$
$= -4y$

c. $3x^2 + 5x^2 - 2 = (3 + 5)x^2 - 2$
$= 8x^2 - 2$

Work Practice 1

Answers

1. a. $-6m$ **b.** $7a$ **c.** $2y^2 + 7$

✓Concept Check Answer

true

Helpful Hint

In this section, we are simplifying expressions. Try not to confuse the two processes below.

Expression: $5y - 8y$ Simplify the expression:	Equation: $8n = -40$ Solve the equation:
$5y - 8y = (5 - 8)y$ $= -3y$	$8n = -40$ $\frac{8n}{8} = \frac{-40}{8}$ Divide both sides by 8. $n = -5$ The solution is -5.

The commutative and associative properties of addition and multiplication can also help us simplify expressions. We presented these properties in Sections 1.3 and 1.5 and state them again using variables.

Properties of Addition and Multiplication

If a, b, and c are numbers, then

$a + b = b + a$ Commutative property of addition

$a \cdot b = b \cdot a$ Commutative property of multiplication

That is, the **order** of adding or multiplying two numbers can be changed without changing their sum or product.

$(a + b) + c = a + (b + c)$ Associative property of addition

$(a \cdot b) \cdot c = a \cdot (b \cdot c)$ Associative property of multiplication

That is, the **grouping** of numbers in addition or multiplication can be changed without changing their sum or product.

Helpful Hint

- Examples of these properties are

$2 + 3 = 3 + 2$ Commutative property of addition

$7 \cdot 9 = 9 \cdot 7$ Commutative property of multiplication

$(1 + 8) + 10 = 1 + (8 + 10)$ Associative property of addition

$(4 \cdot 2) \cdot 3 = 4 \cdot (2 \cdot 3)$ Associative property of multiplication

- These properties are not true for subtraction or division.

Example 2 Simplify: $2y - 6 + 4y + 8$

Solution: We begin by writing subtraction as the opposite of addition.

$2y - 6 + 4y + 8 = 2y + (-6) + 4y + 8$ Apply the commutative property of addition.

$= 2y + 4y + (-6) + 8$

$= (2 + 4)y + (-6) + 8$ Apply the distributive property.

$= 6y + 2$ Simplify.

Work Practice 2

Practice 2

Simplify: $6z + 5 + z - 4$

Answer

2. $7z + 1$

Practice 3–5

Simplify each expression by combining like terms.

3. $6y + 12y - 6$
4. $7y - 5 + y + 8$
5. $-7y + 2 - 2y - 9x + 12 - x$

Examples Simplify each expression by combining like terms.

3. $6x + 2x - 5 = 8x - 5$

4. $4x + 2 - 5x + 3 = 4x + 2 + (-5x) + 3$
$= 4x + (-5x) + 2 + 3$
$= -1x + 5$ or $-x + 5$

5. $2x - 5 + 3y + 4x - 10y + 11$
$= 2x + (-5) + 3y + 4x + (-10y) + 11$
$= 2x + 4x + 3y + (-10y) + (-5) + 11$
$= 6x - 7y + 6$

Work Practice 3–5

As we practice combining like terms, keep in mind that some of the steps may be performed mentally.

Objective B Multiplying Expressions 

We can also use properties of numbers to multiply expressions such as $3(2x)$. By the associative property of multiplication, we can write the product $3(2x)$ as $(3 \cdot 2)x$, which simplifies to $6x$.

Practice 6–7

Multiply.

6. $6(4a)$
7. $-8(9x)$

Examples Multiply.

6. $5(3y) = (5 \cdot 3)y$ Apply the associative property of multiplication.
$= 15y$ Multiply.

7. $-2(4x) = (-2 \cdot 4)x$ Apply the associative property of multiplication.
$= -8x$ Multiply.

Work Practice 6–7

We can use the distributive property to combine like terms, which we have done, and also to multiply expressions such as $2(3 + x)$. By the distributive property, we have

$2(3 + x) = 2 \cdot 3 + 2 \cdot x$ Apply the distributive property.
$= 6 + 2x$ Multiply.

Practice 8

Use the distributive property to multiply: $8(y + 2)$

Example 8 Use the distributive property to multiply: $5(x + 4)$

Solution: By the distributive property,

$5(x + 4) = 5 \cdot x + 5 \cdot 4$ Apply the distributive property.
$= 5x + 20$ Multiply.

Work Practice 8

Answers

3. $18y - 6$ 4. $8y + 3$ 5. $-9y - 10x + 14$ 6. $24a$ 7. $-72x$ 8. $8y + 16$

✓**Concept Check** What's wrong with the following?

$8(a - b) = 8a - b$

✓**Concept Check Answer**

did not distribute the 8; $8(a - b) = 8a - 8b$

Example 9 Multiply: $-3(5a + 2)$

Solution: By the distributive property,

$$-3(5a + 2) = -3(5a) + (-3)(2) \quad \text{Apply the distributive property.}$$
$$= (-3 \cdot 5)a + (-6) \quad \text{Use the associative property and multiply.}$$
$$= -15a - 6 \quad \text{Multiply.}$$

Work Practice 9

Practice 9

Multiply: $3(7a - 5)$

Example 10 Multiply: $8(x - 4)$

Solution:

$$8(x - 4) = 8 \cdot x - 8 \cdot 4$$
$$= 8x - 32$$

Work Practice 10

Practice 10

Multiply: $6(5 - y)$

Objective C Simplifying Expressions

Next we will **simplify** expressions by first using the distributive property to multiply and then **combining** any like terms.

Example 11 Simplify: $2(3 + 7x) - 15$

Solution: First we use the distributive property to remove parentheses.

$$2(3 + 7x) - 15 = 2(3) + 2(7x) - 15 \quad \text{Apply the distributive property.}$$
$$= 6 + 14x - 15 \quad \text{Multiply.}$$
$$= 14x + (-9) \text{ or } 14x - 9 \quad \text{Combine like terms.}$$

Work Practice 11

Practice 11

Simplify: $5(2y - 3) - 8$

Helpful Hint 2 is *not* distributed to the -15 since it is not within the parentheses.

Example 12 Simplify: $-2(x - 5) + 4(2x + 2)$

Solution: First we use the distributive property to remove parentheses.

$$-2(x - 5) + 4(2x + 2) = -2(x) - (-2)(5) + 4(2x) + 4(2) \quad \text{Apply the distributive property.}$$
$$= -2x + 10 + 8x + 8 \quad \text{Multiply.}$$
$$= 6x + 18 \quad \text{Combine like terms.}$$

Work Practice 12

Practice 12

Simplify:
$-7(x - 1) + 5(2x + 3)$

Example 13 Simplify: $-(x + 4) + 5x + 16$

Solution: The expression $-(x + 4)$ means $-1(x + 4)$.

$$-(x + 4) + 5x + 16 = -1(x + 4) + 5x + 16$$
$$= -1 \cdot x + (-1)(4) + 5x + 16 \quad \text{Apply the distributive property.}$$
$$= -x + (-4) + 5x + 16 \quad \text{Multiply.}$$
$$= 4x + 12 \quad \text{Combine like terms.}$$

Work Practice 13

Practice 13

Simplify: $-(y + 1) + 3y - 12$

Answers

9. $21a - 15$ **10.** $30 - 6y$
11. $10y - 23$ **12.** $3x + 22$
13. $2y - 13$

Objective D Finding Perimeter and Area

Practice 14

Find the perimeter of the square.

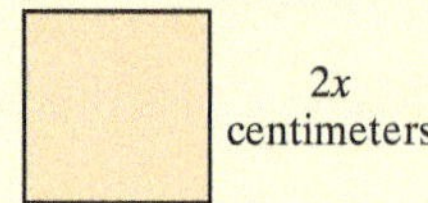

Example 14 Find the perimeter of the triangle.

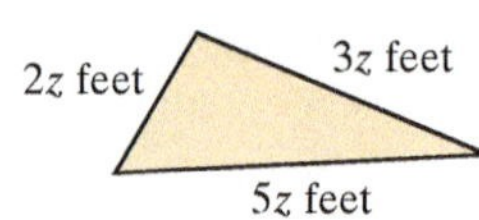

Solution: Recall that the perimeter of a figure is the distance around the figure. To find the perimeter, then, we find the sum of the lengths of the sides. We use the letter P to represent perimeter.

$$P = 2z + 3z + 5z$$
$$= 10z$$

Helpful Hint Don't forget to insert proper units.

The perimeter is $10z$ feet.

Work Practice 14

Practice 15

Find the area of the rectangular garden.

Example 15 Finding the Area of a Basketball Court

Find the area of this YMCA basketball court.

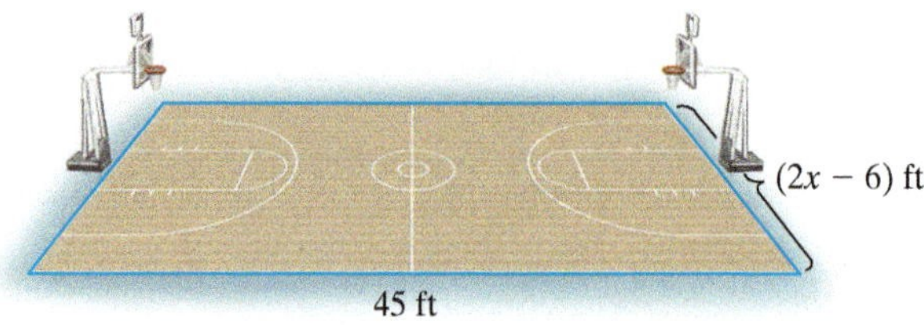

Solution: Recall how to find the area of a rectangle. **Area** = **Length** · **Width**, or if A represents area, l represents length, and w represents width, we have $A = l \cdot w$.

$$A = l \cdot w$$

$$= 45(2x - 6) \quad \text{Let length} = 45 \text{ and width} = (2x - 6).$$
$$= 90x - 270 \quad \text{Multiply.}$$

The area is $(90x - 270)$ *square* feet.

Work Practice 15

Helpful Hint

Don't forget...

Area:	Perimeter:
• surface enclosed	• distance around
• measured in square units	• measured in units

Answers

14. $8x$ cm **15.** $(36y + 27)$ sq yd

Vocabulary, Readiness & Video Check

Use the choices below to fill in each blank. Some choices may be used more than once.

numerical coefficient	combine like terms	like	term	variable	associative
constant	expression	unlike	distributive	commutative	

1. $14y^2 + 2x - 23$ is called a(n) ________ while $14y^2$, $2x$, and -23 are each called a(n) ________.
2. To multiply $3(-7x + 1)$, we use the ________ property.
3. To simplify an expression like $y + 7y$, we ________.
4. By the ________ properties, the *order* of adding or multiplying two numbers can be changed without changing their sum or product.
5. The term $5x$ is called a(n) ________ term while the term 7 is called a(n) ________ term.
6. The term z has an understood ________ of 1.
7. By the ________ properties, the *grouping* of adding or multiplying numbers can be changed without changing their sum or product.
8. The terms $-x$ and $5x$ are ________ terms and the terms $5x$ and $5y$ are ________ terms.
9. For the term $-3x^2y$, -3 is called the ________.

Martin-Gay Interactive Videos *Watch the section lecture video and answer the following questions.*

See Video 3.1

Objective A 10. In Example 2, why can't the expression be simplified after the first step?

Objective B 11. In Example 5, what property is used to multiply?

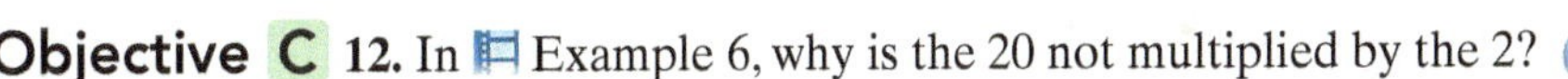

Objective C 12. In Example 6, why is the 20 not multiplied by the 2?

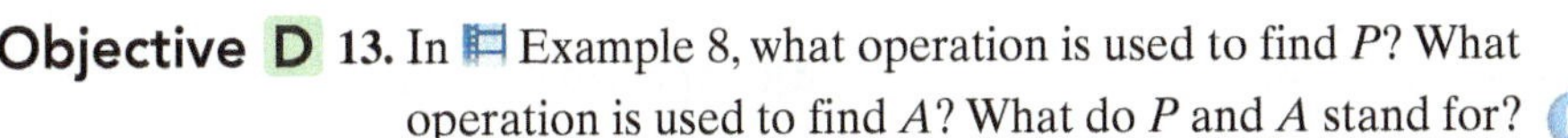

Objective D 13. In Example 8, what operation is used to find P? What operation is used to find A? What do P and A stand for?

3.1 Exercise Set MyMathLab®

Objective A *Simplify each expression by combining like terms. See Examples 1 through 5.*

1. $3x + 5x$
2. $8y + 3y$
3. $2n - 3n$
4. $7z - 10z$
5. $4c + c - 7c$
6. $5b - 8b - b$
7. $4x - 6x + x - 5x$
8. $8y + y - 2y - 8y$
9. $3a + 2a + 7a - 5$
10. $5b - 4b + b - 15$

Objective B *Multiply. See Examples 6 and 7.*

11. $6(7x)$
12. $4(4x)$
13. $-3(11y)$
14. $-3(21z)$
15. $12(6a)$
16. $13(5b)$

Multiply. See Examples 8 through 10.

17. $2(y + 3)$ **18.** $3(x + 1)$ **19.** $3(a - 6)$ **20.** $4(y - 6)$

21. $-4(3x + 7)$ **22.** $-8(8y + 10)$

Objective C *Simplify each expression. First use the distributive property to multiply and remove parentheses. See Examples 11 through 13.*

23. $2(x + 4) - 7$ **24.** $5(6 - y) - 2$ **25.** $8 + 5(3c - 1)$ **26.** $10 + 4(6d - 2)$

27. $-4(6n - 5) + 3n$ **28.** $-3(5 - 2b) - 4b$ **29.** $3 + 6(w + 2) + w$ **30.** $8z + 5(6 + z) + 20$

31. $2(3x + 1) + 5(x - 2)$ **32.** $3(5x - 2) + 2(3x + 1)$ **33.** $-(2y - 6) + 10$ **34.** $-(5x - 1) - 10$

Objectives A B C Mixed Practice *Simplify each expression. See Examples 1 through 13.*

35. $18y - 20y$ **36.** $x + 12x$ **37.** $z - 8z$ **38.** $-12x + 8x$

39. $9d - 3c - d$ **40.** $8r + s - 7s$ **41.** $2y - 6 + 4y - 8$ **42.** $a + 4 - 7a - 5$

43. $5q + p - 6q - p$ **44.** $m - 8n + m + 8n$ **45.** $2(x + 1) + 20$ **46.** $5(x - 1) + 18$

47. $5(x - 7) - 8x$ **48.** $3(x + 2) - 11x$ **49.** $-5(z + 3) + 2z$

50. $-8(1 + v) + 6v$ **51.** $8 - x + 4x - 2 - 9x$ **52.** $5y - 4 + 9y - y + 15$

53. $-7(x + 5) + 5(2x + 1)$ **54.** $-2(x + 4) + 8(3x - 1)$ **55.** $3r - 5r + 8 + r$

56. $6x - 4 + 2x - x + 3$ **57.** $-3(n - 1) - 4n$ **58.** $5(c + 2) + 7c$

59. $4(z - 3) + 5z - 2$ **60.** $8(m + 3) - 20 + m$ **61.** $6(2x - 1) - 12x$

62. $5(2a + 3) - 10a$ **63.** $-(4x - 5) + 5$ **64.** $-(7y - 2) + 6$

65. $-(4x - 10) + 2(3x + 5)$ **66.** $-(12b - 10) + 5(3b - 2)$ **67.** $3a + 4(a + 3)$

68. $b + 2(b - 5)$ **69.** $5y - 2(y - 1) + 3$ **70.** $3x - 4(x + 2) + 1$

Objective D *Find the perimeter of each figure. See Example 14.*

△ **71.** △ **72.** △ **73.**

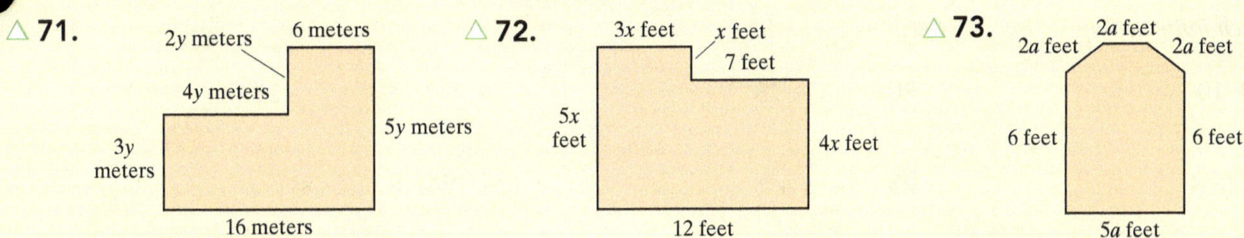

△ **74.** △ **75.** △ **76.**

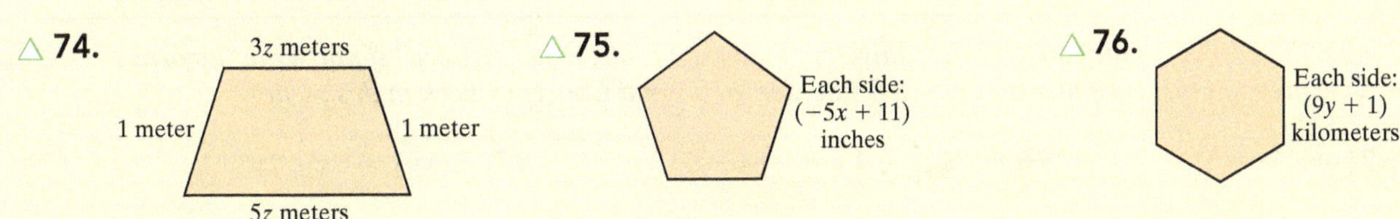

Find the area of each rectangle. See Example 15.

△ **77.** △ **78.** △ **79.**

△ **80.** **81.** △ **82.**

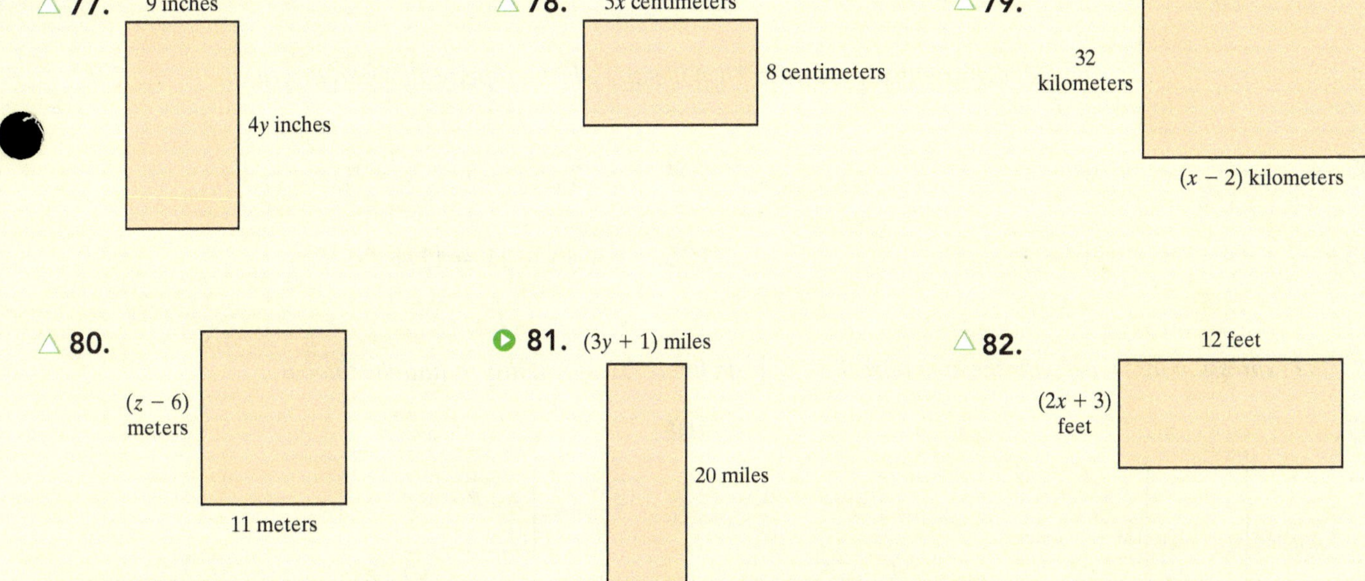

Objectives A B C D Mixed Practice *Solve. See Examples 1 through 15.*

83. Find the area of a regulation NCAA basketball court that is 94 feet long and 50 feet wide.

84. Find the area of a rectangular movie screen that is 50 feet long and 40 feet high.

85. A decorator wishes to put a wallpaper border around a rectangular room that measures 14 feet by 18 feet. Find the room's perimeter.

86. How much fencing will a rancher need for a rectangular cattle lot that measures 80 feet by 120 feet?

87. Find the perimeter of a triangular garden that measures 5 feet by x feet by $(2x + 1)$ feet.

88. Find the perimeter of a triangular picture frame that measures x inches by x inches by $(x - 14)$ inches.

Review

Perform each indicated operation. See Sections 2.2 and 2.3.

89. $-13 + 10$

90. $-15 + 23$

91. $-4 - (-12)$

92. $-7 - (-4)$

93. $-4 + 4$

94. $8 + (-8)$

Concept Extensions

If the expression on the left side of the equal sign is equivalent to the right side, write "correct." If not, write "incorrect" and then write an expression that is equivalent to the left side. See the second Concept Check in this section.

95. $5(3x - 2) \stackrel{?}{=} 15x - 2$

96. $-2(4x - 1) \stackrel{?}{=} -8x - 2$

97. $2(xy) \stackrel{?}{=} 2x \cdot 2y$

98. $-8(ab) \stackrel{?}{=} -8a \cdot (-8b)$

99. $7x - (x + 2) \stackrel{?}{=} 7x - x - 2$

100. $12y - (3y - 1) \stackrel{?}{=} 12y - 3y + 1$

101. $4(y - 3) + 11 \stackrel{?}{=} 4y - 7 + 11$

102. $6(x + 5) + 2 \stackrel{?}{=} 6x + 30 + 12$

Review commutative, associative, and distributive properties. Then identify which property allows us to write the equivalent expression on the right side of the equal sign.

103. $6(2x - 3) + 5 = 12x - 18 + 5$

104. $9 + 7x + (-2) = 7x + 9 + (-2)$

105. $-7 + (4 + y) = (-7 + 4) + y$

106. $(x + y) + 11 = 11 + (x + y)$

Write the expression that represents the area of each composite figure. Then simplify to find the total area.

△ **107.**

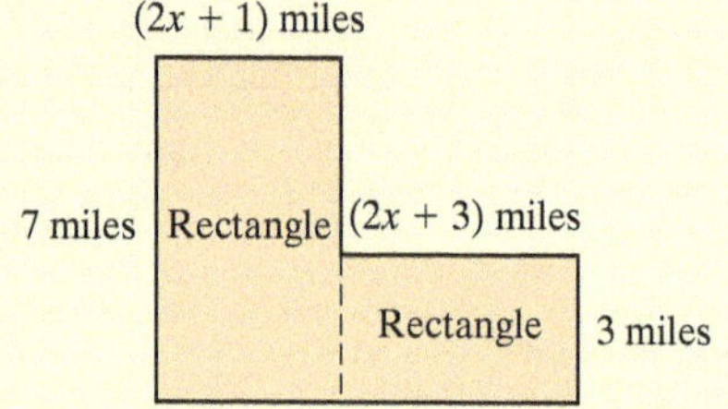

△ **108.**

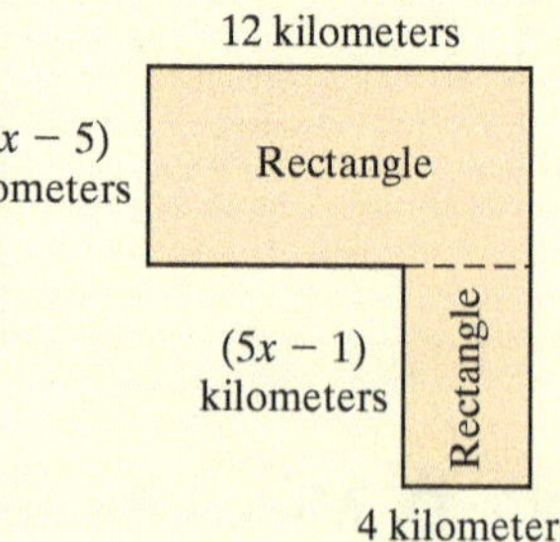

Simplify.

109. $9684q - 686 - 4860q + 12{,}960$

110. $76(268x + 592) - 2960$

111. If x is a whole number, which expression is greater: $2x$ or $5x$? Explain your answer.

112. If x is a whole number, which expression is greater: $-2x$ or $-5x$? Explain your answer.

113. Explain what makes two terms "like terms."

114. Explain how to combine like terms.

3.2 Solving Equations: Review of the Addition and Multiplication Properties

Objective A Using the Addition Property or the Multiplication Property

In this section, we continue solving equations using the properties first introduced in Section 2.6.

First, let's recall the difference between an **expression** and an **equation.** Remember—an equation contains an equal sign and an expression does not.

Equations (equal signs)	Expressions (no equal signs)
$7x = 6x + 4$	$7x - 6x + 4$
$3(3y - 5) = 10y$	$y - 1 + 11y - 21$

Thus far in this text, we have

Solved some equations (Section 2.6)

and

Simplified some expressions (Section 3.1)

As we will see in this section, the ability to simplify expressions will help us as we solve more equations.

The addition and multiplication properties are reviewed below.

Addition Property of Equality

Let a, b, and c represent numbers. Then

$a = b$ and $a + c = b + c$ are equivalent equations.	Also, $a = b$ and $a - c = b - c$ are equivalent equations.

Multiplication Property of Equality

Let a, b, and c represent numbers and let $c \neq 0$. Then

$a = b$ and $a \cdot c = b \cdot c$ are equivalent equations.	Also, $a = b$ and $\frac{a}{c} = \frac{b}{c}$ are equivalent equations.

In other words, the same number may be added to or subtracted from *both* sides of an equation without changing the solution of the equation. Also, *both* sides of an equation may be multiplied or divided by the same nonzero number without changing the solution of the equation.

Many equations in this section will contain expressions that can be simplified. If one or both sides of an equation can be simplified, do that first.

Example 1 Solve: $y - 5 = -2 - 6$

Solution: First we simplify the right side of the equation.

$y - 5 = -2 - 6$

$y - 5 = -8$ Combine like terms.

(*Continued on next page*)

Objectives

A Use the Addition Property or the Multiplication Property to Solve Equations.

B Use Both Properties to Solve Equations.

C Translate Word Phrases to Mathematical Expressions.

Practice 1

Solve: $x + 6 = 1 - 3$

Answer

1. -8

Next we get y alone by using the addition property of equality. We add 5 to both sides of the equation.

$$y - 5 + 5 = -8 + 5 \quad \text{Add 5 to both sides.}$$
$$y = -3 \quad \text{Simplify.}$$

Check: To see that -3 is the solution, replace y with -3 in the original equation.

$$y - 5 = -2 - 6$$
$$-3 - 5 \stackrel{?}{=} -2 - 6 \quad \text{Replace } y \text{ with } -3.$$
$$-8 = -8 \quad \text{True}$$

Since $-8 = -8$ is true, the solution is -3.

Work Practice 1

Practice 2

Solve: $10 = 2m - 4m$

Example 2 Solve: $3y - 7y = 12$

Solution: First, simplify the left side of the equation by combining like terms.

$$3y - 7y = 12$$
$$-4y = 12 \quad \text{Combine like terms.}$$

Next, we get y alone by using the multiplication property of equality and dividing both sides by -4.

$$\frac{-4y}{-4} = \frac{12}{-4} \quad \text{Divide both sides by } -4.$$
$$y = -3 \quad \text{Simplify.}$$

Check: Replace y with -3 in the original equation.

$$3y - 7y = 12$$
$$3(-3) - 7(-3) \stackrel{?}{=} 12$$
$$-9 + 21 \stackrel{?}{=} 12$$
$$12 = 12 \quad \text{True}$$

The solution is -3.

Work Practice 2

✓Concept Check What's wrong with the following solution?

$$4x - 6x = 10$$
$$2x = 10$$
$$\frac{2x}{2} = \frac{10}{2}$$
$$x = 5$$

Practice 3

Solve: $-8 + 6 = \frac{a}{3}$

Example 3 Solve: $\frac{z}{-4} = 11 - 5$

Solution: Simplify the right side of the equation first.

$$\frac{z}{-4} = 11 - 5$$
$$\frac{z}{-4} = 6$$

Answers

2. -5 **3.** -6

✓Concept Check answer

On the left side of the equation, $4x - 6x$ simplifies to $-2x$ *not* $2x$.

Next, to get z alone, multiply both sides by -4.

$$-4 \cdot \frac{z}{-4} = -4 \cdot 6 \quad \text{Multiply both sides by } -4.$$

$$\frac{-4}{-4} \cdot z = -4 \cdot 6$$

$$1z = -24 \quad \text{or} \quad z = -24$$

Check to see that -24 is the solution.

Work Practice 3

Example 4 Solve: $5x + 2 - 4x = 7 - 19$

Solution: First we simplify each side of the equation separately.

$$5x + 2 - 4x = 7 - 19$$

$$5x - 4x + 2 = 7 - 19$$

$$1x + 2 = -12 \quad \text{Combine like terms.}$$

To get x alone on the left side, we subtract 2 from both sides.

$$1x + 2 - 2 = -12 - 2 \quad \text{Subtract 2 from both sides.}$$

$$1x = -14 \quad \text{or} \quad x = -14 \quad \text{Simplify.}$$

Check to see that -14 is the solution.

Work Practice 4

Example 5 Solve: $8x - 9x = 12 - 17$

Solution: First combine like terms on each side of the equation.

$$8x - 9x = 12 - 17$$

$$-x = -5$$

Recall that $-x$ means $-1x$ and divide both sides by -1.

$$\frac{-1x}{-1} = \frac{-5}{-1} \quad \text{Divide both sides by } -1.$$

$$x = 5 \quad \text{Simplify.}$$

Check to see that the solution is 5.

Work Practice 5

Example 6 Solve: $3(3x - 5) = 10x$

Solution: First we multiply on the left side to remove the parentheses.

$$3(3x - 5) = 10x$$

$$3 \cdot 3x - 3 \cdot 5 = 10x \quad \text{Use the distributive property.}$$

$$9x - 15 = 10x$$

Now we subtract $9x$ from both sides.

$$9x - 15 - 9x = 10x - 9x \quad \text{Subtract } 9x \text{ from both sides.}$$

$$-15 = 1x \quad \text{or} \quad x = -15 \quad \text{Simplify.}$$

Work Practice 6

Practice 4

Solve:
$-6y - 1 + 7y = 17 + 2$

Practice 5

Solve: $-4 - 10 = 4y - 5y$

Practice 6

Solve: $13x = 4(3x - 1)$

Answers

4. 20 **5.** 14 **6.** -4

Objective B Using Both Properties to Solve Equations 

We now solve equations in one variable using more than one property of equality. To solve an equation such as $2x - 6 = 18$, we first get the variable term $2x$ alone on one side of the equation.

Practice 7

Solve: $5y + 2 = 17$

Example 7 Solve: $2x - 6 = 18$

Solution: We start by adding 6 to both sides to get the variable term $2x$ alone.

$$2x - 6 = 18$$
$$2x - 6 + 6 = 18 + 6 \quad \text{Add 6 to both sides.}$$
$$2x = 24 \quad \text{Simplify.}$$

To finish solving, we divide both sides by 2.

$$\frac{2x}{2} = \frac{24}{2} \quad \text{Divide both sides by 2.}$$
$$x = 12 \quad \text{Simplify.}$$

Check:

$$2x - 6 = 18$$
$$2(12) - 6 \stackrel{?}{=} 18 \quad \text{Replace } x \text{ with 12 and simplify.}$$
$$24 - 6 \stackrel{?}{=} 18$$
$$18 = 18 \quad \text{True}$$

The solution is 12.

Helpful Hint Don't forget to check the proposed solution in the *original* equation.

Work Practice 7

Don't forget, if one or both sides of an equation can be simplified, do that first.

Practice 8

Solve:
$-4(x + 2) - 60 = 2 - 10$

Example 8 Solve: $2 - 6 = -5(x + 4) - 39$

Solution: First, simplify each side of the equation.

$$2 - 6 = -5(x + 4) - 39$$
$$2 - 6 = -5x - 20 - 39 \quad \text{Use the distributive property.}$$
$$-4 = -5x - 59 \quad \text{Combine like terms on each side.}$$
$$-4 + 59 = -5x - 59 + 59 \quad \text{Add 59 to both sides to get the variable term alone.}$$
$$55 = -5x \quad \text{Simplify.}$$
$$\frac{55}{-5} = \frac{-5x}{-5} \quad \text{Divide both sides by } -5.$$
$$-11 = x \quad \text{or} \quad x = -11 \quad \text{Simplify.}$$

Check to see that -11 is the solution.

Work Practice 8

Answers
7. 3 8. −15

Objective C Translating Word Phrases into Expressions

Section 3.4 in this chapter contains a formal introduction to problem solving. To prepare for this section, let's once again review writing phrases as algebraic expressions using the following key words and phrases as a guide:

Addition	Subtraction	Multiplication	Division
sum	difference	product	quotient
plus	minus	times	divided
added to	subtracted from	multiply	shared equally among
more than	less than	twice	per
increased by	decreased by	of	divided by
total	less	twice/double/triple	divided into

Example 9 Write each phrase as an algebraic expression. Use x to represent "a number."

a. a number increased by -5

b. the product of -7 and a number

c. a number less 20

d. the quotient of -18 and a number

e. a number subtracted from -2

Solution:

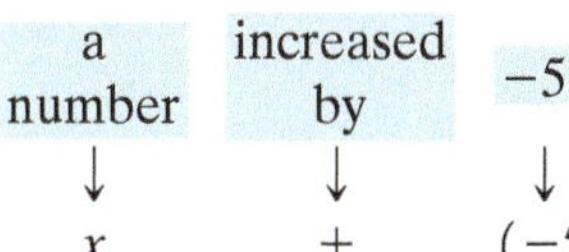

a. In words: a number | increased by | -5

Translate: $x + (-5)$ or $x - 5$

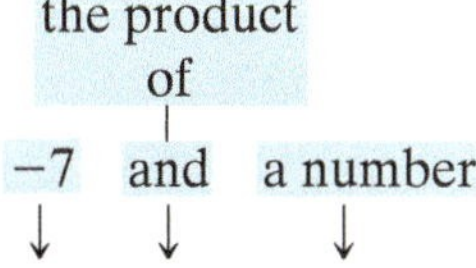

b. In words: the product of -7 and a number

Translate: $-7 \cdot x$ or $-7x$

c. In words: a number | less | 20

Translate: $x - 20$

d. In words: the quotient of -18 and a number

Translate: $-18 \div x$ or $\dfrac{-18}{x}$ or $-\dfrac{18}{x}$

e. In words: a number | subtracted from | -2

Translate: $-2 - x$

Work Practice 9

Practice 9

Write each phrase as an algebraic expression. Use x to represent "a number."

a. the sum of -3 and a number

b. -5 decreased by a number

c. three times a number

d. a number subtracted from 83

e. the quotient of a number and -4

Answers

9. a. $-3 + x$ **b.** $-5 - x$ **c.** $3x$ **d.** $83 - x$ **e.** $\dfrac{x}{-4}$ or $-\dfrac{x}{4}$

Helpful Hint

As we reviewed in Chapter 1, don't forget that order is important when subtracting. Notice the translation order of numbers and variables below.

Phrase	Translation
a number less 9	$x - 9$
a number subtracted from 9	$9 - x$

Practice 10

Translate each phrase into an algebraic expression. Let x be the unknown number.

a. The product of 5 and a number, decreased by 25

b. Twice the sum of a number and 3

c. The quotient of 39 and twice a number

Example 10

Write each phrase as an algebraic expression. Let x be the unknown number.

a. Twice a number, increased by -9

b. Three times the difference of a number and 11

c. The quotient of 5 times a number and 17

Solution:

a. In words: twice a number → increased by → -9

Translate: $2x + (-9)$ or $2x - 9$

b. In words: three times → the difference of (a number and 11)

Translate: $3(x - 11)$

c. In words: the quotient of (5 times a number and 17)

Translate: $5x \div 17$ or $\dfrac{5x}{17}$

Work Practice 10

Answers

10. a. $5x - 25$ **b.** $2(x + 3)$ **c.** $39 \div (2x)$ or $\dfrac{39}{2x}$

Vocabulary, Readiness & Video Check

Use the choices below to fill in each blank.

equation	multiplication	equivalent	expression
solving	addition	simplifying	

1. The equations $-3x = 51$ and $\dfrac{-3x}{-3} = \dfrac{51}{-3}$ are called __________ equations.
2. The difference between an equation and an expression is that a(n) __________ contains an equal sign, while a(n) __________ does not.
3. The process of writing $-3x + 10x$ as $7x$ is called __________ the expression.
4. For the equation $-5x - 1 = -21$, the process of finding that 4 is the solution is called __________ the equation.

5. By the __________ property of equality, $x = -2$ and $x + 7 = -2 + 7$ are equivalent equations.

6. By the __________ property of equality, $y = 8$ and $3 \cdot y = 3 \cdot 8$ are equivalent equations.

Martin-Gay Interactive Videos *Watch the section lecture video and answer the following questions.*

See Video 3.2

Objective A 7. When solving Example 1, what must be done before applying the multiplication property of equality?

8. When solving Example 4, what must be done before applying the addition property of equality?

Objective B 9. In Examples 5 and 6, both the addition and the multiplication properties of equality are used. Which property is used first both times?

Objective C 10. In Example 7, why are we told to be careful when translating a subtraction phrase?

3.2 Exercise Set MyMathLab®

Objective A *Solve each equation. First combine any like terms on each side of the equation. See Examples 1 through 5.*

1. $x - 3 = -1 + 4$
2. $x + 7 = 2 + 3$
3. $-7 + 10 = m - 5$
4. $1 - 8 = n + 2$
5. $2w - 12w = 40$
6. $10y - y = 45$
7. $24 = t + 3t$
8. $100 = 15y + 5y$
9. $2z = 12 - 14$
10. $-3x = 11 - 2$
11. $4 - 10 = \frac{z}{-3}$
12. $20 - 22 = \frac{z}{-4}$
13. $-3x - 3x = 50 - 2$
14. $5y - 9y = -14 + (-14)$
15. $\frac{x}{5} = -26 + 16$
16. $\frac{y}{3} = 32 - 52$
17. $7x + 7 - 6x = 10$
18. $-3 + 5x - 4x = 13$
19. $-8 - 9 = 3x + 5 - 2x$
20. $-7 + 10 = 4x - 6 - 3x$

Solve. First multiply to remove parentheses. See Example 6.

21. $2(5x - 3) = 11x$
22. $6(3x + 1) = 19x$
23. $3y = 2(y + 12)$
24. $17x = 4(4x - 6)$
25. $21y = 5(4y - 6)$
26. $28z = 9(3z - 2)$
27. $-3(-4 - 2z) = 7z$
28. $-2(-1 - 3y) = 7y$

Objective B *Solve each equation. See Examples 7 and 8.*

29. $2x - 8 = 0$
30. $3y - 12 = 0$
31. $7y + 3 = 24$
32. $5m + 1 = 46$
33. $-7 = 2x - 1$
34. $-11 = 3t - 2$
35. $6(6 - 4y) = 12y$
36. $4(3y - 5) = 14y$
37. $11(x - 6) = -4 - 7$
38. $5(x - 6) = -2 - 8$
39. $-3(x + 1) - 10 = 12 + 8$

40. $-2(x + 5) - 2 = -8 - 4$

41. $y - 20 = 6y$

42. $x - 63 = 10x$

43. $22 - 42 = 4(x - 1) - 4$

44. $35 - (-3) = 3(x - 2) + 17$

Objectives A B Mixed Practice *Solve each equation. See Examples 1 through 8.*

45. $-2 - 3 = -4 + x$

46. $7 - (-10) = x - 5$

47. $y + 1 = -3 + 4$

48. $y - 8 = -5 - 1$

49. $3w - 12w = -27$

50. $y - 6y = 20$

51. $-4x = 20 - (-4)$

52. $6x = 5 - 35$

53. $18 - 11 = \frac{x}{-5}$

54. $9 - 14 = \frac{x}{-12}$

55. $9x - 12 = 78$

56. $8x - 8 = 32$

57. $10 = 7t - 12t$

58. $-30 = t + 9t$

59. $5 - 5 = 3x + 2x$

60. $-42 + 20 = -2x + 13x$

61. $50y = 7(7y + 4)$

62. $65y = 8(8y - 9)$

63. $8x = 2(6x + 10)$

64. $10x = 6(2x - 3)$

65. $7x + 14 - 6x = -4 - 10$

66. $-10x + 11x + 5 = 9 - 5$

67. $\frac{x}{-4} = -1 - (-8)$

68. $\frac{y}{-6} = 6 - (-1)$

69. $23x + 8 - 25x = 7 - 9$

70. $8x - 4 - 6x = 12 - 22$

71. $-3(x + 9) - 41 = 4 - 60$

72. $-4(x + 7) - 30 = 3 - 37$

Objective C Translating *Write each phrase as a variable expression. Use x to represent "a number." See Examples 9 and 10.*

73. The sum of -7 and a number

74. Negative eight plus a number

75. Eleven subtracted from a number

76. A number subtracted from twelve

77. The product of -13 and a number

78. Twice a number

79. A number divided by -12

80. The quotient of negative six and a number

81. The product of -11 and a number, increased by 5

82. Negative four times a number, increased by 18

83. Negative ten decreased by 7 times a number

84. Twice a number, decreased by thirty

85. Seven added to the product of 4 and a number

86. The product of 7 and a number, added to 100

87. Twice a number, decreased by 17

88. The difference of -9 times a number, and 1

89. The product of -6 and the sum of a number and 15

90. Twice the sum of a number and -5

91. The quotient of 45 and the product of a number and -5

92. The quotient of ten times a number, and -4

93. The quotient of seventeen and a number, increased by -15

94. The quotient of -20 and a number, decreased by three

Review

This horizontal bar graph shows the top ten states for traveler spending in a recent year. Use this graph to answer Exercises 95 through 98. See Sections 1.2 and 1.3.

95. For what state do travelers spend the most money?

96. For the states shown, which states have traveler amounts less than \$20 billion?

97. What is the combined spending for the neighboring states of Florida and Georgia?

98. What is the combined spending for the two largest states, Texas and California?

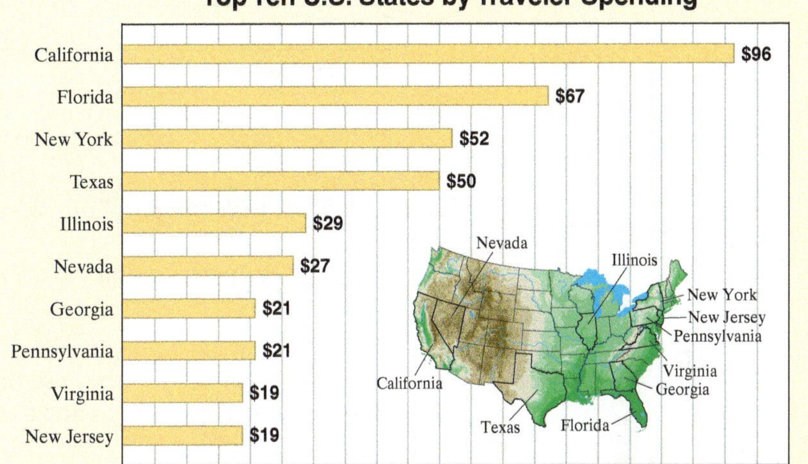

Source: U.S. Travel Association, 2010

Concept Extensions

99. In your own words, explain the addition property of equality.

100. Write an equation that can be solved using the addition property of equality.

101. Are the equations below equivalent? Why or why not?

$$x + 7 = 4 + (-9)$$
$$x + 7 \stackrel{?}{=} 5$$

102. Are the equations below equivalent? Why or why not?

$$3x - 6x = 12$$
$$3x \stackrel{?}{=} 12$$

103. Why does the multiplication property of equality not allow us to divide both sides of an equation by zero?

104. Is the equation $-x = 6$ solved for the variable? Explain why or why not.

Solve.

105. $\frac{y}{72} = -86 - (-1029)$

106. $\frac{x}{-13} = 4^6 - 5^7$

107. $\frac{x}{-2} = 5^2 - |-10| - (-9)$

108. $\frac{y}{10} = (-8)^2 - |20| + (-2)^2$

109. $|-13| + 3^2 = 100y - |-20| - 99y$

110. $4(x - 11) + |90| - |-86| + 2^5 = 5x$

Integrated Review Sections 3.1–3.2

Expressions and Equations

Answers

1. ____________
2. ____________
3. ____________
4. ____________
5. ____________
6. ____________
7. ____________
8. ____________
9. ____________
10. ____________
11. ____________
12. ____________
13. ____________
14. ____________
15. ____________
16. ____________
17. ____________
18. ____________
19. ____________
20. ____________

For the table below, identify each as an expression or an equation.

	Expression or Equation	
1.	$7x - 5y + 14$	
2.	$7x = 35 + 14$	
3.	$3(x - 2) = 5(x + 1) - 17$	
4.	$-9(2x + 1) - 4(x - 2) + 14$	

Fill in each blank with "simplify" or "solve."

5. To ________ an expression, we combine any like terms.

6. To ________ an equation, we use the properties of equality to find any value of the variable that makes the equation a true statement.

Simplify each expression by combining like terms.

7. $7x + x$

8. $6y - 10y$

9. $2a + 5a - 9a - 2$

10. $6a - 12 - a - 14$

Multiply and simplify if possible.

11. $-2(4x + 7)$

12. $-3(2x - 10)$

13. $5(y + 2) - 20$

14. $12x + 3(x - 6) - 13$

△**15.** Find the area of the rectangle.

Rectangle

3 meters

$(4x - 2)$ meters

△**16.** Find the perimeter of the triangle.

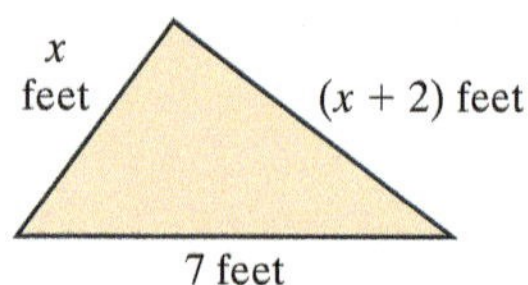

Solve and check.

17. $12 = 11x - 14x$

18. $8y + 7y = -45$

19. $x - 12 = -45 + 23$

20. $6 - (-5) = x + 5$

Solve and check.

21. $\frac{x}{3} = -14 + 9$

22. $\frac{z}{4} = -23 - 7$

23. $-6 + 2 = 4x + 1 - 3x$

24. $5 - 8 = 5x + 10 - 4x$

25. $6(3x - 4) = 19x$

26. $25x = 6(4x - 9)$

27. $-36x - 10 + 37x = -12 - (-14)$

28. $-8 + (-14) = -80y + 20 + 81y$

29. $3x - 16 = -10$

30. $4x - 21 = -13$

31. $-8z - 2z = 26 - (-4)$

32. $-12 + (-13) = 5x - 10x$

33. $-4(x + 8) - 11 = 3 - 26$

34. $-6(x - 2) + 10 = -4 - 10$

Translating *Write each phrase as an algebraic expression. Use x to represent "a number."*

35. The difference of a number and 10

36. The sum of −20 and a number

37. The product of 10 and a number

38. The quotient of 10 and a number

39. Five added to the product of −2 and a number

40. The product of −4 and the difference of a number and 1

21. ____________
22. ____________
23. ____________
24. ____________
25. ____________
26. ____________
27. ____________
28. ____________
29. ____________
30. ____________
31. ____________
32. ____________
33. ____________
34. ____________
35. ____________
36. ____________
37. ____________
38. ____________
39. ____________
40. ____________

3.3 Solving Linear Equations in One Variable

Objectives

A Solve Linear Equations Using the Addition and Multiplication Properties.

B Solve Linear Equations Containing Parentheses.

C Write Numerical Sentences as Equations.

In this chapter, the equations we are solving are called **linear equations in one variable** or **first-degree equations in one variable.** For example, an equation such as $5x - 2 = 6x$ is a linear equation in one variable. It is called linear or first degree because the exponent on each x is 1 and there is no variable below a fraction bar. It is an equation in one variable because it contains one variable, x.

Let's continue solving linear equations in one variable.

Objective A Solving Equations Using the Addition and Multiplication Properties

If an equation contains variable terms on both sides, we use the addition property of equality to get all the variable terms on one side and all the constants or numbers on the other side.

Practice 1

Solve: $7x + 12 = 3x - 4$

Example 1 Solve: $3a - 6 = a + 4$

Solution: Although it makes no difference which side we choose, let's move variable terms to the left side and constants to the right side.

$$3a - 6 = a + 4$$

$$3a - 6 + 6 = a + 4 + 6 \quad \text{Add 6 to both sides.}$$

$$3a = a + 10 \quad \text{Simplify.}$$

$$3a - a = a + 10 - a \quad \text{Subtract } a \text{ from both sides.}$$

$$2a = 10 \quad \text{Simplify.}$$

$$\frac{2a}{2} = \frac{10}{2} \quad \text{Divide both sides by 2.}$$

$$a = 5 \quad \text{Simplify.}$$

Check:

$$3a - 6 = a + 4 \quad \text{Original equation}$$

$$3 \cdot 5 - 6 \stackrel{?}{=} 5 + 4 \quad \text{Replace } a \text{ with 5.}$$

$$15 - 6 \stackrel{?}{=} 9 \quad \text{Simplify.}$$

$$9 = 9 \quad \text{True}$$

The solution is 5.

Work Practice 1

Helpful Hint

Make sure you understand which property to use to solve an equation.

Addition

$$x + 2 = 10$$

To undo addition of 2, we subtract 2 from both sides.

$$x + 2 - 2 = 10 - 2 \quad \text{Use addition property of equality.}$$

$$x = 8$$

Check:

$$x + 2 = 10$$

$$8 + 2 \stackrel{?}{=} 10$$

$$10 = 10 \quad \text{True}$$

Understood multiplication

$$2x = 10$$

To undo multiplication of 2, we divide both sides by 2.

$$\frac{2x}{2} = \frac{10}{2} \quad \text{Use multiplication property of equality.}$$

$$x = 5$$

Check:

$$2x = 10$$

$$2 \cdot 5 \stackrel{?}{=} 10$$

$$10 = 10 \quad \text{True}$$

Answer

1. -4

Example 2 Solve: $17 - 7x + 3 = -3x + 21 - 3x$

Solution: First, simplify both sides of the equation.

$$17 - 7x + 3 = -3x + 21 - 3x$$
$$20 - 7x = -6x + 21 \quad \text{Simplify.}$$

Next, move variable terms to one side of the equation and constants, or numbers, to the other side. To begin, let's add $6x$ to both sides.

$$20 - 7x + 6x = -6x + 21 + 6x \quad \text{Add } 6x \text{ to each side.}$$
$$20 - x = 21 \quad \text{Simplify.}$$
$$20 - x - 20 = 21 - 20 \quad \text{Subtract 20 from both sides.}$$
$$-1x = 1 \quad \text{Simplify. Recall that } -x \text{ means } -1x.$$
$$\frac{-1x}{-1} = \frac{1}{-1} \quad \text{Divide both sides by } -1.$$
$$x = -1 \quad \text{Simplify.}$$

Check:

$$17 - 7x + 3 = -3x + 21 - 3x$$
$$17 - 7(-1) + 3 \stackrel{?}{=} -3(-1) + 21 - 3(-1)$$
$$17 + 7 + 3 \stackrel{?}{=} 3 + 21 + 3$$
$$27 = 27 \quad \text{True}$$

The solution is -1.

Work Practice 2

Practice 2

Solve:
$40 - 5y + 5 = -2y - 10 - 4y$

Objective B Solving Equations Containing Parentheses

Recall from the previous section that if an equation contains parentheses, we will first use the distributive property to remove them.

Example 3 Solve: $7(x - 2) = 9x - 6$

Solution: First we apply the distributive property.

$$7(x - 2) = 9x - 6$$
$$7x - 14 = 9x - 6 \quad \text{Apply the distributive property.}$$

Next we move variable terms to one side of the equation and constants to the other side.

$$7x - 14 - 9x = 9x - 6 - 9x \quad \text{Subtract } 9x \text{ from both sides.}$$
$$-2x - 14 = -6 \quad \text{Simplify.}$$
$$-2x - 14 + 14 = -6 + 14 \quad \text{Add 14 to both sides.}$$
$$-2x = 8 \quad \text{Simplify.}$$
$$\frac{-2x}{-2} = \frac{8}{-2} \quad \text{Divide both sides by } -2.$$
$$x = -4 \quad \text{Simplify.}$$

Check to see that -4 is the solution.

Work Practice 3

Practice 3

Solve: $6(a - 5) = 4a + 4$

✓Concept Check In Example 3, the solution is -4. To check this solution, what equation should we use?

Answers
2. -55 **3.** 17

✓Concept Check Answer
$7(x - 2) = 9x - 6$

Practice 4

Solve: $4(x + 3) + 1 = 13$

You may want to use the following steps to solve equations.

Steps for Solving an Equation

Step 1: If parentheses are present, use the distributive property.

Step 2: Combine any like terms on each side of the equation.

Step 3: Use the addition property of equality to rewrite the equation so that variable terms are on one side of the equation and constant terms are on the other side.

Step 4: Use the multiplication property of equality to divide both sides by the numerical coefficient of the variable to solve.

Step 5: Check the solution in the *original equation.*

Example 4 Solve: $3(2x - 6) + 6 = 0$

Solution: $3(2x - 6) + 6 = 0$

Step 1: $6x - 18 + 6 = 0$ Apply the distributive property.

Step 2: $6x - 12 = 0$ Combine like terms on the left side of the equation.

Step 3: $6x - 12 + 12 = 0 + 12$ Add 12 to both sides.

$6x = 12$ Simplify.

Step 4: $\frac{6x}{6} = \frac{12}{6}$ Divide both sides by 6.

$x = 2$ Simplify.

Check:

Step 5:

$$3(2x - 6) + 6 = 0$$
$$3(2 \cdot 2 - 6) + 6 \stackrel{?}{=} 0$$
$$3(4 - 6) + 6 \stackrel{?}{=} 0$$
$$3(-2) + 6 \stackrel{?}{=} 0$$
$$-6 + 6 \stackrel{?}{=} 0$$
$$0 = 0 \quad \text{True}$$

The solution is 2.

Work Practice 4

Objective C Writing Numerical Sentences as Equations

Next, we practice translating sentences into equations. Below are key words and phrases that translate to an equal sign. (*Note:* For a review of key words and phrases that translate to addition, subtraction, multiplication, and division, see Sections 1.8 and 3.2.)

Key Words or Phrases	Examples	Symbols
equals	3 equals 2 plus 1	$3 = 2 + 1$
gives	the quotient of 10 and −5 gives −2	$\frac{10}{-5} = -2$
is/was	17 minus 12 is 5	$17 - 12 = 5$
yields	11 plus 2 yields 13	$11 + 2 = 13$
amounts to	twice −15 amounts to −30	$2(-15) = -30$
is equal to	−24 is equal to 2 times −12	$-24 = 2(-12)$

Answer
4. 0

Example 5 Translate each sentence into an equation.

a. The product of 7 and 6 is 42.
b. Twice the sum of 3 and 5 is equal to 16.
c. The quotient of -45 and 5 yields -9.

Solution:

a.

In words:	the product of 7 and 6	is	42
	↓	↓	↓
Translate:	$7 \cdot 6$	$=$	42

b.

In words:	twice	the sum of 3 and 5	is equal to	16
	↓	↓	↓	↓
Translate:	2	$(3 + 5)$	$=$	16

c.

In words:	the quotient of -45 and 5	yields	-9
	↓	↓	↓
Translate:	$\frac{-45}{5}$	$=$	-9

Work Practice 5

Practice 5

Translate each sentence into an equation.

a. The difference of 110 and 80 is 30.
b. The product of 3 and the sum of -9 and 11 amounts to 6.
c. The quotient of 24 and -6 yields -4.

Answers

5. a. $110 - 80 = 30$
b. $3(-9 + 11) = 6$ **c.** $\frac{24}{-6} = -4$

Calculator Explorations Checking Equations

A calculator can be used to check possible solutions of equations. To do this, replace the variable by the possible solution and evaluate each side of the equation separately. For example, to see whether 7 is a solution of the equation $52x = 15x + 259$, replace x with 7 and use your calculator to evaluate each side separately.

Equation: $52x = 15x + 259$

$52 \cdot 7 \stackrel{?}{=} 15 \cdot 7 + 259$ Replace x with 7.

Evaluate left side: [52] [×] [7] then [=] or [ENTER].
Display: [364].
Evaluate right side: [15] [×] [7] [+] [259] then [=] or [ENTER].
Display: [364].

Since the left side equals the right side, 7 is a solution of the equation $52x = 15x + 259$.

Use a calculator to determine whether the numbers given are solutions of each equation.

1. $76(x - 25) = -988$; 12
2. $-47x + 862 = -783$; 35
3. $x + 562 = 3x + 900$; -170
4. $55(x + 10) = 75x + 910$; -18
5. $29x - 1034 = 61x - 362$; -21
6. $-38x + 205 = 25x + 120$; 25

Vocabulary, Readiness & Video Check

Use the choices below to fill in each blank. Some choices may be used more than once.

addition	multiplication	combine like terms
$5(2x + 6) - 1 = 39$	$3x - 9 + x - 16$	distributive

1. An example of an expression is ______________ while an example of an equation is ______________.
2. To solve $\frac{x}{-7} = -10$, we use the ______________ property of equality.
3. To solve $x - 7 = -10$, we use the ______________ property of equality.

Use the order of the Steps for Solving an Equation in this section to answer Exercises 4 through 6.

4. To solve $9x - 6x = 10 + 6$, first ______________.
5. To solve $5(x - 1) = 25$, first use the ______________ property.
6. To solve $4x + 3 = 19$, first use the ______________ property of equality.

Martin-Gay Interactive Videos *Watch the section lecture video and answer the following questions.*

See Video 3.3 

Objective A 7. In Example 1, the number 1 is subtracted from the left side of the equation. What property tells us we must also subtract 1 from the right side? Why is it important to do the same thing to both sides?

Objective B 8. From Example 3, what is the first step when solving an equation that contains parentheses? What property do we use to perform this step?

Objective C 9. What word or phrase translates to "equals" in Example 5? In Example 6?

3.3 Exercise Set MyMathLab®

Objective A *Solve each equation. See Examples 1 and 2.*

1. $3x - 7 = 4x + 5$
2. $7x - 1 = 8x + 4$
3. $10x + 15 = 6x + 3$
4. $5x - 3 = 2x - 18$
5. $19 - 3x = 14 + 2x$
6. $4 - 7m = -3m + 4$
7. $-14x - 20 = -12x + 70$
8. $57y + 140 = 54y - 100$
9. $x + 20 + 2x = -10 - 2x - 15$
10. $2x + 10 + 3x = -12 - x - 20$
11. $40 + 4y - 16 = 13y - 12 - 3y$
12. $19x - 2 - 7x = 31 + 6x - 15$

Objective B *Solve each equation. See Examples 3 and 4.*

13. $35 - 17 = 3(x - 2)$
14. $22 - 42 = 4(x - 1)$
15. $3(x - 1) - 12 = 0$
16. $2(x + 5) + 8 = 0$
17. $2(y - 3) = y - 6$
18. $3(z + 2) = 5z + 6$
19. $-2(y + 4) = 2$
20. $-1(y + 3) = 10$
21. $2t - 1 = 3(t + 7)$
22. $-4 + 3c = 4(c + 2)$
23. $3(5c + 1) - 12 = 13c + 3$
24. $4(3t + 4) - 20 = 3 + 5t$

Mixed Practice (*Sections 2.6, 3.2, 3.3*) *Solve each equation.*

25. $-4x = 44$
26. $-3x = 51$
27. $x + 9 = 2$
28. $y - 6 = -11$
29. $8 - b = 13$
30. $7 - z = 15$

31. $-20 - (-50) = \frac{x}{9}$

32. $-2 - 10 = \frac{z}{10}$

33. $3r + 4 = 19$

34. $7y + 3 = 38$

35. $-7c + 1 = -20$

36. $-2b + 5 = -7$

37. $8y - 13y = -20 - 25$

38. $4x - 11x = -14 - 14$

39. $6(7x - 1) = 43x$

40. $5(3y - 2) = 16y$

41. $-4 + 12 = 16x - 3 - 15x$

42. $-9 + 20 = 19x - 4 - 18x$

43. $-10(x + 3) + 28 = -16 - 16$

44. $-9(x + 2) + 25 = -19 - 19$

45. $4x + 3 = 2x + 11$

46. $6y - 8 = 3y + 7$

47. $-2y - 10 = 5y + 18$

48. $7n + 5 = 12n - 10$

49. $-8n + 1 = -6n - 5$

50. $10w + 8 = w - 10$

51. $9 - 3x = 14 + 2x$

52. $4 - 7m = -3m$

53. $9a + 29 + 7 = 0$

54. $10 + 4v + 6 = 0$

55. $7(y - 2) = 4y - 29$

56. $2(z - 2) = 5z + 17$

57. $12 + 5t = 6(t + 2)$

58. $4 + 3c = 2(c + 2)$

59. $3(5c - 1) - 2 = 13c + 3$

60. $4(2t + 5) - 21 = 7t - 6$

61. $10 + 5(z - 2) = 4z + 1$

62. $14 + 4(w - 5) = 6 - 2w$

63. $7(6 + w) = 6(2 + w)$

64. $6(5 + c) = 5(c - 4)$

Objective C Translating *Write each sentence as an equation. See Example 5.*

65. The sum of −42 and 16 is −26.

66. The difference of −30 and 10 equals −40.

67. The product of −5 and −29 gives 145.

68. The quotient of −16 and 2 yields −8.

69. Three times the difference of −14 and 2 amounts to −48.

70. Negative 2 times the sum of 3 and 12 is −30.

71. The quotient of 100 and twice 50 is equal to 1.

72. Seventeen subtracted from −12 equals −29.

Review

The following bar graph shows the number of U.S. federal individual income tax returns that are filed electronically during the years shown. Electronically filed returns include Telefile and online returns. Use this graph to answer Exercises 73 through 76. Write number answers in standard form. See Sections 1.2 and 1.3.

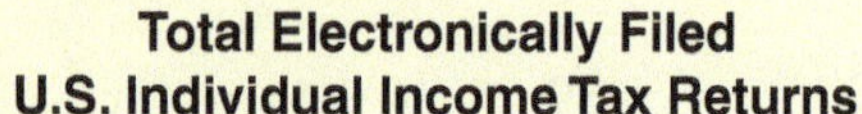

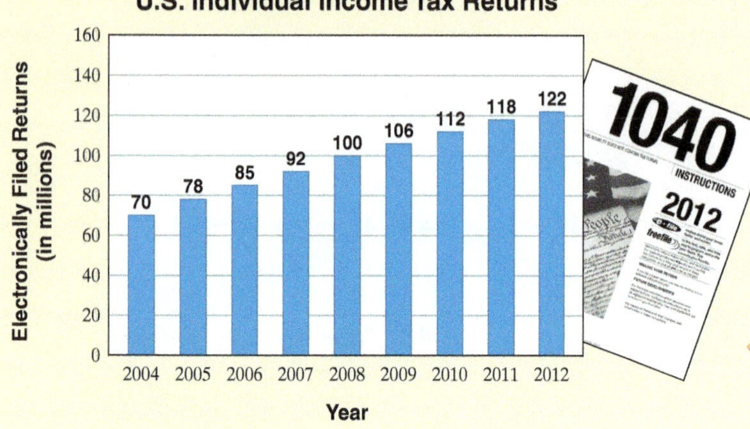

Source: IRS Compliance Research Division

73. Determine the number of electronically filed returns for 2012.

74. Determine the number of electronically filed returns for 2010.

75. By how much did the number of electronically filed returns increase from 2006 to 2012?

76. Describe any trends shown in this graph.

Evaluate each expression for $x = 3$, $y = -1$, and $z = 0$. See Section 2.5.

77. $x^3 - 2xy$

78. $y^3 + 3xyz$

79. $y^5 - 4x^2$

80. $(-y)^3 + 3xyz$

Concept Extensions

Using the Steps for Solving an Equation, choose the next operation for solving the given equation.

81. $2x - 5 = -7$

- **a.** Add 7 to both sides.
- **b.** Add 5 to both sides.
- **c.** Divide both sides by 2.

82. $3x + 2x = -x - 4$

- **a.** Add 4 to both sides.
- **b.** Subtract $2x$ from both sides.
- **c.** Add $3x$ and $2x$.

83. $-3x = -12$

- **a.** Divide both sides by -3.
- **b.** Add 12 to both sides.
- **c.** Add $3x$ to both sides.

84. $9 - 5x = 15$

- **a.** Divide both sides by -5.
- **b.** Subtract 15 from both sides.
- **c.** Subtract 9 from both sides.

A classmate shows you steps for solving an equation. The solution does not check, but the classmate is unable to find the error. For each set of steps, check the solution, find the error, and correct it.

85.
$$\begin{aligned} 2(3x - 5) &= 5x - 7 \\ 6x - 5 &= 5x - 7 \\ 6x - 5 + 5 &= 5x - 7 + 5 \\ 6x &= 5x - 2 \\ 6x - 5x &= 5x - 2 - 5x \\ x &= -2 \end{aligned}$$

86.
$$\begin{aligned} 37x + 1 &= 9(4x - 7) \\ 37x + 1 &= 36x - 7 \\ 37x + 1 - 1 &= 36x - 7 - 1 \\ 37x &= 36x - 8 \\ 37x - 36x &= 36x - 8 - 36x \\ x &= -8 \end{aligned}$$

Solve.

87. $(-8)^2 + 3x = 5x + 4^3$

88. $3^2 \cdot x = (-9)^3$

89. $2^3(x + 4) = 3^2(x + 4)$

90. $x + 45^2 = 54^2$

91. A classmate tries to solve $3x = 39$ by subtracting 3 from both sides of the equation. Will this step solve the equation for x? Why or why not?

92. A classmate tries to solve $2 + x = 20$ by dividing both sides by 2. Will this step solve the equation for x? Why or why not?

3.4 Linear Equations in One Variable and Problem Solving 

Objective A Writing Sentences as Equations

Now that we have practiced solving equations for a variable, we can extend considerably our problem-solving skills. We begin by writing sentences as equations using the following key words and phrases as a guide:

Addition	Subtraction	Multiplication	Division	Equal Sign
sum	difference	product	quotient	equals
plus	minus	times	divide	gives
added to	subtracted from	multiply	shared equally among	is/was
more than	less than	twice	per	yields
increased by	decreased by	of	divided by	amounts to
total	less	double	divided into	is equal to

Objectives

A Write Sentences as Equations.

B Use Problem-Solving Steps to Solve Problems.

Example 1 Write each sentence as an equation. Use x to represent "a number."

a. Twenty increased by a number is 5.
b. Twice a number equals −10.
c. A number minus 11 amounts to 168.
d. Three times the sum of a number and 5 is −30.
e. The quotient of twice a number and 8 is equal to 2.

Solution:

a. In words: twenty | increased by | a number | is | 5

Translate: $20 \quad + \quad x \quad = \quad 5$

b. In words: twice a number | equals | −10

Translate: $2x \quad = \quad -10$

c. In words: a number | minus | 11 | amounts to | 168

Translate: $x \quad - \quad 11 \quad = \quad 168$

d. In words: three times | the sum of a number and 5 | is | −30

Translate: $3 \quad (x + 5) \quad = \quad -30$

(*Continued on next page*)

Practice 1

Write each sentence as an equation. Use x to represent "a number."

a. Four times a number is 20.
b. The sum of a number and −5 yields 32.
c. Fifteen subtracted from a number amounts to −23.
d. Five times the difference of a number and 7 is equal to −8.
e. The quotient of triple a number and 5 gives 1.

Answers

1. a. $4x = 20$ **b.** $x + (-5) = 32$
c. $x - 15 = -23$
d. $5(x - 7) = -8$
e. $\frac{3x}{5} = 1$

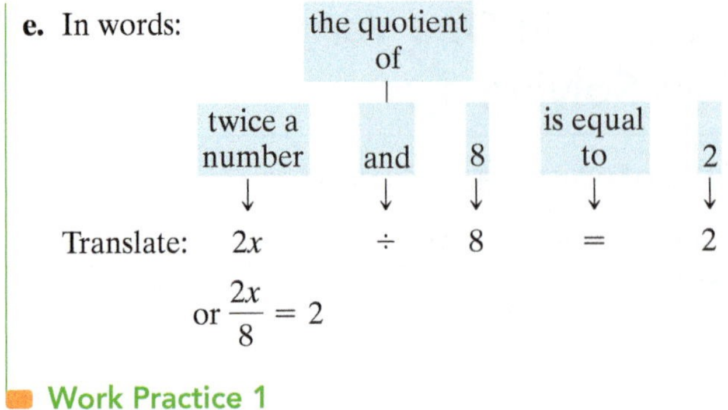

or $\frac{2x}{8} = 2$

Work Practice 1

Objective B Using Problem-Solving Steps to Solve Problems

Our main purpose for studying arithmetic and algebra is to solve problems. In previous sections, we have prepared for problem solving by writing phrases as algebraic expressions and sentences as equations. We now draw upon this experience as we solve problems. The following problem-solving steps will be used throughout this text.

Problem-Solving Steps

1. UNDERSTAND the problem. During this step, become comfortable with the problem. Some ways of doing this are as follows:
   - Read and reread the problem.
   - Construct a drawing.
   - Propose a solution and check. Pay careful attention to how you check your proposed solution. This will help when writing an equation to model the problem.
   - Choose a variable to represent the unknown. Use this variable to represent any other unknowns.
2. TRANSLATE the problem into an equation.
3. SOLVE the equation.
4. INTERPRET the results: *Check* the proposed solution in the stated problem and *state* your conclusion.

The first problem that we solve consists of finding an unknown number.

Practice 2

Translate "the sum of a number and 2 equals 6 added to three times the number" into an equation and solve.

Example 2 Finding an Unknown Number

Twice a number plus 3 is the same as the number minus 6. Find the unknown number.

Solution:

1. UNDERSTAND the problem. To do so, we read and reread the problem.

   Let's propose a solution to help us understand. Suppose the unknown number is 5. Twice this number plus 3 is $2 \cdot 5 + 3$ or 13. Is this the same as the number minus 6, or $5 - 6$, or -1? Since 13 is not the same as -1, we know that 5 is not the solution. However, remember that the purpose of proposing a solution is not to guess correctly, but to better understand the problem.

   Now let's choose a variable to represent the unknown. Let's let

   $x =$ unknown number

Answer

2. -2

2. TRANSLATE the problem into an equation.

In words:	twice a number	plus 3	is the same as	the number minus 6
	↓	↓	↓	↓
Translate:	$2x$	$+3$	$=$	$x - 6$

3. SOLVE the equation. To solve the equation, we first subtract x from both sides.

$$2x + 3 = x - 6$$
$$2x + 3 - x = x - 6 - x$$
$$x + 3 = -6 \quad \text{Simplify.}$$
$$x + 3 - 3 = -6 - 3 \quad \text{Subtract 3 from both sides.}$$
$$x = -9 \quad \text{Simplify.}$$

4. INTERPRET the results. First, *check* the proposed solution in the stated problem. Twice "−9" is −18 and −18 + 3 is −15. This is equal to the number minus 6, or "−9" − 6, or −15. Then *state* your conclusion: The unknown number is −9.

Work Practice 2

✓**Concept Check** Suppose you have solved an equation involving perimeter to find the length of a rectangular table. Explain why you would want to recheck your math if you obtain the result of −5.

Example 3 Determining Distances

The distance by road from Chicago, Illinois, to Los Angeles, California, is 1091 miles *more* than the distance from Chicago to Boston, Massachusetts. If the total of these two distances is 3017 miles, find the distance from Chicago to Boston. (*Source: World Almanac*)

Solution:

1. UNDERSTAND the problem. We read and reread the problem.

Let's propose and check a solution to help us better understand the problem. Suppose the distance from Chicago to Boston is 600 miles. Since the distance from Chicago to Los Angeles is 1091 miles *more*, then this distance is 600 + 1091 = 1691 miles. With these numbers, the total of the distances is 600 + 1691 = 2291 miles. This is less than the given total of 3017 miles, so we are incorrect. But not only do we have a better understanding of this exercise, we also know that the distance from Boston to Chicago is greater than 600 miles since this proposed solution led to a total too small. Now let's choose a variable to represent an unknown. Then we'll use this variable to represent any other unknown quantities. Let

x = distance from Chicago to Boston

Then

$x + 1091$ = distance from Chicago to Los Angeles

Since that distance is 1091 miles more.

2. TRANSLATE the problem into an equation.

In words:	Chicago to Boston distance	+	Chicago to Los Angeles distance	=	total miles
	↓		↓		↓
Translate:	x	+	$x + 1091$	=	3017

(*Continued on next page*)

Practice 3

The distance by road from Cincinnati, Ohio, to Denver, Colorado, is 71 miles *less* than the distance from Denver to San Francisco, California. If the total of these two distances is 2399 miles, find the distance from Denver to San Francisco.

Answer

3. 1235 miles

✓**Concept Check Answer**

Length cannot be negative.

3. SOLVE the equation:

$$
\begin{aligned}
x + x + 1091 &= 3017 \\
2x + 1091 &= 3017 && \text{Combine like terms.} \\
2x + 1091 - 1091 &= 3017 - 1091 && \text{Subtract 1091 from both sides.} \\
2x &= 1926 && \text{Simplify.} \\
\frac{2x}{2} &= \frac{1926}{2} && \text{Divide both sides by 2.} \\
x &= 963 && \text{Simplify.}
\end{aligned}
$$

4. INTERPRET the results. First *check* the proposed solution in the stated problem. Since x represents the distance from Chicago to Boston, this is 963 miles. The distance from Chicago to Los Angeles is $x + 1091 = 963 + 1091 = 2054$ miles. To check, notice that the total number of miles is $963 + 2054 = 3017$ miles, the given total of miles. Also, 2054 is 1091 more miles than 963, so the solution checks. Then, *state* your conclusion: The distance from Chicago to Boston is 963 miles.

Work Practice 3

Practice 4

A woman's $57,000 estate is to be divided so that her husband receives twice as much as her son. How much will each receive?

Example 4 Calculating Separate Costs

A salesperson at an electronics store sold a computer system and software for $2100, receiving four times as much money for the computer system as for the software. Find the price of each.

Solution:

1. UNDERSTAND the problem. We read and reread the problem. Then we choose a variable to represent an unknown. We use this variable to represent any other unknown quantities. We let

$x =$ the software price

Then

$4x =$ the computer system price

2. TRANSLATE the problem into an equation.

In words:	software price	and	computer price	is	2100
	↓	↓	↓	↓	↓
Translate:	x	$+$	$4x$	$=$	2100

3. SOLVE the equation:

$$
\begin{aligned}
x + 4x &= 2100 \\
5x &= 2100 && \text{Combine like terms.} \\
\frac{5x}{5} &= \frac{2100}{5} && \text{Divide both sides by 5.} \\
x &= 420 && \text{Simplify.}
\end{aligned}
$$

4. INTERPRET the results. *Check* the proposed solution in the stated problem. The software sold for $420. The computer system sold for $4x = 4(\$420) = \$1680$. Since $\$420 + \$1680 = \$2100$, the total price, and $1680 is four times $420, the solution checks. *State* your conclusion: The software sold for $420, and the computer system sold for $1680.

Work Practice 4

Answer

4. husband: $38,000; son: $19,000

Vocabulary, Readiness & Video Check

Martin-Gay Interactive Videos *Watch the section lecture video and answer the following questions.*

See Video 3.4 

Objective A 1. In Example 2, why does the left side of the equation translate to $-20 - x$ and not $x - (-20)$?

Objective B 2. Why are parentheses used in the translation of the left side of the equation in Example 4?

3. In Example 5, the solution to the equation is $x = 37$. Why is this not the solution to the application?

3.4 Exercise Set MyMathLab®

Objective A Translating *Write each sentence as an equation. Use x to represent "a number." See Example 1.*

1. A number added to -5 is -7.
2. Five subtracted from a number equals 10.
3. Three times a number yields 27.
4. The quotient of 8 and a number is -2.
5. A number subtracted from -20 amounts to 104.
6. Two added to twice a number gives -14.
7. Twice a number gives 108.
8. Five times a number is equal to -75.
9. The product of 5 and the sum of -3 and a number is -20.
10. Twice the sum of -17 and a number is -14.

Objective B Translating *Translate each sentence into an equation. Then solve the equation. See Example 2.*

11. Three times a number, added to 9, is 33. Find the number.
12. Twice a number, subtracted from 60, is 20. Find the number.
13. The sum of 3, 4, and a number amounts to 16. Find the number.
14. The sum of 7, 9, and a number is 40. Find the number.
15. The difference of a number and 3 is equal to the quotient of 10 and 5. Find the number.
16. Eight decreased by a number equals the quotient of 15 and 5. Find the number.
17. Thirty less a number is equal to the product of 3 and the sum of the number and 6. Find the number.
18. The product of a number and 3 is twice the sum of that number and 5. Find the number.
19. 40 subtracted from five times a number is 8 more than the number. Find the number.
20. Five times the sum of a number and 2 is 11 less than the number times 8. Find the number.

21. Three times the difference of some number and 5 amounts to the quotient of 108 and 12. Find the number.

22. Seven times the difference of some number and 1 gives the quotient of 70 and 10. Find the number.

23. The product of 4 and a number is the same as 30 less twice that same number. Find the number.

24. Twice a number equals 25 less triple that same number. Find the number.

Solve. For Exercises 25 and 26, the solutions have been started for you. See Examples 3 and 4.

25. Florida has 28 fewer electoral votes for president than California. If the total number of electoral votes for these two states is 82, find the number for each state. (*Source: The World Almanac* 2013)

Start the solution:

1. UNDERSTAND the problem. Reread it as many times as needed. Let's let

   x = number of electoral votes for California

   Then

   $x - 28$ = number of electoral votes for Florida
2. TRANSLATE into an equation. (Fill in the blanks below.)

votes for California	+	votes for Florida	= 82
↓		↓	
______	+	______	= 82

Now, you finish with

3. SOLVE the equation.
4. INTERPRET the results.

26. Texas has twice the number of electoral votes for president as Michigan. If the total number of electoral votes for these two states is 51, find the number for each state. (*Source: The World Almanac* 2013)

Start the solution:

1. UNDERSTAND the problem. Reread it as many times as needed. Let's let

   x = number of electoral votes for Michigan

   Then

   $2x$ = number of electoral votes for Texas
2. TRANSLATE into an equation. (Fill in the blanks below.)

votes for Michigan	+	votes for Texas	= 51
↓		↓	
______	+	______	= 51

Now, you finish with

3. SOLVE the equation.
4. INTERPRET the results.

27. A falcon, when diving, can travel five times as fast as a pheasant's top speed. If the total speed for these two birds is 222 miles per hour, find the fastest speed of the falcon and the fastest speed of the pheasant.
(*Source: Fantastic Book of Comparisons*)

28. Norway has had three times as many rulers as Liechtenstein. If the total number of rulers for both countries is 56, find the number of rulers for Norway and the number for Liechtenstein.

29. The U.S. Sunday newspaper with the greatest circulation is *The New York Times,* followed by *The Los Angeles Times*. If the Sunday circulation for *The New York Times* is 384 thousand more than the circulation for *The Los Angeles Times*, and their combined circulation is 2494 thousand, find the circulation for each newspaper. (*Source: The Top Ten of Everything*, 2013)

30. The average life expectancy for an elephant is 24 years longer than the life expectancy for a chimpanzee. If the total of these life expectancies is 130 years, find the life expectancy of each.

31. Coca-Cola was recently named the Most Valuable Global Brand. It has a recognition value worth $43 billion more than Disney's. If the total value of these two global brands is $101 billion, find the value of each of the brands. (See the Chapter 3 Opener.)

32. The global brand Microsoft has a recognition value worth $26 billion more than Apple's. If the total value of these two global brands is $92 billion, find the value of each of the brands. (See the Chapter 3 Opener.)

33. An Xbox 360 game system and several games are sold for $440. The cost of the Xbox 360 is 3 times as much as the cost of the games. Find the cost of the Xbox 360 and the cost of the games.

34. The two top-selling PC games are *Call of Duty: Black Ops 2* and *Madden NFL 13.* A price for *Call of Duty: Black Ops 2* is $13 more than a price for *Madden NFL 13.* If the total of these two prices is $47, find the price of each game.(*Source:* Internet Research)

35. By air, the distance from New York City to London is 2001 miles *less* than the distance from Los Angeles to Tokyo. If the total of these two distances is 8939 miles, find the distance from Los Angeles to Tokyo.

36. By air, the distance from Melbourne, Australia, to Cairo, Egypt, is 2338 miles *more* than the distance from Madrid, Spain, to Bangkok, Thailand. If the total of these distances is 15,012 miles, find the distance from Madrid to Bangkok.

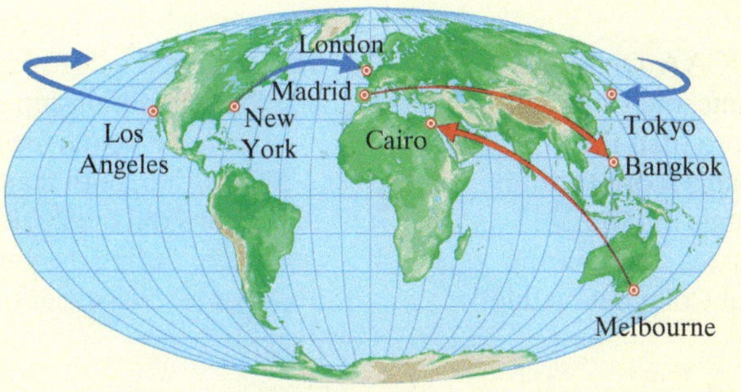

37. The two NCAA stadiums with the largest capacities are Beaver Stadium (Penn State) and Michigan Stadium (Univ. of Michigan). Beaver Stadium has a capacity of 1081 more than Michigan Stadium. If the combined capacity for the two stadiums is 213,483, find the capacity for each stadium. (*Source:* National Collegiate Athletic Association)

38. A National Hot Rod Association (NHRA) top fuel dragster has a top speed of 95 mph faster than an Indy Racing League car. If the top speed for these two cars combined is 565 mph, find the top speed of each car. (*Source: USA Today*)

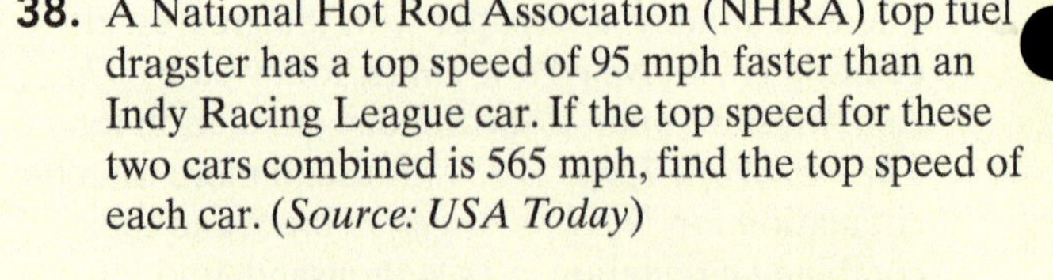

39. In 2020, China is projected to be the country with the greatest number of visiting tourists. This number is twice the number of tourists projected for Spain. If the total number of tourists for these two countries is projected to be 210 million, find the number projected for each. (*Source: The State of the World Atlas* by Dan Smith)

40. California contains the largest state population of native Americans. This population is three times the native American population of Washington state. If the total of these two populations is 412 thousand, find the native American population in each of these two states. (*Source:* U.S. Census Bureau)

41. In Germany, about twice as many cars are manufactured per day than in the United States. If the total number of these cars manufactured per day is 24,258, find the number manufactured in the United States and the number manufactured in Germany. (Based on data from the International Organization of Motor Vehicle Manufacturers)

42. A Toyota Camry is traveling twice as fast as a Dodge truck. If their combined speed is 105 miles per hour, find the speed of the car and find the speed of the truck.

43. A biker sold his used mountain bike and accessories for \$270. If he received five times as much money for the bike as he did for the accessories, find how much money he received for the bike.

44. A tractor and a plow attachment are worth \$1200. The tractor is worth seven times as much money as the plow. Find the value of the tractor and the value of the plow.

45. During the 2013 Women's NCAA Division I basketball championship game, the Connecticut Huskies scored 33 points more than the Louisville Cardinals. Together, both teams scored a total of 153 points. How many points did the 2013 Champion Connecticut Huskies score during this game? (*Source:* National Collegiate Athletic Association)

46. During the 2013 Men's NCAA Division I basketball championship game, the Michigan Wolverines scored 6 points fewer than the Louisville Cardinals. Together, both teams scored 158 points. How many points did the 2013 Champion Louisville Cardinals score during the game? (*Source:* National Collegiate Athletic Association)

47. The USA is the country with the most personal computers in use, followed by China. If the USA has 115,500 thousand more computers than China and the total number of computers for both countries is 505,780 thousand, find the number of computers for each country. (*Source:* Computer Industry Almanac)

48. The total number of personal computers in use for Italy and Russia is 98,240 thousand. If Russia has 8800 thousand more computers than Italy, find the number of computers for each country. (*Source:* Computer Industry Almanac)

Review

Round each number to the given place value. See Section 1.4.

49. 586 to the nearest ten

50. 82 to the nearest ten

51. 1026 to the nearest hundred

52. 52,333 to the nearest thousand

53. 2986 to the nearest thousand

54. 101,552 to the nearest hundred

Concept Extensions

55. Solve Example 3 again, but this time let x be the distance from Chicago to Los Angeles. Did you get the same results? Explain why or why not.

56. Solve Exercise 25 again, but this time let x be the number of electoral votes for Florida. Did you get the same results? Explain why or why not.

In real estate, a house's selling price P is found by adding the real estate agent's commission C to the amount A that the seller of the house receives: $P = A + C$.

57. A house sold for $230,000. The owner's real estate agent received a commission of $13,800. How much did the seller receive? (*Hint:* Substitute the known values into the equation, then solve the equation for the remaining unknown.)

58. A homeowner plans to use a real estate agent to sell his house. He hopes to sell the house for $165,000 and keep $156,750 of that. If everything goes as he has planned, how much will his real estate agent receive as a commission?

In retailing, the retail price P of an item can be computed using the equation $P = C + M$, where C is the wholesale cost of the item and M is the amount of markup.

59. The retail price of a computer system is $999 after a markup of $450. What is the wholesale cost of the computer system? (*Hint:* Substitute the known values into the equation, then solve the equation for the remaining unknown.)

60. Slidell Feed and Seed sells a bag of cat food for $12. If the store paid $7 for the cat food, what is the markup on the cat food?

Chapter 3 Group Activity

Modeling Equation Solving with Addition and Subtraction

Sections 3.1–3.4

We can use positive counters ● and negative counters ● to help us model the equation-solving process. We also need to use an object that represents a variable. We use small slips of paper with the variable name written on them.

Recall that taking a ● and ● together creates a neutral or zero pair. After a neutral pair has been formed, it can be removed from or added to an equation model without changing the overall value. We also need to remember that we can add or remove the same number of positive or negative counters from both sides of an equation without changing the overall value.

We can represent the equation $x + 5 = 2$ as follows:

To get the variable by itself, we must remove 5 black counters from both sides of the model. Because there are only 2 counters on the right side, we must add 5 negative counters to both sides of the model. Then we can remove neutral pairs: 5 from the left side and 2 from the right side.

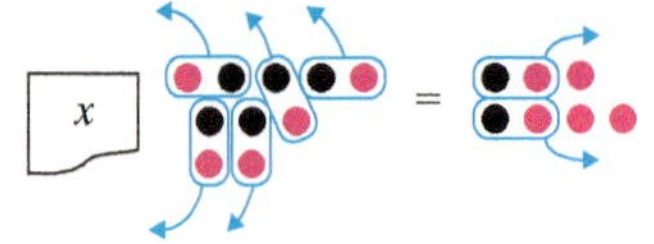

We are left with the following model, which represents the solution, $x = -3$.

Similarly, we can represent the equation $x - 4 = -6$ as follows:

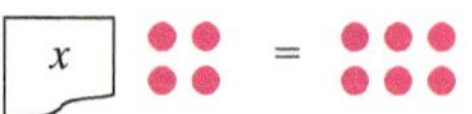

To get the variable by itself, we must remove 4 red counters from both sides of the model.

We are left with the following model, which represents the solution, $x = -2$.

Use the counter model to solve each equation.

1. $x - 3 = -7$
2. $x - 1 = -9$
3. $x + 2 = 8$
4. $x + 4 = 5$
5. $x + 8 = 3$
6. $x - 5 = -1$
7. $x - 2 = 1$
8. $x - 5 = 10$
9. $x + 3 = -7$
10. $x + 8 = -2$

Chapter 3 Vocabulary Check

Fill in each blank with one of the words or phrases listed below.

variable	addition	constant	algebraic expression	equation
terms	simplified	multiplication	evaluating the expression	solution
like	combined	numerical coefficient	distributive	

1. An algebraic expression is ________________ when all like terms have been ________________.
2. Terms that are exactly the same, except that they may have different numerical coefficients, are called ________________ terms.
3. A letter used to represent a number is called a(n) ________________.
4. A combination of operations on variables and numbers is called a(n) ________________.
5. The addends of an algebraic expression are called the ________________ of the expression.
6. The number factor of a variable term is called the ________________.
7. Replacing a variable in an expression by a number and then finding the value of the expression is called ________________ for the variable.
8. A term that is a number only is called a(n) ________________.
9. A(n) ________________ is of the form expression = expression.
10. A(n) ________________ of an equation is a value for the variable that makes the equation a true statement.

11. To multiply $-3(2x + 1)$, we use the ________ property.

12. By the ________ property of equality, we may multiply or divide both sides of an equation by any nonzero number without changing the solution of the equation.

13. By the ________ property of equality, the same number may be added to or subtracted from both sides of an equation without changing the solution of the equation.

Helpful Hint Are you preparing for your test? Don't forget to take the Chapter 3 Test on page 207. Then check your answers at the back of the text and use the Chapter Test Prep Videos to see the fully worked-out solutions to any of the exercises you want to review.

3 Chapter Highlights

Definitions and Concepts	Examples
Section 3.1 Simplifying Algebraic Expressions	
The addends of an algebraic expression are called the **terms** of the expression.	$5x^2 + (-4x) + (-2)$ — 3 terms
The number factor of a variable term is called the **numerical coefficient.**	Term / Numerical Coefficient: $7x$ — 7; $-6y$ — -6; x or $1x$ — 1
Terms that are exactly the same, except that they may have different numerical coefficients, are called **like terms.**	$5x + 11x = (5 + 11)x = 16x$ (like terms)
An algebraic expression is **simplified** when all like terms have been **combined.**	$y - 6y = (1 - 6)y = -5y$
Use the **distributive property** to multiply an algebraic expression by a term.	Simplify: $-4(x + 2) + 3(5x - 7)$ $= -4 \cdot x + (-4) \cdot 2 + 3 \cdot 5x - 3 \cdot 7$ $= -4x + (-8) + 15x + (-21)$ $= 11x + (-29)$ or $11x - 29$
Section 3.2 Solving Equations: Review of the Addition and Multiplication Properties	
Addition Property of Equality Let a, b, and c represent numbers. Then $a = b$ and $a + c = b + c$ are equivalent equations. Also, $a = b$ and $a - c = b - c$ are equivalent equations. In other words, the same number may be added to or subtracted from both sides of an equation without changing the solution of the equation.	Solve for x: $x + 8 = 2 + (-1)$ $x + 8 = 1$ Combine like terms. $x + 8 - 8 = 1 - 8$ Subtract 8 from both sides. $x = -7$ Simplify. The solution is -7.

(continued)

Definitions and Concepts	Examples
Section 3.2 Solving Equations: Review of the Addition and Multiplication Properties *(continued)*	
Multiplication Property of Equality Let a, b, and c represent numbers and let $c \neq 0$. Then $a = b$ and $a \cdot c = b \cdot c$ are equivalent equations. | Also, $a = b$ and $\frac{a}{c} = \frac{b}{c}$ are equivalent equations. In other words, both sides of an equation may be multiplied or divided by the same nonzero number without changing the solution of the equation.	Solve: $y - 7y = 30$ $-6y = 30$ Combine like terms. $\frac{-6y}{-6} = \frac{30}{-6}$ Divide both sides by −6. $y = -5$ Simplify. The solution is −5.
Section 3.3 Solving Linear Equations in One Variable	
Steps for Solving an Equation **Step 1:** If parentheses are present, use the distributive property. **Step 2:** Combine any like terms on each side of the equation. **Step 3:** Use the addition property of equality to rewrite the equation so that variable terms are on one side of the equation and constant terms are on the other side. **Step 4:** Use the multiplication property of equality to divide both sides by the numerical coefficient of the variable to solve. **Step 5:** Check the solution in the *original equation.*	Solve for x: $5(3x - 1) + 15 = -5$ **Step 1:** $15x - 5 + 15 = -5$ Apply the distributive property. **Step 2:** $15x + 10 = -5$ Combine like terms. **Step 3:** $15x + 10 - 10 = -5 - 10$ Subtract 10 from both sides. $15x = -15$ **Step 4:** $\frac{15x}{15} = \frac{-15}{15}$ Divide both sides by 15. $x = -1$ **Step 5:** Check to see that −1 is the solution.
Section 3.4 Linear Equations in One Variable and Problem Solving	
Problem-Solving Steps **1.** UNDERSTAND the problem. Some ways of doing this are Read and reread the problem. Construct a drawing. Choose a variable to represent an unknown in the problem.	The incubation period for a golden eagle is three times the incubation period for a hummingbird. If the total of their incubation periods is 60 days, find the incubation period for each bird. (*Source: Wildlife Fact File,* International Masters Publishers) **1.** UNDERSTAND the problem. Then choose a variable to represent an unknown. Let x = incubation period of a hummingbird Then $3x$ = incubation period of a golden eagle
2. TRANSLATE the problem into an equation.	**2.** TRANSLATE. incubation of hummingbird + incubation of golden eagle is 60 ↓ ↓ ↓ ↓ $x + 3x = 60$

Definitions and Concepts	Examples
Section 3.4 Linear Equations in One Variable and Problem Solving *(continued)*	
3. SOLVE the equation.	**3.** SOLVE: $x + 3x = 60$ $4x = 60$ $\frac{4x}{4} = \frac{60}{4}$ $x = 15$
4. INTERPRET the results. *Check* the proposed solution in the stated problem and *state* your conclusion.	**4.** INTERPRET the solution in the stated problem. The incubation period for a hummingbird is 15 days. The incubation period for a golden eagle is $3x = 3 \cdot 15 = 45$ days. Since 15 days + 45 days = 60 days and 45 is 3(15), the solution checks. *State* your conclusion: The incubation period for a hummingbird is 15 days. The incubation period for a golden eagle is 45 days.

Chapter 3 Review

(3.1) *Simplify each expression by combining like terms.*

1. $3y + 7y - 15$

2. $2y - 10 - 8y$

3. $8a + a - 7 - 15a$

4. $y + 3 - 9y - 1$

Multiply.

5. $2(x + 5)$

6. $-3(y + 8)$

Simplify.

7. $7x + 3(x - 4) + x$

8. $-(3m + 2) - m - 10$

9. $3(5a - 2) - 20a + 10$

10. $6y + 3 + 2(3y - 6)$

11. $6y - 7 + 11y - y + 2$

12. $10 - x + 5x - 12 - 3x$

Find the perimeter of each figure.

△**13.**

2x yards

3 yards

Rectangle

△**14.**

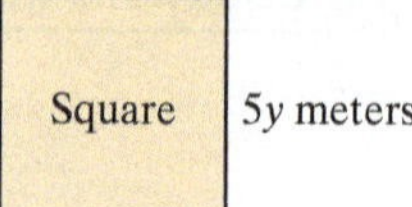

Find the area of each figure.

△**15.**

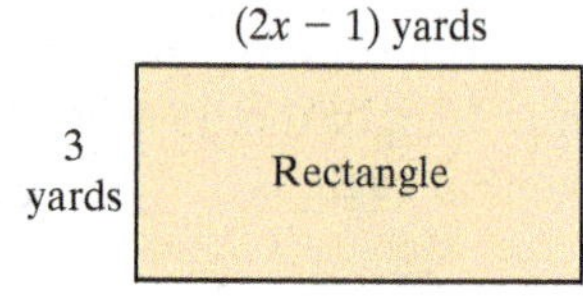

△**16.**

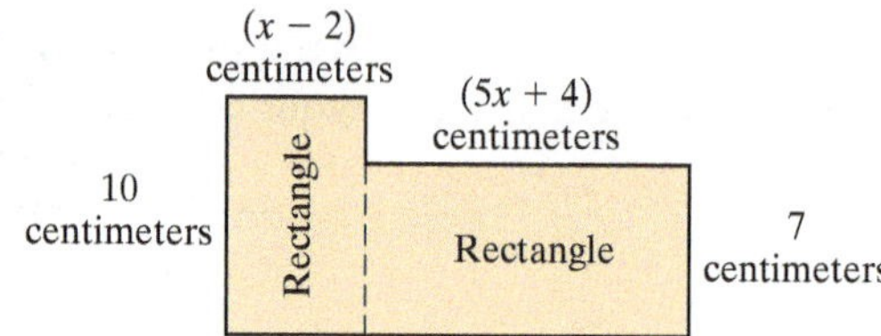

(3.2) *Solve each equation.*

17. $z - 5 = -7$

18. $3x + 10 = 4x$

19. $3y = -21$

20. $-3a = -15$

21. $\frac{x}{-6} = 2$

22. $\frac{y}{-15} = -3$

23. $n + 18 = 10 - (-2)$

24. $c - 5 = -13 + 7$

25. $7x + 5 - 6x = -20$

26. $17x = 2(8x - 4)$

27. $5x + 7 = -3$

28. $-14 = 9y + 4$

29. $\frac{z}{4} = -8 - (-6)$

30. $-1 + (-8) = \frac{x}{5}$

31. $6y - 7y = 100 - 105$

32. $19x - 16x = 45 - 60$

33. $9(2x - 7) = 19x$

34. $-5(3x + 3) = -14x$

35. $3x - 4 = 11$

36. $6y + 1 = 73$

37. $2(x + 4) - 10 = -2(7)$

38. $-3(x - 6) + 13 = 20 - 1$

Translating *Translate each phrase into an algebraic expression. Let x represent "a number."*

39. The product of -5 and a number

40. Three subtracted from a number

41. The sum of -5 and a number

42. The quotient of -2 and a number

43. The product of -5 and a number, decreased by 50

44. Eleven added to twice a number

45. The quotient of 70 and the sum of a number and 6

46. Twice the difference of a number and 13

(3.3) *Solve each equation.*

47. $2x + 5 = 7x - 100$

48. $-6x - 4 = x + 66$

49. $2x + 7 = 6x - 1$

50. $5x - 18 = -4x$

51. $5(n - 3) = 7 + 3n$

52. $7(2 + x) = 4x - 1$

53. $6x + 3 - (-x) = -20 + 5x - 7$

54. $x - 25 + 2x = -5 + 2x - 10$

55. $3(x - 4) = 5x - 8$

56. $4(x - 3) = -2x - 48$

57. $6(2n - 1) + 18 = 0$

58. $7(3y - 2) - 7 = 0$

59. $95x - 14 = 20x - 10 + 10x - 4$

60. $32z + 11 - 28z = 50 + 2z - (-1)$

Translating *Write each sentence as an equation.*

61. The difference of 20 and -8 is 28.

62. Nineteen subtracted from -2 amounts to -21.

63. The quotient of -75 and the sum of 5 and 20 is equal to -3.

64. Five times the sum of 2 and -6 yields -20.

(3.4) Translating *Write each sentence as an equation using x as the variable.*

65. Twice a number minus 8 is 40.

66. The product of a number and 6 is equal to the sum of the number and 20.

67. Twelve subtracted from the quotient of a number and 2 is 10.

68. The difference of a number and 3 is the quotient of 8 and 4.

Solve.

69. Five times a number subtracted from 40 is the same as three times the number. Find the number.

70. The product of a number and 3 is twice the difference of that number and 8. Find the number.

71. In an election, the incumbent received 14,000 votes of the 18,500 votes cast. Of the remaining votes, the Democratic candidate received 272 more than the Independent candidate. Find how many votes the Democratic candidate received.

72. Rajiv Puri has twice as many movies on DVD as he has on Blu-ray Disc. Find the number of DVDs if he has a total of 126 movie recordings.

Mixed Review

Simplify.

73. $9x - 20x$

74. $-5(7x)$

75. $12x + 5(2x - 3) - 4$

76. $-7(x + 6) - 2(x - 5)$

Solve.

77. $c - 5 = -13 + 7$

78. $7x + 5 - 6x = -20$

79. $-7x + 3x = -50 - 2$

80. $-x + 8x = -38 - 4$

81. $9x + 12 - 8x = -6 + (-4)$

82. $-17x + 14 + 20x - 2x = 5 - (-3)$

83. $5(2x - 3) = 11x$

84. $\frac{y}{-3} = -1 - 5$

85. $12y - 10 = -70$

86. $4n - 8 = 2n + 14$

87. $-6(x - 3) = x + 4$

88. $9(3x - 4) + 63 = 0$

89. $-5z + 3z - 7 = 8z - 1 - 6$

90. $4x - 3 + 6x = 5x - 3 - 30$

91. Three times a number added to twelve is 27. Find the number.

92. Twice the sum of a number and four is ten. Find the number.

93. Out of the 50 states, Hawaii has the least number of roadway miles, followed by Delaware. If Delaware has 1931 more roadway miles than Hawaii and the total number of roadway miles for both states is 10,673, find the number of roadway miles for each state. (*Source:* U.S. Federal Highway Administration)

94. North and South Dakota both have over 80,000 roadway miles. North Dakota has 4489 more miles and the total number of roadway miles for both states is 169,197. Find the number of roadway miles for North Dakota and for South Dakota. (*Source:* U.S. Federal Highway Administration)

Step-by-step test solutions are found on the Chapter Test Prep Videos. Where available: MyMathLab® or YouTube™

Test Chapter 3

1. Simplify $7x - 5 - 12x + 10$ by combining like terms.

2. Multiply: $-2(3y + 7)$

3. Simplify: $-(3z + 2) - 5z - 18$

△ 4. Write an expression that represents the perimeter of the equilateral triangle (a triangle with three sides of equal length). Simplify the expression.

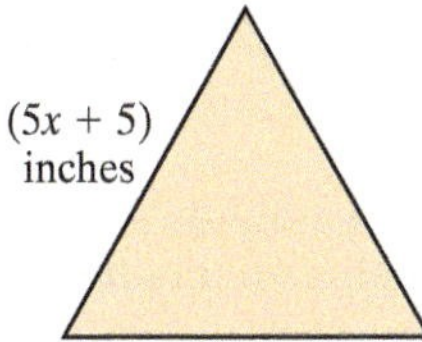

△ 5. Write an expression that represents the area of the rectangle. Simplify the expression.

4 meters

Rectangle

$(3x - 1)$ meters

Solve each equation.

6. $12 = y - 3y$

7. $\frac{x}{2} = -5 - (-2)$

8. $5x + 12 - 4x - 14 = 22$

9. $-4x + 7 = 15$

10. $2(x - 6) = 0$

11. $-4(x - 11) - 34 = 10 - 12$

12. $5x - 2 = x - 10$

13. $4(5x + 3) = 2(7x + 6)$

14. $6 + 2(3n - 1) = 28$

Translate the following phrases into mathematical expressions. If needed, use x to represent "a number."

15. The sum of -23 and a number

16. Three times a number, subtracted from -2

Answers

1. __________

2. __________

3. __________

4. __________

5. __________

6. __________

7. __________

8. __________

9. __________

10. __________

11. __________

12. __________

13. __________

14. __________

15. __________

16. __________

17. ________

18. ________

19. ________

20. ________

21. ________

Translate each sentence into an equation. If needed, use x to represent "a number."

17. The sum of twice 5 and -15 is -5.

18. Six added to three times a number equals -30.

Solve.

19. The difference of three times a number and five times the same number is 4. Find the number.

20. In a championship basketball game, Paula made twice as many free throws as Maria. If the total number of free throws made by both women was 12, find how many free throws Paula made.

21. In a 10-kilometer race, there are 112 more men entered than women. Find the number of female runners if the total number of runners in the race is 600.

Cumulative Review Chapters 1–3

1. Write 308,063,557 in words.

2. Write 276,004 in words.

△ 3. Find the perimeter of the polygon shown.

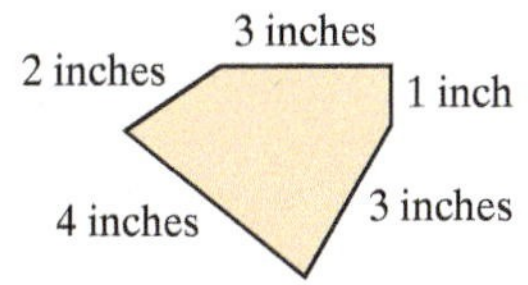

△ 4. Find the perimeter of the rectangle shown.

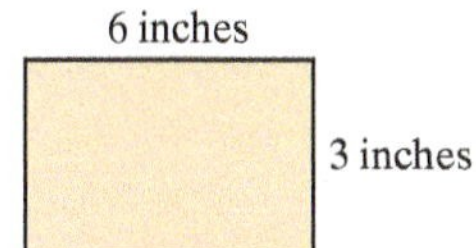

5. Subtract: 900 − 174. Check by adding.

6. Subtract: 17,801 − 8216. Check by adding.

7. Round 248,982 to the nearest hundred.

8. Round 844,497 to the nearest thousand.

9. Multiply: 25×8

10. Multiply: 395×74

11. Divide and check: $1872 \div 9$

12. Divide and check: $3956 \div 46$

13. Simplify: $2 \cdot 4 - 3 \div 3$

14. Simplify: $8 \cdot 4 + 9 \div 3$

15. Evaluate $x^2 + z - 3$ for $x = 5$ and $z = 4$.

16. Evaluate $2a^2 + 5 - c$ for $a = 2$ and $c = 3$.

17. Determine which numbers in the set $\{26, 40, 20\}$ are solutions of the equation $2n - 30 = 10$.

18. Insert < or > to make a true statement.

   a. $-14 \quad 0$

   b. $-(-7) \quad -8$

19. Add using a number line: $5 + (-2)$

20. Add using a number line: $-3 + (-4)$

Answers

1. ____________
2. ____________
3. ____________
4. ____________
5. ____________
6. ____________
7. ____________
8. ____________
9. ____________
10. ____________
11. ____________
12. ____________
13. ____________
14. ____________
15. ____________
16. ____________
17. ____________
18. a. ____________
    b. ____________
19. ____________
20. ____________

21. ______
22. ______
23. ______
24. ______
25. ______
26. ______
27. ______
28. ______
29. ______
30. ______
31. ______
32. ______
33. ______
34. ______
35. ______
36. ______
37. ______
38. ______
39. ______
40. ______
41. ______
42. ______
43. ______
44. ______
45. ______
46. ______
47. ______
48. ______
49. ______
50. ______

Add.

21. $-15 + (-10)$

22. $3 + (-7)$

23. $-2 + 25$

24. $21 + 15 + (-19)$

Subtract.

25. $-4 - 10$

26. $-2 - 3$

27. $6 - (-5)$

28. $19 - (-10)$

29. $-11 - (-7)$

30. $-16 - (-13)$

Divide.

31. $\dfrac{-12}{6}$

32. $\dfrac{-30}{-5}$

33. $-20 \div (-4)$

34. $26 \div (-2)$

35. $\dfrac{48}{-3}$

36. $\dfrac{-120}{12}$

Find the value of each expression.

37. $(-3)^2$

38. -2^5

39. -3^2

40. $(-5)^2$

41. Simplify: $2y - 6 + 4y + 8$

42. Simplify: $6x + 2 - 3x + 7$

43. Determine whether -1 is a solution of the equation $3y + 1 = 3$.

44. Determine whether 2 is a solution of $5x - 3 = 7$.

45. Solve: $-12x = -36$

46. Solve: $-3y = 15$

47. Solve: $2x - 6 = 18$

48. Solve: $3a + 5 = -1$

49. A salesperson at an electronics store sold a computer system and software for \$2100, receiving four times as much money for the computer system as for the software. Find the price of each.

50. Rose Daunis is thinking of a number. Two times the number plus four is the same amount as three times the number minus seven. Find Rose's number.

Fractions and Mixed Numbers

4

The following graph is called a circle graph or a pie chart. Each sector (shaped like a piece of pie) shows the fraction of entering college freshmen who choose to major in each discipline shown. Can you find your current choice of major in this graph? In Section 4.2, Exercises 93–96, we show this same circle graph, but in 3-D design. We simplify some of the fractions in it and also study sector size versus fraction value.

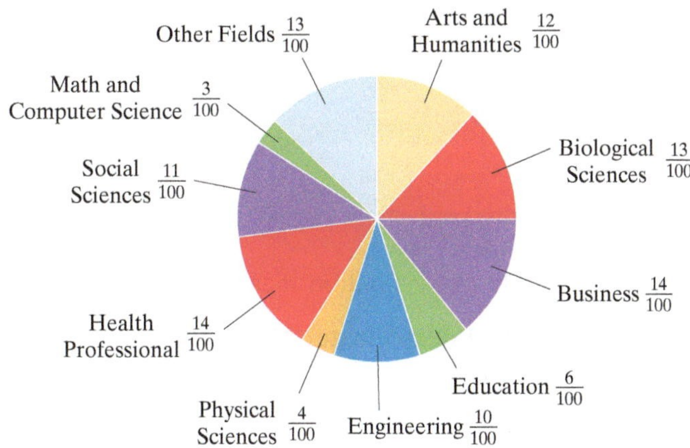

Source: The Higher Education Research Institute

Sections

Check Your Progress

Fractions are numbers and, like whole numbers and integers, they can be added, subtracted, multiplied, and divided. Fractions are very useful and appear frequently in everyday language, in common phrases such as "half an hour," "quarter of a pound," and "third of a cup." This chapter reviews the concepts of fractions and mixed numbers and demonstrates how to add, subtract, multiply, and divide these numbers.

4.1 Introduction to Fractions and Mixed Numbers

Objectives

A Identify the Numerator and the Denominator of a Fraction.

B Write a Fraction to Represent Parts of Figures or Real-Life Data.

C Graph Fractions on a Number Line.

D Review Division Properties of 0 and 1.

E Write Mixed Numbers as Improper Fractions.

F Write Improper Fractions as Mixed Numbers or Whole Numbers.

Objective A Identifying Numerators and Denominators

Whole numbers are used to count whole things or units, such as cars, horses, dollars, and people. To refer to a part of a whole, fractions can be used. Here are some examples of **fractions.** Study these examples for a moment.

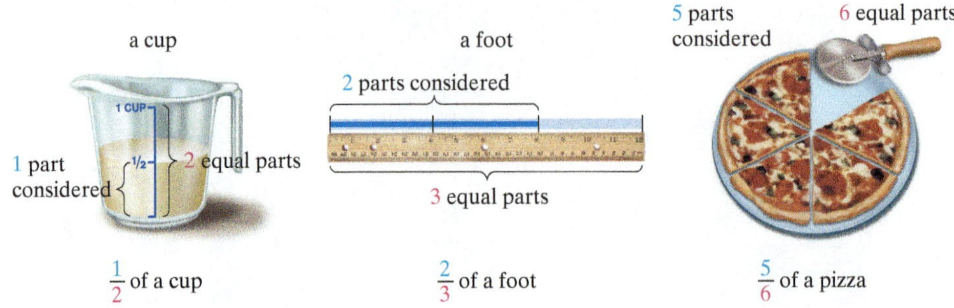

In a fraction, the top number is called the **numerator** and the bottom number is called the **denominator.** The bar between the numbers is called the **fraction bar.**

Name	Fraction	Meaning
numerator ⟶	5	⟵ number of parts being considered
denominator ⟶	6	⟵ number of equal parts in the whole

Examples

Identify the numerator and the denominator of each fraction.

1. $\frac{3}{7}$ ← numerator (3), ← denominator (7)

2. $\frac{13}{5x}$ ← numerator (13), ← denominator ($5x$)

Work Practice 1–2

Practice 1–2

Identify the numerator and the denominator of each fraction.

1. $\frac{11}{2}$

2. $\frac{10y}{17}$

Helpful Hint

$\frac{3}{7}$ ← Remember that the bar in a fraction means division. Since division by 0 is undefined, a fraction with a denominator of 0 is undefined. For example, $\frac{3}{0}$ is undefined.

Objective B Writing Fractions to Represent Parts of Figures or Real-Life Data

One way to become familiar with the concept of fractions is to visualize fractions with shaded figures. We can then write a fraction to represent the shaded area of the figure (or diagram).

Answers

1. numerator $= 11$, denominator $= 2$
2. numerator $= 10y$, denominator $= 17$

Examples Write a fraction to represent the shaded part of each figure.

3. In this figure, 2 of the 5 equal parts are shaded. Thus, the fraction is $\frac{2}{5}$.

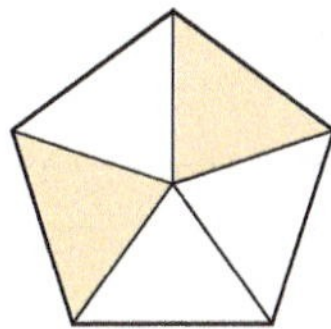

$\frac{2}{5}$ ← number of parts shaded / ← number of equal parts

4. In this figure, 3 of the 10 rectangles are shaded. Thus, the fraction is $\frac{3}{10}$.

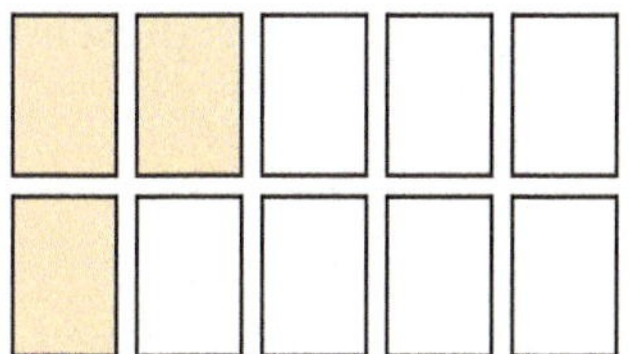

$\frac{3}{10}$ ← number of parts shaded / ← number of equal parts

Work Practice 3–4

Examples Write a fraction to represent the shaded part of the diagram.

5. The fraction is $\frac{3}{10}$.

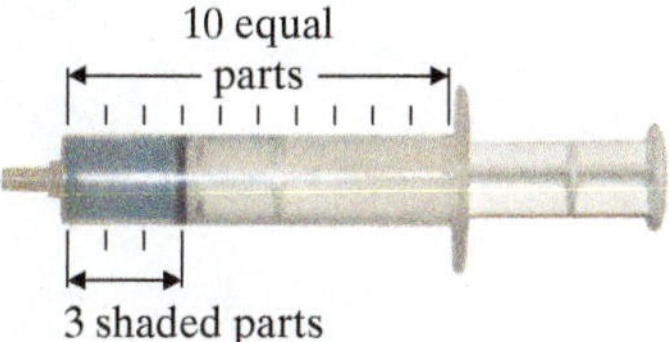

6. The fraction is $\frac{1}{3}$.

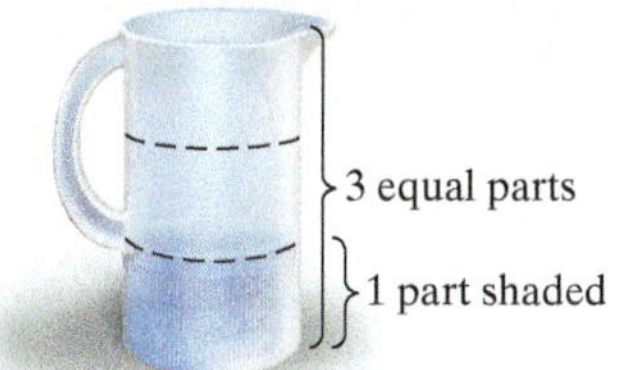

Work Practice 5–6

Examples Draw a figure and then shade a part of it to represent each fraction.

7. $\frac{5}{6}$ of a figure

We will use a geometric figure such as a rectangle. Since the denominator is 6, we divide it into 6 equal parts. Then we shade 5 of the equal parts.

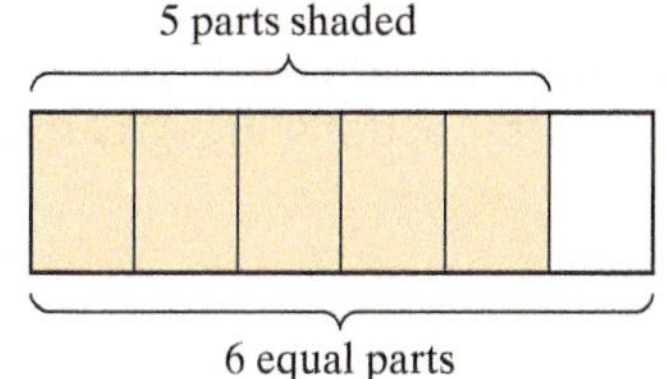

(Continued on next page)

Practice 3–4

Write a fraction to represent the shaded part of each figure.

3.

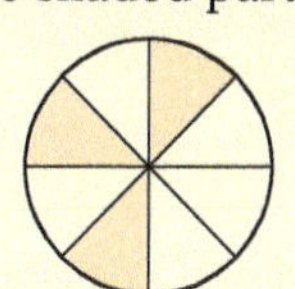

4.

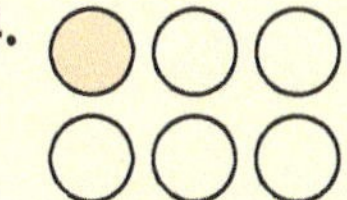

Practice 5–6

Write a fraction to represent the part of the whole shown.

5.

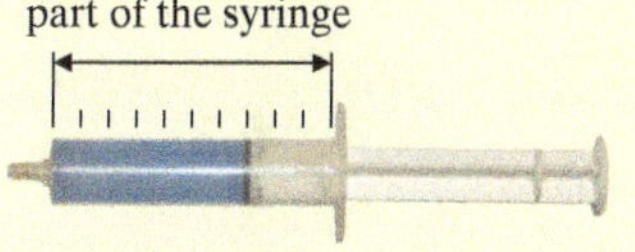

6.

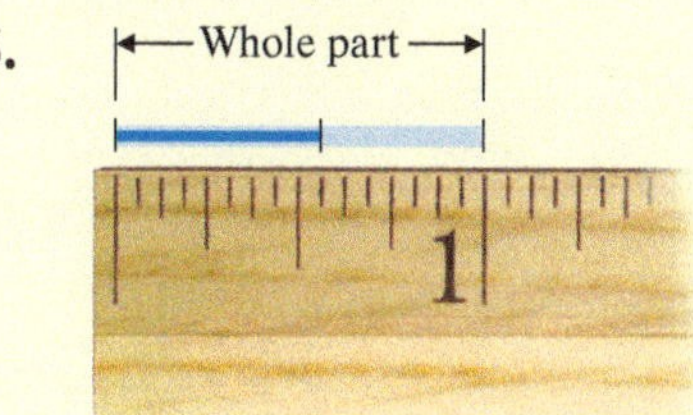

Practice 7

Draw and shade a part of a figure to represent the fraction.

7. $\frac{2}{3}$ of a figure

Answers

3. $\frac{3}{8}$ **4.** $\frac{1}{6}$ **5.** $\frac{7}{10}$ **6.** $\frac{9}{16}$

7. answers may vary; for example,

Practice 8

Draw and shade a part of a figure to represent the fraction.

8. $\frac{7}{11}$ of a figure

8. $\frac{3}{8}$ of a figure

If you'd like, our figure can consist of 8 triangles of the same size. We will shade 3 of the triangles.

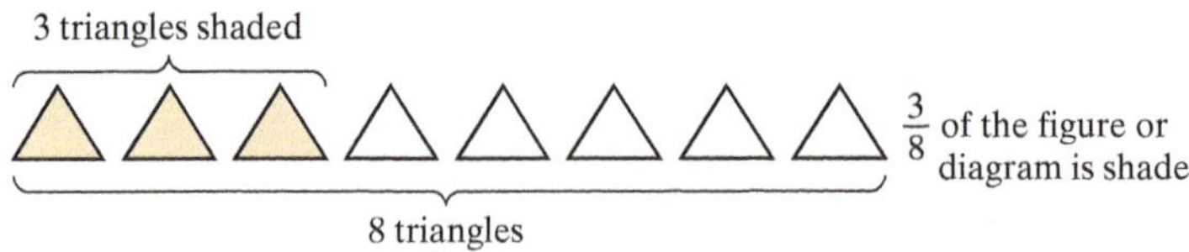

Work Practice 7–8

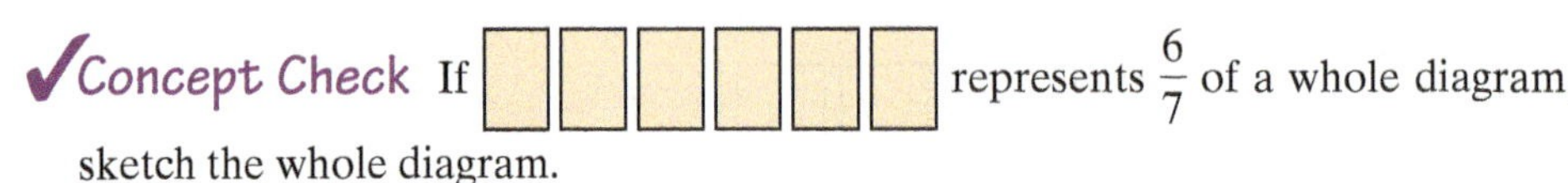

✓**Concept Check** If [6 shaded boxes] represents $\frac{6}{7}$ of a whole diagram, sketch the whole diagram.

Practice 9

Of the eight planets in our solar system, five are farther from the Sun than Earth is. What fraction of the planets are farther from the Sun than Earth is?

Example 9 Writing Fractions from Real-Life Data

Of the eight planets in our solar system (Pluto is now a dwarf planet), three are closer to the Sun than Mars. What fraction of the planets are closer to the Sun than Mars?

Solution: The fraction of planets closer to the Sun than Mars is:

$\frac{3}{8}$ ← number of planets closer / ← number of planets in our solar system

Thus, $\frac{3}{8}$ of the planets in our solar system are closer to the Sun than Mars.

Work Practice 9

The definitions and statements below apply to positive fractions.

A **proper fraction** is a fraction whose numerator is less than its denominator. Proper fractions are less than 1. For example, the shaded portion of the triangle is represented by $\frac{2}{3}$.

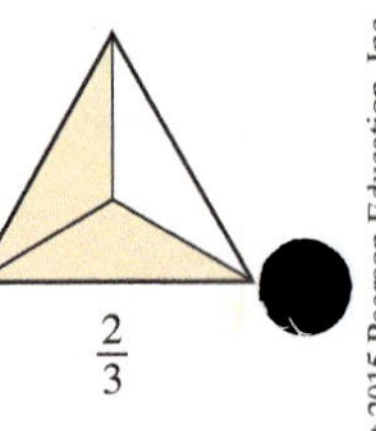

$\frac{2}{3}$

An **improper fraction** is a fraction whose numerator is greater than or equal to its denominator. Improper fractions are greater than or equal to 1.

Answers

8. answers may vary; for example,

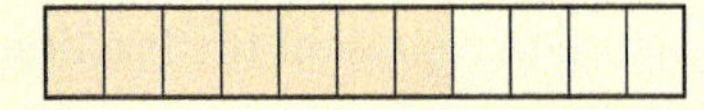

9. $\frac{5}{8}$

✓**Concept Check Answer**

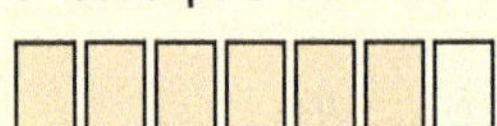

The shaded part of the group of circles below is $\frac{9}{4}$. The shaded part of the rectangle is $\frac{6}{6}$. Recall that $\frac{6}{6}$ simplifies to 1 and notice that the entire rectangle (1 whole figure) is shaded below.

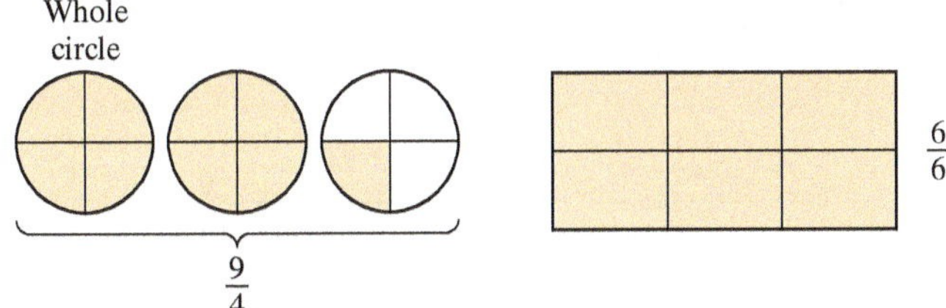

A **mixed number** contains a whole number and a fraction. Mixed numbers are greater than 1. Above, we wrote the shaded part of the group of circles below as the improper fraction $\frac{9}{4}$. Now let's write the shaded part as a mixed number. The shaded part of the group of circles' area is $2\frac{1}{4}$. Read this as "two and one-fourth."

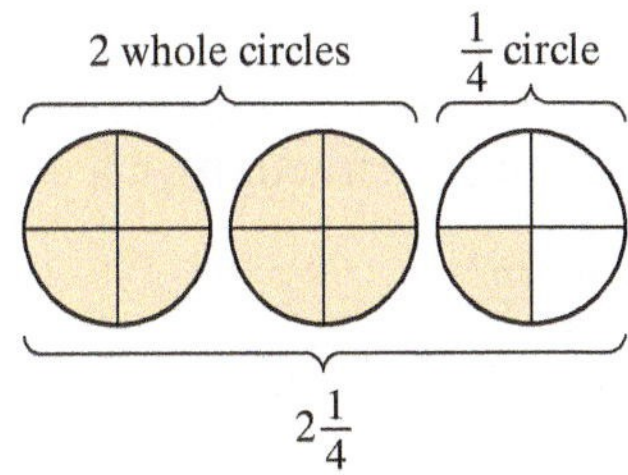

Note: The mixed number $2\frac{1}{4}$, diagrammed above, represents $2 + \frac{1}{4}$.

The mixed number $-3\frac{1}{5}$ represents $-\left(3 + \frac{1}{5}\right)$ or $-3 - \frac{1}{5}$. We review this later in this chapter.

Examples Represent the shaded part of each figure group as both an improper fraction and a mixed number.

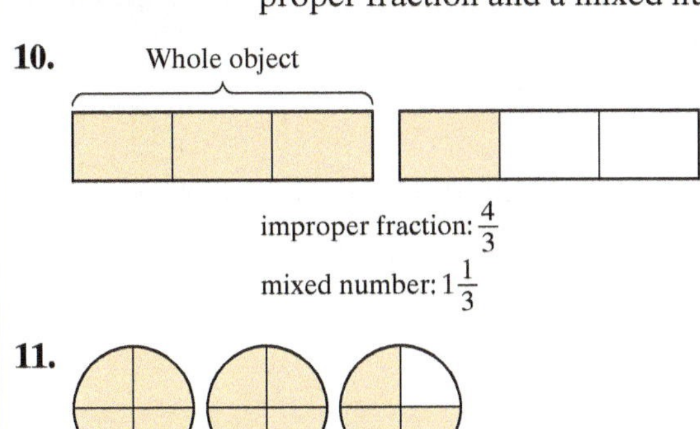

improper fraction: $\frac{4}{3}$

mixed number: $1\frac{1}{3}$

11.

improper fraction: $\frac{11}{4}$

mixed number: $2\frac{3}{4}$

Work Practice 10–11

Concept Check If you were to round $2\frac{3}{4}$, shown in Example 11 above, to the nearest whole number, would you choose 2 or 3? Why?

Practice 10–11

Represent the shaded part of each figure group as both an improper fraction and a mixed number.

10.

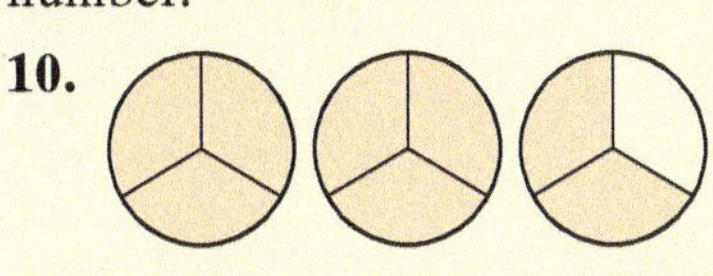

11.

Answers

10. $\frac{8}{3}, 2\frac{2}{3}$ **11.** $\frac{5}{4}, 1\frac{1}{4}$

Concept Check Answer

3; answers may vary

Objective C Graphing Fractions on a Number Line 

Another way to visualize fractions is to graph them on a number line. To do this, think of 1 unit on the number line as a whole. To graph $\frac{2}{5}$, for example, divide the distance from 0 to 1 into 5 equal parts. Then start at 0 and count 2 parts to the right.

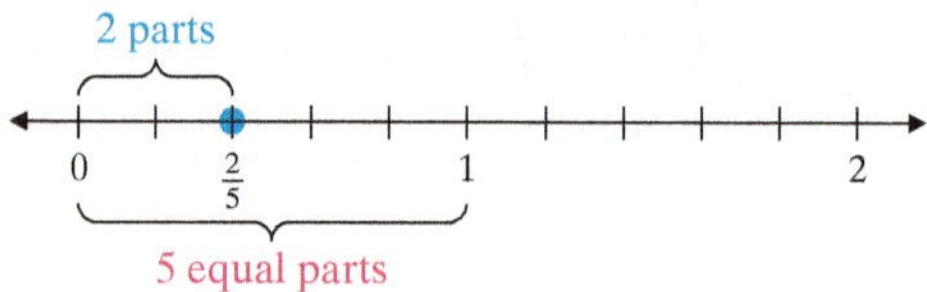

Notice that the graph of $\frac{2}{5}$ lies between 0 and 1. This means

$$0 < \frac{2}{5} \left(\text{or } \frac{2}{5} > 0\right) \text{ and also } \frac{2}{5} < 1$$

Practice 12

Graph each proper fraction on a number line.

a. $\frac{5}{7}$ **b.** $\frac{2}{3}$ **c.** $\frac{4}{6}$

Example 12

Graph each proper fraction on a number line.

a. $\frac{3}{4}$ **b.** $\frac{1}{2}$ **c.** $\frac{3}{6}$

Solution:

a. To graph $\frac{3}{4}$, divide the distance from 0 to 1 into 4 parts. Then start at 0 and count over 3 parts.

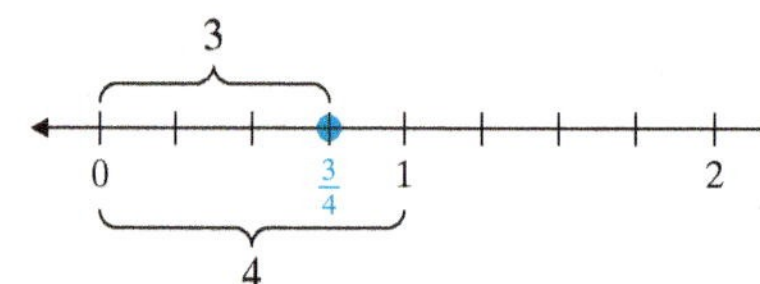

b.

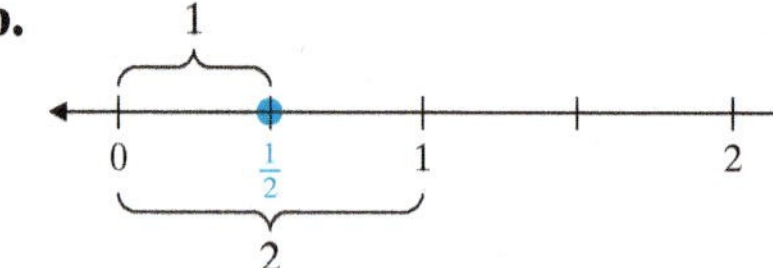

c.

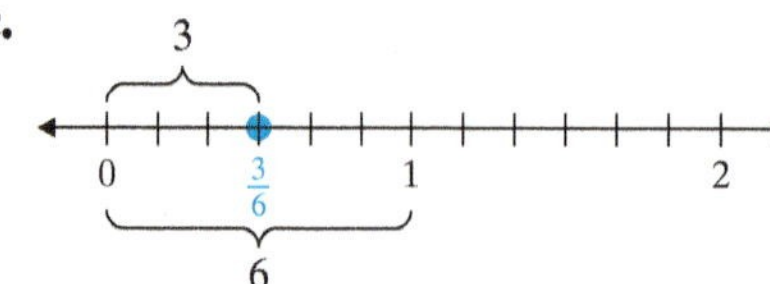

Work Practice 12

The statements below apply to positive fractions.

The fractions in Example 12 are all proper fractions. Notice that the value of each is less than 1. This is always true for proper fractions since the numerator of a proper fraction is less than the denominator.

On the next page, we graph improper fractions. Notice that improper fractions are greater than or equal to 1. This is always true since the numerator of an improper fraction is greater than or equal to the denominator.

Answers

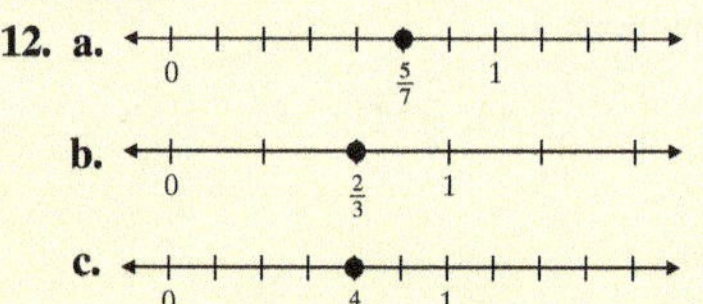

Example 13 Graph each improper fraction on a number line.

a. $\frac{7}{6}$ **b.** $\frac{9}{5}$ **c.** $\frac{6}{6}$ **d.** $\frac{3}{1}$

Solution:

a.

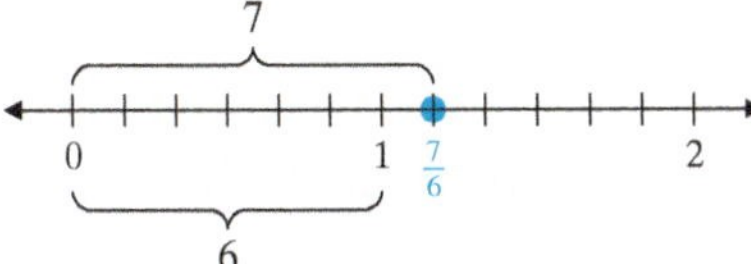

b.

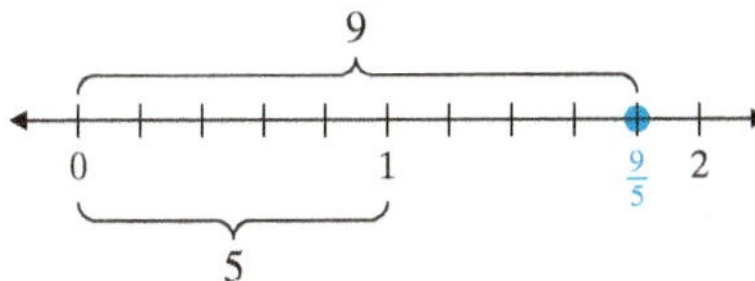

c.

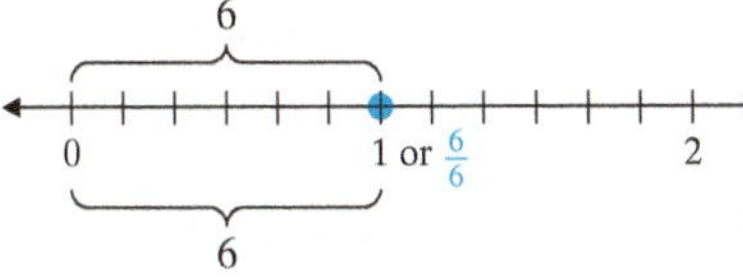

d. Each 1-unit distance has 1 equal part. Count over 3 parts.

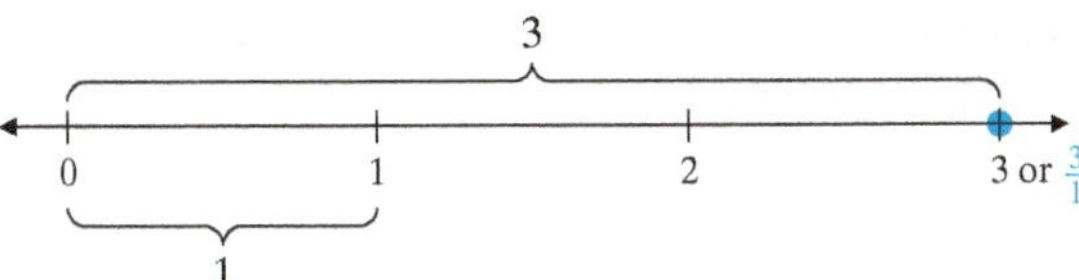

Work Practice 13

Note: We will graph mixed numbers at the end of this chapter.

Practice 13

Graph each improper fraction on a number line.

a. $\frac{8}{3}$ **b.** $\frac{5}{4}$ **c.** $\frac{7}{7}$

Objective D Reviewing Division Properties of 0 and 1 

Before we continue further, don't forget from Section 1.6 that a fraction bar indicates division. Let's review some division properties of 1 and 0.

$\frac{9}{9} = 1$ because $1 \cdot 9 = 9$ $\frac{-11}{1} = -11$ because $-11 \cdot 1 = -11$

$\frac{0}{6} = 0$ because $0 \cdot 6 = 0$ $\frac{6}{0}$ *is undefined* because there is no number that when multiplied by 0 gives 6.

In general, we can say the following.

Let n be any integer except 0.

$\frac{n}{n} = 1$ $\frac{0}{n} = 0$

$\frac{n}{1} = n$ $\frac{n}{0}$ is undefined.

Answers

13. a. **b.**

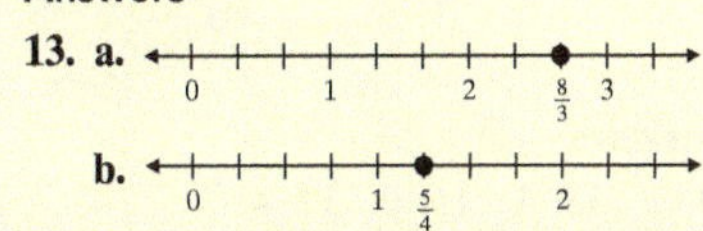

c.

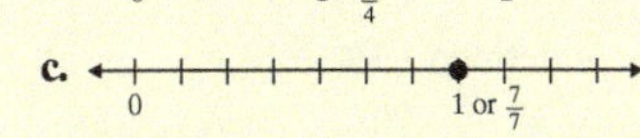

Practice 14–19

Simplify.

14. $\frac{9}{9}$ **15.** $\frac{-6}{-6}$

16. $\frac{0}{-1}$ **17.** $\frac{4}{1}$

18. $\frac{-13}{0}$ **19.** $\frac{-13}{1}$

Examples Simplify.

14. $\frac{5}{5} = 1$ **15.** $\frac{-2}{-2} = 1$ **16.** $\frac{0}{-5} = 0$

17. $\frac{-5}{1} = -5$ **18.** $\frac{41}{1} = 41$ **19.** $\frac{19}{0}$ is undefined

Work Practice 14–19

Notice from Example 17 that we can have negative fractions. In fact,

$$\frac{-5}{1} = -5, \quad \frac{5}{-1} = -5, \quad \text{and} \quad -\frac{5}{1} = -5$$

Because all of the fractions equal -5, we have

$$\frac{-5}{1} = \frac{5}{-1} = -\frac{5}{1}$$

This means that the negative sign in a fraction can be written in the numerator, in the denominator, or in front of the fraction. Remember this as we work with negative fractions.

Helpful Hint Remember, for example, that

$$-\frac{2}{3} = \frac{-2}{3} = \frac{2}{-3}$$

Objective E Writing Mixed Numbers as Improper Fractions

Earlier in this section, mixed numbers and improper fractions were both used to represent the shaded part of figure groups. For example,

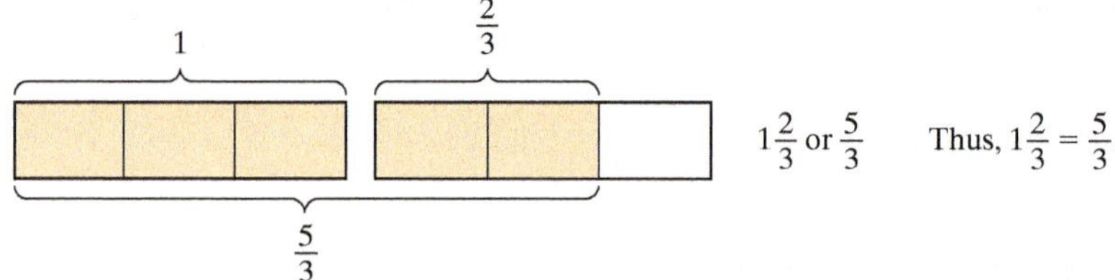

The following steps may be used to write a mixed number as an improper fraction:

Writing a Mixed Number as an Improper Fraction

To write a mixed number as an improper fraction:

Step 1: Multiply the denominator of the fraction by the whole number.

Step 2: Add the numerator of the fraction to the product from Step 1.

Step 3: Write the sum from Step 2 as the numerator of the improper fraction over the original denominator.

For example,

$$1\frac{2}{3} = \frac{\overbrace{3 \cdot 1}^{\text{Step 1}} + \overbrace{2}^{\text{Step 2}}}{\underset{\text{Step 3}}{3}} = \frac{3+2}{3} = \frac{5}{3} \quad \text{or} \quad 1\frac{2}{3} = \frac{5}{3}\text{, as stated above.}$$

Practice 20

Write each as an improper fraction.

a. $5\frac{2}{7}$ **b.** $6\frac{2}{3}$ **c.** $10\frac{9}{10}$ **d.** $4\frac{1}{5}$

Example 20 Write each as an improper fraction.

a. $4\frac{2}{9} = \frac{9 \cdot 4 + 2}{9} = \frac{36+2}{9} = \frac{38}{9}$ **b.** $1\frac{8}{11} = \frac{11 \cdot 1 + 8}{11} = \frac{11+8}{11} = \frac{19}{11}$

Work Practice 20

Answers

14. 1 **15.** 1 **16.** 0 **17.** 4 **18.** undefined **19.** -13

20. a. $\frac{37}{7}$ **b.** $\frac{20}{3}$ **c.** $\frac{109}{10}$ **d.** $\frac{21}{5}$

Objective F Writing Improper Fractions as Mixed Numbers or Whole Numbers

Just as there are times when an improper fraction is preferred, sometimes a mixed or a whole number better suits a situation. To write improper fractions as mixed or whole numbers, we use division. Recall once again from Section 1.6 that the fraction bar means division. This means that the fraction

$\frac{5}{3}$ ← numerator / denominator means $3\overline{)5}$ (3 is the denominator, 5 is the numerator)

Writing an Improper Fraction as a Mixed Number or a Whole Number

To write an improper fraction as a mixed number or a whole number:

Step 1: Divide the denominator into the numerator.

Step 2: The whole number part of the mixed number is the quotient. The fraction part of the mixed number is the remainder over the original denominator.

$$\text{quotient}\frac{\text{remainder}}{\text{original denominator}}$$

For example,

Step 1: $\frac{5}{3}$: $3\overline{)5}$ gives quotient 1; -3; remainder 2

Step 2: $\frac{5}{3} = 1\frac{2}{3}$ ← 2 is the remainder, 3 is the original denominator, 1 is the quotient

Example 21 Write each as a mixed number or a whole number.

a. $\frac{30}{7}$ b. $\frac{16}{15}$ c. $\frac{84}{6}$

Solution:

a. $\frac{30}{7}$: $7\overline{)30}$ gives 4; -28; remainder 2. $\frac{30}{7} = 4\frac{2}{7}$

b. $\frac{16}{15}$: $15\overline{)16}$ gives 1; -15; remainder 1. $\frac{16}{15} = 1\frac{1}{15}$

c. $\frac{84}{6}$: $6\overline{)84}$ gives 14; -6; 24; -24; 0. $\frac{84}{6} = 14$ Since the remainder is 0, the result is the whole number 14.

Helpful Hint When the remainder is 0, the improper fraction is a whole number. For example, $\frac{92}{4} = 23$.

$4\overline{)92}$ gives 23; -8; 12; -12; 0

Work Practice 21

Practice 21

Write each as a mixed number or a whole number.

a. $\frac{9}{5}$ b. $\frac{23}{9}$ c. $\frac{48}{4}$

d. $\frac{62}{13}$ e. $\frac{51}{7}$ f. $\frac{21}{20}$

Answers

21. a. $1\frac{4}{5}$ b. $2\frac{5}{9}$ c. 12 d. $4\frac{10}{13}$ e. $7\frac{2}{7}$ f. $1\frac{1}{20}$

Vocabulary, Readiness & Video Check

Use the choices below to fill in each blank.

improper	fraction	proper	is undefined	mixed number	$= 0$
≥ 1	denominator	$= 1$	< 1	numerator	

1. The number $\frac{17}{31}$ is called a(n) ________. The number 31 is called its ________ and 17 is called its ________.

2. If we simplify each fraction, $\frac{-9}{-9}$ ________, $\frac{0}{-4}$ ________, and we say $\frac{-4}{0}$ ________.

3. The fraction $\frac{8}{3}$ is called a(n) ________ fraction, the fraction $\frac{3}{8}$ is called a(n) ________ fraction, and $10\frac{3}{8}$ is called a(n) ________.

4. The value of an improper fraction is always ________, and the value of a proper fraction is always ________.

Martin-Gay Interactive Videos *Watch the section lecture video and answer the following questions.*

See Video 4.1

Objective A **5.** Complete this statement based on Example 1: When the numerator is greater than or ________ to the denominator, you have a(n) ________ fraction.

Objective B **6.** In Example 4, there are two shapes in the diagram, so why do the representative fractions have a denominator 3?

Objective C **7.** From Examples 6 and 7, when graphing a positive fraction on a number line, how does the denominator help? What does the denominator tell you?

Objective D **8.** From Example 10, what can you conclude about any fraction where the numerator and denominator are the same nonzero number?

Objective E **9.** Complete this statement based on the lecture before Example 13: The operation of ________ is understood in a mixed number notation; for example, $1\frac{1}{3}$ means 1 ________ $\frac{1}{3}$.

Objective F **10.** From the lecture before Example 16, what operation is used to write an improper fraction as a mixed number?

4.1 Exercise Set MyMathLab®

Objectives A B *Identify the numerator and the denominator of each fraction and identify each fraction as proper or improper. See Examples 1, 2, 10, and 11.*

1. $\frac{1}{2}$ **2.** $\frac{1}{4}$ **3.** $\frac{10}{3}$

4. $\frac{53}{21}$ **5.** $\frac{15}{15}$ **6.** $\frac{26}{26}$

Objective B *Write a proper or improper fraction to represent the shaded part of each diagram. If an improper fraction is appropriate, write the shaded part of the diagram as (**a**) an improper fraction and (**b**) a mixed number. (Note to students: In case you know how to simplify fractions, none of the fractions in this section are simplified.) See Examples 3 through 6 and 10 and 11.*

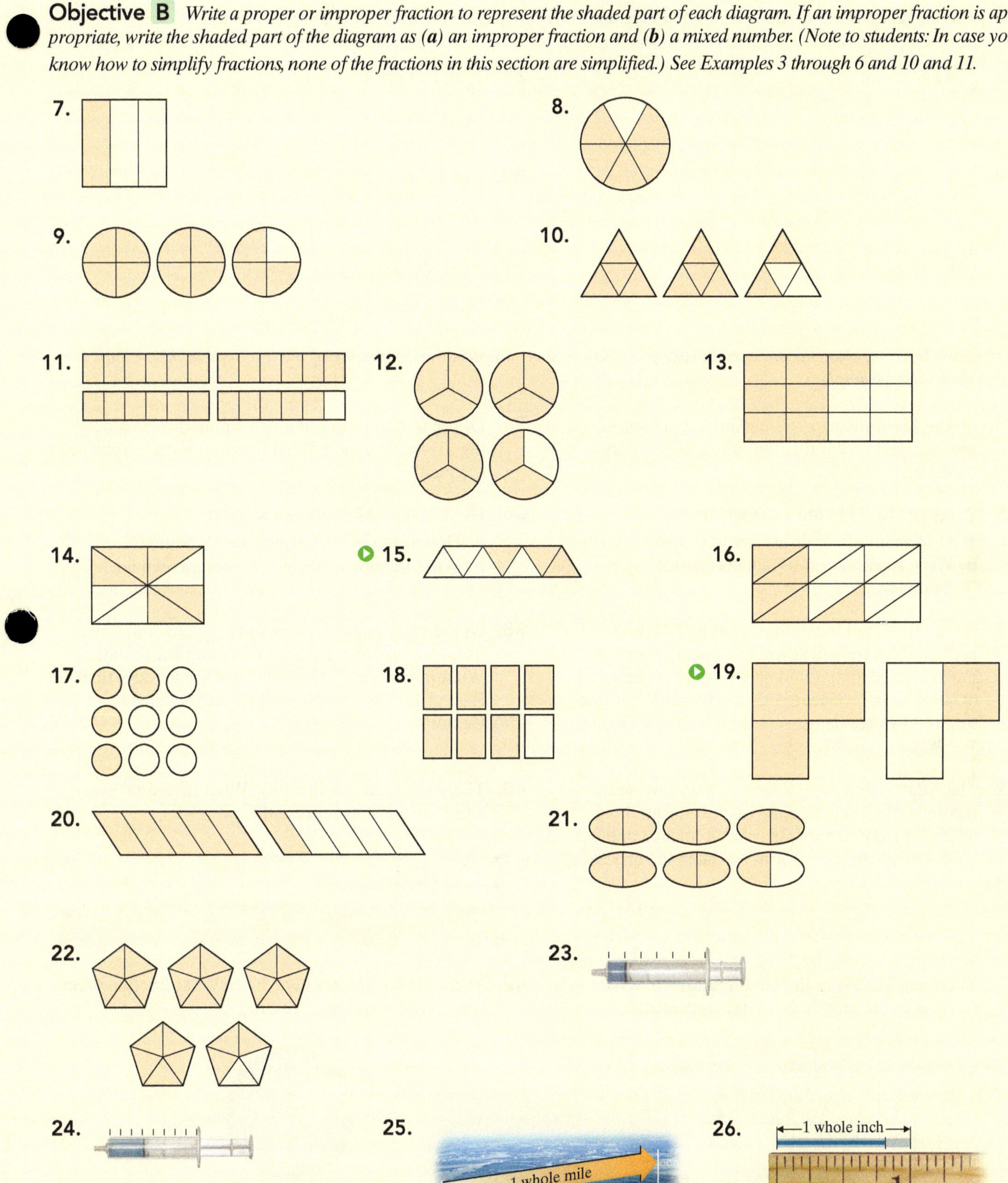

Objective B *Draw and shade a part of a diagram to represent each fraction. See Examples 7 and 8.*

27. $\frac{1}{5}$ of a diagram

28. $\frac{1}{16}$ of a diagram

29. $\frac{6}{7}$ of a diagram

30. $\frac{7}{9}$ of a diagram

31. $\frac{4}{4}$ of a diagram

32. $\frac{6}{6}$ of a diagram

Write each fraction. (Note to students: In case you know how to simplify fractions, none of the fractions in this section are simplified.) See Example 9.

33. Of the 131 students at a small private school, 42 are freshmen. What fraction of the students are freshmen?

34. Of the 63 employees at a new biomedical engineering firm, 22 are men. What fraction of the employees are men?

35. Use Exercise 33 to answer **a** and **b.**

a. How many students are *not* freshmen?
b. What fraction of the students are *not* freshmen?

36. Use Exercise 34 to answer **a** and **b.**

a. How many of the employees are women?
b. What fraction of the employees are women?

37. As of 2013, the United States has had 44 different presidents. A total of seven U.S. presidents were born in the state of Ohio, second only to the state of Virginia in producing U.S. presidents. What fraction of U.S. presidents were born in Ohio? (*Source: World Almanac*, 2013)

38. Of the eight planets in our solar system, four have days that are longer than the 24-hour Earth day. What fraction of the planets have longer days than Earth has? (*Source:* National Space Science Data Center)

39. The Atlantic hurricane season of 2005 rewrote the record books. There were 28 tropical storms, 15 of which turned into hurricanes. What fraction of the 2005 Atlantic tropical storms escalated to hurricanes?

40. There are 12 inches in a foot. What fractional part of a foot do 5 inches represent?

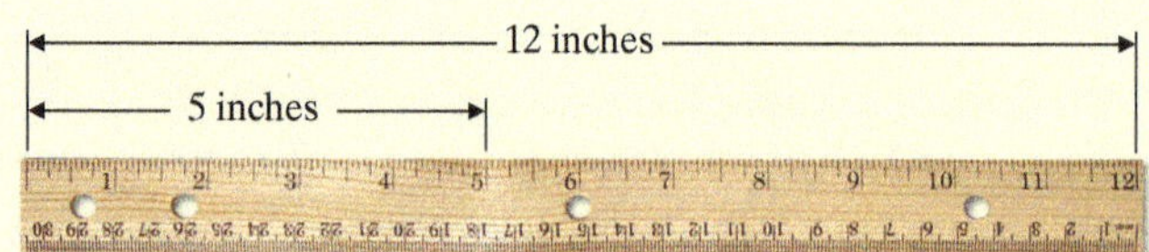

41. There are 31 days in the month of March. What fraction of the month do 11 days represent?

Mon.	Tue.	Wed.	Thu.	Fri.	Sat.	Sun.
					1	2
3	4	5	6	7	8	9
10	11	12	13	14	15	16
17	18	19	20	21	22	23
24	25	26	27	28	29	30
31						

42. There are 60 minutes in an hour. What fraction of an hour do 37 minutes represent?

43. In a prealgebra class containing 31 students, there are 18 freshmen, 10 sophomores, and 3 juniors. What fraction of the class is sophomores?

44. In a sports team with 20 children, there are 9 boys and 11 girls. What fraction of the team is boys?

45. Thirty-three out of the fifty total states in the United States contain federal Indian reservations.

a. What fraction of the states contain federal Indian reservations?

b. How many states do not contain federal Indian reservations?

c. What fraction of the states do not contain federal Indian reservations? (*Source:* Tiller Research, Inc., Albuquerque, NM)

46. Consumer fireworks are legal in 44 out of the 50 total states in the United States.

a. In what fraction of the states are consumer fireworks legal?

b. In how many states are consumer fireworks illegal?

c. In what fraction of the states are consumer fireworks illegal? (*Source:* United States Fireworks Safety Council)

47. A bag contains 50 red or blue marbles. If 21 marbles are blue, answer each question.

a. What fraction of the marbles are blue?

b. How many marbles are red?

c. What fraction of the marbles are red?

48. An art dealer is taking inventory. His shop contains a total of 37 pieces, which are all sculptures, watercolor paintings, or oil paintings. If there are 15 watercolor paintings and 17 oil paintings, answer each question.

a. What fraction of the inventory is watercolor paintings?

b. What fraction of the inventory is oil paintings?

c. How many sculptures are there?

d. What fraction of the inventory is sculptures?

Objective C *Graph each fraction on a number line. See Examples 12 and 13.*

49. $\frac{1}{4}$

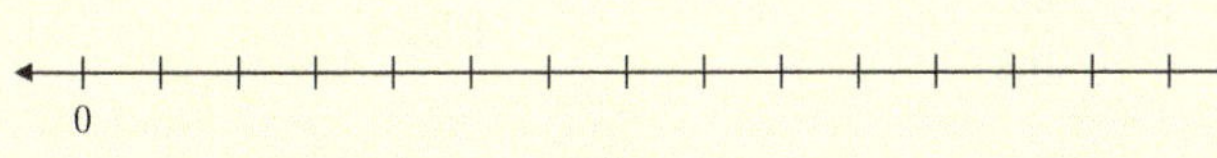

50. 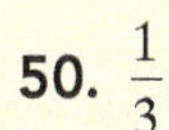 $\frac{1}{3}$

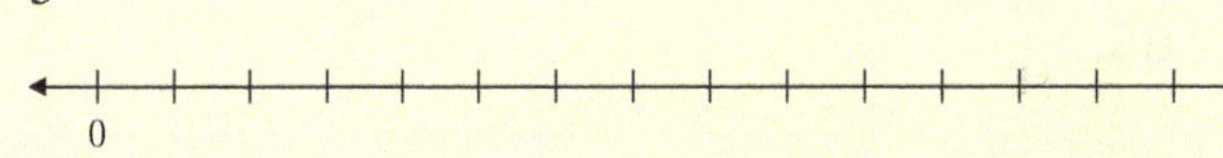

51. $\frac{4}{7}$

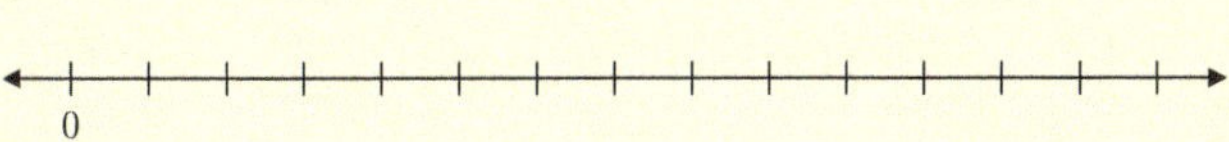

52. $\frac{5}{6}$

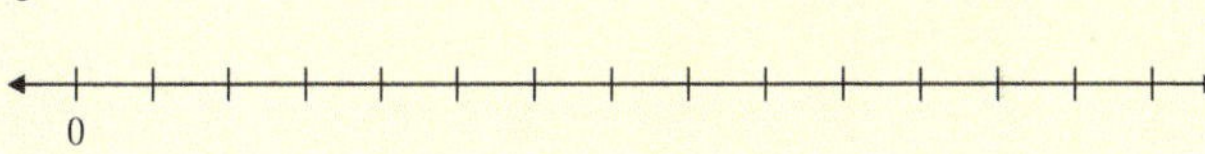

53. $\frac{8}{5}$

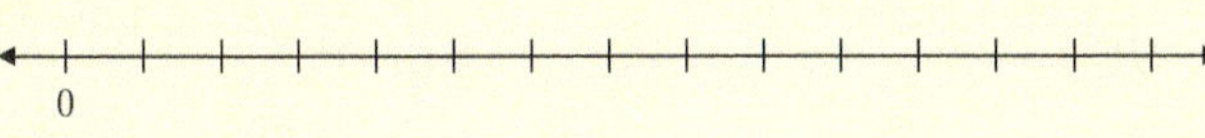

54. $\frac{9}{8}$

0

55. $\frac{7}{3}$

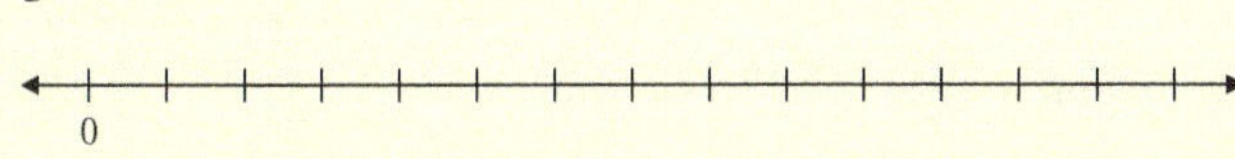

56. $\frac{13}{7}$

0

Objective D *Simplify by dividing. See Examples 14 through 19.*

57. $\frac{12}{12}$

58. $\frac{-3}{-3}$

59. $\frac{-5}{1}$

60. $\frac{-20}{1}$

61. $\frac{0}{-2}$

62. $\frac{0}{-8}$

63. $\frac{-8}{-8}$

64. $\frac{-14}{-14}$

65. $\frac{-9}{0}$

66. $\frac{-7}{0}$

67. $\frac{3}{1}$

68. $\frac{5}{5}$

Objective E *Write each mixed number as an improper fraction. See Example 20.*

69. $2\frac{1}{3}$

70. $1\frac{13}{17}$

71. $3\frac{3}{5}$

72. $2\frac{5}{9}$

73. $6\frac{5}{8}$

74. $7\frac{3}{8}$

75. $11\frac{6}{7}$

76. $12\frac{2}{5}$

77. $9\frac{7}{20}$

78. $10\frac{14}{27}$

79. $166\frac{2}{3}$

80. $114\frac{2}{7}$

Objective F *Write each improper fraction as a mixed number or a whole number. See Example 21.*

81. $\frac{17}{5}$

82. $\frac{13}{7}$

83. $\frac{37}{8}$

84. $\frac{64}{9}$

85. $\frac{47}{15}$

86. $\frac{65}{12}$

87. $\frac{225}{15}$

88. $\frac{196}{14}$

89. $\frac{182}{175}$

90. $\frac{149}{143}$

91. $\frac{737}{112}$

92. $\frac{901}{123}$

Review

Simplify. See Section 1.7.

93. 3^2

94. 4^3

95. 5^3

96. 3^4

Concept Extensions

Write each fraction in two other equivalent ways by inserting the negative sign in different places.

97. $-\frac{11}{2} = \quad = $

98. $-\frac{21}{4} = \quad = $

99. $\frac{-13}{15} = \quad = $

100. $\frac{45}{-57} = \quad = $

101. In your own words, explain why $\frac{0}{10} = 0$ and $\frac{10}{0}$ is undefined.

102. In your own words, explain why $\frac{0}{-3} = 0$ and $\frac{-3}{0}$ is undefined.

Solve. See the Concept Checks in this section.

103. If ○○○○ represents $\frac{4}{9}$ of a whole diagram, sketch the whole diagram.

104. If △△ represents $\frac{1}{3}$ of a whole diagram, sketch the whole diagram.

105. Round the mixed number $7\frac{1}{8}$ to the nearest whole number.

106. Round the mixed number $5\frac{11}{12}$ to the nearest whole number.

107. The Gap Corporation owns stores with six different brand names, with the three most popular shown on the bar graph. What fraction of the total stores shown on the graph are named "Banana Republic"?

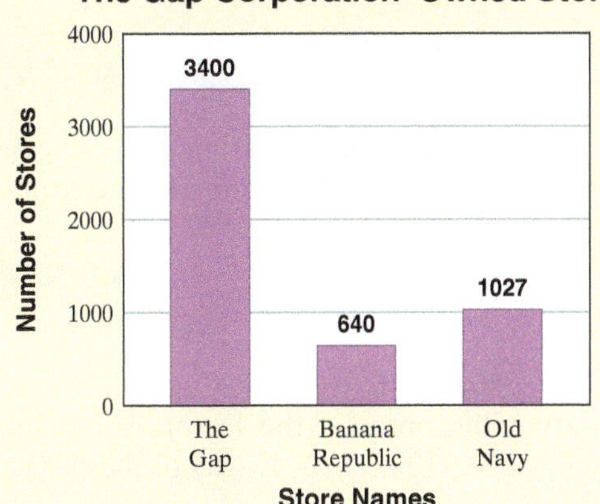

108. The Public Broadcasting Service (PBS) provides programming to the noncommercial public TV stations of the United States. The bar graph shows a breakdown of the public television licensees by type. Each licensee operates one or more PBS member TV stations. What fraction of the public television licensees are universities or colleges? (*Source:* The Public Broadcasting Service)

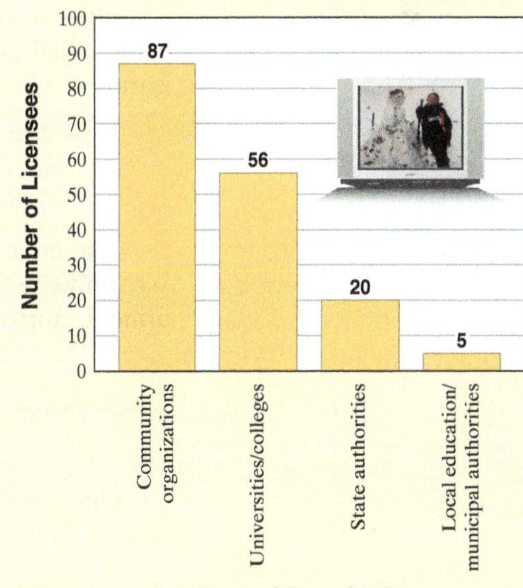

109. Habitat for Humanity is a nonprofit organization that helps provide affordable housing to families in need. Habitat for Humanity does its work of building and renovating houses through 1500 local affiliates in the United States and 80 international affiliates. What fraction of the total Habitat for Humanity affiliates are located in the United States? (*Hint:* First find the total number of affiliates.) (*Source:* Habitat for Humanity International)

110. The United States Marine Corps (USMC) has five principal training centers in California, three in North Carolina, two in South Carolina, one in Arizona, one in Hawaii, and one in Virginia. What fraction of the total USMC principal training centers are located in California? (*Hint:* First find the total number of USMC training centers.) (*Source:* U.S. Department of Defense)

4.2 Factors and Simplest Form 

Objectives

A Write a Number as a Product of Prime Numbers.

B Write a Fraction in Simplest Form.

C Determine Whether Two Fractions Are Equivalent.

D Solve Problems by Writing Fractions in Simplest Form.

Objective A Writing a Number as a Product of Prime Numbers

Recall from Section 1.5 that since $12 = 2 \cdot 2 \cdot 3$, the numbers 2 and 3 are called *factors* of 12. A **factor** is any number that divides a number evenly (with a remainder of 0).

To perform operations on fractions, it is necessary to be able to factor a number. Remember that factoring a number means writing a number as a product. We first practice writing a number as a product of prime numbers.

A **prime number** is a natural number greater than 1 whose only factors are 1 and itself. The first few prime numbers are 2, 3, 5, 7, 11, 13, 17, 19, 23, 29, … .

A **composite number** is a natural number greater than 1 that is not prime.

Helpful Hint

The natural number 1 is neither prime nor composite.

When a composite number is written as a product of prime numbers, this product is called the **prime factorization** of the number. For example, the prime factorization of 12 is $2 \cdot 2 \cdot 3$ because

$$12 = \underbrace{2 \cdot 2 \cdot 3}$$

This product is 12 and each number is a prime number.

Because multiplication is commutative, the order of the factors is not important. We can write the factorization $2 \cdot 2 \cdot 3$ as $2 \cdot 3 \cdot 2$ or $3 \cdot 2 \cdot 2$. Any of these is called the prime factorization of 12.

Every whole number greater than 1 has exactly one prime factorization.

One method for finding the prime factorization of a number is by using a factor tree, as shown in the next example.

Practice 1

Use a factor tree to find the prime factorization of each number.

a. 30 **b.** 56 **c.** 72

Example 1 Write the prime factorization of 45.

Solution: We can begin by writing 45 as the product of two numbers, say, 5 and 9.

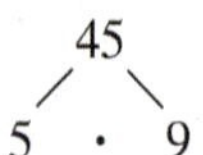

The number 5 is prime but 9 is not, so we write 9 as $3 \cdot 3$.

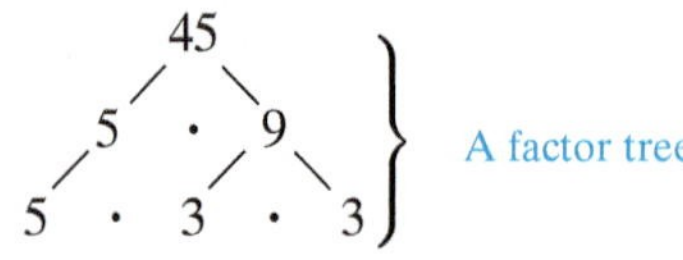

Each factor is now a prime number, so the prime factorization of 45 is $3 \cdot 3 \cdot 5$ or $3^2 \cdot 5$.

Work Practice 1

Answers

1. a. $2 \cdot 3 \cdot 5$ **b.** $2^3 \cdot 7$ **c.** $2^3 \cdot 3^2$

✓**Concept Check** True or false? Two different numbers can have exactly the same prime factorization. Explain your answer.

Example 2 Write the prime factorization of 80.

Solution: Write 80 as a product of two numbers. Continue this process until all factors are prime.

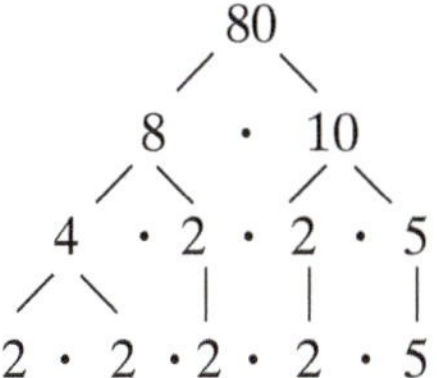

All factors are now prime, so the prime factorization of 80 is

$2 \cdot 2 \cdot 2 \cdot 2 \cdot 5$ or $2^4 \cdot 5$.

Work Practice 2

Practice 2

Write the prime factorization of 60.

Helpful Hint

It makes no difference which factors you start with. The prime factorization of a number will be the same.

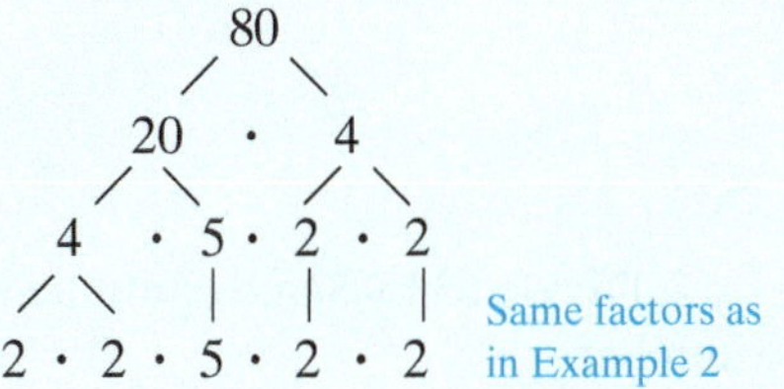

There are a few quick **divisibility tests** to determine whether a number is divisible by the primes 2, 3, or 5. (A number is divisible by 2, for example, if 2 divides it evenly so that the remainder is 0.)

Divisibility Tests

A whole number is divisible by:

- 2 if the last digit is 0, 2, 4, 6, or 8.

  132 is divisible by 2 since the last digit is a 2.
- 3 if the sum of the digits is divisible by 3.

  144 is divisible by 3 since $1 + 4 + 4 = 9$ is divisible by 3.
- 5 if the last digit is 0 or 5.

  1115 is divisible by 5 since the last digit is a 5.

Answer

2. $2^2 \cdot 3 \cdot 5$

✓**Concept Check Answer**

false; answers may vary

Helpful Hint

Here are a few other divisibility tests you may want to use. A whole number is divisible by:

- 4 if its last two digits are divisible by 4.

  1712 is divisible by 4.
- 6 if it's divisible by 2 and 3.

  9858 is divisible by 6.
- 9 if the sum of its digits is divisible by 9.

  5238 is divisible by 9 since $5 + 2 + 3 + 8 = 18$ is divisible by 9.

When finding the prime factorization of larger numbers, you may want to use the procedure shown in Example 3.

Practice 3

Write the prime factorization of 297.

Example 3 Write the prime factorization of 252.

Solution: For this method, we divide prime numbers into the given number. Since the ones digit of 252 is 2, we know that 252 is divisible by 2.

$$2\overline{)252}\ \text{quotient } 126$$

126 is divisible by 2 also.

$$\begin{array}{r} 63 \\ 2\overline{)126} \\ 2\overline{)252} \end{array}$$

63 is not divisible by 2 but is divisible by 3. Divide 63 by 3 and continue in this same manner until the quotient is a prime number.

$$\begin{array}{r} 7 \\ 3\overline{)\ 21} \\ 3\overline{)\ 63} \\ 2\overline{)126} \\ 2\overline{)252} \end{array}$$

Helpful Hint The order of choosing prime numbers does not matter. For consistency, we use the order 2, 3, 5, 7, ...

The prime factorization of 252 is $2 \cdot 2 \cdot 3 \cdot 3 \cdot 7$ or $2^2 \cdot 3^2 \cdot 7$.

Work Practice 3

In this text, we will write the factorization of a number from the smallest factor to the largest factor.

✓**Concept Check** True or false? The prime factorization of 117 is $9 \cdot 13$. Explain your reasoning.

Objective B Writing Fractions in Simplest Form 

Fractions that represent the same portion of a whole or the same point on a number line are called **equivalent fractions.** Study the table on the next page to see two ways to visualize equivalent fractions.

Answer

3. $3^3 \cdot 11$

✓**Concept Check Answer**

False; 9 is not prime.

<table>
<tr><th colspan="2">Equivalent Fractions</th></tr>
<tr><th>Figures</th><th>Number Line</th></tr>
<tr><td>When we shade $\frac{1}{3}$ and $\frac{2}{6}$ on the same-sized figures,
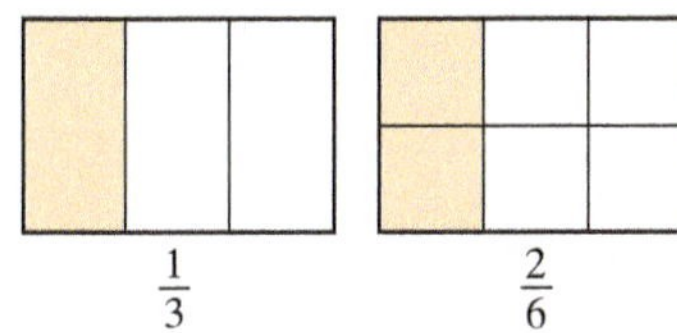
$\frac{1}{3}$ $\frac{2}{6}$
notice that both $\frac{1}{3}$ and $\frac{2}{6}$ represent the same portion of a whole. These fractions are called equivalent fractions and we write $\frac{1}{3} = \frac{2}{6}$.</td><td>When we graph $\frac{1}{3}$ and $\frac{2}{6}$ on a number line,
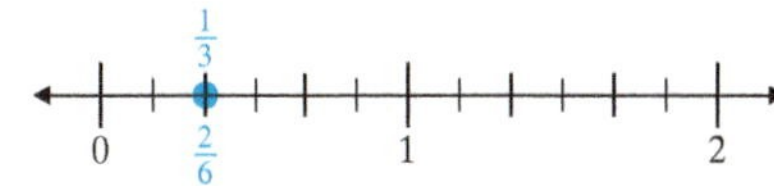

notice that both $\frac{1}{3}$ and $\frac{2}{6}$ correspond to the same point. These fractions are called equivalent fractions, and we write $\frac{1}{3} = \frac{2}{6}$.</td></tr>
<tr><td colspan="2">Thus, $\frac{1}{3} = \frac{2}{6}$ and $\frac{1}{3}$ and $\frac{2}{6}$ are equivalent.</td></tr>
</table>

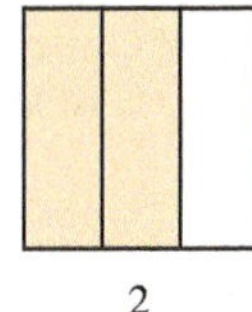

$\frac{2}{3}$

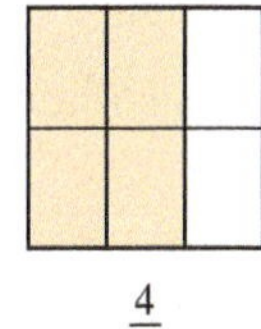

$\frac{4}{6}$

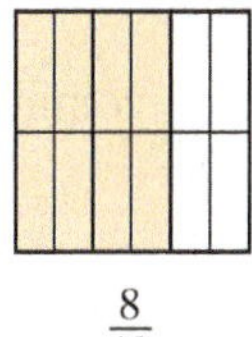

$\frac{8}{12}$

For example, $\frac{2}{3}, \frac{4}{6}$, and $\frac{8}{12}$ all represent the same shaded portion of the rectangle's area, so they are equivalent fractions. To show that these fractions are equivalent, we place an equal sign between them. In other words,

$$\frac{2}{3} = \frac{4}{6} = \frac{8}{12}$$

There are many equivalent forms of a fraction. A special equivalent form of a fraction is called **simplest form.**

Simplest Form of a Fraction

A fraction is written in **simplest form** or **lowest terms** when the numerator and the denominator have no common factors other than 1.

For example, the fraction $\frac{2}{3}$ *is* in simplest form because 2 and 3 have no common factor other than 1. The fraction $\frac{4}{6}$ *is not* in simplest form because 4 and 6 both have a factor of 2. That is, 2 is a common factor of 4 and 6. The process of writing a fraction in simplest form is called **simplifying** the fraction.

To simplify $\frac{4}{6}$ and write it as $\frac{2}{3}$, let's first study a few properties. Recall from Section 4.1 that any nonzero whole number n divided by itself is 1.

Any nonzero number n divided by itself is 1.

$$\frac{5}{5} = 1, \frac{17}{17} = 1, \frac{24}{24} = 1, \text{ or, in general, } \frac{n}{n} = 1$$

Also, in general, if $\frac{a}{b}$ and $\frac{c}{d}$ are fractions (with b and d not 0), the following is true.

$$\frac{a \cdot c}{b \cdot d} = \frac{a}{b} \cdot \frac{c}{d}*$$

Helpful Hint These two properties together are called the Fundamental Property of Fractions

$$\frac{a \cdot c}{b \cdot c} = \frac{a}{b} \cdot \frac{c}{c} = \frac{a}{b}$$

These properties allow us to do the following:

$$\frac{4}{6} = \frac{2 \cdot 2}{2 \cdot 3} = \frac{2}{2} \cdot \frac{2}{3} = 1 \cdot \frac{2}{3} = \frac{2}{3}$$

When 1 is multiplied by a number, the result is the same number. ($\frac{2}{2}$: This is 1.)

Practice 4

Write in simplest form: $\frac{30}{45}$

Example 4 Write in simplest form: $\frac{12}{20}$

Solution: Notice that 12 and 20 have a common factor of 4.

$$\frac{12}{20} = \frac{4 \cdot 3}{4 \cdot 5} = \frac{4}{4} \cdot \frac{3}{5} = 1 \cdot \frac{3}{5} = \frac{3}{5}$$

Since 3 and 5 have no common factors (other than 1), $\frac{3}{5}$ is in simplest form.

Work Practice 4

If you have trouble finding common factors, write the prime factorization of the numerator and the denominator.

Practice 5

Write in simplest form: $\frac{39x}{51}$

Example 5 Write in simplest form: $\frac{42x}{66}$

Solution: Let's write the prime factorizations of 42 and 66. Remember that $42x$ means $42 \cdot x$.

$$\frac{42x}{66} = \frac{2 \cdot 3 \cdot 7 \cdot x}{2 \cdot 3 \cdot 11} = \frac{2}{2} \cdot \frac{3}{3} \cdot \frac{7x}{11} = 1 \cdot 1 \cdot \frac{7x}{11} = \frac{7x}{11}$$

Work Practice 5

In the example above, you may have saved time by noticing that 42 and 66 have a common factor of 6.

$$\frac{42x}{66} = \frac{6 \cdot 7x}{6 \cdot 11} = \frac{6}{6} \cdot \frac{7x}{11} = 1 \cdot \frac{7x}{11} = \frac{7x}{11}$$

Helpful Hint

Writing the prime factorizations of the numerator and the denominator is helpful in finding any common factors.

The method for simplifying negative fractions is the same as for positive fractions.

Practice 6

Write in simplest form: $-\frac{9}{50}$

Example 6 Write in simplest form: $-\frac{10}{27}$

Solution:

$$-\frac{10}{27} = -\frac{2 \cdot 5}{3 \cdot 3 \cdot 3}$$ Prime factorizations of 10 and 27

Since 10 and 27 have no common factors, $-\frac{10}{27}$ is already in simplest form.

Work Practice 6

*Note: We will study this concept further in the next section.

Answers

4. $\frac{2}{3}$ **5.** $\frac{13x}{17}$ **6.** $-\frac{9}{50}$

Example 7 Write in simplest form: $\frac{30}{108}$

Solution:

$$\frac{30}{108} = \frac{2 \cdot 3 \cdot 5}{2 \cdot 2 \cdot 3 \cdot 3 \cdot 3} = \frac{2}{2} \cdot \frac{3}{3} \cdot \frac{5}{2 \cdot 3 \cdot 3} = 1 \cdot 1 \cdot \frac{5}{18} = \frac{5}{18}$$

Work Practice 7

Practice 7

Write in simplest form: $\frac{49}{112}$

We can use a shortcut procedure with common factors when simplifying.

$$\frac{4}{6} = \frac{\overset{1}{\cancel{2}} \cdot 2}{\underset{1}{\cancel{2}} \cdot 3} = \frac{1 \cdot 2}{1 \cdot 3} = \frac{2}{3}$$ Divide out the common factor of 2 in the numerator and denominator.

This procedure is possible because dividing out a common factor in the numerator and denominator is the same as removing a factor of 1 in the product.

Writing a Fraction in Simplest Form

To write a fraction in simplest form, write the prime factorization of the numerator and the denominator and then divide both by all common factors.

Example 8 Write in simplest form: $-\frac{72}{26}$

Solution:

$$-\frac{72}{26} = -\frac{\overset{1}{\cancel{2}} \cdot 2 \cdot 2 \cdot 3 \cdot 3}{\underset{1}{\cancel{2}} \cdot 13} = -\frac{1 \cdot 2 \cdot 2 \cdot 3 \cdot 3}{1 \cdot 13} = -\frac{36}{13}$$

Work Practice 8

Practice 8

Write in simplest form: $-\frac{64}{20}$

✓**Concept Check** Which is the correct way to simplify the fraction $\frac{15}{25}$? Or are both correct? Explain.

a. $\frac{15}{25} = \frac{3 \cdot \overset{1}{\cancel{5}}}{5 \cdot \underset{1}{\cancel{5}}} = \frac{3}{5}$ **b.** $\frac{1\overset{1}{\cancel{5}}}{2\underset{1}{\cancel{5}}} = \frac{11}{21}$

In this chapter, we will simplify and perform operations on fractions containing variables. When the denominator of a fraction contains a variable, such as $\frac{6x^2}{60x^3}$, we will assume that the variable does not represent 0. Recall that the denominator of a fraction cannot be 0.

Example 9 Write in simplest form: $\frac{6x^2}{60x^3}$

Solution: Notice that 6 and 60 have a common factor of 6. Let's also use the definition of an exponent to factor x^2 and x^3.

$$\frac{6x^2}{60x^3} = \frac{\overset{1}{\cancel{6}} \cdot \overset{1}{\cancel{x}} \cdot \overset{1}{\cancel{x}}}{\underset{1}{\cancel{6}} \cdot 10 \cdot \underset{1}{\cancel{x}} \cdot \underset{1}{\cancel{x}} \cdot x} = \frac{1 \cdot 1 \cdot 1}{1 \cdot 10 \cdot 1 \cdot 1 \cdot x} = \frac{1}{10x}$$

Work Practice 9

Practice 9

Write in simplest form: $\frac{7a^3}{56a^2}$

Answers

7. $\frac{7}{16}$ **8.** $-\frac{16}{5}$ **9.** $\frac{a}{8}$

✓**Concept Check Answers**

a. correct **b.** incorrect

Helpful Hint

Be careful when all factors of the numerator or denominator are divided out. In Example 9, the numerator was $1 \cdot 1 \cdot 1 = 1$, so the final result was $\frac{1}{10x}$.

Objective C Determining Whether Two Fractions Are Equivalent

Recall from Objective **B** that two fractions are equivalent if they represent the same part of a whole. One way to determine whether two fractions are equivalent is to see whether they simplify to the same fraction.

Practice 10

Determine whether $\frac{7}{9}$ and $\frac{21}{27}$ are equivalent.

Example 10 Determine whether $\frac{16}{40}$ and $\frac{10}{25}$ are equivalent.

Solution: Simplify each fraction.

$$\frac{16}{40} = \frac{\overset{1}{\cancel{8}} \cdot 2}{\underset{1}{\cancel{8}} \cdot 5} = \frac{1 \cdot 2}{1 \cdot 5} = \frac{2}{5}$$

$$\frac{10}{25} = \frac{2 \cdot \overset{1}{\cancel{5}}}{5 \cdot \underset{1}{\cancel{5}}} = \frac{2 \cdot 1}{5 \cdot 1} = \frac{2}{5}$$

Since these fractions are the same, $\frac{16}{40} = \frac{10}{25}$.

Work Practice 10

There is a shortcut method you may use to check or test whether two fractions are equivalent. In the example above, we learned that the fractions are equivalent, or

$$\frac{16}{40} = \frac{10}{25}$$

In this example above, we call $25 \cdot 16$ and $40 \cdot 10$ **cross products** because they are the products one obtains by multiplying diagonally across the equal sign, as shown below.

Cross Products

$25 \cdot 16$ $\quad$ $40 \cdot 10$

$$\frac{16}{40} = \frac{10}{25}$$

Notice that these cross products are equal:

$$25 \cdot 16 = 400, \quad 40 \cdot 10 = 400$$

In general, this is true for equivalent fractions.

Equality of Fractions

$8 \cdot 6$ $\quad$ $24 \cdot 2$

$$\frac{6}{24} \stackrel{?}{=} \frac{2}{8}$$

Since the cross products ($8 \cdot 6 = 48$ and $24 \cdot 2 = 48$) are equal, the fractions are equal.

Note: If the cross products are not equal, the fractions are not equal.

Answer

10. equivalent

Example 11 Determine whether $\frac{8}{11}$ and $\frac{19}{26}$ are equivalent.

Solution: Let's check cross products.

$$26 \cdot 8 = 208 \qquad \frac{8}{11} \stackrel{?}{=} \frac{19}{26} \qquad 11 \cdot 19 = 209$$

Since $208 \neq 209$, then $\frac{8}{11} \neq \frac{19}{26}$.

Work Practice 11

Practice 11

Determine whether $\frac{4}{13}$ and $\frac{5}{18}$ are equivalent.

Objective D Solving Problems by Writing Fractions in Simplest Form

Many real-life problems can be solved by writing fractions. To make the answers clearer, these fractions should be written in simplest form.

Example 12 Calculating Fraction of Parks in Wyoming

There are currently 58 national parks in the United States. Two of these parks are located in the state of Wyoming. What fraction of the United States' national parks can be found in Wyoming? Write the fraction in simplest form. (*Source: World Almanac*, 2013)

Solution: First we determine the fraction of parks found in Wyoming state.

$\frac{2}{58}$ ← national parks in Wyoming / ← total national parks

Next we simplify the fraction.

$$\frac{2}{58} = \frac{\overset{1}{\cancel{2}}}{\underset{1}{\cancel{2}} \cdot 29} = \frac{1}{1 \cdot 29} = \frac{1}{29}$$

Thus, $\frac{1}{29}$ of the United States' national parks are in Wyoming state.

Work Practice 12

Practice 12

There are eight national parks in Alaska. See Example 12 and determine what fraction of the United States' national parks are located in Alaska. Write the fraction in simplest form.

Answers

11. not equivalent **12.** $\frac{4}{29}$

Calculator Explorations Simplifying Fractions

Scientific Calculator

Many calculators have a fraction key, such as $\boxed{a^b/c}$, that allows you to simplify a fraction on the calculator. For example, to simplify $\frac{324}{612}$, enter

$\boxed{324}$ $\boxed{a^b/c}$ $\boxed{612}$ $\boxed{=}$

The display will read

$\boxed{9 \mid 17}$

which represents $\frac{9}{17}$, the original fraction simplified.

Graphing Calculator

Graphing calculators also allow you to simplify fractions. The fraction option on a graphing calculator may be found under the $\boxed{\text{MATH}}$ menu.

To simplify $\frac{324}{612}$, enter

$\boxed{324}$ $\boxed{\div}$ $\boxed{612}$ $\boxed{\text{MATH}}$ $\boxed{\text{ENTER}}$ $\boxed{\text{ENTER}}$

The display will read

$\boxed{324/612 \blacktriangleright \text{Frac } 9/17}$

Helpful Hint The Calculator Explorations boxes in this chapter provide only an introduction to fraction keys on calculators. Any time you use a calculator, there are both advantages and limitations to its use. Never rely solely on your calculator. It is very important that you understand how to perform all operations on fractions by hand in order to progress through later topics. For further information, talk to your instructor.

Use your calculator to simplify each fraction.

1. $\frac{128}{224}$ **2.** $\frac{231}{396}$ **3.** $\frac{340}{459}$

4. $\frac{999}{1350}$ **5.** $\frac{432}{810}$ **6.** $\frac{225}{315}$

7. $\frac{54}{243}$ **8.** $\frac{455}{689}$

Vocabulary, Readiness & Video Check

Use the choices below to fill in each blank.

cross products	equivalent	composite
simplest form	prime factorization	prime

1. The number 40 equals $2 \cdot 2 \cdot 2 \cdot 5$. Since each factor is prime, we call $2 \cdot 2 \cdot 2 \cdot 5$ the ______________ of 40.

2. A natural number, other than 1, that is not prime is called a(n) ______________ number.

3. A natural number that has exactly two different factors, 1 and itself, is called a(n) ______________ number.

4. In $\frac{11}{48}$, since 11 and 48 have no common factors other than 1, $\frac{11}{48}$ is in ______________.

5. Fractions that represent the same portion of a whole are called ______________ fractions.

6. In the statement $\frac{5}{12} = \frac{15}{36}$, $5 \cdot 36$ and $12 \cdot 15$ are called ______________.

Martin-Gay Interactive Videos *Watch the section lecture video and answer the following questions.*

See Video 4.2

Objective A 7. From Example 1, what two things should you check to make sure your prime factorization of a number is correct?

Objective B 8. From the lecture before Example 3, when you have a common factor in the numerator and denominator of a fraction, essentially you have what?

Objective C 9. Describe another way to solve Example 7 besides using cross products.

Objective D 10. Why isn't $\frac{10}{24}$ the final answer to Example 8? What is the final answer?

4.2 Exercise Set MyMathLab®

Objective A *Write the prime factorization of each number. See Examples 1 through 3.*

1. 20
2. 12
3. 48
4. 75
5. 81
6. 64
7. 162
8. 128
9. 110
10. 130
11. 85
12. 93
13. 240
14. 836
15. 828
16. 504

Objective B *Write each fraction in simplest form. See Examples 4 through 9.*

17. $\frac{3}{12}$
18. $\frac{5}{30}$
19. $\frac{4x}{42}$
20. $\frac{9y}{48}$
21. $\frac{14}{16}$
22. $\frac{22}{34}$
23. $\frac{20}{30}$
24. $\frac{70}{80}$
25. $\frac{35a}{50a}$
26. $\frac{25z}{55z}$
27. $-\frac{63}{81}$
28. $-\frac{21}{49}$
29. $\frac{30x^2}{36x}$
30. $\frac{45b}{80b^2}$
31. $\frac{27}{64}$
32. $\frac{32}{63}$
33. $\frac{25xy}{40y}$
34. $\frac{36y}{42yz}$
35. $-\frac{40}{64}$
36. $-\frac{28}{60}$
37. $\frac{36x^3y^2}{24xy}$
38. $\frac{60a^2b}{36ab^3}$
39. $\frac{90}{120}$
40. $\frac{60}{150}$
41. $\frac{40xy}{64xyz}$

42. $\frac{28abc}{60ac}$ **43.** $\frac{66}{308}$ **44.** $\frac{65}{234}$ **45.** $-\frac{55}{85y}$ **46.** $-\frac{78}{90x}$

47. $\frac{189z}{216z}$ **48.** $\frac{144y}{162y}$ **49.** $\frac{224a^3b^4c^2}{16ab^4c^2}$ **50.** $\frac{270x^4y^3z^3}{15x^3y^3z^3}$

Objective C *Determine whether each pair of fractions is equivalent. See Examples 10 and 11.*

51. $\frac{2}{6}$ and $\frac{4}{12}$ **52.** $\frac{3}{6}$ and $\frac{5}{10}$ **53.** $\frac{7}{11}$ and $\frac{5}{8}$

54. $\frac{2}{5}$ and $\frac{4}{11}$ **55.** $\frac{10}{15}$ and $\frac{6}{9}$ **56.** $\frac{4}{10}$ and $\frac{6}{15}$

57. $\frac{3}{9}$ and $\frac{6}{18}$ **58.** $\frac{2}{8}$ and $\frac{7}{28}$ **59.** $\frac{10}{13}$ and $\frac{13}{15}$

60. $\frac{16}{20}$ and $\frac{9}{16}$ **61.** $\frac{8}{18}$ and $\frac{12}{24}$ **62.** $\frac{6}{21}$ and $\frac{14}{35}$

Objective D *Solve. Write each fraction in simplest form. See Example 12.*

63. A work shift for an employee at Starbucks consists of 8 hours. What fraction of the employee's work shift is represented by 2 hours?

64. Two thousand baseball caps were sold one year at the U.S. Open Golf Tournament. What fractional part of this total do 200 caps represent?

65. There are 5280 feet in a mile. What fraction of a mile is represented by 2640 feet?

66. There are 100 centimeters in 1 meter. What fraction of a meter is 20 centimeters?

67. Sixteen out of the total fifty states in the United States have Ritz-Carlton hotels. (*Source:* Ritz-Carlton Hotel Company, LLC)

a. What fraction of states can claim at least one Ritz-Carlton hotel?

b. How many states do not have a Ritz-Carlton hotel?

c. Write the fraction of states without a Ritz-Carlton hotel.

68. There are 75 national monuments in the United States. Ten of these monuments are located in New Mexico. (*Source: World Almanac*, 2013)

a. What fraction of the national monuments in the United States can be found in New Mexico?

b. How many of the national monuments in the United States are found outside New Mexico?

c. Write the fraction of national monuments found in states other than New Mexico.

69. The outer wall of the Pentagon is 24 inches thick. Ten inches is concrete, 8 inches is brick, and 6 inches is limestone. What fraction of the wall is concrete?

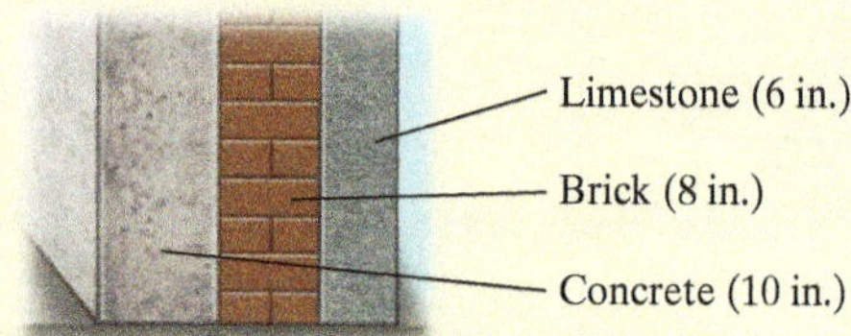

70. There are 35 students in a biology class. If 10 students made an A on the first test, what fraction of the students made an A?

71. As Internet usage grows in the United States, more and more state governments are placing services online. Forty-two out of the total fifty states have Web sites that allow residents to file their federal and state income tax electronically at the same time.
- **a.** How many states do not have this type of Web site?
- **b.** What fraction of states do not have this type of Web site? (*Source:* MeF Federal/State Program)

72. Katy Biagini just bought a brand-new 2013 Toyota Camry hybrid for $24,200. Her old car was traded in for $12,000.
- **a.** How much of her purchase price was not covered by her trade-in?
- **b.** What fraction of the purchase price was not covered by the trade-in?

73. As of this writing, a total of 464 individuals from the United States are or have been astronauts. Of these, 26 were born in Texas. What fraction of U.S. astronauts were born in Texas? (*Source:* Spacefacts Web site)

74. Worldwide, Hallmark employs nearly 12,000 full-time employees. About 3200 employees work at the Hallmark headquarters in Kansas City, Missouri. What fraction of Hallmark employees work in Kansas City? (*Source:* Hallmark Cards, Inc.)

Review

Evaluate each expression using the given replacement numbers. See Section 2.5.

75. $\frac{x^3}{9}$ when $x = -3$

76. $\frac{y^3}{5}$ when $y = -5$

77. $2y$ when $y = -7$

78. $-5a$ when $a = -4$

Concept Extensions

79. In your own words, define equivalent fractions.

80. Given a fraction, say $\frac{15}{40}$, how many fractions are there that are equivalent to it, but in simplest form or lowest terms? Explain your answer.

Write each fraction in simplest form.

81. $\frac{3975}{6625}$

82. $\frac{9506}{12{,}222}$

There are generally considered to be eight basic blood types. The table shows the number of people with the various blood types in a typical group of 100 blood donors. Use the table to answer Exercises 83 through 86. Write each answer in simplest form.

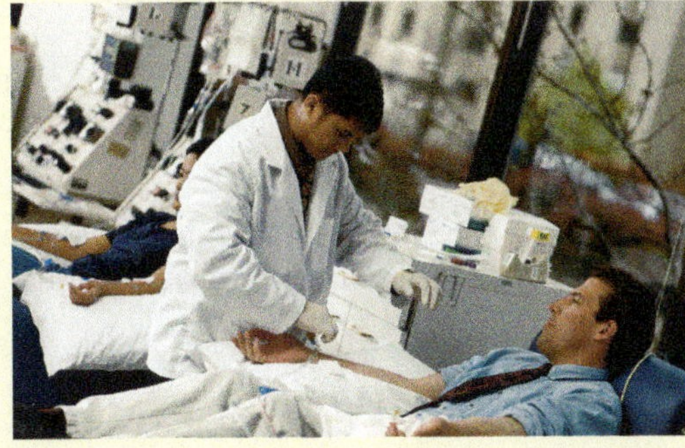

Distribution of Blood Types in Blood Donors

Blood Type	Number of People
O Rh-positive	37
O Rh-negative	7
A Rh-positive	36
A Rh-negative	6
B Rh-positive	9
B Rh-negative	1
AB Rh-positive	3
AB Rh-negative	1

(*Source:* American Red Cross Biomedical Services)

83. What fraction of blood donors have blood type A Rh-positive?

84. What fraction of blood donors have an O blood type?

85. What fraction of blood donors have an AB blood type?

86. What fraction of blood donors have a B blood type?

Find the prime factorization of each number.

87. 34,020

88. 131,625

89. In your own words, define a prime number.

90. The number 2 is a prime number. All other even natural numbers are composite numbers. Explain why.

91. Two students have different prime factorizations for the same number. Is this possible? Explain.

92. Two students work to prime factor 120. One student starts by writing 120 as 12 × 10. The other student writes 120 as 24 × 5. Finish each prime factorization. Are they the same? Why or why not?

The following graph is called a circle graph or pie chart. Each sector (shaped like a piece of pie) shows the fraction of entering college freshmen who choose to major in each discipline shown. The whole circle represents the entire class of college freshmen. Use this graph to answer Exercises 93 through 96. Write each fraction answer in simplest form.

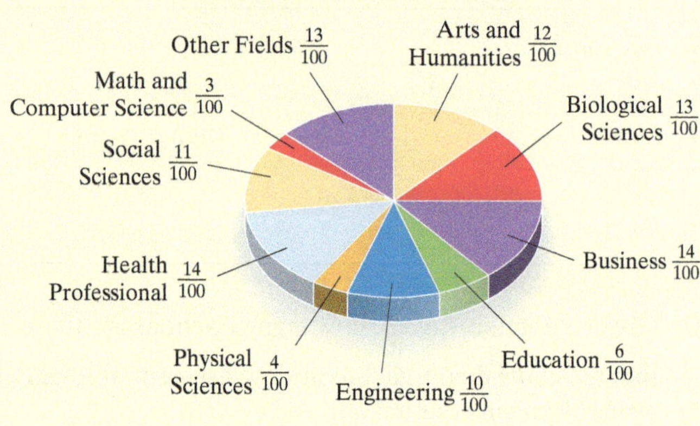

Source: The Higher Education Research Institute

93. What fraction of entering college freshmen plan to major in education?

94. What fraction of entering college freshmen plan to major in engineering?

95. Why is the Business sector the same size as the Health Professional sector?

96. Why is the Physical Sciences sector smaller than the Business sector?

Use this circle graph to answer Exercises 97 through 100. Write each fraction answer in simplest form.

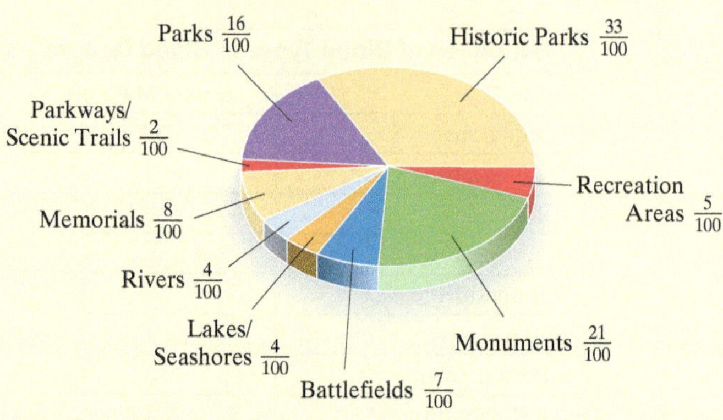

Source: National Park Service

97. What fraction of National Park Service areas are National Memorials?

98. What fraction of National Park Service areas are National Parks?

99. Why is the National Battlefields sector smaller than the National Monuments sector?

100. Why is the National Lakes/National Seashores sector the same size as the National Rivers sector?

Use the following numbers for Exercises 101 through 104.

8691 786 1235 2235 85 105 22 222 900 1470

101. List the numbers divisible by both 2 and 3.

102. List the numbers that are divisible by both 3 and 5.

103. The answers to Exercise 101 are also divisible by what number? Tell why.

104. The answers to Exercise 102 are also divisible by what number? Tell why.

4.3 Multiplying and Dividing Fractions 

Objective A Multiplying Fractions

Let's use a diagram to discover how fractions are multiplied. For example, to multiply $\frac{1}{2}$ and $\frac{3}{4}$, we find $\frac{1}{2}$ of $\frac{3}{4}$. To do this, we begin with a diagram showing $\frac{3}{4}$ of a rectangle's area shaded.

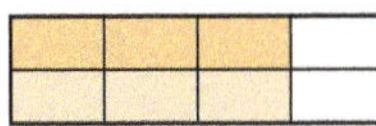

$\frac{3}{4}$ of the rectangle's area is shaded.

To find $\frac{1}{2}$ of $\frac{3}{4}$, we heavily shade $\frac{1}{2}$ of the part that is already shaded.

By counting smaller rectangles, we see that $\frac{3}{8}$ of the larger rectangle is now heavily shaded, so that

$$\frac{1}{2} \text{ of } \frac{3}{4} \text{ is } \frac{3}{8}, \text{ or } \frac{1}{2} \cdot \frac{3}{4} = \frac{3}{8}$$

Notice that $\frac{1}{2} \cdot \frac{3}{4} = \frac{1 \cdot 3}{2 \cdot 4} = \frac{3}{8}$.

Multiplying Fractions

To multiply two fractions, multiply the numerators and multiply the denominators. If a, b, c, and d represent numbers, and b and d are not 0, we have

$$\frac{a}{b} \cdot \frac{c}{d} = \frac{a \cdot c}{b \cdot d}$$

Examples Multiply.

1. $\frac{2}{3} \cdot \frac{5}{11} = \frac{2 \cdot 5}{3 \cdot 11} = \frac{10}{33}$ Multiply numerators. Multiply denominators.

   This fraction is in simplest form since 10 and 33 have no common factors other than 1.

2. $\frac{1}{4} \cdot \frac{1}{2} = \frac{1 \cdot 1}{4 \cdot 2} = \frac{1}{8}$ This fraction is in simplest form.

Work Practice 1–2

Example 3 Multiply and simplify: $\frac{6}{7} \cdot \frac{14}{27}$

Solution:

$$\frac{6}{7} \cdot \frac{14}{27} = \frac{6 \cdot 14}{7 \cdot 27}$$

We can simplify by finding the prime factorizations and using our shortcut procedure of dividing out common factors in the numerator and denominator.

$$\frac{6 \cdot 14}{7 \cdot 27} = \frac{2 \cdot \overset{1}{\cancel{3}} \cdot 2 \cdot \overset{1}{\cancel{7}}}{\underset{1}{\cancel{7}} \cdot \underset{1}{\cancel{3}} \cdot 3 \cdot 3} = \frac{2 \cdot 2}{3 \cdot 3} = \frac{4}{9}$$

Work Practice 3

Objectives

A Multiply Fractions.

B Evaluate Exponential Expressions with Fractional Bases.

C Divide Fractions.

D Multiply and Divide Given Fractional Replacement Values.

E Solve Applications That Require Multiplication of Fractions.

Practice 1–2

Multiply.

1. $\frac{3}{7} \cdot \frac{5}{11}$ 2. $\frac{1}{3} \cdot \frac{1}{9}$

Practice 3

Multiply and simplify: $\frac{6}{77} \cdot \frac{7}{8}$

Answers

1. $\frac{15}{77}$ 2. $\frac{1}{27}$ 3. $\frac{3}{44}$

Helpful Hint

Remember that the shortcut procedure in Example 3 is the same as removing factors of 1 in the product.

$$\frac{6 \cdot 14}{7 \cdot 27} = \frac{2 \cdot 3 \cdot 2 \cdot 7}{7 \cdot 3 \cdot 3 \cdot 3} = \frac{7}{7} \cdot \frac{3}{3} \cdot \frac{2 \cdot 2}{3 \cdot 3} = 1 \cdot 1 \cdot \frac{4}{9} = \frac{4}{9}$$

Practice 4

Multiply and simplify: $\frac{4}{27} \cdot \frac{3}{8}$

Helpful Hint Don't forget that we may identify common factors that are not prime numbers.

Example 4 Multiply and simplify: $\frac{23}{32} \cdot \frac{4}{7}$

Solution: Notice that 4 and 32 have a common factor of 4.

$$\frac{23}{32} \cdot \frac{4}{7} = \frac{23 \cdot 4}{32 \cdot 7} = \frac{23 \cdot \cancel{4}^{1}}{\cancel{4}_{1} \cdot 8 \cdot 7} = \frac{23}{8 \cdot 7} = \frac{23}{56}$$

Work Practice 4

After multiplying two fractions, always check to see whether the product can be simplified.

Practice 5

Multiply.

$\frac{1}{2} \cdot \left(-\frac{11}{28}\right)$

Example 5 Multiply: $-\frac{1}{4} \cdot \frac{1}{2}$

Solution: Recall that the product of a negative number and a positive number is a negative number.

$$-\frac{1}{4} \cdot \frac{1}{2} = -\frac{1 \cdot 1}{4 \cdot 2} = -\frac{1}{8}$$

Work Practice 5

Practice 6–7

Multiply.

6. $\left(-\frac{4}{11}\right)\left(-\frac{33}{16}\right)$

7. $\frac{1}{6} \cdot \frac{3}{10} \cdot \frac{25}{16}$

Examples Multiply.

6. $\left(-\frac{6}{13}\right)\left(-\frac{26}{30}\right) = \frac{6 \cdot 26}{13 \cdot 30} = \frac{\cancel{6}^{1} \cdot \cancel{13}^{1} \cdot 2}{\cancel{13}_{1} \cdot \cancel{6}_{1} \cdot 5} = \frac{2}{5}$ The product of two negative numbers is a positive number.

7. $\frac{1}{3} \cdot \frac{2}{5} \cdot \frac{9}{16} = \frac{1 \cdot 2 \cdot 9}{3 \cdot 5 \cdot 16} = \frac{1 \cdot \cancel{2}^{1} \cdot \cancel{3}^{1} \cdot 3}{\cancel{3}_{1} \cdot 5 \cdot \cancel{2}_{1} \cdot 8} = \frac{3}{40}$

Work Practice 6–7

We multiply fractions in the same way if variables are involved.

Practice 8

Multiply: $\frac{2}{3} \cdot \frac{3y}{2}$

Example 8 Multiply: $\frac{3x}{4} \cdot \frac{8}{5x}$

Solution: Notice that 8 and 4 have a common factor of 4.

$$\frac{3x}{4} \cdot \frac{8}{5x} = \frac{3 \cdot \cancel{x}^{1} \cdot \cancel{4}^{1} \cdot 2}{\cancel{4}_{1} \cdot 5 \cdot \cancel{x}_{1}} = \frac{3 \cdot 1 \cdot 1 \cdot 2}{1 \cdot 5 \cdot 1} = \frac{6}{5}$$

Work Practice 8

Helpful Hint

Recall that when the denominator of a fraction contains a variable, such as $\frac{8}{5x}$, we assume that the variable does not represent 0.

Answers

4. $\frac{1}{18}$ 5. $-\frac{11}{56}$ 6. $\frac{3}{4}$ 7. $\frac{5}{64}$ 8. y

Example 9 Multiply: $\frac{x^2}{y} \cdot \frac{y^3}{x}$

Solution:

$$\frac{x^2}{y} \cdot \frac{y^3}{x} = \frac{x^2 \cdot y^3}{y \cdot x} = \frac{\overset{1}{\cancel{x}} \cdot x \cdot \overset{1}{\cancel{y}} \cdot y \cdot y}{\underset{1}{\cancel{y}} \cdot \underset{1}{\cancel{x}}} = \frac{x \cdot y \cdot y}{1} = xy^2$$

Work Practice 9

Practice 9

Multiply: $\frac{a^3}{b^2} \cdot \frac{b}{a^2}$

Objective B Evaluating Expressions with Fractional Bases

The base of an exponential expression can also be a fraction.

$$\left(\frac{1}{3}\right)^4 = \underbrace{\frac{1}{3} \cdot \frac{1}{3} \cdot \frac{1}{3} \cdot \frac{1}{3}}_{} = \frac{1 \cdot 1 \cdot 1 \cdot 1}{3 \cdot 3 \cdot 3 \cdot 3} = \frac{1}{81}$$

$\frac{1}{3}$ is a factor 4 times.

Example 10 Evaluate.

a. $\left(\frac{2}{5}\right)^4 = \frac{2}{5} \cdot \frac{2}{5} \cdot \frac{2}{5} \cdot \frac{2}{5} = \frac{2 \cdot 2 \cdot 2 \cdot 2}{5 \cdot 5 \cdot 5 \cdot 5} = \frac{16}{625}$

b. $\left(-\frac{1}{4}\right)^2 = \left(-\frac{1}{4}\right) \cdot \left(-\frac{1}{4}\right) = \frac{1 \cdot 1}{4 \cdot 4} = \frac{1}{16}$ The product of two negative numbers is a positive number.

Work Practice 10

Practice 10

Evaluate.

a. $\left(\frac{3}{4}\right)^3$ **b.** $\left(-\frac{4}{5}\right)^2$

Objective C Dividing Fractions

Before we can divide fractions, we need to know how to find the **reciprocal** of a fraction.

Reciprocal of a Fraction

Two numbers are **reciprocals** of each other if their product is 1. The reciprocal of the fraction $\frac{a}{b}$ is $\frac{b}{a}$ because $\frac{a}{b} \cdot \frac{b}{a} = \frac{a \cdot b}{b \cdot a} = 1$.

Helpful Hint

Every number has a reciprocal except 0. The number 0 has no reciprocal because there is no number such that $0 \cdot a = 1$.

For example,

The reciprocal of $\frac{2}{5}$ is $\frac{5}{2}$ because $\frac{2}{5} \cdot \frac{5}{2} = \frac{10}{10} = 1$.

The reciprocal of 5 is $\frac{1}{5}$ because $5 \cdot \frac{1}{5} = \frac{5}{1} \cdot \frac{1}{5} = \frac{5}{5} = 1$.

The reciprocal of $-\frac{7}{11}$ is $-\frac{11}{7}$ because $-\frac{7}{11} \cdot -\frac{11}{7} = \frac{77}{77} = 1$.

Answers

9. $\frac{a}{b}$ **10. a.** $\frac{27}{64}$ **b.** $\frac{16}{25}$

Division of fractions has the same meaning as division of whole numbers. For example,

$10 \div 5$ means: How many 5s are there in 10?

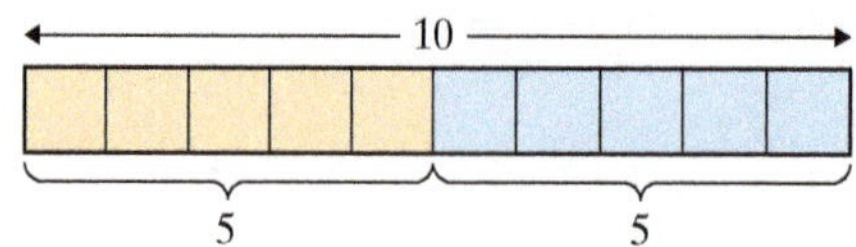

There are two 5s in 10, so $10 \div 5 = 2$.

$\frac{3}{4} \div \frac{1}{8}$ means: How many $\frac{1}{8}$s are there in $\frac{3}{4}$?

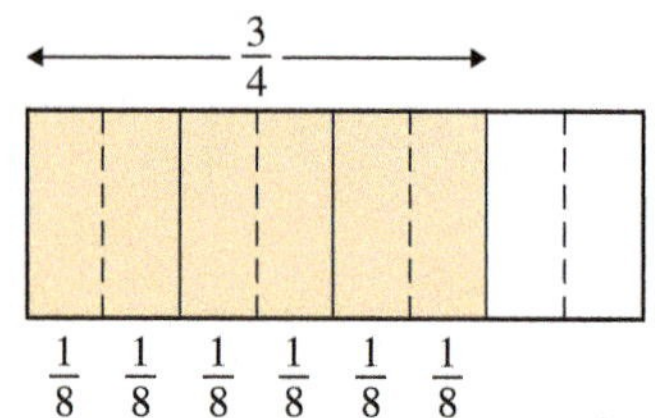

There are six $\frac{1}{8}$s in $\frac{3}{4}$, so $\frac{3}{4} \div \frac{1}{8} = 6$.

We use reciprocals to divide fractions.

Dividing Fractions

To divide two fractions, multiply the first fraction by the reciprocal of the second fraction.

If a, b, c, and d represent numbers, and b, c, and d are not 0, then

$$\frac{a}{b} \div \frac{c}{d} = \frac{a}{b} \cdot \frac{d}{c} = \frac{a \cdot d}{b \cdot c}$$

($\frac{d}{c}$ is the reciprocal of $\frac{c}{d}$)

For example,

multiply by reciprocal

$$\frac{3}{4} \div \frac{1}{8} = \frac{3}{4} \cdot \frac{8}{1} = \frac{3 \cdot 8}{4 \cdot 1} = \frac{3 \cdot 2 \cdot \cancel{4}^{1}}{\cancel{4}_{1} \cdot 1} = \frac{6}{1} \text{ or } 6$$

After dividing fractions, *always* check to see whether the result can be simplified.

Examples Divide and simplify.

11. $\frac{5}{16} \div \frac{3}{4} = \frac{5}{16} \cdot \frac{4}{3} = \frac{5 \cdot 4}{16 \cdot 3} = \frac{5 \cdot \cancel{4}^{1}}{\cancel{4}_{1} \cdot 4 \cdot 3} = \frac{5}{12}$

12. $\frac{2}{5} \div \frac{1}{2} = \frac{2}{5} \cdot \frac{2}{1} = \frac{2 \cdot 2}{5 \cdot 1} = \frac{4}{5}$

Work Practice 11–12

Practice 11–12

Divide and simplify.

11. $\frac{8}{7} \div \frac{2}{9}$ **12.** $\frac{4}{9} \div \frac{1}{2}$

Answers

11. $\frac{36}{7}$ **12.** $\frac{8}{9}$

Helpful Hint

When dividing by a fraction, do not look for common factors to divide out until you rewrite the division as multiplication.

Do not try to divide out these two 2s.

$$\frac{1}{\mathbf{2}} \div \frac{\mathbf{2}}{3} = \frac{1}{2} \cdot \frac{3}{2} = \frac{3}{4}$$

Example 13 Divide: $-\frac{7}{12} \div -\frac{5}{6}$

Solution: Recall that the quotient (or product) of two negative numbers is a positive number.

$$-\frac{7}{12} \div -\frac{5}{6} = -\frac{7}{12} \cdot -\frac{6}{5} = \frac{7 \cdot \overset{1}{\cancel{6}}}{2 \cdot \underset{1}{\cancel{6}} \cdot 5} = \frac{7}{10}$$

Work Practice 13

Practice 13

Divide: $-\frac{10}{4} \div \frac{2}{9}$

Example 14 Divide: $\frac{2x}{3} \div 3x^2$

Solution:

$$\frac{2x}{3} \div 3x^2 = \frac{2x}{3} \div \frac{3x^2}{1} = \frac{2x}{3} \cdot \frac{1}{3x^2} = \frac{2 \cdot \overset{1}{\cancel{x}} \cdot 1}{3 \cdot 3 \cdot \underset{1}{\cancel{x}} \cdot x} = \frac{2}{9x}$$

Work Practice 14

Practice 14

Divide: $\frac{3y}{4} \div 5y^3$

Example 15 Simplify: $\left(\frac{4}{7} \cdot \frac{3}{8}\right) \div -\frac{3}{4}$

Solution: Remember to perform the operations inside the ( ) first.

$$\left(\frac{4}{7} \cdot \frac{3}{8}\right) \div -\frac{3}{4} = \left(\frac{\overset{1}{\cancel{4}} \cdot 3}{7 \cdot 2 \cdot \underset{1}{\cancel{4}}}\right) \div -\frac{3}{4} = \frac{3}{14} \div -\frac{3}{4}$$

Now divide.

$$\frac{3}{14} \div -\frac{3}{4} = \frac{3}{14} \cdot -\frac{4}{3} = -\frac{\overset{1}{\cancel{3}} \cdot \overset{1}{\cancel{2}} \cdot 2}{\underset{1}{\cancel{2}} \cdot 7 \cdot \underset{1}{\cancel{3}}} = -\frac{2}{7}$$

Work Practice 15

Practice 15

Simplify: $\left(-\frac{2}{3} \cdot \frac{9}{14}\right) \div \frac{7}{15}$

✓Concept Check Which is the correct way to divide $\frac{3}{5}$ by $\frac{5}{12}$? Explain.

a. $\frac{3}{5} \div \frac{5}{12} = \frac{5}{3} \cdot \frac{5}{12}$ **b.** $\frac{3}{5} \div \frac{5}{12} = \frac{3}{5} \cdot \frac{12}{5}$

Answers

13. $-\frac{45}{4}$ **14.** $\frac{3}{20y^2}$ **15.** $-\frac{45}{49}$

✓Concept Check Answers

a. incorrect **b.** correct

Objective D Multiplying and Dividing with Fractional Replacement Values

Recall the difference between an expression and an equation. For example, xy and $x \div y$ are expressions. They contain no equal signs. In Example 16, we practice *simplifying* expressions given fractional replacement values.

Practice 16

If $x = -\frac{3}{4}$ and $y = \frac{9}{2}$, evaluate **(a)** xy and **(b)** $x \div y$.

Example 16 If $x = \frac{7}{8}$ and $y = -\frac{1}{3}$, evaluate **(a)** xy and **(b)** $x \div y$.

Solution: Replace x with $\frac{7}{8}$ and y with $-\frac{1}{3}$.

a. $xy = \frac{7}{8} \cdot -\frac{1}{3} = -\frac{7 \cdot 1}{8 \cdot 3} = -\frac{7}{24}$

b. $x \div y = \frac{7}{8} \div -\frac{1}{3} = \frac{7}{8} \cdot -\frac{3}{1} = -\frac{7 \cdot 3}{8 \cdot 1} = -\frac{21}{8}$

Work Practice 16

Practice 17

Is $-\frac{9}{8}$ a solution of the equation $2x = -\frac{9}{4}$?

Example 17 Is $-\frac{2}{3}$ a solution of the equation $-\frac{1}{2}x = \frac{1}{3}$?

Solution: To check whether a number is a solution of an equation, recall that we replace the variable with the given number and see if a true statement results.

$-\frac{1}{2} \cdot x = \frac{1}{3}$ Recall that $-\frac{1}{2}x$ means $-\frac{1}{2} \cdot x$.

$-\frac{1}{2} \cdot -\frac{2}{3} \stackrel{?}{=} \frac{1}{3}$ Replace x with $-\frac{2}{3}$.

$\frac{1 \cdot \overset{1}{\cancel{2}}}{\underset{1}{\cancel{2}} \cdot 3} \stackrel{?}{=} \frac{1}{3}$ The product of two negative numbers is a positive number.

$\frac{1}{3} = \frac{1}{3}$ True

Since we have a true statement, $-\frac{2}{3}$ is a solution.

Work Practice 17

Helpful Hint "of" usually translates to multiplication.

Objective E Solving Problems by Multiplying Fractions

To solve real-life problems that involve multiplying fractions, we use our four problem-solving steps from Chapter 3. In Example 18, a new key word that implies multiplication is used. That key word is "**of.**"

Answers

16. a. $-\frac{27}{8}$ **b.** $-\frac{1}{6}$ **17.** yes

Example 18 Finding the Number of Roller Coasters in an Amusement Park

Cedar Point is an amusement park located in Sandusky, Ohio. Its collection of 72 rides is the largest in the world. Of the rides, $\frac{2}{9}$ are roller coasters. How many roller coasters are in Cedar Point's collection of rides? (*Source:* Wikipedia)

Solution:

1. UNDERSTAND the problem. To do so, read and reread the problem. We are told that $\frac{2}{9}$ of Cedar Point's rides are roller coasters. The word "of" here means multiplication.

2. TRANSLATE.

In words:	number of roller coasters	is	$\frac{2}{9}$	of	total rides at Cedar Point
	↓	↓	↓	↓	↓
Translate:	number of roller coasters	=	$\frac{2}{9}$	·	72

3. SOLVE: Before we solve, let's estimate a reasonable answer. The fraction $\frac{2}{9}$ is less than $\frac{1}{4}$ (draw a diagram, if needed), and $\frac{1}{4}$ of 72 rides is 18 rides, so the number of roller coasters should be less than 18.

$$\frac{2}{9} \cdot 72 = \frac{2}{9} \cdot \frac{72}{1} = \frac{2 \cdot 72}{9 \cdot 1} = \frac{2 \cdot \overset{1}{\cancel{9}} \cdot 8}{\underset{1}{\cancel{9}} \cdot 1} = \frac{16}{1} \quad \text{or} \quad 16$$

4. INTERPRET. *Check* your work. From our estimate, our answer is reasonable. *State* your conclusion: The number of roller coasters at Cedar Point is 16.

Work Practice 18

Practice 18

Hershey Park is an amusement park in Hershey, Pennsylvania. Of its 66 rides, $\frac{1}{6}$ of these are roller coasters. How many roller coasters are in Hershey Park?

Helpful Hint

To help visualize a fractional part of a whole number, look at the diagram below.

$\frac{1}{5}$ of 60 = ?

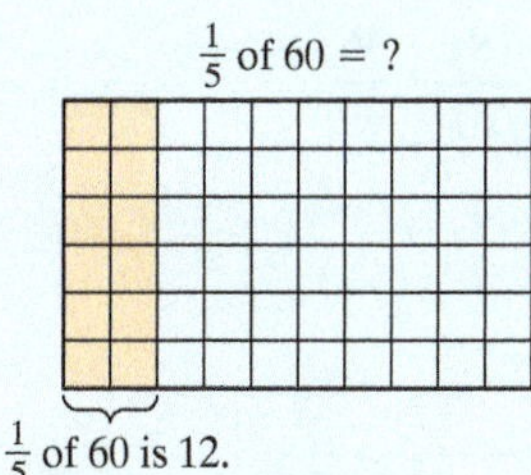

$\frac{1}{5}$ of 60 is 12.

Answer

18. 11 roller coasters

Vocabulary, Readiness & Video Check

Use the choices below to fill in each blank. Not all choices will be used.

multiplication	$\frac{a \cdot d}{b \cdot c}$	$\frac{a \cdot c}{b \cdot d}$	$\frac{2 \cdot 2 \cdot 2}{7}$	$\frac{2}{7} \cdot \frac{2}{7} \cdot \frac{2}{7}$
division	0	reciprocals		

1. To multiply two fractions, we write $\frac{a}{b} \cdot \frac{c}{d} =$ ____________.

2. Two numbers are ____________ of each other if their product is 1.

3. The expression $\frac{2^3}{7} =$ ____________ while $\left(\frac{2}{7}\right)^3 =$ ____________.

4. Every number has a reciprocal except ____________.

5. To divide two fractions, we write $\frac{a}{b} \div \frac{c}{d} =$ ____________.

6. The word "of" indicates ____________.

Martin-Gay Interactive Videos *Watch the section lecture video and answer the following questions.*

See Video 4.3 

Objective A **7.** In Example 2, how do we know that the answer is negative?

Objective B **8.** In Example 4, does the exponent apply to the negative sign? Why or why not?

Objective C **9.** Complete this statement based on Example 5: When writing the reciprocal of a fraction, the denominator becomes the ____________, and the numerator becomes the ____________.

Objective D **10.** In Example 11a, why don't we write out the prime factorizations of 4 and 9 in the numerator?

Objective E **11.** What formula is used to solve Example 12?

4.3 Exercise Set MyMathLab®

Objective A *Multiply. Write the product in simplest form. See Examples 1 through 9.*

1. $\frac{6}{11} \cdot \frac{3}{7}$

2. $\frac{5}{9} \cdot \frac{7}{4}$

3. $-\frac{2}{7} \cdot \frac{5}{8}$

4. $\frac{4}{15} \cdot -\frac{1}{20}$

5. $-\frac{1}{2} \cdot -\frac{2}{15}$

6. $-\frac{3}{11} \cdot -\frac{11}{12}$

7. $\frac{18x}{20} \cdot \frac{36}{99}$

8. $\frac{5}{32} \cdot \frac{64y}{100}$

9. $3a^2 \cdot \frac{1}{4}$

10. $-\frac{2}{3} \cdot 6y^3$

11. $\frac{x^3}{y^3} \cdot \frac{y^2}{x}$

12. $\frac{a}{b^3} \cdot \frac{b}{a^3}$

13. $0 \cdot \frac{8}{9}$

14. $\frac{11}{12} \cdot 0$

15. $-\frac{17y}{20} \cdot \frac{4}{5y}$

16. $-\frac{13x}{20} \cdot \frac{5}{6x}$

17. $\frac{11}{20} \cdot \frac{1}{7} \cdot \frac{5}{22}$

18. $\frac{27}{32} \cdot \frac{10}{13} \cdot \frac{16}{30}$

Objective B *Evaluate. See Example 10.*

19. $\left(\frac{1}{5}\right)^3$
20. $\left(\frac{8}{9}\right)^2$
21. $\left(-\frac{2}{3}\right)^2$
22. $\left(-\frac{1}{2}\right)^4$
23. $\left(-\frac{2}{3}\right)^3 \cdot \frac{1}{2}$
24. $\left(-\frac{3}{4}\right)^3 \cdot \frac{1}{3}$

Objective C *Divide. Write all quotients in simplest form. See Examples 11 through 14.*

25. $\frac{2}{3} \div \frac{5}{6}$
26. $\frac{5}{8} \div \frac{3}{4}$
27. $-\frac{6}{15} \div \frac{12}{5}$
28. $-\frac{4}{15} \div -\frac{8}{3}$
29. $-\frac{8}{9} \div \frac{x}{2}$
30. $\frac{10}{11} \div -\frac{4}{5x}$
31. $\frac{11y}{20} \div \frac{3}{11}$
32. $\frac{9z}{20} \div \frac{2}{9}$
33. $-\frac{2}{3} \div 4$
34. $-\frac{5}{6} \div 10$
35. $\frac{1}{5x} \div \frac{5}{x^2}$
36. $\frac{3}{y^2} \div \frac{9}{y^3}$

Objectives A B C **Mixed Practice** *Perform each indicated operation. See Examples 1 through 15.*

37. $\frac{2}{3} \cdot \frac{5}{9}$
38. $\frac{8}{15} \cdot \frac{5}{32}$
39. $\frac{3x}{7} \div \frac{5}{6x}$
40. $\frac{2}{5y} \div \frac{5y}{11}$
41. $\frac{16}{27y} \div \frac{8}{15y}$
42. $\frac{12y}{21} \div \frac{4y}{7}$
43. $-\frac{5}{28} \cdot \frac{35}{25}$
44. $\frac{24}{45} \cdot -\frac{5}{8}$
45. $\left(-\frac{3}{4}\right)^2$
46. $\left(-\frac{1}{2}\right)^5$
47. $\frac{x^2}{y} \cdot \frac{y^3}{x}$
48. $\frac{b}{a^2} \cdot \frac{a^3}{b^3}$
49. $7 \div \frac{2}{11}$
50. $-100 \div \frac{1}{2}$
51. $-3x \div \frac{x^2}{12}$
52. $-7x^2 \div \frac{14x}{3}$
53. $\left(\frac{2}{7} \div \frac{7}{2}\right) \cdot \frac{3}{4}$
54. $\frac{1}{2} \cdot \left(\frac{5}{6} \div \frac{1}{12}\right)$
55. $-\frac{19}{63y} \cdot 9y^2$
56. $16a^2 \cdot -\frac{31}{24a}$
57. $-\frac{2}{3} \cdot -\frac{6}{11}$
58. $-\frac{1}{5} \cdot -\frac{6}{7}$
59. $\frac{4}{8} \div \frac{3}{16}$
60. $\frac{9}{2} \div \frac{16}{15}$
61. $\frac{21x^2}{10y} \div \frac{14x}{25y}$
62. $\frac{17y^2}{24x} \div \frac{13y}{18x}$
63. $\left(1 \div \frac{3}{4}\right) \cdot \frac{2}{3}$
64. $\left(33 \div \frac{2}{11}\right) \cdot \frac{5}{9}$
65. $\frac{a^3}{2} \div 30a^3$
66. $15c^3 \div \frac{3c^2}{5}$
67. $\frac{ab^2}{c} \cdot \frac{c}{ab}$
68. $\frac{ac}{b} \cdot \frac{b^3}{a^2c}$
69. $\left(\frac{1}{2} \cdot \frac{2}{3}\right) \div \frac{5}{6}$
70. $\left(\frac{3}{4} \cdot \frac{8}{9}\right) \div \frac{2}{5}$
71. $-\frac{4}{7} \div \left(\frac{4}{5} \cdot \frac{3}{7}\right)$
72. $\frac{5}{8} \div \left(\frac{4}{7} \cdot -\frac{5}{16}\right)$

Objective D *Given the following replacement values, evaluate (a) xy and (b) x ÷ y. See Example 16.*

73. $x = \frac{2}{5}$ and $y = \frac{5}{6}$

74. $x = \frac{8}{9}$ and $y = \frac{1}{4}$

75. $x = -\frac{4}{5}$ and $y = \frac{9}{11}$

76. $x = \frac{7}{6}$ and $y = -\frac{1}{2}$

Determine whether the given replacement values are solutions of the given equations. See Example 17.

77. Is $-\frac{5}{18}$ a solution to $3x = -\frac{5}{6}$?

78. Is $\frac{9}{11}$ a solution to $\frac{2}{3}y = \frac{6}{11}$?

79. Is $\frac{2}{5}$ a solution to $-\frac{1}{2}z = \frac{1}{10}$?

80. Is $\frac{3}{5}$ a solution to $5x = \frac{1}{3}$?

Objective E **Translating** *Solve. Write each answer in simplest form. For Exercises 81 through 84, recall that "of" translates to multiplication. See Example 18.*

81. Find $\frac{1}{4}$ of 200.

82. Find $\frac{1}{5}$ of 200.

83. Find $\frac{5}{6}$ of 24.

84. Find $\frac{5}{8}$ of 24.

Solve. For Exercises 85 and 86, the solutions have been started for you. See Example 18.

85. In the United States, $\frac{7}{50}$ of college freshmen major in business. A community college in Pennsylvania has a freshman enrollment of approximately 800 students. How many of these freshmen might we project are majoring in business?

Start the solution:

1. UNDERSTAND the problem. Reread it as many times as needed.
2. TRANSLATE into an equation. (Fill in the blank below.)

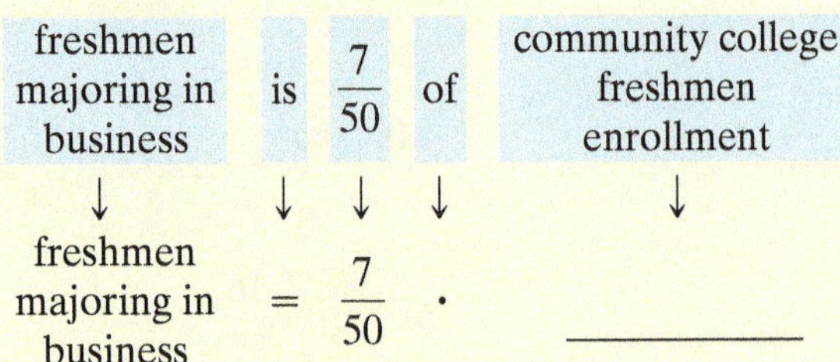

Finish with:

3. SOLVE
4. INTERPRET

86. A patient was told that, at most, $\frac{1}{5}$ of his calories should come from fat. If his diet consists of 3000 calories a day, find the maximum number of calories that can come from fat.

Start the solution:

1. UNDERSTAND the problem. Reread it as many times as needed.
2. TRANSLATE into an equation. (Fill in the blank below.)

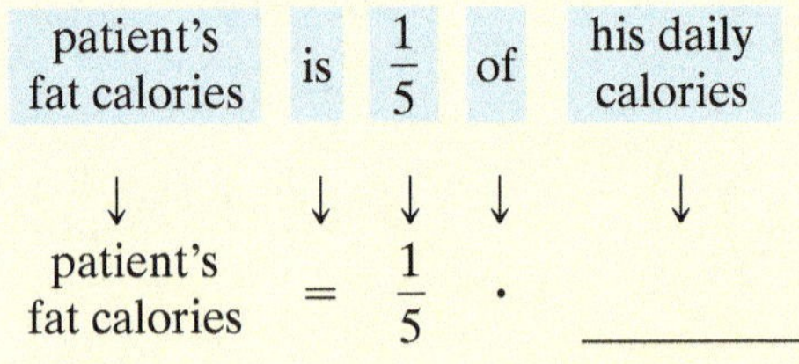

Finish with:

3. SOLVE
4. INTERPRET

87. In 2012, there were approximately 225 million moviegoers in the United States and Canada. Of these, about $\frac{12}{25}$ viewed at least one 3-D movie. Find the approximate number of people who viewed at least one 3-D movie. (*Source:* Motion Picture Association of America)

88. In a recent year, movie theater owners received a total of \$7660 million in movie admission tickets. About $\frac{7}{10}$ of this amount was for R-rated movies. Find the amount of money received from R-rated movies. (*Source:* Motion Picture Association of America)

89. The Oregon National Historic Trail is 2170 miles long. It begins in Independence, Missouri, and ends in Oregon City, Oregon. Manfred Coulon has hiked $\frac{2}{5}$ of the trail before. How many miles has he hiked? (*Source:* National Park Service)

90. Each turn of a screw sinks it $\frac{3}{16}$ of an inch deeper into a piece of wood. Find how deep the screw is after 8 turns.

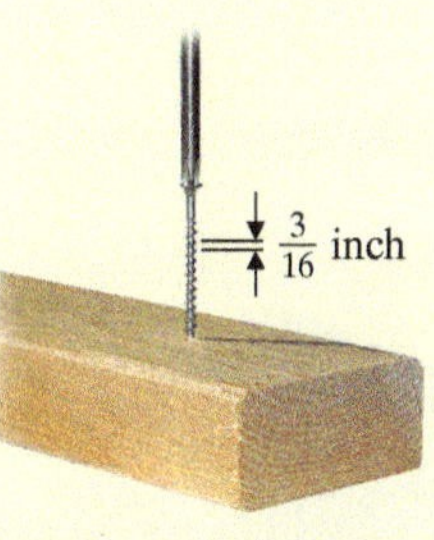

91. The radius of a circle is one-half of its diameter, as shown. If the diameter of a circle is $\frac{3}{8}$ of an inch, what is its radius?

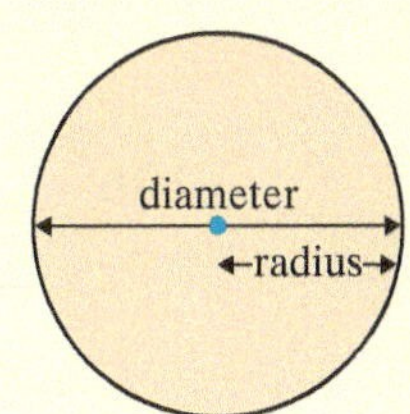

△ **92.** The diameter of a circle is twice its radius, as shown in the Exercise 91 illustration. If the radius of a circle is $\frac{7}{20}$ of a foot, what is its diameter?

93. A special on a cruise to the Bahamas is advertised to be $\frac{2}{3}$ of the regular price. If the regular price is \$2757, what is the sale price?

94. A family recently sold their house for \$102,000, but $\frac{3}{50}$ of this amount goes to the real estate companies that helped them sell their house. How much money does the family pay to the real estate companies?

95. The state of Mississippi houses $\frac{1}{184}$ of the total U.S. libraries. If there are about 9200 libraries in the United States, how many libraries are in Mississippi?

96. There have been about 410 contestants on the reality television show *Survivor* over 27 seasons. Some of these contestants have appeared in multiple seasons. If the number of repeat contestants is $\frac{6}{41}$ of the total number of participants in the first 27 seasons, how many contestants have participated more than once? (*Source:* Survivor.com)

Find the area of each rectangle. Recall that area = length · width.

△ **97.**

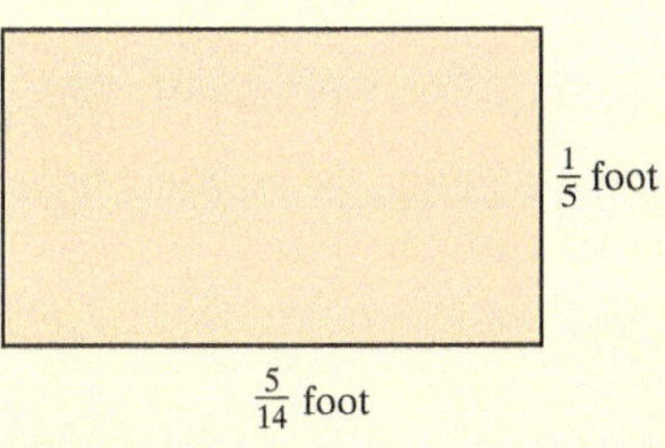

△ **98.**

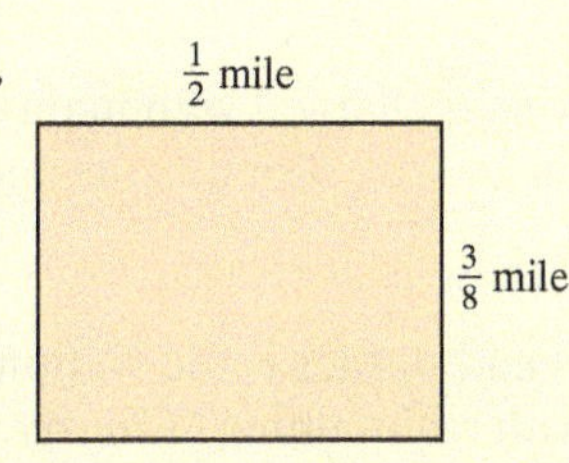

Recall from Section 4.2 that the following graph is called a **circle graph** *or* **pie chart**. *Each sector (shaped like a piece of pie) shows the fractional part of a car's total mileage that falls into a particular category. The whole circle represents a car's total mileage.*

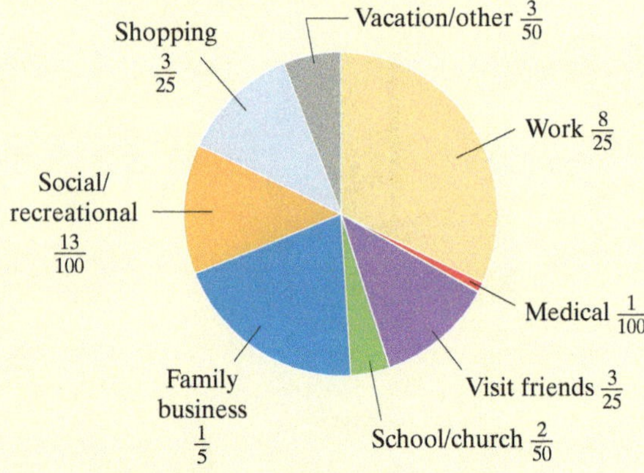

Source: The American Automobile Manufacturers Association and The National Automobile Dealers Association

In one year, a family drove 12,000 miles in the family car. Use the circle graph to determine how many of these miles might be expected to fall in the categories shown in Exercises 99 through 102.

99. Work

100. Shopping

101. Family business

102. Medical

Review

Perform each indicated operation. See Section 1.3.

103. $\begin{array}{r} 27 \\ 76 \\ +\ 98 \\ \hline \end{array}$

104. $\begin{array}{r} 811 \\ 42 \\ +\ 69 \\ \hline \end{array}$

105. $\begin{array}{r} 968 \\ -\ 772 \\ \hline \end{array}$

106. $\begin{array}{r} 882 \\ -\ 773 \\ \hline \end{array}$

Concept Extensions

107. In your own words, describe how to divide fractions.

108. In your own words, explain how to multiply fractions.

Simplify.

109. $\frac{42}{25} \cdot \frac{125}{36} \div \frac{7}{6}$

110. $\left(\frac{8}{13} \cdot \frac{39}{16} \cdot \frac{8}{9}\right)^2 \div \frac{1}{2}$

111. Approximately $\frac{1}{8}$ of the U.S. population lives in the state of California. If the U.S. population is approximately 313,914,000, find the approximate population of California. (*Source:* U.S. Census Bureau)

112. In 2012, there were approximately 11,430 commercial radio stations broadcasting in the United States. Of these, approximately $\frac{9}{51}$ were country stations. How many radio stations were country stations in 2012? (Round to the nearest whole.) (*Source:* Federal Communications Commission)

113. The National Park Service is charged with maintaining 27,000 historic structures. Monuments and statues make up $\frac{63}{200}$ of these historic structures. How many monuments and statues is the National Park Service charged with maintaining? (*Source:* National Park Service)

114. If $\frac{3}{4}$ of 36 students on a first bus are girls and $\frac{2}{3}$ of the 30 students on a second bus are *boys*, how many students on the two buses are girls?

4.4 Adding and Subtracting Like Fractions, Least Common Denominator, and Equivalent Fractions

Fractions with the same denominator are called **like fractions.** Fractions that have different denominators are called **unlike fractions.**

Like Fractions	Unlike Fractions
$\frac{2}{5}$ and $\frac{3}{5}$ ← same denominator	$\frac{2}{5}$ and $\frac{3}{4}$ ← different denominators
$\frac{5}{21}, \frac{16}{21}$, and $\frac{7}{21}$ ← same denominator	$\frac{5}{7}$ and $\frac{5}{9}$ ← different denominators

Objectives

A Add or Subtract Like Fractions.

B Add or Subtract Given Fractional Replacement Values.

C Solve Problems by Adding or Subtracting Like Fractions.

D Find the Least Common Denominator of a List of Fractions.

E Write Equivalent Fractions.

Objective A Adding or Subtracting Like Fractions

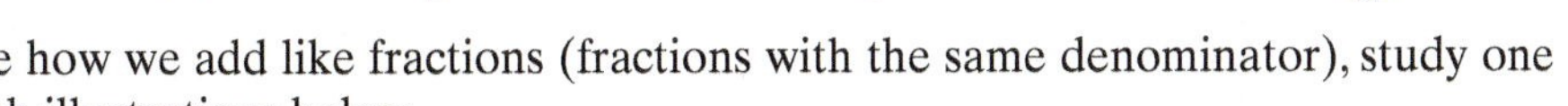

To see how we add like fractions (fractions with the same denominator), study one or both illustrations below.

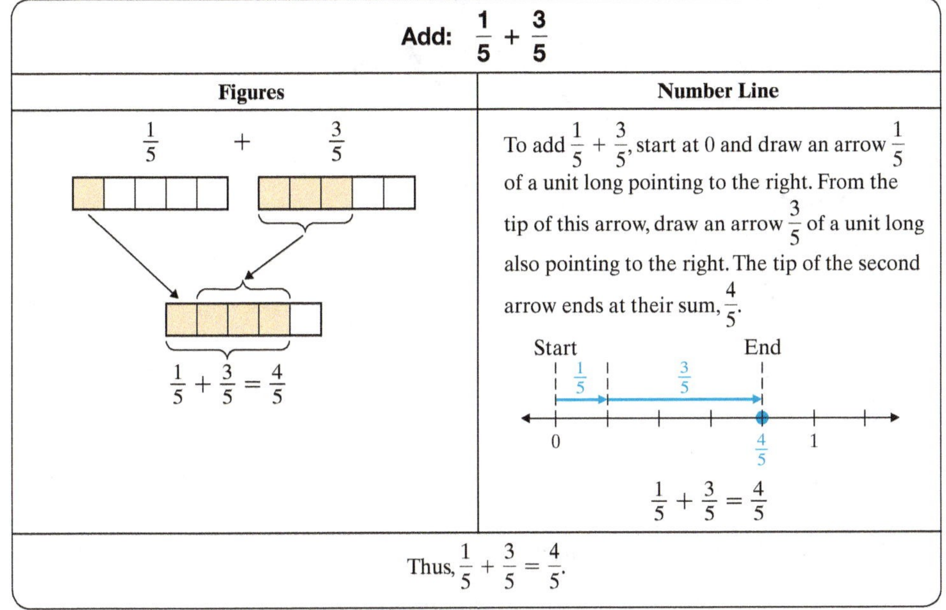

Notice that the numerator of the sum is the sum of the numerators. Also, the denominator of the sum is the **common denominator.** This is how we add fractions. Similar illustrations can be shown for subtracting fractions.

Adding or Subtracting Like Fractions (Fractions with the Same Denominator)

If a, b, and c are numbers and b is not 0, then

$$\frac{a}{b} + \frac{c}{b} = \frac{a+c}{b} \quad \text{and also} \quad \frac{a}{b} - \frac{c}{b} = \frac{a-c}{b}$$

In other words, to add or subtract fractions with the same denominator, add or subtract their numerators and write the sum or difference over the **common** denominator.

For example,

$$\frac{1}{4}+\frac{2}{4}=\frac{1+2}{4}=\frac{3}{4}$$ Add the numerators. Keep the denominator.

$$\frac{4}{5}-\frac{2}{5}=\frac{4-2}{5}=\frac{2}{5}$$ Subtract the numerators. Keep the denominator.

Helpful Hint

As usual, don't forget to write all answers in simplest form.

Examples Add and simplify.

1. $\frac{2}{7}+\frac{3}{7}=\frac{2+3}{7}=\frac{5}{7}$ ← Add the numerators. ← Keep the common denominator.

2. $\frac{3}{16x}+\frac{7}{16x}=\frac{3+7}{16x}=\frac{10}{16x}=\frac{\overset{1}{\cancel{2}}\cdot 5}{\underset{1}{\cancel{2}}\cdot 8\cdot x}=\frac{5}{8x}$

3. $\frac{7}{8}+\frac{6}{8}+\frac{3}{8}=\frac{7+6+3}{8}=\frac{16}{8}$ or 2

Work Practice 1–3

✓Concept Check Find and correct the error in the following:

$$\frac{1}{5}+\frac{1}{5}=\frac{2}{10}$$

Examples Subtract and simplify.

4. $\frac{8}{9}-\frac{1}{9}=\frac{8-1}{9}=\frac{7}{9}$ ← Subtract the numerators. ← Keep the common denominator.

5. $\frac{7}{8}-\frac{5}{8}=\frac{7-5}{8}=\frac{2}{8}=\frac{\overset{1}{\cancel{2}}}{\underset{1}{\cancel{2}}\cdot 4}=\frac{1}{4}$

Work Practice 4–5

From our earlier work, we know that

$$\frac{-12}{6}=\frac{12}{-6}=-\frac{12}{6}$$ since these all simplify to -2.

In general, the following is true:

$$\frac{-a}{b}=\frac{a}{-b}=-\frac{a}{b}$$ as long as b is not 0.

Example 6 Add: $-\frac{11}{8}+\frac{6}{8}$

Solution: $-\frac{11}{8}+\frac{6}{8}=\frac{-11+6}{8}$

$$=\frac{-5}{8} \text{ or } -\frac{5}{8}$$

Work Practice 6

Practice 1–3

Add and simplify.

1. $\frac{6}{13}+\frac{2}{13}$
2. $\frac{5}{8x}+\frac{1}{8x}$
3. $\frac{20}{11}+\frac{6}{11}+\frac{7}{11}$

Practice 4–5

Subtract and simplify.

4. $\frac{11}{12}-\frac{6}{12}$
5. $\frac{7}{15}-\frac{2}{15}$

Practice 6

Add: $-\frac{8}{17}+\frac{4}{17}$

Answers

1. $\frac{8}{13}$ 2. $\frac{3}{4x}$ 3. 3 4. $\frac{5}{12}$ 5. $\frac{1}{3}$
6. $-\frac{4}{17}$

✓Concept Check Answer

We don't add denominators together; correct solution: $\frac{1}{5}+\frac{1}{5}=\frac{2}{5}$.

Example 7 Subtract: $\frac{3x}{4} - \frac{7}{4}$

Solution: $\frac{3x}{4} - \frac{7}{4} = \frac{3x - 7}{4}$

Recall from Section 3.1 that the terms in the numerator are unlike terms and cannot be combined.

Work Practice 7

Example 8 Subtract: $\frac{3}{7} - \frac{6}{7} - \frac{3}{7}$

Solution: $\frac{3}{7} - \frac{6}{7} - \frac{3}{7} = \frac{3 - 6 - 3}{7} = \frac{-6}{7}$ or $-\frac{6}{7}$

Work Practice 8

Helpful Hint

Recall that $\frac{-6}{7} = -\frac{6}{7}$ $\left(\text{Also, } \frac{6}{-7} = -\frac{6}{7}\text{, if needed.}\right)$

Objective B Adding or Subtracting Given Fractional Replacement Values

Example 9 Evaluate $y - x$ if $x = -\frac{3}{10}$ and $y = -\frac{8}{10}$.

Solution: Be very careful when replacing x and y with replacement values.

$$y - x = -\frac{8}{10} - \left(-\frac{3}{10}\right) \quad \text{Replace } x \text{ with } -\frac{3}{10} \text{ and } y \text{ with } -\frac{8}{10}.$$

$$= \frac{-8 - (-3)}{10}$$

$$= \frac{-5}{10} = \frac{-1 \cdot 5}{2 \cdot 5} = \frac{-1}{2} \text{ or } -\frac{1}{2}$$

Work Practice 9

Concept Check Fill in each blank with the best choice given.

expression equation simplified solved

A(n) ________ contains an equal sign and may be ________ for the variable.

A(n) ________ does not contain an equal sign but may be ________.

Objective C Solving Problems by Adding or Subtracting Like Fractions

Many real-life problems involve finding the perimeters of square or rectangular-shaped figures such as pastures, swimming pools, and so on. We can use our knowledge of adding fractions to find perimeters.

Practice 7

Subtract: $\frac{2}{5} - \frac{7y}{5}$

Practice 8

Subtract: $\frac{4}{11} - \frac{6}{11} - \frac{3}{11}$

Practice 9

Evaluate $x + y$ if $x = -\frac{10}{12}$ and $y = \frac{5}{12}$.

Answers

7. $\frac{2 - 7y}{5}$ 8. $-\frac{5}{11}$ 9. $-\frac{5}{12}$

Concept Check Answer

equation; solved; expression; simplified

Practice 10

Find the perimeter of the square.

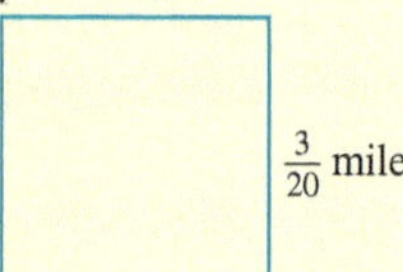

Example 10

Find the perimeter of the rectangle.

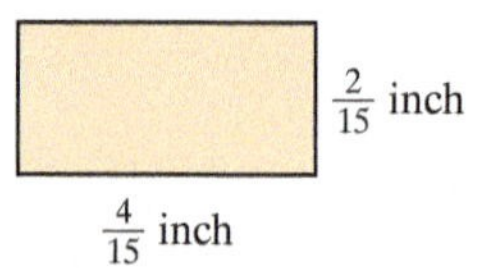

Solution: Recall that perimeter means distance around and that opposite sides of a rectangle are the same length.

$$\text{Perimeter} = \frac{2}{15} + \frac{4}{15} + \frac{2}{15} + \frac{4}{15} = \frac{2+4+2+4}{15}$$

$$= \frac{12}{15} = \frac{\overset{1}{\cancel{3}} \cdot 4}{\underset{1}{\cancel{3}} \cdot 5} = \frac{4}{5}$$

The perimeter of the rectangle is $\frac{4}{5}$ inch.

Work Practice 10

We can combine our skills in adding and subtracting fractions with our four problem-solving steps from Section 3.4 to solve many kinds of real-life problems.

Practice 11

A jogger ran $\frac{13}{4}$ miles on Monday and $\frac{11}{4}$ miles on Wednesday. How much farther did he run on Monday than on Wednesday?

Example 11 Calculating Distance

The distance from home to the World Gym is $\frac{7}{8}$ of a mile and from home to the post office is $\frac{3}{8}$ of a mile. How much farther is it from home to the World Gym than from home to the post office?

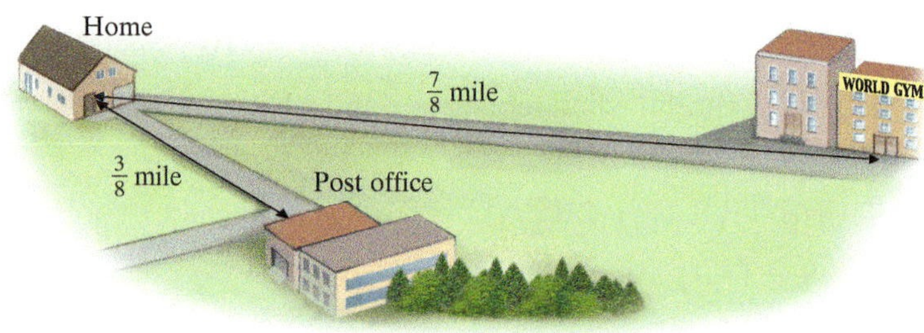

Solution:

1. UNDERSTAND. Read and reread the problem. The phrase "How much farther" tells us to subtract distances.
2. TRANSLATE.

In words:	distance farther	is	home to World Gym distance	minus	home to post office distance
	↓	↓	↓	↓	↓
Translate:	distance farther	=	$\frac{7}{8}$	−	$\frac{3}{8}$

3. SOLVE: $\frac{7}{8} - \frac{3}{8} = \frac{7-3}{8} = \frac{4}{8} = \frac{\overset{1}{\cancel{4}}}{2 \cdot \underset{1}{\cancel{4}}} = \frac{1}{2}$
4. INTERPRET. *Check* your work. *State* your conclusion: The distance from home to the World Gym is $\frac{1}{2}$ mile farther than from home to the post office.

Work Practice 11

Answers

10. $\frac{3}{5}$ mi **11.** $\frac{1}{2}$ mi

Objective D Finding the Least Common Denominator 

In the next section, we will add and subtract fractions that have different, or unlike, denominators. To do so, we first write them as equivalent fractions with a common denominator.

Although any common denominator can be used to add or subtract unlike fractions, we will use the **least common denominator (LCD).** The LCD of a list of fractions is the same as the **least common multiple (LCM)** of the denominators. Why do we use this number as the common denominator? Since the LCD is the *smallest* of all common denominators, operations are usually less tedious with this number.

The **least common denominator (LCD)** of a list of fractions is the smallest positive number divisible by all the denominators in the list. (The least common denominator is also the **least common multiple (LCM) of the denominators.**)

For example, the LCD of $\frac{1}{4}$ and $\frac{3}{10}$ is 20 because 20 is the smallest positive number divisible by both 4 and 10.

Finding the LCD: Method 1

One way to find the LCD is to see whether the larger denominator is divisible by the smaller denominator. If so, the larger number is the LCD. If not, then check consecutive multiples of the larger denominator until the LCD is found.

Method 1: Finding the LCD of a List of Fractions Using Multiples of the Largest Number

Step 1: Write the multiples of the largest denominator (starting with the number itself) until a multiple common to all denominators in the list is found.

Step 2: The multiple found in Step 1 is the LCD.

Example 12 Find the LCD of $\frac{3}{7}$ and $\frac{5}{14}$.

Solution: We write the multiples of 14 until we find one that is also a multiple of 7.

$14 \cdot 1 = 14$ A multiple of 7

The LCD is 14.

Work Practice 12

Practice 12

Find the LCD of $\frac{7}{8}$ and $\frac{11}{16}$.

Example 13 Find the LCD of $\frac{11}{12}$ and $\frac{7}{20}$.

Solution: We write the multiples of 20 until we find one that is also a multiple of 12.

$20 \cdot 1 = 20$ Not a multiple of 12

$20 \cdot 2 = 40$ Not a multiple of 12

$20 \cdot 3 = 60$ A multiple of 12

The LCD is 60.

Work Practice 13

Practice 13

Find the LCD of $\frac{23}{25}$ and $\frac{1}{30}$.

Answers

12. 16 **13.** 150

Method 1 for finding multiples works fine for smaller numbers, but may get tedious for larger numbers. For this reason, let's study a second method, which uses prime factorization.

Finding the LCD: Method 2

For example, to find the LCD of $\frac{11}{12}$ and $\frac{7}{20}$, such as in Example 13, let's look at the prime factorization of each denominator.

$$12 = 2 \cdot 2 \cdot 3$$
$$20 = 2 \cdot 2 \cdot 5$$

Recall that the LCD must be a multiple of both 12 and 20. Thus, to build the LCD, we will circle the greatest number of factors for each different prime number. The LCD is the product of the circled factors.

Prime Number Factors

$$12 = \boxed{2 \cdot 2} \quad \boxed{3}$$
$$20 = 2 \cdot 2 \quad \quad \boxed{5}$$

Circle either pair of 2s, but not both.

$$\text{LCD} = 2 \cdot 2 \cdot 3 \cdot 5 = 60$$

The number 60 is the smallest number that both 12 and 20 divide into evenly. This method is summarized below:

Method 2: Finding the LCD of a List of Denominators Using Prime Factorization

Step 1: Write the prime factorization of each denominator.

Step 2: For each different prime factor in Step 1, circle the *greatest* number of times that factor occurs in any one factorization.

Step 3: The LCD is the product of the circled factors.

Practice 14

Find the LCD of $-\frac{3}{40}$ and $\frac{11}{108}$.

Helpful Hint If you prefer working with exponents, circle the factor with the greatest exponent.

Example 14:

$$72 = \boxed{2^3} \cdot \boxed{3^2}$$
$$60 = 2^2 \cdot 3 \cdot \boxed{5}$$
$$\text{LCD} = 2^3 \cdot 3^2 \cdot 5 = 360$$

Answer

14. 1080

Example 14 Find the LCD of $-\frac{23}{72}$ and $\frac{17}{60}$.

Solution: First we write the prime factorization of each denominator.

$$72 = 2 \cdot 2 \cdot 2 \cdot 3 \cdot 3$$
$$60 = 2 \cdot 2 \cdot 3 \cdot 5$$

For the prime factors shown, we circle the greatest number of factors found in either factorization.

$$72 = \boxed{2 \cdot 2 \cdot 2} \cdot \boxed{3 \cdot 3}$$
$$60 = 2 \cdot 2 \cdot 3 \cdot \boxed{5}$$

The LCD is the product of the circled factors.

$$\text{LCD} = 2 \cdot 2 \cdot 2 \cdot 3 \cdot 3 \cdot 5 = 360$$

The LCD is 360.

Work Practice 14

Helpful Hint

If the number of factors of a prime number is equal, circle either one, but not both. For example,

$$12 = 2 \cdot 2 \cdot 3$$
$$15 = 3 \cdot 5$$

Circle either 3 but not both.

The LCD is $2 \cdot 2 \cdot 3 \cdot 5 = 60$.

Example 15 Find the LCD of $\frac{1}{15}, \frac{5}{18}$, and $\frac{53}{54}$.

Solution:
$$15 = 3 \cdot 5$$
$$18 = 2 \cdot 3 \cdot 3$$
$$54 = 2 \cdot 3 \cdot 3 \cdot 3$$

The LCD is $2 \cdot 3 \cdot 3 \cdot 3 \cdot 5$ or 270.

Work Practice 15

Practice 15

Find the LCD of $\frac{7}{20}, \frac{1}{24}$, and $\frac{13}{45}$.

Example 16 Find the LCD of $\frac{3}{5}, \frac{2}{x}$, and $\frac{7}{x^3}$.

Solution:
$$5 = 5$$
$$x = x$$
$$x^3 = x \cdot x \cdot x$$

$\text{LCD} = 5 \cdot x \cdot x \cdot x = 5x^3$

Work Practice 16

Practice 16

Find the LCD of $\frac{7}{y}$ and $\frac{6}{11}$.

✓Concept Check True or false? The LCD of the fractions $\frac{1}{6}$ and $\frac{1}{8}$ is 48.

Objective E Writing Equivalent Fractions

To add or subtract unlike fractions in the next section, we first write equivalent fractions with the LCD as the denominator.

To write $\frac{1}{3}$ as an equivalent fraction with a denominator of 6, we multiply by 1 in the form of $\frac{2}{2}$. Why? Because $3 \cdot 2 = 6$, so the new denominator will become 6, as shown below.

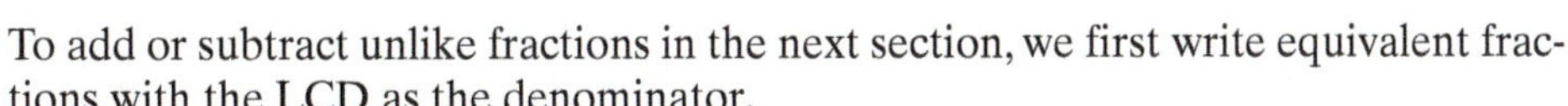

$$\frac{1}{3} = \frac{1}{3} \cdot 1 = \frac{1}{3} \cdot \frac{2}{2} = \frac{1 \cdot 2}{3 \cdot 2} = \frac{2}{6}$$

$\frac{2}{2} = 1$

So $\frac{1}{3} = \frac{2}{6}$.

Helpful Hint Recall from the Helpful Hint on p. 230, that this is also called the Fundamental Property of Fractions.

$$\frac{a}{b} = \frac{a}{b} \cdot \frac{c}{c} = \frac{a \cdot c}{b \cdot c}$$

To write an equivalent fraction,

$$\frac{a}{b} = \frac{a}{b} \cdot \frac{c}{c} = \frac{a \cdot c}{b \cdot c}$$

where a, b, and c are nonzero numbers.

Answers

15. 360 **16.** $11y$

✓Concept Check Answer

false; it is 24

✓**Concept Check** Which of the following is *not* equivalent to $\frac{3}{4}$?

a. $\frac{6}{8}$ **b.** $\frac{18}{24}$ **c.** $\frac{9}{14}$ **d.** $\frac{30}{40}$

Practice 17

Write $\frac{7}{8}$ as an equivalent fraction with a denominator of 56.

$$\frac{7}{8} = \frac{}{56}$$

Example 17

Write $\frac{3}{4}$ as an equivalent fraction with a denominator of 20.

$$\frac{3}{4} = \frac{}{20}$$

Solution: In the denominators, since $4 \cdot 5 = 20$, we will multiply by 1 in the form of $\frac{5}{5}$.

$$\frac{3}{4} = \frac{3}{4} \cdot \frac{5}{5} = \frac{3 \cdot 5}{4 \cdot 5} = \frac{15}{20}$$

Thus, $\frac{3}{4} = \frac{15}{20}$.

Work Practice 17

Helpful Hint

To check Example 17, write $\frac{15}{20}$ in simplest form.

$$\frac{15}{20} = \frac{3 \cdot \overset{1}{\cancel{5}}}{4 \cdot \underset{1}{\cancel{5}}} = \frac{3}{4}, \text{ the original fraction.}$$

If the original fraction is in lowest terms, we can check our work by writing the new, equivalent fraction in simplest form. This should be the original fraction.

✓**Concept Check** True or false? When the fraction $\frac{2}{9}$ is rewritten as an equivalent fraction with 27 as the denominator, the result is $\frac{2}{27}$.

Practice 18

Write an equivalent fraction with the given denominator.

$$\frac{1}{4} = \frac{}{20}$$

Example 18

Write an equivalent fraction with the given denominator.

$$\frac{2}{5} = \frac{}{15}$$

Solution: Since $5 \cdot 3 = 15$, we multiply by 1 in the form of $\frac{3}{3}$.

$$\frac{2}{5} = \frac{2}{5} \cdot \frac{3}{3} = \frac{2 \cdot 3}{5 \cdot 3} = \frac{6}{15}$$

Then $\frac{2}{5}$ is equivalent to $\frac{6}{15}$. They both represent the same part of a whole.

Work Practice 18

Answers

17. $\frac{49}{56}$ **18.** $\frac{5}{20}$

✓**Concept Check Answers**

c

false; the correct result would be $\frac{6}{27}$

Example 19 Write an equivalent fraction with the given denominator.

$$\frac{9x}{11} = \frac{}{44}$$

Solution: Since $11 \cdot 4 = 44$, we multiply by 1 in the form of $\frac{4}{4}$.

$$\frac{9x}{11} = \frac{9x}{11} \cdot \frac{4}{4} = \frac{9x \cdot 4}{11 \cdot 4} = \frac{36x}{44}$$

Then $\frac{9x}{11}$ is equivalent to $\frac{36x}{44}$.

Work Practice 19

Practice 19

Write an equivalent fraction with the given denominator.

$$\frac{3x}{7} = \frac{}{42}$$

Example 20 Write an equivalent fraction with the given denominator.

$$3 = \frac{}{7}$$

Solution: Recall that $3 = \frac{3}{1}$. Since $1 \cdot 7 = 7$, multiply by 1 in the form $\frac{7}{7}$.

$$\frac{3}{1} = \frac{3}{1} \cdot \frac{7}{7} = \frac{3 \cdot 7}{1 \cdot 7} = \frac{21}{7}$$

Work Practice 20

Practice 20

Write an equivalent fraction with the given denominator.

$$4 = \frac{}{6}$$

Don't forget that when the denominator of a fraction contains a variable, such as $\frac{8}{3x}$, we will assume that the variable does not represent 0. Recall that the denominator of a fraction cannot be 0.

Example 21 Write an equivalent fraction with the given denominator.

$$\frac{8}{3x} = \frac{}{24x}$$

Solution: Since $3x \cdot 8 = 24x$, multiply by 1 in the form $\frac{8}{8}$.

$$\frac{8}{3x} = \frac{8}{3x} \cdot \frac{8}{8} = \frac{8 \cdot 8}{3x \cdot 8} = \frac{64}{24x}$$

Work Practice 21

Practice 21

Write an equivalent fraction with the given denominator.

$$\frac{9}{4x} = \frac{}{36x}$$

✓**Concept Check** What is the first step in writing $\frac{3}{10}$ as an equivalent fraction whose denominator is 100?

Answers

19. $\frac{18x}{42}$ **20.** $\frac{24}{6}$ **21.** $\frac{81}{36x}$

✓**Concept Check Answer**

answers may vary

Vocabulary, Readiness & Video Check

Use the choices below to fill in each blank. Not all choices will be used.

least common denominator (LCD)	like	$-\frac{a}{b}$	$\frac{a-c}{b}$	$\frac{a+c}{b}$	$-\frac{a}{-b}$
perimeter	unlike				
equivalent					

1. The fractions $\frac{9}{11}$ and $\frac{13}{11}$ are called ________ fractions while $\frac{3}{4}$ and $\frac{1}{3}$ are called ________ fractions.
2. $\frac{a}{b} + \frac{c}{b} =$ ________ and $\frac{a}{b} - \frac{c}{b} =$ ________.
3. As long as b is not 0, $\frac{-a}{b} = \frac{a}{-b} =$ ________.
4. The distance around a figure is called its ________.
5. The smallest positive number divisible by all the denominators of a list of fractions is called the ________.
6. Fractions that represent the same portion of a whole are called ________ fractions.

Martin-Gay Interactive Videos — *Watch the section lecture video and answer the following questions.*

See Video 4.4

Objective A 7. Complete this statement based on the lecture before Example 1: To add like fractions, we add the ________ and keep the same ________.

Objective B 8. In Example 6, why are we told to be careful when substituting the replacement value for y?

Objective C 9. What is the perimeter equation used to solve Example 7? What is the final answer?

Objective D 10. In Example 8, the LCD is found to be 45. What does this mean in terms of the specific fractions in the problem?

Objective E 11. From Example 10, why can we multiply a fraction by a form of 1 to get an equivalent fraction?

4.4 Exercise Set MyMathLab®

Objective A *Add and simplify. See Examples 1 through 3, and 6.*

1. $\frac{5}{11} + \frac{2}{11}$
2. $\frac{9}{17} + \frac{2}{17}$
3. $\frac{2}{9} + \frac{4}{9}$
4. $\frac{3}{10} + \frac{2}{10}$
5. $-\frac{6}{20} + \frac{1}{20}$
6. $-\frac{3}{8} + \frac{1}{8}$
7. $-\frac{3}{14} + \left(-\frac{4}{14}\right)$
8. $-\frac{5}{24} + \left(-\frac{7}{24}\right)$
9. $\frac{2}{9x} + \frac{4}{9x}$
10. $\frac{3}{10y} + \frac{2}{10y}$
11. $-\frac{7x}{18} + \frac{3x}{18} + \frac{2x}{18}$
12. $-\frac{7z}{15} + \frac{3z}{15} + \frac{1z}{15}$

Subtract and simplify. See Examples 4, 5, 7, and 8.

13. $\frac{10}{11} - \frac{4}{11}$

14. $\frac{9}{13} - \frac{5}{13}$

15. $\frac{7}{8} - \frac{1}{8}$

16. $\frac{5}{6} - \frac{1}{6}$

17. $\frac{1}{y} - \frac{4}{y}$

18. $\frac{4}{z} - \frac{7}{z}$

19. $-\frac{27}{33} - \left(-\frac{8}{33}\right)$

20. $-\frac{37}{45} - \left(-\frac{18}{45}\right)$

21. $\frac{20}{21} - \frac{10}{21} - \frac{17}{21}$

22. $\frac{27}{28} - \frac{5}{28} - \frac{28}{28}$

23. $\frac{7a}{4} - \frac{3}{4}$

24. $\frac{18b}{5} - \frac{3}{5}$

Mixed Practice *Perform the indicated operation. See Examples 1 through 8.*

25. $-\frac{9}{100} + \frac{99}{100}$

26. $-\frac{15}{200} + \frac{85}{200}$

27. $-\frac{13x}{28} - \frac{13x}{28}$

28. $-\frac{15}{26y} - \frac{15}{26y}$

29. $\frac{9x}{15} + \frac{1}{15}$

30. $\frac{2x}{15} + \frac{7}{15}$

31. $\frac{7x}{16} - \frac{15x}{16}$

32. $\frac{3}{16z} - \frac{15}{16z}$

33. $\frac{9}{12} - \frac{7}{12} - \frac{10}{12}$

34. $\frac{1}{8} - \frac{15}{8} + \frac{2}{8}$

35. $\frac{x}{4} + \frac{3x}{4} - \frac{2x}{4} + \frac{x}{4}$

36. $\frac{9y}{8} + \frac{2y}{8} + \frac{5y}{8} - \frac{4y}{8}$

Objective B *Evaluate each expression for the given replacement values. See Example 9.*

37. $x + y; x = \frac{3}{4}, y = \frac{2}{4}$

38. $x - y; x = \frac{7}{8}, y = \frac{9}{8}$

39. $x - y; x = -\frac{1}{5}, y = -\frac{3}{5}$

40. $x + y; x = -\frac{1}{6}, y = \frac{5}{6}$

Objective C *Find the perimeter of each figure. (Hint: Recall that perimeter means distance around.) See Example 10.*

△ **41.** $\frac{4}{20}$ inch, $\frac{7}{20}$ inch, Triangle, $\frac{9}{20}$ inch

△ **42.**

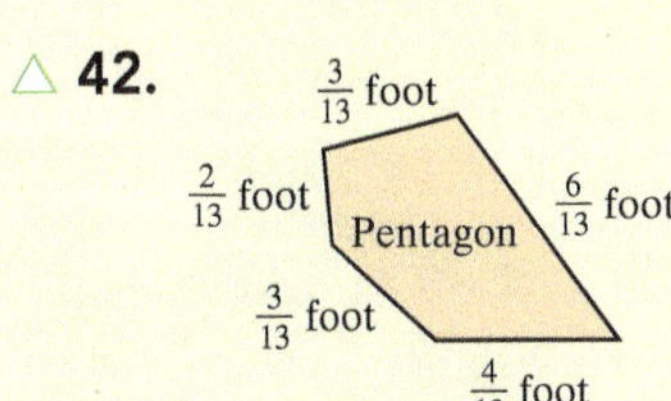

43.

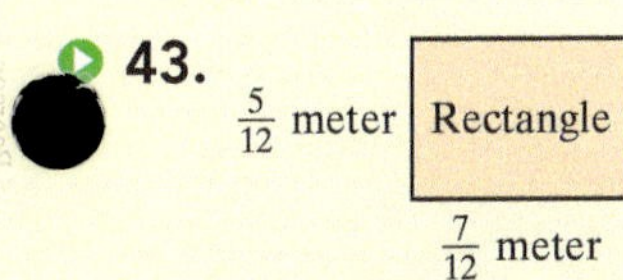

△ **44.**

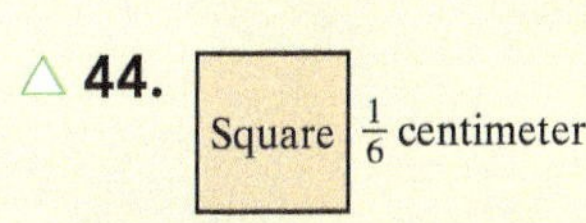

Solve. For Exercises 45 and 46, the solutions have been started for you. Write each answer in simplest form. See Example 11.

45. A railroad inspector must inspect $\frac{19}{20}$ of a mile of railroad track. If she has already inspected $\frac{5}{20}$ of a mile, how much more does she need to inspect?

Start the solution:

1. UNDERSTAND the problem. Reread it as many times as needed.
2. TRANSLATE into an equation. (Fill in the blanks.)

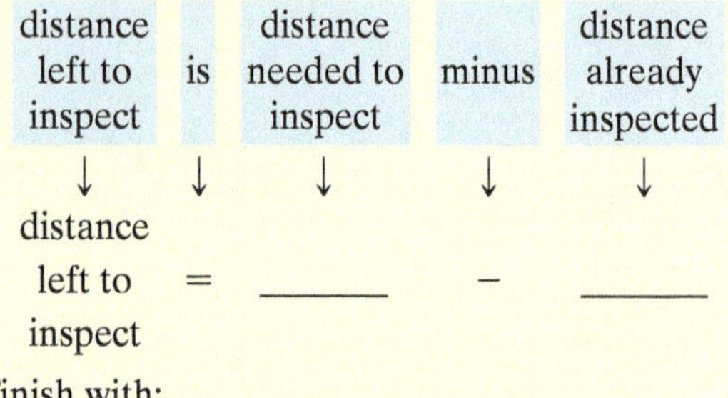

Finish with:

3. SOLVE. and 4. INTERPRET.

46. Scott Davis has run $\frac{11}{8}$ miles already and plans to complete $\frac{16}{8}$ miles. To do this, how much farther must he run?

Start the solution:

1. UNDERSTAND the problem. Reread it as many times as needed.
2. TRANSLATE into an equation. (Fill in the blanks.)

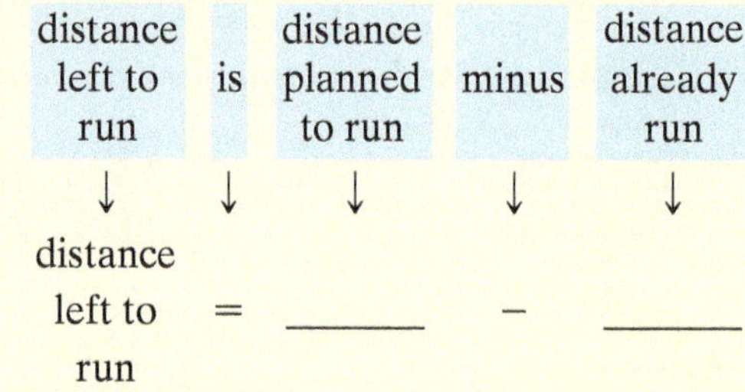

Finish with:

3. SOLVE. and 4. INTERPRET.

47. As of 2013, the fraction of states in the United States with maximum interstate highway speed limits up to and including 70 mph was $\frac{33}{50}$. The fraction of states with 70 mph speed limits was $\frac{20}{50}$. What fraction of states had speed limits that were less than 70 mph? (*Source:* Insurance Institute for Highway Safety)

48. When people take aspirin, $\frac{31}{50}$ of the time it is used to treat some type of pain. Approximately $\frac{7}{50}$ of all aspirin use is for treating headaches. What fraction of aspirin use is for treating pain other than headaches? (*Source:* Bayer Market Research)

The map of the world below shows the fraction of the world's surface land area taken up by each continent. In other words, the continent of Africa, for example, makes up $\frac{20}{100}$ *of the land in the world. Use this map to solve Exercises 49 through 52. Write answers in simplest form.*

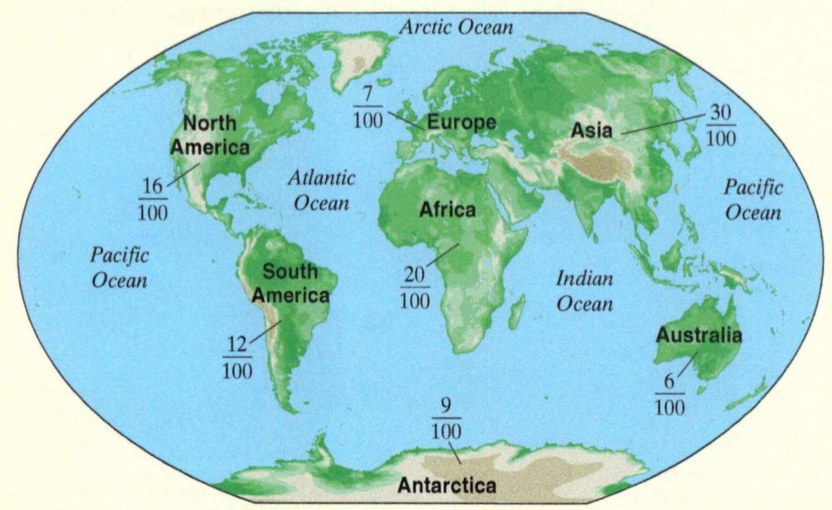

49. Find the fractional part of the world's land area within the continents of North America and South America.

50. Find the fractional part of the world's land area within the continents of Asia and Africa.

51. How much greater is the fractional part of the continent of Antarctica than the fractional part of the continent of Europe?

52. How much greater is the fractional part of the continent of Asia than the continent of Australia?

Objective D *Find the LCD of each list of fractions. See Examples 12 through 16.*

53. $\frac{2}{9}, \frac{6}{15}$

54. $\frac{7}{12}, \frac{3}{20}$

55. $-\frac{1}{36}, \frac{1}{24}$

56. $-\frac{1}{15}, \frac{1}{90}$

57. $\frac{2}{25}, \frac{3}{15}, \frac{5}{6}$

58. $\frac{3}{4}, \frac{1}{14}, \frac{13}{20}$

59. $-\frac{7}{24}, -\frac{5}{x}$

60. $-\frac{11}{y}, -\frac{13}{70}$

61. $\frac{23}{18}, \frac{1}{21}$

62. $\frac{45}{24}, \frac{2}{45}$

63. $\frac{4}{3}, \frac{8}{21}, \frac{3}{56}$

64. $\frac{12}{11}, \frac{20}{33}, \frac{12}{121}$

Objective E *Write each fraction as an equivalent fraction with the given denominator. See Examples 17 through 21.*

65. $\frac{2}{3} = \frac{}{21}$

66. $\frac{5}{6} = \frac{}{24}$

67. $\frac{4}{7} = \frac{}{35}$

68. $\frac{3}{5} = \frac{}{100}$

69. $\frac{1}{2} = \frac{}{50}$

70. $\frac{1}{5} = \frac{}{50}$

71. $\frac{14x}{17} = \frac{}{68}$

72. $\frac{19z}{21} = \frac{}{126}$

73. $\frac{2y}{3} = \frac{}{12}$

74. $\frac{3x}{2} = \frac{}{12}$

75. $\frac{5}{9} = \frac{}{36a}$

76. $\frac{7}{6} = \frac{}{36a}$

The table on the next page shows the fraction of goods sold online by type of goods in a particular year. Use this table to answer Exercises 77 through 80.

77. Complete the table by writing each fraction as an equivalent fraction with a denominator of 100.

78. Which of these types of goods has the largest fraction sold online?

79. Which of these types of goods has the smallest fraction sold online?

80. Which of the types of goods has **more than** $\frac{3}{5}$ of the goods sold online? (*Hint:* Write $\frac{3}{5}$ as an equivalent fraction with a denominator of 100.)

Type of Goods	Fraction of All Goods That Are Sold Online	Equivalent Fraction with a Denominator of 100
books and magazines	$\frac{27}{50}$	
clothing and accessories	$\frac{1}{2}$	
computer hardware	$\frac{23}{50}$	
computer software	$\frac{1}{2}$	
drugs, health and beauty aids	$\frac{3}{20}$	
electronics and appliances	$\frac{13}{20}$	
food, beer, and wine	$\frac{9}{20}$	
home furnishings	$\frac{13}{25}$	
music and videos	$\frac{3}{5}$	
office equipment and supplies	$\frac{61}{100}$	
sporting goods	$\frac{12}{25}$	
toys, hobbies, and games	$\frac{1}{2}$	

(*Source:* Fedstats.gov)

Review

Simplify. See Section 1.7.

81. 3^2

82. 4^3

83. 5^3

84. 3^4

85. 7^2

86. 5^4

87. $2^3 \cdot 3$

88. $4^2 \cdot 5$

Concept Extensions

Find and correct the error. See the first Concept Check in this section.

89. $\frac{2}{7} + \frac{9}{7} = \frac{11}{14}$

90. $\frac{3}{4} - \frac{1}{4} = \frac{2}{8} = \frac{1}{4}$

Solve.

91. In your own words, explain how to add like fractions.

92. In your own words, explain how to subtract like fractions.

93. Use the map of the world for Exercises 49 through 52 and find the sum of all the continents' fractions. Explain your answer.

94. Mike Cannon jogged $\frac{3}{8}$ of a mile from home and then rested. Then he continued jogging farther from home for another $\frac{3}{8}$ of a mile until he discovered his watch had fallen off. He walked back along the same path for $\frac{4}{8}$ of a mile until he found his watch. Find how far he was from his home.

Write each fraction as an equivalent fraction with the indicated denominator.

95. $\frac{37x}{165} = \frac{\quad}{3630}$

96. $\frac{108}{215y} = \frac{\quad}{4085y}$

97. In your own words, explain how to find the LCD of two fractions.

98. In your own words, explain how to write a fraction as an equivalent fraction with a given denominator.

Solve. See the fourth and fifth Concept Checks in this section.

99. Which of the following are equivalent to $\frac{2}{3}$?

a. $\frac{10}{15}$ **b.** $\frac{40}{60}$

c. $\frac{16}{20}$ **d.** $\frac{200}{300}$

100. True or false? When the fraction $\frac{7}{12}$ is rewritten with a denominator of 48, the result is $\frac{11}{48}$. If false, give the correct fraction.

4.5 Adding and Subtracting Unlike Fractions

Objective A Adding and Subtracting Unlike Fractions 

In this section we add and subtract fractions with unlike denominators. To add or subtract these unlike fractions, we first write the fractions as equivalent fractions with a common denominator and then add or subtract the like fractions. Recall from the previous section that the common denominator we use is called the **least common denominator (LCD).**

To begin, let's add the unlike fractions $\frac{3}{4} + \frac{1}{6}$.

The LCD of these fractions is 12. So we write each fraction as an equivalent fraction with a denominator of 12, then add as usual. This addition process is shown next and also illustrated by figures.

Objectives

A Add or Subtract Unlike Fractions.

B Write Fractions in Order.

C Evaluate Expressions Given Fractional Replacement Values.

D Solve Problems by Adding or Subtracting Unlike Fractions.

Add: $\frac{3}{4} + \frac{1}{6}$	The LCD is 12.
Figures	**Algebra**
$\frac{3}{4}$ + $\frac{1}{6}$ $\frac{9}{12}$ + $\frac{2}{12}$ $\frac{9}{12} + \frac{2}{12} = \frac{11}{12}$	$\frac{3}{4} = \frac{3}{4} \cdot \frac{3}{3} = \frac{9}{12}$ and $\frac{1}{6} = \frac{1}{6} \cdot \frac{2}{2} = \frac{2}{12}$ Remember $\frac{3}{3} = 1$ and $\frac{2}{2} = 1$. Now we can add just as we did in Section 4.4. $\frac{3}{4} + \frac{1}{6} = \frac{9}{12} + \frac{2}{12} = \frac{11}{12}$
Thus, the sum is $\frac{11}{12}$.	

Adding or Subtracting Unlike Fractions

Step 1: Find the least common denominator (LCD) of the fractions.

Step 2: Write each fraction as an equivalent fraction whose denominator is the LCD.

Step 3: Add or subtract the like fractions.

Step 4: Write the sum or difference in simplest form.

Practice 1

Add: $\frac{2}{7} + \frac{8}{21}$

Example 1 Add: $\frac{2}{5} + \frac{4}{15}$

Solution:

Step 1: The LCD of the fractions is 15. In later examples, we shall simply say, for example, that the LCD of the denominators 5 and 15 is 15.

Step 2: $\frac{2}{5} = \frac{2}{5} \cdot \frac{3}{3} = \frac{6}{15}$, $\frac{4}{15} = \frac{4}{15}$ ← This fraction already has a denominator of 15.

↑ Multiply by 1 in the form $\frac{3}{3}$.

Step 3: $\frac{2}{5} + \frac{4}{15} = \frac{6}{15} + \frac{4}{15} = \frac{10}{15}$

Step 4: Write in simplest form.

$$\frac{10}{15} = \frac{2 \cdot \overset{1}{\cancel{5}}}{3 \cdot \underset{1}{\cancel{5}}} = \frac{2}{3}$$

Work Practice 1

Answer

1. $\frac{2}{3}$

When the fractions contain variables, we add and subtract the same way.

Example 2 Add: $\frac{2x}{15} + \frac{3x}{10}$

Solution:

Step 1: The LCD of the denominators 15 and 10 is 30.

Step 2: $\frac{2x}{15} = \frac{2x}{15} \cdot \frac{2}{2} = \frac{4x}{30}$ $\quad \frac{3x}{10} = \frac{3x}{10} \cdot \frac{3}{3} = \frac{9x}{30}$

Step 3: $\frac{2x}{15} + \frac{3x}{10} = \frac{4x}{30} + \frac{9x}{30} = \frac{13x}{30}$

Step 4: $\frac{13x}{30}$ is in simplest form.

Work Practice 2

Practice 2

Add: $\frac{5y}{6} + \frac{2y}{9}$

Example 3 Add: $-\frac{1}{6} + \frac{1}{2}$

Solution: The LCD of the denominators 6 and 2 is 6.

$$-\frac{1}{6} + \frac{1}{2} = \frac{-1}{6} + \frac{1 \cdot 3}{2 \cdot 3}$$
$$= \frac{-1}{6} + \frac{3}{6}$$
$$= \frac{2}{6}$$

Helpful Hint

Recall that

$$-\frac{1}{6} = \frac{-1}{6} = \frac{1}{-6}$$

Next, simplify $\frac{2}{6}$.

$$\frac{2}{6} = \frac{\overset{1}{\cancel{2}}}{\underset{1}{\cancel{2}} \cdot 3} = \frac{1}{3}$$

Work Practice 3

Practice 3

Add: $-\frac{1}{5} + \frac{9}{20}$

✓Concept Check Find and correct the error in the following:

$$\frac{2}{9} + \frac{4}{11} = \frac{6}{20} = \frac{3}{10}$$

Example 4 Subtract: $\frac{2}{3} - \frac{10}{11}$

Solution:

Step 1: The LCD of the denominators 3 and 11 is 33.

Step 2: $\frac{2}{3} = \frac{2}{3} \cdot \frac{11}{11} = \frac{22}{33}$ $\quad \frac{10}{11} = \frac{10}{11} \cdot \frac{3}{3} = \frac{30}{33}$

Step 3: $\frac{2}{3} - \frac{10}{11} = \frac{22}{33} - \frac{30}{33} = \frac{-8}{33}$ or $-\frac{8}{33}$

Step 4: $-\frac{8}{33}$ is in simplest form.

Work Practice 4

Practice 4

Subtract: $\frac{5}{7} - \frac{9}{10}$

Answers

2. $\frac{19y}{18}$ **3.** $\frac{1}{4}$ **4.** $-\frac{13}{70}$

✓Concept Check Answer

When adding fractions, we don't add the denominators. Correct solution:

$$\frac{2}{9} + \frac{4}{11} = \frac{22}{99} + \frac{36}{99} = \frac{58}{99}$$

Practice 5

Find: $\frac{5}{8} - \frac{1}{3} - \frac{1}{12}$

Example 5 Find: $-\frac{3}{4} - \frac{1}{14} + \frac{6}{7}$

Solution: The LCD of 4, 14, and 7 is 28.

$$-\frac{3}{4} - \frac{1}{14} + \frac{6}{7} = -\frac{3 \cdot 7}{4 \cdot 7} - \frac{1 \cdot 2}{14 \cdot 2} + \frac{6 \cdot 4}{7 \cdot 4}$$
$$= -\frac{21}{28} - \frac{2}{28} + \frac{24}{28}$$
$$= \frac{1}{28}$$

Work Practice 5

✓**Concept Check** Find and correct the error in the following:

$$\frac{7}{12} - \frac{3}{4} = \frac{4}{8} = \frac{1}{2}$$

Practice 6

Subtract: $5 - \frac{y}{4}$

Example 6 Subtract: $2 - \frac{x}{3}$

Solution: Recall that $2 = \frac{2}{1}$. The LCD of the denominators 1 and 3 is 3.

$$\frac{2}{1} - \frac{x}{3} = \frac{2 \cdot 3}{1 \cdot 3} - \frac{x}{3}$$
$$= \frac{6}{3} - \frac{x}{3}$$
$$= \frac{6 - x}{3}$$

Helpful Hint

The expression $\frac{6 - x}{3}$ from Example 6 *does not simplify* to $2 - x$. The number 3 must be a factor of both terms in the numerator (not just 6) in order to simplify.

The numerator $6 - x$ cannot be simplified further since 6 and $-x$ are unlike terms.

Work Practice 6

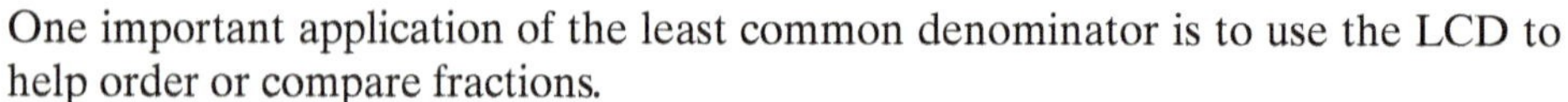

Objective B Writing Fractions in Order

One important application of the least common denominator is to use the LCD to help order or compare fractions.

Practice 7

Insert < or > to form a true sentence.

$\frac{5}{8} \quad \frac{11}{20}$

Example 7 Insert < or > to form a true sentence.

$\frac{3}{4} \quad \frac{9}{11}$

Solution: The LCD for these fractions is 44. Let's write each fraction as an equivalent fraction with a denominator of 44.

$$\frac{3}{4} = \frac{3 \cdot 11}{4 \cdot 11} = \frac{33}{44} \qquad \frac{9}{11} = \frac{9 \cdot 4}{11 \cdot 4} = \frac{36}{44}$$

Answers

5. $\frac{5}{24}$ 6. $\frac{20 - y}{4}$ 7. >

✓**Concept Check Answers**

When adding fractions, we don't add the denominators. Correct solutions:

$\frac{7}{12} - \frac{3}{4} = \frac{7}{12} - \frac{9}{12} = -\frac{2}{12} = -\frac{1}{6}$

Since $33 < 36$, then

$$\frac{33}{44} < \frac{36}{44} \text{ or}$$

$$\frac{3}{4} < \frac{9}{11}$$

Work Practice 7

Example 8 Insert $<$ or $>$ to form a true sentence.

$$-\frac{2}{7} \quad -\frac{1}{3}$$

Solution: The LCD is 21.

$$-\frac{2}{7} = -\frac{2 \cdot 3}{7 \cdot 3} = -\frac{6}{21} \text{ or } \frac{-6}{21} \qquad -\frac{1}{3} = -\frac{1 \cdot 7}{3 \cdot 7} = -\frac{7}{21} \text{ or } \frac{-7}{21}$$

Since $-6 > -7$, then

$$-\frac{6}{21} > -\frac{7}{21} \text{ or}$$

$$-\frac{2}{7} > -\frac{1}{3}$$

Work Practice 8

Practice 8

Insert $<$ or $>$ to form a true sentence.

$$-\frac{17}{20} \quad -\frac{4}{5}$$

Objective C Evaluating Expressions Given Fractional Replacement Values

Example 9 Evaluate $x - y$ if $x = \frac{7}{18}$ and $y = \frac{2}{9}$.

Solution: Replace x with $\frac{7}{18}$ and y with $\frac{2}{9}$ in the expression $x - y$.

$$x - y = \frac{7}{18} - \frac{2}{9}$$

The LCD of the denominators 18 and 9 is 18. Then

$$\frac{7}{18} - \frac{2}{9} = \frac{7}{18} - \frac{2 \cdot 2}{9 \cdot 2}$$

$$= \frac{7}{18} - \frac{4}{18}$$

$$= \frac{3}{18} = \frac{1}{6} \quad \text{Simplified}$$

Work Practice 9

Practice 9

Evaluate $x - y$ if $x = \frac{5}{11}$ and $y = \frac{4}{9}$.

Objective D Solving Problems by Adding or Subtracting Unlike Fractions

Very often, real-world problems involve adding or subtracting unlike fractions.

Answers

8. $<$ **9.** $\frac{1}{99}$

Practice 10

To repair her sidewalk, a homeowner must pour cement in three different locations. She needs $\frac{3}{5}$ of a cubic yard, $\frac{3}{10}$ of a cubic yard, and $\frac{1}{15}$ of a cubic yard for these locations. Find the total amount of cement the homeowner needs.

Example 10 Finding Total Weight

A freight truck has $\frac{1}{4}$ of a ton of computers, $\frac{1}{3}$ of a ton of televisions, and $\frac{3}{8}$ of a ton of small appliances. Find the total weight of its load.

Solution:

1. UNDERSTAND. Read and reread the problem. The phrase "total weight" tells us to add.
2. TRANSLATE.

In words:	total weight	is	weight of computers	plus	weight of televisions	plus	weight of appliances
	↓	↓	↓	↓	↓	↓	↓
Translate:	total weight	=	$\frac{1}{4}$	+	$\frac{1}{3}$	+	$\frac{3}{8}$

3. SOLVE: The LCD is 24.

$$\frac{1}{4}+\frac{1}{3}+\frac{3}{8}=\frac{1}{4}\cdot\frac{6}{6}+\frac{1}{3}\cdot\frac{8}{8}+\frac{3}{8}\cdot\frac{3}{3}$$
$$=\frac{6}{24}+\frac{8}{24}+\frac{9}{24}$$
$$=\frac{23}{24}$$

4. INTERPRET. *Check* the solution. *State* your conclusion: The total weight of the truck's load is $\frac{23}{24}$ ton.

Work Practice 10

Practice 11

Find the difference in length of two boards if one board is $\frac{3}{4}$ of a foot long and the other is $\frac{2}{3}$ of a foot long.

Example 11 Calculating Flight Time

A flight from Tucson to Phoenix, Arizona, requires $\frac{5}{12}$ of an hour. If the plane has been flying $\frac{1}{4}$ of an hour, find how much time remains before landing.

Solution:

1. UNDERSTAND. Read and reread the problem. The phrase "how much time remains" tells us to subtract.

Answers

10. $\frac{29}{30}$ cu yd 11. $\frac{1}{12}$ ft

2. TRANSLATE.

In words:	time remaining	is	flight time from Tucson to Phoenix	minus	flight time already passed
	↓	↓	↓	↓	↓
Translate:	time remaining	$=$	$\frac{5}{12}$	$-$	$\frac{1}{4}$

3. SOLVE: The LCD is 12.

$$\frac{5}{12} - \frac{1}{4} = \frac{5}{12} - \frac{1}{4} \cdot \frac{3}{3}$$

$$= \frac{5}{12} - \frac{3}{12}$$

$$= \frac{2}{12} = \frac{\overset{1}{\cancel{2}}}{\underset{1}{\cancel{2}} \cdot 6} = \frac{1}{6}$$

4. INTERPRET. *Check* the solution. *State* your conclusion: The remaining flight time is $\frac{1}{6}$ of an hour.

Work Practice 11

Calculator Explorations Performing Operations on Fractions

Scientific Calculator

Many calculators have a fraction key, such as [$a^b/_c$], that allows you to enter fractions and perform operations on fractions, and gives the result as a fraction. If your calculator has a fraction key, use it to calculate

$$\frac{3}{5} + \frac{4}{7}$$

Enter the keystrokes

[3] [$a^b/_c$] [5] [+] [4] [$a^b/_c$] [7] [=]

The display should read [1_6 | 35]

which represents the mixed number $1\frac{6}{35}$. Let's write the result as a fraction. To convert from mixed number notation to fractional notation, press

[2nd] [d/c]

The display now reads [41 | 35]

which represents $\frac{41}{35}$, the sum in fractional notation.

Graphing Calculator

Graphing calculators also allow you to perform operations on fractions and will give exact fractional results. The fraction option on a graphing calculator may be found under the [MATH] menu. To perform the addition in the first column, try the keystrokes

[3] [÷] [5] [+] [4] [÷] [7] [MATH] [ENTER] [ENTER]

The display should read

[3/5 + 4/7 ▶ Frac 41/35]

Use a calculator to add the following fractions. Give each sum as a fraction.

1. $\frac{1}{16} + \frac{2}{5}$ **2.** $\frac{3}{20} + \frac{2}{25}$ **3.** $\frac{4}{9} + \frac{7}{8}$

4. $\frac{9}{11} + \frac{5}{12}$ **5.** $\frac{10}{17} + \frac{12}{19}$ **6.** $\frac{14}{31} + \frac{15}{21}$

Vocabulary, Readiness & Video Check

Use the choices below to fill in each blank. Any numerical answers are not listed. Not all choices may be used.

expression	least common denominator	>
equation	equivalent	<

1. To add or subtract unlike fractions, we first write the fractions as ______________ fractions with a common denominator. The common denominator we use is called the ______________.
2. The LCD for $\frac{1}{6}$ and $\frac{5}{8}$ is __________.
3. $\frac{1}{6} + \frac{5}{8} = \frac{1}{6} \cdot \frac{4}{4} + \frac{5}{8} \cdot \frac{3}{3} = \_\_ + \_\_ = \_\_$
4. $\frac{1}{6} - \frac{5}{8} = \frac{1}{6} \cdot \frac{4}{4} - \frac{5}{8} \cdot \frac{3}{3} = \_\_ - \_\_ = \_\_$.
5. $x - y$ is a(n) ______________ while $3x = \frac{1}{5}$ is a(n) ______________.
6. Since $-10 < -1$, we know that $-\frac{10}{13}$ ______________ $-\frac{1}{13}$.

Martin-Gay Interactive Videos

See Video 4.5

Watch the section lecture video and answer the following questions.

Objective A 7. In Example 3, why can't we add the two terms in the numerator?

Objective B 8. In Example 5, when comparing two fractions, how does writing each fraction with the same denominator help?

Objective C 9. In Example 6, if we had chosen to simplify the first fraction before adding, what would our addition problem have become and what would our LCD have been?

Objective D 10. What are the two forms of the answer to Example 7?

4.5 Exercise Set MyMathLab®

Objective A *Add or subtract as indicated. See Examples 1 through 6.*

1. $\frac{2}{3} + \frac{1}{6}$
2. $\frac{5}{6} + \frac{1}{12}$
3. $\frac{1}{2} - \frac{1}{3}$
4. $\frac{2}{3} - \frac{1}{4}$
5. $-\frac{2}{11} + \frac{2}{33}$
6. $-\frac{5}{9} + \frac{1}{3}$
7. $\frac{3}{14} - \frac{3}{7}$
8. $\frac{2}{15} - \frac{2}{5}$
9. $\frac{11x}{35} + \frac{2x}{7}$
10. $\frac{2y}{5} + \frac{3y}{25}$
11. $2 - \frac{y}{12}$
12. $5 - \frac{y}{20}$
13. $\frac{5}{12} - \frac{1}{9}$
14. $\frac{7}{12} - \frac{5}{18}$
15. $-7 + \frac{5}{7}$
16. $-10 + \frac{7}{10}$

17. $\frac{5a}{11} + \frac{4a}{9}$
18. $\frac{7x}{18} + \frac{2x}{9}$
19. $\frac{2y}{3} - \frac{1}{6}$
20. $\frac{5z}{6} - \frac{1}{12}$
21. $\frac{1}{2} + \frac{3}{x}$
22. $\frac{2}{5} + \frac{3}{x}$
23. $-\frac{2}{11} - \frac{2}{33}$
24. $-\frac{5}{9} - \frac{1}{3}$
25. $\frac{9}{14} - \frac{3}{7}$
26. $\frac{4}{5} - \frac{2}{15}$
27. $\frac{11y}{35} - \frac{2}{7}$
28. $\frac{2b}{5} - \frac{3}{25}$
29. $\frac{1}{9} - \frac{5}{12}$
30. $\frac{5}{18} - \frac{7}{12}$
31. $\frac{7}{15} - \frac{5}{12}$
32. $\frac{5}{8} - \frac{3}{20}$
33. $\frac{5}{7} - \frac{1}{8}$
34. $\frac{10}{13} - \frac{7}{10}$
35. $\frac{7}{8} + \frac{3}{16}$
36. $\frac{7}{18} + \frac{2}{9}$
37. $\frac{3}{9} - \frac{5}{9}$
38. $\frac{1}{13} - \frac{4}{13}$
39. $-\frac{2}{5} + \frac{1}{3} - \frac{3}{10}$
40. $-\frac{1}{3} - \frac{1}{4} + \frac{2}{5}$
41. $\frac{5}{11} + \frac{y}{3}$
42. $\frac{5z}{13} + \frac{3}{26}$
43. $-\frac{5}{6} - \frac{3}{7}$
44. $-\frac{1}{2} - \frac{3}{29}$
45. $\frac{x}{2} + \frac{x}{4} + \frac{2x}{16}$
46. $\frac{z}{4} + \frac{z}{8} + \frac{2z}{16}$
47. $\frac{7}{9} - \frac{1}{6}$
48. $\frac{9}{16} - \frac{3}{8}$
49. $\frac{2a}{3} + \frac{6a}{13}$
50. $\frac{3y}{4} + \frac{y}{7}$
51. $\frac{7}{30} - \frac{5}{12}$
52. $\frac{7}{30} - \frac{3}{20}$
53. $\frac{5}{9} + \frac{1}{y}$
54. $\frac{1}{12} - \frac{5}{x}$
55. $\frac{6}{5} - \frac{3}{4} + \frac{1}{2}$
56. $\frac{6}{5} + \frac{3}{4} - \frac{1}{2}$
57. $\frac{4}{5} + \frac{4}{9}$
58. $\frac{11}{12} - \frac{7}{24}$
59. $\frac{5}{9x} + \frac{1}{8}$
60. $\frac{3}{8} + \frac{5}{12x}$
61. $-\frac{9}{12} + \frac{17}{24} - \frac{1}{6}$
62. $-\frac{5}{14} + \frac{3}{7} - \frac{1}{2}$
63. $\frac{3x}{8} + \frac{2x}{7} - \frac{5}{14}$
64. $\frac{9x}{10} - \frac{1}{2} + \frac{x}{5}$

Objective B *Insert $<$ or $>$ to form a true sentence. See Examples 7 and 8.*

65. $\frac{2}{7} \quad \frac{3}{10}$
66. $\frac{5}{9} \quad \frac{6}{11}$
67. $-\frac{5}{6} \quad -\frac{13}{15}$
68. $-\frac{7}{8} \quad -\frac{5}{6}$
69. $-\frac{3}{4} \quad -\frac{11}{14}$
70. $-\frac{2}{9} \quad -\frac{3}{13}$

Objective C *Evaluate each expression if $x = \frac{1}{3}$ and $y = \frac{3}{4}$. See Example 9.*

71. $x + y$
72. $x - y$
73. xy
74. $x \div y$
75. $2y + x$
76. $2x + y$

Objective D *Find the perimeter of each geometric figure. (Hint: Recall that perimeter means distance around.)*

77.

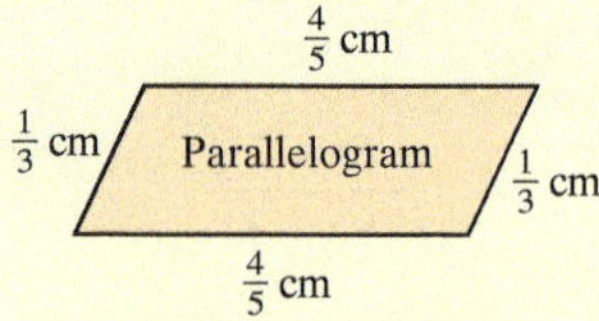

△ **78.**

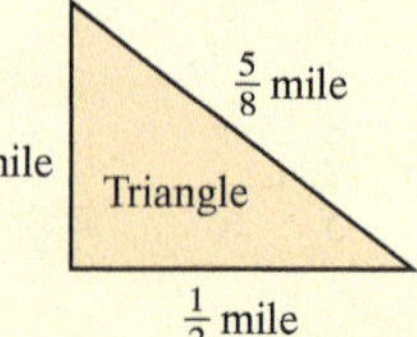

△ **79.**

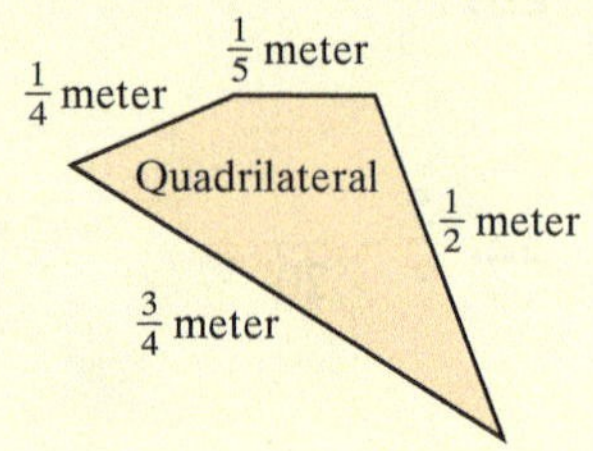

△ **80.**

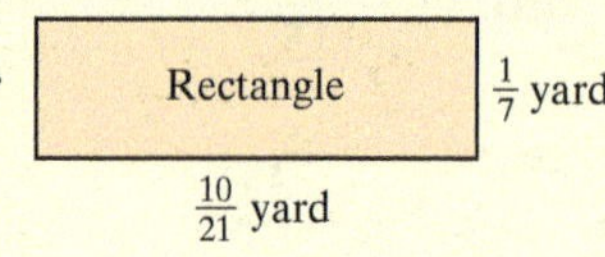

Translating *Translate each phrase into an algebraic expression. Use "x" to represent "a number." See Examples 10 and 11.*

81. The sum of a number and $\frac{1}{2}$

82. A number increased by $-\frac{2}{5}$

83. A number subtracted from $-\frac{3}{8}$

84. The difference of a number and $\frac{7}{20}$

Solve. For Exercises 85 and 86, the solutions have been started for you. See Examples 10 and 11.

85. The slowest mammal is the three-toed sloth from South America. The sloth has an average ground speed of $\frac{1}{10}$ mph. In trees, it can accelerate to $\frac{17}{100}$ mph. How much faster can a sloth travel in trees? (*Source: The Guinness Book of World Records*)

Start the solution:

1. UNDERSTAND the problem. Reread it as many times as needed.
2. TRANSLATE into an equation. (Fill in the blanks.)

how much faster sloth travels in trees	is	sloth speed in trees	minus	sloth speed on ground
↓	↓	↓	↓	↓
how much faster sloth travels in trees	=	____	−	____

Finish with:

3. SOLVE. and
4. INTERPRET.

86. Killer bees have been known to chase people for up to $\frac{1}{4}$ of a mile, while domestic European honeybees will normally chase a person for no more than 100 feet, or $\frac{5}{264}$ of a mile. How much farther will a killer bee chase a person than a domestic honeybee? (*Source:* Coachella Valley Mosquito & Vector Control District)

Start the solution:

1. UNDERSTAND the problem. Reread it as many times as needed.
2. TRANSLATE into an equation. (Fill in the blanks.)

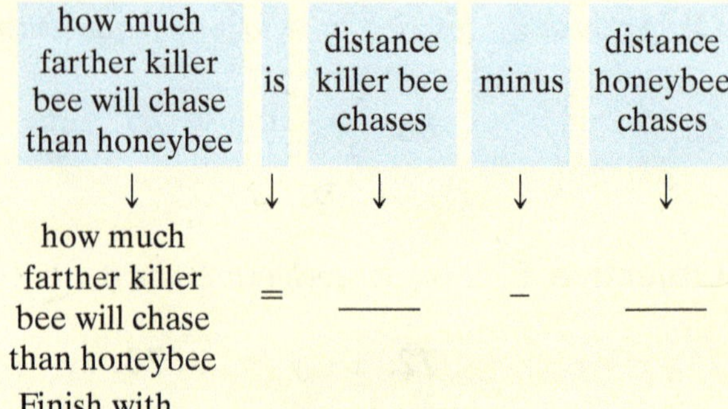

how much farther killer bee will chase than honeybee	is	distance killer bee chases	minus	distance honeybee chases
↓	↓	↓	↓	↓
how much farther killer bee will chase than honeybee	=	____	−	____

Finish with

3. SOLVE. and
4. INTERPRET.

87. Find the inner diameter of the washer.

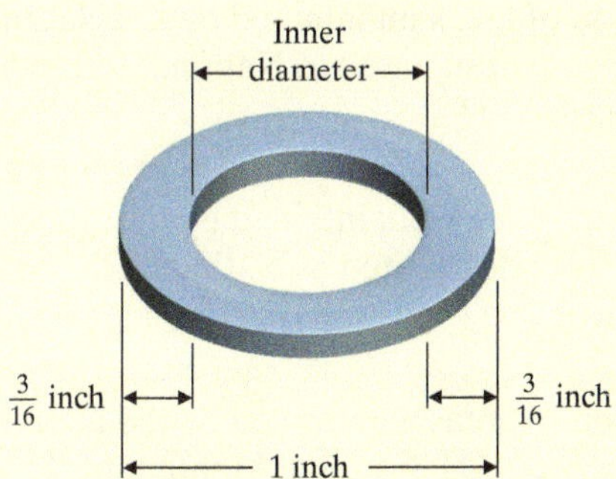

88. Find the inner diameter of the tubing.

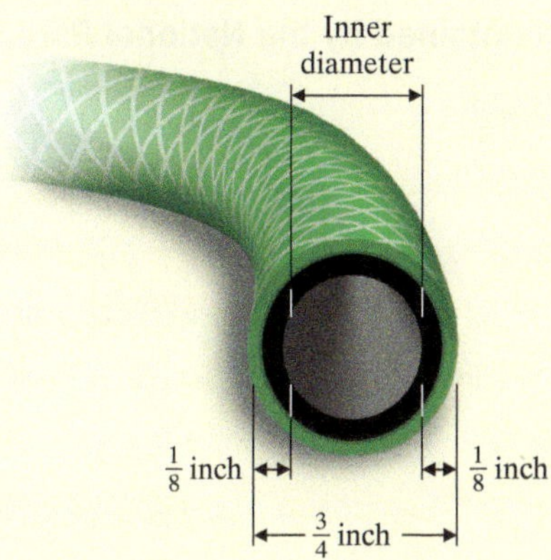

89. About $\frac{13}{20}$ of American students ages 10 to 17 name math, science, or art as their favorite subject in school. Art is the favorite subject for about $\frac{4}{25}$ of the American students ages 10 to 17. For what fraction of students this age is math or science their favorite subject? (*Source:* Peter D. Hart Research Associates for the National Science Foundation)

90. In the 2012 FIA Formula One World Championship, the Red Bull Racing–Renault team won $\frac{7}{20}$ of all races. If the Red Bull driver Sebastian Vettel won $\frac{1}{4}$ of the races, how many did Red Bull driver Mark Webber win? (*Source:* Formula1.com)

91. Given the following diagram, find its total length.

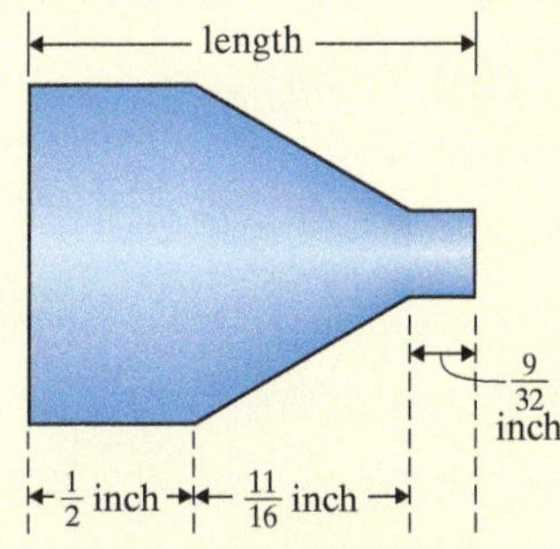

92. Given the following diagram, find its total width.

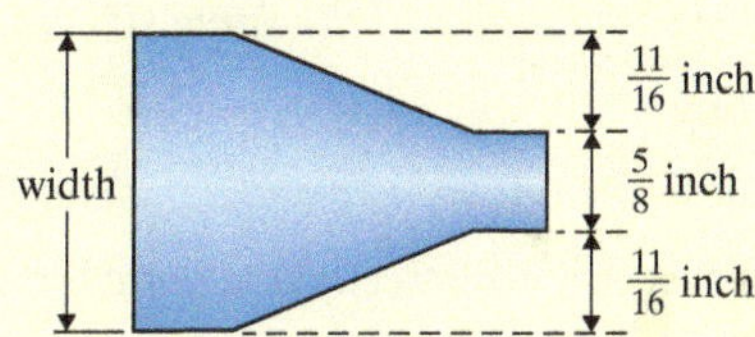

The table below shows the fraction of the Earth's water area taken up by each ocean. Use this table for Exercises 93 and 94.

Fraction of Earth's Water Area per Ocean

Ocean	Fraction
Arctic	$\frac{1}{25}$
Atlantic	$\frac{13}{50}$
Pacific	$\frac{1}{2}$
Indian	$\frac{1}{5}$

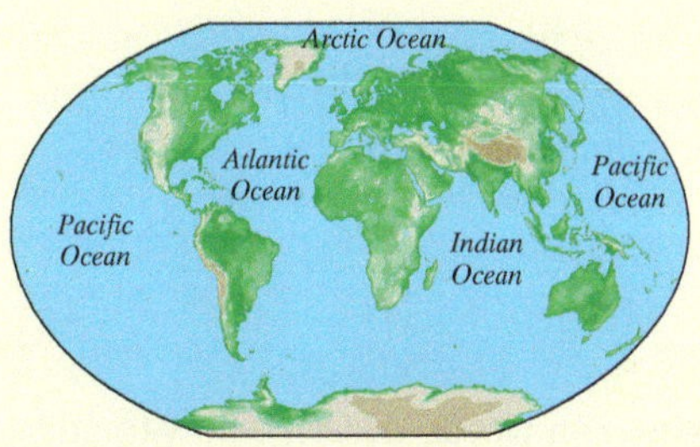

93. What fraction of the world's water surface area is accounted for by the Pacific and Atlantic Oceans?

94. What fraction of the world's water surface area is accounted for by the Arctic and Indian Oceans?

Use this circle graph to answer Exercises 95 through 98.

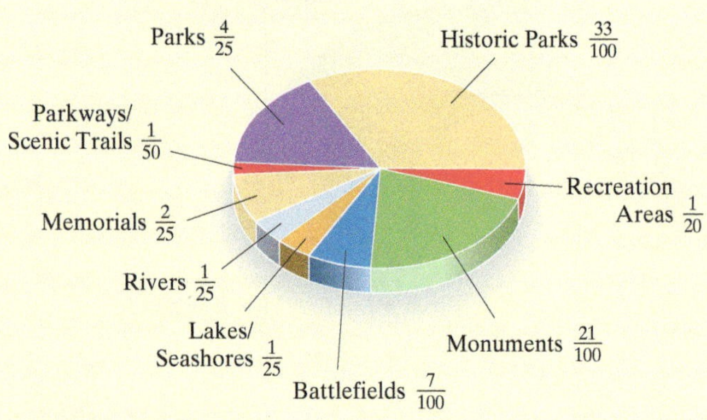

95. What fraction of areas maintained by the National Park Service are designated as National Lakes or National Seashores?

96. What fraction of areas maintained by the National Park Service are designated as National Recreation Areas?

97. What fraction of areas maintained by the National Park Service are NOT National Monuments?

98. What fraction of areas maintained by the National Park Service are NOT National Parkways or Scenic Trails?

Review

Use order of operations to simplify. See Section 2.5.

99. $-50 \div 5 \cdot 2$

100. $8 - 6 \cdot 4 - 7$

101. $(8 - 6) \cdot (4 - 7)$

102. $50 \div (5 \cdot 2)$

Concept Extensions

For Exercises 103 and 104 below, do the following:

a) Draw three rectangles of the same size and represent each fraction in the sum or difference, one fraction per rectangle, by shading.

b) Using these rectangles as estimates, determine whether there is an error in the sum or difference.

c) If there is an error, correctly calculate the sum or difference.

See the Concept Checks in this section.

103. $\frac{3}{5} + \frac{4}{5} \stackrel{?}{=} \frac{7}{10}$

104. $\frac{5}{8} - \frac{3}{4} \stackrel{?}{=} \frac{2}{4}$

Subtract from left to right.

105. $\frac{2}{3} - \frac{1}{4} - \frac{2}{540}$

106. $\frac{9}{10} - \frac{7}{200} - \frac{1}{3}$

Perform each indicated operation.

107. $\frac{30}{55} + \frac{1000}{1760}$

108. $\frac{19}{26} - \frac{968}{1352}$

109. In your own words, describe how to add or subtract two fractions with different denominators.

110. Find the sum of the fractions in the circle graph above. Did the sum surprise you? Why or why not?

111. In 2012, about $\frac{69}{160}$ of the total number of pieces of mail delivered by the United States Postal Service was first-class mail. That same year, about $\frac{1}{2}$ of the total number of pieces of mail delivered by the United States Postal Service was standard mail. Which of these two categories account for a greater portion of the mail handled by volume? (*Source:* United States Postal Service)

Summary on Fractions and Operations on Fractions

Use a fraction to represent the shaded area of each figure. If the fraction is improper, also write the fraction as a mixed number.

1.

2. 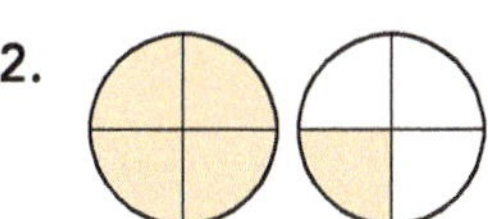

Solve.

3. In a survey, 73 people out of 85 get fewer than 8 hours of sleep each night. What fraction of people in the survey get fewer than 8 hours of sleep?

4. Sketch a diagram to represent $\frac{9}{13}$.

Simplify.

5. $\frac{11}{-11}$ **6.** $\frac{17}{1}$ **7.** $\frac{0}{-3}$ **8.** $\frac{7}{0}$

Write the prime factorization of each composite number. Write any repeated factors using exponents.

9. 65 **10.** 70 **11.** 315 **12.** 441

Write each fraction in simplest form.

13. $\frac{2}{14}$ **14.** $\frac{24}{20}$ **15.** $-\frac{56}{60}$ **16.** $-\frac{72}{80}$

17. $\frac{54x}{135}$ **18.** $\frac{90}{240y}$ **19.** $\frac{165z^3}{210z}$ **20.** $\frac{245ab}{385a^2b^3}$

Determine whether each pair of fractions is equivalent.

21. $\frac{7}{8}$ and $\frac{9}{10}$ **22.** $\frac{10}{12}$ and $\frac{15}{18}$

23. Of the 50 states, 2 states are not adjacent to any other states.

a. What fraction of the states are not adjacent to other states?
b. How many states are adjacent to other states?
c. What fraction of the states are adjacent to other states?

24. There are approximately 90,000 digital movie screens in the world. Of these, about 36,000 are in the U.S./Canada. (*Source:* Motion Picture Association of America)

a. What fraction of digital movie screens are in the U.S./Canada?
b. How many digital movie screens are not in the U.S./Canada?
c. What fraction of digital movie screens are not in the U.S./Canada?

Answers

1. ______
2. ______
3. ______
4. ______
5. ______
6. ______
7. ______
8. ______
9. ______
10. ______
11. ______
12. ______
13. ______
14. ______
15. ______
16. ______
17. ______
18. ______
19. ______
20. ______
21. ______
22. ______
23. a. b. c. ______
24. a. b. c. ______

25. ______ 26. ______
27. ______ 28. ______
29. ______ 30. ______
31. ______ 32. ______
33. ______ 34. ______
35. ______ 36. ______
37. ______
38. ______
39. ______
40. ______
41. ______
42. ______
43. ______
44. ______
45. ______
46. ______
47. ______
48. ______
49. ______
50. ______
51. ______
52. ______
53. ______

Find the LCM of each list of numbers.

25. 5, 6 **26.** 2, 14 **27.** 6, 18, 30

Write each fraction as an equivalent fraction with the indicated denominator.

28. $\frac{7}{9} = \frac{}{36}$ **29.** $\frac{11}{15} = \frac{}{75}$ **30.** $\frac{5}{6} = \frac{}{48}$

The following summary will help you with the following review of operations on fractions.

Operations on Fractions

Let $a, b, c,$ and d be integers.

Addition: $\frac{a}{b} + \frac{c}{b} = \frac{a+c}{b}$ $\quad b \neq 0$ (↑ ↑ common denominator)

Subtraction: $\frac{a}{b} - \frac{c}{b} = \frac{a-c}{b}$ $\quad (b \neq 0)$ (↑ ↑ common denominator)

Multiplication: $\frac{a}{b} \cdot \frac{c}{d} = \frac{a \cdot c}{b \cdot d}$ $\quad (b \neq 0, d \neq 0)$

Division: $\frac{a}{b} \div \frac{c}{d} = \frac{a}{b} \cdot \frac{d}{c} = \frac{a \cdot d}{b \cdot c}$ $\quad (b \neq 0, d \neq 0, c \neq 0)$

Perform each indicated operation.

31. $\frac{1}{5} + \frac{3}{5}$ **32.** $\frac{1}{5} - \frac{3}{5}$ **33.** $\frac{1}{5} \cdot \frac{3}{5}$ **34.** $\frac{1}{5} \div \frac{3}{5}$

35. $\frac{2}{3} \div \frac{5}{6}$ **36.** $\frac{2a}{3} \cdot \frac{5}{6a}$ **37.** $\frac{2}{3y} - \frac{5}{6y}$ **38.** $\frac{2x}{3} + \frac{5x}{6}$

39. $-\frac{1}{7} \cdot -\frac{7}{18}$ **40.** $-\frac{4}{9} \cdot -\frac{3}{7}$ **41.** $-\frac{7z}{8} \div 6z^2$ **42.** $-\frac{9}{10} \div 5$

43. $\frac{7}{8} + \frac{1}{20}$ **44.** $\frac{5}{12} - \frac{1}{9}$ **45.** $\frac{2}{9} + \frac{1}{18} + \frac{1}{3}$ **46.** $\frac{3y}{10} + \frac{y}{5} + \frac{6}{25}$

Translating *Translate each to an expression. Use x to represent "a number."*

47. $\frac{2}{3}$ of a number

48. The quotient of a number and $-\frac{1}{5}$

49. A number subtracted from $-\frac{8}{9}$

50. $\frac{6}{11}$ increased by a number

Solve.

51. Find $\frac{2}{3}$ of 1530.

52. A contractor is using 18 acres of his land to sell $\frac{3}{4}$-acre lots. How many lots can he sell?

53. Suppose that the cross-section of a piece of pipe looks like the diagram shown. What is the inner diameter?

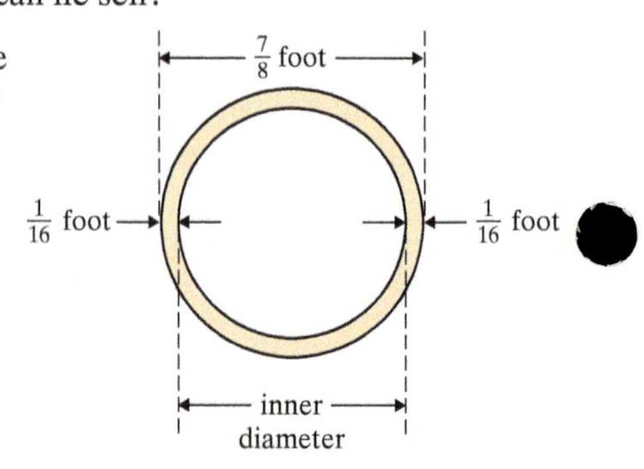

4.6 Complex Fractions and Review of Order of Operations

Objective A Simplifying Complex Fractions

Thus far, we have studied operations on fractions. We now practice simplifying fractions whose numerators or denominators themselves contain fractions. These fractions are called **complex fractions.**

Complex Fraction

A fraction whose numerator or denominator or both numerator and denominator contain fractions is called a **complex fraction.**

Examples of complex fractions are

$$\frac{\frac{x}{4}}{\frac{3}{2}} \qquad \frac{\frac{1}{2}+\frac{3}{8}}{\frac{3}{4}-\frac{1}{6}} \qquad \frac{\frac{y}{5}-2}{\frac{3}{10}} \qquad \frac{-4z}{\frac{3}{5}}$$

Method 1 for Simplifying Complex Fractions

Two methods are presented to simplify complex fractions. The first method makes use of the fact that a fraction bar means division.

Example 1 Simplify: $\dfrac{\frac{x}{4}}{\frac{3}{2}}$

Solution: Since a fraction bar means division, the complex fraction $\dfrac{\frac{x}{4}}{\frac{3}{2}}$ can be written as $\frac{x}{4} \div \frac{3}{2}$. Then divide as usual to simplify.

$$\frac{x}{4} \div \frac{3}{2} = \frac{x}{4}\cdot\frac{2}{3} \quad \text{Multiply by the reciprocal.}$$

$$= \frac{x \cdot \overset{1}{\cancel{2}}}{\underset{1}{\cancel{2}} \cdot 2 \cdot 3}$$

$$= \frac{x}{6}$$

Work Practice 1

Example 2 Simplify: $\dfrac{\frac{1}{2}+\frac{3}{8}}{\frac{3}{4}-\frac{1}{6}}$

Solution: Recall the order of operations. Since the fraction bar is considered a grouping symbol, we simplify the numerator and the denominator of the complex fraction separately. Then we divide.

$$\frac{\frac{1}{2}+\frac{3}{8}}{\frac{3}{4}-\frac{1}{6}} = \frac{\frac{1\cdot 4}{2\cdot 4}+\frac{3}{8}}{\frac{3\cdot 3}{4\cdot 3}-\frac{1\cdot 2}{6\cdot 2}} = \frac{\frac{4}{8}+\frac{3}{8}}{\frac{9}{12}-\frac{2}{12}} = \frac{\frac{7}{8}}{\frac{7}{12}}$$

(Continued on next page)

Objectives

A Simplify Complex Fractions.

B Review the Order of Operations.

C Evaluate Expressions Given Replacement Values.

Practice 1

Simplify: $\dfrac{\frac{7y}{10}}{\frac{1}{5}}$

Practice 2

Simplify: $\dfrac{\frac{1}{2}+\frac{1}{6}}{\frac{3}{4}-\frac{2}{3}}$

Answers

1. $\frac{7y}{2}$ **2.** $\frac{8}{1}$ or 8

Thus,

$$\frac{\frac{1}{2}+\frac{3}{8}}{\frac{3}{4}-\frac{1}{6}}=\frac{\frac{7}{8}}{\frac{7}{12}}$$

$$=\frac{7}{8}\div\frac{7}{12}$$ Rewrite the quotient using the ÷ sign.

$$=\frac{7}{8}\cdot\frac{12}{7}$$ Multiply by the reciprocal.

$$=\frac{\overset{1}{\cancel{7}}\cdot 3\cdot\overset{1}{\cancel{4}}}{2\cdot\underset{1}{\cancel{4}}\cdot\underset{1}{\cancel{7}}}$$ Multiply.

$$=\frac{3}{2}$$ Simplify.

Work Practice 2

Method 2 for Simplifying Complex Fractions

The second method for simplifying complex fractions is to multiply the numerator and the denominator of the complex fraction by the LCD of all the fractions in its numerator and its denominator. This has the effect of leaving sums and differences of integers in the numerator and the denominator, as we shall see in the example below.

Let's use this second method to simplify the complex fraction in Example 2 again.

Practice 3

Use Method 2 to simplify:

$$\frac{\frac{1}{2}+\frac{1}{6}}{\frac{3}{4}-\frac{2}{3}}$$

Example 3 Simplify: $\frac{\frac{1}{2}+\frac{3}{8}}{\frac{3}{4}-\frac{1}{6}}$

Solution: The complex fraction contains fractions with denominators 2, 8, 4, and 6. The LCD is 24. By the fundamental property of fractions (See the Helpful Hints on p. 230 and p. 257), we can multiply the numerator and the denominator of the complex fraction by 24. Notice below that by the distributive property, this means that we multiply each term in the numerator and denominator by 24.

$$\frac{\frac{1}{2}+\frac{3}{8}}{\frac{3}{4}-\frac{1}{6}}=\frac{24\left(\frac{1}{2}+\frac{3}{8}\right)}{24\left(\frac{3}{4}-\frac{1}{6}\right)}$$

$$=\frac{\left(\overset{12}{\cancel{24}}\cdot\frac{1}{\underset{1}{\cancel{2}}}\right)+\left(\overset{3}{\cancel{24}}\cdot\frac{3}{\underset{1}{\cancel{8}}}\right)}{\left(\overset{6}{\cancel{24}}\cdot\frac{3}{\underset{1}{\cancel{4}}}\right)-\left(\overset{4}{\cancel{24}}\cdot\frac{1}{\underset{1}{\cancel{6}}}\right)}$$ Apply the distributive property. Then divide out common factors to aid in multiplying.

$$=\frac{12+9}{18-4}$$ Multiply.

$$=\frac{21}{14}$$

$$=\frac{\overset{1}{\cancel{7}}\cdot 3}{\underset{1}{\cancel{7}}\cdot 2}=\frac{3}{2}$$ Simplify.

Work Practice 3

The simplified result is the same, of course, no matter which method is used.

Answer

3. $\frac{8}{1}$ or 8

Example 4 Simplify: $\dfrac{\frac{y}{5} - 2}{\frac{3}{10}}$

Solution: Use the second method and multiply the numerator and the denominator of the complex fraction by the LCD of all fractions. Recall that $2 = \frac{2}{1}$. The LCD of the denominators 5, 1, and 10 is 10.

$$\frac{\frac{y}{5} - \frac{2}{1}}{\frac{3}{10}} = \frac{10\left(\frac{y}{5} - \frac{2}{1}\right)}{10\left(\frac{3}{10}\right)}$$ Multiply the numerator and denominator by 10.

$$= \frac{\left(\overset{2}{\cancel{10}} \cdot \frac{y}{\underset{1}{\cancel{5}}}\right) - \left(10 \cdot \frac{2}{1}\right)}{\overset{1}{\cancel{10}} \cdot \frac{3}{\underset{1}{\cancel{10}}}}$$ Apply the distributive property. Then divide out common factors to aid in multiplying.

$$= \frac{2y - 20}{3}$$ Multiply.

Work Practice 4

Practice 4

Simplify: $\dfrac{\frac{3}{4}}{\frac{x}{5} - 1}$

Helpful Hint Don't forget to multiply the numerator and the denominator of the complex fraction by the same number—the LCD.

Objective B Reviewing the Order of Operations 

At this time, it is probably a good idea to review the order of operations on expressions containing fractions. Before we do so, let's review how we perform operations on fractions.

Review of Operations on Fractions

Operation	Procedure	Example
Multiply	Multiply the numerators and multiply the denominators.	$\frac{5}{9} \cdot \frac{1}{2} = \frac{5 \cdot 1}{9 \cdot 2} = \frac{5}{18}$
Divide	Multiply the first fraction by the reciprocal of the second fraction.	$\frac{2}{3} \div \frac{11}{13} = \frac{2}{3} \cdot \frac{13}{11} = \frac{2 \cdot 13}{3 \cdot 11} = \frac{26}{33}$
Add or Subtract	**1.** Write each fraction as an equivalent fraction whose denominator is the LCD. **2.** Add or subtract numerators and write the result over the common denominator.	$\frac{3}{4} + \frac{1}{8} = \frac{3}{4} \cdot \frac{2}{2} + \frac{1}{8} = \frac{6}{8} + \frac{1}{8} = \frac{7}{8}$

Now let's review order of operations.

Order of Operations

1. Perform all operations within parentheses ( ), brackets [ ], or other grouping symbols such as fraction bars, starting with the innermost set.
2. Evaluate any expressions with exponents.
3. Multiply or divide in order from left to right.
4. Add or subtract in order from left to right.

Example 5 Simplify: $\left(\frac{4}{5}\right)^2 - 1$

Solution: According to the order of operations, first evaluate $\left(\frac{4}{5}\right)^2$.

$$\left(\frac{4}{5}\right)^2 - 1 = \frac{16}{25} - 1$$ Write $\left(\frac{4}{5}\right)^2$ as $\frac{16}{25}$.

(*Continued on next page*)

Practice 5

Simplify: $\left(\frac{2}{3}\right)^3 - 2$

Answers

4. $\frac{15}{4x - 20}$ **5.** $-\frac{46}{27}$

Next, combine the fractions. The LCD of 25 and 1 is 25.

$$\frac{16}{25} - 1 = \frac{16}{25} - \frac{25}{25} \quad \text{Write 1 as } \frac{25}{25}.$$

$$= \frac{-9}{25} \text{ or } -\frac{9}{25} \quad \text{Subtract.}$$

Work Practice 5

Practice 6

Simplify: $\left(-\frac{1}{2} + \frac{1}{5}\right)\left(\frac{7}{8} + \frac{1}{8}\right)$

Example 6 Simplify: $\left(\frac{1}{4} + \frac{2}{3}\right)\left(\frac{11}{12} + \frac{1}{4}\right)$

Solution: First perform operations inside parentheses. Then multiply.

$$\left(\frac{1}{4} + \frac{2}{3}\right)\left(\frac{11}{12} + \frac{1}{4}\right) = \left(\frac{1\cdot 3}{4\cdot 3} + \frac{2\cdot 4}{3\cdot 4}\right)\left(\frac{11}{12} + \frac{1\cdot 3}{4\cdot 3}\right) \quad \text{Each LCD is 12.}$$

$$= \left(\frac{3}{12} + \frac{8}{12}\right)\left(\frac{11}{12} + \frac{3}{12}\right)$$

$$= \left(\frac{11}{12}\right)\left(\frac{14}{12}\right) \quad \text{Add.}$$

$$= \frac{11\cdot \overset{1}{\cancel{2}}\cdot 7}{\underset{1}{\cancel{2}}\cdot 6\cdot 12} \quad \text{Multiply.}$$

$$= \frac{77}{72} \quad \text{Simplify.}$$

Work Practice 6

Helpful Hint If you find it difficult replacing a variable with a number, try the following. First, replace the variable with a set of parentheses, then place the replacement number between the parentheses.

If $x = \frac{4}{5}$, find $2x + x^2$.

$$2x + x^2 = 2(\ \ ) + (\ \ )^2$$

$$= 2\left(\frac{4}{5}\right) + \left(\frac{4}{5}\right)^2 \ldots$$

then continue simplifying.

✓Concept Check What should be done first to simplify the expression $\frac{1}{5}\cdot\frac{5}{2} - \left(\frac{2}{3} + \frac{4}{5}\right)^2$?

Objective C Evaluating Algebraic Expressions

Practice 7

Evaluate $-\frac{3}{5} - xy$ if $x = \frac{3}{10}$ and $y = \frac{2}{3}$.

Example 7 Evaluate $2x + y^2$ if $x = -\frac{1}{2}$ and $y = \frac{1}{3}$.

Solution: Replace x and y with the given values and simplify.

$$2x + y^2 = 2\left(-\frac{1}{2}\right) + \left(\frac{1}{3}\right)^2 \quad \text{Replace } x \text{ with } -\frac{1}{2} \text{ and } y \text{ with } \frac{1}{3}.$$

$$= 2\left(-\frac{1}{2}\right) + \frac{1}{9} \quad \text{Write } \left(\frac{1}{3}\right)^2 \text{ as } \frac{1}{9}.$$

$$= -1 + \frac{1}{9} \quad \text{Multiply.}$$

$$= -\frac{9}{9} + \frac{1}{9} \quad \text{The LCD is 9.}$$

$$= -\frac{8}{9} \quad \text{Add.}$$

Work Practice 7

Answers

6. $-\frac{3}{10}$ 7. $-\frac{4}{5}$

✓Concept Check Answer

Add inside parentheses.

Vocabulary, Readiness & Video Check

Use the choices below to fill in each blank.

addition	multiplication	evaluate the exponential expression
subtraction	division	complex

1. A fraction whose numerator or denominator or both numerator and denominator contain fractions is called a(n) ______ fraction.
2. To simplify $-\frac{1}{2} + \frac{2}{3} \cdot \frac{7}{8}$, which operation do we perform first? ______
3. To simplify $-\frac{1}{2} \div \frac{2}{3} \cdot \frac{7}{8}$, which operation do we perform first? ______
4. To simplify $\frac{7}{8} \cdot \left(\frac{1}{2} - \frac{2}{3}\right)$, which operation do we perform first? ______
5. To simplify $\frac{1}{3} \div \frac{1}{4} \cdot \left(\frac{9}{11} + \frac{3}{8}\right)^3$, which operation do we perform first? ______
6. To simplify $9 - \left(-\frac{3}{4}\right)^2$, which operation do we perform first? ______

Martin-Gay Interactive Videos

See Video 4.6 

Watch the section lecture video and answer the following questions.

Objective A 7. In Example 2, what property is used to simplify the denominator of the complex fraction?

Objective B 8. In Example 3, why can we add the fractions in the first set of parentheses right away?

Objective C 9. In Example 4, why do we use parentheses when substituting the replacement value for x? What would happen if we didn't use parentheses?

4.6 Exercise Set MyMathLab®

Objective A *Simplify each complex fraction. See Examples 1 through 4.*

1. $\dfrac{\frac{1}{8}}{\frac{3}{4}}$

2. $\dfrac{\frac{5}{12}}{\frac{15}{12}}$

3. $\dfrac{\frac{2}{3}}{\frac{2}{7}}$

4. $\dfrac{\frac{9}{25}}{\frac{6}{25}}$

5. $\dfrac{\frac{2x}{27}}{\frac{4}{9}}$

6. $\dfrac{\frac{3y}{11}}{\frac{1}{2}}$

7. $\dfrac{\frac{3}{4} + \frac{2}{5}}{\frac{1}{2} + \frac{3}{5}}$

8. $\dfrac{\frac{7}{6} + \frac{2}{3}}{\frac{3}{2} - \frac{8}{9}}$

9. $\dfrac{\frac{3x}{4}}{5 - \frac{1}{8}}$

10. $\dfrac{\frac{3}{10} + 2}{\frac{2}{5y}}$

Objective B *Use the order of operations to simplify each expression. See Examples 5 and 6.*

11. $\frac{1}{5}+\frac{1}{3}\cdot\frac{1}{4}$

12. $\frac{1}{2}+\frac{1}{6}\cdot\frac{1}{3}$

13. $\frac{5}{6}\div\frac{1}{3}\cdot\frac{1}{4}$

14. $\frac{7}{8}\div\frac{1}{4}\cdot\frac{1}{7}$

15. $2^2-\left(\frac{1}{3}\right)^2$

16. $3^2-\left(\frac{1}{2}\right)^2$

17. $\left(\frac{2}{9}+\frac{4}{9}\right)\left(\frac{1}{3}-\frac{9}{10}\right)$

18. $\left(\frac{1}{5}-\frac{1}{10}\right)\left(\frac{1}{5}+\frac{1}{10}\right)$

19. $\left(\frac{7}{8}-\frac{1}{2}\right)\div\frac{3}{11}$

20. $\left(-\frac{2}{3}-\frac{7}{3}\right)\div\frac{4}{9}$

21. $2\cdot\left(\frac{1}{4}+\frac{1}{5}\right)+2$

22. $\frac{2}{5}\cdot\left(5-\frac{1}{2}\right)-1$

23. $\left(\frac{3}{4}\right)^2\div\left(\frac{3}{4}-\frac{1}{12}\right)$

24. $\left(\frac{8}{9}\right)^2\div\left(2-\frac{2}{3}\right)$

25. $\left(\frac{2}{5}-\frac{3}{10}\right)^2$

26. $\left(\frac{3}{2}-\frac{4}{3}\right)^3$

27. $\left(\frac{3}{4}+\frac{1}{8}\right)^2-\left(\frac{1}{2}+\frac{1}{8}\right)$

28. $\left(\frac{1}{6}+\frac{1}{3}\right)^3+\left(\frac{2}{5}\cdot\frac{3}{4}\right)^2$

Objective C *Evaluate each expression if* $x=-\frac{1}{3}$, $y=\frac{2}{5}$, *and* $z=\frac{5}{6}$. *See Example 7.*

29. $5y-z$

30. $2z-x$

31. $\frac{x}{z}$

32. $\frac{y+x}{z}$

33. x^2-yz

34. x^2-z^2

35. $(1+x)(1+z)$

36. $(1-x)(1-z)$

Objectives A B Mixed Practice *Simplify the following. See Examples 1 through 6.*

37. $\dfrac{\frac{5a}{24}}{\frac{1}{12}}$

38. $\dfrac{\frac{7}{10}}{\frac{14z}{25}}$

39. $\left(\frac{3}{2}\right)^3+\left(\frac{1}{2}\right)^3$

40. $\left(\frac{5}{21}\div\frac{1}{2}\right)+\left(\frac{1}{7}\cdot\frac{1}{3}\right)$

41. $\left(-\frac{1}{2}\right)^2+\frac{1}{5}$

42. $\left(-\frac{3}{4}\right)^2+\frac{3}{8}$

43. $\dfrac{2 + \dfrac{1}{6}}{1 - \dfrac{4}{3}}$

44. $\dfrac{3 - \dfrac{1}{2}}{4 + \dfrac{1}{5}}$

45. $\left(1 - \dfrac{2}{5}\right)^2$

46. $\left(-\dfrac{1}{2}\right)^2 - \left(\dfrac{3}{4}\right)^2$

47. $\left(\dfrac{3}{4} - 1\right)\left(\dfrac{1}{8} + \dfrac{1}{2}\right)$

48. $\left(\dfrac{1}{10} + \dfrac{3}{20}\right)\left(\dfrac{1}{5} - 1\right)$

49. $\left(-\dfrac{2}{9} - \dfrac{7}{9}\right)^4$

50. $\left(\dfrac{5}{9} - \dfrac{2}{3}\right)^2$

51. $\dfrac{\dfrac{1}{3} - \dfrac{5}{6}}{\dfrac{3}{4} + \dfrac{1}{2}}$

52. $\dfrac{\dfrac{7}{10} + \dfrac{1}{2}}{\dfrac{4}{5} + \dfrac{3}{4}}$

53. $\left(\dfrac{3}{4} \div \dfrac{6}{5}\right) - \left(\dfrac{3}{4} \cdot \dfrac{6}{5}\right)$

54. $\left(\dfrac{1}{2} \cdot \dfrac{2}{7}\right) - \left(\dfrac{1}{2} \div \dfrac{2}{7}\right)$

55. $\dfrac{\dfrac{x}{3} + 2}{5 + \dfrac{1}{3}}$

56. $\dfrac{1 - \dfrac{x}{4}}{2 + \dfrac{3}{8}}$

Review

Perform each indicated operation. If the result is an improper fraction, also write the improper fraction as a mixed number. See Sections 4.1 and 4.5.

57. $3 + \dfrac{1}{2}$

58. $2 + \dfrac{2}{3}$

59. $9 - \dfrac{5}{6}$

60. $4 - \dfrac{1}{5}$

Concept Extensions

61. Calculate $\dfrac{2^3}{3}$ and $\left(\dfrac{2}{3}\right)^3$. Do both of these expressions simplify to the same number? Explain why or why not.

62. Calculate $\left(\dfrac{1}{2}\right)^2 \cdot \left(\dfrac{3}{4}\right)^2$ and $\left(\dfrac{1}{2} \cdot \dfrac{3}{4}\right)^2$. Do both of these expressions simplify to the same number? Explain why or why not.

Recall that to find the average of two numbers, we find their sum and divide by 2. For example, the average of $\frac{1}{2}$ and $\frac{3}{4}$ is $\dfrac{\dfrac{1}{2} + \dfrac{3}{4}}{2}$. Find the average of each pair of numbers.

63. $\dfrac{1}{2}, \dfrac{3}{4}$

64. $\dfrac{3}{5}, \dfrac{9}{10}$

65. $\dfrac{1}{4}, \dfrac{2}{14}$

66. $\dfrac{5}{6}, \dfrac{7}{9}$

67. Two positive numbers, a and b, are graphed below. Where should the graph of their average lie?

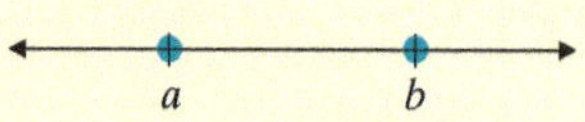

68. Study Exercise 67. Without calculating, can $\frac{1}{3}$ be the average of $\frac{1}{2}$ and $\frac{8}{9}$? Explain why or why not.

Answer true or false for each statement.

69. It is possible for the average of two numbers to be greater than both numbers.

70. It is possible for the average of two numbers to be less than both numbers.

71. The sum of two negative fractions is always a negative number.

72. The sum of a negative fraction and a positive fraction is always a positive number.

73. It is possible for the sum of two fractions to be a whole number.

74. It is possible for the difference of two fractions to be a whole number.

75. What operation should be performed first to simplify

$$\frac{1}{5}\cdot\frac{5}{2} - \left(\frac{2}{3} + \frac{4}{5}\right)^2?$$

Explain your answer.

76. A student is to evaluate $x - y$ when $x = \frac{1}{5}$ and $y = -\frac{1}{7}$. This student is asking you if he should evaluate $\frac{1}{5} - \frac{1}{7}$. What do you tell this student and why?

Each expression contains one addition, one subtraction, one multiplication, and one division. Write the operations in the order that they should be performed. Do not actually simplify. See the Concept Check in this section.

77. $[9 + 3(4 - 2)] \div \frac{10}{21}$

78. $[30 - 4(3 + 2)] \div \frac{5}{2}$

79. $\frac{1}{3} \div \left(\frac{2}{3}\right)\left(\frac{4}{5}\right) - \frac{1}{4} + \frac{1}{2}$

80. $\left(\frac{5}{6} - \frac{1}{3}\right)\cdot\frac{1}{3} + \frac{1}{2} \div \frac{9}{8}$

Evaluate each expression if $x = \frac{3}{4}$ *and* $y = -\frac{4}{7}$.

81. $\frac{2 + x}{y}$

82. $4x + y$

83. $x^2 + 7y$

84. $\frac{\frac{9}{14}}{x + y}$

4.7 Operations on Mixed Numbers 

Objective A Graphing Fractions and Mixed Numbers

Let's review graphing fractions and practice graphing mixed numbers on a number line. This will help us visualize rounding and estimating operations with mixed numbers.

Recall that $5\frac{2}{3}$ means $5 + \frac{2}{3}$ and

$$-4\frac{1}{6} \text{ means } -\left(4 + \frac{1}{6}\right) \quad \text{or} \quad -4 - \frac{1}{6} \quad \text{or} \quad -4 + \left(-\frac{1}{6}\right)$$

Example 1 Graph the numbers on a number line:

$\frac{1}{2}, -\frac{3}{4}, 2\frac{2}{3}, -3, -3\frac{1}{8}$

Solution: Remember that $2\frac{2}{3}$ means $2 + \frac{2}{3}$.

Also, $-3\frac{1}{8}$ means $-3 - \frac{1}{8}$, so $-3\frac{1}{8}$ lies to the left of -3.

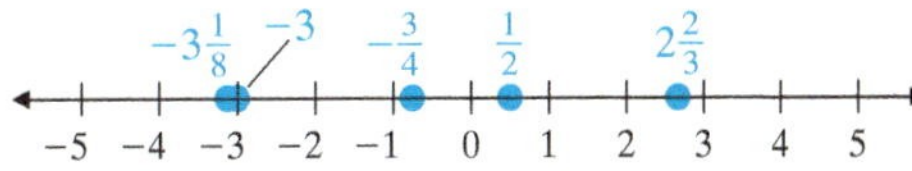

Work Practice 1

✓Concept Check Which of the following is/are equivalent to 9?

a. $7\frac{6}{3}$ **b.** $8\frac{4}{4}$ **c.** $8\frac{9}{9}$ **d.** $\frac{18}{2}$ **e.** all of these

Objective B Multiplying or Dividing with Mixed Numbers or Whole Numbers

When multiplying or dividing a fraction and a mixed or a whole number, remember that mixed and whole numbers can be written as improper fractions.

Multiplying or Dividing Fractions and Mixed Numbers or Whole Numbers

To multiply or divide with mixed numbers or whole numbers, first write any mixed or whole numbers as improper fractions and then multiply or divide as usual.

(*Note:* If an exercise contains a mixed number, we will write the answer as a mixed number, if possible.)

Objectives

A Graph Positive and Negative Fractions and Mixed Numbers.

B Multiply or Divide Mixed or Whole Numbers.

C Add or Subtract Mixed Numbers.

D Solve Problems Containing Mixed Numbers.

E Perform Operations on Negative Mixed Numbers.

Practice 1

Graph the numbers on a number line.

$-5, -4\frac{1}{2}, 2\frac{3}{4}, \frac{1}{8}, -\frac{1}{2}$

−5 −4 −3 −2 −1 0 1 2 3 4 5

Answer

1.

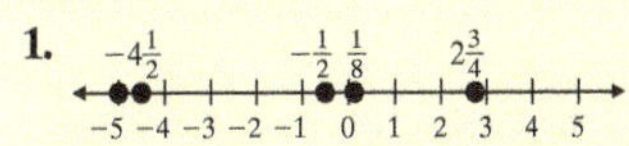

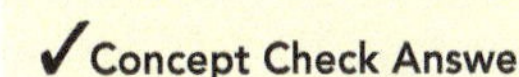

✓Concept Check Answer

e

Practice 2

Multiply and simplify: $1\frac{2}{3} \cdot \frac{11}{15}$

Example 2 Multiply: $3\frac{1}{3} \cdot \frac{7}{8}$

Solution: Recall from Section 4.1 that the mixed number $3\frac{1}{3}$ can be written as the fraction $\frac{10}{3}$. Then

$$3\frac{1}{3} \cdot \frac{7}{8} = \frac{10}{3} \cdot \frac{7}{8} = \frac{\overset{1}{\cancel{2}} \cdot 5 \cdot 7}{3 \cdot \underset{1}{\cancel{2}} \cdot 4} = \frac{35}{12} \text{ or } 2\frac{11}{12}$$

Work Practice 2

Don't forget that a whole number can be written as a fraction by writing the whole number over 1. For example,

$$20 = \frac{20}{1} \quad \text{and} \quad 7 = \frac{7}{1}$$

Practice 3

Multiply: $\frac{5}{6} \cdot 18$

Example 3 Multiply: $\frac{3}{4} \cdot 20$

Solution: $\frac{3}{4} \cdot 20 = \frac{3}{4} \cdot \frac{20}{1} = \frac{3 \cdot 20}{4 \cdot 1} = \frac{3 \cdot \overset{1}{\cancel{4}} \cdot 5}{\underset{1}{\cancel{4}} \cdot 1} = \frac{15}{1} \text{ or } 15$

Work Practice 3

When both numbers to be multiplied are mixed or whole numbers, it is a good idea to estimate the product to see if your answer is reasonable. To do this, we first practice rounding mixed numbers to the nearest whole. If the fraction part of the mixed number is $\frac{1}{2}$ or greater, we round the whole number part up. If the fraction part of the mixed number is less than $\frac{1}{2}$, then we do not round the whole number part up. Study the table below for examples.

Mixed Number	Rounding
$5\frac{1}{4}$ — $\frac{1}{4}$ is less than $\frac{1}{2}$	Thus, $5\frac{1}{4}$ rounds to 5.
$3\frac{9}{16}$ ← 9 is greater than 8. → Half of 16 is 8.	Thus, $3\frac{9}{16}$ rounds to 4.
$1\frac{3}{7}$ ← 3 is less than $3\frac{1}{2}$. → Half of 7 is $3\frac{1}{2}$.	Thus, $1\frac{3}{7}$ rounds to 1.

Practice 4

Multiply. Check by estimating.

$3\frac{1}{5} \cdot 2\frac{3}{4}$

Example 4 Multiply $1\frac{2}{3} \cdot 2\frac{1}{4}$. Check by estimating.

Solution: $1\frac{2}{3} \cdot 2\frac{1}{4} = \frac{5}{3} \cdot \frac{9}{4} = \frac{5 \cdot 9}{3 \cdot 4} = \frac{5 \cdot \overset{1}{\cancel{3}} \cdot 3}{\underset{1}{\cancel{3}} \cdot 4} = \frac{15}{4} \text{ or } 3\frac{3}{4}$ Exact

Let's check by estimating.

$1\frac{2}{3}$ rounds to 2, $2\frac{1}{4}$ rounds to 2, and $2 \cdot 2 = 4$ Estimate

The estimate is close to the exact value, so our answer is reasonable.

Work Practice 4

Answers

2. $1\frac{2}{9}$ 3. 15 4. $8\frac{4}{5}$

Example 5 Multiply: $7 \cdot 2\frac{11}{14}$. Check by estimating.

Solution: $7 \cdot 2\frac{11}{14} = \frac{7}{1} \cdot \frac{39}{14} = \frac{7 \cdot 39}{1 \cdot 14} = \frac{\overset{1}{\cancel{7}} \cdot 39}{1 \cdot 2 \cdot \underset{1}{\cancel{7}}} = \frac{39}{2}$ or $19\frac{1}{2}$ Exact

To estimate,

$2\frac{11}{14}$ rounds to 3 and $7 \cdot 3 = 21$. Estimate

The estimate is close to the exact value, so our answer is reasonable.

Work Practice 5

Practice 5

Multiply. Check by estimating.

$3 \cdot 6\frac{7}{15}$

✓Concept Check Find the error.

$2\frac{1}{4} \cdot \frac{1}{2} = 2\frac{1 \cdot 1}{4 \cdot 2} = 2\frac{1}{8}$

Examples Divide.

6. $\frac{3}{4} \div 5 = \frac{3}{4} \div \frac{5}{1} = \frac{3}{4} \cdot \frac{1}{5} = \frac{3 \cdot 1}{4 \cdot 5} = \frac{3}{20}$

7. $\frac{11}{18} \div 2\frac{5}{6} = \frac{11}{18} \div \frac{17}{6} = \frac{11}{18} \cdot \frac{6}{17} = \frac{11 \cdot 6}{18 \cdot 17} = \frac{11 \cdot \overset{1}{\cancel{6}}}{\underset{1}{\cancel{6}} \cdot 3 \cdot 17} = \frac{11}{51}$

8. $5\frac{2}{3} \div 2\frac{5}{9} = \frac{17}{3} \div \frac{23}{9} = \frac{17}{3} \cdot \frac{9}{23} = \frac{17 \cdot 9}{3 \cdot 23} = \frac{17 \cdot \overset{1}{\cancel{3}} \cdot 3}{\underset{1}{\cancel{3}} \cdot 23} = \frac{51}{23}$ or $2\frac{5}{23}$

Work Practice 6–8

Practice 6–8

Divide.

6. $\frac{4}{9} \div 7$

7. $\frac{8}{15} \div 3\frac{4}{5}$

8. $3\frac{2}{7} \div 2\frac{3}{14}$

Objective C Adding or Subtracting Mixed Numbers

We can add or subtract mixed numbers, too, by first writing each mixed number as an improper fraction. But it is often easier to add or subtract the whole-number parts and add or subtract the proper-fraction parts vertically.

Adding or Subtracting Mixed Numbers

To add or subtract mixed numbers, add or subtract the fraction parts and then add or subtract the whole number parts.

Example 9 Add: $2\frac{1}{3} + 5\frac{3}{8}$. Check by estimating.

Solution: The LCD of the denominators 3 and 8 is 24.

$$\begin{array}{rcr} 2\frac{1 \cdot 8}{3 \cdot 8} & = & 2\frac{8}{24} \\ +\,5\frac{3 \cdot 3}{8 \cdot 3} & = & +\,5\frac{9}{24} \\ \hline & = & 7\frac{17}{24} \end{array}$$

← Add the fractions.

Add the whole numbers.

To check by estimating, we round as usual. The fraction $2\frac{1}{3}$ rounds to 2, $5\frac{3}{8}$ rounds to 5, and $2 + 5 = 7$, our estimate.

Our exact answer is close to 7, so our answer is reasonable.

Work Practice 9

Practice 9

Add: $2\frac{1}{6} + 4\frac{2}{5}$.

Check by estimating.

Answers

5. $19\frac{2}{5}$ 6. $\frac{4}{63}$ 7. $\frac{8}{57}$ 8. $1\frac{15}{31}$

9. $6\frac{17}{30}$

✓Concept Check Answer

forgot to change mixed number to improper fraction

Helpful Hint

When adding or subtracting mixed numbers and whole numbers, it is a good idea to estimate to see if your answer is reasonable.

For the rest of this section, we leave most of the checking by estimating to you.

Practice 10

Add: $3\frac{5}{14} + 2\frac{6}{7}$

Example 10 Add: $3\frac{4}{5} + 1\frac{4}{15}$

Solution: The LCD of the denominators 5 and 15 is 15.

$$\begin{aligned} 3\frac{4}{5} &= 3\frac{12}{15} \\ +1\frac{4}{15} &= +1\frac{4}{15} \quad \text{Add the fractions; then add the whole numbers.} \\ & \quad\ 4\frac{16}{15} \quad \text{Notice that the fraction part is improper.} \end{aligned}$$

Since $\frac{16}{15}$ is $1\frac{1}{15}$ we can write the sum as

$$4\frac{16}{15} = 4 + 1\frac{1}{15} = 5\frac{1}{15}$$

Work Practice 10

✓Concept Check Explain how you could estimate the following sum:

$5\frac{1}{9} + 14\frac{10}{11}$.

Practice 11

Add: $12 + 3\frac{6}{7} + 2\frac{1}{5}$

Example 11 Add: $2\frac{4}{5} + 5 + 1\frac{1}{2}$

Solution: The LCD of the denominators 5 and 2 is 10.

$$\begin{aligned} 2\frac{4}{5} &= 2\frac{8}{10} \\ 5 &= 5 \\ +1\frac{1}{2} &= +1\frac{5}{10} \\ & \quad 8\frac{13}{10} = 8 + 1\frac{3}{10} = 9\frac{3}{10} \end{aligned}$$

Work Practice 11

Practice 12

Subtract: $32\frac{7}{9} - 16\frac{5}{18}$

Example 12 Subtract: $8\frac{3}{7} - 5\frac{2}{21}$. Check by estimating.

Solution: The LCD of the denominators 7 and 21 is 21.

$$\begin{aligned} 8\frac{3}{7} &= 8\frac{9}{21} \quad \leftarrow \text{The LCD of 7 and 21 is 21.} \\ -5\frac{2}{21} &= -5\frac{2}{21} \\ & \quad 3\frac{7}{21} \quad \leftarrow \text{Subtract the fractions.} \end{aligned}$$

↑ Subtract the whole numbers.

Answers

10. $6\frac{3}{14}$ **11.** $18\frac{2}{35}$ **12.** $16\frac{1}{2}$

✓Concept Check Answer

Round each mixed number to the nearest whole number and add. $5\frac{1}{9}$ rounds to 5 and $14\frac{10}{11}$ rounds to 15, and the estimated sum is $5 + 15 = 20$.

Then $3\frac{7}{21}$ simplifies to $3\frac{1}{3}$. The difference is $3\frac{1}{3}$.

To check, $8\frac{3}{7}$ rounds to 8, $5\frac{2}{21}$ rounds to 5, and $8 - 5 = 3$, our estimate.

Our exact answer is close to 3, so our answer is reasonable.

Work Practice 12

When subtracting mixed numbers, borrowing may be needed, as shown in the next example.

Example 13 Subtract: $7\frac{3}{14} - 3\frac{6}{7}$

Solution: The LCD of the denominators 7 and 14 is 14.

$$\begin{array}{rcl} 7\frac{3}{14} & = & 7\frac{3}{14} \\ -3\frac{6}{7} & = & -3\frac{12}{14} \\ \hline \end{array}$$

Notice that we cannot subtract $\frac{12}{14}$ from $\frac{3}{14}$, so we borrow from the whole number 7.

borrow 1 from 7

$$7\frac{3}{14} = 6 + 1\frac{3}{14} = 6 + \frac{17}{14} \text{ or } 6\frac{17}{14}$$

Now subtract.

$$\begin{array}{rcccl} 7\frac{3}{14} & = & 7\frac{3}{14} & = & 6\frac{17}{14} \\ -3\frac{6}{7} & = & -3\frac{12}{14} & = & -3\frac{12}{14} \\ \hline & & & & 3\frac{5}{14} \end{array}$$

$\leftarrow$ Subtract the fractions.

$\uparrow$ Subtract the whole numbers.

Work Practice 13

Practice 13

Subtract: $9\frac{7}{15} - 4\frac{3}{5}$

✓Concept Check In the subtraction problem $5\frac{1}{4} - 3\frac{3}{4}$, $5\frac{1}{4}$ must be rewritten because $\frac{3}{4}$ cannot be subtracted from $\frac{1}{4}$. Why is it incorrect to rewrite $5\frac{1}{4}$ as $5\frac{5}{4}$?

Example 14 Subtract: $14 - 8\frac{3}{7}$

Solution:

$$\begin{array}{rcl} 14 & = & 13\frac{7}{7} \\ -8\frac{3}{7} & = & -8\frac{3}{7} \\ \hline & & 5\frac{4}{7} \end{array}$$

Borrow 1 from 14 and write it as $\frac{7}{7}$.

$\leftarrow$ Subtract the fractions.

$\uparrow$ Subtract the whole numbers.

Work Practice 14

Practice 14

Subtract: $25 - 10\frac{2}{9}$

Answers

13. $4\frac{13}{15}$ **14.** $14\frac{7}{9}$

✓Concept Check Answer

Rewrite $5\frac{1}{4}$ as $4\frac{5}{4}$ by borrowing from the 5.

Objective D Solving Problems Containing Mixed Numbers

Now that we know how to perform operations on mixed numbers, we can solve real-life problems.

Practice 15

The measurement around the trunk of a tree just below shoulder height is called its girth. The largest known American beech tree in the United States has a girth of $23\frac{1}{4}$ feet. The largest known sugar maple tree in the United States has a girth of $19\frac{5}{12}$ feet. How much larger is the girth of the largest known American beech tree than the girth of the largest known sugar maple tree? (*Source: American Forests*)

Example 15 Finding Legal Lobster Size

Lobster fishermen must measure the upper body shells of the lobsters they catch. Lobsters that are too small are thrown back into the ocean. Each state has its own size standard for lobsters to help control the breeding stock. Massachusetts divides its waters into four Lobster Conservation Management Areas, with a different minimum lobster size permitted in each area. In area three, the legal lobster size increased from $3\frac{13}{32}$ inches to $3\frac{1}{2}$ inches. How much of an increase was this? (*Source:* Massachusetts Division of Marine Fisheries)

Solution:

1. UNDERSTAND. Read and reread the problem carefully. The word "increase" found in the problem might make you think that we add to solve the problem. But the phrase "how much of an increase" tells us to subtract to find the increase.

2. TRANSLATE.

In words:	increase	is	new lobster size	minus	old lobster size
	↓	↓	↓	↓	↓
Translate:	increase	=	$3\frac{1}{2}$	−	$3\frac{13}{32}$

3. SOLVE. Before we solve, let's estimate by rounding to the nearest wholes. The fraction $3\frac{1}{2}$ can be rounded up to 4, $3\frac{13}{32}$ rounds to 3, and $4 - 3 = 1$. The increase is not 1, but will be smaller since we rounded $3\frac{1}{2}$ up and rounded $3\frac{13}{32}$ down.

$$\begin{array}{rcr} 3\frac{1}{2} & = & 3\frac{16}{32} \\ -\,3\frac{13}{32} & = & -\,3\frac{13}{32} \\ \hline & & \frac{3}{32} \end{array}$$

4. INTERPRET. *Check* your work. Our estimate tells us that the exact increase of $\frac{3}{32}$ is reasonable. *State* your conclusion: The increase in lobster size was $\frac{3}{32}$ of an inch.

Work Practice 15

Answer

15. $3\frac{5}{6}$ ft

Example 16 Calculating Manufacturing Materials Needed

In a manufacturing process, a metal-cutting machine cuts strips $1\frac{3}{5}$ inches long from a piece of metal stock. How many such strips can be cut from a 48-inch piece of stock?

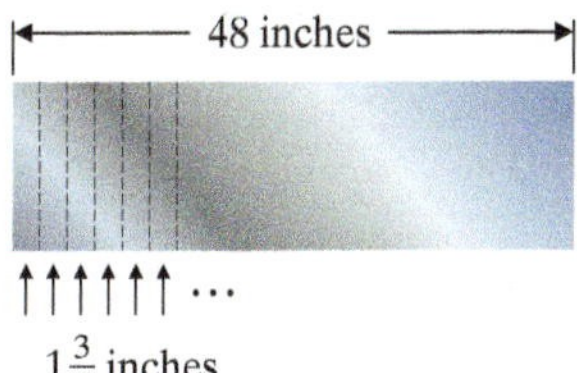

Solution:

1. UNDERSTAND the problem. To do so, read and reread the problem. Then draw a diagram:

   We want to know how many $1\frac{3}{5}$s there are in 48.

2. TRANSLATE.

In words:	number of strips	is	48	divided by	$1\frac{3}{5}$
	↓	↓	↓	↓	↓
Translate:	number of strips	=	48	÷	$1\frac{3}{5}$

3. SOLVE: Let's estimate a reasonable answer. The mixed number $1\frac{3}{5}$ rounds to 2 and 48 ÷ 2 = 24.

$$48 \div 1\frac{3}{5} = 48 \div \frac{8}{5} = \frac{48}{1} \cdot \frac{5}{8} = \frac{48 \cdot 5}{1 \cdot 8} = \frac{\overset{1}{\cancel{8}} \cdot 6 \cdot 5}{1 \cdot \underset{1}{\cancel{8}}} = \frac{30}{1} \text{ or } 30$$

4. INTERPRET. *Check* your work. Since the exact answer of 30 is close to our estimate of 24, our answer is reasonable. *State* your conclusion: Thirty strips can be cut from the 48-inch piece of stock.

Work Practice 16

Practice 16

A designer of women's clothing designs a woman's dress that requires $3\frac{1}{7}$ yards of material. How many dresses can be made from a 44-yard bolt of material?

Objective E Operating on Negative Mixed Numbers

To perform operations on negative mixed numbers, let's first practice writing these numbers as negative fractions and negative fractions as negative mixed numbers.

To understand negative mixed numbers, we simply need to know that, for example,

$$-3\frac{2}{5} \text{ means } -\left(3\frac{2}{5}\right)$$

Thus, to write a negative mixed number as a fraction, we do the following.

$$-3\frac{2}{5} = -\left(3\frac{2}{5}\right) = -\left(\frac{5 \cdot 3 + 2}{5}\right) = -\left(\frac{17}{5}\right) \text{ or } -\frac{17}{5}$$

Examples Write each as a fraction.

17. $-1\frac{7}{8} = -\frac{8 \cdot 1 + 7}{8} = -\frac{15}{8}$ Write $1\frac{7}{8}$ as an improper fraction and keep the negative sign.

18. $-23\frac{1}{2} = -\frac{2 \cdot 23 + 1}{2} = -\frac{47}{2}$ Write $23\frac{1}{2}$ as an improper fraction and keep the negative sign.

Work Practice 17–18

Practice 17–18

Write each as a fraction.

17. $-9\frac{3}{7}$ **18.** $-5\frac{10}{11}$

Answers

16. 14 dresses **17.** $-\frac{66}{7}$ **18.** $-\frac{65}{11}$

To write a negative fraction as a negative mixed number, we use a similar procedure. We simply disregard the negative sign, convert the improper fraction to a mixed number, and then reinsert the negative sign.

Practice 19–20

Write each as a mixed number.

19. $-\frac{37}{8}$ **20.** $-\frac{46}{5}$

Examples Write each as a mixed number.

19. $-\frac{22}{5} = -4\frac{2}{5}$

$$\begin{array}{r} 4 \\ 5\overline{)22} \\ -20 \\ \hline 2 \end{array} \qquad \frac{22}{5} = 4\frac{2}{5}$$

20. $-\frac{9}{4} = -2\frac{1}{4}$

$$\begin{array}{r} 2 \\ 4\overline{)9} \\ -8 \\ \hline 1 \end{array} \qquad \frac{9}{4} = 2\frac{1}{4}$$

Work Practice 19–20

We multiply or divide with negative mixed numbers the same way that we multiply or divide with positive mixed numbers. We first write each mixed number as a fraction.

Practice 21–22

21. $2\frac{3}{4} \cdot \left(-3\frac{3}{5}\right)$

22. $-4\frac{2}{7} \div 1\frac{1}{4}$

Examples Perform the indicated operations.

21. $-4\frac{2}{5} \cdot 1\frac{3}{11} = -\frac{22}{5} \cdot \frac{14}{11} = -\frac{22 \cdot 14}{5 \cdot 11} = -\frac{2 \cdot \overset{1}{\cancel{11}} \cdot 14}{5 \cdot \underset{1}{\cancel{11}}} = -\frac{28}{5}$ or $-5\frac{3}{5}$

22. $-2\frac{1}{3} \div \left(-2\frac{1}{2}\right) = -\frac{7}{3} \div \left(-\frac{5}{2}\right) = -\frac{7}{3} \cdot \left(-\frac{2}{5}\right) = \frac{7 \cdot 2}{3 \cdot 5} = \frac{14}{15}$

Work Practice 21–22

To add or subtract with negative mixed numbers, we must be very careful! Problems arise because recall that

$$-3\frac{2}{5} \text{ means } -\left(3\frac{2}{5}\right)$$

This means that

$$-3\frac{2}{5} = -\left(3\frac{2}{5}\right) = -\left(3 + \frac{2}{5}\right) = -3 - \frac{2}{5}$$ This can sometimes be easily overlooked.

To avoid problems, we will add or subtract negative mixed numbers by rewriting as addition and recalling how to add signed numbers.

Answers

19. $-4\frac{5}{8}$ **20.** $-9\frac{1}{5}$

21. $-9\frac{9}{10}$ **22.** $-3\frac{3}{7}$

Example 23 Add: $6\frac{3}{5} + \left(-9\frac{7}{10}\right)$

Solution: Here we are adding two numbers with different signs. Recall that we then subtract the absolute values and keep the sign of the larger absolute value.

Since $-9\frac{7}{10}$ has the larger absolute value, the answer is negative.

First, subtract absolute values:

$$\begin{array}{rcr} 9\frac{7}{10} & = & 9\frac{7}{10} \\ -\ 6\frac{3\cdot 2}{5\cdot 2} & = & -\ 6\frac{6}{10} \\ \hline & & 3\frac{1}{10} \end{array}$$

Thus,

$$6\frac{3}{5} + \left(-9\frac{7}{10}\right) = -3\frac{1}{10}$$ The result is negative since $-9\frac{7}{10}$ has the larger absolute value.

Work Practice 23

Practice 23

Add: $6\frac{2}{3} + \left(-12\frac{3}{4}\right)$

Example 24 Subtract: $-11\frac{5}{6} - 20\frac{4}{9}$

Solution: Let's write as an equivalent addition: $-11\frac{5}{6} + \left(-20\frac{4}{9}\right)$. Here, we are adding two numbers with like signs. Recall that we add their absolute values and keep the common negative sign.

First, add absolute values:

$$\begin{array}{rcr} 11\frac{5\cdot 3}{6\cdot 3} & = & 11\frac{15}{18} \\ +\ 20\frac{4\cdot 2}{9\cdot 2} & = & +\ 20\frac{8}{18} \\ \hline & & 31\frac{23}{18} \text{ or } 32\frac{5}{18} \end{array}$$

Since $\frac{23}{18} = 1\frac{5}{18}$

Thus,

$$-11\frac{5}{6} - 20\frac{4}{9} = -32\frac{5}{18}$$

Keep the common sign.

Work Practice 24

Practice 24

Subtract: $-9\frac{2}{7} - 30\frac{11}{14}$

Answers

23. $-6\frac{1}{12}$ 24. $-40\frac{1}{14}$

Calculator Explorations Converting Between Mixed Number and Fraction Notation

If your calculator has a fraction key, such as $\boxed{a^b\!/c}$, you can use it to convert between mixed number notation and fraction notation.

To write $13\frac{7}{16}$ as an improper fraction, press

$\boxed{13}$ $\boxed{a^b\!/c}$ $\boxed{7}$ $\boxed{a^b\!/c}$ $\boxed{16}$ $\boxed{2nd}$ $\boxed{d/c}$

The display will read

$\boxed{215|16}$

which represents $\frac{215}{16}$. Thus $13\frac{7}{16} = \frac{215}{16}$.

To convert $\frac{190}{13}$ to a mixed number, press

$\boxed{190}$ $\boxed{a^b\!/c}$ $\boxed{13}$ $=$

The display will read

$\boxed{14\_8/13}$

which represents $14\frac{8}{13}$. Thus $\frac{190}{13} = 14\frac{8}{13}$.

Write each mixed number as a fraction and each fraction as a mixed number.

1. $25\frac{5}{11}$ **2.** $67\frac{14}{15}$ **3.** $107\frac{31}{35}$

4. $186\frac{17}{21}$ **5.** $\frac{365}{14}$ **6.** $\frac{290}{13}$

7. $\frac{2769}{30}$ **8.** $\frac{3941}{17}$

Vocabulary, Readiness & Video Check

Use the choices below to fill in each blank.

round	fraction	whole number
improper	mixed number	

1. The number $5\frac{3}{4}$ is called a(n) ____________.

2. For $5\frac{3}{4}$, the 5 is called the ____________ part and $\frac{3}{4}$ is called the ____________ part.

3. To estimate operations on mixed numbers, we ____________ mixed numbers to the nearest whole number.

4. The mixed number $2\frac{5}{8}$ written as a(n) ____________ fraction is $\frac{21}{8}$.

Martin-Gay Interactive Videos

See Video 4.7

Watch the section lecture video and answer the following questions.

Objective A 5. In Example 1, why is the unit distance between −4 and −3 on the number line split into 5 equal parts?

Objective B 6. Why do we need to know how to multiply fractions to solve Example 2?

Objective C 7. In Example 4, why is the first form of the answer not an appropriate form?

Objective D 8. Why do we need to know how to subtract fractions to solve Example 5?

Objective E 9. In Example 6, how is it determined whether the answer is positive or negative?

4.7 Exercise Set MyMathLab® 

Objective A *Graph each list of numbers on the given number line. See Example 1.*

1. $-2, -2\frac{2}{3}, 0, \frac{7}{8}, -\frac{1}{3}$

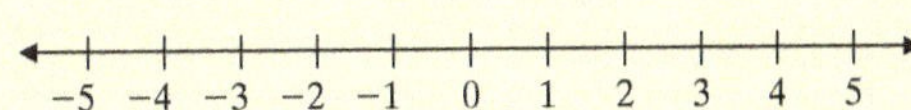

2. $-1, -1\frac{1}{4}, -\frac{1}{4}, 3\frac{1}{4}, 3$

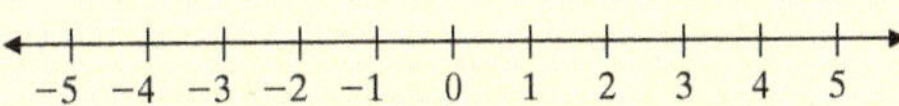

3. $4, \frac{1}{3}, -3, -3\frac{4}{5}, 1\frac{1}{3}$

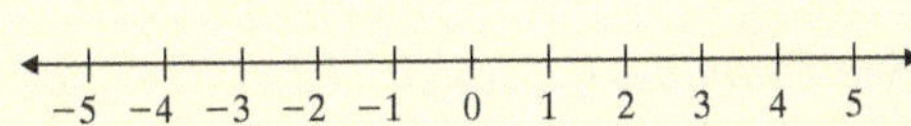

4. $3, \frac{3}{8}, -4, -4\frac{1}{3}, -\frac{9}{10}$

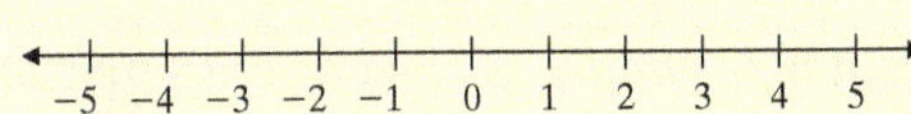

Objective B *Choose the best estimate for each product or quotient. See Examples 4 and 5.*

5. $2\frac{11}{12} \cdot 1\frac{1}{4}$
   a. 2 **b.** 3 **c.** 1 **d.** 12

6. $5\frac{1}{6} \cdot 3\frac{5}{7}$
   a. 9 **b.** 15 **c.** 8 **d.** 20

7. $12\frac{2}{11} \div 3\frac{9}{10}$
   a. 3 **b.** 4 **c.** 36 **d.** 9

8. $20\frac{3}{14} \div 4\frac{8}{11}$
   a. 5 **b.** 80 **c.** 4 **d.** 16

Multiply or divide. For Exercises 13 through 16, find an exact answer and an estimated answer. See Examples 2 through 8.

9. $2\frac{2}{3} \cdot \frac{1}{7}$

10. $\frac{5}{9} \cdot 4\frac{1}{5}$

11. $7 \div 1\frac{3}{5}$

12. $9 \div 1\frac{2}{3}$

13. $2\frac{1}{5} \cdot 3\frac{1}{2}$
    Exact:
    Estimate:

14. $2\frac{1}{4} \cdot 7\frac{1}{8}$
    Exact:
    Estimate:

15. $3\frac{4}{5} \cdot 6\frac{2}{7}$
    Exact:
    Estimate:

16. $5\frac{5}{6} \cdot 7\frac{3}{5}$
    Exact:
    Estimate:

17. $5 \cdot 2\frac{1}{2}$

18. $6 \cdot 3\frac{1}{3}$

19. $3\frac{2}{3} \cdot 1\frac{1}{2}$

20. $2\frac{4}{5} \cdot 2\frac{5}{8}$

21. $2\frac{2}{3} \div \frac{1}{7}$

22. $\frac{5}{9} \div 4\frac{1}{5}$

Objective C *Choose the best estimate for each sum or difference. See Examples 9 and 12.*

23. $3\frac{7}{8} + 2\frac{1}{5}$

a. 6 **b.** 5 **c.** 1 **d.** 2

24. $3\frac{7}{8} - 2\frac{1}{5}$

a. 6 **b.** 5 **c.** 1 **d.** 2

25. $8\frac{1}{3} + 1\frac{1}{2}$

a. 4 **b.** 10 **c.** 6 **d.** 16

26. $8\frac{1}{3} - 1\frac{1}{2}$

a. 4 **b.** 10 **c.** 6 **d.** 16

Add. For Exercises 27 through 30, find an exact sum and an estimated sum. See Examples 9 through 11.

27. $4\frac{7}{12} + 2\frac{1}{12}$

Exact:

Estimate:

28. $7\frac{4}{11} + 3\frac{2}{11}$

Exact:

Estimate:

29. $10\frac{3}{14} + 3\frac{4}{7}$

Exact:

Estimate:

30. $12\frac{5}{12} + 4\frac{1}{6}$

Exact:

Estimate:

31. $9\frac{1}{5} + 8\frac{2}{25}$

32. $6\frac{2}{13} + 8\frac{7}{26}$

33. $12\frac{3}{14}$, 10, $+ 25\frac{5}{12}$

34. $8\frac{2}{9}$, 32, $+ 9\frac{10}{21}$

35. $15\frac{4}{7} + 9\frac{11}{14}$

36. $23\frac{3}{5} + 8\frac{8}{15}$

37. $3\frac{5}{8}$, $2\frac{1}{6}$, $+ 7\frac{3}{4}$

38. $4\frac{1}{3}$, $9\frac{2}{5}$, $+ 3\frac{1}{6}$

Subtract. For Exercises 39 through 42, find an exact difference and an estimated difference. See Examples 12 through 14.

39. $4\frac{7}{10} - 2\frac{1}{10}$

Exact:

Estimate:

40. $7\frac{4}{9} - 3\frac{2}{9}$

Exact:

Estimate:

41. $10\frac{13}{14} - 3\frac{4}{7}$

Exact:

Estimate:

42. $12\frac{5}{12} - 4\frac{1}{6}$

Exact:

Estimate:

43. $\begin{array}{r} 9\frac{1}{5} \\ -8\frac{6}{25} \\ \hline \end{array}$

44. $\begin{array}{r} 5\frac{2}{13} \\ -4\frac{7}{26} \\ \hline \end{array}$

45. $\begin{array}{r} 6 \\ -2\frac{4}{9} \\ \hline \end{array}$

46. $\begin{array}{r} 8 \\ -1\frac{7}{10} \\ \hline \end{array}$

47. $\begin{array}{r} 63\frac{1}{6} \\ -47\frac{5}{12} \\ \hline \end{array}$

48. $\begin{array}{r} 86\frac{2}{15} \\ -27\frac{3}{10} \\ \hline \end{array}$

Objectives B C Mixed Practice *Perform each indicated operation. See Examples 2 through 14.*

49. $\begin{array}{r} 2\frac{3}{4} \\ +1\frac{1}{4} \\ \hline \end{array}$

50. $\begin{array}{r} 5\frac{5}{8} \\ +2\frac{3}{8} \\ \hline \end{array}$

51. $\begin{array}{r} 15\frac{4}{7} \\ -9\frac{11}{14} \\ \hline \end{array}$

52. $\begin{array}{r} 23\frac{8}{15} \\ -8\frac{3}{5} \\ \hline \end{array}$

53. $3\frac{1}{9} \cdot 2$

54. $4\frac{1}{2} \cdot 3$

55. $1\frac{2}{3} \div 2\frac{1}{5}$

56. $5\frac{1}{5} \div 3\frac{1}{4}$

57. $22\frac{4}{9} + 13\frac{5}{18}$

58. $15\frac{3}{25} + 5\frac{2}{5}$

59. $5\frac{2}{3} - 3\frac{1}{6}$

60. $5\frac{3}{8} - 2\frac{3}{16}$

61. $\begin{array}{r} 15\frac{1}{5} \\ 20\frac{3}{10} \\ +37\frac{2}{15} \\ \hline \end{array}$

62. $\begin{array}{r} 3\frac{7}{16} \\ 6\frac{1}{2} \\ +9\frac{3}{8} \\ \hline \end{array}$

63. $6\frac{4}{7} - 5\frac{11}{14}$

64. $47\frac{5}{12} - 23\frac{19}{24}$

65. $4\frac{2}{7} \cdot 1\frac{3}{10}$

66. $6\frac{2}{3} \cdot 2\frac{3}{4}$

67. $\begin{array}{r} 6\frac{2}{11} \\ 3 \\ +4\frac{10}{33} \\ \hline \end{array}$

68. $\begin{array}{r} 7\frac{3}{7} \\ 15 \\ +20\frac{1}{2} \\ \hline \end{array}$

Objective D **Translating** *Translate each phrase into an algebraic expression. Use x to represent "a number." See Examples 15 and 16.*

69. $-5\frac{2}{7}$ decreased by a number

70. The sum of $8\frac{3}{4}$ and a number

71. Multiply $1\frac{9}{10}$ by a number.

72. Divide a number by $-6\frac{1}{11}$.

Solve. For Exercises 73 and 74, the solutions have been started for you. Write each answer in simplest form. See Examples 15 and 16.

73. A heart attack patient in rehabilitation walked on a treadmill $12\frac{3}{4}$ miles over 4 days. How many miles is this per day on average?

Start the solution:

1. UNDERSTAND the problem. Reread it as many times as needed.
2. TRANSLATE into an equation. (Fill in the blanks.)

miles per day	is	total miles	divided by	number of days
↓	↓	↓	↓	↓
miles per day	=	____	÷	____

Finish with:

3. SOLVE and **4.** INTERPRET

74. A local restaurant is selling hamburgers from a booth on Memorial Day. A total of $27\frac{3}{4}$ pounds of hamburger have been ordered. How many quarter-pound hamburgers can this make?

Start the solution:

1. UNDERSTAND the problem. Reread it as many times as needed.
2. TRANSLATE into an equation. (Fill in the blanks.)

how many quarter-pound hamburgers	is	total pounds of hamburger	divided by	a quarter pound
↓	↓	↓	↓	↓
how many quarter-pound hamburgers	=	____	÷	____

Finish with:

3. SOLVE and **4.** INTERPRET

75. The Gauge Act of 1846 set the standard gauge for U.S. railroads at $56\frac{1}{2}$ inches. (See figure.) If the standard gauge in Spain is $65\frac{9}{10}$ inches, how much wider is Spain's standard gauge than the U.S. standard gauge? (*Source:* San Diego Railroad Museum)

76. The standard railroad track gauge (see figure) in Spain is $65\frac{9}{10}$ inches, while in neighboring Portugal it is $65\frac{11}{20}$ inches. Which gauge is wider and by how much? (*Source:* San Diego Railroad Museum)

Track gauge (U.S. $56\frac{1}{2}$ inches)

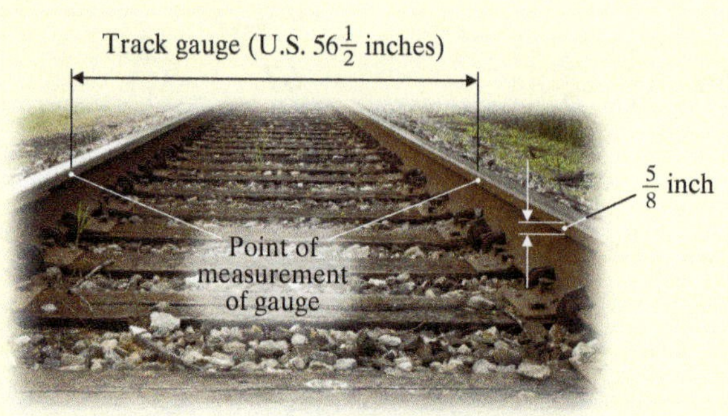

77. If Tucson's average rainfall is $11\frac{1}{4}$ inches and Yuma's is $3\frac{3}{5}$ inches, how much more rain, on the average, does Tucson get than Yuma?

78. A pair of crutches needs adjustment. One crutch is 43 inches and the other is $41\frac{5}{8}$ inches. Find how much the shorter crutch should be lengthened to make both crutches the same length.

For Exercises 79 and 80, find the area of each figure.

△ **79.**

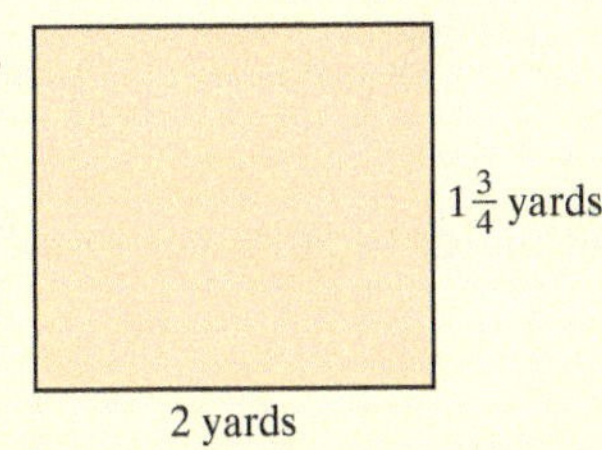

△ **80.**

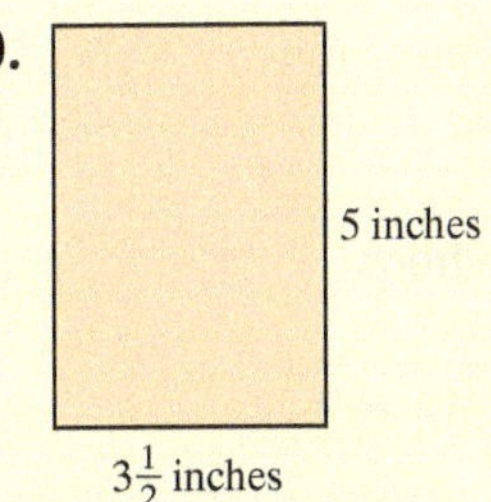

△ **81.** A model for a proposed computer chip measures $\frac{3}{4}$ inch by $1\frac{1}{4}$ inches. Find its area.

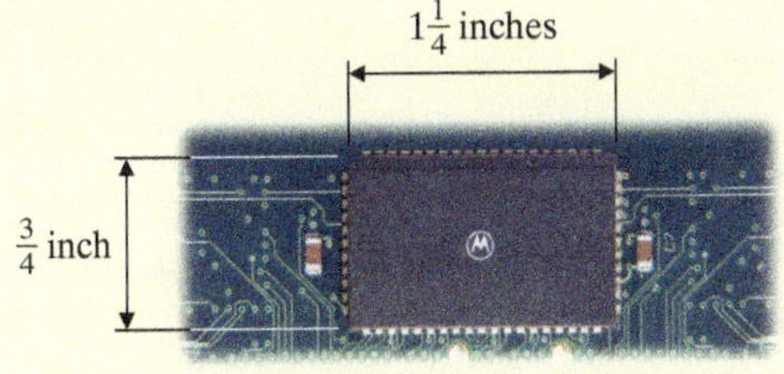

△ **82.** The Saltalamachios are planning to build a deck that measures $4\frac{1}{2}$ yards by $6\frac{1}{3}$ yards. Find the area of their proposed deck.

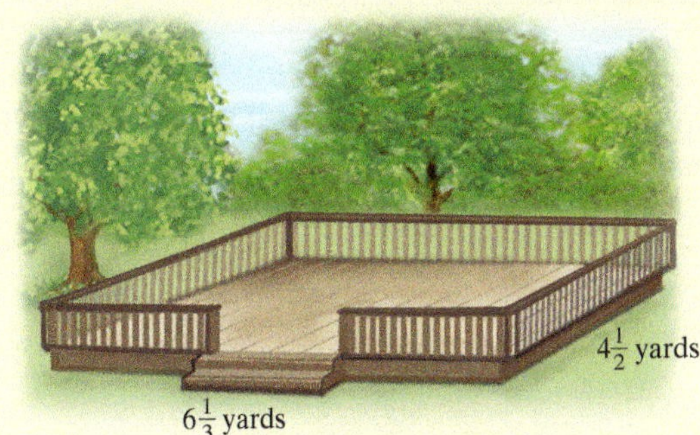

For Exercises 83 and 84, find the perimeter of each figure.

△ **83.**

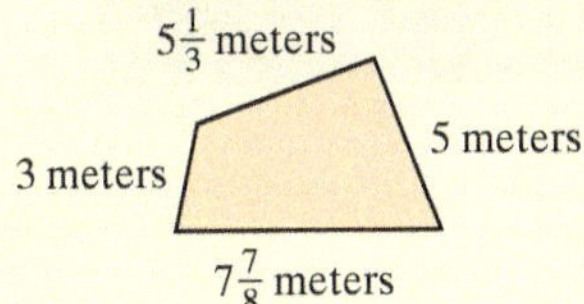

△ **84.**

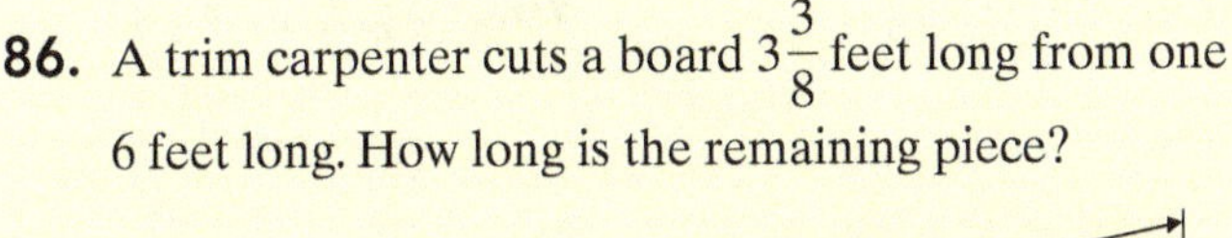

85. A homeowner has $15\frac{2}{3}$ feet of plastic pipe. She cuts off a $2\frac{1}{2}$-foot length and then a $3\frac{1}{4}$-foot length. If she now needs a 10-foot piece of pipe, will the remaining piece do? If not, by how much will the piece be short?

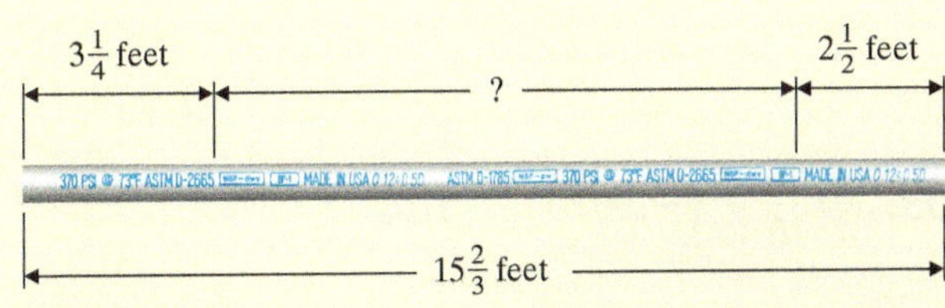

86. A trim carpenter cuts a board $3\frac{3}{8}$ feet long from one 6 feet long. How long is the remaining piece?

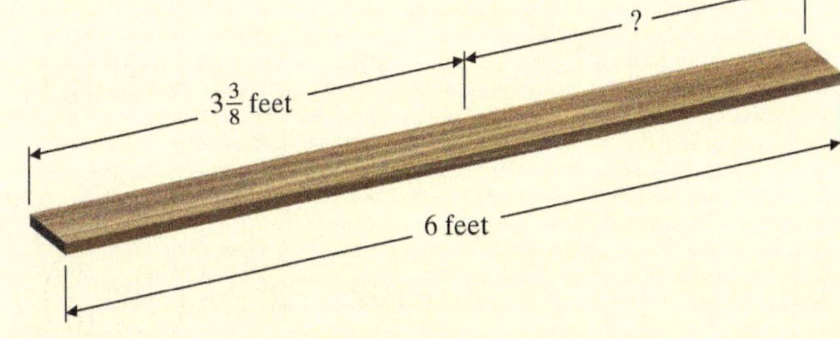

△ **87.** The area of the rectangle below is 12 square meters. If its width is $2\frac{4}{7}$ meters, find its length.

Rectangle

$2\frac{4}{7}$ meters

△ **88.** The perimeter of the square below is $23\frac{1}{2}$ feet. Find the length of each side.

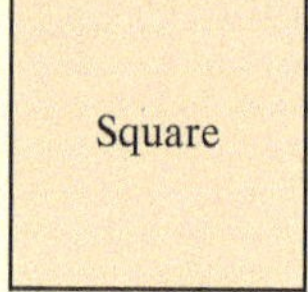

The following table lists three upcoming total eclipses of the Sun that will be visible in North America. The duration of each eclipse is listed in the table. Use the table to answer Exercises 89 through 92.

Total Solar Eclipses Visible from North America

Date of Eclipse	Duration (in minutes)
August 21, 2017	$2\frac{2}{3}$
April 8, 2024	$4\frac{7}{15}$
March 30, 2033	$2\frac{37}{60}$

(*Source:* NASA/Goddard Space Flight Center)

89. What is the total duration for the three eclipses?

90. What is the total duration for the two eclipses occurring in odd-numbered years?

91. How much longer will the April 8, 2024, eclipse be than the August 21, 2017, eclipse?

92. How much longer will the April 8, 2024, eclipse be than the March 30, 2033, eclipse?

Objective E *Perform the indicated operations. See Examples 17 through 24.*

93. $-4\frac{2}{5} \cdot 2\frac{3}{10}$

94. $-3\frac{5}{6} \div \left(-3\frac{2}{3}\right)$

95. $-5\frac{1}{8} - 19\frac{3}{4}$

96. $17\frac{5}{9} + \left(-14\frac{2}{3}\right)$

97. $-31\frac{2}{15} + 17\frac{3}{20}$

98. $-31\frac{7}{8} - \left(-26\frac{5}{12}\right)$

99. $-1\frac{5}{7} \cdot \left(-2\frac{1}{2}\right)$

100. $1\frac{3}{4} \div \left(-3\frac{1}{2}\right)$

101. $11\frac{7}{8} - 13\frac{5}{6}$

102. $-20\frac{2}{5} + \left(-30\frac{3}{10}\right)$

103. $-7\frac{3}{10} \div (-100)$

104. $-4\frac{1}{4} \div 2\frac{3}{8}$

Review

Multiply. See Section 4.3.

105. $\frac{1}{3}(3x)$

106. $\frac{1}{5}(5y)$

107. $\frac{2}{3}\left(\frac{3}{2}a\right)$

108. $-\frac{9}{10}\left(-\frac{10}{9}m\right)$

Concept Extensions

Solve. See the first Concept Check in this section.

109. Which of the following are equivalent to 10?

a. $9\frac{5}{5}$ **b.** $9\frac{100}{100}$ **c.** $6\frac{44}{11}$ **d.** $8\frac{13}{13}$

110. Which of the following are equivalent to $7\frac{3}{4}$?

a. $6\frac{7}{4}$ **b.** $5\frac{11}{4}$ **c.** $7\frac{12}{16}$ **d.** all of them

Solve. See the second Concept Check in this section.

111. A student asked you to check her work below. Is it correct? If not, where is the error?

$$20\frac{2}{3} \div 10\frac{1}{2} \stackrel{?}{=} 2\frac{1}{3}$$

112. A student asked you to check his work below. Is it correct? If not, where is the error?

$$3\frac{2}{3} \cdot 1\frac{1}{7} \stackrel{?}{=} 3\frac{2}{21}$$

113. In your own words, describe how to divide mixed numbers.

114. In your own words, explain how to multiply

a. fractions

b. mixed numbers

Solve. See the third Concept Check in this section.

115. In your own words, explain how to round a mixed number to the nearest whole number.

116. Use rounding to estimate the best sum for $11\frac{19}{20} + 9\frac{1}{10}$.

a. 2 **b.** 3 **c.** 20 **d.** 21

Solve.

117. Explain in your own words why $9\frac{13}{9}$ is equal to $10\frac{4}{9}$.

118. In your own words, explain

a. when to borrow when subtracting mixed numbers, and

b. how to borrow when subtracting mixed numbers.

4.8 Solving Equations Containing Fractions 

Objectives

A Solve Equations Containing Fractions.

B Solve Equations by Multiplying by the LCD.

C Review Adding and Subtracting Fractions.

Objective A Solving Equations Containing Fractions

In Chapter 3, we solved linear equations in one variable. In this section, we practice this skill by solving linear equations containing fractions. To help us solve these equations, let's review the properties of equality.

Addition Property of Equality

Let a, b, and c represent numbers. Then

$a = b$	Also, $a = b$
and $a + c = b + c$	and $a - c = b - c$
are equivalent equations.	are equivalent equations.

Multiplication Property of Equality

Let a, b, and c represent numbers and let $c \neq 0$. Then

$a = b$	Also, $a = b$
and $a \cdot c = b \cdot c$	and $\frac{a}{c} = \frac{b}{c}$
are equivalent equations.	are equivalent equations.

In other words, the same number may be added to or subtracted from both sides of an equation without changing the solution of the equation. Also, both sides of an equation may be multiplied or divided by the same nonzero number without changing the solution.

Also, don't forget that to solve an equation in x, our goal is to use properties of equality to write simpler equations, all equivalent to the original equation, until the final equation has the form

$x =$ **number** or **number** $= x$

Practice 1

Solve: $y - \frac{2}{3} = \frac{5}{12}$

Example 1 Solve: $x - \frac{3}{4} = \frac{1}{20}$

Solution: To get x by itself, add $\frac{3}{4}$ to both sides.

$$x - \frac{3}{4} = \frac{1}{20}$$

$$x - \frac{3}{4} + \frac{3}{4} = \frac{1}{20} + \frac{3}{4} \quad \text{Add } \tfrac{3}{4} \text{ to both sides.}$$

$$x = \frac{1}{20} + \frac{3 \cdot 5}{4 \cdot 5} \quad \text{The LCD of the denominators 20 and 4 is 20.}$$

$$x = \frac{1}{20} + \frac{15}{20}$$

$$x = \frac{16}{20}$$

$$x = \frac{\overset{1}{\cancel{4}} \cdot 4}{\underset{1}{\cancel{4}} \cdot 5} = \frac{4}{5} \quad \text{Write } \tfrac{16}{20} \text{ in simplest form.}$$

Answer

1. $\frac{13}{12}$

Check: To check, replace x with $\frac{4}{5}$ in the original equation.

$$x - \frac{3}{4} = \frac{1}{20}$$

$$\frac{4}{5} - \frac{3}{4} \stackrel{?}{=} \frac{1}{20} \quad \text{Replace } x \text{ with } \frac{4}{5}.$$

$$\frac{4 \cdot 4}{5 \cdot 4} - \frac{3 \cdot 5}{4 \cdot 5} \stackrel{?}{=} \frac{1}{20} \quad \text{The LCD of 5 and 4 is 20.}$$

$$\frac{16}{20} - \frac{15}{20} \stackrel{?}{=} \frac{1}{20}$$

$$\frac{1}{20} = \frac{1}{20} \quad \text{True}$$

Thus $\frac{4}{5}$ is the solution of $x - \frac{3}{4} = \frac{1}{20}$.

Work Practice 1

Example 2 Solve: $\frac{1}{3}x = 7$

Solution: Recall that isolating x means that we want the coefficient of x to be 1. To do so, we use the multiplication property of equality and multiply both sides of the equation by the reciprocal of $\frac{1}{3}$, or 3. Since $\frac{1}{3} \cdot 3 = 1$, we will have isolated x.

$$\frac{1}{3}x = 7$$

$$3 \cdot \frac{1}{3}x = 3 \cdot 7 \quad \text{Multiply both sides by 3.}$$

$$1 \cdot x = 21 \text{ or } x = 21 \quad \text{Simplify.}$$

Check: To check, replace x with 21 in the original equation.

$$\frac{1}{3}x = 7 \quad \text{Original equation}$$

$$\frac{1}{3} \cdot 21 \stackrel{?}{=} 7 \quad \text{Replace } x \text{ with 21.}$$

$$7 = 7 \quad \text{True}$$

Since $7 = 7$ is a true statement, 21 is the solution of $\frac{1}{3}x = 7$.

Work Practice 2

Example 3 Solve: $\frac{3}{5}a = 9$

Solution: Multiply both sides by $\frac{5}{3}$, the reciprocal of $\frac{3}{5}$, so that the coefficient of a is 1.

$$\frac{3}{5}a = 9$$

$$\frac{5}{3} \cdot \frac{3}{5}a = \frac{5}{3} \cdot 9 \quad \text{Multiply both sides by } \frac{5}{3}.$$

$$1a = \frac{5 \cdot 9}{3} \quad \text{Multiply.}$$

$$a = 15 \quad \text{Simplify.}$$

(Continued on next page)

Practice 2

Solve: $\frac{1}{5}y = 2$

Practice 3

Solve: $\frac{5}{7}b = 25$

Answers

2. 10 **3.** 35

Check: To check, replace a with 15 in the original equation.

$$\frac{3}{5}a = 9$$

$$\frac{3}{5} \cdot 15 \stackrel{?}{=} 9 \quad \text{Replace } a \text{ with 15.}$$

$$\frac{3 \cdot 15}{5} \stackrel{?}{=} 9 \quad \text{Multiply.}$$

$$9 = 9 \quad \text{True}$$

Since $9 = 9$ is true, 15 is the solution of $\frac{3}{5}a = 9$.

Work Practice 3

Practice 4

Solve: $-\frac{7}{10}x = \frac{2}{5}$

Example 4 Solve: $\frac{3}{4}x = -\frac{1}{8}$

Solution: Multiply both sides of the equation by $\frac{4}{3}$, the reciprocal of $\frac{3}{4}$.

$$\frac{3}{4}x = -\frac{1}{8}$$

$$\frac{4}{3} \cdot \frac{3}{4}x = \frac{4}{3} \cdot -\frac{1}{8} \quad \text{Multiply both sides by } \frac{4}{3}.$$

$$1x = -\frac{4 \cdot 1}{3 \cdot 8} \quad \text{Multiply.}$$

$$x = -\frac{1}{6} \quad \text{Simplify.}$$

Check: To check, replace x with $-\frac{1}{6}$ in the original equation.

$$\frac{3}{4}x = -\frac{1}{8} \quad \text{Original equation}$$

$$\frac{3}{4} \cdot -\frac{1}{6} \stackrel{?}{=} -\frac{1}{8} \quad \text{Replace } x \text{ with } -\frac{1}{6}.$$

$$-\frac{1}{8} = -\frac{1}{8} \quad \text{True}$$

Since we arrived at a true statement, $-\frac{1}{6}$ is the solution of $\frac{3}{4}x = -\frac{1}{8}$.

Work Practice 4

Practice 5

Solve: $5x = -\frac{3}{4}$

Example 5 Solve: $3y = -\frac{2}{11}$

Solution: We can either divide both sides by 3 or multiply both sides by the reciprocal of 3, which is $\frac{1}{3}$.

$$3y = -\frac{2}{11}$$

$$\frac{1}{3} \cdot 3y = \frac{1}{3} \cdot -\frac{2}{11} \quad \text{Multiply both sides by } \frac{1}{3}.$$

$$1y = -\frac{1 \cdot 2}{3 \cdot 11} \quad \text{Multiply.}$$

$$y = -\frac{2}{33} \quad \text{Simplify.}$$

Check to see that the solution is $-\frac{2}{33}$.

Work Practice 5

Answers

4. $-\frac{4}{7}$ 5. $-\frac{3}{20}$

Objective B Solving Equations by Multiplying by the LCD 

Solving equations with fractions can be tedious. If an equation contains fractions, it is often helpful to first multiply both sides of the equation by the LCD of the fractions. This has the effect of eliminating the fractions in the equation, as shown in the next example.

Let's solve the equation in Example 4 again. This time, we will multiply both sides by the LCD.

Example 6 Solve: $\frac{3}{4}x = -\frac{1}{8}$ (Example 4 solved an alternate way.)

Solution: First, multiply both sides of the equation by the LCD of the fractions $\frac{3}{4}$ and $-\frac{1}{8}$. The LCD of the denominators is 8.

$$\frac{3}{4}x = -\frac{1}{8}$$

$$8 \cdot \frac{3}{4}x = 8 \cdot -\frac{1}{8} \quad \text{Multiply both sides by 8.}$$

$$\frac{8 \cdot 3}{1 \cdot 4}x = -\frac{\overset{1}{\cancel{8}} \cdot 1}{1 \cdot \underset{1}{\cancel{8}}} \quad \text{Multiply the fractions.}$$

$$\frac{2 \cdot \overset{1}{\cancel{4}} \cdot 3}{1 \cdot \underset{1}{\cancel{4}}}x = -\frac{1 \cdot 1}{1 \cdot 1} \quad \text{Simplify.}$$

$$6x = -1$$

$$\frac{6x}{6} = \frac{-1}{6} \quad \text{Divide both sides by 6.}$$

$$x = \frac{-1}{6} \text{ or } -\frac{1}{6} \quad \text{Simplify.}$$

As seen in Example 4, the solution is $-\frac{1}{6}$.

Work Practice 6

Practice 6

Solve: $\frac{11}{15}x = -\frac{3}{5}$

Example 7 Solve: $\frac{x}{6} + 1 = \frac{4}{3}$

Solution:

Solve by multiplying by the LCD:

The LCD of the denominators 6 and 3 is 6.

$$\frac{x}{6} + 1 = \frac{4}{3}$$

$$6\left(\frac{x}{6} + 1\right) = 6\left(\frac{4}{3}\right) \quad \text{Multiply both sides by 6.}$$

$$\overset{1}{\cancel{6}}\left(\frac{x}{\underset{1}{\cancel{6}}}\right) + 6(1) = \overset{2}{\cancel{6}}\left(\frac{4}{\underset{1}{\cancel{3}}}\right) \quad \text{Apply the distributive property.}$$

$$x + 6 = 8 \quad \text{Simplify.}$$

$$x + 6 + (-6) = 8 + (-6) \quad \text{Add } -6 \text{ to both sides.}$$

$$x = 2 \quad \text{Simplify.}$$

Solve with fractions:

$$\frac{x}{6} + 1 = \frac{4}{3}$$

$$\frac{x}{6} + 1 - 1 = \frac{4}{3} - 1 \quad \text{Subtract 1 from both sides.}$$

$$\frac{x}{6} = \frac{4}{3} - \frac{3}{3} \quad \text{Write 1 as } \frac{3}{3}.$$

$$\frac{x}{6} = \frac{1}{3} \quad \text{Subtract.}$$

$$6 \cdot \frac{x}{6} = 6 \cdot \frac{1}{3} \quad \text{Multiply both sides by 6.}$$

$$6 \cdot \frac{1}{6} \cdot x = \frac{\overset{2}{\cancel{6}} \cdot 1}{1 \cdot \underset{1}{\cancel{3}}}$$

$$1x = 2 \text{ or } x = 2$$

(Continued on next page)

Practice 7

Solve: $\frac{y}{8} + \frac{3}{4} = 2$

Answers

6. $-\frac{9}{11}$ 7. 10

Check: To check, replace x with 2 in the original equation.

$$\frac{x}{6} + 1 = \frac{4}{3} \quad \text{Original equation}$$

$$\frac{2}{6} + 1 \stackrel{?}{=} \frac{4}{3} \quad \text{Replace } x \text{ with 2.}$$

$$\frac{1}{3} + \frac{3}{3} \stackrel{?}{=} \frac{4}{3} \quad \text{Simplify } \frac{2}{6}. \text{ The LCD of 3 and 1 is 3.}$$

$$\frac{4}{3} = \frac{4}{3} \quad \text{True}$$

Since we arrived at a true statement, 2 is the solution of $\frac{x}{6} + 1 = \frac{4}{3}$.

Work Practice 7

Let's review the steps for solving equations in x. An extra step is now included to handle equations containing fractions.

Solving an Equation in x

Step 1: If fractions are present, multiply both sides of the equation by the LCD of the fractions.

Step 2: If parentheses are present, use the distributive property.

Step 3: Combine any like terms on each side of the equation.

Step 4: Use the addition property of equality to rewrite the equation so that variable terms are on one side of the equation and constant terms are on the other side.

Step 5: Divide both sides of the equation by the numerical coefficient of x to solve.

Step 6: Check the answer in the **original equation.**

Practice 8

Solve: $\frac{x}{5} - x = \frac{1}{5}$

Helpful Hint Don't forget to multiply *both* sides of the equation by the LCD.

Example 8 Solve: $\frac{z}{5} - \frac{z}{3} = 6$

Solution:

$$\frac{z}{5} - \frac{z}{3} = 6$$

$$15\left(\frac{z}{5} - \frac{z}{3}\right) = 15(6) \quad \text{Multiply both sides by the LCD, 15.}$$

$$\overset{3}{\cancel{15}}\left(\frac{z}{\underset{1}{\cancel{5}}}\right) - \overset{5}{\cancel{15}}\left(\frac{z}{\underset{1}{\cancel{3}}}\right) = 15(6) \quad \text{Apply the distributive property.}$$

$$3z - 5z = 90 \quad \text{Simplify.}$$

$$-2z = 90 \quad \text{Combine like terms.}$$

$$\frac{-2z}{-2} = \frac{90}{-2} \quad \text{Divide both sides by } -2, \text{ the coefficient of } z.$$

$$z = -45 \quad \text{Simplify.}$$

To check, replace z with -45 in the **original equation** to see that a true statement results.

Work Practice 8

Answer

8. $-\frac{1}{4}$

Example 9 Solve: $\frac{x}{2} = \frac{x}{3} + \frac{1}{2}$

Solution: First multiply both sides by the LCD, 6.

$$\frac{x}{2} = \frac{x}{3} + \frac{1}{2}$$

$$6\left(\frac{x}{2}\right) = 6\left(\frac{x}{3} + \frac{1}{2}\right) \quad \text{Multiply both sides by the LCD, 6.}$$

$$\overset{3}{\cancel{6}}\left(\frac{x}{\underset{1}{\cancel{2}}}\right) = \overset{2}{\cancel{6}}\left(\frac{x}{\underset{1}{\cancel{3}}}\right) + \overset{3}{\cancel{6}}\left(\frac{1}{\underset{1}{\cancel{2}}}\right) \quad \text{Apply the distributive property.}$$

$$3x = 2x + 3 \quad \text{Simplify.}$$

$$3x - 2x = 2x + 3 - 2x \quad \text{Subtract } 2x \text{ from both sides.}$$

$$x = 3 \quad \text{Simplify.}$$

To check, replace x with 3 in the original equation to see that a true statement results.

Work Practice 9

Practice 9

Solve: $\frac{y}{2} = \frac{y}{5} + \frac{3}{2}$

Objective C Review of Adding and Subtracting Fractions 

Make sure you understand the difference between **solving an equation** containing fractions and **adding or subtracting two fractions.** To solve an equation containing fractions, we use the multiplication property of equality and multiply both sides by the LCD of the fractions, thus eliminating the fractions. This method does not apply to adding or subtracting fractions. The multiplication property of equality applies only to equations. To add or subtract unlike fractions, we write each fraction as an equivalent fraction using the LCD of the fractions as the denominator. See the next example for a review.

Example 10 Add: $\frac{x}{3} + \frac{2}{5}$

Solution: This is an expression, not an equation. Here, we are adding two unlike fractions. To add unlike fractions, we need to find the LCD. The LCD of the denominators 3 and 5 is 15. Write each fraction as an equivalent fraction with a denominator of 15.

$$\frac{x}{3} + \frac{2}{5} = \frac{x}{3} \cdot \frac{5}{5} + \frac{2}{5} \cdot \frac{3}{3} = \frac{x \cdot 5}{3 \cdot 5} + \frac{2 \cdot 3}{5 \cdot 3}$$

$$= \frac{5x}{15} + \frac{6}{15}$$

$$= \frac{5x + 6}{15}$$

Work Practice 10

Practice 10

Subtract: $\frac{9}{10} - \frac{y}{3}$

✓**Concept Check** Which of the following are equations and which are expressions?

a. $\frac{1}{2} + 3x = 5$ **b.** $\frac{2}{3}x - \frac{x}{5}$

c. $\frac{x}{12} + \frac{5x}{24}$ **d.** $\frac{x}{5} = \frac{1}{10}$

Answers

9. 5 **10.** $\frac{27 - 10y}{30}$

✓**Concept Check Answers**

equations: a, d; expressions: b, c

Vocabulary, Readiness & Video Check

Fill in the blank with the least common denominator (LCD). Do not solve these equations.

1. Equation: $\frac{2}{3} + x = \frac{5}{6}$; LCD = ___

2. Equation: $\frac{x}{21} - 1 = \frac{1}{7}$; LCD = ___

3. Equation: $\frac{y}{5} + \frac{1}{3} = 2$; LCD = ___

4. Equation: $\frac{-2n}{11} + \frac{1}{2} = 5$; LCD = ___

Martin-Gay Interactive Videos

See Video 4.8 

Watch the section lecture video and answer the following questions.

Objective A **5.** In Example 1, what property is used to get x by itself on one side of the equation?

6. Explain how a reciprocal is used to solve Example 2.

Objective B **7.** Why are both sides of the equation multiplied by 12 in Example 3? What effect does this have on the fractions in the equation?

Objective C **8.** In Example 5, why can't we multiply by the LCD of all fractions?

4.8 Exercise Set MyMathLab®

Objective A *Solve each equation. Check your proposed solution. See Example 1.*

1. $x + \frac{1}{3} = -\frac{1}{3}$

2. $x + \frac{1}{9} = -\frac{7}{9}$

3. $y - \frac{3}{13} = -\frac{2}{13}$

4. $z - \frac{5}{14} = \frac{4}{14}$

5. $3x - \frac{1}{5} - 2x = \frac{1}{5} + \frac{2}{5}$

6. $5x + \frac{1}{11} - 4x = \frac{2}{11} - \frac{5}{11}$

7. $x - \frac{1}{12} = \frac{5}{6}$

8. $y - \frac{8}{9} = \frac{1}{3}$

9. $\frac{2}{5} + y = -\frac{3}{10}$

10. $\frac{1}{2} + a = -\frac{3}{8}$

11. $7z + \frac{1}{16} - 6z = \frac{3}{4}$

12. $9x - \frac{2}{7} - 8x = \frac{11}{14}$

13. $-\frac{2}{9} = x - \frac{5}{6}$

14. $-\frac{1}{4} = y - \frac{7}{10}$

Solve each equation. See Examples 2 through 5.

15. $7x = 2$

16. $-5x = 4$

17. $\frac{1}{4}x = 3$

18. $\frac{1}{3}x = 6$

19. $\frac{2}{9}y = -6$

20. $\frac{4}{7}x = -8$

21. $-\frac{4}{9}z = -\frac{3}{2}$

22. $-\frac{11}{10}x = -\frac{2}{7}$

23. $7a = \frac{1}{3}$

24. $2z = -\frac{5}{12}$

25. $-3x = -\frac{6}{11}$

26. $-4z = -\frac{12}{25}$

Objective B *Solve each equation. See Examples 6 through 9.*

27. $\frac{5}{9}x = -\frac{3}{18}$

28. $\frac{3}{5}y = -\frac{7}{20}$

29. $\frac{x}{3} + 2 = \frac{7}{3}$

30. $\frac{x}{5} - 1 = \frac{7}{5}$

31. $\frac{x}{5} - x = -8$

32. $\frac{x}{3} - x = -6$

33. $\frac{1}{2} - \frac{3}{5} = \frac{x}{10}$

34. $\frac{2}{3} - \frac{1}{4} = \frac{x}{12}$

35. $\frac{x}{3} = \frac{x}{5} - 2$

36. $\frac{a}{2} = \frac{a}{7} + \frac{5}{2}$

Objective C *Add or subtract as indicated. See Example 10.*

37. $\frac{x}{7} - \frac{4}{3}$

38. $-\frac{5}{9} + \frac{y}{8}$

39. $\frac{y}{2} + 5$

40. $2 + \frac{7x}{3}$

41. $\frac{3x}{10} + \frac{x}{6}$

42. $\frac{9x}{8} - \frac{5x}{6}$

Objectives A B C Mixed Practice *Solve. If no equation is given, perform the indicated operation. See Examples 1 through 10.*

43. $\frac{3}{8}x = \frac{1}{2}$

44. $\frac{2}{5}y = \frac{3}{10}$

45. $\frac{2}{3} - \frac{x}{5} = \frac{4}{15}$

46. $\frac{4}{5} + \frac{x}{4} = \frac{21}{20}$

47. $\frac{9}{14}z = \frac{27}{20}$

48. $\frac{5}{16}a = \frac{5}{6}$

49. $-3m - 5m = \frac{4}{7}$

50. $30n - 34n = \frac{3}{20}$

51. $\frac{x}{4} + 1 = \frac{1}{4}$

52. $\frac{y}{7} - 2 = \frac{1}{7}$

53. $\frac{5}{9} - \frac{2}{3}$

54. $\frac{8}{11} - \frac{1}{2}$

55. $\frac{1}{5}y = 10$

56. $\frac{1}{4}x = -2$

57. $\frac{5}{7}y = -\frac{15}{49}$

58. $-\frac{3}{4}x = \frac{9}{2}$

59. $\frac{x}{2} - x = -2$

60. $\frac{y}{3} = -4 + y$

61. $-\frac{5}{8}y = \frac{3}{16} - \frac{9}{16}$

62. $-\frac{7}{9}x = -\frac{5}{18} - \frac{4}{18}$

63. $17x - 25x = \frac{1}{3}$

64. $27x - 30x = \frac{4}{9}$

65. $\frac{7}{6}x = \frac{1}{4} - \frac{2}{3}$

66. $\frac{5}{4}y = \frac{1}{2} - \frac{7}{10}$

67. $\frac{b}{4} = \frac{b}{12} + \frac{2}{3}$

68. $\frac{a}{6} = \frac{a}{3} + \frac{1}{2}$

69. $\frac{x}{3} + 2 = \frac{x}{2} + 8$

70. $\frac{y}{5} - 2 = \frac{y}{3} - 4$

Review

Round each number to the given place value. See Section 1.4.

71. 57,236 to the nearest hundred

72. 576 to the nearest hundred

73. 327 to the nearest ten

74. 2333 to the nearest ten

Concept Extensions

75. Explain why the method for eliminating fractions in an **equation** does not apply to simplifying **expressions** containing fractions.

76. Think about which exercise (part **a** or part **b**)may be completed by multiplying by 6. Now complete each exercise.

a. Solve: $\frac{x}{6} - \frac{5}{3} = 2$

b. Subtract: $\frac{x}{6} - \frac{5}{3}$

Solve.

77. $\frac{14}{11} + \frac{3x}{8} = \frac{x}{2}$

78. $\frac{19}{53} = \frac{353x}{1431} + \frac{23}{27}$

79. Find the area and the perimeter of the rectangle. Remember to attach proper units.

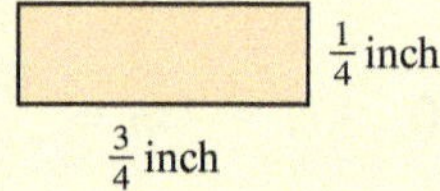

80. The area of the rectangle is $\frac{5}{12}$ square inch. Find its length, x.

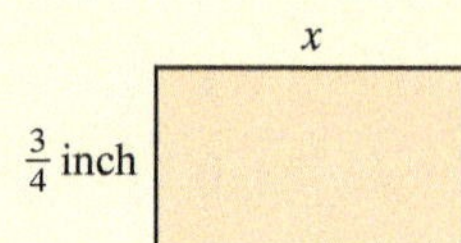

Chapter 4 Group Activity

Lobster Classification

Sections 4.1, 4.7, 4.8

This activity may be completed by working in groups or individually.

Lobsters are normally classified by weight. Use the weight classification table to answer the questions in this activity.

Classification of Lobsters	
Class	**Weight (in pounds)**
Chicken	1 to $1\frac{1}{8}$
Quarter	$1\frac{1}{4}$
Half	$1\frac{1}{2}$ to $1\frac{3}{4}$
Select	$1\frac{3}{4}$ to $2\frac{1}{2}$
Large select	$2\frac{1}{2}$ to $3\frac{1}{2}$
Jumbo	Over $3\frac{1}{2}$
(*Source:* The Maine Lobster Promotion Council)	

1. A lobster fisher has kept four lobsters from a lobster trap. Classify each lobster if they have the following weights:

a. $1\frac{7}{8}$ pounds

b. $1\frac{9}{16}$ pounds

c. $2\frac{3}{4}$ pounds

d. $2\frac{3}{8}$ pounds

2. A recipe requires 5 pounds of lobster. Using the minimum weight for each class, decide whether a chicken, half, and select lobster will be enough for the recipe, and explain your reasoning. If not, suggest a better choice of lobsters to meet the recipe requirements.

3. A lobster market customer has selected two chickens, a select, and a large select. What is the most that these four lobsters could weigh? What is the least that these four lobsters could weigh?

4. A lobster market customer wishes to buy three quarters. If lobsters sell for $7 per pound, how much will the customer owe for her purchase?

5. Why do you think there is no classification for lobsters weighing under 1 pound?

Chapter 4 Vocabulary Check

Fill in each blank with one of the words or phrases listed below.

mixed number	complex fraction	like	numerator	prime factorization
composite number	equivalent	cross products	least common denominator	denominator
prime number	improper fraction	simplest form	undefined	0
reciprocals	proper fraction			

1. Two numbers are ______________ of each other if their product is 1.

2. A(n) ______________ is a natural number greater than 1 that is not prime.

3. Fractions that represent the same portion of a whole are called ______________ fractions.

4. A(n) ______________ is a fraction whose numerator is greater than or equal to its denominator.

5. A(n) ______________ is a natural number greater than 1 whose only factors are 1 and itself.

6. A fraction is in ______________ when the numerator and the denominator have no factors in common other than 1.

7. A(n) ______________ is one whose numerator is less than its denominator.

8. A(n) ______________ contains a whole number part and a fraction part.

9. In the fraction $\frac{7}{9}$, the 7 is called the ______________ and the 9 is called the ______________.

10. The ______________ of a number is the factorization in which all the factors are prime numbers.

11. The fraction $\frac{3}{0}$ is ______________.

12. The fraction $\frac{0}{5}$ = ______________.

13. Fractions that have the same denominator are called ______________ fractions.

14. The LCM of the denominators in a list of fractions is called the ______________.

15. A fraction whose numerator or denominator or both numerator and denominator contain fractions is called a(n) ______________.

16. In $\frac{a}{b} = \frac{c}{d}$, $a \cdot d$ and $b \cdot c$ are called ______________.

Helpful Hint Are you preparing for your test? Don't forget to take the Chapter 4 Test on page 324. Then check your answers at the back of the text and use the Chapter Test Prep Videos to see the fully worked-out solutions to any of the exercises you want to review.

4 Chapter Highlights

Definitions and Concepts	Examples
Section 4.1 Introduction to Fractions and Mixed Numbers	
A **fraction** is of the form $\frac{\text{numerator}}{\text{denominator}}$ ← number of parts being considered / ← number of equal parts in the whole	Write a fraction to represent the shaded part of the figure. 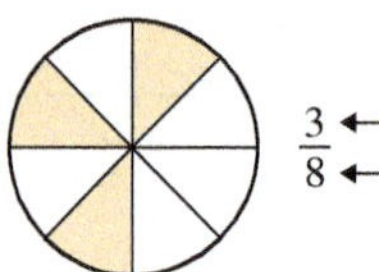 $\frac{3}{8}$ ← number of parts shaded / ← number of equal parts
A fraction is called a **proper fraction** if its numerator is less than its denominator.	Proper Fractions: $\frac{1}{3}, \frac{2}{5}, \frac{7}{8}, \frac{100}{101}$
A fraction is called an **improper fraction** if its numerator is greater than or equal to its denominator.	Improper Fractions: $\frac{5}{4}, \frac{2}{2}, \frac{9}{7}, \frac{101}{100}$
A **mixed number** contains a whole number and a fraction.	Mixed Numbers: $1\frac{1}{2}, 5\frac{7}{8}, 25\frac{9}{10}$
To Write a Mixed Number As an Improper Fraction 1. Multiply the denominator of the fraction by the whole number. 2. Add the numerator of the fraction to the product from Step 1. 3. Write the sum from Step 2 as the numerator of the improper fraction over the original denominator.	$5\frac{2}{7} = \frac{7 \cdot 5 + 2}{7} = \frac{35 + 2}{7} = \frac{37}{7}$
To Write an Improper Fraction As a Mixed Number or a Whole Number 1. Divide the denominator into the numerator. 2. The whole number part of the mixed number is the quotient. The fraction is the remainder over the original denominator. $\text{quotient}\frac{\text{remainder}}{\text{original denominator}}$	$\frac{17}{3} = 5\frac{2}{3}$ $3\overline{)17}$ = 5 -15 2

Definitions and Concepts	Examples
Section 4.2 Factors and Simplest Form	
A **prime number** is a natural number that has exactly two different factors, 1 and itself.	2, 3, 5, 7, 11, 13, 17, . . .
A **composite number** is any natural number other than 1 that is not prime.	4, 6, 8, 9, 10, 12, 14, 15, 16, . . .
The **prime factorization** of a number is the factorization in which all the factors are prime numbers.	Write the prime factorization of 60. $60 = 6 \cdot 10$ $= 2 \cdot 3 \cdot 2 \cdot 5$ or $2^2 \cdot 3 \cdot 5$
Fractions that represent the same portion of a whole are called **equivalent fractions.**	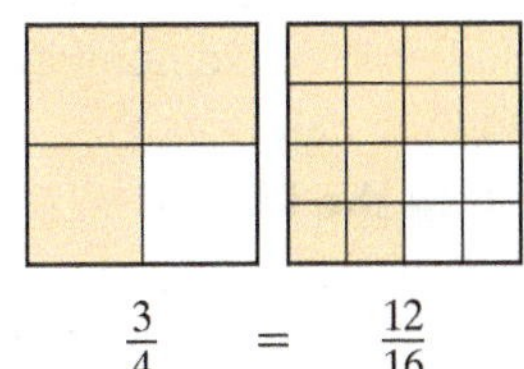 $\frac{3}{4} = \frac{12}{16}$
A fraction is in **simplest form** or **lowest terms** when the numerator and the denominator have no common factors other than 1.	The fraction $\frac{2}{3}$ is in simplest form.
To write a fraction in simplest form, write the prime factorizations of the numerator and the denominator and then divide both by all common factors.	Write in simplest form: $\frac{30}{36}$ $\frac{30}{36} = \frac{2 \cdot 3 \cdot 5}{2 \cdot 2 \cdot 3 \cdot 3} = \frac{2}{2} \cdot \frac{3}{3} \cdot \frac{5}{2 \cdot 3} = 1 \cdot 1 \cdot \frac{5}{6} = \frac{5}{6}$ or $\frac{30}{36} = \frac{\overset{1}{\cancel{2}} \cdot \overset{1}{\cancel{3}} \cdot 5}{\underset{1}{\cancel{2}} \cdot 2 \cdot \underset{1}{\cancel{3}} \cdot 3} = \frac{5}{6}$
Two fractions are equivalent if **Method 1.** They simplify to the same fraction. **Method 2.** Their cross products are equal. $24 \cdot 7 = 168$ $\frac{7}{8} = \frac{21}{24}$ $8 \cdot 21 = 168$ Since $168 = 168$, $\frac{7}{8} = \frac{21}{24}$.	Determine whether $\frac{7}{8}$ and $\frac{21}{24}$ are equivalent. $\frac{7}{8}$ is in simplest form. $\frac{21}{24} = \frac{\overset{1}{\cancel{3}} \cdot 7}{\underset{1}{\cancel{3}} \cdot 8} = \frac{1 \cdot 7}{1 \cdot 8} = \frac{7}{8}$ Since both simplify to $\frac{7}{8}$, then $\frac{7}{8} = \frac{21}{24}$.
Section 4.3 Multiplying and Dividing Fractions	
To multiply two fractions, multiply the numerators and multiply the denominators.	Multiply. $\frac{2x}{3} \cdot \frac{5}{7} = \frac{2x \cdot 5}{3 \cdot 7} = \frac{10x}{21}$ $\frac{3}{4} \cdot \frac{1}{6} = \frac{3 \cdot 1}{4 \cdot 6} = \frac{\overset{1}{\cancel{3}} \cdot 1}{4 \cdot \underset{1}{\cancel{3}} \cdot 2} = \frac{1}{8}$
To find the **reciprocal** of a fraction, interchange its numerator and denominator.	The reciprocal of $\frac{3}{5}$ is $\frac{5}{3}$.
To divide two fractions, multiply the first fraction by the reciprocal of the second fraction.	Divide. $-\frac{3}{10} \div \frac{7}{9} = -\frac{3}{10} \cdot \frac{9}{7} = -\frac{3 \cdot 9}{10 \cdot 7} = -\frac{27}{70}$

Definitions and Concepts	Examples
Section 4.4 Adding and Subtracting Like Fractions, Least Common Denominator, and Equivalent Fractions	
Fractions that have the same denominator are called **like fractions.**	$\frac{1}{3}$ and $\frac{2}{3}$; $\frac{5x}{7}$ and $\frac{6}{7}$
To add or subtract like fractions, combine the numerators and place the sum or difference over the common denominator.	$\frac{2}{7}+\frac{3}{7}=\frac{5}{7}$ ← Add the numerators. ← Keep the common denominator. $\frac{7}{8}-\frac{4}{8}=\frac{3}{8}$ ← Subtract the numerators. ← Keep the common denominator.
The **least common denominator (LCD)** of a list of fractions is the smallest positive number divisible by all the denominators in the list.	The LCD of $\frac{1}{2}$ and $\frac{5}{6}$ is 6 because 6 is the smallest positive number that is divisible by both 2 and 6.
Method 1 for finding the LCD of a list of fractions using multiples	Find the LCD of $\frac{1}{4}$ and $\frac{5}{6}$ using Method 1.
Step 1: Write the multiples of the largest denominator (starting with the number itself) until a multiple common to all denominators in the list is found.	$6 \cdot 1 = 6$ Not a multiple of 4 $6 \cdot 2 = 12$ A multiple of 4
Step 2: The multiple found in Step 1 is the LCD.	The LCD is 12.
Method 2 for finding the LCD of a list of a fractions using prime factorization	Find the LCD of $\frac{5}{6}$ and $\frac{11}{20}$ using Method 2.
Step 1: Write the prime factorization of each denominator.	$6 = 2 \cdot ③$ $20 = \boxed{2 \cdot 2} \cdot ⑤$
Step 2: For each different prime factor in Step 1, circle the greatest number of times that factor occurs in any one factorization.	The LCD is
Step 3: The LCD is the product of the circled factors.	$2 \cdot 2 \cdot 3 \cdot 5 = 60$
Equivalent fractions represent the same portion of a whole.	Write an equivalent fraction with the indicated denominator. $\frac{2}{8}=\frac{}{16}$ $\frac{2 \cdot 2}{8 \cdot 2}=\frac{4}{16}$
Section 4.5 Adding and Subtracting Unlike Fractions	
To add or subtract fractions with unlike denominators	Add: $\frac{3}{20}+\frac{2}{5}$
Step 1: Find the LCD.	**Step 1:** The LCD of the denominators 20 and 5 is 20.
Step 2: Write each fraction as an equivalent fraction whose denominator is the LCD.	**Step 2:** $\frac{3}{20}=\frac{3}{20}$; $\frac{2}{5}=\frac{2}{5}\cdot\frac{4}{4}=\frac{8}{20}$
Step 3: Add or subtract the like fractions.	**Step 3:** $\frac{3}{20}+\frac{2}{5}=\frac{3}{20}+\frac{8}{20}=\frac{11}{20}$
Step 4: Write the sum or difference in simplest form.	**Step 4:** $\frac{11}{20}$ is in simplest form.

Definitions and Concepts	Examples
Section 4.6 Complex Fractions and Review of Order of Operations	
A fraction whose numerator or denominator or both contain fractions is called a **complex fraction.**	Complex Fractions: $\dfrac{\frac{11}{4}}{\frac{7}{10}}, \dfrac{\frac{y}{6}-11}{\frac{4}{3}}$
One method for simplifying complex fractions is to multiply the numerator and the denominator of the complex fraction by the LCD of all fractions in its numerator and its denominator.	$\dfrac{\frac{y}{6}-11}{\frac{4}{3}} = \dfrac{6\left(\frac{y}{6}-11\right)}{6\left(\frac{4}{3}\right)} = \dfrac{6\left(\frac{y}{6}\right)-6(11)}{6\left(\frac{4}{3}\right)}$ $= \dfrac{y-66}{8}$
Section 4.7 Operations on Mixed Numbers	
To multiply with mixed numbers or whole numbers, first write any mixed or whole numbers as improper fractions and then multiply as usual.	$2\frac{1}{3}\cdot\frac{1}{9} = \frac{7}{3}\cdot\frac{1}{9} = \frac{7\cdot 1}{3\cdot 9} = \frac{7}{27}$
To divide with mixed numbers or whole numbers, first write any mixed or whole numbers as fractions and then divide as usual.	$2\frac{5}{8} \div 3\frac{7}{16} = \frac{21}{8} \div \frac{55}{16} = \frac{21}{8}\cdot\frac{16}{55} = \frac{21\cdot 16}{8\cdot 55}$ $= \frac{21\cdot 2\cdot \overset{1}{\cancel{8}}}{\underset{1}{\cancel{8}}\cdot 55} = \frac{42}{55}$
To add or subtract with mixed numbers, add or subtract the fractions and then add or subtract the whole numbers.	Add: $2\frac{1}{2} + 5\frac{7}{8}$ $2\frac{1}{2} = 2\frac{4}{8}$ $+5\frac{7}{8} = +5\frac{7}{8}$ $7\frac{11}{8} = 7 + 1\frac{3}{8} = 8\frac{3}{8}$
Section 4.8 Solving Equations Containing Fractions	
To Solve an Equation in *x*	Solve: $\frac{x}{15} + 2 = \frac{7}{3}$
Step 1: If fractions are present, multiply both sides of the equation by the LCD of the fractions.	$15\left(\frac{x}{15}+2\right) = 15\left(\frac{7}{3}\right)$ Multiply by the LCD 15.
Step 2: If parentheses are present, use the distributive property.	$15\left(\frac{x}{15}\right) + 15\cdot 2 = 15\left(\frac{7}{3}\right)$
Step 3: Combine any like terms on each side of the equation.	$x + 30 = 35$
Step 4: Use the addition property of equality to rewrite the equation so that variable terms are on one side of the equation and constant terms are on the other side.	$x + 30 - 30 = 35 - 30$ $x = 5$
Step 5: Divide both sides by the numerical coefficient of *x* to solve.	
Step 6: Check the answer in the *original equation.*	Check to see that 5 is the solution.

Chapter 4 Review

(4.1) *Write a fraction to represent the shaded area. If the fraction is improper, write the shaded area as a mixed number also. Do not simplify these answers.*

1. 

2. 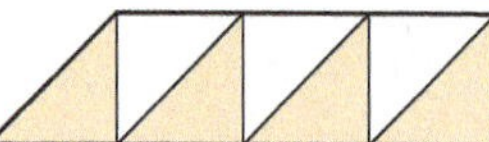

3. 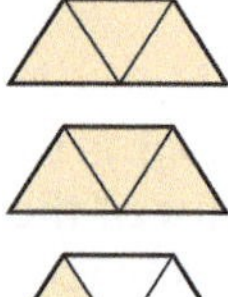

4. 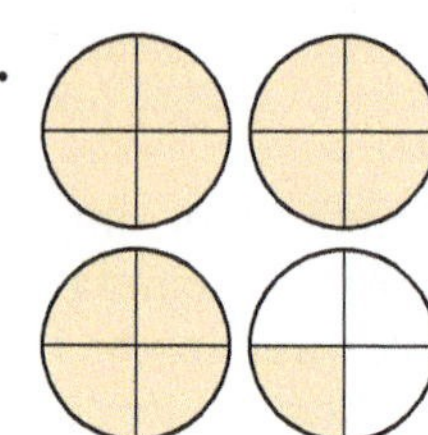

Solve.

5. A basketball player made 11 free throws out of 12 tries during a game. What fraction of free throws did the player make?

6. A new car lot contains 23 blue cars out of a total of 131 cars.

a. How many cars on the lot are not blue?
b. What fraction of cars on the lot are not blue?

Simplify by dividing.

7. $-\frac{3}{3}$

8. $\frac{-20}{-20}$

9. $\frac{0}{-1}$

10. $\frac{4}{0}$

Graph each fraction on a number line.

11. $\frac{7}{9}$

0

12. $\frac{4}{7}$

0

13. $\frac{5}{4}$

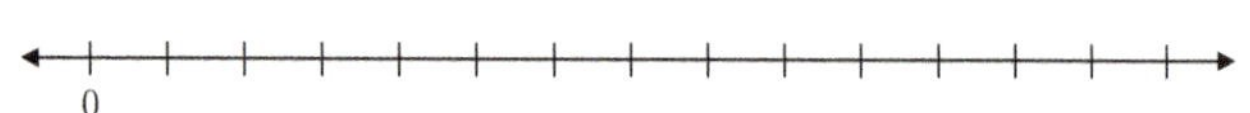

14. $\frac{7}{5}$

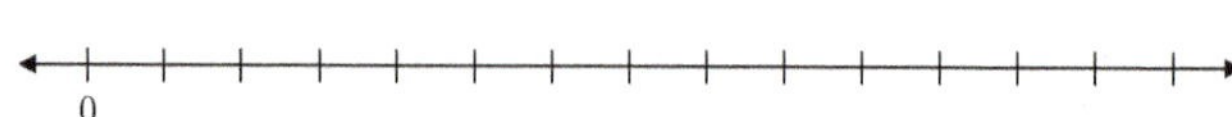

Write each improper fraction as a mixed number or a whole number.

15. $\frac{15}{4}$

16. $\frac{39}{13}$

Write each mixed number as an improper fraction.

17. $2\frac{1}{5}$

18. $3\frac{8}{9}$

(4.2) *Write each fraction in simplest form.*

19. $\frac{12}{28}$

20. $\frac{15}{27}$

21. $-\frac{25x}{75x^2}$

22. $-\frac{36y^3}{72y}$

23. $\frac{29ab}{32abc}$

24. $\frac{18xyz}{23xy}$

25. $\frac{45x^2y}{27xy^3}$

26. $\frac{42ab^2c}{30abc^3}$

27. There are 12 inches in a foot. What fractional part of a foot does 8 inches represent?

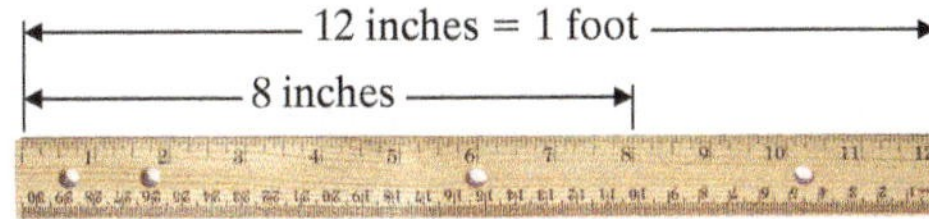

28. Six out of 15 cars are white. What fraction of the cars are *not* white?

Determine whether each two fractions are equivalent.

29. $\frac{10}{34}$ and $\frac{4}{14}$

30. $\frac{30}{50}$ and $\frac{9}{15}$

(4.3) *Multiply.*

31. $\frac{3}{5} \cdot \frac{1}{2}$

32. $-\frac{6}{7} \cdot \frac{5}{12}$

33. $-\frac{24x}{5} \cdot -\frac{15}{8x^3}$

34. $\frac{27y^3}{21} \cdot \frac{7}{18y^2}$

35. $\left(-\frac{1}{3}\right)^3$

36. $\left(-\frac{5}{12}\right)^2$

Divide.

37. $-\frac{3}{4} \div \frac{3}{8}$

38. $\frac{21a}{4} \div \frac{7a}{5}$

39. $-\frac{9}{2} \div -\frac{1}{3}$

40. $-\frac{5}{3} \div 2y$

41. Evaluate $x \div y$ if $x = \frac{9}{7}$ and $y = \frac{3}{4}$.

42. Evaluate ab if $a = -7$ and $b = \frac{9}{10}$.

Find the area of each figure.

△ **43.**

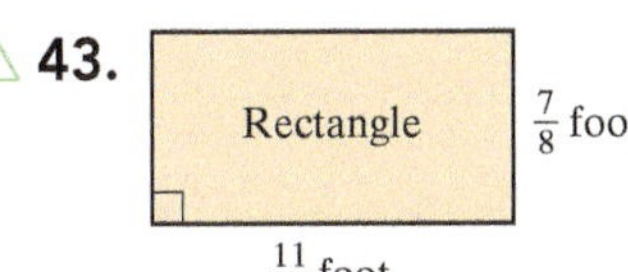

△ **44.**

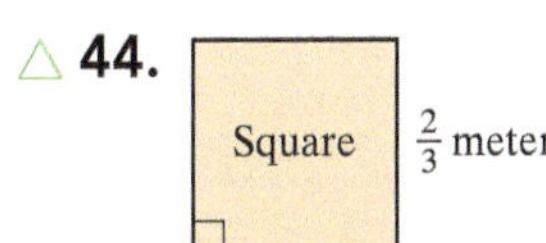

(4.4) *Add or subtract as indicated.*

45. $\frac{7}{11} + \frac{3}{11}$

46. $\frac{4}{9} + \frac{2}{9}$

47. $\frac{1}{12} - \frac{5}{12}$

48. $\frac{11x}{15} + \frac{x}{15}$

49. $\frac{4y}{21} - \frac{3}{21}$

50. $\frac{4}{15} - \frac{3}{15} - \frac{2}{15}$

Find the LCD of each list of fractions.

51. $\frac{2}{3}, \frac{5}{x}$

52. $\frac{3}{4}, \frac{3}{8}, \frac{7}{12}$

Write each fraction as an equivalent fraction with the given denominator.

53. $\frac{2}{3} = \frac{?}{30}$

54. $\frac{5}{8} = \frac{?}{56}$

55. $\frac{7a}{6} = \frac{?}{42}$

56. $\frac{9b}{4} = \frac{?}{20}$

57. $\frac{4}{5x} = \frac{?}{50x}$

58. $\frac{5}{9y} = \frac{?}{18y}$

Solve.

59. One evening Mark Alorenzo did $\frac{3}{8}$ of his homework before supper, another $\frac{2}{8}$ of it while his children did their homework, and $\frac{1}{8}$ after his children went to bed. What part of his homework did he do that evening?

△ **60.** The Simpsons will be fencing in their land, which is in the shape of a rectangle. In order to do this, they need to find its perimeter. Find the perimeter of their land.

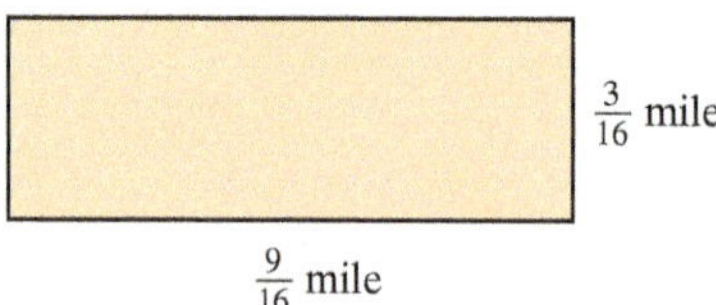

(4.5) *Add or subtract as indicated.*

61. $\frac{7}{18} + \frac{2}{9}$

62. $\frac{4}{13} - \frac{1}{26}$

63. $-\frac{1}{3} + \frac{1}{4}$

64. $-\frac{2}{3} + \frac{1}{4}$

65. $\frac{5x}{11} + \frac{2}{55}$

66. $\frac{4}{15} + \frac{b}{5}$

67. $\frac{5y}{12} - \frac{2y}{9}$

68. $\frac{7x}{18} + \frac{2x}{9}$

69. $\frac{4}{9} + \frac{5}{y}$

70. $-\frac{9}{14} - \frac{3}{7}$

71. $\frac{4}{25} + \frac{23}{75} + \frac{7}{50}$

72. $\frac{2}{3} - \frac{2}{9} - \frac{1}{6}$

Find the perimeter of each figure.

73.

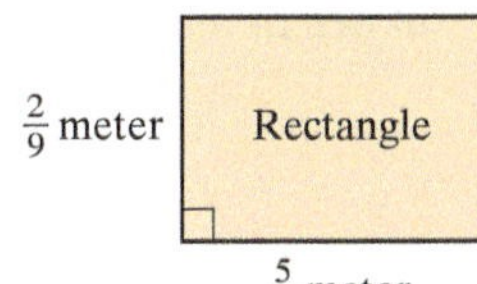

74.

75. In a group of 100 blood donors, typically $\frac{9}{25}$ have type A Rh-positive blood and $\frac{3}{50}$ have type A Rh-negative blood. What fraction have type A blood?

76. Find the difference in length of two scarves if one scarf is $\frac{5}{12}$ of a yard long and the other is $\frac{2}{3}$ of a yard long.

$\frac{2}{3}$ of a yard

$\frac{5}{12}$ of a yard

(4.6) *Simplify each complex fraction.*

77. $\dfrac{\frac{2x}{5}}{\frac{7}{10}}$

78. $\dfrac{\frac{3y}{7}}{\frac{11}{7}}$

79. $\dfrac{\frac{2}{5}-\frac{1}{2}}{\frac{3}{4}-\frac{7}{10}}$

80. $\dfrac{\frac{5}{6}-\frac{1}{4}}{\frac{-1}{12y}}$

Evaluate each expression if $x = \frac{1}{2}$, $y = -\frac{2}{3}$, *and* $z = \frac{4}{5}$.

81. $\dfrac{x}{y+z}$

82. $\dfrac{x+y}{z}$

Evaluate each expression. Use the order of operations to simplify.

83. $\frac{5}{13} \div \frac{1}{2} \cdot \frac{4}{5}$

84. $\frac{2}{27} - \left(\frac{1}{3}\right)^2$

85. $\frac{9}{10} \cdot \frac{1}{3} - \frac{2}{5} \cdot \frac{1}{11}$

86. $-\frac{2}{7} \cdot \left(\frac{1}{5} + \frac{3}{10}\right)$

(4.7) *Perform operations as indicated. Simplify your answers. Estimate where noted.*

87.
$$\begin{array}{r} 7\frac{3}{8} \\ 9\frac{5}{6} \\ +\ 3\frac{1}{12} \\ \hline \end{array}$$

88.
$$\begin{array}{r} 8\frac{1}{5} \\ -\ 5\frac{3}{11} \\ \hline \end{array}$$

Exact:

Estimate:

89. $1\frac{5}{8} \cdot 3\frac{1}{5}$

Exact:

Estimate:

90. $6\frac{3}{4} \div 1\frac{2}{7}$

91. A truck traveled 341 miles on $15\frac{1}{2}$ gallons of gas. How many miles might we expect the truck to travel on 1 gallon of gas?

92. There are $7\frac{1}{3}$ grams of fat in each ounce of hamburger. How many grams of fat are in a 5-ounce hamburger patty?

Find the unknown measurements.

△ **93.**

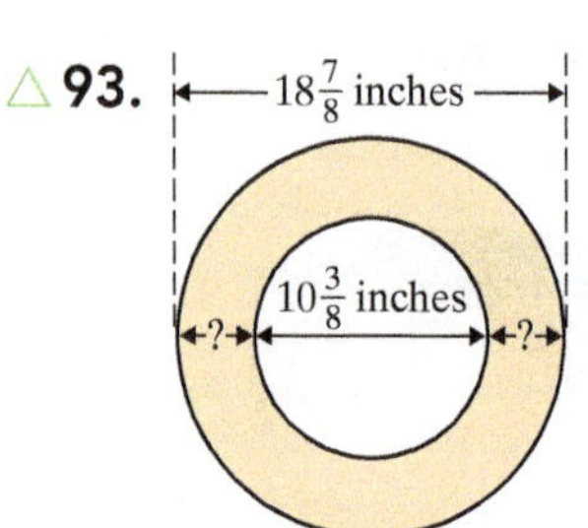

△ **94.**

$\frac{1}{2}$ yard

$\frac{3}{5}$ yard

$1\frac{1}{30}$ yard

$1\frac{3}{10}$ yard

?

$1\frac{8}{15}$ yard

Perform the indicated operations.

95. $-12\frac{1}{7} + \left(-15\frac{3}{14}\right)$

96. $23\frac{7}{8} - 24\frac{7}{10}$

97. $-3\frac{1}{5} \div \left(-2\frac{7}{10}\right)$

98. $-2\frac{1}{4} \cdot 1\frac{3}{4}$

(4.8) *Solve each equation.*

99. $a - \frac{2}{3} = \frac{1}{6}$

100. $9x + \frac{1}{5} - 8x = -\frac{7}{10}$

101. $-\frac{3}{5}x = 6$

102. $\frac{2}{9}y = -\frac{4}{3}$

103. $\frac{x}{7} - 3 = -\frac{6}{7}$

104. $\frac{y}{5} + 2 = \frac{11}{5}$

105. $\frac{1}{6} + \frac{x}{4} = \frac{17}{12}$

106. $\frac{x}{5} - \frac{5}{4} = \frac{x}{2} - \frac{1}{20}$

Mixed Review

Perform the indicated operations. Write each answer in simplest form. Estimate where noted.

107. $\frac{6}{15} \cdot \frac{5}{8}$

108. $\frac{5x^2}{y} \div \frac{10x^3}{y^3}$

109. $\frac{3}{10} - \frac{1}{10}$

110. $\frac{7}{8x} \cdot -\frac{2}{3}$

111. $\frac{2x}{3} + \frac{x}{4}$

112. $-\frac{5}{11} + \frac{2}{55}$

113. $-1\frac{3}{5} \div \frac{1}{4}$

114. $2\frac{7}{8} + 9\frac{1}{2}$

Exact:

Estimate:

115. $12\frac{1}{7} - 9\frac{3}{5}$

Exact:

Estimate:

116. Simplify: $\dfrac{2 + \frac{3}{4}}{1 - \frac{1}{8}}$

117. Evaluate: $-\frac{3}{8} \cdot \left(\frac{2}{3} - \frac{4}{9}\right)$

Solve.

118. $11x - \frac{2}{7} - 10x = -\frac{13}{14}$

119. $-\frac{3}{5}x = \frac{4}{15}$

120. $\frac{x}{12} + \frac{5}{6} = -\frac{3}{4}$

121. A ribbon $5\frac{1}{2}$ yards long is cut from a reel of ribbon with 50 yards on it. Find the length of the piece remaining on the reel.

△ **122.** A slab of natural granite is purchased and a rectangle with length $7\frac{4}{11}$ feet and width $5\frac{1}{2}$ feet is cut from it. Find the area of the rectangle.

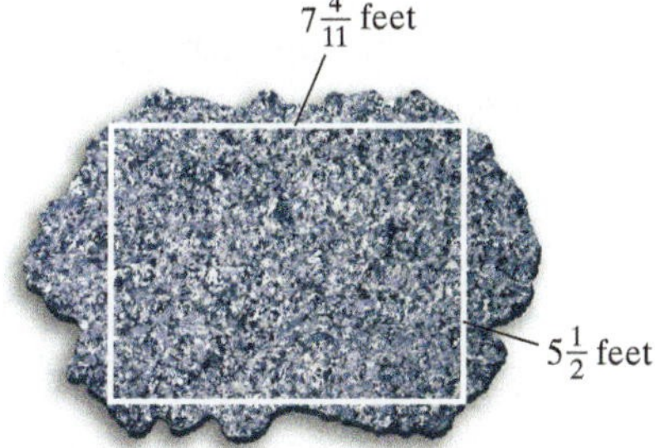

Chapter 4 Test

Step-by-step test solutions are found on the Chapter Test Prep Videos. Where available: MyMathLab® or YouTube

Answers

1. ____
2. ____
3. ____
4. ____
5. ____
6. ____
7. ____
8. ____
9. ____
10. ____
11. ____
12. ____
13. ____
14. ____
15. ____
16. ____
17. ____
18. ____
19. ____
20. ____
21. ____

Write a fraction to represent the shaded area.

1. 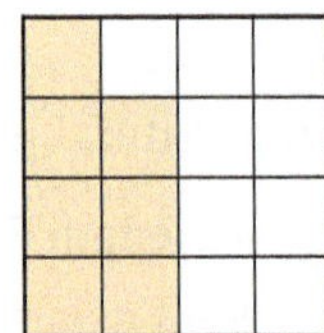

Write the mixed number as an improper fraction.

2. $7\frac{2}{3}$

Write the improper fraction as a mixed number.

3. $\frac{75}{4}$

Write each fraction in simplest form.

4. $\frac{24}{210}$

5. $-\frac{42x}{70}$

Determine whether these fractions are equivalent.

6. $\frac{5}{7}$ and $\frac{8}{11}$

7. $\frac{6}{27}$ and $\frac{14}{63}$

Find the prime factorization of each number.

8. 84

9. 495

Perform each indicated operation and write the answers in simplest form.

10. $\frac{4}{4} \div \frac{3}{4}$

11. $-\frac{4}{3} \cdot \frac{4}{4}$

12. $\frac{7x}{9} + \frac{x}{9}$

13. $\frac{1}{7} - \frac{3}{x}$

14. $\frac{xy^3}{z} \cdot \frac{z}{xy}$

15. $-\frac{2}{3} \cdot -\frac{8}{15}$

16. $\frac{9a}{10} + \frac{2}{5}$

17. $-\frac{8}{15y} - \frac{2}{15y}$

18. $\frac{3a}{8} \cdot \frac{16}{6a^3}$

19. $\frac{11}{12} - \frac{3}{8} + \frac{5}{24}$

20.
$$\begin{array}{r} 3\frac{7}{8} \\ 7\frac{2}{5} \\ +\,2\frac{3}{4} \\ \hline \end{array}$$

21.
$$\begin{array}{r} 19 \\ -2\frac{3}{11} \\ \hline \end{array}$$

22. $-\frac{16}{3} \div -\frac{3}{12}$ **23.** $3\frac{1}{3} \cdot 6\frac{3}{4}$ **24.** $-\frac{2}{7} \cdot \left(6 - \frac{1}{6}\right)$ **25.** $\frac{1}{2} \div \frac{2}{3} \cdot \frac{3}{4}$

26. $\left(-\frac{3}{4}\right)^2 \div \left(\frac{2}{3} + \frac{5}{6}\right)$ **27.** Find the average of $\frac{5}{6}, \frac{4}{3}$, and $\frac{7}{12}$.

Simplify each complex fraction.

28. $\dfrac{\frac{5x}{7}}{\frac{20x^2}{21}}$ **29.** $\dfrac{5 + \frac{3}{7}}{2 - \frac{1}{2}}$

Solve.

30. $-\frac{3}{8}x = \frac{3}{4}$ **31.** $\frac{x}{5} + x = -\frac{24}{5}$ **32.** $\frac{2}{3} + \frac{x}{4} = \frac{5}{12} + \frac{x}{2}$

Evaluate each expression for the given replacement values.

33. $-5x; x = -\frac{1}{2}$ **34.** $x \div y; x = \frac{1}{2}, y = 3\frac{7}{8}$

Solve.

35. A carpenter cuts a piece $2\frac{3}{4}$ feet long from a cedar plank that is $6\frac{1}{2}$ feet long. How long is the remaining piece?

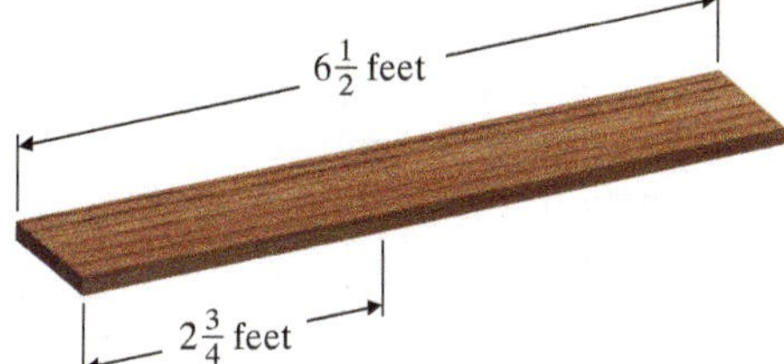

The circle graph below shows us how the average consumer spends money. For example, $\frac{7}{50}$ of spending goes for food. Use this information for Exercises 36 through 38.

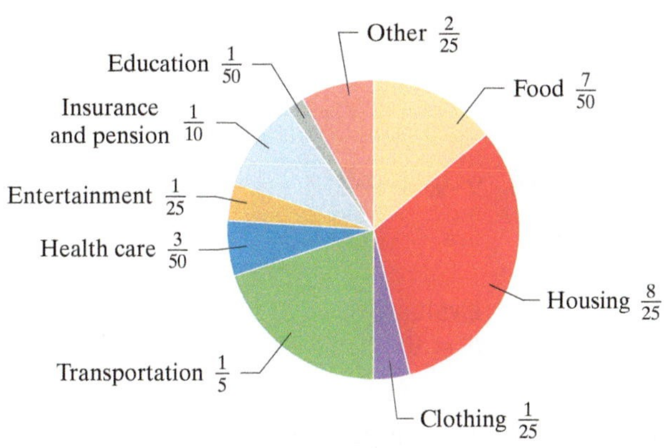

Source: U.S. Bureau of Labor Statistics; based on survey

36. What fraction of spending goes for housing and food combined?

37. What fraction of spending goes for education, transportation, and clothing?

38. Suppose your family spent $47,000 on the items in the graph. How much might we expect was spent on health care?

Find the perimeter and area of the figure.

△ **39.**

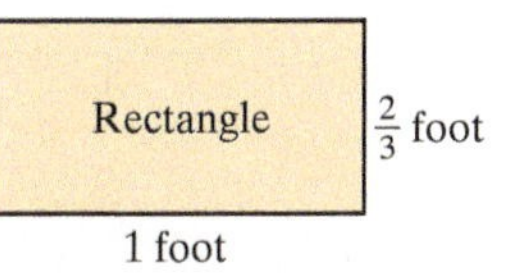

40. During a 258-mile trip, a car used $10\frac{3}{4}$ gallons of gas. How many miles would we expect the car to travel on 1 gallon of gas?

22. ____
23. ____
24. ____
25. ____
26. ____
27. ____
28. ____
29. ____
30. ____
31. ____
32. ____
33. ____
34. ____
35. ____
36. ____
37. ____
38. ____
39. ____
40. ____

Chapters 1–4 Cumulative Review

Answers

1. ______
2. ______
3. ______
4. ______
5. ______
6. ______
7. ______
8. ______
9. ______
10. ______
11. ______
12. ______
13. ______
14. ______
15. ______
16. ______
17. ______
18. ______
19. ______
20. ______
21. ______
22. ______

Write each number in words.

1. 546

2. 115

3. 27,034

4. 6573

5. Add: $46 + 713$

6. Add: $587 + 44$

7. Subtract: $543 - 29$. Check by adding.

8. Subtract: $995 - 62$. Check by adding.

9. Round 278,362 to the nearest thousand.

10. Round 1436 to the nearest ten.

11. A digital video disc (DVD) can hold about 4800 megabytes (MB) of information. How many megabytes can 12 DVDs hold?

12. On a trip across the country, Daniel Daunis travels 435 miles per day. How many total miles does he travel in 3 days?

13. Divide and check: $56{,}717 \div 8$

14. Divide and check: $4558 \div 12$

Write using exponential notation.

15. $7 \cdot 7 \cdot 7$

16. $7 \cdot 7$

17. $3 \cdot 3 \cdot 3 \cdot 3 \cdot 9 \cdot 9 \cdot 9$

18. $9 \cdot 9 \cdot 9 \cdot 9 \cdot 5 \cdot 5$

19. Evaluate $2(x - y)$ for $x = 6$ and $y = 3$.

20. Evaluate $8a + 3(b - 5)$ for $a = 5$ and $b = 9$.

21. The world's deepest cave is Krubera (or Voronja), in the country of Georgia, located by the Black Sea in Asia. It has been explored to a depth of 7188 feet below the surface of Earth. Represent this position using an integer. (*Source:* messagetoeagle.com and Wikipedia)

22. The temperature on a cold day in Minneapolis, MN, was 21°F below zero. Represent this temperature using an integer.

23. Add using a number line: $-7 + 3$

24. Add using a number line: $-3 + 8$

25. Simplify: $7 - 8 - (-5) - 1$

26. Simplify: $6 + (-8) - (-9) + 3$

27. Evaluate: $(-5)^2$

28. Evaluate: -2^4

29. Simplify: $3(4 - 7) + (-2) - 5$

30. Simplify: $(20 - 5^2)^2$

31. Simplify: $2y - 6 + 4y + 8$

32. Simplify: $5x - 1 + x + 10$

Solve.

33. $5x + 2 - 4x = 7 - 19$

34. $9y + 1 - 8y = 3 - 20$

35. $17 - 7x + 3 = -3x + 21 - 3x$

36. $9x - 2 = 7x - 24$

37. Write a fraction to represent the shaded part of the figure.

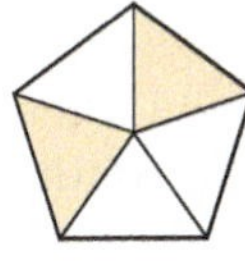

38. Write the prime factorization of 156.

39. Write each as an improper fraction.

a. $4\frac{2}{9}$ **b.** $1\frac{8}{11}$

40. Write $\frac{39}{5}$ as a mixed number.

41. Write in simplest form: $\frac{42x}{66}$

42. Write in simplest form: $\frac{70}{105y}$

43. Multiply: $3\frac{1}{3} \cdot \frac{7}{8}$

44. Multiply: $\frac{2}{3} \cdot 4$

45. Divide and simplify: $\frac{5}{16} \div \frac{3}{4}$

46. Divide: $1\frac{1}{10} \div 5\frac{3}{5}$

23. ______
24. ______
25. ______
26. ______
27. ______
28. ______
29. ______
30. ______
31. ______
32. ______
33. ______
34. ______
35. ______
36. ______
37. ______
38. ______
39. a. ______ b. ______
40. ______
41. ______
42. ______
43. ______
44. ______
45. ______
46. ______

5 Decimals

Sections

Check Your Progress

Decimal numbers represent parts of a whole, just like fractions. For example, one penny is 0.01 or $\frac{1}{100}$ of a dollar. In this chapter, we learn to perform arithmetic operations on decimals and to analyze the relationship between fractions and decimals. We also learn how decimals are used in the real world.

The graph below shows the age group distribution for average daily texting. (Check your age group and see if the data are accurate based on your own experiences.) While we have practiced calculating averages before in this text, an average certainly does not usually simplify to a whole number. While fractions are useful, decimals are also an important system of numbers that can be used to show values between whole numbers. Data are easy to round when they are in the form of a decimal. In Section 5.2, Exercises 83 and 84, we continue to study text messaging and how it is increasing each year.

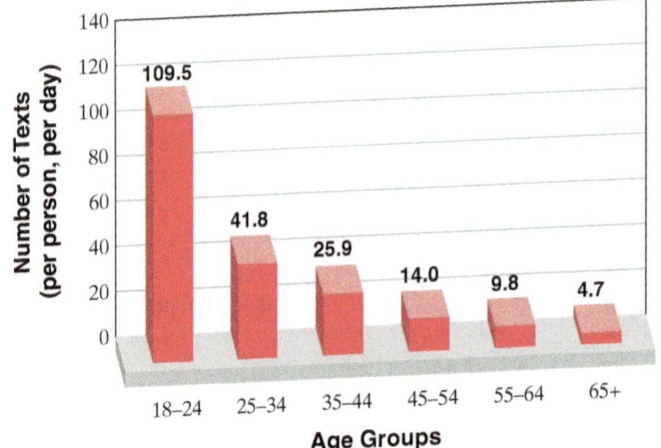

Source: The Pew Research Center, 2011

5.1 Introduction to Decimals 

Objective A Decimal Notation and Writing Decimals in Words

Like fractional notation, decimal notation is used to denote a part of a whole. Numbers written in decimal notation are called **decimal numbers,** or simply **decimals.** The decimal 17.758 has three parts.

```
1 7 . 7 5 8
Whole   Decimal
number  part
part
    Decimal point
```

In Section 1.2, we introduced place value for whole numbers. Place names and place values for the whole number part of a decimal number are exactly the same. Place names and place values for the decimal part are shown below.

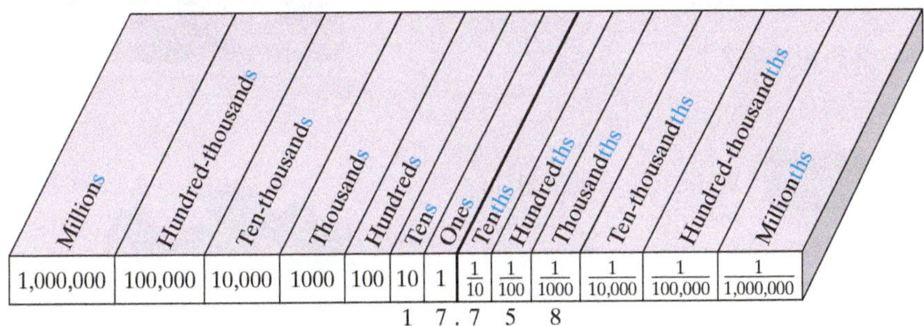

Notice that the value of each place is $\frac{1}{10}$ of the value of the place to its left. For example,

$$1 \cdot \frac{1}{10} = \frac{1}{10} \quad \text{and} \quad \frac{1}{10} \cdot \frac{1}{10} = \frac{1}{100}$$

(ones, tenths; tenths, hundredths)

The decimal number 17.758 means

1 ten + 7 ones + 7 tenths + 5 hundredths + 8 thousandths

$$\text{or} \quad 1 \cdot 10 + 7 \cdot 1 + 7 \cdot \frac{1}{10} + 5 \cdot \frac{1}{100} + 8 \cdot \frac{1}{1000}$$

$$\text{or} \quad 10 + 7 + \frac{7}{10} + \frac{5}{100} + \frac{8}{1000}$$

Writing (or Reading) a Decimal in Words

Step 1: Write the whole number part in words.

Step 2: Write "and" for the decimal point.

Step 3: Write the decimal part in words as though it were a whole number, followed by the place value of the last digit.

Objectives

A Know the Meaning of Place Value for a Decimal Number and Write Decimals in Words.

B Write Decimals in Standard Form.

C Write Decimals as Fractions.

D Compare Decimals.

E Round Decimals to Given Place Values.

Helpful Hint Notice that place values to the left of the decimal point end in "s." Place values to the right of the decimal point end in "ths."

Practice 1

Write each decimal in words.

a. 0.06

b. −200.073

c. 0.0829

Example 1

Write each decimal in words.

a. 0.7 **b.** −50.82 **c.** 21.093

Solution:

a. seven tenths

b. negative fifty and eighty-two hundredths

c. twenty-one and ninety-three thousandths

Work Practice 1

Practice 2

Write the decimal 87.31 in words.

Example 2

Write the decimal in the following sentence in words: The Golden Jubilee Diamond is a 545.67-carat cut diamond. (*Source: The Guinness Book of Records*)

Solution: five hundred forty-five and sixty-seven hundredths

Work Practice 2

Practice 3

Write the decimal 52.1085 in words.

Example 3

Write the decimal in the following sentence in words: The oldest known fragments of the Earth's crust are Zircon crystals; they were discovered in Australia and are thought to be 4.276 billion years old. (*Source: The Guinness Book of Records*)

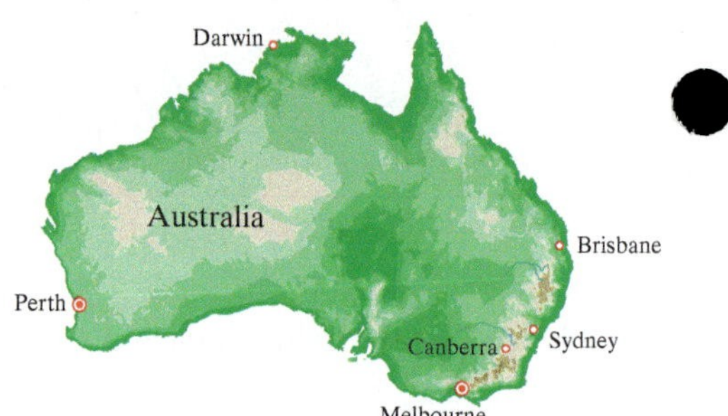

Solution: four and two hundred seventy-six thousandths

Work Practice 3

Suppose that you are paying for a purchase of $368.42 at Circuit City by writing a check. Checks are usually written using the following format.

Elayn Martin-Gay 60-8124/7233 1000613331 1403

DATE (Current date)

PAY TO THE ORDER OF Circuit City $ 368.42

Three hundred sixty-eight and 42/100 —— DOLLARS

FIRST STATE BANK OF FARTHINGTON FARTHINGTON, IL 64422

MEMO Elayn Martin-Gay

⑆621497260⑆ 1000613331⑈ 1403

Answers

1. a. six hundredths **b.** negative two hundred and seventy-three thousandths **c.** eight hundred twenty-nine ten-thousandths

2. eighty-seven and thirty-one hundredths

3. fifty-two and one thousand eighty-five ten-thousandths

Example 4 Fill in the check to Camelot Music to pay for your purchase of \$92.98.

Solution:

Your Preprinted Name
Your Preprinted Address
80-8124/7233 1000613331
1404
DATE (Current date)
PAY TO THE ORDER OF Camelot Music \$ 92.98
Ninety-two and 98/100 DOLLARS
FIRST STATE BANK OF FARTHINGTON
FARTHINGTON, IL 64422
MEMO (Your signature)
⑆621497260⑆ 1000613331⑈ 1404

Work Practice 4

Practice 4

Fill in the check to CLECO (Central Louisiana Electric Company) to pay for your monthly electric bill of \$207.40.

Objective B Writing Decimals in Standard Form

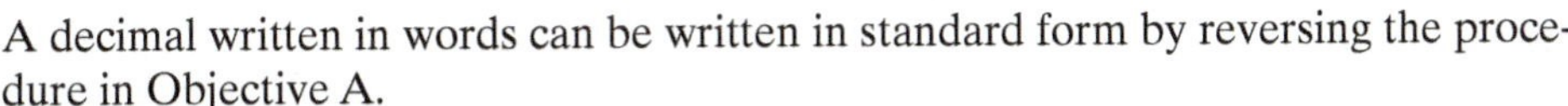

A decimal written in words can be written in standard form by reversing the procedure in Objective A.

Examples Write each decimal in standard form.

5. Forty-eight and twenty-six hundredths is

48.26

hundredths place

6. Six and ninety-five thousandths is

6.095

thousandths place

Work Practice 5–6

Practice 5–6

Write each decimal in standard form.

5. Five hundred and ninety-six hundredths

6. Thirty-nine and forty-two thousandths

Helpful Hint

When converting a decimal from words to decimal notation, make sure the last digit is in the correct place by inserting 0s if necessary. For example,

Two and thirty-eight thousandths is 2.038

thousandths place

Objective C Writing Decimals as Fractions

Once you master reading and writing decimals, writing a decimal as a fraction follows naturally.

Decimal	In Words	Fraction
0.7	seven tenths	$\frac{7}{10}$
0.51	fifty-one hundredths	$\frac{51}{100}$
0.009	nine thousandths	$\frac{9}{1000}$
0.05	five hundredths	$\frac{5}{100} = \frac{1}{20}$

Answers

4.

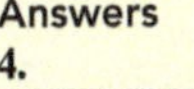

Your Preprinted Name
Your Preprinted Address
1406
DATE (Current date)
PAY TO THE ORDER OF CLECO \$ 207.40
Two hundred seven and 40/100 DOLLARS
FIRST STATE BANK OF FARTHINGTON
MEMO (Your signature)
⑆621497260⑆ 1000613331⑈ 1406

5. 500.96
6. 39.042

Notice that the number of decimal places in a decimal number is the same as the number of zeros in the denominator of the equivalent fraction. We can use this fact to write decimals as fractions.

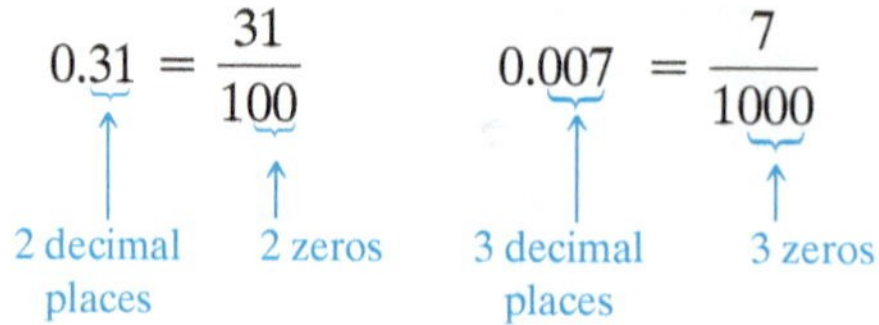

Practice 7

Write 0.051 as a fraction.

Example 7 Write 0.47 as a fraction.

Solution: $0.47 = \frac{47}{100}$

2 decimal places; 2 zeros

Work Practice 7

Practice 8

Write 29.97 as a mixed number.

Example 8 Write 5.9 as a mixed number.

Solution: $5.9 = 5\frac{9}{10}$

1 decimal place; 1 zero

Work Practice 8

Practice 9–11

Write each decimal as a fraction or mixed number. Write your answer in simplest form.

9. 0.12
10. 64.8
11. −209.986

Examples Write each decimal as a fraction or a mixed number. Write your answer in simplest form.

9. $0.125 = \frac{125}{1000} = \frac{\overset{1}{\cancel{125}}}{8 \cdot \underset{1}{\cancel{125}}} = \frac{1}{8}$

10. $43.5 = 43\frac{5}{10} = 43\frac{\overset{1}{\cancel{5}}}{2 \cdot \underset{1}{\cancel{5}}} = 43\frac{1}{2 \cdot 1} = 43\frac{1}{2}$

11. $-105.083 = -105\frac{83}{1000}$

Work Practice 9–11

Later in the chapter, we write fractions as decimals. If you study Examples 7–11, you already know how to write fractions with denominators of 10, 100, 1000, and so on, as decimals.

Objective D Comparing Decimals

One way to compare positive decimals is by comparing digits in corresponding places. To see why this works, let's compare 0.5 or $\frac{5}{10}$ and 0.8 or $\frac{8}{10}$. We know

$\frac{5}{10} < \frac{8}{10}$ since $5 < 8$, so

$0.5 < 0.8$ since $5 < 8$

This leads to the following.

Answers

7. $\frac{51}{1000}$ **8.** $29\frac{97}{100}$ **9.** $\frac{3}{25}$ **10.** $64\frac{4}{5}$ **11.** $-209\frac{493}{500}$

Comparing Two Positive Decimals

Compare digits in the same places from left to right. When two digits are not equal, the number with the larger digit is the larger decimal. If necessary, insert 0s after the last digit to the right of the decimal point to continue comparing.

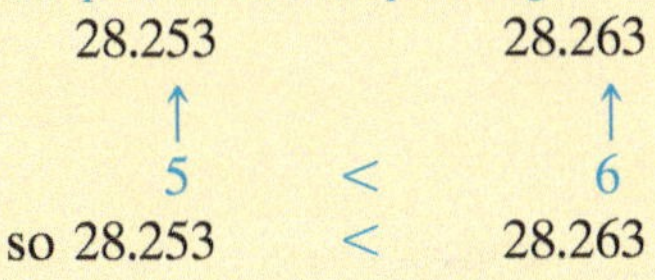

Helpful Hint

For any decimal, writing 0s after the last digit to the right of the decimal point does not change the value of the number.

7.6 = 7.60 = 7.600, and so on

When a whole number is written as a decimal, the decimal point is placed to the right of the ones digit.

25 = 25.0 = 25.00, and so on

Example 12 Insert <, >, or = to form a true statement.

0.378 0.368

Solution:

0.3 78 0.3 68 The tenths places are the same.

0.3 7 8 0.3 6 8 The hundredths places are different.

Since 7 > 6, then 0.378 > 0.368.

Work Practice 12

Practice 12

Insert <, >, or = to form a true statement.

26.208 26.28

Example 13 Insert <, >, or = to form a true statement.

0.052 0.236

Solution: 0. 0 52 < 0. 2 36 0 is smaller than 2 in the tenths place.

Work Practice 13

Practice 13

Insert <, >, or = to form a true statement.

0.12 0.026

We can also use a number line to compare decimals. This is especially helpful when comparing negative decimals. Remember, the number whose graph is to the left is smaller, and the number whose graph is to the right is larger.

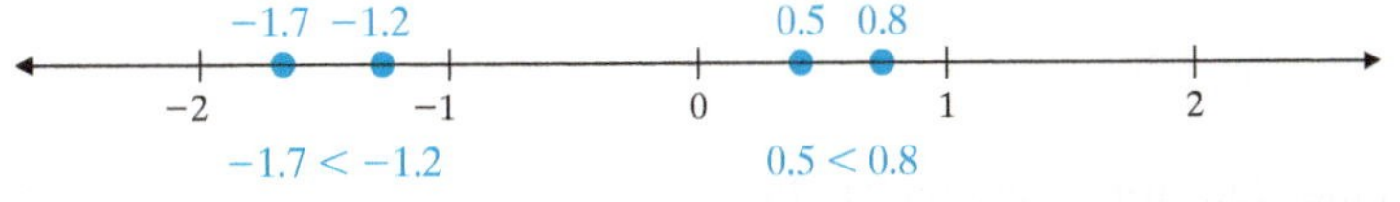

Answers

12. < **13.** >

Helpful Hint

If you have trouble comparing two negative decimals, try the following: Compare their absolute values. Then to correctly compare the negative decimals, reverse the direction of the inequality symbol.

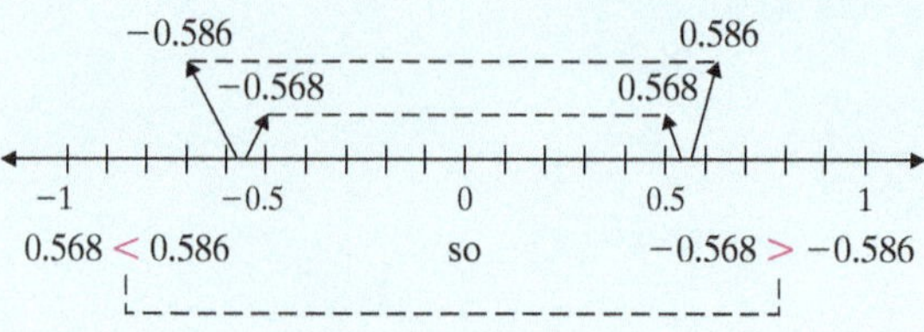

Practice 14

Insert <, >, or = to form a true statement.

$-0.039 \quad -0.0309$

Example 14 Insert <, >, or = to form a true statement.

$-0.0101 \quad -0.00109$

Solution: Since $0.0101 > 0.00109$, then $-0.0101 < -0.00109$.

Work Practice 14

Objective E Rounding Decimals

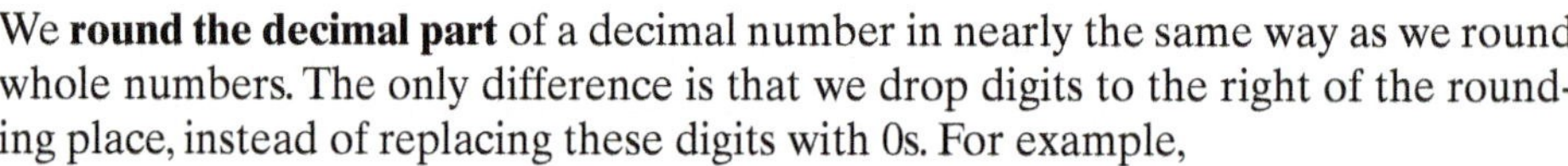

We **round the decimal part** of a decimal number in nearly the same way as we round whole numbers. The only difference is that we drop digits to the right of the rounding place, instead of replacing these digits with 0s. For example,

36.954 rounded to the nearest hundredth is 36.95.

Rounding Decimals to a Place Value to the Right of the Decimal Point

Step 1: Locate the digit to the right of the given place value.

Step 2: If this digit is 5 or greater, add 1 to the digit in the given place value and drop all digits to its right. If this digit is less than 5, drop all digits to the right of the given place.

Practice 15

Round 482.7817 to the nearest thousandth.

Example 15 Round 736.2359 to the nearest tenth.

Solution:

Step 1: We locate the digit to the right of the tenths place.

tenths place

736.2 3 59

digit to the right

Step 2: Since the digit to the right is less than 5, we drop it and all digits to its right.

Thus, 736.2359 rounded to the nearest tenth is 736.2.

Work Practice 15

The same steps for rounding can be used when the decimal is negative.

Answers

14. < **15.** 482.782

Example 16 Round -0.027 to the nearest hundredth.

Solution:

Step 1: Locate the digit to the right of the hundredths place.

-0.027 (hundredths place: 2; digit to the right: 7)

Step 2: Since the digit to the right is 5 or greater, we add 1 to the hundredths digit and drop all digits to its right.

Thus, -0.027 is -0.03 rounded to the nearest hundredth.

Work Practice 16

Practice 16

Round -0.032 to the nearest hundredth.

The following number line illustrates the rounding of negative decimals.

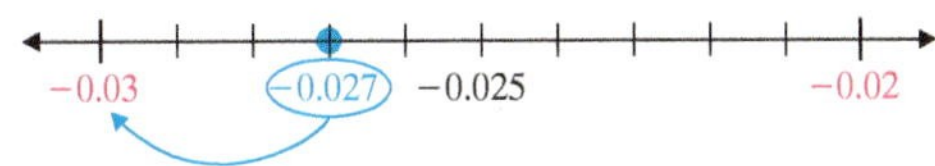

In Section 5.3, we will introduce a formula for the distance around a circle. The distance around a circle is given the special name **circumference.**

The symbol π is the Greek letter pi, pronounced "pie." We use π to denote the following constant:

$$\pi = \frac{\text{circumference of a circle}}{\text{diameter of a circle}}$$

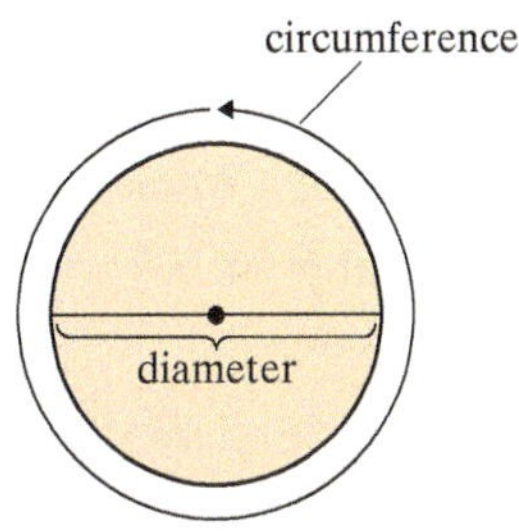

The value π is an **irrational number.** This means if we try to write it as a decimal, it neither ends nor repeats in a pattern.

Example 17 $\pi \approx 3.14159265$. Round π to the nearest hundredth.

Solution:

3.14159265 (hundredths place: 4; 1 is less than 5. Delete these digits.)

Thus, 3.14159265 rounded to the nearest hundredth is 3.14. In other words, $\pi \approx 3.14$.

Work Practice 17

Practice 17

$\pi \approx 3.14159265$. Round π to the nearest ten-thousandth.

Rounding often occurs with money amounts. Since there are 100 cents in a dollar, each cent is $\frac{1}{100}$ of a dollar. This means that if we want to round to the nearest cent, we round to the nearest hundredth of a dollar.

✓**Concept Check** 1756.0894 rounded to the nearest *ten* is

a. 1756.1 **b.** 1760.0894 **c.** 1760 **d.** 1750

Answers

16. -0.03 **17.** $\pi \approx 3.1416$

✓**Concept Check Answer**

c

Practice 18

Water bills in Mexia are always rounded to the nearest dollar. Round a water bill of $24.62 to the nearest dollar.

Answer
18. $25

Example 18 Determining State Taxable Income

A high school teacher's taxable income is $41,567.72. The tax tables in the teacher's state use amounts rounded to the nearest dollar. Round the teacher's income to the nearest whole dollar.

Solution: Rounding to the nearest whole dollar means rounding to the ones place.

once place → $41,567.72 ← 7 is greater than 5.
Add 1. Delete these digits.

Thus, the teacher's income rounded to the nearest dollar is $41,568.

Work Practice 18

Vocabulary, Readiness & Video Check

Use the choices below to fill in each blank.

words	decimals	tenths	after
tens	circumference	and	standard form

1. The number "twenty and eight hundredths" is written in __________ and "20.08" is written in __________.
2. Another name for the distance around a circle is its __________.
3. Like fractions, __________ are used to denote part of a whole.
4. When writing a decimal number in words, the decimal point is written as __________.
5. The place value __________ is to the right of the decimal point while __________ is to the left of the decimal point.
6. The decimal point in a whole number is __________ the last digit.

Martin-Gay Interactive Videos

See Video 5.1

Watch the section lecture video and answer the following questions.

Objective A 7. In Example 1, how is the decimal point written?

Objective B 8. Why is 9.8 not the correct answer to Example 3? What is the correct answer?

Objective C 9. From Example 5, why does reading a decimal number correctly help you write it as an equivalent fraction?

Objective D 10. In Example 7, we compare place value by place value in which direction?

Objective E 11. Example 8 is being rounded to the nearest tenth, so why is the digit 7, which is not in the tenths place, looked at?

5.1 Exercise Set MyMathLab®

Objective A *Write each decimal number in words. See Examples 1 through 3.*

1. 5.62
2. 9.57
3. 16.23
4. 47.65

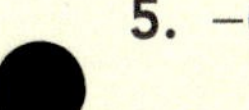

5. −0.205

6. −0.495

7. 167.009

8. 233.056

9. 3000.04

10. 5000.02

11. 105.6

12. 410.3

13. The Akashi Kaikyo Bridge, between Kobe and Awaji-Shima, Japan, is approximately 2.43 miles long.

14. The English Channel Tunnel is a 31.04 miles long undersea rail tunnel connecting England and France. (*Source: Railway Directory & Year Book*)

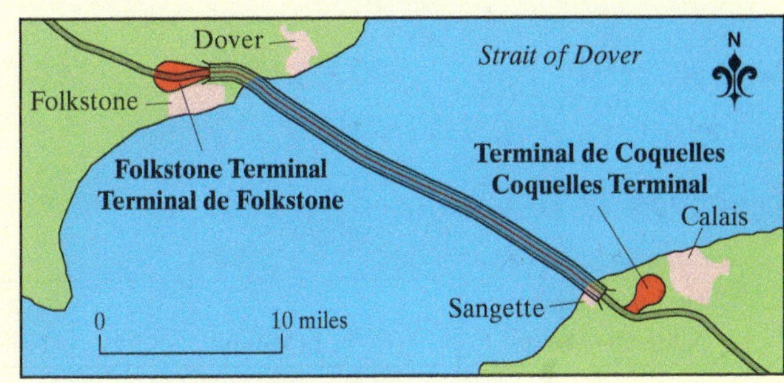

Fill in each check for the described purchase. See Example 4.

15. Your monthly car loan of $321.42 to R. W. Financial.

Your Preprinted Name
Your Preprinted Address
60-8124/7233
1000613331
1407
DATE
PAY TO THE ORDER OF $
DOLLARS
FIRST STATE BANK
OF FARTHINGTON
FARTHINGTON, IL 64422
MEMO
⑆621497260⑆ 1000613331⑈ 1407

16. Your part of the monthly apartment rent, which is $213.70. You pay this to Amanda Dupre.

Your Preprinted Name
Your Preprinted Address
60-8124/7233
1000613331
1408
DATE
PAY TO THE ORDER OF $
DOLLARS
FIRST STATE BANK
OF FARTHINGTON
FARTHINGTON, IL 64422
MEMO
⑆621497260⑆ 1000613331⑈ 1408

17. Your bill of $91.68 to Verizon wireless.

Your Preprinted Name
Your Preprinted Address
60-8124/7233
1000613331
1409
DATE
PAY TO THE ORDER OF $
DOLLARS
FIRST STATE BANK
OF FARTHINGTON
FARTHINGTON, IL 64422
MEMO
⑆621497260⑆ 1000613331⑈ 1409

18. Your grocery bill of $387.49 at Kroger.

Your Preprinted Name
Your Preprinted Address
60-8124/7233
1000613331
1410
DATE
PAY TO THE ORDER OF $
DOLLARS
FIRST STATE BANK
OF FARTHINGTON
FARTHINGTON, IL 64422
MEMO
⑆621497260⑆ 1000613331⑈ 1410

Objective B *Write each decimal number in standard form. See Examples 5 and 6.*

19. Two and eight tenths

20. Five and one tenth

21. Nine and eight hundredths

22. Twelve and six hundredths

23. Negative seven hundred five and six hundred twenty-five thousandths

24. Negative eight hundred four and three hundred ninety-nine thousandths

25. Forty-six ten-thousandths

26. Eighty-three ten-thousandths

Objective C *Write each decimal as a fraction or a mixed number. Write your answer in simplest form. See Examples 7 through 11.*

27. 0.7
28. 0.9
29. 0.27
30. 0.39
31. 0.4
32. 0.8
33. 5.4
34. 6.8
35. −0.058
36. −0.024
37. 7.008
38. 9.005
39. 15.802
40. 11.406
41. 0.3005
42. 0.2006

Objectives A B C Mixed Practice *Fill in the chart. The first row is completed for you. See Examples 1 through 11.*

	Decimal Number in Standard Form	In Words	Fraction
	0.37	thirty-seven hundredths	$\frac{37}{100}$
43.		eight tenths	
44.		five tenths	
45.	0.077		
46.	0.019		

Objective D *Insert <, >, or = between each pair of numbers to form a true statement. See Examples 12 through 14.*

47. 0.15 0.16
48. 0.12 0.15
49. −0.57 −0.54
50. −0.59 −0.52
51. 0.098 0.1
52. 0.0756 0.2
53. 0.54900 0.549
54. 0.98400 0.984
55. 167.908 167.980
56. 519.3405 519.3054
57. −1.062 −1.07
58. −18.1 −18.01
59. −7.052 7.0052
60. 0.01 −0.1
61. −0.023 −0.024
62. −0.562 −0.652

Objective E *Round each decimal to the given place value. See Examples 15 through 18.*

63. 0.57, nearest tenth
64. 0.64, nearest tenth
65. 98,207.23, nearest ten
66. 68,934.543, nearest ten
67. −0.234, nearest hundredth
68. −0.892, nearest hundredth
69. 0.5942, nearest thousandth
70. 63.4523, nearest thousandth

Recall that the number π, written as a decimal, neither ends nor repeats in a pattern. Given that π ≈ 3.14159265, round π to the given place values below. (We study π further in Section 5.3.) See Example 17.

71. tenth **72.** ones **73.** thousandth **74.** hundred-thousandth

Round each monetary amount to the nearest cent or dollar as indicated. See Example 18.

75. $26.95, to the nearest dollar

76. $14,769.52, to the nearest dollar

77. $0.1992, to the nearest cent

78. $0.7633, to the nearest cent

Round each number to the given place value. See Example 18.

79. At the time of this writing, the Apple MacBook Air is the thinnest Mac in production. At its thickest point, it measures 0.68 in. Round this number to the nearest tenth. (*Source:* Apple, Inc.)

13-inch MacBook Air

80. A large tropical cockroach of the family Dictyoptera is the fastest-moving insect. This insect was clocked at a speed of 3.36 miles per hour. Round this number to the nearest tenth. (*Source:* University of California, Berkeley)

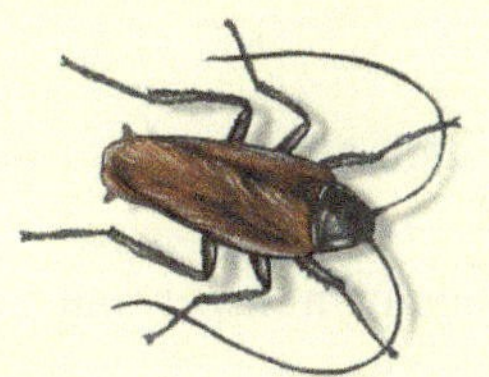

81. Missy Franklin of the United States won the gold medal for the 200 m backstroke in the 2012 London Summer Olympics with a record time of 2.0677 minutes. Round this time to the nearest hundredth of a minute.

82. The population density of the state of Utah is 34.745 people per square mile. Round this population density to the nearest tenth. (*Source:* U.S. Census Bureau)

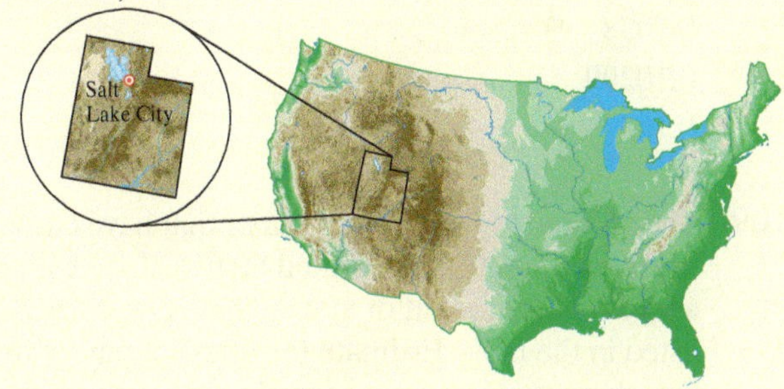

83. A used biology textbook is priced at $67.89. Round this price to the nearest dollar.

84. A used office desk is advertised at $19.95 by Drawley's Office Furniture. Round this price to the nearest dollar.

85. Venus makes a complete orbit around the Sun every 224.695 days. Round this figure to the nearest whole day. (*Source:* National Space Science Data Center)

86. The length of a day on Mars, a full rotation about its axis, is 24.6229 hours. Round this figure to the nearest thousandth. (*Source:* National Space Science Data Center)

Review

Perform each indicated operation. See Section 1.3.

87. 3452 + 2314 **88.** 8945 + 4536 **89.** 82 − 47 **90.** 4002 − 3897

Concept Extensions

Solve. See the Concept Check in this section.

91. 2849.1738 rounded to the nearest hundred is
a. 2849.17 **b.** 2800 **c.** 2850 **d.** 2849.174

92. 146.059 rounded to the nearest ten is
a. 146.0 **b.** 146.1 **c.** 140 **d.** 150

93. 2849.1738 rounded to the nearest hundredth is
a. 2849.17 **b.** 2800 **c.** 2850 **d.** 2849.18

94. 146.059 rounded to the nearest tenth is
a. 146.0 **b.** 146.1 **c.** 140 **d.** 150

95. In your own words, describe how to write a decimal as a fraction or a mixed number.

96. Explain how to identify the value of the 9 in the decimal 486.3297.

97. Write $7\frac{12}{100}$ as a decimal.

98. Write $17\frac{268}{1000}$ as a decimal.

99. Write 0.00026849577 as a fraction.

100. Write 0.00026849576 in words.

101. Write a 5-digit number that rounds to 1.7.

102. Write a 4-digit number that rounds to 26.3.

103. Write a decimal number that is greater than 8 but less than 9.

104. Write a decimal number that is greater than 48.1, but less than 48.2.

105. Which number(s) rounds to 0.26?
0.26559 0.26499 0.25786 0.25186

106. Which number(s) rounds to 0.06?
0.0612 0.066 0.0586 0.0506

Write these numbers from smallest to largest.

107. 0.9
0.1038
0.10299
0.1037

108. 0.01
0.0839
0.09
0.1

109. The all-time top six movies (those that have earned the most money in the United States) along with the approximate amount of money they have earned are listed in the table. Estimate the total amount of money that these movies have earned by first rounding each earning to the nearest hundred-million. (*Source:* The Internet Movie Database)

Top All-Time American Movies

Movie	Gross Domestic Earnings
Avatar (2009)	$760.5 million
Titanic (1997)	$658.7 million
The Avengers (2012)	$623.3 million
The Dark Knight (2008)	$533.3 million
Star Wars: The Phantom Menace (1999)	$474.5 million
Star Wars (1977)	$460.9 million

110. In 2012, there were 1392.2 million singles downloaded at an average price of $1.20 each. Find an estimate of the total revenue from downloaded singles by answering parts **a–c**. (*Source:* Recording Industry Association of America)

a. Round 1392.2 million to the nearest ten million.

b. Multiply the rounded value in part **a** by 12.

c. Move the decimal point in the product from part **b** one place to the left. This number is the total revenue in million dollars.

5.2 Adding and Subtracting Decimals 

Objective A Adding or Subtracting Decimals

Adding or subtracting decimals is similar to adding or subtracting whole numbers. We add or subtract digits in corresponding place values from right to left, carrying or borrowing if necessary. To make sure that digits in corresponding place values are added or subtracted, we line up the decimal points vertically.

Adding or Subtracting Decimals

Step 1: Write the decimals so that the decimal points line up vertically.

Step 2: Add or subtract as with whole numbers.

Step 3: Place the decimal point in the sum or difference so that it lines up vertically with the decimal points in the problem.

In this section, we will insert zeros in decimal numbers so that place-value digits line up neatly. This is shown in Example 1.

Example 1 Add: 23.85 + 1.604

Solution: First we line up the decimal points vertically.

```
  23.850   Insert one 0 so that digits line up neatly.
+  1.604
   ↑
Line up decimal points.
```

Then we add the digits from right to left as for whole numbers.

```
   1
  23.850
+  1.604
  25.454
   ↑___ Place the decimal point in the sum so that all decimal points line up.
```

Work Practice 1

Helpful Hint

Recall that 0s may be placed after the last digit to the right of the decimal point without changing the value of the decimal. This may be used to help line up place values when adding decimals.

```
   3.2     becomes     3.200   Insert two 0s.
  15.567              15.567
+  0.11              + 0.110   Insert one 0.
                      18.877   Add.
```

Objectives

A Add or Subtract Decimals.

B Estimate when Adding or Subtracting Decimals.

C Evaluate Expressions with Decimal Replacement Values.

D Simplify Expressions Containing Decimals.

E Solve Problems That Involve Adding or Subtracting Decimals.

Practice 1

Add.

a. 19.52 + 5.371

b. 40.08 + 17.612

c. 0.125 + 422.8

Answers

1. a. 24.891 **b.** 57.692 **c.** 422.925

Practice 2

Add.

a. 34.567 + 129.43 + 2.8903

b. 11.21 + 46.013 + 362.526

Example 2

Add: 763.7651 + 22.001 + 43.89

Solution: First we line up the decimal points.

```
  1 1 1
  763.7651
   22.0010   Insert one 0.
+  43.8900   Insert two 0s.
  829.6561   Add.
```

Work Practice 2

Helpful Hint

Don't forget that the decimal point in a whole number is positioned after the last digit.

Practice 3

Add: 19 + 26.072

Example 3

Add: 45 + 2.06

Solution:

```
  45.00   Insert a decimal point and two 0s.
+  2.06   Line up decimal points.
  47.06   Add.
```

Work Practice 3

✓**Concept Check** What is wrong with the following calculation of the sum of 7.03, 2.008, 19.16, and 3.1415?

```
    7.03
   2.008
   19.16
 +3.1415
  3.6042
```

Practice 4

Add: 7.12 + (−9.92)

Example 4

Add: 3.62 + (−4.78)

Solution: Recall from Chapter 2 that to add two numbers with different signs, we find the difference of the larger absolute value and the smaller absolute value. The sign of the answer is the same as the sign of the number with the larger absolute value.

```
  4.78
− 3.62
  1.16   Subtract the absolute values.
```

Thus, 3.62 + (−4.78) = −1.16

The sign of the number with the larger absolute value; −4.78 has the larger absolute value.

Work Practice 4

Subtracting decimals is similar to subtracting whole numbers. We line up digits and subtract from right to left, borrowing when needed.

Answers

2. a. 166.8873 b. 419.749
3. 45.072 4. −2.8

✓**Concept Check Answer**

The decimal places are not lined up properly.

Example 5 Subtract: 3.5 − 0.068. Check your answer.

Solution:

$$\begin{array}{r} 3.500 \\ -\,0.068 \\ \hline 3.432 \end{array}$$

Insert two 0s. Line up decimal points. Subtract.

Check: Recall that we can check a subtraction problem by adding.

$$\begin{array}{rl} 3.432 & \text{Difference} \\ +\,0.068 & \text{Subtrahend} \\ \hline 3.500 & \text{Minuend} \end{array}$$

Work Practice 5

Practice 5

Subtract. Check your answers.

a. 6.7 − 3.92

b. 9.72 − 4.068

Example 6 Subtract: 85 − 17.31. Check your answer.

Solution:

$$\begin{array}{r} 85.00 \\ -\,17.31 \\ \hline 67.69 \end{array}$$

Check:

$$\begin{array}{rl} 67.69 & \text{Difference} \\ +\,17.31 & \text{Subtrahend} \\ \hline 85.00 & \text{Minuend} \end{array}$$

Work Practice 6

Practice 6

Subtract. Check your answers.

a. 73 − 29.31

b. 210 − 68.22

Example 7 Subtract 3 from 6.98.

Solution:

$$\begin{array}{r} 6.98 \\ -\,3.00 \\ \hline 3.98 \end{array}$$

Insert two 0s.

Check:

$$\begin{array}{rl} 3.98 & \text{Difference} \\ +\,3.00 & \text{Subtrahend} \\ \hline 6.98 & \text{Minuend} \end{array}$$

Work Practice 7

Practice 7

Subtract 19 from 25.91

Example 8 Subtract: −5.8 − 1.7

Solution: Recall from Chapter 2 that to subtract 1.7, we add the opposite of 1.7, or −1.7. Thus

$-5.8 - 1.7 = -5.8 + (-1.7)$ To subtract, add the opposite of 1.7, which is −1.7.

$= -7.5.$ Add the absolute values. Use the common negative sign.

Work Practice 8

Practice 8

Subtract: −5.4 − 9.6

Example 9 Subtract: −2.56 − (−4.01)

Solution: $-2.56 - (-4.01) = -2.56 + 4.01$ To subtract, add the opposite of −4.01, which is 4.01.

$= 1.45$ Subtract the absolute values. The answer is positive since 4.01 has the larger absolute value.

Work Practice 9

Practice 9

Subtract: −1.05 − (−7.23)

Answers

5. a. 2.78 **b.** 5.652
6. a. 43.69 **b.** 141.78
7. 6.91 **8.** −15 **9.** 6.18

Objective B Estimating when Adding or Subtracting Decimals

To help avoid errors, we can also estimate to see if our answer is reasonable when adding or subtracting decimals. Although only one estimate is needed per operation, we show two for variety.

Practice 10

Add or subtract as indicated. Then estimate to see if the answer is reasonable by rounding the given numbers and adding or subtracting the rounded numbers.

a. 58.1 + 326.97

b. 16.08 − 0.925

Example 10

Add or subtract as indicated. Then estimate to see if the answer is reasonable by rounding the given numbers and adding or subtracting the rounded numbers.

a. 27.6 + 519.25

Exact		Estimate 1		Estimate 2
27.60	rounds to	30		30
+ 519.25	rounds to	+ 500	or	+ 520
546.85		530		550

Since the exact answer is close to either estimate, it is reasonable. (In the first estimate, each number is rounded to the place value of the leftmost digit. In the second estimate, each number is rounded to the nearest ten.)

b. 11.01 − 0.862

Exact		Estimate 1		Estimate 2
11.010	rounds to	10		11
− 0.862	rounds to	− 1	or	− 1
10.148		9		10

In the first estimate, we rounded the first number to the nearest ten and the second number to the nearest one. In the second estimate, we rounded both numbers to the nearest one. Both estimates show us that our answer is reasonable.

Helpful Hint Remember that estimates are used for our convenience to quickly check the reasonableness of an answer.

Work Practice 10

✓**Concept Check** Why shouldn't the sum 21.98 + 42.36 be estimated as 30 + 50 = 80?

Objective C Using Decimals as Replacement Values

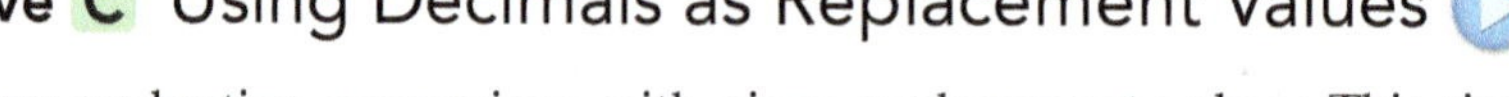

Let's review evaluating expressions with given replacement values. This time the replacement values are decimals.

Practice 11

Evaluate $y - z$ for $y = 11.6$ and $z = 10.8$.

Example 11

Evaluate $x - y$ for $x = 2.8$ and $y = 0.92$.

Solution: Replace x with 2.8 and y with 0.92 and simplify.

$$x - y = 2.8 - 0.92 = 1.88$$

$$\begin{array}{r} 2.80 \\ -0.92 \\ \hline 1.88 \end{array}$$

Work Practice 11

Answers

10. a. Exact: 385.07; an Estimate: 390 **b.** Exact: 15.155; an Estimate: 15 **11.** 0.8

✓**Concept Check Answer**

Each number is rounded incorrectly. The estimate is too high.

Example 12 Is 2.3 a solution of the equation $6.3 = x + 4$?

Solution: Replace x with 2.3 in the equation $6.3 = x + 4$ to see if the result is a true statement.

$$6.3 = x + 4$$
$$6.3 \stackrel{?}{=} 2.3 + 4 \quad \text{Replace } x \text{ with 2.3.}$$
$$6.3 = 6.3 \quad \text{True}$$

Since $6.3 = 6.3$ is a true statement, 2.3 is a solution of $6.3 = x + 4$.

Work Practice 12

Practice 12

Is 12.1 a solution of the equation $y - 4.3 = 7.8$?

Objective D Simplifying Expressions Containing Decimals

Example 13 Simplify by combining like terms:

$$11.1x - 6.3 + 8.9x - 4.6$$

Solution:

$$\begin{aligned} 11.1x - 6.3 + 8.9x - 4.6 &= 11.1x + 8.9x + (-6.3) + (-4.6) \\ &= 20x + (-10.9) \\ &= 20x - 10.9 \end{aligned}$$

Work Practice 13

Practice 13

Simplify by combining like terms:

$$-4.3y + 7.8 - 20.1y + 14.6$$

Objective E Solving Problems by Adding or Subtracting Decimals

Decimals are very common in real-life problems.

Example 14 Calculating the Cost of Owning an Automobile

Find the total monthly cost of owning and operating a certain automobile given the expenses shown.

Monthly car payment:	\$256.63
Monthly insurance cost:	\$47.52
Average gasoline bill per month:	\$195.33

Solution:

1. UNDERSTAND. Read and reread the problem. The phrase "total monthly cost" tells us to add.
2. TRANSLATE.

In words:	total monthly cost	is	car payment	plus	insurance cost	plus	gasoline bill
	↓	↓	↓	↓	↓	↓	↓
Translate:	total monthly cost	=	\$256.63	+	\$47.52	+	\$195.33

(Continued on next page)

Practice 14

Find the total monthly cost of owning and operating a certain automobile given the expenses shown.

Monthly car payment:	\$563.52
Monthly insurance cost:	\$52.68
Average gasoline bill per month:	\$127.50

Answers

12. yes **13.** $-24.4y + 22.4$
14. \$743.70

3. SOLVE: Let's also estimate by rounding each number to the nearest ten.

256.63	rounds to	260	
47.52	rounds to	50	
+ 195.33	rounds to	200	
499.48	Exact	510	Estimate

4. INTERPRET. *Check* your work. Since our estimate is close to our exact answer, our answer is reasonable. *State* your conclusion: The total monthly cost is $499.48.

Work Practice 14

The next bar graph has horizontal bars. To visualize the value represented by a bar, see how far it extends to the right. The value of each bar is labeled, and we will study bar graphs further in a later chapter.

Practice 15

Use the bar graph in Example 15. How much greater is the average height in the Netherlands than the average height in Czechoslovakia?

Example 15 Comparing Average Heights

The bar graph shows the current average heights for adults in various countries. How much greater is the average height in Denmark than the average height in the United States?

Average Adult Height

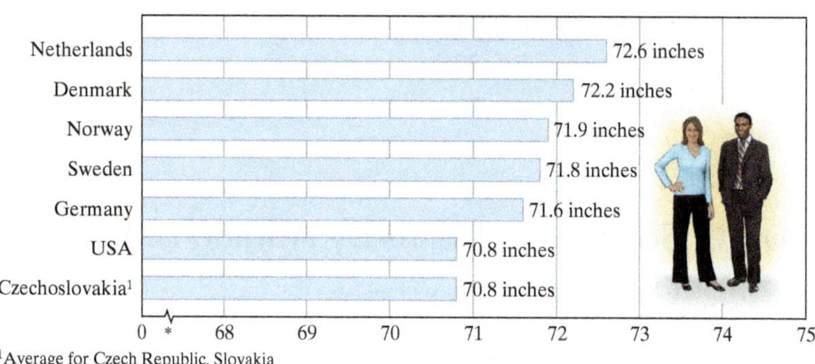

[1]Average for Czech Republic, Slovakia
Source: USA Today

* The -\/- means that some numbers are purposefully missing on the axis.

Solution:

1. UNDERSTAND. Read and reread the problem. Since we want to know "how much greater," we subtract.
2. TRANSLATE.

In words:	How much greater	is	Denmark's average height	minus	U.S. average height
	↓	↓	↓	↓	↓
Translate:	How much greater	=	72.2	−	70.8

3. SOLVE: We estimate by rounding each number to the nearest whole.

72.2	rounds to	72	
− 70.8	rounds to	− 71	
1.4	Exact	1	Estimate

4. INTERPRET. *Check* your work. Since our estimate is close to our exact answer, 1.4 inches is reasonable. *State* your conclusion: The average height in Denmark is 1.4 inches greater than the average U.S. height.

Work Practice 15

Answer
15. 1.8 in.

Calculator Explorations Decimals

Entering Decimal Numbers

To enter a decimal number, find the key marked [·]. To enter the number 2.56, for example, press the keys [2] [·] [56].

The display will read [2.56].

Operations on Decimal Numbers

Operations on decimal numbers are performed in the same way as operations on whole or signed numbers. For example, to find 8.625 − 4.29, press the keys [8.625] [−] [4.29] then [=] or [ENTER].

The display will read [4.335]. (Although entering 8.625, for example, requires pressing more than one key, we group numbers together here for easier reading.)

Use a calculator to perform each indicated operation.

1. 315.782 + 12.96

2. 29.68 + 85.902

3. 6.249 − 1.0076

4. 5.238 − 0.682

5.
$$\begin{array}{r} 12.555 \\ 224.987 \\ 5.2 \\ +622.65 \\ \hline \end{array}$$

6.
$$\begin{array}{r} 47.006 \\ 0.17 \\ 313.259 \\ +139.088 \\ \hline \end{array}$$

Vocabulary, Readiness & Video Check

Use the choices below to fill in each blank. Not all choices will be used.

minuend	vertically	like	true
difference	subtrahend	last	false

1. The decimal point in a whole number is positioned after the ________ digit.
2. In 89.2 − 14.9 = 74.3, the number 74.3 is called the ________, 89.2 is the ________, and 14.9 is the ________.
3. To simplify an expression, we combine any ________ terms.
4. To add or subtract decimals, we line up the decimal points ________.
5. True or false: If we replace x with 11.2 and y with −8.6 in the expression $x - y$, we have 11.2 − 8.6. ________
6. True or false: If we replace x with −9.8 and y with −3.7 in the expression $x + y$, we have −9.8 + 3.7. ________

Martin-Gay Interactive Videos

See Video 5.2

Watch the section lecture video and answer the following questions.

Objective A 7. From Examples 1–3, why do you think we line up decimal points?

Objective B 8. In Example 4, estimating is used to check whether the answer to the subtraction problem is reasonable, but what is the best way to fully check?

Objective C 9. In Example 5, why is the actual subtraction performed to the side?

Objective D 10. How many sets of like terms are there in Example 6?

Objective E 11. In Example 7, to calculate the amount of border material needed, we are actually calculating the ____________ of the triangle.

5.2 Exercise Set MyMathLab® 

Objectives A B Mixed Practice *Add. See Examples 1 through 4 and 10. For those exercises marked, also estimate to see if the answer is reasonable.*

1. 5.6 + 2.1

2. 3.6 + 4.1

3. 8.2 + 2.15

4. 5.17 + 3.7

5. 24.6 + 2.39 + 0.0678

6. 32.4 + 1.58 + 0.0934

7. −2.6 + (−5.97)

8. −18.2 + (−10.8)

9. 18.56 + (−8.23)

10. 4.38 + (−6.05)

11.
$$\begin{array}{r} 234.89 \\ +\,230.67 \\ \hline \end{array}$$
Exact: Estimate:

12.
$$\begin{array}{r} 734.89 \\ +\,640.56 \\ \hline \end{array}$$
Exact: Estimate:

13.
$$\begin{array}{r} 100.009 \\ 6.08 \\ +\quad 9.034 \\ \hline \end{array}$$
Exact: Estimate:

14.
$$\begin{array}{r} 200.89 \\ 7.49 \\ +\;\; 62.83 \\ \hline \end{array}$$
Exact: Estimate:

15. Find the sum of 39, 3.006, and 8.403

16. Find the sum of 65, 5.0903, and 6.9003

Subtract and check. See Examples 5 through 10. For those exercises marked, also estimate to see if the answer is reasonable.

17. 12.6 − 8.2

18. 8.9 − 3.1

19. 18 − 2.7

20. 28 − 3.3

21.
$$\begin{array}{r} 654.9 \\ -\;\; 56.67 \\ \hline \end{array}$$

22.
$$\begin{array}{r} 863.23 \\ -\;\; 39.453 \\ \hline \end{array}$$

23. 5.9 − 4.07
Exact:
Estimate:

24. 6.4 − 3.04
Exact:
Estimate:

25.
$$\begin{array}{r} 1000 \\ -\;\; 123.4 \\ \hline \end{array}$$
Exact:

Estimate:

26.
$$\begin{array}{r} 2000 \\ -\;\; 327.47 \\ \hline \end{array}$$
Exact:

Estimate:

27. 200 − 5.6

28. 800 − 8.9

29. −1.12 − 5.2

30. −8.63 − 5.6

31. 5.21 − 11.36

32. 8.53 − 17.84

33. −2.6 − (−5.7)

34. −9.4 − (−10.4)

35. 3 − 0.0012

36. 7 − 0.097

37. Subtract 6.7 from 23.

38. Subtract 9.2 from 45.

Objective A *Perform the indicated operation. See Examples 1 through 9.*

39. $0.9 + 2.2$
40. $0.7 + 3.4$
41. $-6.06 + 0.44$
42. $-5.05 + 0.88$
43. $500.21 - 136.85$
44. $600.47 - 254.68$
45. $50.2 - 600$
46. $40.3 - 700$
47. Subtract 61.9 from 923.5.
48. Subtract 45.8 from 845.9.
49. Add 100.009 and 6.08 and 9.034.
50. Add 200.89 and 7.49 and 62.83.
51. $-0.003 + 0.091$
52. $-0.004 + 0.085$
53. $-102.4 - 78.04$
54. $-36.2 - 10.02$
55. $-2.9 - (-1.8)$
56. $-6.5 - (-3.3)$

Objective C *Evaluate each expression for $x = 3.6$, $y = 5$, and $z = 0.21$. See Example 11.*

57. $x + z$
58. $y + x$
59. $x - z$
60. $y - z$
61. $y - x + z$
62. $x + y + z$

Determine whether the given values are solutions to the given equations. See Example 12.

63. Is 7 a solution to $x + 2.7 = 9.3$?
64. Is 3.7 a solution to $x + 5.9 = 8.6$?
65. Is -11.4 a solution to $27.4 + y = 16$?
66. Is -22.9 a solution to $45.9 + z = 23$?
67. Is 1 a solution to $2.3 + x = 5.3 - x$?
68. Is 0.9 a solution to $1.9 - x = x + 0.1$?

Objective D *Simplify by combining like terms. See Example 13.*

69. $30.7x + 17.6 - 23.8x - 10.7$
70. $14.2z + 11.9 - 9.6z - 15.2$
71. $-8.61 + 4.23y - 2.36 - 0.76y$
72. $-8.96x - 2.31 - 4.08x + 9.68$

Objective E *Solve. For Exercises 73 and 74, the solutions have been started for you. See Examples 14 and 15.*

73. Ann-Margaret Tober bought a book for \$32.48. If she paid with two \$20 bills, what was her change?

Start the solution:

1. UNDERSTAND the problem. Reread it as many times as needed.
2. TRANSLATE into an equation. (Fill in the blank.)

change	is	two \$20 bills	minus	cost of book
↓	↓	↓	↓	↓
change	=	40	−	____

Finish with

3. SOLVE and 4. INTERPRET

74. Phillip Guillot bought a car part for \$18.26. If he paid with two \$10 bills, what was his change?

Start the solution:

1. UNDERSTAND the problem. Reread it as many times as needed.
2. TRANSLATE into an equation. (Fill in the blank.)

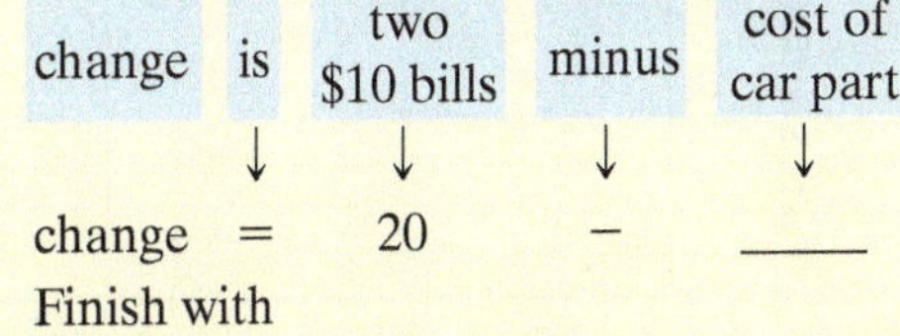

change	is	two \$10 bills	minus	cost of car part
↓	↓	↓	↓	↓
change	=	20	−	____

Finish with

3. SOLVE and 4. INTERPRET

75. Microsoft stock opened the day at \$35.17 per share, and the closing price the same day was \$34.75. By how much did the price of each share change?

76. A pair of eyeglasses costs a total of \$347.89. The frames of the glasses are \$97.23. How much do the lenses of the eyeglasses cost?

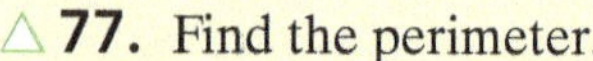

△ **77.** Find the perimeter.

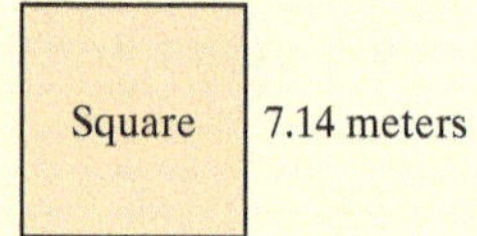

△ **78.** Find the perimeter.

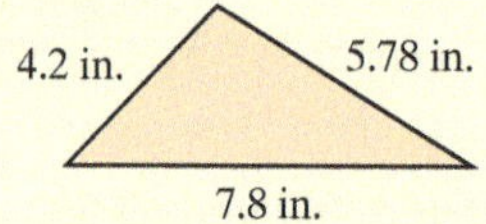

79. The Apple iPhone 5 was released in 2012. It measures 4.87 inches by 2.31 inches. Find the perimeter of this phone. (*Source:* Apple.com)

80. The Google Nexus 4, released in 2012, is the newest Google phone (at this writing). It measures 5.27 inches by 2.7 inches. Find the perimeter of the phone. (*Source:* Google.com)

81. The average wind speed at the weather station on Mt. Washington in New Hampshire is 35.2 miles per hour. The highest speed ever recorded at the station is 231.0 miles per hour. How much faster is the highest speed than the average wind speed? (*Source:* National Climatic Data Center)

82. The average annual rainfall in Omaha, Nebraska, is 30.08 inches. The average annual rainfall in New Orleans, Louisiana, is 64.16 inches. On average, how much more rain does New Orleans receive annually than Omaha? (*Source:* National Climatic Data Center)

This bar graph shows the predicted increase in the total number of text messages per person per day in the United States. Use this graph for Exercises 83 and 84. (Source: Pew Research Center.)

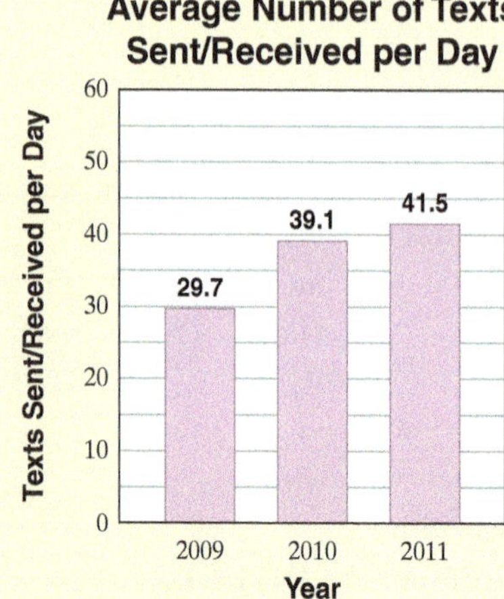

83. Find the increase in the number of texts sent or received per day from 2009 to 2011.

84. Find the increase in the number of texts sent or received per day from 2010 to 2011.

85. As of this writing, the top three U.S. movies that made the most money through movie ticket sales are *Avatar* (2009), $760.5 million; *Titanic* (1997), $658.7 million; and *The Avengers* (2012), $623.3 million. What was the total amount of ticket sales for these three movies? (*Source:* MovieWeb)

86. In 2010, the average credit card late fee was $35. In 2011, the average credit card late fee had decreased by about $11.85. Find the average credit card late fee in 2011. (*Source:* Consumer Financial Protection Bureau)

87. The snowiest city in the United States is Valdez, AK, which receives an average of 110.5 more inches of snow than the second snowiest city. The second snowiest city in the United States is Crested Butte, CO. Crested Butte receives an average of 215.8 inches annually. How much snow does Valdez receive on average each year? (*Source:* The Weather Channel)

88. The driest place in the world is the Atacama Desert in Chile, which receives an average of only 0.004 inch of rain per year. Yuma, Arizona, is the driest city in the United States. Yuma receives an average of 3.006 more inches of rain each year than the Atacama Desert. What is the average annual rainfall in Yuma? (*Source:* National Climatic Data Center)

89. A landscape architect is planning a border for a flower garden shaped like a triangle. The sides of the garden measure 12.4 feet, 29.34 feet, and 25.7 feet. Find the amount of border material needed.

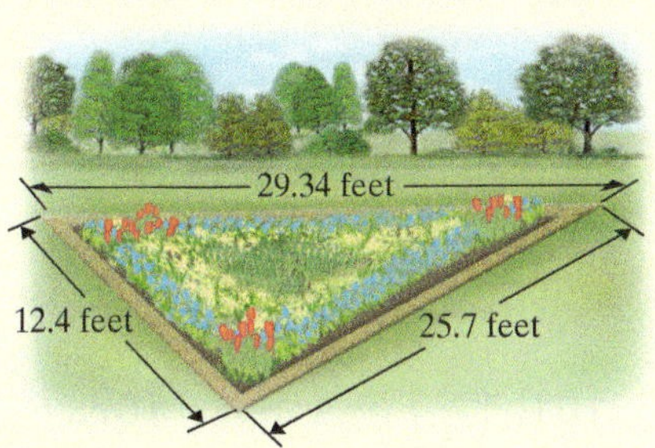

△ **90.** A contractor purchased enough railing to completely enclose the newly built deck shown below. Find the amount of railing purchased.

The table shows the average speeds for the Daytona 500 winners for the years shown. Use this table to answer Exercises 91 and 92. (Source: Daytona International Speedway*)*

Daytona 500 Winners

Year	Winner	Average Speed
1978	Bobby Allison	159.73
1988	Bobby Allison	137.531
1998	Dale Earnhardt	172.712
2008	Ryan Newman	152.672
2013	Jimmie Johnson	159.250

91. How much slower was the average Daytona 500 winning speed in 2013 than in 1998?

92. How much faster was Bobby Allison's average Daytona 500 winning speed in 1978 than his average Daytona 500 winning speed in 1988?

The bar graph shows the top five chocolate-consuming nations in the world. Use this table to answer Exercises 93 through 97.

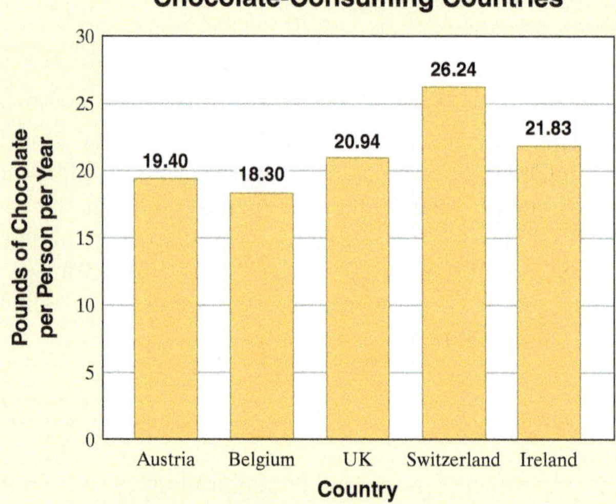

Source: Confectionary News and Leatherhead Food Research

93. Which country in the table has the greatest chocolate consumption per person?

94. Which country in the table has the least chocolate consumption per person?

95. How much more is the greatest chocolate consumption than the least chocolate consumption shown in the table?

96. How much more chocolate does the average Irishman or woman consume per year than the average Austrian?

97. Make a new chart listing the countries and their corresponding chocolate consumptions in order from greatest to least.

Review

Multiply. See Sections 1.5 and 4.3.

98. $23 \cdot 2$ **99.** $46 \cdot 3$ **100.** $39 \cdot 3$ **101.** $\left(\frac{2}{3}\right)^2$ **102.** $\left(\frac{1}{5}\right)^3$

Concept Extensions

A friend asks you to check his calculations for Exercises 103 and 104. Are they correct? If not, explain your friend's errors and correct the calculations. See the first Concept Check in this section.

103.

$$\begin{array}{r} \overset{1}{9}.2 \\ \overset{1}{8}.63 \\ +\ 4.005 \\ \hline 4.960 \end{array}$$

104.

$$\begin{array}{r} \overset{8}{\not9}\overset{9}{\not0}\overset{9}{\not0}.\overset{9}{\not0} \\ -\ 96.4 \\ \hline 803.5 \end{array}$$

Find the unknown length in each figure.

△ **105.**

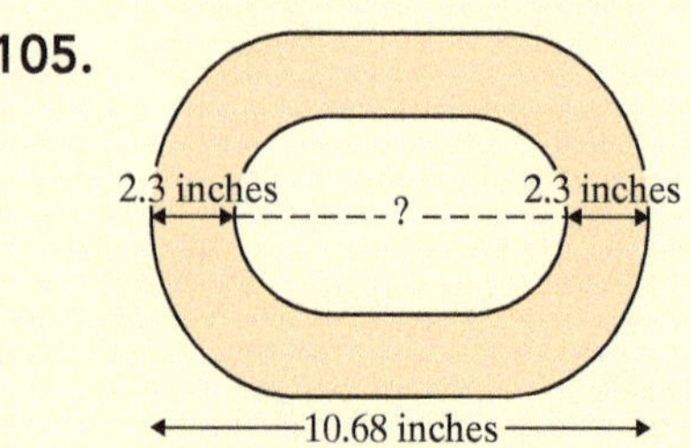

△ **106.**

5.26 meters | 7.82 meters | ? meters

17.67 meters

Let's review the values of these common U.S. coins in order to answer the following exercises.

Penny Nickel Dime Quarter

$0.01 $0.05 $0.10 $0.25

For Exercises 107 and 108, write the value of each group of coins. To do so, it is usually easiest to start with the coin(s) of greatest value and end with the coin(s) of least value.

107. 

108. 

109. Name the different ways that coins can have a value of $0.17 given that you may use no more than 10 coins.

110. Name the different ways that coin(s) can have a value of $0.25 given that there are no pennies.

111. Why shouldn't the sum

$$82.95 + 51.26$$

be estimated as $90 + 60 = 150$?
See the second Concept Check in this section.

112. Laser beams can be used to measure the distance to the moon. One measurement showed the distance to the moon to be 256,435.235 miles. A later measurement showed that the distance is 256,436.012 miles. Find how much farther away the moon is in the second measurement as compared to the first.

113. Explain how adding or subtracting decimals is similar to adding or subtracting whole numbers.

114. Can the sum of two negative decimals ever be a positive decimal? Why or why not?

Combine like terms and simplify.

115. $-8.689 + 4.286x - 14.295 - 12.966x + 30.861x$

116. $14.271 - 8.968x + 1.333 - 201.815x + 101.239x$

5.3 Multiplying Decimals and Circumference of a Circle 

Objectives

A Multiply Decimals.

B Estimate when Multiplying Decimals.

C Multiply Decimals by Powers of 10.

D Evaluate Expressions with Decimal Replacement Values.

E Find the Circumference of Circles.

F Solve Problems by Multiplying Decimals.

Objective A Multiplying Decimals

Multiplying decimals is similar to multiplying whole numbers. The only difference is that we place a decimal point in the product. To discover where a decimal point is placed in a product, let's multiply 0.6 × 0.03. We first write each decimal as an equivalent fraction and then multiply.

$$\underset{\text{1 decimal place}}{0.6} \times \underset{\text{2 decimal places}}{0.03} = \frac{6}{10} \times \frac{3}{100} = \frac{18}{1000} = \underset{\text{3 decimal places}}{0.018}$$

Notice that 1 + 2 = 3, the number of decimal places in the product. Now let's multiply 0.03 × 0.002.

$$\underset{\text{2 decimal place}}{0.03} \times \underset{\text{3 decimal places}}{0.002} = \frac{3}{100} \times \frac{2}{1000} = \frac{6}{100{,}000} = \underset{\text{5 decimal places}}{0.00006}$$

Again, we see that 2 + 3 = 5, the number of decimal places in the product.

Instead of writing decimals as fractions each time we want to multiply, we notice a pattern from these examples and state a rule that we can use:

Multiplying Decimals

Step 1: Multiply the decimals as though they are whole numbers.

Step 2: The decimal point in the product is placed so that the number of decimal places in the product is equal to the *sum* of the number of decimal places in the factors.

Practice 1

Multiply: 34.8 × 0.62

Example 1 Multiply: 23.6 × 0.78

Solution:

```
   23.6   1 decimal place
 × 0.78   2 decimal places
   1888
  16520
 18.408
```

Since 1 + 2 = 3, insert the decimal point in the product so that there are 3 decimal places.

Work Practice 1

Practice 2

Multiply: 0.0641 × 27

Example 2 Multiply: 0.0531 × 16

Solution:

```
 0.0531   4 decimal places
 ×   16   0 decimal places
   3186
   5310
 0.8496
```

4 decimal places (4 + 0 = 4)

Work Practice 2

Answers

1. 21.576 2. 1.7307

✓**Concept Check** True or false? The number of decimal places in the product of 0.261 and 0.78 is 6. Explain.

Example 3 Multiply: $(-2.6)(0.8)$

Solution: Recall that the product of a negative number and a positive number is a negative number.

$$(-2.6)(0.8) = -2.08$$

Work Practice 3

Practice 3

Multiply: $(7.3)(-0.9)$

Objective B Estimating when Multiplying Decimals

Just as for addition and subtraction, we can estimate when multiplying decimals to check the reasonableness of our answer.

Example 4 Multiply: 28.06×1.95. Then estimate to see whether the answer is reasonable by rounding each factor, then multiplying the rounded numbers.

Solution:

Exact	Estimate 1		or	Estimate 2	
28.06	28			30	
× 1.95	× 2	Rounded to ones		× 2	Rounded to one nonzero digit
14030	56			60	
252540					
280600					
54.7170					

The answer 54.7170 or 54.717 is reasonable.

Work Practice 4

Practice 4

Multiply: 30.26×2.89. Then estimate to see whether the answer is reasonable.

As shown in Example 4, estimated results will vary depending on what estimates are used. Notice that estimating results is a good way to see whether the decimal point has been correctly placed.

Objective C Multiplying Decimals by Powers of 10

There are some patterns that occur when we multiply a number by a power of 10 such as 10, 100, 1000, 10,000, and so on.

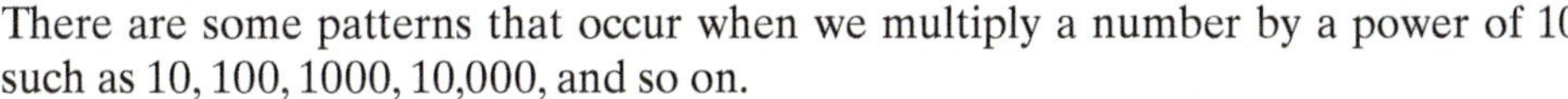

$23.6951 \times 10 = 236.951$ Move the decimal point *1 place* to the *right.*
(10: 1 zero)

$23.6951 \times 100 = 2369.51$ Move the decimal point *2 places* to the *right.*
(100: 2 zeros)

$23.6951 \times 100{,}000 = 2{,}369{,}510.$ Move the decimal point *5 places* to the *right* (insert a 0).
(100,000: 5 zeros)

Notice that we move the decimal point the same number of places as there are zeros in the power of 10.

Answers

3. −6.57 **4.** Exact: 87.4514; Estimate: $30 \cdot 3 = 90$

✓**Concept Check Answer**

false: 3 decimal places and 2 decimal places means 5 decimal places in the product

Multiplying Decimals by Powers of 10 Such as 10, 100, 1000, 10,000

Move the decimal point to the *right* the same number of places as there are *zeros* in the power of 10.

Practice 5–7

Multiply.

5. 46.8×10

6. 203.004×100

7. $(-2.33)(1000)$

Examples Multiply.

5. $7.68 \times 10 = 76.8$ 7.68

6. $23.702 \times 100 = 2370.2$ 23.702

7. $(-76.3)(1000) = -76{,}300$ 76.300

Work Practice 5–7

There are also powers of 10 that are less than 1. The decimals 0.1, 0.01, 0.001, 0.0001, and so on, are examples of powers of 10 less than 1. Notice the pattern when we multiply by these powers of 10:

$569.2 \times 0.1 = 56.92$ Move the decimal point *1 place* to the *left*.

↑ 1 decimal place

$569.2 \times 0.01 = 5.692$ Move the decimal point *2 places* to the *left*.

↑ 2 decimal places

$569.2 \times 0.0001 = 0.05692$ Move the decimal point *4 places* to the *left* (insert one 0).

↑ 4 decimal places

Multiplying Decimals by Powers of 10 Such as 0.1, 0.01, 0.001, 0.0001

Move the decimal point to the *left* the same number of places as there are *decimal places* in the power of 10.

Practice 8–10

Multiply.

8. 6.94×0.1

9. 3.9×0.01

10. $(-7682)(-0.001)$

Examples Multiply.

8. $42.1 \times 0.1 = 4.21$ 42.1

9. $76{,}805 \times 0.01 = 768.05$ 76,805.

10. $(-9.2)(-0.001) = 0.0092$ 0009.2

Work Practice 8–10

Many times we see large numbers written, for example, in the form 297.9 million rather than in the longer standard notation. The next example shows us how to interpret these numbers.

Answers

5. 468 **6.** 20,300.4 **7.** −2330 **8.** 0.694 **9.** 0.039 **10.** 7.682

Example 11 In 2050, the population of the United States is projected to be 420.3 million. Write this number in standard notation. (*Source:* U.S. Census Bureau)

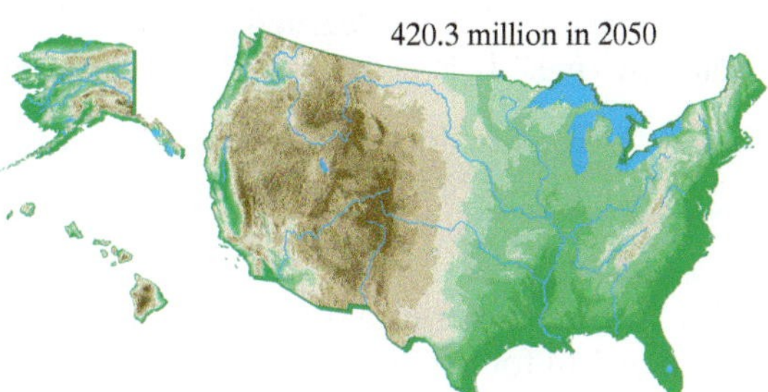

Solution: 420.3 million = 420.3×1 million

$= 420.3 \times 1{,}000{,}000 = 420{,}300{,}000$

Work Practice 11

Practice 11

In 2012, there were 58.9 million married couples in the United States. Write this number in standard notation. (*Source:* U.S. Census Bureau)

Objective D Using Decimals as Replacement Values

Now let's practice working with variables.

Example 12 Evaluate xy for $x = 2.3$ and $y = 0.44$.

Solution: Recall that xy means $x \cdot y$.

$$xy = (2.3)(0.44)$$

$$\begin{array}{r} 2.3 \\ \times\, 0.44 \\ \hline 92 \\ 920 \\ \hline 1.012 \end{array}$$

$$= 1.012$$

Work Practice 12

Practice 12

Evaluate $7y$ for $y = -0.028$.

Example 13 Is -9 a solution of the equation $3.7y = -3.33$?

Solution: Replace y with -9 in the equation $3.7y = -3.33$ to see if a true equation results.

$3.7y = -3.33$

$3.7(-9) \stackrel{?}{=} -3.33$ Replace y with -9.

$-33.3 = -3.33$ False

Since $-33.3 = -3.33$ is a false statement, -9 is **not** a solution of $3.7y = -3.33$.

Work Practice 13

Practice 13

Is -5.5 a solution of the equation $-6x = 33$?

Objective E Finding the Circumference of a Circle

Recall from Section 1.3 that the distance around a polygon is called its perimeter. The distance around a circle is given the special name **circumference,** and this distance depends on the radius or the diameter of the circle.

Circumference of a Circle

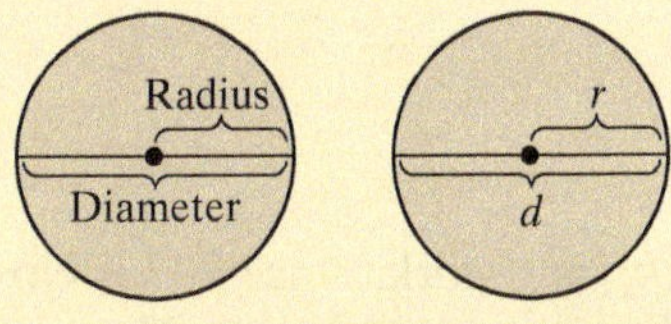

Circumference = $2 \cdot \pi \cdot$ **radius** or **Circumference** = $\pi \cdot$ **diameter**

$$C = 2\pi r \quad \text{or} \quad C = \pi d$$

Answers
11. 58,900,000 **12.** -0.196 **13.** yes

In Section 5.1, we learned about the symbol π as the Greek letter pi, pronounced "pie." It is a constant between 3 and 4. A decimal approximation for π is 3.14. Also, a fraction approximation for π is $\frac{22}{7}$.

Practice 14

Find the circumference of a circle whose radius is 11 meters. Then use the approximation 3.14 for π to approximate this circumference.

Example 14 Circumference of a Circle

Find the circumference of a circle whose radius is 5 inches. Then use the approximation 3.14 for π to approximate the circumference.

Solution: Let $r = 5$ in the formula $C = 2\pi r$.

$$\begin{aligned} C &= 2\pi r \\ &= 2\pi \cdot 5 \\ &= 10\pi \end{aligned}$$

5 inches

Next, replace π with the approximation 3.14.

$$\begin{aligned} C &= 10\pi \\ \text{(is approximately)} \longrightarrow \quad &\approx 10(3.14) \\ &\approx 31.4 \end{aligned}$$

The **exact** circumference or distance around the circle is 10π inches, which is **approximately** 31.4 inches.

Work Practice 14

Objective F Solving Problems by Multiplying Decimals 

The solutions to many real-life problems are found by multiplying decimals. We continue using our four problem-solving steps to solve such problems.

Practice 15

A biology major is fertilizing her personal garden. She uses 5.6 ounces of fertilizer per square yard. The garden measures 60.5 square yards. How much fertilizer does she need?

Example 15 Finding the Total Cost of Materials for a Job

A college student is hired to paint a billboard with paint costing $2.49 per quart. If the job requires 3 quarts of paint, what is the total cost of the paint?

Solution:

1. UNDERSTAND. Read and reread the problem. The phrase "total cost" might make us think addition, but since this problem requires repeated addition, let's multiply.
2. TRANSLATE.

In words:	total cost	is	cost per quart of paint	times	number of quarts
	↓	↓	↓	↓	↓
Translate:	total cost	=	2.49	×	3

3. SOLVE. We can estimate to check our calculations. The number 2.49 rounds to 2 and $2 \times 3 = 6$.

$$\begin{array}{r} 2.49 \\ \times \quad 3 \\ \hline 7.47 \end{array}$$

4. INTERPRET. *Check* your work. Since 7.47 is close to our estimate of 6, our answer is reasonable. *State* your conclusion: The total cost of the paint is $7.47.

Work Practice 15

Answers
14. 22π m $\approx$ 69.08 m **15.** 338.8 oz

Vocabulary, Readiness & Video Check

Use the choices below to fill in each blank.

circumference	left	sum	zeros
decimal places	right	product	factor

1. When multiplying decimals, the number of decimal places in the product is equal to the ____________ of the number of decimal places in the factors.
2. In $8.6 \times 5 = 43$, the number 43 is called the ____________, while 8.6 and 5 are each called a ____________.
3. When multiplying a decimal number by powers of 10 such as 10, 100, 1000, and so on, we move the decimal point in the number to the ____________ the same number of places as there are ____________ in the power of 10.
4. When multiplying a decimal number by powers of 10 such as 0.1, 0.01, and so on, we move the decimal point in the number to the ____________ the same number of places as there are ____________ in the power of 10.
5. The distance around a circle is called its ____________.

Martin-Gay Interactive Videos

See Video 5.3

Watch the section lecture video and answer the following questions.

Objective A 6. From the lecture before Example 1, what's the main difference between multiplying whole numbers and multiplying decimal numbers?

Objective B 7. From Example 3, what does estimating especially help us with?

Objective C 8. Why don't we do any actual multiplying in Example 5?

Objective D 9. In Example 8, once all replacement values are inserted in the variable expression, what is the resulting expression to evaluate?

Objective E 10. Why is 31.4 cm not the exact answer to Example 9?

Objective F 11. In Example 10, why is 24.8 not the complete answer? What is the complete answer?

5.3 Exercise Set MyMathLab®

Objectives A B Mixed Practice *Multiply. See Examples 1 through 4. For those exercises marked, also estimate to see if the answer is reasonable.*

1. 0.17×8
2. 0.23×9
3. 1.2×0.5
4. 6.8×0.3
5. $(-2.3)(7.65)$
6. $(4.7)(-9.02)$
7. $(-5.73)(-9.6)$
8. $(-7.84)(-3.5)$
9. 6.8×4.2
   Exact:
   Estimate:
10. 8.3×2.7
   Exact:
   Estimate:
11. 0.347×0.3
12. 0.864×0.4

13. $\begin{array}{r} 1.0047 \\ \times \quad 8.2 \\ \hline \end{array}$
Exact: Estimate:

14. $\begin{array}{r} 2.0005 \\ \times \quad 5.5 \\ \hline \end{array}$
Exact: Estimate:

15. $\begin{array}{r} 490.2 \\ \times 0.023 \\ \hline \end{array}$

16. $\begin{array}{r} 300.9 \\ \times 0.032 \\ \hline \end{array}$

Objective C *Multiply. See Examples 5 through 10.*

17. 6.5×10

18. 7.2×100

19. 8.3×0.1

20. 23.4×0.1

21. $(-7.093)(1000)$

22. $(-1.123)(1000)$

23. 0.7×100

24. 0.5×100

25. $(-9.83)(-0.01)$

26. $(-4.72)(-0.01)$

27. 25.23×0.001

28. 36.41×0.001

Objectives A B C Mixed Practice *Multiply. See Examples 1 through 10.*

29. 0.123×0.4

30. 0.216×0.3

31. $(147.9)(100)$

32. $(345.2)(100)$

33. 8.6×0.15

34. 0.42×5.7

35. $(937.62)(-0.01)$

36. $(-0.001)(562.01)$

37. 562.3×0.001

38. 993.5×0.001

39. $\begin{array}{r} 6.32 \\ \times 5.7 \\ \hline \end{array}$

40. $\begin{array}{r} 9.21 \\ \times 3.8 \\ \hline \end{array}$

Write each number in standard notation. See Example 11.

41. The cost of the Hubble Space Telescope at launch was \$1.5 billion. (*Source:* NASA)

42. About 56.7 million American households own at least one dog. (*Source:* American Pet Products Manufacturers Association)

43. The Blue Streak is the oldest roller coaster at Cedar Point, an amusement park in Sandusky, Ohio. Since 1964, it has given more than 49.8 million rides. (*Source:* Cedar Fair, L.P.)

44. In 2013, the restaurant industry had projected sales of \$660.5 billion. (*Source:* National Restaurant Association)

Objective D *Evaluate each expression for $x = 3$, $y = -0.2$, and $z = 5.7$. See Example 12.*

45. xy

46. yz

47. $xz - y$

48. $-5y + z$

Determine whether the given value is a solution of each given equation. See Example 13.

49. Is 14.2 a solution of $0.6x = 4.92$?

50. Is 1414 a solution of $100z = 14.14$?

51. Is -4 a solution of $3.5y = -14$?

52. Is -3.6 a solution of $0.7x = -2.52$?

Objective E *Find the circumference of each circle. Then use the approximation 3.14 for π and approximate each circumference. See Example 14.*

53.
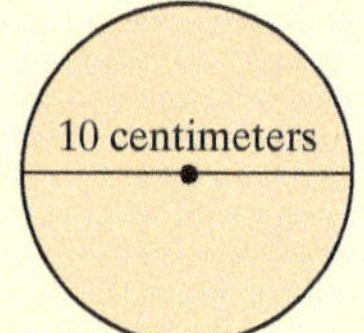

△ **54.**
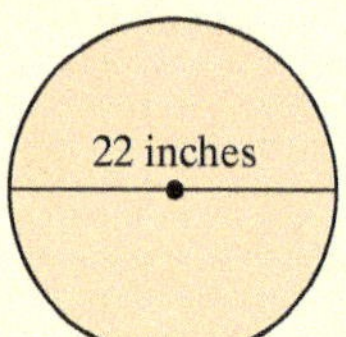

△ **55.**
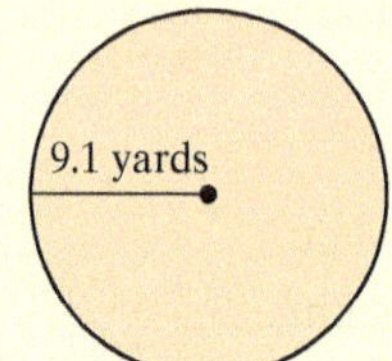

△ **56.**
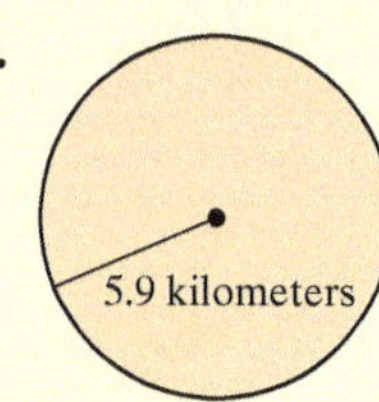

Objectives E F Mixed Practice *Solve. For Exercises 57 and 58, the solutions have been started for you. For circumference applications find the exact circumference and then use 3.14 for π to approximate the circumference. See Examples 14 and 15.*

57. An electrician for Central Power and Light worked 40 hours last week. Calculate his pay before taxes for last week if his hourly wage is $17.88.

Start the solution:

1. UNDERSTAND the problem. Reread it as many times as needed.
2. TRANSLATE into an equation. (Fill in the blanks.)

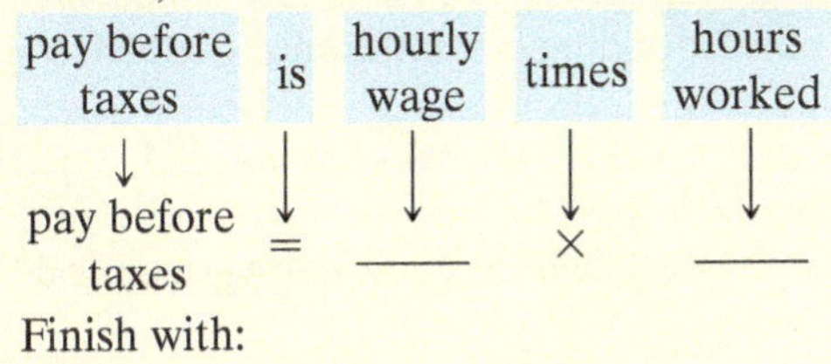

Finish with:

3. SOLVE and 4. INTERPRET.

58. An assembly line worker worked 20 hours last week. Her hourly rate is $19.52 per hour. Calculate her pay before taxes.

Start the solution:

1. UNDERSTAND the problem. Reread it as many times as needed.
2. TRANSLATE into an equation. (Fill in the blanks.)

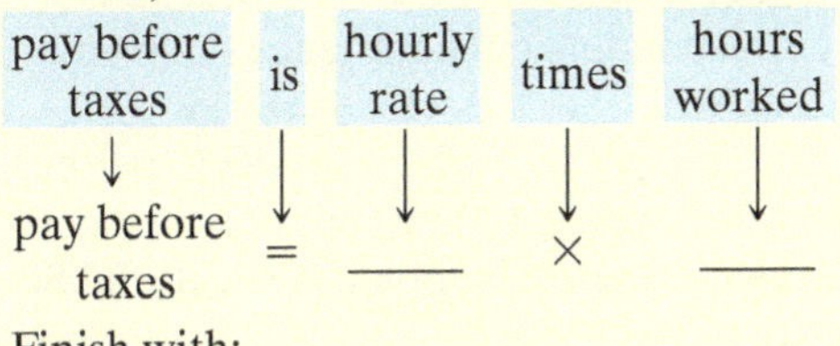

Finish with:

3. SOLVE and 4. INTERPRET.

59. A 1-ounce serving of cream cheese contains 6.2 grams of saturated fat. How much saturated fat is in 4 ounces of cream cheese? (*Source: Home and Garden Bulletin No. 72;* U.S. Department of Agriculture)

60. A 3.5-ounce serving of lobster meat contains 0.1 gram of saturated fat. How much saturated fat do 3 servings of lobster meat contain? (*Source:* The National Institutes of Health)

△ **61.** Recall that the face of the Apple iPhone 5 (see Section 5.2) measures 4.87 inches by 2.3 inches rounded. Find the approximate area of the face of the Apple iPhone 5.

△ **62.** Recall that the face of the Google Nexus 4 (see Section 5.2) measures 5.27 inches by 2.7 inches. Find the area of the face of the Google Nexus 4.

△ **63.** In 1893, the first ride called a Ferris wheel was constructed by Washington Gale Ferris. Its diameter was 250 feet. Find its circumference. Give an exact answer and an approximation using 3.14 for π. (*Source: The Handy Science Answer Book,* Visible Ink Press, 1994)

△ **64.** The radius of Earth is approximately 3950 miles. Find the distance around Earth at the equator. Give an exact answer and an approximation using 3.14 for π. (*Hint:* Find the circumference of a circle with radius 3950 miles.)

△ **65.** The London Eye, built for the millennium celebration in London, resembles a gigantic Ferris wheel with a diameter of 135 meters. If Adam Hawn rides the Eye for one revolution, find how far he travels. Give an exact answer and an approximation using 3.14 for π. (*Source:* Londoneye.com)

△ **66.** The world's longest suspension bridge is the Akashi Kaikyo Bridge in Japan. This bridge has two circular caissons, which are underwater foundations. If the diameter of a caisson is 80 meters, find its circumference. Give an exact answer and an approximation using 3.14 for π. (*Source: Scientific American;* How Things Work Today)

67. A meter is a unit of length in the metric system that is approximately equal to 39.37 inches. Sophia Wagner is 1.65 meters tall. Find her approximate height in inches.

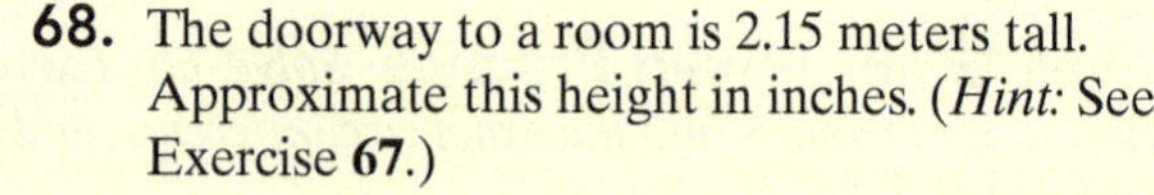

68. The doorway to a room is 2.15 meters tall. Approximate this height in inches. (*Hint:* See Exercise **67**.)

△ **69. a.** Approximate the circumference of each circle.

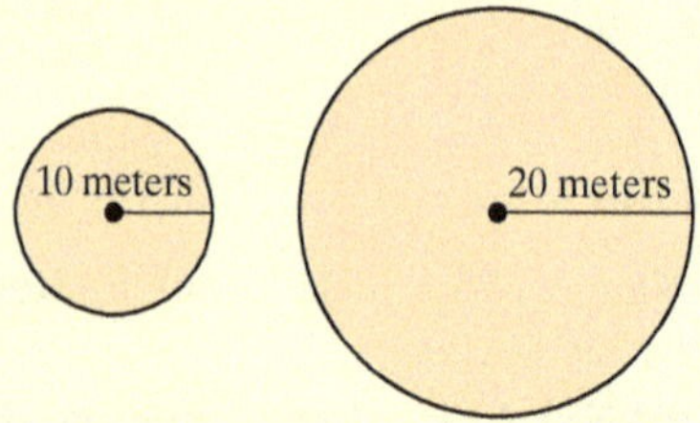

b. If radius of a circle is doubled, is its corresponding circumference doubled?

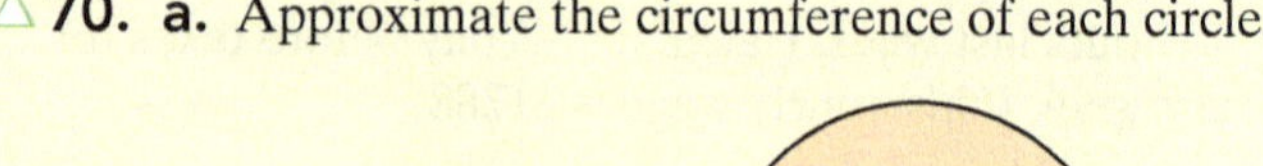

△ **70. a.** Approximate the circumference of each circle.

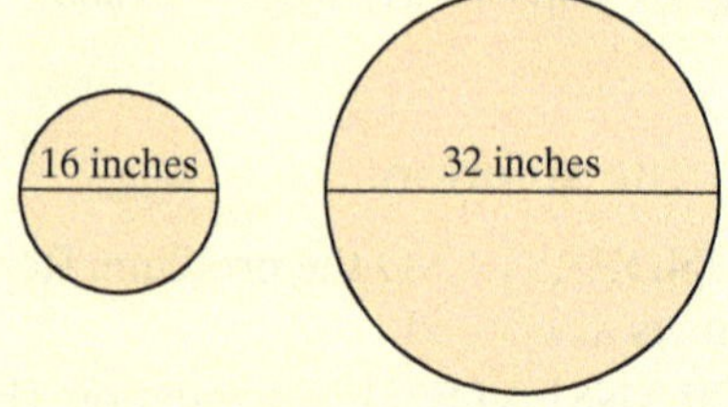

b. If the diameter of a circle is doubled, is its corresponding circumference doubled?

71. In 2011, the price of wheat was $7.30 per bushel. How much would 100 bushels of wheat cost at this price? (*Source:* National Agricultural Statistics Service)

72. In 2011, the price of soybeans was $11.70 per bushel. How much would a company pay for 10,000 bushels of soybeans? (*Source:* National Agricultural Statistics Service)

The table shows currency exchange rates for various countries on August 26, 2013. To find the amount of foreign currency equivalent to an amount of U.S. dollars, multiply the U.S. dollar amount by the exchange rate listed in the table. Use this table to answer Exercises 73 through 76. If needed, round answers to the nearest hundredth.

Foreign Currency Exchange Rates

Country	Exchange Rate
Canadian dollar	1.04920
European Union euro	0.74697
New Zealand dollar	1.28084
Chinese yuan	6.15231
Japanese yen	98.68
Swiss franc	0.92090

73. How many Canadian dollars are equivalent to $750 U.S.?

74. Suppose you wish to exchange 300 American dollars for Chinese yuan. How much money, in Chinese yuan, would you receive?

75. The Scarpulla family is traveling to New Zealand. How many New Zealand dollars can they "buy" with 800 U.S. dollars?

76. A French tourist to the United States spent $130 for souvenirs at the *Head of the Charles Regatta* in Boston. How much money did he spend in euros? Round to the nearest hundredth.

Review

Divide. See Sections 1.6 and 4.3.

77. $2916 \div 6$

78. $2920 \div 365$

79. $-\frac{24}{7} \div \frac{8}{21}$

80. $\frac{162}{25} \div -\frac{9}{75}$

Concept Extensions

Mixed Practice (*Sections 5.2, 5.3*) *Perform the indicated operations.*

81. $3.6 + 0.04$

82. $7.2 + 0.14 + 98.6$

83. $3.6 - 0.04$

84. $100 - 48.6$

85. -0.221×0.5

86. -3.6×0.04

87. Find how far radio waves travel in 20.6 seconds. (Radio waves travel at a speed of 186,000 miles per second.)

88. If it takes radio waves approximately 8.3 minutes to travel from the sun to the earth, find approximately how far it is from the sun to the earth. (*Hint:* See Exercise **87**.)

89. In your own words, explain how to find the number of decimal places in a product of decimal numbers.

90. In your own words, explain how to multiply by a power of 10.

91. Write down two decimal numbers whose product will contain 5 decimal places. Without multiplying, explain how you know your answer is correct.

5.4 Dividing Decimals 

Objective A Dividing Decimals 

Dividing decimal numbers is similar to dividing whole numbers. The only difference is that we place a decimal point in the quotient. If the divisor is a whole number, we place the decimal point in the quotient directly above the decimal point in the dividend, and then divide as with whole numbers. Recall that division can be checked by multiplication.

Example 1 Divide: 270.2 ÷ 7. Check your answer.

Solution: We divide as usual. The decimal point in the quotient is directly above the decimal point in the dividend.

```
                ┌─Write the decimal point
              38.6    ← quotient
divisor →  7)270.2    ← dividend
            −21
              60
             −56
              4 2
             −4 2
                0
```

Check:

```
   64
  38.6
×    7
 270.2
```

The quotient is 38.6.

Work Practice 1

Example 2 Divide: $32\overline{)8.32}$

Solution: We divide as usual. The decimal point in the quotient is directly above the decimal point in the dividend.

```
               0.26   ← quotient
dividend → 32) 8.32   ← divisor
              −64
               192
              −192
                 0
```

Check:

```
  0.26   quotient
× 32     divisor
    52
  7 80
  8.32   dividend
```

Work Practice 2

Objectives

A Divide Decimals.

B Estimate when Dividing Decimals.

C Divide Decimals by Powers of 10.

D Evaluate Expressions with Decimal Replacement Values.

E Solve Problems by Dividing Decimals.

Practice 1

Divide: 370.4 ÷ 8. Check your answer.

Practice 2

Divide: $48\overline{)34.08}$. Check your answer.

Answers

1. 46.3 **2.** 0.71

Sometimes to continue dividing we need to insert zeros after the last digit in the dividend.

Practice 3

Divide and check.

a. $-15.89 \div 14$

b. $-2.808 \div (-104)$

Example 3 Divide: $-5.98 \div 115$

Solution: Recall that a negative number divided by a positive number gives a negative quotient.

$$\begin{array}{r} 0.052 \\ 115\overline{)5.980} \\ -575 \\ \hline 230 \\ -230 \\ \hline 0 \end{array} \quad \leftarrow \text{Insert one 0.}$$

Thus $-5.98 \div 115 = -0.052$.

Work Practice 3

If the divisor is not a whole number, before we divide we need to move the decimal point to the right until the divisor is a whole number.

$$1.5\overline{)64.85}$$

divisor ↑ (1.5) ↑ dividend (64.85)

To understand how this works, let's rewrite

$$1.5\overline{)64.85} \quad \text{as} \quad \frac{64.85}{1.5}$$

and then multiply by 1 in the form of $\frac{10}{10}$. We use the form $\frac{10}{10}$ so that the denominator (divisor) becomes a whole number.

$$\frac{64.85}{1.5} = \frac{64.85}{1.5} \cdot 1 = \frac{64.85}{1.5} \cdot \frac{10}{10} = \frac{64.85 \cdot 10}{1.5 \cdot 10} = \frac{648.5}{15},$$

which can be written as $15.\overline{)648.5}$. Notice that

$1.5\overline{)64.85}$ is equivalent to $15.\overline{)648.5}$

The decimal points in the dividend and the divisor were both moved one place to the right, and the divisor is now a whole number. This procedure is summarized next:

Dividing by a Decimal

Step 1: Move the decimal point in the divisor to the right until the divisor is a whole number.

Step 2: Move the decimal point in the dividend to the right the *same number of places* as the decimal point was moved in Step 1.

Step 3: Divide. Place the decimal point in the quotient directly over the moved decimal point in the dividend.

Answers

3. a. -1.135 **b.** 0.027

Example 4 Divide: 10.764 ÷ 2.3

Solution: We move the decimal points in the divisor and the dividend one place to the right so that the divisor is a whole number.

$2.3\overline{)10.764}$ becomes

```
        4.68
  23.)107.64
      -92
       15 6
      -13 8
        1 84
       -1 84
           0
```

Work Practice 4

Example 5 Divide: 5.264 ÷ 0.32

Solution:

$0.32\overline{)5.264}$ becomes

```
      16.45
  32)526.40   Insert one 0.
     -32
      206
     -192
       14 4
      -12 8
        1 60
       -1 60
           0
```

Work Practice 5

✓**Concept Check** Is it always true that the number of decimal places in a quotient equals the sum of the decimal places in the dividend and divisor?

Example 6 Divide: 17.5 ÷ 0.48. Round the quotient to the nearest hundredth.

Solution: First we move the decimal points in the divisor and the dividend two places. Then we divide and round the quotient to the nearest hundredth.

```
            ┌── hundredths place
        36.458 ≈ 36.46
  48.)1750.000  ↑ "is approximately"
     -144
       310
      -288
        22 0
       -19 2
         2 80
        -2 40
           400
          -384
            16
```

When rounding to the nearest hundredth, carry the division process out to one more decimal place, the thousandths place.

Work Practice 6

Practice 4

Divide: 166.88 ÷ 5.6

Practice 5

Divide: 1.976 ÷ 0.16

Practice 6

Divide: 23.4 ÷ 0.57. Round the quotient to the nearest hundredth.

Answers

4. 29.8 **5.** 12.35 **6.** 41.05

✓**Concept Check Answer**

no

✓**Concept Check** If a quotient is to be rounded to the nearest thousandth, to what place should the division be carried out? (Assume that the division carries out to your answer.)

Objective B Estimating when Dividing Decimals

Just as for addition, subtraction, and multiplication of decimals, we can estimate when dividing decimals to check the reasonableness of our answer.

Practice 7

Divide: 713.7 ÷ 91.5. Then estimate to see whether the proposed answer is reasonable.

Example 7 Divide: 272.356 ÷ 28.4. Then estimate to see whether the proposed result is reasonable.

Solution:

Exact:

$$\begin{array}{r} 9.59 \\ 284.\overline{)2723.56} \\ -2556 \\ \hline 167\;5 \\ -142\;0 \\ \hline 25\;56 \\ -25\;56 \\ \hline 0 \end{array}$$

Estimate 1

$$\begin{array}{r} 9 \\ 30\overline{)270} \end{array}$$

or

Estimate 2

$$\begin{array}{r} 10 \\ 30\overline{)300} \end{array}$$

The estimate is 9 or 10, so 9.59 is reasonable.

Work Practice 7

Objective C Dividing Decimals by Powers of 10

As with multiplication, there are patterns that occur when we divide decimals by powers of 10 such as 10, 100, 1000, and so on.

$$\frac{569.2}{10} = 56.92$$ Move the decimal point *1 place* to the *left*.

(10: 1 zero)

$$\frac{569.2}{10{,}000} = 0.05692$$ Move the decimal point *4 places* to the *left*.

(10,000: 4 zeros)

This pattern suggests the following rule:

> **Dividing Decimals by Powers of 10 Such as 10, 100, or 1000**
>
> Move the decimal point of the dividend to the *left* the same number of places as there are *zeros* in the power of 10.

Practice 8–9

Divide.

8. $\frac{362.1}{1000}$ 9. $-\frac{0.49}{10}$

Examples Divide.

8. $\frac{786.1}{1000} = 0.7861$ Move the decimal point *3 places* to the *left*.

(1000: 3 zeros)

9. $\frac{0.12}{10} = 0.012$ Move the decimal point *1 place* to the *left*.

(10: 1 zero)

Work Practice 8–9

Answers

7. 7.8 8. 0.3621 9. −0.049

✓**Concept Check Answer**

ten-thousandths place

Objective D Using Decimals as Replacement Values 

Example 10 Evaluate $x \div y$ for $x = 2.5$ and $y = 0.05$.

Solution: Replace x with 2.5 and y with 0.05.

$x \div y = 2.5 \div 0.05$ $0.05\overline{)2.5}$ becomes $5\overline{)250}$ with quotient 50

$= 50$

Work Practice 10

Example 11 Is 720 a solution of the equation $\frac{y}{100} = 7.2$?

Solution: Replace y with 720 to see if a true statement results.

$\frac{y}{100} = 7.2$ Original equation

$\frac{720}{100} \stackrel{?}{=} 7.2$ Replace y with 720.

$7.2 = 7.2$ True

Since $7.2 = 7.2$ is a true statement, 720 is a solution of the equation.

Work Practice 11

Objective E Solving Problems by Dividing Decimals

Many real-life problems involve dividing decimals.

Example 12 Calculating Materials Needed for a Job

A gallon of paint covers a 250-square-foot area. If Betty Adkins wishes to paint a wall that measures 1450 square feet, how many gallons of paint does she need? If she can buy only gallon containers of paint, how many gallon containers does she need?

Solution:

1. UNDERSTAND. Read and reread the problem. We need to know how many 250s are in 1450, so we divide.
2. TRANSLATE.

In words:	number of gallons	is	square feet	divided by	square feet per gallon
	↓	↓	↓	↓	↓
Translate:	number of gallons	=	1450	÷	250

3. SOLVE. Let's see if our answer is reasonable by estimating. The dividend 1450 rounds to 1500 and the divisor 250 rounds to 300. Then $1500 \div 300 = 5$.

```
       5.8
250)1450.0
   -1250
     200 0
    -200 0
         0
```

(*Continued on next page*)

Practice 10

Evaluate $x \div y$ for $x = 0.035$ and $y = 0.02$.

Practice 11

Is 39 a solution of the equation $\frac{x}{100} = 3.9$?

Practice 12

A bag of fertilizer covers 1250 square feet of lawn. Tim Parker's lawn measures 14,800 square feet. How many bags of fertilizer does he need? If he can buy only whole bags of fertilizer, how many whole bags does he need?

Answers
10. 1.75 **11.** no
12. 11.84 bags; 12 bags

4. INTERPRET. *Check* your work. Since our estimate is close to our answer of 5, our answer is reasonable. *State* your conclusion: Betty needs 5.8 gallons of paint. If she can buy only gallon containers of paint, she needs 6 gallon containers of paint to complete the job.

Work Practice 12

Calculator Explorations Estimation

Calculator errors can easily be made by pressing an incorrect key or by not pressing a correct key hard enough. Estimation is a valuable tool that can be used to check calculator results.

Example Use estimation to determine whether the calculator result is reasonable or not. (For example, a result that is not reasonable can occur if proper keys are not pressed.)

Divide: 82.064 ÷ 23
Calculator display: 35.68

Solution: Round each number to the nearest 10. Since 80 ÷ 20 = 4, the calculator display 35.68 is not reasonable.

Use estimation to determine whether each result is reasonable or not.

1. 102.62 × 41.8; Result: 428.9516
2. 174.835 ÷ 47.9; Result: 3.65
3. 1025.68 − 125.42; Result: 900.26
4. 562.781 + 2.96; Result: 858.781

Vocabulary, Readiness & Video Check

Use the choices below to fill in each blank. Some choices may be used more than once, and some not used at all.

dividend	divisor	quotient	true
zeros	left	right	false

1. In 6.5 ÷ 5 = 1.3, the number 1.3 is called the ________, 5 is the ________, and 6.5 is the ________.
2. To check a division exercise, we can perform the following multiplication: quotient · ________ = ________.
3. To divide a decimal number by a power of 10 such as 10, 100, 1000, and so on, we move the decimal point in the number to the ________ the same number of places as there are ________ in the power of 10.
4. True or false: If we replace x with −12.6 and y with 0.3 in the expression $y \div x$, we have 0.3 ÷ (−12.6). ________

Martin-Gay Interactive Videos

See Video 5.4

Watch the section lecture video and answer the following questions.

Objective A 5. From the lecture before Example 1, what must we make sure the divisor is before dividing decimals?

Objective B 6. From Example 4, what does estimating especially help us with?

Objective C 7. Why don't we do any actual dividing in Example 5?

Objective D 8. In Example 6, 4 does not divide into 1. How does this affect the quotient?

Objective E 9. In Example 7, why is the division carried to the hundredths place?

5.4 Exercise Set MyMathLab® 

Objectives A B Mixed Practice *Divide. See Examples 1 through 5 and 7. For those exercises marked, also estimate to see if the answer is reasonable.*

1. $6\overline{)27.6}$

2. $4\overline{)23.6}$

3. $5\overline{)0.47}$

4. $6\overline{)0.51}$

5. $0.06\overline{)18}$

6. $0.04\overline{)20}$

7. $0.42\overline{)3.066}$

8. $0.36\overline{)1.764}$

9. $5.5\overline{)36.3}$
Exact:
Estimate:

10. $2.2\overline{)21.78}$
Exact:
Estimate:

11. $7.434 \div 18$

12. $8.304 \div 16$

13. $36 \div (-0.06)$

14. $36 \div (-0.04)$

15. Divide -4.2 by -0.6.

16. Divide -3.6 by -0.9.

17. $0.27\overline{)1.296}$

18. $0.34\overline{)2.176}$

19. $0.02\overline{)42}$

20. $0.03\overline{)24}$

21. $4.756 \div 0.82$

22. $3.312 \div 0.92$

23. $-36.3 \div (-6.6)$

24. $-21.78 \div (-9.9)$

25. $7.2\overline{)70.56}$
Exact:
Estimate:

26. $6.3\overline{)54.18}$
Exact:
Estimate:

27. $5.4\overline{)51.84}$

28. $7.7\overline{)33.88}$

29. $\dfrac{1.215}{0.027}$

30. $\dfrac{3.213}{0.051}$

31. $0.25\overline{)13.648}$

32. $0.75\overline{)49.866}$

33. $3.78\overline{)0.02079}$

34. $2.96\overline{)0.01332}$

Divide. Round the quotients as indicated. See Example 6.

35. Divide: $0.549 \div 0.023$. Round the quotient to the nearest hundredth.

36. Divide: $0.0453 \div 0.98$. Round the quotient to the nearest thousandth.

37. Divide: $68.39 \div 0.6$. Round the quotient to the nearest tenth.

38. Divide: $98.83 \div 3.5$. Round the quotient to the nearest tenth.

Objective C *Divide. See Examples 8 and 9.*

39. $\dfrac{83.397}{100}$

40. $\dfrac{64.423}{100}$

41. $\dfrac{26.87}{10}$

42. $\dfrac{13.49}{10}$

43. $12.9 \div (-1000)$

44. $13.49 \div (-10{,}000)$

Objectives A C Mixed Practice *Divide. See Examples 1 through 5, 8, and 9.*

45. $7\overline{)88.2}$

46. $9\overline{)130.5}$

47. $\dfrac{13.1}{10}$

48. $\dfrac{17.7}{10}$

49. $\dfrac{456.25}{10{,}000}$

50. $\dfrac{986.11}{10{,}000}$

51. $1.239 \div 3$

52. $0.54 \div 12$

53. Divide 4.8 by −0.6. **54.** Divide 4.9 by −0.7. **55.** $-1.224 \div 0.17$ **56.** $-1.344 \div 0.42$

57. Divide 42 by 0.03. **58.** Divide 27 by 0.03. **59.** Divide −18 by −0.6. **60.** Divide 20 by 0.4.

61. Divide 87 by −0.0015. **62.** Divide 35 by −0.0007. **63.** $-1.104 \div 1.6$ **64.** $-2.156 \div 0.98$

65. $-2.4 \div (-100)$ **66.** $-86.79 \div (-1000)$ **67.** $\frac{4.615}{0.071}$ **68.** $\frac{23.8}{0.035}$

Objective D *Evaluate each expression for $x = 5.65$, $y = -0.8$, and $z = 4.52$. See Example 10.*

69. $z \div y$ **70.** $z \div x$ **71.** $x \div y$ **72.** $y \div 2$

Determine whether the given values are solutions of the given equations. See Example 11.

73. $\frac{x}{4} = 3.04$; $x = 12.16$ **74.** $\frac{y}{8} = 0.89$; $y = 7.12$ **75.** $\frac{z}{100} = 0.8$; $z = 8$ **76.** $\frac{x}{10} = 0.23$; $x = 23$

Objective E *Solve. For Exercises 77 and 78, the solutions have been started for you. See Example 12.*

△ **77.** A new homeowner is painting the walls of a room. The walls have a total area of 546 square feet. A quart of paint covers 52 square feet. If the paint is sold in whole quarts only, how many quarts are needed?

Start the solution:

1. UNDERSTAND the problem. Reread it as many times as needed.
2. TRANSLATE into an equation. (Fill in the blanks.)

number of quarts	is	square feet	divided by	square feet per quart
↓	↓	↓	↓	↓
number of quarts	=	____	÷	____

3. SOLVE. Don't forget to round up your quotient.
4. INTERPRET.

78. A shipping box can hold 36 books. If 486 books must be shipped, how many boxes are needed?

Start the solution:

1. UNDERSTAND the problem. Reread it as many times as needed.
2. TRANSLATE into an equation. (Fill in the blanks.)

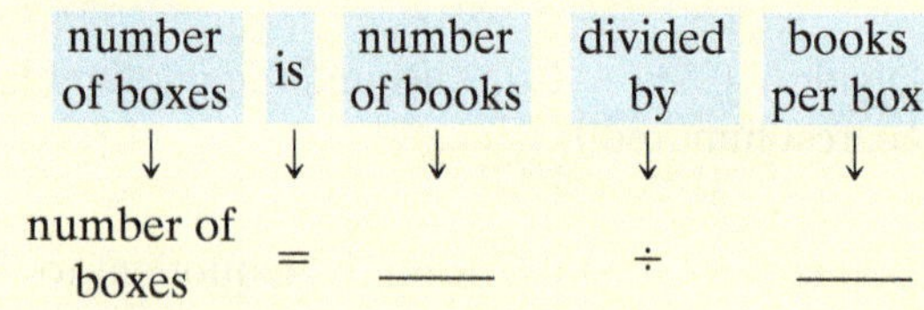

3. SOLVE. Don't forget to round up your quotient.
4. INTERPRET.

79. There are approximately 39.37 inches in 1 meter. How many meters, to the nearest tenth of a meter, are there in 200 inches?

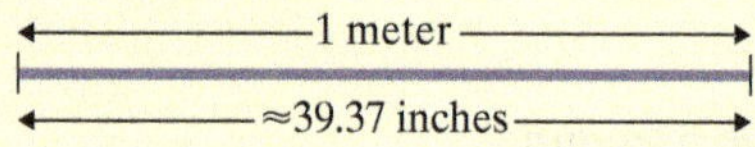

80. There are 2.54 centimeters in 1 inch. How many inches are there in 50 centimeters? Round to the nearest tenth.

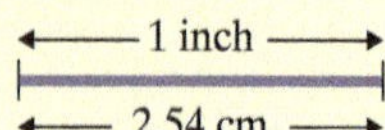

81. In the United States, an average child will wear down 730 crayons by his or her tenth birthday. Find the number of boxes of 64 crayons this is equivalent to. Round to the nearest tenth. (*Source:* Binney & Smith Inc.)

82. In 2011, American farmers received an average of \$68.20 per hundred pounds of turkey. What was the average price per pound for turkeys? Round to the nearest cent. (*Source:* National Agricultural Statistics Service)

A child is to receive a dose of 0.5 teaspoon of cough medicine every 4 hours. If the bottle contains 4 fluid ounces, answer Exercises 83 through 86.

83. A fluid ounce equals 6 teaspoons. How many teaspoons are in 4 fluid ounces?

84. The bottle of medicine contains how many doses for the child? (*Hint:* See Exercise **83**.)

85. If the child takes a dose every four hours, how many days will the medicine last?

86. If the child takes a dose every six hours, how many days will the medicine last?

87. Americans ages 16–19 drive, on average, 7624 miles per year. About how many miles each week is that? Round to the nearest tenth. (*Note:* There are 52 weeks in a year.) (*Source:* Federal Highway Administration)

88. Drake Saucier was interested in the gas mileage on his "new" used car. He filled the tank, drove 423.8 miles, and filled the tank again. When he refilled the tank, it took 19.35 gallons of gas. Calculate the miles per gallon for Drake's car. Round to the nearest tenth.

89. The book *Harry Potter and the Deathly Hallows* was released to the public on July 21, 2007. Booksellers in the United States sold approximately 8292 thousand copies in the first 24 hours after release. If the same number of books was sold each hour, calculate the number of books sold each hour in the United States for that first day.

90. The leading money winner in men's professional golf in 2012 was Rory McIlroy. He earned approximately \$8,048,000. Suppose he had earned this working 40 hours each week for a year. Determine his hourly wage to the nearest cent. (*Note:* There are 52 weeks in a year.) (*Source:* Professional Golfers' Association)

Review

Perform the indicated operation. See Sections 4.3 and 4.5.

91. $\frac{3}{5} \cdot \frac{7}{10}$

92. $\frac{3}{5} \div \frac{7}{10}$

93. $\frac{3}{5} - \frac{7}{10}$

94. $-\frac{3}{4} - \frac{1}{14}$

Concept Extensions

Mixed Practice (*Sections 5.2, 5.3, 5.4*) *Perform the indicated operation.*

95. $1.278 \div 0.3$

96. 1.278×0.3

97. $1.278 + 0.3$

98. $1.278 - 0.3$

99. $(-8.6)(3.1)$

100. $7.2 + 0.05 + 49.1$

101. $\begin{array}{r} 1000 \\ -\ \ 95.71 \\ \hline \end{array}$

102. $\frac{87.2}{-10,000}$

Choose the best estimate.

103. 8.62×41.7
- **a.** 36
- **b.** 32
- **c.** 360
- **d.** 3.6

104. $1.437 + 20.69$
- **a.** 34
- **b.** 22
- **c.** 3.4
- **d.** 2.2

105. $78.6 \div 97$
- **a.** 7.86
- **b.** 0.786
- **c.** 786
- **d.** 7860

106. $302.729 - 28.697$
- **a.** 270
- **b.** 20
- **c.** 27
- **d.** 300

Recall from Section 1.6 that the average of a list of numbers is their total divided by how many numbers there are in the list. Use this procedure to find the average of the test scores listed in Exercises 107 and 108. If necessary, round to the nearest tenth.

107. 86, 78, 91, 87

108. 56, 75, 80

△ **109.** The area of a rectangle is 38.7 square feet. If its width is 4.5 feet, find its length.

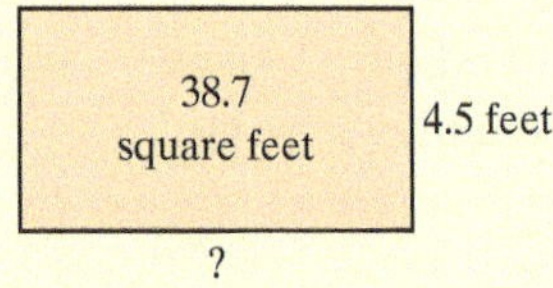

△ **110.** The perimeter of a square is 180.8 centimeters. Find the length of a side.

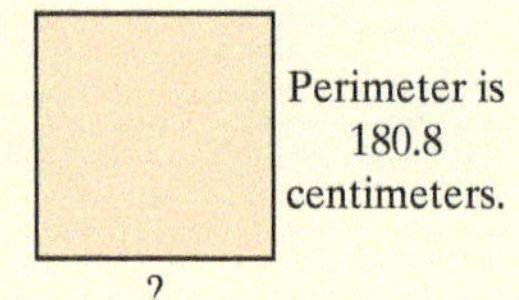

111. When dividing decimals, describe the process you use to place the decimal point in the quotient.

112. In your own words, describe how to quickly divide a number by a power of 10 such as 10, 100, 1000, etc.

To convert wind speeds in miles per hour to knots, divide by 1.15. Use this information and the Saffir-Simpson Hurricane Intensity chart below to answer Exercises 113 and 114. Round to the nearest tenth.

Saffir-Simpson Hurricane Intensity Scale

Category	Wind Speed	Barometric Pressure [inches of mercury (Hg)]	Storm Surge	Damage Potential
1 (Weak)	75–95 mph	≥28.94 in.	4–5 ft	Minimal damage to vegetation
2 (Moderate)	96–110 mph	28.50–28.93 in.	6–8 ft	Moderate damage to houses
3 (Strong)	111–130 mph	27.91–28.49 in.	9–12 ft	Extensive damage to small buildings
4 (Very Strong)	131–155 mph	27.17–27.90 in.	13–18 ft	Extreme structural damage
5 (Devastating)	>155 mph	<27.17 in.	>18 ft	Catastrophic building failures possible

113. The chart gives wind speeds in miles per hour. What is the range of wind speeds for a Category 1 hurricane in knots?

114. What is the range of wind speeds for a Category 4 hurricane in knots?

△ **115.** A rancher is building a horse corral that's shaped like a rectangle with a width of 24.3 meters. He plans to make a four-wire fence; that is, he will string four wires around the corral. If he plans to use all of his 412.8 meters of wire, find the length of the corral he can construct.

116. A college student signed up for a new credit card that guarantees her no interest charges on transferred balances for a year. She transferred over a $2523.86 balance from her old credit card. Her minimum payment is $185.35 per month. If she pays only the minimum, will she pay off her balance before interest charges start again?

Sections 5.1–5.4 **Integrated Review**

Operations on Decimals

Perform the indicated operation.

1. $1.6 + 0.97$

2. $3.2 + 0.85$

3. $9.8 - 0.9$

4. $10.2 - 6.7$

5. $\begin{array}{r} 0.8 \\ \times 0.2 \\ \hline \end{array}$

6. $\begin{array}{r} 0.6 \\ \times 0.4 \\ \hline \end{array}$

7. $8\overline{)2.16}$

8. $6\overline{)3.12}$

9. $(9.6)(-0.5)$

10. $(-8.7)(-0.7)$

11. $\begin{array}{r} 123.6 \\ -\ 48.04 \\ \hline \end{array}$

12. $\begin{array}{r} 325.2 \\ -\ 36.08 \\ \hline \end{array}$

13. $-25 + 0.026$

14. $0.125 + (-44)$

15. $29.24 \div (-3.4)$

16. $-10.26 \div (-1.9)$

17. -2.8×100

18. 1.6×1000

19. $\begin{array}{r} 96.21 \\ 7.028 \\ +\ 121.7 \\ \hline \end{array}$

20. $\begin{array}{r} 0.268 \\ 1.93 \\ +\ 142.881 \\ \hline \end{array}$

21. $-25.76 \div (-46)$

22. $-27.09 \div 43$

23. $\begin{array}{r} 12.004 \\ \times\ \ 2.3 \\ \hline \end{array}$

24. $\begin{array}{r} 28.006 \\ \times\ \ 5.2 \\ \hline \end{array}$

Answers

1. ________
2. ________
3. ________
4. ________
5. ________
6. ________
7. ________
8. ________
9. ________
10. ________
11. ________
12. ________
13. ________
14. ________
15. ________
16. ________
17. ________
18. ________
19. ________
20. ________
21. ________
22. ________
23. ________
24. ________

25. ______
26. ______
27. ______
28. ______
29. ______
30. ______
31. ______
32. ______
33. ______
34. ______
35. ______
36. ______
37. ______
38. ______
39. ______

25. Subtract 4.6 from 10.

26. Subtract 18 from 0.26.

27. $-268.19 - 146.25$

28. $-860.18 - 434.85$

29. $\frac{2.958}{-0.087}$

30. $\frac{-1.708}{0.061}$

31. $160 - 43.19$

32. $120 - 101.21$

33. 15.62×10

34. $15.62 \div 10$

35. $15.62 + 10$

36. $15.62 - 10$

37. Find the distance in miles between Garden City, Kansas, and Wichita, Kansas. Next, estimate the distance by rounding each given distance to the nearest ten.

38. It costs \$7.29 to send a 5-pound package locally via parcel post at a U.S. Post Office. To send the same package Priority Mail costs \$8.10. How much more does it cost to send a package as Priority Mail? (*Source:* United States Postal Service)

39. In 2011, sales of Blu-ray Discs were \$2 billion, but DVD sales dropped to \$6.8 billion. Find the total spent to buy Blu-ray Discs or DVDs in 2011. Write the total in billions of dollars and also in standard notation. (*Source: USA Today*)

5.5 Fractions, Decimals, and Order of Operations 

Objective A Writing Fractions as Decimals

To write a fraction as a decimal, we interpret the fraction bar to mean division and find the quotient.

Writing Fractions as Decimals

To write a fraction as a decimal, divide the numerator by the denominator.

Example 1 Write $\frac{1}{4}$ as a decimal.

Solution: $\frac{1}{4} = 1 \div 4$

$$\begin{array}{r} 0.25 \\ 4\overline{)1.00} \\ -8 \\ \hline 20 \\ -20 \\ \hline 0 \end{array}$$

Thus, $\frac{1}{4}$ written as a decimal is 0.25.

Work Practice 1

Example 2 Write $-\frac{5}{8}$ as a decimal.

Solution: $-\frac{5}{8} = -(5 \div 8) = -0.625$

$$\begin{array}{r} 0.625 \\ 8\overline{)5.000} \\ -4\,8 \\ \hline 20 \\ -16 \\ \hline 40 \\ -40 \\ \hline 0 \end{array}$$

Work Practice 2

Example 3 Write $\frac{2}{3}$ as a decimal.

Solution:

$$\begin{array}{rl} 0.666\ldots & \\ 3\overline{)2.000} & \\ -1\,8 & \\ \hline 20 & \text{Remainder is 2, then 0 is brought down.} \\ -18 & \\ \hline 20 & \text{Remainder is 2, then 0 is brought down.} \\ -18 & \\ \hline 2 & \text{Remainder is 2.} \end{array}$$

(*Continued on next page*)

Objectives

A Write Fractions as Decimals.

B Compare Decimals and Fractions.

C Simplify Expressions Containing Decimals and Fractions Using Order of Operations.

D Solve Area Problems Containing Fractions and Decimals.

E Evaluate Expressions Given Decimal Replacement Values.

Practice 1

a. Write $\frac{2}{5}$ as a decimal.

b. Write $\frac{9}{40}$ as a decimal.

Practice 2

Write $-\frac{3}{8}$ as a decimal.

Practice 3

a. Write $\frac{5}{6}$ as a decimal.

b. Write $\frac{2}{9}$ as a decimal.

Answers

1. a. 0.4 **b.** 0.225 **2.** −0.375
3. a. $0.8\overline{3}$ **b.** $0.\overline{2}$

Notice that the digit 2 keeps occurring as the remainder. This will continue so that the digit 6 will keep repeating in the quotient. We place a bar over the digit 6 to indicate that it repeats.

$$\frac{2}{3} = 0.666\ldots = 0.\overline{6}$$

We can also write a decimal approximation for $\frac{2}{3}$. For example, $\frac{2}{3}$ rounded to the nearest hundredth is 0.67. This can be written as $\frac{2}{3} \approx 0.67$.

Work Practice 3

Practice 4

Write $\frac{28}{13}$ as a decimal. Round to the nearest thousandth.

Example 4 Write $\frac{22}{7}$ as a decimal. (Recall that the fraction $\frac{22}{7}$ is an approximation for π.) Round to the nearest hundredth.

Solution:

$$\begin{array}{r} 3.142 \\ 7\overline{)22.000} \\ -21 \\ \hline 1\,0 \\ -7 \\ \hline 30 \\ -28 \\ \hline 20 \\ -14 \\ \hline 6 \end{array} \approx 3.14$$

Carry the division out to the thousandths place.

The fraction $\frac{22}{7}$ in decimal form is approximately 3.14.

Work Practice 4

Practice 5

Write $3\frac{5}{16}$ as a decimal.

Example 5 Write $2\frac{3}{16}$ as a decimal.

Solution:

Option 1. Write the fractional part only as a decimal.

$$\frac{3}{16} \longrightarrow \begin{array}{r} 0.1875 \\ 16\overline{)3.0000} \\ -1\,6 \\ \hline 1\,40 \\ -1\,28 \\ \hline 120 \\ -112 \\ \hline 80 \\ -80 \\ \hline 0 \end{array}$$

Thus $2\frac{3}{16} = 2.1875$

Option 2. Write $2\frac{3}{16}$ as an improper fraction, and divide.

$$2\frac{3}{16} = \frac{35}{16} \longrightarrow \begin{array}{r} 2.1875 \\ 16\overline{)35.0000} \\ -32 \\ \hline 3\,0 \\ -1\,6 \\ \hline 1\,40 \\ -1\,28 \\ \hline 120 \\ -112 \\ \hline 80 \\ -80 \\ \hline 0 \end{array}$$

Work Practice 5

Answers
4. 2.154 **5.** 3.3125

Some fractions may be written as decimals using our knowledge of decimals. From Section 5.1, we know that if the denominator of a fraction is 10, 100, 1000, or so on, we can immediately write the fraction as a decimal. For example,

$$\frac{4}{10} = 0.4, \quad \frac{12}{100} = 0.12, \text{ and so on}$$

Example 6 Write $\frac{4}{5}$ as a decimal.

Solution: Let's write $\frac{4}{5}$ as an equivalent fraction with a denominator of 10.

$$\frac{4}{5} = \frac{4}{5} \cdot \frac{2}{2} = \frac{8}{10} = 0.8$$

Work Practice 6

Practice 6

Write $\frac{3}{5}$ as a decimal.

Example 7 Write $\frac{1}{25}$ as a decimal.

Solution: $\frac{1}{25} = \frac{1}{25} \cdot \frac{4}{4} = \frac{4}{100} = 0.04$

Work Practice 7

Practice 7

Write $\frac{3}{50}$ as a decimal.

✓**Concept Check** Suppose you are writing the fraction $\frac{9}{16}$ as a decimal. How do you know you have made a mistake if your answer is 1.735?

Objective B Comparing Decimals and Fractions

Now we can compare decimals and fractions by writing fractions as equivalent decimals.

Example 8 Insert <, >, or = to form a true statement.

$\frac{1}{8} \quad 0.12$

Solution: First we write $\frac{1}{8}$ as an equivalent decimal. Then we compare decimal places.

$$\begin{array}{r} 0.125 \\ 8\overline{)1.000} \\ -8 \\ \hline 20 \\ -16 \\ \hline 40 \\ -40 \\ \hline 0 \end{array}$$

Original numbers	$\frac{1}{8}$	0.12
Decimals	0.125	0.120
Compare	0.125 > 0.12	

Thus, $\frac{1}{8} > 0.12$

Work Practice 8

Practice 8

Insert <, >, or = to form a true statement.

$\frac{1}{5} \quad 0.25$

Answers

6. 0.6 **7.** 0.06 **8.** <

✓**Concept Check Answer**

$\frac{9}{16}$ is less than 1 while 1.735 is greater than 1.

Practice 9

Insert <, >, or = to form a true statement.

a. $\frac{1}{2}$ 0.54 **b.** $0.\overline{5}$ $\frac{5}{9}$

c. $\frac{5}{7}$ 0.72

Example 9

Insert <, >, or = to form a true statement.

$0.\overline{7}$ $\frac{7}{9}$

Solution: We write $\frac{7}{9}$ as a decimal and then compare.

$$\begin{array}{r} 0.77\ldots \\ 9\overline{)7.00} \\ -6.3 \\ \hline 70 \\ -63 \\ \hline 7 \end{array} \quad = 0.\overline{7}$$

Original numbers	$0.\overline{7}$	$\frac{7}{9}$
Decimals	$0.\overline{7}$	$0.\overline{7}$
Compare	$0.\overline{7} = 0.\overline{7}$	

Thus, $0.\overline{7} = \frac{7}{9}$

Work Practice 9

Practice 10

Write the numbers in order from smallest to largest.

a. $\frac{1}{3}$, 0.302, $\frac{3}{8}$

b. 1.26, $1\frac{1}{4}$, $1\frac{2}{5}$

c. 0.4, 0.41, $\frac{3}{7}$

Example 10

Write the numbers in order from smallest to largest.

$\frac{9}{20}$, $\frac{4}{9}$, 0.456

Solution:

Original numbers	$\frac{9}{20}$	$\frac{4}{9}$	0.456
Decimals	0.450	0.444 . . .	0.456
Compare in order	2nd	1st	3rd

Written in order, we have

1st 2nd 3rd
↓ ↓ ↓
$\frac{4}{9}$, $\frac{9}{20}$, 0.456

Work Practice 10

Objective C Simplifying Expressions with Decimals and Fractions

In the remaining examples, we will review the order of operations by simplifying expressions that contain decimals.

Order of Operations

1. Perform all operations within parentheses ( ), brackets [ ], or other grouping symbols such as fraction bars.
2. Evaluate any expressions with exponents.
3. Multiply or divide in order from left to right.
4. Add or subtract in order from left to right.

Answers

9. a. < **b.** = **c.** <

10. a. 0.302, $\frac{1}{3}$, $\frac{3}{8}$ **b.** $1\frac{1}{4}$, 1.26, $1\frac{2}{5}$

c. 0.4, 0.41, $\frac{3}{7}$

Example 11 Simplify: $723.6 \div 1000 \times 10$

Solution: Multiply or divide in order from left to right.

$$\begin{aligned} 723.6 \div 1000 \times 10 &= 0.7236 \times 10 && \text{Divide.} \\ &= 7.236 && \text{Multiply.} \end{aligned}$$

Work Practice 11

Practice 11

Simplify: $897.8 \div 100 \times 10$

Example 12 Simplify: $-0.5(8.6 - 1.2)$

Solution: According to the order of operations, we simplify inside the parentheses first.

$$\begin{aligned} -0.5(8.6 - 1.2) &= -0.5(7.4) && \text{Subtract.} \\ &= -3.7 && \text{Multiply.} \end{aligned}$$

Work Practice 12

Practice 12

Simplify: $-8.69(3.2 - 1.8)$

Example 13 Simplify: $(-1.3)^2 + 2.4$

Solution: Recall the meaning of an exponent.

$$\begin{aligned} (-1.3)^2 + 2.4 &= (-1.3)(-1.3) + 2.4 && \text{Use the definition of an exponent.} \\ &= 1.69 + 2.4 && \text{Multiply. The product of two negative numbers is a positive number.} \\ &= 4.09 && \text{Add.} \end{aligned}$$

Work Practice 13

Practice 13

Simplify: $(-0.7)^2 + 2.1$

Example 14 Simplify: $\dfrac{5.68 + (0.9)^2 \div 100}{0.2}$

Solution: First we simplify the numerator of the fraction. Then we divide.

$$\begin{aligned} \frac{5.68 + (0.9)^2 \div 100}{0.2} &= \frac{5.68 + 0.81 \div 100}{0.2} && \text{Simplify } (0.9)^2. \\ &= \frac{5.68 + 0.0081}{0.2} && \text{Divide.} \\ &= \frac{5.6881}{0.2} && \text{Add.} \\ &= 28.4405 && \text{Divide.} \end{aligned}$$

Work Practice 14

Practice 14

Simplify: $\dfrac{20.06 - (1.2)^2 \div 10}{0.02}$

Objective D Solving Area Problems Containing Fractions and Decimals

Sometimes real-life problems contain both fractions and decimals. In the next example, we review the area of a triangle, and when values are substituted, the result may be an expression containing both fractions and decimals.

Answers

11. 89.78 **12.** −12.166
13. 2.59 **14.** 995.8

Practice 15

Find the area of the triangle.

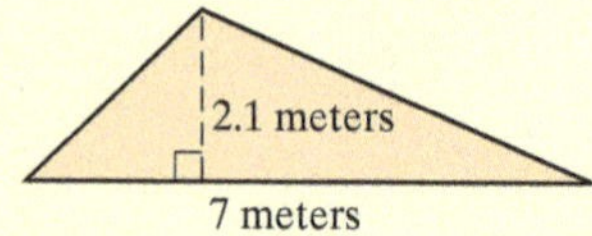

Example 15 The area of a triangle is Area $= \frac{1}{2} \cdot \text{base} \cdot \text{height}$. Find the area of the triangle shown.

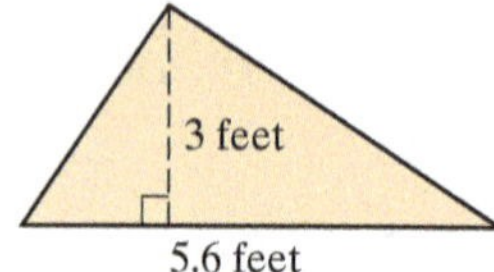

Solution:

$$\begin{aligned} \text{Area} &= \frac{1}{2} \cdot \text{base} \cdot \text{height} \\ &= \frac{1}{2} \cdot 5.6 \cdot 3 \\ &= 0.5 \cdot 5.6 \cdot 3 && \text{Write } \tfrac{1}{2} \text{ as the decimal } 0.5. \\ &= 8.4 \end{aligned}$$

The area of the triangle is 8.4 square feet.

Work Practice 15

Objective E Using Decimals as Replacement Values 

Practice 16

Evaluate $1.7y - 2$ for $y = 2.3$.

Example 16 Evaluate $-2x + 5$ for $x = 3.8$.

Solution: Replace x with 3.8 in the expression $-2x + 5$ and simplify.

$$\begin{aligned} -2x + 5 &= -2(3.8) + 5 && \text{Replace } x \text{ with 3.8.} \\ &= -7.6 + 5 && \text{Multiply.} \\ &= -2.6 && \text{Add.} \end{aligned}$$

Work Practice 16

Answers

15. 7.35 sq m **16.** 1.91

Vocabulary, Readiness & Video Check

Answer each exercise "true" or "false."

1. The number $0.\overline{5}$ means 0.555. _____

2. To write $\frac{9}{19}$ as a decimal, perform the division $19\overline{)9}$. _____

3. $(-1.2)^2$ means $(-1.2)(-1.2)$ or -1.44. _____

4. To simplify $8.6(4.8 - 9.6)$, we first subtract. _____

Martin-Gay Interactive Videos

See Video 5.5

Watch the section lecture video and answer the following questions.

Objective A 5. In Example 2, why is the bar placed over just the 6?

Objective B 6. In Example 3, why do we write the fraction as a decimal rather than the decimal as a fraction?

Objective C 7. In Example 4, besides meaning division, what other purpose does the fraction bar serve?

Objective D 8. What formula is used to solve Example 5? What is the final answer?

Objective E 9. In Example 6, once all replacement values are put into the variable expression, what is the resulting expression to evaluate?

5.5 Exercise Set MyMathLab®

Objective A *Write each number as a decimal. See Examples 1 through 7.*

1. $\frac{1}{5}$
2. $\frac{1}{20}$
3. $\frac{17}{25}$
4. $\frac{13}{25}$
5. $\frac{3}{4}$
6. $\frac{1}{8}$
7. $-\frac{2}{25}$
8. $-\frac{3}{25}$
9. $\frac{9}{4}$
10. $\frac{8}{5}$
11. $\frac{11}{12}$
12. $\frac{5}{12}$
13. $\frac{17}{40}$
14. $\frac{19}{25}$
15. $\frac{18}{40}$
16. $\frac{31}{40}$
17. $-\frac{1}{3}$
18. $-\frac{7}{9}$
19. $\frac{7}{16}$
20. $\frac{9}{16}$
21. $\frac{7}{11}$
22. $\frac{9}{11}$
23. $5\frac{17}{20}$
24. $4\frac{7}{8}$
25. $\frac{78}{125}$
26. $\frac{159}{375}$

Round each number as indicated. See Example 4.

27. Round your decimal answer to Exercise **17** to the nearest hundredth.
28. Round your decimal answer to Exercise **18** to the nearest hundredth.
29. Round your decimal answer to Exercise **19** to the nearest hundredth.
30. Round your decimal answer to Exercise **20** to the nearest hundredth.
31. Round your decimal answer to Exercise **21** to the nearest tenth.
32. Round your decimal answer to Exercise **22** to the nearest tenth.

Write each fraction as a decimal. If necessary, round to the nearest hundredth. See Examples 1 through 7.

33. Of the U.S. mountains that are over 14,000 feet in elevation, $\frac{56}{91}$ are located in Colorado. (*Source:* U.S. Geological Survey)
34. About $\frac{9}{20}$ of all U.S. citizens have type O blood. (*Source:* American Red Cross Biomedical Services)

35. About $\frac{43}{50}$ of Americans are Internet users. (*Source:* Digitalcenter)

36. About $\frac{14}{25}$ of Americans use the Internet through a wireless device. (*Source:* Digitalcenter)

37. When first launched, the Hubble Space Telescope's primary mirror was out of shape on the edges by $\frac{1}{50}$ of a human hair. This very small defect made it difficult to focus faint objects being viewed. Because the HST was in low Earth orbit, it was serviced by a shuttle and the defect was corrected.

38. The two mirrors currently in use in the Hubble Space Telescope were ground so that they do not deviate from a perfect curve by more than $\frac{1}{800{,}000}$ of an inch. Do not round this number.

Objective B *Insert <, >, or = to form a true statement. See Examples 8 and 9.*

39. 0.562 0.569

40. 0.983 0.988

41. 0.215 $\frac{43}{200}$

42. $\frac{29}{40}$ 0.725

43. −0.0932 −0.0923

44. −0.00563 −0.00536

45. $0.\overline{6}$ $\frac{5}{6}$

46. $0.\overline{1}$ $\frac{2}{17}$

47. $\frac{51}{91}$ $0.56\overline{4}$

48. $0.58\overline{3}$ $\frac{6}{11}$

49. $\frac{4}{7}$ 0.14

50. $\frac{5}{9}$ 0.557

51. 1.38 $\frac{18}{13}$

52. 0.372 $\frac{22}{59}$

53. 7.123 $\frac{456}{64}$

54. 12.713 $\frac{89}{7}$

Write the numbers in order from smallest to largest. See Example 10.

55. 0.34, 0.35, 0.32

56. 0.47, 0.42, 0.40

57. 0.49, 0.491, 0.498

58. 0.72, 0.727, 0.728

59. 5.23, $\frac{42}{8}$, 5.34

60. 7.56, $\frac{67}{9}$, 7.562

61. $\frac{5}{8}$, 0.612, 0.649

62. $\frac{5}{6}$, 0.821, 0.849

Objective C *Simplify each expression. See Examples 11 through 14.*

63. $(0.3)^2 + 0.5$

64. $(-2.5)(3) - 4.7$

65. $\frac{1 + 0.8}{-0.6}$

66. $(-0.05)^2 + 3.13$

67. $(-2.3)^2(0.3 + 0.7)$

68. $(8.2)(100) - (8.2)(10)$

69. $(5.6 - 2.3)(2.4 + 0.4)$

70. $\frac{0.222 - 2.13}{12}$

71. $\frac{(4.5)^2}{100}$

72. $0.9(5.6 - 6.5)$

73. $\frac{7 + 0.74}{-6}$

74. $(1.5)^2 + 0.5$

Find the value of each expression. Give the result as a decimal. See Examples 11 through 14.

75. $\frac{1}{5} - 2(7.8)$

76. $\frac{3}{4} - (9.6)(5)$

77. $\frac{1}{4}(-9.6 - 5.2)$

78. $\frac{3}{8}(4.7 - 5.9)$

Objective D *Find the area of each triangle or rectangle. See Example 15.*

△ **79.**

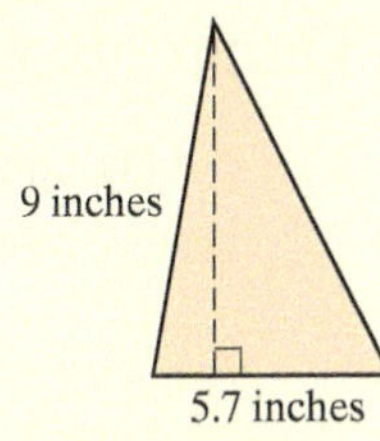

△ **80.**

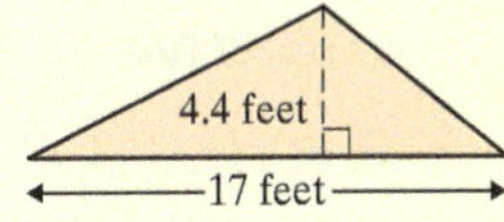

81.

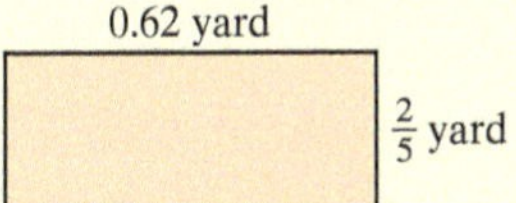

△ **82.**

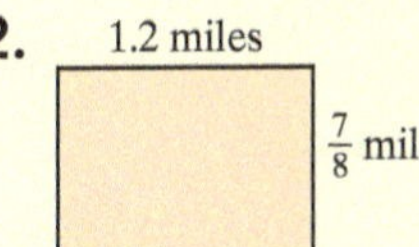

Objective E *Evaluate each expression for $x = 6$, $y = 0.3$, and $z = -2.4$. See Example 16.*

83. z^2 **84.** y^2 **85.** $x - y$ **86.** $x - z$ **87.** $4y - z$ **88.** $\frac{x}{y} + 2z$

Review

Simplify. See Sections 4.3 and 4.5.

89. $\frac{9}{10} + \frac{16}{25}$ **90.** $\frac{4}{11} - \frac{19}{22}$ **91.** $\left(\frac{2}{5}\right)\left(\frac{5}{2}\right)^2$ **92.** $\left(\frac{2}{3}\right)^2\left(\frac{3}{2}\right)^3$

Concept Extensions

Without calculating, describe each number as < 1, $= 1$, or > 1. See the Concept Check in this section.

93. 1.0 **94.** 1.0000 **95.** 1.00001 **96.** $\frac{101}{99}$ **97.** $\frac{99}{100}$ **98.** $\frac{99}{99}$

In 2012, there were 11,434 commercial radio stations in the United States. The most popular formats are listed in the table along with their counts. Use this graph to answer Exercises 99 through 102.

99. Write as a decimal the fraction of radio stations with a classic hits music format. Round to the nearest thousandth.

100. Write as a decimal the fraction of radio stations with a Spanish format. Round to the nearest hundredth.

101. Estimate, by rounding each number in the table to the nearest hundred, the total number of stations with the top six formats in 2012.

102. Use your estimate from Exercise **101** to write the fraction of radio stations accounted for by the top six formats as a decimal. Round to the nearest hundredth.

Top Commercial Radio Station Formats in 2012

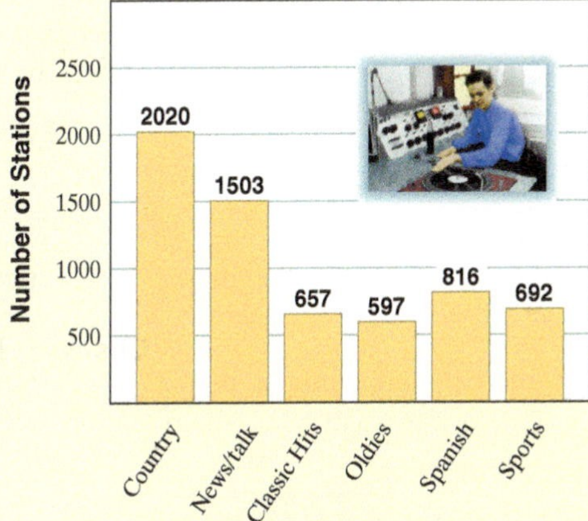

Format (Total stations: 11,434)

103. Describe two ways to write fractions as decimals.

104. Describe two ways to write mixed numbers as decimals.

5.6 Solving Equations Containing Decimals

Objective

A Solve Equations Containing Decimals.

Objective A Solving Equations Containing Decimals

In this section, we continue our work with decimals and algebra by solving equations containing decimals. First, we review the steps given earlier for solving an equation.

Steps for Solving an Equation in x

Step 1: If fractions are present, multiply both sides of the equation by the LCD of the fractions.

Step 2: If parentheses are present, use the distributive property.

Step 3: Combine any like terms on each side of the equation.

Step 4: Use the addition property of equality to rewrite the equation so that variable terms are on one side of the equation and constant terms are on the other side.

Step 5: Divide both sides by the numerical coefficient of x to solve.

Step 6: Check your answer in the **original equation**.

Practice 1

Solve: $z + 0.9 = 1.3$

Example 1 Solve: $x - 1.5 = 8$

Solution: Steps 1 through 3 are not needed for this equation, so we begin with Step 4. To get x alone on one side of the equation, add 1.5 to both sides.

$x - 1.5 = 8$ Original equation

$x - 1.5 + 1.5 = 8 + 1.5$ Add 1.5 to both sides.

$x = 9.5$ Simplify.

Check: To check, replace x with 9.5 in the *original equation*.

$x - 1.5 = 8$ Original equation

$9.5 - 1.5 \stackrel{?}{=} 8$ Replace x with 9.5.

$8 = 8$ True

Since $8 = 8$ is a true statement, 9.5 is a solution of the equation.

Work Practice 1

Practice 2

Solve: $0.17x = -0.34$

Example 2 Solve: $-2y = 6.7$

Solution: Steps 1 through 4 are not needed for this equation, so we begin with Step 5. To solve for y, divide both sides by the coefficient of y, which is -2.

$-2y = 6.7$ Original equation

$\frac{-2y}{-2} = \frac{6.7}{-2}$ Divide both sides by -2.

$y = -3.35$ Simplify.

Check: To check, replace y with -3.35 in the original equation.

$-2y = 6.7$ Original equation

$-2(-3.35) \stackrel{?}{=} 6.7$ Replace y with -3.35.

$6.7 = 6.7$ True

Thus -3.35 is a solution of the equation $-2y = 6.7$.

Work Practice 2

Answers

1. 0.4 **2.** -2

Example 3 Solve: $1.2x + 5.8 = 8.2$

Solution: We begin with Step 4 and get the variable term alone by subtracting 5.8 from both sides.

$$1.2x + 5.8 = 8.2$$
$$1.2x + 5.8 - 5.8 = 8.2 - 5.8 \quad \text{Subtract 5.8 from both sides.}$$
$$1.2x = 2.4 \quad \text{Simplify.}$$
$$\frac{1.2x}{1.2} = \frac{2.4}{1.2} \quad \text{Divide both sides by 1.2.}$$
$$x = 2 \quad \text{Simplify.}$$

To check, replace x with 2 in the original equation. The solution is 2.

Work Practice 3

Practice 3

Solve: $2.9 = 1.7 + 0.3x$

Example 4 Solve: $7x + 3.2 = 4x - 1.6$

Solution: We start with Step 4 to get variable terms on one side and numerical terms on the other.

$$7x + 3.2 = 4x - 1.6$$
$$7x + 3.2 - 3.2 = 4x - 1.6 - 3.2 \quad \text{Subtract 3.2 from both sides.}$$
$$7x = 4x - 4.8 \quad \text{Simplify.}$$
$$7x - 4x = 4x - 4.8 - 4x \quad \text{Subtract } 4x \text{ from both sides.}$$
$$3x = -4.8 \quad \text{Simplify.}$$
$$\frac{3x}{3} = -\frac{4.8}{3} \quad \text{Divide both sides by 3.}$$
$$x = -1.6 \quad \text{Simplify.}$$

Check to see that -1.6 is the solution.

Work Practice 4

Practice 4

Solve: $8x + 4.2 = 10x + 11.6$

Example 5 Solve: $5(x - 0.36) = -x + 2.4$

Solution: First use the distributive property to distribute the factor 5.

$$5(x - 0.36) = -x + 2.4 \quad \text{Original equation}$$
$$5x - 1.8 = -x + 2.4 \quad \text{Apply the distributive property.}$$

Next, get the variable terms alone on the left side of the equation by adding 1.8 to both sides of the equation and then adding x to both sides of the equation.

$$5x - 1.8 + 1.8 = -x + 2.4 + 1.8 \quad \text{Add 1.8 to both sides.}$$
$$5x = -x + 4.2 \quad \text{Simplify.}$$
$$5x + x = -x + 4.2 + x \quad \text{Add } x \text{ to both sides.}$$
$$6x = 4.2 \quad \text{Simplify.}$$
$$\frac{6x}{6} = \frac{4.2}{6} \quad \text{Divide both sides by 6.}$$
$$x = 0.7 \quad \text{Simplify.}$$

To verify that 0.7 is the solution, replace x with 0.7 in the original equation.

Work Practice 5

Practice 5

Solve: $6.3 - 5x = 3(x + 2.9)$

Answers

3. 4 **4.** -3.7 **5.** -0.3

Instead of solving equations with decimals, sometimes it may be easier to first rewrite the equation so that it contains integers only. Recall that multiplying a decimal by a power of 10 such as 10, 100, or 1000 has the effect of moving the decimal point to the right. We can use the multiplication property of equality to multiply both sides of the equation by an appropriate power of 10. The resulting equivalent equation will contain integers only.

Practice 6

Solve: $0.2y + 2.6 = 4$

Example 6 Solve: $0.5y + 2.3 = 1.65$

Solution: Multiply both sides of the equation by 100. This will move the decimal point in each term two places to the right.

$$0.5y + 2.3 = 1.65 \quad \text{Original equation}$$
$$100(0.5y + 2.3) = 100(1.65) \quad \text{Multiply both sides by 100.}$$
$$100(0.5y) + 100(2.3) = 100(1.65) \quad \text{Apply the distributive property.}$$
$$50y + 230 = 165 \quad \text{Simplify.}$$

Now the equation contains integers only. Continue solving by subtracting 230 from both sides.

$$50y + 230 = 165$$
$$50y + 230 - 230 = 165 - 230 \quad \text{Subtract 230 from both sides.}$$
$$50y = -65 \quad \text{Simplify.}$$
$$\frac{50y}{50} = \frac{-65}{50} \quad \text{Divide both sides by 50.}$$
$$y = -1.3 \quad \text{Simplify.}$$

Check to see that -1.3 is the solution by replacing y with -1.3 in the original equation.

Work Practice 6

Answer
6. 7

✓Concept Check By what number would you multiply both sides of the following equation to make calculations easier? Explain your choice.

$$1.7x + 3.655 = -14.2$$

✓Concept Check Answer
Multiply by 1000.

Martin-Gay Interactive Videos

See Video 5.6

Watch the section lecture video and answer the following questions.

Objective A

1. In Example 3, why did we multiply both sides of the equation by 10?

2. In Example 4, we subtracted $3x$ from both sides, but what would have been a potential benefit of subtracting $2x$ instead?

5.6 Exercise Set MyMathLab® 

Objective A *Solve each equation. See Examples 1 and 2.*

1. $x + 1.2 = 7.1$

2. $y - 0.5 = 9$

3. $-5y = 2.15$

4. $-0.4x = 50$

5. $6.2 = y - 4$

6. $9.7 = x + 11.6$

7. $3.1x = -13.95$

8. $3y = -25.8$

Solve each equation. See Examples 3 through 5.

9. $-3.5x + 2.8 = -11.2$

10. $7.1 - 0.2x = 6.1$

11. $6x + 8.65 = 3x + 10$

12. $7x - 9.64 = 5x + 2.32$

13. $2(x - 1.3) = 5.8$

14. $5(x + 2.3) = 19.5$

Solve each equation by first multiplying both sides by an appropriate power of 10 so that the equation contains integers only. See Example 6.

15. $0.4x + 0.7 = -0.9$

16. $0.7x + 0.1 = 1.5$

17. $7x - 10.8 = x$

18. $3y = 7y + 24.4$

19. $2.1x + 5 - 1.6x = 10$

20. $1.5x + 2 - 1.2x = 12.2$

Solve. See Examples 1 through 6.

21. $y - 3.6 = 4$

22. $x + 5.7 = 8.4$

23. $-0.02x = -1.2$

24. $-9y = -0.162$

25. $6.5 = 10x + 7.2$

26. $2x - 4.2 = 8.6$

27. $2.7x - 25 = 1.2x + 5$

28. $9y - 6.9 = 6y - 11.1$

29. $200x - 0.67 = 100x + 0.81$

30. $2.3 + 500x = 600x - 0.2$

31. $3(x + 2.71) = 2x$

32. $7(x + 8.6) = 6x$

33. $8x - 5 = 10x - 8$

34. $24y - 10 = 20y - 17$

35. $1.2 + 0.3x = 0.9$

36. $1.5 = 0.4x + 0.5$

37. $-0.9x + 2.65 = -0.5x + 5.45$

38. $-50x + 0.81 = -40x - 0.48$

39. $4x + 7.6 = 2(3x - 3.2)$

40. $4(2x - 1.6) = 5x - 6.4$

41. $0.7x + 13.8 = x - 2.16$

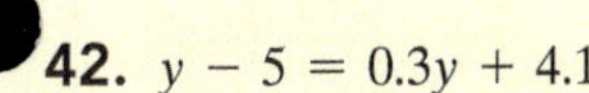

42. $y - 5 = 0.3y + 4.1$

Review

Simplify each expression by combining like terms. See Section 3.1.

43. $2x - 7 + x - 9$

44. $x + 14 - 5x - 17$

Perform the indicated operation. See Sections 4.3, 4.5, and 4.7.

45. $\frac{6x}{5} \cdot \frac{1}{2x^2}$

46. $5\frac{1}{3} \div 9\frac{1}{6}$

47. $\frac{x}{3} + \frac{2x}{7}$

48. $50 - 14\frac{9}{13}$

Concept Extensions

Mixed Practice (*Sections 5.2 and 5.6*) *This section of exercises contains equations and expressions. If the exercise contains an equation, solve it for the variable. If the exercise contains an expression, simplify it by combining any like terms.*

49. $b + 4.6 = 8.3$

50. $y - 15.9 = -3.8$

51. $2x - 0.6 + 4x - 0.01$

52. $-x - 4.1 - x - 4.02$

53. $5y - 1.2 - 7y + 8$

54. $9a - 5.6 - 3a + 6$

55. $2.8 = z - 6.3$

56. $9.7 = x + 4.3$

57. $4.7x + 8.3 = -5.8$

58. $2.8x + 3.4 = -13.4$

59. $7.76 + 8z - 12z + 8.91$

60. $9.21 + x - 4x + 11.33$

61. $5(x - 3.14) = 4x$

62. $6(x + 1.43) = 5x$

63. $2.6y + 8.3 = 4.6y - 3.4$

64. $8.4z - 2.6 = 5.4z + 10.3$

65. $9.6z - 3.2 - 11.7z - 6.9$

66. $-3.2x + 12.6 - 8.9x - 15.2$

67. Explain in your own words the property of equality that allows us to multiply both sides of an equation by a power of 10.

68. By what number would you multiply both sides of $8x - 7.6 = 4.23$ to make calculations easier? Explain your choice.

69. Construct an equation whose solution is 1.4.

70. Construct an equation whose solution is −8.6.

Solve.

71. $-5.25x = -40.33575$

72. $7.68y = -114.98496$

73. $1.95y + 6.834 = 7.65y - 19.8591$

74. $6.11x + 4.683 = 7.51x + 18.235$

5.7 Decimal Applications: Mean, Median, and Mode

Objective A Finding the Mean

Sometimes we want to summarize data by displaying them in a graph, but sometimes it is also desirable to be able to describe a set of data, or a set of numbers, by a single "middle" number. Three such **measures of central tendency** are the **mean,** the **median,** and the **mode.**

The most common measure of central tendency is the mean (sometimes called the "arithmetic mean" or the "average"). Recall that we first introduced finding the average of a list of numbers in Section 1.6.

The **mean (average)** of a set of numbered items is the sum of the items divided by the number of items.

$$\text{mean} = \frac{\text{sum of items}}{\text{number of items}}$$

Example 1 Finding the Mean Time in an Experiment

Seven students in a psychology class conducted an experiment on mazes. Each student was given a pencil and asked to successfully complete the same maze. The timed results are below:

Student	Ann	Thanh	Carlos	Jesse	Melinda	Ramzi	Dayni
Time (Seconds)	13.2	11.8	10.7	16.2	15.9	13.8	18.5

a. Who completed the maze in the shortest time? Who completed the maze in the longest time?

b. Find the mean time.

c. How many students took longer than the mean time? How many students took shorter than the mean time?

Solution:

a. Carlos completed the maze in 10.7 seconds, the shortest time. Dayni completed the maze in 18.5 seconds, the longest time.

b. To find the mean (or average), we find the sum of the items and divide by 7, the number of items.

$$\text{mean} = \frac{13.2 + 11.8 + 10.7 + 16.2 + 15.9 + 13.8 + 18.5}{7}$$

$$= \frac{100.1}{7} = 14.3$$

c. Three students, Jesse, Melinda, and Dayni, had times longer than the mean time. Four students, Ann, Thanh, Carlos, and Ramzi, had times shorter than the mean time.

Work Practice 1

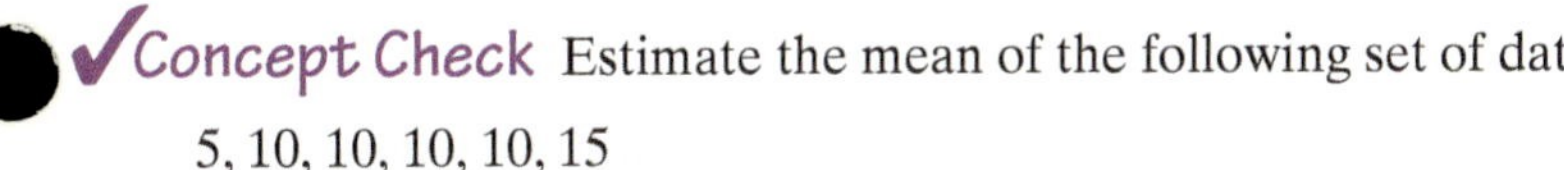

✓Concept Check Estimate the mean of the following set of data:

5, 10, 10, 10, 10, 15

Often in college, the calculation of a **grade point average** (GPA) is a **weighted mean** and is calculated as shown in Example 2.

Objectives

A Find the Mean of a List of Numbers.

B Find the Median of a List of Numbers.

C Find the Mode of a List of Numbers.

Practice 1

Find the mean of the following test scores: 87, 75, 96, 91, and 78.

Answer

1. 85.4

✓Concept Check Answer

10

Practice 2

Find the grade point average if the following grades were earned in one semester.

Grade	Credit Hours
A	2
B	4
C	5
D	2
A	2

Example 2 Calculating Grade Point Average (GPA)

The following grades were earned by a student during one semester. Find the student's grade point average.

Course	Grade	Credit Hours
College mathematics	A	3
Biology	B	3
English	A	3
PE	C	1
Social studies	D	2

Solution: To calculate the grade point average, we need to know the point values for the different possible grades. The point values of grades commonly used in colleges and universities are given below:

A: 4, B: 3, C: 2, D: 1, F: 0

Now, to find the grade point average, we multiply the number of credit hours for each course by the point value of each grade. The grade point average is the sum of these products divided by the sum of the credit hours.

Course	Grade	Point Value of Grade	Credit Hours	Point Value of Credit Hours
College mathematics	A	4	3	12
Biology	B	3	3	9
English	A	4	3	12
PE	C	2	1	2
Social studies	D	1	2	2
		Totals:	12	37

$$\text{grade point average} = \frac{37}{12} \approx 3.08 \text{ rounded to two decimal places}$$

The student earned a grade point average of 3.08.

Work Practice 2

Objective B Finding the Median 

You may have noticed that a very low number or a very high number can affect the mean of a list of numbers. Because of this, you may sometimes want to use another measure of central tendency. A second measure of central tendency is called the **median.** The median of a list of numbers is not affected by a low or high number in the list.

The **median** of a set of numbers in numerical order is the middle number. If the number of items is odd, the median is the middle number. If the number of items is even, the median is the mean of the two middle numbers.

Practice 3

Find the median of the list of numbers: 5, 11, 14, 23, 24, 35, 38, 41, 43

Example 3 Find the median of the following list of numbers:

25, 54, 56, 57, 60, 71, 98

Solution: Because this list is in numerical order, the median is the middle number, 57.

Work Practice 3

Answers
2. 2.67 **3.** 24

Example 4 Find the median of the following list of scores: 67, 91, 75, 86, 55, 91

Solution: First we list the scores in numerical order and then we find the middle number.

55, 67, 75, 86, 91, 91

Since there is an even number of scores, there are two middle numbers, 75 and 86. The median is the mean of the two middle numbers.

$$\text{median} = \frac{75 + 86}{2} = 80.5$$

The median is 80.5.

Helpful Hint Don't forget to write the numbers in order from smallest to largest before finding the median.

Work Practice 4

Practice 4

Find the median of the list of scores:
36, 91, 78, 65, 95, 95, 88, 71

Objective C Finding the Mode

The last common measure of central tendency is called the **mode.**

The **mode** of a set of numbers is the number that occurs most often. (It is possible for a set of numbers to have more than one mode or to have no mode.)

Example 5 Find the mode of the list of numbers:

11, 14, 14, 16, 31, 56, 65, 77, 77, 78, 79

Solution: There are two numbers that occur the most often. They are 14 and 77. This list of numbers has two modes, 14 and 77.

Work Practice 5

Practice 5

Find the mode of the list of numbers:
14, 10, 10, 13, 15, 15, 15, 17, 18, 18, 20

Example 6 Find the median and the mode of the following set of numbers. These numbers were high temperatures for 14 consecutive days in a city in Montana.

76, 80, 85, 86, 89, 87, 82, 77, 76, 79, 82, 89, 89, 92

Solution: First we write the numbers in numerical order.

76, 76, 77, 79, 80, 82, 82, 85, 86, 87, 89, 89, 89, 92

Since there is an even number of items, the median is the mean of the two middle numbers, 82 and 85.

$$\text{median} = \frac{82 + 85}{2} = 83.5$$

The mode is 89, since 89 occurs most often.

Work Practice 6

Practice 6

Find the median and the mode of the list of numbers:
26, 31, 15, 15, 26, 30, 16, 18, 15, 35

Answers

4. 83 **5.** 15 **6.** median: 22; mode: 15

✓**Concept Check** True or false? Every set of numbers *must* have a mean, median, and mode. Explain your answer.

✓**Concept Check Answer**

false; a set of numbers may have no mode

Helpful Hint

Don't forget that it is possible for a list of numbers to have no mode. For example, the list

2, 4, 5, 6, 8, 9

has no mode. There is no number or numbers that occur more often than the others.

Vocabulary, Readiness & Video Check

Use the choices below to fill in each blank. Some choices may be used more than once.

mean	mode	grade point average
median	average	

1. Another word for "mean" is ______________.
2. The number that occurs most often in a set of numbers is called the ______________.
3. The ______________ of a set of number items is $\frac{\text{sum of items}}{\text{number of items}}$.
4. The ______________ of a set of numbers is the middle number. If the number of numbers is even, it is the ______________ of the two middle numbers.
5. An example of weighted mean is a calculation of ______________.

Martin-Gay Interactive Videos

See Video 5.7 

Watch the section lecture video and answer the following questions.

Objective A 6. Why is the ≈ symbol used in Example 1?

Objective B 7. From Example 3, what is always the first step when finding the median of a set of data numbers?

Objective C 8. From Example 4, why do you think it is helpful to have data numbers in numerical order when finding the mode?

5.7 Exercise Set MyMathLab®

Objectives A B C Mixed Practice *For each set of numbers, find the mean, median, and mode. If necessary, round the mean to one decimal place. See Examples 1 and 3 through 6.*

1. 15, 23, 24, 18, 25
2. 45, 36, 28, 46, 52
3. 7.6, 8.2, 8.2, 9.6, 5.7, 9.1
4. 4.9, 7.1, 6.8, 6.8, 5.3, 4.9
5. 0.5, 0.2, 0.2, 0.6, 0.3, 1.3, 0.8, 0.1, 0.5
6. 0.6, 0.6, 0.8, 0.4, 0.5, 0.3, 0.7, 0.8, 0.1
7. 231, 543, 601, 293, 588, 109, 334, 268
8. 451, 356, 478, 776, 892, 500, 467, 780

The ten tallest buildings in the world, completed as of the start of 2012, are listed in the following table. Use this table to answer Exercises 9 through 14. If necessary, round results to one decimal place. See Examples 1 and 3 through 6.

Building	Height (in feet)
Burj Khalifa, Dubai	2717
Makkah Royal Clock Tower Hotel, Saudi Arabia	1972
Taipei 101	1667
Shanghai World Financial Center	1614
International Commerce Centre, Hong Kong	1588
Petronas Tower 1, Kuala Lumpur	1483
Petronas Tower 2, Kuala Lumpur	1483
Zifeng Tower, China	1476
Willis Tower, Chicago	1451
KK 100 Development, China	1449
(*Source:* Council on Tall Buildings and Urban Habitat)	

9. Find the mean height of the five tallest buildings.

10. Find the median height of the five tallest buildings.

11. Find the median height of the eight tallest buildings.

12. Find the mean height of the eight tallest buildings.

13. Given the building heights, explain how you know, without calculating, that the answer to Exercise **10** is greater than the answer to Exercise **11**.

14. Given the building heights, explain how you know, without calculating, that the answer to Exercise **12** is less than the answer to Exercise **9**.

For Exercises 15 through 18, the grades are given for a student for a particular semester. Find the grade point average. If necessary, round the grade point average to the nearest hundredth. See Example 2.

15.

Grade	Credit Hours
B	3
C	3
A	4
C	4

16.

Grade	Credit Hours
D	1
F	1
C	4
B	5

17.

Grade	Credit Hours
A	3
A	3
A	4
B	3
C	1

18.

Grade	Credit Hours
B	2
B	2
C	3
A	3
B	3

During an experiment, the following times (in seconds) were recorded:

7.8, 6.9, 7.5, 4.7, 6.9, 7.0.

19. Find the mean. Round to the nearest tenth.

20. Find the median.

21. Find the mode.

In a mathematics class, the following test scores were recorded for a student: 93, 85, 89, 79, 88, 91.

22. Find the mean. Round to the nearest hundredth.

23. Find the median.

24. Find the mode.

The following pulse rates were recorded for a group of 15 students:

78, 80, 66, 68, 71, 64, 82, 71, 70, 65, 70, 75, 77, 86, 72.

25. Find the mean.

26. Find the median.

27. Find the mode.

28. How many pulse rates were higher than the mean?

29. How many pulse rates were lower than the mean?

Review

Write each fraction in simplest form. See Section 4.2.

30. $\frac{12}{20}$

31. $\frac{6}{18}$

32. $\frac{4x}{36}$

33. $\frac{18}{30y}$

34. $\frac{35a^3}{100a^2}$

35. $\frac{55y^2}{75y^2}$

Concept Extensions

Find the missing numbers in each set of numbers.

36. 16, 18, ________, ________, ________. The mode is 21. The median is 20.

37. ________, ________, ________, 40, ________. The mode is 35. The median is 37. The mean is 38.

38. Write a list of numbers for which you feel the median would be a better measure of central tendency than the mean.

39. Without making any computations, decide whether the median of the following list of numbers will be a whole number. Explain your reasoning.

36, 77, 29, 58, 43

Chapter 5 Group Activity

Maintaining a Checking Account

(Sections 5.1, 5.2, 5.3, 5.4)

This activity may be completed by working in groups or individually.

A checking account is a convenient way of handling money and paying bills. To open a checking account, the bank or savings and loan association requires a customer to make a deposit. Then the customer receives a checkbook that contains checks, deposit slips, and a register for recording checks written and deposits made. It is important to record all payments and deposits that affect the account. It is also important to keep the checkbook balance current by subtracting checks written and adding deposits made.

About once a month, checking customers receive a statement from the bank listing all activity that the account has had in the last month. The statement lists a beginning balance, all checks and deposits, any service charges made against the account, and an ending balance. Because it may take several days for checks that a customer has written to clear the banking system, the check register may list checks that do not appear on the monthly bank statement. These checks are called **outstanding checks.** Deposits that are recorded in the check register but do not appear on the statement are called **deposits in transit.** Because of these differences, it is important to balance, or reconcile, the checkbook against the monthly statement. The steps for doing so are listed below.

Balancing or Reconciling a Checkbook

Step 1: Place a check mark in the checkbook register next to each check and deposit listed on the monthly bank statement. Any entries in the register without a check mark are outstanding checks or deposits in transit.

Step 2: Find the ending checkbook register balance and add to it any outstanding checks and any interest paid on the account.

Step 3: From the total in Step 2, subtract any deposits in transit and any service charges.

Step 4: Compare the amount found in Step 3 with the ending balance listed on the bank statement. If they are the same, the checkbook balances with the bank statement. Be sure to update the check register with service charges and interest.

Step 5: If the checkbook does not balance, recheck the balancing process. Next, make sure that the running checkbook register balance was calculated correctly. Finally, compare the checkbook register with the statement to make sure that each check was recorded for the correct amount.

For the checkbook register and monthly bank statement given:

a. *update the checkbook register* **b.** *list the outstanding checks and their total, and deposits in transit*

c. *balance the checkbook—be sure to update the register with any interest or service fees*

Checkbook Register

#	Date	Description	Payment	✓	Deposit	Balance
						425.86
114	4/1	Market Basket	30.27			
115	4/3	May's Texaco	8.50			
	4/4	Cash at ATM	50.00			
116	4/6	UNO Bookstore	121.38			
	4/7	Deposit			100.00	
117	4/9	MasterCard	84.16			
118	4/10	Redbox	6.12			
119	4/12	Kroger	18.72			
120	4/14	Parking sticker	18.50			
	4/15	Direct deposit			294.36	
121	4/20	Rent	395.00			
122	4/25	Student fees	20.00			
	4/28	Deposit			75.00	

First National Bank Monthly Statement 4/30

BEGINNING BALANCE:		425.86
Date	Number	Amount
CHECKS AND ATM WITHDRAWALS		
4/3	114	30.27
4/4	ATM	50.00
4/11	117	84.16
4/13	115	8.50
4/15	119	18.72
4/22	121	395.00
DEPOSITS		
4/7		100.00
4/15	Direct deposit	294.36
SERVICE CHARGES		
Low balance fee		7.50
INTEREST		
Credited 4/30		1.15
ENDING BALANCE:		227.22

Chapter 5 Vocabulary Check

Fill in each blank with one of the choices listed below. Some choices may be used more than once and some not at all.

vertically	decimal	and	right triangle
standard form	mean	median	circumference
sum	denominator	numerator	mode

1. Like fractional notation, ______________ notation is used to denote a part of a whole.
2. To write fractions as decimals, divide the ______________ by the ______________.
3. To add or subtract decimals, write the decimals so that the decimal points line up ______________.
4. When writing decimals in words, write "______________" for the decimal point.
5. When multiplying decimals, the decimal point in the product is placed so that the number of decimal places in the product is equal to the ______________ of the number of decimal places in the factors.
6. The ______________ of a set of numbers is the number that occurs most often.
7. The distance around a circle is called the ______________.
8. The ______________ of a set of numbers in numerical order is the middle number. If there is an even number of numbers, the median is the ______________ of the two middle numbers.
9. The ______________ of a list of items with number values is $\frac{\text{sum of items}}{\text{number of items}}$.
10. When 2 million is written as 2,000,000, we say it is written in ______________.

Helpful Hint Are you preparing for your test? Don't forget to take the Chapter 5 Test on page 405. Then check your answers at the back of the text and use the Chapter Test Prep Videos to see the fully worked-out solutions to any of the exercises you want to review.

5 Chapter Highlights

Definitions and Concepts	Examples

Section 5.1 Introduction to Decimals

Place-Value Chart

		4	.	2	6	5		
hundreds	tens	ones	↑ decimal point	tenths	hundredths	thousandths	ten-thousandths	hundred-thousandths
100	10	1		$\frac{1}{10}$	$\frac{1}{100}$	$\frac{1}{1000}$	$\frac{1}{10,000}$	$\frac{1}{100,000}$

4.265 means

$$4 \cdot 1 + 2 \cdot \frac{1}{10} + 6 \cdot \frac{1}{100} + 5 \cdot \frac{1}{1000}$$

or

$$4 + \frac{2}{10} + \frac{6}{100} + \frac{5}{1000}$$

Definitions and Concepts	Examples
Section 5.1 Introduction to Decimals *(continued)*	
Writing (or Reading) a Decimal in Words **Step 1:** Write the whole number part in words. **Step 2:** Write "and" for the decimal point. **Step 3:** Write the decimal part in words as though it were a whole number, followed by the place value of the last digit.	Write 3.08 in words. Three and eight hundredths
A decimal written in words can be written in standard form by reversing the above procedure.	Write "negative four and twenty-one thousandths" in standard form. −4.021
To Round a Decimal to a Place Value to the Right of the Decimal Point **Step 1:** Locate the digit to the right of the given place value. **Step 2:** If this digit is 5 or greater, add 1 to the digit in the given place value and drop all digits to its right. If this digit is less than 5, drop all digits to the right of the given place value.	Round 86.1256 to the nearest hundredth. **Step 1:** 86.12 5 6 (hundredths place: 2; digit to the right: 5) **Step 2:** Since the digit to the right is 5 or greater, we add 1 to the digit in the hundredths place and drop all digits to its right. 86.1256 rounded to the nearest hundredth is 86.13.
Section 5.2 Adding and Subtracting Decimals	
To Add or Subtract Decimals **Step 1:** Write the decimals so that the decimal points line up vertically. **Step 2:** Add or subtract as with whole numbers. **Step 3:** Place the decimal point in the sum or difference so that it lines up vertically with the decimal points in the problem.	Add: 4.6 + 0.28 4.60 + 0.28 4.88 Subtract: 2.8 − 1.04 2.8̸0 (7 10) − 1.04 1.76
Section 5.3 Multiplying Decimals and Circumference of a Circle	
To Multiply Decimals **Step 1:** Multiply the decimals as though they are whole numbers. **Step 2:** The decimal point in the product is placed so that the number of decimal places in the product is equal to the *sum* of the number of decimal places in the factors.	Multiply: 1.48 × 5.9 1.48 ← 2 decimal places × 5.9 ← 1 decimal place 1332 7400 8.732 ← 3 decimal places

(continued)

Definitions and Concepts	Examples

Section 5.3 Multiplying Decimals and Circumference of a Circle *(continued)*

The **circumference** of a circle is the distance around the circle.

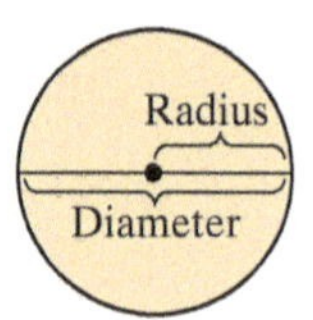

$C = \pi d$ or $C = 2\pi r$

where $\pi \approx 3.14$ or $\pi \approx \frac{22}{7}$.

or

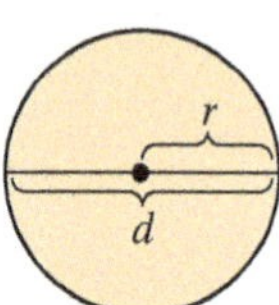

Find the exact circumference of a circle with radius 5 miles and an approximation by using 3.14 for π.

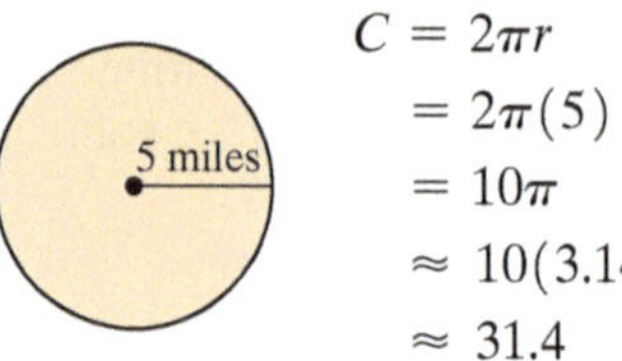

$$\begin{aligned} C &= 2\pi r \\ &= 2\pi(5) \\ &= 10\pi \\ &\approx 10(3.14) \\ &\approx 31.4 \end{aligned}$$

The circumference is exactly 10π miles and approximately 31.4 miles.

Section 5.4 Dividing Decimals

To Divide Decimals

Step 1: If the divisor is not a whole number, move the decimal point in the divisor to the right until the divisor is a whole number.

Step 2: Move the decimal point in the dividend to the right the *same number of places* as the decimal point was moved in Step 1.

Step 3: Divide. The decimal point in the quotient is directly over the moved decimal point in the dividend.

Divide: $1.118 \div 2.6$

$$\begin{array}{r} 0.43 \\ 2.6\overline{)1.118} \\ -104 \\ \hline 78 \\ -78 \\ \hline 0 \end{array}$$

Section 5.5 Fractions, Decimals, and Order of Operations

To **write fractions as decimals,** divide the numerator by the denominator.

Write $\frac{3}{8}$ as a decimal.

$$\begin{array}{r} 0.375 \\ 8\overline{)3.000} \\ -24 \\ \hline 60 \\ -56 \\ \hline 40 \\ -40 \\ \hline 0 \end{array}$$

Order of Operations

1. Perform all operations within parentheses ( ), brackets [ ], or other grouping symbols such as fraction bars.
2. Evaluate any expressions with exponents.
3. Multiply or divide in order from left to right.
4. Add or subtract in order from left to right.

Simplify.

$$\begin{aligned} -1.9(12.8 - 4.1) &= -1.9(8.7) && \text{Subtract.} \\ &= -16.53 && \text{Multiply.} \end{aligned}$$

Definitions and Concepts	Examples
Section 5.6 Solving Equations Containing Decimals	
Steps for Solving an Equation in *x* **Step 1:** If fractions are present, multiply both sides of the equation by the LCD of the fractions. **Step 2:** If parentheses are present, use the distributive property. **Step 3:** Combine any like terms on each side of the equation. **Step 4:** Use the addition property of equality to rewrite the equation so that variable terms are on one side of the equation and constant terms are on the other side. **Step 5:** Divide both sides by the numerical coefficient of x to solve. **Step 6:** Check your answer in the *original equation.*	Solve: $3(x + 2.6) = 10.92$ $3x + 7.8 = 10.92$ Apply the distributive property. $3x + 7.8 - 7.8 = 10.92 - 7.8$ Subtract 7.8 from both sides. $3x = 3.12$ Simplify. $\frac{3x}{3} = \frac{3.12}{3}$ Divide both sides by 3. $x = 1.04$ Simplify. Check 1.04 in the original equation.
Section 5.7 Decimal Applications: Mean, Median, and Mode	
The **mean** (or **average**) of a set of number items is $\text{mean} = \frac{\text{sum of items}}{\text{number of items}}$	Find the mean, median, and mode of the following set of numbers: 33, 35, 35, 43, 68, 68 $\text{mean} = \frac{33 + 35 + 35 + 43 + 68 + 68}{6} = 47$
The **median** of a set of numbers in numerical order is the middle number. If the number of items is even, the median is the mean of the two middle numbers.	The median is the mean of the two middle numbers, 35 and 43 $\text{median} = \frac{35 + 43}{2} = 39$
The **mode** of a set of numbers is the number that occurs most often. (A set of numbers may have no mode or more than one mode.)	There are two modes because there are two numbers that occur twice: 35 and 68

Chapter 5 Review

(5.1) *Determine the place value of the number 4 in each decimal.*

1. 23.45

2. 0.000345

Write each decimal in words.

3. −23.45

4. 0.00345

5. 109.23

6. 200.000032

Write each decimal in standard form.

7. Eight and six hundredths

8. Negative five hundred three and one hundred two thousandths

9. Sixteen thousand twenty-five and fourteen ten-thousandths

10. Fourteen and eleven thousandths

Write each decimal as a fraction or a mixed number.

11. 0.16

12. −12.023

Write each fraction or mixed number as a decimal.

13. $\frac{231}{100,000}$

14. $25\frac{1}{4}$

Insert <, >, or = between each pair of numbers to make a true statement.

15. 0.49 0.43

16. 0.973 0.9730

17. −38.0027 −38.00056

18. −0.230505 −0.23505

Round each decimal to the given place value.

19. 0.623, nearest tenth

20. 0.9384, nearest hundredth

21. −42.895, nearest hundredth

22. 16.34925, nearest thousandth

Write each number in standard notation.

23. Saturn is a distance of about 887 million miles from the Sun.

24. The tail of a comet can be over 600 thousand miles long.

(5.2) *Add.*

25. 8.6 + 9.5

26. 3.9 + 1.2

27. −6.4 + (−0.88)

28. −19.02 + 6.98

29. 200.49 + 16.82 + 103.002

30. 0.00236 + 100.45 + 48.29

Subtract.

31. 4.9 − 3.2

32. 5.23 − 2.74

33. −892.1 − 432.4

34. 0.064 − 10.2

35. 100 − 34.98

36. 200 − 0.00198

Solve.

37. Find the total distance between Grove City and Jerome.

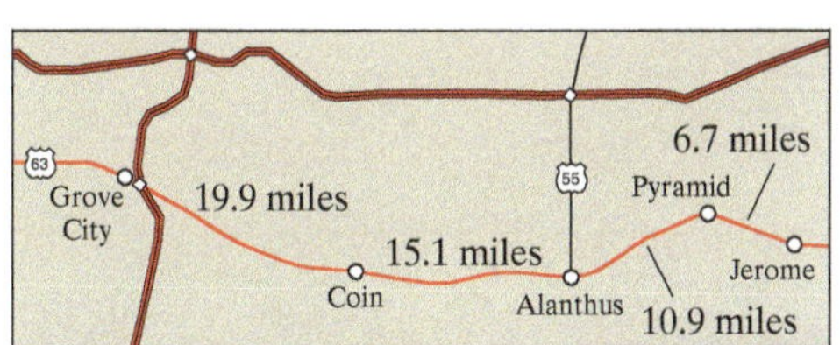

38. Evaluate $x - y$ for $x = 1.2$ and $y = 6.9$.

△ **39.** Find the perimeter.

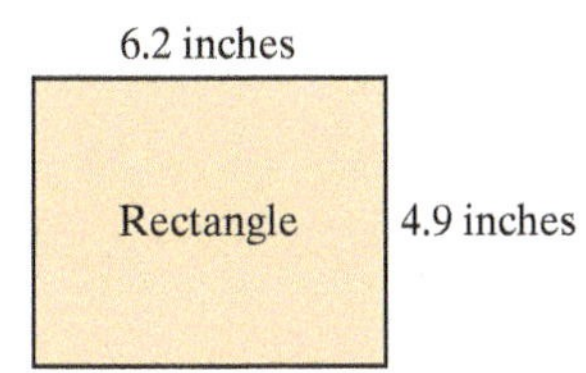

△ **40.** Find the perimeter.

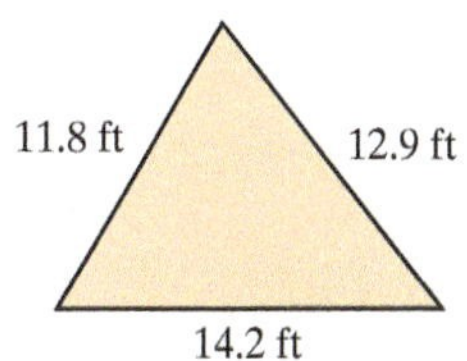

(5.3) *Multiply.*

41. 7.2×10

42. 9.345×1000

43. -34.02×2.3

44. $-839.02 \times (-87.3)$

Find the exact circumference of each circle. Then use the approximation 3.14 for π and approximate the circumference.

△ **45.**

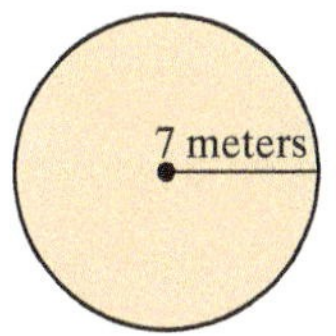

△ **46.**

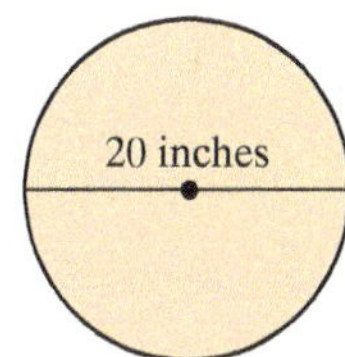

(5.4) *Divide. Round the quotient to the nearest thousandth if necessary.*

47. $3\overline{)0.2631}$

48. $20\overline{)316.5}$

49. $-21 \div (-0.3)$

50. $-0.0063 \div 0.03$

51. $0.34\overline{)2.74}$

52. $19.8\overline{)601.92}$

53. $\frac{23.65}{1000}$

54. $\frac{93}{-10}$

55. There are approximately 3.28 feet in 1 meter. Find how many meters are in 24 feet to the nearest tenth of a meter.

←1 meter→

←≈3.28 feet→

56. George Strait pays \$69.71 per month to pay back a loan of \$3136.95. In how many months will the loan be paid off?

(5.5) *Write each fraction or mixed number as a decimal. Round to the nearest thousandth if necessary.*

57. $\frac{4}{5}$

58. $-\frac{12}{13}$

59. $2\frac{1}{3}$

60. $\frac{13}{60}$

Insert <, >, or = to make a true statement.

61. 0.392 0.39200 **62.** −0.0231 −0.0221 **63.** $\frac{4}{7}$ 0.625 **64.** 0.293 $\frac{5}{17}$

Write the numbers in order from smallest to largest.

65. 0.837, 0.839, 0.832 **66.** 0.685, 0.626, $\frac{5}{8}$ **67.** $\frac{3}{7}$, 0.42, 0.43 **68.** $\frac{18}{11}$, 1.63, $\frac{19}{12}$

Simplify each expression.

69. $-7.6 \times 1.9 + 2.5$ **70.** $(-2.3)^2 - 1.4$ **71.** $0.0726 \div 10 \times 1000$

72. $0.9(6.5 - 5.6)$ **73.** $\frac{(1.5)^2 + 0.5}{0.05}$ **74.** $\frac{7 + 0.74}{-0.06}$

Find each area.

△ **75.**

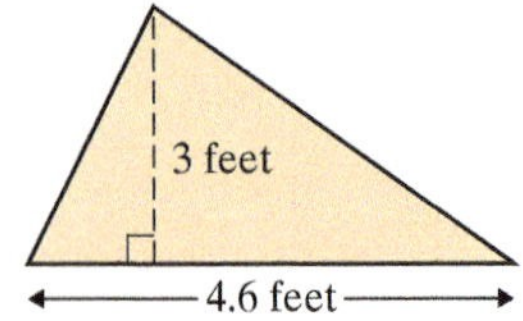

△ **76.**

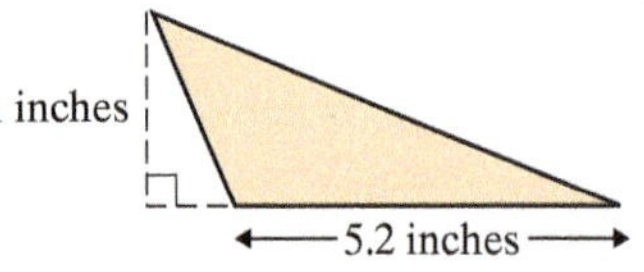

(5.6) *Solve.*

77. $x + 3.9 = 4.2$ **78.** $70 = y - 22.81$ **79.** $2x = 17.2$

80. $-1.1y = 88$ **81.** $3x - 0.78 = 1.2 + 2x$ **82.** $-x + 0.6 - 2x = -4x - 0.9$

83. $-1.3x - 9.4 = -0.4x + 8.6$ **84.** $3(x - 1.1) = 5x - 5.3$

(5.7) *Find the mean, median, and any mode(s) for each list of numbers. If necessary, round to the nearest tenth.*

85. 13, 23, 33, 14, 6 **86.** 45, 86, 21, 60, 86, 64, 45

87. 14,000, 20,000, 12,000, 20,000, 36,000, 45,000 **88.** 560, 620, 123, 400, 410, 300, 400, 780, 430, 450

For Exercises 89 and 90, the grades are given for a student for a particular semester. Find each grade point average. If necessary, round the grade point average to the nearest hundredth.

89.

Grade	Credit Hours
A	3
A	3
C	2
B	3
C	1

90.

Grade	Credit Hours
B	3
B	4
C	2
D	2
B	3

Mixed Review

91. Write 200.0032 in words.

92. Write negative sixteen and nine hundredths in standard form.

93. Write 0.0847 as a fraction or a mixed number.

94. Write the numbers $\frac{6}{7}, \frac{8}{9}$, 0.75 in order from smallest to largest.

Write each fraction as a decimal. Round to the nearest thousandth, if necessary.

95. $-\frac{7}{100}$

96. $\frac{9}{80}$ (Do not round.)

97. $\frac{8935}{175}$

Insert <, >, or = to make a true statement.

98. $-402.000032 \quad -402.00032$

99. $\frac{6}{11} \quad 0.55$

Round each decimal to the given place value.

100. 86.905, nearest hundredth

101. 3.11526, nearest thousandth

Round each money amount to the nearest dollar.

102. $123.46

103. $3645.52

Add or subtract as indicated.

104. $3.2 - 4.9$

105. $9.12 - 3.86$

106. $-102.06 + 89.3$

107. $-4.021 + (-10.83) + (-0.056)$

Multiply or divide as indicated. Round to the nearest thousandth, if necessary.

108. $\begin{array}{r} 2.54 \\ \times\ 3.2 \\ \hline \end{array}$

109. $(-3.45)(2.1)$

110. $0.005\overline{)24.5}$

111. $2.3\overline{)54.98}$

Solve.

△ **112.** Tomaso is going to fertilize his lawn, a rectangle that measures 77.3 feet by 115.9 feet. Approximate the area of the lawn by rounding each measurement to the nearest ten feet.

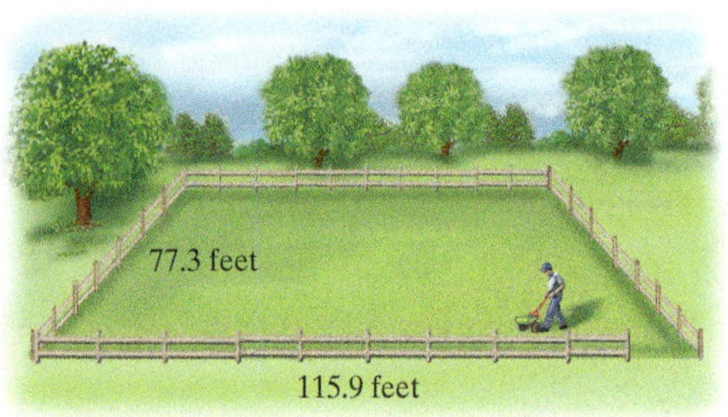

113. Estimate the cost of the items to see whether the groceries can be purchased with a $10 bill.

Simplify each expression.

114. $\dfrac{(3.2)^2}{100}$

115. $(2.6 + 1.4)(4.5 - 3.6)$

Find the mean, median, and any mode(s) for each list of numbers. If needed, round answers to the nearest hundredth.

116. 73, 82, 95, 68, 54

117. 952, 327, 566, 814, 327, 729

Step-by-step test solutions are found on the Chapter Test Prep Videos. Where available: MyMathLab® or YouTube

Chapter 5 Test

Answers

Write each decimal as indicated.

1. 45.092, in words

2. Three thousand and fifty-nine thousandths, in standard form

Perform each indicated operation. Round the result to the nearest thousandth if necessary.

3. $2.893 + 4.21 + 10.492$

4. $-47.92 - 3.28$

5. $9.83 - 30.25$

6. 10.2×4.01

7. $-0.00843 \div (-0.23)$

Round each decimal to the indicated place value.

8. 34.8923, nearest tenth

9. 0.8623, nearest thousandth

Insert $<$, $>$, or $=$ between each pair of numbers to form a true statement.

10. 25.0909 25.9090

11. $\frac{4}{9}$ 0.445

Write each decimal as a fraction or a mixed number.

12. 0.345

13. -24.73

Write each fraction as a decimal. If necessary, round to the nearest thousandth.

14. $-\frac{13}{26}$

15. $\frac{16}{17}$

Simplify.

16. $(-0.6)^2 + 1.57$

17. $\frac{0.23 + 1.63}{-0.3}$

18. $2.4x - 3.6 - 1.9x - 9.8$

1. ______
2. ______
3. ______
4. ______
5. ______
6. ______
7. ______
8. ______
9. ______
10. ______
11. ______
12. ______
13. ______
14. ______
15. ______
16. ______
17. ______
18. ______

19. ______

20. ______

21. ______

22. ______

23. ______

24. ______

25. ______

26. ______

27. a. ______

b. ______

28. ______

Solve.

19. $0.2x + 1.3 = 0.7$

20. $2(x + 5.7) = 6x - 3.4$

Find the mean, median, and mode of each list of numbers.

21. 26, 32, 42, 43, 49

22. 8, 10, 16, 16, 14, 12, 12, 13

Find the grade point average. If necessary, round to the nearest hundredth.

23.

Grade	Credit Hours
A	3
B	3
C	3
B	4
A	1

Solve.

24. At its farthest, Pluto is 4583 million miles from the Sun. Write this number using standard notation.

△ **25.** Find the area.

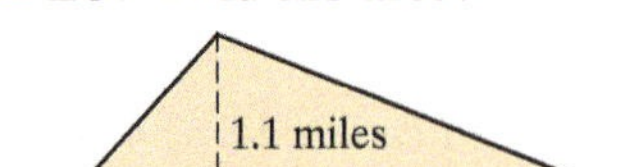

△ **26.** Find the exact circumference of the circle. Then use the approximation 3.14 for π and approximate the circumference.

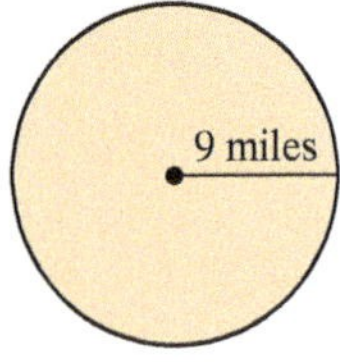

27. Vivian Thomas is going to put insecticide on her lawn to control grubworms. The lawn is a rectangle that measures 123.8 feet by 80 feet. The amount of insecticide required is 0.02 ounce per square foot.

a. Find the area of her lawn.

b. Find how much insecticide Vivian needs to purchase.

28. Find the total distance from Bayette to Center City.

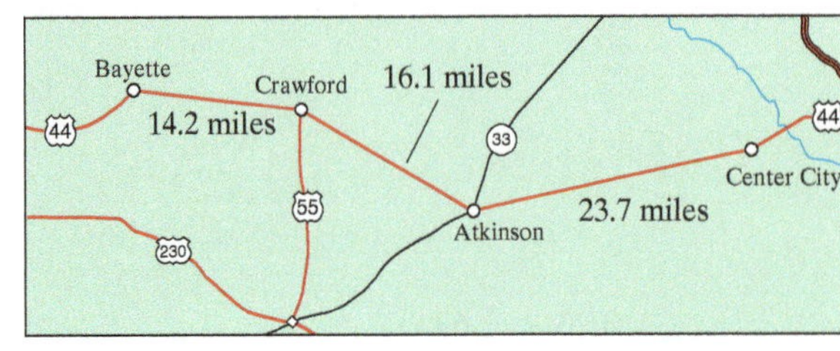

Cumulative Review Chapters 1–5

Write each number in words.

1. 72

2. 107

3. 546

4. 5026

5. Add: $46 + 713$

6. Find the perimeter.

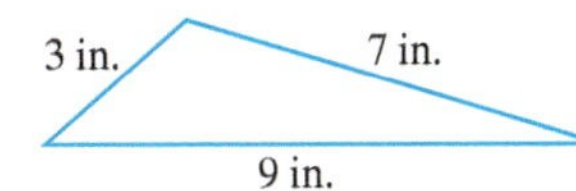

7. Subtract: $543 - 29$. Then check by adding.

8. Divide: $3268 \div 27$

9. Round 278,362 to the nearest thousand.

10. Write the prime factorization of 30.

11. Multiply: 236×86

12. Multiply: $236 \times 86 \times 0$

13. Find each quotient and then check the answer by multiplying.

a. $1\overline{)7}$

b. $12 \div 1$

c. $\frac{6}{6}$

d. $9 \div 9$

e. $\frac{20}{1}$

f. $18\overline{)18}$

14. Find the average of 25, 17, 19, and 39.

15. Simplify: $2 \cdot 4 - 3 \div 3$

16. Simplify: $77 \div 11 \cdot 7$

Evaluate.

17. 9^2

18. 5^3

Answers

1. ______
2. ______
3. ______
4. ______
5. ______
6. ______
7. ______
8. ______
9. ______
10. ______
11. ______
12. ______
13. a. ______
b. ______
c. ______
d. ______
e. ______
f. ______
14. ______
15. ______
16. ______
17. ______
18. ______

19. ______
20. ______
21. ______
22. ______
23. a. ______
b. ______
c. ______
24. a. ______
b. ______
c. ______
25. ______
26. ______
27. ______
28. ______
29. ______
30. ______
31. ______
32. ______
33. ______
34. ______
35. ______
36. ______
37. ______
38. ______

19. 3^4

20. 10^3

21. Evaluate $\dfrac{x - 5y}{y}$ for $x = 35$ and $y = 5$.

22. Evaluate $\dfrac{2a + 4}{c}$ for $a = 7$ and $c = 3$.

23. Find the opposite of each number.

a. 13 **b.** -2 **c.** 0

24. Find the opposite of each number.

a. -7 **b.** 4 **c.** -1

25. Add: $-2 + (-21)$

26. Add: $-7 + (-15)$

Find the value of each expression.

27. $5 \cdot 6^2$

28. $4 \cdot 2^3$

29. -7^2

30. $(-2)^5$

31. $(-5)^2$

32. -3^2

Represent the shaded part as an improper fraction and a mixed number.

33. 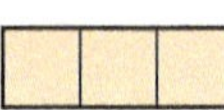

34. 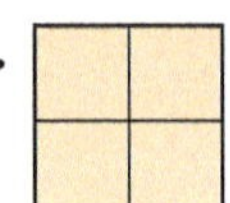 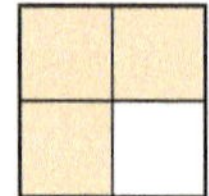

35. 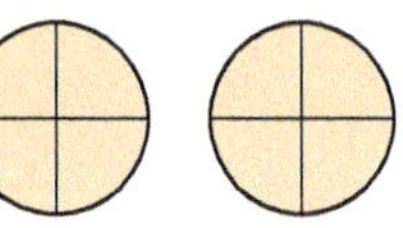

36. 

37. Write the prime factorization of 252.

38. Find the difference of 87 and 25.

39. Write $-\frac{72}{26}$ in simplest form.

40. Write $9\frac{7}{8}$ as an improper fraction.

41. Determine whether $\frac{16}{40}$ and $\frac{10}{25}$ are equivalent.

42. Insert < or > to form a true statement. $\frac{4}{7} \quad \frac{5}{9}$

Multiply.

43. $\frac{2}{3} \cdot \frac{5}{11}$

44. $2\frac{5}{8} \cdot \frac{4}{7}$

45. $\frac{1}{4} \cdot \frac{1}{2}$

46. $7 \cdot 5\frac{2}{7}$

Solve.

47. $\frac{z}{-4} = 11 - 5$

48. $6x - 12 - 5x = -20$

49. Add: $763.7651 + 22.001 + 43.89$

50. Add: $89.27 + 14.361 + 127.2318$

51. Multiply: 23.6×0.78

52. Multiply: 43.8×0.645

39. ____________

40. ____________

41. ____________

42. ____________

43. ____________

44. ____________

45. ____________

46. ____________

47. ____________

48. ____________

49. ____________

50. ____________

51. ____________

52. ____________

6 Ratio, Proportion, and Percent

Sections

Check Your Progress

This chapter is devoted to ratio, proportion, and percent, concepts used virtually every day in ordinary and business life. Understanding percent and using it efficiently depend on understanding ratios, because a percent is a ratio whose denominator is 100. We present techniques to write percents as fractions and as decimals. We then solve percent problems by using equations and also with proportions. A proportion is an equation with two equal ratios. The rest of this chapter is devoted to select applications of percent—including interest rates, sales tax, discounts, and other real-life situations.

Is it "disk" or "disc"? Actually, either spelling can be used unless there is a trademark involved. Disk (or disc) storage is, then, a general category that includes various storage mechanisms where data are recorded and stored by various methods to a surface layer of a rotating disk(s). Disk storage has certainly changed or evolved over the years, and just a few examples are shown here.

In Section 6.1, Exercises 19–22, we find the ratios of diameters of certain disk storage mechanisms.

6.1 Ratio and Proportion 

Objective A Writing Ratios as Fractions

A **ratio** is the quotient of two numbers or two quantities. A ratio, in fact, is no different from a fraction, except that a ratio is sometimes written using notation other than fractional notation. For example, the ratio of 1 to 2 can be written as

$$1 \text{ to } 2 \quad \text{or} \quad \frac{1}{2} \quad \text{or} \quad 1:2$$

fractional notation ($\frac{1}{2}$) colon notation ($1:2$)

These ratios are all read as, "the ratio of 1 to 2."

Ratio

The ratio of a number a to a number b is their quotient. Ways of writing ratios are

$$a \text{ to } b, \qquad a:b, \qquad \text{and} \qquad \frac{a}{b}$$

Whenever possible, we will convert quantities in a ratio to the same unit of measurement.

Example 1 Write a ratio for each phrase. Use fractional notation.

a. The ratio of 2 parts salt to 5 parts water

b. The ratio of 18 inches to 2 feet

Solution:

a. The ratio of 2 parts salt to 5 parts water is $\frac{2}{5}$.

b. First we convert to the same unit of measurement. For example,

$$2 \text{ feet} = 2 \cdot 12 \text{ inches} = 24 \text{ inches}$$

The ratio of 18 inches to 2 feet is then $\frac{18}{24}$, or $\frac{3}{4}$ in lowest terms.

Work Practice 1

If a ratio compares two decimal numbers, we will write the simplified ratio as a ratio of whole numbers.

Example 2 Write the ratio of 2.5 to 3.15 as a fraction in simplest form.

Solution: The ratio is

$$\frac{2.5}{3.15}$$

Now let's clear the ratio of decimals.

$$\frac{2.5}{3.15} = \frac{2.5 \cdot 100}{3.15 \cdot 100} = \frac{250}{315} = \frac{50}{63} \quad \text{Simplest form}$$

Work Practice 2

Objectives

A Write Ratios as Fractions.

B Solve Proportions.

C Solve Problems Modeled by Proportions.

Practice 1

Write a ratio for each phrase. Use fractional notation.

a. The ratio of 3 parts oil to 7 parts gasoline

b. The ratio of 40 minutes to 3 hours

Practice 2

Write the ratio of 1.68 to 4.8 as a fraction in simplest form.

Answers

1. a. $\frac{3}{7}$ **b.** $\frac{2}{9}$ **2.** $\frac{7}{20}$

Practice 3

Given the triangle shown:

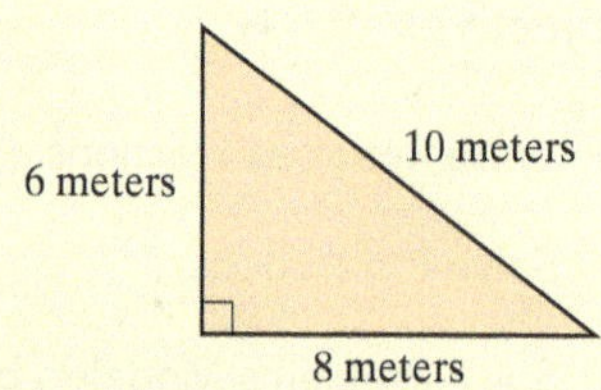

a. Find the ratio of the length of the shortest side to the length of the longest side in simplest form.

b. Find the ratio of the length of the longest side to the perimeter of the triangle in simplest form.

Example 3

Given the rectangle shown:

a. Find the ratio of its width (shorter side) to its length (longer side).

b. Find the ratio of its length to its perimeter.

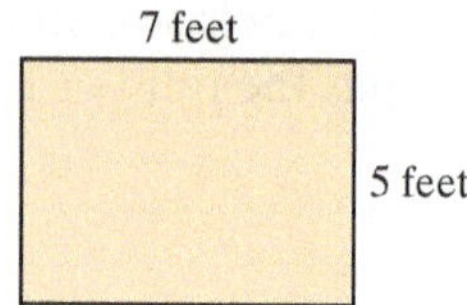

Solution:

a. The ratio of its width to its length is

$$\frac{\text{width}}{\text{length}} = \frac{5 \text{ feet}}{7 \text{ feet}} = \frac{5}{7}$$

b. Recall that the perimeter of a rectangle is the distance around the rectangle: $7 + 5 + 7 + 5 = 24$ feet. The ratio of its length to its perimeter is

$$\frac{\text{length}}{\text{perimeter}} = \frac{7 \text{ feet}}{24 \text{ feet}} = \frac{7}{24}$$

Work Practice 3

✓Concept Check Explain why the answer $\frac{7}{5}$ would be incorrect for part (a) of Example 3.

Objective B Solving Proportions

Ratios can be used to form proportions. A **proportion** is a mathematical statement that two ratios are equal.

For example, the equation

$$\frac{1}{2} = \frac{4}{8}$$

is a proportion that says that the ratios $\frac{1}{2}$ and $\frac{4}{8}$ are equal.

Notice that a proportion contains four numbers. If any three numbers are known, we can solve and find the fourth number. One way to do so is to use cross products. To understand cross products, which were introduced in Section 4.2, let's start with the proportion

$$\frac{a}{b} = \frac{c}{d}$$

and multiply both sides by the LCD, bd.

$$\frac{a}{b} = \frac{c}{d}$$

$$bd\left(\frac{a}{b}\right) = bd\left(\frac{c}{d}\right) \quad \text{Multiply both sides by the LCD, } bd.$$

$$ad = bc \quad \text{Simplify.}$$

(ad: cross product; bc: cross product)

Answers

3. a. $\frac{3}{5}$ **b.** $\frac{5}{12}$

✓Concept Check Answer

$\frac{7}{5}$ is the ratio of the rectangle's length to its width.

Notice why ad and bc are called cross products.

$$ad \qquad \frac{a}{b} = \frac{c}{d} \qquad bc$$

Cross Products

If $\frac{a}{b} = \frac{c}{d}$, then $ad = bc$.

Example 4 Solve for x: $\frac{45}{x} = \frac{5}{7}$

Solution: To solve, we set cross products equal.

$$\frac{45}{x} = \frac{5}{7}$$

$45 \cdot 7 = x \cdot 5$ Set cross products equal.

$315 = 5x$ Multiply.

$\frac{315}{5} = \frac{5x}{5}$ Divide both sides by 5.

$63 = x$ Simplify.

Check: To check, substitute 63 for x in the original proportion. The solution is 63.

Work Practice 4

Practice 4

Solve for x: $\frac{3}{8} = \frac{63}{x}$

Example 5 Solve for x: $\frac{x - 5}{3} = \frac{x + 2}{5}$

Solution:

$$\frac{x - 5}{3} = \frac{x + 2}{5}$$

$5(x - 5) = 3(x + 2)$ Set cross products equal.

$5x - 25 = 3x + 6$ Multiply.

$5x = 3x + 31$ Add 25 to both sides.

$2x = 31$ Subtract $3x$ from both sides.

$\frac{2x}{2} = \frac{31}{2}$ Divide both sides by 2.

$x = \frac{31}{2}$

Check: Verify that $\frac{31}{2}$ is the solution.

Work Practice 5

Practice 5

Solve for x: $\frac{2x + 1}{7} = \frac{x - 3}{5}$

✓**Concept Check** For which of the following equations can we immediately use cross products to solve for x?

a. $\frac{2 - x}{5} = \frac{1 + x}{3}$ **b.** $\frac{2}{5} - x = \frac{1 + x}{3}$

Answers

4. $x = 168$ **5.** $x = -\frac{26}{3}$

✓**Concept Check Answer**

a

Objective C Solving Problems Modeled by Proportions

Writing proportions is a powerful tool for solving problems in almost every field, including business, chemistry, biology, health sciences, and engineering, as well as in daily life. Given a specified ratio (or rate) of two quantities, a proportion can be used to determine an unknown quantity.

Practice 6

On an architect's blueprint, 1 inch corresponds to 4 feet. How long is a wall represented by a $4\frac{1}{4}$-inch line on the blueprint?

Example 6 Determining Distances from a Map

On a Chamber of Commerce map of Abita Springs, 5 miles corresponds to 2 inches. How many miles correspond to 7 inches?

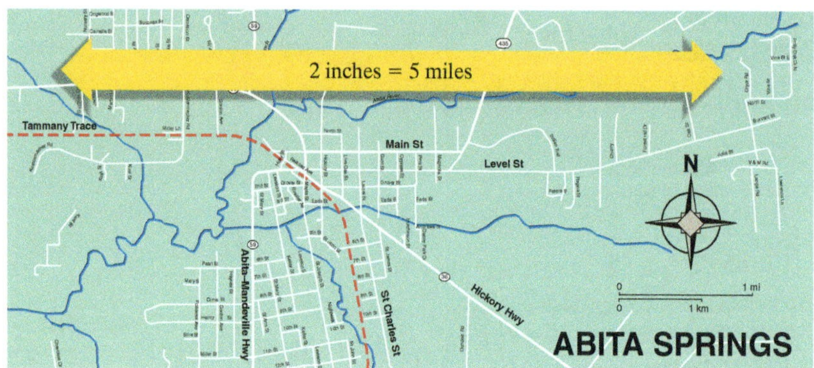

Solution:

1. UNDERSTAND. Read and reread the problem. You may want to draw a diagram.

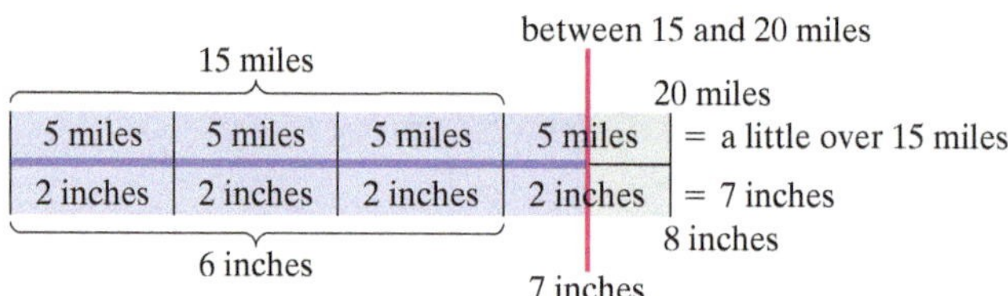

From the diagram we can see that a reasonable solution should be between 15 and 20 miles.

2. TRANSLATE. We will let x be our unknown number. Since 5 miles corresponds to 2 inches as x miles corresponds to 7 inches, we have the proportion

$$\begin{matrix} \text{miles} \rightarrow \\ \text{inches} \rightarrow \end{matrix} \quad \frac{5}{2} = \frac{x}{7} \quad \begin{matrix} \leftarrow \text{miles} \\ \leftarrow \text{inches} \end{matrix}$$

3. SOLVE: In earlier sections, we estimated to obtain a reasonable answer. Notice we did this in Step 1 above.

$$\frac{5}{2} = \frac{x}{7}$$

$5 \cdot 7 = 2 \cdot x$ Set the cross products equal to each other.

$35 = 2x$ Multiply.

$\frac{35}{2} = \frac{2x}{2}$ Divide both sides by 2.

$17\frac{1}{2} = x$ or $x = 17.5$ Simplify.

4. INTERPRET. *Check* your work. This result is reasonable since it is between 15 and 20 miles. *State* your conclusion: 7 inches corresponds to 17.5 miles.

Work Practice 6

Answer

6. 17 ft

Helpful Hint

We can also solve Example 6 by writing the proportion

$$\frac{2 \text{ inches}}{5 \text{ miles}} = \frac{7 \text{ inches}}{x \text{ miles}}$$

Although other proportions may be used to solve Example 6, we will solve by writing proportions so that the numerators have the same unit measures and the denominators have the same unit measures.

Example 7 Finding Medicine Dosage

The standard dose of an antibiotic is 4 cc (cubic centimeters) for every 25 pounds (lb) of body weight. At this rate, find the standard dose for a 140-lb woman.

Solution:

1. UNDERSTAND. Read and reread the problem. You may want to draw a diagram to estimate a reasonable solution.

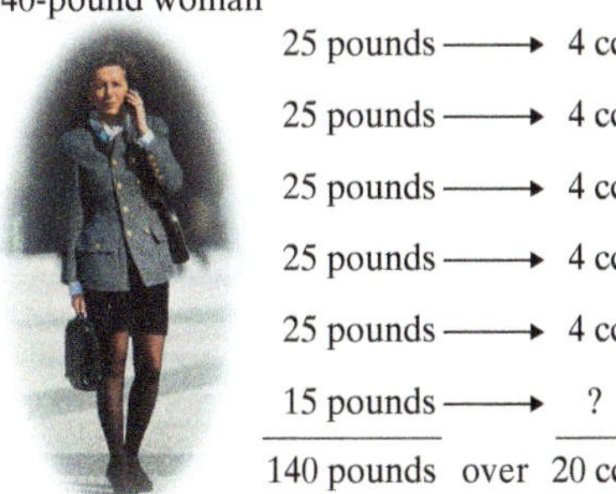

From the diagram, we can see that a reasonable solution is a little over 20 cc.

2. TRANSLATE. We will let x be the unknown number. From the problem, we know that 4 cc is to 25 pounds as x cc is to 140 pounds, or

$$\begin{array}{rcccl} \text{cubic centimeters} & \rightarrow & \frac{4}{25} = \frac{x}{140} & \leftarrow & \text{cubic centimeters} \\ \text{pounds} & \rightarrow & & \leftarrow & \text{pounds} \end{array}$$

3. SOLVE:

$$\frac{4}{25} = \frac{x}{140}$$

$4 \cdot 140 = 25 \cdot x$ Set the cross products equal to each other.

$560 = 25x$ Multiply.

$\frac{560}{25} = \frac{25x}{25}$ Divide both sides by 25.

$22\frac{2}{5} = x$ or $x = 22.4$ Simplify.

4. INTERPRET. *Check* your work. This result is reasonable since it is a little over 20 cc. *State* your conclusion: The standard dose for a 140-lb woman is 22.4 cc.

Work Practice 7

Practice 7

An auto mechanic recommends that 5 ounces of isopropyl alcohol be mixed with a tankful of gas (16 gallons) to increase the octane of the gasoline for better engine performance. At this rate, how many gallons of gas can be treated with an 8-ounce bottle of alcohol?

Answer

7. $25\frac{3}{5}$ or 25.6 gal

Vocabulary, Readiness & Video Check

Use the words and phrases below to fill in each blank.

ratio   cross products   proportion

1. $\frac{4.2}{8.4} = \frac{1}{2}$ is called a ____________ while the fraction $\frac{7}{8}$ may also be called a ____________.

2. In $\frac{a}{b} = \frac{c}{d}$, $a \cdot d$ and $b \cdot c$ are called ____________.

Martin-Gay Interactive Videos

See Video 6.1

Watch the section lecture video and answer the following questions.

Objective A **3.** In Example 2, why are dollars converted to nickels?

Objective B **4.** Based on Examples 3 and 4, can proportions only be solved by using cross products? Explain.

Objective C **5.** In Example 5 we are told there are many ways to set up a correct proportion. Why does this fact make it even more important to check that your solution is reasonable?

6.1 Exercise Set MyMathLab®

Objective A *Write each ratio in fractional notation in lowest terms. See Examples 1 through 3.*

1. 2 megabytes to 15 megabytes

2. 18 disks to 41 disks

3. 10 inches to 12 inches

4. 15 miles to 40 miles

5. 4 nickels to 2 dollars

6. 12 quarters to 2 dollars

7. 190 minutes to 3 hours

8. 60 hours to 2 days

9. 7.7 to 10

10. 8.1 to 10

11. 3.5 to 12.25

12. 1.5 to 9.75

△ **13.** Find the ratio of the length to the width of a regulation size basketball court.

△ **14.** Find the ratio of the base to the height of the triangular mainsail.

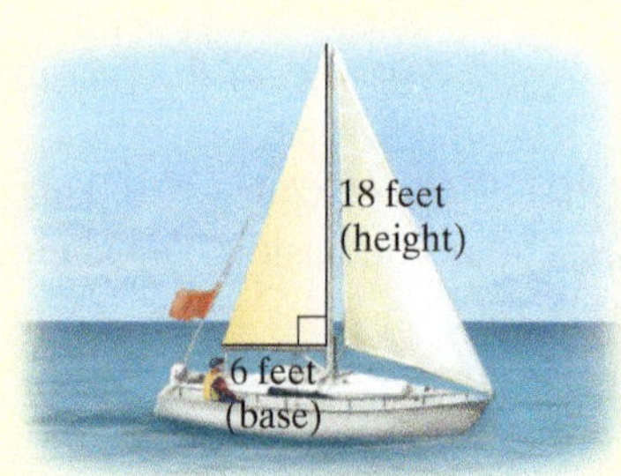

△ **15.** Find the ratio of the longest side to the perimeter of the right-triangular-shaped billboard.

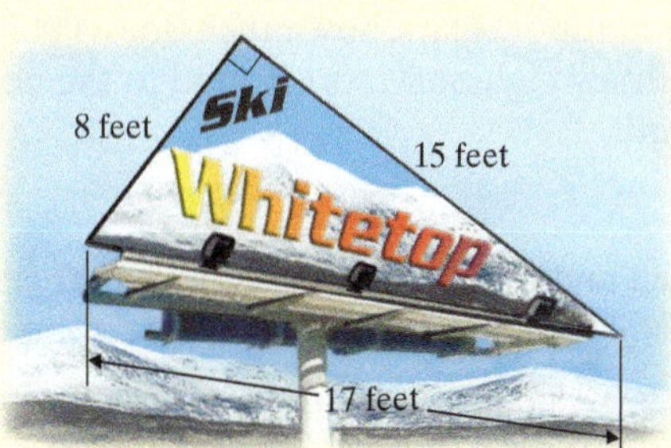

△ **16.** Find the ratio of the width to the perimeter of the rectangular vegetable garden.

Blood contains three types of cells: red blood cells, white blood cells, and platelets. For approximately every 600 red blood cells in healthy humans, there are 40 platelets and 1 white blood cell. Use this for Exercises 17 and 18. (Source: American Red Cross Biomedical Services)

17. Write the ratio of red blood cells to platelet cells.

18. Write the ratio of white blood cells to red blood cells.

Exercises 19–22 have to do with disk storage. See the Chapter Opener for more information on disk storage.

19. A standard DVD has a diameter of 12 centimeters while a miniDVD has a diameter of 8 centimeters. Write the ratio of the miniDVD diameter to the standard DVD diameter.

20. LaserDiscs have a diameter of 12 centimeters and 20 centimeters. Write the ratio of the smaller diameter to the larger diameter.

21. Hard drives come in many diameters. Find the ratio of a $2\frac{1}{2}''$ diameter to a $5\frac{1}{4}''$ diameter hard drive. (Note: If you choose to work with mixed numbers, find $2\frac{1}{2} \div 5\frac{1}{4}$, or write each mixed number as an equivalent decimal number.)

22. Floppy disks come in many diameters. Find the ratio of a $3\frac{1}{2}''$ diameter to a $5\frac{1}{4}''$ diameter floppy disk. (See the note for the previous exercise.)

Objective B *Solve each proportion. See Examples 4 and 5.*

23. $\frac{2}{3} = \frac{x}{6}$

24. $\frac{x}{2} = \frac{16}{6}$

25. $\frac{x}{10} = \frac{5}{9}$

26. $\frac{9}{4x} = \frac{6}{2}$

27. $\frac{4x}{6} = \frac{7}{2}$

28. $\frac{a}{5} = \frac{3}{2}$

29. $\frac{x-3}{x} = \frac{4}{7}$

30. $\frac{y}{y-16} = \frac{5}{3}$

31. $\frac{x+1}{2x+3} = \frac{2}{3}$

32. $\frac{x+1}{x+2} = \frac{5}{3}$

33. $\frac{9}{5} = \frac{12}{3x+2}$

34. $\frac{6}{11} = \frac{27}{3x-2}$

35. $\frac{3}{x+1} = \frac{5}{2x}$

36. $\frac{7}{x-3} = \frac{8}{2x}$

37. $\frac{16}{3x-4} = \frac{5}{x}$

38. $\frac{x}{3} = \frac{2x-5}{7}$

Objective C *Solve. See Examples 6 and 7.*

39. The ratio of the weight of an object on Earth to the weight of the same object on Pluto is 100 to 3. If an elephant weighs 4100 pounds on Earth, find the elephant's weight on Pluto.

40. If a 170-pound person weighs approximately 65 pounds on Mars, how much does a 9000-pound satellite weigh on Mars? Round to the nearest pound.

41. There are 110 calories per 28.4 grams of Crispy Rice cereal. Find how many calories are in 42.6 grams of this cereal.

42. On an architect's blueprint, 1 inch corresponds to 4 feet. Find the length of a wall represented by a line that is $3\frac{7}{8}$ inches long on the blueprint.

43. A 16-oz grande Shaken Sweet Tea at Starbucks has 100 calories. How many calories are there in a 24-oz venti Shaken Sweet Tea? (*Source:* Starbucks Coffee Company)

44. A 16-oz nonfat Skinny Mocha at Starbucks has 140 calories. How many calories are there in a 12-oz nonfat Skinny Mocha? (*Source:* Starbucks Coffee Company)

45. Mosquitos are annoying insects. To eliminate mosquito larvae, a certain granular substance can be applied to standing water in a ratio of 1 tsp per 25 sq ft of standing water.

a. At this rate, find how many teaspoons of granules must be used for 450 square feet.

b. If 3 tsp = 1 tbsp, how many tablespoons of granules must be used?

46. Another type of mosquito control is liquid, where 3 oz of pesticide is mixed with 100 oz of water. This mixture is sprayed on roadsides to control mosquito breeding grounds hidden by tall grass.

a. If one mixture of water with this pesticide can treat 150 feet of roadway, how many ounces of pesticide are needed to treat one mile? (*Hint:* 1 mile = 5280 feet)

b. If 8 liquid ounces equals one cup, write your answer to part **a** in cups. Round to the nearest cup.

47. The daily supply of oxygen for one person is provided by 625 square feet of lawn. A total of 3750 square feet of lawn would provide the daily supply of oxygen for how many people? (*Source:* Professional Lawn Care Association of America)

48. In 2013, approximately 96 million of the 136 million U.S. employees worked in service industries. In a town of 6800 workers, how many would be expected to work in service-industry jobs? (*Source:* U.S. Bureau of Labor Statistics)

49. A student would like to estimate the height of the Statue of Liberty in New York City's harbor. The length of the Statue of Liberty's right arm is 42 feet. The student's right arm is 2 feet long and her height is $5\frac{1}{3}$ feet. Use this information to estimate the height of the Statue of Liberty. How close is your estimate to the statue's actual height of 111 feet, 1 inch from heel to top of head? (*Source:* National Park Service)

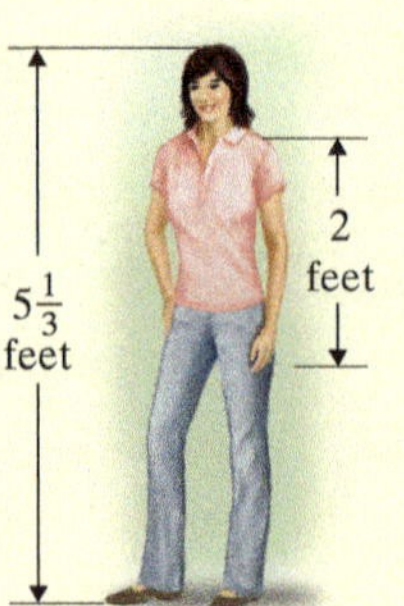

50. The length of the Statue of Liberty's index finger is 8 feet while the height to the top of the head is about 111 feet. Suppose your measurements are proportionaly the same as this statue and that your height is 5 feet.

a. Use this information to find the proposed length of your index finger. Give an exact measurement and then a decimal rounded to the nearest hundredth.

b. Measure your index finger and write it as decimal in feet rounded to the nearest hundredth. How close is the length of your index finger to the answer to **a**? Explain why.

51. There are 72 milligrams of cholesterol in a 3.5-ounce serving of lobster. How much cholesterol is in 5 ounces of lobster? Round to the nearest tenth of a milligram. (*Source:* The National Institutes of Health)

52. There are 76 milligrams of cholesterol in a 3-ounce serving of skinless chicken. How much cholesterol is in 8 ounces of chicken? (*Source:* USDA)

53. Trump World Tower in New York City is 881 feet tall and contains 72 stories. The Empire State Building contains 102 stories. If the Empire State Building has the same number of feet per floor as the Trump World Tower, approximate its height rounded to the nearest foot. (*Source:* Skyscrapers.com)

54. In 2012, 9 out of 10 American preschool-aged children had received a measles vaccine. In a preschool with 130 students, how many students would be expected to have received a measles vaccine? (*Source:* Centers for Disease Control and Prevention)

55. In 2012, nearly one out of every four autos sold in the United States was a crossover utility vehicle (CUV). If an auto dealership sold 4500 vehicles in total that year, how many of these sales would you expect involved a CUV? (*Source:* Alliance of Automobile Manufacturers)

56. In the United States in 2013, 1 in every 17 private-sector jobs depended on the auto industry. In a group of 595 private-sector U.S. workers, how many would you expect to have worked in a job that depended on the auto industry? (*Source:* Alliance of Automobile Manufacturers)

57. In 2012, two out of every three wireless phones sold in the United States were smartphones. Approximately 170 million wireless phones were sold in the United States in 2012. How many of them were smartphones? Round to the nearest million. (*Source:* Consumer Electronics Association)

58. One pound of firmly packed brown sugar yields $2\frac{1}{4}$ cups. How many pounds of brown sugar will be required in a recipe that calls for 6 cups of firmly packed brown sugar? (*Source:* Based on data from *Family Circle* magazine)

When making homemade ice cream in an electric or a hand-cranked freezer, the tub containing the ice cream mix is surrounded by a brine (water/salt) solution. To freeze the ice cream mix rapidly so that smooth and creamy ice cream results, the brine solution should combine crushed ice and rock salt in a ratio of 5 to 1. Use this for Exercises 59 and 60. (Source: White Mountain Freezers, The Rival Company)

59. A small ice cream freezer requires 12 cups of crushed ice. How much rock salt should be mixed with the ice to create the necessary brine solution?

60. A large ice cream freezer requires $18\frac{3}{4}$ cups of crushed ice. How much rock salt will be needed?

61. The gas/oil ratio for a certain chainsaw is 50 to 1.
   a. How much oil (in gallons) should be mixed with 5 gallons of gasoline?
   b. If 1 gallon equals 128 fluid ounces, write the answer to part **a** in fluid ounces. Round to the nearest whole ounce.

62. The gas/oil ratio for a certain tractor mower is 20 to 1.
   a. How much oil (in gallons) should be mixed with 10 gallons of gas?
   b. If 1 gallon equals 4 quarts, write the answer to part **a** in quarts.

63. The adult daily dosage for a certain medicine is 150 mg (milligrams) of medicine for every 20 pounds of body weight.
   a. At this rate, find the daily dose for a man who weighs 275 pounds.
   b. If the man is to receive 500 mg of this medicine every 8 hours, is he receiving the proper dosage?

64. The adult daily dosage for a certain medicine is 80 mg (milligrams) for every 25 pounds of body weight.
   a. At this rate, find the daily dose for a woman who weighs 190 pounds.
   b. If she is to receive this medicine every 6 hours, find the amount to be given every 6 hours.

Review

Find the prime factorization of each number. See Section 4.2.

65. 200

66. 300

67. 32

68. 81

Concept Extensions

69. Suppose someone tells you that the ratio of 11 inches to 2 feet is $\frac{11}{2}$. How would you correct that person and explain the error?

70. Write a real-life example of a ratio that can be written in fractional notation as $\frac{3}{2}$.

As we have seen earlier, proportions are often used in medicine dosage calculations. The exercises below have to do with liquid drug preparations, where the weight of the drug is contained in a volume of solution. The description of mg and ml below will help. We will study metric units further in Chapter 8.

mg means milligrams (A paper clip weighs about a gram. A milligram is about the weight of $\frac{1}{1000}$ *of a paper clip.)*

ml means milliliters (A liter is about a quart. A milliliter is about the amount of liquid in $\frac{1}{1000}$ *of a quart.)*

One way to solve the applications below is to set up the proportion $\frac{\text{mg}}{\text{ml}} = \frac{\text{mg}}{\text{ml}}$.

A solution strength of 15 mg of medicine in 1 ml of solution is available.

71. If a patient needs 12 mg of medicine, how many ml do you administer?

72. If a patient needs 33 mg of medicine, how many ml do you administer?

A solution strength of 8 mg of medicine in 1 ml of solution is available.

73. If a patient needs 10 mg of medicine, how many ml do you administer?

74. If a patient needs 6 mg of medicine, how many ml do you administer?

6.2 Percents, Decimals, and Fractions 

Objectives

A Understand Percent. 

B Write Percents as Decimals or Fractions.

C Write Decimals or Fractions as Percents.

D Applications with Percents, Decimals, and Fractions.

Objective A Understanding Percent

The word **percent** comes from the Latin phrase *per centum,* which means **"per 100."** For example, 53% (53 percent) means 53 per 100. In the square below, 53 of the 100 squares are shaded. Thus, 53% of the figure is shaded.

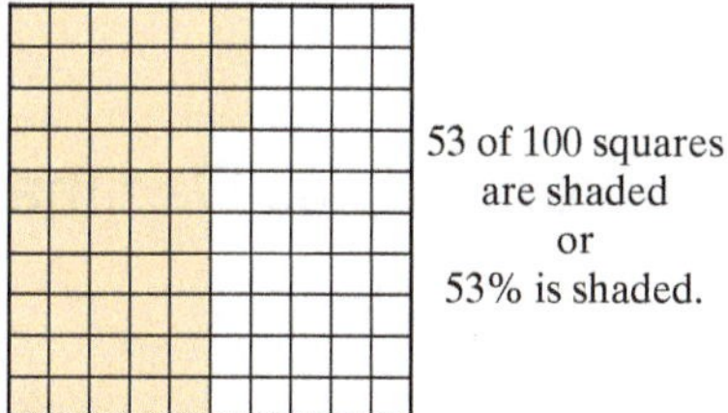

Since 53% means 53 per 100, 53% is the ratio of 53 to 100, or $\frac{53}{100}$.

$$53\% = \frac{53}{100}$$

Also,

$$7\% = \frac{7}{100} \quad \text{7 parts per 100 parts}$$

$$73\% = \frac{73}{100} \quad \text{73 parts per 100 parts}$$

$$109\% = \frac{109}{100} \quad \text{109 parts per 100 parts}$$

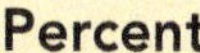

Percent

Percent means **per one hundred.** The "%" symbol is used to denote percent.

Percent is used in a variety of everyday situations. For example,

- 84.1% of the U.S. population uses the Internet.
- The store is having a 20%-off sale.
- For the past two years, the enrollment in community colleges increased 15%.
- The South is the home of 49% of all frequent paintball participants.
- 50% of the cinema screens in the United States are digital non-3D.

Example 1 In a survey of 100 people, 17 people drive blue cars. What percent of people drive blue cars?

Solution: Since 17 people out of 100 drive blue cars, the fraction is $\frac{17}{100}$. Then

$$\frac{17}{100} = 17\%$$

Work Practice 1

Practice 1

Of 100 students in a club, 27 are freshmen. What percent of the students are freshmen?

Answer

1. 27%

Practice 2

31 out of 100 executives are in their forties. What percent of executives are in their forties?

Example 2 45 out of every 100 young college graduates return home to live (at least temporarily). What percent of college graduates is this?  (*Source:* Independent Insurance Agents of America)

Solution:

$$\frac{45}{100} = 45\%$$

Work Practice 2

Objective B Writing Percents as Decimals or Fractions

Since percent means "per hundred," we have that

$$1\% = \frac{1}{100} = 0.01$$

In other words, the percent symbol means "per hundred" or, equivalently, "$\frac{1}{100}$" or "0.01." Thus

Write 87% as a fraction: $87\% = 87 \times \frac{1}{100} = \frac{87}{100}$

or

Write 87% as a decimal: $87\% = 87 \times (0.01) = 0.87$

Results are the same.

Of course, we know that the end results are the same; that is,

$$\frac{87}{100} = 0.87$$

The above gives us two options for converting percents. We can replace the percent symbol, %, by $\frac{1}{100}$ or by 0.01 and then multiply.

For consistency, when we
- convert from a percent to a *decimal,* we will drop the % symbol and multiply by 0.01.
- convert from a percent to a *fraction,* we will drop the % symbol and multiply by $\frac{1}{100}$.

Let's practice writing percents as decimals, then writing percents as fractions.

Writing a Percent as a Decimal

Replace the percent symbol with its decimal equivalent, 0.01; then multiply.

$$43\% = 43(0.01) = 0.43$$

Helpful Hint

If it helps, think of writing a percent as a decimal by

Percent → Remove the % symbol and move the decimal point 2 places to the left. → Decimal

Answer

2. 31%

Examples Write each percent as a decimal.

3. $23\% = 23(0.01) = 0.23$ Replace the percent symbol with 0.01. Then multiply.
4. $4.6\% = 4.6(0.01) = 0.046$ Replace the percent symbol with 0.01. Then multiply.
5. $190\% = 190(0.01) = 1.90$ or 1.9
6. $0.74\% = 0.74(0.01) = 0.0074$
7. $100\% = 100(0.01) = 1.00$ or 1

Helpful Hint We just learned that $100\% = 1$.

Work Practice 3–7

Practice 3–7

Write each percent as a decimal.

3. 49%
4. 3.1%
5. 175%
6. 0.46%
7. 600%

✓**Concept Check** Why is it incorrect to write the percent 0.033% as 3.3 in decimal form?

Now let's write percents as fractions.

Writing a Percent as a Fraction

Replace the percent symbol with its fraction equivalent, $\frac{1}{100}$; then multiply. Don't forget to simplify the fraction if possible.

$$43\% = 43 \cdot \frac{1}{100} = \frac{43}{100}$$

Examples Write each percent as a fraction or mixed number in simplest form.

8. $40\% = 40 \cdot \frac{1}{100} = \frac{40}{100} = \frac{2 \cdot \overset{1}{\cancel{20}}}{5 \cdot \underset{1}{\cancel{20}}} = \frac{2}{5}$

9. $1.9\% = 1.9 \cdot \frac{1}{100} = \frac{1.9}{100}$. We don't want the numerator of the fraction to contain a decimal, so we multiply by 1 in the form of $\frac{10}{10}$.

$$= \frac{1.9}{100} \cdot \frac{10}{10} = \frac{1.9 \cdot 10}{100 \cdot 10} = \frac{19}{1000}$$

10. $125\% = 125 \cdot \frac{1}{100} = \frac{125}{100} = \frac{5 \cdot \overset{1}{\cancel{25}}}{4 \cdot \underset{1}{\cancel{25}}} = \frac{5}{4}$ or $1\frac{1}{4}$

11. $33\frac{1}{3}\% = 33\frac{1}{3} \cdot \frac{1}{100} = \frac{100}{3} \cdot \frac{1}{100} = \frac{\overset{1}{\cancel{100}} \cdot 1}{3 \cdot \underset{1}{\cancel{100}}} = \frac{1}{3}$

Write as an improper fraction.

12. $100\% = 100 \cdot \frac{1}{100} = \frac{100}{100} = 1$

Helpful Hint Just as in Example 7, we confirm that $100\% = 1$.

Work Practice 8–12

Practice 8–12

Write each percent as a fraction or mixed number in simplest form.

8. 50%
9. 2.3%
10. 150%
11. $66\frac{2}{3}\%$
12. 12%

Answers

3. 0.49 4. 0.031 5. 1.75 6. 0.0046 7. 6 8. $\frac{1}{2}$ 9. $\frac{23}{1000}$ 10. $\frac{3}{2}$ or $1\frac{1}{2}$ 11. $\frac{2}{3}$ 12. $\frac{3}{25}$

✓**Concept Check Answers**

To write a percent as a decimal, the decimal point should be moved two places to the left, not to the right. So the correct answer is 0.00033.

Objective C Writing Decimals or Fractions as Percents

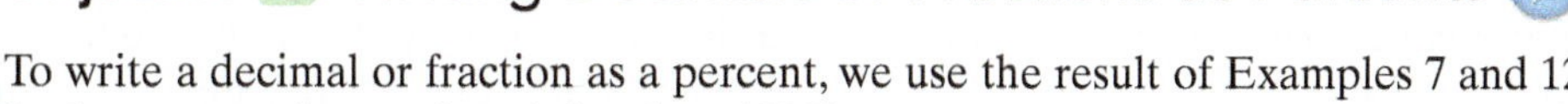

To write a decimal or fraction as a percent, we use the result of Examples 7 and 12. In these examples, we found that $1 = 100\%$.

Write 0.38 as a percent: $0.38 = 0.38(1) = 0.38(100\%) = 38.\%$

Write $\frac{1}{5}$ as a percent: $\frac{1}{5} = \frac{1}{5}(1) = \frac{1}{5} \cdot 100\% = \frac{100}{5}\% = 20\%$

First, let's practice writing decimals as percents.

Writing a Decimal as a Percent

Multiply by 1 in the form of 100%.

$0.27 = 0.27(100\%) = 27.\%$

Helpful Hint

If it helps, think of writing a decimal as a percent by reversing the steps in the Helpful Hint on page 422.

Percent ← | Move the decimal point 2 places to the right and attach a % symbol. | ← Decimal

Practice 13–16

Write each decimal as a percent.

13. 0.14 **14.** 1.75
15. 0.057 **16.** 0.5

Examples Write each decimal as a percent.

13. $0.65 = 0.65(100\%) = 65.\%$ or 65% Multiply by 100%.

14. $1.25 = 1.25(100\%) = 125.\%$ or 125%

15. $0.012 = 0.012(100\%) = 001.2\%$ or 1.2%

16. $0.6 = 0.6(100\%) = 060.\%$ or 60%

Helpful Hint A zero was inserted as a placeholder.

Work Practice 13–16

✓**Concept Check** Why is it incorrect to write the decimal 0.0345 as 34.5% in percent form?

Now let's write fractions as percents.

Writing a Fraction as a Percent

Multiply by 1 in the form of 100%.

$$\frac{1}{8} = \frac{1}{8} \cdot 100\% = \frac{1}{8} \cdot \frac{100}{1}\% = \frac{100}{8}\% = 12\frac{1}{2}\% \quad \text{or} \quad 12.5\%$$

Helpful Hint

From Examples 7 and 12, we know that

$100\% = 1$

Recall that when we multiply a number by 1, we are not changing the value of that number. This means that when we multiply a number by 100%, we are not changing its value but rather writing the number as an equivalent percent.

Answers

13. 14% **14.** 175% **15.** 5.7%
16. 50%

✓**Concept Check Answer**

To change a decimal to a percent, multiply by 100%, or move the decimal point *only* two places to the right. So the correct answer is 3.45%.

Examples Write each fraction or mixed number as a percent.

17. $\frac{7}{20} = \frac{7}{20} \cdot 100\% = \frac{7}{20} \cdot \frac{100}{1}\% = \frac{700}{20}\% = 35\%$

18. $\frac{2}{3} = \frac{2}{3} \cdot 100\% = \frac{2}{3} \cdot \frac{100}{1}\% = \frac{200}{3}\% = 66\frac{2}{3}\%$

19. $2\frac{1}{4} = \frac{9}{4} \cdot 100\% = \frac{9}{4} \cdot \frac{100}{1}\% = \frac{900}{4}\% = 225\%$

Helpful Hint

$\frac{200}{3} = 66.\overline{6}$. Thus, another way to write $\frac{200}{3}\%$ is $66.\overline{6}\%$.

Work Practice 17–19

Practice 17–19

Write each fraction or mixed number as a percent.

17. $\frac{3}{25}$ 18. $\frac{9}{40}$ 19. $5\frac{1}{2}$

✓Concept Check Which digit in the percent 76.4582% represents

a. A tenth percent?
b. A thousandth percent?
c. A hundredth percent?
d. A ten percent?

Example 20 Write $\frac{1}{12}$ as a percent. Round to the nearest hundredth percent.

Solution:

$$\frac{1}{12} = \frac{1}{12} \cdot 100\% = \frac{1}{12} \cdot \frac{100\%}{1} = \frac{100}{12}\% \approx 8.33\%$$

("approximately")

```
      8.333 ≈ 8.33
 12)100.000
   -96
     40
    -36
      40
     -36
       40
      -36
        4
```

Thus, $\frac{1}{12}$ is approximately 8.33%.

Work Practice 20

Practice 20

Write $\frac{3}{17}$ as a percent. Round to the nearest hundredth percent.

Objective D Applications with Percents, Decimals, and Fractions

Let's summarize what we have learned so far about percents, decimals, and fractions:

Summary of Converting Percents, Decimals, and Fractions

- *To write a percent as a decimal,* replace the % symbol with its decimal equivalent, 0.01; then multiply.
- *To write a percent as a fraction,* replace the % symbol with its fraction equivalent, $\frac{1}{100}$; then multiply.
- *To write a decimal or fraction as a percent,* multiply by 100%.

Answers

17. 12% 18. $22\frac{1}{2}\%$ 19. 550%
20. 17.65%

✓Concept Check Answers

a. 4 b. 8 c. 5 d. 7

If we let x represent a number, below we summarize using symbols.

Write a percent as a decimal:	Write a percent as a fraction:	Write a number as a percent:
$x\% = x(0.01)$	$x\% = x \cdot \frac{1}{100}$	$x = x \cdot 100\%$

Practice 21

A family decides to spend no more than 27.5% of its monthly income on rent. Write 27.5% as a decimal and as a fraction.

Example 21 In the last ten years, automobile thefts in the continental United States have decreased 39.8%. Write this percent as a decimal and as a fraction. (*Source:* The American Automobile Manufacturers Association)

Solution:

As a decimal: $39.8\% = 39.8(0.01) = 0.398.$

As a fraction: $39.8\% = 39.8 \cdot \frac{1}{100} = \frac{39.8}{100} = \frac{39.8}{100} \cdot \frac{10}{10} = \frac{398}{1000} = \frac{\overset{1}{\cancel{2}} \cdot 199}{\underset{1}{\cancel{2}} \cdot 500} = \frac{199}{500}.$

Thus, 39.8% written as a decimal is 0.398, and written as a fraction is $\frac{199}{500}$.

Work Practice 21

Practice 22

Provincetown's budget for waste disposal increased by $1\frac{3}{4}$ times over the budget from last year. What percent increase is this?

Example 22 An advertisement for a stereo system reads "$\frac{1}{4}$ off." What percent off is this?

Solution: Write $\frac{1}{4}$ as a percent.

$$\frac{1}{4} = \frac{1}{4} \cdot 100\% = \frac{1}{4} \cdot \frac{100\%}{1} = \frac{100}{4}\% = 25\%$$

Thus, "$\frac{1}{4}$ off" is the same as "25% off."

Work Practice 22

Note: It is helpful to know a few basic percent conversions. Appendix A.2 contains a handy reference of percent, decimal, and fraction equivalencies.

Also, Appendix A.3 shows how to find common percents of a number.

Answers

21. 0.275, $\frac{11}{40}$ **22.** 175%

Vocabulary, Readiness & Video Check

Use the choices below to fill in each blank. Some choices may be used more than once.

$\frac{1}{100}$ 0.01 100% percent

1. ________ means "per hundred."
2. ________ = 1.
3. The % symbol is read as ________.
4. To write a decimal or a fraction as a percent, multiply by 1 in the form of ________.
5. To write a percent as a *decimal,* drop the % symbol and multiply by ________.
6. To write a percent as a *fraction,* drop the % symbol and multiply by ________.

See Video 6.2 

Watch the section lecture video and answer the following questions.

Objective A 7. From the lecture before Example 1, what is the most important thing to remember about percent?

Objective B 8. From Example 4, what is the percent equivalent of 1?

Objective C 9. Complete this statement based on Example 9: Multiplying by 100% is the same as multiplying by ________.

Objective D 10. From Example 15, what is the main difference between writing a percent as a decimal and writing a percent as a fraction?

6.2 Exercise Set MyMathLab®

Objective A *Solve. See Examples 1 and 2.*

1. In a survey of 100 college students, 96 use the Internet. What percent use the Internet?

2. A basketball player makes 81 out of 100 attempted free throws. What percent of free throws are made?

One hundred adults were asked to name their favorite sport, and the results are shown in the circle graph.

3. What sport was preferred by most adults? What percent preferred this sport?

4. What sport was preferred by the least number of adults? What percent preferred this sport?

5. What percent of adults preferred football or soccer?

6. What percent of adults preferred basketball or baseball?

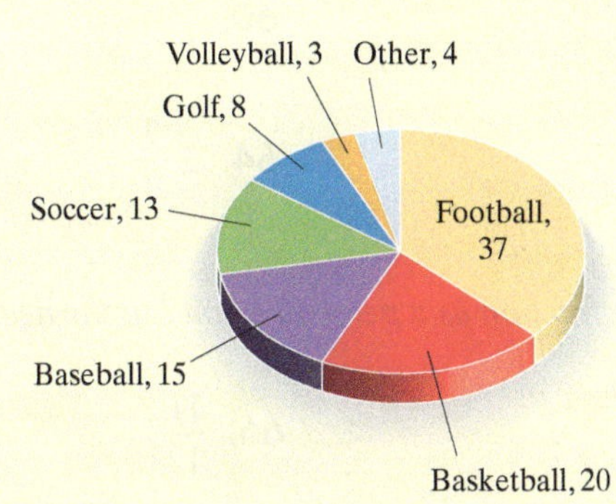

Objective B *Write each percent as a decimal. See Examples 3 through 7.*

7. 41%
8. 62%
9. 6%
10. 3%
11. 100%
12. 136%
13. 73.6%
14. 45.7%
15. 2.8%
16. 1.4%
17. 0.6%
18. 0.9%
19. 300%
20. 500%
21. 32.58%
22. 72.18%

Write each percent as a fraction or mixed number in simplest form. See Examples 8 through 12.

23. 8%
24. 22%
25. 4%
26. 2%
27. 4.5%
28. 7.5%
29. 175%
30. 275%
31. 6.25%
32. 8.75%
33. $10\frac{1}{3}\%$
34. $7\frac{3}{4}\%$
35. $22\frac{3}{8}\%$
36. $21\frac{7}{8}\%$

Objective C *Write each decimal as a percent. See Examples 13 through 16.*

37. 0.22
38. 0.44
39. 0.006
40. 0.008
41. 5.3
42. 2.7
43. 0.056
44. 0.019
45. 0.2228
46. 0.1115
47. 3.00
48. 9.00
49. 0.7
50. 0.8

Write each fraction or mixed number as a percent. See Examples 17 through 19.

51. $\frac{7}{10}$
52. $\frac{9}{10}$
53. $\frac{4}{5}$
54. $\frac{2}{5}$
55. $\frac{34}{50}$
56. $\frac{41}{50}$
57. $\frac{3}{8}$
58. $\frac{5}{16}$
59. $\frac{1}{3}$
60. $\frac{5}{6}$
61. $4\frac{1}{2}$
62. $6\frac{1}{5}$
63. $1\frac{9}{10}$
64. $2\frac{7}{10}$

Write each fraction as a percent. Round to the nearest hundredth percent. See Example 20.

65. $\frac{9}{11}$
66. $\frac{11}{12}$
67. $\frac{4}{15}$
68. $\frac{10}{11}$

Objectives A B C Mixed Practice *Complete each table. See Examples 1 through 20.*

69.

Percent	Decimal	Fraction
60%		
	0.235	
		$\frac{4}{5}$
$33\frac{1}{3}\%$		
		$\frac{7}{8}$
7.5%		

70.

Percent	Decimal	Fraction
	0.525	
		$\frac{3}{4}$
$66\frac{1}{2}\%$		
		$\frac{5}{6}$
100%		
		$\frac{7}{50}$

71.

Percent	Decimal	Fraction
200%		
	2.8	
705%		
		$4\frac{27}{50}$

72.

Percent	Decimal	Fraction
800%		
	3.2	
608%		
		$9\frac{13}{50}$

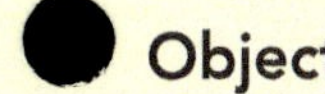

Objective D *Write each percent as a decimal and a fraction. See Examples 21 and 22.*

73. People take aspirin for a variety of reasons. A common use of aspirin is to prevent heart disease, accounting for 38% of all aspirin use. (*Source:* Bayer Market Research)

74. Japan exports 80.5% of all motorcycles manufactured there. (*Source:* Japan Automobile Manufacturers Association)

75. In the United States recently, 35.8% of households had no landlines, just cell phones. (*Source:* CTIA–The Wireless Association)

76. From 2000 to 2011, use of smoked tobacco products in the United States decreased by 27.5%. (*Source:* Centers for Disease Control and Prevention)

77. Approximately 91% of all eighth-grade students in public schools recently reported that they use a computer at home. (*Source:* National Center for Education Statistics)

78. Approximately 15.7% of the American population is not covered by health insurance. (*Source:* U.S. Census Current Population Study)

In Exercises 79 through 82, write the percent from the circle graph as a decimal and a fraction.

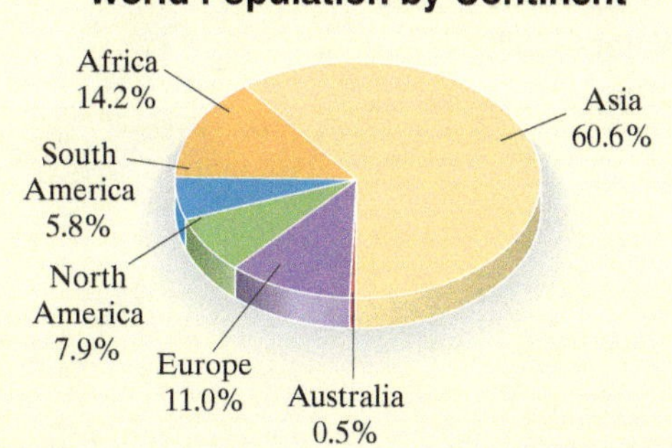

79. Australia: 0.5%

80. Europe: 11%

81. Africa: 14.2%

82. Asia: 60.6%

Solve. See Examples 21 and 22.

83. In a particular year, 0.781 of all electricity produced in France was nuclear generated. Write this decimal as a percent.

84. The United States' share of the total world motor vehicle production is 0.123. Write this decimal as a percent. (*Source:* OICA)

85. The mirrors on the Hubble Space Telescope are able to lock onto a target without deviating more than $\frac{7}{1000}$ of an arc-second. Write this fraction as a percent. (*Source:* NASA)

86. In a recent year, $\frac{1}{4}$ of light trucks sold in the United States were white. Write this fraction as a percent.

87. In 2012, the U.S. Postal Service handled 40% of the world's card and letter mail volume. Write this percent as a decimal. (*Source:* U.S. Postal Service)

88. In 2013, 20% of viewers bought a high-definition television to watch the Super Bowl. Write this percent as a fraction. (*Source:* Nielsen Media)

Review

Perform the indicated operations. See Sections 4.6 and 4.7.

89. $\frac{3}{4} - \frac{1}{2} \cdot \frac{8}{9}$

90. $\left(\frac{2}{11} + \frac{5}{11}\right)\left(\frac{2}{11} - \frac{5}{11}\right)$

91. $6\frac{2}{3} - 4\frac{5}{6}$

92. $6\frac{2}{3} \div 4\frac{5}{6}$

Concept Extensions

Solve. See the Concept Checks in this section.

93. Given the percent 52.8647%, round as indicated.

a. Round to the nearest tenth percent.
b. Round to the nearest hundredth percent.

94. Given the percent 0.5269%, round as indicated.

a. Round to the nearest tenth percent.
b. Round to the nearest hundredth percent.

95. Which of the following are correct?

a. 6.5% = 0.65
b. 7.8% = 0.078
c. 120% = 0.12
d. 0.35% = 0.0035

96. Which of the following are correct?

a. 0.231 = 23.1%
b. 5.12 = 0.0512%
c. 3.2 = 320%
d. 0.0175 = 0.175%

Recall that 1 = 100%. This means that 1 whole is 100%. Use this for Exercises 97 and 98. (Source: Some Body! by Dr. Pete Rowen)

97. The four blood types are A, B, O, and AB. (Each blood type can also be further classified as Rh-positive or Rh-negative depending upon whether your blood contains protein or not.) Given the percent blood types for people in the United States below, calculate the percent of the U.S. population with AB blood type.

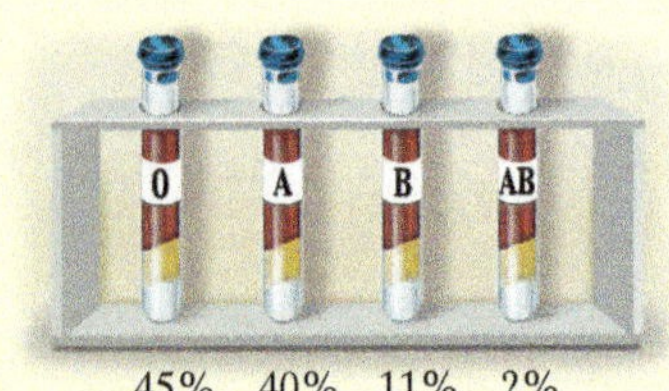

98. The top four components of bone are below. Find the missing percent.

1. Minerals—45%
2. Living tissue—30%
3. Water—20%
4. Other—?

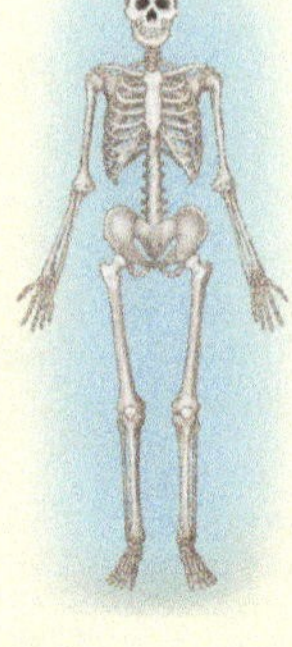

What percent of the figure is shaded?

99.

100.

Fill in the blanks.

101. A fraction written as a percent is greater than 100% when the numerator is ________ than the denominator. (greater/less)

102. A decimal written as a percent is less than 100% when the decimal is ________ than 1. (greater/less)

Write each fraction as a decimal and then write each decimal as a percent. Round the decimal to three decimal places (nearest thousandth) and the percent to the nearest tenth percent.

103. $\frac{21}{79}$

104. $\frac{56}{102}$

The bar graph shows the predicted fastest-growing occupations by percent that require an associate degree or more education. Use this graph for Exercises 105 through 108.

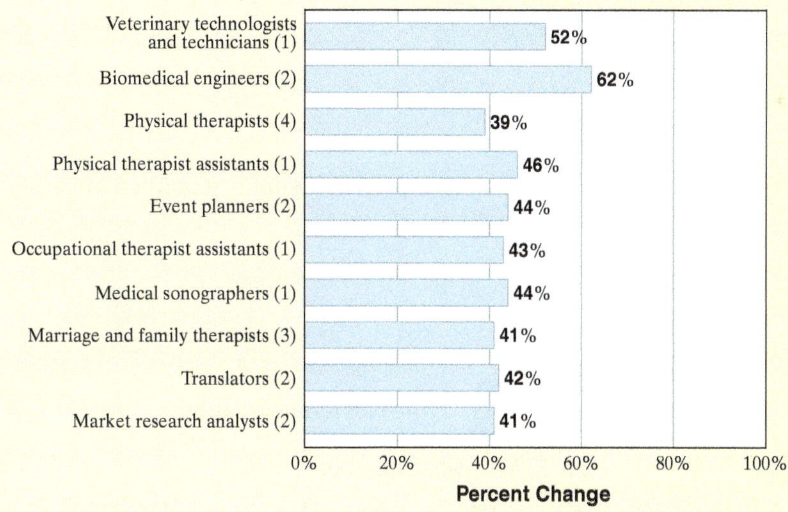

105. What occupation is predicted to be the fastest growing?

106. What occupation is predicted to be the second fastest growing?

107. Write the percent change for physical therapists as a decimal.

108. Write the percent change for event planners as a decimal.

109. In your own words, explain how to write a percent as a decimal.

110. In your own words, explain how to write a decimal as a percent.

6.3 Solving Percent Problems with Equations

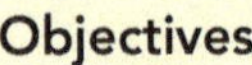

Objectives

A Write Percent Problems as Equations.

B Solve Percent Problems.

Sections 6.3 and 6.4 introduce two methods for solving percent problems. It may not be necessary for you to study both sections. You may want to check with your instructor for further advice.

To solve percent problems in this section, we will translate the problems into mathematical statements, or equations.

Objective A Writing Percent Problems as Equations

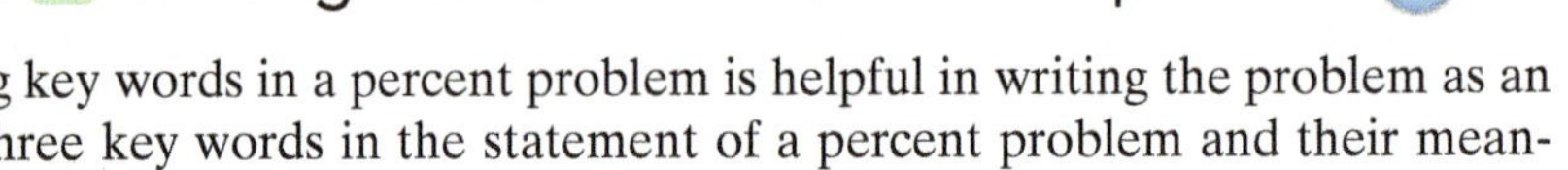

Recognizing key words in a percent problem is helpful in writing the problem as an equation. Three key words in the statement of a percent problem and their meanings are as follows:

of means **multiplication** (·)

is means **equal** (=)

what (or some equivalent) means **the unknown number**

In our examples, we will let the letter x stand for the unknown number.

Practice 1

Translate: 8 is what percent of 48?

Example 1 Translate to an equation:

5 is what percent of 20?

Solution: 5 is what percent of 20?

$$5 = x \cdot 20$$

Work Practice 1

Helpful Hint

Remember that an equation is simply a mathematical statement that contains an equal sign (=).

$$5 = 20x$$

↑ equal sign

Practice 2

Translate: 2.6 is 40% of what number?

Example 2 Translate to an equation:

1.2 is 30% of what number?

Solution: 1.2 is 30% of what number?

$$1.2 = 30\% \cdot x$$

Work Practice 2

Practice 3

Translate: What number is 90% of 0.045?

Example 3 Translate to an equation:

What number is 25% of 0.008?

Solution: What number is 25% of 0.008?

$$x = 25\% \cdot 0.008$$

Work Practice 3

Answers

1. $8 = x \cdot 48$ **2.** $2.6 = 40\% \cdot x$ **3.** $x = 90\% \cdot 0.045$

Examples Translate each question to an equation.

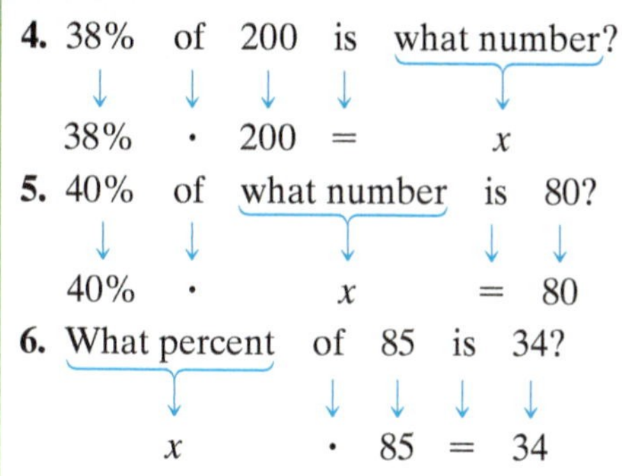

4. 38% of 200 is what number?

$38\% \cdot 200 = x$

5. 40% of what number is 80?

$40\% \cdot x = 80$

6. What percent of 85 is 34?

$x \cdot 85 = 34$

Work Practice 4–6

Practice 4–6

Translate each question to an equation.

4. 56% of 180 is what number?
5. 12% of what number is 21?
6. What percent of 95 is 76?

✓**Concept Check** In the equation $2x = 10$, what step is taken to solve the equation?

Objective B Solving Percent Problems

You may have noticed by now that each percent problem has contained three numbers—in our examples, two are known and one is unknown. Each of these numbers is given a special name.

15%	of	60	is	9
↓	↓	↓	↓	↓
15% percent	·	60 base	=	9 amount

We call this equation the **percent equation.**

Percent Equation

percent · base = amount

Once a percent problem has been written as a percent equation, we can use the equation to find the unknown number, whether it is the percent, the base, or the amount.

Example 7 Solving Percent Equations for the Amount

What number is 35% of 60?

Solution:

$x = 35\% \cdot 60$ Translate to an equation.

$x = 0.35 \cdot 60$ Write 35% as 0.35.

$x = 21$ Multiply:

$$\begin{array}{r} 60 \\ \times\, 0.35 \\ \hline 300 \\ 1800 \\ \hline 21.00 \end{array}$$

Then 21 is 35% of 60. Is this reasonable? To see, round 35% to 40%. Then 40% of 60 or 0.40(60) is 24. Our result is reasonable since 21 is close to 24.

Work Practice 7

Practice 7

What number is 25% of 90?

Answers

4. $56\% \cdot 180 = x$ 5. $12\% \cdot x = 21$ 6. $x \cdot 95 = 76$ 7. 22.5

✓**Concept Check Answer**

Divide both sides of the equation by 2.

Helpful Hint

When solving a percent equation, write the percent as a decimal or fraction.

Practice 8

95% of 400 is what number?

Example 8 85% of 300 is what number?

Solution:

$85\% \cdot 300 = x$ Translate to an equation.

$0.85 \cdot 300 = x$ Write 85% as 0.85.

$255 = x$ Multiply: $0.85 \cdot 300 = 255$.

Then 85% of 300 is 255. Is this result reasonable? To see, round 85% to 90%. Then 90% of 300 or $0.90(300) = 270$, which is close to 255.

Work Practice 8

Practice 9

15% of what number is 2.4?

Example 9 Solving Percent Equations for the Base

12% of what number is 0.6?

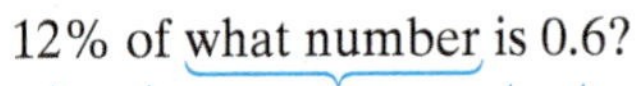

Solution:

$12\% \cdot x = 0.6$ Translate to an equation.

$0.12 \cdot x = 0.6$ Write 12% as 0.12.

$\frac{0.12 \cdot x}{0.12} = \frac{0.6}{0.12}$ Divide both sides by 0.12.

$x = 5$ $\quad 0.12\overline{)0.60}$ gives 5. (60, 0)

Then 12% of 5 is 0.6. Is this reasonable? To see, round 12% to 10%. Then 10% of 5 or $0.10(5) = 0.5$, which is close to 0.6.

Work Practice 9

Practice 10

18 is $4\frac{1}{2}\%$ of what number?

Example 10 13 is $6\frac{1}{2}\%$ of what number?

Solution:

$13 = 6\frac{1}{2}\% \cdot x$ Translate to an equation.

$13 = 0.065 \cdot x$ $\quad 6\frac{1}{2}\% = 6.5\% = 0.065$

$\frac{13}{0.065} = \frac{0.065 \cdot x}{0.065}$ Divide both sides by 0.065. $\quad 0.065\overline{)13.000}$ gives 200. (130, 0)

$200 = x$

Then 13 is $6\frac{1}{2}\%$ of 200. Check to see if this result is reasonable.

Work Practice 10

Answers

8. 380 **9.** 16 **10.** 400

Example 11 Solving Percent Equations for the Percent

What percent of 12 is 9?

Solution:

$$x \cdot 12 = 9 \quad \text{Translate to an equation.}$$

$$\frac{x \cdot 12}{12} = \frac{9}{12} \quad \text{Divide both sides by 12.}$$

$$\text{or } x = 0.75$$

Next, since we are looking for a percent, we can write $\frac{9}{12}$ or 0.75 as a percent.

$x = 75\%$

Then 75% of 12 is 9. To check, see that $75\% \cdot 12 = 9$.

Work Practice 11

Practice 11

What percent of 90 is 27?

Helpful Hint

If your unknown in the percent equation is a percent, don't forget to convert your answer to a percent.

Example 12

78 is what percent of 65?

Solution:

$$78 = x \cdot 65 \quad \text{Translate to an equation.}$$

$$\frac{78}{65} = \frac{x \cdot 65}{65} \quad \text{Divide both sides by 65.}$$

$$1.2 = x$$

$$120\% = x \quad \text{Write 1.2 as a percent.}$$

Then 78 is 120% of 65. Check this result.

Work Practice 12

Practice 12

63 is what percent of 45?

✓Concept Check

Consider these problems

1. 75% of 50 =
   a. 50 **b.** a number greater than 50 **c.** a number less than 50
2. 40% of a number is 10. Is the number
   a. 10? **b.** less than 10? **c.** greater than 10?
3. 800 is 120% of what number? Is the number
   a. 800? **b.** less than 800? **c.** greater than 800?

Helpful Hint

Use the following to see if your answers are reasonable.

(100%) of a number = the number

(a percent greater than 100%) of a number = a number greater than the original number

(a percent less than 100%) of a number = a number less than the original number

Answers

11. 30% **12.** 140%

✓Concept Check Answers

1. c **2.** c **3.** b

Vocabulary, Readiness & Video Check

Use the choices below to fill in each blank.

percent	amount	of	less
base	the number	is	greater

1. The word __________ translates to "=".
2. The word __________ usually translates to "multiplication."
3. In the statement "10% of 90 is 9," the number 9 is called the __________, 90 is called the __________, and 10 is called the __________.
4. 100% of a number = __________.
5. Any "percent greater than 100%" of "a number" = "a number __________ than the original number."
6. Any "percent less than 100%" of "a number" = "a number __________ than the original number."

Martin-Gay Interactive Videos

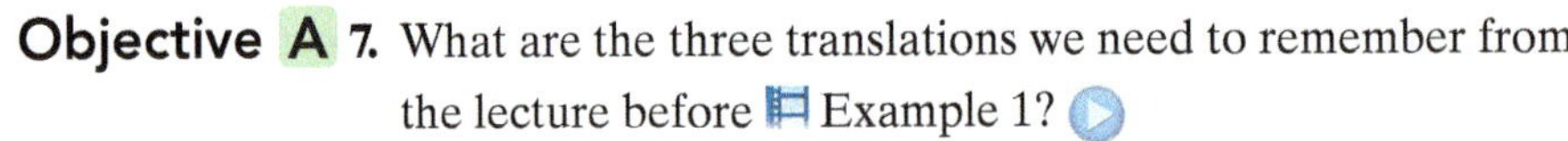

See Video 6.3

Watch the section lecture video and answer the following questions.

Objective A 7. What are the three translations we need to remember from the lecture before Example 1?

Objective B 8. What is different about the translated equation in Example 5?

6.3 Exercise Set MyMathLab®

Objective A Translating *Translate each to an equation. Do not solve. See Examples 1 through 6.*

1. 18% of 81 is what number?
2. 36% of 72 is what number?
3. 20% of what number is 105?
4. 40% of what number is 6?
5. 0.6 is 40% of what number?
6. 0.7 is 20% of what number?
7. What percent of 80 is 3.8?
8. 9.2 is what percent of 92?
9. What number is 9% of 43?
10. What number is 25% of 55?
11. What percent of 250 is 150?
12. What percent of 375 is 300?

Objective B *Solve. See Examples 7 and 8.*

13. 10% of 35 is what number?

14. 25% of 68 is what number?

15. What number is 14% of 205?

16. What number is 18% of 425?

Solve. See Examples 9 and 10.

17. 1.2 is 12% of what number?

18. 0.22 is 44% of what number?

19. $8\frac{1}{2}\%$ of what number is 51?

20. $4\frac{1}{2}\%$ of what number is 45?

Solve. See Examples 11 and 12.

21. What percent of 80 is 88?

22. What percent of 40 is 60?

23. 17 is what percent of 50?

24. 48 is what percent of 50?

Objectives A B Mixed Practice *Solve. See Examples 1 through 12.*

25. 0.1 is 10% of what number?

26. 0.5 is 5% of what number?

27. 150% of 430 is what number?

28. 300% of 56 is what number?

29. 82.5 is $16\frac{1}{2}\%$ of what number?

30. 7.2 is $6\frac{1}{4}\%$ of what number?

31. 2.58 is what percent of 50?

32. 2.64 is what percent of 25?

33. What number is 42% of 60?

34. What number is 36% of 80?

35. What percent of 184 is 64.4?

36. What percent of 120 is 76.8?

37. 120% of what number is 42?

38. 160% of what number is 40?

39. 2.4% of 26 is what number?

40. 4.8% of 32 is what number?

41. What percent of 600 is 3?

42. What percent of 500 is 2?

43. 6.67 is 4.6% of what number?

44. 9.75 is 7.5% of what number?

45. 1575 is what percent of 2500?

46. 2520 is what percent of 3500?

47. 2 is what percent of 50?

48. 2 is what percent of 40?

Review

Find the value of x in each proportion. See Section 6.1.

49. $\frac{27}{x} = \frac{9}{10}$

50. $\frac{35}{x} = \frac{7}{5}$

51. $\frac{x}{5} = \frac{8}{11}$

52. $\frac{x}{3} = \frac{6}{13}$

Write each sentence as a proportion. See Section 6.1.

53. 17 is to 12 as x is to 20.

54. 20 is to 25 as x is to 10.

55. 8 is to 9 as 14 is to x.

56. 5 is to 6 as 15 is to x.

Concept Extensions

For each equation, determine the next step taken to find the value of n. See the first Concept Check in this section.

57. $5 \cdot n = 32$

a. $n = 5 \cdot 32$ **b.** $n = \frac{5}{32}$ **c.** $n = \frac{32}{5}$ **d.** none of these

58. $68 = 8 \cdot n$

a. $n = 8 \cdot 68$ **b.** $n = \frac{68}{8}$ **c.** $n = \frac{8}{68}$ **d.** none of these

59. $0.06 = n \cdot 7$

a. $n = 0.06 \cdot 7$ **b.** $n = \frac{0.06}{7}$ **c.** $n = \frac{7}{0.06}$ **d.** none of these

60. $n = 0.7 \cdot 12$

a. $n = 8.4$ **b.** $n = \frac{12}{0.7}$ **c.** $n = \frac{0.7}{12}$ **d.** none of these

61. Write a word statement for the equation $20\% \cdot x = 18.6$. Use the phrase "what number" for "x."

62. Write a word statement for the equation $x = 33\frac{1}{3}\% \cdot 24$. Use the phrase "what number" for "x."

For each exercise, determine whether the percent, x, is (a) 100%, (b) greater than 100%, or (c) less than 100%. See the second Concept Check in this section.

63. $x\%$ of 20 is 30

64. $x\%$ of 98 is 98

65. $x\%$ of 120 is 85

66. $x\%$ of 35 is 50

For each exercise, determine whether the number, y, is (a) equal to 45, (b) greater than 45, or (c) less than 45.

67. 55% of 45 is y

68. 230% of 45 is y

69. 100% of 45 is y

70. 30% of y is 45

71. 100% of y is 45

72. 180% of y is 45

73. In your own words, explain how to solve a percent equation.

74. Write a percent problem that uses the percent 50%.

Solve.

75. 1.5% of 45,775 is what number?

76. What percent of 75,528 is 27,945.36?

77. 22,113 is 180% of what number?

6.4 Solving Percent Problems with Proportions

There is more than one method that can be used to solve percent problems. (See the note at the beginning of Section 6.3.) In the last section, we used the percent equation. In this section, we will use proportions.

Objectives

A Write Percent Problems as Proportions.

B Solve Percent Problems. 

Objective A Writing Percent Problems as Proportions 

To understand the proportion method, recall that 70% means the ratio of 70 to 100, or $\frac{70}{100}$.

$$70\% = \frac{70}{100} = \frac{7}{10}$$

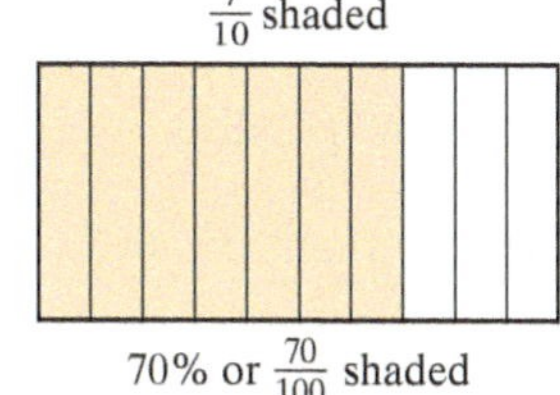

Since the ratio $\frac{70}{100}$ is equal to the ratio $\frac{7}{10}$, we have the proportion

$$\frac{7}{10} = \frac{70}{100}$$

We call this proportion the **percent proportion.** In general, we can name the parts of this proportion as follows:

Percent Proportion

$$\frac{\text{amount}}{\text{base}} = \frac{\text{percent}}{100} \quad \leftarrow \text{always } 100$$

or

$$\begin{matrix} \text{amount} \rightarrow \\ \text{base} \rightarrow \end{matrix} \frac{a}{b} = \frac{p}{100} \leftarrow \text{percent}$$

When we translate percent problems to proportions, the **percent,** p, can be identified by looking for the symbol % or the word *percent.* The **base,** b, usually follows the word *of.* The **amount,** a, is the part compared to the whole.

Helpful Hint

This table may be useful when identifying the parts of a proportion.

Part of Proportion	How It's Identified
Percent	% or *percent*
Base	Appears after *of*
Amount	Part compared to whole

Practice 1

Translate to a proportion.
27% of what number is 54?

Example 1

Translate to a proportion.

12% of what number is 47?

Solution:

percent	base It appears after the word *of*.	amount It is the part compared to the whole.

amount → $\frac{47}{b} = \frac{12}{100}$ ← percent
base →

Work Practice 1

Practice 2

Translate to a proportion.
30 is what percent of 90?

Example 2

Translate to a proportion.

101 is what percent of 200?

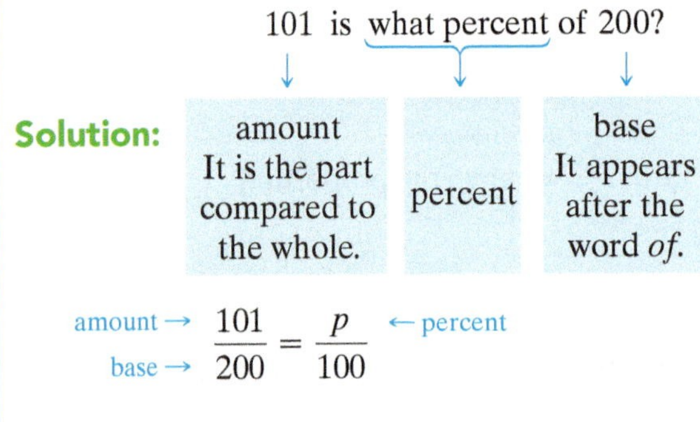

Work Practice 2

Practice 3

Translate to a proportion.
What number is 25% of 116?

Example 3

Translate to a proportion.

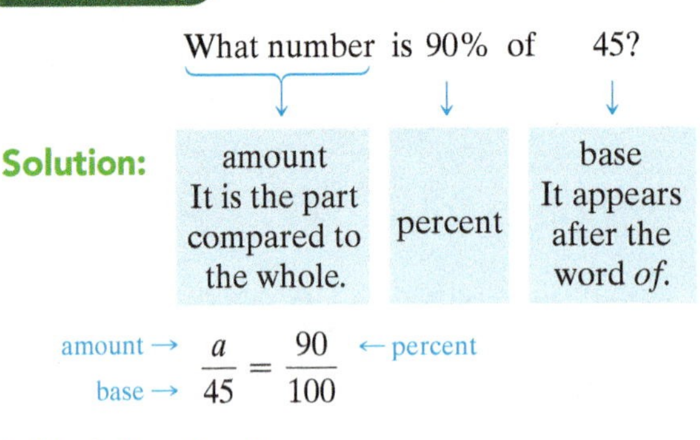

Work Practice 3

Practice 4

Translate to a proportion.
680 is 65% of what number?

Example 4

Translate to a proportion.

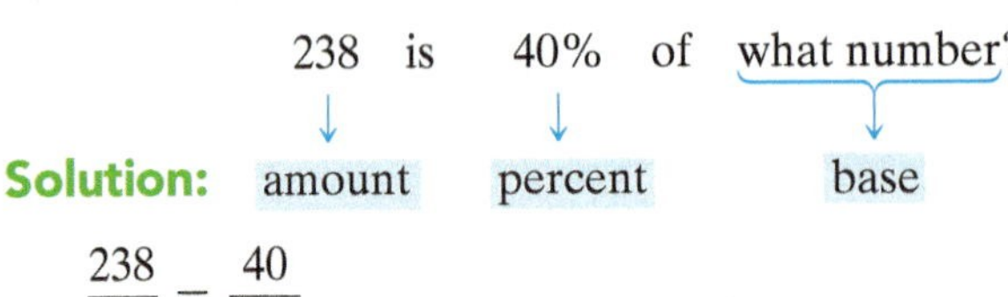

Work Practice 4

Answers

1. $\frac{54}{b} = \frac{27}{100}$ 2. $\frac{30}{90} = \frac{p}{100}$
3. $\frac{a}{116} = \frac{25}{100}$ 4. $\frac{680}{b} = \frac{65}{100}$

Example 5 Translate to a proportion.

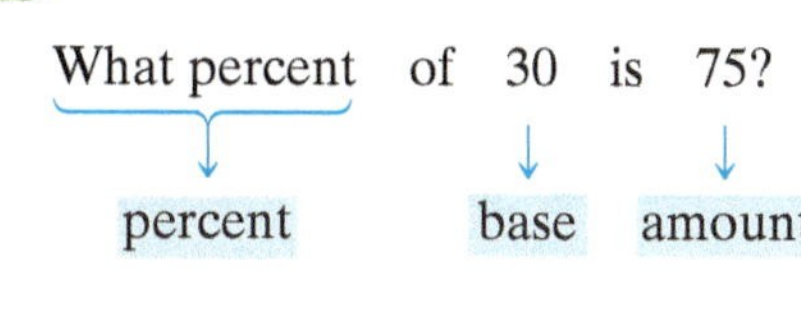

Solution:

$$\frac{75}{30} = \frac{p}{100}$$

Work Practice 5

Practice 5

Translate to a proportion.
What percent of 40 is 75?

Example 6 Translate to a proportion.

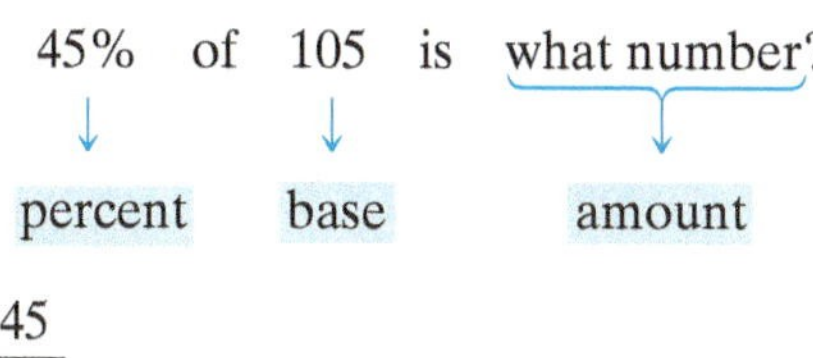

Solution:

$$\frac{a}{105} = \frac{45}{100}$$

Work Practice 6

Practice 6

Translate to a proportion.
46% of 80 is what number?

✓**Concept Check** Consider the statement "78 is what percent of 350?" Which part of the percent proportion is unknown?

a. the amount
b. the base
c. the percent

Consider another statement: "14 is 10% of some number." Which part of the percent proportion is unknown?

a. the amount
b. the base
c. the percent

Objective B Solving Percent Problems 

The proportions that we have written in this section contain three values that can change: the percent, the base, and the amount. If any two of these values are known, we can find the third (the unknown value). To do this, we write a percent proportion and find the unknown value as we did in Section 6.1.

Example 7 Solving Percent Proportions for the Amount

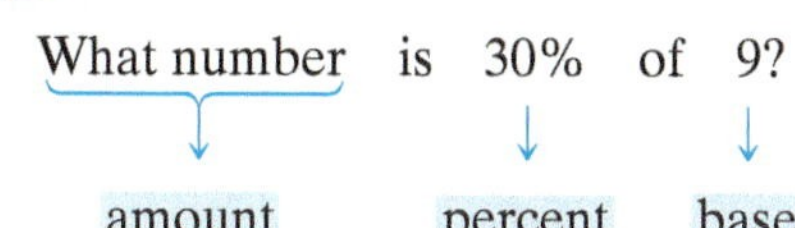

Solution:

$$\frac{a}{9} = \frac{30}{100}$$

Practice 7

What number is 8% of 120?

Answers

5. $\frac{75}{40} = \frac{p}{100}$ **6.** $\frac{a}{80} = \frac{46}{100}$ **7.** 9.6

✓**Concept Check Answers**

c, b

Helpful Hint The proportion in Example 7 contains the ratio $\frac{30}{100}$. A ratio in a proportion may be simplified before solving the proportion. The unknown number in both $\frac{a}{9} = \frac{30}{100}$ and $\frac{a}{9} = \frac{3}{10}$ is 2.7.

To solve, we set cross products equal to each other.

$$\frac{a}{9} = \frac{30}{100}$$

$a \cdot 100 = 9 \cdot 30$ Set cross products equal.

$100a = 270$ Multiply.

$\frac{100a}{100} = \frac{270}{100}$ Divide both sides by 100, the coefficient of a.

$a = 2.7$ Simplify.

Thus, 2.7 is 30% of 9.

Work Practice 7

Practice 8

65% of what number is 52?

Example 8 Solving Percent Problems for the Base

150% of what number is 30?

Solution: 150% → percent; what number → base; 30 → amount

$\frac{30}{b} = \frac{150}{100}$ Write the proportion.

$\frac{30}{b} = \frac{3}{2}$ Simplify $\frac{150}{100}$ and write as $\frac{3}{2}$.

$30 \cdot 2 = b \cdot 3$ Set cross products equal.

$60 = 3b$ Multiply.

$\frac{60}{3} = \frac{3b}{3}$ Divide both sides by 3.

$20 = b$ Simplify.

Thus, 150% of 20 is 30.

Work Practice 8

✓**Concept Check** When solving a percent problem by using a proportion, describe how you can check the result.

Practice 9

15.4 is 5% of what number?

Example 9

20.8 is 40% of what number?

Solution: 20.8 → amount; 40% → percent; what number → base

$\frac{20.8}{b} = \frac{40}{100}$ or $\frac{20.8}{b} = \frac{2}{5}$ Write the proportion and simplify $\frac{40}{100}$.

$20.8 \cdot 5 = b \cdot 2$ Set cross products equal.

$104 = 2b$ Multiply.

$\frac{104}{2} = \frac{2b}{2}$ Divide both sides by 2.

$52 = b$ Simplify.

So, 20.8 is 40% of 52.

Work Practice 9

Answers
8. 80 **9.** 308

✓**Concept Check Answer**
by putting the result into the proportion and checking that the proportion is true

Example 10 Solving Percent Problems for the Percent

What percent of 50 is 8?

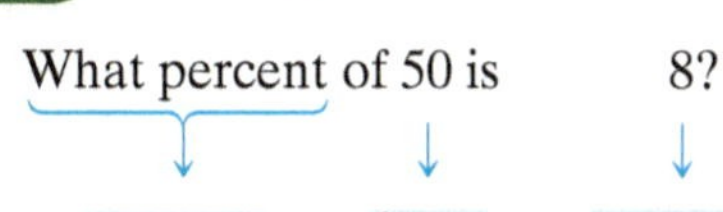

Solution: percent base amount

$$\frac{8}{50} = \frac{p}{100} \quad \text{or} \quad \frac{4}{25} = \frac{p}{100}$$ Write the proportion and simplify $\frac{8}{50}$.

$$4 \cdot 100 = 25 \cdot p$$ Set cross products equal.

$$400 = 25p$$ Multiply.

$$\frac{400}{25} = \frac{25p}{25}$$ Divide both sides by 25.

$$16 = p$$ Simplify.

So, 16% of 50 is 8.

Work Practice 10

Helpful Hint Recall from our percent proportion that this number already is a percent. Just keep the number as is and attach a % symbol.

Example 11

504 is what percent of 360?

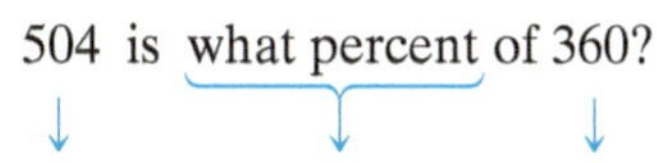

Solution: amount percent base

$$\frac{504}{360} = \frac{p}{100}$$

Let's choose not to simplify the ratio $\frac{504}{360}$.

$$504 \cdot 100 = 360 \cdot p$$ Set cross products equal.

$$50{,}400 = 360p$$ Multiply.

$$\frac{50{,}400}{360} = \frac{360p}{360}$$ Divide both sides by 360.

$$140 = p$$ Simplify.

Notice that by choosing not to simplify $\frac{504}{360}$, we had larger numbers in our equation. Either way, we find that 504 is 140% of 360.

Work Practice 11

Helpful Hint

Use the following to see whether your answers to the above examples and practice problems are reasonable.

100% of a number = the greater

(a percent greater than 100%) of a number = a number greater than the original number

(a percent less than 100%) of a number = a number less than the original number

Practice 10

What percent of 40 is 8?

Practice 11

414 is what percent of 180?

Answers

10. 20% **11.** 230%

Vocabulary, Readiness & Video Check

Use the choices below to fill in each blank. These choices will be used more than once.

amount base percent

1. When translating the statement "20% of 15 is 3" to a proportion, the number 3 is called the ________, 15 is the ________, and 20 is the ________.
2. In the question "50% of what number is 28?", which part of the percent proportion is unknown? ________
3. In the question "What number is 25% of 200?", which part of the percent proportion is unknown? ________
4. In the question "38 is what percent of 380?", which part of the percent proportion is unknown? ________

Martin-Gay Interactive Videos

See Video 6.4 

Watch the section lecture video and answer the following questions.

Objective A 5. In Example 1, how did we identify what part of the percent proportion 45 is?

Objective B 6. From Examples 4–6, what number is *always* part of the cross product equation of a percent proportion?

6.4 Exercise Set MyMathLab®

Objective A **Translating** *Translate each to a proportion. Do not solve. See Examples 1 through 6.*

1. 98% of 45 is what number?
2. 92% of 30 is what number?
3. What number is 4% of 150?
4. What number is 7% of 175?
5. 14.3 is 26% of what number?
6. 1.2 is 47% of what number?
7. 35% of what number is 84?
8. 85% of what number is 520?
9. What percent of 400 is 70?
10. What percent of 900 is 216?
11. 8.2 is what percent of 82?
12. 9.6 is what percent of 96?

Objective B *Solve. See Example 7.*

13. 40% of 65 is what number?
14. 25% of 84 is what number?
15. What number is 18% of 105?
16. What number is 60% of 29?

Solve. See Examples 8 and 9.

17. 15% of what number is 90?

18. 55% of what number is 55?

19. 7.8 is 78% of what number?

20. 1.1 is 44% of what number?

Solve. See Examples 10 and 11.

21. What percent of 35 is 42?

22. What percent of 98 is 147?

23. 14 is what percent of 50?

24. 24 is what percent of 50?

Objectives A B Mixed Practice *Solve. See Examples 1 through 11.*

25. 3.7 is 10% of what number?

26. 7.4 is 5% of what number?

27. 2.4% of 70 is what number?

28. 2.5% of 90 is what number?

29. 160 is 16% of what number?

30. 30 is 6% of what number?

31. 394.8 is what percent of 188?

32. 550.4 is what percent of 172?

33. What number is 89% of 62?

34. What number is 53% of 130?

35. What percent of 6 is 2.7?

36. What percent of 5 is 1.6?

37. 140% of what number is 105?

38. 170% of what number is 221?

39. 1.8% of 48 is what number?

40. 7.8% of 24 is what number?

41. What percent of 800 is 4?

42. What percent of 500 is 3?

43. 3.5 is 2.5% of what number?

44. 9.18 is 6.8% of what number?

45. 20% of 48 is what number?

46. 75% of 14 is what number?

47. 2486 is what percent of 2200?

48. 9310 is what percent of 3800?

Review

Add or subtract the fractions. See Sections 4.4, 4.5, and 4.7.

49. $-\frac{11}{16}+\left(-\frac{3}{16}\right)$

50. $\frac{7}{12}-\frac{5}{8}$

51. $3\frac{1}{2}-\frac{11}{30}$

52. $2\frac{2}{3}+4\frac{1}{2}$

Add or subtract the decimals. See Section 5.2.

53. $\begin{array}{r} 0.41 \\ +0.29 \\ \hline \end{array}$

54. $\begin{array}{r} 10.78 \\ 4.3 \\ +\ 0.21 \\ \hline \end{array}$

55. $\begin{array}{r} 2.38 \\ -0.19 \\ \hline \end{array}$

56. $\begin{array}{r} 16.37 \\ -\ 2.61 \\ \hline \end{array}$

Concept Extensions

57. Write a word statement for the proportion $\frac{x}{28}=\frac{25}{100}$. Use the phrase "what number" for "x."

58. Write a percent statement that translates to $\frac{16}{80}=\frac{20}{100}$.

Suppose you have finished solving four percent problems using proportions that you set up correctly. Check each answer to see if each makes the proportion a true proportion. If any proportion is not true, solve it to find the correct solution. See the Concept Checks in this section.

59. $\frac{a}{64}=\frac{25}{100}$
Is the amount equal to 17?

60. $\frac{520}{b}=\frac{65}{100}$
Is the base equal to 800?

61. $\frac{p}{100}=\frac{13}{52}$
Is the percent equal to 25 (25%)?

62. $\frac{36}{12}=\frac{p}{100}$
Is the percent equal to 50 (50%)?

63. In your own words, describe how to identify the percent, the base, and the amount in a percent problem.

64. In your own words, explain how to use a proportion to solve a percent problem.

Solve. Round to the nearest tenth, if necessary.

65. What number is 22.3% of 53,862?

66. What percent of 110,736 is 88,542?

67. 8652 is 119% of what number?

Ratio, Proportion, and Percent

Write each ratio as a ratio of whole numbers using fractional notation. Write the fraction in simplest form.

1. 18 to 20 **2.** 36 to 100 **3.** 8.6 to 10 **4.** 1.6 to 4.6

Find the ratio described in each problem.

5. Find the ratio of the width to the length of the sign below.

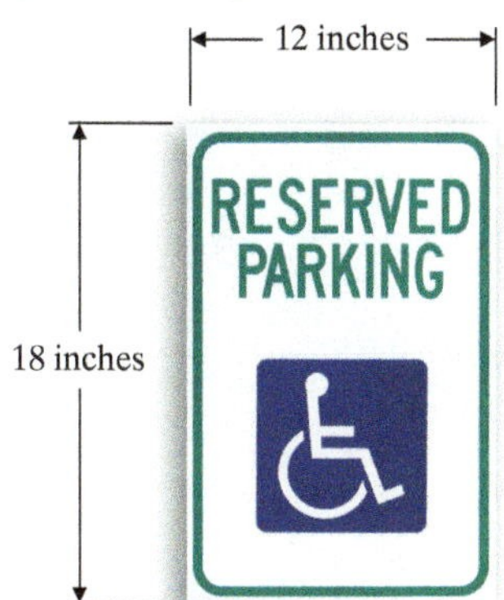

6. At the end of fiscal year 2012, Apple Inc. had approximately $180 billion in assets and approximately $40 billion in debts. Find the ratio of assets to debts. (*Source*: Apple Inc.)

Solve.

7. $\frac{3.5}{12.5} = \frac{7}{z}$

8. $\frac{x + 7}{3} = \frac{2x}{5}$

An office uses 5 boxes of envelopes every 3 weeks.

9. Find how long a gross of envelope boxes is likely to last. (A gross of boxes is 144 boxes.) Round to the nearest week.

10. Find how many boxes should be purchased to last a month. Round to the nearest box.

Write each number as a percent.

11. 0.94 **12.** 0.17 **13.** $\frac{3}{8}$ **14.** $\frac{7}{2}$

15. 4.7 **16.** 8 **17.** $\frac{9}{20}$ **18.** $\frac{53}{50}$

19. $6\frac{3}{4}$ **20.** $3\frac{1}{4}$ **21.** 0.02 **22.** 0.06

Answers

1. ______
2. ______
3. ______
4. ______
5. ______
6. ______
7. ______
8. ______
9. ______
10. ______
11. ______
12. ______
13. ______
14. ______
15. ______
16. ______
17. ______
18. ______
19. ______
20. ______
21. ______
22. ______

23. ______ 24. ______
25. ______ 26. ______
27. ______ 28. ______
29. ______ 30. ______
31. ______
32. ______
33. ______
34. ______
35. ______
36. ______
37. ______
38. ______
39. ______ 40. ______
41. ______ 42. ______
43. ______ 44. ______
45. ______ 46. ______
47. ______ 48. ______
49. ______ 50. ______

Write each percent as a decimal.

23. 71% **24.** 31% **25.** 3% **26.** 4%

27. 224% **28.** 700% **29.** 2.9% **30.** 6.6%

Write each percent as a decimal and as a fraction or mixed number in simplest form. (If necessary when writing as a decimal, round to the nearest thousandth.)

31. 7% **32.** 5% **33.** 6.8% **34.** 11.25%

35. 74% **36.** 45% **37.** $16\frac{1}{3}\%$ **38.** $12\frac{2}{3}\%$

Solve each percent problem.

39. 15% of 90 is what number?

40. 78 is 78% of what number?

41. 297.5 is 85% of what number?

42. 78 is what percent of 65?

43. 23.8 is what percent of 85?

44. 38% of 200 is what number?

45. What number is 40% of 85?

46. What percent of 99 is 128.7?

47. What percent of 250 is 115?

48. What number is 45% of 84?

49. 42% of what number is 63?

50. 95% of what number is 58.9?

6.5 Applications of Percent 

Objective A Solving Applications Involving Percent

Objectives

A Solve Applications Involving Percent.

B Find Percent Increase and Percent Decrease.

Percent is used in a variety of everyday situations. The next examples show just a few ways that percent occurs in real-life settings. (Each of these examples shows two ways of solving these problems. If you studied Section 6.3 only, see *Method 1*. If you studied Section 6.4 only, see *Method 2*.)

The next example has to do with the Appalachian Trail, a hiking trail conceived by a forester in 1921 and diagrammed to the right. (*Note:* The trail mileage changes from year to year as maintenance groups reroute the trail as needed.)

Example 1 The circle graph in the margin shows the Appalachian Trail mileage by state. If the total mileage of the trail is 2174, use the circle graph to determine the number of miles in the state of New York. Round to the nearest whole mile.

Solution: *Method 1.* First, we state the problem in words.

In words:	What number	is	4%	of	2174?
	↓	↓	↓	↓	↓
Translate:	x	$=$	4%	$\cdot$	2174

To solve for x, we find $4\% \cdot 2174$.

$x = 0.04 \cdot 2174$ Write 4% as a decimal.

$x = 86.96$ Multiply.

$x \approx 87$ Round to the nearest whole.

Rounded to the nearest whole mile, we have that approximately 87 miles of the Appalachian Trail are in New York state.

Method 2. State the problem in words; then translate.

In words:	What number	is	4%	of	2174?
	↓		↓		↓
	amount		percent		base

Translate: amount → $\frac{a}{2174} = \frac{4}{100}$ ← percent; base → 2174

Next, we solve for a.

$a \cdot 100 = 2174 \cdot 4$ Set cross products equal.

$100a = 8696$ Multiply.

$\frac{100a}{100} = \frac{8696}{100}$ Divide both sides by 100.

$a = 86.96$ Simplify.

$a \approx 87$ Round to the nearest whole.

Rounded to the nearest whole mile, we have that approximately 87 miles of the Appalachian Trail are in New York state.

Work Practice 1

Practice 1

If the total mileage of the Appalachian Trail is 2174, use the circle graph to determine the number of miles in the state of Virginia.

Appalachian Trail Mileage by State Percent

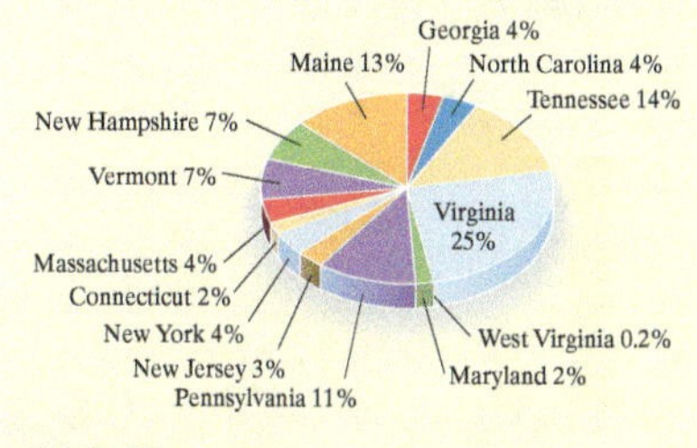

Total miles: 2174

Source: purebound.com

Answer

1. 543.5 mi

Practice 2

From 2010 to 2020, it is projected that the number of employed nurses will grow by 710,000. If the number of nurses employed in 2010 was 2,740,000, find the percent increase in nurses employed from 2010 to 2020. Round to the nearest percent. (*Source:* Bureau of Labor Statistics)

Example 2 Finding Percent of Nursing School Applications Accepted

There continues to be a shortage of nursing school facilities. Recently, of the 256,000 applications for bachelor degree nursing schools, 101,000 of these were accepted. What percent of these applications were accepted? Round to the nearest percent. (*Source:* Bureau of Labor Statistics)

Solution: *Method 1.* First, we state the problem in words.

In words: 101 is what percent of 256?

Translate: $101 = x \cdot 256$

or $101 = 256x$

Next, solve for x.

$\frac{101}{256} = \frac{256x}{256}$ Divide both sides by 256.

$0.39 \approx x$ Divide and round to the nearest hundredth.

$39\% \approx x$ Write as a percent.

About 39% of nursing school applications were accepted.

Method 2.

In words: 101 is what percent of 256?

101 → amount; what percent → percent; 256 → base

Translate: amount → $\frac{101}{256} = \frac{p}{100}$ ← percent; base → 256

Next, solve for p.

$101 \cdot 100 = 256 \cdot p$ Set cross products equal.

$10{,}100 = 256p$ Multiply.

$\frac{10{,}100}{256} = \frac{256p}{256}$ Divide both sides by 256.

$39 \approx p$

About 39% of nursing school applications were accepted.

Work Practice 2

Practice 3

The freshmen class of 864 students is 32% of all students at Euclid University. How many students go to Euclid University?

Example 3 Finding the Base Number of Absences

Mr. Percy, the principal at Slidell High School, counted 31 freshmen absent during a particular day. If this is 4% of the total number of freshmen, how many freshmen are there at Slidell High School?

Solution: *Method 1.* First we state the problem in words; then we translate.

In words: 31 is 4% of what number?

Translate: $31 = 4\% \cdot x$

Next, we solve for x.

$31 = 0.04 \cdot x$ Write 4% as a decimal.

$\frac{31}{0.04} = \frac{0.04x}{0.04}$ Divide both sides by 0.04.

$775 = x$ Simplify.

There are 775 freshmen at Slidell High School.

Answers
2. 26% 3. 2700

Method 2. First we state the problem in words; then we translate.

In words: 31 is 4% of what number?

31 → amount; 4% → percent; what number → base

Translate: $\frac{31}{b} = \frac{4}{100}$ (amount → 31, base → b, 4 ← percent)

Next, we solve for b.

$31 \cdot 100 = b \cdot 4$ Set cross products equal.

$3100 = 4b$ Multiply.

$\frac{3100}{4} = \frac{4b}{4}$ Divide both sides by 4.

$775 = b$ Simplify.

There are 775 freshmen at Slidell High School.

Work Practice 3

Example 4 Finding the Base Increase in Licensed Drivers

From 2000 to 2011, the number of licensed drivers on the road in the United States increased by 11.6%. In 2000, there were about 190 million licensed drivers on the road.

a. Find the increase in licensed drivers from 2000 to 2011.

b. Find the number of licensed drivers on the road in 2011. (*Source:* Federal Highway Administration)

Solution: *Method 1.* First we find the increase in licensed drivers.

In words: What number is 11.6% of 190?

Translate: $x = 11.6\% \cdot 190$

Next, we solve for x.

$x = 0.116 \cdot 190$ Write 11.6 as a decimal.

$x = 22.04$ Multiply.

a. The increase in licensed drivers is 22.04 million.

b. This means that the number of licensed drivers in 2011 was

Number of licensed drivers in 2011 = Number of licensed drivers in 2000 + Increase in number of licensed drivers

= 190 million + 22.04 million

= 212.04 million

(*Continued on next page*)

Practice 4

From 2000 to 2011, the number of registered vehicles on the road in the United States increased by 11.4%. In 2000, the number of vehicles on the road was 220 million.

a. Find the increase in the number of vehicles on the road in 2011.

b. Find the total number of registered vehicles on the road in 2011.

(*Source:* Federal Highway Administration)

Answers

4. a. 25.08 million **b.** 245.08 million

Method 2. First we find the increase in licensed drivers.

In words: What number is 11.6% of 190?

What number → amount; 11.6% → percent; 190 → base

Translate: $\frac{a}{190} = \frac{11.6}{100}$ (amount → a, base → 190, 11.6 ← percent)

Next, we solve for a.

$$a \cdot 100 = 190 \cdot 11.6 \quad \text{Set cross products equal.}$$
$$100a = 2204 \quad \text{Multiply.}$$
$$\frac{100a}{100} = \frac{2204}{100} \quad \text{Divide both sides by 100.}$$
$$a = 22.04 \quad \text{Simplify.}$$

a. The increase in licensed drivers is 22.04 million.

b. This means that the number of licensed drivers in 2011 was

$$\begin{array}{c}\text{Number of}\\ \text{licensed drivers}\\ \text{in 2011}\end{array} = \begin{array}{c}\text{Number of}\\ \text{licensed drivers}\\ \text{in 2000}\end{array} + \begin{array}{c}\text{Increase}\\ \text{in number of}\\ \text{licensed drivers}\end{array}$$
$$= 190 \text{ million} + 22.04 \text{ million}$$
$$= 212.04 \text{ million}$$

Work Practice 4

Objective B Finding Percent Increase and Percent Decrease

We often use percents to show how much an amount has increased or decreased.

Suppose that the population of a town is 10,000 people and then it increases by 2000 people. The **percent of increase** is

$$\frac{2000}{10{,}000} = 0.2 = 20\%$$
(amount of increase → 2000; original amount → 10,000)

In general, we have the following.

Percent of Increase

$$\text{percent of increase} = \frac{\text{amount of increase}}{\text{original amount}}$$

Then write the quotient as a percent.

Practice 5

The number of people attending the local play, *Peter Pan,* increased from 285 on Friday to 333 on Saturday. Find the percent increase in attendance. Round to the nearest tenth percent.

Answer

5. 16.8%

Example 5 Finding Percent Increase

The number of applications for a mathematics scholarship at Yale increased from 34 to 45 in one year. What is the percent increase? Round to the nearest whole percent.

Solution: First we find the amount of increase by subtracting the original number of applicants from the new number of applicants.

$$\text{amount of increase} = 45 - 34 = 11$$

The amount of increase is 11 applicants. To find the percent of increase,

$$\text{percent of increase} = \frac{\text{amount of increase}}{\text{original amount}} = \frac{11}{34} \approx 0.32 = 32\%$$

The number of applications increased by about 32%.

Work Practice 5

Helpful Hint Make sure that this number is the original number and not the new number.

✓Concept Check A student is calculating the percent increase in enrollment from 180 students one year to 200 students the next year. Explain what is wrong with the following calculations:

$$\text{Amount of increase} = 200 - 180 = 20$$

$$\text{Percent of increase} = \frac{20}{200} = 0.1 = 10\%$$

Suppose that your income was \$300 a week and then it decreased by \$30. The **percent of decrease** is

$$\begin{array}{r} \text{amount of decrease} \rightarrow \\ \text{original amount} \rightarrow \end{array} \frac{\$30}{\$300} = 0.1 = 10\%$$

Percent of Decrease

$$\text{percent of decrease} = \frac{\text{amount of decrease}}{\text{original amount}}$$

Then write the quotient as a percent.

Example 6 Finding Percent Decrease

In response to a decrease in sales, a company with 1500 employees reduces the number of employees to 1230. What is the percent decrease?

Solution: First we find the amount of decrease by subtracting 1230 from 1500.

$$\text{amount of decrease} = 1500 - 1230 = 270$$

The amount of decrease is 270. To find the percent of decrease,

$$\text{percent of decrease} = \frac{\text{amount of decrease}}{\text{original amount}} = \frac{270}{1500} = 0.18 = 18\%$$

The number of employees decreased by 18%.

Work Practice 6

✓Concept Check An ice cream stand sold 6000 ice cream cones last summer. This year the same stand sold 5400 cones. Was there a 10% increase, a 10% decrease, or neither? Explain.

Practice 6

A town with a population of 20,200 in 2003 decreased to 18,483 in 2013. What was the percent decrease?

Answer

6. 8.5%

✓Concept Check Answers

To find the percent of increase, you have to divide the amount of increase (20) by the original amount (180); 10% decrease.

Martin-Gay Interactive Videos *Watch the section lecture video and answer the following questions.*

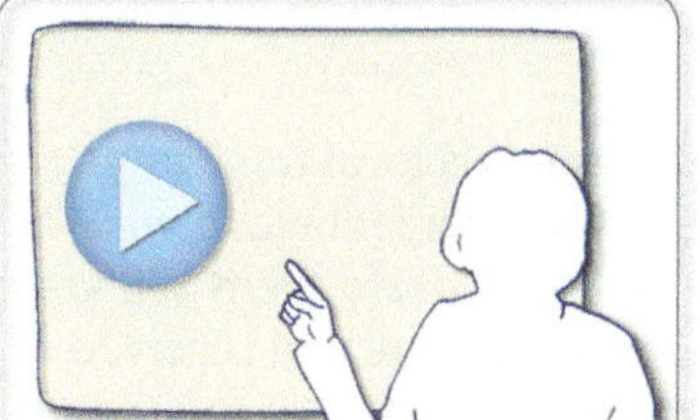

See Video 6.5

Objective A **1.** How do we interpret the answer 175,000 in Example 1?

Objective B **2.** In Example 3, what does the improper fraction tell us?

6.5 Exercise Set MyMathLab®

Objective A *Solve. For Exercises 1 and 2, the solutions have been started for you. See Examples 1 through 4. If necessary, round percents to the nearest tenth and all other answers to the nearest whole.*

1. An inspector found 24 defective bolts during an inspection. If this is 1.5% of the total number of bolts inspected, how many bolts were inspected?

Start the solution:

1. UNDERSTAND the problem. Reread it as many times as needed.

Go to Method 1 or Method 2.

Method 1.

2. TRANSLATE into an equation. (Fill in the boxes.)

24 is 1.5% of what number
↓ ↓ ↓ ↓ ↓
24 □ 1.5% □ x

3. SOLVE for x. (See Example 3, Method 1, for help.)

4. INTERPRET. The total number of bolts inspected was ________.

Method 2.

2. TRANSLATE into a proportion. (Fill in the blanks with "amount" or "base.")

24 is 1.5% of what number
↓ ↓ ↓
_____ percent _____

$$\text{amount} \rightarrow \frac{\_\_\_\_\_}{\_\_\_\_\_} = \frac{1.5}{100} \leftarrow \text{percent}$$
(base → denominator)

3. SOLVE the proportion. (See Example 3, Method 2, for help.)

4. INTERPRET. The total number of bolts inspected was ________.

2. A day care worker found 28 children absent one day during an epidemic of chicken pox. If this was 35% of the total number of children attending the day care center, how many children attend this day care center?

Start the solution:

1. UNDERSTAND the problem. Reread it as many times as needed.

Go to Method 1 or Method 2.

Method 1.

2. TRANSLATE into an equation. (Fill in the boxes.)

28 is 35% of what number
↓ ↓ ↓ ↓ ↓
28 □ 35% □ x

3. SOLVE for x. (See Example 3, Method 1, for help.)

4. INTERPRET. The total number of children attending the day care center is ________.

Method 2.

2. TRANSLATE into a proportion. (Fill in the blanks with "amount" or "base.")

28 is 35% of what number
↓ ↓ ↓
_____ percent _____

$$\text{amount} \rightarrow \frac{\_\_\_\_\_}{\_\_\_\_\_} = \frac{35}{100} \leftarrow \text{percent}$$
(base → denominator)

3. SOLVE the proportion. (See Example 3, Method 2, for help.)

4. INTERPRET. The total number of children attending the day care center is ________.

3. One model of Total Gym® provides weight resistance through adjustments of incline. The minimum weight resistance is 4% of the weight of the person using the Total Gym. Find the minimum weight resistance possible for a 220-pound man. (*Source:* Total Gym)

4. The maximum weight resistance for one model of Total Gym is 60% of the weight of the person using it. Find the maximum weight resistance possible for a 220-pound man. (See Exercise **3** if needed.)

5. A student's cost for last semester at her community college was \$2700. She spent \$378 of that on books. What percent of last semester's college costs was spent on books?

6. Pierre Sampeau belongs to his local food cooperative, where he receives a percentage of what he spends each year as a dividend. He spent \$3850 last year at the food cooperative store and received a dividend of \$154. What percent of his total spending at the food cooperative did he receive as a dividend?

In 2012, there were about 40,000 cinema screens in the United States. Use this information for Exercises 7 and 8.

7. If about 34% of the total screens in the United States are digital 3-D screens, find the number of digital 3-D screens.

8. If about 16% of the total screens in the United States are analog screens, find the number of analog screens.

Solve.

9. Approximately 160,650 of America's 945,000 restaurants are pizza restaurants. Determine the percent of restaurants in America that are pizza restaurants. (*Source:* Pizza Marketplace, National Restaurant Association)

10. Of the 97,100 veterinarians in private practice in the United States, approximately 52,434 are female. Determine the percent of female veterinarians in private practice in the United States. (*Source:* American Veterinary Medical Association)

11. A furniture company currently produces 6200 chairs per month. If production decreases by 8%, find the decrease and the new number of chairs produced each month.

12. The enrollment at a local college decreased by 5% over last year's enrollment of 7640. Find the decrease in enrollment and the current enrollment.

13. From 2010 to 2020, the number of people employed as physician assistants in the United States is expected to increase by 30%. The number of people employed as physician assistants in 2010 was 83,600. Find the predicted number of physician assistants in 2020. (*Source:* Bureau of Labor Statistics)

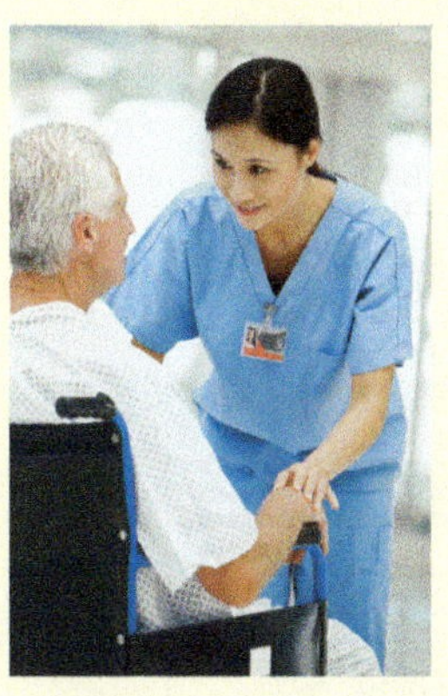

14. From 2007 to 2012, the number of households owning turtles increased by 19.3%. The number of households owning turtles in 2007 was 1,106,000. Find the number of households owning turtles in 2012. (*Source:* American Veterinary Medical Association)

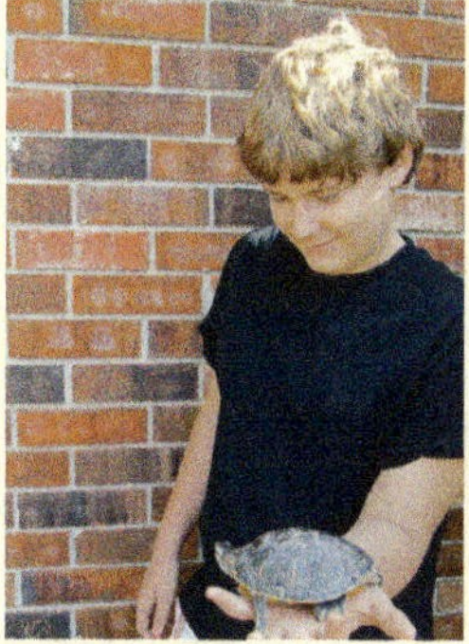

Two states, Michigan and Rhode Island, decreased in population from 2010 to 2012. Their locations are shown on the partial U.S. map below. (Source: U.S. Dept. of Commerce)

15. In 2010, the population of Rhode Island was approximately 1,053,000. If the population decrease was 0.3%, find the population of Rhode Island in 2012.

16. In 2010, the population in Michigan was approximately 9938 thousand. If the population decrease was about 0.6%, find the population of Michigan in 2012.

A popular extreme sport is snowboarding. Ski trails are marked with difficulty levels of easy ●, intermediate ■, difficult ◆, expert ◆◆, and other variations. Use this information for Exercises 17 and 18. Round each percent to the nearest whole.

17. At Keystone ski area in Colorado, approximately 41 of the 135 total ski runs are rated intermediate. What percent of the runs are intermediate?

18. At Telluride ski area in Colorado, about 28 of the 115 total ski runs are rated easy. What percent of the runs are easy?

For each food described, find the percent of total calories from fat. If necessary, round to the nearest tenth percent. See Example 2.

19. Ranch dressing, serving size of 2 tablespoons

	Calories
Total	40
From fat	20

20. Unsweetened cocoa powder, serving size of 1 tablespoon

	Calories
Total	20
From fat	5

21.

Nutrition Facts
Serving Size 1 pouch (20g)
Servings Per Container 6

Amount Per Serving	
Calories	80
Calories from fat	10
	% Daily Value*
Total Fat 1g	2%
Sodium 45mg	2%
Total Carbohydrate 17g	6%
Sugars 9g	
Protein 0g	
Vitamin C	25%

Not a significant source of saturated fat, cholesterol, dietary fiber, vitamin A, calcium and iron.

*Percent Daily Values are based on a 2,000 calorie diet.

Artificial Fruit Snacks

22.

Nutrition Facts
Serving Size $\frac{1}{4}$ cup (33g)
Servings Per Container About 9

Amount Per Serving	
Calories 190 Calories from Fat 130	
	% Daily Value
Total Fat 16g	24%
Saturated Fat 3g	16%
Cholesterol 0mg	0%
Sodium 135mg	6%
Total Carbohydrate 9g	3%
Dietary Fiber 1g	5%
Sugars 2g	
Protein 5g	

Vitamin A 0% • Vitamin C 0%
Calcium 0% • Iron 8%

Peanut Mixture

23.

Nutrition Facts
Serving Size 18 crackers (29g)
Servings Per Container About 9

Amount Per Serving	
Calories 120 Calories from Fat 35	
	% Daily Value*
Total Fat 4g	6%
Saturated Fat 0.5g	3%
Polyunsaturated Fat 0g	
Monounsaturated Fat 1.5g	
Cholesterol 0mg	0%
Sodium 220mg	9%
Total Carbohydrate 21g	7%
Dietary Fiber 2g	7%
Sugars 3g	
Protein 2g	

Vitamin A 0% • Vitamin C 0%
Calcium 2% • Iron 4%
Phosphorus 10%

Snack Crackers

24.

Nutrition Facts
Serving Size 28 crackers (31g)
Servings Per Container About 6

Amount Per Serving	
Calories 130 Calories from Fat 35	
	% Daily Value*
Total Fat 4g	6%
Saturated Fat 2g	10%
Polyunsaturated Fat 1g	
Monounsaturated Fat 1g	
Cholesterol 0mg	0%
Sodium 470mg	20%
Total Carbohydrate 23g	8%
Dietary Fiber 1g	4%
Sugars 4g	
Protein 2g	

Vitamin A 0% • Vitamin C 0%
Calcium 0% • Iron 2%

Snack Crackers

Solve. If necessary, round money amounts to the nearest cent and all other amounts to the nearest tenth. See Examples 1 through 4.

25. A family paid $26,250 as a down payment for a home. If this represents 15% of the price of the home, find the price of the home.

26. A banker learned that $842.40 is withheld from his monthly check for taxes and insurance. If this represents 18% of his total pay, find the total pay.

27. An owner of a repair service company estimates that for every 40 hours a repairperson is on the job, he can bill for only 78% of the hours. The remaining hours, the repairperson is idle or driving to or from a job. Determine the number of hours per 40-hour week the owner can bill for a repairperson.

28. A manufacturer of electronic components expects 1.04% of its products to be defective. Determine the number of defective components expected in a batch of 28,350 components. Round to the nearest whole component.

29. A car manufacturer announced that next year, the price of a certain model of car will increase by 4.5%. This year the price is $19,286. Find the increase in price and the new price.

30. A union contract calls for a 6.5% salary increase for all employees. Determine the increase and the new salary that a worker currently making $58,500 under this contract can expect.

A popular extreme sport is artificial wall climbing. The photo shown is an artificial climbing wall. Exercises 31–32 are about the Footsloggers Climbing Tower in Boone, North Carolina.

31. A climber is resting at a height of 21 feet while on the Footsloggers Climbing Tower. If this is 60% of the tower's total height, find the height of the tower.

32. A group plans to climb the Footsloggers Climbing Tower at the group rate, once they save enough money. Thus far, $126 has been saved. If this is 70% of the total amount needed for the group, find the total price.

33. Tuition for an Ohio resident at the Columbus campus of Ohio State University was $8679 in 2008. The tuition increased by 15.3% during the period from 2008 to 2013. Find the increase, and the tuition for the 2013–2014 school year. Round the increase to the nearest whole dollar. (*Source:* Ohio State University)

34. The population of Americans aged 65 and older was 40 million in 2010. That population is projected to increase by 80.5% by 2030. Find the increase and the projected 2030 population. (*Source:* Bureau of the Census)

35. From 2013–2014 to 2020–2021, the number of associate degrees awarded is projected to increase by 17.4%. If the number of associate degrees awarded in 2013–2014 was 943,000, find the increase and the projected number of associate degrees awarded in the 2020–2021 school year. (*Source:* National Center for Education Statistics)

36. From 2013–2014 to 2020–2021, the number of bachelor degrees awarded is projected to increase by 7.6%. If the number of bachelor degrees awarded in 2013–2014 was 1,836,000, find the increase and the projected number of bachelor degrees awarded in the 2020–2021 school year. (*Source:* National Center for Education Statistics)

Objective B *Find the amount of increase and the percent increase. See Example 5.*

	Original Amount	New Amount	Amount of Increase	Percent Increase
37.	50	80		
38.	8	12		
39.	65	117		
40.	68	170		

Find the amount of decrease and the percent decrease. See Example 6.

	Original Amount	New Amount	Amount of Decrease	Percent Decrease
41.	8	6		
42.	25	20		
43.	160	40		
44.	200	162		

Solve. Round percents to the nearest tenth, if necessary. See Examples 5 and 6.

45. There are 150 calories in a cup of whole milk and only 84 in a cup of skim milk. In switching to skim milk, find the percent decrease in number of calories per cup.

46. In reaction to a slow economy, the number of employees at a soup company decreased from 530 to 477. What was the percent decrease in the number of employees?

47. Before taking a typing course, Geoffry Landers could type 32 words per minute. By the end of the course, he was able to type 76 words per minute. Find the percent increase.

48. The number of cable TV systems recently decreased from 10,845 to 10,700. Find the percent decrease.

49. The number of cell sites in the United States was 178,025 in 2005. By 2012, the number of cell sites had increased to 301,779. What was the percent increase? (*Source:* CTIA—The Wireless Association)

50. The population of Japan is expected to decrease from 127,799 thousand in 2011 to 97,076 thousand in 2050. Find the percent decrease. (*Source:* International Programs Center, Bureau of the Census, U.S. Dept. of Commerce)

51. In 2010, there were 3725 thousand elementary and secondary teachers employed in the United States. This number is expected to increase to 4205 thousand teachers in 2018. What is the percent increase? (*Source:* National Center for Education Statistics)

52. In 2010, approximately 493 thousand correctional officers were employed in the United States. By 2020, this number is expected to increase to 518 thousand correctional officers. What is the percent increase? (*Source:* Bureau of Labor Statistics)

53. In a recent 10-year period, the number of indoor cinema sites in the United States decreased from 5813 to 5331. What is this percent decrease? (*Source:* National Association of Theater Owners)

54. As the largest health care occupation, registered nurses held about 2.7 million jobs in 2010. The number of registered nurses is expected to be 7.1 million by 2020. What is the percent increase? (*Source:* Bureau of Labor Statistics)

Semiconductors are the foundation for solid-state electronics, including digital televisions (DTV). Businesses measure the overall decline in the production of televisions by studying DTV semiconductor revenue. (Examples of semiconductor use in DTVs include the main board for DTVs, power, LED backlighting, flat-panel TVs, and integrated circuits used in remote controls, just to name a few.) Use this graph to answer Exercises 55 and 56.

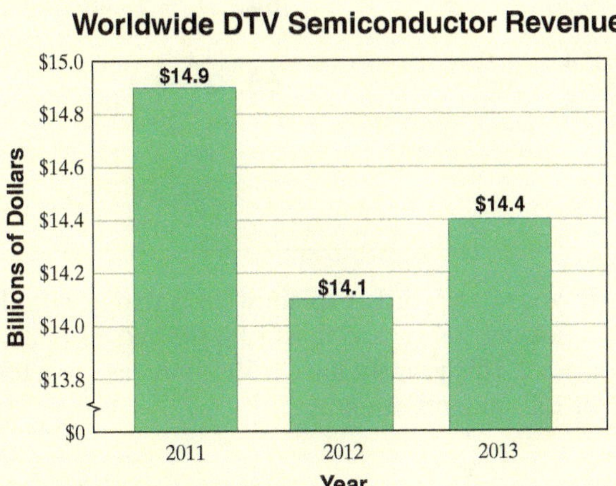

Source: IHS iSuppli Research, 2013

55. Find the percent decrease in DTV semiconductor revenue from 2011 to 2012.

56. Find the percent decrease in DTV semiconductor revenue from 2011 to 2013.

57. In 1940, the average size of a farm in the United States was 174 acres. In a recent year, the average size of a farm in the United States had increased to 420 acres. What is this percent increase? (*Source:* National Agricultural Statistics Service)

58. In 1994, there were 784 deaths from boating accidents in the United States. By 2012, the number of deaths from boating accidents had decreased to 651. What was the percent decrease? (*Source:* U.S. Coast Guard)

In 1999, Napster, a free online file-sharing service, debuted. iTunes, which debuted in 2003, is given credit for getting people to start paying for digital music. This 4-year gap between the starts of these two companies was only the beginning of the decline in music revenue.

59. In 1999, total revenue from U.S. music sales and licensing was $14.6 billion. It was forecasted that this number would continue to drop until it reached $5.5 billion in 2014. Find this percent decrease in music revenue.

60. By comparing prices, a particular music album downloads from a low of $2.99 to a high of $7.99. Find the percent increase from $2.99 to $7.99.

Review

Perform each indicated operation. See Sections 4.3 through 4.5, 4.7, and 5.2 through 5.4.

61. 0.12×38

62. $29.4 \div 0.7$

63. $9.20 + 1.98$

64. $78 - 19.46$

65. $-\frac{3}{8} + \frac{5}{12}$

66. $\left(-\frac{3}{8}\right)\left(-\frac{5}{12}\right)$

67. $2\frac{4}{5} \div 3\frac{9}{10}$

68. $2\frac{4}{5} - 3\frac{9}{10}$

Concept Extensions

69. If a number is increased by 100%, how does the increased number compare with the original number? Explain your answer.

70. In your own words, explain what is wrong with the following statement: "Last year we had 80 students attend. This year we have a 50% increase or a total of 160 students attend."

Use the bar and circle graphs below for Exercises 71 and 72.

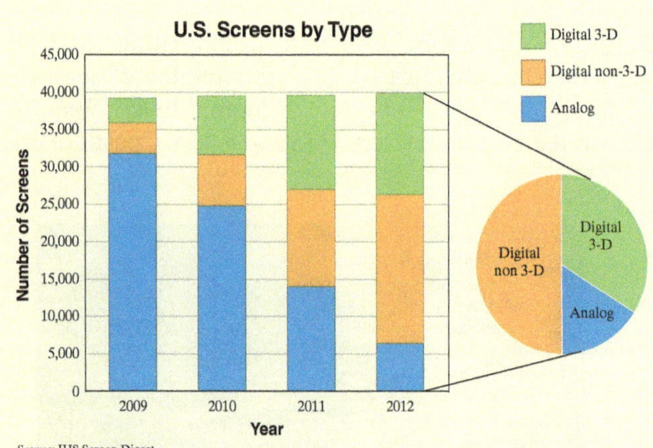

Source: IHS Screen Digest

71. Use the last bar in the bar graph, the 2012 bar. Are your answers for Exercises **7** and **8** reasonable? Explain why or why not.

72. Use the circle graph. If the 40,000 screens represent 100% of this circle graph, do the percents given and your answers in Exercises **7** and **8** seem reasonable? Explain why or why not.

Explain what errors were made by each student when solving percent of increase or decrease problems and then correct the errors. See the Concept Checks in this section. "The population of a certain rural town was 150 in 1990, 180 in 2000, and 150 in 2010."

73. Find the percent of increase in population from 1980 to 1990.

Miranda's solution: Percent of increase $= \frac{30}{180} = 0.1\overline{6} \approx 16.7\%$

74. Find the percent of decrease in population from 1990 to 2000.

Jeremy's solution: Percent of decrease $= \frac{30}{150} = 0.20 = 20\%$

75. The percent of increase from 1980 to 1990 is the same as the percent of decrease from 1990 to 2000. True or false?

Chris's answer: True because they had the same amount of increase as the amount of decrease.

6.6 Percent and Problem Solving: Sales Tax, Commission, and Discount

Objectives

A Calculate Sales Tax and Total Price.

B Calculate Commissions.

C Calculate Discount and Sale Price.

Objective A Calculating Sales Tax and Total Price

Percents are frequently used in the retail trade. For example, most states charge a tax on certain items when purchased. This tax is called a **sales tax,** and retail stores collect it for the state. Sales tax is almost always stated as a percent of the purchase price.

A 9% sales tax rate on a purchase of a \$10 calculator gives a sales tax of

sales tax = 9% of \$10 = 0.09 · \$10.00 = \$0.90

The total price to the customer would be

purchase price plus sales tax
↓ ↓ ↓
\$10.00 + \$0.90 = \$10.90

This example suggests the following equations:

Sales Tax and Total Price

sales tax = tax rate · purchase price
total price = purchase price + sales tax

In this section we round dollar amounts to the nearest cent.

Example 1 Finding Sales Tax and Purchase Price

Find the sales tax and the total price on the purchase of an \$85.50 atlas in a city where the sales tax rate is 7.5%.

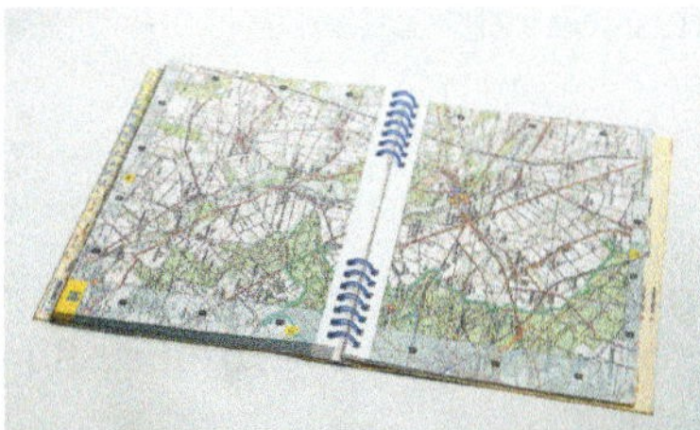

Solution: The purchase price is \$85.50 and the tax rate is 7.5%.

sales tax = tax rate · purchase price
↓ ↓ ↙
sales tax = 7.5% · \$85.50
= 0.075 · \$85.5 Write 7.5% as a decimal.
≈ \$6.41 Rounded to the nearest cent

Thus, the sales tax is \$6.41. Next find the total price.

total price = purchase price + sales tax
↓ ↓ ↙
total price = \$85.50 + \$6.41
= \$91.91

The sales tax on \$85.50 is \$6.41, and the total price is \$91.91.

Work Practice 1

Practice 1

If the sales tax rate is 8.5%, what is the sales tax and the total amount due on a \$59.90 Goodgrip tire? (Round the sales tax to the nearest cent.)

Answer

1. tax: \$5.09; total: \$64.99

✓**Concept Check** The purchase price of a textbook is \$50 and sales tax is 10%. If you are told by the cashier that the total price is \$75, how can you tell that a mistake has been made?

Practice 2

The sales tax on an \$18,500 automobile is \$1665. Find the sales tax rate.

Example 2 Finding a Sales Tax Rate

The sales tax on a \$310 Sony flat-screen, digital 32-inch television is \$26.35. Find the sales tax rate.

Solution: Let r represent the unknown sales tax rate. Then

sales tax = tax rate · purchase price

$$\$26.35 = r \cdot \$310$$

$$\frac{26.35}{310} = \frac{r \cdot 310}{310} \quad \text{Divide both sides by 310.}$$

$$0.085 = r \quad \text{Simplify.}$$

$$8.5\% = r \quad \text{Write 0.085 as a percent.}$$

The sales tax rate is 8.5%.

Work Practice 2

Objective B Calculating Commissions 

A **wage** is payment for performing work. Hourly wage, commissions, and salary are some of the ways wages can be paid. Many people who work in sales are paid a commission. An employee who is paid a **commission** is paid a percent of his or her total sales.

Commission

commission = commission rate · sales

Practice 3

A sales representative for Office Product Copiers sold \$47,632 worth of copy equipment and supplies last month. What is his commission for the month if he is paid a commission of 6.6% of his total sales for the month?

Example 3 Finding the Amount of Commission

Sherry Souter, a real estate broker for Wealth Investments, sold a house for \$214,000 last week. If her commission is 1.5% of the selling price of the home, find the amount of her commission.

Solution:

commission = commission rate · sales

$$\text{commission} = 1.5\% \cdot \$214{,}000$$

$$= 0.015 \cdot \$214{,}000 \quad \text{Write 1.5\% as 0.015.}$$

$$= \$3210 \quad \text{Multiply.}$$

Her commission on the house is \$3210.

Work Practice 3

Answers

2. 9% 3. \$3143.71

✓**Concept Check Answer**

Since $10\% = \frac{1}{10}$, the sales tax is $\frac{\$50}{10} = \$5$. The total price should be \$55.

Example 4 Finding a Commission Rate

A salesperson earned \$1560 for selling \$13,000 worth of electronics equipment. Find the commission rate.

Solution: Let r stand for the unknown commission rate. Then

commission = commission rate · sales

$$\$1560 = r \cdot \$13{,}000$$

$$\frac{1560}{13{,}000} = r \quad \text{Divide 1560 by 13,000, the number multiplied by } r.$$

$$0.12 = r \quad \text{Simplify.}$$

$$12\% = r \quad \text{Write 0.12 as a percent.}$$

The commission rate is 12%.

Work Practice 4

Practice 4

A salesperson earns \$645 for selling \$4300 worth of appliances. Find the commission rate.

Objective C Calculating Discount and Sale Price 

Suppose that an item that normally sells for \$40 is on sale for 25% off. This means that the **original price** of \$40 is reduced, or **discounted,** by 25% of \$40, or \$10. The **discount rate** is 25%, the **amount of discount** is \$10, and the **sale price** is \$40 − \$10, or \$30. Study the diagram below to visualize these terms.

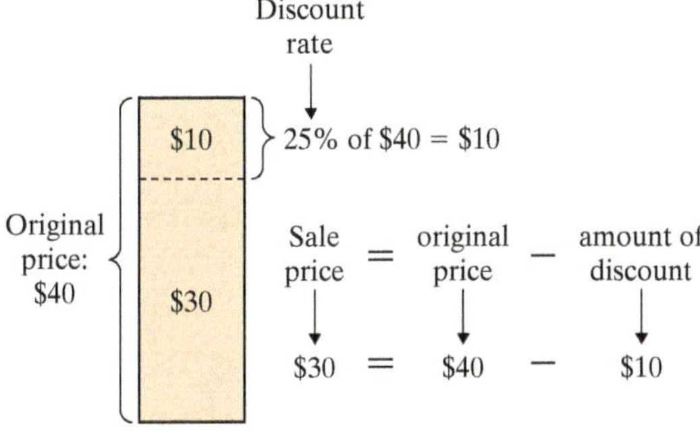

To calculate discounts and sale prices, we can use the following equations:

Discount and Sale Price

amount of discount = discount rate · original price

sale price = original price − amount of discount

Example 5 Finding a Discount and a Sale Price

An electric rice cooker that normally sells for \$65 is on sale for 25% off. What is the amount of discount and what is the sale price?

Solution: First we find the amount of discount, or simply the discount.

amount of discount = discount rate · original price

amount of discount = 25% · \$65

= 0.25 · \$65 Write 25% as 0.25.

= \$16.25 Multiply.

(*Continued on next page*)

Practice 5

A discontinued washer and dryer combo is advertised on sale for 35% off the regular price of \$700. Find the amount of discount and the sale price.

Answers

4. 15% **5.** \$245; \$455

The discount is \$16.25. Next, find the sale price.

$$\text{sale price} = \text{original price} - \text{discount}$$
$$\text{sale price} = \$65 - \$16.25$$
$$= \$48.75 \quad \text{Subtract.}$$

The sale price is \$48.75.

Work Practice 5

Vocabulary, Readiness & Video Check

Use the choices below to fill in each blank.

amount of discount   sale price   sales tax
commission   total price

1. ______________ = tax rate · purchase price
2. ______________ = purchase price + sales tax
3. ______________ = commission rate · sales
4. ______________ = discount rate · original price
5. ______________ = original price − amount of discount

Martin-Gay Interactive Videos *Watch the section lecture video and answer the following questions.*

See Video 6.6 

Objective A 6. In Example 1, what is our first step after translating the problem into an equation?

Objective B 7. What is our final step in solving Example 2?

Objective C 8. In the lecture before Example 3, since both equations shown involve the "amount of discount," how can the two equations be combined into one equation?

6.6 Exercise Set MyMathLab®

Objective A *Solve. See Examples 1 and 2.*

1. What is the sales tax on a jacket priced at \$150 if the sales tax rate is 5%?

2. If the sales tax rate is 6%, find the sales tax on a microwave oven priced at \$188.

3. The purchase price of a camcorder is \$799. What is the total price if the sales tax rate is 7.5%?

4. A stereo system has a purchase price of \$426. What is the total price if the sales tax rate is 8%?

5. A new large-screen television has a purchase price of \$4790. If the sales tax on this purchase is \$335.30, find the sales tax rate.

6. The sales tax on the purchase of a \$6800 used car is \$374. Find the sales tax rate.

7. The sales tax on a table saw is $10.20.
   a. What is the purchase price of the table saw (before tax) if the sales tax rate is 8.5%? (*Hint:* Use the sales tax equation and insert the replacement values.)
   b. Find the total price of the table saw.

8. The sales tax on a one-half-carat diamond ring is $76.
   a. Find the purchase price of the ring (before tax) if the sales tax rate is 9.5%. (See the hint for Exercise **7a.**)
   b. Find the total price of the ring.

9. A gold and diamond bracelet sells for $1800. Find the sales tax and the total price if the sales tax rate is 6.5%.

10. The purchase price of a personal computer is $1890. If the sales tax rate is 8%, what is the sales tax and the total price?

11. The sales tax on the purchase of a futon is $24.25. If the tax rate is 5%, find the purchase price of the futon.

12. The sales tax on the purchase of a TV-DVD combination is $32.85. If the tax rate is 9%, find the purchase price of the TV-DVD.

13. The sales tax is $98.70 on a stereo sound system purchase of $1645. Find the sales tax rate.

14. The sales tax is $103.50 on a necklace purchase of $1150. Find the sales tax rate.

15. A cell phone costs $210, a battery recharger costs $15, and batteries cost $5. What is the sales tax and total price for purchasing these items if the sales tax rate is 7%?

16. Ms. Warner bought a blouse for $35, a skirt for $55, and a blazer for $95. Find the sales tax and the total price she paid, given a sales tax rate of 6.5%.

Objective B *Solve. See Examples 3 and 4.*

17. A sales representative for a large furniture warehouse is paid a commission rate of 4%. Find her commission if she sold $1,329,401 worth of furniture last year.

18. Rosie Davis-Smith is a beauty consultant for a home cosmetic business. She is paid a commission rate of 12.8%. Find her commission if she sold $1638 in cosmetics last month.

19. A salesperson earned a commission of $1380.40 for selling $9860 worth of paper products. Find the commission rate.

20. A salesperson earned a commission of $3575 for selling $32,500 worth of books to various bookstores. Find the commission rate.

21. How much commission will Jack Pruet make on the sale of a $325,900 house if he receives 1.5% of the selling price?

22. Frankie Lopez sold $9638 of jewelry this week. Find her commission for the week if she receives a commission rate of 5.6%.

23. A real estate agent earned a commission of $5565 for selling a house. If his rate is 3%, find the selling price of the house. (*Hint:* Use the commission equation and insert the replacement values.)

24. A salesperson earned $1750 for selling fertilizer. If her commission rate is 7%, find the selling price of the fertilizer. (See the hint for Exercise **23.**)

Objective C *Find the amount of discount and the sale price. See Example 5.*

	Original Price	Discount Rate	Amount of Discount	Sale Price
25.	$89	10%		
26.	$74	20%		
27.	$196.50	50%		
28.	$110.60	40%		
29.	$410	35%		
30.	$370	25%		
31.	$21,700	15%		
32.	$17,800	12%		

33. A $300 fax machine is on sale for 15% off. Find the amount of discount and the sale price.

34. A $4295 designer dress is on sale for 30% off. Find the amount of discount and the sale price.

Objectives A B Mixed Practice *Complete each table.*

	Purchase Price	Tax Rate	Sales Tax	Total Price
35.	$305	9%		
36.	$243	8%		
37.	$56	5.5%		
38.	$65	8.4%		

	Sale	Commission Rate	Commission
39.	$235,800	3%	
40.	$195,450	5%	
41.	$17,900		$1432
42.	$25,600		$2304

Review

Multiply. See Sections 4.3, 5.3, and 5.5.

43. $2000 \cdot \frac{3}{10} \cdot 2$

44. $500 \cdot \frac{2}{25} \cdot 3$

45. $400 \cdot \frac{3}{100} \cdot 11$

46. $1000 \cdot \frac{1}{20} \cdot 5$

47. $600 \cdot 0.04 \cdot \frac{2}{3}$

48. $6000 \cdot 0.06 \cdot \frac{3}{4}$

Concept Extensions

Solve. See the Concept Check in this section.

49. Your purchase price is \$68 and the sales tax rate is 9.5%. Round each amount and use the rounded amounts to estimate the total price. Choose the best estimate.
a. \$105 **b.** \$58 **c.** \$93 **d.** \$77

50. Your purchase price is \$200 and the tax rate is 10%. Choose the best estimate of the total price.
a. \$190 **b.** \$210 **c.** \$220 **d.** \$300

Tipping *One very useful application of percent is mentally calculating a tip. Recall that to find 10% of a number, simply move the decimal point one place to the left. To find 20% of a number, just double 10% of the number. To find 15% of a number, find 10% and then add to that number half of the 10% amount. Mentally fill in the chart below. To do so, start by rounding the bill amount to the nearest dollar.*

Tipping Chart

	Bill Amount	10%	15%	20%
51.	\$40.21			
52.	\$15.89			
53.	\$72.17			
54.	\$9.33			

55. Suppose that the original price of a shirt is \$50. Which is better, a 60% discount or a discount of 30% followed by a discount of 35% of the reduced price? Explain your answer.

56. Which is better, a 30% discount followed by an additional 25% off or a 20% discount followed by an additional 40% off? To see, suppose an item costs \$100 and calculate each discounted price. Explain your answer.

57. A diamond necklace sells for \$24,966. If the tax rate is 7.5%, find the total price.

58. A house recently sold for \$562,560. The commission rate on the sale is 5.5%. If the real estate agent is to receive 60% of the commission, find the amount received by the agent.

6.7 Percent and Problem Solving: Interest 

Objectives

A Calculate Simple Interest.

B Calculate Compound Interest.

Objective A Calculating Simple Interest

Interest is money charged for using other people's money. When you borrow money, you pay interest. When you loan or invest money, you earn interest. The money borrowed, loaned, or invested is called the **principal amount,** or simply **principal.** Interest is normally stated in terms of a percent of the principal for a given period of time. The **interest rate** is the percent used in computing the interest. Unless stated otherwise, *the rate is understood to be per year.* When the interest is computed on the original principal, it is called **simple interest.** Simple interest is calculated using the following equation:

Simple Interest

$$\text{Simple Interest} = \text{Principal} \cdot \text{Rate} \cdot \text{Time}$$
$$I = P \cdot R \cdot T$$

where the rate is understood to be per year and time is in years.

Practice 1

Find the simple interest after 5 years on $875 at an interest rate of 7%.

Practice 2

A student borrowed $1500 for 9 months on her credit card at a simple interest rate of 20%. How much interest did she pay?

Answers
1. $306.25 **2.** $225

✓Concept Check Answer
$16,000 is too much interest

Example 1 Finding Simple Interest

Find the simple interest after 2 years on $500 at an interest rate of 12%.

Solution: In this example, $P = \$500$, $R = 12\%$, and $T = 2$ years. Replace the variables with values in the formula $I = PRT$.

$$I = P \cdot R \cdot T$$
$$I = \$500 \cdot 12\% \cdot 2 \quad \text{Let } P = \$500, R = 12\%, \text{ and } T = 2.$$
$$= \$500 \cdot (0.12) \cdot 2 \quad \text{Write 12\% as a decimal.}$$
$$= \$120 \quad \text{Multiply.}$$

The simple interest is $120.

Work Practice 1

If time is not given in years, we need to convert the given time to years.

Example 2 Finding Simple Interest

A recent college graduate borrowed $2400 at 10% simple interest for 8 months to buy a used Toyota Corolla. Find the simple interest he paid.

Solution: Since there are 12 months in a year, we first find what part of a year 8 months is.

$$8 \text{ months} = \frac{8}{12} \text{ year} = \frac{2}{3} \text{ year}$$

Now we find the simple interest.

$$I = P \cdot R \cdot T$$
$$= \$2400 \cdot (0.10) \cdot \frac{2}{3} \quad \text{Let } P = \$2400, R = 10\% \text{ or } 0.10, \text{ and } T = \frac{2}{3}.$$
$$= \$160$$

The interest on his loan is $160.

Work Practice 2

✓Concept Check Suppose in Example 2 you had obtained an answer of $16,000. How would you know that you had made a mistake in this problem?

When money is borrowed, the borrower pays the original amount borrowed, or the principal, as well as the interest. When money is invested, the investor receives the original amount invested, or the principal, as well as the interest. In either case, the **total amount** is the sum of the principal and the interest.

Finding the Total Amount of a Loan or Investment

total amount (paid or received) = principal + interest

Example 3 Finding the Total Amount of an Investment

An accountant invested \$2000 at a simple interest rate of 10% for 2 years. What total amount of money will she have from her investment in 2 years?

Solution: First we find her interest.

$$I = P \cdot R \cdot T$$
$$= \$2000 \cdot (0.10) \cdot 2 \quad \text{Let } P = \$2000,\ R = 10\% \text{ or } 0.10, \text{ and } T = 2.$$
$$= \$400$$

The interest is \$400.

Next, we add the interest to the principal.

total amount	=	principal	+	interest
↓		↓		↓
total amount	=	\$2000	+	\$400
	=	\$2400		

After 2 years, she will have a total amount of \$2400.

Work Practice 3

Practice 3

If \$2100 is borrowed at a simple interest rate of 13% for 6 months, find the total amount paid.

✓**Concept Check** Which investment would earn more interest: an amount of money invested at 8% interest for 2 years, or the same amount of money invested at 8% for 3 years? Explain.

Objective B Calculating Compound Interest

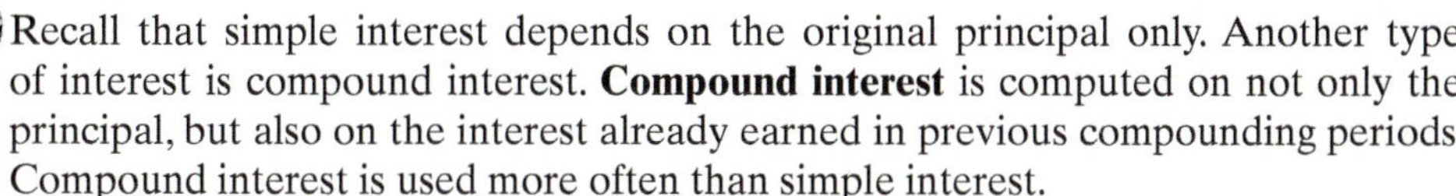

Recall that simple interest depends on the original principal only. Another type of interest is compound interest. **Compound interest** is computed on not only the principal, but also on the interest already earned in previous compounding periods. Compound interest is used more often than simple interest.

Let's see how compound interest differs from simple interest. Suppose that \$2000 is invested at 7% interest **compounded annually** for 3 years. This means that interest is added to the principal at the end of each year and that next year's interest is computed on this new amount. In this section, we round dollar amounts to the nearest cent.

	Amount at Beginning of Year	Principal	·	Rate	·	Time	= Interest	Amount at End of Year
1st year	\$2000	\$2000	·	0.07	·	1	= \$140	\$2000 + 140 = \$2140
2nd year	\$2140	\$2140	·	0.07	·	1	= \$149.80	\$2140 + 149.80 = \$2289.80
3rd year	\$2289.80	\$2289.80	·	0.07	·	1	= \$160.29	\$2289.80 + 160.29 = \$2450.09

The compound interest earned can be found by

total amount	−	original principal	=	compound interest
↓		↓		↓
\$2450.09	−	\$2000	=	\$450.09

The simple interest earned would have been

principal	·	rate	·	time	=	interest
↓		↓		↓		↓
\$2000	·	0.07	·	3	=	\$420

Since compound interest earns "interest on interest," compound interest earns more than simple interest.

Computing compound interest using the method above can be tedious. We can use a calculator and the compound interest formula on the next page to compute compound interest more quickly.

Answer

3. \$2236.50

✓**Concept Check Answer**

8% for 3 years. Since the interest rate is the same, the longer you keep the money invested, the more interest you earn.

Compound Interest Formula

The total amount A in an account is given by

$$A = P\left(1 + \frac{r}{n}\right)^{n \cdot t}$$

where P is the principal, r is the interest rate written as a decimal, t is the length of time in years, and n is the number of times compounded per year.

Practice 4

\$3000 is invested at 4% interest compounded annually. Find the total amount after 6 years.

Example 4 \$1800 is invested at 2% interest compounded annually. Find the total amount after 3 years.

Solution: "Compounded annually" means 1 time a year, so

$n = 1$. Also, $P = \$1800$, $r = 2\% = 0.02$, and $t = 3$ years.

$$\begin{aligned} A &= P\left(1 + \frac{r}{n}\right)^{n \cdot t} \\ &= 1800\left(1 + \frac{0.02}{1}\right)^{1 \cdot 3} \\ &= 1800(1.02)^3 \\ &\approx 1910.17 \quad \text{Round to 2 decimal places.} \end{aligned}$$

Helpful Hint Remember order of operations. **First** evaluate $(1.02)^3$, then multiply by 1800.

The total amount at the end of 3 years is \$1910.17.

Work Practice 4

Practice 5

\$5500 is invested at $6\frac{1}{4}\%$ compounded *daily* for 5 years. Find the total amount at the end of 5 years. (Use 1 year = 365 days.)

Example 5 Finding Total Amount Received from an Investment

\$4000 is invested at 5.3% compounded quarterly for 10 years. Find the total amount at the end of 10 years.

Solution: "Compounded quarterly" means 4 times a year, so

$n = 4$. Also, $P = \$4000$, $r = 5.3\% = 0.053$, and $t = 10$ years.

$$\begin{aligned} A &= P\left(1 + \frac{r}{n}\right)^{n \cdot t} \\ &= 4000\left(1 + \frac{0.053}{4}\right)^{4 \cdot 10} \\ &= 4000(1.01325)^{40} \\ &\approx 6772.12 \end{aligned}$$

The total amount after 10 years is \$6772.12.

Work Practice 5

Answers
4. \$3795.96 **5.** \$7517.41

Calculator Explorations Compound Interest Formula

For a review of using your calculator to evaluate compound interest, see this box.

Let's review the calculator keys pressed to evaluate the expression in Example 5,

$$4000\left(1 + \frac{0.053}{4}\right)^{40}$$

To evaluate, press the keys

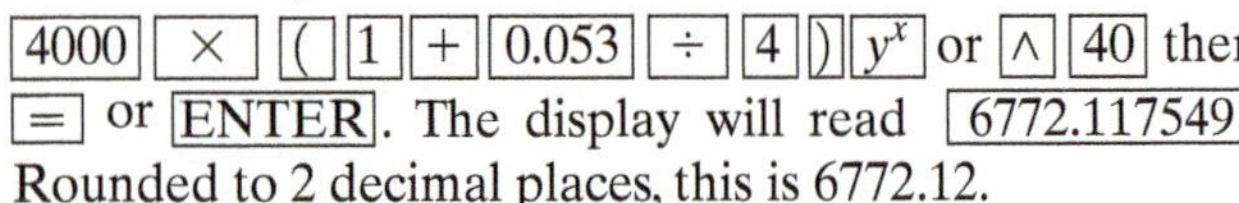

4000 × (1 + 0.053 ÷ 4) y^x or ^ 40 then = or ENTER. The display will read 6772.117549. Rounded to 2 decimal places, this is 6772.12.

Find the compound interest.

1. \$600, 5 years, 9%, compounded quarterly
2. \$10,000, 15 years, 4%, compounded daily
3. \$1200, 20 years, 11%, compounded annually
4. \$5800, 1 year, 7%, compounded semiannually
5. \$500, 4 years, 6%, compounded quarterly
6. \$2500, 19 years, 5%, compounded daily

Vocabulary, Readiness & Video Check

Use the choices below to fill in each blank. Choices may be used more than once.

total amount   simple   principal amount   compound

1. To calculate ________________ interest, use $I = P \cdot R \cdot T$.
2. To calculate ________________ interest, use $A = P\left(1 + \frac{r}{n}\right)^{n \cdot t}$.
3. ________________ interest is computed not only on the original principal, but also on interest already earned in previous compounding periods.
4. When interest is computed on the original principal only, it is called ________________ interest.
5. ________________ (paid or received) = principal + interest.
6. The ________________ is the money borrowed, loaned, or invested.

Martin-Gay Interactive Videos *Watch the section lecture video and answer the following questions.*

See Video 6.7

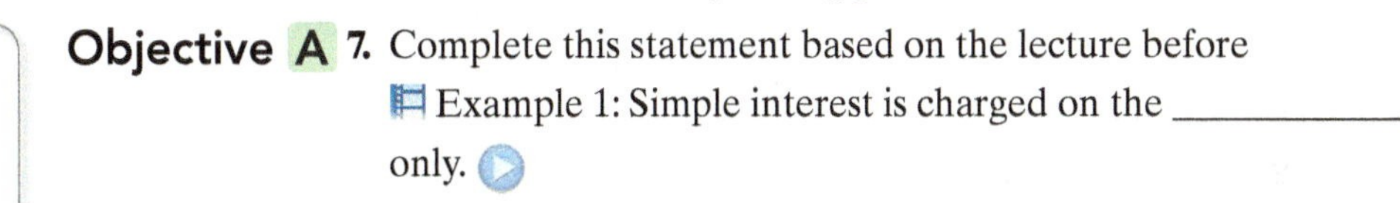

Objective A 7. Complete this statement based on the lecture before Example 1: Simple interest is charged on the ____________ only.

Objective B 8. In Example 2, how often is the interest compounded, and what number does this translate to in the formula?

6.7 Exercise Set MyMathLab® 

Objective A *Find the simple interest. See Examples 1 and 2.*

	Principal	Rate	Time
1.	\$200	8%	2 years
3.	\$160	11.5%	4 years
5.	\$5000	10%	$1\frac{1}{2}$ years
7.	\$375	18%	6 months
9.	\$2500	16%	21 months

	Principal	Rate	Time
2.	\$800	9%	3 years
4.	\$950	12.5%	5 years
6.	\$1500	14%	$2\frac{1}{4}$ years
8.	\$775	15%	8 months
10.	\$1000	10%	18 months

Solve. See Examples 1 through 3.

11. A company borrows \$162,500 for 5 years at a simple interest rate of 12.5%. Find the interest paid on the loan and the total amount paid back.

12. \$265,000 is borrowed to buy a house. If the simple interest rate on the 30-year loan is 8.25%, find the interest paid on the loan and the total amount paid back.

13. A money market fund advertises a simple interest rate of 9%. Find the total amount received on an investment of \$5000 for 15 months.

14. The Real Service Company takes out a 270-day (9-month) short-term, simple interest loan of \$4500 to finance the purchase of some new equipment. If the interest rate is 14%, find the total amount that the company pays back.

15. Marsha borrows \$8500 and agrees to pay it back in 4 years. If the simple interest rate is 17%, find the total amount she pays back.

16. An 18-year-old is given a high school graduation gift of \$2000. If this money is invested at 8% simple interest for 5 years, find the total amount.

Objective B *Find the total amount in each compound interest account. See Examples 4 and 5.*

17. \$6150 is compounded semiannually at a rate of 14% for 15 years.

18. \$2060 is compounded annually at a rate of 15% for 10 years.

19. \$1560 is compounded daily at a rate of 8% for 5 years.

20. \$1450 is compounded quarterly at a rate of 10% for 15 years.

21. \$10,000 is compounded semiannually at a rate of 9% for 20 years.

22. \$3500 is compounded daily at a rate of 8% for 10 years.

23. \$2675 is compounded annually at a rate of 9% for 1 year.

24. \$6375 is compounded semiannually at a rate of 10% for 1 year.

25. \$2000 is compounded annually at a rate of 8% for 5 years.

26. \$2000 is compounded semiannually at a rate of 8% for 5 years.

27. \$2000 is compounded quarterly at a rate of 8% for 5 years.

28. \$2000 is compounded daily at a rate of 8% for 5 years.

Review

Find the perimeter of each figure. See Section 1.3.

△ **29.**

Rectangle

6 yards

10 yards

△ **30.**

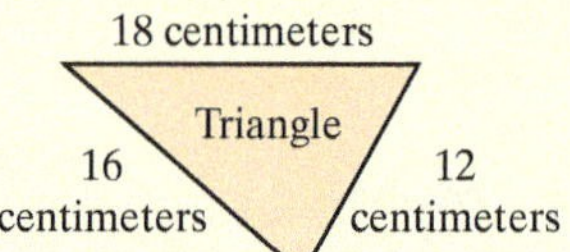

△ **31.**

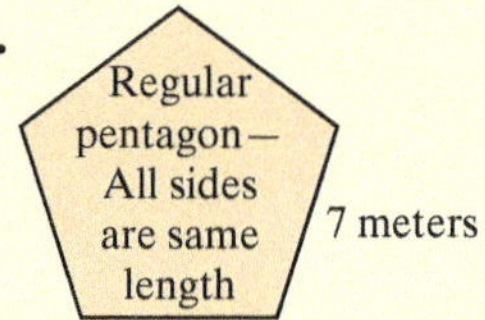

△ **32.**

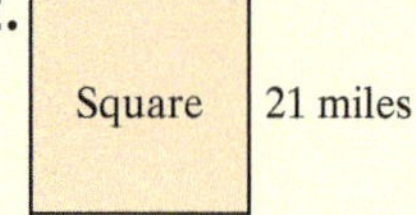

Perform the indicated operations. See Section 4.3 through 4.6.

33. $\frac{x}{4} + \frac{x}{5}$

34. $-\frac{x}{4} \div \left(-\frac{x}{5}\right)$

35. $\left(\frac{2}{3}\right)\left(-\frac{1}{3}\right) - \left(\frac{9}{10}\right)\left(\frac{2}{5}\right)$

36. $\frac{3}{11} \div \frac{9}{22} \cdot \frac{1}{3}$

Concept Extensions

37. Review Exercises **25** through **28**. As the number of compoundings increases per year, how is the total amount affected?

38. Explain how to find the amount of interest in a compounded account.

39. Compare the following accounts: Account 1: \$1000 is invested for 10 years at a simple interest rate of 6%. Account 2: \$1000 is compounded semiannually at a rate of 6% for 10 years. Discuss how the interest is computed for each account. Determine which account earns more interest. Why?

Chapter 6 Group Activity

Investigating Scale Drawings

Sections 6.1

Materials:

- ruler
- tape measure
- grid paper (optional)

This activity may be completed by working in groups or individually.

Scale drawings are used by architects, engineers, interior designers, shipbuilders, and others. In a scale drawing, each unit measurement on the drawing represents a fixed length on the object being drawn. For instance, in an architect's scale drawing, 1 inch on the drawing may represent 10 feet on a building. The scale describes the relationship between the measurements. If the measurements have the same units, the scale can be expressed as a ratio. In this case, the ratio would be 1 : 120, representing 1 inch to 120 inches (or 10 feet).

Use a ruler and the scale drawing of a college building below to answer the following questions.

1. How wide are each of the front doors of the college building?
2. How long is the front of the college building?
3. How tall is the front of the college building?

Now you will draw your own scale floor plan. First choose a room to draw—it can be your math classroom, your living room, your dormitory room, or any room that can be easily measured. Start by using a tape measure to measure the distances around the base of the walls in the room you are drawing.

4. Choose a scale for your floor plan.
5. Convert each measurement in the room you are drawing to the corresponding lengths needed for the scale drawing.
6. Complete your floor plan (you may find it helpful to use grid paper). Mark the locations of doors and windows on your floor plan. Be sure to indicate on the drawing the scale used in your floor plan.

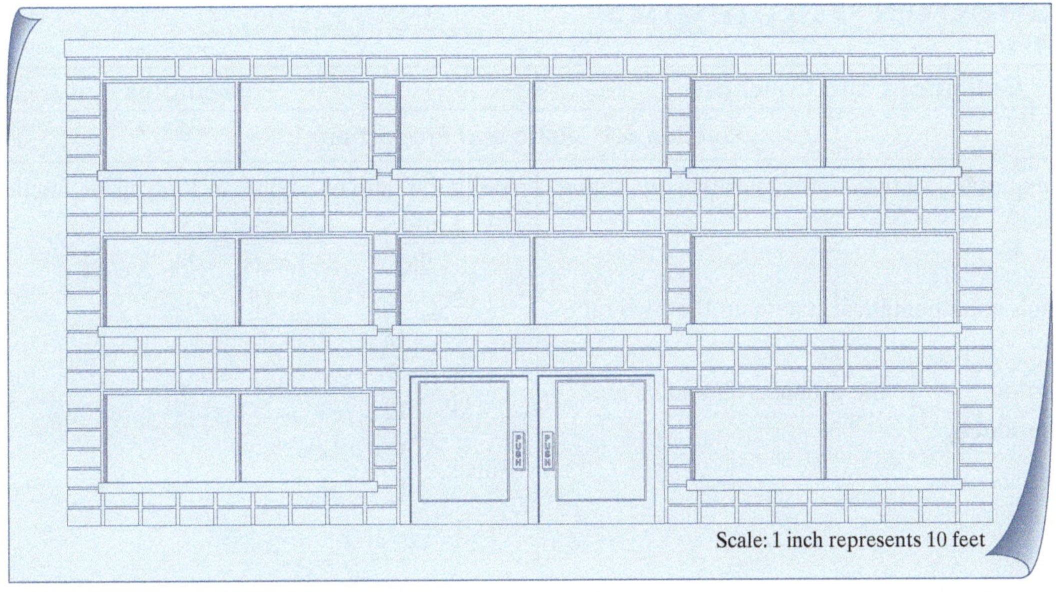

Chapter 6 Vocabulary Check

Fill in each blank with one of the words or phrases listed below. Some words may be used more than once.

percent	sales tax	is	0.01	$\frac{1}{100}$	amount of discount
percent of decrease	total price	ratio	proportion	base	of
amount	100%	compound interest	percent of increase	sale price	commission

1. In a mathematical statement, ________ usually means "multiplication."
2. In a mathematical statement, ________ means "equal."
3. ________ means "per hundred."

4. ______________ is computed not only on the principal, but also on interest already earned in previous compounding periods.

5. In the percent proportion, $\frac{\rule{1cm}{0.4pt}}{\rule{2cm}{0.4pt}} = \frac{\text{percent}}{100}$.

6. To write a decimal or fraction as a percent, multiply by ______________.

7. The decimal equivalent of the % symbol is ______________.

8. The fraction equivalent of the % symbol is ______________.

9. The percent equation is ______________ · percent = ______________.

10. ______________ $= \frac{\text{amount of decrease}}{\text{original amount}}$

11. ______________ $= \frac{\text{amount of increase}}{\text{original amount}}$.

12. ______________ = tax rate · purchase price

13. ______________ = purchase price + sales tax

14. ______________ = commission rate · sales

15. ______________ = discount rate · original price

16. ______________ = original price − amount of discount

17. A(n) ______________ is a mathematical statement that two ratios are equal.

18. A (n) ______________ is the quotient of two numbers or two quantities

Helpful Hint Are you preparing for your test? Don't forget to take the Chapter 6 Test on page 481. Then check your answers at the back of the text and use the Chapter Test Prep Videos to see the fully worked-out solutions to any of the exercises you want to review.

6 Chapter Highlights

Definitions and Concepts	Examples
Section 6.1 Ratio and Proportion	
A **ratio** is the quotient of two numbers or two quantities.	Write the ratio of 5 hours to 1 day using fractional notation. $\frac{5\text{ hours}}{1\text{ day}} = \frac{5\text{ hours}}{24\text{ hours}} = \frac{5}{24}$
A **proportion** is a mathematical statement that two ratios are equal.	$\frac{2}{3} = \frac{8}{12}$ $\quad \frac{x}{7} = \frac{15}{35}$
In the proportion $\frac{a}{b} = \frac{c}{d}$, the products ad and bc are called **cross products.**	$\frac{2}{3} = \frac{8}{12}$ → $3 \cdot 8$ or 24; → $2 \cdot 12$ or 24
If $\frac{a}{b} = \frac{c}{d}$, then $ad = bc$.	Solve: $\frac{3}{4} = \frac{x}{x-1}$ $\frac{3}{4} = \frac{x}{x-1}$ $3(x-1) = 4x$ Set cross products equal. $3x - 3 = 4x$ $-3 = x$
Section 6.2 Percents, Decimals, and Fractions	
Percent means "per hundred." The % symbol denotes percent.	$51\% = \frac{51}{100}$ 51 per 100 $7\% = \frac{7}{100}$ 7 per 100
To write a percent as a decimal, replace the % symbol with its decimal equivalent, 0.01, and multiply.	$32\% = 32(0.01) = 0.32$

Definitions and Concepts	Examples
Section 6.2 Percents, Decimals, and Fractions *(continued)*	
To write a decimal as a percent, multiply by 100%.	$0.08 = 0.08(100\%) = 08.\% = 8\%$
To write a percent as a fraction, replace the % symbol with its fraction equivalent, $\frac{1}{100}$, and multiply.	$25\% = 25 \cdot \frac{1}{100} = \frac{25}{100} = \frac{\overset{1}{\cancel{25}}}{4 \cdot \underset{1}{\cancel{25}}} = \frac{1}{4}$
To write a fraction as a percent, multiply by 100%.	$\frac{1}{6} = \frac{1}{6} \cdot 100\% = \frac{1}{6} \cdot \frac{100}{1}\% = \frac{100}{6}\% = 16\frac{2}{3}\%$
Section 6.3 Solving Percent Problems with Equations	
Three key words in the statement of a percent problem are **of,** which means multiplication (·) **is,** which means equal (=) **what** (or some equivalent word or phrase), which stands for the unknown (x)	Solve: 6 is 12% of what number? ↓ ↓ ↓ ↓ ↓ $6 = 12\% \cdot x$ $6 = 0.12 \cdot x$ Write 12% as a decimal. $\frac{6}{0.12} = \frac{0.12 \cdot x}{0.12}$ Divide both sides by 0.12. $50 = x$ Thus, 6 is 12% of 50.
Section 6.4 Solving Percent Problems with Proportions	
Percent Proportion $\frac{\text{amount}}{\text{base}} = \frac{\text{percent}}{100}$ ← always 100 or amount → $\frac{a}{b} = \frac{p}{100}$ ← percent base →	Solve: 20.4 is what percent of 85? amount percent base amount → $\frac{20.4}{85} = \frac{p}{100}$ ← percent base → $20.4 \cdot 100 = 85 \cdot p$ Set cross products equal. $2040 = 85 \cdot p$ Multiply. $\frac{2040}{85} = \frac{85 \cdot p}{85}$ Divide both sides by 85. $24 = p$ Simplify. Thus, 20.4 is 24% of 85.
Section 6.5 Applications of Percent	
Percent Increase $\text{percent increase} = \frac{\text{amount of increase}}{\text{original amount}}$ **Percent Decrease** $\text{percent decrease} = \frac{\text{amount of decrease}}{\text{original amount}}$	A town with a population of 16,480 decreased to 13,870 over a 12-year period. Find the percent decrease. Round to the nearest whole percent. $\text{amount of decrease} = 16{,}480 - 13{,}870$ $= 2610$ $\text{percent decrease} = \frac{\text{amount of decrease}}{\text{original amount}}$ $= \frac{2610}{16{,}480} \approx 0.16$ $\approx 16\%$ The town's population decreased by about 16%.

(Continued)

Definitions and Concepts	Examples
Section 6.6 Percent and Problem Solving: Sales Tax, Commission, and Discount	
Sales Tax sales tax = sales tax rate · purchase price total price = purchase price + sales tax	Find the sales tax and the total price of a purchase of \$42.00 if the sales tax rate is 9%. sales tax = sales tax rate · purchase price ↓ ↓ ↓ sales tax = 9% · \$42 = 0.09 · \$42 = \$3.78 The total price is total price = purchase price + sales tax ↓ ↓ ↓ total price = \$42.00 + \$3.78 = \$45.78 The total price is \$45.78.
Commission commission = commission rate · sales	A salesperson earns a commission of 3%. Find the commission from sales of \$12,500 worth of appliances. commission = commission rate · sales ↓ ↓ ↓ commission = 3% · \$12,500 = 0.03 · 12,500 = \$375 The commission is \$375.
Discount and Sale Price amount of discount = discount rate · original price sale price = original price − amount of discount	A suit is priced at \$320 and is on sale today for 25% off. What is the sale price? amount of discount = discount rate · original price ↓ ↓ ↓ amount of discount = 25% · \$320 = 0.25 · \$320 = \$80 sale price = original price − amount of discount ↓ ↓ ↓ sale price = \$320 − \$80 = \$240 The sale price is \$240.

Definitions and Concepts	Examples
Section 6.7 Percent and Problem Solving: Interest	
Simple Interest interest = principal · rate · time where the rate is understood to be per year, unless told otherwise.	Find the simple interest after 3 years on \$800 at an interest rate of 5%. interest = principal · rate · time interest = \$800 · 5% · 3 = \$800 · 0.05 · 3 Write 5% as 0.05. = \$120 Multiply. The interest is \$120.
Compound Interest Formula The total amount A in an account is $A = P\left(1 + \frac{r}{n}\right)^{n \cdot t}$ where P is the principal, r is the interest rate written as a decimal, t is time in years, and n is the number of times compounded per year.	\$800 is invested at 5% compounded quarterly for 10 years. Find the total amount at the end of 10 years. "Compounded quarterly" means $n = 4$. Also, $P = \$800$, $r = 5\% = 0.05$, and $t = 10$ years. $A = P\left(1 + \frac{r}{n}\right)^{n \cdot t}$ $= 800\left(1 + \frac{0.05}{4}\right)^{4 \cdot 10}$ $= 800(1.0125)^{40}$ ≈ 1314.90 The total amount is \$1314.90.

Chapter 6 Review

(6.1) *Write each phrase as a ratio in fractional notation.*

1. 20 cents to 1 dollar

2. four parts red to six parts white

Solve each proportion.

3. $\frac{x}{2} = \frac{12}{4}$

4. $\frac{20}{1} = \frac{x}{25}$

5. $\frac{32}{100} = \frac{100}{x}$

6. $\frac{20}{2} = \frac{c}{5}$

7. $\frac{2}{x - 1} = \frac{3}{x + 3}$

8. $\frac{4}{y - 3} = \frac{3}{y + 2}$

9. $\frac{y + 2}{y} = \frac{5}{3}$

10. $\frac{x - 3}{3x + 2} = \frac{2}{5}$

11. A machine can process 300 parts in 20 minutes. Find how many parts can be processed in 45 minutes.

12. As his consulting fee, Mr. Visconti charges \$90.00 per day. Find how much he charges for 3 hours of consulting. Assume an 8-hour workday.

(6.2) *Solve.*

13. In a survey of 100 adults, 37 preferred pepperoni on their pizzas. What percent preferred pepperoni?

14. A basketball player made 77 out 100 attempted free throws. What percent of free throws was made?

Write each percent as a decimal.

15. 26% **16.** 75% **17.** 3.5% **18.** 1.5%

19. 275% **20.** 400% **21.** 47.85% **22.** 85.34%

Write each decimal as a percent.

23. 1.6 **24.** 0.055 **25.** 0.076 **26.** 0.085

27. 0.71 **28.** 0.65 **29.** 6 **30.** 9

Write each percent as a fraction or mixed number in simplest form.

31. 7% **32.** 15% **33.** 25% **34.** 8.5%

35. 10.2% **36.** $16\frac{2}{3}\%$ **37.** $33\frac{1}{3}\%$ **38.** 110%

Write each fraction or mixed number as a percent.

39. $\frac{2}{5}$ **40.** $\frac{7}{10}$ **41.** $\frac{7}{12}$ **42.** $1\frac{2}{3}$

43. $1\frac{1}{4}$ **44.** $\frac{3}{5}$ **45.** $\frac{1}{16}$ **46.** $\frac{5}{8}$

(6.3) Translating *Translate each to an equation and solve.*

47. 1250 is 1.25% of what number?

48. What number is $33\frac{1}{3}\%$ of 24,000?

49. 124.2 is what percent of 540?

50. 22.9 is 20% of what number?

51. What number is 17% of 640?

52. 693 is what percent of 462?

(6.4) Translating *Translate each to a proportion and solve.*

53. 104.5 is 25% of what number?

54. 16.5 is 5.5% of what number?

55. What number is 30% of 532?

56. 63 is what percent of 35?

57. 93.5 is what percent of 85?

58. What number is 33% of 500?

(6.5) *Solve.*

59. In a survey of 2000 people, it was found that 1320 have a microwave oven. Find the percent of people who own microwaves.

60. Of the 12,360 freshmen entering County College, 2000 are enrolled in prealgebra. Find the percent of entering freshmen who are enrolled in prealgebra. Round to the nearest whole percent.

61. The number of violent crimes in a city decreased from 675 to 534. Find the percent decrease. Round to the nearest tenth percent.

62. The current charge for dumping waste in a local landfill is $16 per cubic foot. To cover new environmental costs, the charge will increase to $33 per cubic foot. Find the percent increase.

63. This year the fund drive for a charity collected $215,000. Next year, a 4% decrease is expected. Find how much is expected to be collected in next year's drive.

64. A local union negotiated a new contract that increases the hourly pay 15% over last year's pay. The old hourly rate was $11.50. Find the new hourly rate rounded to the nearest cent.

(6.6) *Solve.*

65. If the sales tax rate is 9.5%, what is the total amount charged for a $250 coat?

66. Find the sales tax paid on a $25.50 purchase if the sales tax rate is 8.5%. Round to the nearest cent.

67. Russ James is a sales representative for a chemical company and is paid a commission rate of 5% on all sales. Find his commission if he sold $100,000 worth of chemicals last month.

68. Carol Sell is a sales clerk in a clothing store. She receives a commission of 7.5% on all sales. Find her commission for the week if her sales for the week were $4005. Round to the nearest cent.

69. A $3000 mink coat is on sale for 30% off. Find the discount and the sale price.

70. A $90 calculator is on sale for 10% off. Find the discount and the sale price.

(6.7) *Solve.*

71. Find the simple interest due on $4000 loaned for 4 months at 12% interest.

72. Find the simple interest due on $6500 loaned for 3 months at 20%.

73. Find the total amount in an account if $5500 is compounded annually at 12% for 15 years.

74. Find the total amount in an account if $6000 is compounded semiannually at 11% for 10 years.

75. Find the total amount in an account if $100 is compounded quarterly at 12% for 5 years.

76. Find the total amount in an account if $1000 is compounded quarterly at 18% for 20 years.

Mixed Review

Write each percent as a decimal.

77. 3.8%

78. 124.5%

Write each decimal as a percent.

79. 0.54

80. 95.2

Write each percent as a fraction or mixed number in simplest form.

81. 47%

82. 5.6%

Write each fraction or mixed number as a percent.

83. $\frac{3}{8}$

84. $\frac{6}{5}$

Translating *Translate each into an equation and solve.*

85. 43 is 16% of what number?

86. 27.5 is what percent of 25?

87. What number is 36% of 1968?

88. 67 is what percent of 50?

Translating *Translate each into a proportion and solve.*

89. 75 is what percent of 25?

90. What number is 16% of 240?

91. 28 is 5% of what number?

92. 52 is what percent of 16?

Solve.

93. The total number of cans in a soft drink machine is 300. If 78 soft drinks have been sold, find the percent of soft drink cans that have been sold.

94. A home valued at $96,950 last year has lost 7% of its value this year. Find the loss in value.

95. A dinette set sells for $568.00. If the sales tax rate is 8.75%, find the total price of the dinette set.

96. The original price of a video game is $23.00. It is on sale for 15% off. What is the amount of the discount?

97. A candy salesman makes a commission of $1.60 from each case of candy he sells. If a case of candy costs $12.80, what is his rate of commission?

98. Find the total amount due on a 6-month loan of $1400 at a simple interest rate of 13%.

99. Find the total amount due on a loan of $5500 for 9 years at 12.5% simple interest.

Step-by-step test solutions are found on the Chapter Test Prep Videos. Where available: MyMathLab® or YouTube

Chapter 6 Test

Answers

Write each percent as a decimal.

1. 85%

2. 500%

3. 0.6%

Write each decimal as a percent.

4. 0.056

5. 6.1

6. 0.35

Write each percent as a fraction or a mixed number in simplest form.

7. 120%

8. 38.5%

9. 0.2%

Write each fraction or mixed number as a percent.

10. $\frac{11}{20}$

11. $\frac{3}{8}$

12. $1\frac{3}{4}$

13. Bottled water accounts for $\frac{1}{5}$ of the total noncarbonated beverage category in convenience stores. Write $\frac{1}{5}$ as a percent. (*Source:* Grocery Manufacturers of America)

14. In small firms in the United States, 64% of full-time employees receive medical insurance benefits. Write 64% as a fraction. (*Source:* U.S. Bureau of Labor Statistics)

Solve.

15. What number is 42% of 80?

16. 0.6% of what number is 7.5?

17. 567 is what percent of 756?

1. ______
2. ______
3. ______
4. ______
5. ______
6. ______
7. ______
8. ______
9. ______
10. ______
11. ______
12. ______
13. ______
14. ______
15. ______
16. ______
17. ______

18. ______

19. ______

20. ______

21. ______

22. ______

23. ______

24. ______

25. ______

26. ______

27. ______

28. ______

29. ______

30. ______

31. ______

Solve. If necessary, round percents to the nearest tenth, dollar amounts to the nearest cent, and all other numbers to the nearest whole.

18. An alloy is 12% copper. How much copper is contained in 320 pounds of this alloy?

19. A farmer in Nebraska estimates that 20% of his potential crop, or \$11,350, has been lost to a hard freeze. Find the total value of his potential crop.

20. If the local sales tax rate is 8.25%, find the total amount charged for a stereo system priced at \$354.

21. A town's population increased from 25,200 to 26,460. Find the percent increase.

22. A \$120 framed picture is on sale for 15% off. Find the discount and the sale price.

23. Randy Nguyen is paid a commission rate of 4% on all sales. Find Randy's commission if his sales were \$9875.

24. A sales tax of \$13.77 is added to an item's price of \$152.99. Find the sales tax rate. Round to the nearest whole percent.

25. Find the simple interest earned on \$2000 saved for $3\frac{1}{2}$ years at an interest rate of 9.25%.

26. \$1365 is compounded annually at 8%. Find the total amount in the account after 5 years.

27. A couple borrowed \$400 from a bank at 13.5% simple interest for 6 months for car repairs. Find the total amount due the bank at the end of the 6-month period.

28. In a recent 5-year period, the number of major crimes reported in New York City decreased from 149,488 to 142,760. Find the percent decrease. (*Source:* New York State Division of Criminal Justice Services)

29. Write the ratio \$75 to \$10 as a fraction in simplest form.

30. Solve: $\frac{5}{y+1} = \frac{4}{y+2}$

31. In a sample of 85 fluorescent bulbs, 3 were found to be defective. At this rate, how many defective bulbs should be found in 510 bulbs?

Cumulative Review Chapters 1–6

1. Multiply: 236×86

2. Multiply: 409×76

3. Subtract 7 from -3.

4. Subtract -2 from 8.

5. Solve: $x - 2 = -1$

6. Solve: $x + 4 = 3$

7. Solve: $3(2x - 6) + 6 = 0$

8. Solve: $5(x - 2) = 3x$

9. Write an equivalent fraction with the given denominator.

   $3 = \frac{}{7}$

10. Write an equivalent fraction with the given denominator.

    $8 = \frac{}{5}$

11. Write in simplest form: $-\frac{10}{27}$

12. Write in simplest form: $\frac{10y}{32}$

13. Divide: $-\frac{7}{12} \div -\frac{5}{6}$

14. Divide: $-\frac{2}{5} \div \frac{7}{10}$

15. Evaluate $y - x$ if $x = -\frac{3}{10}$ and $y = -\frac{8}{10}$.

16. Evaluate $2x + 3y$ if $x = \frac{2}{5}$ and $y = -\frac{1}{5}$.

17. Find: $-\frac{3}{4} - \frac{1}{14} + \frac{6}{7}$

18. Find: $\frac{2}{9} + \frac{7}{15} - \frac{1}{3}$

19. Simplify: $\dfrac{\frac{1}{2} + \frac{3}{8}}{\frac{3}{4} - \frac{1}{6}}$

20. Simplify: $\dfrac{\frac{2}{3} + \frac{1}{6}}{\frac{3}{4} - \frac{3}{5}}$

21. Solve: $\frac{x}{2} = \frac{x}{3} + \frac{1}{2}$

22. Solve: $\frac{x}{2} + \frac{1}{5} = 3 - \frac{x}{5}$

Answers

1. ______
2. ______
3. ______
4. ______
5. ______
6. ______
7. ______
8. ______
9. ______
10. ______
11. ______
12. ______
13. ______
14. ______
15. ______
16. ______
17. ______
18. ______
19. ______
20. ______
21. ______
22. ______

23. a. ____
b. ____
24. a. ____
b. ____
25. ____
26. ____
27. ____
28. ____
29. ____
30. ____
31. ____
32. ____
33. ____
34. ____
35. ____
36. ____
37. ____
38. ____
39. ____
40. ____
41. ____
42. ____
43. ____
44. ____
45. ____
46. ____
47. ____
48. ____
49. ____
50. ____

23. Write each mixed number as an improper fraction.

a. $4\frac{2}{9}$

b. $1\frac{8}{11}$

24. Write each mixed number as an improper fraction.

a. $3\frac{2}{5}$

b. $6\frac{2}{7}$

Write each decimal as a fraction or mixed number. Write your answer in simplest form.

25. 0.125

26. 0.85

27. −105.083

28. 17.015

29. Subtract: $85 - 17.31$

30. Subtract: $38 - 10.06$

Multiply.

31. 7.68×10

32. 12.483×100

33. $(-76.3)(1000)$

34. -853.75×10

35. Evaluate $x \div y$ for $x = 2.5$ and $y = 0.05$.

36. Is 470 a solution of the equation $\frac{x}{100} = 4.75$?

37. Find the median of the scores: 67, 91, 75, 86, 55, 91

38. Find the mean of 36, 40, 86, and 30.

39. Write the ratio of 2.5 to 3.15 as a fraction in simplest form.

40. Write the ratio of 5.8 to 7.6 as a fraction in simplest form.

41. Solve for x: $\frac{45}{x} = \frac{5}{7}$

42. Solve for x: $\frac{x - 1}{3x + 1} = \frac{3}{8}$

43. On a chamber of commerce map of Abita Springs, 5 miles corresponds to 2 inches. How many miles correspond to 7 inches?

44. A student doing math homework can complete 7 problems in about 6 minutes. At this rate, how many problems can be completed in 30 minutes?

Write each percent as a fraction or mixed number in simplest form.

45. 1.9%

46. 2.3%

47. $33\frac{1}{3}\%$

48. 108%

49. What number is 35% of 60?

50. What number is 42% of 85?

7

Graphs and Triangle Applications

Hybrid Electric Vehicle (HEV)

Battery Electric Vehicle (BEV)

The graph below is a broken line graph (or line graph) of hybrid cars sold for the months shown. A graph like this helps us visually see trends and make predictions. In Section 7.2, Exercise 28, we use percent data on hybrid sales to draw a circle graph.

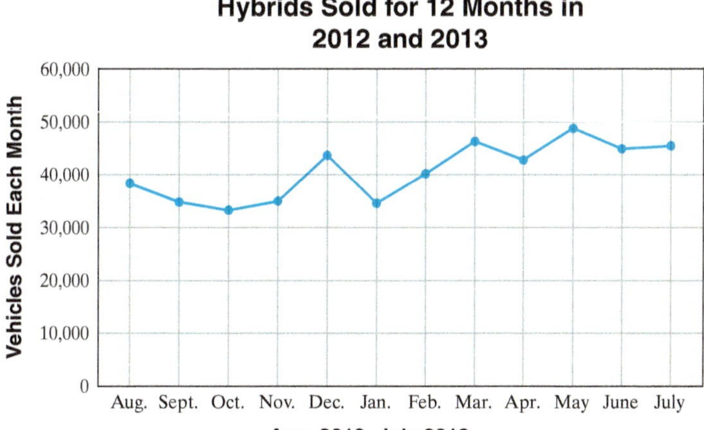

Sections

Check Your Progress

In this chapter, we present data in a usable form on a graph and the basic ideas of probability. Also in this chapter, we introduce important triangle applications, such as the Pythagorean theorem and the concept of congruent and similar triangles.

 Reading Pictographs, Bar Graphs, Histograms, and Line Graphs 

Objectives

A Read Pictographs.

B Read and Construct Bar Graphs.

C Read and Construct Histograms.

D Read Line Graphs.

Often data are presented visually in a graph. In this section, we practice reading several kinds of graphs including pictographs, bar graphs, and line graphs.

Objective A Reading Pictographs

A **pictograph** such as the one below is a graph in which pictures or symbols are used. This type of graph contains a key that explains the meaning of the symbol used. An advantage of using a pictograph to display information is that comparisons can easily be made. A disadvantage of using a pictograph is that it is often hard to tell what fractional part of a symbol is shown. For example, in the pictograph below, Arabic shows a part of a symbol, but it's hard to read with any accuracy what fractional part of a symbol is shown.

Practice 1

Use the pictograph shown in Example 1 to answer the following questions:

a. Approximate the number of people who primarily speak Spanish.

b. Approximate how many more people primarily speak Spanish than Arabic.

Example 1 Calculating Languages Spoken

The following pictograph shows the top eight most-spoken (primary) languages. Use this pictograph to answer the questions.

Top 8 Most-Spoken (Primary) Languages

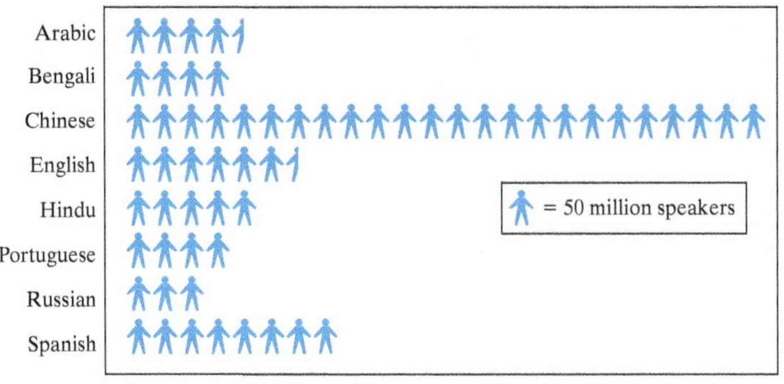

Source: www.ethnologue.com

a. Approximate the number of people who primarily speak Russian.

b. Approximate how many more people primarily speak English than Russian.

Solution:

a. Russian corresponds to 3 symbols, and each symbol represents 50 million speakers. This means that the number of people who primarily speak Russian is approximately $3 \cdot (50 \text{ million})$ or 150 million people.

b. English shows $3\frac{1}{2}$ more symbols than Russian. This means that $3\frac{1}{2} \cdot (50 \text{ million})$ or 175 million more people primarily speak English than Russian.

Work Practice 1

Answers

1. a. 400 million people
b. 175 million people

Objective B Reading and Constructing Bar Graphs

Another way to visually present data is with a **bar graph.** Bar graphs can appear with vertical bars or horizontal bars. Although we have studied bar graphs in previous sections, we now practice reading the height or length of the bars contained in a bar graph. An advantage to using bar graphs is that a scale is usually included for greater accuracy. Care must be taken when reading bar graphs, as well as other types of graphs—they may be misleading, as shown later in this section.

Example 2 Finding the Number of Endangered Species

The following bar graph shows the number of endangered species in the United States in 2013. Use this graph to answer the questions.

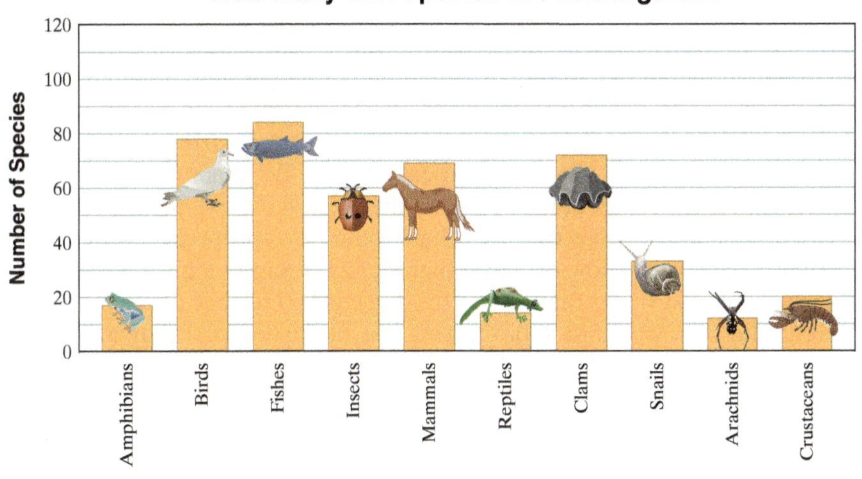

a. Approximate the number of endangered species that are clams.

b. Which category has the most endangered species?

Solution:

a. To approximate the number of endangered species that are clams, we go to the top of the bar that represents clams. From the top of this bar, we move horizontally to the left until the scale is reached. We read the height of the bar on the scale as approximately 72. There are approximately 72 clam species that are endangered, as shown.

How Many U.S. Species Are Endangered?

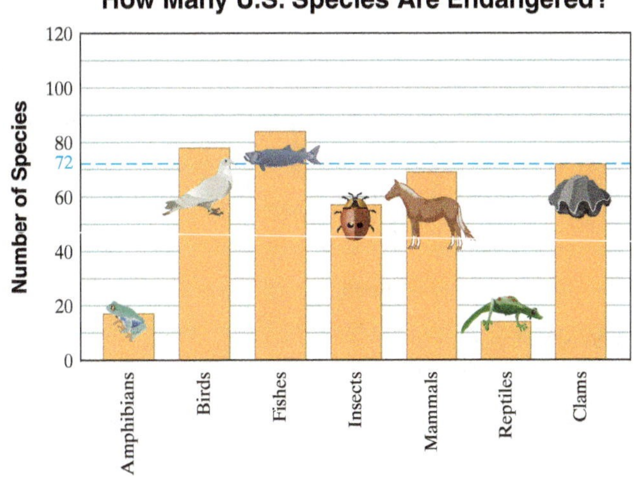

b. The most endangered species is represented by the tallest (longest) bar. The tallest bar corresponds to fishes.

Work Practice 2

Practice 2

Use the bar graph in Example 2 to answer the following questions:

a. Approximate the number of endangered species that are birds.

b. Which category shows the fewest endangered species?

Answers

2. a. 78 **b.** arachnids

Next, we practice constructing a bar graph.

Practice 3

Draw a vertical bar graph using the information in the table about electoral votes for selected states.

Total Electoral Votes by Selected States	
State	**Electoral Votes**
Texas	34
California	55
Florida	27
Nebraska	5
Indiana	11
Georgia	15
(*Source: World Almanac* 2013)	

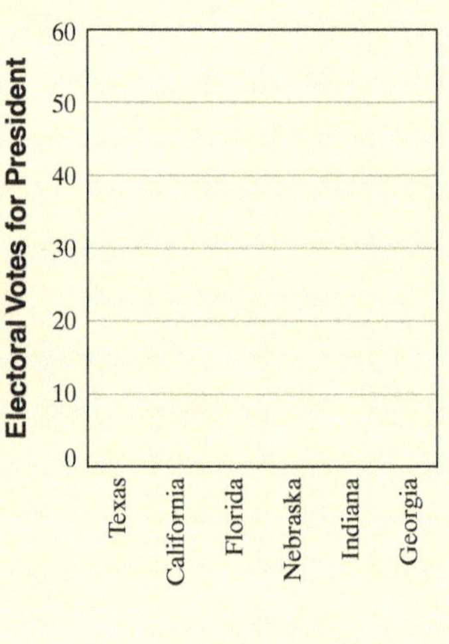

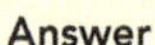
Answer

3.

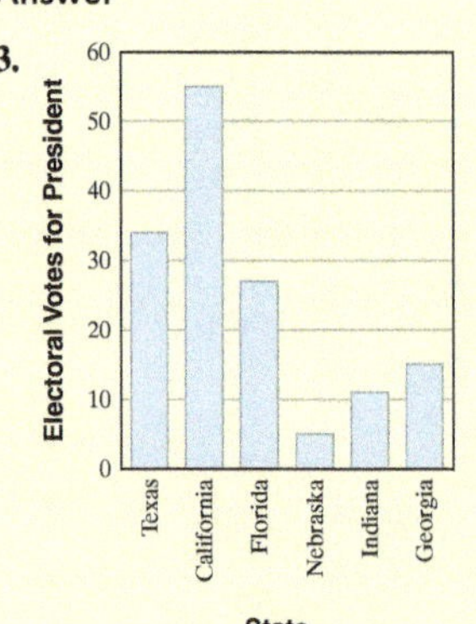

Example 3

Draw a vertical bar graph using the information in the table below, which gives the caffeine content of selected foods.

Average Caffeine Content of Selected Foods			
Food	**Milligrams**	**Food**	**Milligrams**
Brewed coffee (percolator, 8 ounces)	124	Instant coffee (8 ounces)	104
Brewed decaffeinated coffee (8 ounces)	3	Brewed tea (U.S. brands, 8 ounces)	64
Coca-Cola Classic (8 ounces)	31	Mr. Pibb (8 ounces)	27
Dark chocolate (semisweet, $1\frac{1}{2}$ ounces)	30	Milk chocolate (8 ounces)	9
(*Sources:* International Food Information Council and the Coca-Cola Company)			

Solution: We draw and label a vertical line and a horizontal line as shown below on the left. These lines are also called axes. We place the different food categories along the horizontal axis. Along the vertical axis we place a scale.

There are many choices of scales that would be appropriate. Notice that the milligrams range from a low of 3 to a high of 124. From this information, we use a scale that starts at 0 and then shows multiples of 20 so that the scale is not too cluttered. The scale stops at 140, the smallest multiple of 20 that will allow all milligrams to be graphed. It may also be helpful to draw horizontal lines along the scale markings to help draw the vertical bars at the correct height. The finished bar graph is shown below on the right.

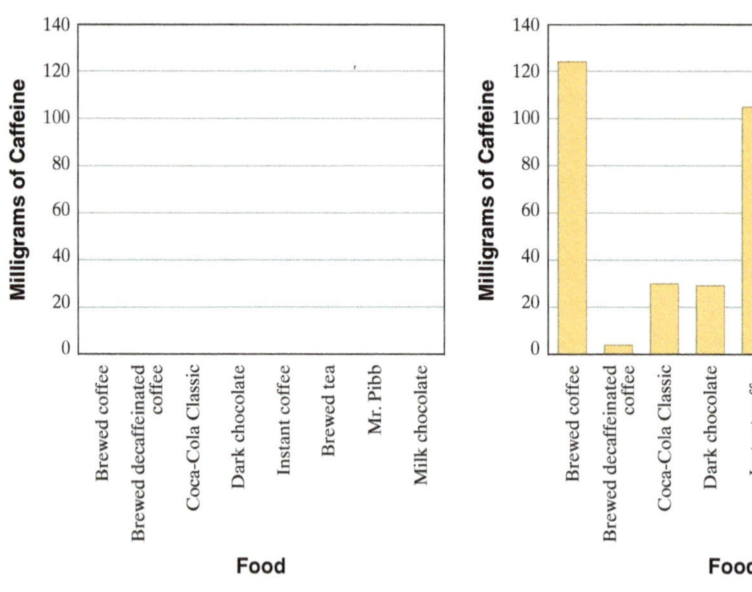

Work Practice 3

As mentioned previously, graphs can be misleading. Both graphs on the next page show the same information, but with different scales. Special care should be taken when forming conclusions from the appearance of a graph.

Notice the ≷ symbol on each vertical scale on the graphs below. This symbol alerts us that numbers are missing from that scale.

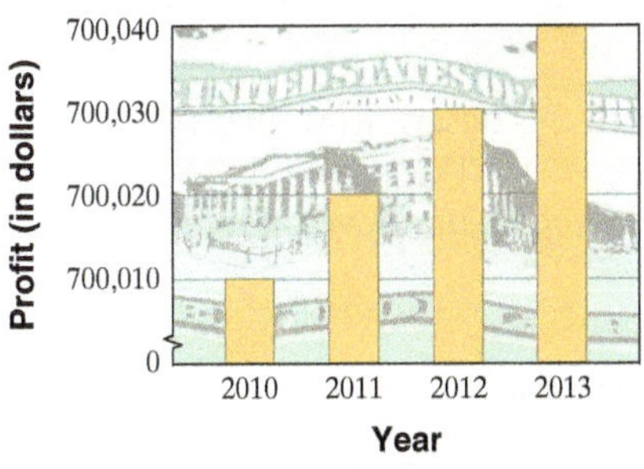

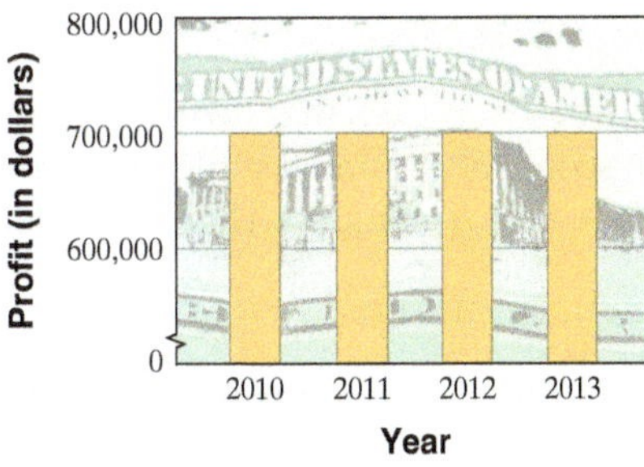

Are the profits shown in the graphs above greatly increasing, or are they remaining about the same?

Objective C Reading and Constructing Histograms 

Suppose that the test scores of 36 students are summarized in the table below:

Student Scores	Frequency (Number of Students)
40–49	1
50–59	3
60–69	2
70–79	10
80–89	12
90–99	8

The results in the table can be displayed in a histogram. A **histogram** is a special bar graph. The width of each bar represents a range of numbers called a **class interval.** The height of each bar corresponds to how many times a number in the class interval occurs and is called the **class frequency.** The bars in a histogram lie side by side with no space between them.

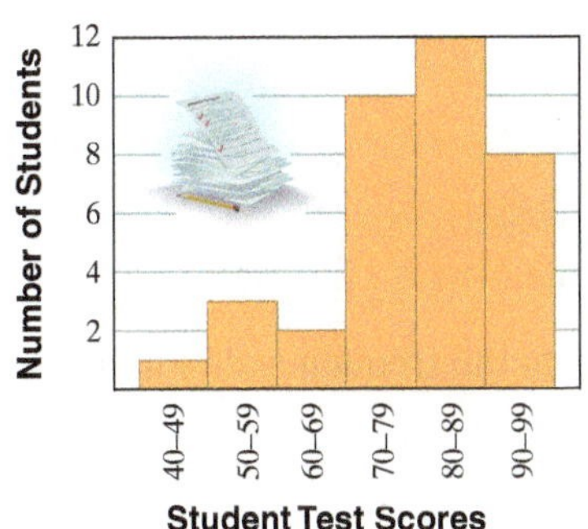

Example 4 Reading a Histogram on Student Test Scores

Use the preceding histogram to determine how many students scored 50–59 on the test.

Solution: We find the bar representing 50–59. The height of this bar is 3, which means 3 students scored 50–59 on the test.

Work Practice 4

Practice 4

Use the histogram for Example 4 to determine how many students scored 80–89 on the test.

Answer

4. 12

Practice 5

Use the histogram for Example 4 to determine how many students scored less than 80 on the test.

Practice 6

Complete the frequency distribution table for the data below. Each number represents a credit card owner's unpaid balance for one month.

0	53	89	125
265	161	37	76
62	201	136	42

Class Intervals (Credit Card Balances)	Tally	Class Frequency (Number of Months)
\$0–\$49	_____	_____
\$50–\$99	_____	_____
\$100–\$149	_____	_____
\$150–\$199	_____	_____
\$200–\$249	_____	_____
\$250–\$299	_____	_____

Practice 7

Construct a histogram from the frequency distribution table for Practice 6.

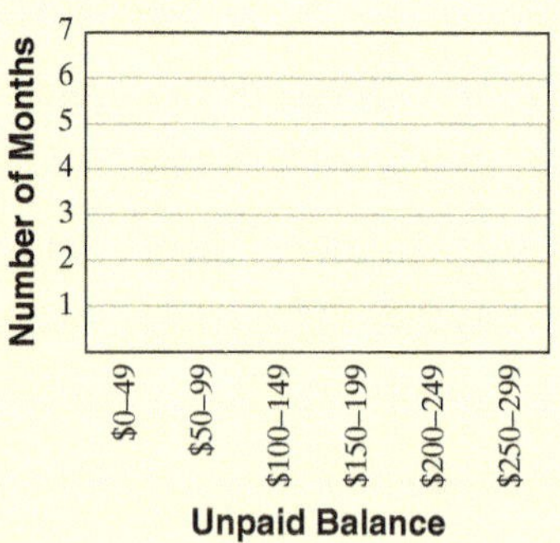

Answers

5. 16

6.

Tally	Class Frequency (Number Months)	Tally	Class Frequency (Number Months)
|||	3	|	1
||||	4	|	1
||	2	|	1

7.

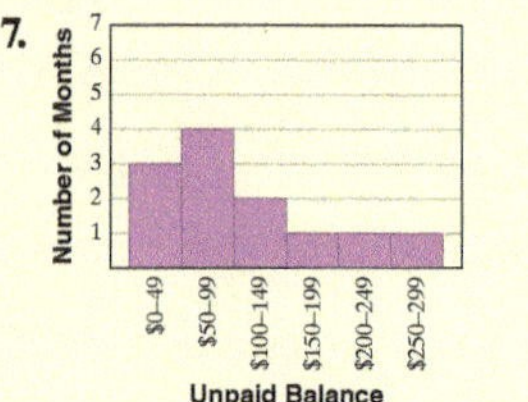

Example 5 Reading a Histogram on Student Test Scores

Use the histogram for Example 4 to determine how many students scored 80 or above on the test.

Solution: We see that two different bars fit this description. There are 12 students who scored 80–89 and 8 students who scored 90–99. The sum of these two categories is 12 + 8 or 20 students. Thus, 20 students scored 80 or above on the test.

Work Practice 5

Now we will look at a way to construct histograms.

The daily high temperatures for 1 month in New Orleans, Louisiana, are recorded in the following list:

85°	90°	95°	89°	88°	94°
87°	90°	95°	92°	95°	94°
82°	92°	96°	91°	94°	92°
89°	89°	90°	93°	95°	91°
88°	90°	88°	86°	93°	89°

The data in this list have not been organized and can be hard to interpret. One way to organize the data is to place them in a **frequency distribution table.** We will do this in Example 6.

Example 6 Completing a Frequency Distribution on Temperature

Complete the frequency distribution table for the preceding temperature data.

Solution: Go through the data and place a tally mark in the second column of the table next to the class interval. Then count the tally marks and write each total in the third column of the table.

Class Intervals (Temperatures)	Tally	Class Frequency (Number of Days)
82°–84°	|	1
85°–87°	|||	3
88°–90°	卌 卌 |	11
91°–93°	卌 ||	7
94°–96°	卌 |||	8

Work Practice 6

Example 7 Constructing a Histogram

Construct a histogram from the frequency distribution table in Example 6.

Solution:

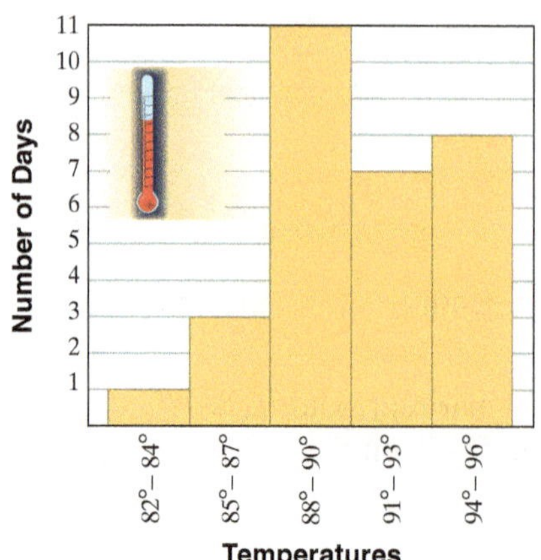

Work Practice 7

✓Concept Check Which of the following sets of data is better suited to representation by a histogram? Explain.

Set 1		Set 2	
Grade on Final	# of Students	Section Number	Avg. Grade on Final
51–60	12	150	78
61–70	18	151	83
71–80	29	152	87
81–90	23	153	73
91–100	25		

Objective D Reading Line Graphs 

Another common way to display information with a graph is by using a **line graph.** An advantage of a line graph is that it can be used to visualize relationships between two quantities. A line graph can also be very useful in showing changes over time.

Example 8 Reading Temperatures from a Line Graph

The following line graph shows the average daily temperature for each month in Omaha, Nebraska. Use this graph to answer the questions below.

Average Daily Temperature for Omaha, Nebraska

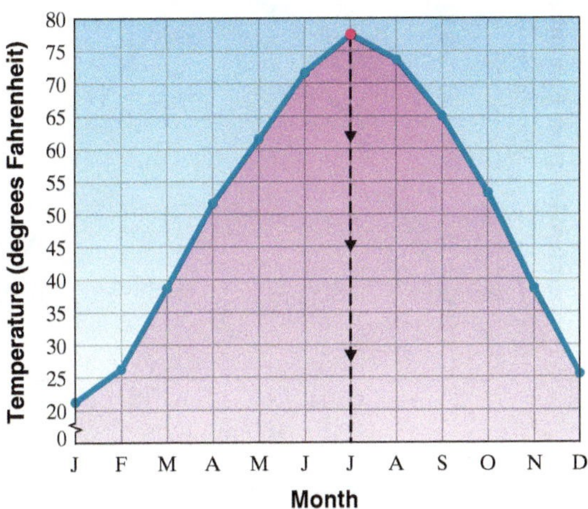

Source: National Climatic Data Center

a. During what month is the average daily temperature the highest?

b. During what month, from July through December, is the average daily temperature 65°F?

c. During what months is the average daily temperature less than 30°F?

Solution:

a. The month with the highest temperature corresponds to the highest point. This is the red point shown on the graph above. We follow this highest point downward to the horizontal month scale and see that this point corresponds to July.

(*Continued on next page*)

Practice 8

Use the temperature graph in Example 8 to answer the following questions:

a. During what month is the average daily temperature the lowest?

b. During what month is the average daily temperature 25°F?

c. During what months is the average daily temperature greater than 70°F?

Answers

8. a. January **b.** December **c.** June, July, and August

✓Concept Check Answer

Set 1; the grades are arranged in ranges of scores.

b. The months July through December correspond to the right side of the graph. We find the 65°F mark on the vertical temperature scale and move to the right until a point on the right side of the graph is reached. From that point, we move downward to the horizontal month scale and read the corresponding month. During the month of September, the average daily temperature is 65°F.

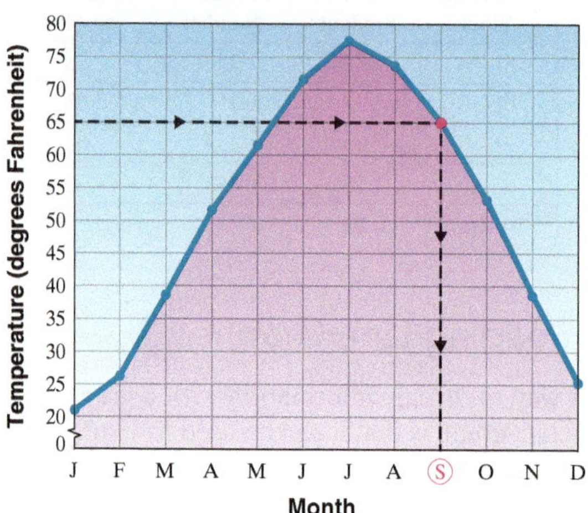

Source: National Climatic Data Center

c. To see what months the temperature is less than 30°F, we find what months correspond to points that fall below the 30°F mark on the vertical scale. These months are January, February, and December.

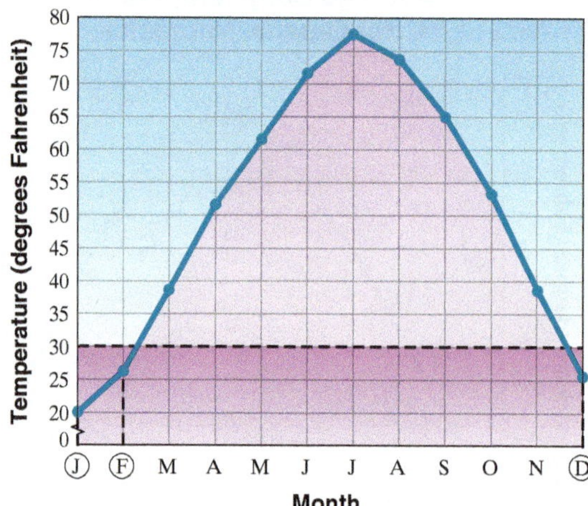

Source: National Climatic Data Center

Work Practice 8

Vocabulary, Readiness & Video Check

Fill in each blank with one of the choices below.

pictograph	bar	class frequency
histogram	line	class interval

1. A __________ graph presents data using vertical or horizontal bars.

2. A __________ is a graph in which pictures or symbols are used to visually present data.

3. A __________ graph displays information with a line that connects data points.

4. A __________ is a special bar graph in which the width of each bar represents a __________ and the height of each bar represents the __________.

Martin-Gay Interactive Videos

See Video 7.1

Watch the section lecture video and answer the following questions.

Objective A 5. From the pictograph in Example 1, how would you approximate the number of wildfires for any given year?

Objective B 6. What is one advantage of displaying data in a bar graph?

Objective C 7. Complete this statement based on the lecture before Example 6: A histogram is a special kind of ____________.

Objective D 8. From the line graph in Examples 10–13, what year averaged the greatest number of goals per game average and what was this average?

7.1 Exercise Set MyMathLab®

Objective A *The following pictograph shows the number of acres devoted to wheat production in selected states. Use this graph to answer Exercises 1 through 8. See Example 1. (Source: U.S. Department of Agriculture)*

1. Which state plants the greatest quantity of acreage in wheat?

2. Which of the states shown plant the least amount of wheat acreage?

3. Approximate the number of acres of wheat planted in Oklahoma.

4. Approximate the number of acres of wheat planted in Kansas.

5. Which state plants about 6,000,000 acres of wheat?

6. Which state plants about 2,000,000 acres of wheat?

7. Which two states together plant about the same acreage of wheat as North Dakota?

Annual Wheat Acreage in Selected Top States

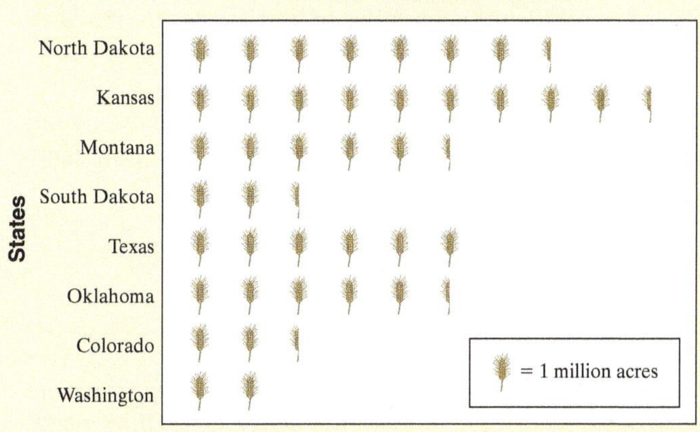

8. Which two states together plant about the same acreage of wheat as Kansas?

The following pictograph shows the average number of wildfires in the United States between 2006 and 2012. Use this graph to answer Exercises 9 through 16. See Example 1. (Source: National Interagency Fire Center)

9. Approximate the number of wildfires in 2008.

10. Approximately how many wildfires were there in 2012?

11. Which year, of the years shown, had the most wildfires?

12. In what years were the number of wildfires greater than 72,000?

Wildfires in the United States

13. What was the amount of decrease in wildfires from 2006 to 2012?

14. What was the amount of decrease in wildfires from 2010 to 2012?

15. What was the average annual number of wildfires from 2006 to 2008? (*Hint:* How do you calculate the average?)

16. Give a possible explanation for the overall decrease in the number of wildfires since 2006.

Objective B *The National Weather Service has exacting definitions for hurricanes; they are tropical storms with winds in excess of 74 mph. The following bar graph shows the number of hurricanes, by month, that have made landfall on the mainland United States between 1851 and 2012. Use this graph to answer Exercises 17 through 22. See Example 2. (Source: National Weather Service: National Hurricane Center)*

17. In which month did the most hurricanes make landfall in the United States?

18. In which month did the fewest hurricanes make landfall in the United States?

19. Approximate the number of hurricanes that made landfall in the United States during the month of August.

20. Approximate the number of hurricanes that made landfall in the United States in September.

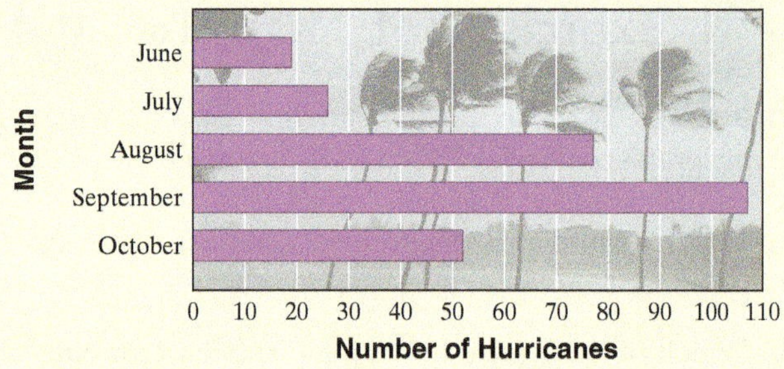

21. In 2008 alone, two hurricanes made landfall during the month of August. What fraction of all the 77 hurricanes that made landfall during August is this?

22. In 2007, only one hurricane made landfall in the United States during the entire season, in the month of September. If there have been 107 hurricanes to make landfall in the month of September since 1851, approximately what percent of these arrived in 2007?

The following horizontal bar graph shows the approximate 2012 population of the world's largest cities (including their suburbs). Use this graph to answer Exercises 23 through 28. See Example 2. (Source: CityPopulation)

23. Name the city with the largest population, and estimate its population.

24. Name the cities whose population is between 19 million and 22 million.

25. Name the city in the United States with the largest population, and estimate its population.

26. Name the two cities that have approximately the same population.

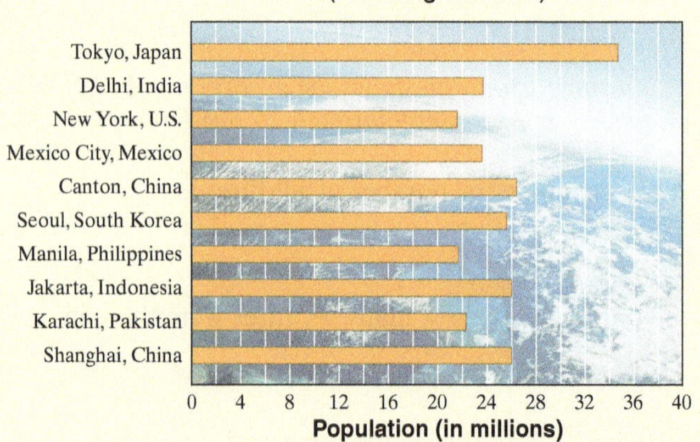

27. How much larger (in terms of population) is Seoul, South Korea, than Delhi, India?

28. How much larger (in terms of population) is, Shanghai, China, than Mexico City, Mexico?

Use the information given to draw a vertical bar graph. Clearly label the bars. See Example 3.

29.

Fiber Content of Selected Foods	
Food	**Grams of Total Fiber**
Kidney beans $\left(\frac{1}{2}\text{ c}\right)$	4.5
Oatmeal, cooked $\left(\frac{3}{4}\text{ c}\right)$	3.0
Peanut butter, chunky (2 tbsp)	1.5
Popcorn (1 c)	1.0
Potato, baked with skin (1 med)	4.0
Whole wheat bread (1 slice)	2.5

(*Sources:* American Dietetic Association and National Center for Nutrition and Dietetics)

Fiber Content of Selected Foods

Food

30.

U.S. Annual Food Sales	
Year	**Sales in Billions of Dollars**
2009	1086
2010	1139
2011	1274
2012	1357

(*Source:* U.S. Department of Agriculture)

U.S. Annual Food Sales

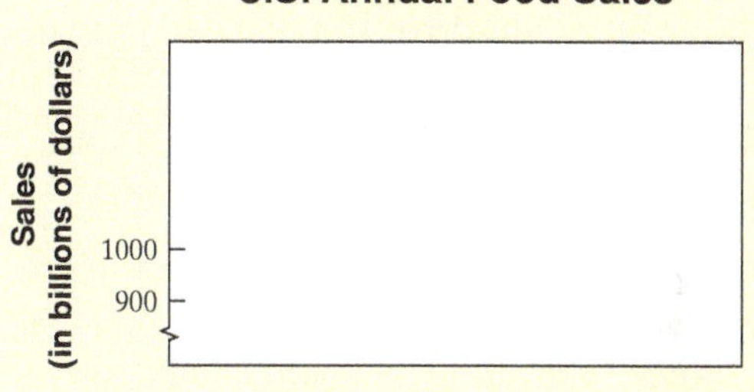

Year

31.

Best-Selling Albums of All Time (U.S. Sales)	
Album	**Estimated Sales (in millions)**
Pink Floyd: *The Wall* (1979)	23
Michael Jackson: *Thriller* (1982)	29
Billy Joel: *Greatest Hits Volumes I + II* (1985)	23
Eagles: *Their Greatest Hits* (1976)	29
Led Zeppelin: *Led Zeppelin IV* (1971)	23

(*Source:* Recording Industry Association of America)

Best-Selling Albums of All Time
(U.S. sales)

Album

32.

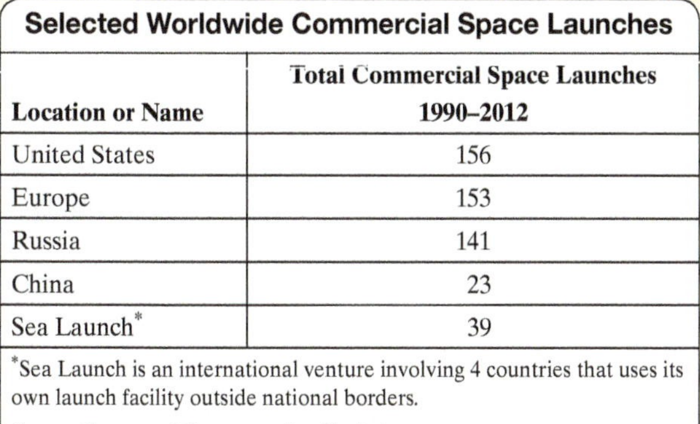

Selected Worldwide Commercial Space Launches	
Location or Name	**Total Commercial Space Launches 1990–2012**
United States	156
Europe	153
Russia	141
China	23
Sea Launch*	39

*Sea Launch is an international venture involving 4 countries that uses its own launch facility outside national borders.

Source: Bureau of Transportation Statistics

Selected Worldwide Commercial Space Launches
(1990–2012)

Number of Commercial Launches

40
20
0

Location or Name

Objective C *The following histogram shows the number of miles that each adult, from a survey of 100 adults, drives per week. Use this histogram to answer Exercises 33 through 42. See Examples 4 and 5.*

33. How many adults drive 100–149 miles per week?

34. How many adults drive 200–249 miles per week?

35. How many adults drive fewer than 150 miles per week?

36. How many adults drive 200 miles or more per week?

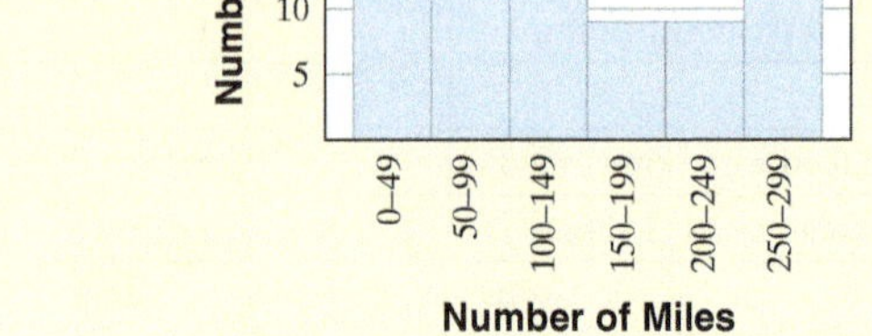

37. How many adults drive 100–199 miles per week?

38. How many adults drive 150–249 miles per week?

39. How many more adults drive 250–299 miles per week than 200–249 miles per week?

40. How many more adults drive 0–49 miles per week than 50–99 miles per week?

41. What is the ratio of adults who drive 150–199 miles per week to the total number of adults surveyed?

42. What is the ratio of adults who drive 50–99 miles per week to the total number of adults surveyed?

The following histogram shows the ages of householders for the year 2010. Use this histogram to answer Exercises 43 through 50. For Exercises 45 through 50, estimate to the nearest whole million. See Examples 4 and 5.

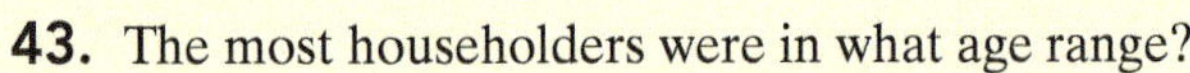

43. The most householders were in what age range?

44. The least number of householders were in what age range?

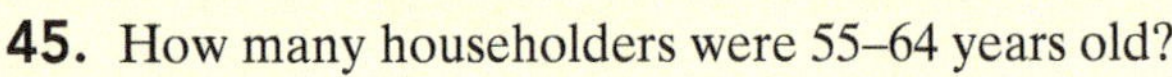

45. How many householders were 55–64 years old?

46. How many householders were 35–44 years old?

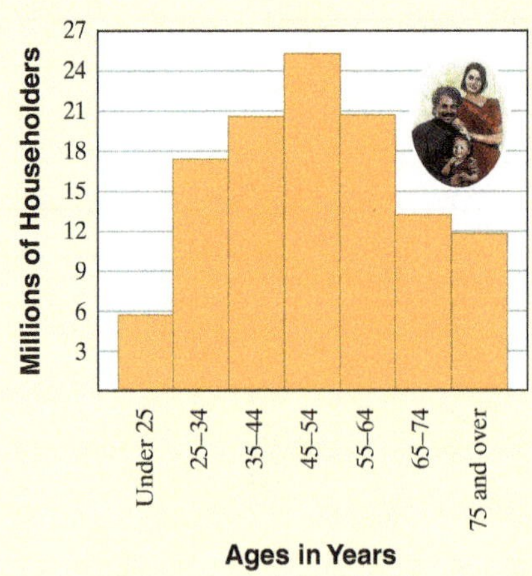

Source: U.S. Bureau of the Census, *Current Population Reports*

47. How many householders were 44 years old or younger?

48. How many householders were 55 years old or older?

49. How many more householders were 45–54 years old than 55–64 years old?

50. How many more householders were 45–54 years old than 75 and over?

The following list shows the golf scores for an amateur golfer. Use this list to complete the frequency distribution table to the right. See Example 6.

78	84	91	93	97
97	95	85	95	96
101	89	92	89	100

	Class Intervals (Scores)	Tally	Class Frequency (Number of Games)
51.	70–79		
52.	80–89		
53.	90–99		
54.	100–109		

Twenty-five people in a survey were asked to give their current checking account balances. Use the balances shown in the following list to complete the frequency distribution table to the right. See Example 6.

$53	$105	$162	$443	$109
$468	$47	$259	$316	$228
$207	$357	$15	$301	$75
$86	$77	$512	$219	$100
$192	$288	$352	$166	$292

	Class Intervals (Account Balances)	Tally	Class Frequency (Number of People)
55.	$0–$99		
56.	$100–$199		
57.	$200–$299		
58.	$300–$399		
59.	$400–$499		
60.	$500–$599		

61. Use the frequency distribution table from Exercises **51** through **54** to construct a histogram. See Example 7.

62. Use the frequency distribution table from Exercises **55** through **60** to construct a histogram. See Example 7.

Account Balances

Objective D *Beach Soccer World Cup is now held every two years. The following line graph shows the World Cup goals per game average for beach soccer during the years shown. Use this graph to answer Exercises 63 through 70. See Example 8.*

63. Find the average number of goals per game in 2011.

64. Find the average number of goals per game in 2009.

65. During what year shown was the average number of goals per game the highest?

66. During what year shown was the average number of goals per game the lowest?

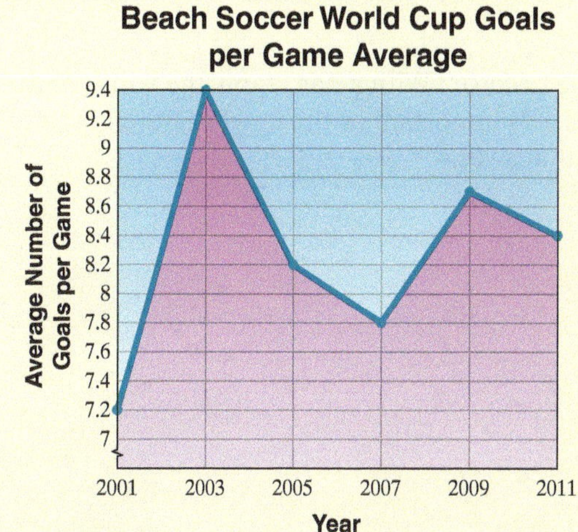

67. From 2007 to 2009, did the average number of goals per game increase or decrease?

68. From 2009 to 2011, did the average number of goals per game increase or decrease?

69. During what year(s) shown were the average goals per game less than 8?

70. During what year(s) shown were the average goals per game greater than 8?

Review

Find each percent. See Sections 6.3 and 6.4.

71. 30% of 12

72. 45% of 120

73. 10% of 62

74. 95% of 50

Write each fraction as a percent. See Section 6.2.

75. $\frac{1}{4}$

76. $\frac{2}{5}$

77. $\frac{17}{50}$

78. $\frac{9}{10}$

Concept Extensions

The following double line graph shows temperature highs and lows for a week. Use this graph to answer Exercises 79 through 84.

79. What was the high temperature reading on Thursday?

80. What was the low temperature reading on Thursday?

81. What day was the temperature the lowest? What was this low temperature?

82. What day of the week was the temperature the highest? What was this high temperature?

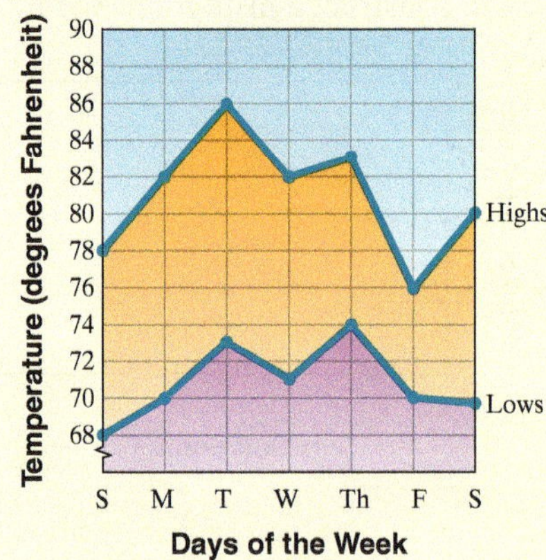

83. On what day of the week was the difference between the high temperature and the low temperature the greatest? What was this difference in temperature?

84. On what day of the week was the difference between the high temperature and the low temperature the least? What was this difference in temperature?

85. True or false? With a bar graph, the width of the bar is just as important as the height of the bar. Explain your answer.

86. Kansas plants about 17% of the wheat acreage in the United States. About how many acres of wheat are planted in the United States, according to the pictograph for Exercises **1** through **8**? Round to the nearest million acre.

7.2 Reading and Drawing Circle Graphs 

Objective A Reading Circle Graphs

In Exercise Set 6.2, the following **circle graph** was shown. This particular graph shows the favorite sport for 100 adults.

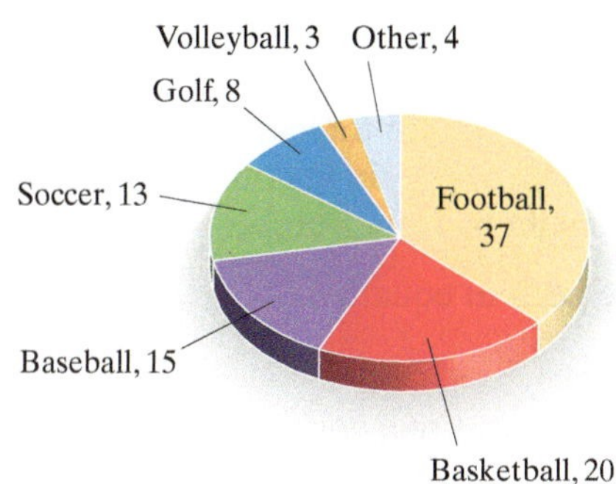

Each sector of the graph (shaped like a piece of pie) shows a category and the relative size of the category. In other words, the most popular sport is football, and it is represented by the largest sector.

Example 1 Find the ratio of adults preferring basketball to total adults. Write the ratio as a fraction in simplest form.

Solution: The ratio is

$$\frac{\text{people preferring basketball}}{\text{total adults}} = \frac{20}{100} = \frac{1}{5}$$

Work Practice 1

A circle graph is often used to show percents in different categories, with the whole circle representing 100%.

Example 2 Using a Circle Graph

The following graph shows the percent of visitors to the United States in a recent year by various regions. Using the circle graph shown, determine the percent of visitors who came to the United States from Mexico or Canada.

Solution: To find this percent, we add the percents corresponding to Mexico and Canada. The percent of visitors to the United States that came from Mexico or Canada is

$$34\% + 21\% = 55\%$$

Work Practice 2

Visitors to U.S. by Region

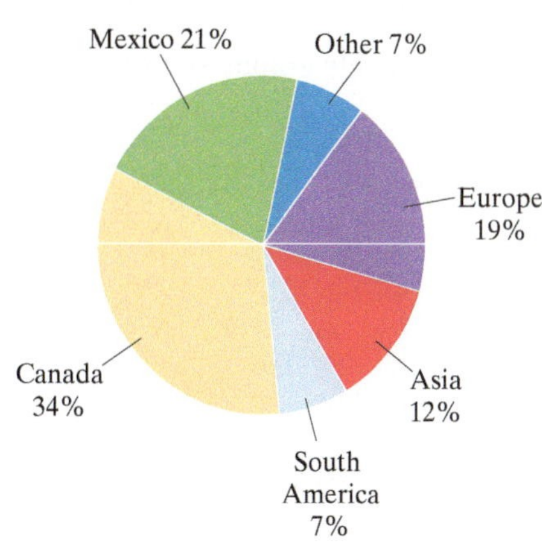

Source: Office of Travel and Tourism Industries, 2012

Objectives

A Read Circle Graphs.

B Draw Circle Graphs.

Practice 1

Find the ratio of adults preferring golf to total adults. Write the ratio as a fraction in simplest form.

Practice 2

Using the circle graph shown in Example 2, determine the percent of visitors to the United States that came from Europe, Asia, and South America.

Answers

1. $\frac{2}{25}$ 2. 38%

Helpful Hint

Since a circle graph represents a whole, the percents should add to 100% or 1. Notice this is true for Example 2.

Practice 3

Use the information in Example 3 and the circle graph from Example 2 to predict the number of tourists from Mexico in 2017.

Example 3 Finding Percent of Population

The U.S. Department of Commerce forecasts 81 million international visitors to the United States in 2017. Use the circle graph from Example 2 and predict the number of tourists that might be from Europe.

Solution: We use the percent equation.

$$\begin{aligned} \text{amount} &= \text{percent} \cdot \text{base} \\ \text{amount} &= 0.19 \cdot 81{,}000{,}000 \\ &= 0.19(81{,}000{,}000) \\ &= 15{,}390{,}000 \end{aligned}$$

Thus, 15,390,000 tourists might come from Europe in 2017.

Work Practice 3

✓Concept Check Can the following data be represented by a circle graph? Why or why not?

Responses to the Question, "In Which Activities Are You Involved?"	
Intramural sports	60%
On-campus job	42%
Fraternity/sorority	27%
Academic clubs	21%
Music programs	14%

Objective B Drawing Circle Graphs 

To draw a circle graph, we use the fact that a whole circle contains 360° (degrees).

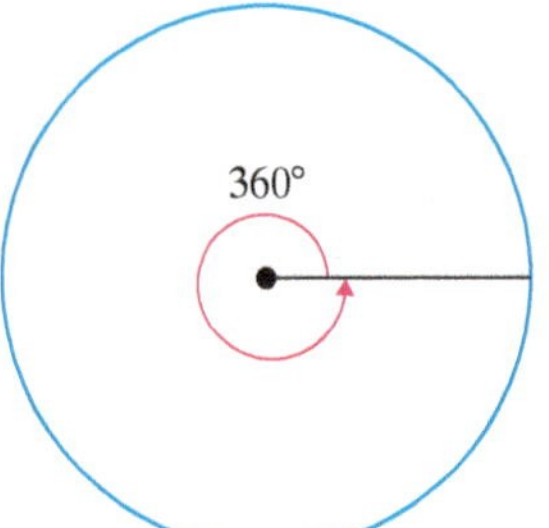

Answer

3. 17,010,000 tourists from Mexico

✓Concept Check Answer

no; the percents add up to more than 100%

Example 4 Drawing a Circle Graph for U.S. Armed Forces Personnel

The following table shows the percent of U.S. armed forces personnel that were in each branch of service in 2008. (*Source:* U.S. Department of Defense)

Branch of Service	Percent
Army	40
Navy	23
Marine Corps	15
Air Force	22
(*Note:* The Coast Guard is now under the Department of Homeland Security.)	

Draw a circle graph showing this data.

Solution: First we find the number of degrees in each sector representing each branch of service. Remember that the whole circle contains 360°. (We will round degrees to the nearest whole.)

Sector	Degrees in Each Sector
Army	$40\% \times 360° = 0.40 \times 360° = 144°$
Navy	$23\% \times 360° = 0.23 \times 360° = 82.8° \approx 83°$
Marine Corps	$15\% \times 360° = 0.15 \times 360° = 54°$
Air Force	$22\% \times 360° = 0.22 \times 360° = 79.2° \approx 79°$

Helpful Hint

Check your calculations by finding the sum of the degrees.

$$144° + 83° + 54° + 79° = 360°$$

The sum should be 360°. (It may vary only slightly because of rounding.)

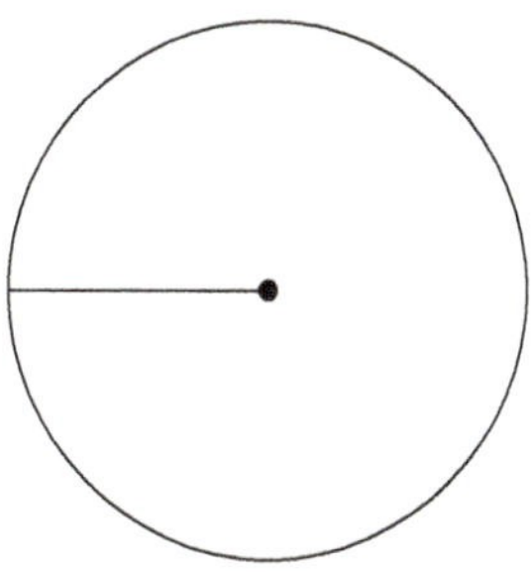

Next we draw a circle and mark its center. Then we draw a line from the center of the circle to the circle itself.

To construct the sectors, we will use a **protractor.** A protractor measures the number of degrees in an angle. We place the hole in the protractor over the center of the circle. Then we adjust the protractor so that 0° on the protractor is aligned with the line that we drew.

It makes no difference which sector we draw first. To construct the "Army" sector, we find 144° on the protractor and mark our circle. Then we remove the protractor and use this mark to draw a second line from the center to the circle itself.

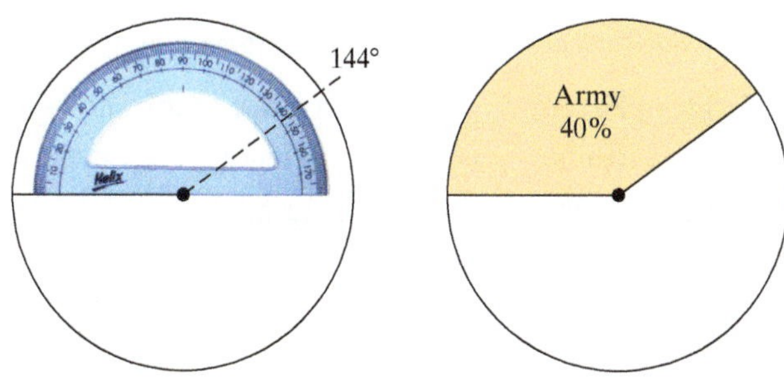

(*Continued on next page*)

Practice 4

Use the data shown to draw a circle graph.

Freshmen	30%
Sophomores	27%
Juniors	25%
Seniors	18%

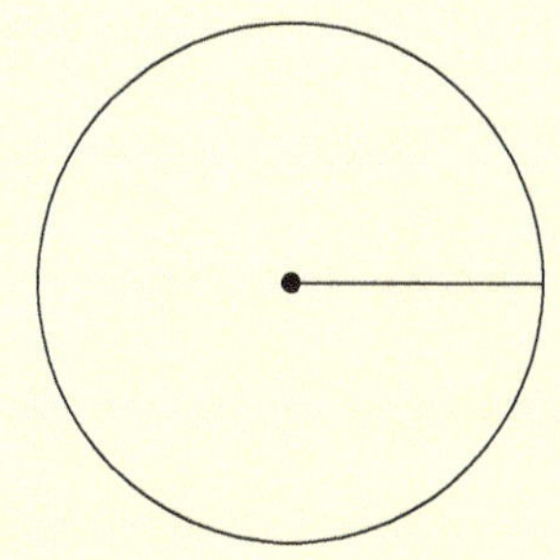

Answer

4.

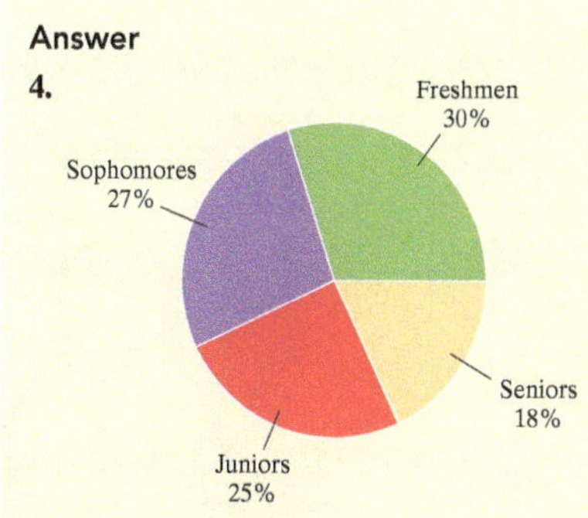

To construct the "Navy" sector, we follow the same procedure as above, except that we line up 0° with the second line we drew and mark the protractor at 83°.

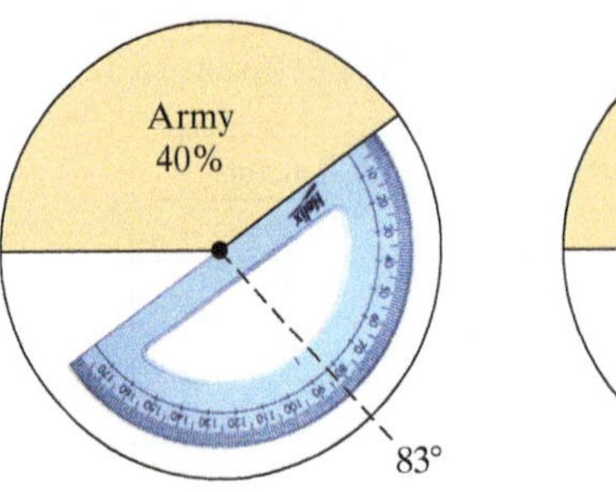

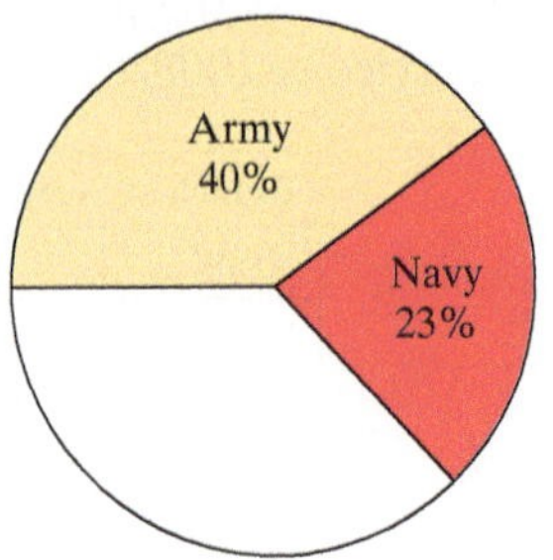

We continue in this manner until the circle graph is complete.

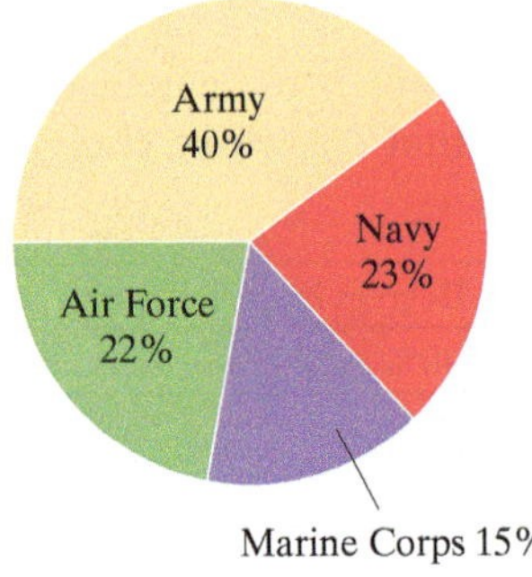

Work Practice 4

✓Concept Check Answer
true

✓Concept Check True or false? The larger a sector in a circle graph, the larger the percent of the total it represents. Explain your answer.

Vocabulary, Readiness & Video Check

Use the choices below to fill in each blank.

sector circle 100 360

1. In a __________ graph, each section (shaped like a piece of pie) shows a category and the relative size of the category.
2. A circle graph contains pie-shaped sections, each called a __________.
3. The number of degrees in a whole circle is __________.
4. If a circle graph has percent labels, the percents should add up to __________.

Martin-Gay Interactive Videos

See Video 7.2

Watch the section lecture video and answer the following questions.

Objective A 5. From Example 3, when a circle graph shows different parts or percents of some whole category, what is the sum of the percents in the whole circle graph?

Objective B 6. From Example 6, when looking at the sector degree measures of a circle graph, the whole circle graph corresponds to what degree measure?

7.2 Exercise Set MyMathLab®

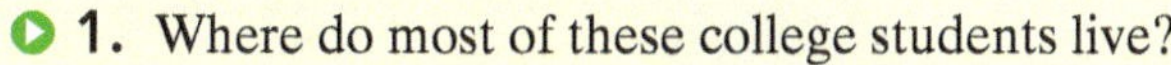

Objective A *The following circle graph is a result of surveying 700 college students. They were asked where they live while attending college. Use this graph to answer Exercises 1 through 6. Write all ratios as fractions in simplest form. See Example 1.*

1. Where do most of these college students live?

2. Besides the category "Other arrangements," where do the fewest of these college students live?

3. Find the ratio of students living in campus housing to total students.

4. Find the ratio of students living in off-campus rentals to total students.

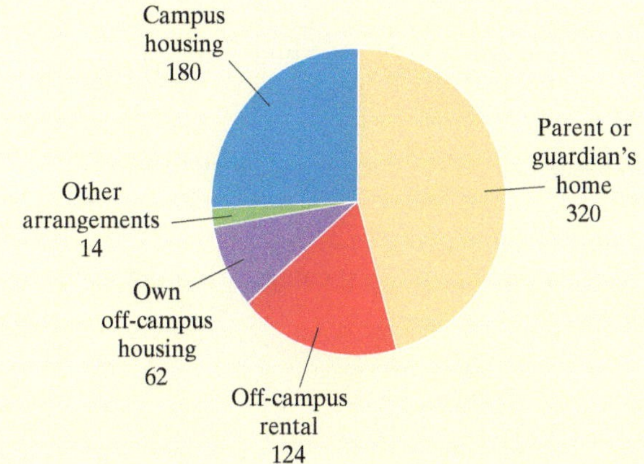

5. Find the ratio of students living in campus housing to students living in a parent or guardian's home.

6. Find the ratio of students living in off-campus rentals to students living in a parent or guardian's home.

The following circle graph shows the percent of the land area of the continents of Earth. Use this graph for Exercises 7 through 14. See Example 2.

7. Which continent is the largest?

8. Which continent is the smallest?

9. What percent of the land on Earth is accounted for by Asia and Europe together?

10. What percent of the land on Earth is accounted for by North and South America?

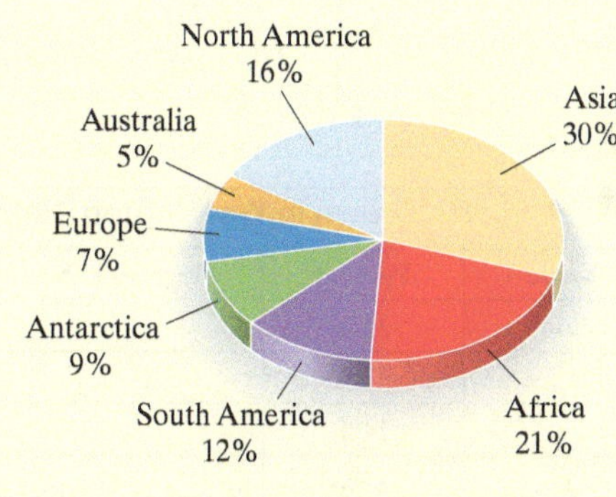

Source: National Geographic Society

The total amount of land from the continents is approximately 57,000,000 square miles. Use the graph to find the area of the continents given in Exercises 11 through 14. See Example 3.

11. Asia **12.** South America **13.** Australia **14.** Europe

The following circle graph shows the percent of the types of books available at Midway Memorial Library. Use this graph for Exercises 15 through 24. See Example 2.

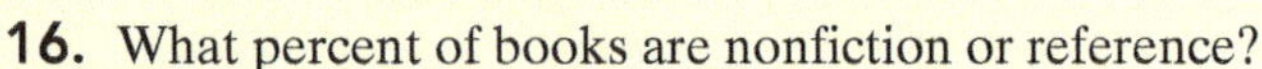

15. What percent of books are classified as some type of fiction?

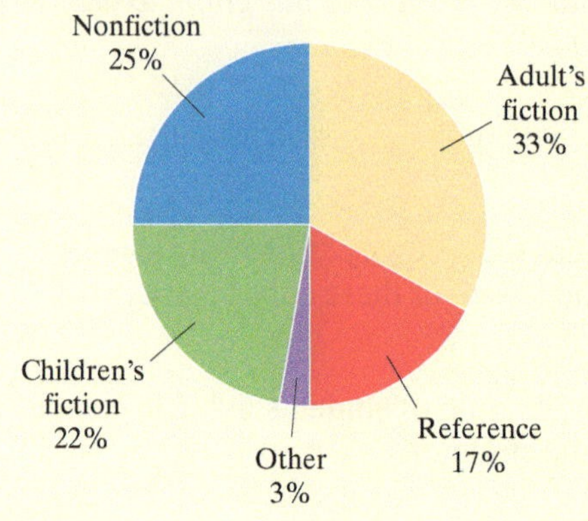

16. What percent of books are nonfiction or reference?

17. What is the second-largest category of books?

18. What is the third-largest category of books?

If this library has 125,600 books, find how many books are in each category given in Exercises 19 through 24. See Example 3.

19. Nonfiction

20. Reference

21. Children's fiction

22. Adult's fiction

23. Reference or other

24. Nonfiction or other

Objective B *Fill in the table. Round to the nearest degree. Then draw a circle graph to represent the information given in each table. (Remember: The total of "Degrees in Sector" column should equal 360° or very close to 360° because of rounding.) See Example 4.*

25.

Types of Apples Grown in Washington State

Type of Apple	Percent	Degrees in Sector
Red Delicious	37%	
Golden Delicious	13%	
Fuji	14%	
Gala	15%	
Granny Smith	12%	
Other varieties	6%	
Braeburn	3%	

(*Source:* U.S. Apple Association)

26.

Color Distribution of M&M's Milk Chocolate

Color	Percent	Degrees in Sector
Blue	24%	
Orange	20%	
Green	16%	
Yellow	14%	
Red	13%	
Brown	13%	

(*Source:* M&M Mars)

27.

Distribution of Large Dams by Continent

Continent	Percent	Degrees in Sector
Europe	19%	
North America	32%	
South America	3%	
Asia	39%	
Africa	5%	
Australia	2%	

(*Source:* International Commission on Large Dams)

28.

2010 Hybrid Sales by Make of Car

Company	Percent	Degrees in Sector
Toyota	69%	
Ford	13%	
Honda	12%	
GM	3%	
Nissan	2%	
Other	1%	

(*Source:* U.S. Department of Energy, 2010 Data)

Review

Write the prime factorization of each number. See Section 4.2.

29. 20 **30.** 25 **31.** 40

32. 16 **33.** 85 **34.** 105

Concept Extensions

The following circle graph shows the relative sizes of the great oceans.

35. Without calculating, determine which ocean is the largest. How can you answer this question by looking at the circle graph?

36. Without calculating, determine which ocean is the smallest. How can you answer this question by looking at the circle graph?

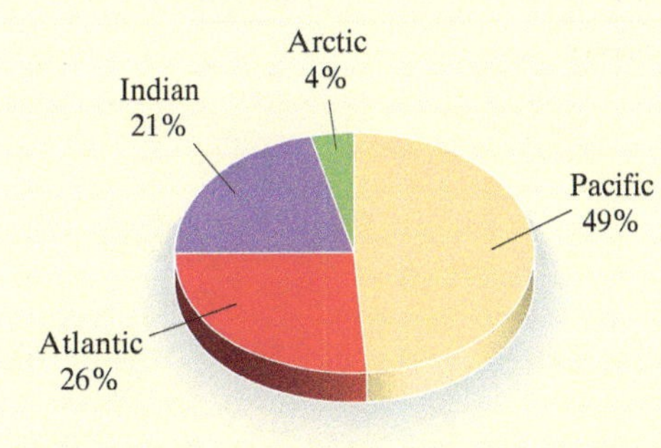

Source: Philip's World Atlas

These oceans together make up 264,489,800 square kilometers of Earth's surface. Find the square kilometers for each ocean.

37. Pacific Ocean **38.** Atlantic Ocean **39.** Indian Ocean **40.** Arctic Ocean

The following circle graph summarizes the results of online spending in America. Let's use these results to make predictions about the online spending behavior of a community of 2800 Internet users age 18 and over. Use this graph for Exercises 41 through 46. Round to the nearest whole. (Note: Because of rounding, these percents do not have a sum of 100%.)

41. How many of the survey respondents said that they spend \$0 online each month?

42. How many of the survey repondents said that they spend \$1–\$100 online each month?

43. How many of the survey respondents said that they spend \$0 to \$100 online each month?

44. How many of the survey respondents said that they spend \$1 to \$1000 online each month?

Online Spending per Month

$101–$1000 14%
> $1000 0.2%
$0 24%
$1–$100 62%

Source: The Digital Future Report, 2013

45. Find the ratio of *number* of respondents who spend \$0 online to *number* of respondents who spend \$1–\$100 online. Write the ratio as a fraction. Simplify the fraction if possible.

46. Find the ratio of *percent* of respondents who spend \$101–\$1000 online to *percent* of those who spend \$1–\$100. Write the ratio as a fraction with integers in the numerator and denominator. Simplify the fraction if possible.

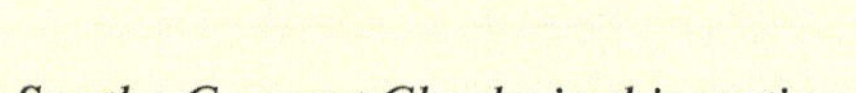

See the Concept Checks in this section.

47. Can the data below be represented by a circle graph? Why or why not?

Responses to the Question, "What Classes Are You Taking?"	
Math	80%
English	72%
History	37%
Biology	21%
Chemistry	14%

48. True or false? The smaller a sector in a circle graph, the smaller the percent of the total it represents. Explain why.

Reading Graphs

The following pictograph shows the six occupations with the largest estimated numerical increase (not salary increase) in employment in the United States between 2010 and 2020. Use this graph to answer Exercises 1 through 4.

Jobs with Projected Highest Numerical Increase: 2010–2020

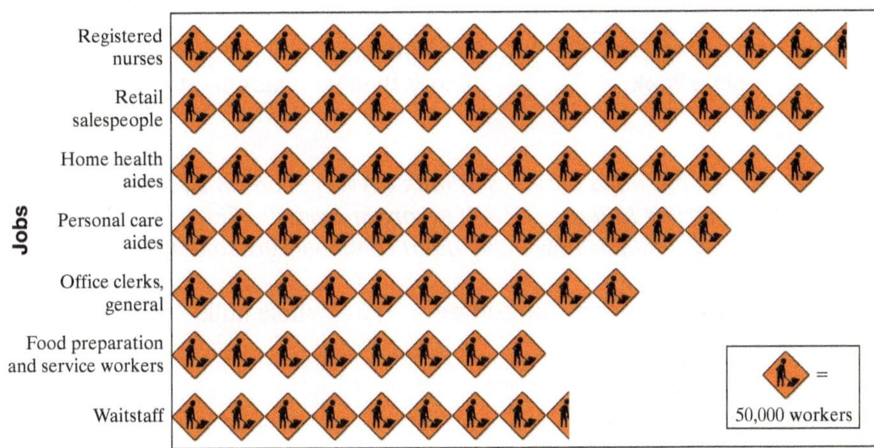

Source: Bureau of Labor Statistics

1. Approximate the increase in the number of retail salespeople from 2010 to 2020.

2. Approximate the increase in the number of registered nurses from 2010 to 2020.

3. Which occupation is expected to show the greatest increase in numbers of employees between the years shown?

4. Which of the listed occupations is expected to show the least increase in numbers of employees between the years shown?

The following bar graph shows the tallest U.S. dams. Use this graph to answer Exercises 5 through 8.

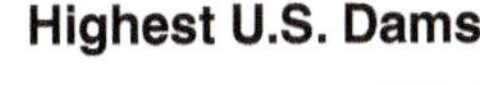

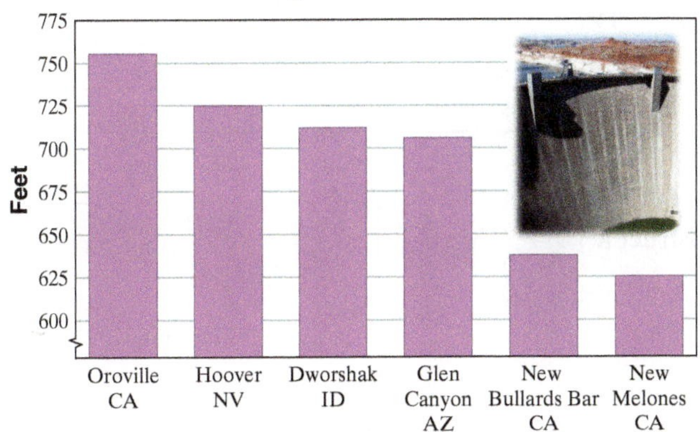

Source: Committee on Register of Dams

5. Name the U.S. dam with the greatest height and estimate its height.

6. Name the U.S. dam whose height is between 625 and 650 feet and estimate its height.

7. Estimate how much taller the Hoover Dam is than the Glen Canyon Dam.

8. How many U.S. dams have heights over 700 feet?

Answers

1. ______
2. ______
3. ______
4. ______
5. ______
6. ______
7. ______
8. ______

9. ______
10. ______
11. ______
12. ______
13. ______
14. ______
15. ______
16. ______
17. ______
18. ______
19. ______
20. ______
21. ______
22. ______

The following line graph shows the daily high temperatures for one week in Annapolis, Maryland. Use this graph to answer Exercises 9 through 12.

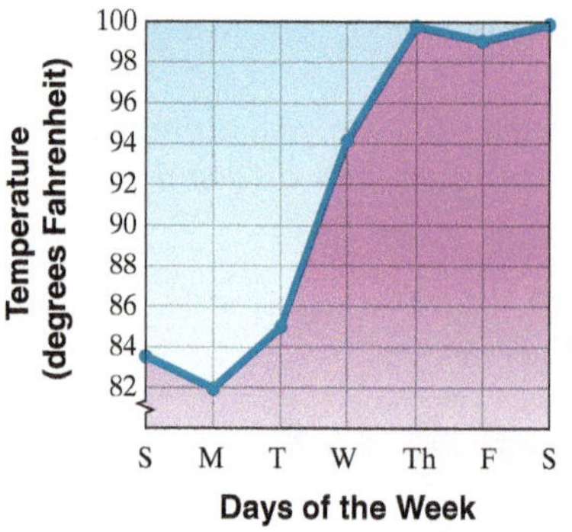

9. Name the day(s) of the week with the highest temperature and give that high temperature.

10. Name the day(s) of the week with the lowest temperature and give that low temperature.

11. On what days of the week was the temperature less than 90° Fahrenheit?

12. On what days of the week was the temperature greater than 90° Fahrenheit?

The following circle graph shows the types of milk beverage consumed in the United States. Use this graph for Exercises 13 through 16. If a store in Kerrville, Texas, sells 200 quart containers of milk per week, estimate how many quart containers are sold in each category below.

Types of Milk Beverage Consumed

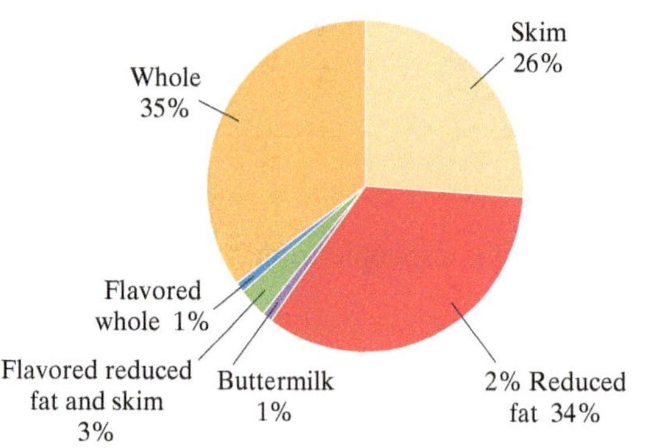

Source: U.S. Department of Agriculture

13. Whole milk

14. Skim milk

15. Buttermilk

16. Flavored reduced fat and skim milk

The following list shows weekly quiz scores for a student in prealgebra. Use this list to complete the frequency distribution table below.

50	80	80	90	88	97
67	89	71	78	93	99
75	53	93	83	86	

	Class Intervals (Scores)	Tally	Class Frequency (Number of Quizzes)
17.	50–59		
18.	60–69		
19.	70–79		
20.	80–89		
21.	90–99		

22. Use the frequency distribution table from Exercises **17** through **21** to construct a histogram.

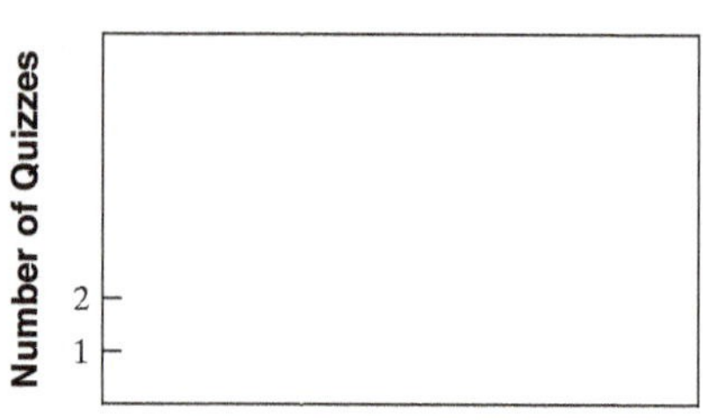

7.3 Square Roots and the Pythagorean Theorem 

In this section, we prepare for work on triangles by studying right triangles and their applications. First, let's practice finding square roots.

Objectives

A Find the Square Root of a Number.

B Approximate Square Roots.

C Use the Pythagorean Theorem.

Objective A Finding Square Roots

The **square** of a number is the number times itself. For example,

The square of 5 is 25 because 5^2 or $5 \cdot 5 = 25$.
The square of -5 is also 25 because $(-5)^2$ or $(-5)(-5) = 25$.

The reverse process of squaring is finding a **square root.** For example,

A square root of 25 is 5 because $5 \cdot 5$ or $5^2 = 25$.
A square root of 25 is also -5 because $(-5)(-5)$ or $(-5)^2 = 25$.

Every positive number has two square roots. We see above that the square roots of 25 are 5 and -5.

We use the symbol $\sqrt{}$, called a **radical sign,** to indicate the positive square root of a nonnegative number. For example,

$\sqrt{25} = 5$ because $5^2 = 25$ and 5 is positive.
$\sqrt{9} = 3$ because $3^2 = 9$ and 3 is positive.

Square Root of a Number

The square root, $\sqrt{}$, of a positive number a is the positive number b whose square is a. In symbols,

$$\sqrt{a} = b, \quad \text{if } b^2 = a$$

Also, $\sqrt{0} = 0$.

Helpful Hint

Remember that the radical sign $\sqrt{}$ is used to indicate the **positive** (or principal) **square root** of a nonnegative number.

Examples Find each square root.

1. $\sqrt{49} = 7$ because $7^2 = 49$.
2. $\sqrt{36} = 6$ because $6^2 = 36$.
3. $\sqrt{1} = 1$ because $1^2 = 1$.
4. $\sqrt{81} = 9$ because $9^2 = 81$.
5. $\sqrt{\frac{1}{36}} = \frac{1}{6}$ because $\left(\frac{1}{6}\right)^2$ or $\frac{1}{6} \cdot \frac{1}{6} = \frac{1}{36}$.
6. $\sqrt{\frac{4}{25}} = \frac{2}{5}$ because $\left(\frac{2}{5}\right)^2$ or $\frac{2}{5} \cdot \frac{2}{5} = \frac{4}{25}$.

Work Practice 1–6

Practice 1–6

Find each square root.

1. $\sqrt{100}$
2. $\sqrt{64}$
3. $\sqrt{169}$
4. $\sqrt{0}$
5. $\sqrt{\frac{1}{4}}$
6. $\sqrt{\frac{9}{16}}$

Answers

1. 10 **2.** 8 **3.** 13 **4.** 0 **5.** $\frac{1}{2}$ **6.** $\frac{3}{4}$

Objective B Approximating Square Roots

Thus far, we have found square roots of perfect squares. Numbers like $\frac{1}{4}$, 36, $\frac{4}{25}$, and 1 are called **perfect squares** because their square root is a whole number or a fraction. A square root such as $\sqrt{5}$ cannot be written as a whole number or a fraction since 5 is not a perfect square.

Although $\sqrt{5}$ cannot be written as a whole number or a fraction, it can be approximated by estimating, by using a table (as in the appendix), or by using a calculator.

Practice 7

Use Appendix A.4 or a calculator to approximate each square root to the nearest thousandth.

a. $\sqrt{10}$ **b.** $\sqrt{62}$

Example 7

Use Appendix A.4 or a calculator to approximate each square root to the nearest thousandth.

a. $\sqrt{43} \approx 6.557$ is approximately

b. $\sqrt{80} \approx 8.944$

Work Practice 7

Helpful Hint

$\sqrt{80}$, above, is *approximately* 8.944. This means that if we multiply 8.944 by 8.944, the product is *close* to 80.

$$8.944 \times 8.944 \approx 79.995$$

It is possible to approximate a square root to the nearest whole number without the use of a calculator or table. To do so, study the number line below and look for patterns.

$\sqrt{25}$ $\sqrt{31}$ $\sqrt{36}$ $\sqrt{40}$ $\sqrt{49}$ $\sqrt{59}$ $\sqrt{64}$ $\sqrt{72}$ $\sqrt{81}$

5 6 7 8 9

Above the number line, notice that as the numbers under the radical signs increase, their value, and thus their placement on the number line, increase also.

Practice 8

Without a calculator or table, approximate $\sqrt{62}$ to the nearest whole.

Example 8

Without a calculator or table:

a. Determine which two whole numbers $\sqrt{78}$ is between.

b. Use part (a) to approximate $\sqrt{78}$ to the nearest whole.

Solution:

a. Review perfect squares and recall that $\sqrt{64} = 8$ and $\sqrt{81} = 9$. Since 78 is between 64 and 81, $\sqrt{78}$ is between $\sqrt{64}$ (or 8) and $\sqrt{81}$ (or 9).

$\sqrt{64}$ $\sqrt{78}$ $\sqrt{81}$

8 9

Thus, $\sqrt{78}$ is between 8 and 9.

b. Since 78 is closer to 81, then (as our number line shows) $\sqrt{78}$ is closer to $\sqrt{81}$, or 9.

Work Practice 8

Answers

7. a. 3.162 **b.** 7.874 **8.** 8

Objective C Using the Pythagorean Theorem 

One important application of square roots has to do with right triangles. Recall that a **right triangle** is a triangle in which one of the angles is a right angle, or measures 90° (degrees). The **hypotenuse** of a right triangle is the side opposite the right angle. The **legs** of a right triangle are the other two sides. These are shown in the following figure. The right angle in the triangle is indicated by the small square drawn in that angle.

The following theorem is true for all right triangles.

Pythagorean Theorem

If a and b are the lengths of the legs of a right triangle and c is the length of the hypotenuse, then

$$a^2 + b^2 = c^2$$

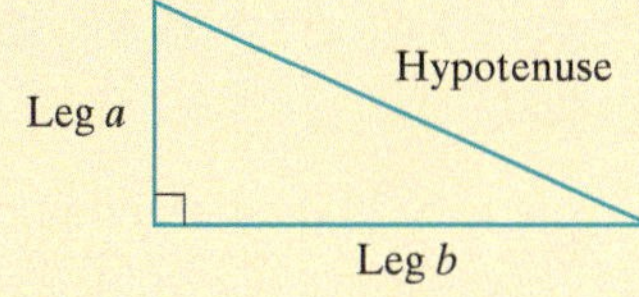

In other words, $(\text{leg})^2 + (\text{other leg})^2 = (\text{hypotenuse})^2$.

Example 9 Find the length of the hypotenuse of the given right triangle.

Solution: Let $a = 6$ and $b = 8$. According to the Pythagorean theorem,

$$a^2 + b^2 = c^2$$
$$6^2 + 8^2 = c^2 \quad \text{Let } a = 6 \text{ and } b = 8.$$
$$36 + 64 = c^2 \quad \text{Evaluate } 6^2 \text{ and } 8^2.$$
$$100 = c^2 \quad \text{Add.}$$

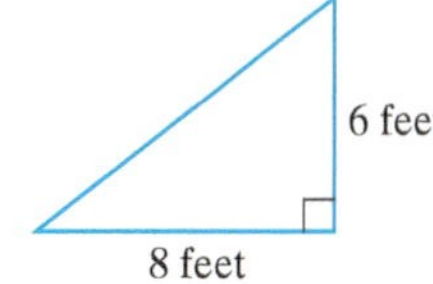

In the equation $c^2 = 100$, the solutions of c are the square roots of 100. Since $10 \cdot 10 = 100$ and $(-10)(-10) = 100$, both 10 and -10 are square roots of 100. Since c represents a length, we are interested only in the positive square root of c^2.

$$c = \sqrt{100}$$
$$= 10$$

The hypotenuse is 10 feet long.

Work Practice 9

Practice 9

Find the length of the hypotenuse of the given right triangle.

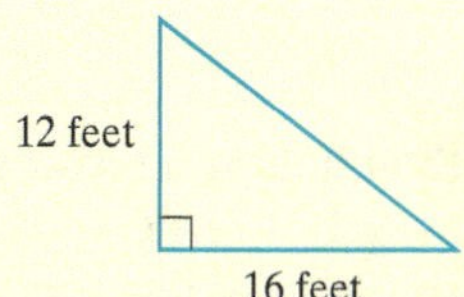

Example 10 Approximate the length of the hypotenuse of the given right triangle. Round the length to the nearest whole unit.

Solution: Let $a = 17$ and $b = 10$.

$$a^2 + b^2 = c^2$$
$$17^2 + 10^2 = c^2$$
$$289 + 100 = c^2$$
$$389 = c^2$$
$$\sqrt{389} = c \text{ or } c \approx 20 \quad \text{From Appendix A.4 or a calculator}$$

10 meters

17 meters

The hypotenuse is exactly $\sqrt{389}$ meters, which is approximately 20 meters.

Work Practice 10

Practice 10

Approximate the length of the hypotenuse of the given right triangle. Round to the nearest whole unit.

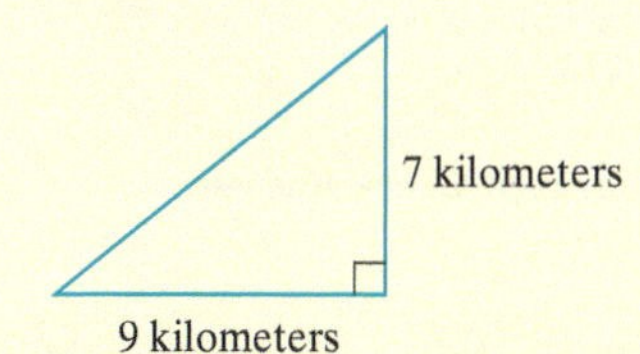

Answers

9. 20 feet **10.** 11 kilometers

Practice 11

Find the length of the leg in the given right triangle. Give the exact length and a two-decimal-place approximation.

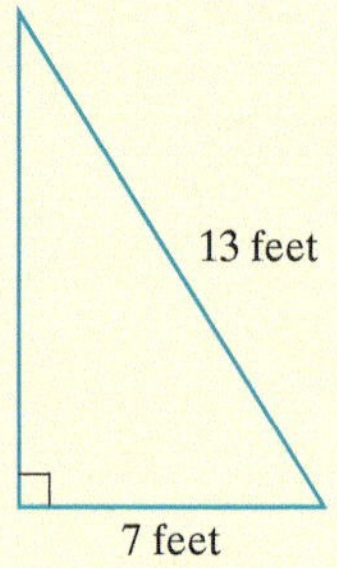

Example 11 Find the length of the leg in the given right triangle. Give the exact length and a two-decimal-place approximation.

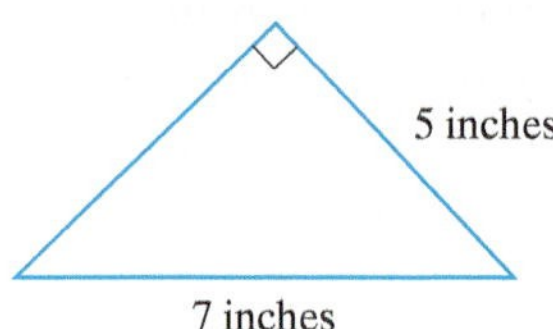

Solution: Notice that the hypotenuse measures 7 inches and that the length of one leg measures 5 inches. Thus, let $c = 7$ and a or b be 5. We will let $a = 5$.

$$a^2 + b^2 = c^2$$

$$5^2 + b^2 = 7^2 \quad \text{Let } a = 5 \text{ and } c = 7.$$

$$25 + b^2 = 49 \quad \text{Evaluate } 5^2 \text{ and } 7^2.$$

$$b^2 = 24 \quad \text{Subtract 25 from both sides.}$$

$$b = \sqrt{24} \approx 4.90$$

The length of the leg is exactly $\sqrt{24}$ inches and approximately 4.90 inches.

Work Practice 11

Concept Check The following lists are the lengths of the sides of two triangles. Which set forms a right triangle?

a. 8, 15, 17 **b.** 24, 30, 40

Practice 12

A football field is a rectangle measuring 100 yards by 53 yards. Draw a diagram and find the length of the diagonal of the football field to the nearest yard.

Example 12 Finding the Dimensions of a Park

An inner-city park is in the shape of a square that measures 300 feet on a side. A sidewalk is to be constructed along the diagonal of the park. Find the length of the sidewalk rounded to the nearest whole foot.

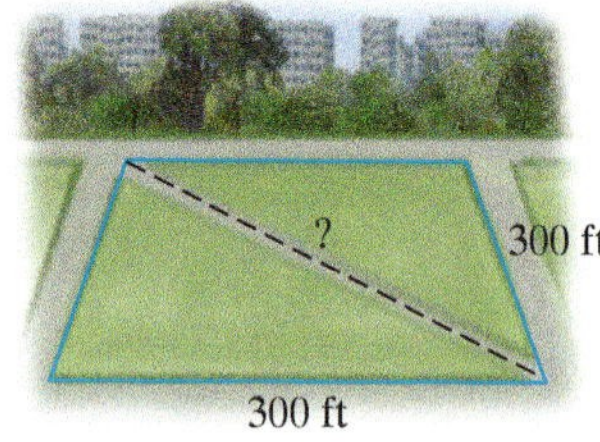

Solution: The diagonal is the hypotenuse of a right triangle, which we label c.

$$a^2 + b^2 = c^2$$

$$300^2 + 300^2 = c^2 \quad \text{Let } a = 300 \text{ and } b = 300.$$

$$90{,}000 + 90{,}000 = c^2 \quad \text{Evaluate } (300)^2.$$

$$180{,}000 = c^2 \quad \text{Add.}$$

$$\sqrt{180{,}000} = c \text{ or } c \approx 424$$

The length of the sidewalk is approximately 424 feet.

Work Practice 12

Answers

11. $\sqrt{120}$ feet $\approx$ 10.95 feet
12. 113 yards

Concept Check Answer

a

Calculator Explorations **Finding and Approximating Square Roots**

To simplify or approximate square roots using a calculator, locate the key marked $\boxed{\sqrt{\ }}$.

To simplify $\sqrt{64}$, for example, press the keys

$\boxed{64}$ $\boxed{\sqrt{\ }}$ or $\boxed{\sqrt{\ }}$ $\boxed{64}$

The display should read $\boxed{\quad 8}$. Then

$\sqrt{64} = 8$

To *approximate* $\sqrt{10}$, press the keys

$\boxed{10}$ $\boxed{\sqrt{\ }}$ or $\boxed{\sqrt{\ }}$ $\boxed{10}$

The display should read $\boxed{3.16227766}$. This is an *approximation* for $\sqrt{10}$. A three-decimal-place approximation is

$\sqrt{10} \approx 3.162$

Is this answer reasonable? Since 10 is between the perfect squares 9 and 16, $\sqrt{10}$ is between $\sqrt{9} = 3$ and $\sqrt{16} = 4$. Our answer is reasonable since 3.162 is between 3 and 4.

Simplify.

1. $\sqrt{1024}$

2. $\sqrt{676}$

Approximate each square root. Round each answer to the nearest thousandth.

3. $\sqrt{15}$

4. $\sqrt{19}$

5. $\sqrt{97}$

6. $\sqrt{56}$

Vocabulary, Readiness & Video Check

Use the choices below to fill in each blank. Some choices will be used more than once.

squaring	Pythagorean theorem	radical	-10	leg
hypotenuse	perfect squares	10	c^2	b^2

1. The square roots of 100 are ________________ and ________________ because $10 \cdot 10 = 100$ and $(-10)(-10) = 100$.

2. $\sqrt{100} =$ ________________ because $10 \cdot 10 = 100$ and 10 is positive.

3. The ________________ sign is used to denote the positive square root of a nonnegative number.

4. The reverse process of ________________ a number is finding a square root of a number.

5. The numbers 9, 1, and $\frac{1}{25}$ are called ________________.

6. Label the parts of the right triangle.

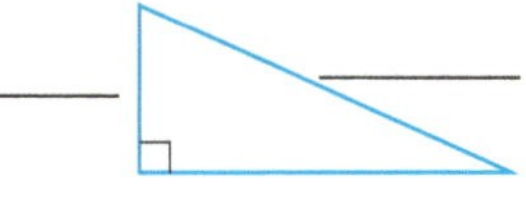

7. In the given triangle, $a^2 +$ __________ $=$ __________.

8. The ________________ can be used for right triangles.

Martin-Gay Interactive Videos *Watch the section lecture video and answer the following questions.*

See Video 7.3

Objective A **9.** From the lecture before Example 1, list the square root(s) of 49. How is this different from finding $\sqrt{49}$?

Objective B **10.** In Example 5, how do we know $\sqrt{15}$ is closer to 4 than to 3?

Objective C **11.** From Example 6, how do we know which side of a right triangle is the hypotenuse?

7.3 Exercise Set MyMathLab®

Objective A *Find each square root. See Examples 1 through 6.*

1. $\sqrt{4}$ **2.** $\sqrt{9}$ **3.** $\sqrt{121}$ **4.** $\sqrt{144}$

5. $\sqrt{\frac{1}{81}}$ **6.** $\sqrt{\frac{1}{64}}$ **7.** $\sqrt{\frac{16}{64}}$ **8.** $\sqrt{\frac{36}{81}}$

Objective B *Use Appendix A.4 or a calculator to approximate each square root. Round the square root to the nearest thousandth. See Example 7.*

9. $\sqrt{3}$ **10.** $\sqrt{5}$ **11.** $\sqrt{15}$ **12.** $\sqrt{17}$

13. $\sqrt{31}$ **14.** $\sqrt{85}$ **15.** $\sqrt{26}$ **16.** $\sqrt{35}$

Determine what two whole numbers each square root is between without using a calculator or table. Then use a calculator or Appendix A.4 to check. See Example 8.

17. $\sqrt{38}$ **18.** $\sqrt{27}$ **19.** $\sqrt{101}$ **20.** $\sqrt{85}$

Objectives A B Mixed Practice *Find each square root. If necessary, round the square root to the nearest thousandth. See Examples 1 through 8.*

21. $\sqrt{256}$ **22.** $\sqrt{625}$ **23.** $\sqrt{92}$ **24.** $\sqrt{18}$

25. $\sqrt{\frac{49}{144}}$ **26.** $\sqrt{\frac{121}{169}}$ **27.** $\sqrt{71}$ **28.** $\sqrt{62}$

Objective C *Find the unknown length in each right triangle. If necessary, approximate the length to the nearest thousandth. See Examples 9 through 12.*

29.

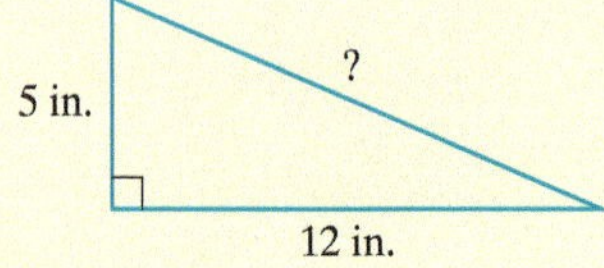

30.

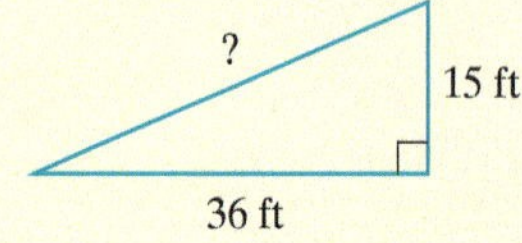

31.

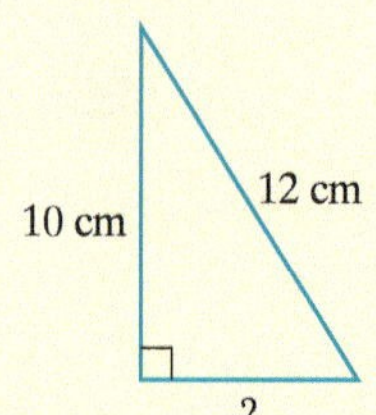

32. 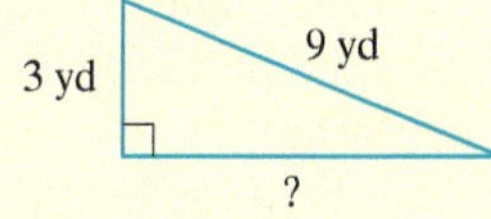

33. 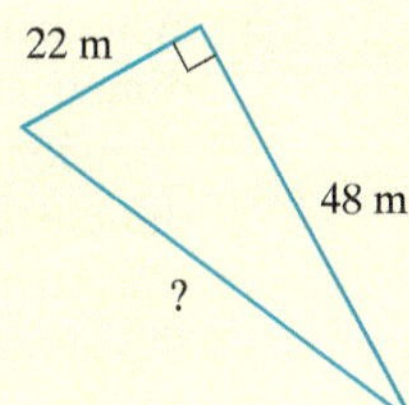

34. 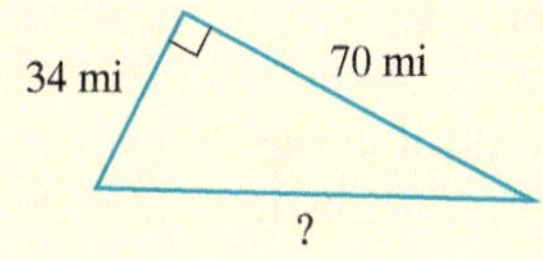

35. 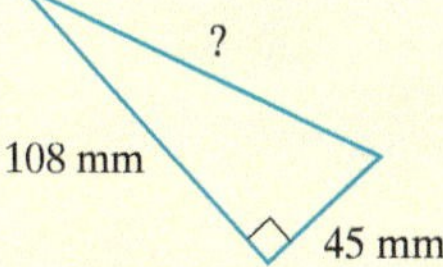

36. 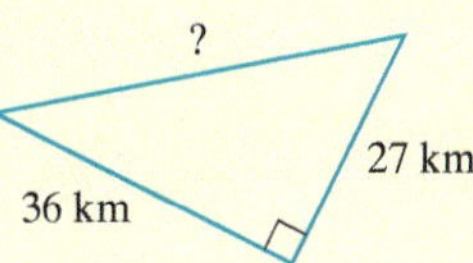

Sketch each right triangle and find the length of the side not given. If necessary, approximate the length to the nearest thousandth. (Each length is in units.) See Examples 9 through 12.

37. leg = 3, leg = 4

38. leg = 9, leg = 12

39. leg = 5, hypotenuse = 13

40. leg = 6, hypotenuse = 10

41. leg = 10, leg = 14

42. leg = 2, leg = 16

43. leg = 35, leg = 28

44. leg = 30, leg = 15

45. leg = 30, leg = 30

46. leg = 21, leg = 21

47. hypotenuse = 2, leg = 1

48. hypotenuse = 9, leg = 8

49. leg = 7.5, leg = 4

50. leg = 12, leg = 22.5

Solve. See Example 12.

51. A standard city block is a square with each side measuring 100 yards. Find the length of the diagonal of a city block to the nearest hundredth yard.

52. A section of land is a square with each side measuring 1 mile. Find the length of the diagonal of the section of land to the nearest thousandth mile.

53. Find the height of the tree. Round the height to one decimal place.

54. Find the height of the antenna. Round the height to one decimal place.

55. The playing field for football is a rectangle that is 300 feet long by 160 feet wide. Find the length of a straight-line run that started at one corner and went diagonally to end at the opposite corner. Round to the nearest foot if necessary.

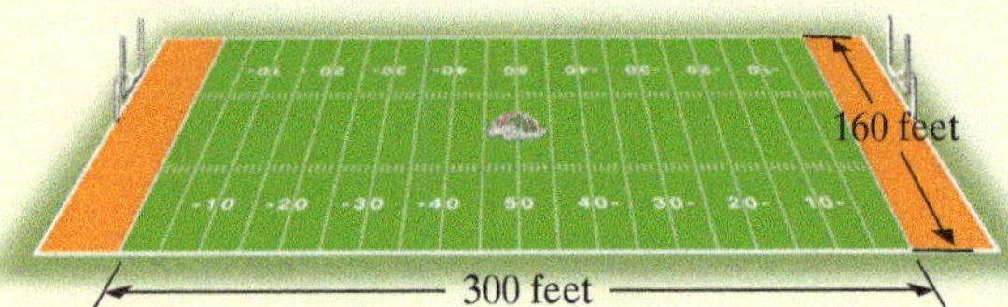

56. A soccer field is in the shape of a rectangle and its dimensions depend on the age of the players. The dimensions of the soccer field below are the minimum dimensions for international play. Find the length of the diagonal of this rectangle. Round the answer to the nearest tenth of a yard.

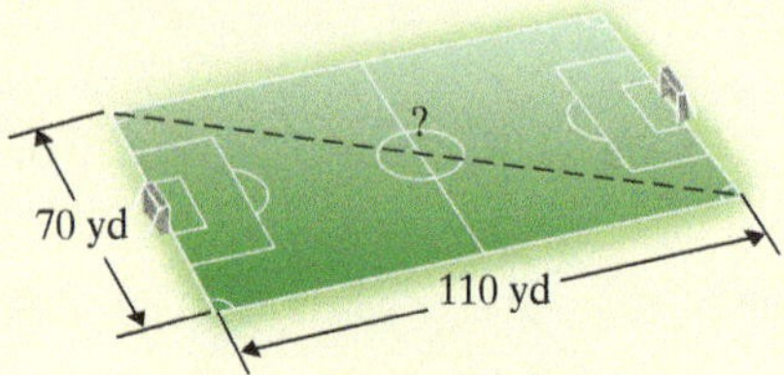

Review

Write each fraction in simplest form. See Section 4.2.

57. $\frac{10}{12}$ **58.** $\frac{10}{15}$ **59.** $\frac{2x}{60}$ **60.** $\frac{35}{75y}$

Perform the indicated operations. See Sections 4.3 and 4.4.

61. $\frac{9}{13y} + \frac{12}{13y}$ **62.** $\frac{3x}{9} - \frac{5}{9}$ **63.** $\frac{9}{8} \cdot \frac{x}{8}$ **64.** $\frac{7x}{11} \div \frac{8x}{11}$

Concept Extensions

*Use the results of Exercises **17–20** and approximate each square root to the nearest whole without using a calculator or table. Then use a calculator or Appendix A.4 to check. See Example 8.*

65. $\sqrt{38}$ **66.** $\sqrt{27}$ **67.** $\sqrt{101}$ **68.** $\sqrt{85}$

69. Without using a calculator, explain how you know that $\sqrt{105}$ is *not* approximately 9.875.

70. Without using a calculator, explain how you know that $\sqrt{27}$ is *not* approximately 3.296.

Does the set form the lengths of the sides of a right triangle? See the Concept Check in this section.

71. 25, 60, 65 **72.** 20, 45, 50

△ **73.** Find the exact length of x. Then give a two-decimal-place approximation.

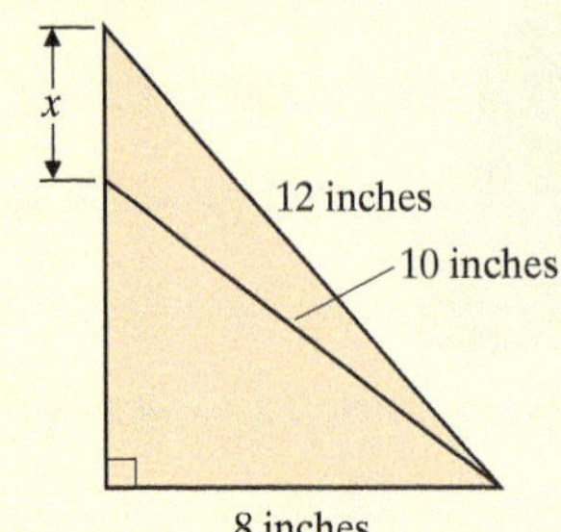

7.4 Congruent and Similar Triangles 

Objectives

A Decide Whether Two Triangles Are Congruent.

B Find the Ratio of Corresponding Sides in Similar Triangles.

C Find Unknown Lengths of Sides in Similar Triangles.

Objective A Deciding Whether Two Triangles Are Congruent

Two triangles are **congruent** when they have the same shape and the same size. In congruent triangles, the measures of corresponding angles are equal and the lengths of corresponding sides are equal. The following triangles are congruent:

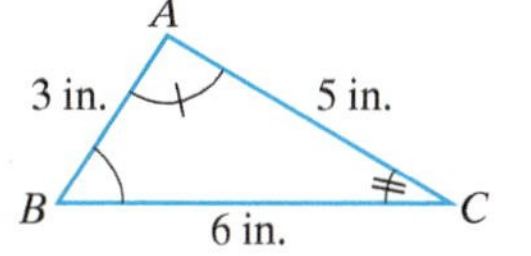

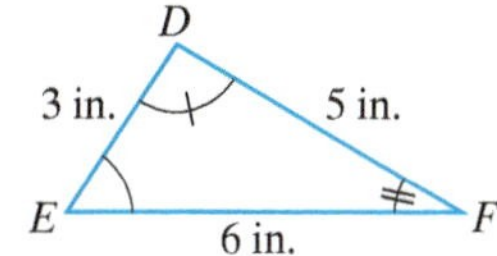

Since these triangles are congruent, the measures of corresponding angles are equal.

Angles with equal measure: $\angle A$ and $\angle D$, $\angle B$ and $\angle E$, $\angle C$ and $\angle F$. Also, the lengths of corresponding sides are equal.

Equal corresponding sides: $\overline{AB}$ and $\overline{DE}$, $\overline{BC}$ and $\overline{EF}$, $\overline{CA}$ and $\overline{FD}$

Any one of the following may be used to determine whether two triangles are congruent:

Congruent Triangles

Angle-Side-Angle (ASA)

If the measures of two angles of a triangle equal the measures of two angles of another triangle, and the lengths of the sides between each pair of angles are equal, the triangles are congruent.

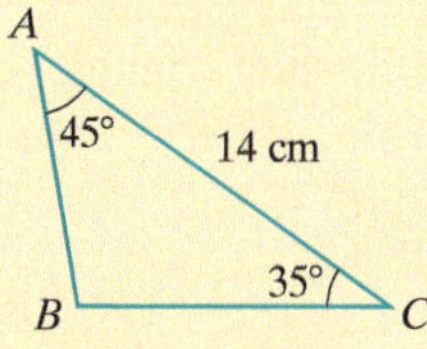

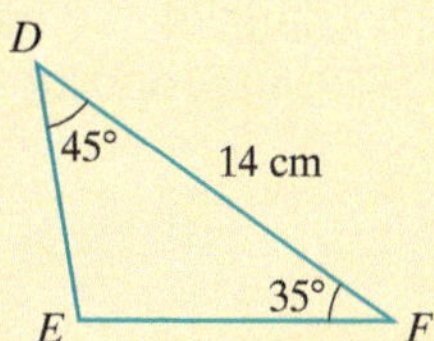

For example, these two triangles are congruent by Angle-Side-Angle.

Side-Side-Side (SSS)

If the lengths of the three sides of a triangle equal the lengths of the corresponding sides of another triangle, the triangles are congruent.

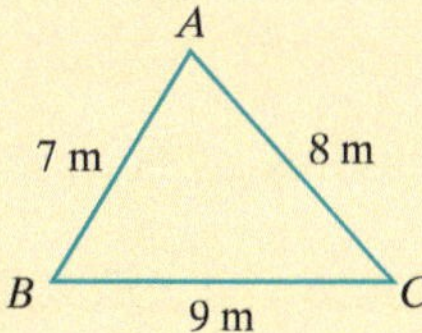

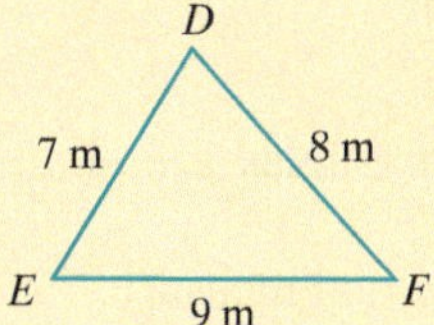

For example, these two triangles are congruent by Side-Side-Side.

Side-Angle-Side (SAS)

If the lengths of two sides of a triangle equal the lengths of corresponding sides of another triangle, and the measures of the angles between each pair of sides are equal, the triangles are congruent.

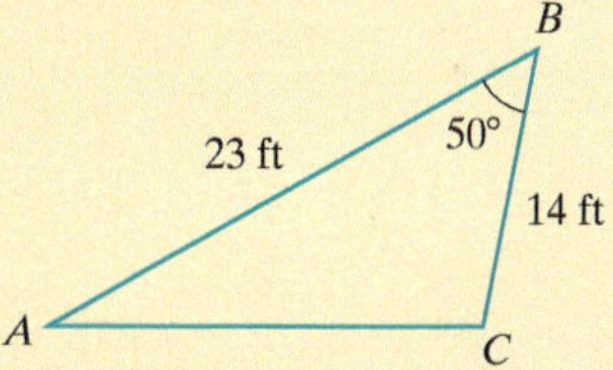

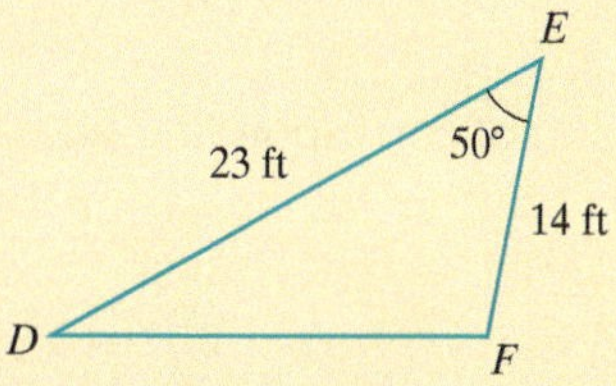

For example, these two triangles are congruent by Side-Angle-Side.

Practice 1

a. Determine whether triangle *MNO* is congruent to triangle *RQS*.

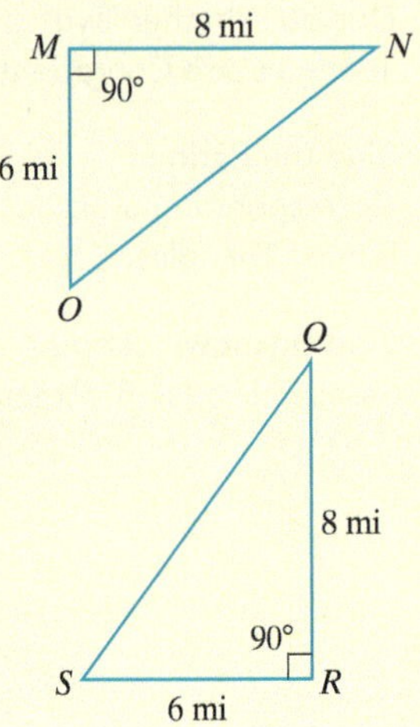

b. Determine whether triangle *GHI* is congruent to triangle *JKL*.

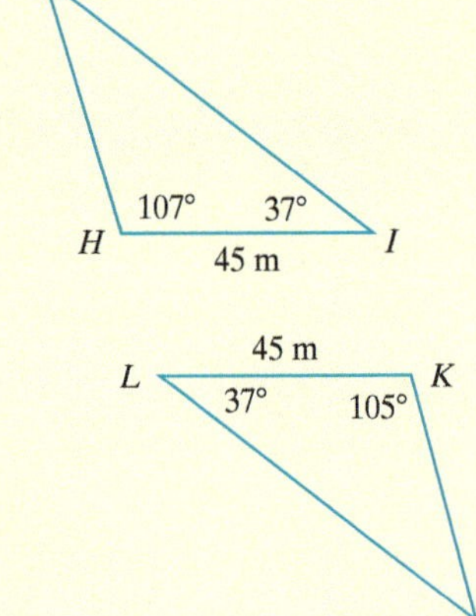

Answers

1. a. congruent **b.** not congruent

Example 1 Determine whether triangle ABC is congruent to triangle DEF.

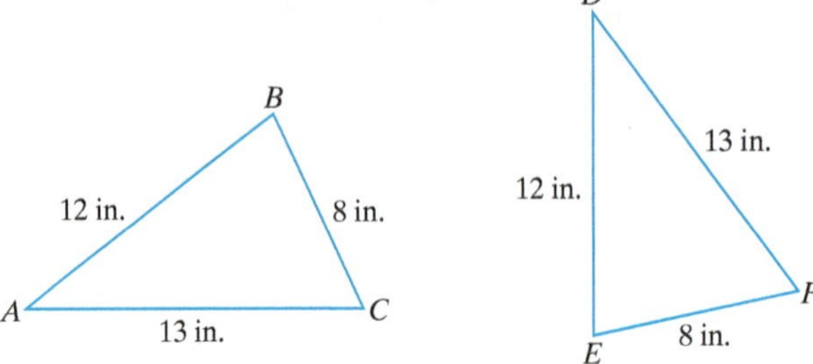

Solution: Since the lengths of all three sides of triangle ABC equal the lengths of all three sides of triangle DEF, the triangles are congruent.

Work Practice 1

In Example 1, notice that as soon as we know that the two triangles are congruent, we know that all three corresponding angles are congruent.

Objective B Finding the Ratios of Corresponding Sides in Similar Triangles

Two triangles are **similar** when they have the same shape but not necessarily the same size. In similar triangles, the measures of corresponding angles are equal and corresponding sides are in proportion. The following triangles are similar:

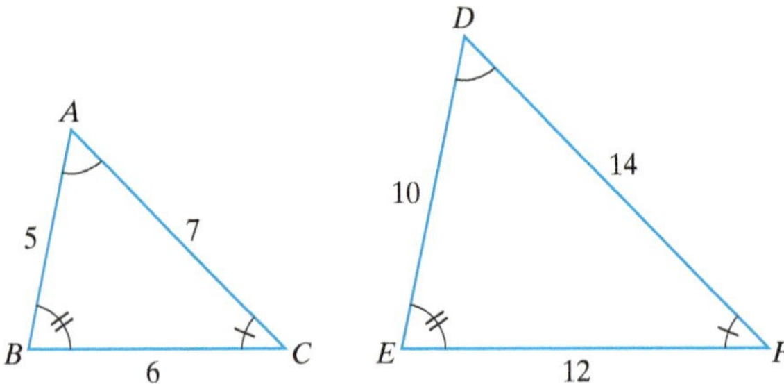

Since these triangles are similar, the measures of corresponding angles are equal.

Angles with equal measure: $\angle A$ and $\angle D$, $\angle B$ and $\angle E$, $\angle C$ and $\angle F$. Also, the lengths of corresponding sides are in proportion.

Sides in proportion: $\frac{AB}{DE} = \frac{BC}{EF} = \frac{CA}{FD}$ or, in this particular case,

$$\frac{AB}{DE} = \frac{5}{10} = \frac{1}{2}, \frac{BC}{EF} = \frac{6}{12} = \frac{1}{2}, \frac{CA}{FD} = \frac{7}{14} = \frac{1}{2}$$

The ratio of corresponding sides is $\frac{1}{2}$.

Example 2 Find the ratio of corresponding sides for the similar triangles ABC and DEF.

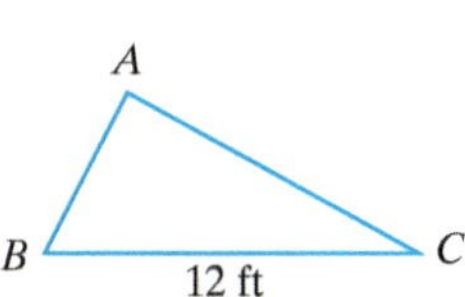

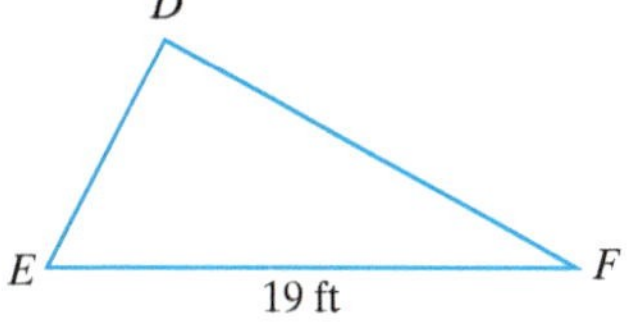

Solution: We are given the lengths of two corresponding sides. Their ratio is

$$\frac{12 \text{ feet}}{19 \text{ feet}} = \frac{12}{19}$$

Work Practice 2

Practice 2

Find the ratio of corresponding sides for the similar triangles QRS and XYZ.

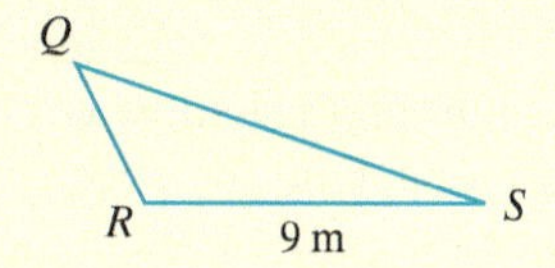

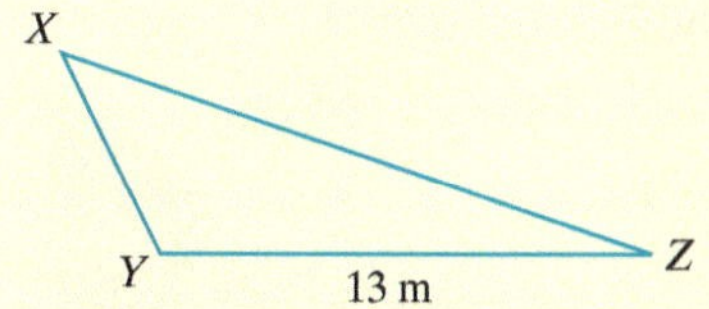

Objective C Finding Unknown Lengths of Sides in Similar Triangles

Because the ratios of lengths of corresponding sides are equal, we can use proportions to find unknown lengths in similar triangles.

Example 3 Given that the triangles are similar, find the missing length y.

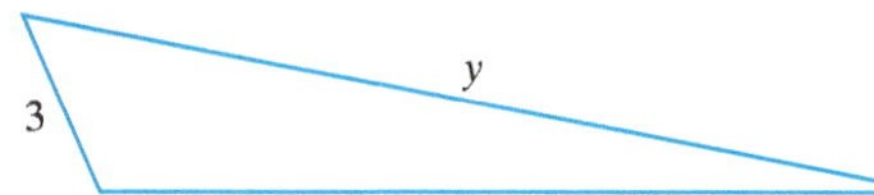

Solution: Since the triangles are similar, corresponding sides are in proportion. Thus, the ratio of 2 to 3 is the same as the ratio of 10 to y, or

$$\frac{2}{3} = \frac{10}{y}$$

To find the unknown length y, we set cross products equal.

$$\frac{2}{3} = \frac{10}{y}$$

$2 \cdot y = 3 \cdot 10$ Set cross products equal.

$2y = 30$ Multiply.

$\frac{2y}{2} = \frac{30}{2}$ Divide both sides by 2.

$y = 15$ Simplify.

The missing length is 15 units.

Work Practice 3

Practice 3

Given that the triangles are similar, find the missing length x.

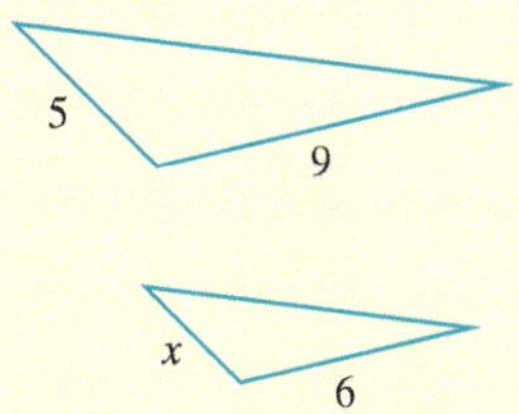

✓Concept Check The following two triangles are similar. Which vertices of the first triangle appear to correspond to which vertices of the second triangle?

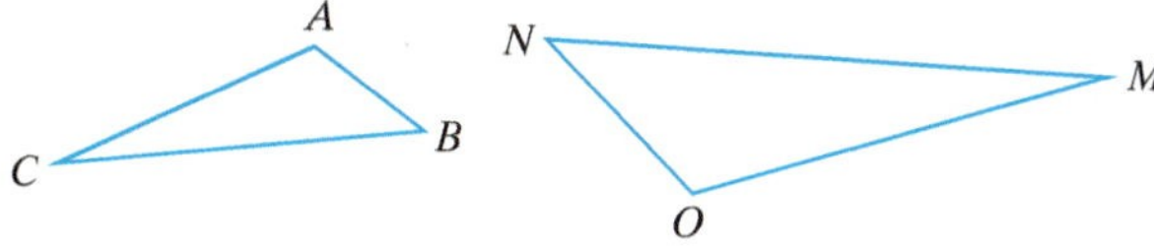

Many applications involve a diagram containing similar triangles. Surveyors, astronomers, and many other professionals continually use similar triangles in their work.

Answers

2. $\frac{9}{13}$ **3.** $x = \frac{10}{3}$ or $3\frac{1}{3}$ units

✓Concept Check Answer

A corresponds to O; B corresponds to N; C corresponds to M

Practice 4

Tammy Shultz, a firefighter, needs to estimate the height of a burning building. She estimates the length of her shadow to be 8 feet long and the length of the building's shadow to be 60 feet long. Find the approximate height of the building if she is 5 feet tall.

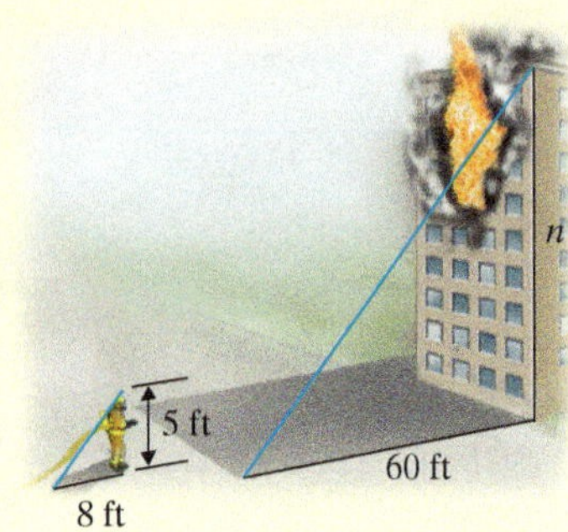

Answer

4. approximately 37.5 ft

Example 4 Finding the Height of a Tree

Mel Wagstaff is a 6-foot-tall park ranger who needs to know the height of a particular tree. He measures the shadow of the tree to be 69 feet long when his own shadow is 9 feet long. Find the height of the tree.

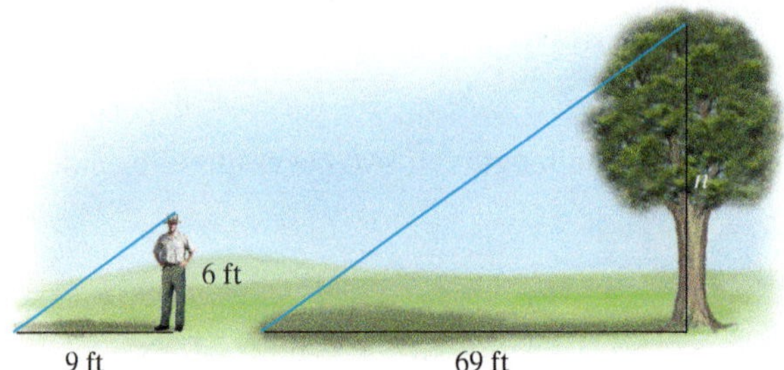

Solution:

1. UNDERSTAND. Read and reread the problem. Notice that the triangle formed by the sun's rays, Mel, and his shadow is similar to the triangle formed by the sun's rays, the tree, and its shadow.
2. TRANSLATE. Write a proportion from the similar triangles formed.

$$\frac{\text{Mel's height}}{\text{height of tree}} \rightarrow \frac{6}{n} = \frac{9}{69} \leftarrow \frac{\text{length of Mel's shadow}}{\text{length of tree's shadow}}$$

$$\text{or } \frac{6}{n} = \frac{3}{23} \quad \text{Simplify } \frac{9}{69} \text{ (ratio in lowest terms).}$$

3. SOLVE for n:

$$\frac{6}{n} = \frac{3}{23}$$

$$6 \cdot 23 = n \cdot 3 \quad \text{Set cross products equal.}$$

$$138 = 3n \quad \text{Multiply.}$$

$$\frac{138}{3} = \frac{3n}{3} \quad \text{Divide both sides by 3.}$$

$$46 = n$$

4. INTERPRET. *Check* to see that replacing n with 46 in the proportion makes the proportion true. *State* your conclusion: The height of the tree is 46 feet.

Work Practice 4

Vocabulary, Readiness & Video Check

Answer each question true or false.

1. Two triangles that have the same shape but not necessarily the same size are congruent. _______
2. Two triangles are congruent if they have the same shape and size. _______
3. Congruent triangles are also similar. _______
4. Similar triangles are also congruent. _______
5. For the two similar triangles, the ratio of corresponding sides is $\frac{5}{6}$. _______

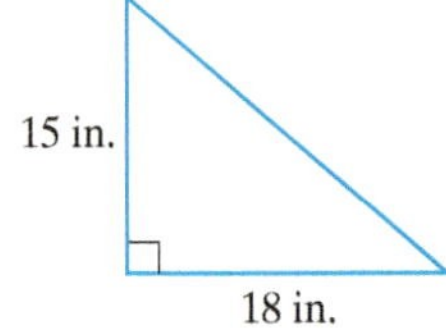

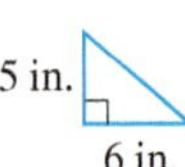

See Video 7.4

Watch the section lecture video and answer the following questions.

Objective A 6. How did we decide which congruency rule to use to determine if the two triangles in Example 1 are congruent?

Objective B 7. From Example 2, what does "corresponding sides are in proportion" mean?

Objective C 8. In Example 4, what is another proportion we could have used to solve the application problem?

7.4 Exercise Set MyMathLab®

Objective A *Determine whether each pair of triangles is congruent. If congruent, state the reason why, such as SSS, SAS, or ASA. See Example 1.*

1.

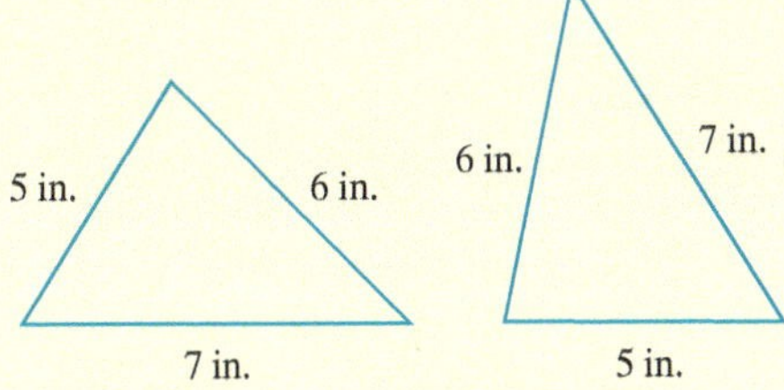

2.

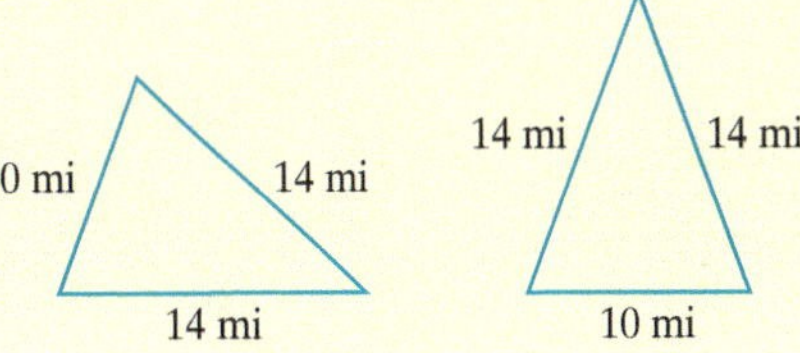

3.

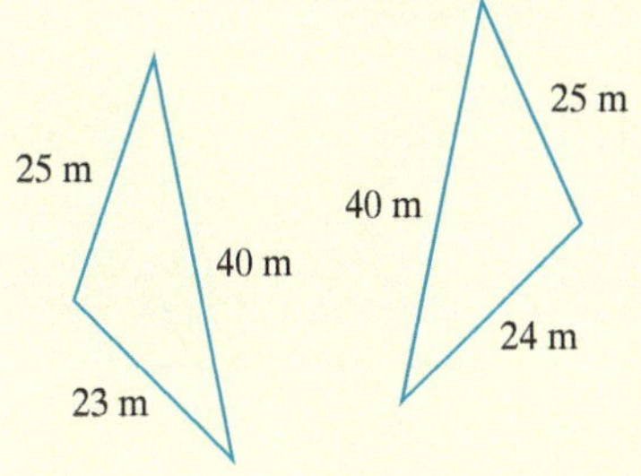

4.

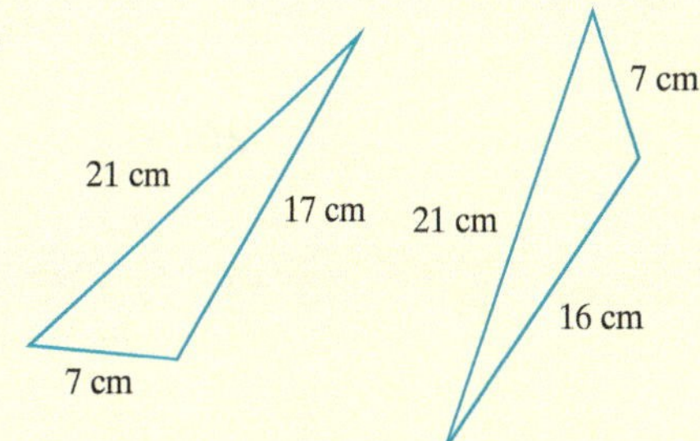

5.

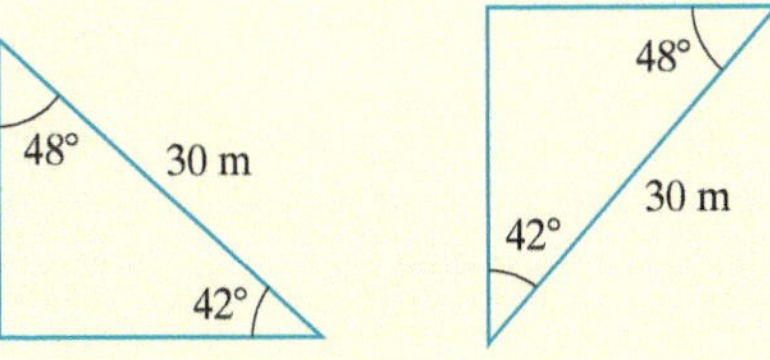

6.

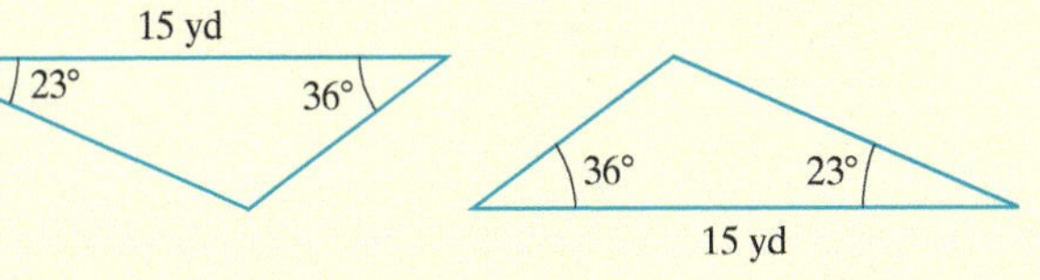

7.

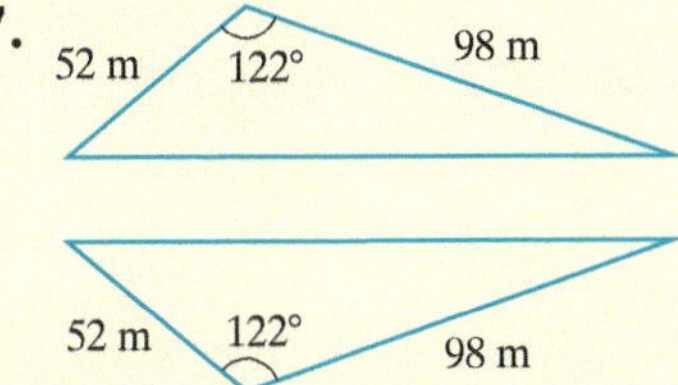

8.

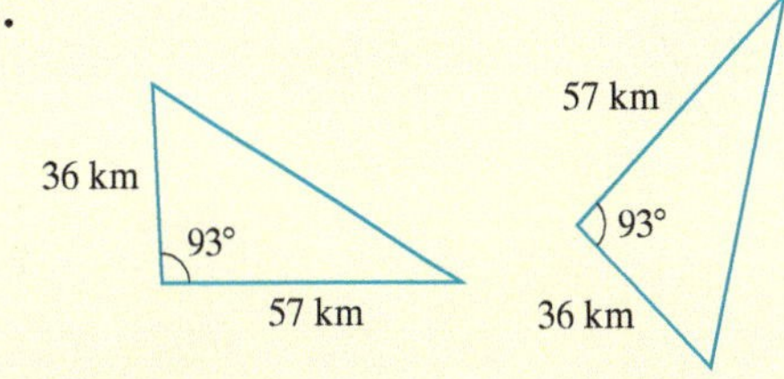

Objective B *Find each ratio of the corresponding sides of the given similar triangles. See Example 2.*

9.

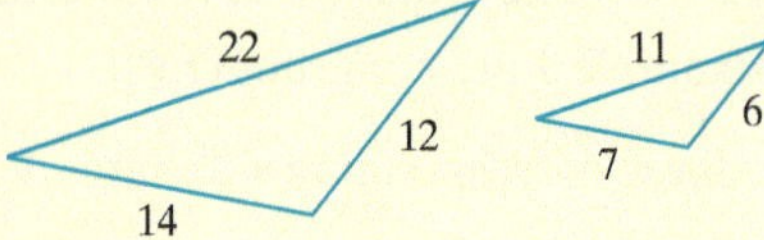

10.

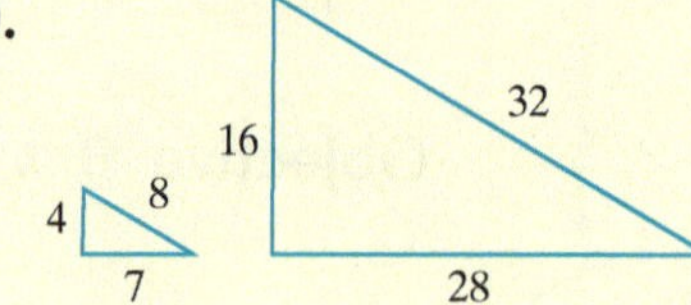

11.

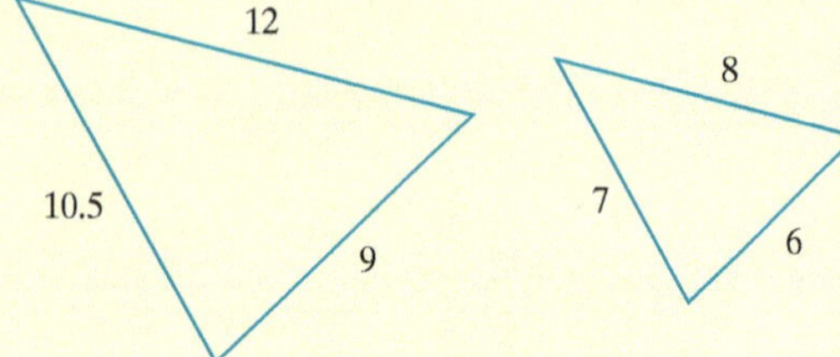

12.

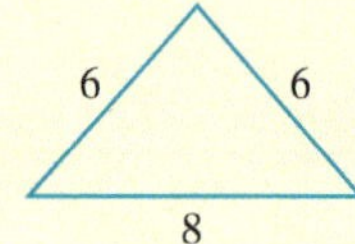

Objective C *Given that the pairs of triangles are similar, find the unknown length of the side labeled with a variable. See Example 3.*

13.

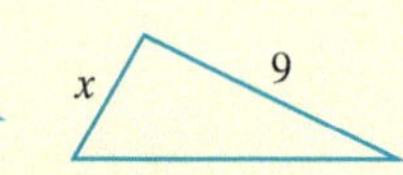

14.

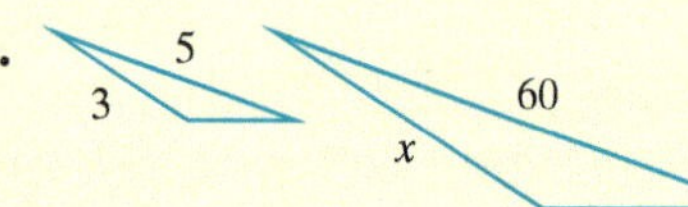

15.

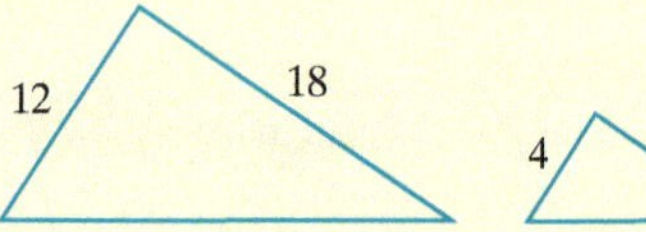

16.

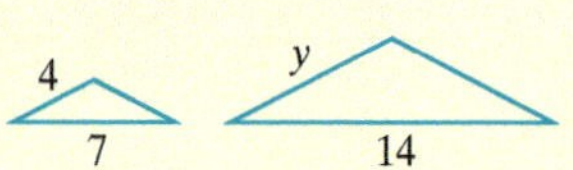

17.

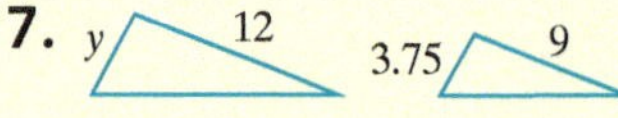

18.

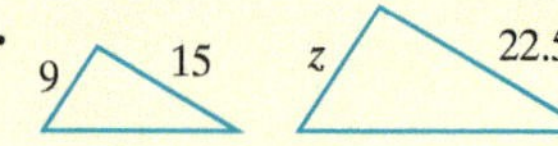

19.

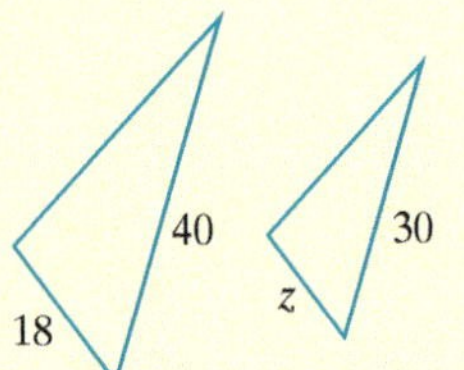

20.

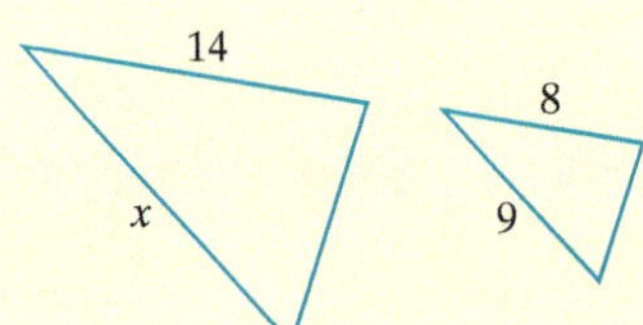

21.

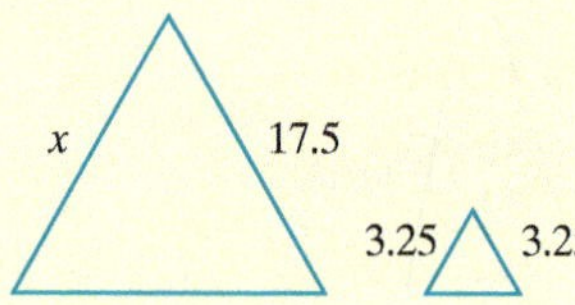

22.

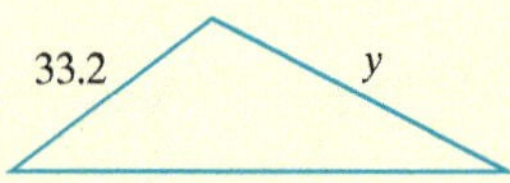

23.

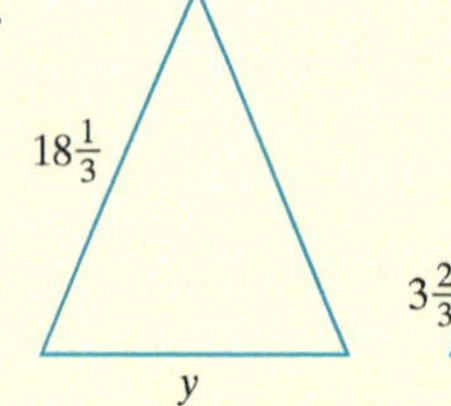

24.

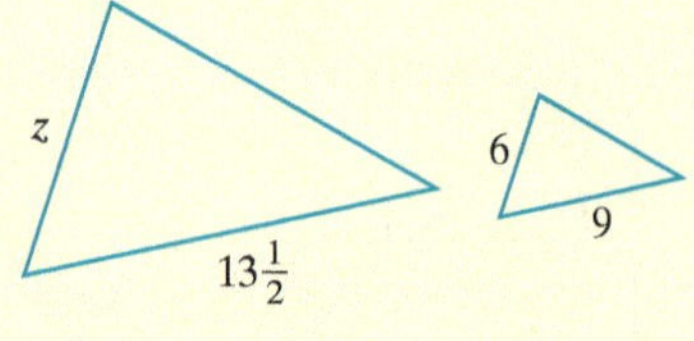

25.

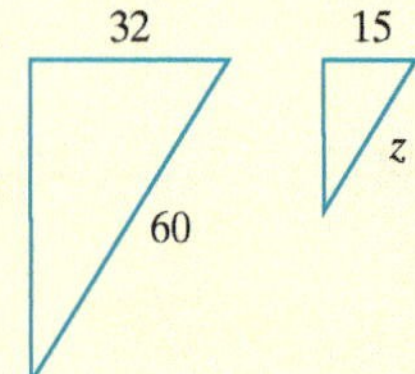

26.

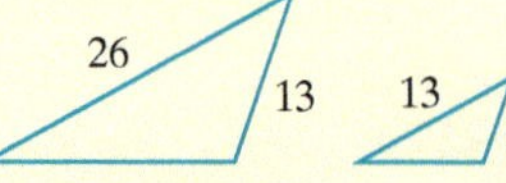

27.

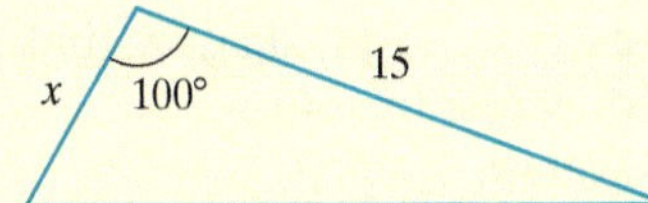

7
100°
$10\frac{1}{2}$

28.

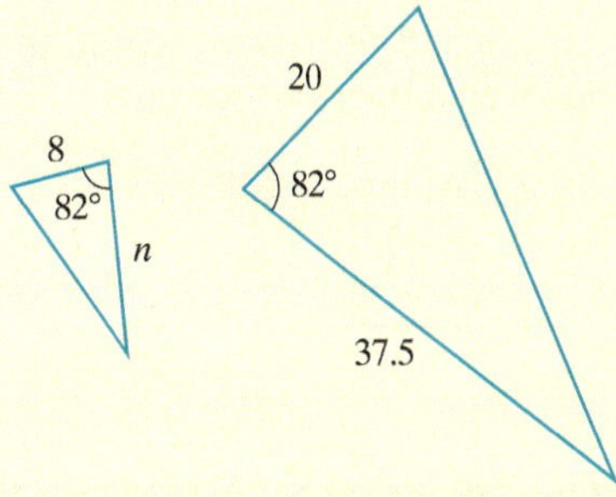

Solve. For Exercises 29 and 30, the solutions have been started for you. See Example 4.

29. Given the following diagram, approximate the height of the observation deck in the Seattle Space Needle in Seattle, Washington. (*Source:* Seattle Space Needle)

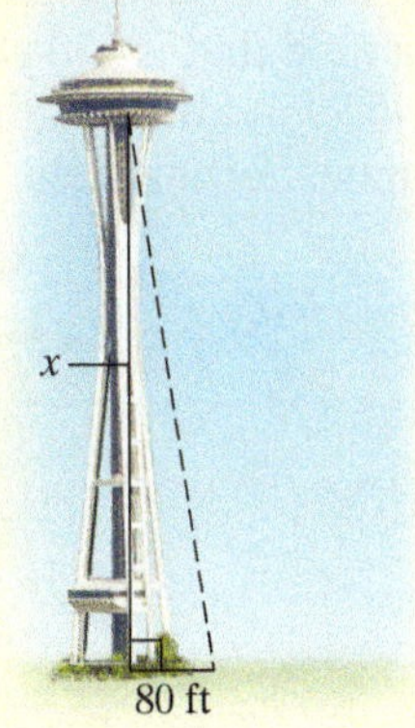

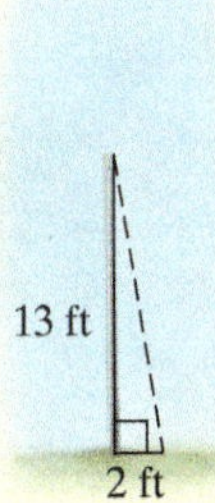

Start the solution:

1. UNDERSTAND the problem. Reread it as many times as needed.
2. TRANSLATE into a proportion using the similar triangles formed. (Fill in the blanks.)

height of observation deck → $\frac{x}{13} = \frac{\_\_}{\_\_}$ ← length of Space Needle shadow
height of pole → ← length of pole shadow

3. SOLVE by setting cross products equal.
4. INTERPRET.

30. Fountain Hills, Arizona, boasts the tallest fountain in the world. The fountain sits in a 28-acre lake and shoots up a column of water every hour. Based on the diagram below, what is the approximate height of the fountain?

x ft
28 ft
5 ft
100 ft

Start the solution:

1. UNDERSTAND the problem. Reread it as many times as needed.
2. TRANSLATE into a proportion using the similar triangles formed. (Fill in the blanks.)

height of tree → $\frac{28}{x} = \frac{\_\_}{\_\_}$ ← length of tree shadow
height of fountain → ← length of fountain shadow

3. SOLVE by setting cross products equal.
4. INTERPRET.

31. Given the following diagram, approximate the height of the Bank One Tower in Oklahoma City, Oklahoma. (*Source: The World Almanac*)

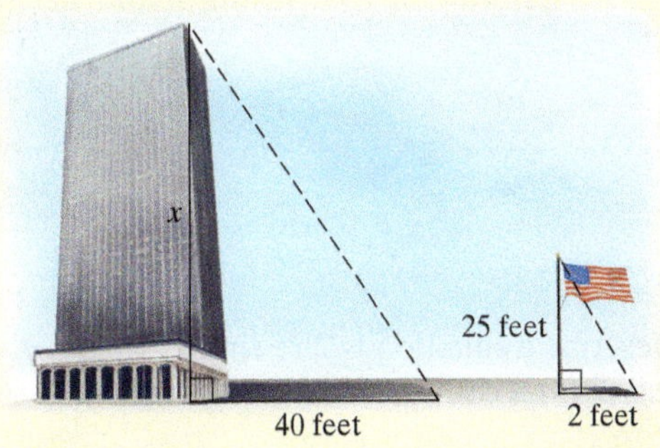

32. The tallest tree currently growing is Hyperion, a redwood located in the Redwood National Park in California. Given the following diagram, approximate its height. (*Source: Guinness World Records*)

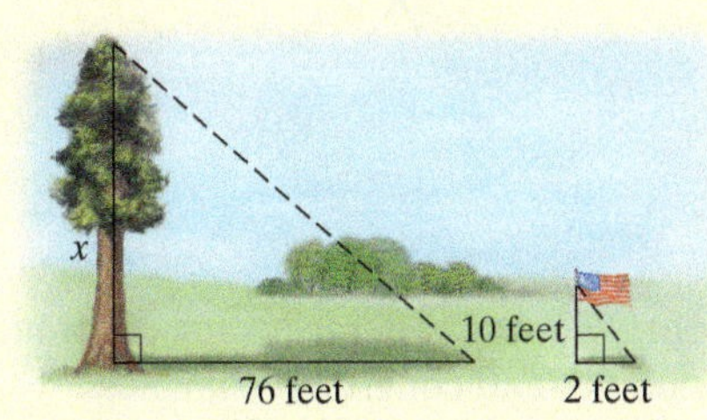

33. If a 30-foot tree casts an 18-foot shadow, find the length of the shadow cast by a 24-foot tree.

34. If a 24-foot flagpole casts a 32-foot shadow, find the length of the shadow cast by a 44-foot antenna. Round to the nearest tenth.

Review

Solve. See Section 6.1.

35. For the health of his fish, Pete's Sea World uses the standard that a 20-gallon tank should house only 19 neon tetras. Find the number of neon tetras that Pete would place into a 55-gallon tank.

36. A local package express deliveryman is traveling the city expressway at 45 mph when he is forced to slow down due to traffic ahead. His truck slows at the rate of 3 mph every 5 seconds. Find his speed 8 seconds after braking.

Solve. See Section 7.3.

37. Launch Umbilical Tower 1 is the name of the gantry used for the *Apollo* launch that took Neil Armstrong and Buzz Aldrin to the moon. Find the height of the gantry to the nearest whole foot.

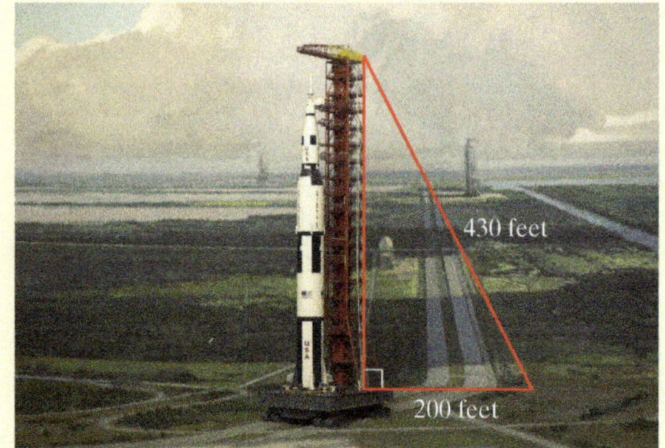

38. Arena polo, popular in the United States and England, is played on a field that is 100 yards long and usually 50 yards wide. Find the length, to the nearest yard, of the diagonal of this field.

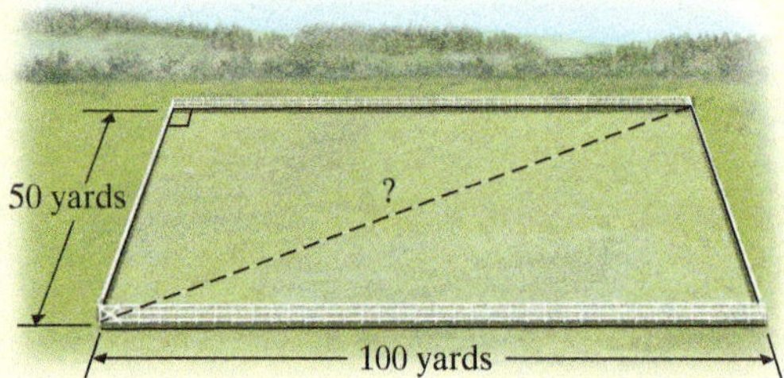

Perform the indicated operation. See Sections 5.2 through 5.4.

39. $3.6 + 0.41$

40. $0.41 - 3.6$

41. $(0.41)(-3)$

42. $-0.48 \div 3$

Concept Extensions

43. The print area on a particular page measures 7 inches by 9 inches. A printing shop is to copy the page and reduce the print area so that its length is 5 inches. What will its width be? Will the print now fit on a 3-by-5-inch index card?

44. The art sample for a banner measures $\frac{1}{3}$ foot in width by $1\frac{1}{2}$ feet in length. If the completed banner is to have a length of 9 feet, find its width.

Given that the pairs of triangles are similar, find the length of the side labeled n. Round your results to one decimal place.

45.

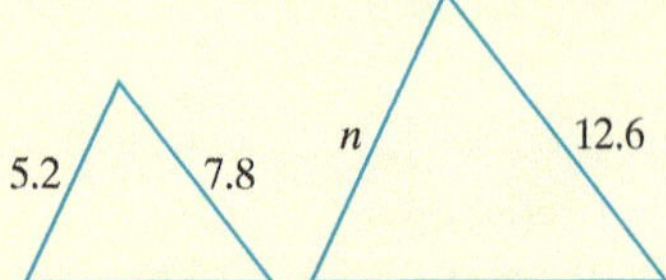

46.

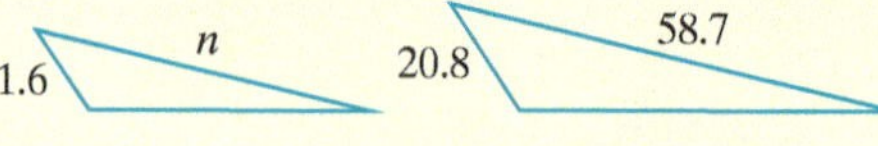

47. In your own words, describe any differences in similar triangles and congruent triangles.

48. Describe a situation where similar triangles would be useful for a contractor building a house.

49. A triangular park is planned and waiting to be approved by the city zoning commission. A drawing of the park shows sides of length 5 inches, $7\frac{1}{2}$ inches, and $10\frac{5}{8}$ inches. If the scale on the drawing is $\frac{1}{4}$ in. = 10 ft, find the actual proposed dimensions of the park.

50. John and Robyn Costello draw a triangular deck on their house plans. Robyn measures the sides of the deck drawing on the plans to be 3 inches, $4\frac{1}{2}$ inches, and 6 inches. If the scale on the drawing is $\frac{1}{4}$ in. = 1 foot, find the lengths of the sides of the deck they want built.

7.5 Counting and Introduction to Probability

Objective A Using a Tree Diagram

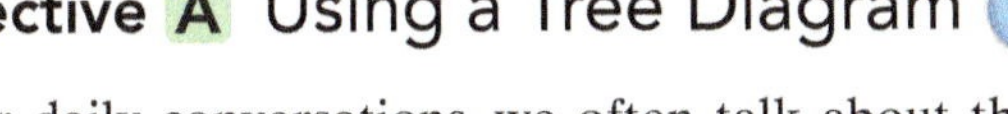

Objectives

A Use a Tree Diagram to Count Outcomes.

B Find the Probability of an Event.

In our daily conversations, we often talk about the likelihood or **probability** of a given result occurring. For example:

The *chance* of thundershowers is 70 percent.

What are the *odds* that the New Orleans Saints will go to the Super Bowl?

What is the *probability* that you will finish cleaning your room today?

Each of these chance happenings—thundershowers, the New Orleans Saints playing in the Super Bowl, and finishing cleaning your room today—is called an **experiment.** The possible results of an experiment are called **outcomes.** For example, flipping a coin is an experiment, and the possible outcomes are heads (H) or tails (T).

One way to picture the outcomes of an experiment is to draw a **tree diagram.** Each outcome is shown on a separate branch. For example, the outcomes of flipping a coin are

Heads

Tails

Example 1 Draw a tree diagram for tossing a coin twice. Then use the diagram to find the number of possible outcomes.

Solution:

First Coin Toss	Second Coin Toss	Outcomes
H	H	H, H
	T	H, T
T	H	T, H
	T	T, T

There are 4 possible outcomes when tossing a coin twice.

Work Practice 1

Practice 1

Draw a tree diagram for tossing a coin three times. Then use the diagram to find the number of possible outcomes.

Answer

1.

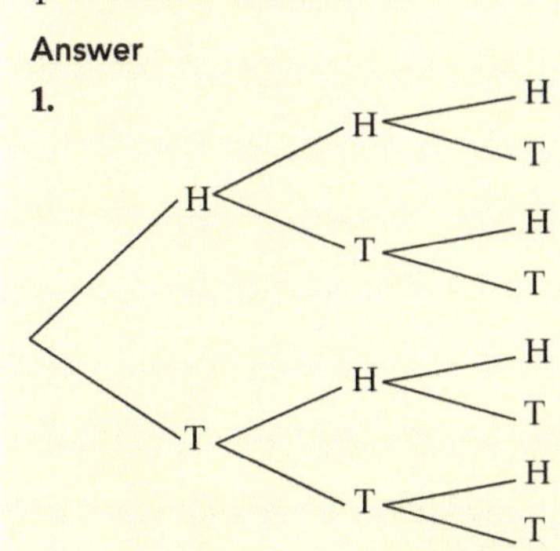

Practice 2

Draw a tree diagram for an experiment consisting of tossing a coin and then rolling a die. Then use the diagram to find the number of possible outcomes.

Example 2 Draw a tree diagram for an experiment consisting of rolling a die and then tossing a coin. Then use the diagram to find the number of possible outcomes.

Die

Solution: Recall that a die has six sides and that each side represents a number, 1 through 6.

Roll a Die	Toss a Coin	Outcomes
1	H	1, H
	T	1, T
2	H	2, H
	T	2, T
3	H	3, H
	T	3, T
4	H	4, H
	T	4, T
5	H	5, H
	T	5, T
6	H	6, H
	T	6, T

There are 12 possible outcomes for rolling a die and then tossing a coin.

Work Practice 2

Any number of outcomes considered together is called an **event.** For example, when tossing a coin twice, H, H is an event. The event is tossing heads first and tossing heads second. Another event would be tossing tails first and then heads (T, H), and so on.

Objective B Finding the Probability of an Event 

As we mentioned earlier, the **probability of an event is a measure of the chance or likelihood of it occurring.** For example, if a coin is tossed, what is the probability that heads occurs? Since one of two equally likely possible outcomes is heads, the probability is $\frac{1}{2}$.

The Probability of an Event

$$\text{probability of an event} = \frac{\text{number of ways that the event can occur}}{\text{number of possible outcomes}}$$

Note from the definition of probability that the probability of an event is always between 0 and 1, inclusive (i.e., including 0 and 1). A probability of 0 means that an event won't occur, and a probability of 1 means that an event is certain to occur.

Answer

2.

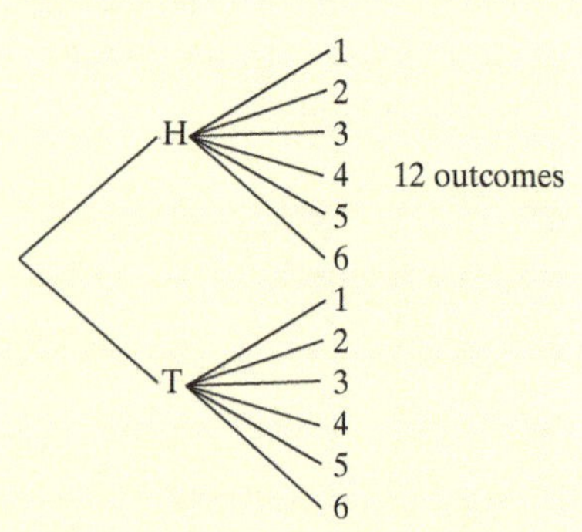

Example 3 If a coin is tossed twice, find the probability of tossing heads on the first toss and then heads again on the second toss (H, H).

Solution: 1 way the event can occur ↓

H, T, H, H, T, H, T, T

4 possible outcomes

$$\text{probability} = \frac{1}{4} \quad \begin{matrix}\text{Number of ways the event can occur} \\ \text{Number of possible outcomes}\end{matrix}$$

The probability of tossing heads and then heads is $\frac{1}{4}$.

Work Practice 3

Practice 3

If a coin is tossed three times, find the probability of tossing tails, then heads, then tails (T, H, T).

Example 4 If a die is rolled one time, find the probability of rolling a 3 or a 4.

Solution: Recall that there are 6 possible outcomes when rolling a die.

2 ways that the event can occur

possible outcomes: 1, 2, 3, 4, 5, 6

6 possible outcomes

$$\text{probability of a 3 or a 4} = \frac{2}{6} \quad \begin{matrix}\text{Number of ways the event can occur} \\ \text{Number of possible outcomes}\end{matrix}$$

$$= \frac{1}{3} \quad \text{Simplest form}$$

Work Practice 4

Practice 4

If a die is rolled one time, find the probability of rolling a 2 or a 5.

✓**Concept Check** Suppose you have calculated a probability of $\frac{11}{9}$. How do you know that you have made an error in your calculation?

Example 5 Find the probability of choosing a red marble from a box containing 1 red, 1 yellow, and 2 blue marbles.

Solution: 1 way that event can occur

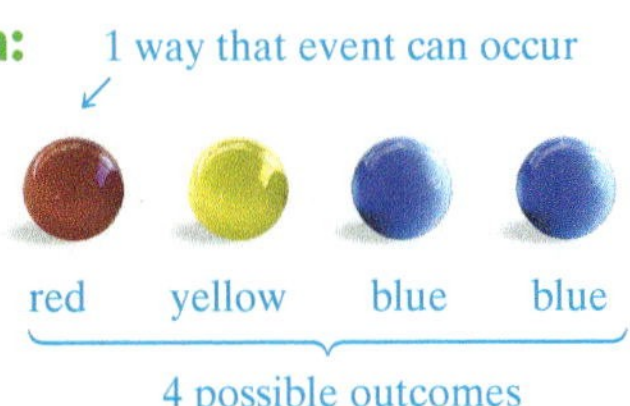

4 possible outcomes

$$\text{probability} = \frac{1}{4}$$

Work Practice 5

Practice 5

Use the diagram and information in Example 5 to find the probability of choosing a blue marble from the box.

Answers

3. $\frac{1}{8}$ **4.** $\frac{1}{3}$ **5.** $\frac{1}{2}$

✓**Concept Check Answer**

The number of ways an event can occur can't be larger than the number of possible outcomes.

Vocabulary, Readiness & Video Check

Use the choices below to fill in each blank. Choices may be used more than once.

0	probability	tree diagram
1	outcome	

1. A possible result of an experiment is called a(n) ___________.

2. A(n) ___________ shows each outcome of an experiment as a separate branch.

3. The ___________ of an event is a measure of the likelihood of it occurring.

4. ___________ is calculated by the number of ways that an event can occur divided by the number of possible outcomes.

5. A probability of ___________ means that an event won't occur.

6. A probability of ___________ means that an event is certain to occur.

Martin-Gay Interactive Videos *Watch the section lecture video and answer the following questions.*

See Video 7.5

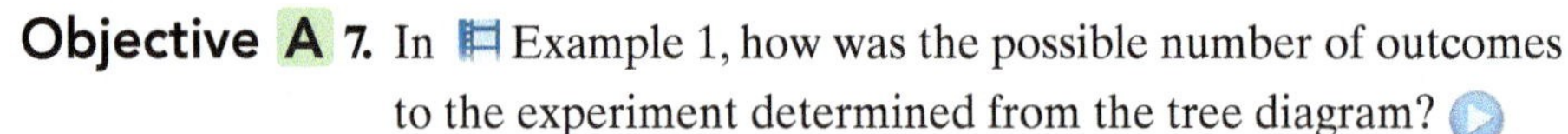

Objective A **7.** In Example 1, how was the possible number of outcomes to the experiment determined from the tree diagram?

Objective B **8.** In Example 2, what is the probability of getting a 7? Explain this result.

7.5 Exercise Set MyMathLab®

Objective A *Draw a tree diagram for each experiment. Then use the diagram to find the number of possible outcomes. See Examples 1 and 2.*

1. Choosing a letter in the word MATH, then a number (1, 2, or 3)

2. Choosing a number (1 or 2) and then a vowel (a, e, i, o, u)

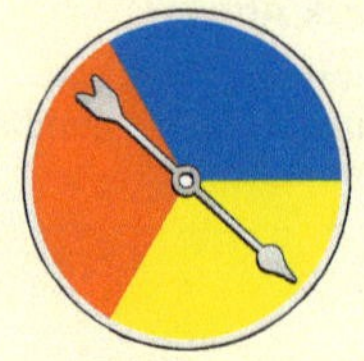

Spinner A

Spinner B

3. Spinning Spinner A once

4. Spinning Spinner B once

5. Spinning Spinner B twice

6. Spinning Spinner A twice

7. Spinning Spinner A and then Spinner B

8. Spinning Spinner B and then Spinner A

9. Tossing a coin and then spinning Spinner B

10. Spinning Spinner A and then tossing a coin

Objective B *If a single die is tossed once, find the probability of each event. See Examples 3 through 5.*

11. A 5

12. A 9

13. A 1 or a 6

14. A 2 or a 3

15. An even number

16. An odd number

17. A number greater than 2

18. A number less than 6

Suppose the spinner shown is spun once. Find the probability of each event. See Examples 3 through 5.

19. The result of the spin is 2.

20. The result of the spin is 3.

21. The result of the spin is 1, 2, or 3.

22. The result of the spin is not 3.

23. The result of the spin is an odd number.

24. The result of the spin is an even number.

If a single choice is made from the bag of marbles shown, find the probability of each event. See Examples 3 through 5.

25. A red marble is chosen.

26. A blue marble is chosen.

27. A yellow marble is chosen.

28. A green marble is chosen.

29. A green or red marble is chosen.

30. A blue or yellow marble is chosen.

A new drug is being tested that is supposed to lower blood pressure. This drug was given to 200 people and the results are shown below.

Lower Blood Pressure	Higher Blood Pressure	Blood Pressure Not Changed
152	38	10

31. If a person is trying this drug, what is the probability that his or her blood pressure will be higher?

32. If a person is trying this drug, what is the probability that his or her blood pressure will be lower?

33. If a person is trying this drug, what is the probability that his or her blood pressure will not change?

34. What is the sum of the answers to Exercises **31**, **32**, and **33**? In your own words, explain why.

Review

Perform each indicated operation. See Sections 4.3 and 4.5.

35. $\frac{1}{2}+\frac{1}{3}$ **36.** $\frac{7}{10}-\frac{2}{5}$ **37.** $\frac{1}{2}\cdot\frac{1}{3}$ **38.** $\frac{7}{10}\div\frac{2}{5}$ **39.** $5\div\frac{3}{4}$ **40.** $\frac{3}{5}\cdot 10$

Concept Extensions

Recall that a deck of cards contains 52 cards. These cards consist of four suits (hearts, spades, clubs, and diamonds) of each of the following: 2, 3, 4, 5, 6, 7, 8, 9, 10, jack, queen, king, and ace. If a card is chosen from a deck of cards, find the probability of each event.

41. The king of hearts **42.** The 10 of spades

43. A king **44.** A 10

45. A heart **46.** A club

47. A card in black ink **48.** A queen or ace

Two dice are tossed. Find the probability of each sum of the dice. (Hint: Draw a tree diagram of the possibilities of two tosses of a die, and then find the sum of the numbers on each branch.)

49. A sum of 6 **50.** A sum of 10 **51.** A sum of 13 **52.** A sum of 2

Solve. See the Concept Check in this section.

53. In your own words, explain why the probability of an event cannot be greater than 1.

54. In your own words, explain when the probability of an event is 0.

Chapter 7 Group Activity

Stem-and-Leaf Displays

Sections 7.1, 7.2

Stem-and-leaf displays are another way to organize data. After data are logically organized, it can be much easier to draw conclusions from them.

Suppose we have collected the following set of data. It could represent the test scores for an algebra class or the pulse rates of a group of small children.

90	73	93	99	79	95	69	78	93	80
89	85	97	78	75	79	72	76	97	88
83	98	72	94	92	79	70	98	85	99

In a stem-and-leaf display, the last digit of each number forms the *leaf,* and the remaining digits to the left form the *stem.* For the first number in the list, 90, 9 is the stem and 0 is the leaf. To make the stem-and-leaf display, we write all of the stems in numerical order in a column. Then we write each leaf on the horizontal line next to its stem, aligning leaves in vertical columns. In this case, because the data range from 69 to 99, we use the stems 6, 7, 8, and 9. Each line of the display represents an interval of data; for instance, the line corresponding to the stem **7** represents all data that fall in the interval **70** to **79,** inclusive. After the data have been divided into stems and leaves on the display as shown in the table below, we simply rearrange the leaves on each line to appear in numerical order, as shown in the table in the rightmost column.

Stem	Leaf		Stem	Leaf
6	9	⟶	6	9
7	39885926290	⟶	7	02235688999
8	095835	⟶	8	035589
9	039537784289	⟶	9	023345778899

Now that the data have been organized into a stem-and-leaf display, it is easy to answer questions about the data such as: What are the least and greatest values in the set of data? Which data interval contains the most items from the data set? How many values fall between 74 and 84? Which data value occurs most frequently in the data set? What patterns or trends do you see in the data?

Make a stem-and-leaf display of the weekend emergency room admission data shown below. Then answer the following questions:

1. What is the difference between the least number and the greatest number of weekend ER admissions?
2. How many weekends had between 125 and 165 ER admissions?
3. What number of weekend ER admissions occurred most frequently?
4. Which interval contains the most weekend ER admissions?

Number of Emergency Room Admissions on Weekends

198	168	117	185	159
160	177	169	112	175
170	188	137	117	145
198	169	154	163	192
167	179	155	133	121
162	188	124	145	146
128	181	198	149	140
122	162	161	180	177

Chapter 7 Vocabulary Check

Fill in each blank with one of the words or phrases listed below. Some items may be used more than once.

outcomes	class interval	experiment	probability	Pythagorean	congruent
pictograph	bar	class frequency	hypotenuse	unit rate	leg
histogram	circle	tree diagram	right	similar	

1. ______________ triangles have the same shape and the same size.

2. ______________ triangles have exactly the same shape but not necessarily the same size.

3–5. Label the sides of the right triangle.

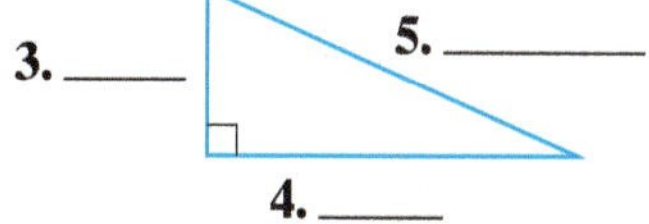

6. A triangle with one right angle is called a(n) ______________ triangle.

7. In the right triangle 

, $a^2 + b^2 = c^2$ is called the ______________ theorem.

8. A(n) ______________ graph presents data using vertical or horizontal bars.

9. The possible results of an experiment are the ______________.

10. A(n) ______________ is a graph in which pictures or symbols are used to visually present data.

11. A(n) ______________ is one way to picture and count outcomes.

12. A(n) ______________ is an activity being considered, such as tossing a coin or rolling a die.

13. In a(n) ______________ graph, each section (shaped like a piece of pie) shows a category and the relative size of the category.

14. The ______________ of an event is $\frac{\text{number of ways that the event can occur}}{\text{number of possible outcomes}}$.

15. A(n) ______________ is a special bar graph in which the width of each bar represents a(n) ______________ and the height of each bar represents the ______________.

Helpful Hint Are you preparing for your test? Don't forget to take the Chapter 7 Test on page 542. Then check your answers at the back of the text and use the Chapter Test Prep Videos to see the fully worked-out solutions to any of the exercises you want to review.

7 Chapter Highlights

Definitions and Concepts	Examples
Section 7.1 Reading Pictographs, Bar Graphs, Histograms, and Line Graphs	
A **pictograph** is a graph in which pictures or symbols are used to visually present data. A **line graph** displays information with a line that connects data points. A **bar graph** presents data using vertical or horizontal bars. The bar graph on the right shows the number of acres of corn harvested in 2012 for leading states.	Corn Production, Selected States 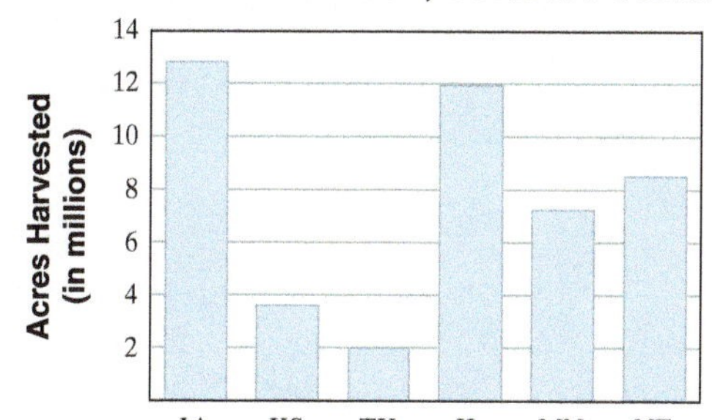 *Source:* U.S. Department of Agriculture

(Continued)

Definitions and Concepts	Examples

Section 7.1 Reading Pictographs, Bar Graphs, Histograms, and Line Graphs *(continued)*

1. Approximately how many acres of corn were harvested in Iowa?

   12,800,000 acres

2. About how many more acres of corn were harvested in Illinois than Nebraska?

$$\begin{array}{r} 12 \text{ million} \\ -\ 8.5 \text{ million} \\ \hline 3.5 \text{ million} \end{array} \quad \text{or } 3{,}500{,}000 \text{ acres}$$

A **histogram** is a special bar graph in which the width of each bar represents a **class interval** and the height of each bar represents the **class frequency.** The histogram on the right shows student quiz scores.

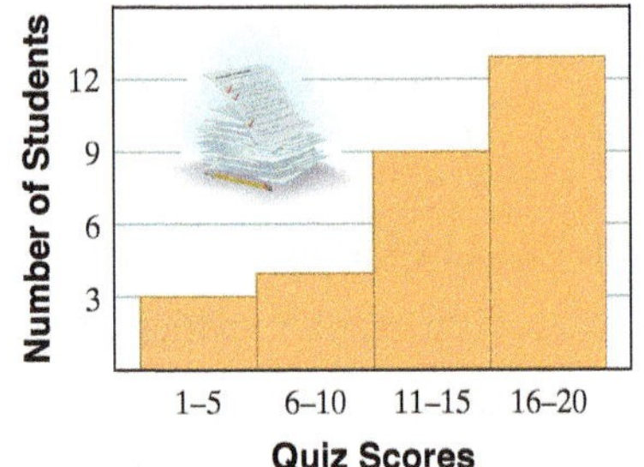

1. How many students received a score of 6–10?

   4 students

2. How many students received a score of 11–20?

   $9 + 13 = 22$ students

Section 7.2 Reading and Drawing Circle Graphs

In a **circle graph,** each section (shaped like a piece of pie) shows a category and the relative size of the category.

The circle graph on the right classifies tornadoes by wind speed.

Tornado Wind Speeds

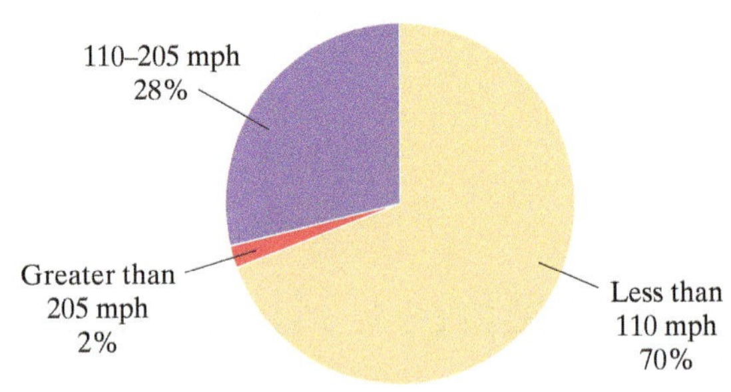

Source: National Oceanic and Atmospheric Administration

1. What percent of tornadoes have wind speeds of 110 mph or greater?

   $28\% + 2\% = 30\%$

2. If there were 939 tornadoes in the United States in 2012, how many of these might we expect to have had wind speeds less than 110 mph? Find 70% of 939.

   $70\%(939) = 0.70(939) = 657.3 \approx 657$

   Around 657 tornadoes would be expected to have had wind speeds of less than 110 mph.

Definitions and Concepts	Examples

Section 7.3 Square Roots and the Pythagorean Theorem

Square Root of a Number

The square root of a positive number a is the positive number b whose square is a. In symbols,

$$\sqrt{a} = b, \quad \text{if} \quad b^2 = a$$

Also, $\sqrt{0} = 0$.

$\sqrt{9} = 3, \quad \sqrt{100} = 10, \quad \sqrt{1} = 1$

Pythagorean Theorem

If a and b are the lengths of the legs of a right triangle and c is the length of the hypotenuse, then

$$a^2 + b^2 = c^2$$

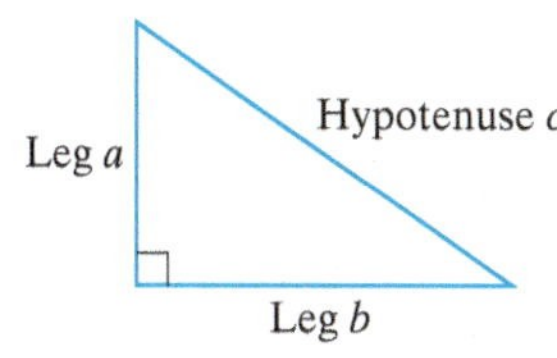

Find c.

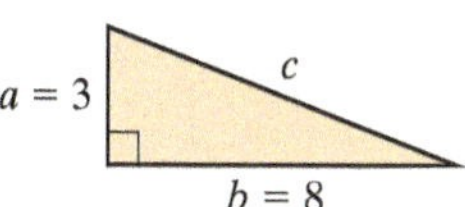

$a^2 + b^2 = c^2$

$3^2 + 8^2 = c^2$ Let $a = 3$ and $b = 8$.

$9 + 64 = c^2$ Multiply.

$73 = c^2$ Simplify.

$\sqrt{73} = c$ or $c \approx 8.5$

Section 7.4 Congruent and Similar Triangles

Congruent triangles have the same shape and the same size. Corresponding angles are equal, and corresponding sides are equal.

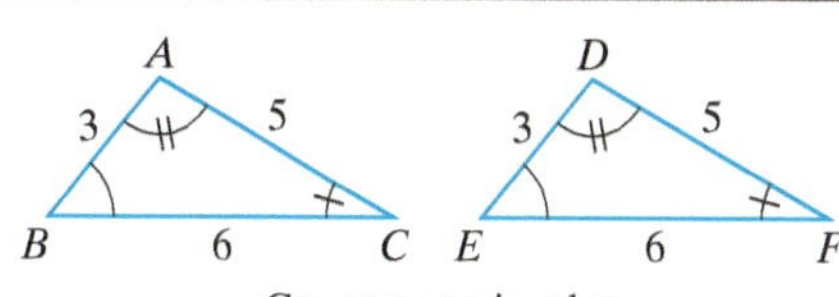

Congruent triangles

Similar triangles have exactly the same shape but not necessarily the same size. Corresponding angles are equal, and the ratios of the lengths of corresponding sides are equal.

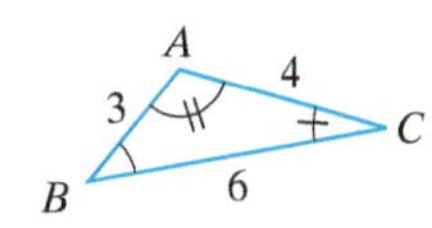

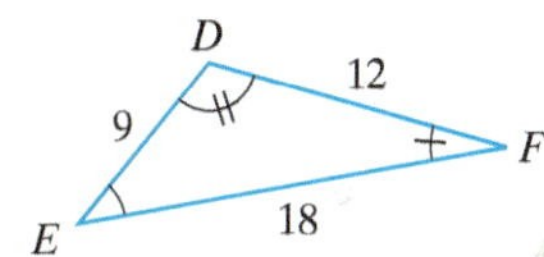

Similar triangles

$\frac{AB}{DE} = \frac{3}{9} = \frac{1}{3}, \frac{BC}{EF} = \frac{6}{18} = \frac{1}{3},$

$\frac{CA}{FD} = \frac{4}{12} = \frac{1}{3}$

Definitions and Concepts	Examples
Section 7.5 Counting and Introduction to Probability	
An **experiment** is an activity being considered, such as tossing a coin or rolling a die. The possible results of an experiment are the **outcomes.** A **tree diagram** is one way to picture and count outcomes.	Draw a tree diagram for tossing a coin and then choosing a number from 1 to 4. 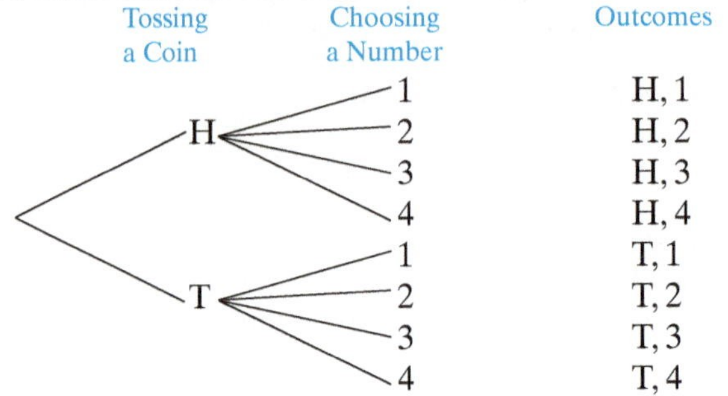
Any number of outcomes considered together is called an **event.** The **probability** of an event is a measure of the chance or likelihood of it occurring. $$\text{probability of an event} = \frac{\text{number of ways that the event can occur}}{\text{number of possible outcomes}}$$	Find the probability of tossing a coin twice and tails occurring each time. 1 way the event can occur: (H, H), (H, T), (T, H), (T, T) — 4 possible outcomes. $\text{probability} = \frac{1}{4}$

Chapter 7 Review

(7.1) *The following pictograph shows the number of new homes constructed from July 2012 to July 2013, by region. Use this graph to answer Exercises 1 through 6.*

New Home Construction July 2012–July 2013

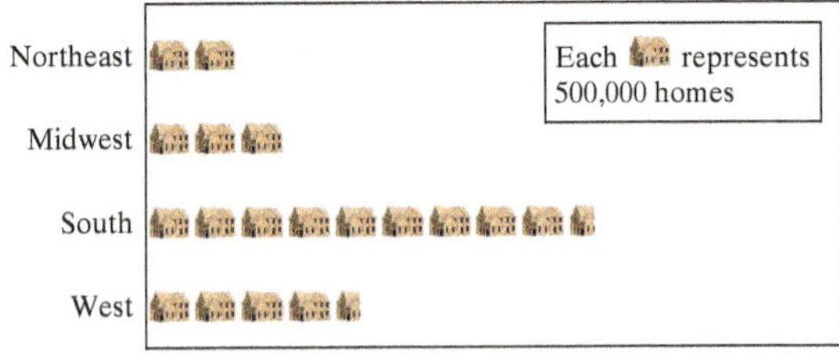

Source: U.S. Census Bureau

1. How many new homes were constructed in the West during the given year?

2. How many new homes were constructed in the South during the given year?

3. Which region had the most new homes constructed?

4. Which region had the fewest new homes constructed?

5. Which region(s) had 2,000,000 or more new homes constructed?

6. Which region(s) had fewer than 2,000,000 new homes constructed?

The following bar graph shows the percent of persons age 25 or over who completed four or more years of college. Use this graph to answer Exercises 7 through 10.

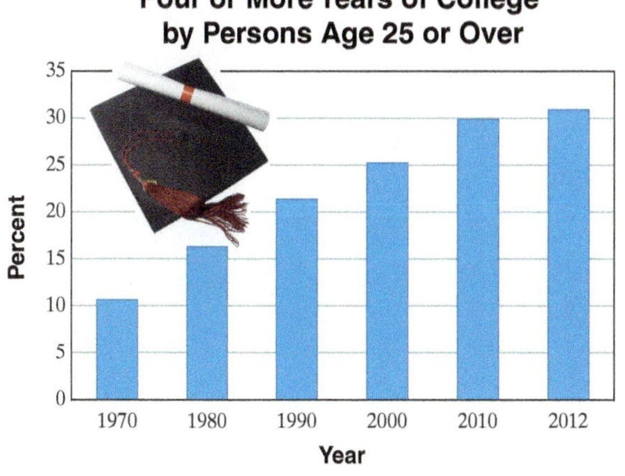

Source: U.S. Census Bureau

7. Approximate the percent of persons who had completed four or more years of college by 2010.

8. What year shown had the greatest percent of persons completing four or more years of college?

9. What years shown had 20% or more of persons completing four or more years of college?

10. Describe any patterns you notice in this graph.

The following line graph shows the total number of Olympic medals awarded during the Summer Olympics since 1992. Use this graph to answer Exercises 11 through 16.

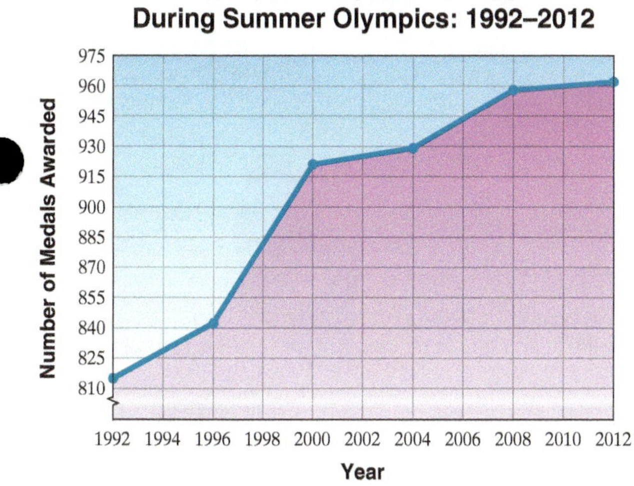

Source: International Olympic Committee

11. Approximate the number of medals awarded during the Summer Olympics of 2012.

12. Approximate the number of medals awarded during the Summer Olympics of 2000.

13. Approximate the number of medals awarded during the Summer Olympics of 2004.

14. Approximate the number of medals awarded during the Summer Olympics of 1992.

15. How many more medals were awarded at the Summer Olympics of 1996 than at the Summer Olympics of 1992?

16. How many more medals were awarded at the Summer Olympics of 2012 than at the Summer Olympics of 1992?

The following histogram shows the hours worked per week by the employees of Southern Star Furniture. Use this histogram to answer Exercises 17 through 20.

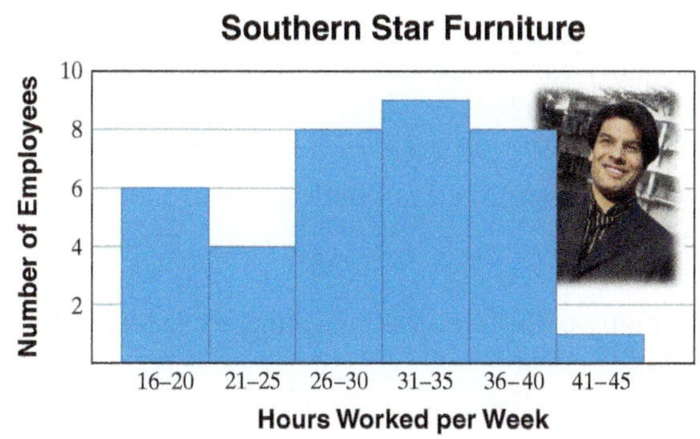

17. How many employees work 41–45 hours per week?

18. How many employees work 21–25 hours per week?

19. How many employees work 30 hours or less per week?

20. How many employees work 36 hours or more per week?

Following is a list of monthly record high temperatures for New Orleans, Louisiana. Use this list to complete the frequency distribution table below.

83	96	101	92
85	100	92	102
89	101	87	84

	Class Intervals (Temperatures)	Tally	Class Frequency (Number of Months)
21.	80°-89°		
22.	90°-99°		
23.	100°-109°		

24. Use the table from Exercises **21–23** to draw a histogram.

(7.2) *The following circle graph shows a family's $4000 monthly budget. Use this graph to answer Exercises 25 through 30. Write all ratios as fractions in simplest form.*

25. What is the largest budget item?

26. What is the smallest budget item?

27. How much money is budgeted for the mortgage payment and utilities?

28. How much money is budgeted for savings and contributions?

29. Find the ratio of the mortage payment to the total monthly budget.

30. Find the ratio of food to the total monthly budget.

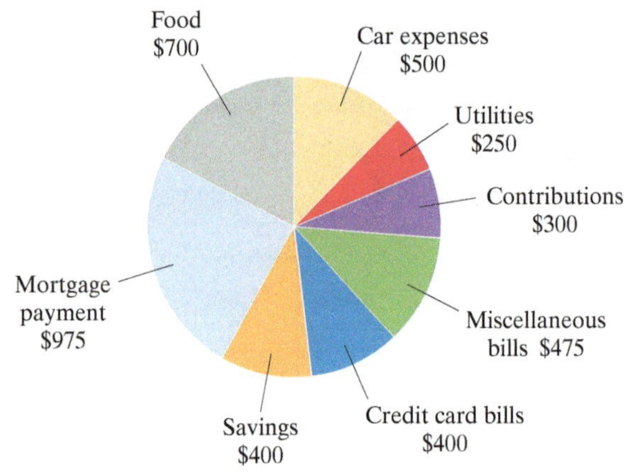

In 2012, there were approximately 62 completed buildings 200 meters or taller in the world. The following circle graph shows the percent of these buildings by area. Use this graph to answer Exercises 31 through 34. Round each answer to the nearest whole.

Percent of Tall Buildings Completed in 2012 200 Meters or Taller by Continent

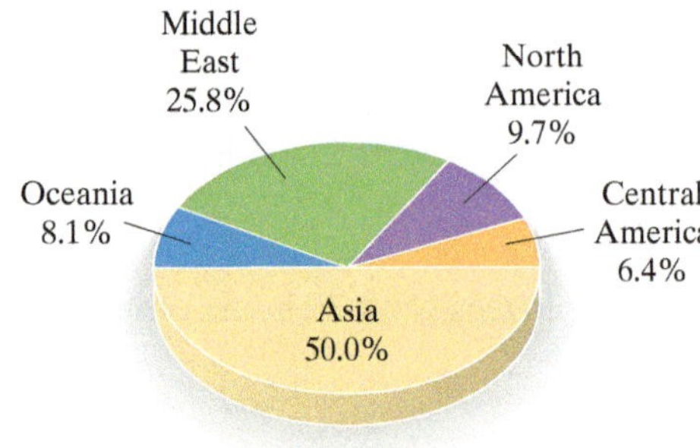

Source: Council on Tall Buildings and Urban Habitats

31. How many completed tall buildings were located in Asia?

32. How many completed tall buildings were located in North America?

33. How many completed tall buildings were located in Oceania?

34. How many completed tall buildings were located in the Middle East?

(7.3) *Find each square root. If necessary, round the square root to the nearest thousandth.*

35. $\sqrt{64}$ **36.** $\sqrt{144}$ **37.** $\sqrt{12}$ **38.** $\sqrt{15}$ **39.** $\sqrt{0}$

40. $\sqrt{1}$ **41.** $\sqrt{50}$ **42.** $\sqrt{65}$ **43.** $\sqrt{\frac{4}{25}}$ **44.** $\sqrt{\frac{1}{100}}$

Find the unknown length in each given right triangle. If necessary, round to the nearest tenth.

△**45.** leg = 12, leg = 5

△**46.** leg = 20, leg = 21

△**47.** leg = 9, hypotenuse = 14

△**48.** leg = 66, hypotenuse = 86

△**49.** Find the length, to the nearest hundredth, of the diagonal of a square that has a side of length 20 centimeters.

△**50.** Find the height of the building rounded to the nearest tenth.

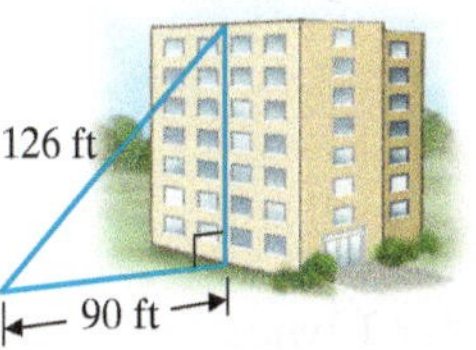

(7.4) *Determine whether each pair of triangles is congruent. If congruent, state the reason why, such as SSS, SAS, or ASA.*

51.

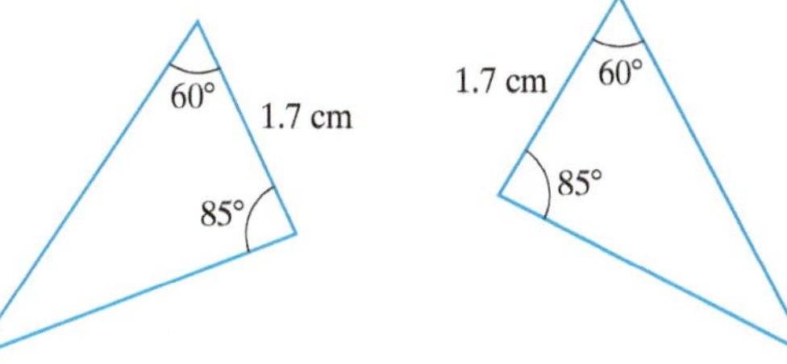

52.

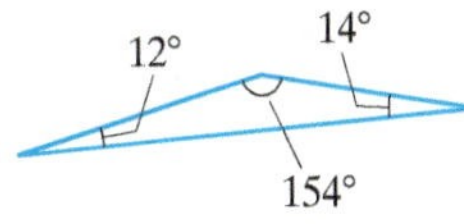

Given that the pairs of triangles are similar, find the unknown length x.

53.

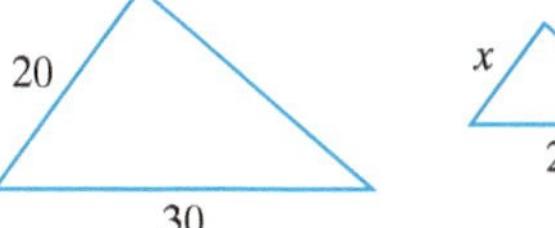

54.

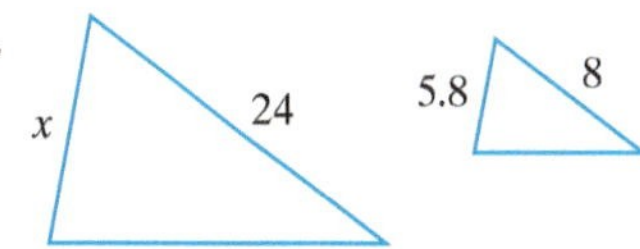

Solve.

55. A housepainter needs to estimate the height of a condominium. He estimates the length of his shadow to be 7 feet long and the length of the building's shadow to be 42 feet long. Find the approximate height of the building if the housepainter is $5\frac{1}{2}$ feet tall.

56. A design company is making a triangular sail for a model sailboat. The model sail is to be the same shape as a life-size sailboat's sail. Use the following diagram to find the unknown lengths x and y.

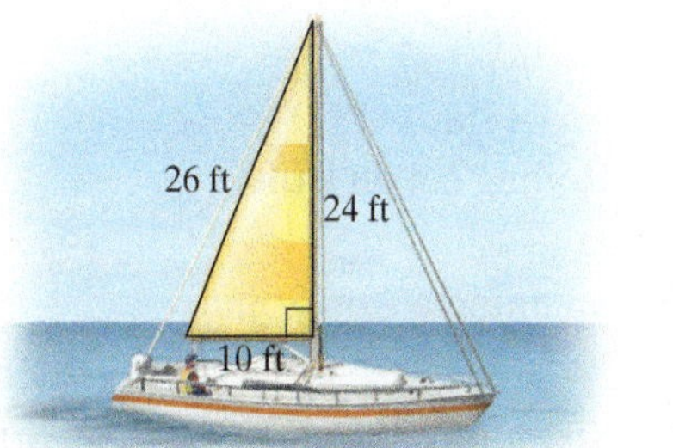

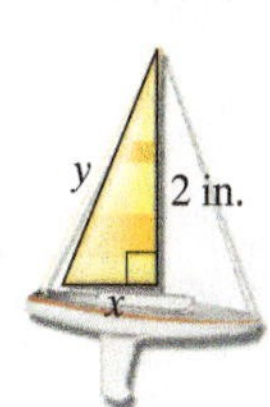

(7.5) *Draw a tree diagram for each experiment. Then use the diagram to determine the number of outcomes.*

Spinner 1

Spinner 2

57. Tossing a coin and then spinning Spinner 1

58. Spinning Spinner 2 and then tossing a coin

59. Spinning Spinner 1 twice

60. Spinning Spinner 1 and then Spinner 2

Find the probability of each event.

Die

61. Rolling a 4 on a die

62. Rolling a 3 on a die

63. Spinning a 4 on Spinner 1

64. Spinning a 3 on Spinner 1

65. Spinning either a 1, 3, or 5 on Spinner 1

66. Spinning either a 2 or a 4 on Spinner 1

Mixed Review

Given a bag containing 2 red marbles, 2 blue marbles, 3 yellow marbles, and 1 green marble, find the following:

67. The probability of choosing a blue marble from the bag

68. The probability of choosing a yellow marble from the bag

69. The probability of choosing a red marble from the bag

70. The probability of choosing a green marble from the bag

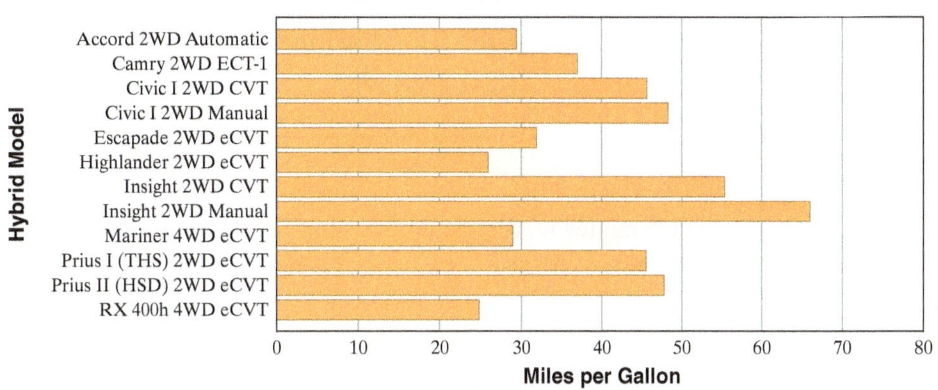

Source: http://www.greenhybrid.com

71. Name the hybrid model with the greatest number of miles per gallon, and estimate the miles per gallon.

72. Name the hybrid model with the least number of miles per gallon, and estimate the miles per gallon.

73. Name the hybrid model whose miles per gallon is between 50 and 60, and estimate the miles per gallon.

74. Name the hybrid model whose miles per gallon is the closest to 40, and estimate the miles per gallon.

Find each square root. If necessary, approximate and round to the nearest thousandth.

75. $\sqrt{36}$

76. $\sqrt{\frac{16}{81}}$

77. $\sqrt{105}$

78. $\sqrt{32}$

Find the unknown length of each given right triangle. If necessary, round to the nearest tenth.

79. leg $= 66$, leg $= 56$

80. leg $= 12$, hypotenuse $= 24$

Given that the pairs of triangles are similar, find the unknown length, n.

81.

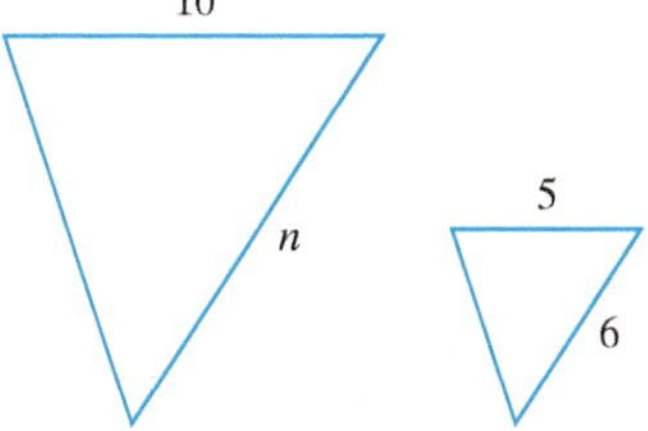

82.

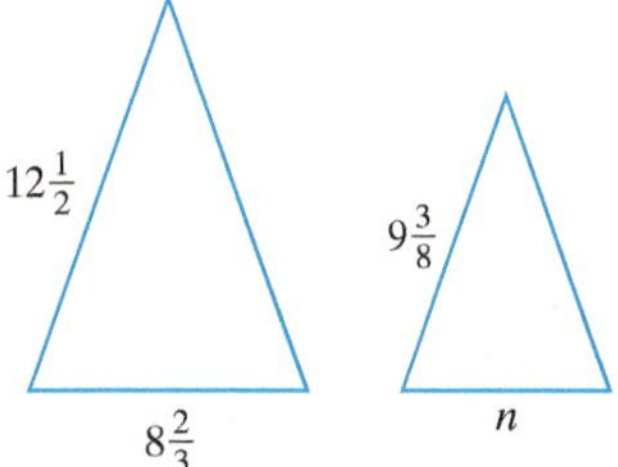

Chapter 7 Test

Step-by-step test solutions are found on the Chapter Test Prep Videos. Where available: MyMathLab® or YouTube

Answers

1. ______

2. ______

3. ______

4. ______

5. ______

The following pictograph shows the money collected each week from a wrapping paper fundraiser. Use this graph to answer Exercises 1 through 3.

Weekly Wrapping Paper Sales

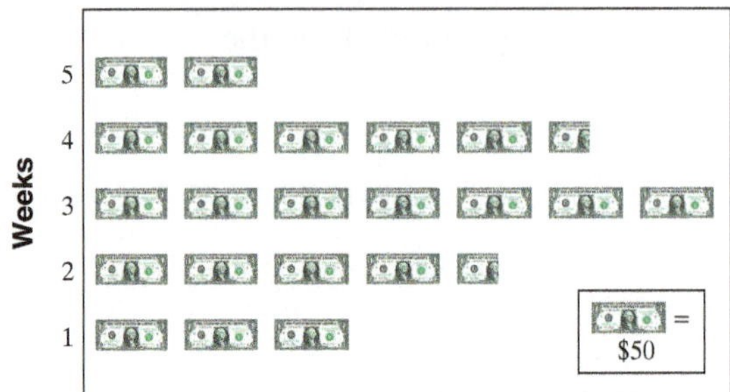

1. How much money was collected during the second week?

2. During which week was the most money collected? How much money was collected during that week?

3. What was the total amount of money collected for the fundraiser?

The bar graph shows the normal monthly precipitation in centimeters for Chicago, Illinois. Use this graph to answer Exercises 4 through 6.

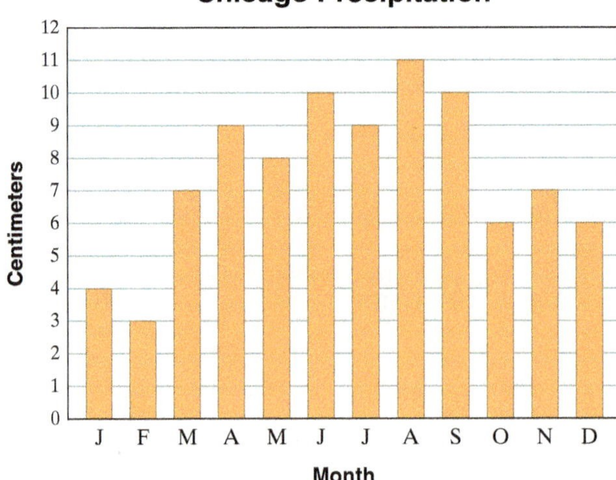

Source: U.S. National Oceanic and Atmospheric Administration, *Climatography of the United States*, No. 81

4. During which month(s) does Chicago normally have more than 9 centimeters of precipitation?

5. During which month does Chicago normally have the least amount of precipitation? How much precipitation occurs during that month?

6. During which month(s) do 7 centimeters of precipitation normally occur?

6. ______

7. Use the information in the table to draw a bar graph. Clearly label each bar.

Most Common Blood Types

Blood Type	% of Population with This Blood Type
O+	38%
A+	34%
B+	9%
O−	7%
A−	6%
AB+	3%
B−	2%
AB−	1%

Most Common Blood Types by Percent in the Population

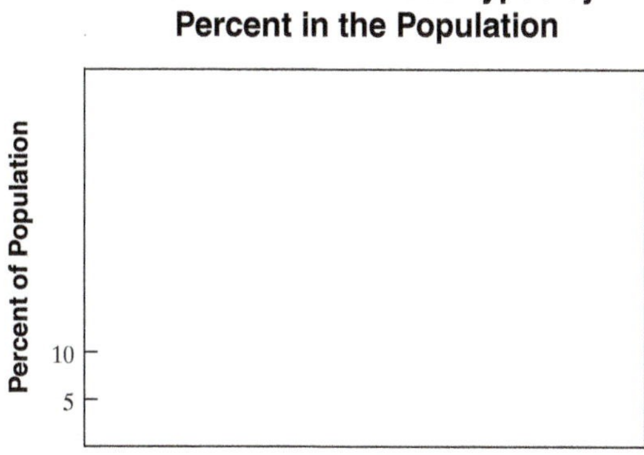

Blood Type

7. ______

8. ______

The following line graph shows the annual inflation rate in the United States for the years 2003–2013. Use this graph to answer Exercises 8 through 10.

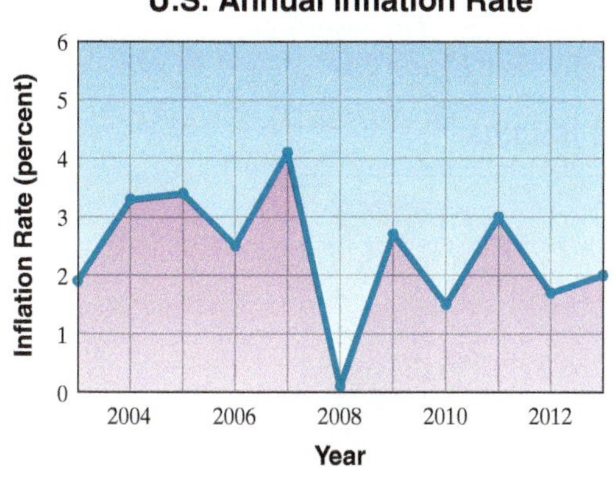

Source: Bureau of Labor Statistics

8. Approximate the annual inflation rate in 2012.

9. ______

9. During which of the years shown was the inflation rate greater than 3%?

10. ______

10. During which sets of years was the inflation rate decreasing?

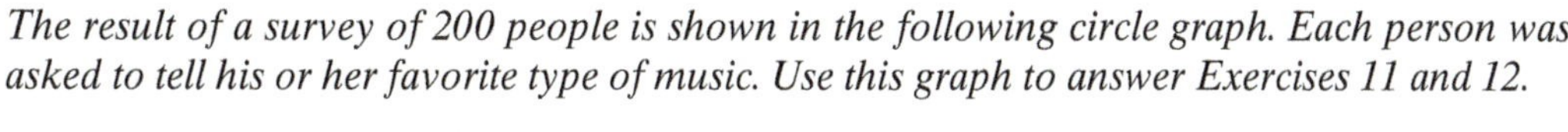

The result of a survey of 200 people is shown in the following circle graph. Each person was asked to tell his or her favorite type of music. Use this graph to answer Exercises 11 and 12.

11. ______

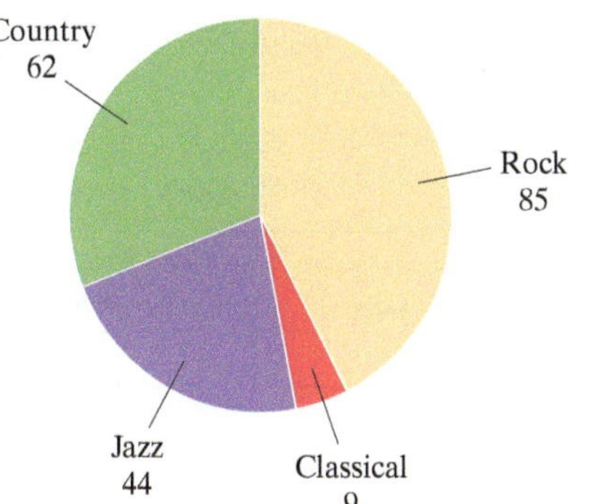

11. Find the ratio of those who prefer rock music to the total number surveyed.

12. Find the ratio of those who prefer country music to those who prefer jazz.

12. ______

13. ______

The following circle graph shows the projected age distribution of the population of the United States in 2015. There are projected to be 326 million people in the United States in 2015. Use the graph to find how many people are expected to be in the age groups given.

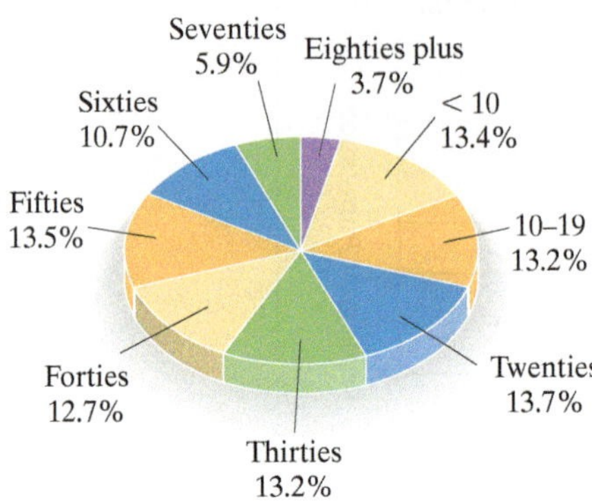

Source: U.S. Census Bureau

14. ______

13. Twenties (Round to nearest whole million.)

14. Eighties plus (Round to nearest whole million.)

A professor measures the heights of the students in her class. The results are shown in the following histogram. Use this histogram to answer Exercises 15 and 16.

15. ______

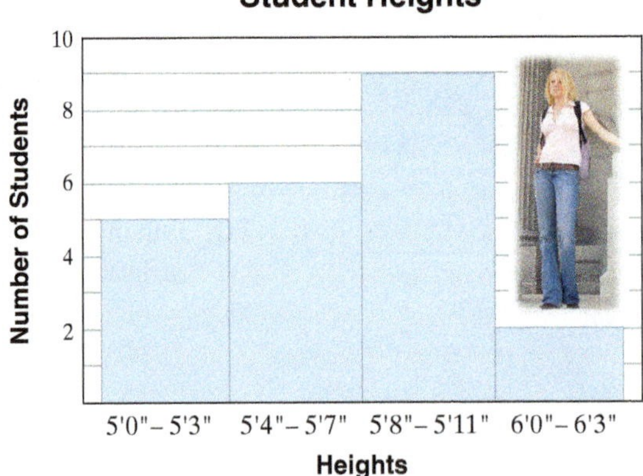

16. ______

15. How many students are 5′8″–5′11″ tall?

16. How many students are 5′7″ or shorter?

17. The history test scores of 25 students are shown below. Use these scores to complete the frequency distribution table.

70	86	81	65	92
43	72	85	69	97
82	51	75	50	68
88	83	85	77	99
77	63	59	84	90

Class Intervals (Scores)	Tally	Class Frequency (Number of Students)
40–49		
50–59		
60–69		
70–79		
80–89		
90–99		

17. ______

18. Use the results of Exercise **17** to draw a histogram.

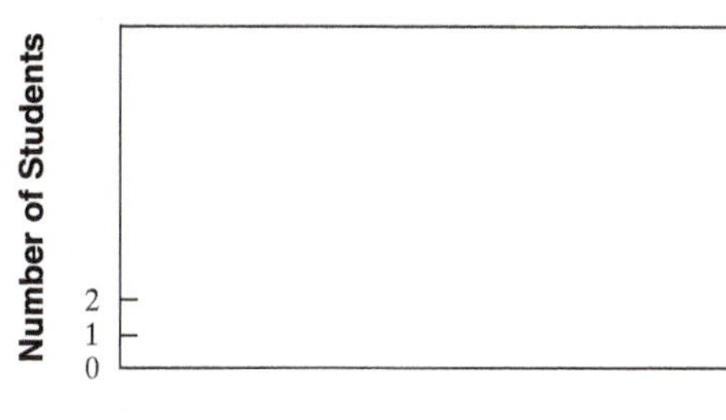

Find each square root and simplify. Round to the nearest thousandth if necessary.

19.  $\sqrt{49}$

20. $\sqrt{157}$

21. $\sqrt{\frac{64}{100}}$

Solve.

△ **22.** Approximate, to the nearest hundredth of a centimeter the length of the missing side of a right triangle with legs of 4 centimeters each.

△ **23.** Given that the following triangles are similar, find the unknown length, n.

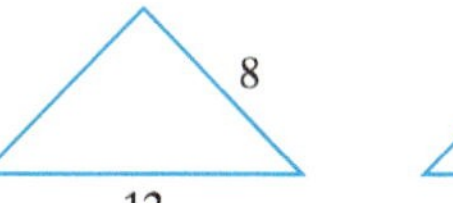

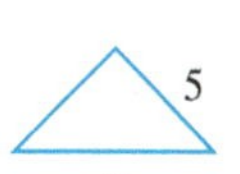

△ **24.** Tamara Watford, surveyor, needs to estimate the height of a tower. She estimates the length of her shadow to be 4 feet long and the length of the tower's shadow to be 48 feet long. Find the height of the tower if she is $5\frac{3}{4}$ feet tall.

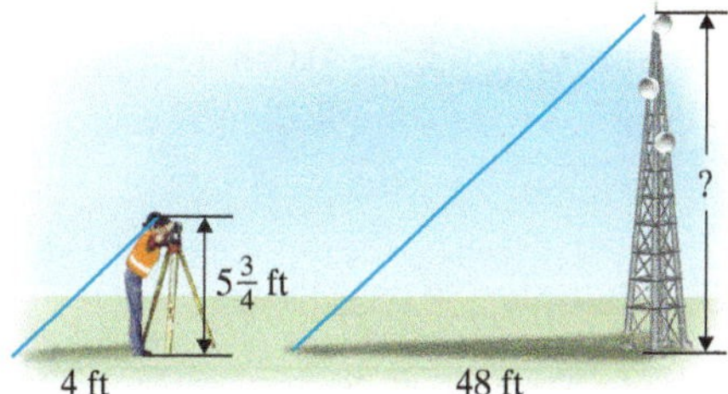

25. Draw a tree diagram for the experiment of spinning the spinner twice. Then use the diagram to determine the number of outcomes.

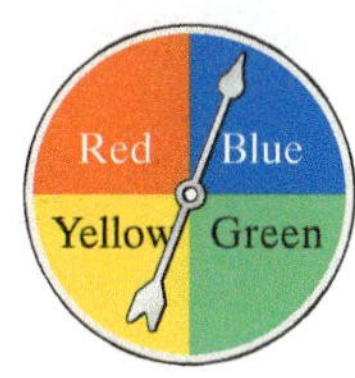

26. Draw a tree diagram for the experiment of tossing a coin twice. Then use the diagram to determine the number of outcomes.

Suppose that the numbers 1 through 10 are each written on same-size pieces of paper and placed in a bag. You then select one piece of paper from the bag.

27. What is the probability of choosing a 6 from the bag?

28. What is the probability of choosing a 3 or a 4 from the bag?

18. ____________
19. ____________
20. ____________
21. ____________
22. ____________
23. ____________
24. ____________
25. ____________
26. ____________
27. ____________
28. ____________

Chapters 1–7 Cumulative Review

Answers

1. ______
2. ______
3. ______
4. ______
5. ______
6. ______
7. ______
8. ______
9. ______
10. ______
11. ______
12. ______
13. ______
14. ______
15. ______
16. ______
17. ______
18. ______
19. ______
20. ______
21. ______
22. ______
23. ______
24. ______
25. ______
26. ______

1. Simplify: $4^3 + [3^2 - (10 \div 2)] - 7 \cdot 3$

2. $7^2 + [5^3 - (6 \div 3)] + 4 \cdot 2$

3. Evaluate $x - y$ for $x = -3$ and $y = 9$.

4. Evaluate $x - y$ for $x = 7$ and $y = -2$.

5. Solve: $3y - 7y = 12$

6. Solve: $2x - 6x = 24$

7. Solve: $\frac{x}{6} + 1 = \frac{4}{3}$

8. Solve: $\frac{7}{2} + \frac{a}{4} = 1$

9. Add: $2\frac{1}{3} + 5\frac{3}{8}$

10. Add: $3\frac{2}{5} + 4\frac{3}{4}$

11. Write 5.9 as a mixed number.

12. Write 2.8 as a mixed number.

13. Subtract: $3.5 - 0.068$

14. Subtract: $7.4 - 0.073$

15. Multiply: 0.0531×16

16. Multiply: 0.147×0.2

17. Divide: $-5.98 \div 115$

18. Divide: $27.88 \div 205$

19. Simplify: $(-1.3)^2 + 2.4$

20. Simplify: $(-2.7)^2$

21. Write $\frac{1}{4}$ as a decimal.

22. Write $\frac{3}{8}$ as a decimal.

23. Solve: $5(x - 0.36) = -x + 2.4$

24. Solve: $4(0.35 - x) = x - 7$

25. Use the appendix or a calculator to approximate $\sqrt{80}$ to the nearest thousandth.

26. Use the appendix or a calculator to approximate $\sqrt{60}$ to the nearest thousandth.

27. Find $\sqrt{\frac{1}{36}}$.

28. Find $\sqrt{\frac{16}{49}}$.

Sketch each right triangle and find the length of the side not given. Round to two decimal places.

29. leg = 5 in., hypotenuse = 7 in.

30. leg = 2, leg = 16

31. Solve $\frac{x-5}{3} = \frac{x+2}{5}$ for x and then check.

32. Solve $\frac{8}{5} = \frac{x}{10}$ for x and then check.

△ **33.** Find the ratio of corresponding sides for the similar triangles ABC and DEF.

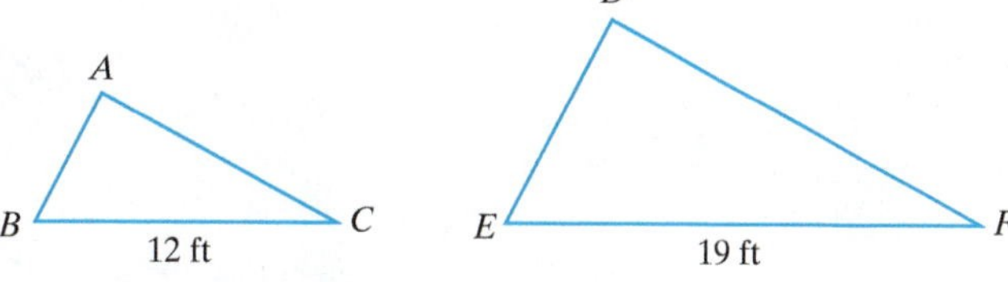

△ **34.** Find the ratio of corresponding sides for the similar triangles GHJ and KLM.

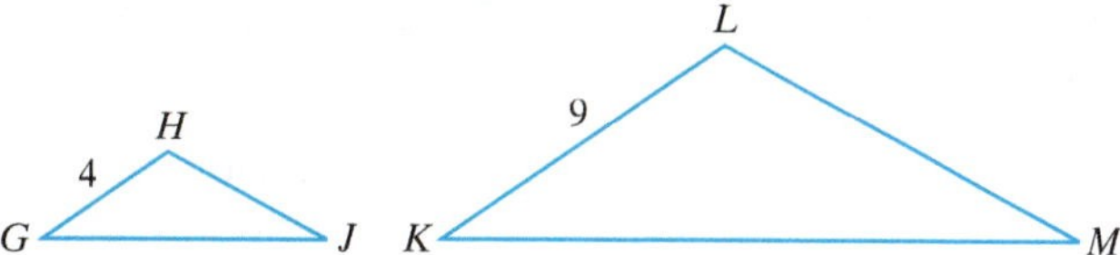

Write each percent as a decimal.

35. 4.6%

36. 32%

37. 0.74%

38. 2.7%

39. What number is 35% of 60?

40. What number is 40% of 36?

41. 20.8 is 40% of what number?

42. 9.5 is 25% of what number?

43. The sales tax on a $310 Sony flat-screen digital 32-inch television is $26.35. Find the sales tax rate.

44. The sales tax on a $2.00 yo-yo is $0.13. Find the sales tax rate.

45. $4000 is invested at 5.3% compounded quarterly for 10 years. Find the total amount at the end of 10 years.

46. Linda Bonnett borrows $1600 for 1 year. If the interest is $128.60, find the monthly payment. (Hint: Find the total amount due and divide by the number of months in a year.)

47. Find the median of the list of numbers: 25, 54, 56, 57, 60, 71, 98

48. Find the median of the list of numbers: 43, 46, 47, 50, 52, 83

49. If a die is rolled one time, find the probability of rolling a 3 or a 4.

50. If a die is rolled one time, find the probability of rolling an even number.

27. ____
28. ____
29. ____
30. ____
31. ____
32. ____
33. ____
34. ____
35. ____
36. ____
37. ____
38. ____
39. ____
40. ____
41. ____
42. ____
43. ____
44. ____
45. ____
46. ____
47. ____
48. ____
49. ____
50. ____

8 Geometry and Measurement

Sections

Check Your Progress

The word *geometry* is formed from the Greek words *geo*, meaning "Earth," and *metron*, meaning "measure." Geometry literally means to measure the Earth. In this chapter we learn about various geometric figures and their properties, such as perimeter, area, and volume. Knowledge of geometry can help us solve practical problems in all types of real-life situations.

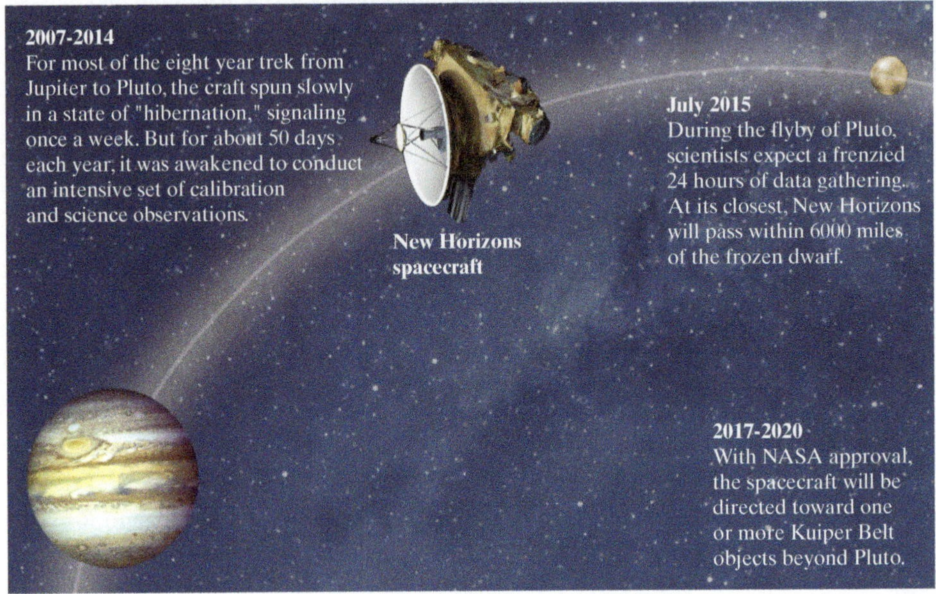

New Horizons is NASA's robotic spacecraft mission, shown above. The spacecraft is about the size and shape of a grand piano with a satellite dish attached. It was launched in January 2006 and is now heading toward Pluto. How do we receive the images of Pluto on Earth when this spacecraft is so far into deep space and how long before we receive the images?

The Deep Space Network (DSN) is a worldwide network of large antennas and communication facilities. When a mission is in deep space, fewer sites are needed for sending and receiving transmissions; thus, the DSN has only three sites—one in California (Goldstone), one in Spain (Madrid), and one in Australia (Canberra). The diagram below shows an overview of Earth from the vantage point of the North Pole and the location of these three sites.

We study some geometry of DSN in Section 8.1, Exercises 65 and 66, and Section 8.7, Exercises 23–26.

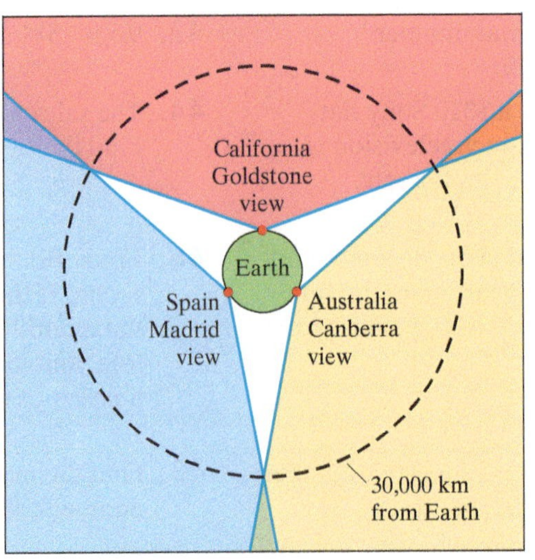

8.1 Lines and Angles 

Objective A Identifying Lines, Line Segments, Rays, and Angles

Objectives

A Identify Lines, Line Segments, Rays, and Angles.

B Classify Angles as Acute, Right, Obtuse, or Straight.

C Identify Complementary and Supplementary Angles.

D Find Measures of Angles. 

Let's begin with a review of two important concepts—space and plane.

Space extends in all directions indefinitely. Examples of objects in space are houses, grains of salt, bushes, your *Algebra Foundations* textbook, and you.

A **plane** is a flat surface that extends indefinitely. Surfaces like a plane are a classroom floor or a blackboard or whiteboard.

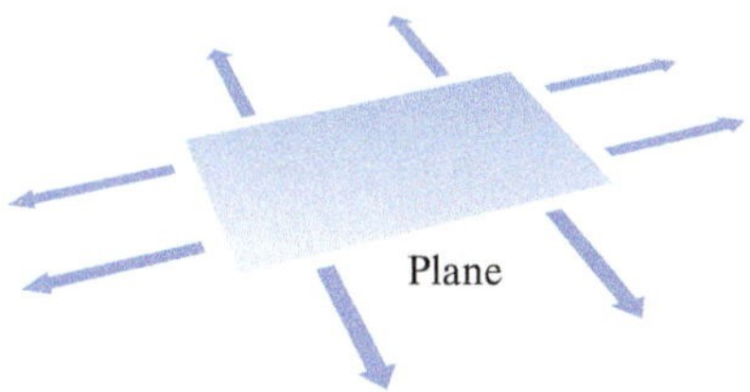

The most basic concept of geometry is the idea of a point in space. A **point** has no length, no width, and no height, but it does have location. We represent a point by a dot, and we usually label points with capital letters.

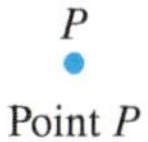

Point P

A **line** is a set of points extending indefinitely in two directions. A line has no width or height, but it does have length. We can name a line by any two of its points or by a single lowercase letter. A **line segment** is a piece of a line with two endpoints.*

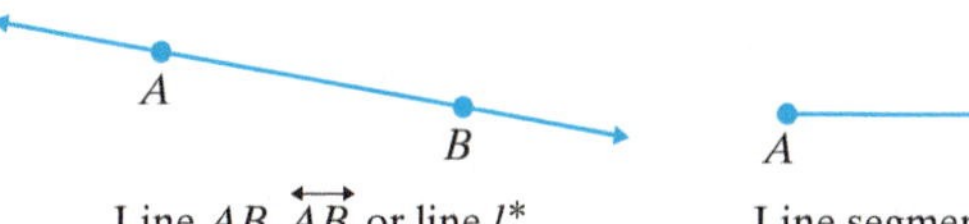

Line AB, $\overleftrightarrow{AB}$, or line l*

Line segment AB or $\overline{AB}$

A **ray** is a part of a line with one endpoint. A ray extends indefinitely in one direction. An **angle** is made up of two rays that share the same endpoint. The common endpoint is called the **vertex.**

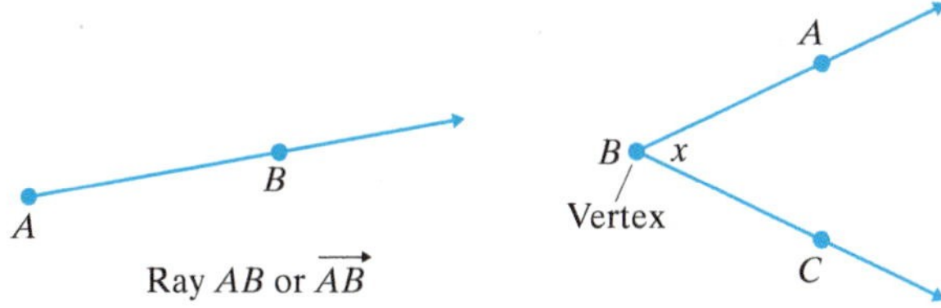

Ray AB or $\overrightarrow{AB}$

The angle in the figure above can be named

$\angle ABC$ $\angle CBA$ $\angle B$ or $\angle x$

The vertex is the middle point.

Rays BA and BC are **sides** of the angle.

*Although line l is also line BA or $\overleftrightarrow{BA}$, we will use only one order of points to name a line or line segment.

Helpful Hint

Naming an Angle

When there is no confusion as to what angle is being named, you may use the vertex alone.

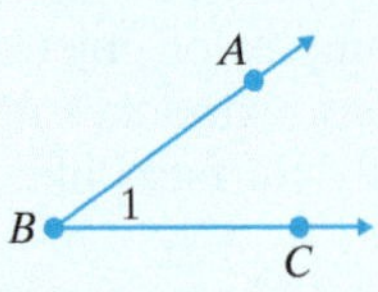

Name of $\angle B$ is all right.
There is no confusion. $\angle B$ means $\angle 1$.

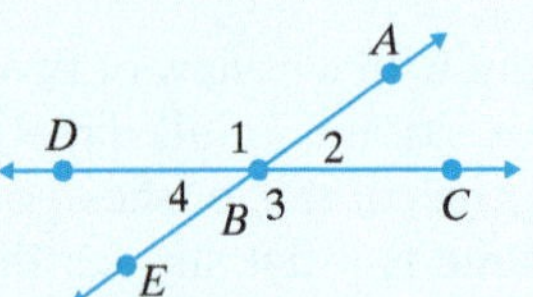

Name of $\angle B$ is *not* all right.
There is confusion. Does $\angle B$ mean $\angle 1$, $\angle 2$, $\angle 3$, or $\angle 4$?

Practice 1

Identify each figure as a line, a ray, a line segment, or an angle. Then name the figure using the given points.

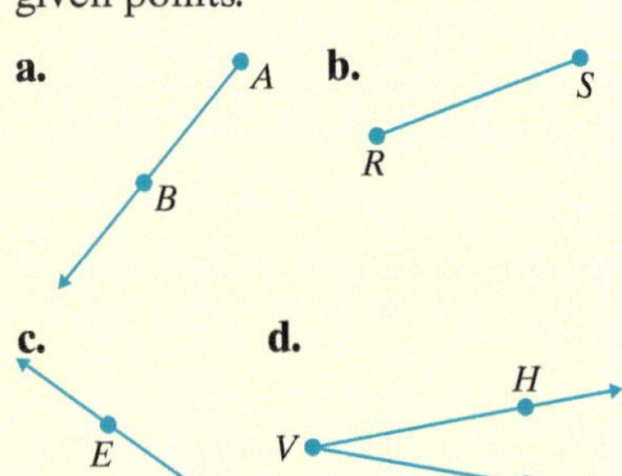

Example 1

Identify each figure as a line, a ray, a line segment, or an angle. Then name the figure using the given points.

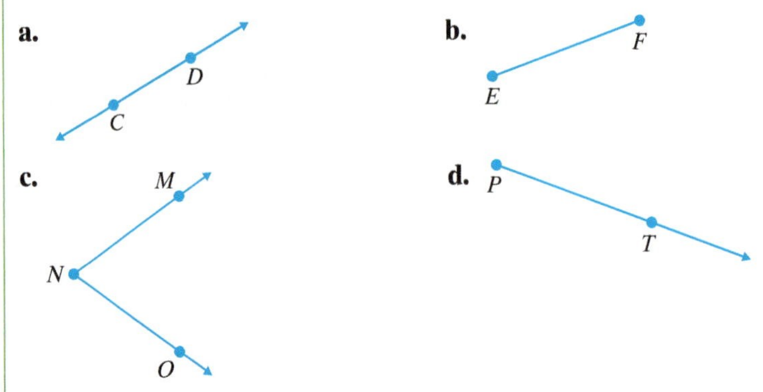

Solution:

Figure (a) extends indefinitely in two directions. It is line *CD* or $\overleftrightarrow{CD}$.

Figure (b) has two endpoints. It is line segment *EF* or $\overline{EF}$.

Figure (c) has two rays with a common endpoint. It is $\angle MNO$, $\angle ONM$, or $\angle N$.

Figure (d) is part of a line with one endpoint. It is ray *PT* or $\overrightarrow{PT}$.

Work Practice 1

Practice 2

Use the figure in Example 2 to list other ways to name $\angle z$.

Example 2

List other ways to name $\angle y$.

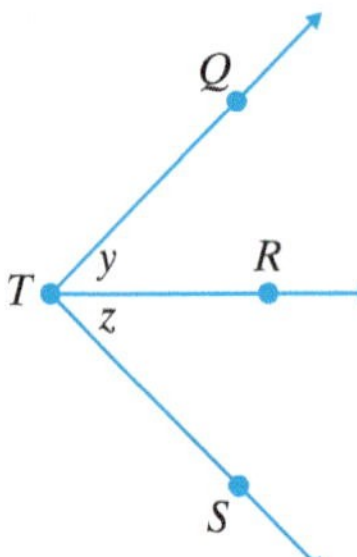

Solution: Two other ways to name $\angle y$ are $\angle QTR$ and $\angle RTQ$. We may *not* use the vertex alone to name this angle because three different angles have *T* as their vertex.

Work Practice 2

Answers

1. a. ray; ray *AB* or $\overrightarrow{AB}$
   b. line segment; line segment *RS* or $\overline{RS}$
   c. line; line *EF* or $\overleftrightarrow{EF}$
   d. angle; $\angle TVH$ or $\angle HVT$ or $\angle V$
2. $\angle RTS$, $\angle STR$

Objective B Classifying Angles as Acute, Right, Obtuse, or Straight

An angle can be measured in **degrees.** The symbol for degrees is a small, raised circle, °. There are 360° in a full revolution, or a full circle.

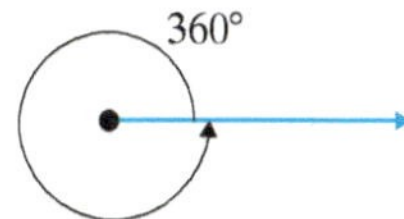

$\frac{1}{2}$ of a revolution measures $\frac{1}{2}(360°) = 180°$. An angle that measures 180° is called a **straight angle.**

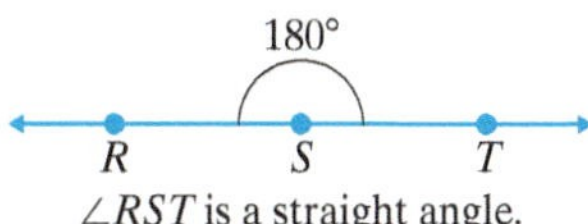

$\angle RST$ is a straight angle.

$\frac{1}{4}$ of a revolution measures $\frac{1}{4}(360°) = 90°$. An angle that measures 90° is called a **right angle.** The symbol ∟ is used to denote a right angle.

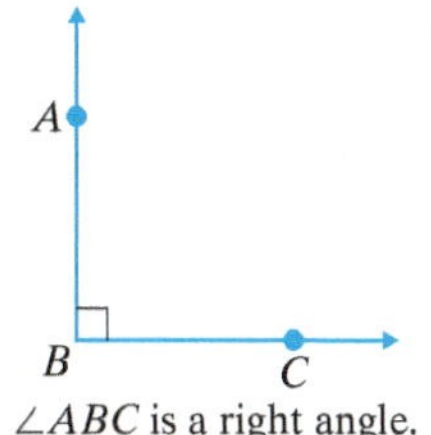

$\angle ABC$ is a right angle.

An angle whose measure is between 0° and 90° is called an **acute angle.**

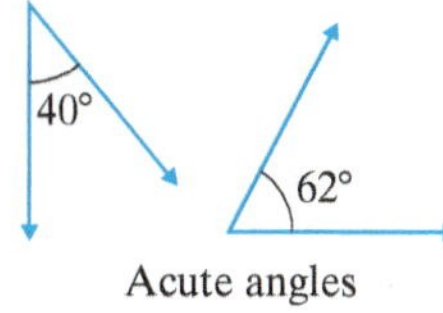

Acute angles

An angle whose measure is between 90° and 180° is called an **obtuse angle.**

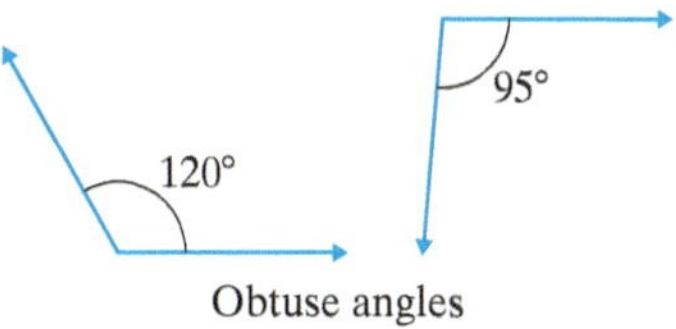

Obtuse angles

Example 3 Classify each angle as acute, right, obtuse, or straight.

a.

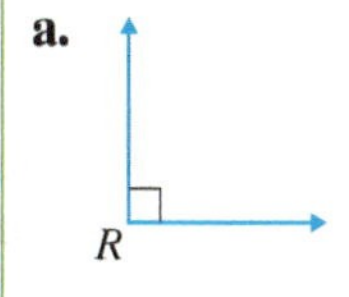

b.

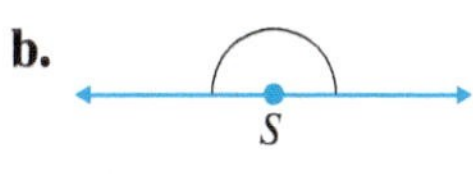

c.

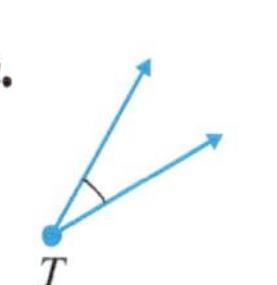

d.

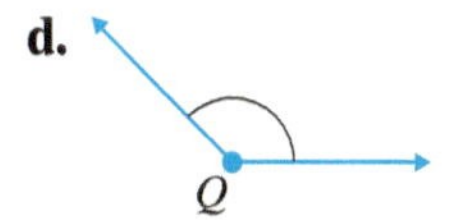

(*Continued on next page*)

Practice 3

Classify each angle as acute, right, obtuse, or straight.

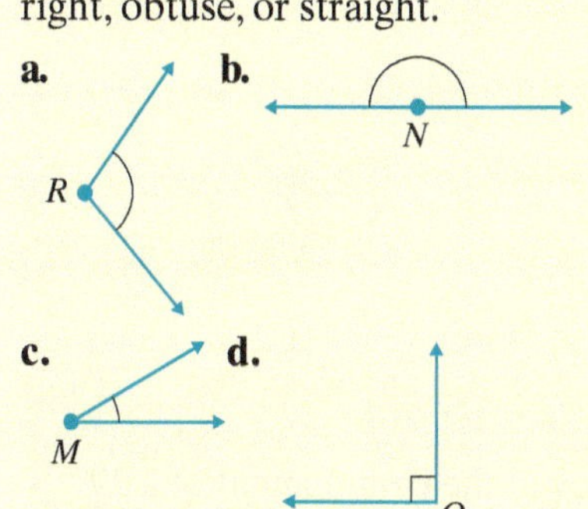

Answers

3. a. obtuse **b.** straight **c.** acute **d.** right

Solution:

a. $\angle R$ is a right angle, denoted by ⦜. It measures 90°.
b. $\angle S$ is a straight angle. It measures 180°.
c. $\angle T$ is an acute angle. It measures between 0° and 90°.
d. $\angle Q$ is an obtuse angle. It measures between 90° and 180°.

Work Practice 3

Let's look at $\angle B$ below, whose measure is 62°.

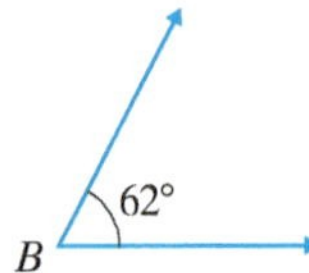

There is a shorthand notation for writing the measure of this angle. To write "The measure of $\angle B$ is 62°, we can write,

$$m\angle B = 62°.$$

By the way, note that $\angle B$ is an acute angle because $m\angle B$ is between 0° and 90°.

Objective C Identifying Complementary and Supplementary Angles

Two angles that have a sum of 90° are called **complementary angles.** We say that each angle is the **complement** of the other.

$\angle R$ and $\angle S$ are complementary angles because

$$m\angle R + m\angle S = 60° + 30° = 90°$$

60° R 30° S

Complementary angles
60° + 30° = 90°

Two angles that have a sum of 180° are called **supplementary angles.** We say that each angle is the **supplement** of the other.

$\angle M$ and $\angle N$ are supplementary angles because

$$m\angle M + m\angle N = 125° + 55° = 180°$$

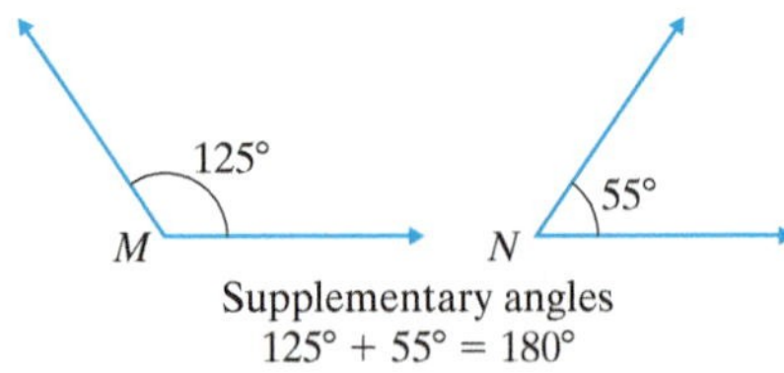

Supplementary angles
125° + 55° = 180°

Practice 4

Find the complement of a 29° angle.

Answer

4. 61°

Example 4 Find the complement of a 48° angle.

Solution: Two angles that have a sum of 90° are complementary. This means that the complement of an angle that measures 48° is an angle that measures $90° - 48° = 42°$.

Work Practice 4

Example 5 Find the supplement of a 107° angle.

Solution: Two angles that have a sum of 180° are supplementary. This means that the supplement of an angle that measures 107° is an angle that measures $180° - 107° = 73°$.

Work Practice 5

Practice 5

Find the supplement of a 67° angle.

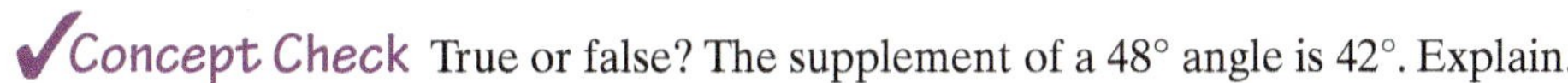

✓Concept Check True or false? The supplement of a 48° angle is 42°. Explain.

Objective D Finding Measures of Angles

Measures of angles can be added or subtracted to find measures of related angles.

Example 6 Find the measure of $\angle x$. Then classify $\angle x$ as an acute, obtuse, or right angle.

Solution:

$$\begin{aligned} m\angle x &= m\angle QTS - m\angle RTS \\ &= 87° - 52° \\ &= 35° \end{aligned}$$

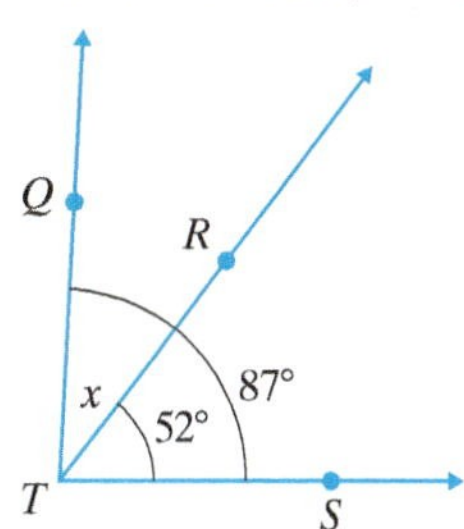

Thus, the measure of $\angle x$ $(m\angle x)$ is 35°.

Since $\angle x$ measures between 0° and 90°, it is an acute angle.

Work Practice 6

Practice 6

a. Find the measure of $\angle y$.

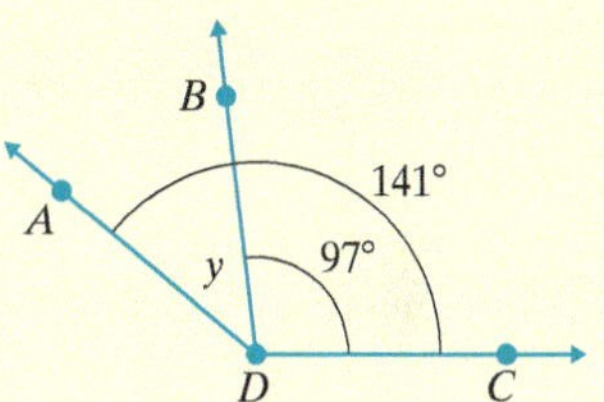

b. Find the measure of $\angle x$.

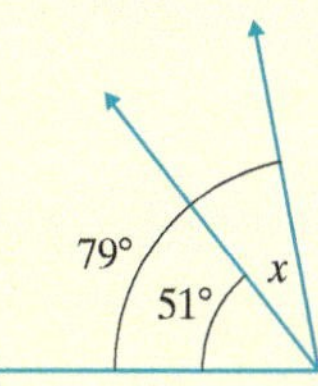

c. Classify $\angle x$ and $\angle y$ as acute, obtuse, or right angles.

Two lines in a plane can be either parallel or intersecting. **Parallel lines** never meet. **Intersecting lines** meet at a point. The symbol $\parallel$ is used to indicate "is parallel to." For example, in the figure, $p \parallel q$.

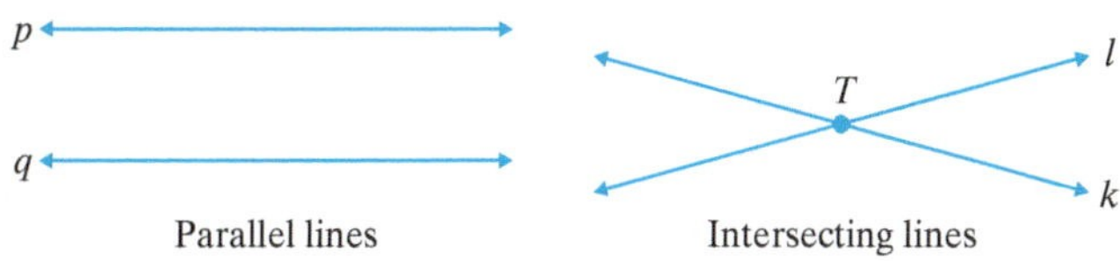

Parallel lines Intersecting lines

Some intersecting lines are perpendicular. Two lines are **perpendicular** if they form right angles when they intersect. The symbol $\perp$ is used to denote "is perpendicular to." For example, in the figure below, $n \perp m$.

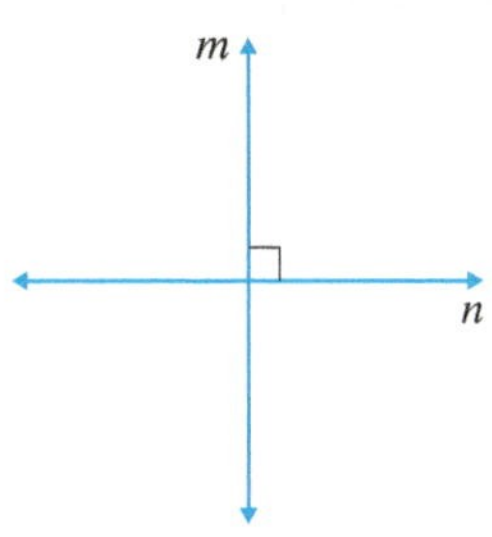

Perpendicular lines

When two lines intersect, four angles are formed. Two angles that are opposite each other are called **vertical angles.** Vertical angles have the same measure.

Two angles that share a common side are called **adjacent angles.** Adjacent angles formed by intersecting lines are supplementary. That is, they have a sum of 180°.

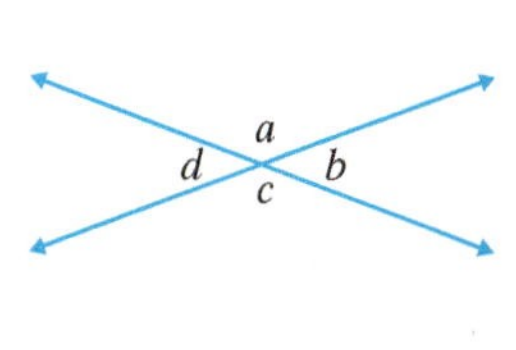

Vertical angles:
$\angle a$ and $\angle c$
$\angle d$ and $\angle b$

Adjacent angles:
$\angle a$ and $\angle b$
$\angle b$ and $\angle c$
$\angle c$ and $\angle d$
$\angle d$ and $\angle a$

Answers

5. 113° **6. a.** 44° **b.** 28° **c.** acute

✓Concept Check Answer

false; the *complement* of a 48° angle is 42°; the *supplement* of a 48° angle is 132°

Here are a few real-life examples of the lines we just discussed.

Parallel lines

Vertical angles

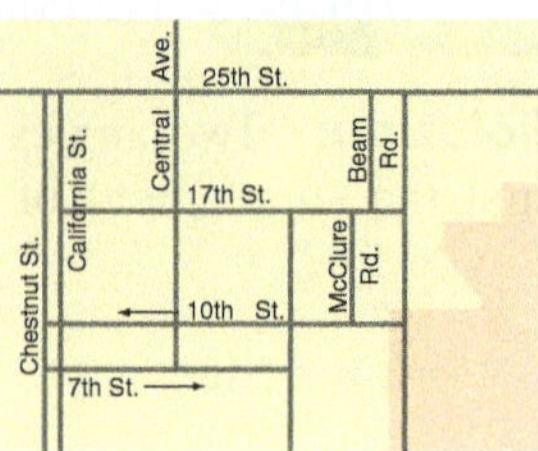

Perpendicular lines

Practice 7

Find the measure of $\angle a$, $\angle b$, and $\angle c$.

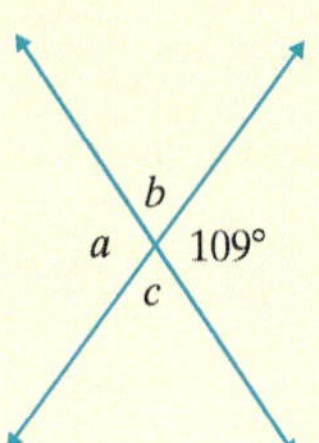

Example 7

Find the measure of $\angle x$, $\angle y$, and $\angle z$ if the measure of $\angle t$ is 42°.

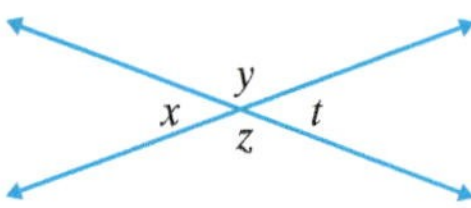

Solution: Since $\angle t$ and $\angle x$ are vertical angles, they have the same measure, so $\angle x$ measures 42°.

Since $\angle t$ and $\angle y$ are adjacent angles, their measures have a sum of 180°. So $\angle y$ measures $180° - 42° = 138°$.

Since $\angle y$ and $\angle z$ are vertical angles, they have the same measure. So $\angle z$ measures 138°.

Work Practice 7

A line that intersects two or more lines at different points is called a **transversal.** Line l is a transversal that intersects lines m and n. The eight angles formed have special names. Some of these names are:

Corresponding angles: $\angle a$ and $\angle e$, $\angle c$ and $\angle g$, $\angle b$ and $\angle f$, $\angle d$ and $\angle h$

Alternate interior angles: $\angle c$ and $\angle f$, $\angle d$ and $\angle e$

When two lines cut by a transversal are *parallel*, the following statement is true:

Parallel Lines Cut by a Transversal

If two parallel lines are cut by a transversal, then the measures of **corresponding angles are equal** and the measures of the **alternate interior angles are equal.**

Practice 8

Given that $m \parallel n$ and that the measure of $\angle w = 45°$, find the measures of all the angles shown.

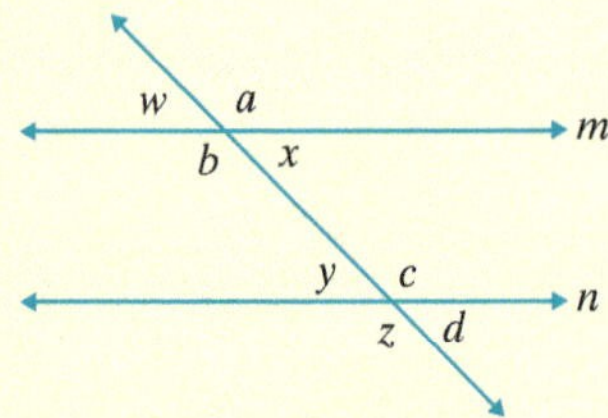

Example 8

Given that $m \parallel n$ and that the measure of $\angle w$ is 100°, find the measures of $\angle x$, $\angle y$, and $\angle z$.

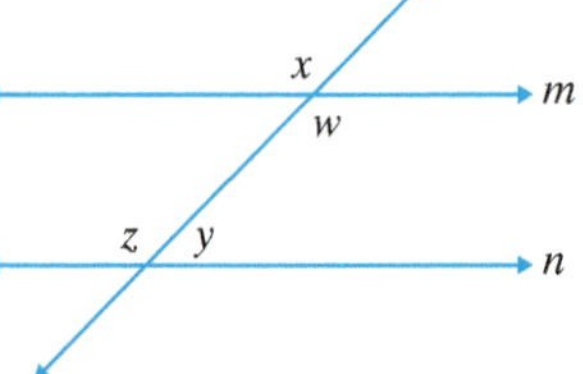

Solution:

$m\angle x = 100°$ — $\angle x$ and $\angle w$ are vertical angles.

$m\angle z = 100°$ — $\angle x$ and $\angle z$ are corresponding angles.

$m\angle y = 180° - 100° = 80°$ — $\angle z$ and $\angle y$ are supplementary angles.

Work Practice 8

Answers

7. $m\angle a = 109°$; $m\angle b = 71°$; $m\angle c = 71°$

8. $m\angle x = 45°$; $m\angle y = 45°$; $m\angle z = 135°$; $m\angle a = 135°$; $m\angle b = 135°$; $m\angle c = 135°$; $m\angle d = 45°$

Vocabulary, Readiness & Video Check

Use the choices below to fill in each blank.

acute	straight	degrees	adjacent	parallel	intersecting
obtuse	space	plane	point	vertical	vertex
right	angle	ray	line	perpendicular	transversal

1. A(n) ____________ is a flat surface that extends indefinitely.
2. A(n) ____________ has no length, no width, and no height.
3. ____________ extends in all directions indefinitely.
4. A(n) ____________ is a set of points extending indefinitely in two directions.
5. A(n) ____________ is part of a line with one endpoint.
6. A(n) ____________ is made up of two rays that share a common endpoint. The common endpoint is called the ____________.
7. A(n) ____________ angle measures 180°.
8. A(n) ____________ angle measures 90°.
9. A(n) ____________ angle measures between 0° and 90°.
10. A(n) ____________ angle measures between 90° and 180°.
11. ____________ lines never meet and ____________ lines meet at a point.
12. Two intersecting lines are ____________ if they form right angles when they intersect.
13. An angle can be measured in ____________.
14. A line that intersects two or more lines at different points is called a(n) ____________.
15. When two lines intersect, four angles are formed, called ____________ angles.
16. Two angles that share a common side are called ____________ angles.

Martin-Gay Interactive Videos

See Video 8.1

Watch the section lecture video and answer the following questions.

Objective A **17.** In the lecture after Example 2, what are the four ways we can name the angle shown?

Objective B **18.** In the lecture before Example 3, what type of angle forms a line? What is its measure?

Objective C **19.** What calculation is used to find the answer to Example 6?

Objective D **20.** In the lecture before Example 7, two lines in a plane that aren't parallel must what?

8.1 Exercise Set MyMathLab®

Objective A *Identify each figure as a line, a ray, a line segment, or an angle. Then name the figure using the given points. See Examples 1 and 2.*

 **1.**

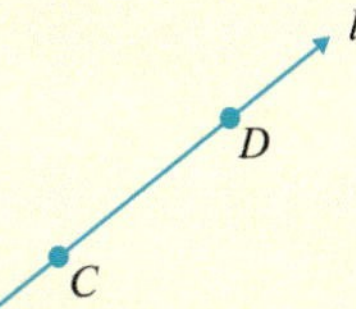

2.

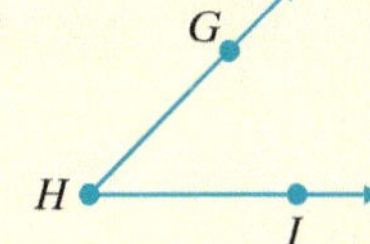

3.

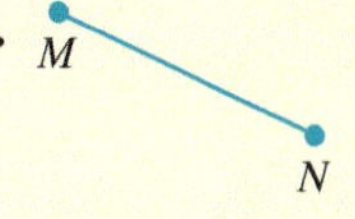

4.

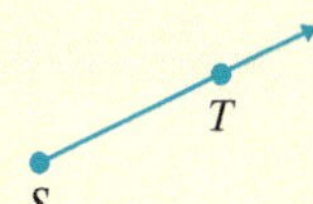

5.

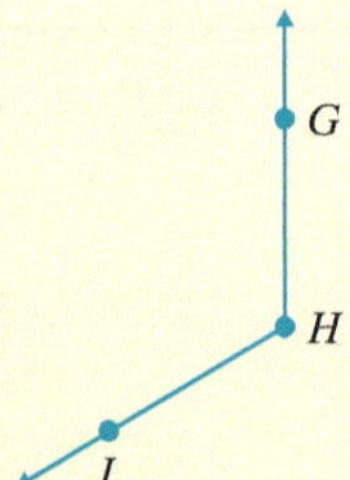

6.

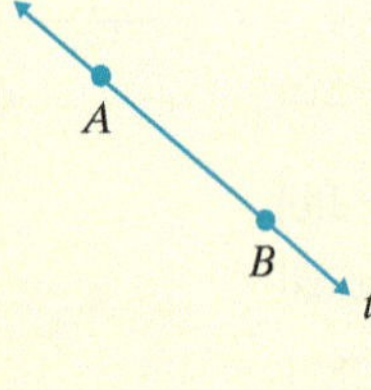

7.

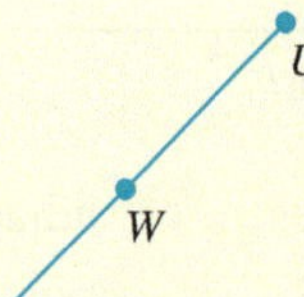

8.

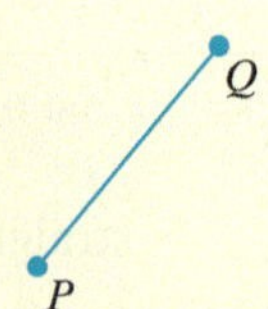

List two other ways to name each angle. See Example 2.

9. $\angle x$

10. $\angle w$

11. $\angle z$

12. $\angle y$

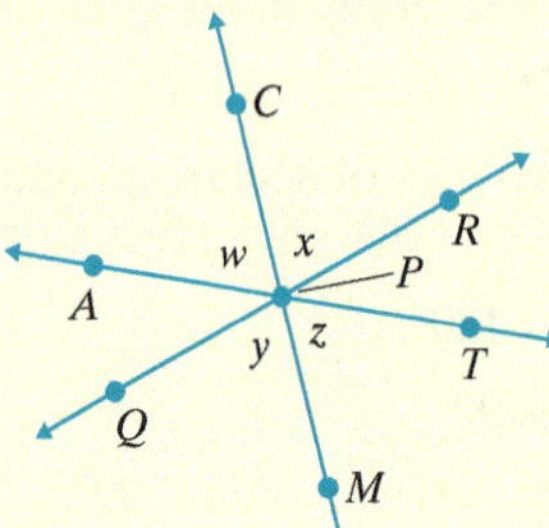

Objective B *Classify each angle as acute, right, obtuse, or straight. See Example 3.*

13.

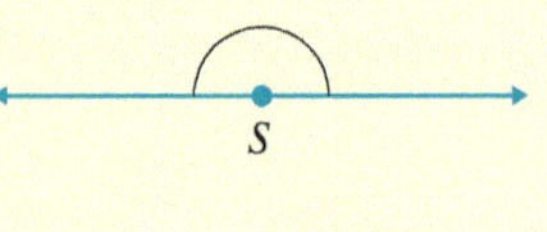

14.

15.

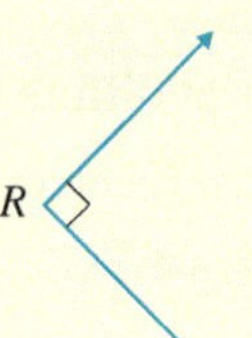

16.

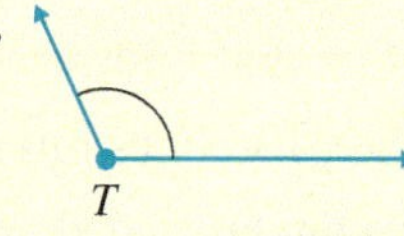

17.

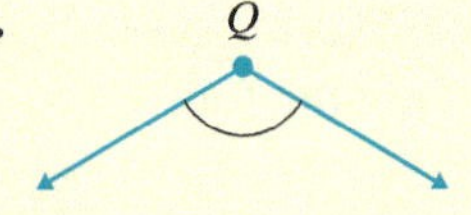

18.

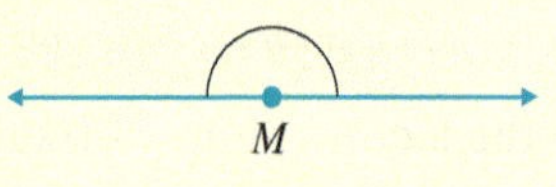

19.

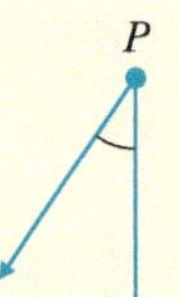

20.

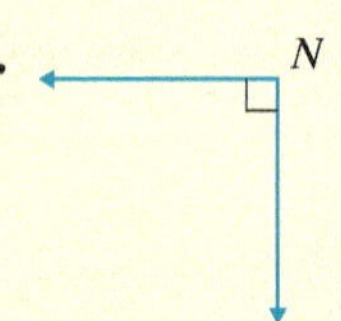

Objective C *Find each complementary or supplementary angle as indicated. See Examples 4 and 5.*

21. Find the complement of a 23° angle.

22. Find the complement of a 77° angle.

23. Find the supplement of a 17° angle.

24. Find the supplement of a 77° angle.

25. Find the complement of a 58° angle.

26. Find the complement of a 22° angle.

27. Find the supplement of a 150° angle.

28. Find the supplement of a 130° angle.

29. Identify the pairs of complementary angles.

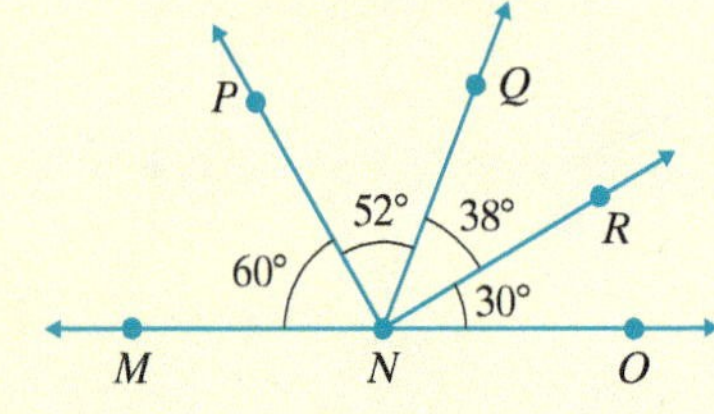

30. Identify the pairs of complementary angles.

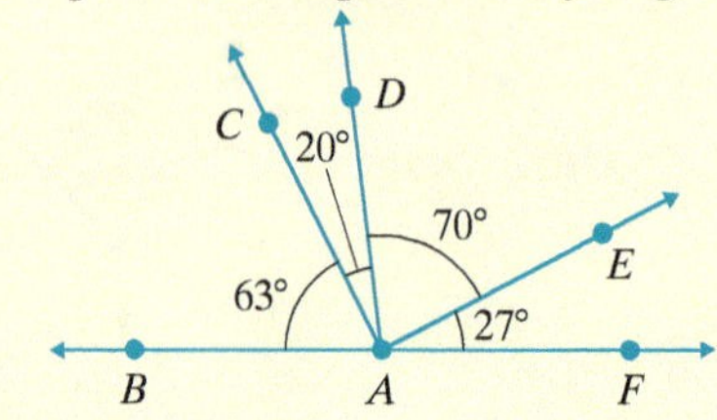

31. Identify the pairs of supplementary angles.

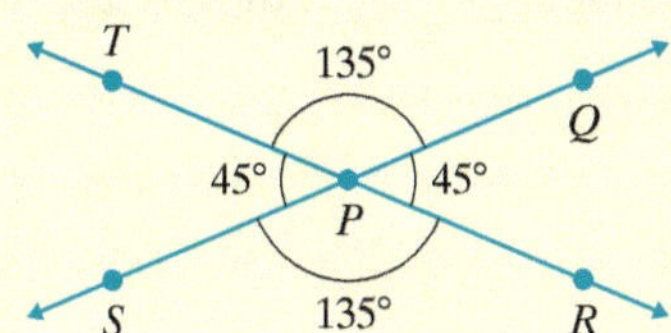

32. Identify the pairs of supplementary angles.

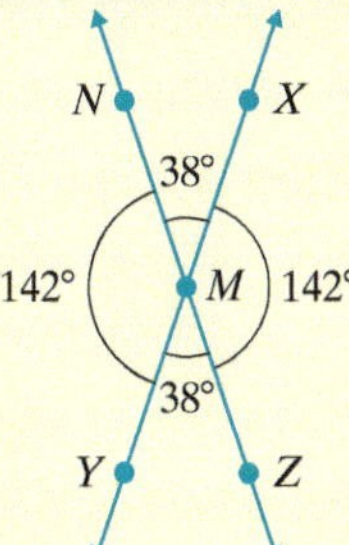

Objective D *Find the measure of ∠x in each figure. See Example 6.*

33.

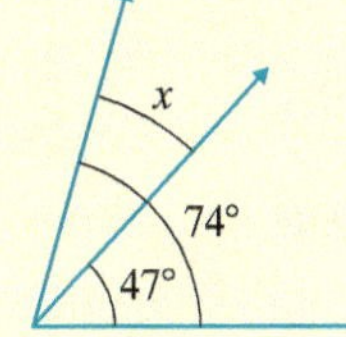

34.

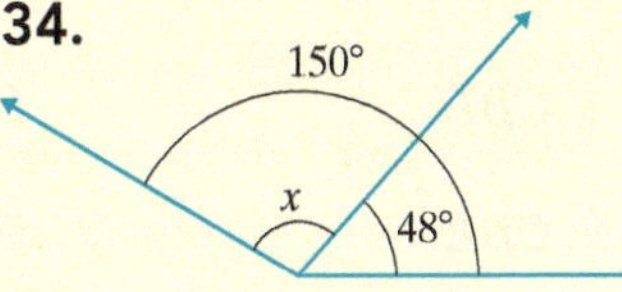

35.

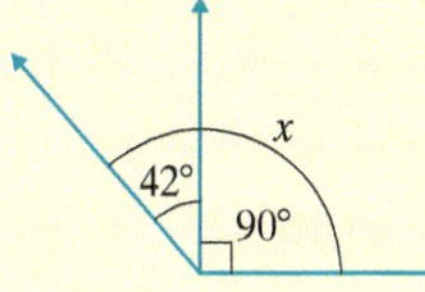

36.

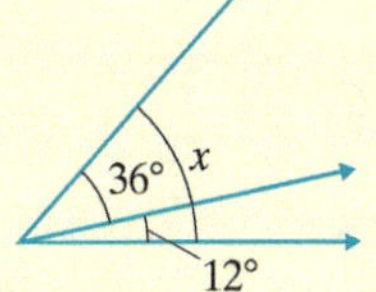

Find the measures of angles x, y, and z in each figure. See Examples 7 and 8.

37.

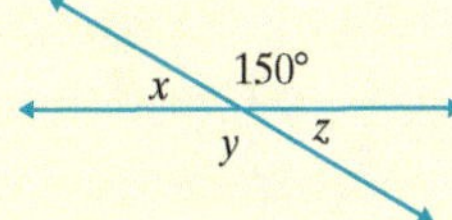

38.

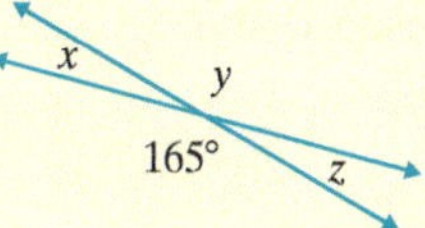

39.

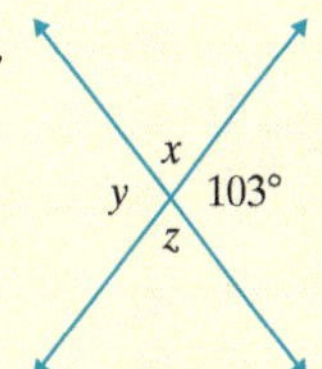

40.

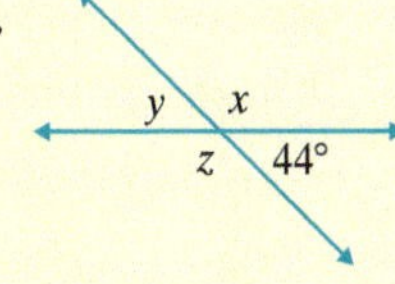

41. $m \parallel n$

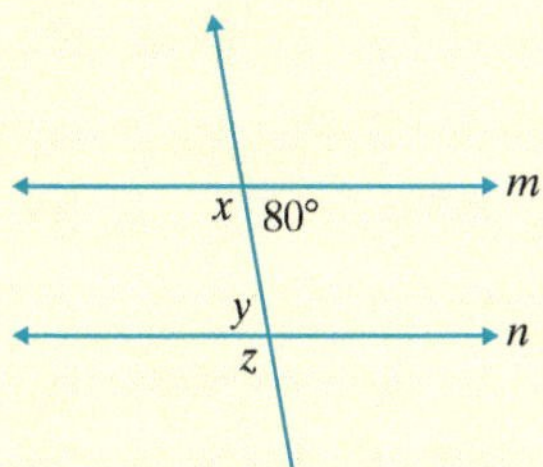

42. $m \parallel n$

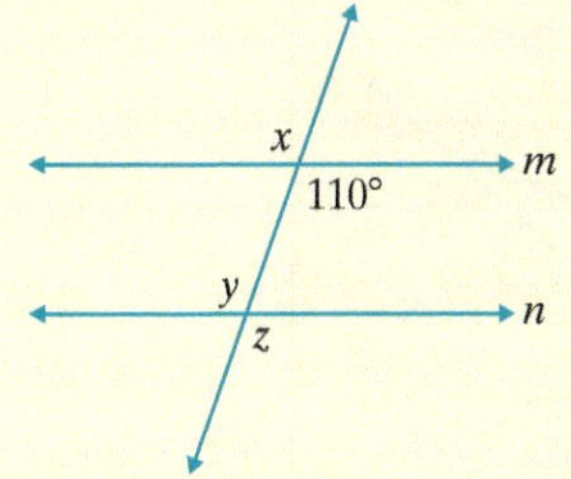

43. $m \parallel n$

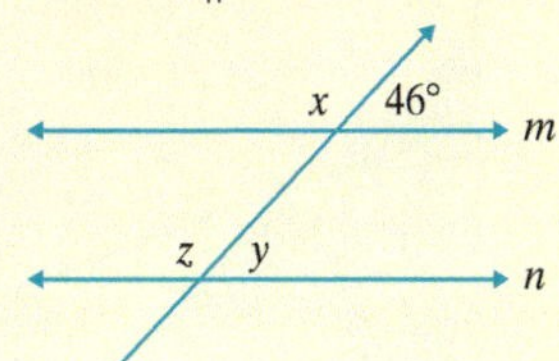

44. $m \parallel n$

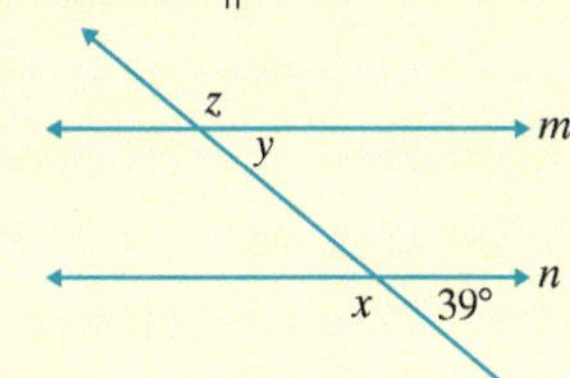

Objectives A D Mixed Practice *Find two other ways of naming each angle. See Example 2.*

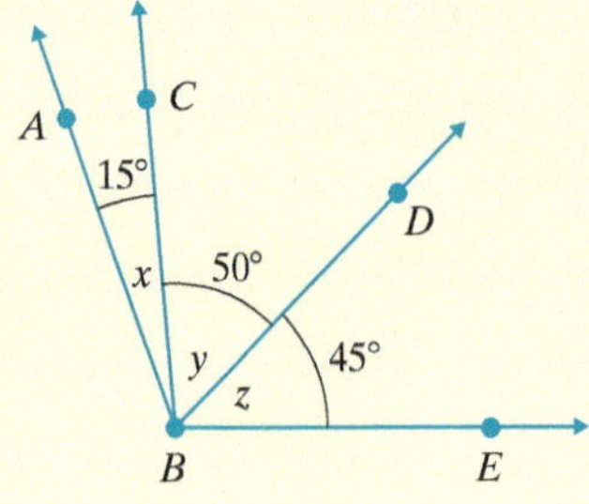

45. $\angle x$

46. $\angle y$

47. $\angle z$

48. $\angle ABE$ (just name one other way)

Find the measure of each angle in the figure above. See Example 6.

49. $\angle ABC$

50. $\angle EBD$

51. $\angle CBD$

52. $\angle CBA$

53. $\angle DBA$

54. $\angle EBC$

55. $\angle CBE$

56. $\angle ABE$

Review

Perform each indicated operation. See Sections 4.3, 4.5, and 4.7.

57. $\frac{7}{8} + \frac{1}{4}$

58. $\frac{7}{8} - \frac{1}{4}$

59. $\frac{7}{8} \cdot \frac{1}{4}$

60. $\frac{7}{8} \div \frac{1}{4}$

61. $3\frac{1}{3} - 2\frac{1}{2}$

62. $3\frac{1}{3} + 2\frac{1}{2}$

63. $3\frac{1}{3} \div 2\frac{1}{2}$

64. $3\frac{1}{3} \cdot 2\frac{1}{2}$

Concept Extensions

Use this North Pole overhead view of the three sites of the Deep Space Network to answer Exercises 65 and 66. (See the Chapter Opener.)

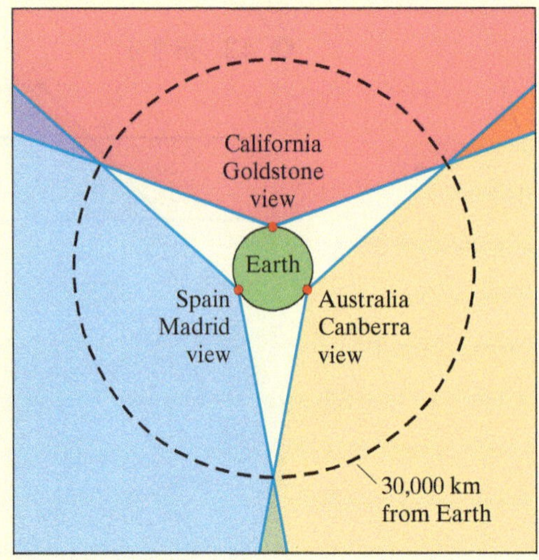

65. How many degrees are there around the Earth at the equator?

66. If the three sites of the Deep Space Network (red dots shown) are about the same number of degrees apart, how many degrees apart are they?

67. The angle between the two walls of the Vietnam Veterans Memorial in Washington, D.C., is 125.2°. Find the supplement of this angle. (*Source:* National Park Service)

68. The faces of Khafre's Pyramid at Giza, Egypt, are inclined at an angle of 53.13°. Find the complement of this angle. (*Source:* PBS *NOVA* Online)

Answer true or false for Exercises 69 through 72. See the Concept Check in this section. If false, explain why.

69. The complement of a 100° angle is an 80° angle.

70. It is possible to find the complement of a 120° angle.

71. It is possible to find the supplement of a 120° angle.

72. The supplement of a 5° angle is a 175° angle.

73. If lines m and n are parallel, find the measures of angles a through e.

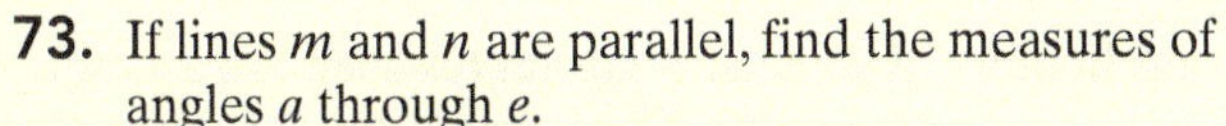

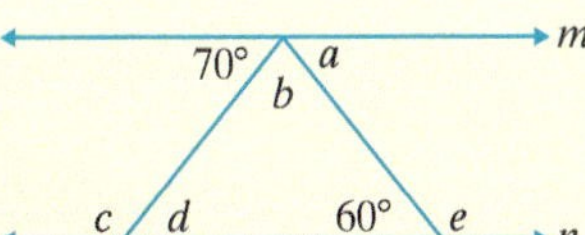

74. Below is a rectangle. List which segments, if extended, would be parallel lines.

75. Can two supplementary angles both be acute? Explain why or why not.

76. In your own words, describe how to find the complement and the supplement of a given angle.

77. Find two complementary angles with the same measure.

78. Is the figure below possible? Why or why not?

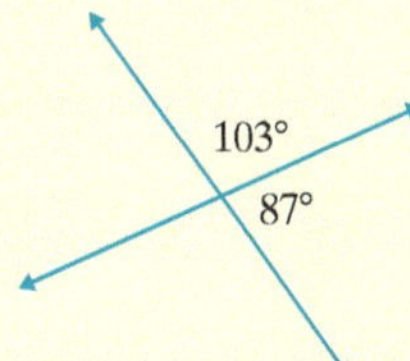

 8.2 Perimeter

Objectives

A Use Formulas to Find Perimeters.

B Use Formulas to Find Circumferences.

Objective A Using Formulas to Find Perimeters 

Recall from Section 1.3 that the perimeter of a polygon is the distance around the polygon. This means that the perimeter of a polygon is the sum of the lengths of its sides.

Practice 1

a. Find the perimeter of the rectangle.

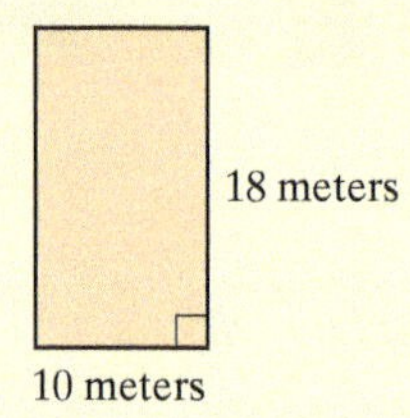

b. Find the perimeter of the rectangular lot shown below:

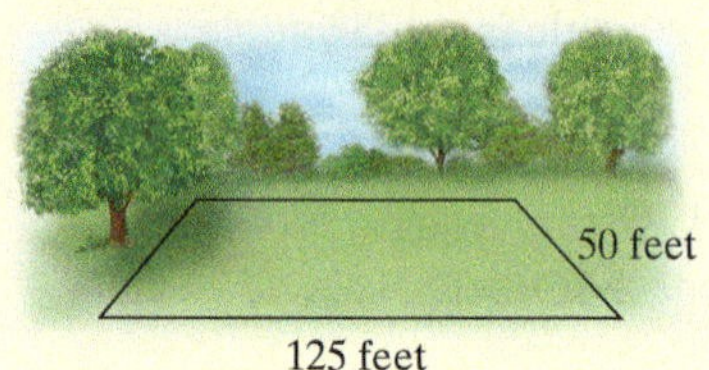

Example 1 Find the perimeter of the rectangle below.

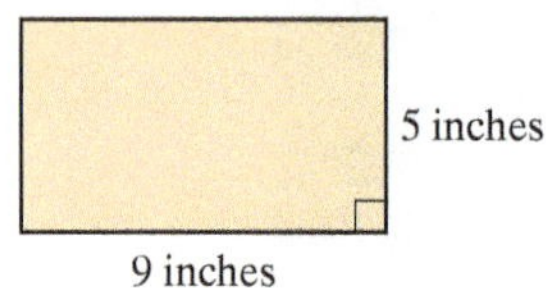

Solution:

$$\text{perimeter} = 9 \text{ inches} + 9 \text{ inches} + 5 \text{ inches} + 5 \text{ inches}$$
$$= 28 \text{ inches}$$

Work Practice 1

Notice that the perimeter of the rectangle in Example 1 can be written as $2 \cdot (9 \text{ inches}) + 2 \cdot (5 \text{ inches})$.

In general, we can say that the perimeter of a rectangle is always

$2 \cdot \text{length} + 2 \cdot \text{width}$

As we have just seen, the perimeters of some special figures such as rectangles form patterns. These patterns are given as **formulas.** The formula for the perimeter of a rectangle is shown next:

Perimeter of a Rectangle

Perimeter = 2 · length + 2 · width

In symbols, this can be written as

$P = 2l + 2w$

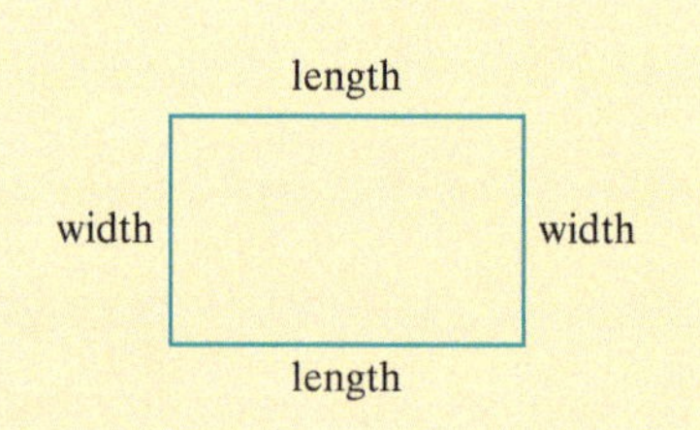

Practice 2

Find the perimeter of a rectangle with a length of 32 centimeters and a width of 15 centimeters.

Example 2 Find the perimeter of a rectangle with a length of 11 inches and a width of 3 inches.

11 in.

3 in.

Solution: We use the formula for perimeter and replace the letters by their known lengths.

$$P = 2l + 2w$$
$$= 2 \cdot 11 \text{ in.} + 2 \cdot 3 \text{ in.} \quad \text{Replace } l \text{ with 11 in. and } w \text{ with 3 in.}$$
$$= 22 \text{ in.} + 6 \text{ in.}$$
$$= 28 \text{ in.}$$

The perimeter is 28 inches.

Work Practice 2

Answers
1. a. 56 m **b.** 350 ft **2.** 94 cm

Recall that a square is a special rectangle with all four sides the same length. The formula for the perimeter of a square is shown next:

Perimeter of a Square

Perimeter = **side** + **side** + **side** + **side**

= 4 · **side**

In symbols,

$P = 4s$

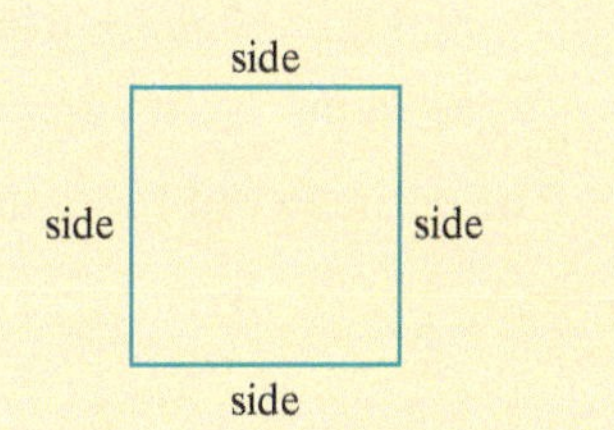

Example 3 Finding the Perimeter of a Field

How much fencing is needed to enclose a square field 50 yards on a side?

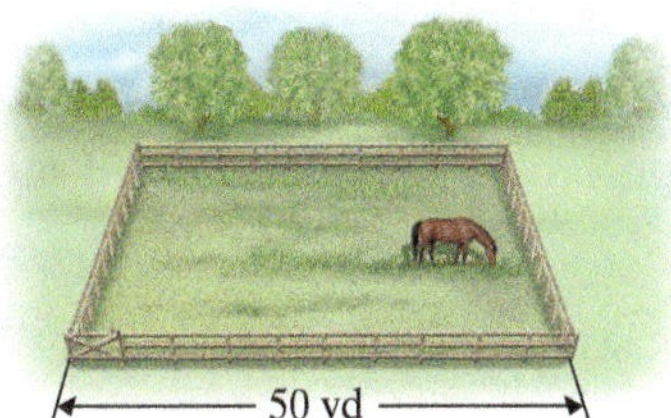

Solution: To find the amount of fencing needed, we find the distance around, or perimeter. The formula for the perimeter of a square is $P = 4s$. We use this formula and replace s by 50 yards.

$$P = 4s$$
$$= 4 \cdot 50 \text{ yd}$$
$$= 200 \text{ yd}$$

The amount of fencing needed is 200 yards.

Work Practice 3

Practice 3

Find the perimeter of a square tabletop if each side is 4 feet long.

The formula for the perimeter of a triangle with sides of lengths a, b, and c is given next:

Perimeter of a Triangle

Perimeter = side **a** + side **b** + side **c**

In symbols,

$P = a + b + c$

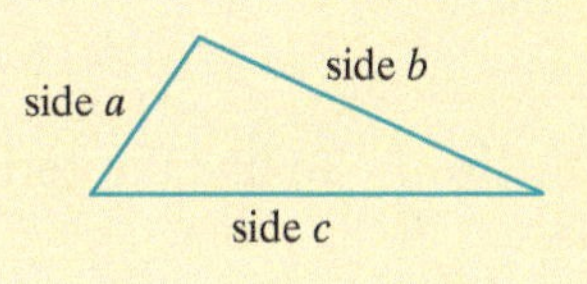

Example 4 Find the perimeter of a triangle if the sides are 3 inches, 7 inches, and 6 inches.

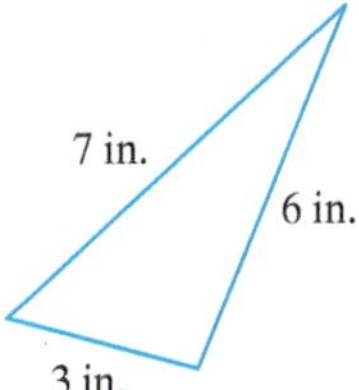

(Continued on next page)

Practice 4

Find the perimeter of a triangle if the sides are 6 centimeters, 10 centimeters, and 8 centimeters in length.

Answers

3. 16 ft **4.** 24 cm

Solution: The formula for the perimeter is $P = a + b + c$, where a, b, and c are the lengths of the sides. Thus,

$$P = a + b + c$$
$$= 3 \text{ in.} + 7 \text{ in.} + 6 \text{ in.}$$
$$= 16 \text{ in.}$$

The perimeter of the triangle is 16 inches.

Work Practice 4

The method for finding the perimeter of any polygon is given next:

Perimeter of a Polygon

The perimeter of a polygon is the sum of the lengths of its sides.

Practice 5

Find the perimeter of the trapezoid shown.

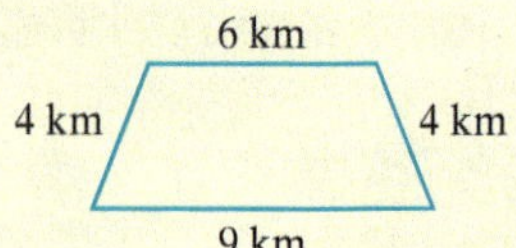

Example 5

Find the perimeter of the trapezoid shown below:

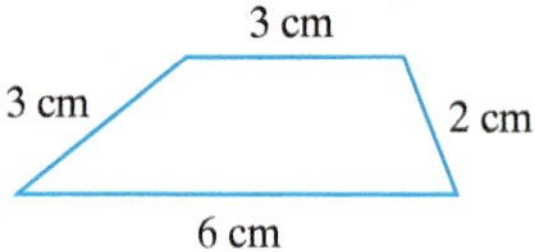

Solution: To find the perimeter, we find the sum of the lengths of its sides.

$$\text{perimeter} = 3 \text{ cm} + 2 \text{ cm} + 6 \text{ cm} + 3 \text{ cm} = 14 \text{ cm}$$

The perimeter is 14 centimeters.

Work Practice 5

Practice 6

Find the perimeter of the room shown.

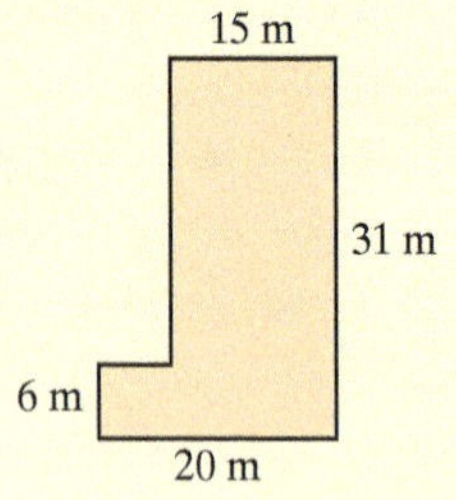

Example 6 Finding the Perimeter of a Room

Find the perimeter of the room shown below:

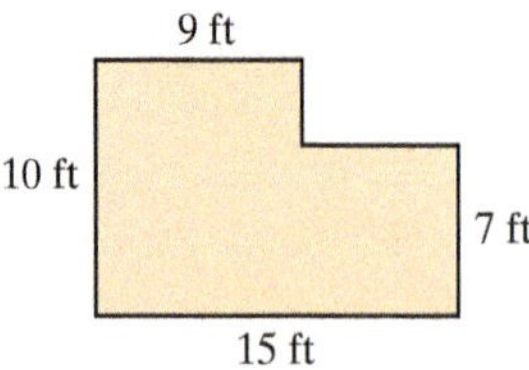

Solution: To find the perimeter of the room, we first need to find the lengths of all sides of the room.

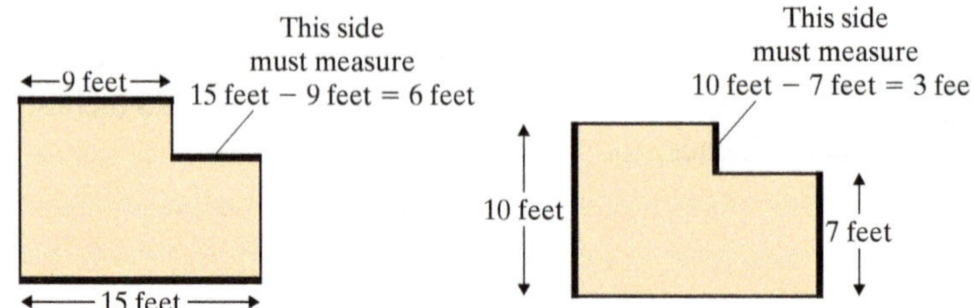

Answers
5. 23 km 6. 102 m

Now that we know the measures of all sides of the room, we can add the measures to find the perimeter.

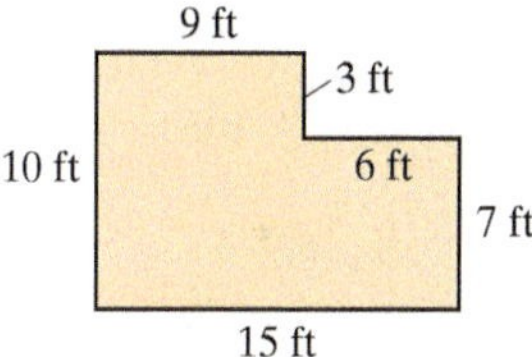

$$\begin{aligned} \text{perimeter} &= 10 \text{ ft} + 9 \text{ ft} + 3 \text{ ft} + 6 \text{ ft} + 7 \text{ ft} + 15 \text{ ft} \\ &= 50 \text{ ft} \end{aligned}$$

The perimeter of the room is 50 feet.

Work Practice 6

Example 7 Calculating the Cost of Wallpaper Border

A rectangular room measures 10 feet by 12 feet. Find the cost to hang a wallpaper border on the walls close to the ceiling if the cost of the wallpaper border is \$1.09 per foot.

Solution: First we find the perimeter of the room.

$$\begin{aligned} P &= 2l + 2w \\ &= 2 \cdot 12 \text{ ft} + 2 \cdot 10 \text{ ft} \quad \text{Replace } l \text{ with 12 feet and } w \text{ with 10 feet.} \\ &= 24 \text{ ft} + 20 \text{ ft} \\ &= 44 \text{ ft} \end{aligned}$$

The cost of the wallpaper is

$$\text{cost} = \$1.09 \cdot 44 \text{ ft} = 47.96$$

The cost of the wallpaper is \$47.96.

Work Practice 7

Practice 7

A rectangular lot measures 60 feet by 120 feet. Find the cost to install fencing around the lot if the cost of fencing is \$1.90 per foot.

Objective B Using Formulas to Find Circumferences 

Recall from Section 5.3 that the distance around a circle is called the **circumference.** This distance depends on the radius or the diameter of the circle.

The formulas for circumference are shown next:

Circumference of a Circle

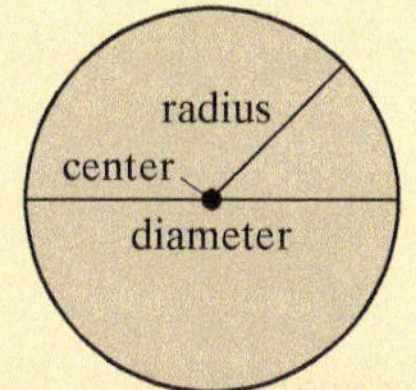

Circumference = $2 \cdot \pi \cdot$ **radius** or Circumference = $\pi \cdot$ **diameter**

In symbols,

$$C = 2\pi r \quad \text{or} \quad C = \pi d$$

where $\pi \approx 3.14$ or $\pi \approx \frac{22}{7}$.

Answer

7. \$684

To better understand circumference and π (pi), try the following experiment. Take any can and measure its circumference and its diameter.

The can in the figure above has a circumference of 23.5 centimeters and a diameter of 7.5 centimeters. Now divide the circumference by the diameter.

$$\frac{\text{circumference}}{\text{diameter}} = \frac{23.5 \text{ cm}}{7.5 \text{ cm}} \approx 3.13$$

Try this with other sizes of cylinders and circles—you should always get a number close to 3.1. The exact ratio of circumference to diameter is π. (Recall that $\pi \approx 3.14$ or $\approx \frac{22}{7}$.)

Practice 8

An irrigation device waters a circular region with a diameter of 20 yards. Find the exact circumference of the watered region, then use $\pi \approx 3.14$ to give an approximation.

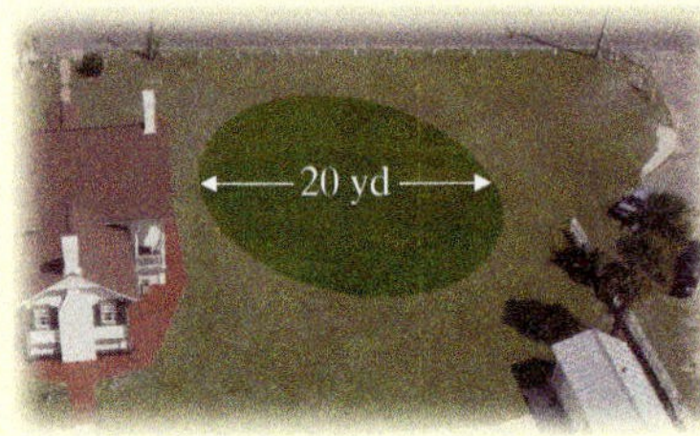

Example 8 Finding Circumference of a Circular Spa

A homeowner plans to install a border of new tiling around the circumference of her circular spa. If her spa has a diameter of 14 feet, find its exact circumference. Then use the approximation 3.14 for π to approximate the circumference.

Solution: Because we are given the diameter, we use the formula $C = \pi d$.

$$\begin{aligned} C &= \pi d \\ &= \pi \cdot 14 \text{ ft} \quad \text{Replace } d \text{ with 14 feet.} \\ &= 14\pi \text{ ft} \end{aligned}$$

The circumference of the spa is *exactly* 14π feet. By replacing π with the *approximation* 3.14, we find that the circumference is *approximately* $14 \text{ feet} \cdot 3.14 = 43.96$ feet.

Work Practice 8

✓Concept Check The distance around which figure is greater: a square with side length 5 inches or a circle with radius 3 inches?

Answer

8. exactly 20π yd ≈ 62.8 yd

✓Concept Check Answer

a square with side length 5 in.

Vocabulary, Readiness & Video Check

Use the choices below to fill in each blank.

circumference	radius	π	$\frac{22}{7}$
diameter	perimeter	3.14	

1. The __________ of a polygon is the sum of the lengths of its sides.

2. The distance around a circle is called the __________.

3. The exact ratio of circumference to diameter is __________.

4. The diameter of a circle is double its __________.

5. Both __________ and __________ are approximations for π.

6. The radius of a circle is half its __________.

Martin-Gay Interactive Videos

See Video 8.2

Watch the section lecture video and answer the following questions.

Objective A **7.** In Example 1, how can the perimeter be found if we forget the formula?

Objective B **8.** From the lecture before Example 6, circumference is a special name for what?

8.2 Exercise Set MyMathLab®

Objective A *Find the perimeter of each figure. (See Appendix A.1 for any unknown geometric figures.) See Examples 1 through 6.*

1.

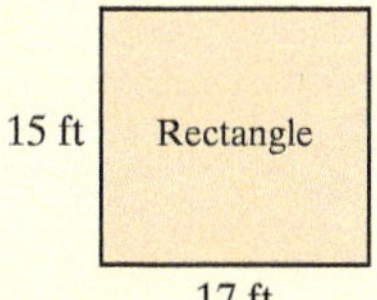

2.

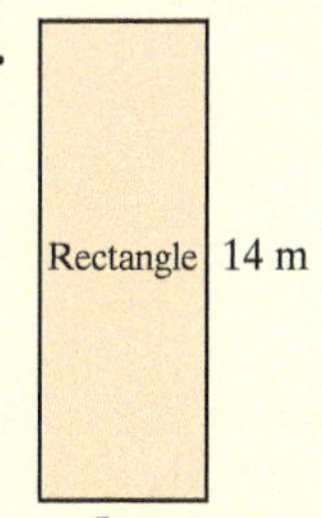

3.

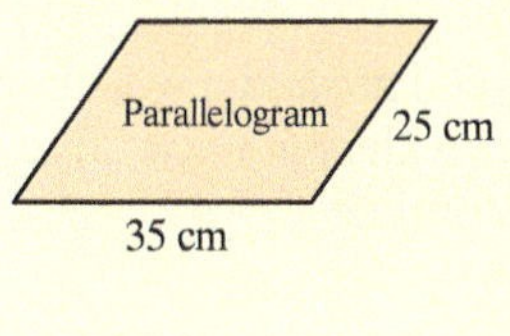

4.

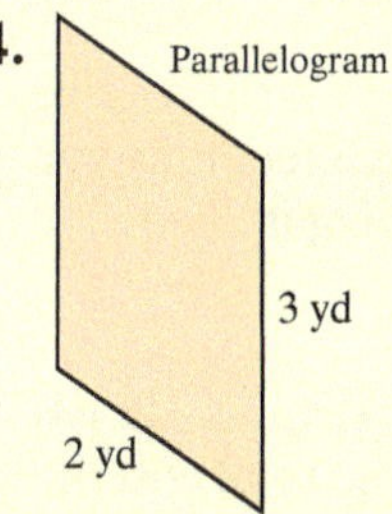

5.

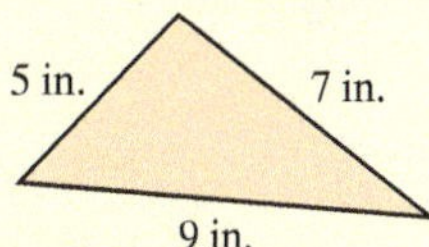

6.

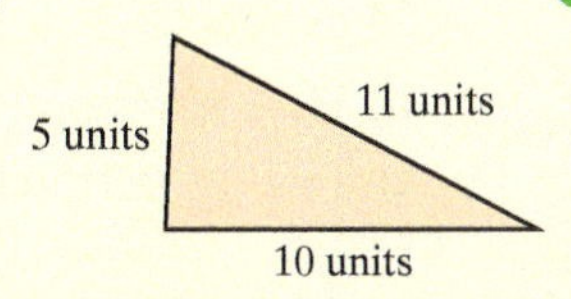

7.

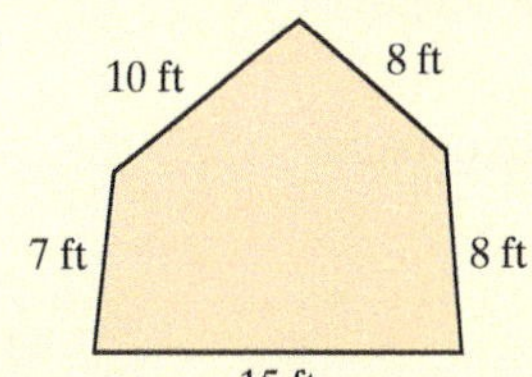

8.

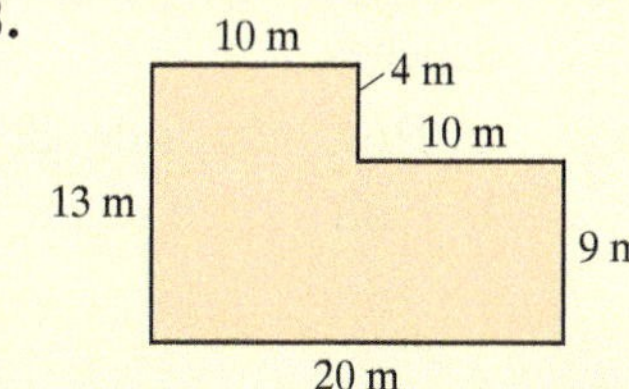

Find the perimeter of each regular polygon. (The sides of a regular polygon have the same length.)

9. 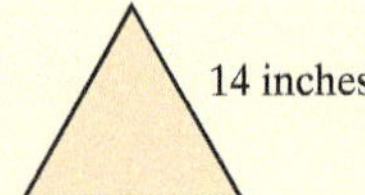

10. 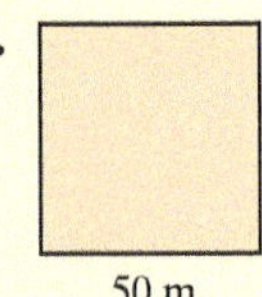

11. 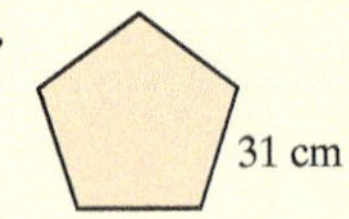

12. 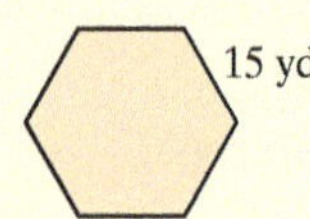

Solve. See Examples 1 through 7.

13. A polygon has sides of length 5 feet, 3 feet, 2 feet, 7 feet, and 4 feet. Find its perimeter.

14. A triangle has sides of length 8 inches, 12 inches, and 10 inches. Find its perimeter.

15. A line-marking machine lays down lime powder to mark both foul lines on a baseball field. If each foul line for this field measures 312 feet, how many feet of lime powder will be deposited?

16. A baseball diamond has 4 sides, with each side length 90 feet. If a baseball player hits a home run, how far does the player run (home plate, around the bases, then back to home plate)?

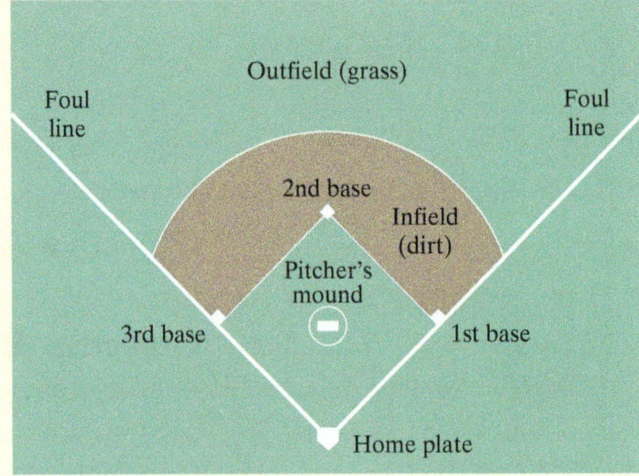

17. If a football field is 53 yards wide and 120 yards long, what is the perimeter?

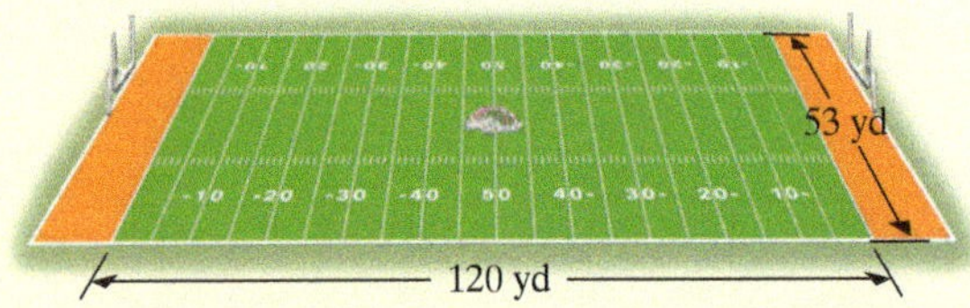

18. A stop sign has eight equal sides of length 12 inches. Find its perimeter.

19. A metal strip is being installed around a workbench that is 8 feet long and 3 feet wide. Find how much stripping is needed for this project.

20. Find how much fencing is needed to enclose a rectangular garden 70 feet by 21 feet.

21. If the stripping in Exercise **19** costs $2.50 per foot, find the total cost of the stripping.

22. If the fencing in Exercise **20** costs $2 per foot, find the total cost of the fencing.

23. A regular octagon has a side length of 9 inches.

a. How many sides does an octagon have?

b. Find its perimeter.

24. A regular pentagon has a side length of 14 meters.

a. How many sides does a pentagon have?

b. Find its perimeter.

25. Find the perimeter of the top of a square compact disc case if the length of one side is 7 inches.

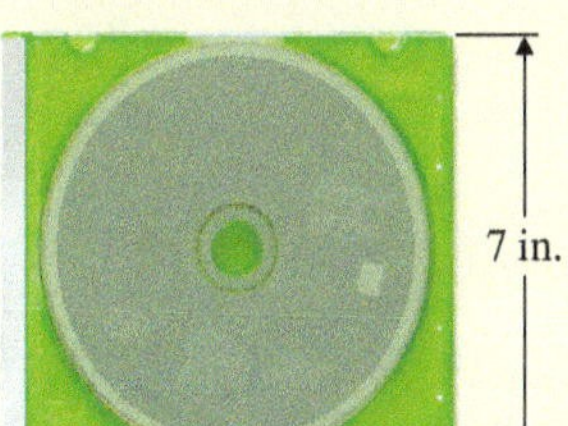

26. Find the perimeter of a square ceramic tile with a side of length 3 inches.

27. A rectangular room measures 10 feet by 11 feet. Find the cost of installing a strip of wallpaper around the room if the wallpaper costs \$0.86 per foot.

28. A rectangular house measures 85 feet by 70 feet. Find the cost of installing gutters around the house if the cost is \$2.36 per foot.

Find the perimeter of each figure. See Example 6.

29.

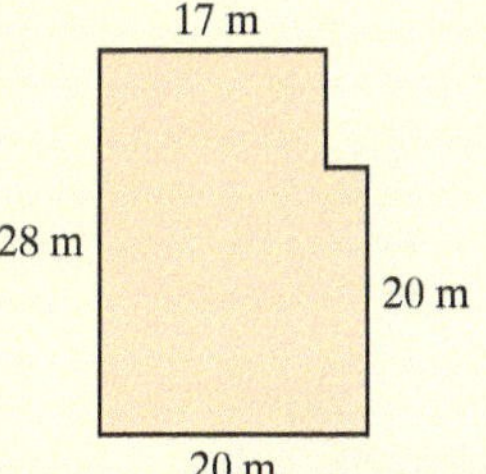

30.

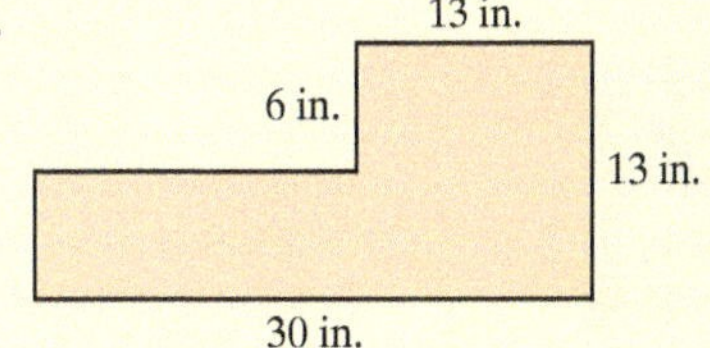

31.

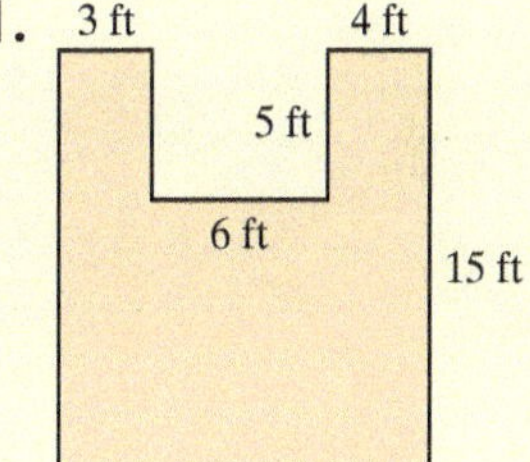

32.

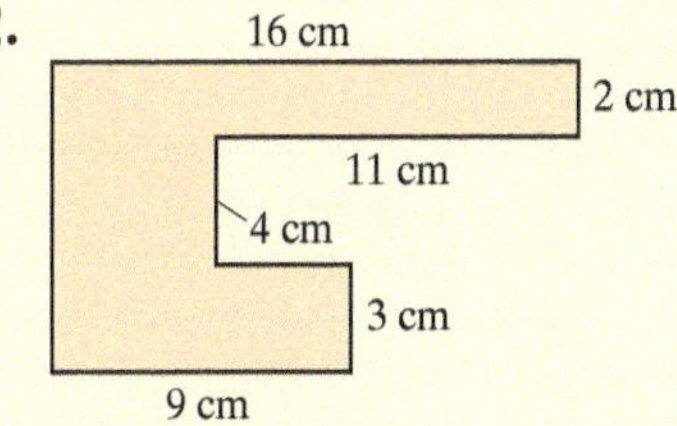

33.

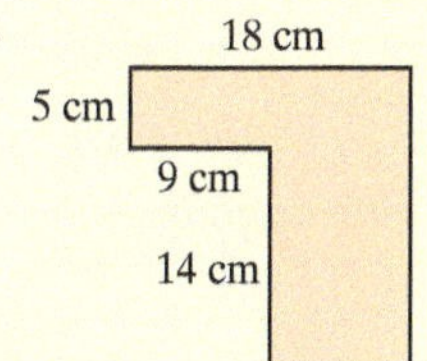

34.

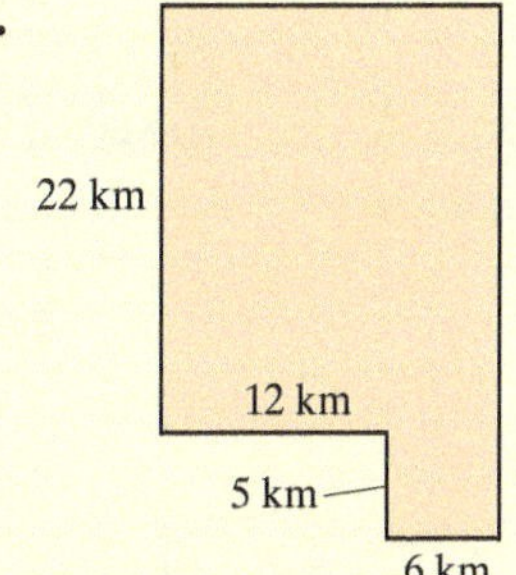

Objective B *Find the circumference of each circle. Give the exact circumference and then an approximation. Use* $\pi \approx 3.14$. *See Example 8.*

35.

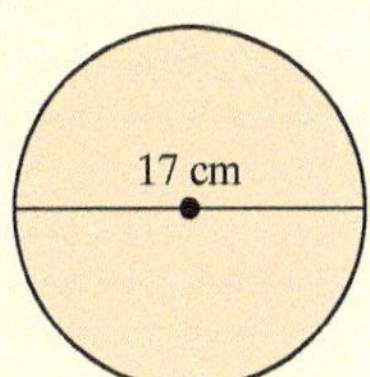

36.

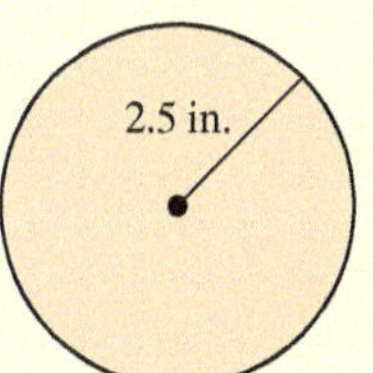

37.

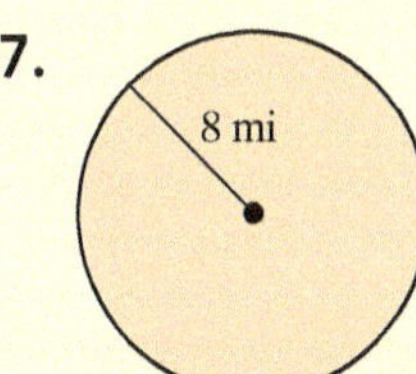

38.

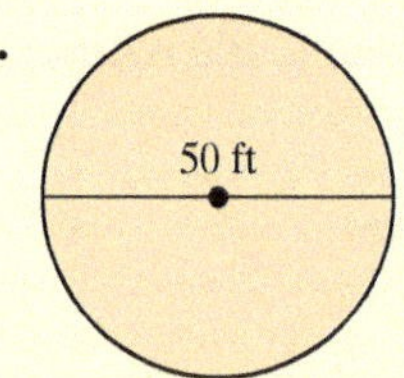

39.

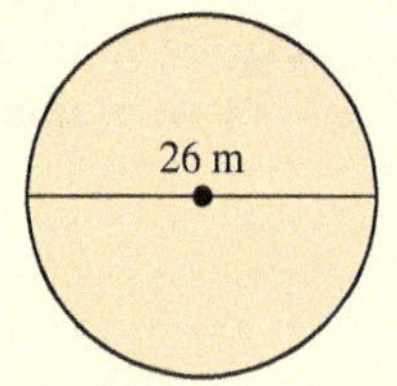

40.

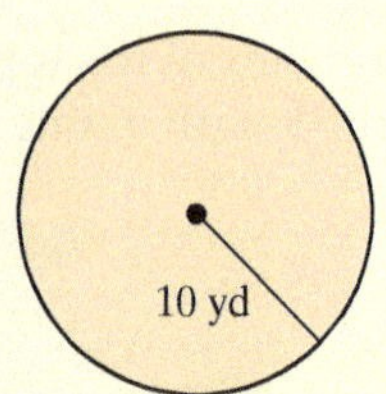

41. Wyley Robinson just bought a trampoline for his children to use. The trampoline has a diameter of 15 feet. If Wyley wishes to buy netting to go around the outside of the trampoline, how many feet of netting does he need?

42. The largest round barn in the world is located at the Marshfield Fairgrounds in Wisconsin. The barn has a diameter of 150 ft. What is the circumference of the barn? (*Source: The Milwaukee Journal Sentinel*)

43. Meteor Crater, near Winslow, Arizona, is 4000 feet in diameter. Approximate the distance around the crater. Use 3.14 for π. (*Source: The Handy Science Answer Book*)

44. The *Pearl of Lao-tze* has a diameter of $5\frac{1}{2}$ inches. Approximate the distance around the pearl. Use $\frac{22}{7}$ for π. (*Source: The Guinness World Records*)

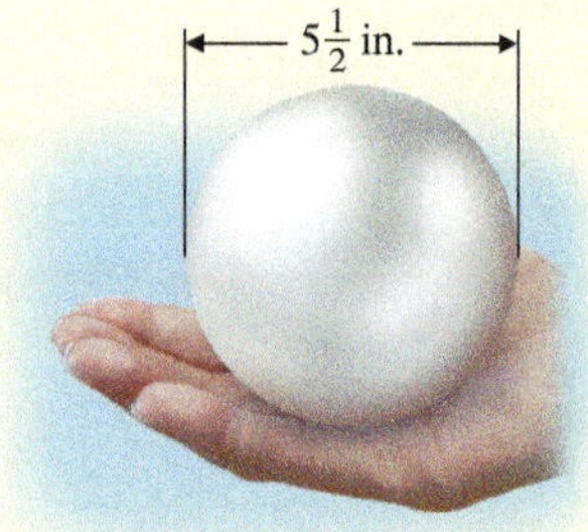

Objectives A B Mixed Practice *Find the distance around each figure. For circles, give the exact circumference and then an approximation. Use* $\pi \approx 3.14$. *See Examples 1 through 8.*

45.

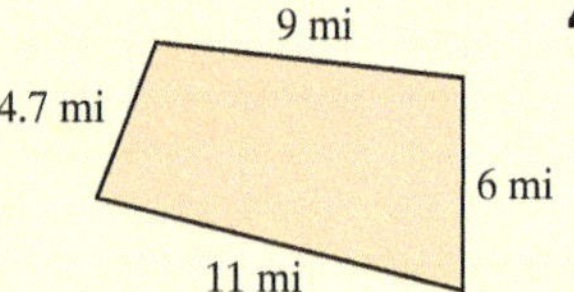

46.

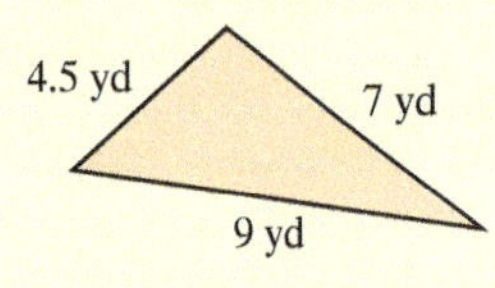

47.

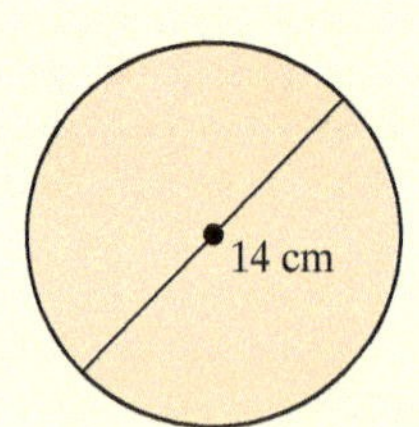

48.

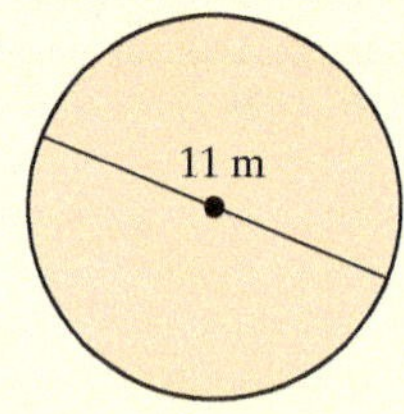

49.

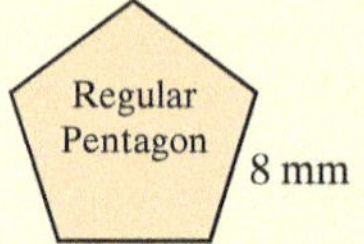

50.

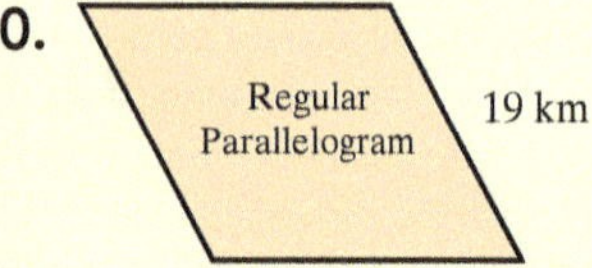

51.

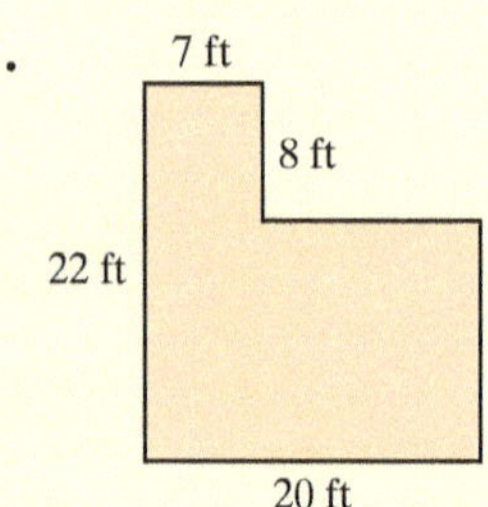

52.

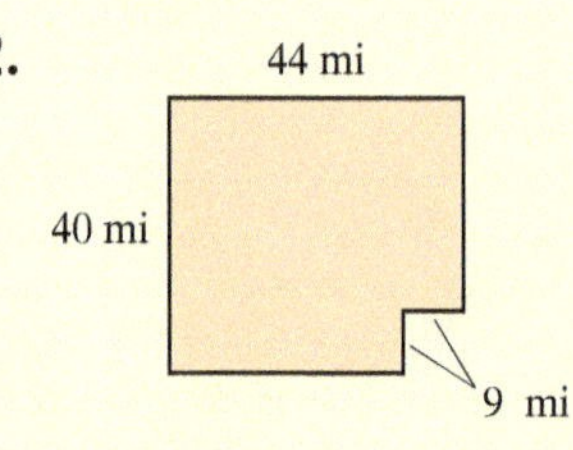

Review

Simplify. See Section 1.7.

53. $5 + 6 \cdot 3$

54. $25 - 3 \cdot 7$

55. $(20 - 16) \div 4$

56. $6 \cdot (8 + 2)$

57. $72 \div (2 \cdot 6)$

58. $(72 \div 2) \cdot 6$

59. $(18 + 8) - (12 + 4)$

60. $4^1 \cdot (2^3 - 8)$

Concept Extensions

There are a number of factors that determine the dimensions of a rectangular soccer field. Use the table below to answer Exercises 61 and 62.

Soccer Field Width and Length

Age	Width Min–Max	Length Min–Max
Under 6/7:	15–20 yards	25–30 yards
Under 8:	20–25 yards	30–40 yards
Under 9:	30–35 yards	40–50 yards
Under 10:	40–50 yards	60–70 yards
Under 11:	40–50 yards	70–80 yards
Under 12:	40–55 yards	100–105 yards
Under 13:	50–60 yards	100–110 yards
International:	70–80 yards	110–120 yards

61. a. Find the minimum length and width of a soccer field for 8-year-old children. (Carefully consider the age.)

b. Find the perimeter of this field.

62. a. Find the maximum length and width of a soccer field for 12-year-old children.

b. Find the perimeter of this field.

Solve. See the Concept Check in this section. Choose the figure that has the greater distance around.

63. a. A square with side length 3 inches

b. A circle with diameter 4 inches

64. a. A circle with diameter 7 inches

b. A square with side length 7 inches

65. a. Find the circumference of each circle. Approximate the circumference by using 3.14 for π.

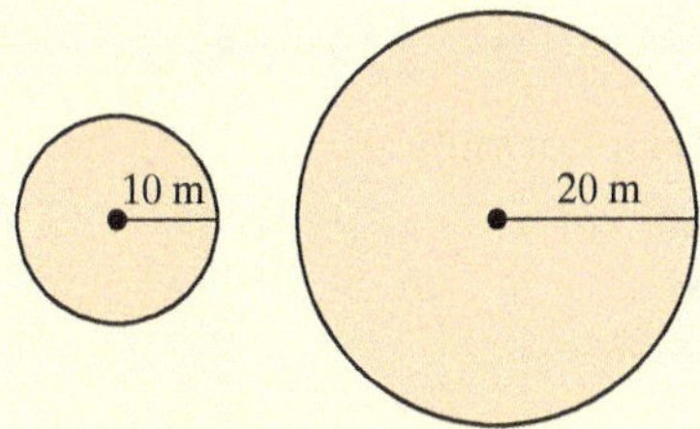

b. If the radius of a circle is doubled, is its corresponding circumference doubled?

66. a. Find the circumference of each circle. Approximate the circumference by using 3.14 for π.

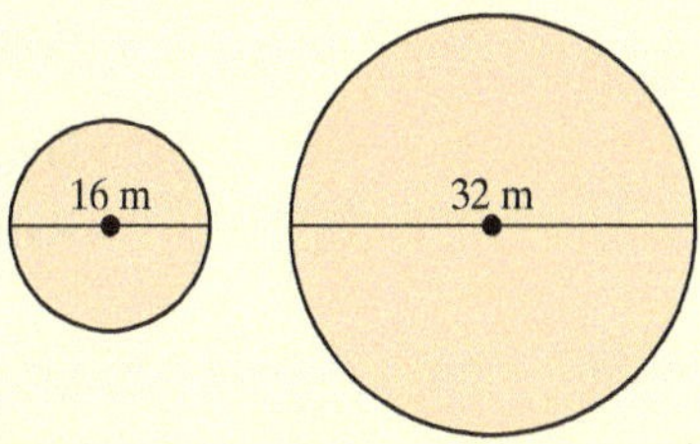

b. If the diameter of a circle is doubled, is its corresponding circumference doubled?

67. In your own words, explain how to find the perimeter of any polygon.

68. In your own words, explain how perimeter and circumference are the same and how they are different.

Find the perimeter. Round your results to the nearest tenth.

69.
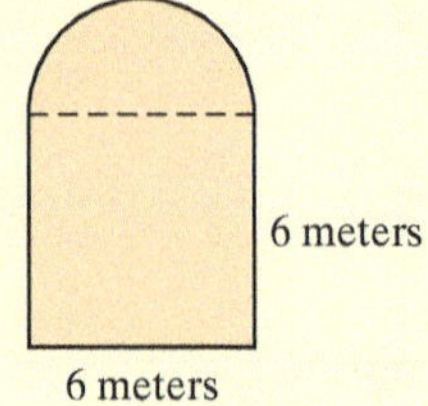

70.
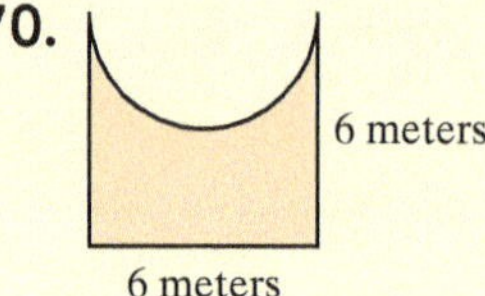

71.
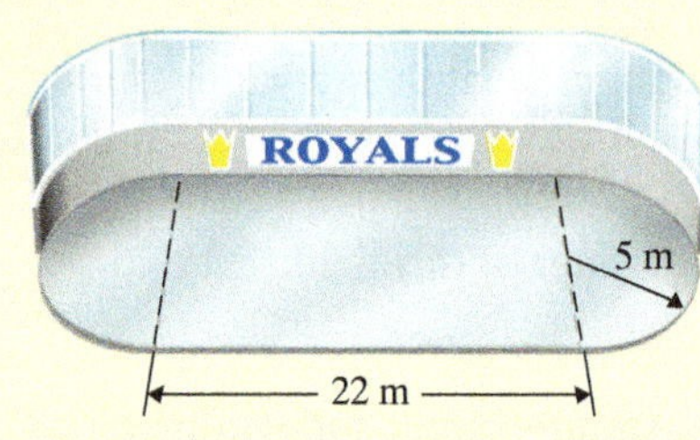

72.
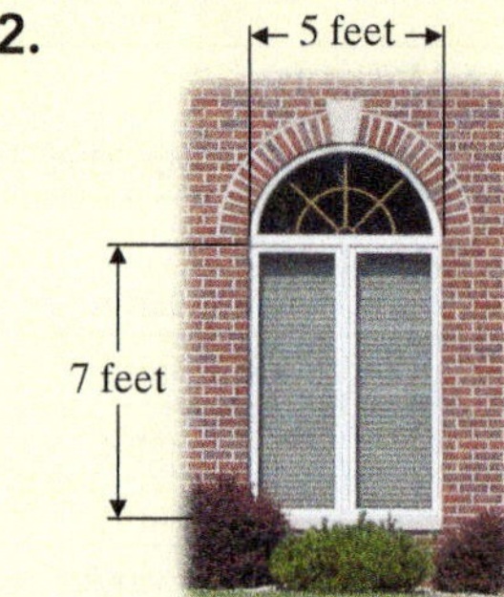

8.3 Area, Volume, and Surface Area 

Objectives

A Find the Area of Plane Regions.

B Find the Volume and Surface Area of Solids.

Objective A Finding Area of Plane Regions 

Recall that area measures the number of square units that cover the surface of a plane region; that is, a region that lies in a plane. Thus far, we know how to find the areas of a rectangle and a square. These formulas, as well as formulas for finding the areas of other common geometric figures, are given next.

Area Formulas of Common Geometric Figures

Geometric Figure	Area Formula
RECTANGLE 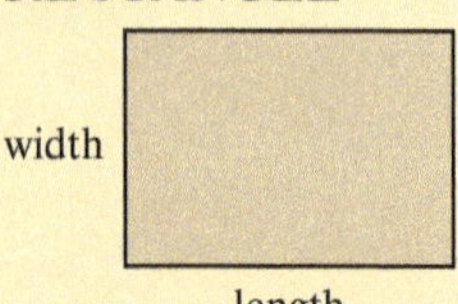	Area of a rectangle: **Area** = **length** · **width**, $A = lw$
SQUARE 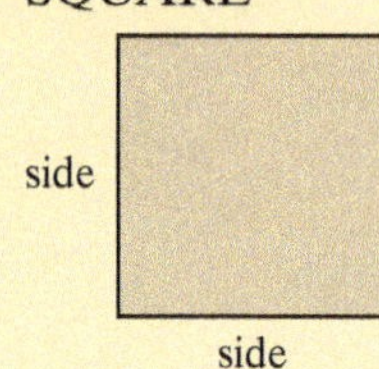	Area of a square: **Area** = **side** · **side**, $A = s \cdot s = s^2$

Geometric Figure	Area Formula
TRIANGLE 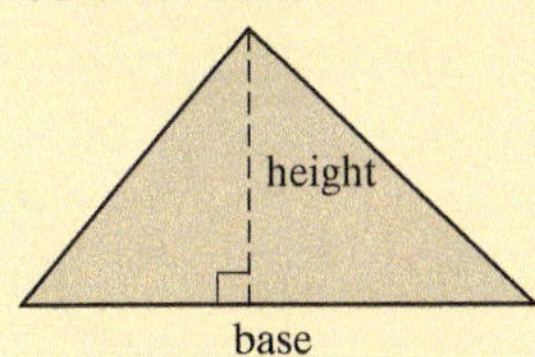	Area of a triangle: **A**rea $= \frac{1}{2} \cdot$ **b**ase $\cdot$ **h**eight; $A = \frac{1}{2}bh$
PARALLELOGRAM 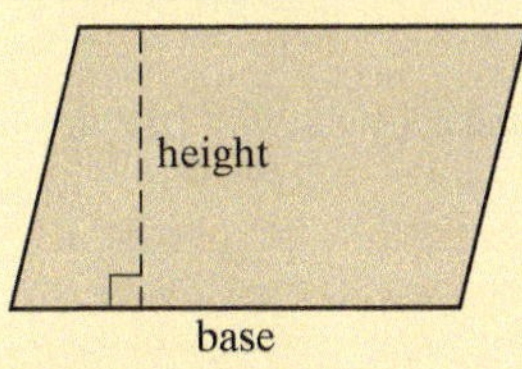	Area of a parallelogram: **A**rea = **b**ase $\cdot$ **h**eight; $A = bh$
TRAPEZOID 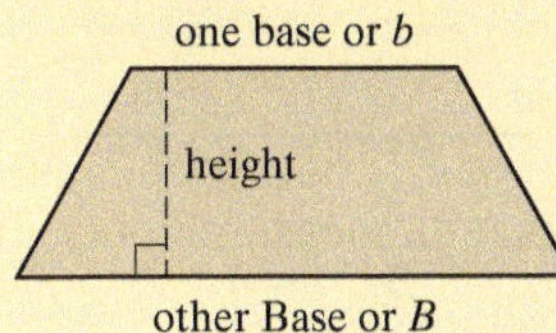	Area of a trapezoid: **A**rea $= \frac{1}{2} \cdot$ (one **b**ase + other **B**ase) $\cdot$ **h**eight; $A = \frac{1}{2}(b + B)h$

Use these formulas for the following examples.

Helpful Hint

Area is always measured in square units.

Example 1 Find the area of the triangle.

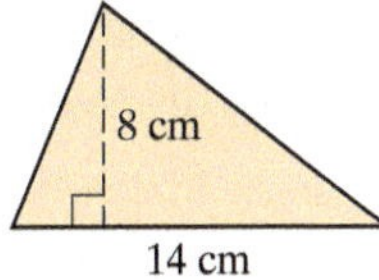

Solution:

$$A = \frac{1}{2}bh$$

$$= \frac{1}{2} \cdot 14 \text{ cm} \cdot 8 \text{ cm}$$ Replace b, base, with 14 cm and h, height, with 8 cm.

$$= \frac{\overset{1}{\cancel{2}} \cdot 7 \cdot 8}{\underset{1}{\cancel{2}}} \text{ square centimeters}$$ Write 14 as $2 \cdot 7$.

$$= 56 \text{ square centimeters}$$

The area is 56 square centimeters.

Work Practice 1

Practice 1

Find the area of the triangle.

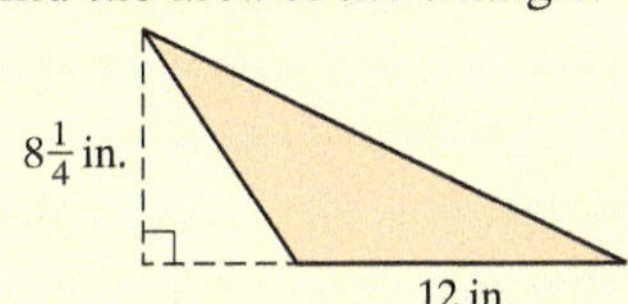

Answer

1. $49\frac{1}{2}$ sq in.

Practice 2

Find the area of the trapezoid.

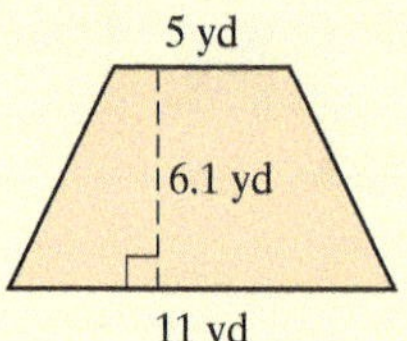

Example 2

Find the area of the parallelogram.

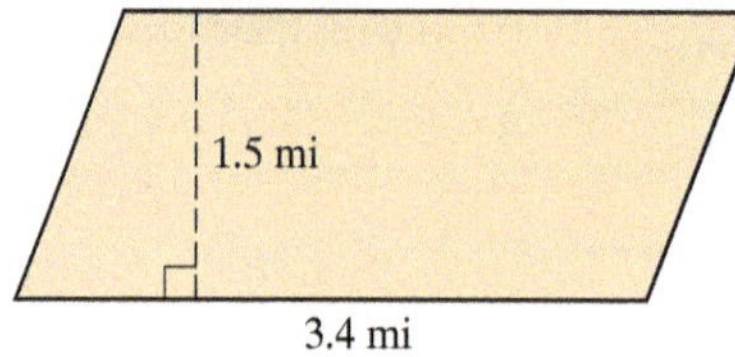

Solution:

$A = bh$

$= 3.4 \text{ miles} \cdot 1.5 \text{ miles}$ Replace b, base, with 3.4 miles and h, height, with 1.5 miles.

$= 5.1 \text{ square miles}$

The area is 5.1 square miles.

Work Practice 2

Helpful Hint

When finding the areas of figures, check to make sure that all measurements are in the same units before calculations are made.

Practice 3

Find the area of the figure.

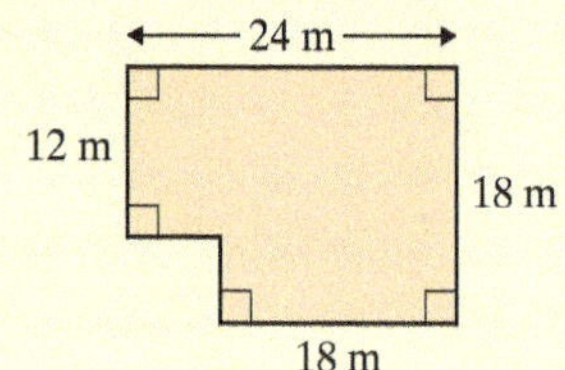

Example 3

Find the area of the figure.

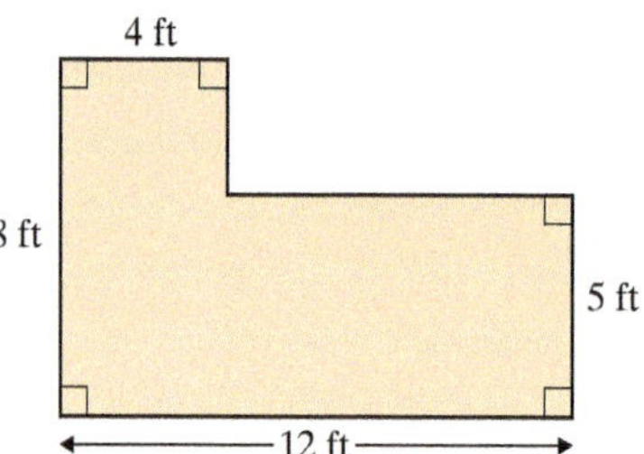

Solution: Split the figure into two rectangles. To find the area of the figure, we find the sum of the areas of the two rectangles.

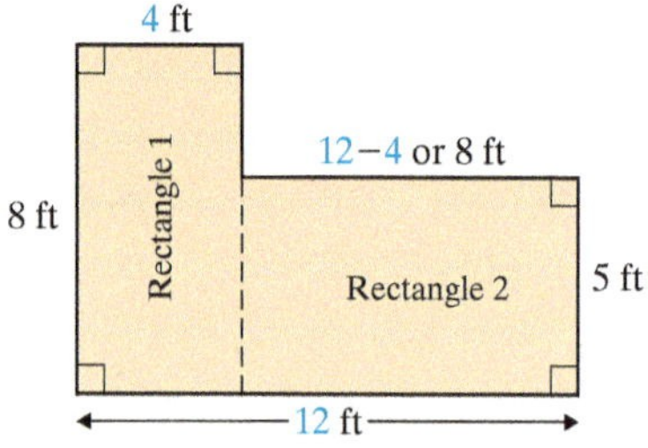

area of Rectangle 1 $= lw$

$= 8 \text{ feet} \cdot 4 \text{ feet}$

$= 32 \text{ square feet}$

Notice that the length of Rectangle 2 is 12 feet − 4 feet or 8 feet.

area of Rectangle 2 $= lw$

$= 8 \text{ feet} \cdot 5 \text{ feet}$

$= 40 \text{ square feet}$

area of the figure = area of Rectangle 1 + area of Rectangle 2

$= 32 \text{ square feet} + 40 \text{ square feet}$

$= 72 \text{ square feet}$

Work Practice 3

Answers

2. 48.8 sq yd **3.** 396 sq m

Helpful Hint

The figure in Example 3 could also be split into two rectangles as shown.

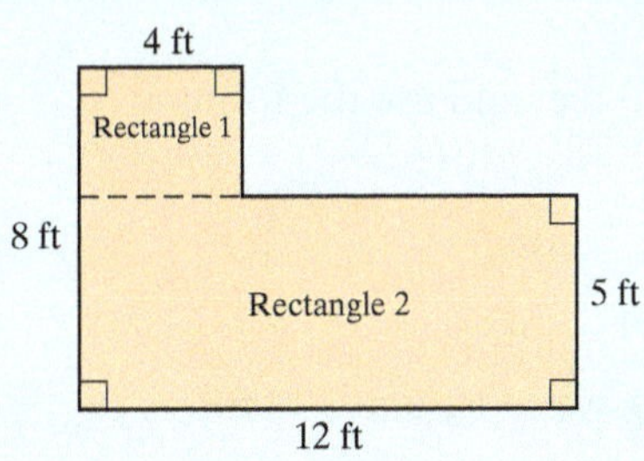

To better understand the formula for area of a circle, try the following. Cut a circle into many pieces, as shown.

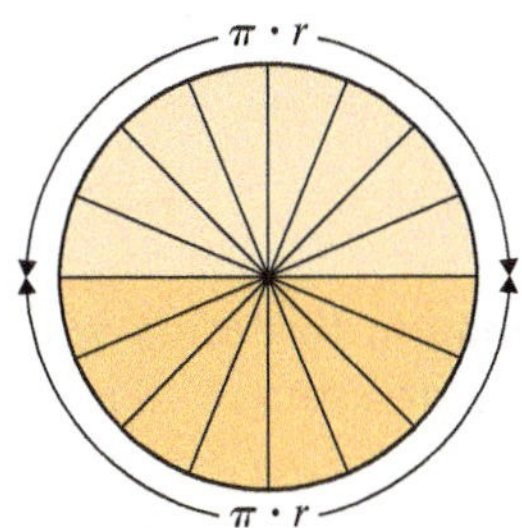

The circumference of a circle is $2\pi r$. This means that the circumference of half a circle is half of $2\pi r$, or πr.

Then unfold the two halves of the circle and place them together, as shown.

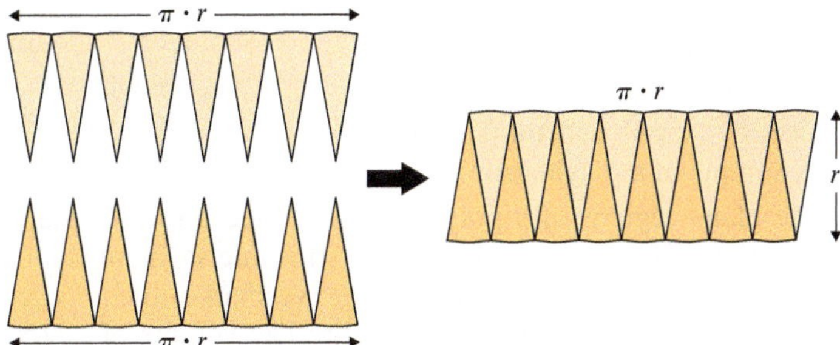

The figure on the right is almost a parallelogram with a base of πr and a height of r. The area is

$$A = \text{base} \cdot \text{height}$$
$$= (\pi r) \cdot r$$
$$= \pi r^2$$

This is the formula for the area of a circle.

Area Formula of a Circle

Circle

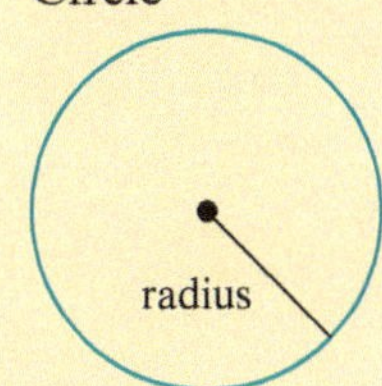

Area of a circle

Area $= \pi \cdot (\text{radius})^2$

$A = \pi r^2$

(A fraction approximation for π is $\frac{22}{7}$.)

(A decimal approximation for π is 3.14.)

Practice 4

Find the area of the given circle. Find the exact area and an approximation. Use 3.14 as an approximation for π.

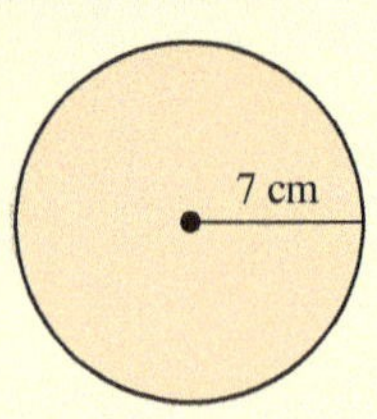

Example 4 Find the area of a circle with a radius of 3 feet. Find the exact area and an approximation. Use 3.14 as an approximation for π.

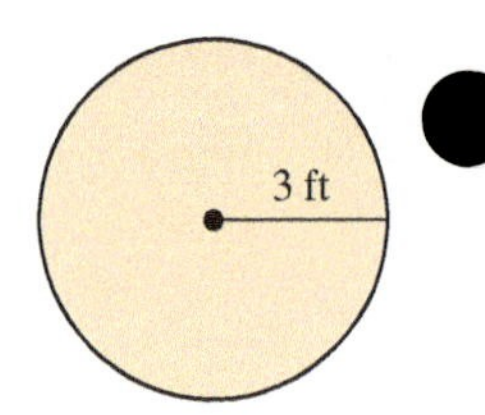

Solution: We let $r = 3$ feet and use the formula

$$
\begin{aligned}
A &= \pi r^2 \\
&= \pi \cdot (3\text{ feet})^2 && \text{Replace } r \text{ with 3 feet.} \\
&= 9 \cdot \pi \text{ square feet} && \text{Replace } (3\text{ feet})^2 \text{ with 9 sq ft.}
\end{aligned}
$$

To approximate this area, we substitute 3.14 for π.

$$
\begin{aligned}
9 \cdot \pi \text{ square feet} &\approx 9 \cdot 3.14 \text{ square feet} \\
&= 28.26 \text{ square feet}
\end{aligned}
$$

The *exact* area of the circle is 9π square feet, which is *approximately* 28.26 square feet.

Work Practice 4

✓**Concept Check** Use estimation to decide which figure would have a larger area: a circle of diameter 10 in. or a square 10 in. long on each side.

Objective B Finding Volume and Surface Area of Solids 

A **convex solid** is a set of points, S, not all in one plane, such that for any two points A and B in S, all points between A and B are also in S. In this section, we will find the volume and surface area of special types of solids called polyhedrons. A solid formed by the intersection of a finite number of planes is called a **polyhedron.** The box to the right is an example of a polyhedron.

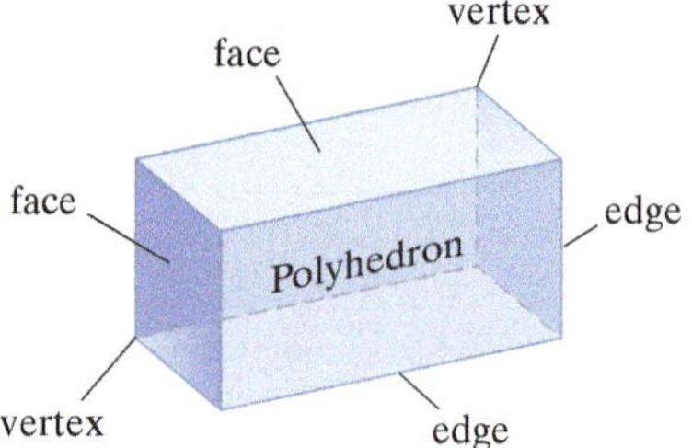

Each of the plane regions of a polyhedron is called a **face** of the polyhedron. If the intersection of two faces is a line segment, this line segment is an **edge** of the polyhedron. The intersections of the edges are the **vertices** of the polyhedron.

Volume is a measure of the space of a region. The volume of a box or can, for example, is the amount of space inside. Volume can be used to describe the amount of juice in a pitcher or the amount of concrete needed to pour a foundation for a house.

The volume of a solid is the number of **cubic units** in the solid. A cubic centimeter and a cubic inch are illustrated.

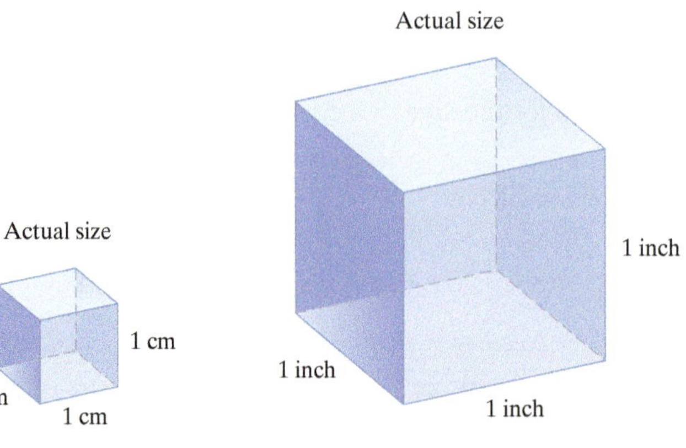

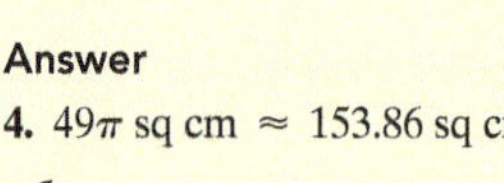

Answer

4. 49π sq cm $\approx$ 153.86 sq cm

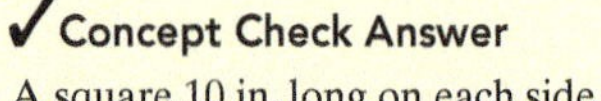

✓**Concept Check Answer**

A square 10 in. long on each side would have a larger area.

The **surface area** of a polyhedron is the sum of the areas of the faces of the polyhedron. For example, each face of the cube to the left on the previous page has an area of 1 square centimeter. Since there are 6 faces of the cube, the sum of the areas of the faces is 6 square centimeters. Surface area can be used to describe the amount of material needed to cover a solid. Surface area is measured in square units.

Formulas for finding the volumes, V, and surface areas, SA, of some common solids are given next. (Note: Spheres, circular cylinders, and cones are not polyhedrons, but they are solids and we will calculate surface areas and volumes of these solids.)

Volume and Surface Area Formulas of Common Solids

Solid	Formulas
RECTANGULAR SOLID height, width, length	$V = lwh$ $SA = 2lh + 2wh + 2lw$ where h = height, w = width, l = length
CUBE side, side, side	$V = s^3$ $SA = 6s^2$ where s = side
SPHERE radius	$V = \frac{4}{3}\pi r^3$ $SA = 4\pi r^2$ where r = radius
CIRCULAR CYLINDER height, radius	$V = \pi r^2 h$ $SA = 2\pi rh + 2\pi r^2$ where h = height, r = radius
CONE height, radius	$V = \frac{1}{3}\pi r^2 h$ $SA = \pi r\sqrt{r^2 + h^2} + \pi r^2$ where h = height, r = radius
SQUARE-BASED PYRAMID slant height, height, side	$V = \frac{1}{3}s^2 h$ $SA = B + \frac{1}{2}pl$ where B = area of base, p = perimeter of base, h = height, s = side, l = slant height

Practice 5

Find the volume and surface area of a rectangular box that is 7 feet long, 3 feet wide, and 4 feet high.

Example 5

Find the volume and surface area of a rectangular box that is 12 inches long, 6 inches wide, and 3 inches high.

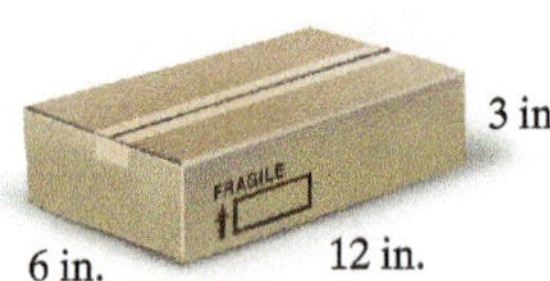

Solution: Let $h = 3$ in., $l = 12$ in., and $w = 6$ in.

$V = lwh$

$V = 12 \text{ inches} \cdot 6 \text{ inches} \cdot 3 \text{ inches} = 216 \text{ cubic inches}$

The volume of the rectangular box is 216 cubic inches.

$$\begin{aligned} SA &= 2lh + 2wh + 2lw \\ &= 2(12 \text{ in.})(3 \text{ in.}) + 2(6 \text{ in.})(3 \text{ in.}) + 2(12 \text{ in.})(6 \text{ in.}) \\ &= 72 \text{ sq in.} + 36 \text{ sq in.} + 144 \text{ sq in.} \\ &= 252 \text{ sq in.} \end{aligned}$$

The surface area of the rectangular box is 252 square inches.

Work Practice 5

✓Concept Check Juan is calculating the volume of the following rectangular solid. Find the error in his calculation.

$$\begin{aligned} \text{Volume} &= l + w + h \\ &= 14 \text{ cm} + 8 \text{ cm} + 5 \text{ cm} \\ &= 27 \text{ cu cm} \end{aligned}$$

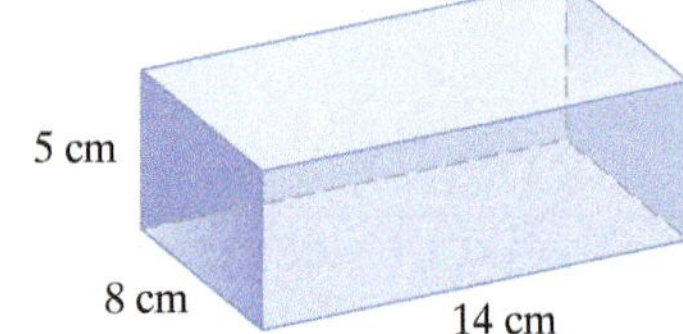

Practice 6

Find the volume and surface area of a ball of radius $\frac{1}{2}$ centimeter. Give the exact volume and surface area. Then use $\frac{22}{7}$ for π and approximate the values.

Example 6

Find the volume and surface area of a ball of radius 2 inches. Give the exact volume and surface area. Then use the approximation $\frac{22}{7}$ for π.

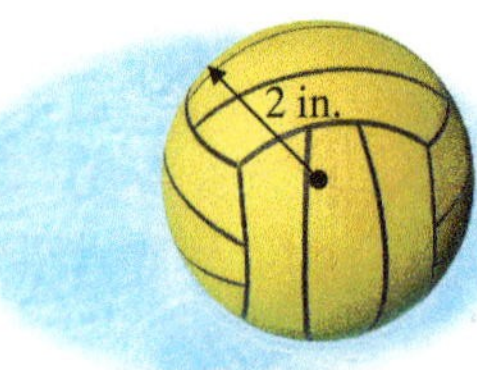

Solution:

$V = \frac{4}{3}\pi r^3$ — Formula for volume of a sphere

$V = \frac{4}{3} \cdot \pi(2 \text{ in.})^3$ — Let $r = 2$ inches.

$= \frac{32}{3}\pi$ cu in. — Exact volume

$\approx \frac{32}{3} \cdot \frac{22}{7}$ cu in. — Approximate π with $\frac{22}{7}$.

$= \frac{704}{21}$ or $33\frac{11}{21}$ cu in. — Approximate volume

Answers

5. $V = 84$ cu ft; $SA = 122$ sq ft

6. $V = \frac{1}{6}\pi$ cu cm $\approx \frac{11}{21}$ cu cm; $SA = \pi$ sq cm $\approx 3\frac{1}{7}$ sq cm

✓Concept Check Answer

$$\begin{aligned} \text{Volume} &= l \cdot w \cdot h \\ &= 14 \text{ cm} \cdot 8 \text{ cm} \cdot 5 \text{ cm} \\ &= 560 \text{ cu cm} \end{aligned}$$

The volume of the sphere is exactly $\frac{32}{3}\pi$ cubic inches or approximately $33\frac{11}{21}$ cubic inches.

$SA = 4\pi r^2$	Formula for surface area
$SA = 4 \cdot \pi(2\text{ in.})^2$	Let $r = 2$ inches.
$= 16\pi$ sq in.	Exact surface area
$\approx 16 \cdot \frac{22}{7}$ sq in.	Approximate π with $\frac{22}{7}$.
$= \frac{352}{7}$ or $50\frac{2}{7}$ sq in.	Approximate surface area

The surface area of the sphere is exactly 16π square inches or approximately $50\frac{2}{7}$ square inches.

Work Practice 6

Example 7 Find the volume of a can that has a $3\frac{1}{2}$-inch radius and a height of 6 inches. Give an exact volume and an approximate volume. Use $\frac{22}{7}$ for π.

Solution: Using the formula for a circular cylinder, we have

$$V = \pi \cdot r^2 \cdot h \qquad 3\frac{1}{2} = \frac{7}{2}$$

$$= \pi \cdot \left(\frac{7}{2}\text{ in.}\right)^2 \cdot 6\text{ in.}$$

$$= \pi \cdot \frac{49}{4}\text{ sq in.} \cdot 6\text{ in.}$$

$$= \frac{\pi \cdot 49 \cdot \overset{1}{\cancel{2}} \cdot 3}{\underset{1}{\cancel{2}} \cdot 2}\text{ cu in.}$$

$$= 73\frac{1}{2}\pi\text{ cu in. or } 73.5\pi\text{ cu in.}$$

This is the exact volume. To approximate the volume, use the approximation $\frac{22}{7}$ for π.

$$V = 73\frac{1}{2}\pi \text{ or } \frac{147}{2} \cdot \frac{22}{7}\text{ cu in.} \qquad \text{Replace } \pi \text{ with } \frac{22}{7}.$$

$$= \frac{21 \cdot \overset{1}{\cancel{7}} \cdot \overset{1}{\cancel{2}} \cdot 11}{\underset{1}{\cancel{2}} \cdot \underset{1}{\cancel{7}}}\text{ cu in.}$$

$$= 231\text{ cu in.}$$

The volume is approximately 231 cubic inches.

Work Practice 7

Practice 7

Find the volume of a cylinder of radius 5 inches and height 9 inches. Give an exact answer and an approximate answer. Use 3.14 for π.

Answer

7. 225π cu in. ≈ 706.5 cu in.

Practice 8

Find the volume of a square-based pyramid that has a 3-meter side and a height of 5.1 meters.

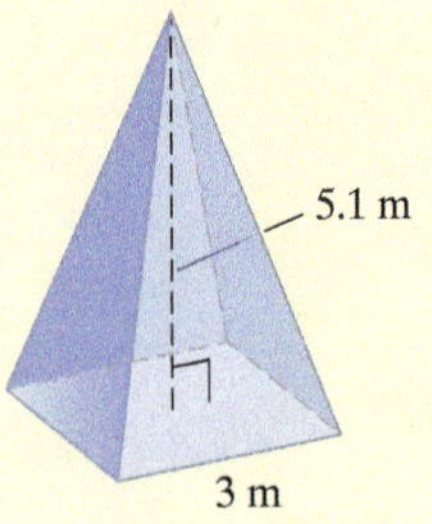

Answer

8. 15.3 cu m

Example 8 Find the volume of a cone that has a height of 14 centimeters and a radius of 3 centimeters. Give an exact answer and an approximate answer. Use 3.14 for π.

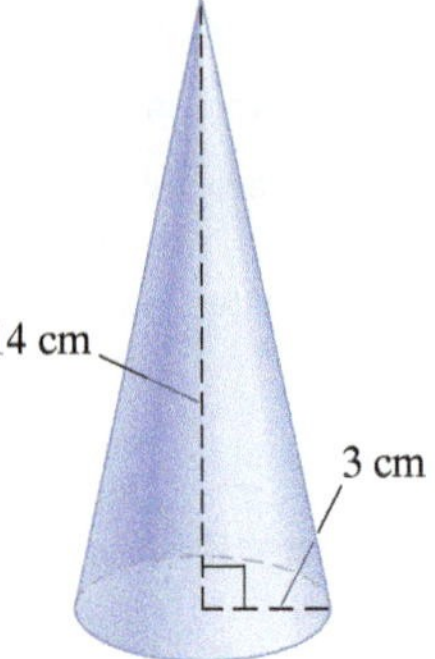

Solution: Using the formula for volume of a cone, we have

$$V = \frac{1}{3} \cdot \pi \cdot r^2 \cdot h$$

$$= \frac{1}{3} \cdot \pi \cdot (3 \text{ cm})^2 \cdot 14 \text{ cm}$$ Replace r with 3 cm and h with 14 cm.

$$= 42\pi \text{ cu cm}$$

Thus, 42π cubic centimeters is the exact volume. To approximate the volume, use the approximation 3.14 for π.

$$V \approx 42 \cdot 3.14 \text{ cu cm}$$ Replace π with 3.14.

$$= 131.88 \text{ cu cm}$$

The volume is approximately 131.88 cubic centimeters.

Work Practice 8

Vocabulary, Readiness & Video Check

Use the choices below to fill in each blank. Some choices may be used more than once.

area	surface area	cubic
volume	square	

1. The ______________ of a polyhedron is the sum of the areas of its faces.
2. The measure of the amount of space inside a solid is its ______________.
3. ______________ measures the amount of surface enclosed by a region.
4. Volume is measured in ______________ units.
5. Area is measured in ______________ units.
6. Surface area is measured in ______________ units.

Watch the section lecture video and answer the following questions.

Objective A 7. Explain why we need to use a formula twice to solve Example 3.

Objective B 8. In Examples 7 and 8, explain the difference in the two volume answers found for each.

8.3 Exercise Set MyMathLab®

Objective A *Find the area of each geometric figure. If the figure is a circle, give an exact area and then use the given approximation for π to approximate the area. See Examples 1 through 4.*

1.

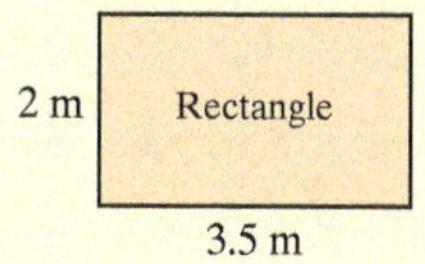

2.

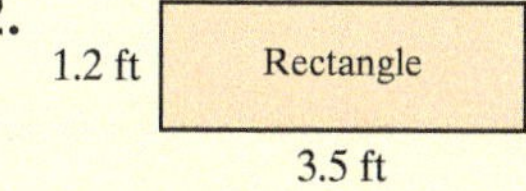

3.

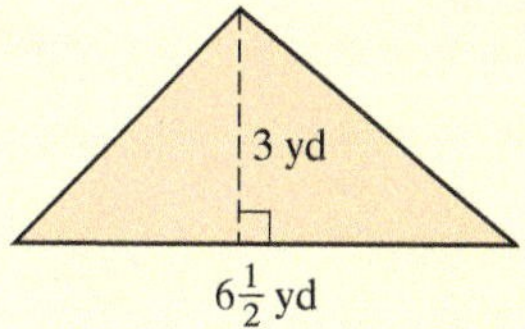

4.

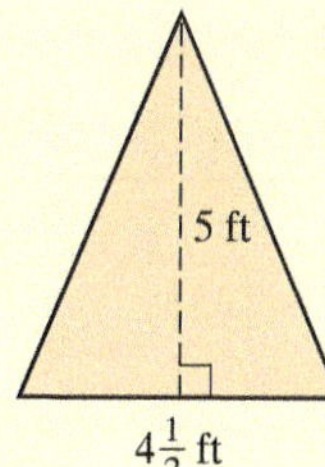

5.

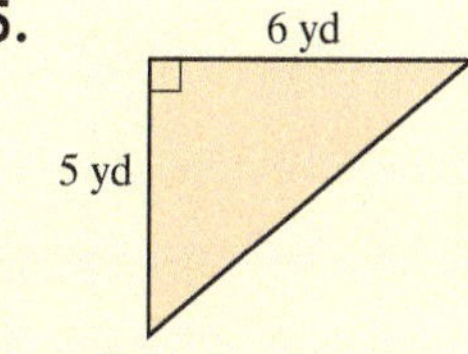

6.

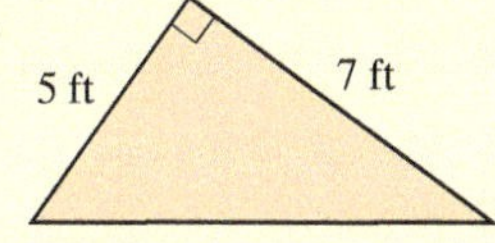

7. Use 3.14 for π.

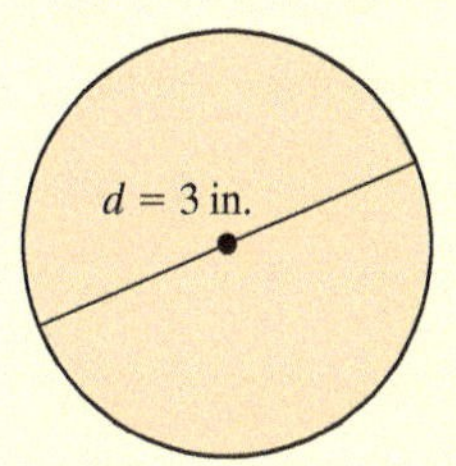

8. Use $\frac{22}{7}$ for π.

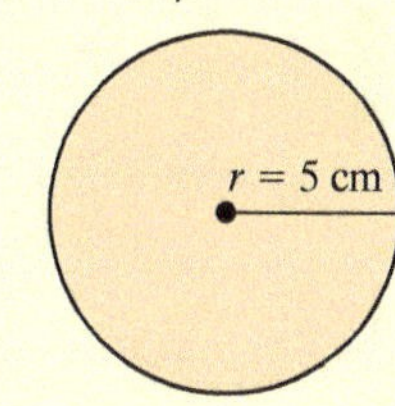

9.

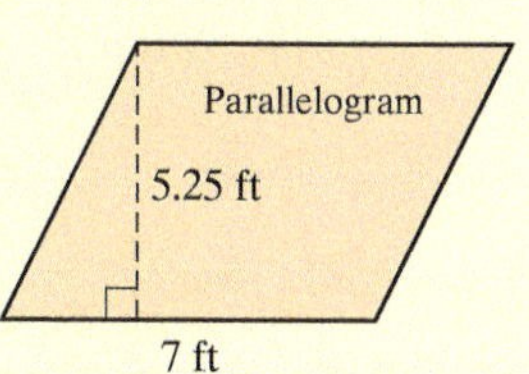

10.

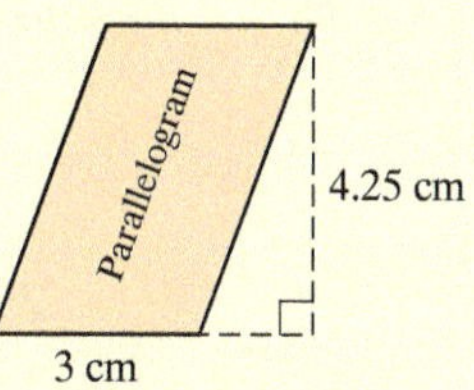

11.

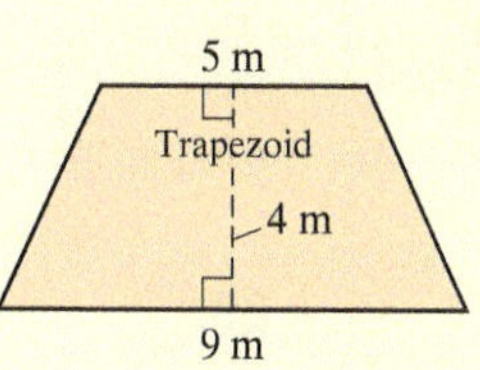

12.

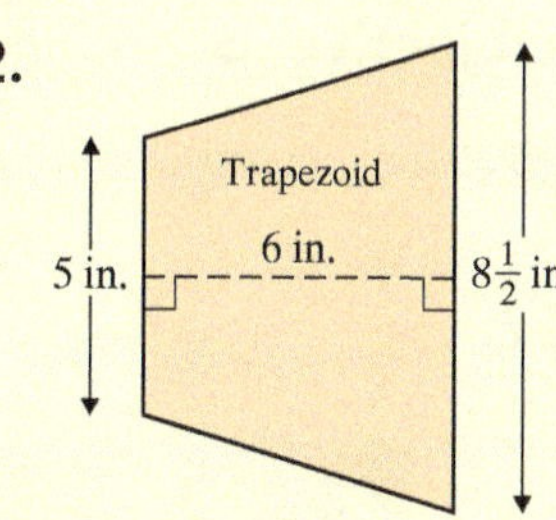

13.

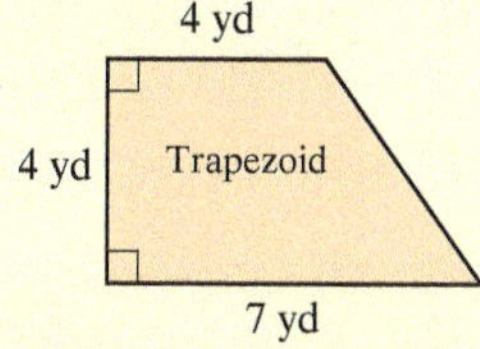

14.

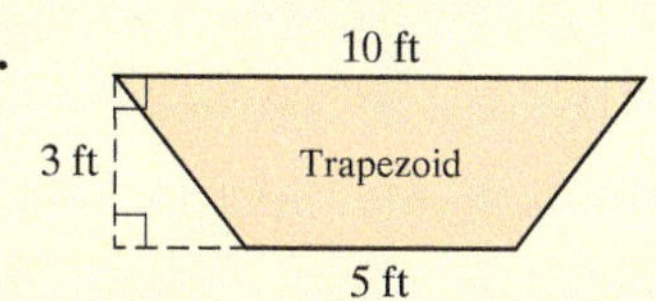

15.

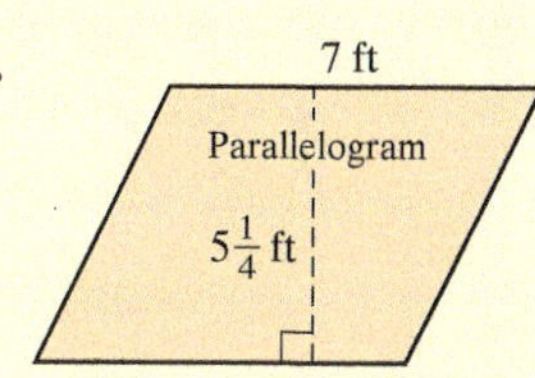

16.
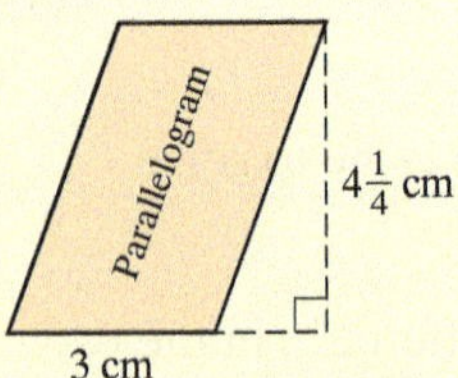

17.
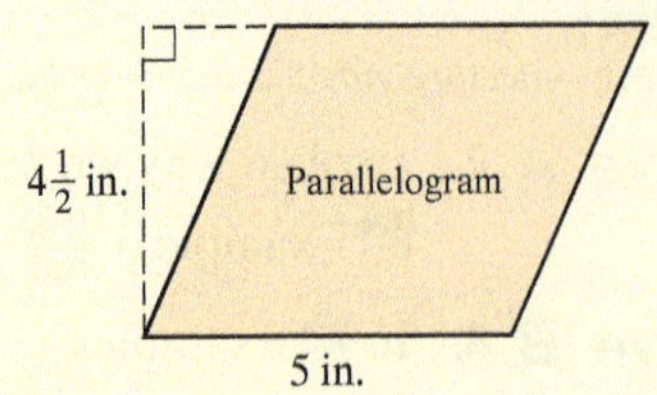

18.
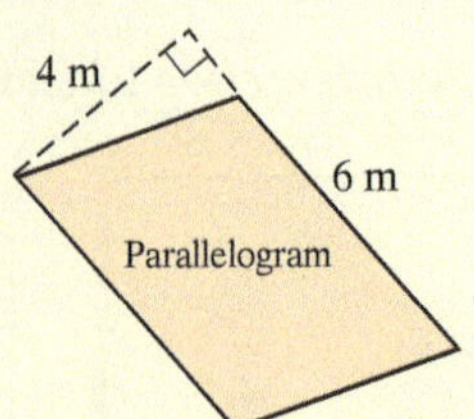

19.
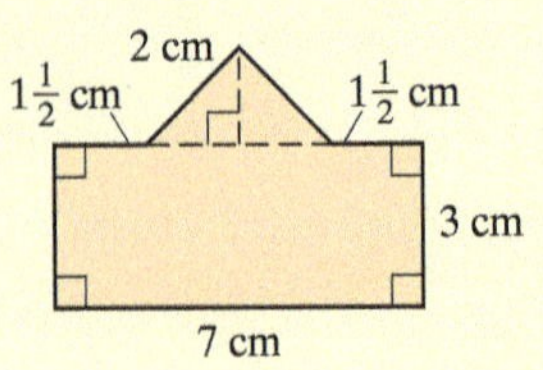

20.
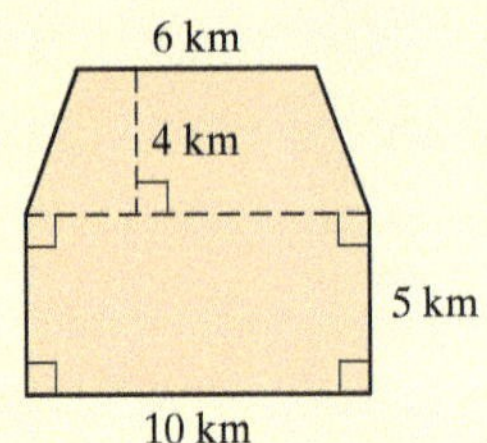

21.
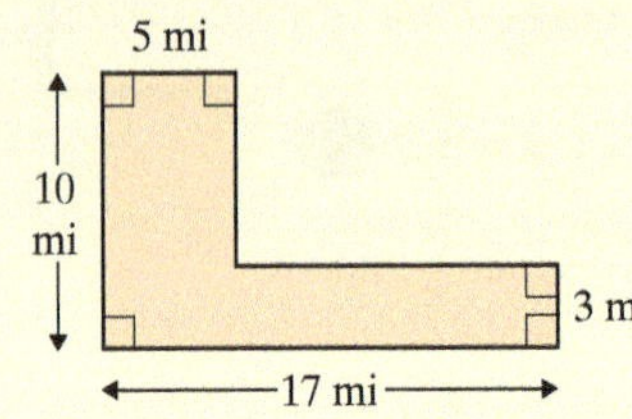

22.
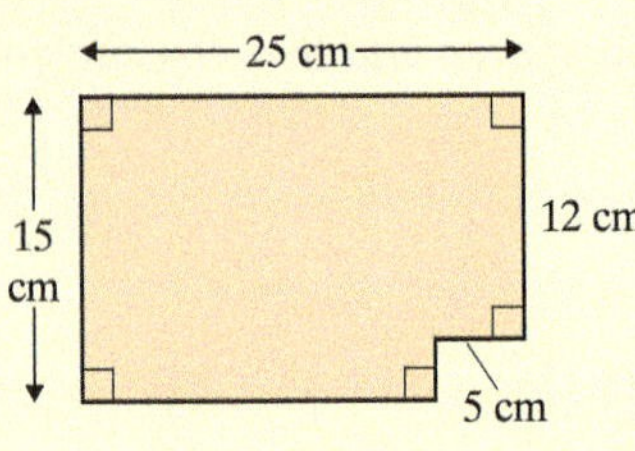

23.
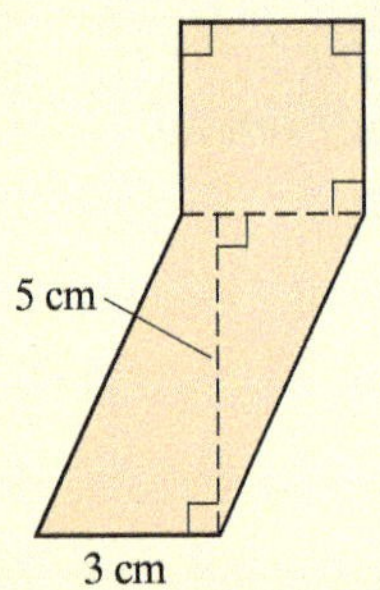

24.
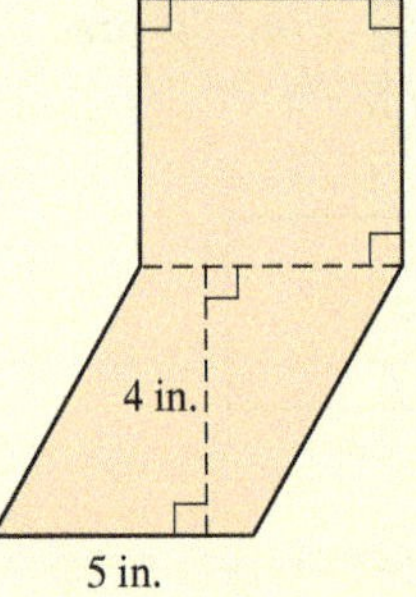

25. Use $\frac{22}{7}$ for π.

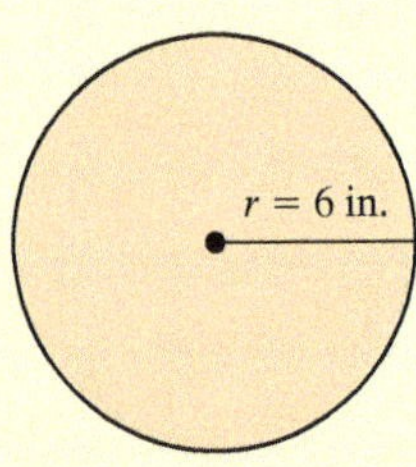

26. Use 3.14 for π.

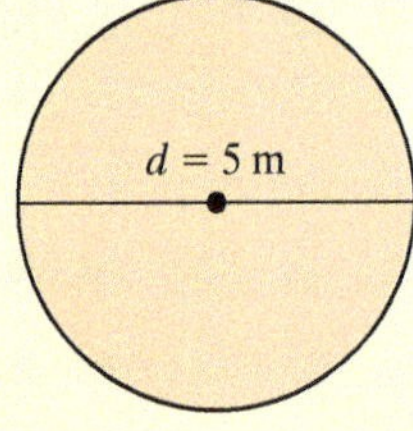

Objective B *Find the volume and surface area of each solid. See Examples 5 through 8. For formulas containing π, give an exact answer and then approximate using $\frac{22}{7}$ for π.*

27.
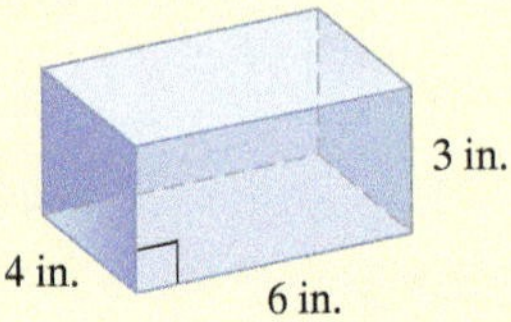

28.
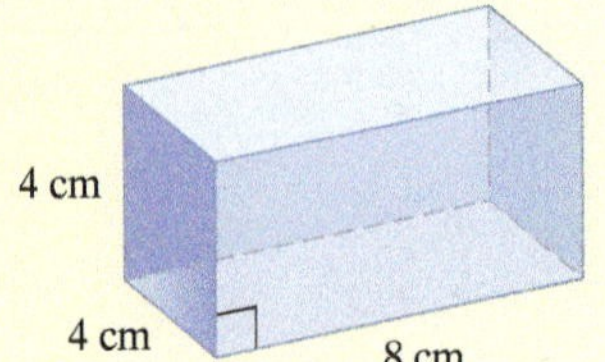

29.
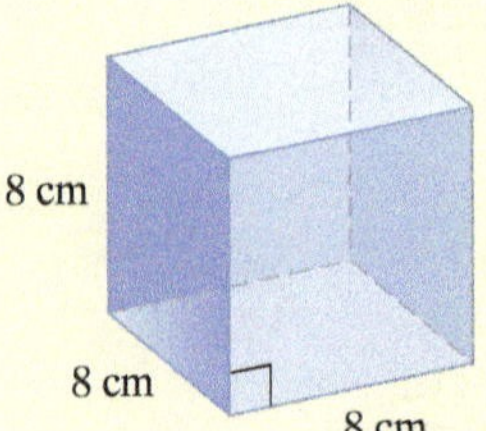

30.

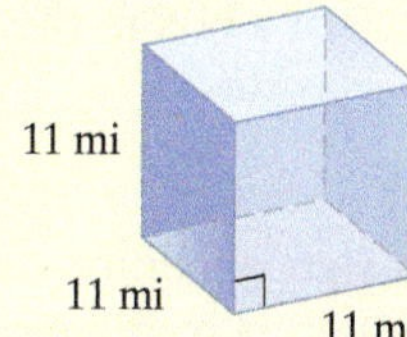

31. For surface area, use $\pi = 3.14$ and round to the nearest hundredth.

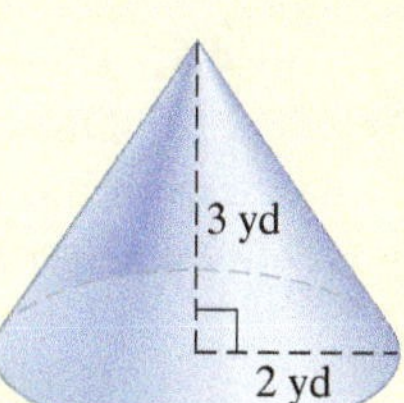

32. For surface area, use $\pi = 3.14$ and round to the nearest hundredth.

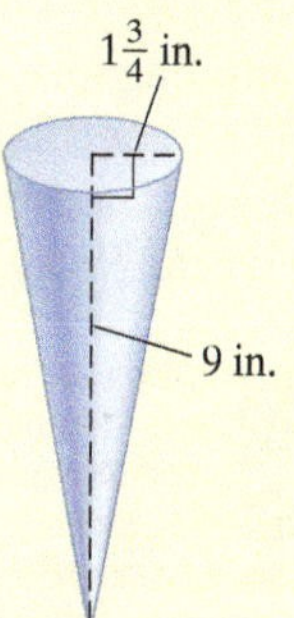

33.

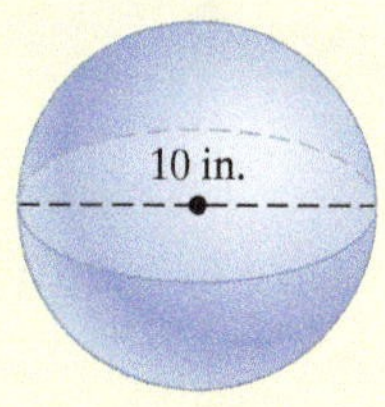

34.

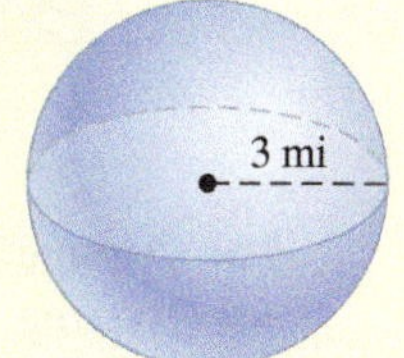

35. Find the volume only.

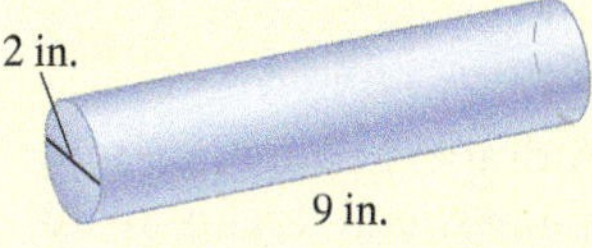

36. Find the volume only.

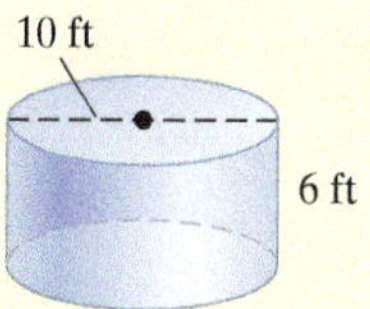

37. Find the volume only.

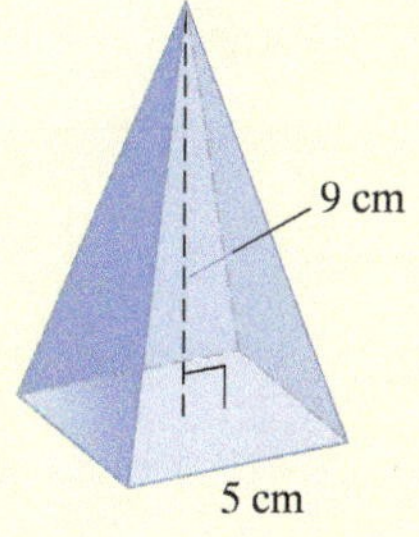

38. Find the volume only.

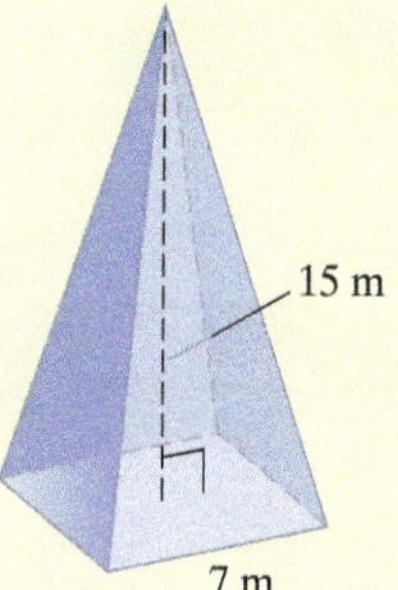

Objectives A B Mixed Practice *Solve. See Examples 1 through 8.*

39. Find the volume of a cube with edges of $1\frac{1}{3}$ inches.

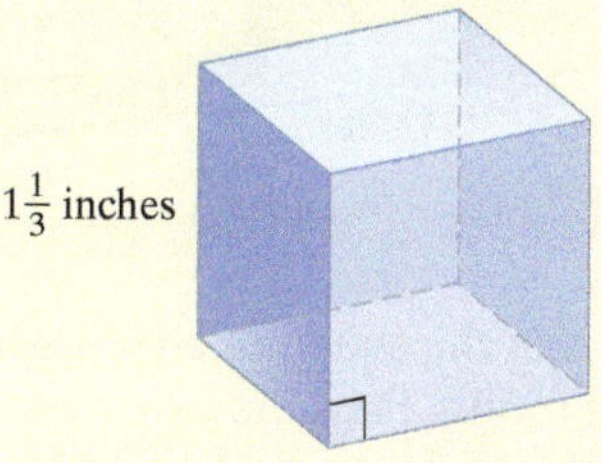

40. A water storage tank is in the shape of a cone with the pointed end down. If the radius is 14 ft and the depth of the tank is 15 ft, approximate the volume of the tank in cubic feet. Use $\frac{22}{7}$ for π.

41. Find the volume and surface area of a rectangular box 2 ft by 1.4 ft by 3 ft.

42. Find the volume and surface area of a box in the shape of a cube that is 5 ft on each side.

43. The largest American flag measures 505 feet by 225 feet. It's the U.S. "Super flag" owned by "Ski" Demski of Long Beach, California. Find its area. (*Source: Guinness World Records*)

44. The largest indoor illuminated sign is a billboard at Dubai Airport. It measures 28 meters by 6.2 meters. Find its area. (*Source: The Guinness Book of World Records*)

45. A drapery panel measures 6 ft by 7 ft. Find how many square feet of material are needed for *four* panels.

46. A page in this book measures 27.6 cm by 5 cm. Find its area.

47. A paperweight is in the shape of a square-based pyramid 20 centimeters tall. If an edge of the base is 12 centimeters, find the volume of the paperweight.

48. A birdbath is made in the shape of a hemisphere (half-sphere). If its radius is 10 inches, approximate its volume. Use $\frac{22}{7}$ for π.

49. Find how many square feet of land are in the following plot:

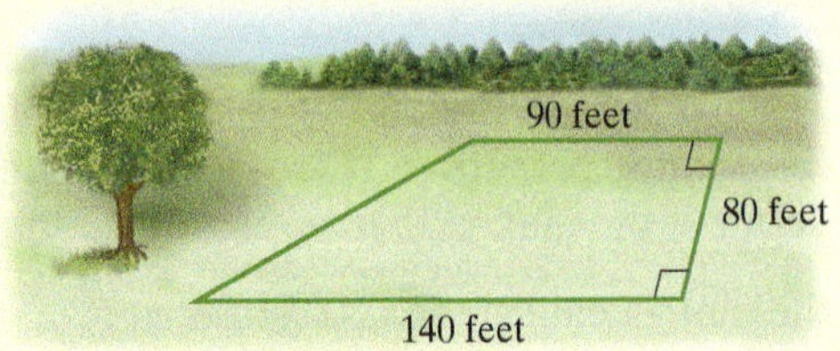

50. For Gerald Gomez to determine how much grass seed he needs to buy, he must know the size of his yard. Use the drawing to determine how many square feet are in his yard.

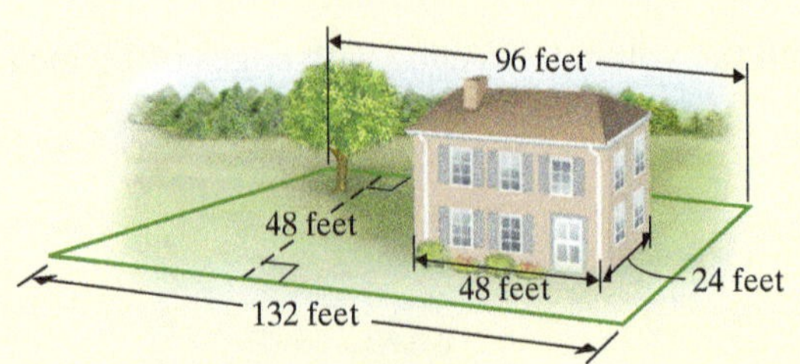

51. Find the exact volume and surface area of a sphere with a radius of 7 inches.

52. A tank is in the shape of a cylinder 8 feet tall and 3 feet in radius. Find the exact volume and surface area of the tank.

53. The outlined part of the roof shown is in the shape of a trapezoid and needs to be shingled. The number of shingles to buy depends on the area.

a. Use the dimensions given to find the area of the outlined part of the roof to the nearest whole square foot.

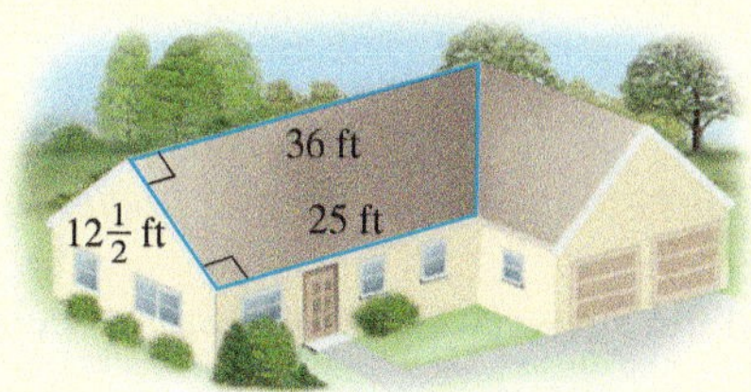

b. Shingles are packaged in a unit called a "square." If a "square" covers 100 square feet, how many whole squares need to be purchased to shingle this part of the roof?

54. The entire side of the building shaded in the drawing is to be bricked. The number of bricks to buy depends on the area.

a. Find the area.

b. If the side area of each brick (including mortar room) is $\frac{1}{6}$ square ft, find the number of bricks that are needed to brick the end of the building.

55. Find the exact volume of a waffle ice cream cone with a 3-in. diameter and a height of 7 inches.

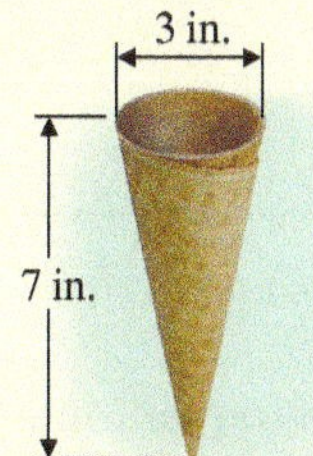

56. A snow globe has a diameter of 6 inches. Find its exact volume. Then approximate its volume using 3.14 for π.

57. Paul Revere's Pizza in the USA will bake and deliver a round pizza with a 4-foot diameter. This pizza is called the "Ultimate Party Pizza" and its current price is $99.99. Find the exact area of the top of the pizza and an approximation. Use 3.14 as an approximation for π.

58. The face of a circular watch has a diameter of 2 centimeters. What is its area? Find the exact area and an approximation. Use 3.14 as an approximation for π.

59. Zorbing is an extreme sport invented by two New Zealanders who joke that they were looking for a way to walk on water. A Zorb is a large sphere inside a second sphere with the space between the spheres pumped full of air. There is a tunnel-like opening so a person can crawl into the inner sphere. You are strapped in and sent down a Zorbing hill. A standard Zorb is approximately 3 m in diameter. Find the exact volume of a Zorb, and approximate the volume using 3.14 for π.

60. Mount Fuji, in Japan, is considered the most beautiful composite volcano in the world. The mountain is in the shape of a cone whose height is about 3.5 kilometers and whose base radius is about 3 kilometers. Approximate the volume of Mt. Fuji in cubic kilometers. Use $\frac{22}{7}$ for π.

61. A $10\frac{1}{2}$-foot by 16-foot concrete wall is to be built using concrete blocks. Find the area of the wall.

62. The floor of Terry's attic is 24 feet by 35 feet. Find how many square feet of insulation are needed to cover the attic floor.

63. Find the volume of a pyramid with a square base 5 inches on a side and a height of $1\frac{3}{10}$ inches.

64. Approximate to the nearest hundredth the volume of a sphere with a radius of 2 centimeters. Use 3.14 for π.

The Space Cube is supposed to be the world's smallest computer, with dimensions of 2 inches by 2 inches by 2.2 inches.

65. Find the volume of the Space Cube.

66. Find the surface area of the Space Cube.

Review

Evaluate. See Section 1.7.

67. 5^2

68. 7^2

69. 3^2

70. 20^2

71. $1^2 + 2^2$

72. $5^2 + 3^2$

73. $4^2 + 2^2$

74. $1^2 + 6^2$

Concept Extensions

Given the following situations, tell whether you are more likely to be concerned with area or perimeter.

75. ordering fencing to fence a yard

76. ordering grass seed to plant in a yard

77. buying carpet to install in a room

78. buying gutters to install on a house

79. ordering paint to paint a wall

80. ordering baseboards to install in a room

81. buying a wallpaper border to go on the walls around a room

82. buying fertilizer for your yard

Solve.

83. A pizza restaurant recently advertised two specials. The first special was a 12-inch pizza for $10. The second special was two 8-inch pizzas for $9. Determine the better buy. (*Hint:* First compare the areas of the two specials and then find a price per square inch for both specials.)

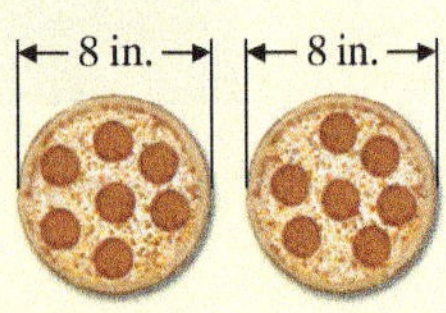

84. Find the approximate area of the state of Utah.

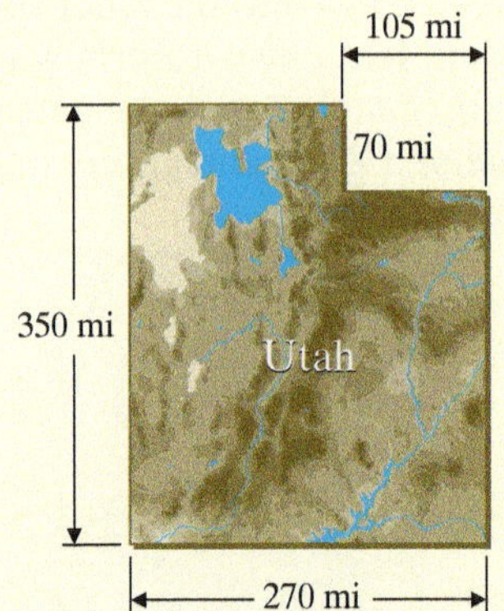

85. The Hayden Planetarium, at the Museum of Natural History in New York City, boasts a dome that has a diameter of 20 m. The dome is a hemisphere, or half a sphere. What is the volume enclosed by the dome at the Hayden Planetarium? Use 3.14 for π and round to the nearest hundredth. (*Source:* Hayden Planetarium)

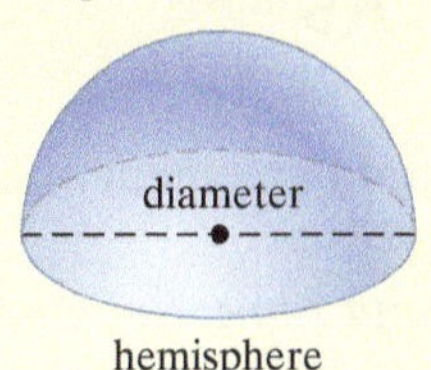

86. The Adler Museum in Chicago has a planetarium, its StarRider Theater, that has a diameter of 55 feet. Find the surface area of its hemispheric (half a sphere) dome. Use 3.14 for π. (*Source:* The Adler Museum)

87. Can you compute the volume of a rectangle? Why or why not?

88. In your own words, explain why perimeter is measured in units and area is measured in square units. (*Hint:* See Section 1.5 for an introduction to the meaning of area.)

89. Find the area of the shaded region. Use the approximation 3.14 for π.

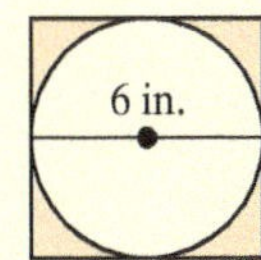

90. The largest pumpkin pie on record was made in New Breman, Ohio, by the New Breman Giant Pumpkin Growers, in September 2010. The pie had a diameter of 240 inches. Find the exact area of the top of the pie and an approximation. Use $\pi \approx 3.14$. (*Source:* World Record Academy)

Find the area of each figure. If needed, use $\pi \approx 3.14$ and round results to the nearest tenth.

91. Find the skating area.

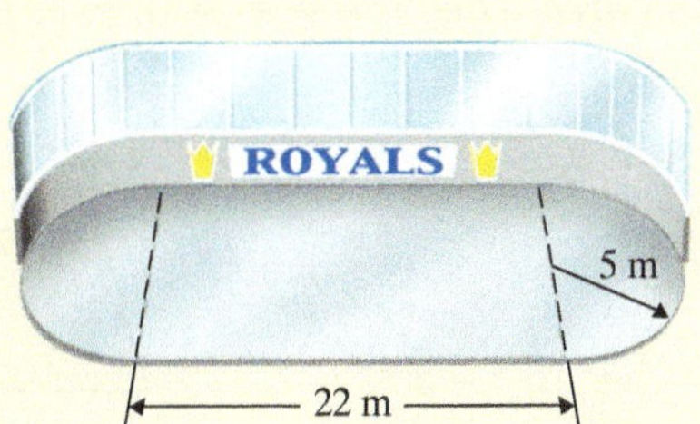

92.

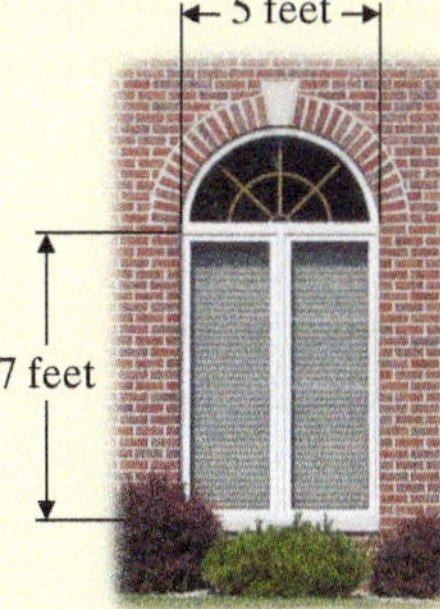

93. Do two rectangles with the same perimeter have the same area? To see, find the perimeter and the area of each rectangle.

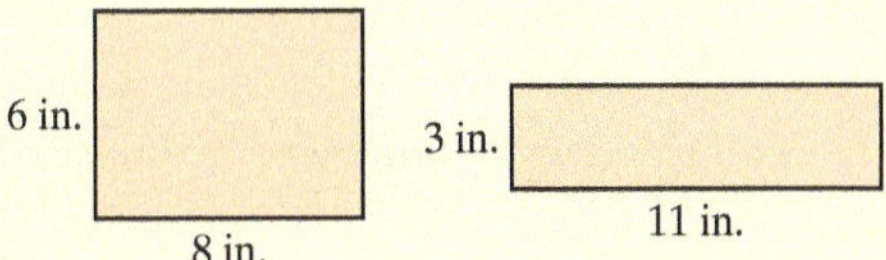

94. Do two rectangular solids with the same volume have the same surface area? To see, find the volume and surface area of each rectangular solid.

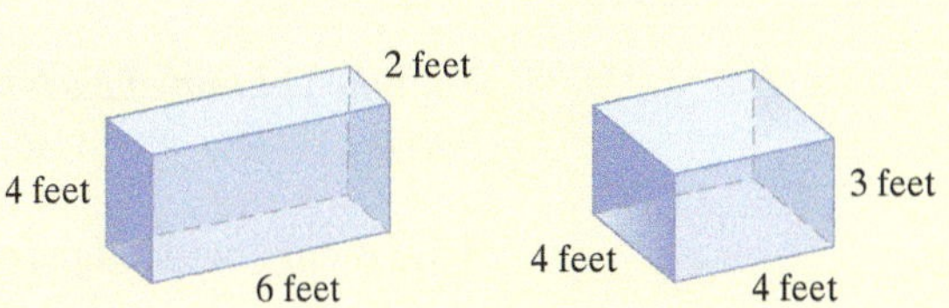

Integrated Review Sections 8.1–8.3

Geometry Concepts

Answers

1. ________
2. ________
3. ________
4. ________
5. ________
6. ________
7. ________
8. ________
9. ________
10. ________
11. ________
12. ________
13. ________
14. ________
15. ________
16. ________

1. Find the supplement and the complement of a 27° angle.

Find the measures of angles x, y, and z in each figure in Exercises 2 and 3.

2.

105°
x z
y

3. $m \| n$

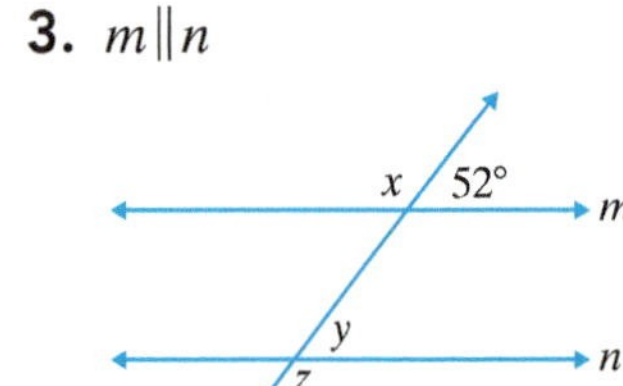

4. Find the measure of $\angle x$. (*Hint:* The sum of the angle measures of a triangle is 180°.)

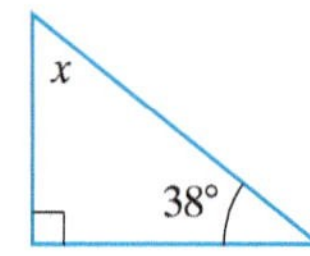

5. Find the diameter.

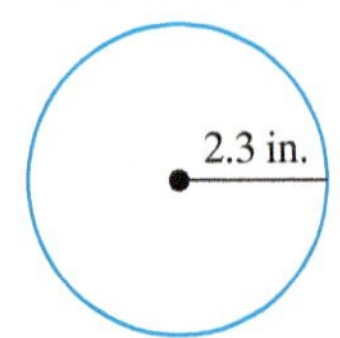

6. Find the radius.

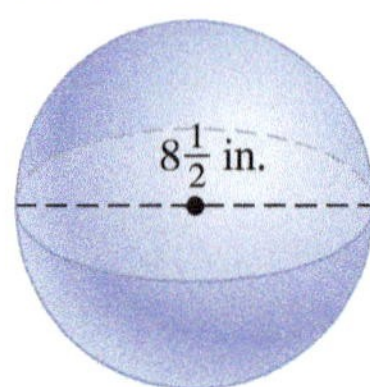

For Exercises 7 through 11, find the perimeter (or circumference) and area of each figure. For the circle, give an exact circumference and area. Then use $\pi \approx 3.14$ to approximate each. Don't forget to attach correct units.

7.

Square 5 m

8.

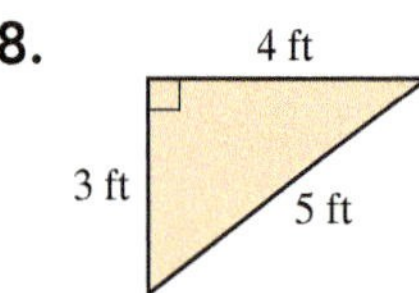

9.

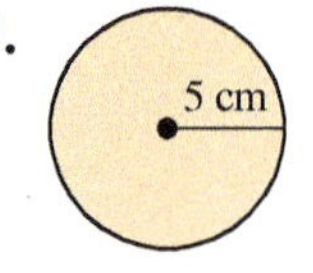

10.

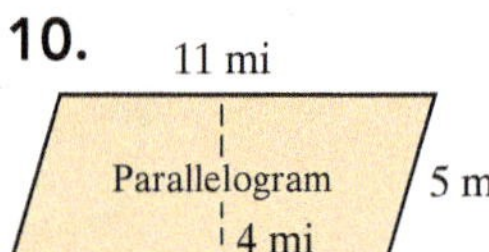

11.

8 cm
3 cm
7 cm
17 cm

12. The smallest cathedral is in Highlandville, Missouri. The rectangular floor of the cathedral measures 14 feet by 17 feet. Find its perimeter and its area. (*Source: The Guinness Book of World Records*)

Find the volume of each solid. Don't forget to attach correct units. For Exercises 13 and 14, find the surface area also.

13. A cube with edges of 4 inches each

14. A rectangular box 2 feet by 3 feet by 5.1 feet

15. A pyramid with a square base 10 centimeters on a side and a height of 12 centimeters

16. A sphere with a diameter of 3 miles. Give the exact volume and then use $\pi \approx \frac{22}{7}$ to approximate.

8.4 Linear Measurement 

Objective A Defining and Converting U.S. System Units of Length

In the United States, two systems of measurement are commonly used. They are the **United States (U.S.), or English, measurement system** and the **metric system.** The U.S. measurement system is familiar to most Americans. Units such as feet, miles, ounces, and gallons are used. However, the metric system is also commonly used in fields such as medicine, sports, international marketing, and certain physical sciences. We are accustomed to buying 2-liter bottles of soft drinks, watching televised coverage of the 100-meter dash at the Olympic Games, or taking a 200-milligram dose of pain reliever.

The U.S. system of measurement uses the **inch, foot, yard,** and **mile** to measure **length.** The following is a summary of equivalencies between units of length:

U.S. Units of Length

12 inches (in.) = 1 foot (ft)
3 feet = 1 yard (yd)
36 inches = 1 yard
5280 feet = 1 mile (mi)

To convert from one unit of length to another, we will use **unit fractions.** We define a unit fraction to be a fraction that is equivalent to 1. Examples of unit fractions are as follows:

Unit Fractions

$$\frac{12 \text{ in.}}{1 \text{ ft}} = 1 \text{ or } \frac{1 \text{ ft}}{12 \text{ in.}} = 1 \text{ (since 12 in.} = 1 \text{ ft)}$$

$$\frac{3 \text{ ft}}{1 \text{ yd}} = 1 \text{ or } \frac{1 \text{ yd}}{3 \text{ ft}} = 1 \text{ (since 3 ft} = 1 \text{ yd)}$$

$$\frac{5280 \text{ ft}}{1 \text{ mi}} = 1 \text{ or } \frac{1 \text{ mi}}{5280 \text{ ft}} = 1 \text{ (since 5280 ft} = 1 \text{ mi)}$$

Remember that multiplying a number by 1 does not change the value of the number.

Example 1 Convert 8 feet to inches.

Solution: We multiply 8 feet by a unit fraction that uses the equality 12 inches = 1 foot. The unit fraction should be in the form $\frac{\text{units to convert to}}{\text{original units}}$ or, in this case, $\frac{12 \text{ inches}}{1 \text{ foot}}$. We do this so that like units will divide out to 1, as shown.

$$8 \text{ ft} = \frac{8 \text{ ft}}{1} \cdot 1 \quad \text{Multiply by 1 in the form of } \frac{12 \text{ in.}}{1 \text{ ft}}.$$

$$= \frac{8 \cancel{\text{ft}}}{1} \cdot \frac{12 \text{ in.}}{1 \cancel{\text{ft}}}$$

$$= 8 \cdot 12 \text{ in.}$$

$$= 96 \text{ in.} \quad \text{Multiply.}$$

(Continued on next page)

Objectives

A Define U.S. Units of Length and Convert from One Unit to Another.

B Use Mixed Units of Length.

C Perform Arithmetic Operations on U.S. Units of Length.

D Define Metric Units of Length and Convert from One Unit to Another.

E Perform Arithmetic Operations on Metric Units of Length.

Practice 1

Convert 6 feet to inches.

Answer

1. 72 in.

Thus, 8 ft = 96 in., as shown in the diagram:

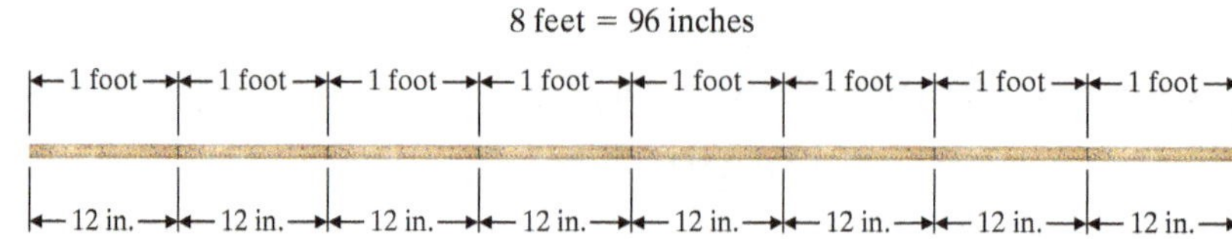

Work Practice 1

Practice 2

Convert 8 yards to feet.

Helpful Hint When converting from one unit to another, select a unit fraction with the properties below:

$$\frac{\text{units you are converting to}}{\text{original units}}$$

By using this unit fraction, the original units will divide out, as wanted.

Example 2 Convert 7 feet to yards.

Solution: We multiply by a unit fraction that compares 1 yard to 3 feet.

$$7 \text{ ft} = \frac{7 \text{ ft}}{1} \cdot 1$$

$$= \frac{7 \cancel{\text{ft}}}{1} \cdot \frac{1 \text{ yd}}{3 \cancel{\text{ft}}} \quad \begin{matrix} \leftarrow \text{Units to convert to} \\ \leftarrow \text{Original units} \end{matrix}$$

$$= \frac{7}{3} \text{ yd}$$

$$= 2\frac{1}{3} \text{ yd} \quad \text{Divide.}$$

Thus, 7 ft = $2\frac{1}{3}$ yd, as shown in the diagram.

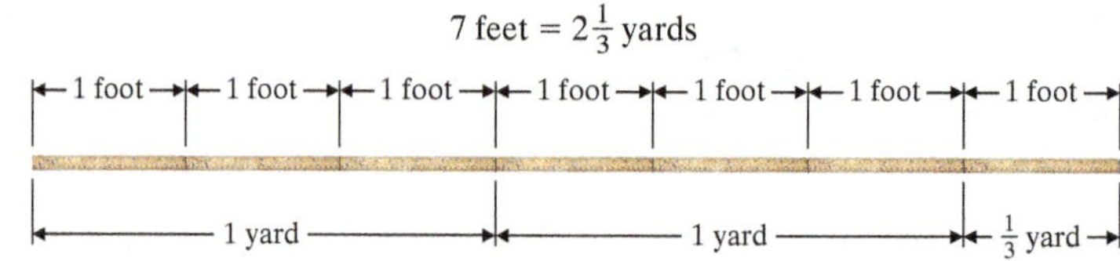

Work Practice 2

Practice 3

Suppose the pelican's bill in the photo measures 18 inches. Convert 18 inches to feet, using decimals.

Example 3 Finding the Length of a Pelican's Bill

The Australian pelican has the longest bill, measuring from 13 to 18.5 inches long. The pelican in the photo has a 15-inch bill. Convert 15 inches to feet, using decimals in your final answer.

Solution:

$$15 \text{ in.} = \frac{15 \cancel{\text{in.}}}{1} \cdot \frac{1 \text{ ft}}{12 \cancel{\text{in.}}} \quad \begin{matrix} \leftarrow \text{Units to convert to} \\ \leftarrow \text{Original units} \end{matrix}$$

$$= \frac{15}{12} \text{ ft}$$

$$= \frac{5}{4} \text{ ft} \quad \text{Simplify } \frac{15}{12}.$$

$$= 1.25 \text{ ft} \quad \text{Divide.}$$

Thus, 15 in. = 1.25 ft, as shown in the diagram.

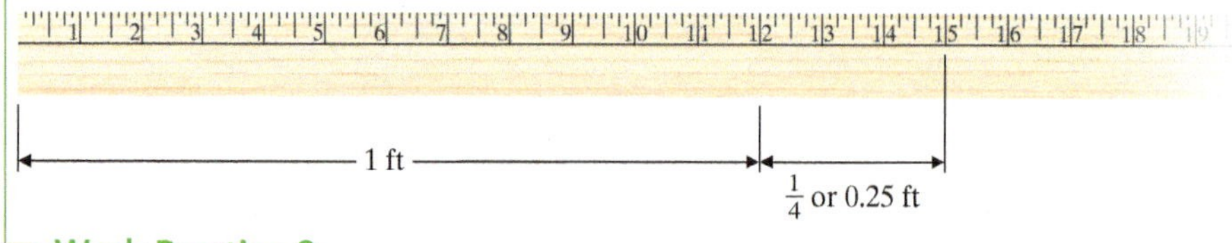

Work Practice 3

Answers
2. 24 ft 3 1.5 ft

Objective B Using Mixed U.S. System Units of Length 

Sometimes it is more meaningful to express a measurement of length with mixed units such as 1 ft and 5 in. We usually condense this and write 1 ft 5 in.

In Example 2, we found that 7 feet is the same as $2\frac{1}{3}$ yards. The measurement can also be written as a mixture of yards and feet. That is,

7 ft = ____ yd ____ ft

Because 3 ft = 1 yd, we divide 3 into 7 to see how many whole yards are in 7 feet. The quotient is the number of yards, and the remainder is the number of feet.

$$\begin{array}{r} 2 \text{ yd } 1 \text{ ft} \\ 3\overline{)7} \quad\quad\; \\ -6 \quad\quad\; \\ \hline 1 \quad\quad\; \end{array}$$

Thus, 7 ft = 2 yd 1 ft, as seen in the diagram:

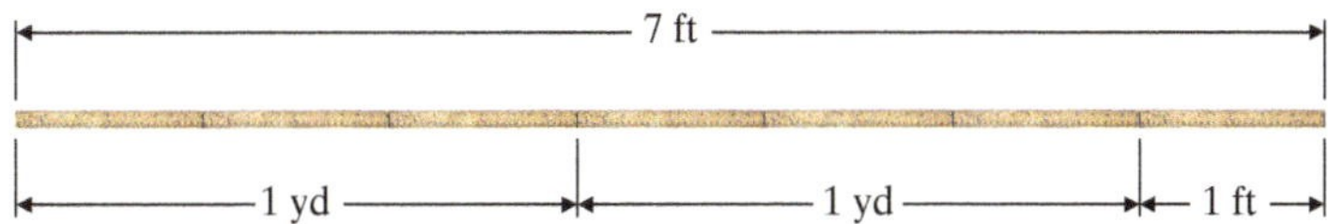

Example 4 Convert: 134 in. = _____ ft _____ in.

Solution: Because 12 in. = 1 ft, we divide 12 into 134. The quotient is the number of feet. The remainder is the number of inches. To see why we divide 12 into 134, notice that

$$134 \text{ in.} = \frac{134 \text{ in.}}{1} \cdot \frac{1 \text{ ft}}{12 \text{ in.}} = \frac{134}{12} \text{ ft}$$

$$\begin{array}{r} 11 \text{ ft } 2 \text{ in.} \\ 12\overline{)134} \quad\quad\quad\; \\ -12 \quad\quad\quad\; \\ \hline 14 \quad\quad\quad\; \\ -12 \quad\quad\quad\; \\ \hline 2 \quad\quad\quad\; \end{array}$$

Thus, 134 in. = 11 ft 2 in.

Work Practice 4

Practice 4

Convert: 68 in. = _____ ft _____ in.

Example 5 Convert 3 feet 7 inches to inches.

Solution: First, we convert 3 feet to inches. Then we add 7 inches.

$$3 \text{ ft} = \frac{3 \text{ ft}}{1} \cdot \frac{12 \text{ in.}}{1 \text{ ft}} = 36 \text{ in.}$$

Then

$$3 \text{ ft } 7 \text{ in.} = 36 \text{ in.} + 7 \text{ in.} = 43 \text{ in.}$$

Work Practice 5

Practice 5

Convert 5 yards 2 feet to feet.

Answers
4. 5 ft 8 in. **5.** 17 ft

Objective C Performing Operations on U.S. System Units of Length

Finding sums or differences of measurements often involves converting units, as shown in the next example. Just remember that, as usual, only like units can be added or subtracted.

Practice 6

Add 4 ft 8 in. to 8 ft 11 in.

Example 6 Add 3 ft 2 in. and 5 ft 11 in.

Solution: To add, we line up the similar units.

$$\begin{array}{r} 3 \text{ ft } \ \ 2 \text{ in.} \\ +\ 5 \text{ ft } 11 \text{ in.} \\ \hline 8 \text{ ft } 13 \text{ in.} \end{array}$$

Since 13 inches is the same as 1 ft 1 in., we have

$$\begin{aligned} 8 \text{ ft } 13 \text{ in.} &= 8 \text{ ft} + 1 \text{ ft } 1 \text{ in.} \\ &= 9 \text{ ft } 1 \text{ in.} \end{aligned}$$

Work Practice 6

✓Concept Check How could you estimate the following sum?

$$\begin{array}{r} 7 \text{ yd } \ \ 4 \text{ in.} \\ +\ 3 \text{ yd } 27 \text{ in.} \\ \hline \end{array}$$

Practice 7

Multiply 4 ft 7 in. by 4.

Example 7 Multiply 8 ft 9 in. by 3.

Solution: By the distributive property, we multiply 8 ft by 3 and 9 in. by 3.

$$\begin{array}{r} 8 \text{ ft } \ \ 9 \text{ in.} \\ \times \qquad\quad 3 \\ \hline 24 \text{ ft } 27 \text{ in.} \end{array}$$

Since 27 in. is the same as 2 ft 3 in., we simplify the product as

$$\begin{aligned} 24 \text{ ft } 27 \text{ in.} &= 24 \text{ ft} + 2 \text{ ft } 3 \text{ in.} \\ &= 26 \text{ ft } 3 \text{ in.} \end{aligned}$$

Work Practice 7

We divide in a similar manner as above.

Practice 8

A carpenter cuts 1 ft 9 in. from a board of length 5 ft 8 in. Find the remaining length of the board.

Example 8 Finding the Length of a Piece of Rope

A rope of length 6 yd 1 ft has 2 yd 2 ft cut from one end. Find the length of the remaining rope.

Solution: Subtract 2 yd 2 ft from 6 yd 1 ft.

$$\begin{array}{rcr} \text{beginning length} & \rightarrow & 6 \text{ yd } 1 \text{ ft} \\ -\quad \text{amount cut} & \rightarrow & -2 \text{ yd } 2 \text{ ft} \\ \hline \text{remaining length} & & \end{array}$$

We cannot subtract 2 ft from 1 ft, so we borrow 1 yd from the 6 yd. One yard is converted to 3 ft and combined with the 1 ft already there.

Answers

6. 13 ft 7 in. 7. 18 ft 4 in.
8. 3 ft 11 in.

✓Concept Check Answer

round each to the nearest yard: 7 yd + 4 yd = 11 yd

Borrow 1 yd = 3 ft

5 yd + 1 yd 3 ft

$$\begin{aligned} \cancel{6\text{ yd}}\ 1\text{ ft} &= 5\text{ yd }4\text{ ft} \\ -2\text{ yd }2\text{ ft} &= -2\text{ yd }2\text{ ft} \\ & 3\text{ yd }2\text{ ft} \end{aligned}$$

The remaining rope is 3 yd 2 ft long.

Work Practice 8

Objective D Defining and Converting Metric System Units of Length

The basic unit of length in the metric system is the **meter.** A meter is slightly longer than a yard. It is approximately 39.37 inches long. Recall that a yard is 36 inches long.

1 yard = 36 inches

1 meter ≈ 39.37 inches

All units of length in the metric system are based on the meter. The following is a summary of the prefixes used in the metric system. Also shown are equivalencies between units of length. Like the decimal system, the metric system uses powers of 10 to define units.

Metric Units of Length
1 **kilo**meter (km) = 1000 meters (m)
1 **hecto**meter (hm) = 100 m
1 **deka**meter (dam) = 10 m
1 meter (m) = 1 m
1 **deci**meter (dm) = 1/10 m or 0.1 m
1 **centi**meter (cm) = 1/100 m or 0.01 m
1 **milli**meter (mm) = 1/1000 m or 0.001 m

The figure below will help you with decimeters, centimeters, and millimeters.

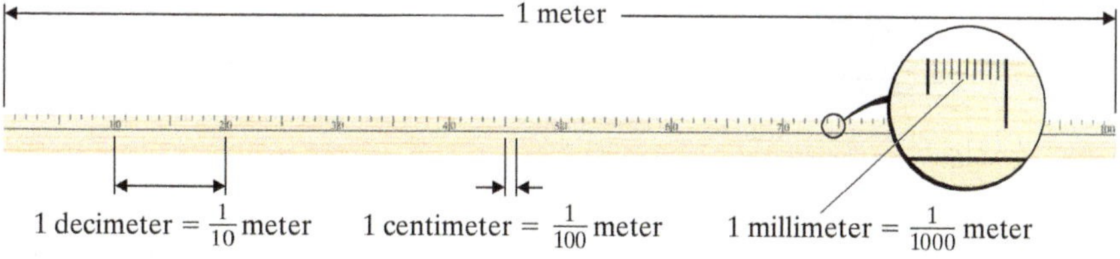

Helpful Hint

Study the figure above for other equivalencies between metric units of length.

10 decimeters = 1 meter
100 centimeters = 1 meter
1000 millimeters = 1 meter
10 millimeters = 1 centimeter
10 centimeters = 1 decimeter

These same prefixes are used in the metric system for mass and capacity. The most commonly used measurements of length in the metric system are the **meter, millimeter, centimeter,** and **kilometer.**

✓Concept Check Is this statement reasonable? "The screen of a home television set has a 30-meter diagonal." Why or why not?

Being comfortable with the metric units of length means gaining a "feeling" for metric lengths, just as you have a "feeling" for the lengths of an inch, a foot, and a mile. To help you accomplish this, study the following examples:

- A millimeter is about the thickness of a large paper clip.

- A centimeter is about the width of a large paper clip.
- A meter is slightly longer than a yard.
- A kilometer is about two-thirds of a mile.

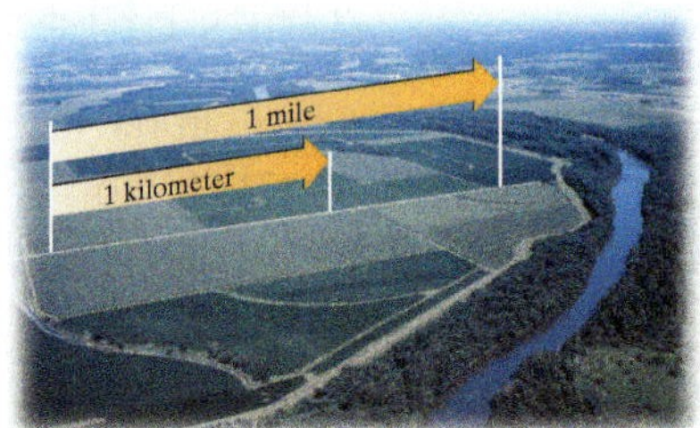

- The width of this book is approximately 21.5 centimeters.
- The distance between New York City and Philadelphia is about 160 kilometers.

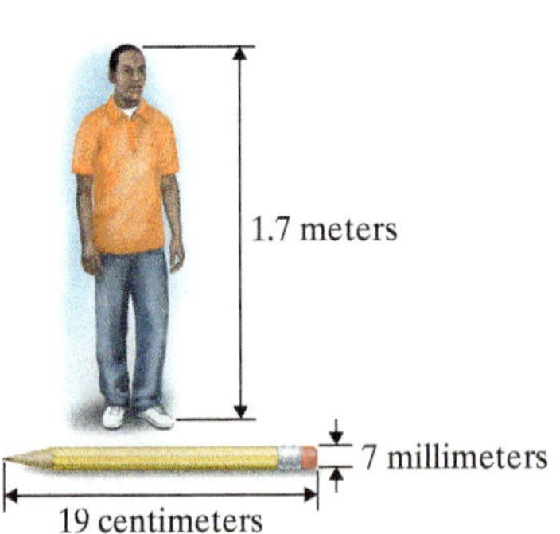

$2\frac{1}{2}$ centimeters is about 1 inch.

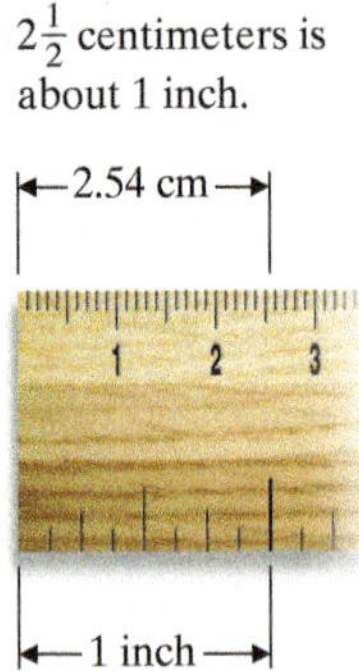

As with the U.S. system of measurement, unit fractions may be used to convert from one unit of length to another. For example, let's convert 1200 meters to kilometers. To do so, we will multiply by 1 in the form of the unit fraction

$$\frac{1\text{ km}}{1000\text{ m}} \quad \begin{array}{l}\leftarrow \text{Units to convert to} \\ \leftarrow \text{Original units}\end{array}$$

$$1200\text{ m} = \frac{1200\text{ m}}{1} \cdot 1 = \frac{1200\text{ }\cancel{\text{m}}}{1} \cdot \overbrace{\frac{1\text{ km}}{1000\text{ }\cancel{\text{m}}}}^{\text{Unit fraction}} = \frac{1200\text{ km}}{1000} = 1.2\text{ km}$$

The metric system does, however, have a distinct advantage over the U.S. system of measurement: the ease of converting from one unit of length to another. Since all units of length are powers of 10 of the meter, converting from one unit of length to another is as simple as moving the decimal point. Listing units of length in

✓Concept Check Answer
no; answers may vary

order from largest to smallest helps to keep track of how many places to move the decimal point when converting.

Let's again convert 1200 meters to kilometers. This time, to convert from meters to kilometers, we move along the chart shown, 3 units to the left, from meters to kilometers. This means that we move the decimal point 3 places to the left.

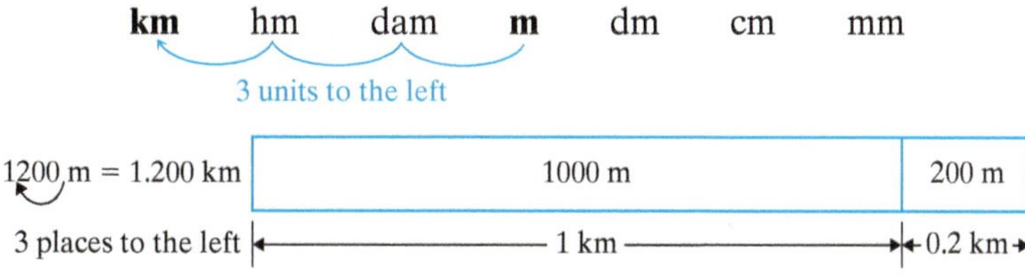

Thus, 1200 m = 1.2 km, as shown in the diagram.

Example 9 Convert 2.3 m to centimeters.

Solution: First we will convert by using a unit fraction.

Unit fraction

$$2.3\text{ m} = \frac{2.3\text{ m}}{1} \cdot \frac{100\text{ cm}}{1\text{ m}} = 230\text{ cm}$$

Now we will convert by listing the units of length in order from left to right and moving from meters to centimeters.

km hm dam m dm cm mm

2 units to the right

2.30 m = 230. cm

2 places to the right

With either method, we get 230 cm.

Work Practice 9

Practice 9

Convert 2.5 m to millimeters.

Example 10 Convert 450,000 mm to meters.

Solution: We list the units of length in order from left to right and move from millimeters to meters.

km hm dam m dm cm mm

3 units to the left

Thus, move the decimal point 3 places to the left.

450,000 mm = 450.000 m or 450 m

Work Practice 10

Practice 10

Convert 3500 m to kilometers.

✓**Concept Check** What is wrong with the following conversion of 150 cm to meters?

150.00 cm = 15,000 m

Objective E Performing Operations on Metric System Units of Length

To add, subtract, multiply, or divide with metric measurements of length, we write all numbers using the same unit of length and then add, subtract, multiply, or divide as with decimals.

Answers

9. 2500 mm **10.** 3.5 km

✓**Concept Check Answer**

decimal point should be moved two places to the left: 1.5 m

Practice 11

Subtract 640 m from 2.1 km.

Practice 12

Multiply 18.3 hm by 5.

Practice 13

Doris Blackwell is knitting a scarf that is currently 0.8 meter long. If she knits an additional 45 centimeters, how long will the scarf be?

Answers

11. 1.46 km or 1460 m **12.** 91.5 hm **13.** 125 cm or 1.25 m

Example 11 Subtract 430 m from 1.3 km.

Solution: First we convert both measurements to kilometers or both to meters.

430 m = 0.43 km or 1.3 km = 1300 m

$$\begin{array}{r} 1.30\text{ km} \\ -0.43\text{ km} \\ \hline 0.87\text{ km} \end{array} \qquad \begin{array}{r} 1300\text{ m} \\ -430\text{ m} \\ \hline 870\text{ m} \end{array}$$

The difference is 0.87 km or 870 m.

Work Practice 11

Example 12 Multiply 5.7 mm by 4.

Solution: Here we simply multiply the two numbers. Note that the unit of measurement remains the same.

$$\begin{array}{r} 5.7\text{ mm} \\ \times\quad 4 \\ \hline 22.8\text{ mm} \end{array}$$

Work Practice 12

Example 13 Finding a Person's Height

Fritz Martinson was 1.2 meters tall on his last birthday. Since then, he has grown 14 centimeters. Find his current height in meters.

Solution:

original height	→	1.20 m	
+ height grown	→	+ 0.14 m	(Since 14 cm = 0.14 m)
current height		1.34 m	

Fritz is now 1.34 meters tall.

Work Practice 13

Vocabulary, Readiness & Video Check

Use the choices below to fill in each blank. Some choices may be used more than once.

inches	yard	unit fraction
feet	meter	

1. The basic unit of length in the metric system is the ________.

2. The expression $\frac{1\text{ foot}}{12\text{ inches}}$ is an example of a(n) ________.

3. A meter is slightly longer than a(n) ________.

4. One foot equals 12 ________.

5. One yard equals 3 ________.

6. One yard equals 36 ________.

7. One mile equals 5280 ________.

Martin-Gay Interactive Videos

See Video 8.4

Watch the section lecture video and answer the following questions.

Objective A 8. In Example 3, what units are used in the denominator of the unit fraction and why was this decided?

Objective B 9. In Example 4, how is a mixed unit similar to a mixed number? Use examples in your answer.

Objective C 10. In Example 5, why is the sum of the addition problem not the final answer? Reference the sum in your answer.

Objective D 11. In the lecture before Example 6, why is it easier to convert metric units than U.S. units?

Objective E 12. What two answers did we get for Example 8? Explain why both answers are correct.

8.4 Exercise Set MyMathLab®

Objective A *Convert each measurement as indicated. See Examples 1 through 3.*

1. 60 in. to feet
2. 84 in. to feet
3. 12 yd to feet
4. 18 yd to feet
5. 42,240 ft to miles
6. 36,960 ft to miles
7. $8\frac{1}{2}$ ft to inches
8. $12\frac{1}{2}$ ft to inches
9. 10 ft to yards
10. 25 ft to yards
11. 6.4 mi to feet
12. 3.8 mi to feet
13. 162 in. to yd (Write answer as a decimal.)
14. 7216 yd to mi (Write answer as a decimal.)
15. 3 in. to ft (Write answer as a decimal.)
16. 129 in. to ft (Write answer as a decimal.)

Objective B *Convert each measurement as indicated. See Examples 4 and 5.*

17. 40 ft = ____ yd ____ ft
18. 100 ft = ____ yd ____ ft
19. 85 in. = ____ ft ____ in.
20. 59 in. = ____ ft ____ in.
21. 10,000 ft = ____ mi ____ ft
22. 25,000 ft = ____ mi ____ ft
23. 5 ft 2 in. = ____ in.
24. 4 ft 11 in. = ____ in.
25. 8 yd 2 ft = ____ ft
26. 4 yd 1 ft = ____ ft
27. 2 yd 1 ft = ____ in.
28. 1 yd 2 ft = ____ in.

Objective C *Perform each indicated operation. Simplify the result if possible. See Examples 6 through 8.*

29. 3 ft 10 in. + 7 ft 4 in.

30. 12 ft 7 in. + 9 ft 11 in.

31. 12 yd 2 ft + 9 yd 2 ft

32. 16 yd 2 ft + 8 yd 2 ft

33. 22 ft 8 in. − 16 ft 3 in.

34. 15 ft 5 in. − 8 ft 2 in.

35. 18 ft 3 in. − 10 ft 9 in.

36. 14 ft 8 in. − 3 ft 11 in.

37. 28 ft 8 in. ÷ 2

38. 34 ft 6 in. ÷ 2

39. 16 yd 2 ft × 5

40. 15 yd 1 ft × 8

Objective D *Convert as indicated. See Examples 9 and 10.*

41. 60 m to centimeters

42. 46 m to centimeters

43. 40 mm to centimeters

44. 14 mm to centimeters

45. 500 m to kilometers

46. 400 m to kilometers

47. 1700 mm to meters

48. 6400 mm to meters

49. 1500 cm to meters

50. 6400 cm to meters

51. 0.42 km to centimeters

52. 0.95 km to centimeters

53. 7 km to meters

54. 5 km to meters

55. 8.3 cm to millimeters

56. 4.6 cm to millimeters

57. 20.1 mm to decimeters

58. 140.2 mm to decimeters

59. 0.04 m to millimeters

60. 0.2 m to millimeters

Objective E *Perform each indicated operation. Remember to insert units when writing your answers. See Examples 11 through 13.*

61. 8.6 m + 0.34 m

62. 14.1 cm + 3.96 cm

63. 2.9 m + 40 mm

64. 30 cm + 8.9 m

65. 24.8 mm − 1.19 cm

66. 45.3 m − 2.16 dam

67. 15 km − 2360 m

68. 14 cm − 15 mm

69. 18.3 m × 3

70. 14.1 m × 4

71. 6.2 km ÷ 4

72. 9.6 m ÷ 5

Objectives A C D E Mixed Practice *Solve. Remember to insert units when writing your answers. For Exercises 73 through 82, complete the charts. See Examples 1 through 13.*

		Yards	Feet	Inches
73.	Chrysler Building in New York City		1046	
74.	4-story building			792
75.	Python length		35	
76.	Ostrich height			108

		Meters	Millimeters	Kilometers	Centimeters
77.	Length of elephant	5			
78.	Height of grizzly bear	3			
79.	Tennis ball diameter				6.5
80.	Golf ball diameter				4.6
81.	Distance from London to Paris			342	
82.	Distance from Houston to Dallas			396	

83. The National Zoo maintains a small patch of bamboo, which it grows as a food supply for its pandas. Two weeks ago, the bamboo was 6 ft 10 in. tall. Since then, the bamboo has grown 3 ft 8 in. How tall is the bamboo now?

84. While exploring in the Marianas Trench, a submarine probe was lowered to a point 1 mile 1400 feet below the ocean's surface. Later it was lowered an additional 1 mile 4000 feet below this point. How far was the probe below the surface of the Pacific?

85. At its deepest point, the Grand Canyon of the Colorado River in Arizona is about 6000 ft. The Grand Canyon of the Yellowstone River, which is in Yellowstone National Park in Wyoming, is at most 900 feet deep. How much deeper is the Grand Canyon of the Colorado River than the Grand Canyon of the Yellowstone River? (*Source:* National Park Service)

86. The Grand Canyon of the Gunnison River, in Colorado, is often called the Black Canyon of the Gunnison because it is so steep that light rarely penetrates the depth of the canyon. The Black Canyon of the Gunnison is only 1150 ft wide at its narrowest point. At its narrowest, the Grand Canyon of the Yellowstone is $\frac{1}{2}$ mile wide. Find the difference in width between the Grand Canyon of the Yellowstone and the Black Canyon of the Gunnison. (*Note:* Notice that the dimensions are different.) (*Source:* National Park Service)

87. The tallest man in the world is recorded as Robert Pershing Wadlow of Alton, Illinois. Born in 1918, he measured 8 ft 11 in. at his tallest. The shortest man in the world is Chandra Bahadur Dangi of Nepal, who measures 21.5 in. How many times taller than Chandra is Robert? Round to one decimal place. (*Source: Guinness World Records*)

88. A 3.4-m rope is attached to a 5.8-m rope. However, when the ropes are tied, 8 cm of length is lost to form the knot. What is the length of the tied ropes?

89. The ice on a pond is 5.33 cm thick. For safe skating, the owner of the pond insists that it be 80 mm thick. How much thicker must the ice be before skating is allowed?

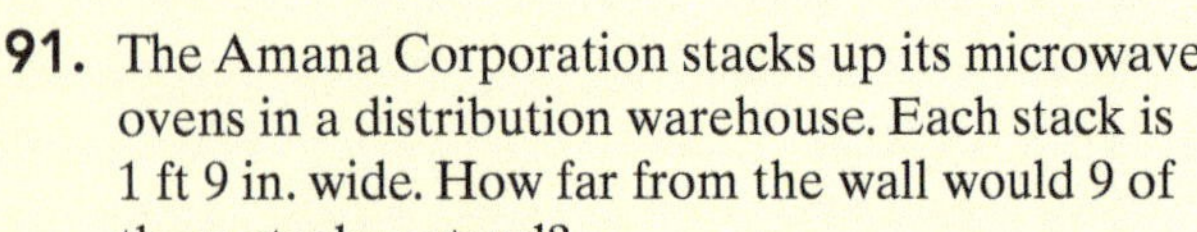

90. The sediment on the bottom of the Towamencin Creek is normally 14 cm thick, but a recent flood washed away 22 mm of sediment. How thick is it now?

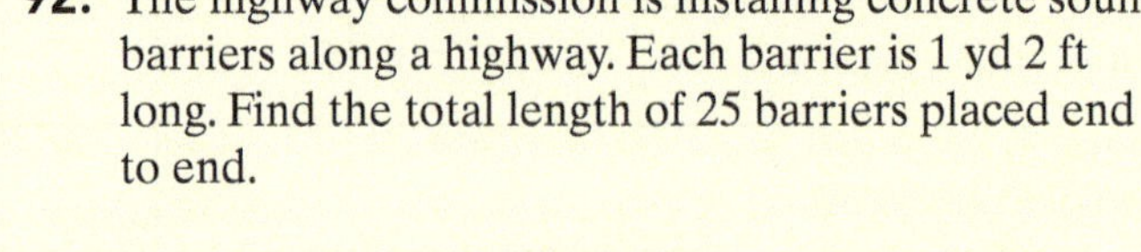

91. The Amana Corporation stacks up its microwave ovens in a distribution warehouse. Each stack is 1 ft 9 in. wide. How far from the wall would 9 of these stacks extend?

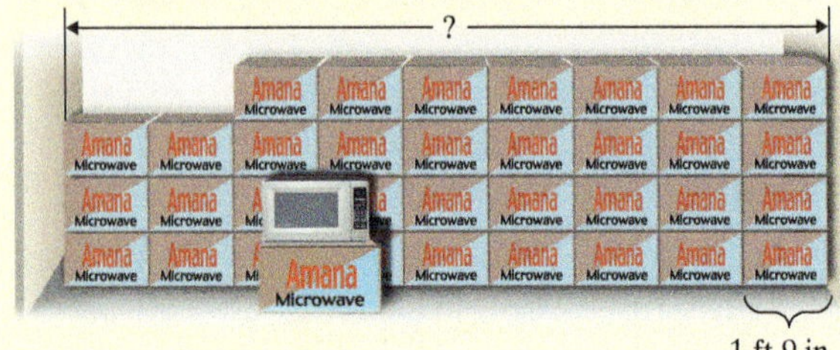

92. The highway commission is installing concrete sound barriers along a highway. Each barrier is 1 yd 2 ft long. Find the total length of 25 barriers placed end to end.

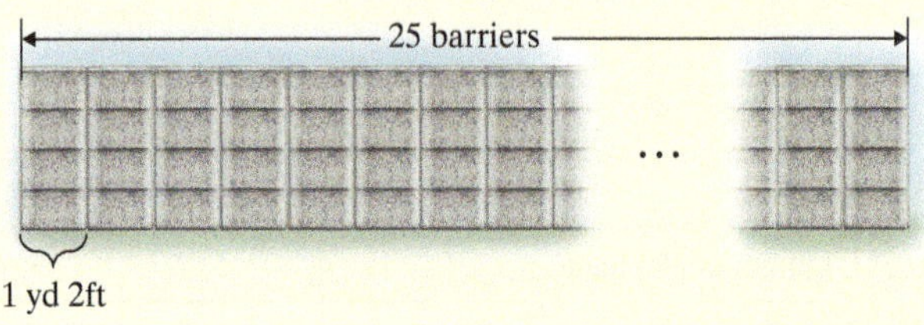

93. A logging firm needs to cut a 67-m-long redwood log into 20 equal pieces before loading it onto a truck for shipment. How long will each piece be?

94. A 112.5-foot-tall dead pinewood tree is removed by starting at the top and cutting off 9-foot-long sections. How many whole sections are removed?

95. The longest truck in the world is operated by Gould Transport in Australia, and is the 182-ft Road Train. How many *yards* long are 2 of these trucks? (*Source: Guinness World Records*)

96. The large Coca-Cola sign in the Tokyo Ginza shopping district is in the shape of a rectangle whose length is 31 yards and whose width is 49 feet. Find the area of the sign in square feet. (*Source:* Coca-Cola Company) (*Hint:* Recall that the area of a rectangle is the product length times width.)

Review

Write each decimal as a fraction and each fraction as a decimal. See Sections 5.1 and 5.5.

97. 0.21

98. 0.86

99. $\frac{13}{100}$

100. $\frac{47}{100}$

101. $\frac{1}{4}$

102. $\frac{3}{20}$

Concept Extensions

Determine whether the measurement in each statement is reasonable.

103. The width of a twin-size bed is 20 meters.

104. A window measures 1 meter by 0.5 meter.

105. A drinking glass is made of glass 2 millimeters thick.

106. A paper clip is 4 kilometers long.

107. The distance across the Colorado River is 50 kilometers.

108. A model's hair is 30 centimeters long.

Estimate each sum or difference. See the first Concept Check in this section.

109.
$$\begin{array}{r} 5\text{ yd}\ \ 2\text{ in.} \\ +\ 7\text{ yd}\ 30\text{ in.} \\ \hline \end{array}$$

110.
$$\begin{array}{r} 45\text{ ft}\ \ 1\text{ in.} \\ -\ 10\text{ ft}\ 11\text{ in.} \\ \hline \end{array}$$

111. Using a unit other than the foot, write a length that is equivalent to 4 feet. (*Hint:* There are many possibilities.)

112. Using a unit other than the meter, write a length that is equivalent to 7 meters. (*Hint:* There are many possibilities.)

113. To convert from meters to centimeters, the decimal point is moved two places to the right. Explain how this relates to the fact that the prefix *centi* means $\frac{1}{100}$.

114. Explain why conversions in the metric system are easier to make than conversions in the U.S. system of measurement.

115. An advertisement sign outside Fenway Park in Boston measures 18.3 m by 18.3 m. What is the area of this sign?

8.5 Weight and Mass 

Objectives

A Define U.S. Units of Weight and Convert from One Unit to Another.

B Perform Arithmetic Operations on U.S. Units of Weight.

C Define Metric Units of Mass and Convert from One Unit to Another.

D Perform Arithmetic Operations on Metric Units of Mass.

Objective A Defining and Converting U.S. System Units of Weight

Whenever we talk about how heavy an object is, we are concerned with the object's **weight.** We discuss weight when we refer to a 12-ounce box of Rice Krispies, a 15-pound tabby cat, or a barge hauling 24 tons of garbage.

12 ounces

15 pounds

24 tons of garbage

The most common units of weight in the U.S. measurement system are the **ounce,** the **pound,** and the **ton.** The following is a summary of equivalencies between units of weight:

U.S. Units of Weight	Unit Fractions
16 ounces (oz) = 1 pound (lb)	$\frac{16\text{ oz}}{1\text{ lb}} = \frac{1\text{ lb}}{16\text{ oz}} = 1$
2000 pounds = 1 ton	$\frac{2000\text{ lb}}{1\text{ ton}} = \frac{1\text{ ton}}{2000\text{ lb}} = 1$

✓**Concept Check** If you were describing the weight of a fully loaded semi-trailer, which type of unit would you use: ounce, pound, or ton? Why?

Unit fractions that equal 1 are used to convert between units of weight in the U.S. system. When converting using unit fractions, recall that the numerator of a unit fraction should contain the units we are converting to and the denominator should contain the original units.

Practice 1

Convert 6500 pounds to tons.

Example 1 Convert 9000 pounds to tons.

Solution: We multiply 9000 lb by a unit fraction that uses the equality 2000 pounds = 1 ton.

Remember, the unit fraction should be $\frac{\text{units to convert to}}{\text{original units}}$ or $\frac{1\text{ ton}}{2000\text{ lb}}$.

$$9000\text{ lb} = \frac{9000\text{ lb}}{1} \cdot 1 = \frac{9000\text{ }\cancel{\text{lb}}}{1} \cdot \frac{1\text{ ton}}{2000\text{ }\cancel{\text{lb}}} = \frac{9000\text{ tons}}{2000} = \frac{9}{2}\text{ tons or } 4\frac{1}{2}\text{ tons}$$

Answer

1. $3\frac{1}{4}$ tons

✓**Concept Check Answer**

ton

Work Practice 1

Example 2 Convert 3 pounds to ounces.

Solution: We multiply by the unit fraction $\frac{16 \text{ oz}}{1 \text{ lb}}$ to convert from pounds to ounces.

$$3 \text{ lb} = \frac{3 \text{ lb}}{1} \cdot 1 = \frac{3 \cancel{\text{lb}}}{1} \cdot \frac{16 \text{ oz}}{1 \cancel{\text{lb}}} = 3 \cdot 16 \text{ oz} = 48 \text{ oz}$$

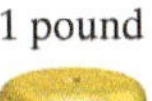

Work Practice 2

Practice 2

Convert 72 ounces to pounds.

As with length, it is sometimes useful to simplify a measurement of weight by writing it in terms of mixed units.

Example 3 Convert: 33 ounces = ______ lb ______ oz

Solution: Because 16 oz = 1 lb, divide 16 into 33 to see how many pounds are in 33 ounces. The quotient is the number of pounds, and the remainder is the number of ounces. To see why we divide 16 into 33, notice that

$$33 \text{ oz} = 33 \cancel{\text{oz}} \cdot \frac{1 \text{ lb}}{16 \cancel{\text{oz}}} = \frac{33}{16} \text{ lb}$$

$$\begin{array}{r} 2 \text{ lb } 1 \text{ oz} \\ 16\overline{)33} \\ -32 \\ \hline 1 \end{array}$$

Thus, 33 ounces is the same as 2 lb 1 oz.

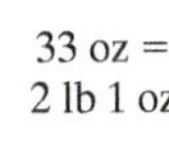

Work Practice 3

Practice 3

Convert:
47 ounces = ______ lb ______ oz

Answers

2. $4\frac{1}{2}$ lb **3.** 2 lb 15 oz

Objective B Performing Operations on U.S. System Units of Weight

Performing arithmetic operations on units of weight works the same way as performing arithmetic operations on units of length.

Practice 4

Subtract 5 tons 1200 lb from 8 tons 100 lb.

Example 4 Subtract 3 tons 1350 lb from 8 tons 1000 lb.

Solution: To subtract, we line up similar units.

$$\begin{array}{r} 8 \text{ tons } 1000 \text{ lb} \\ -\ 3 \text{ tons } 1350 \text{ lb} \\ \hline \end{array}$$

Since we cannot subtract 1350 lb from 1000 lb, we borrow 1 ton from the 8 tons. To do so, we write 1 ton as 2000 lb and combine it with the 1000 lb.

7 tons + 1 ton 2000 lb

$$\begin{array}{rcr} 8 \text{ tons } 1000 \text{ lb} & = & 7 \text{ tons } 3000 \text{ lb} \\ -\ 3 \text{ tons } 1350 \text{ lb} & = & -\ 3 \text{ tons } 1350 \text{ lb} \\ \hline & & 4 \text{ tons } 1650 \text{ lb} \end{array}$$

To check, see that the sum of 4 tons 1650 lb and 3 tons 1350 lb is 8 tons 1000 lb.

Work Practice 4

Practice 5

Divide 5 lb 8 oz by 4.

Example 5 Divide 9 lb 6 oz by 2.

Solution: We divide each of the units by 2.

$$\begin{array}{r} 4 \text{ lb} \quad 11 \text{ oz} \\ 2\overline{)\ 9 \text{ lb} \quad 6 \text{ oz}} \\ -8 \quad\quad\quad \\ \hline 1 \text{ lb} = 16 \text{ oz} \\ 22 \text{ oz} \end{array}$$

Divide 2 into 22 oz to get 11 oz.

To check, multiply 4 pounds 11 ounces by 2. The result is 9 pounds 6 ounces.

Work Practice 5

Practice 6

A 5-lb 14-oz batch of cookies is packed into a 6-oz container before it is mailed. Find the total weight.

Example 6 Finding the Weight of a Child

Bryan weighed 8 lb 8 oz at birth. By the time he was 1 year old, he had gained 11 lb 14 oz. Find his weight at age 1 year.

Solution:

birth weight	→	8 lb 8 oz
+ weight gained	→	+ 11 lb 14 oz
total weight	→	19 lb 22 oz

Since 22 oz equals 1 lb 6 oz,

$$\begin{aligned} 19 \text{ lb } 22 \text{ oz} &= 19 \text{ lb} + 1 \text{ lb } 6 \text{ oz} \\ &= 20 \text{ lb } 6 \text{ oz} \end{aligned}$$

Bryan weighed 20 lb 6 oz on his first birthday.

Work Practice 6

Answers

4. 2 tons 900 lb **5.** 1 lb 6 oz **6.** 6 lb 4 oz

Objective C Defining and Converting Metric System Units of Mass

In scientific and technical areas, a careful distinction is made between **weight** and **mass. Weight** is really a measure of the pull of gravity. The farther from Earth an object gets, the less it weighs. However, **mass** is a measure of the amount of substance in the object and does not change. Astronauts orbiting Earth weigh much less than they weigh on Earth, but they have the same mass in orbit as they do on Earth. Here on Earth, weight and mass are the same, so either term may be used.

The basic unit of mass in the metric system is the **gram.** It is defined as the mass of water contained in a cube 1 centimeter (cm) on each side.

Actual size

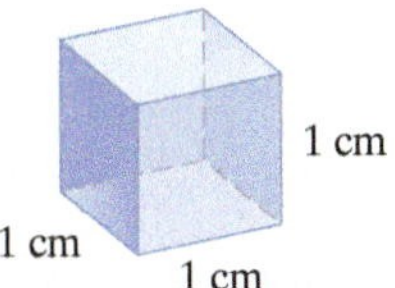

1 cubic centimeter

The following examples may help you get a feeling for metric masses:

- A tablet contains 200 milligrams of ibuprofen.

- A large paper clip weighs approximately 1 gram.

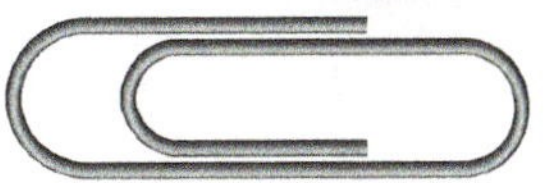

- A box of crackers weighs 453 grams.

A kilogram is slightly over 2 pounds. An adult woman may weigh 60 kilograms.

The prefixes for units of mass in the metric system are the same as for units of length, as shown in the following table:

Metric Units of Mass
1 **kilo**gram (kg) = 1000 grams (g)
1 **hecto**gram (hg) = 100 g
1 **deka**gram (dag) = 10 g
1 gram (g) = 1 g
1 **deci**gram (dg) = 1/10 g or 0.1 g
1 **centi**gram (cg) = 1/100 g or 0.01 g
1 **milli**gram (mg) = 1/1000 g or 0.001 g

✓Concept Check True or false? A decigram is larger than a dekagram. Explain.

The **milligram,** the **gram,** and the **kilogram** are the three most commonly used units of mass in the metric system.

As with lengths, all units of mass are powers of 10 of the gram, so converting from one unit of mass to another only involves moving the decimal point. To convert from one unit of mass to another in the metric system, list the units of mass in order from largest to smallest.

✓Concept Check Answer
false

Let's convert 4300 milligrams to grams. To convert from milligrams to grams, we move along the list 3 units to the left.

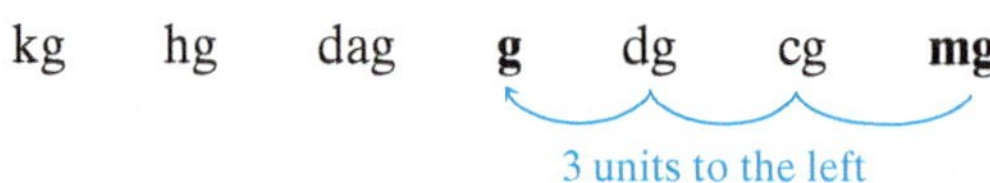

This means that we move the decimal point 3 places to the left to convert from milligrams to grams.

$4300 \text{ mg} = 4.3 \text{ g}$

Don't forget, the same conversion can be done with unit fractions.

$$4300 \text{ mg} = \frac{4300 \text{ mg}}{1} \cdot 1 = \frac{4300 \not{\text{mg}}}{1} \cdot \frac{0.001 \text{ g}}{1 \not{\text{mg}}}$$

$$= 4300 \cdot 0.001 \text{ g}$$

$$= 4.3 \text{ g}$$ To multiply by 0.001, move the decimal point 3 places to the left.

To see that this is reasonable, study the diagram:

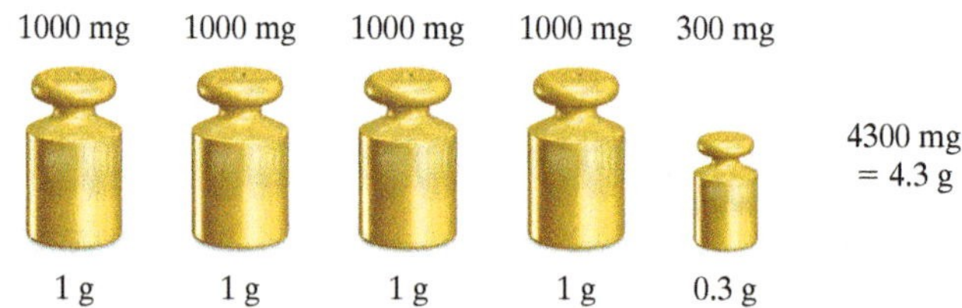

Thus, 4300 mg = 4.3 g.

Practice 7

Convert 3.41 g to milligrams.

Example 7 Convert 3.2 kg to grams.

Solution: First we convert by using a unit fraction.

$$3.2 \text{ kg} = 3.2 \text{ kg} \cdot 1 = 3.2 \not{\text{kg}} \cdot \overbrace{\frac{1000 \text{ g}}{1 \not{\text{kg}}}}^{\text{Unit fraction}} = 3200 \text{ g}$$

Now let's list the units of mass in order from left to right and move from kilograms to grams.

kg hg dag g dg cg mg

3 units to the right

$3.200 \text{ kg} = 3200. \text{ g}$

3 places to the right

1 kg 1 kg 1 kg 0.2 kg

1000 g 1000 g 1000 g 200 g

3.2 kg = 3200 g

Work Practice 7

Answer

7. 3410 mg

Example 8 Convert 2.35 cg to grams.

Solution: We list the units of mass in a chart and move from centigrams to grams.

kg hg dag g dg cg mg

2 units to the left

02.35 cg = 0.0235 g

2 places to the left

Work Practice 8

Practice 8

Convert 56.2 cg to grams.

Objective D Performing Operations on Metric System Units of Mass

Arithmetic operations can be performed with metric units of mass just as we performed operations with metric units of length. We convert each number to the same unit of mass and add, subtract, multiply, or divide as with decimals.

Example 9 Subtract 5.4 dg from 1.6 g.

Solution: We convert both numbers to decigrams or to grams before subtracting.

$5.4\text{ dg} = 0.54\text{ g}$ or $1.6\text{ g} = 16\text{ dg}$

$$\begin{array}{r} 1.60\text{ g} \\ -0.54\text{ g} \\ \hline 1.06\text{ g} \end{array} \qquad \begin{array}{r} 16.0\text{ dg} \\ -5.4\text{ dg} \\ \hline 10.6\text{ dg} \end{array}$$

The difference is 1.06 g or 10.6 dg.

Work Practice 9

Practice 9

Subtract 3.1 dg from 2.5 g.

Example 10 Calculating Allowable Weight in an Elevator

An elevator has a weight limit of 1400 kg. A sign posted in the elevator indicates that the maximum capacity of the elevator is 17 persons. What is the average allowable weight for each passenger, rounded to the nearest kilogram?

Solution: To solve, notice that the total weight of 1400 kilograms ÷ 17 = average weight.

$$\begin{array}{r} 82.3\text{ kg} \approx 82\text{ kg} \\ 17\overline{)1400.0\text{ kg}} \\ -136 \\ \hline 40 \\ -34 \\ \hline 60 \\ -51 \\ \hline 9 \end{array}$$

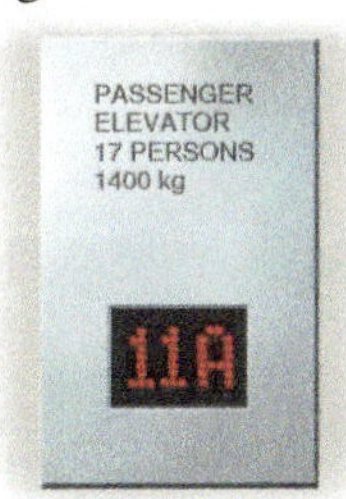

Each passenger can weigh an average of 82 kg. (Recall that a kilogram is slightly over 2 pounds, so 82 kilograms is over 164 pounds.)

Work Practice 10

Practice 10

Twenty-four bags of cement weigh a total of 550 kg. Find the average weight of 1 bag, rounded to the nearest kilogram.

Answers

8. 0.562 g **9.** 2.19 g or 21.9 dg
10. 23 kg

Vocabulary, Readiness & Video Check

Use the choices below to fill in each blank.

mass weight gram

1. __________ is a measure of the amount of substance in an object. This measure does not change.
2. __________ is the measure of the pull of gravity.
3. The basic unit of mass in the metric system is the __________.

Fill in these blanks with the correct number. Choices for these blanks are not shown in the list of terms above.

4. One pound equals __________ ounces.
5. One ton equals __________ pounds.

Martin-Gay Interactive Videos

See Video 8.5 

Watch the section lecture video and answer the following questions.

Objective A 6. In Example 2, what units are used in the numerator of the unit fraction and why was this decided?

Objective B 7. In Example 4, explain the first step taken to solve the problem.

Objective C 8. In Example 5, how many places is the decimal moved and in what direction? What is the final conversion?

Objective D 9. What is the answer to Example 7 in decigrams?

8.5 Exercise Set MyMathLab®

Objective A *Convert as indicated. See Examples 1 through 3.*

1. 2 pounds to ounces
2. 5 pounds to ounces
3. 5 tons to pounds
4. 7 tons to pounds
5. 18,000 pounds to tons
6. 28,000 pounds to tons
7. 60 ounces to pounds
8. 90 ounces to pounds
9. 3500 pounds to tons
10. 11,000 pounds to tons
11. 12.75 pounds to ounces
12. 9.5 pounds to ounces
13. 4.9 tons to pounds
14. 8.3 tons to pounds
15. $4\frac{3}{4}$ pounds to ounces
16. $9\frac{1}{8}$ pounds to ounces
17. 2950 pounds to the nearest tenth of a ton
18. 51 ounces to the nearest tenth of a pound
19. $\frac{4}{5}$ oz to pounds
20. $\frac{1}{4}$ oz to pounds
21. $5\frac{3}{4}$ lb to ounces

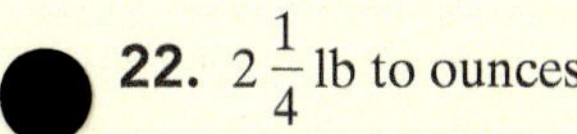

22. $2\frac{1}{4}$ lb to ounces

23. 10 lb 1 oz to ounces

24. 7 lb 6 oz to ounces

25. 89 oz = ____ lb ____ oz

26. 100 oz = ____ lb ____ oz

Objective B *Perform each indicated operation. See Examples 4 through 6.*

27. 34 lb 12 oz + 18 lb 14 oz

28. 6 lb 10 oz + 10 lb 8 oz

29. 3 tons 1820 lb + 4 tons 930 lb

30. 1 ton 1140 lb + 5 tons 1200 lb

31. 5 tons 1050 lb − 2 tons 875 lb

32. 4 tons 850 lb − 1 ton 260 lb

33. 12 lb 4 oz − 3 lb 9 oz

34. 45 lb 6 oz − 26 lb 10 oz

35. 5 lb 3 oz × 6

36. 2 lb 5 oz × 5

37. 6 tons 1500 lb ÷ 5

38. 5 tons 400 lb ÷ 4

Objective C *Convert as indicated. See Examples 7 and 8.*

39. 500 g to kilograms

40. 820 g to kilograms

41. 4 g to milligrams

42. 9 g to milligrams

43. 25 kg to grams

44. 18 kg to grams

45. 48 mg to grams

46. 112 mg to grams

47. 6.3 g to kilograms

48. 4.9 g to kilograms

49. 15.14 g to milligrams

50. 16.23 g to milligrams

51. 6.25 kg to grams

52. 3.16 kg to grams

53. 35 hg to centigrams

54. 4.26 cg to dekagrams

Objective D *Perform each indicated operation. Remember to insert units when writing your answers. See Examples 9 and 10.*

55. 3.8 mg + 9.7 mg

56. 41.6 g + 9.8 g

57. 205 mg + 5.61 g

58. 2.1 g + 153 mg

59. 9 g − 7150 mg

60. 6.13 g − 418 mg

61. 1.61 kg − 250 g

62. 4 kg − 2410 g

63. 5.2 kg × 2.6

64. 4.8 kg × 9.3

65. 17 kg ÷ 8

66. 8.25 g ÷ 6

Objectives A B C D Mixed Practice *Solve. Remember to insert units when writing your answers. For Exercises 67 through 74, complete the chart. See Examples 1 through 10.*

	Object	Tons	Pounds	Ounces
67.	Statue of Liberty—weight of copper sheeting	100		
68.	Statue of Liberty—weight of steel	125		
69.	A 12-inch cube of osmium (heaviest metal)		1345	
70.	A 12-inch cube of lithium (lightest metal)		32	

	Object	Grams	Kilograms	Milligrams	Centigrams
71.	Capsule of amoxicillin (antibiotic)			500	
72.	Tablet of Topamax (epilepsy and migraine uses)			25	
73.	A six-year-old boy		21		
74.	A golf ball	45			

75. A can of 7-Up weighs 336 grams. Find the weight in kilograms of 24 cans.

76. Guy Green normally weighs 73 kg, but he lost 2800 grams after being sick with the flu. Find Guy's new weight.

77. Sudafed is a decongestant that comes in two strengths. Regular strength contains 60 mg of medication. Extra strength contains 0.09 g of medication. How much extra medication is in the extra-strength tablet?

78. A small can of Planters sunflower seeds weighs 177 g. If each can contains 6 servings, find the weight of one serving.

79. Doris Johnson has two open containers of rice. If she combines 1 lb 10 oz from one container with 3 lb 14 oz from the other container, how much total rice does she have?

80. Dru Mizel maintains the records of the amount of coal delivered to his department in the steel mill. In January, 3 tons 1500 lb were delivered. In February, 2 tons 1200 lb were delivered. Find the total amount delivered in these two months.

81. Carla Hamtini was amazed when she grew a 28-lb 10-oz zucchini in her garden, but later she learned that the heaviest zucchini ever grown weighed 64 lb 8 oz in Llanharry, Wales, by B. Lavery in 1990. How far below the record weight was Carla's zucchini? (*Source: Guinness World Records*)

82. The heaviest baby born in good health weighed an incredible 22 lb 8 oz. He was born in Italy in September 1955. How much heavier is this than a 7-lb 12-oz baby? (*Source: Guinness World Records*)

83. The smallest baby born in good health weighed only 8.6 ounces, less than a can of soda. She was born in Chicago in December 2004. How much lighter was she than an average baby, who weighs about 7 lb 8 ounces?

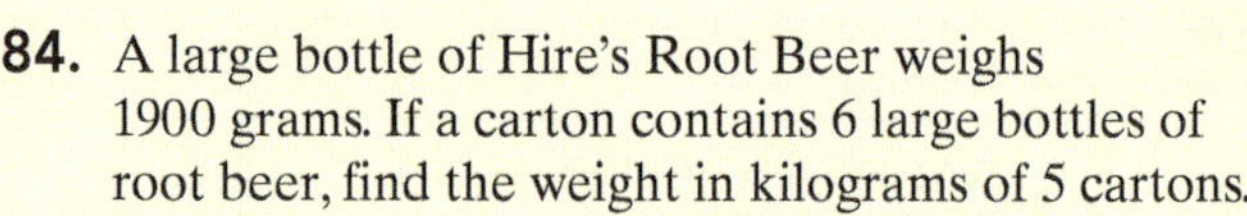

84. A large bottle of Hire's Root Beer weighs 1900 grams. If a carton contains 6 large bottles of root beer, find the weight in kilograms of 5 cartons.

85. Three milligrams of preservatives are added to a 0.5-kg box of dried fruit. How many milligrams of preservatives are in 3 cartons of dried fruit if each carton contains 16 boxes?

86. One box of Swiss Miss Cocoa Mix weighs 0.385 kg, but 39 grams of this weight is the packaging. Find the actual weight of the cocoa in 8 boxes.

87. A carton of 12 boxes of Quaker Oats Oatmeal weighs 6.432 kg. Each box includes 26 grams of packaging material. What is the actual weight of the oatmeal in the carton?

88. The supermarket prepares hamburger in 85-gram market packages. When Leo Gonzalas gets home, he divides the package in half before refrigerating the meat. How much will each package weigh?

89. The Shop 'n Bag supermarket chain ships hamburger meat by placing 10 packages of hamburger in a box, with each package weighing 3 lb 4 oz. How much will 4 boxes of hamburger weigh?

90. The Quaker Oats Company ships its 1-lb 2-oz boxes of oatmeal in cartons containing 12 boxes of oatmeal. How much will 3 such cartons weigh?

91. A carton of Del Monte Pineapple weighs 55 lb 4 oz, but 2 lb 8 oz of this weight is due to packaging. Find the actual weight of the pineapple in 4 cartons.

92. The Hormel Corporation ships cartons of canned ham weighing 43 lb 2 oz each. Of this weight, 3 lb 4 oz is due to packaging. Find the actual weight of the ham found in 3 cartons.

Review

Write each fraction as a decimal. See Section 5.5.

93. $\frac{4}{25}$ **94.** $\frac{3}{5}$ **95.** $\frac{7}{8}$ **96.** $\frac{3}{16}$

Concept Extensions

Determine whether the measurement in each statement is reasonable.

97. The doctor prescribed a pill containing 2 kg of medication.

98. A full-grown cat weighs approximately 15 g.

99. A bag of flour weighs 4.5 kg.

100. A staple weighs 15 mg.

101. A professor weighs less than 150 g.

102. A car weighs 2000 mg.

103. Use a unit other than centigram and write a mass that is equivalent to 25 centigrams. (*Hint:* There are many possibilities.)

104. Use a unit other than pound and write a weight that is equivalent to 4000 pounds. (*Hint:* There are many possibilities.)

True or false? See the second Concept Check in this section.

105. A kilogram is larger than a gram.

106. A decigram is larger than a milligram.

107. Why is the decimal point moved to the right when grams are converted to milligrams?

108. To change 8 pounds to ounces, multiply by 16. Why is this the correct procedure?

8.6 Capacity 

Objectives

A Define U.S. Units of Capacity and Convert from One Unit to Another.

B Perform Arithmetic Operations on U.S. Units of Capacity.

C Define Metric Units of Capacity and Convert from One Unit to Another.

D Perform Arithmetic Operations on Metric Units of Capacity.

Objective A Defining and Converting U.S. System Units of Capacity

Units of **capacity** are generally used to measure liquids. The number of gallons of gasoline needed to fill a gas tank in a car, the number of cups of water needed in a bread recipe, and the number of quarts of milk sold each day at a supermarket are all examples of using units of capacity. The following summary shows equivalencies between units of capacity:

U.S. Units of Capacity

$$\begin{aligned} 8 \text{ fluid ounces (fl oz)} &= 1 \text{ cup (c)} \\ 2 \text{ cups} &= 1 \text{ pint (pt)} \\ 2 \text{ pints} &= 1 \text{ quart (qt)} \\ 4 \text{ quarts} &= 1 \text{ gallon (gal)} \end{aligned}$$

Just as with units of length and weight, we can form unit fractions to convert between different units of capacity. For instance,

$$\frac{2 \text{ c}}{1 \text{ pt}} = \frac{1 \text{ pt}}{2 \text{ c}} = 1 \quad \text{and} \quad \frac{2 \text{ pt}}{1 \text{ qt}} = \frac{1 \text{ qt}}{2 \text{ pt}} = 1$$

Practice 1

Convert 43 pints to quarts.

Example 1 Convert 9 quarts to gallons.

Solution: We multiply by the unit fraction $\frac{1 \text{ gal}}{4 \text{ qt}}$.

$$\begin{aligned} 9 \text{ qt} &= \frac{9 \text{ qt}}{1} \cdot 1 \\ &= \frac{9 \cancel{\text{qt}}}{1} \cdot \frac{1 \text{ gal}}{4 \cancel{\text{qt}}} \\ &= \frac{9 \text{ gal}}{4} \\ &= 2\frac{1}{4} \text{ gal} \end{aligned}$$

Thus, 9 quarts is the same as $2\frac{1}{4}$ gallons, as shown in the diagram:

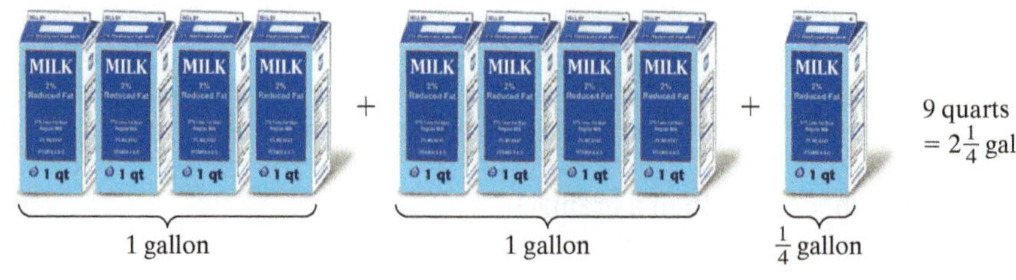

Work Practice 1

Answer

1. $21\frac{1}{2}$ qt

Example 2 Convert 14 cups to quarts.

Solution: Our equivalency table contains no direct conversion from cups to quarts. However, from this table we know that

$$1\text{ qt} = 2\text{ pt} = \frac{2\text{ pt}}{1}\cdot 1 = \frac{2\text{ pt}}{1}\cdot\frac{2\text{ c}}{1\text{ pt}} = 4\text{ c}$$

so 1 qt = 4 c. Now we have the unit fraction $\frac{1\text{ qt}}{4\text{ c}}$. Thus,

$$14\text{ c} = \frac{14\text{ c}}{1}\cdot 1 = \frac{14\text{ c}}{1}\cdot\frac{1\text{ qt}}{4\text{ c}} = \frac{14\text{ qt}}{4} = \frac{7}{2}\text{ qt} \quad\text{or}\quad 3\frac{1}{2}\text{ qt}$$

1 quart + 1 quart + 1 quart + $\frac{1}{2}$ quart; 14 cups $= 3\frac{1}{2}$ qt

Work Practice 2

Practice 2

Convert 26 quarts to cups.

✓**Concept Check** If 50 cups is converted to quarts, will the equivalent number of quarts be less than or greater than 50? Explain.

Objective B Performing Operations on U.S. System Units of Capacity

As is true of units of length and weight, units of capacity can be added, subtracted, multiplied, and divided.

Example 3 Subtract 3 qt from 4 gal 2 qt.

Solution: To subtract, we line up similar units.

$$\begin{array}{r} 4\text{ gal }2\text{ qt} \\ -\quad\quad 3\text{ qt} \\ \hline \end{array}$$

We cannot subtract 3 qt from 2 qt. We need to borrow 1 gallon from the 4 gallons, convert it to 4 quarts, and then combine it with the 2 quarts.

3 gal + 1 gal 4 qt

$$\begin{array}{rcr} 4\text{ gal }2\text{ qt} & = & 3\text{ gal }6\text{ qt} \\ -\quad 3\text{ qt} & = & -\quad 3\text{ qt} \\ \hline & & 3\text{ gal }3\text{ qt} \end{array}$$

To check, see that the sum of 3 gal 3 qt and 3 qt is 4 gal 2 qt.

Work Practice 3

Practice 3

Subtract 2 qt from 1 gal 1 qt.

Example 4 Finding the Amount of Water in an Aquarium

An aquarium contains 6 gal 3 qt of water. If 2 gal 2 qt of water is added, what is the total amount of water in the aquarium?

Solution:

beginning water	→	6 gal 3 qt
+ water added	→	+ 2 gal 2 qt
total water	→	8 gal 5 qt

(*Continued on next page*)

Practice 4

A large oil drum contains 15 gal 3 qt of oil. How much will be in the drum if an additional 4 gal 3 qt of oil is poured into it?

Answers

2. 104 c **3.** 3 qt **4.** 20 gal 2 qt

✓**Concept Check Answer**

less than 50

Since 5 qt = 1 gal 1 qt, we have

$$= \underbrace{8 \text{ gal}}_{8 \text{ gal}} + \underbrace{1 \text{ gal } 1 \text{ qt}}_{5 \text{ qt}}$$

$$= 9 \text{ gal } 1 \text{ qt}$$

The total amount of water is 9 gal 1 qt.

Work Practice 4

Objective C Defining and Converting Metric System Units of Capacity

Thus far, we know that the basic unit of length in the metric system is the meter and that the basic unit of mass in the metric system is the gram. What is the basic unit of capacity? The **liter.** By definition, a **liter** is the capacity or volume of a cube measuring 10 centimeters on each side.

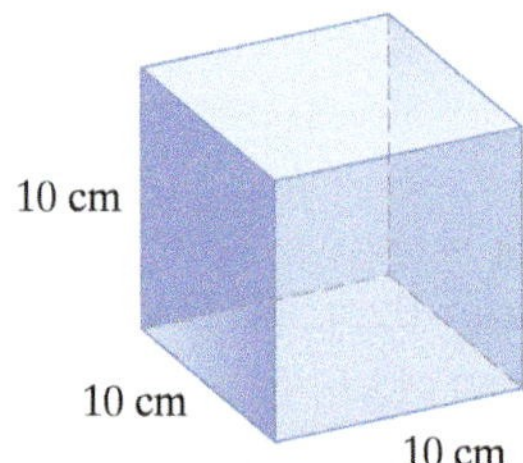

The following examples may help you get a feeling for metric capacities:

- One liter of liquid is slightly more than one quart.

1 liter

1 quart

- Many soft drinks are packaged in 2-liter bottles.

2 liters

The metric system was designed to be a consistent system. Once again, the prefixes for metric units of capacity are the same as for metric units of length and mass, as summarized in the following table:

Metric Units of Capacity
1 **kilo**liter (kl) = 1000 liters (L)
1 **hecto**liter (hl) = 100 L
1 **deka**liter (dal) = 10 L
1 liter (L) = 1 L
1 **deci**liter (dl) = 1/10 L or 0.1 L
1 **centi**liter (cl) = 1/100 L or 0.01 L
1 **milli**liter (ml) = 1/1000 L or 0.001 L

The **milliliter** and the **liter** are the two most commonly used metric units of capacity.

Converting from one unit of capacity to another involves multiplying by powers of 10 or moving the decimal point to the left or to the right. Listing units of capacity in order from largest to smallest helps to keep track of how many places to move the decimal point when converting.

Let's convert 2.6 liters to milliliters. To convert from liters to milliliters, we move along the chart 3 units to the right.

kl hl dal **L** dl cl **ml**

3 units to the right

This means that we move the decimal point 3 places to the right to convert from liters to milliliters.

2.600 L = 2600.ml

This same conversion can be done with unit fractions.

$$\begin{aligned} 2.6 \text{ L} &= \frac{2.6 \text{ L}}{1} \cdot 1 \\ &= \frac{2.6 \text{ \cancel{L}}}{1} \cdot \frac{1000 \text{ ml}}{1 \text{ \cancel{L}}} \\ &= 2.6 \cdot 1000 \text{ ml} \\ &= 2600 \text{ ml} \end{aligned}$$

To multiply by 1000, move the decimal point 3 places to the right.

To visualize the result, study the diagram below:

Thus, 2.6 L = 2600 ml.

Example 5 Convert 3210 ml to liters.

Solution: Let's use the unit fraction method first.

$$3210 \text{ ml} = \frac{3210 \text{ ml}}{1} \cdot 1 = 3210 \text{ \cancel{ml}} \cdot \underbrace{\frac{1 \text{ L}}{1000 \text{ \cancel{ml}}}}_{\text{Unit fraction}} = 3.21 \text{ L}$$

Now let's list the unit measures in order from left to right and move from milliliters to liters.

kl hl dal L dl cl ml

3 units to the left

3210 ml = 3.210 L, the same result as before and shown below in the diagram.

3 places to the left

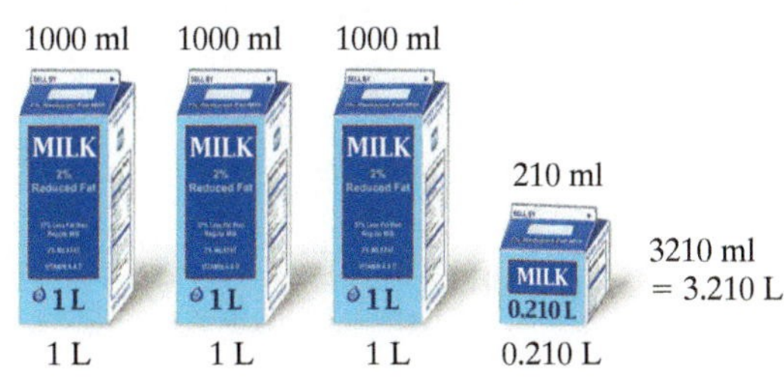

Work Practice 5

Practice 5

Convert 2100 ml to liters.

Answer

5. 2.1 L

Practice 6

Convert 2.13 dal to liters.

Example 6 Convert 0.185 dl to milliliters.

Solution: We list the unit measures in order from left to right and move from deciliters to milliliters.

kl hl dal L dl cl ml

2 units to the right

0.185 dl = 18.5 ml

2 places to the right

Work Practice 6

Objective D Performing Operations on Metric System Units of Capacity

As was true for length and weight, arithmetic operations involving metric units of capacity can also be performed. Make sure that the metric units of capacity are the same before adding, subtracting, multiplying, or dividing.

Practice 7

Add 1250 ml to 2.9 L.

Example 7 Add 2400 ml to 8.9 L.

Solution: We must convert both to liters or both to milliliters before adding the capacities together.

2400 ml = 2.4 L or 8.9 L = 8900 ml

$$\begin{array}{r} 2.4\text{ L} \\ +\ 8.9\text{ L} \\ \hline 11.3\text{ L} \end{array} \qquad \begin{array}{r} 2400\text{ ml} \\ +\ 8900\text{ ml} \\ \hline 11{,}300\text{ ml} \end{array}$$

The total is 11.3 L or 11,300 ml. They both represent the same capacity.

Work Practice 7

✓**Concept Check** How could you estimate the following operation? Subtract 950 ml from 7.5 L.

Practice 8

If 28.6 L of water can be pumped every minute, how much water can be pumped in 85 minutes?

Example 8 Finding the Amount of Medication a Person Has Received

A patient hooked up to an IV unit in the hospital is to receive 12.5 ml of medication every hour. How much medication does the patient receive in 3.5 hours?

Solution: We multiply 12.5 ml by 3.5.

$$\begin{array}{rcr} \text{medication per hour} & \rightarrow & 12.5\text{ ml} \\ \times \qquad\qquad \text{hours} & \rightarrow & \times\ 3.5 \\ \hline \text{total medication} & & 625 \\ & & 3750 \\ \hline & & 43.75\text{ ml} \end{array}$$

The patient receives 43.75 ml of medication.

Work Practice 8

Answers

6. 21.3 L **7.** 4150 ml or 4.15 L
8. 2431 L

✓**Concept Check Answer**

950 ml = 0.95 L; round 0.95 to 1; 7.5 − 1 = 6.5 L

Vocabulary, Readiness & Video Check

Use the choices below to fill in each blank. Some choices may be used more than once.

cups	pints	liter
quarts	fluid ounces	capacity

1. Units of __________ are generally used to measure liquids.
2. The basic unit of capacity in the metric system is the __________.
3. One cup equals 8 __________.
4. One quart equals 2 __________.
5. One pint equals 2 __________.
6. One quart equals 4 __________.
7. One gallon equals 4 __________.

Martin-Gay Interactive Videos

See Video 8.6

Watch the section lecture video and answer the following questions.

Objective A **8.** Complete this statement based on Example 1: When using a unit fraction, we are not changing the __________, we are changing the __________.

Objective B **9.** In Example 4, explain the first step taken to solve the problem.

Objective C **10.** In Example 5, how many places is the decimal moved and in what direction? What is the final conversion?

Objective D **11.** What is the answer to Example 7 in dekaliters?

8.6 Exercise Set MyMathLab®

Objective A *Convert each measurement as indicated. See Examples 1 and 2.*

1. 32 fluid ounces to cups
2. 16 quarts to gallons
3. 8 quarts to pints
4. 9 pints to quarts
5. 14 quarts to gallons
6. 11 cups to pints
7. 80 fluid ounces to pints
8. 18 pints to gallons
9. 2 quarts to cups
10. 3 pints to fluid ounces
11. 120 fluid ounces to quarts
12. 20 cups to gallons
13. 42 cups to quarts
14. 7 quarts to cups
15. $4\frac{1}{2}$ pints to cups
16. $6\frac{1}{2}$ gallons to quarts
17. 5 gal 3 qt to quarts
18. 4 gal 1 qt to quarts
19. $\frac{1}{2}$ cup to pints
20. $\frac{1}{2}$ pint to quarts
21. 58 qt = ____ gal ____ qt
22. 70 qt = ____ gal ____ qt

23. 39 pt = _____ gal _____ qt _____ pt

24. 29 pt = _____ gal _____ qt _____ pt

25. $2\frac{3}{4}$ gallons to pints

26. $3\frac{1}{4}$ quarts to cups

Objective B *Perform each indicated operation. See Examples 3 and 4.*

27. 5 gal 3 qt + 7 gal 3 qt

28. 2 gal 2 qt + 9 gal 3 qt

29. 1 c 5 fl oz + 2 c 7 fl oz

30. 2 c 3 fl oz + 2 c 6 fl oz

31. 3 gal − 1 gal 3 qt

32. 2 pt − 1 pt 1 c

33. 3 gal 1 qt − 1 qt 1 pt

34. 3 qt 1 c − 1 c 4 fl oz

35. 8 gal 2 qt × 2

36. 6 gal 1 pt × 2

37. 9 gal 2 qt ÷ 2

38. 5 gal 6 fl oz ÷ 2

Objective C *Convert as indicated. See Examples 5 and 6.*

39. 5 L to milliliters

40. 8 L to milliliters

41. 0.16 L to kiloliters

42. 0.127 L to kiloliters

43. 5600 ml to liters

44. 1500 ml to liters

45. 3.2 L to centiliters

46. 1.7 L to centiliters

47. 410 L to kiloliters

48. 250 L to kiloliters

49. 64 ml to liters

50. 39 ml to liters

51. 0.16 kl to liters

52. 0.48 kl to liters

53. 3.6 L to milliliters

54. 1.9 L to milliliters

Objective D *Perform each indicated operation. Remember to insert units when writing your answers. See Examples 7 and 8.*

55. 3.4 L + 15.9 L

56. 18.5 L + 4.6 L

57. 2700 ml + 1.8 L

58. 4.6 L + 1600 ml

59. 8.6 L − 190 ml

60. 4.8 L − 283 ml

61. 17,500 ml − 0.9 L

62. 6850 ml − 0.3 L

63. 480 ml × 8

64. 290 ml × 6

65. 81.2 L ÷ 0.5

66. 5.4 L ÷ 3.6

Objectives A B C D Mixed Practice *Solve. Remember to insert units when writing your answers. For Exercises 67 through 70, complete the chart.*

	Capacity	Cups	Gallons	Quarts	Pints
67.	An average-size bath of water		21		
68.	A dairy cow's daily milk yield				38
69.	Your kidneys filter about this amount of blood every minute	4			
70.	The amount of water needed in a punch recipe	2			

71. Mike Schaferkotter drank 410 ml of Mountain Dew from a 2-liter bottle. How much Mountain Dew remains in the bottle?

72. The Werners' Volvo has a 54.5-L gas tank. Only 3.8 liters of gasoline still remain in the tank. How much is needed to fill it?

73. Margie Phitts added 354 ml of Prestone dry gas to the 18.6 L of gasoline in her car's tank. Find the total amount of gasoline in the tank.

74. Chris Peckaitis wishes to share a 2-L bottle of Coca-Cola equally with 7 of his friends. How much will each person get?

75. A garden tool engine requires a 30-to-1 gas-to-oil mixture. This means that $\frac{1}{30}$ of a gallon of oil should be mixed with 1 gallon of gas. Convert $\frac{1}{30}$ gallon to fluid ounces. Round to the nearest tenth.

76. Henning's Supermarket sells homemade soup in 1 qt 1 pt containers. How much soup is contained in three such containers?

77. Can 5 pt 1 c of fruit punch and 2 pt 1 c of ginger ale be poured into a 1-gal container without it overflowing?

78. Three cups of prepared Jell-O are poured into 6 dessert dishes. How many fluid ounces of Jell-O are in each dish?

79. Stanley Fisher paid $14 to fill his car with 44.3 liters of gasoline. Find the price per liter of gasoline to the nearest thousandth of a dollar.

80. A student carelessly misread the scale on a cylinder in the chemistry lab and added 40 cl of water to a mixture instead of 40 ml. Find the excess amount of water.

Review

Write each fraction in simplest form. See Section 4.2.

81. $\frac{20}{25}$ **82.** $\frac{75}{100}$ **83.** $\frac{27}{45}$ **84.** $\frac{56}{60}$ **85.** $\frac{72}{80}$ **86.** $\frac{18}{20}$

Concept Extensions

Determine whether the measurement in each statement is reasonable.

87. Clair took a dose of 2 L of cough medicine to cure her cough.

88. John drank 250 ml of milk for lunch.

89. Jeannie likes to relax in a tub filled with 3000 ml of hot water.

90. Sarah pumped 20 L of gasoline into her car yesterday.

Solve. See the Concept Checks in this section.

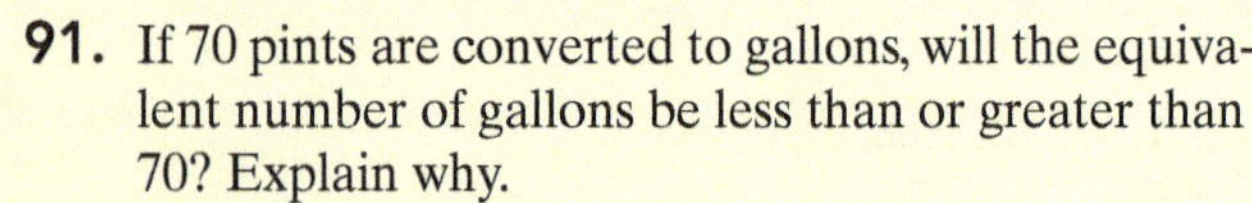

91. If 70 pints are converted to gallons, will the equivalent number of gallons be less than or greater than 70? Explain why.

92. If 30 gallons are converted to quarts, will the equivalent number of quarts be less than or greater than 30? Explain why.

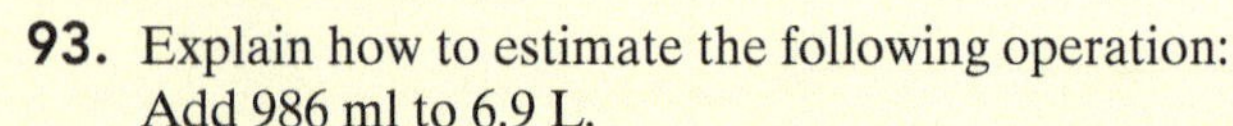

93. Explain how to estimate the following operation: Add 986 ml to 6.9 L.

94. Explain how to borrow in order to subtract 1 gal 2 qt from 3 gal 1 qt.

95. Find the number of fluid ounces in 1 gallon.

96. Find the number of fluid ounces in 1.5 gallons.

A cubic centimeter (cc) is the amount of space that a volume of 1 ml occupies. Because of this, we will say that 1 cc = 1 ml.

A common syringe is one with a capacity of 3 cc. Use the diagram and give the measurement indicated by each arrow.

97. B **98.** A **99.** D **100.** C

In order to measure small dosages, such as for insulin, u-100 syringes are used. For these syringes, 1 cc has been divided into 100 equal units (u). Use the diagram and give the measurement indicated by each arrow in units (u) and then in cubic centimeters. Use 100 u = 1 cc.

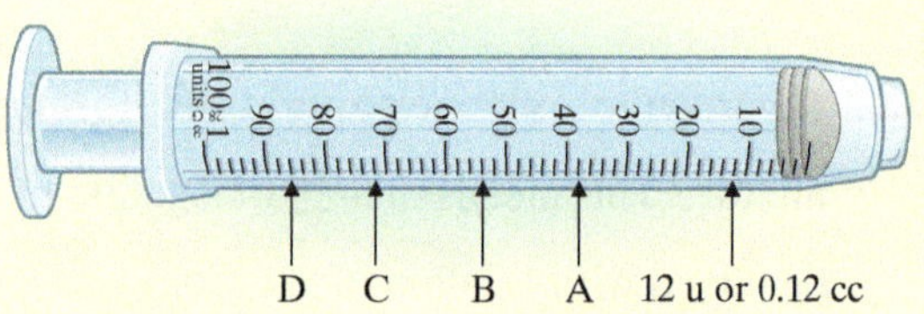

101. B **102.** A

103. D **104.** C

8.7 Temperature and Conversions Between the U.S. and Metric Systems

Objectives

A Convert Between the U.S. and Metric Systems.

B Convert Temperatures from Degrees Celsius to Degrees Fahrenheit.

C Convert Temperatures from Degrees Fahrenheit to Degrees Celsius.

Objective A Converting Between the U.S. and Metric Systems

The metric system probably had its beginnings in France in the 1600s, but it was the Metric Act of 1866 that made the use of this system legal (although not mandatory) in the United States. Other laws have followed that allow for a slow, but deliberate, transfer to the modernized metric system. In April 2001, for example, the U.S. Stock Exchanges completed their change to decimal trading instead of fractions. By the end of 2009, all products sold in Europe (with some exceptions) were required to have only metric units on their labels. (*Source:* U.S. Metric Association and National Institute of Standards and Technology)

You may be surprised at the number of everyday items we use that are already manufactured in metric units. We easily recognize 1 L and 2 L soda bottles, but what about the following?

- Pencil leads (0.5 mm or 0.7 mm)
- Camera film (35 mm)
- Sporting events (5-km or 10-km races)

- Medicines (500-mg capsules)
- Labels on retail goods (dual-labeled since 1994)

Since the United States has not completely converted to the metric system, we need to practice converting from one system to the other. Below is a table of mostly approximate conversions.

Length:		**Capacity:**		**Weight (mass):**	
Metric	U.S. System	Metric	U.S. System	Metric	U.S. System
1 m ≈	1.09 yd	1 L ≈	1.06 qt	1 kg ≈	2.20 lb
1 m ≈	3.28 ft	1 L ≈	0.26 gal	1 g ≈	0.04 oz
1 km ≈	0.62 mi	3.79 L ≈	1 gal	0.45 kg ≈	1 lb
2.54 cm =	1 in.	0.95 L ≈	1 qt	28.35 g ≈	1 oz
0.30 m ≈	1 ft	29.57 ml ≈	1 fl oz		
1.61 km ≈	1 mi				

There are many ways to perform these metric-to-U.S. conversions. We will do so by using unit fractions.

Example 1 Compact Discs

Standard-sized compact discs are 12 centimeters in diameter. Convert this length to inches. Round the result to two decimal places.

Solution: From our length conversion table, we know that 2.54 cm = 1 in. This fact gives us two unit fractions: $\frac{2.54\text{ cm}}{1\text{ in.}}$ and $\frac{1\text{ in.}}{2.54\text{ cm}}$. We use the unit fraction with cm in the denominator so that these units divide out.

1.5 cm

12 cm

$$12\text{ cm} = \frac{12\text{ cm}}{1}\cdot 1 = \frac{12\ \cancel{\text{cm}}}{1}\cdot\overbrace{\frac{1\text{ in.}}{2.54\ \cancel{\text{cm}}}}^{\text{Unit fraction}}\begin{matrix}\leftarrow \text{Units to convert to}\\ \leftarrow \text{Original units}\end{matrix}$$

$$= \frac{12\text{ in.}}{2.54}$$

$$\approx 4.72\text{ in.}\quad \text{Divide.}$$

Thus, the diameter of a standard compact disc is exactly 12 cm or approximately 4.72 inches. For a dimension this size, you can use a ruler to check. Another method is to approximate. Our result, 4.72 in., is close to 5 inches. Since 1 in. is about 2.5 cm, then 5 in. is about $5(2.5\text{ cm}) = 12.5\text{ cm}$, which is close to 12 cm.

Work Practice 1

Practice 1

The center hole of a standard-sized compact disc is 1.5 centimeters in diameter. Convert this length to inches. Round the result to 2 decimal places.

Example 2 Liver

The liver is your largest internal organ. It weighs about 3.5 pounds in a grown man. Convert this weight to kilograms. Round to the nearest tenth. (*Source: Some Body!* by Dr. Pete Rowan)

Solution: $3.5\text{ lb} \approx \frac{3.5\ \cancel{\text{lb}}}{1}\cdot\overbrace{\frac{0.45\text{ kg}}{1\ \cancel{\text{lb}}}}^{\text{Unit fraction}} = 3.5(0.45\text{ kg}) \approx 1.6\text{ kg}$

Thus 3.5 pounds is approximately 1.6 kilograms. From the table of conversions, we know that 1 kg ≈ 2.2 lb. So that means 0.5 kg ≈ 1.1 lb and after adding, we have 1.5 kg ≈ 3.3 lb. Our result is reasonable.

Work Practice 2

Practice 2

A full-grown human heart weighs about 8 ounces. Convert this weight to grams. If necessary, round your result to the nearest tenth of a gram.

Answers

1. 0.59 in. **2.** 226.8 g

Practice 3

Convert 237 ml to fluid ounces. Round to the nearest whole fluid ounce.

Example 3 Postage Stamp

Australia converted to the metric system in 1973. In that year, four postage stamps were issued to publicize this conversion. One such stamp is shown. Let's check the mathematics on the stamp by converting 7 fluid ounces to milliliters. Round to the nearest hundred.

Solution: $7 \text{ fl oz} \approx \frac{7 \cancel{\text{fl oz}}}{1} \cdot \underbrace{\frac{29.57 \text{ ml}}{1 \cancel{\text{fl oz}}}}_{\text{Unit fraction}} = 7(29.57 \text{ ml}) = 206.99 \text{ ml}$

Rounded to the nearest hundred, 7 fl oz ≈ 200 ml.

Work Practice 3

Now that we have practiced converting between two measurement systems, let's practice converting between two temperature scales.

Temperature When Gabriel Fahrenheit and Anders Celsius independently established units for temperature scales, each based his unit on the heat of water the moment it boils compared to the moment it freezes. One degree Celsius is $\frac{1}{100}$ of the difference in heat. One degree Fahrenheit is $\frac{1}{180}$ of the difference in heat. Celsius arbitrarily labeled the temperature at the freezing point at 0°C, making the boiling point 100°C; Fahrenheit labeled the freezing point 32°F, making the boiling point 212°F. Water boils at 212°F or 100°C.

By comparing the two scales in the figure, we see that a 20°C day is as warm as a 68°F day. Similarly, a sweltering 104°F day in the Mojave desert corresponds to a 40°C day.

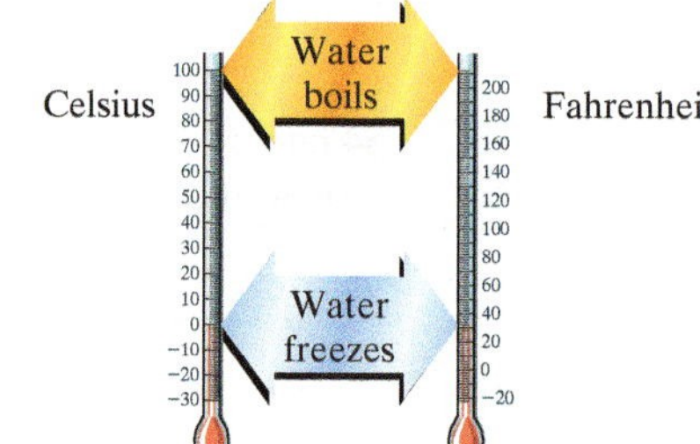

✓Concept Check Which of the following statements is correct? Explain.

a. 6°C is below the freezing point of water.

b. 6°F is below the freezing point of water.

Objective B Converting Degrees Celsius to Degrees Fahrenheit

To convert from Celsius temperatures to Fahrenheit temperatures, see the box below. In this box, we use the symbol F to represent degrees Fahrenheit and the symbol C to represent degrees Celsius.

Converting Celsius to Fahrenheit

$$F = \frac{9}{5}C + 32 \quad \text{or} \quad F = 1.8C + 32$$

(To convert to Fahrenheit temperature, multiply the Celsius temperature by $\frac{9}{5}$ or 1.8, and then add 32.)

Answer

3. 8 fl oz

✓Concept Check Answer

b

Example 4 Convert 15°C to degrees Fahrenheit.

Solution:

$$F = \frac{9}{5}C + 32$$
$$= \frac{9}{5} \cdot 15 + 32 \quad \text{Replace C with 15.}$$
$$= 27 + 32 \quad \text{Simplify.}$$
$$= 59 \quad \text{Add.}$$

Thus, 15°C is equivalent to 59°F.

Work Practice 4

Practice 4

Convert 60°C to degrees Fahrenheit.

Example 5 Convert 29°C to degrees Fahrenheit.

Solution:

$$F = 1.8C + 32$$
$$= 1.8 \cdot 29 + 32 \quad \text{Replace C with 29.}$$
$$= 52.2 + 32 \quad \text{Multiply 1.8 by 29.}$$
$$= 84.2 \quad \text{Add.}$$

Therefore, 29°C is the same as 84.2°F.

Work Practice 5

Practice 5

Convert 32°C to degrees Fahrenheit.

Objective C Converting Degrees Fahrenheit to Degrees Celsius

To convert from Fahrenheit temperatures to Celsius temperatures, see the box below. The symbol C represents degrees Celsius and the symbol F represents degrees Fahrenheit.

Converting Fahrenheit to Celsius

$$C = \frac{5}{9}(F - 32)$$

(To convert to Celsius temperature, subtract 32 from the Fahrenheit temperature, and then multiply by $\frac{5}{9}$.)

Example 6 Convert 59°F to degrees Celsius.

Solution: We evaluate the formula $C = \frac{5}{9}(F - 32)$ when F is 59.

$$C = \frac{5}{9}(F - 32)$$
$$= \frac{5}{9} \cdot (59 - 32) \quad \text{Replace F with 59.}$$
$$= \frac{5}{9} \cdot (27) \quad \text{Subtract inside parentheses.}$$
$$= 15 \quad \text{Multiply.}$$

Therefore, 59°F is the same temperature as 15°C.

Work Practice 6

Practice 6

Convert 68°F to degrees Celsius.

Answers

4. 140°F 5. 89.6°F 6. 20°C

Practice 7

Convert 113°F to degrees Celsius. If necessary, round to the nearest tenth of a degree.

Example 7 Convert 114°F to degrees Celsius. If necessary, round to the nearest tenth of a degree.

Solution:
$$C = \frac{5}{9}(F - 32)$$
$$= \frac{5}{9}(114 - 32) \quad \text{Replace F with 114.}$$
$$= \frac{5}{9}\cdot(82) \quad \text{Subtract inside parentheses.}$$
$$\approx 45.6 \quad \text{Multiply.}$$

Therefore, 114°F is approximately 45.6°C.

Work Practice 7

Practice 8

During a bout with the flu, Albert's temperature reaches 102.8°F. What is his temperature measured in degrees Celsius? Round to the nearest tenth of a degree.

Example 8 Body Temperature

Normal body temperature is 98.6°F. What is this temperature in degrees Celsius?

Solution: We evaluate the formula $C = \frac{5}{9}(F - 32)$ when F is 98.6.

$$C = \frac{5}{9}(F - 32)$$
$$= \frac{5}{9}(98.6 - 32) \quad \text{Replace F with 98.6.}$$
$$= \frac{5}{9}\cdot(66.6) \quad \text{Subtract inside parentheses.}$$
$$= 37 \quad \text{Multiply.}$$

Therefore, normal body temperature is 37°C.

Work Practice 8

✓**Concept Check** Clarissa must convert 40°F to degrees Celsius. What is wrong with her work shown below?

$$F = 1.8\cdot C + 32$$
$$F = 1.8\cdot 40 + 32$$
$$F = 72 + 32$$
$$F = 104$$

Answers
7. 45°C 8. 39.3°C

✓**Concept Check Answer**
She used the conversion for Celsius to Fahrenheit instead of Fahrenheit to Celsius.

Vocabulary, Readiness & Video Check

Martin-Gay Interactive Videos

See Video 8.7

Watch the section lecture video and answer the following questions.

Objective A 1. Write two conversions that may be used to solve Example 2.

2. Why isn't 0.1125 kg the final answer to Example 3?

Objective B 3. Which version of the formula is used to solve Example 4? What is the replacement value for C?

Objective C 4. In Example 5, what is the replacement value for F? What is the final conversion?

8.7 Exercise Set MyMathLab®

Note: Because approximations are used, your answers may vary slightly from the answers given in the back of the book.

Objective A *Convert as indicated. If necessary, round answers to two decimal places. See Examples 1 through 3.*

1. 756 milliliters to fluid ounces
2. 18 liters to quarts
3. 86 inches to centimeters
4. 86 miles to kilometers
5. 1000 grams to ounces
6. 100 kilograms to pounds
7. 93 kilometers to miles
8. 9.8 meters to feet
9. 14.5 liters to gallons
10. 150 milliliters to fluid ounces
11. 30 pounds to kilograms
12. 15 ounces to grams

Fill in the chart. Give exact answers or round to one decimal place. See Examples 1 through 3.

		Meters	Yards	Centimeters	Feet	Inches
13.	The height of a woman				5	
14.	Statue of Liberty length of nose	1.37				
15.	Leaning Tower of Pisa		60			
16.	Blue whale		36			

Solve. If necessary, round answers to two decimal places. See Examples 1 through 3.

17. The balance beam for female gymnasts is 10 centimeters wide. Convert this width to inches.

18. In men's gymnastics, the rings are 250 centimeters from the floor. Convert this height to inches, then to feet.

19. In many states, the maximum speed limit for recreational vehicles is 50 miles per hour. Convert this to kilometers per hour.

20. In some states, the speed limit is 70 miles per hour. Convert this to kilometers per hour.

21. Ibuprofen comes in 200-milligram tablets. Convert this to ounces. (Round your answer to this exercise to 3 decimal places.)

22. Vitamin C tablets come in 500-milligram caplets. Convert this to ounces.

The 70-meter-diameter antenna is the largest and most sensitive Deep Space Network antenna. See the Chapter Opener and answer Exercises 23–26.

70-Meter Antenna

23. Convert 70 meters to feet.

24. The Deep Space Network sites also have a 26-meter antenna. Convert 26 meters to feet.

25. The 70-meter-diameter antenna can track a spacecraft traveling more than 16 billion kilometers from Earth. Convert this distance to miles.

26. The dish reflector and the mount atop the concrete pedestal of the 70-meter antenna weigh nearly 2.7 million kilograms. Convert this number to tons.

27. A stone is a unit in the British customary system. Use the conversion 14 pounds = 1 stone to check the equivalencies in this 1973 Australian stamp. Is 100 kilograms approximately 15 stone 10 pounds?

28. Convert 5 feet 11 inches to centimeters and check the conversion on this 1973 Australian stamp. Is it correct?

29. The Monarch butterfly migrates annually between the northern United States and central Mexico. The trip is about 4500 km long. Convert this to miles.

30. There is a species of African termite that builds nests up to 18 ft high. Convert this to meters.

31. A $3\frac{1}{2}$-inch diskette is not really $3\frac{1}{2}$ inches. To find its actual width, convert this measurement to centimeters, then to millimeters. Round the result to the nearest ten.

32. The average two-year-old is 84 centimeters tall. Convert this to feet and inches.

33. For an average adult, the weight of the right lung is greater than the weight of the left lung. If the right lung weighs 1.5 pounds and the left lung weighs 1.25 pounds, find the difference in grams. (*Source: Some Body!*)

34. The skin of an average adult weighs 9 pounds and is the heaviest organ. Find the weight in grams. (*Source: Some Body!*)

35. A fast sneeze has been clocked at about 167 kilometers per hour. Convert this to miles per hour. Round to the nearest whole.

36. A Boeing 747 has a cruising speed of about 980 kilometers per hour. Convert this to miles per hour. Round to the nearest whole.

37. The General Sherman giant sequoia tree has a diameter of about 8 meters at its base. Convert this to feet. (*Source: Fantastic Book of Comparisons*)

38. The largest crater on the near side of the moon is Billy Crater. It has a diameter of 303 kilometers. Convert this to miles. (*Source: Fantastic Book of Comparisons*)

39. The total length of the track on a CD is about 4.5 kilometers. Convert this to miles. Round to the nearest whole mile.

40. The distance between Mackinaw City, Michigan, and Cheyenne, Wyoming, is 2079 kilometers. Convert this to miles. Round to the nearest whole mile.

41. A doctor orders a dosage of 5 ml of medicine every 4 hours for 1 week. How many fluid ounces of medicine should be purchased? Round up to the next whole fluid ounce.

42. A doctor orders a dosage of 12 ml of medicine every 6 hours for 10 days. How many fluid ounces of medicine should be purchased? Round up to the next whole fluid ounce.

Without actually converting, choose the most reasonable answer.

43. This math book has a height of about__________.
a. 28 mm **b.** 28 cm
c. 28 m **d.** 28 km

44. A mile is __________ a kilometer.
a. shorter than **b.** longer than
c. the same length as

45. A liter has __________ capacity than a quart.
a. less **b.** greater
c. the same

46. A foot is __________ a meter.
a. shorter than **b.** longer than
c. the same length as

47. A kilogram weighs __________ a pound.
a. the same as **b.** less than
c. greater than

48. A football field is 100 yards, which is about__________.
a. 9 m **b.** 90 m
c. 900 m **d.** 9000 m

49. An $8\frac{1}{2}$-ounce glass of water has a capacity of about __________.
a. 250 L **b.** 25 L
c. 2.5 L **d.** 250 ml

50. A 5-gallon gasoline can has a capacity of about __________.
a. 19 L **b.** 1.9 L
c. 19 ml **d.** 1.9 ml

51. The weight of an average man is about __________.
a. 700 kg **b.** 7 kg
c. 0.7 kg **d.** 70 kg

52. The weight of a pill is about __________.
a. 200 kg **b.** 20 kg
c. 2 kg **d.** 200 mg

Objectives B C Mixed Practice *Convert as indicated. When necessary, round to the nearest tenth of a degree. See Examples 4 through 8.*

53. 77°F to degrees Celsius

54. 86°F to degrees Celsius

55. 104°F to degrees Celsius

56. 140°F to degrees Celsius

57. 50°C to degrees Fahrenheit

58. 80°C to degrees Fahrenheit

59. 115°C to degrees Fahrenheit

60. 225°C to degrees Fahrenheit

61. 20°F to degrees Celsius

62. 26°F to degrees Celsius

63. 142.1°F to degrees Celsius

64. 43.4°F to degrees Celsius

65. 92°C to degrees Fahrenheit

66. 75°C to degrees Fahrenheit

67. 12.4°C to degrees Fahrenheit

68. 48.6°C to degrees Fahrenheit

69. The hottest temperature ever recorded in the United States, in Death Valley, was 134°F. Convert this temperature to degrees Celsius. (*Source:* National Climatic Data Center)

70. The hottest temperature ever recorded in the United States in January was 95°F in Los Angeles. Convert this temperature to degrees Celsius. (*Source:* National Climatic Data Center)

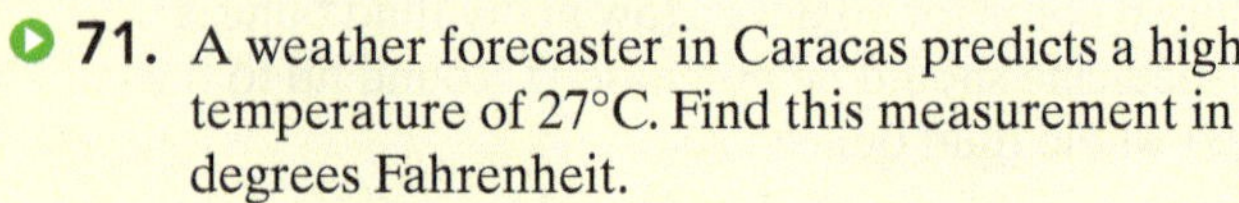

71. A weather forecaster in Caracas predicts a high temperature of 27°C. Find this measurement in degrees Fahrenheit.

72. While driving to work, Alan Olda notices a temperature of 18°C flash on the local bank's temperature display. Find the corresponding temperature in degrees Fahrenheit.

73. At Mack Trucks' headquarters, the room temperature is to be set at 70°F, but the thermostat is calibrated in degrees Celsius. Find the temperature to be set.

74. The computer room at Merck, Sharp, and Dohm is normally cooled to 66°F. Find the corresponding temperature in degrees Celsius.

75. In a European cookbook, a recipe requires the ingredients for caramels to be heated to 118°C, but the cook has access only to a Fahrenheit thermometer. Find the temperature in degrees Fahrenheit that should be used to make the caramels.

76. The ingredients for divinity should be heated to 127°C, but the candy thermometer that Myung Kim has is calibrated to degrees Fahrenheit. Find how hot he should heat the ingredients.

77. The temperature of Earth's core is estimated to be 4000°C. Find the corresponding temperature in degrees Fahrenheit.

78. In 2012, the average temperature of Earth's surface was 58.3°F. Convert this temperature to degrees Celsius. (*Source:* NASA)

Review

Perform the indicated operations. See Section 1.7.

79. $6 \cdot 4 + 5 \div 1$

80. $10 \div 2 + 9(8)$

81. $3[(1 + 5) \cdot (8 - 6)]$

82. $5[(18 - 8) - 9]$

Concept Extensions

Determine whether the measurement in each statement is reasonable.

83. A 72°F room feels comfortable.

84. Water heated to 110°F will boil.

85. Josiah has a fever if a thermometer shows his temperature to be 40°F.

86. An air temperature of 20°F on a Vermont ski slope can be expected in the winter.

87. When the temperature is 30°C outside, an overcoat is needed.

88. An air-conditioned room at 60°C feels quite chilly.

89. Barbara has a fever when a thermometer records her temperature at 40°C.

90. Water cooled to 32°C will freeze.

Body surface area (BSA) is often used to calculate dosages for some drugs. BSA is calculated in square meters using a person's weight and height.

$$\text{BSA} = \sqrt{\frac{(\text{weight in kg}) \times (\text{height in cm})}{3600}}$$

For Exercises 91 through 96, calculate the BSA for each person. Round to the nearest hundredth. You will need to use the square root key on your calculator.

91. An adult whose height is 182 cm and weight is 90 kg

92. An adult whose height is 157 cm and weight is 63 kg

93. A child whose height is 40 in. and weight is 50 kg (*Hint:* Don't forget to first convert inches to centimeters.)

94. A child whose height is 26 in. and weight is 13 kg (*Hint:* Don't forget to first convert inches to centimeters.)

95. An adult whose height is 60 in. and weight is 150 lb

96. An adult whose height is 69 in. and weight is 172 lb

97. In February 2010, at the Brookhaven National Laboratory in Long Island, NY, the highest temperature produced in a laboratory was achieved. This temperature was 7,200,000,000°F. Convert this temperature to degrees Celsius. Round your answer to the nearest billion degrees. (*Source: Guinness World Records*)

98. The hottest-burning substance known is carbon subnitride. Its flame at one atmospheric pressure reaches 9010°F. Convert this temperature to degrees Celsius. (*Source: Guinness World Records*)

99. In your own words, describe how to convert from degrees Celsius to degrees Fahrenheit.

100. In your own words, describe how to convert from degrees Fahrenheit to degrees Celsius.

Chapter 8 Group Activity

Map Reading

Sections 8.1, 8.4, and 8.7

Materials:

- ruler
- string
- calculator

This activity may be completed by working in groups or individually.

Investigate the route you would take from Santa Rosa, New Mexico, to San Antonio, New Mexico. Use the map in the figure to answer the following questions. You may find that using string to match the roads on the map is useful when measuring distances.

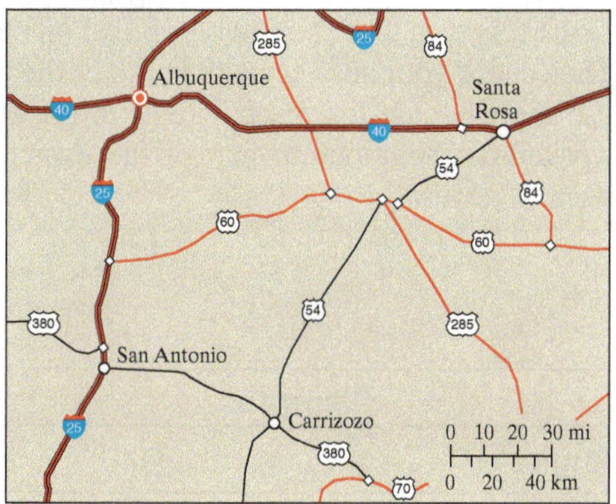

1. How many miles is it from Santa Rosa to San Antonio via Interstate 40 and Interstate 25? Convert this distance to kilometers.
2. How many miles is it from Santa Rosa to San Antonio via U.S. 54 and U.S. 380? Convert this distance to kilometers.
3. Assume that the speed limit on Interstates 40 and 25 is 65 miles per hour. How long would the trip take if you took this route and traveled 65 miles per hour the entire trip?
4. At what average speed would you have to travel on the U.S. routes to make the trip from Santa Rosa to San Antonio in the same amount of time that it would take on the interstate routes? Do you think this speed is reasonable on this route? Explain your reasoning.
5. Discuss in general the factors that might affect your decision between the different routes.
6. Explain which route you would choose in this case and why.

Chapter 8 Vocabulary Check

Fill in each blank with one of the words or phrases listed below.

transversal	line segment	obtuse	straight	adjacent	right	volume	area
acute	perimeter	vertical	supplementary	ray	angle	line	complementary
vertex	mass	unit fractions	gram	weight	meter	liter	surface area

1. ________ is a measure of the pull of gravity.
2. ________ is a measure of the amount of substance in an object. This measure does not change.
3. The basic unit of length in the metric system is the ________.
4. To convert from one unit of length to another, ________ may be used.
5. The ________ is the basic unit of mass in the metric system.
6. The ________ is the basic unit of capacity in the metric system.
7. A(n) ________ is a piece of a line with two endpoints.
8. Two angles that have a sum of 90° are called ________ angles.
9. A(n) ________ is a set of points extending indefinitely in two directions.
10. The ________ of a polygon is the distance around the polygon.
11. A(n) ________ is made up of two rays that share the same endpoint. The common endpoint is called the ________.
12. ________ measures the amount of surface of a region.
13. A(n) ________ is a part of a line with one endpoint. A ray extends indefinitely in one direction.
14. A line that intersects two or more lines at different points is called a(n) ________.
15. An angle that measures 180° is called a(n) ________ angle.
16. The measure of the space of a solid is called its ________.
17. When two lines intersect, four angles are formed. Each pair of angles that are opposite each other are called ________ angles.
18. Two of the angles from Exercise **17** that share a common side are called ________ angles.
19. An angle whose measure is between 90° and 180° is called a(n) ________ angle.
20. An angle that measures 90° is called a(n) ________ angle.
21. An angle whose measure is between 0° and 90° is called a(n) ________ angle.
22. Two angles that have a sum of 180° are called ________ angles.
23. The ________ of a polyhedron is the sum of the areas of the faces of the polyhedron.

Helpful Hint Are you preparing for your test? Don't forget to take the Chapter 8 Test on page 638. Then check your answers at the back of the text and use the Chapter Test Prep Videos to see the fully worked-out solutions to any of the exercises you want to review.

8 Chapter Highlights

Definitions and Concepts	Examples
Section 8.1 Lines and Angles	
A **line** is a set of points extending indefinitely in two directions. A line has no width or height, but it does have length. We name a line by any two of its points.	Line AB or $\overleftrightarrow{AB}$ (A, B)
A **line segment** is a piece of a line with two endpoints.	Line segment AB or $\overline{AB}$ (A, B)

(continued)

Definitions and Concepts	Examples
Section 8.1 Lines and Angles (*continued*)	
A **ray** is a part of a line with one endpoint. A ray extends indefinitely in one direction.	Ray AB or $\overrightarrow{AB}$
An **angle** is made up of two rays that share the same endpoint. The common endpoint is called the **vertex.**	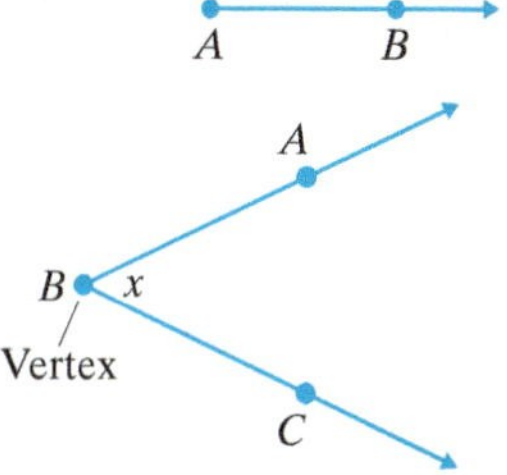
Section 8.2 Perimeter	
Perimeter Formulas **Rectangle:** $P = 2l + 2w$ **Square:** $P = 4s$ **Triangle:** $P = a + b + c$ **Circumference of a Circle:** $C = 2\pi r$ or $C = \pi d$ where $\pi \approx 3.14$ or $\pi \approx \frac{22}{7}$	Find the perimeter of the rectangle. 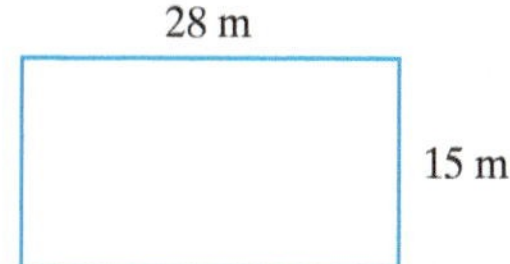 $P = 2l + 2w$ $= 2 \cdot 28 \text{ meters} + 2 \cdot 15 \text{ meters}$ $= 56 \text{ meters} + 30 \text{ meters}$ $= 86 \text{ meters}$ The perimeter is 86 meters.
Section 8.3 Area, Volume, and Surface Area	
Area Formulas **Rectangle:** $A = lw$ **Square:** $A = s^2$ **Triangle:** $A = \frac{1}{2}bh$ **Parallelogram:** $A = bh$ **Trapezoid:** $A = \frac{1}{2}(b + B)h$ **Circle:** $A = \pi r^2$	Find the area of the square.  $A = s^2$ $= (8 \text{ centimeters})^2$ $= 64 \text{ square centimeters}$ The area of the square is 64 square centimeters.
Volume Formulas **Rectangular Solid:** $V = lwh$ **Cube:** $V = s^3$ **Sphere:** $V = \frac{4}{3}\pi r^3$ **Right Circular Cylinder:** $V = \pi r^2 h$ **Cone:** $V = \frac{1}{3}\pi r^2 h$ **Square-Based Pyramid:** $V = \frac{1}{3}s^2 h$ **Surface Area Formulas:** See page 575.	Find the volume of the sphere. Use $\frac{22}{7}$ for π. 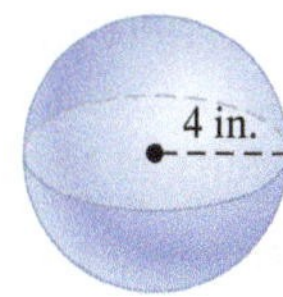 $V = \frac{4}{3}\pi r^3$ $\approx \frac{4}{3} \cdot \frac{22}{7} \cdot (4 \text{ inches})^3$ $= \frac{4 \cdot 22 \cdot 64}{3 \cdot 7} \text{ cubic inches}$ $= \frac{5632}{21}$ or $268\frac{4}{21}$ cubic inches

Definitions and Concepts	Examples

Section 8.4 Linear Measurement

To convert from one unit of length to another, multiply by a **unit fraction** in the form

$$\frac{\text{units to convert to}}{\text{original units}}$$

$$\frac{12 \text{ inches}}{1 \text{ foot}}, \frac{1 \text{ foot}}{12 \text{ inches}}, \frac{3 \text{ feet}}{1 \text{ yard}}$$

Convert 6 feet to inches.

$$6 \text{ ft} = \frac{6 \text{ ft}}{1} \cdot 1$$

$$= \frac{6 \text{ ft}}{1} \cdot \frac{12 \text{ in.}}{1 \text{ ft}} \begin{array}{l} \leftarrow \text{units to convert to} \\ \leftarrow \text{original units} \end{array}$$

$$= 6 \cdot 12 \text{ in.}$$

$$= 72 \text{ in.}$$

Length: U.S. System of Measurement

12 inches (in.) = 1 foot (ft)

3 feet = 1 yard (yd)

5280 feet = 1 mile (mi)

Length: Metric System of Measurement

The basic unit of length in the metric system is the **meter.** A meter is slightly longer than a yard.

Metric Units of Length
1 **kilo**meter (km) = 1000 meters (m)
1 **hecto**meter (hm) = 100 m
1 **deka**meter (dam) = 10 m
1 meter (m) = 1 m
1 **deci**meter (dm) = 1/10 m or 0.1 m
1 **centi**meter (cm) = 1/100 m or 0.01 m
1 **milli**meter (mm) = 1/1000 m or 0.001 m

Convert 3650 centimeters to meters.

$$3650 \text{ cm} = 3650 \text{ cm} \cdot 1$$

$$= \frac{3650 \text{ cm}}{1} \cdot \frac{0.01 \text{ m}}{1 \text{ cm}} = 36.5 \text{ m}$$

or

km hm dam m dm cm mm

2 units to the left

3650 cm = 36.5 m

2 places to the left

Section 8.5 Weight and Mass

Weight is really a measure of the pull of gravity. **Mass** is a measure of the amount of substance in an object and does not change.

Convert 5 pounds to ounces.

$$5 \text{ lb} = 5 \text{ lb} \cdot 1 = \frac{5 \text{ lb}}{1} \cdot \frac{16 \text{ oz}}{1 \text{ lb}} = 80 \text{ oz}$$

Weight: U.S. System of Measurement

16 ounces (oz) = 1 pound (lb)

2000 pounds = 1 ton

Mass: Metric System of Measurement

The **gram** is the basic unit of mass in the metric system. It is the mass of water contained in a cube 1 centimeter on each side. A paper clip weighs about 1 gram.

Metric Units of Mass
1 **kilo**gram (kg) = 1000 grams (g)
1 **hecto**gram (hg) = 100 g
1 **deka**gram (dag) = 10 g
1 gram (g) = 1 g
1 **deci**gram (dg) = 1/10 g or 0.1 g
1 **centi**gram (cg) = 1/100 g or 0.01 g
1 **milli**gram (mg) = 1/1000 g or 0.001 g

Convert 260 grams to kilograms.

$$260 \text{ g} = \frac{260 \text{ g}}{1} \cdot 1 = \frac{260 \text{ g}}{1} \cdot \frac{1 \text{ kg}}{1000 \text{ g}} = 0.26 \text{ kg}$$

or

kg hg dag g dg cg mg

3 units to the left

260 g = 0.26 kg

3 places to the left

Definitions and Concepts	Examples
Section 8.6 Capacity	
Capacity: U.S. System of Measurement 8 fluid ounces (fl oz) = 1 cup (c) 2 cups = 1 pint (pt) 2 pints = 1 quart (qt) 4 quarts = 1 gallon (gal)	Convert 5 pints to gallons. 1 gal = 4 qt = 8 pt $5\text{ pt} = 5\text{ pt} \cdot 1 = \frac{5\text{ pt}}{1} \cdot \frac{1\text{ gal}}{8\text{ pt}} = \frac{5}{8}\text{ gal}$
Capacity: Metric System of Measurement The **liter** is the basic unit of capacity in the metric system. It is the capacity or volume of a cube measuring 10 centimeters on each side. A liter of liquid is slightly more than 1 quart. **Metric Units of Capacity** 1 **kilo**liter (kl) = 1000 liters (L) 1 **hecto**liter (hl) = 100 L 1 **deka**liter (dal) = 10 L 1 liter (L) = 1 L 1 **deci**liter (dl) = 1/10 L or 0.1 L 1 **centi**liter (cl) = 1/100 L or 0.01 L 1 **milli**liter (ml) = 1/1000 L or 0.001 L	Convert 1.5 liters to milliliters. $1.5\text{ L} = \frac{1.5\text{ L}}{1} \cdot 1 = \frac{1.5\text{ L}}{1} \cdot \frac{1000\text{ ml}}{1\text{ L}} = 1500\text{ ml}$ or kl hl dal L dl cl ml 3 units to the left 1.500 L = 1500 ml 3 places to the right
Section 8.7 Temperature and Conversions Between the U.S. and Metric Systems	
To convert between systems, use approximate unit fractions. See page 619.	Convert 7 feet to meters. $7\text{ ft} \approx \frac{7\text{ ft}}{1} \cdot \frac{0.30\text{ m}}{1\text{ ft}} = 2.1\text{ m}$ Convert 8 liters to quarts. $8\text{ L} \approx \frac{8\text{ L}}{1} \cdot \frac{1.06\text{ qt}}{1\text{ L}} = 8.48\text{ qt}$ Convert 363 grams to ounces. $363\text{ g} \approx \frac{363\text{ g}}{1} \cdot \frac{0.04\text{ oz}}{1\text{ g}} = 14.52\text{ oz}$
Celsius to Fahrenheit $F = \frac{9}{5}C + 32$ or $F = 1.8C + 32$	Convert 35°C to degrees Fahrenheit. $F = \frac{9}{5} \cdot 35 + 32 = 63 + 32 = 95$ 35°C = 95°F
Fahrenheit to Celsius $C = \frac{5}{9}(F - 32)$	Convert 50°F to degrees Celsius. $C = \frac{5}{9} \cdot (50 - 32) = \frac{5}{9} \cdot (18) = 10$ 50°F = 10°C

Chapter 8 Review

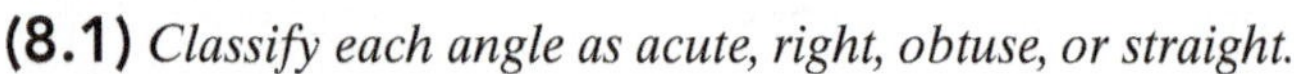

(8.1) *Classify each angle as acute, right, obtuse, or straight.*

1.

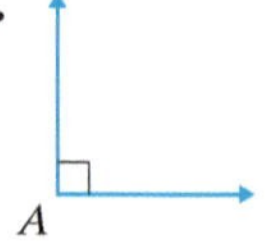

2.

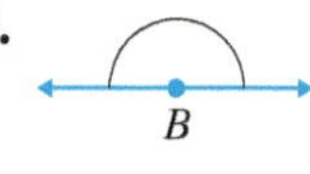

3.

4.

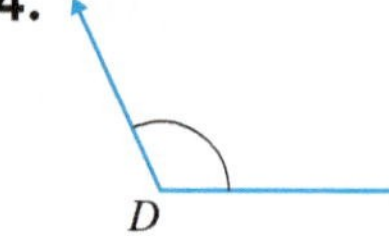

5. Find the complement of a 25° angle.

6. Find the supplement of a 105° angle.

Find the measure of angle x in each figure.

7.

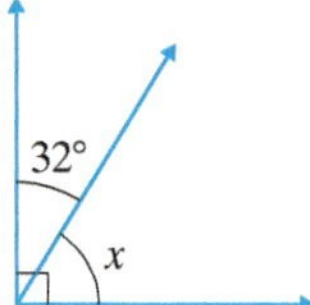

8.

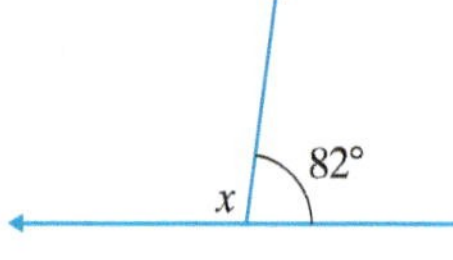

9.

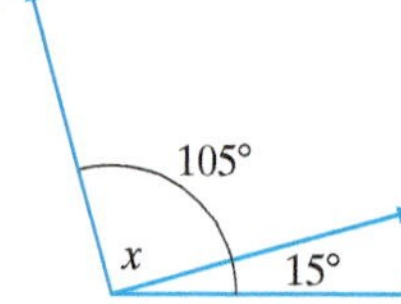

10.

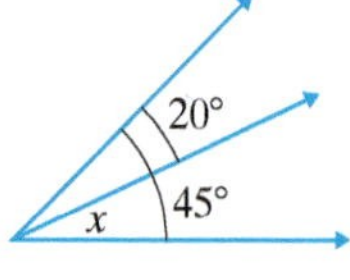

11. Identify the pairs of supplementary angles.

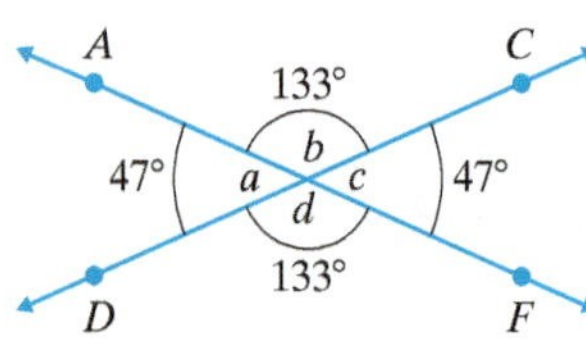

12. Identify the pairs of complementary angles.

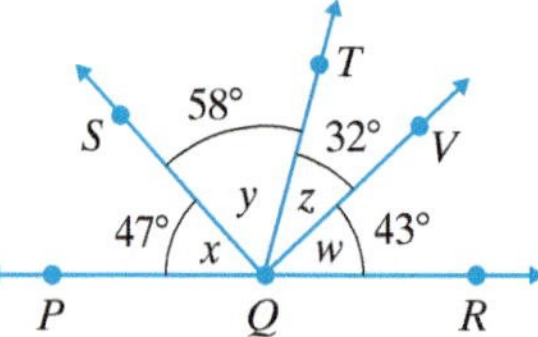

Find the measures of angles x, y, and z in each figure.

13.

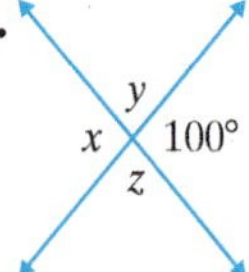

14.

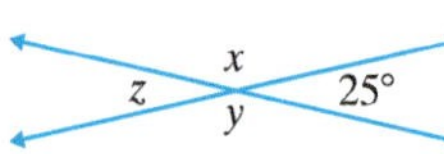

15. Given that $m \parallel n$.

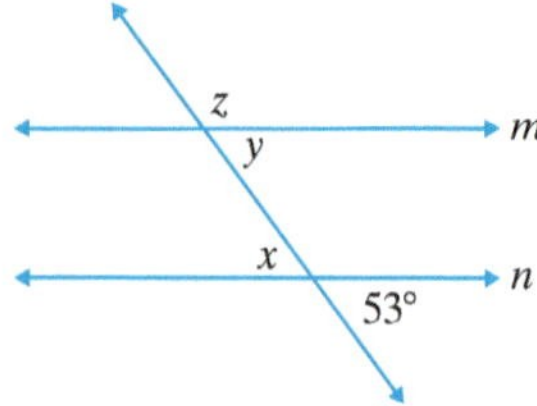

16. Given that $m \parallel n$.

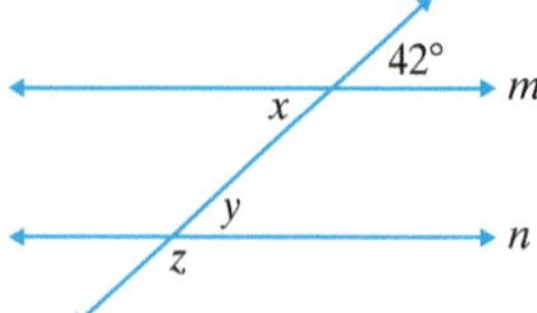

(8.2) *Find the perimeter of each figure.*

17.

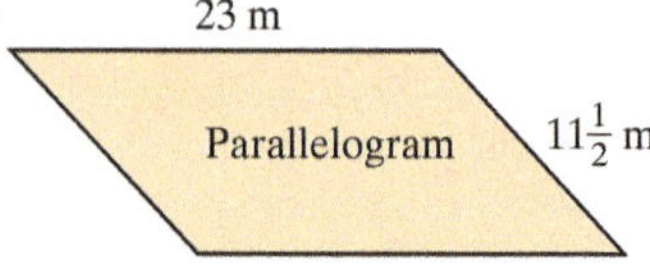

18.

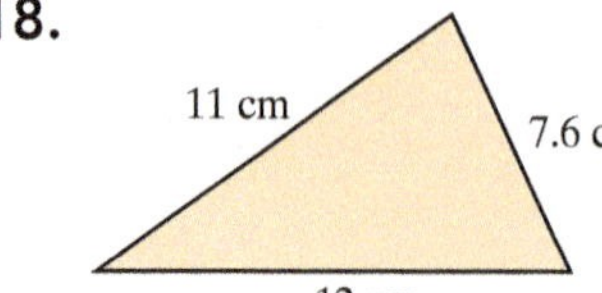

19.

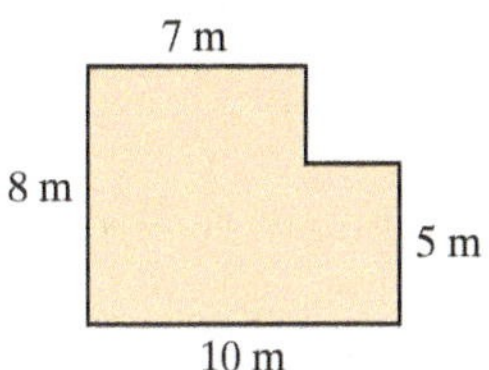

20.

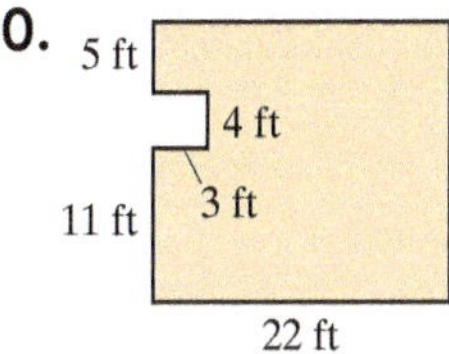

Solve.

21. Find the perimeter of a rectangular sign that measures 6 feet by 10 feet.

22. Find the perimeter of a town square that measures 110 feet on a side.

Find the circumference of each circle. Use $\pi \approx 3.14$.

23.

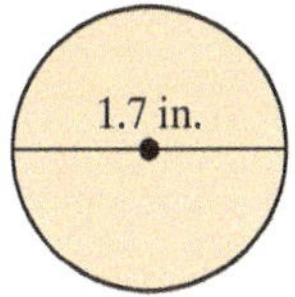

24.

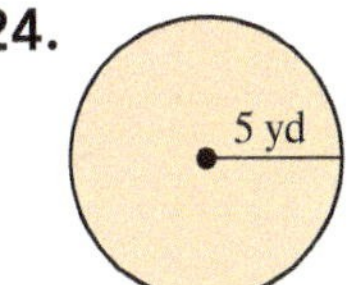

(8.3) *Find the area of each figure. For the circles, find the exact area and then use* $\pi \approx 3.14$ *to approximate the area.*

25.

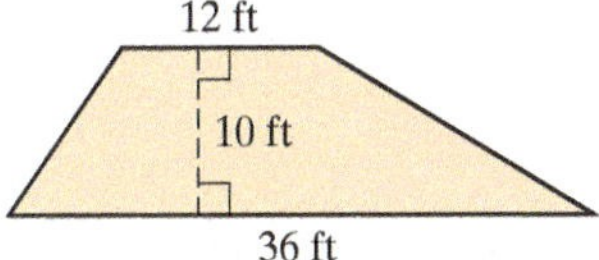

26.

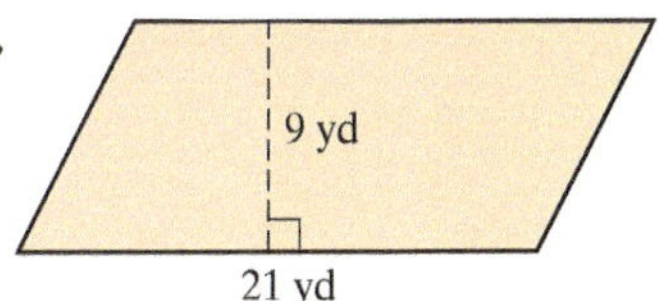

27.

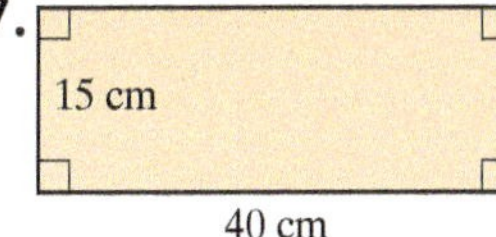

28.

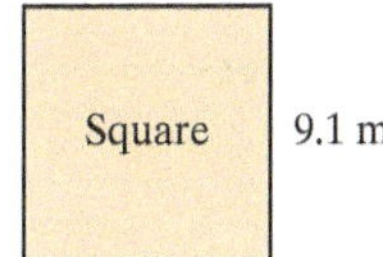

29.

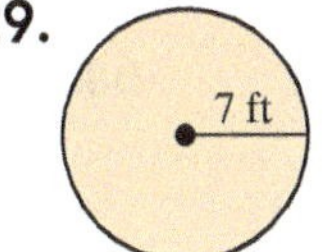

30.

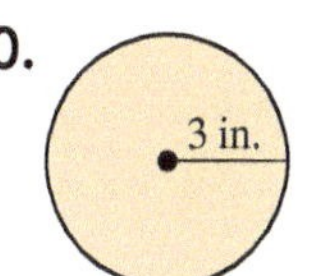

31.

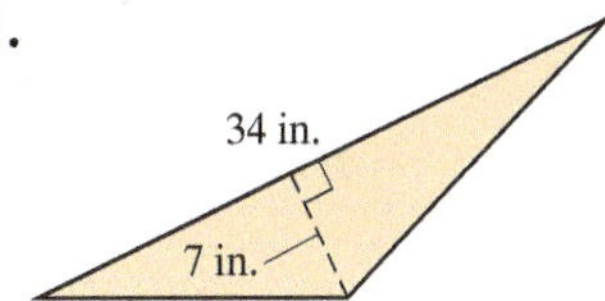

32.

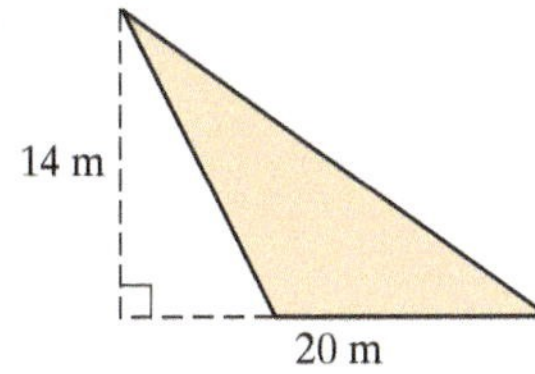

33.

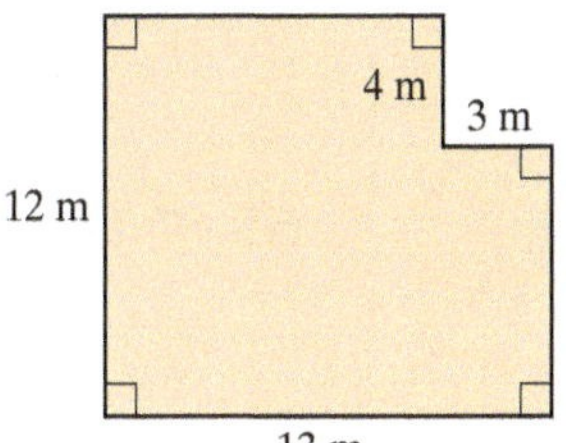

34.

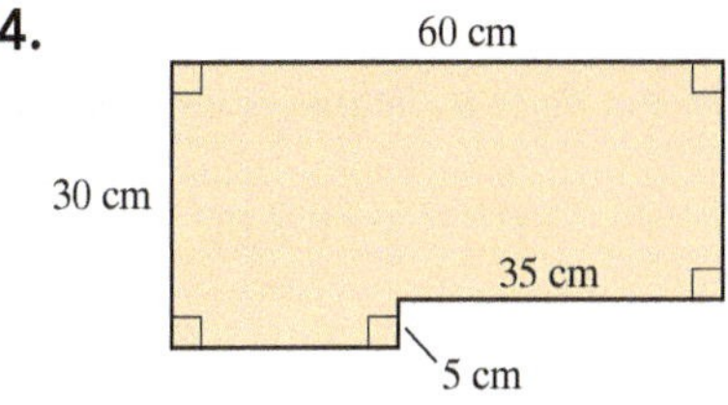

35. The amount of sealer necessary to seal a driveway depends on the area. Find the area of a rectangular driveway 36 feet by 12 feet.

36. Find how much carpet is necessary to cover the floor of the room shown.

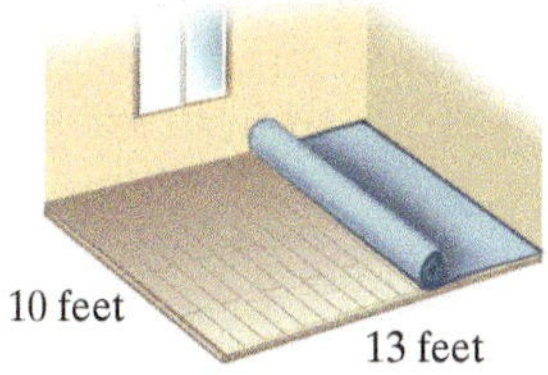

Find the volume and surface area of the solids in Exercises 37 and 38. For Exercises 39 and 40, give an exact volume and an approximation.

37.

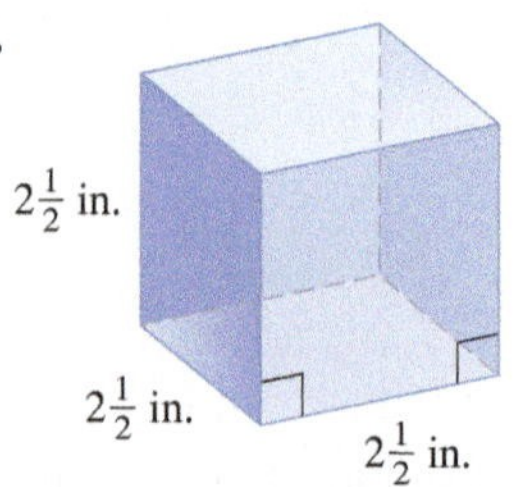

38.

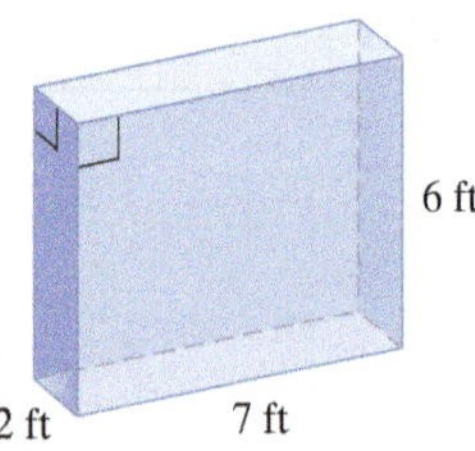

39. Use $\pi \approx 3.14$.

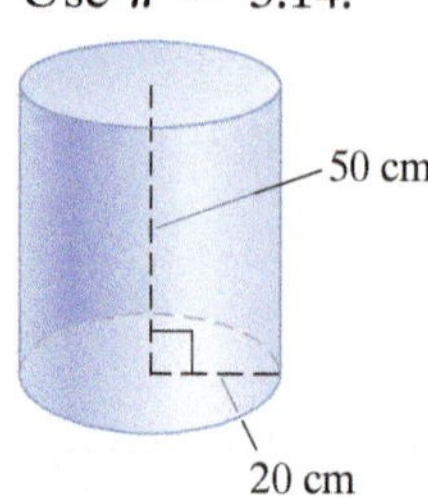

40. Use $\pi \approx \frac{22}{7}$.

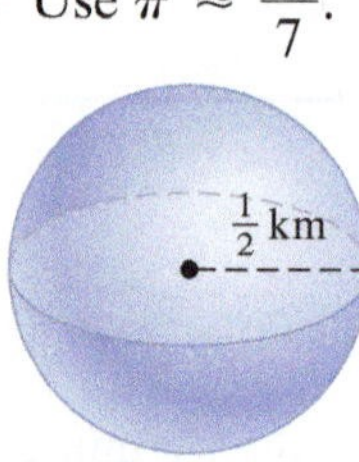

41. Find the volume of a pyramid with a square base 2 feet on a side and a height of 2 feet.

42. Approximate the volume of a tin can 8 inches high and 3.5 inches in radius. Use 3.14 for π.

43. A chest has 3 drawers. If each drawer has inside measurements of $2\frac{1}{2}$ feet by $1\frac{1}{2}$ feet by $\frac{2}{3}$ foot, find the total volume of the 3 drawers.

44. A cylindrical canister for a shop vacuum is 2 feet tall and 1 foot in *diameter*. Find its exact volume.

(8.4) *Convert.*

45. 108 in. to feet

46. 72 ft to yards

47. 1.5 mi to feet

48. $\frac{1}{2}$ yd to inches

49. 52 ft = ______ yd ______ ft

50. 46 in. = ______ ft ______ in.

51. 42 m to centimeters

52. 82 cm to millimeters

53. 12.18 mm to meters

54. 2.31 m to kilometers

Perform each indicated operation.

55. 4 yd 2 ft + 16 yd 2 ft

56. 7 ft 4 in. ÷ 2

57. 8 cm + 15 mm

58. 4 m − 126 cm

Solve.

59. A bolt of cloth contains 333 yd 1 ft of cotton ticking. Find the amount of material that remains after 163 yd 2 ft is removed from the bolt.

60. The student activities club is sponsoring a walk for hunger, and all students who participate will receive a sash with the name of the school to wear on the walk. If each sash requires 5 ft 2 in. of material and there are 50 students participating in the walk, how much material will the student activities club need?

61. The trip from El Paso, TX, to Ontario, CA, is about 1235 km each way. Four friends agree to share the driving equally. How far must each drive on this round-trip vacation?

△ 62. The college has ordered that NO SMOKING signs be placed above the doorway of each classroom. Each sign is 0.8 m long and 30 cm wide. Find the area of each sign. (*Hint:* Recall that the area of a rectangle = width · length.)

(8.5) *Convert.*

63. 66 oz to pounds

64. 2.3 tons to pounds

65. 52 oz = ______ lb ______ oz

66. 10,300 lb = ______ tons ______ lb

67. 27 mg to grams

68. 40 kg to grams

69. 2.1 hg to dekagrams

70. 0.03 mg to decigrams

Perform each indicated operation.

71. 6 lb 5 oz − 2 lb 12 oz

72. 8 lb 6 oz × 4

73. 4.3 mg × 5

74. 4.8 kg − 4200 g

Solve.

75. Donshay Berry ordered 1 lb 12 oz of soft-center candies and 2 lb 8 oz of chewy-center candies for his party. Find the total weight of the candy ordered.

76. Four local townships jointly purchase 38 tons 300 lb of cinders to spread on their roads during an ice storm. Determine the weight of the cinders each township receives if they share the purchase equally.

(8.6) *Convert.*

77. 28 pints to quarts

78. 40 fluid ounces to cups

79. 3 qt 1 pt to pints

80. 18 quarts to cups

81. 9 pt = ______ qt ______ pt

82. 15 qt = ______ gal ______ qt

83. 3.8 L to milliliters

84. 14 hl to kiloliters

85. 30.6 L to centiliters

86. 2.45 ml to liters

Perform each indicated operation.

87. 1 qt 1 pt + 3 qt 1 pt

88. 3 gal 2 qt × 2

89. 0.946 L − 210 ml

90. 6.1 L + 9400 ml

Solve.

91. Carlos Perez prepared 4 gal 2 qt of iced tea for a block party. During the first 30 minutes of the party, 1 gal 3 qt of the tea is consumed. How much iced tea remains?

92. A recipe for soup stock calls for 1 c 4 fl oz of beef broth. How much should be used if the recipe is cut in half?

93. Each bottle of Kiwi liquid shoe polish holds 85 ml of the polish. Find the number of liters of shoe polish contained in 8 boxes if each box contains 16 bottles.

94. Ivan Miller wants to pour three separate containers of saline solution into a single vat with a capacity of 10 liters. Will 6 liters of solution in the first container combined with 1300 milliliters in the second container and 2.6 liters in the third container fit into the larger vat?

(8.7) *Note: Because approximations are used in this section, your answers may vary slightly from the answers given in the back of the book.*

Convert as indicated. If necessary, round to two decimal places.

95. 7 meters to feet

96. 11.5 yards to meters

97. 17.5 liters to gallons

98. 7.8 liters to quarts

99. 15 ounces to grams

100. 23 pounds to kilograms

101. A compact disc is 1.2 mm thick. Find the height (in inches) of 50 discs.

102. If a person weighs 82 kilograms, how many pounds is this?

Convert. Round to the nearest tenth of a degree, if necessary.

103. 42°C to degrees Fahrenheit

104. 160°C to degrees Fahrenheit

105. 41.3°F to degrees Celsius

106. 80°F to degrees Celsius

Solve. Round to the nearest tenth of a degree, if necessary.

107. A sharp dip in the jet stream caused the temperature in New Orleans to drop to 35°F. Find the corresponding temperature in degrees Celsius.

108. A recipe for meat loaf calls for a 165°C oven. Find the setting used if the oven has a Fahrenheit thermometer.

Mixed Review

Find the following.

109. Find the supplement of a 72° angle.

110. Find the complement of a 1° angle.

Find the measure of angle x in each figure.

111.

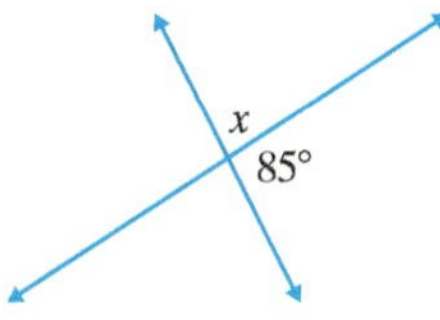

112.

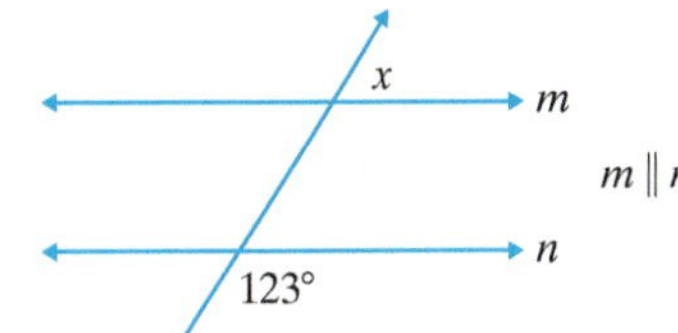

Find the perimeter of each figure.

113.

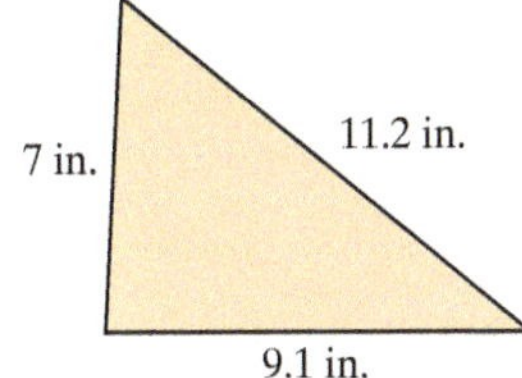

114.

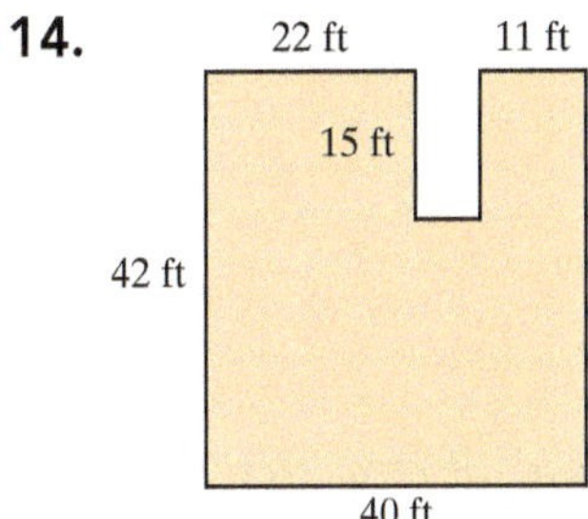

Find the area of each figure. For the circle, find the exact area and then use $\pi \approx 3.14$ to approximate the area.

115.

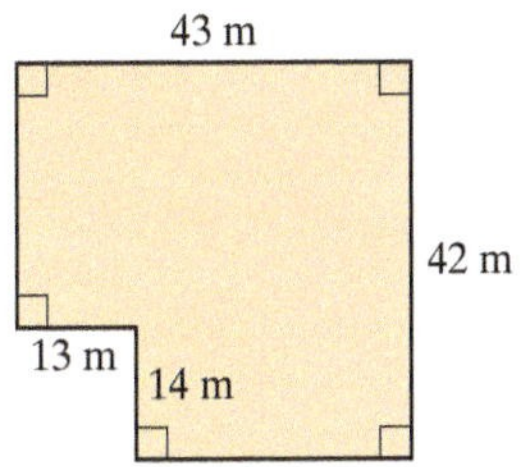

116.

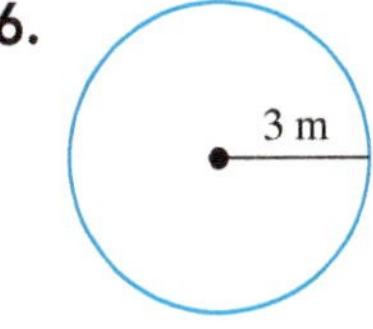

Find the volume of each solid.

117. Give an approximation using $\frac{22}{7}$ for π.

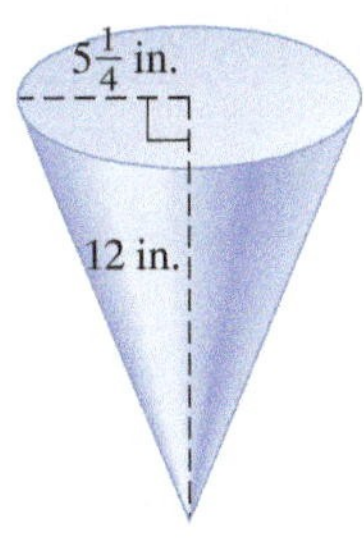

118. Find the surface area also.

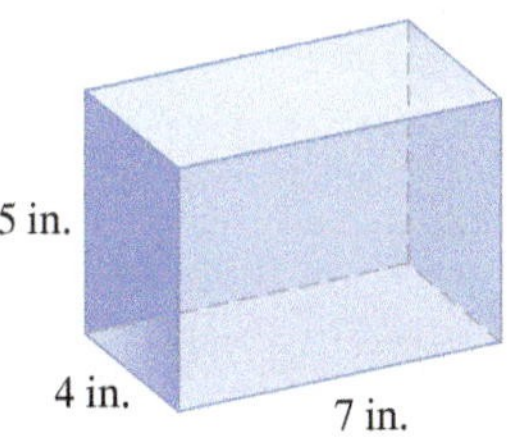

Convert the following.

119. 6.25 ft to inches

120. 8200 lb = ______ tons ______ lb

121. 5 m to centimeters

122. 286 mm to kilometers

123. 1400 mg to grams

124. 6.75 gallons to quarts

125. 86°C to degrees Fahrenheit

126. 51.8°F to degrees Celsius

Perform the indicated operations and simplify.

127. 9.3 km − 183 m

128. 35 L + 700 ml

129. 3 gal 3 qt + 4 gal 2 qt

130. 3.2 kg × 4

Chapter 8 Test

Step-by-step test solutions are found on the Chapter Test Prep Videos. Where available: MyMathLab® or YouTube™

Answers

1. ______
2. ______
3. ______
4. ______
5. ______
6. ______
7. ______
8. ______
9. ______
10. ______

△ **1.** Find the complement of a 78° angle.

△ **2.** Find the supplement of a 124° angle.

△ **3.** Find the measure of $\angle x$.

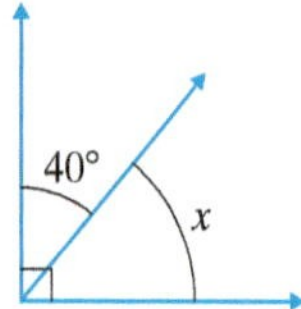

Find the measures of x, y, and z in each figure.

△ **4.**

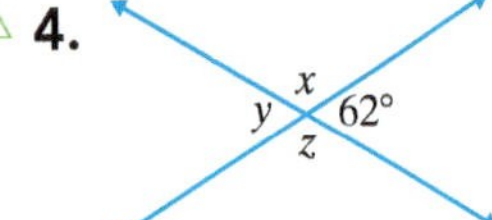

△ **5.** Given that $m \| n$

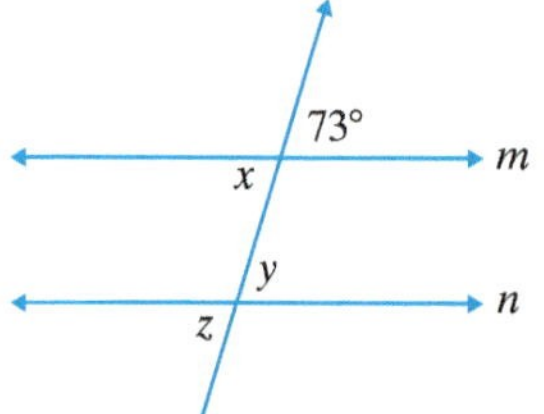

Find the unknown diameter or radius as indicated.

△ **6.**

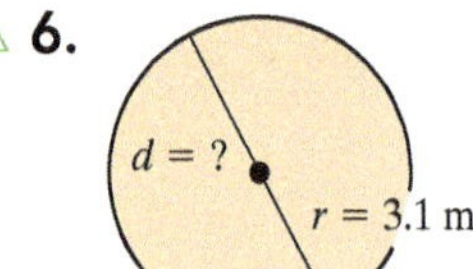

△ **7.**

r = ?
d = 20 in.

Find the perimeter (or circumference) and area of each figure. For the circle, give the exact value and an approximation using $\pi \approx 3.14$.

△ **8.**

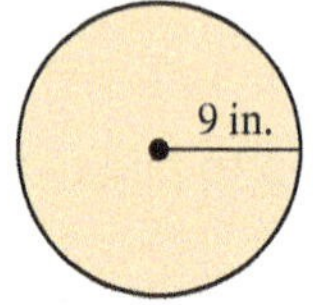

△ **9.**

Rectangle
5.3 yd
7 yd

△ **10.**

6 in.
11 in.
7 in.
23 in.

Find the volume of each solid. For the cylinder, use $\pi \approx \frac{22}{7}$.

△**11.**

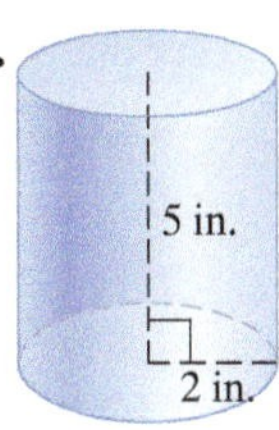

△**12.**

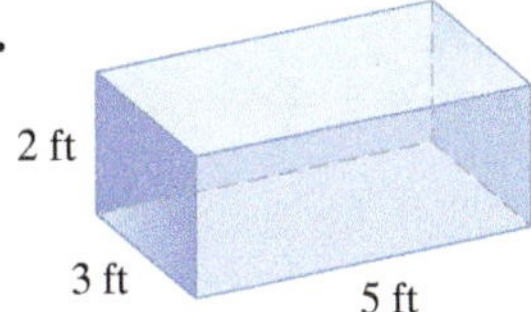

Solve.

△**13.** Find the perimeter of a square photo with a side length of 4 inches.

△**14.** How much soil is needed to fill a rectangular hole 3 feet by 3 feet by 2 feet?

15. Find how much baseboard is needed to go around a rectangular room that measures 18 feet by 13 feet. If baseboard costs \$1.87 per foot, also calculate the total cost needed for materials.

Convert.

16. 280 in. = ______ ft ______ in.

17. $2\frac{1}{2}$ gal to quarts

18. 30 oz to pounds

19. 2.8 tons to pounds

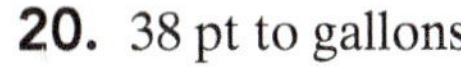

20. 38 pt to gallons

21. 40 mg to grams

22. 2.4 kg to grams

23. 3.6 cm to millimeters

24. 4.3 dg to grams

25. 0.83 L to milliliters

Perform each indicated operation.

26. 3 qt 1 pt + 2 qt 1 pt

27. 8 lb 6 oz − 4 lb 9 oz

28. 2 ft 9 in. × 3

29. 5 gal 2 qt ÷ 2

30. 8 cm − 14 mm

31. 1.8 km + 456 m

Convert. Round to the nearest tenth of a degree, if necessary.

32. 84°F to degrees Celsius

33. 12.6°C to degrees Fahrenheit

34. The sugar maples in front of Bette MacMillan's house are 8.4 meters tall. Because they interfere with the phone lines, the telephone company plans to remove the top third of the trees. How tall will the maples be after they are shortened?

35. A total of 15 gal 1 qt of oil has been removed from a 20-gallon drum. How much oil still remains in the container?

36. The engineer in charge of bridge construction said that the span of a certain bridge would be 88 m. But the actual construction required it to be 340 cm longer. Find the span of the bridge, in meters.

37. If 2 ft 9 in. of material is used to manufacture one scarf, how much material is needed for 6 scarves?

11. ______
12. ______
13. ______
14. ______
15. ______
16. ______
17. ______
18. ______
19. ______
20. ______
21. ______
22. ______
23. ______
24. ______
25. ______
26. ______
27. ______
28. ______
29. ______
30. ______
31. ______
32. ______
33. ______
34. ______
35. ______
36. ______
37. ______

Chapters 1–8 Cumulative Review

Answers

1. ______

2. ______

3. a. ______

b. ______

4. a. ______

b. ______

5. ______

6. ______

7. ______

8. ______

9. ______

10. ______

11. ______

12. ______

13. ______

14. ______

15. ______

16. ______

1. Solve: $3a - 6 = a + 4$

2. Solve: $2x + 1 = 3x - 5$

3. Evaluate:

a. $\left(\frac{2}{5}\right)^4$ **b.** $\left(-\frac{1}{4}\right)^2$

4. Evaluate:

a. $\left(-\frac{1}{3}\right)^3$ **b.** $\left(\frac{3}{7}\right)^2$

5. Add: $2\frac{4}{5} + 5 + 1\frac{1}{2}$

6. Add: $2\frac{1}{3} + 4\frac{2}{5} + 3$

7. Simplify by combining like terms: $11.1x - 6.3 + 8.9x - 4.6$

8. Simplify by combining like terms: $2.5y + 3.7 - 1.3y - 1.9$

9. Simplify: $\dfrac{5.68 + (0.9)^2 \div 100}{0.2}$

10. Simplify: $\dfrac{0.12 + 0.96}{0.5}$

11. Insert <, >, or = to form a true statement.

$0.\overline{7} \quad \frac{7}{9}$

12. Insert <, >, or = to form a true statement.

$0.43 \quad \frac{2}{5}$

13. Solve: $0.5y + 2.3 = 1.65$

14. Solve: $0.4x - 9.3 = 2.7$

△**15.** An inner-city park is in the shape of a square that measures 300 feet on a side. Find the length of the diagonal of the park, rounded to the nearest whole foot.

△**16.** A rectangular field is 200 feet by 125 feet. Find the length of the diagonal of the field, rounded to the nearest whole foot.

△ **17.** Given the rectangle shown:

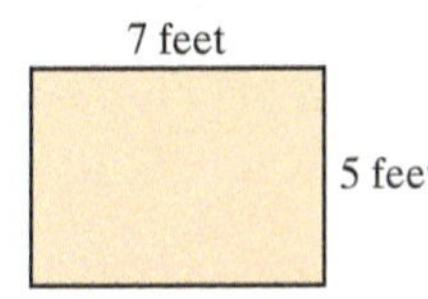

a. Find the ratio of its width to its length.
b. Find the ratio of its length to its perimeter.

△ **18.** A square is 9 inches by 9 inches.

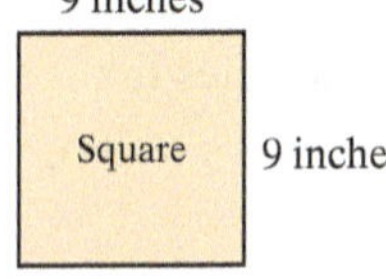

a. Find the ratio of a side to its perimeter.
b. Find the ratio of its perimeter to its area.

19. Determine whether -1 is a solution of the equation $3y + 1 = 3$.

20. Determine whether -5 is a solution of the equation $4(x + 7) = 8$.

21. Solve: $\frac{x}{3} = -2$

22. Solve: $9x = 8x + 1.02$

23. The standard dose of an antibiotic is 4 cc (cubic centimeters) for every 25 pounds (lb) of body weight. At this rate, find the standard dose for a 140-lb woman.

24. A recipe that makes 2 pie crusts calls for 3 cups of flour. How much flour is needed to make 5 pie crusts?

25. In a survey of 100 people, 17 people drive blue cars. What percent of people drive blue cars?

26. Of 100 shoppers surveyed at a mall, 38 paid for their purchases using only cash. What percent of shoppers used only cash to pay for their purchases?

27. 13 is $6\frac{1}{2}\%$ of what number?

28. 54 is $4\frac{1}{2}\%$ of what number?

29. What number is 30% of 9?

30. What number is 42% of 30?

31. The number of applications for a mathematics scholarship at one university increased from 34 to 45 in one year. What is the percent increase? Round to the nearest whole percent.

32. The price of a gallon of paint rose from \$15 to \$19. Find the percent increase, rounded to the nearest whole percent.

17. a. ______
b. ______
18. a. ______
b. ______
19. ______
20. ______
21. ______
22. ______
23. ______
24. ______
25. ______
26. ______
27. ______
28. ______
29. ______
30. ______
31. ______
32. ______

33. ______

34. ______

35. ______

36. ______

37. ______

38. ______

39. ______

40. ______

41. ______

42. ______

43. ______

44. ______

45. ______

46. ______

33. Find the sales tax and the total price on the purchase of an \$85.50 atlas in a city where the sales tax rate is 7.5%.

34. A sofa has a purchase price of \$375. If the sales tax rate is 8%, find the amount of sales tax and the total cost of the sofa.

35. Find the median of the list of numbers: 25, 54, 56, 57, 60, 71, 98

36. Find the median of the list of scores: 60, 95, 89, 72, 83

37. Find the probability of choosing a red marble from a box containing 1 red, 1 yellow, and 2 blue marbles.

38. Find the probability of choosing a nickel at random in a coin purse that contains 2 pennies, 2 nickels, and 3 quarters.

39. Find the complement of a 48° angle.

40. Find the supplement of a 137° angle.

41. Convert 8 feet to inches.

42. Convert 7 yards to feet.

43. Subtract 3 tons 1350 lb from 8 tons 1000 lb.

44. Add 8 lb 15 oz to 9 lb 3 oz.

45. Convert 59°F to degrees Celsius.

46. Convert 86°F to degrees Celsius.

Equations, Inequalities, and Problem Solving

9

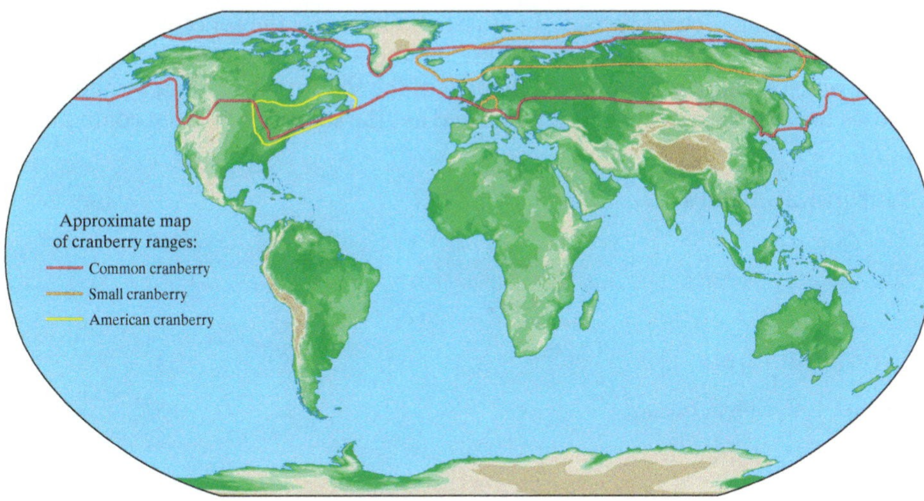

Sections

Check Your Progress

In this chapter, we review solving equations and introduce solving inequalities. Solving equations and inequalities is necessary in order to solve word problems. Of course, problem solving is an integral topic in algebra and its discussion is continued throughout this text.

The cranberry is one of only three fruits that are native to North America (the other two are the blueberry and the Concord grape). Native Americans pounded cranberries into a paste that they mixed with dried meat to create pemmican. American and Canadian sailors took cranberries on long sea voyages to combat scurvy, because of their high vitamin C content.

Americans consume 400 million pounds of cranberries annually, 20% of them during Thanksgiving week. They are available in many forms: fresh, jellied, dried, and in juice. They are high in antioxidants and beneficial to the health of gums and teeth. In Exercises 69–72, Section 9.1, we will explore information about cranberry crops from the top five cranberry-producing states. (*Source:* Cape Cod Cranberry Growers Association)

Cranberry flower—named because of resemblance of Crane

Top Cranberry-Producing States (in millions of pounds)

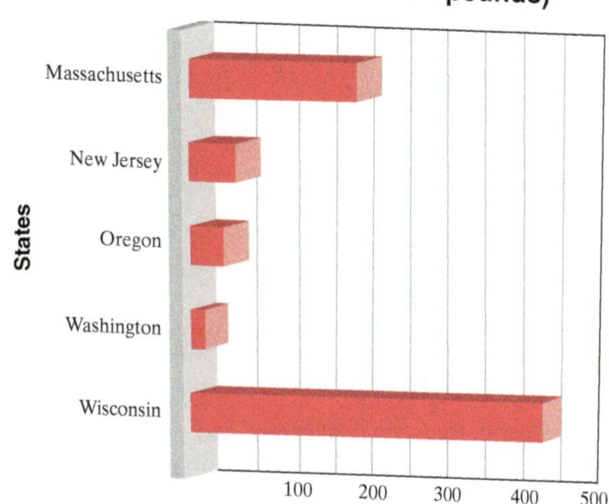

Source: National Agricultural Statistics Service

9.1 Symbols and Sets of Numbers 

Objectives

A Define the Meaning of the Symbols $=$, $\neq$, $<$, $>$, $\leq$, and $\geq$.

B Translate Sentences into Mathematical Statements.

C Identify Integers, Rational Numbers, Irrational Numbers, and Real Numbers.

Throughout the previous chapters, we have studied different sets of numbers. In this section, we review these sets of numbers. We also introduce a few new sets of numbers in order to show the relationships among these common sets of real numbers. We begin with a review of the set of natural numbers and the set of whole numbers and how we use symbols to compare these numbers. A **set** is a collection of objects, each of which is called a **member** or **element** of the set. A pair of brace symbols { } encloses the list of elements and is translated as "the set of" or "the set containing."

Natural numbers

$\{1, 2, 3, 4, 5, 6, \ldots\}$

Whole numbers

$\{0, 1, 2, 3, 4, 5, 6, \ldots\}$

Helpful Hint

The three dots (an ellipsis) at the end of the list of elements of a set means that the list continues in the same manner indefinitely.

Objective A Equality and Inequality Symbols 

Picturing natural numbers and whole numbers on a number line helps us to see the order of the numbers. Symbols can be used to describe in writing the order of two quantities. We will use equality symbols and inequality symbols to compare quantities.

Below is a review of these symbols. The letters a and b are used to represent quantities. Letters such as a and b that are used to represent numbers or quantities are called **variables.**

Equality and Inequality Symbols

		Meaning
Equality symbol:	$a = b$	a is equal to b.
Inequality symbols:	$a \neq b$	a is not equal to b.
	$a < b$	a is less than b.
	$a > b$	a is greater than b.
	$a \leq b$	a is less than or equal to b.
	$a \geq b$	a is greater than or equal to b.

These symbols may be used to form **mathematical statements** such as

$2 = 2$ and $2 \neq 6$

Recall that on a number line, we see that a number **to the right of** another number is **larger.** Similarly, a number **to the left of** another number is **smaller.** For example, 3 is to the left of 5 on the number line, which means that 3 is less than 5, or $3 < 5$. Similarly, 2 is to the right of 0 on the number line, which means that 2 is greater than 0, or $2 > 0$. Since 0 is to the left of 2, we can also say that 0 is less than 2, or $0 < 2$.

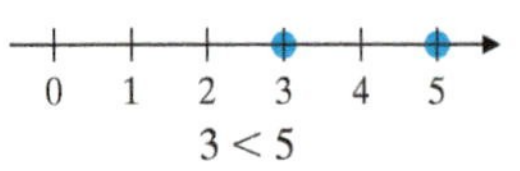

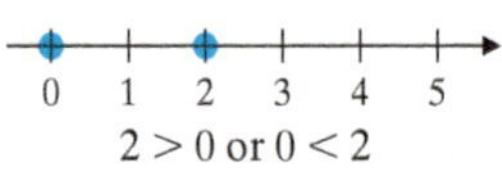

Helpful Hint

Notice that $2 > 0$ has exactly the same meaning as $0 < 2$. Switching the order of the numbers and reversing the "direction" of the inequality symbol does not change the meaning of the statement.

$6 > 4$ has the same meaning as $4 < 6$.

Also notice that when the statement is true, the inequality arrow points to the smaller number.

Our discussion above can be generalized in the order property below.

Order Property for Real Numbers

For any two real numbers a and b, a is less than b if a is to the left of b on a number line.

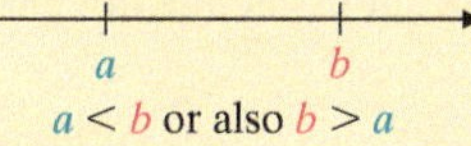

$a < b$ or also $b > a$

Examples Determine whether each statement is true or false.

1. $2 < 3$ True. Since 2 is to the left of 3 on a number line
2. $72 < 27$ False. 72 is to the right of 27 on a number line, so $72 > 27$.
3. $8 \geq 8$ True. Since $8 = 8$ is true
4. $8 \leq 8$ True. Since $8 = 8$ is true
5. $23 \leq 0$ False. Since neither $23 < 0$ nor $23 = 0$ is true
6. $0 \leq 23$ True. Since $0 < 23$ is true

Work Practice 1–6

Practice 1–6

Determine whether each statement is true or false.

1. $8 < 6$
2. $100 > 10$
3. $21 \leq 21$
4. $21 \geq 21$
5. $0 \geq 5$
6. $25 \geq 22$

Helpful Hint

If either $3 < 3$ or $3 = 3$ is true, then $3 \leq 3$ is true.

Objective B Translating Sentences into Mathematical Statements

Now, let's use the symbols discussed on the previous page to translate sentences into mathematical statements.

Example 7 Translate each sentence into a mathematical statement.

a. Nine is less than or equal to eleven.
b. Eight is greater than one.
c. Three is not equal to four.

Solution:

a. nine → 9; is less than or equal to → $\leq$; eleven → 11

b. eight → 8; is greater than → $>$; one → 1

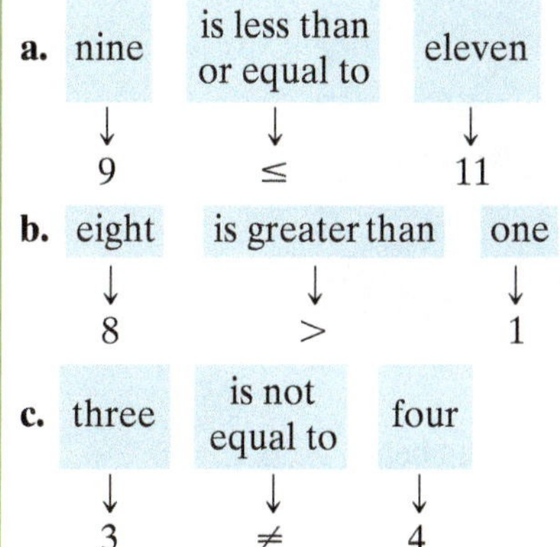

Work Practice 7

Practice 7

Translate each sentence into a mathematical statement.

a. Fourteen is greater than or equal to fourteen.
b. Zero is less than five.
c. Nine is not equal to ten.

Answers

1. false 2. true 3. true 4. true 5. false 6. true 7. a. $14 \geq 14$ b. $0 < 5$ c. $9 \neq 10$

Objective C Identifying Common Sets of Numbers 

Whole numbers are not sufficient to describe many situations in the real world. For example, quantities smaller than zero must sometimes be represented, such as temperatures less than 0 degrees.

Recall that we can place numbers less than zero on a number line as follows: Numbers less than 0 are to the left of 0 and are labeled −1, −2, −3, and so on. The numbers we have labeled on the number line below are called the set of **integers.**

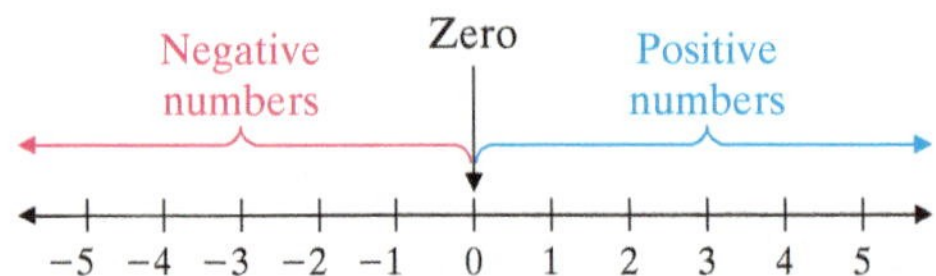

Integers to the left of 0 are called **negative integers;** integers to the right of 0 are called **positive integers.** The integer 0 is neither positive nor negative.

Integers

$\{\ldots, -3, -2, -1, 0, 1, 2, 3, \ldots\}$

Helpful Hint

A − sign, such as the one in −2, tells us that the number is to the left of 0 on a number line.

−2 is read "negative two."

A + sign or no sign tells us that the number lies to the right of 0 on a number line. For example, 3 and +3 both mean positive three.

Practice 8

Use an integer to express the number in the following: "The elevation of the city of New Orleans is 8 feet below sea level." (*Source: The World Almanac*)

Example 8 Use an integer to express the number in the following: "The lowest temperature ever recorded at South Pole Station, Antarctica, occurred during the month of June. The record-low temperature was 117 degrees below zero." (*Source:* The National Oceanic and Atmospheric Administration)

Solution: The integer −117 represents 117 degrees below zero.

Work Practice 8

Answer

8. −8

A problem with integers in real-life settings arises when quantities are smaller than some integer but greater than the next smallest integer. On a number line, these quantities may be visualized by points between integers. Some of these quantities between integers can be represented as quotients of integers. For example,

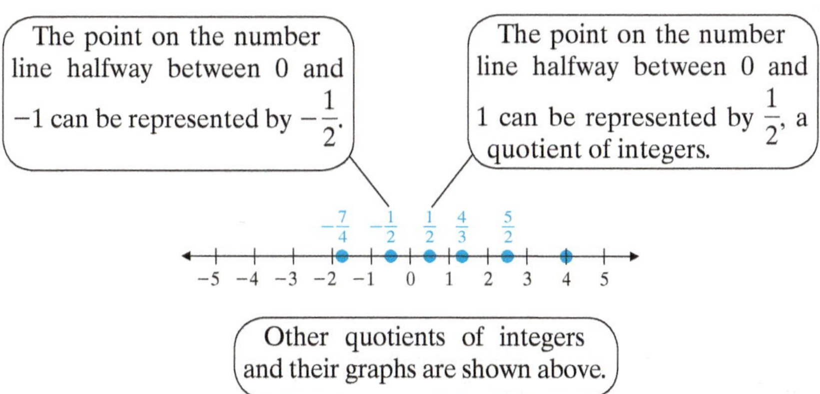

These numbers, each of which can be represented as a quotient of integers, are examples of **rational numbers.** It's not possible to list the set of rational numbers using the notation that we have been using. For this reason, we will use a different notation.

Rational Numbers

$$\left\{\frac{a}{b}\,\middle|\, a \text{ and } b \text{ are integers and } b \neq 0\right\}$$

We read this set as "the set of numbers $\frac{a}{b}$ such that a and b are integers and ***b* is not equal to 0.**"

Helpful Hint

We commonly refer to rational numbers as fractions.

Notice that every integer is also a rational number since each integer can be written as a quotient of integers. For example, the integer 5 is also a rational number since $5 = \frac{5}{1}$. For the rational number $\frac{5}{1}$, recall that the top number, 5, is called the numerator and the bottom number, 1, is called the denominator.

Let's practice **graphing** numbers on a number line.

Example 9 Graph the numbers on a number line.

$$-\frac{4}{3},\ \frac{1}{4},\ \frac{3}{2},\ -2\frac{1}{8},\ 3.5$$

Solution: To help graph the improper fractions in the list, we first write them as mixed numbers.

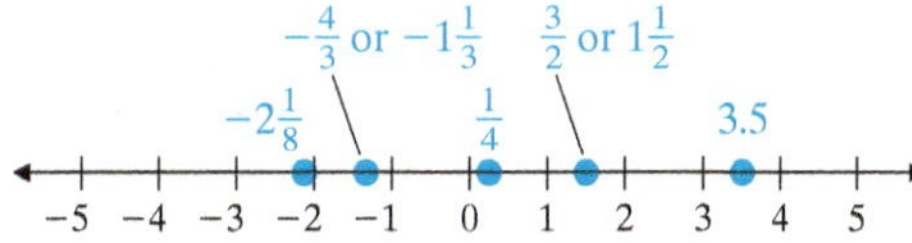

Work Practice 9

Every rational number has a point on the number line that corresponds to it. But not every point on the number line corresponds to a rational number. Those points that do not correspond to rational numbers correspond instead to **irrational numbers.**

Practice 9

Graph the numbers on the number line.

$$-2\frac{1}{2},\ -\frac{2}{3},\ \frac{1}{5},\ \frac{5}{4},\ 2.25$$

−5 −4 −3 −2 −1 0 1 2 3 4 5

Answer

9. $-2\frac{1}{2}$ $-\frac{2}{3}$ $\frac{1}{5}$ $\frac{5}{4}$ 2.25

−5 −4 −3 −2 −1 0 1 2 3 4 5

Irrational Numbers

{Nonrational numbers that correspond to points on a number line}

An irrational number that you have probably seen is π. Also, $\sqrt{2}$, the length of the diagonal of the square shown below, is an irrational number.

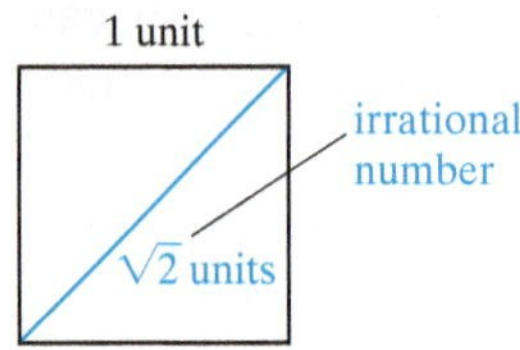

Both rational and irrational numbers can be written as decimal numbers. The decimal equivalent of a rational number will either terminate or repeat in a pattern. For example, upon dividing we find that

$\frac{3}{4} = 0.75$ (Decimal number terminates or ends.)

$\frac{2}{3} = 0.66666\ldots$ (Decimal number repeats in a pattern.)

The decimal representation of an irrational number will neither terminate nor repeat. (For further review of decimals, see Chapter 5.)

The set of numbers, each of which corresponds to a point on a number line, is called the set of **real numbers.** One and only one point on a number line corresponds to each real number.

Real Numbers

{All numbers that correspond to points on a number line}

Several different sets of numbers have been discussed in this section. The following diagram shows the relationships among these sets of real numbers. Notice that, together, the rational numbers and the irrational numbers make up the real numbers.

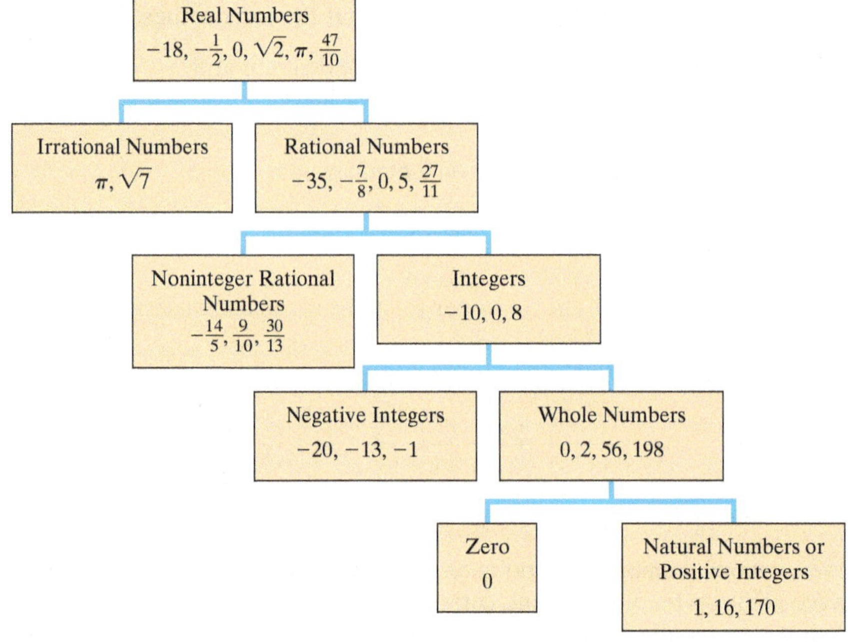

Now that other sets of numbers have been reviewed, let's continue our practice of comparing numbers.

Example 10 Insert $<, >$, or $=$ between the pairs of numbers to form true statements.

a. $-5 \quad -6$ **b.** $3.195 \quad 3.2$ **c.** $\frac{1}{4} \quad \frac{1}{3}$

Solution:

a. $-5 > -6$ since -5 lies to the right of -6 on a number line.

b. By comparing digits in the same place values, we find that $3.195 < 3.2$, since $0.1 < 0.2$.

c. By dividing, we find that $\frac{1}{4} = 0.25$ and $\frac{1}{3} = 0.33\ldots$. Since $0.25 < 0.33\ldots$, $\frac{1}{4} < \frac{1}{3}$.

Work Practice 10

Practice 10

Insert $<, >$, or $=$ between the pairs of numbers to form true statements.

a. $-11 \quad -9$

b. $4.511 \quad 4.151$

c. $\frac{7}{8} \quad \frac{2}{3}$

Example 11 Given the set $\left\{-3, -2, 0, \frac{1}{4}, \sqrt{2}, 11, 112\right\}$, list the numbers in this set that belong to the set of:

a. Natural numbers **b.** Whole numbers **c.** Integers
d. Rational numbers **e.** Irrational numbers **f.** Real numbers

Solution:

a. The natural numbers are 11 and 112.

b. The whole numbers are 0, 11, and 112.

c. The integers are $-3, -2, 0, 11$, and 112.

d. Recall that integers are rational numbers also. The rational numbers are $-3, -2, 0, \frac{1}{4}, 11$, and 112.

e. The only irrational number is $\sqrt{2}$.

f. All numbers in the given set are real numbers.

Work Practice 11

Practice 11

Given the set $\left\{-100, -\frac{2}{5}, 0, \pi, 6, 913\right\}$, list the numbers in this set that belong to the set of:

a. Natural numbers

b. Whole numbers

c. Integers

d. Rational numbers

e. Irrational numbers

f. Real numbers

Answers

10. a. $<$ **b.** $>$ **c.** $>$

11. a. 6, 913 **b.** 0, 6, 913
c. $-100, 0, 6, 913$
d. $-100, -\frac{2}{5}, 0, 6, 913$ **e.** π
f. all numbers in the given set

Vocabulary, Readiness & Video Check

Use the choices below to fill in each blank. Not all choices will be used.

real	natural	whole	$\frac{1}{2}$	$\frac{1}{4}$
rational	inequality	integers	0	1

1. The ____________ numbers are $\{0, 1, 2, 3, 4, \ldots\}$.

2. The ____________ numbers are $\{1, 2, 3, 4, 5, \ldots\}$.

3. The symbols $\neq, \leq$, and $>$ are called ____________ symbols.

4. The ____________ are $\{\ldots, -3, -2, -1, 0, 1, 2, 3, \ldots\}$.

5. The ____________ numbers are {all numbers that correspond to points on a number line}.

6. The ____________ numbers are $\left\{\frac{a}{b} \middle| a \text{ and } b \text{ are integers}, b \neq 0\right\}$.

7. The integer ____________ is neither positive nor negative.

8. The point on a number line halfway between 0 and $\frac{1}{2}$ can be represented by ____________.

Martin-Gay Interactive Videos *Watch the section lecture video and answer the following questions.*

See Video 9.1

Objective A 9. In Example 2, why is the symbol < inserted between the two numbers?

Objective B 10. Write the sentence given in Example 4 and translate it to a mathematical statement using symbols.

Objective C 11. Which sets of numbers does the number in Example 6 belong to? Why is this number not an irrational number?

9.1 Exercise Set MyMathLab®

Objectives A C Mixed Practice *Insert <, >, or = in the space between the paired numbers to make each statement true. See Examples 1 through 6 and 10.*

1. 4 10
2. 8 5
3. 7 3
4. 9 15
5. 6.26 6.26
6. 1.13 1.13
7. 0 7
8. 20 0

9. The freezing point of water is 32° Fahrenheit. The boiling point of water is 212° Fahrenheit. Write an inequality statement using < or > comparing the numbers 32 and 212.

10. The freezing point of water is 0° Celsius. The boiling point of water is 100° Celsius. Write an inequality statement using < or > comparing the numbers 0 and 100.

△ 11. An angle measuring 30° and an angle measuring 45° are shown. Write an inequality statement using ≤ or ≥ comparing the numbers 30 and 45.

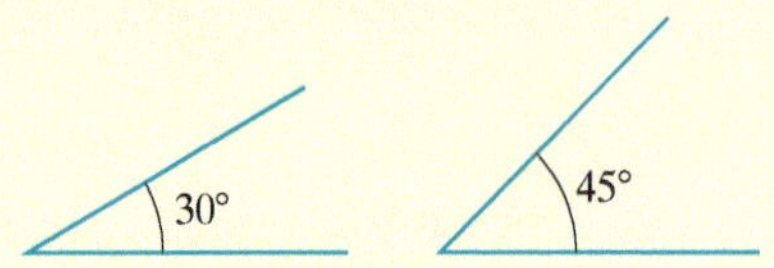

△ 12. The sum of the measures of the angles of a parallelogram is 360°. The sum of the measures of the angles of a triangle is 180°. Write an inequality statement using ≤ or ≥ comparing the numbers 360 and 180.

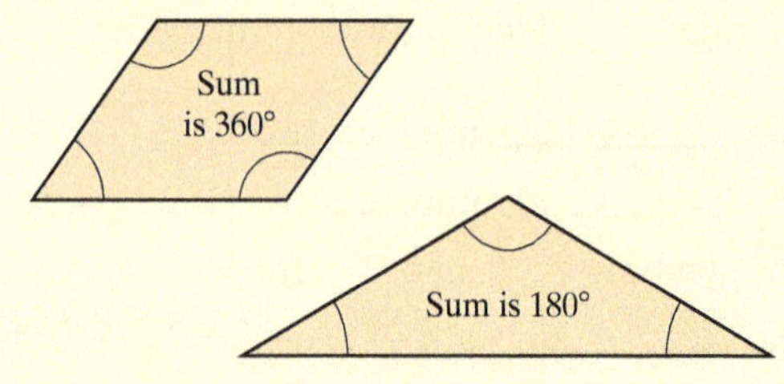

Determine whether each statement is true or false. See Examples 1 through 6 and 10.

13. $11 \le 11$

14. $8 \ge 9$

15. $-11 > -10$

16. $-16 > -17$

17. $5.092 < 5.902$

18. $1.02 > 1.021$

19. $\frac{9}{10} \le \frac{8}{9}$

20. $\frac{4}{5} \le \frac{9}{11}$

Rewrite each inequality so that the inequality symbol points in the opposite direction and the resulting statement has the same meaning as the given one. See Examples 1 through 6 and 10.

21. $25 \ge 20$

22. $-13 \le 13$

23. $0 < 6$

24. $5 > 3$

25. $-10 > -12$

26. $-4 < -2$

Objectives B C Mixed Practice—Translating *Write each sentence as a mathematical statement. See Example 7.*

27. Seven is less than eleven.

28. Twenty is greater than two.

29. Five is greater than or equal to four.

30. Negative ten is less than or equal to thirty-seven.

31. Fifteen is not equal to negative two.

32. Negative seven is not equal to seven.

Use integers to represent the value(s) in each statement. See Example 8.

33. The highest elevation in California is Mt. Whitney, with an altitude of 14,494 feet. The lowest elevation in California is Death Valley, with an altitude of 282 feet below sea level. (*Source:* U.S. Geological Survey)

34. Driskill Mountain, in Louisiana, has an altitude of 535 feet. New Orleans, Louisiana, lies 8 feet below sea level. (*Source:* U.S. Geological Survey)

35. The number of graduate students at the University of Texas at Austin was 27,724 fewer than the number of undergraduate students. (*Source:* University of Texas at Austin, 2012)

36. The number of students admitted to the class of 2011 at UCLA was 56,715 fewer students than the number that had applied. (*Source:* UCLA, 2012)

37. A community college student deposited \$475 in her savings account. She later withdrew \$195.

38. A deep-sea diver ascended 17 feet and later descended 15 feet.

Graph each set of numbers on the number line. See Example 9.

39. $-4, 0, 2, -2$

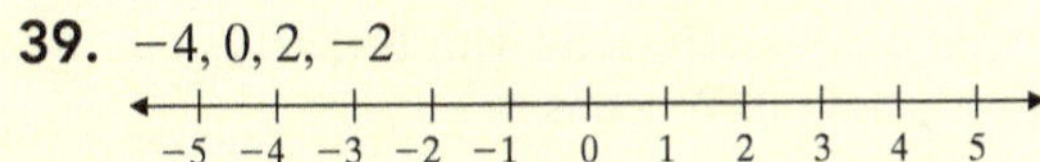

40. $-3, 0, 1, -5$

-5 -4 -3 -2 -1 0 1 2 3 4 5

41. $-2, 4, \frac{1}{3}, -\frac{1}{4}$

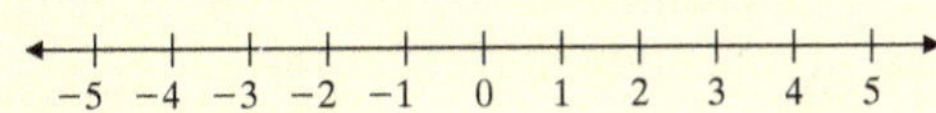

42. $-5, 3, -\frac{1}{3}, \frac{7}{8}$

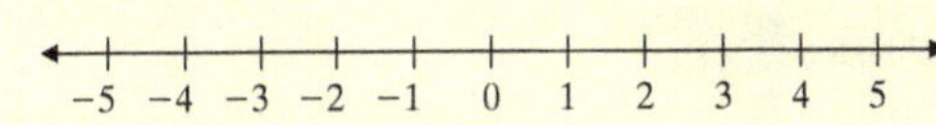

43. $-4.5, \frac{7}{4}, 3.25, -\frac{3}{2}$

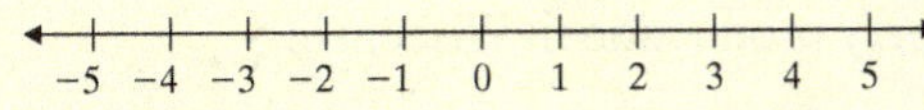

44. $4.5, -\frac{9}{4}, 1.75, -\frac{7}{2}$

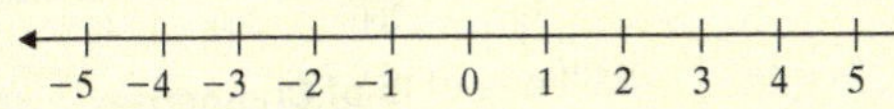

Tell which set or sets each number belongs to: natural numbers, whole numbers, integers, rational numbers, irrational numbers, or real numbers. See Example 11.

45. 0

46. $\frac{1}{4}$

47. -7

48. $-\frac{1}{7}$

49. 265

50. 7941

51. $\frac{2}{3}$

52. $\sqrt{3}$

Determine whether each statement is true or false.

53. Every rational number is also an integer.

54. Every natural number is positive.

55. 0 is a real number.

56. $\frac{1}{2}$ is an integer.

57. Every negative number is also a rational number.

58. Every rational number is also a real number.

59. Every real number is also a rational number.

60. Every whole number is an integer.

Review

Insert <, >, or = in the appropriate space to make each statement true. See Section 2.1.

61. $|-5| \quad -4$

62. $|-12| \quad |0|$

63. $\left|-\frac{5}{8}\right| \quad \left|\frac{5}{8}\right|$

64. $\left|\frac{2}{5}\right| \quad \left|-\frac{2}{5}\right|$

65. $|-2| \quad |-2.7|$

66. $|-5.01| \quad |-5|$

67. $|0| \quad |-8|$

68. $|-12| \quad \frac{-24}{2}$

Concept Extensions

The bar graph shows cranberry production from the top five cranberry-producing states. (Source: National Agricultural Statistics Service)

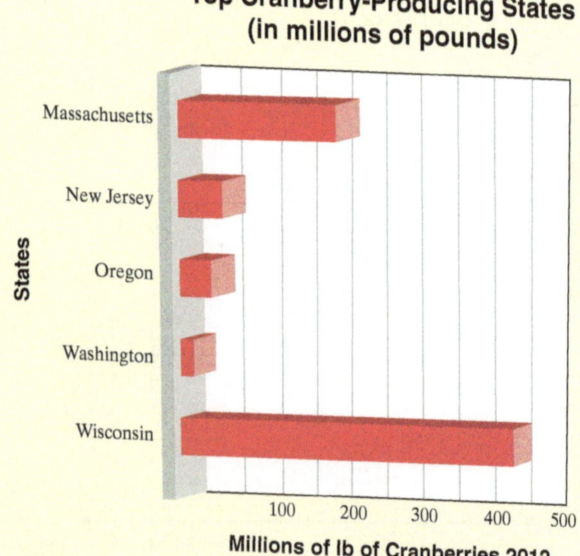

69. Write an inequality comparing the 2012 cranberry production in Oregon with the 2012 cranberry production in Washington.

70. Write an inequality comparing the 2012 cranberry production in Massachusetts with the 2012 cranberry production in Wisconsin.

71. Determine the difference between the 2012 cranberry production in Washington and the 2012 cranberry production in New Jersey.

72. Determine the difference between the 2012 cranberry production in Massachusetts and the 2012 cranberry production in Wisconsin.

The apparent magnitude of a star is the measure of its brightness as seen by someone on Earth. The smaller the apparent magnitude, the brighter the star. Below, the apparent magnitudes of some stars are listed. Use this table to answer Exercises 73 through 78.

Star	Apparent Magnitude	Star	Apparent Magnitude
Arcturus	−0.04	Spica	0.98
Sirius	−1.46	Rigel	0.12
Vega	0.03	Regulus	1.35
Antares	0.96	Canopus	−0.72
Sun (Sol)	−26.7	Hadar	0.61

(*Source: Norton's 2000.0: Star Atlas and Reference Handbook,* 18th ed., Longman Group, UK, 1989)

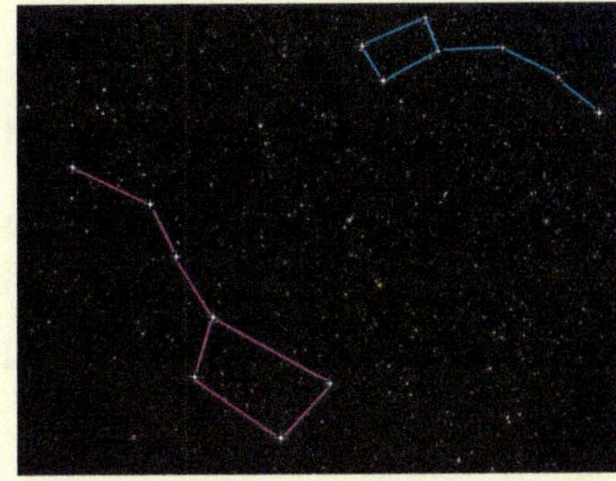

73. The apparent magnitude of the sun is −26.7. The apparent magnitude of the star Arcturus is −0.04. Write an inequality statement comparing the numbers −0.04 and −26.7.

74. The apparent magnitude of Antares is 0.96. The apparent magnitude of Spica is 0.98. Write an inequality statement comparing the numbers 0.96 and 0.98.

75. Which is brighter, the sun or Arcturus?

76. Which is dimmer, Antares or Spica?

77. Which star listed is the brightest?

78. Which star listed is the dimmest?

79. In your own words, explain how to find the absolute value of a number.

80. Give an example of a real-life situation that can be described with integers but not with whole numbers.

9.2 Properties of Real Numbers

Objective A Using the Commutative and Associative Properties

Objectives

A Use the Commutative and Associative Properties.

B Use the Distributive Property.

C Use the Identity and Inverse Properties.

In this section we review names of properties of real numbers with which we are already familiar. Throughout this section, the variables a, b, and c represent real numbers.

We know that order does not matter when adding numbers. For example, we know that $7 + 5$ is the same as $5 + 7$. This property is given a special name—the **commutative property of addition.** We also know that order does not matter when multiplying numbers. For example, we know that $-5(6) = 6(-5)$. This property means that multiplication is commutative also and is called the **commutative property of multiplication.**

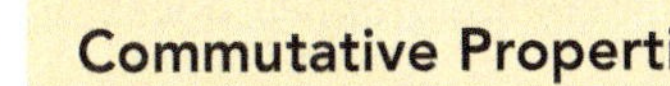

Commutative Properties

Addition: $a + b = b + a$

Multiplication: $a \cdot b = b \cdot a$

These properties state that the *order* in which any two real numbers are added or multiplied does not change their sum or product. For example, if we let $a = 3$ and $b = 5$, then the commutative properties guarantee that

$$3 + 5 = 5 + 3 \quad \text{and} \quad 3 \cdot 5 = 5 \cdot 3$$

Helpful Hint

Is subtraction also commutative? Try an example. Is $3 - 2 = 2 - 3$? **No!** The left side of this statement equals 1; the right side equals -1. There is no commutative property of subtraction. Similarly, there is no commutative property of division. For example, $10 \div 2$ does not equal $2 \div 10$.

Practice 1

Use a commutative property to complete each statement.

a. $7 \cdot y =$ _____

b. $4 + x =$ _____

Example 1 Use a commutative property to complete each statement.

a. $x + 5 =$ _____ **b.** $3 \cdot x =$ _____

Solution:

a. $x + 5 = 5 + x$ By the commutative property of addition

b. $3 \cdot x = x \cdot 3$ By the commutative property of multiplication

Work Practice 1

✓**Concept Check** Which of the following pairs of actions are commutative?

a. "raking the leaves" and "bagging the leaves"

b. "putting on your left glove" and "putting on your right glove"

c. "putting on your coat" and "putting on your shirt"

d. "reading a novel" and "reading a newspaper"

Let's now discuss grouping numbers. When we add three numbers, the way in which they are grouped or associated does not change their sum. For example, we know that $2 + (3 + 4) = 2 + 7 = 9$. This result is the same if we group the numbers differently. In other words, $(2 + 3) + 4 = 5 + 4 = 9$, also. Thus, $2 + (3 + 4) = (2 + 3) + 4$. This property is called the **associative property of addition.**

In the same way, changing the grouping of numbers when multiplying does not change their product. For example, $2 \cdot (3 \cdot 4) = (2 \cdot 3) \cdot 4$ (check it). This is the **associative property of multiplication.**

Associative Properties

Addition: $(a + b) + c = a + (b + c)$

Multiplication: $(a \cdot b) \cdot c = a \cdot (b \cdot c)$

These properties state that the way in which three numbers are *grouped* does not change their sum or their product.

Answers

1. a. $y \cdot 7$ **b.** $x + 4$

✓**Concept Check Answer**

b, d

Example 2 Use an associative property to complete each statement.

a. $5 + (4 + 6) =$ _____ **b.** $(-1 \cdot 2) \cdot 5 =$ _____

c. $(m + n) + 9 =$ _____ **d.** $(xy) \cdot 12 =$ _____

Solution:

a. $5 + (4 + 6) = (5 + 4) + 6$ By the associative property of addition

b. $(-1 \cdot 2) \cdot 5 = -1 \cdot (2 \cdot 5)$ By the associative property of multiplication

c. $(m + n) + 9 = m + (n + 9)$ By the associative property of addition

d. $(xy) \cdot 12 = x \cdot (y \cdot 12)$ Recall that xy means $x \cdot y$.

Work Practice 2

Practice 2

Use an associative property to complete each statement.

a. $5 \cdot (-3 \cdot 6) =$ _____

b. $(-2 + 7) + 3 =$ _____

c. $(q + r) + 17 =$ _____

d. $(ab) \cdot 21 =$ _____

Helpful Hint

Remember the difference between the commutative properties and the associative properties. The commutative properties have to do with the *order* of numbers and the associative properties have to do with the *grouping* of numbers.

Examples Determine whether each statement is true by an associative property or a commutative property.

3. $(7 + 10) + 4 = (10 + 7) + 4$ Since the order of two numbers was changed and their grouping was not, this is true by the commutative property of addition.

4. $2 \cdot (3 \cdot 1) = (2 \cdot 3) \cdot 1$ Since the grouping of the numbers was changed and their order was not, this is true by the associative property of multiplication.

Work Practice 3–4

Practice 3–4

Determine whether each statement is true by an associative property or a commutative property.

3. $5 \cdot (4 \cdot 7) = 5 \cdot (7 \cdot 4)$

4. $-2 + (4 + 9) = (-2 + 4) + 9$

Let's now illustrate how these properties can help us simplify expressions.

Examples Simplify each expression.

5. $10 + (x + 12) = 10 + (12 + x)$ By the commutative property of addition

$= (10 + 12) + x$ By the associative property of addition

$= 22 + x$ Add.

6. $-3(7x) = (-3 \cdot 7)x$ By the associative property of multiplication

$= -21x$ Multiply.

Work Practice 5–6

Practice 5–6

Simplify each expression.

5. $(-3 + x) + 17$ **6.** $4(5x)$

Objective B Using the Distributive Property 

The **distributive property of multiplication over addition** is used repeatedly throughout algebra. It is useful because it allows us to write a product as a sum or a sum as a product.

We know that $7(2 + 4) = 7(6) = 42$. Compare that with

$$7(2) + 7(4) = 14 + 28 = 42$$

Since both original expressions equal 42, they must equal each other, or

$$7(2 + 4) = 7(2) + 7(4)$$

Answers

2. a. $(5 \cdot -3) \cdot 6$ **b.** $-2 + (7 + 3)$ **c.** $q + (r + 17)$ **d.** $a \cdot (b \cdot 21)$ **3.** commutative **4.** associative **5.** $14 + x$ **6.** $20x$

This is an example of the distributive property. The product on the left side of the equal sign is equal to the sum on the right side. We can think of the 7 as being distributed to each number inside the parentheses.

Distributive Property of Multiplication Over Addition

$$a(b + c) = ab + ac$$

Since multiplication is commutative, this property can also be written as

$$(b + c)a = ba + ca$$

The distributive property can also be extended to more than two numbers inside the parentheses. For example,

$$3(x + y + z) = 3(x) + 3(y) + 3(z)$$
$$= 3x + 3y + 3z$$

Since we define subtraction in terms of addition, the distributive property is also true for subtraction. For example,

$$2(x - y) = 2(x) - 2(y)$$
$$= 2x - 2y$$

Practice 7–12

Use the distributive property to write each expression without parentheses. Then simplify the result.

7. $5(x + y)$
8. $-3(2 + 7x)$
9. $4(x + 6y - 2z)$
10. $-1(3 - a)$
11. $-(8 + a - b)$
12. $\frac{1}{2}(2x + 4) + 9$

Answers

7. $5x + 5y$ 8. $-6 - 21x$ 9. $4x + 24y - 8z$ 10. $-3 + a$ 11. $-8 - a + b$ 12. $x + 11$

Examples Use the distributive property to write each expression without parentheses. Then simplify the result.

7. $2(x + y) = 2(x) + 2(y)$
$= 2x + 2y$

8. $-5(-3 + 2z) = -5(-3) + (-5)(2z)$
$= 15 - 10z$

9. $5(x + 3y - z) = 5(x) + 5(3y) - 5(z)$
$= 5x + 15y - 5z$

10. $-1(2 - y) = (-1)(2) - (-1)(y)$
$= -2 + y$

11. $-(3 + x - w) = -1(3 + x - w)$
$= (-1)(3) + (-1)(x) - (-1)(w)$
$= -3 - x + w$

12. $\frac{1}{2}(6x + 14) + 10 = \frac{1}{2}(6x) + \frac{1}{2}(14) + 10$ Apply the distributive property.
$= 3x + 7 + 10$ Multiply.
$= 3x + 17$ Add.

Helpful Hint

Notice in Example 11 that $-(3 + x - w)$ can be rewritten as $-1(3 + x - w)$.

Work Practice 7–12

The distributive property can also be used to write a sum as a product.

Examples Use the distributive property to write each sum as a product.

13. $8 \cdot 2 + 8 \cdot x = 8(2 + x)$
14. $7s + 7t = 7(s + t)$

Work Practice 13–14

Practice 13–14

Use the distributive property to write each sum as a product.

13. $9 \cdot 3 + 9 \cdot y$
14. $4x + 4y$

Objective C Using the Identity and Inverse Properties 

Next, we look at the **identity properties.**

The number 0 is called the identity for addition because when 0 is added to any real number, the result is the same real number. In other words, the *identity* of the real number is not changed.

The number 1 is called the identity for multiplication because when a real number is multiplied by 1, the result is the same real number. In other words, the *identity* of the real number is not changed.

Identities for Addition and Multiplication

0 is the identity element for addition.

$$a + 0 = a \quad \text{and} \quad 0 + a = a$$

1 is the identity element for multiplication.

$$a \cdot 1 = a \quad \text{and} \quad 1 \cdot a = a$$

Notice that 0 is the *only* number that can be added to any real number with the result that the sum is the same real number. Also, 1 is the *only* number that can be multiplied by any real number with the result that the product is the same real number.

Additive inverses or **opposites** were introduced in Section 2.1. Two numbers are called additive inverses or opposites if their sum is 0. The additive inverse or opposite of 6 is -6 because $6 + (-6) = 0$. The additive inverse or opposite of -5 is 5 because $-5 + 5 = 0$.

Reciprocals or **multiplicative inverses** were introduced in Section 4.3. Two nonzero numbers are called reciprocals or multiplicative inverses if their product is 1. The reciprocal or multiplicative inverse of $\frac{2}{3}$ is $\frac{3}{2}$ because $\frac{2}{3} \cdot \frac{3}{2} = 1$. Likewise, the reciprocal of -5 is $-\frac{1}{5}$ because $-5\left(-\frac{1}{5}\right) = 1$.

Additive or Multiplicative Inverses

The numbers a and $-a$ are additive inverses or opposites of each other because their sum is 0; that is,

$$a + (-a) = 0$$

The numbers b and $\frac{1}{b}$ (for $b \neq 0$) are reciprocals or multiplicative inverses of each other because their product is 1; that is,

$$b \cdot \frac{1}{b} = 1$$

Answers
13. $9(3 + y)$ **14.** $4(x + y)$

Practice 15–21

Name the property illustrated by each true statement.

15. $7(a + b) = 7 \cdot a + 7 \cdot b$
16. $12 + y = y + 12$
17. $-4 \cdot (6 \cdot x) = (-4 \cdot 6) \cdot x$
18. $6 + (z + 2) = 6 + (2 + z)$
19. $3\left(\frac{1}{3}\right) = 1$
20. $(x + 0) + 23 = x + 23$
21. $(7 \cdot y) \cdot 10 = y \cdot (7 \cdot 10)$

Answers
15. distributive property
16. commutative property of addition
17. associative property of multiplication
18. commutative property of addition
19. multiplicative inverse property
20. identity element for addition
21. commutative and associative properties of multiplication

✓Concept Check Answers
a. $\frac{3}{10}$ **b.** $-\frac{10}{3}$

✓**Concept Check** Which of the following is the

a. opposite of $-\frac{3}{10}$, and which is the

b. reciprocal of $-\frac{3}{10}$?

$$1, -\frac{10}{3}, \frac{3}{10}, 0, \frac{10}{3}, -\frac{3}{10}$$

Examples Name the property illustrated by each true statement.

15. $3(x + y) = 3 \cdot x + 3 \cdot y$ — Distributive property
16. $(x + 7) + 9 = x + (7 + 9)$ — Associative property of addition (grouping changed)
17. $(b + 0) + 3 = b + 3$ — Identity element for addition
18. $2 \cdot (z \cdot 5) = 2 \cdot (5 \cdot z)$ — Commutative property of multiplication (order changed)
19. $-2 \cdot \left(-\frac{1}{2}\right) = 1$ — Multiplicative inverse property
20. $-2 + 2 = 0$ — Additive inverse property
21. $-6 \cdot (y \cdot 2) = (-6 \cdot 2) \cdot y$ — Commutative and associative properties of multiplication (order and grouping changed)

Work Practice 15–21

Vocabulary, Readiness & Video Check

Use the choices below to fill in each blank.

distributive property	associative property of multiplication	commutative property of addition
opposites or additive inverses	associative property of addition	
reciprocals or multiplicative inverses	commutative property of multiplication	

1. $x + 5 = 5 + x$ is a true statement by the ______________.
2. $x \cdot 5 = 5 \cdot x$ is a true statement by the ______________.
3. $3(y + 6) = 3 \cdot y + 3 \cdot 6$ is true by the ______________.
4. $2 \cdot (x \cdot y) = (2 \cdot x) \cdot y$ is a true statement by the ______________.
5. $x + (7 + y) = (x + 7) + y$ is a true statement by the ______________.
6. The numbers $-\frac{2}{3}$ and $-\frac{3}{2}$ are called ______________.
7. The numbers $-\frac{2}{3}$ and $\frac{2}{3}$ are called ______________.

Martin-Gay Interactive Videos *Watch the section lecture video and answer the following questions.*

See Video 9.2

Objective A **8.** The commutative properties are discussed in Examples 1 and 2 and the associative properties are discussed in Examples 3–7. What's the one word used again and again to describe the commutative property? The associative property?

Objective B **9.** In Example 10, what point is made about the term 2?

Objective C **10.** Complete these statements based on the lecture given before Example 12:

- The identity element for addition is __________ because if we add __________ to any real number, the result is that real number.
- The identity element for multiplication is __________ because any real number times __________ gives a result of that original real number.

9.2 Exercise Set MyMathLab®

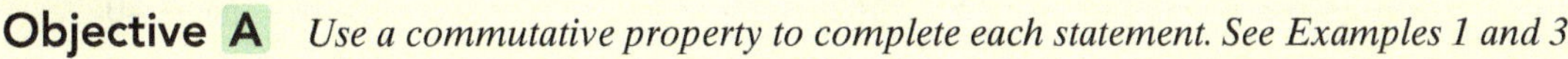

Objective A *Use a commutative property to complete each statement. See Examples 1 and 3.*

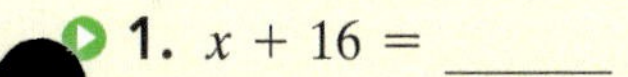

1. $x + 16 =$ _____ **2.** $8 + y =$ _____ **3.** $-4 \cdot y =$ _____ **4.** $-2 \cdot x =$ _____

5. $xy =$ _____ **6.** $ab =$ _____ **7.** $2x + 13 =$ _____ **8.** $19 + 3y =$ _____

Use an associative property to complete each statement. See Examples 2 and 4.

9. $(xy) \cdot z =$ _____ **10.** $3 \cdot (x \cdot y) =$ _____ **11.** $2 + (a + b) =$ _____

12. $(y + 4) + z =$ _____ **13.** $4 \cdot (ab) =$ _____ **14.** $(-3y) \cdot z =$ _____

15. $(a + b) + c =$ _____ **16.** $6 + (r + s) =$ _____

Use the commutative and associative properties to simplify each expression. See Examples 5 and 6.

17. $8 + (9 + b)$ **18.** $(r + 3) + 11$ **19.** $4(6y)$ **20.** $2(42x)$ **21.** $\frac{1}{5}(5y)$

22. $\frac{1}{8}(8z)$ **23.** $(13 + a) + 13$ **24.** $7 + (x + 4)$ **25.** $-9(8x)$ **26.** $-3(12y)$

27. $\frac{3}{4}\left(\frac{4}{3}s\right)$ **28.** $\frac{2}{7}\left(\frac{7}{2}r\right)$ **29.** $-\frac{1}{2}(5x)$ **30.** $-\frac{1}{3}(7x)$

Objective B *Use the distributive property to write each expression without parentheses. Then simplify the result, if possible. See Examples 7 through 12.*

31. $4(x + y)$
32. $7(a + b)$
33. $9(x - 6)$
34. $11(y - 4)$
35. $2(3x + 5)$
36. $5(7 + 8y)$
37. $7(4x - 3)$
38. $3(8x - 1)$
39. $3(6 + x)$
40. $2(x + 5)$
41. $-2(y - z)$
42. $-3(z - y)$
43. $-\frac{1}{3}(3y + 5)$
44. $-\frac{1}{2}(2r + 11)$
45. $5(x + 4m + 2)$
46. $8(3y + z - 6)$
47. $-4(1 - 2m + n) + 4$
48. $-4(4 + 2p + 5) + 16$
49. $-(5x + 2)$
50. $-(9r + 5)$
51. $-(r - 3 - 7p)$
52. $-(q - 2 + 6r)$
53. $\frac{1}{2}(6x + 7) + \frac{1}{2}$
54. $\frac{1}{4}(4x - 2) - \frac{7}{2}$
55. $-\frac{1}{3}(3x - 9y)$
56. $-\frac{1}{5}(10a - 25b)$
57. $3(2r + 5) - 7$
58. $10(4s + 6) - 40$
59. $-9(4x + 8) + 2$
60. $-11(5x + 3) + 10$
61. $-0.4(4x + 5) - 0.5$
62. $-0.6(2x + 1) - 0.1$

Use the distributive property to write each sum as a product. See Examples 13 and 14.

63. $4 \cdot 1 + 4 \cdot y$
64. $14 \cdot z + 14 \cdot 5$
65. $11x + 11y$
66. $9a + 9b$
67. $(-1) \cdot 5 + (-1) \cdot x$
68. $(-3)a + (-3)y$
69. $30a + 30b$
70. $25x + 25y$

Objectives A C Mixed Practice *Name the property illustrated by each true statement. See Examples 15 through 21.*

71. $3 \cdot 5 = 5 \cdot 3$
72. $4(3 + 8) = 4 \cdot 3 + 4 \cdot 8$
73. $2 + (x + 5) = (2 + x) + 5$
74. $9 \cdot (x \cdot 7) = (9 \cdot x) \cdot 7$
75. $(x + 9) + 3 = (9 + x) + 3$
76. $1 \cdot 9 = 9$
77. $(4 \cdot y) \cdot 9 = 4 \cdot (y \cdot 9)$
78. $-4 \cdot (8 \cdot 3) = (8 \cdot 3) \cdot (-4)$

79. $0 + 6 = 6$

80. $(a + 9) + 6 = a + (9 + 6)$

81. $-4(y + 7) = -4 \cdot y + (-4) \cdot 7$

82. $(11 + r) + 8 = (r + 11) + 8$

83. $6 \cdot \frac{1}{6} = 1$

84. $r + 0 = r$

85. $-6 \cdot 1 = -6$

86. $-\frac{3}{4}\left(-\frac{4}{3}\right) = 1$

Review

Evaluate each expression for the given values. See Section 2.5.

87. If $x = -1$ and $y = 3$, find $y - x^2$.

88. If $g = 0$ and $h = -4$, find $gh - h^2$.

89. If $a = 2$ and $b = -5$, find $a - b^2$.

90. If $x = -3$, find $x^3 - x^2 + 4$.

91. If $y = -5$ and $z = 0$, find $yz - y^2$.

92. If $x = -2$, find $x^3 - x^2 - x$.

Concept Extensions

Fill in the table with the opposite (additive inverse), the reciprocal (multiplicative inverse), or the expression. Assume that the value of each expression is not 0.

	93.	94.	95.	96.	97.	98.
Expression	8	$-\frac{2}{3}$	x	$4y$		
Opposite						$7x$
Reciprocal					$\frac{1}{2x}$	

Decide whether each statement is true or false. See the second Concept Check in this section.

99. The opposite of $-\frac{a}{2}$ is $-\frac{2}{a}$.

100. The reciprocal of $-\frac{a}{2}$ is $\frac{a}{2}$.

Determine which pairs of actions are commutative. See the first Concept Check in this section.

101. "taking a test" and "studying for the test"

102. "putting on your shoes" and "putting on your socks"

103. "putting on your left shoe" and "putting on your right shoe"

104. "reading the sports section" and "reading the comics section"

105. "mowing the lawn" and "trimming the hedges"

106. "baking a cake" and "eating the cake"

107. "feeding the dog" and "feeding the cat"

108. "dialing a number" and "turning on the cell phone"

Name the property illustrated by each step.

109. **a.** $\triangle + (\square + \circ) = (\square + \circ) + \triangle$

b. $= (\circ + \square) + \triangle$

c. $= \circ + (\square + \triangle)$

110. **a.** $(x + y) + z = x + (y + z)$

b. $= (y + z) + x$

c. $= (z + y) + x$

111. Explain why 0 is called the identity element for addition.

112. Explain why 1 is called the identity element for multiplication.

113. Write an example that shows that division is not commutative.

114. Write an example that shows that subtraction is not commutative.

9.3 Further Solving Linear Equations 

Objectives

A Apply the General Strategy for Solving a Linear Equation.

B Solve Equations Containing Fractions or Decimals.

C Recognize Identities and Equations with No Solution.

Objective A Solving Linear Equations

Since Chapter 2, we have been solving equations. In this section, we give a definition of a linear equation in one variable and review steps we may use to solve these equations.

Let's begin with a formal definition of a linear equation in one variable.

A **linear equation in one variable** can be written in the form

$$Ax + B = C$$

where A, B, and C are real numbers and $A \neq 0$.

We now combine our knowledge from the previous chapters and review solving linear equations.

To Solve Linear Equations in One Variable

Step 1: If an equation contains fractions or decimals, multiply both sides by the LCD to clear the equation of fractions or decimals.

Step 2: Use the distributive property to remove parentheses if they are present.

Step 3: Simplify each side of the equation by combining like terms.

Step 4: Get all variable terms on one side and all numbers on the other side by using the addition property of equality.

Step 5: Get the variable alone by using the multiplication property of equality.

Step 6: Check the solution by substituting it into the original equation.

Practice 1

Solve:

$5(3x - 1) + 2 = 12x + 6$

Answer

1. $x = 3$

Example 1 Solve: $4(2x - 3) + 7 = 3x + 5$

Solution: There are no fractions, so we begin with Step 2.

$4(2x - 3) + 7 = 3x + 5$

Step 2: $8x - 12 + 7 = 3x + 5$ Use the distributive property.

Step 3: $8x - 5 = 3x + 5$ Combine like terms.

Step 4: Get all variable terms on one side of the equation and all numbers on the other side. One way to do this is by subtracting $3x$ from both sides and then adding 5 to both sides.

$$8x - 5 - 3x = 3x + 5 - 3x \quad \text{Subtract } 3x \text{ from both sides.}$$
$$5x - 5 = 5 \quad \text{Simplify.}$$
$$5x - 5 + 5 = 5 + 5 \quad \text{Add 5 to both sides.}$$
$$5x = 10 \quad \text{Simplify.}$$

Step 5: Use the multiplication property of equality to get x alone.

$$\frac{5x}{5} = \frac{10}{5} \quad \text{Divide both sides by 5.}$$
$$x = 2 \quad \text{Simplify.}$$

Step 6: Check.

$$4(2x - 3) + 7 = 3x + 5 \quad \text{Original equation}$$
$$4[2(2) - 3] + 7 \stackrel{?}{=} 3(2) + 5 \quad \text{Replace } x \text{ with 2.}$$
$$4(4 - 3) + 7 \stackrel{?}{=} 6 + 5$$
$$4(1) + 7 \stackrel{?}{=} 11$$
$$4 + 7 \stackrel{?}{=} 11$$
$$11 = 11 \quad \text{True}$$

The solution is 2.

Work Practice 1

Helpful Hint When checking solutions, use the original equation.

Example 2 Solve: $8(2 - t) = -5t$

Solution: First, we apply the distributive property.

$$8(2 - t) = -5t$$

Step 2: $16 - 8t = -5t$ Use the distributive property.

Step 4: $16 - 8t + 8t = -5t + 8t$ Add $8t$ to both sides.

$$16 = 3t \quad \text{Combine like terms.}$$

Step 5: $\frac{16}{3} = \frac{3t}{3}$ Divide both sides by 3.

$$\frac{16}{3} = t \quad \text{Simplify.}$$

Step 6: Check.

$$8(2 - t) = -5t \quad \text{Original equation}$$
$$8\left(2 - \frac{16}{3}\right) \stackrel{?}{=} -5\left(\frac{16}{3}\right) \quad \text{Replace } t \text{ with } \frac{16}{3}.$$
$$8\left(\frac{6}{3} - \frac{16}{3}\right) \stackrel{?}{=} -\frac{80}{3} \quad \text{The LCD is 3.}$$
$$8\left(-\frac{10}{3}\right) \stackrel{?}{=} -\frac{80}{3} \quad \text{Subtract fractions.}$$
$$-\frac{80}{3} = -\frac{80}{3} \quad \text{True}$$

The solution is $\frac{16}{3}$.

Work Practice 2

Practice 2

Solve: $9(5 - x) = -3x$

Answer

2. $x = \frac{15}{2}$

Objective B Solving Equations Containing Fractions or Decimals

If an equation contains fractions, we can clear the equation of fractions by multiplying both sides by the LCD of all denominators. By doing this, we avoid working with time-consuming fractions.

Practice 3

Solve: $\frac{5}{2}x - 1 = \frac{3}{2}x - 4$

Example 3 Solve: $\frac{x}{2} - 1 = \frac{2}{3}x - 3$

Solution: We begin by clearing fractions. To do this, we multiply both sides of the equation by the LCD, which is 6.

$$\frac{x}{2} - 1 = \frac{2}{3}x - 3$$

Step 1: $6\left(\frac{x}{2} - 1\right) = 6\left(\frac{2}{3}x - 3\right)$ Multiply both sides by the LCD, 6.

Step 2: $6\left(\frac{x}{2}\right) - 6(1) = 6\left(\frac{2}{3}x\right) - 6(3)$ Use the distributive property.

$3x - 6 = 4x - 18$ Simplify.

Helpful Hint Don't forget to multiply *each* term by the LCD.

There are no longer grouping symbols and no like terms on either side of the equation, so we continue with Step 4.

$3x - 6 = 4x - 18$

Step 4: $3x - 6 - 3x = 4x - 18 - 3x$ Subtract $3x$ from both sides.

$-6 = x - 18$ Simplify.

$-6 + 18 = x - 18 + 18$ Add 18 to both sides.

$12 = x$ Simplify.

Step 5: The variable is now alone, so there is no need to apply the multiplication property of equality.

Step 6: Check.

$\frac{x}{2} - 1 = \frac{2}{3}x - 3$ Original equation

$\frac{12}{2} - 1 \stackrel{?}{=} \frac{2}{3} \cdot 12 - 3$ Replace x with 12.

$6 - 1 \stackrel{?}{=} 8 - 3$ Simplify.

$5 = 5$ True

The solution is 12.

Work Practice 3

Practice 4

Solve: $\frac{3(x - 2)}{5} = 3x + 6$

Example 4 Solve: $\frac{2(a + 3)}{3} = 6a + 2$

Solution: We clear the equation of fractions first.

$$\frac{2(a + 3)}{3} = 6a + 2$$

Step 1: $3 \cdot \frac{2(a + 3)}{3} = 3(6a + 2)$ Clear the fraction by multiplying both sides by the LCD, 3.

$2(a + 3) = 3(6a + 2)$ Simplify.

Answers
3. $x = -3$ 4. $x = -3$

Step 2: Next, we use the distributive property to remove parentheses.

$$2a + 6 = 18a + 6$$ Use the distributive property.

Step 4: $2a + 6 - 18a = 18a + 6 - 18a$ Subtract $18a$ from both sides.

$$-16a + 6 = 6$$ Simplify.

$$-16a + 6 - 6 = 6 - 6$$ Subtract 6 from both sides.

$$-16a = 0$$

Step 5: $\frac{-16a}{-16} = \frac{0}{-16}$ Divide both sides by -16.

$$a = 0$$ Simplify.

Step 6: To check, replace a with 0 in the original equation. The solution is 0.

Work Practice 4

Helpful Hint

Remember: When solving an equation, it makes no difference on which side of the equation variable terms lie. Just make sure that constant terms lie on the other side.

When solving a problem about money, you may need to solve an equation containing decimals. If you choose, you may multiply to clear the equation of decimals.

Example 5 Solve: $0.25x + 0.10(x - 3) = 1.1$

Solution: First we clear this equation of decimals by multiplying both sides of the equation by 100. Recall that multiplying a decimal number by 100 has the effect of moving the decimal point 2 places to the right.

$$0.25x + 0.10(x - 3) = 1.1$$

Step 1: $0.25x + 0.10(x - 3) = 1.10$ Multiply both sides by 100.

$$25x + 10(x - 3) = 110$$

Step 2: $25x + 10x - 30 = 110$ Apply the distributive property.

Step 3: $35x - 30 = 110$ Combine like terms.

Step 4: $35x - 30 + 30 = 110 + 30$ Add 30 to both sides.

$$35x = 140$$ Combine like terms.

Step 5: $\frac{35x}{35} = \frac{140}{35}$ Divide both sides by 35.

$$x = 4$$

Step 6: To check, replace x with 4 in the original equation. The solution is 4.

Work Practice 5

Practice 5

Solve:
$0.06x - 0.10(x - 2) = -0.16$

Helpful Hint If you have trouble with this step, try removing parentheses first.

$$0.25x + 0.10(x - 3) = 1.1$$
$$0.25x + 0.10x - 0.3 = 1.1$$
$$0.25x + 0.10x - 0.30 = 1.10$$
$$25x + 10x - 30 = 110$$

Then continue.

Objective C Recognizing Identities and Equations with No Solution

So far, each equation that we have solved has had a single solution. However, not every equation in one variable has a single solution. Some equations have no solution, while others have an infinite number of solutions. For example,

$$x + 5 = x + 7$$

Answer

5. $x = 9$

has **no solution** since no matter which real number we replace x with, the equation is false.

real number + 5 = same real number + 7 FALSE

On the other hand,

$$x + 6 = x + 6$$

has infinitely many solutions since x can be replaced by any real number and the equation will always be true.

real number + 6 = same real number + 6 TRUE

The equation $x + 6 = x + 6$ is called an **identity.** The next two examples illustrate special equations like these.

Practice 6

Solve:
$5(2 - x) + 8x = 3(x - 6)$

Example 6 Solve: $-2(x - 5) + 10 = -3(x + 2) + x$

Solution:

$$-2(x - 5) + 10 = -3(x + 2) + x$$
$$-2x + 10 + 10 = -3x - 6 + x \quad \text{Apply the distributive property on both sides.}$$
$$-2x + 20 = -2x - 6 \quad \text{Combine like terms.}$$
$$-2x + 20 + 2x = -2x - 6 + 2x \quad \text{Add } 2x \text{ to both sides.}$$
$$20 = -6 \quad \text{Combine like terms.}$$

The final equation contains no variable terms, and the result is the false statement $20 = -6$. There is no value for x that makes $20 = -6$ a true equation. Thus, we conclude that there is **no solution** to this equation.

Work Practice 6

Practice 7

Solve:
$-6(2x + 1) - 14 = -10(x + 2) - 2x$

Example 7 Solve: $3(x - 4) = 3x - 12$

Solution:
$$3(x - 4) = 3x - 12$$
$$3x - 12 = 3x - 12 \quad \text{Apply the distributive property.}$$

The left side of the equation is now identical to the right side. Every real number may be substituted for x and a true statement will result. We arrive at the same conclusion if we continue.

$$3x - 12 = 3x - 12$$
$$3x - 12 - 3x = 3x - 12 - 3x \quad \text{Subtract } 3x \text{ from both sides.}$$
$$-12 = -12 \quad \text{Combine like terms.}$$

Again, the final equation contains no variables, but this time the result is the true statement $-12 = -12$. This means that one side of the equation is identical to the other side. Thus, $3(x - 4) = 3x - 12$ is an **identity** and **all real numbers** are solutions.

Work Practice 7

Answers
6. no solution
7. All real numbers are solutions.

✓**Concept Check** Suppose you have simplified several equations and obtained the following results. What can you conclude about the solutions to the original equation?

a. $7 = 7$ **b.** $x = 0$ **c.** $7 = -4$

✓**Concept Check Answers**
a. All real numbers are solutions.
b. The solution is 0.
c. There is no solution.

Calculator Explorations Checking Equations

We can use a calculator to check possible solutions of equations. To do this, replace the variable by the possible solution and evaluate each side of the equation separately.

Equation: $3x - 4 = 2(x + 6)$ Solution: $x = 16$

$3x - 4 = 2(x + 6)$ Original equation

$3(16) - 4 \stackrel{?}{=} 2(16 + 6)$ Replace x with 16.

Now evaluate each side with your calculator.

Evaluate left side: [3] [×] [16] [−] [4] [=] or [ENTER]

Display: [44]

Evaluate right side: [2] [(] [16] [+] [6] [)] [=] or [ENTER]

Display: [44]

Since the left side equals the right side, the equation checks.

Use a calculator to check the possible solutions to each equation.

1. $2x = 48 + 6x;\quad x = -12$
2. $-3x - 7 = 3x - 1;\quad x = -1$
3. $5x - 2.6 = 2(x + 0.8);\quad x = 4.4$
4. $-1.6x - 3.9 = -6.9x - 25.6;\quad x = 5$
5. $\frac{564x}{4} = 200x - 11(649);\quad x = 121$
6. $20(x - 39) = 5x - 432;\quad x = 23.2$

Vocabulary, Readiness & Video Check

Throughout algebra, it is important to be able to distinguish between equations and expressions.

Remember,

- an equation contains an equal sign and
- an expression does not.

Among other things,

- we solve equations and
- we simplify or perform operations on expressions.

Identify each as an equation or an expression.

1. $x = -7$ ____________
2. $x - 7$ ____________
3. $4y - 6 + 9y + 1$ ____________
4. $4y - 6 = 9y + 1$ ____________
5. $\frac{1}{x} - \frac{x - 1}{8}$ ____________
6. $\frac{1}{x} - \frac{x - 1}{8} = 6$ ____________
7. $0.1x + 9 = 0.2x$ ____________
8. $0.1x^2 + 9y - 0.2x^2$ ____________

Martin-Gay Interactive Videos *Watch the section lecture video and answer the following questions.*

See Video 9.3

Objective A **9.** The general strategy for solving linear equations in one variable is discussed after Example 1. How many properties (not steps) are mentioned in this strategy and what are they?

Objective B **10.** In the first step for solving Example 2, both sides of the equation are multiplied by the LCD. Why is the distributive property mentioned?

11. In Example 3, why is the number of decimal places in each term of the equation important?

Objective C **12.** Complete each statement based on Examples 4 and 5. When solving an equation and all variable terms subtract out:

a. If we have a true statement, then the equation has ________ solution(s).

b. If we have a false statement, then the equation has ________ solution(s).

9.3 Exercise Set MyMathLab®

Objective A *Solve each equation. See Examples 1 and 2.*

1. $-4y + 10 = -2(3y + 1)$

2. $-3x + 1 = -2(4x + 2)$

3. $15x - 8 = 10 + 9x$

4. $15x - 5 = 7 + 12x$

5. $-2(3x - 4) = 2x$

6. $-(5x - 10) = 5x$

7. $5(2x - 1) - 2(3x) = 1$

8. $3(2 - 5x) + 4(6x) = 12$

9. $-6(x - 3) - 26 = -8$

10. $-4(n - 4) - 23 = -7$

11. $8 - 2(a + 1) = 9 + a$

12. $5 - 6(2 + b) = b - 14$

13. $4x + 3 = -3 + 2x + 14$

14. $6y - 8 = -6 + 3y + 13$

15. $-2y - 10 = 5y + 18$

16. $-7n + 5 = 8n - 10$

Objective B *Solve each equation. See Examples 3 through 5.*

17. $\frac{2}{3}x + \frac{4}{3} = -\frac{2}{3}$

18. $\frac{4}{5}x - \frac{8}{5} = -\frac{16}{5}$

19. $\frac{3}{4}x - \frac{1}{2} = 1$

20. $\frac{2}{9}x - \frac{1}{3} = 1$

21. $0.50x + 0.15(70) = 35.5$

22. $0.40x + 0.06(30) = 9.8$

23. $\frac{2(x + 1)}{4} = 3x - 2$

24. $\frac{3(y + 3)}{5} = 2y + 6$

25. $x + \frac{7}{6} = 2x - \frac{7}{6}$

26. $\frac{5}{2}x - 1 = x + \frac{1}{4}$

27. $0.12(y - 6) + 0.06y = 0.08y - 0.7$

28. $0.60(z - 300) + 0.05z = 0.70z - 205$

Objective C *Solve each equation. See Examples 6 and 7.*

29. $4(3x + 2) = 12x + 8$

30. $14x + 7 = 7(2x + 1)$

31. $\frac{x}{4} + 1 = \frac{x}{4}$

32. $\frac{x}{3} - 2 = \frac{x}{3}$

33. $3x - 7 = 3(x + 1)$

34. $2(x - 5) = 2x + 10$

35. $-2(6x - 5) + 4 = -12x + 14$

36. $-5(4y - 3) + 2 = -20y + 17$

Objectives A B C Mixed Practice *Solve. See Examples 1 through 7.*

37. $\frac{6(3 - z)}{5} = -z$

38. $\frac{4(5 - w)}{3} = -w$

39. $-3(2t - 5) + 2t = 5t - 4$

40. $-(4a - 7) - 5a = 10 + a$

41. $5y + 2(y - 6) = 4(y + 1) - 2$

42. $9x + 3(x - 4) = 10(x - 5) + 7$

43. $\frac{3(x - 5)}{2} = \frac{2(x + 5)}{3}$

44. $\frac{5(x - 1)}{4} = \frac{3(x + 1)}{2}$

45. $0.7x - 2.3 = 0.5$

46. $0.9x - 4.1 = 0.4$

47. $5x - 5 = 2(x + 1) + 3x - 7$

48. $3(2x - 1) + 5 = 6x + 2$

49. $4(2n + 1) = 3(6n + 3) + 1$

50. $4(4y + 2) = 2(1 + 6y) + 8$

51. $x + \frac{5}{4} = \frac{3}{4}x$

52. $\frac{7}{8}x + \frac{1}{4} = \frac{3}{4}x$

53. $\frac{x}{2} - 1 = \frac{x}{5} + 2$

54. $\frac{x}{5} - 7 = \frac{x}{3} - 5$

55. $2(x + 3) - 5 = 5x - 3(1 + x)$

56. $4(2 + x) + 1 = 7x - 3(x - 2)$

57. $0.06 - 0.01(x + 1) = -0.02(2 - x)$

58. $-0.01(5x + 4) = 0.04 - 0.01(x + 4)$

59. $\frac{9}{2} + \frac{5}{2}y = 2y - 4$

60. $3 - \frac{1}{2}x = 5x - 8$

Review

Translating *Write each algebraic expression described. See Section 3.1.*

△ **61.** A plot of land is in the shape of a triangle. If one side is x meters, a second side is $(2x - 3)$ meters, and a third side is $(3x - 5)$ meters, express the perimeter of the lot as a simplified expression in x.

(3x − 5) meters
(2x − 3) meters
x meters

62. A portion of a board has length x feet. The other part has length $(7x - 9)$ feet. Express the total length of the board as a simplified expression in x.

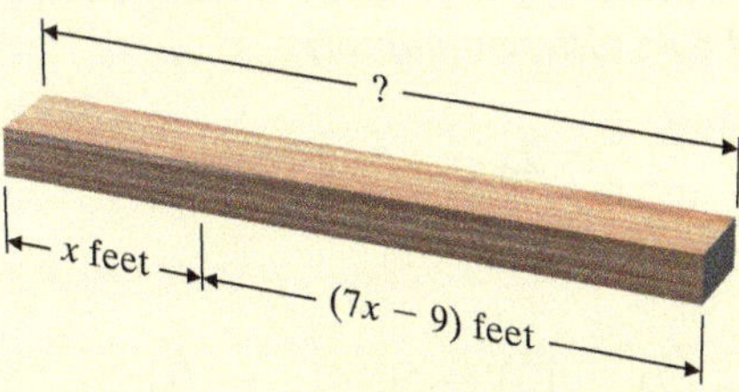

Translating *Write each phrase as an algebraic expression. Use x for the unknown number. See Section 3.2.*

63. A number subtracted from −8

64. Three times a number

65. The sum of −3 and twice a number

66. The difference of 8 and twice a number

67. The product of 9 and the sum of a number and 20

68. The quotient of −12 and the difference of a number and 3

Concept Extensions

See the Concept Check in this section.

69. a. Solve: $x + 3 = x + 3$

b. If you simplify an equation (such as the one in part **a**) and get a true statement such as $3 = 3$ or $0 = 0$, what can you conclude about the solution(s) of the original equation?

c. On your own, construct an equation for which every real number is a solution.

70. a. Solve: $x + 3 = x + 5$

b. If you simplify an equation (such as the one in part **a**) and get a false statement such as $3 = 5$ or $10 = 17$, what can you conclude about the solution(s) of the original equation?

c. On your own, construct an equation that has no solution.

Match each equation in the first column with its solution in the second column. Items in the second column may be used more than once.

71. $5x + 1 = 5x + 1$

72. $3x + 1 = 3x + 2$

73. $2x - 6x - 10 = -4x + 3 - 10$

74. $x - 11x - 3 = -10x - 1 - 2$

75. $9x - 20 = 8x - 20$

76. $-x + 15 = x + 15$

a. all real numbers
b. no solution
c. 0

77. Explain the difference between simplifying an expression and solving an equation.

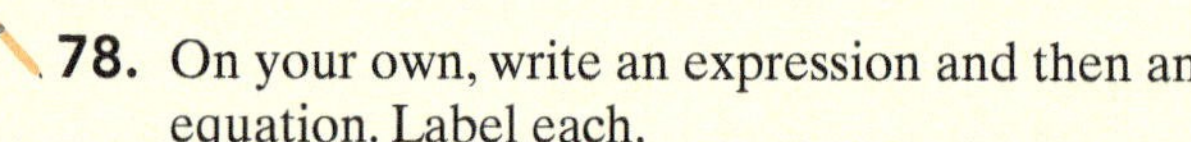

78. On your own, write an expression and then an equation. Label each.

For Exercises 79 and 80, **a.** *Write an equation for perimeter.* **b.** *Solve the equation in part (a).* **c.** *Find the length of each side.*

79. The perimeter of a geometric figure is the sum of the lengths of its sides. If the perimeter of the following pentagon (five-sided figure) is 28 centimeters, find the length of each side.

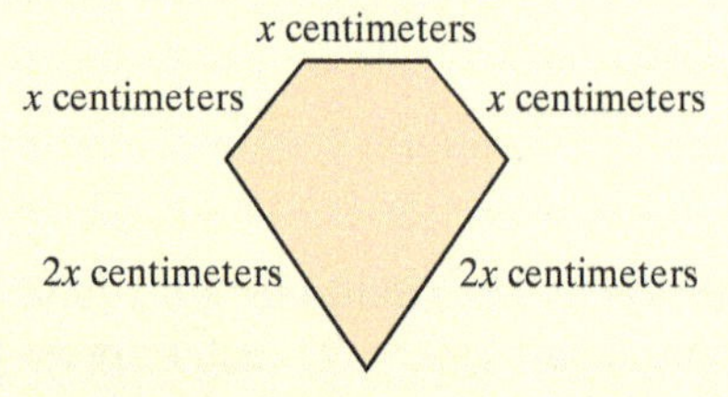

80. The perimeter of the following triangle is 35 meters. Find the length of each side.

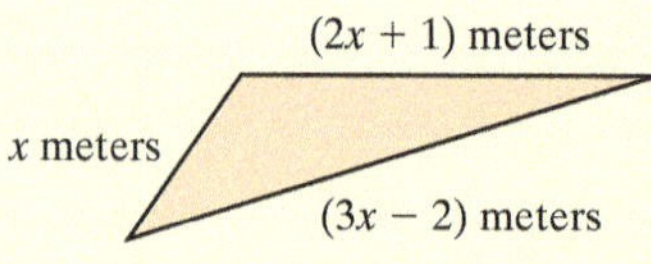

Fill in the blanks with numbers of your choice so that each equation has the given solution. Note: Each blank may be replaced by a different number.

81. $x +$ ____ $= 2x -$ ____; solution: 9

82. $-5x -$ ____ $=$ ____; solution: 2

Solve.

83. $1000(7x - 10) = 50(412 + 100x)$

84. $1000(x + 40) = 100(16 + 7x)$

85. $0.035x + 5.112 = 0.010x + 5.107$

86. $0.127x - 2.685 = 0.027x - 2.38$

Sections 9.1–9.3 **Integrated Review**

Real Numbers and Solving Linear Equations

Tell which set or sets each number belongs to: natural numbers, whole numbers, integers, rational numbers, irrational numbers, or real numbers.

1. 0

2. 143

3. $\frac{3}{8}$

4. 1

5. -13

6. $\frac{9}{10}$

7. $-\frac{1}{9}$

8. $\sqrt{5}$

Remove parentheses and simplify each expression.

9. $7(d - 3) + 10$

10. $9(z + 7) - 15$

11. $-4(3y - 4) + 12y$

12. $-3(2x + 5) - 6x$

Solve. Feel free to use the steps given in Section 9.3.

13. $2x - 7 = 6x - 27$

14. $3 + 8y = 3y - 2$

15. $-3a + 6 + 5a = 7a - 8a$

16. $4b - 8 - b = 10b - 3b$

17. $-\frac{2}{3}x = \frac{5}{9}$

18. $-\frac{3}{8}y = -\frac{1}{16}$

19. $10 = -6n + 16$

20. $-5 = -2m + 7$

Answers

1. ____________
2. ____________
3. ____________
4. ____________
5. ____________
6. ____________
7. ____________
8. ____________
9. ____________
10. ____________
11. ____________
12. ____________
13. ____________
14. ____________
15. ____________
16. ____________
17. ____________
18. ____________
19. ____________
20. ____________

21. ______

22. ______

23. ______

24. ______

25. ______

26. ______

27. ______

28. ______

29. ______

30. ______

31. ______

32. ______

21. $3(5c - 1) - 2 = 13c + 3$

22. $4(3t + 4) - 20 = 3 + 5t$

23. $\frac{2(z + 3)}{3} = 5 - z$

24. $\frac{3(w + 2)}{4} = 2w + 3$

25. $-2(2x - 5) = -3x + 7 - x + 3$

26. $-4(5x - 2) = -12x + 4 - 8x + 4$

27. $0.02(6t - 3) = 0.04(t - 2) + 0.02$

28. $0.03(m + 7) = 0.02(5 - m) + 0.03$

29. $-3y = \frac{4(y - 1)}{5}$

30. $-4x = \frac{5(1 - x)}{6}$

31. $\frac{5}{3}x - \frac{7}{3} = x$

32. $\frac{7}{5}n + \frac{3}{5} = -n$

9.4 Further Problem Solving 

In this section, we review our problem-solving steps first introduced in Section 3.4 and continue to solve problems that are modeled by linear equations.

Objectives

A Solve Problems Involving Direct Translations.

B Solve Problems Involving Relationships Among Unknown Quantities.

C Solve Problems Involving Consecutive Integers.

General Strategy for Problem Solving

1. UNDERSTAND the problem. During this step, become comfortable with the problem. Some ways of doing this are:
   - Read and reread the problem.
   - Choose a variable to represent the unknown.
   - Construct a drawing.
   - Propose a solution and check. Pay careful attention to how you check your proposed solution. This will help when writing an equation to model the problem.
2. TRANSLATE the problem into an equation.
3. SOLVE the equation.
4. INTERPRET the results: *Check* the proposed solution in the stated problem and *state* your conclusion.

Objective A Solving Direct Translation Problems 

Much of problem solving involves a direct translation from a sentence to an equation.

Example 1 Finding an Unknown Number

Twice a number, added to seven, is the same as three subtracted from the number. Find the number.

Solution: Translate the sentence into an equation and solve.

In words:	twice a number	added to	seven	is the same as	three subtracted from the number
	↓	↓	↓	↓	↓
Translate:	$2x$	$+$	7	$=$	$x - 3$

To solve, begin by subtracting x from both sides to isolate the variable term.

$$2x + 7 = x - 3$$

$$2x + 7 - x = x - 3 - x \quad \text{Subtract } x \text{ from both sides.}$$

$$x + 7 = -3 \quad \text{Combine like terms.}$$

$$x + 7 - 7 = -3 - 7 \quad \text{Subtract 7 from both sides.}$$

$$x = -10 \quad \text{Combine like terms.}$$

Check the solution in the problem as it was originally stated. To do so, replace "number" in the sentence with -10. Twice "-10" added to 7 is the same as 3 subtracted from "-10."

$$2(-10) + 7 \stackrel{?}{=} -10 - 3$$

$$-13 = -13$$

The unknown number is -10.

Work Practice 1

Practice 1

Three times a number, minus 6, is the same as two times the number, plus 3. Find the number.

Answer

1. 9

Helpful Hint

When checking solutions, go back to the original stated problem, rather than to your equation, in case errors have been made in translating to an equation.

Practice 2

Three times the difference of a number and 5 is the same as twice the number decreased by 3. Find the number.

Example 2 Finding an Unknown Number

Twice the sum of a number and 4 is the same as four times the number decreased by 12. Find the number.

Solution:

1. UNDERSTAND. Read and reread the problem. If we let x = the unknown number, then

   "the sum of a number and 4" translates to "$x + 4$" and

   "four times the number" translates to "$4x$"

2. TRANSLATE.

twice	sum of a number and 4	is the same as	four times the number	decreased by	12
↓	↓	↓	↓	↓	↓
2	$(x + 4)$	$=$	$4x$	$-$	12

3. SOLVE

$$2(x + 4) = 4x - 12$$
$$2x + 8 = 4x - 12 \quad \text{Apply the distributive property.}$$
$$2x + 8 - 4x = 4x - 12 - 4x \quad \text{Subtract } 4x \text{ from both sides.}$$
$$-2x + 8 = -12$$
$$-2x + 8 - 8 = -12 - 8 \quad \text{Subtract 8 from both sides.}$$
$$-2x = -20$$
$$\frac{-2x}{-2} = \frac{-20}{-2} \quad \text{Divide both sides by } -2.$$
$$x = 10$$

4. INTERPRET.

   Check: Check this solution in the problem as it was originally stated. To do so, replace "number" with 10. Twice the sum of "10" and 4 is 28, which is the same as 4 times "10" decreased by 12.

   State: The number is 10.

Work Practice 2

Objective B Solving Problems Involving Relationships Among Unknown Quantities

Practice 3

An 18-foot wire is to be cut so that the length of the longer piece is 5 times the length of the shorter piece. Find the length of each piece.

Example 3 Finding the Length of a Board

A 10-foot board is to be cut into two pieces so that the length of the longer piece is 4 times the length of the shorter. Find the length of each piece.

Solution:

1. UNDERSTAND the problem. To do so, read and reread the problem. You may also want to propose a solution. For example, if 3 feet represents the length of the shorter piece, then $4(3) = 12$ feet is the length of the longer piece, since it is 4 times the length of the shorter piece.

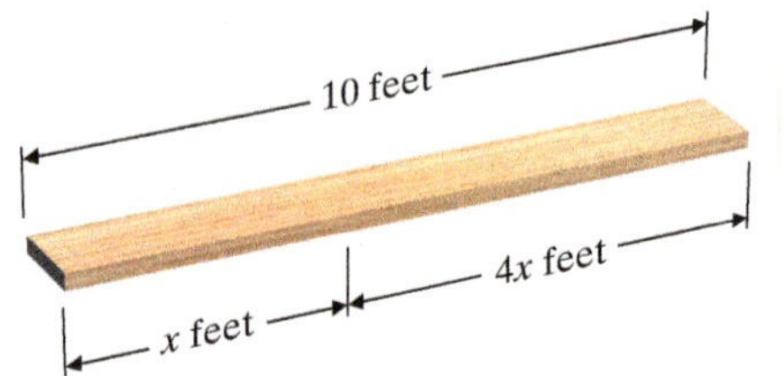

Answers

2. 12

3. shorter piece: 3 feet; longer piece: 15 feet

This guess gives a total board length of 3 feet + 12 feet = 15 feet, which is too long. However, the purpose of proposing a solution is not to guess correctly, but to help better understand the problem and how to model it.

In general, if we let

x = length of shorter piece, then

$4x$ = length of longer piece

2. TRANSLATE the problem. First, we write the equation in words.

length of shorter piece	added to	length of longer piece	equals	total length of board
↓	↓	↓	↓	↓
x	$+$	$4x$	$=$	10

3. SOLVE.

$$x + 4x = 10$$

$$5x = 10 \quad \text{Combine like terms.}$$

$$\frac{5x}{5} = \frac{10}{5} \quad \text{Divide both sides by 5.}$$

$$x = 2$$

4. INTERPRET.

Check: Check the solution in the stated problem. If the length of the shorter piece of board is 2 feet, the length of the longer piece is $4 \cdot (2 \text{ feet}) = 8$ feet and the sum of the lengths of the two pieces is 2 feet + 8 feet = 10 feet.

State: The shorter piece of board is 2 feet and the longer piece of board is 8 feet.

Work Practice 3

Helpful Hint

Make sure that units are included in your answer, if appropriate.

Example 4 Finding the Number of Republican Democratic Senators

As of July 2013, the total number of Democrats and Republicans in the U.S. House of Representatives was 440. There were 28 more Republican representatives than Democratic. Find the number of representatives from each party. (*Source:* Congressional Research Service)

Solution:

1. UNDERSTAND the problem. Read and reread the problem. Let's suppose that there are 200 Democratic representatives. Since there are 28 more Republicans than Democrats, there must be 200 + 28 = 228 Republicans. The total number of Democrats and Republicans is then 200 + 228 = 428. This is incorrect since the total should be 440, but we now have a better understanding of the problem.

In general, if we let

x = number of Democrats, then

$x + 28$ = number of Republicans

(*Continued on next page*)

Practice 4

Through the year 2020, the state of California will have 17 more electoral votes for president than the state of Texas. If the total electoral votes for these two states is 93, find the number of electoral votes for each state.

Answer

4. Texas: 38 electoral votes; California: 55 electoral votes

2. TRANSLATE the problem. First, we write the equation in words.

number of Democrats	added to	number of Republicans	equals	440
↓	↓	↓	↓	↓
x	$+$	$(x + 28)$	$=$	440

3. SOLVE.

$$x + (x + 28) = 440$$
$$2x + 28 = 440 \quad \text{Combine like terms.}$$
$$2x + 28 - 28 = 440 - 28 \quad \text{Subtract 28 from both sides.}$$
$$2x = 412$$
$$\frac{2x}{2} = \frac{412}{2} \quad \text{Divide both sides by 2.}$$
$$x = 206$$

4. INTERPRET.

Check: If there were 206 Democratic representatives, then there were $206 + 28 = 234$ Republican representatives. The total number of representatives is then $206 + 234 = 440$. The results check.

State: There were 206 Democratic and 234 Republican representatives in Congress in July 2013.

Work Practice 4

Practice 5

A car rental agency charges \$28 a day and \$0.15 a mile. If you rent a car for a day and your bill (before taxes) is \$52, how many miles did you drive?

Example 5 Calculating Hours on the Job

A computer science major at a local university has a part-time job working on computers for his clients. He charges \$20 to go to your home or office and then \$25 per hour. During one month he visited 10 homes or offices and his total income was \$575. How many hours did he spend working on computers?

Solution:

1. UNDERSTAND. Read and reread the problem. Let's propose that the student spent 20 hours working on computers. Pay careful attention as to how his income is calculated. For 20 hours and 10 visits, his income is 20(\$25) + 10(\$20) = \$700, which is more than \$575. We now have a better understanding of the problem and know that the time spent working on computers was less than 20 hours.

Let's let

x = hours working on computers. Then

$25x$ = amount of money made while working on computers

2. TRANSLATE.

money made while working on computers	plus	money made for visits	is equal to	575
↓	↓	↓	↓	↓
$25x$	$+$	$10(20)$	$=$	575

3. SOLVE.

$$25x + 200 = 575$$
$$25x + 200 - 200 = 575 - 200 \quad \text{Subtract 200 from both sides.}$$
$$25x = 375 \quad \text{Simplify.}$$
$$\frac{25x}{25} = \frac{375}{25} \quad \text{Divide both sides by 25.}$$
$$x = 15 \quad \text{Simplify.}$$

Answer

5. 160 miles

4. INTERPRET.

Check: If the student works 15 hours and makes 10 visits, his income is $15(\$25) + 10(\$20) = \$575$.

State: The student spent 15 hours working on computers.

Work Practice 5

Example 6 Finding Angle Measures

If the two walls of the Vietnam Veterans Memorial in Washington, D.C., were connected, an isosceles triangle would be formed. The measure of the third angle is 97.5° more than the measure of either of the two equal angles. Find the measure of the third angle. (*Source:* National Park Service)

Solution:

1. UNDERSTAND. Read and reread the problem. We then draw a diagram (recall that an isosceles triangle has two angles with the same measure) and let

$x =$ degree measure of one angle

$x =$ degree measure of the second, equal angle

$x + 97.5 =$ degree measure of the third angle

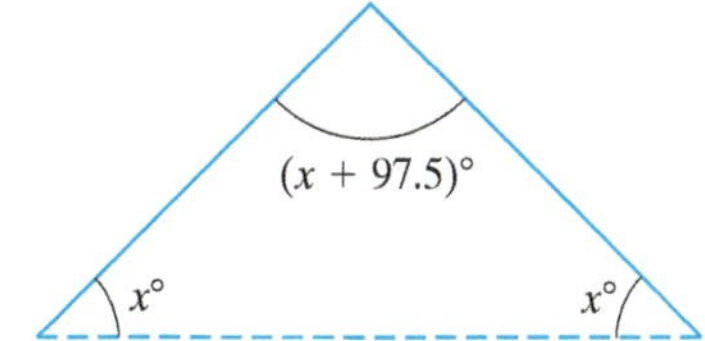

2. TRANSLATE. Recall that the sum of the measures of the angles of a triangle equals 180.

measure of first angle	+	measure of second angle	+	measure of third angle	equals	180
↓		↓		↓	↓	↓
x	+	x	+	$(x + 97.5)$	=	180

3. SOLVE.

$$x + x + (x + 97.5) = 180$$

$$3x + 97.5 = 180 \quad \text{Combine like terms.}$$

$$3x + 97.5 - 97.5 = 180 - 97.5 \quad \text{Subtract 97.5 from both sides.}$$

$$3x = 82.5$$

$$\frac{3x}{3} = \frac{82.5}{3} \quad \text{Divide both sides by 3.}$$

$$x = 27.5$$

4. INTERPRET.

Check: If $x = 27.5$, then the measure of the third angle is $x + 97.5 = 125$. The sum of the angles is then $27.5 + 27.5 + 125 = 180$, the correct sum.

State: The third angle measures 125°.*

Work Practice 6

*The two walls actually meet at an angle of 125 degrees 12 minutes. The measurement of 97.5° given in the problem is an approximation.

Practice 6

The measure of the second angle of a triangle is twice the measure of the smallest angle. The measure of the third angle of the triangle is three times the measure of the smallest angle. Find the measures of the angles.

Answer

6. smallest: 30°; second: 60°; third: 90°

Objective C Solving Consecutive Integer Problems

The next example has to do with consecutive integers. Recall what we have learned thus far about these integers.

	Example	*General Representation*	
Consecutive Integers	11, 12, 13 (+1, +1)	Let x be an integer.	$x,\ x+1,\ x+2$ (+1, +1)
Consecutive Even Integers	38, 40, 42 (+2, +2)	Let x be an even integer.	$x,\ x+2,\ x+4$ (+2, +2)
Consecutive Odd Integers	57, 59, 61 (+2, +2)	Let x be an odd integer.	$x,\ x+2\ \ x+4$ (+2, +2)

Practice 7

The sum of three consecutive even integers is 144. Find the integers.

Example 7

Some states have a single area code for the entire state. Two such states have area codes that are consecutive odd integers. If the sum of these integers is 1208, find the two area codes. (*Source: World Almanac*)

Solution:

1. UNDERSTAND. Read and reread the problem. If we let

$x =$ the first odd integer, then

$x + 2 =$ the next odd integer

Helpful Hint Remember, the 2 here means that odd integers are 2 units apart—for example, the odd integers 13 and $13 + 2 = 15$.

2. TRANSLATE.

first odd integer	the sum of	next odd integer	is	1208
↓	↓	↓		
x	$+$	$(x + 2)$	$=$	1208

3. SOLVE.

$$x + x + 2 = 1208$$
$$2x + 2 = 1208$$
$$2x + 2 - 2 = 1208 - 2$$
$$2x = 1206$$
$$\frac{2x}{2} = \frac{1206}{2}$$
$$x = 603$$

4. INTERPRET.

Check: If $x = 603$, then the next odd integer $x + 2 = 603 + 2 = 605$. Notice their sum, $603 + 605 = 1208$, as needed.

State: The area codes are 603 and 605.

Note: New Hampshire's area code is 603 and South Dakota's area code is 605.

Work Practice 7

Answer
7. 46, 48, 50

Vocabulary, Readiness & Video Check

Fill in the table.

1.	A number: x →	Double the number: →	Double the number, decreased by 31:
2.	A number: x →	Three times the number: →	Three times the number, increased by 17:
3.	A number: x →	The sum of the number and 5: →	Twice the sum of the number and 5:
4.	A number: x →	The difference of the number and 11: →	Seven times the difference of the number and 11:
5.	A number: y →	The difference of 20 and the number: →	The difference of 20 and the number, divided by 3:
6.	A number: y →	The sum of -10 and the number: →	The sum of -10 and the number, divided by 9:

Martin-Gay Interactive Videos

See Video 9.4

Watch the section lecture video and answer the following questions.

Objective A 7. At the end of Example 1, where are we told is the best place to check the solution of an application problem?

Objective B 8. The solution of the equation for Example 3 is $x = 43$. Why is this not the solution to the application?

Objective C 9. What are two things that should be checked to make sure the solution of Example 4 is correct?

9.4 Exercise Set MyMathLab®

Objective A *Solve. See Examples 1 and 2. For Exercises 1 through 4, write each of the following as equations. Then solve.*

1. The sum of twice a number and 7 is equal to the sum of the number and 6. Find the number.

2. The difference of three times a number and 1 is the same as twice the number. Find the number.

3. Three times a number, minus 6, is equal to two times the number, plus 8. Find the number.

4. The sum of 4 times a number and -2 is equal to the sum of 5 times the number and -2. Find the number.

5. Twice the difference of a number and 8 is equal to three times the sum of the number and 3. Find the number.

6. Five times the sum of a number and -1 is the same as 6 times the number. Find the number.

7. The product of twice a number and 3 is the same as the difference of five times the number and $\frac{3}{4}$. Find the number.

8. If the difference of a number and 4 is doubled, the result is $\frac{1}{4}$ less than the number. Find the number.

Objective B *Solve. See Examples 3 and 4.*

9. A 25-inch piece of steel is cut into three pieces so that the second piece is twice as long as the first piece, and the third piece is one inch more than five times the length of the first piece. Find the lengths of the pieces.

10. A 46-foot piece of rope is cut into three pieces so that the second piece is three times as long as the first piece, and the third piece is two feet more than seven times the length of the first piece. Find the lengths of the pieces.

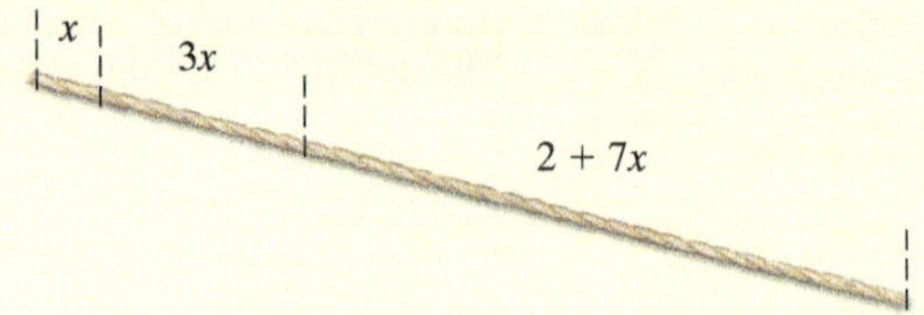

11. A 40-inch board is to be cut into three pieces so that the second piece is twice as long as the first piece and the third piece is 5 times as long as the first piece. If x represents the length of the first piece, find the lengths of all three pieces.

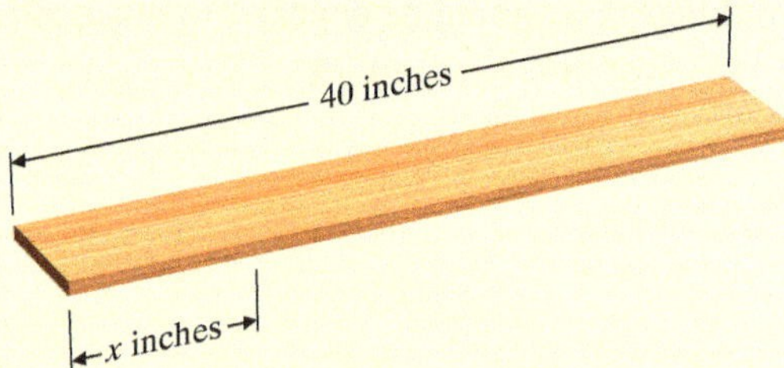

12. A 21-foot beam is to be divided so that the longer piece is 1 foot more than 3 times the length of the shorter piece. If x represents the length of the shorter piece, find the lengths of both pieces.

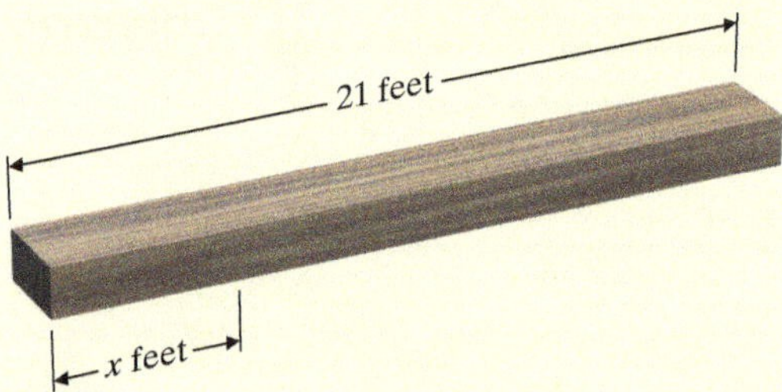

13. In 2012, New York produced 226 million pounds more apples than Pennsylvania. Together, the two states produced 1214 million pounds of apples. Find the amount of apples grown in New York and Pennsylvania in 2012. (*Source:* National Agriculture Statistics Service)

14. In the 2012 Summer Olympics, the U.S. team won 8 more gold medals than the Chinese team. If the total number of gold medals won by both teams was 84, find the number of gold medals won by each team. (*Source:* International Olympic Committee)

Solve. See Example 5.

15. A car rental agency advertised renting a Buick Century for \$24.95 per day and \$0.29 per mile. If you rent this car for 2 days, how many whole miles can you drive on a \$100 budget?

16. A plumber gave an estimate for the renovation of a kitchen. Her hourly pay is \$27 per hour and the plumbing parts will cost \$80. If her total estimate is \$404, how many hours does she expect this job to take?

17. In one U.S. city, the taxi cost is \$3 plus \$0.80 per mile. If you are traveling from the airport, there is an additional charge of \$4.50 for tolls. How far can you travel from the airport by taxi for \$27.50?

18. A professional carpet cleaning service charges \$30 plus \$25.50 per hour to come to your home. If your total bill from this company is \$119.25 before taxes, for how many hours were you charged?

Solve. See Example 6.

△ **19.** The flag of Equatorial Guinea contains an isosceles triangle. (Recall that an isosceles triangle contains two angles with the same measure.) If the measure of the third angle of the triangle is 30° more than twice the measure of either of the other two angles, find the measure of each angle of the triangle. (*Hint:* Recall that the sum of the measures of the angles of a triangle is 180°.)

△ **20.** The flag of Brazil contains a parallelogram. One angle of the parallelogram is 15° less than twice the measure of the angle next to it. Find the measure of each angle of the parallelogram. (*Hint:* Recall that opposite angles of a parallelogram have the same measure and that the sum of the measures of the angles is 360°.)

21. The sum of the measures of the angles of a parallelogram is 360°. In the parallelogram below, angles A and D have the same measure as angles C and B. If the measure of angle C is twice the measure of angle A, find the measure of each angle.

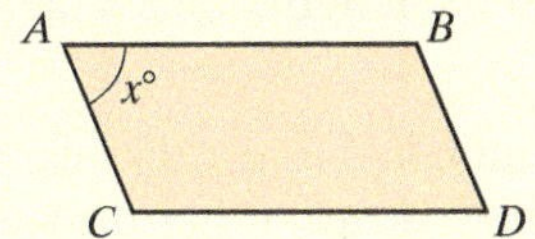

22. Recall that the sum of the measures of the angles of a triangle is 180°. In the triangle below, angle C has the same measure as angle B, and angle A measures 42° less than angle B. Find the measure of each angle.

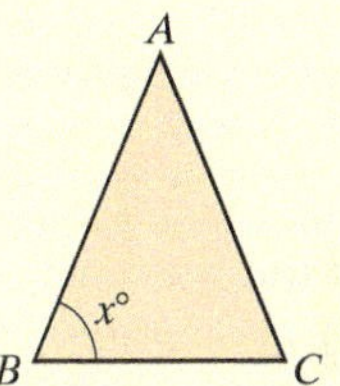

Objective C *Solve. See Example 7. Fill in the table. Most of the first row has been completed for you.*

	First Integer →	*Next Integers*		→	*Indicated Sum*
23. Three consecutive integers:	Integer: x	$x+1$	$x+2$		Sum of the three consecutive integers, simplified:
24. Three consecutive integers:	Integer: x				Sum of the second and third consecutive integers, simplified:
25. Three consecutive even integers:	Even integer: x				Sum of the first and third even consecutive integers, simplified:
26. Three consecutive odd integers:	Odd integer: x				Sum of the three consecutive odd integers, simplified:
27. Four consecutive integers:	Integer: x				Sum of the four consecutive integers, simplified:
28. Four consecutive integers:	Integer: x				Sum of the first and fourth consecutive integers, simplified:
29. Three consecutive odd integers:	Odd integer: x				Sum of the second and third consecutive odd integers, simplified:
30. Three consecutive even integers:	Even integer: x				Sum of the three consecutive even integers, simplified:

Solve. See Example 7.

31. The left and right page numbers of an open book are two consecutive integers whose sum is 469. Find these page numbers.

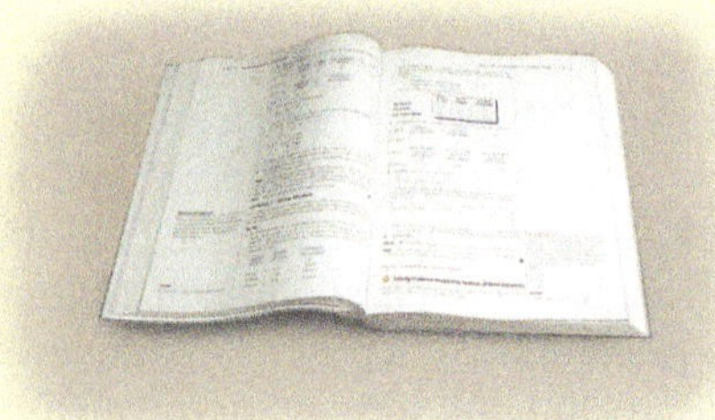

32. The room numbers of two adjacent classrooms are two consecutive even numbers. If their sum is 654, find the classroom numbers.

33. To make an international telephone call, you need the code for the country you are calling. The codes for Belgium, France, and Spain are three consecutive integers whose sum is 99. Find the code for each country. (Source: *The World Almanac and Book of Facts*)

34. The code to unlock a student's combination lock happens to be three consecutive odd integers whose sum is 51. Find the integers.

Objectives A B C Mixed Practice *Solve. See Examples 1 through 7.*

35. A 17-foot piece of string is cut into two pieces so that the longer piece is 2 feet longer than twice the length of the shorter piece. Find the lengths of both pieces.

36. A 25-foot wire is to be cut so that the longer piece is one foot longer than 5 times the length of the shorter piece. Find the length of each piece.

37. Currently, the two fastest trains are the Japanese Maglev and the French TGV. The sum of their fastest speeds is 718.2 miles per hour. If the speed of the Maglev is 3.8 mph faster than the speed of the TGV, find the speeds of each.

38. The Pentagon is the world's largest office building in terms of floor space. It has three times the amount of floor space as the Empire State Building. If the total floor space for these two buildings is approximately 8700 thousand square feet, find the floor space of each building.

39. Two angles are supplementary if their sum is 180°. The larger angle below measures eight degrees more than three times the measure of the smaller angle. If x represents the measure of the smaller angle and these two angles are supplementary, find the measure of each angle.

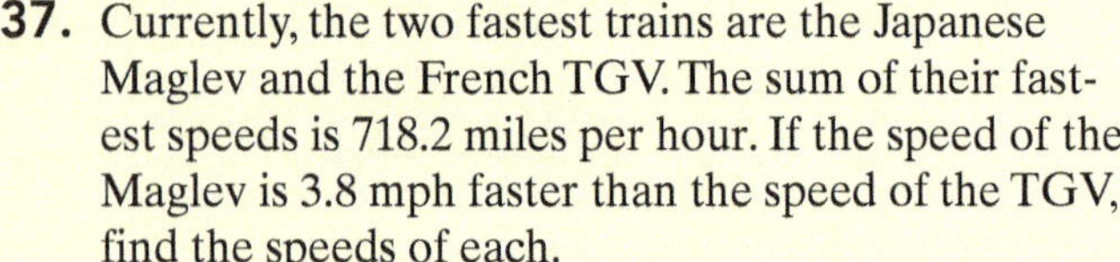

40. Two angles are complementary if their sum is 90°. Given the measures of the complementary angles shown, find the measure of each angle.

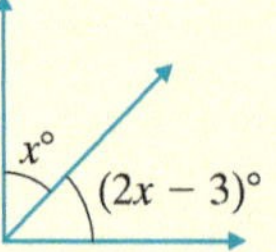

41. The measures of the angles of a triangle are 3 consecutive even integers. Find the measure of each angle.

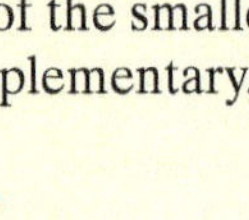

42. A quadrilateral is a polygon with 4 sides. The sum of the measures of the 4 angles in a quadrilateral is 360°. If the measures of the angles of a quadrilateral are consecutive odd integers, find the measures.

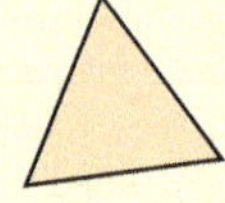

43. The sum of $\frac{1}{5}$ and twice a number is equal to $\frac{4}{5}$ subtracted from three times the number. Find the number.

44. The sum of $\frac{2}{3}$ and four times a number is equal to $\frac{5}{6}$ subtracted from five times the number. Find the number.

45. Hertz Car Rental charges a daily rate of \$39 plus \$0.20 per mile for a certain car. Suppose that you rent that car for a day and your bill (before taxes) is \$95. How many miles did you drive?

46. A woman's \$15,000 estate is to be divided so that her husband receives twice as much as her son. Find the amount of money that her husband receives and the amount of money that her son receives.

47. During the 2013 Rose Bowl, Stanford University beat University of Wisconsin by 6 points. If their combined scores totaled 34, find the individual team scores.

48. In June 2013, there were 10 more Republican governors than Democratic governors in the United States. How many Democrats and how many Republicans held governors' offices after this election? (*Source:* National Governors Association)

49. The number of counties in California and the number of counties in Montana are consecutive even integers whose sum is 114. If California has more counties than Montana, how many counties does each state have? (*Source: The World Almanac and Book of Facts*)

50. A student is building a bookcase with stepped shelves for her dorm room. She buys a 48-inch board and wants to cut the board into three pieces with lengths equal to three consecutive even integers. Find the three board lengths.

51. Scientists are continually updating information about the planets in our solar system, including the number of satellites orbiting each. Uranus is now believed to have 13 more satellites than Neptune. Also, Saturn is now believed to have 8 more than twice the number of satellites of Uranus. If the total number of satellites for these planets is 103, find the number of satellites for each planet. (*Source:* National Space Science Data Center)

52. Apple's iPad Mini tablet computer was launched in 2012. The height of each iPad Mini is 70 millimeters less than twice the width. If the sum of the height and width of an iPad Mini is 335 millimeters, find each dimension. (*Source:* Apple Inc.)

53. If the sum of a number and five is tripled, the result is one less than twice the number. Find the number.

54. Twice the sum of a number and six equals three times the sum of the number and four. Find the number.

55. The area of the Sahara Desert is 7 times the area of the Gobi Desert. If the sum of their areas is 4,000,000 square miles, find the area of each desert.

56. The largest meteorite in the world is the Hoba West, located in Namibia. Its weight is 3 times the weight of the Armanty meteorite, located in Outer Mongolia. If the sum of their weights is 88 tons, find the weight of each.

57. In the 2012 Summer Olympics, New Zealand won more gold medals than Cuba, which won more gold medals than Jamaica. If the number of gold medals won by these three countries is three consecutive integers whose sum is 15, find the number of gold medals won by each. (*Source:* International Olympic Committee)

58. To make an international telephone call, you need the code for the country you are calling. The codes for Mali Republic, Côte d'Ivoire, and Niger are three consecutive odd integers whose sum is 675. Find the code for each country.

59. In the fall of 2012, there were 1580 more female students enrolled at Rutgers University than male students. If the total student enrollment at Rutgers was 58,788 that fall, find the numbers of female students and male students who were enrolled. (*Source:* Rutgers, The State University of New Jersey)

60. In 2012, approximately 1.1 million fewer pickups were sold in the United States than middle-sized cars. If the total number of pickups and middle-sized cars sold was 4.9 million, find the number of vehicles sold in each category. (*Source:* Alliance of Automobile Manufacturers)

61. A geodesic dome, based on the design by Buckminster Fuller, is composed of two different types of triangular panels. One of these is an isosceles triangle. In one geodesic dome, the measure of the third angle is 76.5° more than the measure of either of the two equal angles. Find the measure of the three angles. (*Source:* Buckminster Fuller Institute)

62. The measures of the angles of a particular triangle are such that the second and third angles are each four times the measure of the smallest angle. Find the measures of the angles of this triangle.

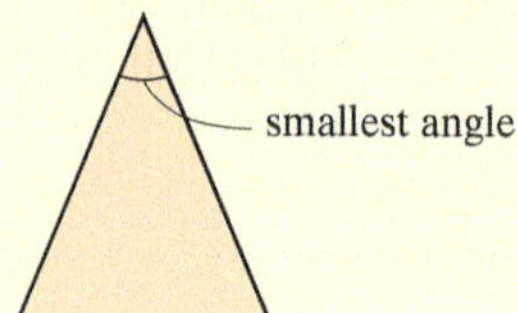

The graph below shows the states with the highest provisional tourism budgets in 2012/2013. Use the graph for Exercises 63 through 66.

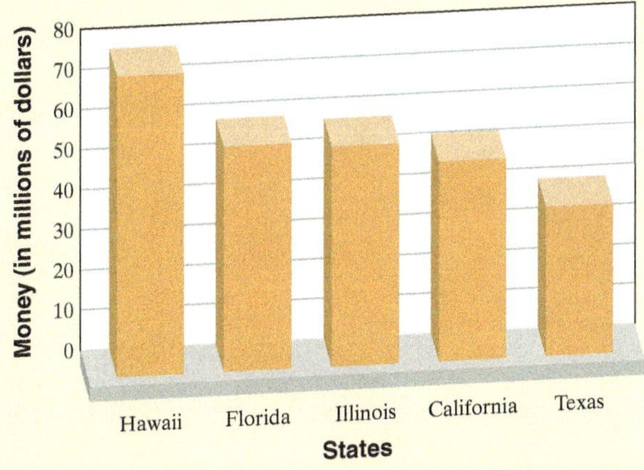

Source: U.S. Travel Association

63. Which state spent the most money on tourism?

64. Which state(s) spent between \$30 and \$40 million on tourism?

65. The states of Florida and California spent a total of \$106.2 million on tourism. The state of Florida spent \$6.2 million more than the state of California. Find the amount that each state spent on tourism.

66. The states of Illinois and Texas spent a total of \$92.6 million on tourism. The state of Illinois spent \$19.0 million less than twice the amount of money that the state of Texas. Find the amount that each state spent on tourism.

Compare the heights of the bars in the graph with your results of the exercises below. Are your answers reasonable?

67. Exercise **65**

68. Exercise **66**

Review

Evaluate each expression for the given values. See Section 1.8.

69. $2W + 2L$; $W = 7$ and $L = 10$

70. $\frac{1}{2}Bh$; $B = 14$ and $h = 22$

71. πr^2; $r = 15$

72. $r \cdot t$; $r = 15$ and $t = 2$

Concept Extensions

△ **73.** A golden rectangle is a rectangle whose length is approximately 1.6 times its width. The early Greeks thought that a rectangle with these dimensions was the most pleasing to the eye, and examples of the golden rectangle are found in many early works of art. For example, the Parthenon in Athens contains many examples of golden rectangles.

Mike Hallahan would like to plant a rectangular garden in the shape of a golden rectangle. If he has 78 feet of fencing available, find the dimensions of the garden.

74. Dr. Dorothy Smith gave the students in her geometry class at the University of New Orleans the following question: Is it possible to construct a triangle such that the second angle of the triangle has a measure that is twice the measure of the first angle and the measure of the third angle is 5 times the measure of the first? If so, find the measure of each angle. (*Hint:* Recall that the sum of the measures of the angles of a triangle is 180°.)

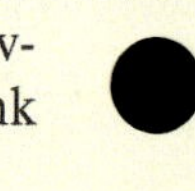

75. Only male crickets chirp. They chirp at different rates depending on their species and the temperature of their environment. Suppose a certain species is currently chirping at a rate of 90 chirps per minute. At this rate, how many chirps occur in one hour? In one 24-hour day? In one year?

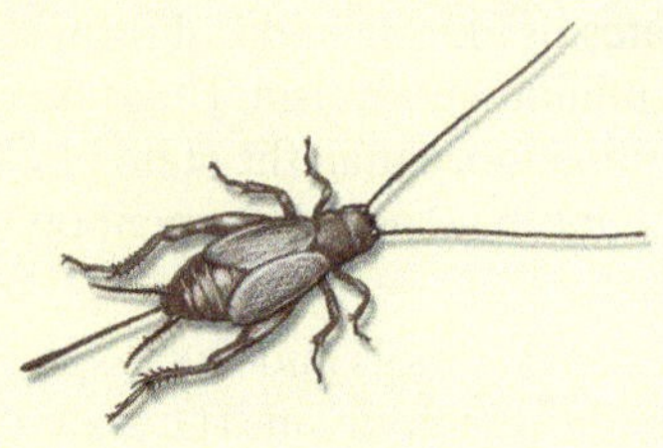

76. The human eye blinks once every 5 seconds on average. How many times does the average eye blink in one hour? In one 16-hour day while awake? In one year while awake?

77. In your own words, explain why a solution of a word problem should be checked using the original wording of the problem and not the equation written from the wording.

78. Give an example of how you recently solved a problem using mathematics.

Recall from Exercise **73** *that a golden rectangle is a rectangle whose length is approximately 1.6 times its width.*

△ **79.** It is thought that for about 75% of adults, a rectangle in the shape of the golden rectangle is the most pleasing to the eye. Draw three rectangles, one in the shape of the golden rectangle, and poll your class. Do the results agree with the percentage given above?

△ **80.** Examples of golden rectangles can be found today in architecture and manufacturing packaging. Find an example of a golden rectangle in your home. A few suggestions: the front face of a book, the floor of a room, the front of a box of food.

For Exercises 81 and 82, measure the dimensions of each rectangle and decide which one best approximates the shape of a golden rectangle.

△ **81.**

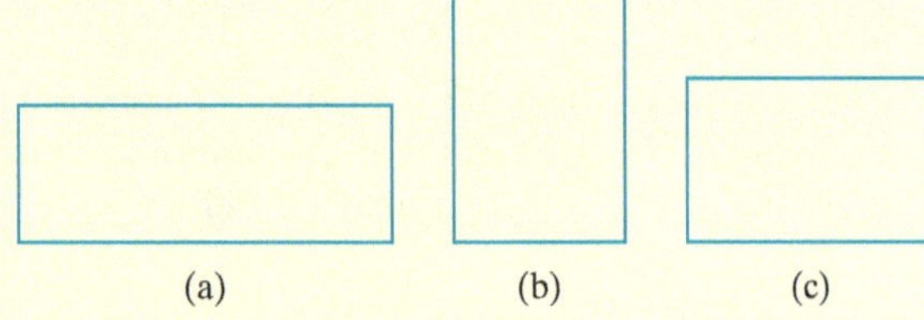

82.

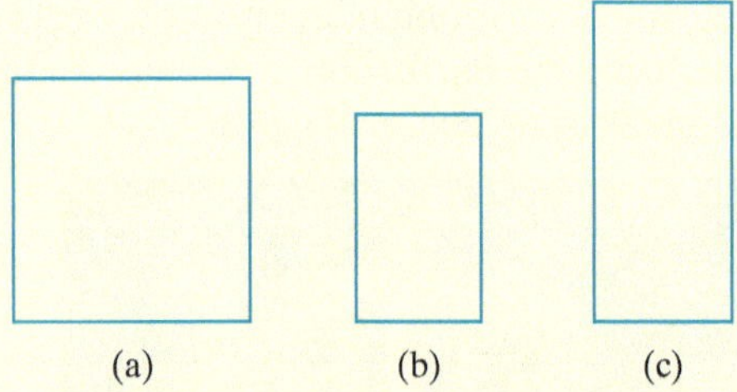

9.5 Formulas and Problem Solving 

Objective A Using Formulas to Solve Problems 

Objectives

A Use Formulas to Solve Problems.

B Solve a Formula or Equation for One of Its Variables.

A **formula** describes a known relationship among quantities. Many formulas are given as equations. For example, the formula

$$d = r \cdot t$$

stands for the relationship

$$\text{distance} = \text{rate} \cdot \text{time}$$

Let's look at one way that we can use this formula.

If we know we traveled a distance of 100 miles at a rate of 40 miles per hour, we can replace the variables d and r in the formula $d = rt$ and find our travel time, t.

$d = rt$ Formula

$100 = 40t$ Replace d with 100 and r with 40.

To solve for t, we divide both sides of the equation by 40.

$\frac{100}{40} = \frac{40t}{40}$ Divide both sides by 40.

$\frac{5}{2} = t$ Simplify.

The travel time was $\frac{5}{2}$ hours, or $2\frac{1}{2}$ hours, or 2.5 hours.

In this section, we solve problems that can be modeled by known formulas. We use the same problem-solving strategy that was used in the previous section.

Example 1 Finding Time Given Rate and Distance

A glacier is a giant mass of rocks and ice that flows downhill like a river. Portage Glacier in Alaska is about 6 miles, or 31,680 *feet,* long and moves 400 *feet* per year. Icebergs are created when the front end of the glacier flows into Portage Lake. How long does it take for ice at the head (beginning) of the glacier to reach the lake?

Solution:

1. UNDERSTAND. Read and reread the problem. The appropriate formula needed to solve this problem is the distance formula, $d = rt$. To become familiar with this formula, let's find the distance that ice traveling at a rate of 400 feet per year travels in 100 years. To do so, we let time t be 100 years and rate r be the given 400 feet per year, and substitute these values into the formula $d = rt$. We then have that distance $d = 400(100) = 40{,}000$ feet. Since we are interested in finding how long it takes ice to travel 31,680 feet, we now know that it is less than 100 years.

(*Continued on next page*)

Practice 1

A family is planning their vacation to visit relatives. They will drive from Cincinnati, Ohio, to Rapid City, South Dakota, a distance of 1180 miles. They plan to average a rate of 50 miles per hour. How much time will they spend driving?

Answer

1. 23.6 hours

Since we are using the formula $d = rt$, we let

t = the time in years for ice to reach the lake
r = rate or speed of ice
d = distance from beginning of glacier to lake

2. TRANSLATE. To translate to an equation, we use the formula $d = rt$ and let distance d = 31,680 feet and rate r = 400 feet per year.

$$d = r \cdot t$$
$$31{,}680 = 400 \cdot t \quad \text{Let } d = 31{,}680 \text{ and } r = 400.$$

3. SOLVE. Solve the equation for t. To solve for t, we divide both sides by 400.

$$\frac{31{,}680}{400} = \frac{400 \cdot t}{400} \quad \text{Divide both sides by 400.}$$
$$79.2 = t \quad \text{Simplify.}$$

4. INTERPRET.

Check: To check, substitute 79.2 for t and 400 for r in the distance formula and check to see that the distance is 31,680 feet.

State: It takes 79.2 years for the ice at the head of Portage Glacier to reach the lake.

Don't forget to include units, if appropriate.

Work Practice 1

Practice 2

A wood deck is being built behind a house. The width of the deck must be 18 feet because of the shape of the house. If there is 450 square feet of decking material, find the length of the deck.

Example 2 Calculating the Length of a Garden

Charles Pecot can afford enough fencing to enclose a rectangular garden with a perimeter of 140 feet. If the width of his garden is to be 30 feet, find the length.

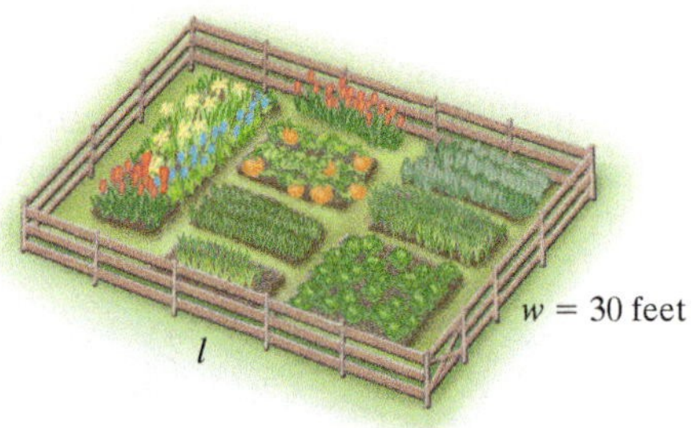

Solution:

1. UNDERSTAND. Read and reread the problem. The formula needed to solve this problem is the formula for the perimeter of a rectangle, $P = 2l + 2w$. Before continuing, let's become familar with this formula.

l = the length of the rectangular garden
w = the width of the rectangular garden
P = perimeter of the garden

2. TRANSLATE. To translate to an equation, we use the formula $P = 2l + 2w$ and let perimeter P = 140 feet and width w = 30 feet.

$$P = 2l + 2w \quad \text{Let } P = 140 \text{ and } w = 30.$$
$$140 = 2l + 2(30)$$

Answer
2. 25 feet

3. SOLVE.

$$140 = 2l + 2(30)$$
$$140 = 2l + 60 \quad \text{Multiply } 2(30).$$
$$140 - 60 = 2l + 60 - 60 \quad \text{Subtract 60 from both sides.}$$
$$80 = 2l \quad \text{Combine like terms.}$$
$$40 = l \quad \text{Divide both sides by 2.}$$

4. INTERPRET.

Check: Substitute 40 for l and 30 for w in the perimeter formula and check to see that the perimeter is 140 feet.

State: The length of the rectangular garden is 40 feet.

Work Practice 2

Example 3 Finding an Equivalent Temperature

The average maximum temperature for January in Algiers, Algeria, is 59° Fahrenheit. Find the equivalent temperature in degrees Celsius.

Solution:

1. UNDERSTAND. Read and reread the problem. A formula that can be used to solve this problem is the formula for converting degrees Celsius to degrees Fahrenheit, $F = \frac{9}{5}C + 32$. Before continuing, become familiar with this formula. Using this formula, we let

C = temperature in degrees Celsius, and

F = temperature in degrees Fahrenheit.

2. TRANSLATE. To translate to an equation, we use the formula $F = \frac{9}{5}C + 32$ and let degrees Fahrenheit $F = 59$.

Formula: $F = \frac{9}{5}C + 32$

Substitute: $59 = \frac{9}{5}C + 32$ Let F = 59.

3. SOLVE.

$$59 = \frac{9}{5}C + 32$$
$$59 - 32 = \frac{9}{5}C + 32 - 32 \quad \text{Subtract 32 from both sides.}$$
$$27 = \frac{9}{5}C \quad \text{Combine like terms.}$$
$$\frac{5}{9} \cdot 27 = \frac{5}{9} \cdot \frac{9}{5}C \quad \text{Multiply both sides by } \frac{5}{9}.$$
$$15 = C \quad \text{Simplify.}$$

4. INTERPRET.

Check: To check, replace C with 15 and F with 59 in the formula and see that a true statement results.

State: Thus, 59° Fahrenheit is equivalent to 15° Celsius.

Work Practice 3

Practice 3

Convert the temperature 5°C to Fahrenheit.

Answer

3. 41°F

In the next example, we use the formula for perimeter of a rectangle as in Example 2. In Example 2, we knew the width of the rectangle. In this example, both the length and width are unknown.

Practice 4

The length of a rectangle is 1 meter more than 4 times its width. Find the dimensions if the perimeter is 52 meters.

Example 4 Finding Road Sign Dimensions

The length of a rectangular road sign is 2 feet less than three times its width. Find the dimensions if the perimeter is 28 feet.

Solution:

1. UNDERSTAND. Read and reread the problem. Recall that the formula for the perimeter of a rectangle is $P = 2l + 2w$. Draw a rectangle and guess the solution. If the width of the rectangular sign is 5 feet, its length is 2 feet less than 3 times the width, or $3(5 \text{ feet}) - 2 \text{ feet} = 13 \text{ feet}$. The perimeter P of the rectangle is then $2(13 \text{ feet}) + 2(5 \text{ feet}) = 36 \text{ feet}$, too large. We now know that the width is less than 5 feet.

   Proposed rectangle:

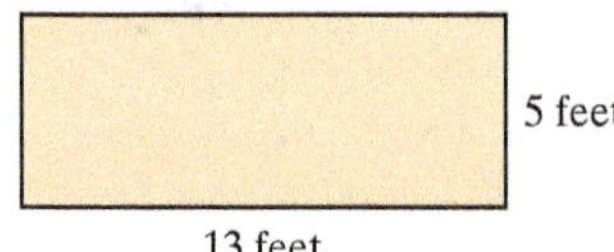

   Let

   $w =$ the width of the rectangular sign; then
   $3w - 2 =$ the length of the sign.

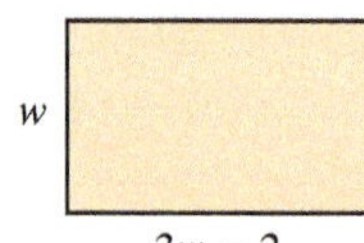

   Draw a rectangle and label it with the assigned variables.

2. TRANSLATE.

   Formula: $P = 2l + 2w$

   Substitute: $28 = 2(3w - 2) + 2w$

3. SOLVE.

$$28 = 2(3w - 2) + 2w$$

$$28 = 6w - 4 + 2w \quad \text{Apply the distributive property.}$$

$$28 = 8w - 4$$

$$28 + 4 = 8w - 4 + 4 \quad \text{Add 4 to both sides.}$$

$$32 = 8w$$

$$\frac{32}{8} = \frac{8w}{8} \quad \text{Divide both sides by 8.}$$

$$4 = w$$

4. INTERPRET.

   Check: If the width of the sign is 4 feet, the length of the sign is $3(4 \text{ feet}) - 2 \text{ feet} = 10 \text{ feet}$. This gives the rectangular sign a perimeter of $P = 2(4 \text{ feet}) + 2(10 \text{ feet}) = 28 \text{ feet}$, the correct perimeter.

   State: The width of the sign is 4 feet and the length of the sign is 10 feet.

Work Practice 4

Answer

4. length: 21 m; width: 5 m

Objective B Solving a Formula for a Variable 

We say that the formula

$$d = rt$$

is solved for d because d is alone on one side of the equation and the other side contains no d's. Suppose that we have a large number of problems to solve where we are given distance d and rate r and asked to find time t. In this case, it may be easier to first solve the formula $d = rt$ for t. To solve for t, we divide both sides of the equation by r.

$$d = rt$$

$$\frac{d}{r} = \frac{rt}{r} \quad \text{Divide both sides by } r.$$

$$\frac{d}{r} = t \quad \text{Simplify.}$$

To solve a formula or an equation for a specified variable, we use the same steps as for solving a linear equation except that we treat the specified variable as the only variable in the equation. These steps are listed next.

Solving Equations for a Specified Variable

Step 1: Multiply on both sides to clear the equation of fractions if they appear.

Step 2: Use the distributive property to remove parentheses if they appear.

Step 3: Simplify each side of the equation by combining like terms.

Step 4: Get all terms containing the specified variable on one side and all other terms on the other side by using the addition property of equality.

Step 5: Get the specified variable alone by using the multiplication property of equality.

Example 5 Solve $V = lwh$ for l.

Solution: This formula is used to find the volume of a box. To solve for l, we divide both sides by wh.

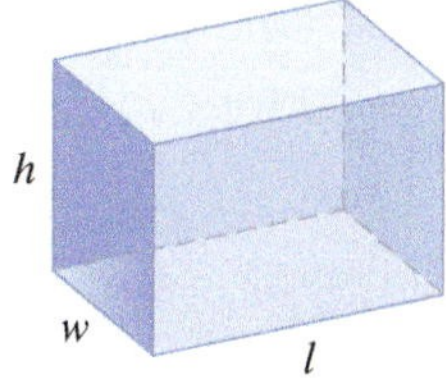

$$V = lwh$$

$$\frac{V}{wh} = \frac{lwh}{wh} \quad \text{Divide both sides by } wh.$$

$$\frac{V}{wh} = l \quad \text{Simplify.}$$

Since we have l alone on one side of the equation, we have solved for l in terms of V, w, and h. Remember that it does not matter on which side of the equation we get the variable alone.

Work Practice 5

Practice 5

Solve $C = 2\pi r$ for r. (This formula is used to find the circumference, C, of a circle given its radius, r.)

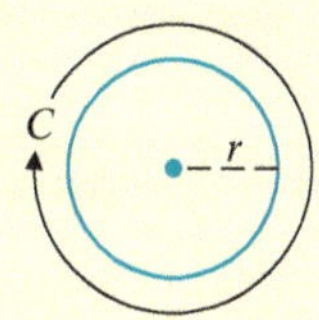

Answer

5. $r = \frac{C}{2\pi}$

Practice 6

Solve $P = 2l + 2w$ for l.

Example 6

Solve $y = mx + b$ for x.

Solution: First we get mx alone by subtracting b from both sides.

$$y = mx + b$$

$$y - b = mx + b - b \quad \text{Subtract } b \text{ from both sides.}$$

$$y - b = mx \quad \text{Combine like terms.}$$

Next we solve for x by dividing both sides by m.

$$\frac{y - b}{m} = \frac{mx}{m}$$

$$\frac{y - b}{m} = x \quad \text{Simplify.}$$

Work Practice 6

✓Concept Check Solve:

a. ● = ■ − ■ for ■

b. ● = ■ · ▲ − ■ for ■

Practice 7

Solve $P = 2a + b - c$ for a.

Example 7

Solve $P = 2l + 2w$ for w.

Solution: This formula relates the perimeter of a rectangle to its length and width. Find the term containing the variable w. To get this term, $2w$, alone, subtract $2l$ from both sides.

$$P = 2l + 2w$$

$$P - 2l = 2l + 2w - 2l \quad \text{Subtract } 2l \text{ from both sides.}$$

$$P - 2l = 2w \quad \text{Combine like terms.}$$

$$\frac{P - 2l}{2} = \frac{2w}{2} \quad \text{Divide both sides by 2.}$$

$$\frac{P - 2l}{2} = w \quad \text{Simplify.}$$

w

l

Helpful Hint

The 2s may *not* be divided out here. Although 2 is a factor of the denominator, 2 is *not* a factor of the numerator since it is not a factor of both terms in the numerator.

Work Practice 7

The next example has an equation containing a fraction. We will first clear the equation of fractions and then solve for the specified variable.

Practice 8

Solve $A = \frac{a + b}{2}$ for b.

Example 8

Solve $F = \frac{9}{5}C + 32$ for C.

Solution:

$$F = \frac{9}{5}C + 32$$

$$5(F) = 5\left(\frac{9}{5}C + 32\right) \quad \text{Clear the fraction by multiplying both sides by the LCD.}$$

$$5F = 9C + 160 \quad \text{Distribute the 5.}$$

$$5F - 160 = 9C + 160 - 160 \quad \text{To get the term containing the variable } C \text{ alone, subtract 160 from both sides.}$$

$$5F - 160 = 9C \quad \text{Combine like terms.}$$

$$\frac{5F - 160}{9} = \frac{9C}{9} \quad \text{Divide both sides by 9.}$$

$$\frac{5F - 160}{9} = C \quad \text{Simplify.}$$

Work Practice 8

Answers

6. $l = \frac{P - 2w}{2}$ 7. $a = \frac{P - b + c}{2}$

8. $b = 2A - a$

✓Concept Check Answers

a.  b. 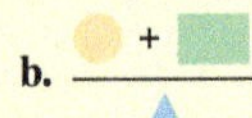

Vocabulary, Readiness & Video Check

Martin-Gay Interactive Videos

See Video 9.5

Watch the section lecture video and answer the following questions.

Objective A **1.** Complete this statement based on the lecture given before Example 1: A formula is an equation that describes known ___________ among quantities.

2. In Example 2, how are the units for the solution determined?

Objective B **3.** In Example 4, what is the equation $5x = 30$ used to show?

9.5 Exercise Set MyMathLab®

Objective A *Substitute the given values into each given formula and solve for the unknown variable. See Examples 1 through 4.*

1. $A = bh$; $A = 45, b = 15$ (Area of a parallelogram)

2. $d = rt$; $d = 195, t = 3$ (Distance formula)

3. $S = 4lw + 2wh$; $S = 102, l = 7, w = 3$ (Surface area of a special rectangular box)

4. $V = lwh$; $l = 14, w = 8, h = 3$ (Volume of a rectangular box)

5. $A = \frac{1}{2}h(B + b)$; $A = 180, B = 11, b = 7$ (Area of a trapezoid)

6. $A = \frac{1}{2}h(B + b)$; $A = 60, B = 7, b = 3$ (Area of a trapezoid)

7. $P = a + b + c$; $P = 30, a = 8, b = 10$ (Perimeter of a triangle)

8. $V = \frac{1}{3}Ah$; $V = 45, h = 5$ (Volume of a pyramid)

9. $C = 2\pi r$; $C = 15.7$ (Circumference of a circle) (Use the approximation 3.14 for π.)

10. $A = \pi r^2$; $r = 4$ (Area of a circle) (Use the approximation 3.14 for π.)

Objective B *Solve each formula for the specified variable. See Examples 5 through 8.*

11. $f = 5gh$ for h

12. $x = 4\pi y$ for y

13. $V = lwh$ for w

14. $T = mnr$ for n

15. $3x + y = 7$ for y

16. $-x + y = 13$ for y

17. $A = P + PRT$ for R

18. $A = P + PRT$ for T

19. $V = \frac{1}{3}Ah$ for A

20. $D = \frac{1}{4}fk$ for k

21. $P = a + b + c$ for a

22. $PR = x + y + z + w$ for z

23. $S = 2\pi rh + 2\pi r^2$ for h

△ **24.** $S = 4lw + 2wh$ for h

Objective A *Solve. See Examples 1 through 4.*

25. For the purpose of purchasing new baseboard and carpet,

a. Find the area and perimeter of the room below (neglecting doors).

b. Identify whether baseboard has to do with area or perimeter and the same with carpet.

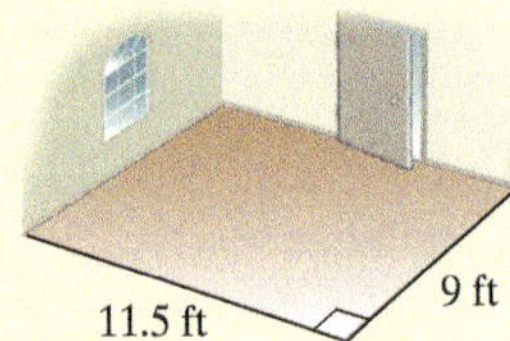

26. For the purpose of purchasing lumber for a new fence and seed to plant grass,

a. Find the area and perimeter of the yard below.

b. Identify whether a fence has to do with area or perimeter and the same with grass seed.

$\left(A = \frac{1}{2}bh\right)$

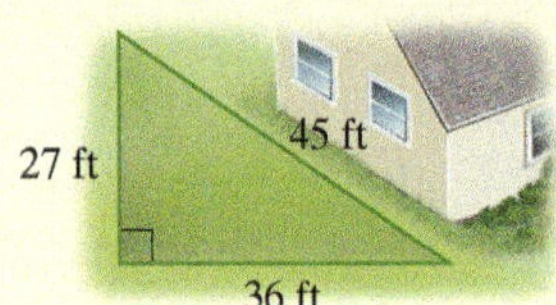

27. A frame shop charges according to both the amount of framing needed to surround the picture and the amount of glass needed to cover the picture.

a. Find the area and perimeter of the picture below.

b. Identify whether the frame has to do with perimeter or area and the same with the glass.

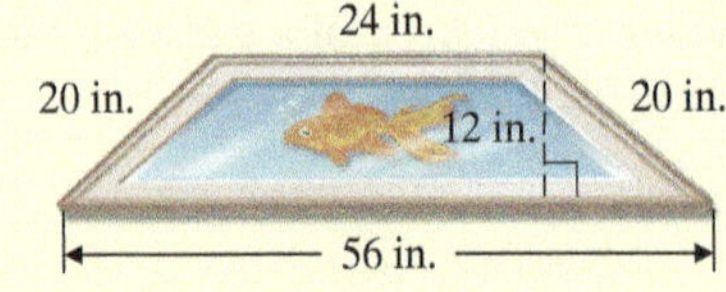

28. A decorator is painting and placing a border completely around a parallelogram-shaped wall.

a. Find the area and perimeter of the wall below.

b. Identify whether the border has to do with perimeter or area and the same with paint.

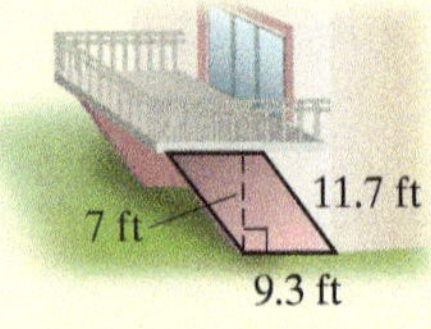

29. The iconic NASDAQ sign in New York's Times Square has a width of 84 feet and an area of 10,080 square feet. Find the height of the sign. (*Source:* livedesignonline.com)

△ **30.** The world's largest sign for Coca-Cola is located in Arica, Chile. The rectangular sign has a length of 400 feet and an area of 52,400 square feet. Find the width of the sign. (*Source:* Fabulous Facts about Coca-Cola, Atlanta, GA)

31. Convert Nome, Alaska's 14°F high temperature to Celsius.

32. Convert Paris, France's low temperature of −5°C to Fahrenheit.

33. The X-30 is a "space plane" that skims the edge of space at 4000 miles per hour. Neglecting altitude, if the circumference of the Earth is approximately 25,000 miles, how long does it take for the X-30 to travel around the Earth?

34. In the United States, a notable hang glider flight was a 303-mile, $8\frac{1}{2}$-hour flight from New Mexico to Kansas. What was the average rate during this flight?

35. An architect designs a rectangular flower garden such that the width is exactly two-thirds of the length. If 260 feet of antique picket fencing is to be used to enclose the garden, find the dimensions of the garden.

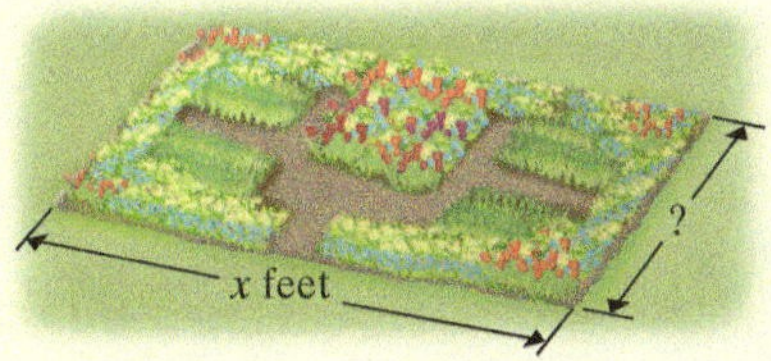

36. If the length of a rectangular parking lot is 10 meters less than twice its width, and the perimeter is 400 meters, find the length of the parking lot.

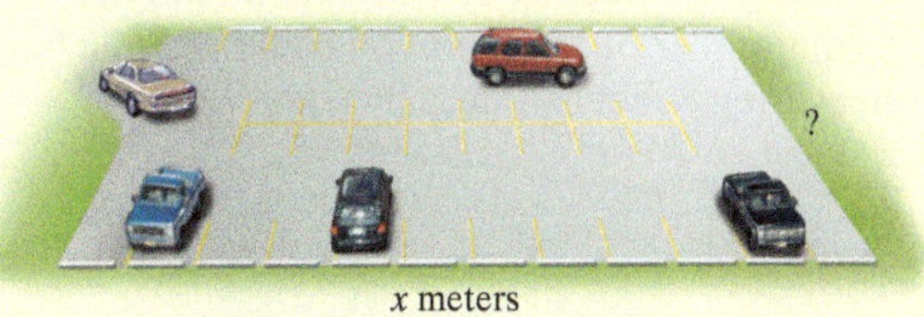

37. A flower bed is in the shape of a triangle with one side twice the length of the shortest side and the third side 30 feet more than the length of the shortest side. Find the dimensions if the perimeter is 102 feet.

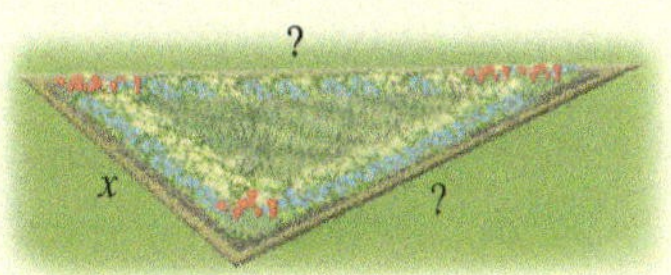

38. The perimeter of a yield sign in the shape of an isosceles triangle is 22 feet. If the shortest side is 2 feet less than the other two sides, find the length of the shortest side. (*Hint:* An isosceles triangle has two sides the same length.)

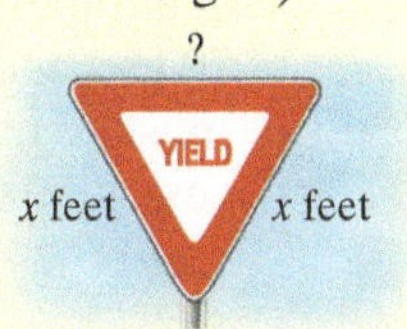

39. The Eurostar is a high-speed train that shuttles passengers between London, England, and Paris, France, via the Channel Tunnel. The Eurostar can make the trip in about $2\frac{1}{4}$ hours at an average speed of 136 mph. About how long is the Eurostar's route connecting London and Paris? (*Source:* Eurostar)

40. A family is planning their vacation to Disney World. They will drive from a small town outside New Orleans, Louisiana, to Orlando, Florida, a distance of 700 miles. They plan to average a rate of 56 mph. How long will this trip take?

Dolbear's Law states the relationship between the rate at which Snowy Tree Crickets chirp and the air temperature of their environment. The formula is

$$T = 50 + \frac{N - 40}{4}, \text{ where } \begin{array}{l} T = \text{temperature in degrees Fahrenheit and} \\ N = \text{number of chirps per minute} \end{array}$$

41. If $N = 86$, find the temperature in degrees Fahrenheit, T.

42. If $N = 94$, find the temperature in degrees Fahrenheit, T.

43. If $T = 55°F$, find the number of chirps per minute.

44. If $T = 65°F$, find the number of chirps per minute.

Use the results of Exercises **41–44** *to complete each sentence with "increases" or "decreases."*

45. As the number of cricket chirps per minute increases, the air temperature of their environment ________.

46. As the air temperature of their environment decreases, the number of cricket chirps per minute ________.

Solve. See Examples 1 through 4.

△ **47.** Piranha fish require 1.5 cubic feet of water per fish to maintain a healthy environment. Find the maximum number of piranhas you could put in a tank measuring 8 feet by 3 feet by 6 feet.

△ **48.** Find the maximum number of goldfish you can put in a cylindrical tank whose diameter is 8 meters and whose height is 3 meters, if each goldfish needs 2 cubic meters of water. $(V = \pi r^2 h)$

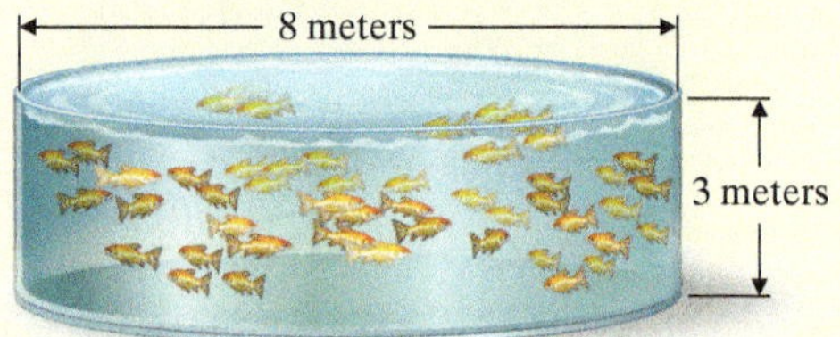

△ **49.** A lawn is in the shape of a trapezoid with a height of 60 feet and bases of 70 feet and 130 feet. How many bags of fertilizer must be purchased to cover the lawn if each bag covers 4000 square feet?

△ **50.** If the area of a right-triangularly shaped sail is 20 square feet and its base is 5 feet, find the height of the sail. $\left(A = \frac{1}{2}bh\right)$

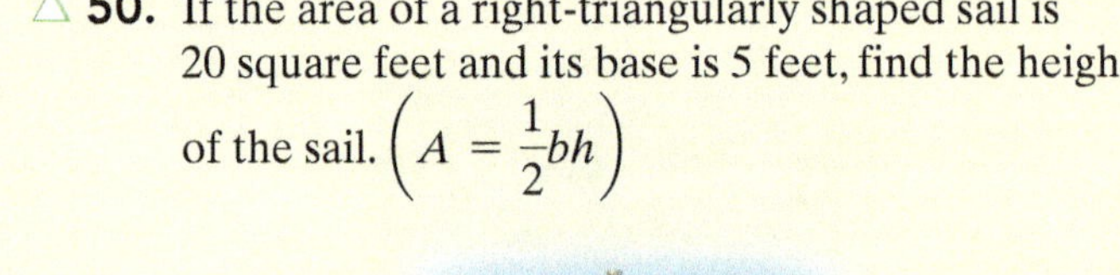

△ **51.** Maria's Pizza sells one 16-inch cheese pizza or two 10-inch cheese pizzas for $9.99. Determine which size gives more pizza.

△ **52.** Find how much rope is needed to wrap around the Earth at the equator, if the radius of the Earth is 4000 miles. (*Hint:* Use 3.14 for π and the formula for circumference.)

△ **53.** The perimeter of an equilateral triangle is 7 inches more than the perimeter of a square, and the side of the triangle is 5 inches longer than the side of the square. Find the side of the triangle. (*Hint:* An equilateral triangle has three sides the same length.)

△ **54.** A square animal pen and a pen shaped like an equilateral triangle have equal perimeters. Find the length of the sides of each pen if the sides of the triangular pen are 15 less than twice a side of the square pen. (*Hint:* An equilateral triangle has three sides the same length.)

55. Find how long it takes Tran Nguyen to drive 135 miles on I-10 if he merges onto I-10 at 10 a.m. and drives nonstop with his cruise control set on 60 mph.

56. Beaumont, Texas, is about 150 miles from Toledo Bend. If Leo Miller leaves Beaumont at 4 a.m. and averages 45 mph, when should he arrive at Toledo Bend?

△ **57.** The longest runway at Los Angeles International Airport has the shape of a rectangle and an area of 1,813,500 square feet. This runway is 150 feet wide. How long is the runway? (*Source:* Los Angeles World Airports)

58. The return stroke of a bolt of lightning can travel at a speed of 87,000 miles per second (almost half the speed of light). At this speed, how many times can an object travel around the world in one second? (See Exercise **52.**) Round to the nearest tenth. (*Source: The Handy Science Answer Book*)

59. The highest temperature ever recorded in Europe was 122°F in Seville, Spain, in August of 1881. Convert this record high temperature to Celsius. (*Source:* National Climatic Data Center)

60. The lowest temperature ever recorded in Oceania was −10°C at the Haleakala Summit in Maui, Hawaii, in January 1961. Convert this record low temperature to Fahrenheit. (*Source:* National Climatic Data Center)

61. The Eurostar high-speed train has a top speed of 186 mph. It travels beneath the English Channel through the Channel Tunnel, which is 31.35 miles long. If the Eurostar were able to travel at its top speed the entire length of the Channel Tunnel, how many *minutes* would the trip take? Round to the nearest tenth of a minute. (*Source:* Eurostar)

62. In 1983, the Hawaiian volcano Kilauea began erupting in a series of episodes still occurring at the time of this writing. At times, the lava flows have advanced at speeds of up to 0.5 kilometer per hour. In 1983 and 1984 lava flows destroyed 16 homes in the Royal Gardens subdivision, about 6 km away from the eruption site. Roughly how long did it take the lava to reach Royal Gardens? (*Source:* U.S. Geological Survey Hawaiian Volcano Observatory)

△ **63.** The IZOD IndyCar series is an open-wheeled race car competition based in the United States. An IndyCar car has a maximum length of 204 inches, a maximum width of 76.5 inches, and a maximum height of 44 inches. When the IZOD IndyCar series travels to another country for a grand prix, teams must ship their cars. Find the volume of the smallest shipping crate needed to ship an IndyCar car of maximum dimensions. (*Source:* Championship Auto Racing Teams, Inc.)

IndyCar Racing Car

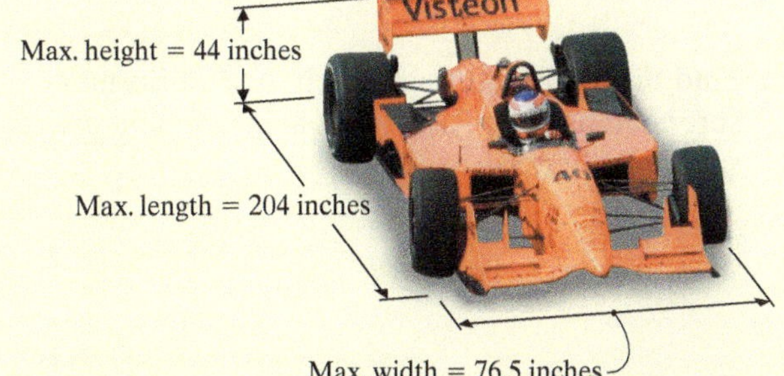

64. During a recent IndyCar road course race, the winner's average speed was 118 mph. Based on this speed, how long would it take an IndyCar driver to travel from Los Angeles to New York City, a distance of about 2810 miles by road, without stopping? Round to the nearest tenth of an hour.

65. The Hoberman Sphere is a toy ball that expands and contracts. When it is completely closed, it has a diameter of 9.5 inches. Find the volume of the Hoberman Sphere when it is completely closed. Use 3.14 for π. Round to the nearest whole cubic inch. (*Hint:* Volume of a sphere $= \frac{4}{3}\pi r^3$.) (*Source:* Hoberman Designs, Inc.)

66. When the Hoberman Sphere (see Exercise **65**) is completely expanded, its diameter is 30 inches. Find the volume of the Hoberman Sphere when it is completely expanded. Use 3.14 for π. (*Source:* Hoberman Designs, Inc.)

67. The average temperature on the planet Mercury is 167°C. Convert this temperature to degrees Fahrenheit. Round to the nearest degree. (*Source:* National Space Science Data Center)

68. The average temperature on the planet Jupiter is −227°F. Convert this temperature to degrees Celsius. Round to the nearest degree. (*Source:* National Space Science Data Center)

Review

Write each percent as a decimal. See Section 6.2.

69. 32%

70. 8%

71. 200%

72. 0.5%

Write each decimal as a percent. See Section 6.2.

73. 0.17

74. 0.03

75. 7.2

76. 5

Concept Extensions

Solve.

77. $N = R + \frac{V}{G}$ for V (Urban forestry: tree plantings per year)

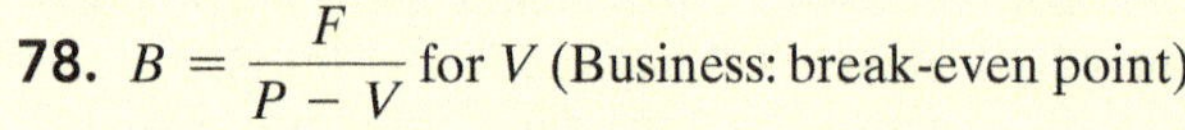

78. $B = \frac{F}{P - V}$ for V (Business: break-even point)

79. The formula $V = lwh$ is used to find the volume of a box. If the length of a box is doubled, the width is doubled, and the height is doubled, how does this affect the volume? Explain your answer.

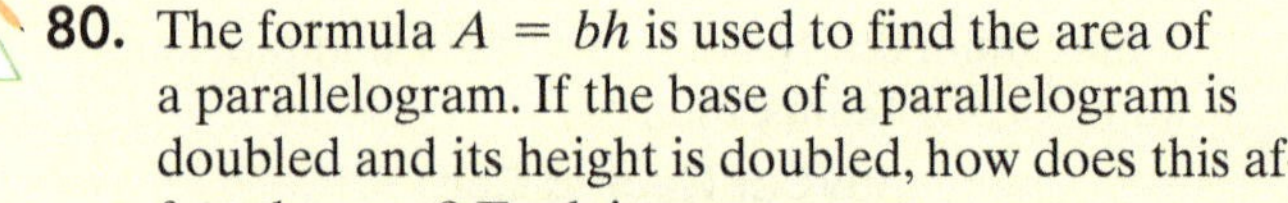

80. The formula $A = bh$ is used to find the area of a parallelogram. If the base of a parallelogram is doubled and its height is doubled, how does this affect the area? Explain your answer.

81. Use the Dolbear's Law formula from Exercises **41–46** and calculate when the number of cricket chirps per minute is the same as the temperature in degrees Fahrenheit. (*Hint:* Replace T with N and solve for N, or replace N with T and solve for T.)

82. Find the temperature at which the Celsius measurement and the Fahrenheit measurement are the same number.

Solve. See the Concept Check in this section.

83. ▲ − ● · ■ = ■ for ●

84. ⬟ · ■ + ▲ = ● for ■

85. Flying fish do not *actually* fly, but glide. They have been known to travel a distance of 1300 feet at a rate of 20 miles per hour. How many seconds would it take to travel this distance? (*Hint:* First convert miles per hour to feet per second. Recall that 1 mile = 5280 feet.) Round to the nearest tenth of a second.

86. A glacier is a giant mass of rocks and ice that flows downhill like a river. Exit Glacier, near Seward, Alaska, moves at a rate of 20 inches a day. Find the distance in feet the glacier moves in a year. (Assume 365 days a year.) Round to two decimal places.

Substitute the given values into each given formula and solve for the unknown variable. If necessary, round to one decimal place.

87. $I = PRT$; $I = 1{,}056{,}000, R = 0.055, T = 6$ (Simple interest formula)

88. $I = PRT$; $I = 3750, P = 25{,}000, R = 0.05$ (Simple interest formula)

89. $V = \frac{4}{3}\pi r^3$; $r = 3$ (Volume of a sphere) (Use a calculator approximation for π.)

90. $V = \frac{1}{3}\pi r^2 h$; $V = 565.2, r = 6$ (Volume of a cone) (Use a calculator approximation for π.)

9.6 Linear Inequalities and Problem Solving

Objectives

A Define Linear Inequality in One Variable, Graph Solution Sets on a Number Line, and Use Interval Notation.

B Use the Addition Property of Inequality to Solve Inequalities.

C Use the Multiplication Property of Inequality to Solve Inequalities.

D Use Both Properties to Solve Inequalities.

E Solve Problems Modeled by Inequalities.

Objective A Graphing Solution Sets to Linear Inequalities and Using Interval Notation

In Section 9.1, we reviewed these inequality symbols and their meanings:

$<$ means "is less than" $\leq$ means "is less than or equal to"

$>$ means "is greater than" $\geq$ means "is greater than or equal to"

Equations	Inequalities
$x = 3$	$x \leq 3$
$5n - 6 = 14$	$5n - 6 < 14$
$12 = 7 - 3y$	$12 \geq 7 - 3y$
$\frac{x}{4} - 6 = 1$	$\frac{x}{4} - 6 > 1$

A linear inequality is similar to a linear equation except that the equality symbol is replaced with an inequality symbol.

Linear Inequality in One Variable

A **linear inequality in one variable** is an inequality that can be written in the form

$$ax + b < c$$

where a, b, and c are real numbers and a is not 0.

This definition and all other definitions, properties, and steps in this section also hold true for the inequality symbols $>$, $\geq$, and $\leq$.

A **solution of an inequality** is a value of the variable that makes the inequality a true statement. The solution set is the set of all solutions. For the inequality $x < 3$, replacing x with any number less than 3—that is, to the left of 3 on a number line—makes the resulting inequality true. This means that any number less than 3 is a solution of the inequality $x < 3$.

Since there are infinitely many such numbers, we cannot list all the solutions of the inequality. We *can* use set notation and write

$\{x \mid x < 3\}$. Recall that this is read

$\{x$ — the set of all x; $\mid$ — such that; $x < 3\}$ — x is less than 3.

We can also picture the solutions on a number line. If we use the open/closed-circle notation, the graph of $\{x|x < 3\}$ looks like the following.

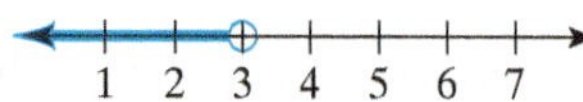

In this text, a convenient notation, called **interval notation,** will be used to write solution sets of inequalities. To help us understand this notation, a different graphing notation will be used. Instead of an open circle, we use a parenthesis; instead of a closed circle, we use a bracket. With this new notation, the graph of $\{x|x < 3\}$ now looks like

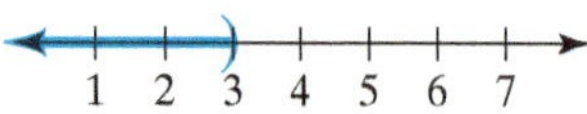

and can be represented in interval notation as $(-\infty, 3)$. The symbol $-\infty$, read as "negative infinity," does not indicate a number but does indicate that the shaded arrow to the left never ends. In other words, the interval $(-\infty, 3)$ includes *all* numbers less than 3.

Helpful Hint

Since we approach $-\infty$ and $+\infty$ (or ∞), but never arrive, we always place parentheses about $-\infty$ and $+\infty$ (or ∞).

Picturing the solutions of an inequality on a number line is called **graphing** the solutions or graphing the inequality, and the picture is called the **graph** of the inequality.

To graph $\{x|x \leq 3\}$ or simply $x \leq 3$, shade the numbers to the left of 3 and place a bracket at 3 on the number line. The bracket indicates that 3 **is** a solution: 3 **is** less than or equal to 3. In interval notation, we write $(-\infty, 3]$.

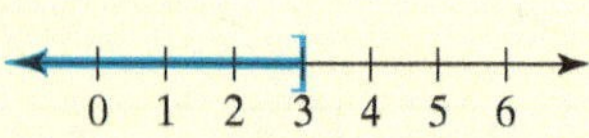

Helpful Hint

When writing an inequality in interval notation, it may be easier to graph the inequality first, and then write it in interval notation. To help, think of the number line as approaching $-\infty$ to the left and $+\infty$ or ∞ to the right. Then simply write the interval notation by following your shading from left to right.

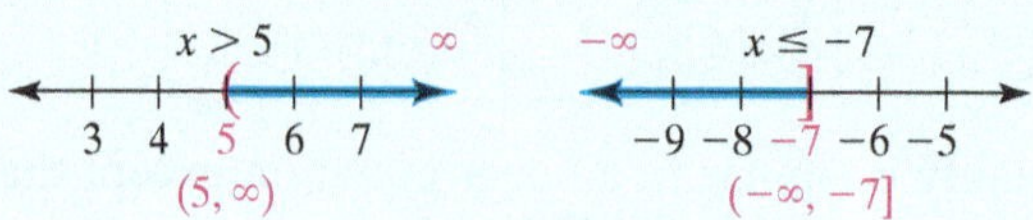

Example 1 Graph: $x \geq -1$. Then write the solutions in interval notation.

Solution: To graph the solutions of $x \geq -1$, we place a bracket at -1 since the inequality symbol is $\geq$ and -1 is greater than or equal to -1. Then we shade to the right of -1.

In interval notation, this is $[-1, \infty)$.

Work Practice 1

Practice 1

Graph: $x \geq -2$. Then write the solutions in interval notation.

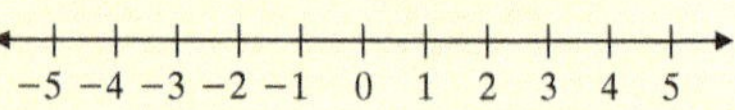

Answer

1. 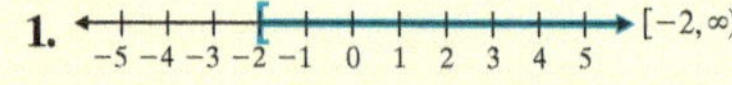 $[-2, \infty)$

Example 2 Graph: $-1 > x$. Then write the solutions in interval notation.

Solution: Recall from Section 9.1 that $-1 > x$ means the same as $x < -1$. The graph of the solutions of $x < -1$ is shown below.

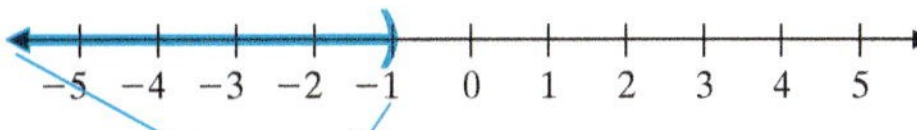

In interval notation, this is $(-\infty, -1)$.

Work Practice 2

Practice 2

Graph: $5 > x$. Then write the solutions in interval notation.

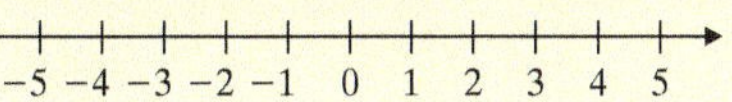

Helpful Hint To write this in interval notation, follow your shading on the number line from left to right.

Example 3 Graph: $-4 < x \le 2$. Then write the solutions in interval notation.

Solution: We read $-4 < x \le 2$ as "-4 is less than x and x is less than or equal to 2," or as "x is greater than -4 and x is less than or equal to 2." To graph the solutions of this inequality, we place parenthesis at -4 (-4 is not part of the graph), a bracket at 2 (2 is part of the graph), and shade all numbers between -4 and 2. Why? All numbers between -4 and 2 are greater than -4 *and* also less than 2.

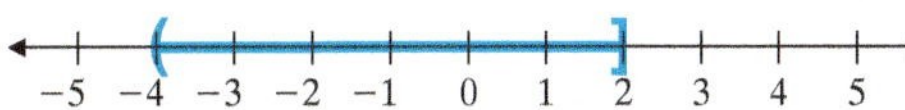

In interval notation, this is $(-4, 2]$.

Work Practice 3

Practice 3

Graph: $-3 \le x < 1$. Then write the solutions in interval notation.

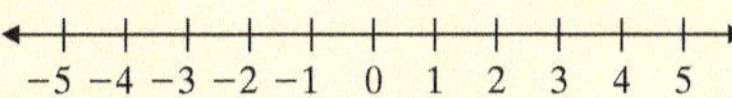

Objective B Using the Addition Property

When solutions of a linear inequality are not immediately obvious, they are found through a process similar to the one used to solve a linear equation. Our goal is to get the variable alone on one side of the inequality. We use properties of inequality similar to properties of equality.

Addition Property of Inequality

If a, b, and c are real numbers, then

$a < b$ and $a + c < b + c$

are equivalent inequalities.

This property also holds true for subtracting values, since subtraction is defined in terms of addition. In other words, adding or subtracting the same quantity from both sides of an inequality does not change the solutions of the inequality.

Example 4 Solve $x + 4 \le -6$. Graph the solution set and write it in interval notation.

Solution: To solve for x, subtract 4 from both sides of the inequality.

$$x + 4 \le -6 \quad \text{Original inequality}$$
$$x + 4 - 4 \le -6 - 4 \quad \text{Subtract 4 from both sides.}$$
$$x \le -10 \quad \text{Simplify.}$$

The graph of the solution set is shown below.

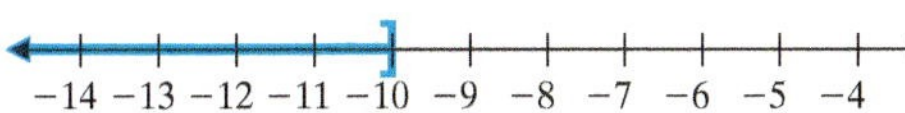

In interval notation, the solution set is $(-\infty, -10]$.

Work Practice 4

Practice 4

Solve $x - 6 \ge -11$. Graph the solution set and write it in interval notation.

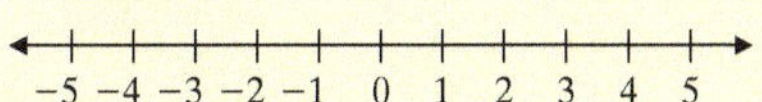

Answers

2. 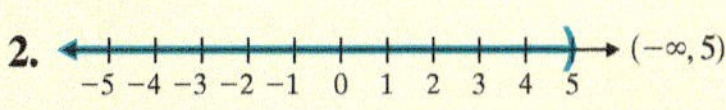

3. 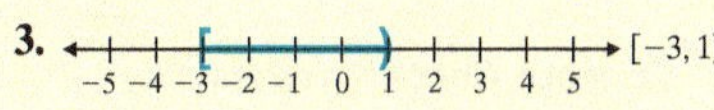

4. 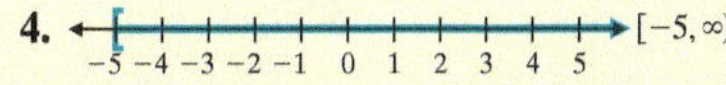

Helpful Hint

Notice that any number less than or equal to -10 is a solution to $x \le -10$. For example, solutions include

$$-10, \quad -200, \quad -11\frac{1}{2}, \quad -\sqrt{130}, \quad \text{and} \quad -50.3$$

Objective C Using the Multiplication Property 

An important difference between solving linear equations and solving linear inequalities is shown when we multiply or divide both sides of an inequality by a nonzero real number. For example, start with the true statement $6 < 8$ and multiply both sides by 2. As we see below, the resulting inequality is also true.

$6 < 8$ True
$2(6) < 2(8)$ Multiply both sides by 2.
$12 < 16$ True

But if we start with the same true statement $6 < 8$ and multiply both sides by -2, the resulting inequality is not a true statement.

$6 < 8$ True
$-2(6) < -2(8)$ Multiply both sides by -2.
$-12 < -16$ False

Notice, however, that if we reverse the direction of the inequality symbol, the resulting inequality is true.

$-12 < -16$ False
$-12 > -16$ True

This demonstrates the multiplication property of inequality.

Multiplication Property of Inequality

1. If a, b, and c are real numbers, and c is **positive,** then
   $a < b$ and $ac < bc$
   are equivalent inequalities.
2. If a, b, and c are real numbers, and c is **negative,** then
   $a < b$ and $ac > bc$
   are equivalent inequalities.

Because division is defined in terms of multiplication, this property also holds true when dividing both sides of an inequality by a nonzero number: If we multiply or divide both sides of an inequality by a negative number, **the direction of the inequality sign must be reversed for the inequalities to remain equivalent.**

✓**Concept Check** Fill in each box with $<$, $>$, $\le$, or $\ge$.

a. Since $-8 < -4$, then $3(-8) \square 3(-4)$.

b. Since $5 \ge -2$, then $\frac{5}{-7} \square \frac{-2}{-7}$.

c. If $a < b$, then $2a \square 2b$.

d. If $a \ge b$, then $\frac{a}{-3} \square \frac{b}{-3}$.

✓**Concept Check Answers**
a. $<$ **b.** $\le$ **c.** $<$ **d.** $\le$

Example 5 Solve $-2x \leq -4$. Graph the solution set and write it in interval notation.

Solution: Remember to reverse the direction of the inequality symbol when dividing by a negative number.

$$-2x \leq -4$$

$$\frac{-2x}{-2} \geq \frac{-4}{-2} \quad \text{Divide both sides by } -2 \text{ and reverse the inequality sign.}$$

$$x \geq 2 \quad \text{Simplify.}$$

The graph of the solution set is shown.

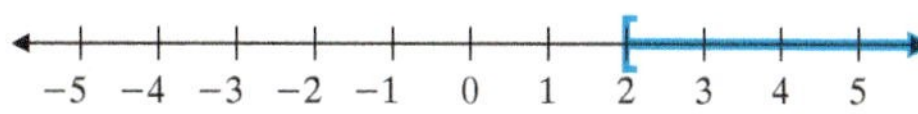

In interval notation, the solution set is $[2, \infty)$.

Work Practice 5

Practice 5

Solve $-3x \leq 12$. Graph the solution set and write it in interval notation.

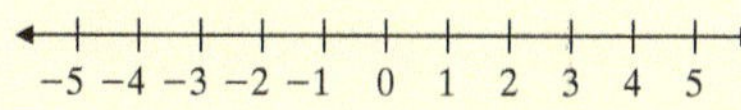

Example 6 Solve $2x < -4$. Graph the solution set and write it in interval notation.

Solution:

$$2x < -4$$

$$\frac{2x}{2} < \frac{-4}{2} \quad \text{Divide both sides by 2. Do not reverse the inequality sign.}$$

$$x < -2 \quad \text{Simplify.}$$

The graph of the solution set is shown.

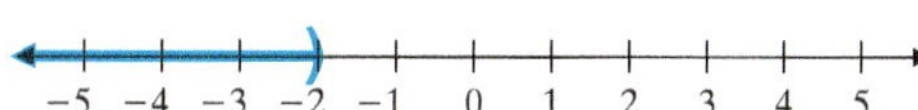

In interval notation, the solution set is $(-\infty, -2)$.

Work Practice 6

Practice 6

Solve $5x > -20$. Graph the solution set and write it in interval notation.

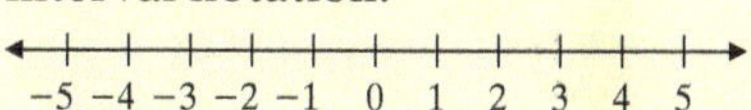

Objective D Using Both Properties of Inequality

The following steps may be helpful when solving inequalities in one variable. Notice that these steps are similar to the ones given in Section 9.3 for solving equations.

To Solve Linear Inequalities in One Variable

Step 1: If an inequality contains fractions or decimals, multiply both sides by the LCD to clear the inequality of fractions or decimals.

Step 2: Use the distributive property to remove parentheses if they appear.

Step 3: Simplify each side of the inequality by combining like terms.

Step 4: Get all variable terms on one side and all numbers on the other side by using the addition property of inequality.

Step 5: Get the variable alone by using the multiplication property of inequality.

Answers

5. 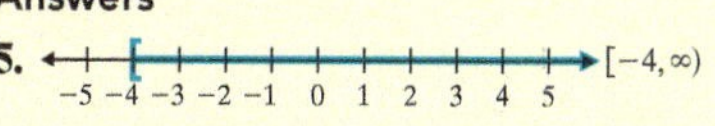 $[-4, \infty)$

6. 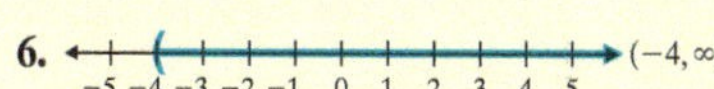 $(-4, \infty)$

Helpful Hint

Don't forget that if both sides of an inequality are multiplied or divided by a negative number, the direction of the inequality sign must be reversed.

Practice 7

Solve $-3x + 11 \le -13$. Graph the solution set and write it in interval notation.

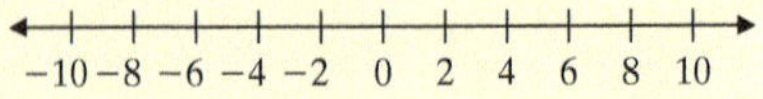
−10 −8 −6 −4 −2 0 2 4 6 8 10

Example 7

Solve $-4x + 7 \ge -9$. Graph the solution set and write it in interval notation.

Solution:

$$-4x + 7 \ge -9$$

$$-4x + 7 - 7 \ge -9 - 7 \quad \text{Subtract 7 from both sides.}$$

$$-4x \ge -16 \quad \text{Simplify.}$$

$$\frac{-4x}{-4} \le \frac{-16}{-4} \quad \text{Divide both sides by } -4 \text{ and reverse the direction of the inequality sign.}$$

$$x \le 4 \quad \text{Simplify.}$$

The graph of the solution set is shown.

−5 −4 −3 −2 −1 0 1 2 3 4 5

In interval notation, the solution set is $(-\infty, 4]$.

Work Practice 7

Practice 8

Solve $2x - 3 > 4(x - 1)$. Graph the solution set and write it in interval notation.

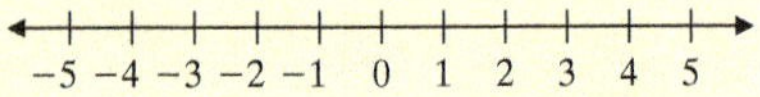
−5 −4 −3 −2 −1 0 1 2 3 4 5

Example 8

Solve $-5x + 7 < 2(x - 3)$. Graph the solution set and write it in interval notation.

Solution:

$$-5x + 7 < 2(x - 3)$$

$$-5x + 7 < 2x - 6 \quad \text{Apply the distributive property.}$$

$$-5x + 7 - 2x < 2x - 6 - 2x \quad \text{Subtract } 2x \text{ from both sides.}$$

$$-7x + 7 < -6 \quad \text{Combine like terms.}$$

$$-7x + 7 - 7 < -6 - 7 \quad \text{Subtract 7 from both sides.}$$

$$-7x < -13 \quad \text{Combine like terms.}$$

$$\frac{-7x}{-7} > \frac{-13}{-7} \quad \text{Divide both sides by } -7 \text{ and reverse the direction of the inequality sign.}$$

$$x > \frac{13}{7} \quad \text{Simplify.}$$

The graph of the solution set is shown.

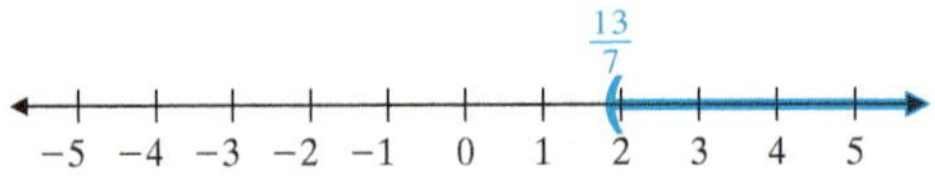
13/7
−5 −4 −3 −2 −1 0 1 2 3 4 5

In interval notation, the solution set is $\left(\frac{13}{7}, \infty\right)$.

Work Practice 8

Answers

7. 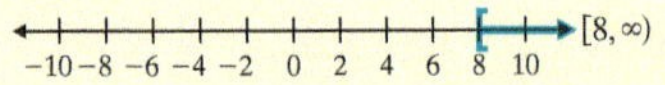
−10 −8 −6 −4 −2 0 2 4 6 8 10
[8, ∞)

8. 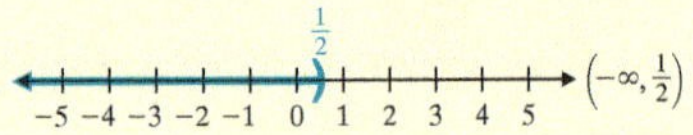
1/2
−5 −4 −3 −2 −1 0 1 2 3 4 5
(−∞, 1/2)

Example 9 Solve: $2(x - 3) - 5 \le 3(x + 2) - 18$. Write the solution set in interval notation.

Solution:

$$2(x - 3) - 5 \le 3(x + 2) - 18$$

$$2x - 6 - 5 \le 3x + 6 - 18 \quad \text{Apply the distributive property.}$$

$$2x - 11 \le 3x - 12 \quad \text{Combine like terms.}$$

$$-x - 11 \le -12 \quad \text{Subtract } 3x \text{ from both sides.}$$

$$-x \le -1 \quad \text{Add 11 to both sides.}$$

$$\frac{-x}{-1} \ge \frac{-1}{-1} \quad \text{Divide both sides by } -1 \text{ and reverse the direction of the inequality sign.}$$

$$x \ge 1 \quad \text{Simplify.}$$

The solution set in interval notation is $[1, \infty)$.

Work Practice 9

Practice 9

Solve:
$3(x + 5) - 1 \ge 5(x - 1) + 7$.
Write the solution set in interval notation.

Objective E Solving Problems Modeled by Inequalities 

Problems containing words such as "at least," "at most," "between," "no more than," and "no less than" usually indicate that an inequality should be solved instead of an equation. In solving applications involving linear inequalities, we use the same procedure we used to solve applications involving linear equations.

Some Inequality Translations

$\ge$	$\le$	$<$	$>$
at least	at most	is less than	is greater than
no less than	no more than		

Example 10 12 subtracted from 3 times a number is less than 21. Find all numbers that make this statement true.

Solution:

1. UNDERSTAND. Read and reread the problem. This is a direct translation problem, and let's let

   $x =$ the unknown number

2. TRANSLATE.

12	subtracted from	3 times a number	is less than	21
$3x$	$-$	12	$<$	21

3. SOLVE. $3x - 12 < 21$

$$3x < 33 \quad \text{Add 12 to both sides.}$$

$$\frac{3x}{3} < \frac{33}{3} \quad \text{Divide both sides by 3 and do not reverse the direction of the inequality sign.}$$

$$x < 11 \quad \text{Simplify.}$$

4. INTERPRET.

   Check: Check the translation; then let's choose a number less than 11 to see if it checks. For example, let's check 10. 12 subtracted from 3 times 10 is 12 subtracted from 30, or 18. Since 18 is less than 21, the number 10 checks.

   State: All numbers less than 11 make the original statement true.

Work Practice 10

Practice 10

Twice a number, subtracted from 35 is greater than 15. Find all numbers that make this true.

Answers

9. $(-\infty, 6]$
10. all numbers less than 10

Practice 11

Alex earns $600 per month plus 4% of all his sales. Find the minimum sales that will allow Alex to earn at least $3000 per month.

Answer

11. $60,000

Example 11 Budgeting for a Wedding

A couple are having their wedding reception at the Gallery reception hall. They may spend at most $1000 for the reception. If the reception hall charges a $100 cleanup fee plus $14 per person, find the greatest number of people that they can invite and still stay within their budget.

Solution:

1. UNDERSTAND. Read and reread the problem. Suppose that 50 people attend the reception. The cost is then $\$100 + \$14(50) = \$100 + \$700 = \$800$.

Let $x =$ the number of people who attend the reception.

2. TRANSLATE.

cleanup fee	+	cost per person	times	number of people	must be less than or equal to	$1000
↓		↓	↓	↓	↓	↓
100	+	14	$\cdot$	x	$\le$	1000

3. SOLVE.

$$100 + 14x \le 1000$$

$$14x \le 900 \quad \text{Subtract 100 from both sides.}$$

$$x \le 64\frac{2}{7} \quad \text{Divide both sides by 14.}$$

4. INTERPRET.

Check: Since x represents the number of people, we round down to the nearest whole, or 64. Notice that if 64 people attend, the cost is $\$100 + \$14(64) = \$996$. If 65 people attend, the cost is $\$100 + \$14(65) = \$1010$, which is more than the given $1000.

State: The couple can invite at most 64 people to the reception.

Work Practice 11

Vocabulary, Readiness & Video Check

Identify each as an equation, expression, or inequality.

1. $6x - 7(x + 9)$ ________

2. $6x = 7(x + 9)$ ________

3. $6x < 7(x + 9)$ ________

4. $5y - 2 \ge -38$ ________

5. $\frac{9}{7} = \frac{x+2}{14}$ ________

6. $\frac{9}{7} - \frac{x+2}{14}$ ________

Decide which number listed is not a solution to each given inequality.

7. $x \ge -3$; $-3, 0, -5, \pi$ ________

8. $x < 6$; $-6, |-6|, 0, -3.2$ ________

9. $x < 4.01$; $4, -4.01, 4.1, -4.1$ ________

10. $x \ge -3$; $-4, -3, -2, -(-2)$ ________

Martin-Gay Interactive Videos *Watch the section lecture video and answer the following questions.*

See Video 9.6

Objective A **11.** From Example 1, when graphing an inequality, what inequality symbol(s) does a parenthesis indicate? What inequality symbol(s) does a bracket indicate?

Objective B **12.** From the lecture before Example 2, which property is the addition property of inequality very similar to?

Objective C **13.** What is the answer to Example 3, written in interval notation?

Objective D **14.** When solving Example 4, why is special attention given to the coefficient of x in the last step?

Objective E **15.** What is the phrase in Example 5 that tells us to translate to an inequality? What does this phrase translate to?

9.6 Exercise Set MyMathLab®

Objective A *Graph each inequality on the number line. Then write the solutions in interval notation. See Examples 1 and 2.*

1. $x \leq -1$

−5 −4 −3 −2 −1 0 1 2 3 4 5

2. $y < 0$

−5 −4 −3 −2 −1 0 1 2 3 4 5

3. $x > \frac{1}{2}$

−5 −4 −3 −2 −1 0 1 2 3 4 5

4. $z \geq -\frac{2}{3}$

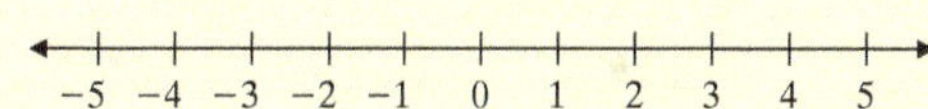

5. $y < 4$

−5 −4 −3 −2 −1 0 1 2 3 4 5

6. $x > 3$

−5 −4 −3 −2 −1 0 1 2 3 4 5

7. $-2 \leq m$

−5 −4 −3 −2 −1 0 1 2 3 4 5

8. $-5 \geq x$

−5 −4 −3 −2 −1 0 1 2 3 4 5

Graph each inequality on the number line. Then write the solutions in interval notation. See Example 3.

9. $-1 < x < 3$

−5 −4 −3 −2 −1 0 1 2 3 4 5

10. $-2 \leq x \leq 3$

−5 −4 −3 −2 −1 0 1 2 3 4 5

11. $0 \leq y < 2$

−5 −4 −3 −2 −1 0 1 2 3 4 5

12. $-4 < x \leq 0$

−5 −4 −3 −2 −1 0 1 2 3 4 5

Objective B *Solve each inequality. Graph the solution set and write it in interval notation. See Example 4.*

13. $x - 2 \geq -7$

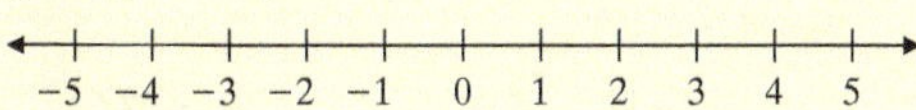

14. $x + 4 \leq 1$

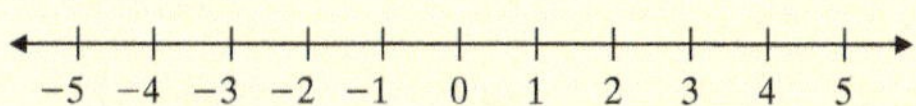

15. $-9 + y < 0$

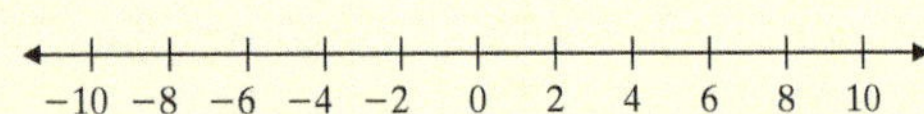

16. $-3 + m > 5$

17. $3x - 5 > 2x - 8$

18. $3 - 7x \geq 10 - 8x$

19. $4x - 1 \leq 5x - 2x$

20. $7x + 3 < 9x - 3x$

Objective C

Solve each inequality. Graph the solution set and write it in interval notation. See Examples 5 and 6.

21. $2x < -6$

22. $3x > -9$

23. $-8x \leq 16$

24. $-5x < 20$

25. $-x > 0$

26. $-y \geq 0$

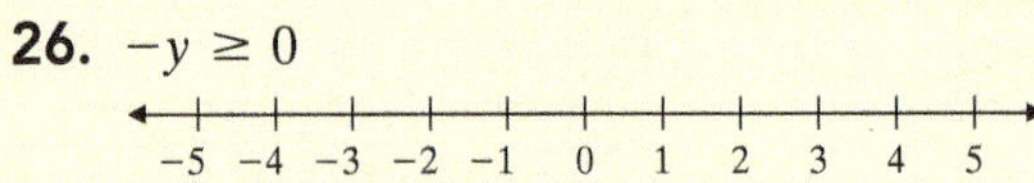

27. $\frac{3}{4}y \geq -2$

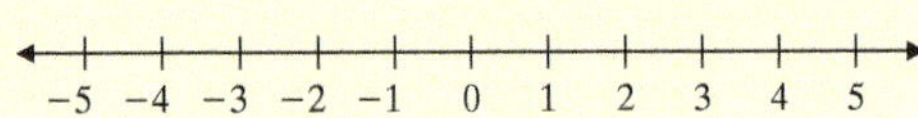

28. $\frac{5}{6}x \leq -8$

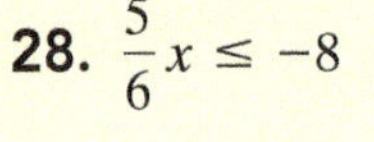

29. $-0.6y < -1.8$

30. $-0.3x > -2.4$

Objectives B C D Mixed Practice

Solve each inequality. Write the solution set in interval notation. See Examples 4 through 9.

31. $-8 < x + 7$

32. $-11 > x + 4$

33. $7(x + 1) - 6x \geq -4$

34. $10(x + 2) - 9x \leq -1$

35. $4x > 1$

36. $6x < 5$

37. $-\frac{2}{3}y \leq 8$

38. $-\frac{3}{4}y \geq 9$

39. $4(2z + 1) < 4$

40. $6(2 - z) \geq 12$

41. $3x - 7 < 6x + 2$

42. $2x - 1 \geq 4x - 5$

43. $5x - 7x \leq x + 2$

44. $4 - x < 8x + 2x$

45. $-6x + 2 \geq 2(5 - x)$

46. $-7x + 4 > 3(4 - x)$

47. $3(x - 5) < 2(2x - 1)$

48. $5(x - 2) \le 3(2x - 1)$

49. $4(3x - 1) \le 5(2x - 4)$

50. $3(5x - 4) \le 4(3x - 2)$

51. $3(x + 2) - 6 > -2(x - 3) + 14$

52. $7(x - 2) + x \le -4(5 - x) - 12$

53. $-5(1 - x) + x \le -(6 - 2x) + 6$

54. $-2(x - 4) - 3x < -(4x + 1) + 2x$

55. $\frac{1}{4}(x + 4) < \frac{1}{5}(2x + 3)$

56. $\frac{1}{2}(x - 5) < \frac{1}{3}(2x - 1)$

57. $-5x + 4 \le -4(x - 1)$

58. $-6x + 2 < -3(x + 4)$

Objective E *Solve the following. See Examples 10 and 11.*

59. Six more than twice a number is greater than negative fourteen. Find all numbers that make this statement true.

60. One more than five times a number is less than or equal to ten. Find all such numbers.

61. The perimeter of a rectangle is to be no greater than 100 centimeters and the width must be 15 centimeters. Find the maximum length of the rectangle.

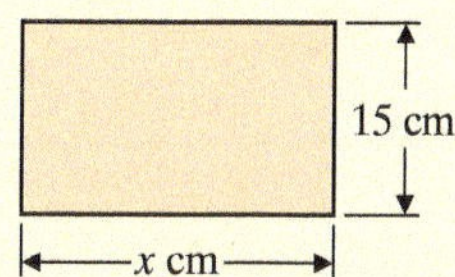

62. One side of a triangle is four times as long as another side, and the third side is 12 inches long. If the perimeter can be no longer than 87 inches, find the maximum lengths of the other two sides.

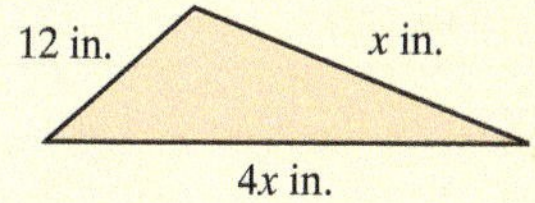

63. Ben Holladay bowled 146 and 201 in his first two games. What must he bowl in his third game to have an average of at least 180? (*Hint:* The average of a list of numbers is their sum divided by the number of numbers in the list.)

64. On an NBA team the two forwards measure 6′8″ and 6′6″ tall and the two guards measure 6′0″ and 5′9″ tall. How tall should the center be if they wish to have a starting team average height of at least 6′5″?

65. Dennis and Nancy Wood are celebrating their 30th wedding anniversary by having a reception at Tiffany Oaks reception hall. They have budgeted $3000 for their reception. If the reception hall charges a $50 cleanup fee plus $34 per person, find the greatest number of people that they may invite and still stay within their budget.

66. A surprise retirement party is being planned for Pratap Puri. A total of $860 has been collected for the event, which is to be held at a local reception hall. This reception hall charges a cleanup fee of $40 and $15 per person for drinks and light snacks. Find the greatest number of people that may be invited and still stay within the $860 budget.

67. A 150-pound person uses 5.8 calories per minute when walking at a speed of 4 mph. How long must a person walk at this speed to use at least 200 calories? Round up to the nearest minute. (*Source: Home & Garden Bulletin* No. 72)

68. A 170-pound person uses 5.3 calories per minute when bicycling at a speed of 5.5 mph. How long must a person ride a bike at this speed in order to use at least 200 calories? Round up to the nearest minute. (*Source:* Same as Exercise **67**)

Review

Evaluate each expression. See Sections 1.7 and 4.3.

69. 3^4 **70.** 4^3 **71.** 1^8 **72.** 0^7 **73.** $\left(\frac{7}{8}\right)^2$ **74.** $\left(\frac{2}{3}\right)^3$

The graph shows the number of recreation visitors to Pecos National Historical Park in New Mexico from 2005 to 2012. Use this graph to answer Exercises 75 through 80. See Section 7.1.

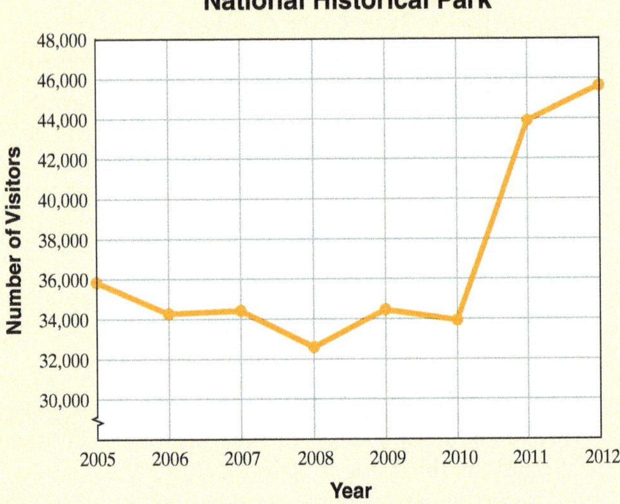

Source: National Park Service

75. How many people visited Pecos National Historical Park in 2006?

76. How many people visited Pecos National Historical Park in 2011?

77. Between which two years did the greatest increase in the number of visitors to Pecos National Historical Park occur?

78. In what year were there approximately 35,800 visitors to Pecos National Historical Park?

79. In which year did the number of visitors to Pecos National Historical Park first fall below 34,000?

80. In which year did the number of visitors to Pecos National Historical Park first rise above 45,000?

Concept Extensions

Fill in the box with $<$, $>$, $\leq$, or $\geq$. See the Concept Check in this section.

81. Since $3 < 5$, then $3(-4) \square 5(-4)$.

82. If $m \leq n$, then $2m \square 2n$.

83. If $m \leq n$, then $-2m \square -2n$.

84. If $-x < y$, then $x \square -y$.

85. When solving an inequality, when must you reverse the direction of the inequality symbol?

86. If both sides of the inequality $-3x < -30$ are divided by 3, do you reverse the direction of the inequality symbol? Why or why not?

Solve.

87. A history major has scores of 75, 83, and 85 on his history tests. Use an inequality to find the scores he can get on his final exam to receive a B in the class. The final exam counts as **two** tests, and a B is received if the final course average is greater than or equal to 80.

88. A mathematics major has scores of 85, 95, and 92 on her college geometry tests. Use an inequality to find the scores she can get on her final exam to receive an A in the course. The final exam counts as **three** tests, and an A is received if the final course average is greater than or equal to 90. Round to one decimal place.

Chapter 9 Group Activity

Investigating Averages

Sections 9.1–9.6

Materials:

- small rubber ball or crumpled paper ball
- bucket or waste can

This activity may be completed by working in groups or individually.

1. Try shooting the ball into the bucket or waste can 5 times. Record your results below.

Shots Made	**Shots Missed**

2. Find your shooting percent for the 5 shots (that is, the percent of the shots you actually made out of the number you tried).
3. Suppose you are going to try an additional 5 shots. How many of the next 5 shots will you have to make to have a 50% shooting percent for all 10 shots? An 80% shooting percent?
4. Did you solve an equation in Question **3**? If so, explain what you did. If not, explain how you could use an equation to find the answers.
5. Suppose instead of trying 5 additional shots, you try 22 additional shots. How many of the next 22 shots will you have to make to have at least a 50% shooting percent for all 27 shots? At least a 70% shooting percent?
6. Choose one of the sports played at your college that is currently in season. How many regular-season games are scheduled? What is the team's current percent of games won?
7. Suppose the team has a goal of finishing the season with a winning percent better than 110% of their current wins. At least how many of the remaining games must they win to achieve their goal?

Chapter 9 Vocabulary Check

Fill in each blank with one of the words or phrases listed below.

no solution	all real numbers	linear equation in one variable
reciprocals	formula	reversed
equivalent inequalities	opposites	
linear inequality in one variable	the same	

1. A(n) ______________ can be written in the form $ax + b = c$.
2. Inequalities that have the same solution are called ______________.
3. An equation that describes a known relationship among quantities is called a(n) ______________.
4. A(n) ______________ can be written in the form $ax + b < c$ (or $>, \leq, \geq$).
5. The solution(s) to the equation $x + 5 = x + 5$ is/are ______________.
6. The solution(s) to the equation $x + 5 = x + 4$ is/are ______________.
7. If both sides of an inequality are multiplied or divided by the same positive number, the direction of the inequality symbol is ______________.

8. If both sides of an inequality are multiplied by the same negative number, the direction of the inequality symbol is ______________.

9. Two numbers whose sum is 0 are called ______________.

10. Two numbers whose product is 1 are called ______________.

Helpful Hint

Are you preparing for your test? Don't forget to take the Chapter 9 Test on page 719. Then check your answers at the back of the text and use the Chapter Test Prep Videos to see the fully worked-out solutions to any of the exercises you want to review.

9 Chapter Highlights

Definitions and Concepts	Examples
Section 9.1 Symbols and Sets of Numbers	
A **set** is a collection of objects, called **elements,** enclosed in braces. A line used to picture numbers is called a **number line.**	$\{a, c, e\}$ 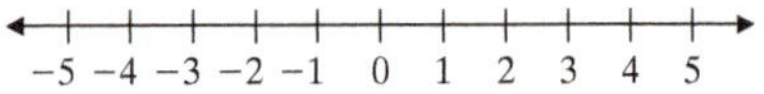
Natural numbers: $\{1, 2, 3, 4, \ldots\}$ **Whole numbers:** $\{0, 1, 2, 3, 4, \ldots\}$ **Integers:** $\{\ldots, -3, -2, -1, 0, 1, 2, 3, \ldots\}$ **Rational numbers:** {real numbers that can be expressed as quotients of integers} **Irrational numbers:** {real numbers that cannot be expressed as quotients of integers} **Real numbers:** {all numbers that correspond to points on a number line}	Given the set $\left\{-3.4, \sqrt{3}, 0, \frac{2}{3}, 5, -4\right\}$, list the numbers that belong to the set of Natural numbers 5 Whole numbers $0, 5$ Integers $-4, 0, 5$ Rational numbers $-4, -3.4, 0, \frac{2}{3}, 5$ Irrational numbers $\sqrt{3}$ Real numbers all numbers in the given set
Symbols: $=$ is equal to $\neq$ is not equal to $>$ is greater than $<$ is less than $\leq$ is less than or equal to $\geq$ is greater than or equal to	$-7 = -7$ $3 \neq -3$ $4 > 1$ $1 < 4$ $6 \leq 6$ $18 \geq -\frac{1}{3}$
Section 9.2 Properties of Real Numbers	
Commutative Properties Addition: $a + b = b + a$ Multiplication: $a \cdot b = b \cdot a$	$3 + (-7) = -7 + 3$ $-8 \cdot 5 = 5 \cdot (-8)$
Associative Properties Addition: $(a + b) + c = a + (b + c)$ Multiplication: $(a \cdot b) \cdot c = a \cdot (b \cdot c)$	$(5 + 10) + 20 = 5 + (10 + 20)$ $(-3 \cdot 2) \cdot 11 = -3 \cdot (2 \cdot 11)$

(continued)

Definitions and Concepts	Examples
Section 9.2 Properties of Real Numbers (*continued*)	
Two numbers whose product is 1 are called **multiplicative inverses** or **reciprocals.** The reciprocal of a nonzero number a is $\frac{1}{a}$ because $a \cdot \frac{1}{a} = 1$.	The reciprocal of 3 is $\frac{1}{3}$. The reciprocal of $-\frac{2}{5}$ is $-\frac{5}{2}$.
Distributive Property $a(b + c) = a \cdot b + a \cdot c$	$5(6 + 10) = 5 \cdot 6 + 5 \cdot 10$ $-2(3 + x) = -2 \cdot 3 + (-2)(x)$
Identities $a + 0 = a \quad 0 + a = a$ $a \cdot 1 = a \quad 1 \cdot a = a$	$5 + 0 = 5 \quad 0 + (-2) = -2$ $-14 \cdot 1 = -14 \quad 1 \cdot 27 = 27$
Inverses Additive or opposite: $a + (-a) = 0$ Multiplicative or reciprocal: $b \cdot \frac{1}{b} = 1, \quad b \neq 0$	$7 + (-7) = 0$ $3 \cdot \frac{1}{3} = 1$
Section 9.3 Further Solving Linear Equations	
To Solve Linear Equations	*Solve:* $\frac{5(-2x + 9)}{6} + 3 = \frac{1}{2}$
1. Clear the equation of fractions.	1. $6 \cdot \frac{5(-2x + 9)}{6} + 6 \cdot 3 = 6 \cdot \frac{1}{2}$
2. Remove any grouping symbols such as parentheses.	2. $5(-2x + 9) + 18 = 3$ Apply the distributive property. $-10x + 45 + 18 = 3$
3. Simplify each side by combining like terms.	3. $-10x + 63 = 3$ Combine like terms.
4. Get all variable terms on one side and all numbers on the other side by using the addition property of equality.	4. $-10x + 63 - 63 = 3 - 63$ Subtract 63. $-10x = -60$
5. Get the variable alone by using the multiplication property of equality.	5. $\frac{-10x}{-10} = \frac{-60}{-10}$ Divide by -10. $x = 6$
6. Check the solution by substituting it into the original equation.	
Section 9.4 Further Problem Solving	
Problem-Solving Steps	The height of the Hudson volcano in Chile is twice the height of the Kiska volcano in the Aleutian Islands. If the sum of their heights is 12,870 feet, find the height of each.
1. UNDERSTAND the problem.	1. Read and reread the problem. Guess a solution and check your guess. Let x be the height of the Kiska volcano. Then $2x$ is the height of the Hudson volcano. 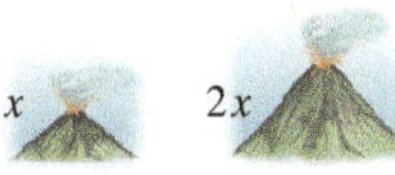

(*continued*)

Definitions and Concepts	Examples
Section 9.4 Further Problem Solving (*continued*)	
2. TRANSLATE the problem.	**2.** height of Kiska → x; added to → $+$; height of Hudson → $2x$; is → $=$; 12,870 → 12,870
3. SOLVE the equation.	**3.** $x + 2x = 12{,}870$ $3x = 12{,}870$ $x = 4290$
4. INTERPRET the results.	**4.** *Check:* If x is 4290, then $2x$ is 2(4290) or 8580. Their sum is 4290 + 8580 or 12,870, the required amount. *State:* The Kiska volcano is 4290 feet tall, and the Hudson volcano is 8580 feet tall.
Section 9.5 Formulas and Problem Solving	
An equation that describes a known relationship among quantities is called a **formula.** **To solve a formula for a specified variable,** use the same steps as for solving a linear equation. Treat the specified variable as the only variable of the equation.	$A = lw$ (area of a rectangle) $I = PRT$ (simple interest) *Solve:* $P = 2l + 2w$ for l $P = 2l + 2w$ $P - 2w = 2l + 2w - 2w$ Subtract $2w$. $P - 2w = 2l$ $\frac{P - 2w}{2} = \frac{2l}{2}$ Divide by 2. $\frac{P - 2w}{2} = l$
Section 9.6 Linear Inequalities and Problem Solving	
Properties of inequalities are similar to properties of equations. However, if you multiply or divide both sides of an inequality by the same *negative* number, you must reverse the direction of the inequality symbol.	*Solve:* $-2x \leq 4$ $\frac{-2x}{-2} \geq \frac{4}{-2}$ Divide by −2; reverse the inequality symbol. $x \geq -2$  In interval notation, the solution set is $[-2, \infty)$.
To Solve Linear Inequalities **1.** Clear the inequality of fractions. **2.** Remove grouping symbols. **3.** Simplify each side by combining like terms. **4.** Write all variable terms on one side and all numbers on the other side using the addition property of inequality. **5.** Get the variable alone by using the multiplication property of inequality.	*Solve:* $3(x + 2) \leq -2 + 8$. Graph the solution set and write it in interval notation. **1.** $3(x + 2) \leq -2 + 8$ No fractions to clear. **2.** $3x + 6 \leq -2 + 8$ Apply the distributive property. **3.** $3x + 6 \leq 6$ Combine like terms. **4.** $3x + 6 - 6 \leq 6 - 6$ Subtract 6. $3x \leq 0$ **5.** $\frac{3x}{3} \leq \frac{0}{3}$ Divide by 3. $x \leq 0$ The solution set is $(-\infty, 0]$. 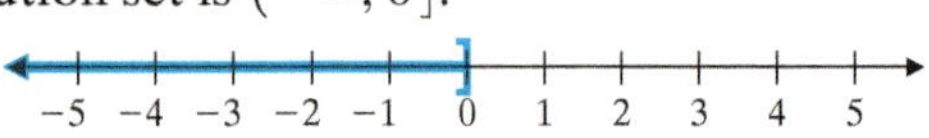

Chapter 9 Review

(9.1) *Insert <, >, or = in the appropriate space to make each statement true. (See Section 2.1 for a review of absolute value.)*

1. $8 \quad 10$ **2.** $7 \quad 2$ **3.** $-4 \quad -5$ **4.** $\frac{12}{2} \quad -8$

5. $|-7| \quad |-8|$ **6.** $|-9| \quad -9$ **7.** $-|-1| \quad -1$ **8.** $|-14| \quad -(-14)$

9. $1.2 \quad 1.02$ **10.** $-\frac{3}{2} \quad -\frac{3}{4}$

Translating *Translate each statement into symbols.*

11. Four is greater than or equal to negative three.

12. Six is not equal to five.

13. 0.03 is less than 0.3.

14. For fiscal year 2012–2013, Massachusetts had a \$10.9 million tourism budget, and Louisiana had a \$13.2 million tourism budget. Write an inequality statement comparing the numbers 10.9 million and 13.2 million. (*Source:* U.S. Travel Association)

Given the sets of numbers below, list the numbers in each set that also belong to the set of:

a. Natural numbers **b.** Whole numbers
c. Integers **d.** Rational numbers
e. Irrational numbers **f.** Real numbers

15. $\left\{-6, 0, 1, 1\frac{1}{2}, 3, \pi, 9.62\right\}$

16. $\left\{-3, -1.6, 2, 5, \frac{11}{2}, 15.1, \sqrt{5}, 2\pi\right\}$

The following chart shows the gains and losses in dollars of Density Oil and Gas stock for a particular week. Use this chart to answer Exercises 17 and 18.

Day	Gain or Loss (in dollars)
Monday	+1
Tuesday	−2
Wednesday	+5
Thursday	+1
Friday	−4

17. Which day showed the greatest loss?

18. Which day showed the greatest gain?

(9.2) *Name the property illustrated in each equation.*

19. $-6 + 5 = 5 + (-6)$ **20.** $6 \cdot 1 = 6$ **21.** $3(8 - 5) = 3 \cdot 8 + 3 \cdot (-5)$

22. $4 + (-4) = 0$

23. $2 + (3 + 9) = (2 + 3) + 9$

24. $2 \cdot 8 = 8 \cdot 2$

25. $6(8 + 5) = 6 \cdot 8 + 6 \cdot 5$

26. $(3 \cdot 8) \cdot 4 = 3 \cdot (8 \cdot 4)$

27. $4 \cdot \frac{1}{4} = 1$

28. $8 + 0 = 8$

(9.3) *Solve each equation.*

29. $\frac{5}{3}x + 4 = \frac{2}{3}x$

30. $\frac{7}{8}x + 1 = \frac{5}{8}x$

31. $-(5x + 1) = -7x + 3$

32. $-4(2x + 1) = -5x + 5$

33. $-6(2x - 5) = -3(9 + 4x)$

34. $3(8y - 1) = 6(5 + 4y)$

35. $\frac{3(2 - z)}{5} = z$

36. $\frac{4(n + 2)}{5} = -n$

37. $0.5(2n - 3) - 0.1 = 0.4(6 + 2n)$

38. $1.72y - 0.04y = 0.42$

39. $\frac{5(c + 1)}{6} = 2c - 3$

40. $-9 - 5a = 3(6a - 1)$

(9.4) *Solve each of the following.*

41. The height of the Washington Monument is 50.5 inches more than 10 times the length of a side of its square base. If the sum of these two dimensions is 7327 inches, find the height of the Washington Monument. (*Source:* National Park Service)

42. A 12-foot board is to be divided into two pieces so that one piece is twice as long as the other. If x represents the length of the shorter piece, find the length of each piece.

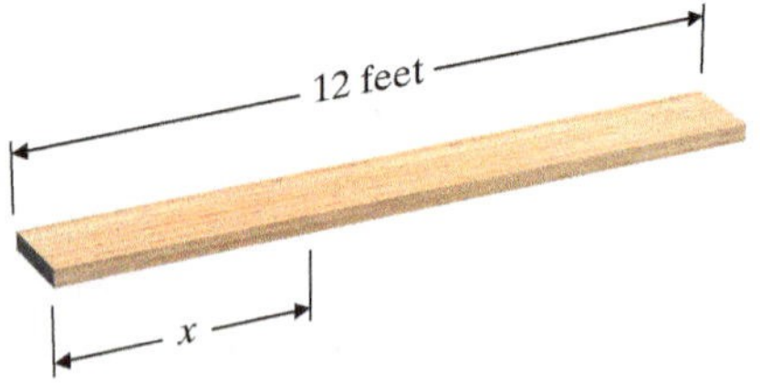

43. The National Park System in the United States includes a variety of park unit types. In 2013, there were a total of 40 national battlefields and national memorials. The number of national memorials is four less than three times the number of national battlefields. How many of each park unit were there? (*Source:* National Park System)

44. Find three consecutive integers whose sum is -114.

45. The quotient of a number and 3 is the same as the difference of the number and 2. Find the number.

46. Double the sum of a number and 6 is the opposite of the number. Find the number.

(9.5) *Substitute the given values into the given formulas and solve for the unknown variable.*

47. $P = 2l + 2w$; $P = 46, l = 14$

48. $V = lwh$; $V = 192, l = 8, w = 6$

Solve each equation for the indicated variable.

49. $y = mx + b$ for m

50. $r = vst - 5$ for s

51. $2y - 5x = 7$ for x

52. $3x - 6y = -2$ for y

△ **53.** $C = \pi d$ for d

△ **54.** $C = 2\pi r$ for r

△ **55.** A swimming pool holds 900 cubic meters of water. If its length is 20 meters and its height is 3 meters, find its width.

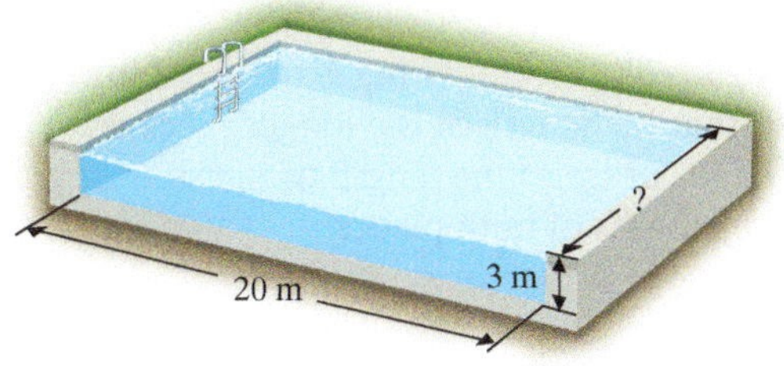

56. The perimeter of a rectangular billboard is 60 feet, and it has a length 6 feet longer than its width. Find the dimensions of the billboard.

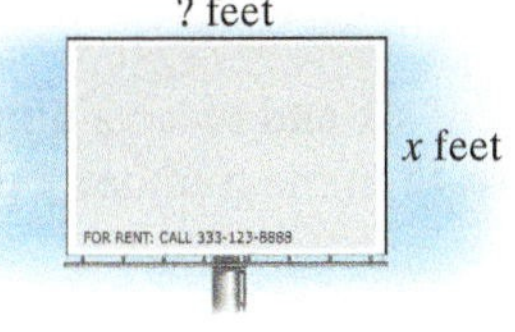

57. A charity 10K race is given annually to benefit a local hospice organization. How long would it take to run/walk a 10K race (10 kilometers or 10,000 meters) if your average pace is 125 **meters** per minute? Give your time in hours and minutes.

58. On September 14, 2013, the highest temperature recorded in the United States was 113°F, which occurred in Death Valley, California. Convert this temperature to degrees Celsius. (*Source:* National Weather Service)

(9.6) *Graph on the given number line and write the solution set in interval notation.*

59. $x \leq -2$

−5 −4 −3 −2 −1 0 1 2 3 4 5

60. $0 < x \leq 5$

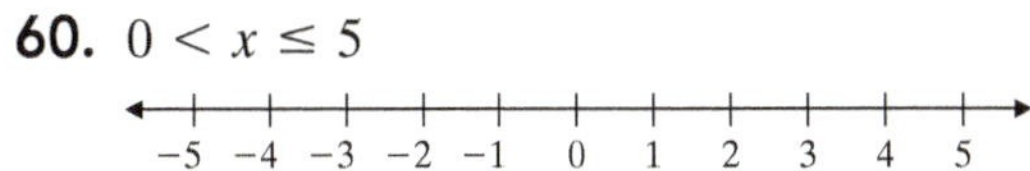

Solve each inequality. Write the solution set in interval notation.

61. $x - 5 \leq -4$

62. $x + 7 > 2$

63. $-2x \geq -20$

64. $-3x > 12$

65. $5x - 7 > 8x + 5$

66. $x + 4 \geq 6x - 16$

67. $\frac{2}{3}y > 6$

68. $-0.5y \le 7.5$

69. $-2(x - 5) > 2(3x - 2)$

70. $4(2x - 5) \le 5x - 1$

71. A part-time salesperson earns $175 per week plus a 5% commission on all her sales. Find the minimum amount of sales she must make to ensure that she earns at least $300 per week.

72. An amateur golfer shot rounds of 76, 82, and 79 golfing. What must he shoot on his next round so that his average will be below 80?

Mixed Review

Solve each equation.

73. $6x + 2x - 1 = 5x + 11$

74. $2(3y - 4) = 6 + 7y$

75. $4(3 - a) - (6a + 9) = -12a$

76. $\frac{x}{3} - 2 = 5$

77. $2(y + 5) = 2y + 10$

78. $7x - 3x + 2 = 2(2x - 1)$

Solve.

79. The sum of six and twice a number is equal to seven less than the number. Find the number.

80. A 23-inch piece of string is to be cut into two pieces so that the length of the longer piece is three more than four times the length of the shorter piece. Find the lengths of both pieces.

Solve for the specified variable.

81. $V = \frac{1}{3}Ah$ for h

Solve each inequality. Graph the solution set and write it in interval notation.

82. $4x - 7 > 3x + 2$

−10 −8 −6 −4 −2 0 2 4 6 8 10

83. $-5x < 20$

−5 −4 −3 −2 −1 0 1 2 3 4 5

84. $-3(1 + 2x) + x \ge -(3 - x)$

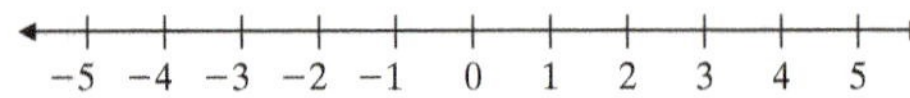

Step-by-step test solutions are found on the Chapter Test Prep Videos. Where available: MyMathLab® or YouTube

Chapter 9 Test

Answers

Translate each statement into symbols.

1. The absolute value of negative seven is greater than five.

2. The sum of nine and five is greater than or equal to four.

3. Given

$\{-5, -1, \frac{1}{4}, 0, 1, 7, 11.6, \sqrt{7}, 3\pi\}$, list the numbers in this set that also belong to the set of:

a. Natural numbers **b.** Whole numbers
c. Integers **d.** Rational numbers
e. Irrational numbers **f.** Real numbers

Identify the property illustrated by each expression.

4. $8 + (9 + 3) = (8 + 9) + 3$

5. $6 \cdot 8 = 8 \cdot 6$

6. $-6(2 + 4) = -6 \cdot 2 + (-6) \cdot 4$

7. $\frac{1}{6}(6) = 1$

8. Find the opposite of -9.

9. Find the reciprocal of $-\frac{1}{3}$.

Use the distributive property to write each expression without parentheses. Then simplify if possible.

10. $7 + 2(5y - 3)$

11. $4(x - 2) - 3(2x - 6)$

Solve each equation.

12. $4(n - 5) = -(4 - 2n)$

13. $-2(x - 3) = x + 5 - 3x$

14. $4z + 1 - z = 1 + z$

15. $\frac{2(x + 6)}{3} = x - 5$

16. $\frac{1}{2} - x + \frac{3}{2} = x - 4$

17. $-0.3(x - 4) + x = 0.5(3 - x)$

18. $-4(a + 1) - 3a = -7(2a - 3)$

1. ______
2. ______
3. a. ______
b. ______
c. ______
d. ______
e. ______
f. ______
4. ______
5. ______
6. ______
7. ______
8. ______
9. ______
10. ______
11. ______
12. ______
13. ______
14. ______
15. ______
16. ______
17. ______
18. ______

19. ______

20. ______

21. ______

22. ______

23. ______

24. ______

25. ______

26. ______

27. ______

28. ______

Solve each application.

19. A number increased by two-thirds of the number is 35. Find the number.

△ **20.** A gallon of water seal covers 200 square feet. How many gallons are needed to apply two coats of water seal on a deck that measures 20 feet by 35 feet?

21. Find the value of x if $y = -14$, $m = -2$, and $b = -2$ in the formula $y = mx + b$.

Solve the equation for the indicated variable.

22. $V = \pi r^2 h$ for h

23. $3x - 4y = 10$ for y

Solve the inequality. Graph the solution set and write it in interval notation.

24. $3x - 5 > 7x + 3$

−5 −4 −3 −2 −1 0 1 2 3 4 5

Solve each inequality. Write each answer using interval notation.

25. $-0.3x \geq 2.4$

26. $-5(x - 1) + 6 \leq -3(x + 4) + 1$

27. $\dfrac{2(5x + 1)}{3} > 2$

28. California has more public libraries than any other state. It has 387 more public libraries than Ohio. If the total number of public libraries for these states is 1827, find the number of public libraries in California and the number in Ohio. (*Source:* Institute of Museum and Library Services)

Cumulative Review Chapters 1–9

Simplify.

1. a. $-(-4)$ **b.** $-|-5|$ **c.** $-|6|$

2. Insert $<$, $>$, or $=$ in the appropriate space to make each statement true.

a. $|0| \quad 2$ **b.** $|-5| \quad 5$
c. $|-3| \quad |-2|$ **d.** $|5| \quad |6|$
e. $|-7| \quad |6|$

3. Simplify: $-2[-3 + 2(-1 + 6)] - 5$

4. Simplify: $\dfrac{3 + |4 - 3| + 2^2}{6 - 4}$

Add.

5. $-18 + 10$ **6.** $(-8) + (-11)$ **7.** $12 + (-8)$

8. $(-2) + 10$ **9.** $(-3) + 4 + (-11)$ **10.** $0.2 + (-0.5)$

Perform the indicated operation.

11. $\dfrac{2}{3} - \dfrac{10}{11}$ **12.** $\dfrac{2}{5} - \dfrac{39}{40}$ **13.** $2 - \dfrac{x}{3}$ **14.** $5 + \dfrac{x}{2}$

Evaluate. Let $x = -\dfrac{1}{2}$ *and* $y = \dfrac{1}{3}$.

15. $2x + y^2$ **16.** $2y + x^2$

Perform the indicated operation.

17. $-4\frac{2}{5} \cdot 1\frac{3}{11}$ **18.** $-2\frac{1}{2} \cdot \left(-2\frac{1}{2}\right)$ **19.** $-2\frac{1}{3} \div \left(-2\frac{1}{2}\right)$

20. $3\frac{3}{5} \div \left(-3\frac{1}{3}\right)$ **21.** $(-1.3)^2 + 2.4$ **22.** $1.2(7.3 - 9.3)$

Answers

1. a. ______
b. ______
c. ______
2. a. ______
b. ______
c. ______
d. ______
e. ______
3. ______
4. ______
5. ______
6. ______
7. ______
8. ______
9. ______
10. ______
11. ______
12. ______
13. ______
14. ______
15. ______
16. ______
17. ______
18. ______
19. ______
20. ______
21. ______
22. ______

23. ______
24. ______
25. ______
26. ______
27. ______
28. ______
29. ______
30. ______
31. ______
32. ______
33. ______
34. ______
35. ______
36. ______
37. ______
38. ______
39. ______
40. ______
41. ______
42. ______
43. ______
44. ______
45. ______
46. ______
47. ______
48. ______
49. ______
50. ______

23. $\sqrt{49}$ **24.** $\sqrt{64}$ **25.** $\sqrt{\frac{4}{25}}$ **26.** $\sqrt{\frac{9}{100}}$

Solve.

27. $1.2x + 5.8 = 8.2$ **28.** $1.3x - 2.6 = -9.1$

Determine whether each statement is true or false.

29. $8 \geq 8$ **30.** $-4 \leq -4$ **31.** $8 \leq 8$ **32.** $-4 \geq -4$

33. $23 \leq 0$ **34.** $-8 \leq 0$ **35.** $0 \leq 23$ **36.** $0 \leq -8$

Use the distributive property to write each expression without parentheses. Then simplify the result.

37. $-5(-3 + 2z)$ **38.** $-4(2x - 1)$ **39.** $\frac{1}{2}(6x + 14) + 10$ **40.** $9 + 2(5x + 6)$

Solve.

41. $\frac{2(a + 3)}{3} = 6a + 2$ **42.** $\frac{x}{2} + \frac{x}{5} = 3$

43. As of July 2013, the total number of Democrats and Republicans in the U.S. House of Representatives was 440. There were 28 more Republican representatives than Democratic. Find the number of representatives from each party.

44. Three times the sum of a number and 2 is the same as 9 times the number. Find the number.

45. A glacier is a giant mass of rocks and ice that flows downhill like a river. Portage Glacier in Alaska is about 6 miles, or 31,680 feet, long and moves 400 feet per year. Icebergs are created when the front end of the glacier flows into Portage Lake. How long does it take for ice at the head (beginning) of the glacier to reach the lake?

46. Solve $V = lwh$ for w.

47. Graph $-1 > x$ and write the solution set in interval notation.

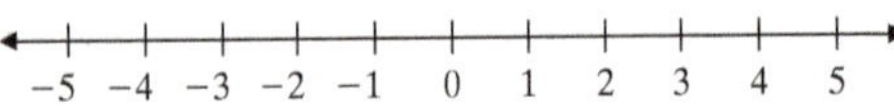

For Exercises 48–50, write the solution set in interval notation.

48. Solve: $-3x < -30$

49. Solve: $2(x - 3) - 5 \leq 3(x + 2) - 18$

50. Solve: $10 + x < 6x - 10$

10

Graphing

Tourism 2020 Vision is the World Tourism Organization's long-term forecast of world tourism through 2020. The broken-line graph below shows the forecasts. In Section 10.1, Exercises 1 through 6, we read a bar graph showing the top tourist destinations by country.

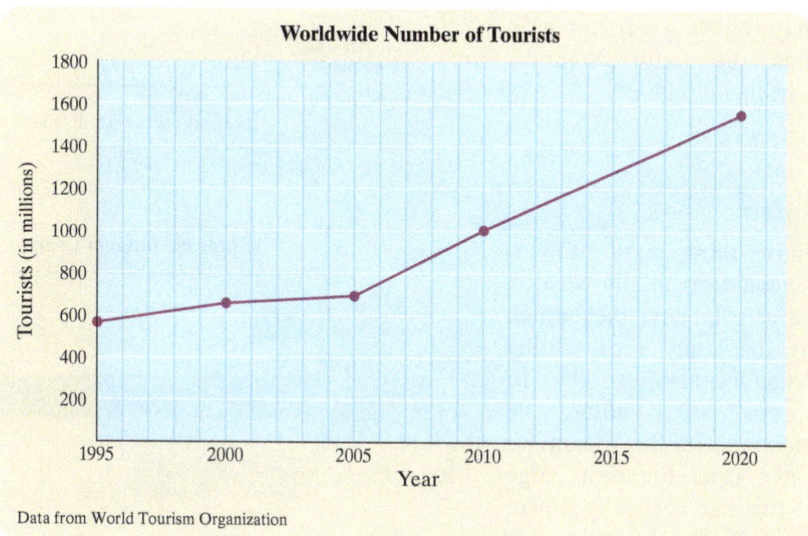

Sections

Check Your Progress

In the previous chapter, we learned to solve and graph the solutions of linear equations and inequalities in one variable. Now we define and present techniques for solving and graphing linear equations and inequalities in two variables.

10.1 Reading Graphs and the Rectangular Coordinate System

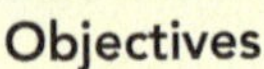

Objectives

1 Read Bar and Line Graphs.

2 Define the Rectangular Coordinate System and Plot Ordered Pairs of Numbers.

3 Graph Paired Data to Create a Scatter Diagram.

4 Determine Whether an Ordered Pair Is a Solution of an Equation in Two Variables.

5 Find the Missing Coordinate of an Ordered Pair Solution, Given One Coordinate of the Pair.

In today's world, where the exchange of information must be fast and entertaining, graphs are becoming increasingly popular. They provide a quick way of making comparisons, drawing conclusions, and approximating quantities.

Objective 1 Reading Bar and Line Graphs 

A **bar graph** consists of a series of bars arranged vertically or horizontally. The bar graph in Example 1 shows a comparison of worldwide Internet users by region. The names of the regions are listed vertically and a bar is shown for each region. Corresponding to the length of the bar for each region is a number along a horizontal axis. These horizontal numbers are numbers of Internet users in millions.

Example 1 The following bar graph shows the estimated number of Internet users worldwide by region as of a recent year.

a. Find the region that has the most Internet users and approximate the number of users.

b. How many more users are in the North America region than the Latin America/Caribbean region?

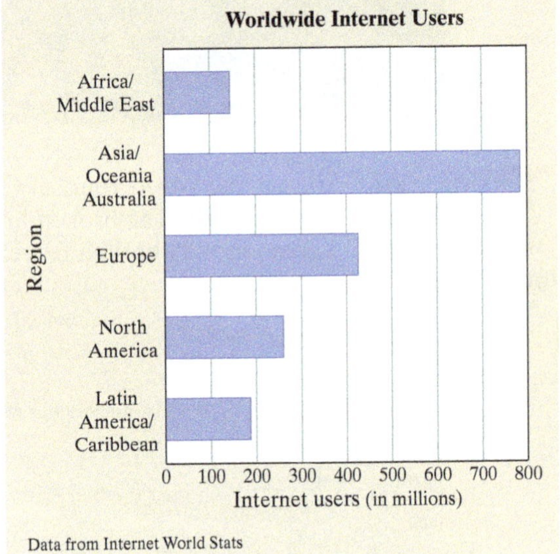

Solution:

a. Since these bars are arranged horizontally, we look for the longest bar, which is the bar representing Asia/Oceania/Australia. To approximate the number associated with this region, we move from the right edge of this bar vertically downward to the Internet user axis. This region has approximately 785 million Internet users.

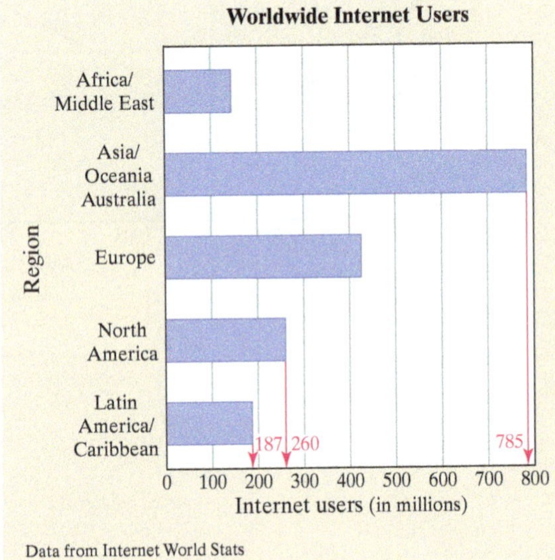

b. The North America region has approximately 260 million Internet users. The Latin America/Caribbean region has approximately 187 million Internet users. To find how many more users are in the North America region, we subtract $260 - 187 = 73$ million more Internet users.

Work Practice 1

Practice 1

Use the graph from Example 1 to answer the following.

a. Find the region with the fewest Internet users and approximate the number of users.

b. How many more users are in the Asia/Oceania/Australia region than in the Africa/Middle East region?

Answers

1. a. Africa/Middle East region; 145 million Internet users **b.** 640 million more Internet users

A **line graph** consists of a series of points connected by a line. The next graph is an example of a line graph. It is also sometimes called a **broken-line graph.**

Example 2 The line graph shows the relationship between time spent smoking a cigarette and pulse rate. Time is recorded along the horizontal axis in minutes, with 0 minutes being the moment a smoker lights a cigarette. Pulse is recorded along the vertical axis in heartbeats per minute.

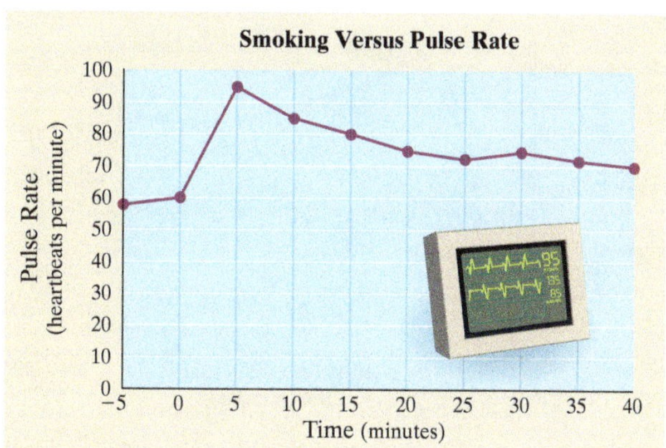

a. What is the pulse rate 15 minutes after a cigarette is lit?

b. When is the pulse rate the lowest?

c. When does the pulse rate show the greatest change?

Solution:

a. We locate the number 15 along the time axis and move vertically upward until the line is reached. From this point on the line, we move horizontally to the left until the pulse rate axis is reached. Reading the number of beats per minute, we find that the pulse rate is 80 beats per minute 15 minutes after a cigarette is lit.

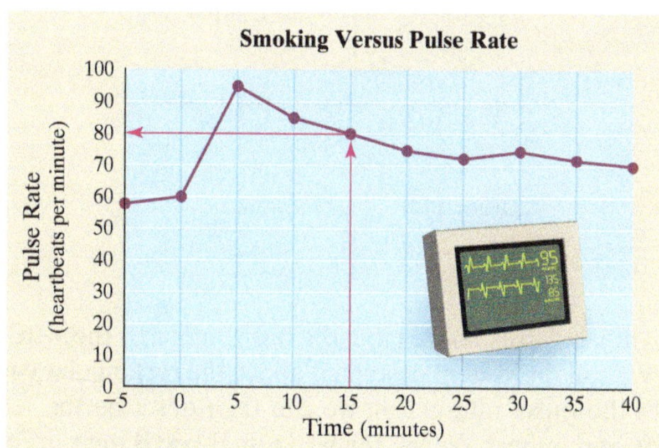

b. We find the lowest point of the line graph, which represents the lowest pulse rate. From this point, we move vertically downward to the time axis. We find that the pulse rate is the lowest at −5 minutes, which means 5 minutes *before* lighting a cigarette.

c. The pulse rate shows the greatest change during the 5 minutes between 0 and 5. Notice that the line graph is *steepest* between 0 and 5 minutes.

Work Practice 2

Practice 2

Use the graph from Example 2 to answer the following.

a. What is the pulse rate 40 minutes after lighting a cigarette?

b. What is the pulse rate when the cigarette is being lit?

c. When is the pulse rate the highest?

Answers

2. a. 70 beats per minute **b.** 60 beats per minute **c.** 5 minutes after lighting

Objective 2 Defining the Rectangular Coordinate System and Plotting Ordered Pairs of Numbers

Notice in the previous graph that two numbers are associated with each point of the graph. For example, we discussed earlier that 15 minutes after lighting a cigarette, the pulse rate is 80 beats per minute. If we agree to write the time first and the pulse rate second, we can say there is a point on the graph corresponding to the **ordered pair** of numbers $(15, 80)$. A few more ordered pairs are listed alongside their corresponding points.

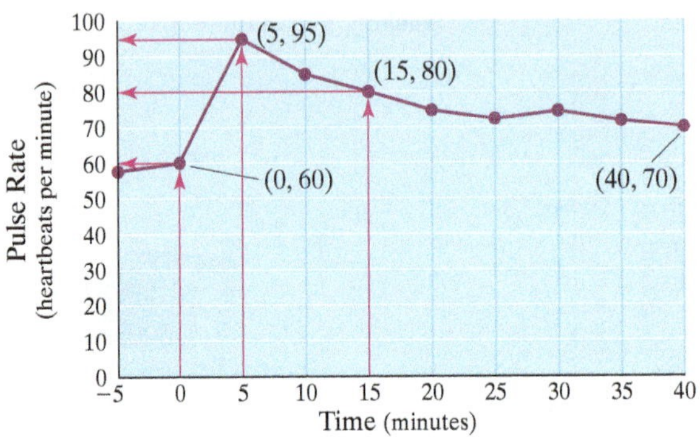

In general, we use this same ordered pair idea to describe the location of a point in a plane (such as a piece of paper). We start with a horizontal and a vertical axis. Each axis is a number line and, for the sake of consistency, we construct our axes to intersect at the 0 coordinate of both. This point of intersection is called the **origin.** Notice that these two number lines or axes divide the plane into four regions called **quadrants.** The quadrants are usually numbered with Roman numerals as shown. The axes are not considered to be in any quadrant.

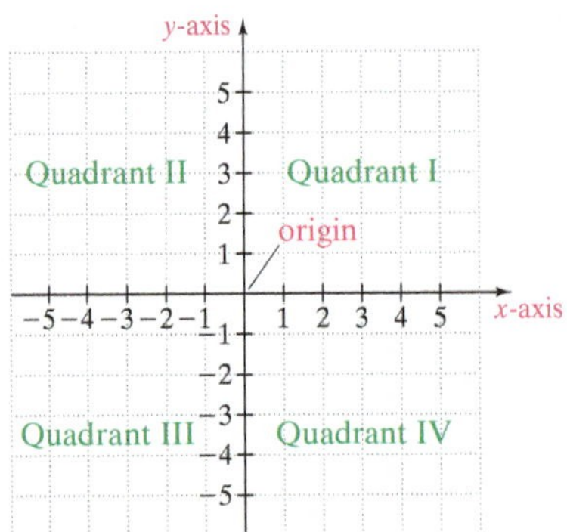

It is helpful to label axes, so we label the horizontal axis the ***x*-axis** and the vertical axis the ***y*-axis.** We call the system described above the **rectangular coordinate system.**

Just as with the pulse rate graph, we can then describe the locations of points by ordered pairs of numbers. We list the horizontal ***x*-axis** measurement first and the vertical ***y*-axis** measurement second.

To plot or graph the point corresponding to the ordered pair

$$(a, b)$$

we start at the origin. We then move a units left or right (right if a is positive, left if a is negative). From there, we move b units up or down (up if b is positive, down if b is negative). For example, to plot the point corresponding to the ordered pair $(3, 2)$, we start at the origin, move 3 units right, and from there move 2 units up. (See the figure to the left.) The x-value, 3, is called the ***x*-coordinate** and the y-value, 2, is called the ***y*-coordinate.** From now on, we will call the point with coordinates $(3, 2)$ simply the point $(3, 2)$. The point $(-2, 5)$ is graphed to the left also.

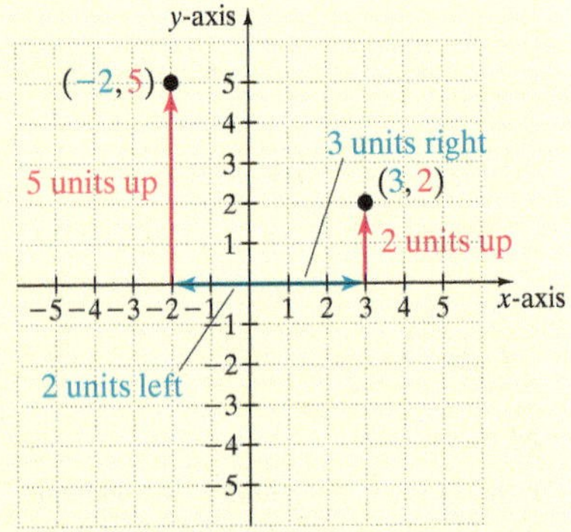

Does the order in which the coordinates are listed matter? Yes! Notice below that the point corresponding to the ordered pair $(2, 3)$ is in a different location than the point corresponding to $(3, 2)$. These two ordered pairs of numbers describe two different points of the plane.

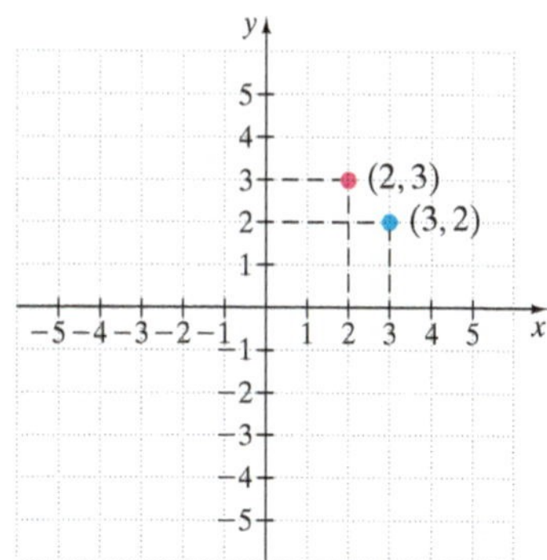

✓**Concept Check** Is the graph of the point $(-5, 1)$ in the same location as the graph of the point $(1, -5)$? Explain.

Helpful Hint

Don't forget that **each ordered pair corresponds to exactly one point in the plane and that each point in the plane corresponds to exactly one ordered pair.**

Example 3 On a single coordinate system, plot each ordered pair. State in which quadrant, if any, each point lies.

a. $(5, 3)$ **b.** $(-5, 3)$ **c.** $(-2, -4)$ **d.** $(1, -2)$ **e.** $(0, 0)$

f. $(0, 2)$ **g.** $(-5, 0)$ **h.** $\left(0, -5\frac{1}{2}\right)$ **i.** $\left(4\frac{2}{3}, -3\right)$

Solution:

a. Point $(5, 3)$ lies in quadrant I.
b. Point $(-5, 3)$ lies in quadrant II.
c. Point $(-2, -4)$ lies in quadrant III.
d. Point $(1, -2)$ lies in quadrant IV.
e.–h. Points $(0, 0)$, $(0, 2)$, $(-5, 0)$, and $\left(0, -5\frac{1}{2}\right)$ lie on axes, so they are not in any quadrant.
i. Point $\left(4\frac{2}{3}, -3\right)$ lies in quadrant IV.

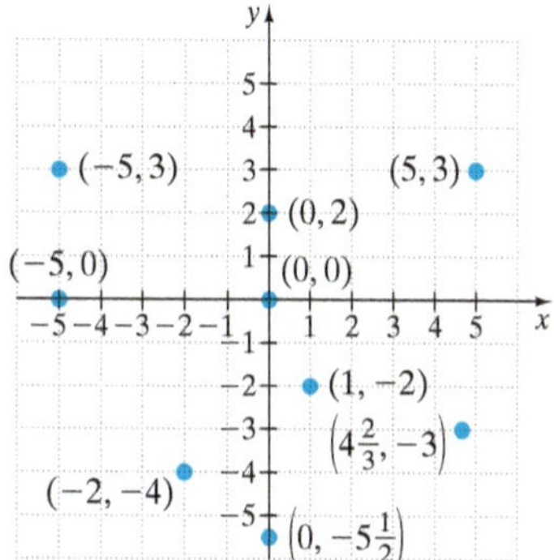

Work Practice 3

From Example 3, notice that the y-coordinate of any point on the x-axis is 0. For example, the point $(-5, 0)$ lies on the x-axis. Also, the x-coordinate of any point on the y-axis is 0. For example, the point $(0, 2)$ lies on the y-axis.

Practice 3

On a single coordinate system, plot each ordered pair. State in which quadrant, if any, each point lies.

a. $(4, -3)$ **b.** $(-3, 5)$
c. $(0, 4)$ **d.** $(-6, 1)$
e. $(-2, 0)$ **f.** $(5, 5)$
g. $\left(3\frac{1}{2}, 1\frac{1}{2}\right)$ **h.** $(-4, -5)$

Answers

3.

$(5, 5)$ and $\left(3\frac{1}{2}, 1\frac{1}{2}\right)$ are in quadrant I; $(-3, 5)$ and $(-6, 1)$ are in quadrant II; $(-4, -5)$ is in quadrant III; $(4, -3)$ is in quadrant IV; $(0, 4)$ is on the y-axis; $(-2, 0)$ is on the x-axis

✓**Concept Check Answer**

The graph of point $(-5, 1)$ lies in quadrant II and the graph of point $(1, -5)$ lies in quadrant IV. They are *not* in the same location.

✓Concept Check For each description of a point in the rectangular coordinate system, write an ordered pair that represents it.

a. Point A is located three units to the left of the y-axis and five units above the x-axis.

b. Point B is located six units below the origin.

Objective 3 Graphing Paired Data

Data that can be represented as an ordered pair is called **paired data.** Many types of data collected from the real world are paired data. For instance, the annual measurement of a child's height can be written as an ordered pair of the form (year, height in inches) and is paired data. The graph of paired data as points in the rectangular coordinate system is called a **scatter diagram.** Scatter diagrams can be used to look for patterns and trends in paired data.

Example 4 The table gives the annual net sales (in billions of dollars) for Target stores for the years shown. (Data from www.retailsales.com)

Year	Target Net Sales (in billions of dollars)
2006	58
2007	61
2008	63
2009	63
2010	66

a. Write this paired data as a set of ordered pairs of the form (year, sales in billions of dollars).

b. Create a scatter diagram of the paired data.

c. What trend in the paired data does the scatter diagram show?

Solution:

a. The ordered pairs are $(2006, 58)$, $(2007, 61)$, $(2008, 63)$, $(2009, 63)$, $(2010, 66)$.

b. We begin by plotting the ordered pairs. Because the x-coordinate in each ordered pair is a year, we label the x-axis "Year" and mark the horizontal axis with the years given. Then we label the y-axis or vertical axis "Net Sales (in billions of dollars)." In this case we can mark the vertical axis in multiples of 2. Since no net sale is less than 58, we use the notation ⚡ to skip to 58, then proceed by multiples of 2.

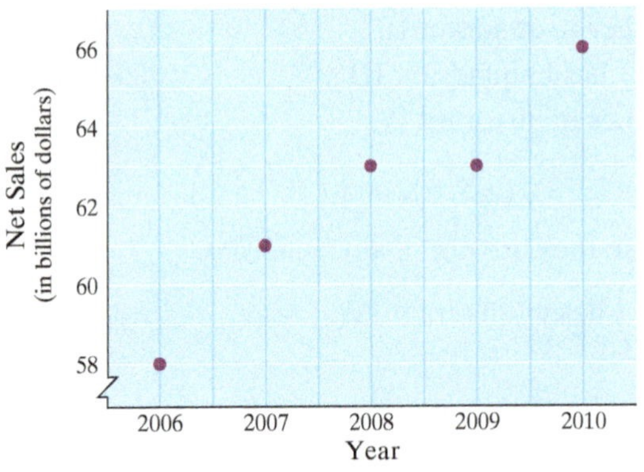

c. The scatter diagram shows that Target net sales were constant or steadily increased over the years 2006–2010.

Work Practice 4

Practice 4

The table gives the approximate annual number of wildfires (in thousands) that have occurred in the United States for the years shown. (Data from National Interagency Fire Center)

Year	Wildfires (in thousands)
2004	65
2005	67
2006	96
2007	86
2008	79
2009	79
2010	72

a. Write this paired data as a set of ordered pairs of the form (year, number of wildfires in thousands).

b. Create a scatter diagram of the paired data.

Answers

4. a. $(2004, 65), (2005, 67), (2006, 96), (2007, 86), (2008, 79), (2009, 79), (2010, 72)$

b.

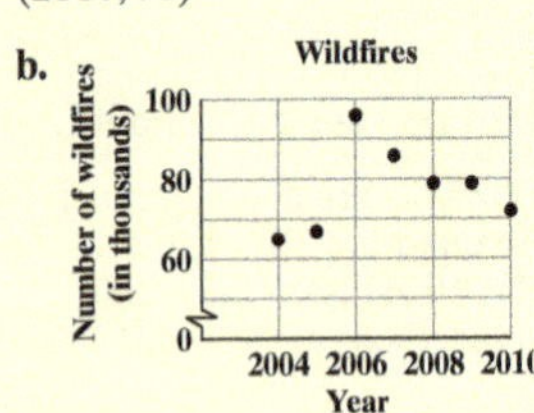

✓Concept Check Answers

a. $(-3, 5)$ **b.** $(0, -6)$

Objective 4 Determining Whether an Ordered Pair Is a Solution

Let's see how we can use ordered pairs to record solutions of equations containing two variables. An equation in one variable such as $x + 1 = 5$ has one solution, which is 4: The number 4 is the value of the variable x that makes the equation true.

An equation in two variables, such as $2x + y = 8$, has solutions consisting of two values, one for x and one for y. For example, $x = 3$ and $y = 2$ is a solution of $2x + y = 8$ because, if x is replaced with 3 and y with 2, we get a true statement.

$$2x + y = 8$$
$$2(3) + 2 = 8$$
$$8 = 8 \quad \text{True}$$

The solution $x = 3$ and $y = 2$ can be written as $(3, 2)$, an **ordered pair** of numbers. The first number, 3, is the x-value and the second number, 2, is the y-value.

In general, an ordered pair is a **solution** of an equation in two variables if replacing the variables by the values of the ordered pair results in a true statement.

Example 5

Determine whether each ordered pair is a solution of the equation $x - 2y = 6$.

a. $(6, 0)$ **b.** $(0, 3)$ **c.** $\left(1, -\frac{5}{2}\right)$

Solution:

a. Let $x = 6$ and $y = 0$ in the equation $x - 2y = 6$.

$$x - 2y = 6$$
$$6 - 2(0) = 6 \quad \text{Replace } x \text{ with 6 and } y \text{ with 0.}$$
$$6 - 0 = 6 \quad \text{Simplify.}$$
$$6 = 6 \quad \text{True}$$

$(6, 0)$ is a solution, since $6 = 6$ is a true statement.

b. Let $x = 0$ and $y = 3$.

$$x - 2y = 6$$
$$0 - 2(3) = 6 \quad \text{Replace } x \text{ with 0 and } y \text{ with 3.}$$
$$0 - 6 = 6$$
$$-6 = 6 \quad \text{False}$$

$(0, 3)$ is *not* a solution, since $-6 = 6$ is a false statement.

c. Let $x = 1$ and $y = -\frac{5}{2}$ in the equation.

$$x - 2y = 6$$
$$1 - 2\left(-\frac{5}{2}\right) = 6 \quad \text{Replace } x \text{ with 1 and } y \text{ with } -\frac{5}{2}.$$
$$1 + 5 = 6$$
$$6 = 6 \quad \text{True}$$

$\left(1, -\frac{5}{2}\right)$ is a solution, since $6 = 6$ is a true statement.

Work Practice 5

Practice 5

Determine whether each ordered pair is a solution of the equation $x + 3y = 6$.

a. $(3, 1)$

b. $(6, 0)$

c. $\left(-2, \frac{2}{3}\right)$

Answers

5. a. yes **b.** yes **c.** no

Objective 5 Completing Ordered Pair Solutions

If one value of an ordered pair solution of an equation is known, the other value can be determined. To find the unknown value, replace one variable in the equation by its known value. Doing so results in an equation with just one variable that can be solved for the variable using the methods of Chapter 9.

Practice 6

Complete the following ordered pair solutions for the equation $2x - y = 8$.

a. $(0, \ )$
b. $(\ , 4)$
c. $(-3, \ )$

Example 6 Complete the following ordered pair solutions for the equation $3x + y = 12$.

a. $(0, \ )$ **b.** $(\ , 6)$ **c.** $(-1, \ )$

Solution:

a. In the ordered pair $(0, \ )$, the x-value is 0. Let $x = 0$ in the equation and solve for y.

$$3x + y = 12$$
$$3(0) + y = 12 \quad \text{Replace } x \text{ with } 0.$$
$$0 + y = 12$$
$$y = 12$$

The completed ordered pair is $(0, 12)$.

b. In the ordered pair $(\ , 6)$, the y-value is 6. Let $y = 6$ in the equation and solve for x.

$$3x + y = 12$$
$$3x + 6 = 12 \quad \text{Replace } y \text{ with } 6.$$
$$3x = 6 \quad \text{Subtract 6 from both sides.}$$
$$x = 2 \quad \text{Divide both sides by 3.}$$

The ordered pair is $(2, 6)$.

c. In the ordered pair $(-1, \ )$, the x-value is -1. Let $x = -1$ in the equation and solve for y.

$$3x + y = 12$$
$$3(-1) + y = 12 \quad \text{Replace } x \text{ with } -1.$$
$$-3 + y = 12$$
$$y = 15 \quad \text{Add 3 to both sides.}$$

The ordered pair is $(-1, 15)$.

Work Practice 6

Solutions of equations in two variables can also be recorded in a **table of values,** as shown in the next example.

Practice 7

Complete the table for the equation $y = -4x$.

	x	y
a.	-2	
b.		-12
c.	0	

Example 7 Complete the table for the equation $y = 3x$.

	x	y
a.	-1	
b.		0
c.		-9

Solution:

a. Replace x with -1 in the equation and solve for y.

$$y = 3x$$
$$y = 3(-1) \quad \text{Let } x = -1.$$
$$y = -3$$

The ordered pair is $(-1, -3)$.

Answers

6. a. $(0, -8)$ **b.** $(6, 4)$ **c.** $(-3, -14)$

7.

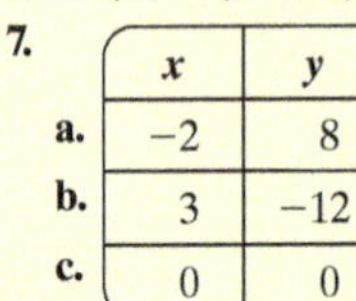

	x	y
a.	-2	8
b.	3	-12
c.	0	0

b. Replace y with 0 in the equation and solve for x.

$$y = 3x$$
$$0 = 3x \quad \text{Let } y = 0.$$
$$0 = x \quad \text{Divide both sides by 3.}$$

The ordered pair is $(0, 0)$.

c. Replace y with -9 in the equation and solve for x.

$$y = 3x$$
$$-9 = 3x \quad \text{Let } y = -9.$$
$$-3 = x \quad \text{Divide both sides by 3.}$$

The ordered pair is $(-3, -9)$. The completed table is shown to the right.

x	y
−1	−3
0	0
−3	−9

Work Practice 7

Example 8 Complete the table for the equation $y = \frac{1}{2}x - 5$.

	x	y
a.	−2	
b.	0	
c.		0

Solution:

a. Let $x = -2$.

$$y = \frac{1}{2}x - 5$$
$$y = \frac{1}{2}(-2) - 5$$
$$y = -1 - 5$$
$$y = -6$$

b. Let $x = 0$.

$$y = \frac{1}{2}x - 5$$
$$y = \frac{1}{2}(0) - 5$$
$$y = 0 - 5$$
$$y = -5$$

c. Let $y = 0$.

$$y = \frac{1}{2}x - 5$$
$$0 = \frac{1}{2}x - 5 \quad \text{Now, solve for } x.$$
$$5 = \frac{1}{2}x \quad \text{Add 5.}$$
$$10 = x \quad \text{Multiply by 2.}$$

Ordered Pairs: $(-2, -6)$ $(0, -5)$ $(10, 0)$

The completed table is

x	y
−2	−6
0	−5
10	0

Work Practice 8

Practice 8

Complete the table for the equation $y = \frac{1}{5}x - 2$.

	x	y
a.	−10	
b.	0	
c.		0

Example 9 Finding the Value of a Computer

A computer was recently purchased for a small business for \$2000. The business manager predicts that the computer will be used for 5 years and the value in dollars y of the computer in x years is $y = -300x + 2000$. Complete the table.

x	0	1	2	3	4	5
y						

Solution:

To find the value of y when x is 0, replace x with 0 in the equation. We use this same procedure to find y when x is 1 and when x is 2.

(*Continued on next page*)

Practice 9

A college student purchased a used car for \$12,000. The student predicted that she would need to use the car for four years and the value in dollars y of the car in x years is $y = -1800x + 12{,}000$. Complete this table.

x	0	1	2	3	4
y					

Answers

8.

	x	y
a.	−10	−4
b.	0	−2
c.	10	0

9.

x	0	1	2	3	4
y	12,000	10,200	8400	6600	4800

When $x = 0$,	**When $x = 1$,**	**When $x = 2$,**
$y = -300x + 2000$	$y = -300x + 2000$	$y = -300x + 2000$
$y = -300 \cdot 0 + 2000$	$y = -300 \cdot 1 + 2000$	$y = -300 \cdot 2 + 2000$
$y = 0 + 2000$	$y = -300 + 2000$	$y = -600 + 2000$
$y = 2000$	$y = 1700$	$y = 1400$

We have the ordered pairs (0, 2000), (1, 1700), and (2, 1400). This means that in 0 years, the value of the computer is \$2000, in 1 year the value of the computer is \$1700, and in 2 years the value is \$1400. To complete the table of values, we continue the procedure for $x = 3$, $x = 4$, and $x = 5$.

When $x = 3$,	**When $x = 4$,**	**When $x = 5$,**
$y = -300x + 2000$	$y = -300x + 2000$	$y = -300x + 2000$
$y = -300 \cdot 3 + 2000$	$y = -300 \cdot 4 + 2000$	$y = -300 \cdot 5 + 2000$
$y = -900 + 2000$	$y = -1200 + 2000$	$y = -1500 + 2000$
$y = 1100$	$y = 800$	$y = 500$

The completed table is

x	0	1	2	3	4	5
y	2000	1700	1400	1100	800	500

Work Practice 9

The ordered pair solutions recorded in the completed table for the example above are graphed below. Notice that the graph gives a visual picture of the decrease in value of the computer.

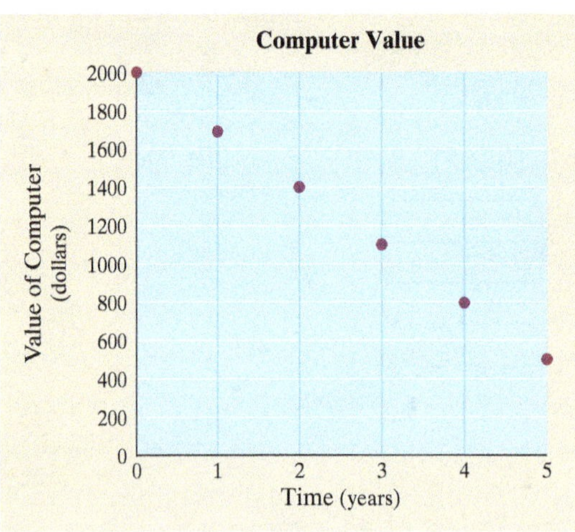

x	y
0	2000
1	1700
2	1400
3	1100
4	800
5	500

Vocabulary, Readiness & Video Check

Use the choices below to fill in each blank. The exercises below all have to do with the rectangular coordinate system.

origin	x-coordinate	x-axis	one	four
quadrants	y-coordinate	y-axis	solution	

1. The horizontal axis is called the __________ and the vertical axis is called the __________.
2. The intersection of the horizontal axis and the vertical axis is a point called the __________.
3. The axes divide the plane into regions called __________. There are __________ of these regions.

4. In the ordered pair of numbers $(-2, 5)$, the number -2 is called the __________ and the number 5 is called the __________.

5. Each ordered pair of numbers corresponds to __________ point in the plane.

6. An ordered pair is a(n) __________ of an equation in two variables if replacing the variables by the coordinates of the ordered pair results in a true statement.

Martin-Gay Interactive Videos

See Video 10.1

Watch the section lecture video and answer the following questions.

Objective 1 7. Examples 1–3 ask you to answer questions about a bar graph. What information is provided on the horizontal axis of this bar graph? On the vertical axis?

Objective 2 8. Several points are plotted in Examples 4–11. Where do you always start when plotting a point? How does the 1st coordinate tell you to move? How does the 2nd coordinate tell you to move?

Objective 3 9. In the lecture before Example 12, what connection is made between data and graphing?

Objective 4 10. An ordered pair is a solution of an equation if, when the variables are replaced with their values, a true statement results. In Example 13, three ordered pairs are tested. What are the last two points to be tested? What lesson can be learned by the results of testing these two points and why?

Objective 5 11. In Example 14, when one variable of a linear equation in two variables is replaced by a replacement value, what type of equation results?

10.1 Exercise Set MyMathLab®

The following bar graph shows the top 10 tourist destinations and the number of tourists that visit each country per year. Use this graph to answer Exercises 1 through 6. See Example 1.

Top Tourist Destinations

Number of Arrivals (in millions): 0, 10, 20, 30, 40, 50, 60, 70, 80, 90

Country: France, U.S., Spain, Italy, U.K., China, Ukraine, Mexico, Turkey, Germany

Data from MapXL (www.mapsofworld.com)

1. Which country shown is the most popular tourist destination?

2. Which country shown is the least popular tourist destination?

3. Which countries shown have more than 40 million tourists per year?

4. Which countries shown have fewer than 30 million tourists per year?

5. Estimate the number of tourists per year whose destination is Italy.

6. Estimate the number of tourists per year whose destination is the U.K.

The following line graph shows the paid attendance at each Super Bowl game from 2005 through 2011. Use this graph to answer Exercises 7 through 10. See Example 2.

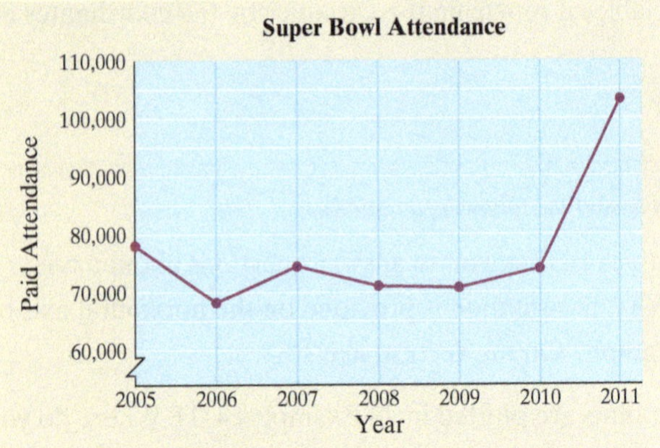

7. Estimate the Super Bowl attendance in 2009.
8. Estimate the Super Bowl attendance in 2010.
9. Find the year on the graph with the greatest Super Bowl attendance and approximate that attendance.
10. Find the year on the graph with the least Super Bowl attendance and approximate that attendance.

The line graph below shows the number of students per teacher in U.S. public elementary and secondary schools. Use this graph for Exercises 11 through 16. See Example 2.

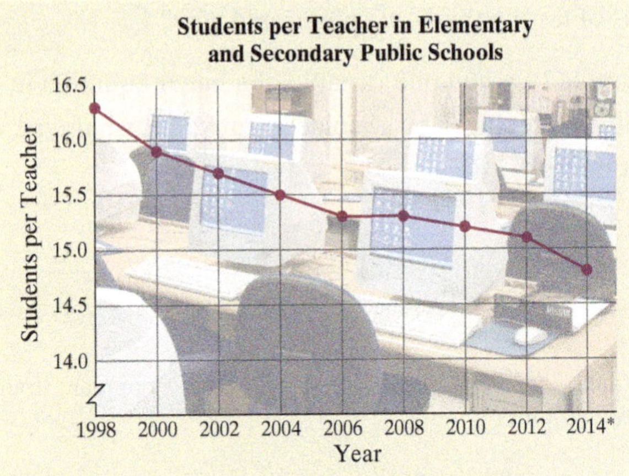

11. Approximate the number of students per teacher in 2000.
12. Approximate the number of students per teacher in 2012.
13. Between what years shown did the greatest decrease in number of students per teacher occur?
14. What was the first year shown that the number of students per teacher fell below 16?
15. What was the first year shown that the number of students per teacher fell below 15?
16. Discuss any trends shown by this line graph.

Plot each ordered pair. State in which quadrant or on which axis each point lies. See Example 3.

17. **a.** $(1, 5)$ **b.** $(-5, -2)$
c. $(-3, 0)$ **d.** $(0, -1)$
e. $(2, -4)$ **f.** $\left(-1, 4\frac{1}{2}\right)$
g. $(3.7, 2.2)$ **h.** $\left(\frac{1}{2}, -3\right)$

18. **a.** $(2, 4)$ **b.** $(0, 2)$
c. $(-2, 1)$ **d.** $(-3, -3)$
e. $\left(3\frac{3}{4}, 0\right)$ **f.** $(5, -4)$
g. $(-3.4, 4.8)$ **h.** $\left(\frac{1}{3}, -5\right)$

Find the x- and y-coordinates of each labeled point. See Example 3.

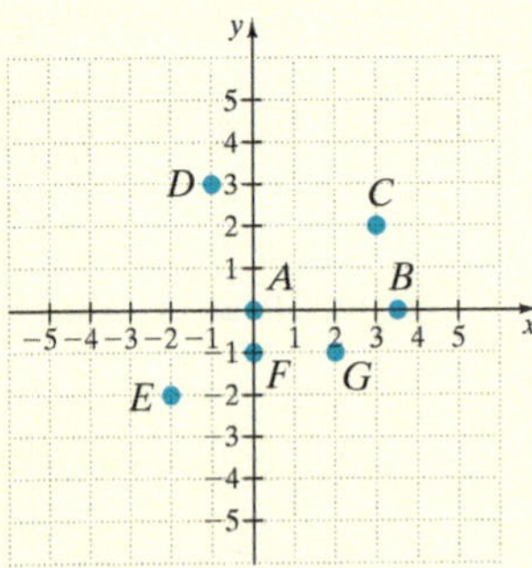

19. *A*
20. *B*
21. *C*
22. *D*
23. *E*
24. *F*
25. *G*

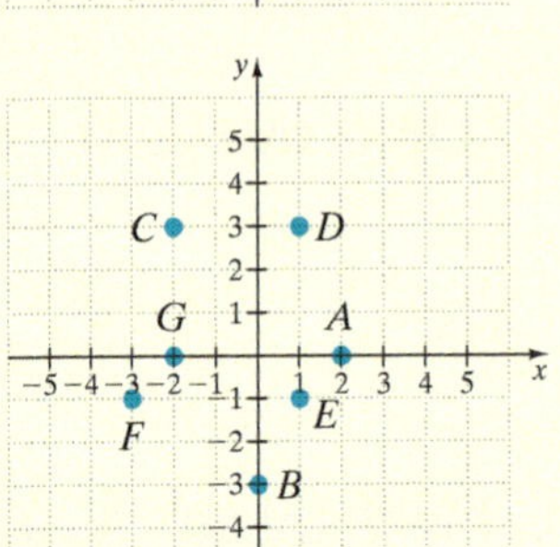

26. *A*
27. *B*
28. *C*
29. *D*
30. *E*
31. *F*
32. *G*

Solve. See Example 4.

33. The table shows the worldwide box office (in billions of dollars) for the movie industry during the years shown. (Data from Motion Picture Association of America)

Year	Box Office (in billions of dollars)
2006	25.5
2007	26.3
2008	27.7
2009	29.4
2010	31.8

a. Write this paired data as a set of ordered pairs of the form (year, box office).

b. In your own words, write the meaning of the ordered pair (2010, 31.8).

c. Create a scatter diagram of the paired data. Be sure to label the axes appropriately.

d. What trend in the paired data does the scatter diagram show?

34. The table shows the amount of money (in billions of dollars) Americans spent on their pets for the years shown. (Data from American Pet Products Manufacturers Association)

Year	Pet-Related Expenditures (in billions of dollars)
2008	43.2
2009	45.5
2010	48.4
2011	50.8

a. Write this paired data as a set of ordered pairs of the form (year, pet-related expenditures).

b. In your own words, write the meaning of the ordered pair (2011, 50.8).

c. Create a scatter diagram of the paired data. Be sure to label the axes appropriately.

d. What trend in the paired data does the scatter diagram show?

35. Minh, a psychology student, kept a record of how much time she spent studying for each of her 20-point psychology quizzes and her score on each quiz.

Hours Spent Studying	0.50	0.75	1.00	1.25	1.50	1.50	1.75	2.00
Quiz Score	10	12	15	16	18	19	19	20

a. Write the data as ordered pairs of the form (hours spent studying, quiz score).

b. In your own words, write the meaning of the ordered pair (1.25, 16).

c. Create a scatter diagram of the paired data. Be sure to label the axes appropriately.

d. What might Minh conclude from the scatter diagram?

36. A local lumberyard uses quantity pricing. The table shows the price per board for different amounts of lumber purchased.

Price per Board (in dollars)	Number of Boards Purchased
8.00	1
7.50	10
6.50	25
5.00	50
2.00	100

a. Write the data as ordered pairs of the form (price per board, number of boards purchased).

b. In your own words, write the meaning of the ordered pair (2.00, 100).

c. Create a scatter diagram of the paired data. Be sure to label the axes appropriately.

d. What trend in the paired data does the scatter diagram show?

37. The table shows the distance from the equator (in miles) and the average annual snowfall (in inches) for each of eight selected U.S. cities. (Data from National Climatic Data Center, Wake Forest University Albatross Project)

City	Distance from Equator (in miles)	Average Annual Snowfall (in inches)
1. Atlanta, GA	2313	2
2. Austin, TX	2085	1
3. Baltimore, MD	2711	21
4. Chicago, IL	2869	39
5. Detroit, MI	2920	42
6. Juneau, AK	4038	99
7. Miami, FL	1783	0
8. Winston-Salem, NC	2493	9

a. Write this paired data as a set of ordered pairs of the form (distance from equator, average annual snowfall).

b. Create a scatter diagram of the paired data. Be sure to label the axes appropriately.

c. What trend in the paired data does the scatter diagram show?

38. The table shows the average farm size (in acres) in the United States during the years shown. (Data from National Agricultural Statistics Service)

Year	*Average Farm Size (in acres)*
2001	438
2002	440
2003	441
2004	443
2005	445
2006	446

a. Write this paired data as a set of ordered pairs of the form (year, average farm size).

b. Create a scatter diagram of the paired data. Be sure to label the axes appropriately.

Determine whether each ordered pair is a solution of the given linear equation. See Example 5.

39. $2x + y = 7$; (3, 1), (7, 0), (0, 7)

40. $3x + y = 8$; (2, 3), (0, 8), (8, 0)

41. $x = -\frac{1}{3}y$; (0, 0), (3, −9)

42. $y = -\frac{1}{2}x$; (0, 0), (4, 2)

43. $x = 5$; (4, 5), (5, 4), (5, 0)

44. $y = -2$; (−2, 2), (2, −2), (0, −2)

Complete each ordered pair so that it is a solution of the given linear equation. See Examples 6 through 8.

45. $x - 4y = 4$; (, −2), (4,)

46. $x - 5y = -1$; (, −2), (4,)

47. $y = \frac{1}{4}x - 3$; (−8,), (, 1)

48. $y = \frac{1}{5}x - 2$; (−10,), (, 1)

Complete the table of ordered pairs for each linear equation. See Examples 6 through 8.

49. $y = -7x$

x	y
0	
−1	
	2

50. $y = -9x$

x	y
	0
−3	
	2

51. $y = -x + 2$

x	y
0	
	0
−3	

52. $x = -y + 4$

x	y
	0
0	
	−3

53. $y = \frac{1}{2}x$

x	y
0	
−6	
	1

54. $y = \frac{1}{3}x$

x	y
0	
−6	
	1

55. $x + 3y = 6$

x	y
0	
	0
	1

56. $2x + y = 4$

x	y
	4
2	
	2

57. $y = 2x - 12$

x	y
0	
	−2
3	

58. $y = 5x + 10$

x	y
	0
	5
0	

59. $2x + 7y = 5$

x	y
0	
	0
	1

60. $x - 6y = 3$

x	y
0	
1	
	−1

MIXED PRACTICE

Complete the table of ordered pairs for each equation. Then plot the ordered pair solutions. See Examples 1 through 7.

61. $x = -5y$

x	y
	0
	1
10	

62. $y = -3x$

x	y
0	
−2	
	9

63. $y = \frac{1}{3}x + 2$

x	y
0	
−3	
	0

64. $y = \frac{1}{2}x + 3$

x	y
0	
−4	
	0

Solve. See Example 9.

65. The cost in dollars y of producing x computer desks is given by $y = 80x + 5000$.

a. Complete the table.

x	100	200	300
y			

b. Find the number of computer desks that can be produced for \$8600. (*Hint:* Find x when $y = 8600$.)

66. The hourly wage y of an employee at a certain production company is given by $y = 0.25x + 9$ where x is the number of units produced by the employee in an hour.

a. Complete the table.

x	0	1	5	10
y				

b. Find the number of units that an employee must produce each hour to earn an hourly wage of \$12.25. (*Hint:* Find x when $y = 12.25$.)

67. The average annual cinema admission price y (in dollars) from 2001 through 2010 is given by $y = 0.24x + 5.28$. In this equation, x represents the number of years after 2000. (*Source:* Motion Picture Association of America)

a. Complete the table.

x	1	3	5
y			

b. Find the year in which the average cinema admission price was approximately \$7.50. (*Hint:* Find x when $y = 7.50$ and round to the nearest whole number.)

c. Use the given equation to predict when the cinema admission price might be \$9.00. (Use the hint for part b.)

d. In your own words, write the meaning of the ordered pair (5, 6.48).

68. The average amount of money y spent per person on recorded music from 2002 to 2009 is given by $y = -4.22x + 65.70$. In this equation, x represents the number of years after 2000. (*Source:* edgeqld.org)

a. Complete the table.

x	1	5	9
y			

b. Find the year in which the yearly average amount of money per person spent on recorded music was approximately \$30. (*Hint:* Find x when $y = 30$ and round to the nearest whole number.)

The graph below shows the number of U.S. Walmart stores for each year. Use this graph to answer Exercises 69 through 72.

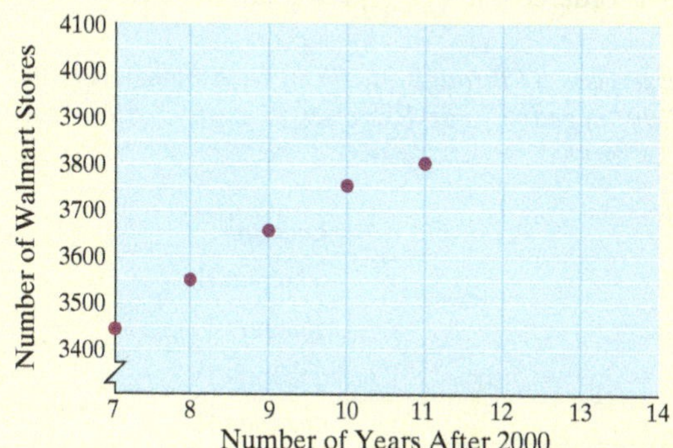

Data from Walmart

69. The ordered pair (10, 3755) is a point of the graph. Write a sentence describing the meaning of this ordered pair.

70. The ordered pair (11, 3804) is a point of the graph. Write a sentence describing the meaning of this ordered pair.

71. Estimate the increase in Walmart stores for years 8, 9, and 10.

72. Use a straightedge or ruler and this graph to estimate the number of Walmart stores in the year 2014.

73. Describe what is similar about the coordinates of points whose graph lies on the x-axis.

74. Describe what is similar about the coordinates of points whose graph lies on the y-axis.

REVIEW AND PREVIEW

Solve each equation for y. See Section 9.5.

75. $x + y = 5$

76. $x - y = 3$

77. $2x + 4y = 5$

78. $5x + 2y = 7$

79. $10x = -5y$

80. $4y = -8x$

81. $x - 3y = 6$

82. $2x - 9y = -20$

CONCEPT EXTENSIONS

Answer each exercise with true or false.

83. Point $(-1, 5)$ lies in quadrant IV.

84. Point $(3, 0)$ lies on the y-axis.

85. For the point $\left(-\frac{1}{2}, 1.5\right)$, the first value, $-\frac{1}{2}$, is the x-coordinate and the second value, 1.5, is the y-coordinate.

86. The ordered pair $\left(2, \frac{2}{3}\right)$ is a solution of $2x - 3y = 6$.

For Exercises 87 through 91, fill in each blank with "0," "positive," or "negative." For Exercises 92 and 93, fill in each blank with "x" or "y."

	Point	Location
87.	(______, ______)	quadrant III
88.	(______, ______)	quadrant I
89.	(______, ______)	quadrant IV
90.	(______, ______)	quadrant II
91.	(______, ______)	origin
92.	(number, 0)	_____-axis
93.	(0, number)	_____-axis

94. Give an example of an ordered pair whose location is in (or on)

a. quadrant I
b. quadrant II
c. quadrant III
d. quadrant IV
e. x-axis
f. y-axis

Solve. See the first Concept Check in this section.

95. Is the graph of $(3, 0)$ in the same location as the graph of $(0, 3)$? Explain why or why not.

96. Give the coordinates of a point such that if the coordinates are reversed, their location is the same.

97. In general, what points can have coordinates reversed and still have the same location?

98. In your own words, describe how to plot or graph an ordered pair of numbers.

Write an ordered pair for each point described. See the second Concept Check in this section.

99. Point C is four units to the right of the y-axis and seven units below the x-axis.

100. Point D is three units to the left of the origin.

Solve.

△ **101.** Find the perimeter of the rectangle whose vertices are the points with coordinates $(-1, 5)$, $(3, 5)$, $(3, -4)$, and $(-1, -4)$.

△ **102.** Find the area of the rectangle whose vertices are the points with coordinates $(5, 2)$, $(5, -6)$, $(0, -6)$, and $(0, 2)$.

103. Three vertices of a rectangle are $(-2, -3)$, $(-7, -3)$, and $(-7, 6)$.

a. Find the coordinates of the fourth vertex of the rectangle.
b. Find the perimeter of the rectangle.
c. Find the area of the rectangle.

104. Three vertices of a square are $(-4, -1)$, $(-4, 8)$, and $(5, 8)$.

a. Find the coordinates of the fourth vertex of the square.
b. Find the perimeter of the square.
c. Find the area of the square.

10.2 Graphing Linear Equations

Objectives

1 Identify Linear Equations.

2 Graph a Linear Equation by Finding and Plotting Ordered Pair Solutions.

Objective 1 Identifying Linear Equations

In the previous section, we found that equations in two variables may have more than one solution. For example, both $(6, 0)$ and $(2, -2)$ are solutions of the equation $x - 2y = 6$. In fact, this equation has an infinite number of solutions. Other solutions include $(0, -3)$, $(4, -1)$, and $(-2, -4)$. If we graph these solutions, notice that a pattern appears.

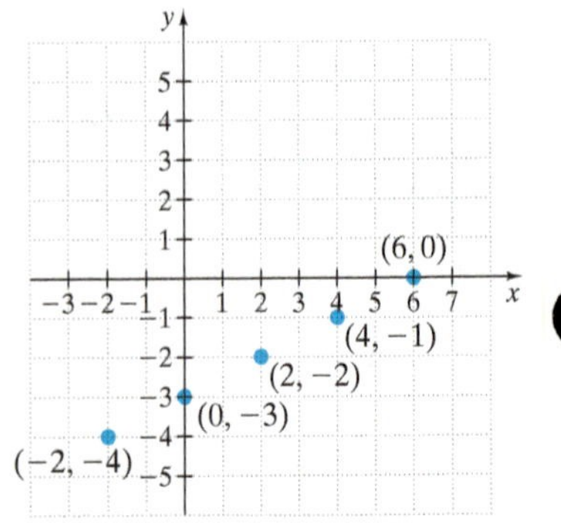

These solutions all appear to lie on the same line, which has been filled in below. It can be shown that every ordered pair solution of the equation corresponds to a point on this line, and every point on this line corresponds to an ordered pair solution. Thus, we say that this line is the **graph of the equation** $x - 2y = 6$.

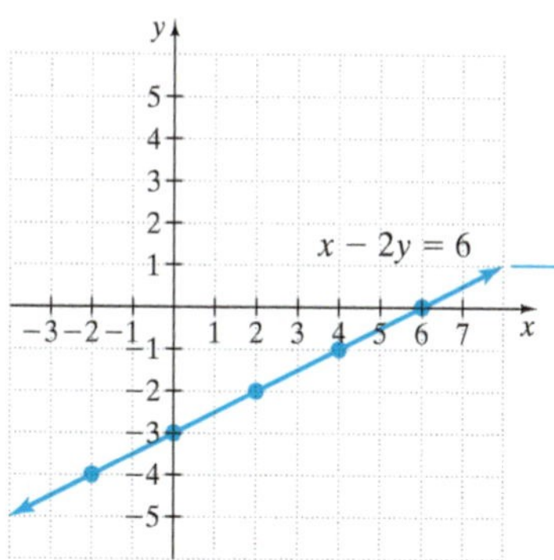

Helpful Hint Notice that we can only show a part of a line on a graph. The arrowheads on each end of the line remind us that the line actually extends indefinitely in both directions.

The equation $x - 2y = 6$ is called a **linear equation in two variables** and **the graph of every linear equation in two variables is a line.**

Linear Equation in Two Variables

A linear equation in two variables is an equation that can be written in the form

$$Ax + By = C$$

where A, B, and C are real numbers and A and B are not both 0. **The graph of a linear equation in two variables is a straight line.**

The form $Ax + By = C$ is called **standard form.**

Helpful Hint

Notice in the form $Ax + By = C$, the understood exponent on both x and y is 1.

Examples of Linear Equations in Two Variables

$2x + y = 8$ (Standard Form) $\quad -2x = 7y \quad y = \frac{1}{3}x + 2 \quad y = 7$

Before we graph linear equations in two variables, let's practice identifying these equations.

Example 1 Determine whether each equation is a linear equation in two variables.

a. $x - 1.5y = -1.6$ **b.** $y = -2x$ **c.** $x + y^2 = 9$ **d.** $x = 5$

Solution:

a. This is a linear equation in two variables because it is written in the form $Ax + By = C$ with $A = 1, B = -1.5$, and $C = -1.6$.

b. This is a linear equation in two variables because it can be written in the form $Ax + By = C$.

$$y = -2x$$
$$2x + y = 0 \quad \text{Add } 2x \text{ to both sides.}$$

(*Continued on next page*)

Practice 1

Determine whether each equation is a linear equation in two variables.

a. $3x + 2.7y = -5.3$
b. $x^2 + y = 8$
c. $y = 12$
d. $5x = -3y$

Answers

1. a. yes **b.** no **c.** yes **d.** yes

c. This is *not* a linear equation in two variables because y is squared.

d. This is a linear equation in two variables because it can be written in the form $Ax + By = C$.

$$x = 5$$
$$x + 0y = 5 \quad \text{Add } 0 \cdot y.$$

Work Practice 1

Objective 2 Graphing Linear Equations by Plotting Ordered Pair Solutions

From geometry, we know that a straight line is determined by just two points. Graphing a linear equation in two variables, then, requires that we find just two of its infinitely many solutions. Once we do so, we plot the solution points and draw the line connecting the points. Usually, we find a third solution as well, as a check.

Practice 2

Graph the linear equation $x + 3y = 9$.

Example 2 Graph the linear equation $2x + y = 5$.

Solution:

Find three ordered pair solutions of $2x + y = 5$. To do this, choose a value for one variable, x or y, and solve for the other variable. For example, let $x = 1$. Then $2x + y = 5$ becomes

$$2x + y = 5$$
$$2(1) + y = 5 \quad \text{Replace } x \text{ with 1.}$$
$$2 + y = 5 \quad \text{Multiply.}$$
$$y = 3 \quad \text{Subtract 2 from both sides.}$$

Since $y = 3$ when $x = 1$, the ordered pair $(1, 3)$ is a solution of $2x + y = 5$. Next, let $x = 0$.

$$2x + y = 5$$
$$2(0) + y = 5 \quad \text{Replace } x \text{ with 0.}$$
$$0 + y = 5$$
$$y = 5$$

The ordered pair $(0, 5)$ is a second solution.

The two solutions found so far allow us to draw the straight line that is the graph of all solutions of $2x + y = 5$. However, we find a third ordered pair as a check. Let $y = -1$.

$$2x + y = 5$$
$$2x + (-1) = 5 \quad \text{Replace } y \text{ with } -1.$$
$$2x - 1 = 5$$
$$2x = 6 \quad \text{Add 1 to both sides.}$$
$$x = 3 \quad \text{Divide both sides by 2.}$$

The third solution is $(3, -1)$. These three ordered pair solutions are listed in table form as shown on the next page. The graph of $2x + y = 5$ is the line through the three points.

Answer

2.
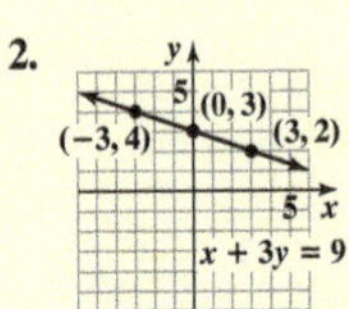

x	y
1	3
0	5
3	−1

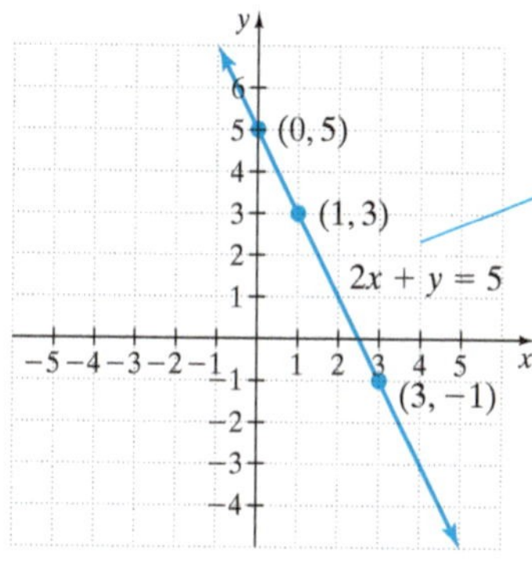

Helpful Hint All three points should fall on the same straight line. If not, check your ordered pair solutions for a mistake.

■ Work Practice 2

Example 3

Graph the linear equation $-5x + 3y = 15$.

Solution:

Find three ordered pair solutions of $-5x + 3y = 15$.

Let $x = 0$.

$$-5x + 3y = 15$$
$$-5 \cdot 0 + 3y = 15$$
$$0 + 3y = 15$$
$$3y = 15$$
$$y = 5$$

Let $y = 0$.

$$-5x + 3y = 15$$
$$-5x + 3 \cdot 0 = 15$$
$$-5x + 0 = 15$$
$$-5x = 15$$
$$x = -3$$

Let $x = -2$.

$$-5x + 3y = 15$$
$$-5(-2) + 3y = 15$$
$$10 + 3y = 15$$
$$3y = 5$$
$$y = \frac{5}{3}$$

The ordered pairs are $(0, 5)$, $(-3, 0)$, and $\left(-2, \frac{5}{3}\right)$. The graph of $-5x + 3y = 15$ is the line through the three points.

x	y
0	5
−3	0
−2	$\frac{5}{3} = 1\frac{2}{3}$

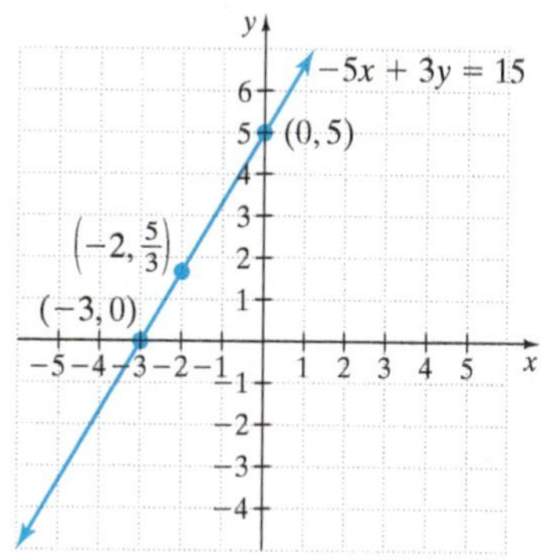

■ Work Practice 3

Example 4

Graph the linear equation $y = 3x$.

Solution:

To graph this linear equation, we find three ordered pair solutions. Since this equation is solved for y, choose three x values.

If $x = 2$, $y = 3 \cdot 2 = 6$.
If $x = 0$, $y = 3 \cdot 0 = 0$.
If $x = -1$, $y = 3 \cdot -1 = -3$.

x	y
2	6
0	0
−1	−3

(*Continued on next page*)

Practice 3

Graph the linear equation $3x - 4y = 12$.

Practice 4

Graph the linear equation $y = -2x$.

Answers

3.

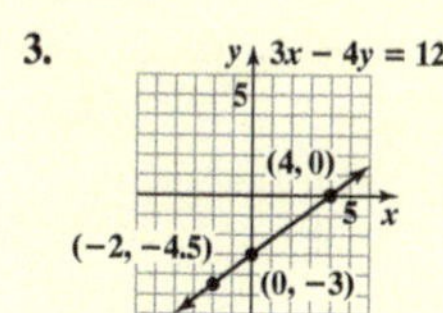

4.

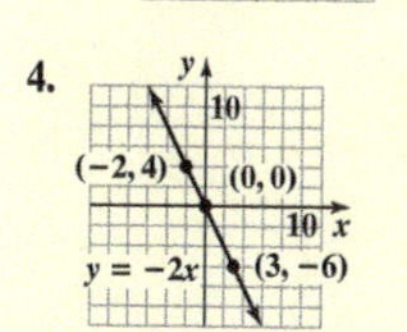

Next, graph the ordered pair solutions listed in the table on the previous page and draw a line through the plotted points as shown below. The line is the graph of $y = 3x$. Every point on the graph represents an ordered pair solution of the equation and every ordered pair solution is a point on this line.

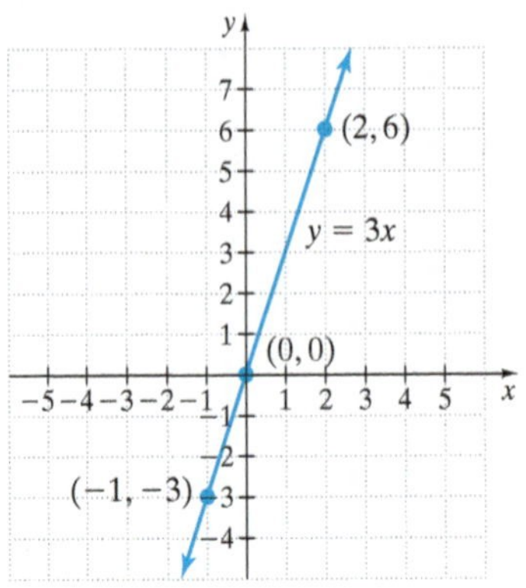

Work Practice 4

Helpful Hint

When graphing a linear equation in two variables, if it is

- solved for y, it may be easier to find ordered pair solutions by choosing x-values. If it is
- solved for x, it may be easier to find ordered pair solutions by choosing y-values.

Practice 5

Graph the linear equation $y = \frac{1}{2}x + 3$.

Example 5 Graph the linear equation $y = -\frac{1}{3}x + 2$.

Solution:

Find three ordered pair solutions, graph the solutions, and draw a line through the plotted solutions. To avoid fractions, choose x values that are multiples of 3 to substitute in the equation. When a multiple of 3 is multiplied by $-\frac{1}{3}$, the result is an integer. See the calculations shown above the table.

If $x = 6$, then $y = -\frac{1}{3} \cdot 6 + 2 = -2 + 2 = 0$.

If $x = 0$, then $y = -\frac{1}{3} \cdot 0 + 2 = 0 + 2 = 2$.

If $x = -3$, then $y = -\frac{1}{3} \cdot -3 + 2 = 1 + 2 = 3$.

x	y
6	0
0	2
−3	3

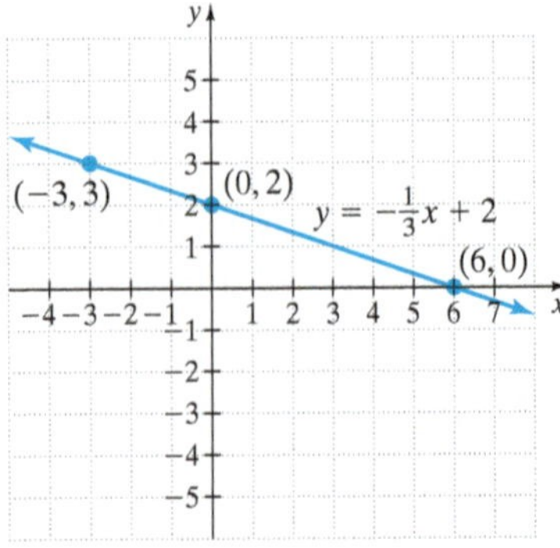

Work Practice 5

Answer

5.

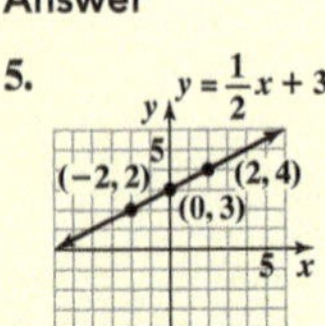

Let's compare the graphs in Examples 4 and 5. The graph of $y = 3x$ tilts upward (as we follow the line from left to right) and the graph of $y = -\frac{1}{3}x + 2$ tilts downward (as we follow the line from left to right). We will learn more about the tilt, or slope, of a line in Section 10.4.

Example 6 Graph the linear equation $y = -2$.

Solution:

The equation $y = -2$ can be written in standard form as $0x + y = -2$. No matter what value we replace x with, y is always -2.

x	y
0	−2
3	−2
−2	−2

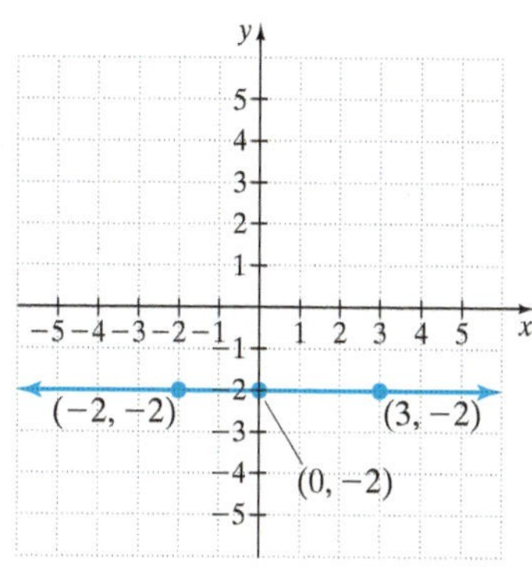

Notice that the graph of $y = -2$ is a horizontal line.

Work Practice 6

Practice 6

Graph the linear equation $x = -2$.

Example 7 Graph the linear equation $y = 3x + 6$ and compare this graph with the graph of $y = 3x$ in Example 4.

Solution:

Find ordered pair solutions, graph the solutions, and draw a line through the plotted solutions. We choose x values and substitute in the equation $y = 3x + 6$.

If $x = -3$, then $y = 3(-3) + 6 = -3$.
If $x = 0$, then $y = 3(0) + 6 = 6$.
If $x = 1$, then $y = 3(1) + 6 = 9$.

x	y
−3	−3
0	6
1	9

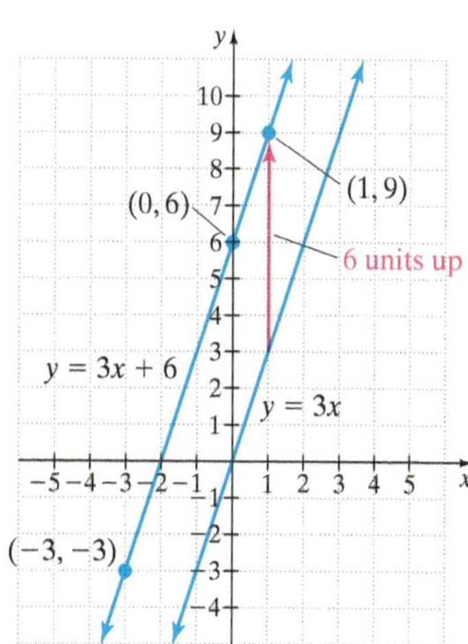

(Continued on next page)

Practice 7

Graph the linear equation $y = -2x + 3$ and compare this graph with the graph of $y = -2x$ in Practice 4.

Answers

6.

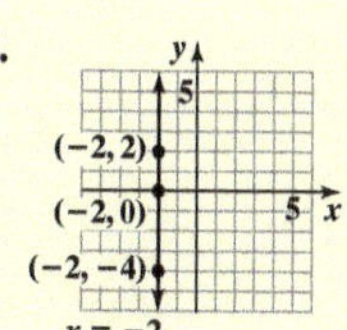

7. The graph of $y = -2x + 3$ is the same as the graph of $y = -2x$ except that the graph of $y = -2x + 3$ is moved 3 units upward.

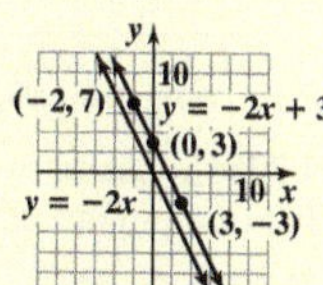

The most startling similarity is that both graphs appear to have the same upward tilt as we move from left to right. Also, the graph of $y = 3x$ crosses the y-axis at the origin, while the graph of $y = 3x + 6$ crosses the y-axis at 6. In fact, the graph of $y = 3x + 6$ is the same as the graph of $y = 3x$ moved vertically upward 6 units.

Work Practice 7

Notice that the graph of $y = 3x + 6$ crosses the y-axis at 6. This happens because when $x = 0$, $y = 3x + 6$ becomes $y = 3 \cdot 0 + 6 = 6$. The graph contains the point $(0, 6)$, which is on the y-axis.

In general, if a linear equation in two variables is solved for y, we say that it is written in the form $y = mx + b$. The graph of this equation contains the point $(0, b)$ because when $x = 0$, $y = mx + b$ is $y = m \cdot 0 + b = b$.

The graph of $y = mx + b$ crosses the y-axis at $(0, b)$.

We will review this again in Section 10.5.

Linear equations are often used to model real data as seen in the next example.

Example 8 Estimating the Number of Registered Nurses

The occupation expected to have the most employment growth in the next few years is registered nurse. The number of people y (in thousands) employed as registered nurses in the United States can be estimated by the linear equation $y = 58.1x + 2619$, where x is the number of years after the year 2008. (Data from U.S. Bureau of Labor Statistics)

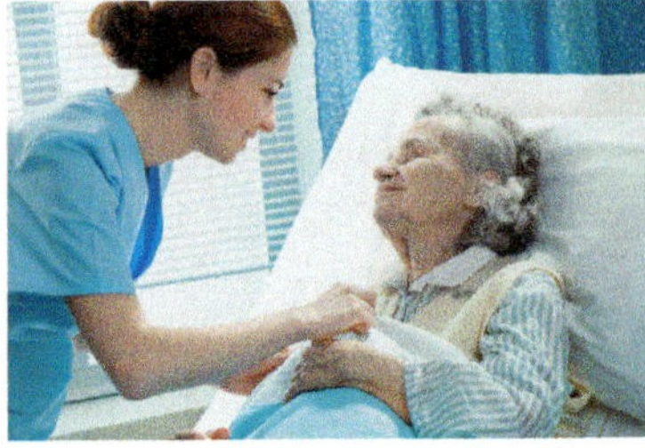

a. Graph the equation.

b. Use the graph to predict the number of registered nurses in the year 2018.

Solution:

a. To graph $y = 58.1x + 2619$, choose x-values and substitute in the equation.

If $x = 0$, then $y = 58.1(0) + 2619 = 2619$.
If $x = 2$, then $y = 58.1(2) + 2619 = 2735.2$.
If $x = 5$, then $y = 58.1(5) + 2619 = 2909.5$.

x	y
0	2619
2	2735.2
5	2909.5

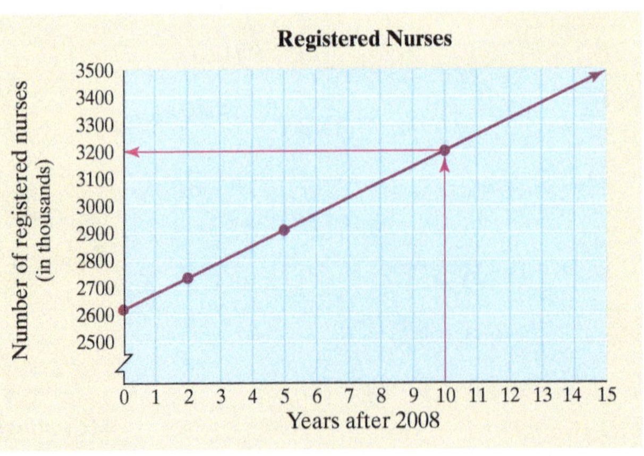

Practice 8

One of the occupations expected to have a large growth in employment in the next few years is computer software application engineers. The number of people y (in thousands) employed as computer software application engineers in the United States can be estimated by the linear equation $y = 17.5x + 515$, where x is the number of years after 2008. (*Source:* Based on data from the Bureau of Labor Statistics)

a. Graph the equation.

b. Use the graph to predict the number of computer software application engineers in the year 2020.

Answers

8. a.

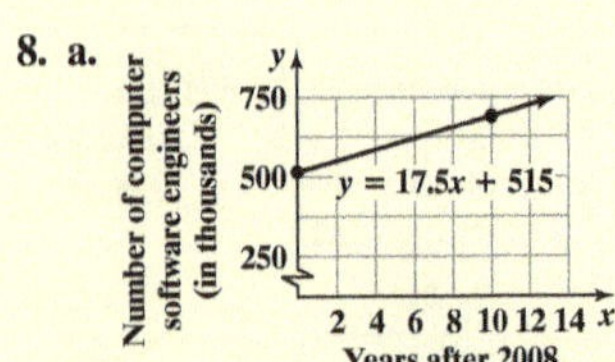

b. We predict 725 thousand computer software application engineers in the year 2020.

b. To use the graph to *predict* the number of registered nurses in the year 2018, we need to find the y-coordinate that corresponds to $x = 10$. (10 years after 2008 is the year 2018.) To do so, find 10 on the x-axis. Move vertically upward to the graphed line and then horizontally to the left. We approximate the number on the y-axis to be 3200. Thus, in the year 2018, we predict that there will be 3200 thousand registered nurses. (The actual value, using 10 for x, is 3200.)

Work Practice 8

Helpful Hint

Make sure you understand that models are mathematical approximations of the data for the known years. (For example, see the model in Example 8.) Any number of unknown factors can affect future years, so be cautious when using models to predict.

Graphing Calculator Explorations

In this section, we begin an optional study of graphing calculators and graphing software packages for computers. These graphers use the same point plotting technique that was introduced in this section. The advantage of this graphing technology is, of course, that graphing calculators and computers can find and plot ordered pair solutions much faster than we can. Note, however, that the features described in these boxes may not be available on all graphing calculators.

The rectangular screen where a portion of the rectangular coordinate system is displayed is called a **window.** We call it a **standard window** for graphing when both the x- and y-axes show coordinates between -10 and 10. This information is often displayed in the window menu on a graphing calculator as

Xmin = −10
Xmax = 10
Xscl = 1 The scale on the x-axis is one unit per tick mark.
Ymin = −10
Ymax = 10
Yscl = 1 The scale on the y-axis is one unit per tick mark.

To use a graphing calculator to graph the equation $y = 2x + 3$, press the [Y =] key and enter the keystrokes [2] [x] [+] [3]. The top row should now read $Y_1 = 2x + 3$. Next press the [GRAPH] key, and the display should look like this:

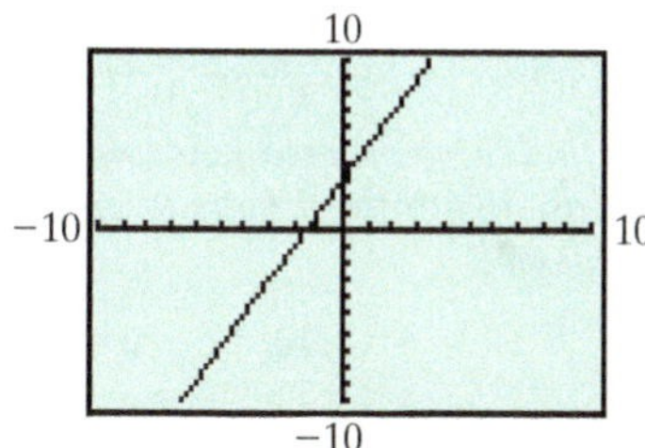

Use a standard window and graph the following linear equations. (Unless otherwise stated, use a standard window when graphing.)

1. $y = -3x + 7$

2. $y = -x + 5$

3. $y = 2.5x - 7.9$

4. $y = -1.3x + 5.2$

5. $y = -\frac{3}{10}x + \frac{32}{5}$

6. $y = \frac{2}{9}x - \frac{22}{3}$

Vocabulary, Readiness & Video Check

Martin-Gay Interactive Videos

See Video 10.2

Watch the section lecture video and answer the following questions.

Objective 1 1. Exponents aren't mentioned in the definition of a linear equation in two variables. However, in determining whether Example 3 is a linear equation in two variables, the exponents or powers on the variables are discussed. Explain.

Objective 2 2. In the lecture before Example 5, it's mentioned that you need only two points to determine a line. Why then are three ordered pair solutions found in Examples 5–7?

Objective 3 3. What does a graphed line represent as discussed at the end of Examples 5 and 7?

10.2 Exercise Set MyMathLab®

Determine whether each equation is a linear equation in two variables. See Example 1.

1. $-x = 3y + 10$
2. $y = x - 15$
3. $x = y$
4. $x = y^3$
5. $x^2 + 2y = 0$
6. $0.01x - 0.2y = 8.8$
7. $y = -1$
8. $x = 25$

For each equation, find three ordered pair solutions by completing the table. Then use the ordered pairs to graph the equation. See Examples 2 through 7.

9. $x - y = 6$

x	y
	0
4	
	−1

10. $x - y = 4$

x	y
0	
	2
−1	

11. $y = -4x$

x	y
1	
0	
−1	

12. $y = -5x$

x	y
1	
0	
−1	

13. $y = \frac{1}{3}x$

x	y
0	
6	
−3	

14. $y = \frac{1}{2}x$

x	y
0	
−4	
2	

15. $y = -4x + 3$

x	y
0	
1	
2	

16. $y = -5x + 2$

x	y
0	
1	
2	

MIXED PRACTICE

Graph each linear equation. See Examples 2 through 7.

17. $x + y = 1$
18. $x + y = 7$
19. $x - y = -2$
20. $-x + y = 6$
21. $x - 2y = 6$
22. $-x + 5y = 5$
23. $y = 6x + 3$
24. $y = -2x + 7$
25. $x = -4$
26. $y = 5$
27. $y = 3$
28. $x = -1$
29. $y = x$
30. $y = -x$
31. $x = -3y$
32. $x = -5y$
33. $x + 3y = 9$
34. $2x + y = 2$
35. $y = \frac{1}{2}x + 2$
36. $y = \frac{1}{4}x + 3$
37. $3x - 2y = 12$
38. $2x - 7y = 14$
39. $y = -3.5x + 4$
40. $y = -1.5x - 3$

Graph each pair of linear equations on the same set of axes. Discuss how the graphs are similar and how they are different. See Example 7.

41. $y = 5x; y = 5x + 4$
42. $y = 2x; y = 2x + 5$
43. $y = -2x; y = -2x - 3$
44. $y = x; y = x - 7$
45. $y = \frac{1}{2}x; y = \frac{1}{2}x + 2$
46. $y = -\frac{1}{4}x; y = -\frac{1}{4}x + 3$

The graph of $y = 5x$ is given below as well as Figures a–d. For Exercises 47 through 50, match each equation with its graph. Hint: Recall that if an equation is written in the form $y = mx + b$, its graph crosses the y-axis at $(0, b)$.

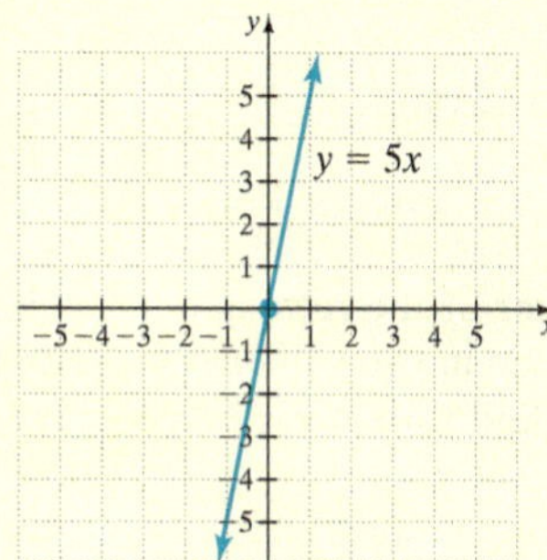

47. $y = 5x + 5$

48. $y = 5x - 4$

49. $y = 5x - 1$

50. $y = 5x + 2$

a.

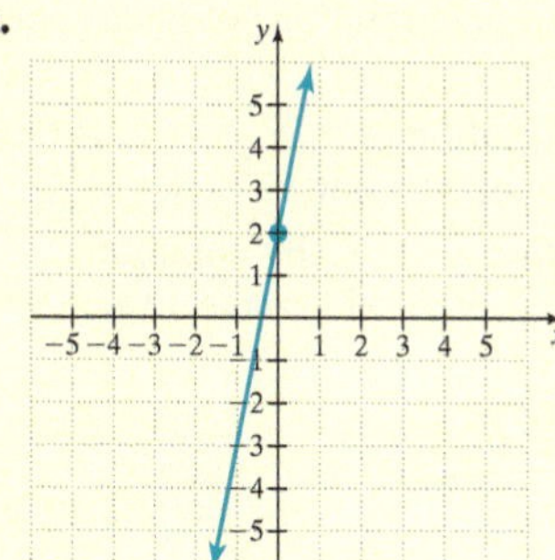

b.

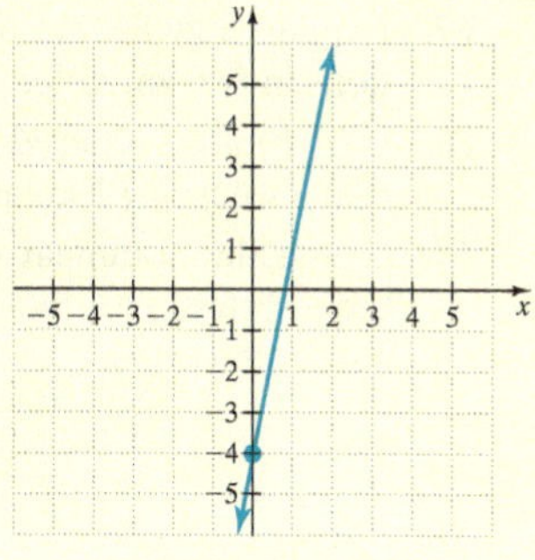

c.

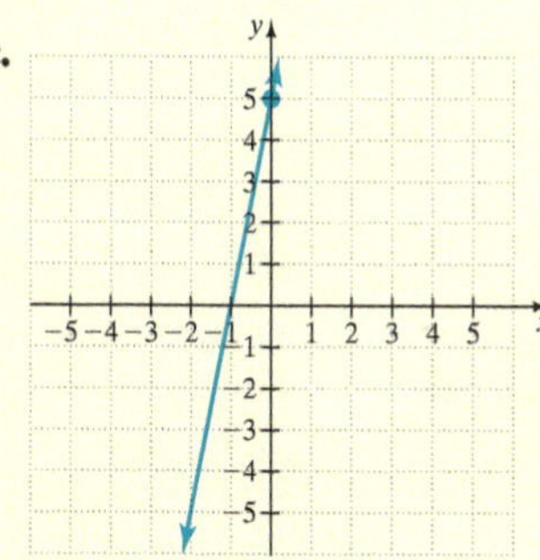

d.

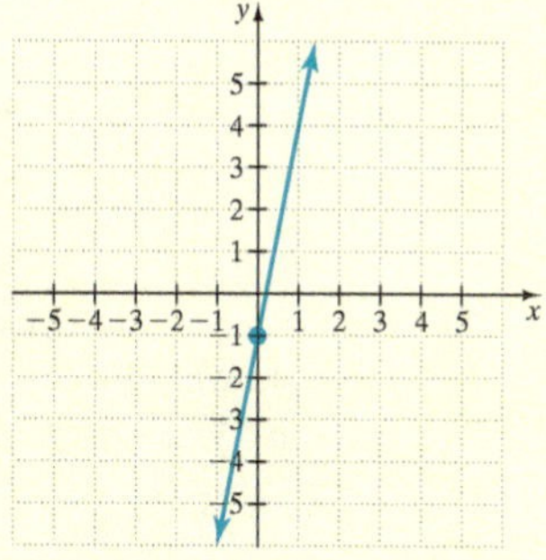

Solve. See Example 8.

51. Jogging is one of the few sports that has been consistently increasing over the past few years. The number of people jogging (in millions) from the years 2000 to 2009 is given by the equation $y = x + 23$, where x is the number of years after 2000. (*Source:* Based on data from the National Sporting Goods Association)

a. Use this equation or a graph of it to complete the ordered pair (8,).

b. Write a sentence explaining the meaning of the answer to part (a).

c. If this trend continues, how many joggers will there be in 2017?

52. The revenue y (in billions of dollars) for Home Depot stores during the years 2006 through 2010 is given by the equation $y = -3x + 79$, where x is the number of years after 2006. (*Source:* Based on data from Home Depot stores)

a. Use this equation or a graph of it to complete the ordered pair (4,).

b. Write a sentence explaining the meaning of the answer to part (a).

c. If this trend continues, predict the revenue for Home Depot stores for the year 2015.

53. One American rite of passage is a driver's license. The number of people y (in millions) who have a driver's license can be estimated by the linear equation $y = 2.2x + 190$, where x is the number of years after 2000. (Data from Federal Highway Administration)

a. Use this equation to complete the ordered pair (8,).

b. Write a sentence explaining the meaning of the ordered pair in part (a).

c. If this trend continues, predict the number of people with driver's licenses in 2016.

54. The percent of U.S. households y with at least one computer can be approximated by the linear equation $y = 4.3x + 51$, where x is the number of years since 2000. (Data from *Statistical Abstract of the United States*)

a. Use the equation to complete the ordered pair (9,).
b. Write a sentence explaining the meaning of the ordered pair found in part (a).
c. If this trend continued, estimate the percent of U.S. households that had at least one computer in 2012.
d. Explain any issues with your answer to part (c).

REVIEW AND PREVIEW

△ **55.** The coordinates of three vertices of a rectangle are $(-2, 5)$, $(4, 5)$, and $(-2, -1)$. Find the coordinates of the fourth vertex. See Section 10.1.

△ **56.** The coordinates of two vertices of a square are $(-3, -1)$ and $(2, -1)$. Find the coordinates of two pairs of points possible for the third and fourth vertices. See Section 10.1.

Complete each table. See Section 10.1.

57. $x - y = -3$

x	y
0	
	0

58. $y - x = 5$

x	y
0	
	0

59. $y = 2x$

x	y
0	
	0

60. $x = -3y$

x	y
0	
	0

CONCEPT EXTENSIONS

Write each statement as an equation in two variables. Then graph the equation.

61. The y-value is 5 more than the x-value.

62. The y-value is twice the x-value.

63. Two times the x-value, added to three times the y-value, is 6.

64. Five times the x-value, added to twice the y-value, is -10.

△ **65.** The perimeter of the trapezoid below is 22 centimeters. Write a linear equation in two variables for the perimeter. Find y if x is 3 cm.

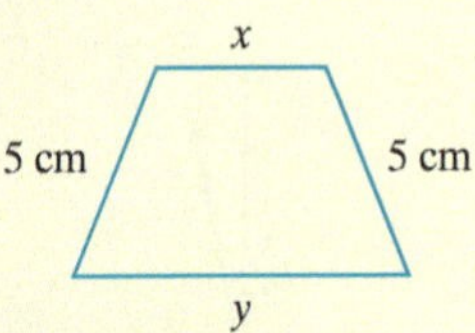

△ **66.** The perimeter of the rectangle below is 50 miles. Write a linear equation in two variables for this perimeter. Use this equation to find x when y is 20.

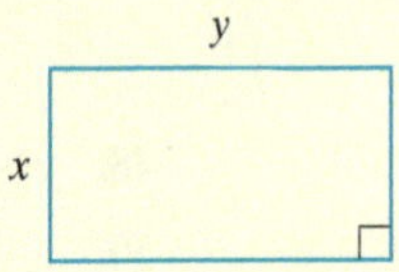

67. Explain how to find ordered pair solutions of linear equations in two variables.

68. If (a, b) is an ordered pair solution of $x + y = 5$, is (b, a) also a solution? Explain why or why not.

69. Graph the nonlinear equation $y = x^2$ by completing the table shown. Plot the ordered pairs and connect them with a smooth curve.

x	y
0	
1	
−1	
2	
−2	

70. Graph the nonlinear equation $y = |x|$ by completing the table shown. Plot the ordered pairs and connect them. This curve is "V" shaped.

$y = |x|$

x	y
0	
1	
−1	
2	
−2	

10.3 Intercepts 

Objective 1 Identifying Intercepts

In this section, we graph linear equations in two variables by identifying intercepts. For example, the graph of $y = 4x - 8$ is shown below. Notice that this graph crosses the y-axis at the point $(0, -8)$. This point is called the **y-intercept.** Likewise, the graph crosses the x-axis at $(2, 0)$, and this point is called the **x-intercept.**

The intercepts are $(2, 0)$ and $(0, -8)$.

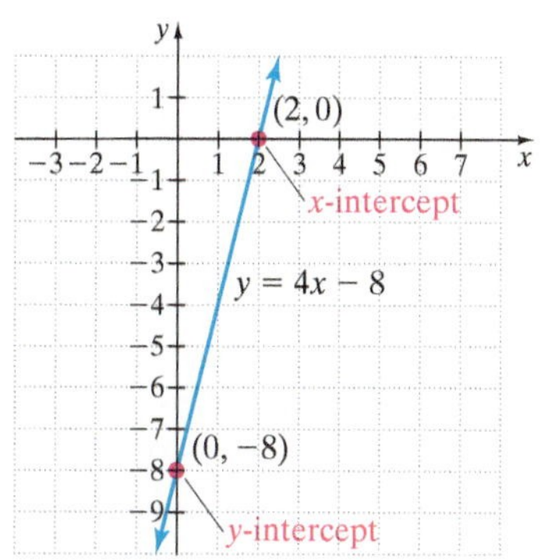

Helpful Hint

If a graph crosses the x-axis at $(-3, 0)$ and the y-axis at $(0, 7)$, then

$(-3, 0)$ ← x-intercept $(0, 7)$ ← y-intercept

Notice that for the y-intercept, the x-value is 0 and for the x-intercept, the y-value is 0. **Note:** Sometimes in mathematics, you may see just the number 7 stated as the y-intercept, and -3 stated as the x-intercept.

Examples

Identify the x- and y-intercepts.

1.

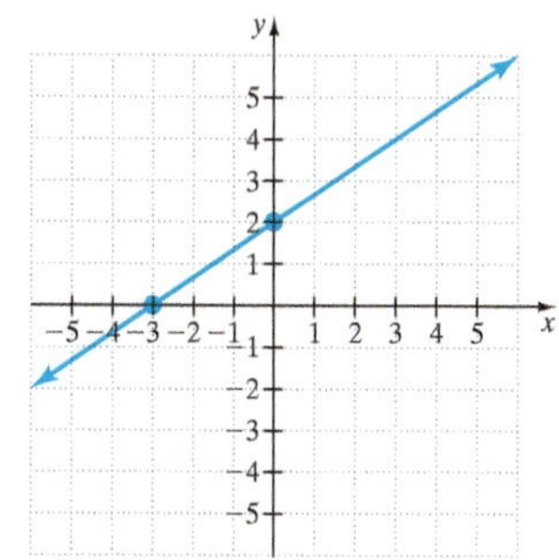

Solution:

x-intercept: $(-3, 0)$
y-intercept: $(0, 2)$

2.

Solution:

x-intercepts: $(-4, 0)$, $(-1, 0)$
y-intercept: $(0, 1)$

(*Continued on next page*)

Objectives

1 Identify Intercepts of a Graph.

2 Graph a Linear Equation by Finding and Plotting Intercepts.

3 Identify and Graph Vertical and Horizontal Lines.

Practice 1–5

Identify the x- and y-intercepts.

1.

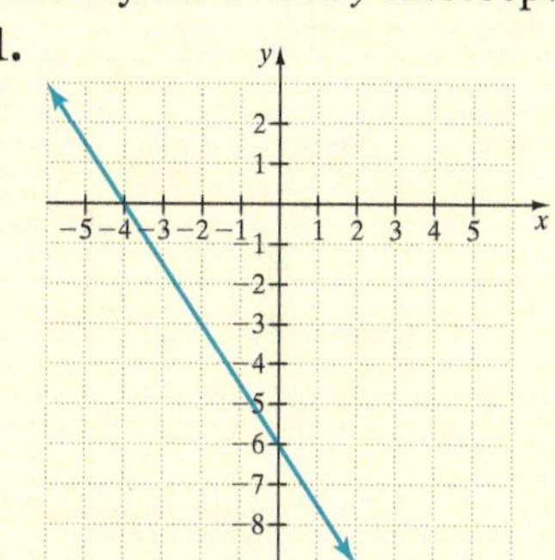

2.

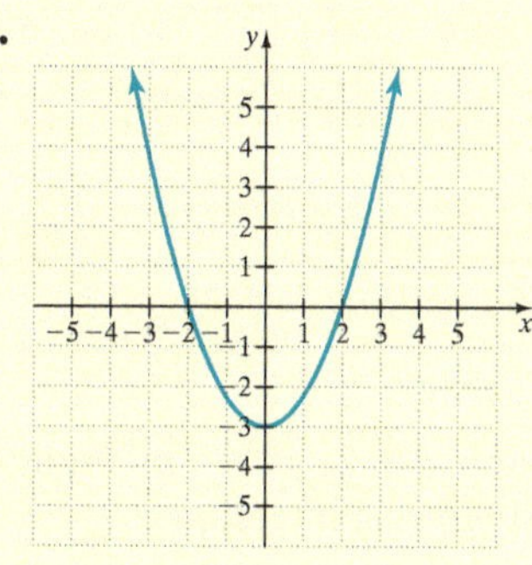

Answers

1. x-intercept: $(-4, 0)$; y-intercept: $(0, -6)$

2. x-intercepts: $(-2, 0)$, $(2, 0)$; y-intercept: $(0, -3)$

3.

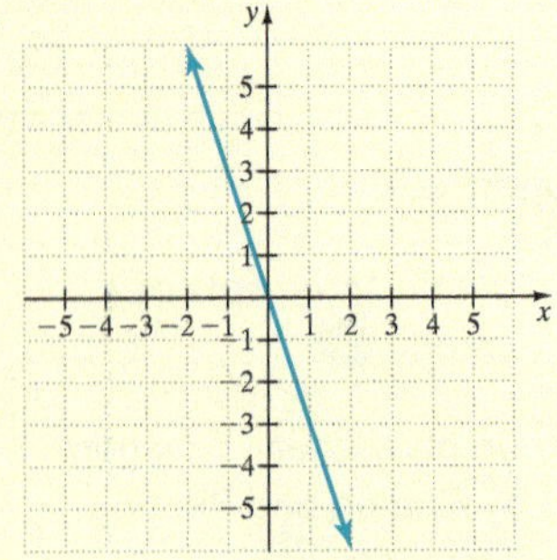

4.

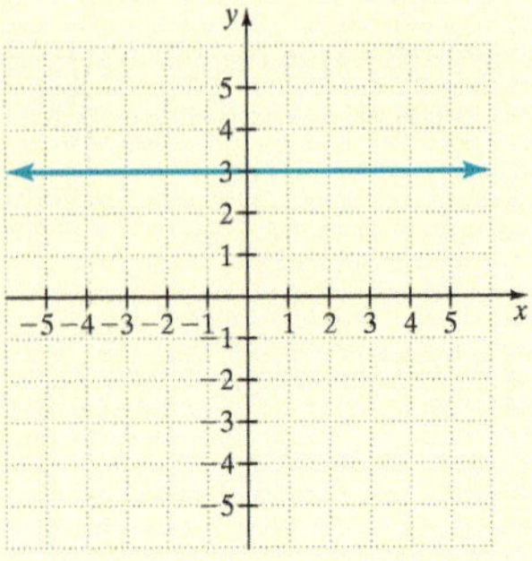

5.

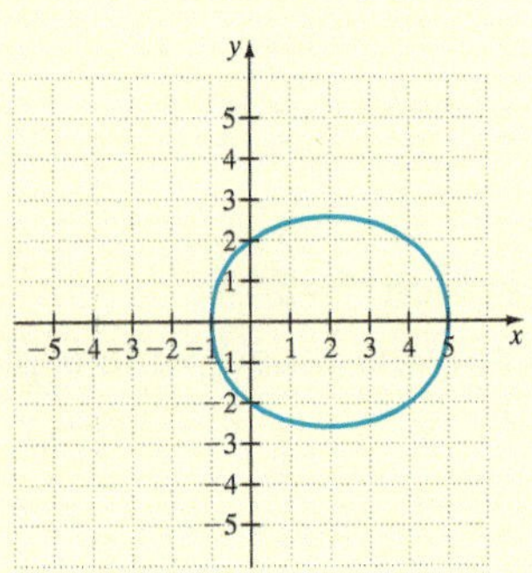

3.

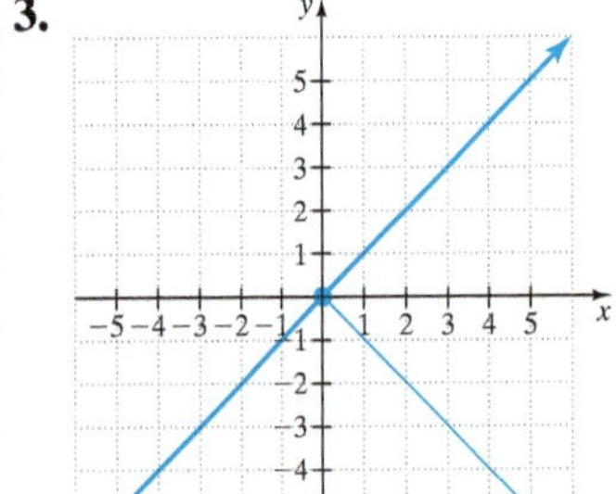

Solution:

x-intercept: $(0, 0)$
y-intercept: $(0, 0)$

4.

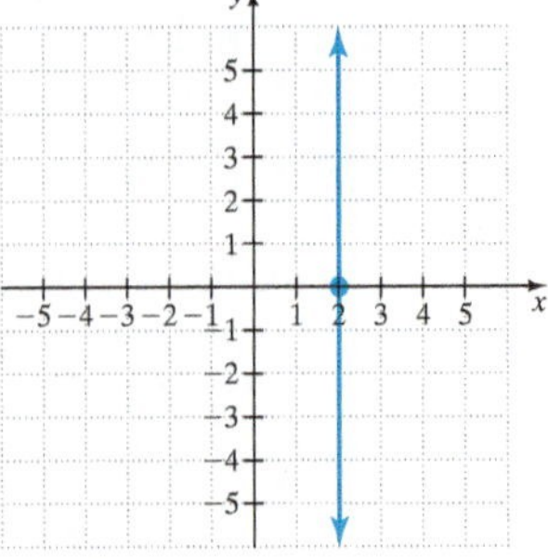

Solution:

x-intercept: $(2, 0)$
y-intercept: none

5.

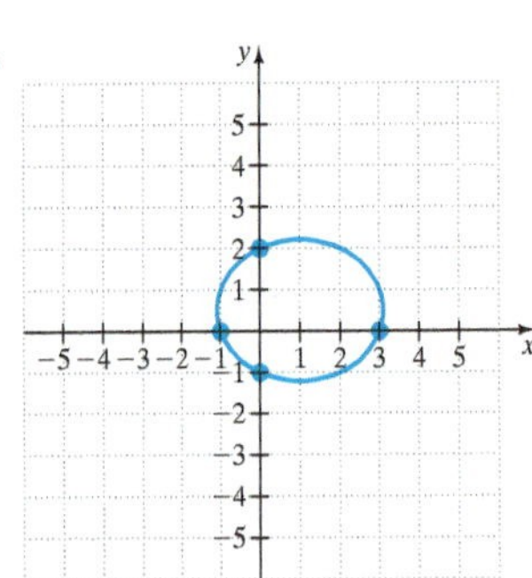

Solution:

x-intercepts: $(-1, 0)$, $(3, 0)$
y-intercepts: $(0, 2)$, $(0, -1)$

Helpful Hint Notice that any time $(0, 0)$ is a point of a graph, then it is an x-intercept and a y-intercept. Why? It is the only point that lies on both axes.

Work Practice 1–5

Objective 2 Using Intercepts to Graph a Linear Equation

Given the equation of a line, intercepts are usually easy to find since one coordinate is 0.

One way to find the y-intercept of a line, given its equation, is to let $x = 0$, since a point on the y-axis has an x-coordinate of 0. To find the x-intercept of a line, let $y = 0$, since a point on the x-axis has a y-coordinate of 0.

Finding x- and y-intercepts

To find the x-intercept, let $y = 0$ and solve for x.
To find the y-intercept, let $x = 0$ and solve for y.

Example 6 Graph $x - 3y = 6$ by finding and plotting intercepts.

Solution:

Let $y = 0$ to find the x-intercept and let $x = 0$ to find the y-intercept.

Let $y = 0$	Let $x = 0$
$x - 3y = 6$	$x - 3y = 6$
$x - 3(0) = 6$	$0 - 3y = 6$
$x - 0 = 6$	$-3y = 6$
$x = 6$	$y = -2$

Practice 6

Graph $x + 2y = -4$ by finding and plotting intercepts.

Answers

3. x-intercept and y-intercept: $(0, 0)$
4. no x-intercept; y-intercept: $(0, 3)$
5. x-intercepts: $(-1, 0)$, $(5, 0)$; y-intercepts: $(0, 2)$, $(0, -2)$
6.

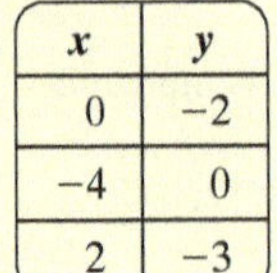

x	y
0	−2
−4	0
2	−3

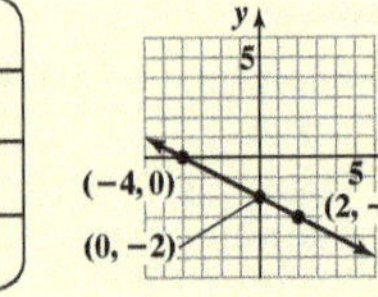

The x-intercept is $(6, 0)$ and the y-intercept is $(0, -2)$. We find a third ordered pair solution to check our work. If we let $y = -1$, then $x = 3$. Plot the points $(6, 0)$, $(0, -2)$, and $(3, -1)$. The graph of $x - 3y = 6$ is the line drawn through these points, as shown.

x	y
6	0
0	−2
3	−1

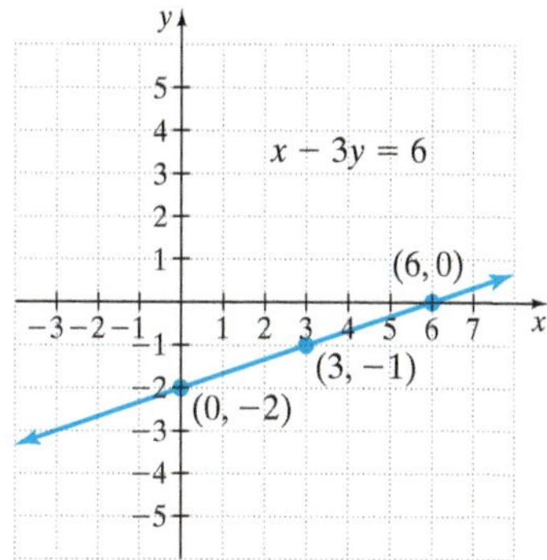

Work Practice 6

Example 7

Graph $x = -2y$ by plotting intercepts.

Solution:

Let $y = 0$ to find the x-intercept and $x = 0$ to find the y-intercept.

Let $y = 0$	Let $x = 0$
$x = -2y$	$x = -2y$
$x = -2(0)$	$0 = -2y$
$x = 0$	$0 = y$

Both the x-intercept and y-intercept are $(0, 0)$. In other words, when $x = 0$, then $y = 0$, which gives the ordered pair $(0, 0)$. Also, when $y = 0$, then $x = 0$, which gives the same ordered pair $(0, 0)$. This happens when the graph passes through the origin. Since two points are needed to determine a line, we must find at least one more ordered pair that satisfies $x = -2y$. We will let $y = -1$ to find a second ordered pair solution and let $y = 1$ as a checkpoint.

Let $y = -1$	Let $y = 1$
$x = -2(-1)$	$x = -2(1)$
$x = 2$	$x = -2$

The ordered pairs are $(0, 0)$, $(2, -1)$, and $(-2, 1)$. Plot these points to graph $x = -2y$.

x	y
0	0
2	−1
−2	1

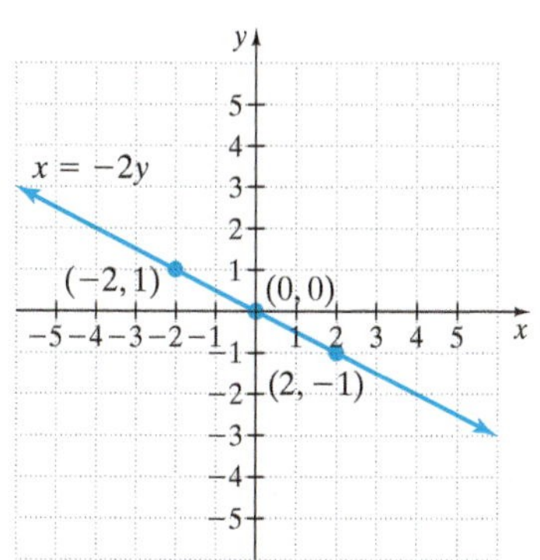

Work Practice 7

Practice 7

Graph $x = 3y$ by plotting intercepts.

Answer

7.

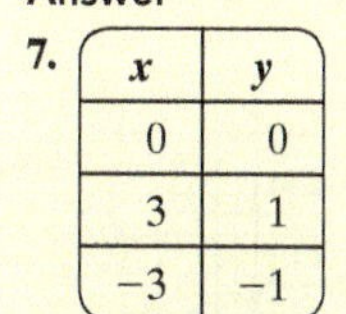

x	y
0	0
3	1
−3	−1

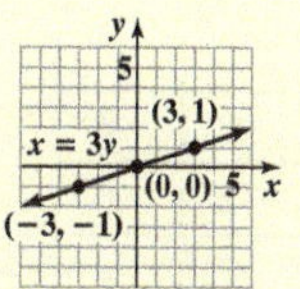

Practice 8

Graph $3x = 2y + 4$.

Example 8 Graph $4x = 3y - 9$.

Solution:

Find the x- and y-intercepts, and then choose $x = 2$ to find a checkpoint.

Let $y = 0$	Let $x = 0$	Let $x = 2$
$4x = 3(0) - 9$	$4 \cdot 0 = 3y - 9$	$4(2) = 3y - 9$
$4x = -9$	$9 = 3y$	$8 = 3y - 9$
Solve for x.	Solve for y.	Solve for y.
$x = -\frac{9}{4}$ or $-2\frac{1}{4}$	$3 = y$	$17 = 3y$
		$\frac{17}{3} = y$ or $y = 5\frac{2}{3}$

The ordered pairs are $\left(-2\frac{1}{4}, 0\right)$, $(0, 3)$, and $\left(2, 5\frac{2}{3}\right)$. The equation $4x = 3y - 9$ is graphed as follows.

x	y
$-2\frac{1}{4}$	0
0	3
2	$5\frac{2}{3}$

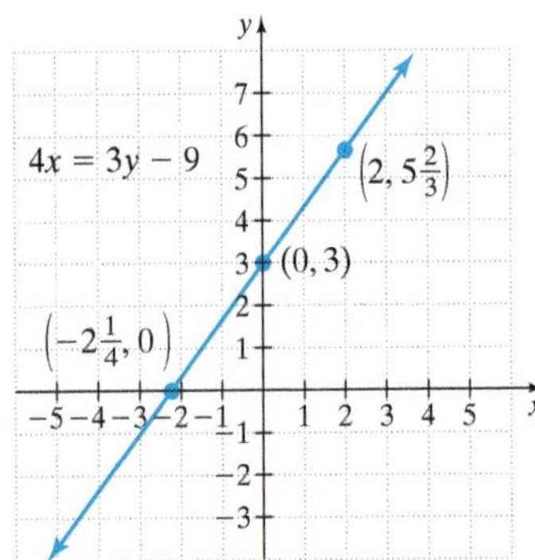

Work Practice 8

Objective 3 Graphing Vertical and Horizontal Lines 

The equation $x = c$, where c is a real number constant, is a linear equation in two variables because it can be written in the form $x + 0y = c$. The graph of this equation is a vertical line as shown in the next example.

Practice 9

Graph $x = -2$.

Example 9 Graph $x = 2$.

Solution:

The equation $x = 2$ can be written as $x + 0y = 2$. For any y-value chosen, notice that x is 2. No other value for x satisfies $x + 0y = 2$. Any ordered pair whose x-coordinate is 2 is a solution of $x + 0y = 2$. We will use the ordered pair solutions $(2, 3)$, $(2, 0)$, and $(2, -3)$ to graph $x = 2$.

x	y
2	3
2	0
2	−3

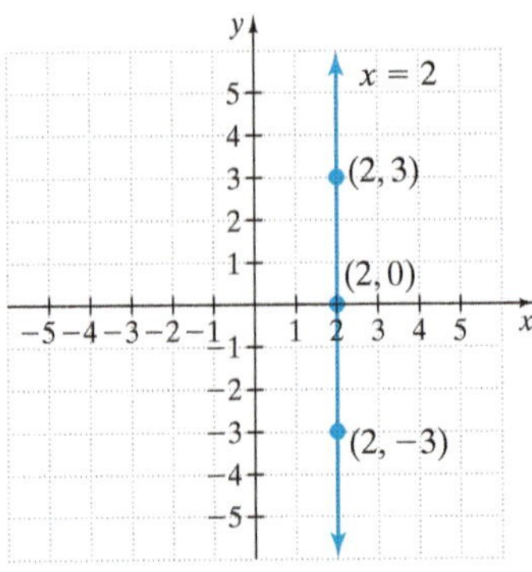

Answers

8.

x	y
0	−2
$\frac{4}{3}$	0
2	1

9.

x	y
−2	4
−2	0
−2	−4

The graph is a vertical line with x-intercept $(2, 0)$. Note that this graph has no y-intercept because x is never 0.

Work Practice 9

Vertical Lines

The graph of $x = c$, where c is a real number, is a vertical line with x-intercept $(c, 0)$.

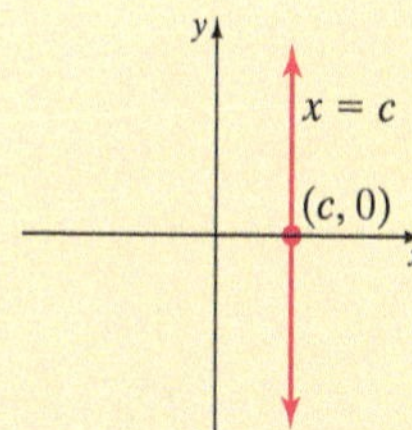

Example 10 Graph $y = -3$.

Solution:

The equation $y = -3$ can be written as $0x + y = -3$. For any x-value chosen, y is -3. If we choose 4, 1, and -2 as x-values, the ordered pair solutions are $(4, -3)$, $(1, -3)$, and $(-2, -3)$. Use these ordered pairs to graph $y = -3$. The graph is a horizontal line with y-intercept $(0, -3)$ and no x-intercept.

x	y
4	-3
1	-3
-2	-3

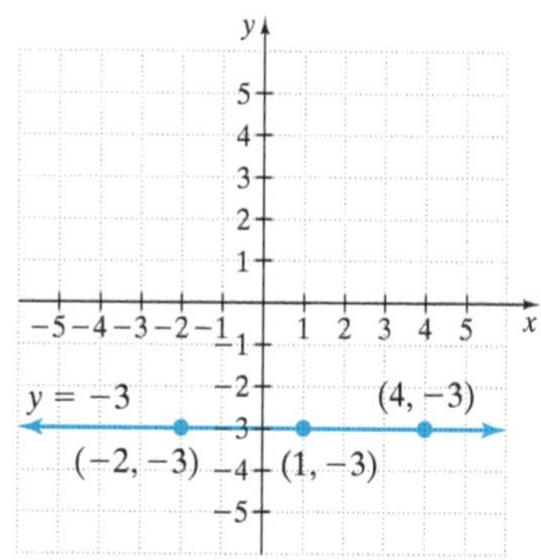

Work Practice 10

Horizontal Lines

The graph of $y = c$, where c is a real number, is a horizontal line with y-intercept $(0, c)$.

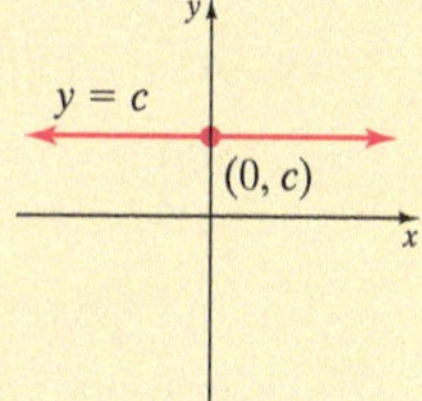

Practice 10

Graph $y = 2$.

Answer

10.

x	y
-5	2
0	2
5	2

Graphing Calculator Explorations

You may have noticed that to use the $\boxed{Y=}$ key on a grapher to graph an equation, the equation must be solved for y. For example, to graph $2x + 3y = 7$, we solve this equation for y.

$$2x + 3y = 7$$

$$3y = -2x + 7 \quad \text{Subtract } 2x \text{ from both sides.}$$

$$\frac{3y}{3} = -\frac{2x}{3} + \frac{7}{3} \quad \text{Divide both sides by 3.}$$

$$y = -\frac{2}{3}x + \frac{7}{3} \quad \text{Simplify.}$$

To graph $2x + 3y = 7$ or $y = -\frac{2}{3}x + \frac{7}{3}$, press the $\boxed{Y=}$ key and enter

$$Y_1 = -\frac{2}{3}x + \frac{7}{3}$$

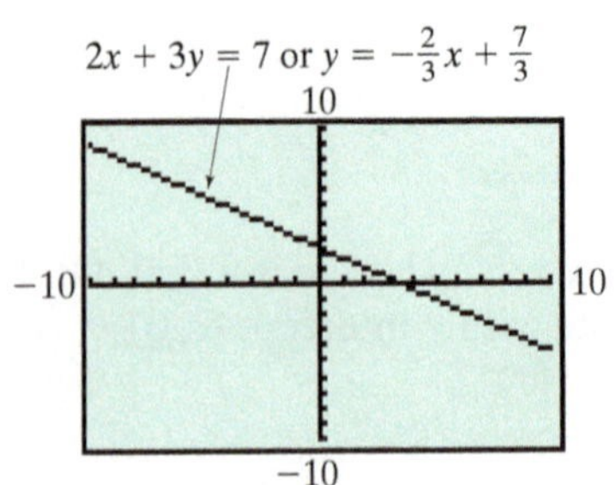

Graph each linear equation.

1. $x = 3.78y$
2. $-2.61y = x$
3. $3x + 7y = 21$
4. $-4x + 6y = 21$
5. $-2.2x + 6.8y = 15.5$
6. $5.9x - 0.8y = -10.4$

Vocabulary, Readiness & Video Check

Use the choices below to fill in each blank. Some choices may be used more than once. Exercises 1 and 2 come from Section 10.2.

x	vertical	x-intercept	linear
y	horizontal	y-intercept	standard

1. An equation that can be written in the form $Ax + By = C$ is called a(n) ________ equation in two variables.
2. The form $Ax + By = C$ is called ________ form.
3. The graph of the equation $y = -1$ is a(n) ________ line.
4. The graph of the equation $x = 5$ is a(n) ________ line.
5. A point where a graph crosses the y-axis is called a(n) ________.
6. A point where a graph crosses the x-axis is called a(n) ________.
7. Given an equation of a line, to find the x-intercept (if there is one), let ______ = 0 and solve for ______.
8. Given an equation of a line, to find the y-intercept (if there is one), let ______ = 0 and solve for ______.

Martin-Gay Interactive Videos

See Video 10.3

Watch the section lecture video and answer the following questions.

Objective 1 9. At the end of Example 2, patterns are discussed. What reason is given for why x-intercepts have y-values of 0? For why y-intercepts have x-values of 0?

Objective 2 10. In Example 3, the goal is to use the x- and y-intercepts to graph a line. Yet once the two intercepts are found, a third point is also found before the line is graphed. Why do think this practice of finding a third point is continued?

Objective 3 11. From Examples 5 and 6, what can you say about the coefficient of x when the equation of a horizontal line is written as $Ax + By = C$? What about the coefficient of y when the equation of a vertical line is written as $Ax + By = C$?

10.3 Exercise Set MyMathLab® 

Identify the intercepts. See Examples 1 through 5.

1.

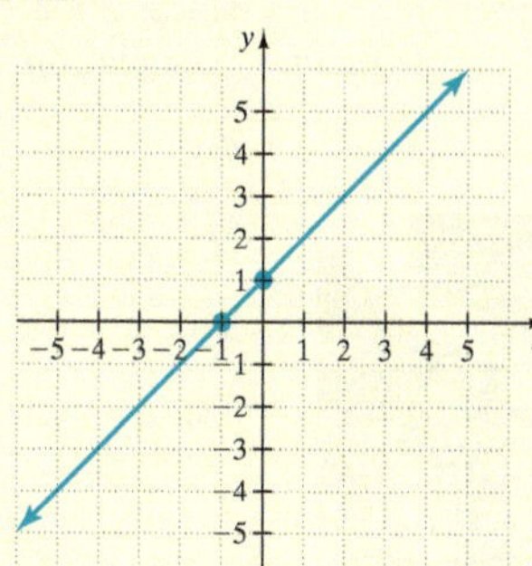

2.

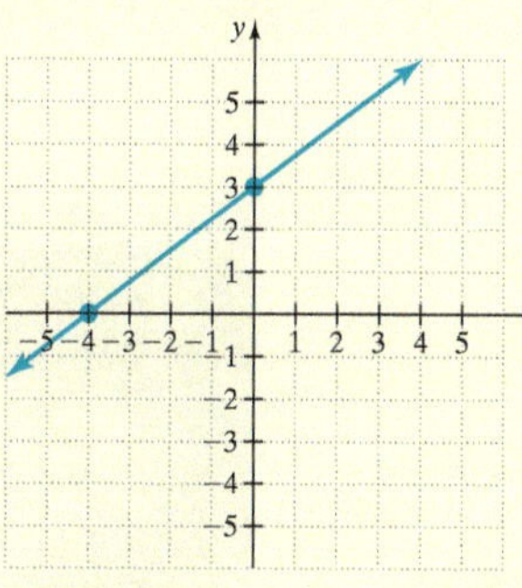

3.

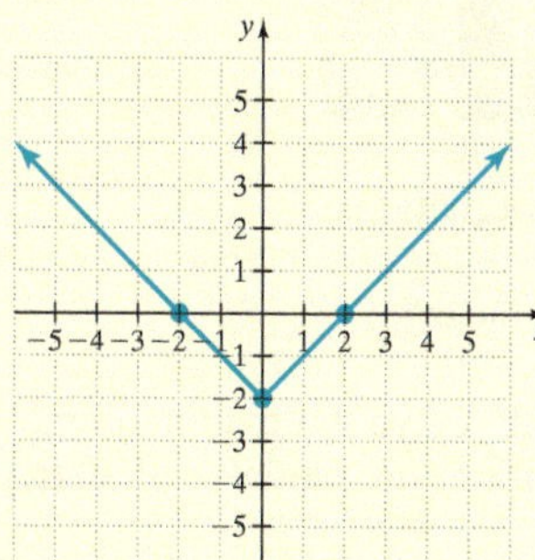

4.

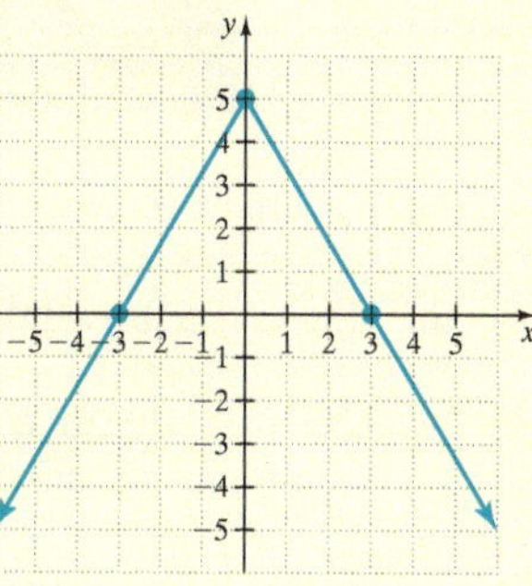

5.

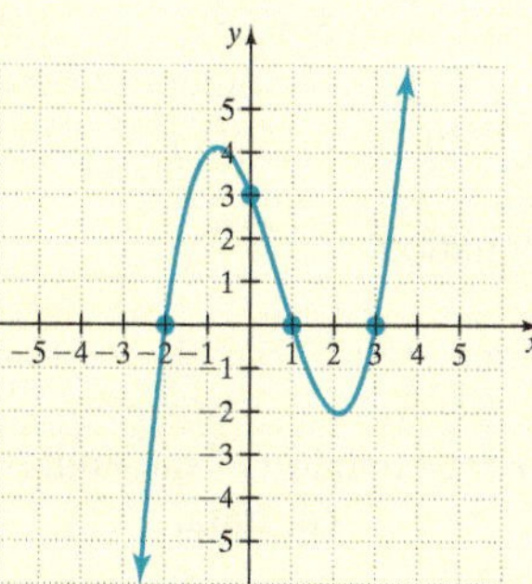

6.

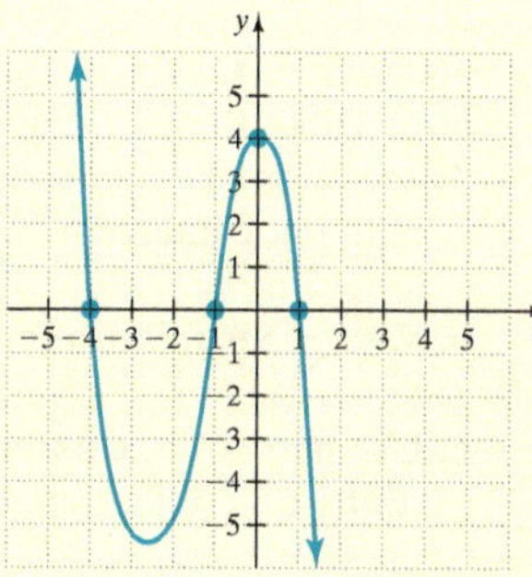

7.

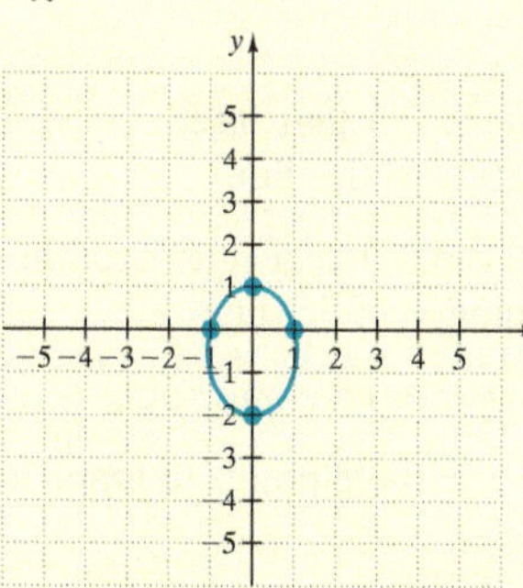

8.

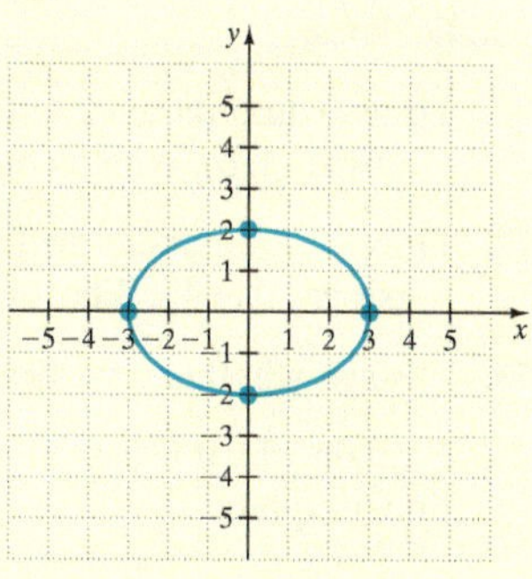

Solve. See Examples 1 through 5.

9. What is the greatest number of intercepts for a line?
10. What is the least number of intercepts for a line?
11. What is the least number of intercepts for a circle?
12. What is the greatest number of intercepts for a circle?

Graph each linear equation by finding and plotting its intercepts. See Examples 6 through 8.

13. $x - y = 3$
14. $x - y = -4$
15. $x = 5y$
16. $x = 2y$
17. $-x + 2y = 6$
18. $x - 2y = -8$
19. $2x - 4y = 8$
20. $2x + 3y = 6$
21. $y = 2x$
22. $y = -2x$
23. $y = 3x + 6$
24. $y = 2x + 10$

Graph each linear equation. See Examples 9 and 10.

25. $x = -1$
26. $y = 5$
27. $y = 0$
28. $x = 0$
29. $y + 7 = 0$
30. $x - 2 = 0$
31. $x + 3 = 0$
32. $y - 6 = 0$

MIXED PRACTICE

Graph each linear equation. See Examples 6 through 10.

33. $x = y$
34. $x = -y$
35. $x + 8y = 8$
36. $x + 3y = 9$
37. $5 = 6x - y$
38. $4 = x - 3y$
39. $-x + 10y = 11$
40. $-x + 9y = 10$
41. $x = -4\frac{1}{2}$
42. $x = -1\frac{3}{4}$
43. $y = 3\frac{1}{4}$
44. $y = 2\frac{1}{2}$
45. $y = -\frac{2}{3}x + 1$
46. $y = -\frac{3}{5}x + 3$
47. $4x - 6y + 2 = 0$
48. $9x - 6y + 3 = 0$

For Exercises 49 through 54, match each equation with its graph. See Graphs A.–F. below and on the next page.

49. $y = 3$
50. $y = 2x + 2$
51. $x = -1$
52. $x = 3$
53. $y = 2x + 3$
54. $y = -2x$

A.

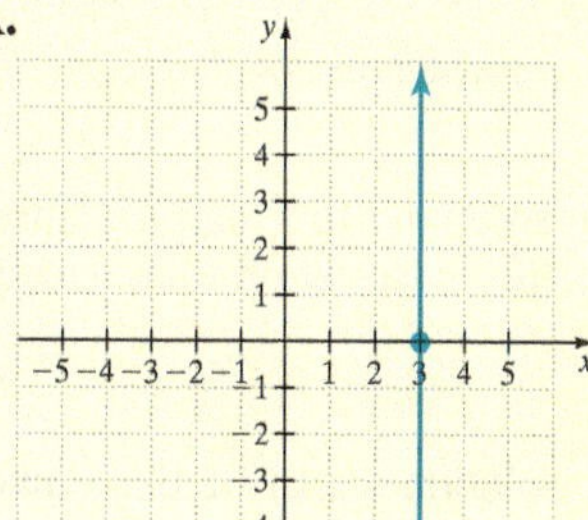

B.

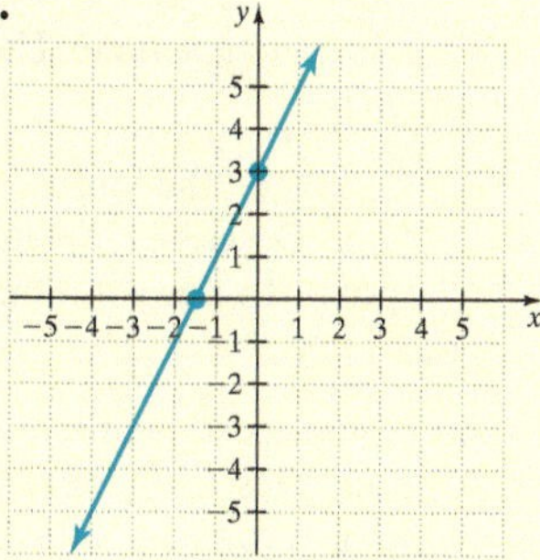

C.

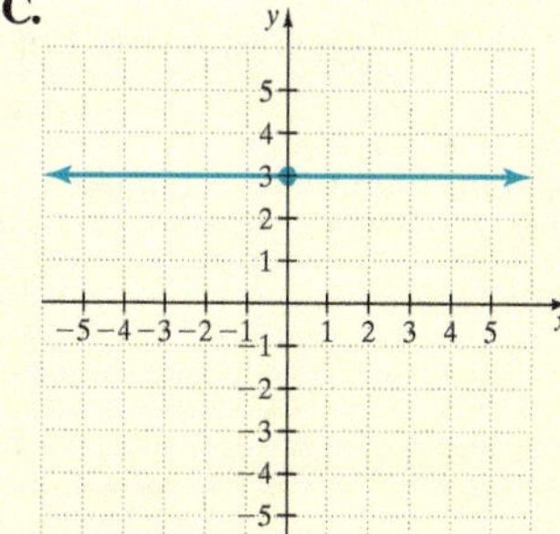

D.

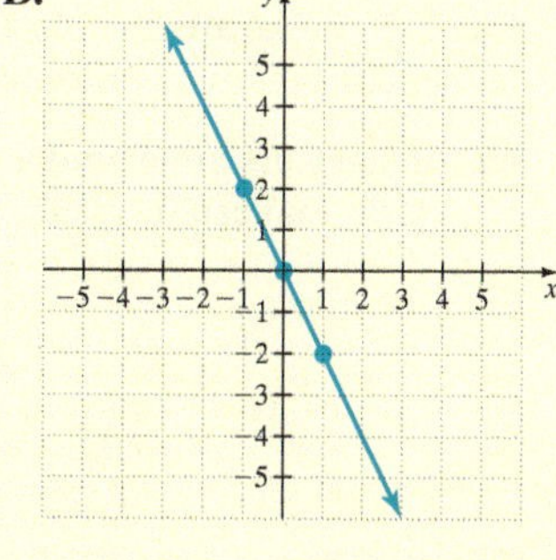

E.

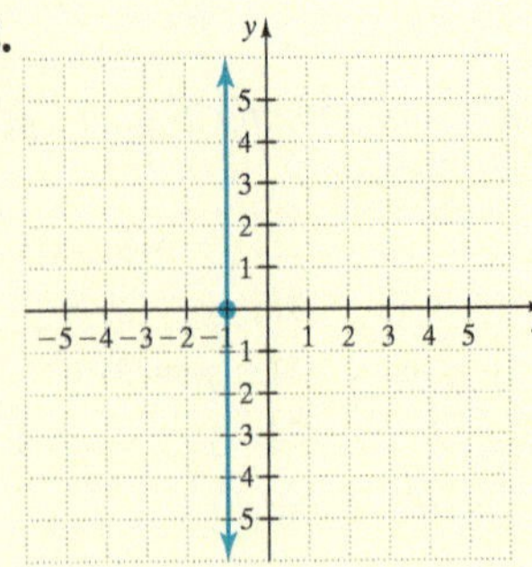

F.

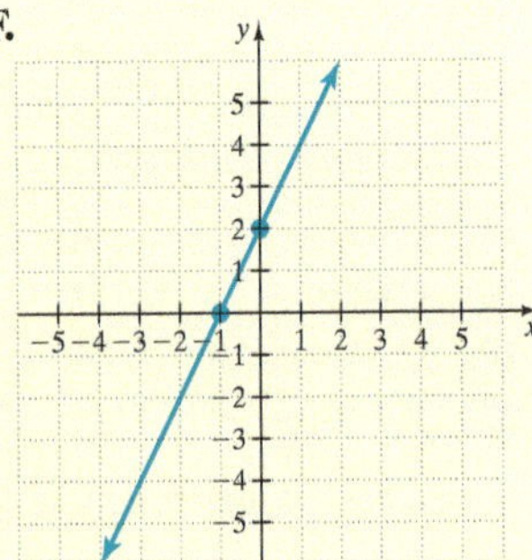

REVIEW AND PREVIEW

Simplify. See Section 4.6.

55. $\frac{-6-3}{2-8}$

56. $\frac{4-5}{-1-0}$

57. $\frac{-8-(-2)}{-3-(-2)}$

58. $\frac{12-3}{10-9}$

59. $\frac{0-6}{5-0}$

60. $\frac{2-2}{3-5}$

CONCEPT EXTENSIONS

Answer the following true or false.

61. All lines have an x-intercept *and* a y-intercept.

62. The graph of $y = 4x$ contains the point $(0, 0)$.

63. The graph of $x + y = 5$ has an x-intercept of $(5, 0)$ and a y-intercept of $(0, 5)$.

64. The graph of $y = 5x$ contains the point $(5, 1)$.

The production supervisor at Alexandra's Office Products finds that it takes 3 hours to manufacture a particular office chair and 6 hours to manufacture an office desk. A total of 1200 hours is available to produce office chairs and desks of this style. The linear equation that models this situation is $3x + 6y = 1200$, where x represents the number of chairs produced and y the number of desks manufactured. Use this information for Exercises 65 through 68.

65. Complete the ordered pair solution (0,) of this equation. Describe the manufacturing situation that corresponds to this solution.

66. Complete the ordered pair solution (, 0) of this equation. Describe the manufacturing situation that corresponds to this solution.

67. If 50 desks are manufactured, find the greatest number of chairs that can be made.

68. If 50 chairs are manufactured, find the greatest number of desks that can be made.

69. Since 2002, admissions at movie theaters in the U.S. and Canada have mostly been in a decline. The number of people y (in billions) who go to movie theaters each year can be estimated by the equation $y = -0.025x + 1.55$, where x represents the number of years since 2002. (*Source:* Based on data from Motion Picture Association of America)

a. Find the x-intercept of this equation.

b. What does this x-intercept mean?

c. Use part (b) to comment on your opinion of the limitations of using equations to model real data.

70. The price of admission to a movie theater has been steadily increasing. The price of regular admission y (in dollars) to a movie theater may be represented by the equation $y = 0.24x + 5.28$, where x is the number of years after 2000. (*Source:* Based on data from Motion Picture Association of America)

a. Find the x-intercept of this equation.

b. What does this x-intercept mean?

c. Use part (b) to comment on your opinion of the limitations of using equations to model real data.

Two lines in the same plane that do not intersect are called **parallel lines.**

△ **71.** Draw a line parallel to the line $x = 5$ that intersects the x-axis at $(1, 0)$. What is the equation of this line?

△ **72.** Draw a line parallel to the line $y = -1$ that intersects the y-axis at $(0, -4)$. What is the equation of this line?

73. Discuss whether a vertical line ever has a y-intercept.

74. Explain why it is a good idea to use three points to graph a linear equation.

75. Discuss whether a horizontal line ever has an x-intercept.

76. Explain how to find intercepts.

10.4 Slope and Rate of Change 

Objectives

1 Find the Slope of a Line Given Two Points of the Line.

2 Find the Slope of a Line Given Its Equation.

3 Find the Slopes of Horizontal and Vertical Lines.

4 Compare the Slopes of Parallel and Perpendicular Lines.

5 Interpret Slope as a Rate of Change.

Objective 1 Finding the Slope of a Line Given Two Points of the Line

Thus far, much of this chapter has been devoted to graphing lines. You have probably noticed by now that a key feature of a line is its slant or steepness. In mathematics, the slant or steepness of a line is formally known as its **slope.** We measure the slope of a line by the ratio of vertical change to the corresponding horizontal change as we move along the line.

On the line below, for example, suppose that we begin at the point (1, 2) and move to the point (4, 6). The vertical change is the change in y-coordinates: $6 - 2$ or 4 units. The corresponding horizontal change is the change in x-coordinates: $4 - 1 = 3$ units. The ratio of these changes is

$$\text{slope} = \frac{\text{change in } y \text{ (vertical change)}}{\text{change in } x \text{ (horizontal change)}} = \frac{4}{3}$$

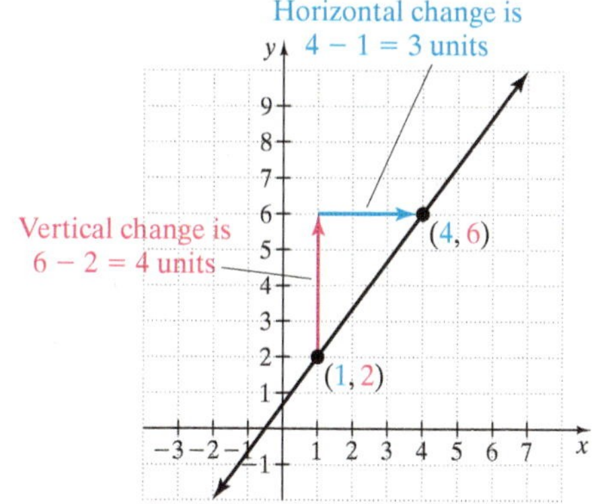

The slope of this line, then, is $\frac{4}{3}$. This means that for every 4 units of change in y-coordinates, there is a corresponding change of 3 units in x-coordinates.

Helpful Hint

It makes no difference what two points of a line are chosen to find its slope. The slope of a line is the same everywhere on the line.

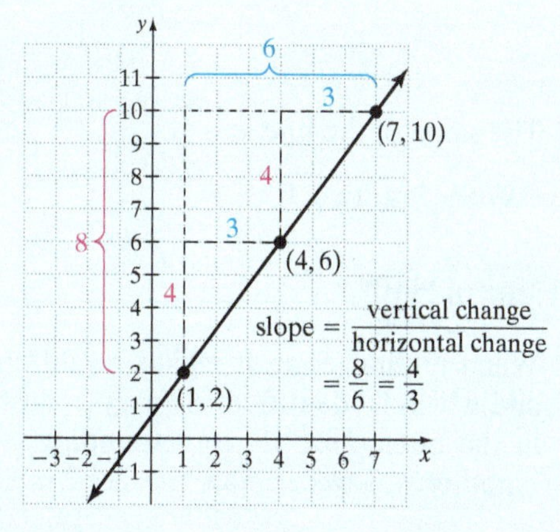

To find the slope of a line, then, choose two points of the line. Label the two x-coordinates of two points x_1 and x_2 (read "x sub one" and "x sub two"), and label the corresponding y-coordinates y_1 and y_2.

The vertical change or **rise** between these points is the difference in the y-coordinates: $y_2 - y_1$. The horizontal change or **run** between the points is the difference of the x-coordinates: $x_2 - x_1$. The slope of the line is the ratio of $y_2 - y_1$ to $x_2 - x_1$, and we traditionally use the letter m to denote slope: $m = \frac{y_2 - y_1}{x_2 - x_1}$.

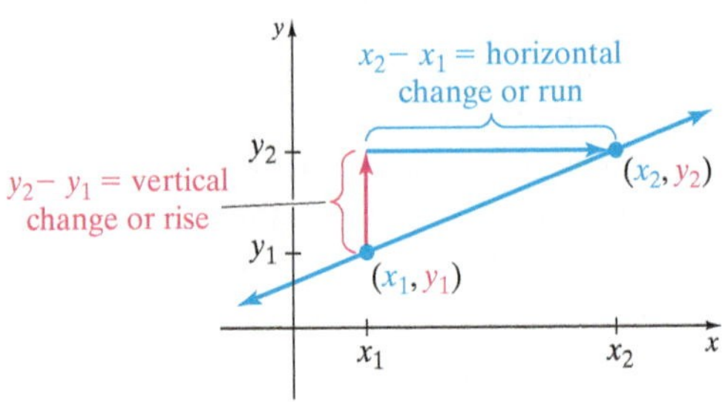

Slope of a Line

The slope m of the line containing the points (x_1, y_1) and (x_2, y_2) is given by

$$m = \frac{\text{rise}}{\text{run}} = \frac{\text{change in } y}{\text{change in } x} = \frac{y_2 - y_1}{x_2 - x_1}, \quad \text{as long as } x_2 \neq x_1$$

Practice 1

Find the slope of the line through $(-4, 11)$ and $(2, 5)$.

Example 1 Find the slope of the line through $(-1, 5)$ and $(2, -3)$. Graph the line.

Solution: If we let (x_1, y_1) be $(-1, 5)$, then $x_1 = -1$ and $y_1 = 5$. Also, let (x_2, y_2) be $(2, -3)$ so that $x_2 = 2$ and $y_2 = -3$. Then, by the definition of slope,

$$m = \frac{y_2 - y_1}{x_2 - x_1} = \frac{-3 - 5}{2 - (-1)} = \frac{-8}{3} = -\frac{8}{3}$$

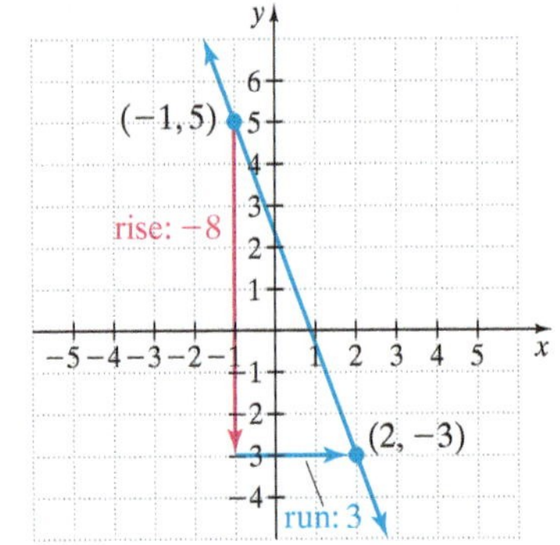

The slope of the line is $-\frac{8}{3}$.

Work Practice 1

Helpful Hint

When finding slope, it makes no difference which point is identified as (x_1, y_1) and which is identified as (x_2, y_2). Just remember that whatever y-value is first in the numerator, its corresponding x-value must be first in the denominator. Another way to calculate the slope in Example 1 is:

$$m = \frac{y_2 - y_1}{x_2 - x_1} = \frac{5 - (-3)}{-1 - 2} = \frac{8}{-3} \text{ or } -\frac{8}{3} \quad \leftarrow \text{Same slope as found in Example 1.}$$

✓Concept Check The points $(-2, -5)$, $(0, -2)$, $(4, 4)$, and $(10, 13)$ all lie on the same line. Work with a partner and verify that the slope is the same no matter which points are used to find slope.

Answer

1. -1

✓Concept Check Answer

$m = \frac{3}{2}$

Example 2 Find the slope of the line through $(-1, -2)$ and $(2, 4)$. Graph the line.

Solution: Let (x_1, y_1) be $(2, 4)$ and (x_2, y_2) be $(-1, -2)$.

$$m = \frac{y_2 - y_1}{x_2 - x_1}$$

$$= \frac{-2 - 4}{-1 - 2} \quad \begin{matrix}\text{y-value} \\ \text{corresponding x-value}\end{matrix}$$

$$= \frac{-6}{-3} = 2$$

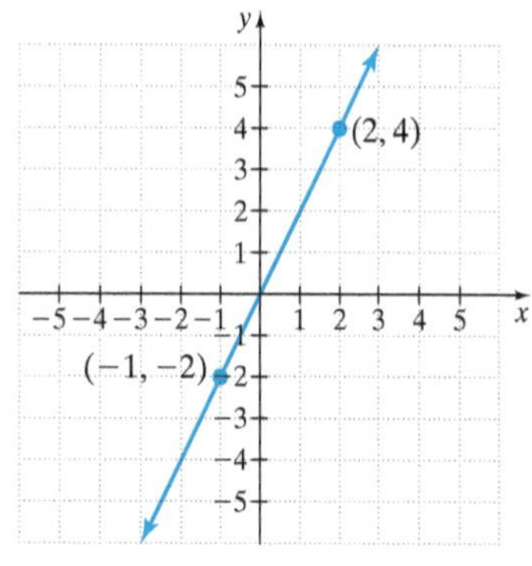

Work Practice 2

Practice 2

Find the slope of the line through $(-3, -1)$ and $(3, 1)$.

Helpful Hint

The slope for Example 2 is the same if we let (x_1, y_1) be $(-1, -2)$ and (x_2, y_2) be $(2, 4)$.

y-value

$$m = \frac{4 - (-2)}{2 - (-1)} = \frac{6}{3} = 2$$

corresponding x-value

✓**Concept Check** What is wrong with the following slope calculation for the points $(3, 5)$ and $(-2, 6)$?

$$m = \frac{5 - 6}{-2 - 3} = \frac{-1}{-5} = \frac{1}{5}$$

Notice that the slope of the line in Example 1 is negative, whereas the slope of the line in Example 2 is positive. Let your eye follow the line with negative slope from left to right and notice that the line "goes down." Following the line with positive slope from left to right, notice that the line "goes up." This is true in general.

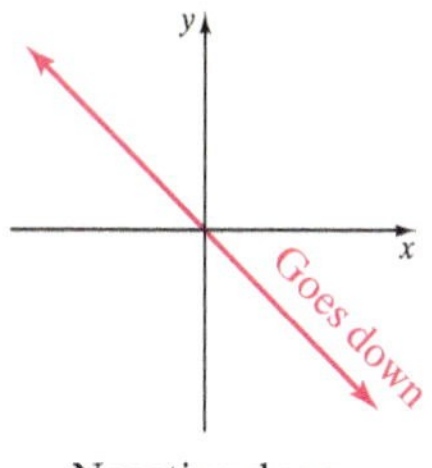

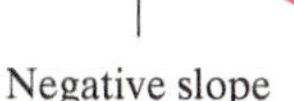

Negative slope

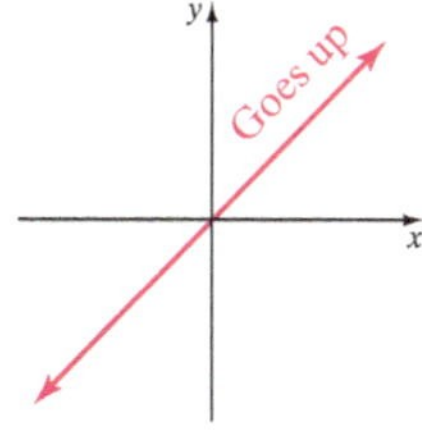

Positive slope

Helpful Hint

To decide whether a line "goes up" or "goes down," always follow the line from left to right.

Objective 2 Finding the Slope of a Line Given Its Equation

As we have seen, the slope of a line is defined by two points on the line. Thus, if we know the equation of a line, we can find its slope by finding two of its points. For example, let's find the slope of the line

$$y = 3x + 2$$

To find two points, we can choose two values for x and substitute to find corresponding y-values. If $x = 0$, for example, $y = 3 \cdot 0 + 2$ or $y = 2$. If $x = 1$, $y = 3 \cdot 1 + 2$ or $y = 5$. This gives the ordered pairs $(0, 2)$ and $(1, 5)$. Using the definition for slope, we have

$$m = \frac{5 - 2}{1 - 0} = \frac{3}{1} = 3 \quad \text{The slope is 3.}$$

Notice that the slope, 3, is the same as the coefficient of x in the equation $y = 3x + 2$.

Also, recall from Section 10.2 that the graph of an equation of the form $y = mx + b$ has y-intercept $(0, b)$.

Answer

2. $\frac{1}{3}$

✓**Concept Check Answer**

The order in which the x- and y-values are used must be the same.

$$m = \frac{5 - 6}{3 - (-2)} = \frac{-1}{5} = -\frac{1}{5}$$

This means that the y-intercept of the graph of $y = 3x + 2$ is $(0, 2)$. This is true in general and the form $y = mx + b$ is appropriately called the **slope-intercept form.**

$$y = \underset{\text{slope}}{\underset{\uparrow}{m}}x + \underset{\substack{y\text{-intercept}\\(0,\,b)}}{\underset{\uparrow}{b}}$$

Slope-Intercept Form

When a linear equation in two variables is written in slope-intercept form,

$$y = mx + b$$

m is the slope of the line and $(0, b)$ is the y-intercept of the line.

Practice 3

Find the slope and y-intercept of the line whose equation is $y = \frac{2}{3}x - 2$.

Example 3 Find the slope and y-intercept of the line whose equation is $y = \frac{3}{4}x + 6$

Solution: The equation is in slope-intercept form, $y = mx + b$.

$$y = \frac{3}{4}x + 6$$

The coefficient of x, $\frac{3}{4}$, is the slope and the constant term, 6, is the y-value of the y-intercept, $(0, 6)$.

Work Practice 3

Practice 4

Find the slope and y-intercept of the line whose equation is $-y = -6x + 5$.

Example 4 Find the slope and the y-intercept of the line whose equation is $-y = 5x - 2$.

Solution: Remember, the equation must be solved for y (not $-y$) in order for it to be written in slope-intercept form.

To solve for y, let's divide both sides of the equation by -1.

$$-y = 5x - 2$$
$$\frac{-y}{-1} = \frac{5x}{-1} - \frac{2}{-1} \quad \text{Divide both sides by } -1.$$
$$y = -5x + 2 \quad \text{Simplify.}$$

The coefficient of x, -5, is the slope and the constant term, 2, is the y-value of the y-intercept, $(0, 2)$.

Work Practice 4

Practice 5

Find the slope and the y-intercept of the line whose equation is $5x + 2y = 8$.

Example 5 Find the slope and the y-intercept of the line whose equation is $3x - 4y = 4$.

Solution: Write the equation in slope-intercept form by solving for y.

$$3x - 4y = 4$$
$$-4y = -3x + 4 \quad \text{Subtract } 3x \text{ from both sides.}$$
$$\frac{-4y}{-4} = \frac{-3x}{-4} + \frac{4}{-4} \quad \text{Divide both sides by } -4.$$
$$y = \frac{3}{4}x - 1 \quad \text{Simplify.}$$

The coefficient of x, $\frac{3}{4}$, is the slope, and the y-intercept is $(0, -1)$.

Work Practice 5

Answers

3. $m = \frac{2}{3}$; y-intercept: $(0, -2)$

4. $m = 6$; y-intercept: $(0, -5)$

5. $m = -\frac{5}{2}$; y-intercept: $(0, 4)$

Objective 3 Finding Slopes of Horizontal and Vertical Lines

Recall that if a line tilts upward from left to right, its slope is positive. If a line tilts downward from left to right, its slope is negative. Let's now find the slopes of two special lines, horizontal and vertical lines.

Example 6 Find the slope of the line $y = -1$.

Solution: Recall that $y = -1$ is a horizontal line with y-intercept $(0, -1)$. To find the slope, find two ordered pair solutions of $y = -1$. Solutions of $y = -1$ must have a y-value of -1. Let's use points $(2, -1)$ and $(-3, -1)$, which are on the line.

$$m = \frac{y_2 - y_1}{x_2 - x_1} = \frac{-1 - (-1)}{-3 - 2} = \frac{0}{-5} = 0$$

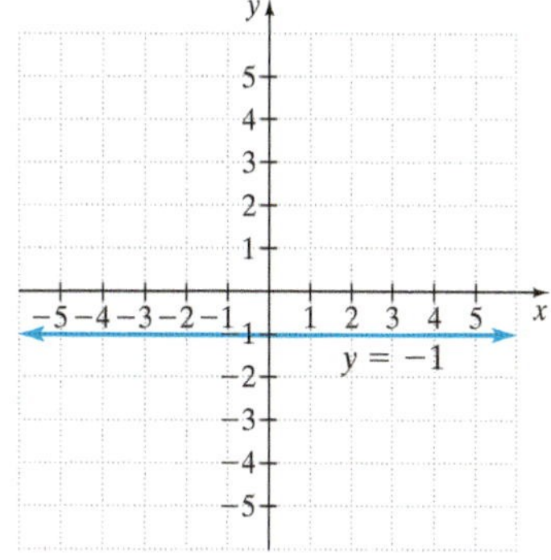

The slope of the line $y = -1$ is 0 and its graph is shown.

Work Practice 6

Practice 6

Find the slope of the line $y = 3$.

Any two points of a horizontal line will have the same y-values. This means that the y-values will always have a difference of 0 for all horizontal lines. Thus, **all horizontal lines have a slope of 0.**

Example 7 Find the slope of the line $x = 5$.

Solution: Recall that the graph of $x = 5$ is a vertical line with x-intercept $(5, 0)$.

To find the slope, find two ordered pair solutions of $x = 5$. Solutions of $x = 5$ must have an x-value of 5. Let's use points $(5, 0)$ and $(5, 4)$, which are on the line.

$$m = \frac{y_2 - y_1}{x_2 - x_1} = \frac{4 - 0}{5 - 5} = \frac{4}{0}$$

Since $\frac{4}{0}$ is undefined, we say the slope of the vertical line $x = 5$ is undefined, and its graph is shown.

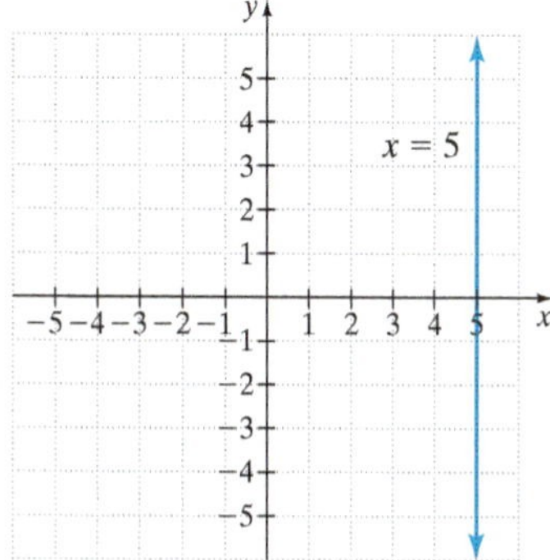

Work Practice 7

Practice 7

Find the slope of the line $x = -4$.

Any two points of a vertical line will have the same x-values. This means that the x-values will always have a difference of 0 for all vertical lines. Thus **all vertical lines have undefined slope.**

Helpful Hint

Slope of 0 and undefined slope are not the same. Vertical lines have undefined slope or no slope, while horizontal lines have a slope of 0.

Answers

6. $m = 0$ **7.** slope is undefined

Here is a general review of slope.

Summary of Slope

Slope m of the line through (x_1, y_1) and (x_2, y_2) is given by the equation $m = \frac{y_2 - y_1}{x_2 - x_1}$.

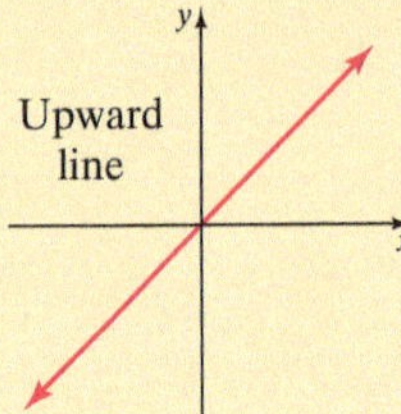

Positive slope: $m > 0$

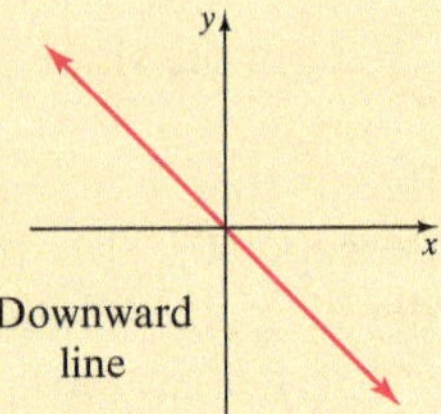

Negative slope: $m < 0$

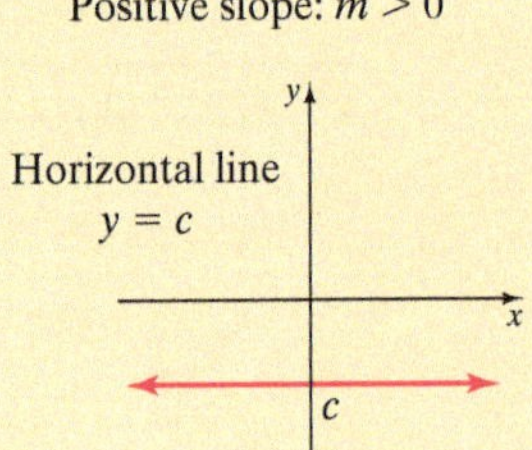

Zero slope: $m = 0$

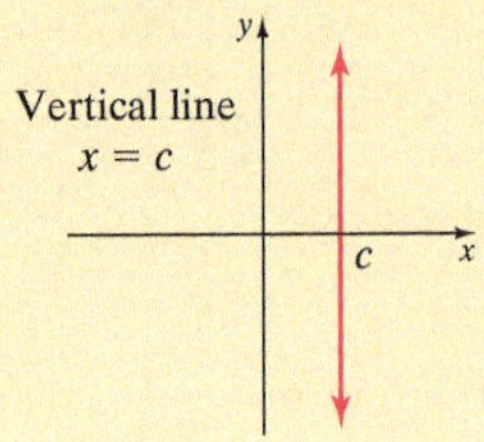

Undefined slope or no slope

Objective 4 Slopes of Parallel and Perpendicular Lines

Two lines in the same plane are **parallel** if they do not intersect. Slopes of lines can help us determine whether lines are parallel. Parallel lines have the same steepness, so it follows that they have the same slope.

For example, the graphs of

$$y = -2x + 4$$

and

$$y = -2x - 3$$

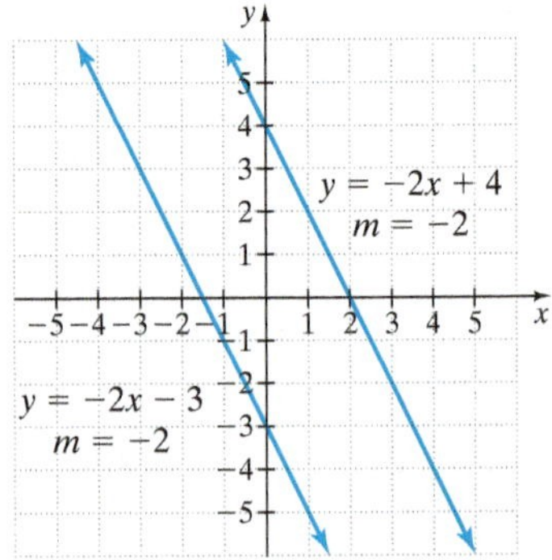

are shown. These lines have the same slope, -2. They also have different y-intercepts, so the lines are distinct and parallel. (If the y-intercepts were the same also, the lines would be the same.)

Parallel Lines

Nonvertical parallel lines have the same slope and different y-intercepts.

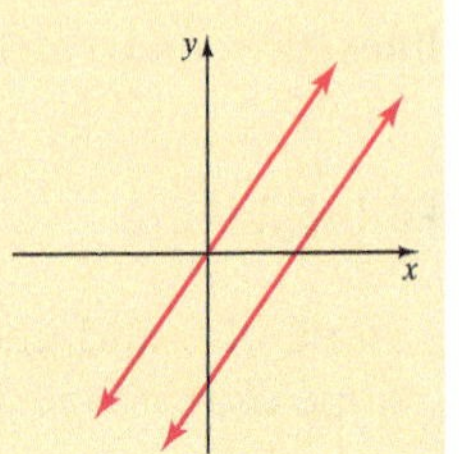

Two lines are **perpendicular** if they lie in the same plane and meet at a 90° (right) angle. How do the slopes of perpendicular lines compare? The product of the slopes of two perpendicular lines is −1.

For example, the graphs of

$$y = 4x + 1$$

and

$$y = -\frac{1}{4}x - 3$$

are shown. The slopes of the lines are 4 and $-\frac{1}{4}$. Their product is $4\left(-\frac{1}{4}\right) = -1$, so the lines are perpendicular.

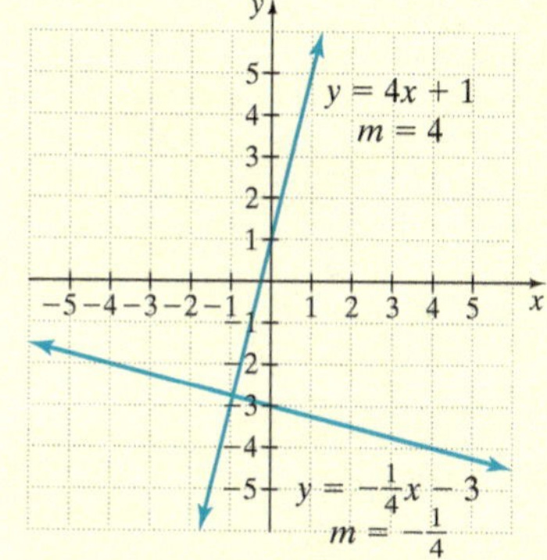

Perpendicular Lines

If the product of the slopes of two lines is −1, then the lines are perpendicular.

(Two nonvertical lines are perpendicular if the slope of one is the negative reciprocal of the slope of the other.)

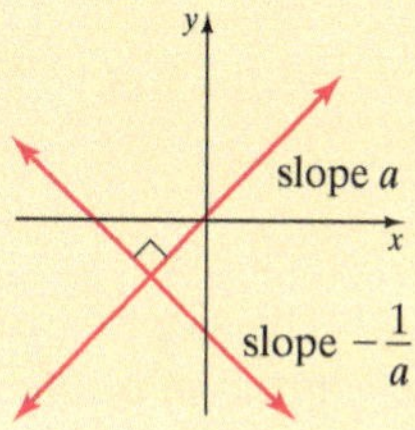

Helpful Hint

Here are examples of numbers that are negative (opposite) reciprocals.

Number	Negative Reciprocal	Their Product Is −1.
$\frac{2}{3}$	$-\frac{3}{2}$	$\frac{2}{3} \cdot -\frac{3}{2} = -\frac{6}{6} = -1$
-5 or $-\frac{5}{1}$	$\frac{1}{5}$	$-5 \cdot \frac{1}{5} = -\frac{5}{5} = -1$

Helpful Hint

Here are a few important facts about vertical and horizontal lines.

- Two distinct vertical lines are parallel.
- Two distinct horizontal lines are parallel.
- A horizontal line and a vertical line are always perpendicular.

Example 8 Determine whether each pair of lines is parallel, perpendicular, or neither.

a. $y = -\frac{1}{5}x + 1$
$2x + 10y = 3$

b. $x + y = 3$
$-x + y = 4$

c. $3x + y = 5$
$2x + 3y = 6$

(Continued on next page)

Practice 8

Determine whether each pair of lines is parallel, perpendicular, or neither.

a. $y = -5x + 1$
$x - 5y = 10$

b. $x + y = 11$
$2x + y = 11$

c. $2x + 3y = 21$
$6y = -4x - 2$

Answers

8. a. perpendicular **b.** neither **c.** parallel

Solution:

a. The slope of the line $y = -\frac{1}{5}x + 1$ is $-\frac{1}{5}$. We find the slope of the second line by solving its equation for y.

$$2x + 10y = 3$$
$$10y = -2x + 3 \quad \text{Subtract } 2x \text{ from both sides.}$$
$$y = \frac{-2}{10}x + \frac{3}{10} \quad \text{Divide both sides by 10.}$$
$$y = -\frac{1}{5}x + \frac{3}{10} \quad \text{Simplify.}$$

The slope of this line is $-\frac{1}{5}$ also. Since the lines have the same slope and different y-intercepts, they are parallel, as shown in the figure below.

b. To find each slope, we solve each equation for y.

$$x + y = 3 \qquad\qquad -x + y = 4$$
$$y = -x + 3 \qquad\qquad y = x + 4$$

The slope is −1. The slope is 1.

The slopes are not the same, so the lines are not parallel. Next we check the product of the slopes: $(-1)(1) = -1$. Since the product is −1, the lines are perpendicular, as shown in the figure.

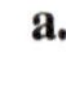

a.

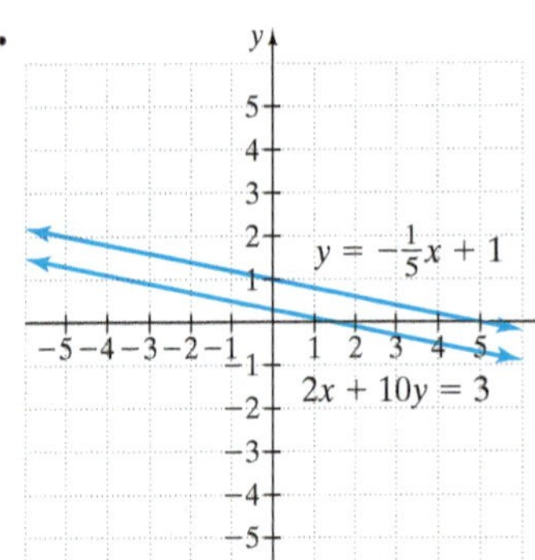

b.

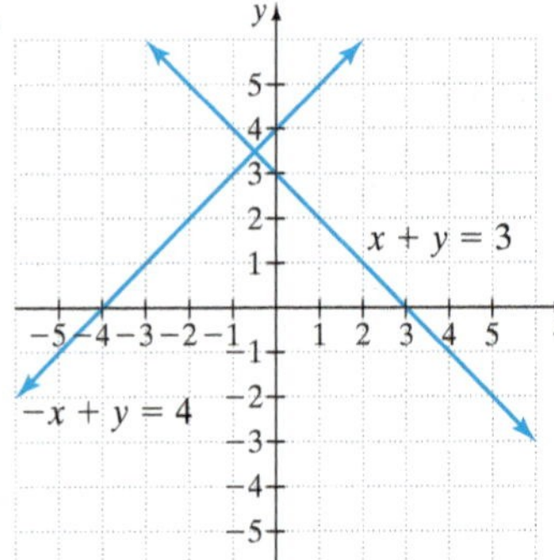

c. We solve each equation for y to find each slope. The slopes are −3 and $-\frac{2}{3}$. The slopes are not the same and their product is not −1. Thus, the lines are neither parallel nor perpendicular.

Work Practice 8

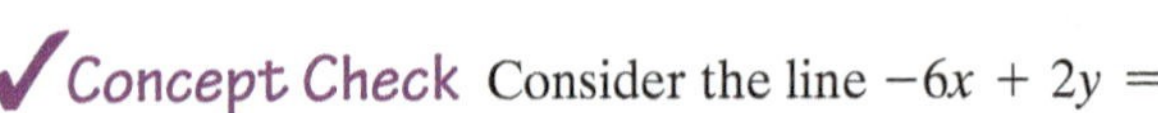

✓Concept Check Consider the line $-6x + 2y = 1$.

a. Write the equations of two lines parallel to this line.

b. Write the equations of two lines perpendicular to this line.

Objective 5 Slope as a Rate of Change

Slope can also be interpreted as a rate of change. In other words, slope tells us how fast y is changing with respect to x. To see this, let's look at a few of the many real-world applications of slope. For example, the pitch of a roof, used by builders and architects, is its slope. The pitch of the roof on the left is $\frac{7}{10}\left(\frac{\text{rise}}{\text{run}}\right)$. This means that the roof rises vertically 7 feet for every horizontal 10 feet. The rate of change for the roof is 7 vertical feet (y) per 10 horizontal feet (x).

$\frac{7}{10}$ pitch, 7 feet, 10 feet

✓Concept Check Answers

a. any two lines with $m = 3$ and y-intercept not $\left(0, \frac{1}{2}\right)$

b. any two lines with $m = -\frac{1}{3}$

The grade of a road is its slope written as a percent. A 7% grade, as shown below, means that the road rises (or falls) 7 feet for every horizontal 100 feet. $\left(\text{Recall that } 7\% = \frac{7}{100}.\right)$ Here, the slope of $\frac{7}{100}$ gives us the rate of change. The road rises (in our diagram) 7 vertical feet (y) for every 100 horizontal feet (x).

Example 9 Finding the Grade of a Road

At one part of the road to the summit of Pikes Peak, the road rises at a rate of 15 vertical feet for a horizontal distance of 250 feet. Find the grade of the road.

Solution: Recall that the grade of a road is its slope written as a percent.

$$\text{grade} = \frac{\text{rise}}{\text{run}} = \frac{15}{250} = 0.06 = 6\%$$

15 feet

250 feet

The grade is 6%.

Work Practice 9

Example 10 Finding the Slope of a Line

The graph on the right shows annual food and drink sales y (in billions of dollars) for year x. Find the slope of the line and attach the proper units for the rate of change. Then write a sentence explaining the meaning of the slope for this application.

Solution: Use (1990, 240) and (2011, 597) to calculate slope.

$$m = \frac{597 - 240}{2011 - 1990} = \frac{357}{21} = \frac{17 \text{ billion dollars}}{1 \text{ year}}$$

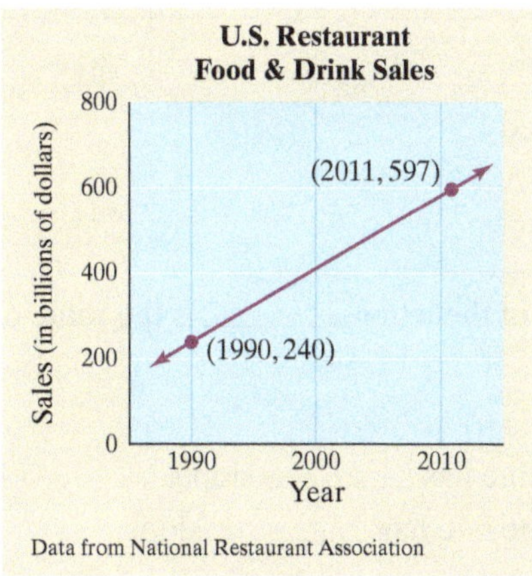

This means that the rate of change of restaurant food and drink sales increases by 17 billion dollars every 1 year, or $17 billion per year.

Work Practice 10

Practice 9

One part of the Mt. Washington (New Hampshire) cog railway rises about 1794 feet over a horizontal distance of 7176 feet. Find the grade of this part of the railway.

Practice 10

The graph below shows the cost y (in dollars) of having laundry done at the Wash-n-Fold, where x is the number of pounds of laundry. Find the slope of the line, and attach the proper units for the rate of change.

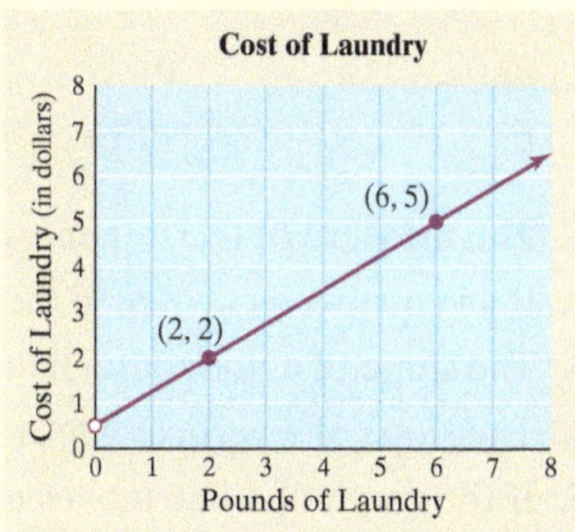

Answers

9. 25%

10. $m = \frac{0.75 \text{ dollar}}{1 \text{ pound}}$; The Wash-n-Fold charges $0.75 per pound of laundry.

Graphing Calculator Explorations

It is possible to use a grapher to sketch the graph of more than one equation on the same set of axes. This feature can be used to confirm our findings from Section 10.2 when we learned that the graph of an equation written in the form $y = mx + b$ has a y-intercept of b. For example, graph the equations $y = \frac{2}{5}x$, $y = \frac{2}{5}x + 7$, and $y = \frac{2}{5}x - 4$ on the same set of axes. To do so, press the $\boxed{Y=}$ key and enter the equations on the first three lines.

$$Y_1 = \left(\frac{2}{5}\right)x$$

$$Y_2 = \left(\frac{2}{5}\right)x + 7$$

$$Y_3 = \left(\frac{2}{5}\right)x - 4$$

The screen should look like:

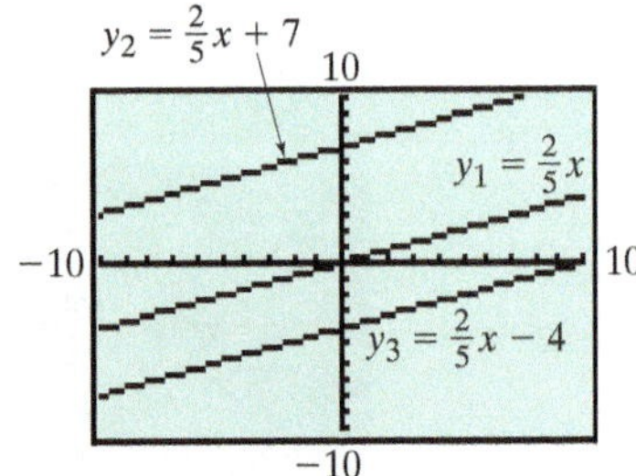

Notice that all three graphs appear to have the same positive slope. The graph of $y = \frac{2}{5}x + 7$ is the graph of $y = \frac{2}{5}x$ moved 7 units upward with a y-intercept of 7. Also, the graph of $y = \frac{2}{5}x - 4$ is the graph of $y = \frac{2}{5}x$ moved 4 units downward with a y-intercept of −4.

Graph the equations on the same set of axes. Describe the similarities and differences in their graphs.

1. $y = 3.8x$, $y = 3.8x - 3$, $y = 3.8x + 9$
2. $y = -4.9x$, $y = -4.9x + 1$, $y = -4.9x + 8$
3. $y = \frac{1}{4}x$; $y = \frac{1}{4}x + 5$, $y = \frac{1}{4}x - 8$
4. $y = -\frac{3}{4}x$, $y = -\frac{3}{4}x - 5$, $y = -\frac{3}{4}x + 6$

Vocabulary, Readiness & Video Check

Use the choices below to fill in each blank. Not all choices will be used.

m	x	0	positive	undefined
b	y	slope	negative	

1. The measure of the steepness or tilt of a line is called ________.
2. If an equation is written in the form $y = mx + b$, the value of the letter ________ is the value of the slope of the graph.
3. The slope of a horizontal line is ________.
4. The slope of a vertical line is ________.
5. If the graph of a line moves upward from left to right, the line has ________ slope.
6. If the graph of a line moves downward from left to right, the line has ________ slope.
7. Given two points of a line, slope $= \dfrac{\text{change in } \_\_}{\text{change in } \_\_}$.

Martin-Gay Interactive Videos

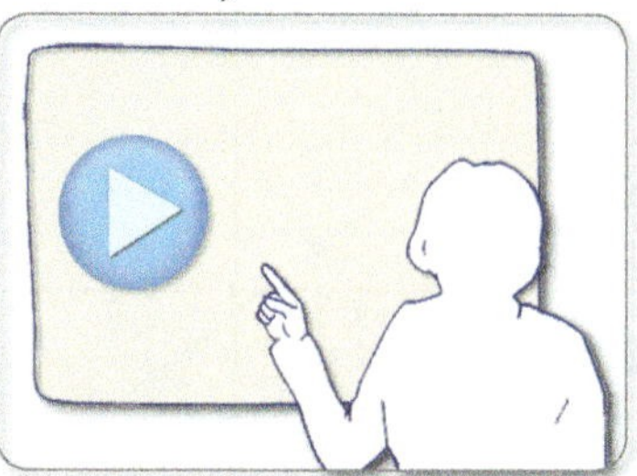

See Video 10.4

Watch the section lecture video and answer the following questions.

Objective 1 8. What important point is made during Example 1 having to do with the order of the points in the slope formula?

Objective 2 9. From Example 5, how do you write an equation in "slope-intercept form"? Once the equation is in slope-intercept form, how do you identify the slope?

Objective 3 10. In the lecture after Example 8, different slopes are summarized. What is the difference between zero slope and undefined slope? What does "no slope" mean?

Objective 4 11. From Example 10, what form of the equations is best to determine if two lines are parallel or perpendicular? Why?

Objective 5 12. Writing the slope as a rate of change in Example 11 gave real-life meaning to the slope. What step in the general strategy for problem solving does this correspond to?

10.4 Exercise Set MyMathLab®

Find the slope of the line that passes through the given points. See Examples 1 and 2.

1. $(-1, 5)$ and $(6, -2)$
2. $(3, 1)$ and $(2, 6)$
3. $(-4, 3)$ and $(-4, 5)$
4. $(6, -6)$ and $(6, 2)$
5. $(-2, 8)$ and $(1, 6)$
6. $(4, -3)$ and $(2, 2)$
7. $(5, 1)$ and $(-2, 1)$
8. $(0, 13)$ and $(-4, 13)$

Find the slope of each line if it exists. See Examples 1 and 2.

9.

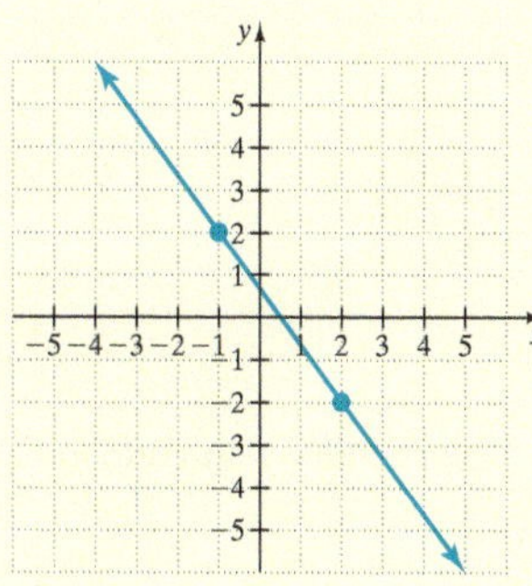

10.

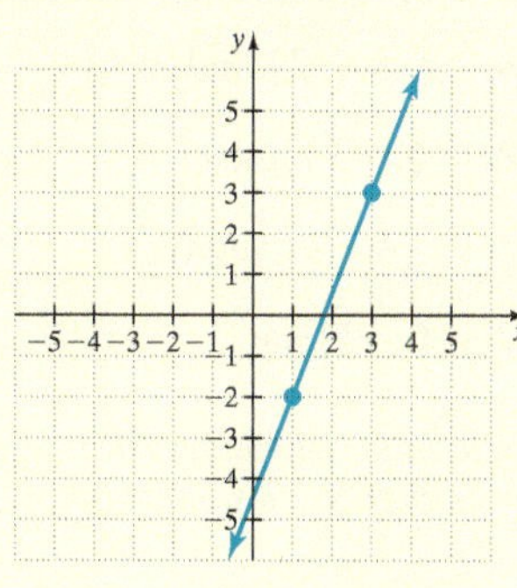

11.

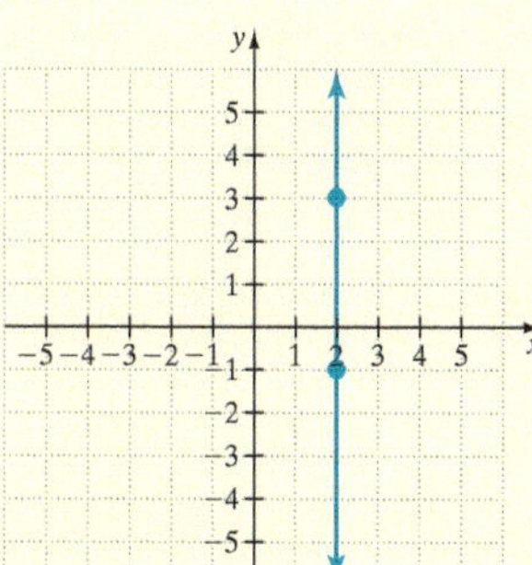

12.

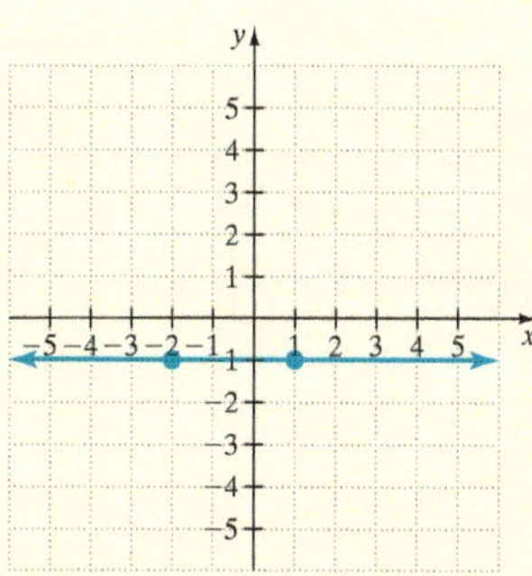

13.

14.

State whether the slope of the line is positive, negative, 0, or undefined. See the top box on p. 762.

15.

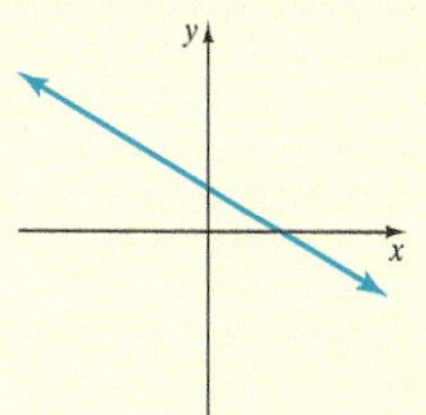

16.

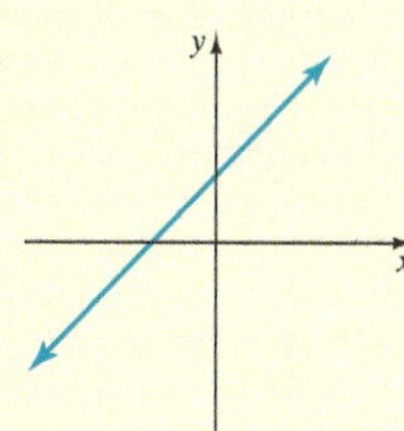

17.

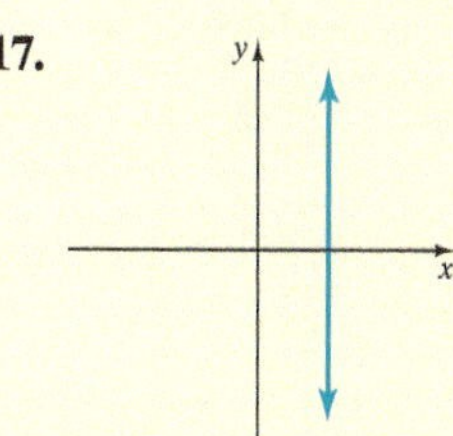

18.

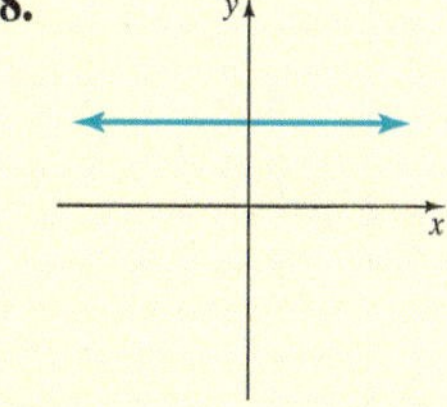

Decide whether a line with the given slope is upward, downward, horizontal, or vertical. See the top box on p. 762.

19. $m = \frac{7}{6}$ ________ **20.** $m = -3$ ________

21. $m = 0$ ________ **22.** m is undefined. ________

For each graph, determine which line has the greater slope. See the top box on p. 762.

23.

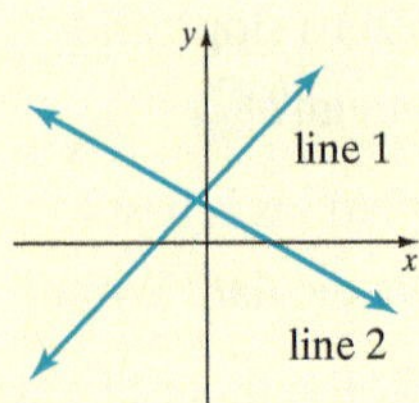

24.

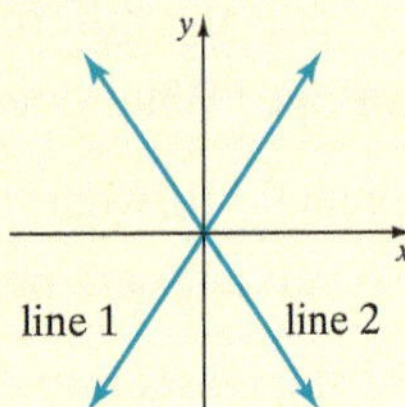

25.

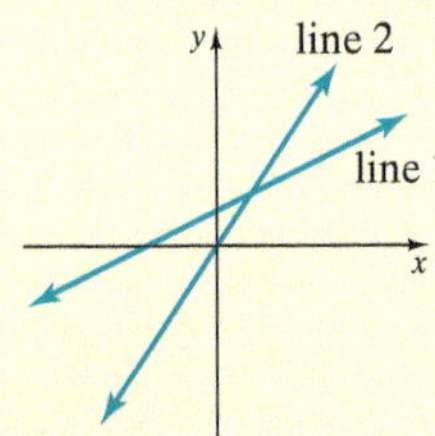

26.

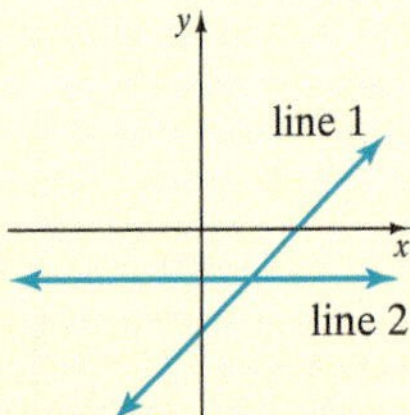

In Exercises 27 through 32, match each line with its slope. See Examples 1 and 2 and the top box on p. 762.

A. $m = 0$ **B.** undefined slope **C.** $m = 3$

D. $m = 1$ **E.** $m = -\frac{1}{2}$ **F.** $m = -\frac{3}{4}$

27.

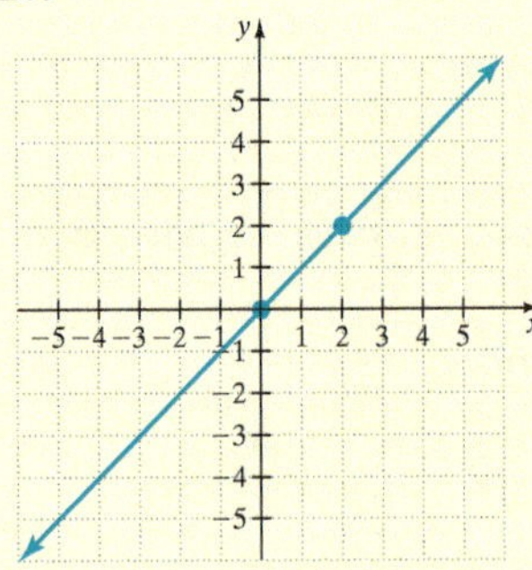

28.

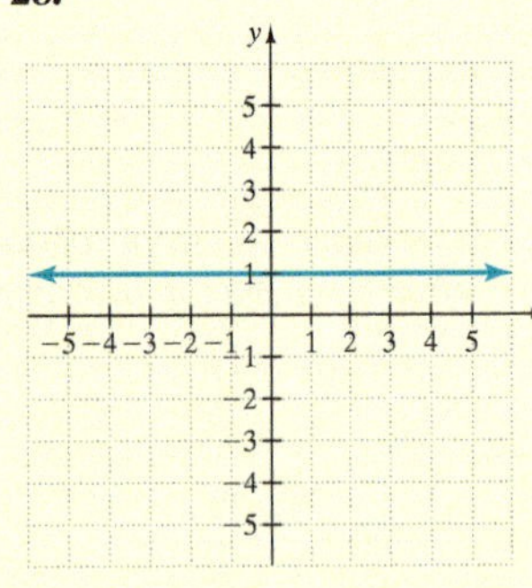

29.

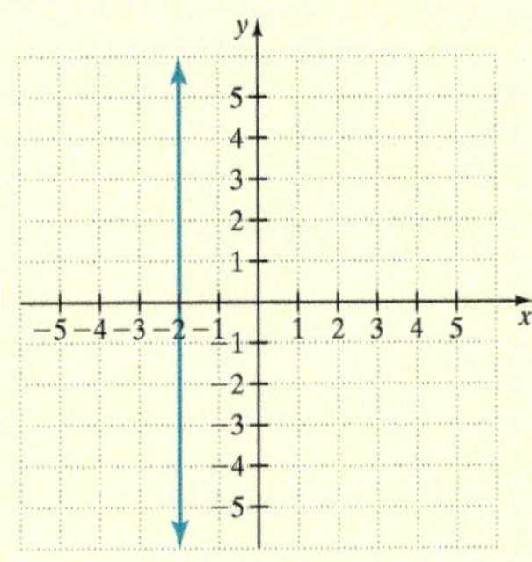

30.

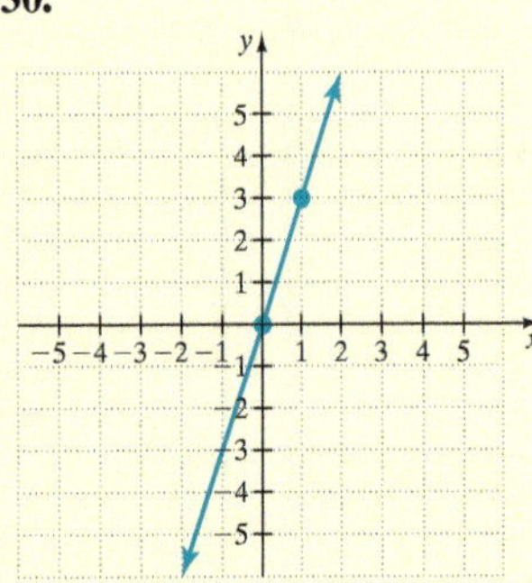

31.

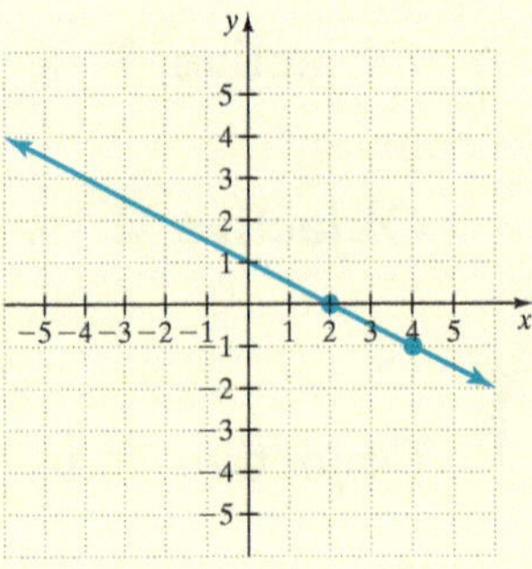

32.

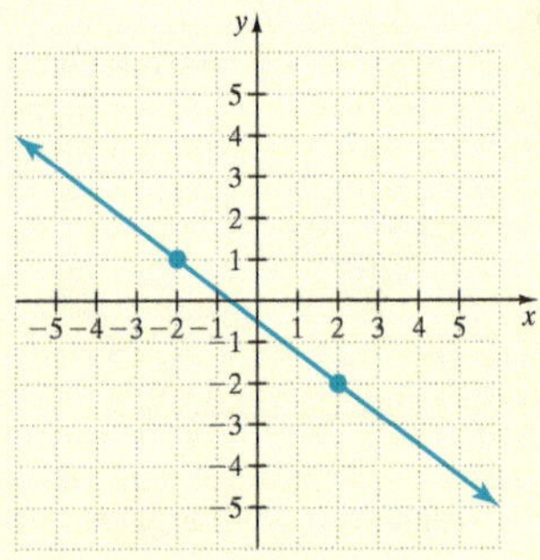

Find the slope of each line. See Examples 6 and 7.

33. $x = 6$ **34.** $y = 4$

35. $y = -4$ **36.** $x = 2$

37. $x = -3$ **38.** $y = -11$

39. $y = 0$ **40.** $x = 0$

MIXED PRACTICE

Find the slope of each line. See Examples 3 through 7.

41. $y = 5x - 2$ **42.** $y = -2x + 6$

43. $y = -0.3x + 2.5$ **44.** $y = -7.6x - 0.1$

45. $2x + y = 7$ **46.** $-5x + y = 10$

47. $2x - 3y = 10$ **48.** $3x - 5y = 1$

49. $x = 1$ **50.** $y = -2$

51. $x = 2y$ **52.** $x = -4y$

53. $y = -3$ **54.** $x = 5$

55. $-3x - 4y = 6$ **56.** $-4x - 7y = 9$

57. $20x - 5y = 1.2$ **58.** $24x - 3y = 5.7$

△ *Find the slope of the line that is **(a)** parallel and **(b)** perpendicular to the line through each pair of points. See Example 8.*

59. $(-3, -3)$ and $(0, 0)$

60. $(6, -2)$ and $(1, 4)$

61. $(-8, -4)$ and $(3, 5)$

62. $(6, -1)$ and $(-4, -10)$

△ *Determine whether each pair of lines is parallel, perpendicular, or neither. See Example 8.*

63. $y = \frac{2}{9}x + 3$
$y = -\frac{2}{9}x$

64. $y = \frac{1}{5}x + 20$
$y = -\frac{1}{5}x$

65. $x - 3y = -6$
$y = 3x - 9$

66. $y = 4x - 2$
$4x + y = 5$

67. $6x = 5y + 1$
$-12x + 10y = 1$

68. $-x + 2y = -2$
$2x = 4y + 3$

69. $6 + 4x = 3y$
$3x + 4y = 8$

70. $10 + 3x = 5y$
$5x + 3y = 1$

The pitch of a roof is its slope. Find the pitch of each roof shown. See Example 9.

71.

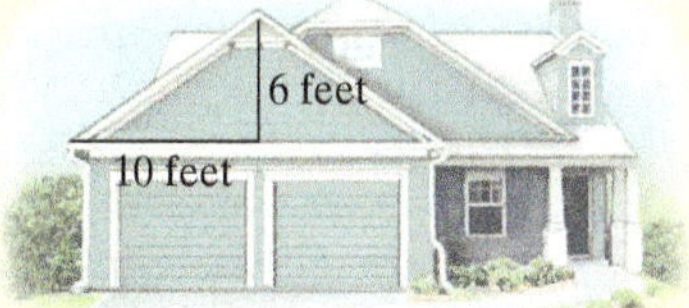

72.

The grade of a road is its slope written as a percent. Find the grade of each road shown. See Example 9.

73.

74.

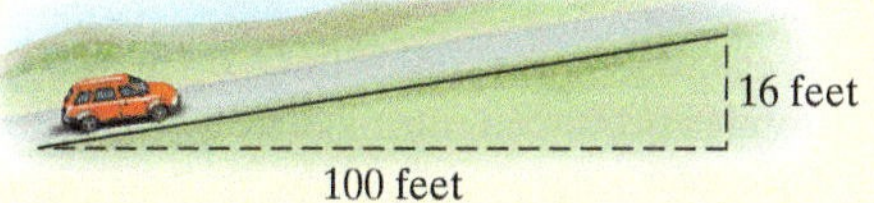

75. One of Japan's superconducting "bullet" trains is researched and tested at the Yamanashi Maglev Test Line near Otsuki City. The steepest section of the track has a rise of 2580 meters for a horizontal distance of 6450 meters. What is the grade of this section of track? (*Source:* Japan Railways Central Co.)

76. Professional plumbers suggest that a sewer pipe should rise 0.25 inch for every horizontal foot. Find the recommended slope for a sewer pipe. Round to the nearest hundredth.

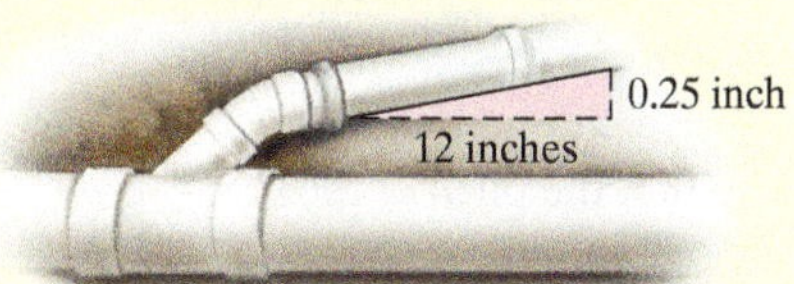

77. There has been controversy over the past few years about the world's steepest street. The *Guinness Book of Records* actually listed Baldwin Street, in Dunedin, New Zealand, as the world's steepest street, but Canton Avenue in the Pittsburgh neighborhood of Beechview may be steeper. Calculate each grade to the nearest percent.

		Grade (%)
Canton Avenue	for every 30 meters of horizontal distance, the vertical change is 11 meters	
Baldwin Street	for every 2.86 meters of horizontal distance, the vertical change is 1 meter	

78. According to federal regulations, a wheelchair ramp should rise no more than 1 foot for a horizontal distance of 12 feet. Write the slope as a grade. Round to the nearest tenth of a percent.

Find the slope of each line and write the slope as a rate of change. Don't forget to attach the proper units. See Example 10.

79. This graph approximates the number of U.S. households that have televisions y (in millions) for year x.

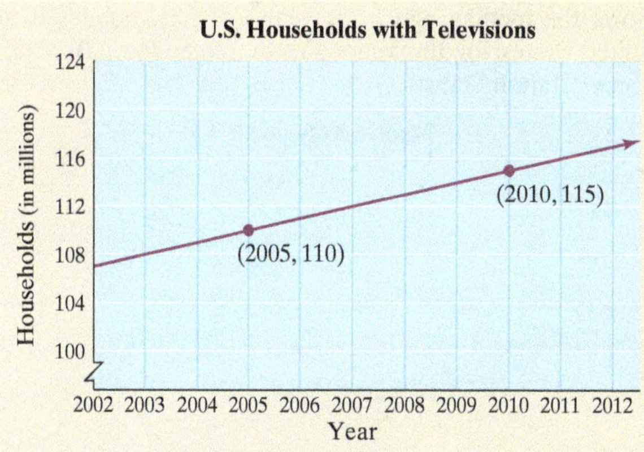

80. The graph approximates the amount of money y (in billions of dollars) spent worldwide on tourism for year x. (*Data from* World Tourism Organization)

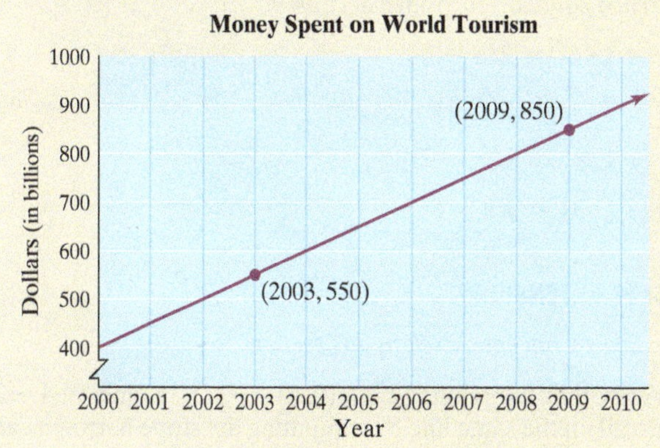

For Exercises 81 and 82, write the slope as a decimal.

81. The graph below shows the total cost y (in dollars) of owning and operating a compact car where x is the number of miles driven.

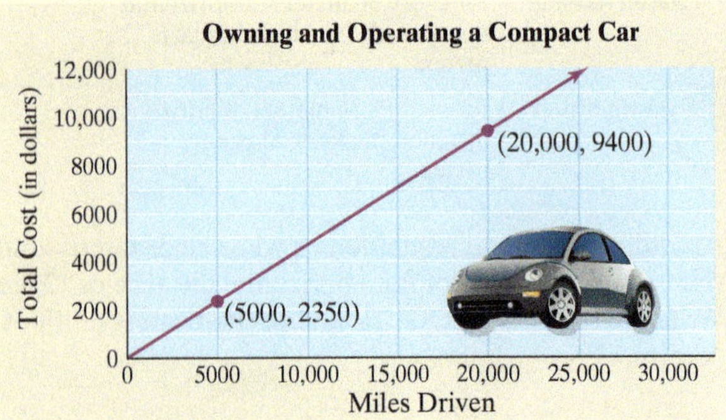

82. Americans are keeping their cars longer. The graph below shows the median age y (in years) of automobiles in the United States for the years shown. (*Data from* Bureau of Transportation Statistics)

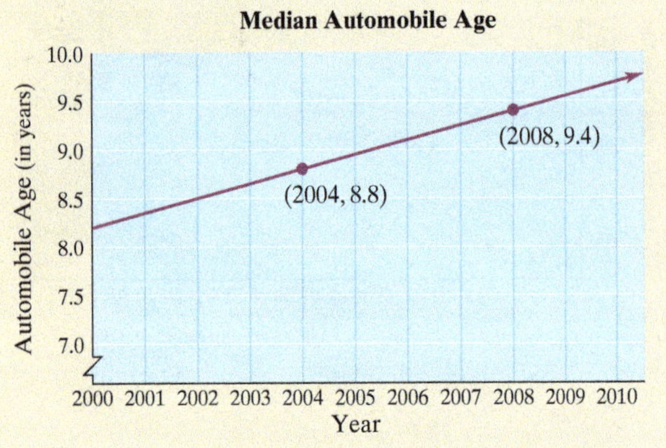

REVIEW AND PREVIEW

Solve each equation for y. See Section 9.5.

83. $y - (-6) = 2(x - 4)$

84. $y - 7 = -9(x - 6)$

85. $y - 1 = -6(x - (-2))$

86. $y - (-3) = 4(x - (-5))$

CONCEPT EXTENSIONS

Solve. See a Concept Check in this section.

87. Verify that the points $(2, 1)$, $(0, 0)$, $(-2, -1)$, and $(-4, -2)$ are all on the same line by computing the slope between each pair of points. (See the first Concept Check.)

88. Given the points $(2, 3)$ and $(-5, 1)$, can the slope of the line through these points be calculated by $\frac{1-3}{2-(-5)}$? Why or why not? (See the second Concept Check.)

89. Write the equations of three lines parallel to $10x - 5y = -7$. (See the third Concept Check.)

90. Write the equations of two lines perpendicular to $10x - 5y = -7$. (See the third Concept Check.)

The following line graph shows the average fuel economy (in miles per gallon) of passenger automobiles produced during each of the model years shown. Use this graph to answer Exercises 91 through 96.

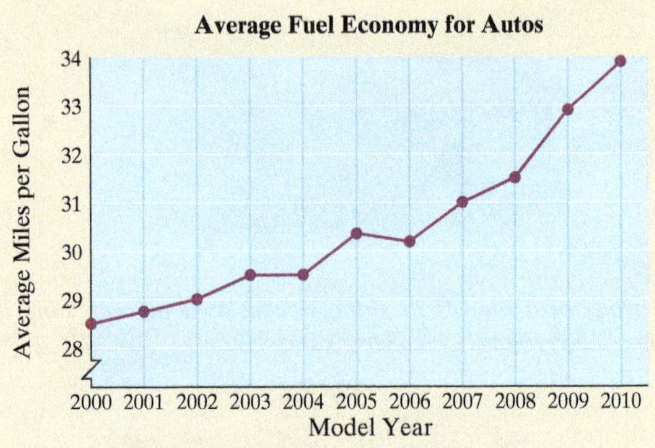

91. Between what two years shown was there a decrease in average fuel economy for automobiles?

92. What was the average fuel economy (in miles per gallon) for automobiles produced during 2008?

93. During which of the model years shown was average fuel economy the lowest?
What was the average fuel economy for that year?

94. During which of the model years shown was average fuel economy the highest?
What was the average fuel economy for that year?

95. Of the following line segments, which has the greatest slope: from 2002 to 2003, from 2006 to 2007, or from 2008 to 2009?

96. Which line segment has a slope of 0?

Solve.

97. Find x so that the pitch of the roof is $\frac{1}{3}$.

98. Find x so that the pitch of the roof is $\frac{2}{5}$.

99. Approximately 2207 heart transplants were performed in the United States in 2007. In 2010, the number of heart transplants in the United States rose to 2333. (*Source:* Organ Procurement and Transplantation Network)

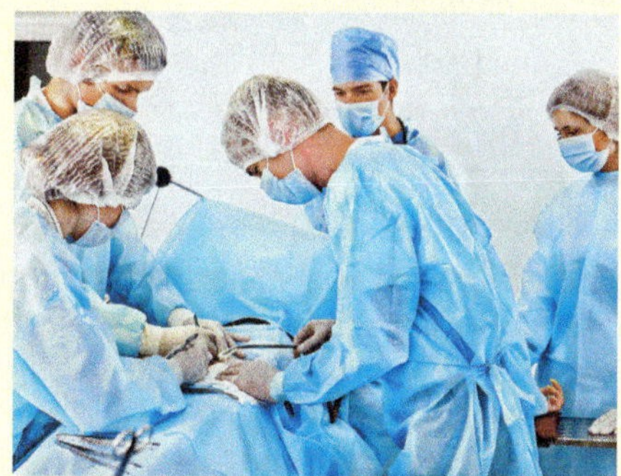

a. Write two ordered pairs of the form (year, number of heart transplants).

b. Find the slope of the line between the two points.

c. Write a sentence explaining the meaning of the slope as a rate of change.

100. The average price of an acre of U.S. farmland was \$1210 in 2002. In 2008, the price of an acre rose to \$2350. (*Source:* National Agricultural Statistics Services)

a. Write two ordered pairs of the form (year, price of an acre).

b. Find the slope of the line through the two points.

c. Write a sentence explaining the meaning of the slope as a rate of change.

101. Show that a triangle with vertices at the points $(1, 1)$, $(-4, 4)$, and $(-3, 0)$ is a right triangle.

102. Show that the quadrilateral with vertices $(1, 3)$, $(2, 1)$, $(-4, 0)$, and $(-3, -2)$ is a parallelogram.

Find the slope of the line through the given points.

103. $(2.1, 6.7)$ and $(-8.3, 9.3)$

104. $(-3.8, 1.2)$ and $(-2.2, 4.5)$

105. $(2.3, 0.2)$ and $(7.9, 5.1)$

106. $(14.3, -10.1)$ and $(9.8, -2.9)$

107. The graph of $y = -\frac{1}{3}x + 2$ has a slope of $-\frac{1}{3}$. The graph of $y = -2x + 2$ has a slope of -2. The graph of $y = -4x + 2$ has a slope of -4. Graph all three equations on a single coordinate system. As the absolute value of the slope becomes larger, how does the steepness of the line change?

108. The graph of $y = \frac{1}{2}x$ has a slope of $\frac{1}{2}$. The graph of $y = 3x$ has a slope of 3. The graph of $y = 5x$ has a slope of 5. Graph all three equations on a single coordinate system. As slope becomes larger, how does the steepness of the line change?

Integrated Review Sections 10.1–10.4

Summary on Slope and Graphing Linear Equations

Find the slope of each line.

1.

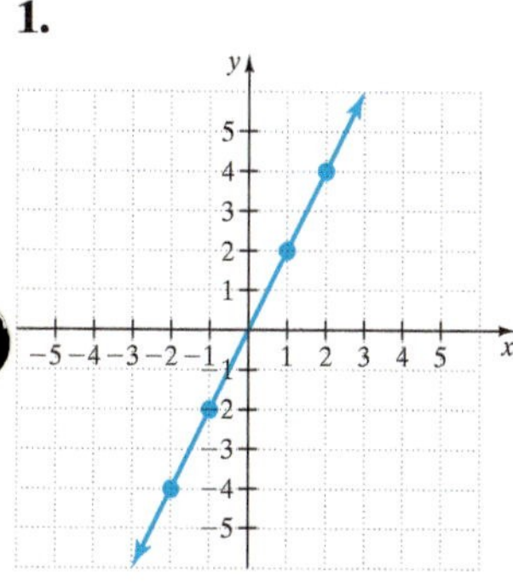

2.

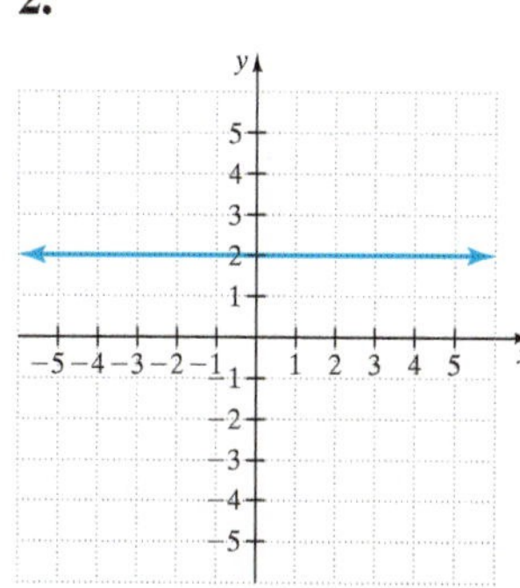

3.

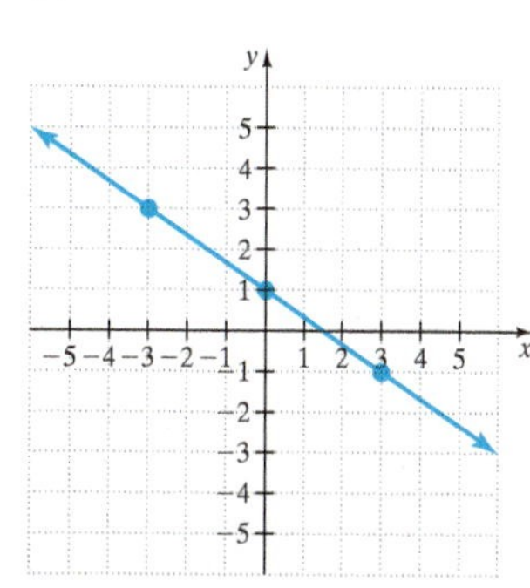

4.

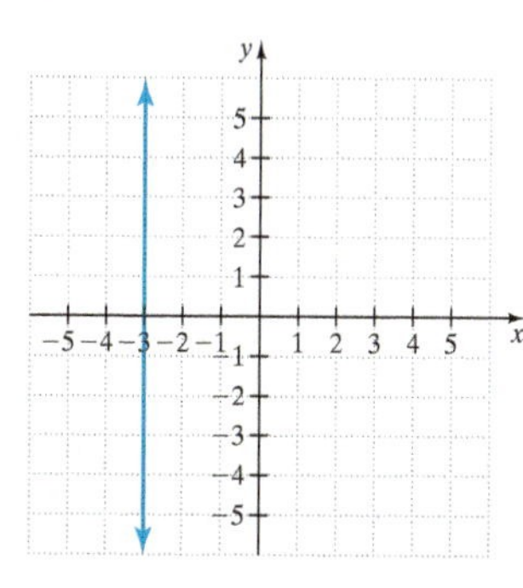

Graph each linear equation.

5. $y = -2x$
6. $x + y = 3$
7. $x = -1$
8. $y = 4$
9. $x - 2y = 6$
10. $y = 3x + 2$
11. $5x + 3y = 15$
12. $2x - 4y = 8$

Determine whether the lines through the points are parallel, perpendicular, or neither.

13. $y = -\frac{1}{5}x + \frac{1}{3}$
$3x = -15y$

14. $x - y = \frac{1}{2}$
$3x - y = \frac{1}{2}$

15. In the years 2000 through 2009 the number of bridges on public roads (in thousands) in the United States can be modeled by the linear equation $y = 1.7x + 587$ where x is years after 2000 and y is number of bridges (in thousands). (*Source:* U.S. Dept. of Transportation)

a. Find the y-intercept of this line.

b. Write a sentence explaining the meaning of this intercept.

c. Find the slope of this line.

d. Write a sentence explaining the meaning of the slope as a rate of change.

16. Online advertising is a means of promoting products and services using the Internet. The revenue (in billions of dollars) for online advertising for the years 2003 through 2010 is given by $y = 3.3x - 3.1$, where x is the number of years after 2000.

a. Use this equation to complete the ordered pair (9,   ).

b. Write a sentence explaining the meaning of the answer to part (a).

10.5 Equations of Lines 

Objectives

1 Use the Slope-Intercept Form to Graph a Linear Equation.

2 Use the Slope-Intercept Form to Write an Equation of a Line.

3 Use the Point-Slope Form to Find an Equation of a Line Given Its Slope and a Point of the Line.

4 Use the Point-Slope Form to Find an Equation of a Line Given Two Points of the Line.

5 Find Equations of Vertical and Horizontal Lines.

6 Use the Point-Slope Form to Solve Problems.

Recall that the form $y = mx + b$ is appropriately called the *slope-intercept form* of a linear equation.

(m: slope; y-intercept is $(0, b)$)

Slope-Intercept Form

When a linear equation in two variables is written in **slope-intercept form,**

$$y = mx + b$$

(m: slope; $(0, b)$, y-intercept)

then m is the slope of the line and $(0, b)$ is the y-intercept of the line.

Objective 1 Using the Slope-Intercept Form to Graph an Equation

We can use the slope-intercept form of the equation of a line to graph a linear equation.

Example 1 Use the slope-intercept form to graph the equation

$$y = \frac{3}{5}x - 2$$

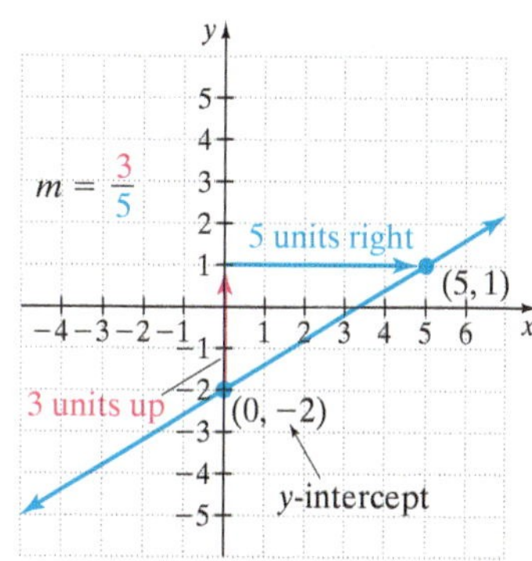

Solution: Since the equation $y = \frac{3}{5}x - 2$ is written in slope-intercept form $y = mx + b$, the slope of its graph is $\frac{3}{5}$ and the y-intercept is $(0, -2)$. To graph this equation, we begin by plotting the point $(0, -2)$. From this point, we can find another point of the graph by using the slope $\frac{3}{5}$ and recalling that slope is $\frac{\text{rise}}{\text{run}}$. We start at the y-intercept and move 3 units up since the numerator of the slope is 3; then we move 5 units to the right since the denominator of the slope is 5. We stop at the point $(5, 1)$. The line through $(0, -2)$ and $(5, 1)$ is the graph of $y = \frac{3}{5}x - 2$.

Work Practice 1

Practice 1

Graph $y = \frac{2}{3}x - 5$.

Example 2 Use the slope-intercept form to graph the equation $4x + y = 1$.

Solution: First we write the given equation in slope-intercept form.

$$4x + y = 1$$
$$y = -4x + 1$$

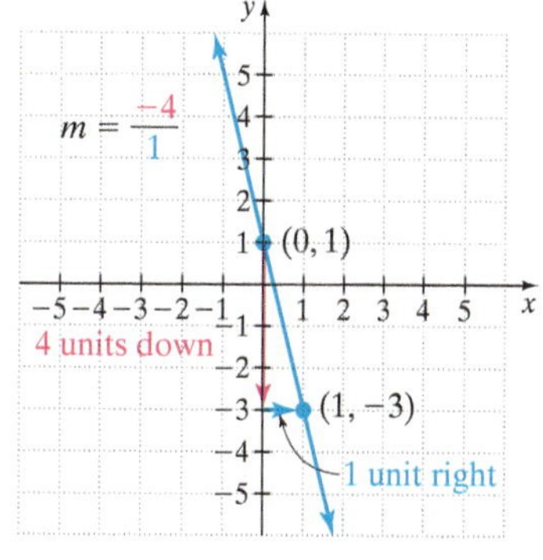

The graph of this equation will have slope -4 and y-intercept $(0, 1)$. To graph this line, we first plot the point $(0, 1)$. To find another point of the graph, we use the slope -4, which can be written as $\frac{-4}{1}$ $\left(\frac{4}{-1} \text{ could also be used}\right)$. We start at the point $(0, 1)$ and move 4 units down (since the numerator of the slope is -4) and then 1 unit to the right (since the denominator of the slope is 1).

We arrive at the point $(1, -3)$. The line through $(0, 1)$ and $(1, -3)$ is the graph of $4x + y = 1$.

Work Practice 2

Practice 2

Use the slope-intercept form to graph the equation $3x - y = 2$.

Helpful Hint

In Example 2, if we interpret the slope of -4 as $\frac{4}{-1}$, we arrive at $(-1, 5)$ for a second point. Notice that this point is also on the line.

Objective 2 Using the Slope-Intercept Form to Write an Equation

The slope-intercept form can also be used to write the equation of a line when we know its slope and y-intercept.

Answers

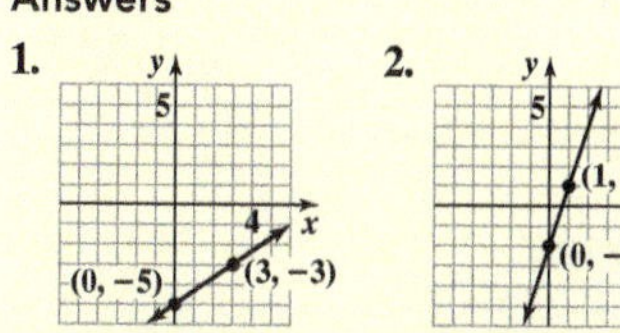

Practice 3

Find an equation of the line with y-intercept $(0, 7)$ and slope of $\frac{1}{2}$.

Example 3 Find an equation of the line with y-intercept $(0, -3)$ and slope of $\frac{1}{4}$.

Solution: We are given the slope and the y-intercept. We let $m = \frac{1}{4}$ and $b = -3$ and write the equation in slope-intercept form, $y = mx + b$.

$$y = mx + b$$

$$y = \frac{1}{4}x + (-3) \quad \text{Let } m = \frac{1}{4} \text{ and } b = -3.$$

$$y = \frac{1}{4}x - 3 \quad \text{Simplify.}$$

Work Practice 3

Objective 3 Writing an Equation Given Slope and a Point

Thus far, we have seen that we can write an equation of a line if we know its slope and y-intercept. We can also write an equation of a line if we know its slope and any point on the line. To see how we do this, let m represent slope and (x_1, y_1) represent the point on the line. Then if (x, y) is any other point on the line, we have that

$$\frac{y - y_1}{x - x_1} = m$$

$$y - y_1 = m(x - x_1) \quad \text{Multiply both sides by } (x - x_1).$$

(↑ slope)

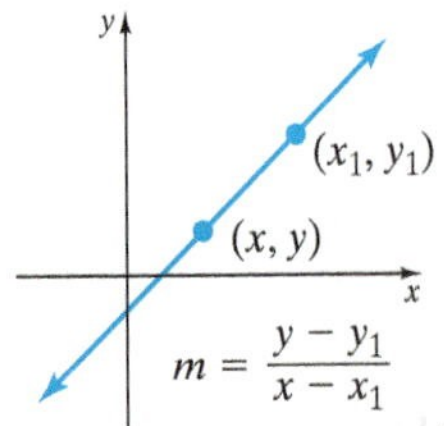

This is the *point-slope form* of the equation of a line.

Point-Slope Form of the Equation of a Line

The **point-slope form** of the equation of a line is

$$y - y_1 = m(x - x_1)$$

(slope: m; (x_1, y_1) point on the line)

where m is the slope of the line and (x_1, y_1) is a point on the line.

Practice 4

Find an equation of the line passing through $(2, 3)$ with slope 4. Write the equation in standard form: $Ax + By = C$.

Example 4 Find an equation of the line with slope -2 that passes through $(-1, 5)$. Write the equation in slope-intercept form, $y = mx + b$, and in standard form, $Ax + By = C$.

Solution: Since the slope and a point on the line are given, we use point-slope form $y - y_1 = m(x - x_1)$ to write the equation. Let $m = -2$ and $(-1, 5) = (x_1, y_1)$.

$$y - y_1 = m(x - x_1)$$

$$y - 5 = -2[x - (-1)] \quad \text{Let } m = -2 \text{ and } (x_1, y_1) = (-1, 5).$$

$$y - 5 = -2(x + 1) \quad \text{Simplify.}$$

$$y - 5 = -2x - 2 \quad \text{Use the distributive property.}$$

To write the equation in slope-intercept form, $y = mx + b$, we simply solve the equation for y. To do this, we add 5 to both sides.

$$y - 5 = -2x - 2$$

$$y = -2x + 3 \quad \text{Slope-intercept form.}$$

$$2x + y = 3 \quad \text{Add } 2x \text{ to both sides and we have standard form.}$$

Work Practice 4

Answers

3. $y = \frac{1}{2}x + 7$ 4. $4x - y = 5$

Objective 4 Writing an Equation Given Two Points 

We can also find an equation of a line when we are given any two points of the line.

Example 5 Find an equation of the line through (2, 5) and (−3, 4). Write the equation in standard form.

Solution: First, use the two given points to find the slope of the line.

$$m = \frac{4 - 5}{-3 - 2} = \frac{-1}{-5} = \frac{1}{5}$$

Next we use the slope $\frac{1}{5}$ and either one of the given points to write the equation in point-slope form. We use (2, 5). Let $x_1 = 2$, $y_1 = 5$, and $m = \frac{1}{5}$.

$y - y_1 = m(x - x_1)$	Use point-slope form.
$y - 5 = \frac{1}{5}(x - 2)$	Let $x_1 = 2$, $y_1 = 5$, and $m = \frac{1}{5}$.
$5(y - 5) = 5 \cdot \frac{1}{5}(x - 2)$	Multiply both sides by 5 to clear fractions.
$5y - 25 = x - 2$	Use the distributive property and simplify.
$-x + 5y - 25 = -2$	Subtract x from both sides.
$-x + 5y = 23$	Add 25 to both sides.

Work Practice 5

Practice 5

Find an equation of the line through (−1, 6) and (3, 1). Write the equation in standard form.

Helpful Hint

Multiply both sides of the equation $-x + 5y = 23$ by −1, and it becomes $x - 5y = -23$. Both $-x + 5y = 23$ and $x - 5y = -23$ are in standard form, and they are equations of the same line.

Objective 5 Finding Equations of Vertical and Horizontal Lines

Recall from Section 10.3 that:

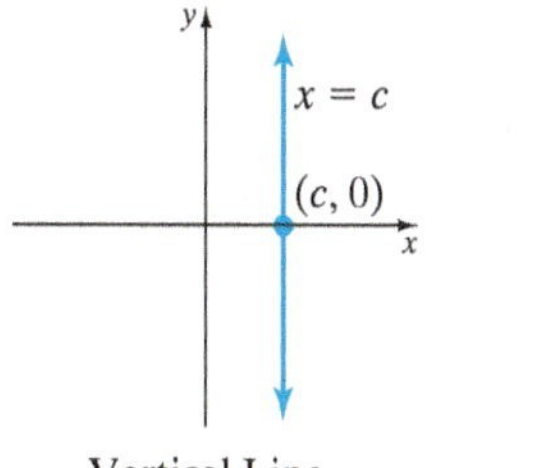

Vertical Line

Horizontal Line

Example 6 Find an equation of the vertical line through (−1, 5).

Solution: The equation of a vertical line can be written in the form $x = c$, so an equation for a vertical line passing through (−1, 5) is $x = -1$.

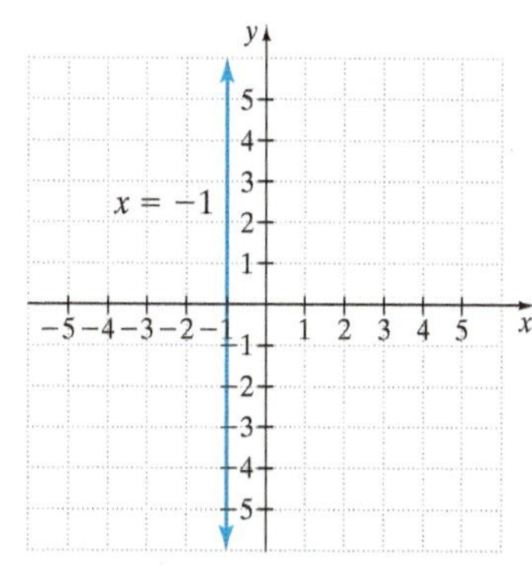

Work Practice 6

Practice 6

Find an equation of the vertical line through (3, −2).

Answers

5. $5x + 4y = 19$ **6.** $x = 3$

Practice 7

Find an equation of the line parallel to the line $y = -2$ and passing through $(4, 3)$.

Example 7

Find an equation of the line parallel to the line $y = 5$ and passing through $(-2, -3)$.

Solution: Since the graph of $y = 5$ is a horizontal line, any line parallel to it is also horizontal. The equation of a horizontal line can be written in the form $y = c$. An equation for the horizontal line passing through

$$(-2, -3) \text{ is } y = -3.$$

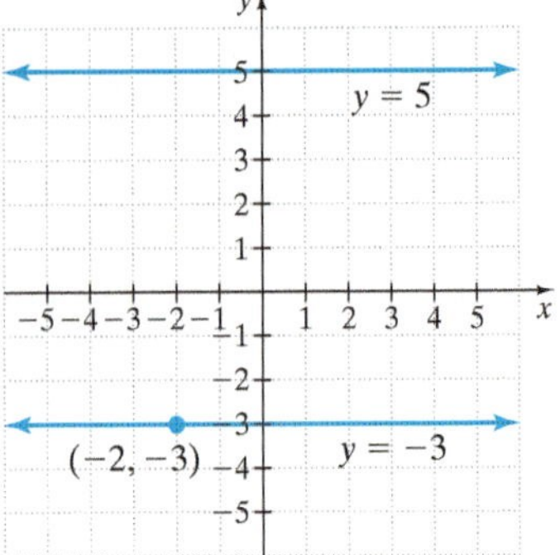

Work Practice 7

Objective 6 Using the Point-Slope Form to Solve Problems

Problems occurring in many fields can be modeled by linear equations in two variables. The next example is from the field of marketing and shows how consumer demand of a product depends on the price of the product.

Practice 8

The new *Camelot* condos were selling at a rate of 30 per month when they were priced at $150,000 each. Lowering the price to $120,000 caused the sales to rise to 50 condos per month.

a. Assume that the relationship between number of condos sold and price is linear, and write an equation describing this relationship. Write the equation in slope-intercept form.

b. How should the condos be priced if the developer wishes to sell 60 condos per month?

Example 8 Predicting the Sales of T-Shirts

A web-based T-shirt company has learned that by pricing a clearance-sale T-shirt at $6, sales will reach 2000 T-shirts per day. Raising the price to $8 will cause the sales to fall to 1500 T-shirts per day.

a. Assume that the relationship between sales price and number of T-shirts sold is linear and write an equation describing this relationship. Write the equation in slope-intercept form.

b. Predict the daily sales of T-shirts if the price is $7.50.

Solution:

a. First, use the given information and write two ordered pairs. Ordered pairs will be in the form (sales price, number sold) so that our ordered pairs are $(6, 2000)$ and $(8, 1500)$. Use the point-slope form to write an equation. To do so, we find the slope of the line that contains these points.

$$m = \frac{2000 - 1500}{6 - 8} = \frac{500}{-2} = -250$$

Next, use the slope and either one of the points to write the equation in point-slope form. We use $(6, 2000)$.

$$y - y_1 = m(x - x_1) \quad \text{Use point-slope form.}$$
$$y - 2000 = -250(x - 6) \quad \text{Let } x_1 = 6,\ y_1 = 2000, \text{ and } m = -250.$$
$$y - 2000 = -250x + 1500 \quad \text{Use the distributive property.}$$
$$y = -250x + 3500 \quad \text{Write in slope-intercept form.}$$

b. To predict the sales if the price is $7.50, we find y when $x = 7.50$.

$$y = -250x + 3500$$
$$y = -250(7.50) + 3500 \quad \text{Let } x = 7.50.$$
$$y = -1875 + 3500$$
$$y = 1625$$

If the price is $7.50, sales will reach 1625 T-shirts per day.

Work Practice 8

Answers
7. $y = 3$ **8. a.** $y = -1500x + 195,000$ **b.** $105,000

The preceding example may also be solved by using ordered pairs of the form (number sold, sales price).

Forms of Linear Equations

$Ax + By = C$	**Standard form** of a linear equation. A and B are not both 0.
$y = mx + b$	**Slope-intercept form** of a linear equation. The slope is m and the y-intercept is $(0, b)$.
$y - y_1 = m(x - x_1)$	**Point-slope form** of a linear equation. The slope is m and (x_1, y_1) is a point on the line.
$y = c$	**Horizontal line** The slope is 0 and the y-intercept is $(0, c)$.
$x = c$	**Vertical line** The slope is undefined and the x-intercept is $(c, 0)$.

Parallel and Perpendicular Lines

Nonvertical parallel lines have the same slope.
The product of the slopes of two nonvertical perpendicular lines is -1.

Graphing Calculator Explorations

A grapher is a very useful tool for discovering patterns. To discover the change in the graph of a linear equation caused by a change in slope, try the following. Use a standard window and graph a linear equation in the form $y = mx + b$. Recall that the graph of such an equation will have slope m and y-intercept b.

First, graph $y = x + 3$. To do so, press the $\boxed{Y =}$ key and enter $Y_1 = x + 3$. Notice that this graph has slope 1 and that the y-intercept is 3. Next, on the same set of axes, graph $y = 2x + 3$ and $y = 3x + 3$ by pressing $\boxed{Y =}$ and entering $Y_2 = 2x + 3$ and $Y_3 = 3x + 3$.

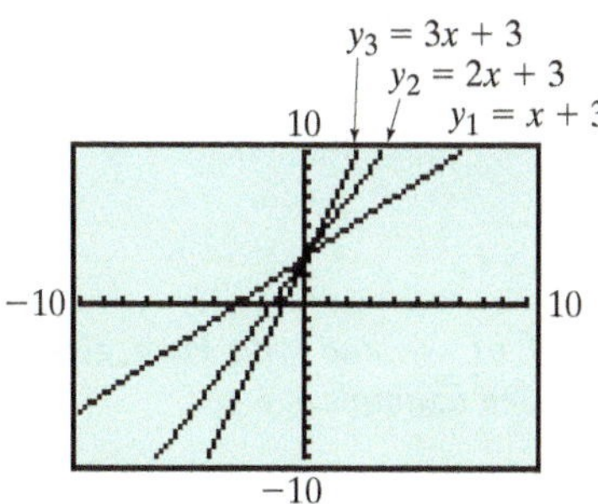

Notice the difference in the graph of each equation as the slope changes from 1 to 2 to 3. How would the graph of $y = 5x + 3$ appear? To see the change in the graph caused by a change in negative slope, try graphing $y = -x + 3$, $y = -2x + 3$, and $y = -3x + 3$ on the same set of axes.

Use a grapher to graph the following equations. For each exercise, graph the first equation and use its graph to predict the appearance of the other equations. Then graph the other equations on the same set of axes and check your prediction.

1. $y = x; y = 6x, y = -6x$
2. $y = -x; y = -5x, y = -10x$
3. $y = \frac{1}{2}x + 2; y = \frac{3}{4}x + 2, y = x + 2$
4. $y = x + 1; y = \frac{5}{4}x + 1, y = \frac{5}{2}x + 1$
5. $y = -7x + 5; y = 7x + 5$
6. $y = 3x - 1; y = -3x - 1$

Vocabulary, Readiness & Video Check

Use the choices below to fill in each blank. Some choices may be used more than once and some not at all.

b	(y_1, x_1)	point-slope	vertical	standard
m	(x_1, y_1)	slope-intercept	horizontal	

1. The form $y = mx + b$ is called ______________ form. When a linear equation in two variables is written in this form, ______________ is the slope of its graph and (0, ______________) is its y-intercept.
2. The form $y - y_1 = m(x - x_1)$ is called ______________ form. When a linear equation in two variables is written in this form, ______________ is the slope of its graph and ______________ is a point on the graph.

Martin-Gay Interactive Videos

See Video 10.5

Watch the section lecture video and answer the following questions.

Objective 1 3. We can use the slope-intercept form to graph a line. Complete these statements based on Example 1. Start by graphing the ______________.
Find another point by applying the slope to this point—rewrite the slope as a ______________ if necessary.

Objective 2 4. In Example 3, what is the y-intercept?

Objective 3 5. In Example 4 we use the point-slope form to find the equation of a line given the slope and a point. How do we then write this equation in standard form?

Objective 4 6. The lecture before Example 5 discusses how to find the equation of a line given two points. Is there any circumstance when you might want to use the slope-intercept form to find the equation of a line given two points? If so, when?

Objective 5 7. Solve Examples 6 and 7 again, this time using the point $(-2, -3)$ in each example.

Objective 6 8. From Example 8, we are told to use ordered pairs of the form (time, speed). Explain why it is important to keep track of how you define your ordered pairs and/or define your variables.

10.5 Exercise Set MyMathLab®

Use the slope-intercept form to graph each equation. See Examples 1 and 2.

1. $y = 2x + 1$
2. $y = -4x - 1$
3. $y = \frac{2}{3}x + 5$
4. $y = \frac{1}{4}x - 3$
5. $y = -5x$
6. $y = -6x$
7. $4x + y = 6$
8. $-3x + y = 2$
9. $4x - 7y = -14$
10. $3x - 4y = 4$
11. $x = \frac{5}{4}y$
12. $x = \frac{3}{2}y$

Write an equation of the line with each given slope, m, and y-intercept, (0, b). See Example 3.

13. $m = 5, b = 3$
14. $m = -3, b = -3$
15. $m = -4, b = -\frac{1}{6}$
16. $m = 2, b = \frac{3}{4}$
17. $m = \frac{2}{3}, b = 0$

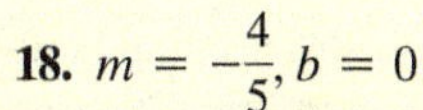

18. $m = -\frac{4}{5}, b = 0$
19. $m = 0, b = -8$
20. $m = 0, b = -2$
21. $m = -\frac{1}{5}, b = \frac{1}{9}$
22. $m = \frac{1}{2}, b = -\frac{1}{3}$

Find an equation of each line with the given slope that passes through the given point. Write the equation in the form $Ax + By = C$. See Example 4.

23. $m = 6;\ (2, 2)$
24. $m = 4;\ (1, 3)$
25. $m = -8;\ (-1, -5)$
26. $m = -2;\ (-11, -12)$
27. $m = \frac{3}{2};\ (5, -6)$
28. $m = \frac{2}{3};\ (-8, 9)$
29. $m = -\frac{1}{2};\ (-3, 0)$
30. $m = -\frac{1}{5};\ (4, 0)$

Find an equation of the line passing through each pair of points. Write the equation in the form $Ax + By = C$. See Example 5.

31. $(3, 2)$ and $(5, 6)$
32. $(6, 2)$ and $(8, 8)$
33. $(-1, 3)$ and $(-2, -5)$
34. $(-4, 0)$ and $(6, -1)$
35. $(2, 3)$ and $(-1, -1)$
36. $(7, 10)$ and $(-1, -1)$
37. $(0, 0)$ and $\left(-\frac{1}{8}, \frac{1}{13}\right)$
38. $(0, 0)$ and $\left(-\frac{1}{2}, \frac{1}{3}\right)$

Find an equation of each line. See Example 6.

39. Vertical line through $(0, 2)$
40. Horizontal line through $(1, 4)$
41. Horizontal line through $(-1, 3)$
42. Vertical line through $(-1, 3)$
43. Vertical line through $\left(-\frac{7}{3}, -\frac{2}{5}\right)$
44. Horizontal line through $\left(\frac{2}{7}, 0\right)$

Find an equation of each line. See Example 7.

45. Parallel to $y = 5$, through $(1, 2)$
46. Perpendicular to $y = 5$, through $(1, 2)$
47. Perpendicular to $x = -3$, through $(-2, 5)$
48. Parallel to $y = -4$, through $(0, -3)$
49. Parallel to $x = 0$, through $(6, -8)$
50. Perpendicular to $x = 7$, through $(-5, 0)$

MIXED PRACTICE

See Examples 1 through 7. Find an equation of each line described. Write each equation in slope-intercept form (solved for y), when possible.

51. With slope $-\frac{1}{2}$, through $\left(0, \frac{5}{3}\right)$
52. With slope $\frac{5}{7}$, through $(0, -3)$
53. Through $(10, 7)$ and $(7, 10)$
54. Through $(5, -6)$ and $(-6, 5)$
55. With undefined slope, through $\left(-\frac{3}{4}, 1\right)$
56. With slope 0, through $(6.7, 12.1)$
57. Slope 1, through $(-7, 9)$
58. Slope 5, through $(6, -8)$
59. Slope -5, y-intercept $(0, 7)$
60. Slope -2; y-intercept $(0, -4)$
61. Through $(6, 7)$, parallel to the x-axis
62. Through $(1, -5)$, parallel to the y-axis
63. Through $(2, 3)$ and $(0, 0)$
64. Through $(4, 7)$ and $(0, 0)$
65. Through $(-2, -3)$, perpendicular to the y-axis
66. Through $(0, 12)$, perpendicular to the x-axis
67. Slope $-\frac{4}{7}$, through $(-1, -2)$
68. Slope $-\frac{3}{5}$, through $(4, 4)$

Solve. Assume each exercise describes a linear relationship. Write the equations in slope-intercept form. See Example 8.

69. A rock is dropped from the top of a 400-foot cliff. After 1 second, the rock is traveling 32 feet per second. After 3 seconds, the rock is traveling 96 feet per second.

a. Assume that the relationship between time and speed is linear and write an equation describing this relationship. Use ordered pairs of the form (time, speed).
b. Use this equation to determine the speed of the rock 4 seconds after it was dropped.

70. A Hawaiian fruit company is studying the sales of a pineapple sauce to see if this product is to be continued. At the end of its first year, profits from this product amounted to \$30,000. At the end of the fourth year, profits were \$66,000.

a. Assume that the relationship between years on the market and profit is linear and write an equation describing this relationship. Use ordered pairs of the form (years on the market, profit).

b. Use this equation to predict the profit at the end of 7 years.

71. Sales of hybrid (gas-electric) vehicles have been decreasing for the past few years. In 2007, 356 thousand new gasoline-electric hybrid vehicles were sold in the United States. In 2009, only 290 thousand were sold. (*Source:* U.S. Energy Information Administration)

a. Write an equation describing the relationship between time and number of registered gasoline-electric hybrid vehicles sold. Use ordered pairs of the form (years past 2007, number of hybrids sold).

b. Use this equation to predict the number of gasoline-electric hybrids sold in the year 2015.

72. In 2011, there were 960 thousand eating establishments in the United States. In 2006, there were 935 thousand eating establishments. (*Source:* National Restaurant Association)

a. Write an equation describing the relationship between time and number of eating establishments. Use ordered pairs of the form (years past 2006, number of eating establishments in thousands).

b. Use this equation to predict the number of eating establishments in 2018.

73. In 2006, the U.S. population per square mile of land area was 85. In 2011, the person per square mile population was 88.

a. Write an equation describing the relationship between year and persons per square mile. Use ordered pairs of the form (years past 2006, persons per square mile).

b. Use this equation to predict the person per square mile population in 2020.

74. In 2005, there were a total of 150 thousand apparel and accessory stores. In 2009, there were a total of 157 thousand apparel and accessory stores. (*Source:* U.S. Bureau of the Census. *County Business Patterns*)

a. Write an equation describing this relationship. Use ordered pairs of the form (years past 2005, numbers of stores in thousands).

b. Use this equation to predict the number of apparel and accessory stores in 2015.

75. It has been said that newspapers are disappearing, replaced by various electronic media. In 2001, newspaper circulation (number of copies it distributed in a day) was about 60 million. In 2009, this circulation dropped to 46 million. (*Source:* Newspaper Asso. of America)

a. Write an equation describing the relationship between time and circulation. Use ordered pairs of the form (years past 2001, number of newspapers circulated).

b. Use this equation to predict the newspaper circulation in 2017.

76. A certain chain of book stores is slowly closing down stores. Suppose that in 2006 there were 3991 stores and in 2010 there were 3200 stores.

a. Write an equation describing the relationship between time and number of store locations. Use ordered pairs of the form (years past 2006, number of stores).

b. Use this equation to predict the number of stores in 2018.

77. The Pool Fun Company has learned that, by pricing a newly released Fun Noodle at \$3, sales will reach 10,000 Fun Noodles per day during the summer. Raising the price to \$5 will cause sales to fall to 8000 Fun Noodles per day.

a. Assume that the relationship between price and number of Fun Noodles sold is linear and write an equation describing this relationship. Use ordered pairs of the form (price, number sold).

b. Predict the daily sales of Fun Noodles if the price is \$3.50.

78. The value of a building bought in 2000 may be depreciated (or decreased) as time passes for income tax purposes. Seven years after the building was bought, this value was \$225,000 and 12 years after it was bought, this value was \$195,000.

a. If the relationship between number of years past 2000 and the depreciated value of the building is linear, write an equation describing this relationship. Use ordered pairs of the form (years past 2000, value of building).

b. Use this equation to estimate the depreciated value of the building in 2018.

REVIEW AND PREVIEW

Find the value of $x^2 - 3x + 1$ *for each given value of x. See Section 2.5.*

79. 2 **80.** 5

81. -1 **82.** -3

For each graph, determine whether any x-values correspond to two or more y-values. See Section 10.1.

83.

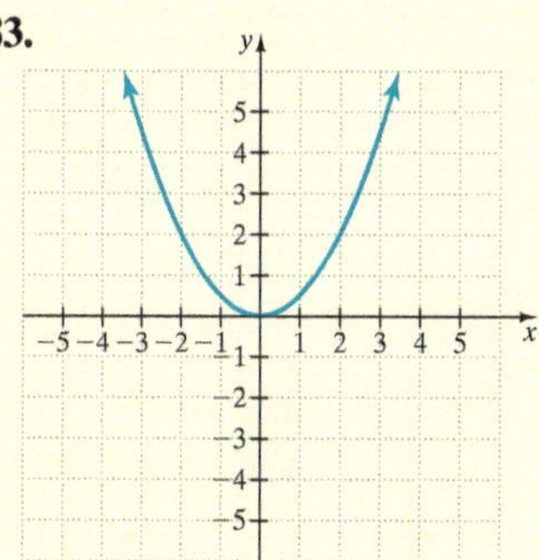

84.

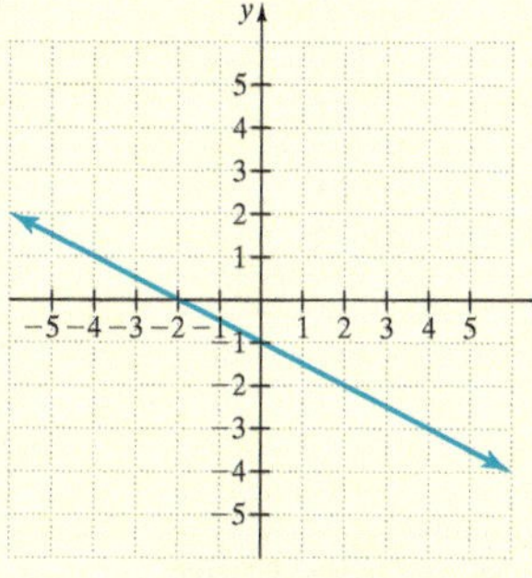

85. **86.**

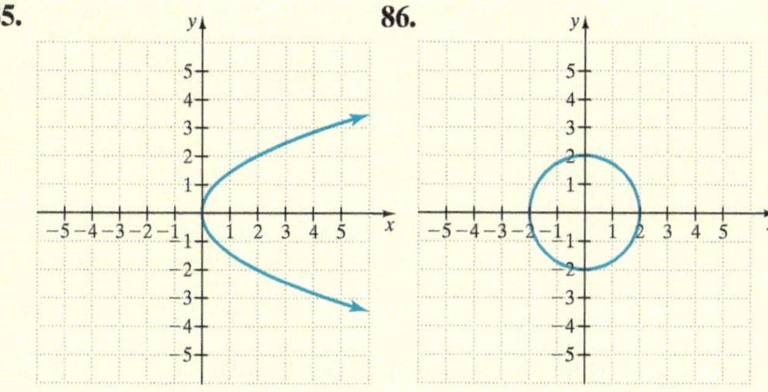

CONCEPT EXTENSIONS

For Exercises 87 through 90, identify the form that the linear equation in two variables is written in. For Exercises 91 and 92, identify the appearance of the graph of the equation.

87. $y - 7 = 4(x + 3)$;__________ form

88. $5x - 9y = 11$;__________ form

89. $y = \frac{3}{4}x - \frac{1}{3}$;__________ form

90. $y + 2 = \frac{-1}{3}(x - 2)$;__________ form

91. $y = \frac{1}{2}$;__________ line **92.** $x = -17$;__________ line

93. Given the equation of a nonvertical line, explain how to find the slope without finding two points on the line.

94. Given two points on a nonvertical line, explain how to use the point-slope form to find the equation of the line.

△ **95.** Write an equation in standard form of the line that contains the point $(-1, 2)$ and is

a. parallel to the line $y = 3x - 1$.
b. perpendicular to the line $y = 3x - 1$.

△ **96.** Write an equation in standard form of the line that contains the point $(4, 0)$ and is

a. parallel to the line $y = -2x + 3$.
b. perpendicular to the line $y = -2x + 3$.

△ **97.** Write an equation in standard form of the line that contains the point $(3, -5)$ and is

a. parallel to the line $3x + 2y = 7$.
b. perpendicular to the line $3x + 2y = 7$.

△ **98.** Write an equation in standard form of the line that contains the point $(-2, 4)$ and is

a. parallel to the line $x + 3y = 6$.
b. perpendicular to the line $x + 3y = 6$.

10.6 Functions 

Objectives

1 Identify Relations, Domains, and Ranges.

2 Identify Functions.

3 Use the Vertical Line Test.

4 Use Function Notation.

Objective 1 Identifying Relations, Domains, and Ranges

In previous sections, we have discussed the relationships between two quantities. For example, the relationship between the length of the side of a square x and its area y is described by the equation $y = x^2$. Ordered pairs can be used to write down solutions of this equation. For example, (2, 4) is a solution of $y = x^2$, and this notation tells us that the x-value 2 is related to the y-value 4 for this equation. In other words, when the length of the side of a square is 2 units, its area is 4 square units.

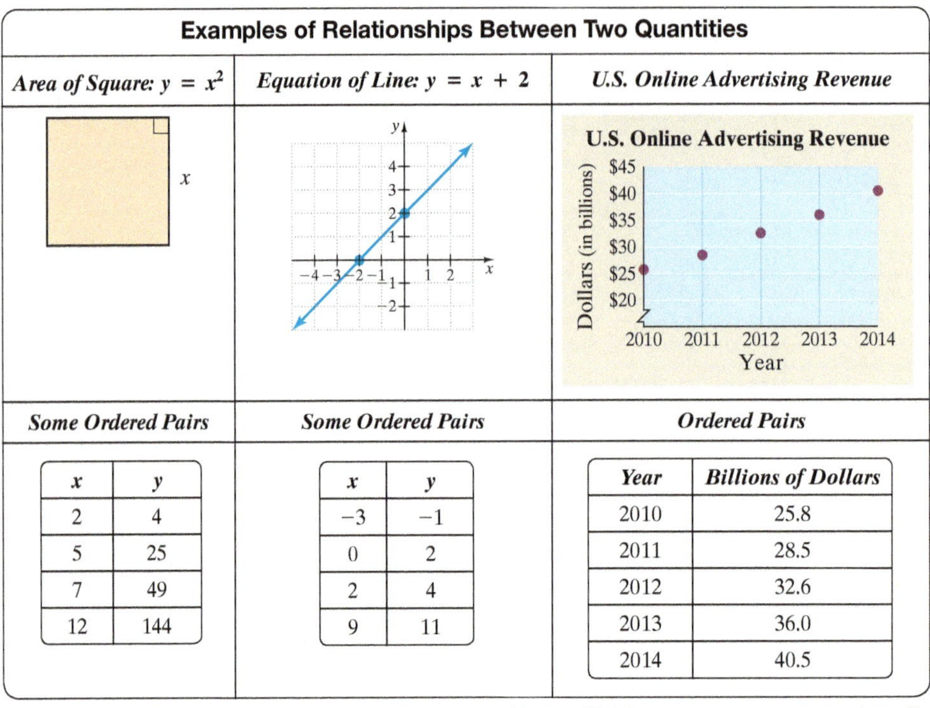

Examples of Relationships Between Two Quantities

Area of Square: $y = x^2$ — *Some Ordered Pairs*

x	y
2	4
5	25
7	49
12	144

Equation of Line: $y = x + 2$ — *Some Ordered Pairs*

x	y
−3	−1
0	2
2	4
9	11

U.S. Online Advertising Revenue — *Ordered Pairs*

Year	*Billions of Dollars*
2010	25.8
2011	28.5
2012	32.6
2013	36.0
2014	40.5

(*Source:* GRABstats.com; some years are estimated)

A set of ordered pairs is called a **relation.** The set of all x-coordinates is called the **domain** of a relation, and the set of all y-coordinates is called the **range** of a relation. Equations such as $y = x^2$ are also called relations since equations in two variables define a set of ordered pair solutions.

Practice 1

Find the domain and the range of the relation $\{(1, 3), (5, 0), (0, -2), (5, 4)\}$.

Example 1 Find the domain and the range of the relation $\{(0, 2), (3, 3), (-1, 0), (3, -2)\}$.

Solution: The domain is the set of all x-values or $\{-1, 0, 3\}$, and the range is the set of all y-values, or $\{-2, 0, 2, 3\}$.

Work Practice 1

Objective 2 Identifying Functions 

Some relations are also functions.

Function

A function is a set of ordered pairs that assigns to each x-value exactly one y-value.

In other words, a function cannot have two ordered pairs with the same x-coordinate but different y-coordinates.

Practice 2

Determine whether each relation is also a function.

a. $\{(4, 1), (3, -2), (8, 5), (-5, 3)\}$

b. $\{(1, 2), (-4, 3), (0, 8), (1, 4)\}$

Example 2 Determine whether each relation is also a function.

a. $\{(-1, 1), (2, 3), (7, 3), (8, 6)\}$

b. $\{(0, -2), (1, 5), (0, 3), (7, 7)\}$

Solution:

a. Although the ordered pairs (2, 3) and (7, 3) have the same y-value, each x-value is assigned to only one y-value, so this set of ordered pairs is a function.

b. The x-value 0 is assigned to two y-values, −2 and 3, so this set of ordered pairs is not a function.

Work Practice 2

Answers

1. Domain: $\{0, 1, 5\}$; Range: $\{-2, 0, 3, 4\}$

2. a. function **b.** not a function

Relations and functions can be described by a graph of their ordered pairs.

Example 3 Determine whether each graph is the graph of a function.

a.

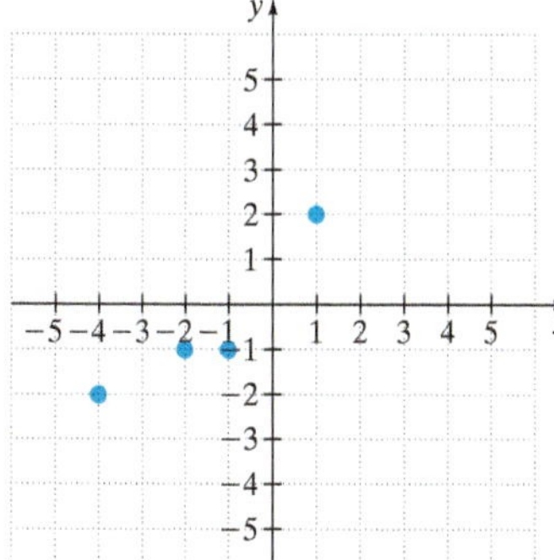

b.

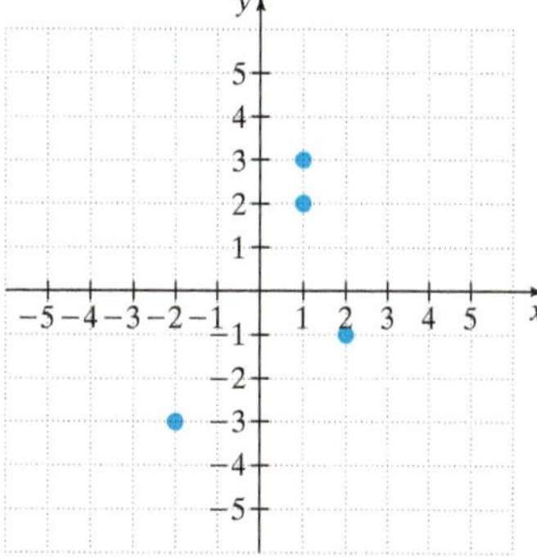

Solution:

a. This is the graph of the relation $\{(-4, -2), (-2, -1), (-1, -1), (1, 2)\}$. Each x-coordinate has exactly one y-coordinate, so this is the graph of a function.

b. This is the graph of the relation $\{(-2, -3), (1, 2), (1, 3), (2, -1)\}$. The x-coordinate 1 is paired with two y-coordinates, 2 and 3, so this is not the graph of a function.

Work Practice 3

Practice 3

Determine whether each graph is the graph of a function.

a.

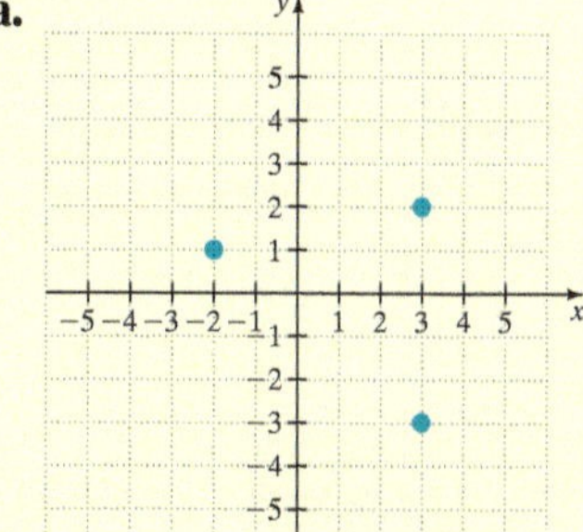

b.

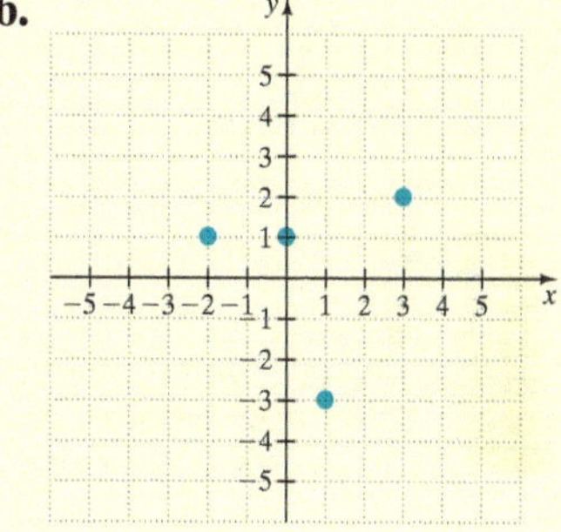

Objective 3 Using the Vertical Line Test 

The graph in Example 3(b) was not the graph of a function because the x-coordinate 1 was paired with two y-coordinates, 2 and 3. Notice that when an x-coordinate is paired with more than one y-coordinate, a vertical line can be drawn that will intersect the graph at more than one point. We can use this fact to determine whether a relation is also a function. We call this the **vertical line test.**

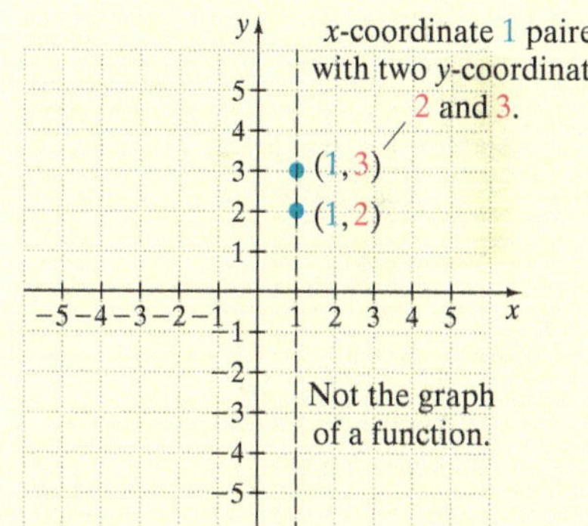

Vertical Line Test

If a vertical line can be drawn so that it intersects a graph more than once, the graph is not the graph of a function.

This vertical line test works for all types of graphs on the rectangular coordinate system.

Example 4 Use the vertical line test to determine whether each graph is the graph of a function.

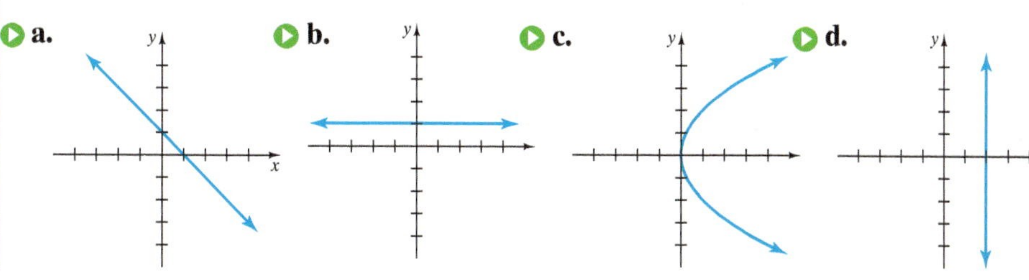

(Continued on next page)

Answers

3. a. not a function b. function

Practice 4

Use the vertical line test to determine whether each graph is the graph of a function.

a.

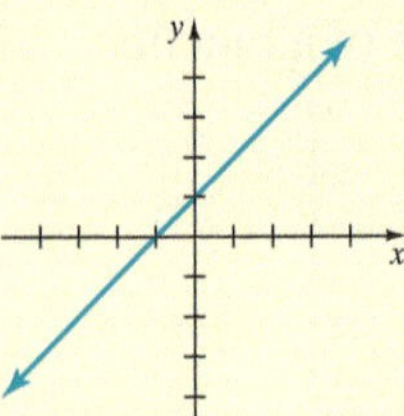

b.

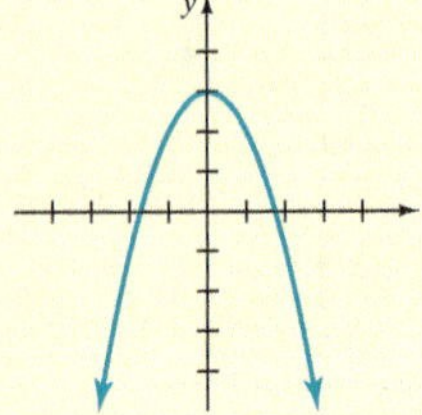

c.

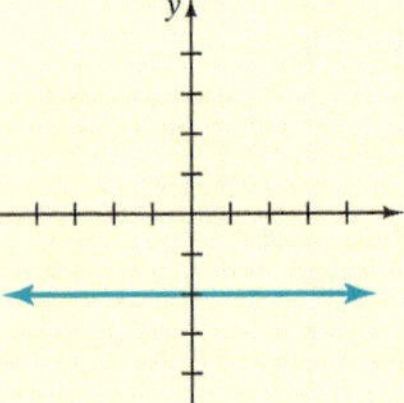

d.

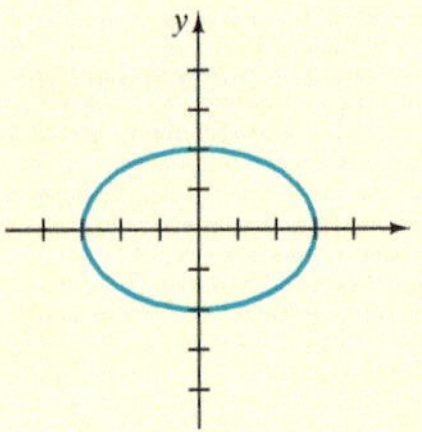

Practice 5

Decide whether the equation describes a function.

a. $y = 2x$ b. $y = -3x - 1$

c. $y = 8$ d. $x = 2$

Answers

4. a. function b. function
c. function d. not a function
5. a. function b. function
c. function d. not a function

Solution:

a. This graph is the graph of a function since no vertical line will intersect this graph more than once.

b. This graph is also the graph of a function; no vertical line will intersect it more than once.

c. This graph is not the graph of a function. Vertical lines can be drawn that intersect the graph in two points. An example of one is shown.

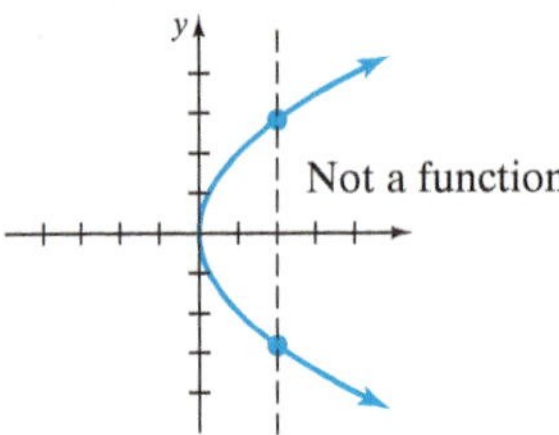

d. This graph is not the graph of a function. A vertical line can be drawn that intersects this line at every point.

Work Practice 4

Recall that the graph of a linear equation is a line, and a line that is not vertical will pass the vertical line test. **Thus, all linear equations are functions except those of the form $x = c$, which are vertical lines.**

Example 5 Decide whether the equation describes a function.

a. $y = x$ b. $y = 2x + 1$ c. $y = 5$ d. $x = -1$

Solution: **a, b,** and **c** are functions because their graphs are nonvertical lines. **d** is not a function because its graph is a vertical line.

Work Practice 5

Examples of functions can often be found in magazines, newspapers, books, and other printed material in the form of tables or graphs such as that in Example 6.

Example 6 The graph shows the sunrise time for Indianapolis, Indiana, for the year. Use this graph to answer the questions.

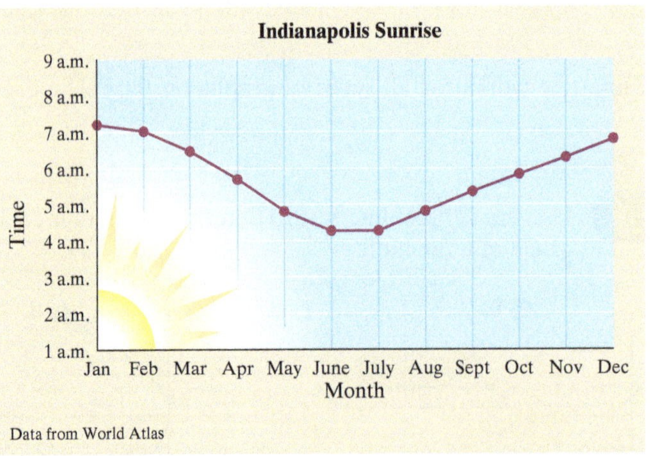

a. Approximate the time of sunrise on February 1.

b. Approximately when does the sun rise at 5 a.m.?

c. Is this the graph of a function?

Solution:

a. To approximate the time of sunrise on February 1, we find the mark on the horizontal axis that corresponds to February 1. From this mark, we move vertically upward until the graph is reached. From that point on the graph, we move horizontally to the left until the vertical axis is reached. The vertical axis there reads 7 a.m.

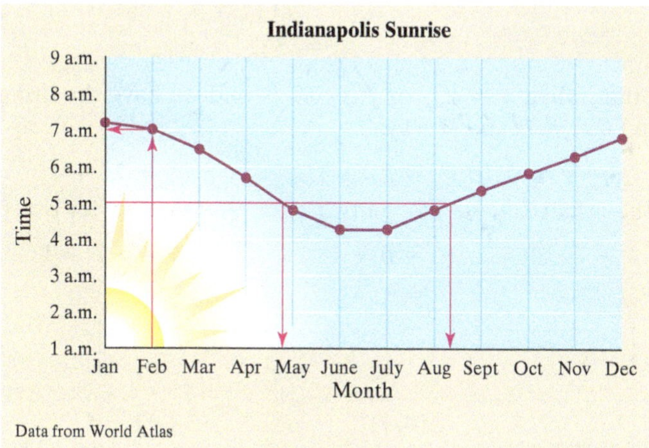

b. To approximate when the sun rises at 5 a.m., we find 5 a.m. on the time axis and move horizontally to the right. Notice that we will reach the graph twice, corresponding to two dates for which the sun rises at 5 a.m. We follow both points on the graph vertically downward until the horizontal axis is reached. The sun rises at 5 a.m. at approximately the end of the month of April and the middle of the month of August.

c. The graph is the graph of a function since it passes the vertical line test. In other words, for every day of the year in Indianapolis, there is exactly one sunrise time.

Work Practice 6

Practice 6

The graph shows the average monthly temperature for Chicago, Illinois, for the year. Use this graph to answer the questions.

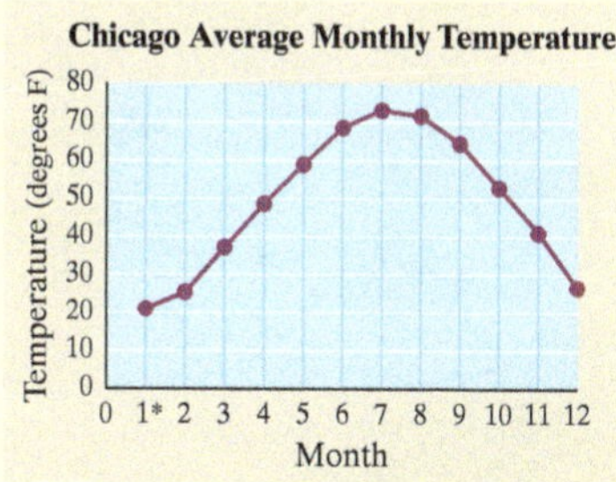

* (1 is Jan.; 12 is Dec.)

a. Approximate the average monthly temperature for June.

b. For what month is the average monthly temperature 26°?

c. Is this the graph of a function?

Objective 4 Using Function Notation

The graph of the linear equation $y = 2x + 1$ passes the vertical line test, so we say that $y = 2x + 1$ is a function. In other words, $y = 2x + 1$ gives us a rule for writing ordered pairs where every x-coordinate is paired with one and only one y-coordinate.

The variable y is a function of the variable x. For each value of x, there is only one value of y. Thus, we say the variable x is the **independent variable** because any value in the domain can be assigned to x. The variable y is the **dependent variable** because its value depends on x.

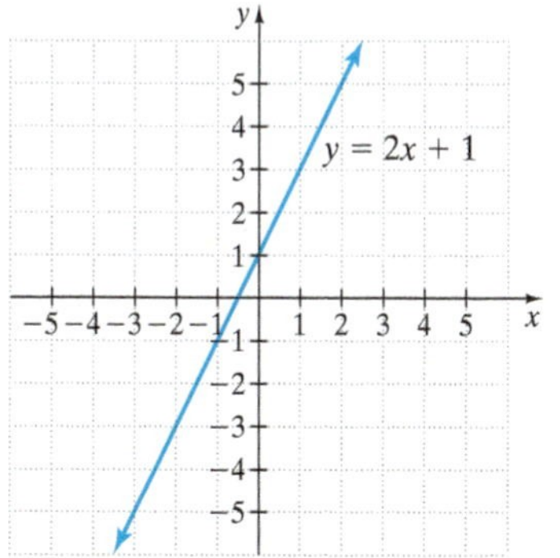

Answers

6. a. 69°F **b.** February **c.** yes

We often use letters such as f, g, and h to name functions. For example, the symbol $f(x)$ means *function of* x and is read "f of x." This notation is called **function notation.** The equation $y = 2x + 1$ can be written as $f(x) = 2x + 1$ using function notation, and these equations mean the same thing. In other words, $y = f(x)$.

The notation $f(1)$ means to replace x with 1 and find the resulting y or function value. Since

$$f(x) = 2x + 1$$

then

$$\begin{aligned} f(1) &= 2(1) + 1 \\ &= 3 \end{aligned}$$

This means that, when $x = 1$, y or $f(x) = 3$, and we have the ordered pair $(1, 3)$. Now let's find $f(2)$, $f(0)$, and $f(-1)$.

$f(x) = 2x + 1$	$f(x) = 2x + 1$	$f(x) = 2x + 1$
$f(2) = 2(2) + 1$	$f(0) = 2(0) + 1$	$f(-1) = 2(-1) + 1$
$= 4 + 1$	$= 0 + 1$	$= -2 + 1$
$= 5$	$= 1$	$= -1$
Ordered Pair: $(2, 5)$	$(0, 1)$	$(-1, -1)$

Helpful Hint Note that, for example, if $f(2) = 5$, the corresponding ordered pair is $(2, 5)$.

Helpful Hint

Note that $f(x)$ is a special symbol in mathematics used to denote a function. The symbol $f(x)$ is read "f of x." It does **not** mean $f \cdot x$ (f times x).

Practice 7

Given $h(x) = x^2 + 5$, find the following. Then write the corresponding ordered pairs generated.

a. $h(2)$ **b.** $h(-5)$ **c.** $h(0)$

Example 7 Given $g(x) = x^2 - 3$, find the following. Then write the corresponding ordered pairs generated.

a. $g(2)$ **b.** $g(-2)$ **c.** $g(0)$

Solution:

a. $\begin{aligned} g(x) &= x^2 - 3 \\ g(2) &= 2^2 - 3 \\ &= 4 - 3 \\ &= 1 \end{aligned}$

b. $\begin{aligned} g(x) &= x^2 - 3 \\ g(-2) &= (-2)^2 - 3 \\ &= 4 - 3 \\ &= 1 \end{aligned}$

c. $\begin{aligned} g(x) &= x^2 - 3 \\ g(0) &= 0^2 - 3 \\ &= 0 - 3 \\ &= -3 \end{aligned}$

Ordered Pairs:	$g(2) = 1$ gives $(2, 1)$	$g(-2) = 1$ gives $(-2, 1)$	$g(0) = -3$ gives $(0, -3)$

Work Practice 7

We now practice finding the domain and the range of a function. The domains of our functions will be the set of all possible real numbers that x can be replaced by. The range is the set of corresponding y-values.

Answers

7. **a.** $h(2) = 9$; $(2, 9)$ **b.** $h(-5) = 30$; $(-5, 30)$ **c.** $h(0) = 5$; $(0, 5)$

Example 8 Find the domain of each function.

a. $g(x) = \frac{1}{x}$ **b.** $f(x) = 2x + 1$

Solution:

a. Recall that we cannot divide by 0, so the domain of $g(x)$ is the set of all real numbers except 0. In interval notation, we can write $(-\infty, 0) \cup (0, \infty)$.

b. In this function, x can be any real number. The domain of $f(x)$ is the set of all real numbers, or $(-\infty, \infty)$ in interval notation.

Work Practice 8

Practice 8

Find the domain of each function.

a. $h(x) = 6x + 3$

b. $f(x) = \frac{1}{x^2}$

✓**Concept Check** Suppose that the value of f is -7 when the function is evaluated at 2. Write this situation in function notation.

Example 9 Find the domain and the range of each function graphed. Use interval notation.

a.

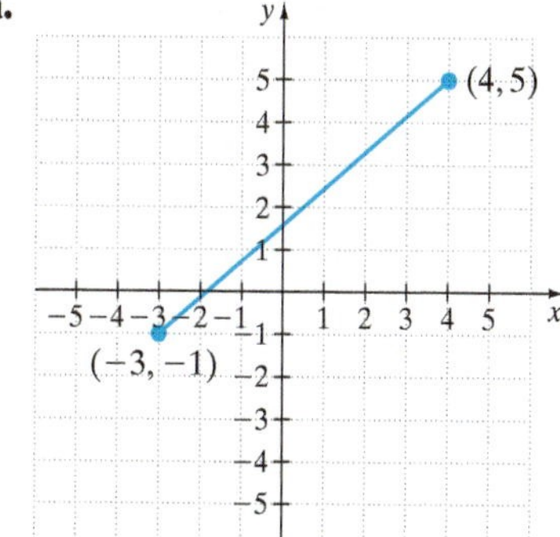

b.

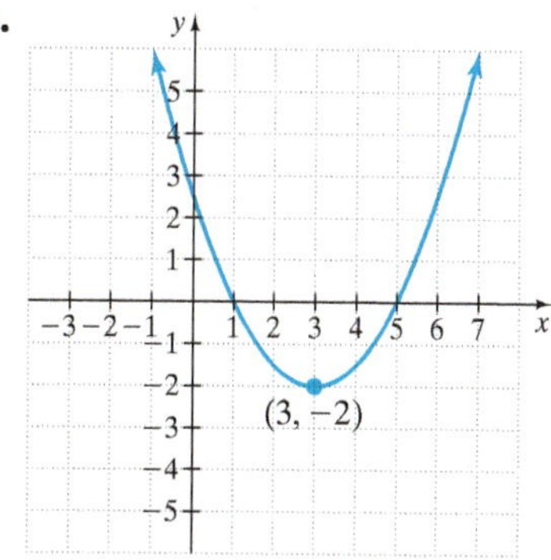

Solution:

a.

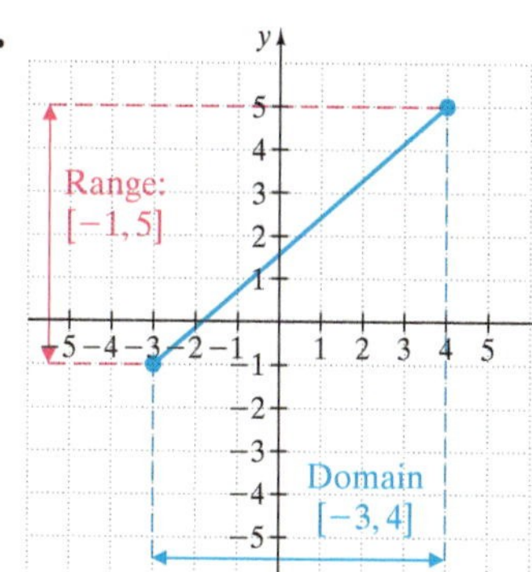

b.

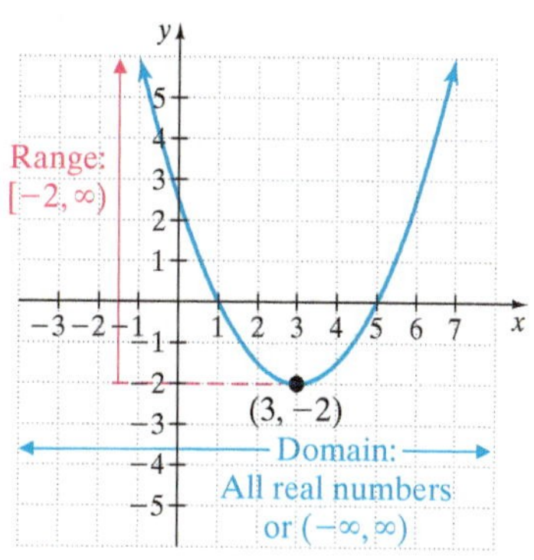

Work Practice 9

Practice 9

Find the domain and the range of each function graphed. Use interval notation.

a.

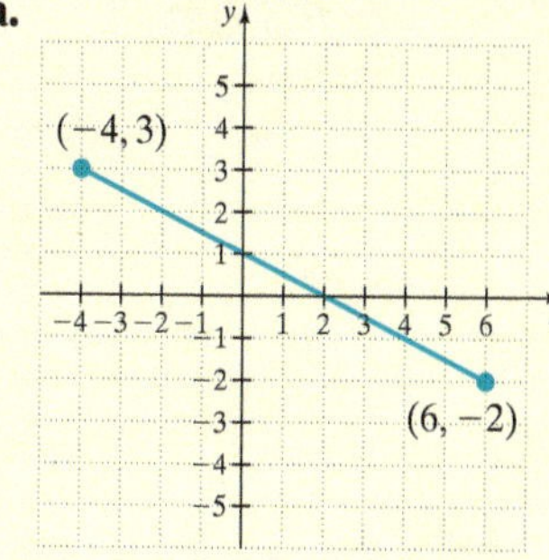

b.

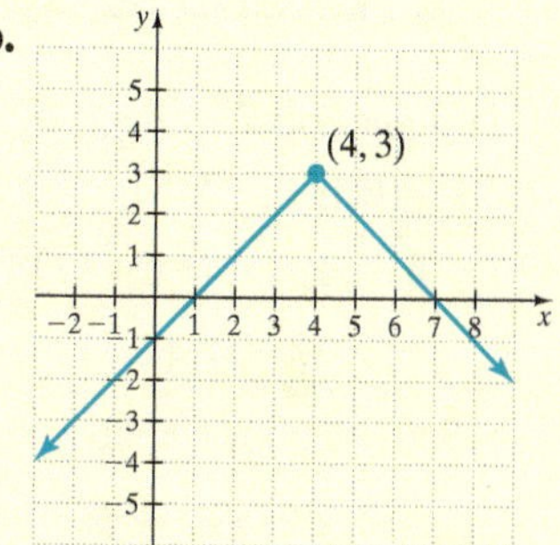

Answers

8. a. Domain: $(-\infty, \infty)$
b. Domain: $(-\infty, 0) \cup (0, \infty)$

9. a. Domain: $[-4, 6]$; range: $[-2, 3]$
b. Domain: $(-\infty, \infty)$; range: $(-\infty, 3]$

✓**Concept Check Answer**

$f(2) = -7$

Vocabulary, Readiness & Video Check

Use the choices below to fill in each blank. Some choices may not be used.

$x = c$	horizontal	domain	relation	(7, 3)
$y = c$	vertical	range	function	(3, 7)

1. A set of ordered pairs is called a(n) ________.
2. A set of ordered pairs that assigns to each x-value exactly one y-value is called a(n) ________.

3. The set of all y-coordinates of a relation is called the ________.
4. The set of all x-coordinates of a relation is called the ________.
5. All linear equations are functions except those whose graphs are ________ lines.
6. All linear equations are functions except those whose equations are of the form ________.

Martin-Gay Interactive Videos *Watch the section lecture video and answer the following questions.*

See Video 10.6

Objective 1 7. In the lecture before Example 1, relations are discussed. Why can an equation in two variables define a relation?

Objective 2 8. Based on Examples 2 and 3, can a set of ordered pairs with no repeated x-values, but with repeated y-values, be a function? For example: $\{(0, 4), (-3, 4), (2, 4)\}$.

Objective 3 9. After reviewing Example 8, explain why the vertical line test works.

Objective 4 10. Using Example 10, write the three function values found and their corresponding ordered pairs. For example, $f(0) = 2$ corresponds to $(0, 2)$.

10.6 Exercise Set MyMathLab®

Find the domain and the range of each relation. See Example 1.

1. $\{(2,4), (0,0), (-7,10), (10,-7)\}$
2. $\{(3,-6), (1,4), (-2,-2)\}$
3. $\{(0,-2), (1,-2), (5,-2)\}$
4. $\{(5,0), (5,-3), (5,4), (5,3)\}$

Determine whether each relation is also a function. See Example 2.

5. $\{(1,1), (2,2), (-3,-3), (0,0)\}$
6. $\{(11,6), (-1,-2), (0,0), (3,-2)\}$
7. $\{(-1,0), (-1,6), (-1,8)\}$
8. $\{(1,2), (3,2), (1,4)\}$

MIXED PRACTICE

Determine whether each graph is the graph of a function. See Examples 3 and 4.

9.

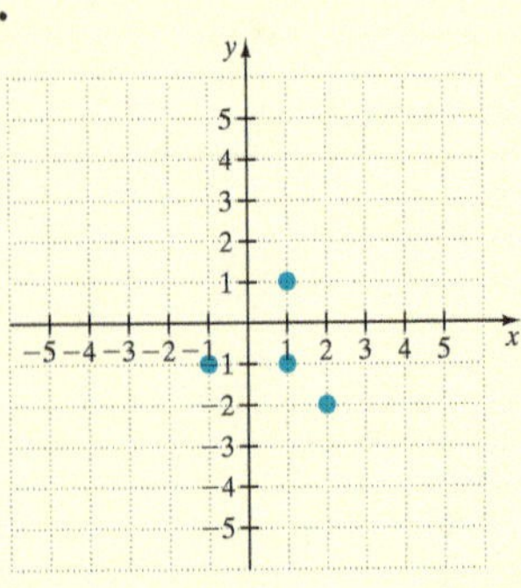

10.

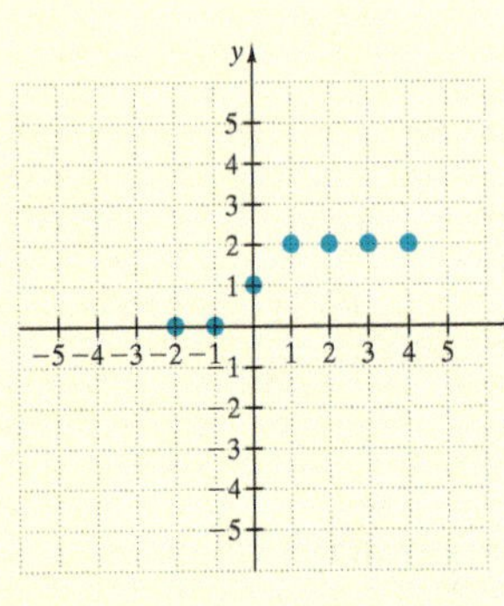

11.

12.

13.

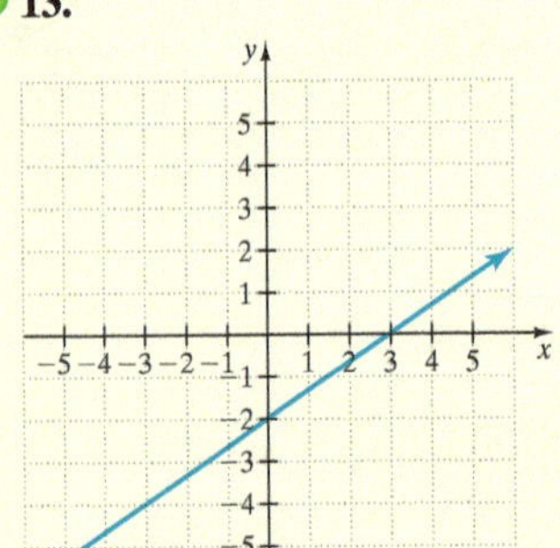

14.

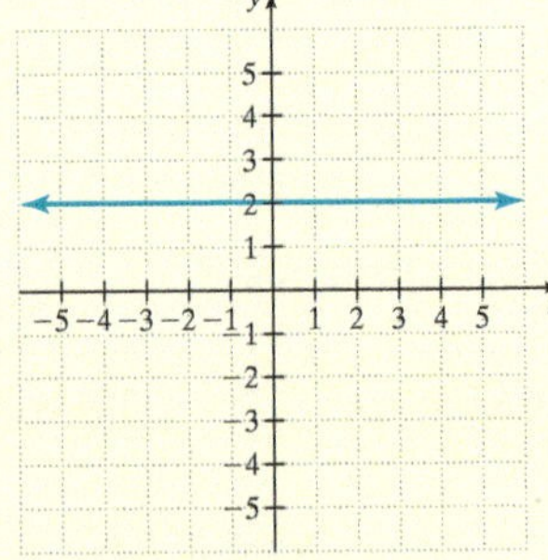

15.

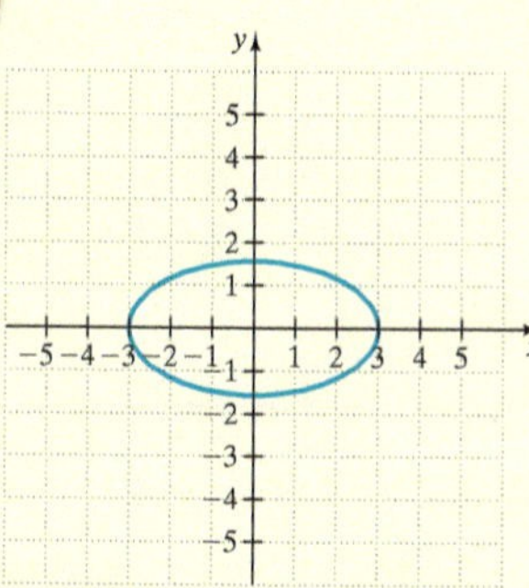

16.

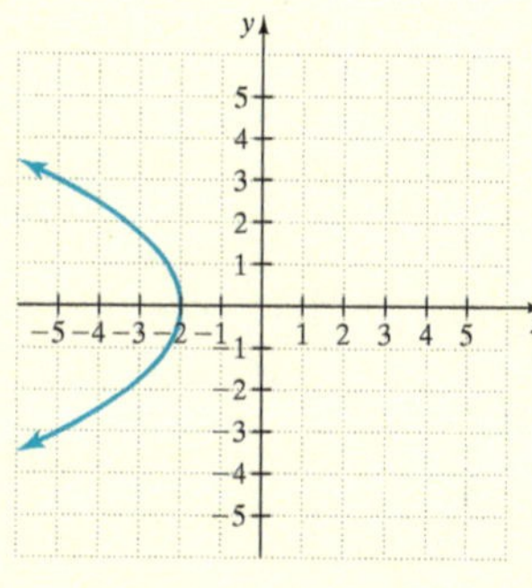

Decide whether the equation describes a function. See Example 5.

17. $y = x + 1$

18. $y = x - 1$

19. $y - x = 7$

20. $2x - 3y = 9$

21. $y = 6$

22. $x = 3$

23. $x = -2$

24. $y = -9$

25. $x = y^2$

26. $y = x^2 - 3$

The graph shows the sunset times for Seward, Alaska. Use this graph to answer Exercises 27 through 32. See Example 6.

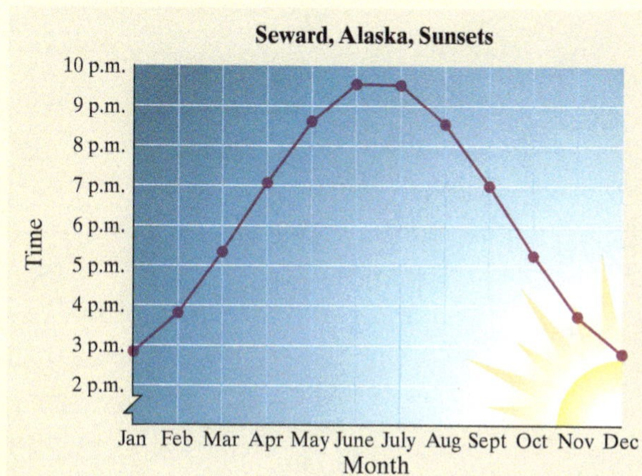

27. Approximate the time of sunset on June 1.

28. Approximate the time of sunset on November 1.

29. Approximate when the sunset is at 3 p.m.

30. Approximate when the sunset is at 9 p.m.

31. Is this graph the graph of a function? Why or why not?

32. Do you think a graph of sunset times for any location will always be a function? Why or why not?

This graph shows the U.S. hourly minimum wage for each year shown. Use this graph to answer Exercises 33 through 38. See Example 6.

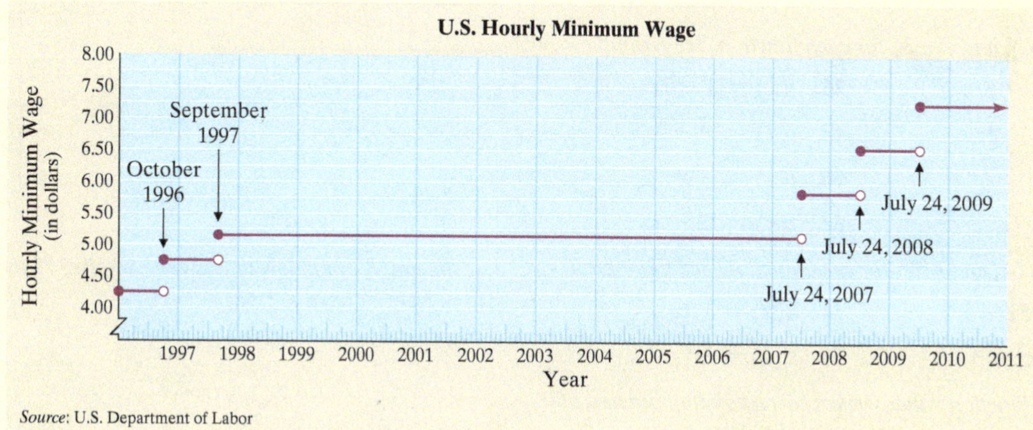

33. Approximate the minimum wage before October 1996.

34. Approximate the minimum wage in 2006.

35. Approximate the year when the minimum wage increased to over \$7.00 per hour.

36. According to the graph, what hourly wage was in effect for the greatest number of years?

37. Is this graph the graph of a function? Why or why not?

38. Do you think that a similar graph of your hourly wage on January 1 of every year (whether you are working or not) will be the graph of a function? Why or why not?

This graph shows the cost of mailing a large envelope through the U.S. Postal Service by weight. (Note: These costs will increase.) Use this graph to answer Exercises 39 through 44. See Example 6.

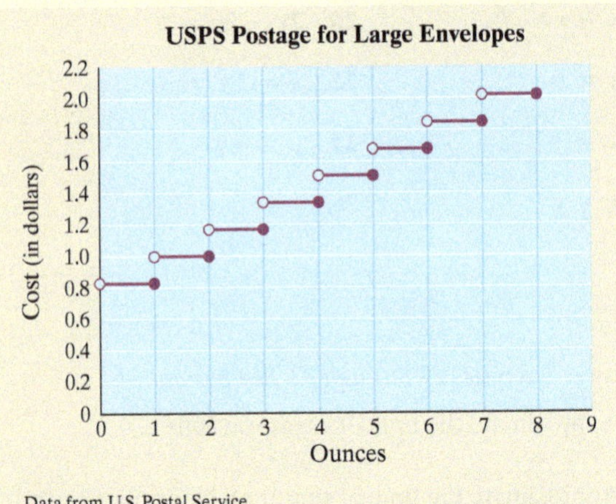

39. Approximate the postage to mail a large envelope weighing more than 4 ounces but not more than 5 ounces.

40. Approximate the postage to mail a large envelope weighing more than 7 ounces but not more than 8 ounces.

41. Give the weight of a large envelope that costs \$1 to mail.?

42. If you have \$1.25, what is the weight of the largest envelope you can mail for that amount of money?

43. Is this graph a function? Why or why not?

44. Do you think that a similar graph of postage to mail a first-class letter will be the graph of a function? Why or why not?

Find $f(-2)$, $f(0)$, and $f(-3)$ for each function. See Example 7.

45. $f(x) = 2x - 5$

46. $f(x) = 3 - 7x$

47. $f(x) = x^2 + 2$

48. $f(x) = x^2 - 4$

49. $f(x) = 3x$

50. $f(x) = -3x$

51. $f(x) = |x|$

52. $f(x) = |2 - x|$

Find $h(-1)$, $h(0)$, and $h(4)$ for each function. See Example 7.

53. $h(x) = -5x$

54. $h(x) = -3x$

55. $h(x) = 2x^2 + 3$

56. $h(x) = 3x^2$

For each given function value, write a corresponding ordered pair.

57. $f(3) = 6$

58. $f(7) = -2$

59. $g(0) = -\frac{1}{2}$

60. $g(0) = -\frac{7}{8}$

61. $h(-2) = 9$

62. $h(-10) = 1$

Find the domain of each function. See Example 8.

63. $f(x) = 3x - 7$

64. $g(x) = 5 - 2x$

65. $h(x) = \frac{1}{x + 5}$

66. $f(x) = \frac{1}{x - 6}$

67. $g(x) = |x + 1|$

68. $h(x) = |2x|$

Find the domain and the range of each relation graphed. See Example 9.

69.

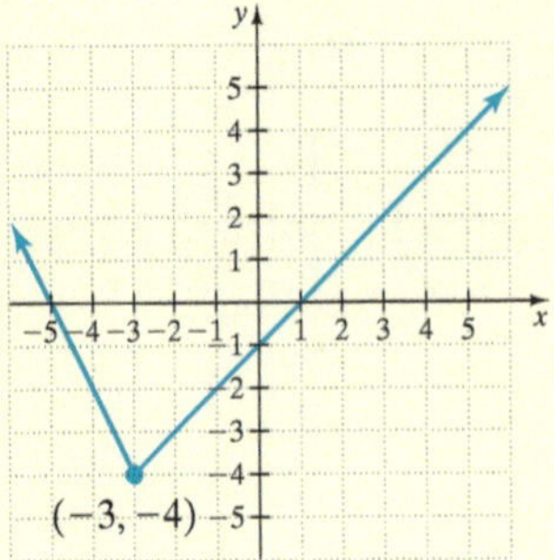

70.

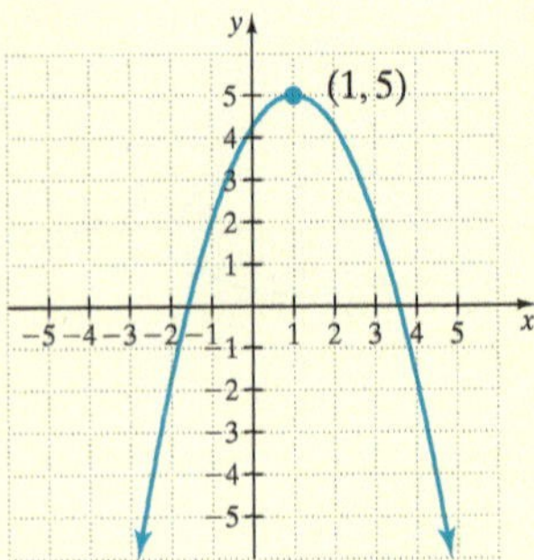

71.

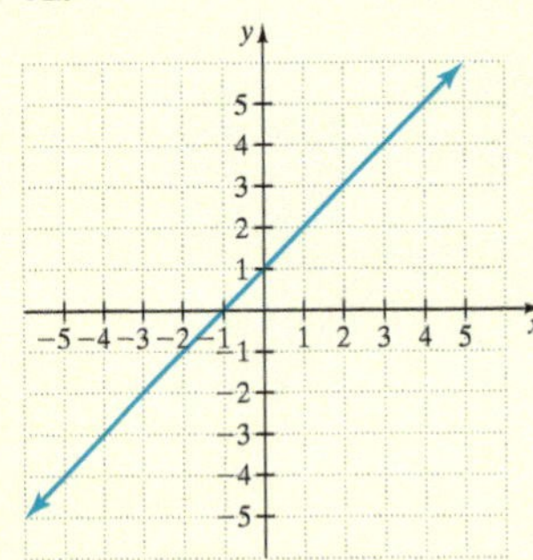

72.

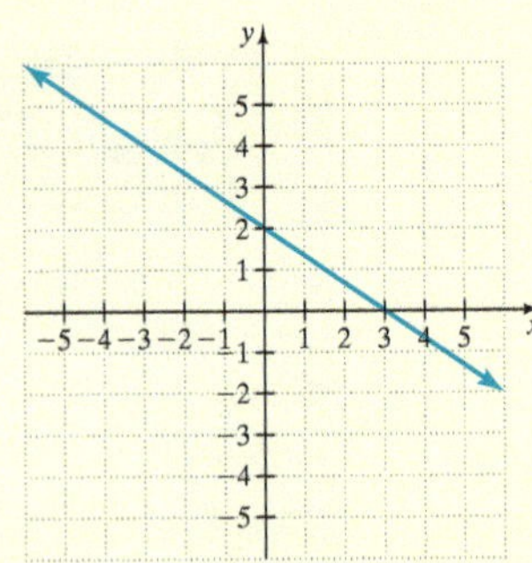

73.

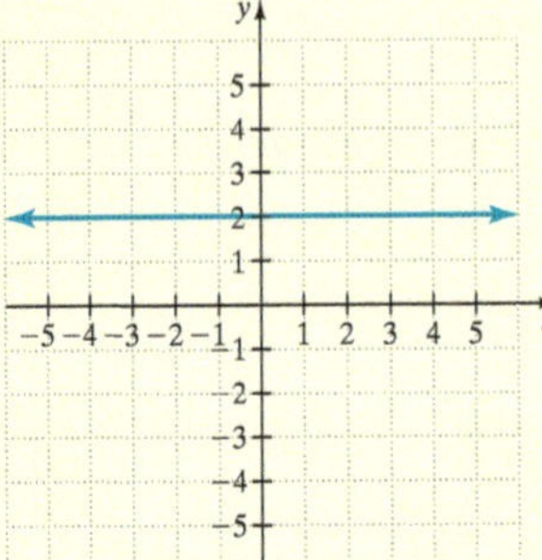

74.

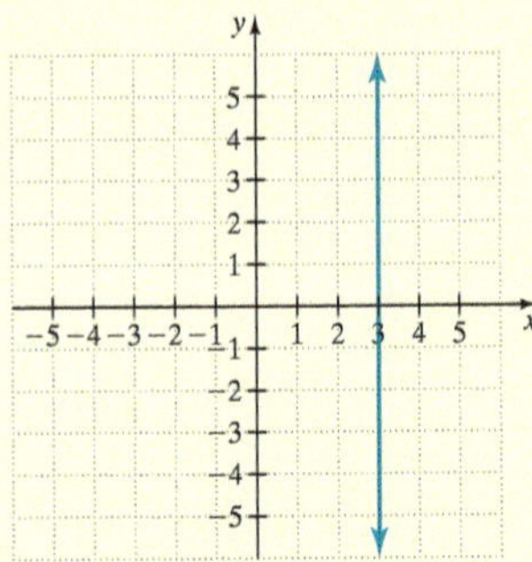

Use the graph of f below to answer Exercises 75 through 80.

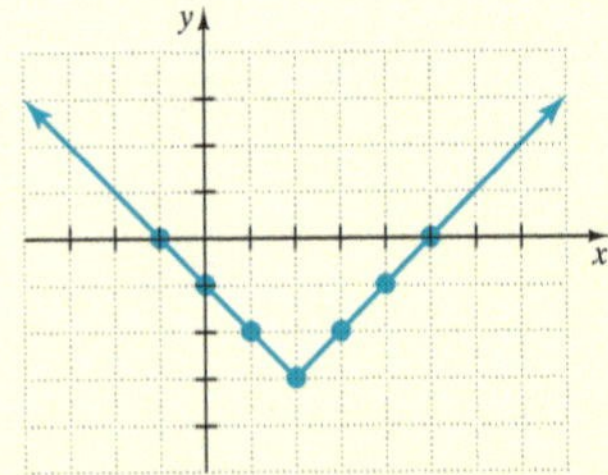

75. Complete the ordered pair solution for f. (0,)

76. Complete the ordered pair solution for f. (3,)

77. $f(0) =$ ____?

78. $f(3) =$ ____?

79. If $f(x) = 0$, find the value(s) of x.

80. If $f(x) = -1$, find the value(s) of x.

REVIEW AND PREVIEW

Find the coordinates of the point of intersection. See Section 10.1.

81.

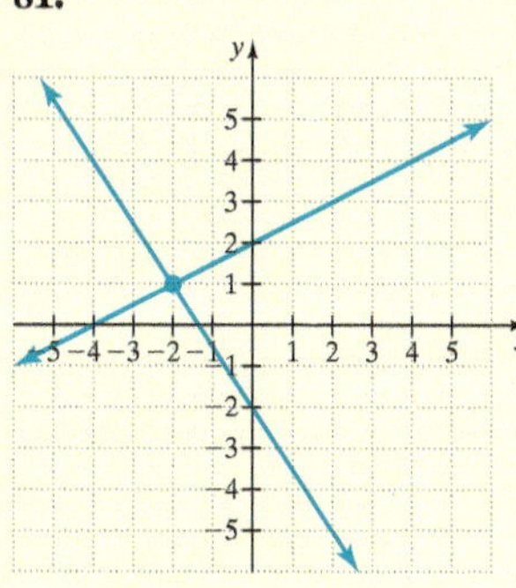

82.

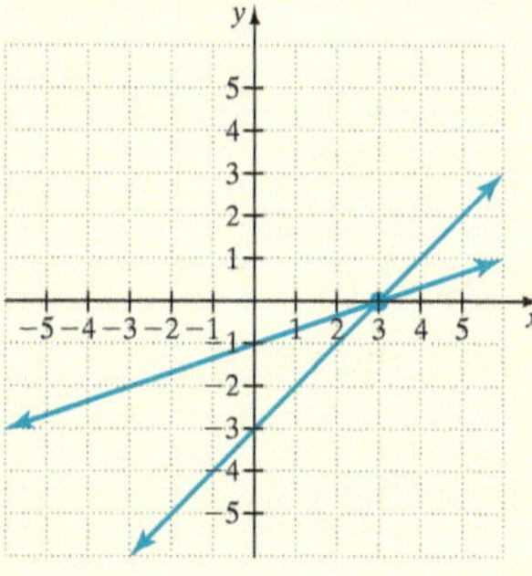

83.

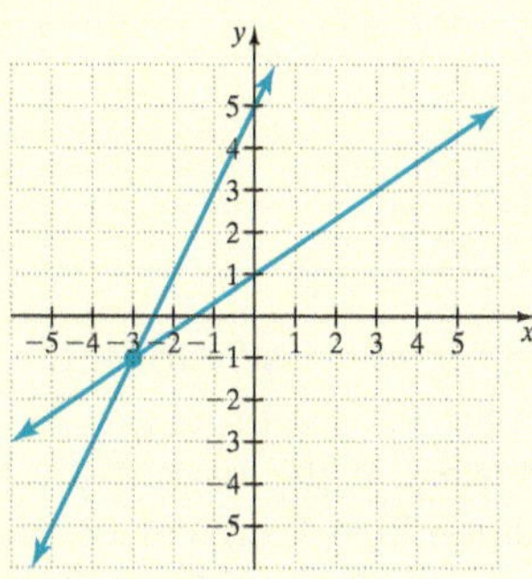

84.

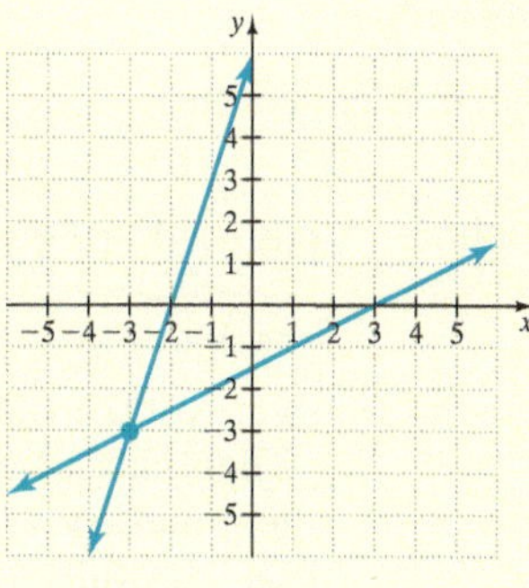

CONCEPT EXTENSIONS

Solve. See the Concept Check in this section.

85. If a function f is evaluated at -5, the value of the function is 12. Write this situation using function notation.

86. Suppose (9, 20) is an ordered pair solution for the function g. Write this situation using function notation.

The graph of the function, f, is below. Use this graph to answer Exercises 87 through 90.

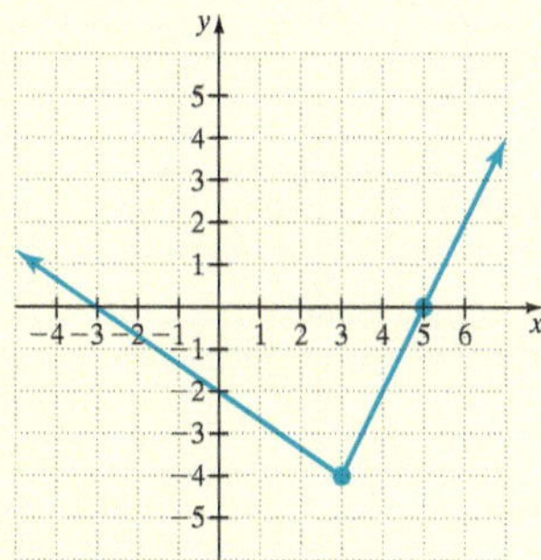

87. Write the coordinates of the lowest point of the graph.

88. Write the answer to Exercise 87 in function notation.

89. An x-intercept of this graph is (5, 0). Write this using function notation.

90. Write the other x-intercept of this graph (see Exercise 89) using function notation.

91. Forensic scientists use the function

$$H(x) = 2.59x + 47.24$$

to estimate the height of a woman in centimeters given the length x of her femur bone.

a. Estimate the height of a woman whose femur measures 46 centimeters.

b. Estimate the height of a woman whose femur measures 39 centimeters.

92. The dosage in milligrams D of Ivermectin, a heartworm preventive for a dog who weighs x pounds, is given by the function

$$D(x) = \frac{136}{25}x$$

a. Find the proper dosage for a dog that weighs 35 pounds.

b. Find the proper dosage for a dog that weighs 70 pounds.

93. In your own words, define **(a)** function; **(b)** domain; **(c)** range.

94. Explain the vertical line test and how it is used.

95. Since $y = x + 7$ is a function, rewrite the equation using function notation.

96. Since $y = 3$ is a function, rewrite this equation using function notation.

See the example below for Exercises 97 through 100.

Example

If $f(x) = x^2 + 2x + 1$, find $f(\pi)$.

Solution:

$$f(x) = x^2 + 2x + 1$$
$$f(\pi) = \pi^2 + 2\pi + 1$$

Given the following functions, find the indicated values.

97. $f(x) = 2x + 7$
a. $f(2)$ 11 **b.** $f(a)$

98. $g(x) = -3x + 12$
a. $g(s)$ **b.** $g(r)$

99. $h(x) = x^2 + 7$
a. $h(3)$ **b.** $h(a)$

100. $f(x) = x^2 - 12$
a. $f(12)$ **b.** $f(a)$

Chapter 10 Vocabulary Check

Fill in each blank with one of the words listed below.

relation	function	domain	range	standard	slope-intercept	x-intercept	x
y-axis	x-axis	solution	linear	slope	point-slope	y-intercept	y

1. An ordered pair is a(n) ____________ of an equation in two variables if replacing the variables by the coordinates of the ordered pair results in a true statement.
2. The vertical number line in the rectangular coordinate system is called the ____________.
3. A(n) ____________ equation can be written in the form $Ax + By = C$.
4. A(n) ____________ is a point of the graph where the graph crosses the x-axis.
5. The form $Ax + By = C$ is called ____________ form.
6. A(n) ____________ is a point of the graph where the graph crosses the y-axis.
7. The equation $y = 7x - 5$ is written in ____________ form.
8. The equation $y + 1 = 7(x - 2)$ is written in ____________ form.
9. To find an x-intercept of a graph, let ____________ = 0.
10. The horizontal number line in the rectangular coordinate system is called the ____________.
11. To find a y-intercept of a graph, let ____________ = 0.
12. The ____________ of a line measures the steepness or tilt of a line.
13. A set of ordered pairs that assigns to each x-value exactly one y-value is called a(n) ____________.
14. The set of all x-coordinates of a relation is called the ____________ of the relation.
15. The set of all y-coordinates of a relation is called the ____________ of the relation.
16. A set of ordered pairs is called a(n) ____________.

Helpful Hint

Are you preparing for your test? Don't forget to take the Chapter 10 Test on page 799. Then check your answers at the back of the text and use the Chapter Test Prep Videos to see the fully worked-out solutions to any of the exercises you want to review.

10 Chapter Highlights

Definitions and Concepts	Examples
Section 10.1 Reading Graphs and the Rectangular Coordinate System	
The **rectangular coordinate system** consists of a plane and a vertical and a horizontal number line intersecting at their 0 coordinates. The vertical number line is called the **y-axis** and the horizontal number line is called the **x-axis.** The point of intersection of the axes is called the **origin.**	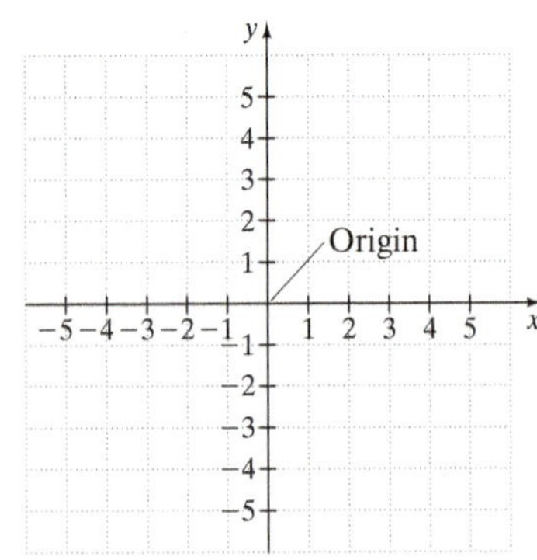
To **plot** or **graph** an ordered pair means to find its corresponding point on a rectangular coordinate system. To plot or graph an ordered pair such as $(3, -2)$, start at the origin. Move 3 units to the right and, from there, 2 units down. To plot or graph $(-3, 4)$, start at the origin. Move 3 units to the left and, from there, 4 units up.	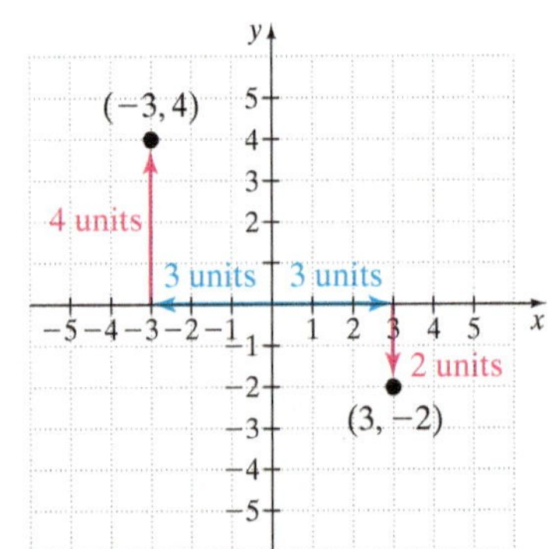

Definitions and Concepts	Examples
Section 10.1 Reading Graphs and the Rectangular Coordinate System *(continued)*	
An ordered pair is a **solution** of an equation in two variables if replacing the variables by the coordinates of the ordered pair results in a true statement.	Determine whether $(-1, 5)$ is a solution of $2x + 3y = 13$. $2x + 3y = 13$ $2(-1) + 3 \cdot 5 = 13$ Let $x = -1, y = 5$ $-2 + 15 = 13$ $13 = 13$ True
If one coordinate of an ordered pair solution is known, the other value can be determined by substitution.	Complete the ordered pair solution (0,) for the equation $x - 6y = 12$. $x - 6y = 12$ $0 - 6y = 12$ Let $x = 0$. $\frac{-6y}{-6} = \frac{12}{-6}$ Divide by -6. $y = -2$ The ordered pair solution is $(0, -2)$.
Section 10.2 Graphing Linear Equations	
A **linear equation in two variables** is an equation that can be written in the form $Ax + By = C$ where A and B are not both 0. The form $Ax + By = C$ is called **standard form.**	***Linear Equations*** $3x + 2y = -6$ $x = -5$ $y = 3$ $y = -x + 10$ $x + y = 10$ is in standard form.
To graph a linear equation in two variables, find three ordered pair solutions. Plot the solution points and draw the line connecting the points.	Graph $x - 2y = 5$. 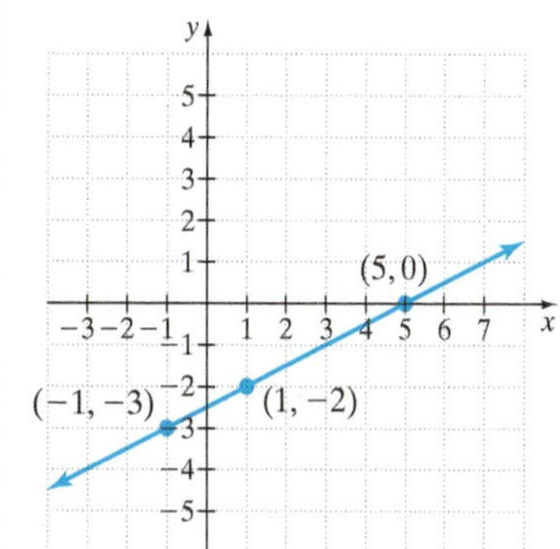 x: 5, 1, −1; y: 0, −2, −3
Section 10.3 Intercepts	
An **intercept** of a graph is a point where the graph intersects an axis. If a graph intersects the x-axis at a, then $(a, 0)$ is the ***x*-intercept.** If a graph intersects the y-axis at b, then $(0, b)$ is the ***y*-intercept.**	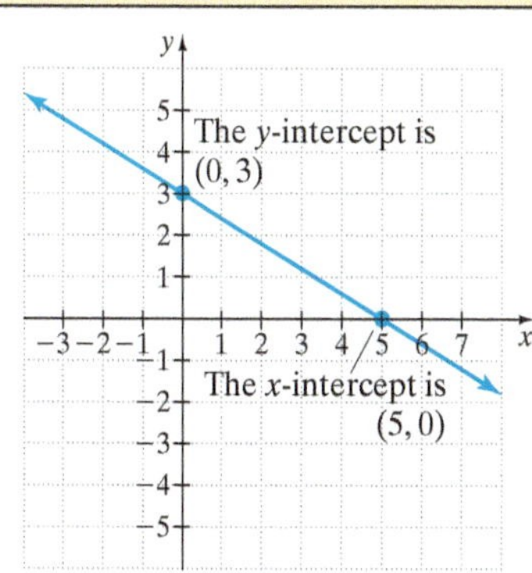

x	y
5	0
1	−2
−1	−3

(continued)

Definitions and Concepts	Examples

Section 10.3 Intercepts (continued)

To find the *x*-intercept, let $y = 0$ and solve for x.

To find the *y*-intercept, let $x = 0$ and solve for y.

Graph $2x - 5y = -10$ by finding intercepts.

If $y = 0$, then

$$2x - 5 \cdot 0 = -10$$
$$2x = -10$$
$$\frac{2x}{2} = \frac{-10}{2}$$
$$x = -5$$

If $x = 0$, then

$$2 \cdot 0 - 5y = -10$$
$$-5y = -10$$
$$\frac{-5y}{-5} = \frac{-10}{-5}$$
$$y = 2$$

The x-intercept is $(-5, 0)$. The y-intercept is $(0, 2)$.

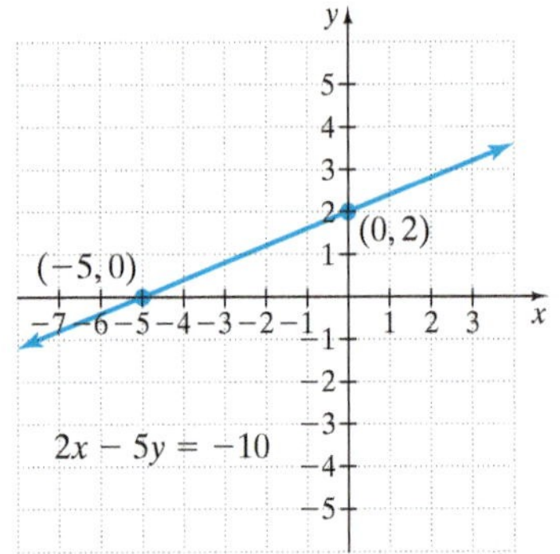

The graph of $x = c$ is a vertical line with x-intercept $(c, 0)$.

The graph of $y = c$ is a horizontal line with y-intercept $(0, c)$.

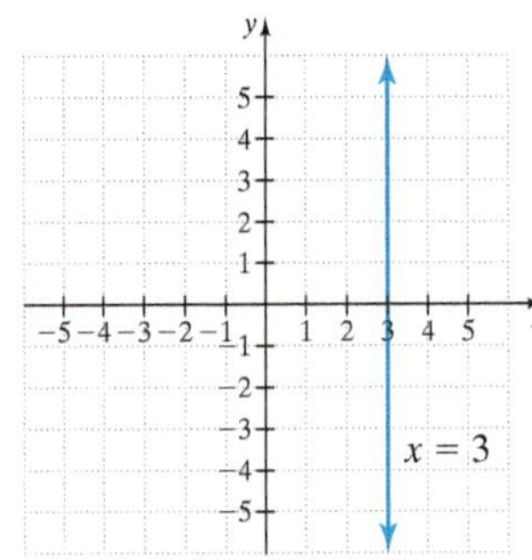

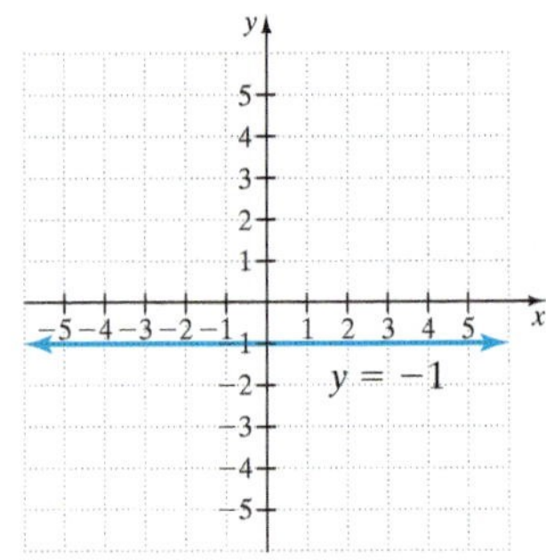

Section 10.4 Slope and Rate of Change

The **slope *m*** of the line through points (x_1, y_1) and (x_2, y_2) is given by

$$m = \frac{y_2 - y_1}{x_2 - x_1} \quad \text{as long as } x_2 \neq x_1$$

The slope of the line through points $(-1, 6)$ and $(-5, 8)$ is

$$m = \frac{y_2 - y_1}{x_2 - x_1} = \frac{8 - 6}{-5 - (-1)} = \frac{2}{-4} = -\frac{1}{2}$$

A horizontal line has slope 0.
The slope of a vertical line is undefined.

The slope of the line $y = -5$ is 0.
The line $x = 3$ has undefined slope.

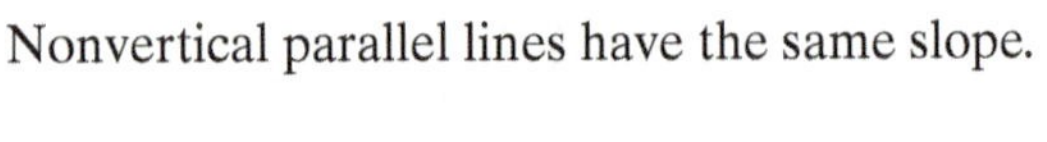

Nonvertical parallel lines have the same slope.

Two nonvertical lines are perpendicular if the slope of one is the negative reciprocal of the slope of the other.

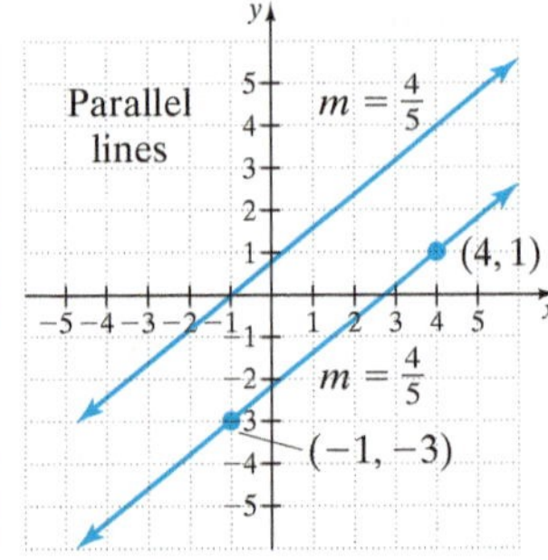

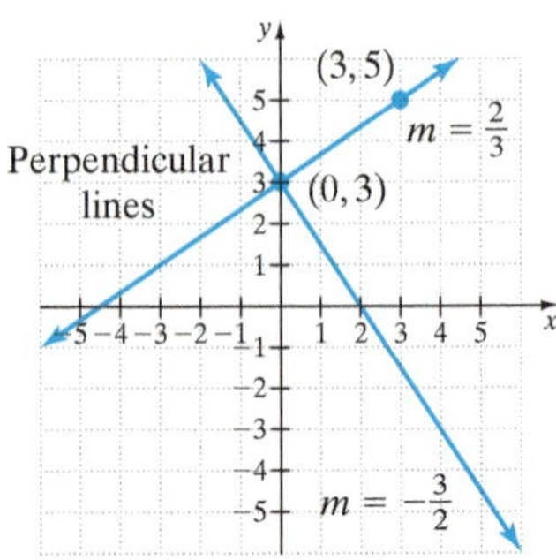

Definitions and Concepts	Examples
Section 10.5 Equations of Lines	
Slope-Intercept Form $y = mx + b$ m is the slope of the line. $(0, b)$ is the y-intercept.	Find the slope and the y-intercept of the line whose equation is $2x + 3y = 6$. Solve for y: $2x + 3y = 6$ $3y = -2x + 6$ Subtract $2x$. $y = -\frac{2}{3}x + 2$ Divide by 3. The slope of the line is $-\frac{2}{3}$ and the y-intercept is $(0, 2)$. Find an equation of the line with slope 3 and y-intercept $(0, -1)$.
Point-Slope Form $y - y_1 = m(x - x_1)$ m is the slope. (x_1, y_1) is a point on the line.	The equation is $y = 3x - 1$. Find an equation of the line with slope $\frac{3}{4}$ that contains the point $(-1, 5)$. $y - 5 = \frac{3}{4}[x - (-1)]$ $4(y - 5) = 3(x + 1)$ Multiply by 4. $4y - 20 = 3x + 3$ Distribute. $-3x + 4y = 23$ Subtract $3x$ and add 20.
Section 10.6 Functions	
A set of ordered pairs is a **relation.** The set of all x-coordinates is called the **domain** of the relation, and the set of all y-coordinates is called the **range** of the relation.	The domain of the relation $\{(0, 5), (2, 5), (4, 5), (5, -2)\}$ is $\{0, 2, 4, 5\}$. The range is $\{-2, 5\}$.
A **function** is a set of ordered pairs that assigns to each x-value exactly one y-value. ***Vertical Line Test*** If a vertical line can be drawn so that it intersects a graph more than once, the graph is not the graph of a function.	Which are graphs of functions? 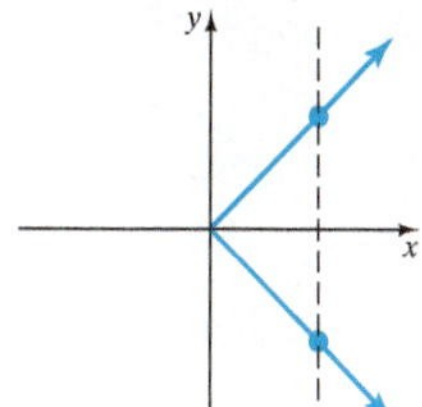 This graph is not the graph of a function. 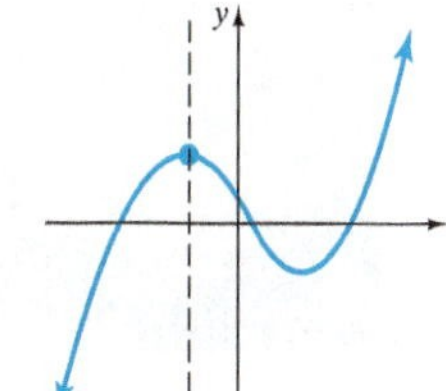 This graph is the graph of a function.
The symbol $f(x)$ means **function of x.** This notation is called **function** notation.	If $f(x) = 2x^2 + 6x - 1$, find $f(3)$. $f(3) = 2(3)^2 + 6 \cdot 3 - 1$ $= 2 \cdot 9 + 18 - 1$ $= 18 + 18 - 1$ $= 35$

Chapter 10 Review

(10.1) Plot the following ordered pairs on a Cartesian coordinate system.

1. $(-7, 0)$
2. $\left(0, 4\frac{4}{5}\right)$
3. $(-2, -5)$
4. $(1, -3)$
5. $(0.7, 0.7)$
6. $(-6, 4)$

7. A local office supply store uses quantity pricing. The table shows the price per box of #10 security envelopes for different numbers of envelopes in a box purchased.

Price per Box of Envelopes (in dollars)	*Number of Envelopes in Box*
5.00	50
8.50	100
20.00	250
27.00	500

a. Write each paired data as an ordered pair of the form (price per box of envelopes, number of envelopes in box).
b. Create a scatter diagram of the paired data. Be sure to label the axes appropriately.

8. The table shows the annual overnight stays in national parks. (Data from National Park Service)

Year	*Overnight Stays in National Parks (in millions)*
2005	13.8
2006	13.6
2007	14.1
2008	13.9
2009	14.6
2010	14.6

a. Write each paired data as an ordered pair of the form (year, number of overnight stays).

b. Create a scatter diagram of the paired data. Be sure to label the axes properly.

Determine whether each ordered pair is a solution of the given equation.

9. $7x - 8y = 56$; $(0, 56), (8, 0)$
10. $-2x + 5y = 10$; $(-5, 0), (1, 1)$
11. $x = 13$; $(13, 5), (13, 13)$
12. $y = 2$; $(7, 2), (2, 7)$

Complete the ordered pairs so that each is a solution of the given equation.

13. $-2 + y = 6x$; $(7, \quad)$
14. $y = 3x + 5$; $\left(\quad , -8\right)$

Complete the table of values for each given equation; then plot the ordered pairs. Use a single coordinate system for each exercise.

15. $9 = -3x + 4y$

x	y
	0
	3
9	

16. $x = 2y$

x	y
	0
	5
	−5

The cost in dollars of producing x compact disc holders is given by $y = 5x + 2000$. Use this equation for Exercises 17 and 18.

17. Complete the following table.

x	y
1	
100	
1000	

18. Find the number of compact disc holders that can be produced for $6430.

(10.2) Graph each linear equation.

19. $x - y = 1$
20. $x + y = 6$
21. $x - 3y = 12$
22. $5x - y = -8$
23. $x = 3y$
24. $y = -2x$
25. $2x - 3y = 6$
26. $4x - 3y = 12$

(10.3) Identify the intercepts.

27.

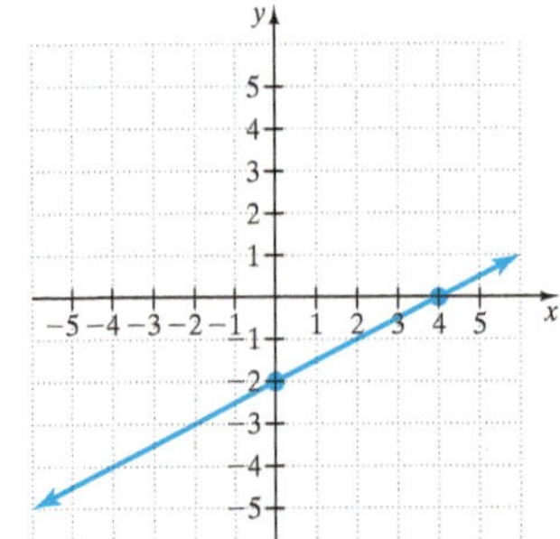

28.

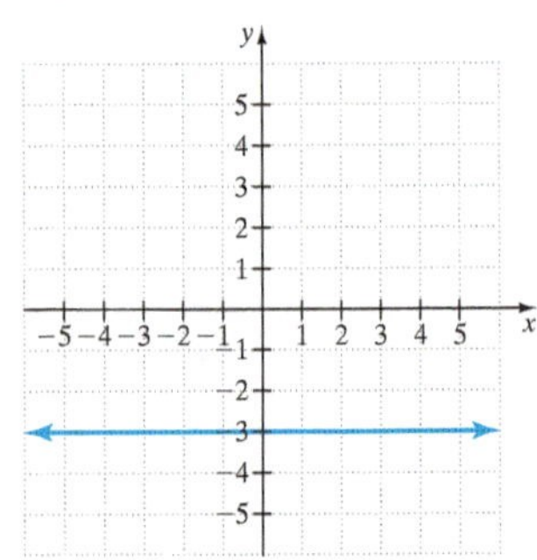

29.

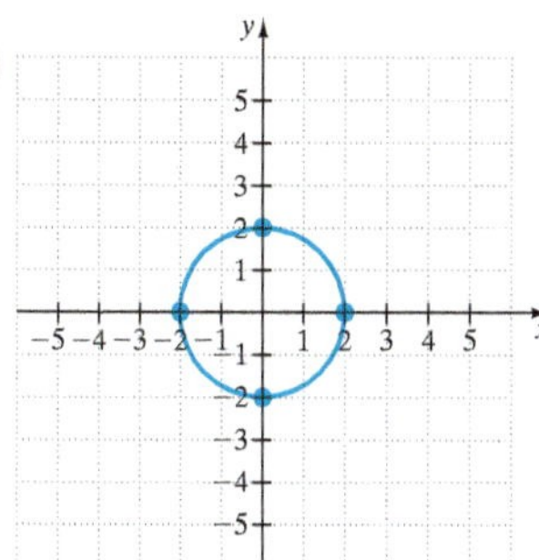

30.

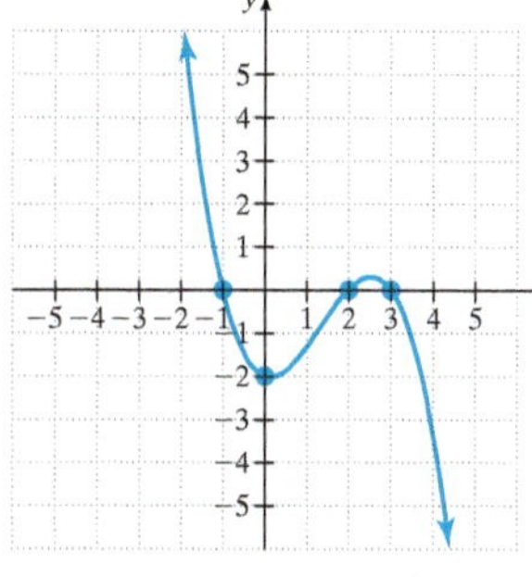

Graph each linear equation by finding its intercepts.

31. $x - 3y = 12$

32. $-4x + y = 8$

33. $y = -3$

34. $x = 5$

35. $y = -3x$

36. $x = 5y$

37. $x - 2 = 0$

38. $y + 6 = 0$

(10.4) *Find the slope of each line.*

39.

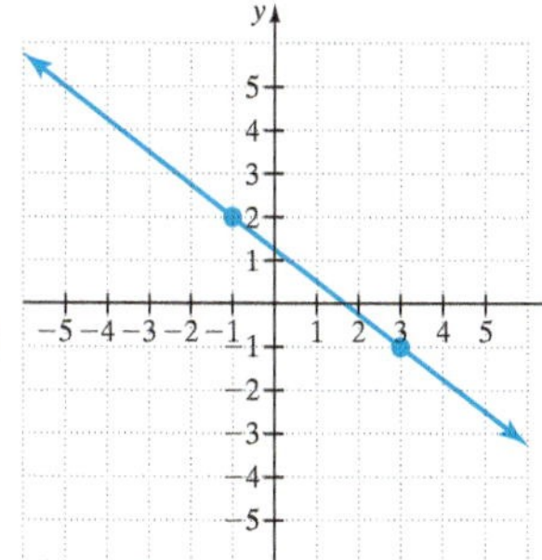

40.

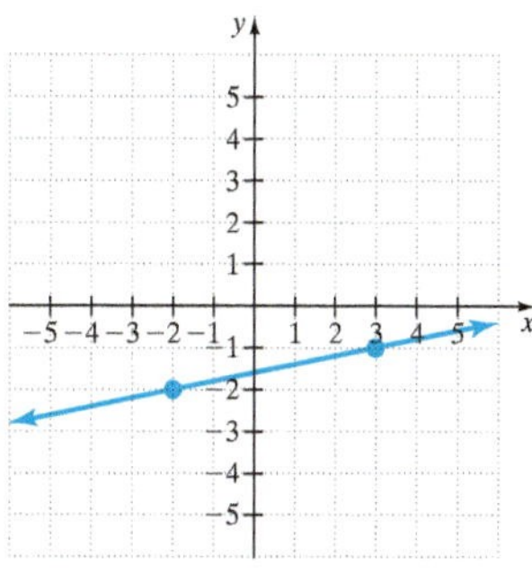

In Exercises 41 through 44, match each line with its slope.

a.

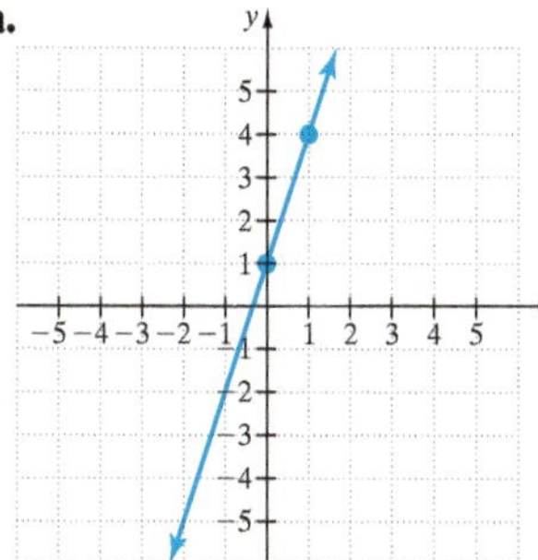

b.

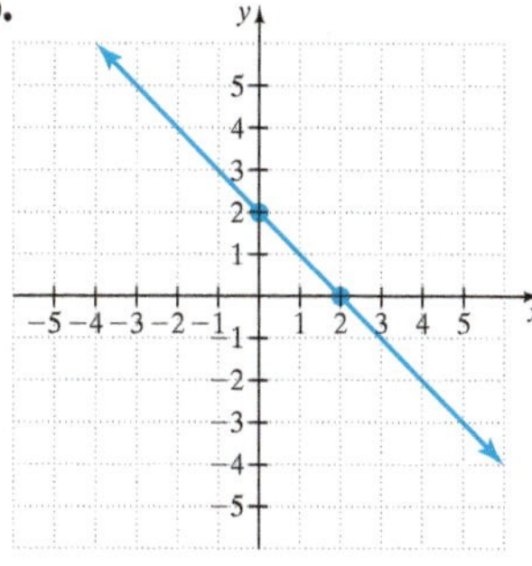

c.

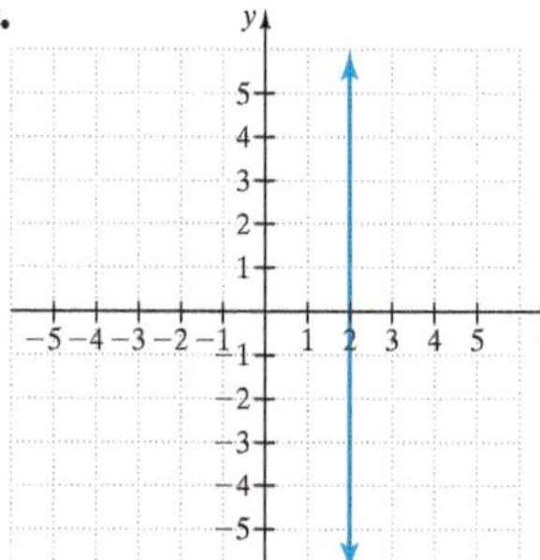

d.

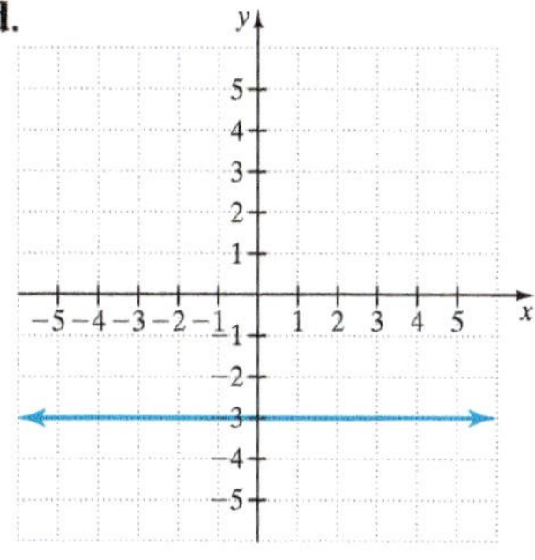

41. $m = 0$

42. $m = -1$

43. undefined slope

44. $m = 3$

Find the slope of the line that goes through the given points.

45. $(2, 5)$ and $(6, 8)$

46. $(4, 7)$ and $(1, 2)$

47. $(1, 3)$ and $(-2, -9)$

48. $(-4, 1)$ and $(3, -6)$

Find the slope of each line.

49. $y = 3x + 7$

50. $x - 2y = 4$

51. $y = -2$

52. $x = 0$

Determine whether each pair of lines is parallel, perpendicular, or neither.

53. $x - y = -6$
$x + y = 3$

54. $3x + y = 7$
$-3x - y = 10$

55. $y = 4x + \frac{1}{2}$
$4x + 2y = 1$

56. $x = 4$
$y = -2$

Find the slope of each line and write the slope as a rate of change. Don't forget to attach the proper units.

57. The graph below approximates the number of U.S. college students (in thousands) earning an associate's degree for each year x.

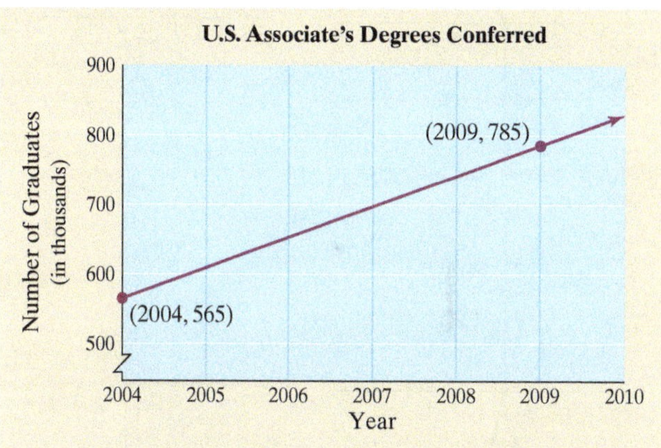

58. The graph below approximates the number of kidney transplants y in the United States for year x.

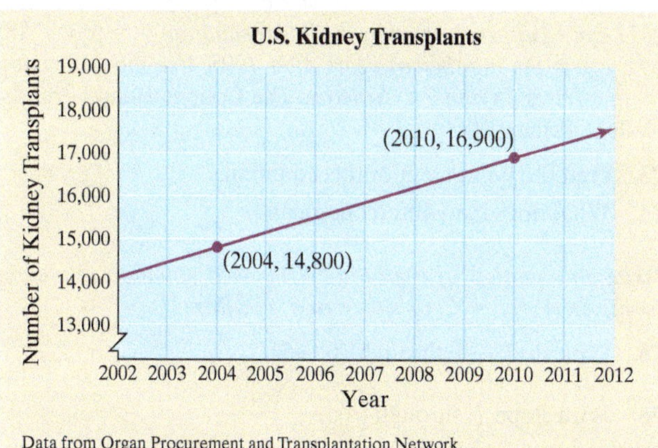

(10.5) Determine the slope and the y-intercept of the graph of each equation.

59. $3x + y = 7$ **60.** $x - 6y = -1$

61. $y = 2$ **62.** $x = -5$

Use the slope-intercept form to graph each equation.

63. $y = 3x - 1$ **64.** $y = -3x$

65. $5x - 3y = 15$ **66.** $-x + 2y = 8$

Write an equation of each line in slope-intercept form.

67. slope -5; y-intercept $\left(0, \frac{1}{2}\right)$

68. slope $\frac{2}{3}$; y-intercept $(0, 6)$

Match each equation with its graph.

69. $y = -4x$ **70.** $y = 2x$

71. $y = 2x - 1$ **72.** $y = -2x + 1$

a.

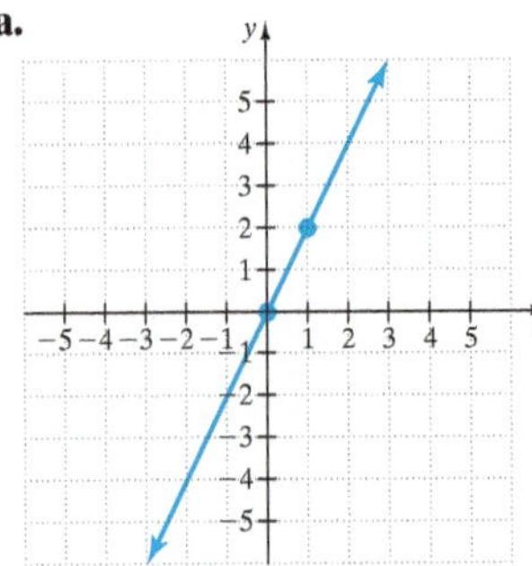

b.

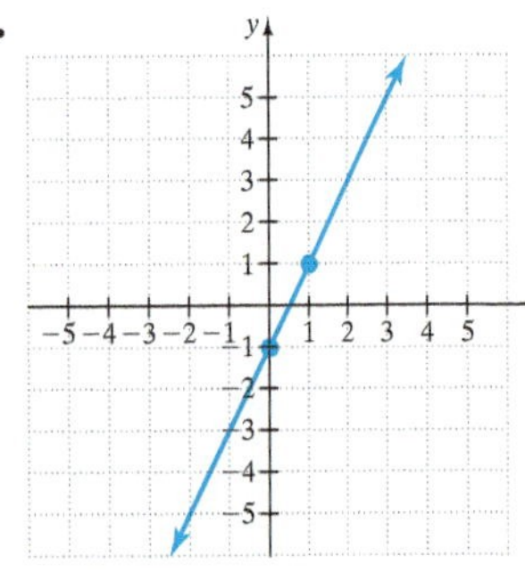

c.

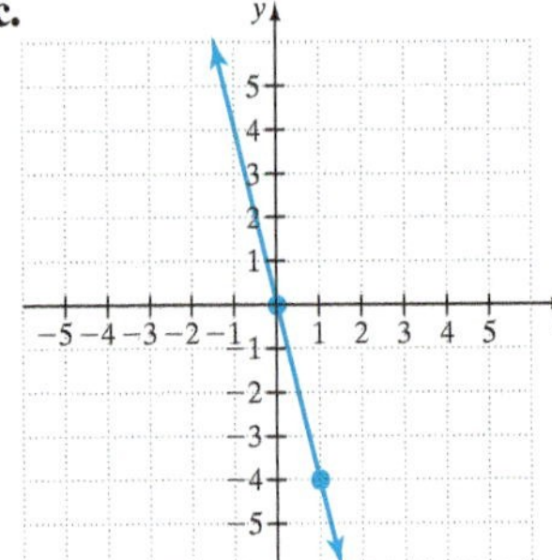

d.

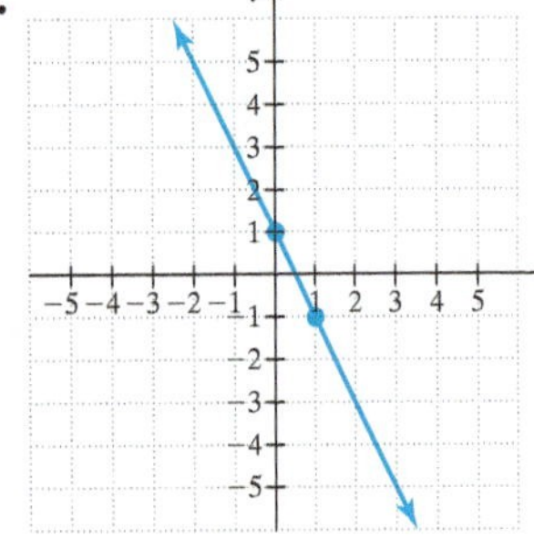

College is getting more expensive every year. The average cost for tuition and fees at a public two-year college y from 1995 through 2011 can be approximated by the linear equation $y = 56x + 1859$, where x is the number of years after 1995. Use this information for Exercises 73 and 74. (Source: The College Board: Trends in College Pricing 2010*)*

73. Find the y-intercept of this equation.

74. What does the y-intercept mean?

Write an equation of each line. Write each equation in standard form, $Ax + By = C$, or $x = c$ or $y = c$ form.

75. With slope -3, through $(0, -5)$

76. With slope $\frac{1}{2}$, through $\left(0, -\frac{7}{2}\right)$

77. With slope 0, through $(-2, -3)$

78. With 0 slope, through the origin

79. With slope -6, through $(2, -1)$

80. With slope 12, through $\left(\frac{1}{2}, 5\right)$

81. Through $(0, 6)$ and $(6, 0)$

82. Through $(0, -4)$ and $(-8, 0)$

83. Vertical line, through $(5, 7)$

84. Horizontal line, through $(-6, 8)$

85. Through $(6, 0)$, perpendicular to $y = 8$

86. Through $(10, 12)$, perpendicular to $x = -2$

(10.6) Determine whether each of the following are functions.

87. $\{(7, 1), (7, 5), (2, 6)\}$

88. $\{(0, -1), (5, -1), (2, 2)\}$

89. $7x - 6y = 1$ **90.** $y = 7$

91. $x = 2$ **92.** $y = x^3$

93.

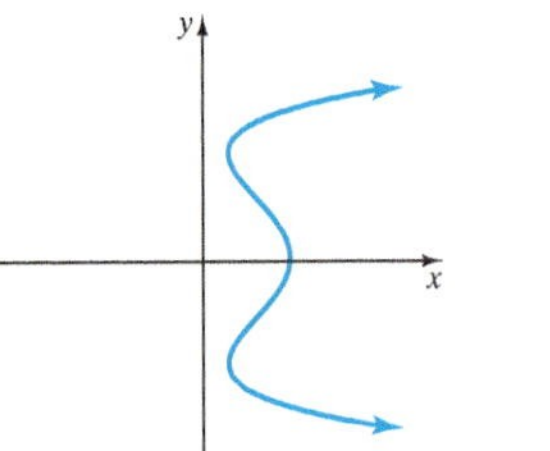

94.

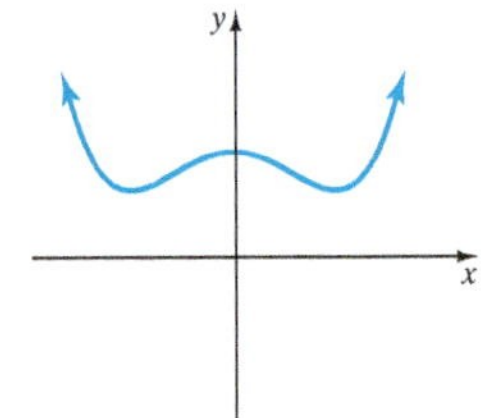

Given the following functions, find the indicated function values.

95. Given $f(x) = -2x + 6$, find

a. $f(0)$ **b.** $f(-2)$ **c.** $f\left(\frac{1}{2}\right)$

96. Given $h(x) = -5 - 3x$, find

a. $h(2)$ **b.** $h(-3)$ **c.** $h(0)$

97. Given $g(x) = x^2 + 12x$, find

a. $g(3)$ **b.** $g(-5)$ **c.** $g(0)$

98. Given $h(x) = 6 - |x|$, find

a. $h(-1)$ **b.** $h(1)$ **c.** $h(-4)$

Find the domain of each function.

99. $f(x) = 2x + 7$ **100.** $g(x) = \frac{7}{x - 2}$

Find the domain and the range of each function graphed.

101.

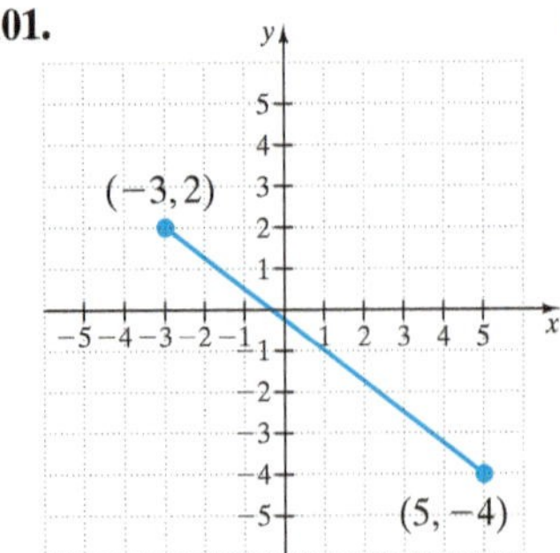

102.

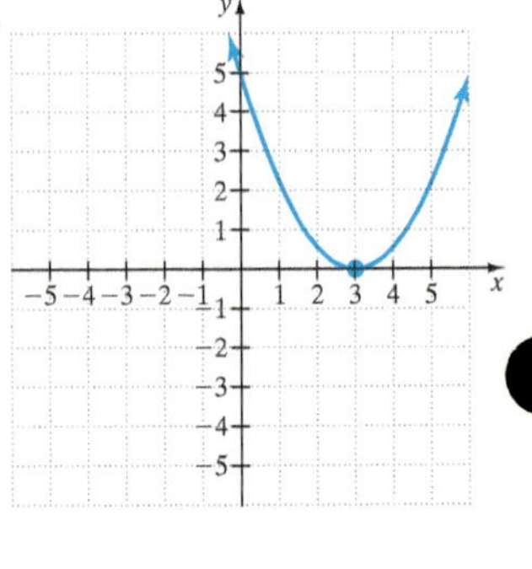

103.

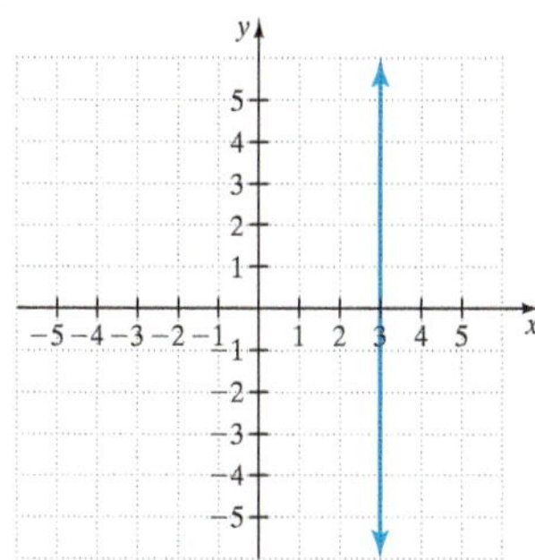

104.

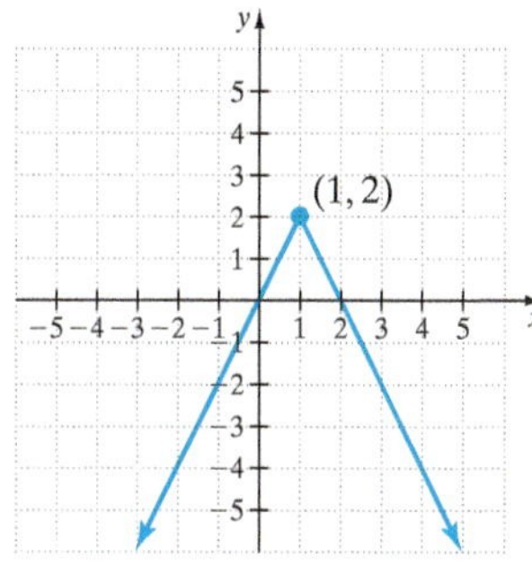

MIXED REVIEW

Complete the table of values for each given equation.

105. $2x - 5y = 9$

x	y
	1
2	
	−3

106. $x = -3y$

x	y
0	
	1
6	

Find the intercepts for each equation.

107. $2x - 3y = 6$

108. $-5x + y = 10$

Graph each linear equation.

109. $x - 5y = 10$

110. $x + y = 4$

111. $y = -4x$

112. $2x + 3y = -6$

113. $x = 3$

114. $y = -2$

Find the slope of the line that passes through each pair of points.

115. $(3, -5)$ and $(-4, 2)$

116. $(1, 3)$ and $(-6, -8)$

Find the slope of each line.

117.

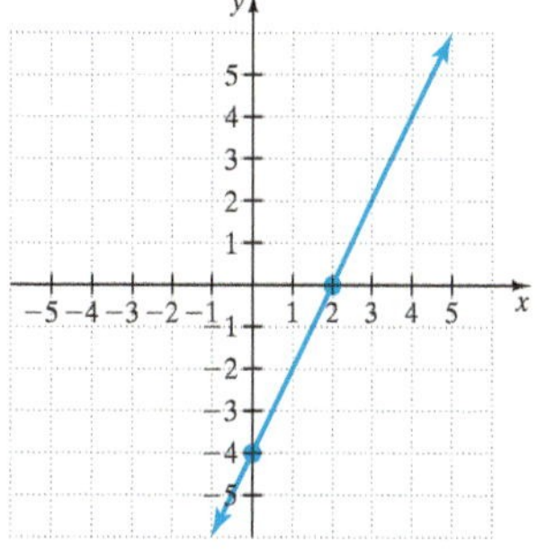

118.

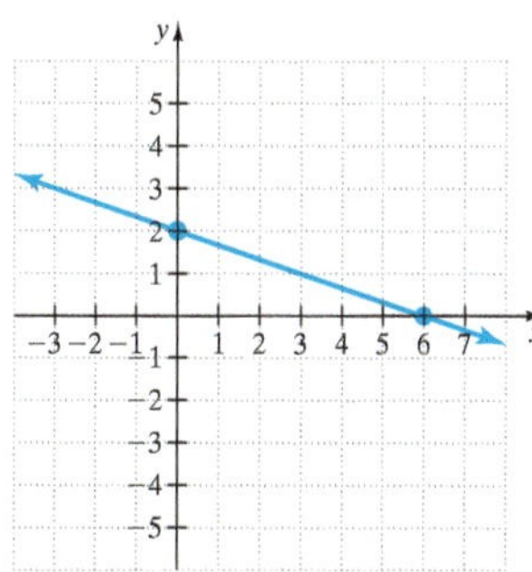

Determine the slope and y-intercept of the graph of each equation.

119. $-2x + 3y = -15$

120. $6x + y - 2 = 0$

Write an equation of the line with the given slope that passes through the given point. Write the equation in the form $Ax + By = C$.

121. $m = -5$; $(3, -7)$

122. $m = 3$; $(0, 6)$

Write an equation of the line passing through each pair of points. Write the equation in the form $Ax + By = C$.

123. $(-3, 9)$ and $(-2, 5)$

124. $(3, 1)$ and $(5, -9)$

Yogurt is an ever more popular food item. In 2002, American Dairy affiliates produced 2134 million pounds of yogurt. In 2009, this number rose to 3800 million pounds of yogurt. Use this information for Exercises 125 and 126.

125. Assume that the relationship between time and yogurt produced is linear and write an equation describing this relationship. Use ordered pairs of the form (years past 2002, millions of pounds of yogurt produced).

126. Use this equation to estimate yogurt production in the year 2014.

Chapter 10 Test

Step-by-step test solutions are found on the Chapter Test Prep Videos. Where available: MyMathLab® or

Graph the following.

1. $y = \frac{1}{2}x$

2. $2x + y = 8$

3. $5x - 7y = 10$

4. $y = -1$

5. $x - 3 = 0$

For Exercises 6 through 10, find the slopes of the lines.

6.

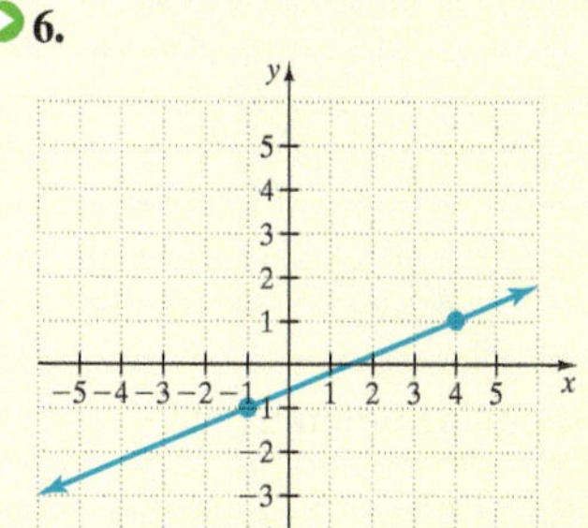

7.

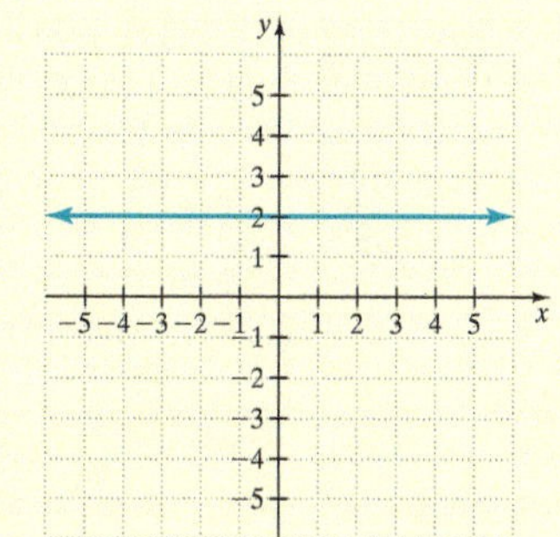

8. Through $(6, -5)$ and $(-1, 2)$

9. $-3x + y = 5$

10. $x = 6$

11. Determine the slope and the y-intercept of the graph of $7x - 3y = 2$.

12. Determine whether the graphs of $y = 2x - 6$ and $-4x = 2y$ are parallel lines, perpendicular lines, or neither.

Find equations of the following lines. Write the equation in standard form.

13. With slope of $-\frac{1}{4}$, through $(2, 2)$

14. Through the origin and $(6, -7)$

15. Through $(2, -5)$ and $(1, 3)$

16. Through $(-5, -1)$ and parallel to $x = 7$

17. With slope $\frac{1}{8}$ and y-intercept $(0, 12)$

Determine whether each graph is the graph of a function.

18.  **19.**

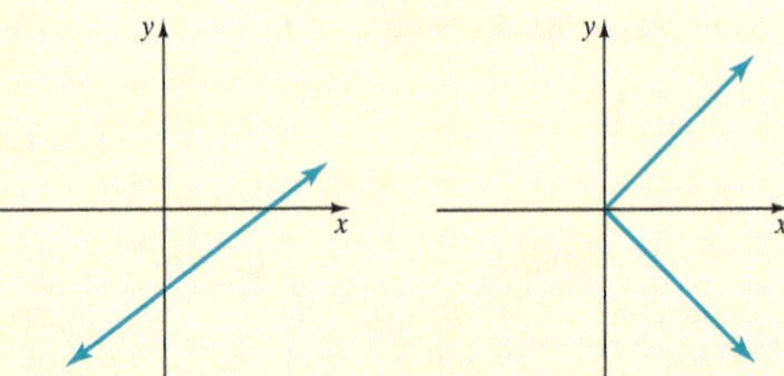

Given the following functions, find the indicated function values.

20. $h(x) = x^3 - x$

a. $h(-1)$ **b.** $h(0)$ **c.** $h(4)$

21. Find the domain of $y = \frac{1}{x + 1}$.

For Exercises 22 and 23,

a. *Identify the x- and y-intercepts.*

b. *Find the domain and the range of each function graphed.*

22. **23.**

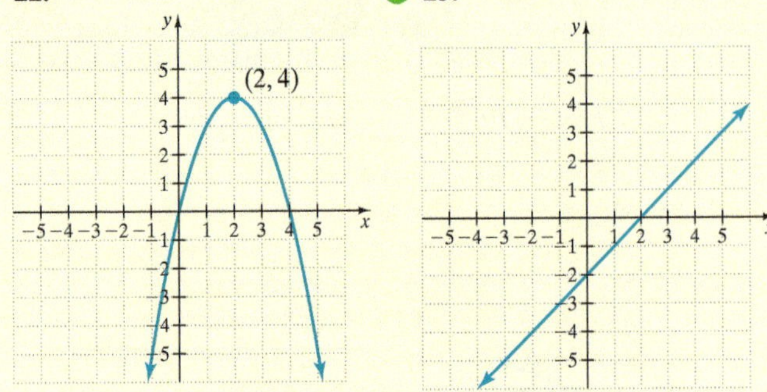

24. If $f(7) = 20$, write the corresponding ordered pair.

Use the bar graph below to answer Exercises 25 and 26.

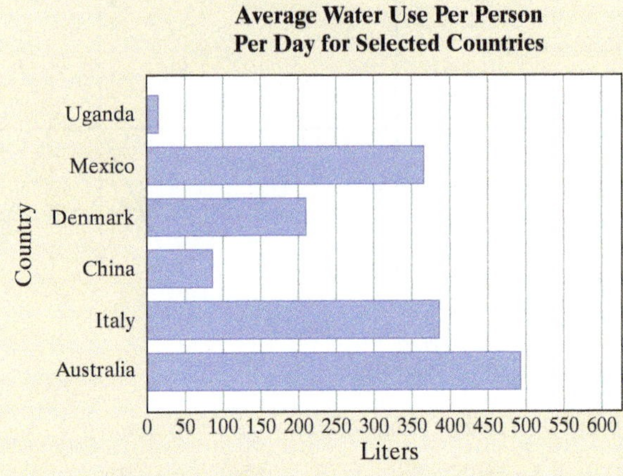

Data from United Nations Development Programme

25. Estimate the average water use per person per day in Denmark.

26. Estimate the average water use per person per day in Australia.

Use this graph to answer Exercises 27 through 29.

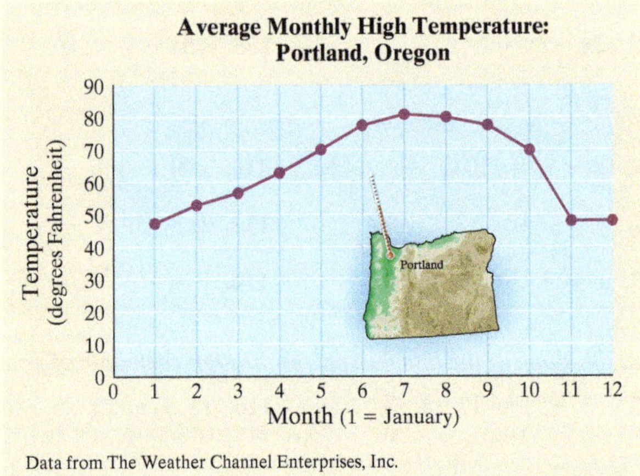

Data from The Weather Channel Enterprises, Inc.

27. During what month is the average high temperature the greatest?

28. Approximate the average high temperature for the month of April.

29. During what month(s) is the average high temperature below 60°F?

30. The table gives the number of basic cable TV subscribers (in millions) for the years shown. (Data from National Cable and Telecommunications Association)

Year	Basic Cable TV Subscribers (in millions)
2003	66.0
2004	65.4
2005	65.4
2006	65.6
2007	64.9
2008	63.7
2009	62.1

a. Write this data as a set of ordered pairs of the form (year, number of basic cable TV subscribers in millions).

b. Create a scatter diagram of the data. Be sure to label the axes properly.

31. This graph approximates the number of movie ticket sales y (in millions) for the year x.

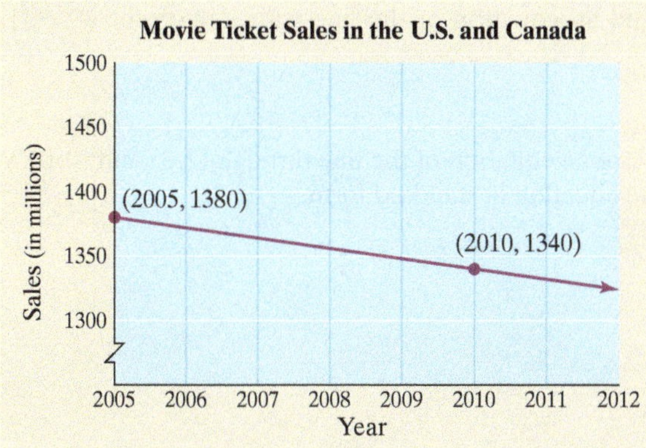

a. Find the slope of the line and write the slope as a rate of change. Don't forget to attach the proper units.

b. Write two ordered pairs of the form (years past 2005, number of tickets sold in millions).

c. Use the two ordered pairs from part (b) to write a linear equation. Write the equation in slope-intercept form.

d. Use the equation from part (c) to predict the number of movie tickets sold in 2015.

Chapters 1–10 Cumulative Review

1. Simplify the expression: $10 + (x + 12)$

2. Simplify the expression: $(y + 5) + 6$

3. Insert $<$, $>$, or $=$ between the pairs of numbers to form true statements.

   a. $-5 \quad -6$ b. $3.195 \quad 3.2$

   c. $\frac{1}{4} \quad \frac{1}{3}$

4. Determine whether the statement is true or false: $0.405 < 0.0405$

5. Solve: $4(2x - 3) + 7 = 3x + 5$

6. Solve: $5(3 + z) - (8z + 9) = -4$

7. Solve: $-2(x - 5) + 10 = -3(x + 2) + x$

8. Solve: $\frac{x}{3} - 2 = \frac{x}{3}$

9. Solve: $0.25x + 0.10(x - 3) = 1.1$

10. Solve: $\frac{x}{3} - 1 = \frac{x}{6} + 1$

11. Some states have a single area code for the entire state. Two such states have area codes that are consecutive odd integers. If the sum of these integers is 1208, find the two area codes. (*Source: World Almanac*)

12. The sum of three consecutive even integers is 48. Find the integers.

13. A 10-foot board is to be cut into two pieces so that the length of the longer piece is 4 times the length of the shorter. Find the length of each piece.

14. Solve $5(x + 4) \geq 4(2x + 3)$. Write the solution set in interval notation.

15. Charles Pecot can afford enough fencing to enclose a rectangular garden with a perimeter of 140 feet. If the width of his garden is to be 30 feet, find the length.

16. Solve $-4x + 3 < 3(x - 5)$. Write the solution set in interval notation.

17. Solve $y = mx + b$ for x.

18. Complete the table for $y = -5x$.

x	y
0	
−1	
	−10

19. 12 subtracted from 3 times a number is less than 21. Find all numbers that make this statement true.

20. Graph: $y = -3x + 5$

21. Graph $x \geq -1$. Then write the solutions in interval notation.

22. Find the x- and y-intercepts of $2x + 4y = -8$.

23. Solve $-2x \leq -4$. Graph the solution set.

24. Graph $x = 2$ on a rectangular coordinate system.

25. Determine whether each ordered pair is a solution of the equation $x - 2y = 6$.

a. $(6, 0)$ b. $(0, 3)$

c. $\left(1, -\frac{5}{2}\right)$

26. Find the slope of the line through $(0, 5)$ and $(-5, 4)$.

27. Determine whether each equation is a linear equation in two variables.

a. $x - 1.5y = -1.6$ b. $y = -2x$

c. $x + y^2 = 9$ d. $x = 5$

28. Find the slope of the line $x = -10$.

29. Find the slope of the line $y = -1$.

30. Find the slope and y-intercept of the line whose equation is $2x - 5y = 10$.

31. Find an equation of the line with y-intercept $(0, -3)$ and slope of $\frac{1}{4}$.

32. Write an equation of the line through $(2, 3)$ and $(0, 0)$. Write the equation in standard form.

Solving Systems of Linear Equations

11

According to the Bureau of Labor Statistics, workers age 55 years and over are projected to have the greatest increase in the labor force during the years 2008 to 2018. These workers are anticipated to increase from 18.1% to 23.9% of the labor force during this period. Workers age 25 to 34 years are projected to increase also, but not at the same rate. In Section 11.3, Exercise 73, we will predict when these two age groups will have the same percent of the labor force.

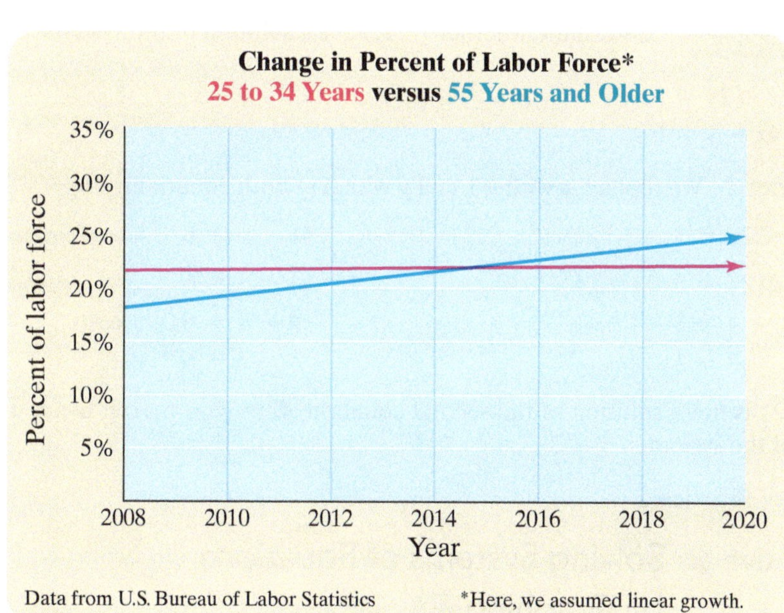

Sections

Check Your Progress

In Chapter 10, we graphed equations containing two variables. Equations like these are often needed to represent relationships between two different values. There are also many real-life opportunities to compare and contrast two such equations, called a system of equations. This chapter presents linear systems and ways we solve these systems and apply them to real-life situations.

11.1 Solving Systems of Linear Equations by Graphing

Objectives

1. Determine If an Ordered Pair Is a Solution of a System of Equations in Two Variables.
2. Solve a System of Linear Equations by Graphing.
3. Without Graphing, Determine the Number of Solutions of a System.

A **system of linear equations** consists of two or more linear equations. In this section, we focus on solving systems of linear equations containing two equations in two variables. Examples of such linear systems are

$$\begin{cases} 3x - 3y = 0 \\ x = 2y \end{cases} \quad \begin{cases} x - y = 0 \\ 2x + y = 10 \end{cases} \quad \begin{cases} y = 7x - 1 \\ y = 4 \end{cases}$$

Objective 1 Deciding Whether an Ordered Pair Is a Solution

A **solution** of a system of two equations in two variables is an ordered pair of numbers that is a solution of both equations in the system.

Example 1 Determine whether (12, 6) is a solution of the system.

$$\begin{cases} 2x - 3y = 6 \\ x = 2y \end{cases}$$

Solution: To determine whether (12, 6) is a solution of the system, we replace x with 12 and y with 6 in both equations.

$2x - 3y = 6$	First equation	$x = 2y$	Second equation
$2(12) - 3(6) \stackrel{?}{=} 6$	Let $x = 12$ and $y = 6$.	$12 \stackrel{?}{=} 2(6)$	Let $x = 12$ and $y = 6$.
$24 - 18 \stackrel{?}{=} 6$	Simplify.	$12 = 12$	True
$6 = 6$	True		

Since (12, 6) is a solution of both equations, it is a solution of the system.

Work Practice 1

Practice 1

Determine whether (4, 12) is a solution of the system.

$$\begin{cases} 4x - y = 2 \\ y = 3x \end{cases}$$

Example 2 Determine whether $(-1, 2)$ is a solution of the system.

$$\begin{cases} x + 2y = 3 \\ 4x - y = 6 \end{cases}$$

Solution: We replace x with -1 and y with 2 in both equations.

$x + 2y = 3$	First equation	$4x - y = 6$	Second equation
$-1 + 2(2) \stackrel{?}{=} 3$	Let $x = -1$ and $y = 2$.	$4(-1) - 2 \stackrel{?}{=} 6$	Let $x = -1$ and $y = 2$.
$-1 + 4 \stackrel{?}{=} 3$	Simplify.	$-4 - 2 \stackrel{?}{=} 6$	Simplify.
$3 = 3$	True	$-6 = 6$	False

$(-1, 2)$ is not a solution of the second equation, $4x - y = 6$, so it is not a solution of the system.

Work Practice 2

Practice 2

Determine whether $(-4, 1)$ is a solution of the system.

$$\begin{cases} x - 3y = -7 \\ 2x + 9y = 1 \end{cases}$$

Objective 2 Solving Systems of Equations by Graphing

Since a solution of a system of two equations in two variables is a solution common to both equations, it is also a point common to the graphs of both equations. Let's practice finding solutions of both equations in a system—that is, solutions of a system—by graphing and identifying points of intersection.

Answers
1. no **2.** yes

Example 3 Solve the system of equations by graphing.

$$\begin{cases} -x + 3y = 10 \\ x + y = 2 \end{cases}$$

Solution: On a single set of axes, graph each linear equation.

$-x + 3y = 10$

x	y
0	$\frac{10}{3}$
−4	2
2	4

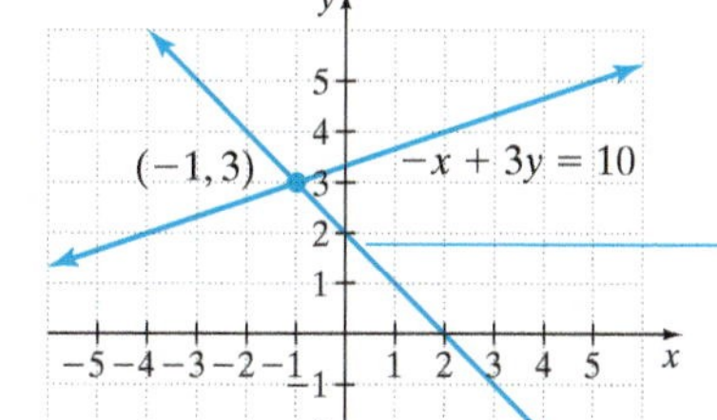

$x + y = 2$

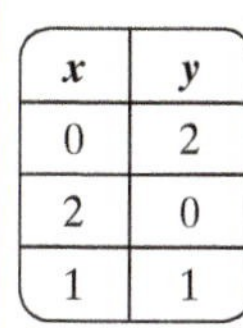

x	y
0	2
2	0
1	1

Helpful Hint The point of intersection gives the solution of the system.

The two lines appear to intersect at the point $(-1, 3)$. To check, we replace x with -1 and y with 3 in both equations.

$-x + 3y = 10$	First equation	$x + y = 2$	Second equation
$-(-1) + 3(3) \stackrel{?}{=} 10$	Let $x = -1$ and $y = 3$.	$-1 + 3 \stackrel{?}{=} 2$	Let $x = -1$ and $y = 3$.
$1 + 9 \stackrel{?}{=} 10$	Simplify.	$2 = 2$	True
$10 = 10$	True		

$(-1, 3)$ checks, so it is the solution of the system and the solution set is $\{(-1, 3)\}$. (*Note:* For single ordered-pair solutions, we shall list the ordered pair only.)

Work Practice 3

Practice 3

Solve the system of equations by graphing:

$$\begin{cases} x - y = 3 \\ x + 2y = 18 \end{cases}$$

Helpful Hint

Neatly drawn graphs can help when you are estimating the solution of a system of linear equations by graphing.

In the example above, notice that the two lines intersect in a point. This means that the system has 1 solution.

Example 4 Solve the system of equations by graphing.

$$\begin{cases} 2x + 3y = -2 \\ x = 2 \end{cases}$$

Solution: We graph each linear equation on a single set of axes.

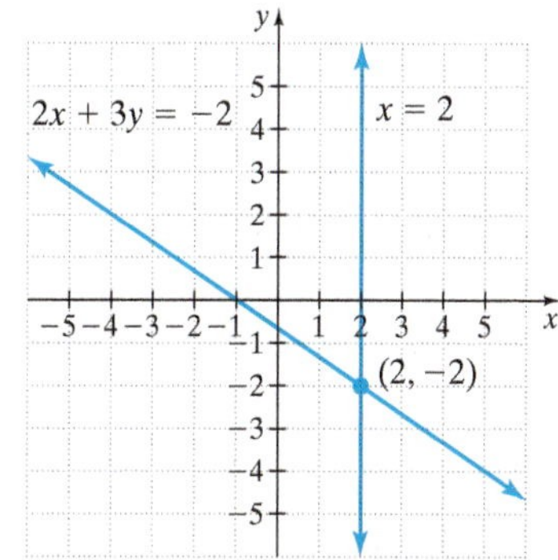

(Continued on next page)

Practice 4

Solve the system of equations by graphing.

$$\begin{cases} -4x + 3y = -3 \\ y = -5 \end{cases}$$

Answers

3. (8, 5)

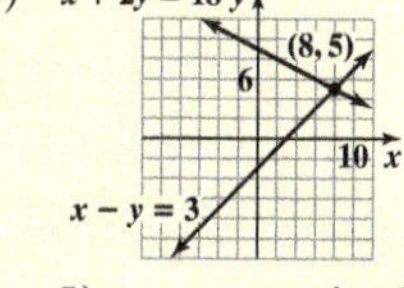

4. (−3, −5)

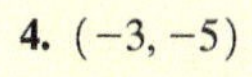

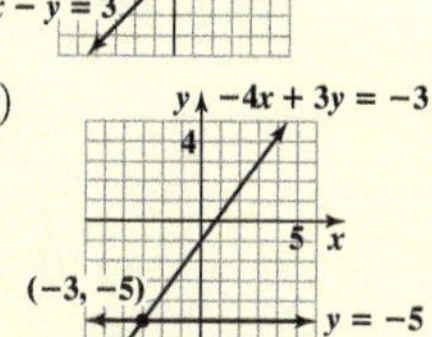

The two lines appear to intersect at the point $(2, -2)$. To determine whether $(2, -2)$ is the solution, we replace x with 2 and y with -2 in both equations.

$2x + 3y = -2$	First equation	$x = 2$	Second equation
$2(2) + 3(-2) \stackrel{?}{=} -2$	Let $x = 2$ and $y = -2$.	$2 \stackrel{?}{=} 2$	Let $x = 2$.
$4 + (-6) \stackrel{?}{=} -2$	Simplify.	$2 = 2$	True
$-2 = -2$	True		

Since a true statement results in both equations, $(2, -2)$ is the solution of the system.

Work Practice 4

A system of equations that has at least one solution as in Examples 3 and 4 is said to be a **consistent system.** A system that has no solution is said to be an **inconsistent system.**

Practice 5

Solve the system of equations by graphing.

$$\begin{cases} 3y = 9x \\ 6x - 2y = 12 \end{cases}$$

Example 5 Solve the following system of equations by graphing.

$$\begin{cases} 2x + y = 7 \\ 2y = -4x \end{cases}$$

Solution: Graph the two lines in the system.

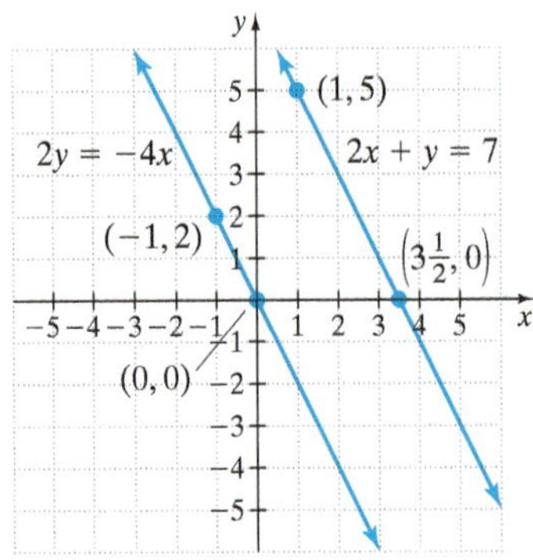

The lines **appear** to be parallel. To confirm this, write both equations in slope-intercept form by solving each equation for y.

$2x + y = 7$	First equation	$2y = -4x$	Second equation
$\mathbf{y = -2x + 7}$	Subtract $2x$ from both sides.	$\frac{2y}{2} = \frac{-4x}{2}$	Divide both sides by 2.
		$\mathbf{y = -2x}$	

Recall that when an equation is written in slope-intercept form, the coefficient of x is the slope. Since both equations have the same slope, -2, but different y-intercepts, the lines are parallel and have no points in common. Thus, there is no solution of the system and the system is inconsistent. To indicate this, we can say the system has no solution or the solution set is $\{\ \}$ or $\varnothing$.

Work Practice 5

In Examples 3, 4, and 5, the graphs of the two linear equations of each system are different. When this happens, we call these equations **independent equations.** If the graphs of the two equations in a system are identical, we call the equations **dependent equations.**

Answer

5. no solution, inconsistent system; $\{\ \}$ or $\varnothing$

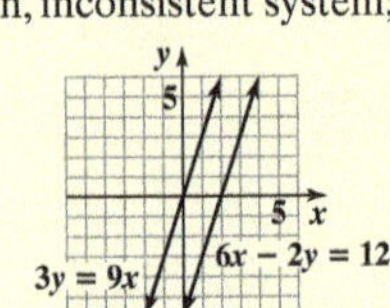

Example 6 Solve the system of equations by graphing.

$$\begin{cases} x - y = 3 \\ -x + y = -3 \end{cases}$$

Solution: Graph each line.

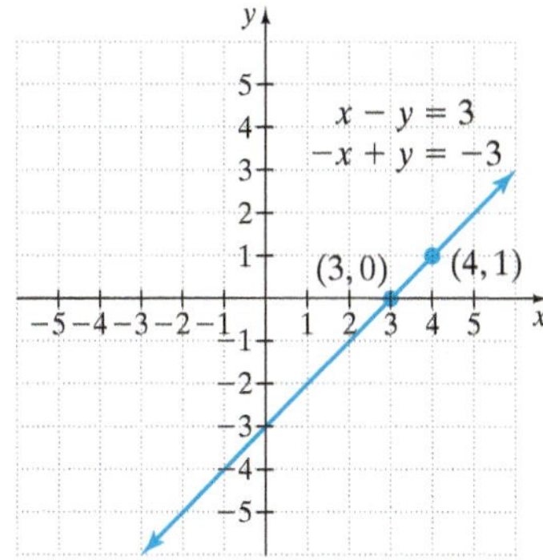

These graphs **appear** to be identical. To confirm this, write each equation in slope-intercept form.

$x - y = 3$ First equation

$-y = -x + 3$ Subtract x from both sides.

$\frac{-y}{-1} = \frac{-x}{-1} + \frac{3}{-1}$ Divide both sides by -1.

$\mathbf{y = x - 3}$

$-x + y = -3$ Second equation

$\mathbf{y = x - 3}$ Add x to both sides.

The equations are identical and so must be their graphs. The lines have an infinite number of points in common. Thus, there is an infinite number of solutions of the system and this is a consistent system. The equations are dependent equations. Here, we can say that there is an infinite number of solutions or the solution set is $\{(x, y) | x - y = 3\}$ or equivalently $\{(x, y) | -x + y = -3\}$ because the equations describe identical ordered pairs. The second set is read "the set of all ordered pairs (x, y) such that $-x + y = -3$.

Work Practice 6

As we have seen, three different situations can occur when graphing the two lines associated with the two equations in a linear system:

One point of intersection: one solution

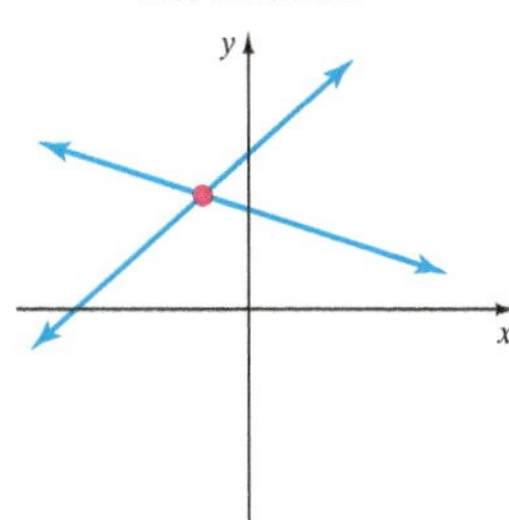

Consistent system (at least one solution)
Independent equations (graphs of equations differ)

Parallel lines: no solution

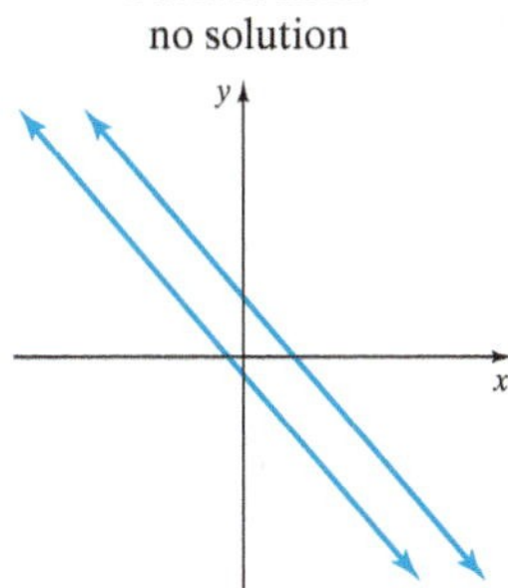

Inconsistent system (no solution)
Independent equations (graphs of equations differ)

Same line: infinite number of solutions

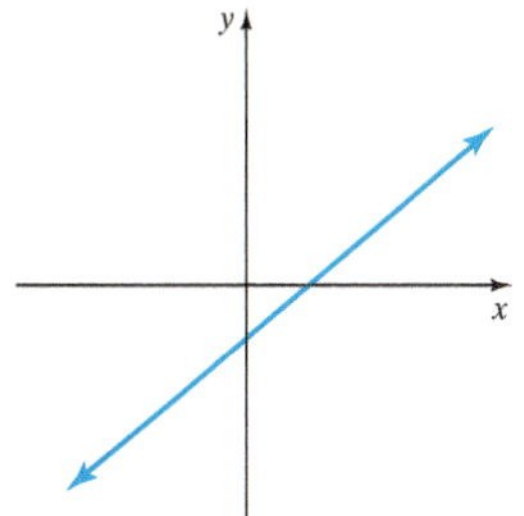

Consistent system (at least one solution)
Dependent equations (graphs of equations identical)

Practice 6

Solve the system of equations by graphing.

$$\begin{cases} x - y = 4 \\ -2x + 2y = -8 \end{cases}$$

Answer

6. infinite number of solutions, consistent system, $\{(x, y) | x - y = 4\}$ or $\{(x, y) | -2x + 2y = -8\}$

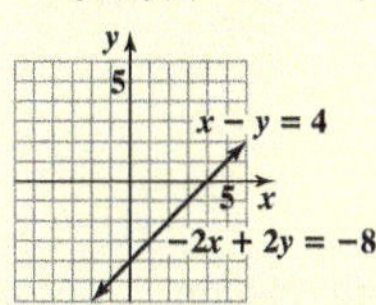

Objective 3 Finding the Number of Solutions of a System Without Graphing

You may have suspected by now that graphing alone is not an accurate way to solve a system of linear equations. For example, a solution of $\left(\frac{1}{2}, \frac{2}{9}\right)$ is unlikely to be read correctly from a graph. The next two sections present two accurate methods of solving these systems. In the meantime, we can decide how many solutions a system has by writing each equation in the slope-intercept form.

Practice 7

Without graphing, determine the number of solutions of the system.

$$\begin{cases} 5x + 4y = 6 \\ x - y = 3 \end{cases}$$

Example 7 Without graphing, determine the number of solutions of the system.

$$\begin{cases} \frac{1}{2}x - y = 2 \\ x = 2y + 5 \end{cases}$$

Solution: First write each equation in slope-intercept form.

$\frac{1}{2}x - y = 2$	First equation	$x = 2y + 5$	Second equation
		$x - 5 = 2y$	Subtract 5 from both sides.
$\frac{1}{2}x = y + 2$	Add y to both sides.	$\frac{x}{2} - \frac{5}{2} = \frac{2y}{2}$	Divide both sides by 2.
$\frac{1}{2}x - 2 = y$	Subtract 2 from both sides.	$\frac{1}{2}x - \frac{5}{2} = y$	Simplify.

The slope of each line is $\frac{1}{2}$, but they have different y-intercepts. This tells us that the lines representing these equations are parallel. Since the lines are parallel, the system has no solution and is inconsistent.

Work Practice 7

Practice 8

Without graphing, determine the number of solutions of the system.

$$\begin{cases} -\frac{2}{3}x + y = 6 \\ 3y = 2x + 5 \end{cases}$$

Example 8 Without graphing, determine the number of solutions of the system.

$$\begin{cases} 3x - y = 4 \\ x + 2y = 8 \end{cases}$$

Solution: Once again, the slope-intercept form helps determine how many solutions this system has.

$3x - y = 4$	First equation	$x + 2y = 8$	Second equation
$3x = y + 4$	Add y to both sides.	$x = -2y + 8$	Subtract $2y$ from both sides.
$3x - 4 = y$	Subtract 4 from both sides.	$x - 8 = -2y$	Subtract 8 from both sides.
		$\frac{x}{-2} - \frac{8}{-2} = \frac{-2y}{-2}$	Divide both sides by -2.
		$-\frac{1}{2}x + 4 = y$	Simplify.

The slope of the second line is $-\frac{1}{2}$, whereas the slope of the first line is 3. Since the slopes are not equal, the two lines are neither parallel nor identical and must intersect. Therefore, this system has one solution and is consistent.

Work Practice 8

Answers

7. one solution 8. no solution

Graphing Calculator Explorations

A graphing calculator may be used to approximate solutions of systems of equations. For example, to approximate the solution of the system

$$\begin{cases} y = -3.14x - 1.35 \\ y = 4.88x + 5.25, \end{cases}$$

first graph each equation on the same set of axes. Then use the intersect feature of your calculator to approximate the point of intersection.

The approximate point of intersection is $(-0.82, 1.23)$.

Solve each system of equations. Approximate the solutions to two decimal places.

1. $\begin{cases} y = -2.68x + 1.21 \\ y = 5.22x - 1.68 \end{cases}$

2. $\begin{cases} y = 4.25x + 3.89 \\ y = -1.88x + 3.21 \end{cases}$

3. $\begin{cases} 4.3x - 2.9y = 5.6 \\ 8.1x + 7.6y = -14.1 \end{cases}$

4. $\begin{cases} -3.6x - 8.6y = 10 \\ -4.5x + 9.6y = -7.7 \end{cases}$

Vocabulary, Readiness & Video Check

Fill in each blank with one of the words or phrases listed below.

system of linear equations	solution	consistent
dependent	inconsistent	independent

1. In a system of linear equations in two variables, if the graphs of the equations are the same, the equations are ________________ equations.

2. Two or more linear equations are called a(n) ________________.

3. A system of equations that has at least one solution is called a(n) ________________ system.

4. A(n) ________________ of a system of two equations in two variables is an ordered pair of numbers that is a solution of both equations in the system.

5. A system of equations that has no solution is called a(n) ________________ system.

6. In a system of linear equations in two variables, if the graphs of the equations are different, the equations are ________________ equations.

Martin-Gay Interactive Videos

See Video 11.1

Watch the section lecture video and answer the following questions.

Objective 1 **7.** In Example 1, the first ordered pair is a solution of the first equation of the system. Why is this not enough to determine whether the first ordered pair is a solution of the system?

Objective 2 **8.** From Examples 2 and 3, why is finding the solution of a system of equations from a graph considered "guessing" and this proposed solution checked algebraically?

Objective 3 **9.** From Examples 5–7, explain how the slope-intercept form tells us how many solutions a system of equations has.

11.1 Exercise Set MyMathLab®

Each rectangular coordinate system shows the graph of the equations in a system of equations. Use each graph to determine the number of solutions for each associated system. If the system has only one solution, give its coordinates.

1.

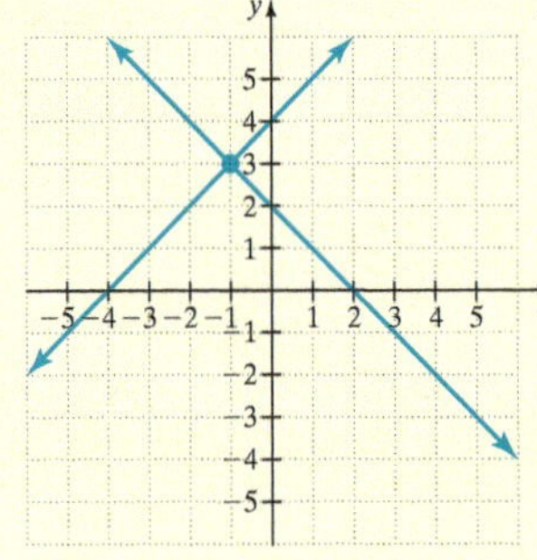

2.

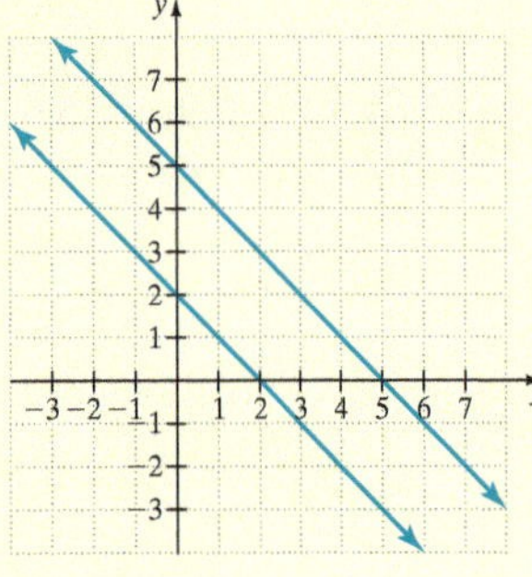

3.

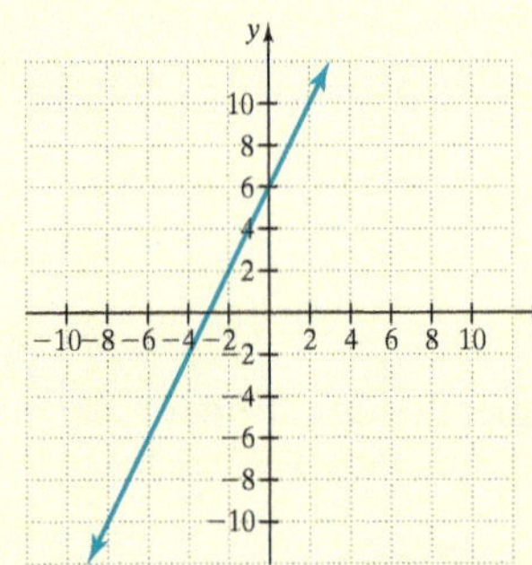

4.

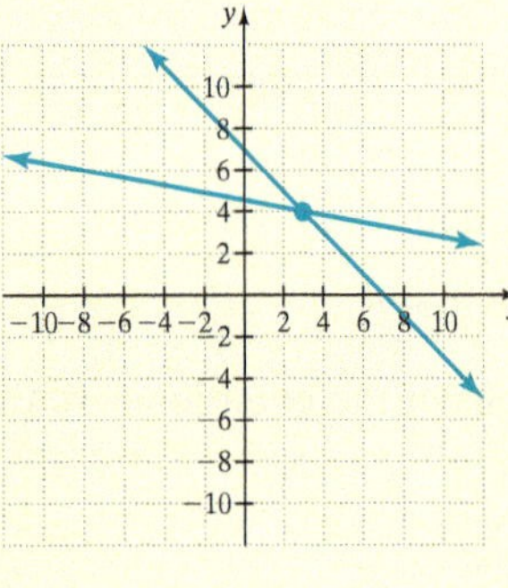

Determine whether each ordered pair is a solution of the system of linear equations. See Examples 1 and 2.

5. $\begin{cases} x + y = 8 \\ 3x + 2y = 21 \end{cases}$
a. $(2, 4)$
b. $(5, 3)$

6. $\begin{cases} 2x + y = 5 \\ x + 3y = 5 \end{cases}$
a. $(5, 0)$
b. $(2, 1)$

7. $\begin{cases} 3x - y = 5 \\ x + 2y = 11 \end{cases}$
a. $(3, 4)$
b. $(0, -5)$

8. $\begin{cases} 2x - 3y = 8 \\ x - 2y = 6 \end{cases}$
a. $(-2, -4)$
b. $(7, 2)$

9. $\begin{cases} 2y = 4x + 6 \\ 2x - y = -3 \end{cases}$
a. $(-3, -3)$
b. $(0, 3)$

10. $\begin{cases} x + 5y = -4 \\ -2x = 10y + 8 \end{cases}$
a. $(-4, 0)$
b. $(6, -2)$

11. $\begin{cases} -2 = x - 7y \\ 6x - y = 13 \end{cases}$
a. $(-2, 0)$
b. $\left(\frac{1}{2}, \frac{5}{14}\right)$

12. $\begin{cases} 4x = 1 - y \\ x - 3y = -8 \end{cases}$
a. $(0, 1)$
b. $\left(\frac{1}{6}, \frac{1}{3}\right)$

MIXED PRACTICE

Solve each system of linear equations by graphing. See Examples 3 through 6.

13. $\begin{cases} x + y = 4 \\ x - y = 2 \end{cases}$

14. $\begin{cases} x + y = 3 \\ x - y = 5 \end{cases}$

15. $\begin{cases} x + y = 6 \\ -x + y = -6 \end{cases}$

16. $\begin{cases} x + y = 1 \\ -x + y = -3 \end{cases}$

17. $\begin{cases} y = 2x \\ 3x - y = -2 \end{cases}$

18. $\begin{cases} y = -3x \\ 2x - y = -5 \end{cases}$

19. $\begin{cases} y = x + 1 \\ y = 2x - 1 \end{cases}$

20. $\begin{cases} y = 3x - 4 \\ y = x + 2 \end{cases}$

21. $\begin{cases} 2x + y = 0 \\ 3x + y = 1 \end{cases}$

22. $\begin{cases} 2x + y = 1 \\ 3x + y = 0 \end{cases}$

23. $\begin{cases} y = -x - 1 \\ y = 2x + 5 \end{cases}$

24. $\begin{cases} y = x - 1 \\ y = -3x - 5 \end{cases}$

25. $\begin{cases} x + y = 5 \\ x + y = 6 \end{cases}$

26. $\begin{cases} x - y = 4 \\ x - y = 1 \end{cases}$

27. $\begin{cases} 2x - y = 6 \\ y = 2 \end{cases}$

28. $\begin{cases} x + y = 5 \\ x = 4 \end{cases}$

29. $\begin{cases} x - 2y = 2 \\ 3x + 2y = -2 \end{cases}$

30. $\begin{cases} x + 3y = 7 \\ 2x - 3y = -4 \end{cases}$

31. $\begin{cases} 2x + y = 4 \\ 6x = -3y + 6 \end{cases}$

32. $\begin{cases} y + 2x = 3 \\ 4x = 2 - 2y \end{cases}$

33. $\begin{cases} y - 3x = -2 \\ 6x - 2y = 4 \end{cases}$

34. $\begin{cases} x - 2y = -6 \\ -2x + 4y = 12 \end{cases}$

35. $\begin{cases} x = 3 \\ y = -1 \end{cases}$

36. $\begin{cases} x = -5 \\ y = 3 \end{cases}$

37. $\begin{cases} y = x - 2 \\ y = 2x + 3 \end{cases}$

38. $\begin{cases} y = x + 5 \\ y = -2x - 4 \end{cases}$

39. $\begin{cases} 2x - 3y = -2 \\ -3x + 5y = 5 \end{cases}$

40. $\begin{cases} 4x - y = 7 \\ 2x - 3y = -9 \end{cases}$

41. $\begin{cases} 6x - y = 4 \\ \frac{1}{2}y = -2 + 3x \end{cases}$

42. $\begin{cases} 3x - y = 6 \\ \frac{1}{3}y = -2 + x \end{cases}$

Without graphing, decide:

a. Are the graphs of the equations identical lines, parallel lines, or lines intersecting at a single point?

b. How many solutions does the system have? See Examples 7 and 8.

43. $\begin{cases} 4x + y = 24 \\ x + 2y = 2 \end{cases}$

44. $\begin{cases} 3x + y = 1 \\ 3x + 2y = 6 \end{cases}$

45. $\begin{cases} 2x + y = 0 \\ 2y = 6 - 4x \end{cases}$

46. $\begin{cases} 3x + y = 0 \\ 2y = -6x \end{cases}$

47. $\begin{cases} 6x - y = 4 \\ \frac{1}{2}y = -2 + 3x \end{cases}$

48. $\begin{cases} 3x - y = 2 \\ \frac{1}{3}y = -2 + 3x \end{cases}$

49. $\begin{cases} x = 5 \\ y = -2 \end{cases}$

50. $\begin{cases} y = 3 \\ x = -4 \end{cases}$

51. $\begin{cases} 3y - 2x = 3 \\ x + 2y = 9 \end{cases}$

52. $\begin{cases} 2y = x + 2 \\ y + 2x = 3 \end{cases}$

53. $\begin{cases} 6y + 4x = 6 \\ 3y - 3 = -2x \end{cases}$

54. $\begin{cases} 8y + 6x = 4 \\ 4y - 2 = 3x \end{cases}$

55. $\begin{cases} x + y = 4 \\ x + y = 3 \end{cases}$

56. $\begin{cases} 2x + y = 0 \\ y = -2x + 1 \end{cases}$

REVIEW AND PREVIEW

Solve each equation. See Section 9.3.

57. $5(x - 3) + 3x = 1$
58. $-2x + 3(x + 6) = 17$
59. $4\left(\frac{y + 1}{2}\right) + 3y = 0$
60. $-y + 12\left(\frac{y - 1}{4}\right) = 3$
61. $8a - 2(3a - 1) = 6$
62. $3z - (4z - 2) = 9$

CONCEPT EXTENSIONS

63. Draw a graph of two linear equations whose associated system has the solution $(-1, 4)$.
64. Draw a graph of two linear equations whose associated system has the solution $(3, -2)$.
65. Draw a graph of two linear equations whose associated system has no solution.
66. Draw a graph of two linear equations whose associated system has an infinite number of solutions.
67. The ordered pair $(-2, 3)$ is a solution of all three independent equations:

$$x + y = 1$$
$$2x - y = -7$$
$$x + 3y = 7$$

Describe the graph of all three equations on the same axes.

68. Explain how to use a graph to determine the number of solutions of a system.

The double-line graph below shows the number of pounds of fishery products from U.S. domestic catch and from imports. Use this graph to answer Exercises 69 and 70.

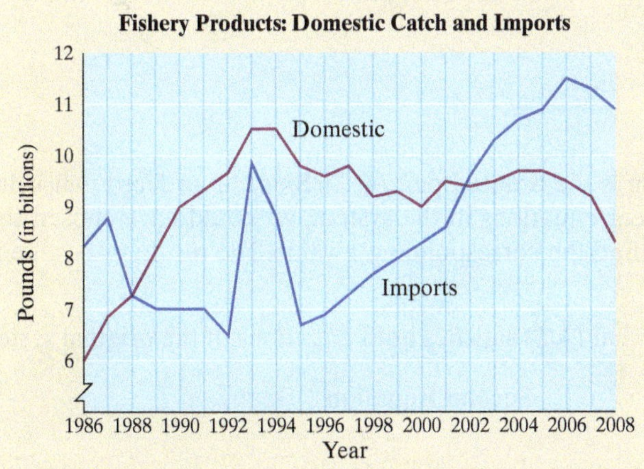

Data from *Statistical Abstract of the United States*, 2011

69. Between what pairs of years did the number of pounds of imported fishery products equal the number of pounds of domestic catch?
70. For what year(s) was the number of pounds of imported fishery products less than the number of pounds of domestic catch?

The double-line graph below shows the average attendance per game for the years shown for the Minnesota Twins and the Texas Rangers baseball teams. Use this for Exercises 71 and 72.

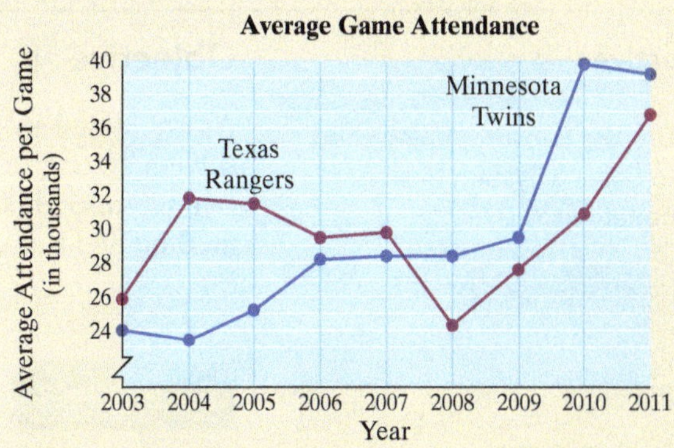

Data from *Baseball Almanac*

71. In what year(s) was the average attendance per game for the Texas Rangers greater than the average attendance per game for the Minnesota Twins?
72. In what year was the average attendance per game for the Texas Rangers closest to the average attendance per game for the Minnesota Twins, 2006 or 2009?
73. Construct a system of two linear equations that has $(1, 3)$ as a solution.
74. Construct a system of two linear equations that has $(0, 7)$ as a solution.
75. Below are tables of values for two linear equations.
   a. Find a solution of the corresponding system.
   b. Graph several ordered pairs from each table and sketch the two lines.

x	y
1	3
2	5
3	7
4	9
5	11

x	y
1	6
2	7
3	8
4	9
5	10

   c. Does your graph confirm the solution from part (a)?

76. Tables of values for two linear equations are shown.
   a. Find a solution of the corresponding system.
   b. Graph several ordered pairs from each table and sketch the two lines.

x	y
−3	5
−1	1
0	−1
1	−3
2	−5

x	y
−3	7
−1	1
0	−2
1	−5
2	−8

   c. Does your graph confirm the solution from part (a)?

77. Explain how writing each equation in a linear system in slope-intercept form helps determine the number of solutions of a system.
78. Is it possible for a system of two linear equations in two variables to be inconsistent but have dependent equations? Why or why not?

11.2 Solving Systems of Linear Equations by Substitution

Objective

1 Use the Substitution Method to Solve a System of Linear Equations.

Objective 1 Using the Substitution Method

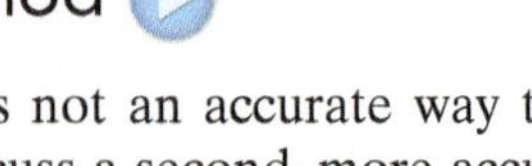

As we stated in the preceding section, graphing alone is not an accurate way to solve a system of linear equations. In this section, we discuss a second, more accurate method for solving systems of equations. This method is called the **substitution method** and is introduced in the next example.

Practice 1

Solve the system:

$$\begin{cases} 2x - y = 9 \\ x = y + 1 \end{cases}$$

Example 1 Solve the system:

$$\begin{cases} 2x + y = 10 & \text{First equation} \\ x = y + 2 & \text{Second equation} \end{cases}$$

Solution: The second equation in this system is $x = y + 2$. This tells us that x and $y + 2$ have the same value. This means that we may substitute $y + 2$ for x in the first equation.

$$\begin{aligned} 2x + y &= 10 && \text{First equation} \\ 2(y + 2) + y &= 10 && \text{Substitute } y + 2 \text{ for } x \text{ since } x = y + 2. \end{aligned}$$

Notice that this equation now has one variable, y. Let's now solve this equation for y.

Helpful Hint Don't forget the distributive property.

$$\begin{aligned} 2(y + 2) + y &= 10 \\ 2y + 4 + y &= 10 && \text{Use the distributive property.} \\ 3y + 4 &= 10 && \text{Combine like terms.} \\ 3y &= 6 && \text{Subtract 4 from both sides.} \\ y &= 2 && \text{Divide both sides by 3.} \end{aligned}$$

Now we know that the y-value of the ordered pair solution of the system is 2. To find the corresponding x-value, we replace y with 2 in the equation $x = y + 2$ and solve for x.

$$\begin{aligned} x &= y + 2 \\ x &= 2 + 2 && \text{Let } y = 2. \\ x &= 4 \end{aligned}$$

The solution of the system is the ordered pair $(4, 2)$. Since an ordered pair solution must satisfy both linear equations in the system, we could have chosen the equation $2x + y = 10$ to find the corresponding x-value. The resulting x-value is the same.

Check: We check to see that $(4, 2)$ satisfies both equations of the original system.

First Equation	**Second Equation**
$2x + y = 10$	$x = y + 2$
$2(4) + 2 \stackrel{?}{=} 10$	$4 \stackrel{?}{=} 2 + 2$ Let $x = 4$ and $y = 2$.
$10 = 10$ True	$4 = 4$ True

The solution of the system is $(4, 2)$.

Answer

1. $(8, 7)$

A graph of the two equations shows the two lines intersecting at the point (4, 2).

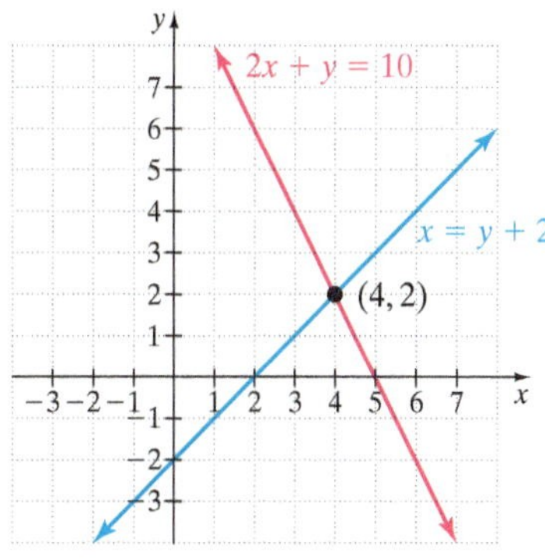

Work Practice 1

Example 2 Solve the system:

$$\begin{cases} 5x - y = -2 \\ y = 3x \end{cases}$$

Solution: The second equation is solved for y in terms of x. We substitute $3x$ for y in the first equation.

$5x - y = -2$ First equation

$5x - (3x) = -2$ Substitute $3x$ for y.

Now we solve for x.

$5x - 3x = -2$

$2x = -2$ Combine like terms.

$x = -1$ Divide both sides by 2.

The x-value of the ordered pair solution is -1. To find the corresponding y-value, we replace x with -1 in the second equation, $y = 3x$.

$y = 3x$ Second equation

$y = 3(-1)$ Let $x = -1$.

$y = -3$

Check to see that the solution of the system is $(-1, -3)$.

Work Practice 2

To solve a system of equations by substitution, we first need an equation solved for one of its variables, as in Examples 1 and 2. If neither equation in a system is solved for x or y, this will be our first step.

Example 3 Solve the system:

$$\begin{cases} x + 2y = 7 \\ 2x + 2y = 13 \end{cases}$$

Solution: We choose one of the equations and solve for x or y. We will solve the first equation for x by subtracting $2y$ from both sides.

$x + 2y = 7$ First equation

$x = 7 - 2y$ Subtract $2y$ from both sides.

(Continued on next page)

Practice 2

Solve the system:

$$\begin{cases} 7x - y = -15 \\ y = 2x \end{cases}$$

Practice 3

Solve the system:

$$\begin{cases} x + 3y = 6 \\ 2x + 3y = 10 \end{cases}$$

Answers

2. $(-3, -6)$ **3.** $\left(4, \frac{2}{3}\right)$

Helpful Hint Don't forget to insert parentheses when substituting $7 - 2y$ for x.

Since $x = 7 - 2y$, we now substitute $7 - 2y$ for x in the second equation and solve for y.

$$\begin{aligned} 2x + 2y &= 13 && \text{Second equation} \\ 2(7 - 2y) + 2y &= 13 && \text{Let } x = 7 - 2y. \\ 14 - 4y + 2y &= 13 && \text{Use the distributive property.} \\ 14 - 2y &= 13 && \text{Simplify.} \\ -2y &= -1 && \text{Subtract 14 from both sides.} \\ y &= \frac{1}{2} && \text{Divide both sides by } -2. \end{aligned}$$

To find x, we let $y = \frac{1}{2}$ in the equation $x = 7 - 2y$.

$$\begin{aligned} x &= 7 - 2y \\ x &= 7 - 2\left(\frac{1}{2}\right) && \text{Let } y = \frac{1}{2}. \\ x &= 7 - 1 \\ x &= 6 \end{aligned}$$

Helpful Hint To find x, any equation in two variables equivalent to the original equations of the system may be used. We used this equation since it is solved for x.

The solution is $\left(6, \frac{1}{2}\right)$. Check the solution in both equations of the original system.

Work Practice 3

The following steps may be used to solve a system of equations by the substitution method.

Solving a System of Two Linear Equations by the Substitution Method

Step 1: Solve one of the equations for one of its variables.

Step 2: Substitute the expression for the variable found in Step 1 into the other equation.

Step 3: Solve the equation from Step 2 to find the value of one variable.

Step 4: Substitute the value found in Step 3 in any equation containing both variables to find the value of the other variable.

Step 5: Check the proposed solution in the original system.

✓**Concept Check** As you solve the system $\begin{cases} 2x + y = -5 \\ x - y = 5 \end{cases}$ you find that $y = -5$. Is this the solution of the system?

Practice 4

Solve the system:

$$\begin{cases} 5x + 3y = -9 \\ -2x + y = 8 \end{cases}$$

Example 4 Solve the system:

$$\begin{cases} 7x - 3y = -14 \\ -3x + y = 6 \end{cases}$$

Solution: To avoid introducing fractions, we will solve the second equation for y.

$$\begin{aligned} -3x + y &= 6 && \text{Second equation} \\ y &= 3x + 6 \end{aligned}$$

Answer

4. $(-3, 2)$

✓**Concept Check Answer**

No, the solution will be an ordered pair.

Next, substitute $3x + 6$ for y in the first equation.

$$7x - 3y = -14 \quad \text{First equation}$$
$$7x - 3(3x + 6) = -14 \quad \text{Let } y = 3x + 6.$$
$$7x - 9x - 18 = -14 \quad \text{Use the distributive property.}$$
$$-2x - 18 = -14 \quad \text{Simplify.}$$
$$-2x = 4 \quad \text{Add 18 to both sides.}$$
$$x = -2 \quad \text{Divide both sides by } -2.$$

To find the corresponding y-value, substitute -2 for x in the equation $y = 3x + 6$. Then $y = 3(-2) + 6$ or $y = 0$. The solution of the system is $(-2, 0)$. Check this solution in both equations of the system.

Work Practice 4

Helpful Hint

When solving a system of equations by the substitution method, begin by solving an equation for one of its variables. If possible, solve for a variable that has a coefficient of 1 or -1. This way, we avoid working with time-consuming fractions.

✓Concept Check To avoid fractions, which of the equations below would you use to solve for x?

a. $3x - 4y = 15$ **b.** $14 - 3y = 8x$ **c.** $7y + x = 12$

Example 5 Solve the system:

$$\begin{cases} \frac{1}{2}x - y = 3 \\ x = 6 + 2y \end{cases}$$

Solution: The second equation is already solved for x in terms of y. Thus we substitute $6 + 2y$ for x in the first equation and solve for y.

$$\frac{1}{2}x - y = 3 \quad \text{First equation}$$
$$\frac{1}{2}(6 + 2y) - y = 3 \quad \text{Let } x = 6 + 2y.$$
$$3 + y - y = 3 \quad \text{Use the distributive property.}$$
$$3 = 3 \quad \text{Simplify.}$$

Arriving at a true statement such as $3 = 3$ indicates that the two linear equations in the original system are equivalent. This means that their graphs are identical and there is an infinite number of solutions of the system. Any solution of one equation is also a solution of the other. For the solution, we can write "infinite number of solutions" or, in set notation, $\left\{(x, y) \middle| \frac{1}{2}x - y = 3\right\}$ or $\{(x, y) | x = 6 + 2y\}$.

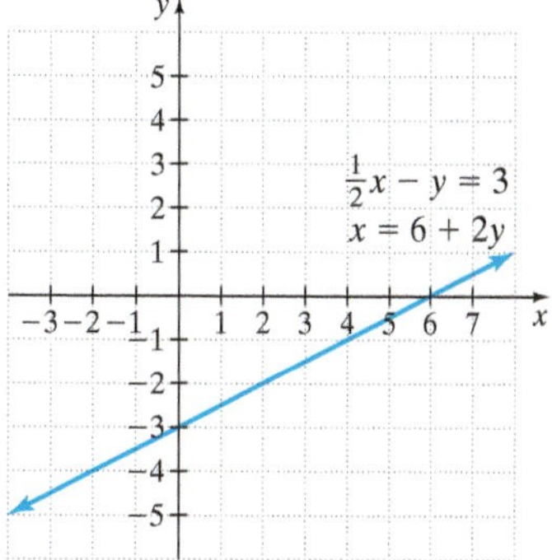

Work Practice 5

Practice 5

Solve the system:

$$\begin{cases} \frac{1}{4}x - y = 2 \\ x = 4y + 8 \end{cases}$$

Answer

5. infinite number of solutions; $\left\{(x, y) \middle| \frac{1}{4}x - y = 2\right\}$ or $\{(x, y) | x = 4y + 8\}$

✓Concept Check Answer

c

Practice 6

Use substitution to solve the system.

$$\begin{cases} 4x - 3y = 12 \\ -8x + 6y = -30 \end{cases}$$

Example 6 Use substitution to solve the system.

$$\begin{cases} 6x + 12y = 5 \\ -4x - 8y = 0 \end{cases}$$

Solution: Choose the second equation and solve for y.

$-4x - 8y = 0$ Second equation

$-8y = 4x$ Add $4x$ to both sides.

$\frac{-8y}{-8} = \frac{4x}{-8}$ Divide both sides by -8.

$y = -\frac{1}{2}x$ Simplify.

Now replace y with $-\frac{1}{2}x$ in the first equation.

$6x + 12y = 5$ First equation

$6x + 12\left(-\frac{1}{2}x\right) = 5$ Let $y = -\frac{1}{2}x$.

$6x + (-6x) = 5$ Simplify.

$0 = 5$ Combine like terms.

The false statement $0 = 5$ indicates that this system has no solution and is inconsistent. The graph of the linear equations in the system is a pair of parallel lines. For the solution, we can write "no solution" or, in set notation, write { } or ∅.

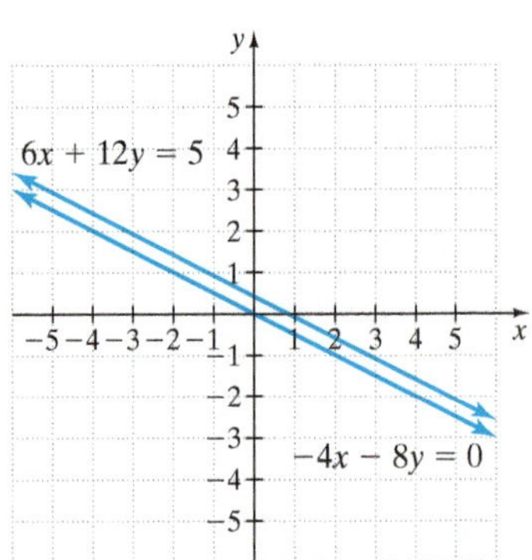

Work Practice 6

✓**Concept Check** Describe how the graphs of the equations in a system appear if the system has

a. no solution **b.** one solution **c.** an infinite number of solutions

Answer

6. no solution; { } or ∅

✓**Concept Check Answers**

a. parallel lines
b. intersect at one point
c. identical graphs

Vocabulary, Readiness & Video Check

Give the solution of each system. If the system has no solution or an infinite number of solutions, say so. If the system has one solution, find it.

1. $\begin{cases} y = 4x \\ -3x + y = 1 \end{cases}$
When solving, you obtain $x = 1$

2. $\begin{cases} 4x - y = 17 \\ -8x + 2y = 0 \end{cases}$
When solving, you obtain $0 = 34$

3. $\begin{cases} 4x - y = 17 \\ -8x + 2y = -34 \end{cases}$
When solving, you obtain $0 = 0$

4. $\begin{cases} 5x + 2y = 25 \\ x = y + 5 \end{cases}$
When solving, you obtain $y = 0$

5. $\begin{cases} x + y = 0 \\ 7x - 7y = 0 \end{cases}$
When solving, you obtain $x = 0$

6. $\begin{cases} y = -2x + 5 \\ 4x + 2y = 10 \end{cases}$
When solving, you obtain $0 = 0$

Martin-Gay Interactive Videos *Watch the section lecture video and answer the following question.*

See Video 11.2

Objective 1 7. The systems in Examples 2–4 all need one of their equations solved for a variable as a first step. What important part of the substitution method is emphasized in each example?

11.2 Exercise Set MyMathLab®

Solve each system of equations by the substitution method. See Examples 1 and 2.

1. $\begin{cases} x + y = 3 \\ x = 2y \end{cases}$

2. $\begin{cases} x + y = 20 \\ x = 3y \end{cases}$

3. $\begin{cases} x + y = 6 \\ y = -3x \end{cases}$

4. $\begin{cases} x + y = 6 \\ y = -4x \end{cases}$

5. $\begin{cases} y = 3x + 1 \\ 4y - 8x = 12 \end{cases}$

6. $\begin{cases} y = 2x + 3 \\ 5y - 7x = 18 \end{cases}$

7. $\begin{cases} y = 2x + 9 \\ y = 7x + 10 \end{cases}$

8. $\begin{cases} y = 5x - 3 \\ y = 8x + 4 \end{cases}$

MIXED PRACTICE

Solve each system of equations by the substitution method. See Examples 1 through 6.

9. $\begin{cases} 3x - 4y = 10 \\ y = x - 3 \end{cases}$

10. $\begin{cases} 4x - 3y = 10 \\ y = x - 5 \end{cases}$

11. $\begin{cases} x + 2y = 6 \\ 2x + 3y = 8 \end{cases}$

12. $\begin{cases} x + 3y = -5 \\ 2x + 2y = 6 \end{cases}$

13. $\begin{cases} 3x + 2y = 16 \\ x = 3y - 2 \end{cases}$

14. $\begin{cases} 2x + 3y = 18 \\ x = 2y - 5 \end{cases}$

15. $\begin{cases} 2x - 5y = 1 \\ 3x + y = -7 \end{cases}$

16. $\begin{cases} 3y - x = 6 \\ 4x + 12y = 0 \end{cases}$

17. $\begin{cases} 4x + 2y = 5 \\ -2x = y + 4 \end{cases}$

18. $\begin{cases} 2y = x + 2 \\ 6x - 12y = 0 \end{cases}$

19. $\begin{cases} 4x + y = 11 \\ 2x + 5y = 1 \end{cases}$

20. $\begin{cases} 3x + y = -14 \\ 4x + 3y = -22 \end{cases}$

21. $\begin{cases} x + 2y + 5 = -4 + 5y - x \\ 2x + x = y + 4 \end{cases}$
(*Hint:* First simplify each equation.)

22. $\begin{cases} 5x + 4y - 2 = -6 + 7y - 3x \\ 3x + 4x = y + 3 \end{cases}$
(*Hint:* See Exercise 21.)

23. $\begin{cases} 6x - 3y = 5 \\ x + 2y = 0 \end{cases}$

24. $\begin{cases} 10x - 5y = -21 \\ x + 3y = 0 \end{cases}$

25. $\begin{cases} 3x - y = 1 \\ 2x - 3y = 10 \end{cases}$

26. $\begin{cases} 2x - y = -7 \\ 4x - 3y = -11 \end{cases}$

27. $\begin{cases} -x + 2y = 10 \\ -2x + 3y = 18 \end{cases}$

28. $\begin{cases} -x + 3y = 18 \\ -3x + 2y = 19 \end{cases}$

29. $\begin{cases} 5x + 10y = 20 \\ 2x + 6y = 10 \end{cases}$

30. $\begin{cases} 6x + 3y = 12 \\ 9x + 6y = 15 \end{cases}$

31. $\begin{cases} 3x + 6y = 9 \\ 4x + 8y = 16 \end{cases}$

32. $\begin{cases} 2x + 4y = 6 \\ 5x + 10y = 16 \end{cases}$

33. $\begin{cases} \frac{1}{3}x - y = 2 \\ x - 3y = 6 \end{cases}$

34. $\begin{cases} \frac{1}{4}x - 2y = 1 \\ x - 8y = 4 \end{cases}$

35. $\begin{cases} x = \frac{3}{4}y - 1 \\ 8x - 5y = -6 \end{cases}$

36. $\begin{cases} x = \frac{5}{6}y - 2 \\ 12x - 5y = -9 \end{cases}$

Solve each system by the substitution method. First, simplify each equation by combining like terms.

37. $\begin{cases} -5y + 6y = 3x + 2(x - 5) - 3x + 5 \\ 4(x + y) - x + y = -12 \end{cases}$

38. $\begin{cases} 5x + 2y - 4x - 2y = 2(2y + 6) - 7 \\ 3(2x - y) - 4x = 1 + 9 \end{cases}$

REVIEW AND PREVIEW

Write equivalent equations by multiplying both sides of the given equation by the given nonzero number. See Section 3.2.

39. $3x + 2y = 6$ by -2
40. $-x + y = 10$ by 5
41. $-4x + y = 3$ by 3
42. $5a - 7b = -4$ by -4

Add the expressions by combining any like terms. See Section 3.1.

43. $\begin{array}{r} 3n + 6m \\ \underline{2n - 6m} \end{array}$

44. $\begin{array}{r} -2x + 5y \\ \underline{2x + 11y} \end{array}$

45. $\begin{array}{r} -5a - 7b \\ \underline{5a - 8b} \end{array}$

46. $\begin{array}{r} 9q + p \\ \underline{-9q - p} \end{array}$

CONCEPT EXTENSIONS

47. Explain how to identify a system with no solution when using the substitution method.

48. Occasionally, when using the substitution method, we obtain the equation $0 = 0$. Explain how this result indicates that the graphs of the equations in the system are identical.

Solve. See a Concept Check in this section.

49. As you solve the system $\begin{cases} 3x - y = -6 \\ -3x + 2y = 7 \end{cases}$, you find that $y = 1$. Is this the solution to the system?

50. As you solve the system $\begin{cases} x = 5y \\ y = 2x \end{cases}$, you find that $x = 0$ and $y = 0$. What is the solution to this system?

51. To avoid fractions, which of the equations below would you use if solving for y? Explain why.
   a. $\frac{1}{2}x - 4y = \frac{3}{4}$
   b. $8x - 5y = 13$
   c. $7x - y = 19$

52. Give the number of solutions for a system if the graphs of the equations in the system are
   a. lines intersecting in one point
   b. parallel lines
   c. same line

53. The number of men and women receiving bachelor's degrees each year has been steadily increasing. For the years 1970 through the estimation of 2014, the number of men receiving degrees (in thousands) is given by the equation $y = 3.9x + 443$, and for women, the equation is $y = 14.2x + 314$, where x is the number of years after 1970. (*Source:* National Center for Education Statistics)
   a. Use the substitution method to solve this system of equations. (Round your final results to the nearest whole numbers.)
   b. Explain the meaning of your answer to part (a).
   c. Sketch a graph of the system of equations. Write a sentence describing the trends for men and women receiving bachelor's degrees.

54. The number of Adult Contemporary Music radio stations in the United States from 2000 to 2009 is given by the equation $y = -11x + 734$, where x is the number of years after 2000. The number of Spanish radio stations is given by $y = 31x + 542$ for the same time period. (*Source:* M Street Corporation)
   a. Use the substitution method to solve this system of equations. (Round your numbers to the nearest tenth.)
   b. Explain the meaning of your answer to part (a).
   c. Sketch a graph of the system of equations. Write a sentence describing the trends in the popularity of these two types of music formats.

Solve each system by substitution. When necessary, round answers to the nearest hundredth.

55. $\begin{cases} y = 5.1x + 14.56 \\ y = -2x - 3.9 \end{cases}$

56. $\begin{cases} y = 3.1x - 16.35 \\ y = -9.7x + 28.45 \end{cases}$

57. $\begin{cases} 3x + 2y = 14.05 \\ 5x + y = 18.5 \end{cases}$

58. $\begin{cases} x + y = -15.2 \\ -2x + 5y = -19.3 \end{cases}$

11.3 Solving Systems of Linear Equations by Addition

Objective

1 Use the Addition Method to Solve a System of Linear Equations.

Objective 1 Using the Addition Method

We have seen that substitution is an accurate way to solve a linear system. Another method for solving a system of equations accurately is the **addition** or **elimination method.** The addition method is based on the addition property of equality: adding equal quantities to both sides of an equation does not change the solution of the equation. In symbols,

$$\text{if } A = B \text{ and } C = D, \text{ then } A + C = B + D.$$

Example 1 Solve the system:

$$\begin{cases} x + y = 7 \\ x - y = 5 \end{cases}$$

Solution: Since the left side of each equation is equal to the right side, we add equal quantities by adding the left sides of the equations together and the right sides of the equations together. This adding eliminates the variable y and gives us an equation in one variable, x. We can then solve for x.

$$\begin{aligned} x + y &= 7 && \text{First equation} \\ x - y &= 5 && \text{Second equation} \\ 2x \quad &= 12 && \text{Add the equations.} \\ x &= 6 && \text{Divide both sides by 2.} \end{aligned}$$

The x-value of the solution is 6. To find the corresponding y-value, let $x = 6$ in either equation of the system. We will use the first equation.

$$\begin{aligned} x + y &= 7 && \text{First equation} \\ 6 + y &= 7 && \text{Let } x = 6. \\ y &= 7 - 6 && \text{Solve for } y. \\ y &= 1 && \text{Simplify.} \end{aligned}$$

Check: The solution is (6, 1). Check this in both equations.

First Equation	Second Equation	
$x + y = 7$	$x - y = 5$	
$6 + 1 \stackrel{?}{=} 7$	$6 - 1 \stackrel{?}{=} 5$	Let $x = 6$ and $y = 1$.
$7 = 7$ True	$5 = 5$ True	

Thus, the solution of the system is (6, 1) and the graphs of the two equations intersect at the point (6, 1) as shown next.

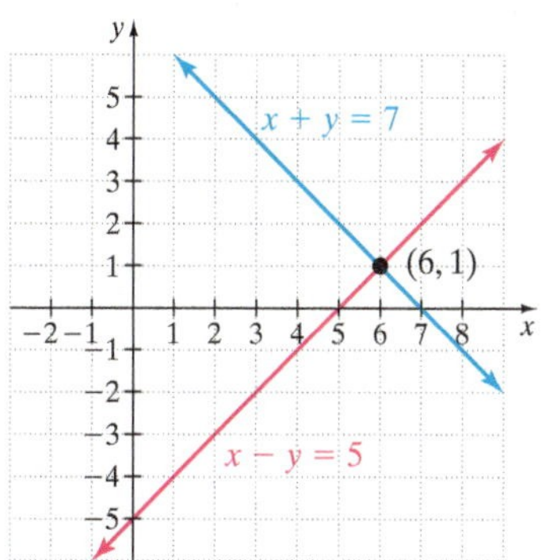

Work Practice 1

Practice 1

Solve the system:

$$\begin{cases} x - y = 2 \\ x + y = 8 \end{cases}$$

Helpful Hint Notice in Example 1 that our goal when solving a system of equations by the addition method is to eliminate a variable when adding the equations.

Example 2 Solve the system: $\begin{cases} -2x + y = 2 \\ -x + 3y = -4 \end{cases}$

Solution: If we simply add the two equations, the result is still an equation in two variables. However, remember from Example 1 that our goal is to eliminate one of the variables. Notice what happens if we multiply *both sides* of the first equation by −3, which we are allowed to do by the multiplication property of equality. The system

$$\begin{cases} -3(-2x + y) = -3(2) \\ -x + 3y = -4 \end{cases} \quad \text{simplifies to} \quad \begin{cases} 6x - 3y = -6 \\ -x + 3y = -4 \end{cases}$$

(Continued on next page)

Practice 2

Solve the system:

$$\begin{cases} x - 2y = 11 \\ 3x - y = 13 \end{cases}$$

Answers

1. (5, 3) **2.** (3, −4)

Now add the resulting equations and the y-variable is eliminated.

$$\begin{aligned} 6x - 3y &= -6 \\ -x + 3y &= -4 \\ \hline 5x \quad &= -10 \quad \text{Add.} \\ x &= -2 \quad \text{Divide both sides by 5.} \end{aligned}$$

To find the corresponding y-value, let $x = -2$ in any of the preceding equations containing both variables. We use the first equation of the original system.

$$\begin{aligned} -2x + y &= 2 \quad \text{First equation} \\ -2(-2) + y &= 2 \quad \text{Let } x = -2. \\ 4 + y &= 2 \\ y &= -2 \quad \text{Subtract 4 from both sides.} \end{aligned}$$

The solution is $(-2, -2)$. Check this ordered pair in both equations of the original system.

Work Practice 2

Helpful Hint When finding the second value of an ordered pair solution, any equation equivalent to one of the original equations in the system may be used.

In Example 2, the decision to multiply the first equation by -3 was no accident. **To eliminate a variable** when adding two equations, **the coefficient of the variable in one equation must be the opposite of its coefficient in the other equation.**

Helpful Hint

Be sure to multiply *both sides* of an equation by a chosen number when solving by the addition method. A common mistake is to multiply only the side containing the variables.

Practice 3

Solve the system:
$\begin{cases} x - 3y = 5 \\ 2x - 6y = -3 \end{cases}$

Example 3 Solve the system: $\begin{cases} 2x - y = 7 \\ 8x - 4y = 1 \end{cases}$

Solution: Multiply both sides of the first equation by -4 and the resulting coefficient of x is -8, the opposite of 8, the coefficient of x in the second equation. The system becomes

$$\begin{cases} -4(2x - y) = -4(7) \\ 8x - 4y = 1 \end{cases} \quad \text{simplifies to} \quad \begin{cases} -8x + 4y = -28 \\ 8x - 4y = 1 \end{cases}$$

Helpful Hint Don't forget to multiply **both** sides by -4.

Now add the resulting equations.

$$\begin{aligned} -8x + 4y &= -28 \\ 8x - 4y &= 1 \quad \text{Add the equations.} \\ \hline 0 &= -27 \quad \text{False} \end{aligned}$$

When we add the equations, both variables are eliminated and we have $0 = -27$, a false statement. This means that the system has no solution. The graphs of these equations are parallel lines. For the solution, we can write "no solution" or, in set notation, { } or $\varnothing$.

Work Practice 3

Practice 4

Solve the system:
$\begin{cases} 4x - 3y = 5 \\ -8x + 6y = -10 \end{cases}$

Example 4 Solve the system: $\begin{cases} 3x - 2y = 2 \\ -9x + 6y = -6 \end{cases}$

Solution: First we multiply both sides of the first equation by 3, then we add the resulting equations.

$$\begin{cases} 3(3x - 2y) = 3(2) \\ -9x + 6y = -6 \end{cases} \quad \text{simplifies to} \quad \begin{aligned} 9x - 6y &= 6 \\ -9x + 6y &= -6 \quad \text{Add the equations.} \\ \hline 0 &= 0 \quad \text{True} \end{aligned}$$

Answers

3. no solution; { } or $\varnothing$ **4.** infinite number of solutions; $\{(x, y) \mid 4x - 3y = 5\}$ or $\{(x, y) \mid -8x + 6y = -10\}$

Both variables are eliminated and we have $0 = 0$, a true statement. Whenever you eliminate a variable and get the equation $0 = 0$, the system has an infinite number of solutions. The graphs of these equations are identical. The solution is "infinite number of solutions" or, in set notation, $\{(x, y) | 3x - 2y = 2\}$ or $\{(x, y) | -9x + 6y = -6\}$.

Work Practice 4

✓Concept Check Suppose you are solving the system

$$\begin{cases} 3x + 8y = -5 \\ 2x - 4y = 3 \end{cases}$$

You decide to use the addition method and begin by multiplying both sides of the first equation by -2. In which of the following was the multiplication performed correctly? Explain.

a. $-6x - 16y = -5$ **b.** $-6x - 16y = 10$

Example 5 Solve the system: $\begin{cases} 3x + 4y = 13 \\ 5x - 9y = 6 \end{cases}$

Solution: We can eliminate the variable y by multiplying the first equation by 9 and the second equation by 4.

$$\begin{cases} 9(3x + 4y) = 9(13) \\ 4(5x - 9y) = 4(6) \end{cases} \text{ simplifies to } \begin{cases} 27x + 36y = 117 \\ 20x - 36y = 24 \end{cases}$$

$$47x = 141 \quad \text{Add the equations.}$$

$$x = 3 \quad \text{Divide both sides by 47.}$$

To find the corresponding y-value, we let $x = 3$ in any equation in this example containing two variables. Doing so in any of these equations will give $y = 1$. The solution to this system is $(3, 1)$. Check to see that $(3, 1)$ satisfies each equation in the original system.

Work Practice 5

If we had decided to eliminate x instead of y in Example 5, the first equation could have been multiplied by 5 and the second by -3. Try solving the original system this way to check that the solution is $(3, 1)$.

The following steps summarize how to solve a system of linear equations by the addition method.

Solving a System of Two Linear Equations by the Addition Method

Step 1: Rewrite each equation in standard form $Ax + By = C$.

Step 2: If necessary, multiply one or both equations by a nonzero number so that the coefficients of a chosen variable in the system are opposites.

Step 3: Add the equations.

Step 4: Find the value of one variable by solving the resulting equation from Step 3.

Step 5: Find the value of the second variable by substituting the value found in Step 4 into either of the original equations.

Step 6: Check the proposed solution in the original system.

Practice 5

Solve the system:
$$\begin{cases} 4x + 3y = 14 \\ 3x - 2y = 2 \end{cases}$$

Answer

5. $(2, 2)$

✓ Concept Check Answer

b

✓Concept Check Suppose you are solving the system

$$\begin{cases} -4x + 7y = 6 \\ x + 2y = 5 \end{cases}$$

by the addition method.

a. What step(s) should you take if you wish to eliminate x when adding the equations?

b. What step(s) should you take if you wish to eliminate y when adding the equations?

Practice 6

Solve the system:

$$\begin{cases} -2x + \frac{3y}{2} = 5 \\ -\frac{x}{2} - \frac{y}{4} = \frac{1}{2} \end{cases}$$

Example 6 Solve the system: $\begin{cases} -x - \frac{y}{2} = \frac{5}{2} \\ \frac{x}{6} - \frac{y}{2} = 0 \end{cases}$

Solution: We begin by clearing each equation of fractions. To do so, we multiply both sides of the first equation by the LCD, 2, and both sides of the second equation by the LCD, 6. Then the system

$$\begin{cases} 2\left(-x - \frac{y}{2}\right) = 2\left(\frac{5}{2}\right) \\ 6\left(\frac{x}{6} - \frac{y}{2}\right) = 6(0) \end{cases} \quad \text{simplifies to} \quad \begin{cases} -2x - y = 5 \\ x - 3y = 0 \end{cases}$$

We can now eliminate the variable x by multiplying the second equation by 2.

$$\begin{cases} -2x - y = 5 \\ 2(x - 3y) = 2(0) \end{cases} \quad \text{simplifies to} \quad \begin{cases} -2x - y = 5 \\ 2x - 6y = 0 \end{cases}$$

$$-7y = 5 \qquad \text{Add the equations.}$$

$$y = -\frac{5}{7} \qquad \text{Solve for } y.$$

To find x, we could replace y with $-\frac{5}{7}$ in one of the equations with two variables. Instead, let's go back to the simplified system and multiply by appropriate factors to eliminate the variable y and solve for x. To do this, we multiply the first equation by -3. Then the system

$$\begin{cases} -3(-2x - y) = -3(5) \\ x - 3y = 0 \end{cases} \quad \text{simplifies to} \quad \begin{cases} 6x + 3y = -15 \\ x - 3y = 0 \end{cases}$$

$$7x = -15 \qquad \text{Add the equations.}$$

$$x = -\frac{15}{7} \qquad \text{Solve for } x.$$

Check the ordered pair $\left(-\frac{15}{7}, -\frac{5}{7}\right)$ in both equations of the original system. The solution is $\left(-\frac{15}{7}, -\frac{5}{7}\right)$.

Work Practice 6

Answer

6. $\left(-\frac{8}{5}, \frac{6}{5}\right)$

✓ Concept Check Answers

a. multiply the second equation by 4

b. possible answer: multiply the first equation by -2 and the second equation by 7

Vocabulary, Readiness & Video Check

Given the system $\begin{cases} 3x - 2y = -9 \\ x + 5y = 14 \end{cases}$ *read each row (Step 1, Step 2, and Result). Then answer whether the result is true or false.*

	Step 1	Step 2	Result	True or False?
1.	Multiply 2nd equation through by -3.	Add the resulting equation to the 1st equation.	The y's are eliminated.	
2.	Multiply 2nd equation through by -3.	Add the resulting equation to the 1st equation.	The x's are eliminated.	
3.	Multiply 1st equation by 5 and 2nd equation by 2.	Add the two new equations.	The y's are eliminated.	
4.	Multiply 1st equation by 5 and 2nd equation by -2.	Add the two new equations.	The y's are eliminated.	

Martin-Gay Interactive Videos

See Video 11.3 

Watch the section lecture video and answer the following question.

Objective 1 5. For the addition/elimination methods, sometimes you need to multiply an equation through by a nonzero number so that the coefficients of a variable are opposites, as is shown in Example 2. What property allows us to do this? What important reminder is made at this step?

11.3 Exercise Set MyMathLab®

Solve each system of equations by the addition method. See Example 1.

1. $\begin{cases} 3x + y = 5 \\ 6x - y = 4 \end{cases}$

2. $\begin{cases} 4x + y = 13 \\ 2x - y = 5 \end{cases}$

3. $\begin{cases} x - 2y = 8 \\ -x + 5y = -17 \end{cases}$

4. $\begin{cases} x - 2y = -11 \\ -x + 5y = 23 \end{cases}$

Solve each system of equations by the addition method. If a system contains fractions or decimals, you may want to clear each equation of fractions or decimals first. See Examples 2 through 6.

5. $\begin{cases} 3x + y = -11 \\ 6x - 2y = -2 \end{cases}$

6. $\begin{cases} 4x + y = -13 \\ 6x - 3y = -15 \end{cases}$

7. $\begin{cases} 3x + 2y = 11 \\ 5x - 2y = 29 \end{cases}$

8. $\begin{cases} 4x + 2y = 2 \\ 3x - 2y = 12 \end{cases}$

9. $\begin{cases} x + 5y = 18 \\ 3x + 2y = -11 \end{cases}$

10. $\begin{cases} x + 4y = 14 \\ 5x + 3y = 2 \end{cases}$

11. $\begin{cases} x + y = 6 \\ x - y = 6 \end{cases}$

12. $\begin{cases} x - y = 1 \\ -x + 2y = 0 \end{cases}$

13. $\begin{cases} 2x + 3y = 0 \\ 4x + 6y = 3 \end{cases}$

14. $\begin{cases} 3x + y = 4 \\ 9x + 3y = 6 \end{cases}$

15. $\begin{cases} -x + 5y = -1 \\ 3x - 15y = 3 \end{cases}$

16. $\begin{cases} 2x + y = 6 \\ 4x + 2y = 12 \end{cases}$

17. $\begin{cases} 3x - 2y = 7 \\ 5x + 4y = 8 \end{cases}$

18. $\begin{cases} 6x - 5y = 25 \\ 4x + 15y = 13 \end{cases}$

19. $\begin{cases} 8x = -11y - 16 \\ 2x + 3y = -4 \end{cases}$

20. $\begin{cases} 10x + 3y = -12 \\ 5x = -4y - 16 \end{cases}$

21. $\begin{cases} 4x - 3y = 7 \\ 7x + 5y = 2 \end{cases}$

22. $\begin{cases} -2x + 3y = 10 \\ 3x + 4y = 2 \end{cases}$

23. $\begin{cases} 4x - 6y = 8 \\ 6x - 9y = 16 \end{cases}$

24. $\begin{cases} 9x - 3y = 12 \\ 12x - 4y = 18 \end{cases}$

25. $\begin{cases} 2x - 5y = 4 \\ 3x - 2y = 4 \end{cases}$

26. $\begin{cases} 6x - 5y = 7 \\ 4x - 6y = 7 \end{cases}$

27. $\begin{cases} \frac{x}{3} + \frac{y}{6} = 1 \\ \frac{x}{2} - \frac{y}{4} = 0 \end{cases}$

28. $\begin{cases} \frac{x}{2} + \frac{y}{8} = 3 \\ x - \frac{y}{4} = 0 \end{cases}$

29. $\begin{cases} \frac{10}{3}x + 4y = -4 \\ 5x + 6y = -6 \end{cases}$

30. $\begin{cases} \frac{3}{2}x + 4y = 1 \\ 9x + 24y = 6 \end{cases}$

31. $\begin{cases} x - \dfrac{y}{3} = -1 \\ -\dfrac{x}{2} + \dfrac{y}{8} = \dfrac{1}{4} \end{cases}$

32. $\begin{cases} 2x - \dfrac{3y}{4} = -3 \\ x + \dfrac{y}{9} = \dfrac{13}{3} \end{cases}$

33. $\begin{cases} -4(x + 2) = 3y \\ 2x - 2y = 3 \end{cases}$

34. $\begin{cases} -9(x + 3) = 8y \\ 3x - 3y = 8 \end{cases}$

35. $\begin{cases} \dfrac{x}{3} - y = 2 \\ -\dfrac{x}{2} + \dfrac{3y}{2} = -3 \end{cases}$

36. $\begin{cases} \dfrac{x}{2} + \dfrac{y}{4} = 1 \\ -\dfrac{x}{4} - \dfrac{y}{8} = 1 \end{cases}$

37. $\begin{cases} \dfrac{3}{5}x - y = -\dfrac{4}{5} \\ 3x + \dfrac{y}{2} = -\dfrac{9}{5} \end{cases}$

38. $\begin{cases} 3x + \dfrac{7}{2}y = \dfrac{3}{4} \\ -\dfrac{x}{2} + \dfrac{5}{3}y = -\dfrac{5}{4} \end{cases}$

39. $\begin{cases} 3.5x + 2.5y = 17 \\ -1.5x - 7.5y = -33 \end{cases}$

40. $\begin{cases} -2.5x - 6.5y = 47 \\ 0.5x - 4.5y = 37 \end{cases}$

41. $\begin{cases} 0.02x + 0.04y = 0.09 \\ -0.1x + 0.3y = 0.8 \end{cases}$

42. $\begin{cases} 0.04x - 0.05y = 0.105 \\ 0.2x - 0.6y = 1.05 \end{cases}$

MIXED PRACTICE

Solve each system by either the addition method or the substitution method.

43. $\begin{cases} 2x - 3y = -11 \\ y = 4x - 3 \end{cases}$

44. $\begin{cases} 4x - 5y = 6 \\ y = 3x - 10 \end{cases}$

45. $\begin{cases} x + 2y = 1 \\ 3x + 4y = -1 \end{cases}$

46. $\begin{cases} x + 3y = 5 \\ 5x + 6y = -2 \end{cases}$

47. $\begin{cases} 2y = x + 6 \\ 3x - 2y = -6 \end{cases}$

48. $\begin{cases} 3y = x + 14 \\ 2x - 3y = -16 \end{cases}$

49. $\begin{cases} y = 2x - 3 \\ y = 5x - 18 \end{cases}$

50. $\begin{cases} y = 6x - 5 \\ y = 4x - 11 \end{cases}$

51. $\begin{cases} x + \dfrac{1}{6}y = \dfrac{1}{2} \\ 3x + 2y = 3 \end{cases}$

52. $\begin{cases} x + \dfrac{1}{3}y = \dfrac{5}{12} \\ 8x + 3y = 4 \end{cases}$

53. $\begin{cases} \dfrac{x + 2}{2} = \dfrac{y + 11}{3} \\ \dfrac{x}{2} = \dfrac{2y + 16}{6} \end{cases}$

54. $\begin{cases} \dfrac{2x + 5}{2} = \dfrac{4y + 14}{4} \\ \dfrac{x}{3} = \dfrac{2y + 2}{6} \end{cases}$

55. $\begin{cases} 2x + 3y = 14 \\ 3x - 4y = -69.1 \end{cases}$

56. $\begin{cases} 5x - 2y = -19.8 \\ -3x + 5y = -3.7 \end{cases}$

REVIEW AND PREVIEW

Translating *Rewrite the following sentences using mathematical symbols. Do not solve the equations. See Sections 3.3 and 3.4.*

57. Twice a number, added to 6, is 3 less than the number.

58. The sum of three consecutive integers is 66.

59. Three times a number, subtracted from 20, is 2.

60. Twice the sum of 8 and a number is the difference of the number and 20.

61. The product of 4 and the sum of a number and 6 is twice the number.

62. If the quotient of twice a number and 7 is subtracted from the reciprocal of the number, the result is 2.

CONCEPT EXTENSIONS

Solve. See a Concept Check in this section.

63. To solve this system by the addition method and eliminate the variable y,

$$\begin{cases} 4x + 2y = -7 \\ 3x - y = -12 \end{cases}$$

by what value would you multiply the second equation? What do you get when you complete the multiplication?

64. Given the system of linear equations $\begin{cases} 3x - y = -8 \\ 5x + 3y = 2 \end{cases}$, use the addition method and

a. Solve the system by eliminating x.
b. Solve the system by eliminating y.

65. Suppose you are solving the system $\begin{cases} 3x + 8y = -5 \\ 2x - 4y = 3. \end{cases}$ You decide to use the addition method by multiplying both sides of the second equation by 2. In which of the following was the multiplication performed correctly? Explain.

a. $4x - 8y = 3$
b. $4x - 8y = 6$

66. Suppose you are solving the system $\begin{cases} -2x - y = 0 \\ -2x + 3y = 6. \end{cases}$

You decide to use the addition method by multiplying both sides of the first equation by 3, then adding the resulting equation to the second equation. Which of the following is the correct sum? Explain.

a. $-8x = 6$
b. $-8x = 9$

67. When solving a system of equations by the addition method, how do we know when the system has no solution?

68. Explain why the addition method might be preferred over the substitution method for solving the system

$$\begin{cases} 2x - 3y = 5 \\ 5x + 2y = 6. \end{cases}$$

69. Use the system of linear equations below to answer the questions.

$$\begin{cases} x + y = 5 \\ 3x + 3y = b \end{cases}$$

a. Find the value of b so that the system has an infinite number of solutions.
b. Find a value of b so that there are no solutions to the system.

70. Use the system of linear equations below to answer the questions.

$$\begin{cases} x + y = 4 \\ 2x + by = 8 \end{cases}$$

a. Find the value of b so that the system has an infinite number of solutions.

b. Find a value of b so that the system has a single solution.

Solve each system by the addition method.

71. $\begin{cases} 1.2x + 3.4y = 27.6 \\ 7.2x - 1.7y = -46.56 \end{cases}$

72. $\begin{cases} 5.1x - 2.4y = 3.15 \\ -15.3x + 1.2y = 27.75 \end{cases}$

73. Two age groups predicted to increase in percent of the workforce are those age 25–34 and those age 55 and older. The percent of workers age 25–34 predicted for 2008 through 2018 can be approximated by $0.05x - y = -21.6$. The percent of workers age 55 and older for the same years can be approximated by $0.58x - y = -18.1$. For both equations, x is the number of years since 2008 and y is the percent of the workforce. (*Source:* Bureau of Labor Statistics)

a. Use the addition method to solve the system:

$$\begin{cases} 0.05x - y = -21.6 \\ 0.58x - y = -18.1 \end{cases}$$

(Round answer to the nearest whole.)

b. Use your result from part (a) and estimate the year in which both percents were the same.

c. Use your results from part (a) and part (b) to estimate the percent of workers age 25–34 and those age 55 and older.

74. Two occupations predicted to greatly increase in the number of jobs are pharmacy technicians and network system analysts. The number of pharmacy technician jobs predicted for 2006 through 2016 can be approximated by $9.1x - y = -295$. The number of network system analyst jobs predicted for 2006 through 2016 can be approximated by $14x - y = -262$. For both equations, x is the number of years since 2006, and y is the number of jobs in the thousands. (*Source:* Bureau of Labor Statistics)

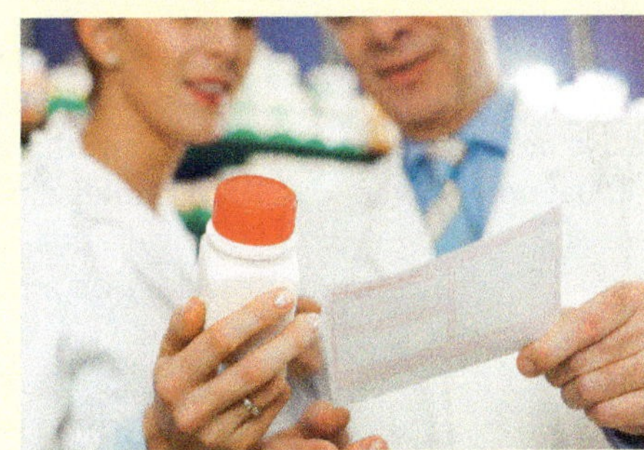

a. Use the addition method to solve this system of equations.

$$\begin{cases} 9.1x - y = -295 \\ 14x - y = -262 \end{cases}$$

(Eliminate y first and solve for x. Round this result to the nearest whole.)

b. Interpret your solution from part (a).

c. Using the year in your answer to part (b), estimate the number of pharmacy technician jobs and network system analyst jobs in that year.

Integrated Review Sections 11.1–11.3

Solving Systems of Equations

Solve each system by either the addition method or the substitution method.

1. $\begin{cases} 2x - 3y = -11 \\ y = 4x - 3 \end{cases}$

2. $\begin{cases} 4x - 5y = 6 \\ y = 3x - 10 \end{cases}$

3. $\begin{cases} x + y = 3 \\ x - y = 7 \end{cases}$

4. $\begin{cases} x - y = 20 \\ x + y = -8 \end{cases}$

5. $\begin{cases} x + 2y = 1 \\ 3x + 4y = -1 \end{cases}$

6. $\begin{cases} x + 3y = 5 \\ 5x + 6y = -2 \end{cases}$

7. $\begin{cases} y = x + 3 \\ 3x - 2y = -6 \end{cases}$

8. $\begin{cases} y = -2x \\ 2x - 3y = -16 \end{cases}$

9. $\begin{cases} y = 2x - 3 \\ y = 5x - 18 \end{cases}$

10. $\begin{cases} y = 6x - 5 \\ y = 4x - 11 \end{cases}$

11. $\begin{cases} x + \frac{1}{6}y = \frac{1}{2} \\ 3x + 2y = 3 \end{cases}$

12. $\begin{cases} x + \frac{1}{3}y = \frac{5}{12} \\ 8x + 3y = 4 \end{cases}$

13. $\begin{cases} x - 5y = 1 \\ -2x + 10y = 3 \end{cases}$

14. $\begin{cases} -x + 2y = 3 \\ 3x - 6y = -9 \end{cases}$

15. $\begin{cases} 0.2x - 0.3y = -0.95 \\ 0.4x + 0.1y = 0.55 \end{cases}$

16. $\begin{cases} 0.08x - 0.04y = -0.11 \\ 0.02x - 0.06y = -0.09 \end{cases}$

17. $\begin{cases} x = 3y - 7 \\ 2x - 6y = -14 \end{cases}$

18. $\begin{cases} y = \dfrac{x}{2} - 3 \\ 2x - 4y = 0 \end{cases}$

19. $\begin{cases} 2x + 5y = -1 \\ 3x - 4y = 33 \end{cases}$

20. $\begin{cases} 7x - 3y = 2 \\ 6x + 5y = -21 \end{cases}$

21. Which method, substitution or addition, would you prefer to use to solve the system below? Explain your reasoning.

$$\begin{cases} 3x + 2y = -2 \\ y = -2x \end{cases}$$

22. Which method, substitution or addition, would you prefer to use to solve the system below? Explain your reasoning.

$$\begin{cases} 3x - 2y = -3 \\ 6x + 2y = 12 \end{cases}$$

11.4 Solving Systems of Linear Equations in Three Variables

Objective

1 Solve a System of Three Linear Equations in Three Variables.

In this section, the algebraic methods of solving systems of two linear equations in two variables are extended to systems of three linear equations in three variables. We call the equation $3x - y + z = -15$, for example, a **linear equation in three variables** since there are three variables and each variable is raised only to the power 1. A solution of this equation is an **ordered triple (x, y, z)** that makes the equation a true statement. For example, the ordered triple $(2, 0, -21)$ is a solution of $3x - y + z = -15$ since replacing x with 2, y with 0, and z with -21 yields the true statement $3(2) - 0 + (-21) = -15$. The graph of this equation is a plane in three-dimensional space, just as the graph of a linear equation in two variables is a line in two-dimensional space.

Although we will not discuss the techniques for graphing equations in three variables, visualizing the possible patterns of intersecting planes gives us insight into the possible patterns of solutions of a system of three three-variable linear equations. There are four possible patterns.

1. Three planes have a single point in common. This point represents the single solution of the system. This system is **consistent.**

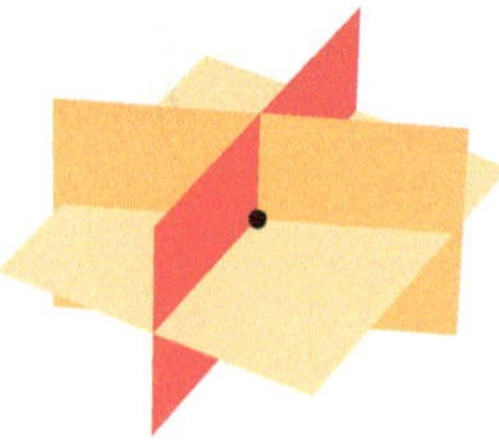

2. Three planes intersect at no point common to all three. This system has no solution. A few ways that this can occur are shown. This system is **inconsistent.**

3. Three planes intersect at all the points of a single line. The system has infinitely many solutions. This system is **consistent.**

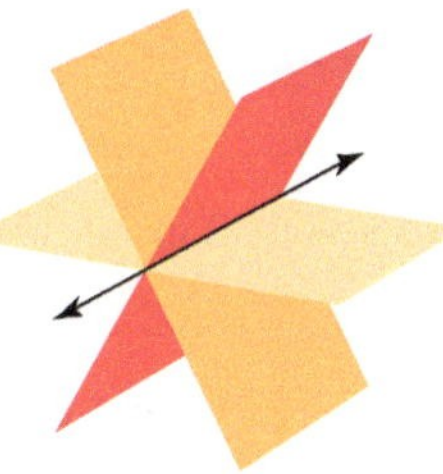

4. Three planes coincide at all points on the plane. The system is consistent, and the equations are **dependent.**

Objective 1 Solving a System of Three Linear Equations in Three Variables

Just as with systems of two equations in two variables, we can use the elimination or substitution method to solve a system of three equations in three variables. To use the elimination method, we eliminate a variable and obtain a system of two equations in two variables. Then we use the methods we learned in the previous two sections to solve the system of two equations.

Example 1 Solve the system.

$$\begin{cases} 3x - y + z = -15 & \text{Equation (1)} \\ x + 2y - z = 1 & \text{Equation (2)} \\ 2x + 3y - 2z = 0 & \text{Equation (3)} \end{cases}$$

Solution: Add equations (1) and (2) to eliminate z.

$$\begin{array}{l} 3x - y + z = -15 \\ \underline{x + 2y - z = 1} \\ 4x + y \quad = -14 \quad \text{Equation (4)} \end{array}$$

Next, add two *other* equations and *eliminate z again*. To do so, multiply both sides of equation (1) by 2 and add this resulting equation to equation (3). Then

$$\begin{cases} 2(3x - y + z) = 2(-15) \\ 2x + 3y - 2z = 0 \end{cases} \text{ simplifies to } \begin{cases} 6x - 2y + 2z = -30 \\ \underline{2x + 3y - 2z = 0} \end{cases}$$

$$8x + y \quad = -30 \quad \text{Equation (5)}$$

Now solve equations (4) and (5) for x and y. To solve by elimination, multiply both sides of equation (4) by -1 and add this resulting equation to equation (5). Then

$$\begin{cases} -1(4x + y) = -1(-14) \\ 8x + y = -30 \end{cases} \text{ simplifies to } \begin{cases} -4x - y = 14 \\ \underline{8x + y = -30} \end{cases}$$

$$4x = -16 \quad \text{Add the equations.}$$

$$x = -4 \quad \text{Solve for } x.$$

(Continued on next page)

Practice 1

Solve the system.

$$\begin{cases} 3x + 2y - z = 0 \\ x - y + 5z = 2 \\ 2x + 3y + 3z = 7 \end{cases}$$

Helpful Hint Don't forget to add two other equations besides equations (1) and (2) *and* to **eliminate the same variable.**

Answer

1. $(-1, 2, 1)$

Replace x with -4 in equation (4) or (5).

$$4x + y = -14 \quad \text{Equation (4)}$$
$$4(-4) + y = -14 \quad \text{Let } x = -4.$$
$$y = 2 \quad \text{Solve for } y.$$

Finally, replace x with -4 and y with 2 in equation (1), (2), or (3).

$$x + 2y - z = 1 \quad \text{Equation (2)}$$
$$-4 + 2(2) - z = 1 \quad \text{Let } x = -4 \text{ and } y = 2.$$
$$-4 + 4 - z = 1$$
$$-z = 1$$
$$z = -1$$

The solution is $(-4, 2, -1)$. To check, let $x = -4$, $y = 2$, and $z = -1$ in all three original equations of the system.

Equation (1)

$$3x - y + z = -15$$
$$3(-4) - 2 + (-1) \stackrel{?}{=} -15$$
$$-12 - 2 - 1 \stackrel{?}{=} -15$$
$$-15 = -15 \quad \text{True}$$

Equation (2)

$$x + 2y - z = 1$$
$$-4 + 2(2) - (-1) \stackrel{?}{=} 1$$
$$-4 + 4 + 1 \stackrel{?}{=} 1$$
$$1 = 1 \quad \text{True}$$

Equation (3)

$$2x + 3y - 2z = 0$$
$$2(-4) + 3(2) - 2(-1) \stackrel{?}{=} 0$$
$$-8 + 6 + 2 \stackrel{?}{=} 0$$
$$0 = 0 \quad \text{True}$$

All three statements are true, so the solution is $(-4, 2, -1)$.

Work Practice 1

Practice 2

Solve the system.

$$\begin{cases} 6x - 3y + 12z = 4 \\ -6x + 4y - 2z = 7 \\ -2x + y - 4z = 3 \end{cases}$$

Example 2

Solve the system.

$$\begin{cases} 2x - 4y + 8z = 2 & (1) \\ -x - 3y + z = 11 & (2) \\ x - 2y + 4z = 0 & (3) \end{cases}$$

Solution: Add equations (2) and (3) to eliminate x, and the new equation is

$$-5y + 5z = 11 \quad (4)$$

To eliminate x again, multiply both sides of equation (2) by 2 and add the resulting equation to equation (1). Then

$$\begin{cases} 2x - 4y + 8z = 2 \\ 2(-x - 3y + z) = 2(11) \end{cases} \text{ simplifies to } \begin{cases} 2x - 4y + 8z = 2 \\ -2x - 6y + 2z = 22 \end{cases}$$
$$-10y + 10z = 24 \quad (5)$$

Next, solve for y and z using equations (4) and (5). Multiply both sides of equation (4) by -2 and add the resulting equation to equation (5).

$$\begin{cases} -2(-5y + 5z) = -2(11) \\ -10y + 10z = 24 \end{cases} \text{ simplifies to } \begin{cases} 10y - 10z = -22 \\ -10y + 10z = 24 \end{cases}$$
$$0 = 2 \quad \text{False}$$

Since the statement is false, this system is inconsistent and has no solution. The solution set is the empty set $\{\ \}$ or $\varnothing$.

Work Practice 2

Answer

2. $\{\ \}$ or $\varnothing$

The elimination method is summarized next.

Solving a System of Three Linear Equations by the Elimination Method

Step 1: Write each equation in standard form $Ax + By + Cz = D$.

Step 2: Choose a pair of equations and use the equations to eliminate a variable.

Step 3: Choose any **other** pair of equations and eliminate the **same variable** as in Step 2.

Step 4: Two equations in two variables should be obtained from Step 2 and Step 3. Use methods from Section 11.2 or 11.3 to solve this system for both variables.

Step 5: To solve for the third variable, substitute the values of the variables found in Step 4 into any of the original equations containing the third variable.

Step 6: Check the ordered triple solution in *all three* original equations.

Make sure you read closely and follow Step 3.

✓Concept Check In the system

$$\begin{cases} x + y + z = 6 & \text{Equation (1)} \\ 2x - y + z = 3 & \text{Equation (2)} \\ x + 2y + 3z = 14 & \text{Equation (3)} \end{cases}$$

equations (1) and (2) are used to eliminate y. Which action could be used to finish solving best? Why?

a. Use (1) and (2) to eliminate z. **b.** Use (2) and (3) to eliminate y.

c. Use (1) and (3) to eliminate x.

Example 3 Solve the system.

$$\begin{cases} 2x + 4y = 1 & (1) \\ 4x - 4z = -1 & (2) \\ y - 4z = -3 & (3) \end{cases}$$

Solution: Notice that equation (2) has no term containing the variable y. Let us eliminate y using equations (1) and (3). Multiply both sides of equation (3) by -4 and add the resulting equation to equation (1). Then

$$\begin{cases} 2x + 4y = 1 \\ -4(y - 4z) = -4(-3) \end{cases} \text{ simplifies to } \begin{cases} 2x + 4y = 1 \\ -4y + 16z = 12 \end{cases}$$

$$2x + 16z = 13 \quad (4)$$

Next, solve for z using equations (4) and (2). Multiply both sides of equation (4) by -2 and add the resulting equation to equation (2).

$$\begin{cases} -2(2x + 16z) = -2(13) \\ 4x - 4z = -1 \end{cases} \text{ simplifies to } \begin{cases} -4x - 32z = -26 \\ 4x - 4z = -1 \end{cases}$$

$$-36z = -27$$

$$z = \frac{3}{4}$$

(Continued on next page)

Practice 3

Solve the system.

$$\begin{cases} 3x + 4y = 0 \\ 9x - 4z = 6 \\ -2y + 7z = 1 \end{cases}$$

Answer

3. $\left(\frac{2}{3}, -\frac{1}{2}, 0\right)$

✓Concept Check Answer

b

Replace z with $\frac{3}{4}$ in equation (3) and solve for y.

$$y - 4\left(\frac{3}{4}\right) = -3 \quad \text{Let } z = \frac{3}{4} \text{ in equation (3).}$$
$$y - 3 = -3$$
$$y = 0$$

Replace y with 0 in equation (1) and solve for x.

$$2x + 4(0) = 1$$
$$2x = 1$$
$$x = \frac{1}{2}$$

The solution is $\left(\frac{1}{2}, 0, \frac{3}{4}\right)$. Check to see that this solution satisfies all three equations of the system.

Work Practice 3

Practice 4

Solve the system.

$$\begin{cases} 2x + y - 3z = 6 \\ x + \frac{1}{2}y - \frac{3}{2}z = 3 \\ -4x - 2y + 6z = -12 \end{cases}$$

Example 4 Solve the system.

$$\begin{cases} x - 5y - 2z = 6 & (1) \\ -2x + 10y + 4z = -12 & (2) \\ \frac{1}{2}x - \frac{5}{2}y - z = 3 & (3) \end{cases}$$

Solution: Multiply both sides of equation (3) by 2 to eliminate fractions and multiply both sides of equation (2) by $-\frac{1}{2}$ so that the coefficient of x is 1. The resulting system is then

$$\begin{cases} x - 5y - 2z = 6 & (1) \\ x - 5y - 2z = 6 & \text{Multiply (2) by } -\frac{1}{2}. \\ x - 5y - 2z = 6 & \text{Multiply (3) by 2.} \end{cases}$$

All three equations are identical, and therefore equations (1), (2), and (3) are all equivalent. There are infinitely many solutions of this system. The equations are dependent. The solution set can be written as $\{(x, y, z) \mid x - 5y - 2z = 6\}$.

Work Practice 4

As mentioned earlier, we can also use the substitution method to solve a system of linear equations in three variables.

Practice 5

Solve the system.

$$\begin{cases} x + 2y + 4z = 16 \\ x + 2z = -4 \\ y - 3z = 30 \end{cases}$$

Example 5 Solve the system:

$$\begin{cases} x - 4y - 5z = 35 & (1) \\ x - 3y = 0 & (2) \\ -y + z = -55 & (3) \end{cases}$$

Solution: Notice in equations (2) and (3) that a variable is missing. Also notice that both equations contain the variable y. Let's use the substitution method by solving equation (2) for x and equation (3) for z and substituting the results in equation (1).

Answers

4. $\{(x, y, z) \mid 2x + y - 3z = 6\}$
5. $(6, 15, -5)$

$x - 3y = 0$ (2)

$x = 3y$ Solve equation (2) for x.

$-y + z = -55$ (3)

$z = y - 55$ Solve equation (3) for z.

Now substitute $3y$ for x and $y - 55$ for z in equation (1).

$x - 4y - 5z = 35$ (1)

$3y - 4y - 5(y - 55) = 35$ Let $x = 3y$ and $z = y - 55$.

$3y - 4y - 5y + 275 = 35$ Use the distributive law and multiply.

$-6y + 275 = 35$ Combine like terms.

$-6y = -240$ Subtract 275 from both sides.

$y = 40$ Solve.

Do not forget to distribute.

To find x, recall that $x = 3y$ and substitute 40 for y. Then $x = 3y$ becomes $x = 3 \cdot 40 = 120$. To find z, recall that $z = y - 55$ and substitute 40 for y, also. Then $z = y - 55$ becomes $z = 40 - 55 = -15$. The solution is $(120, 40, -15)$.

Work Practice 5

Vocabulary, Readiness & Video Check

Solve.

1. Choose the equation(s) that has $(-1, 3, 1)$ as a solution.

   a. $x + y + z = 3$ **b.** $-x + y + z = 5$ **c.** $-x + y + 2z = 0$ **d.** $x + 2y - 3z = 2$

2. Choose the equation(s) that has $(2, 1, -4)$ as a solution.

   a. $x + y + z = -1$ **b.** $x - y - z = -3$ **c.** $2x - y + z = -1$ **d.** $-x - 3y - z = -1$

3. Use the result of Exercise 1 to determine whether $(-1, 3, 1)$ is a solution of the system below. Explain your answer.

$$\begin{cases} x + y + z = 3 \\ -x + y + z = 5 \\ x + 2y - 3z = 2 \end{cases}$$

4. Use the result of Exercise 2 to determine whether $(2, 1, -4)$ is a solution of the system below. Explain your answer.

$$\begin{cases} x + y + z = -1 \\ x - y - z = -3 \\ 2x - y + z = -1 \end{cases}$$

Martin-Gay Interactive Videos *Watch the section lecture video and answer the following question.*

See Video 11.4

Objective 1 5. From Example 1 and the lecture before, why does Step 3 stress that the same variable be eliminated from two other equations?

11.4 Exercise Set MyMathLab® 

Solve each system. See Examples 1 through 5.

1. $\begin{cases} x - y + z = -4 \\ 3x + 2y - z = 5 \\ -2x + 3y - z = 15 \end{cases}$

2. $\begin{cases} x + y - z = -1 \\ -4x - y + 2z = -7 \\ 2x - 2y - 5z = 7 \end{cases}$

3. $\begin{cases} x + y = 3 \\ 2y = 10 \\ 3x + 2y - 3z = 1 \end{cases}$

4. $\begin{cases} 5x = 5 \\ 2x + y = 4 \\ 3x + y - 4z = -15 \end{cases}$

5. $\begin{cases} 2x + 2y + z = 1 \\ -x + y + 2z = 3 \\ x + 2y + 4z = 0 \end{cases}$

6. $\begin{cases} 2x - 3y + z = 5 \\ x + y + z = 0 \\ 4x + 2y + 4z = 4 \end{cases}$

7. $\begin{cases} x - 2y + z = -5 \\ -3x + 6y - 3z = 15 \\ 2x - 4y + 2z = -10 \end{cases}$

8. $\begin{cases} 3x + y - 2z = 2 \\ -6x - 2y + 4z = -4 \\ 9x + 3y - 6z = 6 \end{cases}$

9. $\begin{cases} 4x - y + 2z = 5 \\ 2y + z = 4 \\ 4x + y + 3z = 10 \end{cases}$

10. $\begin{cases} 5y - 7z = 14 \\ 2x + y + 4z = 10 \\ 2x + 6y - 3z = 30 \end{cases}$

11. $\begin{cases} x + 5z = 0 \\ 5x + y = 0 \\ y - 3z = 0 \end{cases}$

12. $\begin{cases} x - 5y = 0 \\ x - z = 0 \\ -x + 5z = 0 \end{cases}$

13. $\begin{cases} 6x - 5z = 17 \\ 5x - y + 3z = -1 \\ 2x + y = -41 \end{cases}$

14. $\begin{cases} x + 2y = 6 \\ 7x + 3y + z = -33 \\ x - z = 16 \end{cases}$

15. $\begin{cases} x + y + z = 8 \\ 2x - y - z = 10 \\ x - 2y - 3z = 22 \end{cases}$

16. $\begin{cases} 5x + y + 3z = 1 \\ x - y + 3z = -7 \\ -x + y = 1 \end{cases}$

17. $\begin{cases} x + 2y - z = 5 \\ 6x + y + z = 7 \\ 2x + 4y - 2z = 5 \end{cases}$

18. $\begin{cases} 4x - y + 3z = 10 \\ x + y - z = 5 \\ 8x - 2y + 6z = 10 \end{cases}$

19. $\begin{cases} 2x - 3y + z = 2 \\ x - 5y + 5z = 3 \\ 3x + y - 3z = 5 \end{cases}$

20. $\begin{cases} 4x + y - z = 8 \\ x - y + 2z = 3 \\ 3x - y + z = 6 \end{cases}$

21. $\begin{cases} -2x - 4y + 6z = -8 \\ x + 2y - 3z = 4 \\ 4x + 8y - 12z = 16 \end{cases}$

22. $\begin{cases} -6x + 12y + 3z = -6 \\ 2x - 4y - z = 2 \\ -x + 2y + \frac{z}{2} = -1 \end{cases}$

23. $\begin{cases} 2x + 2y - 3z = 1 \\ y + 2z = -14 \\ 3x - 2y = -1 \end{cases}$

24. $\begin{cases} 7x + 4y = 10 \\ x - 4y + 2z = 6 \\ y - 2z = -1 \end{cases}$

25. $\begin{cases} x + 2y - z = 5 \\ -3x - 2y - 3z = 11 \\ 4x + 4y + 5z = -18 \end{cases}$

26. $\begin{cases} 3x - 3y + z = -1 \\ 3x - y - z = 3 \\ -6x + 4y + 3z = -8 \end{cases}$

27. $\begin{cases} \frac{3}{4}x - \frac{1}{3}y + \frac{1}{2}z = 9 \\ \frac{1}{6}x + \frac{1}{3}y - \frac{1}{2}z = 2 \\ \frac{1}{2}x - y + \frac{1}{2}z = 2 \end{cases}$

28. $\begin{cases} \frac{1}{3}x - \frac{1}{4}y + z = -9 \\ \frac{1}{2}x - \frac{1}{3}y - \frac{1}{4}z = -6 \\ x - \frac{1}{2}y - z = -8 \end{cases}$

REVIEW AND PREVIEW

Translating Solve. *See Section 3.4.*

29. The sum of two numbers is 45 and one number is twice the other. Find the numbers.

30. The difference of two numbers is 5. Twice the smaller number added to five times the larger number is 53. Find the numbers.

Solve. See Section 9.3.

31. $2(x - 1) - 3x = x - 12$

32. $7(2x - 1) + 4 = 11(3x - 2)$

33. $-y - 5(y + 5) = 3y - 10$

34. $z - 3(z + 7) = 6(2z + 1)$

CONCEPT EXTENSIONS

35. Write a single linear equation in three variables that has $(-1, 2, -4)$ as a solution. (There are many possibilities.) Explain the process you used to write the equation.

36. Write a system of three linear equations in three variables that has $(2, 1, 5)$ as a solution. (There are many possibilities.) Explain the process you used to write the system.

37. Write a system of linear equations in three variables that has the solution $(-1, 2, -4)$. Explain the process you used to write your system.

38. When solving a system of three equations in three unknowns, explain how to determine that the system has no solution.

39. The fraction $\frac{1}{24}$ can be written as the following sum:

$$\frac{1}{24} = \frac{x}{8} + \frac{y}{4} + \frac{z}{3}$$

where the numbers x, y, and z are solutions of

$$\begin{cases} x + y + z = 1 \\ 2x - y + z = 0 \\ -x + 2y + 2z = -1 \end{cases}$$

Solve the system and see that the sum of the fractions is $\frac{1}{24}$.

40. The fraction $\frac{1}{18}$ can be written as the following sum:

$$\frac{1}{18} = \frac{x}{2} + \frac{y}{3} + \frac{z}{9}$$

where the numbers x, y, and z are solutions of

$$\begin{cases} x + 3y + z = -3 \\ -x + y + 2z = -14 \\ 3x + 2y - z = 12 \end{cases}$$

Solve the system and see that the sum of the fractions is $\frac{1}{18}$.

Solving systems involving more than three variables can be accomplished with methods similar to those encountered in this section. Apply what you already know to solve each system of equations in four variables.

41. $\begin{cases} x + y - w = 0 \\ y + 2z + w = 3 \\ x - z = 1 \\ 2x - y - w = -1 \end{cases}$

42. $\begin{cases} 5x + 4y = 29 \\ y + z - w = -2 \\ 5x + z = 23 \\ y - z + w = 4 \end{cases}$

43. $\begin{cases} x + y + z + w = 5 \\ 2x + y + z + w = 6 \\ x + y + z = 2 \\ x + y = 0 \end{cases}$

44. $\begin{cases} 2x - z = -1 \\ y + z + w = 9 \\ y - 2w = -6 \\ x + y = 3 \end{cases}$

45. Write a system of three linear equations in three variables that are dependent equations.

46. What is the solution to the system in Exercise 45?

11.5 Systems of Linear Equations and Problem Solving

Objectives

1. Solve Problems That Can Be Modeled by a System of Two Linear Equations.
2. Solve Problems with Cost and Revenue Functions.
3. Solve Problems That Can Be Modeled by a System of Three Linear Equations.

Objective 1 Solving Problems Modeled by Systems of Two Equations

Thus far, we have solved problems by writing one-variable equations and solving for the variable. Some of these problems can be solved, perhaps more easily, by writing a system of equations, as illustrated in this section.

Example 1 Predicting Equal Consumption of Red Meat and Poultry

America's consumption of red meat has decreased most years since 2000, while consumption of poultry has increased. The function $y = -0.56x + 113.6$ approximates the annual pounds of red meat consumed per capita, where x is the number of years since 2000. The function $y = 0.76x + 68.57$ approximates the annual pounds of poultry consumed per capita, where x is also the number of years since 2000. If this trend continues, determine the year when the annual consumption of red meat and poultry will be equal. (*Source:* USDA: Economic Research Service)

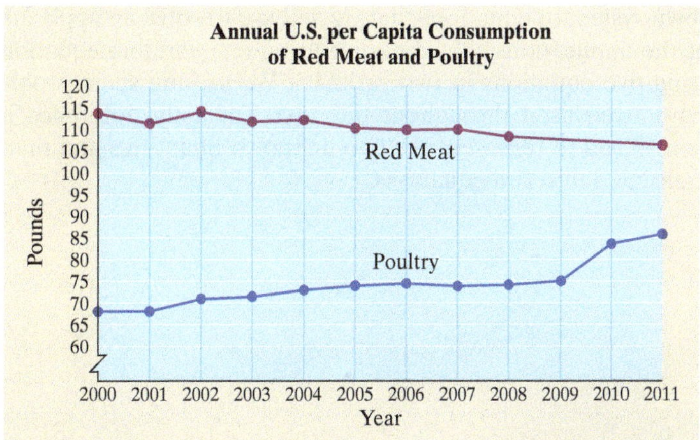

(Continued on next page)

Practice 1

Read Example 1. If we use the years 2005, 2006, 2007, and 2008 only to write functions approximating the consumption of red meat and poultry, we have the following:

Red meat: $y = -0.54x + 110.6$

Poultry: $y = -0.36x + 74.1$

where x is the number of years since 2005 and y is pounds per year consumed.

a. Assuming this trend continues, predict the year when consumption of red meat and poultry will be the same. Round to the nearest year.

b. Does your answer differ from the answer to Example 1? Why or why not?

Solution:

1. UNDERSTAND. Read and reread the problem and guess a year. Let's guess the year 2020. This year is 20 years since 2000, so $x = 20$. Now let $x = 20$ in each given function.

Red meat: $y = -0.56x + 113.6 = -0.56(20) + 113.6 = 102.4$ pounds

Poultry: $y = 0.76x + 68.57 = 0.76(20) + 68.57 = 83.77$ pounds

Since the projected pounds in 2020 for red meat and poultry are not the same, we guessed incorrectly, but we do have a better understanding of the problem. We know that the year will be later than 2020.

2. TRANSLATE. We are already given the system of equations.

3. SOLVE. We want to know the year x in which pounds y are the same, so we solve the system:

$$\begin{cases} y = -0.56x + 113.6 \\ y = 0.76x + 68.57 \end{cases}$$

Since both equations are solved for y, one way to solve is to use the substitution method.

$$y = 0.76x + 68.57 \quad \text{Second equation}$$

$$-0.56x + 113.6 = 0.76x + 68.57 \quad \text{Let } y = -0.56x + 113.6.$$

$$-1.32x = -45.03$$

$$x = \frac{-45.03}{-1.32} \approx 34.11$$

4. INTERPRET. Since we are asked to find only the year, we need only solve for x.

Check: To check, see whether $x \approx 34.11$ gives approximately the same number of pounds of red meat and poultry.

Red meat: $y = -0.56x + 113.6 = -0.56(34.11) + 113.6 \approx 94.4984$ pounds

Poultry: $y = 0.76x + 68.57 = 0.76(34.11) + 68.57 \approx 94.4936$ pounds

Since we rounded the number of years, the numbers of pounds do differ slightly. They differ only by 0.0048, so we can assume we solved correctly.

State: The consumption of red meat and poultry will be the same about 34.11 years after 2000, or 2034.11. Thus, in the year 2034, we predict the consumption will be the same.

Work Practice 1

For Example 1, the equations in the system were given to us. Let's now practice writing our own system of equations that we will use to solve an application.

Many of the applications solved earlier using one-variable equations can also be solved using two equations in **two** variables. We use the same problem-solving steps that have been used throughout this text. The only difference is that two variables are assigned to represent the two unknown quantities and that the stated problem is translated into **two** equations.

Answers

1. a. 2208 b. yes; answers may vary

Problem-Solving Steps

Step 1: UNDERSTAND the problem. During this step, become comfortable with the problem. Some ways of doing this are to

Read and reread the problem.

Choose two variables to represent the two unknowns.

Construct a drawing if possible.

Propose a solution and check. Pay careful attention to how you check your proposed solution. This will help when writing equations to model the problem.

Step 2: TRANSLATE the problem into two equations.

Step 3: SOLVE the system of equations.

Step 4: INTERPRET the results: **Check** the proposed solution in the stated problem and **state** your conclusion.

Example 2 Finding Unknown Numbers

Find two numbers whose sum is 37 and whose difference is 21.

Solution:

1. UNDERSTAND. Read and reread the problem. Suppose that one number is 20. If their sum is 37, the other number is 17 because $20 + 17 = 37$. Is their difference 21? No; $20 - 17 = 3$. Our proposed solution is incorrect, but we now have a better understanding of the problem.

Since we are looking for two numbers, we let

$x =$ first number

$y =$ second number

2. TRANSLATE. Since we have assigned two variables to this problem, we translate our problem into two equations.

In words:	two numbers whose sum	is	37
	↓	↓	↓
Translate:	$x + y$	$=$	37

In words:	two numbers whose difference	is	21
	↓	↓	↓
Translate:	$x - y$	$=$	21

3. SOLVE. Now we solve the system

$$\begin{cases} x + y = 37 \\ x - y = 21 \end{cases}$$

Notice that the coefficients of the variable y are opposites. Let's then solve by the addition method and begin by adding the equations.

$$\begin{aligned} x + y &= 37 \\ x - y &= 21 \\ \hline 2x \quad &= 58 \quad \text{Add the equations.} \\ x &= \frac{58}{2} = 29 \quad \text{Divide both sides by 2.} \end{aligned}$$

(Continued on next page)

Practice 2

Find two numbers whose sum is 30 and whose difference is 6.

Answer

2. 18, 12

Now we let $x = 29$ in the first equation to find y.

$$x + y = 37 \quad \text{First equation}$$
$$29 + y = 37$$
$$y = 8 \quad \text{Subtract 29 from both sides.}$$

4. INTERPRET. The solution of the system is (29, 8).

Check: Notice that the sum of 29 and 8 is $29 + 8 = 37$, the required sum. Their difference is $29 - 8 = 21$, the required difference.

State: The numbers are 29 and 8.

Work Practice 2

Practice 3

A first number is 5 more than a second number. Twice the first number is 2 less than 3 times the second number. Find the numbers.

Example 3 Finding Unknown Numbers

A first number is 4 less than a second number. Four times the first number is 6 more than twice the second. Find the numbers.

Solution:

1. UNDERSTAND. Read and reread the problem and guess a solution. If a first number is 10 and this is 4 less than a second number, the second number is 14. Four times the first number is $4(10)$, or 40. This is not equal to 6 more than twice the second number, which is $2(14) + 6$ or 34. Although we guessed incorrectly, we now have a better understanding of the problem.

Since we are looking for two numbers, we will let

x = first number

y = second number

2. TRANSLATE. Since we have assigned two variables to this problem, we will translate the given facts into two equations. For the first statement we have

In words:	the first number	is	4 less than the second number
	↓	↓	↓
Translate:	x	$=$	$y - 4$

Next we translate the second statement into an equation.

In words:	four times the first number	is	6 more than twice the second number
	↓	↓	↓
Translate:	$4x$	$=$	$2y + 6$

3. SOLVE. Here we solve the system

$$\begin{cases} x = y - 4 \\ 4x = 2y + 6 \end{cases}$$

Since the first equation expresses x in terms of y, we will use substitution. We substitute $y - 4$ for x in the second equation and solve for y.

$$4x = 2y + 6 \quad \text{Second equation}$$
$$4(y - 4) = 2y + 6$$
$$4y - 16 = 2y + 6 \quad \text{Let } x = y - 4.$$
$$2y = 22$$
$$y = 11$$

Answer

3. 17 and 12

Now we replace y with 11 in the equation $x = y - 4$ and solve for x. Then $x = y - 4$ becomes $x = 11 - 4 = 7$. The ordered pair solution of the system is $(7, 11)$.

4. INTERPRET. Since the solution of the system is $(7, 11)$, then the first number we are looking for is 7 and the second number is 11.

Check: Notice that 7 *is* 4 less than 11, and 4 times 7 *is* 6 more than twice 11. The proposed numbers, 7 and 11, are correct.

State: The numbers are 7 and 11.

Work Practice 3

Example 4 Solving a Problem About Prices

The Cirque du Soleil show Varekai is performing locally. Matinee admission for 4 adults and 2 children is \$374, while admission for 2 adults and 3 children is \$285.

a. What is the price of an adult's ticket?

b. What is the price of a child's ticket?

c. Suppose that a special rate of \$1000 is offered for groups of 20 persons. Should a group of 4 adults and 16 children use the group rate? Why or why not?

Solution:

1. UNDERSTAND. Read and reread the problem and guess a solution. Let's suppose that the price of an adult's ticket is \$50 and the price of a child's ticket is \$40. To check our proposed solution, let's see if admission for 4 adults and 2 children is \$374. Admission for 4 adults is 4(\$50) or \$200 and admission for 2 children is 2(\$40) or \$80. This gives a total admission of \$200 + \$80 = \$280, not the required \$374. Again, though, we have accomplished the purpose of this process: We have a better understanding of the problem. To continue, we let

A = the price of an adult's ticket and

C = the price of a child's ticket

2. TRANSLATE. We translate the problem into two equations using both variables.

In words:	admission for 4 adults	and	admission for 2 children	is	\$374
	↓	↓	↓	↓	↓
Translate:	$4A$	$+$	$2C$	$=$	374
In words:	admission for 2 adults	and	admission for 3 children	is	\$285
	↓	↓	↓	↓	↓
Translate:	$2A$	$+$	$3C$	$=$	285

(Continued on next page)

Practice 4

It is considered a premium game when the Red Sox or the Yankees go to Ameriquest Park in Texas to play the Rangers. Admission for one of these games for three adults and three children under 14 is \$75, while admission for two adults and four children is \$62. (*Source:* MLB.com, Texas Rangers)

a. What is the price of an adult admission at Ameriquest Park?

b. What is the price of a child's admission?

c. Suppose that a special rate of \$200 is offered for groups of 20 persons. Should a group of 5 adults and 15 children use the group rate? Why or why not?

Answers

4. a. Adult: \$19 **b.** Child: \$6 **c.** No, the regular rates are less than the group rate.

3. SOLVE. We solve the system.

$$\begin{cases} 4A + 2C = 374 \\ 2A + 3C = 285 \end{cases}$$

Since both equations are written in standard form, we solve by the addition method. First we multiply the second equation by -2 so that when we add the equations, we eliminate the variable A. Then the system

$$\begin{cases} 4A + 2C = 374 \\ -2(2A + 3C) = -2(285) \end{cases} \quad \text{simplifies to} \quad \begin{cases} 4A + 2C = 374 \\ -4A - 6C = -570 \end{cases}$$

Add the equations.

$$-4C = -196$$

$C = 49$ or \$49, the children's ticket price.

To find A, we replace C with 49 in the first equation.

$$4A + 2C = 374 \quad \text{First equation}$$
$$4A + 2(49) = 374 \quad \text{Let } C = 49.$$
$$4A + 98 = 374$$
$$4A = 276$$

$A = 69$ or \$69, the adult's ticket price

4. INTERPRET.

Check: Notice that 4 adults and 2 children will pay $4(\$69) + 2(\$49) = \$276 + \$98 = \$374$, the required amount. Also, the price for 2 adults and 3 children is $2(\$69) + 3(\$49) = \$138 + \$147 = \$285$, the required amount.

State: Answer the three original questions.

a. Since $A = 69$, the price of an adult's ticket is \$69.

b. Since $C = 49$, the price of a child's ticket is \$49.

c. The regular admission price for 4 adults and 16 children is

$$4(\$69) + 16(\$49) = \$276 + \$784$$
$$= \$1060$$

This is \$60 more than the special group rate of \$1000, so they should request the group rate.

Work Practice 4

Practice 5

In 2007, the French train TGV V150 became the fastest conventional rail train in the world. It broke the 1990 record of the next fastest conventional rail train, the French TGV Atlantique. Assume the V150 and the Atlantique leave the same station in Paris, with one heading west and one heading east. After 2 hours, they are 2150 kilometers apart. If the V150 is 75 kph faster than the Atlantique, what is the speed of each?

Answer

5. Atlantique: 500 kph; V150: 575 kph

Example 5 Finding the Rate of Speed

Two cars leave Indianapolis, one traveling east and the other west. After 3 hours, they are 297 miles apart. If one car is traveling 5 mph faster than the other, what is the speed of each?

Solution:

1. UNDERSTAND. Read and reread the problem. Let's guess a solution and use the formula $d = rt$ (distance = rate · time) to check. Suppose that one car is traveling at a rate of 55 miles per hour. This means that the other car is traveling at a rate of 50 miles per hour since we are told that one car is traveling 5 mph faster than the other. To find the distance apart after 3 hours, we will first find the distance traveled by each car. One car's distance is rate · time $= 55(3) = 165$ miles. The other car's distance is rate · time $= 50(3) = 150$ miles. Since one car is traveling east and the other west, their distance apart is the sum of their distances, or 165 miles + 150 miles = 315 miles. Although this distance apart is not the required distance of 297 miles, we now have a better understanding of the problem.

150 + 165 = 315 mi

50(3) = 150 mi Indianapolis 55(3) = 165 mi

Let's model the problem with a system of equations. We will let

x = speed of one car
y = speed of the other car

We summarize the information on the following chart. Both cars have traveled 3 hours. Since distance = rate · time, their distances are $3x$ and $3y$ miles, respectively.

	Rate ·	Time =	Distance
One Car	x	3	$3x$
Other Car	y	3	$3y$

2. TRANSLATE. We can now translate the stated conditions into two equations.

In words:	one car's distance	added to	the other car's distance	is	297
	↓	↓	↓	↓	↓
Translate:	$3x$	$+$	$3y$	$=$	297

In words:	one car's speed	is	5 mph faster than the other's
	↓	↓	↓
Translate:	x	$=$	$y + 5$

3. SOLVE. Here we solve the system

$$\begin{cases} 3x + 3y = 297 \\ x = y + 5 \end{cases}$$

Again, the substitution method is appropriate. We replace x with $y + 5$ in the first equation and solve for y.

$$3x + 3y = 297 \quad \text{First equation}$$
$$3(y + 5) + 3y = 297 \quad \text{Let } x = y + 5.$$
$$3y + 15 + 3y = 297$$
$$6y = 282$$
$$y = 47$$

To find x, we replace y with 47 in the equation $x = y + 5$. Then $x = 47 + 5 = 52$. The ordered pair solution of the system is $(52, 47)$.

4. INTERPRET. The solution $(52, 47)$ means that the cars are traveling at 52 mph and 47 mph, respectively.

Check: Notice that one car is traveling 5 mph faster than the other. Also, if one car travels 52 mph for 3 hours, the distance is $3(52) = 156$ miles. The other car traveling for 3 hours at 47 mph travels a distance of $3(47) = 141$ miles. The sum of the distances $156 + 141$ is 297 miles, the required distance.

State: The cars are traveling at 52 mph and 47 mph.

Helpful Hint Don't forget to attach units if appropriate.

Work Practice 5

Practice 6

Keith Robinson is a chemistry teacher who needs 1 liter of a solution of 5% hydrochloric acid to carry out an experiment. If he only has a stock solution of 99% hydrochloric acid, how much water (0% acid) and how much stock solution (99%) of HCL must he mix to get 1 liter of 5% solution? Round answers to the nearest hundredth of a liter.

Example 6 Mixing Solutions

Lynn Pike, a pharmacist, needs 70 liters of a 50% alcohol solution. She has available a 30% alcohol solution and an 80% alcohol solution. How many liters of each solution should she mix to obtain 70 liters of a 50% alcohol solution?

Solution:

1. UNDERSTAND. Read and reread the problem. Next, guess the solution. Suppose that we need 20 liters of the 30% solution. Then we need $70 - 20 = 50$ liters of the 80% solution. To see if this gives us 70 liters of a 50% alcohol solution, let's find the amount of pure alcohol in each solution.

number of liters	×	alcohol strength	=	amount of pure alcohol
↓		↓		↓
20 liters	×	0.30	=	6 liters
50 liters	×	0.80	=	40 liters
70 liters	×	0.50	=	35 liters

Since 6 liters + 40 liters = 46 liters and not 35 liters, our guess is incorrect, but we have gained some insight as to how to model and check this problem.

We will let

x = amount of 30% solution, in liters

y = amount of 80% solution, in liters

and use a table to organize the given data.

	Number of Liters	Alcohol Strength	Amount of Pure Alcohol
30% Solution	x	30%	$0.30x$
80% Solution	y	80%	$0.80y$
50% Solution Needed	70	50%	$(0.50)(70)$

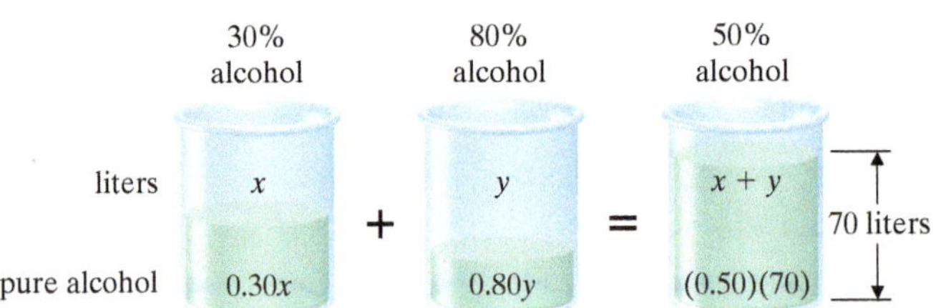

2. TRANSLATE. We translate the stated conditions into two equations.

In words:	amount of 30% solution	+	amount of 80% solution	=	70
	↓		↓		↓
Translate:	x	+	y	=	70
In words:	amount of pure alcohol in 30% solution	+	amount of pure alcohol in 80% solution	=	amount of pure alcohol in 50% solution
	↓		↓		↓
Translate:	$0.30x$	+	$0.80y$	=	$(0.50)(70)$

3. SOLVE. Here we solve the system

$$\begin{cases} x + y = 70 \\ 0.30x + 0.80y = (0.50)(70) \end{cases}$$

Answer

6. 0.95 liter of water; 0.05 liter of 99% HCL

To solve this system, we use the elimination method. We multiply both sides of the first equation by -3 and both sides of the second equation by 10. Then

$$\begin{cases} -3(x + y) = -3(70) \\ 10(0.30x + 0.80y) = 10(0.50)(70) \end{cases} \text{ simplifies to } \begin{cases} -3x - 3y = -210 \\ 3x + 8y = 350 \end{cases}$$

$$5y = 140$$

$$y = 28$$

Now we replace y with 28 in the equation $x + y = 70$ and find that $x + 28 = 70$, or $x = 42$.

The ordered pair solution of the system is (42, 28).

4. INTERPRET.

Check: Check the solution in the same way that we checked our guess.

State: The pharmacist needs to mix 42 liters of 30% solution and 28 liters of 80% solution to obtain 70 liters of 50% solution.

Work Practice 6

✓Concept Check Suppose you mix an amount of 25% acid solution with an amount of 60% acid solution. You then calculate the acid strength of the resulting acid mixture. For which of the following results should you suspect an error in your calculation? Why?

a. 14% **b.** 32% **c.** 55%

Objective 2 Solving Problems with Cost and Revenue Functions

Recall that businesses are often computing cost and revenue functions or equations to predict sales, to determine whether prices need to be adjusted, and to see whether the company is making or losing money. Recall also that the value at which revenue equals cost is called the break-even point. When revenue is less than cost, the company is losing money; when revenue is greater than cost, the company is making money.

Example 7 Finding a Break-Even Point

A manufacturing company recently purchased $3000 worth of new equipment to offer new personalized stationery to its customers. The cost of producing a package of personalized stationery is $3.00, and it is sold for $5.50. Find the number of packages that must be sold for the company to break even.

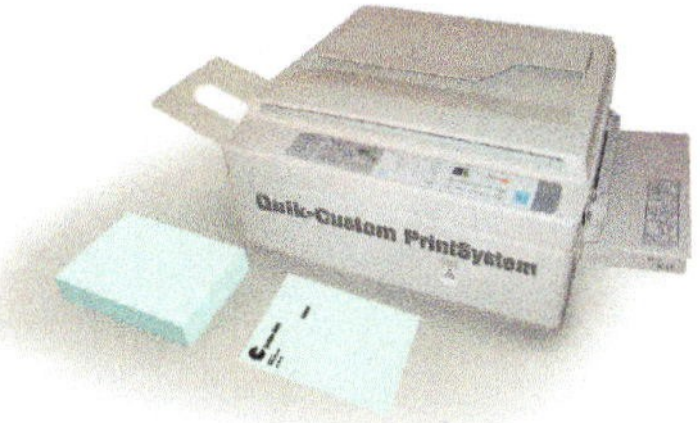

Solution:

1. UNDERSTAND. Read and reread the problem. Notice that the cost to the company will include a one-time cost of $3000 for the equipment and then $3.00 per package produced. The revenue will be $5.50 per package sold.

To model this problem, we will let

x = number of packages of personalized stationery

$C(x)$ = total cost of producing x packages of stationery

$R(x)$ = total revenue from selling x packages of stationery

(Continued on next page)

Practice 7

An online-only electronics firm recently purchased $3000 worth of new equipment to create shock-proof packaging for its products. The cost of producing one shock-proof package is $2.50, and the firm charges the customer $4.50 for the packaging. Find the number of packages that must be sold for the company to break even.

Answer

7. 1500 packages

✓Concept Check Answer

a; answers may vary

2. TRANSLATE. The revenue equation is

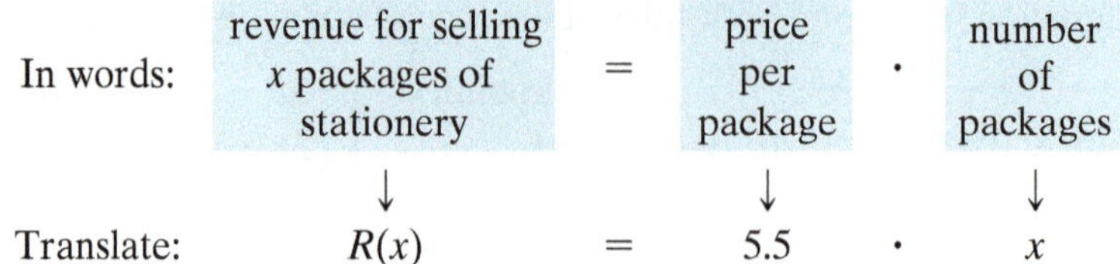

In words:	revenue for selling x packages of stationery	=	price per package	·	number of packages
	↓		↓		↓
Translate:	$R(x)$	=	5.5	·	x

The cost equation is

In words:	cost for producing x packages of stationery	=	cost per package	·	number of packages	+	cost for equipment
	↓		↓		↓		↓
Translate:	$C(x)$	=	3	·	x	+	3000

Since the break-even point is when $R(x) = C(x)$, we solve the equation

$$5.5x = 3x + 3000$$

3. SOLVE.

$$5.5x = 3x + 3000$$

$$2.5x = 3000 \quad \text{Subtract } 3x \text{ from both sides.}$$

$$x = 1200 \quad \text{Divide both sides by 2.5.}$$

4. INTERPRET.

Check: To see whether the break-even point occurs when 1200 packages are produced and sold, see if revenue equals cost when $x = 1200$. When $x = 1200$, $R(x) = 5.5x = 5.5(1200) = 6600$ and $C(x) = 3x + 3000 = 3(1200) + 3000 = 6600$. Since $R(1200) = C(1200) = 6600$, the break-even point is 1200.

State: The company must sell 1200 packages of stationery to break even. The graph of this system is shown.

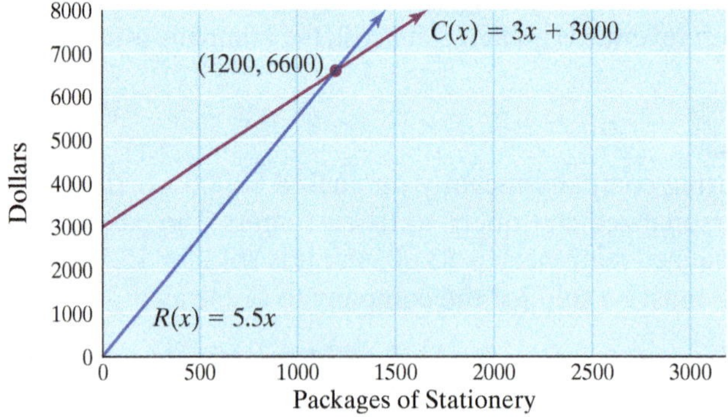

Work Practice 7

Objective 3 Solving Problems Modeled by Systems of Three Equations

To introduce problem solving by writing a system of three linear equations in three variables, we solve a problem about triangles.

Example 8 Finding Angle Measures

The measure of the largest angle of a triangle is 80° more than the measure of the smallest angle, and the measure of the remaining angle is 10° more than the measure of the smallest angle. Find the measure of each angle.

Solution:

1. UNDERSTAND. Read and reread the problem. Recall that the sum of the measures of the angles of a triangle is 180°. Then guess a solution. If the smallest angle measures 20°, the measure of the largest angle is 80° more, or 20° + 80° = 100°. The measure of the remaining angle is 10° more than the measure of the smallest angle, or 20° + 10° = 30°. The sum of these three angles is 20° + 100° + 30° = 150°, not the required 180°. We now know that the measure of the smallest angle is greater than 20°.

   To model this problem, we will let

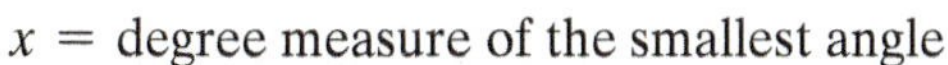

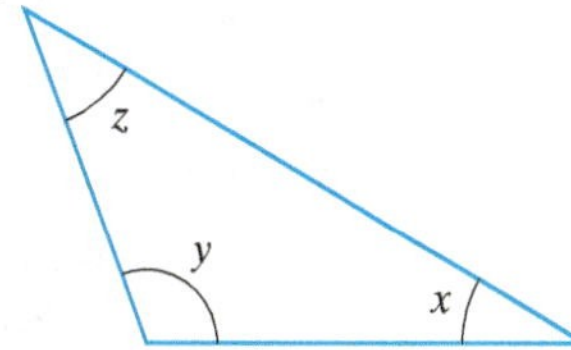

   x = degree measure of the smallest angle
   y = degree measure of the largest angle
   z = degree measure of the remaining angle

2. TRANSLATE. We translate the given information into three equations.

In words:	the sum of the measures	=	180
	↓		↓
Translate:	$x + y + z$	=	180
In words:	the largest angle	is	80 more than the smallest angle
	↓	↓	↓
Translate:	y	=	$x + 80$
In words:	the remaining angle	is	10 more than the smallest angle
	↓	↓	↓
Translate:	z	=	$x + 10$

3. SOLVE. We solve the system

$$\begin{cases} x + y + z = 180 \\ y = x + 80 \\ z = x + 10 \end{cases}$$

   Since y and z are both expressed in terms of x, we will solve using the substitution method. We substitute $y = x + 80$ and $z = x + 10$ in the first equation. Then

$$x + y + z = 180$$

$$\begin{aligned} x + (x + 80) + (x + 10) &= 180 && \text{First equation} \\ 3x + 90 &= 180 && \text{Let } y = x + 80 \text{ and } z = x + 10. \\ 3x &= 90 \\ x &= 30 \end{aligned}$$

   Then $y = x + 80 = 30 + 80 = 110$, and $z = x + 10 = 30 + 10 = 40$. The ordered triple solution is (30, 110, 40).

(Continued on next page)

Practice 8

The measure of the largest angle of a triangle is 40° more than the measure of the smallest angle, and the measure of the remaining angle is 20° more than the measure of the smallest angle. Find the measure of each angle.

Answer

8. 40°, 60°, 80°

4. INTERPRET.

Check: Notice that $30° + 40° + 110° = 180°$. Also, the measure of the largest angle, 110°, is 80° more than the measure of the smallest angle, 30°. The measure of the remaining angle, 40°, is 10° more than the measure of the smallest angle, 30°.

Work Practice 8

Vocabulary, Readiness & Video Check

Martin-Gay Interactive Videos

See Video 11.5

Watch the section lecture video and answer the following questions.

Objective 1 1. In Example 1 and the lecture before, the problem-solving steps for solving applications are mentioned. What is the difference here from when we've used these steps in the past?

Objective 2 2. Based on Example 6, explain the meaning of a break-even point. How do you find the break-even point algebraically?

Objective 3 3. In Example 7, why is the ordered triple not the final stated solution to the application?

11.5 Exercise Set MyMathLab®

Without actually solving each problem, choose each correct solution by deciding which choice satisfies the given conditions.

△ **1.** The length of a rectangle is 3 feet longer than the width. The perimeter is 30 feet. Find the dimensions of the rectangle.

a. length = 8 feet; width = 5 feet
b. length = 8 feet; width = 7 feet
c. length = 9 feet; width = 6 feet

△ **2.** An isosceles triangle, a triangle with two sides of equal length, has a perimeter of 20 inches. Each of the equal sides is one inch longer than the third side. Find the lengths of the three sides.

a. 6 inches, 6 inches, and 7 inches
b. 7 inches, 7 inches, and 6 inches
c. 6 inches, 7 inches, and 8 inches

3. Two computer disks and three notebooks cost $17. However, five computer disks and four notebooks cost $32. Find the price of each.

a. notebook = $4; computer disk = $3
b. notebook = $3; computer disk = $4
c. notebook = $5; computer disk = $2

4. Two music CDs and four music cassette tapes cost a total of $40. However, three music CDs and five cassette tapes cost $55. Find the price of each.

a. CD = $12; cassette = $4
b. CD = $15; cassette = $2
c. CD = $10; cassette = $5

5. Kesha has a total of 100 coins, all of which are either dimes or quarters. The total value of the coins is $13.00. Find the number of each type of coin.

a. 80 dimes; 20 quarters
b. 20 dimes; 44 quarters
c. 60 dimes; 40 quarters

6. Samuel has 28 gallons of saline solution available in two large containers at his pharmacy. One container holds three times as much as the other container. Find the capacity of each container.

a. 15 gallons; 5 gallons
b. 20 gallons; 8 gallons
c. 21 gallons; 7 gallons

TRANSLATING

Write a system of equations in x and y describing each situation. Do not solve the system. See Example 2.

7. A smaller number and a larger number add up to 15 and have a difference of 7. (Let x be the larger number.)

8. The total of two numbers is 16. The first number plus 2 more than 3 times the second equals 18. (Let x be the first number.)

9. Keiko has a total of $6500, which she has invested in two accounts. The larger account is $800 greater than the smaller account. (Let x be the amount of money in the larger account.)

10. Dominique has four times as much money in his savings account as in his checking account. The total amount is $2300. (Let x be the amount of money in his checking account.)

MIXED PRACTICE

Solve. See Examples 1 through 6. For Exercises 13 and 14, the solutions have been started for you.

11. Two numbers total 83 and have a difference of 17. Find the two numbers.

12. The sum of two numbers is 76 and their difference is 52. Find the two numbers.

13. One number is two more than a second number. Twice the first is 4 less than 3 times the second. Find the numbers.

Start the solution:

1. UNDERSTAND the problem. Since we are looking for two numbers, let

x = one number

y = second number

2. TRANSLATE. Since we have assigned two variables, we will translate the facts into two equations. (Fill in the blanks.)

First equation:

In words:	One number	is	two	more than	second number
	↓	↓			
Translate:	x	$=$	___	___	___

Second equation:

In words:	Twice the first number	is	4	less than	3 times the second number
	↓	↓			
Translate:	$2x$	$=$	___	___	___

Finish with:

3. SOLVE the system and
4. INTERPRET the results.

14. Three times one number minus a second number is 8, and the sum of the numbers is 12. Find the numbers.

Start the solution:

1. UNDERSTAND the problem. Since we are looking for two numbers, let

x = one number

y = second number

2. TRANSLATE. Since we have assigned two variables, we will translate the facts into two equations. (Fill in the blanks.)

First equation:

In words:	Three times one number	minus	a second number	is	8
	↓	↓	↓	↓	↓
Translate:	$3x$	___	___	$=$	8

Second equation:

In words:	The sum of the numbers	is	12
	↓	↓	↓
Translate:	x + ___	___	12

Finish with:

3. SOLVE the system and
4. INTERPRET the results.

15. A first number plus twice a second number is 8. Twice the first number, plus the second, totals 25. Find the numbers.

16. One number is 4 more than twice the second number. Their total is 25. Find the numbers.

17. Miguel Cabrera of the Detroit Tigers led Major League Baseball in runs batted in for the 2010 regular season. Alex Rodriguez of the New York Yankees, who came in second to Cabrera, had 1 fewer run batted in for the 2010 regular season. Together, these two players brought home 251 runs during the 2010 regular season. How many runs batted in did each player account for? (*Source:* Major League Baseball)

18. The highest scorer during the WNBA 2010 regular season was Cappie Pondexter of the New York Liberty. Over the season, Pondexter scored 13 more points than the second-highest scorer, Angel McCoughtry of the Atlanta Dream. Together, Pondexter and McCoughtry scored 1445 points during the 2010 regular season. How many points did each player score over the course of the season? (*Source:* Women's National Basketball Association)

19. Ann Marie Jones has been pricing Amtrak train fares for a group trip to New York. Three adults and four children must pay \$159. Two adults and three children must pay \$112. Find the price of an adult's ticket and find the price of a child's ticket.

20. Last month, Jerry Papa purchased two DVDs and five CDs at Wall-to-Wall Sound for \$65. This month, he bought four DVDs and three CDs for \$81. Find the price of each DVD and find the price of each CD.

21. Johnston and Betsy Waring have a jar containing 80 coins, all of which are either quarters or nickels. The total value of the coins is \$14.60. How many of each type of coin do they have?

22. Sarah and Keith Robinson purchased 40 stamps, a mixture of 44¢ and 20¢ stamps. Find the number of each type of stamp if they spent \$16.40.

23. Norman and Suzanne Scarpulla own 35 shares of McDonald's stock and 69 shares of The Ohio Art Company stock (makers of Etch A Sketch and other toys). On a particular day in 2011, their stock portfolio consisting of these two stocks was worth \$2814. The McDonald's stock was \$70 more per share than The Ohio Art Company stock. What was the price of each stock on that day? (*Source:* Yahoo finance)

24. Saralee Rose has investments in Google and Nintendo stock. During a particular day in 2011, Google stock was at \$540 per share, and Nintendo stock was at \$30 per share. Saralee's portfolio made up of these two stocks was worth \$20,610 at that time. If Saralee owns 16 more shares of Google stock than she owns of Nintendo stock, how many shares of each type of stock does she own?

25. Twice last month, Judy Carter rented a car from Enterprise in Fresno, California, and traveled around the Southwest on business. Enterprise rents this car for a daily fee plus an additional charge per mile driven. Judy recalls that her first trip lasted 4 days, she drove 450 miles, and the rental cost her \$240.50. On her second business trip, she drove the same model of car a distance of 200 miles in 3 days and paid \$146.00 for the rental. Find the daily fee and the mileage charge.

26. Joan Gundersen rented the same car model twice from Hertz, which rents this car model for a daily fee plus an additional charge per mile driven. Joan recalls that the car rented for 5 days and driven for 300 miles cost her \$178, while the same model car rented for 4 days and driven for 500 miles cost \$197. Find the daily fee and find the mileage charge.

27. Pratap Puri rowed 18 miles down the Delaware River in 2 hours, but the return trip took him $4\frac{1}{2}$ hours. Find the rate Pratap can row in still water and find the rate of the current.

Let x = rate Pratap can row in still water and
y = rate of the current

	d	$= r$	$\cdot\ t$
Downstream		$x + y$	
Upstream		$x - y$	

28. The Jonathan Schultz family took a canoe 10 miles down the Allegheny River in $1\frac{1}{4}$ hours. After lunch, it took them 4 hours to return. Find the rate of the current.

Let x = rate the family can row in still water and
y = rate of the current

	d	$= r$	$\cdot\ t$
Downstream		$x + y$	
Upstream		$x - y$	

29. Dave and Sandy Hartranft are frequent flyers with Delta Airlines. They often fly from Philadelphia to Chicago, a distance of 780 miles. On one particular trip, they fly into the wind, and the flight takes 2 hours. The return trip, with the wind behind them, only takes $1\frac{1}{2}$ hours. If the wind speed is the same on each trip, find the speed of the wind and find the speed of the plane in still air.

30. With a strong wind behind it, a United Airlines jet flies 2400 miles from Los Angeles to Orlando in $4\frac{3}{4}$ hours. The return trip takes 6 hours because the plane flies into the wind. If the wind speed is the same on each trip, find the speed of the plane in still air and find the wind speed to the nearest tenth of a mile per hour.

31. Kevin Briley began a 114-mile bicycle trip to build up stamina for a triathlete competition. Unfortunately, his bicycle chain broke, so he finished the trip walking. The whole trip took 6 hours. If Kevin walks at a rate of 4 miles per hour and rides at 24 miles per hour, find the amount of time he spent on the bicycle.

32. In Canada, eastbound and westbound trains travel along the same track, with sidings to pull onto to avoid accidents. Two trains are now 150 miles apart, with the westbound train traveling twice as fast as the eastbound train. A warning must be issued to pull one train onto a siding or else the trains will crash in $1\frac{1}{4}$ hours. Find the speed of the eastbound train and the speed of the westbound train.

33. Doreen Schmidt is a chemist with Gemco Pharmaceutical. She needs to prepare 12 liters of a 9% hydrochloric acid solution. Find the amount of a 4% solution and the amount of a 12% solution she should mix to get this solution.

Concentration Rate	Liters of Solution	Liters of Pure Acid
0.04	x	$0.04x$
0.12	y	?
0.09	12	?

34. Elise Everly is preparing 15 liters of a 25% saline solution. Elise has two other saline solutions, with strengths of 40% and 10%. Find the amount of 40% solution and the amount of 10% solution she should mix to get 15 liters of a 25% solution.

Concentration Rate	Liters of Solution	Liters of Pure Salt
0.40	x	$0.40x$
0.10	y	?
0.25	15	?

35. Wayne Osby blends coffee for a local coffee café. He needs to prepare 200 pounds of blended coffee beans selling for \$3.95 per pound. He intends to do this by blending together a high-quality bean costing \$4.95 per pound and a cheaper bean costing \$2.65 per pound. To the nearest pound, find how much high-quality coffee bean and how much cheaper coffee bean he should blend.

36. Macadamia nuts cost an astounding \$16.50 per pound, but research by an independent firm says that mixed nuts sell better if macadamias are included. The standard mix costs \$9.25 per pound. Find how many pounds of macadamias and how many pounds of the standard mix should be combined

to produce 40 pounds that will cost $10 per pound. Find the amounts to the nearest tenth of a pound.

37. Recall that two angles are complementary if the sum of their measures is 90°. Find the measures of two complementary angles if one angle is twice the other.

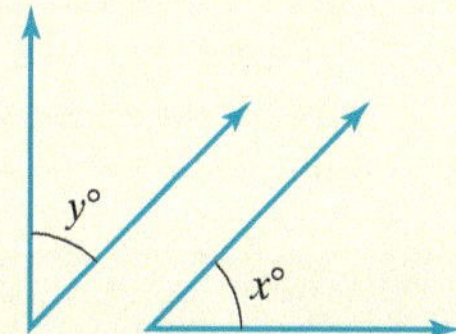

38. Recall that two angles are supplementary if the sum of their measures is 180°. Find the measures of two supplementary angles if one angle is 20° more than four times the other.

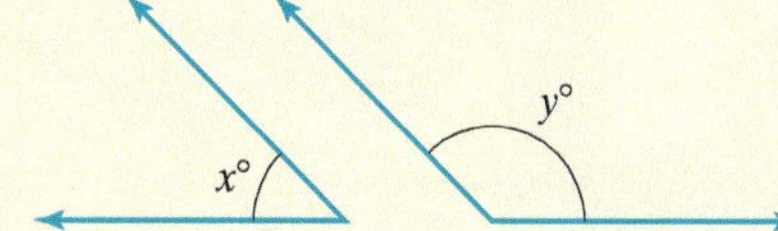

39. Find the measures of two complementary angles if one angle is 10° more than three times the other.

40. Find the measures of two supplementary angles if one angle is 18° more than twice the other.

41. Kathi and Robert Hawn had a pottery stand at the annual Skippack Craft Fair. They sold some of their pottery at the original price of $9.50 each but later decreased the price of each by $2. If they sold all 90 pieces and took in $721, find how many they sold at the original price and how many they sold at the reduced price.

42. A charity fundraiser consisted of a spaghetti supper where a total of 387 people were fed. They charged $6.80 for adults and half price for children. If they took in $2444.60, find how many adults and how many children attended the supper.

43. The Santa Fe National Historic Trail is approximately 1200 miles between Old Franklin, Missouri, and Santa Fe, New Mexico. Suppose that a group of hikers start from each town and walk the trail toward each other. They meet after a total hiking time of 240 hours. If one group travels $\frac{1}{2}$ mile per hour slower than the other group, find the rate of each group. (*Source:* National Park Service)

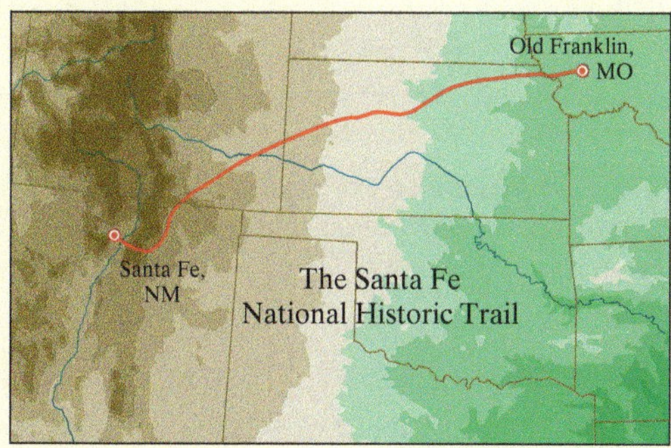

44. California 1 South is a historic highway that stretches 123 miles along the coast from Monterey to Morro Bay. Suppose that two antique cars start driving this highway, one from each town. They meet after 3 hours. Find the rate of each car if one car travels 1 mile per hour faster than the other car. (*Source: National Geographic*)

45. A 30% solution of fertilizer is to be mixed with a 60% solution of fertilizer to get 150 gallons of a 50% solution. How many gallons of the 30% solution and 60% solution should be mixed?

46. A 10% acid solution is to be mixed with a 50% acid solution to get 120 ounces of a 20% acid solution. How many ounces of the 10% solution and 50% solution should be mixed?

47. Traffic signs are regulated by the *Manual on Uniform Traffic Control Devices* (MUTCD). According to this manual, if the sign below is placed on a freeway, its perimeter must be 144 inches. Also, its length is 12 inches longer than its width. Find the dimensions of this sign.

48. According to the MUTCD (see Exercise 47), this sign must have a perimeter of 60 inches. Also, its length must be 6 inches longer than its width. Find the dimensions of this sign.

49. In the United States, the percent of adult blogging has changed within the various age ranges. For 2007 to 2009, the function $y = -4.5x + 24$ can be used to estimate the percent of adults under 30 who blogged, and the function $y = 2x + 7$ can be used to estimate the percent of adults over 30 who blogged. For both functions, x is the number of years after 2007. (*Source:* Pew Internet & American Life Project)

a. If this trend continued, estimate the year in which the percent of adults under 30 and the percent of adults over 30 who blogged was the same.

b. Use these equations to predict the percent of adults under 30 who blog and the percent of adults over 30 who blog for the current year.

50. The rate of fatalities per 100 million vehicle-miles has been decreasing for both automobiles and light trucks (pickups, sport-utility vehicles, and minivans). For the years 2001 through 2009, the function $y = -0.06x + 1.7$ can be used to estimate the rate of fatalities per 100 million vehicle-miles for automobiles during this period, and the function $y = -0.08x + 2.1$ can be used to estimate the rate of fatalities per 100 million vehicle-miles for light trucks during this period. For both functions, x is the number of years since 2000. (*Source:* Bureau of Statistics, U.S. Department of Transportation)

a. If this trend continues, predict the year in which the fatality rate for automobiles equals the fatality rate for light trucks.

b. Use these equations to predict the fatality rate per million vehicle-miles for automobiles and light trucks for the current year.

51. The annual U.S. per capita consumption of cheddar cheese has remained about the same since the millennium, while the consumption of mozzarella cheese has increased. For the years 2000–2010, the function $y = 0.06x + 9.7$ approximates the annual U.S. per capita consumption of cheddar cheese in pounds, and the function $y = 0.21x + 9.3$ approximates the annual U.S. per capita consumption of mozzarella cheese in pounds. For both functions, x is the number of years after 2000.

a. Explain how the given function verifies that the consumption of cheddar cheese has remained the same, while the given function verifies that the consumption of mozzarella cheese has increased.

b. Based on this information, determine the year in which the pounds of cheddar cheese consumed equaled the pounds of mozzarella cheese consumed.
(*Source:* Based on data from the U.S. Department of Agriculture)

52. Two of the major job categories defined by the U.S. Department of Labor are manufacturing jobs and jobs in the service sector. Jobs in the manufacturing sector have decreased nearly every year since the 1960s. During the same time period, service sector jobs have been steadily increasing. For the years from 1988 through 2009, the function $y = -0.225x + 16.1$ approximates the percent of jobs in the U.S. economy that are manufacturing jobs, while the function $y = 0.45x + 21.7$ approximates the percent of jobs that are service sector jobs. (*Source:* Based on data from the U.S. Department of Labor)

a. Explain how the decrease in manufacturing jobs can be verified by the given function, and the increase of service sector jobs can be verified by the given function.

b. Based on this information, determine the year when the percent of manufacturing jobs and the percent of service sector jobs were the same.

△ **53.** In the figure, line l and line m are parallel lines cut by transversal t. Find the values of x and y.

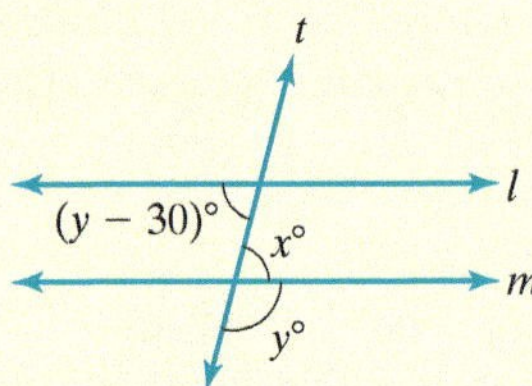

△ **54.** Find the values of x and y in the following isosceles triangle.

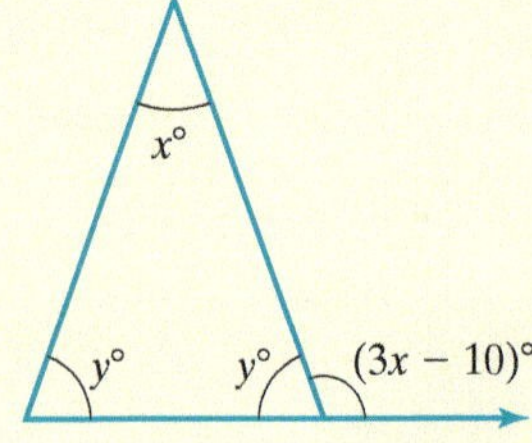

Given the cost function $C(x)$ and the revenue function $R(x)$, find the number of units x that must be sold to break even. See Example 7.

55. $C(x) = 30x + 10{,}000;\ R(x) = 46x$

56. $C(x) = 12x + 15{,}000;\ R(x) = 32x$

57. $C(x) = 1.2x + 1500;\ R(x) = 1.7x$

58. $C(x) = 0.8x + 900;\ R(x) = 2x$

59. $C(x) = 75x + 160{,}000;\ R(x) = 200x$

60. $C(x) = 105x + 70{,}000;\ R(x) = 245x$

61. The planning department of Abstract Office Supplies has been asked to determine whether the company should introduce a new computer desk next year. The department estimates that \$6000 of new manufacturing equipment will need to be purchased and that the cost of constructing each desk will be \$200. The department also estimates that the revenue from each desk will be \$450.

 a. Determine the revenue function $R(x)$ from the sale of x desks.
 b. Determine the cost function $C(x)$ for manufacturing x desks.
 c. Find the break-even point.

62. Baskets, Inc., is planning to introduce a new woven basket. The company estimates that \$500 worth of new equipment will be needed to manufacture this new type of basket and that it will cost \$15 per basket to manufacture. The company also estimates that the revenue from each basket will be \$31.

 a. Determine the revenue function $R(x)$ from the sale of x baskets.
 b. Determine the cost function $C(x)$ for manufacturing x baskets.
 c. Find the break-even point. Round up to the nearest whole basket.

Solve. See Example 8.

63. Rabbits in a lab are to be kept on a strict daily diet that includes 30 grams of protein, 16 grams of fat, and 24 grams of carbohydrates. The scientist has only three food mixes available with the following grams of nutrients per unit.

	Protein	Fat	Carbohydrate
Mix A	4	6	3
Mix B	6	1	2
Mix C	4	1	12

Find how many units of each mix are needed daily to meet each rabbit's dietary need.

64. Gerry Gundersen mixes different solutions with concentrations of 25%, 40%, and 50% to get 200 liters of a 32% solution. If he uses twice as much of the 25% solution as of the 40% solution, find how many liters of each kind he uses.

65. The perimeter of a quadrilateral (four-sided polygon) is 29 inches. The longest side is twice as long as the shortest side. The other two sides are equally long and are 2 inches longer than the shortest side. Find the lengths of all four sides.

66. The measure of the largest angle of a triangle is 90° more than the measure of the smallest angle, and the measure of the remaining angle is 30° more than the measure of the smallest angle. Find the measure of each angle.

67. The sum of three numbers is 40. The first number is five more than the second number. It is also twice the third. Find the numbers.

68. The sum of the digits of a three-digit number is 15. The tens-place digit is twice the hundreds-place digit, and the ones-place digit is 1 less than the hundreds-place digit. Find the three-digit number.

69. During the 2010–2011 regular NBA season, the top-scoring player was Kevin Durant of the Oklahoma City Thunder. Durant scored a total of 2161 points during the regular season. The number of free throws (each worth one point) he made was 14 more than four times the number of three-point field goals he made. The number of two-point field goals that Durant made was 28 less than the number of free throws he made. How many free throws, two-point field goals, and three-point field goals did Kevin Durant make during the 2010–2011 NBA season? (*Source:* National Basketball Association)

70. For 2010, the WNBA's top scorer was Diana Taurasi of the Phoenix Mercury. She scored a total of 745 points during the regular season. The number of two-point field goals Taurasi made was 36 fewer than two times the number of three-point field goals she made. The number of free throws (each worth one point) she made was 61 more than the number of two-point field goals she made. Find how many free throws, two-point field goals, and three-point field goals Diana Taurasi made during the 2010 regular season. (*Source:* Women's National Basketball Association)

71. Find the values of x, y, and z in the following triangle.

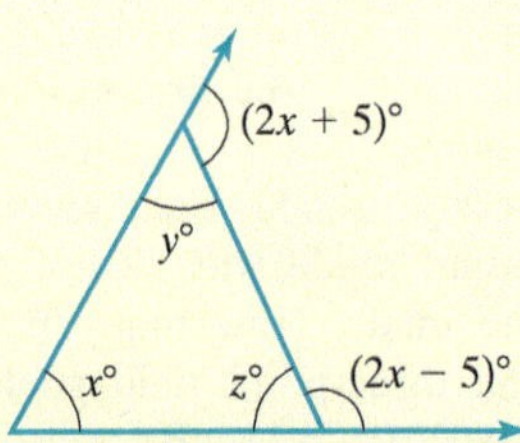

72. The sum of the measures of the angles of a quadrilateral is 360°. Find the values of x, y, and z in the following quadrilateral.

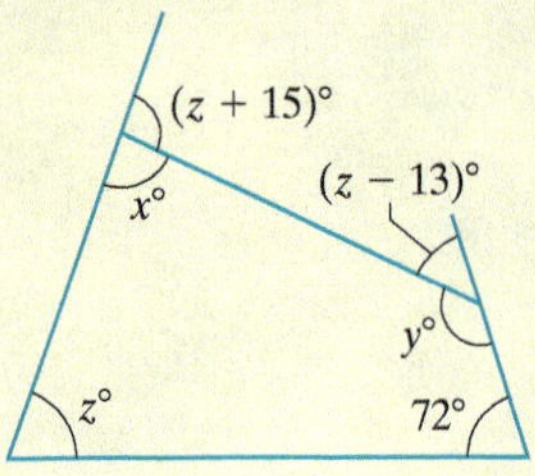

REVIEW AND PREVIEW

Solve each linear inequality. Write your solution in interval notation. See Section 9.6.

73. $-3x < -9$

74. $2x - 7 \leq 5x + 11$

75. $4(2x - 1) \geq 0$

76. $\frac{2}{3}x < \frac{1}{3}$

CONCEPT EXTENSIONS

Solve. See the Concept Check in this section.

77. Suppose you mix an amount of candy costing \$0.49 a pound with candy costing \$0.65 a pound. Which of the following costs per pound could result?

a. \$0.58 **b.** \$0.72 **c.** \$0.29

78. Suppose you mix a 50% acid solution with pure acid (100%). Which of the following acid strengths are possible for the resulting acid mixture?

a. 25% **b.** 150% **c.** 62% **d.** 90%

79. Dale and Sharon Mahnke have decided to fence off a garden plot behind their house, using their house as the "fence" along one side of the garden. The length (which runs parallel to the house) is 3 feet less than twice the width. Find the dimensions if 33 feet of fencing is used along the three sides requiring it.

80. Judy McElroy plans to erect 152 feet of fencing around her rectangular horse pasture. A river bank serves as one side length of the rectangle. If each width is 4 feet longer than half the length, find the dimensions.

81. Find the values of $a, b,$ and c such that the equation $y = ax^2 + bx + c$ has ordered pair solutions $(1, 6)$, $(-1, -2)$, and $(0, -1)$. To do so, substitute each ordered pair solution into the equation. Each time, the result is an equation in three unknowns: a, b, and c. Then solve the resulting system of three linear equations in three unknowns, a, b, and c.

82. Find the values of $a, b,$ and c such that the equation $y = ax^2 + bx + c$ has ordered pair solutions $(1, 2)$, $(2, 3)$, and $(-1, 6)$. (*Hint:* See Exercise 81.)

83. Data (x, y) for the total number (in thousands) of college-bound students who took the ACT assessment in the year x are approximately (3, 927), (11, 1179), and (19, 1495), where $x = 3$ represents 1993 and $x = 11$ represents 2001. Find the values a, b, and c such that the equation $y = ax^2 + bx + c$ models these data. According to your model, how many students took the ACT in 2015? (*Source:* ACT, Inc.)

84. Monthly normal rainfall data (x, y) for Portland, Oregon, are $(4, 2.47)$, $(7, 0.58)$, $(8, 1.07)$, where x represents time in months (with $x = 1$ representing January) and y represents rainfall in inches. Find the values of a, b, and c rounded to 2 decimal places such that the equation $y = ax^2 + bx + c$ models this data. According to your model, how much rain should Portland expect during September? (*Source:* National Climatic Data Center)

The function $f(x) = -8.6x + 275$ represents the U.S. average number of monthly calls (sent or received) per wireless subscriber and the function $f(x) = 204.9x - 1217$ represents the average number of text messages (sent or received) per wireless subscriber. For both functions, x is the number of years since 2000, and these functions are good for the years 2006–2010.

85. Solve the system formed by these functions. Round each coordinate to the nearest whole number.

86. Use your answer to Exercise 85 to estimate the year in which the monthly calls and text messages are/were the same.

Chapter 11 Vocabulary Check

Fill in each blank with one of the words or phrases listed below.

system of linear equations	solution	consistent	independent
dependent	inconsistent	substitution	addition

1. In a system of linear equations in two variables, if the graphs of the equations are the same, the equations are ____________ equations.
2. Two or more linear equations are called a(n) ____________.
3. A system of equations that has at least one solution is called a(n) ____________ system.
4. A(n) ____________ of a system of two equations in two variables is an ordered pair of numbers that is a solution of both equations in the system.
5. Two algebraic methods for solving systems of equations are ____________ and ____________.
6. A system of equations that has no solution is called a(n) ____________ system.
7. In a system of linear equations in two variables, if the graphs of the equations are different, the equations are ____________ equations.

Helpful Hint

Are you preparing for your test? Don't forget to take the Chapter 11 Test on page 856. Then check your answers at the back of the text and use the Chapter Test Prep Videos to see the fully worked-out solutions to any of the exercises you want to review.

11 Chapter Highlights

Definitions and Concepts	Examples
Section 11.1 Solving Systems of Linear Equations by Graphing	
A **solution** of a system of two equations in two variables is an ordered pair of numbers that is a solution of both equations in the system.	Determine whether $(-1, 3)$ is a solution of the system: $\begin{cases} 2x - y = -5 \\ x = 3y - 10 \end{cases}$ Replace x with -1 and y with 3 in both equations. $2x - y = -5$; $2(-1) - 3 \stackrel{?}{=} -5$; $-5 = -5$ True. $x = 3y - 10$; $-1 \stackrel{?}{=} 3 \cdot 3 - 10$; $-1 = -1$ True. $(-1, 3)$ is a solution of the system.
Graphically, a solution of a system is a point common to the graphs of both equations.	Solve by graphing. $\begin{cases} 3x - 2y = -3 \\ x + y = 4 \end{cases}$ 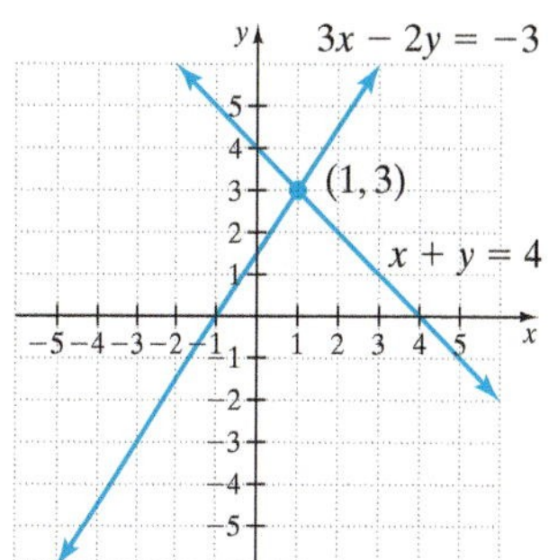

(continued)

Definitions and Concepts	Examples
Section 11.1 Solving Systems of Linear Equations by Graphing (*continued*)	
A system of equations with at least one solution is a **consistent system.** A system that has no solution is an **inconsistent system.** If the graphs of two linear equations are identical, the equations are **dependent.** If their graphs are different, the equations are **independent.**	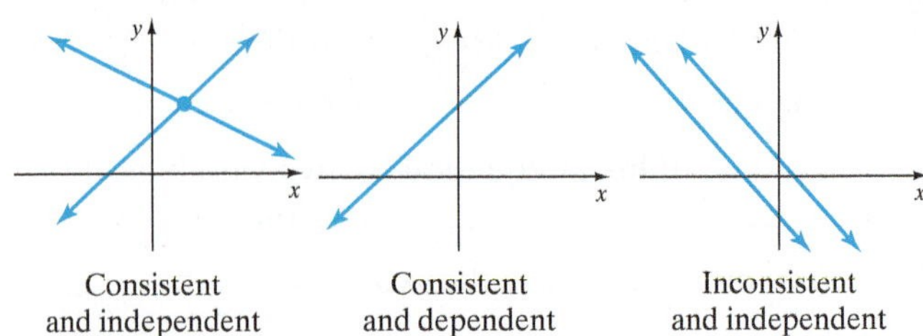
Section 11.2 Solving Systems of Linear Equations by Substitution	
To solve a system of linear equations by the substitution method: **Step 1:** Solve one equation for a variable. **Step 2:** Substitute the expression for the variable into the other equation. **Step 3:** Solve the equation from Step 2 to find the value of one variable. **Step 4:** Substitute the value from Step 3 in either original equation to find the value of the other variable. **Step 5:** Check the solution in both equations.	Solve by substitution. $\begin{cases} 3x + 2y = 1 \\ x = y - 3 \end{cases}$ Substitute $y - 3$ for x in the first equation. $3x + 2y = 1$ $3(y - 3) + 2y = 1$ $3y - 9 + 2y = 1$ $5y = 10$ $y = 2$ Divide by 5. To find x, substitute 2 for y in $x = y - 3$ so that $x = 2 - 3$ or -1. The solution $(-1, 2)$ checks.
Section 11.3 Solving Systems of Linear Equations by Addition	
To solve a system of linear equations by the addition method: **Step 1:** Rewrite each equation in standard form $Ax + By = C$. **Step 2:** Multiply one or both equations by a nonzero number so that the coefficients of a variable are opposites. **Step 3:** Add the equations. **Step 4:** Find the value of one variable by solving the resulting equation. **Step 5:** Substitute the value from Step 4 into either original equation to find the value of the other variable. **Step 6:** Check the solution in both equations. If solving a system of linear equations by substitution or addition yields a true statement such as $-2 = -2$, then the graphs of the equations in the system are identical and there is an infinite number of solutions of the system.	Solve by addition. $\begin{cases} x - 2y = 8 \\ 3x + y = -4 \end{cases}$ Multiply both sides of the first equation by -3. $\begin{cases} -3x + 6y = -24 \\ 3x + y = -4 \end{cases}$ $7y = -28$ Add. $y = -4$ Divide by 7. To find x, let $y = -4$ in an original equation. $x - 2(-4) = 8$ First equation $x + 8 = 8$ $x = 0$ The solution $(0, -4)$ checks. Solve: $\begin{cases} 2x - 6y = -2 \\ x = 3y - 1 \end{cases}$ Substitute $3y - 1$ for x in the first equation. $2(3y - 1) - 6y = -2$ $6y - 2 - 6y = -2$ $-2 = -2$ True The system has an infinite number of solutions. In set notation, we write $\{(x, y) \mid 2x - 6y = -2\}$ or $\{(x, y) \mid x = 3y - 1\}$.

Definitions and Concepts	Examples
Section 11.4 Solving Systems of Linear Equations in Three Variables	

A **solution** of an equation in three variables x, y, and z is an **ordered triple** (x, y, z) that makes the equation a true statement.

Verify that $(-2, 1, 3)$ is a solution of $2x + 3y - 2z = -7$. Replace x with -2, y with 1, and z with 3.

$$\begin{aligned} 2(-2) + 3(1) - 2(3) &\stackrel{?}{=} -7 \\ -4 + 3 - 6 &\stackrel{?}{=} -7 \\ -7 &= -7 \quad \text{True} \end{aligned}$$

$(-2, 1, 3)$ is a solution.

Solving a System of Three Linear Equations by the Elimination Method

Step 1: Write each equation in standard form, $Ax + By + Cz = D$.

Step 2: Choose a pair of equations and use them to eliminate a variable.

Step 3: Choose any other pair of equations and eliminate the same variable.

Step 4: Solve the system of two equations in two variables from Steps 2 and 3.

Step 5: Solve for the third variable by substituting the values of the variables from Step 4 into any of the original equations.

Step 6: Check the solution in all three original equations.

Solve:

$$\begin{cases} 2x + y - z = 0 & (1) \\ x - y - 2z = -6 & (2) \\ -3x - 2y + 3z = -22 & (3) \end{cases}$$

1. Each equation is written in standard form.

2.
$$\begin{aligned} 2x + y - z &= 0 && (1) \\ x - y - 2z &= -6 && (2) \\ \hline 3x - 3z &= -6 && (4) \quad \text{Add.} \end{aligned}$$

3. Eliminate y from equations (1) and (3) also.

$$\begin{aligned} 4x + 2y - 2z &= 0 && \text{Multiply equation} \\ -3x - 2y + 3z &= -22 && (3) \quad \text{(1) by 2.} \\ \hline x + z &= -22 && (5) \quad \text{Add.} \end{aligned}$$

4. Solve.

$$\begin{cases} 3x - 3z = -6 & (4) \\ x + z = -22 & (5) \end{cases}$$

$$\begin{aligned} x - z &= -2 && \text{Divide equation (4) by 3.} \\ x + z &= -22 \\ \hline 2x &= -24 && \text{Add.} \\ x &= -12 \end{aligned}$$

To find z, use equation (5).

$$\begin{aligned} x + z &= -22 \\ -12 + z &= -22 \\ z &= -10 \end{aligned}$$

5. To find y, use equation (1).

$$\begin{aligned} 2x + y - z &= 0 \\ 2(-12) + y - (-10) &= 0 \\ -24 + y + 10 &= 0 \\ y &= 14 \end{aligned}$$

6. The solution $(-12, 14, -10)$ checks.

Definitions and Concepts	Examples

Section 11.5 Systems of Linear Equations and Problem Solving

Problem-solving steps

1. UNDERSTAND. Read and reread the problem.

Two angles are supplementary if their sum is 180°.

The larger of two supplementary angles is three times the smaller, decreased by twelve. Find the measure of each angle. Let

$x =$ measure of smaller angle
$y =$ measure of larger angle

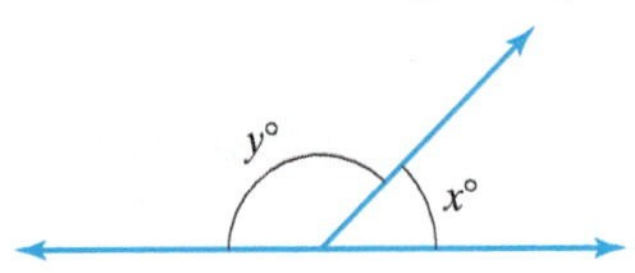

2. TRANSLATE.

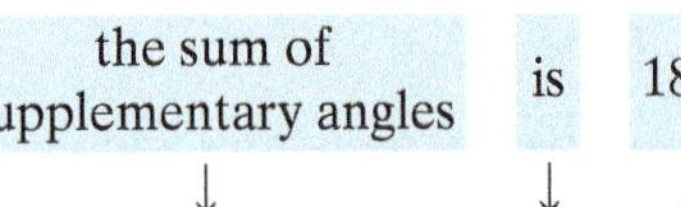

In words:	the sum of supplementary angles	is	180°
	↓	↓	↓
Translate:	$x + y$	$=$	180

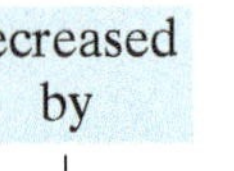

In words:	larger angle	is	3 times smaller	decreased by	12
	↓	↓	↓	↓	↓
Translate:	y	$=$	$3x$	$-$	12

3. SOLVE.

Solve the system:

$$\begin{cases} x + y = 180 \\ y = 3x - 12 \end{cases}$$

Use the substitution method and replace y with $3x - 12$ in the first equation.

$$\begin{aligned} x + y &= 180 \\ x + (3x - 12) &= 180 \\ 4x &= 192 \\ x &= 48 \end{aligned}$$

Since $y = 3x - 12$, then $y = 3 \cdot 48 - 12$ or 132.

4. INTERPRET.

The solution checks. The smaller angle measures 48° and the larger angle measures 132°.

Chapter 11 Review

***(11.1)** Determine whether any of the following ordered pairs satisfy the system of linear equations.*

1. $\begin{cases} 2x - 3y = 12 \\ 3x + 4y = 1 \end{cases}$

   a. $(12, 4)$ **b.** $(3, -2)$ **c.** $(-3, 6)$

2. $\begin{cases} 4x + y = 0 \\ -8x - 5y = 9 \end{cases}$

   a. $\left(\frac{3}{4}, -3\right)$ **b.** $(-2, 8)$ **c.** $\left(\frac{1}{2}, -2\right)$

3. $\begin{cases} 5x - 6y = 18 \\ 2y - x = -4 \end{cases}$

   a. $(-6, -8)$ **b.** $\left(3, \frac{5}{2}\right)$ **c.** $\left(3, -\frac{1}{2}\right)$

4. $\begin{cases} 2x + 3y = 1 \\ 3y - x = 4 \end{cases}$

   a. $(2, 2)$ **b.** $(-1, 1)$ **c.** $(2, -1)$

Solve each system of equations by graphing.

5. $\begin{cases} x + y = 5 \\ x - y = 1 \end{cases}$

6. $\begin{cases} x + y = 3 \\ x - y = -1 \end{cases}$

7. $\begin{cases} x = 5 \\ y = -1 \end{cases}$

8. $\begin{cases} x = -3 \\ y = 2 \end{cases}$

9. $\begin{cases} 2x + y = 5 \\ x = -3y \end{cases}$

10. $\begin{cases} 3x + y = -2 \\ y = -5x \end{cases}$

11. $\begin{cases} y = 3x \\ -6x + 2y = 6 \end{cases}$

12. $\begin{cases} x - 2y = 2 \\ -2x + 4y = -4 \end{cases}$

***(11.2)** Solve each system of equations by the substitution method.*

13. $\begin{cases} y = 2x + 6 \\ 3x - 2y = -11 \end{cases}$

14. $\begin{cases} y = 3x - 7 \\ 2x - 3y = 7 \end{cases}$

15. $\begin{cases} x + 3y = -3 \\ 2x + y = 4 \end{cases}$

16. $\begin{cases} 3x + y = 11 \\ x + 2y = 12 \end{cases}$

17. $\begin{cases} 4y = 2x + 6 \\ x - 2y = -3 \end{cases}$

18. $\begin{cases} 9x = 6y + 3 \\ 6x - 4y = 2 \end{cases}$

19. $\begin{cases} x + y = 6 \\ y = -x - 4 \end{cases}$

20. $\begin{cases} -3x + y = 6 \\ y = 3x + 2 \end{cases}$

***(11.3)** Solve each system of equations by the addition method.*

21. $\begin{cases} 2x + 3y = -6 \\ x - 3y = -12 \end{cases}$

22. $\begin{cases} 4x + y = 15 \\ -4x + 3y = -19 \end{cases}$

23. $\begin{cases} 2x - 3y = -15 \\ x + 4y = 31 \end{cases}$

24. $\begin{cases} x - 5y = -22 \\ 4x + 3y = 4 \end{cases}$

25. $\begin{cases} 2x - 6y = -1 \\ -x + 3y = \frac{1}{2} \end{cases}$

26. $\begin{cases} 0.6x - 0.3y = -1.5 \\ 0.04x - 0.02y = -0.1 \end{cases}$

27. $\begin{cases} \frac{3}{4}x + \frac{2}{3}y = 2 \\ x + \frac{y}{3} = 6 \end{cases}$

28. $\begin{cases} 10x + 2y = 0 \\ 3x + 5y = 33 \end{cases}$

***(11.4)** Solve each system of equations in three variables.*

29. $\begin{cases} x + z = 4 \\ 2x - y = 4 \\ x + y - z = 0 \end{cases}$

30. $\begin{cases} 2x + 5y = 4 \\ x - 5y + z = -1 \\ 4x - z = 11 \end{cases}$

31. $\begin{cases} 4y + 2z = 5 \\ 2x + 8y = 5 \\ 6x + 4z = 1 \end{cases}$

32. $\begin{cases} 5x + 7y = 9 \\ 14y - z = 28 \\ 4x + 2z = -4 \end{cases}$

33. $\begin{cases} 3x - 2y + 2z = 5 \\ -x + 6y + z = 4 \\ 3x + 14y + 7z = 20 \end{cases}$

34. $\begin{cases} x + 2y + 3z = 11 \\ y + 2z = 3 \\ 2x + 2z = 10 \end{cases}$

35. $\begin{cases} 7x - 3y + 2z = 0 \\ 4x - 4y - z = 2 \\ 5x + 2y + 3z = 1 \end{cases}$

36. $\begin{cases} x - 3y - 5z = -5 \\ 4x - 2y + 3z = 13 \\ 5x + 3y + 4z = 22 \end{cases}$

***(11.5)** Solve each problem by writing and solving a system of linear equations.*

37. The sum of two numbers is 16. Three times the larger number decreased by the smaller number is 72. Find the two numbers.

38. The Forrest Theater can seat a total of 360 people. They take in $15,150 when every seat is sold. If orchestra section tickets cost $45 and balcony tickets cost $35, find the number of seats in the orchestra section and the number of seats in the balcony.

39. A riverboat can head 340 miles upriver in 19 hours, but the return trip takes only 14 hours. Find the current of the river and find the speed of the riverboat in still water to the nearest tenth of a mile.

	d =	r ·	t
Upriver	340	$x - y$	19
Downriver	340	$x + y$	14

40. Find the amount of a 6% acid solution and the amount of a 14% acid solution Pat Mayfield should combine to prepare 50 cc (cubic centimeters) of a 12% solution.

41. A deli charges $3.80 for a breakfast of three eggs and four strips of bacon. The charge is $2.75 for two eggs and three strips of bacon. Find the cost of each egg and the cost of each strip of bacon.

42. An exercise enthusiast alternates between jogging and walking. He traveled 15 miles during the past 3 hours. He jogs at a rate of 7.5 miles per hour and walks at a rate of 4 miles per hour. Find how much time, to the nearest hundredth of an hour, he actually spent jogging and how much time he spent walking.

43. Chris Kringler has $2.77 in her coin jar—all in pennies, nickels, and dimes. If she has 53 coins in all and four more nickels than dimes, find how many of each type of coin she has.

44. An employee at See's Candy Store needs a special mixture of candy. She has creme-filled chocolates that sell for $3.00 per pound, chocolate-covered nuts that sell for $2.70 per pound, and chocolate-covered raisins that sell for $2.25 per pound. She wants to have twice as many raisins as nuts in the mixture. Find how many pounds of each she should use to make 45 pounds worth $2.80 per pound.

45. The perimeter of an isosceles (two sides equal) triangle is 73 centimeters. If the unequal side is 7 centimeters longer than the two equal sides, find the lengths of the three sides.

46. The sum of three numbers is 295. One number is five more than a second and twice the third. Find the numbers.

MIXED REVIEW

Solve each system of equations by graphing.

47. $\begin{cases} x - 2y = 1 \\ 2x + 3y = -12 \end{cases}$

48. $\begin{cases} 3x - y = -4 \\ 6x - 2y = -8 \end{cases}$

Solve each system of equations.

49. $\begin{cases} x + 4y = 11 \\ 5x - 9y = -3 \end{cases}$

50. $\begin{cases} x + 9y = 16 \\ 3x - 8y = 13 \end{cases}$

51. $\begin{cases} y = -2x \\ 4x + 7y = -15 \end{cases}$

52. $\begin{cases} 3y = 2x + 15 \\ -2x + 3y = 21 \end{cases}$

53. $\begin{cases} 3x - y = 4 \\ 4y = 12x - 16 \end{cases}$

54. $\begin{cases} x + y = 19 \\ x - y = -3 \end{cases}$

55. $\begin{cases} x - 3y = -11 \\ 4x + 5y = -10 \end{cases}$

56. $\begin{cases} -x - 15y = 44 \\ 2x + 3y = 20 \end{cases}$

57. $\begin{cases} x - 3y + 2z = 0 \\ 9y - 9z = 22 \\ 5x + 3z = 10 \end{cases}$

58. $\begin{cases} x - 4y = 4 \\ \frac{1}{8}x - \frac{1}{2}y = 3 \end{cases}$

Solve each problem by writing and solving a system of linear equations.

59. The sum of two numbers is 12. Three times the smaller number increased by the larger number is 20. Find the numbers.

60. The difference of two numbers is −18. Twice the smaller decreased by the larger is −23. Find the two numbers.

61. Emma Hodges has a jar containing 65 coins, all of which are either nickels or dimes. The total value of the coins is \$5.30. How many of each type does she have?

62. Sarah and Owen Hebert purchased 26 stamps, a mixture of 17¢ and 20¢ stamps. Find the number of each type of stamp if they spent \$4.93.

63. The perimeter of a triangle is 126 units. The length of one side is twice the length of the shortest side. The length of the third side is fourteen more than the length of the shortest side. Find the lengths of the sides of the triangle.

Chapter 11 Test

Answer each question true or false.

1. A system of two linear equations in two variables can have exactly two solutions.

2. Although (1, 4) is not a solution of $x + 2y = 6$, it can still be a solution of the system $\begin{cases} x + 2y = 6 \\ x + y = 5 \end{cases}$.

3. If the two equations in a system of linear equations are added and the result is $3 = 0$, the system has no solution.

4. If the two equations in a system of linear equations are added and the result is $3x = 0$, the system has no solution.

Is the ordered pair a solution of the given linear system?

5. $\begin{cases} 2x - 3y = 5 \\ 6x + y = 1 \end{cases}$; (1, −1)

6. $\begin{cases} 4x - 3y = 24 \\ 4x + 5y = -8 \end{cases}$; (3, −4)

7. Use graphing to find the solutions of the system $\begin{cases} y - x = 6 \\ y + 2x = -6 \end{cases}$

8. Use the substitution method to solve the system $\begin{cases} 3x - 2y = -14 \\ x + 3y = -1 \end{cases}$

9. Use the substitution method to solve the system $\begin{cases} \frac{1}{2}x + 2y = -\frac{15}{4} \\ 4x = -y \end{cases}$

10. Use the addition method to solve the system $\begin{cases} 3x + 5y = 2 \\ 2x - 3y = 14 \end{cases}$

11. Use the addition method to solve the system $\begin{cases} 4x - 6y = 7 \\ -2x + 3y = 0 \end{cases}$

Solve each system using the substitution method or the addition method.

12. $\begin{cases} 3x + y = 7 \\ 4x + 3y = 1 \end{cases}$

13. $\begin{cases} 3(2x + y) = 4x + 20 \\ x - 2y = 3 \end{cases}$

14. $\begin{cases} \frac{x - 3}{2} = \frac{2 - y}{4} \\ \frac{7 - 2x}{3} = \frac{y}{2} \end{cases}$

Solve each problem by writing and using a system of linear equations.

15. Two numbers have a sum of 124 and a difference of 32. Find the numbers.

16. Find the amount of a 12% saline solution a lab assistant should add to 80 cc (cubic centimeters) of a 22% saline solution to have a 16% solution.

17. Texas and Missouri are the states with the most farms. Texas has 140 thousand more farms than Missouri and the total number of farms for these two states is 356 thousand. Find the number of farms for each state.

18. Two hikers start at opposite ends of the St. Tammany Trail and walk toward each other. The trail is 36 miles long and they meet in 4 hours. If one hiker is twice as fast as the other, find both hiking speeds.

19. $\begin{cases} 2x - 3y = 4 \\ 3y + 2z = 2 \\ x - z = -5 \end{cases}$

20. $\begin{cases} 3x - 2y - z = -1 \\ 2x - 2y = 4 \\ 2x - 2z = -12 \end{cases}$

21. The measure of the largest angle of a triangle is three less than 5 times the measure of the smallest angle. The measure of the remaining angle is 1 less than twice the measure of the smallest angle. Find the measure of each angle.

Chapters 1–11 Cumulative Review

1. Insert $<$, $>$, or $=$ between the pairs of numbers to form true statements.

   a. $-5 \quad -6$ **b.** $3.195 \quad 3.2$ **c.** $\frac{1}{4} \quad \frac{1}{3}$

2. Evaluate.

   a. 5^2 **b.** 2^5

3. Name the property or properties illustrated by each true statement.

   a. $3(x+y) = 3 \cdot x + 3 \cdot y$

   b. $(x+7)+9 = x+(7+9)$

   c. $(b+0)+3 = b+3$

   d. $2 \cdot (z \cdot 5) = 2 \cdot (5 \cdot z)$

   e. $-2 \cdot \left(-\frac{1}{2}\right) = 1$

   f. $-2+2=0$

   g. $-6 \cdot (y \cdot 2) = (-6 \cdot 2) \cdot y$

4. Evaluate $y^2 - 3x$ for $x = 8$ and $y = 5$.

5. Simplify: $2y - 6 + 4y + 8$

6. Simplify: $7 - 12 + (-5) - 2 + (-2)$

7. Solve: $3(3x - 5) = 10x$

8. Evaluate $2y^2 - x^2$ for $x = -7$ and $y = -3$.

9. Solve: $\frac{z}{-4} = 11 - 5$

10. Simplify: $6(2 + 5x) - 18$

11. Solve: $\frac{x}{2} - 1 = \frac{2}{3}x - 3$

12. Solve: $7(x - 2) - 6(x + 1) = 20$

13. Twice the sum of a number and 4 is the same as four times the number decreased by 12. Find the number.

14. Solve: $5(y - 5) = 5y + 10$

15. Solve $y = mx + b$ for x.

16. Five times the sum of a number and -1 is the same as 6 times the number. Find the number.

17. Solve $-2x \le -4$. Write the solution set in interval notation.

18. Solve $P = a + b + c$ for b.

19. Graph $x = -2y$ by plotting intercepts.

20. Solve $3x + 7 \ge x - 9$. Write the solution set in interval notation.

21. Find the slope of the line through $(-1, 5)$ and $(2, -3)$.

22. Complete the table of values for $x - 3y = 3$.

x	y
	-1
3	
	2

23. Find the slope and y-intercept of the line whose equation is $y = \frac{3}{4}x + 6$.

24. Find the slope of a line parallel to the line passing through $(-1, 3)$ and $(2, -8)$.

25. Find the slope and the y-intercept of the line whose equation is $3x - 4y = 4$.

26. Find the slope and y-intercept of the line whose equation is $y = 7x$.

27. Find an equation of the line with slope –2 that passes through $(-1, 5)$. Write the equation in slope-intercept form, $y = mx + b$, and in standard form, $Ax + By = C$.

28. Determine whether the lines are parallel, perpendicular, or neither.

$$y = 4x - 5$$
$$-4x + y = 7$$

29. Find an equation of the vertical line through $(-1, 5)$.

30. Write an equation of the line with slope -5 through $(-2, 3)$.

31. Find the domain and the range of the relation $\{(0, 2), (3, 3), (-1, 0), (3, -2)\}$.

32. If $f(x) = 5x^2 - 6$, find $f(0)$ and $f(-2)$.

33. Determine whether each relation is also a function.

a. $\{(-1, 1), (2, 3), (7, 3), (8, 6)\}$

b. $\{(0, -2), (1, 5), (0, 3), (7, 7)\}$

34. Determine whether each graph is also the graph of a function.

a.

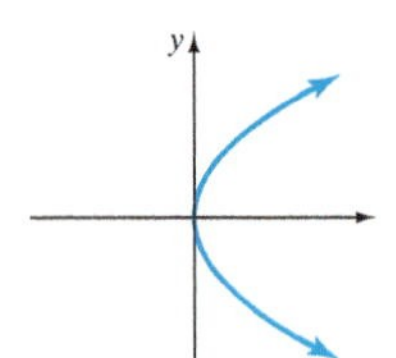

b.

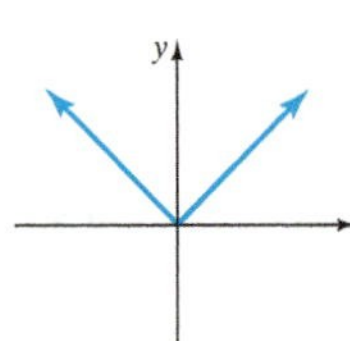

c.

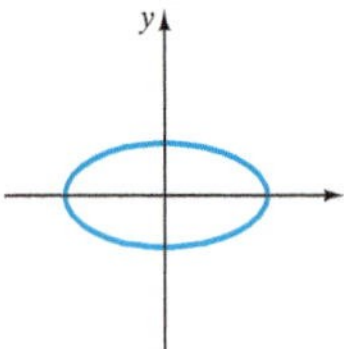

35. Without graphing, determine the number of solutions of the system. $\begin{cases} 3x - y = 4 \\ x + 2y = 8 \end{cases}$

36. Determine whether each ordered pair is a solution of the given system.

$$\begin{cases} 2x - y = 6 \\ 3x + 2y = -5 \end{cases}$$

a. $(1, -4)$ **b.** $(0, 6)$ **c.** $(3, 0)$

Solve each system.

37. $\begin{cases} x + 2y = 7 \\ 2x + 2y = 13 \end{cases}$

38. $\begin{cases} 3x - 4y = 10 \\ y = 2x \end{cases}$

39. $\begin{cases} x + y = 7 \\ x - y = 5 \end{cases}$

40. $\begin{cases} x = 5y - 3 \\ x = 8y + 4 \end{cases}$

41. Solve the system.

$$\begin{cases} 3x - y + z = -15 \\ x + 2y - z = 1 \\ 2x + 3y - 2z = 0 \end{cases}$$

42. Solve the system.

$$\begin{cases} x - 2y + z = 0 \\ 3x - y - 2z = -15 \\ 2x - 3y + 3z = 7 \end{cases}$$

43. A first number is 4 less than a second number. Four times the first number is 6 more than twice the second. Find the numbers.

44. Find two numbers whose sum is 25 and whose difference is 9.

12

Exponents and Polynomials

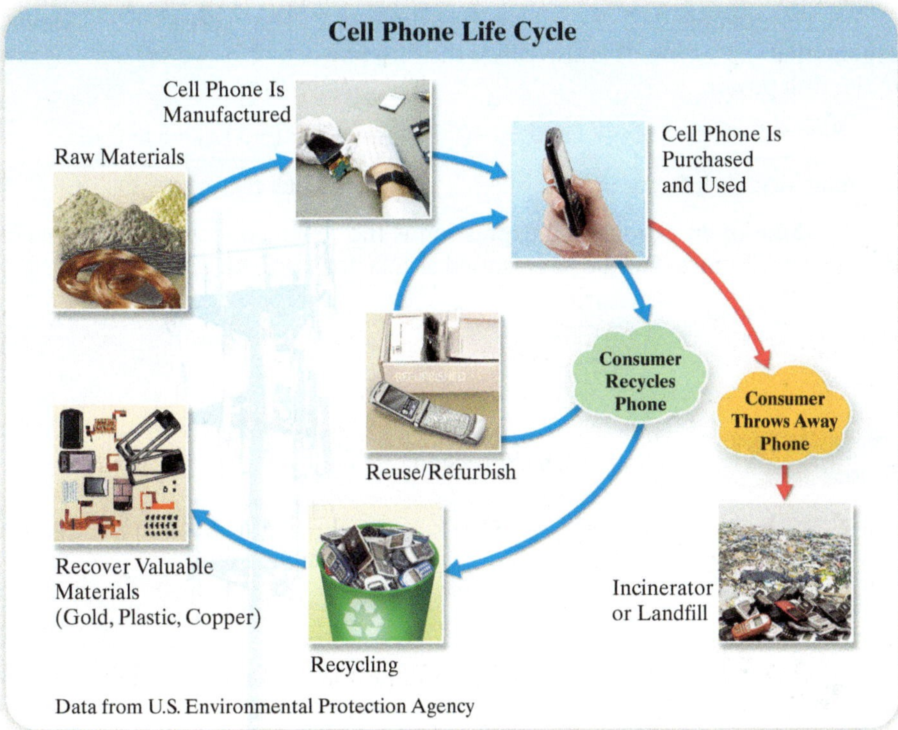

Data from U.S. Environmental Protection Agency

According to a recent survey, the average American adult owns 2.4 cell phones, and over 56% of those surveyed still have their old cell phones. In fact, it is estimated that 130 million cell phones will be retired this year. With new cell phone models having new features and increased technologies, this number of retired cell phones will probably only increase. The good news is that the number of cell phones being recycled is increasing. Below is a graph showing the growth of cell phones recycled by the largest recycler. In Exercises 99 and 100 of Section 12.2, you will explore some information about recycling cell phones.

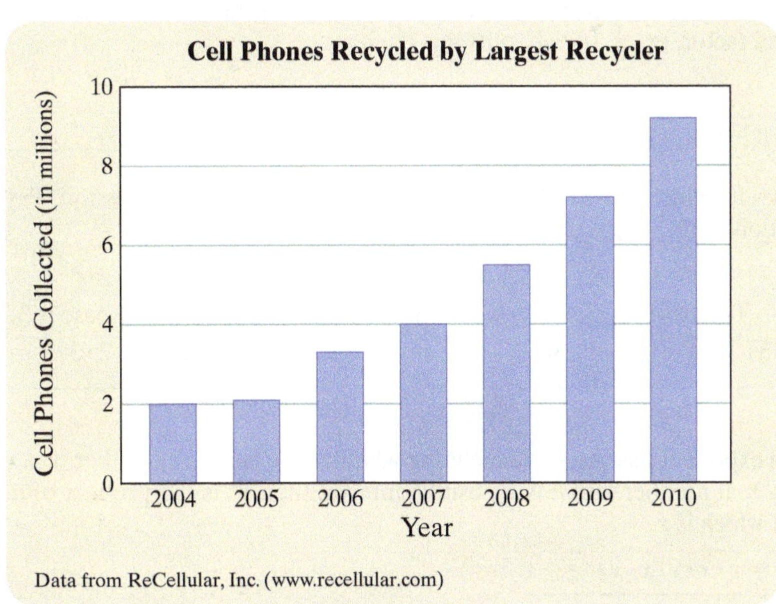

Data from ReCellular, Inc. (www.recellular.com)

Sections

Check Your Progress

Recall from Chapter 1 that an exponent is a shorthand notation for repeated factors. This chapter explores additional concepts about exponents and exponential expressions. An especially useful type of exponential expression is a polynomial. Polynomials model many real-world phenomena. In this chapter, we focus on polynomials and operations on polynomials.

12.1 Exponents 

Objectives

1. Evaluate Exponential Expressions.
2. Use the Product Rule for Exponents.
3. Use the Power Rule for Exponents.
4. Use the Power Rules for Products and Quotients.
5. Use the Quotient Rule for Exponents, and Define a Number Raised to the 0 Power.
6. Decide Which Rule(s) to Use to Simplify an Expression.

Objective 1 Evaluating Exponential Expressions

As we reviewed in Section 1.7, an exponent is a shorthand notation for repeated factors. For example, $2 \cdot 2 \cdot 2 \cdot 2 \cdot 2$ can be written as 2^5. The expression 2^5 is called an **exponential expression.** It is also called the fifth **power** of 2, or we say that 2 is **raised** to the fifth power.

$$5^6 = \underbrace{5 \cdot 5 \cdot 5 \cdot 5 \cdot 5 \cdot 5}_{6 \text{ factors; each factor is } 5} \quad \text{and} \quad (-3)^4 = \underbrace{(-3) \cdot (-3) \cdot (-3) \cdot (-3)}_{4 \text{ factors; each factor is } -3}$$

The **base** of an exponential expression is the repeated factor. The **exponent** is the number of times that the base is used as a factor.

Example 1

Evaluate each expression.

a. 2^3 **b.** 3^1 **c.** $(-4)^2$ **d.** -4^2 **e.** $\left(\frac{1}{2}\right)^4$ **f.** $(0.5)^3$ **g.** $4 \cdot 3^2$

Solution:

a. $2^3 = 2 \cdot 2 \cdot 2 = 8$

b. To raise 3 to the first power means to use 3 as a factor only once. Therefore, $3^1 = 3$. Also, when no exponent is shown, the exponent is assumed to be 1.

c. $(-4)^2 = (-4)(-4) = 16$ **d.** $-4^2 = -(4 \cdot 4) = -16$

e. $\left(\frac{1}{2}\right)^4 = \frac{1}{2} \cdot \frac{1}{2} \cdot \frac{1}{2} \cdot \frac{1}{2} = \frac{1}{16}$ **f.** $(0.5)^3 = (0.5)(0.5)(0.5) = 0.125$

g. $4 \cdot 3^2 = 4 \cdot 9 = 36$

Work Practice 1

Practice 1

Evaluate each expression.

a. 3^3 **b.** 4^1 **c.** $(-8)^2$ **d.** -8^2 **e.** $\left(\frac{3}{4}\right)^3$ **f.** $(0.3)^4$ **g.** $3 \cdot 5^2$

Notice how similar -4^2 is to $(-4)^2$ in the example above. The difference between the two is the parentheses. In $(-4)^2$, the parentheses tell us that the base, or repeated factor, is -4. In -4^2, only 4 is the base.

Helpful Hint

Be careful when identifying the base of an exponential expression. Pay close attention to the use of parentheses.

$(-3)^2$	-3^2	$2 \cdot 3^2$
The base is -3.	The base is 3.	The base is 3.
$(-3)^2 = (-3)(-3) = 9$	$-3^2 = -(3 \cdot 3) = -9$	$2 \cdot 3^2 = 2 \cdot 3 \cdot 3 = 18$

An exponent has the same meaning whether the base is a number or a variable. If x is a real number and n is a positive integer, then x^n is the product of n factors, each of which is x.

$$x^n = \underbrace{x \cdot x \cdot x \cdot x \cdot x \cdot \cdots \cdot x}_{n \text{ factors of } x}$$

Answers

1. a. 27 **b.** 4 **c.** 64 **d.** -64 **e.** $\frac{27}{64}$ **f.** 0.0081 **g.** 75

Example 2 Evaluate each expression for the given value of x.

a. $2x^3$; x is 5 **b.** $\frac{9}{x^2}$; x is -3

Solution: **a.** If x is 5, $2x^3 = 2 \cdot (5)^3$
$= 2 \cdot (5 \cdot 5 \cdot 5)$
$= 2 \cdot 125$
$= 250$

b. If x is -3, $\frac{9}{x^2} = \frac{9}{(-3)^2}$
$= \frac{9}{(-3)(-3)}$
$= \frac{9}{9}$
$= 1$

Work Practice 2

Practice 2

Evaluate each expression for the given value of x.

a. $3x^4$; x is 3

b. $\frac{6}{x^2}$; x is -4

Objective 2 Using the Product Rule

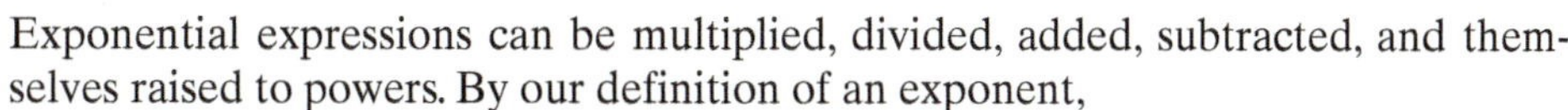

Exponential expressions can be multiplied, divided, added, subtracted, and themselves raised to powers. By our definition of an exponent,

$$5^4 \cdot 5^3 = \underbrace{(5 \cdot 5 \cdot 5 \cdot 5)}_{4 \text{ factors of } 5} \cdot \underbrace{(5 \cdot 5 \cdot 5)}_{3 \text{ factors of } 5}$$
$$= \underbrace{5 \cdot 5 \cdot 5 \cdot 5 \cdot 5 \cdot 5 \cdot 5}_{7 \text{ factors of } 5}$$
$$= 5^7$$

Also,

$$x^2 \cdot x^3 = (x \cdot x) \cdot (x \cdot x \cdot x)$$
$$= x \cdot x \cdot x \cdot x \cdot x$$
$$= x^5$$

In both cases, notice that the result is exactly the same if the exponents are added.

$$5^4 \cdot 5^3 = 5^{4+3} = 5^7 \quad \text{and} \quad x^2 \cdot x^3 = x^{2+3} = x^5$$

This suggests the following rule.

Product Rule for Exponents

If m and n are positive integers and a is a real number, then

$a^m \cdot a^n = a^{m+n}$ ← Add exponents.
↑ Keep common base.

For example, $3^5 \cdot 3^7 = 3^{5+7} = 3^{12}$ ← Add exponents.
↑ Keep common base.

Helpful Hint

Don't forget that

$3^5 \cdot 3^7 \neq 9^{12}$ ← Add exponents.
↑ **Common base *not* kept.**

$$3^5 \cdot 3^7 = \underbrace{3 \cdot 3 \cdot 3 \cdot 3 \cdot 3}_{5 \text{ factors of } 3} \cdot \underbrace{3 \cdot 3 \cdot 3 \cdot 3 \cdot 3 \cdot 3 \cdot 3}_{7 \text{ factors of } 3}$$
$= 3^{12}$ 12 factors of 3, *not* 9

In other words, to multiply two exponential expressions with the **same base,** we keep the base and add the exponents. We call this **simplifying** the exponential expression.

Answers

2. a. 243 **b.** $\frac{3}{8}$

Practice 3

Use the product rule to simplify.

a. $3^4 \cdot 3^6$ **b.** $y^3 \cdot y^2$ **c.** $z \cdot z^4$
d. $x^3 \cdot x^2 \cdot x^6$ **e.** $(-2)^5 \cdot (-2)^3$
f. $b^3 \cdot t^5$

Example 3 Use the product rule to simplify.

a. $4^2 \cdot 4^5$ **b.** $x^4 \cdot x^6$ **c.** $y^3 \cdot y$ **d.** $y^3 \cdot y^2 \cdot y^7$ **e.** $(-5)^7 \cdot (-5)^8$ **f.** $a^2 \cdot b^2$

Solution:

a. $4^2 \cdot 4^5 = 4^{2+5} = 4^7$ ← Add exponents. **Keep** common base.

b. $x^4 \cdot x^6 = x^{4+6} = x^{10}$

c. $y^3 \cdot y = y^3 \cdot y^1$
$= y^{3+1}$
$= y^4$

Helpful Hint Don't forget that if no exponent is written, it is assumed to be 1.

d. $y^3 \cdot y^2 \cdot y^7 = y^{3+2+7} = y^{12}$

e. $(-5)^7 \cdot (-5)^8 = (-5)^{7+8} = (-5)^{15}$

f. $a^2 \cdot b^2$ Cannot be simplified because a and b are different bases.

Work Practice 3

✓**Concept Check** Where possible, use the product rule to simplify the expression.

a. $z^2 \cdot z^{14}$ **b.** $x^2 \cdot y^{14}$ **c.** $9^8 \cdot 9^3$ **d.** $9^8 \cdot 2^7$

Practice 4

Use the product rule to simplify $(-5y^3)(-3y^4)$.

Example 4 Use the product rule to simplify $(2x^2)(-3x^5)$.

Solution: Recall that $2x^2$ means $2 \cdot x^2$ and $-3x^5$ means $-3 \cdot x^5$.

$(2x^2)(-3x^5) = 2 \cdot x^2 \cdot -3 \cdot x^5$ Remove parentheses.
$= 2 \cdot -3 \cdot x^2 \cdot x^5$ Group factors with common bases.
$= -6x^7$ Simplify.

Work Practice 4

Practice 5

Simplify.

a. $(y^7z^3)(y^5z)$
b. $(-m^4n^4)(7mn^{10})$

Example 5 Simplify.

a. $(x^2y)(x^3y^2)$ **b.** $(-a^7b^4)(3ab^9)$

Solution:

a. $(x^2y)(x^3y^2) = (x^2 \cdot x^3) \cdot (y^1 \cdot y^2)$ Group like bases and write y as y^1.
$= x^5 \cdot y^3$ or x^5y^3 Multiply.

b. $(-a^7b^4)(3ab^9) = (-1 \cdot 3) \cdot (a^7 \cdot a^1) \cdot (b^4 \cdot b^9)$
$= -3a^8b^{13}$

Work Practice 5

Helpful Hint

These examples will remind you of the difference between adding and multiplying terms.

Addition

$5x^3 + 3x^3 = (5 + 3)x^3 = 8x^3$ By the distributive property
$7x + 4x^2 = 7x + 4x^2$ Cannot be combined

Multiplication

$(5x^3)(3x^3) = 5 \cdot 3 \cdot x^3 \cdot x^3 = 15x^{3+3} = 15x^6$ By the product rule
$(7x)(4x^2) = 7 \cdot 4 \cdot x \cdot x^2 = 28x^{1+2} = 28x^3$ By the product rule

Answers

3. a. 3^{10} **b.** y^5 **c.** z^5 **d.** x^{11} **e.** $(-2)^8$ **f.** $b^3 \cdot t^5$ **4.** $15y^7$ **5. a.** $y^{12}z^4$ **b.** $-7m^5n^{14}$

✓**Concept Check Answers**

a. z^{16} **b.** cannot be simplified **c.** 9^{11} **d.** cannot be simplified

Objective 3 Using the Power Rule 

Exponential expressions can themselves be raised to powers. Let's try to discover a rule that simplifies an expression like $(x^2)^3$. By definition,

$$(x^2)^3 = \underbrace{(x^2)(x^2)(x^2)}_{3 \text{ factors of } x^2}$$

which can be simplified by the product rule for exponents.

$$(x^2)^3 = (x^2)(x^2)(x^2) = x^{2+2+2} = x^6$$

Notice that the result is exactly the same if we multiply the exponents.

$$(x^2)^3 = x^{2 \cdot 3} = x^6$$

The following property states this result.

Power Rule for Exponents

If m and n are positive integers and a is a real number, then

$$(a^m)^n = a^{mn} \leftarrow \text{Multiply exponents.}$$

$\uparrow$ Keep common base.

For example, $(7^2)^5 = 7^{2 \cdot 5} = 7^{10}$ ← Multiply exponents.

↑ Keep common base.

To raise a power to a power, keep the base and multiply the exponents.

Example 6 Use the power rule to simplify.

a. $(y^8)^2$ **b.** $(8^4)^5$ **c.** $[(-5)^3]^7$

Solution:

a. $(y^8)^2 = y^{8 \cdot 2} = y^{16}$ **b.** $(8^4)^5 = 8^{4 \cdot 5} = 8^{20}$ **c.** $[(-5)^3]^7 = (-5)^{21}$

Work Practice 6

Practice 6

Use the power rule to simplify.

a. $(z^3)^7$ **b.** $(4^9)^2$ **c.** $[(-2)^3]^5$

Helpful Hint

Take a moment to make sure that you understand when to apply the product rule and when to apply the power rule.

Product Rule → Add Exponents	Power Rule → Multiply Exponents
$x^5 \cdot x^7 = x^{5+7} = x^{12}$	$(x^5)^7 = x^{5 \cdot 7} = x^{35}$
$y^6 \cdot y^2 = y^{6+2} = y^8$	$(y^6)^2 = y^{6 \cdot 2} = y^{12}$

Answers

6. a. z^{21} **b.** 4^{18} **c.** $(-2)^{15}$

Objective 4 Using the Power Rules for Products and Quotients

When the base of an exponential expression is a product, the definition of x^n still applies. To simplify $(xy)^3$, for example,

$$
\begin{aligned}
(xy)^3 &= (xy)(xy)(xy) && (xy)^3 \text{ means 3 factors of } (xy). \\
&= x \cdot x \cdot x \cdot y \cdot y \cdot y && \text{Group factors with common bases.} \\
&= x^3y^3 && \text{Simplify.}
\end{aligned}
$$

Notice that to simplify the expression $(xy)^3$, we raise each factor within the parentheses to a power of 3.

$$(xy)^3 = x^3y^3$$

In general, we have the following rule.

Power of a Product Rule

If n is a positive integer and a and b are real numbers, then

$$(ab)^n = a^nb^n$$

For example, $(3x)^5 = 3^5x^5$.

In other words, to raise a product to a power, we raise each factor to the power.

Practice 7

Simplify each expression.

a. $(pr)^5$ **b.** $(6b)^2$

c. $\left(\frac{1}{4}x^2y\right)^3$ **d.** $(-3a^3b^4c)^4$

Example 7 Simplify each expression.

a. $(st)^4$ **b.** $(2a)^3$ **c.** $\left(\frac{1}{3}mn^3\right)^2$ **d.** $(-5x^2y^3z)^2$

Solution:

a. $(st)^4 = s^4 \cdot t^4 = s^4t^4$ Use the power of a product rule.

b. $(2a)^3 = 2^3 \cdot a^3 = 8a^3$ Use the power of a product rule.

c. $\left(\frac{1}{3}mn^3\right)^2 = \left(\frac{1}{3}\right)^2 \cdot (m)^2 \cdot (n^3)^2 = \frac{1}{9}m^2n^6$ Use the power of a product rule.

d. $(-5x^2y^3z)^2 = (-5)^2 \cdot (x^2)^2 \cdot (y^3)^2 \cdot (z^1)^2$ Use the power of a product rule.

$= 25x^4y^6z^2$ Use the power rule for exponents.

Work Practice 7

Let's see what happens when we raise a quotient to a power. To simplify $\left(\frac{x}{y}\right)^3$, for example,

$$
\begin{aligned}
\left(\frac{x}{y}\right)^3 &= \left(\frac{x}{y}\right)\left(\frac{x}{y}\right)\left(\frac{x}{y}\right) && \left(\frac{x}{y}\right)^3 \text{ means 3 factors of } \left(\frac{x}{y}\right). \\
&= \frac{x \cdot x \cdot x}{y \cdot y \cdot y} && \text{Multiply fractions.} \\
&= \frac{x^3}{y^3} && \text{Simplify.}
\end{aligned}
$$

Answers

7. a. p^5r^5 **b.** $36b^2$ **c.** $\frac{1}{64}x^6y^3$ **d.** $81a^{12}b^{16}c^4$

Notice that to simplify the expression $\left(\frac{x}{y}\right)^3$, we raise both the numerator and the denominator to a power of 3.

$$\left(\frac{x}{y}\right)^3 = \frac{x^3}{y^3}$$

In general, we have the following.

Power of a Quotient Rule

If n is a positive integer and a and c are real numbers, then

$$\left(\frac{a}{c}\right)^n = \frac{a^n}{c^n}, \quad c \neq 0$$

For example, $\left(\frac{y}{7}\right)^4 = \frac{y^4}{7^4}$.

In other words, to raise a quotient to a power, we raise both the numerator and the denominator to the power.

Example 8 Simplify each expression.

a. $\left(\frac{m}{n}\right)^7$ **b.** $\left(\frac{x^3}{3y^5}\right)^4$

Solution:

a. $\left(\frac{m}{n}\right)^7 = \frac{m^7}{n^7}, n \neq 0$ Use the power of a quotient rule.

b. $\left(\frac{x^3}{3y^5}\right)^4 = \frac{(x^3)^4}{3^4 \cdot (y^5)^4}, y \neq 0$ Use the power of a product and quotient rule.

$= \frac{x^{12}}{81y^{20}}$ Use the power rule for exponents.

Work Practice 8

Practice 8

Simplify each expression.

a. $\left(\frac{x}{y^2}\right)^5$ **b.** $\left(\frac{2a^4}{b^3}\right)^5$

Objective 5 Using the Quotient Rule and Defining the Zero Exponent

Another pattern for simplifying exponential expressions involves quotients.

To simplify an expression like $\frac{x^5}{x^3}$, in which the numerator and the denominator have a common base, we can apply the fundamental principle of fractions and divide the numerator and the denominator by the common base factors. Assume for the remainder of this section that denominators are not 0.

$$\frac{x^5}{x^3} = \frac{x \cdot x \cdot x \cdot x \cdot x}{x \cdot x \cdot x}$$

$$= \frac{x \cdot x \cdot x \cdot x \cdot x}{x \cdot x \cdot x}$$

$$= x \cdot x$$

$$= x^2$$

Answers

8. a. $\frac{x^5}{y^{10}}$ **b.** $\frac{32a^{20}}{b^{15}}$

Notice that the result is exactly the same if we subtract exponents of the common bases.

$$\frac{x^5}{x^3} = x^{5-3} = x^2$$

The quotient rule for exponents states this result in a general way.

Quotient Rule for Exponents

If m and n are positive integers and a is a real number, then

$$\frac{a^m}{a^n} = a^{m-n}$$

as long as a is not 0.

For example, $\frac{x^6}{x^2} = x^{6-2} = x^4$.

In other words, to divide one exponential expression by another with a common base, keep the base and subtract exponents.

Practice 9

Simplify each quotient.

a. $\frac{z^8}{z^4}$ **b.** $\frac{(-5)^5}{(-5)^3}$ **c.** $\frac{8^8}{8^6}$

d. $\frac{q^5}{t^2}$ **e.** $\frac{6x^3y^7}{xy^5}$

Example 9

Simplify each quotient.

a. $\frac{x^5}{x^2}$ **b.** $\frac{4^7}{4^3}$ **c.** $\frac{(-3)^5}{(-3)^2}$ **d.** $\frac{s^2}{t^3}$ **e.** $\frac{2x^5y^2}{xy}$

Solution:

a. $\frac{x^5}{x^2} = x^{5-2} = x^3$ Use the quotient rule.

b. $\frac{4^7}{4^3} = 4^{7-3} = 4^4 = 256$ Use the quotient rule.

c. $\frac{(-3)^5}{(-3)^2} = (-3)^3 = -27$ Use the quotient rule.

d. $\frac{s^2}{t^3}$ Cannot be simplified because s and t are different bases.

e. Begin by grouping common bases.

$$\frac{2x^5y^2}{xy} = 2 \cdot \frac{x^5}{x^1} \cdot \frac{y^2}{y^1}$$

$$= 2 \cdot (x^{5-1}) \cdot (y^{2-1}) \quad \text{Use the quotient rule.}$$

$$= 2x^4y^1 \quad \text{or} \quad 2x^4y$$

Work Practice 9

✓**Concept Check** Suppose you are simplifying each expression. Tell whether you would *add* the exponents, *subtract* the exponents, *multiply* the exponents, *divide* the exponents, or *none of these*.

a. $(x^{63})^{21}$ **b.** $\frac{y^{15}}{y^3}$ **c.** $z^{16} + z^8$ **d.** $w^{45} \cdot w^9$

Answers

9. a. z^4 **b.** 25 **c.** 64 **d.** $\frac{q^5}{t^2}$ **e.** $6x^2y^2$

✓**Concept Check Answers**

a. multiply **b.** subtract
c. none of these **d.** add

Let's now give meaning to an expression such as x^0. To do so, we will simplify $\frac{x^3}{x^3}$ in two ways and compare the results.

$\frac{x^3}{x^3} = x^{3-3} = x^0$ Apply the quotient rule.

$\frac{x^3}{x^3} = \frac{x \cdot x \cdot x}{x \cdot x \cdot x} = 1$ Apply the fundamental principle of fractions.

Since $\frac{x^3}{x^3} = x^0$ and $\frac{x^3}{x^3} = 1$, we define that $x^0 = 1$ as long as x is not 0.

Zero Exponent

$a^0 = 1$, as long as a is not 0.

In other words, any base raised to the 0 power is 1 as long as the base is not 0.

Example 10 Simplify each expression.

a. 3^0 **b.** $(5x^3y^2)^0$ **c.** $(-5)^0$ **d.** -5^0 **e.** $\left(\frac{3}{100}\right)^0$ **f.** $4x^0$

Solution:

a. $3^0 = 1$

b. Assume that neither x nor y is zero.

$(5x^3y^2)^0 = 1$

c. $(-5)^0 = 1$

d. $-5^0 = -1 \cdot 5^0 = -1 \cdot 1 = -1$

e. $\left(\frac{3}{100}\right)^0 = 1$

f. $4x^0 = 4 \cdot x^0 = 4 \cdot 1 = 4$

Work Practice 10

Practice 10

Simplify the following expressions.

a. -3^0 **b.** $(-3)^0$ **c.** 8^0

d. $(0.2)^0$ **e.** $(7a^2y^4)^0$ **f.** $7y^0$

Objective 6 Deciding Which Rule to Use

Let's practice deciding which rule(s) to use to simplify. We will continue this discussion with more examples in Section 12.5.

Example 11 Simplify each expression.

a. $x^7 \cdot x^4$ **b.** $\left(\frac{t}{2}\right)^4$ **c.** $(9y^5)^2$

Solution:

a. Here, we have a product, so we use the product rule to simplify.

$x^7 \cdot x^4 = x^{7+4} = x^{11}$

b. This is a quotient raised to a power, so we use the power of a quotient rule.

$\left(\frac{t}{2}\right)^4 = \frac{t^4}{2^4} = \frac{t^4}{16}$

c. This is a product raised to a power, so we use the power of a product rule.

$(9y^5)^2 = 9^2(y^5)^2 = 81y^{10}$

Work Practice 11

Practice 11

Simplify each expression.

a. $\left(\frac{z}{12}\right)^2$ **b.** $(4x^6)^3$ **c.** $y^{10} \cdot y^3$

Answers

10. a. -1 **b.** 1 **c.** 1 **d.** 1 **e.** 1 **f.** 7

11. a. $\frac{z^2}{144}$ **b.** $64x^{18}$ **c.** y^{13}

Practice 12

Simplify each expression.

a. $8^2 - 8^0$ **b.** $(z^0)^6 + (4^0)^5$

c. $\left(\frac{5x^3}{15y^4}\right)^2$ **d.** $\frac{(2z^8x^5)^4}{-16z^2x^{20}}$

Answers

12. a. 63 **b.** 2 **c.** $\frac{x^6}{9y^8}$ **d.** $-z^{30}$

Example 12

Simplify each expression.

a. $4^2 - 4^0$ **b.** $(x^0)^3 + (2^0)^5$ **c.** $\left(\frac{3y^7}{6x^5}\right)^2$ **d.** $\frac{(2a^3b^4)^3}{-8a^9b^2}$

Solution:

a. $4^2 - 4^0 = 16 - 1 = 15$ Remember that $4^0 = 1$.

b. $(x^0)^3 + (2^0)^5 = 1^3 + 1^5 = 1 + 1 = 2$

c. $\left(\frac{3y^7}{6x^5}\right)^2 = \frac{3^2(y^7)^2}{6^2(x^5)^2} = \frac{9 \cdot y^{14}}{36 \cdot x^{10}} = \frac{y^{14}}{4x^{10}}$

d. $\frac{(2a^3b^4)^3}{-8a^9b^2} = \frac{2^3(a^3)^3(b^4)^3}{-8a^9b^2} = \frac{8a^9b^{12}}{-8a^9b^2} = -1 \cdot (a^{9-9}) \cdot (b^{12-2})$

$= -1 \cdot a^0 \cdot b^{10} = -1 \cdot 1 \cdot b^{10} = -b^{10}$

Work Practice 12

Vocabulary, Readiness & Video Check

Use the choices below to fill in each blank. Some choices may be used more than once.

0	base	add
1	exponent	multiply

1. Repeated multiplication of the same factor can be written using a(n) __________.

2. In 5^2, the 2 is called the __________ and the 5 is called the __________.

3. To simplify $x^2 \cdot x^7$, keep the base and __________ the exponents.

4. To simplify $(x^3)^6$, keep the base and __________ the exponents.

5. The understood exponent on the term y is __________.

6. If $x^{\square} = 1$, the exponent is __________.

Martin-Gay Interactive Videos

See Video 12.1

Watch the section lecture video and answer the following questions.

Objective 1 **7.** Examples 3 and 4 illustrate how to find the base of an exponential expression both with and without parentheses. Explain how identifying the base of Example 7 is similar to identifying the base of Example 4.

Objective 2 **8.** Why were the commutative and associative properties applied in Example 12? Are these properties used in another example?

Objective 3 **9.** What point is made at the end of Example 15?

Objective 4 **10.** Although it's not especially emphasized in Example 20, what is helpful to remind yourself about the −2 in the problem?

Objective 5 **11.** In Example 24, which exponent rule is used to show that any nonzero base raised to zero is 1?

Objective 6 **12.** When simplifying an exponential expression that's a fraction, will you always use the quotient rule? Refer to Example 30 for this objective to support your answer.

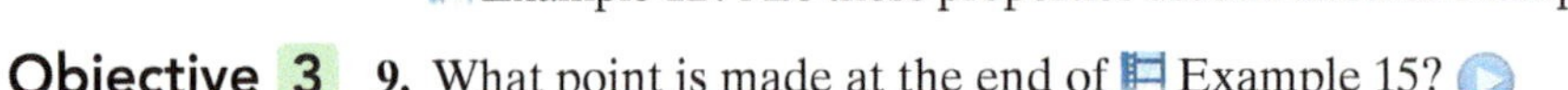

12.1 Exercise Set MyMathLab®

For each of the following expressions, state the exponent shown and its corresponding base.

1. 3^2
2. $(-3)^6$
3. -4^2
4. $5 \cdot 3^4$
5. $5x^2$
6. $(5x)^2$

Evaluate each expression. See Example 1.

7. 7^2
8. -3^2
9. $(-5)^1$
10. $(-3)^2$
11. -2^4
12. -4^3
13. $(-2)^4$
14. $(-4)^3$
15. $(0.1)^5$
16. $(0.2)^5$
17. $\left(\frac{1}{3}\right)^4$
18. $\left(-\frac{1}{9}\right)^2$
19. $7 \cdot 2^5$
20. $9 \cdot 1^7$
21. $-2 \cdot 5^3$
22. $-4 \cdot 3^3$

Evaluate each expression given the replacement values for x. See Example 2.

23. $x^2; x = -2$
24. $x^3; x = -2$
25. $5x^3; x = 3$
26. $4x^2; x = -1$
27. $2xy^2; x = 3$ and $y = 5$
28. $-4x^2y^3; x = 2$ and $y = -1$
29. $\frac{2z^4}{5}; z = -2$
30. $\frac{10}{3y^3}; y = 5$

Use the product rule to simplify each expression. Write the results using exponents. See Examples 3 through 5.

31. $x^2 \cdot x^5$
32. $y^2 \cdot y$
33. $(-3)^3 \cdot (-3)^9$
34. $(-5)^7 \cdot (-5)^6$
35. $(5y^4)(3y)$
36. $(-2z^3)(-2z^2)$
37. $(x^9y)(x^{10}y^5)$
38. $(a^2b)(a^{13}b^{17})$
39. $(-8mn^6)(9m^2n^2)$
40. $(-7a^3b^3)(7a^{19}b)$
41. $(4z^{10})(-6z^7)(z^3)$
42. $(12x^5)(-x^6)(x^4)$

△ 43. The rectangle below has width $4x^2$ feet and length $5x^3$ feet. Find its area as an expression in x. ($A = l \cdot w$)

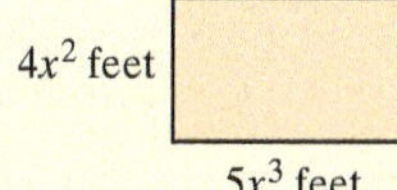

△ 44. The parallelogram below has base length $9y^7$ meters and height $2y^{10}$ meters. Find its area as an expression in y. ($A = b \cdot h$)

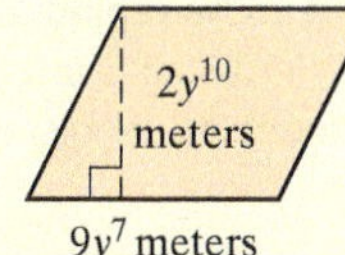

MIXED PRACTICE

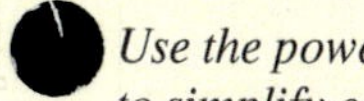

Use the power rule and the power of a product or quotient rule to simplify each expression. See Examples 6 through 8.

45. $(x^9)^4$
46. $(y^7)^5$
47. $(pq)^8$
48. $(ab)^6$
49. $(2a^5)^3$
50. $(4x^6)^2$
51. $(x^2y^3)^5$
52. $(a^4b)^7$
53. $(-7a^2b^5c)^2$
54. $(-3x^7yz^2)^3$
55. $\left(\frac{r}{s}\right)^9$
56. $\left(\frac{q}{t}\right)^{11}$
57. $\left(\frac{mp}{n}\right)^5$
58. $\left(\frac{xy}{7}\right)^2$
59. $\left(\frac{-2xz}{y^5}\right)^2$
60. $\left(\frac{xy^4}{-3z^3}\right)^3$

△ 61. The square shown has sides of length $8z^5$ decimeters. Find its area. ($A = s^2$)

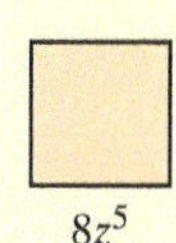

△ 62. Given the circle below with radius $5y$ centimeters, find its area. Do not approximate π. ($A = \pi r^2$)

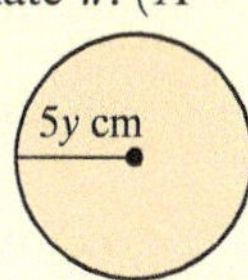

△ 63. The vault below is in the shape of a cube. If each side is $3y^4$ feet, find its volume. ($V = s^3$)

△ 64. The silo shown is in the shape of a cylinder. If its radius is $4x$ meters and its height is $5x^3$ meters, find its volume. Do not approximate π. ($V = \pi r^2 h$)

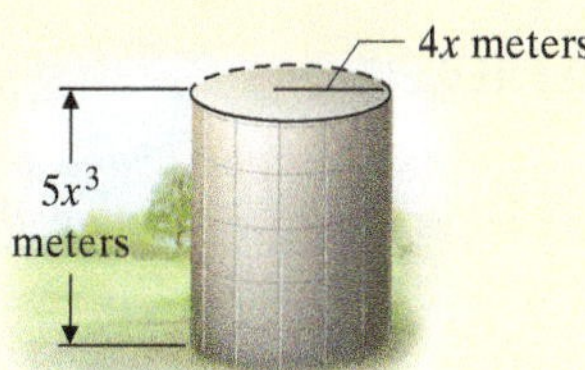

Use the quotient rule and simplify each expression. See Example 9.

65. $\frac{x^3}{x}$
66. $\frac{y^{10}}{y^9}$
67. $\frac{(-4)^6}{(-4)^3}$
68. $\frac{(-6)^{13}}{(-6)^{11}}$
69. $\frac{p^7q^{20}}{pq^{15}}$
70. $\frac{x^8y^6}{xy^5}$
71. $\frac{7x^2y^6}{14x^2y^3}$
72. $\frac{9a^4b^7}{27ab^2}$

Simplify each expression. See Example 10.

73. 7^0
74. 23^0
75. $(2x)^0$
76. $(4y)^0$
77. $-7x^0$
78. $-2x^0$
79. $5^0 + y^0$
80. $-3^0 + 4^0$

MIXED PRACTICE

Simplify each expression. See Examples 1 through 12.

81. -9^2
82. $(-9)^2$
83. $\left(\frac{1}{4}\right)^3$
84. $\left(\frac{2}{3}\right)^3$
85. b^4b^2
86. y^4y
87. $a^2a^3a^4$
88. $x^2x^{15}x^9$
89. $(2x^3)(-8x^4)$
90. $(3y^4)(-5y)$
91. $(a^7b^{12})(a^4b^8)$
92. $(y^2z^2)(y^{15}z^{13})$
93. $(-2mn^6)(-13m^8n)$
94. $(-3s^5t)(-7st^{10})$
95. $(z^4)^{10}$
96. $(t^5)^{11}$
97. $(4ab)^3$
98. $(2ab)^4$
99. $(-6xyz^3)^2$
100. $(-3xy^2a^3)^3$
101. $\frac{3x^5}{x^4}$
102. $\frac{5x^9}{x^3}$
103. $(9xy)^2$
104. $(2ab)^5$
105. $2^3 + 2^0$
106. $7^2 - 7^0$
107. $\left(\frac{3y^5}{6x^4}\right)^3$
108. $\left(\frac{2ab}{6yz}\right)^4$
109. $\frac{2x^3y^2z}{xyz}$
110. $\frac{x^{12}y^{13}}{x^5y^7}$
111. $(5^0)^3 + (y^0)^7$
112. $(9^0)^4 + (z^0)^5$
113. $\left(\frac{5x^9}{10y^{11}}\right)^2$
114. $\left(\frac{3a^4}{9b^5}\right)^2$
115. $\frac{(2a^5b^3)^4}{-16a^{20}b^7}$
116. $\frac{(2x^6y^2)^5}{-32x^{20}y^{10}}$

REVIEW AND PREVIEW

Simplify each expression by combining any like terms. Use the distributive property to remove any parentheses. See Section 3.1.

117. $y - 10 + y$
118. $-6z + 20 - 3z$
119. $7x + 2 - 8x - 6$
120. $10y - 14 - y - 14$
121. $2(x - 5) + 3(5 - x)$
122. $-3(w + 7) + 5(w + 1)$

CONCEPT EXTENSIONS

Solve. See the Concept Checks in this section. For Exercises 123 through 126, match the expression with the operation needed to simplify it. A letter may be used more than once and a letter may not be used at all.

123. $(x^{14})^{23}$
124. $x^{14} \cdot x^{23}$
125. $x^{14} + x^{23}$
126. $\frac{x^{35}}{x^{17}}$

a. Add the exponents.
b. Subtract the exponents.
c. Multiply the exponents.
d. Divide the exponents.
e. None of these

Fill in the boxes so that each statement is true. (More than one answer is possible for each exercise.)

127. $x^{\square} \cdot x^{\square} = x^{12}$
128. $(x^{\square})^{\square} = x^{20}$
129. $\frac{y^{\square}}{y^{\square}} = y^7$
130. $(y^{\square})^{\square} \cdot (y^{\square})^{\square} = y^{30}$

131. The formula $V = x^3$ can be used to find the volume V of a cube with side length x. Find the volume of a cube with side length 7 meters. (Volume is measured in cubic units.)

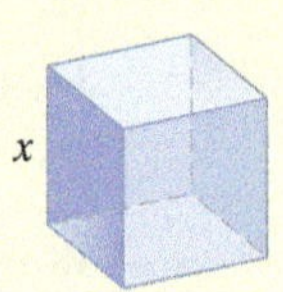

132. The formula $S = 6x^2$ can be used to find the surface area S of a cube with side length x. Find the surface area of a cube with side length 5 meters. (Surface area is measured in square units.)
133. To find the amount of water that a swimming pool in the shape of a cube can hold, do we use the formula for volume of the cube or surface area of the cube? (See Exercises 131 and 132.)
134. To find the amount of material needed to cover an ottoman in the shape of a cube, do we use the formula for volume of the cube or surface area of the cube? (See Exercises 131 and 132.)
135. Explain why $(-5)^4 = 625$, while $-5^4 = -625$.
136. Explain why $5 \cdot 4^2 = 80$, while $(5 \cdot 4)^2 = 400$.
137. In your own words, explain why $5^0 = 1$.
138. In your own words, explain when $(-3)^n$ is positive and when it is negative.

Simplify each expression. Assume that variables represent positive integers.

139. $x^{5a}x^{4a}$
140. $b^{9a}b^{4a}$
141. $(a^b)^5$
142. $(2a^{4b})^4$
143. $\frac{x^{9a}}{x^{4a}}$
144. $\frac{y^{15b}}{y^{6b}}$

Solve. Round money amounts to 2 decimal places.

145. Suppose you borrow money for 6 months. If the interest is compounded monthly, the formula $A = P\left(1 + \frac{r}{12}\right)^6$ gives the total amount A to be repaid at the end of 6 months. For a loan of $P = \$1000$ and interest rate of 9% $(r = 0.09)$, how much money is needed to pay off the loan?
146. Suppose you borrow money for 3 years. If the interest is compounded quarterly, the formula $A = P\left(1 + \frac{r}{4}\right)^{12}$ gives the total amount A to be repaid at the end of 3 years. For a loan of \$10,000 and interest rate of 8% $(r = 0.08)$, how much money is needed to pay off the loan in 3 years?

12.2 Polynomial Functions and Adding and Subtracting Polynomials 

Objective 1 Defining Polynomial, Monomial, Binomial, Trinomial, and Degree

In this section, we introduce a special algebraic expression called a polynomial. Let's first review some definitions presented in Section 3.1.

Recall that a **term** is a number or the product of a number and variables raised to powers. The terms of the expression $4x^2 + 3x$ are $4x^2$ and $3x$.

The terms of the expression $9x^4 - 7x - 1$ are $9x^4$, $-7x$, and -1.

Expression	Terms
$4x^2 + 3x$	$4x^2, 3x$
$9x^4 - 7x - 1$	$9x^4, -7x, -1$
$7y^3$	$7y^3$
5	5

The **numerical coefficient** of a term, or simply the **coefficient**, is the numerical factor of each term. If no numerical factor appears in the term, then the coefficient is understood to be 1. If the term is a number only, it is called a **constant term** or simply a **constant.**

Term	Coefficient
x^5	1
$3x^2$	3
$-4x$	-4
$-x^2y$	-1
3 (constant)	3

Now we are ready to define a polynomial.

Polynomial

A **polynomial in *x*** is a finite sum of terms of the form ax^n, where a is a real number and n is a whole number.

For example,

$$x^5 - 3x^3 + 2x^2 - 5x + 1$$

is a polynomial. Notice that this polynomial is written in **descending powers** of x because the powers of x decrease from left to right. (Recall that the term 1 can be thought of as $1x^0$.)

On the other hand,

$$x^{-5} + 2x - 3$$

is **not** a polynomial because it contains an exponent, -5, that is not a whole number. (We study negative exponents in Section 12.5 of this chapter.)

Some polynomials are given special names.

Types of Polynomials

A **monomial** is a polynomial with exactly one term.

A **binomial** is a polynomial with exactly two terms.

A **trinomial** is a polynomial with exactly three terms.

Objectives

1 Define Polynomial, Monomial, Binomial, Trinomial, and Degree.

2 Define Polynomial Functions.

3 Simplify a Polynomial by Combining Like Terms.

4 Add and Subtract Polynomials.

The following are examples of monomials, binomials, and trinomials. Each of these examples is also a polynomial.

POLYNOMIALS			
Monomials	**Binomials**	**Trinomials**	**More than Three Terms**
ax^2	$x + y$	$x^2 + 4xy + y^2$	$5x^3 - 6x^2 + 3x - 6$
$-3z$	$3p + 2$	$x^5 + 7x^2 - x$	$-y^5 + y^4 - 3y^3 - y^2 + y$
4	$4x^2 - 7$	$-q^4 + q^3 - 2q$	$x^6 + x^4 - x^3 + 1$

Each term of a polynomial has a **degree.**

Degree of a Term

The degree of a term is the sum of the exponents on the variables contained in the term.

Practice 1

Find the degree of each term.

a. $5y^3$
b. $10xy$
c. z
d. $-3a^2b^5c$
e. 8

Example 1 Find the degree of each term.

a. $3x^2$ **b.** -2^3x^5 **c.** y **d.** $12x^2yz^3$ **e.** 5

Solution:

a. The exponent on x is 2, so the degree of the term is 2.
b. The exponent on x is 5, so the degree of the term is 5. (Recall that the degree is the sum of the exponents on only the *variables.*)
c. The degree of y, or y^1, is 1.
d. The degree is the sum of the exponents on the variables, or $2 + 1 + 3 = 6$.
e. The degree of 5, which can be written as $5x^0$, is 0.

Work Practice 1

From the preceding, we can say that **the degree of a constant is 0.**

Each polynomial also has a degree.

Degree of a Polynomial

The degree of a polynomial is the greatest degree of any term of the polynomial.

Practice 2

Find the degree of each polynomial and tell whether the polynomial is a monomial, binomial, trinomial, or none of these.

a. $5b^2 - 3b + 7$
b. $7t + 3$
c. $5x^2 + 3x - 6x^3 + 4$

Example 2 Find the degree of each polynomial and tell whether the polynomial is a monomial, binomial, trinomial, or none of these.

a. $-2t^2 + 3t + 6$ **b.** $15x - 10$ **c.** $7x + 3x^3 + 2x^2 - 1$

Solution:

a. The degree of the trinomial $-2t^2 + 3t + 6$ is 2, the greatest degree of any of its terms.
b. The degree of the binomial $15x - 10$ or $15x^1 - 10$ is 1.
c. The degree of the polynomial $7x + 3x^3 + 2x^2 - 1$ is 3.

Work Practice 2

Answers

1. a. degree 3 **b.** degree 2 **c.** degree 1 **d.** degree 8 **e.** degree 0 **2. a.** trinomial, degree 2 **b.** binomial, degree 1 **c.** none of these, degree 3

Example 3 Complete the table for the polynomial

$$7x^2y - 6xy + x^2 - 3y + 7$$

Use the table to give the degree of the polynomial.

Solution:

Term	Numerical Coefficient	Degree of Term
$7x^2y$	7	3
$-6xy$	-6	2
x^2	1	2
$-3y$	-3	1
7	7	0

The degree of the polynomial is 3.

Work Practice 3

Practice 3

Complete the table for the polynomial $-3x^3y^2 + 4xy^2 - y^2 + 3x - 2$. Use the table to give the degree of the polynomial.

Term	Numerical Coefficient	Degree of Term
$-3x^3y^2$		
$4xy^2$		
$-y^2$		
$3x$		
-2		

Objective 2 Defining Polynomial Functions 

At times, it is convenient to use function notation to represent polynomials. For example, we may write $P(x)$ to represent the polynomial $3x^2 - 2x - 5$. In symbols, this is

$$P(x) = 3x^2 - 2x - 5$$

This function is called a **polynomial function** because the expression $3x^2 - 2x - 5$ is a polynomial.

Helpful Hint

Recall that the symbol $P(x)$ **does not mean** P times x. It is a special symbol used to denote a function.

Example 4 If $P(x) = 3x^2 - 2x - 5$, find the following.

a. $P(1)$ **b.** $P(-2)$

Solution:

a. Substitute 1 for x in $P(x) = 3x^2 - 2x - 5$ and simplify.

$$P(x) = 3x^2 - 2x - 5$$
$$P(1) = 3(1)^2 - 2(1) - 5 = -4$$

b. Substitute -2 for x in $P(x) = 3x^2 - 2x - 5$ and simplify.

$$P(x) = 3x^2 - 2x - 5$$
$$P(-2) = 3(-2)^2 - 2(-2) - 5 = 11$$

Work Practice 4

Practice 4

If $P(x) = -2x^2 - x + 7$, find

a. $P(1)$

b. $P(-4)$

Many real-world phenomena are modeled by polynomial functions. If the polynomial function model is given, we can often find the solution of a problem by evaluating the function at a certain value.

Answers

3.

Term	Numerical Coefficient	Degree of Term
$-3x^3y^2$	-3	5
$4xy^2$	4	3
$-y^2$	-1	2
$3x$	3	1
-2	-2	0

degree of polynomial: 5

4. a. 4 **b.** -21

Practice 5

The cliff divers of Acapulco dive 130 feet into La Quebrada several times a day for the entertainment of the tourists. If a tourist is standing near the diving platform and drops his camera off the cliff, the height of the camera above the water at time t seconds is given by the polynomial function $P(t) = -16t^2 + 130$. Find the height of the camera when $t = 1$ second and when $t = 2$ seconds.

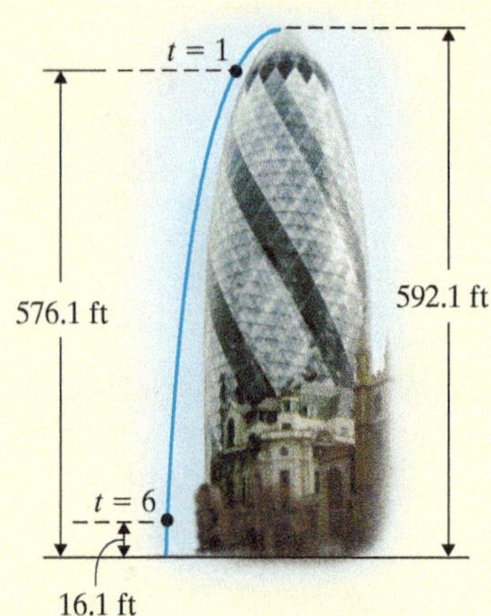

Example 5 Finding the Height of a Dropped Object

The Swiss Re Building, in London, is a unique building. Londoners often refer to it as the "pickle building." The building is 592.1 feet tall. An object is dropped from the highest point of this building. Neglecting air resistance, the height in feet of the object above ground at time t seconds is given by the polynomial function $P(t) = -16t^2 + 592.1$. Find the height of the object when $t = 1$ second, and when $t = 6$ seconds.

Solution: To find each height, we find $P(1)$ and $P(6)$.

$$P(t) = -16t^2 + 592.1$$
$$P(1) = -16(1)^2 + 592.1 \quad \text{Replace } t \text{ with 1.}$$
$$P(1) = -16 + 592.1$$
$$P(1) = 576.1$$

The height of the object at 1 second is 576.1 feet.

$$P(t) = -16t^2 + 592.1$$
$$P(6) = -16(6)^2 + 592.1 \quad \text{Replace } t \text{ with 6.}$$
$$P(6) = -576 + 592.1$$
$$P(6) = 16.1$$

The height of the object at 6 seconds is 16.1 feet.

Work Practice 5

Objective 3 Simplifying Polynomials by Combining Like Terms

Polynomials with like terms can be simplified by combining the like terms. Recall that like terms are terms that contain exactly the same variables raised to exactly the same powers.

Like Terms	Unlike Terms
$5x^2, -7x^2$	$3x, 3y$
$y, 2y$	$-2x^2, -5x$
$\frac{1}{2}a^2b, -a^2b$	$6st^2, 4s^2t$

Only like terms can be combined. We combine like terms by applying the distributive property.

Practice 6

Simplify each polynomial by combining any like terms.

a. $-4y + 2y$

b. $z + 5z^3$

c. $15x^3 - x^3$

d. $7a^2 - 5 - 3a^2 - 7$

e. $\frac{3}{8}x^3 - x^2 + \frac{5}{6}x^4 + \frac{1}{12}x^3 - \frac{1}{2}x^4$

Example 6 Simplify each polynomial by combining any like terms.

a. $-3x + 7x$

b. $x + 3x^2$

c. $9x^3 + x^3$

d. $11x^2 + 5 + 2x^2 - 7$

e. $\frac{2}{5}x^4 + \frac{2}{3}x^3 - x^2 + \frac{1}{10}x^4 - \frac{1}{6}x^3$

Solution:

a. $-3x + 7x = (-3 + 7)x = 4x$

b. $x + 3x^2$ These terms cannot be combined because x and $3x^2$ are not like terms.

c. $9x^3 + x^3 = 9x^3 + 1x^3 = 10x^3$

d. $11x^2 + 5 + 2x^2 - 7 = 11x^2 + 2x^2 + 5 - 7$
$= 13x^2 - 2$ Combine like terms.

Answers

5. 114 ft; 66 ft 6. a. $-2y$ b. $z + 5z^3$ c. $14x^3$ d. $4a^2 - 12$ e. $\frac{1}{3}x^4 + \frac{11}{24}x^3 - x^2$

e. $\frac{2}{5}x^4 + \frac{2}{3}x^3 - x^2 + \frac{1}{10}x^4 - \frac{1}{6}x^3$

$$= \left(\frac{2}{5} + \frac{1}{10}\right)x^4 + \left(\frac{2}{3} - \frac{1}{6}\right)x^3 - x^2$$
$$= \left(\frac{4}{10} + \frac{1}{10}\right)x^4 + \left(\frac{4}{6} - \frac{1}{6}\right)x^3 - x^2$$
$$= \frac{5}{10}x^4 + \frac{3}{6}x^3 - x^2$$
$$= \frac{1}{2}x^4 + \frac{1}{2}x^3 - x^2$$

Work Practice 6

✓Concept Check When combining like terms in the expression $5x - 8x^2 - 8x$, which of the following is the proper result?

a. $-11x^2$ **b.** $-8x^2 - 3x$ **c.** $-11x$ **d.** $-11x^4$

Example 7 Combine like terms to simplify.

$-9x^2 + 3xy - 5y^2 + 7yx$

Solution:

$$-9x^2 + 3xy - 5y^2 + 7yx = -9x^2 + (3 + 7)xy - 5y^2$$
$$= -9x^2 + 10xy - 5y^2$$

Work Practice 7

This term can be written as $7yx$ or $7xy$.

Practice 7

Combine like terms to simplify: $9xy - 3x^2 - 4yx + 5y^2$.

Example 8 Write a polynomial that describes the total area of the squares and rectangles. Then simplify the polynomial.

Solution:

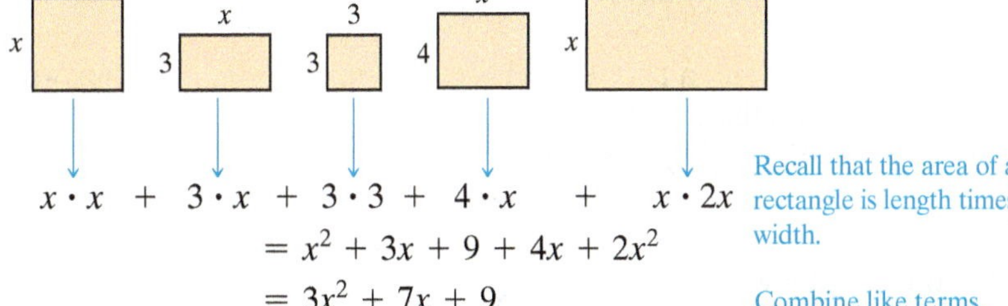

Area: $x \cdot x + 3 \cdot x + 3 \cdot 3 + 4 \cdot x + x \cdot 2x$ — Recall that the area of a rectangle is length times width.

$= x^2 + 3x + 9 + 4x + 2x^2$

$= 3x^2 + 7x + 9$ — Combine like terms.

Work Practice 8

Practice 8

Write a polynomial that describes the total area of the squares and rectangles shown below. Then simplify the polynomial.

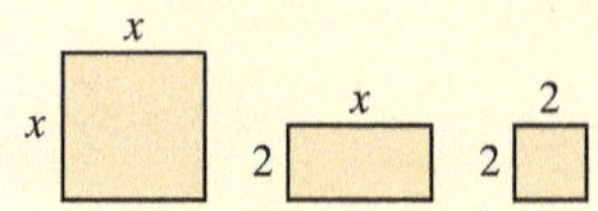

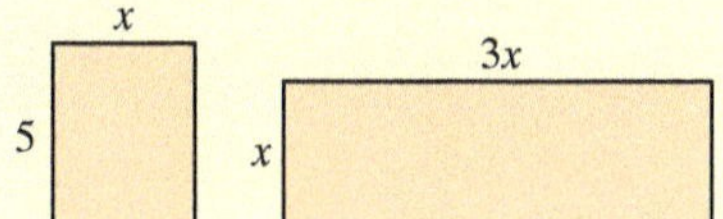

Objective 4 Adding and Subtracting Polynomials

We now practice adding and subtracting polynomials.

Adding Polynomials

To add polynomials, combine all like terms.

Answers

7. $-3x^2 + 5xy + 5y^2$

8. $x^2 + 2x + 4 + 5x + 3x^2$; $4x^2 + 7x + 4$

✓Concept Check Answer

b

Practice 9

Add.

a. $(4y^2 + x - 3y - 7) + (x + y^2 - 2)$

b. $(-8a^2b - ab^2 + 10) + (-2ab^2 - 10)$

Example 9

Add.

a. $(7x^3y - xy^3 + 11) + (6x^3y - 4)$ b. $(3a^3 - b + 2a - 5) + (a + b + 5)$

Solution:

a. To add, remove the parentheses and group like terms.

$$(7x^3y - xy^3 + 11) + (6x^3y - 4)$$
$$= 7x^3y - xy^3 + 11 + 6x^3y - 4$$
$$= 7x^3y + 6x^3y - xy^3 + 11 - 4 \quad \text{Group like terms.}$$
$$= 13x^3y - xy^3 + 7 \quad \text{Combine like terms.}$$

b. $(3a^3 - b + 2a - 5) + (a + b + 5)$
$$= 3a^3 - b + 2a - 5 + a + b + 5$$
$$= 3a^3 - b + b + 2a + a - 5 + 5 \quad \text{Group like terms.}$$
$$= 3a^3 + 3a \quad \text{Combine like terms.}$$

Work Practice 9

Practice 10

Add $(3x^2 - 9x + 11)$ and $(-3x^2 + 7x^3 + 3x - 4)$

Example 10

Add $(11x^3 - 12x^2 + x - 3)$ and $(x^3 - 10x + 5)$.

Solution:

$$(11x^3 - 12x^2 + x - 3) + (x^3 - 10x + 5)$$
$$= 11x^3 + x^3 - 12x^2 + x - 10x - 3 + 5 \quad \text{Group like terms.}$$
$$= 12x^3 - 12x^2 - 9x + 2 \quad \text{Combine like terms.}$$

Work Practice 10

Sometimes it is more convenient to add polynomials vertically. To do this, line up like terms beneath one another and add like terms.

The definition of subtraction of real numbers can be extended to apply to polynomials. To subtract a number, we add its opposite.

$$a - b = a + (-b)$$

Likewise, to subtract a polynomial, we add its opposite. In other words, if P and Q are polynomials, then

$$P - Q = P + (-Q)$$

The polynomial $-Q$ is the **opposite,** or **additive inverse,** of the polynomial Q. We can find $-Q$ by writing the opposite of each term of Q.

Subtracting Polynomials

To subtract two polynomials, change the signs of the terms of the polynomial being subtracted and then add.

✓Concept Check Which polynomial is the opposite of $16x^3 - 5x + 7$

a. $-16x^3 - 5x + 7$ b. $-16x^3 + 5x - 7$

c. $16x^3 + 5x + 7$ d. $-16x^3 + 5x + 7$

Answers

9. a. $5y^2 - 3y + 2x - 9$ b. $-8a^2b - 3ab^2$ 10. $7x^3 - 6x + 7$

✓Concept Check Answer

b

Example 11 Subtract: $(2x^3 + 8x^2 - 6x) - (2x^3 - x^2 + 1)$.

Solution: First, change the sign of each term of the second polynomial and then add.

$$\begin{aligned}(2x^3 + 8x^2 - 6x) - (2x^3 - x^2 + 1) &= (2x^3 + 8x^2 - 6x) + (-2x^3 + x^2 - 1)\\ &= 2x^3 - 2x^3 + 8x^2 + x^2 - 6x - 1\\ &= 9x^2 - 6x - 1 \quad \text{Combine like terms.}\end{aligned}$$

Work Practice 11

Practice 11

Subtract: $(3x^3 - 5x^2 + 4x) - (x^3 - x^2 + 6)$.

Notice the sign of each term is changed.

Example 12 Subtract $(5z - 7)$ from the sum of $(8z + 11)$ and $(9z - 2)$.

Solution: Notice that $(5z - 7)$ is to be subtracted **from** a sum. The translation is

$$\begin{aligned}&[(8z + 11) + (9z - 2)] - (5z - 7)\\ &= 8z + 11 + 9z - 2 - 5z + 7 \quad \text{Remove grouping symbols.}\\ &= 8z + 9z - 5z + 11 - 2 + 7 \quad \text{Group like terms.}\\ &= 12z + 16 \quad \text{Combine like terms.}\end{aligned}$$

Work Practice 12

Practice 12

Subtract $(3x + 5)$ from the sum of $(8x - 11)$ and $(2x + 5)$.

Example 13 Add or subtract as indicated.

a. $(3x^2 - 6xy + 5y^2) + (-2x^2 + 8xy - y^2)$

b. $(9a^2b^2 + 6ab - 3ab^2) - (5b^2a + 2ab - 3 - 9b^2)$

Solution:

a. $(3x^2 - 6xy + 5y^2) + (-2x^2 + 8xy - y^2)$

$= 3x^2 - 6xy + 5y^2 - 2x^2 + 8xy - y^2$

$= x^2 + 2xy + 4y^2$ Combine like terms.

b. $(9a^2b^2 + 6ab - 3ab^2) - (5b^2a + 2ab - 3 - 9b^2)$

$= 9a^2b^2 + 6ab - 3ab^2 - 5b^2a - 2ab + 3 + 9b^2$ Change the sign of each term of the polynomial being subtracted.

$= 9a^2b^2 + 4ab - 8ab^2 + 3 + 9b^2$ Combine like terms.

Work Practice 13

Practice 13

Add or subtract as indicated.

a. $(3a^2 - 4ab + 7b^2) + (-8a^2 + 3ab - b^2)$

b. $(5x^2y^2 - 6xy - 4xy^2) - (2x^2y^2 + 4xy - 5 + 6y^2)$

✓**Concept Check** If possible, simplify each expression by performing the indicated operation.

a. $2y + y$ **b.** $2y \cdot y$ **c.** $-2y - y$ **d.** $(-2y)(-y)$ **e.** $2x + y$

To add or subtract polynomials vertically, just remember to line up like terms. For example, perform the subtraction $(10x^3y^2 - 7x^2y^2) - (4x^3y^2 - 3x^2y^2 + 2y^2)$ vertically.

Add the opposite of the second polynomial.

$$\begin{array}{r}10x^3y^2 - 7x^2y^2\\ \underline{-(4x^3y^2 - 3x^2y^2 + 2y^2)}\end{array} \quad \text{is equivalent to} \quad \begin{array}{r}10x^3y^2 - 7x^2y^2\\ \underline{-4x^3y^2 + 3x^2y^2 - 2y^2}\\ 6x^3y^2 - 4x^2y^2 - 2y^2\end{array}$$

Answers

11. $2x^3 - 4x^2 + 4x - 6$ **12.** $7x - 11$

13. a. $-5a^2 - ab + 6b^2$

b. $3x^2y^2 - 10xy - 4xy^2 + 5 - 6y^2$

✓**Concept Check Answer**

a. $3y$ **b.** $2y^2$ **c.** $-3y$ **d.** $2y^2$

e. cannot be simplified

✓Concept Check Why is the following subtraction incorrect?

$$(7z - 5) - (3z - 4)$$
$$= 7z - 5 - 3z - 4$$
$$= 4z - 9$$

Polynomial functions, like polynomials, can be added, subtracted, multiplied, and divided. For example, if

$$P(x) = x^2 + x + 1$$

then

$$2P(x) = 2(x^2 + x + 1) = 2x^2 + 2x + 2 \quad \text{Use the distributive property.}$$

Also, if $Q(x) = 5x^2 - 1$, then $P(x) + Q(x) = (x^2 + x + 1) + (5x^2 - 1) = 6x^2 + x$.

A useful business and economics application of subtracting polynomial functions is finding the profit function $P(x)$ when given a revenue function $R(x)$ and a cost function $C(x)$. In business, it is true that

profit = revenue − cost, or
$$P(x) = R(x) - C(x)$$

For example, if the revenue function is $R(x) = 7x$ and the cost function is $C(x) = 2x + 5000$, then the profit function is

$$P(x) = R(x) - C(x)$$

or

$$P(x) = 7x - (2x + 5000) \quad \text{Substitute } R(x) = 7x \text{ and } C(x) = 2x + 5000.$$
$$P(x) = 5x - 5000$$

✓Concept Check Answer
With parentheses removed, the expression should be
$7z - 5 - 3z + 4 = 4z - 1$

Graphing Calculator Explorations

A graphing calculator may be used to visualize addition and subtraction of polynomials in one variable. For example, to visualize the following polynomial subtraction statement

$$(3x^2 - 6x + 9) - (x^2 - 5x + 6) = 2x^2 - x + 3$$

graph both

$$Y_1 = (3x^2 - 6x + 9) - (x^2 - 5x + 6) \quad \text{Left side of equation}$$

and

$$Y_2 = 2x^2 - x + 3 \quad \text{Right side of equation}$$

on the same screen and see that their graphs coincide. (*Note:* If the graphs do not coincide, we can be sure that a mistake has been made in combining polynomials or in calculator keystrokes. If the graphs appear to coincide, we cannot be sure that our work is correct. This is because it is possible for the graphs to differ so slightly that we do not notice it.)

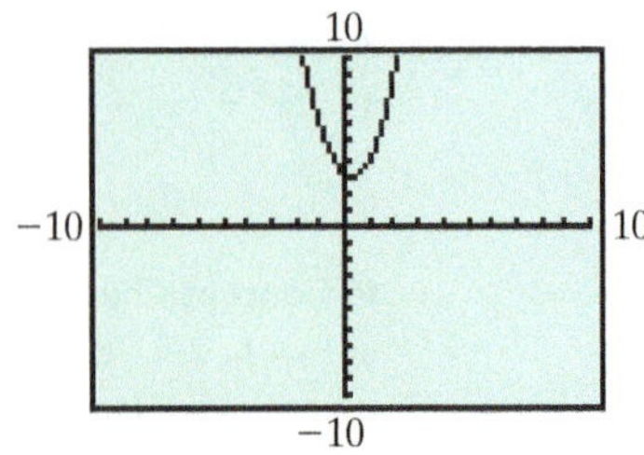

The graphs of Y_1 and Y_2 are shown. The graphs appear to coincide, so the subtraction statement

$$(3x^2 - 6x + 9) - (x^2 - 5x + 6) = 2x^2 - x + 3$$

appears to be correct.

Perform the indicated operations. Then visualize by using the procedure described above.

1. $(2x^2 + 7x + 6) + (x^3 - 6x^2 - 14)$

2. $(-14x^3 - x + 2) + (-x^3 + 3x^2 + 4x)$

3. $(1.8x^2 - 6.8x - 1.7) - (3.9x^2 - 3.6x)$

4. $(-4.8x^2 + 12.5x - 7.8) - (3.1x^2 - 7.8x)$

5. $(1.29x - 5.68) + (7.69x^2 - 2.55x + 10.98)$

6. $(-0.98x^2 - 1.56x + 5.57) + (4.36x - 3.71)$

Vocabulary, Readiness & Video Check

Use the choices below to fill in each blank. Not all choices will be used.

least	monomial	trinomial	coefficient
greatest	binomial	constant	

1. A ________ is a polynomial with exactly 2 terms.
2. A ________ is a polynomial with exactly one term.
3. A ________ is a polynomial with exactly three terms.
4. The numerical factor of a term is called the ________.
5. A number term is also called a ________.
6. The degree of a polynomial is the ________ degree of any term of the polynomial.

Martin-Gay Interactive Videos

See Video 12.2

Watch the section lecture video and answer the following questions.

Objective 1 7. For Example 2, why is the degree of each **term** found when the example asks for the degree of the **polynomial** only?

Objective 2 8. From Example 3, how do you find the value of a polynomial function $Q(x)$ given a value for x?

Objective 3 9. When combining any like terms in a polynomial, as in Examples 4–6, what are we doing to the polynomial?

Objective 4 10. From Example 7, when we simply remove parentheses and combine the like terms of two polynomials, what operation do we perform? Is this true of Examples 9–11?

12.2 Exercise Set MyMathLab®

Find the degree of each of the following polynomials and determine whether it is a monomial, binomial, trinomial, or none of these. See Examples 1 through 3.

1. $x + 2$
2. $-6y + 4$
3. $9m^3 - 5m^2 + 4m - 8$
4. $a + 5a^2 + 3a^3 - 4a^4$
5. $12x^4y - x^2y^2 - 12x^2y^4$
6. $7r^2s^2 + 2rs - 3rs^5$
7. $3 - 5x^8$
8. $5y^7 + 2$

In the second column, write the degree of the polynomial in the first column. See Examples 1 through 3.

	Polynomial	Degree
9.	$3xy^2 - 4$	
10.	$8x^2y^2$	
11.	$5a^2 - 2a + 1$	
12.	$4z^6 + 3z^2$	

If $P(x) = x^2 + x + 1$ and $Q(x) = 5x^2 - 1$, find the following. See Examples 4 and 5.

13. $P(7)$
14. $Q(4)$
15. $Q(-10)$
16. $P(-4)$
17. $P(0)$
18. $Q(0)$
19. $Q\left(\frac{1}{4}\right)$
20. $P\left(\frac{1}{2}\right)$

The CN Tower in Toronto, Ontario, is 1821 feet tall and is the world's tallest self-supporting structure. An object is dropped from the Skypod of the Tower, which is at 1150 feet. Neglecting air resistance, the height of the object at time t seconds is given by the polynomial function $P(t) = -16t^2 + 1150$. Find the height of the object at the given times.

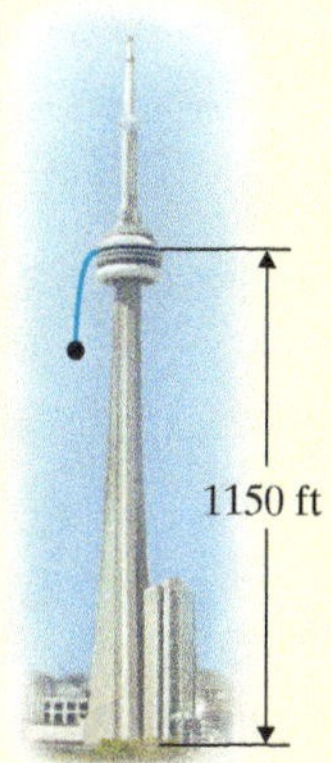

	Time, t (in seconds)	Height $P(t) = -16t^2 + 1150$
21.	1	
22.	7	
23.	3	
24.	6	

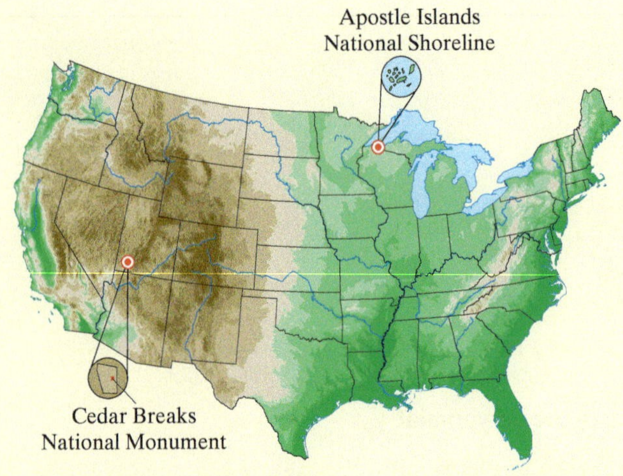

25. The polynomial $P(x) = -7.5x^2 + 93x - 100$ models the yearly number of visitors (in thousands) x years after 2000 at Apostle Islands National Park. Use this polynomial to estimate the number of visitors to the park in 2008 ($P(8)$).

26. The polynomial $P(x) = 8x^2 - 90.6x + 752$ models the yearly number of visitors (in thousands) x years after 2000 at Cedar Breaks National Park. Use this polynomial to estimate the number of visitors to the park in 2007 ($P(7)$).

Simplify each of the following by combining like terms. See Examples 6 and 7.

27. $14x^2 + 9x^2$
28. $18x^3 - 4x^3$
29. $15x^2 - 3x^2 - y$
30. $12k^3 - 9k^3 + 11$
31. $8s - 5s + 4s$
32. $5y + 7y - 6y$
33. $0.1y^2 - 1.2y^2 + 6.7 - 1.9$
34. $7.6y + 3.2y^2 - 8y - 2.5y^2$
35. $\frac{2}{5}x^2 - \frac{1}{3}x^3 + x^2 - \frac{1}{4}x^3 + 6$
36. $\frac{1}{6}x^4 - \frac{1}{7}x^2 + 5 - \frac{1}{2}x^4 - \frac{3}{7}x^2 + \frac{1}{3}$
37. $6a^2 - 4ab + 7b^2 - a^2 - 5ab + 9b^2$
38. $x^2y + xy - y + 10x^2y - 2y + xy$

Perform the indicated operations. See Examples 9 through 13.

39. $(-7x + 5) + (-3x^2 + 7x + 5)$
40. $(3x - 8) + (4x^2 - 3x + 3)$
41. $(2x^2 + 5) - (3x^2 - 9)$
42. $(5x^2 + 4) - (-2y^2 + 4)$
43. $3x - (5x - 9)$
44. $4 - (-y - 4)$
45. $(2x^2 + 3x - 9) - (-4x + 7)$
46. $(-7x^2 + 4x + 7) - (-8x + 2)$
47. $\begin{array}{r} 3t^2 + 4 \\ \underline{+5t^2 - 8} \end{array}$

48. $\begin{array}{r} 7x^3 + 3 \\ \underline{+\ 2x^3 + 1} \end{array}$

49. $\begin{array}{r} 4z^2 - 8z + 3 \\ \underline{-(6z^2 + 8z - 3)} \end{array}$

50. $\begin{array}{r} 5u^5 - 4u^2 + 3u - 7 \\ \underline{-(3u^5 + 6u^2 - 8u + 2)} \end{array}$

51. $\begin{array}{r} 5x^3 - 4x^2 + 6x - 2 \\ \underline{-(3x^3 - 2x^2 - x - 4)} \end{array}$

52. $\begin{array}{r} 7a^2 - 9a + 6 \\ \underline{-(11a^2 - 4a + 2)} \end{array}$

MIXED PRACTICE

Perform the indicated operations.

53. $(-3y^2 - 4y) + (2y^2 + y - 1)$
54. $(7x^2 + 2x - 9) + (-3x^2 + 5)$
55. $(5x + 8) - (-2x^2 - 6x + 8)$
56. $(-6y^2 + 3y - 4) - (9y^2 - 3y)$
57. $(-8x^4 + 7x) + (-8x^4 + x + 9)$
58. $(6y^5 - 6y^3 + 4) + (-2y^5 - 8y^3 - 7)$
59. $(3x^2 + 5x - 8) + (5x^2 + 9x + 12) - (x^2 - 14)$
60. $(-a^2 + 1) - (a^2 - 3) + (5a^2 - 6a + 7)$

TRANSLATING

Perform each indicated operation.

61. Subtract $4x$ from $(7x - 3)$.
62. Subtract y from $(y^2 - 4y + 1)$.
63. Add $(4x^2 - 6x + 1)$ and $(3x^2 + 2x + 1)$.
64. Add $(-3x^2 - 5x + 2)$ and $(x^2 - 6x + 9)$.
65. Subtract $(19x^2 + 5)$ from $(81x^2 + 10)$.
66. Subtract $(2x + xy)$ from $(3x - 9xy)$.
67. Subtract $(5x + 7)$ from $(7x^2 + 3x + 9)$.
68. Subtract $(5y^2 + 8y + 2)$ from $(7y^2 + 9y - 8)$.
69. Subtract $(2x + 2)$ from the sum of $(8x + 1)$ and $(6x + 3)$.
70. Subtract $(-12x - 3)$ from the sum of $(-5x - 7)$ and $(12x + 3)$.
71. Subtract $(4y^2 - 6y - 3)$ from the sum of $(8y^2 + 7)$ and $(6y + 9)$.
72. Subtract $(4x^2 - 2x + 2)$ from the sum of $(x^2 + 7x + 1)$ and $(7x + 5)$.

Find the area of each figure. Write a polynomial that describes the total area of the rectangles and squares shown in Exercises 73–74. Then simplify the polynomial. See Example 8.

△ 73.

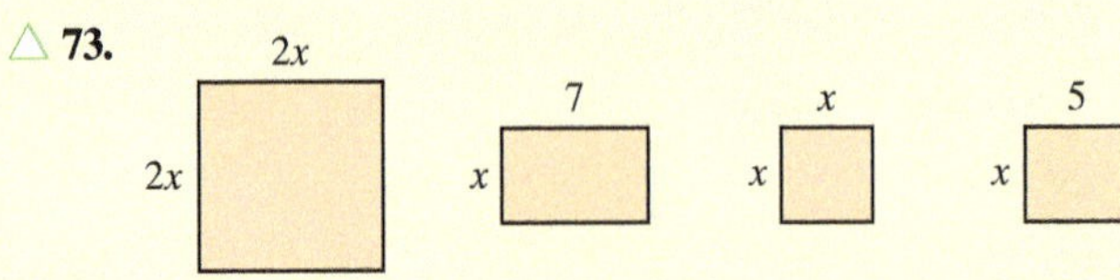

△ 74.

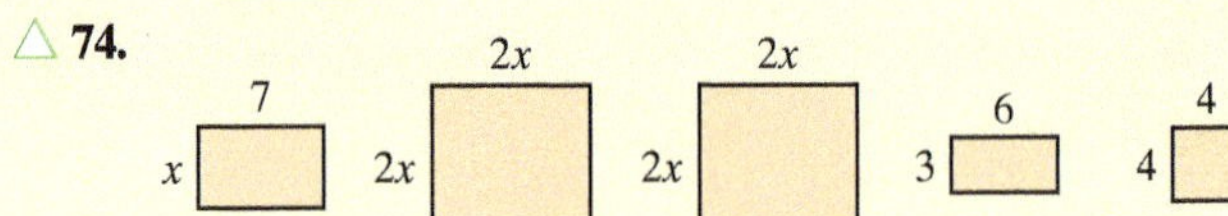

Add or subtract as indicated. See Examples 9 through 13.

75. $(9a + 6b - 5) + (-11a - 7b + 6)$

76. $(3x - 2 + 6y) + (7x - 2 - y)$

77. $(4x^2 + y^2 + 3) - (x^2 + y^2 - 2)$

78. $(7a^2 - 3b^2 + 10) - (-2a^2 + b^2 - 12)$

79. $(x^2 + 2xy - y^2) + (5x^2 - 4xy + 20y^2)$

80. $(a^2 - ab + 4b^2) + (6a^2 + 8ab - b^2)$

81. $(11r^2s + 16rs - 3 - 2r^2s^2) - (3sr^2 + 5 - 9r^2s^2)$

82. $(3x^2y - 6xy + x^2y^2 - 5) - (11x^2y^2 - 1 + 5yx^2)$

Simplify each polynomial by combining like terms.

83. $7.75x + 9.16x^2 - 1.27 - 14.58x^2 - 18.34$

84. $1.85x^2 - 3.76x + 9.25x^2 + 10.76 - 4.21x$

Perform each indicated operation.

85. $[(7.9y^4 - 6.8y^3 + 3.3y) + (6.1y^3 - 5)] - (4.2y^4 + 1.1y - 1)$

86. $[(1.2x^2 - 3x + 9.1) - (7.8x^2 - 3.1 + 8)] + (1.2x - 6)$

REVIEW AND PREVIEW

Multiply. See Section 12.1.

87. $3x(2x)$

88. $-7x(x)$

89. $(12x^3)(-x^5)$

90. $6r^3(7r^{10})$

91. $10x^2(20xy^2)$

92. $-z^2y(11zy)$

CONCEPT EXTENSIONS

Recall that the perimeter of a figure is the sum of the lengths of its sides. For Exercises 93 through 96, find the perimeter of each polynomial.

△ **93.**

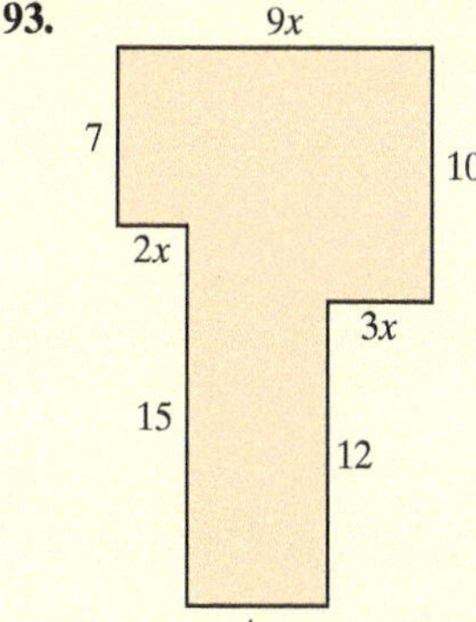

△ **94.**

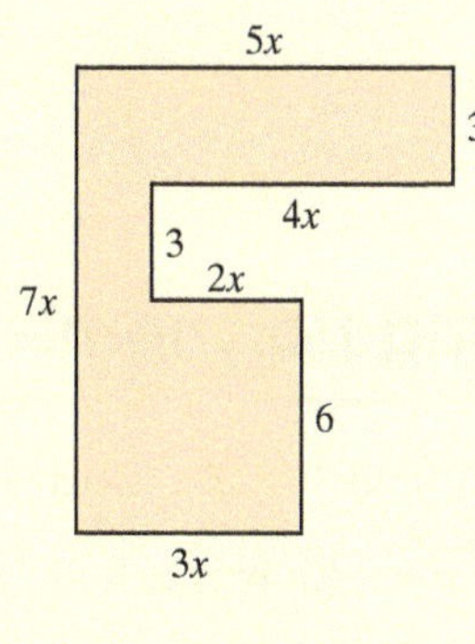

△ **95.**

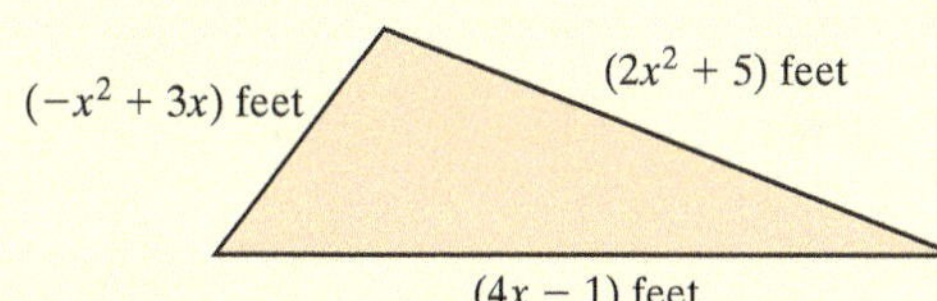

△ **96.**

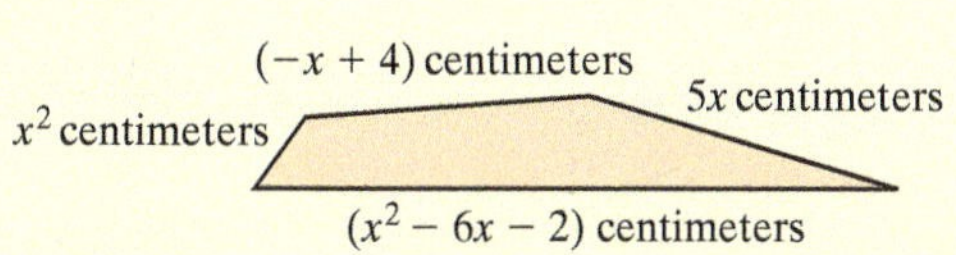

△ **97.** A wooden beam is $(4y^2 + 4y + 1)$ meters long. If a piece $(y^2 - 10)$ meters is cut, express the length of the remaining piece of beam as a polynomial in y.

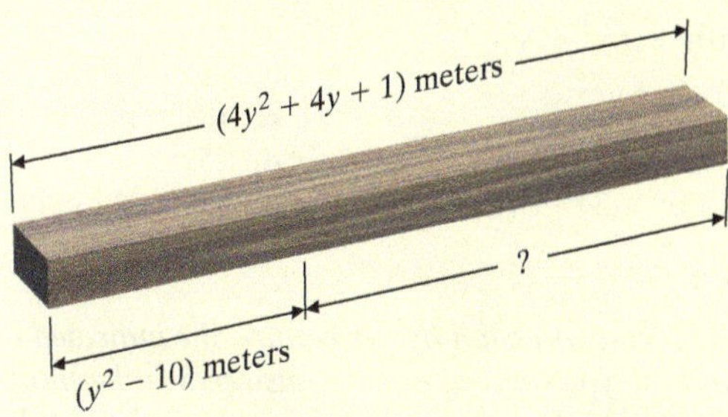

△ **98.** A piece of quarter-round molding is $(13x - 7)$ inches long. If a piece $(2x + 2)$ inches is removed, express the length of the remaining piece of molding as a polynomial in x.

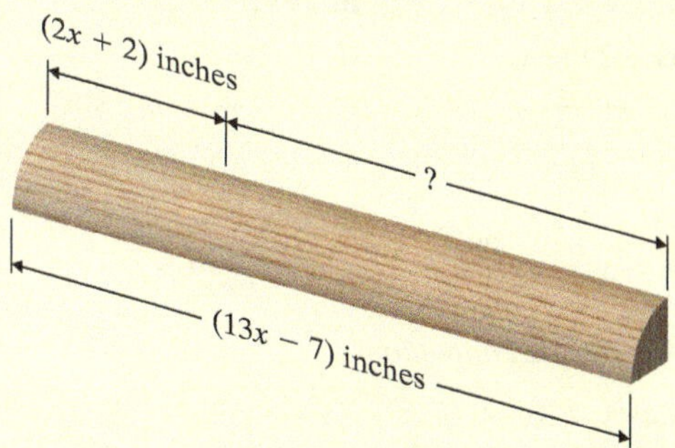

The number of cell phones recycled by the largest recycler x years after 2004 is given by the polynomial function $P(x) = 0.16x^2 + 0.29x + 1.88$ for the years 2004 through 2010. Use this polynomial function for Exercises 99 and 100.

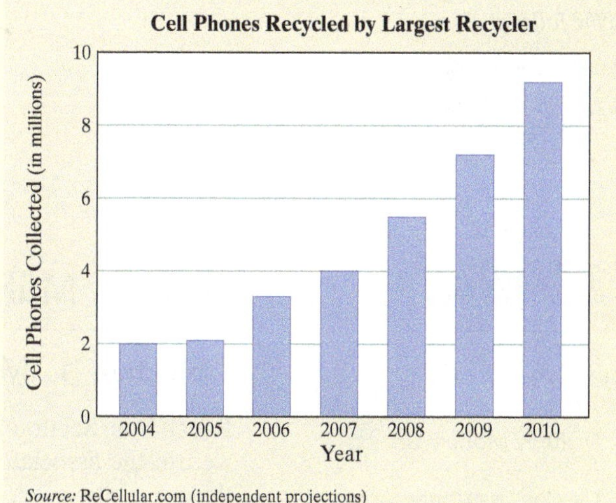

99. Estimate the number of cell phones recycled by this company in 2009.

100. Use the given polynomial function to estimate the number of cell phones recycled in 2014.

101. Describe how to find the degree of a term.

102. Describe how to find the degree of a polynomial.

103. Explain why xyz is a monomial while $x + y + z$ is a trinomial.

104. Explain why the degree of the term $5y^3$ is 3 and the degree of the polynomial $2y + y + 2y$ is 1.

Match each expression on the left with its simplification on the right. Not all letters on the right must be used, and a letter may be used more than once.

105. $10y - 6y^2 - y$

106. $5x + 5x$

107. $(5x - 3) + (5x - 3)$

108. $(15x - 3) - (5x - 3)$

a. $3y$
b. $9y - 6y^2$
c. $10x$
d. $25x^2$
e. $10x - 6$
f. none of these

Simplify each expression by performing the indicated operation. Explain how you arrived at each answer. See the third Concept Check in this section.

109. **a.** $z + 3z$ **b.** $z \cdot 3z$ **c.** $-z - 3z$ **d.** $(-z)(-3z)$

110. **a.** $2y + y$ **b.** $2y \cdot y$ **c.** $-2y - y$ **d.** $(-2y)(-y)$

111. **a.** $m \cdot m \cdot m$ **b.** $m + m + m$ **c.** $(-m)(-m)(-m)$ **d.** $-m - m - m$

112. **a.** $x + x$ **b.** $x \cdot x$ **c.** $-x - x$ **d.** $(-x)(-x)$

Perform indicated operations.

113. $(4x^{2a} - 3x^a + 0.5) - (x^{2a} - 5x^a - 0.2)$

114. $(9y^{5a} - 4y^{3a} + 1.5y) - (6y^{5a} - y^{3a} + 4.7y)$

115. $(8x^{2y} - 7x^y + 3) + (-4x^{2y} + 9x^y - 14)$

116. $(14z^{5x} + 3z^{2x} + z) - (2z^{5x} - 10z^{2x} + 3z)$

If $P(x) = 3x + 3$, $Q(x) = 4x^2 - 6x + 3$, and $R(x) = 5x^2 - 7$, find the following.

117. $P(x) + Q(x)$

118. $R(x) + P(x)$

119. $Q(x) - R(x)$

120. $P(x) - Q(x)$

121. $2[Q(x)] - R(x)$

122. $-5[P(x)] - Q(x)$

If $P(x)$ is the polynomial given, find **a.** *$P(a)$,* **b.** *$P(-x)$, and* **c.** *$P(x + h)$.*

123. $P(x) = 2x - 3$

124. $P(x) = 8x + 3$

125. $P(x) = 4x$

126. $P(x) = -4x$

Fill in the squares so that each is a true statement.

127. $3x^{\square} + 4x^2 = 7x^{\square}$

128. $9y^7 + 3y^{\square} = 12y^7$

129. $2x^{\square} + 3x^{\square} - 5x^{\square} + 4x^{\square} = 6x^4 - 2x^3$

130. $3y^{\square} + 7y^{\square} - 2y^{\square} - y^{\square} = 10y^5 - 3y^2$

Write a polynomial that describes the surface area of each figure. (Recall that the surface area of a solid is the sum of the areas of the faces or sides of the solid.)

△ **131.**

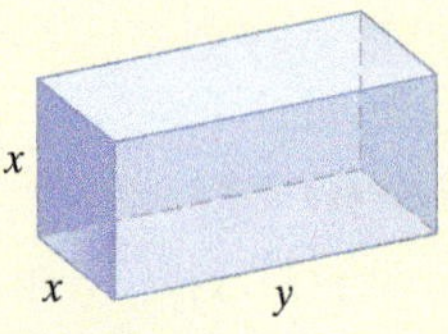

△ **132.**

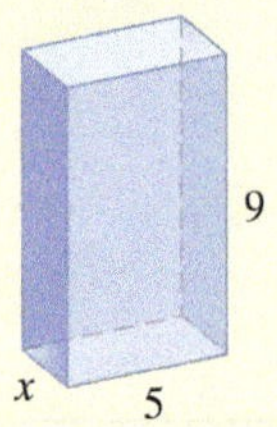

12.3 Multiplying Polynomials

Objectives

1. Multiply Monomials.
2. Use the Distributive Property to Multiply Polynomials.
3. Multiply Polynomials Vertically.

Objective 1 Multiplying Monomials

Recall from Section 12.1 that to multiply two monomials such as $(-5x^3)$ and $(-2x^4)$, we use the associative and commutative properties and regroup. Remember, also, that to multiply exponential expressions with a common base, we use the product rule for exponents and add exponents.

$$(-5x^3)(-2x^4) = (-5)(-2)(x^3)(x^4) = 10x^7$$

Examples Multiply.

1. $6x \cdot 4x = (6 \cdot 4)(x \cdot x)$ Use the commutative and associative properties.
$= 24x^2$ Multiply.

2. $-7x^2 \cdot 0.2x^5 = (-7 \cdot 0.2)(x^2 \cdot x^5)$
$= -1.4x^7$

Practice 1–3

Multiply.

1. $5y \cdot 2y$

2. $(5z^3) \cdot (-0.4z^5)$

3. $\left(-\frac{1}{9}b^6\right)\left(-\frac{7}{8}b^3\right)$

3. $\left(-\frac{1}{3}x^5\right)\left(-\frac{2}{9}x\right) = \left(-\frac{1}{3}\cdot -\frac{2}{9}\right)\cdot (x^5 \cdot x)$

$= \frac{2}{27}x^6$

Work Practice 1–3

✓Concept Check Simplify.

a. $3x \cdot 2x$ **b.** $3x + 2x$

Objective 2 Using the Distributive Property to Multiply Polynomials

To multiply polynomials that are not monomials, use the distributive property.

Example 4 Use the distributive property to find each product.

a. $5x(2x^3 + 6)$ **b.** $-3x^2(5x^2 + 6x - 1)$

Solution:

a. $5x(2x^3 + 6) = 5x(2x^3) + 5x(6)$ Use the distributive property.

$= 10x^4 + 30x$ Multiply.

b. $-3x^2(5x^2 + 6x - 1)$

$= (-3x^2)(5x^2) + (-3x^2)(6x) + (-3x^2)(-1)$ Use the distributive property.

$= -15x^4 - 18x^3 + 3x^2$ Multiply.

Work Practice 4

Practice 4

Use the distributive property to find each product.

a. $3x(9x^5 + 11)$

b. $-6x^3(2x^2 - 9x + 2)$

We also use the distributive property to multiply two binomials. To multiply $(x + 3)$ by $(x + 1)$, distribute the factor $(x + 3)$ first.

$(x + 3)(x + 1) = x(x + 1) + 3(x + 1)$ Distribute $(x + 3)$.

$= x(x) + x(1) + 3(x) + 3(1)$ Apply the distributive property a second time.

$= x^2 + x + 3x + 3$ Multiply.

$= x^2 + 4x + 3$ Combine like terms.

This idea can be expanded so that we can multiply any two polynomials.

To Multiply Two Polynomials

Multiply each term of the first polynomial by each term of the second polynomial and then combine like terms.

Answers

1. $10y^2$ **2.** $-2z^8$ **3.** $\frac{7}{72}b^9$

4. a. $27x^6 + 33x$

b. $-12x^5 + 54x^4 - 12x^3$

✓Concept Check Answers

a. $6x^2$ **b.** $5x$

Practice 5

Multiply $(5x - 2)(2x + 3)$.

Example 5 Multiply $(3x + 2)(2x - 5)$.

Solution: Multiply each term of the first binomial by each term of the second.

$$\begin{aligned}(3x + 2)(2x - 5) &= 3x(2x) + 3x(-5) + 2(2x) + 2(-5)\\ &= 6x^2 - 15x + 4x - 10 && \text{Multiply.}\\ &= 6x^2 - 11x - 10 && \text{Combine like terms.}\end{aligned}$$

Work Practice 5

Practice 6

Multiply $(5x - 3y)^2$.

Example 6 Multiply $(2x - y)^2$.

Solution: Recall that $a^2 = a \cdot a$, so $(2x - y)^2 = (2x - y)(2x - y)$. Multiply each term of the first polynomial by each term of the second.

$$\begin{aligned}(2x - y)(2x - y) &= 2x(2x) + 2x(-y) + (-y)(2x) + (-y)(-y)\\ &= 4x^2 - 2xy - 2xy + y^2 && \text{Multiply.}\\ &= 4x^2 - 4xy + y^2 && \text{Combine like terms.}\end{aligned}$$

Work Practice 6

✓**Concept Check** Square where indicated. Simplify if possible.

a. $(4a)^2 + (3b)^2$

b. $(4a + 3b)^2$

Practice 7

Multiply $(y + 4)$ by $(2y^2 - 3y + 5)$.

Example 7 Multiply $(t + 2)$ by $(3t^2 - 4t + 2)$.

Solution: Multiply each term of the first polynomial by each term of the second.

$$\begin{aligned}(t + 2)(3t^2 - 4t + 2) &= t(3t^2) + t(-4t) + t(2) + 2(3t^2) + 2(-4t) + 2(2)\\ &= 3t^3 - 4t^2 + 2t + 6t^2 - 8t + 4\\ &= 3t^3 + 2t^2 - 6t + 4 && \text{Combine like terms.}\end{aligned}$$

Work Practice 7

Practice 8

Multiply $(s + 2t)^3$.

Example 8 Multiply $(3a + b)^3$.

Solution: Write $(3a + b)^3$ as $(3a + b)(3a + b)(3a + b)$.

$$\begin{aligned}(3a + b)(3a + b)(3a + b) &= (9a^2 + 3ab + 3ab + b^2)(3a + b)\\ &= (9a^2 + 6ab + b^2)(3a + b)\\ &= 9a^2(3a + b) + 6ab(3a + b) + b^2(3a + b)\\ &= 27a^3 + 9a^2b + 18a^2b + 6ab^2 + 3ab^2 + b^3\\ &= 27a^3 + 27a^2b + 9ab^2 + b^3\end{aligned}$$

Work Practice 8

Answers

5. $10x^2 + 11x - 6$
6. $25x^2 - 30xy + 9y^2$
7. $2y^3 + 5y^2 - 7y + 20$
8. $s^3 + 6s^2t + 12st^2 + 8t^3$

✓**Concept Check Answers**

a. $16a^2 + 9b^2$
b. $16a^2 + 24ab + 9b^2$

Objective 3 Multiplying Polynomials Vertically 

Another convenient method for multiplying polynomials is to use a vertical format similar to the format used to multiply real numbers. We demonstrate this method by multiplying $(3y^2 - 4y + 1)$ by $(y + 2)$.

Example 9 Multiply $(3y^2 - 4y + 1)(y + 2)$. Use a vertical format.

Solution:

$$\begin{array}{rrrrl}
 & 3y^2 & -\ 4y & +\ 1 & \\
\times & & y & +\ 2 & \text{1st, Multiply } 3y^2 - 4y + 1 \text{ by } 2. \\
\hline
 & 6y^2 & -\ 8y & +\ 2 & \text{2nd, Multiply } 3y^2 - 4y + 1 \text{ by } y. \\
3y^3 & -\ 4y^2 & +\ y & & \text{Line up like terms.} \\
\hline
3y^3 & +\ 2y^2 & -\ 7y & +\ 2 & \text{3rd, Combine like terms.}
\end{array}$$

Thus, $(3y^2 - 4y + 1)(y + 2) = 3y^3 + 2y^2 - 7y + 2$.

Work Practice 9

Practice 9

Multiply $(5x^2 - 3x + 5)(x - 4)$.

Helpful Hint Make sure like terms are lined up.

When multiplying vertically, be careful if a power is missing; you may want to leave space in the partial products and take care that like terms are lined up.

Example 10 Multiply $(2x^3 - 3x + 4)(x^2 + 1)$. Use a vertical format.

Solution:

$$\begin{array}{rrrrrl}
 & 2x^3 & & -\ 3x & +\ 4 & \\
\times & & & x^2 & +\ 1 & \\
\hline
 & 2x^3 & & -\ 3x & +\ 4 & \text{Leave space for missing powers of } x. \\
2x^5 & -\ 3x^3 & +\ 4x^2 & & & \leftarrow \text{Line up like terms.} \\
\hline
2x^5 & -\ x^3 & +\ 4x^2 & -\ 3x & +\ 4 & \text{Combine like terms.}
\end{array}$$

Work Practice 10

Practice 10

Multiply $(x^3 - 2x^2 + 1)(x^2 + 2)$ using a vertical format.

Example 11 Find the product of $(2x^2 - 3x + 4)$ and $(x^2 + 5x - 2)$ using a vertical format.

Solution: First, we arrange the polynomials in a vertical format. Then we multiply each term of the second polynomial by each term of the first polynomial.

$$\begin{array}{rrrrrl}
 & & 2x^2 & -\ 3x & +\ 4 & \\
\times & & x^2 & +\ 5x & -\ 2 & \\
\hline
 & & -4x^2 & +\ 6x & -\ 8 & \text{Multiply } 2x^2 - 3x + 4 \text{ by } -2. \\
 & 10x^3 & -\ 15x^2 & +\ 20x & & \text{Multiply } 2x^2 - 3x + 4 \text{ by } 5x. \\
2x^4 & -\ 3x^3 & +\ 4x^2 & & & \text{Multiply } 2x^2 - 3x + 4 \text{ by } x^2. \\
\hline
2x^4 & +\ 7x^3 & -\ 15x^2 & +\ 26x & -\ 8 & \text{Combine like terms.}
\end{array}$$

Work Practice 11

Practice 11

Find the product of $(5x^2 + 2x - 2)$ and $(x^2 - x + 3)$ using a vertical format.

Answers

9. $5x^3 - 23x^2 + 17x - 20$
10. $x^5 - 2x^4 + 2x^3 - 3x^2 + 2$
11. $5x^4 - 3x^3 + 11x^2 + 8x - 6$

Vocabulary, Readiness & Video Check

Fill in each blank with the correct choice.

1. The expression $5x(3x + 2)$ equals $5x \cdot 3x + 5x \cdot 2$ by the ____________ property.

a. commutative **b.** associative **c.** distributive

2. The expression $(x + 4)(7x - 1)$ equals $x(7x - 1) + 4(7x - 1)$ by the ____________ property.

a. commutative **b.** associative **c.** distributive

3. The expression $(5y - 1)^2$ equals ____________.

a. $2(5y - 1)$ **b.** $(5y - 1)(5y + 1)$ **c.** $(5y - 1)(5y - 1)$

4. The expression $9x \cdot 3x$ equals ____________.

a. $27x$ **b.** $27x^2$ **c.** $12x$ **d.** $12x^2$

Martin-Gay Interactive Videos *Watch the section lecture video and answer the following questions.*

See Video 12.3

Objective 1 **5.** For Example 1, we use the product property to multiply the monomials. Is it possible to add the same two monomials? Why or why not?

Objective 2 **6.** What property and what exponent rule are used in Examples 2–6?

Objective 3 **7.** Would you say the vertical format used in Example 7 also applies the distributive property? Explain.

12.3 Exercise Set MyMathLab®

Multiply. See Examples 1 through 3.

1. $-4n^3 \cdot 7n^7$

2. $9t^6(-3t^5)$

3. $(-3.1x^3)(4x^9)$

4. $(-5.2x^4)(3x^4)$

5. $\left(-\frac{1}{3}y^2\right)\left(\frac{2}{5}y\right)$

6. $\left(-\frac{3}{4}y^7\right)\left(\frac{1}{7}y^4\right)$

7. $(2x)(-3x^2)(4x^5)$

8. $(x)(5x^4)(-6x^7)$

Multiply. See Example 4.

9. $3x(2x + 5)$

10. $2x(6x + 3)$

11. $-2a(a + 4)$

12. $-3a(2a + 7)$

13. $3x(2x^2 - 3x + 4)$

14. $4x(5x^2 - 6x - 10)$

15. $-2a^2(3a^2 - 2a + 3)$

16. $-4b^2(3b^3 - 12b^2 - 6)$

17. $-y(4x^3 - 7x^2y + xy^2 + 3y^3)$

18. $-x(6y^3 - 5xy^2 + x^2y - 5x^3)$

19. $\frac{1}{2}x^2(8x^2 - 6x + 1)$

20. $\frac{1}{3}y^2(9y^2 - 6y + 1)$

Multiply. See Examples 5 and 6.

21. $(x + 4)(x + 3)$

22. $(x + 2)(x + 9)$

23. $(a + 7)(a - 2)$

24. $(y - 10)(y + 11)$

25. $\left(x + \frac{2}{3}\right)\left(x - \frac{1}{3}\right)$

26. $\left(x + \frac{3}{5}\right)\left(x - \frac{2}{5}\right)$

27. $(3x^2 + 1)(4x^2 + 7)$

28. $(5x^2 + 2)(6x^2 + 2)$

29. $(2y - 4)^2$

30. $(6x - 7)^2$

31. $(4x - 3)(3x - 5)$

32. $(8x - 3)(2x - 4)$

33. $(3x^2 + 1)^2$

34. $(x^2 + 4)^2$

35. Perform the indicated operations.

a. $4y^2(-y^2)$

b. $4y^2 - y^2$

c. Explain the difference between the two expressions.

36. Perform the indicated operations.

a. $9x^2(-10x^2)$

b. $9x^2 - 10x^2$

c. Explain the difference between the two expressions.

Multiply. See Example 7.

37. $(x - 2)(x^2 - 3x + 7)$
38. $(x + 3)(x^2 + 5x - 8)$
39. $(x + 5)(x^3 - 3x + 4)$
40. $(a + 2)(a^3 - 3a^2 + 7)$
41. $(2a - 3)(5a^2 - 6a + 4)$
42. $(3 + b)(2 - 5b - 3b^2)$

Multiply. See Example 8.

43. $(x + 2)^3$
44. $(y - 1)^3$
45. $(2y - 3)^3$
46. $(3x + 4)^3$

Multiply vertically. See Examples 9 through 11.

47. $(2x - 11)(6x + 1)$
48. $(4x - 7)(5x + 1)$
49. $(5x + 1)(2x^2 + 4x - 1)$
50. $(4x - 5)(8x^2 + 2x - 4)$
51. $(x^2 + 5x - 7)(2x^2 - 7x - 9)$
52. $(3x^2 - x + 2)(x^2 + 2x + 1)$

MIXED PRACTICE

Multiply. See Examples 1 through 11.

53. $-1.2y(-7y^6)$
54. $-4.2x(-2x^5)$
55. $-3x(x^2 + 2x - 8)$
56. $-5x(x^2 - 3x + 10)$
57. $(x + 19)(2x + 1)$
58. $(3y + 4)(y + 11)$
59. $\left(x + \frac{1}{7}\right)\left(x - \frac{3}{7}\right)$
60. $\left(m + \frac{2}{9}\right)\left(m - \frac{1}{9}\right)$
61. $(3y + 5)^2$
62. $(7y + 2)^2$
63. $(a + 4)(a^2 - 6a + 6)$
64. $(t + 3)(t^2 - 5t + 5)$
65. $(2x - 5)^3$
66. $(3y - 1)^3$
67. $(4x + 5)(8x^2 + 2x - 4)$
68. $(5x + 4)(x^2 - x + 4)$
69. $(3x^2 + 2x - 4)(2x^2 - 4x + 3)$
70. $(a^2 + 3a - 2)(2a^2 - 5a - 1)$

Express as the product of polynomials. Then multiply.

△ 71. Find the area of the rectangle.

(2x + 5) yards

(2x − 5) yards

△ 72. Find the area of the square field.

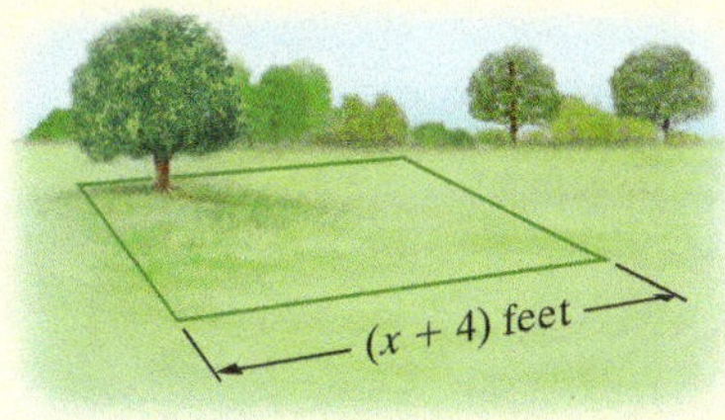

△ 73. Find the area of the triangle.
(Triangle Area $= \frac{1}{2} \cdot \text{base} \cdot \text{height}$)

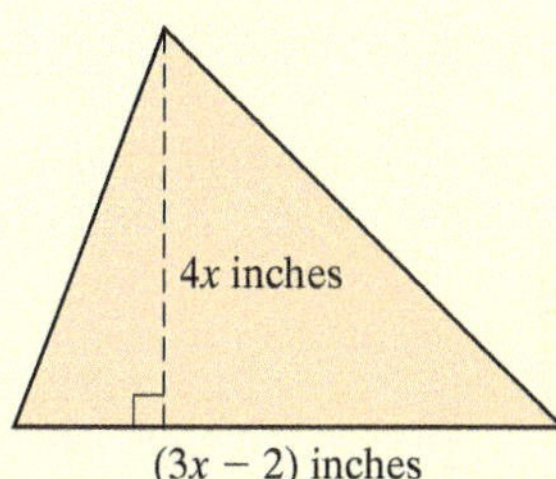

△ 74. Find the volume of the cube.
(Volume = length · width · height)

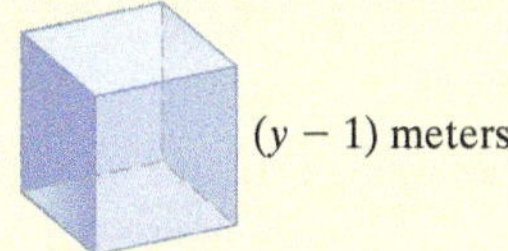

REVIEW AND PREVIEW

In this section, we review operations on monomials. Study the box below, then proceed. See Sections 3.1, 12.1, and 12.2. (continued on next page)

Operations on Monomials	
Multiply	Review the product rule for exponents.
Divide	Review the quotient rule for exponents.
Add or Subtract	Remember, we may only combine like terms.

Perform the operations on the monomials if possible. The first two rows have been completed for you.

	Monomials	Add	Subtract	Multiply	Divide
	$6x, 3x$	$6x + 3x = 9x$	$6x - 3x = 3x$	$6x \cdot 3x = 18x^2$	$\frac{6x}{3x} = 2$
	$-12x^2, 2x$	$-12x^2 + 2x$, can't be simplified	$-12x^2 - 2x$, can't be simplified	$-12x^2 \cdot 2x = -24x^3$	$\frac{-12x^2}{2x} = -6x$
75.	$5a, 15a$				
76.	$4y^7, 4y^3$				
77.	$-3y^5, 9y^4$				
78.	$-14x^2, 2x^2$				

CONCEPT EXTENSIONS

79. Perform each indicated operation. Explain the difference between the two expressions.

a. $(3x + 5) + (3x + 7)$
b. $(3x + 5)(3x + 7)$

80. Perform each indicated operation. Explain the difference between the two expressions.

a. $(8x - 3) - (5x - 2)$
b. $(8x - 3)(5x - 2)$

MIXED PRACTICE

Perform the indicated operations. See Sections 12.1 and 12.2.

81. $(3x - 1) + (10x - 6)$

82. $(2x - 1) + (10x - 7)$

83. $(3x - 1)(10x - 6)$

84. $(2x - 1)(10x - 7)$

85. $(3x - 1) - (10x - 6)$

86. $(2x - 1) - (10x - 7)$

CONCEPT EXTENSIONS

△ **87.** The area of the larger rectangle on the right is $x(x + 3)$. Find another expression for this area by finding the sum of the areas of the two smaller rectangles.

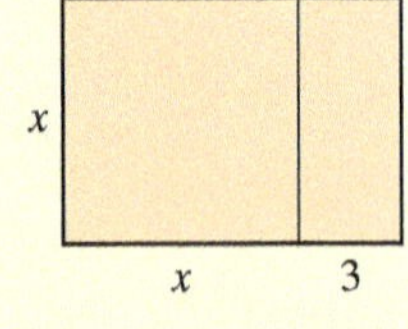

△ **88.** Write an expression for the area of the larger rectangle on the right in two different ways.

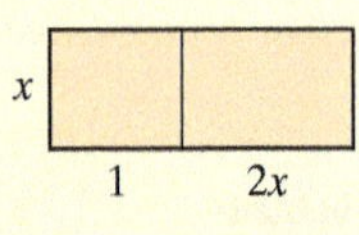

△ **89.** The area of the figure on the right is $(x + 2)(x + 3)$. Find another expression for this area by finding the sum of the areas of the four smaller rectangles.

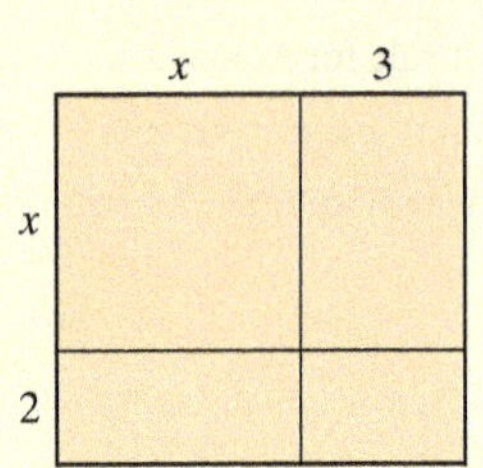

△ **90.** Write an expression for the area of the figure in two different ways.

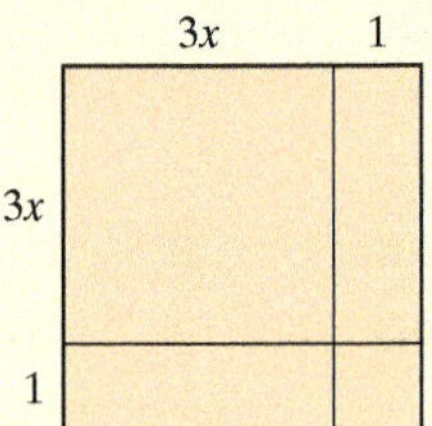

Simplify. See the Concept Checks in this section.

91. $5a + 6a$

92. $5a \cdot 6a$

Square where indicated. Simplify if possible.

93. $(5x)^2 + (2y)^2$

94. $(5x + 2y)^2$

95. Multiply each of the following polynomials.

a. $(a + b)(a - b)$
b. $(2x + 3y)(2x - 3y)$
c. $(4x + 7)(4x - 7)$
d. Can you make a general statement about all products of the form $(x + y)(x - y)$?

96. Evaluate each of the following.

a. $(2 + 3)^2; 2^2 + 3^2$
b. $(8 + 10)^2; 8^2 + 10^2$
c. Does $(a + b)^2 = a^2 + b^2$ no matter what the values of a and b are? Why or why not?

△ **97.** Write a polynomial that describes the area of the shaded region. (Find the area of the larger square minus the area of the smaller square.)

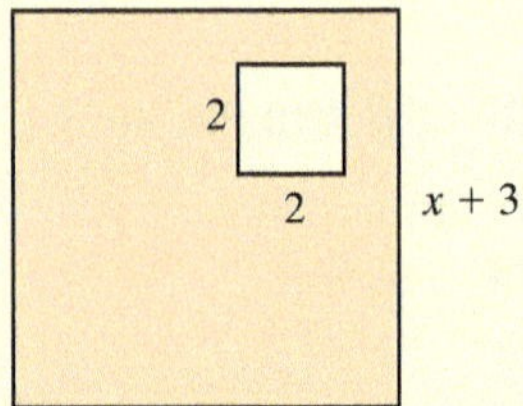

△ **98.** Write a polynomial that describes the area of the shaded region. (See Exercise 97.)

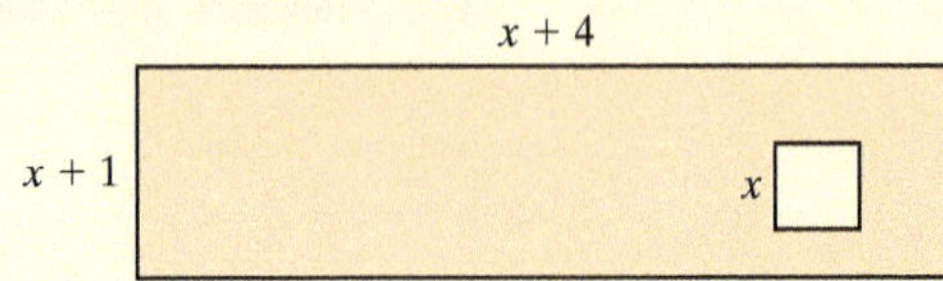

12.4 Special Products 

Objective 1 Using the FOIL Method

In this section, we multiply binomials using special products. First, a special order for multiplying binomials called the FOIL order or method is introduced. This method is demonstrated by multiplying $(3x + 1)$ by $(2x + 5)$ as shown below.

The FOIL Method

F stands for the product of the **First** terms. $(3x + 1)(2x + 5)$

$(3x)(2x) = 6x^2$ **F**

O stands for the product of the **Outer** terms. $(3x + 1)(2x + 5)$

$(3x)(5) = 15x$ **O**

I stands for the product of the **Inner** terms. $(3x + 1)(2x + 5)$

$(1)(2x) = 2x$ **I**

L stands for the product of the **Last** terms. $(3x + 1)(3x + 5)$

$(1)(5) = 5$ **L**

$$(3x + 1)(2x + 5) = \overset{F}{6x^2} + \overset{O}{15x} + \overset{I}{2x} + \overset{L}{5}$$

$$= 6x^2 + 17x + 5 \quad \text{Combine like terms.}$$

✓**Concept Check** Multiply $(3x + 1)(2x + 5)$ using methods from the last section. Show that the product is still $6x^2 + 17x + 5$.

Example 1 Multiply $(x - 3)(x + 4)$ by the FOIL method.

Solution:

$$(x - 3)(x + 4) = \overset{F}{(x)(x)} + \overset{O}{(x)(4)} + \overset{I}{(-3)(x)} + \overset{L}{(-3)(4)}$$

$$= x^2 + 4x - 3x - 12$$

$$= x^2 + x - 12 \quad \text{Combine like terms.}$$

Work Practice 1

Example 2 Multiply $(5x - 7)(x - 2)$ by the FOIL method.

Solution:

$$(5x - 7)(x - 2) = \overset{F}{5x(x)} + \overset{O}{5x(-2)} + \overset{I}{(-7)(x)} + \overset{L}{(-7)(-2)}$$

$$= 5x^2 - 10x - 7x + 14$$

$$= 5x^2 - 17x + 14 \quad \text{Combine like terms.}$$

Work Practice 2

Objectives

1 Multiply Two Binomials Using the FOIL Method.

2 Square a Binomial.

3 Multiply the Sum and Difference of Two Terms.

4 Use Special Products to Multiply Binomials.

Practice 1

Multiply $(x + 2)(x - 5)$ by the FOIL method.

Helpful Hint Remember that the FOIL order for multiplying can be used only for the product of 2 binomials.

Practice 2

Multiply $(4x - 9)(x - 1)$ by the FOIL method.

Answers

1. $x^2 - 3x - 10$ **2.** $4x^2 - 13x + 9$

✓**Concept Check Answer**

Multiply and simplify: $3x(2x + 5) + 1(2x + 5)$

Practice 3

Multiply $3(x + 5)(3x - 1)$.

Example 3 Multiply $2(y + 6)(2y - 1)$.

Solution:

$$2(y+6)(2y-1) = 2(\overset{F}{2y^2} - \overset{O}{1y} + \overset{I}{12y} - \overset{L}{6})$$
$$= 2(2y^2 + 11y - 6) \quad \text{Simplify inside parentheses.}$$
$$= 4y^2 + 22y - 12 \quad \text{Now use the distributive property.}$$

Work Practice 3

Objective 2 Squaring Binomials

Now, try squaring a binomial using the FOIL method.

Practice 4

Multiply $(4x - 1)^2$.

Example 4 Multiply $(3y + 1)^2$.

Solution: $(3y + 1)^2 = (3y + 1)(3y + 1)$

$$= \overset{F}{(3y)(3y)} + \overset{O}{(3y)(1)} + \overset{I}{1(3y)} + \overset{L}{1(1)}$$
$$= 9y^2 + 3y + 3y + 1$$
$$= 9y^2 + 6y + 1$$

Work Practice 4

Notice the pattern that appears in Example 4.

$(3y + 1)^2 = 9y^2 + 6y + 1$

$9y^2$ is the first term of the binomial squared. $(3y)^2 = 9y^2$.

$6y$ is 2 times the product of both terms of the binomial. $(2)(3y)(1) = 6y$.

1 is the second term of the binomial squared. $(1)^2 = 1$.

This pattern leads to the following, which can be used when squaring a binomial. We call these **special products.**

Squaring a Binomial

A binomial squared is equal to the square of the first term plus or minus twice the product of both terms plus the square of the second term.

$$(a + b)^2 = a^2 + 2ab + b^2$$
$$(a - b)^2 = a^2 - 2ab + b^2$$

This product can be visualized geometrically.

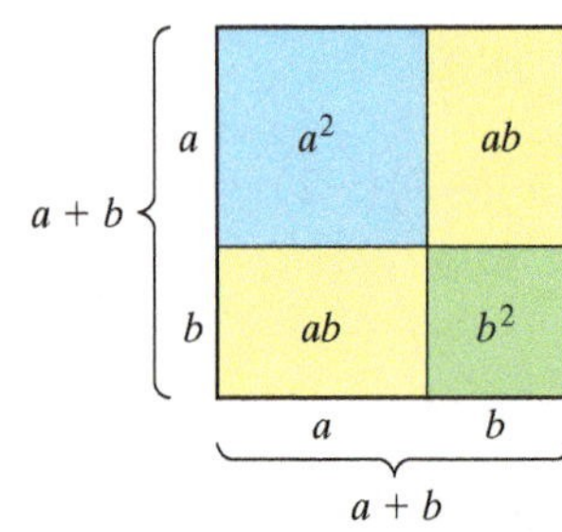

The area of the large square is side · side.

$$\text{Area} = (a + b)(a + b) = (a + b)^2$$

The area of the large square is also the sum of the areas of the smaller rectangles.

$$\text{Area} = a^2 + ab + ab + b^2 = a^2 + 2ab + b^2$$

Thus, $(a + b)^2 = a^2 + 2ab + b^2$.

Answers

3. $9x^2 + 42x - 15$ **4.** $16x^2 - 8x + 1$

Example 5 Use a special product to square each binomial.

a. $(t + 2)^2$ **b.** $(p - q)^2$ **c.** $(2x + 5)^2$ **d.** $(x^2 - 7y)^2$

Solution:

		first term squared	plus or minus	twice the product of the terms	plus	second term squared
a. $(t + 2)^2$	=	t^2	+	$2(t)(2)$	+	$2^2 = t^2 + 4t + 4$
b. $(p - q)^2$	=	p^2	−	$2(p)(q)$	+	$q^2 = p^2 - 2pq + q^2$
c. $(2x + 5)^2$	=	$(2x)^2$	+	$2(2x)(5)$	+	$5^2 = 4x^2 + 20x + 25$
d. $(x^2 - 7y)^2$	=	$(x^2)^2$	−	$2(x^2)(7y)$	+	$(7y^2) = x^4 - 14x^2y + 49y^2$

Work Practice 5

Practice 5

Use a special product to square each binomial.

a. $(b + 3)^2$ **b.** $(x - y)^2$

c. $(3y + 2)^2$ **d.** $(a^2 - 5b)^2$

Helpful Hint

Notice that

$(a + b)^2 \neq a^2 + b^2$ The middle term $2ab$ is missing.

$(a + b)^2 = (a + b)(a + b) = a^2 + 2ab + b^2$

Likewise,

$(a - b)^2 \neq a^2 - b^2$

$(a - b)^2 = (a - b)(a - b) = a^2 - 2ab + b^2$

Objective 3 Multiplying the Sum and Difference of Two Terms

Another special product is the product of the sum and difference of the same two terms, such as $(x + y)(x - y)$. Finding this product by the FOIL method, we see a pattern emerge.

$$\begin{aligned}(x + y)(x - y) &= \overset{F}{x^2} - \overset{O}{xy} + \overset{I}{xy} - \overset{L}{y^2} \\ &= x^2 - y^2\end{aligned}$$

Notice that the middle two terms subtract out. This is because the **O**uter product is the opposite of the **I**nner product. Only the **difference of squares** remains.

Multiplying the Sum and Difference of Two Terms

The product of the sum and difference of two terms is the square of the first term minus the square of the second term.

$(a + b)(a - b) = a^2 - b^2$

Answers

5. a. $b^2 + 6b + 9$ **b.** $x^2 - 2xy + y^2$ **c.** $9y^2 + 12y + 4$ **d.** $a^4 - 10a^2b + 25b^2$

Practice 6

Use a special product to multiply.

a. $3(x + 5)(x - 5)$

b. $(4b - 3)(4b + 3)$

c. $\left(x + \frac{2}{3}\right)\left(x - \frac{2}{3}\right)$

d. $(5s + t)(5s - t)$

e. $(2y - 3z^2)(2y + 3z^2)$

Example 6 Use a special product to multiply.

a. $4(x + 4)(x - 4)$ **b.** $(6t + 7)(6t - 7)$ **c.** $\left(x - \frac{1}{4}\right)\left(x + \frac{1}{4}\right)$

d. $(2p - q)(2p + q)$ **e.** $(3x^2 - 5y)(3x^2 + 5y)$

Solution:

first term squared ↓ minus ↓ second term squared ↙

a. $4(x + 4)(x - 4) = 4(x^2 - 4^2) = 4(x^2 - 16) = 4x^2 - 64$

b. $(6t + 7)(6t - 7) = (6t)^2 - 7^2 = 36t^2 - 49$

c. $\left(x - \frac{1}{4}\right)\left(x + \frac{1}{4}\right) = x^2 - \left(\frac{1}{4}\right)^2 = x^2 - \frac{1}{16}$

d. $(2p - q)(2p + q) = (2p)^2 - q^2 = 4p^2 - q^2$

e. $(3x^2 - 5y)(3x^2 + 5y) = (3x^2)^2 - (5y)^2 = 9x^4 - 25y^2$

Work Practice 6

✓**Concept Check** Match expression number 1 and expression number 2 to the equivalent expression or expressions in the list below.

1. $(a + b)^2$ **2.** $(a + b)(a - b)$

a. $(a + b)(a + b)$ **b.** $a^2 - b^2$ **c.** $a^2 + b^2$

d. $a^2 - 2ab + b^2$ **e.** $a^2 + 2ab + b^2$

Objective 4 Using Special Products

Let's now practice multiplying polynomials in general. If possible, use a special product.

Practice 7

Use a special product to multiply, if possible.

a. $(4x + 3)(x - 6)$

b. $(7b - 2)^2$

c. $(x + 0.4)(x - 0.4)$

d. $\left(x^2 - \frac{3}{7}\right)\left(3x^4 + \frac{2}{7}\right)$

e. $(x + 1)(x^2 + 5x - 2)$

Example 7 Use a special product to multiply, if possible.

a. $(x - 5)(3x + 4)$ **b.** $(7x + 4)^2$ **c.** $(y - 0.6)(y + 0.6)$

d. $\left(y^4 + \frac{2}{5}\right)\left(3y^2 - \frac{1}{5}\right)$ **e.** $(a - 3)(a^2 + 2a - 1)$

Solution:

a. $(x - 5)(3x + 4) = 3x^2 + 4x - 15x - 20$ FOIL
$= 3x^2 - 11x - 20$

b. $(7x + 4)^2 = (7x)^2 + 2(7x)(4) + 4^2$ Squaring a binomial
$= 49x^2 + 56x + 16$

c. $(y - 0.6)(y + 0.6) = y^2 - (0.6)^2 = y^2 - 0.36$ Multiplying the sum and difference of 2 terms

d. $\left(y^4 + \frac{2}{5}\right)\left(3y^2 - \frac{1}{5}\right) = 3y^6 - \frac{1}{5}y^4 + \frac{6}{5}y^2 - \frac{2}{25}$ FOIL

Answers

6. a. $3x^2 - 75$ **b.** $16b^2 - 9$ **c.** $x^2 - \frac{4}{9}$ **d.** $25s^2 - t^2$ **e.** $4y^2 - 9z^4$

7. a. $4x^2 - 21x - 18$ **b.** $49b^2 - 28b + 4$ **c.** $x^2 - 0.16$ **d.** $3x^6 - \frac{9}{7}x^4 + \frac{2}{7}x^2 - \frac{6}{49}$ **e.** $x^3 + 6x^2 + 3x - 2$

✓**Concept Check Answer**

1. a and e **2.** b

e. I've inserted this product as a reminder that since it is not a binomial times a binomial, the FOIL order may not be used.

$$\begin{aligned}(a - 3)(a^2 + 2a - 1) &= a(a^2 + 2a - 1) - 3(a^2 + 2a - 1)\\ &= a^3 + 2a^2 - a - 3a^2 - 6a + 3\\ &= a^3 - a^2 - 7a + 3\end{aligned}$$

Multiplying each term of the binomial by each term of the trinomial

Work Practice 7

Helpful Hint

- When multiplying two binomials, you may always use the FOIL order or method.
- When multiplying any two polynomials, you may always use the distributive property to find the product.

Vocabulary, Readiness & Video Check

Answer each exercise true or false.

1. $(x + 4)^2 = x^2 + 16$

2. For $(x + 6)(2x - 1)$ the product of the first terms is $2x^2$.

3. $(x + 4)(x - 4) = x^2 + 16$

4. The product $(x - 1)(x^3 + 3x - 1)$ is a polynomial of degree 5.

Martin-Gay Interactive Videos

See Video 12.4 

Watch the section lecture video and answer the following questions.

Objective 1 **5.** From Examples 1–3, for what type of multiplication problem is the FOIL order of multiplication used?

Objective 2 **6.** Name at least one other method you can use to multiply Example 4.

Objective 3 **7.** From Example 5, why does multiplying the sum and difference of the same two terms always give you a binomial answer?

Objective 4 **8.** Why was the FOIL method not used for Example 10?

12.4 Exercise Set MyMathLab®

Multiply using the FOIL method. See Examples 1 through 3.

1. $(x + 3)(x + 4)$

2. $(x + 5)(x - 1)$

3. $(x - 5)(x + 10)$

4. $(y - 12)(y + 4)$

5. $(5x - 6)(x + 2)$

6. $(3y - 5)(2y - 7)$

7. $5(y - 6)(4y - 1)$

8. $2(x - 11)(2x - 9)$

9. $(2x + 5)(3x - 1)$

10. $(6x + 2)(x - 2)$

11. $\left(x - \frac{1}{3}\right)\left(x + \frac{2}{3}\right)$

12. $\left(x - \frac{2}{5}\right)\left(x + \frac{1}{5}\right)$

Multiply. See Examples 4 and 5.

13. $(x + 2)^2$

14. $(x + 7)^2$

15. $(2x - 1)^2$

16. $(7x - 3)^2$

17. $(3a - 5)^2$

18. $(5a + 2)^2$

19. $(5x + 9)^2$

20. $(6s - 2)^2$

Multiply. See Example 6.

21. $(a - 7)(a + 7)$

22. $(b + 3)(b - 3)$

23. $(3x - 1)(3x + 1)$

24. $(4x - 5)(4x + 5)$

25. $\left(3x - \frac{1}{2}\right)\left(3x + \frac{1}{2}\right)$

26. $\left(10x + \frac{2}{7}\right)\left(10x - \frac{2}{7}\right)$

27. $(9x + y)(9x - y)$

28. $(2x - y)(2x + y)$

29. $(2x + 0.1)(2x - 0.1)$

30. $(5x - 1.3)(5x + 1.3)$

MIXED PRACTICE

Multiply. See Example 7.

31. $(a + 5)(a + 4)$

32. $(a - 5)(a - 7)$

33. $(a + 7)^2$

34. $(b - 2)^2$

35. $(4a + 1)(3a - 1)$

36. $(6a + 7)(6a + 5)$

37. $(x + 2)(x - 2)$

38. $(x - 10)(x + 10)$

39. $(3a + 1)^2$

40. $(4a - 2)^2$

41. $(x^2 + y)(4x - y^4)$

42. $(x^3 - 2)(5x + y)$

43. $(x + 3)(x^2 - 6x + 1)$

44. $(x - 2)(x^2 - 4x + 2)$

45. $(2a - 3)^2$

46. $(5b - 4x)^2$

47. $(5x - 6z)(5x + 6z)$

48. $(11x - 7y)(11x + 7y)$

49. $(x^5 - 3)(x^5 - 5)$

50. $(a^4 + 5)(a^4 + 6)$

51. $(x + 0.8)(x - 0.8)$

52. $(y - 0.9)(y + 0.9)$

53. $(a^3 + 11)(a^4 - 3)$

54. $(x^5 + 5)(x^2 - 8)$

55. $3(x - 2)^2$

56. $2(3b + 7)^2$

57. $(3b + 7)(2b - 5)$

58. $(3y - 13)(y - 3)$

59. $(7p - 8)(7p + 8)$

60. $(3s - 4)(3s + 4)$

61. $\left(\frac{1}{3}a^2 - 7\right)\left(\frac{1}{3}a^2 + 7\right)$

62. $\left(\frac{2}{3}a - b^2\right)\left(\frac{2}{3}a - b^2\right)$

63. $5x^2(3x^2 - x + 2)$

64. $4x^3(2x^2 + 5x - 1)$

65. $(2r - 3s)(2r + 3s)$

66. $(6r - 2x)(6r + 2x)$

67. $(3x - 7y)^2$

68. $(4s - 2y)^2$

69. $(4x + 5)(4x - 5)$

70. $(3x + 5)(3x - 5)$

71. $(8x + 4)^2$

72. $(3x + 2)^2$

73. $\left(a - \frac{1}{2}y\right)\left(a + \frac{1}{2}y\right)$

74. $\left(\frac{a}{2} + 4y\right)\left(\frac{a}{2} - 4y\right)$

75. $\left(\frac{1}{5}x - y\right)\left(\frac{1}{5}x + y\right)$

76. $\left(\frac{y}{6} - 8\right)\left(\frac{y}{6} + 8\right)$

77. $(a + 1)(3a^2 - a + 1)$

78. $(b + 3)(2b^2 + b - 3)$

Express each as a product of polynomials in x. Then multiply and simplify.

△ **79.** Find the area of the square rug shown if its side is $(2x + 1)$ feet.

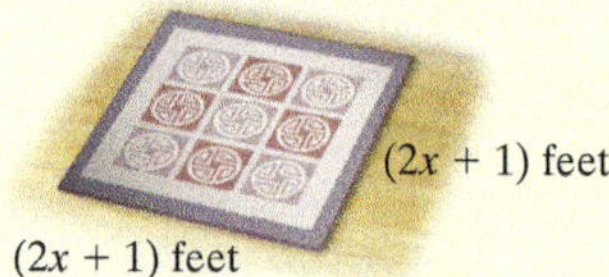

△ **80.** Find the area of the rectangular canvas if its length is $(3x - 2)$ inches and its width is $(x - 4)$ inches.

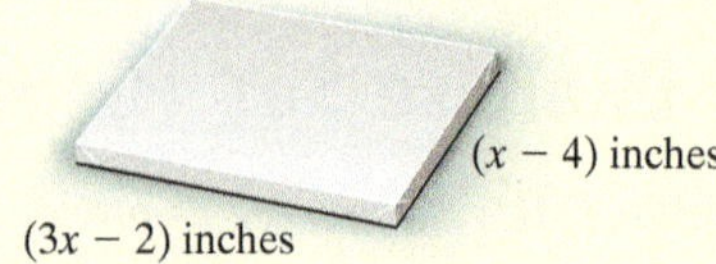

REVIEW AND PREVIEW

Simplify each expression. See Section 12.1.

81. $\frac{50b^{10}}{70b^5}$

82. $\frac{x^3y^6}{xy^2}$

83. $\frac{8a^{17}b^{15}}{-4a^7b^{10}}$

84. $\frac{-6a^8y}{3a^4y}$

85. $\frac{2x^4y^{12}}{3x^4y^4}$

86. $\frac{-48ab^6}{32ab^3}$

Find the slope of each line. See Section 10.4.

87.

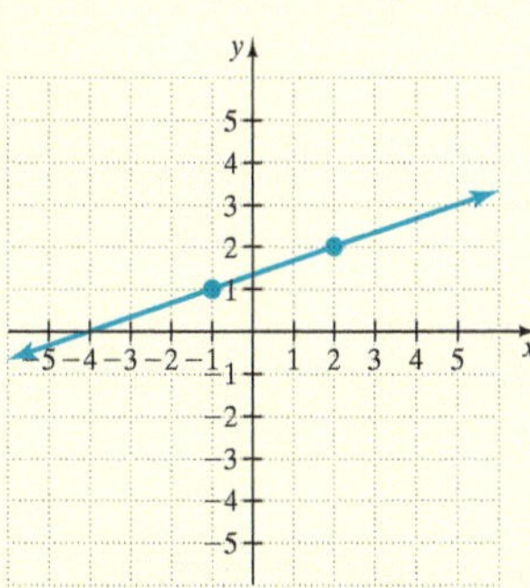

88.

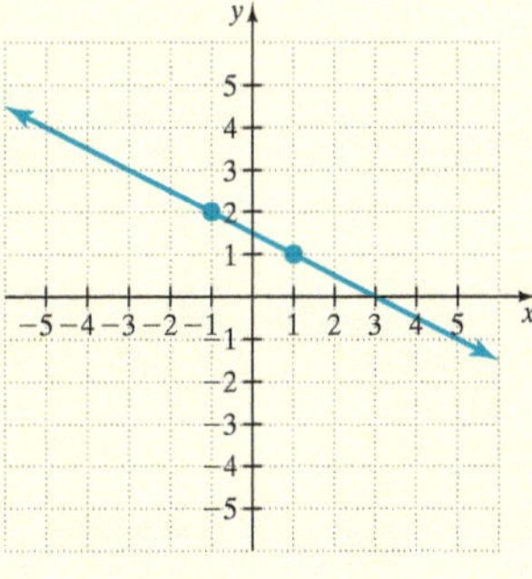

89.

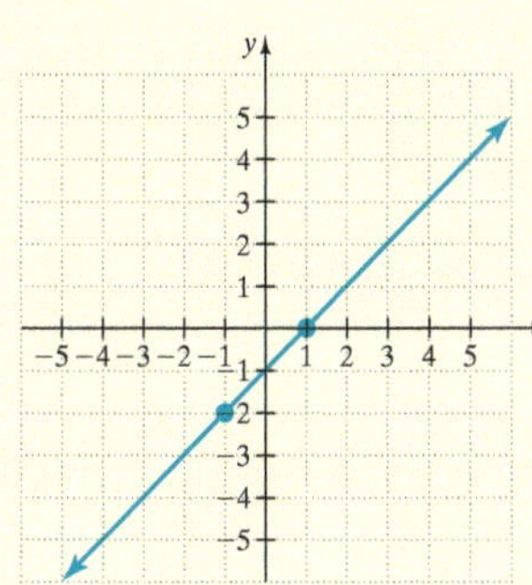

90.

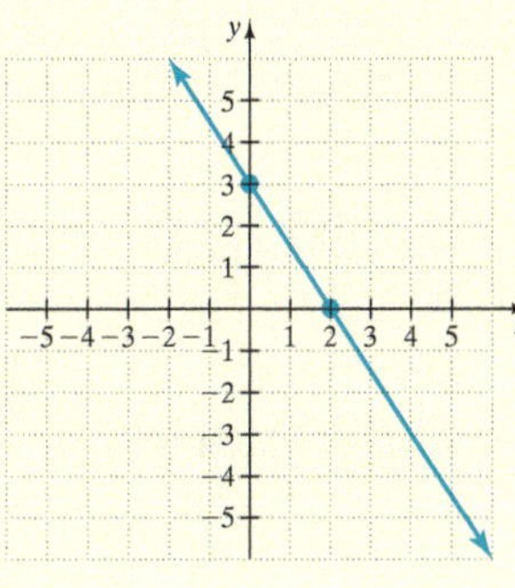

CONCEPT EXTENSIONS

Match each expression on the left to the equivalent expression on the right. See the second Concept Check in this section.

91. $(a - b)^2$ **a.** $a^2 - b^2$

92. $(a - b)(a + b)$ **b.** $a^2 + b^2$

93. $(a + b)^2$ **c.** $a^2 - 2ab + b^2$

94. $(a + b)^2(a - b)^2$ **d.** $a^2 + 2ab + b^2$

e. none of these

Fill in the squares so that a true statement forms.

95. $(x^{\square} + 7)(x^{\square} + 3) = x^4 + 10x^2 + 21$

96. $(5x^{\square} - 2)^2 = 25x^6 - 20x^3 + 4$

Find the area of the shaded figure. To do so, subtract the area of the smaller square(s) from the area of the larger geometric figure.

△ **97.**

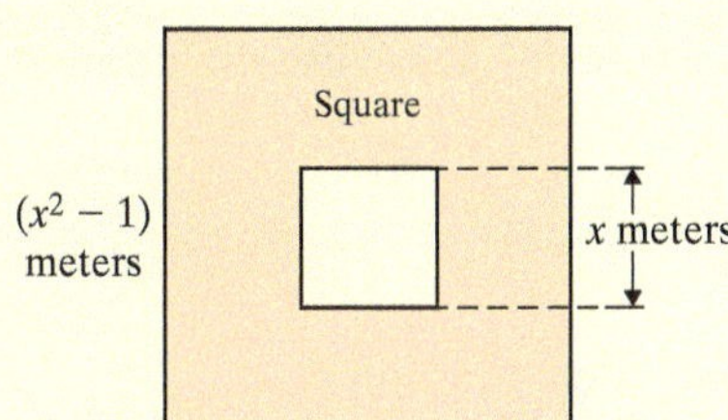

△ **98.**

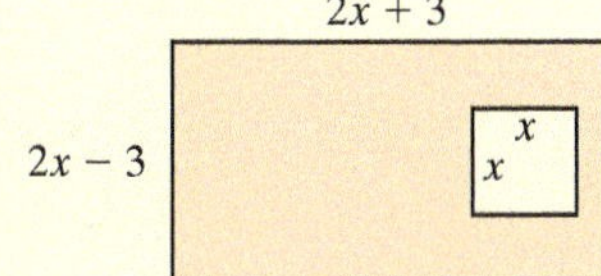

△ **99.**

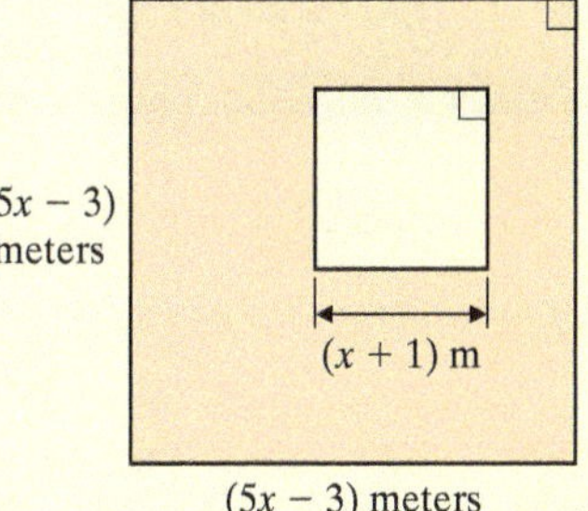

△ **100.**

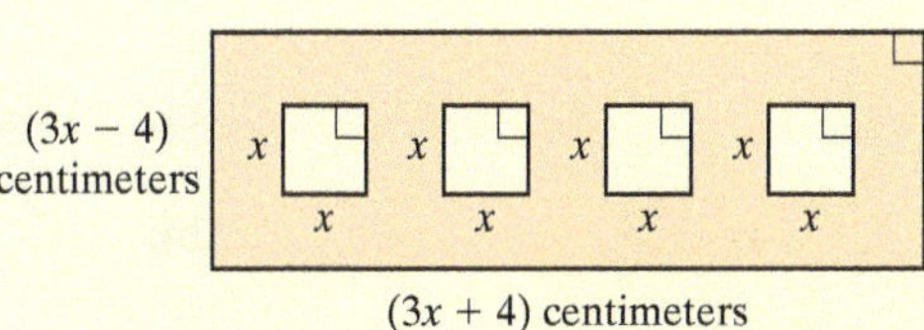

For Exercises 101 and 102, find the area of the shaded figure.

△ **101.**

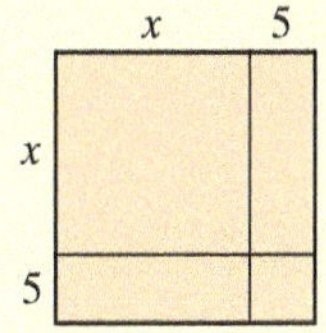

△ **102.**

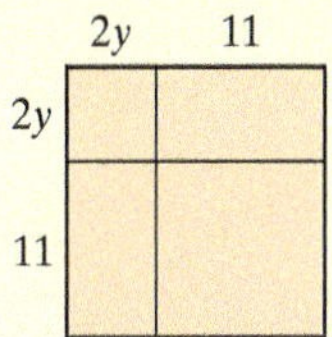

103. In your own words, describe the different methods that can be used to find the product: $(2x - 5)(3x + 1)$.

104. In your own words, describe the different methods that can be used to find the product: $(5x + 1)^2$.

105. Suppose that a classmate asked you why $(2x + 1)^2$ is **not** $(4x^2 + 1)$. Write down your response to this classmate.

106. Suppose that a classmate asked you why $(2x + 1)^2$ **is** $(4x^2 + 4x + 1)$. Write down your response to this classmate.

107. Using your own words, explain how to square a binomial such as $(a + b)^2$.

108. Explain how to find the product of two binomials using the FOIL method.

Find each product. For example,

$$[(a + b) - 2][(a + b) + 2] = (a + b)^2 - 2^2$$
$$= a^2 + 2ab + b^2 - 4$$

109. $[(x + y) - 3][(x + y) + 3]$

110. $[(a + c) - 5][(a + c) + 5]$

111. $[(a - 3) + b][(a - 3) - b]$

112. $[(x - 2) + y][(x - 2) - y]$

Integrated Review Sections 12.1–12.4

Exponents and Operations on Polynomials

Perform the indicated operations and simplify.

1. $(5x^2)(7x^3)$
2. $(4y^2)(8y^7)$
3. -4^2
4. $(-4)^2$
5. $(x-5)(2x+1)$
6. $(3x-2)(x+5)$
7. $(x-5)+(2x+1)$
8. $(3x-2)+(x+5)$
9. $\dfrac{7x^9y^{12}}{x^3y^{10}}$
10. $\dfrac{20a^2b^8}{14a^2b^2}$
11. $(12m^7n^6)^2$
12. $(4y^9z^{10})^3$
13. $3(4y-3)(4y+3)$
14. $2(7x-1)(7x+1)$
15. $(x^7y^5)^9$
16. $(3^1x^9)^3$
17. $(7x^2-2x+3)-(5x^2+9)$
18. $(10x^2+7x-9)-(4x^2-6x+2)$
19. $0.7y^2-1.2+1.8y^2-6y+1$
20. $7.8x^2-6.8x+3.3+0.6x^2-9$
21. $(x+4y)^2$
22. $(y-9z)^2$
23. $(x+4y)+(x+4y)$
24. $(y-9z)+(y-9z)$
25. $7x^2-6xy+4(y^2-xy)$
26. $5a^2-3ab+6(b^2-a^2)$
27. $(x-3)(x^2+5x-1)$
28. $(x+1)(x^2-3x-2)$
29. $(2x^3-7)(3x^2+10)$
30. $(5x^3-1)(4x^4+5)$
31. $(2x-7)(x^2-6x+1)$
32. $(5x-1)(x^2+2x-3)$

Perform these exercises and simplify if possible.

33. $5x^3+5y^3$
34. $(5x^3)(5y^3)$
35. $(5x^3)^3$
36. $\dfrac{5x^3}{5y^3}$
37. $x+x$
38. $x \cdot x$

12.5 Negative Exponents and Scientific Notation

Objectives

1. Simplify Expressions Containing Negative Exponents.
2. Use All the Rules and Definitions for Exponents to Simplify Exponential Expressions.
3. Write Numbers in Scientific Notation.
4. Convert Numbers from Scientific Notation to Standard Form.
5. Perform Operations on Numbers Written in Scientific Notation.

Objective 1 Simplifying Expressions Containing Negative Exponents

Our work with exponential expressions so far has been limited to exponents that are positive integers or 0. Here we expand this idea to give meaning to an expression like x^{-3}.

Suppose that we wish to simplify the expression $\frac{x^2}{x^5}$. If we use the quotient rule for exponents, we subtract exponents:

$$\frac{x^2}{x^5} = x^{2-5} = x^{-3}, \quad x \neq 0$$

But what does x^{-3} mean? Let's simplify $\frac{x^2}{x^5}$ using the definition of x^n.

$$\frac{x^2}{x^5} = \frac{x \cdot x}{x \cdot x \cdot x \cdot x \cdot x}$$

$$= \frac{x \cdot x}{x \cdot x \cdot x \cdot x \cdot x}$$ Divide numerator and denominator by common factors by applying the fundamental principle for fractions.

$$= \frac{1}{x^3}$$

If the quotient rule is to hold true for negative exponents, then x^{-3} must equal $\frac{1}{x^3}$.

From this example, we state the definition for negative exponents.

Negative Exponents

If a is a real number other than 0 and n is an integer, then

$$a^{-n} = \frac{1}{a^n}$$

For example, $x^{-3} = \frac{1}{x^3}$.

In other words, another way to write a^{-n} is to take its reciprocal and change the sign of its exponent.

Example 1 Simplify by writing each expression with positive exponents only.

a. 3^{-2} **b.** $2x^{-3}$ **c.** $2^{-1} + 4^{-1}$ **d.** $(-2)^{-4}$ **e.** y^{-4}

Solution:

a. $3^{-2} = \frac{1}{3^2} = \frac{1}{9}$ Use the definition of negative exponents.

b. $2x^{-3} = 2 \cdot \frac{1}{x^3} = \frac{2}{x^3}$ Use the definition of negative exponents.

c. $2^{-1} + 4^{-1} = \frac{1}{2} + \frac{1}{4} = \frac{2}{4} + \frac{1}{4} = \frac{3}{4}$

d. $(-2)^{-4} = \frac{1}{(-2)^4} = \frac{1}{(-2)(-2)(-2)(-2)} = \frac{1}{16}$

e. $y^{-4} = \frac{1}{y^4}$

Work Practice 1

Practice 1

Simplify by writing each expression with positive exponents only.

a. 5^{-3} **b.** $3y^{-4}$

c. $3^{-1} + 2^{-1}$ **d.** $(-5)^{-2}$

e. x^{-5}

Helpful Hint

Don't forget that since there are no parentheses, only x is the base for the exponent -3.

Helpful Hint

A negative exponent *does not affect* the sign of its base.

Remember: Another way to write a^{-n} is to take its reciprocal and change the sign of its exponent: $a^{-n} = \frac{1}{a^n}$. For example,

$$x^{-2} = \frac{1}{x^2}, \qquad 2^{-3} = \frac{1}{2^3} \text{ or } \frac{1}{8}$$

$$\frac{1}{y^{-4}} = \frac{1}{\frac{1}{y^4}} = y^4, \qquad \frac{1}{5^{-2}} = 5^2 \text{ or } 25$$

From the preceding Helpful Hint, we know that $x^{-2} = \frac{1}{x^2}$ and $\frac{1}{y^{-4}} = y^4$. We can use this to include another statement in our definition of negative exponents.

Negative Exponents

If a is a real number other than 0 and n is an integer, then

$$a^{-n} = \frac{1}{a^n} \quad \text{and} \quad \frac{1}{a^{-n}} = a^n$$

Answers

1. a. $\frac{1}{125}$ **b.** $\frac{3}{y^4}$ **c.** $\frac{5}{6}$ **d.** $\frac{1}{25}$ **e.** $\frac{1}{x^5}$

Practice 2

Simplify each expression. Write results using positive exponents only.

a. $\frac{1}{s^{-5}}$ **b.** $\frac{1}{2^{-3}}$

c. $\frac{x^{-7}}{y^{-5}}$ **d.** $\frac{4^{-3}}{3^{-2}}$

Example 2 Simplify each expression. Write results using positive exponents only.

a. $\frac{1}{x^{-3}}$ **b.** $\frac{1}{3^{-4}}$ **c.** $\frac{p^{-4}}{q^{-9}}$ **d.** $\frac{5^{-3}}{2^{-5}}$

Solution:

a. $\frac{1}{x^{-3}} = \frac{x^3}{1} = x^3$ **b.** $\frac{1}{3^{-4}} = \frac{3^4}{1} = 81$ **c.** $\frac{p^{-4}}{q^{-9}} = \frac{q^9}{p^4}$ **d.** $\frac{5^{-3}}{2^{-5}} = \frac{2^5}{5^3} = \frac{32}{125}$

Work Practice 2

Practice 3

Simplify each expression. Write answers with positive exponents.

a. $\frac{x^{-3}}{x^2}$ **b.** $\frac{5}{y^{-7}}$ **c.** $\frac{z}{z^{-4}}$ **d.** $\left(\frac{5}{9}\right)^{-2}$

Example 3 Simplify each expression. Write answers with positive exponents.

a. $\frac{y}{y^{-2}}$ **b.** $\frac{3}{x^{-4}}$ **c.** $\frac{x^{-5}}{x^7}$ **d.** $\left(\frac{2}{3}\right)^{-3}$

Solution:

a. $\frac{y}{y^{-2}} = \frac{y^1}{y^{-2}} = y^{1-(-2)} = y^3$ Remember that $\frac{a^m}{a^n} = a^{m-n}$.

b. $\frac{3}{x^{-4}} = 3 \cdot \frac{1}{x^{-4}} = 3 \cdot x^4$ or $3x^4$

c. $\frac{x^{-5}}{x^7} = x^{-5-7} = x^{-12} = \frac{1}{x^{12}}$

d. $\left(\frac{2}{3}\right)^{-3} = \frac{2^{-3}}{3^{-3}} = \frac{3^3}{2^3} = \frac{27}{8}$

Work Practice 3

Objective 2 Simplifying Exponential Expressions 

All the previously stated rules for exponents apply for negative exponents also. Here is a summary of the rules and definitions for exponents.

Summary of Exponent Rules

If m and n are integers and a, b, and c are real numbers, then:

Product rule for exponents: $a^m \cdot a^n = a^{m+n}$

Power rule for exponents: $(a^m)^n = a^{m \cdot n}$

Power of a product: $(ab)^n = a^n b^n$

Power of a quotient: $\left(\frac{a}{c}\right)^n = \frac{a^n}{c^n}, \quad c \neq 0$

Quotient rule for exponents: $\frac{a^m}{a^n} = a^{m-n}, \quad a \neq 0$

Zero exponent: $a^0 = 1, \quad a \neq 0$

Negative exponent: $a^{-n} = \frac{1}{a^n}, \quad a \neq 0$

Answers

2. a. s^5 **b.** 8 **c.** $\frac{y^5}{x^7}$ **d.** $\frac{9}{64}$

3. a. $\frac{1}{x^5}$ **b.** $5y^7$ **c.** z^5 **d.** $\frac{81}{25}$

Example 4 Simplify the following expressions. Write each result using positive exponents only.

a. $(y^{-3}z^6)^{-6}$ **b.** $\dfrac{(2x^3)^4x}{x^7}$ **c.** $\left(\dfrac{3a^2}{b}\right)^{-3}$ **d.** $\dfrac{4^{-1}x^{-3}y}{4^{-3}x^2y^{-6}}$ **e.** $\left(\dfrac{-2x^3y}{xy^{-1}}\right)^3$

Solution:

a. $(y^{-3}z^6)^{-6} = y^{18} \cdot z^{-36} = \dfrac{y^{18}}{z^{36}}$

b. $\dfrac{(2x^3)^4x}{x^7} = \dfrac{2^4 \cdot x^{12} \cdot x}{x^7} = \dfrac{16 \cdot x^{12+1}}{x^7} = \dfrac{16x^{13}}{x^7} = 16x^{13-7} = 16x^6$ Use the power rule.

c. $\left(\dfrac{3a^2}{b}\right)^{-3} = \dfrac{3^{-3}(a^2)^{-3}}{b^{-3}}$ Raise each factor in the numerator and the denominator to the −3 power.

$= \dfrac{3^{-3}a^{-6}}{b^{-3}}$ Use the power rule.

$= \dfrac{b^3}{3^3a^6}$ Use the negative exponent rule.

$= \dfrac{b^3}{27a^6}$ Write 3^3 as 27.

d. $\dfrac{4^{-1}x^{-3}y}{4^{-3}x^2y^{-6}} = 4^{-1-(-3)}x^{-3-2}y^{1-(-6)} = 4^2x^{-5}y^7 = \dfrac{4^2y^7}{x^5} = \dfrac{16y^7}{x^5}$

e. $\left(\dfrac{-2x^3y}{xy^{-1}}\right)^3 = \dfrac{(-2)^3x^9y^3}{x^3y^{-3}} = \dfrac{-8x^9y^3}{x^3y^{-3}} = -8x^{9-3}y^{3-(-3)} = -8x^6y^6$

Work Practice 4

Practice 4

Simplify the following expressions. Write each result using positive exponents only.

a. $(a^4b^{-3})^{-5}$ **b.** $\dfrac{x^2(x^5)^3}{x^7}$

c. $\left(\dfrac{5p^8}{q}\right)^{-2}$ **d.** $\dfrac{6^{-2}x^{-4}y^{-7}}{6^{-3}x^3y^{-9}}$

e. $\left(\dfrac{-3x^4y}{x^2y^{-2}}\right)^3$

Objective 3 Writing Numbers in Scientific Notation

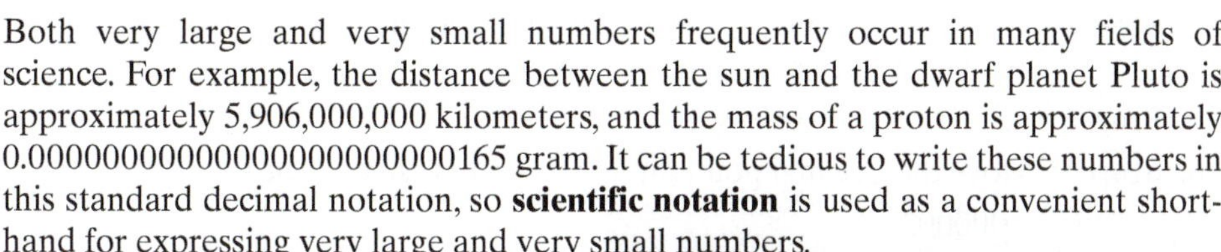

Both very large and very small numbers frequently occur in many fields of science. For example, the distance between the sun and the dwarf planet Pluto is approximately 5,906,000,000 kilometers, and the mass of a proton is approximately 0.00000000000000000000000165 gram. It can be tedious to write these numbers in this standard decimal notation, so **scientific notation** is used as a convenient shorthand for expressing very large and very small numbers.

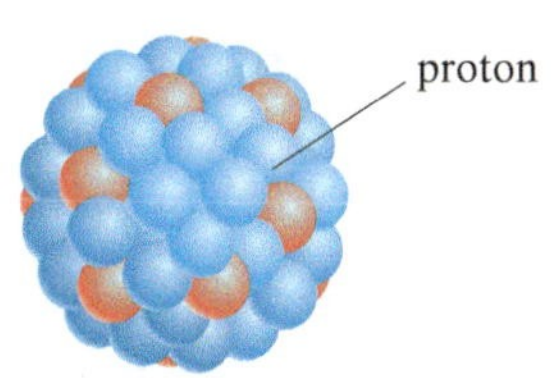

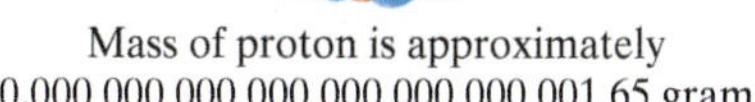

Mass of proton is approximately 0.000 000 000 000 000 000 000 001 65 gram

Scientific Notation

A positive number is written in scientific notation if it is written as the product of a number a, where $1 \le a < 10$, and an integer power r of 10:

$$a \times 10^r$$

Answers

4. a. $\dfrac{b^{15}}{a^{20}}$ **b.** x^{10} **c.** $\dfrac{q^2}{25p^{16}}$ **d.** $\dfrac{6y^2}{x^7}$ **e.** $-27x^6y^9$

The numbers below are written in scientific notation. The $\times$ sign for multiplication is used as part of the notation.

2.03×10^2 7.362×10^7 5.906×10^9 (Distance between the sun and Pluto)

1×10^{-3} 8.1×10^{-5} 1.65×10^{-24} (Mass of a proton)

The following steps are useful when writing numbers in scientific notation.

To Write a Number in Scientific Notation

Step 1: Move the decimal point in the original number to the left or right so that the new number has a value between 1 and 10.

Step 2: Count the number of decimal places the decimal point is moved in Step 1. If the original number is 10 or greater, the count is positive. If the original number is less than 1, the count is negative.

Step 3: Multiply the new number in Step 1 by 10 raised to an exponent equal to the count found in Step 2.

Practice 5

Write each number in scientific notation.

a. 0.000007

b. 20,700,000

c. 0.0043

d. 812,000,000

Example 5 Write each number in scientific notation.

a. 367,000,000 **b.** 0.000003 **c.** 20,520,000,000 **d.** 0.00085

Solution:

a. Step 1: Move the decimal point until the number is between 1 and 10.

367,000,000.

8 places

Step 2: The decimal point is moved 8 places, and the original number is 10 or greater, so the count is positive 8.

Step 3: $367{,}000{,}000 = 3.67 \times 10^8$

b. Step 1: Move the decimal point until the number is between 1 and 10.

0.000003

6 places

Step 2: The decimal point is moved 6 places, and the original number is less than 1, so the count is -6.

Step 3: $0.000003 = 3.0 \times 10^{-6}$

c. $20{,}520{,}000{,}000 = 2.052 \times 10^{10}$

d. $0.00085 = 8.5 \times 10^{-4}$

■ **Work Practice 5**

Objective 4 Converting Numbers to Standard Form

A number written in scientific notation can be rewritten in standard form. For example, to write 8.63×10^3 in standard form, recall that $10^3 = 1000$.

$$8.63 \times 10^3 = 8.63(1000) = 8630$$

Notice that the exponent on the 10 is positive 3, and we moved the decimal point 3 places to the right.

Answers

5. a. 7×10^{-6} **b.** 2.07×10^7 **c.** 4.3×10^{-3} **d.** 8.12×10^8

To write 7.29×10^{-3} in standard form, recall that $10^{-3} = \frac{1}{10^3} = \frac{1}{1000}$.

$$7.29 \times 10^{-3} = 7.29\left(\frac{1}{1000}\right) = \frac{7.29}{1000} = 0.00729$$

The exponent on the 10 is negative 3, and we moved the decimal point to the left 3 places.

In general, **to write a scientific notation number in standard form,** move the decimal point the same number of places as the exponent on 10. If the exponent is positive, move the decimal point to the right; if the exponent is negative, move the decimal point to the left.

Example 6 Write each number in standard notation, without exponents.

a. 1.02×10^5 **b.** 7.358×10^{-3} **c.** 8.4×10^7 **d.** 3.007×10^{-5}

Solution:

a. Move the decimal point 5 places to the right.

$1.02 \times 10^5 = 102{,}000.$

b. Move the decimal point 3 places to the left.

$7.358 \times 10^{-3} = 0.007358$

c. $8.4 \times 10^7 = 84{,}000{,}000.$ 7 places to the right

d. $3.007 \times 10^{-5} = 0.00003007$ 5 places to the left

Work Practice 6

Practice 6

Write each number in standard notation, without exponents.

a. 3.67×10^{-4}

b. 8.954×10^6

c. 2.009×10^{-5}

d. 4.054×10^3

✓**Concept Check** Which number in each pair is larger?

a. 7.8×10^3 or 2.1×10^5

b. 9.2×10^{-2} or 2.7×10^4

c. 5.6×10^{-4} or 6.3×10^{-5}

Objective 5 Performing Operations with Scientific Notation

Performing operations on numbers written in scientific notation uses the rules and definitions for exponents.

Example 7 Perform each indicated operation. Write each result in standard decimal notation.

a. $(8 \times 10^{-6})(7 \times 10^3)$ **b.** $\frac{12 \times 10^2}{6 \times 10^{-3}}$

Solution:

a.
$$\begin{aligned}(8 \times 10^{-6})(7 \times 10^3) &= (8 \cdot 7) \times (10^{-6} \cdot 10^3) \\ &= 56 \times 10^{-3} \\ &= 0.056\end{aligned}$$

b.
$$\frac{12 \times 10^2}{6 \times 10^{-3}} = \frac{12}{6} \times 10^{2-(-3)} = 2 \times 10^5 = 200{,}000$$

Work Practice 7

Practice 7

Perform each indicated operation. Write each result in standard decimal notation.

a. $(5 \times 10^{-4})(8 \times 10^6)$

b. $\frac{64 \times 10^3}{32 \times 10^{-7}}$

Answers

6. a. 0.000367 **b.** 8,954,000 **c.** 0.00002009 **d.** 4054

7. a. 4000 **b.** 20,000,000,000

✓**Concept Check Answers**

a. 2.1×10^5 **b.** 2.7×10^4 **c.** 5.6×10^{-4}

Graphing Calculator Explorations Scientific Notation

To enter a number written in scientific notation on a scientific calculator, locate the scientific notation key, which may be marked [EE] or [EXP]. To enter 3.1×10^7, press [3.1] [EE] [7]. The display should read [3.1 07].

Enter each number written in scientific notation on your calculator.

1. 5.31×10^3
2. -4.8×10^{14}
3. 6.6×10^{-9}
4. -9.9811×10^{-2}

Multiply each of the following on your calculator. Notice the form of the result.

5. $3{,}000{,}000 \times 5{,}000{,}000$
6. $230{,}000 \times 1000$

Multiply each of the following on your calculator. Write the product in scientific notation.

7. $(3.26 \times 10^6)(2.5 \times 10^{13})$
8. $(8.76 \times 10^{-4})(1.237 \times 10^9)$

Vocabulary, Readiness & Video Check

Fill in each blank with the correct choice.

1. The expression x^{-3} equals ______________.
   a. $-x^3$ b. $\frac{1}{x^3}$ c. $\frac{-1}{x^3}$ d. $\frac{1}{x^{-3}}$
2. The expression 5^{-4} equals ______________.
   a. -20 b. -625 c. $\frac{1}{20}$ d. $\frac{1}{625}$
3. The number 3.021×10^{-3} is written in ______________.
   a. standard form b. expanded form c. scientific notation
4. The number 0.0261 is written in ______________.
   a. standard form b. expanded form c. scientific notation

Martin-Gay Interactive Videos

See Video 12.5

Watch the section lecture video and answer the following questions.

Objective 1 5. What important reminder is made at the end of Example 1?

Objective 2 6. Name all the rules and definitions used to simplify Example 8.

Objective 3 7. From Examples 9 and 10, explain how the movement of the decimal point in Step 1 suggests the sign of the exponent on the number 10.

Objective 4 8. From Example 11, what part of a number written in scientific notation is key in telling you how to write the number in standard form?

Objective 5 9. For Example 13, what exponent rule(s) were needed to evaluate?

12.5 Exercise Set MyMathLab®

Simplify each expression. Write each result using positive exponents only. See Examples 1 through 3.

1. 4^{-3} **2.** 6^{-2} **3.** $(-3)^{-4}$

4. $(-3)^{-5}$ **5.** $7x^{-3}$ **6.** $(7x)^{-3}$

7. $\left(\frac{1}{2}\right)^{-5}$ **8.** $\left(\frac{1}{8}\right)^{-2}$ **9.** $\left(-\frac{1}{4}\right)^{-3}$

10. $\left(-\frac{1}{8}\right)^{-2}$ **11.** $3^{-1} + 5^{-1}$ **12.** $4^{-1} + 4^{-2}$

13. $\frac{1}{p^{-3}}$ **14.** $\frac{1}{q^{-5}}$ **15.** $\frac{p^{-5}}{q^{-4}}$

16. $\frac{r^{-5}}{s^{-2}}$ **17.** $\frac{x^{-2}}{x}$ **18.** $\frac{y}{y^{-3}}$

19. $\frac{z^{-4}}{z^{-7}}$ **20.** $\frac{x^{-4}}{x^{-1}}$ **21.** $3^{-2} + 3^{-1}$

22. $4^{-2} - 4^{-3}$ **23.** $\frac{-1}{p^{-4}}$ **24.** $\frac{-1}{y^{-6}}$

25. $-2^0 - 3^0$ **26.** $5^0 + (-5)^0$

MIXED PRACTICE

Simplify each expression. Write each result using positive exponents only. See Examples 1 through 4.

27. $\frac{x^2x^5}{x^3}$ **28.** $\frac{y^4y^5}{y^6}$ **29.** $\frac{p^2p}{p^{-1}}$

30. $\frac{y^3y}{y^{-2}}$ **31.** $\frac{(m^5)^4m}{m^{10}}$ **32.** $\frac{(x^2)^8x}{x^9}$

33. $\frac{r}{r^{-3}r^{-2}}$ **34.** $\frac{p}{p^{-3}q^{-5}}$ **35.** $(x^5y^3)^{-3}$

36. $(z^5x^5)^{-3}$ **37.** $\frac{(x^2)^3}{x^{10}}$ **38.** $\frac{(y^4)^2}{y^{12}}$

39. $\frac{(a^5)^2}{(a^3)^4}$ **40.** $\frac{(x^2)^5}{(x^4)^3}$ **41.** $\frac{8k^4}{2k}$

42. $\frac{27r^4}{3r^6}$ **43.** $\frac{-6m^4}{-2m^3}$

44. $\frac{15a^4}{-15a^5}$ **45.** $\frac{-24a^6b}{6ab^2}$

46. $\frac{-5x^4y^5}{15x^4y^2}$ **47.** $(-2x^3y^{-4})(3x^{-1}y)$

48. $(-5a^4b^{-7})(-a^{-4}b^3)$ **49.** $(a^{-5}b^2)^{-6}$

50. $(4^{-1}x^5)^{-2}$ **51.** $\left(\frac{x^{-2}y^4}{x^3y^7}\right)^2$

52. $\left(\frac{a^5b}{a^7b^{-2}}\right)^{-3}$ **53.** $\frac{4^2z^{-3}}{4^3z^{-5}}$

54. $\frac{3^{-1}x^4}{3^3x^{-7}}$ **55.** $\frac{2^{-3}x^{-4}}{2^2x}$

56. $\frac{5^{-1}z^7}{5^{-2}z^9}$ **57.** $\frac{7ab^{-4}}{7^{-1}a^{-3}b^2}$

58. $\frac{6^{-5}x^{-1}y^2}{6^{-2}x^{-4}y^4}$ **59.** $\left(\frac{a^{-5}b}{ab^3}\right)^{-4}$

60. $\left(\frac{r^{-2}s^{-3}}{r^{-4}s^{-3}}\right)^{-3}$ **61.** $\frac{(xy^3)^5}{(xy)^{-4}}$

62. $\frac{(rs)^{-3}}{(r^2s^3)^2}$ **63.** $\frac{(-2xy^{-3})^{-3}}{(xy^{-1})^{-1}}$

64. $\frac{(-3x^2y^2)^{-2}}{(xyz)^{-2}}$ **65.** $\frac{6x^2y^3}{-7xy^5}$

66. $\frac{-8xa^2b}{-5xa^5b}$ **67.** $\frac{(a^4b^{-7})^{-5}}{(5a^2b^{-1})^{-2}}$

68. $\frac{(a^6b^{-2})^4}{(4a^{-3}b^{-3})^3}$

Write each number in scientific notation. See Example 5.

69. 78,000 **70.** 9,300,000,000

71. 0.00000167 **72.** 0.00000017

73. 0.00635 **74.** 0.00194

75. 1,160,000 **76.** 700,000

77. More than 2,000,000,000 pencils are manufactured in the United States annually. Write this number in scientific notation. (*Source*: AbsoluteTrivia.com)

78. The temperature at the interior of the Earth is 20,000,000 degrees Celsius. Write 20,000,000 in scientific notation.

79. As of this writing, the world's largest optical telescope is the Gran Telescopio Canaris, located in La Palma, Canary Islands, Spain. The elevation of this telescope is 2400 meters above sea level. Write 2400 in scientific notation.

80. In January 2009, the twin Mars rovers, Spirit and Opportunity, celebrated their fifth anniversary of landing on Mars. These rovers, which were expected to last about 90 days, have defied all expectations and have been transmitting signals back to Earth from as far away as 250,000,000 miles. Write 250,000,000 in scientific notation. (*Source:* NASA)

Write each number in standard notation. See Example 6.

81. 8.673×10^{-10}

82. 9.056×10^{-4}

83. 3.3×10^{-2}

84. 4.8×10^{-6}

85. 2.032×10^{4}

86. 9.07×10^{10}

87. Each second, the Sun converts 7.0×10^{8} tons of hydrogen into helium and energy in the form of gamma rays. Write this number in standard notation. (*Source:* Students for the Exploration and Development of Space)

88. In chemistry, Avogadro's number is the number of atoms in one mole of an element. Avogadro's number is $6.02214199 \times 10^{23}$. Write this number in standard notation. (*Source:* National Institute of Standards and Technology)

89. The distance light travels in 1 year is 9.460×10^{12} kilometers. Write this number in standard notation.

90. The population of the world is 6.95×10^{9}. Write this number in standard notation. (*Source:* U.S. Bureau of the Census)

MIXED PRACTICE

See Examples 5 and 6. Below are some interesting facts about selected countries' national debts during a certain time period. If a number is written in standard form, write it in scientific notation. If a number is written in scientific notation, write it in standard form. (Data from CIA World Factbook)

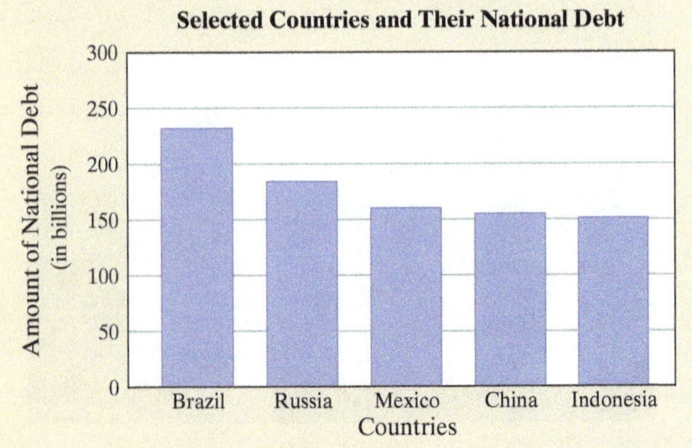

91. The national debt of Russia during a certain time period was $184,000,000,000.

92. The amount by which Russia's debt is greater than Mexico's debt is $24,000,000,000.

93. At a certain time period, China's national debt was $\$1.55 \times 10^{11}$.

94. At a certain time period, the national debt of the United States was $\$1.1 \times 10^{13}$.

95. At a certain time period, the estimated per person share of the United States' national debt was $\$3.5 \times 10^{4}$.

96. The bar graph shows the national debt of five countries. Estimate the height of the tallest bar and the shortest bar in standard notation. Then write each number in scientific notation.

Evaluate each expression using exponential rules. Write each result in standard notation. See Example 7.

97. $(1.2 \times 10^{-3})(3 \times 10^{-2})$

98. $(2.5 \times 10^{6})(2 \times 10^{-6})$

99. $(4 \times 10^{-10})(7 \times 10^{-9})$

100. $(5 \times 10^{6})(4 \times 10^{-8})$

101. $\dfrac{8 \times 10^{-1}}{16 \times 10^{5}}$

102. $\dfrac{25 \times 10^{-4}}{5 \times 10^{-9}}$

103. $\dfrac{1.4 \times 10^{-2}}{7 \times 10^{-8}}$

104. $\dfrac{0.4 \times 10^{5}}{0.2 \times 10^{11}}$

REVIEW AND PREVIEW

Simplify the following. See Section 12.1.

105. $\dfrac{5x^7}{3x^4}$

106. $\dfrac{27y^{14}}{3y^7}$

107. $\dfrac{15z^4y^3}{21zy}$

108. $\dfrac{18a^7b^{17}}{30a^7b}$

Use the distributive property and multiply. See Section 12.3.

109. $\dfrac{1}{y}(5y^2 - 6y + 5)$

110. $\dfrac{2}{x}(3x^5 + x^4 - 2)$

CONCEPT EXTENSIONS

△ 111. Find the volume of the cube.

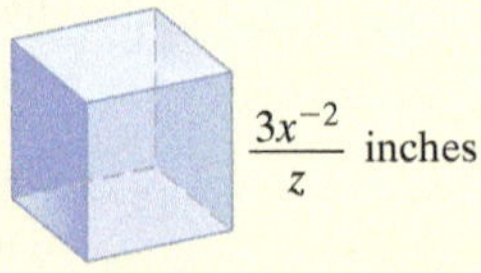

△ 112. Find the area of the triangle.

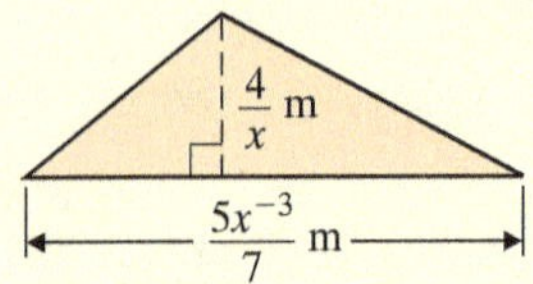

Simplify.

113. $(2a^3)^3a^4 + a^5a^8$

114. $(2a^3)^3a^{-3} + a^{11}a^{-5}$

Fill in the boxes so that each statement is true. (More than one answer may be possible for these exercises.)

115. $x^{\square} = \frac{1}{x^5}$

116. $7^{\square} = \frac{1}{49}$

117. $z^{\square} \cdot z^{\square} = z^{-10}$

118. $(x^{\square})^{\square} = x^{-15}$

119. Which is larger? See the Concept Check in this section.

a. 9.7×10^{-2} or 1.3×10^1
b. 8.6×10^5 or 4.4×10^7
c. 6.1×10^{-2} or 5.6×10^{-4}

120. Determine whether each statement is true or false.

a. $5^{-1} < 5^{-2}$
b. $\left(\frac{1}{5}\right)^{-1} < \left(\frac{1}{5}\right)^{-2}$
c. $a^{-1} < a^{-2}$ for all nonzero numbers.

121. It was stated earlier that for an integer n,

$$x^{-n} = \frac{1}{x^n}, \quad x \neq 0.$$

Explain why x may not equal 0.

122. The quotient rule states that

$$\frac{a^m}{a^n} = a^{m-n}, a \neq 0.$$

Explain why a may not equal 0.

Simplify each expression. Assume that variables represent positive integers.

123. $a^{-4m} \cdot a^{5m}$

124. $(x^{-3s})^3$

125. $(3y^{2z})^3$

126. $a^{4m+1} \cdot a^4$

Simplify each expression. Write each result in standard notation.

127. $(2.63 \times 10^{12})(-1.5 \times 10^{-10})$

128. $(6.785 \times 10^{-4})(4.68 \times 10^{10})$

Light travels at a rate of 1.86×10^5 miles per second. Use this information and the distance formula $d = r \cdot t$ to answer Exercises 129 and 130.

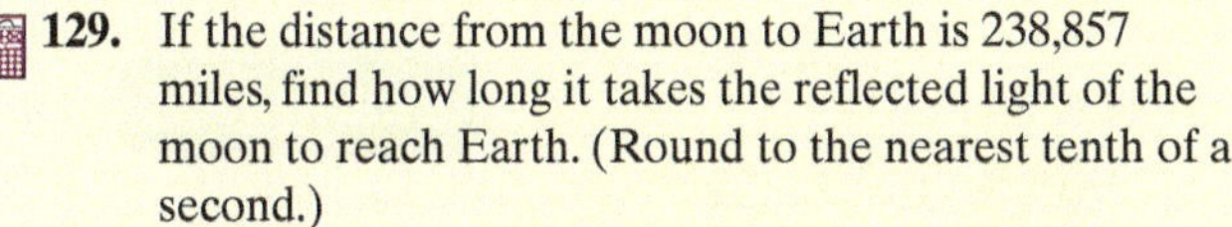

129. If the distance from the moon to Earth is 238,857 miles, find how long it takes the reflected light of the moon to reach Earth. (Round to the nearest tenth of a second.)

130. If the distance from the sun to Earth is 93,000,000 miles, find how long it takes the light of the sun to reach Earth.

12.6 Dividing Polynomials

Objective 1 Dividing by a Monomial

Objectives

1 Divide a Polynomial by a Monomial.

2 Use Long Division to Divide a Polynomial by Another Polynomial.

Now that we know how to add, subtract, and multiply polynomials, we practice dividing polynomials.

To divide a polynomial by a monomial, recall addition of fractions. Fractions that have a common denominator are added by adding the numerators:

$$\frac{a}{c} + \frac{b}{c} = \frac{a + b}{c}$$

If we read this equation from right to left and let a, b, and c be monomials, $c \neq 0$, we have the following:

Dividing a Polynomial by a Monomial

Divide each term of the polynomial by the monomial.

$$\frac{a + b}{c} = \frac{a}{c} + \frac{b}{c}, \quad c \neq 0$$

Throughout this section, we assume that denominators are not 0.

Practice 1

Divide $8t^3 + 4t^2$ by $4t^2$.

Example 1 Divide $6m^2 + 2m$ by $2m$.

Solution: We begin by writing the quotient in fraction form. Then we divide each term of the polynomial $6m^2 + 2m$ by the monomial $2m$.

$$\frac{6m^2 + 2m}{2m} = \frac{6m^2}{2m} + \frac{2m}{2m}$$

$$= 3m + 1 \quad \text{Simplify.}$$

Check: We know that if $\frac{6m^2 + 2m}{2m} = 3m + 1$, then $2m \cdot (3m + 1)$ must equal $6m^2 + 2m$. Thus, to check, we multiply.

$$2m(3m + 1) = 2m(3m) + 2m(1) = 6m^2 + 2m$$

The quotient $3m + 1$ checks.

Work Practice 1

Practice 2

Divide $\frac{16x^6 + 20x^3 - 12x}{4x^2}$.

Example 2 Divide $\frac{9x^5 - 12x^2 + 3x}{3x^2}$.

Solution:

$$\frac{9x^5 - 12x^2 + 3x}{3x^2} = \frac{9x^5}{3x^2} - \frac{12x^2}{3x^2} + \frac{3x}{3x^2} \quad \text{Divide each term by } 3x^2.$$

$$= 3x^3 - 4 + \frac{1}{x} \quad \text{Simplify.}$$

Notice that the quotient is not a polynomial because of the term $\frac{1}{x}$. This expression is called a rational expression—we will study rational expressions further in Chapter 14. Although the quotient of two polynomials is not always a polynomial, we may still check by multiplying.

Check: $$3x^2\left(3x^3 - 4 + \frac{1}{x}\right) = 3x^2(3x^3) - 3x^2(4) + 3x^2\left(\frac{1}{x}\right)$$

$$= 9x^5 - 12x^2 + 3x$$

Work Practice 2

Practice 3

Divide $\frac{15x^4y^4 - 10xy + y}{5xy}$.

Example 3 Divide $\frac{8x^2y^2 - 16xy + 2x}{4xy}$.

Solution:

$$\frac{8x^2y^2 - 16xy + 2x}{4xy} = \frac{8x^2y^2}{4xy} - \frac{16xy}{4xy} + \frac{2x}{4xy} \quad \text{Divide each term by } 4xy.$$

$$= 2xy - 4 + \frac{1}{2y} \quad \text{Simplify.}$$

Check: $$4xy\left(2xy - 4 + \frac{1}{2y}\right) = 4xy(2xy) - 4xy(4) + 4xy\left(\frac{1}{2y}\right)$$

$$= 8x^2y^2 - 16xy + 2x$$

Work Practice 3

✓**Concept Check** In which of the following is $\frac{x + 5}{5}$ simplified correctly?

a. $\frac{x}{5} + 1$ **b.** x **c.** $x + 1$

Answers

1. $2t + 1$ **2.** $4x^4 + 5x - \frac{3}{x}$

3. $3x^3y^3 - 2 + \frac{1}{5x}$

✓**Concept Check Answer**

a

Objective 2 Using Long Division to Divide by a Polynomial

To divide a polynomial by a polynomial other than a monomial, we use a process known as long division. Polynomial long division is similar to number long division, so we review long division by dividing 13 into 3660.

$$\begin{array}{r l}
281 & \\
13\overline{)3660} & \\
\underline{26}\downarrow & 2\cdot 13 = 26 \\
106 & \text{Subtract and bring down the next digit in the dividend.} \\
\underline{104}\downarrow & 8\cdot 13 = 104 \\
20 & \text{Subtract and bring down the next digit in the dividend.} \\
\underline{13} & 1\cdot 13 = 13 \\
7 & \text{Subtract. There are no more digits to bring down, so the remainder is 7.}
\end{array}$$

Helpful Hint Recall that 3660 is called the dividend.

The quotient is 281 R 7, which can be written as $281\frac{7}{13}$. ← remainder, ← divisor

Recall that division can be checked by multiplication. To check a division problem such as this one, we see that

$$13\cdot 281 + 7 = 3660$$

Now we demonstrate long division of polynomials.

Example 4 Divide $x^2 + 7x + 12$ by $x + 3$ using long division.

Solution:

$$\begin{array}{r l}
x & \text{How many times does } x \text{ divide } x^2?\ \frac{x^2}{x} = x. \\
x + 3\overline{)x^2 + 7x + 12} & \\
\underline{\overset{-}{}x^2 \overset{-}{\cancel{+}} 3x}\quad\downarrow & \text{Multiply: } x(x + 3). \\
4x + 12 & \text{Subtract and bring down the next term.}
\end{array}$$

To subtract, change the signs of these terms and add.

Now we repeat this process.

$$\begin{array}{r l}
x + 4 & \text{How many times does } x \text{ divide } 4x?\ \frac{4x}{x} = 4. \\
x + 3\overline{)x^2 + 7x + 12} & \\
\underline{\overset{-}{}x^2 \overset{-}{\cancel{+}} 3x} & \\
4x + 12 & \\
\underline{\overset{-}{}4x \overset{-}{\cancel{+}} 12} & \text{Multiply: } 4(x + 3). \\
0 & \text{Subtract. The remainder is 0.}
\end{array}$$

To subtract, change the signs of these terms and add.

The quotient is $x + 4$.

Check: We check by multiplying.

divisor	·	quotient	+	remainder	=	dividend
↓		↓		↓		↓
$(x + 3)$	·	$(x + 4)$	+	0	=	$x^2 + 7x + 12$

The quotient checks.

Work Practice 4

Practice 4

Divide $x^2 + 5x + 6$ by $x + 2$ using long division.

Answer

4. $x + 3$

Practice 5

Divide $4x^2 + 8x - 7$ by $2x + 1$ using long division.

Example 5 Divide $6x^2 + 10x - 5$ by $3x - 1$ using long division.

Solution:

$$\begin{array}{r|l} 2x + 4 & \\ 3x - 1\overline{)6x^2 + 10x - 5} & \frac{6x^2}{3x} = 2x\text{, so } 2x \text{ is a term of the quotient.} \\ -6x^2 + 2x \quad\downarrow & \text{Multiply } 2x(3x - 1). \\ 12x - 5 & \text{Subtract and bring down the next term.} \\ -12x + 4 & \frac{12x}{3x} = 4\text{, multiply } 4(3x - 1) \\ -1 & \text{Subtract. The remainder is } -1. \end{array}$$

Thus $(6x^2 + 10x - 5)$ divided by $(3x - 1)$ is $(2x + 4)$ with a remainder of -1. This can be written as

$$\frac{6x^2 + 10x - 5}{3x - 1} = 2x + 4 + \frac{-1}{3x - 1} \begin{array}{l} \leftarrow \text{remainder} \\ \leftarrow \text{divisor} \end{array}$$

Check: To check, we multiply $(3x - 1)(2x + 4)$. Then we add the remainder, -1, to this product.

$$\begin{aligned} (3x - 1)(2x + 4) + (-1) &= (6x^2 + 12x - 2x - 4) - 1 \\ &= 6x^2 + 10x - 5 \end{aligned}$$

The quotient checks.

Work Practice 5

In Example 5, the degree of the divisor, $3x - 1$, is 1 and the degree of the remainder, -1, is 0. The division process is continued until the degree of the remainder polynomial is less than the degree of the divisor polynomial.

Writing the dividend and divisor in a form with descending order of powers and with no missing terms is helpful when dividing polynomials.

Practice 6

Divide $\frac{11x - 3 + 9x^3}{3x + 2}$.

Example 6 Divide $\frac{4x^2 + 7 + 8x^3}{2x + 3}$.

Solution: Before we begin the division process, we rewrite

$4x^2 + 7 + 8x^3$ as $8x^3 + 4x^2 + 0x + 7$

Notice that we have written the polynomial in descending order and have represented the missing x-term by $0x$.

$$\begin{array}{r} 4x^2 - 4x + 6 \\ 2x + 3\overline{)8x^3 + 4x^2 + 0x + 7} \\ -8x^3 - 12x^2 \qquad\qquad \\ -8x^2 + 0x \qquad \\ +8x^2 + 12x \qquad \\ 12x + 7 \\ -12x - 18 \\ -11 \end{array} \quad \text{Remainder}$$

Thus, $\frac{4x^2 + 7 + 8x^3}{2x + 3} = 4x^2 - 4x + 6 + \frac{-11}{2x + 3}$.

Work Practice 6

Answers

5. $2x + 3 + \frac{-10}{2x + 1}$ or $2x + 3 - \frac{10}{2x + 1}$

6. $3x^2 - 2x + 5 + \frac{-13}{3x + 2}$ or $3x^2 - 2x + 5 - \frac{13}{3x - 2}$

Example 7 Divide $\frac{2x^4 - x^3 + 3x^2 + x - 1}{x^2 + 1}$.

Solution: Before dividing, rewrite the divisor polynomial

$x^2 + 1$ as $x^2 + 0x + 1$

The $0x$ term represents the missing x^1-term in the divisor.

$$\begin{array}{r} 2x^2 - x + 1 \\ x^2 + 0x + 1\overline{)2x^4 - x^3 + 3x^2 + x - 1} \\ \underline{2x^4 + 0x^3 + 2x^2} \\ -x^3 + x^2 + x \\ \underline{-x^3 - 0x^2 - x} \\ x^2 + 2x - 1 \\ \underline{x^2 + 0x + 1} \\ 2x - 2 \end{array} \quad \text{Remainder}$$

Thus, $\frac{2x^4 - x^3 + 3x^2 + x - 1}{x^2 + 1} = 2x^2 - x + 1 + \frac{2x - 2}{x^2 + 1}$.

Work Practice 7

Practice 7

Divide $\frac{3x^4 - 2x^3 - 3x^2 + x + 4}{x^2 + 2}$.

Example 8 Divide $x^3 - 8$ by $x - 2$.

Solution: Notice that the polynomial $x^3 - 8$ is missing an x^2-term and an x-term. We'll represent these terms by inserting $0x^2$ and $0x$.

$$\begin{array}{r} x^2 + 2x + 4 \\ x - 2\overline{)x^3 + 0x^2 + 0x - 8} \\ \underline{x^3 - 2x^2} \\ 2x^2 + 0x \\ \underline{2x^2 - 4x} \\ 4x - 8 \\ \underline{4x - 8} \\ 0 \end{array}$$

Thus, $\frac{x^3 - 8}{x - 2} = x^2 + 2x + 4$.

Check: To check, see that $(x^2 + 2x + 4)(x - 2) = x^3 - 8$.

Work Practice 8

Practice 8

Divide $x^3 + 27$ by $x + 3$.

Answers

7. $3x^2 - 2x - 9 + \frac{5x + 22}{x^2 + 2}$

8. $x^2 - 3x + 9$

Vocabulary, Readiness & Video Check

Use the choices below to fill in each blank. Choices may be used more than once.

dividend divisor quotient

1. In $6\overline{)18}^{\,3}$, the 18 is the ________, the 3 is the ________ and the 6 is the ________.

2. In $x + 1\overline{)x^2 + 3x + 2}^{\,x + 2}$, the $x + 1$ is the ________, the $x^2 + 3x + 2$ is the ________ and the $x + 2$ is the ________.

Simplify each expression mentally.

3. $\frac{a^6}{a^4}$ **4.** $\frac{p^8}{p^3}$ **5.** $\frac{y^2}{y}$ **6.** $\frac{a^3}{a}$

Martin-Gay Interactive Videos

See Video 12.6

Watch the section lecture video and answer the following questions.

Objective 1 7. The lecture before Example 1 begins with adding two fractions with the same denominator. From there, the lecture continues to a method for dividing a polynomial by a monomial. What role does the monomial play in the fraction example?

Objective 2 8. In Example 5, you're told that although you don't have to fill in missing powers in the divisor and the dividend, it really is a good idea to do so. Why?

12.6 Exercise Set MyMathLab®

Perform each division. See Examples 1 through 3.

1. $\dfrac{12x^4 + 3x^2}{x}$
2. $\dfrac{15x^2 - 9x^5}{x}$
3. $\dfrac{20x^3 - 30x^2 + 5x + 5}{5}$
4. $\dfrac{8x^3 - 4x^2 + 6x + 2}{2}$
5. $\dfrac{15p^3 + 18p^2}{3p}$
6. $\dfrac{14m^2 - 27m^3}{7m}$
7. $\dfrac{-9x^4 + 18x^5}{6x^5}$
8. $\dfrac{6x^5 + 3x^4}{3x^4}$
9. $\dfrac{-9x^5 + 3x^4 - 12}{3x^3}$
10. $\dfrac{6a^2 - 4a + 12}{-2a^2}$
11. $\dfrac{4x^4 - 6x^3 + 7}{-4x^4}$
12. $\dfrac{-12a^3 + 36a - 15}{3a}$

Find each quotient using long division. See Examples 4 and 5.

13. $\dfrac{x^2 + 4x + 3}{x + 3}$
14. $\dfrac{x^2 + 7x + 10}{x + 5}$
15. $\dfrac{2x^2 + 13x + 15}{x + 5}$
16. $\dfrac{3x^2 + 8x + 4}{x + 2}$
17. $\dfrac{2x^2 - 7x + 3}{x - 4}$
18. $\dfrac{3x^2 - x - 4}{x - 1}$
19. $\dfrac{9a^3 - 3a^2 - 3a + 4}{3a + 2}$
20. $\dfrac{4x^3 + 12x^2 + x - 14}{2x + 3}$
21. $\dfrac{8x^2 + 10x + 1}{2x + 1}$
22. $\dfrac{3x^2 + 17x + 7}{3x + 2}$
23. $\dfrac{2x^3 + 2x^2 - 17x + 8}{x - 2}$
24. $\dfrac{4x^3 + 11x^2 - 8x - 10}{x + 3}$

Find each quotient using long division. Don't forget to write the polynomials in descending order and fill in any missing terms. See Examples 6 through 8.

25. $\dfrac{x^2 - 36}{x - 6}$
26. $\dfrac{a^2 - 49}{a - 7}$
27. $\dfrac{x^3 - 27}{x - 3}$
28. $\dfrac{x^3 + 64}{x + 4}$
29. $\dfrac{1 - 3x^2}{x + 2}$
30. $\dfrac{7 - 5x^2}{x + 3}$
31. $\dfrac{-4b + 4b^2 - 5}{2b - 1}$
32. $\dfrac{-3y + 2y^2 - 15}{2y + 5}$

MIXED PRACTICE

Divide. If the divisor contains 2 or more terms, use long division. See Examples 1 through 8.

33. $\dfrac{a^2b^2 - ab^3}{ab}$
34. $\dfrac{m^3n^2 - mn^4}{mn}$
35. $\dfrac{8x^2 + 6x - 27}{2x - 3}$
36. $\dfrac{18w^2 + 18w - 8}{3w + 4}$
37. $\dfrac{2x^2y + 8x^2y^2 - xy^2}{2xy}$

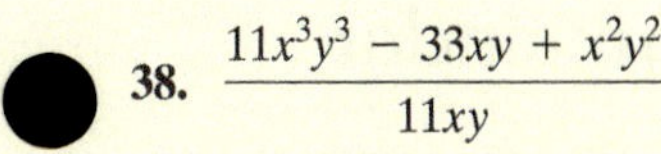

38. $\dfrac{11x^3y^3 - 33xy + x^2y^2}{11xy}$

39. $\dfrac{2b^3 + 9b^2 + 6b - 4}{b + 4}$

40. $\dfrac{2x^3 + 3x^2 - 3x + 4}{x + 2}$

41. $\dfrac{5x^2 + 28x - 10}{x + 6}$

42. $\dfrac{2x^2 + x - 15}{x + 3}$

43. $\dfrac{10x^3 - 24x^2 - 10x}{10x}$

44. $\dfrac{2x^3 + 12x^2 + 16}{4x^2}$

45. $\dfrac{6x^2 + 17x - 4}{x + 3}$

46. $\dfrac{2x^2 - 9x + 15}{x - 6}$

47. $\dfrac{30x^2 - 17x + 2}{5x - 2}$

48. $\dfrac{4x^2 - 13x - 12}{4x + 3}$

49. $\dfrac{3x^4 - 9x^3 + 12}{-3x}$

50. $\dfrac{8y^6 - 3y^2 - 4y}{4y}$

51. $\dfrac{x^3 + 6x^2 + 18x + 27}{x + 3}$

52. $\dfrac{x^3 - 8x^2 + 32x - 64}{x - 4}$

53. $\dfrac{y^3 + 3y^2 + 4}{y - 2}$

54. $\dfrac{3x^3 + 11x + 12}{x + 4}$

55. $\dfrac{5 - 6x^2}{x - 2}$

56. $\dfrac{3 - 7x^2}{x - 3}$

Divide.

57. $\dfrac{x^5 + x^2}{x^2 + x}$

58. $\dfrac{x^6 - x^4}{x^3 + 1}$

REVIEW AND PREVIEW

Multiply each expression. See Section 12.3.

59. $2a(a^2 + 1)$

60. $-4a(3a^2 - 4)$

61. $2x(x^2 + 7x - 5)$

62. $4y(y^2 - 8y - 4)$

63. $-3xy(xy^2 + 7x^2y + 8)$

64. $-9xy(4xyz + 7xy^2z + 2)$

65. $9ab(ab^2c + 4bc - 8)$

66. $-7sr(6s^2r + 9sr^2 + 9rs + 8)$

CONCEPT EXTENSIONS

67. The perimeter of a square is $(12x^3 + 4x - 16)$ feet. Find the length of its side.

Perimeter is $(12x^3 + 4x - 16)$ feet

68. The volume of the swimming pool shown is $(36x^5 - 12x^3 + 6x^2)$ cubic feet. If its height is $2x$ feet and its width is $3x$ feet, find its length.

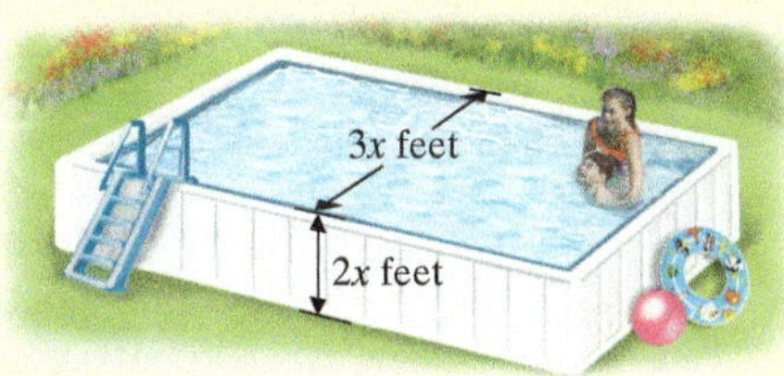

69. In which of the following is $\dfrac{a + 7}{7}$ simplified correctly? See the Concept Check in this section.

a. $a + 1$ **b.** a **c.** $\dfrac{a}{7} + 1$

70. In which of the following is $\dfrac{5x + 15}{5}$ simplified correctly? See the Concept Check in this section.

a. $x + 15$ **b.** $x + 3$ **c.** $x + 1$

71. Explain how to check a polynomial long division result when the remainder is 0.

72. Explain how to check a polynomial long division result when the remainder is not 0.

73. The area of the following parallelogram is $(10x^2 + 31x + 15)$ square meters. If its base is $(5x + 3)$ meters, find its height.

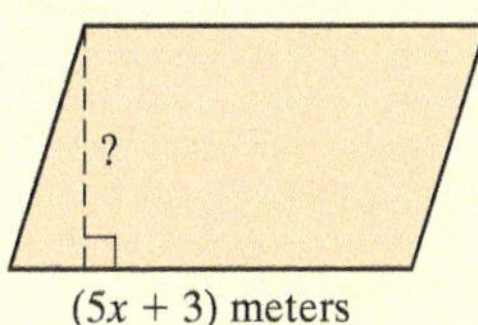

74. The area of the top of the Ping-Pong table is $(49x^2 + 70x - 200)$ square inches. If its length is $(7x + 20)$ inches, find its width.

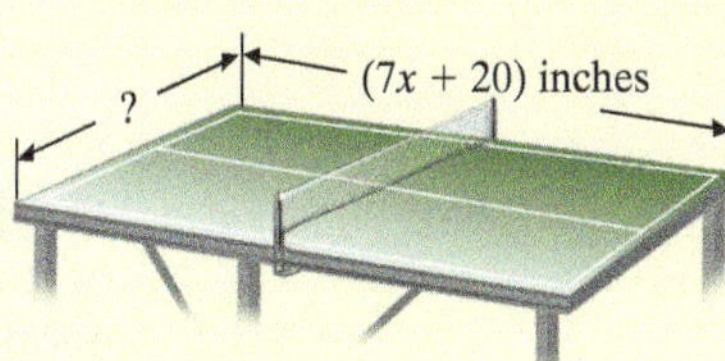

75. $(18x^{10a} - 12x^{8a} + 14x^{5a} - 2x^{3a}) \div 2x^{3a}$

76. $(25y^{11b} + 5y^{6b} - 20y^{3b} + 100y^{b}) \div 5y^{b}$

12.7 Synthetic Division and the Remainder Theorem

Objectives

1. Use Synthetic Division to Divide a Polynomial by a Binomial.
2. Use the Remainder Theorem to Evaluate Polynomials.

Objective 1 Using Synthetic Division 

When a polynomial is to be divided by a binomial of the form $x - c$, a shortcut process called **synthetic division** may be used. On the left is an example of long division, and on the right, the same example showing the coefficients of the variables only.

$$
\begin{array}{r}
2x^2 + 5x + 2 \\
x - 3\overline{)2x^3 - x^2 - 13x + 1} \\
\underline{2x^3 - 6x^2} \qquad\qquad \\
5x^2 - 13x \qquad \\
\underline{5x^2 - 15x} \qquad \\
2x + 1 \\
\underline{2x - 6} \\
7
\end{array}
\qquad
\begin{array}{r}
2 \quad 5 \quad 2 \\
1 - 3\overline{)2 - 1 - 13 + 1} \\
\underline{2 - 6} \qquad\quad \\
5 - 13 \qquad \\
\underline{5 - 15} \qquad \\
2 + 1 \\
\underline{2 - 6} \\
7
\end{array}
$$

Notice that as long as we keep coefficients of powers of x in the same column, we can perform division of polynomials by performing algebraic operations on the coefficients only. This shortcut process of dividing with coefficients only in a special format is called synthetic division. To find $(2x^3 - x^2 - 13x + 1) \div (x - 3)$ by synthetic division, follow the next example.

Practice 1

Use synthetic division to divide $4x^3 - 3x^2 + 6x + 5$ by $x - 1$.

Example 1 Use synthetic division to divide $2x^3 - x^2 - 13x + 1$ by $x - 3$.

Solution: To use synthetic division, the divisor must be in the form $x - c$. Since we are dividing by $x - 3$, c is 3. Write down 3 and the coefficients of the dividend.

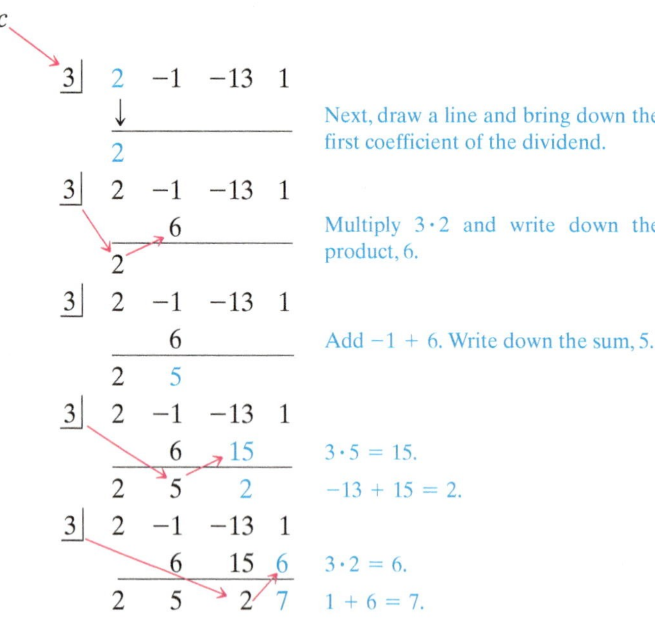

The quotient is found in the bottom row. The numbers 2, 5, and 2 are the coefficients of the quotient polynomial, and the number 7 is the remainder. The degree of the quotient polynomial is one less than the degree of the dividend. In our example, the degree of the dividend is 3, so the degree of the quotient polynomial is 2. As we found when we performed the long division, the quotient is

Answer

1. $4x^2 + x + 7 + \frac{12}{x - 1}$

$2x^2 + 5x + 2$, remainder 7

or

$$2x^2 + 5x + 2 + \frac{7}{x - 3}.$$

Work Practice 1

Example 2 Use synthetic division to divide $x^4 - 2x^3 - 11x^2 + 5x + 34$ by $x + 2$.

Solution: The divisor is $x + 2$, which we write in the form $x - c$ as $x - (-2)$. Thus, c is -2. The dividend coefficients are 1, -2, -11, 5, and 34.

$$\begin{array}{r|rrrrr} -2 & 1 & -2 & -11 & 5 & 34 \\ & & -2 & 8 & 6 & -22 \\ \hline & 1 & -4 & -3 & 11 & 12 \end{array}$$

The dividend is a fourth-degree polynomial, so the quotient polynomial is a third-degree polynomial. The quotient is $x^3 - 4x^2 - 3x + 11$ with a remainder of 12. Thus,

$$\frac{x^4 - 2x^3 - 11x^2 + 5x + 34}{x + 2} = x^3 - 4x^2 - 3x + 11 + \frac{12}{x + 2}.$$

Work Practice 2

Practice 2

Use synthetic division to divide $x^4 + 3x^3 - 5x^2 + 6x + 12$ by $x + 3$.

✓Concept Check Which division problems are candidates for the synthetic division process?

a. $(3x^2 + 5) \div (x + 4)$ **b.** $(x^3 - x^2 + 2) \div (3x^3 - 2)$

c. $(y^4 + y - 3) \div (x^2 + 1)$ **d.** $x^5 \div (x - 5)$

Helpful Hint

Before dividing by synthetic division, write the dividend in descending order of variable exponents. Any "missing powers" of the variable should be represented by 0 times the variable raised to the missing power.

Example 3 If $P(x) = 2x^3 - 4x^2 + 5$,

a. Find $P(2)$ by substitution.

b. Use synthetic division to find the remainder when $P(x)$ is divided by $x - 2$.

Solution:

a. $P(x) = 2x^3 - 4x^2 + 5$

$P(2) = 2(2)^3 - 4(2)^2 + 5$

$= 2(8) - 4(4) + 5 = 16 - 16 + 5 = 5$

Thus, $P(2) = 5$.

(*Continued on next page*)

Practice 3

If $P(x) = x^3 - 5x - 2$,

a. Find $P(2)$ by substitution.

b. Use synthetic division to find the remainder when $P(x)$ is divided by $x - 2$.

Answers

2. $x^3 - 5x + 21 - \frac{51}{x + 3}$

3. a. -4 **b.** -4

✓Concept Check Answer

a and d

b. The coefficients of $P(x)$ are 2, −4, 0, and 5. The number 0 is a coefficient of the missing power of x^1. The divisor is $x - 2$, so c is 2.

c → 2	2	−4	0	5	
		4	0	0	
	2	0	0	5	remainder

The remainder when $P(x)$ is divided by $x - 2$ is 5.

Work Practice 3

Objective 2 Using the Remainder Theorem 

Notice in the preceding example that $P(2) = 5$ and that the remainder when $P(x)$ is divided by $x - 2$ is 5. This is no accident. This illustrates the **remainder theorem.**

Remainder Theorem

If a polynomial $P(x)$ is divided by $x - c$, then the remainder is $P(c)$.

Practice 4

Use the remainder theorem and synthetic division to find $P(3)$ if $P(x) = 2x^5 - 18x^4 + 90x^2 + 59x$.

Example 4 Use the remainder theorem and synthetic division to find $P(4)$ if

$$P(x) = 4x^6 - 25x^5 + 35x^4 + 17x^2.$$

Solution: To find $P(4)$ by the remainder theorem, we divide $P(x)$ by $x - 4$. The coefficients of $P(x)$ are 4, −25, 35, 0, 17, 0, and 0. Also, c is 4.

c → 4	4	−25	35	0	17	0	0	
		16	−36	−4	−16	4	16	
	4	−9	−1	−4	1	4	16	remainder

Thus, $P(4) = 16$, the remainder.

Work Practice 4

Answer

4. 15

Vocabulary, Readiness & Video Check

Martin-Gay Interactive Videos

See Video 12.7

Watch the section lecture video and answer the following questions.

Objective 1 **1.** From Example 1, once you've completed the synthetic division, what does the bottom row of numbers mean? What is the degree of the quotient?

Objective 2 **2.** From Example 4, given a polynomial function $P(x)$, under what circumstances might it be easier/faster to use the remainder theorem to find $P(c)$ rather than substituting the value c for x and then simplifying?

12.7 Exercise Set MyMathLab® 

Use synthetic division to divide. See Examples 1 and 2.

1. $(x^2 + 3x - 40) \div (x - 5)$
2. $(x^2 - 14x + 24) \div (x - 2)$
3. $(x^2 + 5x - 6) \div (x + 6)$
4. $(x^2 + 12x + 32) \div (x + 4)$
5. $(x^3 - 7x^2 - 13x + 5) \div (x - 2)$
6. $(x^3 + 6x^2 + 4x - 7) \div (x + 5)$
7. $(4x^2 - 9) \div (x - 2)$
8. $(3x^2 - 4) \div (x - 1)$

For the given polynomial P(x) and the given c, find P(c) by **(a)** *direct substitution and* **(b)** *the remainder theorem. See Examples 3 and 4.*

9. $P(x) = 3x^2 - 4x - 1; P(2)$
10. $P(x) = x^2 - x + 3; P(5)$
11. $P(x) = 4x^4 + 7x^2 + 9x - 1; P(-2)$
12. $P(x) = 8x^5 + 7x + 4; P(-3)$
13. $P(x) = x^5 + 3x^4 + 3x - 7; P(-1)$
14. $P(x) = 5x^4 - 4x^3 + 2x - 1; P(-1)$

MIXED PRACTICE

Use synthetic division to divide.

15. $(x^3 - 3x^2 + 2) \div (x - 3)$
16. $(x^2 + 12) \div (x + 2)$
17. $(6x^2 + 13x + 8) \div (x + 1)$
18. $(x^3 - 5x^2 + 7x - 4) \div (x - 3)$
19. $(2x^4 - 13x^3 + 16x^2 - 9x + 20) \div (x - 5)$
20. $(3x^4 + 5x^3 - x^2 + x - 2) \div (x + 2)$
21. $(3x^2 - 15) \div (x + 3)$
22. $(3x^2 + 7x - 6) \div (x + 4)$
23. $(3x^3 - 6x^2 + 4x + 5) \div \left(x - \frac{1}{2}\right)$
24. $(8x^3 - 6x^2 - 5x + 3) \div \left(x + \frac{3}{4}\right)$
25. $(3x^3 + 2x^2 - 4x + 1) \div \left(x - \frac{1}{3}\right)$
26. $(9y^3 + 9y^2 - y + 2) \div \left(y + \frac{2}{3}\right)$
27. $(7x^2 - 4x + 12 + 3x^3) \div (x + 1)$
28. $(x^4 + 4x^3 - x^2 - 16x - 4) \div (x - 2)$
29. $(x^3 - 1) \div (x - 1)$
30. $(y^3 - 8) \div (y - 2)$
31. $(2x^3 + 12x^2 - 3x - 20) \div (x + 6)$
32. $(4x^3 + 12x^2 + x - 12) \div (x + 3)$

For the given polynomial P(x) and the given c, use the remainder theorem to find P(c).

33. $P(x) = x^3 + 3x^2 - 7x + 4; 1$
34. $P(x) = x^3 + 5x^2 - 4x - 6; 2$
35. $P(x) = 3x^3 - 7x^2 - 2x + 5; -3$
36. $P(x) = 4x^3 + 5x^2 - 6x - 4; -2$
37. $P(x) = 4x^4 + x^2 - 2; -1$
38. $P(x) = x^4 - 3x^2 - 2x + 5; -2$
39. $P(x) = 2x^4 - 3x^2 - 2; \frac{1}{3}$
40. $P(x) = 4x^4 - 2x^3 + x^2 - x - 4; \frac{1}{2}$
41. $P(x) = x^5 + x^4 - x^3 + 3; \frac{1}{2}$
42. $P(x) = x^5 - 2x^3 + 4x^2 - 5x + 6; \frac{2}{3}$
43. Explain an advantage of using the remainder theorem instead of direct substitution.
44. Explain an advantage of using synthetic division instead of long division.

REVIEW AND PREVIEW

Solve each equation for x. See Section 9.3.

45. $7x + 2 = x - 3$
46. $4 - 2x = 17 - 5x$
47. $\frac{x}{3} - 5 = 13$
48. $\frac{2x}{9} + 1 = \frac{7}{9}$

Evaluate. See Section 12.1.

49. 2^3
50. 3^4
51. $(-2)^5$
52. -2^5
53. $3 \cdot 4^2$
54. $4 \cdot 3^3$

Evaluate each expression for the given replacement value. See Section 12.1.

55. x^2 if x is -5
56. x^3 if x is -5
57. $2x^3$ if x is -1
58. $3x^2$ if x is -1

CONCEPT EXTENSIONS

Determine whether each division problem is a candidate for the synthetic division process. See the Concept Check in this section.

59. $(5x^2 - 3x + 2) \div (x + 2)$
60. $(x^4 - 6) \div (x^3 + 3x - 1)$
61. $(x^7 - 2) \div (x^5 + 1)$
62. $(3x^2 + 7x - 1) \div \left(x - \frac{1}{3}\right)$

△ **63.** If the area of a parallelogram is $(x^4 - 23x^2 + 9x - 5)$ square centimeters and its base is $(x + 5)$ centimeters, find its height.

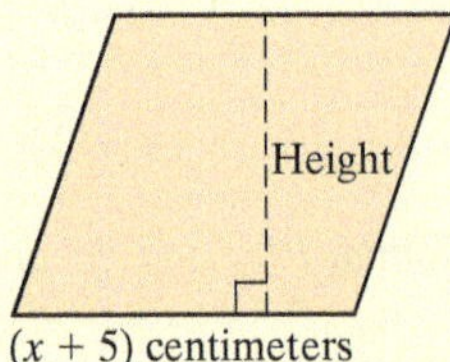

△ **64.** If the volume of a box is $(x^4 + 6x^3 - 7x^2)$ cubic meters, its height is x^2 meters, and its length is $(x + 7)$ meters, find its width.

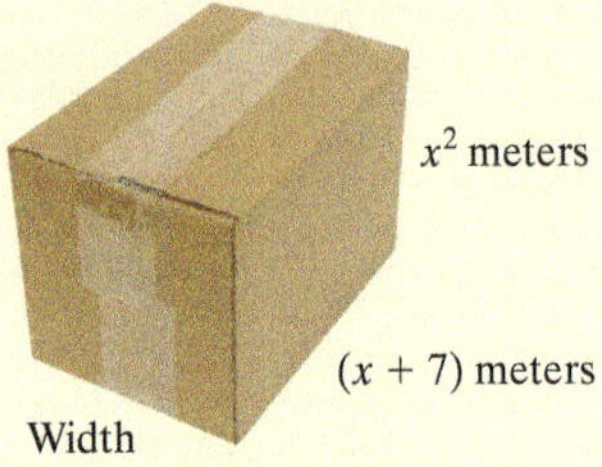

Divide.

65. $\left(x^4 + \frac{2}{3}x^3 + x\right) \div (x - 1)$

66. $\left(2x^3 + \frac{9}{2}x^2 - 4x - 10\right) \div (x + 2)$

We say that 2 is a factor of 8 because 2 divides 8 evenly, or with a remainder of 0. In the same manner, the polynomial $x - 2$ is a factor of the polynomial $x^3 - 14x^2 + 24x$ because the remainder is 0 when $x^3 - 14x^2 + 24x$ is divided by $x - 2$. Use this information for Exercises 67 through 69.

67. Use synthetic division to show that $x + 3$ is a factor of $x^3 + 3x^2 + 4x + 12$.

68. Use synthetic division to show that $x - 2$ is a factor of $x^3 - 2x^2 - 3x + 6$.

69. From the remainder theorem, the polynomial $x - c$ is a factor of a polynomial function $P(x)$ if $P(c)$ is what value?

70. If a polynomial is divided by $x - 5$, the quotient is $2x^2 + 5x - 6$ and the remainder is 3. Find the original polynomial.

71. If a polynomial is divided by $x + 3$, the quotient is $x^2 - x + 10$ and the remainder is -2. Find the original polynomial.

Chapter 12 Vocabulary Check

Fill in each blank with one of the words or phrases listed below.

term	coefficient	monomial	binomial	trinomial
polynomials	degree of a term	distributive	FOIL	degree of a polynomial

1. A ____________ is a number or the product of numbers and variables raised to powers.

2. The ____________ method may be used when multiplying two binomials.

3. A polynomial with exactly 3 terms is called a ____________.

4. The ____________ is the greatest degree of any term of the polynomial.

5. A polynomial with exactly 2 terms is called a ____________.

6. The ____________ of a term is its numerical factor.

7. The ____________ is the sum of the exponents on the variables in the term.

8. A polynomial with exactly 1 term is called a ____________.

9. Monomials, binomials, and trinomials are all examples of ____________.

10. The ____________ property is used to multiply $2x(x - 4)$.

Helpful Hint Are you preparing for your test? Don't forget to take the Chapter 12 Test on page 923. Then check your answers at the back of the text and use the Chapter Test Prep Videos to see the fully worked-out solutions to any of the exercises you want to review.

12 Chapter Highlights

Definitions and Concepts	Examples
Section 12.1 Exponents	
a^n means the product of n factors, each of which is a.	$3^2 = 3 \cdot 3 = 9$ $(-5)^3 = (-5)(-5)(-5) = -125$ $\left(\frac{1}{2}\right)^4 = \frac{1}{2}\cdot\frac{1}{2}\cdot\frac{1}{2}\cdot\frac{1}{2} = \frac{1}{16}$
If m and n are integers and no denominators are 0,	
Product Rule: $a^m \cdot a^n = a^{m+n}$	$x^2 \cdot x^7 = x^{2+7} = x^9$
Power Rule: $(a^m)^n = a^{mn}$	$(5^3)^8 = 5^{3\cdot 8} = 5^{24}$
Power of a Product Rule: $(ab)^n = a^n b^n$	$(7y)^4 = 7^4 y^4$
Power of a Quotient Rule: $\left(\frac{a}{b}\right)^n = \frac{a^n}{b^n}$	$\left(\frac{x}{8}\right)^3 = \frac{x^3}{8^3}$
Quotient Rule: $\frac{a^m}{a^n} = a^{m-n}$	$\frac{x^9}{x^4} = x^{9-4} = x^5$
Zero Exponent: $a^0 = 1, a \neq 0.$	$5^0 = 1, x^0 = 1, x \neq 0$
Section 12.2 Polynomial Functions and Adding and Subtracting Polynomials	
	Terms
A **term** is a number or the product of numbers and variables raised to powers.	$-5x, 7a^2b, \frac{1}{4}y^4, 0.2$
	Term — **Coefficient**
The **numerical coefficient** or **coefficient** of a term is its numerical factor.	$7x^2$ — 7 y — 1 $-a^2b$ — -1
	Polynomials
A **polynomial** is a finite sum of terms in which all variables have exponents that are nonnegative integers and no variables appear in the denominator.	$1.3x^2$ (monomial) $-\frac{1}{3}y + 5$ (binomial) $6z^2 - 5z + 7$ (trinomial)
A function P is a **polynomial function** if $P(x)$ is a polynomial.	For the polynomial function $P(x) = -x^2 + 6x - 12$, find $P(-2)$. $P(-2) = -(-2)^2 + 6(-2) - 12 = -28$

(continued)

Definitions and Concepts	Examples
Section 12.2 Polynomial Functions and Adding and Subtracting Polynomials *(continued)*	
The **degree of a term** is the sum of the exponents on the variables in the term.	**Term** / **Degree** $-5x^3$ — 3 3 (or $3x^0$) — 0 $2a^2b^2c$ — $2 + 2 + 1 = 5$
The **degree of a polynomial** is the greatest degree of any term of the polynomial.	**Polynomial** / **Degree** $5x^2 - 3x + 2$ — 2 $7y + 8y^2z^3 - 12$ — $2 + 3 = 5$
To add polynomials, add or combine like terms.	Add: $(7x^2 - 3x + 2) + (-5x - 6) = 7x^2 - 3x + 2 - 5x - 6$ $= 7x^2 - 8x - 4$
To subtract two polynomials, change the signs of the terms of the second polynomial, then add.	Subtract: $(17y^2 - 2y + 1) - (-3y^3 + 5y - 6)$ $= (17y^2 - 2y + 1) + (3y^3 - 5y + 6)$ $= 17y^2 - 2y + 1 + 3y^3 - 5y + 6$ $= 3y^3 + 17y^2 - 7y + 7$
Section 12.3 Multiplying Polynomials	
To multiply two polynomials, multiply each term of one polynomial by each term of the other polynomial, and then combine like terms.	Multiply: $(2x + 1)(5x^2 - 6x + 2)$ $= 2x(5x^2 - 6x + 2) + 1(5x^2 - 6x + 2)$ $= 10x^3 - 12x^2 + 4x + 5x^2 - 6x + 2$ $= 10x^3 - 7x^2 - 2x + 2$
Section 12.4 Special Products	
The **FOIL method** may be used when multiplying two binomials.	Multiply: $(5x - 3)(2x + 3)$ First, Last, Outer, Inner $(5x - 3)(2x + 3)$ F O I L $= (5x)(2x) + (5x)(3) + (-3)(2x) + (-3)(3)$ $= 10x^2 + 15x - 6x - 9$ $= 10x^2 + 9x - 9$
Squaring a Binomial $(a + b)^2 = a^2 + 2ab + b^2$ $(a - b)^2 = a^2 - 2ab + b^2$	Square each binomial. $(x + 5)^2 = x^2 + 2(x)(5) + 5^2$ $= x^2 + 10x + 25$ $(3x - 2y)^2 = (3x)^2 - 2(3x)(2y) + (2y)^2$ $= 9x^2 - 12xy + 4y^2$
Multiplying the Sum and Difference of Two Terms $(a + b)(a - b) = a^2 - b^2$	Multiply: $(6y + 5)(6y - 5) = (6y)^2 - 5^2$ $= 36y^2 - 25$

Definitions and Concepts	Examples
Section 12.5 Negative Exponents and Scientific Notation	
If $a \neq 0$ and n is an integer, $a^{-n} = \frac{1}{a^n}$	$3^{-2} = \frac{1}{3^2} = \frac{1}{9}$; $5x^{-2} = \frac{5}{x^2}$
Rules for exponents are true for positive and negative integers.	Simplify: $\left(\frac{x^{-2}y}{x^5}\right)^{-2} = \frac{x^4y^{-2}}{x^{-10}} = x^{4-(-10)}y^{-2} = \frac{x^{14}}{y^2}$
A positive number is written in scientific notation if it is written as the product of a number a, $1 \le a < 10$, and an integer power r of 10. $a \times 10^r$	Write each number in scientific notation. $12{,}000 = 1.2 \times 10^4$ $0.00000568 = 5.68 \times 10^{-6}$
Section 12.6 Dividing Polynomials	
To divide a polynomial by a monomial: $\frac{a+b}{c} = \frac{a}{c} + \frac{b}{c}$	Divide: $\frac{15x^5 - 10x^3 + 5x^2 - 2x}{5x^2} = \frac{15x^5}{5x^2} - \frac{10x^3}{5x^2} + \frac{5x^2}{5x^2} - \frac{2x}{5x^2} = 3x^3 - 2x + 1 - \frac{2}{5x}$
To divide a polynomial by a polynomial other than a monomial, use long division.	$2x + 3\overline{)10x^2 + 13x - 7}$ quotient $5x - 1 + \frac{-4}{2x+3}$; $\underline{10x^2 + 15x}$ (signs changed: $-10x^2 - 15x$); $-2x - 7$; $\underline{-2x - 3}$ (signs changed: $+2x + 3$); -4
Section 12.7 Synthetic Division and the Remainder Theorem	
A shortcut method called **synthetic division** may be used to divide a polynomial by a binomial of the form $x - c$.	Use synthetic division to divide $2x^3 - x^2 - 8x - 1$ by $x - 2$. $\begin{array}{r\|rrrr} 2 & 2 & -1 & -8 & -1 \\ & \downarrow & 4 & 6 & -4 \\ \hline & 2 & 3 & -2 & -5 \end{array}$ The quotient is $2x^2 + 3x - 2 - \frac{5}{x-2}$.

Chapter 12 Review

(12.1) *State the base and the exponent for each expression.*

1. 7^9

2. $(-5)^4$

3. -5^4

4. x^6

Evaluate each expression.

5. 8^3

6. $(-6)^2$

7. -6^2

8. $-4^3 - 4^0$

9. $(3b)^0$

10. $\dfrac{8b}{8b}$

Simplify each expression.

11. $y^2 \cdot y^7$

12. $x^9 \cdot x^5$

13. $(2x^5)(-3x^6)$

14. $(-5y^3)(4y^4)$

15. $(x^4)^2$

16. $(y^3)^5$

17. $(3y^6)^4$

18. $(2x^3)^3$

19. $\dfrac{x^9}{x^4}$

20. $\dfrac{z^{12}}{z^5}$

21. $\dfrac{a^5b^4}{ab}$

22. $\dfrac{x^4y^6}{xy}$

23. $\dfrac{3x^4y^{10}}{12xy^6}$

24. $\dfrac{2x^7y^8}{8xy^2}$

25. $5a^7(2a^4)^3$

26. $(2x)^2(9x)$

27. $(-5a)^0 + 7^0 + 8^0$

28. $8x^0 + 9^0$

Simplify the given expression and choose the correct result.

29. $\left(\dfrac{3x^4}{4y}\right)^3$

a. $\dfrac{27x^{64}}{64y^3}$ **b.** $\dfrac{27x^{12}}{64y^3}$ **c.** $\dfrac{9x^{12}}{12y^3}$ **d.** $\dfrac{3x^{12}}{4y^3}$

30. $\left(\dfrac{5a^6}{b^3}\right)^2$

a. $\dfrac{10a^{12}}{b^6}$ **b.** $\dfrac{25a^{36}}{b^9}$ **c.** $\dfrac{25a^{12}}{b^6}$ **d.** $25a^{12}b^6$

(12.2) *Find the degree of each term.*

31. $-5x^4y^3$

32. $10x^3y^2z$

33. $35a^5bc^2$

34. $95xyz$

Find the degree of each polynomial.

35. $y^5 + 7x - 8x^4$

36. $9y^2 + 30y + 25$

37. $-14x^2y - 28x^2y^3 - 42x^2y^2$

38. $6x^2y^2z^2 + 5x^2y^3 - 12xyz$

39. The Glass Bridge Skywalk is suspended 4000 feet over the Colorado River at the very edge of the Grand Canyon. Neglecting air resistance, the height of an object dropped from the Skywalk at time t seconds is given by the polynomial function $P(t) = -16t^2 + 4000$. Find the height of the object at the given times.

t	0 seconds	1 second	3 seconds	5 seconds
$P(t) = -16t^2 + 4000$				

△ **40.** The surface area of a box with a square base and a height of 5 units is given by the polynomial function $P(x) = 2x^2 + 20x$. Fill in the table below by evaluating the function for the given values of x.

x	1	3	5.1	10
$P(x) = 2x^2 + 20x$				

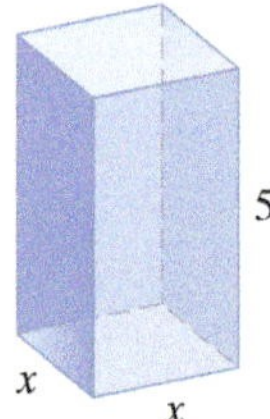

Combine like terms in each expression.

41. $6a^2 + 4a + 9a^2$

42. $21x^2 + 3x + x^2 + 6$

43. $4a^2b - 3b^2 - 8q^2 - 10a^2b + 7q^2$

44. $2s^{14} + 3s^{13} + 12s^{12} - s^{10}$

Add or subtract as indicated.

45. $(3x^2 + 2x + 6) + (5x^2 + x)$

46. $(2x^5 + 3x^4 + 4x^3 + 5x^2) + (4x^2 + 7x + 6)$

47. $(-5y^2 + 3) - (2y^2 + 4)$

48. $(3x^2 - 7xy + 7y^2) - (4x^2 - xy + 9y^2)$

TRANSLATING

Perform the indicated operations.

49. Subtract $(3x - y)$ from $(7x - 14y)$.

50. Subtract $(4x^2 + 8x - 7)$ from the sum of $(x^2 + 7x + 9)$ and $(x^2 + 4)$.

If $P(x) = 9x^2 - 7x + 8$, *find the following.*

51. $P(6)$ **52.** $P(-2)$

53. Find the perimeter of the rectangle.

$x^2y + 5$ cm

$2x^2y - 6x + 1$ cm

54. With the ownership of computers growing rapidly, the market for new software is also increasing. The revenue for software publishers (in millions of dollars) in the United States from 2001 to 2006 can be represented by the polynomial function $f(x) = 754x^2 - 228x + 80{,}134$ where x is the number of years since 2001. Use this model to find the revenues from software sales in 2009. (*Source:* Software & Information Industry Association)

(12.3) *Multiply each expression.*

55. $4(2a + 7)$

56. $9(6a - 3)$

57. $-7x(x^2 + 5)$

58. $-8y(4y^2 - 6)$

59. $(3a^3 - 4a + 1)(-2a)$

60. $(6b^3 - 4b + 2)(7b)$

61. $(2x + 2)(x - 7)$

62. $(2x - 5)(3x + 2)$

63. $(x - 9)^2$

64. $(x - 12)^2$

65. $(4a - 1)(a + 7)$

66. $(6a - 1)(7a + 3)$

67. $(5x + 2)^2$

68. $(3x + 5)^2$

69. $(x + 7)(x^3 + 4x - 5)$

70. $(x + 2)(x^5 + x + 1)$

71. $(x^2 + 2x + 4)(x^2 + 2x - 4)$

72. $(x^3 + 4x + 4)(x^3 + 4x - 4)$

73. $(x + 7)^3$

74. $(2x - 5)^3$

(12.4) *Use special products to multiply each of the following.*

75. $(x + 7)^2$

76. $(x - 5)^2$

77. $(3x - 7)^2$

78. $(4x + 2)^2$

79. $(5x - 9)^2$

80. $(5x + 1)(5x - 1)$

81. $(7x + 4)(7x - 4)$

82. $(a + 2b)(a - 2b)$

83. $(2x - 6)(2x + 6)$

84. $(4a^2 - 2b)(4a^2 + 2b)$

Express each as a product of polynomials in x. Then multiply and simplify.

△ **85.** Find the area of the square if its side is $(3x - 1)$ meters.

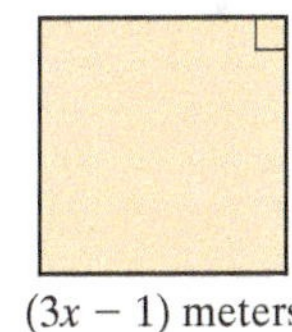

△ **86.** Find the area of the rectangle.

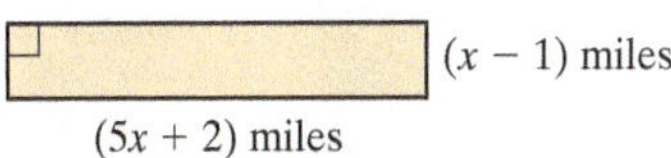

(12.5) *Simplify each expression.*

87. 7^{-2} **88.** -7^{-2} **89.** $2x^{-4}$

90. $(2x)^{-4}$ **91.** $\left(\frac{1}{5}\right)^{-3}$ **92.** $\left(\frac{-2}{3}\right)^{-2}$

93. $2^0 + 2^{-4}$ **94.** $6^{-1} - 7^{-1}$

Simplify each expression. Write each answer using positive exponents only.

95. $\frac{x^5}{x^{-3}}$ **96.** $\frac{z^4}{z^{-4}}$ **97.** $\frac{r^{-3}}{r^{-4}}$

98. $\frac{y^{-2}}{y^{-5}}$ **99.** $\left(\frac{bc^{-2}}{bc^{-3}}\right)^4$ **100.** $\left(\frac{x^{-3}y^{-4}}{x^{-2}y^{-5}}\right)^{-3}$

101. $\frac{x^{-4}y^{-6}}{x^2y^7}$ **102.** $\frac{a^5b^{-5}}{a^{-5}b^5}$ **103.** $a^{6m}a^{5m}$

104. $\frac{(x^{5+h})^3}{x^5}$ **105.** $(3xy^{2z})^3$ **106.** $a^{m+2}a^{m+3}$

Write each number in scientific notation.

107. 0.00027 **108.** 0.8868

109. 80,800,000 **110.** 868,000

111. Google.com is an Internet search engine that handles 91,000,000 searches every day. Write 91,000,000 in scientific notation. (*Source:* Google, Inc.)

112. The approximate diameter of the Milky Way galaxy is 150,000 light years. Write this number in scientific notation. (*Source:* NASA IMAGE/POETRY Education and Public Outreach Program)

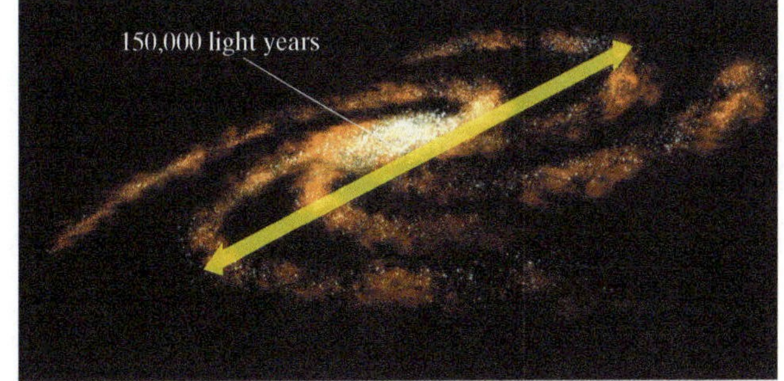

Write each number in standard form.

113. 8.67×10^5

114. 3.86×10^{-3}

115. 8.6×10^{-4}

116. 8.936×10^5

117. The volume of the planet Jupiter is 1.43128×10^{15} cubic kilometers. Write this number in standard notation. (*Source:* National Space Science Data Center)

118. An angstrom is a unit of measure, equal to 1×10^{-10} meter, used for measuring wavelengths or the diameters of atoms. Write this number in standard notation. (*Source:* National Institute of Standards and Technology)

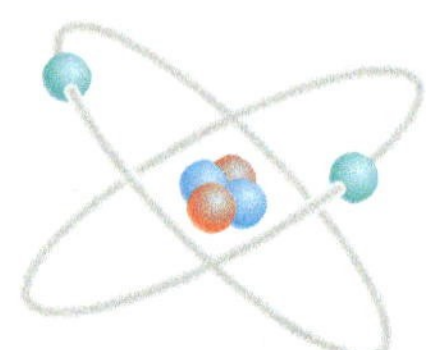

Simplify. Express each result in standard form.

119. $(8 \times 10^4)(2 \times 10^{-7})$

120. $\frac{8 \times 10^4}{2 \times 10^{-7}}$

(12.6) *Divide.*

121. $\frac{x^2 + 21x + 49}{7x^2}$

122. $\frac{5a^3b - 15ab^2 + 20ab}{-5ab}$

123. $(a^2 - a + 4) \div (a - 2)$

124. $(4x^2 + 20x + 7) \div (x + 5)$

125. $\frac{a^3 + a^2 + 2a + 6}{a - 2}$

126. $\frac{9b^3 - 18b^2 + 8b - 1}{3b - 2}$

127. $\frac{4x^4 - 4x^3 + x^2 + 4x - 3}{2x - 1}$

128. $\frac{-10x^2 - x^3 - 21x + 18}{x - 6}$

△ **129.** The area of the rectangle below is $(15x^3 - 3x^2 + 60)$ square feet. If its length is $3x^2$ feet, find its width.

Area is $(15x^3 - 3x^2 + 60)$ sq feet

△ **130.** The perimeter of the equilateral triangle below is $(21a^3b^6 + 3a - 3)$ units. Find the length of a side.

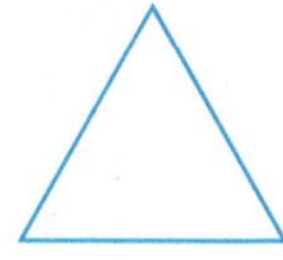

Perimeter is $(21a^3b^6 + 3a - 3)$ units

(12.7) *Use synthetic division to find each quotient.*

131. $(3x^3 + 12x - 4) \div (x - 2)$

132. $(3x^3 + 2x^2 - 4x - 1) \div \left(x + \frac{3}{2}\right)$

133. $(x^5 - 1) \div (x + 1)$

134. $(x^3 - 81) \div (x - 3)$

135. $(x^3 - x^2 + 3x^4 - 2) \div (x - 4)$

136. $(3x^4 - 2x^2 + 10) \div (x + 2)$

If $P(x) = 3x^5 - 9x + 7$, use the remainder theorem to find the following.

137. $P(4)$

138. $P(-5)$

MIXED REVIEW

Evaluate.

139. $\left(-\frac{1}{2}\right)^3$

Simplify each expression. Write each answer using positive exponents only.

140. $(4xy^2)(x^3y^5)$

141. $\frac{18x^9}{27x^3}$

142. $\left(\frac{3a^4}{b^2}\right)^3$

143. $(2x^{-4}y^3)^{-4}$

144. $\frac{a^{-3}b^6}{9^{-1}a^{-5}b^{-2}}$

Perform the indicated operations and simplify.

145. $(6x + 2) + (5x - 7)$

146. $(-y^2 - 4) + (3y^2 - 6)$

147. $(8y^2 - 3y + 1) - (3y^2 + 2)$

148. $(5x^2 + 2x - 6) - (-x - 4)$

149. $4x(7x^2 + 3)$

150. $(2x + 5)(3x - 2)$

151. $(x - 3)(x^2 + 4x - 6)$

152. $(7x - 2)(4x - 9)$

Use special products to multiply.

153. $(5x + 4)^2$

154. $(6x + 3)(6x - 3)$

Divide.

155. $\frac{8a^4 - 2a^3 + 4a - 5}{2a^3}$

156. $\frac{x^2 + 2x + 10}{x + 5}$

157. $\frac{4x^3 + 8x^2 - 11x + 4}{2x - 3}$

Chapter 12 Test

Step-by-step test solutions are found on the Chapter Test Prep Videos. Where available: MyMathLab® or YouTube

Evaluate each expression.

1. 2^5 **2.** $(-3)^4$ **3.** -3^4 **4.** 4^{-3}

Simplify each exponential expression. Write the result using only positive exponents.

5. $(3x^2)(-5x^9)$

6. $\frac{y^7}{y^2}$

7. $\frac{r^{-8}}{r^{-3}}$

8. $\left(\frac{x^2y^3}{x^3y^{-4}}\right)^2$

9. $\frac{6^2x^{-4}y^{-1}}{6^3x^{-3}y^7}$

Express each number in scientific notation.

10. 563,000 **11.** 0.0000863

Write each number in standard form.

12. 1.5×10^{-3} **13.** 6.23×10^4

14. Simplify. Write the answer in standard form.

$$(1.2 \times 10^5)(3 \times 10^{-7})$$

15. a. Complete the table for the polynomial $4xy^2 + 7xyz + x^3y - 2$.

Term	Numerical Coefficient	Degree of Term
$4xy^2$		
$7xyz$		
x^3y		
-2		

b. What is the degree of the polynomial?

16. Simplify by combining like terms.

$$5x^2 + 4xy - 7x^2 + 11 + 8xy$$

Perform each indicated operation.

17. $(8x^3 + 7x^2 + 4x - 7) + (8x^3 - 7x - 6)$

18. $5x^3 + x^2 + 5x - 2 - (8x^3 - 4x^2 + x - 7)$

19. Subtract $(4x + 2)$ from the sum of $(8x^2 + 7x + 5)$ and $(x^3 - 8)$.

Multiply.

20. $(3x + 7)(x^2 + 5x + 2)$

21. $3x^2(2x^2 - 3x + 7)$

22. $(x + 7)(3x - 5)$

23. $\left(3x - \frac{1}{5}\right)\left(3x + \frac{1}{5}\right)$

24. $(4x - 2)^2$

25. $(8x + 3)^2$

26. $(x^2 - 9b)(x^2 + 9b)$

Solve.

27. The height of the Bank of China in Hong Kong is 1001 feet. Neglecting air resistance, the height of an object dropped from this building at time t seconds is given by the polynomial function $P(t) = -16t^2 + 1001$. Find the height of the object at the given times below.

t	0 seconds	1 second	3 seconds	5 seconds
$P(t) = -16t^2 + 1001$				

28. △ Find the area of the top of the table. Express the area as a product, then multiply and simplify.

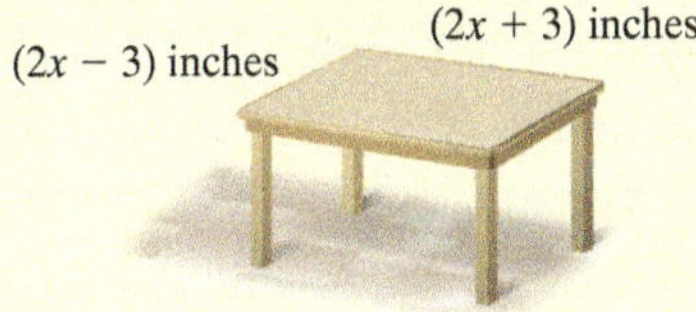

Divide.

29. $\frac{4x^2 + 24xy - 7x}{8xy}$

30. $(x^2 + 7x + 10) \div (x + 5)$

31. $\frac{27x^3 - 8}{3x + 2}$

32. A pebble is hurled upward from the top of the Canada Trust Tower, which is 880 feet tall, with an initial velocity of 96 feet per second. Neglecting air resistance, the height $h(t)$ of the pebble after t seconds is given by the polynomial function

$$h(t) = -16t^2 + 96t + 880$$

a. Find the height of the pebble when $t = 1$.

b. Find the height of the pebble when $t = 5.1$.

33. Use synthetic division to divide $(4x^4 - 3x^3 - x - 1)$ by $(x + 3)$.

34. If $P(x) = 4x^4 + 7x^2 - 2x - 5$, use the remainder theorem to find $P(-2)$.

Chapters 1–12 Cumulative Review

1. Determine whether each statement is true or false.
   a. $8 \geq 8$ b. $8 \leq 8$
   c. $23 \leq 0$ d. $23 \geq 0$
2. Find the absolute value of each number.
   a. $|-7.2|$ b. $|0|$ c. $\left|-\frac{1}{2}\right|$
3. Divide and simplify. $\frac{5}{16} \div \frac{3}{4}$
4. Multiply and simplify.
   $\frac{3}{4} \cdot \frac{7}{21}$
5. Evaluate.
   a. 9^2 b. 6^1 c. 3^4
   d. $5 \cdot 6^2$
6. Evaluate $\frac{2x - 7y}{y^2}$ for $x = 5$ and $y = 1$.
7. Add.
   a. $-18 + 10$ b. $12 + (-8)$
   c. $0 + (-5)$
8. Simplify: $8 + 3(2 \cdot 6 - 1)$
9. Subtract 7 from -3.
10. Is $x = 1$ a solution of $5x^2 + 2 = x - 8$?
11. Subtract.
    a. $8 - 5$ b. $-4 - 10$
    c. $6 - (-5)$ d. $-11 - (-7)$
12. Subtract:
    a. $7 - 40$ b. $-5 - (-10)$
13. Use an associative property to complete each statement.
    a. $5 + (4 + 6) =$ _____
    b. $(-1 \cdot 2) \cdot 5 =$ _____
    c. $(m + n) + 9 =$ _____
    d. $(xy) \cdot 12 =$ _____
14. Multiply: $-3(-6)$
15. Simplify each expression.
    a. $10 + (x + 12)$
    b. $-3(7x)$
16. Use the distributive property to write $-2(x + 3y - z)$ without parentheses.
17. Use the distributive property to multiply: $5(x + 4)$
18. Simplify: $2(6x - 1) - (x - 7)$
19. Solve $x - 2 = -1$ for x.
20. Write the phrase as an algebraic expression: double a number, subtracted from the sum of the number and seven
21. Solve: $\frac{z}{-4} = 11 - 5$
22. Solve: $2x + 1 = x - 3$
23. Twice a number plus 3 is the same as the number minus 6. Find the unknown number.
24. Solve: $-10 = 5x$
25. Twice the sum of a number and 4 is the same as four times the number decreased by 12. Find the number.
26. Solve: $\frac{7x + 5}{3} = x + 3$
27. The length of a rectangular road sign is 2 feet less than three times its width. Find the dimensions if the perimeter is 28 feet.
28. Graph $x < 5$ and write in interval notation.
29. Solve $F = \frac{9}{5}C + 32$ for C.
30. Find the slope of each line.
    a. $x = -1$
    b. $y = 7$
31. Graph $-4 < x \leq 2$. Then write the solutions in interval notation.
32. Recall that the grade of a road is its slope written as a percent. Find the grade of the road shown.

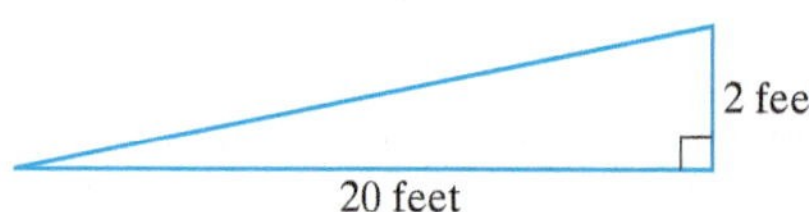

33. Complete the following ordered pair solutions for the equation $3x + y = 12$.
    a. $(0, \ )$ b. $(\ , 6)$
    c. $(-1, \ )$
34. Solve the system: $\begin{cases} 3x + 2y = -8 \\ 2x - 6y = -9 \end{cases}$
35. Graph the linear equation $2x + y = 5$.
36. Solve the system: $\begin{cases} x = -3y + 3 \\ 2x + 9y = 5 \end{cases}$
37. Graph $x = 2$.
38. Evaluate.
    a. $(-5)^2$ b. -5^2 c. $2 \cdot 5^2$
39. Find the slope of the line $x = 5$.
40. Simplify: $\frac{(z^2)^3 \cdot z^7}{z^9}$
41. Subtract: $(2x^3 + 8x^2 - 6x) - (2x^3 - x^2 + 1)$.
42. Subtract $(5y^2 - 6) - (y^2 + 2)$.
43. Use the product rule to simplify $(2x^2)(-3x^5)$.
44. Find the value of $-x^2$ when
    a. $x = 2$ b. $x = -2$
45. Add $(11x^3 - 12x^2 + x - 3)$ and $(x^3 - 10x + 5)$.
46. Multiply $(10x^2 - 3)(10x^2 + 3)$.
47. Multiply $(2x - y)^2$.
48. Multiply $(10x^2 + 3)^2$.
49. Divide $6m^2 + 2m$ by $2m$.
50. Evaluate.
    a. 5^{-1} b. 7^{-2}

13

Factoring Polynomials

As shown on the circle graph below, games and music are the leading moneymakers in the digital market. Record companies' revenue from digital music channels grew 6% in a recent year to total $4.6 billion in the United States alone. From 2004 to 2010, the digital music market increased 1000%.

In Section 13.1, Exercise 109, we explore digital music downloads further.

Global Digital Revenues

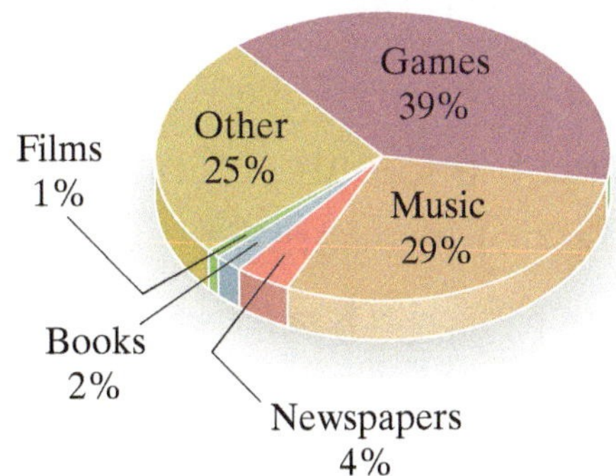

Data from NPD Group; Pollstar; Recording Industry Association of America (RIAA); U.S. Department of Labor; and Tera Consultants

Sections

Check Your Progress

In Chapter 12, you learned how to multiply polynomials. This chapter deals with an operation that is the reverse process of multiplying, called *factoring*. Factoring is an important algebraic skill because this process allows us to write a sum as a product.

At the end of this chapter, we use factoring to help us solve equations other than linear equations, and in Chapter 14, we use factoring to simplify and perform arithmetic operations on rational expressions.

13.1 The Greatest Common Factor and Factoring by Grouping 

Objectives

1 Find the Greatest Common Factor of a List of Integers.

2 Find the Greatest Common Factor of a List of Terms.

3 Factor Out the Greatest Common Factor from a Polynomial.

4 Factor a Polynomial by Grouping.

In the product $2 \cdot 3 = 6$, the numbers 2 and 3 are called **factors** of 6, and $2 \cdot 3$ is a **factored form** of 6. This is true of polynomials also. Since $(x + 2)(x + 3) = x^2 + 5x + 6$, $(x + 2)$ and $(x + 3)$ are factors of $x^2 + 5x + 6$, and $(x + 2)(x + 3)$ is a factored form of the polynomial.

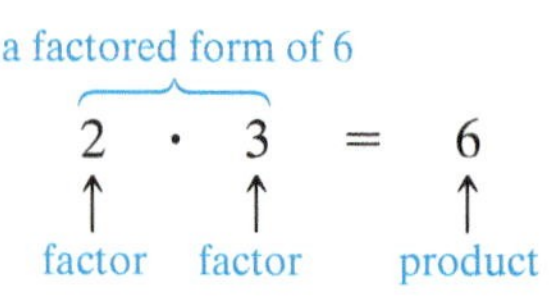

a factored form of x^5

$$x^2 \cdot x^3 = x^5$$

factor factor product

a factored form of $x^2 + 5x + 6$

$$(x + 2)(x + 3) = x^2 + 5x + 6$$

factor factor product

The process of writing a polynomial as a product is called **factoring** the polynomial.

Do you see that factoring is the reverse process of multiplying?

$$x^2 + 5x + 6 = (x + 2)(x + 3)$$

factoring →; ← multiplying

✓Concept Check Multiply: $2(x - 4)$
What do you think the result of factoring $2x - 8$ would be? Why?

Objective 1 Finding the Greatest Common Factor of a List of Integers

The first step in factoring a polynomial is to see whether the terms of the polynomial have a common factor. If there is one, we can write the polynomial as a product by **factoring out** the common factor. We will usually factor out the **greatest common factor (GCF).**

The GCF of a list of integers is the largest integer that is a factor of all the integers in the list. For example, the GCF of 12 and 20 is 4 because 4 is the largest integer that is a factor of both 12 and 20. With large integers, the GCF may not be found easily by inspection. When this happens, use the following steps.

Finding the GCF of a List of Integers

Step 1: Write each number as a product of prime numbers.

Step 2: Identify the common prime factors.

Step 3: The product of all common prime factors found in Step 2 is the greatest common factor. If there are no common prime factors, the greatest common factor is 1.

Recall from Section 4.2 that a prime number is a whole number other than 1 whose only factors are 1 and itself.

✓Concept Check Answers
$2x - 8$; The result would be $2(x - 4)$ because factoring is the reverse process of multiplying.

Example 1 Find the GCF of each list of numbers.

a. 28 and 40 **b.** 55 and 21 **c.** 15, 18, and 66

Solution:

a. Write each number as a product of primes.

$$28 = 2 \cdot 2 \cdot 7 = 2^2 \cdot 7$$
$$40 = 2 \cdot 2 \cdot 2 \cdot 5 = 2^3 \cdot 5$$

There are two common factors, each of which is 2, so the GCF is

$$\text{GCF} = 2 \cdot 2 = 4$$

b. $55 = 5 \cdot 11$

$21 = 3 \cdot 7$

There are no common prime factors; thus, the GCF is 1.

c. $15 = 3 \cdot 5$

$18 = 2 \cdot 3 \cdot 3 = 2 \cdot 3^2$

$66 = 2 \cdot 3 \cdot 11$

The only prime factor common to all three numbers is 3, so the GCF is

$$\text{GCF} = 3$$

Work Practice 1

Practice 1

Find the GCF of each list of numbers.

a. 36 and 42

b. 35 and 44

c. 12, 16, and 40

Objective 2 Finding the Greatest Common Factor of a List of Terms

The greatest common factor of a list of variables raised to powers is found in a similar way. For example, the GCF of x^2, x^3, and x^5 is x^2 because each term contains a factor of x^2 and no higher power of x is a factor of each term.

$$x^2 = x \cdot x$$
$$x^3 = x \cdot x \cdot x$$
$$x^5 = x \cdot x \cdot x \cdot x \cdot x$$

There are two common factors, each of which is x, so the GCF $= x \cdot x$ or x^2.

From this example, we see that **the GCF of a list of common variables raised to powers is the variable raised to the smallest exponent in the list.**

Example 2 Find the GCF of each list of terms.

a. x^3, x^7, and x^5 **b.** y, y^4, and y^7

Solution:

a. The GCF is x^3, since 3 is the smallest exponent to which x is raised.

b. The GCF is y^1 or y, since 1 is the smallest exponent on y.

Work Practice 2

Practice 2

Find the GCF of each list of terms.

a. y^7, y^4, and y^6

b. x, x^4, and x^2

In general, the **greatest common factor (GCF) of a list of terms** is the product of the GCF of the numerical coefficients and the GCF of the variable factors.

$$20x^2y^2 = 2 \cdot 2 \cdot 5 \cdot x \cdot x \cdot y \cdot y$$
$$6xy^3 = 2 \cdot 3 \cdot x \cdot y \cdot y \cdot y$$
$$\text{GCF} = 2 \cdot x \cdot y \cdot y = 2xy^2$$

Answers

1. a. 6 **b.** 1 **c.** 4 **2. a.** y^4 **b.** x

Helpful Hint

Remember that the GCF of a list of terms contains the smallest exponent on each common variable.

The GCF of x^5y^6, x^2y^7, and x^3y^4 is x^2y^4. — Smallest exponent on x — Smallest exponent on y

Practice 3

Find the GCF of each list of terms.

a. $5y^4$, $15y^2$, and $-20y^3$

b. $4x^2$, x^3, and $3x^8$

c. a^4b^2, a^3b^5, and a^2b^3

Example 3

Find the GCF of each list of terms.

a. $6x^2$, $10x^3$, and $-8x$ **b.** $-18y^2$, $-63y^3$, and $27y^4$ **c.** a^3b^2, a^5b, and a^6b^2

Solution:

a.
$$6x^2 = 2 \cdot 3 \cdot x^2$$
$$10x^3 = 2 \cdot 5 \cdot x^3$$
$$-8x = -1 \cdot 2 \cdot 2 \cdot 2 \cdot x^1$$
→ The GCF of x^2, x^3, and x^1 is x^1 or x.
$$\text{GCF} = 2 \cdot x^1 \quad \text{or} \quad 2x$$

b.
$$-18y^2 = -1 \cdot 2 \cdot 3 \cdot 3 \cdot y^2$$
$$-63y^3 = -1 \cdot 3 \cdot 3 \cdot 7 \cdot y^3$$
$$27y^4 = 3 \cdot 3 \cdot 3 \cdot y^4$$
→ The GCF of y^2, y^3, and y^4 is y^2.
$$\text{GCF} = 3 \cdot 3 \cdot y^2 \quad \text{or} \quad 9y^2$$

c. The GCF of a^3, a^5, and a^6 is a^3.

The GCF of b^2, b, and b^2 is b.

Thus, the GCF of a^3b^2, a^5b, and a^6b^2 is a^3b.

Work Practice 3

Objective 3 Factoring Out the Greatest Common Factor

The first step in factoring a polynomial is to find the GCF of its terms. Once we do so, we can write the polynomial as a product by **factoring out** the GCF.

The polynomial $8x + 14$, for example, contains two terms: $8x$ and 14. The GCF of these terms is 2. We factor out 2 from each term by writing each term as a product of 2 and the term's remaining factors.

$$8x + 14 = 2 \cdot 4x + 2 \cdot 7$$

Using the distributive property, we can write

$$8x + 14 = 2 \cdot 4x + 2 \cdot 7$$
$$= 2(4x + 7)$$

Thus, a factored form of $8x + 14$ is $2(4x + 7)$. We can check by multiplying:

$$2(4x + 7) = 2 \cdot 4x + 2 \cdot 7 = 8x + 14.$$

Helpful Hint

A factored form of $8x + 14$ is *not*

$$2 \cdot 4x + 2 \cdot 7$$

Although the *terms* have been factored (written as a product), the *polynomial* $8x + 14$ has not been factored (written as a product).

A factored form of $8x + 14$ is the *product* $2(4x + 7)$.

Answers

3. a. $5y^2$ **b.** x^2 **c.** a^2b^2

✓**Concept Check** Which of the following is/are factored form(s) of $7t + 21$?

a. 7 **b.** $7 \cdot t + 7 \cdot 3$ **c.** $7(t + 3)$ **d.** $7(t + 21)$

Example 4 Factor each polynomial by factoring out the GCF.

a. $6t + 18$ **b.** $y^5 - y^7$

Solution:

a. The GCF of terms $6t$ and 18 is 6.

$$6t + 18 = 6 \cdot t + 6 \cdot 3$$
$$= 6(t + 3) \quad \text{Apply the distributive property.}$$

Our work can be checked by multiplying 6 and $(t + 3)$.

$$6(t + 3) = 6 \cdot t + 6 \cdot 3 = 6t + 18, \text{ the original polynomial.}$$

b. The GCF of y^5 and y^7 is y^5. Thus,

$$y^5 - y^7 = y^5(1) - y^5(y^2)$$
$$= y^5(1 - y^2)$$

Don't forget the 1.

Work Practice 4

Practice 4

Factor each polynomial by factoring out the GCF.

a. $4t + 12$

b. $y^8 + y^4$

Example 5 Factor: $-9a^5 + 18a^2 - 3a$

Solution:

$$-9a^5 + 18a^2 - 3a = (3a)(-3a^4) + (3a)(6a) + (3a)(-1)$$
$$= 3a(-3a^4 + 6a - 1)$$

Helpful Hint

Don't forget the -1.

Work Practice 5

Practice 5

Factor $-8b^6 + 16b^4 - 8b^2$.

In Example 5, we could have chosen to factor out $-3a$ instead of $3a$. If we factor out $-3a$, we have

$$-9a^5 + 18a^2 - 3a = (-3a)(3a^4) + (-3a)(-6a) + (-3a)(1)$$
$$= -3a(3a^4 - 6a + 1)$$

Helpful Hint

Notice the changes in signs when factoring out $-3a$.

Examples Factor.

6. $6a^4 - 12a = 6a(a^3 - 2)$

7. $\frac{3}{7}x^4 + \frac{1}{7}x^3 - \frac{5}{7}x^2 = \frac{1}{7}x^2(3x^2 + x - 5)$

8. $15p^2q^4 + 20p^3q^5 + 5p^3q^3 = 5p^2q^3(3q + 4pq^2 + p)$

Work Practice 6–8

Practice 6–8

Factor.

6. $5x^4 - 20x$

7. $\frac{5}{9}z^5 + \frac{1}{9}z^4 - \frac{2}{9}z^3$

8. $8a^2b^4 - 20a^3b^3 + 12ab^3$

Answers

4. a. $4(t + 3)$ **b.** $y^4(y^4 + 1)$

5. $8b^2(-b^4 + 2b^2 - 1)$ or $-8b^2(b^4 - 2b^2 + 1)$

6. $5x(x^3 - 4)$

7. $\frac{1}{9}z^3(5z^2 + z - 2)$

8. $4ab^3(2ab - 5a^2 + 3)$

✓**Concept Check Answer**

c

Practice 9

Factor $8(y - 2) + x(y - 2)$.

Example 9 Factor: $5(x + 3) + y(x + 3)$

Solution: The binomial $(x + 3)$ is the greatest common factor. Use the distributive property to factor out $(x + 3)$.

$$5(x + 3) + y(x + 3) = (x + 3)(5 + y)$$

Work Practice 9

Practice 10

Factor $7xy^3(p + q) - (p + q)$.

Example 10 Factor: $3m^2n(a + b) - (a + b)$

Solution: The greatest common factor is $(a + b)$.

$$3m^2n(a + b) - 1(a + b) = (a + b)(3m^2n - 1)$$

Work Practice 10

Objective 4 Factoring by Grouping

Once the GCF is factored out, we can often continue to factor the polynomial, using a variety of techniques. We discuss here a technique for factoring polynomials called **factoring by grouping.**

Practice 11

Factor $xy + 3y + 4x + 12$ by grouping. Check by multiplying.

Example 11 Factor $xy + 2x + 3y + 6$ by grouping. Check by multiplying.

Solution: The GCF of the first two terms is x, and the GCF of the last two terms is 3.

$$xy + 2x + 3y + 6 = (xy + 2x) + (3y + 6) \quad \text{Group terms.}$$
$$= x(y + 2) + 3(y + 2) \quad \text{Factor out GCF from each grouping.}$$

Notice that this form, $x(y + 2) + 3(y + 2)$, is *not* a factored form of the original polynomial. It is a sum, not a product.

Next we factor out the common binomial factor, $(y + 2)$.

$$x(y + 2) + 3(y + 2) = (y + 2)(x + 3)$$

Now the result is a factored form because it is a product. We were able to write the polynomial as a product because of the common binomial factor, $(y + 2)$, that appeared. If this does not happen, try rearranging the terms of the original polynomial.

Check: Multiply $(y + 2)$ by $(x + 3)$.

$$(y + 2)(x + 3) = xy + 2x + 3y + 6,$$

the original polynomial.
Thus, the factored form of $xy + 2x + 3y + 6$ is the product $(y + 2)(x + 3)$.

Work Practice 11

You may want to try these steps when factoring by grouping.

To Factor a Four-Term Polynomial by Grouping

Step 1: Group the terms in two groups of two terms so that each group has a common factor.

Step 2: Factor out the GCF from each group.

Step 3: If there is now a common binomial factor in the groups, factor it out.

Step 4: If not, rearrange the terms and try these steps again.

Answers

9. $(y - 2)(8 + x)$
10. $(p + q)(7xy^3 - 1)$
11. $(x + 3)(y + 4)$

Examples Factor by grouping.

12. $15x^3 - 10x^2 + 6x - 4$

$= (15x^3 - 10x^2) + (6x - 4)$ Group the terms.

$= 5x^2(3x - 2) + 2(3x - 2)$ Factor each group.

$= (3x - 2)(5x^2 + 2)$ Factor out the common factor, $(3x - 2)$.

13. $3x^2 + 4xy - 3x - 4y$

$= (3x^2 + 4xy) + (-3x - 4y)$

$= x(3x + 4y) - 1(3x + 4y)$ Factor each group. A -1 is factored from the second pair of terms so that there is a common factor, $(3x + 4y)$.

$= (3x + 4y)(x - 1)$ Factor out the common factor, $(3x + 4y)$.

14. $2a^2 + 5ab + 2a + 5b$

$= (2a^2 + 5ab) + (2a + 5b)$

$= a(2a + 5b) + 1(2a + 5b)$ Factor each group. An understood 1 is written before $(2a+5b)$ to help remember that $(2a+5b)$ is $1(2a+5b)$.

$= (2a + 5b)(a + 1)$ Factor out the common factor, $(2a + 5b)$.

Work Practice 12–14

Practice 12–14

12. Factor $40x^3 - 24x^2 + 15x - 9$ by grouping.

13. Factor $2xy + 3y^2 - 2x - 3y$ by grouping.

14. Factor $7a^3 + 5a^2 + 7a + 5$ by grouping.

Helpful Hint Notice the factor of 1 is written when $(2a + 5b)$ is factored out.

Examples Factor by grouping.

15. $3xy + 2 - 3x - 2y$

Notice that the first two terms have no common factor other than 1. However, if we rearrange these terms, a grouping emerges that does lead to a common factor.

$3xy + 2 - 3x - 2y$

$= (3xy - 3x) + (-2y + 2)$

$= 3x(y - 1) - 2(y - 1)$ Factor -2 from the second group so that there is a common factor, $(y - 1)$.

$= (y - 1)(3x - 2)$ Factor out the common factor, $(y - 1)$.

16. $5x - 10 + x^3 - x^2 = 5(x - 2) + x^2(x - 1)$

There is no common binomial factor that can now be factored out. No matter how we rearrange the terms, no grouping will lead to a common factor. Thus, this polynomial is not factorable by grouping.

Work Practice 15–16

Practice 15–16

15. Factor $4xy + 15 - 12x - 5y$ by grouping.

16. Factor $9y - 18 + y^3 - 4y^2$ by grouping.

Helpful Hint

One more reminder: When **factoring** a polynomial, make sure the polynomial is written as a **product.** For example, it is true that

$$3x^2 + 4xy - 3x - 4y = \underbrace{x(3x + 4y) - 1(3x + 4y)}_{\text{but is not a factored form}},$$

since it is a **sum (difference),** not a **product.** A factored form of $3x^2 + 4xy - 3x - 4y$ is the product $(3x + 4y)(x - 1)$.

Answers

12. $(5x - 3)(8x^2 + 3)$
13. $(2x + 3y)(y - 1)$
14. $(7a + 5)(a^2 + 1)$
15. $(y - 3)(4x - 5)$
16. Cannot be factored by grouping

Practice 17

Factor $3xy - 3ay - 6ax + 6a^2$.

Helpful Hint Throughout this chapter, we will be factoring polynomials. Even when the instructions do not so state, it is always a good idea to check your answers by multiplying.

Answer

17. $3(x - a)(y - 2a)$

Factoring out a greatest common factor first makes factoring by any method easier, as we see in the next example.

Example 17 Factor: $4ax - 4ab - 2bx + 2b^2$

Solution: First, factor out the common factor 2 from all four terms.

$4ax - 4ab - 2bx + 2b^2$

$= 2(2ax - 2ab - bx + b^2)$ Factor out 2 from all four terms.

$= 2[2a(x - b) - b(x - b)]$ Factor each pair of terms. A "$-b$" is factored from the second pair so that there is a common factor, $x - b$.

$= 2(x - b)(2a - b)$ Factor out the common binomial.

Work Practice 17

Vocabulary, Readiness & Video Check

Use the choices below to fill in each blank. Some choices may be used more than once and some may not be used at all.

greatest common factor factors factoring true false least greatest

1. Since $5 \cdot 4 = 20$, the numbers 5 and 4 are called ________________ of 20.
2. The ________________ of a list of integers is the largest integer that is a factor of all the integers in the list.
3. The greatest common factor of a list of common variables raised to powers is the variable raised to the ________________ exponent in the list.
4. The process of writing a polynomial as a product is called ________________.
5. True or false: A factored form of $7x + 21 + xy + 3y$ is $7(x + 3) + y(x + 3)$. ________________
6. True or false: A factored form of $3x^3 + 6x + x^2 + 2$ is $3x(x^2 + 2)$. ________________

Martin-Gay Interactive Videos

See Video 13.1

Watch the section lecture video and answer the following questions.

Objective 1 7. Based on Example 1, give a general definition for the greatest common factor (GCF) of a list of numbers.

Objective 2 8. In Example 3, why are the numbers factored out, but not the variables?

Objective 3 9. From Example 5, how can the number of terms in the other factor, once you factor out the GCF, help you determine if your factorization is correct?

Objective 4 10. In Examples 7 and 8, what are you reminded to always do first when factoring a polynomial? Also, explain how a polynomial looks if it suggests it might be factored by grouping.

13.1 Exercise Set MyMathLab®

Find the GCF for each list. See Examples 1 through 3.

1. 32, 36
2. 36, 90
3. 18, 42, 84
4. 30, 75, 135
5. 24, 14, 21
6. 15, 25, 27
7. y^2, y^4, y^7
8. x^3, x^2, x^5
9. z^7, z^9, z^{11}
10. y^8, y^{10}, y^{12}
11. $x^{10}y^2, xy^2, x^3y^3$
12. p^7q, p^8q^2, p^9q^3
13. $14x, 21$
14. $20y, 15$
15. $12y^4, 20y^3$
16. $32x^5, 18x^2$
17. $-10x^2, 15x^3$
18. $-21x^3, 14x$
19. $12x^3, -6x^4, 3x^5$
20. $15y^2, 5y^7, -20y^3$
21. $-18x^2y, 9x^3y^3, 36x^3y$
22. $7x^3y^3, -21x^2y^2, 14xy^4$
23. $20a^6b^2c^8, 50a^7b$
24. $40x^7y^2z, 64x^9y$

Factor out the GCF from each polynomial. See Examples 4 through 10.

25. $3a + 6$
26. $18a + 12$
27. $30x - 15$
28. $42x - 7$
29. $x^3 + 5x^2$
30. $y^5 + 6y^4$
31. $6y^4 + 2y^3$
32. $5x^2 + 10x^6$
33. $4x - 8y + 4$
34. $7x + 21y - 7$
35. $6x^3 - 9x^2 + 12x$
36. $12x^3 + 16x^2 - 8x$
37. $a^7b^6 - a^3b^2 + a^2b^5 - a^2b^2$
38. $x^9y^6 + x^3y^5 - x^4y^3 + x^3y^3$
39. $8x^5 + 16x^4 - 20x^3 + 12$
40. $9y^6 - 27y^4 + 18y^2 + 6$
41. $\frac{1}{3}x^4 + \frac{2}{3}x^3 - \frac{4}{3}x^5 + \frac{1}{3}x$
42. $\frac{2}{5}y^7 - \frac{4}{5}y^5 + \frac{3}{5}y^2 - \frac{2}{5}y$
43. $y(x^2 + 2) + 3(x^2 + 2)$
44. $x(y^2 + 1) - 3(y^2 + 1)$
45. $z(y + 4) - 3(y + 4)$
46. $8(x + 2) - y(x + 2)$
47. $r(z^2 - 6) + (z^2 - 6)$
48. $q(b^3 - 5) + (b^3 - 5)$

Factor a negative number or a GCF with a negative coefficient from each polynomial. See Example 5.

49. $-2x - 14$
50. $-7y - 21$
51. $-2x^5 + x^7$
52. $-5y^3 + y^6$
53. $-6a^4 + 9a^3 - 3a^2$
54. $-5m^6 + 10m^5 - 5m^3$

Factor each four-term polynomial by grouping. If this is not possible, write "not factorable by grouping." See Examples 11 through 17.

55. $x^3 + 2x^2 + 5x + 10$
56. $x^3 + 4x^2 + 3x + 12$
57. $5x + 15 + xy + 3y$
58. $xy + y + 2x + 2$
59. $6x^3 - 4x^2 + 15x - 10$
60. $16x^3 - 28x^2 + 12x - 21$
61. $5m^3 + 6mn + 5m^2 + 6n$
62. $8w^2 + 7wv + 8w + 7v$
63. $2y - 8 + xy - 4x$
64. $6x - 42 + xy - 7y$
65. $2x^3 - x^2 + 8x - 4$
66. $2x^3 - x^2 - 10x + 5$
67. $3x - 3 + x^3 - 4x^2$
68. $7x - 21 + x^3 - 2x^2$
69. $4x^2 - 8xy - 3x + 6y$
70. $5xy - 15x - 6y + 18$
71. $5q^2 - 4pq - 5q + 4p$
72. $6m^2 - 5mn - 6m + 5n$
73. $2x^4 + 5x^3 + 2x^2 + 5x$
74. $4y^4 + y^2 + 20y^3 + 5y$
75. $12x^2y - 42x^2 - 4y + 14$
76. $90 + 15y^2 - 18x - 3xy^2$

MIXED PRACTICE

Factor. See Examples 4 through 17.

77. $32xy - 18x^2$
78. $10xy - 15x^2$
79. $y(x + 2) - 3(x + 2)$
80. $z(y - 4) + 3(y - 4)$
81. $14x^3y + 7x^2y - 7xy$
82. $5x^3y - 15x^2y + 10xy$
83. $28x^3 - 7x^2 + 12x - 3$
84. $15x^3 + 5x^2 - 6x - 2$
85. $-40x^8y^6 - 16x^9y^5$
86. $-21x^3y - 49x^2y^2$
87. $6a^2 + 9ab^2 + 6ab + 9b^3$
88. $16x^2 + 4xy^2 + 8xy + 2y^3$

REVIEW AND PREVIEW

Multiply. See Section 12.3.

89. $(x + 2)(x + 5)$
90. $(y + 3)(y + 6)$
91. $(b + 1)(b - 4)$
92. $(x - 5)(x + 10)$

Fill in the chart by finding two numbers that have the given product and sum. The first column is filled in for you.

		93.	94.	95.	96.	97.	98.
Two Numbers	4, 7						
Their Product	28	12	20	8	16	−10	−24
Their Sum	11	8	9	−9	−10	3	−5

CONCEPT EXTENSIONS

See the Concept Checks in this section.

99. Which of the following is/are factored form(s) of $8a - 24$?

a. $8 \cdot a - 24$ **b.** $8(a - 3)$

c. $4(2a - 12)$ **d.** $8 \cdot a - 2 \cdot 12$

100. Which of the following is/are factored form(s) of $-2x + 14$?

a. $-2(x + 7)$ **b.** $-2 \cdot x + 14$

c. $-2(x - 14)$ **d.** $-2(x - 7)$

Determine whether the following expressions are factored.

101. $(a + 6)(a + 2)$

102. $(x + 5)(x + y)$

103. $5(2y + z) - b(2y + z)$

104. $3x(a + 2b) + 2(a + 2b)$

105. Construct a binomial whose greatest common factor is $5a^3$. (*Hint:* Multiply $5a^3$ by a binomial whose terms contain no common factor other than 1. $5a^3(\square + \square)$.)

106. Construct a trinomial whose greatest common factor is $2x^2$. See the hint for Exercise 105.

107. Explain how you can tell whether a polynomial is written in factored form.

108. Construct a four-term polynomial that can be factored by grouping.

109. The number (in millions) of single digital downloads annually in the United States each year during 2004–2010 can be modeled by the polynomial $-20x^2 + 300x + 120$, where x is the number of years since 2004. (*Source:* Recording Industry Association of America)

a. Find the number of single digital downloads in 2010. To do so, let $x = 6$ and evaluate $-20x^2 + 300x + 120$.

b. Use this expression to estimate the number of single digital downloads in 2014.

c. Factor the polynomial $-20x^2 + 300x + 120$ by factoring -20 from each term.

110. The number (in thousands) of students who graduated from U.S. public high schools each year during 2000–2009 can be modeled by $0.8x^2 + 44x + 2552$, where x is the number of years since 2000. (*Source:* Digest of Education Statistics)

a. Find the number of students who graduated from U.S. high schools in 2005. To do so, let $x = 5$ and evaluate $0.8x^2 + 44x + 2552$.

b. Use this expression to estimate the number of students who graduated from U.S. high schools in 2015.

c. Factor the polynomial $0.8x^2 + 44x + 2552$ by factoring out 0.8 from each term.

Write an expression for the area of each shaded region. Then write the expression as a factored polynomial.

△ **111.**

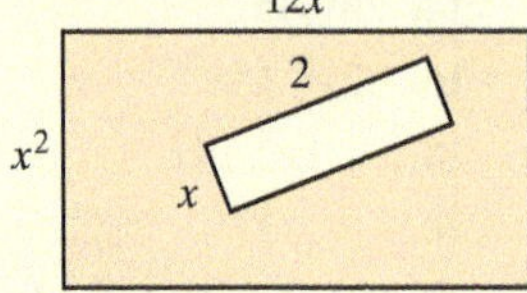

△ **112.**

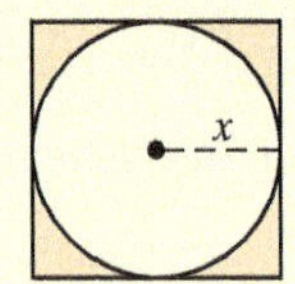

*Write an expression for the length of each rectangle. (**Hint:** Factor the area binomial and recall that Area = width · length.)*

△ **113.**

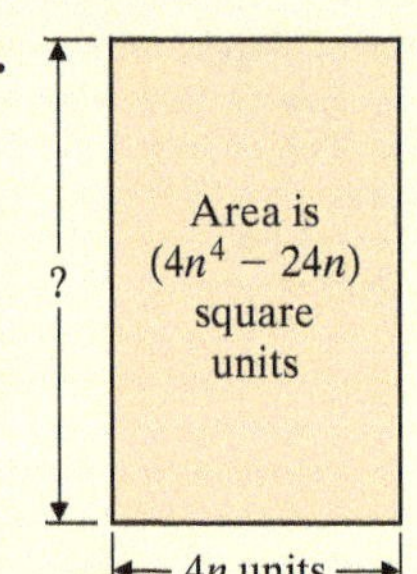

△ **114.**

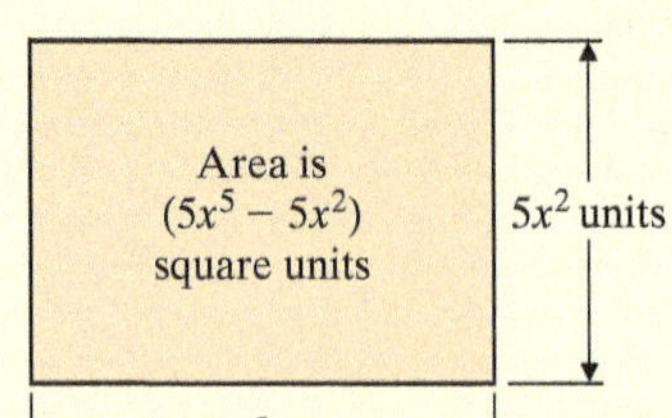

Factor each polynomial by grouping.

115. $x^{2n} + 2x^n + 3x^n + 6$

(*Hint:* Don't forget that $x^{2n} = x^n \cdot x^n$.)

116. $x^{2n} + 6x^n + 10x^n + 60$

117. $3x^{2n} + 21x^n - 5x^n - 35$

118. $12x^{2n} - 10x^n - 30x^n + 25$

13.2 Factoring Trinomials of the Form $x^2 + bx + c$ 

Objective 1 Factoring Trinomials of the Form $x^2 + bx + c$

Objectives

1. Factor Trinomials of the Form $x^2 + bx + c$.
2. Factor Out the Greatest Common Factor and Then Factor a Trinomial of the Form $x^2 + bx + c$.

In this section, we factor trinomials of the form $x^2 + bx + c$, such as

$$x^2 + 4x + 3, \quad x^2 - 8x + 15, \quad x^2 + 4x - 12, \quad r^2 - r - 42$$

Notice that for these trinomials, the coefficient of the squared variable is 1.

Recall that factoring means to write as a product and that factoring and multiplying are reverse processes. Using the FOIL method of multiplying binomials, we have that

$$\begin{aligned} & \quad\ \ \text{F} \quad \text{O} \quad \text{I} \quad \text{L} \\ (x + 3)(x + 1) &= x^2 + 1x + 3x + 3 \\ &= x^2 + 4x + 3 \end{aligned}$$

Thus, a factored form of $x^2 + 4x + 3$ is $(x + 3)(x + 1)$.

Notice that the product of the first terms of the binomials is $x \cdot x = x^2$, the first term of the trinomial. Also, the product of the last two terms of the binomials is $3 \cdot 1 = 3$, the third term of the trinomial. The sum of these same terms is $3 + 1 = 4$, the coefficient of the middle term, x, of the trinomial.

The product of these numbers is 3.

$$x^2 + 4x + 3 = (x + 3)(x + 1)$$

The sum of these numbers is 4.

Many trinomials, such as the one above, factor into two binomials. To factor $x^2 + 7x + 10$, let's assume that it factors into two binomials and begin by writing two pairs of parentheses. The first term of the trinomial is x^2, so we use x and x as the first terms of the binomial factors.

$$x^2 + 7x + 10 = (x + \square)(x + \square)$$

To determine the last term of each binomial factor, we look for two integers whose product is 10 and whose sum is 7. Since our numbers must have a positive product and a positive sum, we list pairs of positive integer factors of 10 only.

Positive Factors of 10	Sum of Factors
1, 10	$1 + 10 = 11$
2, 5	$2 + 5 = 7$

The correct pair of numbers is 2 and 5 because their product is 10 and their sum is 7. Now we can fill in the last terms of the binomial factors.

$$x^2 + 7x + 10 = (x + 2)(x + 5)$$

Check: To see if we have factored correctly, multiply.

$$\begin{aligned} (x + 2)(x + 5) &= x^2 + 5x + 2x + 10 \\ &= x^2 + 7x + 10 \quad \text{Combine like terms.} \end{aligned}$$

Helpful Hint

Since multiplication is commutative, the factored form of $x^2 + 7x + 10$ can be written as either $(x + 2)(x + 5)$ or $(x + 5)(x + 2)$.

Factoring a Trinomial of the Form $x^2 + bx + c$

The factored form of $x^2 + bx + c$ is

The product of these numbers is c.

$$x^2 + bx + c = (x + \square)(x + \square)$$

The sum of these numbers is b.

Practice 1

Factor $x^2 + 5x + 6$.

Example 1 Factor: $x^2 + 7x + 12$

Solution: We begin by writing the first terms of the binomial factors.

$(x + \square)(x + \square)$

Next we look for two numbers whose product is 12 and whose sum is 7. Since our numbers must have a positive product and a positive sum, we look at pairs of positive factors of 12 only.

Positive Factors of 12	Sum of Factors
1, 12	13
2, 6	8
3, 4	7

Correct sum, so the numbers are 3 and 4.

Thus, $x^2 + 7x + 12 = (x + 3)(x + 4)$

Check: $(x + 3)(x + 4) = x^2 + 4x + 3x + 12 = x^2 + 7x + 12$.

Work Practice 1

Practice 2

Factor $x^2 - 17x + 70$.

Example 2 Factor: $x^2 - 12x + 35$

Solution: Again, we begin by writing the first terms of the binomials.

$(x + \square)(x + \square)$

Now we look for two numbers whose product is 35 and whose sum is -12. Since our numbers must have a positive product and a negative sum, we look at pairs of negative factors of 35 only.

Negative Factors of 35	Sum of Factors
−1, −35	−36
−5, −7	−12

Correct sum, so the numbers are −5 and −7.

Thus, $x^2 - 12x + 35 = (x - 5)(x - 7)$

Check: To check, multiply $(x - 5)(x - 7)$.

Work Practice 2

Practice 3

Factor $x^2 + 5x - 14$.

Answers

1. $(x + 2)(x + 3)$
2. $(x - 10)(x - 7)$
3. $(x + 7)(x - 2)$

Example 3 Factor: $x^2 + 4x - 12$

Solution:

$$x^2 + 4x - 12 = (x + \square)(x + \square)$$

We look for two numbers whose product is -12 and whose sum is 4. Since our numbers must have a negative product, we look at pairs of factors with opposite signs.

Factors of −12	Sum of Factors
−1, 12	11
1, −12	−11
−2, 6	4
2, −6	−4
−3, 4	1
3, −4	−1

Correct sum, so the numbers are −2 and 6.

Thus, $x^2 + 4x - 12 = (x - 2)(x + 6)$

Work Practice 3

Example 4 Factor: $r^2 - r - 42$

Solution: Because the variable in this trinomial is r, the first term of each binomial factor is r.

$$r^2 - r - 42 = (r + \square)(r + \square)$$

Now we look for two numbers whose product is −42 and whose sum is −1, the numerical coefficient of r. The numbers are 6 and −7. Therefore,

$$r^2 - r - 42 = (r + 6)(r - 7)$$

Work Practice 4

Practice 4

Factor $p^2 - 2p - 63$.

Example 5 Factor: $a^2 + 2a + 10$

Solution: Look for two numbers whose product is 10 and whose sum is 2. Neither 1 and 10 nor 2 and 5 give the required sum, 2. We conclude that $a^2 + 2a + 10$ is not factorable with integers. A polynomial such as $a^2 + 2a + 10$ is called a **prime polynomial.**

Work Practice 5

Practice 5

Factor $b^2 + 5b + 1$.

Example 6 Factor: $x^2 + 7xy + 6y^2$

Solution:

$$x^2 + 7xy + 6y^2 = (x + \square)(x + \square)$$

Recall that the middle term, $7xy$, is the same as $7yx$. Thus, we can see that $7y$ is the "coefficient" of x. We then look for two terms whose product is $6y^2$ and whose sum is $7y$. The terms are $6y$ and $1y$ or $6y$ and y because $6y \cdot y = 6y^2$ and $6y + y = 7y$. Therefore,

$$x^2 + 7xy + 6y^2 = (x + 6y)(x + y)$$

Work Practice 6

Practice 6

Factor $x^2 + 7xy + 12y^2$.

Example 7 Factor: $x^4 + 5x^2 + 6$

Solution: As usual, we begin by writing the first terms of the binomials. Since the greatest power of x in this polynomial is x^4, we write

$$(x^2 + \square)(x^2 + \square) \quad \text{since } x^2 \cdot x^2 = x^4$$

Now we look for two factors of 6 whose sum is 5. The numbers are 2 and 3. Thus,

$$x^4 + 5x^2 + 6 = (x^2 + 2)(x^2 + 3)$$

Work Practice 7

Practice 7

Factor $x^4 + 13x^2 + 12$.

Answers

4. $(p - 9)(p + 7)$
5. prime polynomial
6. $(x + 3y)(x + 4y)$
7. $(x^2 + 12)(x^2 + 1)$

If the terms of a polynomial are not written in descending powers of the variable, you may want to do so before factoring.

Practice 8

Factor $48 - 14x + x^2$.

Example 8 Factor: $40 - 13t + t^2$

Solution: First, we rearrange terms so that the trinomial is written in descending powers of t.

$$40 - 13t + t^2 = t^2 - 13t + 40$$

Next, try to factor.

$$t^2 - 13t + 40 = (t + \square)(t + \square)$$

Now we look for two factors of 40 whose sum is -13. The numbers are -8 and -5. Thus,

$$t^2 - 13t + 40 = (t - 8)(t - 5)$$

Work Practice 8

The following sign patterns may be useful when factoring trinomials.

Helpful Hint

A positive constant in a trinomial tells us to look for two numbers with the same sign. The sign of the coefficient of the middle term tells us whether the signs are both positive or both negative.

both positive ↓ same sign ↙

$$x^2 + 10x + 16 = (x + 2)(x + 8)$$

both negative ↓ same sign ↙

$$x^2 - 10x + 16 = (x - 2)(x - 8)$$

A negative constant in a trinomial tells us to look for two numbers with opposite signs.

opposite signs ↓

$$x^2 + 6x - 16 = (x + 8)(x - 2)$$

opposite signs ↓

$$x^2 - 6x - 16 = (x - 8)(x + 2)$$

Objective 2 Factoring Out the Greatest Common Factor

Remember that the first step in factoring any polynomial is to factor out the greatest common factor (if there is one other than 1 or -1).

Practice 9

Factor $4x^2 - 24x + 36$.

Example 9 Factor: $3m^2 - 24m - 60$

Solution: First we factor out the greatest common factor, 3, from each term.

$$3m^2 - 24m - 60 = 3(m^2 - 8m - 20)$$

Now we factor $m^2 - 8m - 20$ by looking for two factors of -20 whose sum is -8. The factors are -10 and 2. Therefore, the complete factored form is

$$3m^2 - 24m - 60 = 3(m + 2)(m - 10)$$

Helpful Hint Remember to write the common factor 3 as part of the factored form.

Work Practice 9

Answers

8. $(x - 6)(x - 8)$
9. $4(x - 3)(x - 3)$

Example 10 Factor: $2x^4 - 26x^3 + 84x^2$

Solution:

$$2x^4 - 26x^3 + 84x^2 = 2x^2(x^2 - 13x + 42) \quad \text{Factor out common factor, } 2x^2.$$
$$= 2x^2(x - 6)(x - 7) \quad \text{Factor } x^2 - 13x + 42.$$

Work Practice 10

Practice 10

Factor $3y^4 - 18y^3 - 21y^2$.

Answer

10. $3y^2(y - 7)(y + 1)$

Vocabulary, Readiness & Video Check

Fill in each blank with "true" or "false."

1. To factor $x^2 + 7x + 6$, we look for two numbers whose product is 6 and whose sum is 7. ______
2. We can write the factorization $(y + 2)(y + 4)$ also as $(y + 4)(y + 2)$. ______
3. The factorization $(4x - 12)(x - 5)$ is completely factored. ______
4. The factorization $(x + 2y)(x + y)$ may also be written as $(x + 2y)^2$. ______

Complete each factored form.

5. $x^2 + 9x + 20 = (x + 4)(x \quad)$
6. $x^2 + 12x + 35 = (x + 5)(x \quad)$
7. $x^2 - 7x + 12 = (x - 4)(x \quad)$
8. $x^2 - 13x + 22 = (x - 2)(x \quad)$
9. $x^2 + 4x + 4 = (x + 2)(x \quad)$
10. $x^2 + 10x + 24 = (x + 6)(x \quad)$

Martin-Gay Interactive Videos

See Video 13.2

Watch the section lecture video and answer the following questions.

Objective 1 11. In Example 2, why are only negative factors of 15 considered?

Objective 2 12. In Example 5, we know we need a positive and a negative factor of -10. How do we determine which factor is negative?

13.2 Exercise Set MyMathLab®

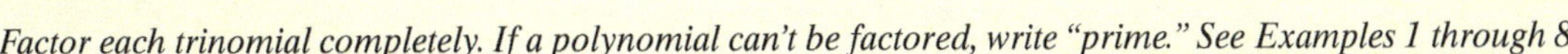

Factor each trinomial completely. If a polynomial can't be factored, write "prime." See Examples 1 through 8.

1. $x^2 + 7x + 6$
2. $x^2 + 6x + 8$
3. $y^2 - 10y + 9$
4. $y^2 - 12y + 11$
5. $x^2 - 6x + 9$
6. $x^2 - 10x + 25$
7. $x^2 - 3x - 18$
8. $x^2 - x - 30$
9. $x^2 + 3x - 70$
10. $x^2 + 4x - 32$
11. $x^2 + 5x + 2$
12. $x^2 - 7x + 5$
13. $x^2 + 8xy + 15y^2$
14. $x^2 + 6xy + 8y^2$
15. $a^4 - 2a^2 - 15$
16. $y^4 - 3y^2 - 70$
17. $13 + 14m + m^2$
18. $17 + 18n + n^2$
19. $10t - 24 + t^2$
20. $6q - 27 + q^2$

21. $a^2 - 10ab + 16b^2$
22. $a^2 - 9ab + 18b^2$

MIXED PRACTICE

Factor each trinomial completely. Some of these trinomials contain a greatest common factor (other than 1). Don't forget to factor out the GCF first. See Examples 1 through 10.

23. $2z^2 + 20z + 32$
24. $3x^2 + 30x + 63$
25. $2x^3 - 18x^2 + 40x$
26. $3x^3 - 12x^2 - 36x$
27. $x^2 - 3xy - 4y^2$
28. $x^2 - 4xy - 77y^2$
29. $x^2 + 15x + 36$
30. $x^2 + 19x + 60$
31. $x^2 - x - 2$
32. $x^2 - 5x - 14$
33. $r^2 - 16r + 48$
34. $r^2 - 10r + 21$
35. $x^2 + xy - 2y^2$
36. $x^2 - xy - 6y^2$
37. $3x^2 + 9x - 30$
38. $4x^2 - 4x - 48$
39. $3x^2 - 60x + 108$
40. $2x^2 - 24x + 70$
41. $x^2 - 18x - 144$
42. $x^2 + x - 42$
43. $r^2 - 3r + 6$
44. $x^2 + 4x - 10$
45. $x^2 - 8x + 15$
46. $x^2 - 9x + 14$
47. $6x^3 + 54x^2 + 120x$
48. $3x^3 + 3x^2 - 126x$
49. $4x^2y + 4xy - 12y$
50. $3x^2y - 9xy + 45y$
51. $x^2 - 4x - 21$
52. $x^2 - 4x - 32$
53. $x^2 + 7xy + 10y^2$
54. $x^2 - 2xy - 15y^2$
55. $64 + 24t + 2t^2$
56. $50 + 20t + 2t^2$
57. $x^3 - 2x^2 - 24x$
58. $x^3 - 3x^2 - 28x$
59. $2t^5 - 14t^4 + 24t^3$
60. $3x^6 + 30x^5 + 72x^4$
61. $5x^3y - 25x^2y^2 - 120xy^3$
62. $7a^3b - 35a^2b^2 + 42ab^3$
63. $162 - 45m + 3m^2$
64. $48 - 20n + 2n^2$
65. $-x^2 + 12x - 11$ (Factor out -1 first.)
66. $-x^2 + 8x - 7$ (Factor out -1 first.)
67. $\frac{1}{2}y^2 - \frac{9}{2}y - 11$ (Factor out $\frac{1}{2}$ first.)
68. $\frac{1}{3}y^2 - \frac{5}{3}y - 8$ (Factor out $\frac{1}{3}$ first.)
69. $x^3y^2 + x^2y - 20x$
70. $a^2b^3 + ab^2 - 30b$

REVIEW AND PREVIEW

Multiply. See Section 12.3.

71. $(2x + 1)(x + 5)$
72. $(3x + 2)(x + 4)$
73. $(5y - 4)(3y - 1)$
74. $(4z - 7)(7z - 1)$
75. $(a + 3b)(9a - 4b)$
76. $(y - 5x)(6y + 5x)$

CONCEPT EXTENSIONS

77. Write a polynomial that factors as $(x - 3)(x + 8)$.
78. To factor $x^2 + 13x + 42$, think of two numbers whose __________ is 42 and whose __________ is 13.

Complete each sentence in your own words.

79. If $x^2 + bx + c$ is factorable and c is negative, then the signs of the last-term factors of the binomials are opposite because...
80. If $x^2 + bx + c$ is factorable and c is positive, then the signs of the last-term factors of the binomials are the same because...

Remember that perimeter means distance around. Write the perimeter of each rectangle as a simplified polynomial. Then factor the polynomial.

81.

82.

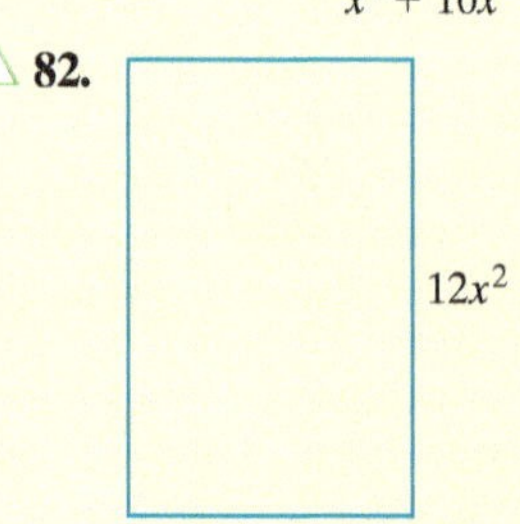

83. An object is thrown upward from the top of an 80-foot building with an initial velocity of 64 feet per second. Neglecting air resistance, the height of the object

after t seconds is given by $-16t^2 + 64t + 80$. Factor this polynomial.

84. An object is thrown upward from the top of a 112-foot building with an initial velocity of 96 feet per second. Neglecting air resistance, the height of the object after t seconds is given by $-16t^2 + 96t + 112$. Factor this polynomial.

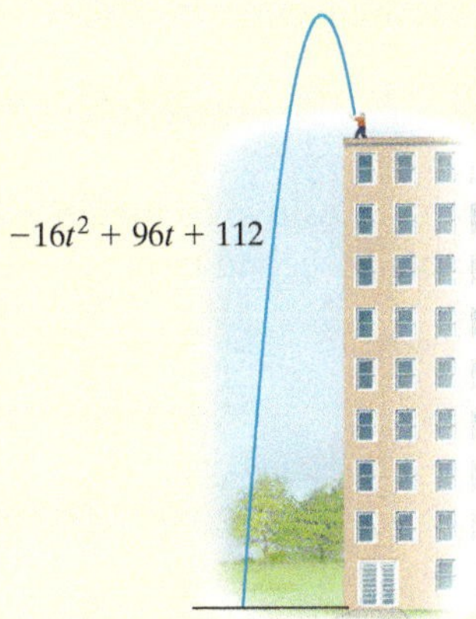

Factor each trinomial completely.

85. $x^2 + \frac{1}{2}x + \frac{1}{16}$

86. $x^2 + x + \frac{1}{4}$

87. $z^2(x + 1) - 3z(x + 1) - 70(x + 1)$

88. $y^2(x + 1) - 2y(x + 1) - 15(x + 1)$

*Factor each trinomial. (**Hint:** Notice that $x^{2n} + 4x^n + 3$ factors as $(x^n + 1)(x^n + 3)$. **Remember:** $x^n \cdot x^n = x^{n+n}$ or x^{2n}.)*

89. $x^{2n} + 8x^n - 20$

90. $x^{2n} + 5x^n + 6$

Find a positive value of c so that each trinomial is factorable.

91. $x^2 + 6x + c$

92. $t^2 + 8t + c$

93. $y^2 - 4y + c$

94. $n^2 - 16n + c$

Find a positive value of b so that each trinomial is factorable.

95. $x^2 + bx + 15$

96. $y^2 + by + 20$

97. $m^2 + bm - 27$

98. $x^2 + bx - 14$

13.3 Factoring Trinomials of the Form $ax^2 + bx + c$ and Perfect Square Trinomials

Objectives

1. Factor Trinomials of the Form $ax^2 + bx + c$, Where $a \neq 1$.
2. Factor Out a GCF Before Factoring a Trinomial of the Form $ax^2 + bx + c$.
3. Factor Perfect Square Trinomials.

Objective 1 Factoring Trinomials of the Form $ax^2 + bx + c$

In this section, we factor trinomials of the form $ax^2 + bx + c$, such as

$$3x^2 + 11x + 6, \quad 8x^2 - 22x + 5, \quad \text{and} \quad 2x^2 + 13x - 7$$

Notice that the coefficient of the squared variable in these trinomials is a number other than 1. We will factor these trinomials using a trial-and-check method based on our work in the last section.

To begin, let's review the relationship between the numerical coefficients of the trinomial and the numerical coefficients of its factored form. For example, since

$$(2x + 1)(x + 6) = 2x^2 + 13x + 6,$$

a factored form of $2x^2 + 13x + 6$ is $(2x + 1)(x + 6)$

Notice that $2x$ and x are factors of $2x^2$, the first term of the trinomial. Also, 6 and 1 are factors of 6, the last term of the trinomial, as shown:

$$2x^2 + 13x + 6 = (2x + 1)(x + 6)$$

$2x \cdot x$

$1 \cdot 6$

Also notice that $13x$, the middle term, is the sum of the following products:

$$2x^2 + 13x + 6 = (2x + 1)(x + 6)$$

$$\begin{array}{r} 1x \\ +12x \\ \hline 13x \end{array}$$ Middle term

Let's use this pattern to factor $5x^2 + 7x + 2$. First, we find factors of $5x^2$. Since all numerical coefficients in this trinomial are positive, we will use factors with positive numerical coefficients only. Thus, the factors of $5x^2$ are $5x$ and x. Let's try these factors as first terms of the binomials. Thus far, we have

$$5x^2 + 7x + 2 = (5x + \square)(x + \square)$$

Next, we need to find positive factors of 2. Positive factors of 2 are 1 and 2. Now we try possible combinations of these factors as second terms of the binomials until we obtain a middle term of $7x$.

$$(5x + 1)(x + 2) = 5x^2 + 11x + 2$$

$$\begin{array}{r} 1x \\ +10x \\ \hline 11x \end{array}$$ → Incorrect middle term

Let's try switching factors 2 and 1.

$$(5x + 2)(x + 1) = 5x^2 + 7x + 2$$

$$\begin{array}{r} 2x \\ +5x \\ \hline 7x \end{array}$$ → Correct middle term

Thus a factored form of $5x^2 + 7x + 2$ is $(5x + 2)(x + 1)$. To check, we multiply $(5x + 2)$ and $(x + 1)$. The product is $5x^2 + 7x + 2$.

Practice 1

Factor $2x^2 + 11x + 15$.

Example 1 Factor: $3x^2 + 11x + 6$

Solution: Since all numerical coefficients are positive, we use factors with positive numerical coefficients. We first find factors of $3x^2$.

Factors of $3x^2$: $3x^2 = 3x \cdot x$

If factorable, the trinomial will be of the form

$$3x^2 + 11x + 6 = (3x + \square)(x + \square)$$

Next we factor 6.

Factors of 6: $6 = 1 \cdot 6, \quad 6 = 2 \cdot 3$

Answer

1. $(2x + 5)(x + 3)$

Now we try combinations of factors of 6 until a middle term of $11x$ is obtained. Let's try 1 and 6 first.

$$(3x + 1)(x + 6) = 3x^2 + 19x + 6$$

$1x$
$+18x$
$19x$ → Incorrect middle term

Now let's next try 6 and 1.

$$(3x + 6)(x + 1)$$

Before multiplying, notice that the terms of the factor $3x + 6$ have a common factor of 3. The terms of the original trinomial $3x^2 + 11x + 6$ have no common factor other than 1, so the terms of its factors will also contain no common factor other than 1. This means that $(3x + 6)(x + 1)$ is not a factored form.

Next let's try 2 and 3 as last terms.

$$(3x + 2)(x + 3) = 3x^2 + 11x + 6$$

$2x$
$+9x$
$11x$ → Correct middle term

Thus a factored form of $3x^2 + 11x + 6$ is $(3x + 2)(x + 3)$.

Work Practice 1

Helpful Hint

If the terms of a trinomial have no common factor (other than 1), then neither of the terms of its binomial factors will contain a common factor (other than 1).

✓Concept Check Do the terms of $3x^2 + 29x + 18$ have a common factor? Without multiplying, decide which of the following factored forms could *not* be a factored form of $3x^2 + 29x + 18$.

a. $(3x + 18)(x + 1)$ **b.** $(3x + 2)(x + 9)$
c. $(3x + 6)(x + 3)$ **d.** $(3x + 9)(x + 2)$

Example 2 Factor: $8x^2 - 22x + 5$

Solution: Factors of $8x^2$: $8x^2 = 8x \cdot x$, $8x^2 = 4x \cdot 2x$

We'll try $8x$ and x.

$$8x^2 - 22x + 5 = (8x + \square)(x + \square)$$

Since the middle term, $-22x$, has a negative numerical coefficient, we factor 5 into negative factors.

Factors of 5: $5 = -1 \cdot -5$

Let's try -1 and -5.

$$(8x - 1)(x - 5) = 8x^2 - 41x + 5$$

$-1x$
$+(-40x)$
$-41x$ → Incorrect middle term

(Continued on next page)

Practice 2

Factor $15x^2 - 22x + 8$.

Answer

2. $(5x - 4)(3x - 2)$

✓Concept Check Answers

no; a, c, d

Now let's try -5 and -1.

$$(8x - 5)(x - 1) = 8x^2 - 13x + 5$$

$-5x$
$+(-8x)$
$-13x$ → Incorrect middle term

Don't give up yet! We can still try other factors of $8x^2$. Let's try $4x$ and $2x$ with -1 and -5.

$$(4x - 1)(2x - 5) = 8x^2 - 22x + 5$$

$-2x$
$+(-20x)$
$-22x$ → Correct middle term

A factored form of $8x^2 - 22x + 5$ is $(4x - 1)(2x - 5)$.

Work Practice 2

Practice 3

Factor $4x^2 + 11x - 3$.

Example 3 Factor: $2x^2 + 13x - 7$

Solution: Factors of $2x^2$: $2x^2 = 2x \cdot x$

Factors of -7: $-7 = -1 \cdot 7$, $-7 = 1 \cdot -7$

We try possible combinations of these factors:

$(2x + 1)(x - 7) = 2x^2 - 13x - 7$ Incorrect middle term

$(2x - 1)(x + 7) = 2x^2 + 13x - 7$ Correct middle term

A factored form of $2x^2 + 13x - 7$ is $(2x - 1)(x + 7)$.

Work Practice 3

Practice 4

Factor $21x^2 + 11xy - 2y^2$.

Example 4 Factor: $10x^2 - 13xy - 3y^2$

Solution: Factors of $10x^2$: $10x^2 = 10x \cdot x$, $10x^2 = 2x \cdot 5x$

Factors of $-3y^2$: $-3y^2 = -3y \cdot y$, $-3y^2 = 3y \cdot -y$

We try some combinations of these factors:

Correct ↓ (first term) Correct ↓ (last term)

$(10x - 3y)(x + y) = 10x^2 + 7xy - 3y^2$

$(x + 3y)(10x - y) = 10x^2 + 29xy - 3y^2$

$(5x + 3y)(2x - y) = 10x^2 + xy - 3y^2$

$(2x - 3y)(5x + y) = 10x^2 - 13xy - 3y^2$ Correct middle term

A factored form of $10x^2 - 13xy - 3y^2$ is $(2x - 3y)(5x + y)$.

Work Practice 4

Answers

3. $(4x - 1)(x + 3)$
4. $(7x - y)(3x + 2y)$

Example 5 Factor: $3x^4 - 5x^2 - 8$

Solution: Factors of $3x^4$: $3x^4 = 3x^2 \cdot x^2$

Factors of -8: $-8 = -2 \cdot 4, -8 = 2 \cdot -4, -8 = -1 \cdot 8, -8 = 1 \cdot -8$

Try combinations of these factors:

$$(3x^2 - 2)(x^2 + 4) = 3x^4 + 10x^2 - 8$$
Correct (first and last terms)
$$(3x^2 + 4)(x^2 - 2) = 3x^4 - 2x^2 - 8$$
$$(3x^2 + 8)(x^2 - 1) = 3x^4 + 5x^2 - 8$$
Incorrect sign on middle term, so switch signs in binomial factors.
$$(3x^2 - 8)(x^2 + 1) = 3x^4 - 5x^2 - 8$$
Correct middle term.

A factored form of $3x^4 - 5x^2 - 8$ is $(3x^2 - 8)(x^2 + 1)$.

Work Practice 5

Practice 5

Factor $2x^4 - 5x^2 - 7$.

Helpful Hint

Study the last two lines of Example 5. If a factoring attempt gives you a middle term whose numerical coefficient is the opposite of the desired numerical coefficient, try switching the signs of the last terms in the binomials.

Switched signs

$$(3x^2 + 8)(x^2 - 1) = 3x^4 + 5x^2 - 8$$ Middle term: $+5x^2$

$$(3x^2 - 8)(x^2 + 1) = 3x^4 - 5x^2 - 8$$ Middle term: $-5x^2$

Objective 2 Factoring Out the Greatest Common Factor

Don't forget that the first step in factoring any polynomial is to look for a common factor to factor out.

Example 6 Factor: $24x^4 + 40x^3 + 6x^2$

Solution: Notice that all three terms have a common factor of $2x^2$. Thus we factor out $2x^2$ first.

$$24x^4 + 40x^3 + 6x^2 = 2x^2(12x^2 + 20x + 3)$$

Next we factor $12x^2 + 20x + 3$.

Factors of $12x^2$: $12x^2 = 4x \cdot 3x$, $12x^2 = 12x \cdot x$, $12x^2 = 6x \cdot 2x$

Since all terms in the trinomial have positive numerical coefficients, we factor 3 using positive factors only.

Factors of 3: $3 = 1 \cdot 3$

We try some combinations of the factors.

$$2x^2(4x + 3)(3x + 1) = 2x^2(12x^2 + 13x + 3)$$
$$2x^2(12x + 1)(x + 3) = 2x^2(12x^2 + 37x + 3)$$
$$2x^2(2x + 3)(6x + 1) = 2x^2(12x^2 + 20x + 3)$$ Correct middle term

A factored form of $24x^4 + 40x^3 + 6x^2$ is $2x^2(2x + 3)(6x + 1)$.

Work Practice 6

Practice 6

Factor $3x^3 + 17x^2 + 10x$.

Helpful Hint Don't forget to include the common factor in the factored form.

Answers

5. $(2x^2 - 7)(x^2 + 1)$
6. $x(3x + 2)(x + 5)$

When the term containing the squared variable has a negative coefficient, you may want to first factor out a common factor of -1.

Practice 7

Factor $-8x^2 + 2x + 3$.

Example 7 Factor: $-6x^2 - 13x + 5$

Solution: We begin by factoring out a common factor of -1.

$$-6x^2 - 13x + 5 = -1(6x^2 + 13x - 5) \quad \text{Factor out } -1.$$
$$= -1(3x - 1)(2x + 5) \quad \text{Factor } 6x^2 + 13x - 5.$$

Work Practice 7

Objective 3 Factoring Perfect Square Trinomials 

A trinomial that is the square of a binomial is called a **perfect square trinomial.** For example,

$$(x + 3)^2 = (x + 3)(x + 3)$$
$$= x^2 + 6x + 9$$

Thus $x^2 + 6x + 9$ is a perfect square trinomial.

In Chapter 12, we discovered special product formulas for squaring binomials.

$$(a + b)^2 = a^2 + 2ab + b^2 \quad \text{and} \quad (a - b)^2 = a^2 - 2ab + b^2$$

Because multiplication and factoring are reverse processes, we can now use these special products to help us factor perfect square trinomials. If we reverse these equations, we have the following.

Factoring Perfect Square Trinomials

$$a^2 + 2ab + b^2 = (a + b)^2$$
$$a^2 - 2ab + b^2 = (a - b)^2$$

Helpful Hint

Notice that for both given forms of a perfect square trinomial, the last term is positive. This is because the last term is a square.

To use these equations to help us factor, we must first be able to recognize a perfect square trinomial. A trinomial is a perfect square when

1. two terms, a^2 and b^2, are squares and
2. the remaining term is $2 \cdot a \cdot b$ or $-2 \cdot a \cdot b$. That is, this term is twice the product of a and b, or its opposite.

When a trinomial fits this description, its factored form is $(a + b)^2$ or $(a - b)^2$.

Practice 8

Factor $x^2 + 14x + 49$.

Example 8 Factor: $x^2 + 12x + 36$

Solution: First, is this a perfect square trinomial?

$$x^2 + 12x + 36$$

1. $x^2 = (x)^2$ and $36 = 6^2$.
2. Is the middle term $2 \cdot x \cdot 6$? Yes, $2 \cdot x \cdot 6 = 12x$, the middle term.

Thus, $x^2 + 12x + 36$ factors as $(x + 6)^2$.

Work Practice 8

Answers

7. $-1(4x - 3)(2x + 1)$ 8. $(x + 7)^2$

Example 9 Factor: $25x^2 + 25xy + 4y^2$

Solution: Is this a perfect square trinomial?

$$25x^2 + 25xy + 4y^2$$

1. $25x^2 = (5x)^2$ and $4y^2 = (2y)^2$.
2. Is the middle term $2 \cdot 5x \cdot 2y$? **No,** $2 \cdot 5x \cdot 2y = 20xy$, **not** the middle term $25xy$.

Therefore, $25x^2 + 25xy + 4y^2$ is not a perfect square trinomial. It is factorable, though. Using earlier techniques, we find that $25x^2 + 25xy + 4y^2$ factors as $(5x + 4y)(5x + y)$.

Work Practice 9

Practice 9

Factor $4x^2 + 20xy + 9y^2$.

Helpful Hint A perfect square trinomial can also be factored by other methods.

Example 10 Factor: $4m^4 - 4m^2 + 1$

Solution: Is this a perfect square trinomial?

$$4m^4 - 4m^2 + 1$$

1. $4m^4 = (2m^2)^2$ and $1 = 1^2$.
2. Is the middle term $2 \cdot 2m^2 \cdot 1$ or $-2 \cdot 2m^2 \cdot 1$? Yes, $-2 \cdot 2m^2 \cdot 1 = -4m^2$, the middle term.

Thus, $4m^4 - 4m^2 + 1$ factors as $(2m^2 - 1)^2$.

Work Practice 10

Practice 10

Factor $36n^4 - 12n^2 + 1$.

Example 11 Factor: $162x^3 - 144x^2 + 32x$

Solution: Don't forget to first look for a common factor. There is a greatest common factor of $2x$ in this trinomial.

$$\begin{aligned} 162x^3 - 144x^2 + 32x &= 2x(81x^2 - 72x + 16) \\ &= 2x[(9x)^2 - 2 \cdot 9x \cdot 4 + 4^2] \\ &= 2x(9x - 4)^2 \end{aligned}$$

Work Practice 11

Practice 11

Factor $12x^3 - 84x^2 + 147x$.

Answers

9. $(2x + 9y)(2x + y)$ **10.** $(6n^2 - 1)^2$ **11.** $3x(2x - 7)^2$

Vocabulary, Readiness & Video Check

Use the choices below to fill in each blank. Some choices will be used more than once and some not used at all.

$5y^2$	$(x + 5y)^2$	perfect square trinomial
$(5y)^2$	$(x - 5y)^2$	perfect square binomial

1. A ______________ is a trinomial that is the square of a binomial.
2. The term $25y^2$ written as a square is ______________.
3. The expression $x^2 + 10xy + 25y^2$ is called a ______________.
4. The factorization $(x + 5y)(x + 5y)$ may also be written as ______________.

Complete each factorization.

5. $2x^2 + 5x + 3$ factors as $(2x + 3)(\ ?\ )$.

a. $(x + 3)$ **b.** $(2x + 1)$ **c.** $(3x + 4)$ **d.** $(x + 1)$

6. $7x^2 + 9x + 2$ factors as $(7x + 2)(\ ?\ )$.

a. $(3x + 1)$ **b.** $(x + 1)$ **c.** $(x + 2)$ **d.** $(7x + 1)$

Martin-Gay Interactive Videos *Watch the section lecture video and answer the following questions.*

See Video 13.3 

Objective 1 **7.** From Example 1, explain in general terms how you would go about factoring a trinomial with a first-term coefficient $\neq 1$.

Objective 2 **8.** From Examples 3 and 5, how can factoring the GCF from a trinomial help you save time when trying to factor the remaining trinomial?

Objective 3 **9.** Describe in words the special patterns that the trinomials in Examples 7 and 8 have that identify them as perfect square trinomials.

13.3 Exercise Set MyMathLab®

Complete each factored form. See Examples 1 through 5, and 8 through 10.

1. $5x^2 + 22x + 8 = (5x + 2)(\quad)$
2. $2y^2 + 27y + 25 = (2y + 25)(\quad)$
3. $50x^2 + 15x - 2 = (5x + 2)(\quad)$
4. $6y^2 + 11y - 10 = (2y + 5)(\quad)$
5. $25x^2 - 20x + 4 = (5x - 2)(\quad)$
6. $4y^2 - 20y + 25 = (2y - 5)(\quad)$

Factor completely. See Examples 1 through 5.

7. $2x^2 + 13x + 15$
8. $3x^2 + 8x + 4$
9. $8y^2 - 17y + 9$
10. $21x^2 - 31x + 10$
11. $2x^2 - 9x - 5$
12. $36r^2 - 5r - 24$
13. $20r^2 + 27r - 8$
14. $3x^2 + 20x - 63$
15. $10x^2 + 31x + 3$
16. $12x^2 + 17x + 5$
17. $2m^2 + 17m + 10$
18. $3n^2 + 20n + 5$
19. $6x^2 - 13xy + 5y^2$
20. $8x^2 - 14xy + 3y^2$
21. $15m^2 - 16m - 15$
22. $25n^2 - 5n - 6$

Factor completely. See Examples 1 through 7.

23. $12x^3 + 11x^2 + 2x$
24. $8a^3 + 14a^2 + 3a$
25. $21b^2 - 48b - 45$
26. $12x^2 - 14x - 10$
27. $7z + 12z^2 - 12$
28. $16t + 15t^2 - 15$
29. $6x^2y^2 - 2xy^2 - 60y^2$
30. $8x^2y + 34xy - 84y$
31. $4x^2 - 8x - 21$
32. $6x^2 - 11x - 10$
33. $-x^2 + 2x + 24$
34. $-x^2 + 4x + 21$
35. $4x^3 - 9x^2 - 9x$
36. $6x^3 - 31x^2 + 5x$
37. $24x^2 - 58x + 9$
38. $36x^2 + 55x - 14$

Factor each perfect square trinomial completely. See Examples 8 through 11.

39. $x^2 + 22x + 121$
40. $x^2 + 18x + 81$
41. $x^2 - 16x + 64$
42. $x^2 - 12x + 36$
43. $16a^2 - 24a + 9$

44. $25x^2 - 20x + 4$
45. $x^4 + 4x^2 + 4$
46. $m^4 + 10m^2 + 25$
47. $2n^2 - 28n + 98$
48. $3y^2 - 6y + 3$
49. $16y^2 + 40y + 25$
50. $9y^2 + 48y + 64$

MIXED PRACTICE

Factor each trinomial completely. See Examples 1 through 11 and Section 13.2.

51. $2x^2 - 7x - 99$
52. $2x^2 + 7x - 72$
53. $24x^2 + 41x + 12$
54. $24x^2 - 49x + 15$
55. $3a^2 + 10ab + 3b^2$
56. $2a^2 + 11ab + 5b^2$
57. $-9x + 20 + x^2$
58. $-7x + 12 + x^2$
59. $p^2 + 12pq + 36q^2$
60. $m^2 + 20mn + 100n^2$
61. $x^2y^2 - 10xy + 25$
62. $x^2y^2 - 14xy + 49$
63. $40a^2b + 9ab - 9b$
64. $24y^2x + 7yx - 5x$
65. $30x^3 + 38x^2 + 12x$
66. $6x^3 - 28x^2 + 16x$
67. $6y^3 - 8y^2 - 30y$
68. $12x^3 - 34x^2 + 24x$
69. $10x^4 + 25x^3y - 15x^2y^2$
70. $42x^4 - 99x^3y - 15x^2y^2$
71. $-14x^2 + 39x - 10$
72. $-15x^2 + 26x - 8$
73. $16p^4 - 40p^3 + 25p^2$
74. $9q^4 - 42q^3 + 49q^2$
75. $x + 3x^2 - 2$
76. $y + 8y^2 - 9$
77. $8x^2 + 6xy - 27y^2$
78. $54a^2 + 39ab - 8b^2$
79. $1 + 6x^2 + x^4$
80. $1 + 16x^2 + x^4$
81. $9x^2 - 24xy + 16y^2$
82. $25x^2 - 60xy + 36y^2$
83. $18x^2 - 9x - 14$
84. $42a^2 - 43a + 6$
85. $-27t + 7t^2 - 4$
86. $-3t + 4t^2 - 7$
87. $49p^2 - 7p - 2$
88. $3r^2 + 10r - 8$
89. $m^3 + 18m^2 + 81m$
90. $y^3 + 12y^2 + 36y$
91. $5x^2y^2 + 20xy + 1$
92. $3a^2b^2 + 12ab + 1$
93. $6a^5 + 37a^3b^2 + 6ab^4$
94. $5m^5 + 26m^3h^2 + 5mh^4$

REVIEW AND PREVIEW

Multiply the following. See Sections 12.3 and 12.4.

95. $(x - 2)(x + 2)$
96. $(y^2 + 3)(y^2 - 3)$
97. $(a + 3)(a^2 - 3a + 9)$
98. $(z - 2)(z^2 + 2z + 4)$

The following graph shows the percent of text message users in each age group. See Section 10.1.

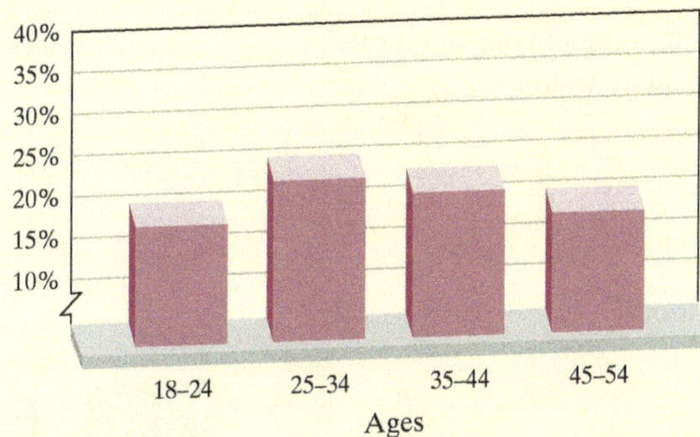

Data from CellSigns, Inc.

99. What range of ages shown has the highest percent of text message users?
100. What range of ages shown has the lowest percent of text message users?
101. Describe any trend you see.
102. Why don't the percents shown in the graph add to 100%?

CONCEPT EXTENSIONS

See the Concept Check in this section.

103. Do the terms of $4x^2 + 19x + 12$ have a common factor (other than 1)?
104. Without multiplying, decide which of the following factored forms is not a factored form of $4x^2 + 19x + 12$.

a. $(2x + 4)(2x + 3)$ **b.** $(4x + 4)(x + 3)$
c. $(4x + 3)(x + 4)$ **d.** $(2x + 2)(2x + 6)$

105. Describe a perfect square trinomial.

106. Write the perfect square trinomial that factors as $(x + 3y)^2$.

Write the perimeter of each figure as a simplified polynomial. Then factor the polynomial.

107.

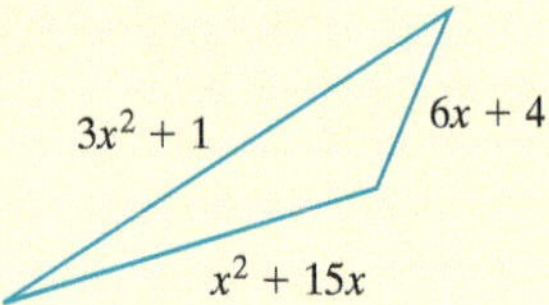

108.

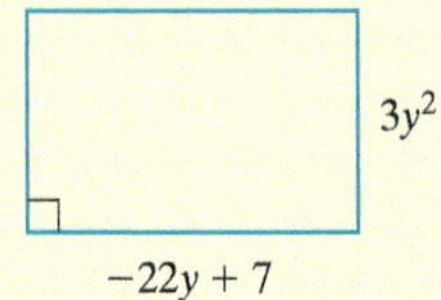

Factor each trinomial completely.

109. $4x^2 + 2x + \frac{1}{4}$

110. $27x^2 + 2x - \frac{1}{9}$

111. $4x^2(y - 1)^2 + 10x(y - 1)^2 + 25(y - 1)^2$

112. $3x^2(a + 3)^3 - 10x(a + 3)^3 + 25(a + 3)^3$

113. Fill in the blank so that $x^2 +$ _____ $x + 16$ is a perfect square trinomial.

114. Fill in the blank so that $9x^2 +$ _____ $x + 25$ is a perfect square trinomial.

The area of the largest square in the figure is $(a + b)^2$. Use this figure to answer Exercises 115 and 116.

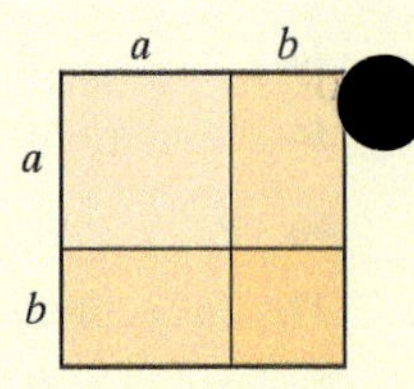

△ **115.** Write the area of the largest square as the sum of the areas of the smaller squares and rectangles.

△ **116.** What factoring formula from this section is visually represented by this square?

Find a positive value of b so that each trinomial is factorable.

117. $3x^2 + bx - 5$

118. $2y^2 + by + 3$

Find a positive value of c so that each trinomial is factorable.

119. $5x^2 + 7x + c$

120. $11y^2 - 40y + c$

Factor completely. Don't forget to first factor out the greatest common factor.

121. $-12x^3y^2 + 3x^2y^2 + 15xy^2$

122. $-12r^3x^2 + 38r^2x^2 + 14rx^2$

123. $4x^2(y - 1)^2 + 20x(y - 1)^2 + 25(y - 1)^2$

124. $3x^2(a + 3)^3 - 28x(a + 3)^3 + 25(a + 3)^3$

Factor.

125. $3x^{2n} + 17x^n + 10$

126. $2x^{2n} + 5x^n - 12$

127. In your own words, describe the steps you will use to factor a trinomial.

13.4 Factoring Trinomials of the Form $ax^2 + bx + c$ by Grouping

Objective

1 Use the Grouping Method to Factor Trinomials of the Form $ax^2 + bx + c$.

Objective 1 Using the Grouping Method

There is an alternative method that can be used to factor trinomials of the form $ax^2 + bx + c, a \neq 1$. This method is called the **grouping method** because it uses factoring by grouping as we learned in Section 13.1.

To see how this method works, recall from Section 13.2 that to factor a trinomial such as $x^2 + 11x + 30$, we find two numbers such that

Product is 30.

$$x^2 + 11x + 30$$

Sum is 11.

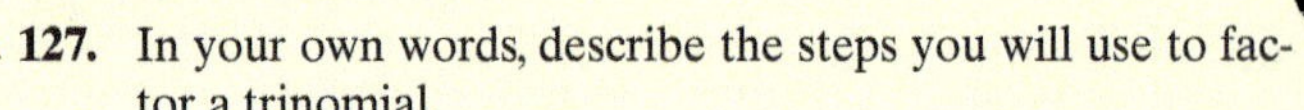

To factor a trinomial such as $2x^2 + 11x + 12$ by grouping, we use an extension of the method in Section 13.2. Here we look for two numbers such that

Product is $2 \cdot 12 = 24$

$$2x^2 + 11x + 12$$

Sum is 11.

This time, we use the two numbers to write

$$2x^2 + 11x + 12 \text{ as}$$
$$= 2x^2 + \square x + \square x + 12$$

Then we factor by grouping. Since we want a positive product, 24, and a positive sum, 11, we consider pairs of positive factors of 24 only.

Factors of 24	Sum of Factors
1, 24	25
2, 12	14
3, 8	11

Correct sum

The factors are 3 and 8. Now we use these factors to write the middle term $11x$ as $3x + 8x$ (or $8x + 3x$). We replace $11x$ with $3x + 8x$ in the original trinomial and then we can factor by grouping.

$$\begin{aligned} 2x^2 + 11x + 12 &= 2x^2 + 3x + 8x + 12 \\ &= (2x^2 + 3x) + (8x + 12) && \text{Group the terms.} \\ &= x(2x + 3) + 4(2x + 3) && \text{Factor each group.} \\ &= (2x + 3)(x + 4) && \text{Factor out } (2x + 3). \end{aligned}$$

In general, we have the following procedure.

To Factor Trinomials by Grouping

Step 1: Factor out a greatest common factor if there is one other than 1.

Step 2: For the resulting trinomial $ax^2 + bx + c$, find two numbers whose product is $a \cdot c$ and whose sum is b.

Step 3: Write the middle term, bx, using the factors found in Step 2.

Step 4: Factor by grouping.

Example 1 Factor $3x^2 + 31x + 10$ by grouping.

Solution:

Step 1: The terms of this trinomial contain no greatest common factor other than 1 (or -1).

Step 2: In $3x^2 + 31x + 10$, $a = 3$, $b = 31$, and $c = 10$.
Let's find two numbers whose product is $a \cdot c$ or $3(10) = 30$ and whose sum is b or 31. The numbers are 1 and 30.

Factors of 30	Sum of Factors
5, 6	11
3, 10	13
2, 15	17
1, 30	31

Correct sum

(Continued on next page)

Practice 1

Factor $5x^2 + 61x + 12$ by grouping.

Answer

1. $(5x + 1)(x + 12)$

Step 3: Write $31x$ as $1x + 30x$ so that $3x^2 + 31x + 10 = 3x^2 + 1x + 30x + 10$.

Step 4: Factor by grouping.

$$3x^2 + 1x + 30x + 10 = x(3x + 1) + 10(3x + 1)$$
$$= (3x + 1)(x + 10)$$

Work Practice 1

Practice 2

Factor $12x^2 - 19x + 5$ by grouping.

Example 2 Factor $8x^2 - 14x + 5$ by grouping.

Solution:

Step 1: The terms of this trinomial contain no greatest common factor other than 1.

Step 2: This trinomial is of the form $ax^2 + bx + c$ with $a = 8, b = -14$, and $c = 5$. Find two numbers whose product is $a \cdot c$ or $8 \cdot 5 = 40$, and whose sum is b or -14.

The numbers are -4 and -10.

Factors of 40	Sum of Factors
−40, −1	−41
−20, −2	−22
−10, −4	−14

Correct sum

Step 3: Write $-14x$ as $-4x - 10x$ so that

$$8x^2 - 14x + 5 = 8x^2 - 4x - 10x + 5$$

Step 4: Factor by grouping.

$$8x^2 - 4x - 10x + 5 = 4x(2x - 1) - 5(2x - 1)$$
$$= (2x - 1)(4x - 5)$$

Work Practice 2

Practice 3

Factor $30x^2 - 14x - 4$ by grouping.

Example 3 Factor $6x^2 - 2x - 20$ by grouping.

Solution:

Step 1: First factor out the greatest common factor, 2.

$$6x^2 - 2x - 20 = 2(3x^2 - x - 10)$$

Step 2: Next, notice that $a = 3, b = -1$, and $c = -10$ in the resulting trinomial. Find two numbers whose product is $a \cdot c$ or $3(-10) = -30$ and whose sum is b, -1. The numbers are -6 and 5.

Step 3: $3x^2 - x - 10 = 3x^2 - 6x + 5x - 10$

Step 4: $3x^2 - 6x + 5x - 10 = 3x(x - 2) + 5(x - 2)$
$$= (x - 2)(3x + 5)$$

The factored form of $6x^2 - 2x - 20 = 2(x - 2)(3x + 5)$.

Don't forget to include the common factor of 2.

Work Practice 3

Practice 4

Factor $40m^4 + 5m^3 - 35m^2$ by grouping.

Answers

2. $(4x - 5)(3x - 1)$
3. $2(5x + 1)(3x - 2)$
4. $5m^2(8m - 7)(m + 1)$

Example 4 Factor $18y^4 + 21y^3 - 60y^2$ by grouping.

Solution:

Step 1: First factor out the greatest common factor, $3y^2$.

$$18y^4 + 21y^3 - 60y^2 = 3y^2(6y^2 + 7y - 20)$$

Step 2: Notice that $a = 6, b = 7$, and $c = -20$ in the resulting trinomial. Find two numbers whose product is $a \cdot c$ or $6(-20) = -120$ and whose sum is 7. It may help to factor -120 as a product of primes and -1.

$-120 = 2 \cdot 2 \cdot 2 \cdot 3 \cdot 5 \cdot (-1)$

Then choose pairings of factors until you have a pairing whose sum is 7.

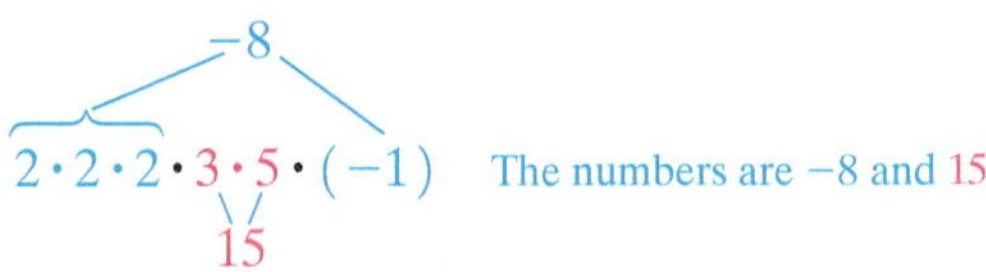

Step 3: $6y^2 + 7y - 20 = 6y^2 - 8y + 15y - 20$

Step 4: $6y^2 - 8y + 15y - 20 = 2y(3y - 4) + 5(3y - 4)$
$= (3y - 4)(2y + 5)$

The factored form of $18y^4 + 21y^3 - 60y^2$ is $3y^2(3y - 4)(2y + 5)$.

Don't forget to include the common factor of $3y^2$.

Work Practice 4

Example 5 Factor $4x^2 + 20x + 25$ by grouping.

Solution:

Step 1: The terms of this trinomial contain no greatest common factor other than 1 (or -1).

Step 2: In $4x^2 + 20x + 25$, $a = 4$, $b = 20$, and $c = 25$. Find two numbers whose product is $a \cdot c$ or $4 \cdot 25 = 100$ and whose sum is 20. The numbers are 10 and 10.

Step 3: Write $20x$ as $10x + 10x$ so that

$4x^2 + 20x + 25 = 4x^2 + 10x + 10x + 25$

Step 4: Factor by grouping.

$4x^2 + 10x + 10x + 25 = 2x(2x + 5) + 5(2x + 5)$
$= (2x + 5)(2x + 5)$

The factored form of $4x^2 + 20x + 25$ is $(2x + 5)(2x + 5)$ or $(2x + 5)^2$.

Work Practice 5

Practice 5

Factor $16x^2 + 24x + 9$ by grouping.

A trinomial that is the square of a binomial, such as the trinomial in Example 5, is called a **perfect square trinomial.** From Chapter 12, there are special product formulas we can use to help us recognize and factor these trinomials. To study these formulas further, see Section 13.3, Objective 3. **Remember:** A perfect square trinomial, such as the one in Example 5, may be factored by special product formulas or by other methods of factoring trinomials, such as by grouping.

Answer
5. $(4x + 3)^2$

Vocabulary, Readiness & Video Check

For each trinomial $ax^2 + bx + c$, choose two numbers whose product is $a \cdot c$ and whose sum is b.

1. $x^2 + 6x + 8$

a. 4, 2 **b.** 7, 1 **c.** 6, 2 **d.** 6, 8

2. $x^2 + 11x + 24$

a. 6, 4 **b.** 24, 1 **c.** 8, 3 **d.** 2, 12

3. $2x^2 + 13x + 6$

a. 2, 6 b. 12, 1 c. 13, 1 d. 3, 4

4. $4x^2 + 8x + 3$

a. 4, 3 b. 4, 4 c. 12, 1 d. 2, 6

Martin-Gay Interactive Videos

See Video 13.4

Watch the section lecture video and answer the following question.

Objective 1 5. In the lecture following Example 1, why does writing a term as the sum or difference of two terms suggest we'd then try to factor by grouping?

13.4 Exercise Set MyMathLab®

Factor each polynomial by grouping. Notice that Step 3 has already been done in these exercises. See Examples 1 through 5.

1. $x^2 + 3x + 2x + 6$
2. $x^2 + 5x + 3x + 15$
3. $y^2 + 8y - 2y - 16$
4. $z^2 + 10z - 7z - 70$
5. $8x^2 - 5x - 24x + 15$
6. $4x^2 - 9x - 32x + 72$
7. $5x^4 - 3x^2 + 25x^2 - 15$
8. $2y^4 - 10y^2 + 7y^2 - 35$

MIXED PRACTICE

Factor each trinomial by grouping. Exercises 9–12 are broken into parts to help you get started. See Examples 1 through 5.

9. $6x^2 + 11x + 3$
   a. Find two numbers whose product is $6 \cdot 3 = 18$ and whose sum is 11.
   b. Write $11x$ using the factors from part (a).
   c. Factor by grouping.
10. $8x^2 + 14x + 3$
   a. Find two numbers whose product is $8 \cdot 3 = 24$ and whose sum is 14.
   b. Write $14x$ using the factors from part (a).
   c. Factor by grouping.
11. $15x^2 - 23x + 4$
   a. Find two numbers whose product is $15 \cdot 4 = 60$ and whose sum is -23.
   b. Write $-23x$ using the factors from part (a).
   c. Factor by grouping.
12. $6x^2 - 13x + 5$
   a. Find two numbers whose product is $6 \cdot 5 = 30$ and whose sum is -13.
   b. Write $-13x$ using the factors from part (a).
   c. Factor by grouping.
13. $21y^2 + 17y + 2$
14. $15x^2 + 11x + 2$
15. $7x^2 - 4x - 11$
16. $8x^2 - x - 9$
17. $10x^2 - 9x + 2$
18. $30x^2 - 23x + 3$
19. $2x^2 - 7x + 5$
20. $2x^2 - 7x + 3$
21. $12x + 4x^2 + 9$
22. $20x + 25x^2 + 4$
23. $4x^2 - 8x - 21$
24. $6x^2 - 11x - 10$
25. $10x^2 - 23x + 12$
26. $21x^2 - 13x + 2$
27. $2x^3 + 13x^2 + 15x$
28. $3x^3 + 8x^2 + 4x$
29. $16y^2 - 34y + 18$
30. $4y^2 - 2y - 12$
31. $-13x + 6 + 6x^2$
32. $-25x + 12 + 12x^2$
33. $54a^2 - 9a - 30$
34. $30a^2 + 38a - 20$
35. $20a^3 + 37a^2 + 8a$

36. $10a^3 + 17a^2 + 3a$
37. $12x^3 - 27x^2 - 27x$
38. $30x^3 - 155x^2 + 25x$
39. $3x^2y + 4xy^2 + y^3$
40. $6r^2t + 7rt^2 + t^3$
41. $20z^2 + 7z + 1$
42. $36z^2 + 6z + 1$
43. $5x^2 + 50xy + 125y^2$
44. $3x^2 + 42xy + 147y^2$
45. $24a^2 - 6ab - 30b^2$
46. $30a^2 + 5ab - 25b^2$
47. $15p^4 + 31p^3q + 2p^2q^2$
48. $20s^4 + 61s^3t + 3s^2t^2$
49. $162a^4 - 72a^2 + 8$
50. $32n^4 - 112n^2 + 98$
51. $35 + 12x + x^2$
52. $33 + 14x + x^2$
53. $6 - 11x + 5x^2$
54. $5 - 12x + 7x^2$

REVIEW AND PREVIEW

Multiply. See Sections 12.3 and 12.4.

55. $(x - 2)(x + 2)$
56. $(y - 5)(y + 5)$
57. $(y + 4)(y + 4)$
58. $(x + 7)(x + 7)$
59. $(9z + 5)(9z - 5)$
60. $(8y + 9)(8y - 9)$
61. $(x - 3)(x^2 + 3x + 9)$
62. $(2z - 1)(4z^2 + 2z + 1)$

CONCEPT EXTENSIONS

Write the perimeter of each figure as a simplified polynomial. Then factor the polynomial.

63.
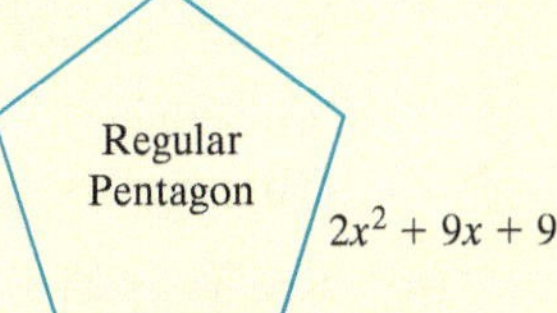

64.
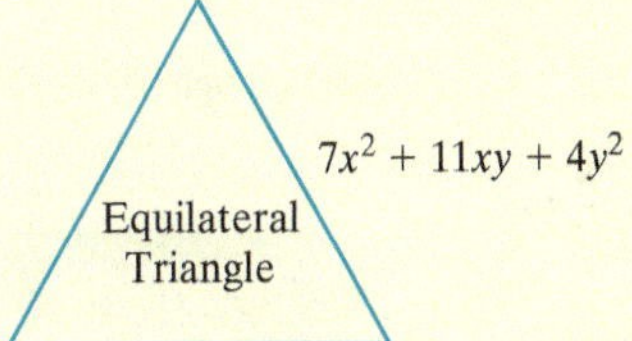

Factor each polynomial by grouping.

65. $x^{2n} + 2x^n + 3x^n + 6$
(***Hint:*** Don't forget that $x^{2n} = x^n \cdot x^n$.)
66. $x^{2n} + 6x^n + 10x^n + 60$
67. $3x^{2n} + 16x^n - 35$
68. $12x^{2n} - 40x^n + 25$
69. In your own words, explain how to factor a trinomial by grouping.

13.5 Factoring Binomials 

Objective 1 Factoring the Difference of Two Squares 

Objectives

1. Factor the Difference of Two Squares.
2. Factor the Sum or Difference of Two Cubes.

When learning to multiply binomials in Chapter 12, we studied a special product, the product of the sum and difference of two terms, a and b:

$$(a + b)(a - b) = a^2 - b^2$$

For example, the product of $x + 3$ and $x - 3$ is

$$(x + 3)(x - 3) = x^2 - 9$$

The binomial $x^2 - 9$ is called a **difference of squares.** In this section, we reverse the pattern for the product of a sum and difference to factor the binomial difference of squares.

Factoring the Difference of Two Squares

$$a^2 - b^2 = (a + b)(a - b)$$

Helpful Hint

Since multiplication is commutative, remember that the order of factors does not matter.

In other words,

$$a^2 - b^2 = (a + b)(a - b) \text{ or } (a - b)(a + b)$$

Practice 1

Factor $x^2 - 81$.

Example 1 Factor: $x^2 - 25$

Solution: $x^2 - 25$ is the difference of two squares since $x^2 - 25 = x^2 - 5^2$. Therefore,

$$x^2 - 25 = x^2 - 5^2 = (x + 5)(x - 5)$$

Multiply to check.

Work Practice 1

Practice 2

Factor each difference of squares.

a. $9x^2 - 1$

b. $36a^2 - 49b^2$

c. $p^2 - \frac{25}{36}$

Example 2 Factor each difference of squares.

a. $4x^2 - 1$ **b.** $25a^2 - 9b^2$ **c.** $y^2 - \frac{4}{9}$

Solution:

a. $4x^2 - 1 = (2x)^2 - 1^2 = (2x + 1)(2x - 1)$

b. $25a^2 - 9b^2 = (5a)^2 - (3b)^2 = (5a + 3b)(5a - 3b)$

c. $y^2 - \frac{4}{9} = y^2 - \left(\frac{2}{3}\right)^2 = \left(y + \frac{2}{3}\right)\left(y - \frac{2}{3}\right)$

Work Practice 2

Practice 3

Factor $p^4 - q^{10}$.

Example 3 Factor: $x^4 - y^6$

Solution: This is a difference of squares since $x^4 = (x^2)^2$ and $y^6 = (y^3)^2$. Thus,

$$x^4 - y^6 = (x^2)^2 - (y^3)^2 = (x^2 + y^3)(x^2 - y^3)$$

Work Practice 3

Practice 4

Factor each binomial.

a. $z^4 - 81$

b. $m^2 + 49$

Example 4 Factor each binomial.

a. $y^4 - 16$ **b.** $x^2 + 4$

Solution:

a. $y^4 - 16 = (y^2)^2 - 4^2$

$= (y^2 + 4)(y^2 - 4)$ Factor the difference of two squares. This binomial can be factored further since it is the difference of two squares.

$= (y^2 + 4)(y + 2)(y - 2)$ Factor the difference of two squares.

b. $x^2 + 4$

Note that the binomial $x^2 + 4$ is the *sum* of two squares since we can write $x^2 + 4$ as $x^2 + 2^2$. We might try to factor using $(x + 2)(x + 2)$ or $(x - 2)(x - 2)$. But when we multiply to check, we find that neither factoring is correct.

Answers

1. $(x + 9)(x - 9)$
2. a. $(3x - 1)(3x + 1)$
b. $(6a - 7b)(6a + 7b)$
c. $\left(p + \frac{5}{6}\right)\left(p - \frac{5}{6}\right)$
3. $(p^2 - q^5)(p^2 + q^5)$
4. a. $(z^2 + 9)(z + 3)(z - 3)$
b. prime polynomial

$(x + 2)(x + 2) = x^2 + 4x + 4$
$(x - 2)(x - 2) = x^2 - 4x + 4$

In both cases, the product is a trinomial, not the required binomial. In fact, $x^2 + 4$ is a prime polynomial.

Work Practice 4

Helpful Hint

When factoring, don't forget:

- See whether the terms have a greatest common factor (GCF) (other than 1) that can be factored out.
- Other than a GCF, the **sum** of two squares cannot be factored using real numbers.
- Factor completely. Always check to see whether any factors can be factored further.

Examples Factor each binomial.

5. $4x^3 - 49x = x(4x^2 - 49)$ Factor out the common factor, x.
$= x[(2x)^2 - 7^2]$
$= x(2x + 7)(2x - 7)$ Factor the difference of two squares.

6. $162x^4 - 2 = 2(81x^4 - 1)$ Factor out the common factor, 2.
$= 2(9x^2 + 1)(9x^2 - 1)$ Factor the difference of two squares.
$= 2(9x^2 + 1)(3x + 1)(3x - 1)$ Factor the difference of two squares.

Work Practice 5–6

Practice 5–6

Factor each binomial.

5. $36y^3 - 25y$ **6.** $80y^4 - 5$

Example 7 Factor: $-49x^2 + 16$

Solution: Factor as is, or, if you like, rearrange terms.

Factor as is: $-49x^2 + 16 = -1(49x^2 - 16)$ Factor out -1.
$= -1(7x + 4)(7x - 4)$ Factor the difference of two squares.

Rewrite binomial: $-49x^2 + 16 = 16 - 49x^2 = 4^2 - (7x)^2$
$= (4 + 7x)(4 - 7x)$

Both factorizations are correct and are equal. To see this, factor -1 from $(4 - 7x)$ in the second factorization.

Work Practice 7

Practice 7

Factor $-9x^2 + 100$.

Helpful Hint

When rearranging terms, keep in mind that the sign of a term is in front of the term.

Objective 2 Factoring the Sum or Difference of Two Cubes

Although the sum of two squares usually does not factor, the sum or difference of two cubes can be factored and reveals factoring patterns. The pattern for the sum of cubes is illustrated by multiplying the binomial $x + y$ and the trinomial $x^2 - xy + y^2$.

Answers

5. $y(6y + 5)(6y - 5)$
6. $5(4y^2 + 1)(2y + 1)(2y - 1)$
7. $-1(3x + 10)(3x - 10)$ or $(10 + 3x)(10 - 3x)$

$$\begin{array}{r} x^2 - xy + y^2 \\ x + y \\ \hline x^2y - xy^2 + y^3 \\ x^3 - x^2y + xy^2 \quad \\ \hline x^3 \qquad\qquad + y^3 \end{array}$$

Thus, $(x + y)(x^2 - xy + y^2) = x^3 + y^3$ Sum of cubes

The pattern for the difference of two cubes is illustrated by multiplying the binomial $x - y$ by the trinomial $x^2 + xy + y^2$. The result is

$(x - y)(x^2 + xy + y^2) = x^3 - y^3$ Difference of cubes

Factoring the Sum or Difference of Two Cubes

$$a^3 + b^3 = (a + b)(a^2 - ab + b^2)$$
$$a^3 - b^3 = (a - b)(a^2 + ab + b^2)$$

Recall that "factor" means "to write as a product." Above are patterns for writing sums and differences as products.

Practice 8

Factor $x^3 + 64$.

Example 8 Factor: $x^3 + 8$

Solution: First, write the binomial in the form $a^3 + b^3$.

$x^3 + 8 = x^3 + 2^3$ Write in the form $a^3 + b^3$.

If we replace a with x and b with 2 in the formula above, we have

$$\begin{aligned} x^3 + 2^3 &= (x + 2)[x^2 - (x)(2) + 2^2] \\ &= (x + 2)(x^2 - 2x + 4) \end{aligned}$$

Work Practice 8

Helpful Hint

When factoring sums or differences of cubes, notice the sign patterns.

same sign

$x^3 + y^3 = (x + y)(x^2 - xy + y^2)$

opposite signs always positive

same sign

$x^3 - y^3 = (x - y)(x^2 + xy + y^2)$

opposite signs always positive

Practice 9

Factor $x^3 - 125$.

Example 9 Factor: $y^3 - 27$

Solution:

$$\begin{aligned} y^3 - 27 &= y^3 - 3^3 \quad \text{Write in the form } a^3 - b^3. \\ &= (y - 3)[y^2 + (y)(3) + 3^2] \\ &= (y - 3)(y^2 + 3y + 9) \end{aligned}$$

Work Practice 9

Answers

8. $(x + 4)(x^2 - 4x + 16)$
9. $(x - 5)(x^2 + 5x + 25)$

Example 10 Factor: $64x^3 + 1$

Solution:

$$64x^3 + 1 = (4x)^3 + 1^3$$
$$= (4x + 1)[(4x)^2 - (4x)(1) + 1^2]$$
$$= (4x + 1)(16x^2 - 4x + 1)$$

Work Practice 10

Practice 10

Factor $27y^3 + 1$.

Example 11 Factor: $54a^3 - 16b^3$

Solution: Remember to factor out common factors first before using other factoring methods.

$$54a^3 - 16b^3 = 2(27a^3 - 8b^3) \quad \text{Factor out the GCF, 2.}$$
$$= 2[(3a)^3 - (2b)^3] \quad \text{Difference of two cubes}$$
$$= 2(3a - 2b)[(3a)^2 + (3a)(2b) + (2b)^2]$$
$$= 2(3a - 2b)(9a^2 + 6ab + 4b^2)$$

Work Practice 11

Practice 11

Factor $32x^3 - 500y^3$.

Answers

10. $(3y + 1)(9y^2 - 3y + 1)$
11. $4(2x - 5y)(4x^2 + 10xy + 25y^2)$

Graphing Calculator Explorations Graphing

A graphing calculator is a convenient tool for evaluating an expression at a given replacement value. For example, let's evaluate $x^2 - 6x$ when $x = 2$. To do so, store the value 2 in the variable x and then enter and evaluate the algebraic expression.

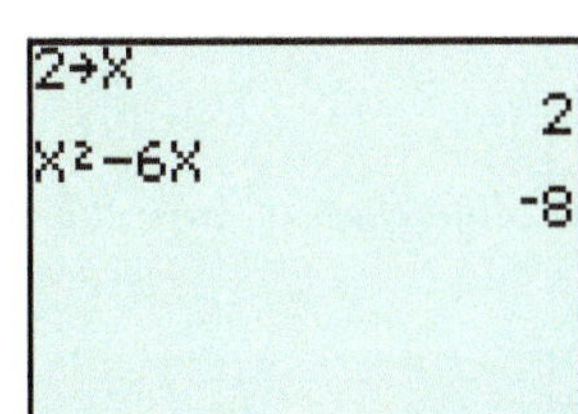

The value of $x^2 - 6x$ when $x = 2$ is -8. You may want to use this method for evaluating expressions as you explore the following.

We can use a graphing calculator to explore factoring patterns numerically. Use your calculator to evaluate $x^2 - 2x + 1$, $x^2 - 2x - 1$, and $(x - 1)^2$ for each value of x given in the table. What do you observe?

	$x^2 - 2x + 1$	$x^2 - 2x - 1$	$(x - 1)^2$
$x = 5$			
$x = -3$			
$x = 2.7$			
$x = -12.1$			
$x = 0$			

Notice in each case that $x^2 - 2x - 1 \neq (x - 1)^2$. Because for each x in the table the value of $x^2 - 2x + 1$ and the value of $(x - 1)^2$ are the same, we might guess that $x^2 - 2x + 1 = (x - 1)^2$. We can verify our guess algebraically with multiplication:

$$(x - 1)(x - 1) = x^2 - x - x + 1 = x^2 - 2x + 1$$

Vocabulary, Readiness & Video Check

Use the choices below to fill in each blank. Some choices may be used more than once and some choices may not be used at all.

true	difference of two squares	sum of two cubes
false	difference of two cubes	

1. The expression $x^3 - 27$ is called a ______________________.

2. The expression $x^2 - 49$ is called a ______________________.

3. The expression $z^3 + 1$ is called a ______________________.

4. True or false: The binomial $y^2 + 9$ factors as $(y + 3)^2$. ______________________

Write each term as a square.

5. $49x^2$ **6.** $25y^4$

Write each term as a cube.

7. $8y^3$ **8.** x^6

Martin-Gay Interactive Videos

See Video 13.5 

Watch the section lecture video and answer the following questions.

Objective 1 **9.** In Examples 1 and 2, what are two reasons the original binomial? is rewritten so that each term is a square?

Objective 2 **10.** From Example 3, what is a prime polynomial?

Objective 3 **11.** In Examples 6–8, what tips are given to remember how to factor the sum or difference of two cubes rather than memorizing the formulas?

13.5 Exercise Set MyMathLab®

Factor each binomial completely. See Examples 1 through 7.

1. $x^2 - 4$

2. $x^2 - 36$

3. $81p^2 - 1$

4. $49m^2 - 1$

5. $25y^2 - 9$

6. $49a^2 - 16$

7. $121m^2 - 100n^2$

8. $169a^2 - 49b^2$

9. $x^2y^2 - 1$

10. $a^2b^2 - 16$

11. $x^2 - \frac{1}{4}$

12. $y^2 - \frac{1}{16}$

13. $-4r^2 + 1$

14. $-9t^2 + 1$

15. $16r^2 + 1$

16. $49y^2 + 1$

17. $-36 + x^2$

18. $-1 + y^2$

19. $m^4 - 1$

20. $n^4 - 16$

21. $m^4 - n^{18}$

22. $n^4 - r^6$

Factor the sum or difference of two cubes. See Examples 8 through 11.

23. $x^3 + 125$

24. $p^3 + 1$

25. $8a^3 - 1$

26. $27y^3 - 1$

27. $m^3 + 27n^3$

28. $y^3 + 64z^3$

29. $5k^3 + 40$

30. $6r^3 + 162$

31. $x^3y^3 - 64$

32. $a^3b^3 - 8$

33. $250r^3 - 128t^3$

34. $24x^3 - 81y^3$

MIXED PRACTICE

Factor each binomial completely. See Examples 1 through 11.

35. $r^2 - 64$

36. $q^2 - 121$

37. $x^2 - 169y^2$

38. $x^2 - 225y^2$

39. $27 - t^3$

40. $125 - r^3$

41. $18r^2 - 8$

42. $32t^2 - 50$

43. $9xy^2 - 4x$

44. $36x^2y - 25y$

45. $8m^3 + 64$

46. $2x^3 + 54$

47. $xy^3 - 9xyz^2$

48. $x^3y - 4xy^3$

49. $36x^2 - 64y^2$

50. $225a^2 - 81b^2$

51. $144 - 81x^2$

52. $12x^2 - 27$

53. $x^3y^3 - z^6$

54. $a^3b^3 - c^9$

55. $49 - \frac{9}{25}m^2$

56. $100 - \frac{4}{81}n^2$

57. $t^3 + 343$

58. $s^3 + 216$

59. $n^3 + 49n$

60. $y^3 + 64y$

61. $x^6 - 81x^2$

62. $n^9 - n^5$

63. $64p^3q - 81pq^3$

64. $100x^3y - 49xy^3$

65. $27x^2y^3 + xy^2$

66. $8x^3y^3 + x^3y$

67. $125a^4 - 64ab^3$

68. $64m^4 - 27mn^3$

69. $16x^4 - 64x^2$

70. $25y^4 - 100y^2$

REVIEW AND PREVIEW

Solve each equation. See Section 3.2.

71. $x - 6 = 0$

72. $y + 5 = 0$

73. $2m + 4 = 0$

74. $3x - 9 = 0$

75. $5z - 10 = 0$

76. $4a + 16 = 0$

CONCEPT EXTENSIONS

Factor each expression completely.

77. $(x + 2)^2 - y^2$

78. $(y - 6)^2 - z^2$

79. $a^2(b - 4) - 16(b - 4)$

80. $m^2(n + 8) - 9(n + 8)$

81. $(x^2 + 6x + 9) - 4y^2$ (*Hint:* Factor the trinomial in parentheses first.)

82. $(x^2 + 2x + 1) - 36y^2$

83. $x^{2n} - 100$

84. $x^{2n} - 81$

85. What binomial multiplied by $(x - 6)$ gives the difference of two squares?

86. What binomial multiplied by $(5 + y)$ gives the difference of two squares?

87. In your own words, explain how to tell whether a binomial is a difference of squares. Then explain how to factor a difference of squares.

88. In your own words, explain how to tell whether a binomial is a sum of cubes. Then explain how to factor a sum of cubes.

89. The Toroweap Overlook, on the North Rim of the Grand Canyon, lies 3000 vertical feet above the Colorado River. The view is spectacular, and the sheer drop is dramatic. A film crew creating a documentary about the Grand Canyon has suspended a camera platform 296 feet below the Overlook. A camera filter comes loose and falls to the river below. The height of the filter above the river after t seconds is given by the expression $2704 - 16t^2$.

a. Find the height of the filter above the river after 3 seconds.

b. Find the height of the filter above the river after 7 seconds.

c. To the nearest whole second, estimate when the filter lands in the river.

d. Factor $2704 - 16t^2$.

90. An object is dropped from the top of Pittsburgh's USX Tower, which is 841 feet tall. (*Source: World Almanac* research) The height of the object after t seconds is given by the expression $841 - 16t^2$.

a. Find the height of the object after 2 seconds.

b. Find the height of the object after 5 seconds.

c. To the nearest whole second, estimate when the object hits the ground.

d. Factor $841 - 16t^2$.

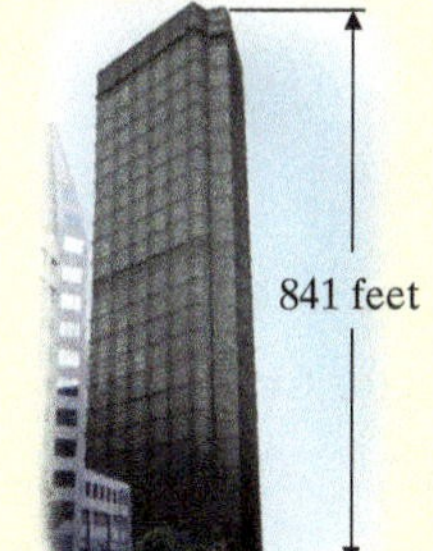

91. At this writing, the world's second tallest building is the Taipei 101 in Taipei, Taiwan, at a height of 1671 feet. (*Source:* Council on Tall Buildings and Urban Habitat) Suppose a worker is suspended 71 feet below the top of the pinnacle atop the building, at a height of 1600 feet above the ground. If the worker accidentally drops a bolt, the height of the bolt after t seconds is given by the expression $1600 - 16t^2$.

a. Find the height of the bolt after 3 seconds.

b. Find the height of the bolt after 7 seconds.

c. To the nearest whole second, estimate when the bolt hits the ground.

d. Factor $1600 - 16t^2$.

92. A performer with the Moscow Circus is planning a stunt involving a free fall from the top of the Moscow State University building, which is 784 feet tall. (*Source:* Council on Tall Buildings and Urban Habitat) Neglecting air resistance, the performer's height above gigantic cushions positioned at ground level after t seconds is given by the expression $784 - 16t^2$.

a. Find the performer's height after 2 seconds.

b. Find the performer's height after 5 seconds.

c. To the nearest whole second, estimate when the performer reaches the cushions positioned at ground level.

d. Factor $784 - 16t^2$.

Integrated Review Sections 13.1–13.5

Choosing a Factoring Strategy

The following steps may be helpful when factoring polynomials.

Factoring a Polynomial

Step 1: Are there any common factors? If so, factor out the GCF.

Step 2: How many terms are in the polynomial?

a. If there are **two** terms, decide if one of the following can be applied.

i. Difference of two squares: $a^2 - b^2 = (a + b)(a - b)$.

ii. Difference of two cubes: $a^3 - b^3 = (a - b)(a^2 + ab + b^2)$.

iii. Sum of two cubes: $a^3 + b^3 = (a + b)(a^2 - ab + b^2)$.

b. If there are **three** terms, try one of the following.

i. Perfect square trinomial: $a^2 + 2ab + b^2 = (a + b)^2$
$a^2 - 2ab + b^2 = (a - b)^2$.

ii. If not a perfect square trinomial, factor using the methods presented in Sections 13.2 through 13.4.

c. If there are **four** or more terms, try factoring by grouping.

Step 3: See if any factors in the factored polynomial can be factored further.

Step 4: Check by multiplying.

Study the next five examples to help you use the steps above.

Practice 1

Factor $6x^2 - 11x + 3$.

Example 1 Factor $10t^2 - 17t + 3$.

Solution:

Step 1: The terms of this polynomial have no common factor (other than 1).

Step 2: There are three terms, so this polynomial is a trinomial. This trinomial is not a perfect square trinomial, so factor using methods from earlier sections.

Factors of $10t^2$: $10t^2 = 2t \cdot 5t$, $\quad 10t^2 = t \cdot 10t$

Answer

1. $(3x - 1)(2x - 3)$

Since the middle term, $-17t$, has a negative numerical coefficient, find negative factors of 3.

Factors of 3: $3 = -1 \cdot -3$

Try different combinations of these factors. The correct combination is

$$(2t - 3)(5t - 1) = 10t^2 - 17t + 3$$

$-15t$
$-2t$
$-17t$ Correct middle term

Step 3: No factor can be factored further, so we have factored completely.

Step 4: To check, multiply $2t - 3$ and $5t - 1$.

$$(2t - 3)(5t - 1) = 10t^2 - 2t - 15t + 3 = 10t^2 - 17t + 3$$

The factored form of $10t^2 - 17t + 3$ is $(2t - 3)(5t - 1)$.

Work Practice 1

Example 2 Factor $2x^3 + 3x^2 - 2x - 3$.

Practice 2

Factor $3x^3 + x^2 - 12x - 4$.

Solution:

Step 1: There are no factors common to all terms.

Step 2: Try factoring by grouping since this polynomial has four terms.

$2x^3 + 3x^2 - 2x - 3 = x^2(2x + 3) - 1(2x + 3)$ Factor out the greatest common factor for each pair of terms.

$= (2x + 3)(x^2 - 1)$ Factor out $2x + 3$.

Step 3: The binomial $x^2 - 1$ can be factored further. It is the difference of two squares.

$= (2x + 3)(x + 1)(x - 1)$ Factor $x^2 - 1$ as a difference of squares.

Step 4: Check by finding the product of the three binomials. The polynomial factored completely is $(2x + 3)(x + 1)(x - 1)$.

Work Practice 2

Example 3 Factor $12m^2 - 3n^2$.

Practice 3

Factor $27x^2 - 3y^2$.

Solution:

Step 1: The terms of this binomial contain a greatest common factor of 3.

$12m^2 - 3n^2 = 3(4m^2 - n^2)$ Factor out the greatest common factor.

Step 2: The binomial $4m^2 - n^2$ is a difference of squares.

$= 3(2m + n)(2m - n)$ Factor the difference of squares.

Step 3: No factor can be factored further.

Step 4: We check by multiplying.

$$3(2m + n)(2m - n) = 3(4m^2 - n^2) = 12m^2 - 3n^2$$

The factored form of $12m^2 - 3n^2$ is $3(2m + n)(2m - n)$.

Work Practice 3

Answers

2. $(3x + 1)(x - 2)(x + 2)$
3. $3(3x - y)(3x + y)$

Practice 4

Factor $8a^3 + b^3$.

Example 4 Factor $x^3 + 27y^3$.

Solution:

Step 1: The terms of this binomial contain no common factor (other than 1).

Step 2: This binomial is the sum of two cubes.

$$\begin{aligned} x^3 + 27y^3 &= (x)^3 + (3y)^3 \\ &= (x + 3y)[x^2 - x(3y) + (3y)^2] \\ &= (x + 3y)(x^2 - 3xy + 9y^2) \end{aligned}$$

Step 3: No factor can be factored further.

Step 4: We check by multiplying.

$$\begin{aligned} (x + 3y)(x^2 - 3xy + 9y^2) &= x(x^2 - 3xy + 9y^2) + 3y(x^2 - 3xy + 9y^2) \\ &= x^3 - 3x^2y + 9xy^2 + 3x^2y - 9xy^2 + 27y^3 \\ &= x^3 + 27y^3 \end{aligned}$$

Thus, $x^3 + 27y^3$ factored completely is $(x + 3y)(x^2 - 3xy + 9y^2)$.

Work Practice 4

Practice 5

Factor $60x^3y^2 - 66x^2y^2 - 36xy^2$.

Example 5 Factor $30a^2b^3 + 55a^2b^2 - 35a^2b$.

Solution:

Step 1: $30a^2b^3 + 55a^2b^2 - 35a^2b = 5a^2b(6b^2 + 11b - 7)$ Factor out the GCF.

Step 2: $= 5a^2b(2b - 1)(3b + 7)$ Factor the resulting trinomial.

Step 3: No factor can be factored further.

Step 4: Check by multiplying.

The trinomial factored completely is $5a^2b(2b - 1)(3b + 7)$.

Work Practice 5

Answers

4. $(2a + b)(4a^2 - 2ab + b^2)$
5. $6xy^2(5x + 2)(2x - 3)$

Factor the following completely.

1. $x^2 + 2xy + y^2$
2. $x^2 - 2xy + y^2$
3. $a^2 + 11a - 12$
4. $a^2 - 11a + 10$
5. $a^2 - a - 6$
6. $a^2 - 2a + 1$
7. $x^2 + 2x + 1$
8. $x^2 + x - 2$
9. $x^2 + 4x + 3$
10. $x^2 + x - 6$
11. $x^2 + 7x + 12$
12. $x^2 + x - 12$
13. $x^2 + 3x - 4$
14. $x^2 - 7x + 10$
15. $x^2 + 2x - 15$
16. $x^2 + 11x + 30$
17. $x^2 - x - 30$
18. $x^2 + 11x + 24$
19. $2x^2 - 98$
20. $3x^2 - 75$
21. $x^2 + 3x + xy + 3y$
22. $3y - 21 + xy - 7x$
23. $x^2 + 6x - 16$
24. $x^2 - 3x - 28$
25. $4x^3 + 20x^2 - 56x$
26. $6x^3 - 6x^2 - 120x$
27. $12x^2 + 34x + 24$
28. $8a^2 + 6ab - 5b^2$
29. $4a^2 - b^2$
30. $28 - 13x - 6x^2$
31. $20 - 3x - 2x^2$
32. $x^2 - 2x + 4$
33. $a^2 + a - 3$
34. $6y^2 + y - 15$
35. $4x^2 - x - 5$
36. $x^2y - y^3$
37. $4t^2 + 36$
38. $x^2 + x + xy + y$
39. $ax + 2x + a + 2$
40. $18x^3 - 63x^2 + 9x$
41. $12a^3 - 24a^2 + 4a$
42. $x^2 + 14x - 32$
43. $x^2 - 14x - 48$
44. $16a^2 - 56ab + 49b^2$
45. $25p^2 - 70pq + 49q^2$
46. $7x^2 + 24xy + 9y^2$
47. $125 - 8y^3$
48. $64x^3 + 27$
49. $-x^2 - x + 30$
50. $-x^2 + 6x - 8$
51. $14 + 5x - x^2$

52. $3 - 2x - x^2$
53. $3x^4y + 6x^3y - 72x^2y$
54. $2x^3y + 8x^2y^2 - 10xy^3$
55. $5x^3y^2 - 40x^2y^3 + 35xy^4$
56. $4x^4y - 8x^3y - 60x^2y$
57. $12x^3y + 243xy$
58. $6x^3y^2 + 8xy^2$
59. $4 - x^2$
60. $9 - y^2$
61. $3rs - s + 12r - 4$
62. $x^3 - 2x^2 + 3x - 6$
63. $4x^2 - 8xy - 3x + 6y$
64. $4x^2 - 2xy - 7yz + 14xz$
65. $6x^2 + 18xy + 12y^2$
66. $12x^2 + 46xy - 8y^2$
67. $xy^2 - 4x + 3y^2 - 12$
68. $x^2y^2 - 9x^2 + 3y^2 - 27$
69. $5(x + y) + x(x + y)$
70. $7(x - y) + y(x - y)$
71. $14t^2 - 9t + 1$
72. $3t^2 - 5t + 1$
73. $3x^2 + 2x - 5$
74. $7x^2 + 19x - 6$
75. $x^2 + 9xy - 36y^2$
76. $3x^2 + 10xy - 8y^2$
77. $1 - 8ab - 20a^2b^2$
78. $1 - 7ab - 60a^2b^2$
79. $9 - 10x^2 + x^4$
80. $36 - 13x^2 + x^4$
81. $x^4 - 14x^2 - 32$
82. $x^4 - 22x^2 - 75$
83. $x^2 - 23x + 120$
84. $y^2 + 22y + 96$
85. $6x^3 - 28x^2 + 16x$
86. $6y^3 - 8y^2 - 30y$
87. $27x^3 - 125y^3$
88. $216y^3 - z^3$
89. $x^3y^3 + 8z^3$
90. $27a^3b^3 + 8$
91. $2xy - 72x^3y$
92. $2x^3 - 18x$
93. $x^3 + 6x^2 - 4x - 24$
94. $x^3 - 2x^2 - 36x + 72$
95. $6a^3 + 10a^2$
96. $4n^2 - 6n$
97. $a^2(a + 2) + 2(a + 2)$
98. $a - b + x(a - b)$
99. $x^3 - 28 + 7x^2 - 4x$
100. $a^3 - 45 - 9a + 5a^2$

CONCEPT EXTENSIONS

Factor.

101. $(x - y)^2 - z^2$
102. $(x + 2y)^2 - 9$
103. $81 - (5x + 1)^2$
104. $b^2 - (4a + c)^2$
105. Explain why it makes good sense to factor out the GCF first, before using other methods of factoring.
106. The sum of two squares usually does not factor. Is the sum of two squares $9x^2 + 81y^2$ factorable?
107. Which of the following are equivalent to $(x + 10)(x - 7)$?
 a. $(x - 7)(x + 10)$ b. $-1(x + 10)(x - 7)$
 c. $-1(x + 10)(7 - x)$ d. $-1(-x - 10)(7 - x)$
108. Which of the following are equivalent to $(x - 2)(x - 5)$?
 a. $-1(x + 2)(x + 5)$ b. $(x - 5)(x - 2)$
 c. $(5 - x)(2 - x)$ d. $-1(x + 2)(x - 5)$

13.6 Solving Quadratic Equations by Factoring

Objectives

1 Solve Quadratic Equations by Factoring.

2 Solve Equations with Degree Greater than 2 by Factoring.

3 Find the x-Intercepts of the Graph of a Quadratic Equation in Two Variables.

In this section, we introduce a new type of equation—the **quadratic equation.**

Quadratic Equation

A quadratic equation is one that can be written in the form

$$ax^2 + bx + c = 0$$

where a, b, and c are real numbers and $a \neq 0$.

Some examples of quadratic equations are shown below.

$$x^2 - 9x - 22 = 0 \qquad 4x^2 - 28 = -49 \qquad x(2x - 7) = 4$$

The form $ax^2 + bx + c = 0$ is called the **standard form** of a quadratic equation. The quadratic equation $x^2 - 9x - 22 = 0$ is the only equation on the previous page that is in standard form.

Quadratic equations model many real-life situations. For example, let's suppose we want to know how long before a person diving from a 144-foot cliff reaches the ocean. The answer to this question is found by solving the quadratic equation $-16t^2 + 144 = 0$. (See Example 1 in Section 13.7.)

Objective 1 Solving Quadratic Equations by Factoring

Some quadratic equations can be solved by making use of factoring and the **zero factor property.**

Zero Factor Theorem

If a and b are real numbers and if $ab = 0$, then $a = 0$ or $b = 0$.

This theorem states that if the product of two numbers is 0, then at least one of the numbers must be 0.

Practice 1

Solve: $(x + 4)(x - 5) = 0$.

Example 1 Solve: $(x - 3)(x + 1) = 0$

Solution: If this equation is to be a true statement, then either the factor $x - 3$ must be 0 or the factor $x + 1$ must be 0. In other words, either

$$x - 3 = 0 \quad \text{or} \quad x + 1 = 0$$

If we solve these two linear equations, we have

$$x = 3 \quad \text{or} \quad x = -1$$

Thus, 3 and -1 are both solutions of the equation $(x - 3)(x + 1) = 0$. To check, we replace x with 3 in the original equation. Then we replace x with -1 in the original equation.

Check:

Let $x = 3$.		Let $x = -1$.	
$(x - 3)(x + 1) = 0$		$(x - 3)(x + 1) = 0$	
$(3 - 3)(3 + 1) \stackrel{?}{=} 0$	Replace x with 3.	$(-1 - 3)(-1 + 1) \stackrel{?}{=} 0$	Replace x with -1.
$0(4) = 0$	True	$(-4)(0) = 0$	True

The solutions are 3 and -1, or we say that the solution set is $\{-1, 3\}$.

Work Practice 1

Helpful Hint

The zero factor property says that *if a product is 0, then a factor is 0.*

If $a \cdot b = 0$, then $a = 0$ or $b = 0$.

If $x(x + 5) = 0$, then $x = 0$ or $x + 5 = 0$.

If $(x + 7)(2x - 3) = 0$, then $x + 7 = 0$ or $2x - 3 = 0$.

Use this property only when the product is 0.

For example, if $a \cdot b = 8$, we do not know the value of a or b. The values may be $a = 2, b = 4$ or $a = 8, b = 1$, or any other two numbers whose product is 8.

Answer

1. $-4, 5$

Example 2 Solve: $(x-5)(2x+7)=0$

Solution: The product is 0. By the zero factor property, this is true only when a factor is 0. To solve, we set each factor equal to 0 and solve the resulting linear equations.

$$(x-5)(2x+7)=0$$
$$x-5=0 \quad \text{or} \quad 2x+7=0$$
$$x=5 \qquad\qquad 2x=-7$$
$$x=-\frac{7}{2}$$

Check: Let $x=5$.

$$(x-5)(2x+7)=0$$
$$(5-5)(2\cdot 5+7)\stackrel{?}{=}0 \quad \text{Replace } x \text{ with } 5.$$
$$0\cdot 17\stackrel{?}{=}0$$
$$0=0 \quad \text{True}$$

Let $x=-\frac{7}{2}$.

$$(x-5)(2x+7)=0$$
$$\left(-\frac{7}{2}-5\right)\left(2\left(-\frac{7}{2}\right)+7\right)\stackrel{?}{=}0 \quad \text{Replace } x \text{ with } -\frac{7}{2}.$$
$$\left(-\frac{17}{2}\right)(-7+7)\stackrel{?}{=}0$$
$$\left(-\frac{17}{2}\right)\cdot 0\stackrel{?}{=}0$$
$$0=0 \quad \text{True}$$

The solutions are 5 and $-\frac{7}{2}$.

Work Practice 2

Practice 2

Solve: $(x-12)(4x+3)=0$.

Example 3 Solve: $x(5x-2)=0$

Solution:

$$x(5x-2)=0$$
$$x=0 \quad \text{or} \quad 5x-2=0 \quad \text{Use the zero factor property.}$$
$$5x=2$$
$$x=\frac{2}{5}$$

Check: Let $x=0$.

$$x(5x-2)=0$$
$$0(5\cdot 0-2)\stackrel{?}{=}0 \quad \text{Replace } x \text{ with } 0.$$
$$0(-2)\stackrel{?}{=}0$$
$$0=0 \quad \text{True}$$

Let $x=\frac{2}{5}$.

$$x(5x-2)=0$$
$$\frac{2}{5}\left(5\cdot\frac{2}{5}-2\right)\stackrel{?}{=}0 \quad \text{Replace } x \text{ with } \frac{2}{5}.$$
$$\frac{2}{5}(2-2)\stackrel{?}{=}0$$
$$\frac{2}{5}(0)\stackrel{?}{=}0$$
$$0=0 \quad \text{True}$$

The solutions are 0 and $\frac{2}{5}$.

Work Practice 3

Practice 3

Solve: $x(7x-6)=0$.

Answers

2. $12, -\frac{3}{4}$ **3.** $0, \frac{6}{7}$

Practice 4

Solve: $x^2 - 8x - 48 = 0$.

Example 4 Solve: $x^2 - 9x - 22 = 0$

Solution: One side of the equation is 0. However, to use the zero factor property, one side of the equation must be 0 *and* the other side must be written as a product (must be factored). Thus, we must first factor this polynomial.

$$x^2 - 9x - 22 = 0$$
$$(x - 11)(x + 2) = 0 \quad \text{Factor.}$$

Now we can apply the zero factor property.

$$x - 11 = 0 \quad \text{or} \quad x + 2 = 0$$
$$x = 11 \qquad\qquad x = -2$$

Check: Let $x = 11$.

$$x^2 - 9x - 22 = 0$$
$$11^2 - 9 \cdot 11 - 22 \stackrel{?}{=} 0$$
$$121 - 99 - 22 \stackrel{?}{=} 0$$
$$22 - 22 \stackrel{?}{=} 0$$
$$0 = 0 \quad \text{True}$$

Let $x = -2$.

$$x^2 - 9x - 22 = 0$$
$$(-2)^2 - 9(-2) - 22 \stackrel{?}{=} 0$$
$$4 + 18 - 22 \stackrel{?}{=} 0$$
$$22 - 22 \stackrel{?}{=} 0$$
$$0 = 0 \quad \text{True}$$

The solutions are 11 and −2.

Work Practice 4

Practice 5

Solve: $9x^2 - 24x = -16$.

Example 5 Solve: $4x^2 - 28x = -49$

Solution: First we rewrite the equation in standard form so that one side is 0. Then we factor the polynomial.

$$4x^2 - 28x = -49$$
$$4x^2 - 28x + 49 = 0 \quad \text{Write in standard form by adding 49 to both sides.}$$
$$(2x - 7)(2x - 7) = 0 \quad \text{Factor.}$$

Next we use the zero factor property and set each factor equal to 0. Since the factors are the same, the related equations will give the same solution.

$$2x - 7 = 0 \quad \text{or} \quad 2x - 7 = 0 \quad \text{Set each factor equal to 0.}$$
$$2x = 7 \qquad\qquad 2x = 7 \quad \text{Solve.}$$
$$x = \frac{7}{2} \qquad\qquad x = \frac{7}{2}$$

Check: Although $\frac{7}{2}$ occurs twice, there is a single solution. Check this solution in the original equation. The solution is $\frac{7}{2}$.

Work Practice 5

Answers

4. $-4, 12$ **5.** $\frac{4}{3}$

The following steps may be used to solve a quadratic equation by factoring.

To Solve Quadratic Equations by Factoring

Step 1: Write the equation in standard form so that one side of the equation is 0.

Step 2: Factor the quadratic expression completely.

Step 3: Set each factor containing a variable equal to 0.

Step 4: Solve the resulting equations.

Step 5: Check each solution in the original equation.

Since it is not always possible to factor a quadratic polynomial, not all quadratic equations can be solved by factoring. Other methods of solving quadratic equations are presented in Chapter 18.

Example 6 Solve: $x(2x - 7) = 4$

Solution: First we write the equation in standard form; then we factor.

$$x(2x - 7) = 4$$
$$2x^2 - 7x = 4 \quad \text{Multiply.}$$
$$2x^2 - 7x - 4 = 0 \quad \text{Write in standard form.}$$
$$(2x + 1)(x - 4) = 0 \quad \text{Factor.}$$
$$2x + 1 = 0 \quad \text{or} \quad x - 4 = 0 \quad \text{Set each factor equal to zero.}$$
$$2x = -1 \qquad x = 4 \quad \text{Solve.}$$
$$x = -\frac{1}{2}$$

Check the solutions in the original equation. The solutions are $-\frac{1}{2}$ and 4.

Work Practice 6

Practice 6

Solve: $x(3x + 7) = 6$.

Helpful Hint To solve the equation $x(2x - 7) = 4$, **do not** set each factor equal to 4. Remember that to apply the zero factor property, one side of the equation must be 0 and the other side of the equation must be in factored form.

✓Concept Check Explain the error and solve the equation correctly.

$$(x - 3)(x + 1) = 5$$
$$x - 3 = 5 \quad \text{or} \quad x + 1 = 5$$
$$x = 8 \quad \text{or} \quad x = 4$$

Example 7 Solve: $-2x^2 - 4x + 30 = 0$.

Solution: The equation is in standard form, so we begin by factoring out a common factor of -2.

$$-2x^2 - 4x + 30 = 0$$
$$-2(x^2 + 2x - 15) = 0 \quad \text{Factor out } -2.$$
$$-2(x + 5)(x - 3) = 0 \quad \text{Factor the quadratic.}$$

Next, set each factor **containing a variable** equal to 0.

$$x + 5 = 0 \quad \text{or} \quad x - 3 = 0 \quad \text{Set each factor containing a variable equal to 0.}$$
$$x = -5 \quad \text{or} \quad x = 3 \quad \text{Solve.}$$

Note: The factor -2 is a constant term containing no variables and can never equal 0. The solutions are -5 and 3.

Work Practice 7

Practice 7

Solve: $-3x^2 - 6x + 72 = 0$.

Answers

6. $-3, \frac{2}{3}$ **7.** $-6, 4$

✓Concept Check Answer

To use the zero factor property, one side of the equation must be 0, not 5. Correctly, $(x - 3)(x + 1) = 5$, $x^2 - 2x - 3 = 5$, $x^2 - 2x - 8 = 0$, $(x - 4)(x + 2) = 0$, $x - 4 = 0$ or $x + 2 = 0$, $x = 4$ or $x = -2$.

Objective 2 Solving Equations with Degree Greater than Two by Factoring

Some equations involving polynomials of degree higher than 2 may also be solved by factoring and then applying the zero factor theorem.

Practice 8

Solve: $7x^3 - 63x = 0$.

Example 8 Solve: $3x^3 - 12x = 0$.

Solution: Factor the left side of the equation. Begin by factoring out the common factor of $3x$.

$$3x^3 - 12x = 0$$

$$3x(x^2 - 4) = 0 \quad \text{Factor out the GCF, } 3x.$$

$$3x(x + 2)(x - 2) = 0 \quad \text{Factor } x^2 - 4 \text{, a difference of squares.}$$

$$3x = 0 \quad \text{or} \quad x + 2 = 0 \quad \text{or} \quad x - 2 = 0 \quad \text{Set each factor equal to 0.}$$

$$x = 0 \quad \text{or} \quad x = -2 \quad \text{or} \quad x = 2 \quad \text{Solve.}$$

Thus, the equation $3x^3 - 12x = 0$ has three solutions: 0, −2, and 2. To check, replace x with each solution in the original equation.

Let x = 0.	***Let x = −2.***	***Let x = 2.***
$3(0)^3 - 12(0) \stackrel{?}{=} 0$	$3(-2)^3 - 12(-2) \stackrel{?}{=} 0$	$3(2)^3 - 12(2) \stackrel{?}{=} 0$
$0 = 0$	$3(-8) + 24 \stackrel{?}{=} 0$	$3(8) - 24 \stackrel{?}{=} 0$
	$0 = 0$	$0 = 0$

Substituting 0, −2, or 2 into the original equation results each time in a true equation. The solutions are 0, −2, and 2.

Work Practice 8

Practice 9

Solve: $(3x - 2)(2x^2 - 13x + 15) = 0$.

Example 9 Solve: $(5x - 1)(2x^2 + 15x + 18) = 0$.

Solution:

$$(5x - 1)(2x^2 + 15x + 18) = 0$$

$$(5x - 1)(2x + 3)(x + 6) = 0 \quad \text{Factor the trinomial.}$$

$$5x - 1 = 0 \quad \text{or} \quad 2x + 3 = 0 \quad \text{or} \quad x + 6 = 0 \quad \text{Set each factor equal to 0.}$$

$$5x = 1 \quad \text{or} \quad 2x = -3 \quad \text{or} \quad x = -6 \quad \text{Solve.}$$

$$x = \frac{1}{5} \quad \text{or} \quad x = -\frac{3}{2}$$

The solutions are $\frac{1}{5}$, $-\frac{3}{2}$, and −6. Check by replacing x with each solution in the original equation. The solutions are −6, $-\frac{3}{2}$, and $\frac{1}{5}$.

Work Practice 9

Answers

8. −3, 0, 3 **9.** $\frac{2}{3}, \frac{3}{2}, 5$

Example 10 Solve: $2x^3 - 4x^2 - 30x = 0$.

Solution: Begin by factoring out the GCF, $2x$.

$$2x^3 - 4x^2 - 30x = 0$$

$$2x(x^2 - 2x - 15) = 0 \quad \text{Factor out the GCF, } 2x.$$

$$2x(x - 5)(x + 3) = 0 \quad \text{Factor the quadratic.}$$

$$2x = 0 \quad \text{or} \quad x - 5 = 0 \quad \text{or} \quad x + 3 = 0 \quad \text{Set each factor containing a variable equal to 0.}$$

$$x = 0 \quad \text{or} \quad x = 5 \quad \text{or} \quad x = -3 \quad \text{Solve.}$$

Check by replacing x with each solution in the cubic equation. The solutions are -3, 0, and 5.

Work Practice 10

Practice 10

Solve: $5x^3 + 5x^2 - 30x = 0$.

Objective 3 Finding x-Intercepts of the Graph of a Quadratic Equation

In Chapter 10, we graphed linear equations in two variables, such as $y = 5x - 6$. Recall that to find the x-intercept of the graph of a linear equation, let $y = 0$ and solve for x. This is also how to find the x-intercepts of the graph of a **quadratic equation in two variables,** such as $y = x^2 - 5x + 4$.

Example 11 Find the x-intercepts of the graph of $y = x^2 - 5x + 4$.

Solution: Let $y = 0$ and solve for x.

$$y = x^2 - 5x + 4$$

$$0 = x^2 - 5x + 4 \quad \text{Let } y = 0.$$

$$0 = (x - 1)(x - 4) \quad \text{Factor.}$$

$$x - 1 = 0 \quad \text{or} \quad x - 4 = 0 \quad \text{Set each factor equal to 0.}$$

$$x = 1 \quad \text{or} \quad x = 4 \quad \text{Solve.}$$

The x-intercepts of the graph of $y = x^2 - 5x + 4$ are $(1, 0)$ and $(4, 0)$.

The graph of $y = x^2 - 5x + 4$ is shown in the margin.

Work Practice 11

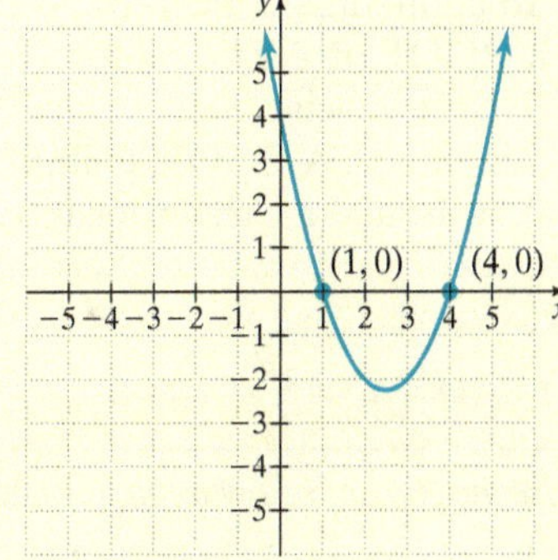

Practice 11

Find the x-intercepts of the graph of $y = x^2 - 6x + 8$.

In general, a quadratic equation in two variables is one that can be written in the form $y = ax^2 + bx + c$ where $a \neq 0$. The graph of such an equation is called a **parabola** and will open up or down depending on the sign of a.

Notice that the x-intercepts of the graph of $y = ax^2 + bx + c$ are the real number solutions of $0 = ax^2 + bx + c$. Also, the real number solutions of $0 = ax^2 + bx + c$ are the x-intercepts of the graph of $y = ax^2 + bx + c$. We study more about graphs of quadratic equations in two variables in Chapter 18.

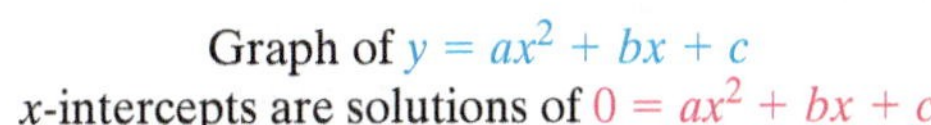

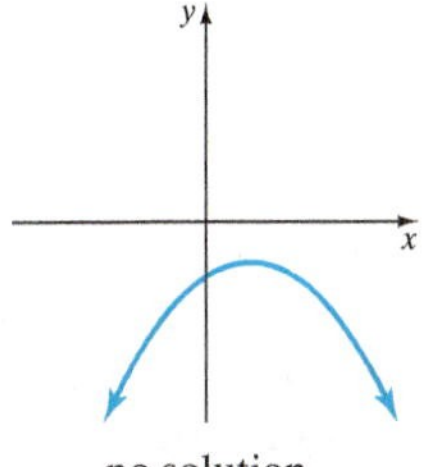
no solution

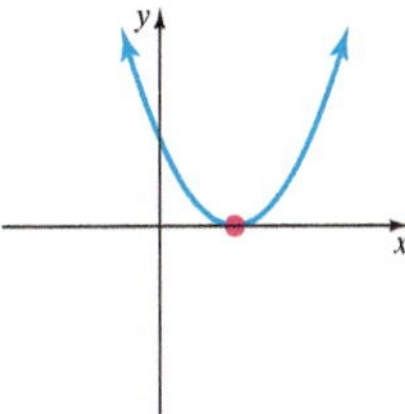
1 solution

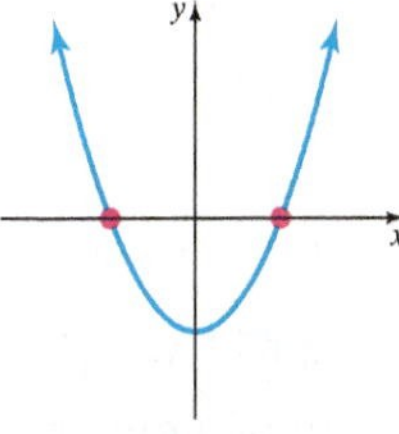
2 solutions

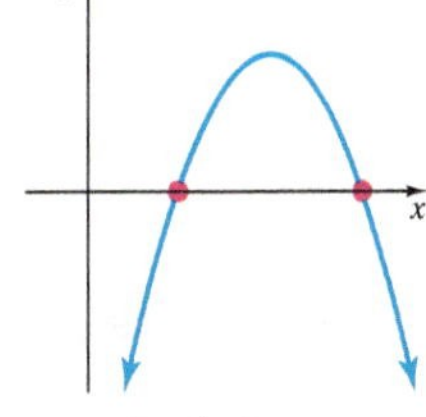
2 solutions

Answers

10. $-3, 0, 2$

11. The x-intercepts are $(2, 0)$ and $(4, 0)$.

Graphing Calculator Explorations

A grapher may be used to find solutions of a quadratic equation whether the related quadratic polynomial is factorable or not. For example, let's use a grapher to approximate the solutions of $0 = x^2 + 4x - 3$. To do so, graph $y_1 = x^2 + 4x - 3$. Recall that the x-intercepts of this graph are the solutions of $0 = x^2 + 4x - 3$.

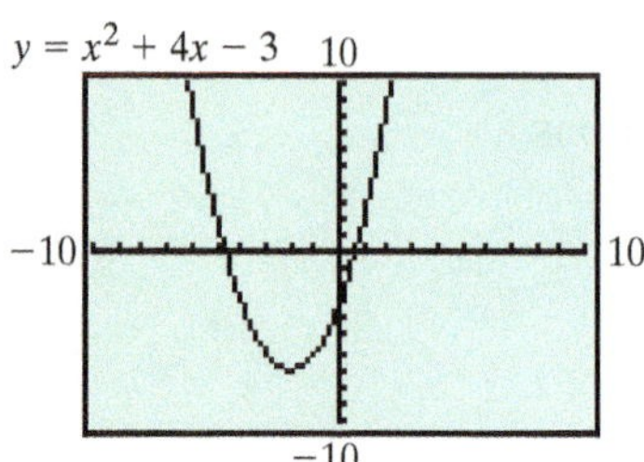

Notice that the graph appears to have an x-intercept between -5 and -4 and one between 0 and 1. Many graphers contain a TRACE feature. This feature activates a graph cursor that can be used to *trace* along a graph while the corresponding x- and y-coordinates are shown on the screen. Use the TRACE feature to confirm that x-intercepts lie between -5 and -4 and between 0 and 1. To approximate the x-intercepts to the nearest tenth, use a ROOT or a ZOOM feature on your grapher or redefine the viewing window. (A ROOT feature calculates the x-intercept. A ZOOM feature magnifies the viewing window around a specific location such as the graph cursor.)

If we redefine the window to [0, 1] on the x-axis and [−1, 1] on the y-axis, the graph to the left is generated.

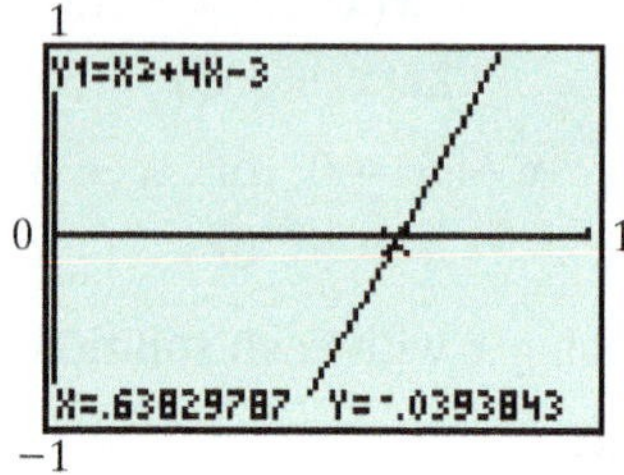

By using the TRACE feature, we can conclude that one x-intercept is approximately 0.6 to the nearest tenth. By repeating these steps for the other x-intercept, we find that it is approximately -4.6.

Use a grapher to approximate the real number solutions to the nearest tenth. If an equation has no real number solution, state so.

1. $3x^2 - 4x - 6 = 0$
2. $x^2 - x - 9 = 0$
3. $2x^2 + x + 2 = 0$
4. $-4x^2 - 5x - 4 = 0$
5. $-x^2 + x + 5 = 0$
6. $10x^2 + 6x - 3 = 0$

Vocabulary, Readiness & Video Check

Use the choices below to fill in each blank. Not all choices will be used.

$-3, 5$	$a = 0$ or $b = 0$	0	linear
$3, -5$	quadratic	1	

1. An equation that can be written in the form $ax^2 + bx + c = 0$, with $a \neq 0$, is called a ______________ equation.
2. If the product of two numbers is 0, then at least one of the numbers must be ______________.
3. The solutions to $(x - 3)(x + 5) = 0$ are ______________.
4. If $a \cdot b = 0$, then ______________.

Martin-Gay Interactive Videos

See Video 13.6

Watch the section lecture video and answer the following questions.

Objective 1 5. As shown in Examples 1–3, what two things have to be true in order to use the zero factor theorem?

Objective 2 6. Example 4 implies that the zero factor theorem can be used with any number of factors on one side of the equation so long as the other side of the equation is zero. Why do you think this is true?

Objective 3 7. From Example 5, how does finding the x-intercepts of the graph of a quadratic equation in two variables lead to solving a quadratic equation?

13.6 Exercise Set MyMathLab® 

Solve each equation. See Examples 1 through 3.

1. $(x - 6)(x - 7) = 0$
2. $(x - 10)(x - 5) = 0$
3. $(x - 2)(x + 1) = 0$
4. $(x + 4)(x - 10) = 0$
5. $(x + 9)(x + 17) = 0$
6. $(x + 11)(x + 1) = 0$
7. $x(x + 6) = 0$
8. $x(x - 7) = 0$
9. $3x(x - 8) = 0$
10. $2x(x + 12) = 0$
11. $(2x + 3)(4x - 5) = 0$
12. $(3x - 2)(5x + 1) = 0$
13. $(2x - 7)(7x + 2) = 0$
14. $(9x + 1)(4x - 3) = 0$
15. $\left(x - \frac{1}{2}\right)\left(x + \frac{1}{3}\right) = 0$
16. $\left(x + \frac{2}{9}\right)\left(x - \frac{1}{4}\right) = 0$
17. $(x + 0.2)(x + 1.5) = 0$
18. $(x + 1.7)(x + 2.3) = 0$

Solve. See Examples 4 through 7.

19. $x^2 - 13x + 36 = 0$
20. $x^2 + 2x - 63 = 0$
21. $x^2 + 2x - 8 = 0$
22. $x^2 - 5x + 6 = 0$
23. $x^2 - 7x = 0$
24. $x^2 - 3x = 0$
25. $x^2 - 4x = 32$
26. $x^2 - 5x = 24$
27. $x^2 = 16$
28. $x^2 = 9$
29. $(x + 4)(x - 9) = 4x$
30. $(x + 3)(x + 8) = x$
31. $x(3x - 1) = 14$
32. $x(4x - 11) = 3$
33. $-3x^2 + 75 = 0$
34. $-2y^2 + 72 = 0$
35. $24x^2 + 44x = 8$
36. $6x^2 + 57x = 30$

Solve each equation. See Examples 8 through 10.

37. $x^3 - 12x^2 + 32x = 0$
38. $x^3 - 14x^2 + 49x = 0$
39. $(4x - 3)(16x^2 - 24x + 9) = 0$
40. $(2x + 5)(4x^2 + 20x + 25) = 0$
41. $4x^3 - x = 0$
42. $4y^3 - 36y = 0$
43. $32x^3 - 4x^2 - 6x = 0$
44. $15x^3 + 24x^2 - 63x = 0$

MIXED PRACTICE

Solve each equation. See Examples 1 through 10. (A few exercises are linear equations.)

45. $(x + 3)(x - 2) = 0$
46. $(x - 6)(x + 7) = 0$
47. $x^2 + 20x = 0$
48. $x^2 + 15x = 0$
49. $4(x - 7) = 6$
50. $5(3 - 4x) = 9$
51. $4y^2 - 1 = 0$
52. $4y^2 - 81 = 0$
53. $(2x + 3)(2x^2 - 5x - 3) = 0$
54. $(2x - 9)(x^2 + 5x - 36) = 0$
55. $x^2 - 15 = -2x$
56. $x^2 - 26 = -11x$
57. $30x^2 - 11x - 30 = 0$
58. $12x^2 + 7x - 12 = 0$
59. $5x^2 - 6x - 8 = 0$
60. $9x^2 + 7x = 2$
61. $6y^2 - 22y - 40 = 0$
62. $3x^2 - 6x - 9 = 0$
63. $(y - 2)(y + 3) = 6$
64. $(y - 5)(y - 2) = 28$
65. $3x^3 + 19x^2 - 72x = 0$
66. $36x^3 + x^2 - 21x = 0$
67. $x^2 + 14x + 49 = 0$
68. $x^2 + 22x + 121 = 0$
69. $12y = 8y^2$
70. $9y = 6y^2$
71. $7x^3 - 7x = 0$
72. $3x^3 - 27x = 0$
73. $3x^2 + 8x - 11 = 13 - 6x$
74. $2x^2 + 12x - 1 = 4 + 3x$
75. $3x^2 - 20x = -4x^2 - 7x - 6$
76. $4x^2 - 20x = -5x^2 - 6x - 5$

Find the x-intercepts of the graph of each equation. See Example 11.

77. $y = (3x + 4)(x - 1)$
78. $y = (5x - 3)(x - 4)$
79. $y = x^2 - 3x - 10$
80. $y = x^2 + 7x + 6$
81. $y = 2x^2 + 11x - 6$
82. $y = 4x^2 + 11x + 6$

For Exercises 83 through 88, match each equation with its graph. See Example 11.

83. $y = (x + 2)(x - 1)$ **84.** $y = (x - 5)(x + 2)$

85. $y = x(x + 3)$ **86.** $y = x(x - 4)$

87. $y = 2x^2 - 8$ **88.** $y = 2x^2 - 2$

a.

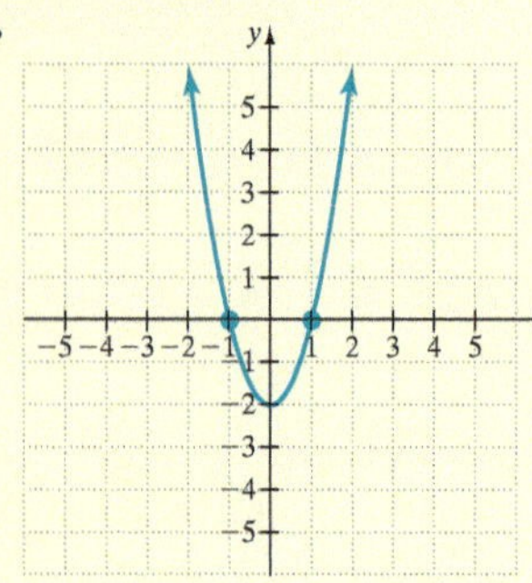

b.

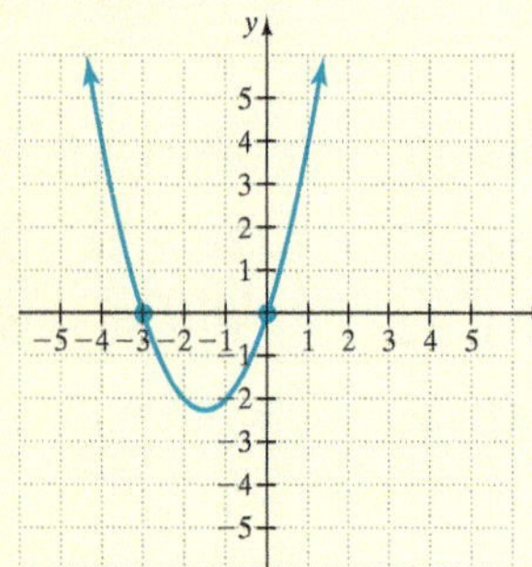

c.

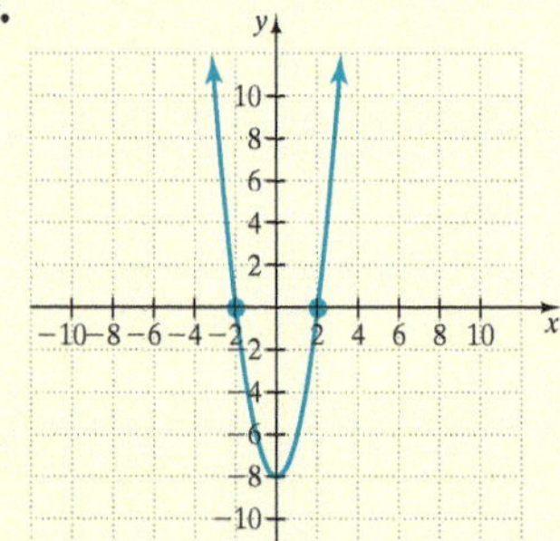

d.

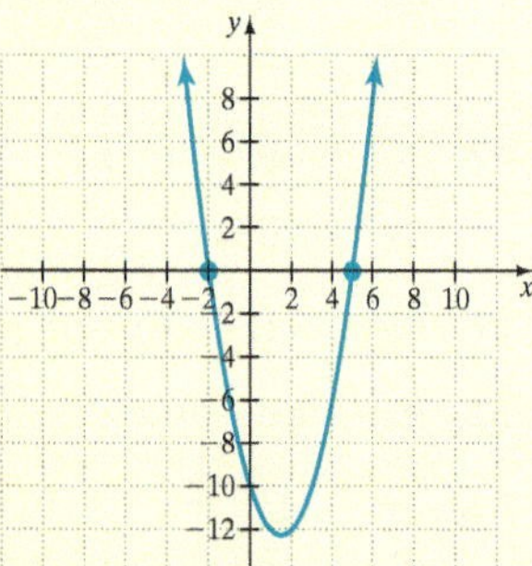

e.

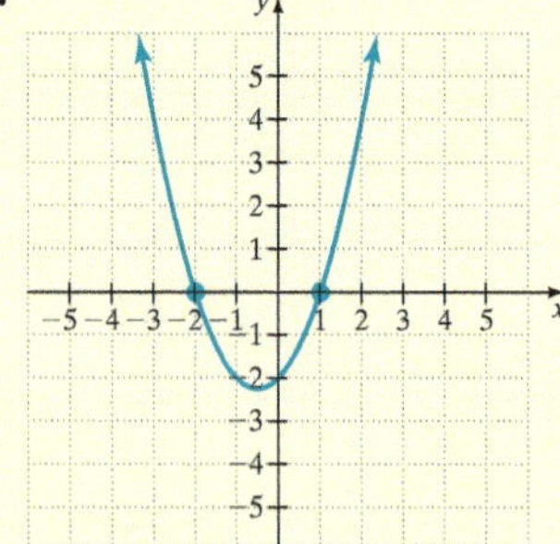

f.

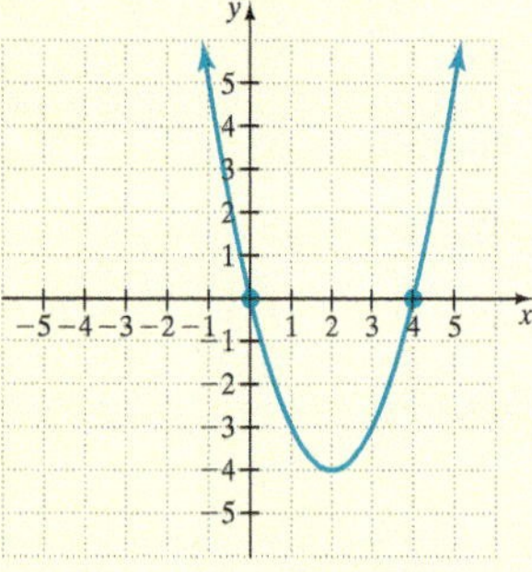

REVIEW AND PREVIEW

Perform the following operations. Write all results in lowest terms. See Sections 4.3, 4.4, and 4.5.

89. $\frac{3}{5} + \frac{4}{9}$ **90.** $\frac{2}{3} + \frac{3}{7}$

91. $\frac{7}{10} - \frac{5}{12}$ **92.** $\frac{5}{9} - \frac{5}{12}$

93. $\frac{7}{8} \div \frac{7}{15}$ **94.** $\frac{5}{12} - \frac{3}{10}$

95. $\frac{4}{5} \cdot \frac{7}{8}$ **96.** $\frac{3}{7} \cdot \frac{12}{17}$

CONCEPT EXTENSIONS

For Exercises 97 and 98, see the Concept Check in this section.

97. Explain the error and solve correctly:

$$x(x - 2) = 8$$
$$x = 8 \quad \text{or} \quad x - 2 = 8$$
$$x = 10$$

98. Explain the error and solve correctly:

$$(x - 4)(x + 2) = 0$$
$$x = -4 \quad \text{or} \quad x = 2$$

99. Write a quadratic equation that has two solutions, 6 and −1. Leave the polynomial in the equation in factored form.

100. Write a quadratic equation that has two solutions, 0 and −2. Leave the polynomial in the equation in factored form.

101. Write a quadratic equation in standard form that has two solutions, 5 and 7.

102. Write an equation that has three solutions, 0, 1, and 2.

103. A compass is accidentally thrown upward and out of an air balloon at a height of 300 feet. The height, y, of the compass at time x in seconds is given by the equation

$$y = -16x^2 + 20x + 300$$

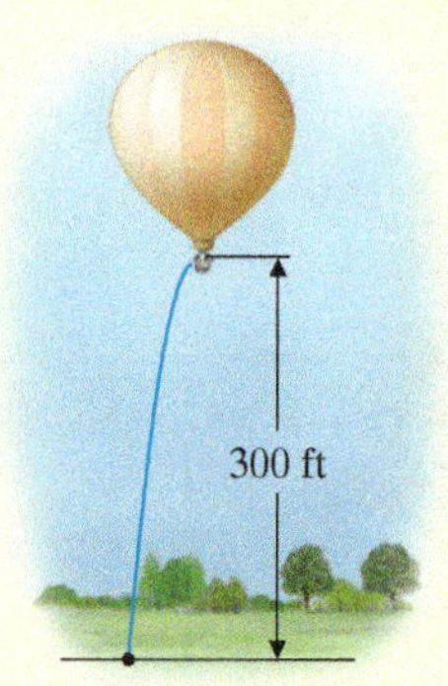

a. Find the height of the compass at the given times by filling in the table below.

time, x	0	1	2	3	4	5	6
height, y							

b. Use the table to determine when the compass strikes the ground.

c. Use the table to approximate the maximum height of the compass.

d. Plot the points (x, y) on a rectangular coordinate system and connect them with a smooth curve. Explain your results.

104. A rocket is fired upward from the ground with an initial velocity of 100 feet per second. The height, y, of the rocket at any time x is given by the equation

$$y = -16x^2 + 100x$$

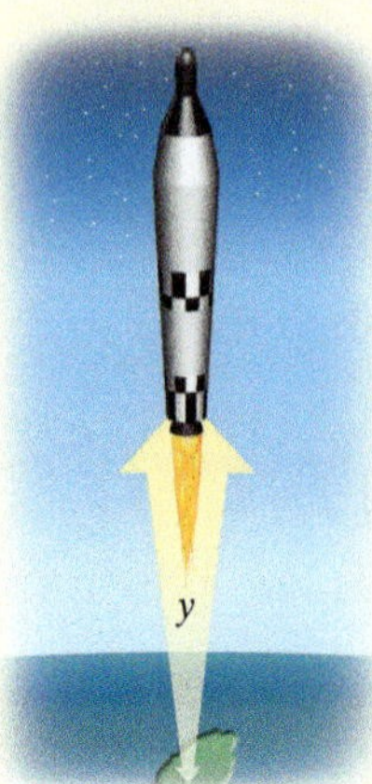

a. Find the height of the rocket at the given times by filling in the table below.

time, x	0	1	2	3	4	5	6	7
height, y								

b. Use the table to approximate when the rocket strikes the ground to the nearest second.

c. Use the table to approximate the maximum height of the rocket.

d. Plot the points (x, y) on a rectangular coordinate system and connect them with a smooth curve. Explain your results.

Solve each equation. First, multiply the binomial.

To solve $(x - 6)(2x - 3) = (x + 2)(x + 9)$, see below.

$$(x - 6)(2x - 3) = (x + 2)(x + 9)$$
$$2x^2 - 15x + 18 = x^2 + 11x + 18$$
$$x^2 - 26x = 0$$
$$x(x - 26) = 0$$
$$x = 0 \quad \text{or} \quad x - 26 = 0$$
$$x = 26$$

105. $(x - 3)(3x + 4) = (x + 2)(x - 6)$

106. $(2x - 3)(x + 6) = (x - 9)(x + 2)$

107. $(2x - 3)(x + 8) = (x - 6)(x + 4)$

108. $(x + 6)(x - 6) = (2x - 9)(x + 4)$

13.7 Quadratic Equations and Problem Solving

Objective

1 Solve Problems That Can Be Modeled by Quadratic Equations.

Objective 1 Solving Problems Modeled by Quadratic Equations

Some problems may be modeled by quadratic equations. To solve these problems, we use the same problem-solving steps that were introduced in Section 3.4. When solving these problems, keep in mind that a solution of an equation that models a problem may not be a solution to the problem. For example, a person's age or the length of a rectangle is always a positive number. Discard solutions that do not make sense as solutions of the problem.

Example 1 Finding Free-Fall Time

Since the 1940s, one of the top tourist attractions in Acapulco, Mexico, is watching the cliff divers off La Quebrada. The divers' platform is about 144 feet above the sea. These divers must time their descent just right, since they land in the crashing Pacific, in an inlet that is at most $9\frac{1}{2}$ feet deep. Neglecting air resistance, the height h in feet of a cliff diver above the ocean after t seconds is given by the quadratic equation $h = -16t^2 + 144$.

Find out how long it takes the diver to reach the ocean.

Solution:

1. UNDERSTAND. Read and reread the problem. Then draw a picture of the problem.

The equation $h = -16t^2 + 144$ models the height of the falling diver at time t. Familiarize yourself with this equation by finding the height of the diver at time $t = 1$ second and $t = 2$ seconds.

144 feet

When $t = 1$ second, the height of the diver is $h = -16(1)^2 + 144 = 128$ feet.

When $t = 2$ seconds, the height of the diver is $h = -16(2)^2 + 144 = 80$ feet.

(Continued on next page)

Practice 1

Cliff divers also frequent the falls at Waimea Falls Park in Oahu, Hawaii. One of the popular diving spots is 64 feet high. Neglecting air resistance, the height of a diver above the pool after t seconds is $h = -16t^2 + 64$. Find how long it takes a diver to reach the pool.

Answer

1. 2 sec

2. TRANSLATE. To find out how long it takes the diver to reach the ocean, we want to know the value of t for which $h = 0$.
3. SOLVE.

$$0 = -16t^2 + 144$$
$$0 = -16(t^2 - 9) \quad \text{Factor out } -16.$$
$$0 = -16(t - 3)(t + 3) \quad \text{Factor completely.}$$
$$t - 3 = 0 \quad \text{or} \quad t + 3 = 0 \quad \text{Set each factor containing a variable equal to 0.}$$
$$t = 3 \quad \text{or} \quad t = -3 \quad \text{Solve.}$$

4. INTERPRET. Since the time t cannot be negative, the proposed solution is 3 seconds.

Check: Verify that the height of the diver when t is 3 seconds is 0.

When $t = 3$ seconds, $h = -16(3)^2 + 144 = -144 + 144 = 0$.

State: It takes the diver 3 seconds to reach the ocean.

Work Practice 1

Practice 2

The square of a number minus eight times the number is equal to forty-eight. Find the number.

Example 2 Finding an Unknown Number

The square of a number plus three times the number is 70. Find the number.

Solution:

1. UNDERSTAND. Read and reread the problem. Suppose that the number is 5. The square of 5 is 5^2 or 25. Three times 5 is 15. Then $25 + 15 = 40$, not 70, so the number is not 5. Remember, the purpose of proposing a number, such as 5, is to understand the problem better. Now that we do, we will let $x =$ the number.
2. TRANSLATE.

the square of a number	plus	three times the number	is	70
↓	↓	↓	↓	↓
x^2	$+$	$3x$	$=$	70

3. SOLVE.

$$x^2 + 3x = 70$$
$$x^2 + 3x - 70 = 0 \quad \text{Subtract 70 from both sides.}$$
$$(x + 10)(x - 7) = 0 \quad \text{Factor.}$$
$$x + 10 = 0 \quad \text{or} \quad x - 7 = 0 \quad \text{Set each factor equal to 0.}$$
$$x = -10 \quad \quad x = 7 \quad \text{Solve.}$$

4. INTERPRET.

Check: The square of -10 is $(-10)^2$, or 100. Three times -10 is $3(-10)$ or -30. Then $100 + (-30) = 70$, the correct sum, so -10 checks.

The square of 7 is 7^2 or 49. Three times 7 is 3(7), or 21. Then $49 + 21 = 70$, the correct sum, so 7 checks.

State: There are two numbers. They are -10 and 7.

Work Practice 2

Answer

2. There are two numbers. They are -4 and 12.

Example 3 Finding the Dimensions of a Sail

The height of a triangular sail is 2 meters less than twice the length of the base. If the sail has an area of 30 square meters, find the length of its base and the height.

Solution:

1. UNDERSTAND. Read and reread the problem. Since we are finding the length of the base and the height, we let

$x =$ the length of the base

and since the height is 2 meters less than twice the base,

$2x - 2 =$ the height

An illustration is shown to the right.

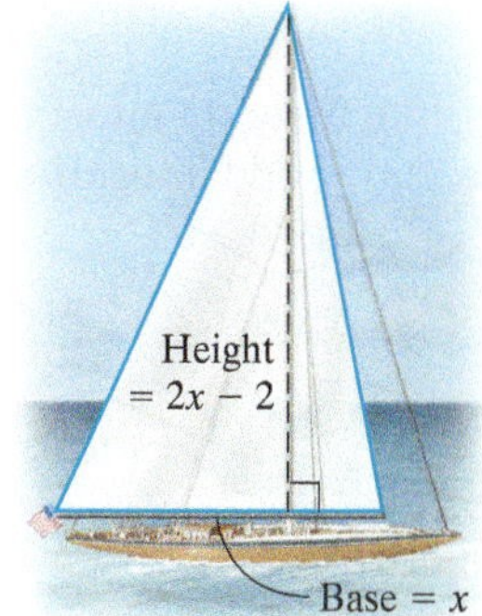

2. TRANSLATE. We are given that the area of the triangle is 30 square meters, so we use the formula for area of a triangle.

area of triangle	=	$\frac{1}{2}$	·	base	·	height
↓		↓		↓		↓
30	=	$\frac{1}{2}$	·	x	·	$(2x - 2)$

3. SOLVE. Now we solve the quadratic equation.

$$30 = \frac{1}{2}x(2x - 2)$$

$$30 = x^2 - x \quad \text{Multiply.}$$

$$x^2 - x - 30 = 0 \quad \text{Write in standard form.}$$

$$(x - 6)(x + 5) = 0 \quad \text{Factor.}$$

$$x - 6 = 0 \text{ or } x + 5 = 0 \quad \text{Set each factor equal to 0.}$$

$$x = 6 \qquad x = -5$$

4. INTERPRET. Since x represents the length of the base, we discard the solution -5. The base of a triangle cannot be negative. The base is then 6 meters and the height is $2(6) - 2 = 10$ meters.

Check: To check this problem, we recall that $\frac{1}{2}\text{base} \cdot \text{height} = \text{area}$, or

$$\frac{1}{2}(6)(10) = 30 \quad \text{The required area}$$

State: The base of the triangular sail is 6 meters and the height is 10 meters.

Work Practice 3

Practice 3

An engineering team from Georgia Tech earned second place in a flight competition, with their triangular-shaped paper hang glider. The base of their prize-winning entry was 1 foot less than three times the height. If the area of the triangular glider wing was 210 square feet, find the dimensions of the wing. *(Source: The Technique* [Georgia Tech's newspaper], April 18, 2003)

Answer

3. base: 35 ft; height: 12 ft

Study the following diagrams for a review of consecutive integers.

Examples

If x is the first integer, then consecutive integers are $x, x + 1, x + 2, \ldots$

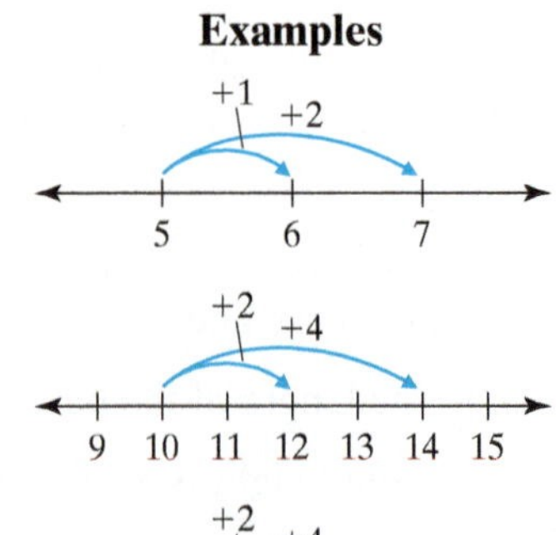

If x is the first even integer, then consecutive even integers are $x, x + 2, x + 4, \ldots$

If x is the first odd integer, then consecutive odd integers are $x, x + 2, x + 4, \ldots$

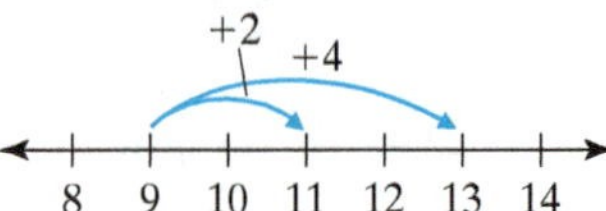

Practice 4

Find two consecutive integers whose product is 41 more than their sum.

Example 4 Finding Consecutive Even Integers

Find two consecutive even integers whose product is 34 more than their sum.

Solution:

1. UNDERSTAND. Read and reread the problem. Let's just choose two consecutive even integers to help us better understand the problem. Let's choose 10 and 12. Their product is $10(12) = 120$ and their sum is $10 + 12 = 22$. The product is $120 - 22$, or 98 greater than the sum. Thus our guess is incorrect, but we have a better understanding of this example.

   Let's let x and $x + 2$ be the consecutive even integers.

2. TRANSLATE.

Product of integers	is	34	more than	sum of integers
↓	↓	↓	↓	↓
$x(x+2)$	$=$	34	$+$	$x + (x+2)$

3. SOLVE. Now we solve the equation.

$$x(x + 2) = 34 + x + (x + 2)$$

$$x^2 + 2x = 34 + x + x + 2 \quad \text{Multiply.}$$

$$x^2 + 2x = 2x + 36 \quad \text{Combine like terms.}$$

$$x^2 - 36 = 0 \quad \text{Write in standard form.}$$

$$(x + 6)(x - 6) = 0 \quad \text{Factor.}$$

$$x + 6 = 0 \quad \text{or} \quad x - 6 = 0 \quad \text{Set each factor equal to 0.}$$

$$x = -6 \qquad x = 6 \quad \text{Solve.}$$

4. INTERPRET. If $x = -6$, then $x + 2 = -6 + 2$, or -4.
   If $x = 6$, then $x + 2 = 6 + 2$, or 8.

Check: $-6, -4$

$$-6(-4) \stackrel{?}{=} 34 + (-6) + (-4)$$

$$24 \stackrel{?}{=} 34 + (-10)$$

$$24 = 24 \quad \text{True}$$

$6, 8$

$$6(8) \stackrel{?}{=} 34 + 6 + 8$$

$$48 \stackrel{?}{=} 34 + 14$$

$$48 = 48 \quad \text{True}$$

State: The two consecutive even integers are -6 and -4 or 6 and 8.

Work Practice 4

Answer

4. 7 and 8 or -6 and -5

The next example uses the **Pythagorean theorem** and consecutive integers. Before we review this theorem, recall that a **right triangle** is a triangle that contains a 90° or right angle. The **hypotenuse** of a right triangle is the side opposite the right angle and is the longest side of the triangle. The **legs** of a right triangle are the other sides of the triangle.

Pythagorean Theorem

In a right triangle, the sum of the squares of the lengths of the two legs is equal to the square of the length of the hypotenuse.

$$(\text{leg})^2 + (\text{leg})^2 = (\text{hypotenuse})^2 \quad \text{or} \quad a^2 + b^2 = c^2$$

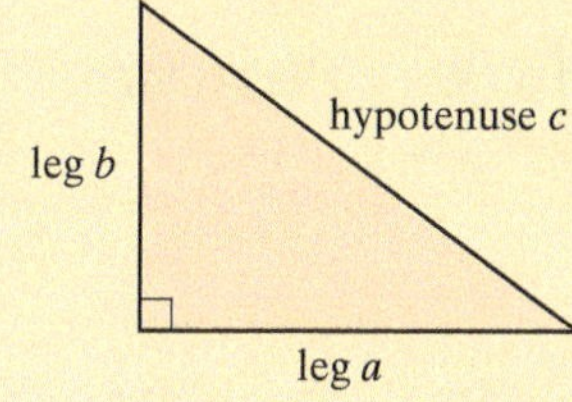

Helpful Hint If you use this formula, don't forget that c represents the length of the hypotenuse.

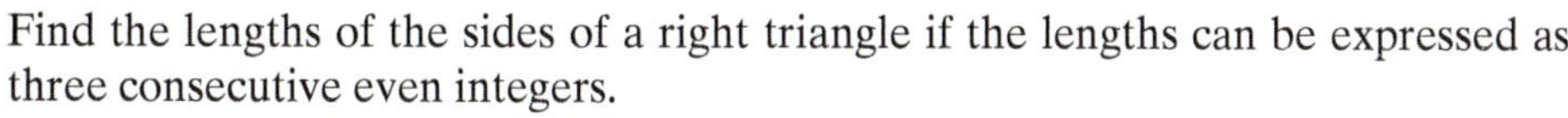

Example 5 Finding the Dimensions of a Triangle

Find the lengths of the sides of a right triangle if the lengths can be expressed as three consecutive even integers.

Solution:

1. UNDERSTAND. Read and reread the problem. Let's suppose that the length of one leg of the right triangle is 4 units. Then the other leg is the next even integer, or 6 units, and the hypotenuse of the triangle is the next even integer, or 8 units. Remember that the hypotenuse is the longest side. Let's see if a triangle with sides of these lengths forms a right triangle. To do this, we check to see whether the Pythagorean theorem holds true.

$$4^2 + 6^2 \stackrel{?}{=} 8^2$$
$$16 + 36 \stackrel{?}{=} 64$$
$$52 = 64 \quad \text{False}$$

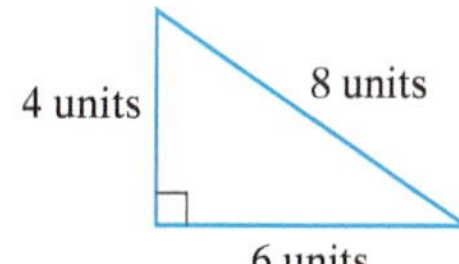

Our proposed numbers do not check, but we now have a better understanding of the problem.

We let x, $x + 2$, and $x + 4$ be three consecutive even integers. Since these integers represent lengths of the sides of a right triangle, we have the following.

$$x = \text{one leg}$$
$$x + 2 = \text{other leg}$$
$$x + 4 = \text{hypotenuse (longest side)}$$

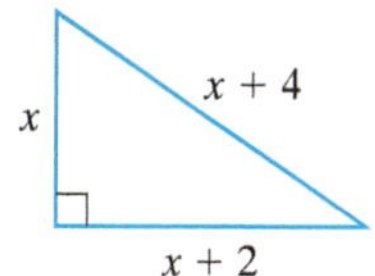

(Continued on next page)

Practice 5

Find the dimensions of a right triangle where the second leg is 1 unit less than double the first leg, and the hypotenuse is 1 unit more than double the length of the first leg.

Answer

5. leg: 8 units; leg: 15 units; hypotenuse: 17 units

2. TRANSLATE. By the Pythagorean theorem, we have that

$$(\text{leg})^2 + (\text{leg})^2 = (\text{hypotenuse})^2$$
$$(x)^2 + (x+2)^2 = (x+4)^2$$

3. SOLVE. Now we solve the equation.

$$x^2 + (x+2)^2 = (x+4)^2$$
$$x^2 + x^2 + 4x + 4 = x^2 + 8x + 16 \quad \text{Multiply.}$$
$$2x^2 + 4x + 4 = x^2 + 8x + 16 \quad \text{Combine like terms.}$$
$$x^2 - 4x - 12 = 0 \quad \text{Write in standard form.}$$
$$(x-6)(x+2) = 0 \quad \text{Factor.}$$
$$x - 6 = 0 \quad \text{or} \quad x + 2 = 0 \quad \text{Set each factor equal to 0.}$$
$$x = 6 \qquad x = -2$$

4. INTERPRET. We discard $x = -2$ since length cannot be negative. If $x = 6$, then $x + 2 = 8$ and $x + 4 = 10$.

Check: Verify that

$$(\text{leg})^2 + (\text{leg})^2 = (\text{hypotenuse})^2$$
$$6^2 + 8^2 \stackrel{?}{=} 10^2$$
$$36 + 64 \stackrel{?}{=} 100$$
$$100 = 100 \quad \text{True}$$

State: The sides of the right triangle have lengths 6 units, 8 units, and 10 units.

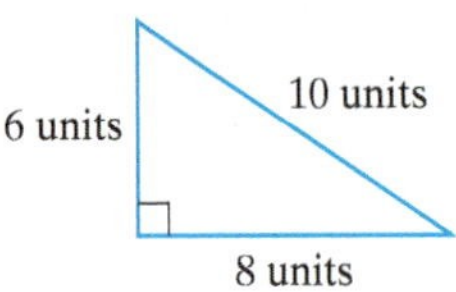

Work Practice 5

Vocabulary, Readiness & Video Check

Martin-Gay Interactive Videos

See Video 13.7

Watch the section lecture video and answer the following question.

Objective 1 **1.** In each of Examples 1–3, why aren't both solutions of the translated equation accepted as solutions to the application?

13.7 Exercise Set MyMathLab®

MIXED PRACTICE

See Examples 1 through 5 for all exercises.

TRANSLATING

For Exercises 1 through 6, represent each given condition using a single variable, x.

△ **1.** The length and width of a rectangle whose length is 4 centimeters more than its width

△ **2.** The length and width of a rectangle whose length is twice its width

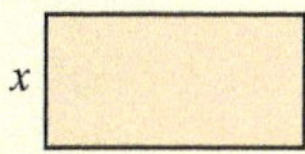

3. Two consecutive odd integers

4. Two consecutive even integers

△ **5.** The base and height of a triangle whose height is one more than four times its base

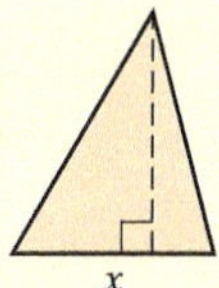

△ **6.** The base and height of a trapezoid whose base is three less than five times its height

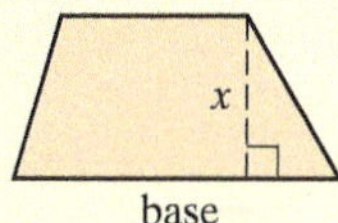

Use the information given to find the dimensions of each figure.

△ **7.** The *area* of the square is 121 square units. Find the length of its sides.

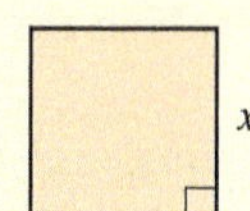

△ **8.** The *area* of the rectangle is 84 square inches. Find its length and width.

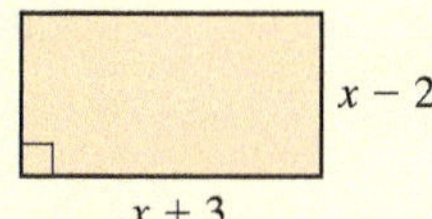

△ **9.** The *perimeter* of the quadrilateral is 120 centimeters. Find the lengths of the sides.

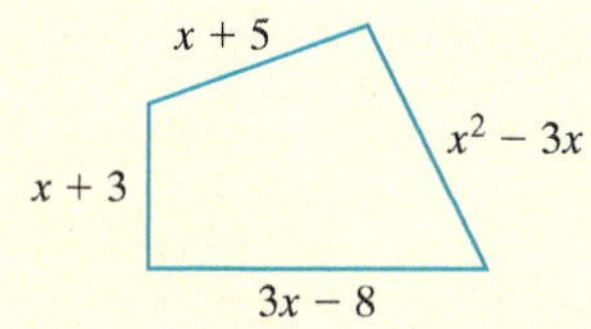

△ **10.** The *perimeter* of the triangle is 85 feet. Find the lengths of its sides.

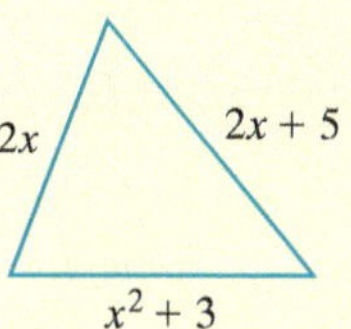

△ **11.** The *area* of the parallelogram is 96 square miles. Find its base and height.

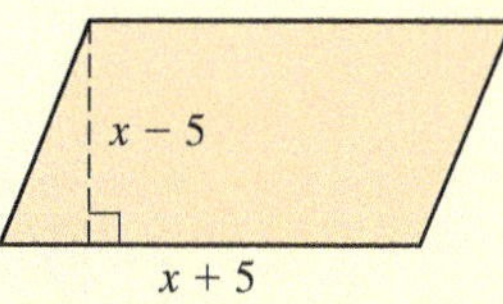

△ **12.** The *area* of the circle is 25π square kilometers. Find its radius.

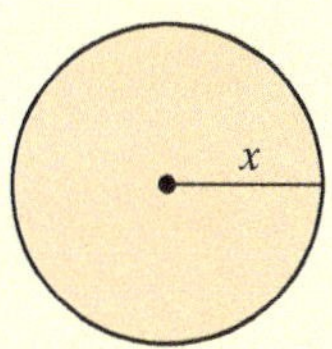

Solve.

13. An object is thrown upward from the top of an 80-foot building with an initial velocity of 64 feet per second. The height h of the object after t seconds is given by the quadratic equation $h = -16t^2 + 64t + 80$. When will the object hit the ground?

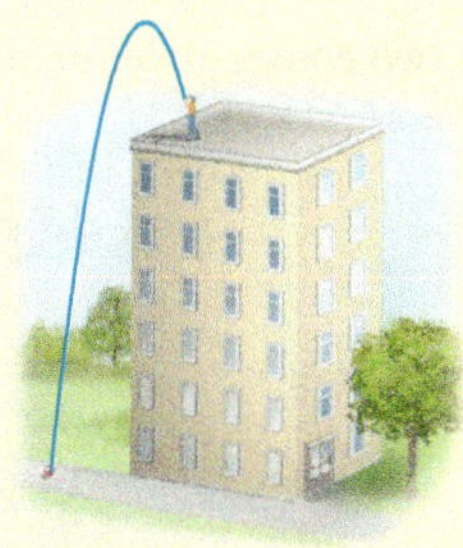

14. A hang glider pilot accidentally drops her compass from the top of a 400-foot cliff. The height h of the compass after t seconds is given by the quadratic equation $h = -16t^2 + 400$. When will the compass hit the ground?

△ **15.** The width of a rectangle is 7 centimeters less than twice its length. Its area is 30 square centimeters. Find the dimensions of the rectangle.

16. The length of a rectangle is 9 inches more than its width. Its area is 112 square inches. Find the dimensions of the rectangle.

The equation $D = \frac{1}{2}n(n - 3)$ *gives the number of diagonals* D *for a polygon with* n *sides. For example, a polygon with 6 sides has* $D = \frac{1}{2} \cdot 6(6 - 3)$ *or* $D = 9$ *diagonals. (See if you can count all 9 diagonals. Some are shown in the figure.) Use this equation,* $D = \frac{1}{2}n(n - 3)$, *for Exercises 17 through 20.*

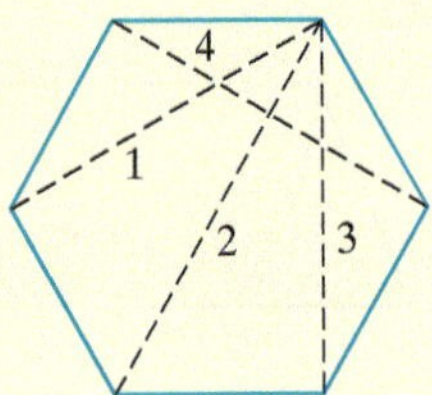

17. Find the number of diagonals for a polygon that has 12 sides.

18. Find the number of diagonals for a polygon that has 15 sides.

19. Find the number of sides n for a polygon that has 35 diagonals.

20. Find the number of sides n for a polygon that has 14 diagonals.

Solve.

21. The sum of a number and its square is 132. Find the number(s).

22. The sum of a number and its square is 182. Find the number(s).

23. The product of two consecutive room numbers is 210. Find the room numbers.

24. The product of two consecutive page numbers is 420. Find the page numbers.

25. A ladder is leaning against a building so that the distance from the ground to the top of the ladder is one foot less than the length of the ladder. Find the length of the ladder if the distance from the bottom of the ladder to the building is 5 feet.

26. Use the given figure to find the length of the guy wire.

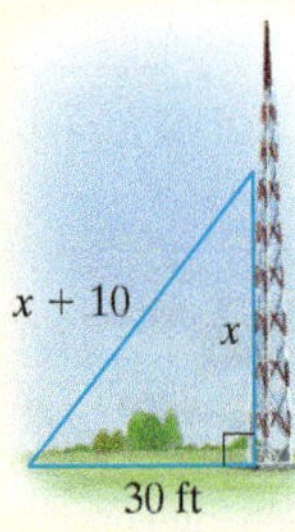

27. If the sides of a square are increased by 3 inches, the area becomes 64 square inches. Find the length of the sides of the original square.

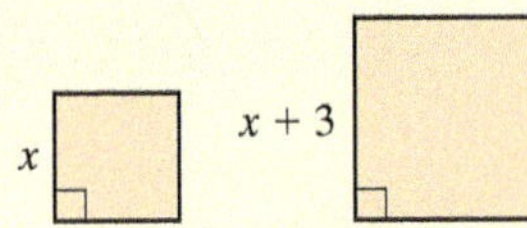

28. If the sides of a square are increased by 5 meters, the area becomes 100 square meters. Find the length of the sides of the original square.

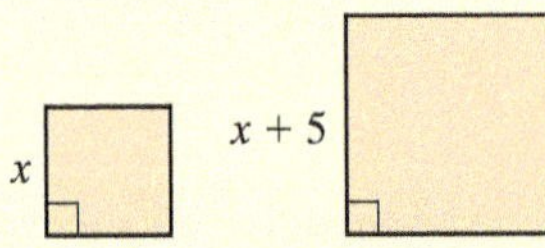

29. One leg of a right triangle is 4 millimeters longer than the smaller leg and the hypotenuse is 8 millimeters longer than the smaller leg. Find the lengths of the sides of the triangle.

30. One leg of a right triangle is 9 centimeters longer than the other leg and the hypotenuse is 45 centimeters. Find the lengths of the legs of the triangle.

31. The length of the base of a triangle is twice its height. If the area of the triangle is 100 square kilometers, find the height.

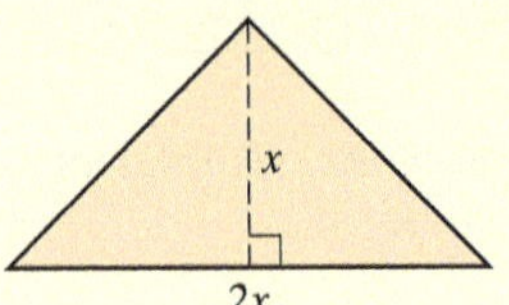

△ **32.** The height of a triangle is 2 millimeters less than the base. If the area is 60 square millimeters, find the base.

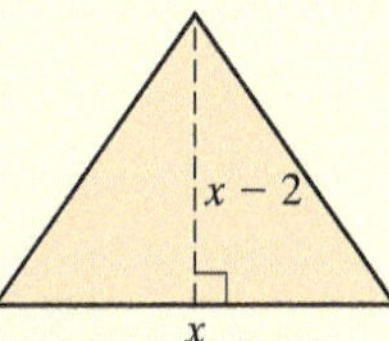

△ **33.** Find the length of the shorter leg of a right triangle if the longer leg is 12 feet more than the shorter leg and the hypotenuse is 12 feet less than twice the shorter leg.

△ **34.** Find the length of the shorter leg of a right triangle if the longer leg is 10 miles more than the shorter leg and the hypotenuse is 10 miles less than twice the shorter leg.

35. An object is dropped from 39 feet below the tip of the pinnacle atop one of the 1483-foot-tall Petronas Twin Towers in Kuala Lumpur, Malaysia. (*Source:* Council on Tall Buildings and Urban Habitat) The height h of the object after t seconds is given by the equation $h = -16t^2 + 1444$. Find how many seconds pass before the object reaches the ground.

36. An object is dropped from the top of 311 South Wacker Drive, a 961-foot-tall office building in Chicago. (*Source:* Council on Tall Buildings and Urban Habitat) The height h of the object after t seconds is given by the equation $h = -16t^2 + 961$. Find how many seconds pass before the object reaches the ground.

37. At the end of 2 years, P dollars invested at an interest rate r compounded annually increases to an amount, A dollars, given by

$$A = P(1 + r)^2$$

Find the interest rate if \$100 increased to \$144 in 2 years. Write your answer as a percent.

38. At the end of 2 years, P dollars invested at an interest rate r compounded annually increases to an amount, A dollars, given by

$$A = P(1 + r)^2$$

Find the interest rate if \$2000 increased to \$2420 in 2 years. Write your answer as a percent.

△ **39.** Find the dimensions of a rectangle whose width is 7 miles less than its length and whose area is 120 square miles.

△ **40.** Find the dimensions of a rectangle whose width is 2 inches less than half its length and whose area is 160 square inches.

41. If the cost, C, for manufacturing x units of a certain product is given by $C = x^2 - 15x + 50$, find the number of units manufactured at a cost of \$9500.

42. If a switchboard handles n telephones, the number C of telephone connections it can make simultaneously is given by the equation $C = \frac{n(n-1)}{2}$. Find how many telephones are handled by a switchboard making 120 telephone connections simultaneously.

REVIEW AND PREVIEW

The following double-line graph shows a comparison of the number of annual visitors (in millions) to Glacier National Park and Gettysburg National Military Park for the years shown. Use this graph to answer Exercises 43 through 50. See Section 10.1.

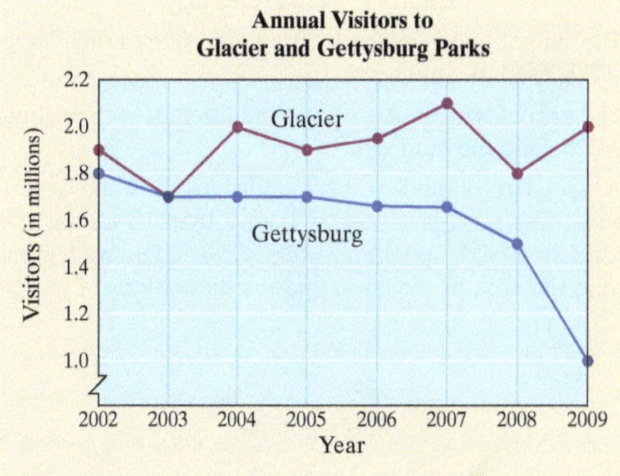

43. Approximate the number of visitors to Glacier National Park in 2009.

44. Approximate the number of visitors to Gettysburg National Military Park in 2009.

45. Approximate the number of visitors to Glacier National Park in 2005.

46. Approximate the number of visitors to Gettysburg National Military Park in 2005.

47. Determine the year that the colored lines in this graph intersect.

48. For what years on the graph is the number of visitors to Glacier Park greater than the number of visitors to Gettysburg Park?

49. In your own words, explain the meaning of the point of intersection in the graph.

50. Describe the trends shown in this graph and speculate as to why these trends have occurred.

Write each fraction in simplest form. See Section 4.2.

51. $\frac{20}{35}$ **52.** $\frac{24}{32}$ **53.** $\frac{27}{18}$

54. $\frac{15}{27}$ **55.** $\frac{14}{42}$ **56.** $\frac{45}{50}$

CONCEPT EXTENSIONS

△ **57.** Two boats travel at right angles to each other after leaving the same dock at the same time. One hour later, the boats are 17 miles apart. If one boat travels 7 miles per hour faster than the other boat, find the rate of each boat.

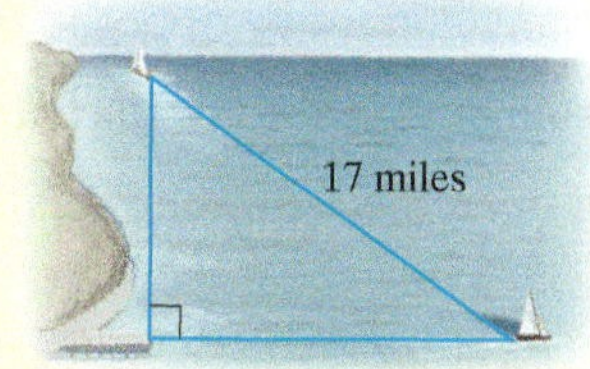

△ **58.** The side of a square equals the width of a rectangle. The length of the rectangle is 6 meters longer than its width. The sum of the areas of the square and the rectangle is 176 square meters. Find the side of the square.

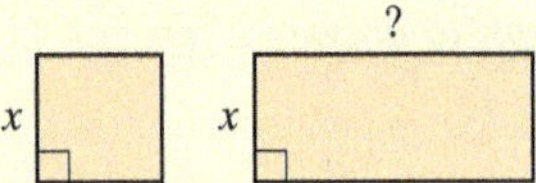

59. The sum of two numbers is 20, and the sum of their squares is 218. Find the numbers.

60. The sum of two numbers is 25, and the sum of their squares is 325. Find the numbers.

△ **61.** A rectangular pool is surrounded by a walk 4 meters wide. The pool is 6 meters longer than its width. If the total area of the pool and walk is 576 square meters more than the area of the pool, find the dimensions of the pool.

△ **62.** A rectangular garden is surrounded by a walk of uniform width. The area of the garden is 180 square yards. If the dimensions of the garden plus the walk are 16 yards by 24 yards, find the width of the walk.

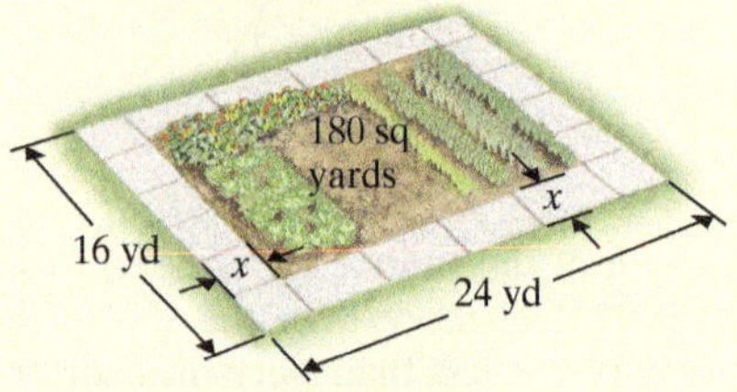

63. Write down two numbers whose sum is 10. Square each number and find the sum of the squares. Use this work to write a word problem like Exercise 59. Then give the word problem to a classmate to solve.

64. Write down two numbers whose sum is 12. Square each number and find the sum of the squares. Use this work to write a word problem like Exercise 60. Then give the word problem to a classmate to solve.

Chapter 13 Vocabulary Check

Fill in each blank with one of the words or phrases listed below. Not all choices will be used and some choices may be used more than once.

factoring	quadratic equation	perfect square trinomial	0
greatest common factor	hypotenuse	sum of two cubes	1
difference of two cubes	difference of two squares	triangle	leg

1. An equation that can be written in the form $ax^2 + bx + c = 0$ (with a not 0) is called a ________________.

2. ________________ is the process of writing an expression as a product.

3. The ________________ of a list of terms is the product of all common factors.

4. A trinomial that is the square of some binomial is called a ________________.

5. The expression $a^2 - b^2$ is called a ________________.

6. The expression $a^3 - b^3$ is called a ________________.

7. The expression $a^3 + b^3$ is called a ________________.

8. By the zero factor property, if the product of two numbers is 0, then at least one of the numbers must be ________________.

9. In a right triangle, the side opposite the right angle is called the ________________.

10. In a right triangle, each side adjacent to the right angle is called a ________________.

11. The Pythagorean theorem states that $(\text{leg})^2 + (\text{leg})^2 = (\_\_\_\_\_\_\_\_)^2$.

Helpful Hint Are you preparing for your test? Don't forget to take the Chapter 13 Test on page 990. Then check your answers at the back of the text and use the Chapter Test Prep Videos to see the fully worked-out solutions to any of the exercises you want to review.

13 Chapter Highlights

Definitions and Concepts	Examples
Section 13.1 The Greatest Common Factor and Factoring by Grouping	
Factoring is the process of writing an expression as a product.	Factor: $6 = 2 \cdot 3$ $x^2 + 5x + 6 = (x + 2)(x + 3)$
To Find the GCF of a List of Integers **Step 1:** Write each number as a product of primes. **Step 2:** Identify the common prime factors. **Step 3:** The product of all common factors is the greatest common factor. If there are no common prime factors, the GCF is 1.	Find the GCF of 12, 36, and 48. $12 = 2 \cdot 2 \cdot 3$ $36 = 2 \cdot 2 \cdot 3 \cdot 3$ $48 = 2 \cdot 2 \cdot 2 \cdot 2 \cdot 3$ $\text{GCF} = 2 \cdot 2 \cdot 3 = 12$
The GCF of a list of common variables raised to powers is the variable raised to the smallest exponent in the list.	The GCF of z^5, z^3, and z^{10} is z^3.
The GCF of a list of terms is the product of all common factors.	Find the GCF of $8x^2y$, $10x^3y^2$, and $26x^2y^3$. The GCF of 8, 10, and 26 is 2. The GCF of x^2, x^3, and x^2 is x^2. The GCF of y, y^2, and y^3 is y. The GCF of the terms is $2x^2y$.
To Factor by Grouping **Step 1:** Arrange the terms so that the first two terms have a common factor and the last two have a common factor. **Step 2:** For each pair of terms, factor out the pair's GCF. **Step 3:** If there is now a common binomial factor, factor it out. **Step 4:** If there is no common binomial factor, begin again, rearranging the terms differently. If no rearrangement leads to a common binomial factor, the polynomial cannot be factored.	Factor $10ax + 15a - 6xy - 9y$. **Step 1:** $10ax + 15a - 6xy - 9y$ **Step 2:** $5a(2x + 3) - 3y(2x + 3)$ **Step 3:** $(2x + 3)(5a - 3y)$
Section 13.2 Factoring Trinomials of the Form $x^2 + bx + c$	
The product of these numbers is c. $x^2 + bx + c = (x + \square)(x + \square)$ The sum of these numbers is b.	Factor: $x^2 + 7x + 12$ $3 + 4 = 7 \quad 3 \cdot 4 = 12$ $x^2 + 7x + 12 = (x + 3)(x + 4)$

Definitions and Concepts	Examples
Section 13.3 Factoring Trinomials of the Form $ax^2 + bx + c$ and Perfect Square Trinomials	
To factor $ax^2 + bx + c$, try various combinations of factors of ax^2 and c until a middle term of bx is obtained when checking.	Factor: $3x^2 + 14x - 5$ Factors of $3x^2$: $3x, x$ Factors of -5: $-1, 5$ and $1, -5$. $(3x - 1)(x + 5)$ $-1x$ $+15x$ Correct middle term $14x$
A **perfect square trinomial** is a trinomial that is the square of some binomial.	Perfect square trinomial = square of binomial $x^2 + 4x + 4 = (x + 2)^2$ $25x^2 - 10x + 1 = (5x - 1)^2$
Factoring Perfect Square Trinomials: $a^2 + 2ab + b^2 = (a + b)^2$ $a^2 - 2ab + b^2 = (a - b)^2$	Factor: $x^2 + 6x + 9 = x^2 + 2 \cdot x \cdot 3 + 3^2 = (x + 3)^2$ $4x^2 - 12x + 9 = (2x)^2 - 2 \cdot 2x \cdot 3 + 3^2 = (2x - 3)^2$
Section 13.4 Factoring Trinomials of the Form $ax^2 + bx + c$ by Grouping	
To Factor $ax^2 + bx + c$ by Grouping	Factor: $3x^2 + 14x - 5$
Step 1: Find two numbers whose product is $a \cdot c$ and whose sum is b.	**Step 1:** Find two numbers whose product is $3 \cdot (-5)$ or -15 and whose sum is 14. They are 15 and -1.
Step 2: Rewrite bx, using the factors found in Step 1.	**Step 2:** $3x^2 + 14x - 5$ $= 3x^2 + 15x - 1x - 5$
Step 3: Factor by grouping.	**Step 3:** $= 3x(x + 5) - 1(x + 5)$ $= (x + 5)(3x - 1)$
Section 13.5 Factoring Binomials	
Difference of Squares $a^2 - b^2 = (a + b)(a - b)$	Factor: $x^2 - 9 = x^2 - 3^2 = (x + 3)(x - 3)$
Sum or Difference of Cubes $a^3 + b^3 = (a + b)(a^2 - ab + b^2)$ $a^3 - b^3 = (a - b)(a^2 + ab + b^2)$	$y^3 + 8 = y^3 + 2^3 = (y + 2)(y^2 - 2y + 4)$ $125z^3 - 1 = (5z)^3 - 1^3 = (5z - 1)(25z^2 + 5z + 1)$
Integrated Review—Choosing a Factoring Strategy	
To Factor a Polynomial, **Step 1:** Factor out the GCF. **Step 2:** **a.** If two terms, **i.** $a^2 - b^2 = (a + b)(a - b)$ **ii.** $a^3 - b^3 = (a - b)(a^2 + ab + b^2)$ **iii.** $a^3 + b^3 = (a + b)(a^2 - ab + b^2)$ **b.** If three terms, **i.** $a^2 + 2ab + b^2 = (a + b)^2$ **ii.** Methods in Sections 13.2 and 13.3 **c.** If four or more terms, try factoring by grouping.	Factor: $2x^4 - 6x^2 - 8$ **Step 1:** $2x^4 - 6x^2 - 8 = 2(x^4 - 3x^2 - 4)$ **Step 2:** **b. ii.** $= 2(x^2 + 1)(x^2 - 4)$

Definitions and Concepts	Examples
Integrated Review—Choosing a Factoring Strategy (*continued*)	
Step 3: See if any factors can be factored further.	**Step 3:** $= 2(x^2 + 1)(x + 2)(x - 2)$
Step 4: Check by multiplying.	**Step 4:** Check by multiplying. $2(x^2 + 1)(x + 2)(x - 2) = 2(x^2 + 1)(x^2 - 4)$ $= 2(x^4 - 3x^2 - 4)$ $= 2x^4 - 6x^2 - 8$
Section 13.6 Solving Quadratic Equations by Factoring	
A **quadratic equation** is an equation that can be written in the form $ax^2 + bx + c = 0$ with a not 0. The form $ax^2 + bx + c = 0$ is called the **standard form** of a quadratic equation.	**Quadratic Equation** / **Standard Form**: $x^2 = 16$ / $x^2 - 16 = 0$; $y = -2y^2 + 5$ / $2y^2 + y - 5 = 0$
Zero Factor Theorem If a and b are real numbers and if $ab = 0$, then $a = 0$ or $b = 0$.	If $(x + 3)(x - 1) = 0$, then $x + 3 = 0$ or $x - 1 = 0$
To solve quadratic equations by factoring,	Solve: $3x^2 = 13x - 4$
Step 1: Write the equation in standard form: $ax^2 + bx + c = 0$.	**Step 1:** $3x^2 - 13x + 4 = 0$
Step 2: Factor the quadratic.	**Step 2:** $(3x - 1)(x - 4) = 0$
Step 3: Set each factor containing a variable equal to 0.	**Step 3:** $3x - 1 = 0$ or $x - 4 = 0$
Step 4: Solve the equations.	**Step 4:** $3x = 1$ or $x = 4$; $x = \frac{1}{3}$
Step 5: Check in the original equation.	**Step 5:** Check both $\frac{1}{3}$ and 4 in the original equation.
Section 13.7 Quadratic Equations and Problem Solving	
Problem-Solving Steps	A garden is in the shape of a rectangle whose length is two feet more than its width. If the area of the garden is 35 square feet, find its dimensions.
1. UNDERSTAND the problem.	1. Read and reread the problem. Guess a solution and check your guess. Let x be the width of the rectangular garden. Then $x + 2$ is the length. 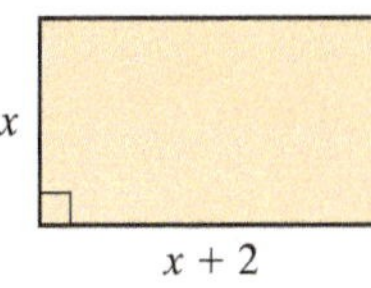
2. TRANSLATE.	2. In words: length · width = area; Translate: $(x + 2) \cdot x = 35$

(continued)

Definitions and Concepts	Examples
Section 13.7 Quadratic Equations and Problem Solving (*continued*)	
3. SOLVE.	**3.** $(x+2)x = 35$ $x^2 + 2x - 35 = 0$ $(x-5)(x+7) = 0$ $x - 5 = 0$ or $x + 7 = 0$ $x = 5$ or $x = -7$
4. INTERPRET.	**4.** Discard the solution of -7 since x represents width. *Check:* If x is 5 feet, then $x + 2 = 5 + 2 = 7$ feet. The area of a rectangle whose width is 5 feet and whose length is 7 feet is (5 feet)(7 feet) or 35 square feet. *State:* The garden is 5 feet by 7 feet.

Chapter 13 Review

(13.1) *Complete the factoring.*

1. $6x^2 - 15x = 3x(\quad)$

2. $2x^3y + 6x^2y^2 + 8xy^3 = 2xy(\quad)$

Factor the GCF from each polynomial.

3. $20x^2 + 12x$

4. $6x^2y^2 - 3xy^3$

5. $3x(2x+3) - 5(2x+3)$

6. $5x(x+1) - (x+1)$

Factor each polynomial by grouping.

7. $3x^2 - 3x + 2x - 2$

8. $3a^2 + 9ab + 3b^2 + ab$

9. $10a^2 + 5ab + 7b^2 + 14ab$

10. $6x^2 + 10x - 3x - 5$

(13.2) *Factor each trinomial.*

11. $x^2 + 6x + 8$

12. $x^2 - 11x + 24$

13. $x^2 + x + 2$

14. $x^2 - x + 2$

15. $x^2 + 4xy - 12y^2$

16. $x^2 + 8xy + 15y^2$

17. $72 - 18x - 2x^2$

18. $32 + 12x - 4x^2$

19. $10a^3 - 110a^2 + 100a$

20. $5y^3 - 50y^2 + 120y$

21. To factor $x^2 + 2x - 48$, think of two numbers whose product is ________ and whose sum is ________.

22. What is the first step in factoring $3x^2 + 15x + 30$?

(13.3) or (13.4) *Factor each trinomial.*

23. $2x^2 + 13x + 6$

24. $4x^2 + 4x - 3$

25. $6x^2 + 5xy - 4y^2$

26. $18x^2 - 9xy - 20y^2$

27. $10y^3 + 25y^2 - 60y$

28. $60y^3 - 39y^2 + 6y$

29. $18x^2 - 60x + 50$

30. $4x^2 - 28xy + 49y^2$

(13.5) *Factor each binomial.*

31. $4x^2 - 9$

32. $9t^2 - 25s^2$

33. $16x^2 + y^2$

34. $x^3 - 8y^3$

35. $8x^3 + 27$

36. $2x^3 + 8x$

37. $54 - 2x^3y^3$

38. $9x^2 - 4y^2$

39. $16x^4 - 1$

40. $x^4 + 16$

(13.6) *Solve the following equations.*

41. $(x+6)(x-2) = 0$

42. $3x(x+1)(7x-2) = 0$

43. $4(5x+1)(x+3) = 0$

44. $x^2 + 8x + 7 = 0$

45. $x^2 - 2x - 24 = 0$

46. $x^2 + 10x = -25$

47. $x(x-10) = -16$

48. $(3x-1)(9x^2 - 6x + 1) = 0$

49. $56x^2 - 5x - 6 = 0$

50. $20x^2 - 7x - 6 = 0$

51. $5(3x+2) = 4$

52. $6x^2 - 3x + 8 = 0$

53. $12 - 5t = -3$

54. $5x^3 + 20x^2 + 20x = 0$

55. $4t^3 - 5t^2 - 21t = 0$

56. Write a quadratic equation that has the two solutions 4 and 5.

(13.7) Use the given information to choose the correct dimensions.

△ 57. The perimeter of a rectangle is 24 inches. The length is twice the width. Find the dimensions of the rectangle.

a. 5 inches by 7 inches
b. 5 inches by 10 inches
c. 4 inches by 8 inches
d. 2 inches by 10 inches

△ 58. The area of a rectangle is 80 square meters. The length is one more than three times the width. Find the dimensions of the rectangle.

a. 8 meters by 10 meters
b. 4 meters by 13 meters
c. 4 meters by 20 meters
d. 5 meters by 16 meters

Use the given information to find the dimensions of each figure.

△ 59. The *area* of the square is 81 square units. Find the length of a side.

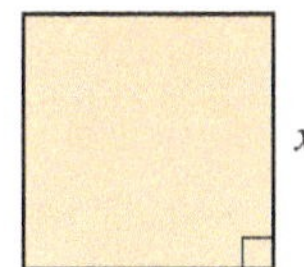

△ 60. The *perimeter* of the quadrilateral is 47 units. Find the lengths of the sides.

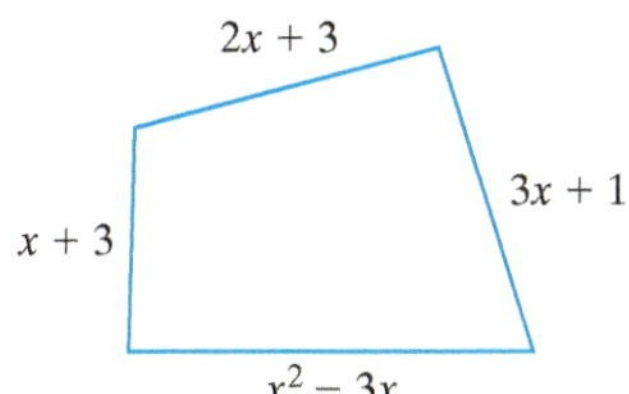

△ 61. A flag for a local organization is in the shape of a rectangle whose length is 15 inches less than twice its width. If the area of the flag is 500 square inches, find its dimensions.

△ 62. The base of a triangular sail is four times its height. If the area of the triangle is 162 square yards, find the base.

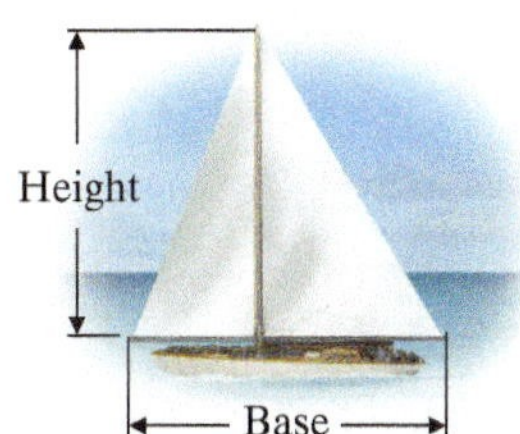

63. Find two consecutive positive integers whose product is 380.

64. Find two consecutive positive even integers whose product is 440.

65. A rocket is fired from the ground with an initial velocity of 440 feet per second. Its height h after t seconds is given by the equation

$$h = -16t^2 + 440t$$

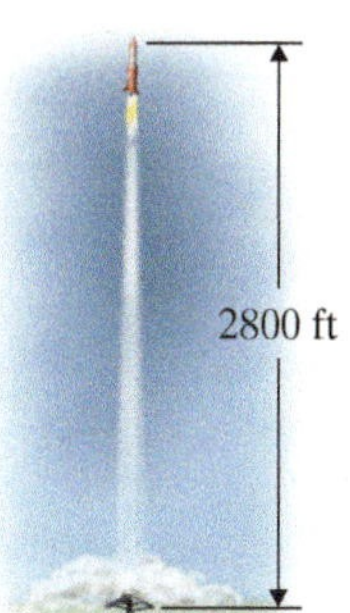

a. Find how many seconds pass before the rocket reaches a height of 2800 feet. Explain why two answers are obtained.
b. Find how many seconds pass before the rocket reaches the ground again.

△ 66. An architect's squaring instrument is in the shape of a right triangle. Find the length of the longer leg of the right triangle if the hypotenuse is 8 centimeters longer than the longer leg and the shorter leg is 8 centimeters shorter than the longer leg.

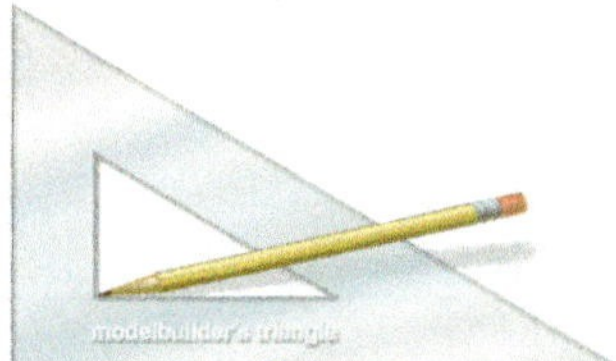

MIXED REVIEW

Factor completely.

67. $7x - 63$

68. $11x(4x - 3) - 6(4x - 3)$

69. $m^2 - \frac{4}{25}$

70. $3x^3 - 4x^2 + 6x - 8$

71. $xy + 2x - y - 2$

72. $2x^2 + 2x - 24$

73. $3x^3 - 30x^2 + 27x$

74. $4x^2 - 81$

75. $2x^2 - 18$

76. $16x^2 - 24x + 9$

77. $5x^2 + 20x + 20$

78. $2x^2 + 5x - 12$

79. $4x^2y - 6xy^2$

80. $125x^3 + 27$

81. $24x^2 - 3x - 18$

82. $(x + 7)^2 - y^2$

83. $x^2(x + 3) - 4(x + 3)$

84. $54a^3b - 2b$

Write the perimeter of each figure as a simplified polynomial. Then factor each polynomial.

△ **85.**

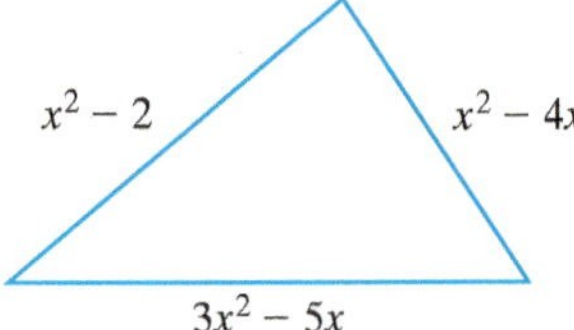

△ **86.**

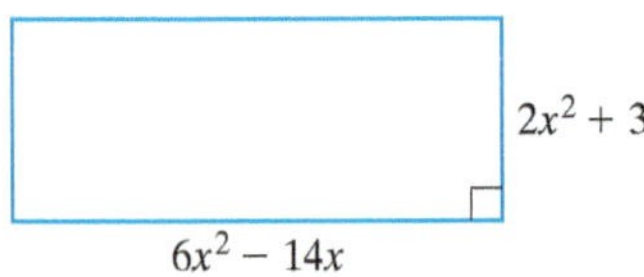

Solve.

87. $2x^2 - x - 28 = 0$

88. $x^2 - 2x = 15$

89. $2x(x + 7)(x + 4) = 0$

90. $x(x - 5) = -6$

91. $x^2 = 16x$

Solve.

△ **92.** The perimeter of the following triangle is 48 inches. Find the lengths of its sides.

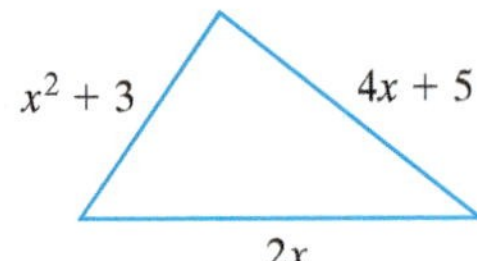

93. The width of a rectangle is 4 inches less than its length. Its area is 12 square inches. Find the dimensions of the rectangle.

94. A 6-foot-tall person drops an object from the top of the Westin Peachtree Plaza in Atlanta, Georgia. The Westin building is 723 feet tall. (*Source: World Almanac* research) The height h of the object after t seconds is given by the equation $h = -16t^2 + 729$. Find how many seconds pass before the object reaches the ground.

Write an expression for the area of the shaded region. Then write the expression as a factored polynomial.

△ **95.**

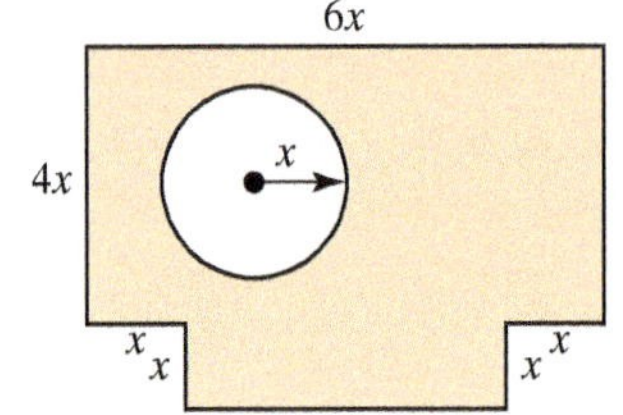

Chapter 13 Test

Step-by-step test solutions are found on the Chapter Test Prep Videos. Where available: MyMathLab® or YouTube

Factor each polynomial completely. If a polynomial cannot be factored, write "prime."

1. $x^2 + 11x + 28$

2. $49 - m^2$

3. $y^2 + 22y + 121$

4. $4(a + 3) - y(a + 3)$

5. $x^2 + 4$

6. $y^2 - 8y - 48$

7. $x^2 + x - 10$

8. $9x^3 + 39x^2 + 12x$

9. $3a^2 + 3ab - 7a - 7b$

10. $3x^2 - 5x + 2$

11. $x^2 + 14xy + 24y^2$

12. $180 - 5x^2$

13. $6t^2 - t - 5$

14. $xy^2 - 7y^2 - 4x + 28$

15. $x - x^5$

16. $-xy^3 - x^3y$

17. $64x^3 - 1$

18. $8y^3 - 64$

Solve each equation.

19. $(x - 3)(x + 9) = 0$

20. $x^2 + 5x = 14$

21. $x(x + 6) = 7$

22. $3x(2x - 3)(3x + 4) = 0$

23. $5t^3 - 45t = 0$

24. $t^2 - 2t - 15 = 0$

25. $6x^2 = 15x$

Solve each problem.

△ **26.** A deck for a home is in the shape of a triangle. The length of the base of the triangle is 9 feet longer than its altitude. If the area of the triangle is 68 square feet, find the length of the base.

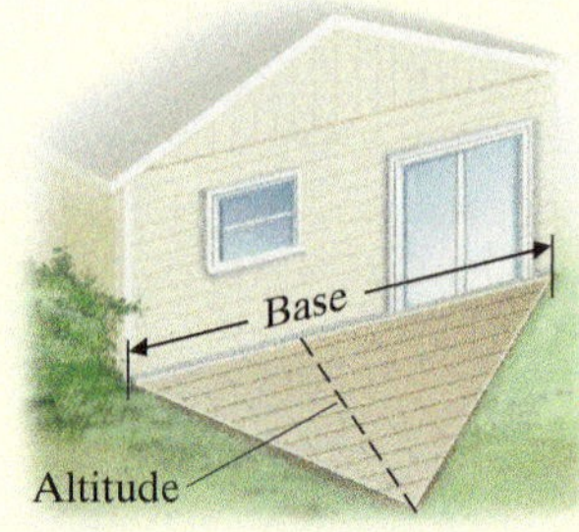

27. The sum of two numbers is 17 and the sum of their squares is 145. Find the numbers.

28. An object is dropped from the top of the Woolworth Building on Broadway in New York City. The height h of the object after t seconds is given by the equation

$$h = -16t^2 + 784$$

Find how many seconds pass before the object reaches the ground.

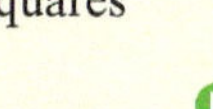

29. Find the lengths of the sides of a right triangle if the hypotenuse is 10 centimeters longer than the shorter leg and 5 centimeters longer than the longer leg.

Chapters 1–13 Cumulative Review

1. Translate each sentence into a mathematical statement.
   a. Nine is less than or equal to eleven.
   b. Eight is greater than one.
   c. Three is not equal to four.
2. Insert $<$ or $>$ in the space to make each statement true.
   a. $-5 \quad -3$
   b. $0 \quad -2$
3. Write in simplest form: $\frac{12}{20}$
4. Write the prime factorization of 60.
5. Simplify: $\left(\frac{4}{5}\right)^2 - 1$
6. Evaluate $\frac{x}{y}$ if $x = \frac{1}{2}$ and $y = \frac{3}{8}$.
7. Add: $(-3) + 4 + (-11)$
8. Evaluate $y + 5x$ if $x = -20$ and $y = -10$.
9. Multiply.
   a. $-7 \cdot 3$
   b. $-3(-5)$
   c. $0 \cdot (-4)$
10. Simplify: $5 - 2(3x - 7)$
11. Simplify each expression by combining like terms.
    a. $4x + 6x$
    b. $y - 5y$
    c. $3x^2 + 5x^2 - 2$
12. Solve: $8y + 2(y - 1) = 18$

Solve.

13. $3y - 7y = 12$
14. $\frac{x}{-7} = -4$
15. $3(3x - 5) = 10x$
16. $-\frac{2}{3}x = -22$
17. $7(x - 2) = 9x - 6$
18. $-3(x - 2) = 5x - 10$
19. A 10-foot board is to be cut into two pieces so that the length of the longer piece is 4 times the length of the shorter. Find the length of each piece.
20. Solve $3x + 9 \le 5(x - 1)$. Write the solution set using interval notation.
21. Graph the linear equation $y = -\frac{1}{3}x + 2$.
22. Is the ordered pair $(-1, 2)$ a solution of $-7x - 8y = -9$?
23. Find the slope and the y-intercept of the line whose equation is $3x - 4y = 4$.
24. Find the slope of the line through $(5, -6)$ and $(5, 2)$.
25. Evaluate each expression for the given value of x.
    a. $2x^3$; x is 5
    b. $\frac{9}{x^2}$; x is -3
26. Find the slope and y-intercept of the line whose equation is $7x - 3y = 2$.
27. Find the degree of each term.
    a. $3x^2$
    b. -2^3x^5
    c. y
    d. $12x^2yz^3$
    e. 5
28. Find an equation of the vertical line through $(0, 7)$.
29. Subtract: $(2x^3 + 8x^2 - 6x) - (2x^3 - x^2 + 1)$
30. Find an equation of the line with slope 4 and y-intercept $\left(0, \frac{1}{2}\right)$. Write the equation in standard form.
31. Multiply $(3x + 2)(2x - 5)$.
32. Write an equation of the line through $(-4, 0)$ and $(6, -1)$. Write the equation in standard form.
33. Multiply $(3y + 1)^2$.
34. Solve the system: $\begin{cases} -x + 3y = 18 \\ -3x + 2y = 19 \end{cases}$
35. Simplify by writing each expression with positive exponents only.
    a. 3^{-2}
    b. $2x^{-3}$
    c. $2^{-1} + 4^{-1}$
    d. $(-2)^{-4}$
    e. y^{-4}
36. Simplify: $\frac{(5a^7)^2}{a^5}$
37. Write each number in scientific notation.
    a. 367,000,000
    b. 0.000003
    c. 20,520,000,000
    d. 0.00085

38. Multiply $(3x - 7y)^2$.

39. Divide $x^2 + 7x + 12$ by $x + 3$ using long division.

40. Simplify: $\dfrac{(xy)^{-3}}{(x^5y^6)^3}$

41. Find the GCF of each list of terms.
a. x^3, x^7, and x^5 **b.** y, y^4, and y^7

Factor.

42. $z^3 + 7z + z^2 + 7$

43. $x^2 + 7x + 12$

44. $2x^3 + 2x^2 - 84x$

45. $8x^2 - 22x + 5$

46. $-4x^2 - 23x + 6$

47. $25a^2 - 9b^2$

48. $9xy^2 - 16x$

49. Solve: $(x - 3)(x + 1) = 0$

50. Solve $x^2 - 13x = -36$.

14

Rational Expressions

Many sport statistics are calculated using formulas containing rational expressions. Below are a few examples:

Sport	Formula	Explanation
Baseball	$S = \frac{h + d + 2t + 3r}{b}$	A baseball player's slugging percentage S, where h = number of hits, d = number of doubles, t = number of triples, r = number of home runs, and b = number of at bats
NCAA Football	$R_{NCAA} = \frac{100C + 330T - 200I + 8.4Y}{A}$	A quarterback's rating in NCAA football R, where C = the number of completed passes, A = the number of attempted passes, T = the number of touchdown passes, Y = the number of yards in the completed passes, and I = the number of interceptions
NFL Football	$R_{NFL} = \frac{a + b + c + d}{6} \times 100$	A quarterback's rating in NFL football R, where a, b, c, and d are each formulas containing rational expressions.

(*Source:* Wikipedia.org.)

In Section 14.1, Exercises 65 and 66, you will have the opportunity to calculate some sport statistics.

Sections

Check Your Progress

In this chapter, we expand our knowledge of algebraic expressions to include algebraic fractions, called *rational expressions.* We explore the operations of addition, subtraction, multiplication, and division using principles similar to the principles for numerical fractions.

14.1 Rational Functions and Simplifying Rational Expressions

Objectives

1 Find the Domain of a Rational Function.

2 Simplify or Write Rational Expressions in Lowest Terms.

3 Write Equivalent Rational Expressions of the Form $-\frac{a}{b} = \frac{-a}{b} = \frac{a}{-b}$.

4 Use Rational Functions in Applications.

As we reviewed in Chapter 9, a rational number is a number that can be written as a quotient of integers. A **rational expression** is also a quotient; it is a quotient of polynomials.

Rational Expression

A rational expression is an expression that can be written in the form

$$\frac{P}{Q},$$

where P and Q are polynomials and $Q \neq 0$.

Rational Expressions

$$-\frac{2}{7} \qquad \frac{3y^3}{8} \qquad \frac{-4p}{p^3 + 2p + 1} \qquad \frac{5x^2 - 3x + 2}{3x + 7}$$

The first rational expression (or fraction) above is $-\frac{2}{7}$. For a negative fraction such as $-\frac{2}{7}$, recall from Section 4.1 that

$$-\frac{2}{7} = \frac{2}{-7} = \frac{-2}{7}$$

In general, for any fraction,

$$\frac{-a}{b} = \frac{a}{-b} = -\frac{a}{b}, \qquad b \neq 0$$

This is also true for rational expressions. For example,

$$\frac{-(x + 2)}{x} = \frac{x + 2}{-x} = -\frac{x + 2}{x}$$

Notice the parentheses.

Rational expressions are sometimes used to describe functions. For example, we call the function $f(x) = \frac{x^2 + 2}{x - 3}$ a **rational function** since $\frac{x^2 + 2}{x - 3}$ is a rational expression.

Objective 1 Finding the Domain of a Rational Function 

As with fractions, a rational expression is **undefined** if the denominator is 0. If a variable in a rational expression is replaced with a number that makes the denominator 0, we say that the rational expression is **undefined** for this value of the variable. For example, the rational expression $\frac{x^2 + 2}{x - 3}$ is undefined when x is 3, because replacing x with 3 results in a denominator of 0. For this reason, we must exclude 3 from the domain of the function $f(x) = \frac{x^2 + 2}{x - 3}$.

The domain of f is then

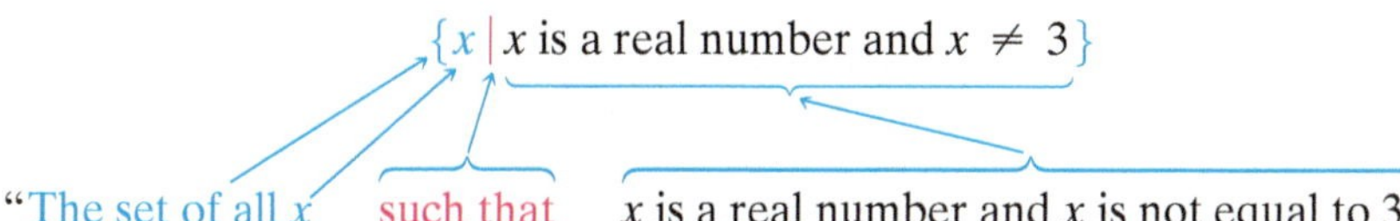

In this section, we will use this set builder notation to write domains. Unless told otherwise, we assume that the domain of a function described by an equation is the set of all real numbers for which the equation is defined.

Example 1 Find the domain of each rational function.

a. $f(x) = \dfrac{8x^3 + 7x^2 + 20}{2}$ **b.** $g(x) = \dfrac{5x^2 - 3}{x - 1}$ **c.** $f(x) = \dfrac{7x - 2}{x^2 - 2x - 15}$

Solution: The domain of each function will contain all real numbers except those values that make the denominator 0.

a. No matter what the value of x, the denominator of $f(x) = \dfrac{8x^3 + 7x^2 + 20}{2}$ is never 0, so the domain of f is $\{x \mid x \text{ is a real number}\}$.

b. To find the values of x that make the denominator of $g(x)$ equal to 0, we solve the equation "denominator = 0":

$$x - 1 = 0, \quad \text{or} \quad x = 1$$

The domain must exclude 1 since the rational expression is undefined when x is 1. The domain of g is $\{x \mid x \text{ is a real number and } x \neq 1\}$.

c. We find the domain by setting the denominator equal to 0.

$$x^2 - 2x - 15 = 0 \qquad \text{Set the denominator equal to 0 and solve.}$$
$$(x - 5)(x + 3) = 0$$
$$x - 5 = 0 \quad \text{or} \quad x + 3 = 0$$
$$x = 5 \quad \text{or} \quad x = -3$$

If x is replaced with 5 or with -3, the rational expression is undefined. The domain of f is $\{x \mid x \text{ is a real number and } x \neq 5, x \neq -3\}$.

Work Practice 1

Practice 1

Find the domain of each rational function.

a. $f(x) = \dfrac{4x^5 - 3x^2 + 2}{-6}$

b. $g(x) = \dfrac{6x^2 + 1}{x + 3}$

c. $h(x) = \dfrac{8x - 3}{x^2 - 5x + 6}$

✓**Concept Check** For which of these values (if any) is the rational expression $\dfrac{x - 3}{x^2 + 2}$ undefined?

a. 2 **b.** 3 **c.** -2 **d.** 0 **e.** None of these

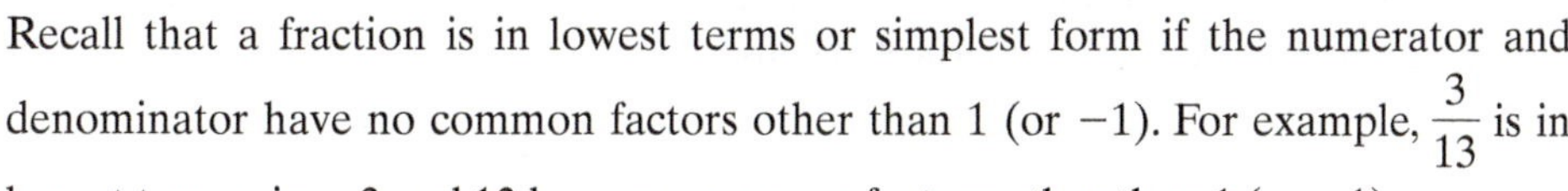

Objective 2 Simplifying Rational Expressions

Recall that a fraction is in lowest terms or simplest form if the numerator and denominator have no common factors other than 1 (or -1). For example, $\dfrac{3}{13}$ is in lowest terms since 3 and 13 have no common factors other than 1 (or -1).

To **simplify** a rational expression, or to write it in lowest terms, we use a method similar to simplifying a fraction.

Answers

1. a. $\{x \mid x \text{ is a real number}\}$
b. $\{x \mid x \text{ is a real number and } x \neq -3\}$
c. $\{x \mid x \text{ is a real number and } x \neq 2, x \neq 3\}$

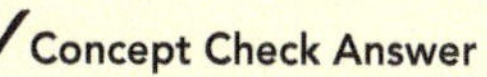

✓**Concept Check Answer**

e

Recall that to simplify a fraction, we essentially "remove factors of 1." Our ability to do this comes from these facts:

- If $c \neq 0$, then $\frac{c}{c} = 1$. For example, $\frac{7}{7} = 1$ and $\frac{-8.65}{-8.65} = 1$.
- $n \cdot 1 = n$. For example, $-5 \cdot 1 = -5$, $126.8 \cdot 1 = 126.8$, and $\frac{a}{b} \cdot 1 = \frac{a}{b}$, $b \neq 0$.

In other words, we have the following:

$$\frac{a \cdot c}{b \cdot c} = \frac{a}{b} \cdot \frac{c}{c} = \frac{a}{b}$$

Since $\frac{a}{b} \cdot 1 = \frac{a}{b}$

Let's practice simplifying a fraction by simplifying $\frac{15}{65}$.

$$\frac{15}{65} = \frac{3 \cdot 5}{13 \cdot 5} = \frac{3}{13} \cdot \frac{5}{5} = \frac{3}{13} \cdot 1 = \frac{3}{13}$$

Let's use the same technique and simplify the rational expression $\frac{x^2 - 9}{x^2 + x - 6}$.

$$\frac{x^2 - 9}{x^2 + x - 6} = \frac{(x - 3)(x + 3)}{(x - 2)(x + 3)}$$ Factor the numerator and the denominator.

$$= \frac{(x - 3)(x + 3)}{(x - 2)(x + 3)}$$ Look for common factors.

$$= \frac{x - 3}{x - 2} \cdot \frac{x + 3}{x + 3}$$

$$= \frac{x - 3}{x - 2} \cdot 1$$ Write $\frac{x + 3}{x + 3}$ as 1.

$$= \frac{x - 3}{x - 2}$$ Multiply to remove a factor of 1.

This "removing a factor of 1" is stated in the principle below:

Fundamental Principle of Rational Expressions

For any rational expression $\frac{P}{Q}$ and any polynomial R, where $R \neq 0$,

$$\frac{PR}{QR} = \frac{P}{Q} \cdot \frac{R}{R} = \frac{P}{Q} \cdot 1 = \frac{P}{Q}$$

or, simply,

$$\frac{PR}{QR} = \frac{P}{Q}$$

In general, the following steps may be used to simplify rational expressions or to write a rational expression in lowest terms.

Simplifying or Writing a Rational Expression in Lowest Terms

Step 1: Completely factor the numerator and denominator of the rational expression.

Step 2: Divide out factors common to the numerator and denominator. (This is the same as "removing a factor of 1.")

For now, we assume that variables in a rational expression do not represent values that make the denominator 0.

Example 2 Simplify each rational expression.

a. $\dfrac{2x^2}{10x^3 - 2x^2}$ **b.** $\dfrac{9x^2 + 13x + 4}{8x^2 + x - 7}$

Solution:

a. $\dfrac{2x^2}{10x^3 - 2x^2} = \dfrac{2x^2 \cdot 1}{2x^2(5x - 1)} = 1 \cdot \dfrac{1}{5x - 1} = \dfrac{1}{5x - 1}$

b. $\dfrac{9x^2 + 13x + 4}{8x^2 + x - 7} = \dfrac{(9x + 4)(x + 1)}{(8x - 7)(x + 1)}$ Factor the numerator and denominator.

$= \dfrac{9x + 4}{8x - 7} \cdot 1$ Since $\dfrac{x + 1}{x + 1} = 1$

$= \dfrac{9x + 4}{8x - 7}$ Simplest form

Work Practice 2

Practice 2

Simplify each rational expression.

a. $\dfrac{5z^4}{10z^5 - 5z^4}$

b. $\dfrac{5x^2 + 13x + 6}{6x^2 + 7x - 10}$

Just as for numerical fractions, we can use a shortcut notation. Remember that as long as exact factors in both the numerator and denominator are divided out, we are "removing a factor of 1." We will use the following notation to show this:

$\dfrac{x^2 - 9}{x^2 + x - 6} = \dfrac{(x - 3)(x + 3)}{(x - 2)(x + 3)}$ A factor of 1 is identified by the shading.

$= \dfrac{x - 3}{x - 2}$ Remove a factor of 1.

Thus, the rational expression $\dfrac{x^2 - 9}{x^2 + x - 6}$ has the same value as the rational expression $\dfrac{x - 3}{x - 2}$ for all values of x except 2 and -3. (Remember that when x is 2, the denominator of both rational expressions is 0 and when x is -3, the original rational expression has a denominator of 0.)

As we simplify rational expressions, we will assume that the simplified rational expression is equal to the original rational expression for all real numbers except those for which the original denominator is 0.

Example 3 Simplify each rational expression.

a. $\dfrac{2 + x}{x + 2}$ **b.** $\dfrac{2 - x}{x - 2}$

Solution:

a. $\dfrac{2 + x}{x + 2} = \dfrac{x + 2}{x + 2} = 1$ By the commutative property of addition, $2 + x = x + 2$.

b. $\dfrac{2 - x}{x - 2}$

The terms in the numerator of $\dfrac{2 - x}{x - 2}$ differ by sign from the terms of the denominator, so the polynomials are opposites of each other and the expression simplifies to -1. To see this, we factor out -1 from the numerator or the denominator. If -1 is factored from the numerator, then

$$\frac{2 - x}{x - 2} = \frac{-1(-2 + x)}{x - 2} = \frac{-1(x - 2)}{x - 2} = \frac{-1}{1} = -1$$

(Continued on next page)

Practice 3

Simplify each rational expression.

a. $\dfrac{x + 3}{3 + x}$ **b.** $\dfrac{3 - x}{x - 3}$

Answers

2. a. $\dfrac{1}{2z - 1}$ **b.** $\dfrac{5x + 3}{6x - 5}$

3. a. 1 **b.** -1

Helpful Hint When the numerator and the denominator of a rational expression are opposites of each other, the expression simplifies to -1.

If -1 is factored from the denominator, the result is the same.

$$\frac{2-x}{x-2} = \frac{2-x}{-1(-x+2)} = \frac{2-x}{-1(2-x)} = \frac{1}{-1} = -1$$

Work Practice 3

Practice 4

Simplify $\dfrac{20-5x^2}{x^2+x-6}$.

Example 4

Simplify $\dfrac{18-2x^2}{x^2-2x-3}$.

Solution:

$$\frac{18-2x^2}{x^2-2x-3} = \frac{2(9-x^2)}{(x+1)(x-3)}$$ Factor.

$$= \frac{2(3+x)(3-x)}{(x+1)(x-3)}$$ Factor completely.

$$= \frac{2(3+x)\cdot -1(x-3)}{(x+1)(x-3)}$$ Notice the opposites $3-x$ and $x-3$. Write $3-x$ as $-1(x-3)$ and simplify.

$$= -\frac{2(3+x)}{x+1}$$

Work Practice 4

Helpful Hint

When simplifying a rational expression, we look for **common *factors*, not common *terms*.**

$$\frac{x\cdot(x+2)}{x\cdot x} = \frac{x+2}{x}$$

Common factors. These can be divided out.

$$\frac{x+2}{x}$$

Common terms. There is no factor of 1 that can be generated.

✓Concept Check Recall that we can only remove *factors* of 1. Which of the following are *not* true? Explain why.

a. $\dfrac{3-1}{3+5}$ simplifies to $-\dfrac{1}{5}$.

b. $\dfrac{2x+10}{2}$ simplifies to $x+5$.

c. $\dfrac{37}{72}$ simplifies to $\dfrac{3}{2}$.

d. $\dfrac{2x+3}{2}$ simplifies to $x+3$.

Practice 5

Simplify each rational expression.

a. $\dfrac{x^3+64}{4+x}$

b. $\dfrac{5z^2+10}{z^3-3z^2+2z-6}$

Example 5

Simplify each rational expression.

a. $\dfrac{x^3+8}{2+x}$ **b.** $\dfrac{2y^2+2}{y^3-5y^2+y-5}$

Solution:

a. $$\frac{x^3+8}{2+x} = \frac{(x+2)(x^2-2x+4)}{x+2}$$ Factor the sum of the two cubes.

$$= x^2-2x+4$$ Divide out common factors.

Answers

4. $-\dfrac{5(2+x)}{x+3}$

5. a. $x^2-4x+16$ **b.** $\dfrac{5}{z-3}$

✓Concept Check Answers

a, c, d

b. $\dfrac{2y^2 + 2}{y^3 - 5y^2 + y - 5} = \dfrac{2(y^2 + 1)}{(y^3 - 5y^2) + (y - 5)}$ Factor the numerator; group the denominator.

$= \dfrac{2(y^2 + 1)}{y^2(y - 5) + 1(y - 5)}$ Factor the denominator by grouping.

$= \dfrac{2(y^2 + 1)}{(y - 5)(y^2 + 1)}$

$= \dfrac{2}{y - 5}$ Divide out common factors.

Work Practice 5

✓Concept Check Does $\dfrac{n}{n + 2}$ simplify to $\dfrac{1}{2}$? Why or why not?

Objective 3 Writing Equivalent Forms of Rational Expressions

From Example 3, we have

$$\frac{2 + x}{x + 2} = \frac{x + 2}{x + 2} = 1 \quad \text{and} \quad \frac{2 - x}{x - 2} = \frac{2 - x}{-1(2 - x)} = \frac{1}{-1} = -1.$$

When performing operations on rational expressions, equivalent forms of answers often result. For this reason, it is very important to be able to recognize equivalent answers.

Example 6 List some equivalent forms of $-\dfrac{5x - 1}{x + 9}$.

Solution: To do so, recall that $-\dfrac{a}{b} = \dfrac{-a}{b} = \dfrac{a}{-b}$. Thus

$$-\frac{5x - 1}{x + 9} = \frac{-(5x - 1)}{x + 9} = \frac{-5x + 1}{x + 9} \quad \text{or} \quad \frac{1 - 5x}{x + 9}$$

Also,

$$-\frac{5x - 1}{x + 9} = \frac{5x - 1}{-(x + 9)} = \frac{5x - 1}{-x - 9} \quad \text{or} \quad \frac{5x - 1}{-9 - x}$$

Thus $-\dfrac{5x - 1}{x + 9} = \dfrac{-(5x - 1)}{x + 9} = \dfrac{-5x + 1}{x + 9} = \dfrac{5x - 1}{-(x + 9)} = \dfrac{5x - 1}{-x - 9}$

Work Practice 6

Keep in mind that many rational expressions may look different, but in fact be equivalent.

Practice 6

List some equivalent forms of $-\dfrac{x + 3}{6x - 11}$.

Helpful Hint Remember, a negative sign in front of a fraction or rational expression may be moved to the numerator or the denominator, but *not* both.

Answer

6. $\dfrac{-(x + 3)}{6x - 11}; \dfrac{-x - 3}{6x - 11}; \dfrac{x + 3}{-(6x - 11)}; \dfrac{x + 3}{-6x + 11}; \dfrac{x + 3}{11 - 6x}$

✓Concept Check Answers

no; answers may vary.

Objective 4 Using Rational Functions in Applications

Rational functions occur often in real-life situations.

Example 7 Cost for Pressing Compact Discs

For the ICL Production Company, the rational function $C(x) = \frac{2.6x + 10{,}000}{x}$ describes the company's cost per disc of pressing x compact discs. Find the cost per disc for pressing:

a. 100 compact discs **b.** 1000 compact discs

Solution:

a. $C(100) = \frac{2.6(100) + 10{,}000}{100} = \frac{10{,}260}{100} = 102.6$

The cost per disc for pressing 100 compact discs is \$102.60.

b. $C(1000) = \frac{2.6(1000) + 10{,}000}{1000} = \frac{12{,}600}{1000} = 12.6$

The cost per disc for pressing 1000 compact discs is \$12.60. Notice that as more compact discs are produced, the cost per disc decreases.

Work Practice 7

Practice 7

A company's cost per T-shirt for silk screening x T-shirts is given by the rational function $C(x) = \frac{3.2x + 400}{x}$. Find the cost per T-shirt for printing:

a. 100 T-shirts

b. 1000 T-shirts

Answers

7. a. \$7.20 **b.** \$3.60

Graphing Calculator Explorations

(Note: The information below about *connected* mode and *dot* mode may not apply to your graphing calculator.)

Recall that since the rational expression $\frac{7x - 2}{(x - 2)(x + 5)}$ is not defined when $x = 2$ or when $x = -5$, we say that the domain of the rational function $f(x) = \frac{7x - 2}{(x - 2)(x + 5)}$ is all real numbers except 2 and -5. This domain can be written as $\{x|x \text{ is a real number and } x \neq 2, x \neq -5\}$. This means that the graph of $f(x)$ should not cross the vertical lines $x = 2$ and $x = -5$. The graph of $f(x)$ in *connected* mode is below to the left. In connected mode the graphing calculator tries to connect all dots of the graph so that the result is a smooth curve. This is what has happened in the graph. Notice that the graph appears to contain vertical lines at $x = 2$ and at $x = -5$. We know that this cannot happen because the function is not defined at $x = 2$ and at $x = -5$. We also know that this cannot happen because the graph of this function would not pass the vertical line test.

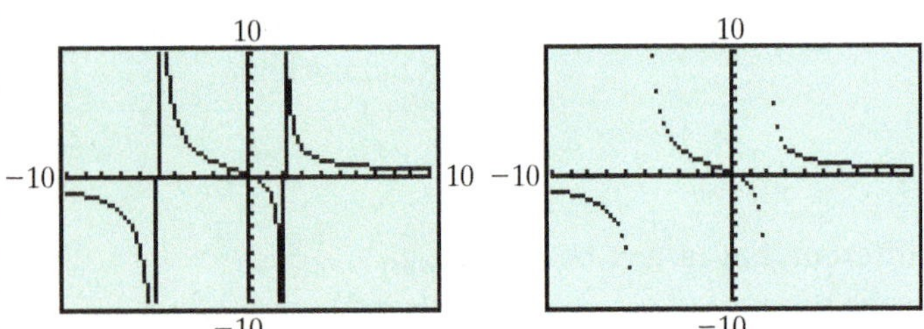

The graph of $f(x)$ in *dot* mode is above to the right. In dot mode the graphing calculator will not connect dots with a smooth curve. Notice that the vertical lines have disappeared, and we have a better picture of the graph. The graph, however, actually appears more like the hand-drawn graph below. By using a Table feature, a Calculate Value feature, or by tracing, we can see that the function is not defined at $x = 2$ and at $x = -5$.

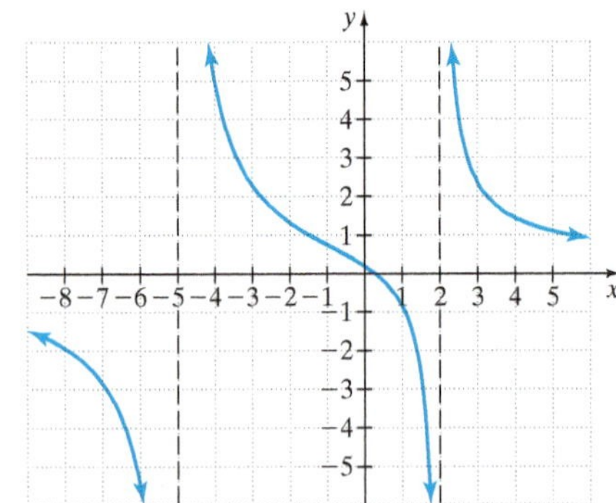

Find the domain of each rational function. Then graph each rational function and use the graph to confirm the domain.

1. $f(x) = \frac{x + 1}{x^2 - 4}$

2. $g(x) = \frac{5x}{x^2 - 9}$

3. $h(x) = \frac{x^2}{2x^2 + 7x - 4}$

4. $f(x) = \frac{3x + 2}{4x^2 - 19x - 5}$

Vocabulary, Readiness & Video Check

Use the choices below to fill in each blank. Some choices may not be used.

1	true	rational	simplified	$\frac{-a}{-b}$	$\frac{-a}{b}$	$\frac{a}{-b}$
-1	false	domain	0			

1. A ________ expression is an expression that can be written as the quotient $\frac{P}{Q}$ of two polynomials P and Q as long as $Q \neq 0$.

2. A rational expression is undefined if the denominator is ________.

3. The ________ of the rational function $f(x) = \frac{2}{x}$ is $\{x | x \text{ is a real number and } x \neq 0\}$.

4. A rational expression is ________ if the numerator and denominator have no common factors other than 1 or -1.

5. The expression $\frac{x^2 + 2}{2 + x^2}$ simplifies to ________.

6. The expression $\frac{y - z}{z - y}$ simplifies to ________.

7. For a rational expression, $-\frac{a}{b} = \underline{\quad} = \underline{\quad}$.

8. True or false: $\frac{a - 6}{a + 2} = \frac{-(a - 6)}{-(a + 2)} = \frac{-a + 6}{-a - 2}$. ________

Martin-Gay Interactive Videos

See Video 14.1

Watch the section lecture video and answer the following questions.

Objective 1 **9.** Why can't the denominators of rational expressions be zero? How can we find the domain of a rational function?

Objective 2 **10.** In Example 6, why isn't a factor of x divided out of the expression at the end?

Objective 3 **11.** From Example 8, if we move a negative sign from in front of a rational expression to either the numerator or denominator, when do we insert parentheses and why?

Objective 4 **12.** From Example 9, why do we subtract parts a and b to find the answer to part c?

14.1 Exercise Set MyMathLab®

Find the domain of each rational expression. See Example 1.

1. $f(x) = \frac{5x - 7}{4}$

2. $g(x) = \frac{4 - 3x}{2}$

3. $s(t) = \frac{t^2 + 1}{2t}$

4. $v(t) = -\frac{5t + t^2}{3t}$

5. $f(x) = \frac{3x}{7 - x}$

6. $f(x) = \frac{-4x}{-2 + x}$

7. $f(x) = \frac{x}{3x - 1}$

8. $g(x) = \frac{-2}{2x + 5}$

9. $R(x) = \frac{3 + 2x}{x^3 + x^2 - 2x}$

10. $h(x) = \dfrac{5 - 3x}{2x^2 - 14x + 20}$

11. $C(x) = \dfrac{x + 3}{x^2 - 4}$

12. $R(x) = \dfrac{5}{x^2 - 7x}$

Study Example 6. Then list four equivalent forms for each rational expression.

13. $-\dfrac{x - 10}{x + 8}$

14. $-\dfrac{x + 11}{x - 4}$

15. $-\dfrac{5y - 3}{y - 12}$

16. $-\dfrac{8y - 1}{y - 15}$

MIXED PRACTICE

Simplify each expression. See Examples 2 through 5.

17. $\dfrac{x + 7}{7 + x}$

18. $\dfrac{y + 9}{9 + y}$

19. $\dfrac{x - 7}{7 - x}$

20. $\dfrac{y - 9}{9 - y}$

21. $\dfrac{2}{8x + 16}$

22. $\dfrac{3}{9x + 6}$

23. $\dfrac{-5a - 5b}{a + b}$

24. $\dfrac{-4x - 4y}{x + y}$

25. $\dfrac{7x + 35}{x^2 + 5x}$

26. $\dfrac{9x + 99}{x^2 + 11x}$

27. $\dfrac{x + 5}{x^2 - 4x - 45}$

28. $\dfrac{x - 3}{x^2 - 6x + 9}$

29. $\dfrac{5x^2 + 11x + 2}{x + 2}$

30. $\dfrac{12x^2 + 4x - 1}{2x + 1}$

31. $\dfrac{x^3 + 7x^2}{x^2 + 5x - 14}$

32. $\dfrac{x^4 - 10x^3}{x^2 - 17x + 70}$

33. $\dfrac{2x^2 - 8}{4x - 8}$

34. $\dfrac{5x^2 - 500}{35x + 350}$

35. $\dfrac{4 - x^2}{x - 2}$

36. $\dfrac{49 - y^2}{y - 7}$

37. $\dfrac{11x^2 - 22x^3}{6x - 12x^2}$

38. $\dfrac{24y^2 - 8y^3}{15y - 5y^2}$

39. $\dfrac{x^2 + xy + 2x + 2y}{x + 2}$

40. $\dfrac{ab + ac + b^2 + bc}{b + c}$

41. $\dfrac{x^3 + 8}{x + 2}$

42. $\dfrac{x^3 + 64}{x + 4}$

43. $\dfrac{x^3 - 1}{1 - x}$

44. $\dfrac{3 - x}{x^3 - 27}$

45. $\dfrac{2xy + 5x - 2y - 5}{3xy + 4x - 3y - 4}$

46. $\dfrac{2xy + 2x - 3y - 3}{2xy + 4x - 3y - 6}$

47. $\dfrac{3x^2 - 5x - 2}{6x^3 + 2x^2 + 3x + 1}$

48. $\dfrac{2x^2 - x - 3}{2x^3 - 3x^2 + 2x - 3}$

49. $\dfrac{9x^2 - 15x + 25}{27x^3 + 125}$

50. $\dfrac{8x^3 - 27}{4x^2 + 6x + 9}$

MIXED PRACTICE

Simplify each expression. Then determine whether the given answer is correct. See Examples 3 through 6.

51. $\dfrac{9 - x^2}{x - 3}$; Answer: $-3 - x$

52. $\dfrac{100 - x^2}{x - 10}$; Answer: $-10 - x$

53. $\dfrac{7 - 34x - 5x^2}{25x^2 - 1}$; Answer: $\dfrac{x + 7}{-5x - 1}$

54. $\dfrac{2 - 15x - 8x^2}{64x^2 - 1}$; Answer: $\dfrac{x + 2}{-8x - 1}$

Find each function value. See Example 7.

55. If $f(x) = \dfrac{x + 8}{2x - 1}$, find $f(2)$, $f(0)$, and $f(-1)$.

56. If $f(x) = \dfrac{x - 2}{-5 + x}$, find $f(-5)$, $f(0)$, and $f(10)$.

57. If $g(x) = \dfrac{x^2 + 8}{x^3 - 25x}$, find $g(3)$, $g(-2)$, and $g(1)$.

58. If $s(t) = \dfrac{t^3 + 1}{t^2 + 1}$, find $s(-1)$, $s(1)$, and $s(2)$.

Solve. See Example 7.

59. The total revenue from the sale of a popular book is approximated by the rational function $R(x) = \dfrac{1000x^2}{x^2 + 4}$, where x is the number of years since publication and $R(x)$ is the total revenue in millions of dollars.
   a. Find the total revenue at the end of the first year.
   b. Find the total revenue at the end of the second year.
   c. Find the revenue during the second year only.
   d. Find the domain of function R.

60. The function $f(x) = \dfrac{100{,}000x}{100 - x}$ models the cost in dollars for removing x percent of the pollutants from a bayou in which a nearby company dumped creosol.
   a. Find the cost of removing 20% of the pollutants from the bayou. [*Hint:* Find $f(20)$.]
   b. Find the cost of removing 60% of the pollutants and then 80% of the pollutants.
   c. Find $f(90)$, then $f(95)$, and then $f(99)$. What happens to the cost as x approaches 100%?
   d. Find the domain of function f.

61. The dose of medicine prescribed for a child depends on the child's age A in years and the adult dose D for the medication. Young's Rule is a formula used by pediatricians that gives a child's dose C as

$$C = \frac{DA}{A + 12}$$

Suppose that an 8-year-old child needs medication, and the normal adult dose is 1000 mg. What size dose should the child receive?

62. Calculating body-mass index is a way to gauge whether a person should lose weight. Doctors recommend that body-mass index values fall between 18.5 and 25. The formula for body-mass index B is

$$B = \frac{703w}{h^2}$$

where w is weight in pounds and h is height in inches. Should a 148-pound person who is 5 feet 6 inches tall lose weight?

63. Anthropologists and forensic scientists use a measure called the cephalic index to help classify skulls. The cephalic index of a skull with width W and length L from front to back is given by the formula

$$C = \frac{100W}{L}$$

A long skull has an index value less than 75, a medium skull has an index value between 75 and 85, and a broad skull has an index value over 85. Find the cephalic index of a skull that is 5 inches wide and 6.4 inches long. Classify the skull.

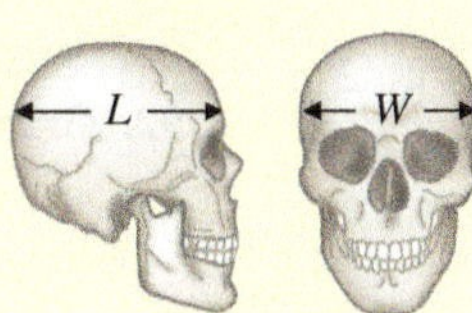

64. A company's gross profit margin P can be computed with the formula $P = \frac{R - C}{R}$, where R = the company's revenue and C = cost of goods sold. During a recent fiscal year, computer company Apple had revenues of \$32.5 billion and cost of goods sold \$21.3 billion. (*Source:* Apple, Inc.) What was Apple's gross profit margin in this year? Express the answer as a percent, rounded to the nearest tenth of a percent.

65. A baseball player's slugging percentage S can be calculated with the following formula:

$$S = \frac{h + d + 2t + 3r}{b}, \text{ where } h = \text{number of hits},$$

d = number of doubles, t = number of triples, r = number of home runs, and b = number of at bats. In 2010, Albert Pujols of the St. Louis Cardinals led Major League Baseball in slugging percent. During the 2010 season, Pujols had 587 at bats, 183 hits, 39 doubles, 1 triple, and 42 home runs. (*Source:* Major League Baseball) Calculate Pujols' 2010 slugging percent. Round to the nearest tenth of a percent.

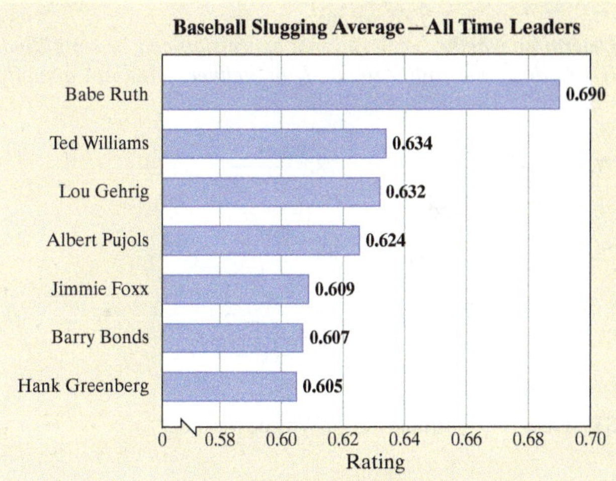

66. To calculate a quarterback's rating in NCAA football, you may use the formula $\frac{100C + 330T - 200I + 8.4Y}{A}$, where C = the number of completed passes, A = the number of attempted passes, T = the number of touchdown passes, Y = the number of yards in the completed passes, and I = the number of interceptions. Cam Newton of Auburn University was selected as the 2010 winner of the Heisman Memorial Trophy as the Most Outstanding Football Player. Newton, a junior quarterback, ended the season with 280 attempts, 185 completions, 2854 yards, 30 touchdowns, and only 7 interceptions. Calculate Newton's quarterback rating for the 2010 season. (*Source:* NCAA) Round the answer to the nearest tenth.

REVIEW AND PREVIEW

Perform each indicated operation. See Section 4.3.

67. $\frac{1}{3} \cdot \frac{9}{11}$

68. $\frac{5}{27} \cdot \frac{2}{5}$

69. $\frac{1}{3} \div \frac{1}{4}$

70. $\frac{7}{8} \div \frac{1}{2}$

71. $\frac{13}{20} \div \frac{2}{9}$

72. $\frac{8}{15} \div \frac{5}{8}$

CONCEPT EXTENSIONS

Which of the following are incorrect and why? See the second Concept Check in this section.

73. $\frac{5a - 15}{5}$ simplifies to $a - 3$?

74. $\frac{7m - 9}{7}$ simplifies to $m - 9$?

75. $\frac{1 + 2}{1 + 3}$ simplifies to $\frac{2}{3}$?

76. $\frac{46}{54}$ simplifies to $\frac{6}{5}$?

Determine whether each rational expression can be simplified. If yes, does it simplify to 1, –1, or neither? (Do not actually simplify.)

77. $\frac{x}{x + 7}$

78. $\frac{x + 9}{x - 9}$

79. $\frac{3 + x}{x + 3}$

80. $\frac{8 + x}{x + 8}$

81. $\frac{5 - x}{x - 5}$

82. $\frac{x - 7}{-x + 7}$

83. Does $\frac{x}{x + 5}$ simplify to $\frac{1}{5}$? Why or why not?

84. Does $\frac{x + 7}{x}$ simplify to 7? Why or why not?

85. In your own words, explain how to simplify a rational? expression.

86. In your own words, explain how to find the domain of a rational function.

87. Graph a portion of the function $f(x) = \frac{20x}{100 - x}$. To do so, complete the given table, plot the points, and then connect the plotted points with a smooth curve. (Note: The domain of this function is all real numbers except 100. We are graphing just a portion of this function.)

x	0	10	30	50	70	90	95	99
y or $f(x)$								

88. The domain of the function $f(x) = \frac{1}{x}$ is all real numbers except 0. This means that the graph of this function will be in two pieces: one piece corresponding to x values less than 0 and one piece corresponding to x values greater than 0. Graph the function by completing the following tables, separately plotting the points, and connecting each set of plotted points with a smooth curve.

x	$\frac{1}{4}$	$\frac{1}{2}$	1	2	4
y or $f(x)$					

x	−4	−2	−1	$-\frac{1}{2}$	$-\frac{1}{4}$
y or $f(x)$					

How does the graph of $y = \frac{x^2 - 9}{x - 3}$ compare to the graph of $y = x + 3$? Recall that $\frac{x^2 - 9}{x - 3} = \frac{(x + 3)(x - 3)}{x - 3} = x + 3$ as long as x is not 3. This means that the graph of $y = \frac{x^2 - 9}{x - 3}$ is the same as the graph of $y = x + 3$ with $x \neq 3$. To graph $y = \frac{x^2 - 9}{x - 3}$, then, graph the linear equation $y = x + 3$ and place an open dot on the graph at 3. This open dot or interruption of the line at 3 means $x \neq 3$.

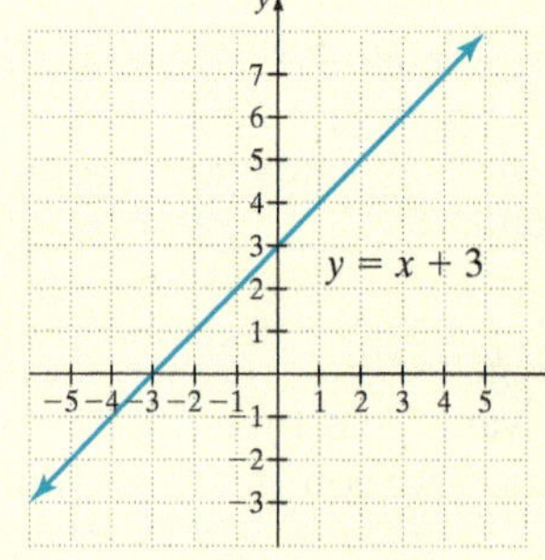

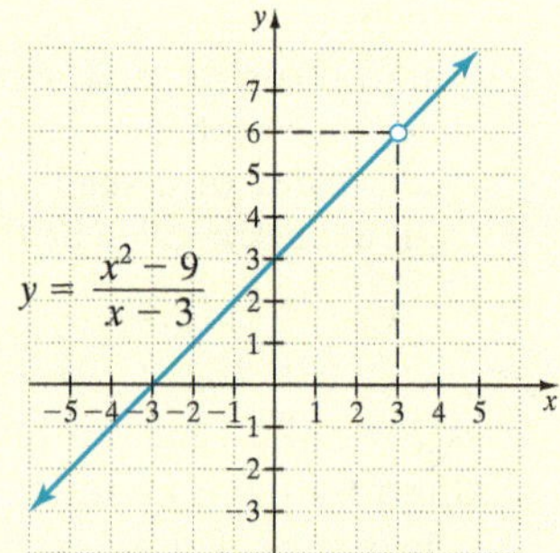

89. Graph $y = \frac{x^2 - 16}{x - 4}$.

90. Graph $y = \frac{x^2 - 25}{x + 5}$.

91. Graph $y = \frac{x^2 - 6x + 8}{x - 2}$.

92. Graph $y = \frac{x^2 + x - 12}{x + 4}$.

14.2 Multiplying and Dividing Rational Expressions

Objective 1 Multiplying Rational Expressions

Just as simplifying rational expressions is similar to simplifying number fractions, multiplying and dividing rational expressions is similar to multiplying and dividing number fractions.

Fractions	**Rational Expressions**
Multiply: $\frac{3}{5} \cdot \frac{10}{11}$	Multiply: $\frac{x-3}{x+5} \cdot \frac{2x+10}{x^2-9}$

Multiply numerators and multiply denominators.

$\frac{3}{5} \cdot \frac{10}{11} = \frac{3 \cdot 10}{5 \cdot 11}$	$\frac{x-3}{x+5} \cdot \frac{2x+10}{x^2-9} = \frac{(x-3)\cdot(2x+10)}{(x+5)\cdot(x^2-9)}$

Simplify by factoring numerators and denominators.

$= \frac{3 \cdot 2 \cdot 5}{5 \cdot 11}$	$= \frac{(x-3)\cdot 2(x+5)}{(x+5)(x+3)(x-3)}$

Apply the fundamental principle.

$= \frac{3 \cdot 2}{11}$ or $\frac{6}{11}$	$= \frac{2}{x+3}$

Multiplying Rational Expressions

If $\frac{P}{Q}$ and $\frac{R}{S}$ are rational expressions, then

$$\frac{P}{Q} \cdot \frac{R}{S} = \frac{PR}{QS}$$

To multiply rational expressions, multiply the numerators and multiply the denominators.

Note: Recall that for Sections 14.1 through 14.4, we assume variables in rational expressions have only those replacement values for which the expressions are defined.

Example 1 Multiply.

a. $\frac{25x}{2} \cdot \frac{1}{y^3}$ **b.** $\frac{-7x^2}{5y} \cdot \frac{3y^5}{14x^2}$

Solution: To multiply rational expressions, multiply the numerators and multiply the denominators of both expressions. Then simplify if possible.

a. $\frac{25x}{2} \cdot \frac{1}{y^3} = \frac{25x \cdot 1}{2 \cdot y^3} = \frac{25x}{2y^3}$

The expression $\frac{25x}{2y^3}$ is in simplest form.

b. $\frac{-7x^2}{5y} \cdot \frac{3y^5}{14x^2} = \frac{-7x^2 \cdot 3y^5}{5y \cdot 14x^2}$ Multiply.

(Continued on next page)

Objectives

1. Multiply Rational Expressions.
2. Divide Rational Expressions.
3. Multiply or Divide Rational Expressions.
4. Convert Between Units of Measure.

Practice 1

Multiply.

a. $\frac{4a}{5} \cdot \frac{3}{b^2}$ **b.** $\frac{-3p^4}{q^2} \cdot \frac{2q^3}{9p^4}$

Answers

1. a. $\frac{12a}{5b^2}$ **b.** $-\frac{2q}{3}$

The expression $\frac{-7x^2 \cdot 3y^5}{5y \cdot 14x^2}$ is not in simplest form, so we factor the numerator and the denominator and divide out common factors.

$$= \frac{-1 \cdot 7 \cdot 3 \cdot x^2 \cdot y \cdot y^4}{5 \cdot 2 \cdot 7 \cdot x^2 \cdot y}$$

$$= -\frac{3y^4}{10}$$

Work Practice 1

When multiplying rational expressions, it is usually best to factor each numerator and denominator. This will help us when we divide out common factors to write the product in lowest terms.

Practice 2

Multiply: $\frac{x^2 - x}{5x} \cdot \frac{15}{x^2 - 1}$

Example 2 Multiply: $\frac{x^2 + x}{3x} \cdot \frac{6}{5x + 5}$

Solution:

$$\frac{x^2 + x}{3x} \cdot \frac{6}{5x + 5} = \frac{x(x + 1)}{3x} \cdot \frac{2 \cdot 3}{5(x + 1)}$$ Factor numerators and denominators.

$$= \frac{x(x + 1) \cdot 2 \cdot 3}{3x \cdot 5\,(x + 1)}$$ Multiply.

$$= \frac{2}{5}$$ Simplify by dividing out common factors.

Work Practice 2

The following steps may be used to multiply rational expressions.

Multiplying Rational Expressions

Step 1: Completely factor numerators and denominators.

Step 2: Multiply numerators and multiply denominators.

Step 3: Simplify or write the product in lowest terms by dividing out common factors.

✓**Concept Check** Which of the following is a true statement?

a. $\frac{1}{3} \cdot \frac{1}{2} = \frac{1}{5}$ **b.** $\frac{2}{x} \cdot \frac{5}{x} = \frac{10}{x}$ **c.** $\frac{3}{x} \cdot \frac{1}{2} = \frac{3}{2x}$ **d.** $\frac{x}{7} \cdot \frac{x + 5}{4} = \frac{2x + 5}{28}$

Practice 3

Multiply:
$\frac{6 - 3x}{6x + 6x^2} \cdot \frac{3x^2 - 2x - 5}{x^2 - 4}$.

Example 3 Multiply: $\frac{3x + 3}{5x - 5x^2} \cdot \frac{2x^2 + x - 3}{4x^2 - 9}$

Solution:

$$\frac{3x + 3}{5x - 5x^2} \cdot \frac{2x^2 + x - 3}{4x^2 - 9} = \frac{3(x + 1)}{5x(1 - x)} \cdot \frac{(2x + 3)(x - 1)}{(2x - 3)(2x + 3)}$$ Factor.

$$= \frac{3(x + 1)(2x + 3)(x - 1)}{5x(1 - x)(2x - 3)(2x + 3)}$$ Multiply.

$$= \frac{3(x + 1)(x - 1)}{5x(1 - x)(2x - 3)}$$ Divide out common factors.

Answers

2. $\frac{3}{x + 1}$ **3.** $-\frac{3x - 5}{2x(x + 2)}$

✓**Concept Check Answer**

c

Next, recall that $x - 1$ and $1 - x$ are opposites so that $x - 1 = -1(1 - x)$.

$$= \frac{3(x+1)(-1)(1-x)}{5x(1-x)(2x-3)} \quad \text{Write } x - 1 \text{ as } -1(1-x).$$

$$= \frac{-3(x+1)}{5x(2x-3)} \quad \text{or} \quad -\frac{3(x+1)}{5x(2x-3)} \quad \text{Divide out common factors.}$$

Work Practice 3

Objective 2 Dividing Rational Expressions

We can divide by a rational expression in the same way we divide by a fraction. To divide by a fraction, multiply by its reciprocal.

Helpful Hint

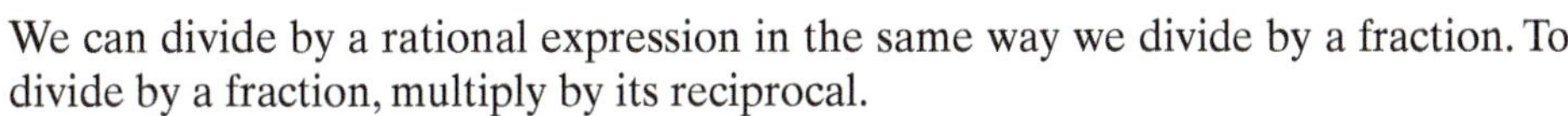

Don't forget how to find reciprocals. The reciprocal of $\frac{a}{b}$ is $\frac{b}{a}$, $a \neq 0, b \neq 0$.

For example, to divide $\frac{3}{2}$ by $\frac{7}{8}$, multiply $\frac{3}{2}$ by $\frac{8}{7}$.

$$\frac{3}{2} \div \frac{7}{8} = \frac{3}{2} \cdot \frac{8}{7} = \frac{3 \cdot 4 \cdot 2}{2 \cdot 7} = \frac{12}{7}$$

Dividing Rational Expressions

If $\frac{P}{Q}$ and $\frac{R}{S}$ are rational expressions and $\frac{R}{S}$ is not 0, then

$$\frac{P}{Q} \div \frac{R}{S} = \frac{P}{Q} \cdot \frac{S}{R} = \frac{PS}{QR}$$

To divide two rational expressions, multiply the first rational expression by the reciprocal of the second rational expression.

Example 4 Divide: $\frac{3x^3y^7}{40} \div \frac{4x^3}{y^2}$

Solution:

$$\frac{3x^3y^7}{40} \div \frac{4x^3}{y^2} = \frac{3x^3y^7}{40} \cdot \frac{y^2}{4x^3} \quad \text{Multiply by the reciprocal of } \frac{4x^3}{y^2}.$$

$$= \frac{3x^3y^9}{160x^3}$$

$$= \frac{3y^9}{160} \quad \text{Simplify.}$$

Work Practice 4

Practice 4

Divide: $\frac{5a^3b^2}{24} \div \frac{10a^5}{6}$

Answer

4. $\frac{b^2}{8a^2}$

Practice 5

Divide $\frac{(x-5)^2}{3}$ by $\frac{4x-20}{9}$.

Helpful Hint Remember, **to Divide by a Rational Expression,** multiply by its reciprocal.

Example 5

Divide: $\frac{(x+2)^2}{10} \div \frac{2x+4}{5}$

Solution:

$$\frac{(x+2)^2}{10} \div \frac{2x+4}{5} = \frac{(x+2)^2}{10} \cdot \frac{5}{2x+4}$$ Multiply by the reciprocal of $\frac{2x+4}{5}$.

$$= \frac{(x+2)(x+2) \cdot 5}{5 \cdot 2 \cdot 2 \cdot (x+2)}$$ Factor and multiply.

$$= \frac{x+2}{4}$$ Simplify.

Work Practice 5

The following may be used to divide by a rational expression.

Dividing by a Rational Expression

Multiply by its reciprocal.

Practice 6

Divide $\frac{10x-2}{x^2-9} \div \frac{5x^2-x}{x+3}$.

Example 6

Divide: $\frac{6x+2}{x^2-1} \div \frac{3x^2+x}{x-1}$

Solution:

$$\frac{6x+2}{x^2-1} \div \frac{3x^2+x}{x-1} = \frac{6x+2}{x^2-1} \cdot \frac{x-1}{3x^2+x}$$ Multiply by the reciprocal.

$$= \frac{2(3x+1)(x-1)}{(x+1)(x-1) \cdot x(3x+1)}$$ Factor and multiply.

$$= \frac{2}{x(x+1)}$$ Simplify.

Work Practice 6

Practice 7

Divide

$\frac{3x^2-11x-4}{2x-8} \div \frac{9x+3}{6}$.

Example 7

Divide: $\frac{2x^2-11x+5}{5x-25} \div \frac{4x-2}{10}$

Solution:

$$\frac{2x^2-11x+5}{5x-25} \div \frac{4x-2}{10} = \frac{2x^2-11x+5}{5x-25} \cdot \frac{10}{4x-2}$$ Multiply by the reciprocal.

$$= \frac{(2x-1)(x-5) \cdot 2 \cdot 5}{5(x-5) \cdot 2(2x-1)}$$ Factor and multiply.

$$= \frac{1}{1} \text{ or } 1$$ Simplify.

Work Practice 7

Answers

5. $\frac{3(x-5)}{4}$ **6.** $\frac{2}{x(x-3)}$ **7.** 1

Objective 3 Multiplying or Dividing Rational Expressions

Let's make sure we understand the difference between multiplying and dividing rational expressions.

Rational Expressions	
Multiplication	Multiply the numerators and multiply the denominators.
Division	Multiply by the reciprocal of the divisor.

Example 8 Multiply or divide as indicated.

a. $\frac{x-4}{5} \cdot \frac{x}{x-4}$ **b.** $\frac{x-4}{5} \div \frac{x}{x-4}$ **c.** $\frac{x^2-4}{2x+6} \cdot \frac{x^2+4x+3}{2-x}$

Solution:

a. $\frac{x-4}{5} \cdot \frac{x}{x-4} = \frac{(x-4) \cdot x}{5 \cdot (x-4)} = \frac{x}{5}$

b. $\frac{x-4}{5} \div \frac{x}{x-4} = \frac{x-4}{5} \cdot \frac{x-4}{x} = \frac{(x-4)^2}{5x}$

c. $\frac{x^2-4}{2x+6} \cdot \frac{x^2+4x+3}{2-x} = \frac{(x-2)(x+2) \cdot (x+1)(x+3)}{2(x+3) \cdot (2-x)}$ Factor and multiply.

$= \frac{(x-2)(x+2) \cdot (x+1)(x+3)}{2(x+3) \cdot (2-x)}$

$= \frac{-1(x+2)(x+1)}{2}$ Divide out common factors. Recall that $\frac{x-2}{2-x} = -1$.

$= -\frac{(x+2)(x+1)}{2}$

Work Practice 8

Practice 8

Multiply or divide as indicated.

a. $\frac{y+9}{8x} \cdot \frac{y+9}{2x}$

b. $\frac{y+9}{8x} \div \frac{y+9}{2}$

c. $\frac{35x-7x^2}{x^2-25} \cdot \frac{x^2+3x-10}{x^2+4x}$

Objective 4 Converting Between Units of Measure 

How many square inches are in 1 square foot?
How many cubic feet are in a cubic yard?

If you have trouble answering these questions, this section will be helpful to you.

Now that we know how to multiply fractions and rational expressions, we can use this knowledge to help us convert between units of measure. To do so, we will use **unit fractions.** A unit fraction is a fraction that equals 1. For example, since 12 in. = 1 ft, we have the unit fractions

$$\frac{12 \text{ in.}}{1 \text{ ft}} = 1 \quad \text{and} \quad \frac{1 \text{ ft}}{12 \text{ in.}} = 1$$

Answers

8. a. $\frac{(y+9)^2}{16x^2}$ **b.** $\frac{1}{4x}$ **c.** $-\frac{7(x-2)}{x+4}$

Practice 9

288 square inches = _____ square feet

Example 9 18 square feet = _____ square yards

Solution: Let's multiply 18 square feet by a unit fraction that has square feet in the denominator and square yards in the numerator. From the diagram, you can see that

1 square yard = 9 square feet

Thus,

$$18 \text{ sq ft} = \frac{18 \text{ sq ft}}{1} \cdot 1 = \frac{\overset{2}{\cancel{18}} \cancel{\text{sq ft}}}{1} \cdot \frac{1 \text{ sq yd}}{\underset{1}{\cancel{9}} \cancel{\text{sq ft}}}$$

$$= \frac{2 \cdot 1}{1 \cdot 1} \text{ sq yd} = 2 \text{ sq yd}$$

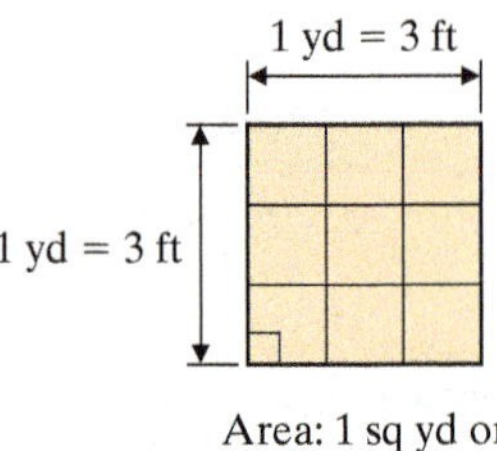

Area: 1 sq yd or 9 sq ft

Thus, 18 sq ft = 2 sq yd.

Draw a diagram of 18 sq ft to help you see that this is reasonable.

Work Practice 9

Practice 10

3.5 square feet = _____ square inches

Example 10 5.2 square yards = _____ square feet

Solution:

$$5.2 \text{ sq yd} = \frac{5.2 \text{ sq yd}}{1} \cdot 1 = \frac{5.2 \cancel{\text{sq yd}}}{1} \cdot \frac{9 \text{ sq ft}}{1 \cancel{\text{sq yd}}}$$ ← Units converting to / ← Units given

$$= \frac{5.2 \cdot 9}{1 \cdot 1} \text{ sq ft}$$

$$= 46.8 \text{ sq ft}$$

Thus, 5.2 sq yd = 46.8 sq ft.

Draw a diagram to see that this is reasonable.

Work Practice 10

Practice 11

The largest casino in the world is the Venetian, in Macau, on the southern tip of China. The gaming area for this casino is approximately 61,000 *square yards*. Find the size of the gaming area in *square feet*. (*Source: USA Today*)

Example 11 Converting from Cubic Feet to Cubic Yards

The largest building in the world by volume is The Boeing Company's Everett, Washington, factory complex, where Boeing's wide-body jetliners, the 747, 767, and 777, are built. The volume of this factory complex is 472,370,319 cubic feet. Find the volume of this Boeing facility in cubic yards. (*Source:* The Boeing Company)

Solution: There are 27 cubic feet in 1 cubic yard. (See the diagram.)

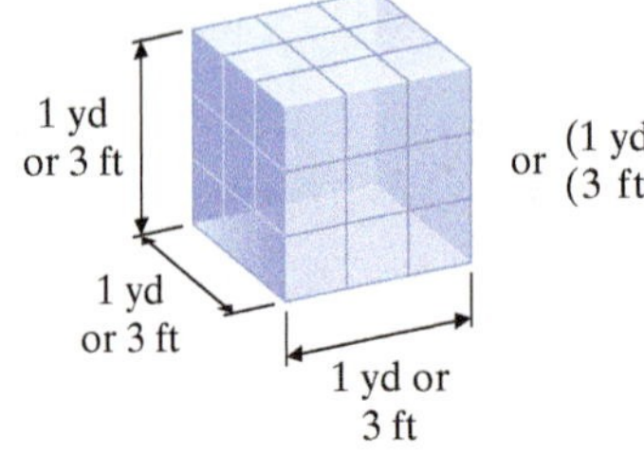

or (1 yd)(1 yd)(1 yd) = 1 cubic yard
(3 ft)(3 ft)(3 ft) = 27 cubic feet

Answers

9. 2 sq ft **10.** 504 sq in. **11.** 549,000 sq ft

$$472{,}370{,}319 \text{ cu ft} = 472{,}370{,}319 \cancel{\text{cu ft}} \cdot \frac{1 \text{ cu yd}}{27 \cancel{\text{cu ft}}}$$

$$= \frac{472{,}370{,}319}{27} \text{ cu yd}$$

$$= 17{,}495{,}197 \text{ cu yd}$$

Work Practice 11

Helpful Hint

When converting among units of measurement, if possible write the unit fraction so that **the numerator contains the units you are converting to** and **the denominator contains the original units.**

Unit fraction

$$48 \text{ in.} = \frac{48 \cancel{\text{in.}}}{1} \cdot \frac{1 \text{ ft}}{12 \cancel{\text{in.}}} \quad \begin{matrix} \leftarrow \text{Units converting to} \\ \leftarrow \text{Original units} \end{matrix}$$

$$= \frac{48}{12} \text{ ft} = 4 \text{ ft}$$

Example 12

At the 2012 Summer Olympics, Jamaican athlete Usain Bolt won the gold medal in the men's 100-meter track event. He ran the distance at an average speed of 34.1 feet per second. Convert this speed to miles per hour. (*Source:* London 2012 Olympics Committee)

Solution: Recall that 1 mile = 5280 feet and 1 hour = 3600 seconds $(60 \cdot 60)$.

Unit fractions

$$34.1 \text{ feet/second} = \frac{34.1 \text{ feet}}{1 \text{ second}} \cdot \frac{3600 \text{ seconds}}{1 \text{ hour}} \cdot \frac{1 \text{ mile}}{5280 \text{ feet}}$$

$$= \frac{34.1 \cdot 3600}{5280} \text{ miles/hour}$$

$$\approx 23.3 \text{ miles/hour (rounded to the nearest tenth)}$$

Work Practice 12

Practice 12

The cheetah is the fastest land animal, being clocked at about 102.7 feet per second. Convert this to miles per hour. Round to the nearest tenth. (*Source: World Almanac and Book of Facts*)

Answer

12. 70.0 miles per hour

Vocabulary, Readiness & Video Check

Use one of the choices below to fill in the blank.

opposites reciprocals

1. The expressions $\frac{x}{2y}$ and $\frac{2y}{x}$ are called ________.

Multiply or divide as indicated.

2. $\frac{a}{b} \cdot \frac{c}{d} =$ ________ **3.** $\frac{a}{b} \div \frac{c}{d} =$ ________ **4.** $\frac{x}{7} \cdot \frac{x}{6} =$ ________ **5.** $\frac{x}{7} \div \frac{x}{6} =$ ________

Martin-Gay Interactive Videos

See Video 14.2

Watch the section lecture video and answer the following questions.

Objective 1 **6.** Would you say a person needs to be quite comfortable with factoring polynomials in order to be successful with multiplying rational expressions? Explain, referencing Example 2 in your answer.

Objective 2 **7.** Based on the lecture before Example 3, complete the following statements. Dividing rational expressions is exactly like dividing ________. Therefore, to divide by a rational expression, multiply by its ________.

Objective 3 **8.** In Examples 4 and 5, determining the operation is the first step in deciding how to simplify. Why do you think this is so?

Objective 4 **9.** When converting between units of measurement, a unit fraction may be used. What units are used in the numerator and what units are used in the denominator of your unit fraction?

14.2 Exercise Set MyMathLab®

Find each product and simplify if possible. See Examples 1 through 3.

1. $\frac{3x}{y^2} \cdot \frac{7y}{4x}$

2. $\frac{9x^2}{y} \cdot \frac{4y}{3x^3}$

3. $\frac{8x}{2} \cdot \frac{x^5}{4x^2}$

4. $\frac{6x^2}{10x^3} \cdot \frac{5x}{12}$

5. $-\frac{5a^2b}{30a^2b^2} \cdot b^3$

6. $-\frac{9x^3y^2}{18xy^5} \cdot y^3$

7. $\frac{x}{2x-14} \cdot \frac{x^2-7x}{5}$

8. $\frac{4x-24}{20x} \cdot \frac{5}{x-6}$

9. $\frac{6x+6}{5} \cdot \frac{10}{36x+36}$

10. $\frac{x^2+x}{8} \cdot \frac{16}{x+1}$

11. $\frac{(m+n)^2}{m-n} \cdot \frac{m}{m^2+mn}$

12. $\frac{(m-n)^2}{m+n} \cdot \frac{m}{m^2-mn}$

13. $\frac{x^2-25}{x^2-3x-10} \cdot \frac{x+2}{x}$

14. $\frac{a^2-4a+4}{a^2-4} \cdot \frac{a+3}{a-2}$

15. $\frac{x^2+6x+8}{x^2+x-20} \cdot \frac{x^2+2x-15}{x^2+8x+16}$

16. $\frac{x^2+9x+20}{x^2-15x+44} \cdot \frac{x^2-11x+28}{x^2+12x+35}$

Find each quotient and simplify. See Examples 4 through 7.

17. $\frac{5x^7}{2x^5} \div \frac{15x}{4x^3}$

18. $\frac{9y^4}{6y} \div \frac{y^2}{3}$

19. $\frac{8x^2}{y^3} \div \frac{4x^2y^3}{6}$

20. $\frac{7a^2b}{3ab^2} \div \frac{21a^2b^2}{14ab}$

21. $\frac{(x-6)(x+4)}{4x} \div \frac{2x-12}{8x^2}$

22. $\frac{(x+3)^2}{5} \div \frac{5x+15}{25}$

23. $\frac{3x^2}{x^2-1} \div \frac{x^5}{(x+1)^2}$

24. $\frac{9x^5}{a^2-b^2} \div \frac{27x^2}{3b-3a}$

25. $\frac{m^2-n^2}{m+n} \div \frac{m}{m^2+nm}$

26. $\frac{(m-n)^2}{m+n} \div \frac{m^2-mn}{m}$

27. $\frac{x+2}{7-x} \div \frac{x^2-5x+6}{x^2-9x+14}$

28. $\frac{x-3}{2-x} \div \frac{x^2+3x-18}{x^2+2x-8}$

29. $\frac{x^2+7x+10}{x-1} \div \frac{x^2+2x-15}{x-1}$

30. $\frac{x+1}{(x+1)(2x+3)} \div \frac{20x+100}{2x+3}$

MIXED PRACTICE

Multiply or divide as indicated. See Examples 1 through 8.

31. $\frac{5x-10}{12} \div \frac{4x-8}{8}$

32. $\frac{6x+6}{5} \div \frac{9x+9}{10}$

33. $\frac{x^2+5x}{8} \cdot \frac{9}{3x+15}$

34. $\frac{3x^2+12x}{6} \cdot \frac{9}{2x+8}$

35. $\frac{7}{6p^2+q} \div \frac{14}{18p^2+3q}$

36. $\frac{3x+6}{20} \div \frac{4x+8}{8}$

37. $\frac{3x+4y}{x^2+4xy+4y^2} \cdot \frac{x+2y}{2}$

38. $\frac{x^2-y^2}{3x^2+3xy} \cdot \frac{3x^2+6x}{3x^2-2xy-y^2}$

39. $\frac{(x+2)^2}{x-2} \div \frac{x^2-4}{2x-4}$

40. $\frac{x+3}{x^2-9} \div \frac{5x+15}{(x-3)^2}$

41. $\frac{x^2-4}{24x} \div \frac{2-x}{6xy}$

42. $\frac{3y}{3-x} \div \frac{12xy}{x^2-9}$

43. $\frac{a^2+7a+12}{a^2+5a+6} \cdot \frac{a^2+8a+15}{a^2+5a+4}$

44. $\frac{b^2+2b-3}{b^2+b-2} \cdot \frac{b^2-4}{b^2+6b+8}$

45. $\frac{5x-20}{3x^2+x} \cdot \frac{3x^2+13x+4}{x^2-16}$

46. $\frac{9x+18}{4x^2-3x} \cdot \frac{4x^2-11x+6}{x^2-4}$

47. $\frac{8n^2-18}{2n^2-5n+3} \div \frac{6n^2+7n-3}{n^2-9n+8}$

48. $\frac{36n^2-64}{3n^2+10n+8} \div \frac{3n^2-13n+12}{n^2-5n-14}$

49. Find the quotient of $\frac{x^2-9}{2x}$ and $\frac{x+3}{8x^4}$.

50. Find the quotient of $\frac{4x^2+4x+1}{4x+2}$ and $\frac{4x+2}{16}$.

Multiply or divide as indicated. Some of these expressions contain 4-term polynomials and sums and differences of cubes. See Examples 1 through 8.

51. $\frac{a^2+ac+ba+bc}{a-b} \div \frac{a+c}{a+b}$

52. $\frac{x^2+2x-xy-2y}{x^2-y^2} \div \frac{2x+4}{x+y}$

53. $\frac{3x^2+8x+5}{x^2+8x+7} \cdot \frac{x+7}{x^2+4}$

54. $\frac{16x^2+2x}{16x^2+10x+1} \cdot \frac{1}{4x^2+2x}$

55. $\frac{x^3+8}{x^2-2x+4} \cdot \frac{4}{x^2-4}$

56. $\frac{9y}{3y-3} \cdot \frac{y^3-1}{y^3+y^2+y}$

57. $\frac{a^2-ab}{6a^2+6ab} \div \frac{a^3-b^3}{a^2-b^2}$

58. $\frac{x^3+27y^3}{6x} \div \frac{x^2-9y^2}{x^2-3xy}$

Convert as indicated. See Examples 9 through 12.

59. 10 square feet = ________ square inches.

60. 1008 square inches = ________ square feet.

61. 45 square feet = _____ square yards.

62. 2 square yards = ________ square inches.

63. 3 cubic yards = ________ cubic feet.

64. 2 cubic yards = ________ cubic inches.

65. 50 miles per hour = ________ feet per second (round to the nearest whole).

66. 10 feet per second = ________ miles per hour (round to the nearest tenth).

67. 6.3 square yards = ________ square feet.

68. 3.6 square yards = ________ square feet.

69. In January 2010, the Burj Khalifa Tower officially became the tallest building in the world. This tower has a curtain wall (the exterior skin of the building) that is approximately 133,500 square yards. Convert this to square feet. (*Source:* Burj Khalifa)

70. The Pentagon, headquarters for the Department of Defense, contains 3,705,793 square feet of office and storage space. Convert this to square yards. Round to the nearest square yard. (*Source:* U.S. Department of Defense)

71. On October 9, 2007, Russ Wicks set a new stock car world speed record of 359.2 feet/second on the Bonneville Salt Flats in Utah. Convert this speed to miles/hour. Round to the nearest tenth. (*Source:* RussWicks.com)

72. On October 4, 2004, the rocket plane *SpaceShipOne* shot to an altitude of more than 100 km for the second time inside a week to claim the $10 million Ansari X-Prize. At one point in its flight, *SpaceShipOne* was traveling past Mach 1, about 930 miles per hour. Find this speed in feet per second. (*Source:* Space.com)

REVIEW AND PREVIEW

Perform each indicated operation. See Section 4.4.

73. $\frac{1}{5}+\frac{4}{5}$

74. $\frac{3}{15}+\frac{6}{15}$

75. $\frac{9}{9} - \frac{19}{9}$

76. $\frac{4}{3} - \frac{8}{3}$

77. $\frac{6}{5} + \left(\frac{1}{5} - \frac{8}{5}\right)$

78. $-\frac{3}{2} + \left(\frac{1}{2} - \frac{3}{2}\right)$

Graph each linear equation. See Section 10.2.

79. $x - 2y = 6$

80. $5x - y = 10$

CONCEPT EXTENSIONS

Identify each statement as true or false. If false, correct the multiplication. See the Concept Check in this section.

81. $\frac{4}{a} \cdot \frac{1}{b} = \frac{4}{ab}$

82. $\frac{2}{3} \cdot \frac{2}{4} = \frac{2}{7}$

83. $\frac{x}{5} \cdot \frac{x+3}{4} = \frac{2x+3}{20}$

84. $\frac{7}{a} \cdot \frac{3}{a} = \frac{21}{a}$

85. Find the area of the rectangle.

$\frac{2x}{x^2 - 25}$ feet

$\frac{x+5}{9x}$ feet

86. Find the area of the square.

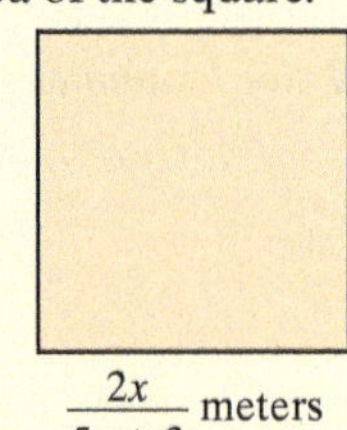

$\frac{2x}{5x+3}$ meters

Multiply or divide as indicated.

87. $\left(\frac{x^2 - y^2}{x^2 + y^2} \div \frac{x^2 - y^2}{3x}\right) \cdot \frac{x^2 + y^2}{6}$

88. $\left(\frac{x^2 - 9}{x^2 - 1} \cdot \frac{x^2 + 2x + 1}{2x^2 + 9x + 9}\right) \div \frac{2x + 3}{1 - x}$

89. $\left(\frac{2a + b}{b^2} \cdot \frac{3a^2 - 2ab}{ab + 2b^2}\right) \div \frac{a^2 - 3ab + 2b^2}{5ab - 10b^2}$

90. $\left(\frac{x^2y^2 - xy}{4x - 4y} \div \frac{3y - 3x}{8x - 8y}\right) \cdot \frac{y - x}{8}$

91. In your own words, explain how you multiply rational expressions.

92. Explain how dividing rational expressions is similar to dividing rational numbers.

14.3 Adding and Subtracting Rational Expressions with Common Denominators and Least Common Denominator

Objectives

1. Add and Subtract Rational Expressions with the Same Denominator.
2. Find the Least Common Denominator of a List of Rational Expressions.
3. Write a Rational Expression as an Equivalent Expression Whose Denominator Is Given.

Objective 1 Adding and Subtracting Rational Expressions with the Same Denominator 

Like multiplication and division, addition and subtraction of rational expressions is similar to addition and subtraction of rational numbers. In this section, we add and subtract rational expressions with a common (or the same) denominator.

Add: $\frac{6}{5} + \frac{2}{5}$ | Add: $\frac{9}{x+2} + \frac{3}{x+2}$

Add the numerators and place the sum over the common denominator.

$$\frac{6}{5} + \frac{2}{5} = \frac{6+2}{5}$$
$$= \frac{8}{5} \quad \text{Simplify.}$$

$$\frac{9}{x+2} + \frac{3}{x+2} = \frac{9+3}{x+2}$$
$$= \frac{12}{x+2} \quad \text{Simplify.}$$

Adding and Subtracting Rational Expressions with Common Denominators

If $\frac{P}{R}$ and $\frac{Q}{R}$ are rational expressions, then

$$\frac{P}{R} + \frac{Q}{R} = \frac{P+Q}{R} \quad \text{and} \quad \frac{P}{R} - \frac{Q}{R} = \frac{P-Q}{R}$$

To add or subtract rational expressions, add or subtract the numerators and place the sum or difference over the common denominator.

Example 1 Add: $\frac{5m}{2n} + \frac{m}{2n}$

Solution:

$$\frac{5m}{2n} + \frac{m}{2n} = \frac{5m+m}{2n}$$ Add the numerators.

$$= \frac{6m}{2n}$$ Simplify the numerator by combining like terms.

$$= \frac{3m}{n}$$ Simplify by applying the fundamental principle.

Work Practice 1

Practice 1

Add: $\frac{7a}{4b} + \frac{a}{4b}$

Example 2 Subtract: $\frac{2y}{2y-7} - \frac{7}{2y-7}$

Solution:

$$\frac{2y}{2y-7} - \frac{7}{2y-7} = \frac{2y-7}{2y-7}$$ Subtract the numerators.

$$= \frac{1}{1} \text{ or } 1$$ Simplify.

Work Practice 2

Practice 2

Subtract: $\frac{3x}{3x-2} - \frac{2}{3x-2}$

Example 3 Subtract: $\frac{3x^2+2x}{x-1} - \frac{10x-5}{x-1}$

Solution:

$$\frac{3x^2+2x}{x-1} - \frac{10x-5}{x-1} = \frac{(3x^2+2x)-(10x-5)}{x-1}$$ Subtract the numerators. Notice the parentheses.

$$= \frac{3x^2+2x-10x+5}{x-1}$$ Use the distributive property.

$$= \frac{3x^2-8x+5}{x-1}$$ Combine like terms.

$$= \frac{(x-1)(3x-5)}{x-1}$$ Factor.

$$= 3x-5$$ Simplify.

Work Practice 3

Practice 3

Subtract: $\frac{4x^2+15x}{x+3} - \frac{8x+15}{x+3}$

Helpful Hint

Parentheses are inserted so the entire numerator, $10x - 5$, can be subtracted.

Answers

1. $\frac{2a}{b}$ **2.** 1 **3.** $4x - 5$

Helpful Hint

Notice how the numerator $10x - 5$ has been subtracted in Example 3.

This − sign applies to the entire numerator of $10x - 5$.

So parentheses are inserted here to indicate this.

$$\frac{3x^2 + 2x}{x - 1} - \frac{10x - 5}{x - 1} = \frac{3x^2 + 2x - (10x - 5)}{x - 1}$$

Objective 2 Finding the Least Common Denominator

To add and subtract fractions with **unlike** denominators, first find a least common denominator (LCD) and then write all fractions as equivalent fractions with the LCD.

For example, suppose we add $\frac{8}{3}$ and $\frac{2}{5}$. The LCD of denominators 3 and 5 is 15, since 15 is the least common multiple (LCM) of 3 and 5. That is, 15 is the smallest number that both 3 and 5 divide into evenly.

Next, rewrite each fraction so that its denominator is 15.

$$\frac{8}{3} + \frac{2}{5} = \frac{8(5)}{3(5)} + \frac{2(3)}{5(3)} = \frac{40}{15} + \frac{6}{15} = \frac{40 + 6}{15} = \frac{46}{15}$$

We are multiplying by 1.

To add or subtract rational expressions with unlike denominators, we also first find an LCD and then write all rational expressions as equivalent expressions with the LCD. The **least common denominator (LCD) of a list of rational expressions** is a polynomial of least degree whose factors include all the factors of the denominators in the list.

Finding the Least Common Denominator (LCD)

Step 1: Factor each denominator completely.

Step 2: The least common denominator (LCD) is the product of all unique factors found in Step 1, each raised to a power equal to the greatest number of times that the factor appears in any one factored denominator.

Practice 4

Find the LCD for each pair.

a. $\frac{3}{14}, \frac{5}{21}$ **b.** $\frac{4}{9y}, \frac{11}{15y^3}$

Example 4 Find the LCD for each pair.

a. $\frac{1}{8}, \frac{3}{22}$ **b.** $\frac{7}{5x}, \frac{6}{15x^2}$

Solution:

a. Start by finding the prime factorization of each denominator.

$$8 = 2 \cdot 2 \cdot 2 = 2^3 \quad \text{and}$$
$$22 = 2 \cdot 11$$

Next, write the product of all the unique factors, each raised to a power equal to the greatest number of times that the factor appears in any denominator.

The greatest number of times that the factor 2 appears is 3.
The greatest number of times that the factor 11 appears is 1.

$$\text{LCD} = 2^3 \cdot 11^1 = 8 \cdot 11 = 88$$

Answers

4. a. 42 **b.** $45y^3$

b. Factor each denominator.

$$5x = 5 \cdot x \quad \text{and}$$
$$15x^2 = 3 \cdot 5 \cdot x^2$$

The greatest number of times that the factor 5 appears is 1.
The greatest number of times that the factor 3 appears is 1.
The greatest number of times that the factor x appears is 2.

$$\text{LCD} = 3^1 \cdot 5^1 \cdot x^2 = 15x^2$$

Work Practice 4

Example 5 Find the LCD of

a. $\frac{7x}{x+2}$ and $\frac{5x^2}{x-2}$ **b.** $\frac{3}{x}$ and $\frac{6}{x+4}$

Solution:

a. The denominators $x + 2$ and $x - 2$ are completely factored already. The factor $x + 2$ appears once and the factor $x - 2$ appears once.

$$\text{LCD} = (x + 2)(x - 2)$$

b. The denominators x and $x + 4$ cannot be factored further. The factor x appears once and the factor $x + 4$ appears once.

$$\text{LCD} = x(x + 4)$$

Work Practice 5

Example 6 Find the LCD of $\frac{6m^2}{3m+15}$ and $\frac{2}{(m+5)^2}$.

Solution: We factor each denominator.

$$3m + 15 = 3(m + 5)$$
$$(m + 5)^2 = (m + 5)^2 \quad \text{This denominator is already factored.}$$

The greatest number of times that the factor 3 appears is 1.
The greatest number of times that the factor $m + 5$ appears *in any one denominator* is 2.

$$\text{LCD} = 3(m + 5)^2$$

Work Practice 6

✓**Concept Check** Choose the correct LCD of $\frac{x}{(x+1)^2}$ and $\frac{5}{x+1}$.

a. $x + 1$ **b.** $(x + 1)^2$ **c.** $(x + 1)^3$ **d.** $5x(x + 1)^2$

Example 7 Find the LCD of $\frac{t-10}{t^2-t-6}$ and $\frac{t+5}{t^2+3t+2}$.

Solution: Start by factoring each denominator.

$$t^2 - t - 6 = (t - 3)(t + 2)$$
$$t^2 + 3t + 2 = (t + 1)(t + 2)$$
$$\text{LCD} = (t - 3)(t + 2)(t + 1)$$

Work Practice 7

Practice 5

Find the LCD of

a. $\frac{16}{y-5}$ and $\frac{3y^3}{y-4}$

b. $\frac{8}{a}$ and $\frac{5}{a+2}$

Practice 6

Find the LCD of $\frac{2x^3}{(2x-1)^2}$ and $\frac{5x}{6x-3}$.

Practice 7

Find the LCD of $\frac{x-5}{x^2+5x+4}$ and $\frac{x+8}{x^2-16}$.

Answers

5. a. $(y - 5)(y - 4)$ **b.** $a(a + 2)$
6. $3(2x - 1)^2$
7. $(x + 1)(x + 4)(x - 4)$

✓**Concept Check Answer**

b

Practice 8

Find the LCD of $\frac{5}{3-x}$ and $\frac{4}{x-3}$.

Example 8 Find the LCD of $\frac{2}{x-2}$ and $\frac{10}{2-x}$.

Solution: The denominators $x-2$ and $2-x$ are opposites. That is, $2-x=-1(x-2)$. Use $x-2$ or $2-x$ as the LCD.

$$\text{LCD} = x-2 \quad \text{or} \quad \text{LCD} = 2-x$$

Work Practice 8

Objective 3 Writing Equivalent Rational Expressions 

Next we practice writing a rational expression as an equivalent rational expression with a given denominator. To do this, we multiply by a form of 1. Recall that multiplying an expression by 1 produces an equivalent expression. In other words,

$$\frac{P}{Q} = \frac{P}{Q}\cdot 1 = \frac{P}{Q}\cdot\frac{R}{R} = \frac{PR}{QR}.$$

Practice 9

Write each rational expression as an equivalent fraction with the given denominator.

a. $\frac{3x}{5y} = \frac{\quad}{35xy^2}$

b. $\frac{9x}{4x+7} = \frac{\quad}{8x+14}$

Example 9 Write each rational expression as an equivalent rational expression with the given denominator.

a. $\frac{4b}{9a} = \frac{\quad}{27a^2b}$ **b.** $\frac{7x}{2x+5} = \frac{\quad}{6x+15}$

Solution:

a. We can ask ourselves: "What do we multiply $9a$ by to get $27a^2b$?" The answer is $3ab$, since $9a(3ab) = 27a^2b$. So we multiply by 1 in the form of $\frac{3ab}{3ab}$.

$$\frac{4b}{9a} = \frac{4b}{9a}\cdot 1$$

$$= \frac{4b}{9a}\cdot\frac{3ab}{3ab}$$

$$= \frac{4b(3ab)}{9a(3ab)} = \frac{12ab^2}{27a^2b}$$

b. First, factor the denominator on the right.

$$\frac{7x}{2x+5} = \frac{\quad}{3(2x+5)}$$

To obtain the denominator on the right from the denominator on the left, we multiply by 1 in the form of $\frac{3}{3}$.

$$\frac{7x}{2x+5} = \frac{7x}{2x+5}\cdot\frac{3}{3} = \frac{7x\cdot 3}{(2x+5)\cdot 3} = \frac{21x}{3(2x+5)} \text{ or } \frac{21x}{6x+15}$$

Work Practice 9

Answers

8. $3-x$ or $x-3$

9. a. $\frac{21x^2y}{35xy^2}$ **b.** $\frac{18x}{8x+14}$

Example 10 Write the rational expression as an equivalent rational expression with the given denominator.

$$\frac{5}{x^2 - 4} = \frac{\qquad}{(x - 2)(x + 2)(x - 4)}$$

Solution: First, factor the denominator $x^2 - 4$ as $(x - 2)(x + 2)$.

If we multiply the original denominator $(x - 2)(x + 2)$ by $x - 4$, the result is the new denominator, $(x - 2)(x + 2)(x - 4)$. Thus, we multiply by 1 in the form of $\frac{x - 4}{x - 4}$.

$$\frac{5}{x^2 - 4} = \frac{5}{(x - 2)(x + 2)} = \frac{5}{(x - 2)(x + 2)} \cdot \frac{x - 4}{x - 4}$$

Factored denominator

$$= \frac{5(x - 4)}{(x - 2)(x + 2)(x - 4)}$$

$$= \frac{5x - 20}{(x - 2)(x + 2)(x - 4)}$$

Work Practice 10

Practice 10

Write the rational expression as an equivalent rational expression with the given denominator.

$$\frac{3}{x^2 - 2x - 15} = \frac{\qquad}{(x - 2)(x + 3)(x - 5)}$$

Answer

10. $3x - 6$

Vocabulary, Readiness & Video Check

Use the choices below to fill in each blank. Not all choices will be used.

$\frac{9}{22}$ $\quad \frac{5}{22}$ $\quad \frac{9}{11}$ $\quad \frac{5}{11}$ $\quad \frac{ac}{b}$ $\quad \frac{a - c}{b}$ $\quad \frac{a + c}{b}$ $\quad \frac{5 - 6 + x}{x}$ $\quad \frac{5 - (6 + x)}{x}$

1. $\frac{7}{11} + \frac{2}{11} =$ ______

2. $\frac{7}{11} - \frac{2}{11} =$ ______

3. $\frac{a}{b} + \frac{c}{b} =$ ______

4. $\frac{a}{b} - \frac{c}{b} =$ ______

5. $\frac{5}{x} - \frac{6 + x}{x} =$ ______

Martin-Gay Interactive Videos

See Video 14.3

Watch the section lecture video and answer the following questions.

Objective 1 **6.** In Example 3, why is it important to place parentheses around the second numerator when writing as one expression?

Objective 2 **7.** In Examples 4 and 5, we factor the denominators completely. How does this help determine the LCD?

Objective 3 **8.** Based on Example 6, complete the following statements. To write an equivalent rational expression, you multiply the ______ of a rational expression by the same expression as the denominator. This means you're multiplying the original rational expression by a factor of ______ and therefore not changing the ______ of the original expression.

14.3 Exercise Set MyMathLab®

Add or subtract as indicated. Simplify the result if possible. See Examples 1 through 3.

1. $\frac{a+1}{13} + \frac{8}{13}$
2. $\frac{x+1}{7} + \frac{6}{7}$
3. $\frac{4m}{3n} + \frac{5m}{3n}$
4. $\frac{3p}{2q} + \frac{11p}{2q}$
5. $\frac{4m}{m-6} - \frac{24}{m-6}$
6. $\frac{8y}{y-2} - \frac{16}{y-2}$
7. $\frac{9}{3+y} + \frac{y+1}{3+y}$
8. $\frac{9}{y+9} + \frac{y-5}{y+9}$
9. $\frac{5x^2+4x}{x-1} - \frac{6x+3}{x-1}$
10. $\frac{x^2+9x}{x+7} - \frac{4x+14}{x+7}$
11. $\frac{4a}{a^2+2a-15} - \frac{12}{a^2+2a-15}$
12. $\frac{3y}{y^2+3y-10} - \frac{6}{y^2+3y-10}$
13. $\frac{2x+3}{x^2-x-30} - \frac{x-2}{x^2-x-30}$
14. $\frac{3x-1}{x^2+5x-6} - \frac{2x-7}{x^2+5x-6}$
15. $\frac{2x+1}{x-3} + \frac{3x+6}{x-3}$
16. $\frac{4p-3}{2p+7} + \frac{3p+8}{2p+7}$
17. $\frac{2x^2}{x-5} - \frac{25+x^2}{x-5}$
18. $\frac{6x^2}{2x-5} - \frac{25+2x^2}{2x-5}$
19. $\frac{5x+4}{x-1} - \frac{2x+7}{x-1}$
20. $\frac{7x+1}{x-4} - \frac{2x+21}{x-4}$

Find the LCD for each list of rational expressions. See Examples 4 through 8.

21. $\frac{19}{2x}, \frac{5}{4x^3}$
22. $\frac{17x}{4y^5}, \frac{2}{8y}$
23. $\frac{9}{8x}, \frac{3}{2x+4}$
24. $\frac{1}{6y}, \frac{3x}{4y+12}$
25. $\frac{2}{x+3}, \frac{5}{x-2}$
26. $\frac{-6}{x-1}, \frac{4}{x+5}$
27. $\frac{x}{x+6}, \frac{10}{3x+18}$
28. $\frac{12}{x+5}, \frac{x}{4x+20}$
29. $\frac{8x^2}{(x-6)^2}, \frac{13x}{5x-30}$
30. $\frac{9x^2}{7x-14}, \frac{6x}{(x-2)^2}$
31. $\frac{1}{3x+3}, \frac{8}{2x^2+4x+2}$
32. $\frac{19x+5}{4x-12}, \frac{3}{2x^2-12x+18}$
33. $\frac{5}{x-8}, \frac{3}{8-x}$
34. $\frac{2x+5}{3x-7}, \frac{5}{7-3x}$
35. $\frac{5x+1}{x^2+3x-4}, \frac{3x}{x^2+2x-3}$
36. $\frac{4}{x^2+4x+3}, \frac{4x-2}{x^2+10x+21}$
37. $\frac{2x}{3x^2+4x+1}, \frac{7}{2x^2-x-1}$
38. $\frac{3x}{4x^2+5x+1}, \frac{5}{3x^2-2x-1}$
39. $\frac{1}{x^2-16}, \frac{x+6}{2x^3-8x^2}$
40. $\frac{5}{x^2-25}, \frac{x+9}{3x^3-15x^2}$

Rewrite each rational expression as an equivalent rational expression with the given denominator. See Examples 9 and 10.

41. $\frac{3}{2x} = \frac{\quad}{4x^2}$
42. $\frac{3}{9y^5} = \frac{\quad}{72y^9}$
43. $\frac{6}{3a} = \frac{\quad}{12ab^2}$
44. $\frac{5}{4y^2x} = \frac{\quad}{32y^3x^2}$
45. $\frac{9}{2x+6} = \frac{\quad}{2y(x+3)}$
46. $\frac{4x+1}{3x+6} = \frac{\quad}{3y(x+2)}$
47. $\frac{9a+2}{5a+10} = \frac{\quad}{5b(a+2)}$
48. $\frac{5+y}{2x^2+10} = \frac{\quad}{4(x^2+5)}$
49. $\frac{x}{x^3+6x^2+8x} = \frac{\quad}{x(x+4)(x+2)(x+1)}$
50. $\frac{5x}{x^3+2x^2-3x} = \frac{\quad}{x(x-1)(x-5)(x+3)}$
51. $\frac{9y-1}{15x^2-30} = \frac{\quad}{30x^2-60}$
52. $\frac{6m-5}{3x^2-9} = \frac{\quad}{12x^2-36}$

MIXED PRACTICE

Perform the indicated operations.

53. $\frac{5x}{7} + \frac{9x}{7}$
54. $\frac{5x}{7} \cdot \frac{9x}{7}$
55. $\frac{x+3}{4} \div \frac{2x-1}{4}$
56. $\frac{x+3}{4} - \frac{2x-1}{4}$
57. $\frac{x^2}{x-6} - \frac{5x+6}{x-6}$
58. $\frac{x^2+5x}{x^2-25} \cdot \frac{3x-15}{x^2}$
59. $\frac{-2x}{x^3-8x} + \frac{3x}{x^3-8x}$
60. $\frac{-2x}{x^3-8x} \div \frac{3x}{x^3-8x}$

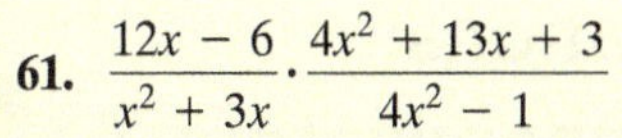

61. $\frac{12x - 6}{x^2 + 3x} \cdot \frac{4x^2 + 13x + 3}{4x^2 - 1}$

62. $\frac{x^3 + 7x^2}{3x^3 - x^2} \div \frac{5x^2 + 36x + 7}{9x^2 - 1}$

REVIEW AND PREVIEW

Perform each indicated operation. See Section 4.5.

63. $\frac{2}{3} + \frac{5}{7}$ **64.** $\frac{9}{10} - \frac{3}{5}$ **65.** $\frac{1}{6} - \frac{3}{4}$

66. $\frac{11}{15} + \frac{5}{9}$ **67.** $\frac{1}{12} + \frac{3}{20}$ **68.** $\frac{7}{30} + \frac{3}{18}$

CONCEPT EXTENSIONS

For Exercises 69 and 70, see the Concept Check in this section.

69. Choose the correct LCD of $\frac{11a^3}{4a - 20}$ and $\frac{15a^3}{(a - 5)^2}$.

a. $4a(a - 5)(a + 5)$ **b.** $a - 5$ **c.** $(a - 5)^2$
d. $4(a - 5)^2$ **e.** $(4a - 20)(a - 5)^2$

70. Choose the correct LCD of $\frac{5}{14x^2}$ and $\frac{y}{6x^3}$.

a. $84x^5$ **b.** $84x^3$ **c.** $42x^3$ **d.** $42x^5$

For Exercises 71 and 72, an algebra student approaches you with each incorrect solution. Find the error and correct the work shown below.

71. $\frac{2x - 6}{x - 5} - \frac{x + 4}{x - 5}$
$= \frac{2x - 6 - x + 4}{x - 5}$
$= \frac{x - 2}{x - 5}$

72. $\frac{x}{x + 3} + \frac{2}{x + 3}$
$= \frac{x + 2}{x + 3}$
$= \frac{2}{3}$

Multiple choice. Select the correct result.

73. $\frac{3}{x} + \frac{y}{x} =$

a. $\frac{3 + y}{x^2}$ **b.** $\frac{3 + y}{2x}$ **c.** $\frac{3 + y}{x}$

74. $\frac{3}{x} - \frac{y}{x} =$

a. $\frac{3 - y}{x^2}$ **b.** $\frac{3 - y}{2x}$ **c.** $\frac{3 - y}{x}$

75. $\frac{3}{x} \cdot \frac{y}{x} =$

a. $\frac{3y}{x}$ **b.** $\frac{3y}{x^2}$ **c.** $3y$

76. $\frac{3}{x} \div \frac{y}{x} =$

a. $\frac{3}{y}$ **b.** $\frac{y}{3}$ **c.** $\frac{3}{x^2y}$

Write each rational expression as an equivalent expression with a denominator of $x - 2$.

77. $\frac{5}{2 - x}$ **78.** $\frac{8y}{2 - x}$

79. $-\frac{7 + x}{2 - x}$ **80.** $\frac{x - 3}{-(x - 2)}$

△ **81.** A square has a side of length $\frac{5}{x - 2}$ meters. Express its perimeter as a rational expression.

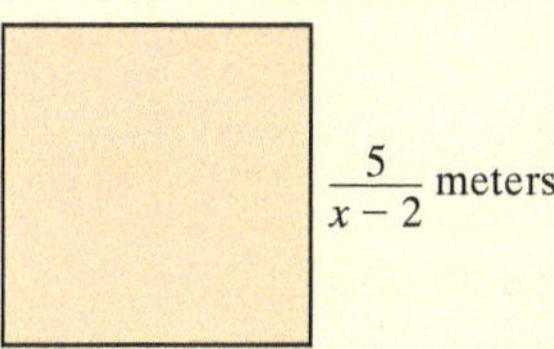

△ **82.** A trapezoid has sides of the indicated lengths. Find its perimeter.

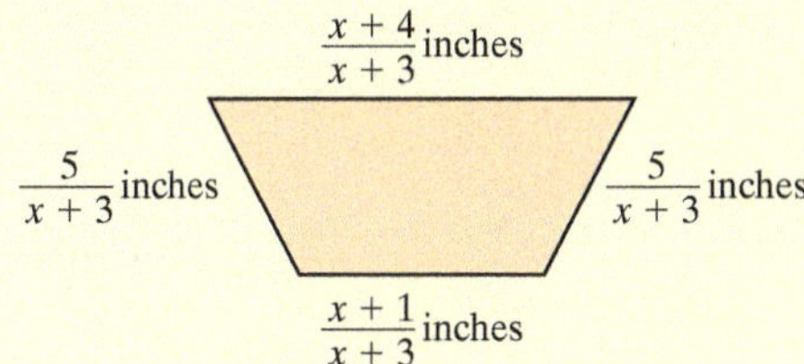

83. Write two rational expressions with the same denominator whose sum is $\frac{5}{3x - 1}$.

84. Write two rational expressions with the same denominator whose difference is $\frac{x - 7}{x^2 + 1}$.

85. The planet Mercury revolves around the sun in 88 Earth days. It takes Jupiter 4332 Earth days to make one revolution around the sun. (*Source:* National Space Science Data Center) If the two planets are aligned as shown in the figure, how long will it take for them to align again?

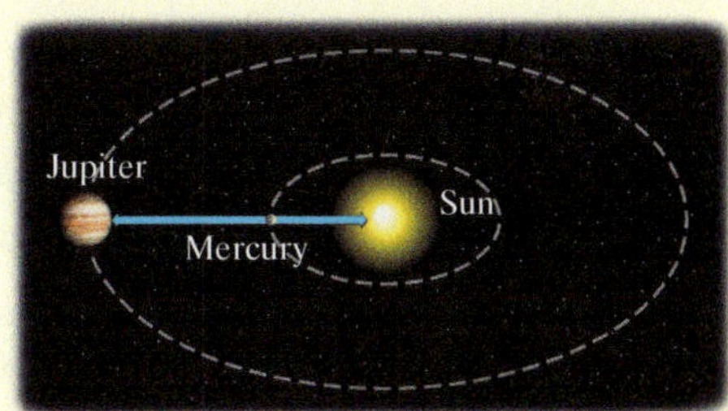

86. You are throwing a barbecue and you want to make sure that you purchase the same number of hot dogs as hot dog buns. Hot dogs come 8 to a package and hot dog buns come 12 to a package. What is the least number of each type of package you should buy?

87. Write some instructions to help a friend who is having difficulty finding the LCD of two rational expressions.

88. Explain why the LCD of the rational expressions $\frac{7}{x + 1}$ and $\frac{9x}{(x + 1)^2}$ is $(x + 1)^2$ and not $(x + 1)^3$.

89. In your own words, describe how to add or subtract two rational expressions with the same denominators.

90. Explain the similarities between subtracting $\frac{3}{8}$ from $\frac{7}{8}$ and subtracting $\frac{6}{x + 3}$ from $\frac{9}{x + 3}$.

14.4 Adding and Subtracting Rational Expressions with Unlike Denominators

Objective

1 Add and Subtract Rational Expressions with Unlike Denominators.

Objective 1 Adding and Subtracting Rational Expressions with Unlike Denominators

Let's add $\frac{3}{8}$ and $\frac{1}{6}$. From the previous section, the LCD of 8 and 6 is 24. Now let's write equivalent fractions with denominator 24 by multiplying by different forms of 1.

$$\frac{3}{8} = \frac{3}{8} \cdot 1 = \frac{3}{8} \cdot \frac{3}{3} = \frac{3 \cdot 3}{8 \cdot 3} = \frac{9}{24}$$

$$\frac{1}{6} = \frac{1}{6} \cdot 1 = \frac{1}{6} \cdot \frac{4}{4} = \frac{1 \cdot 4}{6 \cdot 4} = \frac{4}{24}$$

Now that the denominators are the same, we may add.

$$\frac{3}{8} + \frac{1}{6} = \frac{9}{24} + \frac{4}{24} = \frac{9+4}{24} = \frac{13}{24}$$

We add or subtract rational expressions the same way. You may want to use the steps below.

Adding or Subtracting Rational Expressions with Unlike Denominators

Step 1: Find the LCD of the rational expressions.

Step 2: Rewrite each rational expression as an equivalent expression whose denominator is the LCD found in Step 1.

Step 3: Add or subtract numerators and write the sum or difference over the common denominator.

Step 4: Simplify or write the rational expression in simplest form.

Practice 1

Perform each indicated operation.

a. $\frac{2x}{5} - \frac{6x}{15}$ **b.** $\frac{7}{8a} + \frac{5}{12a^2}$

Example 1 Perform each indicated operation.

a. $\frac{a}{4} - \frac{2a}{8}$ **b.** $\frac{3}{10x^2} + \frac{7}{25x}$

Solution:

a. First, we must find the LCD. Since $4 = 2^2$ and $8 = 2^3$, the LCD $= 2^3 = 8$. Next we write each fraction as an equivalent fraction with the denominator 8, then we subtract.

$$\frac{a}{4} - \frac{2a}{8} = \frac{a(2)}{4(2)} - \frac{2a}{8} = \frac{2a}{8} - \frac{2a}{8} = \frac{2a - 2a}{8} = \frac{0}{8} = 0$$

Multiplying the numerator and denominator by 2 is the same as multiplying by $\frac{2}{2}$ or 1.

b. Since $10x^2 = 2 \cdot 5 \cdot x \cdot x$ and $25x = 5 \cdot 5 \cdot x$, the LCD $= 2 \cdot 5^2 \cdot x^2 = 50x^2$. We write each fraction as an equivalent fraction with a denominator of $50x^2$.

$$\frac{3}{10x^2} + \frac{7}{25x} = \frac{3(5)}{10x^2(5)} + \frac{7(2x)}{25x(2x)}$$

$$= \frac{15}{50x^2} + \frac{14x}{50x^2}$$

$$= \frac{15 + 14x}{50x^2}$$ Add numerators. Write the sum over the common denominator.

Work Practice 1

Answers

1. a. 0 **b.** $\frac{21a+10}{24a^2}$

Example 2 Subtract: $\frac{6x}{x^2 - 4} - \frac{3}{x + 2}$

Solution: Since $x^2 - 4 = (x + 2)(x - 2)$, the LCD $= (x - 2)(x + 2)$. We write equivalent expressions with the LCD as denominators.

$$\frac{6x}{x^2 - 4} - \frac{3}{x + 2} = \frac{6x}{(x - 2)(x + 2)} - \frac{3(x - 2)}{(x + 2)(x - 2)}$$

$$= \frac{6x - 3(x - 2)}{(x + 2)(x - 2)}$$ Subtract numerators. Write the difference over the common denominator.

$$= \frac{6x - 3x + 6}{(x + 2)(x - 2)}$$ Apply the distributive property in the numerator.

$$= \frac{3x + 6}{(x + 2)(x - 2)}$$ Combine like terms in the numerator.

Next we factor the numerator to see if this rational expression can be simplified.

$$= \frac{3(x + 2)}{(x + 2)(x - 2)}$$ Factor.

$$= \frac{3}{x - 2}$$ Divide out common factors to simplify.

Work Practice 2

Practice 2

Subtract: $\frac{12x}{x^2 - 25} - \frac{6}{x + 5}$

Example 3 Add: $\frac{2}{3t} + \frac{5}{t + 1}$

Solution: The LCD is $3t(t + 1)$. We write each rational expression as an equivalent rational expression with a denominator of $3t(t + 1)$.

$$\frac{2}{3t} + \frac{5}{t + 1} = \frac{2(t + 1)}{3t(t + 1)} + \frac{5(3t)}{(t + 1)(3t)}$$

$$= \frac{2(t + 1) + 5(3t)}{3t(t + 1)}$$ Add numerators. Write the sum over the common denominator.

$$= \frac{2t + 2 + 15t}{3t(t + 1)}$$ Apply the distributive property in the numerator.

$$= \frac{17t + 2}{3t(t + 1)}$$ Combine like terms in the numerator.

Work Practice 3

Practice 3

Add: $\frac{3}{5y} + \frac{2}{y + 1}$

Example 4 Subtract: $\frac{7}{x - 3} - \frac{9}{3 - x}$

Solution: To find a common denominator, we notice that $x - 3$ and $3 - x$ are opposites. That is, $3 - x = -(x - 3)$. We write the denominator $3 - x$ as $-(x - 3)$ and simplify.

$$\frac{7}{x - 3} - \frac{9}{3 - x} = \frac{7}{x - 3} - \frac{9}{-(x - 3)}$$

$$= \frac{7}{x - 3} - \frac{-9}{x - 3}$$ Apply $\frac{a}{-b} = \frac{-a}{b}$.

$$= \frac{7 - (-9)}{x - 3}$$ Subtract numerators. Write the difference over the common denominator.

$$= \frac{16}{x - 3}$$

Work Practice 4

Practice 4

Subtract: $\frac{6}{x - 5} - \frac{7}{5 - x}$

Answers

2. $\frac{6}{x - 5}$ 3. $\frac{13y + 3}{5y(y + 1)}$ 4. $\frac{13}{x - 5}$

Practice 5

Add: $2 + \frac{b}{b+3}$

Example 5 Add: $1 + \frac{m}{m+1}$

Solution: Recall that 1 is the same as $\frac{1}{1}$. The LCD of $\frac{1}{1}$ and $\frac{m}{m+1}$ is $m + 1$.

$$1 + \frac{m}{m+1} = \frac{1}{1} + \frac{m}{m+1}$$ Write 1 as $\frac{1}{1}$.

$$= \frac{1(m+1)}{1(m+1)} + \frac{m}{m+1}$$ Multiply both the numerator and the denominator of $\frac{1}{1}$ by $m + 1$.

$$= \frac{m+1+m}{m+1}$$ Add numerators. Write the sum over the common denominator.

$$= \frac{2m+1}{m+1}$$ Combine like terms in the numerator.

Work Practice 5

Practice 6

Subtract: $\frac{5}{2x^2+3x} - \frac{3x}{4x+6}$

Example 6 Subtract: $\frac{3}{2x^2+x} - \frac{2x}{6x+3}$

Solution: First, we factor the denominators.

$$\frac{3}{2x^2+x} - \frac{2x}{6x+3} = \frac{3}{x(2x+1)} - \frac{2x}{3(2x+1)}$$

The LCD is $3x(2x + 1)$. We write equivalent expressions with denominators of $3x(2x + 1)$.

$$= \frac{3(3)}{x(2x+1)(3)} - \frac{2x(x)}{3(2x+1)(x)}$$

$$= \frac{9-2x^2}{3x(2x+1)}$$ Subtract numerators. Write the difference over the common denominator.

Work Practice 6

Practice 7

Add: $\frac{2x}{x^2+7x+12} + \frac{3x}{x^2-9}$

Example 7 Add: $\frac{2x}{x^2+2x+1} + \frac{x}{x^2-1}$

Solution: First we factor the denominators.

$$\frac{2x}{x^2+2x+1} + \frac{x}{x^2-1} = \frac{2x}{(x+1)(x+1)} + \frac{x}{(x+1)(x-1)}$$

Now we write the rational expressions as equivalent expressions with denominators of $(x + 1)(x + 1)(x - 1)$, the LCD.

$$= \frac{2x(x-1)}{(x+1)(x+1)(x-1)} + \frac{x(x+1)}{(x+1)(x-1)(x+1)}$$

$$= \frac{2x(x-1)+x(x+1)}{(x+1)^2(x-1)}$$ Add numerators. Write the sum over the common denominator.

$$= \frac{2x^2-2x+x^2+x}{(x+1)^2(x-1)}$$ Apply the distributive property in the numerator.

$$= \frac{3x^2-x}{(x+1)^2(x-1)} \text{ or } \frac{x(3x-1)}{(x+1)^2(x-1)}$$

Work Practice 7

The numerator was factored as a last step to see if the rational expression could be simplified further. Since there are no factors common to the numerator and the denominator, we can't simplify further.

Answers

5. $\frac{3b+6}{b+3}$ or $\frac{3(b+2)}{b+3}$
6. $\frac{10-3x^2}{2x(2x+3)}$
7. $\frac{x(5x+6)}{(x+4)(x+3)(x-3)}$

Vocabulary, Readiness & Video Check

Match each exercise with the first step needed to perform the operation. Do not actually perform the operation.

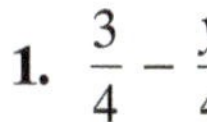

1. $\frac{3}{4} - \frac{y}{4}$ **2.** $\frac{2}{a} \cdot \frac{3}{(a+6)}$ **3.** $\frac{x+1}{x} \div \frac{x-1}{x}$ **4.** $\frac{9}{x-2} - \frac{x}{x+2}$

a. Multiply the first rational expression by the reciprocal of the second rational expression.

b. Find the LCD. Write each expression as an equivalent expression with the LCD as denominator.

c. Multiply numerators and multiply denominators.

d. Subtract numerators. Place the difference over a common denominator.

Martin-Gay Interactive Videos *Watch the section lecture video and answer the following question.*

See Video 14.4

Objective 1 **5.** What special case is shown in Example 2 and what's the purpose of presenting it?

14.4 Exercise Set MyMathLab®

MIXED PRACTICE

Perform each indicated operation. Simplify if possible. See Examples 1 through 7.

1. $\frac{4}{2x} + \frac{9}{3x}$

2. $\frac{15}{7a} + \frac{8}{6a}$

3. $\frac{15a}{b} - \frac{6b}{5}$

4. $\frac{4c}{d} - \frac{8d}{5}$

5. $\frac{3}{x} + \frac{5}{2x^2}$

6. $\frac{14}{3x^2} + \frac{6}{x}$

7. $\frac{6}{x+1} + \frac{10}{2x+2}$

8. $\frac{8}{x+4} - \frac{3}{3x+12}$

9. $\frac{3}{x+2} - \frac{2x}{x^2-4}$

10. $\frac{5}{x-4} + \frac{4x}{x^2-16}$

11. $\frac{3}{4x} + \frac{8}{x-2}$

12. $\frac{5}{y^2} - \frac{y}{2y+1}$

13. $\frac{6}{x-3} + \frac{8}{3-x}$

14. $\frac{15}{y-4} + \frac{20}{4-y}$

15. $\frac{9}{x-3} + \frac{9}{3-x}$

16. $\frac{5}{a-7} + \frac{5}{7-a}$

17. $\frac{-8}{x^2-1} - \frac{7}{1-x^2}$

18. $\frac{-9}{25x^2-1} + \frac{7}{1-25x^2}$

19. $\frac{5}{x} + 2$

20. $\frac{7}{x^2} - 5x$

21. $\frac{5}{x-2} + 6$

22. $\frac{6y}{y+5} + 1$

23. $\frac{y+2}{y+3} - 2$

24. $\frac{7}{2x-3} - 3$

25. $\frac{-x+2}{x} - \frac{x-6}{4x}$

26. $\frac{-y+1}{y} - \frac{2y-5}{3y}$

27. $\frac{5x}{x+2} - \frac{3x-4}{x+2}$

28. $\frac{7x}{x-3} - \frac{4x+9}{x-3}$

29. $\frac{3x^4}{7} - \frac{4x^2}{21}$

30. $\frac{5x}{6} + \frac{11x^2}{2}$

31. $\frac{1}{x+3} - \frac{1}{(x+3)^2}$

32. $\frac{5x}{(x-2)^2} - \frac{3}{x-2}$

33. $\frac{4}{5b} + \frac{1}{b-1}$

34. $\frac{1}{y+5} + \frac{2}{3y}$

35. $\frac{2}{m} + 1$

36. $\frac{6}{x} - 1$

37. $\frac{2x}{x-7} - \frac{x}{x-2}$

38. $\frac{9x}{x-10} - \frac{x}{x-3}$

39. $\frac{6}{1-2x} - \frac{4}{2x-1}$

40. $\frac{10}{3n-4} - \frac{5}{4-3n}$

41. $\frac{7}{(x+1)(x-1)} + \frac{8}{(x+1)^2}$

42. $\frac{5}{(x+1)(x+5)} - \frac{2}{(x+5)^2}$

43. $\frac{x}{x^2-1} - \frac{2}{x^2-2x+1}$

44. $\frac{x}{x^2-4} - \frac{5}{x^2-4x+4}$

45. $\frac{3a}{2a+6} - \frac{a-1}{a+3}$

46. $\frac{1}{x+y} - \frac{y}{x^2-y^2}$

47. $\frac{y-1}{2y+3} + \frac{3}{(2y+3)^2}$

48. $\frac{x-6}{5x+1} + \frac{6}{(5x+1)^2}$

49. $\frac{5}{2-x} + \frac{x}{2x-4}$

50. $\frac{-1}{a-2} + \frac{4}{4-2a}$

51. $\frac{15}{x^2+6x+9} + \frac{2}{x+3}$

52. $\frac{2}{x^2+4x+4} + \frac{1}{x+2}$

53. $\frac{13}{x^2-5x+6} - \frac{5}{x-3}$

54. $\frac{-7}{y^2-3y+2} - \frac{2}{y-1}$

55. $\frac{70}{m^2-100} + \frac{7}{2(m+10)}$

56. $\frac{27}{y^2-81} + \frac{3}{2(y+9)}$

57. $\frac{x+8}{x^2-5x-6} + \frac{x+1}{x^2-4x-5}$

58. $\frac{x+4}{x^2+12x+20} + \frac{x+1}{x^2+8x-20}$

59. $\frac{5}{4n^2-12n+8} - \frac{3}{3n^2-6n}$

60. $\frac{6}{5y^2-25y+30} - \frac{2}{4y^2-8y}$

MIXED PRACTICE

Perform the indicated operations. Addition, subtraction, multiplication, and division of rational expressions are included here.

61. $\frac{15x}{x+8} \cdot \frac{2x+16}{3x}$

62. $\frac{9z+5}{15} \cdot \frac{5z}{81z^2-25}$

63. $\frac{8x+7}{3x+5} - \frac{2x-3}{3x+5}$

64. $\frac{2z^2}{4z-1} - \frac{z-2z^2}{4z-1}$

65. $\frac{5a+10}{18} \div \frac{a^2-4}{10a}$

66. $\frac{9}{x^2-1} \div \frac{12}{3x+3}$

67. $\frac{5}{x^2-3x+2} + \frac{1}{x-2}$

68. $\frac{4}{2x^2+5x-3} + \frac{2}{x+3}$

REVIEW AND PREVIEW

Solve the following linear and quadratic equations. See Sections 9.3 and 13.6.

69. $3x + 5 = 7$

70. $5x - 1 = 8$

71. $2x^2 - x - 1 = 0$

72. $4x^2 - 9 = 0$

73. $4(x + 6) + 3 = -3$

74. $2(3x + 1) + 15 = -7$

CONCEPT EXTENSIONS

Perform each indicated operation.

75. $\frac{3}{x} - \frac{2x}{x^2-1} + \frac{5}{x+1}$

76. $\frac{5}{x-2} + \frac{7x}{x^2-4} - \frac{11}{x}$

77. $\frac{5}{x^2-4} + \frac{2}{x^2-4x+4} - \frac{3}{x^2-x-6}$

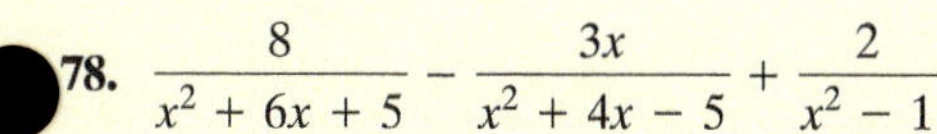

78. $\frac{8}{x^2 + 6x + 5} - \frac{3x}{x^2 + 4x - 5} + \frac{2}{x^2 - 1}$

79. $\frac{9}{x^2 + 9x + 14} - \frac{3x}{x^2 + 10x + 21} + \frac{x + 4}{x^2 + 5x + 6}$

80. $\frac{x + 10}{x^2 - 3x - 4} - \frac{8}{x^2 + 6x + 5} - \frac{9}{x^2 + x - 20}$

81. A board of length $\frac{3}{x + 4}$ inches was cut into two pieces. If one piece is $\frac{1}{x - 4}$ inches, express the length of the other piece as a rational expression.

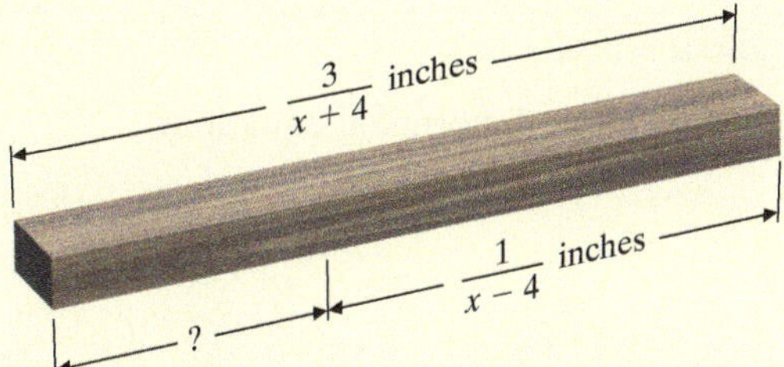

△ **82.** The length of a rectangle is $\frac{3}{y - 5}$ feet, while its width is $\frac{2}{y}$ feet. Find its perimeter and then find its area.

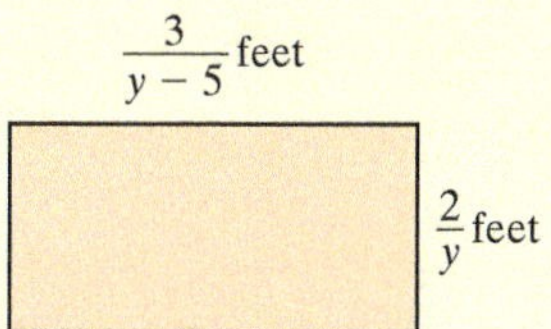

83. In ice hockey, penalty killing percentage is a statistic calculated as $1 - \frac{G}{P}$, where G = opponent's power play goals and P = opponent's power play opportunities. Simplify this expression.

84. The dose of medicine prescribed for a child depends on the child's age A in years and the adult dose D for the medication. Two expressions that give a child's dose are Young's Rule, $\frac{DA}{A + 12}$, and Cowling's Rule, $\frac{D(A + 1)}{24}$. Find an expression for the difference in the doses given by these expressions.

85. Explain when the LCD of the rational expressions in a sum is the product of the denominators.

86. Explain when the LCD is the same as one of the denominators of a rational expression to be added or subtracted.

△ **87.** Two angles are said to be complementary if the sum of their measures is 90°. If one angle measures $\frac{40}{x}$ degrees, find the measure of its complement.

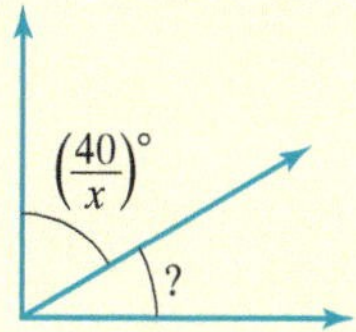

△ **88.** Two angles are said to be supplementary if the sum of their measures is 180°. If one angle measures $\frac{x + 2}{x}$ degrees, find the measure of its supplement.

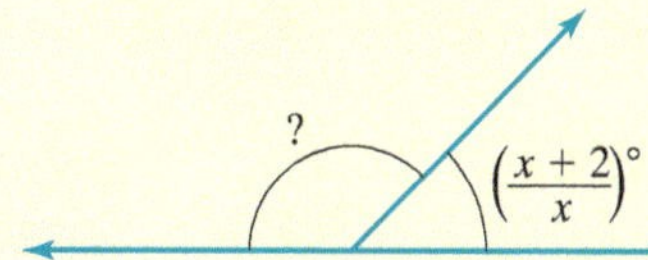

89. In your own words, explain how to add two rational expressions with different denominators.

90. In your own words, explain how to subtract two rational xpressions with different denominators.

14.5 Solving Equations Containing Rational Expressions

Objective 1 Solving Equations Containing Rational Expressions

Objectives

1. Solve Equations Containing Rational Expressions.
2. Solve Equations Containing Rational Expressions for a Specified Variable.

In Chapter 4, we solved equations containing fractions. In this section, we continue the work we began in Chapter 4 by solving equations containing rational expressions.

Examples of Equations Containing Rational Expressions

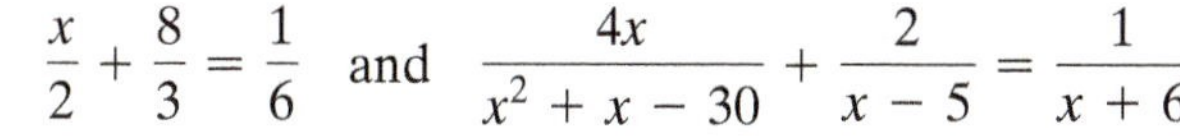

$\frac{x}{2} + \frac{8}{3} = \frac{1}{6}$ and $\frac{4x}{x^2 + x - 30} + \frac{2}{x - 5} = \frac{1}{x + 6}$

To solve equations such as these, use the multiplication property of equality to clear the equation of fractions by multiplying both sides of the equation by the LCD.

Practice 1

Solve: $\frac{x}{3} + \frac{4}{5} = \frac{2}{15}$

Example 1 Solve: $\frac{x}{2} + \frac{8}{3} = \frac{1}{6}$

Solution: The LCD of denominators 2, 3, and 6 is 6, so we multiply both sides of the equation by 6.

$$6\left(\frac{x}{2} + \frac{8}{3}\right) = 6\left(\frac{1}{6}\right)$$

$$6\left(\frac{x}{2}\right) + 6\left(\frac{8}{3}\right) = 6\left(\frac{1}{6}\right) \quad \text{Use the distributive property.}$$

Helpful Hint Make sure that *each* term is multiplied by the LCD, 6.

$$3 \cdot x + 16 = 1 \quad \text{Multiply and simplify.}$$

$$3x = -15 \quad \text{Subtract 16 from both sides.}$$

$$x = -5 \quad \text{Divide both sides by 3.}$$

Check: To check, we replace x with -5 in the original equation.

$$\frac{x}{2} + \frac{8}{3} = \frac{1}{6}$$

$$\frac{-5}{2} + \frac{8}{3} \stackrel{?}{=} \frac{1}{6} \quad \text{Replace } x \text{ with } -5.$$

$$\frac{1}{6} = \frac{1}{6} \quad \text{True}$$

This number checks, so the solution is -5.

Work Practice 1

Practice 2

Solve: $\frac{x+4}{4} - \frac{x-3}{3} = \frac{11}{12}$

Example 2 Solve: $\frac{t-4}{2} - \frac{t-3}{9} = \frac{5}{18}$

Solution: The LCD of denominators 2, 9, and 18 is 18, so we multiply both sides of the equation by 18.

$$18\left(\frac{t-4}{2} - \frac{t-3}{9}\right) = 18\left(\frac{5}{18}\right)$$

$$18\left(\frac{t-4}{2}\right) - 18\left(\frac{t-3}{9}\right) = 18\left(\frac{5}{18}\right) \quad \text{Use the distributive property.}$$

Helpful Hint Multiply *each* term by 18.

$$9(t-4) - 2(t-3) = 5 \quad \text{Simplify.}$$

$$9t - 36 - 2t + 6 = 5 \quad \text{Use the distributive property.}$$

$$7t - 30 = 5 \quad \text{Combine like terms.}$$

$$7t = 35$$

$$t = 5 \quad \text{Solve for } t.$$

Check: $\frac{t-4}{2} - \frac{t-3}{9} = \frac{5}{18}$

$$\frac{5-4}{2} - \frac{5-3}{9} \stackrel{?}{=} \frac{5}{18} \quad \text{Replace } t \text{ with 5.}$$

$$\frac{1}{2} - \frac{2}{9} \stackrel{?}{=} \frac{5}{18} \quad \text{Simplify.}$$

$$\frac{5}{18} = \frac{5}{18} \quad \text{True}$$

The solution is 5.

Work Practice 2

Answers
1. -2 **2.** 13

Recall from Section 14.1 that a rational expression is defined for all real numbers except those that make the denominator of the expression 0. This means that if an equation contains *rational expressions with variables in the denominator*, we must be certain that the proposed solution does not make the denominator 0. If replacing the variable with the proposed solution makes the denominator 0, the rational expression is undefined and this proposed solution must be rejected.

Example 3 Solve: $3 - \frac{6}{x} = x + 8$

Solution: In this equation, 0 cannot be a solution because if x is 0, the rational expression $\frac{6}{x}$ is undefined. The LCD is x, so we multiply both sides of the equation by x.

$$x\left(3 - \frac{6}{x}\right) = x(x + 8)$$

$$x(3) - x\left(\frac{6}{x}\right) = x \cdot x + x \cdot 8 \quad \text{Use the distributive property.}$$

$$3x - 6 = x^2 + 8x \quad \text{Simplify.}$$

Multiply *each* term by x.

Now we write the quadratic equation in standard form and solve for x.

$$0 = x^2 + 5x + 6$$

$$0 = (x + 3)(x + 2) \quad \text{Factor.}$$

$$x + 3 = 0 \quad \text{or} \quad x + 2 = 0 \quad \text{Set each factor equal to 0 and solve.}$$

$$x = -3 \qquad x = -2$$

Notice that neither -3 nor -2 makes the denominator in the original equation equal to 0.

Check: To check these solutions, we replace x in the original equation by -3, and then by -2.

If $x = -3$:

$$3 - \frac{6}{x} = x + 8$$

$$3 - \frac{6}{-3} \stackrel{?}{=} -3 + 8$$

$$3 - (-2) \stackrel{?}{=} 5$$

$$5 = 5 \quad \text{True}$$

If $x = -2$:

$$3 - \frac{6}{x} = x + 8$$

$$3 - \frac{6}{-2} \stackrel{?}{=} -2 + 8$$

$$3 - (-3) \stackrel{?}{=} 6$$

$$6 = 6 \quad \text{True}$$

Both -3 and -2 are solutions.

Work Practice 3

Practice 3

Solve: $8 + \frac{7}{x} = x + 2$

The following steps may be used to solve an equation containing rational expressions.

Solving an Equation Containing Rational Expressions

Step 1: Multiply both sides of the equation by the LCD of all rational expressions in the equation.

Step 2: Remove any grouping symbols and solve the resulting equation.

Step 3: Check the solution in the original equation.

Answer

3. $-1, 7$

Practice 4

Solve:
$$\frac{6x}{x^2 - 5x - 14} - \frac{3}{x + 2} = \frac{1}{x - 7}$$

Example 4 Solve: $\frac{4x}{x^2 + x - 30} + \frac{2}{x - 5} = \frac{1}{x + 6}$

Solution: The denominator $x^2 + x - 30$ factors as $(x + 6)(x - 5)$. The LCD is then $(x + 6)(x - 5)$, so we multiply both sides of the equation by this LCD.

$$(x + 6)(x - 5)\left(\frac{4x}{x^2 + x - 30} + \frac{2}{x - 5}\right) = (x + 6)(x - 5)\left(\frac{1}{x + 6}\right)$$ Multiply by the LCD.

$$(x + 6)(x - 5) \cdot \frac{4x}{x^2 + x - 30} + (x + 6)(x - 5) \cdot \frac{2}{x - 5}$$ Apply the distributive property.

$$= (x + 6)(x - 5) \cdot \frac{1}{x + 6}$$

$$4x + 2(x + 6) = x - 5$$ Simplify.

$$4x + 2x + 12 = x - 5$$ Apply the distributive property.

$$6x + 12 = x - 5$$ Combine like terms.

$$5x = -17$$

$$x = -\frac{17}{5}$$ Divide both sides by 5.

Check: Check by replacing x with $-\frac{17}{5}$ in the original equation. The solution is $-\frac{17}{5}$.

Work Practice 4

Practice 5

Solve: $\frac{7}{x - 2} = \frac{3}{x - 2} + 4$

Example 5 Solve: $\frac{2x}{x - 4} = \frac{8}{x - 4} + 1$

Solution: Multiply both sides by the LCD, $x - 4$.

$$(x - 4)\left(\frac{2x}{x - 4}\right) = (x - 4)\left(\frac{8}{x - 4} + 1\right)$$ Multiply by the LCD. Notice that 4 cannot be a solution.

$$(x - 4) \cdot \frac{2x}{x - 4} = (x - 4) \cdot \frac{8}{x - 4} + (x - 4) \cdot 1$$ Use the distributive property.

$$2x = 8 + (x - 4)$$ Simplify.

$$2x = 4 + x$$

$$x = 4$$

Notice that 4 makes the denominators 0 in the original equation. Therefore, 4 is *not* a solution.

This equation has *no solution*.

Work Practice 5

Helpful Hint

As we can see from Example 5, it is important to check the proposed solution(s) in the *original* equation.

✓Concept Check When can we clear fractions by multiplying through by the LCD?

a. When adding or subtracting rational expressions

b. When solving an equation containing rational expressions

c. Both of these

d. Neither of these

Answers

4. $-\frac{19}{2}$ **5.** 3

✓Concept Check Answer

b

Example 6 Solve: $x + \frac{14}{x - 2} = \frac{7x}{x - 2} + 1$

Solution: Notice the denominators in this equation. We can see that 2 can't be a solution. The LCD is $x - 2$, so we multiply both sides of the equation by $x - 2$.

$$(x - 2)\left(x + \frac{14}{x - 2}\right) = (x - 2)\left(\frac{7x}{x - 2} + 1\right)$$

$$(x - 2)(x) + (x - 2)\left(\frac{14}{x - 2}\right) = (x - 2)\left(\frac{7x}{x - 2}\right) + (x - 2)(1)$$

$$x^2 - 2x + 14 = 7x + x - 2 \quad \text{Simplify.}$$

$$x^2 - 2x + 14 = 8x - 2 \quad \text{Combine like terms.}$$

$$x^2 - 10x + 16 = 0 \quad \text{Write the quadratic equation in standard form.}$$

$$(x - 8)(x - 2) = 0 \quad \text{Factor.}$$

$$x - 8 = 0 \quad \text{or} \quad x - 2 = 0 \quad \text{Set each factor equal to 0.}$$

$$x = 8 \qquad x = 2 \quad \text{Solve.}$$

As we have already noted, 2 can't be a solution of the original equation. So we need only replace x with 8 in the original equation. We find that 8 is a solution; the only solution is 8.

Work Practice 6

Practice 6

Solve: $x + \frac{x}{x - 5} = \frac{5}{x - 5} - 7$

Objective 2 Solving Equations for a Specified Variable 

The last example in this section is an equation containing several variables, and we are directed to solve for one of the variables. The steps used in the preceding examples can be applied to solve equations for a specified variable as well.

Example 7 Solve $\frac{1}{a} + \frac{1}{b} = \frac{1}{x}$ for x.

Solution: (This type of equation often models a work problem, as we shall see in Section 14.6.) The LCD is abx, so we multiply both sides by abx.

$$abx\left(\frac{1}{a} + \frac{1}{b}\right) = abx\left(\frac{1}{x}\right)$$

$$abx\left(\frac{1}{a}\right) + abx\left(\frac{1}{b}\right) = abx \cdot \frac{1}{x}$$

$$bx + ax = ab \quad \text{Simplify.}$$

$$x(b + a) = ab \quad \text{Factor out } x \text{ from each term on the left side.}$$

$$\frac{x(b + a)}{b + a} = \frac{ab}{b + a} \quad \text{Divide both sides by } b + a.$$

$$x = \frac{ab}{b + a} \quad \text{Simplify.}$$

This equation is now solved for x.

Work Practice 7

Practice 7

Solve: $\frac{1}{a} + \frac{1}{b} = \frac{1}{x}$ for b

Answers

6. -8 **7.** $b = \frac{ax}{a - x}$

Graphing Calculator Explorations

A graphing calculator may be used to check solutions of equations containing rational expressions. For example, to check the solution of Example 1, $\frac{x}{2} + \frac{8}{3} = \frac{1}{6}$, graph $y_1 = \frac{x}{2} + \frac{8}{3}$ and $y_2 = \frac{1}{6}$.

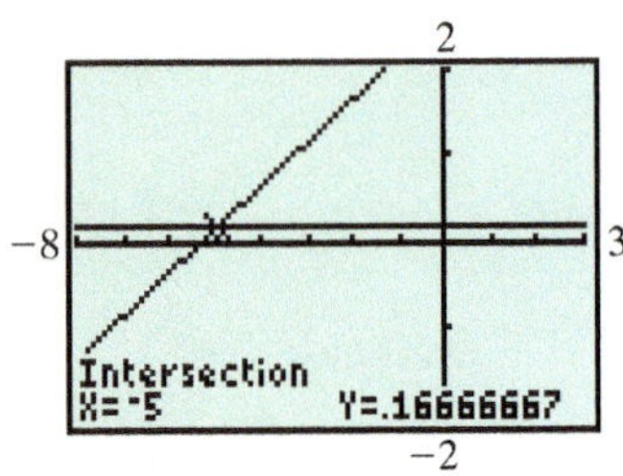

Use TRACE and ZOOM, or use INTERSECT, to find the point of intersection. The point of intersection has an x-value of -5, so the solution of the equation is -5.

Use a graphing calculator to check the examples of this section.

1. Example 2
2. Example 3
3. Example 5
4. Example 6

Vocabulary, Readiness & Video Check

Multiple choice. Choose the correct response.

1. Multiply both sides of the equation $\frac{3x}{2} + 5 = \frac{1}{4}$ by 4. The result is:

   a. $3x + 5 = 1$ b. $6x + 5 = 1$ c. $6x + 20 = 1$ d. $6x + 9 = 1$

2. Multiply both sides of the equation $\frac{1}{x} - \frac{3}{5x} = 2$ by $5x$. The result is:

   a. $1 - 3 = 10x$ b. $5 - 3 = 10x$ c. $1 - 3 = 7x$ d. $5 - 3 = 7x$

Choose the correct LCD for the fractions in each equation.

3. Equation: $\frac{9}{x} + \frac{3}{4} = \frac{1}{12}$; LCD: ____________

   a. $4x$ b. $12x$ c. $48x$ d. x

4. Equation: $\frac{8}{3x} - \frac{1}{x} = \frac{7}{9}$; LCD: ____________

   a. x b. $3x$ c. $27x$ d. $9x$

5. Equation: $\frac{9}{x-1} = \frac{7}{(x-1)^2}$; LCD: ____________

   a. $(x-1)^2$ b. $x - 1$ c. $(x-1)^3$ d. 63

6. Equation: $\frac{1}{x-2} - \frac{3}{x^2-4} = 8$; LCD: ____________

   a. $(x-2)$ b. $x + 2$ c. $x^2 - 4$ d. $(x-2)(x^2-4)$

Martin-Gay Interactive Videos

See Video 14.5

Watch the section lecture video and answer the following questions.

Objective 1 7. After multiplying through by the LCD and then simplifying, why is it important to take a moment and determine whether you have a linear or a quadratic equation before you finish solving the problem?

Objective 2 8. From Examples 2–5, what extra step is needed when checking solutions to an equation containing rational expressions?

Objective 3 9. The steps for solving Example 6 for a specified variable are the same as what other steps? How do you treat this specified variable?

14.5 Exercise Set MyMathLab®

Solve each equation and check each solution. See Examples 1 through 3.

1. $\frac{x}{5} + 3 = 9$
2. $\frac{x}{5} - 2 = 9$
3. $\frac{x}{2} + \frac{5x}{4} = \frac{x}{12}$
4. $\frac{x}{6} + \frac{4x}{3} = \frac{x}{18}$
5. $2 - \frac{8}{x} = 6$
6. $5 + \frac{4}{x} = 1$
7. $2 + \frac{10}{x} = x + 5$
8. $6 + \frac{5}{y} = y - \frac{2}{y}$
9. $\frac{a}{5} = \frac{a-3}{2}$
10. $\frac{b}{5} = \frac{b+2}{6}$
11. $\frac{x-3}{5} + \frac{x-2}{2} = \frac{1}{2}$
12. $\frac{a+5}{4} + \frac{a+5}{2} = \frac{a}{8}$

Solve each equation and check each proposed solution. See Examples 4 through 6.

13. $\frac{3}{2a-5} = -1$
14. $\frac{6}{4-3x} = -3$
15. $\frac{4y}{y-4} + 5 = \frac{5y}{y-4}$
16. $\frac{2a}{a+2} - 5 = \frac{7a}{a+2}$
17. $2 + \frac{3}{a-3} = \frac{a}{a-3}$
18. $\frac{2y}{y-2} - \frac{4}{y-2} = 4$
19. $\frac{1}{x+3} + \frac{6}{x^2-9} = 1$
20. $\frac{1}{x+2} + \frac{4}{x^2-4} = 1$
21. $\frac{2y}{y+4} + \frac{4}{y+4} = 3$
22. $\frac{5y}{y+1} - \frac{3}{y+1} = 4$
23. $\frac{2x}{x+2} - 2 = \frac{x-8}{x-2}$
24. $\frac{4y}{y-3} - 3 = \frac{3y-1}{y+3}$

MIXED PRACTICE

Solve each equation. See Examples 1 through 6.

25. $\frac{2}{y} + \frac{1}{2} = \frac{5}{2y}$
26. $\frac{6}{3y} + \frac{3}{y} = 1$
27. $\frac{a}{a-6} = \frac{-2}{a-1}$
28. $\frac{5}{x-6} = \frac{x}{x-2}$
29. $\frac{11}{2x} + \frac{2}{3} = \frac{7}{2x}$
30. $\frac{5}{3} - \frac{3}{2x} = \frac{3}{2}$
31. $\frac{2}{x-2} + 1 = \frac{x}{x+2}$
32. $1 + \frac{3}{x+1} = \frac{x}{x-1}$
33. $\frac{x+1}{3} - \frac{x-1}{6} = \frac{1}{6}$
34. $\frac{3x}{5} - \frac{x-6}{3} = -\frac{2}{5}$
35. $\frac{t}{t-4} = \frac{t+4}{6}$
36. $\frac{15}{x+4} = \frac{x-4}{x}$
37. $\frac{y}{2y+2} + \frac{2y-16}{4y+4} = \frac{2y-3}{y+1}$
38. $\frac{1}{x+2} = \frac{4}{x^2-4} - \frac{1}{x-2}$
39. $\frac{4r-4}{r^2+5r-14} + \frac{2}{r+7} = \frac{1}{r-2}$
40. $\frac{3}{x+3} = \frac{12x+19}{x^2+7x+12} - \frac{5}{x+4}$
41. $\frac{x+1}{x+3} = \frac{x^2-11x}{x^2+x-6} - \frac{x-3}{x-2}$
42. $\frac{2t+3}{t-1} - \frac{2}{t+3} = \frac{5-6t}{t^2+2t-3}$

Solve each equation for the indicated variable. See Example 7.

43. $R = \frac{E}{I}$ for I (Electronics: resistance of a circuit)
44. $T = \frac{V}{Q}$ for Q (Water purification: settling time)
45. $T = \frac{2U}{B+E}$ for B (Merchandising: stock turnover rate)
46. $i = \frac{A}{t+B}$ for t (Hydrology: rainfall intensity)
47. $B = \frac{705w}{h^2}$ for w (Health: body-mass index)
48. $\frac{A}{W} = L$ for W (Geometry: area of a rectangle)
49. $N = R + \frac{V}{G}$ for G (Urban forestry: tree plantings per year)

50. $C = \dfrac{D(A + 1)}{24}$ for A (Medicine: Cowling's Rule for child's dose)

△ **51.** $\dfrac{C}{\pi r} = 2$ for r (Geometry: circumference of a circle)

52. $W = \dfrac{CE^2}{2}$ for C (Electronics: energy stored in a capacitor)

53. $\dfrac{1}{y} + \dfrac{1}{3} = \dfrac{1}{x}$ for x

54. $\dfrac{1}{5} + \dfrac{2}{y} = \dfrac{1}{x}$ for x

REVIEW AND PREVIEW

TRANSLATING

Write each phrase as an expression.

55. The reciprocal of x

56. The reciprocal of $x + 1$

57. The reciprocal of x, added to the reciprocal of 2

58. The reciprocal of x, subtracted from the reciprocal of 5

Answer each question.

59. If a tank is filled in 3 hours, what fractional part of the tank is filled in 1 hour?

60. If a strip of beach is cleaned in 4 hours, what fractional part of the beach is cleaned in 1 hour?

Identify the x- and y-intercepts. See Section 10.3.

61.

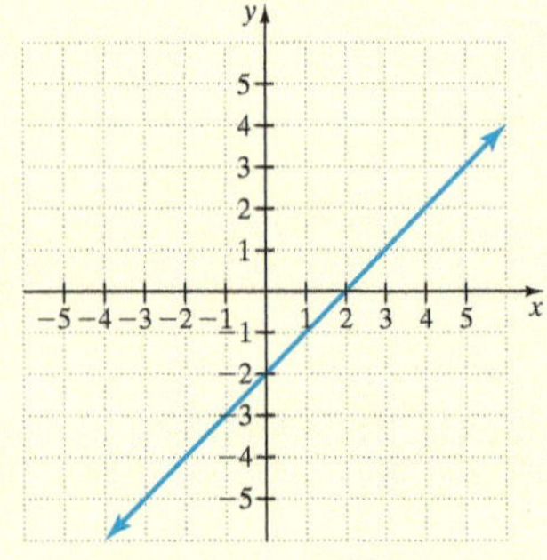

62.

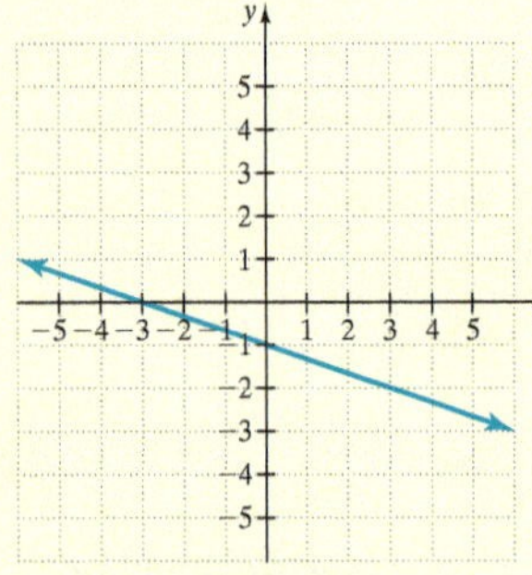

63.

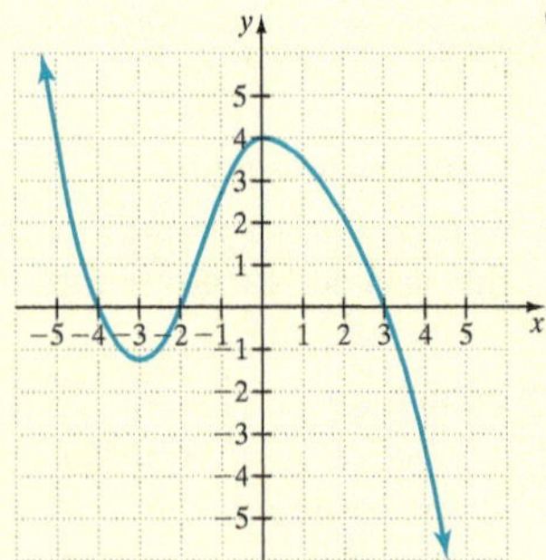

64.

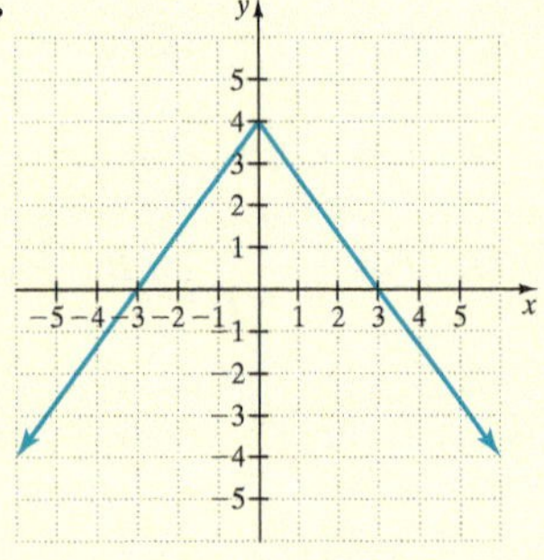

CONCEPT EXTENSIONS

65. Explain the difference between solving an equation such as $\dfrac{x}{2} + \dfrac{3}{4} = \dfrac{x}{4}$ for x and performing an operation such as adding $\dfrac{x}{2} + \dfrac{3}{4}$.

66. When solving an equation such as $\dfrac{y}{4} = \dfrac{y}{2} - \dfrac{1}{4}$, we may multiply all terms by 4. When subtracting two rational expressions such as $\dfrac{y}{2} - \dfrac{1}{4}$, we may not. Explain why.

Determine whether each of the following is an equation or an expression. If it is an equation, then solve it for its variable. If it is an expression, perform the indicated operation.

△ **67.** $\dfrac{1}{x} + \dfrac{5}{9}$

68. $\dfrac{1}{x} + \dfrac{5}{9} = \dfrac{2}{3}$

69. $\dfrac{5}{x - 1} - \dfrac{2}{x} = \dfrac{5}{x(x - 1)}$

70. $\dfrac{5}{x - 1} - \dfrac{2}{x}$

Recall that two angles are supplementary if the sum of their measures is 180°. Find the measures of the following supplementary angles.

△ **71.**

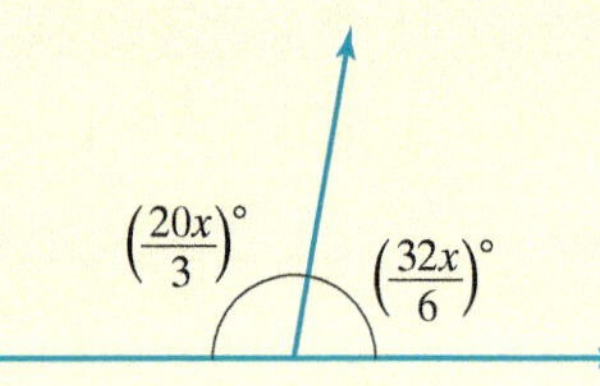

△ **72.**

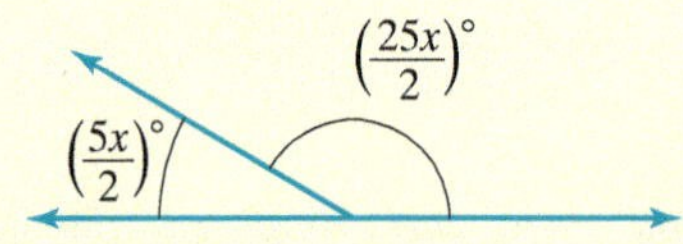

Recall that two angles are complementary if the sum of their measures is 90°. Find the measures of the following complementary angles.

△ **73.**

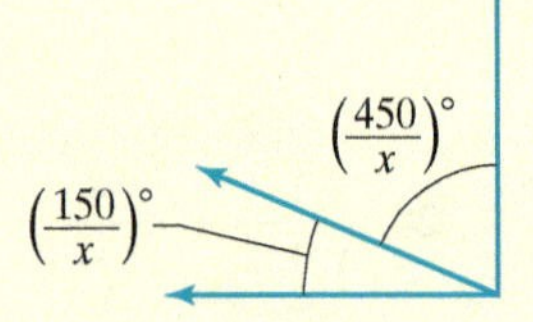

△ **74.**

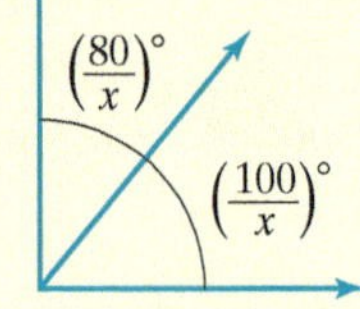

Solve each equation.

75. $\dfrac{5}{a^2 + 4a + 3} + \dfrac{2}{a^2 + a - 6} - \dfrac{3}{a^2 - a - 2} = 0$

76. $\dfrac{-2}{a^2 + 2a - 8} + \dfrac{1}{a^2 + 9a + 20} = \dfrac{-4}{a^2 + 3a - 10}$

Integrated Review Sections 14.1–14.5

Summary on Rational Expressions

It is important to know the difference between performing operations with rational expressions and solving an equation containing rational expressions. Study the examples below.

Performing Operations with Rational Expressions

Adding: $\frac{1}{x} + \frac{1}{x+5} = \frac{1 \cdot (x+5)}{x(x+5)} + \frac{1 \cdot x}{x(x+5)} = \frac{x+5+x}{x(x+5)} = \frac{2x+5}{x(x+5)}$

Subtracting: $\frac{3}{x} - \frac{5}{x^2y} = \frac{3 \cdot xy}{x \cdot xy} - \frac{5}{x^2y} = \frac{3xy-5}{x^2y}$

Multiplying: $\frac{2}{x} \cdot \frac{5}{x-1} = \frac{2 \cdot 5}{x(x-1)} = \frac{10}{x(x-1)}$

Dividing: $\frac{4}{2x+1} \div \frac{x-3}{x} = \frac{4}{2x+1} \cdot \frac{x}{x-3} = \frac{4x}{(2x+1)(x-3)}$

Solving an Equation Containing Rational Expressions

To solve an equation containing rational expressions, we clear the equation of fractions by multiplying both sides by the LCD.

$$\frac{3}{x} - \frac{5}{x-1} = \frac{1}{x(x-1)}$$ Note that x can't be 0 or 1.

$$x(x-1)\left(\frac{3}{x}\right) - x(x-1)\left(\frac{5}{x-1}\right) = x(x-1) \cdot \frac{1}{x(x-1)}$$ Multiply both sides by the LCD.

$$3(x-1) - 5x = 1$$ Simplify.

$$3x - 3 - 5x = 1$$ Use the distributive property.

$$-2x - 3 = 1$$ Combine like terms.

$$-2x = 4$$ Add 3 to both sides.

$$x = -2$$ Divide both sides by -2.

Determine whether each of the following is an equation or an expression. If it is an equation, solve it for its variable. If it is an expression, perform the indicated operation.

1. $\frac{1}{x} + \frac{2}{3}$
2. $\frac{3}{a} + \frac{5}{6}$
3. $\frac{1}{x} + \frac{2}{3} = \frac{3}{x}$
4. $\frac{3}{a} + \frac{5}{6} = 1$
5. $\frac{2}{x-1} - \frac{1}{x}$
6. $\frac{4}{x-3} - \frac{1}{x}$
7. $\frac{2}{x+1} - \frac{1}{x} = 1$
8. $\frac{4}{x-3} - \frac{1}{x} = \frac{6}{x(x-3)}$
9. $\frac{15x}{x+8} \cdot \frac{2x+16}{3x}$
10. $\frac{9z+5}{15} \cdot \frac{5z}{81z^2-25}$
11. $\frac{2x+1}{x-3} + \frac{3x+6}{x-3}$
12. $\frac{4p-3}{2p+7} + \frac{3p+8}{2p+7}$
13. $\frac{x+5}{7} = \frac{8}{2}$
14. $\frac{1}{2} = \frac{x-1}{8}$
15. $\frac{5a+10}{18} \div \frac{a^2-4}{10a}$
16. $\frac{9}{x^2-1} + \frac{12}{3x+3}$
17. $\frac{x+2}{3x-1} + \frac{5}{(3x-1)^2}$
18. $\frac{4}{(2x-5)^2} + \frac{x+1}{2x-5}$
19. $\frac{x-7}{x} - \frac{x+2}{5x}$
20. $\frac{10x-9}{x} - \frac{x-4}{3x}$

21. $\frac{3}{x+3} = \frac{5}{x^2-9} - \frac{2}{x-3}$

22. $\frac{9}{x^2-4} + \frac{2}{x+2} = \frac{-1}{x-2}$

23. Explain the difference between solving an equation such as $\frac{x}{5} + \frac{3}{10} = \frac{x}{10}$ for x, and performing an operation such as adding $\frac{x}{5} + \frac{3}{10}$.

24. When solving an equation such as $\frac{y}{10} = \frac{y}{5} - \frac{1}{10}$, we may multiply all terms by 4. When subtracting two rational expressions such as $\frac{y}{5} - \frac{1}{10}$, we may not. Explain why.

14.6 Problem Solving with Proportions and Rational Equations

Objectives

1. Use Proportions to Solve Problems.
2. Solve Problems About Numbers.
3. Solve Problems About Work.
4. Solve Problems About Distance.

Objective 1 Using Proportions to Solve Problems

Proportions can be used to model and solve many real-life problems. When using proportions in this way, it is important to judge whether the solution is reasonable. Doing so helps us decide if the proportion has been formed correctly. We use the same problem-solving steps that were introduced in Section 3.4.

Before we solve an application, recall from Section 6.1 that a **proportion** is a mathematical statement (equation) that two ratios are equal.

A proportion, such as $\frac{a}{b} = \frac{c}{d}$, contains four numbers and/or expressions. To solve a proportion for an unknown, we may either set cross products equal or multiply both sides by the least common denominator, LCD.

Recall below why the products ad and bc are called cross products.

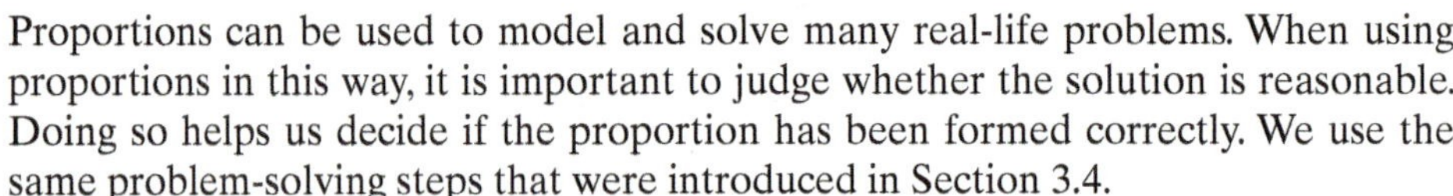

$$\frac{a}{b} = \frac{c}{d}$$

Helpful Hint In a true proportion, we may set cross products equal because that is the simplified result after multiplying both sides of the proportion (equation) by the LCD.

Cross Products

If $\frac{a}{b} = \frac{c}{d}$, then $ad = bc$.

Practice 1

Four 2-liter bottles of Diet Pepsi cost \$5.16. How much will seven 2-liter bottles cost?

Example 1 Calculating the Cost of Recordable Compact Discs

Three boxes of CD-Rs (recordable compact discs) cost \$37.47. How much should 5 boxes cost?

Solution:

1. UNDERSTAND. Read and reread the problem. We know that the cost of 5 boxes is more than the cost of 3 boxes, or \$37.47, and less than the cost of 6 boxes, which is double the cost of 3 boxes, or 2(\$37.47) = \$74.94. Let's suppose that 5 boxes cost \$60.00. To check, we see if 3 boxes is to 5 boxes as the *price* of 3 boxes is to the *price* of 5 boxes. In other words, we see if

$$\frac{3 \text{ boxes}}{5 \text{ boxes}} = \frac{\text{price of 3 boxes}}{\text{price of 5 boxes}}$$

or

$$\frac{3}{5} = \frac{37.47}{60.00}$$

Answer

1. \$9.03

$$3(60.00) = 5(37.47) \quad \text{Set cross products equal.}$$

or

$$180.00 = 187.35 \quad \text{Not a true statement.}$$

Thus, \$60 is not correct, but we now have a better understanding of the problem.

Let x = price of 5 boxes of CD-Rs.

2. TRANSLATE.

$$\frac{3 \text{ boxes}}{5 \text{ boxes}} = \frac{\text{price of 3 boxes}}{\text{price of 5 boxes}}$$

$$\frac{3}{5} = \frac{37.47}{x}$$

3. SOLVE.

$$\frac{3}{5} = \frac{37.47}{x}$$

$$3x = 5(37.47) \quad \text{Set cross products equal.}$$

$$3x = 187.35$$

$$x = 62.45 \quad \text{Divide both sides by } 3.$$

4. INTERPRET.

Check: Verify that 3 boxes is to 5 boxes as \$37.47 is to \$62.45. Also, notice that our solution is a reasonable one as discussed in Step 1.

State: Five boxes of CD-Rs cost \$62.45.

Work Practice 1

Helpful Hint

The proportion $\frac{5 \text{ boxes}}{3 \text{ boxes}} = \frac{\text{price of 5 boxes}}{\text{price of 3 boxes}}$ could also have been used to solve Example 1. Notice that the cross products are the same.

Similar triangles have the same shape but not necessarily the same size. In similar triangles, the measures of corresponding angles are equal, and corresponding sides are in proportion.

If triangle ABC and triangle XYZ shown are similar, then we know that the measure of angle A = the measure of angle X, the measure of angle B = the measure of angle Y, and the measure of angle C = the measure of angle Z. We also know that corresponding sides are in proportion: $\frac{a}{x} = \frac{b}{y} = \frac{c}{z}$.

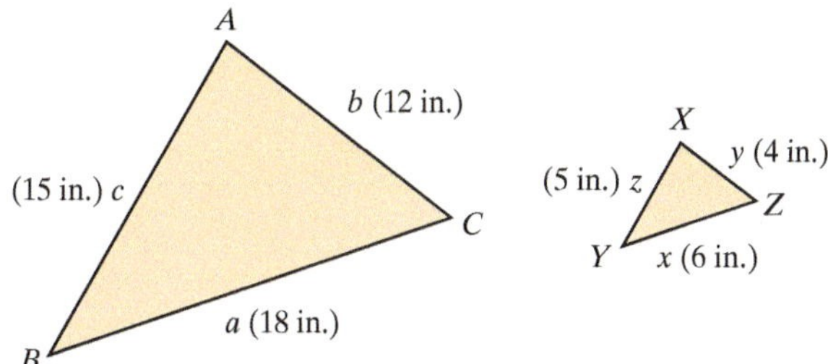

In this section, we will position similar triangles so that they have the same orientation.

To show that corresponding sides are in proportion for the triangles above, we write the ratios of the corresponding sides.

$$\frac{a}{x} = \frac{18}{6} = 3 \quad \frac{b}{y} = \frac{12}{4} = 3 \quad \frac{c}{z} = \frac{15}{5} = 3$$

Practice 2

If the following two triangles are similar, find x.

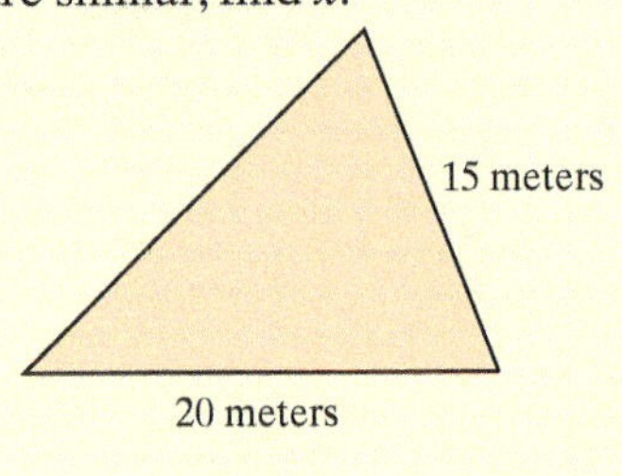

x meters
8 meters

Example 2 Finding the Length of a Side of a Triangle

If the following two triangles are similar, find the missing length, x.

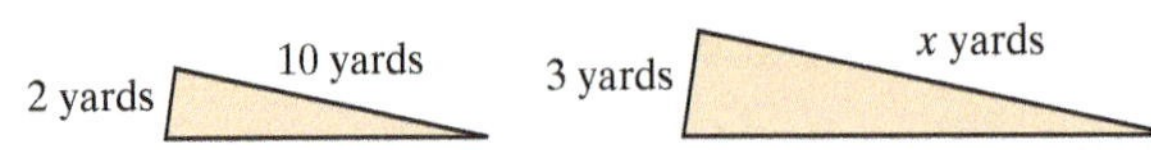

Solution:

1. UNDERSTAND. Read the problem and study the figure.
2. TRANSLATE. Since the triangles are similar, their corresponding sides are in proportion and we have

$$\frac{2}{3} = \frac{10}{x}$$

3. SOLVE. To solve, we multiply both sides by the LCD, $3x$, or cross multiply.

$$2x = 30$$

$$x = 15 \quad \text{Divide both sides by 2.}$$

4. INTERPRET.

Check: To check, replace x with 15 in the original proportion and see that a true statement results.

State: The missing length is 15 yards.

Work Practice 2

Objective 2 Solving Problems About Numbers 

Let's continue to solve problems. The remaining problems are all modeled by rational equations.

Practice 3

The quotient of a number and 5, minus $\frac{3}{2}$, is the quotient of the number and 10. Find the number.

Example 3 Finding an Unknown Number

The quotient of a number and 6, minus $\frac{5}{3}$, is the quotient of the number and 2. Find the number.

Solution:

1. UNDERSTAND. Read and reread the problem. Suppose that the unknown number is 2; then we see if the quotient of 2 and 6, or $\frac{2}{6}$, minus $\frac{5}{3}$ is equal to the quotient of 2 and 2, or $\frac{2}{2}$.

$$\frac{2}{6} - \frac{5}{3} = \frac{1}{3} - \frac{5}{3} = -\frac{4}{3}, \text{ not } \frac{2}{2}$$

Don't forget that the purpose of a proposed solution is to better understand the problem.

Let $x =$ the unknown number.

2. TRANSLATE.

In words:	the quotient of x and 6	minus	$\frac{5}{3}$	is	the quotient of x and 2
	↓	↓	↓	↓	↓
Translate:	$\frac{x}{6}$	$-$	$\frac{5}{3}$	$=$	$\frac{x}{2}$

Answers
2. 6 **3.** 15

3. SOLVE. Here, we solve the equation $\frac{x}{6} - \frac{5}{3} = \frac{x}{2}$. We begin by multiplying both sides of the equation by the LCD, 6.

$$6\left(\frac{x}{6} - \frac{5}{3}\right) = 6\left(\frac{x}{2}\right)$$

$$6\left(\frac{x}{6}\right) - 6\left(\frac{5}{3}\right) = 6\left(\frac{x}{2}\right) \quad \text{Apply the distributive property.}$$

$$x - 10 = 3x \quad \text{Simplify.}$$

$$-10 = 2x \quad \text{Subtract } x \text{ from both sides.}$$

$$\frac{-10}{2} = \frac{2x}{2} \quad \text{Divide both sides by 2.}$$

$$-5 = x \quad \text{Simplify.}$$

4. INTERPRET.

Check: To check, we verify that "the quotient of −5 and 6 minus $\frac{5}{3}$ is the quotient of −5 and 2," or $-\frac{5}{6} - \frac{5}{3} = -\frac{5}{2}$.

State: The unknown number is −5.

Work Practice 3

Objective 3 Solving Problems About Work

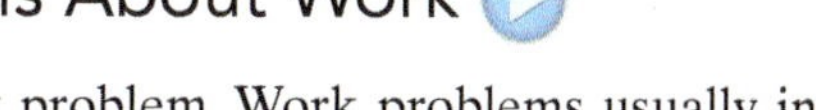

The next example is often called a work problem. Work problems usually involve people or machines doing a certain task.

Example 4 Finding Work Rates

Sam Waterton and Frank Schaffer work in a plant that manufactures automobiles. Sam can complete a quality control tour of the plant in 3 hours, while his assistant, Frank, needs 7 hours to complete the same job. The regional manager is coming to inspect the plant facilities, so both Sam and Frank are directed to complete a quality control tour together. How long will this take?

Solution:

1. UNDERSTAND. Read and reread the problem. The key idea here is the relationship between the **time** (hours) it takes to complete the job and the **part of the job** completed in 1 unit of time (hour). For example, if the **time** it takes Sam to complete the job is 3 hours, the **part of the job** he can complete in 1 hour is $\frac{1}{3}$. Similarly, Frank can complete $\frac{1}{7}$ of the job in 1 hour.

Let x = the **time** in hours it takes Sam and Frank to complete the job together. Then $\frac{1}{x}$ = the **part of the job** they can complete in 1 hour.

	Hours to Complete Total Job	Part of Job Completed in 1 Hour
Sam	3	$\frac{1}{3}$
Frank	7	$\frac{1}{7}$
Together	x	$\frac{1}{x}$

(Continued on next page)

Practice 4

Cindy Liu and Mary Beckwith own a landscaping company. Cindy can complete a certain garden planting in 3 hours, while Mary takes 4 hours to complete the same job. If both of them work together, how long will it take to plant the garden?

Answer

4. $1\frac{5}{7}$ hr

2. TRANSLATE.

In words:	part of job Sam completes in 1 hour	added to	part of job Frank completes in 1 hour	is equal to	part of job they complete together in 1 hour
	↓	↓	↓	↓	↓
Translate:	$\frac{1}{3}$	$+$	$\frac{1}{7}$	$=$	$\frac{1}{x}$

3. SOLVE. Here, we solve the equation $\frac{1}{3} + \frac{1}{7} = \frac{1}{x}$. We begin by multiplying both sides of the equation by the LCD, $21x$.

$$21x\left(\frac{1}{3}\right) + 21x\left(\frac{1}{7}\right) = 21x\left(\frac{1}{x}\right)$$

$$7x + 3x = 21 \quad \text{Simplify.}$$

$$10x = 21$$

$$x = \frac{21}{10} \quad \text{or} \quad 2\frac{1}{10} \text{ hours}$$

4. INTERPRET.

Check: Our proposed solution is $2\frac{1}{10}$ hours. This proposed solution is reasonable since $2\frac{1}{10}$ hours is more than half of Sam's time and less than half of Frank's time.

Check this solution in the *originally stated* problem.

State: Sam and Frank can complete the quality control tour in $2\frac{1}{10}$ hours.

Work Practice 4

Solve $E = mc^2$

a. for m. **b.** for c^2.

Objective 4 Solving Problems About Distance

Next we look at a problem solved by the distance formula,

$$d = r \cdot t$$

Practice 5

A bus travels 180 miles in the same time that a car travels 240 miles. If the car's speed is 15 mph faster than the speed of the bus, find the speed of the car and the speed of the bus.

Example 5 Finding Speeds of Vehicles

A car travels 180 miles in the same time that a truck travels 120 miles. If the car's speed is 20 miles per hour faster than the truck's, find the car's speed and the truck's speed.

Solution:

1. UNDERSTAND. Read and reread the problem. Suppose that the truck's speed is 45 miles per hour. Then the car's speed is 20 miles per hour more, or 65 miles per hour.

Answer

5. bus: 45 mph; car: 60 mph

Concept Check Answers

a. $m = \frac{E}{c^2}$ **b.** $c^2 = \frac{E}{m}$

We are given that the car travels 180 miles in the same time that the truck travels 120 miles. To find the time it takes the car to travel 180 miles, remember that since $d = rt$, we know that $\frac{d}{r} = t$.

Car's Time

$$t = \frac{d}{r} = \frac{180}{65} = 2\frac{50}{65} = 2\frac{10}{13} \text{ hours}$$

Truck's Time

$$t = \frac{d}{r} = \frac{120}{45} = 2\frac{30}{45} = 2\frac{2}{3} \text{ hours}$$

Since the times are not the same, our proposed solution is not correct. But we have a better understanding of the problem.

Let $x =$ the speed of the truck.

Since the car's speed is 20 miles per hour faster than the truck's, then

$x + 20 =$ the speed of the car

Use the formula $d = r \cdot t$ or **d**istance = **r**ate · **t**ime. Prepare a chart to organize the information in the problem.

Helpful Hint

If $d = r \cdot t$, then $t = \frac{d}{r}$ or *time* $= \frac{\textit{distance}}{\textit{rate}}$.

	Distance	=	Rate	·	Time
Truck	120		x		$\frac{120}{x}$ ← distance / ← rate
Car	180		$x + 20$		$\frac{180}{x + 20}$ ← distance / ← rate

2. TRANSLATE. Since the car and the truck travel the same amount of time, we have that

In words: car's time = truck's time

Translate: $\frac{180}{x + 20} = \frac{120}{x}$

3. SOLVE. We begin by multiplying both sides of the equation by the LCD, $x(x + 20)$, or cross multiplying.

$$\frac{180}{x + 20} = \frac{120}{x}$$

$180x = 120(x + 20)$

$180x = 120x + 2400$ Use the distributive property.

$60x = 2400$ Subtract $120x$ from both sides.

$x = 40$ Divide both sides by 60.

4. INTERPRET. The speed of the truck is 40 miles per hour. The speed of the car must then be $x + 20$ or 60 miles per hour.

Check: Find the time it takes the car to travel 180 miles and the time it takes the truck to travel 120 miles.

Car's Time

$$t = \frac{d}{r} = \frac{180}{60} = 3 \text{ hours}$$

Truck's Time

$$t = \frac{d}{r} = \frac{120}{40} = 3 \text{ hours}$$

Since both travel the same amount of time, the proposed solution is correct.

State: The car's speed is 60 miles per hour and the truck's speed is 40 miles per hour.

Work Practice 5

Vocabulary, Readiness & Video Check

Without solving algebraically, select the best choice for each exercise.

1. One person can complete a job in 7 hours. A second person can complete the same job in 5 hours. How long will it take them to complete the job if they work together?
a. more than 7 hours **b.** between 5 and 7 hours
c. less than 5 hours

2. One inlet pipe can fill a pond in 30 hours. A second inlet pipe can fill the same pond in 25 hours. How long before the pond is filled if both inlet pipes are on?
a. less than 25 hours **b.** between 25 and 30 hours
c. more than 30 hours

TRANSLATING

Given the variable in the first column, use the phrase in the second column to translate to an expression and then continue to the phrase in the third column to translate to another expression.

3.	A number: x	The reciprocal of the number:	The reciprocal of the number, decreased by 3:
4.	A number: y	The reciprocal of the number:	The reciprocal of the number, increased by 2:
5.	A number: z	The sum of the number and 5:	The reciprocal of the sum of the number and 5:
6.	A number: x	The difference of the number and 1:	The reciprocal of the difference of the number and 1:
7.	A number: y	Twice the number:	Eleven divided by twice the number:
8.	A number: z	Triple the number:	Negative ten divided by triple the number:

Martin-Gay Interactive Videos

See Video 14.6

Watch the section lecture video and answer the following questions.

Objective 1 **9.** In Example 1, we are told there are many ways to set up a correct proportion. Why does this fact make it even more important to check that your solution is reasonable?

Objective 2 **10.** What words or phrases in Example 3 told you to translate to an equation containing rational expressions?

Objective 3 **11.** From Example 4, how can you determine a somewhat reasonable answer to a work problem before you even begin to solve it?

Objective 4 **12.** The following problem is worded like Example 5 in the video, but uses different quantities.

A car travels 325 miles in the same time that a motorcycle travels 290 miles. If the car's speed is 7 miles per hour more than the motorcycle's, find the speed of the car and the speed of the motorcycle. Fill in the table and set up an equation based on this problem (do not solve). Use Example 7 in the video as a model for your work.

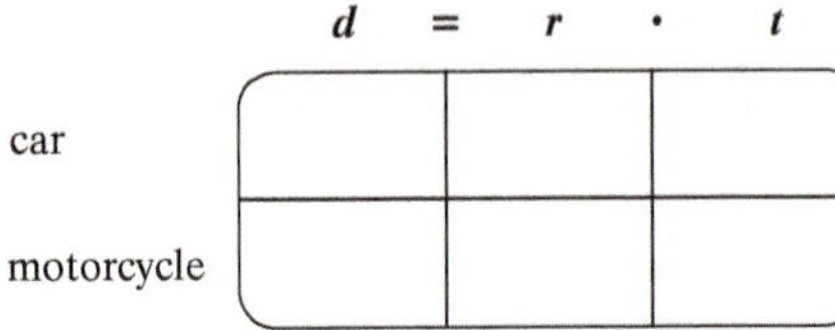

	d	$= r$	$\cdot\ t$
car			
motorcycle			

14.6 Exercise Set MyMathLab®

Solve. See Example 1.

1. The ratio of the weight of an object on Earth to the weight of the same object on Pluto is 100 to 3. If an elephant weighs 4100 pounds on Earth, find the elephant's weight on Pluto.
2. If a 170-pound person weighs approximately 65 pounds on Mars, about how much does a 9000-pound satellite weigh? Round your answer to the nearest pound.
3. There are 110 calories per 28.8 grams of Frosted Flakes cereal. Find how many calories are in 43.2 grams of this cereal.
4. On an architect's blueprint, 1 inch corresponds to 4 feet. Find the length of a wall represented by a line that is $3\frac{7}{8}$ inches long on the blueprint.

Find the unknown length x or y in the following pairs of similar triangles. See Example 2.

5.
6.
7.
8.

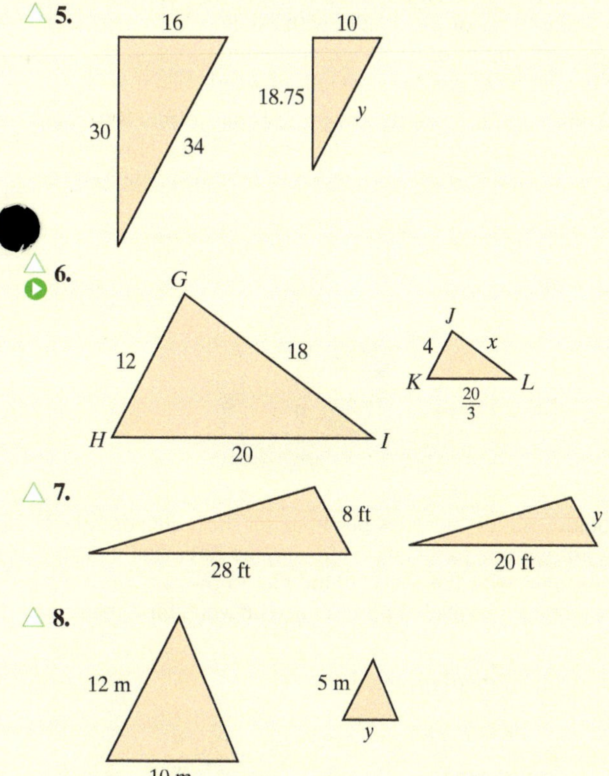

Solve the following. See Example 3.

9. Three times the reciprocal of a number equals 9 times the reciprocal of 6. Find the number.
10. Twelve divided by the sum of x and 2 equals the quotient of 4 and the difference of x and 2. Find x.
11. If twice a number added to 3 is divided by the number plus 1, the result is three halves. Find the number.
12. A number added to the product of 6 and the reciprocal of the number equals -5. Find the number.

See Example 4.

13. Smith Engineering found that an experienced surveyor surveys a roadbed in 4 hours. An apprentice surveyor needs 5 hours to survey the same stretch of road. If the two work together, find how long it takes them to complete the job.
14. An experienced bricklayer constructs a small wall in 3 hours. The apprentice completes the job in 6 hours. Find how long it takes if they work together.
15. In 2 minutes, a conveyor belt moves 300 pounds of recyclable aluminum from the delivery truck to a storage area. A smaller belt moves the same quantity of cans the same distance in 6 minutes. If both belts are used, find how long it takes to move the cans to the storage area.
16. Find how long it takes the conveyor belts described in Exercise 15 to move 1200 pounds of cans. (*Hint:* Think of 1200 pounds as four 300-pound jobs.)

See Example 5.

17. A jogger begins her workout by jogging to the park, a distance of 12 miles. She then jogs home at the same speed but along a different route. This return trip is 18 miles and her time is one hour longer. Find her jogging speed. Complete the accompanying chart and use it to find her jogging speed.

	Distance	=	Rate	·	Time
Trip to Park	12				
Return Trip	18				

18. A boat can travel 9 miles upstream in the same amount of time it takes to travel 11 miles downstream. If the current of the river is 3 miles per hour, complete the chart below and use it to find the speed of the boat in still water.

	Distance	=	Rate	·	Time
Upstream	9		$r - 3$		
Downstream	11		$r + 3$		

19. A cyclist rode the first 20-mile portion of his workout at a constant speed. For the 16-mile cooldown portion of his workout, he reduced his speed by 2 miles per hour. Each portion of the workout took the same time. Find the cyclist's speed during the first portion and find his speed during the cooldown portion.
20. A semi-truck travels 300 miles through the flatland in the same amount of time that it travels 180 miles through mountains. The rate of the truck is 20 miles per hour slower in the mountains than in the flatland. Find both the flatland rate and mountain rate.

MIXED PRACTICE

Solve the following. See Examples 1 through 5. (Note: *Some exercises can be modeled by equations without rational expressions.*)

21. A human factors expert recommends that there be at least 9 square feet of floor space in a college classroom for every student in the class. Find the minimum floor space that 40 students need.

22. Due to space problems at a local university, a 20-foot by 12-foot conference room is converted into a classroom. Find the maximum number of students the room can accommodate. (See Exercise 21.)

23. One-fourth equals the quotient of a number and 8. Find the number.

24. Four times a number added to 5 is divided by 6. The result is $\frac{7}{2}$. Find the number.

25. Marcus and Tony work for Lombardo's Pipe and Concrete. Mr. Lombardo is preparing an estimate for a customer. He knows that Marcus lays a slab of concrete in 6 hours. Tony lays the same size slab in 4 hours. If both work on the job and the cost of labor is \$45.00 per hour, decide what the labor estimate should be.

26. Mr. Dodson can paint his house by himself in 4 days. His son needs an additional day to complete the job if he works by himself. If they work together, find how long it takes to paint the house.

27. A pilot can travel 400 miles with the wind in the same amount of time as 336 miles against the wind. Find the speed of the wind if the pilot's speed in still air is 230 miles per hour.

28. A fisherman on Pearl River rows 9 miles downstream in the same amount of time he rows 3 miles upstream. If the current is 6 miles per hour, find how long it takes him to cover the 12 miles.

29. Find the unknown length y.

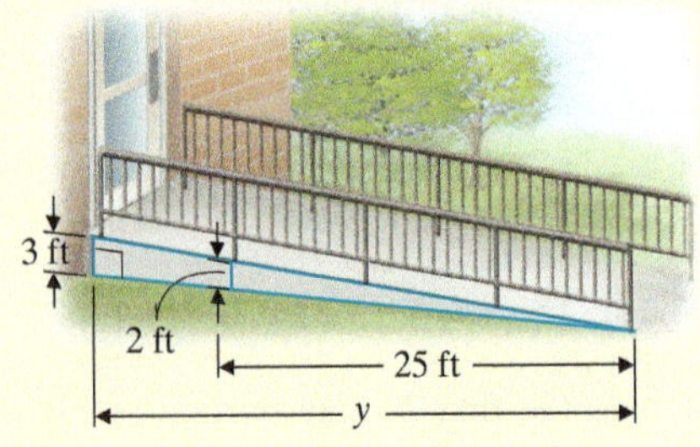

△ **30.** Find the unknown length y.

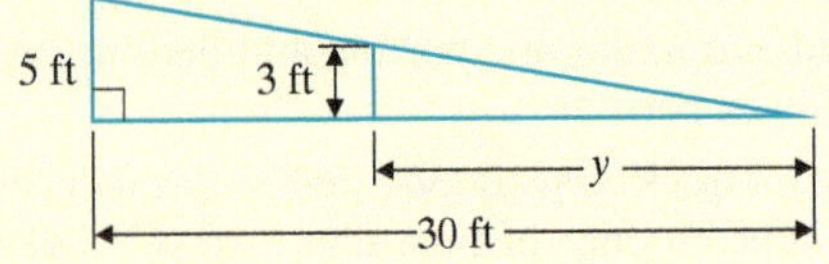

31. Suppose two trains leave Holbrook, Arizona, at the same time, traveling in opposite directions. One train travels 10 mph faster than the other. In 3.5 hours, the trains are 322 miles apart. Find the speed of each train.

32. Suppose two cars leave Brinkley, Arkansas, at the same time, traveling in opposite directions. One car travels 8 mph faster than the other car. In 2.5 hours, the cars are 280 miles apart. Find the speed of each car.

33. Two divided by the difference of a number and 3, minus 4 divided by the number plus 3, equals 8 times the reciprocal of the difference of the number squared and 9. What is the number?

34. If 15 times the reciprocal of a number is added to the ratio of 9 times the number minus 7 and the number plus 2, the result is 9. What is the number?

35. A pilot flies 630 miles with a tailwind of 35 miles per hour. Against the wind, he flies only 455 miles in the same amount of time. Find the rate of the plane in still air.

36. A marketing manager travels 1080 miles in a corporate jet and then an additional 240 miles by car. If the car ride takes one hour longer than the jet ride takes, and if the rate of the jet is 6 times the rate of the car, find the time the manager travels by jet and find the time the manager travels by car.

37. To mix weed killer with water correctly, it is necessary to mix 8 teaspoons of weed killer with 2 gallons of water. Find how many gallons of water are needed to mix with the entire box if it contains 36 teaspoons of weed killer.

38. The directions for a certain bug spray concentrate is to mix 3 ounces of concentrate with 2 gallons of water. How many ounces of concentrate are needed to mix with 5 gallons of water?

39. A boater travels 16 miles per hour on the water on a still day. During one particular windy day, he finds that he travels 48 miles with the wind behind him in the same amount of time that he travels 16 miles into the wind. Find the rate of the wind.

Let x be the rate of the wind.

	r	×	t	=	d
with wind	$16 + x$				48
into wind	$16 - x$				16

40. The current on a portion of the Mississippi River is 3 miles per hour. A barge can go 6 miles upstream in the same amount of time it takes to go 10 miles downstream. Find the speed of the boat in still water.

Let x be the speed of the boat in still water.

	r	×	t	=	d
upstream	$x - 3$				6
downstream	$x + 3$				10

41. Two hikers are 11 miles apart and walking toward each other. They meet in 2 hours. Find the rate of each hiker if one hiker walks 1.1 mph faster than the other.

42. On a 255-mile trip, Gary Alessandrini traveled at an average speed of 70 mph, got a speeding ticket, and then traveled at 60 mph for the remainder of the trip. If the entire trip took 4.5 hours and the speeding ticket stop took 30 minutes, how long did Gary speed before getting stopped?

43. One custodian cleans a suite of offices in 3 hours. When a second worker is asked to join the regular custodian, the job takes only $1\frac{1}{2}$ hours. How long does it take the second worker to do the same job alone?

44. One person proofreads copy for a small newspaper in 4 hours. If a second proofreader is also employed, the job can be done in $2\frac{1}{2}$ hours. How long does it take for the second proofreader to do the same job alone?

△ **45.** An architect is completing the plans for a triangular deck. Use the diagram below to find the missing dimension.

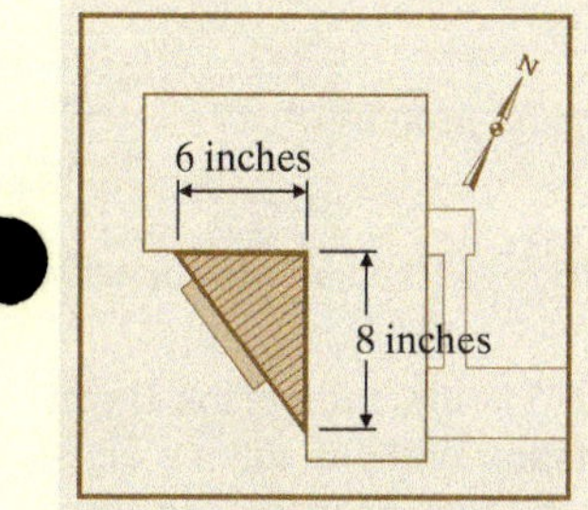

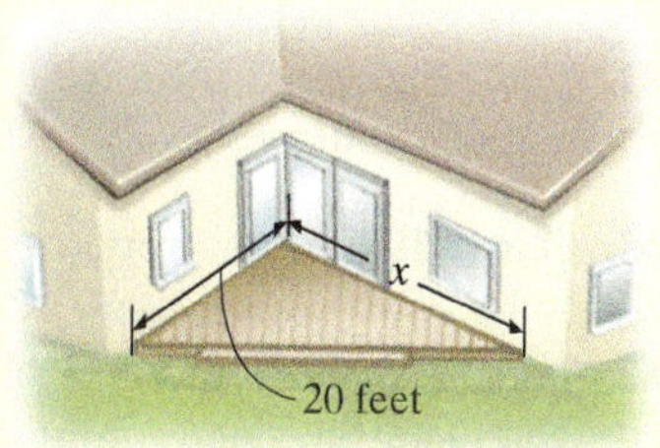

△ **46.** A student wishes to make a small model of a triangular mainsail to study the effects of wind on the sail. The smaller model will be the same shape as a regular-size sailboat's mainsail. Use the following diagram to find the missing dimensions.

47. The manufacturers of cans of salted mixed nuts state that the ratio of peanuts to other nuts is 3 to 2. If 324 peanuts are in a can, find how many other nuts should also be in the can.

48. There are 1280 calories in a 14-ounce portion of Eagle Brand Milk. Find how many calories are in 2 ounces of Eagle Brand Milk.

49. A jet plane traveling at 500 mph overtakes a propeller plane traveling at 200 mph that had a 2-hour head start. How far from the starting point are the planes?

50. How long will it take a bus traveling at 60 miles per hour to overtake a car traveling at 40 mph if the car had a 1.5-hour head start?

51. One pipe fills a storage pool in 20 hours. A second pipe fills the same pool in 15 hours. When a third pipe is added and all three are used to fill the pool, it takes only 6 hours. Find how long it takes the third pipe to do the job.

52. One pump fills a tank in 9 hours. A second pump fills the same tank in 6 hours. When a third pump is added and all three are used to fill the tank, it takes only 3 hours. Find how long it takes the third pump to fill the tank.

53. A car travels 280 miles in the same time that a motorcycle travels 240 miles. If the car's speed is 10 miles per hour more than the motorcycle's, find the speed of the car and the speed of the motorcycle.

54. A bus traveled on a level road for 3 hours at an average speed 20 miles per hour faster than it traveled on a winding road. The time spent on the winding road was 4 hours. Find the average speed on the level road if the entire trip was 305 miles.

55. In 6 hours, an experienced cook prepares enough pies to supply a local restaurant's daily order. Another cook prepares the same number of pies in 7 hours. Together with a third cook, they prepare the pies in 2 hours. Find how long it takes the third cook to prepare the pies alone.

56. Mrs. Smith balances the company books in 8 hours. It takes her assistant 12 hours to do the same job. If they work together, find how long it takes them to balance the books.

57. The quotient of a number and 3, minus 1, equals $\frac{5}{3}$. Find the number.

58. The quotient of a number and 5, minus 1, equals $\frac{7}{5}$. Find the number.

59. Currently, the Toyota Corolla is the best-selling car in the world. Suppose that during a test drive of two Corollas, one car travels 224 miles in the same time that the second car travels 175 miles. If the speed of the first car is 14 miles per hour faster than the speed of the second car, find the speed of both cars. (*Source: Top Ten of Everything*)

60. The second best-selling car is the Volkswagen Golf. A driver of this car took a day trip around the California coastline driving at two speeds. He drove 70 miles at a slower speed and 300 miles at a speed 40 miles per hour faster. If the time spent driving the faster speed was twice that spent at the slower speed, find the two speeds during the trip. (*Source: Top Ten of Everything*)

61. A pilot can fly an MD-11 2160 miles with the wind in the same time she can fly 1920 miles against the wind. If the speed of the wind is 30 mph, find the speed of the plane in still air. (*Source:* Air Transport Association of America)

62. A pilot can fly a DC-10 1365 miles against the wind in the same time he can fly 1575 miles with the wind. If the speed of the plane in still air is 490 miles per hour, find the speed of the wind. (*Source:* Air Transport Association of America)

Given that the following pairs of triangles are similar, find each missing length.

63.

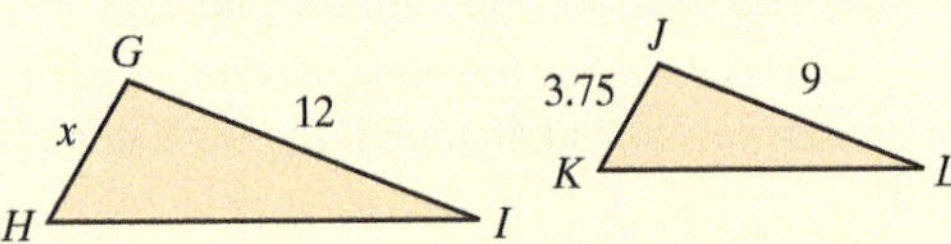

64.

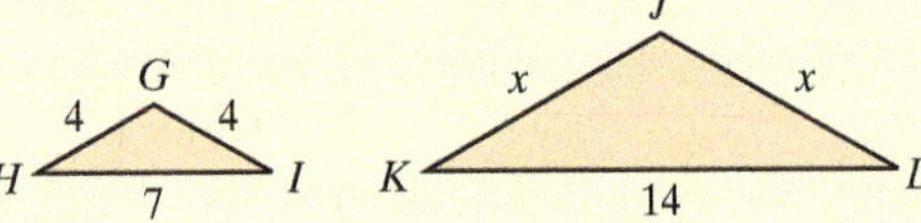

65.

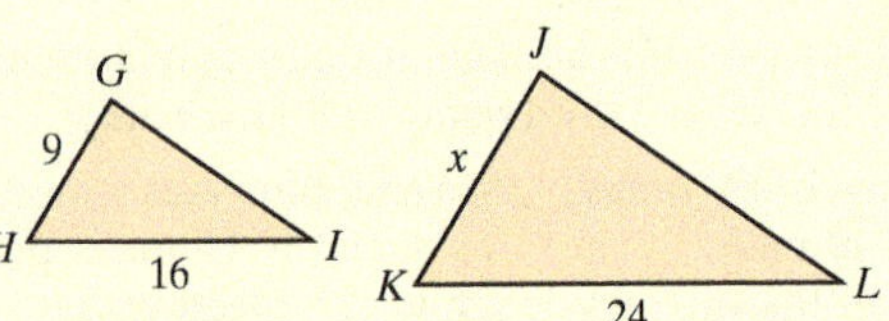

66.

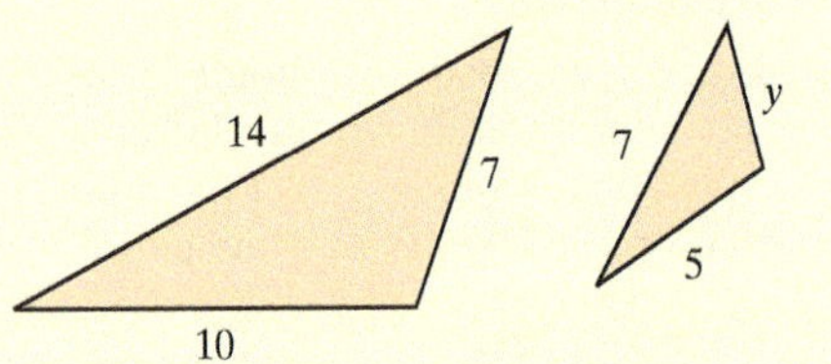

REVIEW AND PREVIEW

Find the slope of the line through each pair of points. Use the slope to determine whether the line is vertical, horizontal, or moves upward or downward from left to right. See Section 10.4.

67. $(-2, 5), (4, -3)$

68. $(0, 4), (2, 10)$

69. $(-3, -6), (1, 5)$

70. $(-2, 7), (3, -2)$

71. $(3, 7), (3, -2)$

72. $(0, -4), (2, -4)$

CONCEPT EXTENSIONS

The following bar graph shows the capacity of the United States to generate electricity from the wind in the years shown. Use this graph for Exercises 73 and 74.

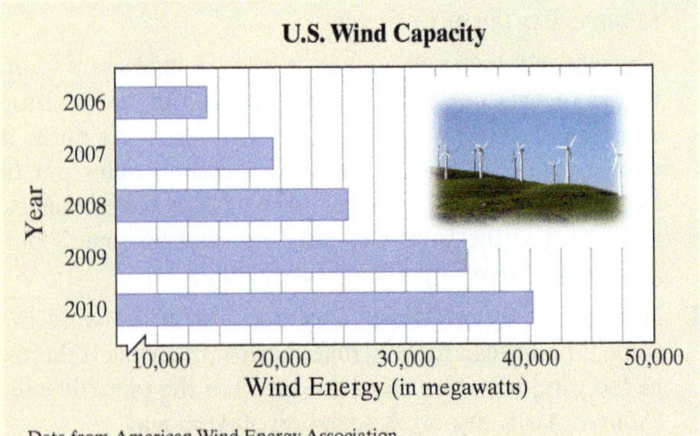

73. Find the approximate megawatt capacity in 2010.

74. Find the approximate megawatt capacity in 2009.

In general, 1000 megawatts will serve the average electricity needs of 560,000 people. Use this fact and the preceding graph to answer Exercises 75 and 76.

75. In 2010, the number of megawatts that were generated from wind served the electricity needs of how many people? (Round to the nearest ten-thousand.)

76. How many megawatts of electricity are needed to serve the city or town in which you live?

For Exercises 77 and 78 decide whether we can immediately use cross products to solve for x. Do not actually solve.

77. $\frac{2 - x}{5} = \frac{1 + x}{3}$

78. $\frac{2}{5} - x = \frac{1 + x}{3}$

Solve.

79. One pump fills a tank 3 times as fast as another pump. If the pumps work together, they fill the tank in 21 minutes. How long does it take for each pump to fill the tank?

80. It takes 9 hours for pump A to fill a tank alone. Pump B takes 15 hours to fill the same tank alone. If pumps A, B, and C are used, the tank fills in 5 hours. How long does it take pump C to fill the tank alone?

81. For what value of x is $\frac{x}{x - 1}$ in proportion to $\frac{x + 1}{x}$? Explain your result.

82. If x is 10, is $\frac{2}{x}$ in proportion to $\frac{x}{50}$? Explain why or why not.

83. Person A can complete a job in 5 hours, and person B can complete the same job in 3 hours. Without solving algebraically, discuss reasonable and unreasonable answers for how long it would take them to complete the job together.

84. A hyena spots a giraffe 0.5 mile away and begins running toward it. The giraffe starts running away from the hyena just as the hyena begins running toward it. A hyena can run at a speed of 40 mph and a giraffe can run at 32 mph. How long will it take for the hyena to overtake the giraffe? (*Source: World Almanac and Book of Facts*)

Solve. See the Concept Check in this section.

Solve $D = RT$

85. for R.

86. for T.

14.7 Simplifying Complex Fractions 

A rational expression whose numerator, denominator, or both contain one or more rational expressions is called a **complex rational expression** or a **complex fraction.**

Complex Fractions

$$\frac{\frac{1}{a}}{\frac{b}{2}} \qquad \frac{\frac{x}{2y^2}}{\frac{6x-2}{9y}} \qquad \frac{x+\frac{1}{y}}{y+1}$$

The parts of a complex fraction are

$$\frac{\left.\frac{x}{y+2}\right\}}{\left.7+\frac{1}{y}\right\}}$$

← Numerator of complex fraction
← Main fraction bar
← Denominator of complex fraction

Objectives

1. Simplify Complex Fractions by Simplifying the Numerator and Denominator and Then Dividing.
2. Simplify Complex Fractions by Multiplying by a Common Denominator.
3. Simplify Expressions with Negative Exponents.

Our goal in this section is to simplify complex fractions. A complex fraction is simplified when it is in the form $\frac{P}{Q}$, where P and Q are polynomials that have no common factors. Two methods of simplifying complex fractions are introduced. The first method evolves from the definition of a fraction as a quotient.

Objective 1 Simplifying Complex Fractions: Method 1 

Simplifying a Complex Fraction: Method I

Step 1: Simplify the numerator and the denominator of the complex fraction so that each is a single fraction.

Step 2: Perform the indicated division by multiplying the numerator of the complex fraction by the reciprocal of the denominator of the complex fraction.

Step 3: Simplify if possible.

Example 1 Simplify each complex fraction.

a. $\dfrac{\frac{2x}{27y^2}}{\frac{6x^2}{9}}$ **b.** $\dfrac{\frac{5x}{x+2}}{\frac{10}{x-2}}$ **c.** $\dfrac{\frac{x}{y^2}+\frac{1}{y}}{\frac{y}{x^2}+\frac{1}{x}}$

Solution:

a. The numerator of the complex fraction is already a single fraction, and so is the denominator. Perform the indicated division by multiplying the numerator, $\frac{2x}{27y^2}$, by the reciprocal of the denominator, $\frac{6x^2}{9}$. Then simplify.

(Continued on next page)

Practice 1

Simplify each complex fraction.

a. $\dfrac{\frac{5k}{36m}}{\frac{15k}{9}}$ **b.** $\dfrac{\frac{8x}{x-4}}{\frac{3}{x+4}}$

c. $\dfrac{\frac{5}{a}+\frac{b}{a^2}}{\frac{5a}{b^2}+\frac{1}{b}}$

Answers

1. a. $\frac{1}{12m}$ **b.** $\frac{8x(x+4)}{3(x-4)}$ **c.** $\frac{b^2}{a^2}$

$$\frac{\frac{2x}{27y^2}}{\frac{6x^2}{9}} = \frac{2x}{27y^2} \div \frac{6x^2}{9}$$

$$= \frac{2x}{27y^2} \cdot \frac{9}{6x^2} \quad \text{Multiply by the reciprocal of } \frac{6x^2}{9}.$$

$$= \frac{2x \cdot 9}{27y^2 \cdot 6x^2}$$

$$= \frac{1}{9xy^2}$$

Helpful Hint Both the numerator and denominator are single fractions, so we perform the indicated division.

b. $$\frac{\frac{5x}{x+2}}{\frac{10}{x-2}} = \frac{5x}{x+2} \div \frac{10}{x-2} = \frac{5x}{x+2} \cdot \frac{x-2}{10} \quad \text{Multiply by the reciprocal of } \frac{10}{x-2}.$$

$$= \frac{5x(x-2)}{2 \cdot 5(x+2)}$$

$$= \frac{x(x-2)}{2(x+2)} \quad \text{Simplify.}$$

c. First simplify the numerator and the denominator of the complex fraction separately so that each is a single fraction. Then perform the indicated division.

$$\frac{\frac{x}{y^2} + \frac{1}{y}}{\frac{y}{x^2} + \frac{1}{x}} = \frac{\frac{x}{y^2} + \frac{1 \cdot y}{y \cdot y}}{\frac{y}{x^2} + \frac{1 \cdot x}{x \cdot x}} \quad \begin{array}{l}\text{Simplify the numerator. The LCD is } y^2.\\ \text{Simplify the denominator. The LCD is } x^2.\end{array}$$

$$= \frac{\frac{x+y}{y^2}}{\frac{y+x}{x^2}} \quad \text{Add.}$$

$$= \frac{x+y}{y^2} \cdot \frac{x^2}{y+x} \quad \text{Multiply by the reciprocal of } \frac{y+x}{x^2}.$$

$$= \frac{x^2(x+y)}{y^2(y+x)}$$

$$= \frac{x^2}{y^2} \quad \text{Simplify.}$$

Work Practice 1

✓Concept Check Which of the following are equivalent to $\frac{\frac{5}{y}}{\frac{2}{z}}$?

a. $\frac{5}{y} \div \frac{2}{z}$ **b.** $\frac{5}{y} \cdot \frac{z}{2}$ **c.** $\frac{5}{y} \div \frac{z}{2}$

Objective 2 Simplifying Complex Fractions: Method 2

Next we look at another method of simplifying complex fractions. With this method, we multiply the numerator and the denominator of the complex fraction by the LCD of all fractions in the complex fraction.

✓Concept Check Answer
a and b

Simplifying a Complex Fraction: Method 2

Step 1: Multiply the numerator and the denominator of the complex fraction by the LCD of the fractions in both the numerator and the denominator.

Step 2: Simplify.

Example 2 Simplify each complex fraction.

a. $\dfrac{\dfrac{5x}{x+2}}{\dfrac{10}{x-2}}$ **b.** $\dfrac{\dfrac{x}{y^2}+\dfrac{1}{y}}{\dfrac{y}{x^2}+\dfrac{1}{x}}$

Solution:

a. Notice we are reworking Example 1b using method 2. The least common denominator of $\frac{5x}{x+2}$ and $\frac{10}{x-2}$ is $(x+2)(x-2)$. Multiply both the numerator, $\frac{5x}{x+2}$, and the denominator, $\frac{10}{x-2}$, by the LCD.

$$\frac{\frac{5x}{x+2}}{\frac{10}{x-2}} = \frac{\left(\frac{5x}{x+2}\right)\cdot(x+2)(x-2)}{\left(\frac{10}{x-2}\right)\cdot(x+2)(x-2)}$$ Multiply numerator and denominator by the LCD.

$$= \frac{5x\cdot(x-2)}{2\cdot 5\cdot(x+2)}$$ Simplify.

$$= \frac{x(x-2)}{2(x+2)}$$ Simplify.

b. Here, we are reworking Example 1c using method 2. The least common denominator of $\frac{x}{y^2}, \frac{1}{y}, \frac{y}{x^2}$, and $\frac{1}{x}$ is x^2y^2.

$$\frac{\frac{x}{y^2}+\frac{1}{y}}{\frac{y}{x^2}+\frac{1}{x}} = \frac{\left(\frac{x}{y^2}+\frac{1}{y}\right)\cdot x^2y^2}{\left(\frac{y}{x^2}+\frac{1}{x}\right)\cdot x^2y^2}$$ Multiply the numerator and denominator by the LCD.

$$= \frac{\frac{x}{y^2}\cdot x^2y^2 + \frac{1}{y}\cdot x^2y^2}{\frac{y}{x^2}\cdot x^2y^2 + \frac{1}{x}\cdot x^2y^2}$$ Use the distributive property.

$$= \frac{x^3 + x^2y}{y^3 + xy^2}$$ Simplify.

$$= \frac{x^2(x+y)}{y^2(y+x)}$$ Factor.

$$= \frac{x^2}{y^2}$$ Simplify.

Work Practice 2

Practice 2

Use method 2 to simplify:

a. $\dfrac{\dfrac{8x}{x-4}}{\dfrac{3}{x+4}}$ **b.** $\dfrac{\dfrac{b}{a^2}+\dfrac{1}{a}}{\dfrac{a}{b^2}+\dfrac{1}{b}}$

Answers

2. a. $\frac{8x(x+4)}{3(x-4)}$ **b.** $\frac{b^2}{a^2}$

Objective 3 Simplifying Expressions with Negative Exponents

If an expression contains negative exponents, write the expression as an equivalent expression with positive exponents.

Practice 3

Simplify: $\dfrac{3x^{-1} + x^{-2}y^{-1}}{y^{-2} + xy^{-1}}$

Example 3 Simplify.

$$\frac{x^{-1} + 2xy^{-1}}{x^{-2} - x^{-2}y^{-1}}$$

Solution: This fraction does not appear to be a complex fraction. If we write it by using only positive exponents, however, we see that it is a complex fraction.

$$\frac{x^{-1} + 2xy^{-1}}{x^{-2} - x^{-2}y^{-1}} = \frac{\frac{1}{x} + \frac{2x}{y}}{\frac{1}{x^2} - \frac{1}{x^2y}}$$

The LCD of $\frac{1}{x}, \frac{2x}{y}, \frac{1}{x^2}$, and $\frac{1}{x^2y}$ is x^2y. Multiply both the numerator and denominator by x^2y.

$$= \frac{\left(\frac{1}{x} + \frac{2x}{y}\right) \cdot x^2y}{\left(\frac{1}{x^2} - \frac{1}{x^2y}\right) \cdot x^2y}$$

$$= \frac{\frac{1}{x} \cdot x^2y + \frac{2x}{y} \cdot x^2y}{\frac{1}{x^2} \cdot x^2y - \frac{1}{x^2y} \cdot x^2y} \quad \text{Apply the distributive property.}$$

$$= \frac{xy + 2x^3}{y - 1} \text{ or } \frac{x(y + 2x^2)}{y - 1} \quad \text{Simplify.}$$

Work Practice 3

Practice 4

Simplify: $\dfrac{(3x)^{-1} - 2}{5x^{-1} + 2}$

Example 4 Simplify: $\dfrac{(2x)^{-1} + 1}{2x^{-1} - 1}$

Solution:

$$\frac{(2x)^{-1} + 1}{2x^{-1} - 1} = \frac{\frac{1}{2x} + 1}{\frac{2}{x} - 1} \quad \text{Write using positive exponents.}$$

$$= \frac{\left(\frac{1}{2x} + 1\right) \cdot 2x}{\left(\frac{2}{x} - 1\right) \cdot 2x} \quad \text{The LDC of } \frac{1}{2x} \text{ and } \frac{2}{x} \text{ is } 2x.$$

$$= \frac{\frac{1}{2x} \cdot 2x + 1 \cdot 2x}{\frac{2}{x} \cdot 2x - 1 \cdot 2x} \quad \text{Use distributive property.}$$

$$= \frac{1 + 2x}{4 - 2x} \text{ or } \frac{1 + 2x}{2(2 - x)} \quad \text{Simplify.}$$

Work Practice 4

Helpful Hint

Don't forget that $(2x)^{-1} = \frac{1}{2x}$, but $2x^{-1} = 2 \cdot \frac{1}{x} = \frac{2}{x}$.

Answers

3. $\dfrac{y(3xy + 1)}{x^2(1 + xy)}$

4. $\dfrac{1 - 6x}{15 + 6x}$ or $\dfrac{1 - 6x}{3(5 + 2x)}$

Vocabulary, Readiness & Video Check

Complete the steps by writing the simplified complex fraction.

1. $\dfrac{\frac{7}{x}}{\frac{1}{x}+\frac{z}{x}} = \dfrac{x\left(\frac{7}{x}\right)}{x\left(\frac{1}{x}\right)+x\left(\frac{z}{x}\right)} =$ ______

2. $\dfrac{\frac{x}{4}}{\frac{x^2}{2}+\frac{1}{4}} = \dfrac{4\left(\frac{x}{4}\right)}{4\left(\frac{x^2}{2}\right)+4\left(\frac{1}{4}\right)} =$ ______

Write with positive exponents.

3. $x^{-2} =$ ______ **4.** $y^{-3} =$ ______ **5.** $2x^{-1} =$ ______

6. $(2x)^{-1} =$ ______ **7.** $(9y)^{-1} =$ ______ **8.** $9y^{-2} =$ ______

Martin-Gay Interactive Videos

See Video 14.7 

Watch the section lecture video and answer the following questions.

Objective 1 **9.** From Example 2, before you can rewrite the complex fraction as division, describe how it must appear.

Objective 2 **10.** How does finding an LCD in method 2, as in Example 3, differ from finding an LCD in method 1? In your answer, mention the purpose of the LCD in each method.

Objective 3 **11.** Based on Example 4, what connection is there between negative exponents and complex fractions?

14.7 Exercise Set MyMathLab®

Simplify each complex fraction. See Examples 1 and 2.

1. $\dfrac{\frac{10}{3x}}{\frac{5}{6x}}$

2. $\dfrac{\frac{15}{2x}}{\frac{5}{6x}}$

3. $\dfrac{1+\frac{2}{5}}{2+\frac{3}{5}}$

4. $\dfrac{2+\frac{1}{7}}{3-\frac{4}{7}}$

5. $\dfrac{\frac{4}{x-1}}{\frac{x}{x-1}}$

6. $\dfrac{\frac{x}{x+2}}{\frac{2}{x+2}}$

7. $\dfrac{1-\frac{2}{x}}{x+\frac{4}{9x}}$

8. $\dfrac{5-\frac{3}{x}}{x+\frac{2}{3x}}$

9. $\dfrac{\frac{4x^2-y^2}{xy}}{\frac{2}{y}-\frac{1}{x}}$

10. $\dfrac{\frac{x^2-9y^2}{xy}}{\frac{1}{y}-\frac{3}{x}}$

11. $\dfrac{\frac{x+1}{3}}{\frac{2x-1}{6}}$

12. $\dfrac{\frac{x+3}{12}}{\frac{4x-5}{15}}$

13. $\dfrac{\frac{2}{x}+\frac{3}{x^2}}{\frac{4}{x^2}-\frac{9}{x}}$

14. $\dfrac{\frac{2}{x^2}+\frac{1}{x}}{\frac{4}{x^2}-\frac{1}{x}}$

15. $\dfrac{\frac{1}{x}+\frac{2}{x^2}}{x+\frac{8}{x^2}}$

16. $\dfrac{\frac{1}{y}+\frac{3}{y^2}}{y+\frac{27}{y^2}}$

17. $\dfrac{\frac{4}{5-x}+\frac{5}{x-5}}{\frac{2}{x}+\frac{3}{x-5}}$

18. $\dfrac{\frac{3}{x-4}-\frac{2}{4-x}}{\frac{2}{x-4}-\frac{2}{x}}$

19. $\dfrac{\frac{x+2}{x}-\frac{2}{x-1}}{\frac{x+1}{x}+\frac{x+1}{x-1}}$

20. $\dfrac{\frac{5}{a+2}-\frac{1}{a-2}}{\frac{3}{2+a}+\frac{6}{2-a}}$

21. $\dfrac{\frac{2}{x}+3}{\frac{4}{x^2}-9}$

22. $\dfrac{2+\frac{1}{x}}{4x-\frac{1}{x}}$

23. $\dfrac{1-\frac{x}{y}}{\frac{x^2}{y^2}-1}$

24. $\dfrac{1-\frac{2}{x}}{x-\frac{4}{x}}$

25. $\dfrac{\frac{-2x}{x-y}}{\frac{y}{x^2}}$

26. $\dfrac{\frac{7y}{x^2+xy}}{\frac{y^2}{x^2}}$

27. $\dfrac{\frac{2}{x}+\frac{1}{x^2}}{\frac{y}{x^2}}$

28. $\dfrac{\frac{5}{x^2}-\frac{2}{x}}{\frac{1}{x}+2}$

29. $\dfrac{\frac{x}{9}-\frac{1}{x}}{1+\frac{3}{x}}$

30. $\dfrac{\frac{x}{4}-\frac{4}{x}}{1-\frac{4}{x}}$

31. $\dfrac{\frac{x-1}{x^2-4}}{1+\frac{1}{x-2}}$

32. $\dfrac{\frac{x+3}{x^2-9}}{1+\frac{1}{x-3}}$

33. $\dfrac{\frac{2}{x+5}+\frac{4}{x+3}}{\frac{3x+13}{x^2+8x+15}}$

34. $\dfrac{\frac{2}{x+2}+\frac{6}{x+7}}{\frac{4x+13}{x^2+9x+14}}$

Simplify. See Examples 3 and 4.

35. $\dfrac{x^{-1}}{x^{-2}+y^{-2}}$

36. $\dfrac{a^{-3}+b^{-1}}{a^{-2}}$

37. $\dfrac{2a^{-1}+3b^{-2}}{a^{-1}-b^{-1}}$

38. $\dfrac{x^{-1}+y^{-1}}{3x^{-2}+5y^{-2}}$

39. $\dfrac{1}{x-x^{-1}}$

40. $\dfrac{x^{-2}}{x+3x^{-1}}$

41. $\dfrac{a^{-1}+1}{a^{-1}-1}$

42. $\dfrac{a^{-1}-4}{4+a^{-1}}$

43. $\dfrac{3x^{-1}+(2y)^{-1}}{x^{-2}}$

44. $\dfrac{5x^{-2}-3y^{-1}}{x^{-1}+y^{-1}}$

45. $\dfrac{2a^{-1}+(2a)^{-1}}{a^{-1}+2a^{-2}}$

46. $\dfrac{a^{-1}+2a^{-2}}{2a^{-1}+(2a)^{-1}}$

47. $\dfrac{5x^{-1}+2y^{-1}}{x^{-2}y^{-2}}$

48. $\dfrac{x^{-2}y^{-2}}{5x^{-1}+2y^{-1}}$

49. $\dfrac{5x^{-1}-2y^{-1}}{25x^{-2}-4y^{-2}}$

50. $\dfrac{3x^{-1}+3y^{-1}}{4x^{-2}-9y^{-2}}$

REVIEW AND PREVIEW

Simplify. See Section 12.1.

51. $\dfrac{3x^3y^2}{12x}$

52. $\dfrac{-36xb^3}{9xb^2}$

53. $\dfrac{144x^5y^5}{-16x^2y}$

54. $\dfrac{48x^3y^2}{-4xy}$

Solve the following. See Section 10.6.

55. If $P(x) = -x^2$, find $P(-3)$.

56. If $f(x) = x^2 - 6$, find $f(-1)$.

CONCEPT EXTENSIONS

Solve. See the Concept Check in this section.

57. Which of the following are equivalent to $\dfrac{\frac{x+1}{9}}{\frac{y-2}{5}}$?

a. $\frac{x+1}{9} \div \frac{y-2}{5}$ b. $\frac{x+1}{9} \cdot \frac{y-2}{5}$ c. $\frac{x+1}{9} \cdot \frac{5}{y-2}$

58. Which of the following are equivalent to $\dfrac{\frac{a}{7}}{\frac{b}{13}}$?

a. $\frac{a}{7} \cdot \frac{b}{13}$ b. $\frac{a}{7} \div \frac{b}{13}$ c. $\frac{a}{7} \div \frac{13}{b}$ d. $\frac{a}{7} \cdot \frac{13}{b}$

59. When the source of a sound is traveling toward a listener, the pitch that the listener hears due to the Doppler effect is given by the complex rational expression $\dfrac{a}{1-\frac{s}{770}}$, where a is the actual pitch of the sound and s is the speed of the sound source. Simplify this expression.

60. In baseball, the earned run average (ERA) statistic gives the average number of earned runs scored on a pitcher per game. It is computed with the following expression: $\dfrac{E}{\frac{I}{9}}$, where E is the number of earned runs scored on a pitcher and I is the total number of innings pitched by the pitcher. Simplify this expression.

61. Which of the following is/are equivalent to $\dfrac{\frac{1}{x}}{\frac{3}{y}}$?

a. $\frac{1}{x} \div \frac{3}{y}$ b. $\frac{1}{x} \cdot \frac{y}{3}$ c. $\frac{1}{x} \div \frac{y}{3}$

62. Which of the following is/are equivalent to $\dfrac{5}{\frac{2}{a}}$?

a. $\frac{5}{1} \div \frac{2}{a}$ b. $\frac{1}{5} \div \frac{2}{a}$ c. $\frac{5}{1} \cdot \frac{2}{a}$

63. In your own words, explain one method for simplifying a complex fraction.

64. Explain your favorite method for simplifying a complex fraction and why.

Simplify.

65. $\dfrac{1}{1 + (1 + x)^{-1}}$

66. $\dfrac{(x + 2)^{-1} + (x - 2)^{-1}}{(x^2 - 4)^{-1}}$

67. $\dfrac{x}{1 - \dfrac{1}{1 + \dfrac{1}{x}}}$

68. $\dfrac{x}{1 - \dfrac{1}{1 - \dfrac{1}{x}}}$

69. $\dfrac{\dfrac{2}{y^2} - \dfrac{5}{xy} - \dfrac{3}{x^2}}{\dfrac{2}{y^2} + \dfrac{7}{xy} + \dfrac{3}{x^2}}$

70. $\dfrac{\dfrac{2}{x^2} - \dfrac{1}{xy} - \dfrac{1}{y^2}}{\dfrac{1}{x^2} - \dfrac{3}{xy} + \dfrac{2}{y^2}}$

71. $\dfrac{3(a + 1)^{-1} + 4a^{-2}}{(a^3 + a^2)^{-1}}$

72. $\dfrac{9x^{-1} - 5(x - y)^{-1}}{4(x - y)^{-1}}$

In the study of calculus, the difference quotient $\dfrac{f(a + h) - f(a)}{h}$ *is often found and simplified. Find and simplify this quotient for each function f(x) by following steps **a** through **d**.*

a. *Find* $(a + h)$.

b. *Find* $f(a)$.

c. *Use steps **a** and **b** to find* $\dfrac{f(a + h) - f(a)}{h}$

d. *Simplify the result of step **c**.*

73. $f(x) = \dfrac{1}{x}$

74. $f(x) = \dfrac{5}{x}$

75. $f(x) = \dfrac{3}{x + 1}$

76. $f(x) = \dfrac{2}{x^2}$

Chapter 14 Vocabulary Check

Fill in each blank with one of the words or phrases listed below. Not all choices will be used.

least common denominator	simplifying	reciprocals	numerator	$\dfrac{-a}{b}$ $\dfrac{a}{-b}$
cross products	ratio	proportion		
rational expression	domain	complex fraction	denominator	$\dfrac{-a}{-b}$

1. A(n) ______________________ is an expression that can be written in the form $\dfrac{P}{Q}$, where P and Q are polynomials and Q is not 0.

2. In a(n) ______________________, the numerator or denominator or both may contain fractions.

3. For a rational expression, $-\dfrac{a}{b} =$ ______________________ $=$ ______________________.

4. A rational expression is undefined when the ______________________ is 0.

5. The process of writing a rational expression in lowest terms is called ______________________.

6. The expressions $\dfrac{2x}{7}$ and $\dfrac{7}{2x}$ are called ______________________.

7. The ______________________ of a list of rational expressions is a polynomial of least degree whose factors include all factors of the denominators in the list.

8. A(n) ______________________ is the quotient of two numbers.

9. $\dfrac{x}{2} = \dfrac{7}{16}$ is an example of a(n) ______________________.

10. If $\dfrac{a}{b} = \dfrac{c}{d}$, then ad and bc are called ______________________.

11. The ______________________ of the rational function $f(x) = \dfrac{1}{x - 3}$ is $\{x | x \text{ is a real number}, x \neq 3\}$.

Helpful Hint Are you preparing for your test? Don't forget to take the Chapter 14 Test on page 1060. Then check your answers at the back of the text and use the Chapter Test Prep Videos to see the fully worked-out solutions to any of the exercises you want to review.

14 Chapter Highlights

Definitions and Concepts | **Examples**

Section 14.1 Rational Functions and Simplifying Rational Expressions

A **rational expression** is an expression that can be written in the form $\frac{P}{Q}$, where P and Q are polynomials and Q does not equal 0.

$$\frac{7y^3}{4}, \frac{x^2+6x+1}{x-3}, \frac{-5}{s^3+8}$$

To find values for which a rational expression is undefined, find values for which the denominator is 0.

Find any values for which the expression $\frac{5y}{y^2-4y+3}$ is undefined.

$$y^2 - 4y + 3 = 0 \quad \text{Set the denominator equal to 0.}$$

$$(y-3)(y-1) = 0 \quad \text{Factor.}$$

$$y - 3 = 0 \quad \text{or} \quad y - 1 = 0 \quad \text{Set each factor equal to 0.}$$

$$y = 3 \qquad\qquad y = 1 \quad \text{Solve.}$$

The expression is undefined when y is 3 and when y is 1.

To Simplify a Rational Expression

Step 1: Factor the numerator and denominator.

Step 2: Divide out factors common to the numerator and denominator. (This is the same as removing a factor of 1.)

Simplify: $\frac{4x+20}{x^2-25}$

$$\frac{4x+20}{x^2-25} = \frac{4(x+5)}{(x+5)(x-5)} = \frac{4}{x-5}$$

A **rational function** is a function described by a rational expression.

$$f(x) = \frac{2x-6}{7}, h(t) = \frac{t^2-3t+5}{t-1}$$

Section 14.2 Multiplying and Dividing Rational Expressions

To Multiply Rational Expressions,

Step 1: Factor numerators and denominators.

Step 2: Multiply numerators and multiply denominators.

Step 3: Write the product in simplest form.

$$\frac{P}{Q}\cdot\frac{R}{S} = \frac{PR}{QS}$$

Multiply: $\frac{4x+4}{2x-3}\cdot\frac{2x^2+x-6}{x^2-1}$

$$\frac{4x+4}{2x-3}\cdot\frac{2x^2+x-6}{x^2-1} = \frac{4(x+1)}{2x-3}\cdot\frac{(2x-3)(x+2)}{(x+1)(x-1)}$$

$$= \frac{4(x+1)(2x-3)(x+2)}{(2x-3)(x+1)(x-1)}$$

$$= \frac{4(x+2)}{x-1}$$

To divide by a rational expression, multiply by the reciprocal.

$$\frac{P}{Q} \div \frac{R}{S} = \frac{P}{Q}\cdot\frac{S}{R} = \frac{PS}{QR}$$

Divide: $\frac{15x+5}{3x^2-14x-5} \div \frac{15}{3x-12}$

$$\frac{15x+5}{3x^2-14x-5} \div \frac{15}{3x-12} = \frac{5(3x+1)}{(3x+1)(x-5)}\cdot\frac{3(x-4)}{3\cdot 5}$$

$$= \frac{x-4}{x-5}$$

Definitions and Concepts	Examples
Section 14.3 Adding and Subtracting Rational Expressions with Common Denominators and Least Common Denominator	
To add or subtract rational expressions with the same denominator, add or subtract numerators and place the sum or difference over a common denominator. $\frac{P}{R} + \frac{Q}{R} = \frac{P + Q}{R}$ $\frac{P}{R} - \frac{Q}{R} = \frac{P - Q}{R}$	Perform indicated operations. $\frac{5}{x + 1} + \frac{x}{x + 1} = \frac{5 + x}{x + 1}$ $\frac{2y + 7}{y^2 - 9} - \frac{y + 4}{y^2 - 9} = \frac{(2y + 7) - (y + 4)}{y^2 - 9} = \frac{2y + 7 - y - 4}{y^2 - 9} = \frac{y + 3}{(y + 3)(y - 3)} = \frac{1}{y - 3}$
To Find the Least Common Denominator (LCD), **Step 1:** Factor the denominators. **Step 2:** The LCD is the product of all unique factors, each raised to a power equal to the greatest number of times that it appears in any one factored denominator.	Find the LCD for $\frac{7x}{x^2 + 10x + 25}$ and $\frac{11}{3x^2 + 15x}$ $x^2 + 10x + 25 = (x + 5)(x + 5)$ $3x^2 + 15x = 3x(x + 5)$ LCD is $3x(x + 5)(x + 5)$ or $3x(x + 5)^2$
Section 14.4 Adding and Subtracting Rational Expressions with Unlike Denominators	
To Add or Subtract Rational Expressions with Unlike Denominators, **Step 1:** Find the LCD. **Step 2:** Rewrite each rational expression as an equivalent expression whose denominator is the LCD. **Step 3:** Add or subtract numerators and place the sum or difference over the common denominator. **Step 4:** Write the result in simplest form.	Perform the indicated operation. $\frac{9x + 3}{x^2 - 9} - \frac{5}{x - 3} = \frac{9x + 3}{(x + 3)(x - 3)} - \frac{5}{x - 3}$ LCD is $(x + 3)(x - 3)$. $= \frac{9x + 3}{(x + 3)(x - 3)} - \frac{5(x + 3)}{(x - 3)(x + 3)} = \frac{9x + 3 - 5(x + 3)}{(x + 3)(x - 3)} = \frac{9x + 3 - 5x - 15}{(x + 3)(x - 3)} = \frac{4x - 12}{(x + 3)(x - 3)} = \frac{4(x - 3)}{(x + 3)(x - 3)} = \frac{4}{x + 3}$
Section 14.5 Solving Equations Containing Rational Expressions	
To Solve an Equation Containing Rational Expressions, **Step 1:** Multiply both sides of the equation by the LCD of all rational expressions in the equation.	Solve: $\frac{5x}{x + 2} + 3 = \frac{4x - 6}{x + 2}$

(continued)

Definitions and Concepts	Examples
Section 14.5 Solving Equations Containing Rational Expressions (*continued*)	
Step 2: Remove any grouping symbols and solve the resulting equation. **Step 3:** Check the solution in the original equation.	$(x+2)\left(\frac{5x}{x+2}+3\right)=(x+2)\left(\frac{4x-6}{x+2}\right)$ $(x+2)\left(\frac{5x}{x+2}\right)+(x+2)(3)=(x+2)\left(\frac{4x-6}{x+2}\right)$ $5x+3x+6=4x-6$ $4x=-12$ $x=-3$ The solution checks and the solution is -3.
Section 14.6 Problem Solving with Proportions and Rational Equations	
A **proportion** is a mathematical statement that two ratios are equal.	**Proportions** $\frac{2}{3}=\frac{8}{12} \qquad \frac{x}{7}=\frac{15}{35}$
Cross products: If $\frac{a}{b}=\frac{c}{d}$, then $ad=bc$.	**Cross Products** $2\cdot 12$ or 24 $\quad \frac{2}{3}=\frac{8}{12} \quad$ $3\cdot 8$ or 24 Solve: $\frac{3}{4}=\frac{x}{x-1}$ $\frac{3}{4}=\frac{x}{x-1}$ $3(x-1)=4x$ Set cross products equal. $3x-3=4x$ $-3=x$
Problem-Solving Steps **1.** UNDERSTAND. Read and reread the problem.	A small plane and a car leave Kansas City, Missouri, and head for Minneapolis, Minnesota, a distance of 450 miles. The speed of the plane is 3 times the speed of the car, and the plane arrives 6 hours ahead of the car. Find the speed of the car. Let $x=$ the speed of the car. Then $3x=$ the speed of the plane.

	Distance	= Rate	· Time
Car	450	x	$\frac{450}{x}\left(\frac{\text{distance}}{\text{rate}}\right)$
Plane	450	$3x$	$\frac{450}{3x}\left(\frac{\text{distance}}{\text{rate}}\right)$

Definitions and Concepts	Examples
Section 14.6 Problem Solving with Proportions and Rational Equations (*continued*)	
2. TRANSLATE.	In words: plane's time + 6 hours = car's time Translate: $\frac{450}{3x} + 6 = \frac{450}{x}$
3. SOLVE.	$\frac{450}{3x} + 6 = \frac{450}{x}$ $3x\left(\frac{450}{3x}\right) + 3x(6) = 3x\left(\frac{450}{x}\right)$ $450 + 18x = 1350$ $18x = 900$ $x = 50$
4. INTERPRET.	**Check** this solution in the originally stated problem. **State** the conclusion: The speed of the car is 50 miles per hour.
Section 14.7 Simplifying Complex Fractions	
Method 1: Simplify the numerator and the denominator so that each is a single fraction. Then perform the indicated division and simplify if possible.	Simplify $\dfrac{\frac{x+2}{x}}{x - \frac{4}{x}}$. Method 1: $\dfrac{\frac{x+2}{x}}{\frac{x \cdot x}{1 \cdot x} - \frac{4}{x}} = \dfrac{\frac{x+2}{x}}{\frac{x^2-4}{x}}$ $= \frac{x+2}{x} \cdot \frac{x}{(x+2)(x-2)} = \frac{1}{x-2}$
Method 2: Multiply the numerator and the denominator of the complex fraction by the LCD of the fractions in both the numerator and the denominator. Then simplify if possible.	Method 2: $\dfrac{\left(\frac{x+2}{x}\right) \cdot x}{\left(x - \frac{4}{x}\right) \cdot x} = \dfrac{x+2}{x \cdot x - \frac{4}{x} \cdot x}$ $= \frac{x+2}{x^2-4} = \frac{x+2}{(x+2)(x-2)} = \frac{1}{x-2}$

Chapter 14 Review

(14.1) *Find the domain of each rational function.*

1. $f(x) = \frac{3 - 5x}{7}$

2. $g(x) = \frac{2x + 4}{11}$

3. $f(x) = \frac{-3x^2}{x - 5}$

4. $h(x) = \frac{4x}{3x - 12}$

5. $f(x) = \frac{x^3 + 2}{x^2 + 8x}$

6. $G(x) = \frac{20}{3x^2 - 48}$

Simplify each rational expression.

7. $\frac{x + 12}{12 - x}$

8. $\frac{2x}{2x^2 - 2x}$

9. $\frac{x + 7}{x^2 - 49}$

10. $\frac{2x^2 + 4x - 30}{x^2 + x - 20}$

Simplify each expression. This section contains four-term polynomials and sums and differences of two cubes.

11. $\dfrac{x^2 + xa + xb + ab}{x^2 - xc + bx - bc}$

12. $\dfrac{x^2 + 5x - 2x - 10}{x^2 - 3x - 2x + 6}$

13. $\dfrac{4 - x}{x^3 - 64}$

14. $\dfrac{x^2 - 4}{x^3 + 8}$

The average cost (per bookcase) of manufacturing x bookcases is given by the rational function

$$C(x) = \frac{35x + 4200}{x}$$

15. Find the average cost per bookcase of manufacturing 50 bookcases.

16. Find the average cost per bookcase of manufacturing 100 bookcases.

(14.2) *Perform each indicated operation and simplify.*

17. $\dfrac{15x^3y^2}{z} \cdot \dfrac{z}{5xy^3}$

18. $\dfrac{-y^3}{8} \cdot \dfrac{9x^2}{y^3}$

19. $\dfrac{x^2 - 9}{x^2 - 4} \cdot \dfrac{x - 2}{x + 3}$

20. $\dfrac{2x + 5}{x - 6} \cdot \dfrac{2x}{-x + 6}$

21. $\dfrac{x^2 - 5x - 24}{x^2 - x - 12} \div \dfrac{x^2 - 10x + 16}{x^2 + x - 6}$

22. $\dfrac{4x + 4y}{xy^2} \div \dfrac{3x + 3y}{x^2y}$

23. $\dfrac{x^2 + x - 42}{x - 3} \cdot \dfrac{(x - 3)^2}{x + 7}$

24. $\dfrac{2a + 2b}{3} \cdot \dfrac{a - b}{a^2 - b^2}$

25. $\dfrac{2x^2 - 9x + 9}{8x - 12} \div \dfrac{x^2 - 3x}{2x}$

26. $\dfrac{x^2 - y^2}{x^2 + xy} \div \dfrac{3x^2 - 2xy - y^2}{3x^2 + 6x}$

27. $\dfrac{x - y}{4} \div \dfrac{y^2 - 2y - xy + 2x}{16x + 24}$

28. $\dfrac{5 + x}{7} \div \dfrac{xy + 5y - 3x - 15}{7y - 35}$

(14.3) *Perform each indicated operation and simplify.*

29. $\dfrac{x}{x^2 + 9x + 14} + \dfrac{7}{x^2 + 9x + 14}$

30. $\dfrac{x}{x^2 + 2x - 15} + \dfrac{5}{x^2 + 2x - 15}$

31. $\dfrac{4x - 5}{3x^2} - \dfrac{2x + 5}{3x^2}$

32. $\dfrac{9x + 7}{6x^2} - \dfrac{3x + 4}{6x^2}$

Find the LCD of each pair of rational expressions.

33. $\dfrac{x + 4}{2x}, \dfrac{3}{7x}$

34. $\dfrac{x - 2}{x^2 - 5x - 24}, \dfrac{3}{x^2 + 11x + 24}$

Rewrite each rational expression as an equivalent expression whose denominator is the given polynomial.

35. $\dfrac{5}{7x} = \dfrac{}{14x^3y}$

36. $\dfrac{9}{4y} = \dfrac{}{16y^3x}$

37. $\dfrac{x + 2}{x^2 + 11x + 18} = \dfrac{}{(x + 2)(x - 5)(x + 9)}$

38. $\dfrac{3x - 5}{x^2 + 4x + 4} = \dfrac{}{(x + 2)^2(x + 3)}$

(14.4) *Perform each indicated operation and simplify.*

39. $\dfrac{4}{5x^2} - \dfrac{6}{y}$

40. $\dfrac{2}{x - 3} - \dfrac{4}{x - 1}$

41. $\dfrac{4}{x + 3} - 2$

42. $\dfrac{3}{x^2 + 2x - 8} + \dfrac{2}{x^2 - 3x + 2}$

43. $\dfrac{2x - 5}{6x + 9} - \dfrac{4}{2x^2 + 3x}$

44. $\dfrac{x - 1}{x^2 - 2x + 1} - \dfrac{x + 1}{x - 1}$

Find the perimeter and the area of each figure.

△ **45.**

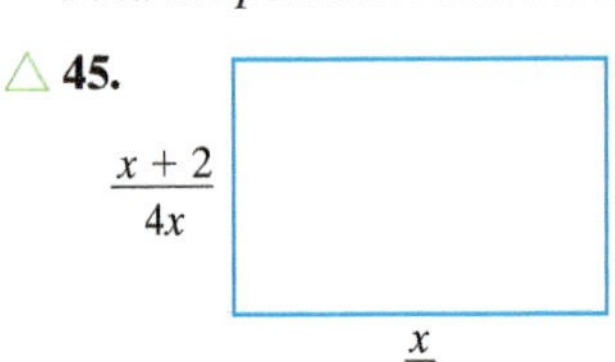

46.

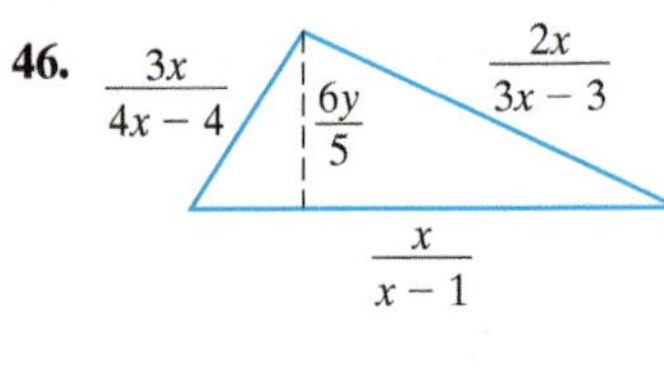

(14.5) *Solve each equation.*

47. $\dfrac{n}{10} = 9 - \dfrac{n}{5}$

48. $\dfrac{2}{x + 1} - \dfrac{1}{x - 2} = -\dfrac{1}{2}$

49. $\dfrac{y}{2y + 2} + \dfrac{2y - 16}{4y + 4} = \dfrac{y - 3}{y + 1}$

50. $\dfrac{2}{x - 3} - \dfrac{4}{x + 3} = \dfrac{8}{x^2 - 9}$

51. $\dfrac{x - 3}{x + 1} - \dfrac{x - 6}{x + 5} = 0$

52. $x + 5 = \dfrac{6}{x}$

Solve the equation for the indicated variable.

53. $\dfrac{4A}{5b} = x^2$, for b

54. $\dfrac{x}{7} + \dfrac{y}{8} = 10$, for y

(14.6) *Solve.*

55. A machine can process 300 parts in 20 minutes. Find how many parts can be processed in 45 minutes.

56. As his consulting fee, Mr. Visconti charges $90.00 per day. Find how much he charges for 3 hours of consulting. Assume an 8-hour work day.

57. Five times the reciprocal of a number equals the sum of $\frac{3}{2}$ the reciprocal of the number and $\frac{7}{6}$. What is the number?

58. The reciprocal of a number equals the reciprocal of the difference of 4 and the number. Find the number.

59. A car travels 90 miles in the same time that another car traveling 10 miles per hour slower travels 60 miles. Find the speed of each car.

60. The current in a bayou near Lafayette, Louisiana, is 4 miles per hour. A paddle boat travels 48 miles upstream in the same amount of time it takes to travel 72 miles downstream. Find the speed of the boat in still water.

61. When Mark and Maria manicure Mr. Stergeon's lawn, it takes them 5 hours. If Mark works alone, it takes 7 hours. Find how long it takes Maria alone.

62. It takes pipe A 20 days to fill a fish pond. Pipe B takes 15 days. Find how long it takes both pipes together to fill the pond.

Given that the pairs of triangles are similar, find each missing length x.

 63.

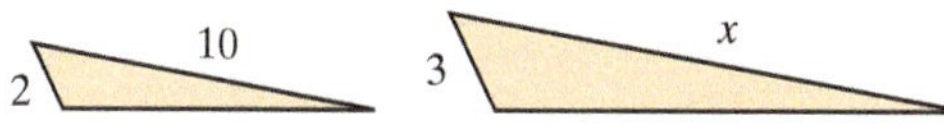

 64.

(14.7) *Simplify each complex fraction.*

65. $\dfrac{\frac{5x}{27}}{-\frac{10xy}{21}}$

66. $\dfrac{\frac{3}{5}+\frac{2}{7}}{\frac{1}{5}+\frac{5}{6}}$

67. $\dfrac{3-\frac{1}{y}}{2-\frac{1}{y}}$

68. $\dfrac{\frac{6}{x+2}+4}{\frac{8}{x+2}-4}$

69. $\dfrac{\frac{x-3}{x+3}+\frac{x+3}{x-3}}{\frac{x-3}{x+3}-\frac{x+3}{x-3}}$

70. $\dfrac{\frac{3}{x-1}-\frac{2}{1-x}}{\frac{2}{x-1}-\frac{2}{x}}$

71. $\dfrac{x+y^{-1}}{\frac{x}{y}}$

72. $\dfrac{x-xy^{-1}}{\frac{1+x}{y}}$

MIXED REVIEW

Simplify each rational expression.

73. $\dfrac{4x+12}{8x^2+24x}$

74. $\dfrac{x^3-6x^2+9x}{x^2+4x-21}$

Perform the indicated operations and simplify.

75. $\dfrac{x^2+9x+20}{x^2-25}\cdot\dfrac{x^2-9x+20}{x^2+8x+16}$

76. $\dfrac{x^2-x-72}{x^2-x-30}\div\dfrac{x^2+6x-27}{x^2-9x+18}$

77. $\dfrac{x}{x^2-36}+\dfrac{6}{x^2-36}$

78. $\dfrac{5x-1}{4x}-\dfrac{3x-2}{4x}$

79. $\dfrac{4}{3x^2+8x-3}+\dfrac{2}{3x^2-7x+2}$

80. $\dfrac{3x}{x^2+9x+14}-\dfrac{6x}{x^2+4x-21}$

Solve.

81. $\dfrac{4}{a-1}+2=\dfrac{3}{a-1}$

82. $\dfrac{x}{x+3}+4=\dfrac{x}{x+3}$

Solve.

83. The quotient of twice a number and three, minus one-sixth, is the quotient of the number and two. Find the number.

84. Mr. Crocker can paint his house by himself in three days. His son will need an additional day to complete the job if he works alone. If they work together, find how long it takes to paint the house.

Given that the following pairs of triangles are similar, find each missing length.

85.

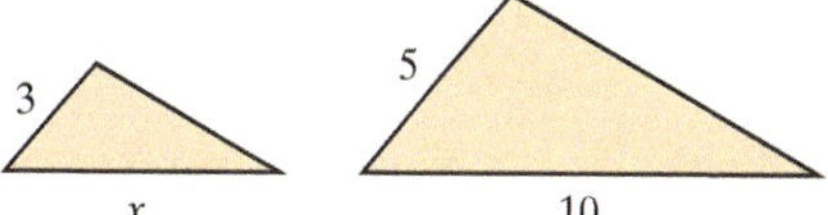

86.

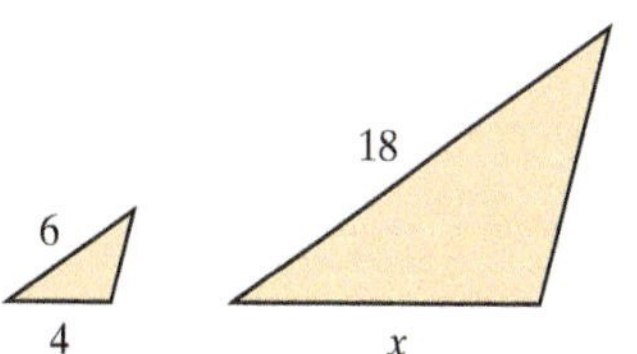

Simplify each complex fraction.

87. $\dfrac{\frac{1}{4}}{\frac{1}{3}+\frac{1}{2}}$

88. $\dfrac{4+\frac{2}{x}}{6+\frac{3}{x}}$

89. $\dfrac{y^{-2}}{1-y^{-2}}$

90. $\dfrac{4+x^{-1}}{3+x^{-1}}$

Chapter 14 Test

Step-by-step test solutions are found on the Chapter Test Prep Videos. Where available: MyMathLab® or 

1. Find the domain of the rational function

$$g(x) = \frac{9x^2 - 9}{x^2 + 4x + 3}$$

2. For a certain computer desk, the average cost C (in dollars) per desk manufactured is

$$C = \frac{100x + 3000}{x}$$

where x is the number of desks manufactured.

a. Find the average cost per desk when manufacturing 200 computer desks.

b. Find the average cost per desk when manufacturing 1000 computer desks.

Simplify each rational expression.

3. $\frac{3x - 6}{5x - 10}$

4. $\frac{x + 6}{x^2 + 12x + 36}$

5. $\frac{x + 3}{x^3 + 27}$

6. $\frac{2m^3 - 2m^2 - 12m}{m^2 - 5m + 6}$

7. $\frac{ay + 3a + 2y + 6}{ay + 3a + 5y + 15}$

8. $\frac{y - x}{x^2 - y^2}$

Perform the indicated operation and simplify if possible.

9. $\frac{3}{x - 1} \cdot (5x - 5)$

10. $\frac{y^2 - 5y + 6}{2y + 4} \cdot \frac{y + 2}{2y - 6}$

11. $\frac{15x}{2x + 5} - \frac{6 - 4x}{2x + 5}$

12. $\frac{5a}{a^2 - a - 6} - \frac{2}{a - 3}$

13. $\frac{6}{x^2 - 1} + \frac{3}{x + 1}$

14. $\frac{x^2 - 9}{x^2 - 3x} \div \frac{xy + 5x + 3y + 15}{2x + 10}$

15. $\frac{x + 2}{x^2 + 11x + 18} + \frac{5}{x^2 - 3x - 10}$

Solve each equation.

16. $\frac{4}{y} - \frac{5}{3} = \frac{-1}{5}$

17. $\frac{5}{y + 1} = \frac{4}{y + 2}$

18. $\frac{a}{a - 3} = \frac{3}{a - 3} - \frac{3}{2}$

19. $x - \frac{14}{x - 1} = 4 - \frac{2x}{x - 1}$

20. $\frac{10}{x^2 - 25} = \frac{3}{x + 5} + \frac{1}{x - 5}$

Simplify each complex fraction.

21. $\dfrac{\frac{5x^2}{yz^2}}{\frac{10x}{z^3}}$

22. $\dfrac{5 - \frac{1}{y^2}}{\frac{1}{y} + \frac{2}{y^2}}$

23. $\dfrac{\frac{b}{a} - \frac{a}{b}}{\frac{1}{b} + \frac{1}{a}}$

24. In a sample of 85 fluorescent bulbs, 3 were found to be defective. At this rate, how many defective bulbs should be found in 510 bulbs?

25. One number plus five times its reciprocal is equal to six. Find the number.

26. A pleasure boat traveling down the Red River takes the same time to go 14 miles upstream as it takes to go 16 miles downstream. If the current of the river is 2 miles per hour, find the speed of the boat in still water.

27. An inlet pipe can fill a tank in 12 hours. A second pipe can fill the tank in 15 hours. If both pipes are used, find how long it takes to fill the tank.

28. Given that the two triangles are similar, find x.

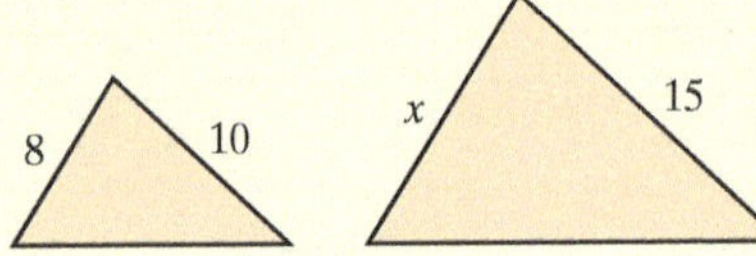

Chapters 1–14 Cumulative Review

TRANSLATING

1. Translate each sentence into an equation.
   a. The product of 7 and 6 is 42.
   b. Twice the sum of 3 and 5 is equal to 16.
   c. The quotient of −45 and 5 yields −9.

2. Translate each sentence into an equation.
   a. The difference of 20 and −15 equals 35.
   b. The product of 12 and 18 gives 216.
   c. Five times the sum of −2 and 4 yields 10.

3. Write each sentence as an equation. Use x to represent "a number."
   a. Twenty increased by a number is 5.
   b. Twice a number equals -10.
   c. A number minus 11 amounts to 168.
   d. Three times the sum of a number and 5 is -30.
   e. The quotient of twice a number and 8 is equal to 2.
4. The sum of a number and 9 is the same as 5 times the number minus 3. Find the unknown number.
5. Graph $x - 3y = 6$ by finding and plotting intercepts.
6. Find the slope of the line whose equation is $7x + 2y = 9$.
7. Use the product rule to simplify.
   a. $4^2 \cdot 4^5$ b. $x^4 \cdot x^6$
   c. $y^3 \cdot y$ d. $y^3 \cdot y^2 \cdot y^7$
   e. $(-5)^7 \cdot (-5)^8$ f. $a^2 \cdot b^2$
8. Simplify.
   a. $\frac{x^9}{x^7}$ b. $\frac{x^{19}y^5}{xy}$
   c. $(x^5y^2)^3$ d. $(-3a^2b)(5a^3b)$
9. Subtract $(5z - 7)$ from the sum of $(8z + 11)$ and $(9z - 2)$.
10. Subtract $(9x^2 - 6x + 2)$ from $(x + 1)$.
11. Multiply: $(3a + b)^3$
12. Multiply: $(2x + 1)(5x^2 - x + 2)$
13. Use a special product to square each binomial.
    a. $(t + 2)^2$
    b. $(p - q)^2$
    c. $(2x + 5)^2$
    d. $(x^2 - 7y)^2$
14. Multiply.
    a. $(x + 9)^2$
    b. $(2x + 1)(2x - 1)$
    c. $8x(x^2 + 1)(x^2 - 1)$
15. Simplify each expression. Write results using positive exponents only.
    a. $\frac{1}{x^{-3}}$ b. $\frac{1}{3^{-4}}$ c. $\frac{p^{-4}}{q^{-9}}$ d. $\frac{5^{-3}}{2^{-5}}$
16. Simplify. Write results with positive exponents only.
    a. 5^{-3} b. $\frac{9}{x^{-7}}$ c. $\frac{11^{-1}}{7^{-2}}$
17. Divide: $\frac{4x^2 + 7 + 8x^3}{2x + 3}$
18. Divide $(4x^3 - 9x + 2)$ by $(x - 4)$.
19. Find the GCF of each list of numbers.
    a. 28 and 40 b. 55 and 21
    c. 15, 18, and 66
20. Find the GCF of $9x^2$, $6x^3$, and $21x^5$.

Factor.

21. $-9a^5 + 18a^2 - 3a$
22. $7x^6 - 7x^5 + 7x^4$
23. $3m^2 - 24m - 60$
24. $-2a^2 + 10a + 12$
25. $3x^2 + 11x + 6$
26. $10m^2 - 7m + 1$
27. $x^2 + 12x + 36$
28. $4x^2 + 12x + 9$
29. $x^2 + 4$
30. $x^2 - 4$
31. $x^3 + 8$
32. $27y^3 - 1$
33. $2x^3 + 3x^2 - 2x - 3$
34. $3x^3 + 5x^2 - 12x - 20$
35. $12m^2 - 3n^2$
36. $x^5 - x$
37. Solve: $x(2x - 7) = 4$
38. Solve: $3x^2 + 5x = 2$
39. Find the x-intercepts of the graph of $y = x^2 - 5x + 4$.
40. Find the x-intercepts of the graph of $y = x^2 - x - 6$.
41. The height of a triangular sail is 2 meters less than twice the length of the base. If the sail has an area of 30 square meters, find the length of its base and the height.
42. The height of a parallelogram is 5 feet more than three times its base. If the area of the parallelogram is 182 square feet, find the length of its base and height.
43. Simplify: $\frac{18 - 2x^2}{x^2 - 2x - 3}$
44. Simplify: $\frac{2x^2 - 50}{4x^4 - 20x^3}$
45. Divide: $\frac{6x + 2}{x^2 - 1} \div \frac{3x^2 + x}{x - 1}$
46. Multiply: $\frac{6x^2 - 18x}{3x^2 - 2x} \cdot \frac{15x - 10}{x^2 - 9}$
47. Simplify: $\frac{(2x)^{-1} + 1}{2x^{-1} - 1}$
48. Simplify: $\frac{\frac{m}{3} + \frac{n}{6}}{\frac{m + n}{12}}$

15 More on Functions and Graphs

Sections

Check Your Progress

In Section 10.6, we introduced the notion of relation and the notion of function, perhaps the single most important and useful concepts in all of mathematics. In this chapter, we explore the concept of functions further.

We define online courses as courses in which at least 80% of the content is delivered online. Although there are many types of course delivery used by instructors, the bar graph below shows the increase in percent of students taking at least one online course. Notice that the two functions, $f(x)$ and $g(x)$, both approximate the percent of students taking at least one online course. Also, for both functions, x is the number of years since 2000. In Section 15.1, Exercises 69–74, we use these functions to predict the growth of online courses.

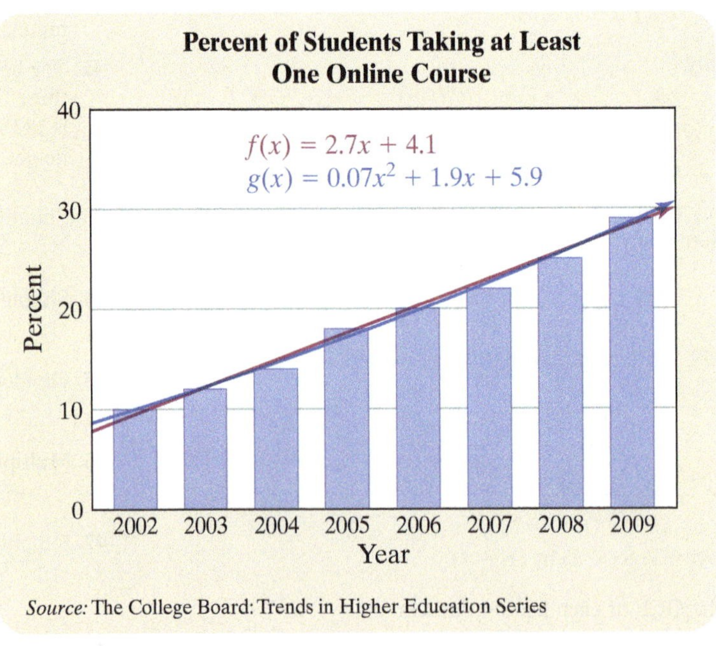

Source: The College Board: Trends in Higher Education Series

15.1 Graphing and Writing Linear Functions 

Objective 1 Graphing Linear Functions

Objectives

1. Graph Linear Functions.
2. Write an Equation of a Line Using Function Notation.
3. Find Equations of Parallel and Perpendicular Lines.

In this section, we identify and graph linear functions. By the vertical line test, Section 10.6, we know that all linear equations except those whose graphs are vertical lines are functions. Thus, all linear equations except those of the form $x = c$ (vertical lines) are linear functions. For example, we know from Section 10.2 that $y = 2x$ is a linear equation in two variables. Its graph is shown.

x	$y = 2x$
1	2
0	0
−1	−2

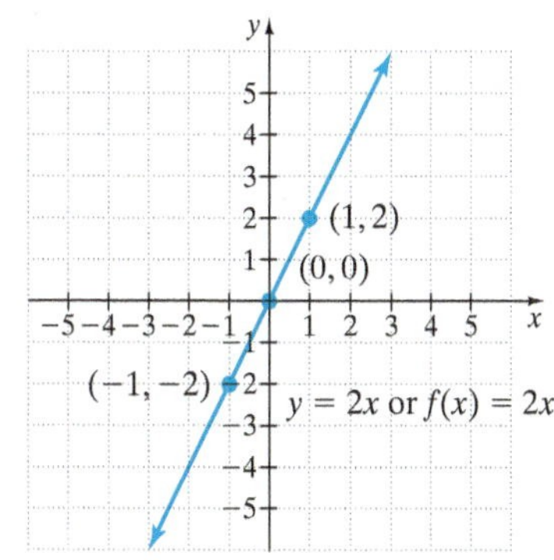

Because this graph passes the vertical line test, we know that $y = 2x$ is a function. If we want to emphasize that this equation describes a function, we may write $y = 2x$ as $f(x) = 2x$.

Example 1 Graph $g(x) = 2x + 1$. Compare this graph with the graph of $f(x) = 2x$.

Solution: To graph $g(x) = 2x + 1$, find three ordered pair solutions.

add 1

x	$f(x) = 2x$	$g(x) = 2x + 1$
0	0	1
−1	−2	−1
1	2	3

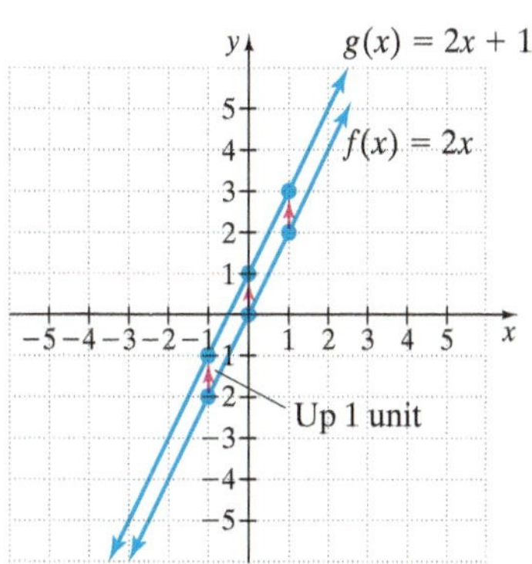

Notice that y-values for the graph of $g(x) = 2x + 1$ are obtained by adding 1 to each y-value of each corresponding point of the graph of $f(x) = 2x$. The graph of $g(x) = 2x + 1$ is the same as the graph of $f(x) = 2x$ shifted upward 1 unit.

Work Practice 1

Practice 1

Graph $g(x) = 4x - 3$ and $f(x) = 4x$ on the same axes.

If a linear function is solved for y, we can easily use function notation to describe it by replacing y with $f(x)$. Recall the slope-intercept form of a linear equation, $y = mx + b$, where m is the slope of the line and $(0, b)$ is the y-intercept. Since this form is solved for y, we use it to define a linear function.

In general, a **linear function** is a function that can be written in the form $f(x) = mx + b$. For example, $g(x) = 2x + 1$ is in this form, with $m = 2$ and $b = 1$. Thus, the slope of the linear function $g(x)$ is 2 and the y-intercept is $(0, 1)$.

Answer

1.

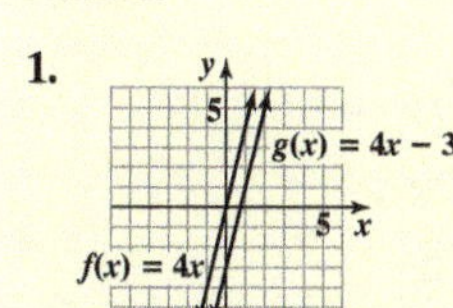

Practice 2

Graph the linear functions $f(x) = -2x$ and $g(x) = -2x + 5$ on the same set of axes.

Example 2 Graph the linear functions $f(x) = -3x$ and $g(x) = -3x - 6$ on the same set of axes.

Solution: To graph $f(x)$ and $g(x)$, find ordered pair solutions.

subtract 6

x	$f(x) = -3x$	$g(x) = -3x - 6$
0	0	−6
1	−3	−9
−1	3	−3
−2	6	0

subtract 6

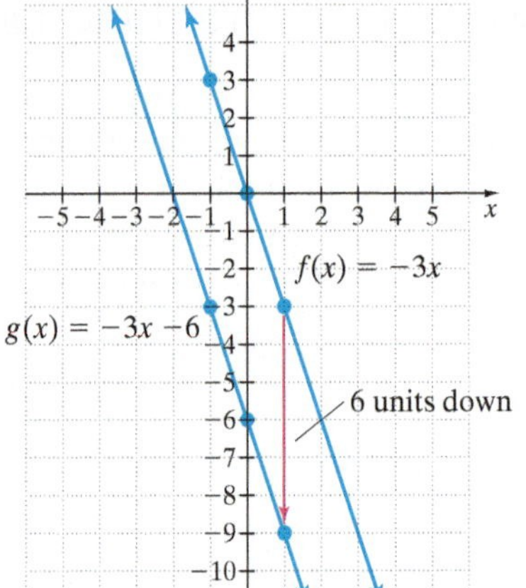

Each y-value for the graph of $g(x) = -3x - 6$ is obtained by subtracting 6 from the y-value of the corresponding point of the graph of $f(x) = -3x$. The graph of $g(x) = -3x - 6$ is the same as the graph of $f(x) = -3x$ shifted down 6 units.

Work Practice 2

Objective 2 Writing Equations of Lines Using Function Notation

We now practice writing linear functions.

This means the graph is a line that passes the vertical line test.

Below is a review of some tools we can use.

$y = mx + b$	**Slope-intercept form** of a linear equation. The slope is m, and the y-intercept is $(0, b)$.
$y - y_1 = m(x - x_1)$	**Point-slope form** of a linear equation. The slope is m, and (x_1, y_1) is a point on the line.
$y = c$	**Horizontal line** The slope is 0, and the y-intercept is $(0, c)$.

Note: $x = c$, whose graph is a vertical line, is not included above, as these equations do **not** define functions.

Practice 3

Find an equation of the line with slope −4 and y-intercept $(0, -3)$. Write the equation using function notation.

Example 3 Find an equation of the line with slope −3 and y-intercept $(0, -5)$. Write the equation using function notation.

Solution: Because we know the slope and the y-intercept, we use the slope-intercept form with $m = -3$ and $b = -5$.

$y = mx + b$ Slope-intercept form

$y = -3 \cdot x + (-5)$ Let $m = -3$ and $b = -5$.

$y = -3x - 5$ Simplify.

This equation is solved for y. To write using function notation, we replace y with $f(x)$.

$f(x) = -3x - 5$

Work Practice 3

Answers

2. 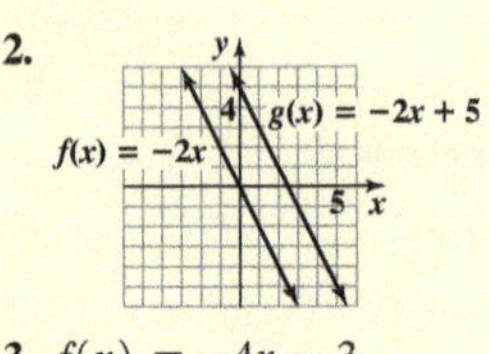

3. $f(x) = -4x - 3$

Example 4 Find an equation of the line through points $(4, 0)$ and $(-4, -5)$. Write the equation using function notation.

Solution: First, find the slope of the line.

$$m = \frac{-5 - 0}{-4 - 4} = \frac{-5}{-8} = \frac{5}{8}$$

Next, make use of the point-slope form. Replace (x_1, y_1) by either $(4, 0)$ or $(-4, -5)$ in the point-slope equation. We will choose the point $(4, 0)$. The line through $(4, 0)$ with slope $\frac{5}{8}$ is

$y - y_1 = m(x - x_1)$ Point-slope form

$y - 0 = \frac{5}{8}(x - 4)$ Let $m = \frac{5}{8}$ and $(x_1, y_1) = (4, 0)$.

$8y = 5(x - 4)$ Multiply both sides by 8.

$8y = 5x - 20$ Apply the distributive property.

To write the equation using function notation, we solve for y, then replace y with $f(x)$.

$8y = 5x - 20$

$y = \frac{5}{8}x - \frac{20}{8}$ Divide both sides by 8.

$f(x) = \frac{5}{8}x - \frac{5}{2}$ Write using function notation.

Work Practice 4

Practice 4

Find an equation of the line through points $(-1, 2)$ and $(2, 0)$. Write the equation using function notation.

Helpful Hint

If two points of a line are given, either one may be used with the point-slope form to write an equation of the line.

Example 5 Find an equation of the horizontal line containing the point $(2, 3)$. Write the equation using function notation.

Solution: A horizontal line has an equation of the form $y = c$. Since the line contains the point $(2, 3)$, the equation is $y = 3$, as shown to the right.

Using function notation, the equation is

$f(x) = 3.$

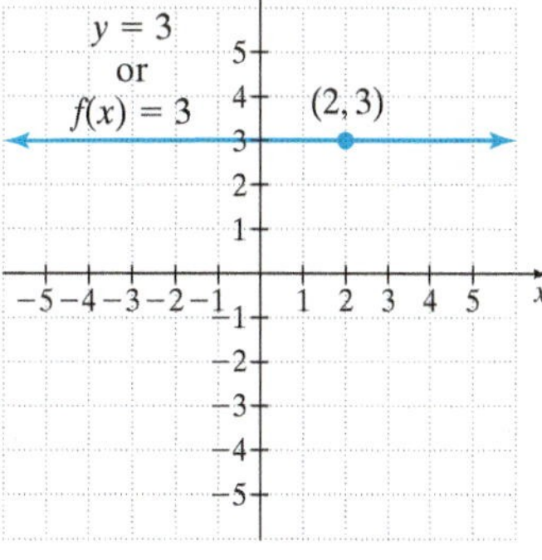

Work Practice 5

Practice 5

Find an equation of the horizontal line containing the point $(6, -2)$. Use function notation.

Answers

4. $f(x) = -\frac{2}{3}x + \frac{4}{3}$ **5.** $f(x) = -2$

Objective 3 Finding Equations of Parallel and Perpendicular Lines

Next, we find equations of parallel and perpendicular lines.

Practice 6

Find an equation of the line containing the point $(8, -3)$ and parallel to the line $3x + 4y = 1$. Write the equation in standard form.

Example 6 Find an equation of the line containing the point $(4, 4)$ and parallel to the line $2x + 3y = -6$. Write the equation in standard form.

Solution: Because the line we want to find is *parallel* to the line $2x + 3y = -6$, the two lines must have equal slopes. Find the slope of $2x + 3y = -6$ by writing it in the form $y = mx + b$. In other words, solve the equation for y.

$$2x + 3y = -6$$

$$3y = -2x - 6 \quad \text{Subtract } 2x \text{ from both sides.}$$

$$y = \frac{-2x}{3} - \frac{6}{3} \quad \text{Divide by 3.}$$

$$y = -\frac{2}{3}x - 2 \quad \text{Write in slope-intercept form.}$$

The slope of this line is $-\frac{2}{3}$. Thus, a line parallel to this line will also have a slope of $-\frac{2}{3}$. The equation we are asked to find describes a line containing the point $(4, 4)$ with a slope of $-\frac{2}{3}$. We use the point-slope form.

$$y - y_1 = m(x - x_1)$$

$$y - 4 = -\frac{2}{3}(x - 4) \quad \text{Let } m = -\frac{2}{3}, x_1 = 4, \text{ and } y_1 = 4.$$

$$3(y - 4) = -2(x - 4) \quad \text{Multiply both sides by 3.}$$

$$3y - 12 = -2x + 8 \quad \text{Apply the distributive property.}$$

$$2x + 3y = 20 \quad \text{Write in standard form.}$$

Work Practice 6

Helpful Hint Multiply both sides of the equation $2x + 3y = 20$ by -1 and it becomes $-2x - 3y = -20$. Both equations are in standard form, and their graphs are the same line.

Practice 7

Write a function that describes the line containing the point $(8, -3)$ and perpendicular to the line $3x + 4y = 1$.

Example 7 Write a function that describes the line containing the point $(4, 4)$ and perpendicular to the line $2x + 3y = -6$.

Solution: In the previous example, we found that the slope of the line $2x + 3y = -6$ is $-\frac{2}{3}$. A line perpendicular to this line will have a slope that is the negative reciprocal of $-\frac{2}{3}$, or $\frac{3}{2}$. From the point-slope equation, we have

$$y - y_1 = m(x - x_1)$$

$$y - 4 = \frac{3}{2}(x - 4) \quad \text{Let } x_1 = 4, y_1 = 4, \text{ and } m = \frac{3}{2}.$$

$$2(y - 4) = 3(x - 4) \quad \text{Multiply both sides by 2.}$$

$$2y - 8 = 3x - 12 \quad \text{Apply the distributive property.}$$

$$2y = 3x - 4 \quad \text{Add 8 to both sides.}$$

$$y = \frac{3}{2}x - 2 \quad \text{Divide both sides by 2.}$$

$$f(x) = \frac{3}{2}x - 2 \quad \text{Write using function notation.}$$

Work Practice 7

Answers

6. $3x + 4y = 12$ 7. $f(x) = \frac{4}{3}x - \frac{41}{3}$

Graphing Calculator Explorations

You may have noticed by now that to use the $\boxed{Y =}$ key on a graphing calculator to graph an equation, the equation must be solved for y.

Graph each function by first solving the function for y.

1. $x = 3.5y$
2. $-2.7y = x$
3. $5.78x + 2.31y = 10.98$
4. $-7.22x + 3.89y = 12.57$
5. $y - |x| = 3.78$
6. $3y - 5x^2 = 6x - 4$
7. $y - 5.6x^2 = 7.7x + 1.5$
8. $y + 2.6|x| = -3.2$

Vocabulary, Readiness & Video Check

Use the choices given to fill in each blank. Some choices may not be used.

linear	$(0, b)$	m
quadratic	$(b, 0)$	mx

1. A ______ function can be written in the form $f(x) = mx + b$.
2. In the form $f(x) = mx + b$, the y-intercept is ______ and the slope is ______.

State the slope and the y-intercept of the graph of each function.

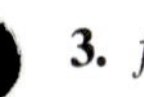

3. $f(x) = -4x + 12$
4. $g(x) = \frac{2}{3}x - \frac{7}{2}$
5. $g(x) = 5x$
6. $f(x) = -x$

Decide whether the lines are parallel, perpendicular, or neither.

7. $y = 12x + 6$
   $y = 12x - 2$
8. $y = -5x + 8$
   $y = -5x - 8$
9. $y = -9x + 3$
   $y = \frac{3}{2}x - 7$
10. $y = 2x - 12$
    $y = \frac{1}{2}x - 6$

Martin-Gay Interactive Videos

See Video 15.1

Watch the section lecture video and answer the following questions.

Objective 1 11. Based on the lecture before Example 1, in what form can a linear function be written?

Objective 2 12. From Example 2, given a y-intercept, how do you know which value to use for b in the slope-intercept form?

13. Example 4 discusses how to find an equation of a line given two points. Under what circumstances might the slope-intercept form be chosen over the point-slope form to find an equation?

Objective 3 14. Solve Example 6 again, this time writing the equation of the line in function notation *parallel* to the given line through the given point.

15.1 Exercise Set MyMathLab®

Graph each linear function. See Examples 1 and 2.

1. $f(x) = -2x$
2. $f(x) = 2x$
3. $f(x) = -2x + 3$
4. $f(x) = 2x + 6$
5. $f(x) = \frac{1}{2}x$
6. $f(x) = \frac{1}{3}x$
7. $f(x) = \frac{1}{2}x - 4$
8. $f(x) = \frac{1}{3}x - 2$

The graph of $f(x) = 5x$ follows. Use this graph to match each linear function with its graph. See Examples 1 and 2.

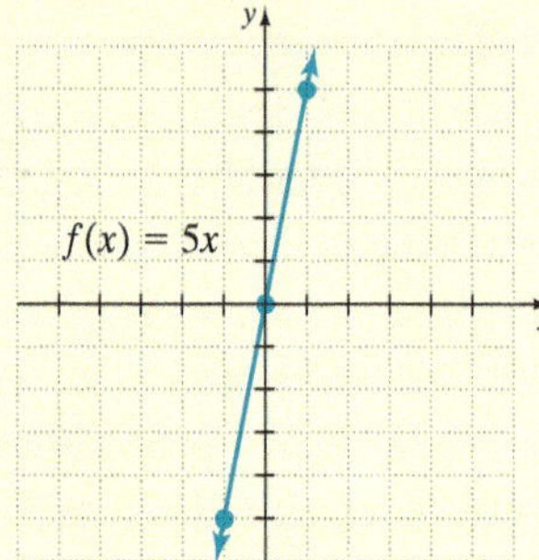

A

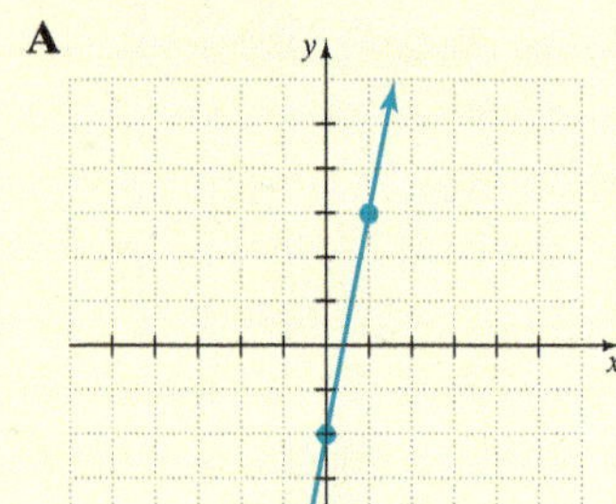

B

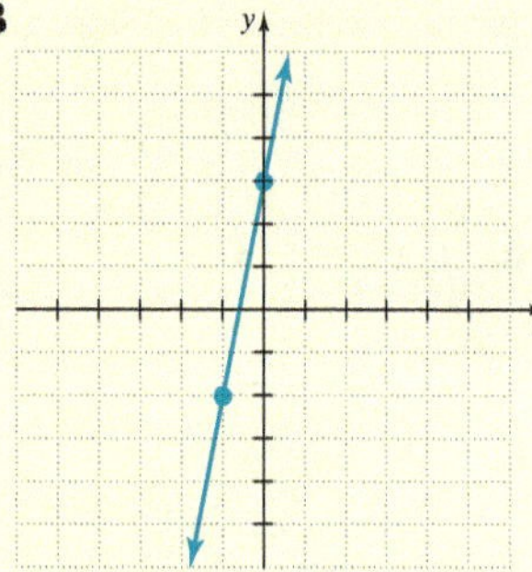

C

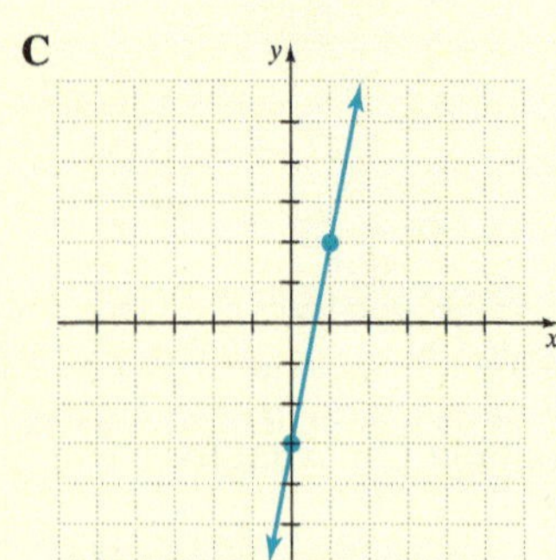

D

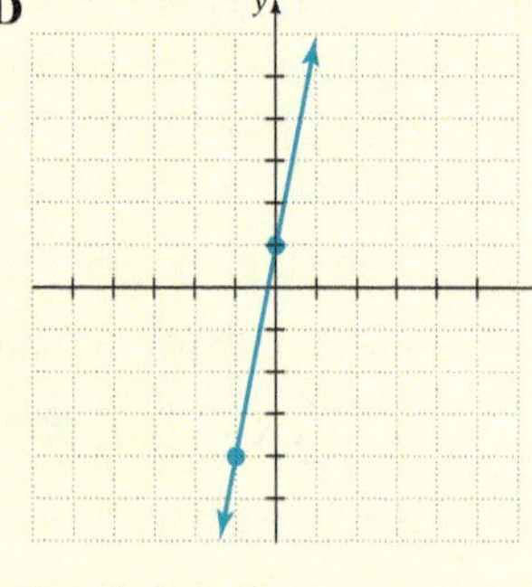

9. $f(x) = 5x - 3$
10. $f(x) = 5x - 2$
11. $f(x) = 5x + 1$
12. $f(x) = 5x + 3$

Use function notation to write the equation of each line with the given slope and y-intercept. See Example 3.

13. Slope -1; y-intercept $(0, 1)$
14. Slope $\frac{1}{2}$; y-intercept $(0, -6)$
15. Slope 2; y-intercept $\left(0, \frac{3}{4}\right)$
16. Slope -3; y-intercept $\left(0, -\frac{1}{5}\right)$
17. Slope $\frac{2}{7}$; y-intercept $(0, 0)$
18. Slope $-\frac{4}{5}$; y-intercept $(0, 0)$

Find an equation of the line with the given slope and containing the given point. Write the equation using function notation. See Example 3.

19. Slope 3; through $(1, 2)$
20. Slope 4; through $(5, 1)$
21. Slope -2; through $(1, -3)$
22. Slope -4; through $(2, -4)$
23. Slope $\frac{1}{2}$; through $(-6, 2)$
24. Slope $\frac{2}{3}$; through $(-9, 4)$
25. Slope $-\frac{9}{10}$; through $(-3, 0)$
26. Slope $-\frac{1}{5}$; through $(4, -6)$

Find an equation of the line passing through the given points. Use function notation to write the equation. See Example 4.

27. $(2, 0), (4, 6)$
28. $(3, 0), (7, 8)$
29. $(-2, 5), (-6, 13)$
30. $(7, -4), (2, 6)$
31. $(-2, -4), (-4, -3)$
32. $(-9, -2), (-3, 10)$
33. $(-3, -8), (-6, -9)$
34. $(8, -3), (4, -8)$
35. $\left(\frac{3}{5}, \frac{2}{5}\right)$ and $\left(-\frac{1}{5}, \frac{7}{10}\right)$
36. $\left(\frac{1}{2}, -\frac{1}{4}\right)$ and $\left(\frac{3}{2}, \frac{3}{4}\right)$

Write an equation of each line using function notation. See Example 5.

37. Slope 0; through $(-2, -4)$
38. Horizontal; through $(-3, 1)$
39. Horizontal; through $(0, 5)$
40. Slope 0; through $(-10, 23)$

Find an equation of each line. Write the equation using function notation. See Examples 6 and 7.

41. Through $(3, 8)$; parallel to $f(x) = 4x - 2$
42. Through $(1, 5)$; parallel to $f(x) = 3x - 4$
43. Through $(2, -5)$; perpendicular to $3y = x - 6$
44. Through $(-4, 8)$; perpendicular to $2x - 3y = 1$
45. Through $(-2, -3)$; parallel to $3x + 2y = 5$
46. Through $(-2, -3)$; perpendicular to $3x + 2y = 5$

MIXED PRACTICE

Find the equation of each line. Write the equation using standard notation unless indicated otherwise. See Examples 3 through 7.

47. Slope 2; through $(-2, 3)$
48. Slope 3; through $(-4, 2)$
49. Through $(1, 6)$ and $(5, 2)$; use function notation
50. Through $(2, 9)$ and $(8, 6)$; use function notation
51. With slope $-\frac{1}{2}$; y-intercept 11; use function notation
52. With slope -4; y-intercept $\frac{2}{9}$; use function notation
53. Through $(-7, -4)$ and $(0, -6)$
54. Through $(2, -8)$ and $(-4, -3)$
55. Slope $-\frac{4}{3}$; through $(-5, 0)$
56. Slope $-\frac{3}{5}$; through $(4, -1)$
57. Horizontal line; through $(-2, -10)$; use function notation
58. Horizontal line; through $(1, 0)$; use function notation
59. Through $(6, -2)$; parallel to the line $2x + 4y = 9$
60. Through $(8, -3)$; parallel to the line $6x + 2y = 5$
61. Slope 0; through $(-9, 12)$; use function notation
62. Slope 0; through $(10, -8)$; use function notation
63. Through $(6, 1)$; parallel to the line $8x - y = 9$
64. Through $(3, 5)$; perpendicular to the line $2x - y = 8$
65. Through $(5, -6)$; perpendicular to $y = 9$
66. Through $(-3, -5)$; parallel to $y = 9$
67. Through $(2, -8)$ and $(-6, -5)$; use function notation
68. Through $(-4, -2)$ and $(-6, 5)$; use function notation

REVIEW AND PREVIEW

From the Chapter 15 opener, we have two functions to describe the percent of college students taking at least one online course. For both functions, x is the number of years since 2000 and y (or f(x) or g(x)) is the percent of students taking at least one online course.

$$f(x) = 2.7x + 4.1 \quad \text{or} \quad g(x) = 0.07x^2 + 1.9x + 5.9$$

Use this for Exercises 69–74. See Section 10.6.

69. Find $f(9)$ and describe in words what this means.
70. Find $g(9)$ and describe in words what this means.
71. Assume the trend of $g(x)$ continues. Find $g(16)$ and describe in words what this means.
72. Assume the trend of $f(x)$ continues. Find $f(16)$ and describe in words what this means.
73. Use Exercises 69–72 and compare $f(9)$ and $g(9)$, then $f(16)$ and $g(16)$. As x increases, are the function values staying about the same or not? Explain your answer.

74. Use the Chapter 15 opener graph and study the graphs of $f(x)$ and $g(x)$. Use these graphs to answer Exercise 73. Explain your answer.

CONCEPT EXTENSIONS

Find an equation of each line graphed. Write the equation using function notation. (Hint: Use each graph to write 2 ordered pair solutions. Find the slope of each line, and then refer to Examples 3 or 4 to complete.)

75.

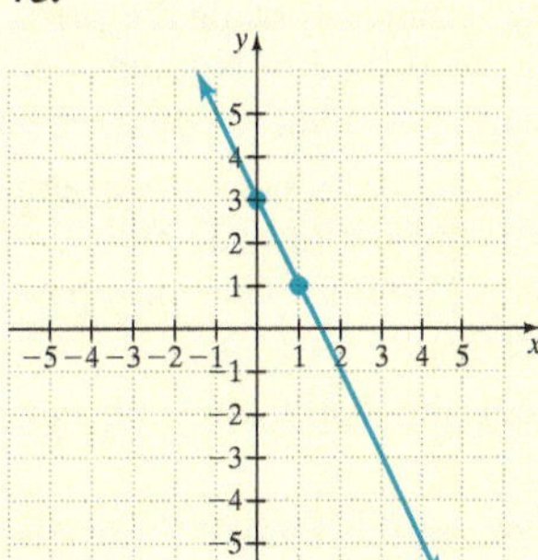

76.

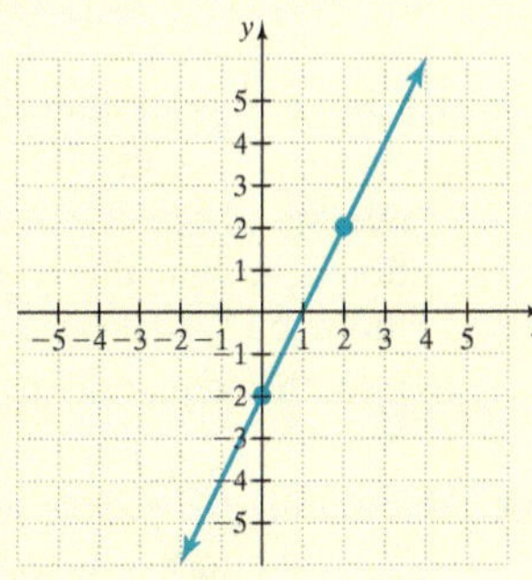

77.

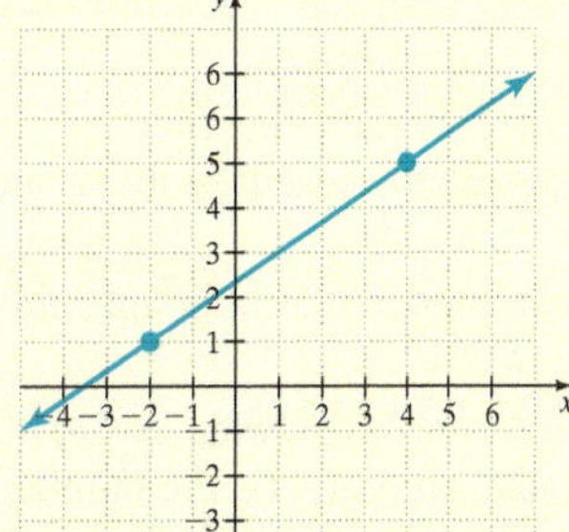

78.

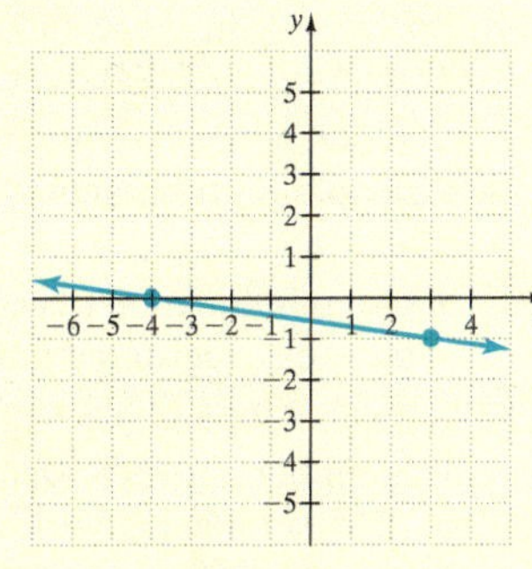

Solve.

79. A rock is dropped from the top of a 400-foot building. After 1 second, the rock is traveling 32 feet per second. After 3 seconds, the rock is traveling 96 feet per second. Let y be the rate of descent and x be the number of seconds since the rock was dropped.
    a. Write a linear equation that relates time x to rate y. [*Hint:* Use the ordered pairs $(1, 32)$ and $(3, 96)$.]
    b. Use this equation to determine the rate of travel of the rock 4 seconds after it was dropped.
80. A fruit company recently released a new applesauce. By the end of its first year, profits on this product amounted to \$30,000. The anticipated profit for the end of the fourth year is \$66,000. The ratio of change in time to change in profit is constant. Let x be years and y be profit.
    a. Write a linear equation that relates profit and time. [*Hint:* Use the ordered pairs $(1, 30{,}000)$ and $(4, 66{,}000)$.]
    b. Use this equation to predict the company's profit at the end of the seventh year.
    c. Predict when the profit should reach \$126,000.
81. The Whammo Company has learned that by pricing a newly released Frisbee at \$6, sales will reach 2000 per day. Raising the price to \$8 will cause the sales to fall to

1500 per day. Assume that the ratio of change in price to change in daily sales is constant and let x be the price of the Frisbee and y be number of sales.

a. Find the linear equation that models the price–sales relationship for this Frisbee. [*Hint:* The line must pass through $(6, 2000)$ and $(8, 1500)$.]

b. Use this equation to predict the daily sales of Frisbees if the price is set at $7.50.

82. The Pool Fun Company has learned that, by pricing a newly released Fun Noodle at $3, sales will reach 10,000 Fun Noodles per day during the summer. Raising the price to $5 will cause the sales to fall to 8000 Fun Noodles per day. Let x be price and y be the number sold.

a. Assume that the relationship between sales price and number of Fun Noodles sold is linear and write an equation describing this relationship. [*Hint:* The line must pass through $(3, 10{,}000)$ and $(5, 8000)$.]

b. Use this equation to predict the daily sales of Fun Noodles if the price is $3.50.

83. The number of people employed in the United States as registered nurses was 2619 thousand in 2008. By 2018, this number is expected to rise to 3200 thousand. Let y be the number of registered nurses (in thousands) employed in the United States in the year x, where $x = 0$ represents 2008. (*Source:* U.S. Bureau of Labor Statistics)

a. Write a linear equation that models the number of people (in thousands) employed as registered nurses in year x.

b. Use this equation to estimate the number of people employed as registered nurses in 2012.

84. In 2008, IBM had 398,500 employees worldwide. By 2010, this number had increased to 426,751. Let y be the number of IBM employees worldwide in the year x, where $x = 0$ represents 2008. (*Source:* IBM Corporation)

a. Write a linear equation that models the growth in the number of IBM employees worldwide, in terms of the year x.

b. Use this equation to estimate the number of IBM employees worldwide in 2013.

85. In 2010, the average price of a new home sold in the United States was $272,900. In 2005, the average price of a new home in the United States was $297,000. Let y be the average price of a new home in the year x, where $x = 0$ represents the year 2005. (*Source:* Based on data from U.S. census)

a. Write a linear equation that models the average price of a new home in terms of the year x. [*Hint:* The line must pass through the points $(0, 297{,}000)$ and $(5, 272{,}900)$.]

b. Use this equation to estimate the average price of a new home in 2013.

86. The number of McDonald's restaurants worldwide in 2010 was 32,737. In 2005, there were 31,046 McDonald's restaurants worldwide. Let y be the number of McDonald's restaurants in the year x, where $x = 0$ represents the year 2005. (*Source:* McDonald's Corporation)

a. Write a linear equation that models the growth in the number of McDonald's restaurants worldwide in terms of the year x. [*Hint:* The line must pass through the points $(0, 31{,}046)$ and $(5, 32{,}737)$.]

b. Use this equation to estimate the number of McDonald's restaurants worldwide in 2013.

Example:

Find an equation of the perpendicular bisector of the line segment whose endpoints are $(2, 6)$ *and* $(0, -2)$.

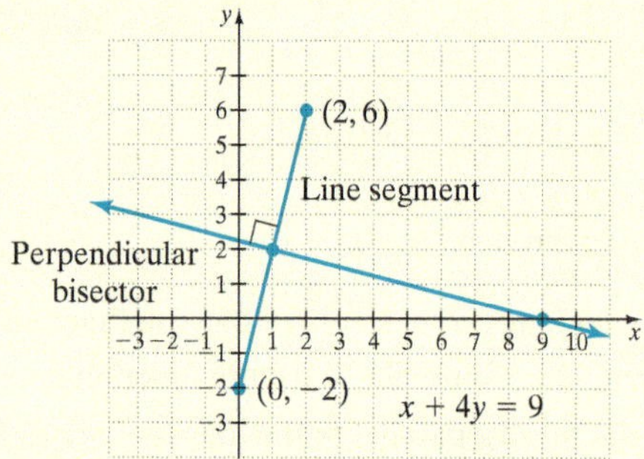

Solution:

A perpendicular bisector is a line that contains the midpoint of the given segment and is perpendicular to the segment.

Step 1: The midpoint of the segment with endpoints $(2, 6)$ and $(0, -2)$ is $(1, 2)$.

Step 2: The slope of the segment containing points $(2, 6)$ and $(0, -2)$ is 4.

Step 3: A line perpendicular to this line segment will have slope of $-\frac{1}{4}$.

Step 4: The equation of the line through the midpoint $(1, 2)$ with a slope of $-\frac{1}{4}$ will be the equation of the perpendicular bisector. This equation in standard form is $x + 4y = 9$.

Find an equation of the perpendicular bisector of the line segment whose endpoints are given. See the previous example.

△ **87.** $(3, -1); (-5, 1)$

△ **88.** $(-6, -3); (-8, -1)$

△ **89.** $(-2, 6); (-22, -4)$

△ **90.** $(5, 8); (7, 2)$

△ **91.** $(2, 3); (-4, 7)$

△ **92.** $(-6, 8); (-4, -2)$

93. Describe how to check to see if the graph of $2x - 4y = 7$ passes through the points $(1.4, -1.05)$ and $(0, -1.75)$. Then follow your directions and check these points.

15.2 Reviewing Function Notation and Graphing Nonlinear Functions

In the previous section, we studied linear equations that describe functions. Not all equations in two variables are linear equations, and not all graphs of equations in two variables are lines. In Chapter 10, we saw graphs of nonlinear equations, some of which were functions since they passed the vertical line test. In this section, we study the functions whose graphs may not be lines. First, let's review function notation.

Objectives

1. Review Function Notation.
2. Find Square Roots of Numbers.
3. Graph Nonlinear Functions.

Objective 1 Reviewing Function Notation

Suppose we have a function f such that $f(2) = -1$. Recall this means that when $x = 2$, $f(x)$, or y, $= -1$. Thus, the graph of f passes through $(2, -1)$.

Helpful Hint

Remember that $f(x)$ is a special symbol in mathematics used to denote a function. The symbol $f(x)$ is read "f of x." It does *not* mean $f \cdot x$ (f times x).

✓Concept Check Suppose $y = f(x)$ and we are told that $f(3) = 9$. Which is not true?

a. When $x = 3$, $y = 9$.

b. A possible function is $f(x) = x^2$.

c. A point on the graph of the function is $(3, 9)$.

d. A possible function is $f(x) = 2x + 4$.

If it helps, think of a function, f, as a machine that has been programmed with a certain correspondence or rule. An input value (a member of the domain) is then fed into the machine, the machine does the correspondence or rule, and the result is the output (a member of the range).

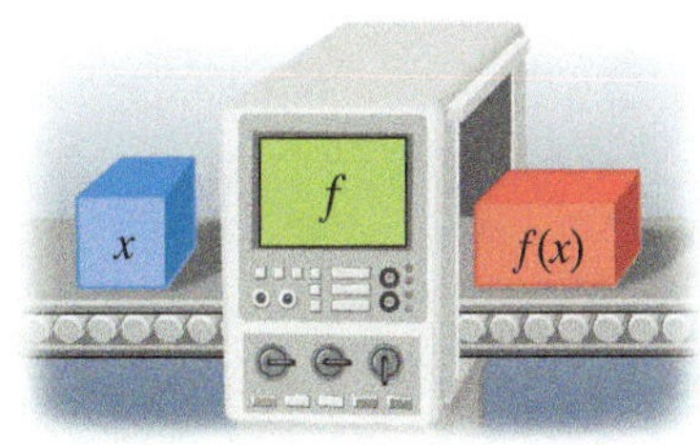

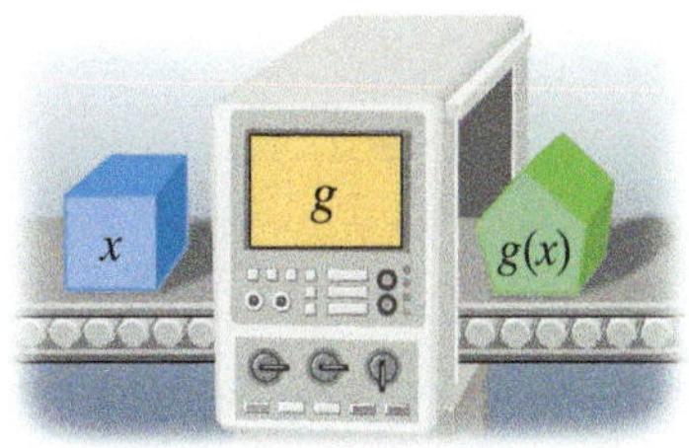

Example 1 Given the graphs of the functions f and g, find each function value by inspecting the graphs.

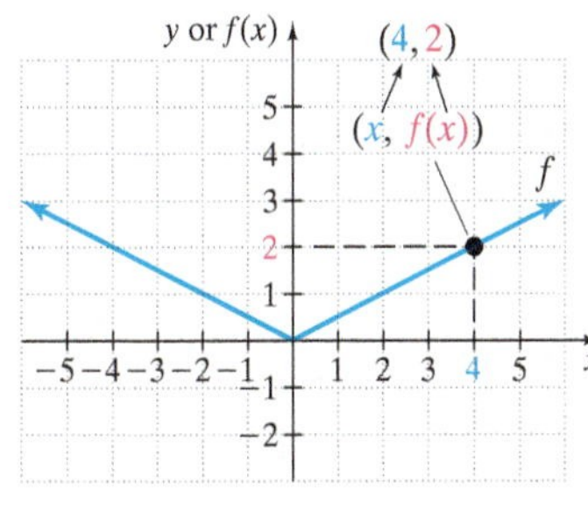

a. $f(4)$ **b.** $f(-2)$ **c.** $g(5)$ **d.** $g(0)$

e. Find all x-values such that $f(x) = 1$. **f.** Find all x-values such that $g(x) = 0$.

(*Continued on next page*)

Practice 1

Given the graphs of the functions f and g, find each function value by inspecting the graphs.

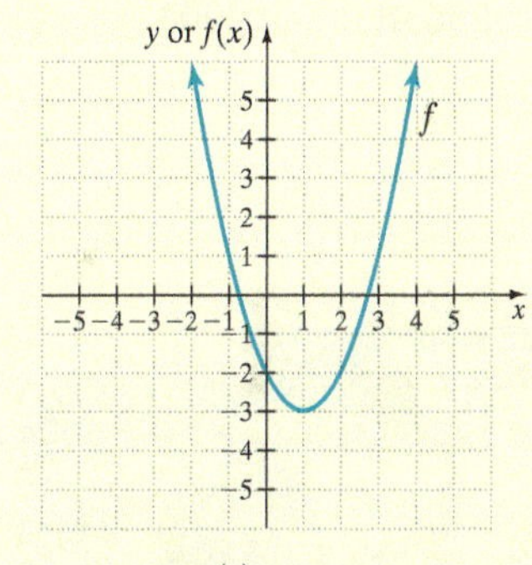

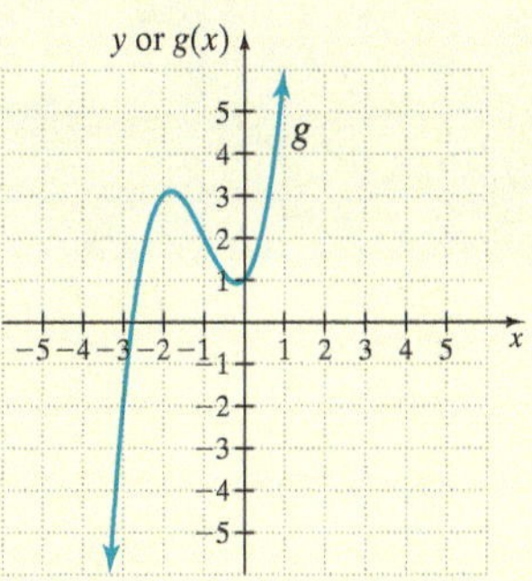

a. $f(1)$ **b.** $f(0)$

c. $g(-2)$ **d.** $g(0)$

e. Find all x-values such that $f(x) = 1$.

f. Find all x-values such that $g(x) = -2$.

Answers

1. a. -3 **b.** -2 **c.** 3 **d.** 1 **e.** -1 and 3 **f.** -3

✓Concept Check Answer

d

Solution:

a. To find $f(4)$, find the y-value when $x = 4$. We see from the graph that when $x = 4$, y or $f(x) = 2$. Thus, $f(4) = 2$.

b. $f(-2) = 1$ from the ordered pair $(-2, 1)$.

c. $g(5) = 3$ from the ordered pair $(5, 3)$.

d. $g(0) = 0$ from the ordered pair $(0, 0)$.

e. To find x-values such that $f(x) = 1$, we are looking for any ordered pairs on the graph of f whose $f(x)$, or y-, value is 1. They are $(2, 1)$ and $(-2, 1)$. Thus $f(2) = 1$ and $f(-2) = 1$. The x-values are 2 and -2.

f. Find ordered pairs on the graph of g whose $g(x)$ or y-value is 0. They are (3, 0), (0, 0), and $(-4, 0)$. Thus $g(3) = 0, g(0) = 0$, and $g(-4) = 0$. The x-values are 3, 0, and -4.

Work Practice 1

Many types of real-world paired data form functions. The broken-line graphs below and on the next page show the total and online enrollment in postsecondary institutions.

Practice 2

Use the graph in Example 2 and approximate the total enrollment in fall 2003.

Example 2 The following graph shows the total and online enrollments in postsecondary institutions as functions of time.

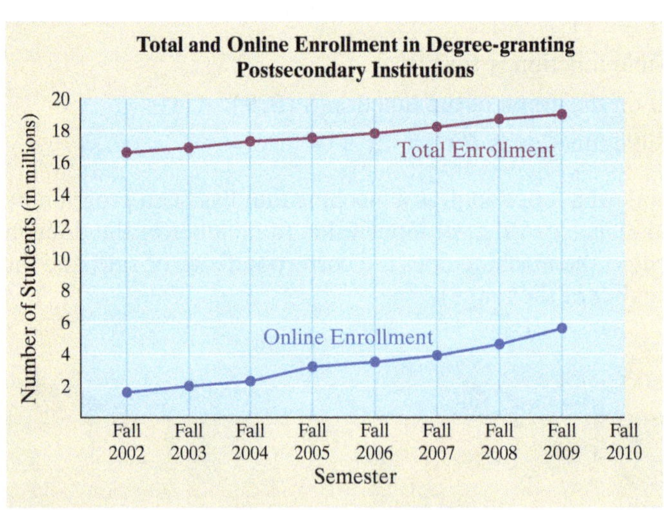

Source: Projections of Education Statistics to 2018, National Center for Education Statistics

a. Approximate the total enrollment in fall 2009.

b. In fall 2002, the total enrollment was 16.6 million students. Find the increase in total enrollment from fall 2002 to fall 2009.

Solution:

a. Find the semester fall 2009 and move upward until you reach the top broken-line graph. From the point on the graph, move horizontally to the left until the vertical axis is reached. In fall 2009, approximately 19 million students, or 19,000,000 students, were enrolled in degree-granting postsecondary institutions.

b. The increase from fall 2002 to fall 2009 was 19 million − 16.6 million = 2.4 million or 2,400,000 students.

Work Practice 2

Answer

2. 17 million or 17,000,000 students

Notice that each graph separately in Example 2 is the graph of a function since for each semester there is only one total enrollment and only one online enrollment. Also notice that each graph resembles the graph of a line. Often, businesses depend on equations that closely fit data-defined functions like this one to model the data and predict future trends. For example, by a method called **least squares,** the function $f(x) = 0.34x + 16$ approximates the data for the red graph, and the function $f(x) = 0.55x + 0.3$ approximates the data for the blue graph. For each function, x is the number of years since 2000, and $f(x)$ is the number of students (in millions). The graphs and the data functions are shown next.

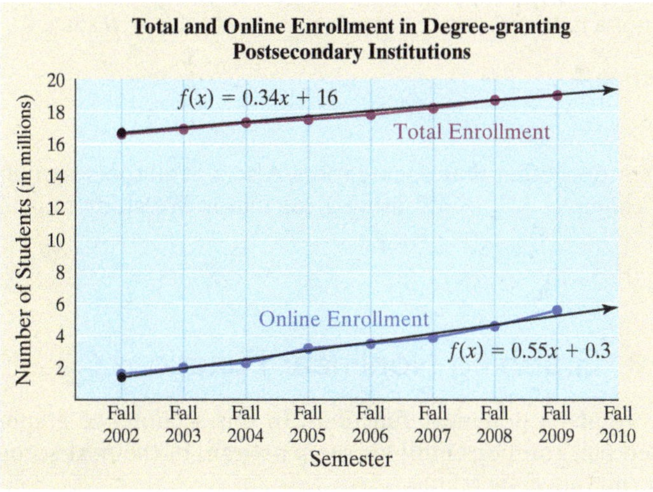

Source: Projections of Education Statistics to 2018, National Center for Education Statistics

Helpful Hint Each function graphed is the graph of a function and passes the vertical line test.

Example 3 Use the function $f(x) = 0.34x + 16$ and the discussion following Example 2 to estimate the total enrollment in degree-granting postsecondary institutions for fall 2010.

Solution: To estimate the total enrollment in fall 2010, remember that x represents the number of years since 2000, so $x = 2010 - 2000 = 10$. Use $f(x) = 0.34x + 16$ and find $f(10)$.

$$f(x) = 0.34x + 16$$
$$f(10) = 0.34(10) + 16$$
$$= 19.4$$

We estimate that in the fall 2010 semester, the total enrollment was 19.4 million, or 19,400,000 students.

Work Practice 3

Practice 3

Use $f(x) = 0.55x + 0.3$ to approximate the online enrollment in fall 2010.

Objective 2 Finding Square Roots of Numbers 

Later in this section, we graph the square root function, $f(x) = \sqrt{x}$. To prepare for this graph, let's review finding square roots of numbers.

The opposite of squaring a number is taking the **square root** of a number. For example, since the square of 4, or 4^2, is 16, we say that a square root of 16 is 4. The notation $\sqrt{a}$ is used to denote the **positive,** or **principal, square root** of a nonnegative number a. We then have in symbols that $\sqrt{16} = 4$. The negative square root of 16 is written $-\sqrt{16} = -4$. The square root of a negative number, such as $\sqrt{-16}$, is not a real number. Why? There is no real number that, when squared, gives a negative number.

Answer

3. 5.8 million or 5,800,000 students

Practice 4

Find the square roots.

a. $\sqrt{121}$ **b.** $\sqrt{\frac{1}{16}}$

c. $-\sqrt{64}$ **d.** $\sqrt{-64}$

e. $\sqrt{100}$

Example 4 Find the square roots.

a. $\sqrt{9}$ **b.** $\sqrt{25}$ **c.** $\sqrt{\frac{1}{4}}$ **d.** $-\sqrt{36}$ **e.** $\sqrt{-36}$ **f.** $\sqrt{0}$

Solution:

a. $\sqrt{9} = 3$ since 3 is positive and $3^2 = 9$.

b. $\sqrt{25} = 5$ since $5^2 = 25$.

c. $\sqrt{\frac{1}{4}} = \frac{1}{2}$ since $\left(\frac{1}{2}\right)^2 = \frac{1}{4}$.

d. $-\sqrt{36} = -6$

e. $\sqrt{-36}$ is not a real number.

f. $\sqrt{0} = 0$ since $0^2 = 0$.

Work Practice 4

We can find roots other than square roots. Also, not all roots simplify to rational numbers. For example, $\sqrt{3} \approx 1.7$ using a calculator. We study radicals further in Chapter 17.

Objective 3 Graphing Nonlinear Functions 

Let's practice graphing nonlinear functions. In this section, we graph by plotting enough ordered pair solutions until we see a pattern. In the next section, we learn about shifting and reflecting graphs.

Practice 5

Graph $f(x) = 2x^2$.

Example 5 Graph $f(x) = x^2$.

Solution: This equation is not linear because the x^2 term does not allow us to write the equation in the form $Ax + By = C$. Its graph is not a line. We begin by finding ordered pair solutions. Because $f(x) = y$, feel free to think of this equation as $f(x) = x^2$ or $y = x^2$. This graph is solved for y, so we choose x-values and find corresponding y-values.

If $x = -3$, then $f(-3) = (-3)^2$, or 9.

If $x = -2$, then $f(-2) = (-2)^2$, or 4.

If $x = -1$, then $f(-1) = (-1)^2$, or 1.

If $x = 0$, then $f(0) = 0^2$, or 0.

If $x = 1$, then $f(1) = 1^2$, or 1.

If $x = 2$, then $f(2) = 2^2$, or 4.

If $x = 3$, then $f(3) = 3^2$, or 9.

x	y or $f(x)$
−3	9
−2	4
−1	1
0	0
1	1
2	4
3	9

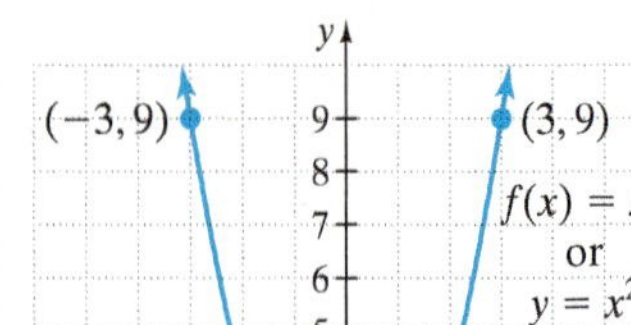
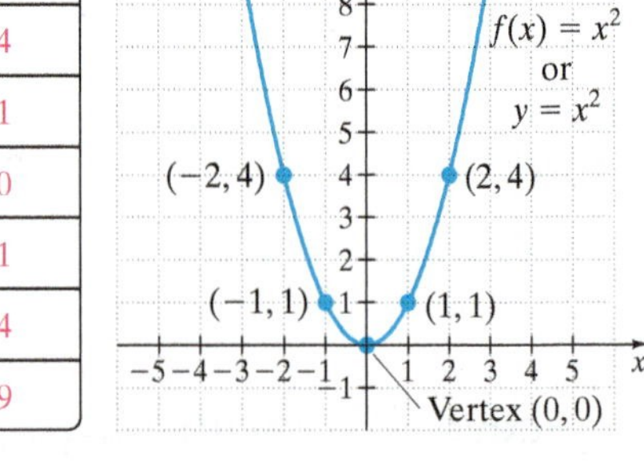

Study the table a moment and look for patterns. Notice that the ordered pair solution $(0, 0)$ contains the smallest y-value because any other x-value squared will give a positive result. This means that the point $(0, 0)$ will be the lowest point on the graph. Also notice that all other y-values correspond to two different x-values. For example, $3^2 = 9$ and also $(-3)^2 = 9$. This means that the graph will be a

Answers

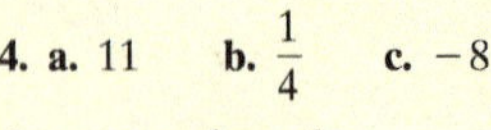

4. a. 11 **b.** $\frac{1}{4}$ **c.** -8 **d.** not a real number **e.** 10

5.

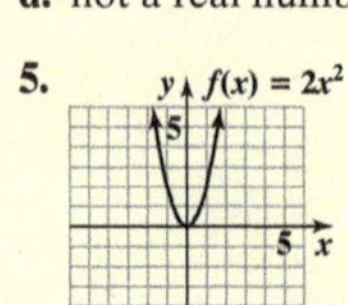

mirror image of itself across the y-axis. Connect the plotted points with a smooth curve to sketch the graph.

This curve is given a special name, a **parabola.** We will study more about parabolas in later chapters.

Work Practice 5

Example 6 Graph the nonlinear function $f(x) = |x|$.

Solution: This is not a linear equation since it cannot be written in the form $Ax + By = C$. Its graph is not a line. Because we do not know the shape of this graph, we find many ordered pair solutions. We will choose x-values and substitute to find corresponding y-values.

If $x = -3$, then $f(-3) = |-3|$, or 3.

If $x = -2$, then $f(-2) = |-2|$, or 2.

If $x = -1$, then $f(-1) = |-1|$, or 1.

If $x = 0$, then $f(0) = |0|$, or 0.

If $x = 1$, then $f(1) = |1|$, or 1.

If $x = 2$, then $f(2) = |2|$, or 2.

If $x = 3$, then $f(3) = |3|$, or 3.

x	y or $f(x)$
−3	3
−2	2
−1	1
0	0
1	1
2	2
3	3

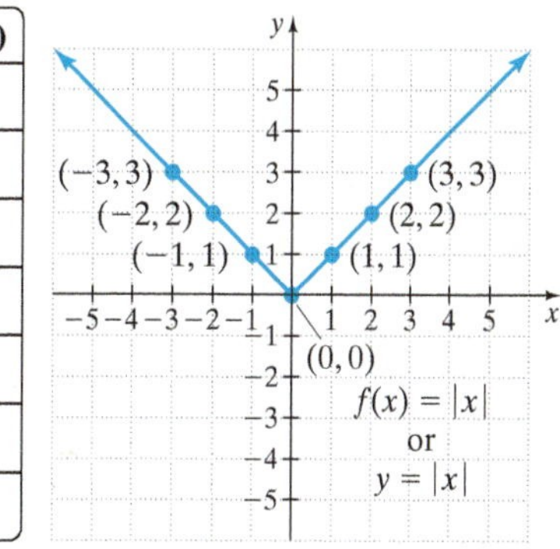

Again, study the table of values for a moment and notice any patterns.

From the plotted ordered pairs, we see that the graph of this absolute value equation is V-shaped.

Work Practice 6

Practice 6

Graph $f(x) = -|x|$.

Example 7 Graph the nonlinear function $f(x) = \sqrt{x}$.

Solution: To graph this square root function, we identify the domain, evaluate the function for several values of x, plot the resulting points, and connect the points with a smooth curve. Since $\sqrt{x}$ represents the nonnegative square root of x, the domain of this function is the set of all nonnegative numbers, $\{x|x \geq 0\}$, or $[0, \infty)$. We have approximated $\sqrt{3}$ below to help us locate the point corresponding to $(3, \sqrt{3})$.

If $x = 0$, then $f(0) = \sqrt{0}$, or 0.

If $x = 1$, then $f(1) = \sqrt{1}$, or 1.

If $x = 3$, then $f(3) = \sqrt{3}$, or 1.7.

If $x = 4$, then $f(4) = \sqrt{4}$, or 2.

If $x = 9$, then $f(9) = \sqrt{9}$, or 3.

x	y or $f(x)$
0	0
1	1
3	$\sqrt{3} \approx 1.7$
4	2
9	3

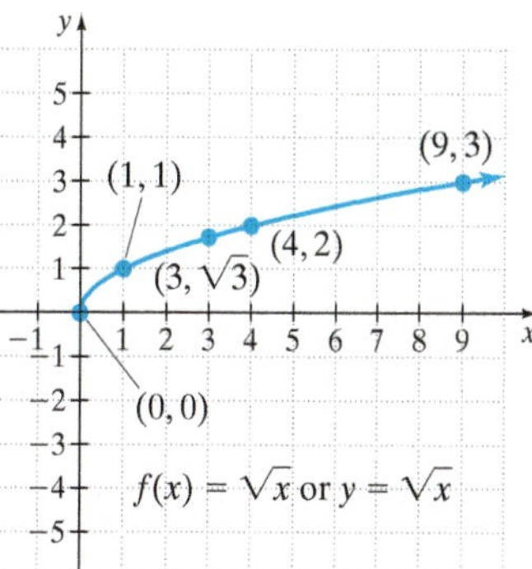

Work Practice 7

Practice 7

Graph $f(x) = \sqrt{x} + 1$.

Answers

6. 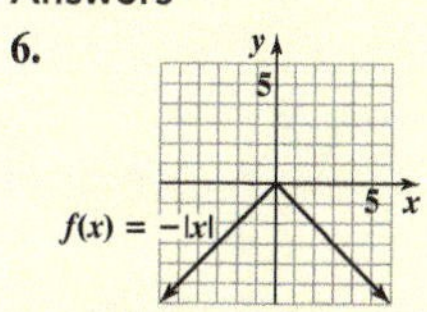

7. 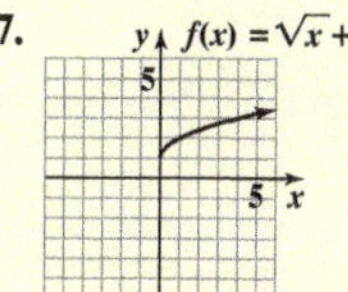

Graphing Calculator Explorations

It is possible to use a graphing calculator to sketch the graph of more than one equation on the same set of axes. For example, graph the functions $f(x) = x^2$ and $g(x) = x^2 + 4$ on the same set of axes.

To graph on the same set of axes, press the $\boxed{Y=}$ key and enter the equations on the first two lines.

$$Y_1 = x^2$$
$$Y_2 = x^2 + 4$$

Then press the $\boxed{\text{GRAPH}}$ key as usual. The screen should look like this.

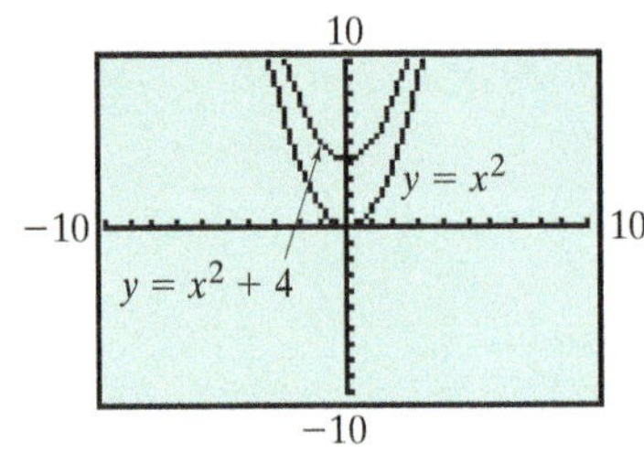

Notice that the graph of y or $g(x) = x^2 + 4$ is the graph of $y = x^2$ moved 4 units upward.

Graph each pair of functions on the same set of axes.

1. $f(x) = |x|$
   $g(x) = |x| + 1$
2. $f(x) = x^2$
   $h(x) = x^2 - 5$
3. $f(x) = x$
   $H(x) = x - 6$
4. $f(x) = |x|$
   $G(x) = |x| + 3$
5. $f(x) = -x^2$
   $F(x) = -x^2 + 7$
6. $f(x) = x$
   $F(x) = x + 2$

Vocabulary, Readiness & Video Check

Use the choices below to fill in each blank. Some choices may not be used.

(1.7, −2)	line	parabola	−6	−9
(−2, 1.7)	V-shaped	6	9	

1. The graph of $y = |x|$ looks ________.
2. The graph of $y = x^2$ is a ________.
3. If $f(-2) = 1.7$, the corresponding ordered pair is ________.
4. If $f(x) = x^2$, then $f(-3) =$ ________.

Martin-Gay Interactive Videos

See Video 15.2

Watch the section lecture video and answer the following questions.

Objective 1 5. From Examples 1 and 2, what is the connection between function notation to evaluate a function at certain values and ordered pair solutions of the function?

Objective 2 6. Explain why Example 6 does not simplify to a real number.

Objective 3 7. Based on Examples 7, 8, and 9, complete the following statements. When graphing a nonlinear equation, first recognize it as a nonlinear equation and know that the graph is ________ a line. If you don't know the ________ of the graph, plot enough points until you see a pattern.

15.2 Exercise Set MyMathLab® 

Use the graph of the following function $f(x)$ to find each value. See Examples 1 and 2.

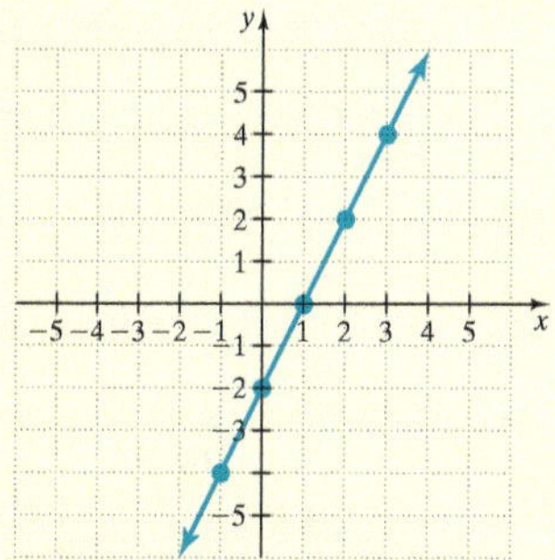

1. $f(1)$
2. $f(0)$
3. $f(-1)$
4. $f(2)$
5. Find x such that $f(x) = 4$.
6. Find x such that $f(x) = -6$.

Use the graph of the functions below to answer Exercises 7 through 18. See Examples 1 and 2.

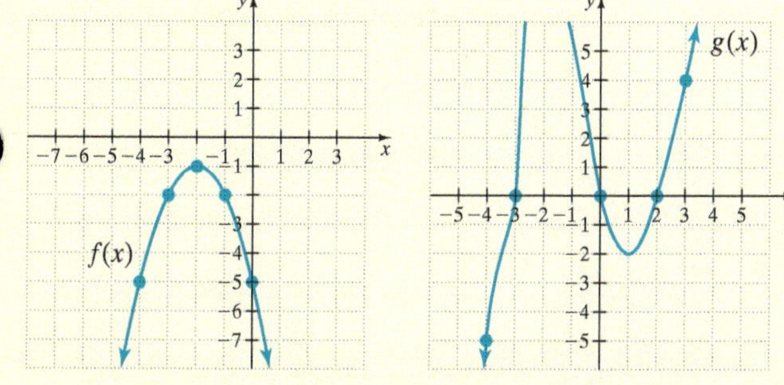

7. If $f(1) = -10$, write the corresponding ordered pair.
8. If $f(-5) = -10$, write the corresponding ordered pair.
9. If $g(4) = 56$, write the corresponding ordered pair.
10. If $g(-2) = 8$, write the corresponding ordered pair.
11. Find $f(-1)$.
12. Find $f(-2)$.
13. Find $g(2)$.
14. Find $g(-4)$.
15. Find all values of x such that $f(x) = -5$.
16. Find all values of x such that $f(x) = -2$.
17. Find all positive values of x such that $g(x) = 4$.
18. Find all values of x such that $g(x) = 0$.

Find the following roots. See Example 4.

19. $\sqrt{49}$
20. $\sqrt{144}$
21. $-\sqrt{\frac{4}{9}}$
22. $-\sqrt{\frac{4}{25}}$
23. $\sqrt{64}$
24. $\sqrt{4}$
25. $\sqrt{81}$
26. $\sqrt{1}$
27. $\sqrt{-100}$
28. $\sqrt{-25}$

MIXED PRACTICE

Graph each function by finding and plotting ordered pair solutions. See Examples 5 through 7.

29. $f(x) = x^2 + 3$
30. $g(x) = (x + 2)^2$
31. $h(x) = |x| - 2$
32. $f(x) = |x - 2|$
33. $g(x) = 2x^2$
34. $h(x) = 5x^2$
35. $f(x) = 5x - 1$
36. $g(x) = -3x + 2$
37. $f(x) = \sqrt{x + 1}$
38. $f(x) = \sqrt{x} - 1$
39. $g(x) = -2|x|$
40. $g(x) = -3|x|$
41. $h(x) = \sqrt{x} + 2$
42. $h(x) = \sqrt{x + 2}$

Use the graph on page 1072 to answer the following. Also see Example 3.

43. **a.** Use the graph to approximate the online enrollment in fall 2009.
    b. The function $f(x) = 0.55x + 0.3$ approximates the online enrollment (in millions). Use this function to approximate the online enrollment in fall 2009.
44. **a.** Use the graph to approximate the total enrollment in fall 2008.
    b. The function $f(x) = 0.34x + 16$ approximates the total enrollment (in millions). Use this function to approximate the total enrollment in fall 2008.

The function $f(x) = 0.42x + 10.5$ can be used to predict diamond production. For this function, x is the number of years after 2000, and $f(x)$ is the value (in billions of dollars) of diamond production.

45. Use the function in the directions above to estimate diamond production in 2012.
46. Use the function in the directions above to estimate diamond production in 2015.

The function $A(r) = \pi r^2$ may be used to find the area of a circle if we are given its radius.

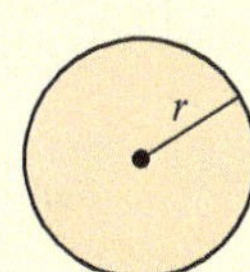

△ 47. Find the area of a circle whose radius is 5 centimeters. (Do not approximate π.)
△ 48. Find the area of a circular garden whose radius is 8 feet. (Do not approximate π.)

The function $V(x) = x^3$ may be used to find the volume of a cube if we are given the length x of a side.

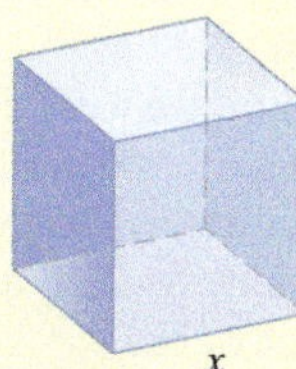

49. Find the volume of a cube whose side is 14 inches.
50. Find the volume of a die whose side is 1.7 centimeters.

Forensic scientists use the following functions to find the height of a woman if they are given the height of her femur bone f or her tibia bone t in centimeters.

$$H(f) = 2.59f + 47.24$$
$$H(t) = 2.72t + 61.28$$

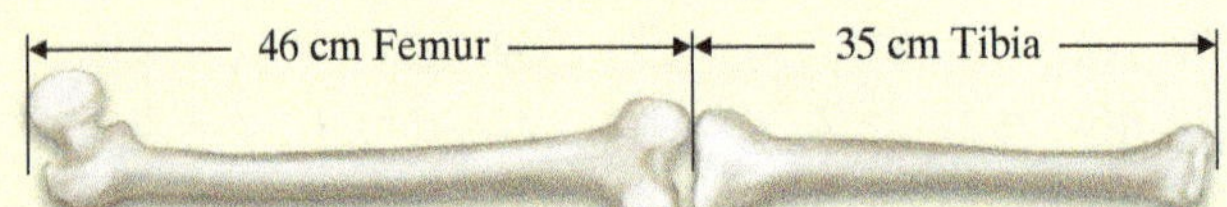

51. Find the height of a woman whose femur measures 46 centimeters.

52. Find the height of a woman whose tibia measures 35 centimeters.

The dosage in milligrams D of Ivermectin, a heartworm preventive, for a dog who weighs x pounds is given by

$$D(x) = \frac{136}{25}x$$

53. Find the proper dosage for a dog that weighs 30 pounds.

54. Find the proper dosage for a dog that weighs 50 pounds.

55. What is the greatest number of x-intercepts that a function may have? Explain your answer.

56. What is the greatest number of y-intercepts that a function may have? Explain your answer.

REVIEW AND PREVIEW

Solve the following equations. See Section 9.3.

57. $3(x - 2) + 5x = 6x - 16$

58. $5 + 7(x + 1) = 12 + 10x$

59. $3x + \frac{2}{5} = \frac{1}{10}$

60. $\frac{1}{6} + 2x = \frac{2}{3}$

CONCEPT EXTENSIONS

For Exercises 61 through 64, match each description with the graph that best illustrates it.

61. Moe worked 40 hours per week until the fall semester started. He quit and didn't work again until he worked 60 hours a week during the holiday season starting mid-December.

62. Kawana worked 40 hours a week for her father during the summer. She slowly cut back her hours to not working at all during the fall semester. During the holiday season in December, she started working again and increased her hours to 60 hours per week.

63. Wendy worked from July through February, never quitting. She worked between 10 and 30 hours per week.

64. Bartholomew worked from July through February. During the holiday season between mid-November and the beginning of January, he worked 40 hours per week. The rest of the time, he worked between 10 and 40 hours per week.

a.

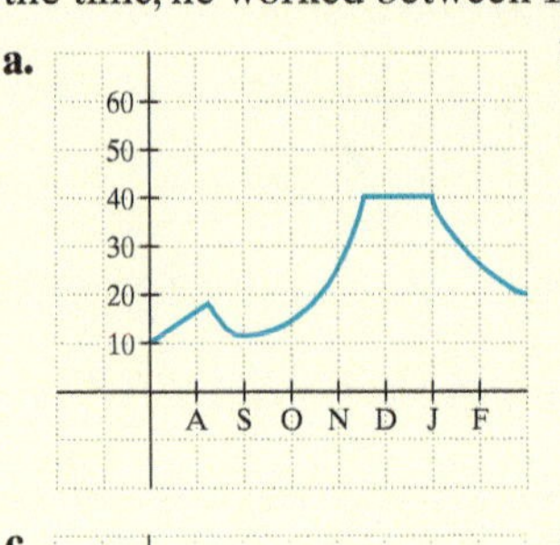

b.

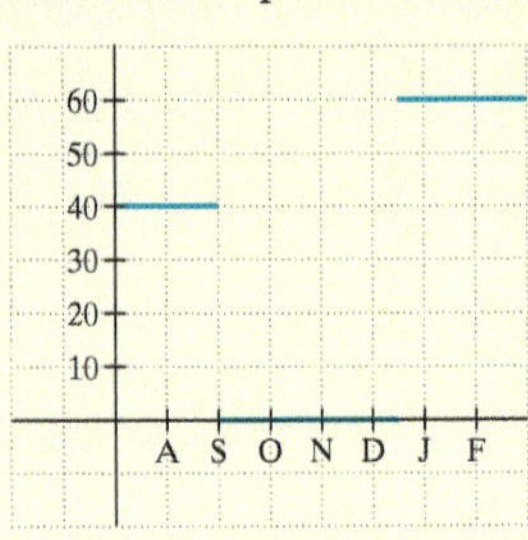

c.

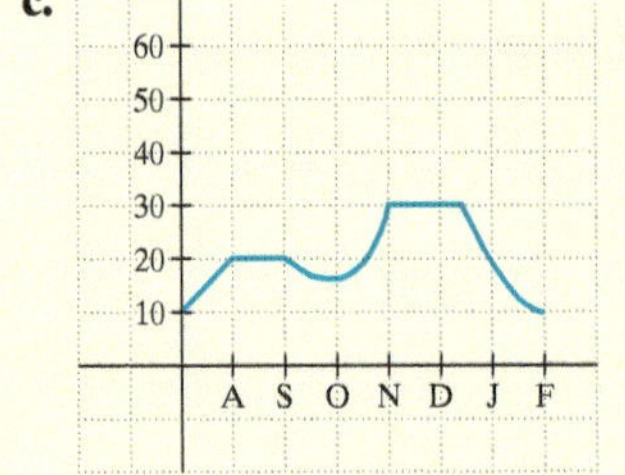

d.

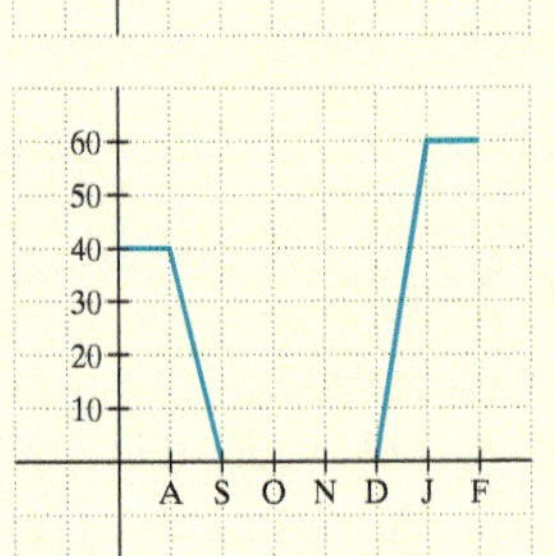

This broken-line graph shows the hourly minimum wage and the years it increased. Use this graph for Exercises 65 through 68.

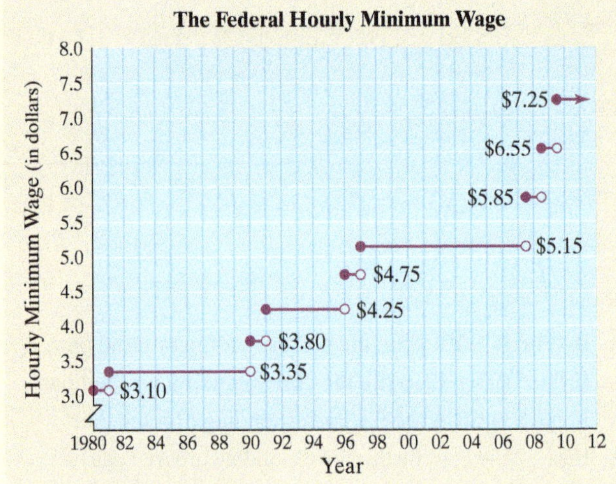

65. What was the first year that the minimum hourly wage rose above $5.00?

66. What was the first year that the minimum hourly wage rose above $6.00?

67. Why do you think that this graph is shaped the way it is?

68. The federal hourly minimum wage started in 1938 at $0.25. How much had it increased by 2011?

69. Graph $y = x^2 - 4x + 7$. Let $x = 0, 1, 2, 3, 4$ to generate ordered pair solutions.

70. Graph $y = x^2 + 2x + 3$. Let $x = -3, -2, -1, 0, 1$ to generate ordered pair solutions.

71. The function $f(x) = [x]$ is called the greatest integer function and its graph is similar to the graph above. The value of $[x]$ is the greatest integer less than or equal to x. For example, $f(1.5) = [1.5] = 1$ since the greatest integer ≤ 1.5 is 1. Sketch the graph of $f(x) = [x]$.

Integrated Review Sections 15.1–15.2

Summary on Functions and Equations of Lines

Find the slope and y-intercept of the graph of each function.

1. $f(x) = 3x - 5$

2. $f(x) = \frac{5}{2}x - \frac{7}{2}$

Determine whether each pair of lines is parallel, perpendicular, or neither.

3. $f(x) = 8x - 6$
$g(x) = 8x + 6$

4. $f(x) = \frac{2}{3}x + 1$
$2y + 3x = 1$

Find the equation of each line. Write the equation using function notation.

5. Through $(1, 6)$ and $(5, 2)$

6. Through $(2, -8)$ and $(-6, -5)$

7. Through $(-1, -5)$; parallel to $3x - y = 5$

8. Through $(0, 4)$; perpendicular to $4x - 5y = 10$

9. Through $(2, -3)$; perpendicular to $4x + y = \frac{2}{3}$

10. Through $(-1, 0)$; parallel to $5x + 2y = 2$

Determine whether each function is linear or not. Then graph the function.

11. $f(x) = 4x - 2$

12. $f(x) = 6x - 5$

13. $g(x) = |x| + 3$

14. $h(x) = |x| + 2$

15. $f(x) = 2x^2$

16. $F(x) = 3x^2$

17. $h(x) = x^2 - 3$

18. $G(x) = x^2 + 3$

19. $F(x) = -2x$

20. $H(x) = -3x$

21. $G(x) = |x + 2|$

22. $g(x) = |x - 1|$

23. $f(x) = \frac{1}{3}x - 1$

24. $f(x) = \frac{1}{2}x - 3$

25. $g(x) = -\frac{3}{2}x + 1$

26. $G(x) = -\frac{2}{3}x + 1$

15.3 Graphing Piecewise-Defined Functions and Shifting and Reflecting Graphs of Functions

Objective 1 Graphing Piecewise-Defined Functions 

Thus far in Chapter 15, we have graphed functions. There are many special functions. In this objective, we study functions defined by two or more expressions. The expression used to complete the function varies with and depends upon the value of x. Before we actually graph these piecewise-defined functions, let's practice finding function values.

Objectives

1 Graph Piecewise-Defined Functions.

2 Vertical and Horizontal Shifts.

3 Reflect Graphs.

Practice 1

Evaluate $f(4), f(-2)$, and $f(0)$ for the function

$$f(x) = \begin{cases} -4x - 2 & \text{if } x \le 0 \\ x + 1 & \text{if } x > 0 \end{cases}$$

Then write your results in ordered pair form.

Practice 2

Graph

$$f(x) = \begin{cases} -4x - 2 & \text{if } x \le 0 \\ x + 1 & \text{if } x > 0 \end{cases}$$

Answers

1. $f(4) = 5; f(-2) = 6; f(0) = -2;$ $(4, 5), (-2, 6), (0, -2)$

2. 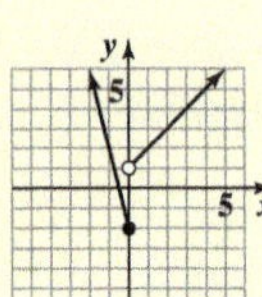

Example 1 Evaluate $f(2), f(-6)$, and $f(0)$ for the function

$$f(x) = \begin{cases} 2x + 3 & \text{if } x \le 0 \\ -x - 1 & \text{if } x > 0 \end{cases}$$

Then write your results in ordered pair form.

Solution: Take a moment and study this function. It is a single function defined by two expressions depending on the value of x. From above, if $x \le 0$, use $f(x) = 2x + 3$. If $x > 0$, use $f(x) = -x - 1$. Thus

$f(2) = -(2) - 1$	$f(-6) = 2(-6) + 3$	$f(0) = 2(0) + 3$
$= -3$ since $2 > 0$	$= -9$ since $-6 \le 0$	$= 3$ since $0 \le 0$
$f(2) = -3$	$f(-6) = -9$	$f(0) = 3$
Ordered pairs: $(2, -3)$	$(-6, -9)$	$(0, 3)$

Work Practice 1

Now, let's graph a piecewise-defined function.

Example 2 Graph $f(x) = \begin{cases} 2x + 3 & \text{if } x \le 0 \\ -x - 1 & \text{if } x > 0 \end{cases}$

Solution: Let's graph each piece.

If $x \le 0$,
$f(x) = 2x + 3$

Values ≤ 0:

x	$f(x) = 2x + 3$
0	3 Closed circle
−1	1
−2	−1

If $x > 0$,
$f(x) = -x - 1$

Values > 0:

x	$f(x) = -x - 1$
1	−2
2	−3
3	−4

The graph of the first part of $f(x)$ listed will look like a ray with a closed-circle endpoint at $(0, 3)$. The graph of the second part of $f(x)$ listed will look like a ray with an open-circle endpoint. To find the exact location of the open-circle endpoint, use $f(x) = -x - 1$ and find $f(0)$. Since $f(0) = -0 - 1 = -1$, we graph the values from the second table and place an open circle at $(0, -1)$.

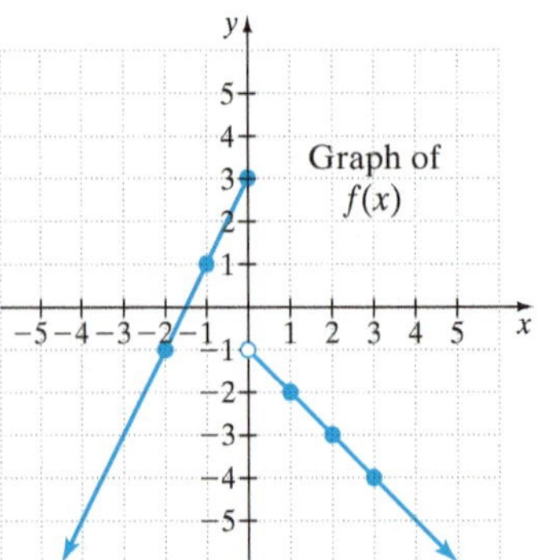

Notice that this graph is the graph of a function because it passes the vertical line test. The domain of this function is $(-\infty, \infty)$ and the range is $(-\infty, 3]$.

Work Practice 2

Objective 2 Vertical and Horizontal Shifting 

Review of Common Graphs

We now take common graphs and learn how more complicated graphs are actually formed by shifting and reflecting these common graphs. These shifts and reflections are called transformations, and it is possible to combine transformations. A knowledge of these transformations will help make graphing simpler.

Let's begin with a review of the graphs of four common functions. We graphed many of these functions in earlier sections. (Much of this review can be found in the previous section.)

First, **let's graph the linear function $f(x) = x$,** or **$y = x$.** Ordered pair solutions of this graph consist of ordered pairs whose x- and y-values are the same.

x	y or $f(x) = x$
-3	-3
0	0
1	1
4	4

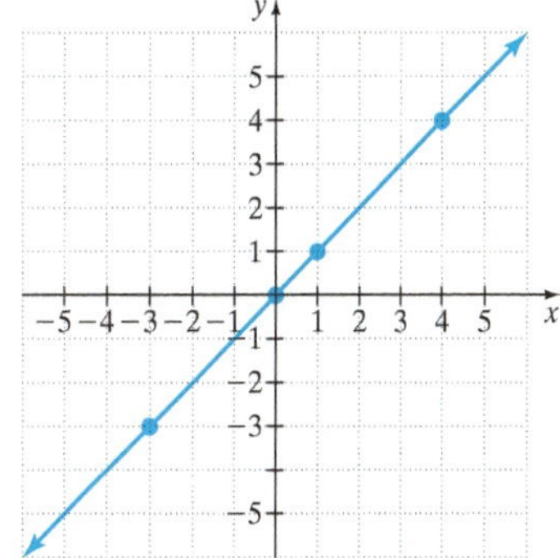

Next, **let's graph the nonlinear function $f(x) = x^2$ or $y = x^2$.**

This equation is not linear because the x^2 term does not allow us to write the equation in the form $Ax + By = C$. Its graph is not a line. We begin by finding ordered pair solutions. Because this graph is solved for $f(x)$, or y, we choose x-values and find corresponding $f(x)$, or y-values.

If $x = -3$, then $y = (-3)^2$, or 9.

If $x = -2$, then $y = (-2)^2$, or 4.

If $x = -1$, then $y = (-1)^2$, or 1.

If $x = 0$, then $y = 0^2$, or 0.

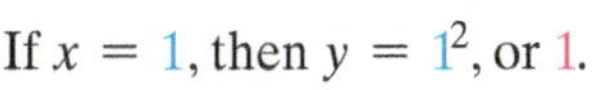

If $x = 1$, then $y = 1^2$, or 1.

If $x = 2$, then $y = 2^2$, or 4.

If $x = 3$, then $y = 3^2$, or 9.

x	$f(x)$ or y
-3	9
-2	4
-1	1
0	0
1	1
2	4
3	9

(−3, 9) (3, 9) $f(x) = x^2$ or $y = x^2$ (−2, 4) (2, 4) (−1, 1) (1, 1) Vertex (0, 0)

Study the table for a moment and look for patterns. Notice that the ordered pair solution (0, 0) contains the smallest y-value because any other x-value squared will give a positive result. This means that the point (0, 0) will be the lowest point on the graph. Also notice that all other y-values correspond to two different x-values, for example, $3^2 = 9$ and $(-3)^2 = 9$. This means that the graph will be a mirror image of itself across the y-axis. Connect the plotted points with a smooth curve to sketch its graph.

This curve is given a special name, a **parabola.** We will study more about parabolas in later chapters.

Next, **let's graph another nonlinear function, $f(x) = |x|$ or $y = |x|$.**

This is not a linear equation since it cannot be written in the form $Ax + By = C$. Its graph is not a line. Because we do not know the shape of this graph, we find many ordered pair solutions. We will choose x-values and substitute to find corresponding y-values.

If $x = -3$, then $y = |-3|$, or 3.

If $x = -2$, then $y = |-2|$, or 2.

If $x = -1$, then $y = |-1|$, or 1.

If $x = 0$, then $y = |0|$, or 0.

If $x = 1$, then $y = |1|$, or 1.

If $x = 2$, then $y = |2|$, or 2.

If $x = 3$, then $y = |3|$, or 3.

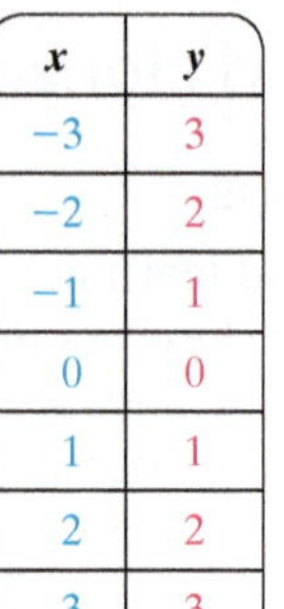

x	y
-3	3
-2	2
-1	1
0	0
1	1
2	2
3	3

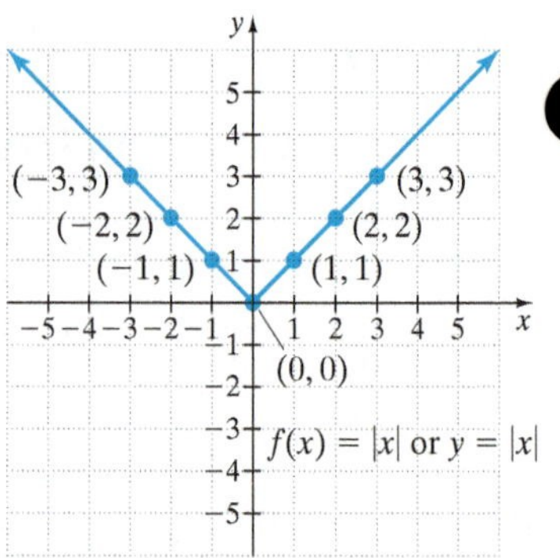

Again, study the table of values for a moment and notice any patterns.

From the plotted ordered pairs, we see that the graph of this absolute value equation is V-shaped.

Finally, a fourth common function is $f(x) = \sqrt{x}$ or $y = \sqrt{x}$. For this graph, you need to recall basic facts about square roots and use your calculator to approximate some square roots to help locate points. Recall also that the square root of a negative number is not a real number, so be careful when finding your domain.

Now **let's graph the square root function $f(x) = \sqrt{x}$, or $y = \sqrt{x}$.**

To graph, we identify the domain, evaluate the function for several values of x, plot the resulting points, and connect the points with a smooth curve. Since $\sqrt{x}$ represents the nonnegative square root of x, the domain of this function is the set of all nonnegative numbers, $\{x | x \geq 0\}$, or $[0, \infty)$. We have approximated $\sqrt{3}$ to help us locate the point corresponding to $(3, \sqrt{3})$.

If $x = 0$, then $y = \sqrt{0}$, or 0.

If $x = 1$, then $y = \sqrt{1}$, or 1.

If $x = 3$, then $y = \sqrt{3}$, or 1.7.

If $x = 4$, then $y = \sqrt{4}$, or 2.

If $x = 9$, then $y = \sqrt{9}$, or 3.

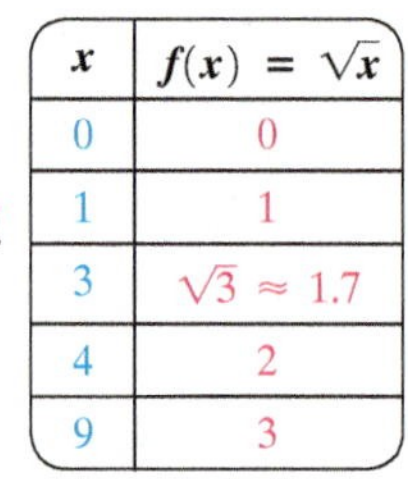

x	$f(x) = \sqrt{x}$
0	0
1	1
3	$\sqrt{3} \approx 1.7$
4	2
9	3

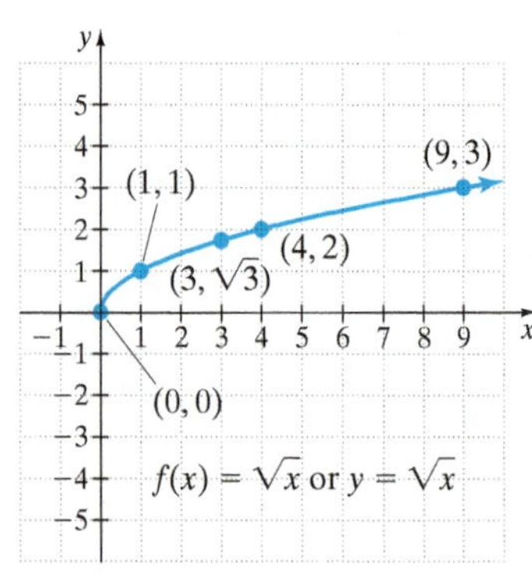

Notice that the graph of this function passes the vertical line test, as expected.

Below is a summary of our four common graphs. Take a moment and study these graphs. Your success in the rest of this section depends on your knowledge of these graphs.

Common Graphs

$f(x) = x$

$f(x) = x^2$

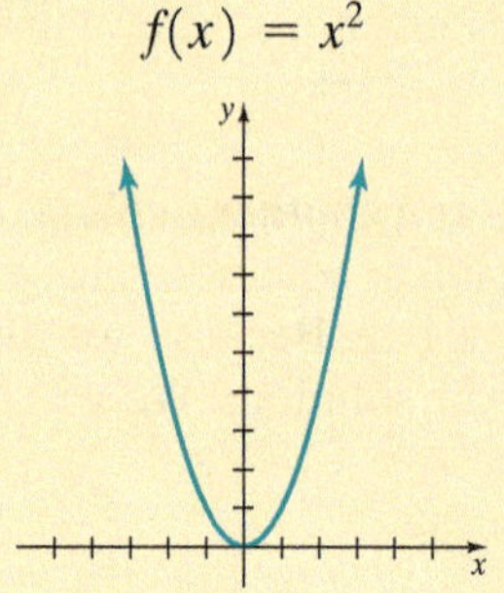

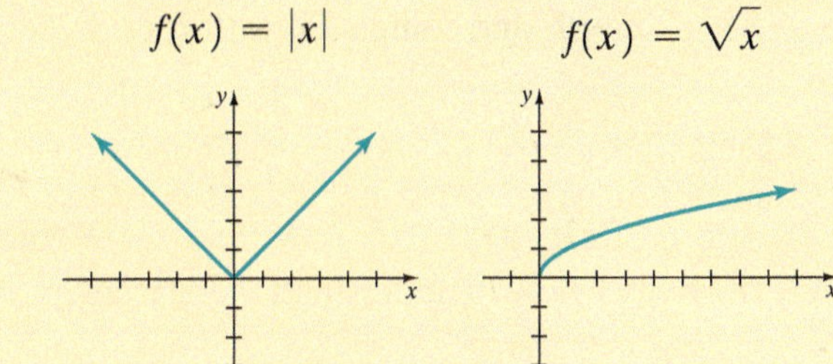

Your knowledge of the slope-intercept form, $f(x) = mx + b$, will help you understand simple shifting of transformations such as vertical shifts. For example, what is the difference between the graphs of $f(x) = x$ and $g(x) = x + 3$?

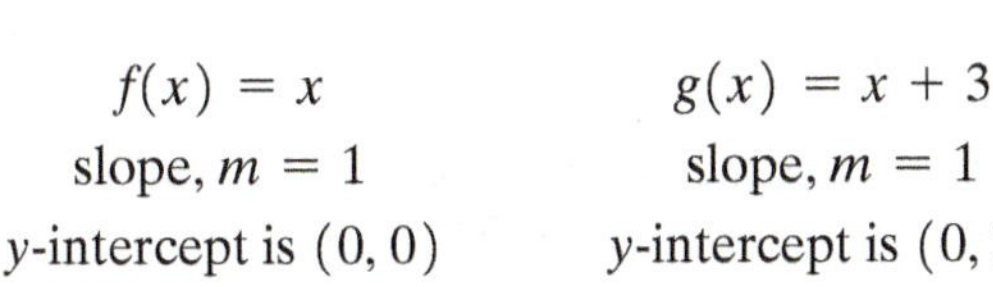

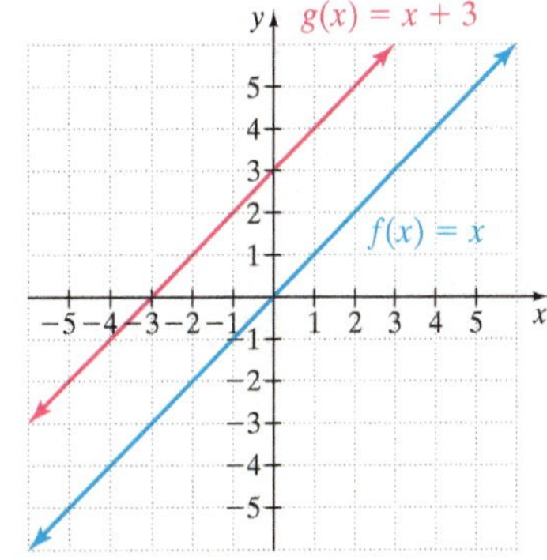

Notice that the graph of $g(x) = x + 3$ is the same as the graph of $f(x) = x$, but moved upward 3 units. This is an example of a **vertical shift** and is true for graphs in general.

Vertical Shifts (Upward or Downward)
Let *k* Be a Positive Number

Graph of	Same as	Moved
$g(x) = f(x) + k$	$f(x)$	k units upward
$g(x) = f(x) - k$	$f(x)$	k units downward

Examples Without plotting points, sketch the graph of each pair of functions on the same set of axes.

3. $f(x) = x^2$ and $g(x) = x^2 + 2$

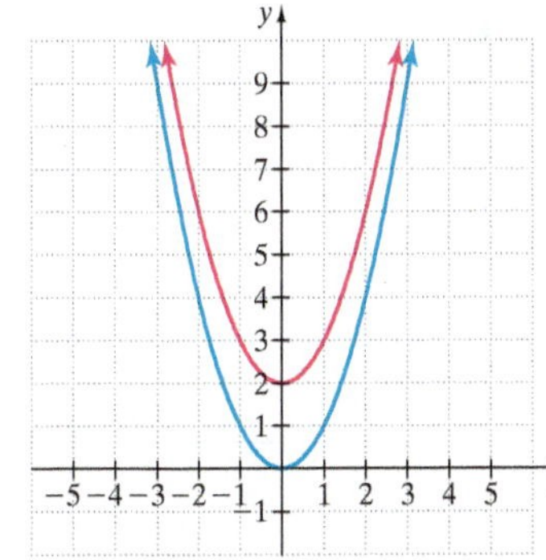

4. $f(x) = \sqrt{x}$ and $g(x) = \sqrt{x} - 3$

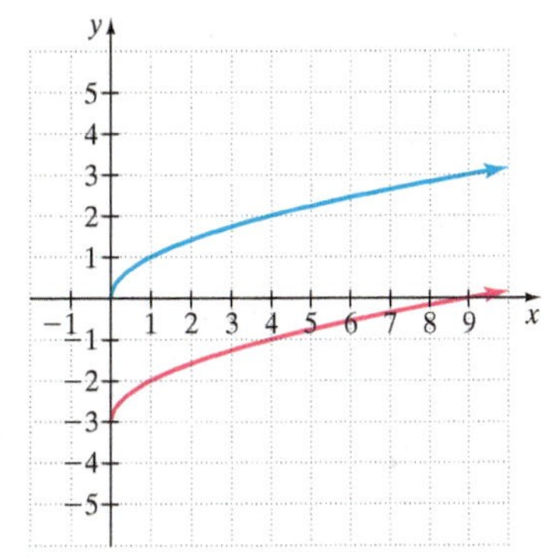

Work Practice 3–4

Practice 3–4

Without plotting points, sketch the graphs of each pair of functions on the same set of axes.

3. $f(x) = x^2$ and $g(x) = x^2 - 3$

4. $f(x) = \sqrt{x}$ and $g(x) = \sqrt{x} + 1$

Answers

3. **4.**

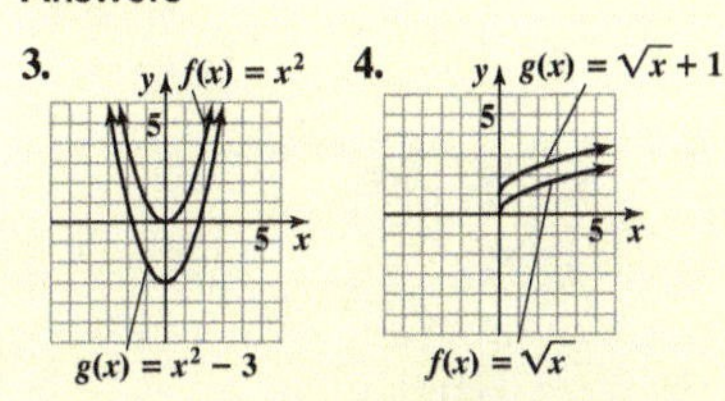

A horizontal shift to the left or right may be slightly more difficult to understand. Let's graph $g(x) = |x - 2|$ and compare it with $f(x) = |x|$.

Practice 5

Sketch the graphs of $f(x) = |x|$ and $g(x) = |x - 3|$ on the same set of axes.

Example 5 Sketch the graphs of $f(x) = |x|$ and $g(x) = |x - 2|$ on the same set of axes.

Solution: Study the table below to understand the placement of both graphs.

x	$f(x) = \lvert x\rvert$	$g(x) = \lvert x - 2\rvert$
−3	3	5
−2	2	4
−1	1	3
0	0	2
1	1	1
2	2	0
3	3	1

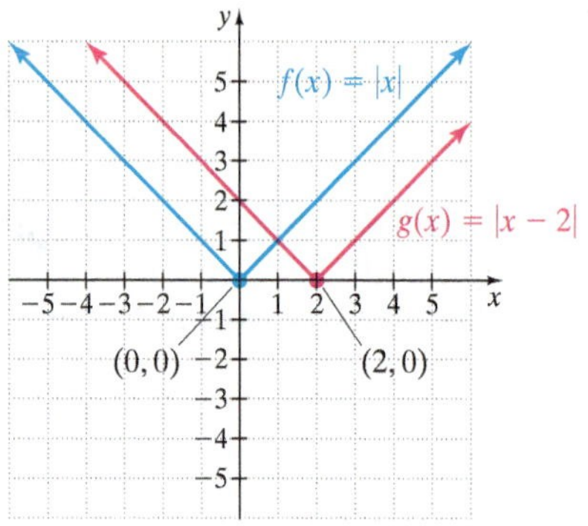

Work Practice 5

The graph of $g(x) = |x - 2|$ is the same as the graph of $f(x) = |x|$, but moved 2 units to the right. This is an example of a **horizontal shift** and is true for graphs in general.

Horizontal Shift (To the Left or Right) Let h Be a Positive Number

Graph of	Same as	Moved
$g(x) = f(x - h)$	$f(x)$	h units to the right
$g(x) = f(x + h)$	$f(x)$	h units to the left

Helpful Hint

Notice that $f(x - h)$ corresponds to a shift to the right and $f(x + h)$ corresponds to a shift to the left.

Vertical and horizontal shifts can be combined.

Practice 6

Sketch the graphs of $f(x) = |x|$ and $g(x) = |x - 2| + 3$ on the same set of axes.

Example 6 Sketch the graphs of $f(x) = x^2$ and $g(x) = (x - 2)^2 + 1$ on the same set of axes.

Solution: The graph of $g(x)$ is the same as the graph of $f(x)$ shifted 2 units to the right and 1 unit up.

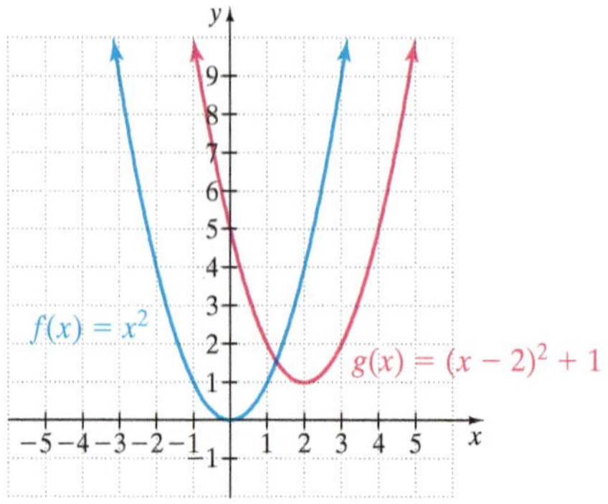

Work Practice 6

Answers

5. 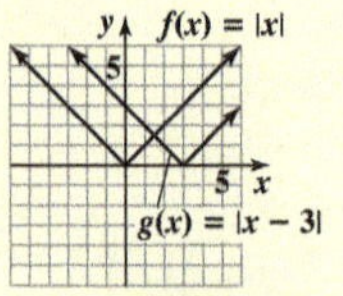

6. 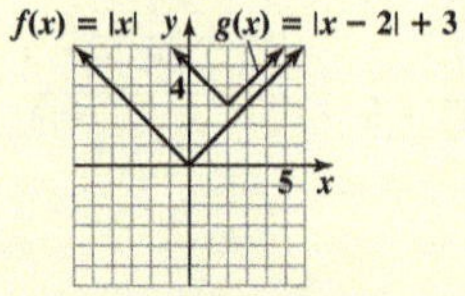

Objective 3 Reflecting Graphs 

Another type of transformation is called a **reflection.** In this section, we will study reflections (mirror images) about the x-axis only. For example, take a moment and study these two graphs. The graph of $g(x) = -x^2$ can be verified, as usual, by plotting points.

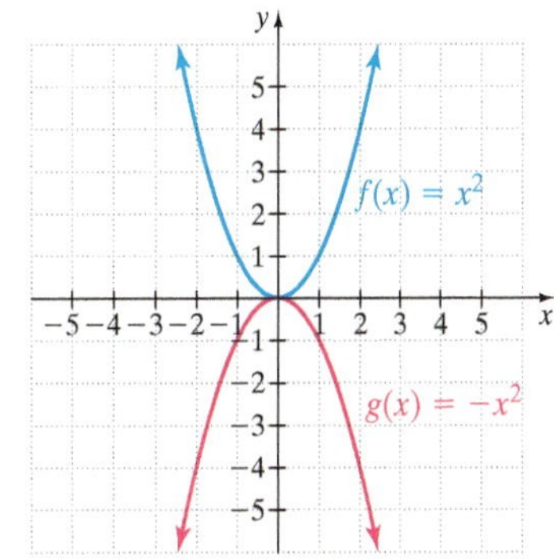

Reflection about the x-axis

The graph of $g(x) = -f(x)$ is the graph of $f(x)$ reflected about the x-axis.

Example 7 Sketch the graph of $h(x) = -|x - 3| + 2$.

Solution: The graph of $h(x) = -|x - 3| + 2$ is the same as the graph of $f(x) = |x|$ reflected about the x-axis, then moved three units to the right and two units upward.

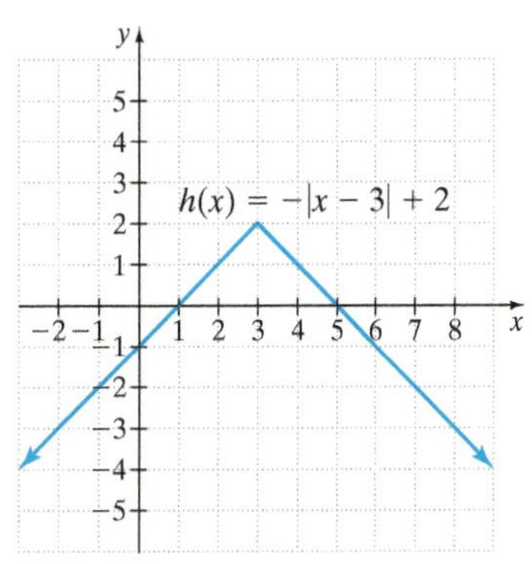

Work Practice 7

Practice 7

Sketch the graph of $h(x) = -(x + 2)^2 - 1$.

Answer

7.

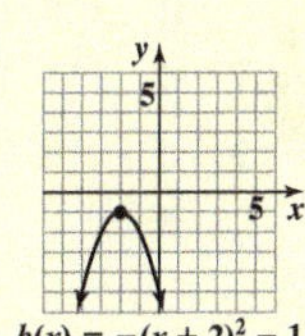

There are other transformations, such as stretching, that won't be covered in this section. For a review of this transformation, see Appendix C.

Vocabulary, Readiness & Video Check

Match each equation with its graph.

1. $y = \sqrt{x}$ **2.** $y = x^2$ **3.** $y = x$ **4.** $y = |x|$

A

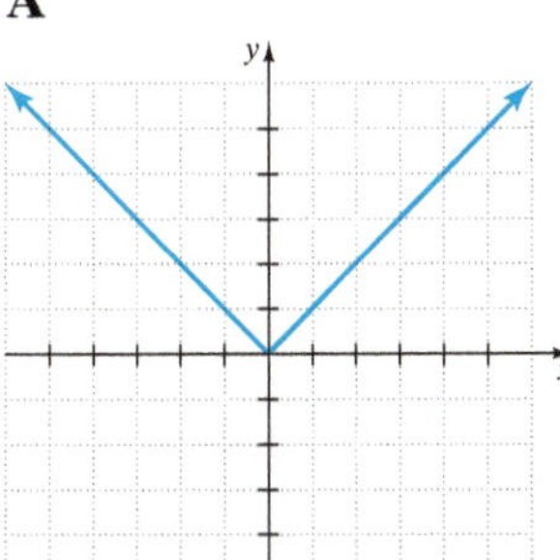

B

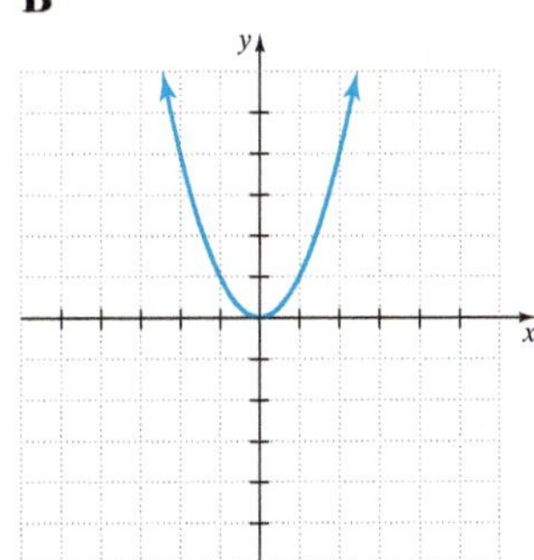

C

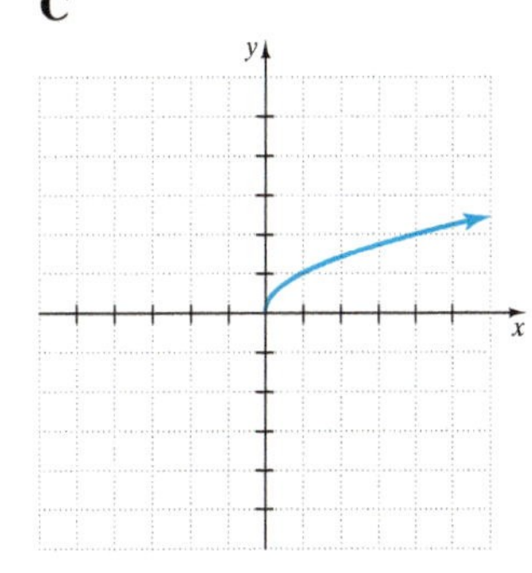

D

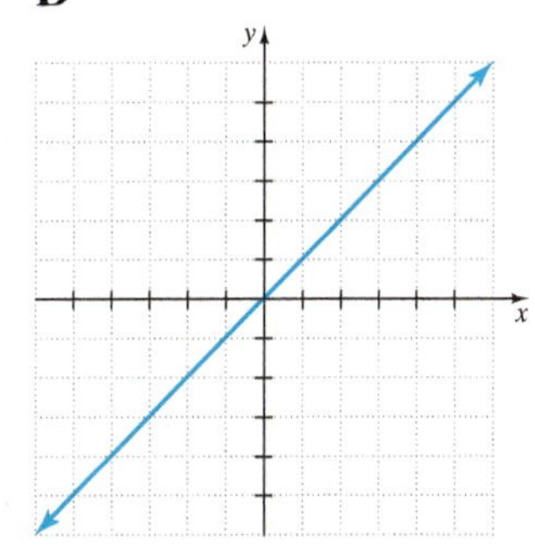

Martin-Gay Interactive Videos

See Video 15.3

Watch the section lecture video and answer the following questions.

Objective 1 **5.** In Example 1, only one piece of the function is defined for the value $x = -1$. Why do we find $f(-1)$ for $f(x) = x + 3$?

Objective 2 **6.** For Examples 2–8, why is it helpful to be familiar with common graphs and their basic shapes?

Objective 3 **7.** Based on the lecture before Example 9, complete the following statement. The graph of $f(x) = -\sqrt{x+6}$ has the same shape as the graph of $f(x) = \sqrt{x+6}$ but it is reflected about the ____________.

15.3 Exercise Set MyMathLab®

Graph each piecewise-defined function. See Examples 1 and 2.

1. $f(x) = \begin{cases} 2x + & \text{if } x < 0 \\ x + 1 & \text{if } x \ge 0 \end{cases}$

2. $f(x) = \begin{cases} 3x & \text{if } x < 0 \\ x + 2 & \text{if } x \ge 0 \end{cases}$

3. $f(x) = \begin{cases} 4x + 5 & \text{if } x \le 0 \\ \frac{1}{4}x + 2 & \text{if } x > 0 \end{cases}$

4. $f(x) = \begin{cases} 5x + 4 & \text{if } x \le 0 \\ \frac{1}{3}x - 1 & \text{if } x > 0 \end{cases}$

5. $g(x) = \begin{cases} -x & \text{if } x \le 1 \\ 2x + 1 & \text{if } x > 1 \end{cases}$

6. $g(x) = \begin{cases} 3x - 1 & \text{if } x \le 2 \\ -x & \text{if } x > 2 \end{cases}$

7. $f(x) = \begin{cases} 5 & \text{if } x < -2 \\ 3 & \text{if } x \ge -2 \end{cases}$

8. $f(x) = \begin{cases} 4 & \text{if } x < -3 \\ -2 & \text{if } x \ge -3 \end{cases}$

MIXED PRACTICE

Graph each piecewise-defined function. Use the graph to determine the domain and range of the function. See Examples 1 and 2.

9. $f(x) = \begin{cases} -2x & \text{if } x \le 0 \\ 2x + 1 & \text{if } x > 0 \end{cases}$

10. $g(x) = \begin{cases} -3x & \text{if } x \le 0 \\ 3x + 2 & \text{if } x > 0 \end{cases}$

11. $h(x) = \begin{cases} 5x - 5 & \text{if } x < 2 \\ -x + 3 & \text{if } x \ge 2 \end{cases}$

12. $f(x) = \begin{cases} 4x - 4 & \text{if } x < 2 \\ -x + 1 & \text{if } x \ge 2 \end{cases}$

13. $f(x) = \begin{cases} x + 3 & \text{if } x < -1 \\ -2x + 4 & \text{if } x \ge -1 \end{cases}$

14. $h(x) = \begin{cases} x + 2 & \text{if } x < 1 \\ 2x + 1 & \text{if } x \ge 1 \end{cases}$

15. $g(x) = \begin{cases} -2 & \text{if } x \le 0 \\ -4 & \text{if } x \ge 1 \end{cases}$

16. $f(x) = \begin{cases} -1 & \text{if } x \le 0 \\ -3 & \text{if } x \ge 2 \end{cases}$

MIXED PRACTICE

Sketch the graph of each function. See Examples 3 through 6.

17. $f(x) = |x| + 3$

18. $f(x) = |x| - 2$

19. $f(x) = \sqrt{x} - 2$

20. $f(x) = \sqrt{x} + 3$

21. $f(x) = |x - 4|$

22. $f(x) = |x + 3|$

23. $f(x) = \sqrt{x + 2}$

24. $f(x) = \sqrt{x - 2}$

25. $y = (x - 4)^2$

26. $y = (x + 4)^2$

27. $f(x) = x^2 + 4$

28. $f(x) = x^2 - 4$

29. $f(x) = \sqrt{x - 2} + 3$

30. $f(x) = \sqrt{x - 1} + 3$

31. $f(x) = |x - 1| + 5$

32. $f(x) = |x - 3| + 2$

33. $f(x) = \sqrt{x + 1} + 1$

34. $f(x) = \sqrt{x + 3} + 2$

35. $f(x) = |x + 3| - 1$

36. $f(x) = |x + 1| - 4$

37. $g(x) = (x - 1)^2 - 1$

38. $h(x) = (x + 2)^2 + 2$

39. $f(x) = (x + 3)^2 - 2$

40. $f(x) = (x + 2)^2 + 4$

Sketch the graph of each function. See Examples 3 through 7.

41. $f(x) = -(x - 1)^2$

42. $g(x) = -(x + 2)^2$

43. $h(x) = -\sqrt{x} + 3$

44. $f(x) = -\sqrt{x + 3}$

45. $h(x) = -|x + 2| + 3$

46. $g(x) = -|x + 1| + 1$

47. $f(x) = (x - 3) + 2$

48. $f(x) = (x - 1) + 4$

REVIEW AND PREVIEW

Match each equation with its graph on the next page. See Section 10.3.

49. $y = -1$

50. $x = -1$

51. $x = 3$

52. $y = 3$

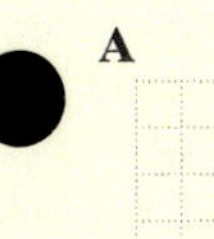

A

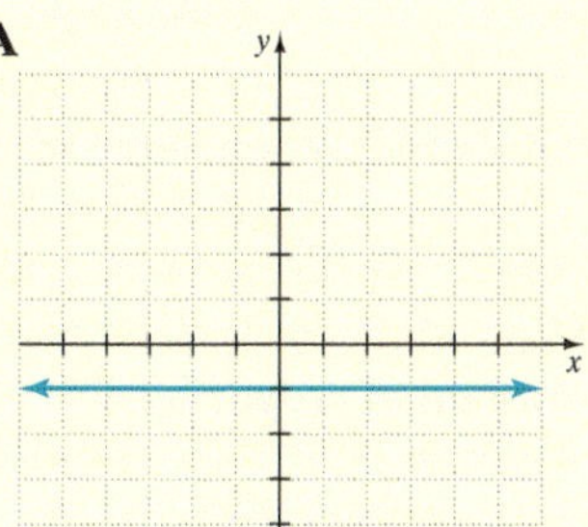

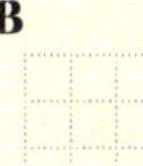

B

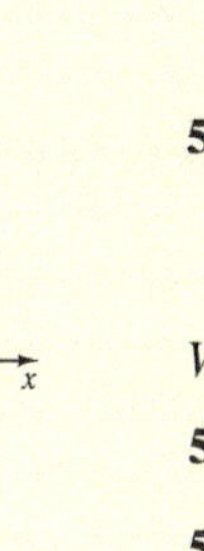

C

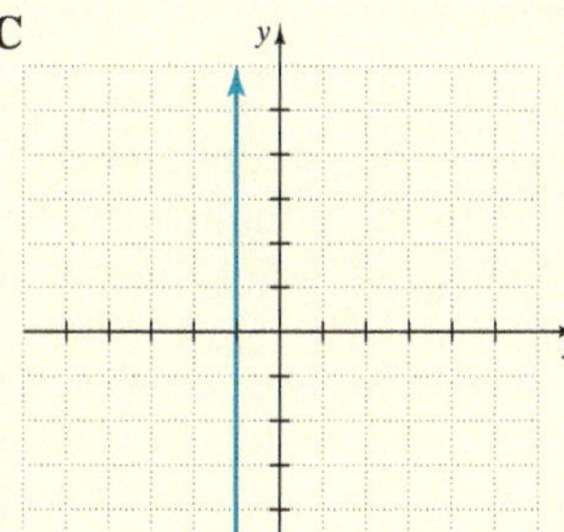

D

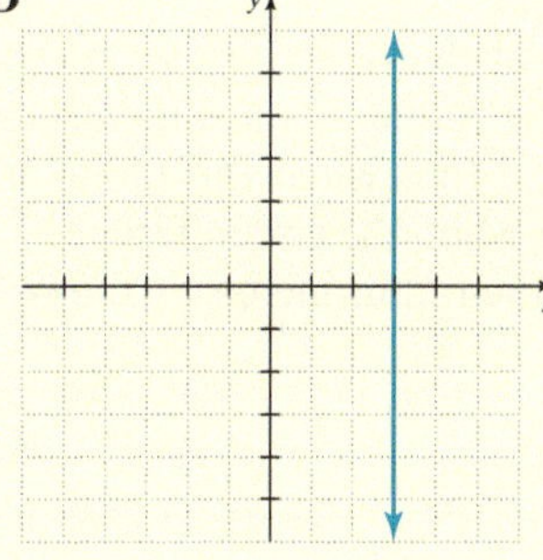

CONCEPT EXTENSIONS

53. Draw a graph whose domain is $(-\infty, 5]$ and whose range is $[2, \infty)$.

54. In your own words, describe how to graph a piecewise-defined function.

55. Graph: $f(x) = \begin{cases} -\frac{1}{2}x & \text{if } x \le 0 \\ x + 1 & \text{if } 0 < x \le 2 \\ 2x - 1 & \text{if } x > 2 \end{cases}$

56. Graph: $f(x) = \begin{cases} -\frac{1}{3}x & \text{if } x \le 0 \\ x + 2 & \text{if } 0 < x \le 4 \\ 3x - 4 & \text{if } x > 4 \end{cases}$

Write the domain and range of the following exercises.

57. Exercise 29

58. Exercise 30

59. Exercise 45

60. Exercise 46

Without graphing, find the domain of each function.

61. $f(x) = 5\sqrt{x - 20} + 1$

62. $g(x) = -3\sqrt{x + 5}$

63. $h(x) = 5|x - 20| + 1$

64. $f(x) = -3|x + 5.7|$

65. $g(x) = 9 - \sqrt{x + 103}$

66. $h(x) = \sqrt{x - 17} - 3$

Sketch the graph of each piecewise-defined function. Write the domain and range of each function.

67. $f(x) = \begin{cases} |x| & \text{if } x \le 0 \\ x^2 & \text{if } x > 0 \end{cases}$

68. $f(x) = \begin{cases} x^2 & \text{if } x < 0 \\ \sqrt{x} & \text{if } x \ge 0 \end{cases}$

69. $g(x) = \begin{cases} |x - 2| & \text{if } x < 0 \\ -x^2 & \text{if } x \ge 0 \end{cases}$

70. $g(x) = \begin{cases} -|x + 1| - 1 & \text{if } x < -2 \\ \sqrt{x + 2} - 4 & \text{if } x \ge -2 \end{cases}$

15.4 Variation and Problem Solving

Objectives

1 Solve Problems Involving Direct Variation.

2 Solve Problems Involving Inverse Variation.

3 Solve Problems Involving Joint Variation.

4 Solve Problems Involving Combined Variation.

Objective 1 Solving Problems Involving Direct Variation

A very familiar example of direct variation is the relationship of the circumference C of a circle to its radius r. The formula $C = 2\pi r$ expresses that the circumference is always 2π times the radius. In other words, C is always a constant multiple (2π) of r. Because it is, we say that **C varies directly as r,** that **C varies directly with r,** or that **C is directly proportional to r.**

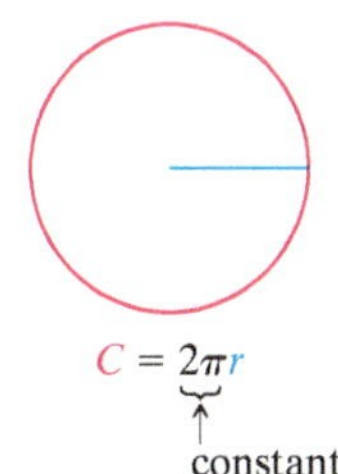

$C = 2\pi r$

constant

Direct Variation

y varies directly as x, or y is directly proportional to x, if there is a nonzero constant k such that

$$y = kx$$

The number k is called the **constant of variation** or the **constant of proportionality.**

In the definition above, the relationship described between x and y is a linear one. In other words, the graph of $y = kx$ is a line. The slope of the line is k, and the line passes through the origin.

For example, the graph of the direct variation equation $C = 2\pi r$ is shown. The horizontal axis represents the radius r, and the vertical axis is the circumference C. From the graph, we can read that when the radius is 6 units, the circumference is approximately 38 units. Also, when the circumference is 45 units, the radius is between 7 and 8 units. Notice that as the radius increases, the circumference increases.

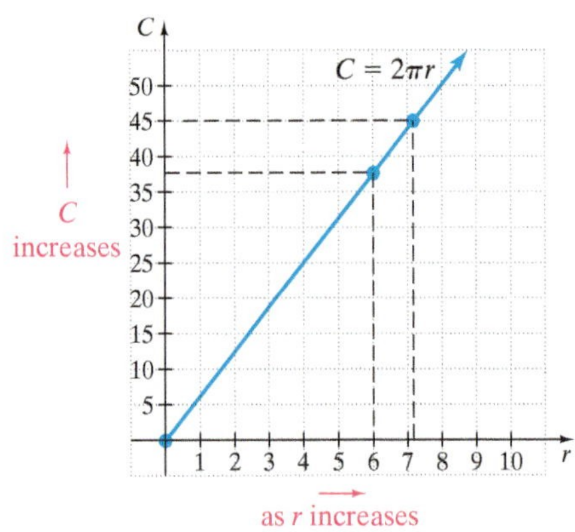

Practice 1

Suppose that y varies directly as x. If y is 20 when x is 15, find the constant of variation and the direct variation equation.

Example 1 Suppose that y varies directly as x. If y is 5 when x is 30, find the constant of variation and the direct variation equation.

Solution: Since y varies directly as x, we write $y = kx$. If $y = 5$ when $x = 30$, we have that

$$y = kx$$

$$5 = k(30) \quad \text{Replace } y \text{ with 5 and } x \text{ with 30.}$$

$$\frac{1}{6} = k \quad \text{Solve for } k.$$

The constant of variation is $\frac{1}{6}$.

After finding the constant of variation k, the direct variation equation can be written as $y = \frac{1}{6}x$.

Work Practice 1

Practice 2

Use Hooke's law as stated in Example 2. If a 36-pound weight attached to a spring stretches the spring 9 inches, find the distance that a 75-pound weight attached to the spring stretches the spring.

Example 2 Using Direct Variation and Hooke's Law

Hooke's law states that the distance a spring stretches is directly proportional to the weight attached to the spring. If a 40-pound weight attached to the spring stretches the spring 5 inches, find the distance that a 65-pound weight attached to the spring stretches the spring.

Answers

1. $k = \frac{4}{3}; y = \frac{4}{3}x$ **2.** $18\frac{3}{4}$ in.

Solution:

1. UNDERSTAND. Read and reread the problem. Notice that we are given that the distance a spring stretches is **directly proportional** to the weight attached. We let

   d = the distance stretched

   w = the weight attached

   The constant of variation is represented by k.

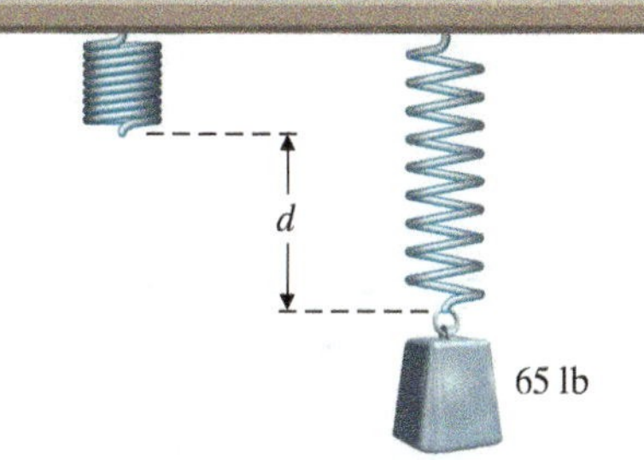

2. TRANSLATE. Because d is directly proportional to w, we write

   $d = kw$

3. SOLVE. When a weight of 40 pounds is attached, the spring stretches 5 inches. That is, when $w = 40$, $d = 5$.

   $d = kw$

   $5 = k(40)$ Replace d with 5 and w with 40.

   $\frac{1}{8} = k$ Solve for k.

   Now when we replace k with $\frac{1}{8}$ in the equation $d = kw$, we have

   $d = \frac{1}{8}w$

   To find the stretch when a weight of 65 pounds is attached, we replace w with 65 to find d.

   $$d = \frac{1}{8}(65)$$

   $$= \frac{65}{8} = 8\frac{1}{8} \quad \text{or} \quad 8.125$$

4. INTERPRET.

   Check: Check the proposed solution of 8.125 inches in the original problem.

   State: The spring stetches 8.125 inches when a 65-pound weight is attached.

Work Practice 2

Objective 2 Solving Problems Involving Inverse Variation 

When y is proportional to the **reciprocal** of another variable x, we say that **y varies inversely as x,** or that **y is inversely proportional to x.** An example of the inverse variation relationship is the relationship between the pressure that a gas exerts and the volume of its container. As the volume of a container decreases, the pressure of the gas it contains increases.

Inverse Variation

y varies inversely as x, or y is inversely proportional to x, if there is a nonzero constant k such that

$$y = \frac{k}{x}$$

The number k is called the **constant of variation** or the **constant of proportionality.**

Notice that $y = \frac{k}{x}$ is a rational equation. Its graph for $k > 0$ and $x > 0$ is shown. From the graph, we can see that as x increases, y decreases.

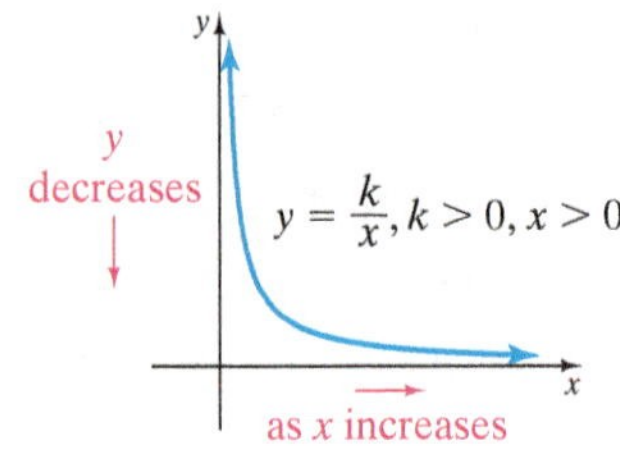

Example 3 Suppose that u varies inversely as w. If u is 3 when w is 5, find the constant of variation and the inverse variation equation.

Solution: Since u varies inversely as w, we have $u = \frac{k}{w}$. We let $u = 3$ and $w = 5$, and we solve for k.

$$u = \frac{k}{w}$$

$$3 = \frac{k}{5} \quad \text{Let } u = 3 \text{ and } w = 5.$$

$$15 = k \quad \text{Multiply both sides by 5.}$$

The constant of variation k is 15. This gives the inverse variation equation

$$u = \frac{15}{w}$$

Work Practice 3

Practice 3

Suppose that b varies inversely as a. If b is 5 when a is 9, find the constant of variation and the inverse variation equation.

Example 4 Using Inverse Variation and Boyle's Law

Boyle's law says that if the temperature stays the same, the pressure P of a gas is inversely proportional to the volume V. If a cylinder in a steam engine has a pressure of 960 kilopascals when the volume is 1.4 cubic meters, find the pressure when the volume increases to 2.5 cubic meters.

Solution:

1. UNDERSTAND. Read and reread the problem. Notice that we are given that the pressure of a gas is *inversely proportional* to the volume. We will let $P =$ the pressure and $V =$ the volume. The constant of variation is represented by k.
2. TRANSLATE. Because P is inversely proportional to V, we write

$$P = \frac{k}{V}$$

When $P = 960$ kilopascals, the volume $V = 1.4$ cubic meters. We use this information to find k.

$$960 = \frac{k}{1.4} \quad \text{Let } P = 960 \text{ and } V = 1.4.$$

$$1344 = k \quad \text{Multiply both sides by 1.4.}$$

Thus, the value of k is 1344. Replacing k with 1344 in the variation equation, we have

$$P = \frac{1344}{V}$$

Next we find P when V is 2.5 cubic meters.

Practice 4

Use Boyle's law as stated in Example 4. If $P = 350$ kilopascals when $V = 2.8$ cubic meters, find the pressure when the volume decreases to 1.5 cubic meters.

Answers

3. $k = 45$; $b = \frac{45}{a}$ 4. $653\frac{1}{3}$ kilopascals

3. SOLVE.

$$P = \frac{1344}{2.5} \quad \text{Let } V = 2.5.$$
$$= 537.6$$

4. INTERPRET.

Check: Check the proposed solution in the original problem.

State: When the volume is 2.5 cubic meters, the pressure is 537.6 kilopascals.

Work Practice 4

Objective 3 Solving Problems Involving Joint Variation

Sometimes the ratio of a variable to the product of many other variables is constant. For example, the ratio of distance traveled to the product of speed and time traveled is always 1.

$$\frac{d}{rt} = 1 \quad \text{or} \quad d = rt$$

Such a relationship is called **joint variation.**

Joint Variation

If the ratio of a variable y to the product of two or more variables is constant, then **y varies jointly as,** or **is jointly proportional to,** the other variables. If

$$y = kxz$$

then the number k is the **constant of variation** or the **constant of proportionality.**

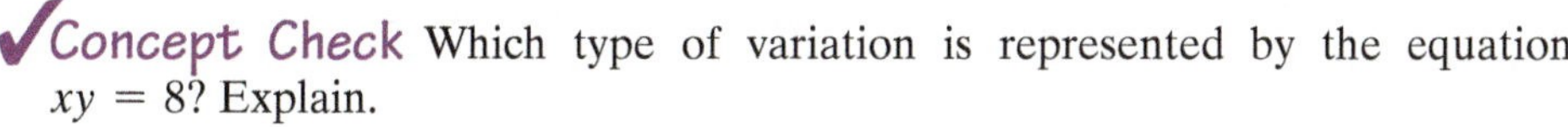

✓**Concept Check** Which type of variation is represented by the equation $xy = 8$? Explain.

a. Direct variation **b.** Inverse variation **c.** Joint variation

Example 5 Expressing Surface Area

The lateral surface area of a cylinder varies jointly as its radius and height. Express this surface area S in terms of radius r and height h.

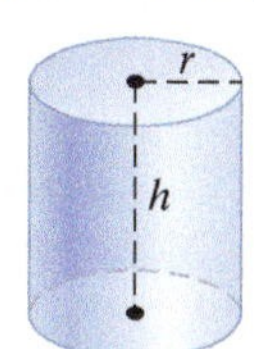

Solution: Because the surface area varies jointly as the radius r and the height h, we equate S to a constant multiple of r and h.

$$S = krh$$

In the equation $S = krh$, it can be determined that the constant k is 2π, and we then have the formula $S = 2\pi rh$. (The lateral surface area formula does not include the areas of the two circular bases.)

Work Practice 5

Practice 5

The area of a regular polygon varies jointly as its apothem and its perimeter. Express the area in terms of the apothem a and the perimeter p.

Answer

5. $A = kpa$

✓**Concept Check Answers**

b; answers may vary

Objective 4 Solving Problems Involving Combined Variation

There are many examples of variation in which y may vary directly or inversely as a *power* of x. Also, there are combinations of direct, inverse, and joint variation. We will call these variations **combined variation.**

Practice 6

Suppose that y varies inversely as the cube of x. If y is $\frac{1}{2}$ when x is 2, find the constant of variation and the variation equation.

Example 6 Suppose that y varies directly as the square of x. If y is 24 when x is 2, find the constant of variation and the variation equation.

Solution: Since y varies directly as the square of x, we have

$$y = kx^2$$

Now let $y = 24$ and $x = 2$ and solve for k.

$$y = kx^2$$
$$24 = k \cdot 2^2$$
$$24 = 4k$$
$$6 = k$$

The constant of variation is 6, so the variation equation is

$$y = 6x^2$$

Work Practice 6

Practice 7

Suppose that y varies directly as z and inversely as the cube of x. If y is 15 when $z = 5$ and $x = 3$, find the constant of variation and the variation equation.

Example 7 Finding Column Weight

The maximum weight that a circular column can support is directly proportional to the fourth power of its diameter and is inversely proportional to the square of its height. A 2-meter-diameter column that is 8 meters in height can support 1 ton. Find the weight that a 1-meter-diameter column that is 4 meters in height can support.

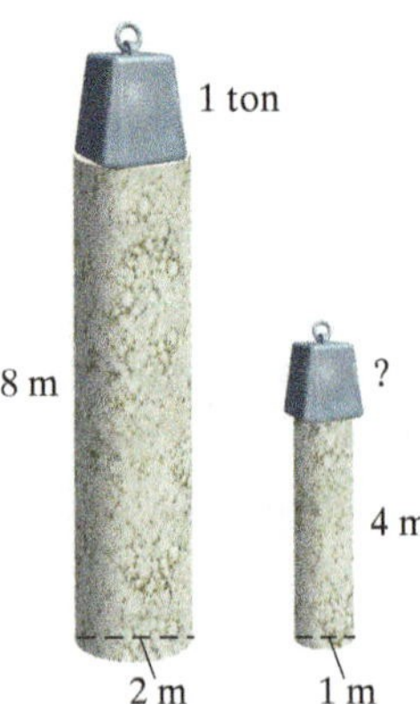

Solution:

1. UNDERSTAND. Read and reread the problem. Let w = weight, d = diameter, h = height, and k = the constant of variation.
2. TRANSLATE. Since w is directly proportional to d^4 and inversely proportional to h^2, we have

$$w = \frac{kd^4}{h^2}$$

Answers

6. $k = 4; y = \frac{4}{x^3}$ 7. $k = 81; y = \frac{81z}{x^3}$

3. SOLVE. To find k, we are given that a 2-meter-diameter column that is 8 meters in height can support 1 ton. That is, $w = 1$ when $d = 2$ and $h = 8$, or

$$1 = \frac{k \cdot 2^4}{8^2} \quad \text{Let } w = 1, d = 2, \text{ and } h = 8.$$

$$1 = \frac{k \cdot 16}{64}$$

$$4 = k \quad \text{Solve for } k.$$

Now replace k with 4 in the equation $w = \frac{kd^4}{h^2}$ and we have

$$w = \frac{4d^4}{h^2}$$

To find weight w for a 1-meter-diameter column that is 4 meters in height, let $d = 1$ and $h = 4$.

$$w = \frac{4 \cdot 1^4}{4^2}$$

$$w = \frac{4}{16} = \frac{1}{4}$$

4. INTERPRET.

Check: Check the proposed solution in the original problem.

State: The 1-meter-diameter column that is 4 meters in height can support $\frac{1}{4}$ ton of weight.

Work Practice 7

Vocabulary, Readiness & Video Check

State whether each equation represents direct, inverse, or joint variation.

1. $y = 5x$

2. $y = \frac{700}{x}$

3. $y = 5xz$

4. $y = \frac{1}{2}abc$

5. $y = \frac{9.1}{x}$

6. $y = 2.3x$

7. $y = \frac{2}{3}x$

8. $y = 3.1st$

Martin-Gay Interactive Videos *Watch the section lecture video and answer the following questions.*

See Video 15.4

Objective 1 **9.** Based on the lecture before Example 1, what kind of equation is a direct variation equation? What does k, the constant of variation, represent in this equation?

Objective 2 **10.** In Example 3, why is it not necessary to replace the given values of x and y in the inverse variation equation in order to find k?

Objective 3 **11.** Based on Example 5 and the lecture before, what is the variation equation for "y varies jointly as the square of a and the fifth power of b"?

Objective 4 **12.** From Example 6, what kind of variation does a combined variation application involve?

15.4 Exercise Set MyMathLab®

If y varies directly as x, find the constant of variation and the direct variation equation for each situation. See Example 1.

1. $y = 4$ when $x = 20$
2. $y = 9$ when $x = 54$
3. $y = 6$ when $x = 4$
4. $y = 12$ when $x = 8$
5. $y = 7$ when $x = \frac{1}{2}$
6. $y = 11$ when $x = \frac{1}{3}$
7. $y = 0.2$ when $x = 0.8$
8. $y = 0.4$ when $x = 2.5$

Solve. See Example 2.

9. The weight of a synthetic ball varies directly with the cube of its radius. A ball with a radius of 2 inches weighs 1.20 pounds. Find the weight of a ball of the same material with a 3-inch radius.

10. At sea, the distance to the horizon is directly proportional to the square root of the elevation of the observer. If a person who is 36 feet above the water can see 7.4 miles, find how far a person 64 feet above the water can see. Round to the nearest tenth of a mile.
11. The amount P of pollution varies directly with the population N of people. Kansas City has a population of 460,000 and produces about 270,000 tons of pollutants. Find how many tons of pollution we should expect St. Louis to produce if we know that its population is 319,000. Round to the nearest whole ton. (*Source:* Wikipedia)
12. Charles's law states that if the pressure P stays the same, the volume V of a gas is directly proportional to its temperature T. If a balloon is filled with 20 cubic meters of a gas at a temperature of 300 K, find the new volume if the temperature rises to 360 K while the pressure stays the same.

If y varies inversely as x, find the constant of variation and the inverse variation equation for each situation. See Example 3.

13. $y = 6$ when $x = 5$
14. $y = 20$ when $x = 9$
15. $y = 100$ when $x = 7$
16. $y = 63$ when $x = 3$
17. $y = \frac{1}{8}$ when $x = 16$
18. $y = \frac{1}{10}$ when $x = 40$
19. $y = 0.2$ when $x = 0.7$
20. $y = 0.6$ when $x = 0.3$

Solve. See Example 4.

21. Pairs of markings a set distance apart are made on highways so that police can detect drivers exceeding the speed limit. Over a fixed distance, the speed R varies inversely with the time T. In one particular pair of markings, R is 45 mph when T is 6 seconds. Find the speed of a car that travels the given distance in 5 seconds.
22. The weight of an object on or above the surface of Earth varies inversely as the square of the distance between the object and Earth's center. If a person weighs 160 pounds on Earth's surface, find the individual's weight if he moves 200 miles above Earth. Round to the nearest whole pound. (Assume that Earth's radius is 4000 miles.)

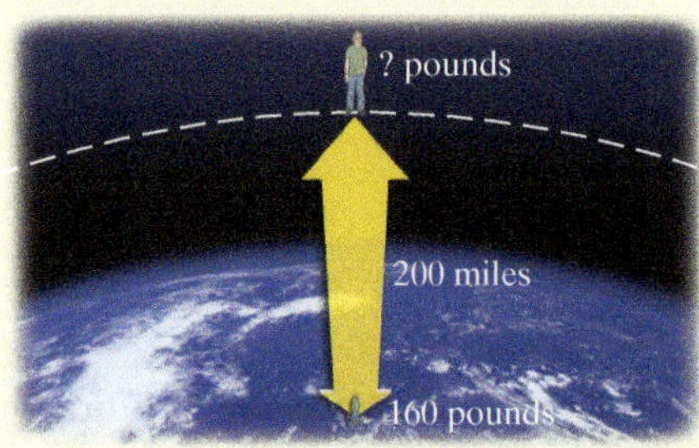

23. If the voltage V in an electric circuit is held constant, the current I is inversely proportional to the resistance R. If the current is 40 amperes when the resistance is 270 ohms, find the current when the resistance is 150 ohms.
24. Because it is more efficient to produce larger numbers of items, the cost of producing a certain computer DVD is inversely proportional to the number produced. If 4000 can be produced at a cost of \$1.20 each, find the cost per DVD when 6000 are produced.

25. The intensity I of light varies inversely as the square of the distance d from the light source. If the distance from the light source is doubled (see the figure), determine what happens to the intensity of light at the new location.

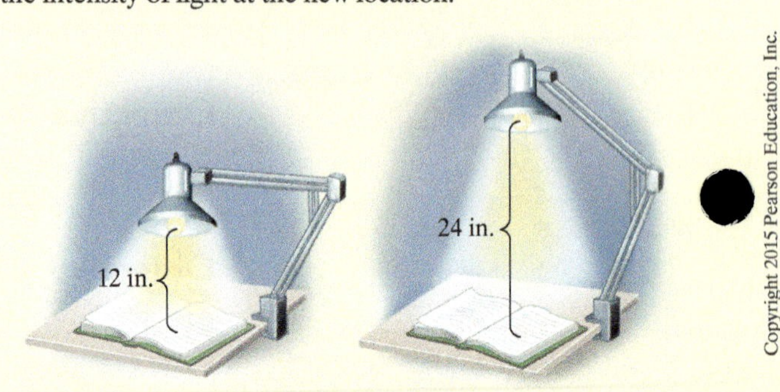

26. The maximum weight that a circular column can hold is inversely proportional to the square of its height. If an 8-foot column can hold 2 tons, find how much weight a 10-foot column can hold.

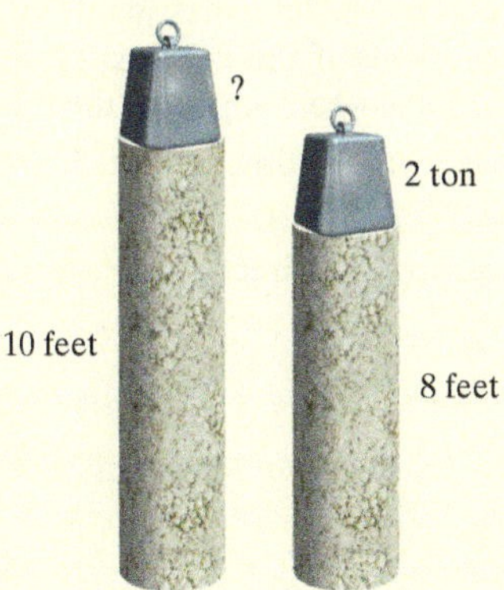

MIXED PRACTICE

Write each statement as an equation. Use k as the constant of variation. See Example 5.

27. x varies jointly as y and z
28. P varies jointly as R and the square of S
29. r varies jointly as s and the cube of t
30. a varies jointly as b and c

For each statement, find the constant of variation and the variation equation. See Examples 5 and 6.

31. y varies directly as the cube of x; $y = 9$ when $x = 3$
32. y varies directly as the cube of x; $y = 32$ when $x = 4$
33. y varies directly as the square root of x; $y = 0.4$ when $x = 4$
34. y varies directly as the square root of x; $y = 2.1$ when $x = 9$
35. y varies inversely as the square of x; $y = 0.052$ when $x = 5$
36. y varies inversely as the square of x; $y = 0.011$ when $x = 10$
37. y varies jointly as x and the cube of z; $y = 120$ when $x = 5$ and $z = 2$
38. y varies jointly as x and the square of z; $y = 360$ when $x = 4$ and $z = 3$

Solve. See Example 7.

39. The maximum weight that a rectangular beam can support varies jointly as its width and the square of its height and inversely as its length. If a beam $\frac{1}{2}$ foot wide, $\frac{1}{3}$ foot high, and 10 feet long can support 12 tons, find how much a similar beam can support if the beam is $\frac{2}{3}$ foot wide, $\frac{1}{2}$ foot high, and 16 feet long.

40. The number of cars manufactured on an assembly line at a General Motors plant varies jointly as the number of workers and the time they work. If 200 workers can produce 60 cars in 2 hours, find how many cars 240 workers should be able to make in 3 hours.

41. The volume of a cone varies jointly as its height and the square of its radius. If the volume of a cone is 32π cubic inches when the radius is 4 inches and the height is 6 inches, find the volume of a cone when the radius is 3 inches and the height is 5 inches.

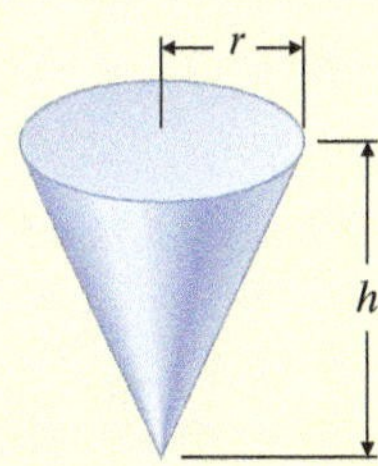

42. When a wind blows perpendicularly against a flat surface, its force is jointly proportional to the surface area and the speed of the wind. A sail whose surface area is 12 square feet experiences a 20-pound force when the wind speed is 10 miles per hour. Find the force on an 8-square-foot sail if the wind speed is 12 miles per hour.

43. The intensity of light (in foot-candles) varies inversely as the square of x, the distance in feet from the light source. The intensity of light 2 feet from the source is 80 foot-candles. How far away is the source if the intensity of light is 5 foot-candles?

44. The horsepower that can be safely transmitted to a shaft varies jointly as the shaft's angular speed of rotation (in revolutions per minute) and the cube of its diameter. A 2-inch shaft making 120 revolutions per minute safely transmits 40 horsepower. Find how much horsepower can be safely transmitted by a 3-inch shaft making 80 revolutions per minute.

MIXED PRACTICE

Write an equation to describe each variation. Use k for the constant of proportionality. See Examples 1 through 7.

45. y varies directly as x
46. p varies directly as q
47. a varies inversely as b
48. y varies inversely as x
49. y varies jointly as x and z
50. y varies jointly as q, r, and t
51. y varies inversely as x^3
52. y varies inversely as a^4
53. y varies directly as x and inversely as p^2
54. y varies directly as a^5 and inversely as b

REVIEW AND PREVIEW

Find the exact circumference and area of each circle. See the inside cover for a list of geometric formulas.

55. 56.

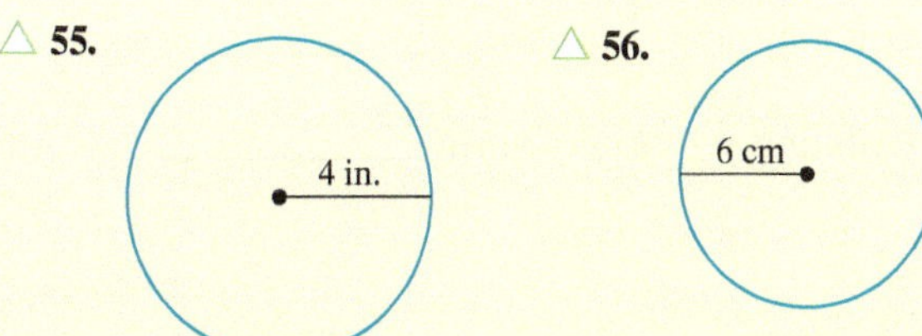

△ **57.**

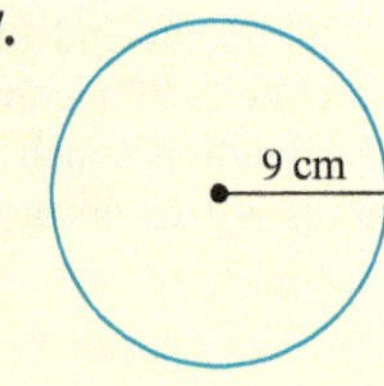

△ **58.**

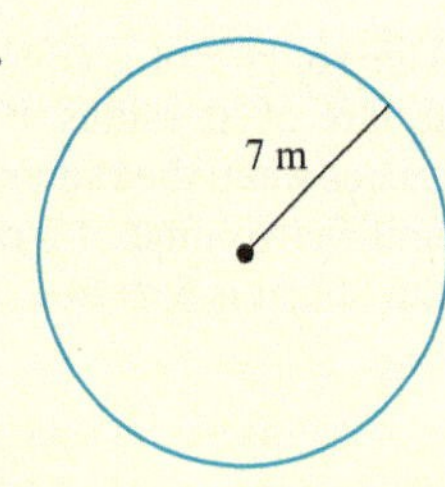

Simplify. See Sections 2.1, 4.3, 4.5, and 5.2.

59. $|-1.2|$

60. $|-3|$

61. $-|7|$

62. $|0|$

63. $-\left|-\frac{1}{2}\right|$

64. $-\left|\frac{1}{5}\right|$

65. $\left(\frac{2}{3}\right)^3$

66. $\left(\frac{5}{11}\right)^2$

CONCEPT EXTENSIONS

Solve. See the Concept Check in this section. Choose the type of variation that each equation represents. **a.** *Direct variation* **b.** *Inverse variation* **c.** *Joint variation*

67. $y = \frac{2}{3}x$

68. $y = \frac{0.6}{x}$

69. $y = 9ab$

70. $xy = \frac{2}{11}$

71. The horsepower required to drive a boat varies directly as the cube of the speed of the boat. If the speed of the boat is to double, determine the corresponding increase in horsepower required.

72. The volume of a cylinder varies jointly as the height and the square of the radius. If the height is halved and the radius is doubled, determine what happens to the volume.

73. Suppose that y varies directly as x. If x is doubled, what is the effect on y?

74. Suppose that y varies directly as x^2. If x is doubled, what is the effect on y?

Complete the following table for the inverse variation $y = \frac{k}{x}$ over each given value of k. Plot the points on a rectangular coordinate system.

x	$\frac{1}{4}$	$\frac{1}{2}$	1	2	4
$y = \frac{k}{x}$					

75. $k = 3$

76. $k = 1$

77. $k = \frac{1}{2}$

78. $k = 5$

Chapter 15 Vocabulary Check

Fill in each blank with one of the words or phrases listed below.

slope-intercept	directly	slope
jointly	parallel	perpendicular
function	inversely	linear function

1. ______________ lines have the same slope and different y-intercepts.

2. ______________ form of a linear equation in two variables is $y = mx + b$.

3. A(n) ______________ is a relation in which each first component in the ordered pairs corresponds to exactly one second component.

4. In the equation $y = 4x - 2$, the coefficient of x is the ______________ of its corresponding graph.

5. Two lines are ______________ if the product of their slopes is -1.

6. A(n) ______________ is a function that can be written in the form $f(x) = mx + b$.

7. In the equation $y = kx$, y varies ______________ as x.

8. In the equation $y = \frac{k}{x}$, y varies ______________ as x.

9. In the equation $y = kxz$, y varies ______________ as x and z.

Helpful Hint Are you preparing for your test? Don't forget to take the Chapter 15 Test on page 1100. Then check your answers at the back of the text and use the Chapter Test Prep Videos to see the fully worked-out solutions to any of the exercises you want to review.

15 Chapter Highlights

Definitions and Concepts	Examples

Section 15.1 Graphing and Writing Linear Functions

A **linear function** is a function that can be written in the form $f(x) = mx + b$.

Linear functions

$$f(x) = -3,\ g(x) = 5x,\ h(x) = -\frac{1}{3}x - 7$$

To graph a linear function, find three ordered pair solutions. Graph the solutions and draw a line through the plotted points.

Graph $f(x) = -2x$.

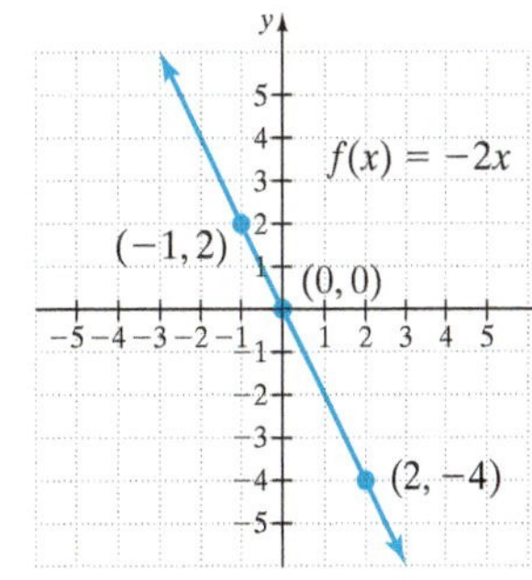

x	y or $f(x)$
−1	2
0	0
2	−4

The point-slope form of the equation of a line is $y - y_1 = m(x - x_1)$, where m is the slope of the line and (x_1, y_1) is a point on the line.

Find an equation of the line parallel to $g(x) = 2x - 1$ and containing the point $(1, -4)$. Write the equation using function notation.

Since we want a parallel line, use the same slope of $g(x)$, which is 2.

$$y - y_1 = m(x - x_1)$$
$$y - (-4) = 2(x - 1)$$
$$y + 4 = 2x - 2$$
$$y = 2x - 6 \quad \text{Solve for } y.$$
$$f(x) = 2x - 6 \quad \text{Let } y = f(x).$$

Section 15.2 Reviewing Function Notation and Graphing Nonlinear Functions

The graph of $y = mx + b$ is the same as the graph of $y = mx$, but shifted $|b|$ units up if b is positive and $|b|$ units down if b is negative.

Graph $g(x) = -2x + 3$.

This is the same as the graph of $f(x) = -2x$ shifted 3 units up.

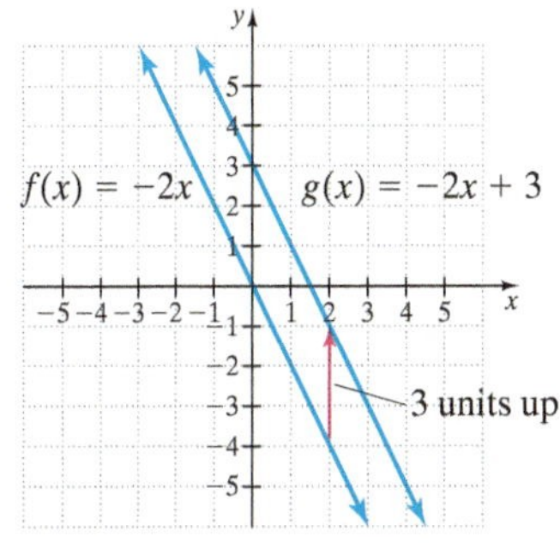

(continued)

Definitions and Concepts	Examples

Section 15.2 Reviewing Function Notation and Graphing Nonlinear Functions (*continued*)

To graph a function that is not linear, find a sufficient number of ordered pair solutions so that a pattern may be discovered.

Graph $f(x) = x^2 + 2$.

x	y or $f(x)$
−2	6
−1	3
0	2
1	3
2	6

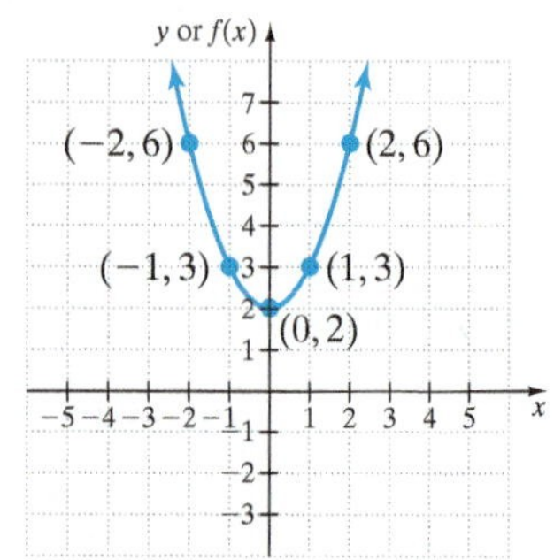

Section 15.3 Graphing Piecewise-Defined Functions and Shifting and Reflecting Graphs of Functions

Vertical shifts (upward or downward):
Let k be a positive number.

Graph of	Same as	Moved
$g(x) = f(x) + k$	$f(x)$	k units upward
$g(x) = f(x) - k$	$f(x)$	k units downward

Horizontal shift (to the left or right):
Let h be a positive number.

Graph of	Same as	Moved
$g(x) = f(x - h)$	$f(x)$	h units to the right
$g(x) = f(x + h)$	$f(x)$	h units to the left

Reflection about the x-axis

The graph of $g(x) = -f(x)$ is the graph of $f(x)$ reflected about the x-axis.

The graph of $h(x) = -|x - 3| + 1$ is the same as the graph of $f(x) = |x|$, reflected about the x-axis, shifted 3 units right, then 1 unit up.

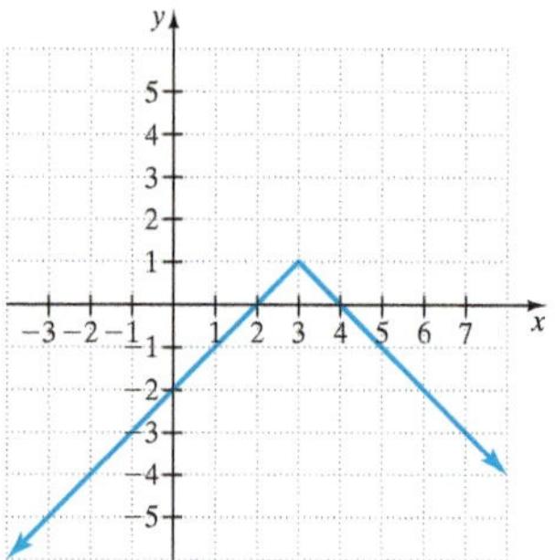

Section 15.4 Variation and Problem Solving

y **varies directly as** x, or y is **directly proportional to** x, if there is a nonzero constant k such that

$$y = kx$$

The circumference of a circle C varies directly as its radius r.

$$C = \underbrace{2\pi}_{k} r$$

y **varies inversely as** x, or y is **inversely proportional to** x, if there is a nonzero constant k such that

$$y = \frac{k}{x}$$

Pressure P varies inversely with volume V.

$$P = \frac{k}{V}$$

y **varies jointly as** x and z or y is **jointly proportional to** x and z if there is a nonzero constant k such that

$$y = kxz$$

The lateral surface area S of a cylinder varies jointly as its radius r and height h.

$$S = \underbrace{2\pi}_{k} rh$$

Chapter 15 Review

(15.1) *Graph each linear function.*

1. $f(x) = x$

2. $f(x) = -\frac{1}{3}x$

3. $g(x) = 4x - 1$

4. $F(x) = -\frac{2}{3}x + 2$

The graph of $f(x) = 3x$ is sketched below. Use this graph to match each linear function with its graph.

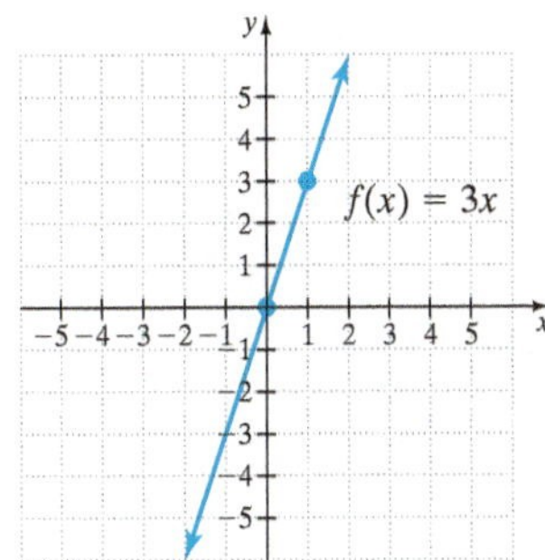

A

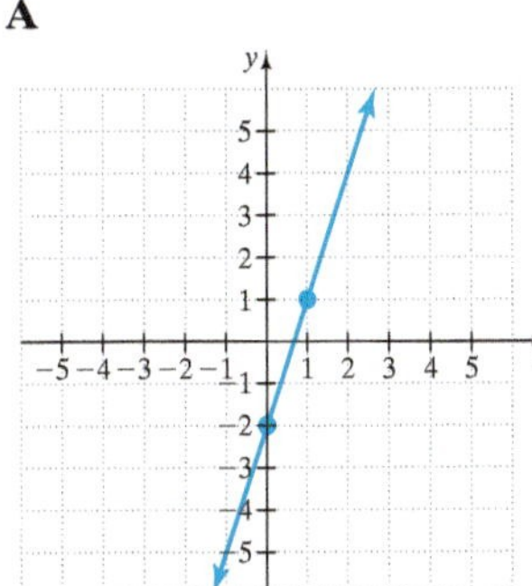

B

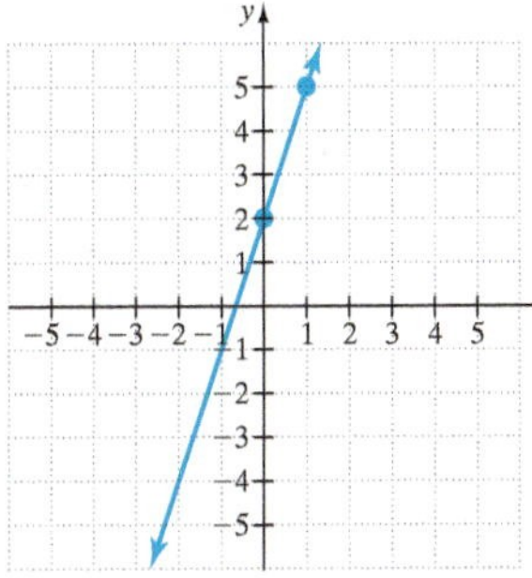

C

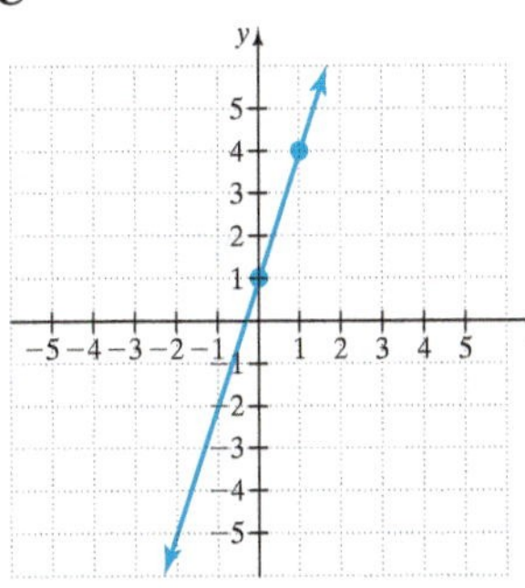

D

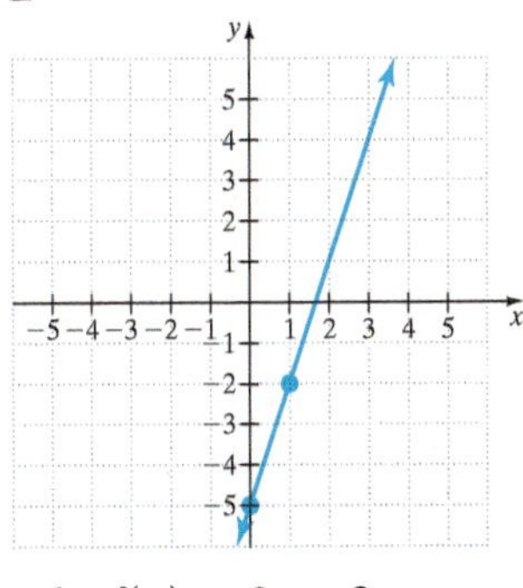

5. $f(x) = 3x + 1$

6. $f(x) = 3x - 2$

7. $f(x) = 3x + 2$

8. $f(x) = 3x - 5$

Find the slope and y-intercept of each function.

9. $f(x) = \frac{2}{5}x - \frac{4}{3}$

10. $f(x) = -\frac{2}{7}x + \frac{3}{2}$

Find the standard form equation of each line that satisfies the given conditions.

11. Slope 2; through $(5, -2)$

12. Through $(-3, 5)$; slope 3

13. Through $(-5, 3)$ and $(-4, -8)$

14. Through $(-6, -1)$ and $(-4, -2)$

15. Through $(-2, -5)$; parallel to $y = 8$

16. Through $(-2, 3)$; perpendicular to $x = 4$

Find the equation of each line that satisfies the given conditions. Write each equation using function notation.

17. Horizontal; through $(3, -1)$

18. Slope $-\frac{2}{3}$; y-intercept $(0, 4)$

19. Slope -1; y-intercept $(0, -2)$

△ **20.** Through $(2, -6)$; parallel to $6x + 3y = 5$

△ **21.** Through $(-4, -2)$; parallel to $3x + 2y = 8$

△ **22.** Through $(-6, -1)$; perpendicular to $4x + 3y = 5$

23. Through $(-4, 5)$; perpendicular to $2x - 3y = 6$

24. The value of an automobile bought in 2006 continues to decrease as time passes. Two years after the car was bought, it was worth \$17,500; four years after it was bought, it was worth \$14,300.

a. Assuming that this relationship between the number of years past 2006 and the value of the car is linear, write an equation describing this relationship. [*Hint:* Use ordered pairs of the form (years past 2006, value of the automobile).]

b. Use this equation to estimate the value of the automobile in 2012.

△ **25.** The value of a building bought in 2000 continues to increase as time passes. Seven years after the building was bought, it was worth \$210,000; 12 years after it was bought, it was worth \$270,000.

a. Assuming that this relationship between the number of years past 2000 and the value of the building is linear, write an equation describing this relationship. [*Hint:* Use ordered pairs of the form (years past 2000, value of the building).]

b. Use this equation to estimate the value of the building in 2018.

26. Decide whether the lines are parallel, perpendicular, or neither.

$$-x + 3y = 2$$
$$6x - 18y = 3$$

(15.2) *Use the graph of the function on the next page to answer Exercises 27 through 30.*

27. Find $f(-1)$.

28. Find $f(1)$.

29. Find all values of x such that $f(x) = 1$.

30. Find all values of x such that $f(x) = -1$.

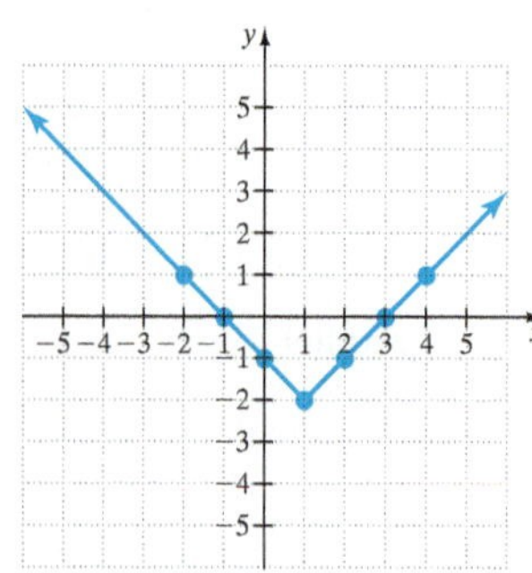

Determine whether each function is linear or not. Then graph the function.

31. $f(x) = 3x$

32. $f(x) = 5x$

33. $g(x) = |x| + 4$

34. $h(x) = x^2 + 4$

35. $F(x) = -\frac{1}{2}x + 2$

36. $G(x) = -x + 5$

37. $y = -1.36x$

38. $y = 2.1x + 5.9$

39. $H(x) = (x - 2)^2$

40. $f(x) = -|x - 3|$

(15.3) *Graph each function.*

41. $g(x) = \begin{cases} -\frac{1}{5}x & \text{if } x \le -1 \\ -4x + 2 & \text{if } x > -1 \end{cases}$

42. $f(x) = \begin{cases} -3x & \text{if } x < 0 \\ x - 3 & \text{if } x \ge 0 \end{cases}$

Graph each function.

43. $f(x) = \sqrt{x - 4}$

44. $y = \sqrt{x} - 4$

45. $h(x) = -(x + 3)^2 - 1$

46. $g(x) = |x - 2| - 2$

(15.4) *Solve each variation problem.*

47. A is directly proportional to B. If $A = 6$ when $B = 14$, find A when $B = 21$.

48. C is inversely proportional to D. If $C = 12$ when $D = 8$, find C when $D = 24$.

49. According to Boyle's law, the pressure exerted by a gas is inversely proportional to the volume, as long as the temperature stays the same. If a gas exerts a pressure of 1250 pounds per square inch when the volume is 2 cubic feet, find the volume when the pressure is 800 pounds per square inch.

△ **50.** The surface area of a sphere varies directly as the square of its radius. If the surface area is 36π square inches when the radius is 3 inches, find the surface area when the radius is 4 inches.

MIXED REVIEW

Write an equation of the line that satisfies each set of conditions. Write the equation in the form $f(x) = mx + b$.

51. Slope 0; through $\left(-4, \frac{9}{2}\right)$

52. Slope $\frac{3}{4}$; through $(-8, -4)$

53. Through $(-3, 8)$ and $(-2, 3)$

54. Through $(-6, 1)$; parallel to $y = -\frac{3}{2}x + 11$

55. Through $(-5, 7)$; perpendicular to $5x - 4y = 10$

Graph each piecewise-defined function.

56. $g(x) = \begin{cases} 4x - 3 & \text{if } x \le 1 \\ 2x & \text{if } x > 1 \end{cases}$

57. $f(x) = \begin{cases} x - 2 & \text{if } x \le 0 \\ -\frac{x}{3} & \text{if } x \ge 3 \end{cases}$

Graph each function.

58. $f(x) = |x + 1| - 3$

59. $f(x) = \sqrt{x - 2}$

60. y is inversely proportional to x. If $y = 14$ when $x = 6$, find y when $x = 21$.

Chapter 15 Test

Step-by-step test solutions are found on the Chapter Test Prep Videos. Where available: MyMathLab® or YouTube

Use the graph of the function f to find each value.

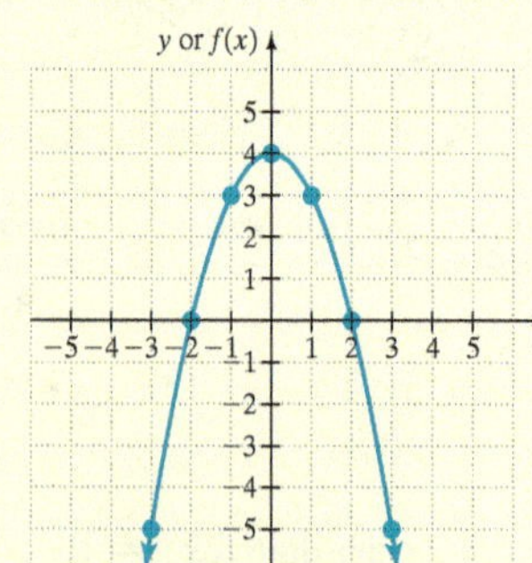

1. Find $f(1)$.

2. Find $f(-3)$.

3. Find all values of x such that $f(x) = 0$.

4. Find all values of x such that $f(x) = 4$.

Graph each line.

5. $2x - 3y = -6$

6. $f(x) = \frac{2}{3}x$

Find an equation of each line that satisfies the given conditions. Write Exercises 7–9 in standard form. Write Exercises 10–12 using function notation.

7. Horizontal; through $(2, -8)$

8. Through $(4, -1)$; slope -3

9. Through $(0, -2)$; slope 5

10. Through $(4, -2)$ and $(6, -3)$

△ **11.** Through $(-1, 2)$; perpendicular to $3x - y = 4$

△ **12.** Parallel to $2y + x = 3$; through $(3, -2)$

13. Line L_1 has the equation $2x - 5y = 8$. Line L_2 passes through the points $(1, 4)$ and $(-1, -1)$. Determine whether these lines are parallel lines, perpendicular lines, or neither.

Find the domain and range of each relation. Also determine whether the relation is a function.

14.

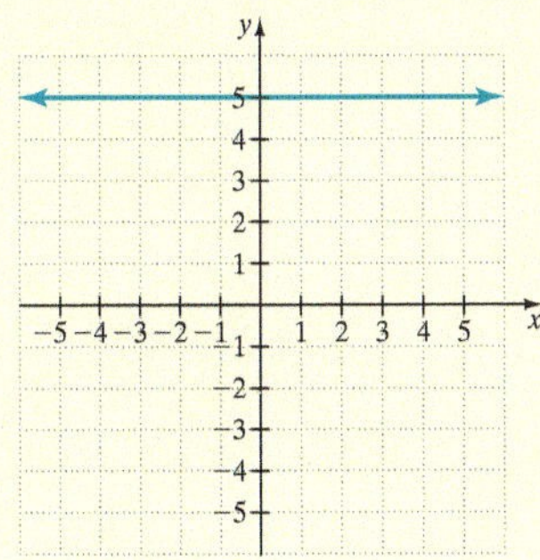

15.

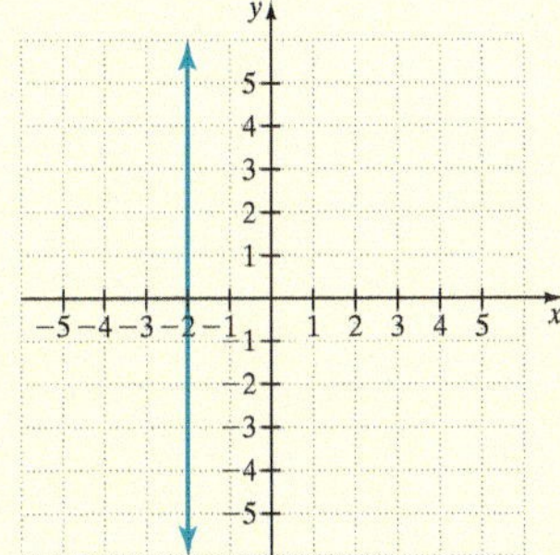

△ **16.**

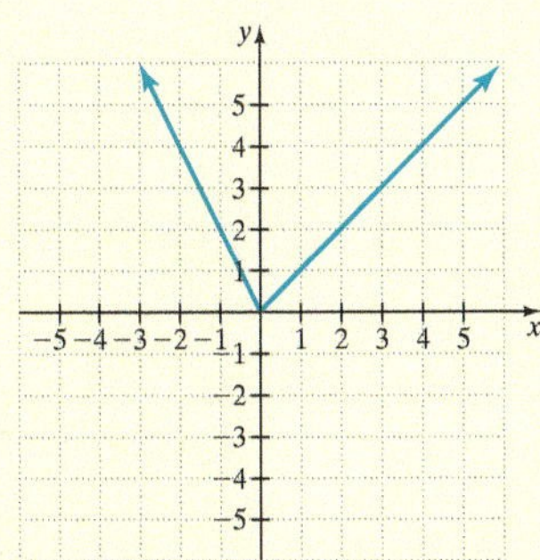

△ **17.**

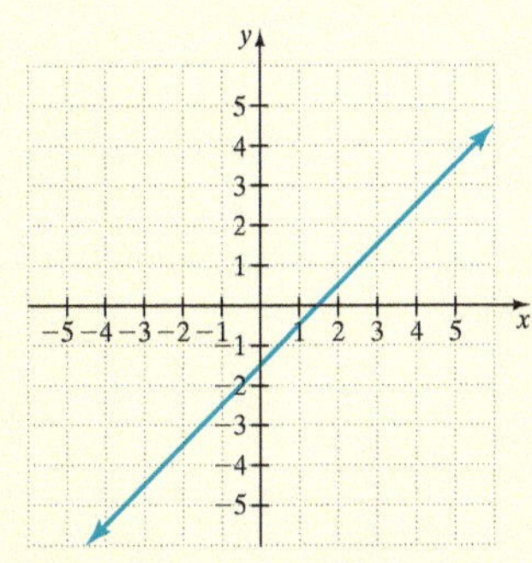

18. For the 2009 Major League Baseball season, the following linear function describes the relationship between a team's payroll x (in millions of dollars) and the number of games y that team won during the regular season.

$$f(x) = 0.096x + 72.81$$

Round to the nearest whole. (*Sources:* Based on data from Major League Baseball and *USA Today*)

a. According to this equation, how many games would have been won during the 2009 season by a team with a payroll of \$90 million?

b. The Baltimore Orioles had a payroll of \$67 million in 2009. According to this equation, how many games would they have won during the season?

c. According to this equation, what payroll would have been necessary in 2009 to win 95 games during the season?

d. Find and interpret the slope of the equation.

Graph each function. For Exercises 19 and 21, state the domain and the range of the function.

19. $f(x) = \begin{cases} -\frac{1}{2}x & \text{if } x \le 0 \\ 2x - 3 & \text{if } x > 0 \end{cases}$

20. $f(x) = (x - 4)^2$

21. $g(x) = -|x + 2| - 1$

22. $h(x) = \sqrt{x} - 1$

23. Suppose that W is inversely proportional to V. If $W = 20$ when $V = 12$, find W when $V = 15$.

24. Suppose that Q is jointly proportional to R and the square of S. If $Q = 24$ when $R = 3$ and $S = 4$, find Q when $R = 2$ and $S = 3$.

25. When an anvil is dropped into a gorge, the speed with which it strikes the ground is directly proportional to the square root of the distance it falls. An anvil that falls 400 feet hits the ground at a speed of 160 feet per second. Find the height of a cliff over the gorge if a dropped anvil hits the ground at a speed of 128 feet per second.

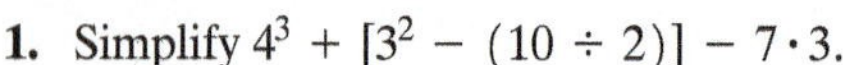

Chapters 1–15 Cumulative Review

1. Simplify $4^3 + [3^2 - (10 \div 2)] - 7 \cdot 3$.
2. Simplify $5^2 + [8 \div (6 - 2)] + 4 \cdot 3$.

Evaluate each expression for $x = -4$.

3. $7 - x^2$
4. $2x^3 - 7$

Solve.

5. $-5x = 15$
6. $\frac{2}{3}y = 7$
7. $8(2 - t) = -5t$
8. $5x - 9 = 5x - 29$
9. Solve $y = mx + b$ for x.
10. Solve $y = 7x - 2$ for x.
11. $-4x + 7 \geq -9$ Write the solution using interval notation.
12. $-5x - 6 < 3x + 1$ Write the solution using interval notation.
13. Find the slope of the line $y = -1$.
14. Find the slope of a line parallel to the line passing through the points $(0, 7)$ and $(-1, 0)$.
15. Given $g(x) = x^2 - 3$, find the following. Then write the corresponding ordered pairs generated.
    a. $g(2)$ **b.** $g(-2)$
    c. $g(0)$
16. Given $f(x) = 3 - x^2$, find the following. Then write the corresponding ordered pairs generated.
    a. $f(2)$ **b.** $f(-2)$
    c. $f(0)$
17. Solve the system: $\begin{cases} 2x + y = 10 \\ x = y + 2 \end{cases}$
18. Solve the system: $\begin{cases} 3y = x + 10 \\ 2x + 5y = 24 \end{cases}$
19. Solve the system: $\begin{cases} -x - \frac{y}{2} = \frac{5}{2} \\ \frac{x}{6} - \frac{y}{2} = 0 \end{cases}$
20. Solve the system: $\begin{cases} \frac{x}{2} + y = \frac{5}{6} \\ 2x - y = \frac{5}{6} \end{cases}$
21. Divide $x^2 + 7x + 12$ by $x + 3$ using long division.
22. Divide: $\frac{5x^2y - 6xy + 2}{6xy}$

Factor out the GCF (greatest common factor).

23. **a.** $6t + 18$
    b. $y^5 - y^7$
24. **a.** $5y - 20$
    b. $z^{10} - z^3$

Factor completely.

25. $x^2 + 4x - 12$
26. $x^2 - 10x + 21$
27. $10x^2 - 13xy - 3y^2$
28. $12a^2 + 5ab - 2b^2$
29. $x^3 + 8$
30. $y^3 - 27$

Solve.

31. $x^2 - 9x - 22 = 0$
32. $y^2 - 5y = -6$

Simplify.

33. **a.** $\frac{2x^2}{10x^3 - 2x^2}$
    b. $\frac{9x^2 + 13x + 4}{8x^2 + x - 7}$
34. **a.** $\frac{33x^4y^2}{3xy}$
    b. $\frac{9y}{90y^2 + 9y}$

Perform the indicated operations.

35. $\frac{3x + 3}{5x - 5x^2} \cdot \frac{2x^2 + x - 3}{4x^2 - 9}$
36. $\frac{2x}{x - 6} - \frac{x + 6}{x - 6}$
37. $\frac{3x^2 + 2x}{x - 1} - \frac{10x - 5}{x - 1}$
38. $\frac{9}{y^2} - 4y$

Solve.

39. $3 - \frac{6}{x} = x + 8$
40. $\frac{x}{2} + \frac{x}{5} = \frac{x - 7}{20}$

Find an equation of the line through the given points. Write the equation using function notation.

41. $(4, 0)$ and $(-4, -5)$
42. $(-1, 3)$ and $(-2, 7)$

16

Inequalities and Absolute Value

Sections

Check Your Progress

Today, it seems that most people in the world want to stay connected most of the time. In fact, 86% of U.S. citizens own cell phones. Also, computers with Internet access are just as important in our lives. Thus, the merging of these two into Wi-Fi-enabled cell phones might be the next big technological explosion. In Section 16.3, Exercises 83 and 84, you will find the estimated increase in the number of Wi-Fi-enabled cell phones in the United States as well as the percent increase. (*Source:* Techcrunchies.com)

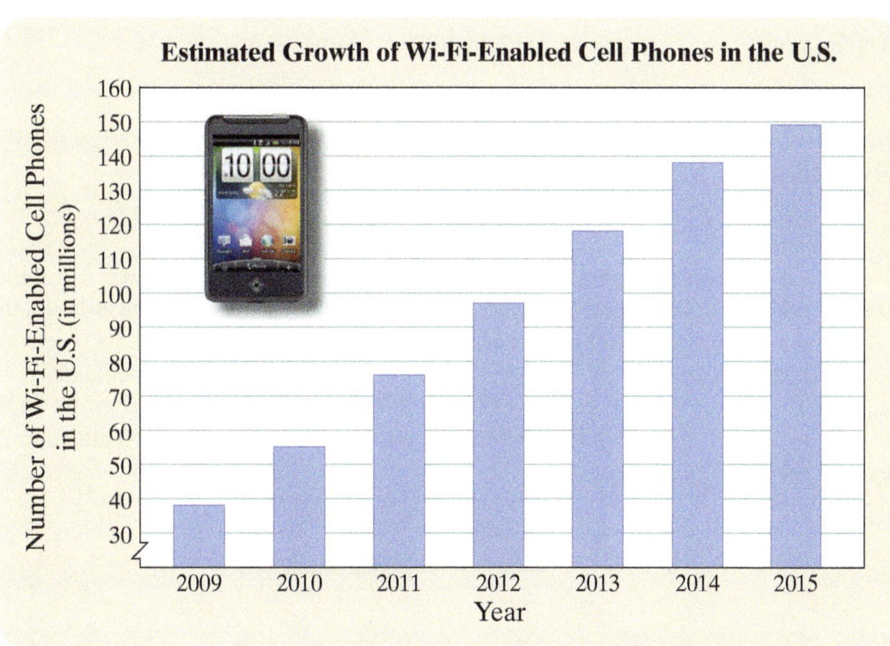

Mathematics is a tool for solving problems in such diverse fields as transportation, engineering, economics, medicine, business, and biology. We solve problems using mathematics by modeling real-world phenomena with mathematical equations or inequalities. Our ability to solve problems using mathematics, then, depends in part on our ability to solve different types of equations and inequalities. This chapter includes solving absolute value equations and inequalities and other types of inequalities.

16.1 Compound Inequalities

Objectives

1. Find the Intersection of Two Sets.
2. Solve Compound Inequalities Containing **and**.
3. Find the Union of Two Sets.
4. Solve Compound Inequalities Containing **or**.

Two inequalities joined by the words **and** or **or** are called **compound inequalities.**

Compound Inequalities

$x + 3 < 8 \quad and \quad x > 2$

$\frac{2x}{3} \geq 5 \quad or \quad -x + 10 < 7$

Objective 1 Finding the Intersection of Two Sets

The solution set of a compound inequality formed by the word **and** is the **intersection** of the solution sets of the two inequalities. We use the symbol $\cap$ to represent "intersection."

Intersection of Two Sets

The intersection of two sets, A and B, is the set of all elements common to both sets. A intersect B is denoted by $A \cap B$.

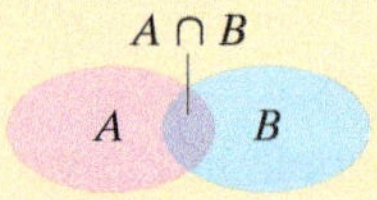

Example 1 If $A = \{x|x$ is an even number greater than 0 and less than 10$\}$ and $B = \{3, 4, 5, 6\}$, find $A \cap B$.

Solution: Let's list the elements in set A.

$A = \{2, 4, 6, 8\}$

The numbers 4 and 6 are in sets A and B. The intersection is $\{4, 6\}$.

Work Practice 1

Practice 1

If $A = \{x \mid x$ is an odd number greater than 0 and less than 10$\}$ and $B = \{1, 2, 3, 4\}$, find $A \cap B$.

Objective 2 Solving Compound Inequalities Containing "and"

A value is a solution of a compound inequality formed by the word **and** if it is a solution of *both* inequalities. For example, the solution set of the compound inequality $x \leq 5$ and $x \geq 3$ contains all values of x that make the inequality $x \leq 5$ a true statement **and** the inequality $x \geq 3$ a true statement. The first graph shown below is the graph of $x \leq 5$, the second graph is the graph of $x \geq 3$, and the third graph shows the intersection of the two graphs. The third graph is the graph of $x \leq 5$ **and** $x \geq 3$.

$\{x|x \leq 5\}$

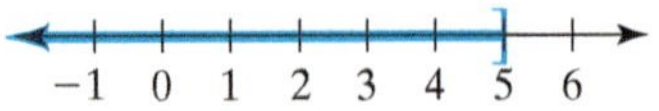

$(-\infty, 5]$

$\{x|x \geq 3\}$

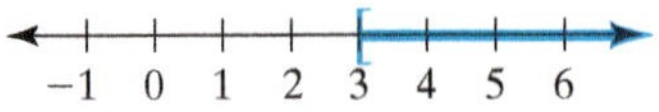

$[3, \infty)$

$\{x|x \leq 5 \text{ and } x \geq 3\}$
also $\{x|3 \leq x \leq 5\}$
(see below)

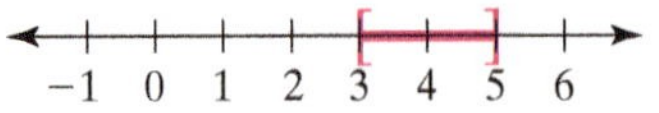

$[3, 5]$

Since $x \geq 3$ is the same as $3 \leq x$, the compound inequality $3 \leq x$ and $x \leq 5$ can be written in a more compact form as $3 \leq x \leq 5$. The solution set $\{x \mid 3 \leq x \leq 5\}$

Answer
1. $\{1, 3\}$

includes all numbers that are greater than or equal to 3 and at the same time are less than or equal to 5.

In interval notation, the set $\{x \mid x \le 5 \text{ and } x \ge 3\}$ or the set $\{x \mid 3 \le x \le 5\}$ is written as $[3, 5]$.

Helpful Hint

Don't forget that some compound inequalities containing "and" can be written in a more compact form.

Compound Inequality	Compact Form	Interval Notation
$2 \le x$ *and* $x \le 6$	$2 \le x \le 6$	$[2, 6]$
Graph:	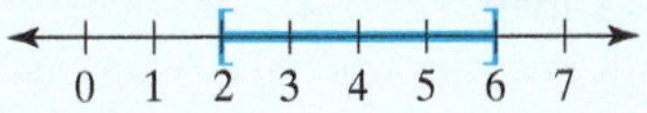	

Example 2 Solve: $x - 7 < 2$ *and* $2x + 1 < 9$.

Solution: First we solve each inequality separately.

$$\begin{aligned} x - 7 &< 2 \quad \text{and} \quad 2x + 1 < 9 \\ x &< 9 \quad \text{and} \quad 2x < 8 \\ x &< 9 \quad \text{and} \quad x < 4 \end{aligned}$$

Now we can graph the two intervals on two number lines and find their intersection. Their intersection is shown on the third number line.

$\{x \mid x < 9\}$ $(-\infty, 9)$

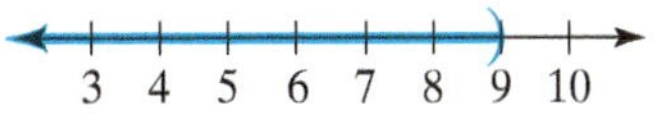

$\{x \mid x < 4\}$ $(-\infty, 4)$

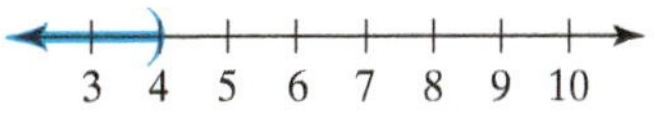

$\{x \mid x < 9 \text{ and } x < 4\} = \{x \mid x < 4\}$ $(-\infty, 4)$

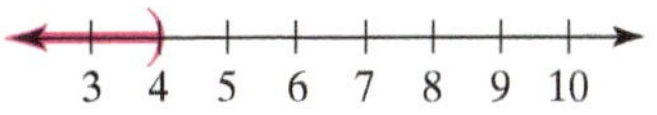

The solution set is $(-\infty, 4)$.

Work Practice 2

Practice 2

Solve: $x + 3 < 8$ *and* $2x - 1 < 3$. Write the solution set in interval notation.

Example 3 Solve: $2x \ge 0$ *and* $4x - 1 \le -9$.

Solution: First we solve each inequality separately.

$$\begin{aligned} 2x &\ge 0 \quad \text{and} \quad 4x - 1 \le -9 \\ x &\ge 0 \quad \text{and} \quad 4x \le -8 \\ x &\ge 0 \quad \text{and} \quad x \le -2 \end{aligned}$$

Now we can graph the two intervals and find their intersection.

$\{x \mid x \ge 0\}$ 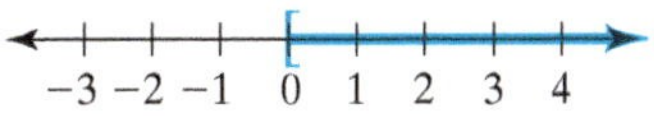 $[0, \infty)$

$\{x \mid x \le -2\}$ 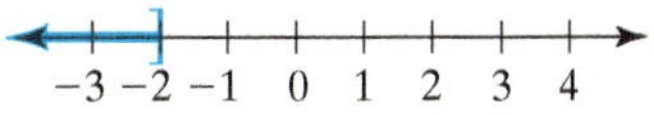 $(-\infty, -2]$

$\{x \mid x \ge 0 \text{ and } x \le -2\} = \varnothing$  $\varnothing$

There is no number that is greater than or equal to 0 *and* less than or equal to -2. The solution set is $\varnothing$.

Work Practice 3

Practice 3

Solve: $4x \le 0$ *and* $3x + 2 > 8$. Write the solution set in interval notation.

Answers

2. $(-\infty, 2)$ **3.** $\{\ \}$ or $\varnothing$

Helpful Hint

Example 3 shows that some compound inequalities have no solution. Also, some have all real numbers as solutions.

To solve a compound inequality written in a compact form, such as $2 < 4 - x < 7$, we get x alone in the "middle part." Since a compound inequality is really two inequalities in one statement, we must perform the same operations on all three parts of the inequality. For example:

$$2 < 4 - x < 7 \quad \text{means} \quad 2 < 4 - x \quad \textit{and} \quad 4 - x < 7.$$

Practice 4

Solve: $3 < 5 - x < 9$. Write the solution set in interval notation.

Example 4

Solve: $2 < 4 - x < 7$.

Solution: To get x alone, we first subtract 4 from all three parts.

$$2 < 4 - x < 7$$

$$2 - 4 < 4 - x - 4 < 7 - 4 \qquad \text{Subtract 4 from all three parts.}$$

$$-2 < -x < 3 \qquad \text{Simplify.}$$

$$\frac{-2}{-1} > \frac{-x}{-1} > \frac{3}{-1} \qquad \text{Divide all three parts by } -1 \text{ and reverse the inequality symbols.}$$

$$2 > x > -3$$

Helpful Hint Don't forget to reverse both inequality symbols.

This is equivalent to $-3 < x < 2$.

The solution set in interval notation is $(-3, 2)$, and its graph is shown.

−4 −3 −2 −1 0 1 2 3

Work Practice 4

Practice 5

Solve: $-4 \le \frac{x}{2} - 1 \le 3$.

Write the solution set in interval notation.

Example 5

Solve: $-1 \le \frac{2x}{3} + 5 \le 2$.

Solution: First, clear the inequality of fractions by multiplying all three parts by the LCD, 3.

$$-1 \le \frac{2x}{3} + 5 \le 2$$

$$3(-1) \le 3\left(\frac{2x}{3} + 5\right) \le 3(2) \qquad \text{Multiply all three parts by the LCD, 3.}$$

$$-3 \le 2x + 15 \le 6 \qquad \text{Use the distributive property and multiply.}$$

$$-3 - 15 \le 2x + 15 - 15 \le 6 - 15 \qquad \text{Subtract 15 from all three parts.}$$

$$-18 \le 2x \le -9 \qquad \text{Simplify.}$$

$$\frac{-18}{2} \le \frac{2x}{2} \le \frac{-9}{2} \qquad \text{Divide all three parts by 2.}$$

$$-9 \le x \le -\frac{9}{2} \qquad \text{Simplify.}$$

The graph of the solution is shown.

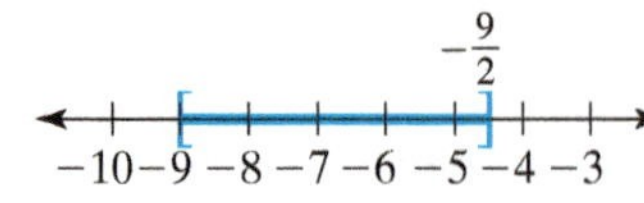

The solution set in interval notation is $\left[-9, -\frac{9}{2}\right]$.

Work Practice 5

Answers

4. $(-4, 2)$ 5. $[-6, 8]$

Objective 3 Finding the Union of Two Sets

The solution set of a compound inequality formed by the word **or** is the **union** of the solution sets of the two inequalities. We use the symbol $\cup$ to denote "union."

Union of Two Sets

The **union** of two sets, A and B, is the set of elements that belong to *either* of the sets. A union B is denoted by $A \cup B$.

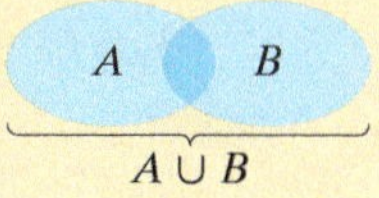

Helpful Hint The word *either* in this definition means "one or the other or both."

Example 6 If $A = \{x \mid x$ is an even number greater than 0 and less than $10\}$ and $B = \{3, 4, 5, 6\}$, find $A \cup B$.

Solution: Recall from Example 1 that $A = \{2, 4, 6, 8\}$. The numbers that are in either set or both sets are $\{2, 3, 4, 5, 6, 8\}$. This set is the union.

Work Practice 6

Practice 6

If $A = \{x \mid x$ is an odd number greater than 0 and less than $10\}$ and $B = \{2, 3, 4, 5, 6\}$, find $A \cup B$.

Objective 4 Solving Compound Inequalities Containing "or"

A value is a solution of a compound inequality formed by the word **or** if it is a solution of **either** inequality. For example, the solution set of the compound inequality $x \le 1$ **or** $x \ge 3$ contains all numbers that make the inequality $x \le 1$ a true statement **or** the inequality $x \ge 3$ a true statement.

$\{x \mid x \le 1\}$

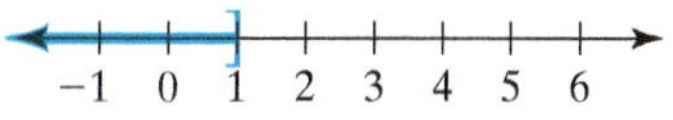

$(-\infty, 1]$

$\{x \mid x \ge 3\}$

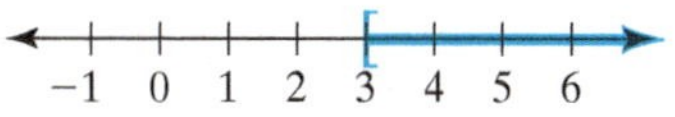

$[3, \infty)$

$\{x \mid x \le 1 \text{ or } x \ge 3\}$

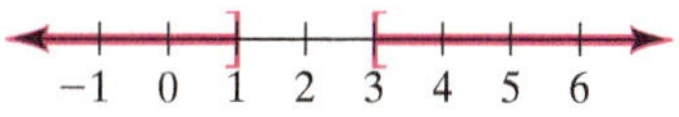

$(-\infty, 1] \cup [3, \infty)$

In interval notation, the set $\{x \mid x \le 1 \text{ or } x \ge 3\}$ is written as $(-\infty, 1] \cup [3, \infty)$.

Example 7 Solve: $5x - 3 \le 10$ *or* $x + 1 \ge 5$.

Solution: First we solve each inequality separately.

$$5x - 3 \le 10 \quad or \quad x + 1 \ge 5$$
$$5x \le 13 \quad or \quad x \ge 4$$
$$x \le \frac{13}{5} \quad or \quad x \ge 4$$

Now we can graph each interval and find their union.

$\left\{x \,\middle|\, x \le \frac{13}{5}\right\}$

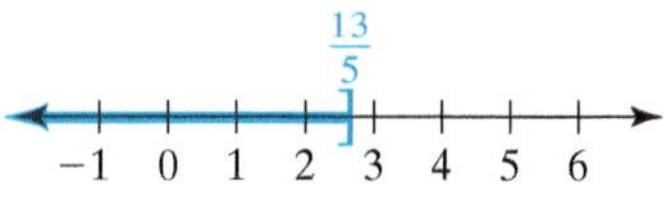

$\left(-\infty, \frac{13}{5}\right]$

$\{x \mid x \ge 4\}$

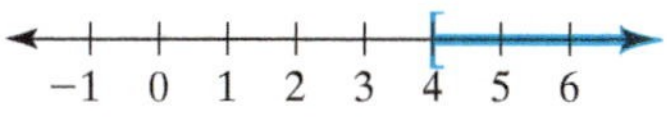

$[4, \infty)$

(*Continued on next page*)

Practice 7

Solve: $8x + 5 \le 8$ or $x - 1 \ge 2$. Write the solution set in interval notation.

Answers

6. $\{1, 2, 3, 4, 5, 6, 7, 9\}$

7. $\left(-\infty, \frac{3}{8}\right] \cup [3, \infty)$

$\left\{x \middle| x \le \frac{13}{5} \text{ or } x \ge 4\right\}$ $\left(-\infty, \frac{13}{5}\right] \cup [4, \infty)$

The solution set is $\left(-\infty, \frac{13}{5}\right] \cup [4, \infty)$.

Work Practice 7

Practice 8

Solve: $-3x - 2 > -8$ *or* $5x > 0$. Write the solution set in interval notation.

Example 8 Solve: $-2x - 5 < -3$ *or* $6x < 0$.

Solution: First we solve each inequality separately.

$$-2x - 5 < -3 \quad \text{or} \quad 6x < 0$$
$$-2x < 2 \quad \text{or} \quad x < 0$$
$$x > -1 \quad \text{or} \quad x < 0$$

Now we can graph each interval and find their union.

$\{x|x > -1\}$ $(-1, \infty)$

$\{x|x < 0\}$ $(-\infty, 0)$

$\{x|x > -1 \text{ or } x < 0\}$ = all real numbers $(-\infty, \infty)$

The solution set is $(-\infty, \infty)$.

Work Practice 8

Answer

8. $(-\infty, \infty)$

✓Concept Check Answer

b is not correct

✓Concept Check Which of the following is *not* a correct way to represent the set of all numbers between −3 and 5?

a. $\{x \mid -3 < x < 5\}$

b. $-3 < x$ *or* $x < 5$

c. $(-3, 5)$

d. $x > -3$ *and* $x < 5$

Vocabulary, Readiness & Video Check

Use the choices below to fill in each blank.

or	$\cup$	$\varnothing$
and	$\cap$	compound

1. Two inequalities joined by the words "and" or "or" are called ________ inequalities.

2. The word ________ means intersection.

3. The word ________ means union.

4. The symbol ________ represents intersection.

5. The symbol ________ represents union.

6. The symbol ________ is the empty set.

Martin-Gay Interactive Videos — *Watch the section lecture video and answer the following questions.*

See Video 16.1

Objective 1 7. Based on Example 1 and the lecture before, complete the following statement. For an element to be in the intersection of sets A and B, the element must be in set A __________ in set B.

Objective 2 8. In Example 2, how can using three number lines help us find the solution to this "and" compound inequality?

Objective 3 9. Based on Example 4 and the lecture before, complete the following statement. For an element to be in the union of sets A and B, the element must be in set A __________ in set B.

Objective 4 10. In Example 5, how can using three number lines help us find the solution to this "or" compound inequality?

16.1 Exercise Set MyMathLab®

MIXED PRACTICE

If $A = \{x|x$ is an even integer$\}$, $B = \{x|x$ *is an odd integer*$\}$, $C = \{2, 3, 4, 5\}$, *and* $D = \{4, 5, 6, 7\}$, *list the elements of each set. See Examples 1 and 6.*

1. $C \cup D$
2. $C \cap D$
3. $A \cap D$
4. $A \cup D$
5. $A \cup B$
6. $A \cap B$
7. $B \cap D$
8. $B \cup D$
9. $B \cup C$
10. $B \cap C$
11. $A \cap C$
12. $A \cup C$

Solve each compound inequality. Graph the solution set and write it in interval notation. See Examples 2 and 3.

13. $x < 1$ *and* $x > -3$
14. $x \le 0$ *and* $x \ge -2$
15. $x \le -3$ *and* $x \ge -2$
16. $x < 2$ *and* $x > 4$
17. $x < -1$ *and* $x < 1$
18. $x \ge -4$ *and* $x > 1$

Solve each compound inequality. Write solutions in interval notation. See Examples 2 and 3.

19. $x + 1 \ge 7$ *and* $3x - 1 \ge 5$
20. $x + 2 \ge 3$ *and* $5x - 1 \ge 9$
21. $4x + 2 \le -10$ *and* $2x \le 0$
22. $2x + 4 > 0$ *and* $4x > 0$
23. $-2x < -8$ *and* $x - 5 < 5$
24. $-7x \le -21$ *and* $x - 20 \le -15$

Solve each compound inequality. See Examples 4 and 5.

25. $5 < x - 6 < 11$
26. $-2 \le x + 3 \le 0$
27. $-2 \le 3x - 5 \le 7$
28. $1 < 4 + 2x < 7$
29. $1 \le \frac{2}{3}x + 3 \le 4$
30. $-2 < \frac{1}{2}x - 5 < 1$
31. $-5 \le \frac{-3x + 1}{4} \le 2$
32. $-4 \le \frac{-2x + 5}{3} \le 1$

Solve each compound inequality. Graph the solution set and write it in interval notation. See Examples 7 and 8.

33. $x < 4$ *or* $x < 5$
34. $x \ge -2$ *or* $x \le 2$
35. $x \le -4$ *or* $x \ge 1$
36. $x < 0$ *or* $x < 1$
37. $x > 0$ *or* $x < 3$
38. $x \ge -3$ *or* $x \le -4$

Solve each compound inequality. Write solutions in interval notation. See Examples 7 and 8.

39. $-2x \le -4$ *or* $5x - 20 \ge 5$
40. $-5x \le 10$ *or* $3x - 5 \ge 1$
41. $x + 4 < 0$ *or* $6x > -12$
42. $x + 9 < 0$ *or* $4x > -12$
43. $3(x - 1) < 12$ *or* $x + 7 > 10$
44. $5(x - 1) \ge -5$ *or* $5 + x \le 11$

MIXED PRACTICE

Solve each compound inequality. Write solutions in interval notation. See Examples 1 through 8.

45. $x < \frac{2}{3}$ *and* $x > -\frac{1}{2}$
46. $x < \frac{5}{7}$ *and* $x < 1$
47. $x < \frac{2}{3}$ *or* $x > -\frac{1}{2}$

48. $x < \frac{5}{7}$ *or* $x < 1$

49. $0 \le 2x - 3 \le 9$

50. $3 < 5x + 1 < 11$

51. $\frac{1}{2} < x - \frac{3}{4} < 2$

52. $\frac{2}{3} < x + \frac{1}{2} < 4$

53. $x + 3 \ge 3$ *and* $x + 3 \le 2$

54. $2x - 1 \ge 3$ *and* $-x > 2$

55. $3x \ge 5$ *or* $-\frac{5}{8}x - 6 > 1$

56. $\frac{3}{8}x + 1 \le 0$ *or* $-2x < -4$

57. $0 < \frac{5 - 2x}{3} < 5$

58. $-2 < \frac{-2x - 1}{3} < 2$

59. $-6 < 3(x - 2) \le 8$

60. $-5 < 2(x + 4) < 8$

61. $-x + 5 > 6$ *and* $1 + 2x \le -5$

62. $5x \le 0$ and $-x + 5 < 8$

63. $3x + 2 \le 5$ *or* $7x > 29$

64. $-x < 7$ *or* $3x + 1 < -20$

65. $5 - x > 7$ *and* $2x + 3 \ge 13$

66. $-2x < -6$ *or* $1 - x > -2$

67. $-\frac{1}{2} \le \frac{4x - 1}{6} < \frac{5}{6}$

68. $-\frac{1}{2} \le \frac{3x - 1}{10} < \frac{1}{2}$

69. $\frac{1}{15} < \frac{8 - 3x}{15} < \frac{4}{5}$

70. $-\frac{1}{4} < \frac{6 - x}{12} < -\frac{1}{6}$

71. $0.3 < 0.2x - 0.9 < 1.5$

72. $-0.7 \le 0.4x + 0.8 < 0.5$

REVIEW AND PREVIEW

Evaluate the following. See Sections 2.2 and 2.3.

73. $|-7| - |19|$

74. $|-7 - 19|$

75. $-(-6) - |-10|$

76. $|-4| - (-4) + |-20|$

Find by inspection all values for x that make each equation true.

77. $|x| = 7$

78. $|x| = 5$

79. $|x| = 0$

80. $|x| = -2$

CONCEPT EXTENSIONS

Use the graph to answer Exercises 81 and 82.

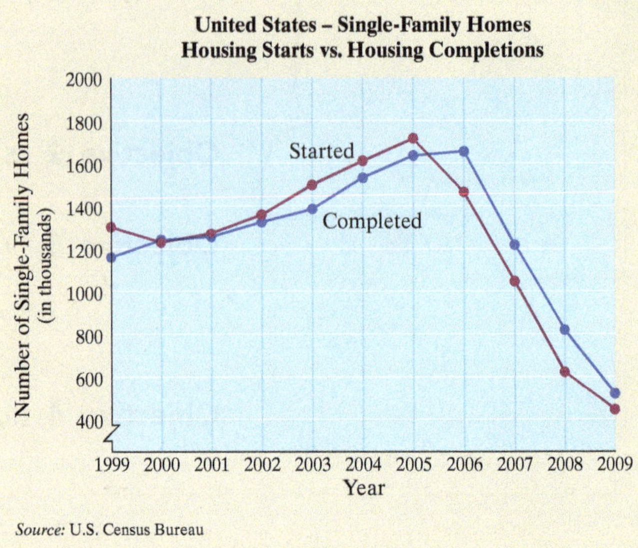

81. For which years were the number of single-family housing starts greater than 1500 and the number of single-family home completions greater than 1500?

82. For which years were the number of single-family housing starts less than 1000 or the number of single-family housing completions greater than 1500?

83. In your own words, describe how to find the union of two sets.

84. In your own words, describe how to find the intersection of two sets.

*Solve each compound inequality for x. See the example below. To solve $x - 6 < 3x < 2x + 5$, notice that this inequality contains a variable not only in the middle but also on the left and the right. When this occurs, we solve by rewriting the inequality using the word **and**.*

$$x - 6 < 3x \quad and \quad 3x < 2x + 5$$
$$-6 < 2x \quad and \quad x < 5$$
$$-3 < x$$
$$x > -3 \quad and \quad x < 5$$

−4 −3 −2 −1 0 1 2 3 4 5 6

$x > -3$

−4 −3 −2 −1 0 1 2 3 4 5 6

$x < 5$

−4 −3 −2 −1 0 1 2 3 4 5 6

$-3 < x < 5$ or $(-3, 5)$

85. $2x - 3 < 3x + 1 < 4x - 5$

86. $x + 3 < 2x + 1 < 4x + 6$

87. $-3(x - 2) \le 3 - 2x \le 10 - 3x$
88. $7x - 1 \le 7 + 5x \le 3(1 + 2x)$
89. $5x - 8 < 2(2 + x) < -2(1 + 2x)$
90. $1 + 2x < 3(2 + x) < 1 + 4x$

The formula for converting Fahrenheit temperatures to Celsius temperatures is $C = \frac{5}{9}(F - 32)$. Use this formula for Exercises 91 and 92.

91. During a recent year, the temperatures in Chicago ranged from −29°C to 35°C. Use a compound inequality to convert these temperatures to Fahrenheit temperatures.
92. In Oslo, the average temperature ranges from −10° to 18° Celsius. Use a compound inequality to convert these temperatures to the Fahrenheit scale.

Solve.

93. Christian D'Angelo has scores of 68, 65, 75, and 78 on his algebra tests. Use a compound inequality to find the scores he can make on his final exam to receive a C in the course. The final exam counts as two tests, and a C is received if the final course average is from 70 to 79.
94. Wendy Wood has scores of 80, 90, 82, and 75 on her chemistry tests. Use a compound inequality to find the range of scores she can make on her final exam to receive a B in the course. The final exam counts as two tests, and a B is received if the final course average is from 80 to 89.

16.2 Absolute Value Equations 

Objective

1 Solve Absolute Value Equations.

Objective 1 Solving Absolute Value Equations

In Chapter 2, we defined the absolute value of a number as its distance from 0 on a number line.

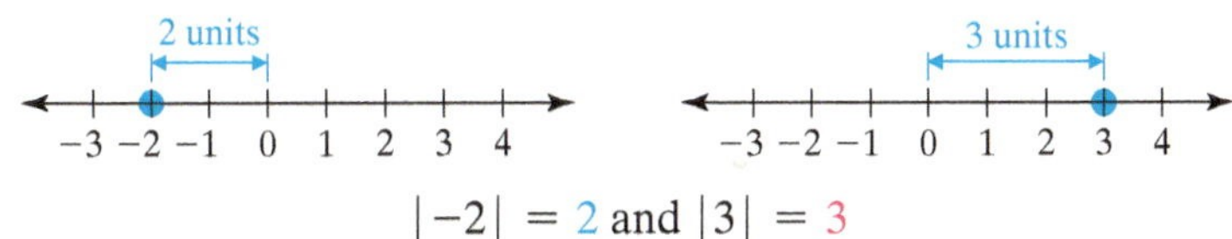

$|-2| = 2$ and $|3| = 3$

In this section, we concentrate on solving equations containing the absolute value of a variable or a variable expression. Examples of absolute value equations are

$$|x| = 3 \qquad -5 = |2y + 7| \qquad |z - 6.7| = |3z + 1.2|$$

Since distance and absolute value are so closely related, absolute value equations and inequalities (see Section 16.3) are extremely useful in solving distance-type problems such as calculating the possible error in a measurement.

For the absolute value equation $|x| = 3$, its solution set will contain all numbers whose distance from 0 is 3 units. Two numbers are 3 units away from 0 on the number line: 3 and −3.

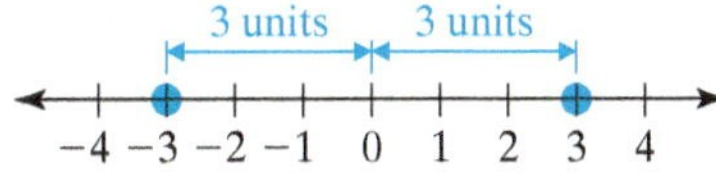

Thus, the solution set of the equation $|x| = 3$ is $\{3, -3\}$. This suggests the following:

Solving Equations of the Form $|X| = a$

If a is a positive number, then $|X| = a$ is equivalent to $X = a$ or $X = -a$.

Practice 1

Solve: $|q| = 3$.

Example 1 Solve: $|p| = 2$.

Solution: Since 2 is positive, $|p| = 2$ is equivalent to $p = 2$ or $p = -2$.

To check, let $p = 2$ and then $p = -2$ in the original equation.

$\lvert p\rvert = 2$	Original equation	$\lvert p\rvert = 2$	Original equation
$\lvert 2\rvert = 2$	Let $p = 2$.	$\lvert -2\rvert = 2$	Let $p = -2$.
$2 = 2$	True	$2 = 2$	True

The solutions are 2 and -2 or the solution set is $\{2, -2\}$.

Work Practice 1

If the expression inside the absolute value bars is more complicated than a single variable, we can still apply the absolute value property.

Helpful Hint

For the equation $|X| = a$ in the box on the previous page, X can be a single variable or a variable expression.

Practice 2

Solve: $|2x - 3| = 5$.

Example 2 Solve: $|5w + 3| = 7$.

Solution: Here the expression inside the absolute value bars is $5w + 3$. If we think of the expression $5w + 3$ as X in the absolute value property, we see that $|X| = 7$ is equivalent to

$$X = 7 \quad \text{or} \quad X = -7$$

Then substitute $5w + 3$ for X, and we have

$$5w + 3 = 7 \quad \text{or} \quad 5w + 3 = -7$$

Solve these two equations for w.

$$\begin{aligned} 5w + 3 &= 7 \quad \text{or} \quad & 5w + 3 &= -7 \\ 5w &= 4 \quad \text{or} \quad & 5w &= -10 \\ w &= \frac{4}{5} \quad \text{or} \quad & w &= -2 \end{aligned}$$

Check: To check, let $w = -2$ and then $w = \frac{4}{5}$ in the original equation.

Let $w = -2$		Let $w = \frac{4}{5}$	
$\lvert 5(-2) + 3\rvert = 7$		$\left\lvert 5\left(\frac{4}{5}\right) + 3\right\rvert = 7$	
$\lvert -10 + 3\rvert = 7$		$\lvert 4 + 3\rvert = 7$	
$\lvert -7\rvert = 7$		$\lvert 7\rvert = 7$	
$7 = 7$	True	$7 = 7$	True

Both solutions check, and the solutions are -2 and $\frac{4}{5}$ or the solution set is $\left\{-2, \frac{4}{5}\right\}$.

Work Practice 2

Answers

1. $-3, 3$ 2. $-1, 4$

Example 3 Solve: $\left|\frac{x}{2} - 1\right| = 11.$

Solution: $\left|\frac{x}{2} - 1\right| = 11$ is equivalent to

$$\frac{x}{2} - 1 = 11 \quad \text{or} \quad \frac{x}{2} - 1 = -11$$

$$2\left(\frac{x}{2} - 1\right) = 2(11) \quad \text{or} \quad 2\left(\frac{x}{2} - 1\right) = 2(-11) \qquad \text{Clear fractions.}$$

$$x - 2 = 22 \quad \text{or} \quad x - 2 = -22 \qquad \text{Apply the distributive property.}$$

$$x = 24 \quad \text{or} \quad x = -20$$

The solutions are 24 and -20.

Work Practice 3

Practice 3

Solve: $\left|\frac{x}{5} + 1\right| = 15.$

To apply the absolute value property, first make sure that the absolute value expression is isolated.

Helpful Hint

If the equation has a single absolute value expression containing variables, isolate the absolute value expression first.

Example 4 Solve: $|2x| + 5 = 7.$

Solution: We want the absolute value expression alone on one side of the equation, so begin by subtracting 5 from both sides. Then apply the absolute value property.

$$|2x| + 5 = 7$$

$$|2x| = 2 \qquad \text{Subtract 5 from both sides.}$$

$$2x = 2 \quad \text{or} \quad 2x = -2$$

$$x = 1 \quad \text{or} \quad x = -1$$

The solutions are -1 and 1.

Work Practice 4

Practice 4

Solve: $|3x| + 8 = 14.$

Example 5 Solve: $|y| = 0.$

Solution: We are looking for all numbers whose distance from 0 is zero units. The only number is 0. The solution is 0.

Work Practice 5

Practice 5

Solve: $|z| = 0.$

The next two examples illustrate a special case for absolute value equations. This special case occurs when an isolated absolute value is equal to a negative number.

Answers

3. $-80, 70$ **4.** $-2, 2$ **5.** 0

Practice 6

Solve: $3|z| + 9 = 7$.

Example 6 Solve: $2|x| + 25 = 23$.

Solution: First, isolate the absolute value.

$$2|x| + 25 = 23$$
$$2|x| = -2 \quad \text{Subtract 25 from both sides.}$$
$$|x| = -1 \quad \text{Divide both sides by 2.}$$

The absolute value of a number is never negative, so this equation has no solution. The solution set is { } or ∅.

Work Practice 6

Practice 7

Solve: $\left|\frac{5x + 3}{4}\right| = -8$.

Example 7 Solve: $\left|\frac{3x + 1}{2}\right| = -2$.

Solution: Again, the absolute value of any expression is never negative, so no solution exists. The solution set is { } or ∅.

Work Practice 7

Given two absolute value expressions, we might ask, when are the absolute values of two expressions equal? To see the answer, notice that

$|2| = |2|$, $|-2| = |-2|$, $|-2| = |2|$, and $|2| = |-2|$

same, same, opposites, opposites

Two absolute value expressions are equal when the expressions inside the absolute value bars are equal to or are opposites of each other.

Practice 8

Solve: $|2x + 4| = |3x - 1|$.

Example 8 Solve: $|3x + 2| = |5x - 8|$.

Solution: This equation is true if the expressions inside the absolute value bars are equal to or are opposites of each other.

$$3x + 2 = 5x - 8 \quad \text{or} \quad 3x + 2 = -(5x - 8)$$

Next, solve each equation.

$$3x + 2 = 5x - 8 \quad \text{or} \quad 3x + 2 = -5x + 8$$
$$-2x + 2 = -8 \quad \text{or} \quad 8x + 2 = 8$$
$$-2x = -10 \quad \text{or} \quad 8x = 6$$
$$x = 5 \quad \text{or} \quad x = \frac{3}{4}$$

The solutions are $\frac{3}{4}$ and 5.

Work Practice 8

Answers

6. { } or ∅ 7. { } or ∅

8. $-\frac{3}{5}, 5$

Example 9 Solve: $|x - 3| = |5 - x|$.

Solution:

$$\begin{aligned} x - 3 &= 5 - x \quad \text{or} \quad & x - 3 &= -(5 - x) \\ 2x - 3 &= 5 \quad \text{or} \quad & x - 3 &= -5 + x \\ 2x &= 8 \quad \text{or} \quad & x - 3 - x &= -5 + x - x \\ x &= 4 \quad \text{or} \quad & -3 &= -5 \quad \text{False} \end{aligned}$$

Recall from Section 9.3 that when an equation simplifies to a false statement, the equation has no solution. Thus, the only solution for the original absolute value equation is 4.

Work Practice 9

Practice 9

Solve: $|x - 2| = |8 - x|$.

✓Concept Check True or false? Absolute value equations always have two solutions. Explain your answer.

The following box summarizes the methods shown for solving absolute value equations.

Absolute Value Equations

$|X| = a$
- If a is positive, then solve $X = a$ or $X = -a$.
- If a is 0, solve $X = 0$.
- If a is negative, the equation $|X| = a$ has no solution.

$|X| = |Y|$ Solve $X = Y$ or $X = -Y$.

Answer

9. 5

✓Concept Check Answer

false; answers may vary

Vocabulary, Readiness & Video Check

Match each absolute value equation with an equivalent statement.

1. $|x - 2| = 5$

2. $|x - 2| = 0$

3. $|x - 2| = |x + 3|$

4. $|x + 3| = 5$

5. $|x + 3| = -5$

A. $x - 2 = 0$

B. $x - 2 = x + 3$ or $x - 2 = -(x + 3)$

C. $x - 2 = 5$ or $x - 2 = -5$

D. $\varnothing$

E. $x + 3 = 5$ or $x + 3 = -5$

Martin-Gay Interactive Videos

See Video 16.2

Watch the section lecture video and answer the following question.

Objective 1 **6.** As explained in Example 3, why is a positive in the rule "$|X| = a$ is equivalent to $X = a$ or $X = -a$"?

16.2 Exercise Set MyMathLab® 

Solve each absolute value equation. See Examples 1 through 7.

1. $|x| = 7$
2. $|y| = 15$
3. $|3x| = 12.6$
4. $|6n| = 12.6$
5. $|2x - 5| = 9$
6. $|6 + 2n| = 4$
7. $\left|\frac{x}{2} - 3\right| = 1$
8. $\left|\frac{n}{3} + 2\right| = 4$
9. $|z| + 4 = 9$
10. $|x| + 1 = 3$
11. $|3x| + 5 = 14$
12. $|2x| - 6 = 4$
13. $|2x| = 0$
14. $|7z| = 0$
15. $|4n + 1| + 10 = 4$
16. $|3z - 2| + 8 = 1$
17. $|5x - 1| = 0$
18. $|3y + 2| = 0$

Solve. See Examples 8 and 9.

19. $|5x - 7| = |3x + 11|$
20. $|9y + 1| = |6y + 4|$
21. $|z + 8| = |z - 3|$
22. $|2x - 5| = |2x + 5|$

MIXED PRACTICE

Solve each absolute value equation. See Examples 1 through 9.

23. $|x| = 4$
24. $|x| = 1$
25. $|y| = 0$
26. $|y| = 8$
27. $|z| = -2$
28. $|y| = -9$
29. $|7 - 3x| = 7$
30. $|4m + 5| = 5$
31. $|6x| - 1 = 11$
32. $|7z| + 1 = 22$
33. $|4p| = -8$
34. $|5m| = -10$
35. $|x - 3| + 3 = 7$
36. $|x + 4| - 4 = 1$
37. $\left|\frac{z}{4} + 5\right| = -7$
38. $\left|\frac{c}{5} - 1\right| = -2$
39. $|9v - 3| = -8$
40. $|1 - 3b| = -7$
41. $|8n + 1| = 0$
42. $|5x - 2| = 0$
43. $|1 + 6c| - 7 = -3$
44. $|2 + 3m| - 9 = -7$
45. $|5x + 1| = 11$
46. $|8 - 6c| = 1$
47. $|4x - 2| = |-10|$
48. $|3x + 5| = |-4|$
49. $|5x + 1| = |4x - 7|$
50. $|3 + 6n| = |4n + 11|$
51. $|6 + 2x| = -|-7|$
52. $|4 - 5y| = -|-3|$
53. $|2x - 6| = |10 - 2x|$
54. $|4n + 5| = |4n + 3|$
55. $\left|\frac{2x - 5}{3}\right| = 7$
56. $\left|\frac{1 + 3n}{4}\right| = 4$
57. $2 + |5n| = 17$
58. $8 + |4m| = 24$
59. $\left|\frac{2x - 1}{3}\right| = |-5|$
60. $\left|\frac{5x + 2}{2}\right| = |-6|$
61. $|2y - 3| = |9 - 4y|$
62. $|5z - 1| = |7 - z|$
63. $\left|\frac{3n + 2}{8}\right| = |-1|$
64. $\left|\frac{2r - 6}{5}\right| = |-2|$
65. $|x + 4| = |7 - x|$
66. $|8 - y| = |y + 2|$
67. $\left|\frac{8c - 7}{3}\right| = -|-5|$
68. $\left|\frac{5d + 1}{6}\right| = -|-9|$

REVIEW AND PREVIEW

The circle graph shows the types of cheese produced in the United States in 2010. Use this graph to answer Exercises 69 through 72. See Section 7.2.

U.S. Cheese[1] Production by Variety, 2010

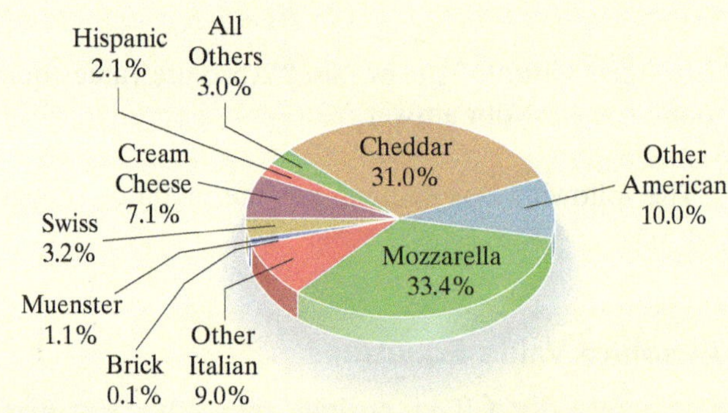

[1]Excludes Cottage Cheese

Source: USDA, *Dairy Products Annual Survey*

69. In 2010, cheddar cheese made up what percent of U.S. cheese production?
70. Which cheese had the highest U.S. production in 2010?
71. A circle contains 360°. Find the number of degrees found in the 9% sector for Other Italian Cheese.
72. In 2010, the total production of cheese in the United States was 10,109,293,000 pounds. Find the amount of cream cheese produced during that year.

List five integer solutions of each inequality. See Section 9.1.

73. $|x| \leq 3$
74. $|x| \geq -2$
75. $|y| > -10$
76. $|y| < 0$

CONCEPT EXTENSIONS

Without going through a solution procedure, determine the solution of each absolute value equation or inequality.

77. $|x - 7| = -4$
78. $|x - 7| < -4$
79. Write an absolute value equation representing all numbers x whose distance from 0 is 5 units.
80. Write an absolute value equation representing all numbers x whose distance from 0 is 2 units.
81. Explain why some absolute value equations have two solutions.
82. Explain why some absolute value equations have one solution.

83. Write an absolute value equation representing all numbers x whose distance from 1 is 5 units.

84. Write an absolute value equation representing all numbers x whose distance from 7 is 2 units.

85. Describe how solving an absolute value equation such as $|2x - 1| = 3$ is similar to solving an absolute value equation such as $|2x - 1| = |x - 5|$.

86. Describe how solving an absolute value equation such as $|2x - 1| = 3$ is different from solving an absolute value equation such as $|2x - 1| = |x - 5|$.

Write each as an equivalent absolute value equation.

87. $x = 6$ or $x = -6$

88. $2x - 1 = 4$ or $2x - 1 = -4$

89. $x - 2 = 3x - 4$ or $x - 2 = -(3x - 4)$

90. For what value(s) of c will an absolute value equation of the form $|ax + b| = c$ have

a. one solution?
b. no solution?
c. two solutions?

16.3 Absolute Value Inequalities 

Objective 1 Solving Absolute Value Inequalities of the Form $|X| < a$

The solution set of an absolute value inequality such as $|x| < 2$ contains all numbers whose distance from 0 is less than 2 units, as shown below.

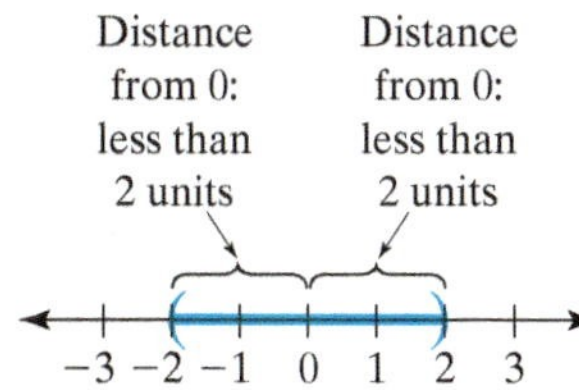

The solution set is $\{x \mid -2 < x < 2\}$, or $(-2, 2)$ in interval notation.

Example 1 Solve: $|x| \le 3$ and graph the solution set.

Solution: The solution set of this inequality contains all numbers whose distance from 0 is less than or equal to 3. Thus 3, −3, and all numbers between 3 and −3 are in the solution set.

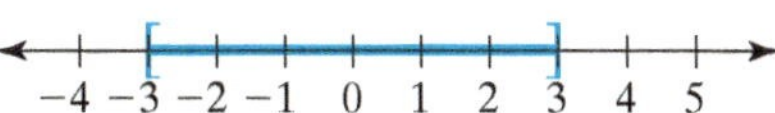

The solution set is $[-3, 3]$.

Work Practice 1

In general, we have the following.

> **Solving Absolute Value Inequalities of the Form $|X| < a$**
>
> If a is a positive number, then $|X| < a$ is equivalent to $-a < X < a$.

This property also holds true for the inequality symbol $\le$.

Objectives

1. Solve Absolute Value Inequalities of the Form $|X| < a$.
2. Solve Absolute Value Inequalities of the Form $|X| > a$.

Practice 1

Solve: $|x| < 5$ and graph the solution set.

Answer

1. $(-5, 5)$ 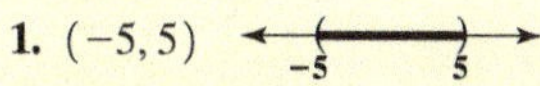

Practice 2

Solve for b: $|b + 1| < 3$. Graph the solution set.

Example 2

Solve for m: $|m - 6| < 2$. Graph the solution set.

Solution: Replace X with $m - 6$ and a with 2 in the preceding property (on the previous page), and we see that

$$|m - 6| < 2 \quad \text{is equivalent to} \quad -2 < m - 6 < 2$$

Solve this compound inequality for m by adding 6 to all three parts.

$$-2 < m - 6 < 2$$
$$-2 + 6 < m - 6 + 6 < 2 + 6 \quad \text{Add 6 to all three parts.}$$
$$4 < m < 8 \quad \text{Simplify.}$$

The solution set is $(4, 8)$, and its graph is shown.

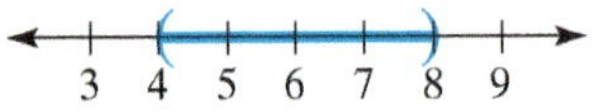

Work Practice 2

Helpful Hint

Before using an absolute value inequality property, isolate the absolute value expression on one side of the inequality.

Practice 3

Solve for x: $|3x - 2| + 5 \le 9$. Graph the solution set.

Example 3

Solve for x: $|5x + 1| + 1 \le 10$. Graph the solution set.

Solution: First, isolate the absolute value expression by subtracting 1 from both sides.

$$|5x + 1| + 1 \le 10$$
$$|5x + 1| \le 10 - 1 \quad \text{Subtract 1 from both sides.}$$
$$|5x + 1| \le 9 \quad \text{Simplify.}$$

Since 9 is positive, we apply the absolute value property for $|X| \le a$.

$$-9 \le 5x + 1 \le 9$$
$$-9 - 1 \le 5x + 1 - 1 \le 9 - 1 \quad \text{Subtract 1 from all three parts.}$$
$$-10 \le 5x \le 8 \quad \text{Simplify.}$$
$$-2 \le x \le \frac{8}{5} \quad \text{Divide all three parts by 5.}$$

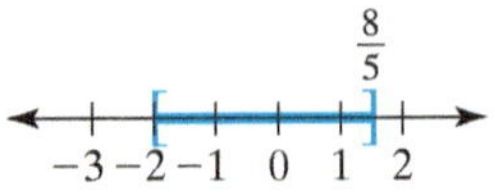

The solution set is $\left[-2, \frac{8}{5}\right]$, and the graph is shown above.

Work Practice 3

Practice 4

Solve for x: $\left|3x + \frac{5}{8}\right| < -4$.

Example 4

Solve for x: $\left|2x - \frac{1}{10}\right| < -13$.

Solution: The absolute value of a number is always nonnegative and can never be less than −13. Thus this absolute value inequality has no solution. The solution set is { } or ∅.

Work Practice 4

Answers

2. $(-4, 2)$

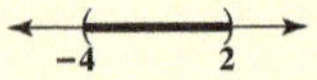

3. $\left[-\frac{2}{3}, 2\right]$

4. { } or ∅

Example 5 Solve for x: $\left|\frac{2(x+1)}{3}\right| \le 0$.

Solution: Recall that "$\le$" means "is less than or equal to." The absolute value of any expression will never be less than 0, but it may be equal to 0. Thus, to solve $\left|\frac{2(x+1)}{3}\right| \le 0$, we solve $\left|\frac{2(x+1)}{3}\right| = 0$.

$$\frac{2(x+1)}{3} = 0$$

$$3\left[\frac{2(x+1)}{3}\right] = 3(0) \quad \text{Clear the equation of fractions.}$$

$$2x + 2 = 0 \quad \text{Apply the distributive property.}$$

$$2x = -2 \quad \text{Subtract 2 from both sides.}$$

$$x = -1 \quad \text{Divide both sides by 2.}$$

The solution set is $\{-1\}$.

Work Practice 5

Practice 5

Solve for x: $\left|\frac{3(x-2)}{5}\right| \le 0$.

Objective 2 Solving Absolute Value Inequalities of the Form $|X| > a$

Let us now solve an absolute value inequality of the form $|X| > a$, such as $|x| \ge 3$. The solution set contains all numbers whose distance from 0 is 3 or more units. Thus the graph of the solution set contains 3 and all points to the right of 3 on the number line or -3 and all points to the left of -3 on the number line.

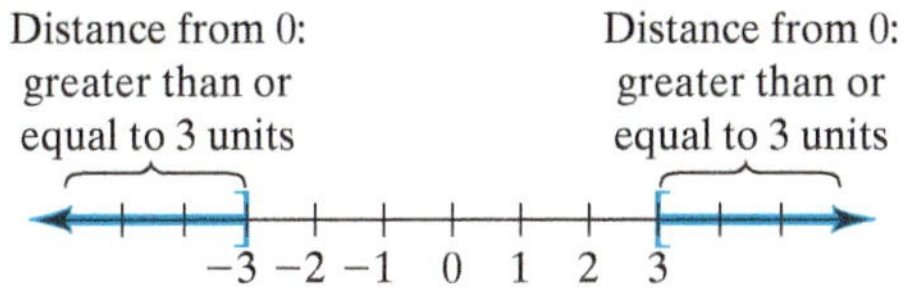

This solution set is written as $\{x \mid x \le -3 \text{ or } x \ge 3\}$. In interval notation, the solution is $(-\infty, -3] \cup [3, \infty)$, since "or" means "union." In general, we have the following.

Solving Absolute Value Inequalities of the Form $|X| > a$

If a is a positive number, then $|X| > a$ is equivalent to $X < -a$ or $X > a$.

This property also holds true for the inequality symbol $\ge$.

Example 6 Solve for y: $|y - 3| > 7$.

Solution: Since 7 is positive, we apply the property for $|X| > a$.

$|y - 3| > 7$ is equivalent to $y - 3 < -7$ or $y - 3 > 7$

Next, solve the compound inequality.

$$y - 3 < -7 \quad \text{or} \quad y - 3 > 7$$

$$y - 3 + 3 < -7 + 3 \quad \text{or} \quad y - 3 + 3 > 7 + 3 \quad \text{Add 3 to both sides.}$$

$$y < -4 \quad \text{or} \quad y > 10 \quad \text{Simplify.}$$

(Continued on next page)

Practice 6

Solve for y: $|y + 4| \ge 6$.

Answers

5. $\{2\}$

6. $(-\infty, -10] \cup [2, \infty)$ 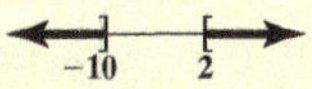

The solution set is $(-\infty, -4) \cup (10, \infty)$, and its graph is shown.

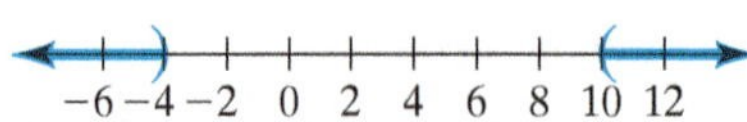

Work Practice 6

Example 7 illustrates another special case of absolute value inequalities when an isolated absolute value expression is less than, less than or equal to, greater than, or greater than or equal to a negative number or 0.

Practice 7

Solve: $|4x + 3| + 5 > 3$. Graph the solution set.

Example 7 Solve: $|2x + 9| + 5 > 3$.

Solution: First isolate the absolute value expression by subtracting 5 from both sides.

$$|2x + 9| + 5 > 3$$

$$|2x + 9| + 5 - 5 > 3 - 5 \quad \text{Subtract 5 from both sides.}$$

$$|2x + 9| > -2 \quad \text{Simplify.}$$

The absolute value of any number is always nonnegative and thus is always greater than -2. This inequality and the original inequality are true for all values of x. The solution set is $\{x|x \text{ is a real number}\}$ or $(-\infty, \infty)$, and its graph is shown.

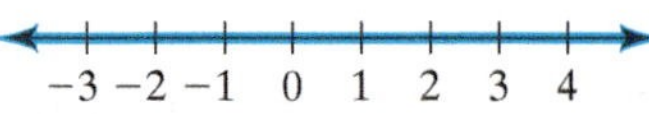

Work Practice 7

✓**Concept Check** Without taking any solution steps, how do you know that the absolute value inequality $|3x - 2| > -9$ has a solution? What is its solution?

Practice 8

Solve: $\left|\frac{x}{2} - 3\right| - 5 > -2$.

Graph the solution set.

Example 8 Solve: $\left|\frac{x}{3} - 1\right| - 7 \geq -5$.

Solution: First, isolate the absolute value expression by adding 7 to both sides.

$$\left|\frac{x}{3} - 1\right| - 7 \geq -5$$

$$\left|\frac{x}{3} - 1\right| - 7 + 7 \geq -5 + 7 \quad \text{Add 7 to both sides.}$$

$$\left|\frac{x}{3} - 1\right| \geq 2 \quad \text{Simplify.}$$

Next, write the absolute value inequality as an equivalent compound inequality and solve.

$$\frac{x}{3} - 1 \leq -2 \quad \text{or} \quad \frac{x}{3} - 1 \geq 2$$

$$3\left(\frac{x}{3} - 1\right) \leq 3(-2) \quad \text{or} \quad 3\left(\frac{x}{3} - 1\right) \geq 3(2) \quad \text{Clear the inequalities of fractions.}$$

$$x - 3 \leq -6 \quad \text{or} \quad x - 3 \geq 6 \quad \text{Apply the distributive property.}$$

$$x \leq -3 \quad \text{or} \quad x \geq 9 \quad \text{Add 3 to both sides.}$$

The solution set is $(-\infty, -3] \cup [9, \infty)$, and its graph is shown.

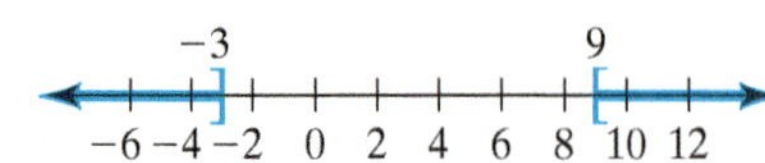

Work Practice 8

Answers

7. $(-\infty, \infty)$

8. $(-\infty, 0) \cup (12, \infty)$

✓**Concept Check Answer**

$(-\infty, \infty)$ since the absolute value is always nonnegative

The following box summarizes the types of absolute value equations and inequalities.

Solving Absolute Value Equations and Inequalities with $a > 0$

Algebraic Solution	Solution Graph
$\lvert X \rvert = a$ is equivalent to $X = a$ or $X = -a$.	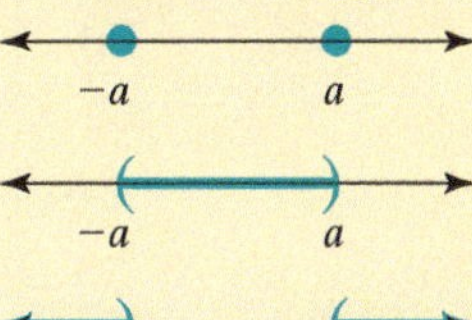
$\lvert X \rvert < a$ is equivalent to $-a < X < a$.	
$\lvert X \rvert > a$ is equivalent to $X < -a$ or $X > a$.	

Vocabulary, Readiness & Video Check

Match each absolute value statement with an equivalent statement.

1. $|2x + 1| = 3$
2. $|2x + 1| \le 3$
3. $|2x + 1| < 3$
4. $|2x + 1| \ge 3$
5. $|2x + 1| > 3$

A. $2x + 1 > 3$ or $2x + 1 < -3$
B. $2x + 1 \ge 3$ or $2x + 1 \le -3$
C. $-3 < 2x + 1 < 3$
D. $2x + 1 = 3$ or $2x + 1 = -3$
E. $-3 \le 2x + 1 \le 3$

Martin-Gay Interactive Videos

See Video 16.3

Watch the section lecture video and answer the following questions.

Objective 1 6. In Example 3, how can you reason that the inequality has no solution even if you don't know the rule?

Objective 2 7. In Example 4, why is the union symbol used when the solution is written in interval notation?

16.3 Exercise Set MyMathLab®

Solve each inequality. Then graph the solution set and write it in interval notation. See Examples 1 through 4.

1. $|x| \le 4$
2. $|x| < 6$
3. $|x - 3| < 2$
4. $|y - 7| \le 5$
5. $|x + 3| < 2$
6. $|x + 4| < 6$
7. $|2x + 7| \le 13$
8. $|5x - 3| \le 18$
9. $|x| + 7 \le 12$
10. $|x| + 6 \le 7$
11. $|3x - 1| < -5$
12. $|8x - 3| < -2$
13. $|x - 6| - 7 \le -1$
14. $|z + 2| - 7 < -3$

Solve each inequality. Graph the solution set and write it in interval notation. See Examples 6 through 8.

15. $|x| > 3$
16. $|y| \ge 4$
17. $|x + 10| \ge 14$
18. $|x - 9| \ge 2$
19. $|x| + 2 > 6$
20. $|x| - 1 > 3$
21. $|5x| > -4$

22. $|4x - 11| > -1$

23. $|6x - 8| + 3 > 7$

24. $|10 + 3x| + 1 > 2$

Solve each inequality. Graph the solution set and write it in interval notation. See Example 5.

25. $|x| \le 0$

26. $|x| \ge 0$

27. $|8x + 3| > 0$

28. $|5x - 6| < 0$

MIXED PRACTICE

Solve each inequality. Graph the solution set and write it in interval notation. See Examples 1 through 8.

29. $|x| \le 2$

30. $|z| < 8$

31. $|y| > 1$

32. $|x| \ge 10$

33. $|x - 3| < 8$

34. $|-3 + x| \le 10$

35. $|0.6x - 3| > 0.6$

36. $|1 + 0.3x| \ge 0.1$

37. $5 + |x| \le 2$

38. $8 + |x| < 1$

39. $|x| > -4$

40. $|x| \le -7$

41. $|2x - 7| \le 11$

42. $|5x + 2| < 8$

43. $|x + 5| + 2 \ge 8$

44. $|-1 + x| - 6 > 2$

45. $|x| > 0$

46. $|x| < 0$

47. $9 + |x| > 7$

48. $5 + |x| \ge 4$

49. $6 + |4x - 1| \le 9$

50. $-3 + |5x - 2| \le 4$

51. $\left|\frac{2}{3}x + 1\right| > 1$

52. $\left|\frac{3}{4}x - 1\right| \ge 2$

53. $|5x + 3| < -6$

54. $|4 + 9x| \ge -6$

55. $\left|\frac{8x - 3}{4}\right| \le 0$

56. $\left|\frac{5x + 6}{2}\right| \le 0$

57. $|1 + 3x| + 4 < 5$

58. $|7x - 3| - 1 \le 10$

59. $\left|\frac{x + 6}{3}\right| > 2$

60. $\left|\frac{7 + x}{2}\right| \ge 4$

61. $-15 + |2x - 7| \le -6$

62. $-9 + |3 + 4x| < -4$

63. $\left|2x + \frac{3}{4}\right| - 7 \le -2$

64. $\left|\frac{3}{5} + 4x\right| - 6 < -1$

MIXED PRACTICE

Solve each equation or inequality for x. See Sections 16.2 and 16.3.

65. $|2x - 3| < 7$

66. $|2x - 3| > 7$

67. $|2x - 3| = 7$

68. $|5 - 6x| = 29$

69. $|x - 5| \ge 12$

70. $|x + 4| \ge 20$

71. $|9 + 4x| = 0$

72. $|9 + 4x| \ge 0$

73. $|2x + 1| + 4 < 7$

74. $8 + |5x - 3| \ge 11$

75. $|3x - 5| + 4 = 5$

76. $|5x - 3| + 2 = 4$

77. $|x + 11| = -1$

78. $|4x - 4| = -3$

79. $\left|\frac{2x - 1}{3}\right| = 6$

80. $\left|\frac{6 - x}{4}\right| = 5$

81. $\left|\frac{3x - 5}{6}\right| > 5$

82. $\left|\frac{4x - 7}{5}\right| < 2$

MIXED PRACTICE

Solve. See Section 9.4.

Many companies predict the growth or decline of various technologies. The following data is based on information from Techcrunchies, a technological information site. Notice that the first table is the estimated increase in the number of Wi-Fi-enabled cell phones (in millions), and the second is the estimated percent increase in the number of Wi-Fi-enabled cell phones in the United States.

83. Use the middle column in the table to find the estimated number of Wi-Fi-enabled cell phones for each year.

Year	Increase in Wi-Fi-Enabled Cell Phones	Estimated Number
2010	$2x - 21$	
2012	$\frac{5}{2}x + 2$	
2014	$3x + 24$	
Total	290 million	

84. Use the middle column in the table to find the estimated percent increase in the number of Wi-Fi-enabled cell phones for each year.

Year	Percent Increase in Wi-Fi-Enabled Cell Phones since 2009	Estimated Percent Increase
2010	x	
2011	$2x + 10$	
2012	$4x - 25$	
	300%	

Consider the equation $3x - 4y = 12$. For each value of x or y given, find the corresponding value of the other variable that makes the statement true. See Section 10.1.

85. If $x = 2$, find y.

86. If $y = -1$, find x.

87. If $y = -3$, find x.

88. If $x = 4$, find y.

CONCEPT EXTENSIONS

89. Write an absolute value inequality representing all numbers x whose distance from 0 is less than 7 units.

90. Write an absolute value inequality representing all numbers x whose distance from 0 is greater than 4 units.

91. Write $-5 \le x \le 5$ as an equivalent inequality containing an absolute value.

92. Write $x > 1$ or $x < -1$ as an equivalent inequality containing an absolute value.

93. Describe how solving $|x - 3| = 5$ is different from solving $|x - 3| < 5$.

94. Describe how solving $|x + 4| = 0$ is similar to solving $|x + 4| \le 0$.

The expression $|x_T - x|$ is defined to be the absolute error in x, where x_T is the true value of a quantity and x is the measured value or the value as stored in a computer.

95. If the true value of a quantity is 3.5 and the absolute error must be less than 0.05, find the acceptable measured values.

96. If the true value of a quantity is 0.2 and the approximate value stored in a computer is $\frac{51}{256}$, find the absolute error.

Integrated Review Sections 16.1–16.3

Solving Compound Inequalities and Absolute Value Equations and Inequalities

Solve each equation or inequality. Write inequality solution sets in interval notation. For inequalities containing "and" or "or," also graph the solution set and write it in interval notation.

1. $x < 7$ and $x > -5$
2. $x < 7$ or $x > -5$
3. $|4x - 3| = 1$
4. $|2x + 1| < 5$
5. $|6x| - 9 \ge -3$
6. $|x - 7| = |2x + 11|$
7. $-5 \le \frac{3x - 8}{2} \le 2$
8. $|9x - 1| = -3$
9. $3x + 2 \le 5$ or $-3x \ge 0$
10. $3x + 2 \le 5$ and $-3x \ge 0$
11. $|3 - x| - 5 \le -2$
12. $\left|\frac{4x + 1}{5}\right| = |-1|$

Match each equation or inequality on the left with an equivalent statement on the right.

13. $|2x + 1| = 5$ A. $2x + 1 > 5$ or $2x + 1 < -5$
14. $|2x + 1| < 5$ B. $2x + 1 = 5$ or $2x + 1 = -5$
15. $|2x + 1| > 5$ C. $x < 5$
16. $x < 3$ or $x < 5$ D. $x < 3$
17. $x < 3$ and $x < 5$ E. $-5 < 2x + 1 < 5$

16.4 Graphing Linear Inequalities in Two Variables and Systems of Linear Inequalities

In this section, we first learn to graph a single linear inequality in two variables. Then we solve systems of linear inequalities.

Recall that a linear equation in two variables is an equation that can be written in the form $Ax + By = C$ where A, B, and C are real numbers and A and B are not both 0. The definition of a linear inequality is the same except that the equal sign is replaced with an inequality sign.

A **linear inequality in two variables** is an inequality that can be written in one of the forms:

$$Ax + By < C \quad Ax + By \le C$$
$$Ax + By > C \quad Ax + By \ge C$$

Objectives

1 Graph a Linear Inequality in Two Variables.

2 Solve a System of Linear Inequalities.

where A, B, and C are real numbers and A and B are not both 0. Just as for linear equations in x and y, an ordered pair is a **solution** of an inequality in x and y if replacing the variables by coordinates of the ordered pair results in a true statement.

Objective 1 Graphing Linear Inequalities in Two Variables

The linear equation $x - y = 1$ is graphed next. Recall that all points on the line correspond to ordered pairs that satisfy the equation $x - y = 1$.

Notice the line defined by $x - y = 1$ divides the rectangular coordinate system plane into 2 sides. All points on one side of the line satisfy the inequality $x - y < 1$ and all points on the other side satisfy the inequality $x - y > 1$. The graph below shows a few examples of this.

Check	$x - y < 1$
(1, 3)	$1 - 3 < 1$ True
(−2, 1)	$-2 - 1 < 1$ True
(−4, −4)	$-4 - (-4) < 1$ True

Check	$x - y > 1$
(4, 1)	$4 - 1 > 1$ True
(2, −2)	$2 - (-2) > 1$ True
(0, −4)	$0 - (-4) > 1$ True

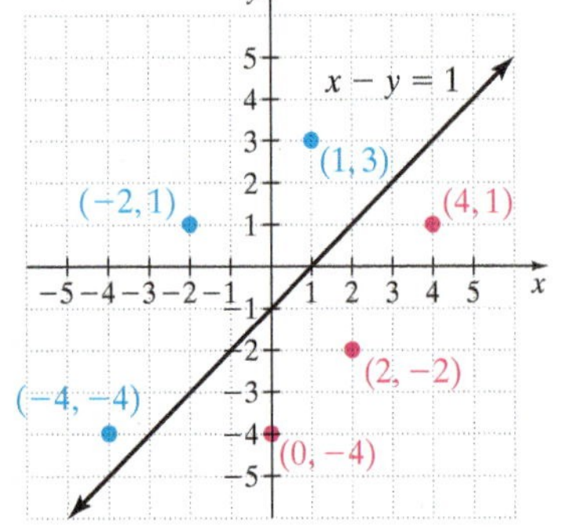

The graph of $x - y < 1$ is the region shaded blue, and the graph of $x - y > 1$ is the region shaded red below.

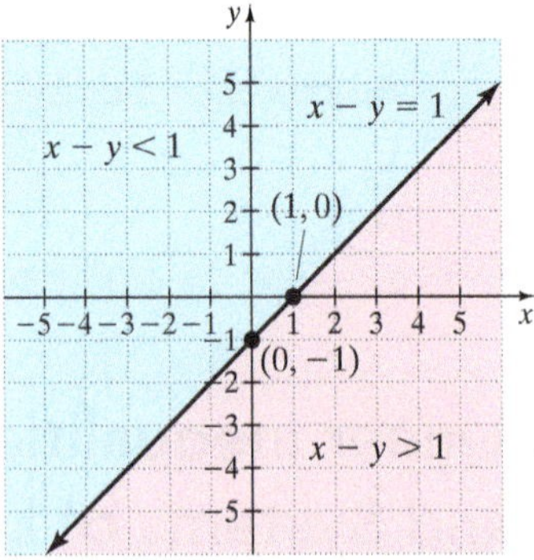

The region to the left of the line and the region to the right of the line are called **half-planes.** Every line divides the plane (similar to a sheet of paper extending indefinitely in all directions) into two half-planes; the line is called the **boundary.**

Recall that the inequality $x - y \le 1$ means

$$x - y = 1 \quad \text{or} \quad x - y < 1$$

Thus, the graph of $x - y \le 1$ is the half-plane $x - y < 1$ along with the boundary line $x - y = 1$.

Graphing a Linear Inequality in Two Variables

Step 1: Graph the boundary line found by replacing the inequality sign with an equal sign. If the inequality sign is $>$ or $<$, graph a dashed boundary line (indicating that the points on the line are not solutions of the inequality). If the inequality sign is $\geq$ or $\leq$, graph a solid boundary line (indicating that the points on the line are solutions of the inequality).

Step 2: Choose a point *not* on the boundary line as a test point. Substitute the coordinates of this test point into the *original* inequality.

Step 3: If a true statement is obtained in Step 2, shade the half-plane that contains the test point. If a false statement is obtained, shade the half-plane that does not contain the test point.

Example 1 Graph: $x + y < 7$

Solution:

Step 1: First we graph the boundary line by graphing the equation $x + y = 7$. We graph this boundary as a dashed line because the inequality sign is $<$, and thus the points on the line are not solutions of the inequality $x + y < 7$.

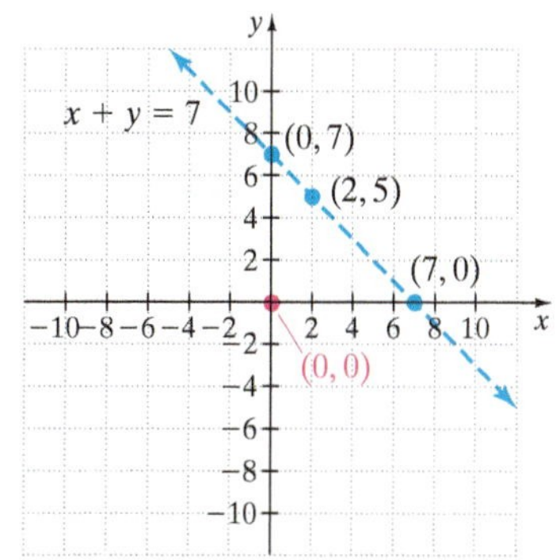

Step 2: Next, we choose a test point, being careful not to choose a point on the boundary line. We choose (0, 0). Substitute the coordinates of (0, 0) into $x + y < 7$.

$x + y < 7$ Original inequality

$0 + 0 \overset{?}{<} 7$ Replace x with 0 and y with 0.

$0 < 7$ True

Step 3: Since the result is a true statement, (0, 0) is a solution of $x + y < 7$, and every point in the same half-plane as (0, 0) is also a solution. To indicate this, shade the entire half-plane containing (0, 0), as shown.

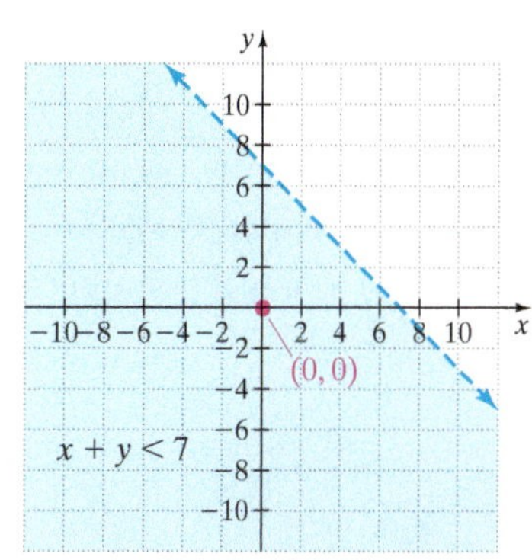

Work Practice 1

Practice 1

Graph: $x + y > 5$

Answer

1.

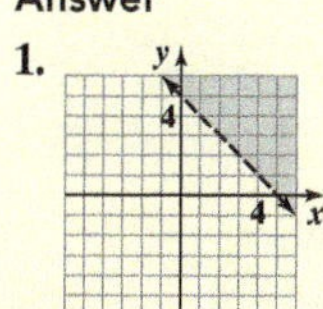

✓Concept Check Determine whether (0, 0) is included in the graph of

a. $y \geq 2x + 3$ **b.** $x < 7$ **c.** $2x - 3y < 6$

Practice 2

Graph: $3x - y \geq 4$

Example 2 Graph: $2x - y \geq 3$

Solution:

Step 1: We graph the boundary line by graphing $2x - y = 3$. We draw this line as a solid line because the inequality sign is $\geq$, and thus the points on the line are solutions of $2x - y \geq 3$.

Step 2: Once again, (0, 0) is a convenient test point since it is not on the boundary line. We substitute 0 for x and 0 for y into the original inequality.

$$2x - y \geq 3$$
$$2(0) - 0 \geq 3 \quad \text{Let } x = 0 \text{ and } y = 0.$$
$$0 \geq 3 \quad \text{False}$$

Step 3: Since the statement is false, no point in the half-plane containing (0, 0) is a solution. Therefore, we shade the half-plane that does not contain (0, 0). Every point in the shaded half-plane and every point on the boundary line is a solution of $2x - y \geq 3$.

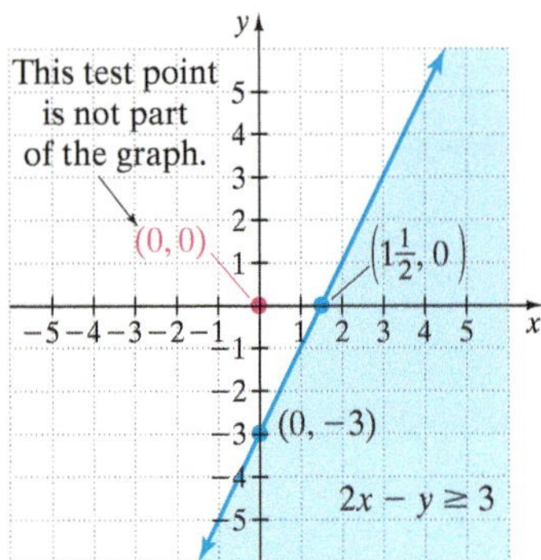

Work Practice 2

Helpful Hint

When graphing an inequality, make sure the test point is substituted into the **original inequality.** For Example 2, we substituted the test point (0, 0) into the **original inequality** $2x - y \geq 3$, *not* $2x - y = 3$.

Practice 3

Graph: $x > 3y$

Example 3 Graph: $x > 2y$

Solution:

Step 1: We find the boundary line by graphing $x = 2y$. The boundary line is a dashed line since the inequality symbol is $>$.

Step 2: We cannot use (0, 0) as a test point because it is a point on the boundary line. We choose (0, 2) instead.

$$x > 2y$$
$$0 > 2(2) \quad \text{Let } x = 0 \text{ and } y = 2.$$
$$0 > 4 \quad \text{False}$$

Answers

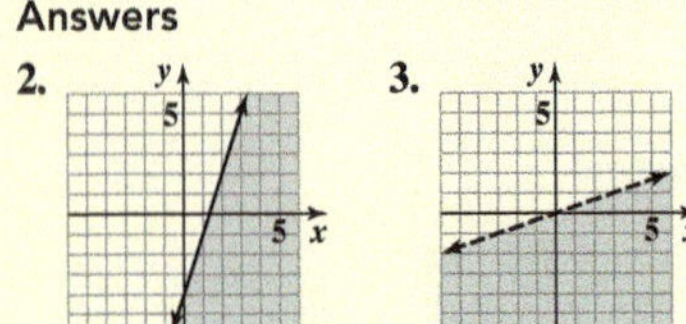

Concept Check Answers

a. no **b.** yes **c.** yes

Step 3: Since the statement is false, we shade the half-plane that does not contain the test point (0, 2), as shown.

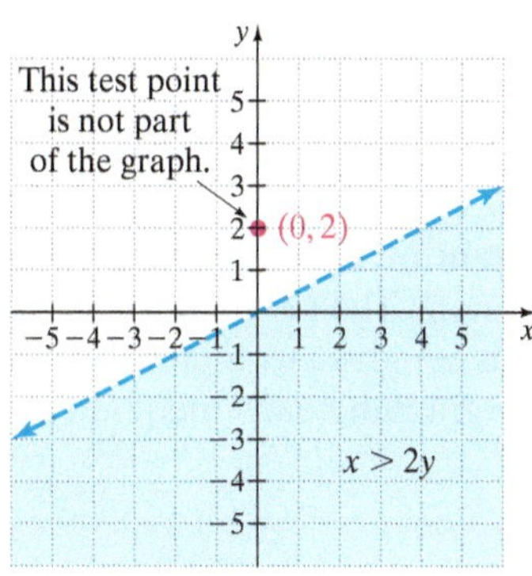

Work Practice 3

Example 4 Graph: $5x + 4y \le 20$

Solution: We graph the solid boundary line $5x + 4y = 20$ and choose (0, 0) as the test point.

$$5x + 4y \le 20$$
$$5(0) + 4(0) \stackrel{?}{\le} 20 \quad \text{Let } x = 0 \text{ and } y = 0.$$
$$0 \le 20 \quad \text{True}$$

We shade the half-plane that contains (0, 0), as shown.

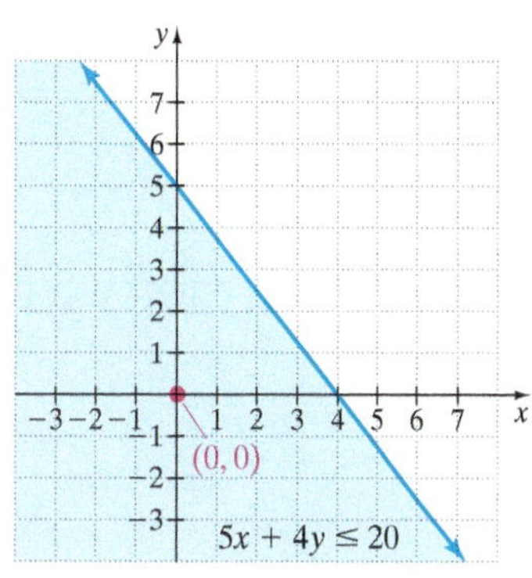

Work Practice 4

Practice 4

Graph: $3x + 4y \ge 12$

Example 5 Graph: $y > 3$

Solution: We graph the dashed boundary line $y = 3$ and choose (0, 0) as the test point. (Recall that the graph of $y = 3$ is a horizontal line with y-intercept 3.)

$$y > 3$$
$$0 \stackrel{?}{>} 3 \quad \text{Let } y = 0.$$
$$0 > 3 \quad \text{False}$$

We shade the half-plane that does not contain (0, 0), as shown.

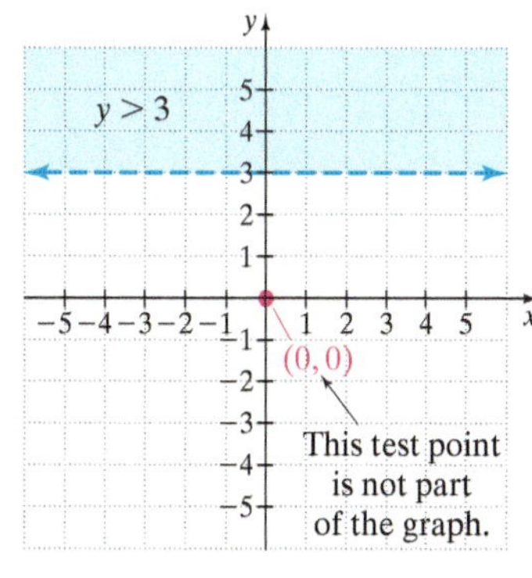

Work Practice 5

Practice 5

Graph: $x > 3$

Answers

4. 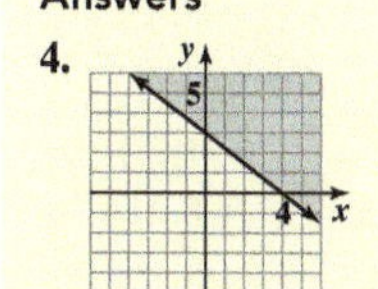

5. 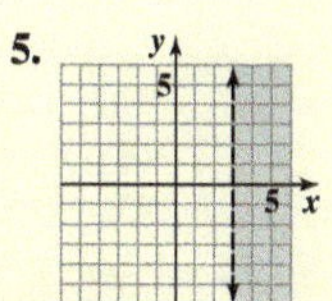

Objective 2 Solving Systems of Linear Inequalities

Just as two linear equations make a system of linear equations, two linear inequalities make a **system of linear inequalities.** Systems of inequalities are very important in a process called linear programming. Many businesses use linear programming to find the most profitable way to use limited resources such as employees, machines, or buildings.

A **solution of a system of linear inequalities** is an ordered pair that satisfies each inequality in the system. The set of all such ordered pairs is the solution set of the system. Graphing this set gives us a picture of the solution set. We can graph a system of inequalities by graphing each inequality in the system and identifying the region of overlap.

Practice 6

Graph the solution of the system: $\begin{cases} 4x \le y \\ x + 3y \ge 9 \end{cases}$

Example 6 Graph the solution of the system: $\begin{cases} 3x \ge y \\ x + 2y \le 8 \end{cases}$

Solution: We begin by graphing each inequality on the same set of axes. The graph of the solution region of the system is the region contained in the graphs of both inequalities. It is their intersection.

First, graph $3x \ge y$. The boundary line is the graph of $3x = y$. Sketch a solid boundary line since the inequality $3x \ge y$ means $3x > y$ or $3x = y$. The test point $(1, 0)$ satisfies the inequality, so shade the half-plane that includes $(1, 0)$.

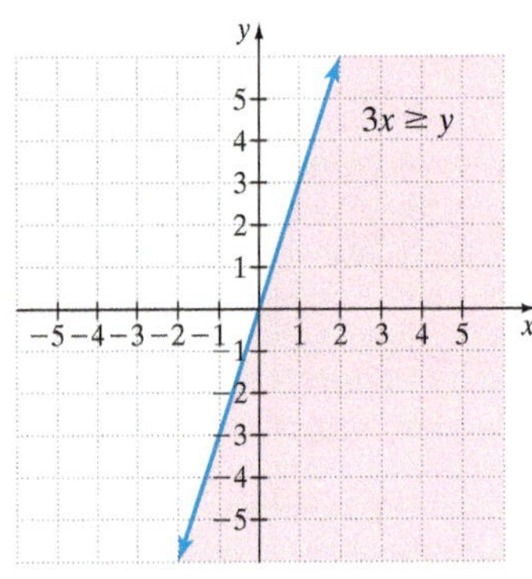

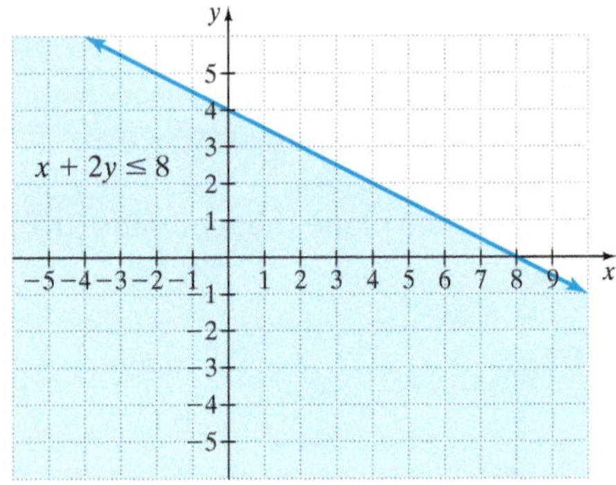

Next, sketch a solid boundary line $x + 2y = 8$ on the same set of axes. The test point $(0, 0)$ satisfies the inequality $x + 2y \le 8$, so shade the half-plane that includes $(0, 0)$. (For clarity, the graph of $x + 2y \le 8$ is shown on a separate set of axes.)

An ordered pair solution of the system must satisfy both inequalities. These solutions are points that lie in both shaded regions. The solution region of the system is the purple shaded region as seen below. This solution region includes parts of both boundary lines.

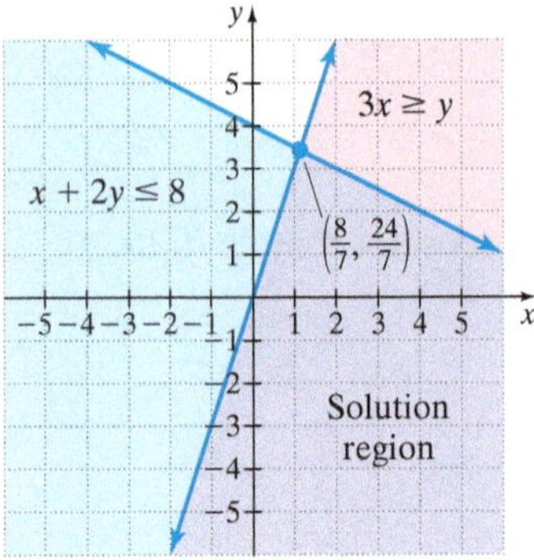

Work Practice 6

Answer

6. 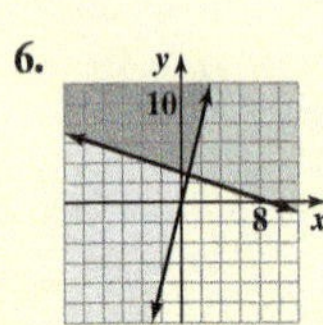

In linear programming, it is sometimes necessary to find the coordinates of the **corner point:** the point at which the two boundary lines intersect. To find the point of intersection, solve the related linear system

$$\begin{cases} 3x = y \\ x + 2y = 8 \end{cases}$$

by the substitution method or the addition method. The lines intersect at $\left(\frac{8}{7}, \frac{24}{7}\right)$, the corner point of the graph.

Graphing the Solution Region of a System of Linear Inequalities

Step 1: Graph each inequality in the system on the same set of axes.

Step 2: The solutions (or solution region) of the system are the points common to the graphs of all the inequalities in the system.

Example 7 Graph the solution of the system: $\begin{cases} x - y < 2 \\ x + 2y > -1 \end{cases}$

Solution: Graph both inequalities on the same set of axes. Both boundary lines are dashed lines since the inequality symbols are $<$ and $>$. The solution region of the system is the region shown by the purple shading. In this example, the boundary lines are not a part of the solution.

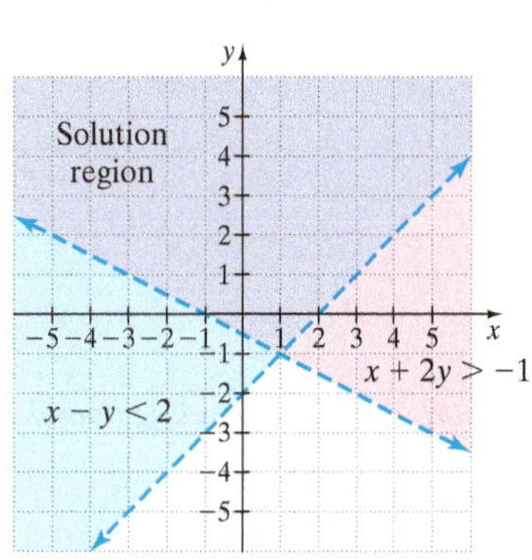

Work Practice 7

Practice 7

Graph the solution of the system: $\begin{cases} x - y > 4 \\ x + 3y < -4 \end{cases}$

Example 8 Graph the solution of the system: $\begin{cases} -3x + 4y < 12 \\ x \geq 2 \end{cases}$

Solution: Graph both inequalities on the same set of axes.

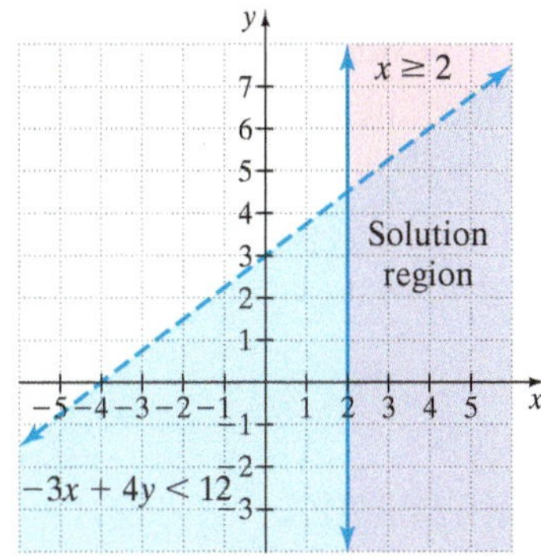

The solution region of the system is the purple shaded region, including a portion of the line $x = 2$.

Work Practice 8

Practice 8

Graph the solution of the system: $\begin{cases} y \leq 6 \\ -2x + 5y > 10 \end{cases}$

Answers

7. 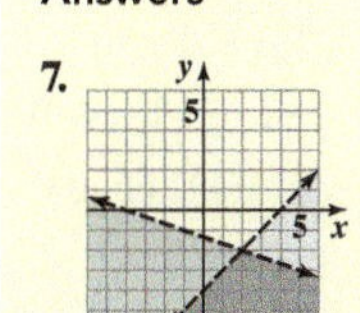

8. 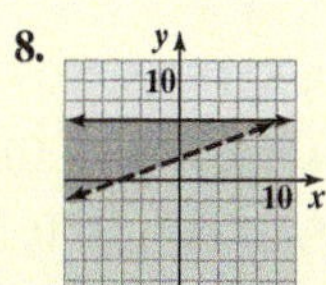

Vocabulary, Readiness & Video Check

Use the choices below to fill in each blank. Some choices may be used more than once and some not at all.

true	$x < 3$	$y < 3$	half-planes	yes
false	$x \leq 3$	$y \leq 3$	linear inequality in two variables	no

1. The statement $5x - 6y < 7$ is an example of a(n) ________________.
2. A boundary line divides a plane into two regions called ________________.
3. True or false: The graph of $5x - 6y < 7$ includes its corresponding boundary line. ________________
4. True or false: When graphing a linear inequality, to determine which side of the boundary line to shade, choose a point *not* on the boundary line. ________________
5. True or false: The boundary line for the inequality $5x - 6y < 7$ is the graph of $5x - 6y = 7$.

____

6. The graph of ________________ is

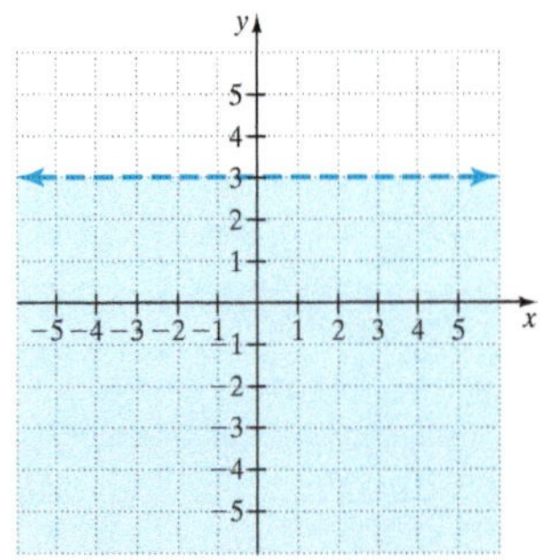

State whether the graph of each inequality includes its corresponding boundary line. Answer yes or no.

7. $y \geq x + 4$
8. $x - y > -7$
9. $y \geq x$
10. $x > 0$

Martin-Gay Interactive Videos *Watch the section lecture video and answer the following questions.*

See Video 16.4 

Objective 1 11. From Example 1, how do you find the equation of the boundary line? How do you determine if the points on the boundary line are solutions of the inequality?

Objective 2 12. In Example 2, did the graph of the first inequality of the system limit where we could choose the test point for the second inequality? Why or why not?

16.4 Exercise Set MyMathLab®

Determine whether the ordered pairs given are solutions of the linear inequality in two variables.

1. $x - y > 3$; $(2, -1)$, $(5, 1)$
2. $y - x < -2$; $(2, 1)$, $(5, -1)$
3. $3x - 5y \leq -4$; $(-1, -1)$, $(4, 0)$
4. $2x + y \geq 10$; $(-1, -4)$, $(5, 0)$
5. $x < -y$; $(0, 2)$, $(-5, 1)$
6. $y > 3x$; $(0, 0)$, $(-1, -4)$

MIXED PRACTICE

Graph each inequality. See Examples 1 through 5.

7. $x + y \leq 1$
8. $x + y \geq -2$
9. $2x + y > -4$
10. $x + 3y \leq 3$
11. $x + 6y \leq -6$
12. $7x + y > -14$
13. $2x + 5y > -10$
14. $5x + 2y \leq 10$
15. $x + 2y \leq 3$
16. $2x + 3y > -5$

17. $2x + 7y > 5$
18. $3x + 5y \le -2$
19. $x - 2y \ge 3$
20. $4x + y \le 2$
21. $5x + y < 3$
22. $x + 2y > -7$
23. $4x + y < 8$
24. $9x + 2y \ge -9$
25. $y \ge 2x$
26. $x < 5y$
27. $x \ge 0$
28. $y \le 0$
29. $y \le -3$
30. $x > -\frac{2}{3}$
31. $2x - 7y > 0$
32. $5x + 2y \le 0$
33. $3x - 7y \ge 0$
34. $-2x - 9y > 0$
35. $x > y$
36. $x \le -y$
37. $x - y \le 6$
38. $x - y > 10$
39. $-\frac{1}{4}y + \frac{1}{3}x > 1$
40. $\frac{1}{2}x - \frac{1}{3}y \le -1$
41. $-x < 0.4y$
42. $0.3x \ge 0.1y$

In Exercises 43 through 48, match each inequality with its graph.

a. $x > 2$ b. $y < 2$ c. $y < 2x$
d. $y \le -3x$ e. $2x + 3y < 6$ f. $3x + 2y > 6$

43. 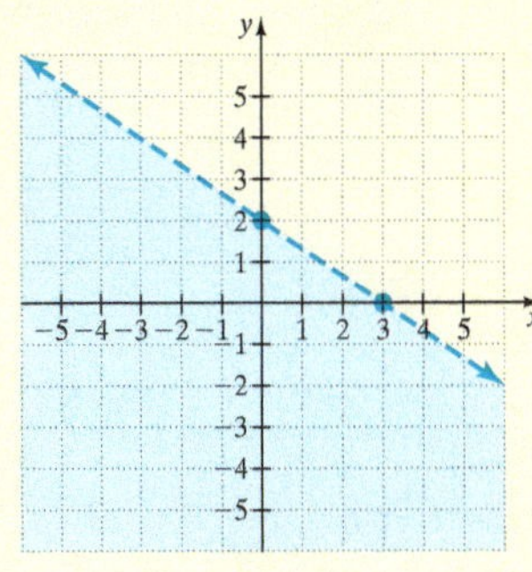

44. 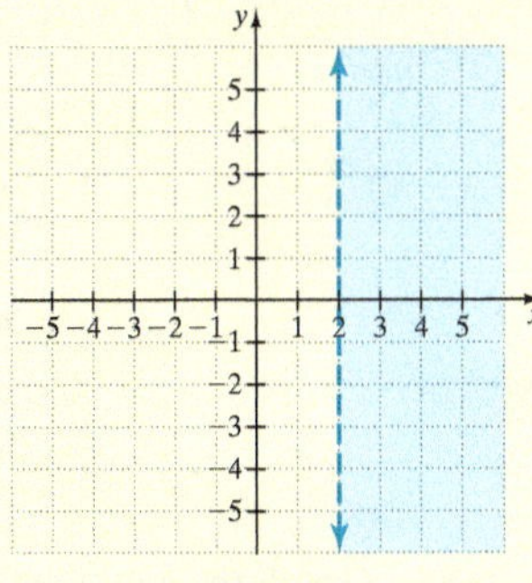

45. 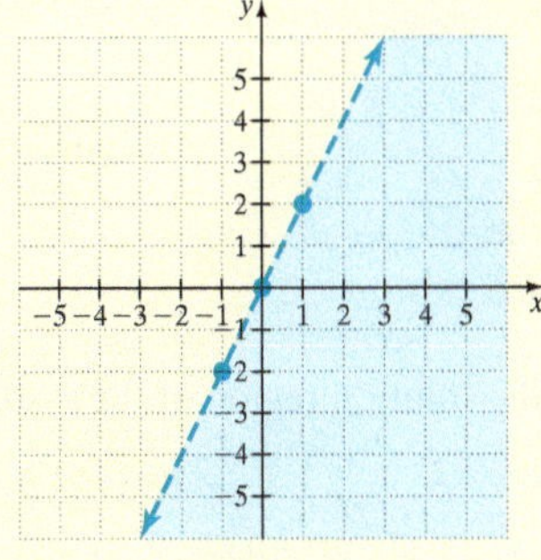

46. 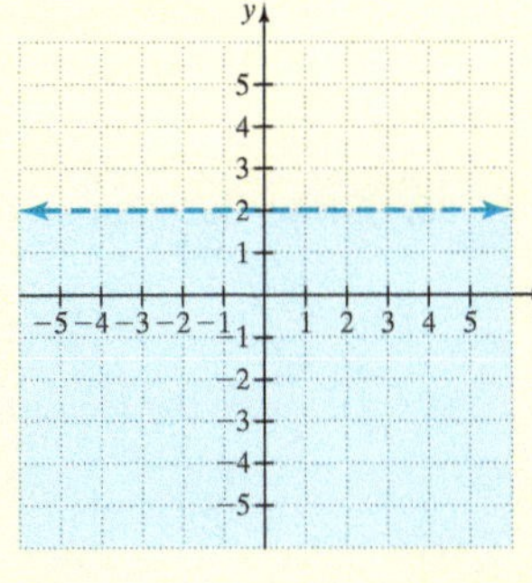

47. 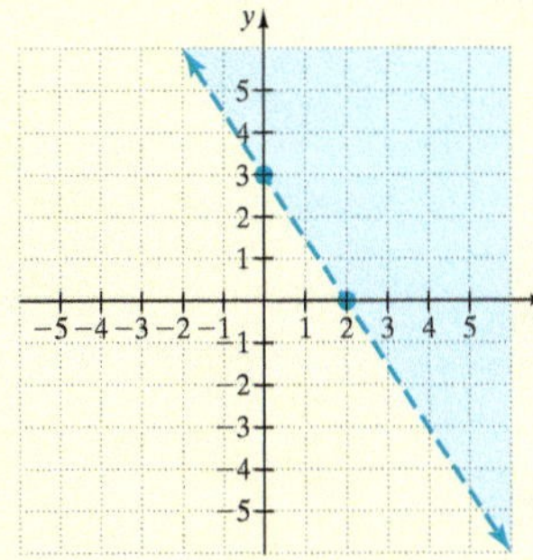

48. 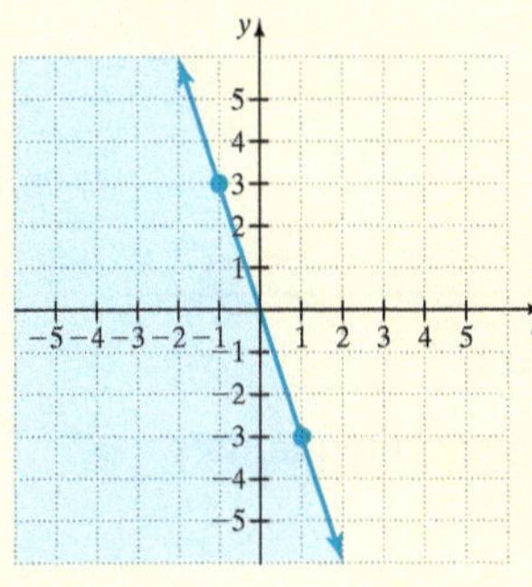

Graph the solution of each system of linear inequalities. See Examples 6 through 8.

49. $\begin{cases} y \ge x + 1 \\ y \ge 3 - x \end{cases}$
50. $\begin{cases} y \ge x - 3 \\ y \ge -1 - x \end{cases}$
51. $\begin{cases} y < 3x - 4 \\ y \le x + 2 \end{cases}$
52. $\begin{cases} y \le 2x + 1 \\ y > x + 2 \end{cases}$
53. $\begin{cases} y \le -2x - 2 \\ y \ge x + 4 \end{cases}$
54. $\begin{cases} y \le 2x + 4 \\ y \ge -x - 5 \end{cases}$
55. $\begin{cases} y \ge -x + 2 \\ y \le 2x + 5 \end{cases}$
56. $\begin{cases} y \ge x - 5 \\ y \le -3x + 3 \end{cases}$
57. $\begin{cases} x \ge 3y \\ x + 3y \le 6 \end{cases}$
58. $\begin{cases} -2x < y \\ x + 2y < 3 \end{cases}$
59. $\begin{cases} y + 2x \ge 0 \\ 5x - 3y \le 12 \end{cases}$
60. $\begin{cases} y + 2x \le 0 \\ 5x + 3y \ge -2 \end{cases}$
61. $\begin{cases} 3x - 4y \ge -6 \\ 2x + y \le 7 \end{cases}$
62. $\begin{cases} 4x - y \ge -2 \\ 2x + 3y \le -8 \end{cases}$
63. $\begin{cases} x \le 2 \\ y \ge -3 \end{cases}$
64. $\begin{cases} x \ge -3 \\ y \ge -2 \end{cases}$
65. $\begin{cases} y \ge 1 \\ x < -3 \end{cases}$
66. $\begin{cases} y > 2 \\ x \ge -1 \end{cases}$
67. $\begin{cases} 2x + 3y < -8 \\ x \ge -4 \end{cases}$
68. $\begin{cases} 3x + 2y \le 6 \\ x < 2 \end{cases}$
69. $\begin{cases} 2x - 5y \le 9 \\ y \le -3 \end{cases}$
70. $\begin{cases} 2x + 5y \le -10 \\ y \ge 1 \end{cases}$
71. $\begin{cases} y \ge \frac{1}{2}x + 2 \\ y \le \frac{1}{2}x - 3 \end{cases}$
72. $\begin{cases} y \ge -\frac{3}{2}x + 3 \\ y < -\frac{3}{2}x + 6 \end{cases}$

REVIEW AND PREVIEW

Evaluate each expression for the given replacement value. See Section 12.1.

73. x^2 if x is -5
74. x^3 if x is -5
75. $2x^3$ if x is -1
76. $3x^2$ if x is -1

CONCEPT EXTENSIONS

Determine whether (1, 1) is included in each graph. See the Concept Check in this section.

77. $3x + 4y < 8$
78. $y > 5x$
79. $y \ge -\frac{1}{2}x$
80. $x > 3$
81. Write an inequality whose solutions are all pairs of numbers x and y whose sum is at least 13. Graph the inequality.
82. Write an inequality whose solutions are all the pairs of numbers x and y whose sum is at most -4. Graph the inequality.
83. Explain why a point on the boundary line should not be chosen as the test point.
84. Describe the graph of a linear inequality.
85. The price for a taxi cab in a small city is \$2.50 per mile, x, while traveling and \$0.25 every minute, y, while waiting. If you have \$20 to spend on a cab ride, the inequality

$$2.5x + 0.25y \le 20$$

represents your situation. Graph this inequality in the first quadrant only.

86. A word processor charges \$22 per hour, x, for typing a first draft and \$15 per hour, y, for making changes and typing a second draft. If you need a document typed and have \$100, the inequality

$$22x + 15y \le 100$$

represents your situation. Graph the inequality in the first quadrant only.

87. In Exercises 85 and 86, why were you instructed to graph each inequality in the first quadrant only?

88. Scott Sambracci and Sara Thygeson are planning their wedding. They have calculated that they want the cost of their wedding ceremony, x, plus the cost of their reception, y, to be no more than \$5000.

a. Write an inequality describing this relationship.

b. Graph this inequality.

c. Why should we be interested in only quadrant I of this graph?

89. It's the end of the budgeting period for Dennis Fernandes, and he has \$500 left in his budget for car rental expenses. He plans to spend this budget on a sales trip throughout southern Texas. He will rent a car that costs \$30 per day and \$0.15 per mile, and he can spend no more than \$500.

a. Write an inequality describing this situation. Let x = number of days and let y = number of miles.

b. Graph this inequality.

c. Why should we be interested in only quadrant I of this graph?

90. Explain how to decide which region to shade to show the solution region of the following system.

$$\begin{cases} x \ge 3 \\ y \ge -2 \end{cases}$$

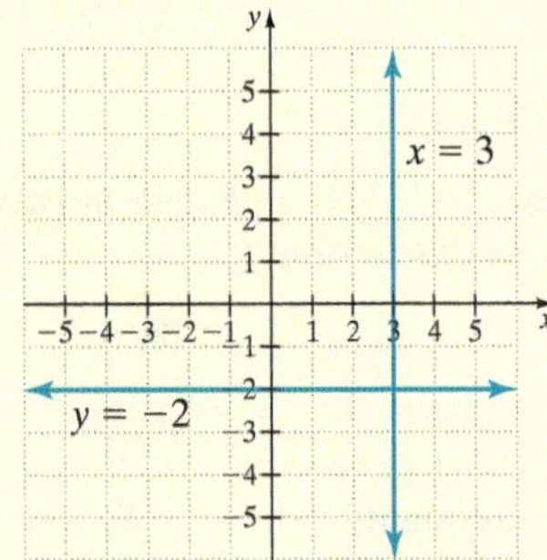

For each system of inequalities, choose the corresponding graph.

91. $\begin{cases} y < 5 \\ x > 3 \end{cases}$

92. $\begin{cases} y > 5 \\ x < 3 \end{cases}$

93. $\begin{cases} y \le 5 \\ x < 3 \end{cases}$

94. $\begin{cases} y > 5 \\ x \ge 3 \end{cases}$

A.

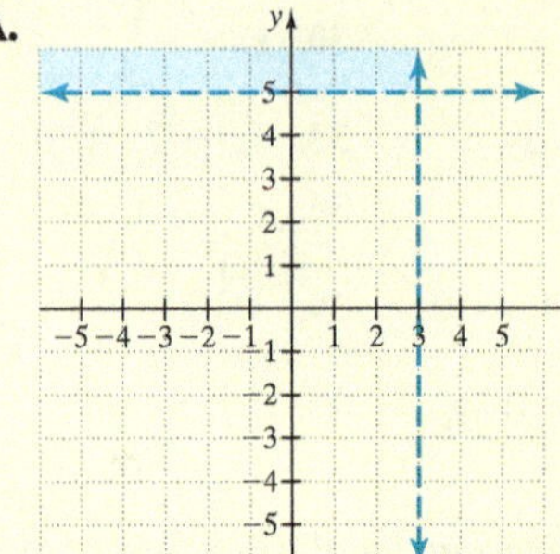

B.

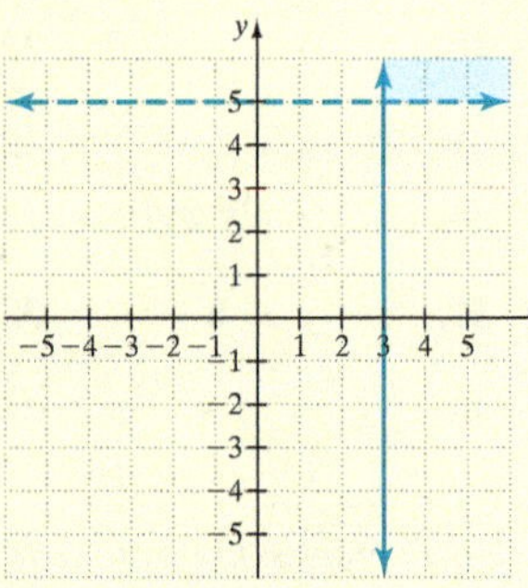

C.

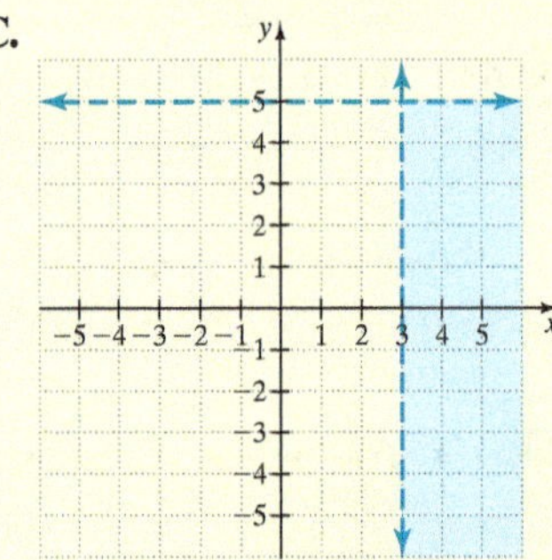

D.

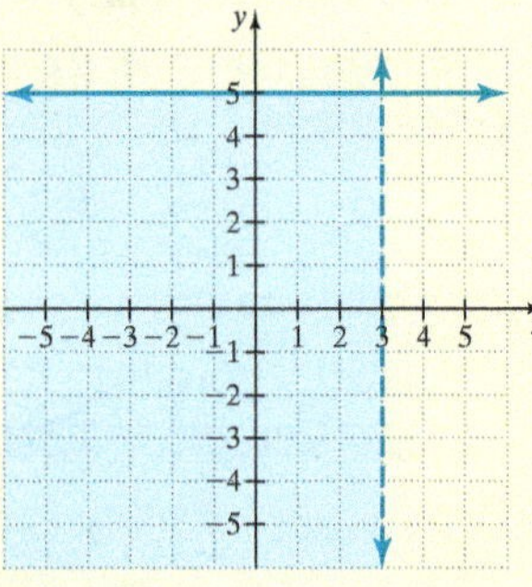

95. Graph the solution of $\begin{cases} 2x - y \le 6 \\ x \ge 3 \\ y > 2 \end{cases}$

96. Graph the solution of $\begin{cases} x + y < 5 \\ y < 2x \\ x \ge 0 \\ y \ge 0 \end{cases}$

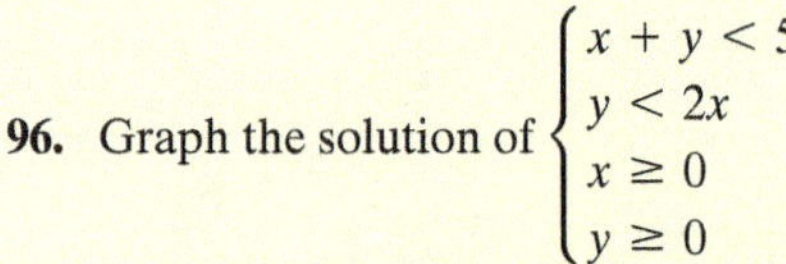

97. Describe the location of the solution region of the system $\begin{cases} x > 0 \\ y > 0 \end{cases}$.

Chapter 16 Vocabulary Check

Fill in each blank with one of the words or phrases listed below.

compound inequality	solution	system of linear inequalities
absolute value	union	intersection

1. The statement "$x < 5$ or $x > 7$" is called a(n) ________________.

2. The ________________ of two sets is the set of all elements common to both sets.

3. The ________________ of two sets is the set of all elements that belong to either of the sets.

4. A number's distance from 0 is called its ________________.

5. When a variable in an equation is replaced by a number and the resulting equation is true, then that number is called a(n) ______________ of the equation.

6. Two or more linear inequalities are called a(n) ______________.

Helpful Hint

Are you preparing for your test? Don't forget to take the Chapter 16 Test on page 1136. Then check your answers at the back of the text and use the Chapter Test Prep Videos to see the fully worked-out solutions to any of the exercises you want to review.

16 Chapter Highlights

Definitions and Concepts	Examples
Section 16.1 Compound Inequalities	
Two inequalities joined by the words **and** or **or** are called **compound inequalities.**	Compound inequalities: $x - 7 \le 4$ *and* $x \ge -21$ $2x + 7 > x - 3$ *or* $5x + 2 > -3$
The solution set of a compound inequality formed by the word **and** is the **intersection,** ∩, of the solution sets of the two inequalities.	Solve for x: $x < 5$ *and* $x < 3$ $\{x \mid x < 5\}$ 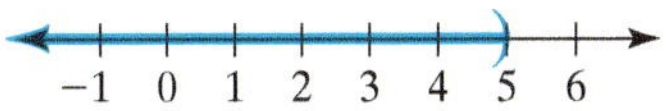 $(-\infty, 5)$ $\{x \mid x < 3\}$ 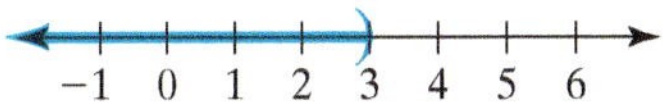 $(-\infty, 3)$ $\{x \mid x < 3$ and $x < 5\}$ 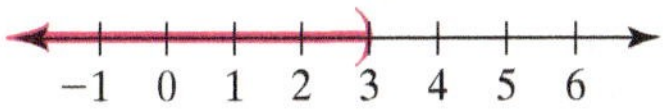 $(-\infty, 3)$
The solution set of a compound inequality formed by the word **or** is the **union,** ∪, of the solution sets of the two inequalities.	Solve for x: $x - 2 \ge -3$ or $2x \le -4$ $x \ge -1$ or $x \le -2$ $\{x \mid x \ge -1\}$ 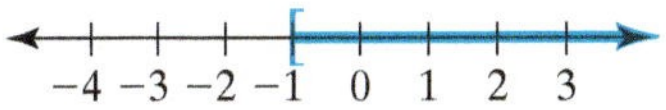 $[-1, \infty)$ $\{x \mid x \le -2\}$ 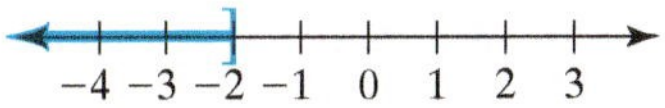 $(-\infty, -2]$ $\{x \mid x \le -2$ or $x \ge -1\}$ 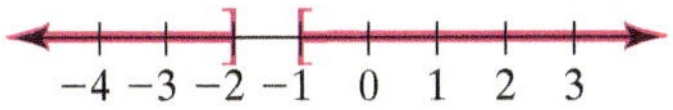 $(-\infty, -2] \cup [-1, \infty)$
Section 16.2 Absolute Value Equations	
If a is a positive number, then $\lvert x \rvert = a$ is equivalent to $x = a$ or $x = -a$.	Solve for y: $\lvert 5y - 1 \rvert - 7 = 4$ $\lvert 5y - 1 \rvert = 11$ Add 7. $5y - 1 = 11$ or $5y - 1 = -11$ $5y = 12$ or $5y = -10$ Add 1. $y = \frac{12}{5}$ or $y = -2$ Divide by 5. The solutions are -2 and $\frac{12}{5}$.

(continued)

Definitions and Concepts	Examples
Section 16.2 Absolute Value Equations (*continued*)	
If a is negative, then $\lvert x \rvert = a$ has no solution.	Solve for x: $\left\lvert \frac{x}{2} - 7 \right\rvert = -1$ The solution set is $\{\ \}$ or $\varnothing$.
If an absolute value equation is of the form $\lvert x \rvert = \lvert y \rvert$, solve $x = y$ or $x = -y$.	Solve for x: $\lvert x - 7 \rvert = \lvert 2x + 1 \rvert$ $x - 7 = 2x + 1$ or $x - 7 = -(2x + 1)$ $x = 2x + 8$ $\quad$ $x - 7 = -2x - 1$ $-x = 8$ $\quad$ $x = -2x + 6$ $x = -8$ or $3x = 6$ $x = 2$ The solutions are -8 and 2.
Section 16.3 Absolute Value Inequalities	
If a is a positive number, then $\lvert x \rvert < a$ is equivalent to $-a < x < a$.	Solve for y: $\lvert y - 5 \rvert \le 3$ $-3 \le y - 5 \le 3$ $-3 + 5 \le y - 5 + 5 \le 3 + 5$ Add 5. $2 \le y \le 8$ The solution set is $[2, 8]$.
If a is a positive number, then $\lvert x \rvert > a$ is equivalent to $x < -a$ or $x > a$.	Solve for x: $\left\lvert \frac{x}{2} - 3 \right\rvert > 7$ $\frac{x}{2} - 3 < -7$ or $\frac{x}{2} - 3 > 7$ $x - 6 < -14$ or $x - 6 > 14$ Multiply by 2. $x < -8$ or $x > 20$ Add 6. The solution set is $(-\infty, -8) \cup (20, \infty)$.
Section 16.4 Graphing Linear Inequalities in Two Variables and Systems of Linear Inequalities	
A **linear inequality in two variables** is an inequality that can be written in one of the forms: $Ax + By < C$ $\quad$ $Ax + By \le C$ $Ax + By > C$ $\quad$ $Ax + By \ge C$ To graph a linear inequality: **1.** Graph the boundary line by graphing the related equation. Draw the line solid if the inequality symbol is $\le$ or $\ge$. Draw the line dashed if the inequality symbol is $<$ or $>$. **2.** Choose a test point not on the line. Substitute its coordinates into the original inequality.	**Linear Inequalities** $2x - 5y < 6$ $\quad$ $x \ge -5$ $y > -8x$ $\quad$ $y \le 2$ Graph $2x - y \le 4$. **1.** Graph $2x - y = 4$. Draw a solid line because the inequality symbol is $\le$. **2.** Check the test point $(0, 0)$ in the inequality $2x - y \le 4$. $2 \cdot 0 - 0 \le 4$ Let $x = 0$ and $y = 0$. $0 \le 4$ True

Definitions and Concepts	Examples
Section 16.4 Graphing Linear Inequalities in Two Variables and Systems of Linear Inequalities (*continued*)	
3. If the resulting inequality is true, shade the half-plane that contains the test point. If the inequality is not true, shade the half-plane that does not contain the test point.	**3.** The inequality is true, so we shade the half-plane containing (0, 0). 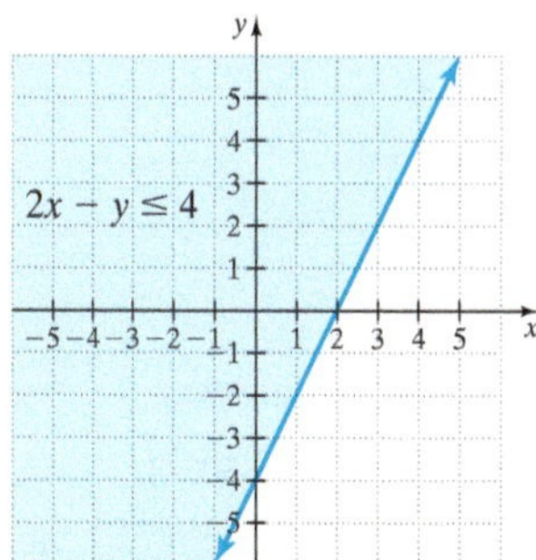
A system of linear inequalities consists of two or more linear inequalities.	System of Linear Inequalities: $\begin{cases} x - y \ge 3 \\ y \le -2x \end{cases}$
To graph a system of inequalities, graph each inequality in the system. The overlapping region is the solution of the system.	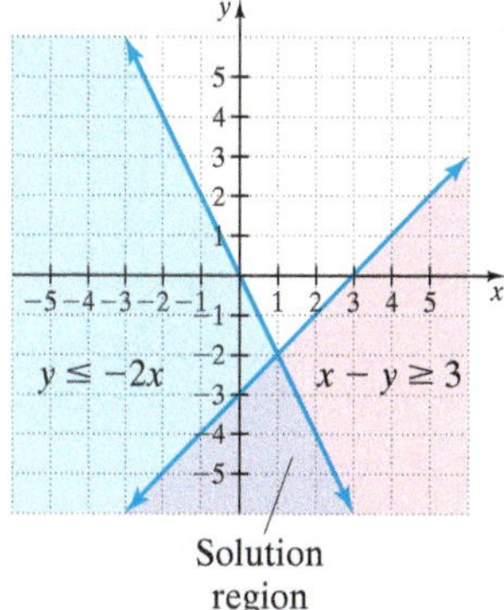

Chapter 16 Review

(16.1) *Solve each inequality. Write your answers in interval notation.*

1. $-3 < 4(2x - 1) < 12$

2. $-2 \le 8 + 5x < -1$

3. $\frac{1}{6} < \frac{4x - 3}{3} \le \frac{4}{5}$

4. $-6 < x - (3 - 4x) < -3$

5. $3x - 5 > 6$ *or* $-x < -5$

6. $x \le 2$ *and* $x > -5$

(16.2) *Solve each absolute value equation.*

7. $|8 - x| = 3$

8. $|x - 7| = 9$

9. $|-3x + 4| = 7$

10. $|2x + 9| = 9$

11. $5 + |6x + 1| = 5$

12. $|3x - 2| + 6 = 10$

13. $|5 - 6x| + 8 = 3$

14. $-5 = |4x - 3|$

15. $\left|\frac{3x - 7}{4}\right| = 2$

16. $-4 = \left|\frac{x - 3}{2}\right| - 5$

17. $|6x + 1| = |15 + 4x|$

18. $|x - 3| = |x + 5|$

(16.3) *Solve each absolute value inequality. Graph the solution set and write it in interval notation.*

19. $|5x - 1| < 9$

20. $|6 + 4x| \ge 10$

21. $|3x| - 8 > 1$

22. $9 + |5x| < 24$

23. $|6x - 5| \le -1$

24. $|6x - 5| \leq 5$

25. $\left|3x + \frac{2}{5}\right| \geq 4$

26. $|5x - 3| > 2$

27. $\left|\frac{x}{3} + 6\right| - 8 > -5$

28. $\left|\frac{4(x - 1)}{7}\right| + 10 < 2$

(16.4) Graph the following inequalities.

29. $3x - 4y \leq 0$

30. $3x - 4y \geq 0$

31. $x + 6y < 6$

32. $y \leq -4$

33. $y \geq -7$

34. $x \geq -y$

Graph the solution region of the following systems of linear inequalities.

35. $\begin{cases} y \geq 2x - 3 \\ y \leq -2x + 1 \end{cases}$

36. $\begin{cases} y \leq -3x - 3 \\ y \leq 2x + 7 \end{cases}$

37. $\begin{cases} x + 2y > 0 \\ x - y \leq 6 \end{cases}$

38. $\begin{cases} 4x - y \leq 0 \\ 3x - 2y \geq -5 \end{cases}$

39. $\begin{cases} 3x - 2y \leq 4 \\ 2x + y \geq 5 \end{cases}$

40. $\begin{cases} -2x + 3y > -7 \\ x \geq -2 \end{cases}$

MIXED REVIEW

Solve. If an inequality, write your solutions in interval notation.

41. $0 \leq \frac{2(3x + 4)}{5} \leq 3$

42. $x \leq 2$ or $x > -5$

43. $-2x \leq 6$ and $-2x + 3 < -7$

44. $|7x| - 26 = -5$

45. $\left|\frac{9 - 2x}{5}\right| = -3$

46. $|x - 3| = |7 + 2x|$

47. $|6x - 5| \geq -1$

48. $\left|\frac{4x - 3}{5}\right| < 1$

Graph the solutions.

49. $-x \leq y$

50. $x + y > -2$

51. $\begin{cases} -3x + 2y > -1 \\ y < -2 \end{cases}$

52. $\begin{cases} x - 2y \geq 7 \\ x + y \leq -5 \end{cases}$

Chapter 16 Test

Step-by-step test solutions are found on the Chapter Test Prep Videos. Where available: MyMathLab® or YouTube™

Solve each equation or inequality.

1. $|6x - 5| - 3 = -2$

2. $|8 - 2t| = -6$

3. $|x - 5| = |x + 2|$

4. $-3 < 2(x - 3) \leq 4$

5. $|3x + 1| > 5$

6. $|x - 5| - 4 < -2$

7. $x \leq -2$ and $x \leq -5$

8. $x \leq -2$ or $x \leq -5$

9. $-x > 1$ and $3x + 3 \geq x - 3$

10. $6x + 1 > 5x + 4$ or $1 - x > -4$

11. $\left|\frac{5x - 7}{2}\right| = 4$

12. $\left|17x - \frac{1}{5}\right| > -2$

13. $-1 \leq \frac{2x - 5}{3} < 2$

Graph each linear equality.

14. $y > -4x$

15. $2x - 3y > -6$

Graph the solutions of the following systems of linear inequalities.

16. $\begin{cases} y + 2x \leq 4 \\ y \geq 2 \end{cases}$

17. $\begin{cases} 2y - x \geq 1 \\ x + y \geq -4 \end{cases}$

Chapter 16 Cumulative Review

1. Find the value of each expression when $x = \frac{7}{8}$ and $y = -\frac{1}{3}$.

   a. xy b. $x \div y$

2. Find the value of each expression when $x = -\frac{1}{2}$ and $y = \frac{2}{3}$.

   a. xy b. $x \div y$

3. Simplify: $\frac{12 - 16}{-1 + 3}$

4. Simplify each expression.

   a. $\frac{4(-3) - (-6)}{-8 + 4}$ b. $\frac{3 + (-3)(-2)^3}{-1 - (-4)}$

5. Simplify each expression by combining like terms.

   a. $4x + 6x$ b. $y - 5y$

   c. $3x^2 + 5x^2 - 2$

6. Simplify each expression by combining like terms.

   a. $4x - 3 + 7 - 5x$

   b. $-6y + 3y - 8 + 8y$

   c. $2 + 8.1a + a - 6$

   d. $2x^2 - 2x$

7. Solve $5x + 2 - 4x = 7 - 19$.

8. Solve $6y - 11 + 4 + 2y = 8 + 15y - 8y$.

9. Complete the table for the equation $y = 3x$.

x	y
-1	
	0
	-9

10. Complete the table for the equation $2x + y = 6$.

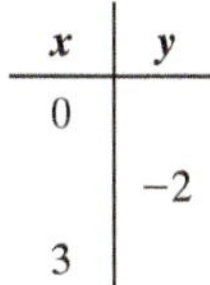

x	y
0	
	-2
3	

11. Identify the x- and y-intercepts.

a.

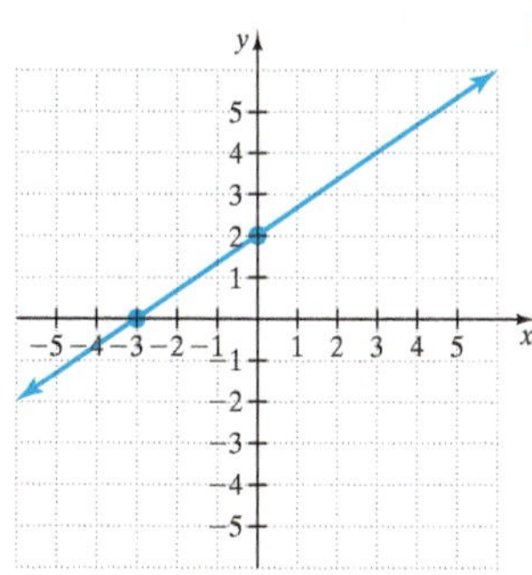

b.

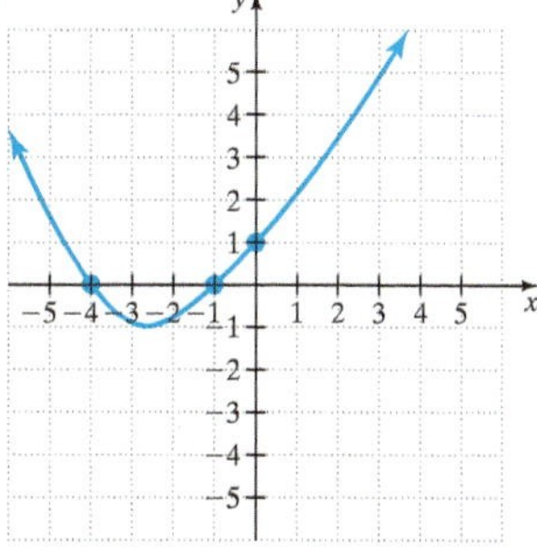

c.

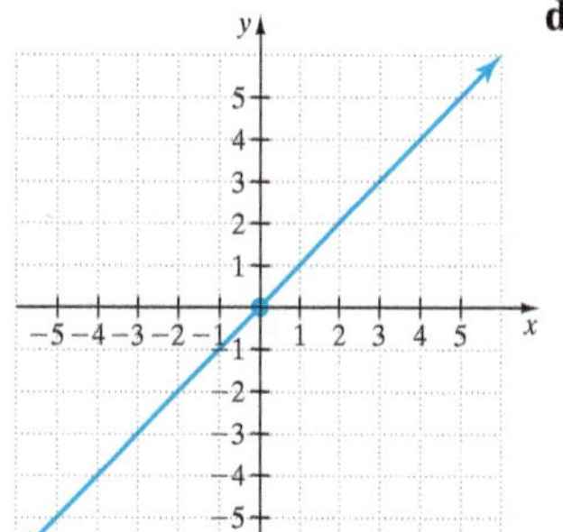

d.

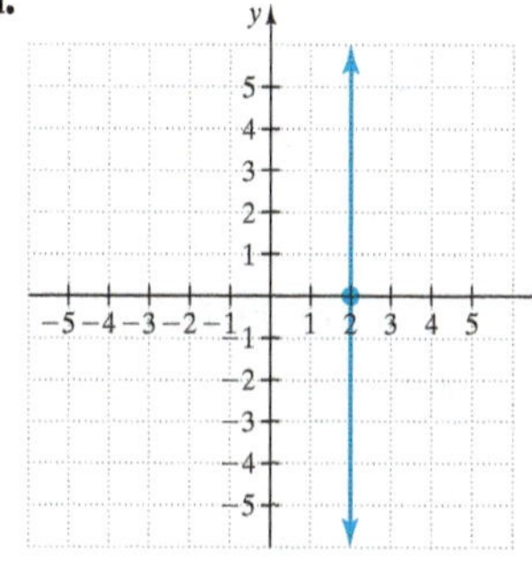

e.

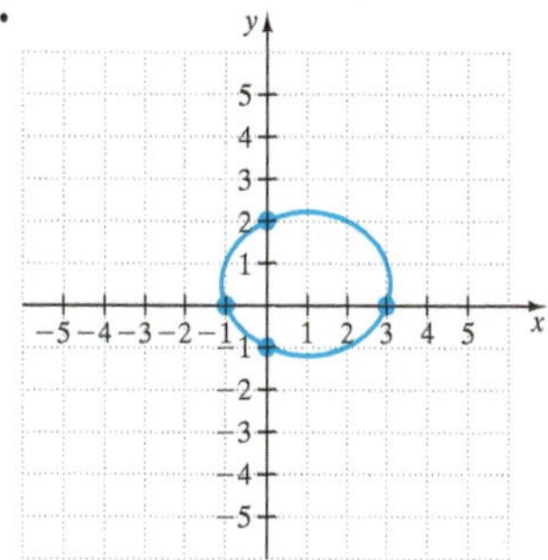

12. Identify the x- and y-intercepts.

a.

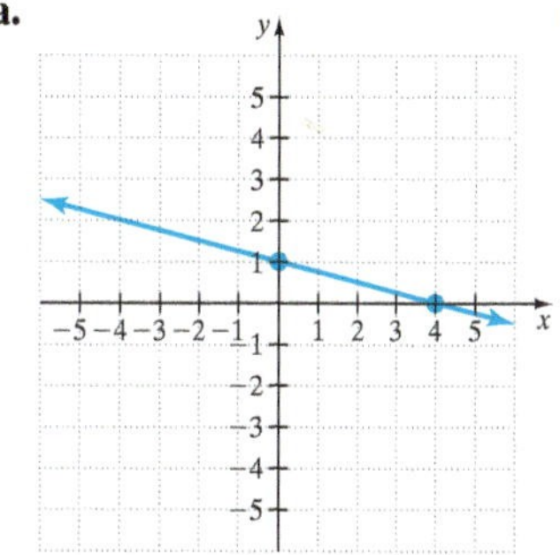

b.

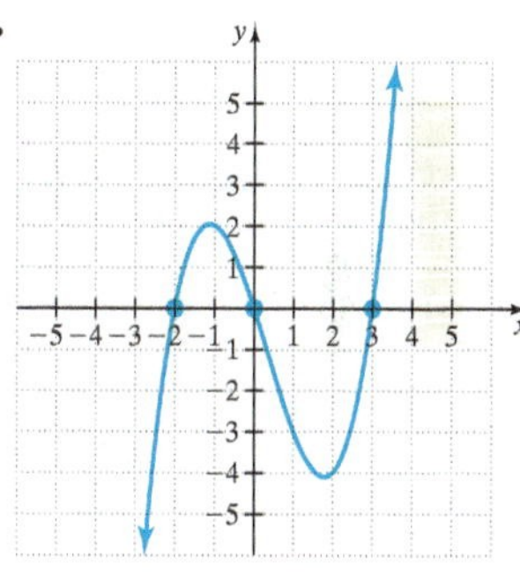

c.

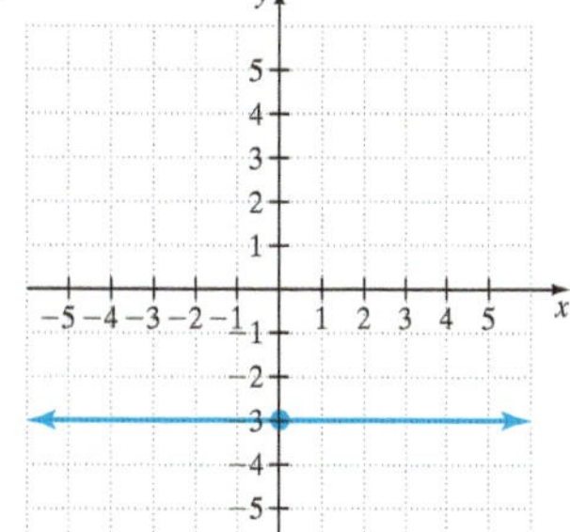

d.

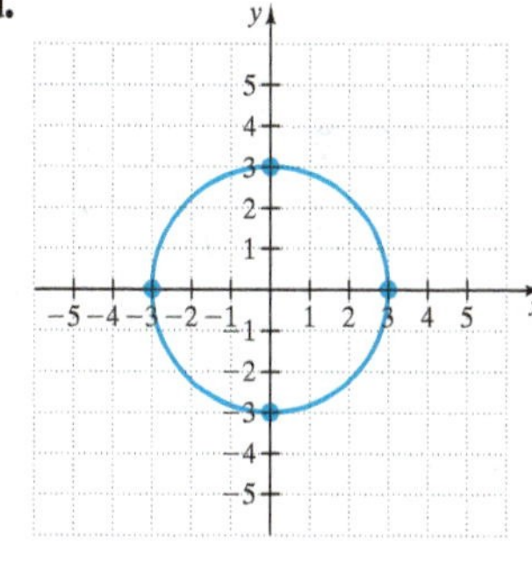

13. Determine whether the graphs of $y = -\frac{1}{5}x + 1$ and $2x + 10y = 3$ are parallel lines, perpendicular lines, or neither.

14. Determine whether the graphs of $y = 3x + 7$ and $x + 3y = -15$ are parallel lines, perpendicular lines, or neither.

15. Find an equation of the line with y-intercept $(0, -3)$ and slope of $\frac{1}{4}$.

16. Find an equation of the line with y-intercept $(0, 4)$ and slope of -2.

17. Find an equation of the line parallel to the line $y = 5$ and passing through $(-2, -3)$.

18. Find an equation of the line perpendicular to $y = 2x + 4$ and passing through $(1, 5)$.

19. Which of the following linear equations are functions?

a. $y = x$
b. $y = 2x + 1$
c. $y = 5$
d. $x = -1$

20. Which of the following linear equations are functions?

a. $2x + 3 = y$ **b.** $x + 4 = 0$

c. $\frac{1}{2}y = 2x$ **d.** $y = 0$

21. Determine whether $(12, 6)$ is a solution of the given system.

$$\begin{cases} 2x - 3y = 6 \\ x = 2y \end{cases}$$

22. Which of the following ordered pairs is a solution of the given system?

$$\begin{cases} 2x + y = 4 \\ x + y = 2 \end{cases}$$

a. $(1, 1)$
b. $(2, 0)$

23. Add $(11x^3 - 12x^2 + x - 3)$ and $(x^3 - 10x + 5)$.

24. Combine like terms to simplify.
$4a^2 + 3a - 2a^2 + 7a - 5$

25. Factor $x^2 + 7yx + 6y^2$.

26. Factor $3x^2 + 15x + 18$.

27. Divide $\frac{3x^3y^7}{40} \div \frac{4x^3}{y^2}$.

28. Divide $\frac{12x^2y^3}{5} \div \frac{3y^2}{x}$.

29. Subtract $\frac{2y}{2y - 7} - \frac{7}{2y - 7}$.

30. Subtract $\frac{-4x^2}{x + 1} - \frac{4x}{x + 1}$.

31. Add $\frac{2x}{x^2 + 2x + 1} + \frac{x}{x^2 - 1}$.

32. Add $\frac{3x}{x^2 + 5x + 6} + \frac{1}{x^2 + 2x - 3}$.

33. Solve $\frac{x}{2} + \frac{8}{3} = \frac{1}{6}$.

34. Solve $\frac{1}{21} + \frac{x}{7} = \frac{5}{3}$.

35. Solve the following system of equations by graphing.

$$\begin{cases} 2x + y = 7 \\ 2y = -4x \end{cases}$$

36. Solve the following system by graphing.

$$\begin{cases} y = x + 2 \\ 2x + y = 5 \end{cases}$$

37. Solve the system.

$$\begin{cases} 7x - 3y = -14 \\ -3x + y = 6 \end{cases}$$

38. Solve the system.

$$\begin{cases} 5x + y = 3 \\ y = -5x \end{cases}$$

39. Solve the system.

$$\begin{cases} 3x - 2y = 2 \\ -9x + 6y = -6 \end{cases}$$

40. Solve the system.

$$\begin{cases} -2x + y = 7 \\ 6x - 3y = -21 \end{cases}$$

41. Graph the solution of the system.

$$\begin{cases} -3x + 4y < 12 \\ x \geq 2 \end{cases}$$

42. Graph the solution of the system.

$$\begin{cases} 2x - y \leq 6 \\ y \geq 2 \end{cases}$$

43. Simplify each expression.

a. $x^7 \cdot x^4$ **b.** $\left(\frac{t}{2}\right)^4$

c. $(9y^5)^2$

44. Simplify.

a. $\left(\frac{-6x}{y^3}\right)^3$ **b.** $\frac{(2b^2)^5}{a^2b^7}$

c. $\frac{(3y)^2}{y^2}$ **d.** $\frac{(x^2y^4)^2}{xy^3}$

45. Solve $(5x - 1)(2x^2 + 15x + 18) = 0$.

46. Solve $(x + 1)(2x^2 - 3x - 5) = 0$.

47. Three boxes of CD-Rs (recordable compact discs) cost \$37.47. How much should 5 boxes cost?

48. If the sum of twice a number and 5 is divided by the number plus 1, the result is five-halves. Find the number.

Rational Exponents, Radicals, and Complex Numbers

17

Sections

Check Your Progress

In this chapter, radical notation is reviewed, and then rational exponents are introduced. As the name implies, rational exponents are exponents that are rational numbers. We present an interpretation of rational exponents that is consistent with the meaning and rules already established for integer exponents, and we present two forms of notation for roots: radical and exponent. We conclude this chapter with complex numbers, a natural extension of the real number system.

The Google Lunar X PRIZE is an international competition to safely land a robot on the surface of the moon, travel 500 meters over the lunar surface, and send images and data back to Earth. Teams needed to be registered by December 31, 2010. There are multiple prizes and bonuses, but the first team to land on the moon and complete the mission objectives will be awarded $20 million. The deadline for winning the competition is December 31, 2015, and 18 teams are competing for the prize.

To reach the moon, these vehicles must first leave the gravity of Earth. In Exercises 115 and 116 of Section 17.1, you will calculate the escape velocity of Earth and the moon, the minimum speed an object must reach to escape the pull of a planet's gravity. (*Source:* X PRIZE Foundation)

Composition of the Moon

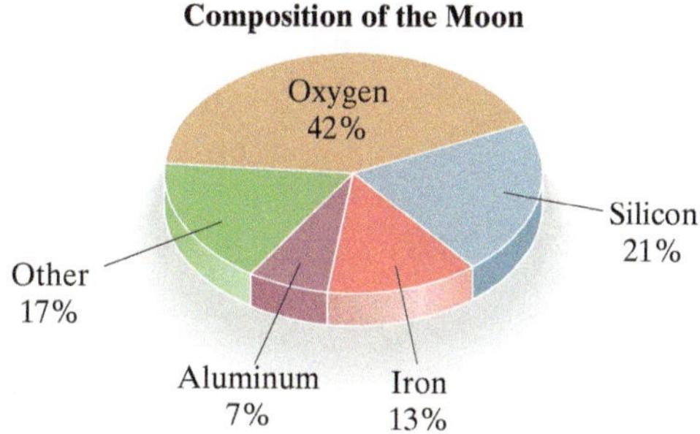

Why the moon? To name a few reasons:

- It is closest to Earth (1.3 seconds for light or radio), so lunar machines can be directly controlled from Earth.
- It is also the closest source of materials to use for any other space project, and it is 22 times easier to launch from the moon than from Earth.
- The moon is 42% oxygen by weight (see the circle graph to the left) and oxygen is the main ingredient of rocket fuel.
- We can collect energy from the moon's surface and transmit it to Earth.

17.1 Radicals and Radical Functions 

Objectives

1. Find Square Roots.
2. Approximate Roots.
3. Find Cube Roots.
4. Find *n*th Roots.
5. Find $\sqrt[n]{a^n}$ Where *a* Is a Real Number.
6. Graph Square and Cube Root Functions.

Objective 1 Finding Square Roots

Recall from Section 15.2 that to find a **square root** of a number *a*, we find a number that was squared to get *a*.

Thus, because

$5^2 = 25$ and $(-5)^2 = 25,$

both 5 and -5 are square roots of 25.

Recall that we denote the **nonnegative,** or **principal, square root** with the **radical sign.**

$\sqrt{25} = 5$

We denote the **negative square root** with the **negative radical sign.**

$-\sqrt{25} = -5$

An expression containing a radical sign is called a **radical expression.** An expression within, or "under," a radical sign is called a **radicand.**

radical expression: 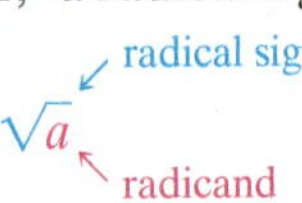

Principal and Negative Square Roots

If *a* is a nonnegative number, then

$\sqrt{a}$ is the **principal,** or **nonnegative, square root** of *a*

$-\sqrt{a}$ is the **negative square root** of *a*

Practice 1

Simplify. Assume that all variables represent positive numbers.

a. $\sqrt{49}$ **b.** $\sqrt{\frac{0}{1}}$ **c.** $\sqrt{\frac{16}{81}}$

d. $\sqrt{0.64}$ **e.** $\sqrt{z^8}$ **f.** $\sqrt{16b^4}$

g. $-\sqrt{36}$ **h.** $\sqrt{-36}$

Example 1

Simplify. Assume that all variables represent positive numbers.

a. $\sqrt{36}$ **b.** $\sqrt{0}$ **c.** $\sqrt{\frac{4}{49}}$ **d.** $\sqrt{0.25}$

e. $\sqrt{x^6}$ **f.** $\sqrt{9x^{12}}$ **g.** $-\sqrt{81}$ **h.** $\sqrt{-81}$

Solution:

a. $\sqrt{36} = 6$ because $6^2 = 36$ and 6 is not negative.

b. $\sqrt{0} = 0$ because $0^2 = 0$ and 0 is not negative.

c. $\sqrt{\frac{4}{49}} = \frac{2}{7}$ because $\left(\frac{2}{7}\right)^2 = \frac{4}{49}$ and $\frac{2}{7}$ is not negative.

d. $\sqrt{0.25} = 0.5$ because $(0.5)^2 = 0.25$.

e. $\sqrt{x^6} = x^3$ because $(x^3)^2 = x^6$.

f. $\sqrt{9x^{12}} = 3x^6$ because $(3x^6)^2 = 9x^{12}$.

g. $-\sqrt{81} = -9$. The negative in front of the radical indicates the negative square root of 81.

h. $\sqrt{-81}$ is not a real number.

Work Practice 1

Recall from Section 15.2 our discussion of the square root of a negative number. For example, can we simplify $\sqrt{-4}$? That is, can we find a real number whose

Answers

1. a. 7 **b.** 0 **c.** $\frac{4}{9}$ **d.** 0.8 **e.** z^4 **f.** $4b^2$ **g.** -6 **h.** not a real number

square is -4? No, there is no real number whose square is -4, and we say that $\sqrt{-4}$ is not a real number. In general:

The square root of a negative number is not a real number.

Helpful Hint

- Remember: $\sqrt{0} = 0$.
- Don't forget that the square root of a negative number is not a real number. For example,

  $\sqrt{-9}$ is not a real number

because there is no real number that when multiplied by itself would give a product of -9. In Section 17.7, we will see what kind of a number $\sqrt{-9}$ is.

Objective 2 Approximating Roots

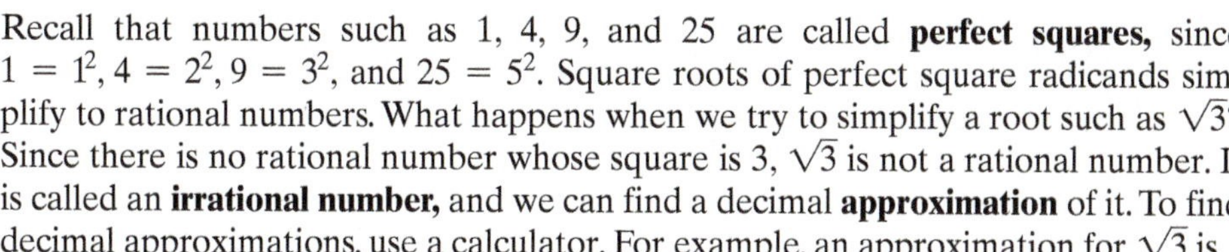

Recall that numbers such as 1, 4, 9, and 25 are called **perfect squares,** since $1 = 1^2, 4 = 2^2, 9 = 3^2$, and $25 = 5^2$. Square roots of perfect square radicands simplify to rational numbers. What happens when we try to simplify a root such as $\sqrt{3}$? Since there is no rational number whose square is 3, $\sqrt{3}$ is not a rational number. It is called an **irrational number,** and we can find a decimal **approximation** of it. To find decimal approximations, use a calculator. For example, an approximation for $\sqrt{3}$ is

$$\sqrt{3} \approx 1.732$$

↑ approximation symbol

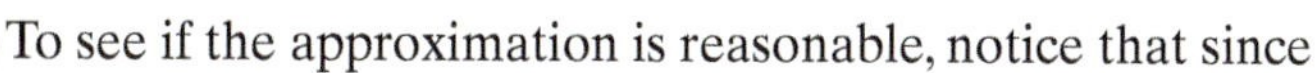

To see if the approximation is reasonable, notice that since

$$1 < 3 < 4,$$
$$\sqrt{1} < \sqrt{3} < \sqrt{4}, \text{ or}$$
$$1 < \sqrt{3} < 2.$$

We found $\sqrt{3} \approx 1.732$, a number between 1 and 2, so our result is reasonable.

Example 2 Use a calculator to approximate $\sqrt{20}$. Round the approximation to 3 decimal places and check to see that your approximation is reasonable.

$$\sqrt{20} \approx 4.472$$

Solution: Is this reasonable? Since $16 < 20 < 25$, $\sqrt{16} < \sqrt{20} < \sqrt{25}$, or $4 < \sqrt{20} < 5$. The approximation is between 4 and 5 and thus is reasonable.

Work Practice 2

Practice 2

Use a calculator to approximate $\sqrt{45}$. Round the approximation to three decimal places and check to see that your approximation is reasonable.

Objective 3 Finding Cube Roots 

Finding roots can be extended to other roots such as cube roots. For example, since $2^3 = 8$, we call 2 the **cube root** of 8. In symbols, we write

$$\sqrt[3]{8} = 2$$

Cube Root

The **cube root** of a real number a is written as $\sqrt[3]{a}$, and

$$\sqrt[3]{a} = b \text{ only if } b^3 = a$$

Answer

2. 6.708

From this definition, we have

$$\sqrt[3]{64} = 4 \text{ since } 4^3 = 64$$
$$\sqrt[3]{-27} = -3 \text{ since } (-3)^3 = -27$$
$$\sqrt[3]{x^3} = x \text{ since } x^3 = x^3$$

Notice that, unlike with square roots, *it is possible to have a negative radicand when finding a cube root*. This is so because the *cube* of a negative number is a negative number. Therefore, the *cube root* of a negative number is a negative number.

Practice 3

Find the cube roots.

a. $\sqrt[3]{-1}$ b. $\sqrt[3]{27}$

c. $\sqrt[3]{\frac{27}{64}}$ d. $\sqrt[3]{x^{12}}$

e. $\sqrt[3]{-8x^3}$

Example 3 Find the cube roots.

a. $\sqrt[3]{1}$ **b.** $\sqrt[3]{-64}$ **c.** $\sqrt[3]{\frac{8}{125}}$ **d.** $\sqrt[3]{x^6}$ **e.** $\sqrt[3]{-27x^9}$

Solution:

a. $\sqrt[3]{1} = 1$ because $1^3 = 1$.

b. $\sqrt[3]{-64} = -4$ because $(-4)^3 = -64$.

c. $\sqrt[3]{\frac{8}{125}} = \frac{2}{5}$ because $\left(\frac{2}{5}\right)^3 = \frac{8}{125}$.

d. $\sqrt[3]{x^6} = x^2$ because $(x^2)^3 = x^6$.

e. $\sqrt[3]{-27x^9} = -3x^3$ because $(-3x^3)^3 = -27x^9$.

Work Practice 3

Objective 4 Finding *n*th Roots

Just as we can raise a real number to powers other than 2 or 3, we can find roots other than square roots and cube roots. In fact, we can find the ***n*th root** of a number, where n is any natural number. In symbols, the nth root of a is written as $\sqrt[n]{a}$, where n is called the **index.** The index 2 is usually omitted for square roots.

Helpful Hint

If the index is even, such as $\sqrt{\ }$, $\sqrt[4]{\ }$, $\sqrt[6]{\ }$, and so on, the radicand must be nonnegative for the root to be a real number. For example,

$$\sqrt[4]{16} = 2, \text{ but } \sqrt[4]{-16} \text{ is not a real number.}$$
$$\sqrt[6]{64} = 2, \text{ but } \sqrt[6]{-64} \text{ is not a real number.}$$

If the index is odd, such as $\sqrt[3]{\ }$, $\sqrt[5]{\ }$, and so on, the radicand may be any real number. For example,

$$\sqrt[3]{64} = 4 \quad \text{and} \quad \sqrt[3]{-64} = -4$$
$$\sqrt[5]{32} = 2 \quad \text{and} \quad \sqrt[5]{-32} = -2$$

✓**Concept Check** Which one is not a real number?

a. $\sqrt[3]{-15}$ **b.** $\sqrt[4]{-15}$ **c.** $\sqrt[5]{-15}$ **d.** $\sqrt{(-15)^2}$

Answers

3. a. -1 b. 3 c. $\frac{3}{4}$ d. x^4 e. $-2x$

✓**Concept Check Answer**

b

Example 4 Simplify the following expressions.

a. $\sqrt[4]{81}$ **b.** $\sqrt[5]{-243}$ **c.** $-\sqrt{25}$ **d.** $\sqrt[4]{-81}$ **e.** $\sqrt[3]{64x^3}$

Solution:

a. $\sqrt[4]{81} = 3$ because $3^4 = 81$ and 3 is positive.

b. $\sqrt[5]{-243} = -3$ because $(-3)^5 = -243$.

c. $-\sqrt{25} = -5$ because -5 is the opposite of $\sqrt{25}$.

d. $\sqrt[4]{-81}$ is not a real number. There is no real number that, when raised to the fourth power, is -81.

e. $\sqrt[3]{64x^3} = 4x$ because $(4x)^3 = 64x^3$.

Work Practice 4

Practice 4

Simplify the following expressions.

a. $\sqrt[4]{10{,}000}$ **b.** $\sqrt[5]{-1}$

c. $-\sqrt{81}$ **d.** $\sqrt[4]{-625}$

e. $\sqrt[3]{27x^9}$

Objective 5 Finding $\sqrt[n]{a^n}$ Where *a* Is a Real Number

Recall that the notation $\sqrt{a^2}$ indicates the positive square root of a^2 only. For example,

$$\sqrt{(-7)^2} = \sqrt{49} = 7$$

When variables are present in the radicand and it is *unclear whether the variable represents a positive number or a negative number*, absolute value bars are sometimes needed to ensure that the result is a positive number. For example,

$$\sqrt{x^2} = |x|$$

This ensures that the result is positive. This same situation may occur when the index is any *even* positive integer. When the index is any *odd* positive integer, absolute value bars are not necessary.

Finding $\sqrt[n]{a^n}$

If n is an *even* positive integer, then $\sqrt[n]{a^n} = |a|$.

If n is an *odd* positive integer, then $\sqrt[n]{a^n} = a$.

Example 5 Simplify.

a. $\sqrt{(-3)^2}$ **b.** $\sqrt{x^2}$ **c.** $\sqrt[4]{(x-2)^4}$ **d.** $\sqrt[3]{(-5)^3}$

e. $\sqrt[5]{(2x-7)^5}$ **f.** $\sqrt{25x^2}$ **g.** $\sqrt{x^2+2x+1}$

Solution:

a. $\sqrt{(-3)^2} = |-3| = 3$ When the index is even, the absolute value bars ensure that our result is not negative.

b. $\sqrt{x^2} = |x|$

c. $\sqrt[4]{(x-2)^4} = |x-2|$

d. $\sqrt[3]{(-5)^3} = -5$

e. $\sqrt[5]{(2x-7)^5} = 2x - 7$ Absolute value bars are not needed when the index is odd.

f. $\sqrt{25x^2} = 5|x|$

g. $\sqrt{x^2+2x+1} = \sqrt{(x+1)^2} = |x+1|$

Work Practice 5

Practice 5

Simplify.

a. $\sqrt{(-4)^2}$ **b.** $\sqrt{x^{14}}$

c. $\sqrt[4]{(x+7)^4}$ **d.** $\sqrt[3]{(-7)^3}$

e. $\sqrt[5]{(3x-5)^5}$ **f.** $\sqrt{49x^2}$

g. $\sqrt{x^2+16x+64}$

Answers

4. a. 10 **b.** -1 **c.** -9 **d.** not a real number **e.** $3x^3$

5. a. 4 **b.** $|x^7|$ **c.** $|x+7|$ **d.** -7 **e.** $3x-5$ **f.** $7|x|$ **g.** $|x+8|$

Objective 6 Graphing Square and Cube Root Functions

Recall that an equation in x and y describes a function if each x-value is paired with exactly one y-value. With this in mind, does the equation

$$y = \sqrt{x}$$

describe a function? First, notice that replacement values for x must be nonnegative real numbers, since $\sqrt{x}$ is not a real number if $x < 0$. The notation $\sqrt{x}$ denotes the principal square root of x, so for every nonnegative number x, there is exactly one number, $\sqrt{x}$. Therefore, $y = \sqrt{x}$ describes a function, and we may write it as

$$f(x) = \sqrt{x}$$

In general, radical functions are functions of the form

$$f(x) = \sqrt[n]{x}.$$

Recall that the domain of a function in x is the set of all possible replacement values of x. This means that if n is even, the domain is the set of all nonnegative numbers, or $\{x|x \geq 0\}$ or $[0, \infty)$. If n is odd, the domain is the set of all real numbers, or $(-\infty, \infty)$. Keep this in mind as we find function values.

Practice 6

If $f(x) = \sqrt{x+5}$ and $g(x) = \sqrt[3]{x-3}$, find each function value.

a. $f(11)$ **b.** $f(-1)$
c. $g(11)$ **d.** $g(-6)$

Example 6 If $f(x) = \sqrt{x-4}$ and $g(x) = \sqrt[3]{x+2}$, find each function value.

a. $f(8)$ **b.** $f(6)$ **c.** $g(-1)$ **d.** $g(1)$

Solution:

a. $f(8) = \sqrt{8-4} = \sqrt{4} = 2$ **b.** $f(6) = \sqrt{6-4} = \sqrt{2}$

c. $g(-1) = \sqrt[3]{-1+2} = \sqrt[3]{1} = 1$ **d.** $g(1) = \sqrt[3]{1+2} = \sqrt[3]{3}$

Work Practice 6

Helpful Hint

Notice that for the function $f(x) = \sqrt{x-4}$, the domain includes all real numbers that make the radicand ≥ 0. To see what numbers these are, solve $x - 4 \geq 0$ and find that $x \geq 4$. The domain is $\{x|x \geq 4\}$, or $[4, \infty)$.

The domain of the cube root function $g(x) = \sqrt[3]{x+2}$ is the set of real numbers, or $(-\infty, \infty)$.

Practice 7

Graph the square root function $h(x) = \sqrt{x+2}$.

Example 7 Graph the square root function $f(x) = \sqrt{x}$.

Solution: To graph, we identify the domain, evaluate the function for several values of x, plot the resulting points, and connect the points with a smooth curve. Since $\sqrt{x}$ represents the nonnegative square root of x, the domain of this function is the set of all nonnegative numbers, $\{x|x \geq 0\}$ or $[0, \infty)$. We have approximated $\sqrt{3}$ in the table on the next page to help us locate the point corresponding to $(3, \sqrt{3})$.

Answers

6. a. 4 **b.** 2 **c.** 2 **d.** $\sqrt[3]{-9}$

7.

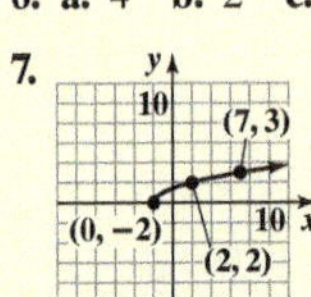

x	$f(x) = \sqrt{x}$
0	0
1	1
3	$\sqrt{3} \approx 1.7$
4	2
9	3

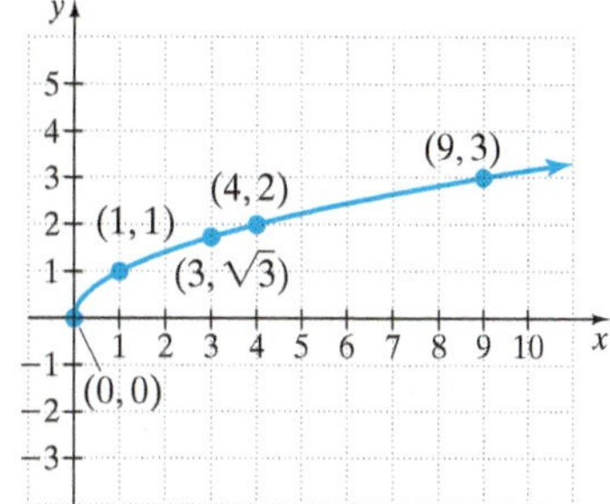

Notice that the graph of this function passes the vertical line test, as expected.

Work Practice 7

The equation $f(x) = \sqrt[3]{x}$ also describes a function. Here, x may be any real number, so the domain of this function is the set of all real numbers, or $(-\infty, \infty)$. A few function values are given next.

$$f(0) = \sqrt[3]{0} = 0$$
$$f(1) = \sqrt[3]{1} = 1$$
$$f(-1) = \sqrt[3]{-1} = -1$$
$$f(6) = \sqrt[3]{6}$$
$$f(-6) = \sqrt[3]{-6}$$

Here, there is no rational number whose cube is 6. Thus, the radicals do not simplify to rational numbers.

$$f(8) = \sqrt[3]{8} = 2$$
$$f(-8) = \sqrt[3]{-8} = -2$$

Example 8 Graph the function $f(x) = \sqrt[3]{x}$.

Solution: To graph, we identify the domain, plot points, and connect the points with a smooth curve. The domain of this function is the set of all real numbers. The table comes from the function values obtained earlier. We have approximated $\sqrt[3]{6}$ and $\sqrt[3]{-6}$ for graphing purposes.

x	$f(x) = \sqrt[3]{x}$
0	0
1	1
−1	−1
6	$\sqrt[3]{6} \approx 1.8$
−6	$\sqrt[3]{-6} \approx -1.8$
8	2
−8	−2

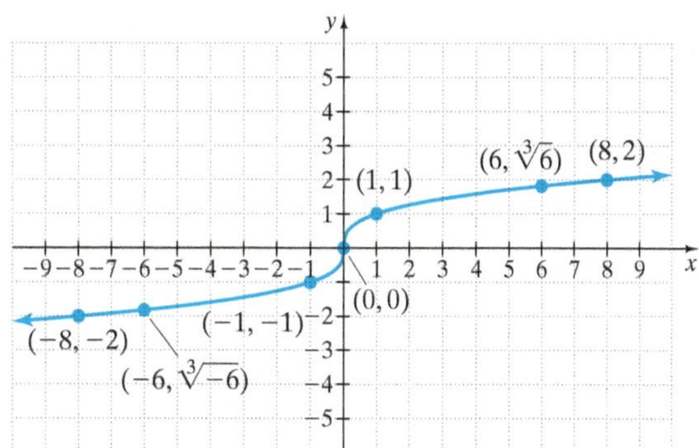

The graph of this function passes the vertical line test, as expected.

Work Practice 8

Practice 8

Graph the function $f(x) = \sqrt[3]{x} - 4$.

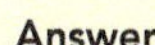

Answer

8.

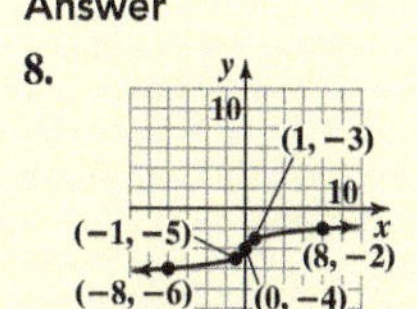

Vocabulary, Readiness & Video Check

Use the choices below to fill in each blank. Not all choices will be used.

is	cubes	$-\sqrt{a}$	radical sign	index
is not	squares	$\sqrt{-a}$	radicand	

1. In the expression $\sqrt[n]{a}$, the n is called the ________, the $\sqrt{\ }$ is called the ________, and a is called the ________.
2. If $\sqrt{a}$ is the positive square root of a, $a \neq 0$, then ________ is the negative square root of a.
3. The square root of a negative number ________ a real number.
4. Numbers such as 1, 4, 9, and 25 are called perfect ________, whereas numbers such as 1, 8, 27, and 125 are called perfect ________.

Fill in the blank.

5. The domain of the function $f(x) = \sqrt{x}$ is ________.
6. The domain of the function $f(x) = \sqrt[3]{x}$ is ________.
7. If $f(16) = 4$, the corresponding ordered pair is ________.
8. If $g(-8) = -2$, the corresponding ordered pair is ________.

Martin-Gay Interactive Videos *Watch the section lecture video and answer the following questions.*

See Video 17.1

Objective 1 9. From Examples 5 and 6, when simplifying radicals containing variables with exponents, describe a shortcut you can use.

Objective 2 10. From Example 9, how can you determine a reasonable approximation for a non–perfect square root without using a calculator?

Objective 3 11. From Example 11, what is an important difference between the square root and the cube root of a negative number?

Objective 4 12. From Example 12, what conclusion is made about the even root of a negative number?

Objective 5 13. From the lecture before Example 17, why do you think no absolute value bars are used when n is odd?

Objective 6 14. In Example 19, the domain is found by looking at the graph. How can the domain be found by looking at the function?

17.1 Exercise Set MyMathLab®

Simplify. Assume that variables represent positive real numbers. See Example 1.

1. $\sqrt{100}$
2. $\sqrt{400}$
3. $\sqrt{\frac{1}{4}}$
4. $\sqrt{\frac{9}{25}}$
5. $\sqrt{0.0001}$
6. $\sqrt{0.04}$
7. $-\sqrt{36}$
8. $-\sqrt{9}$
9. $\sqrt{x^{10}}$
10. $\sqrt{x^{16}}$
11. $\sqrt{16y^6}$
12. $\sqrt{64y^{20}}$

Use a calculator to approximate each square root to 3 decimal places. Check to see that each approximation is reasonable. See Example 2.

13. $\sqrt{7}$
14. $\sqrt{11}$
15. $\sqrt{38}$
16. $\sqrt{56}$
17. $\sqrt{200}$
18. $\sqrt{300}$

Find each cube root. See Example 3.

19. $\sqrt[3]{64}$
20. $\sqrt[3]{27}$
21. $\sqrt[3]{\frac{1}{8}}$
22. $\sqrt[3]{\frac{27}{64}}$
23. $\sqrt[3]{-1}$
24. $\sqrt[3]{-125}$
25. $\sqrt[3]{x^{12}}$
26. $\sqrt[3]{x^{15}}$
27. $\sqrt[3]{-27x^9}$
28. $\sqrt[3]{-64x^6}$

Find each root. Assume that all variables represent nonnegative real numbers. See Example 4.

29. $-\sqrt[4]{16}$
30. $\sqrt[5]{-243}$
31. $\sqrt[4]{-16}$
32. $\sqrt{-16}$
33. $\sqrt[5]{-32}$
34. $\sqrt[5]{-1}$
35. $\sqrt[5]{x^{20}}$
36. $\sqrt[4]{x^{20}}$
37. $\sqrt[6]{64x^{12}}$
38. $\sqrt[5]{-32x^{15}}$
39. $\sqrt{81x^4}$
40. $\sqrt[4]{81x^4}$
41. $\sqrt[4]{256x^8}$
42. $\sqrt{256x^8}$

Simplify. Assume that the variables represent any real number. See Example 5.

43. $\sqrt{(-8)^2}$
44. $\sqrt{(-7)^2}$
45. $\sqrt[3]{(-8)^3}$
46. $\sqrt[5]{(-7)^5}$
47. $\sqrt{4x^2}$
48. $\sqrt[4]{16x^4}$
49. $\sqrt[3]{x^3}$
50. $\sqrt[5]{x^5}$
51. $\sqrt{(x-5)^2}$
52. $\sqrt{(y-6)^2}$
53. $\sqrt{x^2+4x+4}$
(*Hint:* Factor the polynomial first.)
54. $\sqrt{x^2-8x+16}$
(*Hint:* Factor the polynomial first.)

MIXED PRACTICE

Simplify each radical. Assume that all variables represent positive real numbers.

55. $-\sqrt{121}$
56. $-\sqrt[3]{125}$
57. $\sqrt[3]{8x^3}$
58. $\sqrt{16x^8}$
59. $\sqrt{y^{12}}$
60. $\sqrt[3]{y^{12}}$
61. $\sqrt{25a^2b^{20}}$
62. $\sqrt{9x^4y^6}$
63. $\sqrt[3]{-27x^{12}y^9}$
64. $\sqrt[3]{-8a^{21}b^6}$
65. $\sqrt[4]{a^{16}b^4}$
66. $\sqrt[4]{x^8y^{12}}$
67. $\sqrt[5]{-32x^{10}y^5}$
68. $\sqrt[5]{-243x^5z^{15}}$
69. $\sqrt{\frac{25}{49}}$
70. $\sqrt{\frac{4}{81}}$
71. $\sqrt{\frac{x^{20}}{4y^2}}$
72. $\sqrt{\frac{y^{10}}{9x^6}}$
73. $-\sqrt[3]{\frac{z^{21}}{27x^3}}$
74. $-\sqrt[3]{\frac{64a^3}{b^9}}$
75. $\sqrt[4]{\frac{x^4}{16}}$
76. $\sqrt[4]{\frac{y^4}{81x^4}}$

If $f(x) = \sqrt{2x+3}$ and $g(x) = \sqrt[3]{x-8}$, find the following function values. See Example 6.

77. $f(0)$
78. $g(0)$
79. $g(7)$
80. $f(-1)$
81. $g(-19)$
82. $f(3)$
83. $f(2)$
84. $g(1)$

Identify the domain and then graph each function. See Example 7.

85. $f(x) = \sqrt{x} + 2$
86. $f(x) = \sqrt{x} - 2$
87. $f(x) = \sqrt{x-3}$; use the following table.

x	$f(x)$
3	
4	
7	
12	

88. $f(x) = \sqrt{x+1}$; use the following table.

x	$f(x)$
−1	
0	
3	
8	

Identify the domain and then graph each function. See Example 8.

89. $f(x) = \sqrt[3]{x} + 1$
90. $f(x) = \sqrt[3]{x} - 2$
91. $g(x) = \sqrt[3]{x-1}$; use the following table.

x	$g(x)$
1	
2	
0	
9	
−7	

92. $g(x) = \sqrt[3]{x+1}$; use the following table.

x	$g(x)$
−1	
0	
−2	
7	
−9	

REVIEW AND PREVIEW

Simplify each exponential expression. See Sections 12.1 and 12.5.

93. $(-2x^3y^2)^5$

94. $(4y^6z^7)^3$

95. $(-3x^2y^3z^5)(20x^5y^7)$

96. $(-14a^5bc^2)(2abc^4)$

97. $\dfrac{7x^{-1}y}{14(x^5y^2)^{-2}}$

98. $\dfrac{(2a^{-1}b^2)^3}{(8a^2b)^{-2}}$

CONCEPT EXTENSIONS

Determine whether the following are real numbers. See the Concept Check in this section.

99. $\sqrt{-17}$

100. $\sqrt[3]{-17}$

101. $\sqrt[10]{-17}$

102. $\sqrt[15]{-17}$

Choose the correct letter or letters. No pencil is needed, just think your way through these.

103. Which radical is not a real number?

a. $\sqrt{3}$ **b.** $-\sqrt{11}$ **c.** $\sqrt[3]{-10}$ **d.** $\sqrt{-10}$

104. Which radical(s) simplify to 3?

a. $\sqrt{9}$ **b.** $\sqrt{-9}$ **c.** $\sqrt[3]{27}$ **d.** $\sqrt[3]{-27}$

105. Which radical(s) simplify to -3?

a. $\sqrt{9}$ **b.** $\sqrt{-9}$ **c.** $\sqrt[3]{27}$ **d.** $\sqrt[3]{-27}$

106. Which radical does not simplify to a whole number?

a. $\sqrt{64}$ **b.** $\sqrt[3]{64}$ **c.** $\sqrt{8}$ **d.** $\sqrt[3]{8}$

For Exercises 107 through 110, do not use a calculator.

107. $\sqrt{160}$ is closest to

a. 10 **b.** 13 **c.** 20 **d.** 40

108. $\sqrt{1000}$ is closest to

a. 10 **b.** 30 **c.** 100 **d.** 500

△ **109.** The perimeter of the triangle is closest to

a. 12 **b.** 18 **c.** 66 **d.** 132

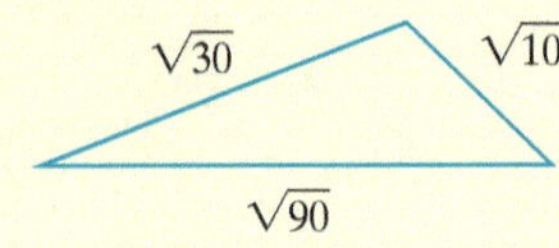

110. The length of the bent wire is closest to

a. 5 **b.** $\sqrt{28}$ **c.** 7 **d.** 14

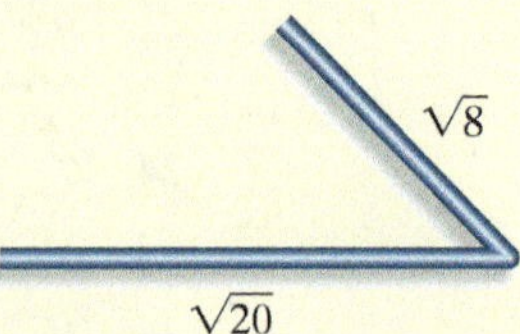

111. Explain why $\sqrt{-64}$ is not a real number.

112. Explain why $\sqrt[3]{-64}$ is a real number.

The Mosteller formula for calculating adult body surface area is $B = \sqrt{\dfrac{hw}{3131}}$, where B is an individual's body surface area in square meters, h is the individual's height in inches, and w is the individual's weight in pounds. Use this information to answer Exercises 113 and 114. Round answers to 2 decimal places.

△ **113.** Find the body surface area of an individual who is 66 inches tall and who weighs 135 pounds.

△ **114.** Find the body surface area of an individual who is 74 inches tall and who weighs 225 pounds.

115. Escape velocity is the minimum speed that an object must reach to escape the pull of a planet's gravity. Escape velocity v is given by the equation $v = \sqrt{\dfrac{2Gm}{r}}$, where m is the mass of the planet, r is its radius, and G is the universal gravitational constant, which has a value of $G = 6.67 \times 10^{-11}\ \text{m}^3/\text{kg}\cdot\text{s}^2$. The mass of Earth is 5.97×10^{24} kg, and its radius is 6.37×10^6 m. Use this information to find the escape velocity for Earth in meters per second. Round to the nearest whole number. (*Source:* National Space Science Data Center)

116. Use the formula from Exercise 115 to determine the escape velocity for the moon. The mass of the moon is 7.35×10^{22} kg, and its radius is 1.74×10^6 m. Round to the nearest whole number. (*Source:* National Space Science Data Center)

117. Suppose a classmate tells you that $\sqrt{13} \approx 5.7$. Without a calculator, how can you convince your classmate that he or she must have made an error?

118. Suppose a classmate tells you that $\sqrt[3]{10} \approx 3.2$. Without a calculator, how can you convince your friend that he or she must have made an error?

Use a graphing calculator to verify the domain of each function and its graph.

119. Exercise 85

120. Exercise 86

121. Exercise 89

122. Exercise 90

17.2 Rational Exponents 

Objective 1 Understanding the Meaning of $a^{1/n}$ 

So far in this text, we have not defined expressions with rational exponents such as $3^{1/2}$, $x^{2/3}$, and $-9^{-1/4}$. We will define these expressions so that the rules for exponents will apply to these rational exponents as well.

Suppose that $x = 5^{1/3}$. Then

$$x^3 = (5^{1/3})^3 = 5^{1/3 \cdot 3} = 5^1 \text{ or } 5$$

using rules for exponents

Since $x^3 = 5$, x is the number whose cube is 5, or $x = \sqrt[3]{5}$. Notice that we also know that $x = 5^{1/3}$. This means

$$5^{1/3} = \sqrt[3]{5}$$

Definition of $a^{1/n}$

If n is a positive integer greater than 1 and $\sqrt[n]{a}$ is a real number, then

$$a^{1/n} = \sqrt[n]{a}$$

Notice that the denominator of the rational exponent corresponds to the index of the radical.

Example 1 Use radical notation to write the following. Simplify if possible.

a. $4^{1/2}$ **b.** $64^{1/3}$ **c.** $x^{1/4}$ **d.** $0^{1/6}$ **e.** $-9^{1/2}$ **f.** $(81x^8)^{1/4}$ **g.** $5y^{1/3}$

Solution:

a. $4^{1/2} = \sqrt{4} = 2$

b. $64^{1/3} = \sqrt[3]{64} = 4$

c. $x^{1/4} = \sqrt[4]{x}$

d. $0^{1/6} = \sqrt[6]{0} = 0$

e. $-9^{1/2} = -\sqrt{9} = -3$

f. $(81x^8)^{1/4} = \sqrt[4]{81x^8} = 3x^2$

g. $5y^{1/3} = 5\sqrt[3]{y}$

Work Practice 1

Objective 2 Understanding the Meaning of $a^{m/n}$ 

As we expand our use of exponents to include $\frac{m}{n}$, we define their meaning so that rules for exponents still hold true. For example, by properties of exponents,

$$8^{2/3} = (8^{1/3})^2 = (\sqrt[3]{8})^2 \quad \text{or}$$
$$8^{2/3} = (8^2)^{1/3} = \sqrt[3]{8^2}$$

Definition of $a^{m/n}$

If m and n are positive integers greater than 1 with $\frac{m}{n}$ in simplest form, then

$$a^{m/n} = \sqrt[n]{a^m} = (\sqrt[n]{a})^m$$

as long as $\sqrt[n]{a}$ is a real number.

Objectives

1. Understand the Meaning of $a^{1/n}$.
2. Understand the Meaning of $a^{m/n}$.
3. Understand the Meaning of $a^{-m/n}$.
4. Use Rules for Exponents to Simplify Expressions That Contain Rational Exponents.
5. Use Rational Exponents to Simplify Radical Expressions.

Practice 1

Use radical notation to write the following. Simplify if possible.

a. $36^{1/2}$ **b.** $1000^{1/3}$
c. $x^{1/3}$ **d.** $1^{1/4}$
e. $-64^{1/2}$ **f.** $(125x^9)^{1/3}$
g. $3x^{1/4}$

Answers

1. a. 6 **b.** 10 **c.** $\sqrt[3]{x}$ **d.** 1 **e.** -8 **f.** $5x^3$ **g.** $3\sqrt[4]{x}$

Notice that the denominator n of the rational exponent corresponds to the index of the radical. The numerator m of the rational exponent indicates that the base is to be raised to the mth power. This means

$$8^{2/3} = \sqrt[3]{8^2} = \sqrt[3]{64} = 4 \quad \text{or}$$
$$8^{2/3} = (\sqrt[3]{8})^2 = 2^2 = 4$$

From simplifying $8^{2/3}$, can you see that it doesn't matter whether you raise to a power first and then take the nth root or you take the nth root first and then raise to a power?

Helpful Hint

Most of the time, $(\sqrt[n]{a})^m$ will be easier to calculate than $\sqrt[n]{a^m}$.

Practice 2

Use radical notation to write the following. Simplify if possible.

a. $16^{3/2}$ **b.** $-1^{3/5}$

c. $-(81)^{3/4}$ **d.** $\left(\frac{1}{25}\right)^{3/2}$

e. $(3x + 2)^{5/9}$

Example 2 Use radical notation to write the following. Then simplify if possible.

a. $4^{3/2}$ **b.** $-16^{3/4}$ **c.** $(-27)^{2/3}$ **d.** $\left(\frac{1}{9}\right)^{3/2}$ **e.** $(4x - 1)^{3/5}$

Solution:

a. $4^{3/2} = (\sqrt{4})^3 = 2^3 = 8$

b. $-16^{3/4} = -(\sqrt[4]{16})^3 = -(2)^3 = -8$

c. $(-27)^{2/3} = (\sqrt[3]{-27})^2 = (-3)^2 = 9$

d. $\left(\frac{1}{9}\right)^{3/2} = \left(\sqrt{\frac{1}{9}}\right)^3 = \left(\frac{1}{3}\right)^3 = \frac{1}{27}$

e. $(4x - 1)^{3/5} = \sqrt[5]{(4x - 1)^3}$

Work Practice 2

Helpful Hint

The *denominator* of a rational exponent is the index of the corresponding radical. For example, $x^{1/5} = \sqrt[5]{x}$ and $z^{2/3} = \sqrt[3]{z^2}$, or $z^{2/3} = (\sqrt[3]{z})^2$.

Objective 3 Understanding the Meaning of $a^{-m/n}$

The rational exponents we have given meaning to exclude negative rational numbers. To complete the set of definitions, we define $a^{-m/n}$.

Definition of $a^{-m/n}$

$$a^{-m/n} = \frac{1}{a^{m/n}}$$

as long as $a^{m/n}$ is a nonzero real number.

Practice 3

Write each expression with a positive exponent; then simplify.

a. $9^{-3/2}$ **b.** $(-64)^{-2/3}$

Example 3 Write each expression with a positive exponent, and then simplify.

a. $16^{-3/4}$ **b.** $(-27)^{-2/3}$

Solution:

a. $16^{-3/4} = \frac{1}{16^{3/4}} = \frac{1}{(\sqrt[4]{16})^3} = \frac{1}{2^3} = \frac{1}{8}$

b. $(-27)^{-2/3} = \frac{1}{(-27)^{2/3}} = \frac{1}{(\sqrt[3]{-27})^2} = \frac{1}{(-3)^2} = \frac{1}{9}$

Work Practice 3

Answers

2. a. 64 **b.** −1 **c.** −27 **d.** $\frac{1}{125}$ **e.** $\sqrt[9]{(3x+2)^5}$ **3. a.** $\frac{1}{27}$ **b.** $\frac{1}{16}$

Helpful Hint

If an expression contains a negative rational exponent, such as $9^{-3/2}$, you may want to first write the expression with a positive exponent and then interpret the rational exponent. Notice that the sign of the base is not affected by the sign of its exponent. For example,

$$9^{-3/2} = \frac{1}{9^{3/2}} = \frac{1}{(\sqrt{9})^3} = \frac{1}{27}$$

Also,

$$(-27)^{-1/3} = \frac{1}{(-27)^{1/3}} = -\frac{1}{3}$$

✓Concept Check Which one is correct?

a. $-8^{2/3} = \frac{1}{4}$ **b.** $8^{-2/3} = -\frac{1}{4}$ **c.** $8^{-2/3} = -4$ **d.** $-8^{-2/3} = -\frac{1}{4}$

Objective 4 Using Rules for Exponents to Simplify Expressions

It can be shown that the properties of integer exponents hold for rational exponents. By using these properties and definitions, we can now simplify expressions that contain rational exponents.

These rules are repeated here for review.

Note: For the remainder of this chapter, we will assume that variables represent positive real numbers. Since this is so, we need not insert absolute value bars when we simplify even roots.

Summary of Exponent Rules

If m and n are rational numbers, and a, b, and c are numbers for which the expressions below exist, then

Product rule for exponents: $a^m \cdot a^n = a^{m+n}$

Power rule for exponents: $(a^m)^n = a^{m \cdot n}$

Power rules for products and quotients: $(ab)^n = a^n b^n$ and $\left(\frac{a}{c}\right)^n = \frac{a^n}{c^n}, c \neq 0$

Quotient rule for exponents: $\frac{a^m}{a^n} = a^{m-n}, a \neq 0$

Zero exponent: $a^0 = 1, a \neq 0$

Negative exponent: $a^{-n} = \frac{1}{a^n}, a \neq 0$

Example 4 Use properties of exponents to simplify. Write results with only positive exponents.

a. $b^{1/3} \cdot b^{5/3}$ **b.** $x^{1/2}x^{1/3}$ **c.** $\frac{7^{1/3}}{7^{4/3}}$

d. $y^{-4/7} \cdot y^{6/7}$ **e.** $\frac{(2x^{2/5}y^{-1/3})^5}{x^2y}$

(*Continued on next page*)

Practice 4

Use properties of exponents to simplify.

a. $y^{2/3} \cdot y^{8/3}$ **b.** $x^{3/5} \cdot x^{1/4}$

c. $\frac{9^{2/7}}{9^{9/7}}$ **d.** $b^{4/9} \cdot b^{-2/9}$

e. $\frac{(3x^{1/4}y^{-2/3})^4}{x^4y}$

Answers

4. a. $y^{10/3}$ **b.** $x^{17/20}$ **c.** $\frac{1}{9}$ **d.** $b^{2/9}$ **e.** $\frac{81}{x^3y^{11/3}}$

✓Concept Check Answer

d

Solution:

a. $b^{1/3} \cdot b^{5/3} = b^{(1/3+5/3)} = b^{6/3} = b^2$

b. $x^{1/2}x^{1/3} = x^{(1/2+1/3)} = x^{3/6+2/6} = x^{5/6}$ Use the product rule.

c. $\dfrac{7^{1/3}}{7^{4/3}} = 7^{1/3-4/3} = 7^{-3/3} = 7^{-1} = \dfrac{1}{7}$ Use the quotient rule.

d. $y^{-4/7} \cdot y^{6/7} = y^{-4/7+6/7} = y^{2/7}$ Use the product rule.

e. We begin by using the power rule $(ab)^m = a^m b^m$ to simplify the numerator.

$$\frac{(2x^{2/5}y^{-1/3})^5}{x^2y} = \frac{2^5(x^{2/5})^5(y^{-1/3})^5}{x^2y} = \frac{32x^2y^{-5/3}}{x^2y}$$ Use the power rule and simplify.

$$= 32x^{2-2}y^{-5/3-3/3}$$ Apply the quotient rule.

$$= 32x^0y^{-8/3}$$

$$= \frac{32}{y^{8/3}}$$

Work Practice 4

Practice 5

Multiply.

a. $x^{3/5}(x^{1/3} - x^2)$

b. $(x^{1/2} + 6)(x^{1/2} - 2)$

Example 5 Multiply.

a. $z^{2/3}(z^{1/3} - z^5)$ **b.** $(x^{1/3} - 5)(x^{1/3} + 2)$

Solution:

a. $z^{2/3}(z^{1/3} - z^5) = z^{2/3}z^{1/3} - z^{2/3}z^5$ Apply the distributive property.

$= z^{(2/3+1/3)} - z^{(2/3+5)}$ Use the product rule.

$= z^{3/3} - z^{(2/3+15/3)}$

$= z - z^{17/3}$

b. $(x^{1/3} - 5)(x^{1/3} + 2) = x^{2/3} + 2x^{1/3} - 5x^{1/3} - 10$

$= x^{2/3} - 3x^{1/3} - 10$

Think of $(x^{1/3} - 5)$ and $(x^{1/3} + 2)$ as 2 binomials, and use FOIL.

Work Practice 5

Practice 6

Factor $x^{-1/5}$ from the expression $2x^{-1/5} - 7x^{4/5}$.

Example 6 Factor $x^{-1/2}$ from the expression $3x^{-1/2} - 7x^{5/2}$. Assume that all variables represent positive numbers.

Solution:

$$3x^{-1/2} - 7x^{5/2} = (x^{-1/2})(3) - (x^{-1/2})(7x^{6/2})$$

$$= x^{-1/2}(3 - 7x^3)$$

To check, multiply $x^{-1/2}(3 - 7x^3)$ to see that the product is $3x^{-1/2} - 7x^{5/2}$.

Work Practice 6

Objective 5 Using Rational Exponents to Simplify Radical Expressions

Some radical expressions are easier to simplify when we first write them with rational exponents. Use properties of exponents to simplify, and then convert back to radical notation.

Answers

5. a. $x^{14/15} - x^{13/5}$ **b.** $x + 4x^{1/2} - 12$

6. $x^{-1/5}(2 - 7x)$

Example 7 Use rational exponents to simplify. Assume that variables represent positive numbers.

a. $\sqrt[8]{x^4}$ **b.** $\sqrt[6]{25}$ **c.** $\sqrt[4]{r^2s^6}$

Solution:

a. $\sqrt[8]{x^4} = x^{4/8} = x^{1/2} = \sqrt{x}$

b. $\sqrt[6]{25} = 25^{1/6} = (5^2)^{1/6} = 5^{2/6} = 5^{1/3} = \sqrt[3]{5}$

c. $\sqrt[4]{r^2s^6} = (r^2s^6)^{1/4} = r^{2/4}s^{6/4} = r^{1/2}s^{3/2} = (rs^3)^{1/2} = \sqrt{rs^3}$

Work Practice 7

Practice 7

Use rational exponents to simplify. Assume that the variables represent positive numbers.

a. $\sqrt[9]{x^3}$ **b.** $\sqrt[4]{36}$

c. $\sqrt[8]{a^4b^2}$

Example 8 Use rational exponents to write as a single radical.

a. $\sqrt{x}\cdot\sqrt[4]{x}$ **b.** $\dfrac{\sqrt{x}}{\sqrt[3]{x}}$ **c.** $\sqrt[3]{3}\cdot\sqrt{2}$

Solution:

a. $\sqrt{x}\cdot\sqrt[4]{x} = x^{1/2}\cdot x^{1/4} = x^{1/2+1/4}$
$= x^{3/4} = \sqrt[4]{x^3}$

b. $\dfrac{\sqrt{x}}{\sqrt[3]{x}} = \dfrac{x^{1/2}}{x^{1/3}} = x^{1/2-1/3} = x^{3/6-2/6}$
$= x^{1/6} = \sqrt[6]{x}$

c. $\sqrt[3]{3}\cdot\sqrt{2} = 3^{1/3}\cdot 2^{1/2}$ Write with rational exponents.
$= 3^{2/6}\cdot 2^{3/6}$ Write the exponents so that they have the same denominator.
$= (3^2\cdot 2^3)^{1/6}$ Use $a^nb^n = (ab)^n$.
$= \sqrt[6]{3^2\cdot 2^3}$ Write with radical notation.
$= \sqrt[6]{72}$ Multiply $3^2\cdot 2^3$.

Work Practice 8

Practice 8

Use rational expressions to write each of the following as a single radical.

a. $\sqrt[3]{x}\cdot\sqrt[4]{x}$ **b.** $\dfrac{\sqrt[3]{y}}{\sqrt[5]{y}}$

c. $\sqrt[3]{5}\cdot\sqrt{3}$

Answers

7. a. $\sqrt[3]{x}$ **b.** $\sqrt{6}$ **c.** $\sqrt[4]{a^2b}$

8. a. $\sqrt[12]{x^7}$ **b.** $\sqrt[15]{y^2}$ **c.** $\sqrt[6]{675}$

Vocabulary, Readiness & Video Check

Answer each true or false.

1. $9^{-1/2}$ is a positive number. ________

2. $9^{-1/2}$ is a whole number. ________

3. $\dfrac{1}{a^{-m/n}} = a^{m/n}$ (where $a^{m/n}$ is a nonzero real number). ________

Fill in the blank with the correct choice.

4. To simplify $x^{2/3}\cdot x^{1/5}$, ________ the exponents.

a. add **b.** subtract **c.** multiply **d.** divide

5. To simplify $(x^{2/3})^{1/5}$, ________ the exponents.

a. add **b.** subtract **c.** multiply **d.** divide

6. To simplify $\dfrac{x^{2/3}}{x^{1/5}}$, ________ the exponents.

a. add **b.** subtract **c.** multiply **d.** divide

Martin-Gay Interactive Videos *Watch the section lecture video and answer the following questions.*

See Video 17.2

Objective 1 **7.** From looking at Example 2, what is $-(3x)^{1/5}$ in radical notation?

Objective 2 **8.** From Examples 3 and 4, in a fractional exponent, what do the numerator and denominator each represent in radical form?

Objective 3 **9.** Based on Example 5, complete the following statement. A negative fractional exponent will move a base from the numerator to the ____________ with the fractional exponent becoming ____________.

Objective 4 **10.** Based on Examples 7–9, complete the following statements. Assume you have an expression with fractional exponents. If applying the product rule of exponents, you ____________ the exponents. If applying the quotient rule of exponents, you ____________ the exponents. If applying the power rule of exponents, you ____________ the exponents.

Objective 5 **11.** From Example 10, describe a way to simplify a radical of a variable raised to a power if the index and the exponent have a common factor.

17.2 Exercise Set MyMathLab®

Use radical notation to write each expression. Simplify if possible. See Example 1.

1. $49^{1/2}$ **2.** $64^{1/3}$

3. $27^{1/3}$ **4.** $8^{1/3}$

5. $\left(\frac{1}{16}\right)^{1/4}$ **6.** $\left(\frac{1}{64}\right)^{1/2}$

7. $169^{1/2}$ **8.** $81^{1/4}$

9. $2m^{1/3}$ **10.** $(2m)^{1/3}$

11. $(9x^4)^{1/2}$ **12.** $(16x^8)^{1/2}$

13. $(-27)^{1/3}$ **14.** $-64^{1/2}$

15. $-16^{1/4}$ **16.** $(-32)^{1/5}$

Use radical notation to write each expression. Simplify if possible. See Example 2.

17. $16^{3/4}$ **18.** $4^{5/2}$

19. $(-64)^{2/3}$ **20.** $(-8)^{4/3}$

21. $(-16)^{3/4}$ **22.** $(-9)^{3/2}$

23. $(2x)^{3/5}$ **24.** $2x^{3/5}$

25. $(7x+2)^{2/3}$ **26.** $(x-4)^{3/4}$

27. $\left(\frac{16}{9}\right)^{3/2}$ **28.** $\left(\frac{49}{25}\right)^{3/2}$

Write with positive exponents. Simplify if possible. See Example 3.

29. $8^{-4/3}$ **30.** $64^{-2/3}$

31. $(-64)^{-2/3}$ **32.** $(-8)^{-4/3}$

33. $(-4)^{-3/2}$ **34.** $(-16)^{-5/4}$

35. $x^{-1/4}$ **36.** $y^{-1/6}$

37. $\frac{1}{a^{-2/3}}$ **38.** $\frac{1}{n^{-8/9}}$

39. $\frac{5}{7x^{-3/4}}$ **40.** $\frac{2}{3y^{-5/7}}$

Use the properties of exponents to simplify each expression. Write with positive exponents. See Example 4.

41. $a^{2/3}a^{5/3}$ **42.** $b^{9/5}b^{8/5}$

43. $x^{-2/5} \cdot x^{7/5}$ **44.** $y^{4/3} \cdot y^{-1/3}$

45. $3^{1/4} \cdot 3^{3/8}$ **46.** $5^{1/2} \cdot 5^{1/6}$

47. $\frac{y^{1/3}}{y^{1/6}}$ **48.** $\frac{x^{3/4}}{x^{1/8}}$

49. $(4u^2)^{3/2}$ **50.** $(32^{1/5}x^{2/3})^3$

51. $\frac{b^{1/2}b^{3/4}}{-b^{1/4}}$ **52.** $\frac{a^{1/4}a^{-1/2}}{a^{2/3}}$

53. $\frac{(x^3)^{1/2}}{x^{7/2}}$ **54.** $\frac{y^{11/3}}{(y^5)^{1/3}}$

55. $\frac{(3x^{1/4})^3}{x^{1/12}}$ **56.** $\frac{(2x^{1/5})^4}{x^{3/10}}$

57. $\frac{(y^3z)^{1/6}}{y^{-1/2}z^{1/3}}$ **58.** $\frac{(m^2n)^{1/4}}{m^{-1/2}n^{5/8}}$

59. $\frac{(x^3y^2)^{1/4}}{(x^{-5}y^{-1})^{-1/2}}$ **60.** $\frac{(a^{-2}b^3)^{1/8}}{(a^{-3}b)^{-1/4}}$

Multiply. See Example 5.

61. $y^{1/2}(y^{1/2} - y^{2/3})$
62. $x^{1/2}(x^{1/2} + x^{3/2})$
63. $x^{2/3}(x - 2)$
64. $3x^{1/2}(x + y)$
65. $(2x^{1/3} + 3)(2x^{1/3} - 3)$
66. $(y^{1/2} + 5)(y^{1/2} + 5)$

Factor the given factor from the expression. See Example 6.

67. $x^{8/3}; x^{8/3} + x^{10/3}$
68. $x^{3/2}; x^{5/2} - x^{3/2}$
69. $x^{1/5}; x^{2/5} - 3x^{1/5}$
70. $x^{2/7}; x^{3/7} - 2x^{2/7}$
71. $x^{-1/3}; 5x^{-1/3} + x^{2/3}$
72. $x^{-3/4}; x^{-3/4} + 3x^{1/4}$

Use rational exponents to simplify each radical. Assume that all variables represent positive numbers. See Example 7.

73. $\sqrt[6]{x^3}$
74. $\sqrt[9]{a^3}$
75. $\sqrt[6]{4}$
76. $\sqrt[4]{36}$
77. $\sqrt[4]{16x^2}$
78. $\sqrt[8]{4y^2}$
79. $\sqrt[8]{x^4y^4}$
80. $\sqrt[9]{y^6z^3}$
81. $\sqrt[12]{a^8b^4}$
82. $\sqrt[10]{a^5b^5}$
83. $\sqrt[4]{(x + 3)^2}$
84. $\sqrt[8]{(y + 1)^4}$

Use rational expressions to write as a single radical expression. See Example 8.

85. $\sqrt[3]{y} \cdot \sqrt[5]{y^2}$
86. $\sqrt[3]{y^2} \cdot \sqrt[6]{y}$
87. $\dfrac{\sqrt[3]{b^2}}{\sqrt[4]{b}}$
88. $\dfrac{\sqrt[4]{a}}{\sqrt[5]{a}}$
89. $\sqrt[3]{x} \cdot \sqrt[4]{x} \cdot \sqrt[8]{x^3}$
90. $\sqrt[6]{y} \cdot \sqrt[3]{y} \cdot \sqrt[5]{y^2}$
91. $\dfrac{\sqrt[3]{a^2}}{\sqrt[6]{a}}$
92. $\dfrac{\sqrt[5]{b^2}}{\sqrt[10]{b^3}}$
93. $\sqrt{3} \cdot \sqrt[3]{4}$
94. $\sqrt[3]{5} \cdot \sqrt{2}$
95. $\sqrt[5]{7} \cdot \sqrt[3]{y}$
96. $\sqrt[4]{5} \cdot \sqrt[3]{x}$
97. $\sqrt{5r} \cdot \sqrt[3]{s}$
98. $\sqrt[3]{b} \cdot \sqrt[5]{4a}$

REVIEW AND PREVIEW

Write each integer as a product of two integers such that one of the factors is a perfect square. For example, write 18 as $9 \cdot 2$ because 9 is a perfect square.

99. 75
100. 20
101. 48
102. 45

Write each integer as a product of two integers such that one of the factors is a perfect cube. For example, write 24 as $8 \cdot 3$ because 8 is a perfect cube.

103. 16
104. 56
105. 54
106. 80

CONCEPT EXTENSIONS

Choose the correct letter for each exercise. Letters will be used more than once. No pencil is needed. Just think about the meaning of each expression.

A = 2, B = −2, C = not a real number

107. $4^{1/2}$ _____
108. $-4^{1/2}$ _____
109. $(-4)^{1/2}$ _____
110. $8^{1/3}$ _____
111. $-8^{1/3}$ _____
112. $(-8)^{1/3}$ _____

Basal metabolic rate (BMR) is the number of calories per day a person needs to maintain life. A person's basal metabolic rate $B(w)$ in calories per day can be estimated with the function $B(w) = 70w^{3/4}$, where w is the person's weight in kilograms. Use this information to answer Exercises 113 and 114.

113. Estimate the BMR for a person who weighs 60 kilograms. Round to the nearest calorie. (*Note:* 60 kilograms is approximately 132 pounds.)

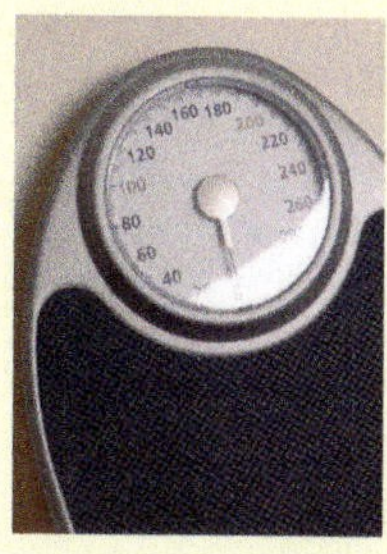

114. Estimate the BMR for a person who weighs 90 kilograms. Round to the nearest calorie. (*Note:* 90 kilograms is approximately 198 pounds.)

The number of cell telephone subscribers in the United States from 1995–2010 can be modeled by $f(x) = 25x^{23/25}$, where $f(x)$ is the number of cellular telephone subscriptions in millions, x years after 1995. (Source: CTIA-Wireless Association, 1995–2010) Use this information to answer Exercises 115 and 116.

115. Use this model to estimate the number of cellular subscriptions in 2010. Round to the nearest tenth of a million.

116. Estimate the number of cellular telephone subscriptions in 2015. Round to the nearest tenth of a million.

117. Explain how writing x^{-7} with positive exponents is similar to writing $x^{-1/4}$ with positive exponents.

118. Explain how writing $2x^{-5}$ with positive exponents is similar to writing $2x^{-3/4}$ with positive exponents.

Fill in each box with the correct expression.

119. $\square \cdot a^{2/3} = a^{3/3}$, or a

120. $\square \cdot x^{1/8} = x^{4/8}$, or $x^{1/2}$

121. $\dfrac{\square}{x^{-2/5}} = x^{3/5}$

122. $\dfrac{\square}{y^{-3/4}} = y^{4/4}$, or y

Use a calculator to write a four-decimal-place approximation of each number.

123. $8^{1/4}$
124. $20^{1/5}$
125. $18^{3/5}$
126. $76^{5/7}$

127. In physics, the speed of a wave traveling over a stretched string with tension t and density u is given by the expression $\dfrac{\sqrt{t}}{\sqrt{u}}$. Write this expression with rational exponents.

128. In electronics, the angular frequency of oscillations in a certain type of circuit is given by the expression $(LC)^{-1/2}$. Use radical notation to write this expression.

17.3 Simplifying Radical Expressions 

Objectives

1. Use the Product Rule for Radicals.
2. Use the Quotient Rule for Radicals.
3. Simplify Radicals.
4. Use the Distance and Midpoint Formulas.

Objective 1 Using the Product Rule

It is possible to simplify some radicals that do not evaluate to rational numbers. To do so, we use a product rule and a quotient rule for radicals. To discover the product rule, notice the following pattern.

$$\sqrt{9} \cdot \sqrt{4} = 3 \cdot 2 = 6$$
$$\sqrt{9 \cdot 4} = \sqrt{36} = 6$$

Since both expressions simplify to 6, it is true that

$$\sqrt{9} \cdot \sqrt{4} = \sqrt{9 \cdot 4}$$

This pattern suggests the following product rule for radicals.

Product Rule for Radicals

If $\sqrt[n]{a}$ and $\sqrt[n]{b}$ are real numbers, then

$$\sqrt[n]{a} \cdot \sqrt[n]{b} = \sqrt[n]{ab}$$

Notice that the product rule is the relationship $a^{1/n} \cdot b^{1/n} = (ab)^{1/n}$ stated in radical notation.

Practice 1

Multiply.

a. $\sqrt{5} \cdot \sqrt{7}$ b. $\sqrt{13} \cdot \sqrt{z}$
c. $\sqrt[4]{125} \cdot \sqrt[4]{5}$ d. $\sqrt[3]{5y} \cdot \sqrt[3]{3x^2}$
e. $\sqrt{\dfrac{5}{m}} \cdot \sqrt{\dfrac{t}{2}}$

Example 1 Multiply.

a. $\sqrt{3} \cdot \sqrt{5}$ b. $\sqrt{21} \cdot \sqrt{x}$ c. $\sqrt[3]{4} \cdot \sqrt[3]{2}$
d. $\sqrt[4]{5y^2} \cdot \sqrt[4]{2x^3}$ e. $\sqrt{\dfrac{2}{a}} \cdot \sqrt{\dfrac{b}{3}}$

Solution:

a. $\sqrt{3} \cdot \sqrt{5} = \sqrt{3 \cdot 5} = \sqrt{15}$
b. $\sqrt{21} \cdot \sqrt{x} = \sqrt{21x}$
c. $\sqrt[3]{4} \cdot \sqrt[3]{2} = \sqrt[3]{4 \cdot 2} = \sqrt[3]{8} = 2$
d. $\sqrt[4]{5y^2} \cdot \sqrt[4]{2x^3} = \sqrt[4]{5y^2 \cdot 2x^3} = \sqrt[4]{10y^2x^3}$
e. $\sqrt{\dfrac{2}{a}} \cdot \sqrt{\dfrac{b}{3}} = \sqrt{\dfrac{2}{a} \cdot \dfrac{b}{3}} = \sqrt{\dfrac{2b}{3a}}$

Work Practice 1

Objective 2 Using the Quotient Rule

To discover a quotient rule for radicals, notice the following pattern.

$$\sqrt{\frac{4}{9}} = \frac{2}{3}$$
$$\frac{\sqrt{4}}{\sqrt{9}} = \frac{2}{3}$$

Answers

1. a. $\sqrt{35}$ b. $\sqrt{13z}$ c. 5 d. $\sqrt[3]{15x^2y}$ e. $\sqrt{\dfrac{5t}{2m}}$

Since both expressions simplify to $\frac{2}{3}$, it is true that

$$\sqrt{\frac{4}{9}} = \frac{\sqrt{4}}{\sqrt{9}}$$

This pattern suggests the following quotient rule for radicals.

Quotient Rule for Radicals

If $\sqrt[n]{a}$ and $\sqrt[n]{b}$ are real numbers and $\sqrt[n]{b}$ is not zero, then

$$\sqrt[n]{\frac{a}{b}} = \frac{\sqrt[n]{a}}{\sqrt[n]{b}}$$

Notice that the quotient rule is the relationship $\left(\frac{a}{b}\right)^{1/n} = \frac{a^{1/n}}{b^{1/n}}$ stated in radical notation. We can use the quotient rule to simplify radical expressions by reading the rule from left to right or to divide radicals by reading the rule from right to left.

For example,

$$\sqrt{\frac{x}{16}} = \frac{\sqrt{x}}{\sqrt{16}} = \frac{\sqrt{x}}{4} \qquad \text{Using } \sqrt[n]{\frac{a}{b}} = \frac{\sqrt[n]{a}}{\sqrt[n]{b}}$$

$$\frac{\sqrt{75}}{\sqrt{3}} = \sqrt{\frac{75}{3}} = \sqrt{25} = 5 \qquad \text{Using } \frac{\sqrt[n]{a}}{\sqrt[n]{b}} = \sqrt[n]{\frac{a}{b}}$$

Note: *Recall that from Section 17.2 on, we assume that variables represent positive real numbers. Since this is so, we need not insert absolute value bars when we simplify even roots.*

Example 2 Use the quotient rule to simplify.

a. $\sqrt{\frac{25}{49}}$ **b.** $\sqrt{\frac{x}{9}}$ **c.** $\sqrt[3]{\frac{8}{27}}$ **d.** $\sqrt[4]{\frac{3}{16y^4}}$

Solution:

a. $\sqrt{\frac{25}{49}} = \frac{\sqrt{25}}{\sqrt{49}} = \frac{5}{7}$

b. $\sqrt{\frac{x}{9}} = \frac{\sqrt{x}}{\sqrt{9}} = \frac{\sqrt{x}}{3}$

c. $\sqrt[3]{\frac{8}{27}} = \frac{\sqrt[3]{8}}{\sqrt[3]{27}} = \frac{2}{3}$

d. $\sqrt[4]{\frac{3}{16y^4}} = \frac{\sqrt[4]{3}}{\sqrt[4]{16y^4}} = \frac{\sqrt[4]{3}}{2y}$

Work Practice 2

Practice 2

Use the quotient rule to simplify.

a. $\sqrt{\frac{36}{49}}$ **b.** $\sqrt{\frac{z}{16}}$

c. $\sqrt[3]{\frac{125}{8}}$ **d.** $\sqrt[4]{\frac{5}{81x^8}}$

Objective 3 Simplifying Radicals

Both the product and quotient rules can be used to simplify a radical. If the product rule is read from right to left, we have that

$$\sqrt[n]{ab} = \sqrt[n]{a} \cdot \sqrt[n]{b}.$$

This is used to simplify the following radicals.

Answers

2. a. $\frac{6}{7}$ **b.** $\frac{\sqrt{z}}{4}$ **c.** $\frac{5}{2}$ **d.** $\frac{\sqrt[4]{5}}{3x^2}$

Practice 3

Simplify the following.

a. $\sqrt{98}$ **b.** $\sqrt[3]{54}$
c. $\sqrt{35}$ **d.** $\sqrt[4]{243}$

Example 3 Simplify the following.

a. $\sqrt{50}$ **b.** $\sqrt[3]{24}$ **c.** $\sqrt{26}$ **d.** $\sqrt[4]{32}$

Solution:

a. Factor 50 such that one factor is the largest perfect square that divides 50. The largest perfect square factor of 50 is 25, so we write 50 as $25 \cdot 2$ and use the product rule for radicals to simplify.

$$\sqrt{50} = \sqrt{25 \cdot 2} = \sqrt{25} \cdot \sqrt{2} = 5\sqrt{2}$$

The largest perfect square factor of 50

Helpful Hint Don't forget that, for example, $5\sqrt{2}$ means $5 \cdot \sqrt{2}$.

b. $\sqrt[3]{24} = \sqrt[3]{8 \cdot 3} = \sqrt[3]{8} \cdot \sqrt[3]{3} = 2\sqrt[3]{3}$

The largest perfect cube factor of 24

c. $\sqrt{26}$ The largest perfect square factor of 26 is 1, so $\sqrt{26}$ cannot be simplified further.

d. $\sqrt[4]{32} = \sqrt[4]{16 \cdot 2} = \sqrt[4]{16} \cdot \sqrt[4]{2} = 2\sqrt[4]{2}$

The largest fourth power factor of 32

Work Practice 3

After simplifying a radical such as a square root, always check the radicand to see that it contains no other perfect square factors. It may, if the largest perfect square factor of the radicand was not originally recognized. For example,

$$\sqrt{200} = \sqrt{4 \cdot 50} = \sqrt{4} \cdot \sqrt{50} = 2\sqrt{50}$$

Notice that the radicand 50 still contains the perfect square factor 25. This is because 4 is not the largest perfect square factor of 200. We continue as follows.

$$2\sqrt{50} = 2\sqrt{25 \cdot 2} = 2 \cdot \sqrt{25} \cdot \sqrt{2} = 2 \cdot 5 \cdot \sqrt{2} = 10\sqrt{2}$$

The radical is now simplified since 2 contains no perfect square factors (other than 1).

Helpful Hint

To help you recognize largest perfect power factors of a radicand, it will help if you are familiar with some perfect powers. A few are listed below.

Perfect Squares	1,	4,	9,	16,	25,	36,	49,	64,	81,	100,	121,	144
	1^2	2^2	3^2	4^2	5^2	6^2	7^2	8^2	9^2	10^2	11^2	12^2
Perfect Cubes	1,	8,	27,	64,	125							
	1^3	2^3	3^3	4^3	5^3							
Perfect Fourth Powers	1,	16,	81,	256								
	1^4	2^4	3^4	4^4								

In general, we say that a radicand of the form $\sqrt[n]{a}$ is simplified when the radicand a contains no factors that are perfect nth powers (other than 1 or -1).

Answers

3. a. $7\sqrt{2}$ **b.** $3\sqrt[3]{2}$ **c.** $\sqrt{35}$ **d.** $3\sqrt[4]{3}$

Example 4 Use the product rule to simplify.

a. $\sqrt{25x^3}$ **b.** $\sqrt[3]{54x^6y^8}$ **c.** $\sqrt[4]{81z^{11}}$

Solution:

a. $\sqrt{25x^3} = \sqrt{25x^2 \cdot x}$ Find the largest perfect square factor.

$= \sqrt{25x^2} \cdot \sqrt{x}$ Apply the product rule.

$= 5x\sqrt{x}$ Simplify.

b. $\sqrt[3]{54x^6y^8} = \sqrt[3]{27 \cdot 2 \cdot x^6 \cdot y^6 \cdot y^2}$ Factor the radicand and identify perfect cube factors.

$= \sqrt[3]{27x^6y^6 \cdot 2y^2}$

$= \sqrt[3]{27x^6y^6} \cdot \sqrt[3]{2y^2}$ Apply the product rule.

$= 3x^2y^2\sqrt[3]{2y^2}$ Simplify.

c. $\sqrt[4]{81z^{11}} = \sqrt[4]{81 \cdot z^8 \cdot z^3}$ Factor the radicand and identify perfect fourth power factors.

$= \sqrt[4]{81z^8} \cdot \sqrt[4]{z^3}$ Apply the product rule.

$= 3z^2\sqrt[4]{z^3}$ Simplify.

Work Practice 4

Practice 4

Use the product rule to simplify.

a. $\sqrt{36z^7}$ **b.** $\sqrt[3]{32p^4q^7}$

c. $\sqrt[4]{16x^{15}}$

Example 5 Use the quotient rule to divide, and simplify if possible.

a. $\dfrac{\sqrt{20}}{\sqrt{5}}$ **b.** $\dfrac{\sqrt{50x}}{2\sqrt{2}}$ **c.** $\dfrac{7\sqrt[3]{48x^4y^8}}{\sqrt[3]{6y^2}}$ **d.** $\dfrac{2\sqrt[4]{32a^8b^6}}{\sqrt[4]{a^{-1}b^2}}$

Solution:

a. $\dfrac{\sqrt{20}}{\sqrt{5}} = \sqrt{\dfrac{20}{5}}$ Apply the quotient rule.

$= \sqrt{4}$ Simplify.

$= 2$ Simplify.

b. $\dfrac{\sqrt{50x}}{2\sqrt{2}} = \dfrac{1}{2} \cdot \sqrt{\dfrac{50x}{2}}$ Apply the quotient rule.

$= \dfrac{1}{2} \cdot \sqrt{25x}$ Simplify.

$= \dfrac{1}{2} \cdot \sqrt{25} \cdot \sqrt{x}$ Factor $25x$.

$= \dfrac{1}{2} \cdot 5 \cdot \sqrt{x}$ Simplify.

$= \dfrac{5}{2}\sqrt{x}$

c. $\dfrac{7\sqrt[3]{48x^4y^8}}{\sqrt[3]{6y^2}} = 7 \cdot \sqrt[3]{\dfrac{48x^4y^8}{6y^2}}$ Apply the quotient rule.

$= 7 \cdot \sqrt[3]{8x^4y^6}$ Simplify.

$= 7\sqrt[3]{8x^3y^6 \cdot x}$ Factor.

$= 7 \cdot \sqrt[3]{8x^3y^6} \cdot \sqrt[3]{x}$ Apply the product rule.

$= 7 \cdot 2xy^2 \cdot \sqrt[3]{x}$ Simplify.

$= 14xy^2\sqrt[3]{x}$

(*Continued on next page*)

Practice 5

Use the quotient rule to divide and simplify.

a. $\dfrac{\sqrt{80}}{\sqrt{5}}$ **b.** $\dfrac{\sqrt{98z}}{3\sqrt{2}}$

c. $\dfrac{5\sqrt[3]{40x^5y^7}}{\sqrt[3]{5y}}$ **d.** $\dfrac{3\sqrt[5]{64x^9y^8}}{\sqrt[5]{x^{-1}y^2}}$

Answers

4. a. $6z^3\sqrt{z}$ **b.** $2pq^2\sqrt[3]{4pq}$ **c.** $2x^3\sqrt[4]{x^3}$

5. a. 4 **b.** $\dfrac{7}{3}\sqrt{z}$ **c.** $10xy^2\sqrt[3]{x^2}$

d. $6x^2y\sqrt[5]{2y}$

d. $\dfrac{2\sqrt[4]{32a^8b^6}}{\sqrt[4]{a^{-1}b^2}} = 2\sqrt[4]{\dfrac{32a^8b^6}{a^{-1}b^2}} = 2\sqrt[4]{32a^9b^4} = 2\sqrt[4]{16\cdot a^8\cdot b^4\cdot 2\cdot a}$

$= 2\sqrt[4]{16a^8b^4}\cdot\sqrt[4]{2a} = 2\cdot 2a^2b\cdot\sqrt[4]{2a} = 4a^2b\sqrt[4]{2a}$

Work Practice 5

✓Concept Check Find and correct the error:

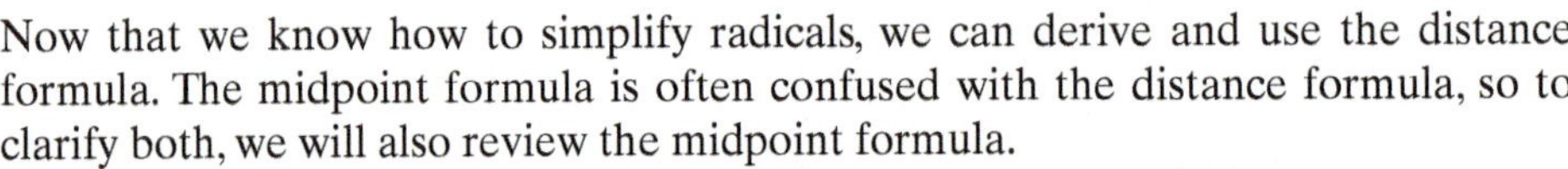

$$\frac{\sqrt[3]{27}}{\sqrt{9}} = \sqrt[3]{\frac{27}{9}} = \sqrt[3]{3}$$

Objective 4 Using the Distance and Midpoint Formulas

Now that we know how to simplify radicals, we can derive and use the distance formula. The midpoint formula is often confused with the distance formula, so to clarify both, we will also review the midpoint formula.

The Cartesian coordinate system helps us visualize a distance between points. To find the distance between two points, we use the distance formula, which is derived from the Pythagorean theorem.

To find the distance d between two points (x_1, y_1) and (x_2, y_2) as shown to the left, notice that the length of leg a is $x_2 - x_1$ and that the length of leg b is $y_2 - y_1$.

Thus, the Pythagorean theorem tells us that

$$d^2 = a^2 + b^2$$

or

$$d^2 = (x_2 - x_1)^2 + (y_2 - y_1)^2$$

or

$$d = \sqrt{(x_2 - x_1)^2 + (y_2 - y_1)^2}$$

This formula gives us the distance between any two points on the real plane.

Distance Formula

The distance d between two points (x_1, y_1) and (x_2, y_2) is given by

$$d = \sqrt{(x_2 - x_1)^2 + (y_2 - y_1)^2}$$

Practice 6

Find the distance between $(-3, 7)$ and $(-2, 3)$. Give an exact distance and a three-decimal-place approximation.

Example 6 Find the distance between $(2, -5)$ and $(1, -4)$. Give an exact distance and a three-decimal-place approximation.

Solution: To use the distance formula, it makes no difference which point we call (x_1, y_1) and which point we call (x_2, y_2). We will let $(x_1, y_1) = (2, -5)$ and $(x_2, y_2) = (1, -4)$.

$$\begin{aligned} d &= \sqrt{(x_2 - x_1)^2 + (y_2 - y_1)^2} \\ &= \sqrt{(1-2)^2 + [-4-(-5)]^2} \\ &= \sqrt{(-1)^2 + (1)^2} \\ &= \sqrt{1+1} \\ &= \sqrt{2} \approx 1.414 \end{aligned}$$

The distance between the two points is exactly $\sqrt{2}$ units, or approximately 1.414 units.

Work Practice 6

Answer

6. $\sqrt{17}$ units $\approx$ 4.123 units

✓Concept Check Answer

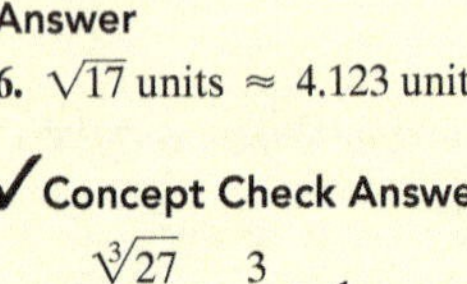

$$\frac{\sqrt[3]{27}}{\sqrt{9}} = \frac{3}{3} = 1$$

The **midpoint** of a line segment is the **point** located exactly halfway between the two endpoints of the line segment. On the graph to the left, the point M is the midpoint of line segment PQ. Thus, the distance between M and P equals the distance between M and Q.

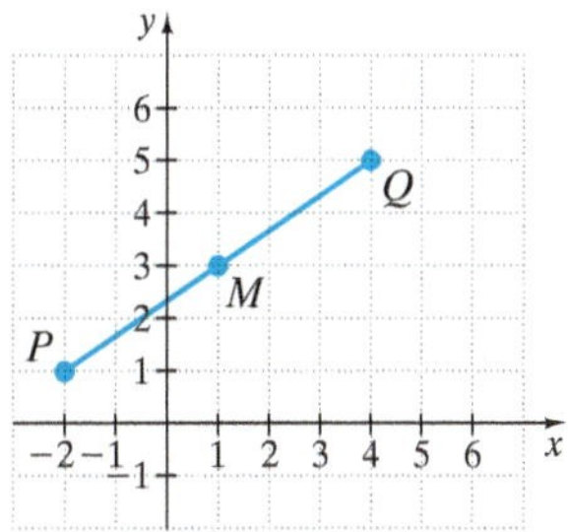

Note: We usually need no knowledge of roots to calculate the midpoint of a line segment. We review midpoint here only because it is often confused with the distance between two points.

The x-coordinate of M is at half the distance between the x-coordinates of P and Q, and the y-coordinate of M is at half the distance between the y-coordinates of P and Q. That is, the x-coordinate of M is the average of the x-coordinates of P and Q; the y-coordinate of M is the average of the y-coordinates of P and Q.

Midpoint Formula

The midpoint of the line segment whose endpoints are (x_1, y_1) and (x_2, y_2) is the point with coordinates

$$\left(\frac{x_1 + x_2}{2}, \frac{y_1 + y_2}{2}\right)$$

Example 7 Find the midpoint of the line segment that joins points $P(-3, 3)$ and $Q(1, 0)$.

Solution: Use the midpoint formula. It makes no difference which point we call (x_1, y_1) or which point we call (x_2, y_2). Let $(x_1, y_1) = (-3, 3)$ and $(x_2, y_2) = (1, 0)$.

$$\begin{aligned} \text{midpoint} &= \left(\frac{x_1 + x_2}{2}, \frac{y_1 + y_2}{2}\right) \\ &= \left(\frac{-3 + 1}{2}, \frac{3 + 0}{2}\right) \\ &= \left(\frac{-2}{2}, \frac{3}{2}\right) \\ &= \left(-1, \frac{3}{2}\right) \end{aligned}$$

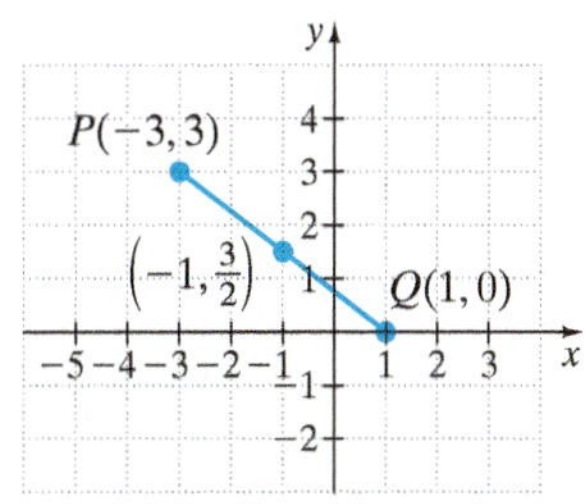

The midpoint of the segment is $\left(-1, \frac{3}{2}\right)$.

Work Practice 7

Practice 7

Find the midpoint of the line segment that joins points $P(5, -2)$ and $Q(8, -6)$.

Answer

7. $\left(\frac{13}{2}, -4\right)$

Helpful Hint

The distance between two points is a distance. The midpoint of a line segment is the point halfway between the endpoints of the segment.

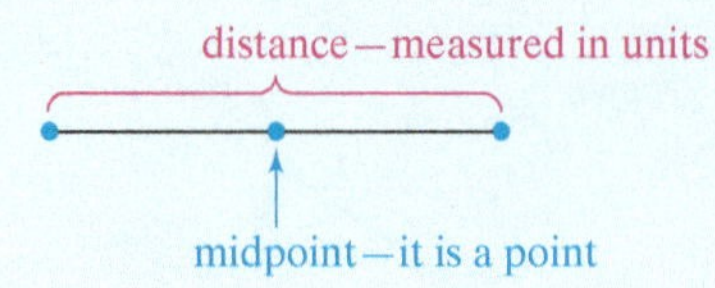

Vocabulary, Readiness & Video Check

Use the choices below to fill in each blank. Some choices may be used more than once.

distance midpoint point

1. The __________ of a line segment is a __________ exactly halfway between the two endpoints of the line segment.
2. The __________ between two points is a distance, measured in units.
3. The __________ formula is $d = \sqrt{(x_2 - x_1)^2 + (y_2 - y_1)^2}$.
4. The __________ formula is $\left(\frac{x_1 + x_2}{2}, \frac{y_1 + y_2}{2}\right)$.

Martin-Gay Interactive Videos *Watch the section lecture video and answer the following questions.*

See Video 17.3

Objective 1 5. From Example 1 and the lecture before, in order to apply the product rule for radicals, what must be true about the indexes of the radicals being multiplied?

Objective 2 6. From Examples 2–6, when might you apply the quotient rule (in either direction) in order to simplify a fractional radical expression?

Objective 3 7. From Example 8, we know that an even power of a variable is a perfect square factor of the variable, leaving no factor in the radicand once simplified. Therefore, what must be true about the power of any variable left in the radicand of a simplified square root? Explain.

Objective 4 8. From Example 10, the formula uses the coordinates of two points similar to the slope formula. What caution should you take when replacing values in the formula?

Objective 5 9. Based on Example 11, complete the following statement. The x-value of the midpoint is the ____________ of the x-values of the endpoints and the y-value of the midpoint is the ____________ of the y-values of the endpoints.

17.3 Exercise Set MyMathLab®

Use the product rule to multiply. See Example 1.

1. $\sqrt{7}\cdot\sqrt{2}$
2. $\sqrt{11}\cdot\sqrt{10}$
3. $\sqrt[4]{8}\cdot\sqrt[4]{2}$
4. $\sqrt[4]{27}\cdot\sqrt[4]{3}$
5. $\sqrt[3]{4}\cdot\sqrt[3]{9}$
6. $\sqrt[3]{10}\cdot\sqrt[3]{5}$
7. $\sqrt{2}\cdot\sqrt{3x}$
8. $\sqrt{3y}\cdot\sqrt{5x}$
9. $\sqrt{\frac{7}{x}}\cdot\sqrt{\frac{2}{y}}$
10. $\sqrt{\frac{6}{m}}\cdot\sqrt{\frac{n}{5}}$
11. $\sqrt[4]{4x^3}\cdot\sqrt[4]{5}$
12. $\sqrt[4]{ab^2}\cdot\sqrt[4]{27ab}$

Use the quotient rule to simplify. See Examples 2 and 3.

13. $\sqrt{\frac{6}{49}}$
14. $\sqrt{\frac{8}{81}}$
15. $\sqrt{\frac{2}{49}}$
16. $\sqrt{\frac{5}{121}}$
17. $\sqrt[4]{\frac{x^3}{16}}$
18. $\sqrt[4]{\frac{y}{81x^4}}$
19. $\sqrt[3]{\frac{4}{27}}$
20. $\sqrt[3]{\frac{3}{64}}$
21. $\sqrt[4]{\frac{8}{x^8}}$
22. $\sqrt[4]{\frac{a^3}{81}}$
23. $\sqrt[3]{\frac{2x}{81y^{12}}}$
24. $\sqrt[3]{\frac{3}{8x^6}}$
25. $\sqrt{\frac{x^2y}{100}}$
26. $\sqrt{\frac{y^2z}{36}}$
27. $\sqrt{\frac{5x^2}{4y^2}}$
28. $\sqrt{\frac{y^{10}}{9x^6}}$
29. $-\sqrt[3]{\frac{z^7}{27x^3}}$
30. $-\sqrt[3]{\frac{64a}{b^9}}$

Simplify. See Examples 3 and 4.

31. $\sqrt{32}$
32. $\sqrt{27}$
33. $\sqrt[3]{192}$
34. $\sqrt[3]{108}$
35. $5\sqrt{75}$
36. $3\sqrt{8}$
37. $\sqrt{24}$
38. $\sqrt{20}$
39. $\sqrt{100x^5}$
40. $\sqrt{64y^9}$
41. $\sqrt[3]{16y^7}$
42. $\sqrt[3]{64y^9}$
43. $\sqrt[4]{a^8b^7}$
44. $\sqrt[5]{32z^{12}}$
45. $\sqrt{y^5}$
46. $\sqrt[3]{y^5}$
47. $\sqrt{25a^2b^3}$
48. $\sqrt{9x^5y^7}$
49. $\sqrt[5]{-32x^{10}y}$
50. $\sqrt[5]{-243z^9}$
51. $\sqrt[3]{50x^{14}}$
52. $\sqrt[3]{40y^{10}}$
53. $-\sqrt{32a^8b^7}$
54. $-\sqrt{20ab^6}$
55. $\sqrt{9x^7y^9}$
56. $\sqrt{12r^9s^{12}}$
57. $\sqrt[3]{125r^9s^{12}}$
58. $\sqrt[3]{8a^6b^9}$
59. $\sqrt[4]{32x^{12}y^5}$
60. $\sqrt[4]{162x^7y^{20}}$

Use the quotient rule to divide. Then simplify if possible. See Example 5.

61. $\frac{\sqrt{14}}{\sqrt{7}}$
62. $\frac{\sqrt{45}}{\sqrt{9}}$
63. $\frac{\sqrt[3]{24}}{\sqrt[3]{3}}$
64. $\frac{\sqrt[3]{10}}{\sqrt[3]{2}}$
65. $\frac{5\sqrt[4]{48}}{\sqrt[4]{3}}$
66. $\frac{7\sqrt[4]{162}}{\sqrt[4]{2}}$
67. $\frac{\sqrt{x^5y^3}}{\sqrt{xy}}$
68. $\frac{\sqrt{a^7b^6}}{\sqrt{a^3b^2}}$
69. $\frac{8\sqrt[3]{54m^7}}{\sqrt[3]{2m}}$
70. $\frac{\sqrt[3]{128x^3}}{-3\sqrt[3]{2x}}$
71. $\frac{3\sqrt{100x^2}}{2\sqrt{2x^{-1}}}$
72. $\frac{\sqrt{270y^2}}{5\sqrt{3y^{-4}}}$
73. $\frac{\sqrt[4]{96a^{10}b^3}}{\sqrt[4]{3a^2b^3}}$
74. $\frac{\sqrt[4]{160x^{10}y^5}}{\sqrt[4]{2x^2y^2}}$
75. $\frac{\sqrt[5]{64x^{10}y^3}}{\sqrt[5]{2x^3y^{-7}}}$
76. $\frac{\sqrt[5]{192x^6y^{12}}}{\sqrt[5]{2x^{-1}y^{-3}}}$

Find the distance between each pair of points. Give an exact distance and a three-decimal-place approximation. See Example 6.

77. $(5, 1)$ and $(8, 5)$
78. $(2, 3)$ and $(14, 8)$
79. $(-3, 2)$ and $(1, -3)$
80. $(3, -2)$ and $(-4, 1)$
81. $(-9, 4)$ and $(-8, 1)$
82. $(-5, -2)$ and $(-6, -6)$
83. $(0, -\sqrt{2})$ and $(\sqrt{3}, 0)$
84. $(-\sqrt{5}, 0)$ and $(0, \sqrt{7})$
85. $(1.7, -3.6)$ and $(-8.6, 5.7)$
86. $(9.6, 2.5)$ and $(-1.9, -3.7)$

Find the midpoint of the line segment whose endpoints are given. See Example 7.

87. $(6, -8), (2, 4)$
88. $(3, 9), (7, 11)$
89. $(-2, -1), (-8, 6)$
90. $(-3, -4), (6, -8)$
91. $(7, 3), (-1, -3)$
92. $(-2, 5), (-1, 6)$
93. $\left(\frac{1}{2}, \frac{3}{8}\right), \left(-\frac{3}{2}, \frac{5}{8}\right)$
94. $\left(-\frac{2}{5}, \frac{7}{15}\right), \left(-\frac{2}{5}, -\frac{4}{15}\right)$
95. $(\sqrt{2}, 3\sqrt{5}), (\sqrt{2}, -2\sqrt{5})$
96. $(\sqrt{8}, -\sqrt{12}), (3\sqrt{2}, 7\sqrt{3})$
97. $(4.6, -3.5), (7.8, -9.8)$
98. $(-4.6, 2.1), (-6.7, 1.9)$

REVIEW AND PREVIEW

Perform each indicated operation. See Sections 3.1, 12.3, and 12.4.

99. $6x + 8x$
100. $(6x)(8x)$

101. $(2x + 3)(x - 5)$

102. $(2x + 3) + (x - 5)$

103. $9y^2 - 8y^2$

104. $(9y^2)(-8y^2)$

105. $-3(x + 5)$

106. $-3 + x + 5$

107. $(x - 4)^2$

108. $(2x + 1)^2$

CONCEPT EXTENSIONS

Answer true or false. Assume all radicals represent nonzero real numbers.

109. $\sqrt[n]{a} \cdot \sqrt[n]{b} = \sqrt[n]{ab}$, ______

110. $\sqrt[3]{7} \cdot \sqrt[3]{11} = \sqrt[3]{18}$, ______

111. $\sqrt[3]{7} \cdot \sqrt{11} = \sqrt{77}$, ______

112. $\sqrt{x^7 y^8} = \sqrt{x^7} \cdot \sqrt{y^8}$, ______

113. $\dfrac{\sqrt[n]{a}}{\sqrt[n]{b}} = \sqrt[n]{\dfrac{a}{b}}$, ______

114. $\dfrac{\sqrt[3]{12}}{\sqrt[3]{4}} = \sqrt[3]{8}$, ______

Find and correct the error. See the Concept Check in this section.

115. $\dfrac{\sqrt[3]{64}}{\sqrt{64}} = \sqrt[3]{\dfrac{64}{64}} = \sqrt[3]{1} = 1$

116. $\dfrac{\sqrt[4]{16}}{\sqrt{4}} = \sqrt[4]{\dfrac{16}{4}} = \sqrt[4]{4}$

Simplify. See the Concept Check in this section. Assume variables represent positive numbers.

117. $\sqrt[5]{x^{35}}$

118. $\sqrt[6]{y^{48}}$

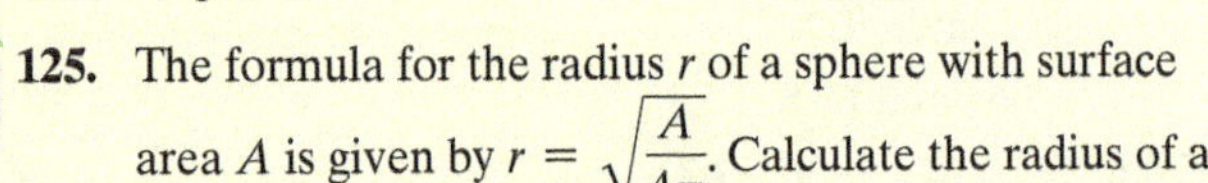

119. $\sqrt[4]{a^{12}b^4c^{20}}$

120. $\sqrt[3]{a^9b^{21}c^3}$

121. $\sqrt[3]{z^{32}}$

122. $\sqrt[5]{x^{49}}$

123. $\sqrt[7]{q^{17}r^{40}s^7}$

124. $\sqrt[4]{p^{11}q^4r^{45}}$

125. The formula for the radius r of a sphere with surface area A is given by $r = \sqrt{\dfrac{A}{4\pi}}$. Calculate the radius of a standard Zorb, whose outside surface area is 32.17 sq m. Round to the nearest tenth. (A Zorb is a large inflated ball within a ball in which a person, strapped inside, may choose to roll down a hill. *Source:* Zorb, Ltd.)

126. The owner of Knightime Classic Movie Rentals has determined that the demand equation for renting older released DVDs is $F(x) = 0.6\sqrt{49 - x^2}$, where x is the price in dollars per two-day rental and $F(x)$ is the number of times the DVD is demanded per week.

a. Approximate to one decimal place the demand per week of an older released DVD if the rental price is \$3 per two-day rental.

b. Approximate to one decimal place the demand per week of an older released DVD if the rental price is \$5 per two-day rental.

c. Explain how the owner of the store can use this equation to predict the number of copies of each DVD that should be in stock.

127. The formula for the lateral surface area A of a cone with height h and radius r is given by

$$A = \pi r\sqrt{r^2 + h^2}$$

a. Find the lateral surface area of a cone whose height is 3 centimeters and whose radius is 4 centimeters.

b. Approximate to two decimal places the lateral surface area of a cone whose height is 7.2 feet and whose radius is 6.8 feet.

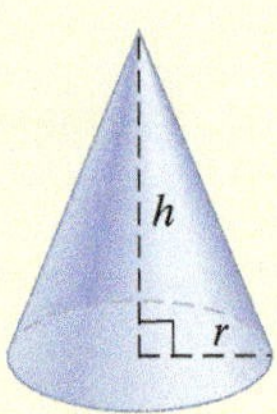

128. Before Mount Vesuvius, a volcano in Italy, erupted violently in 79 C.E., its height was 4190 feet. Vesuvius was roughly cone-shaped, and its base had a radius of approximately 25,200 feet. Use the formula for the lateral surface area of a cone, given in Exercise 127, to approximate the surface area this volcano had before it erupted. (*Source:* Global Volcanism Network)

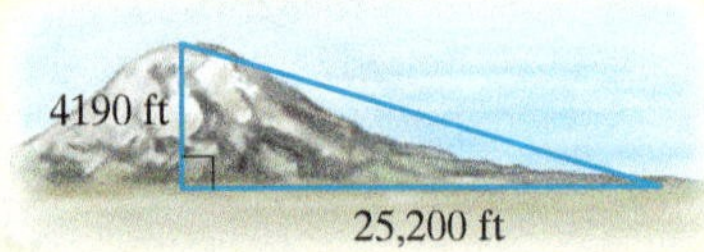

17.4 Adding, Subtracting, and Multiplying Radical Expressions

Objectives

1. Add or Subtract Radical Expressions.
2. Multiply Radical Expressions.

Objective 1 Adding or Subtracting Radical Expressions

We have learned that sums or differences of like terms can be simplified. To simplify these sums or differences, we use the distributive property. For example,

$$2x + 3x = (2 + 3)x = 5x \quad \text{and} \quad 7x^2y - 4x^2y = (7 - 4)x^2y = 3x^2y$$

The distributive property can also be used to add **like radicals.**

Like Radicals

Radicals with the same index and the same radicand are like radicals.

For example, $2\sqrt{7} + 3\sqrt{7} = (2 + 3)\sqrt{7} = 5\sqrt{7}$. Also,

Like radicals

$$5\sqrt{3x} - 7\sqrt{3x} = (5 - 7)\sqrt{3x} = -2\sqrt{3x}$$

The expression $2\sqrt{7} + 2\sqrt[3]{7}$ cannot be simplified further since $2\sqrt{7}$ and $2\sqrt[3]{7}$ are not like radicals.

Unlike radicals

Example 1 Add or subtract as indicated. Assume all variables represent positive real numbers.

a. $4\sqrt{11} + 8\sqrt{11}$ **b.** $5\sqrt[3]{3x} - 7\sqrt[3]{3x}$ **c.** $4\sqrt{5} + 4\sqrt[3]{5}$

Solution:

a. $4\sqrt{11} + 8\sqrt{11} = (4 + 8)\sqrt{11} = 12\sqrt{11}$

b. $5\sqrt[3]{3x} - 7\sqrt[3]{3x} = (5 - 7)\sqrt[3]{3x} = -2\sqrt[3]{3x}$

c. $4\sqrt{5} + 4\sqrt[3]{5}$

This expression cannot be simplified since $4\sqrt{5}$ and $4\sqrt[3]{5}$ do not contain like radicals.

Work Practice 1

Practice 1

Add or subtract as indicated.

a. $3\sqrt{17} + 5\sqrt{17}$

b. $7\sqrt[3]{5z} - 12\sqrt[3]{5z}$

c. $3\sqrt{2} + 5\sqrt[3]{2}$

When adding or subtracting radicals, always check first to see whether any radicals can be simplified.

Concept Check True or false?

$$\sqrt{a} + \sqrt{b} = \sqrt{a + b}$$

Explain.

Example 2 Add or subtract. Assume that variables represent positive real numbers.

a. $\sqrt{20} + 2\sqrt{45}$ **b.** $\sqrt[3]{54} - 5\sqrt[3]{16} + \sqrt[3]{2}$ **c.** $\sqrt{27x} - 2\sqrt{9x} + \sqrt{72x}$

d. $\sqrt[3]{98} + \sqrt{98}$ **e.** $\sqrt[3]{48y^4} + \sqrt[3]{6y^4}$

(Continued on next page)

Practice 2

Add or subtract.

a. $\sqrt{24} + 3\sqrt{54}$

b. $\sqrt[3]{24} - 4\sqrt[3]{81} + \sqrt[3]{3}$

c. $\sqrt{75x} - 3\sqrt{27x} + \sqrt{12x}$

d. $\sqrt{40} + \sqrt[3]{40}$

e. $\sqrt[3]{81x^4} + \sqrt[3]{3x^4}$

Answers

1. a. $8\sqrt{17}$ **b.** $-5\sqrt[3]{5z}$ **c.** $3\sqrt{2} + 5\sqrt[3]{2}$

2. a. $11\sqrt{6}$ **b.** $-9\sqrt[3]{3}$ **c.** $-2\sqrt{3x}$ **d.** $2\sqrt{10} + 2\sqrt[3]{5}$ **e.** $4x\sqrt[3]{3x}$

Concept Check Answer

false; answers may vary

Solution: First, simplify each radical. Then add or subtract any like radicals.

a. $\sqrt{20} + 2\sqrt{45} = \sqrt{4 \cdot 5} + 2\sqrt{9 \cdot 5}$ — Factor 20 and 45.
$= \sqrt{4} \cdot \sqrt{5} + 2 \cdot \sqrt{9} \cdot \sqrt{5}$ — Use the product rule.
$= 2 \cdot \sqrt{5} + 2 \cdot 3 \cdot \sqrt{5}$ — Simplify $\sqrt{4}$ and $\sqrt{9}$.
$= 2\sqrt{5} + 6\sqrt{5}$
$= 8\sqrt{5}$ — Add like radicals.

b. $\sqrt[3]{54} - 5\sqrt[3]{16} + \sqrt[3]{2}$
$= \sqrt[3]{27} \cdot \sqrt[3]{2} - 5 \cdot \sqrt[3]{8} \cdot \sqrt[3]{2} + \sqrt[3]{2}$ — Factor and use the product rule.
$= 3 \cdot \sqrt[3]{2} - 5 \cdot 2 \cdot \sqrt[3]{2} + \sqrt[3]{2}$ — Simplify $\sqrt[3]{27}$ and $\sqrt[3]{8}$.
$= 3\sqrt[3]{2} - 10\sqrt[3]{2} + \sqrt[3]{2}$ — Write $5 \cdot 2$ as 10.
$= -6\sqrt[3]{2}$ — Combine like radicals.

c. $\sqrt{27x} - 2\sqrt{9x} + \sqrt{72x}$
$= \sqrt{9} \cdot \sqrt{3x} - 2 \cdot \sqrt{9} \cdot \sqrt{x} + \sqrt{36} \cdot \sqrt{2x}$ — Factor and use the product rule.
$= 3 \cdot \sqrt{3x} - 2 \cdot 3 \cdot \sqrt{x} + 6 \cdot \sqrt{2x}$ — Simplify $\sqrt{9}$ and $\sqrt{36}$.
$= 3\sqrt{3x} - 6\sqrt{x} + 6\sqrt{2x}$ — Write $2 \cdot 3$ as 6.

Helpful Hint None of these terms contain like radicals. We can simplify no further.

d. $\sqrt[3]{98} + \sqrt{98} = \sqrt[3]{98} + \sqrt{49} \cdot \sqrt{2}$ — Factor and use the product rule.
$= \sqrt[3]{98} + 7\sqrt{2}$ — No further simplification is possible.

e. $\sqrt[3]{48y^4} + \sqrt[3]{6y^4} = \sqrt[3]{8y^3} \cdot \sqrt[3]{6y} + \sqrt[3]{y^3} \cdot \sqrt[3]{6y}$ — Factor and use the product rule.
$= 2y\sqrt[3]{6y} + y\sqrt[3]{6y}$ — Simplify $\sqrt[3]{8y^3}$ and $\sqrt[3]{y^3}$.
$= 3y\sqrt[3]{6y}$ — Combine like radicals.

Work Practice 2

Let's continue to assume that variables represent positive real numbers.

Practice 3

Add or subtract as indicated.

a. $\dfrac{\sqrt{28}}{3} - \dfrac{\sqrt{7}}{4}$

b. $\sqrt[3]{\dfrac{6y}{64}} + 3\sqrt[3]{6y}$

Example 3 Add or subtract as indicated.

a. $\dfrac{\sqrt{45}}{4} - \dfrac{\sqrt{5}}{3}$ **b.** $\sqrt[3]{\dfrac{7x}{8}} + 2\sqrt[3]{7x}$

Solution:

a. $\dfrac{\sqrt{45}}{4} - \dfrac{\sqrt{5}}{3} = \dfrac{3\sqrt{5}}{4} - \dfrac{\sqrt{5}}{3}$ — To subtract, notice that the LCD is 12.
$= \dfrac{3\sqrt{5} \cdot 3}{4 \cdot 3} - \dfrac{\sqrt{5} \cdot 4}{3 \cdot 4}$ — Write each expression as an equivalent expression with a denominator of 12.
$= \dfrac{9\sqrt{5}}{12} - \dfrac{4\sqrt{5}}{12}$ — Multiply factors in the numerator and the denominator.
$= \dfrac{5\sqrt{5}}{12}$ — Subtract.

b. $\sqrt[3]{\dfrac{7x}{8}} + 2\sqrt[3]{7x} = \dfrac{\sqrt[3]{7x}}{\sqrt[3]{8}} + 2\sqrt[3]{7x}$ — Apply the quotient rule for radicals.
$= \dfrac{\sqrt[3]{7x}}{2} + 2\sqrt[3]{7x}$ — Simplify.
$= \dfrac{\sqrt[3]{7x}}{2} + \dfrac{2\sqrt[3]{7x} \cdot 2}{2}$ — Write each expression as an equivalent expression with a denominator of 2.
$= \dfrac{\sqrt[3]{7x}}{2} + \dfrac{4\sqrt[3]{7x}}{2}$
$= \dfrac{5\sqrt[3]{7x}}{2}$ — Add.

Work Practice 3

Answers

3. a. $\dfrac{5\sqrt{7}}{12}$ **b.** $\dfrac{13\sqrt[3]{6y}}{4}$

Objective 2 Multiplying Radical Expressions

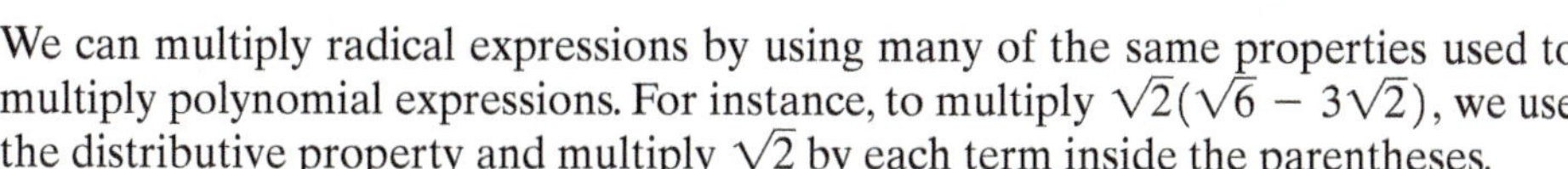

We can multiply radical expressions by using many of the same properties used to multiply polynomial expressions. For instance, to multiply $\sqrt{2}(\sqrt{6} - 3\sqrt{2})$, we use the distributive property and multiply $\sqrt{2}$ by each term inside the parentheses.

$$\begin{aligned}\sqrt{2}(\sqrt{6} - 3\sqrt{2}) &= \sqrt{2}(\sqrt{6}) - \sqrt{2}(3\sqrt{2}) && \text{Use the distributive property.}\\ &= \sqrt{2\cdot 6} - 3\sqrt{2\cdot 2}\\ &= \sqrt{2\cdot 2\cdot 3} - 3\cdot 2 && \text{Use the product rule for radicals.}\\ &= 2\sqrt{3} - 6\end{aligned}$$

Example 4 Multiply.

a. $\sqrt{3}(5 + \sqrt{30})$ **b.** $(\sqrt{5} - \sqrt{6})(\sqrt{7} + 1)$ **c.** $(7\sqrt{x} + 5)(3\sqrt{x} - \sqrt{5})$
d. $(4\sqrt{3} - 1)^2$ **e.** $(\sqrt{2x} - 5)(\sqrt{2x} + 5)$ **f.** $(\sqrt{x - 3} + 5)^2$

Solution:

a. $$\begin{aligned}\sqrt{3}(5 + \sqrt{30}) &= \sqrt{3}(5) + \sqrt{3}(\sqrt{30})\\ &= 5\sqrt{3} + \sqrt{3\cdot 30}\\ &= 5\sqrt{3} + \sqrt{3\cdot 3\cdot 10}\\ &= 5\sqrt{3} + 3\sqrt{10}\end{aligned}$$

b. To multiply, we can use the FOIL method.

$$\begin{aligned}(\sqrt{5} - \sqrt{6})(\sqrt{7} + 1) &= \overset{\text{First}}{\sqrt{5}\cdot\sqrt{7}} + \overset{\text{Outer}}{\sqrt{5}\cdot 1} - \overset{\text{Inner}}{\sqrt{6}\cdot\sqrt{7}} - \overset{\text{Last}}{\sqrt{6}\cdot 1}\\ &= \sqrt{35} + \sqrt{5} - \sqrt{42} - \sqrt{6}\end{aligned}$$

c. $$\begin{aligned}(7\sqrt{x} + 5)(3\sqrt{x} - \sqrt{5}) &= 7\sqrt{x}(3\sqrt{x}) - 7\sqrt{x}(\sqrt{5}) + 5(3\sqrt{x}) - 5(\sqrt{5})\\ &= 21x - 7\sqrt{5x} + 15\sqrt{x} - 5\sqrt{5}\end{aligned}$$

d. $$\begin{aligned}(4\sqrt{3} - 1)^2 &= (4\sqrt{3} - 1)(4\sqrt{3} - 1)\\ &= 4\sqrt{3}(4\sqrt{3}) - 4\sqrt{3}(1) - 1(4\sqrt{3}) - 1(-1)\\ &= 16\cdot 3 - 4\sqrt{3} - 4\sqrt{3} + 1\\ &= 48 - 8\sqrt{3} + 1\\ &= 49 - 8\sqrt{3}\end{aligned}$$

e. $$\begin{aligned}(\sqrt{2x} - 5)(\sqrt{2x} + 5) &= \sqrt{2x}\cdot\sqrt{2x} + 5\sqrt{2x} - 5\sqrt{2x} - 5\cdot 5\\ &= 2x - 25\end{aligned}$$

f. $$\begin{aligned}(\underbrace{\sqrt{x - 3}}_{a} + \underbrace{5}_{b})^2 &= \underbrace{(\sqrt{x - 3})^2}_{a^2} \underbrace{+ 2\cdot}_{+\,2\cdot} \underbrace{\sqrt{x - 3}}_{a}\underbrace{\cdot 5}_{\cdot b} \underbrace{+ 5^2}_{+\,b^2}\\ &= x - 3 + 10\sqrt{x - 3} + 25 && \text{Simplify.}\\ &= x + 22 + 10\sqrt{x - 3} && \text{Combine like terms.}\end{aligned}$$

Work Practice 4

Practice 4

Multiply.

a. $\sqrt{5}(2 + \sqrt{15})$
b. $(\sqrt{2} - \sqrt{5})(\sqrt{6} + 2)$
c. $(3\sqrt{z} - 4)(2\sqrt{z} + 3)$
d. $(\sqrt{6} - 3)^2$
e. $(\sqrt{5x} + 3)(\sqrt{5x} - 3)$
f. $(\sqrt{x + 2} + 3)^2$

Answers

4. a. $2\sqrt{5} + 5\sqrt{3}$
b. $2\sqrt{3} + 2\sqrt{2} - \sqrt{30} - 2\sqrt{5}$
c. $6z + \sqrt{z} - 12$ **d.** $-6\sqrt{6} + 15$
e. $5x - 9$ **f.** $6\sqrt{x + 2} + x + 11$

Vocabulary, Readiness & Video Check

Complete the table with "Like" or "Unlike."

	Terms	Like or Unlike Radical Terms?
1.	$\sqrt{7}, \sqrt[3]{7}$	
2.	$\sqrt[3]{x^2y}, \sqrt[3]{yx^2}$	
3.	$\sqrt[3]{abc}, \sqrt[3]{cba}$	
4.	$2x\sqrt{5}, 2x\sqrt{10}$	

Simplify. Assume that all variables represent positive real numbers.

5. $2\sqrt{3} + 4\sqrt{3} =$ _____ **6.** $5\sqrt{7} + 3\sqrt{7} =$ _____ **7.** $8\sqrt{x} - \sqrt{x} =$ _____

8. $3\sqrt{y} - \sqrt{y} =$ _____ **9.** $7\sqrt[3]{x} + \sqrt[3]{x} =$ _____ **10.** $8\sqrt[3]{z} + \sqrt[3]{z} =$ _____

Martin-Gay Interactive Videos *Watch the section lecture video and answer the following questions.*

See Video 17.4

Objective 1 **11.** From Examples 1 and 2, why should you always check to see if all terms in your expression are simplified before attempting to add or subtract radicals?

Objective 2 **12.** In Example 4, what are you told to remember about the square root of a positive number?

17.4 Exercise Set MyMathLab® 

Add or subtract. See Examples 1 through 3.

1. $\sqrt{8} - \sqrt{32}$

2. $\sqrt{27} - \sqrt{75}$

3. $2\sqrt{2x^3} + 4x\sqrt{8x}$

4. $3\sqrt{45x^3} + x\sqrt{5x}$

5. $2\sqrt{50} - 3\sqrt{125} + \sqrt{98}$

6. $4\sqrt{32} - \sqrt{18} + 2\sqrt{128}$

7. $\sqrt[3]{16x} - \sqrt[3]{54x}$

8. $2\sqrt[3]{3a^4} - 3a\sqrt[3]{81a}$

9. $\sqrt{9b^3} - \sqrt{25b^3} + \sqrt{49b^3}$

10. $\sqrt{4x^7} + 9x^2\sqrt{x^3} - 5x\sqrt{x^5}$

11. $\dfrac{5\sqrt{2}}{3} + \dfrac{2\sqrt{2}}{5}$

12. $\dfrac{\sqrt{3}}{2} + \dfrac{4\sqrt{3}}{3}$

13. $\sqrt[3]{\dfrac{11}{8}} - \dfrac{\sqrt[3]{11}}{6}$

14. $\dfrac{2\sqrt[3]{4}}{7} - \dfrac{\sqrt[3]{4}}{14}$

15. $\dfrac{\sqrt{20x}}{9} + \sqrt{\dfrac{5x}{9}}$

16. $\dfrac{3x\sqrt{7}}{5} + \sqrt{\dfrac{7x^2}{100}}$

17. $7\sqrt{9} - 7 + \sqrt{3}$

18. $\sqrt{16} - 5\sqrt{10} + 7$

19. $2 + 3\sqrt{y^2} - 6\sqrt{y^2} + 5$

20. $3\sqrt{7} - \sqrt[3]{x} + 4\sqrt{7} - 3\sqrt[3]{x}$

21. $3\sqrt{108} - 2\sqrt{18} - 3\sqrt{48}$

22. $-\sqrt{75} + \sqrt{12} - 3\sqrt{3}$

23. $-5\sqrt[3]{625} + \sqrt[3]{40}$

24. $-2\sqrt[3]{108} - \sqrt[3]{32}$

25. $a^3\sqrt{9ab^3} - \sqrt{25a^7b^3} + \sqrt{16a^7b^3}$

26. $\sqrt{4x^7y^5} + 9x^2\sqrt{x^3y^5} - 5xy\sqrt{x^5y^3}$

27. $5y\sqrt{8y} + 2\sqrt{50y^3}$

28. $3\sqrt{8x^2y^3} - 2x\sqrt{32y^3}$

29. $\sqrt[3]{54xy^3} - 5\sqrt[3]{2xy^3} + y\sqrt[3]{128x}$

30. $2\sqrt[3]{24x^3y^4} + 4x\sqrt[3]{81y^4}$

31. $6\sqrt[3]{11} + 8\sqrt{11} - 12\sqrt{11}$

32. $3\sqrt[3]{5} + 4\sqrt{5} - 8\sqrt{5}$

33. $-2\sqrt[4]{x^7} + 3\sqrt[4]{16x^7} - x\sqrt[4]{x^3}$

34. $6\sqrt[3]{24x^3} - 2\sqrt[3]{81x^3} - x\sqrt[3]{3}$

35. $\dfrac{4\sqrt{3}}{3} - \dfrac{\sqrt{12}}{3}$

36. $\dfrac{\sqrt{45}}{10} + \dfrac{7\sqrt{5}}{10}$

37. $\dfrac{\sqrt[3]{8x^4}}{7} + \dfrac{3x\sqrt[3]{x}}{7}$

38. $\dfrac{\sqrt[4]{48}}{5x} - \dfrac{2\sqrt[4]{3}}{10x}$

39. $\sqrt{\dfrac{28}{x^2}} + \sqrt{\dfrac{7}{4x^2}}$

40. $\dfrac{\sqrt{99}}{5x} - \sqrt{\dfrac{44}{x^2}}$

41. $\sqrt[3]{\dfrac{16}{27}} - \dfrac{\sqrt[3]{54}}{6}$

42. $\dfrac{\sqrt[3]{3}}{10} + \sqrt[3]{\dfrac{24}{125}}$

43. 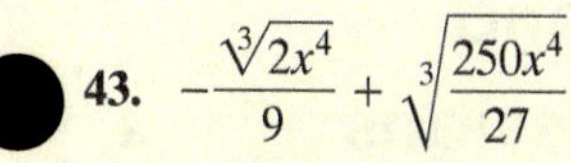 $-\frac{\sqrt[3]{2x^4}}{9} + \sqrt[3]{\frac{250x^4}{27}}$

44. $\frac{\sqrt[3]{y^5}}{8} + \frac{5y\sqrt[3]{y^2}}{4}$

△ 45. Find the perimeter of the trapezoid.

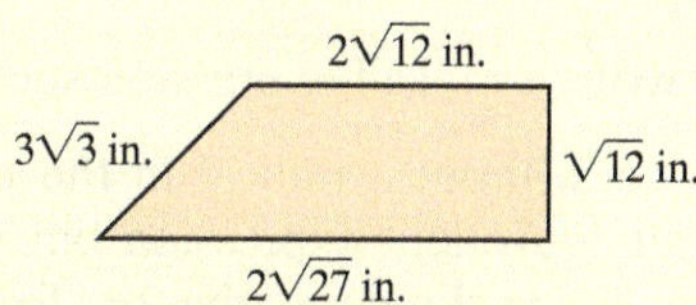

△ 46. Find the perimeter of the triangle.

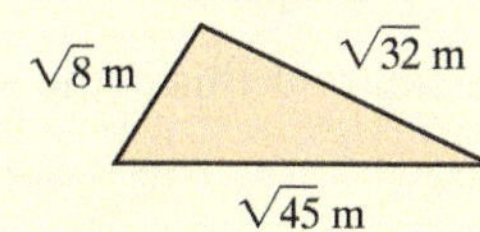

Multiply and then simplify if possible. See Example 4.

47. $\sqrt{7}(\sqrt{5} + \sqrt{3})$

48. $\sqrt{5}(\sqrt{15} - \sqrt{35})$

49. $(\sqrt{5} - \sqrt{2})^2$

50. $(3x - \sqrt{2})(3x - \sqrt{2})$

51. $\sqrt{3x}(\sqrt{3} - \sqrt{x})$

52. $\sqrt{5y}(\sqrt{y} + \sqrt{5})$

53. $(2\sqrt{x} - 5)(3\sqrt{x} + 1)$

54. $(8\sqrt{y} + z)(4\sqrt{y} - 1)$

55. $(\sqrt[3]{a} - 4)(\sqrt[3]{a} + 5)$

56. $(\sqrt[3]{a} + 2)(\sqrt[3]{a} + 7)$

57. $6(\sqrt{2} - 2)$

58. $\sqrt{5}(6 - \sqrt{5})$

59. $\sqrt{2}(\sqrt{2} + x\sqrt{6})$

60. $\sqrt{3}(\sqrt{3} - 2\sqrt{5x})$

61. $(2\sqrt{7} + 3\sqrt{5})(\sqrt{7} - 2\sqrt{5})$

62. $(\sqrt{6} - 4\sqrt{2})(3\sqrt{6} + \sqrt{2})$

63. $(\sqrt{x} - y)(\sqrt{x} + y)$

64. $(\sqrt{3x} + 2)(\sqrt{3x} - 2)$

65. $(\sqrt{3} + x)^2$

66. $(\sqrt{y} - 3x)^2$

67. $(\sqrt{5x} - 2\sqrt{3x})(\sqrt{5x} - 3\sqrt{3x})$

68. $(5\sqrt{7x} - \sqrt{2x})(4\sqrt{7x} + 6\sqrt{2x})$

69. $(\sqrt[3]{4} + 2)(\sqrt[3]{2} - 1)$

70. $(\sqrt[3]{3} + \sqrt[3]{2})(\sqrt[3]{9} - \sqrt[3]{4})$

71. $(\sqrt[3]{x} + 1)(\sqrt[3]{x^2} - \sqrt[3]{x} + 1)$

72. $(\sqrt[3]{3x} + 2)(\sqrt[3]{9x^2} - 2\sqrt[3]{3x} + 4)$

73. $(\sqrt{x - 1} + 5)^2$

74. $(\sqrt{3x + 1} + 2)^2$

75. $(\sqrt{2x + 5} - 1)^2$

76. $(\sqrt{x - 6} - 7)^2$

REVIEW AND PREVIEW

Factor each numerator and denominator. Then simplify if possible. See Section 14.1.

77. $\frac{2x - 14}{2}$

78. $\frac{8x - 24y}{4}$

79. $\frac{7x - 7y}{x^2 - y^2}$

80. $\frac{x^3 - 8}{4x - 8}$

81. $\frac{6a^2b - 9ab}{3ab}$

82. $\frac{14r - 28r^2s^2}{7rs}$

83. $\frac{-4 + 2\sqrt{3}}{6}$

84. $\frac{-5 + 10\sqrt{7}}{5}$

CONCEPT EXTENSIONS

△ 85. Find the perimeter and area of the rectangle.

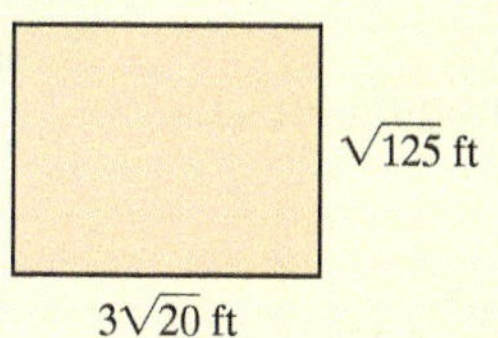

△ 86. Find the area and perimeter of the trapezoid. (*Hint:* The area of a trapezoid is the product of half the height $6\sqrt{3}$ meters and the sum of the bases $2\sqrt{63}$ and $7\sqrt{7}$ meters.)

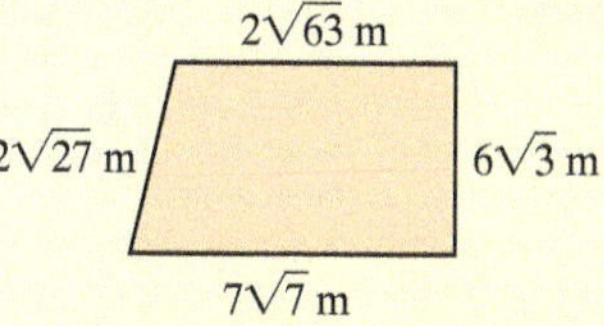

87. a. Add: $\sqrt{3} + \sqrt{3}$
b. Multiply: $\sqrt{3} \cdot \sqrt{3}$
c. Describe the differences in parts (a) and (b).

88. a. Add: $2\sqrt{5} + \sqrt{5}$
b. Multiply: $2\sqrt{5} \cdot \sqrt{5}$
c. Describe the differences in parts (a) and (b).

89. Multiply: $(\sqrt{2} + \sqrt{3} - 1)^2$

90. Multiply: $(\sqrt{5} - \sqrt{2} + 1)^2$

91. Explain how simplifying $2x + 3x$ is similar to simplifying $2\sqrt{x} + 3\sqrt{x}$.

92. Explain how multiplying $(x - 2)(x + 3)$ is similar to multiplying $(\sqrt{x} - \sqrt{2})(\sqrt{x} + 3)$.

17.5 Rationalizing Denominators and Numerators of Radical Expressions

Objectives

1 Rationalize Denominators.

2 Rationalize Denominators Having Two Terms.

3 Rationalize Numerators.

Objective 1 Rationalizing Denominators of Radical Expressions

Often in mathematics, it is helpful to write a radical expression such as $\frac{\sqrt{3}}{\sqrt{2}}$ either without a radical in the denominator or without a radical in the numerator. The process of writing this expression as an equivalent expression but without a radical in the denominator is called **rationalizing the denominator.** To rationalize the denominator of $\frac{\sqrt{3}}{\sqrt{2}}$, we use the fundamental principle of fractions and multiply the numerator and the denominator by $\sqrt{2}$. Recall that this is the same as multiplying by $\frac{\sqrt{2}}{\sqrt{2}}$, which simplifies to 1.

$$\frac{\sqrt{3}}{\sqrt{2}} = \frac{\sqrt{3}\cdot\sqrt{2}}{\sqrt{2}\cdot\sqrt{2}} = \frac{\sqrt{6}}{\sqrt{4}} = \frac{\sqrt{6}}{2}$$

In this section, we continue to assume that variables represent positive real numbers.

Practice 1

Rationalize the denominator of each expression.

a. $\frac{5}{\sqrt{3}}$ **b.** $\frac{3\sqrt{25}}{\sqrt{4x}}$ **c.** $\sqrt[3]{\frac{2}{9}}$

Example 1 Rationalize the denominator of each expression.

a. $\frac{2}{\sqrt{5}}$ **b.** $\frac{2\sqrt{16}}{\sqrt{9x}}$ **c.** $\sqrt[3]{\frac{1}{2}}$

Solution:

a. To rationalize the denominator, we multiply the numerator and denominator by a factor that makes the radicand in the denominator a perfect square.

$$\frac{2}{\sqrt{5}} = \frac{2\cdot\sqrt{5}}{\sqrt{5}\cdot\sqrt{5}} = \frac{2\sqrt{5}}{5}$$ The denominator is now rationalized.

b. First, we simplify the radicals and then rationalize the denominator.

$$\frac{2\sqrt{16}}{\sqrt{9x}} = \frac{2(4)}{3\sqrt{x}} = \frac{8}{3\sqrt{x}}$$

To rationalize the denominator, multiply the numerator and denominator by $\sqrt{x}$. Then

$$\frac{8}{3\sqrt{x}} = \frac{8\cdot\sqrt{x}}{3\sqrt{x}\cdot\sqrt{x}} = \frac{8\sqrt{x}}{3x}$$

c. $\sqrt[3]{\frac{1}{2}} = \frac{\sqrt[3]{1}}{\sqrt[3]{2}} = \frac{1}{\sqrt[3]{2}}$. Now we rationalize the denominator. Since $\sqrt[3]{2}$ is a cube root, we want to multiply by a value that will make the radicand 2 a perfect cube. If we multiply $\sqrt[3]{2}$ by $\sqrt[3]{2^2}$, we get $\sqrt[3]{2^3} = \sqrt[3]{8} = 2$.

$$\frac{1\cdot\sqrt[3]{2^2}}{\sqrt[3]{2}\cdot\sqrt[3]{2^2}} = \frac{\sqrt[3]{4}}{\sqrt[3]{2^3}} = \frac{\sqrt[3]{4}}{2}$$ Multiply the numerator and denominator by $\sqrt[3]{2^2}$ and then simplify.

Work Practice 1

Answers

1. a. $\frac{5\sqrt{3}}{3}$ **b.** $\frac{15\sqrt{x}}{2x}$ **c.** $\frac{\sqrt[3]{6}}{3}$

✓**Concept Check** Determine by which number both the numerator and denominator can be multiplied to rationalize the denominator of the radical expression.

a. $\dfrac{1}{\sqrt[3]{7}}$ **b.** $\dfrac{1}{\sqrt[4]{8}}$

Example 2 Rationalize the denominator of $\sqrt{\dfrac{7x}{3y}}$.

Solution:

$\sqrt{\dfrac{7x}{3y}} = \dfrac{\sqrt{7x}}{\sqrt{3y}}$ Use the quotient rule. No radical may be simplified further.

$= \dfrac{\sqrt{7x}\cdot\sqrt{3y}}{\sqrt{3y}\cdot\sqrt{3y}}$ Multiply numerator and denominator by $\sqrt{3y}$ so that the radicand in the denominator is a perfect square.

$= \dfrac{\sqrt{21xy}}{3y}$ Use the product rule in the numerator and denominator. Remember that $\sqrt{3y}\cdot\sqrt{3y} = 3y$.

Work Practice 2

Practice 2

Rationalize the denominator of $\sqrt{\dfrac{3z}{5y}}$.

Example 3 Rationalize the denominator of $\dfrac{\sqrt[4]{x}}{\sqrt[4]{81y^5}}$.

Solution: First, simplify each radical if possible.

$\dfrac{\sqrt[4]{x}}{\sqrt[4]{81y^5}} = \dfrac{\sqrt[4]{x}}{\sqrt[4]{81y^4}\cdot\sqrt[4]{y}}$ Use the product rule in the denominator.

$= \dfrac{\sqrt[4]{x}}{3y\sqrt[4]{y}}$ Write $\sqrt[4]{81y^4}$ as $3y$.

$= \dfrac{\sqrt[4]{x}\cdot\sqrt[4]{y^3}}{3y\sqrt[4]{y}\cdot\sqrt[4]{y^3}}$ Multiply numerator and denominator by $\sqrt[4]{y^3}$ so that the radicand in the denominator is a perfect fourth power.

$= \dfrac{\sqrt[4]{xy^3}}{3y\sqrt[4]{y^4}}$ Use the product rule in the numerator and denominator.

$= \dfrac{\sqrt[4]{xy^3}}{3y^2}$ In the denominator, $\sqrt[4]{y^4} = y$ and $3y\cdot y = 3y^2$.

Work Practice 3

Practice 3

Rationalize the denominator of $\dfrac{\sqrt[3]{z^2}}{\sqrt[3]{27x^4}}$.

Objective 2 Rationalizing Denominators Having Two Terms

Remember the product of the sum and difference of two terms?

$$(a + b)(a - b) = a^2 - b^2$$

These two expressions are called **conjugates** of each other.

To rationalize a numerator or denominator that is a sum or difference of two terms, we use conjugates. To see how and why this works, let's rationalize the denominator of the expression $\dfrac{5}{\sqrt{3} - 2}$. To do so, we multiply both the numerator and the denominator by $\sqrt{3} + 2$, the **conjugate** of the denominator $\sqrt{3} - 2$, and see what happens.

Answers

2. $\dfrac{\sqrt{15yz}}{5y}$ **3.** $\dfrac{\sqrt[3]{z^2x^2}}{3x^2}$

✓**Concept Check Answers**

a. $\sqrt[3]{7^2}$ or $\sqrt[3]{49}$ **b.** $\sqrt[4]{2}$

$$\frac{5}{\sqrt{3}-2} = \frac{5(\sqrt{3}+2)}{(\sqrt{3}-2)(\sqrt{3}+2)}$$

$$= \frac{5(\sqrt{3}+2)}{(\sqrt{3})^2 - 2^2}$$ Multiply the sum and difference of two terms: $(a+b)(a-b) = a^2 - b^2$.

$$= \frac{5(\sqrt{3}+2)}{3-4}$$

$$= \frac{5(\sqrt{3}+2)}{-1}$$

$$= -5(\sqrt{3}+2) \quad \text{or} \quad -5\sqrt{3}-10$$

Notice in the denominator that the product of $(\sqrt{3}-2)$ and its conjugate, $(\sqrt{3}+2)$, is -1. In general, the product of an expression and its conjugate will contain no radical terms. This is why, when rationalizing a denominator or a numerator containing two terms, we multiply by its conjugate. Examples of conjugates are

$\sqrt{a}-\sqrt{b}$ and $\sqrt{a}+\sqrt{b}$

$x+\sqrt{y}$ and $x-\sqrt{y}$

Practice 4

Rationalize the denominator.

a. $\dfrac{5}{3\sqrt{5}+2}$ b. $\dfrac{\sqrt{2}+5}{\sqrt{3}-\sqrt{5}}$

c. $\dfrac{3\sqrt{x}}{2\sqrt{x}+\sqrt{y}}$

Example 4 Rationalize each denominator.

a. $\dfrac{2}{3\sqrt{2}+4}$ b. $\dfrac{\sqrt{6}+2}{\sqrt{5}-\sqrt{3}}$ c. $\dfrac{2\sqrt{m}}{3\sqrt{x}+\sqrt{m}}$

Solution:

a. Multiply the numerator and denominator by the conjugate of the denominator, $3\sqrt{2}+4$.

$$\frac{2}{3\sqrt{2}+4} = \frac{2(3\sqrt{2}-4)}{(3\sqrt{2}+4)(3\sqrt{2}-4)}$$

$$= \frac{2(3\sqrt{2}-4)}{(3\sqrt{2})^2 - 4^2}$$

$$= \frac{2(3\sqrt{2}-4)}{18-16}$$

$$= \frac{2(3\sqrt{2}-4)}{2}, \quad \text{or} \quad 3\sqrt{2}-4$$

It is often useful to leave a numerator in factored form to help determine whether the expression can be simplified.

b. Multiply the numerator and denominator by the conjugate of $\sqrt{5}-\sqrt{3}$.

$$\frac{\sqrt{6}+2}{\sqrt{5}-\sqrt{3}} = \frac{(\sqrt{6}+2)(\sqrt{5}+\sqrt{3})}{(\sqrt{5}-\sqrt{3})(\sqrt{5}+\sqrt{3})}$$

$$= \frac{\sqrt{6}\sqrt{5}+\sqrt{6}\sqrt{3}+2\sqrt{5}+2\sqrt{3}}{(\sqrt{5})^2-(\sqrt{3})^2}$$

$$= \frac{\sqrt{30}+\sqrt{18}+2\sqrt{5}+2\sqrt{3}}{5-3}$$

$$= \frac{\sqrt{30}+3\sqrt{2}+2\sqrt{5}+2\sqrt{3}}{2}$$

Answers

4. a. $\dfrac{5(3\sqrt{5}-2)}{41}$ b. $\dfrac{\sqrt{6}+5\sqrt{3}+\sqrt{10}+5\sqrt{5}}{-2}$ c. $\dfrac{6x-3\sqrt{xy}}{4x-y}$

c. Multiply by the conjugate of $3\sqrt{x} + \sqrt{m}$ to eliminate the radicals from the denominator.

$$\frac{2\sqrt{m}}{3\sqrt{x}+\sqrt{m}} = \frac{2\sqrt{m}(3\sqrt{x}-\sqrt{m})}{(3\sqrt{x}+\sqrt{m})(3\sqrt{x}-\sqrt{m})} = \frac{6\sqrt{mx}-2m}{(3\sqrt{x})^2-(\sqrt{m})^2}$$
$$= \frac{6\sqrt{mx}-2m}{9x-m}$$

Work Practice 4

Objective 3 Rationalizing Numerators 

As mentioned earlier, it is also often helpful to write an expression such as $\frac{\sqrt{3}}{\sqrt{2}}$ as an equivalent expression without a radical in the numerator. This process is called **rationalizing the numerator.** To rationalize the numerator of $\frac{\sqrt{3}}{\sqrt{2}}$, we multiply the numerator and the denominator by $\sqrt{3}$.

$$\frac{\sqrt{3}}{\sqrt{2}} = \frac{\sqrt{3}\cdot\sqrt{3}}{\sqrt{2}\cdot\sqrt{3}} = \frac{\sqrt{9}}{\sqrt{6}} = \frac{3}{\sqrt{6}}$$

Example 5 Rationalize the numerator of $\frac{\sqrt{7}}{\sqrt{45}}$.

Solution: First we simplify $\sqrt{45}$.

$$\frac{\sqrt{7}}{\sqrt{45}} = \frac{\sqrt{7}}{\sqrt{9\cdot 5}} = \frac{\sqrt{7}}{3\sqrt{5}}$$

Next we rationalize the numerator by multiplying the numerator and the denominator by $\sqrt{7}$.

$$\frac{\sqrt{7}}{3\sqrt{5}} = \frac{\sqrt{7}\cdot\sqrt{7}}{3\sqrt{5}\cdot\sqrt{7}} = \frac{7}{3\sqrt{5\cdot 7}} = \frac{7}{3\sqrt{35}}$$

Work Practice 5

Practice 5

Rationalize the numerator of $\frac{\sqrt{32}}{\sqrt{80}}$.

Example 6 Rationalize the numerator of $\frac{\sqrt[3]{2x^2}}{\sqrt[3]{5y}}$.

Solution: The numerator and the denominator of this expression are already simplified. To rationalize the numerator, $\sqrt[3]{2x^2}$, we multiply the numerator and denominator by a factor that will make the radicand a perfect cube. If we multiply $\sqrt[3]{2x^2}$ by $\sqrt[3]{4x}$, we get $\sqrt[3]{8x^3} = 2x$.

$$\frac{\sqrt[3]{2x^2}}{\sqrt[3]{5y}} = \frac{\sqrt[3]{2x^2}\cdot\sqrt[3]{4x}}{\sqrt[3]{5y}\cdot\sqrt[3]{4x}} = \frac{\sqrt[3]{8x^3}}{\sqrt[3]{20xy}} = \frac{2x}{\sqrt[3]{20xy}}$$

Work Practice 6

Practice 6

Rationalize the numerator of $\frac{\sqrt[3]{5b}}{\sqrt[3]{2a}}$.

Answers

5. $\frac{2}{\sqrt{10}}$ **6.** $\frac{5b}{\sqrt[3]{50ab^2}}$

Practice 7

Rationalize the numerator of $\frac{\sqrt{x} - 3}{4}$.

Answer

7. $\frac{x - 9}{4(\sqrt{x} + 3)}$

Example 7 Rationalize the numerator of $\frac{\sqrt{x} + 2}{5}$.

Solution: We multiply the numerator and the denominator by the conjugate of the numerator, $\sqrt{x} + 2$.

$$\frac{\sqrt{x} + 2}{5} = \frac{(\sqrt{x} + 2)(\sqrt{x} - 2)}{5(\sqrt{x} - 2)}$$ Multiply by $\sqrt{x} - 2$, the conjugate of $\sqrt{x} + 2$.

$$= \frac{(\sqrt{x})^2 - 2^2}{5(\sqrt{x} - 2)}$$ $(a + b)(a - b) = a^2 - b^2$

$$= \frac{x - 4}{5(\sqrt{x} - 2)}$$

Work Practice 7

Vocabulary, Readiness & Video Check

Use the choices below to fill in each blank. Not all choices will be used.

rationalizing the numerator	conjugate	$\frac{\sqrt{3}}{\sqrt{3}}$
rationalizing the denominator	$\frac{5}{5}$	

1. The ______________ of $a + b$ is $a - b$.
2. The process of writing an equivalent expression, but without a radical in the denominator, is called ______________.
3. The process of writing an equivalent expression, but without a radical in the numerator, is called ______________.
4. To rationalize the denominator of $\frac{5}{\sqrt{3}}$, we multiply by ______________.

Martin-Gay Interactive Videos

See Video 17.5

Watch the section lecture video and answer the following questions.

Objective 1 5. From Examples 1–3, what is the goal of rationalizing a denominator?

Objective 2 6. From Example 4, why will multiplying a denominator by its conjugate always rationalize the denominator?

Objective 3 7. From Example 5, is the process of rationalizing a numerator any different from rationalizing a denominator?

17.5 Exercise Set MyMathLab®

Rationalize each denominator. See Examples 1 through 3.

1. $\frac{\sqrt{2}}{\sqrt{7}}$
2. $\frac{\sqrt{3}}{\sqrt{2}}$
3. $\sqrt{\frac{1}{5}}$
4. $\sqrt{\frac{1}{2}}$
5. $\sqrt{\frac{4}{x}}$
6. $\sqrt{\frac{25}{y}}$
7. $\frac{4}{\sqrt[3]{3}}$
8. $\frac{6}{\sqrt[3]{9}}$
9. $\frac{3}{\sqrt{8x}}$
10. $\frac{5}{\sqrt{27a}}$
11. $\frac{3}{\sqrt[3]{4x^2}}$
12. $\frac{5}{\sqrt[3]{3y}}$
13. $\frac{9}{\sqrt{3a}}$
14. $\frac{x}{\sqrt{5}}$
15. $\frac{3}{\sqrt[3]{2}}$
16. $\frac{5}{\sqrt[3]{9}}$
17. $\frac{2\sqrt{3}}{\sqrt{7}}$
18. $\frac{-5\sqrt{2}}{\sqrt{11}}$
19. $\sqrt{\frac{2x}{5y}}$
20. $\sqrt{\frac{13a}{2b}}$
21. $\sqrt[3]{\frac{3}{5}}$
22. $\sqrt[3]{\frac{7}{10}}$
23. $\sqrt{\frac{3x}{50}}$
24. $\sqrt{\frac{11y}{45}}$
25. $\frac{1}{\sqrt{12z}}$
26. $\frac{1}{\sqrt{32x}}$
27. $\frac{\sqrt[3]{2y^2}}{\sqrt[3]{9x^2}}$
28. $\frac{\sqrt[3]{3x}}{\sqrt[3]{4y^4}}$
29. $\sqrt[4]{\frac{81}{8}}$
30. $\sqrt[4]{\frac{1}{9}}$
31. $\sqrt[4]{\frac{16}{9x^7}}$
32. $\sqrt[5]{\frac{32}{m^6n^{13}}}$
33. $\frac{5a}{\sqrt[5]{8a^9b^{11}}}$
34. $\frac{9y}{\sqrt[4]{4y^9}}$

Write the conjugate of each expression.

35. $\sqrt{2} + x$
36. $\sqrt{3} + y$
37. $5 - \sqrt{a}$
38. $6 - \sqrt{b}$
39. $-7\sqrt{5} + 8\sqrt{x}$
40. $-9\sqrt{2} - 6\sqrt{y}$

Rationalize each denominator. See Example 4.

41. $\frac{6}{2 - \sqrt{7}}$
42. $\frac{3}{\sqrt{7} - 4}$
43. $\frac{-7}{\sqrt{x} - 3}$
44. $\frac{-8}{\sqrt{y} + 4}$
45. $\frac{\sqrt{2} - \sqrt{3}}{\sqrt{2} + \sqrt{3}}$
46. $\frac{\sqrt{3} + \sqrt{4}}{\sqrt{2} - \sqrt{3}}$
47. $\frac{\sqrt{a} + 1}{2\sqrt{a} - \sqrt{b}}$
48. $\frac{2\sqrt{a} - 3}{2\sqrt{a} + \sqrt{b}}$
49. $\frac{8}{1 + \sqrt{10}}$
50. $\frac{-3}{\sqrt{6} - 2}$
51. $\frac{\sqrt{x}}{\sqrt{x} + \sqrt{y}}$
52. $\frac{2\sqrt{a}}{2\sqrt{x} - \sqrt{y}}$
53. $\frac{2\sqrt{3} + \sqrt{6}}{4\sqrt{3} - \sqrt{6}}$
54. $\frac{4\sqrt{5} + \sqrt{2}}{2\sqrt{5} - \sqrt{2}}$

Rationalize each numerator. See Examples 5 and 6.

55. $\sqrt{\frac{5}{3}}$
56. $\sqrt{\frac{3}{2}}$
57. $\sqrt{\frac{18}{5}}$
58. $\sqrt{\frac{12}{7}}$
59. $\frac{\sqrt{4x}}{7}$
60. $\frac{\sqrt{3x^5}}{6}$
61. $\frac{\sqrt[3]{5y^2}}{\sqrt[3]{4x}}$
62. $\frac{\sqrt[3]{4x}}{\sqrt[3]{z^4}}$
63. $\sqrt{\frac{2}{5}}$
64. $\sqrt{\frac{3}{7}}$
65. $\frac{\sqrt{2x}}{11}$
66. $\frac{\sqrt{y}}{7}$
67. $\sqrt[3]{\frac{7}{8}}$
68. $\sqrt[3]{\frac{25}{2}}$
69. $\frac{\sqrt[3]{3x^5}}{10}$
70. $\sqrt[3]{\frac{9y}{7}}$
71. $\sqrt{\frac{18x^4y^6}{3z}}$
72. $\sqrt{\frac{8x^5y}{2z}}$

Rationalize each numerator. See Example 7.

73. $\frac{2 - \sqrt{11}}{6}$
74. $\frac{\sqrt{15} + 1}{2}$
75. $\frac{2 - \sqrt{7}}{-5}$
76. $\frac{\sqrt{5} + 2}{\sqrt{2}}$
77. $\frac{\sqrt{x} + 3}{\sqrt{x}}$
78. $\frac{5 + \sqrt{2}}{\sqrt{2x}}$
79. $\frac{\sqrt{2} - 1}{\sqrt{2} + 1}$
80. $\frac{\sqrt{8} - \sqrt{3}}{\sqrt{2} + \sqrt{3}}$
81. $\frac{\sqrt{x} + 1}{\sqrt{x} - 1}$
82. $\frac{\sqrt{x} + \sqrt{y}}{\sqrt{x} - \sqrt{y}}$

REVIEW AND PREVIEW

Solve each equation. See Sections 9.3 and 13.6.

83. $2x - 7 = 3(x - 4)$
84. $9x - 4 = 7(x - 2)$
85. $(x - 6)(2x + 1) = 0$
86. $(y + 2)(5y + 4) = 0$
87. $x^2 - 8x = -12$
88. $x^3 = x$

CONCEPT EXTENSIONS

△ **89.** The formula of the radius r of a sphere with surface area A is

$$r = \sqrt{\frac{A}{4\pi}}$$

Rationalize the denominator of the radical expression in this formula.

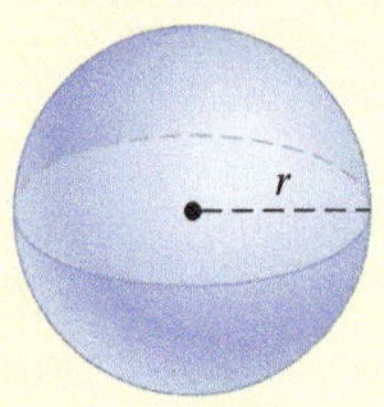

△ **90.** The formula for the radius r of a cone with height 7 centimeters and volume V is

$$r = \sqrt{\frac{3V}{7\pi}}$$

Rationalize the numerator of the radical expression in this formula.

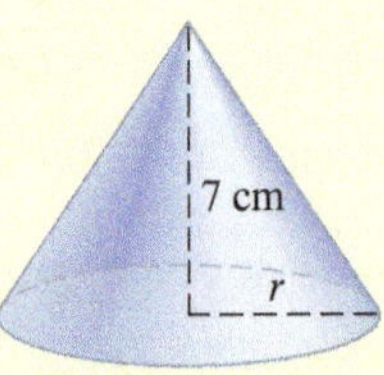

91. Given $\frac{\sqrt{5y^3}}{\sqrt{12x^3}}$, rationalize the denominator by following parts (a) and (b).

a. Multiply the numerator and denominator by $\sqrt{12x^3}$.

b. Multiply the numerator and denominator by $\sqrt{3x}$.

c. What can you conclude from parts (a) and (b)?

92. Given $\frac{\sqrt[3]{5y}}{\sqrt[3]{4}}$, rationalize the denominator by following parts (a) and (b).

a. Multiply the numerator and denominator by $\sqrt[3]{16}$.

b. Multiply the numerator and denominator by $\sqrt[3]{2}$.

c. What can you conclude from parts (a) and (b)?

Determine the smallest number both the numerator and denominator should be multiplied by to rationalize the denominator of the radical expression. See the Concept Check in this section.

93. $\frac{9}{\sqrt[3]{5}}$

94. $\frac{5}{\sqrt{27}}$

95. When rationalizing the denominator of $\frac{\sqrt{5}}{\sqrt{7}}$, explain why both the numerator and the denominator must be multiplied by $\sqrt{7}$.

96. When rationalizing the numerator of $\frac{\sqrt{5}}{\sqrt{7}}$, explain why both the numerator and the denominator must be multiplied by $\sqrt{5}$.

97. Explain why rationalizing the denominator does not change the value of the original expression.

98. Explain why rationalizing the numerator does not change the value of the original expression.

Integrated Review Sections 17.1–17.5

Radicals and Rational Exponents

Throughout this review, assume that all variables represent positive real numbers.

Find each root.

1. $\sqrt{81}$

2. $\sqrt[3]{-8}$

3. $\sqrt[4]{\frac{1}{16}}$

4. $\sqrt{x^6}$

5. $\sqrt[3]{y^9}$

6. $\sqrt{4y^{10}}$

7. $\sqrt[5]{-32y^5}$

8. $\sqrt[4]{81b^{12}}$

Use radical notation to write each expression. Simplify if possible.

9. $36^{1/2}$

10. $(3y)^{1/4}$

11. $64^{-2/3}$

12. $(x + 1)^{3/5}$

Use the properties of exponents to simplify each expression. Write with positive exponents.

13. $y^{-1/6} \cdot y^{7/6}$

14. $\frac{(2x^{1/3})^4}{x^{5/6}}$

15. $\frac{x^{1/4}x^{3/4}}{x^{-1/4}}$

16. $4^{1/3} \cdot 4^{2/5}$

Use rational exponents to simplify each radical.

17. $\sqrt[3]{8x^6}$ **18.** $\sqrt[12]{a^9b^6}$

Use rational exponents to write each as a single radical expression.

19. $\sqrt[4]{x}\cdot\sqrt{x}$ **20.** $\sqrt{5}\cdot\sqrt[3]{2}$

Simplify.

21. $\sqrt{40}$ **22.** $\sqrt[4]{16x^7y^{10}}$ **23.** $\sqrt[3]{54x^4}$ **24.** $\sqrt[5]{-64b^{10}}$

Multiply or divide. Then simplify if possible.

25. $\sqrt{5}\cdot\sqrt{x}$ **26.** $\sqrt[3]{8x}\cdot\sqrt[3]{8x^2}$ **27.** $\dfrac{\sqrt{98y^6}}{\sqrt{2y}}$ **28.** $\dfrac{\sqrt[4]{48a^9b^3}}{\sqrt[4]{ab^3}}$

Perform each indicated operation.

29. $\sqrt{20}-\sqrt{75}+5\sqrt{7}$ **30.** $\sqrt[3]{54y^4}-y\sqrt[3]{16y}$ **31.** $\sqrt{3}(\sqrt{5}-\sqrt{2})$

32. $(\sqrt{7}+\sqrt{3})^2$ **33.** $(2x-\sqrt{5})(2x+\sqrt{5})$ **34.** $(\sqrt{x+1}-1)^2$

Rationalize each denominator.

35. $\sqrt{\dfrac{7}{3}}$ **36.** $\dfrac{5}{\sqrt[3]{2x^2}}$ **37.** $\dfrac{\sqrt{3}-\sqrt{7}}{2\sqrt{3}+\sqrt{7}}$

Rationalize each numerator.

38. $\sqrt{\dfrac{7}{3}}$ **39.** $\sqrt[3]{\dfrac{9y}{11}}$ **40.** $\dfrac{\sqrt{x}-2}{\sqrt{x}}$

17.6 Radical Equations and Problem Solving 

Objectives

1. Solve Equations That Contain Radical Expressions.
2. Use the Pythagorean Theorem to Model Problems.

Objective 1 Solving Equations That Contain Radical Expressions

In this section, we present techniques to solve equations containing radical expressions such as

$$\sqrt{2x-3}=9$$

We use the power rule to help us solve these radical equations.

Power Rule

If both sides of an equation are raised to the same power, **all** solutions of the original equation are **among** the solutions of the new equation.

This property *does not* say that raising both sides of an equation to a power yields an equivalent equation. A solution of the new equation *may or may not* be a solution of the original equation. For example, $(-2)^2 = 2^2$, but $-2 \neq 2$. Thus, *each solution of the new equation must be checked* to make sure it is a solution of the original equation. Recall that a proposed solution that is not a solution of the original equation is called an **extraneous solution.**

Practice 1

Solve: $\sqrt{3x - 5} = 7$.

Example 1 Solve: $\sqrt{2x - 3} = 9$.

Solution: We use the power rule to square both sides of the equation to eliminate the radical.

$$\begin{aligned} \sqrt{2x - 3} &= 9 \\ (\sqrt{2x - 3})^2 &= 9^2 \\ 2x - 3 &= 81 \\ 2x &= 84 \\ x &= 42 \end{aligned}$$

Now we check the solution in the original equation.

Check:

$$\begin{aligned} \sqrt{2x - 3} &= 9 \\ \sqrt{2(42) - 3} &\stackrel{?}{=} 9 \quad \text{Let } x = 42. \\ \sqrt{84 - 3} &\stackrel{?}{=} 9 \\ \sqrt{81} &\stackrel{?}{=} 9 \\ 9 &= 9 \quad \text{True} \end{aligned}$$

The solution checks, so we conclude that the solution is 42, or the solution set is $\{42\}$.

Work Practice 1

To solve a radical equation, first isolate a radical on one side of the equation.

Practice 2

Solve: $\sqrt{16x - 3} - 4x = 0$.

Example 2 Solve: $\sqrt{-10x - 1} + 3x = 0$.

Solution: First, isolate the radical on one side of the equation. To do this, we subtract $3x$ from both sides.

$$\begin{aligned} \sqrt{-10x - 1} + 3x &= 0 \\ \sqrt{-10x - 1} + 3x - 3x &= 0 - 3x \\ \sqrt{-10x - 1} &= -3x \end{aligned}$$

Next we use the power rule to eliminate the radical.

$$\begin{aligned} (\sqrt{-10x - 1})^2 &= (-3x)^2 \\ -10x - 1 &= 9x^2 \end{aligned}$$

Since this is a quadratic equation, we can set the equation equal to 0 and try to solve by factoring.

$$\begin{aligned} 9x^2 + 10x + 1 &= 0 \\ (9x + 1)(x + 1) &= 0 \quad \text{Factor.} \\ 9x + 1 = 0 \quad \text{or} \quad x + 1 &= 0 \quad \text{Set each factor equal to 0.} \\ x = -\frac{1}{9} \quad \text{or} \quad x &= -1 \end{aligned}$$

Answers

1. 18 **2.** $\frac{1}{4}, \frac{3}{4}$

Check: Let $x = -\frac{1}{9}$.

$$\sqrt{-10x - 1} + 3x = 0$$
$$\sqrt{-10\left(-\frac{1}{9}\right) - 1} + 3\left(-\frac{1}{9}\right) \stackrel{?}{=} 0$$
$$\sqrt{\frac{10}{9} - \frac{9}{9}} - \frac{3}{9} \stackrel{?}{=} 0$$
$$\sqrt{\frac{1}{9}} - \frac{1}{3} \stackrel{?}{=} 0$$
$$\frac{1}{3} - \frac{1}{3} = 0 \quad \text{True}$$

Let $x = -1$.

$$\sqrt{-10x - 1} + 3x = 0$$
$$\sqrt{-10(-1) - 1} + 3(-1) \stackrel{?}{=} 0$$
$$\sqrt{10 - 1} - 3 \stackrel{?}{=} 0$$
$$\sqrt{9} - 3 \stackrel{?}{=} 0$$
$$3 - 3 = 0 \quad \text{True}$$

Both solutions check. The solutions are $-\frac{1}{9}$ and -1, or the solution set is $\left\{-\frac{1}{9}, -1\right\}$.

Work Practice 2

The following steps may be used to solve a radical equation.

Solving a Radical Equation

Step 1: Isolate one radical on one side of the equation.

Step 2: Raise each side of the equation to a power equal to the index of the radical and simplify.

Step 3: If the equation still contains a radical term, repeat Steps 1 and 2. If not, solve the equation.

Step 4: Check all proposed solutions in the original equation.

Example 3 Solve: $\sqrt[3]{x + 1} + 5 = 3$.

Solution: First we isolate the radical by subtracting 5 from both sides of the equation.

$$\sqrt[3]{x + 1} + 5 = 3$$
$$\sqrt[3]{x + 1} = -2$$

Next we raise both sides of the equation to the third power to eliminate the radical.

$$\left(\sqrt[3]{x + 1}\right)^3 = (-2)^3$$
$$x + 1 = -8$$
$$x = -9$$

The solution checks in the original equation, so the solution is -9.

Work Practice 3

Practice 3

Solve: $\sqrt[3]{x - 2} + 1 = 3$.

Answer

3. 10

Practice 4

Solve: $\sqrt{16 + x} = x - 4$.

Example 4 Solve: $\sqrt{4 - x} = x - 2$.

Solution:

$$\sqrt{4 - x} = x - 2$$
$$(\sqrt{4 - x})^2 = (x - 2)^2$$
$$4 - x = x^2 - 4x + 4$$
$$x^2 - 3x = 0 \quad \text{Write the quadratic equation in standard form.}$$
$$x(x - 3) = 0 \quad \text{Factor.}$$
$$x = 0 \quad \text{or} \quad x - 3 = 0 \quad \text{Set each factor equal to 0.}$$
$$x = 3$$

Check:

$$\sqrt{4 - x} = x - 2 \qquad \sqrt{4 - x} = x - 2$$
$$\sqrt{4 - 0} \stackrel{?}{=} 0 - 2 \quad \text{Let } x = 0. \qquad \sqrt{4 - 3} \stackrel{?}{=} 3 - 2 \quad \text{Let } x = 3.$$
$$2 = -2 \quad \text{False} \qquad 1 = 1 \quad \text{True}$$

The proposed solution 3 checks, but 0 does not. Since 0 is an extraneous solution, the only solution is 3.

Work Practice 4

Helpful Hint

In Example 4, notice that $(x - 2)^2 = x^2 - 4x + 4$. Make sure binomials are squared correctly.

✓**Concept Check** How can you immediately tell that the equation $\sqrt{2y + 3} = -4$ has no real solution?

Practice 5

Solve: $\sqrt{8x + 1} + \sqrt{3x} = 2$.

Example 5 Solve: $\sqrt{2x + 5} + \sqrt{2x} = 3$.

Solution: We get one radical alone by subtracting $\sqrt{2x}$ from both sides.

$$\sqrt{2x + 5} + \sqrt{2x} = 3$$
$$\sqrt{2x + 5} = 3 - \sqrt{2x}$$

Now we use the power rule to begin eliminating the radicals. First we square both sides.

$$(\sqrt{2x + 5})^2 = (3 - \sqrt{2x})^2$$
$$2x + 5 = 9 - 6\sqrt{2x} + 2x \quad \text{Multiply } (3 - \sqrt{2x})(3 - \sqrt{2x}).$$

There is still a radical in the equation, so we get a radical alone again. Then we square both sides.

$$2x + 5 = 9 - 6\sqrt{2x} + 2x$$
$$6\sqrt{2x} = 4 \quad \text{Get the radical alone.}$$
$$36(2x) = 16 \quad \text{Square both sides of the equation to eliminate the radical.}$$
$$72x = 16 \quad \text{Multiply.}$$
$$x = \frac{16}{72} \quad \text{Solve.}$$
$$x = \frac{2}{9} \quad \text{Simplify.}$$

The proposed solution, $\frac{2}{9}$, checks in the original equation. The solution is $\frac{2}{9}$.

Work Practice 5

Answers

4. 9 **5.** $\frac{3}{25}$

✓**Concept Check Answer**

answers may vary

Helpful Hint

Make sure expressions are squared correctly. In Example 5, we squared $(3 - \sqrt{2x})$ as

$$\begin{aligned}(3 - \sqrt{2x})^2 &= (3 - \sqrt{2x})(3 - \sqrt{2x}) \\ &= 3\cdot 3 - 3\sqrt{2x} - 3\sqrt{2x} + \sqrt{2x}\cdot\sqrt{2x} \\ &= 9 - 6\sqrt{2x} + 2x\end{aligned}$$

✓**Concept Check** What is wrong with the following solution?

$$\begin{aligned}\sqrt{2x+5} + \sqrt{4-x} &= 8 \\ (\sqrt{2x+5} + \sqrt{4-x})^2 &= 8^2 \\ (2x+5) + (4-x) &= 64 \\ x + 9 &= 64 \\ x &= 55\end{aligned}$$

Objective 2 Using the Pythagorean Theorem

Recall that the Pythagorean theorem states that in a right triangle, the length of the hypotenuse squared equals the sum of the lengths of each of the legs squared.

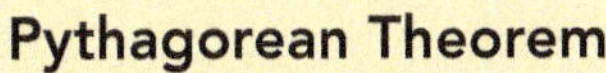

Pythagorean Theorem

If a and b are the lengths of the legs of a right triangle and c is the length of the hypotenuse, then $a^2 + b^2 = c^2$.

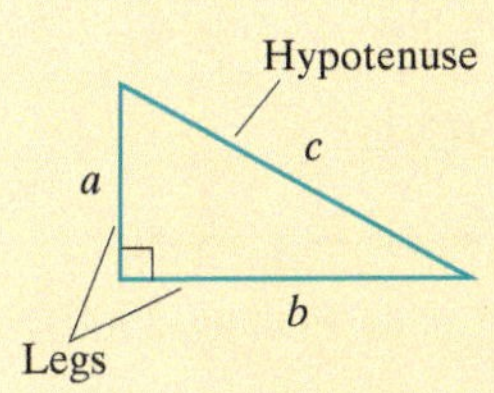

Example 6 Find the length of the unknown leg of the right triangle.

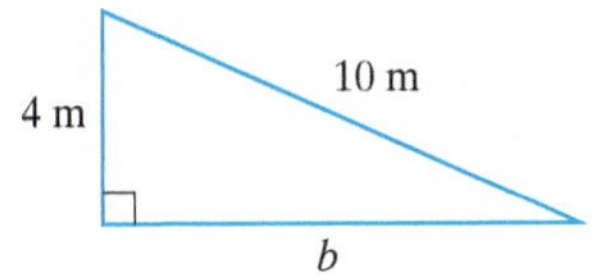

Solution: In the formula $a^2 + b^2 = c^2$, c is the hypotenuse. Here, $c = 10$, the length of the hypotenuse, and $a = 4$. We solve for b. Then $a^2 + b^2 = c^2$ becomes

$$\begin{aligned}4^2 + b^2 &= 10^2 \\ 16 + b^2 &= 100 \\ b^2 &= 84 \quad \text{Subtract 16 from both sides.} \\ b &= \pm\sqrt{84} = \pm\sqrt{4\cdot 21} = \pm 2\sqrt{21}\end{aligned}$$

Since b is a length and thus is positive, we will use the positive value only. The unknown leg of the triangle is $2\sqrt{21}$ meters long.

Work Practice 6

Practice 6

Find the length of the unknown leg of the right triangle.

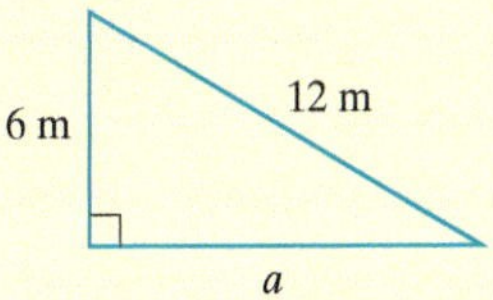

Answer

6. $6\sqrt{3}$ m

✓ **Concept Check Answer**

$(\sqrt{2x+5} + \sqrt{4-x})^2$ is not $(2x+5) + (4-x)$.

Practice 7

Keith Robinson buys two Siamese fighting fish, but when he gets home, he finds he only has one rectangular tank that is 12 in. long, 7 in. wide, and 5 in. deep. Since the fish must be kept separated, he needs to insert a plastic divider in the diagonal of the tank. He already has a piece that is 5 in. in one dimension, but how long must it be to fit corner to corner in the tank?

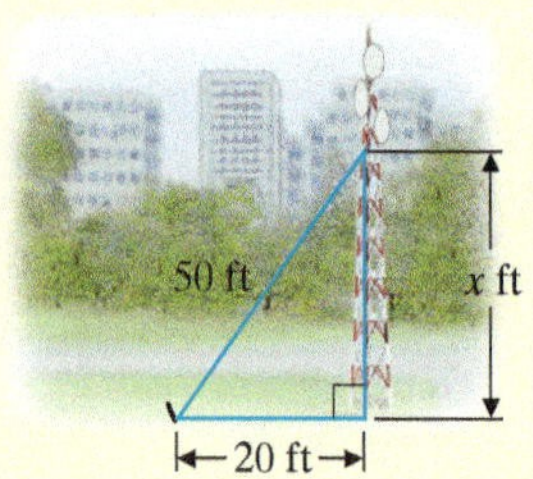

Answer

7. $\sqrt{193}$ in. $\approx$ 13.89 in.

Example 7 Calculating Placement of a Wire

A 50-foot supporting wire is to be attached to a 75-foot antenna. Because of surrounding buildings, sidewalks, and roadways, the wire must be anchored exactly 20 feet from the base of the antenna.

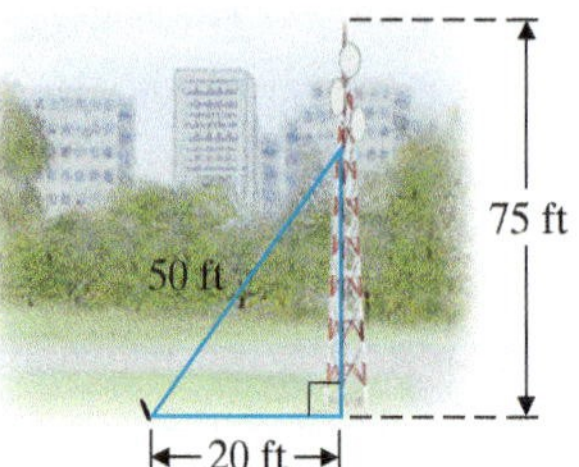

a. How high from the base of the antenna is the wire attached?

b. Local regulations require that a supporting wire be attached at a height no less than $\frac{3}{5}$ of the total height of the antenna. From part (a), have local regulations been met?

Solution:

1. UNDERSTAND. Read and reread the problem. From the diagram, we notice that a right triangle is formed with hypotenuse 50 feet and one leg 20 feet. Let x be the height from the base of the antenna to the attached wire.
2. TRANSLATE. Use the Pythagorean theorem.

$$a^2 + b^2 = c^2$$
$$20^2 + x^2 = 50^2 \quad a = 20, c = 50$$

3. SOLVE.

$$20^2 + x^2 = 50^2$$
$$400 + x^2 = 2500$$
$$x^2 = 2100 \quad \text{Subtract 400 from both sides.}$$
$$x = \pm\sqrt{2100}$$
$$= \pm 10\sqrt{21}$$

4. INTERPRET. *Check* the work and *state* the solution.

Check: We will use only the positive value, $x = 10\sqrt{21}$, because x represents length. The wire is attached exactly $10\sqrt{21}$ feet from the base of the pole, or approximately 45.8 feet.

State: The supporting wire must be attached at a height no less than $\frac{3}{5}$ of the total height of the antenna. This height is $\frac{3}{5}$(75 feet), or 45 feet. Since we know from part (a) that the wire is to be attached at a height of approximately 45.8 feet, local regulations have been met.

Work Practice 7

Graphing Calculator Explorations

We can use a graphing calculator to solve radical equations. For example, to use a graphing calculator to approximate the solutions of the equation solved in Example 4, we graph the following.

$$Y_1 = \sqrt{4 - x} \quad \text{and} \quad Y_2 = x - 2$$

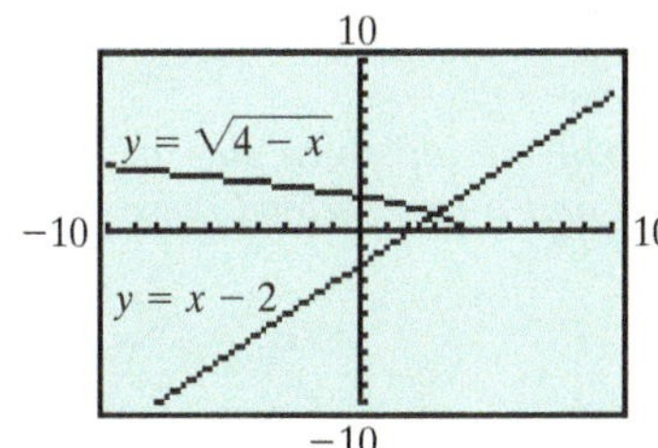

The x-value of the point of intersection is the solution. Use the Intersect feature or the Zoom and Trace features of your graphing calculator to see that the solution is 3.

Use a graphing calculator to solve each radical equation. Round all solutions to the nearest hundredth.

1. $\sqrt{x + 7} = x$
2. $\sqrt{3x + 5} = 2x$
3. $\sqrt{2x + 1} = \sqrt{2x + 2}$
4. $\sqrt{10x - 1} = \sqrt{-10x + 10} - 1$
5. $1.2x = \sqrt{3.1x + 5}$
6. $\sqrt{1.9x^2 - 2.2} = -0.8x + 3$

Vocabulary, Readiness & Video Check

Use the choices below to fill in each blank. Not all choices will be used.

hypotenuse	right	$x^2 + 25$	$16 - 8\sqrt{7x} + 7x$
extraneous solution	legs	$x^2 - 10x + 25$	$16 + 7x$

1. A proposed solution that is not a solution of the original equation is called a(n) ________.
2. The Pythagorean theorem states that $a^2 + b^2 = c^2$, where a and b are the lengths of the ________ of a(n) ________ triangle and c is the length of the ________.
3. The square of $x - 5$, or $(x - 5)^2 =$ ________.
4. The square of $4 - \sqrt{7x}$, or $\left(4 - \sqrt{7x}\right)^2 =$ ________.

Martin-Gay Interactive Videos

See Video 17.6

Watch the section lecture video and answer the following questions.

Objective 1 5. From Examples 1–4, why must you be careful and check your proposed solution(s) in the original equation?

Objective 2 6. From Example 5, when solving problems using the Pythagorean theorem, what two things must you remember?

Objective 3 7. What important reminder is given as the final answer to Example 5 is being found?

17.6 Exercise Set MyMathLab®

Solve. See Examples 1 and 2.

1. $\sqrt{2x} = 4$
2. $\sqrt{3x} = 3$
3. $\sqrt{x-3} = 2$
4. $\sqrt{x+1} = 5$
5. $\sqrt{2x} = -4$
6. $\sqrt{5x} = -5$
7. $\sqrt{4x-3} - 5 = 0$
8. $\sqrt{x-3} - 1 = 0$
9. $\sqrt{2x-3} - 2 = 1$
10. $\sqrt{3x+3} - 4 = 8$

Solve. See Example 3.

11. $\sqrt[3]{6x} = -3$
12. $\sqrt[3]{4x} = -2$
13. $\sqrt[3]{x-2} - 3 = 0$
14. $\sqrt[3]{2x-6} - 4 = 0$

Solve. See Examples 4 and 5.

15. $\sqrt{13-x} = x - 1$
16. $\sqrt{2x-3} = 3 - x$
17. $x - \sqrt{4-3x} = -8$
18. $2x + \sqrt{x+1} = 8$
19. $\sqrt{y+5} = 2 - \sqrt{y-4}$
20. $\sqrt{x+3} + \sqrt{x-5} = 3$
21. $\sqrt{x-3} + \sqrt{x+2} = 5$
22. $\sqrt{2x-4} - \sqrt{3x+4} = -2$

MIXED PRACTICE

Solve. See Examples 1 through 5.

23. $\sqrt{3x-2} = 5$
24. $\sqrt{5x-4} = 9$
25. $-\sqrt{2x} + 4 = -6$
26. $-\sqrt{3x+9} = -12$
27. $\sqrt{3x+1} + 2 = 0$
28. $\sqrt{3x+1} - 2 = 0$
29. $\sqrt[4]{4x+1} - 2 = 0$
30. $\sqrt[4]{2x-9} - 3 = 0$
31. $\sqrt{4x-3} = 7$
32. $\sqrt{3x+9} = 6$
33. $\sqrt[3]{6x-3} - 3 = 0$
34. $\sqrt[3]{3x} + 4 = 7$
35. $\sqrt[3]{2x-3} - 2 = -5$
36. $\sqrt[3]{x-4} - 5 = -7$
37. $\sqrt{x+4} = \sqrt{2x-5}$
38. $\sqrt{3y+6} = \sqrt{7y-6}$
39. $x - \sqrt{1-x} = -5$
40. $x - \sqrt{x-2} = 4$
41. $\sqrt[3]{-6x-1} = \sqrt[3]{-2x-5}$
42. $\sqrt[3]{-4x-3} = \sqrt[3]{-x-15}$
43. $\sqrt{5x-1} - \sqrt{x} + 2 = 3$
44. $\sqrt{2x-1} - 4 = -\sqrt{x-4}$
45. $\sqrt{2x-1} = \sqrt{1-2x}$
46. $\sqrt{7x-4} = \sqrt{4-7x}$
47. $\sqrt{3x+4} - 1 = \sqrt{2x+1}$
48. $\sqrt{x-2} + 3 = \sqrt{4x+1}$
49. $\sqrt{y+3} - \sqrt{y-3} = 1$
50. $\sqrt{x+1} - \sqrt{x-1} = 2$

Find the length of the unknown side of each triangle. See Example 6.

51.

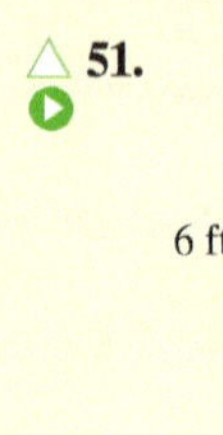

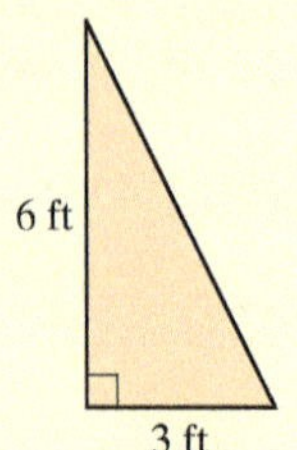

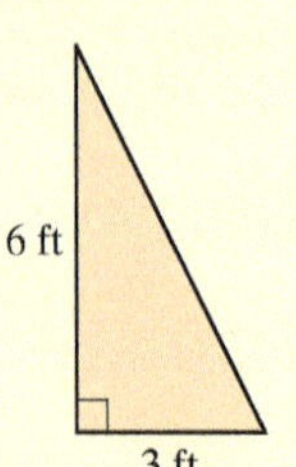

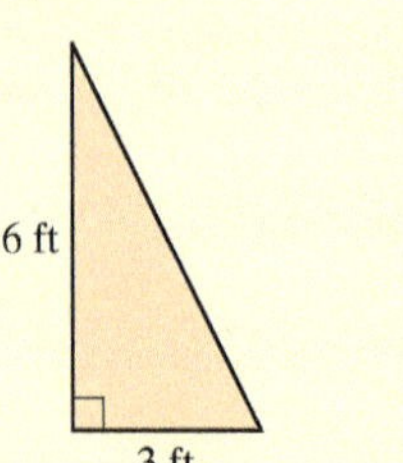

52.

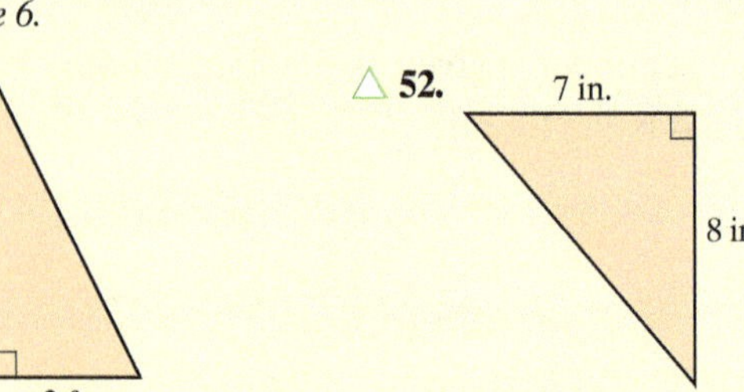

53. 3 m, 7 m

54. 4 cm, 7 cm

Find the length of the unknown side of each triangle. Give the exact length and a one-decimal-place approximation. See Example 6.

55. 9 m, $11\sqrt{5}$ m

56.

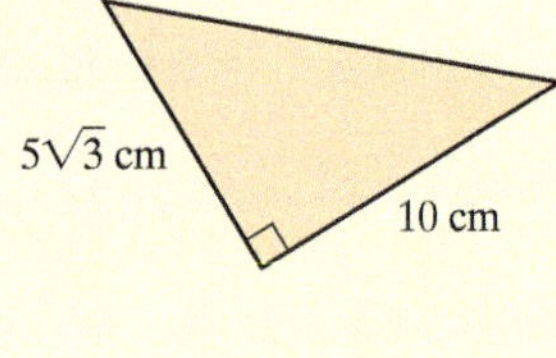

57.

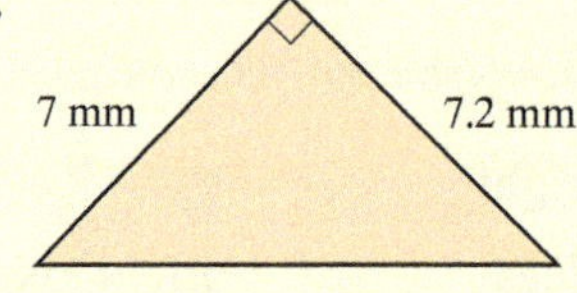

58.

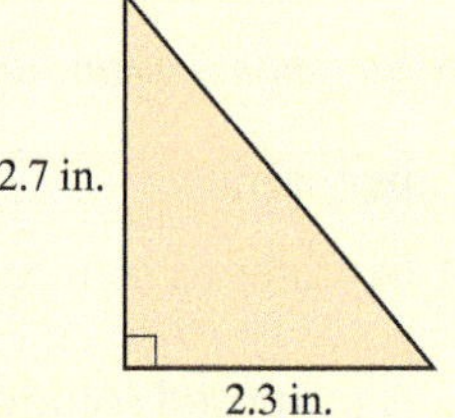

Solve. Give exact answers and two-decimal-place approximations where appropriate. For Exercises 59 and 60, the solutions have been started for you. See Example 7.

59. A wire is needed to support a vertical pole 15 feet tall. The cable will be anchored to a stake 8 feet from the base of the pole. How much cable is needed?

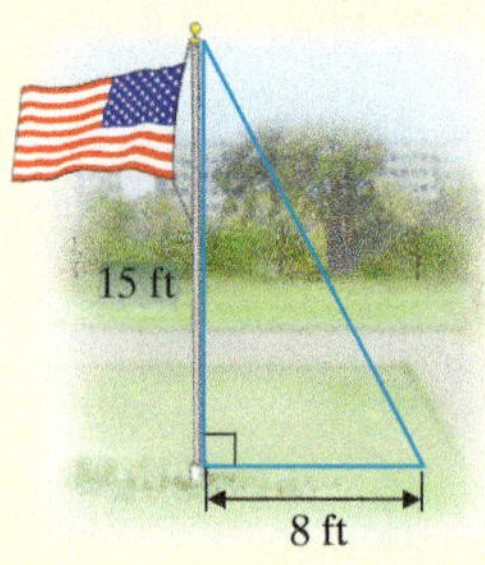

Start the solution:

1. UNDERSTAND the problem. Reread it as many times as needed. Notice that a right triangle is formed with legs of length 8 ft and 15 ft.

   Since we are looking for how much cable is needed, let

   x = amount of cable needed

2. TRANSLATE into an equation. We use the Pythagorean theorem. (Fill in the blanks below.)

$$a^2 + b^2 = c^2$$
$$\downarrow \qquad \downarrow$$
$$\_\_\_^2 + \_\_\_^2 = x^2$$

Finish with:

3. SOLVE and 4. INTERPRET

60. The tallest structure in the United States is a TV tower in Blanchard, North Dakota. Its height is 2063 feet. A 2382-foot length of wire is to be used as a guy wire attached to the top of the tower. Approximate to the nearest foot how far from the base of the tower the guy wire must be anchored. (*Source:* U.S. Geological Survey)

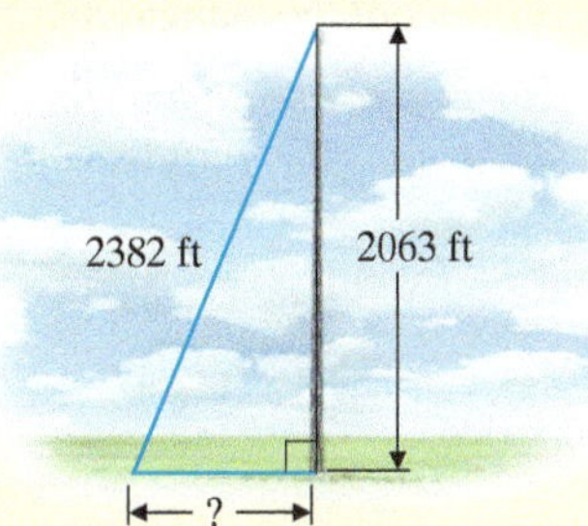

Start the solution:

1. UNDERSTAND the problem. Reread it as many times as needed. Notice that a right triangle is formed with hypotenuse 2382 ft and one leg 2063 ft.

   Since we are looking for how far from the base of the tower the guy wire is anchored, let

   x = distance from base of tower to where guy wire is anchored.

2. TRANSLATE into an equation. We use the Pythagorean theorem. (Fill in the blanks below.)

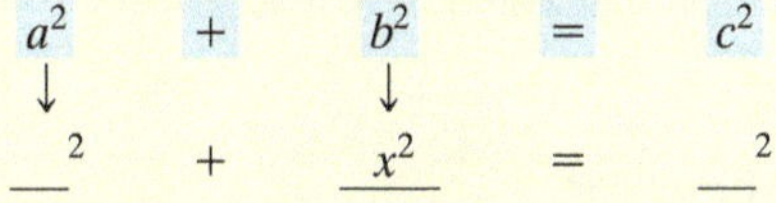

$$\begin{array}{ccccc} a^2 & + & b^2 & = & c^2 \\ \downarrow & & \downarrow & & \\ \_\_^2 & + & \underline{x^2} & = & \_\_^2 \end{array}$$

Finish with:

3. SOLVE and 4. INTERPRET

△ **61.** A spotlight is mounted on the eaves of a house 12 feet above the ground. A flower bed runs between the house and the sidewalk, so the closest a ladder can be placed to the house is 5 feet. How long of a ladder is needed so that an electrician can reach the place where the light is mounted?

△ **62.** A wire is to be attached to support a telephone pole. Because of surrounding buildings, sidewalks, and roadways, the wire must be anchored exactly 15 feet from the base of the pole. Telephone company workers have only 30 feet of cable, and 2 feet of that must be used to attach the cable to the pole and to the stake on the ground. How high from the base of the pole can the wire be attached?

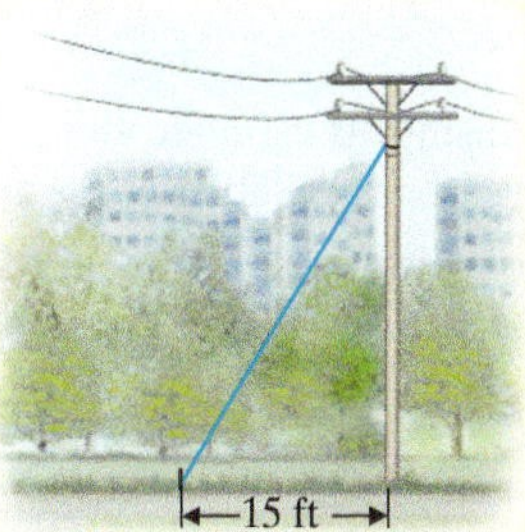

△ **63.** The radius of the moon is 1080 miles. Use the formula for the radius r of a sphere given its surface area A,

$$r = \sqrt{\frac{A}{4\pi}}$$

to find the surface area of the moon. Round to the nearest square mile. (*Source:* National Space Science Data Center)

64. Police departments find it very useful to be able to approximate the speed of a car when they are given the distance that the car skidded before it came to a stop. If the road surface is wet concrete, the function $S(x) = \sqrt{10.5x}$ is used, where $S(x)$ is the speed of the car in miles per hour and x is the distance skidded in feet. Find how fast a car was moving if it skidded 280 feet on wet concrete.

65. The formula $v = \sqrt{2gh}$ gives the velocity v, in feet per second, of an object when it falls h feet accelerated by gravity g, in feet per second squared. If g is approximately 32 feet per second squared, find how far an object has fallen if its velocity is 80 feet per second.

66. Two tractors are pulling a tree stump from a field. If two forces A and B pull at right angles (90°) to each other, the size of the resulting force R is given by the formula $R = \sqrt{A^2 + B^2}$. If tractor A is exerting 600 pounds of force and the resulting force is 850 pounds, find how much force tractor B is exerting.

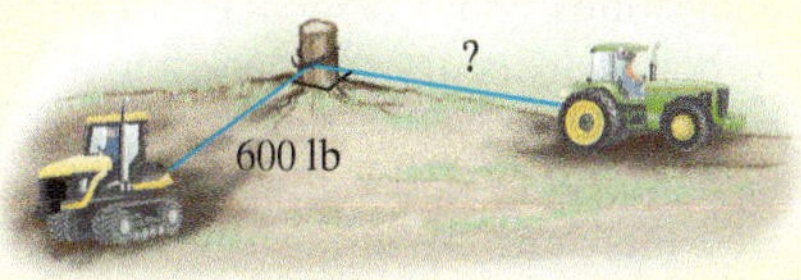

In psychology, it has been suggested that the number S of nonsense syllables that a person can repeat consecutively depends on his or her IQ score I according to the equation $S = 2\sqrt{I} - 9$.

67. Use this relationship to estimate the IQ of a person who can repeat 11 nonsense syllables consecutively.

68. Use this relationship to estimate the IQ of a person who can repeat 15 nonsense syllables consecutively.

*The **period** of a pendulum is the time it takes for the pendulum to make one full back-and-forth swing. The period of a pendulum depends on the length of the pendulum. The formula for the period P, in seconds, is* $P = 2\pi\sqrt{\frac{l}{32}}$, *where l is the length of the pendulum in feet. Use this formula for Exercises 69 through 74.*

69. Find the period of a pendulum whose length is 2 feet. Give an exact answer and a two-decimal-place approximation.

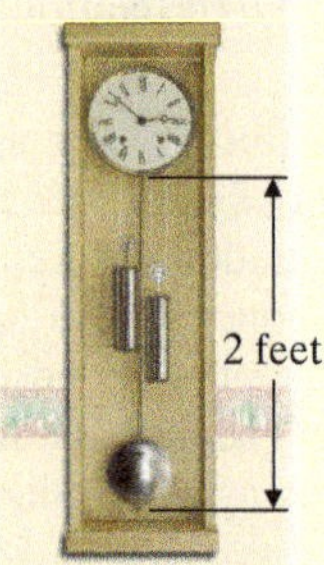

70. Klockit sells a 43-inch lyre pendulum. Find the period of this pendulum. Round your answer to 2 decimal places. (*Hint:* First convert inches to feet.)

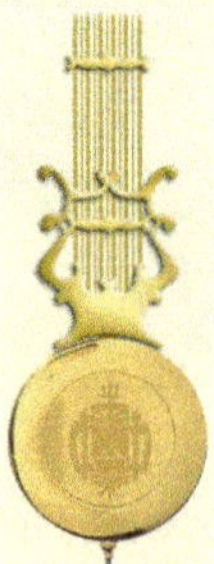

71. Find the length of a pendulum whose period is 4 seconds. Round your answer to 2 decimal places.

72. Find the length of a pendulum whose period is 3 seconds. Round your answer to 2 decimal places.

73. Study the relationship between period and pendulum length in Exercises 69 through 72 and make a conjecture about this relationship.

74. Galileo experimented with pendulums. He supposedly made conjectures about pendulums of equal length with different bob weights. Try this experiment. Make two pendulums 3 feet long. Attach a heavy weight (lead) to one and a light weight (a cork) to the other. Pull both pendulums back the same angle measure and release. Make a conjecture from your observations.

If the three lengths of the sides of a triangle are known, Heron's formula can be used to find its area. If a, b, and c are the lengths of the three sides, Heron's formula for area is

$$A = \sqrt{s(s-a)(s-b)(s-c)}$$

where s is half the perimeter of the triangle, or $s = \frac{1}{2}(a+b+c)$. *Use this formula to find the area of each triangle. Give an exact answer and then a two-decimal-place approximation.*

75. **76.**

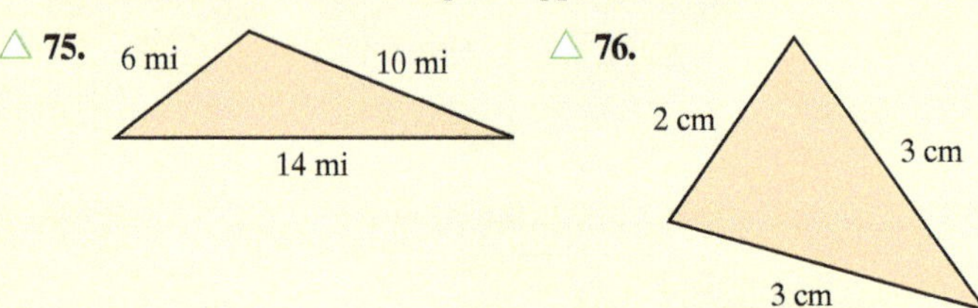

77. Describe when Heron's formula might be useful.

78. In your own words, explain why you think s in Heron's formula is called the *semiperimeter*.

The maximum distance D(h) in kilometers that a person can see from a height h kilometers above the ground is given by the function $D(h) = 111.7\sqrt{h}$. *Use this function for Exercises 79 and 80. Round your answers to two decimal places.*

79. Find the height that would allow a person to see 80 kilometers.

80. Find the height that would allow a person to see 40 kilometers.

REVIEW AND PREVIEW

Use the vertical line test to determine whether each graph represents the graph of a function. See Section 10.6.

81. **82.**

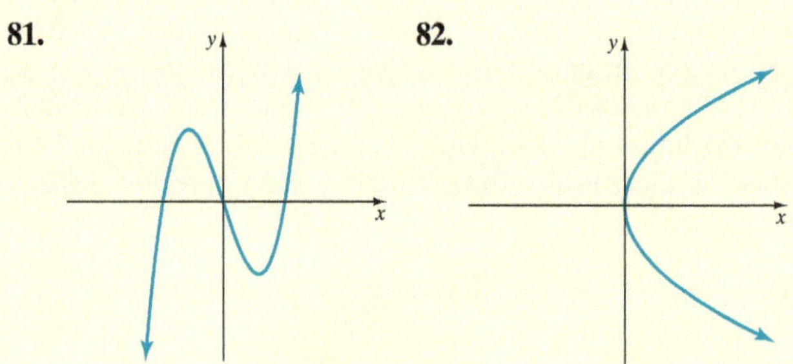

83. **84.**

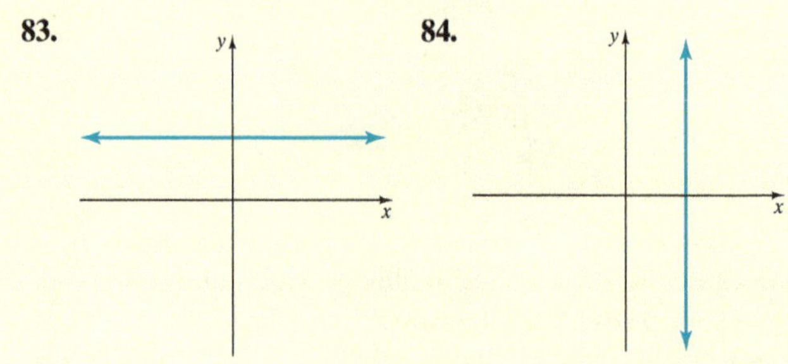

85.

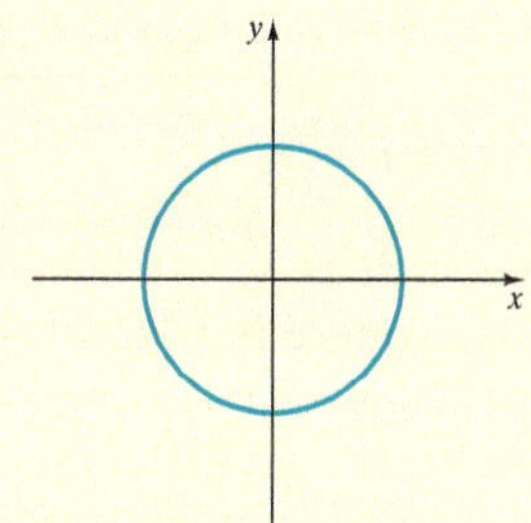

86.

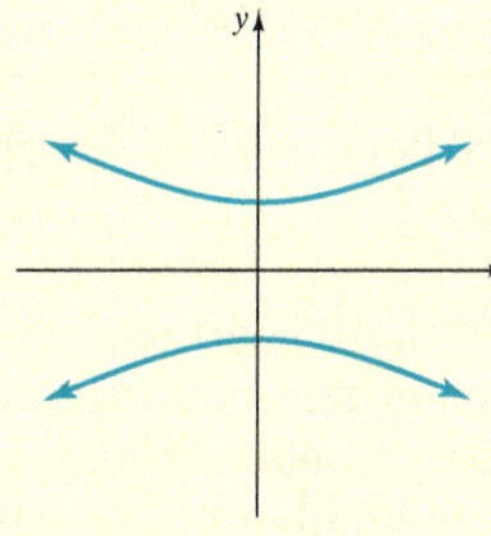

Simplify. See Section 14.7.

87. $\dfrac{\frac{x}{6}}{\frac{2x}{3}+\frac{1}{2}}$

88. $\dfrac{\frac{1}{y}+\frac{4}{5}}{-\frac{3}{20}}$

89. $\dfrac{\frac{z}{5}+\frac{1}{10}}{\frac{z}{20}-\frac{z}{5}}$

90. $\dfrac{\frac{1}{y}+\frac{1}{x}}{\frac{1}{y}-\frac{1}{x}}$

CONCEPT EXTENSIONS

Find the error in each solution and correct it. See the second Concept Check in this section.

91.
$$\begin{aligned} \sqrt{5x-1}+4&=7\\ (\sqrt{5x-1}+4)^2&=7^2\\ 5x-1+16&=49\\ 5x&=34\\ x&=\frac{34}{5} \end{aligned}$$

92.
$$\begin{aligned} \sqrt{2x+3}+4&=1\\ \sqrt{2x+3}&=5\\ (\sqrt{2x+3})^2&=5^2\\ 2x+3&=25\\ 2x&=22\\ x&=11 \end{aligned}$$

93. Solve: $\sqrt{\sqrt{x+3}+\sqrt{x}}=\sqrt{3}$

94. The cost $C(x)$ in dollars per day to operate a small delivery service is given by $C(x)=80\sqrt[3]{x}+500$, where x is the number of deliveries per day. In July, the manager decides that it is necessary to keep delivery costs below \$1620.00. Find the greatest number of deliveries this company can make per day and still keep overhead below \$1620.00.

95. Consider the equations $\sqrt{2x}=4$ and $\sqrt[3]{2x}=4$.

a. Explain the difference in solving these equations.
b. Explain the similarity in solving these equations.

96. Explain why proposed solutions of radical equations must be checked.

Example

For Exercises 97 through 100, see the example below.

Solve $(t^2-3t)-2\sqrt{t^2-3t}=0$.

Solution

Substitution can be used to make this problem somewhat simpler. Since t^2-3t occurs more than once, let $x=t^2-3t$.

$$\begin{aligned} (t^2-3t)-2\sqrt{t^2-3t}&=0\\ x-2\sqrt{x}&=0\\ x&=2\sqrt{x}\\ x^2&=(2\sqrt{x})^2\\ x^2&=4x\\ x^2-4x&=0\\ x(x-4)&=0\\ x=0 \quad\text{or}\quad x-4&=0\\ x&=4 \end{aligned}$$

Now we "undo" the substitution.

$x=0$ Replace x with t^2-3t.

$$\begin{aligned} t^2-3t&=0\\ t(t-3)&=0\\ t=0 \quad\text{or}\quad t-3&=0\\ t&=3 \end{aligned}$$

$x=4$ Replace x with t^2-3t.

$$\begin{aligned} t^2-3t&=4\\ t^2-3t-4&=0\\ (t-4)(t+1)&=0\\ t-4=0 \quad&\text{or}\quad t+1=0\\ t=4 \quad&\qquad t=-1 \end{aligned}$$

In this problem, we have four possible solutions: 0, 3, 4, and −1. All four solutions check in the original equation, so the solutions are −1, 0, 3, 4.

Solve. See the preceding example.

97. $3\sqrt{x^2-8x}=x^2-8x$

98. $\sqrt{(x^2-x)+7}=2(x^2-x)-1$

99. $7-(x^2-3x)=\sqrt{(x^2-3x)+5}$

100. $x^2+6x=4\sqrt{x^2+6x}$

17.7 Complex Numbers 

Objectives

1. Write Square Roots of Negative Numbers in the Form *bi*.
2. Add or Subtract Complex Numbers.
3. Multiply Complex Numbers.
4. Divide Complex Numbers.
5. Raise *i* to Powers.

Objective 1 Writing Numbers in the Form *bi*

Our work with radical expressions has excluded expressions such as $\sqrt{-16}$ because $\sqrt{-16}$ is not a real number; there is no real number whose square is -16. In this section, we discuss a number system that includes roots of negative numbers. This number system is the **complex number system,** and it includes the set of real numbers as a subset. The complex number system allows us to solve equations such as $x^2 + 1 = 0$ that have no real number solutions. The set of complex numbers includes the **imaginary unit**.

Imaginary Unit

The imaginary unit, written i, is the number whose square is -1. That is,

$$i^2 = -1 \quad \text{and} \quad i = \sqrt{-1}$$

To write the square root of a negative number in terms of i, use the property that if a is a positive number, then

$$\sqrt{-a} = \sqrt{-1}\cdot\sqrt{a} = i\cdot\sqrt{a}$$

Using i, we can write $\sqrt{-16}$ as

$$\sqrt{-16} = \sqrt{-1\cdot 16} = \sqrt{-1}\cdot\sqrt{16} = i\cdot 4, \text{ or } 4i$$

Practice 1

Write with i notation.

a. $\sqrt{-4}$ **b.** $\sqrt{-7}$ **c.** $-\sqrt{-18}$

Example 1 Write with i notation.

a. $\sqrt{-36}$ **b.** $\sqrt{-5}$ **c.** $-\sqrt{-20}$

Solution:

a. $\sqrt{-36} = \sqrt{-1\cdot 36} = \sqrt{-1}\cdot\sqrt{36} = i\cdot 6$, or $6i$

b. $\sqrt{-5} = \sqrt{-1(5)} = \sqrt{-1}\cdot\sqrt{5} = i\sqrt{5}$.

c. $-\sqrt{-20} = -\sqrt{-1\cdot 20} = -\sqrt{-1}\cdot\sqrt{4\cdot 5} = -i\cdot 2\sqrt{5} = -2i\sqrt{5}$

Helpful Hint Since $\sqrt{5}i$ can easily be confused with $\sqrt{5i}$, we write $\sqrt{5}i$ as $i\sqrt{5}$.

Work Practice 1

The product rule for radicals does not necessarily hold true for imaginary numbers. *To multiply square roots of negative numbers, first we write each number in terms of the imaginary unit i.* For example, to multiply $\sqrt{-4}$ and $\sqrt{-9}$, we first write each number in the form *bi*.

$$\sqrt{-4}\sqrt{-9} = 2i(3i) = 6i^2 = 6(-1) = -6 \quad \text{Correct}$$

We will also use this method to simplify quotients of square roots of negative numbers. Why? The product rule does not work for this example. In other words,

$$\sqrt{-4}\cdot\sqrt{-9} = \sqrt{(-4)(-9)} = \sqrt{36} = 6 \quad \text{Incorrect}$$

Practice 2

Multiply or divide as indicated.

a. $\sqrt{-5}\cdot\sqrt{-6}$ **b.** $\sqrt{-9}\cdot\sqrt{-1}$

c. $\sqrt{125}\cdot\sqrt{-5}$ **d.** $\dfrac{\sqrt{-27}}{\sqrt{3}}$

Example 2 Multiply or divide as indicated.

a. $\sqrt{-3}\cdot\sqrt{-5}$ **b.** $\sqrt{-36}\cdot\sqrt{-1}$ **c.** $\sqrt{8}\cdot\sqrt{-2}$ **d.** $\dfrac{\sqrt{-125}}{\sqrt{5}}$

Solution:

a. $\sqrt{-3}\cdot\sqrt{-5} = i\sqrt{3}(i\sqrt{5}) = i^2\sqrt{15} = -1\sqrt{15} = -\sqrt{15}$

Answers

1. a. $2i$ **b.** $i\sqrt{7}$ **c.** $-3i\sqrt{2}$

2. a. $-\sqrt{30}$ **b.** -3 **c.** $25i$ **d.** $3i$

b. $\sqrt{-36} \cdot \sqrt{-1} = 6i(i) = 6i^2 = 6(-1) = -6$

c. $\sqrt{8} \cdot \sqrt{-2} = 2\sqrt{2}(i\sqrt{2}) = 2i(\sqrt{2}\sqrt{2}) = 2i(2) = 4i$

d. $\frac{\sqrt{-125}}{\sqrt{5}} = \frac{i\sqrt{125}}{\sqrt{5}} = i\sqrt{25} = 5i$

Work Practice 2

Now that we have practiced working with the imaginary unit, we define complex numbers.

Complex Numbers

A **complex number** is a number that can be written in the form $a + bi$, where a and b are real numbers.

Notice that the set of real numbers is a subset of the complex numbers since any real number can be written in the form of a complex number. For example,

$$16 = 16 + 0i$$

In general, a complex number $a + bi$ is a real number if $b = 0$. Also, a complex number is called a **pure imaginary number** or an imaginary number if $a = 0$ and $b \neq 0$. For example,

$$3i = 0 + 3i \quad \text{and} \quad i\sqrt{7} = 0 + i\sqrt{7}$$

are pure imaginary numbers.

The following diagram shows the relationship between complex numbers and their subsets.

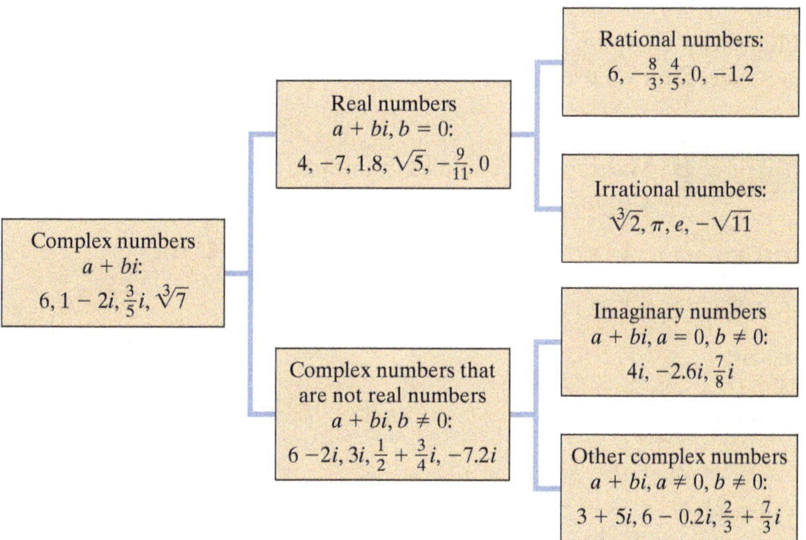

✓Concept Check True or false? Every complex number is also a real number.

Objective 2 Adding or Subtracting Complex Numbers

Two complex numbers $a + bi$ and $c + di$ are equal if and only if $a = c$ and $b = d$. Complex numbers can be added or subtracted by adding or subtracting their real parts and then adding or subtracting their imaginary parts.

✓**Concept Check Answer**
false

Sum or Difference of Complex Numbers

If $a + bi$ and $c + di$ are complex numbers, then their sum is

$$(a + bi) + (c + di) = (a + c) + (b + d)i$$

Their difference is

$$(a + bi) - (c + di) = a + bi - c - di = (a - c) + (b - d)i$$

Practice 3

Add or subtract the complex numbers. Write the sum or difference in the form $a + bi$.

a. $(3 - 5i) + (-4 + i)$
b. $4i - (3 - i)$
c. $(-5 - 2i) - (-8)$

Example 3 Add or subtract the complex numbers. Write the sum or difference in the form $a + bi$.

a. $(2 + 3i) + (-3 + 2i)$ **b.** $5i - (1 - i)$ **c.** $(-3 - 7i) - (-6)$

Solution:

a. $(2 + 3i) + (-3 + 2i) = (2 - 3) + (3 + 2)i = -1 + 5i$

b. $5i - (1 - i) = 5i - 1 + i$
$= -1 + (5 + 1)i$
$= -1 + 6i$

c. $(-3 - 7i) - (-6) = -3 - 7i + 6$
$= (-3 + 6) - 7i$
$= 3 - 7i$

Work Practice 3

Objective 3 Multiplying Complex Numbers

To multiply two complex numbers of the form $a + bi$, we multiply as though they are binomials. Then we use the relationship $i^2 = -1$ to simplify.

Practice 4

Multiply the complex numbers. Write the product in the form $a + bi$.

a. $-4i \cdot 5i$ **b.** $5i(2 + i)$
c. $(2 + 3i)(6 - i)$
d. $(3 - i)^2$
e. $(9 + 2i)(9 - 2i)$

Example 4 Multiply the complex numbers. Write the product in the form $a + bi$.

a. $-7i \cdot 3i$ **b.** $3i(2 - i)$ **c.** $(2 - 5i)(4 + i)$
d. $(2 - i)^2$ **e.** $(7 + 3i)(7 - 3i)$

Solution:

a. $-7i \cdot 3i = -21i^2$
$= -21(-1)$ Replace i^2 with -1.
$= 21 + 0i$

b. $3i(2 - i) = 3i \cdot 2 - 3i \cdot i$ Use the distributive property.
$= 6i - 3i^2$ Multiply.
$= 6i - 3(-1)$ Replace i^2 with -1.
$= 6i + 3$
$= 3 + 6i$ Use the FOIL order (First, Outer, Inner, Last) below.

c. $(2 - 5i)(4 + i) = 2(4) + 2(i) - 5i(4) - 5i(i)$
F O I L
$= 8 + 2i - 20i - 5i^2$
$= 8 - 18i - 5(-1)$ $i^2 = -1$
$= 8 - 18i + 5$
$= 13 - 18i$

Answers

3. a. $-1 - 4i$ **b.** $-3 + 5i$ **c.** $3 - 2i$
4. a. $20 + 0i$ **b.** $-5 + 10i$ **c.** $15 + 16i$
d. $8 - 6i$ **e.** $85 + 0i$

d. $(2 - i)^2 = (2 - i)(2 - i)$
$= 2(2) - 2(i) - 2(i) + i^2$
$= 4 - 4i + (-1)$ $\quad i^2 = -1$
$= 3 - 4i$

e. $(7 + 3i)(7 - 3i) = 7(7) - 7(3i) + 3i(7) - 3i(3i)$
$= 49 - 21i + 21i - 9i^2$
$= 49 - 9(-1)$ $\quad i^2 = -1$
$= 49 + 9$
$= 58 + 0i$

Work Practice 4

Notice that if you add, subtract, or multiply two complex numbers, just like real numbers, the result is a complex number.

Objective 4 Dividing Complex Numbers

From Example 4e, notice that the product of $7 + 3i$ and $7 - 3i$ is a real number. These two complex numbers are called **complex conjugates** of one another. In general, we have the following definition.

Complex Conjugates

The complex numbers $(a + bi)$ and $(a - bi)$ are called **complex conjugates** of each other, and

$$(a + bi)(a - bi) = a^2 + b^2.$$

To see that the product of a complex number $a + bi$ and its conjugate $a - bi$ is the real number $a^2 + b^2$, we multiply.

$$(a + bi)(a - bi) = a^2 - abi + abi - b^2i^2$$
$$= a^2 - b^2(-1)$$
$$= a^2 + b^2$$

We use complex conjugates to divide by a complex number.

Example 5 Divide. Write in the form $a + bi$.

a. $\dfrac{2 + i}{1 - i}$ **b.** $\dfrac{7}{3i}$

Solution:

a. Multiply the numerator and denominator by the complex conjugate of $1 - i$ to eliminate the imaginary number in the denominator.

$$\frac{2 + i}{1 - i} = \frac{(2 + i)(1 + i)}{(1 - i)(1 + i)}$$
$$= \frac{2(1) + 2(i) + 1(i) + i^2}{1^2 - i^2}$$
$$= \frac{2 + 3i - 1}{1 + 1} \quad \text{Here, } i^2 = -1.$$
$$= \frac{1 + 3i}{2} \quad \text{or} \quad \frac{1}{2} + \frac{3}{2}i$$

(*Continued on next page*)

Practice 5

Divide. Write in the form $a + bi$.

a. $\dfrac{4 - i}{3 + i}$ **b.** $\dfrac{5}{2i}$

Answers

5. a. $\dfrac{11}{10} - \dfrac{7}{10}i$ **b.** $0 - \dfrac{5}{2}i$

b. Multiply the numerator and denominator by the conjugate of $3i$. Note that $3i = 0 + 3i$, so its conjugate is $0 - 3i$ or $-3i$.

$$\frac{7}{3i} = \frac{7(-3i)}{(3i)(-3i)} = \frac{-21i}{-9i^2} = \frac{-21i}{-9(-1)} = \frac{-21i}{9} = \frac{-7i}{3} \text{ or } 0 - \frac{7}{3}i$$

Work Practice 5

Helpful Hint

Recall that division can be checked by multiplication.

To check that $\frac{2+i}{1-i} = \frac{1}{2} + \frac{3}{2}i$, in Example 5a, multiply $\left(\frac{1}{2} + \frac{3}{2}i\right)(1 - i)$ to verify that the product is $2 + i$.

Objective 5 Finding Powers of i

We can use the fact that $i^2 = -1$ to find higher powers of i. To find i^3, we rewrite it as the product of i^2 and i.

$$i^3 = i^2 \cdot i = (-1)i = -i$$
$$i^4 = i^2 \cdot i^2 = (-1) \cdot (-1) = 1$$

We continue this process and use the fact that $i^4 = 1$ and $i^2 = -1$ to simplify i^5 and i^6.

$$i^5 = i^4 \cdot i = 1 \cdot i = i$$
$$i^6 = i^4 \cdot i^2 = 1 \cdot (-1) = -1$$

If we continue finding powers of i, we generate the following pattern. Notice that the values i, -1, $-i$, and 1 repeat as i is raised to higher and higher powers.

$i^1 = i$	$i^5 = i$	$i^9 = i$
$i^2 = -1$	$i^6 = -1$	$i^{10} = -1$
$i^3 = -i$	$i^7 = -i$	$i^{11} = -i$
$i^4 = 1$	$i^8 = 1$	$i^{12} = 1$

This pattern allows us to find other powers of i. To do so, we will use the fact that $i^4 = 1$ and rewrite a power of i in terms of i^4. For example,

$$i^{22} = i^{20} \cdot i^2 = (i^4)^5 \cdot i^2 = 1^5 \cdot (-1) = 1 \cdot (-1) = -1.$$

Practice 6

Find the following powers of i.

a. i^9 **b.** i^{16} **c.** i^{34} **d.** i^{-24}

Example 6 Find the following powers of i.

a. i^7 **b.** i^{20} **c.** i^{46} **d.** i^{-12}

Solution:

a. $i^7 = i^4 \cdot i^3 = 1(-i) = -i$

b. $i^{20} = (i^4)^5 = 1^5 = 1$

c. $i^{46} = i^{44} \cdot i^2 = (i^4)^{11} \cdot i^2 = 1^{11}(-1) = -1$

d. $i^{-12} = \frac{1}{i^{12}} = \frac{1}{(i^4)^3} = \frac{1}{(1)^3} = \frac{1}{1} = 1$

Work Practice 6

Answers

6. **a.** i **b.** 1 **c.** -1 **d.** 1

Vocabulary, Readiness & Video Check

Use the choices below to fill in each blank. Not all choices will be used.

-1	$\sqrt{-1}$	real	imaginary unit
1	$\sqrt{1}$	complex	pure imaginary

1. A(n) ________ number is one that can be written in the form $a + bi$, where a and b are real numbers.

2. In the complex number system, i denotes the ________.

3. $i^2 =$ ________

4. $i =$ ________

5. A complex number, $a + bi$, is a(n) ________ number if $b = 0$.

6. A complex number, $a + bi$, is a(n) ________ number if $a = 0$ and $b \neq 0$.

Martin-Gay Interactive Videos *Watch the section lecture video and answer the following questions.*

See Video 17.7

Objective 1 **7.** From Example 4, with what rule must you be especially careful when working with imaginary numbers and why?

Objective 2 **8.** In Examples 5 and 6, what is the process of adding and subtracting complex numbers compared to? What important reminder is given about i?

Objective 3 **9.** In Examples 7 and 8, what part of the definition of the imaginary unit i may be used during the multiplication of complex numbers to help simplify products?

Objective 4 **10.** In Example 9, using complex conjugates to divide complex numbers is compared to what process?

Objective 5 **11.** From the lecture before Example 10, what are the first four powers of i whose values keep repeating?

17.7 Exercise Set MyMathLab®

Simplify. See Example 1.

1. $\sqrt{-81}$ **2.** $\sqrt{-49}$ **3.** $\sqrt{-7}$

4. $\sqrt{-3}$ **5.** $-\sqrt{16}$ **6.** $-\sqrt{4}$

7. $\sqrt{-64}$ **8.** $\sqrt{-100}$

Write in terms of i. See Example 1.

9. $\sqrt{-24}$ **10.** $\sqrt{-32}$

11. $-\sqrt{-36}$ **12.** $-\sqrt{-121}$

13. $8\sqrt{-63}$ **14.** $4\sqrt{-20}$

15. $-\sqrt{54}$ **16.** $\sqrt{-63}$

Multiply or divide. See Example 2.

17. $\sqrt{-2}\cdot\sqrt{-7}$ **18.** $\sqrt{-11}\cdot\sqrt{-3}$

19. $\sqrt{-5}\cdot\sqrt{-10}$ **20.** $\sqrt{-2}\cdot\sqrt{-6}$

21. $\sqrt{16}\cdot\sqrt{-1}$ **22.** $\sqrt{3}\cdot\sqrt{-27}$

23. $\dfrac{\sqrt{-9}}{\sqrt{3}}$ **24.** $\dfrac{\sqrt{49}}{\sqrt{-10}}$

25. $\dfrac{\sqrt{-80}}{\sqrt{-10}}$ **26.** $\dfrac{\sqrt{-40}}{\sqrt{-8}}$

Add or subtract. Write the sum or difference in the form a + bi. See Example 3.

27. $(4 - 7i) + (2 + 3i)$ **28.** $(2 - 4i) - (2 - i)$

29. $(6 + 5i) - (8 - i)$ **30.** $(8 - 3i) + (-8 + 3i)$

31. $6 - (8 + 4i)$ **32.** $(9 - 4i) - 9$

Multiply. Write the product in the form a + bi. See Example 4.

33. $-10i \cdot -4i$ **34.** $-2i \cdot -11i$

35. $6i(2 - 3i)$ **36.** $5i(4 - 7i)$

37. $(\sqrt{3} + 2i)(\sqrt{3} - 2i)$ **38.** $(\sqrt{5} - 5i)(\sqrt{5} + 5i)$

39. $(4 - 2i)^2$ **40.** $(6 - 3i)^2$

Write each quotient in the form a + bi. See Example 5.

41. $\dfrac{4}{i}$ **42.** $\dfrac{5}{6i}$

43. $\frac{7}{4 + 3i}$

44. $\frac{9}{1 - 2i}$

45. $\frac{3 + 5i}{1 + i}$

46. $\frac{6 + 2i}{4 - 3i}$

47. $\frac{5 - i}{3 - 2i}$

48. $\frac{6 - i}{2 + i}$

MIXED PRACTICE

Perform each indicated operation. Write the result in the form a + bi.

49. $(7i)(-9i)$

50. $(-6i)(-4i)$

51. $(6 - 3i) - (4 - 2i)$

52. $(-2 - 4i) - (6 - 8i)$

53. $-3i(-1 + 9i)$

54. $-5i(-2 + i)$

55. $\frac{4 - 5i}{2i}$

56. $\frac{6 + 8i}{3i}$

57. $(4 + i)(5 + 2i)$

58. $(3 + i)(2 + 4i)$

59. $(6 - 2i)(3 + i)$

60. $(2 - 4i)(2 - i)$

61. $(8 - 3i) + (2 + 3i)$

62. $(7 + 4i) + (4 - 4i)$

63. $(1 - i)(1 + i)$

64. $(6 + 2i)(6 - 2i)$

65. $\frac{16 + 15i}{-3i}$

66. $\frac{2 - 3i}{-7i}$

67. $(9 + 8i)^2$

68. $(4 - 7i)^2$

69. $\frac{2}{3 + i}$

70. $\frac{5}{3 - 2i}$

71. $(5 - 6i) - 4i$

72. $(6 - 2i) + 7i$

73. $\frac{2 - 3i}{2 + i}$

74. $\frac{6 + 5i}{6 - 5i}$

75. $(2 + 4i) + (6 - 5i)$

76. $(5 - 3i) + (7 - 8i)$

77. $(\sqrt{6} + i)(\sqrt{6} - i)$

78. $(\sqrt{14} - 4i)(\sqrt{14} + 4i)$

79. $4(2 - i)^2$

80. $9(2 - i)^2$

Find each power of i. See Example 6.

81. i^8

82. i^{10}

83. i^{21}

84. i^{15}

85. i^{11}

86. i^{40}

87. i^{-6}

88. i^{-9}

89. $(2i)^6$

90. $(5i)^4$

91. $(-3i)^5$

92. $(-2i)^7$

REVIEW AND PREVIEW

Recall that the sum of the measures of the angles of a triangle is 180°. Find the unknown angle in each triangle.

93.

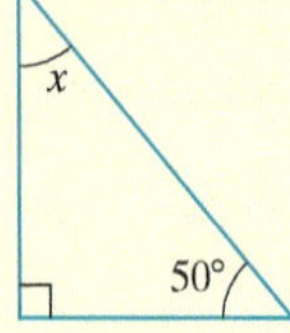

94.

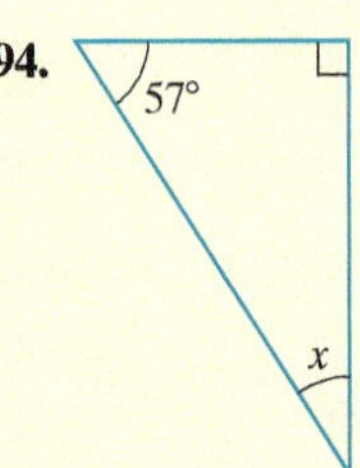

Use synthetic division to divide the following. See Section 12.7.

95. $(x^3 - 6x^2 + 3x - 4) \div (x - 1)$

96. $(5x^4 - 3x^2 + 2) \div (x + 2)$

Thirty people were recently polled about the average monthly balance in their checking accounts. The results of this poll are shown in the following histogram. Use this graph to answer Exercises 97 through 102. See Section 10.1.

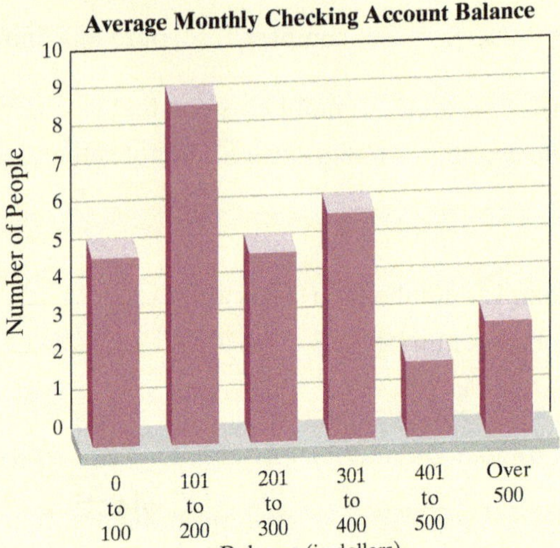

97. How many people polled reported an average checking account balance of \$201 to \$300?

98. How many people polled reported an average checking account balance of \$0 to \$100?

99. How many people polled reported an average checking account balance of \$200 or less?

100. How many people polled reported an average checking account balance of \$301 or more?

101. What percent of people polled reported an average checking account balance of \$201 to \$300? Round to the nearest tenth of a percent.

102. What percent of people polled reported an average checking account balance of \$0 to \$100? Round to the nearest tenth of a percent.

CONCEPT EXTENSIONS

Write in the form a + bi.

103. $i^3 - i^4$

104. $i^8 - i^7$

105. $i^6 + i^8$

106. $i^4 + i^{12}$

107. $2 + \sqrt{-9}$

108. $5 - \sqrt{-16}$

109. $\frac{6 + \sqrt{-18}}{3}$

110. $\frac{4 - \sqrt{-8}}{2}$

111. $\frac{5 - \sqrt{-75}}{10}$

112. $\frac{7 + \sqrt{-98}}{14}$

113. Describe how to find the conjugate of a complex number.

114. Explain why the product of a complex number and its complex conjugate is a real number.

Simplify.

115. $(8 - \sqrt{-3}) - (2 + \sqrt{-12})$

116. $(8 - \sqrt{-4}) - (2 + \sqrt{-16})$

117. Determine whether $2i$ is a solution of $x^2 + 4 = 0$.

118. Determine whether $-1 + i$ is a solution of $x^2 + 2x = -2$.

Chapter 17 Vocabulary Check

Fill in each blank with one of the words or phrases listed below.

index	rationalizing	conjugate	principal square root	cube root	midpoint
complex number	like radicals	radicand	imaginary unit	distance	

1. The ________________ of $\sqrt{3} + 2$ is $\sqrt{3} - 2$.

2. The ________________ of a nonnegative number a is written as $\sqrt{a}$.

3. The process of writing a radical expression as an equivalent expression but without a radical in the denominator is called ________________ the denominator.

4. The ________________, written i, is the number whose square is -1.

5. The ________________ of a number is written as $\sqrt[3]{a}$.

6. In the notation $\sqrt[n]{a}$, n is called the ________________ and a is called the ________________.

7. Radicals with the same index and the same radicand are called ________________.

8. A(n) ________________ is a number that can be written in the form $a + bi$, where a and b are real numbers.

9. The ________________ formula is $d = \sqrt{(x_2 - x_1)^2 + (y_2 - y_1)^2}$.

10. The ________________ formula is $\left(\frac{x_1 + x_2}{2}, \frac{y_1 + y_2}{2}\right)$.

Helpful Hint

Are you preparing for your test? Don't forget to take the Chapter 17 Test on page 1201. Then check your answers at the back of the text and use the Chapter Test Prep Videos to see the fully worked-out solutions to any of the exercises you want to review.

17 Chapter Highlights

Definitions and Concepts	Examples
Section 17.1 Radicals and Radical Functions	
The **positive,** or **principal, square root** of a nonnegative number a is written as $\sqrt{a}$.	$\sqrt{36} = 6$ $\quad \sqrt{\frac{9}{100}} = \frac{3}{10}$
$\sqrt{a} = b$ only if $b^2 = a$ and $b \geq 0$	$-\sqrt{36} = -6$ $\quad -\sqrt{0.04} = -0.2$
The **negative square root of** a is written as $-\sqrt{a}$. The **cube root** of a real number a is written as $\sqrt[3]{a}$.	$\sqrt[3]{27} = 3$ $\quad \sqrt[3]{-\frac{1}{8}} = -\frac{1}{2}$
$\sqrt[3]{a} = b$ only if $b^3 = a$	$\sqrt[3]{y^6} = y^2$ $\quad \sqrt[3]{64x^9} = 4x^3$
If n is an even positive integer, then $\sqrt[n]{a^n} = \lvert a \rvert$.	$\sqrt{(-3)^2} = \lvert -3 \rvert = 3$
If n is an odd positive integer, then $\sqrt[n]{a^n} = a$.	$\sqrt[3]{(-7)^3} = -7$
A **radical function** in x is a function defined by an expression containing a root of x.	If $f(x) = \sqrt{x} + 2$, $f(1) = \sqrt{(1)} + 2 = 1 + 2 = 3$ $f(3) = \sqrt{(3)} + 2 \approx 3.73$

Definitions and Concepts	Examples
Section 17.2 Rational Exponents	
$a^{1/n} = \sqrt[n]{a}$ if $\sqrt[n]{a}$ is a real number.	$81^{1/2} = \sqrt{81} = 9$ $(-8x^3)^{1/3} = \sqrt[3]{-8x^3} = -2x$
If m and n are positive integers greater than 1 with $\frac{m}{n}$ in lowest terms and $\sqrt[n]{a}$ is a real number, then $a^{m/n} = (a^{1/n})^m = (\sqrt[n]{a})^m$	$4^{5/2} = (\sqrt{4})^5 = 2^5 = 32$ $27^{2/3} = (\sqrt[3]{27})^2 = 3^2 = 9$
$a^{-m/n} = \frac{1}{a^{m/n}}$ as long as $a^{m/n}$ is a nonzero number.	$16^{-3/4} = \frac{1}{16^{3/4}} = \frac{1}{(\sqrt[4]{16})^3} = \frac{1}{2^3} = \frac{1}{8}$
Exponent rules are true for rational exponents.	$x^{2/3} \cdot x^{-5/6} = x^{2/3-5/6} = x^{-1/6} = \frac{1}{x^{1/6}}$ $(8^4)^{1/2} = 8^2 = 64$ $\frac{a^{4/5}}{a^{-2/5}} = a^{4/5-(-2/5)} = a^{6/5}$
Section 17.3 Simplifying Radical Expressions	
Product and Quotient Rules If $\sqrt[n]{a}$ and $\sqrt[n]{b}$ are real numbers, $\sqrt[n]{a} \cdot \sqrt[n]{b} = \sqrt[n]{a \cdot b}$ $\frac{\sqrt[n]{a}}{\sqrt[n]{b}} = \sqrt[n]{\frac{a}{b}}$, provided $\sqrt[n]{b} \neq 0$	Multiply or divide as indicated: $\sqrt{11} \cdot \sqrt{3} = \sqrt{33}$ $\frac{\sqrt[3]{40x}}{\sqrt[3]{5x}} = \sqrt[3]{8} = 2$
A radical of the form $\sqrt[n]{a}$ is **simplified** when a contains no factors that are perfect nth powers.	$\sqrt{40} = \sqrt{4 \cdot 10} = 2\sqrt{10}$ $\sqrt{36x^5} = \sqrt{36x^4 \cdot x} = 6x^2\sqrt{x}$ $\sqrt[3]{24x^7y^3} = \sqrt[3]{8x^6y^3 \cdot 3x} = 2x^2y\sqrt[3]{3x}$
Distance Formula The distance d between two points (x_1, y_1) and (x_2, y_2) is given by $d = \sqrt{(x_2 - x_1)^2 + (y_2 - y_1)^2}$	Find the distance between points $(-1, 6)$ and $(-2, -4)$. Let $(x_1, y_1) = (-1, 6)$ and $(x_2, y_2) = (-2, -4)$. $d = \sqrt{(x_2 - x_1)^2 + (y_2 - y_1)^2}$ $= \sqrt{(-2-(-1))^2 + (-4-6)^2}$ $= \sqrt{1 + 100} = \sqrt{101}$
Midpoint Formula The midpoint of the line segment whose endpoints are (x_1, y_1) and (x_2, y_2) is the point with coordinates $\left(\frac{x_1 + x_2}{2}, \frac{y_1 + y_2}{2}\right)$	Find the midpoint of the line segment whose endpoints are $(-1, 6)$ and $(-2, -4)$. $\left(\frac{-1 + (-2)}{2}, \frac{6 + (-4)}{2}\right)$ The midpoint is $\left(-\frac{3}{2}, 1\right)$.

Definitions and Concepts	Examples
Section 17.4 Adding, Subtracting, and Multiplying Radical Expressions	
Radicals with the same index and the same radicand are **like radicals.**	$5\sqrt{6} + 2\sqrt{6} = (5 + 2)\sqrt{6} = 7\sqrt{6}$
The distributive property can be used to add like radicals.	$\sqrt[3]{3x} - 10\sqrt[3]{3x} + 3\sqrt[3]{10x}$ $= (-1 - 10)\sqrt[3]{3x} + 3\sqrt[3]{10x}$ $= -11\sqrt[3]{3x} + 3\sqrt[3]{10x}$
Radical expressions are multiplied by using many of the same properties used to multiply polynomials.	Multiply: $(\sqrt{5} - \sqrt{2x})(\sqrt{2} + \sqrt{2x})$ $= \sqrt{10} + \sqrt{10x} - \sqrt{4x} - 2x$ $= \sqrt{10} + \sqrt{10x} - 2\sqrt{x} - 2x$ $(2\sqrt{3} - \sqrt{8x})(2\sqrt{3} + \sqrt{8x})$ $= 4(3) - 8x = 12 - 8x$
Section 17.5 Rationalizing Denominators and Numerators of Radical Expressions	
The **conjugate** of $a + b$ is $a - b$.	The conjugate of $\sqrt{7} + \sqrt{3}$ is $\sqrt{7} - \sqrt{3}$.
The process of writing the denominator of a radical expression without a radical is called **rationalizing the denominator.**	Rationalize each denominator: $\frac{\sqrt{5}}{\sqrt{3}} = \frac{\sqrt{5}\cdot\sqrt{3}}{\sqrt{3}\cdot\sqrt{3}} = \frac{\sqrt{15}}{3}$ $\frac{6}{\sqrt{7} + \sqrt{3}} = \frac{6(\sqrt{7} - \sqrt{3})}{(\sqrt{7} + \sqrt{3})(\sqrt{7} - \sqrt{3})}$ $= \frac{6(\sqrt{7} - \sqrt{3})}{7 - 3}$ $= \frac{6(\sqrt{7} - \sqrt{3})}{4} = \frac{3(\sqrt{7} - \sqrt{3})}{2}$
The process of writing the numerator of a radical expression without a radical is called **rationalizing the numerator.**	Rationalize each numerator: $\frac{\sqrt[3]{9}}{\sqrt[3]{5}} = \frac{\sqrt[3]{9}\cdot\sqrt[3]{3}}{\sqrt[3]{5}\cdot\sqrt[3]{3}} = \frac{\sqrt[3]{27}}{\sqrt[3]{15}} = \frac{3}{\sqrt[3]{15}}$ $\frac{\sqrt{9} + \sqrt{3x}}{12} = \frac{(\sqrt{9} + \sqrt{3x})(\sqrt{9} - \sqrt{3x})}{12(\sqrt{9} - \sqrt{3x})}$ $= \frac{9 - 3x}{12(\sqrt{9} - \sqrt{3x})}$ $= \frac{3(3 - x)}{3\cdot 4(3 - \sqrt{3x})} = \frac{3 - x}{4(3 - \sqrt{3x})}$

Definitions and Concepts	Examples
Section 17.6 Radical Equations and Problem Solving	
To Solve a Radical Equation	Solve: $x = \sqrt{4x + 9} + 3$.
Step 1: Write the equation so that one radical is by itself on one side of the equation.	**1.** $x - 3 = \sqrt{4x + 9}$
Step 2: Raise each side of the equation to a power equal to the index of the radical and simplify.	**2.** $(x - 3)^2 = (\sqrt{4x + 9})^2$ $x^2 - 6x + 9 = 4x + 9$
Step 3: If the equation still contains a radical, repeat Steps 1 and 2. If not, solve the equation.	**3.** $x^2 - 10x = 0$ $x(x - 10) = 0$ $x = 0$ or $x = 10$
Step 4: Check all proposed solutions in the original equation.	**4.** The proposed solution 10 checks, but 0 does not. The solution is 10.
Section 17.7 Complex Numbers	
$i^2 = -1$ and $i = \sqrt{-1}$	Simplify: $\sqrt{-9}$. $\sqrt{-9} = \sqrt{-1 \cdot 9} = \sqrt{-1} \cdot \sqrt{9} = i \cdot 3$ or $3i$
A **complex number** is a number that can be written in the form $a + bi$, where a and b are real numbers.	**Complex Numbers** — **Written in Form $a + bi$** 12 — $12 + 0i$ $-5i$ — $0 + (-5)i$ $-2 - 3i$ — $-2 + (-3)i$ Multiply. $\sqrt{-3} \cdot \sqrt{-7} = i\sqrt{3} \cdot i\sqrt{7}$ $= i^2\sqrt{21}$ $= -\sqrt{21}$
To add or subtract complex numbers, add or subtract their real parts and then add or subtract their imaginary parts.	Perform each indicated operation. $(-3 + 2i) - (7 - 4i) = -3 + 2i - 7 + 4i$ $= -10 + 6i$
To multiply complex numbers, multiply as though they are binomials.	$(-7 - 2i)(6 + i) = -42 - 7i - 12i - 2i^2$ $= -42 - 19i - 2(-1)$ $= -42 - 19i + 2$ $= -40 - 19i$
The complex numbers $(a + bi)$ and $(a - bi)$ are called **complex conjugates.**	The complex conjugate of $(3 + 6i)$ is $(3 - 6i)$. Their product is a real number: $(3 - 6i)(3 + 6i) = 9 - 36i^2$ $= 9 - 36(-1) = 9 + 36 = 45$
To divide complex numbers, multiply the numerator and the denominator by the conjugate of the denominator.	Divide. $\frac{4}{2 - i} = \frac{4(2 + i)}{(2 - i)(2 + i)}$ $= \frac{4(2 + i)}{4 - i^2}$ $= \frac{4(2 + i)}{5}$ $= \frac{8 + 4i}{5} = \frac{8}{5} + \frac{4}{5}i$

Chapter 17 Review

(17.1) *Find the root. Assume that all variables represent positive numbers.*

1. $\sqrt{81}$
2. $\sqrt[4]{81}$
3. $\sqrt[3]{-8}$
4. $\sqrt[4]{-16}$
5. $-\sqrt{\frac{1}{49}}$
6. $\sqrt{x^{64}}$
7. $-\sqrt{36}$
8. $\sqrt[3]{64}$
9. $\sqrt[3]{-a^6b^9}$
10. $\sqrt{16a^4b^{12}}$
11. $\sqrt[5]{32a^5b^{10}}$
12. $\sqrt[5]{-32x^{15}y^{20}}$
13. $\sqrt{\frac{x^{12}}{36y^2}}$
14. $\sqrt[3]{\frac{27y^3}{z^{12}}}$

Simplify. Use absolute value bars when necessary.

15. $\sqrt{(-x)^2}$
16. $\sqrt[4]{(x^2-4)^4}$
17. $\sqrt[3]{(-27)^3}$
18. $\sqrt[5]{(-5)^5}$
19. $-\sqrt[5]{x^5}$
20. $-\sqrt[3]{x^3}$
21. $\sqrt[4]{16(2y+z)^4}$
22. $\sqrt{25(x-y)^2}$
23. $\sqrt[5]{y^5}$
24. $\sqrt[6]{x^6}$

25. Let $f(x) = \sqrt{x} + 3$.
 a. Find $f(0)$ and $f(9)$.
 b. Find the domain of $f(x)$.
 c. Graph $f(x)$.

26. Let $g(x) = \sqrt[3]{x} - 3$.
 a. Find $g(11)$ and $g(20)$.
 b. Find the domain of $g(x)$.
 c. Graph $g(x)$.

(17.2) *Evaluate.*

27. $\left(\frac{1}{81}\right)^{1/4}$
28. $\left(-\frac{1}{27}\right)^{1/3}$
29. $(-27)^{-1/3}$
30. $(-64)^{-1/3}$
31. $-9^{3/2}$
32. $64^{-1/3}$
33. $(-25)^{5/2}$
34. $\left(\frac{25}{49}\right)^{-3/2}$
35. $\left(\frac{8}{27}\right)^{-2/3}$
36. $\left(-\frac{1}{36}\right)^{-1/4}$

Write with rational exponents.

37. $\sqrt[3]{x^2}$
38. $\sqrt[5]{5x^2y^3}$

Write using radical notation.

39. $y^{4/5}$
40. $5(xy^2z^5)^{1/3}$
41. $(x+2)^{-1/3}$
42. $(x+2y)^{-1/2}$

Simplify each expression. Assume that all variables represent positive real numbers. Write with only positive exponents.

43. $a^{1/3}a^{4/3}a^{1/2}$
44. $\frac{b^{1/3}}{b^{4/3}}$
45. $(a^{1/2}a^{-2})^3$
46. $(x^{-3}y^6)^{1/3}$
47. $\left(\frac{b^{3/4}}{a^{-1/2}}\right)^8$
48. $\frac{x^{1/4}x^{-1/2}}{x^{2/3}}$
49. $\left(\frac{49c^{5/3}}{a^{-1/4}b^{5/6}}\right)^{-1}$
50. $a^{-1/4}(a^{5/4} - a^{9/4})$

Use a calculator and write a three-decimal-place approximation of each number.

51. $\sqrt{20}$
52. $\sqrt[3]{-39}$
53. $\sqrt[4]{726}$
54. $56^{1/3}$
55. $-78^{3/4}$
56. $105^{-2/3}$

Use rational exponents to write each as a single radical.

57. $\sqrt[3]{2}\cdot\sqrt{7}$
58. $\sqrt[3]{3}\cdot\sqrt[4]{x}$

(17.3) *Perform each indicated operation and then simplify if possible. Assume that all variables represent positive real numbers.*

59. $\sqrt{3}\cdot\sqrt{8}$
60. $\sqrt[3]{7y}\cdot\sqrt[3]{x^2z}$
61. $\frac{\sqrt{44x^3}}{\sqrt{11x}}$
62. $\frac{\sqrt[4]{a^6b^{13}}}{\sqrt[4]{a^2b}}$

Simplify.

63. $\sqrt{60}$
64. $-\sqrt{75}$
65. $\sqrt[3]{162}$
66. $\sqrt[3]{-32}$
67. $\sqrt{36x^7}$
68. $\sqrt[3]{24a^5b^7}$
69. $\sqrt{\frac{p^{17}}{121}}$
70. $\sqrt[3]{\frac{y^5}{27x^6}}$
71. $\sqrt[4]{\frac{xy^6}{81}}$
72. $\sqrt{\frac{2x^3}{49y^4}}$

△ *The formula for the radius r of a circle of area A is* $r = \sqrt{\frac{A}{\pi}}$. *Use this for Exercises 73 and 74.*

73. Find the exact radius of a circle whose area is 25 square meters.

74. Approximate to two decimal places the radius of a circle whose area is 104 square inches.

Find the distance between each pair of points. Give an exact value and a three-decimal-place approximation.

75. $(-6, 3)$ and $(8, 4)$
76. $(-4, -6)$ and $(-1, 5)$
77. $(-1, 5)$ and $(2, -3)$
78. $(-\sqrt{2}, 0)$ and $(0, -4\sqrt{6})$
79. $(-\sqrt{5}, -\sqrt{11})$ and $(-\sqrt{5}, -3\sqrt{11})$
80. $(7.4, -8.6)$ and $(-1.2, 5.6)$

Find the midpoint of each line segment whose endpoints are given.

81. $(2, 6)$; $(-12, 4)$
82. $(-6, -5)$; $(-9, 7)$
83. $(4, -6)$; $(-15, 2)$
84. $\left(0, -\frac{3}{8}\right)$; $\left(\frac{1}{10}, 0\right)$

85. $\left(\frac{3}{4}, -\frac{1}{7}\right); \left(-\frac{1}{4}, -\frac{3}{7}\right)$

86. $(\sqrt{3}, -2\sqrt{6}); (\sqrt{3}, -4\sqrt{6})$

(17.4) *Perform each indicated operation. Assume that all variables represent positive real numbers.*

87. $\sqrt{20} + \sqrt{45} - 7\sqrt{5}$

88. $x\sqrt{75x} - \sqrt{27x^3}$

89. $\sqrt[3]{128} + \sqrt[3]{250}$

90. $3\sqrt[4]{32a^5} - a\sqrt[4]{162a}$

91. $\frac{5}{\sqrt{4}} + \frac{\sqrt{3}}{3}$

92. $\sqrt{\frac{8}{x^2}} - \sqrt{\frac{50}{16x^2}}$

93. $2\sqrt{50} - 3\sqrt{125} + \sqrt{98}$

94. $2a\sqrt[4]{32b^5} - 3b\sqrt[4]{162a^4b} + \sqrt[4]{2a^4b^5}$

Multiply and then simplify if possible. Assume that all variables represent positive real numbers.

95. $\sqrt{3}(\sqrt{27} - \sqrt{3})$

96. $(\sqrt{x} - 3)^2$

97. $(\sqrt{5} - 5)(2\sqrt{5} + 2)$

98. $(2\sqrt{x} - 3\sqrt{y})(2\sqrt{x} + 3\sqrt{y})$

99. $(\sqrt{a} + 3)(\sqrt{a} - 3)$

100. $(\sqrt[3]{a} + 2)^2$

101. $(\sqrt[3]{5x} + 9)(\sqrt[3]{5x} - 9)$

102. $(\sqrt[3]{a} + 4)(\sqrt[3]{a^2} - 4\sqrt[3]{a} + 16)$

(17.5) *Rationalize each denominator. Assume that all variables represent positive real numbers.*

103. $\frac{3}{\sqrt{7}}$

104. $\sqrt{\frac{x}{12}}$

105. $\frac{5}{\sqrt[3]{4}}$

106. $\sqrt{\frac{24x^5}{3y}}$

107. $\sqrt[3]{\frac{15x^6y^7}{z^2}}$

108. $\sqrt[4]{\frac{81}{8x^{10}}}$

109. $\frac{3}{\sqrt{y} - 2}$

110. $\frac{\sqrt{2} - \sqrt{3}}{\sqrt{2} + \sqrt{3}}$

Rationalize each numerator. Assume that all variables represent positive real numbers.

111. $\frac{\sqrt{11}}{3}$

112. $\sqrt{\frac{18}{y}}$

113. $\frac{\sqrt[3]{9}}{7}$

114. $\sqrt{\frac{24x^5}{3y^2}}$

115. $\sqrt[3]{\frac{xy^2}{10z}}$

116. $\frac{\sqrt{x} + 5}{-3}$

(17.6) *Solve each equation.*

117. $\sqrt{y - 7} = 5$

118. $\sqrt{2x} + 10 = 4$

119. $\sqrt[3]{2x - 6} = 4$

120. $\sqrt{x + 6} = \sqrt{x + 2}$

121. $2x - 5\sqrt{x} = 3$

122. $\sqrt{x + 9} = 2 + \sqrt{x - 7}$

Find each unknown length.

△ **123.** △ **124.**

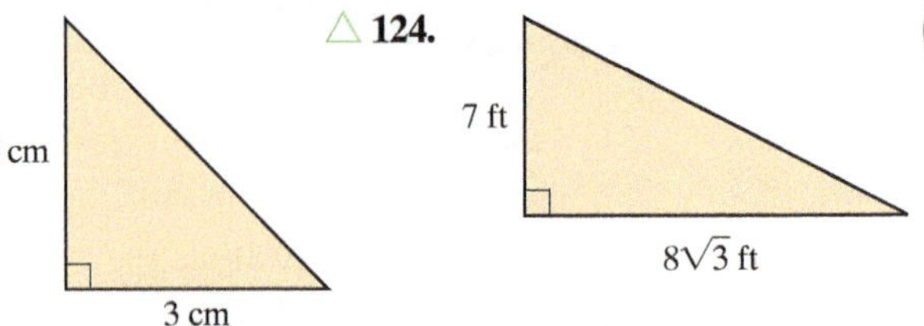

△ **125.** Craig and Daniel Cantwell want to determine the distance x across a pond on their property. They are able to measure the distances shown on the following diagram. Find how wide the pond is at the crossing point indicated by the triangle to the nearest tenth of a foot.

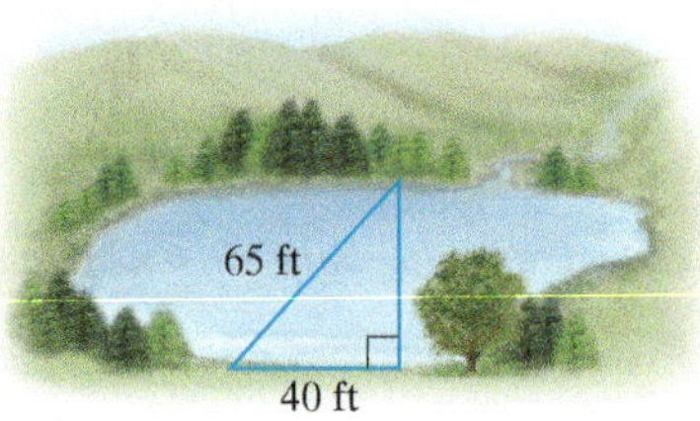

△ **126.** Andrea Roberts, a pipefitter, needs to connect two underground pipelines that are offset by 3 feet, as pictured in the diagram. Neglecting the joints needed to join the pipes, find the length of the shortest possible connecting pipe rounded to the nearest hundredth of a foot.

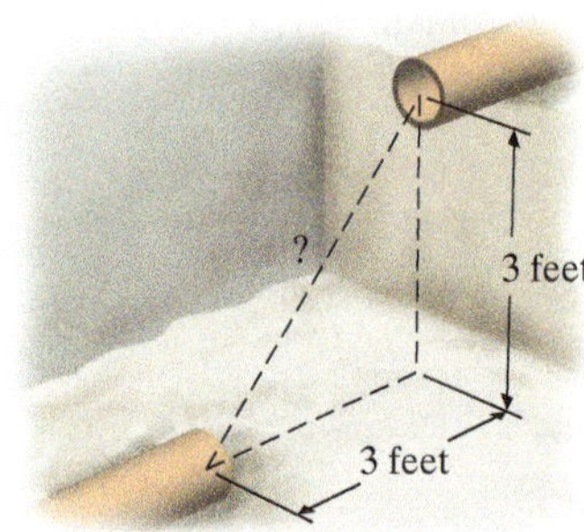

(17.7) *Perform each indicated operation and simplify. Write the results in the form $a + bi$.*

127. $\sqrt{-8}$

128. $-\sqrt{-6}$

129. $\sqrt{-4} + \sqrt{-16}$

130. $\sqrt{-2} \cdot \sqrt{-5}$

131. $(12 - 6i) + (3 + 2i)$

132. $(-8 - 7i) - (5 - 4i)$

133. $(2i)^6$

134. $(3i)^4$

135. $-3i(6 - 4i)$

136. $(3 + 2i)(1 + i)$

137. $(2 - 3i)^2$

138. $(\sqrt{6} - 9i)(\sqrt{6} + 9i)$

139. $\frac{2 + 3i}{2i}$

140. $\frac{1 + i}{-3i}$

MIXED REVIEW

Simplify. Use absolute value bars when necessary.

141. $\sqrt[3]{x^3}$

142. $\sqrt{(x + 2)^2}$

Simplify. Assume that all variables represent positive real numbers. If necessary, write answers with positive exponents only.

143. $-\sqrt{100}$

144. $\sqrt[3]{-x^{12}y^3}$

145. $\sqrt[4]{\dfrac{y^{20}}{16x^{12}}}$

146. $9^{1/2}$

147. $64^{-1/2}$

148. $\left(\dfrac{27}{64}\right)^{-2/3}$

149. $\dfrac{(x^{2/3}x^{-3})^3}{x^{-1/2}}$

150. $\sqrt{200x^9}$

151. $\sqrt{\dfrac{3n^3}{121m^{10}}}$

152. $3\sqrt{20} - 7x\sqrt[3]{40} + 3\sqrt[3]{5x^3}$

153. $(2\sqrt{x} - 5)^2$

154. Find the distance between $(-3, 5)$ and $(-8, 9)$.

155. Find the midpoint of the line segment joining $(-3, 8)$ and $(11, 24)$.

Rationalize each denominator.

156. $\dfrac{7}{\sqrt{13}}$

157. $\dfrac{2}{\sqrt{x} + 3}$

Solve.

158. $\sqrt{x} + 2 = x$

159. $\sqrt{2x - 1} + 2 = x$

Chapter 17 Test

Step-by-step test solutions are found on the Chapter Test Prep Videos. Where available: MyMathLab® or 

Raise to the power or find the root. Assume that all variables represent positive numbers. Write with only positive exponents.

1. $\sqrt{216}$

2. $-\sqrt[4]{x^{64}}$

3. $\left(\dfrac{1}{125}\right)^{1/3}$

4. $\left(\dfrac{1}{125}\right)^{-1/3}$

5. $\left(\dfrac{8x^3}{27}\right)^{2/3}$

6. $\sqrt[3]{-a^{18}b^9}$

7. $\left(\dfrac{64c^{4/3}}{a^{-2/3}b^{5/6}}\right)^{1/2}$

8. $a^{-2/3}(a^{5/4} - a^3)$

Find the root. Use absolute value bars when necessary.

9. $\sqrt[4]{(4xy)^4}$

10. $\sqrt[3]{(-27)^3}$

Rationalize the denominator. Assume that all variables represent positive numbers.

11. $\sqrt{\dfrac{9}{y}}$

12. $\dfrac{4 - \sqrt{x}}{4 + 2\sqrt{x}}$

13. $\dfrac{\sqrt[3]{ab}}{\sqrt[3]{ab^2}}$

14. Rationalize the numerator of $\dfrac{\sqrt{6} + x}{8}$ and simplify.

Perform the indicated operations. Assume that all variables represent positive numbers.

15. $\sqrt{125x^3} - 3\sqrt{20x^3}$

16. $\sqrt{3}(\sqrt{16} - \sqrt{2})$

17. $(\sqrt{x} + 1)^2$

18. $(\sqrt{2} - 4)(\sqrt{3} + 1)$

19. $(\sqrt{5} + 5)(\sqrt{5} - 5)$

Use a calculator to approximate each to three decimal places.

20. $\sqrt{561}$

21. $386^{-2/3}$

Solve.

22. $x = \sqrt{x - 2} + 2$

23. $\sqrt{x^2 - 7} + 3 = 0$

24. $\sqrt[3]{x + 5} = \sqrt[3]{2x - 1}$

Perform the indicated operation and simplify. Write the result in the form $a + bi$.

25. $\sqrt{-2}$

26. $-\sqrt{-8}$

27. $(12 - 6i) - (12 - 3i)$

28. $(6 - 2i)(6 + 2i)$

29. $(4 + 3i)^2$

30. $\dfrac{1 + 4i}{1 - i}$

31. Find x.

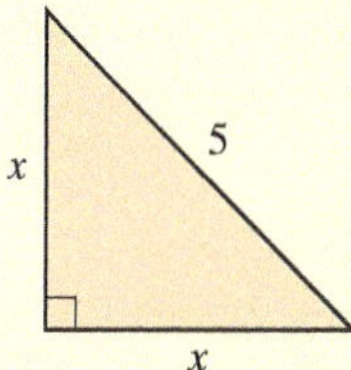

32. Identify the domain of $g(x)$. Then complete the accompanying table and graph $g(x)$.

$$g(x) = \sqrt{x + 2}$$

x	-2	-1	2	7
$g(x)$				

33. Find the distance between the points $(-6, 3)$ and $(-8, -7)$.

34. Find the distance between the points $(-2\sqrt{5}, \sqrt{10})$ and $(-\sqrt{5}, 4\sqrt{10})$.

35. Find the midpoint of the line segment whose endpoints are $(-2, -5)$ and $(-6, 12)$.

36. Find the midpoint of the line segment whose endpoints are $\left(-\dfrac{2}{3}, -\dfrac{1}{5}\right)$ and $\left(-\dfrac{1}{3}, \dfrac{4}{5}\right)$.

Solve.

37. The function $V(r) = \sqrt{2.5r}$ can be used to estimate the maximum safe velocity V in miles per hour at which a car can travel if it is driven along a curved road with a *radius of curvature* r in feet. To the nearest whole number, find the maximum safe speed if a cloverleaf exit on an expressway has a radius of curvature of 300 feet.

38. Use the formula from Exercise 37 to find the radius of curvature if the safe velocity is 30 mph.

Chapters 1–17 Cumulative Review

1. Simplify: $-2[-3 + 2(-1 + 6)] - 5$
2. Simplify each expression.
   a. $2(x - 3) + (5x + 3)$
   b. $4(3x + 2) - 3(5x - 1)$
   c. $7x + 2(x - 7) - 3x$
3. Solve: $\frac{x}{2} - 1 = \frac{2}{3}x - 3$
4. Solve: $\frac{a - 1}{2} + a = 2 - \frac{2a + 7}{8}$
5. A 10-foot board is to be cut into two pieces so that the length of the longer piece is 4 times the length of the shorter. Find the length of each piece.
6. The Smith family owns a lake house 121.5 miles from home. If it takes them $4\frac{1}{2}$ hours to drive round-trip from their house to their lake house, find their average speed.
7. Without graphing, determine the number of solutions of the system.
   $\begin{cases} 3x - y = 4 \\ x + 2y = 8 \end{cases}$
8. Solve: $|3x - 2| + 5 = 5$
9. Solve the system: $\begin{cases} x + 2y = 7 \\ 2x + 2y = 13 \end{cases}$
10. Solve: $\left|\frac{x}{2} - 1\right| \le 0$.
11. Solve the system: $\begin{cases} 2x - y = 7 \\ 8x - 4y = 1 \end{cases}$
12. Graph $y = |x - 2|$.
13. Lynn Pike, a pharmacist, needs 70 liters of a 50% alcohol solution. She has available a 30% alcohol solution and an 80% alcohol solution. How many liters of each solution should she mix to obtain 70 liters of a 50% alcohol solution?
14. Find the domain and the range of each relation. Use the vertical line test to determine whether each graph is the graph of a function.
   a.
   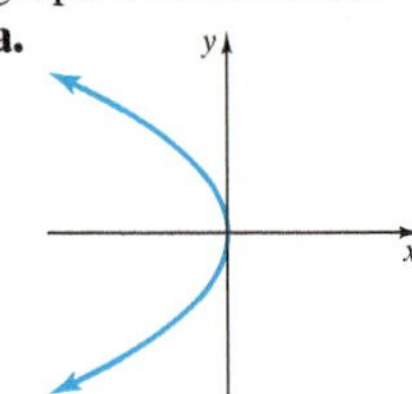

   b.
   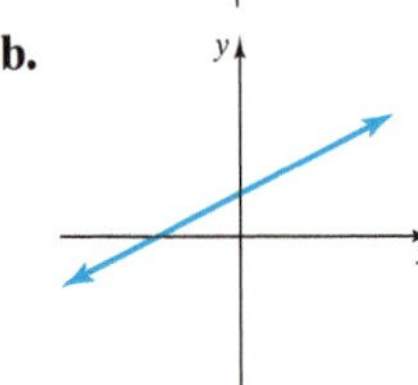

   c.
   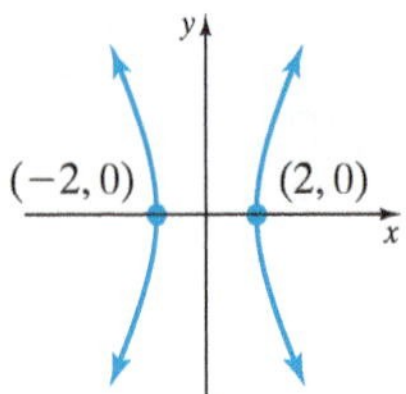

15. If $P(x) = 3x^2 - 2x - 5$, find the following.
   a. $P(1)$
   b. $P(-2)$
16. Graph $f(x) = -2$.
17. Divide $6m^2 + 2m$ by $2m$.
18. Find the slope of $y = -3$.
19. Use synthetic division to divide $2x^3 - x^2 - 13x + 1$ by $x - 3$.
20. Solve the system.
   $\begin{cases} \frac{x}{6} - \frac{y}{2} = 1 \\ \frac{x}{3} - \frac{y}{4} = 2 \end{cases}$
21. Factor $40 - 13t + t^2$.
22. At a seasonal clearance sale, Nana Long spent \$33.75. She paid \$3.50 for T-shirts and \$4.25 for shorts. If she bought 9 items, how many of each item did she buy?
23. Simplify each rational expression.
   a. $\frac{x^3 + 8}{2 + x}$
   b. $\frac{2y^2 + 2}{y^3 - 5y^2 + y - 5}$
24. Use scientific notation to simplify and write the answer in scientific notation. $\frac{0.0000035 \times 4000}{0.28}$
25. Solve: $|x - 3| = |5 - x|$
26. Subtract $(2x - 5)$ from the sum of $(5x^2 - 3x + 6)$ and $(4x^2 + 5x - 3)$.
27. Subtract: $\frac{3x^2 + 2x}{x - 1} - \frac{10x - 5}{x - 1}$
28. Multiply and simplify the product if possible.
   a. $(y - 2)(3y + 4)$
   b. $(3y - 1)(2y^2 + 3y - 1)$
29. Add: $1 + \frac{m}{m + 1}$
30. Factor. $x^3 - x^2 + 4x - 4$

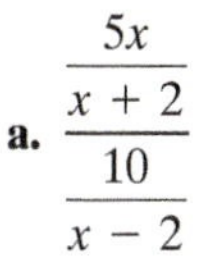

31. Simply each complex fraction.

a. $\dfrac{\dfrac{5x}{x+2}}{\dfrac{10}{x-2}}$

b. $\dfrac{\dfrac{x}{y^2}+\dfrac{1}{y}}{\dfrac{y}{x^2}+\dfrac{1}{x}}$

32. Simplify each rational expression.

a. $\dfrac{a^3-8}{2-a}$

b. $\dfrac{3a^2-3}{a^3+5a^2-a-5}$

33. Solve: $|5x+1|+1 \le 10$

34. Perform the indicated operations.

a. $\dfrac{3}{xy^2}-\dfrac{2}{3x^2y}$

b. $\dfrac{5x}{x+3}-\dfrac{2x}{x-3}$

c. $\dfrac{x}{x-2}-\dfrac{5}{2-x}$

35. If the following two triangles are similar, find the missing length, x.

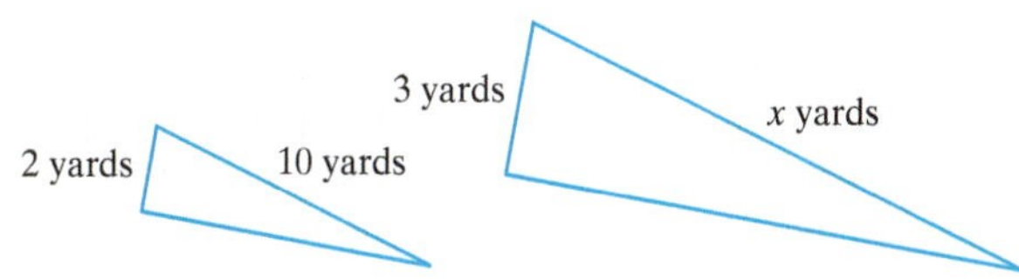

36. Simplify each complex fraction.

a. $\dfrac{\dfrac{y-2}{16}}{\dfrac{2y+3}{12}}$

b. $\dfrac{\dfrac{x}{16}-\dfrac{1}{x}}{1-\dfrac{4}{x}}$

37. Find the cube roots.

a. $\sqrt[3]{1}$

b. $\sqrt[3]{-64}$

c. $\sqrt[3]{\dfrac{8}{125}}$

d. $\sqrt[3]{x^6}$

e. $\sqrt[3]{-27x^9}$

38. Divide x^3-2x^2+3x-6 by $x-2$.

39. Write each expression with a positive exponent, and then simplify.

a. $16^{-3/4}$

b. $(-27)^{-2/3}$

40. Use synthetic division to divide $4y^3-12y^2-y+12$ by $y-3$.

41. Rationalize the numerator of $\dfrac{\sqrt{x}+2}{5}$.

42. Solve: $\dfrac{28}{9-a^2}=\dfrac{2a}{a-3}+\dfrac{6}{a+3}$

43. Suppose that u varies inversely as w. If u is 3 when w is 5, find the constant of variation and the inverse variation equation.

44. Suppose that y varies directly as x. If $y = 0.51$ when $x = 3$, find the constant of variation and the direct variation equation.

18 Quadratic Equations and Functions

Sections

Check Your Progress

An important part of the study of algebra is learning to model and solve problems. Often, the model of a problem is a quadratic equation or a function containing a second-degree polynomial. In this chapter, we continue the work begun in Chapter 13, when we solved polynomial equations in one variable by factoring. Two additional methods of solving quadratic equations are analyzed as are methods of solving nonlinear inequalities in one variable.

Man has always desired to reach the stars, and some buildings seem to be trying to do just that. As populations expand and land becomes scarcer, ever taller and more spectacular buildings are being constructed. As of 2010, the tallest building in the world was the Burj Khalifa, in Dubai. In Exercise 80, Section 18.1, you will explore the height of the Burj Khalifa. (*Source:* Council on Tall Buildings and Urban Habitat, *Fast Company*)

Snapshot of Selected Tall Structures (with completion dates)

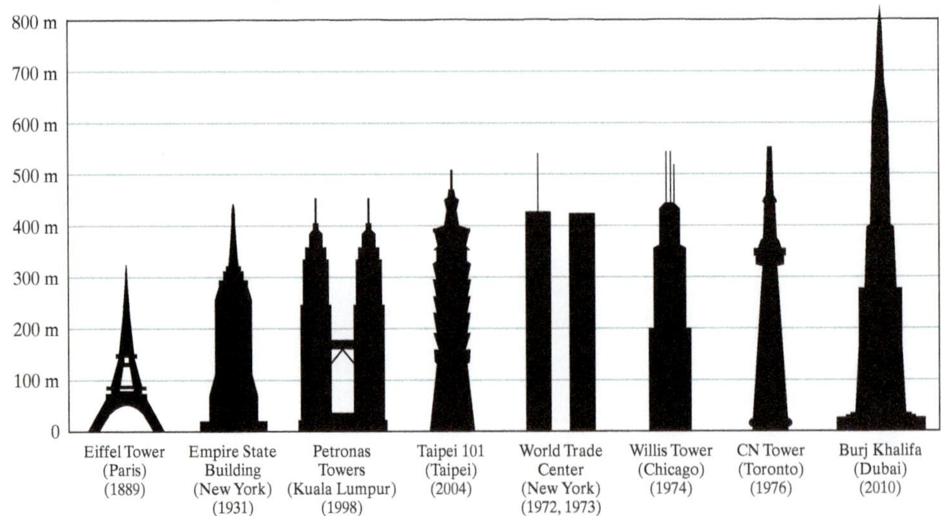

Source: Council on Tall Buildings and Urban Habitat, (*Fast Company*)

18.1 Solving Quadratic Equations by Completing the Square 

Objective 1 Using the Square Root Property

In Chapter 13, we solved quadratic equations by factoring. Recall that a **quadratic,** or **second-degree, equation** is an equation that can be written in the form $ax^2 + bx + c = 0$, where a, b, and c are real numbers and a is not 0. To solve a quadratic equation such as $x^2 = 9$ by factoring, we use the zero factor theorem. To use the zero factor theorem, the equation must first be written in standard form, $ax^2 + bx + c = 0$.

$$x^2 = 9$$
$$x^2 - 9 = 0 \quad \text{Subtract 9 from both sides.}$$
$$(x + 3)(x - 3) = 0 \quad \text{Factor.}$$
$$x + 3 = 0 \quad \text{or} \quad x - 3 = 0 \quad \text{Set each factor equal to 0.}$$
$$x = -3 \qquad x = 3 \quad \text{Solve.}$$

The solution set is $\{-3, 3\}$, the positive and negative square roots of 9. Not all quadratic equations can be solved by factoring, so we need to explore other methods. Notice that the solutions of the equation $x^2 = 9$ are two numbers whose square is 9.

$$3^2 = 9 \quad \text{and} \quad (-3)^2 = 9$$

Thus, we can solve the equation $x^2 = 9$ by taking the square root of both sides. Be sure to include both $\sqrt{9}$ and $-\sqrt{9}$ as solutions since both $\sqrt{9}$ and $-\sqrt{9}$ are numbers whose square is 9.

$$x^2 = 9$$
$$\sqrt{x^2} = \pm\sqrt{9} \quad \text{The notation } \pm\sqrt{9} \text{ (read as "plus or minus } \sqrt{9}\text{") indicates}$$
$$x = \pm 3 \quad \text{the pair of numbers } +\sqrt{9} \text{ and } -\sqrt{9}.$$

This illustrates the square root property.

Square Root Property

If b is a real number and if $a^2 = b$, then $a = \pm\sqrt{b}$.

Helpful Hint

The notation ± 3, for example, is read as "plus or minus 3." It is a shorthand notation for the pair of numbers +3 and −3.

Example 1 Use the square root property to solve $x^2 = 50$.

Solution:

$$x^2 = 50$$
$$x = \pm\sqrt{50} \quad \text{Use the square root property.}$$
$$x = \pm 5\sqrt{2} \quad \text{Simplify the radical.}$$

Check:

Let $x = 5\sqrt{2}$.	Let $x = -5\sqrt{2}$.
$x^2 = 50$	$x^2 = 50$
$(5\sqrt{2})^2 \stackrel{?}{=} 50$	$(-5\sqrt{2})^2 \stackrel{?}{=} 50$
$25 \cdot 2 \stackrel{?}{=} 50$	$25 \cdot 2 \stackrel{?}{=} 50$
$50 = 50$ True	$50 = 50$ True

The solutions are $5\sqrt{2}$ and $-5\sqrt{2}$, or the solution set is $\{-5\sqrt{2}, 5\sqrt{2}\}$.

Work Practice 1

Objectives

1 Use the Square Root Property to Solve Quadratic Equations.

2 Solve Quadratic Equations by Completing the Square.

3 Use Quadratic Equations to Solve Problems.

Practice 1

Use the square root property to solve $x^2 = 32$.

Answer

1. $-4\sqrt{2}, 4\sqrt{2}$

Practice 2

Use the square root property to solve $5x^2 - 50 = 0$.

Example 2 Use the square root property to solve $2x^2 - 14 = 0$.

Solution: First we get the squared variable alone on one side of the equation.

$$2x^2 - 14 = 0$$

$$2x^2 = 14 \quad \text{Add 14 to both sides.}$$

$$x^2 = 7 \quad \text{Divide both sides by 2.}$$

$$x = \pm\sqrt{7} \quad \text{Use the square root property.}$$

Check to see that the solutions are $\sqrt{7}$ and $-\sqrt{7}$, or the solution set is $\{-\sqrt{7}, \sqrt{7}\}$.

Work Practice 2

Practice 3

Use the square root property to solve $(x + 3)^2 = 20$.

Helpful Hint

Don't forget that $-1 \pm 2\sqrt{3}$, for example, means $-1 + 2\sqrt{3}$ and $-1 - 2\sqrt{3}$. In other words, the equation in Example 3 has two solutions.

Example 3 Use the square root property to solve $(x + 1)^2 = 12$.

Solution:

$$(x + 1)^2 = 12$$

$$x + 1 = \pm\sqrt{12} \quad \text{Use the square root property.}$$

$$x + 1 = \pm 2\sqrt{3} \quad \text{Simplify the radical.}$$

$$x = -1 \pm 2\sqrt{3} \quad \text{Subtract 1 from both sides.}$$

Check: Below is a check for $-1 + 2\sqrt{3}$. The check for $-1 - 2\sqrt{3}$ is almost the same and is left for you to do on your own.

$$(x + 1)^2 = 12$$

$$(-1 + 2\sqrt{3} + 1)^2 \stackrel{?}{=} 12$$

$$(2\sqrt{3})^2 \stackrel{?}{=} 12$$

$$4 \cdot 3 \stackrel{?}{=} 12$$

$$12 = 12 \quad \text{True}$$

The solutions are $-1 + 2\sqrt{3}$ and $-1 - 2\sqrt{3}$.

Work Practice 3

Practice 4

Use the square root property to solve $(5x - 2)^2 = -9$.

Example 4 Use the square root property to solve $(2x - 5)^2 = -16$.

Solution:

$$(2x - 5)^2 = -16$$

$$2x - 5 = \pm\sqrt{-16} \quad \text{Use the square root property.}$$

$$2x - 5 = \pm 4i \quad \text{Simplify the radical.}$$

$$2x = 5 \pm 4i \quad \text{Add 5 to both sides.}$$

$$x = \frac{5 \pm 4i}{2} \quad \text{Divide both sides by 2.}$$

The solutions are $\frac{5 + 4i}{2}$ and $\frac{5 - 4i}{2}$, or in standard form $a + bi$ for complex numbers, $\frac{5}{2} + 2i$ and $\frac{5}{2} - 2i$.

Work Practice 4

Answers

2. $-\sqrt{10}, \sqrt{10}$
3. $-3 - 2\sqrt{5}, -3 + 2\sqrt{5}$
4. $\frac{2 + 3i}{5}, \frac{2 - 3i}{5}$ or $\frac{2}{5} \pm \frac{3}{5}i$

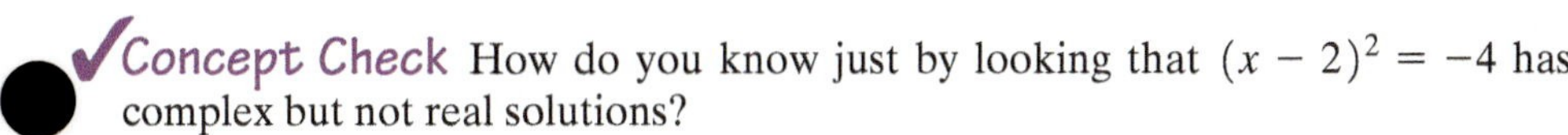

Concept Check How do you know just by looking that $(x - 2)^2 = -4$ has complex but not real solutions?

Objective 2 Solving by Completing the Square 

Notice from Examples 3 and 4 that, if we write a quadratic equation so that one side is the square of a binomial, we can solve by using the square root property. To write the square of a binomial, we write perfect square trinomials. Recall that a perfect square trinomial is a trinomial that can be factored into two identical binomial factors.

Perfect Square Trinomials	Factored Form
$x^2 + 8x + 16$	$(x + 4)^2$
$x^2 - 6x + 9$	$(x - 3)^2$
$x^2 + 3x + \frac{9}{4}$	$\left(x + \frac{3}{2}\right)^2$

Notice that for each perfect square trinomial in x, **the constant term of the trinomial is the square of half the coefficient of the x-term.** For example,

$$x^2 + 8x + 16 \qquad\qquad x^2 - 6x + 9$$

$$\frac{1}{2}(8) = 4 \text{ and } 4^2 = 16 \qquad \frac{1}{2}(-6) = -3 \text{ and } (-3)^2 = 9$$

The process of writing a quadratic equation so that one side is a perfect square trinomial is called **completing the square.**

Example 5 Solve $p^2 + 2p = 4$ by completing the square.

Solution: First, add the square of half the coefficient of p to both sides so that the resulting trinomial will be a perfect square trinomial. The coefficient of p is 2.

$$\frac{1}{2}(2) = 1 \quad \text{and} \quad 1^2 = 1$$

Add 1 to both sides of the original equation.

$p^2 + 2p = 4$

$p^2 + 2p + 1 = 4 + 1$ Add 1 to both sides.

$(p + 1)^2 = 5$ Factor the trinomial; simplify the right side.

We may now use the square root property and solve for p.

$p + 1 = \pm\sqrt{5}$ Use the square root property.

$p = -1 \pm \sqrt{5}$ Subtract 1 from both sides.

Notice that there are two solutions: $-1 + \sqrt{5}$ and $-1 - \sqrt{5}$.

Work Practice 5

Practice 5

Solve $b^2 + 4b = 3$ by completing the square.

Example 6 Solve $m^2 - 7m - 1 = 0$ for m by completing the square.

Solution: First, add 1 to both sides of the equation so that the left side has no constant term.

$m^2 - 7m - 1 = 0$

$m^2 - 7m = 1$

(*Continued on next page*)

Practice 6

Solve $p^2 - 3p + 1 = 0$ by completing the square.

Answers

5. $-2 - \sqrt{7}, -2 + \sqrt{7}$

6. $\frac{3 - \sqrt{5}}{2}, \frac{3 + \sqrt{5}}{2}$

✓Concept Check Answer

answers may vary

Now find the constant term that makes the left side a perfect square trinomial by squaring half the coefficient of m. Add this constant to both sides of the equation.

$$\frac{1}{2}(-7) = -\frac{7}{2} \quad \text{and} \quad \left(-\frac{7}{2}\right)^2 = \frac{49}{4}$$

$m^2 - 7m + \frac{49}{4} = 1 + \frac{49}{4}$ Add $\frac{49}{4}$ to both sides of the equation.

$\left(m - \frac{7}{2}\right)^2 = \frac{53}{4}$ Factor the perfect square trinomial and simplify the right side.

$m - \frac{7}{2} = \pm\sqrt{\frac{53}{4}}$ Apply the square root property.

$m = \frac{7}{2} \pm \frac{\sqrt{53}}{2}$ Add $\frac{7}{2}$ to both sides and simplify $\sqrt{\frac{53}{4}}$.

$m = \frac{7 \pm \sqrt{53}}{2}$ Simplify.

The solutions are $\frac{7 + \sqrt{53}}{2}$ and $\frac{7 - \sqrt{53}}{2}$.

Work Practice 6

The following steps may be used to solve a quadratic equation such as $ax^2 + bx + c = 0$ by completing the square. This method may be used whether or not the polynomial $ax^2 + bx + c$ is factorable.

Solving a Quadratic Equation in x by Completing the Square

Step 1: If the coefficient of x^2 is 1, go to Step 2. Otherwise, divide both sides of the equation by the coefficient of x^2.

Step 2: Isolate all variable terms on one side of the equation.

Step 3: Complete the square for the resulting binomial by adding the square of half of the coefficient of x to both sides of the equation.

Step 4: Factor the resulting perfect square trinomial and write it as the square of a binomial.

Step 5: Use the square root property to solve for x.

Practice 7

Solve: $3x^2 - 12x + 1 = 0$.

Example 7 Solve: $2x^2 - 8x + 3 = 0$.

Solution: Our procedure for finding the constant term to complete the square works only if the coefficient of the squared variable term is 1. Therefore, to solve this equation, the first step is to divide both sides by 2, the coefficient of x^2.

$$2x^2 - 8x + 3 = 0$$

Step 1: $x^2 - 4x + \frac{3}{2} = 0$ Divide both sides by 2.

Step 2: $x^2 - 4x = -\frac{3}{2}$ Subtract $\frac{3}{2}$ from both sides.

Next find the square of half of -4.

$$\frac{1}{2}(-4) = -2 \quad \text{and} \quad (-2)^2 = 4$$

Add 4 to both sides of the equation to complete the square.

Step 3: $x^2 - 4x + 4 = -\frac{3}{2} + 4$

Answer

7. $\frac{6 - \sqrt{33}}{3}, \frac{6 + \sqrt{33}}{3}$

Step 4: $(x - 2)^2 = \frac{5}{2}$ Factor the perfect square and simplify the right side.

Step 5: $x - 2 = \pm\sqrt{\frac{5}{2}}$ Apply the square root property.

$x - 2 = \pm\frac{\sqrt{10}}{2}$ Rationalize the denominator.

$x = 2 \pm \frac{\sqrt{10}}{2}$ Add 2 to both sides.

$= \frac{4}{2} \pm \frac{\sqrt{10}}{2}$ Find a common denominator.

$= \frac{4 \pm \sqrt{10}}{2}$ Simplify.

The solutions are $\frac{4 + \sqrt{10}}{2}$ and $\frac{4 - \sqrt{10}}{2}$.

Work Practice 7

Example 8 Solve $3x^2 - 9x + 8 = 0$ by completing the square.

Solution: $3x^2 - 9x + 8 = 0$

Step 1: $x^2 - 3x + \frac{8}{3} = 0$ Divide both sides of the equation by 3.

Step 2: $x^2 - 3x = -\frac{8}{3}$ Subtract $\frac{8}{3}$ from both sides.

Since $\frac{1}{2}(-3) = -\frac{3}{2}$ and $\left(-\frac{3}{2}\right)^2 = \frac{9}{4}$, we add $\frac{9}{4}$ to both sides of the equation.

Step 3: $x^2 - 3x + \frac{9}{4} = -\frac{8}{3} + \frac{9}{4}$

Step 4: $\left(x - \frac{3}{2}\right)^2 = -\frac{5}{12}$ Factor the perfect square trinomial.

Step 5: $x - \frac{3}{2} = \pm\sqrt{-\frac{5}{12}}$ Apply the square root property.

$x - \frac{3}{2} = \pm\frac{i\sqrt{5}}{2\sqrt{3}}$ Simplify the radical.

$x - \frac{3}{2} = \pm\frac{i\sqrt{15}}{6}$ Rationalize the denominator.

$x = \frac{3}{2} \pm \frac{i\sqrt{15}}{6}$ Add $\frac{3}{2}$ to both sides.

$= \frac{9}{6} \pm \frac{i\sqrt{15}}{6}$ Find a common denominator.

$= \frac{9 \pm i\sqrt{15}}{6}$ Simplify.

The solutions are $\frac{9 + i\sqrt{15}}{6}$ and $\frac{9 - i\sqrt{15}}{6}$, or in $a + bi$ form, $\frac{3}{2} + \frac{\sqrt{15}}{6}i$ and $\frac{3}{2} - \frac{\sqrt{15}}{6}i$.

Work Practice 8

Practice 8

Solve $2x^2 - 5x + 7 = 0$ by completing the square.

Answer

8. $\frac{5 - i\sqrt{31}}{4}, \frac{5 + i\sqrt{31}}{4}$ or $\frac{5}{4} \pm \frac{\sqrt{31}}{4}i$

Objective 3 Solving Problems Modeled by Quadratic Equations

Recall the **simple interest** formula $I = Prt$, where I is the interest earned, P is the principal, r is the rate of interest, and t is time in years. If \$100 is invested at a simple interest rate of 5% annually, at the end of 3 years the total interest I earned is

$$I = P \cdot r \cdot t$$

or

$$I = 100 \cdot 0.05 \cdot 3 = \$15$$

and the new principal is

$$\$100 + \$15 = \$115$$

Most of the time, the interest computed on money borrowed or money deposited is **compound interest.** Compound interest, unlike simple interest, is computed on original principal *and* on interest already earned. To see the difference between simple interest and compound interest, suppose that \$100 is invested at a rate of 5% compounded annually. To find the total amount of money at the end of 3 years, we calculate as follows.

$$I = P \cdot r \cdot t$$

First year: Interest $= \$100 \cdot 0.05 \cdot 1 = \$5.00$
New principal $= \$100.00 + \$5.00 = \$105.00$

Second year: Interest $= \$105.00 \cdot 0.05 \cdot 1 = \$5.25$
New principal $= \$105.00 + \$5.25 = \$110.25$

Third year: Interest $= \$110.25 \cdot 0.05 \cdot 1 \approx \$5.51$
New principal $= \$110.25 + \$5.51 = \$115.76$

At the end of the third year, the total compound interest earned is \$15.76, whereas the total simple interest earned is \$15.

It is tedious to calculate compound interest as we did above, so we use a compound interest formula. The formula for calculating the total amount of money when interest is compounded annually is

$$A = P(1 + r)^t$$

where P is the original investment, r is the interest rate per compounding period, and t is the number of periods. For example, the amount of money A at the end of 3 years if \$100 is invested at 5% compounded annually is

$$A = \$100(1 + 0.05)^3 \approx \$100(1.1576) = \$115.76$$

as we previously calculated.

Practice 9

Use the formula from Example 9 to find the interest rate r if \$5000 compounded annually grows to \$5618 in 2 years.

Example 9 Finding Interest Rates

Use the formula $A = P(1 + r)^t$ to find the interest rate r if \$2000 compounded annually grows to \$2420 in 2 years.

Solution:

1. UNDERSTAND the problem. Since the \$2000 is compounded annually, we use the compound interest formula. For this example, make sure that you understand the formula for compounding interest annually.
2. TRANSLATE. We substitute the given values into the formula.

$$A = P(1 + r)^t$$

$$2420 = 2000(1 + r)^2 \quad \text{Let } A = 2420,\ P = 2000, \text{ and } t = 2.$$

Answer

9. 6%

3. SOLVE. Solve the equation for r.

$$2420 = 2000(1 + r)^2$$

$$\frac{2420}{2000} = (1 + r)^2 \quad \text{Divide both sides by 2000.}$$

$$\frac{121}{100} = (1 + r)^2 \quad \text{Simplify the fraction.}$$

$$\pm\sqrt{\frac{121}{100}} = 1 + r \quad \text{Use the square root property.}$$

$$\pm\frac{11}{10} = 1 + r \quad \text{Simplify.}$$

$$-1 \pm \frac{11}{10} = r$$

$$-\frac{10}{10} \pm \frac{11}{10} = r$$

$$\frac{1}{10} = r \quad \text{or} \quad -\frac{21}{10} = r$$

4. INTERPRET. The rate cannot be negative, so we reject $-\frac{21}{10}$.

Check: $\frac{1}{10} = 0.10 = 10\%$ per year. If we invest \$2000 at 10% compounded annually, in 2 years the amount in the account would be $2000(1 + 0.10)^2 = 2420$ dollars, the desired amount.

State: The interest rate is 10% compounded annually.

Work Practice 9

Graphing Calculator Explorations

In Section 13.6, we showed how we can use a grapher to approximate real number solutions of a quadratic equation written in standard form. We can also use a grapher to solve a quadratic equation when it is not written in standard form. For example, to solve $(x + 1)^2 = 12$, the quadratic equation in Example 3, we graph the following on the same set of axes. Use Xmin $= -10$, Xmax $= 10$, Ymin $= -13$, and Ymax $= 13$.

$$Y_1 = (x + 1)^2 \quad \text{and} \quad Y_2 = 12$$

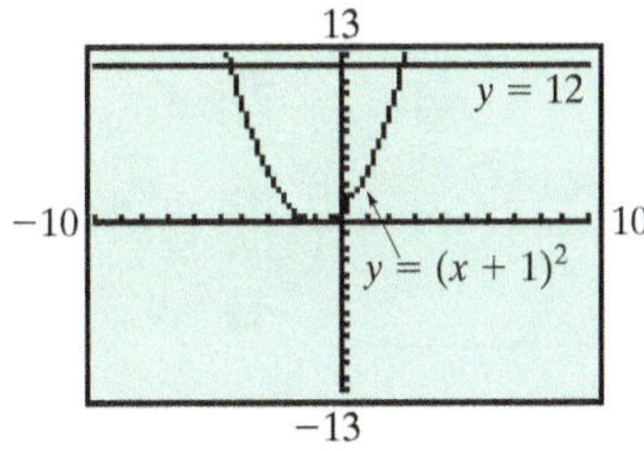

Use the Intersect feature or the Zoom and Trace features to locate the points of intersection of the graphs. (See your manuals for specific instructions.) The x-values of these points are the solutions of $(x + 1)^2 = 12$. The solutions, rounded to two decimal places, are 2.46 and -4.46.

Check to see that these numbers are approximations of the exact solutions $-1 \pm 2\sqrt{3}$.

Use a grapher to solve each quadratic equation. Round all solutions to the nearest hundredth.

1. $x(x - 5) = 8$

2. $x(x + 2) = 5$

3. $x^2 + 0.5x = 0.3x + 1$

4. $x^2 - 2.6x = -2.2x + 3$

5. Use a grapher and solve $(2x - 5)^2 = -6$, Example 4 in this section, using the window

Xmin $= -20$
Xmax $= 20$
Xscl $= 1$
Ymin $= -20$
Ymax $= 20$
Yscl $= 1$

Explain the results. Compare your results with the solution found in Example 4.

6. What are the advantages and disadvantages of using a grapher to solve quadratic equations?

Vocabulary, Readiness & Video Check

Use the choices below to fill in each blank. Not all choices will be used.

binomial	$\sqrt{b}$	$\pm\sqrt{b}$	b^2	9	25	completing the square
quadratic	$-\sqrt{b}$	$\frac{b}{2}$	$\left(\frac{b}{2}\right)^2$	3	5	

1. By the square root property, if b is a real number, and $a^2 = b$, then $a =$ ________________.

2. A ________________ equation can be written in the form $ax^2 + bx + c = 0, a \neq 0$.

3. The process of writing a quadratic equation so that one side is a perfect square trinomial is called ________________.

4. A perfect square trinomial is one that can be factored as a ________________ squared.

5. To solve $x^2 + 6x = 10$ by completing the square, add ________________ to both sides.

6. To solve $x^2 + bx = c$ by completing the square, add ________________ to both sides.

Martin-Gay Interactive Videos *Watch the section lecture video and answer the following questions.*

See Video 18.1

Objective 1 **7.** From Examples 2 and 3, explain a step you can perform so that you may easily apply the square root property to $2x^2 = 16$. Explain why you perform this step.

Objective 2 **8.** In Example 5, why is the equation first divided through by 3?

Objective 3 **9.** In Example 6, why is the negative solution not considered?

18.1 Exercise Set MyMathLab®

Use the square root property to solve each equation. These equations have real number solutions. See Examples 1 through 3.

1. $x^2 = 16$
2. $x^2 = 49$
3. $x^2 - 7 = 0$
4. $x^2 - 11 = 0$
5. $x^2 = 18$
6. $y^2 = 20$
7. $3z^2 - 30 = 0$
8. $2x^2 - 4 = 0$
9. $(x + 5)^2 = 9$
10. $(y - 3)^2 = 4$
11. $(z - 6)^2 = 18$
12. $(y + 4)^2 = 27$
13. $(2x - 3)^2 = 8$
14. $(4x + 9)^2 = 6$

Use the square root property to solve each equation. See Examples 1 through 4.

15. $x^2 + 9 = 0$
16. $x^2 + 4 = 0$
17. $x^2 - 6 = 0$
18. $y^2 - 10 = 0$
19. $2z^2 + 16 = 0$
20. $3p^2 + 36 = 0$
21. $(3x - 1)^2 = -16$
22. $(4y + 2)^2 = -25$
23. $(z + 7)^2 = 5$
24. $(x + 10)^2 = 11$
25. $(x + 3)^2 + 8 = 0$
26. $(y - 4)^2 + 18 = 0$

Add the proper constant to each binomial so that the resulting trinomial is a perfect square trinomial. Then factor the trinomial.

27. $x^2 + 16x +$ ____________
28. $y^2 + 2y +$ ____________
29. $z^2 - 12z +$ ____________
30. $x^2 - 8x +$ ____________
31. $p^2 + 9p +$ ____________
32. $n^2 + 5n +$ ____________
33. $x^2 + x +$ ____________
34. $y^2 - y +$ ____________

MIXED PRACTICE

Solve each equation by completing the square. These equations have real number solutions. See Examples 5 through 7.

35. $x^2 + 8x = -15$
36. $y^2 + 6y = -8$

37. $x^2 + 6x + 2 = 0$

38. $x^2 - 2x - 2 = 0$

39. $x^2 + x - 1 = 0$

40. $x^2 + 3x - 2 = 0$

41. $x^2 + 2x - 5 = 0$

42. $x^2 - 6x + 3 = 0$

43. $y^2 + y - 7 = 0$

44. $x^2 - 7x - 1 = 0$

45. $3p^2 - 12p + 2 = 0$

46. $2x^2 + 14x - 1 = 0$

47. $4y^2 - 2 = 12y$

48. $6x^2 - 3 = 6x$

49. $2x^2 + 7x = 4$

50. $3x^2 - 4x = 4$

51. $x^2 + 8x + 1 = 0$

52. $x^2 - 10x + 2 = 0$

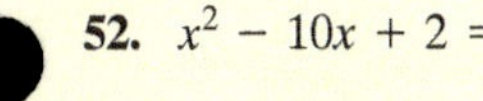

53. $3y^2 + 6y - 4 = 0$

54. $2y^2 + 12y + 3 = 0$

55. $2x^2 - 3x - 5 = 0$

56. $5x^2 + 3x - 2 = 0$

Solve each equation by completing the square. See Examples 5 through 8.

57. $y^2 + 2y + 2 = 0$

58. $x^2 + 4x + 6 = 0$

59. $y^2 + 6y - 8 = 0$

60. $y^2 + 10y - 26 = 0$

61. $2a^2 + 8a = -12$

62. $3x^2 + 12x = -14$

63. $5x^2 + 15x - 1 = 0$

64. $16y^2 + 16y - 1 = 0$

65. $2x^2 - x + 6 = 0$

66. $4x^2 - 2x + 5 = 0$

67. $x^2 + 10x + 28 = 0$

68. $y^2 + 8y + 18 = 0$

69. $z^2 + 3z - 4 = 0$

70. $y^2 + y - 2 = 0$

71. $2x^2 - 4x = -3$

72. $9x^2 - 36x = -40$

73. $3x^2 + 3x = 5$

74. $10y^2 - 30y = 2$

Use the formula $A = P(1 + r)^t$ to solve Exercises 75 through 78. See Example 9.

75. Find the rate r at which \$3000 compounded annually grows to \$4320 in 2 years.

76. Find the rate r at which \$800 compounded annually grows to \$882 in 2 years.

77. Find the rate at which \$15,000 compounded annually grows to \$16,224 in 2 years.

78. Find the rate at which \$2000 compounded annually grows to \$2880 in 2 years.

Neglecting air resistance, the distance $s(t)$ in feet traveled by a freely falling object is given by the function $s(t) = 16t^2$, where t is time in seconds. Use this formula to solve Exercises 79 through 82. Round answers to two decimal places.

79. The Petronas Towers in Kuala Lumpur, completed in 1998, are the tallest buildings in Malaysia. Each tower is 1483 feet tall. How long would it take an object to fall to the ground from the top of one of the towers? (*Source:* Council on Tall Buildings and Urban Habitat, Lehigh University)

80. The Burj Khalifa, the tallest building in the world, was completed in 2010 in Dubai. It is estimated to be 2717 feet tall. How long would it take an object to fall to the ground from the top of the building? (*Source:* Council on Tall Buildings and Urban Habitat)

81. When completed, the Rogun Dam in Tajikistan (part of the former USSR that borders Afghanistan) will be the tallest dam in the world at 1100 feet. How long would it take an object to fall from the top to the base of the dam? (*Source:* U.S. Committee on Large Dams of the International Commission on Large Dams)

82. The Hoover Dam, located on the Colorado River on the border of Nevada and Arizona near Las Vegas, is 725 feet tall. How long would it take an object to fall from the top to the base of the dam? (*Source:* U.S. Committee on Large Dams of the International Commission on Large Dams)

Solve.

△ **83.** The area of a square room is 225 square feet. Find the dimensions of the room.

△ **84.** The area of a circle is 36π square inches. Find the radius of the circle.

△ **85.** An isosceles right triangle has legs of equal length. If the hypotenuse is 20 centimeters long, find the length of each leg.

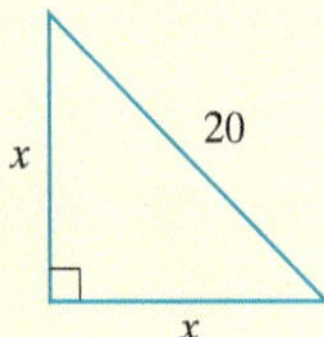

△ **86.** The top of a square coffee table has a diagonal that measures 30 inches. Find the length of each side of the top of the coffee table.

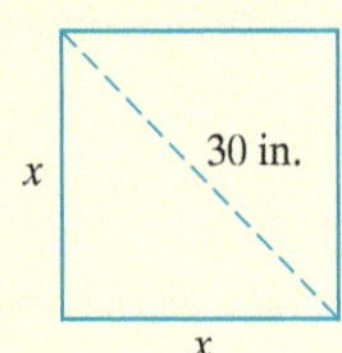

REVIEW AND PREVIEW

Simplify each expression. See Section 17.1.

87. $\frac{1}{2} - \sqrt{\frac{9}{4}}$

88. $\frac{9}{10} - \sqrt{\frac{49}{100}}$

Simplify each expression. See Section 17.5.

89. $\frac{6 + 4\sqrt{5}}{2}$

90. $\frac{10 - 20\sqrt{3}}{2}$

91. $\frac{3 - 9\sqrt{2}}{6}$

92. $\frac{12 - 8\sqrt{7}}{16}$

Evaluate $\sqrt{b^2 - 4ac}$ for each set of values. See Section 17.3.

93. $a = 2, b = 4, c = -1$

94. $a = 1, b = 6, c = 2$

95. $a = 3, b = -1, c = -2$

96. $a = 1, b = -3, c = -1$

CONCEPT EXTENSIONS

Without solving, determine whether the solutions of each equation are real numbers or complex but not real numbers. See the Concept Check in this section.

97. $(x + 1)^2 = -1$

98. $(y - 5)^2 = -9$

99. $3z^2 = 10$

100. $4x^2 = 17$

101. $(2y - 5)^2 + 7 = 3$

102. $(3m + 2)^2 + 4 = 1$

Find two possible missing terms so that each is a perfect square trinomial.

103. $x^2 + \quad + 16$

104. $y^2 + \quad + 9$

105. $z^2 + \quad + \frac{25}{4}$

106. $x^2 + \quad + \frac{1}{4}$

107. In your own words, explain how to calculate the number that will complete the square on an expression such as $x^2 - 5x$.

108. In your own words, what is the difference between simple interest and compound interest?

109. If you are depositing money in an account that pays 4%, would you prefer the interest to be simple or compound? Explain your answer.

110. If you are borrowing money at a rate of 10%, would you prefer the interest to be simple or compound? Explain your answer.

A common equation used in business is a demand equation. It expresses the relationship between the unit price of some commodity and the quantity demanded. For Exercises 111 and 112, p represents the unit price and x represents the quantity demanded in thousands.

111. A manufacturing company has found that the demand equation for a certain type of scissors is given by the equation $p = -x^2 + 47$. Find the demand for the scissors if the price is $11 per pair.

112. Acme, Inc., sells desk lamps and has found that the demand equation for a certain style of desk lamp is given by the equation $p = -x^2 + 15$. Find the demand for the desk lamp if the price is $7 per lamp.

18.2 Solving Quadratic Equations by the Quadratic Formula

Objectives

1 Solve Quadratic Equations by Using the Quadratic Formula.

2 Determine the Number and Type of Solutions of a Quadratic Equation by Using the Discriminant.

3 Solve Problems Modeled by Quadratic Equations.

Objective 1 Solving Quadratic Equations by Using the Quadratic Formula

Any quadratic equation can be solved by completing the square. Since the same sequence of steps is repeated each time we complete the square, let's complete the square for a general quadratic equation, $ax^2 + bx + c = 0, a \neq 0$. By doing so, we find a pattern for the solutions of a quadratic equation known as the **quadratic formula.**

Recall that to complete the square for an equation such as $ax^2 + bx + c = 0$, we first divide both sides by the coefficient of x^2.

$$ax^2 + bx + c = 0$$

$$x^2 + \frac{b}{a}x + \frac{c}{a} = 0 \quad \text{Divide both sides by } a\text{, the coefficient of } x^2.$$

$$x^2 + \frac{b}{a}x = -\frac{c}{a} \quad \text{Subtract the constant } \frac{c}{a} \text{ from both sides.}$$

Next, find the square of half $\frac{b}{a}$, the coefficient of x.

$$\frac{1}{2}\left(\frac{b}{a}\right) = \frac{b}{2a} \quad \text{and} \quad \left(\frac{b}{2a}\right)^2 = \frac{b^2}{4a^2}$$

Add this result to both sides of the equation.

$$x^2 + \frac{b}{a}x + \frac{b^2}{4a^2} = -\frac{c}{a} + \frac{b^2}{4a^2} \quad \text{Add } \frac{b^2}{4a^2} \text{ to both sides.}$$

$$x^2 + \frac{b}{a}x + \frac{b^2}{4a^2} = \frac{-c \cdot 4a}{a \cdot 4a} + \frac{b^2}{4a^2} \quad \text{Find a common denominator on the right side.}$$

$$x^2 + \frac{b}{a}x + \frac{b^2}{4a^2} = \frac{b^2 - 4ac}{4a^2} \quad \text{Simplify the right side.}$$

$$\left(x + \frac{b}{2a}\right)^2 = \frac{b^2 - 4ac}{4a^2} \quad \text{Factor the perfect square trinomial on the left side.}$$

$$x + \frac{b}{2a} = \pm\sqrt{\frac{b^2 - 4ac}{4a^2}} \quad \text{Apply the square root property.}$$

$$x + \frac{b}{2a} = \pm\frac{\sqrt{b^2 - 4ac}}{2a} \quad \text{Simplify the radical.}$$

$$x = -\frac{b}{2a} \pm \frac{\sqrt{b^2 - 4ac}}{2a} \quad \text{Subtract } \frac{b}{2a} \text{ from both sides.}$$

$$x = \frac{-b \pm \sqrt{b^2 - 4ac}}{2a} \quad \text{Simplify.}$$

This equation identifies the solutions of the general quadratic equation in standard form and is called the quadratic formula. It can be used to solve any equation written in standard form $ax^2 + bx + c = 0$ as long as a is not 0.

Quadratic Formula

A quadratic equation written in the form $ax^2 + bx + c = 0$ has the solutions

$$x = \frac{-b \pm \sqrt{b^2 - 4ac}}{2a}$$

Practice 1

Solve $3x^2 - 5x - 2 = 0$ for x.

Example 1 Solve $3x^2 + 16x + 5 = 0$ for x.

Solution: This equation is in standard form, so $a = 3, b = 16$, and $c = 5$. Substitute these values into the quadratic formula.

$$x = \frac{-b \pm \sqrt{b^2 - 4ac}}{2a} \quad \text{Quadratic formula}$$

$$= \frac{-16 \pm \sqrt{16^2 - 4(3)(5)}}{2 \cdot 3} \quad \text{Use } a = 3, b = 16, \text{ and } c = 5.$$

$$= \frac{-16 \pm \sqrt{256 - 60}}{6}$$

$$= \frac{-16 \pm \sqrt{196}}{6} = \frac{-16 \pm 14}{6}$$

$$x = \frac{-16 + 14}{6} = -\frac{1}{3} \quad \text{or} \quad x = \frac{-16 - 14}{6} = -\frac{30}{6} = -5$$

The solutions are $-\frac{1}{3}$ and -5, or the solution set is $\left\{-\frac{1}{3}, -5\right\}$.

Work Practice 1

Practice 2

Solve: $3x^2 - 8x = 2$.

Helpful Hint

To replace a, b, and c correctly in the quadratic formula, write the quadratic equation in standard form $ax^2 + bx + c = 0$.

Example 2 Solve: $2x^2 - 4x = 3$.

Solution: First write the equation in standard form by subtracting 3 from both sides.

$$2x^2 - 4x - 3 = 0$$

Now $a = 2, b = -4$, and $c = -3$. Substitute these values into the quadratic formula.

$$x = \frac{-b \pm \sqrt{b^2 - 4ac}}{2a}$$

$$= \frac{-(-4) \pm \sqrt{(-4)^2 - 4(2)(-3)}}{2 \cdot 2}$$

$$= \frac{4 \pm \sqrt{16 + 24}}{4}$$

$$= \frac{4 \pm \sqrt{40}}{4} = \frac{4 \pm 2\sqrt{10}}{4}$$

$$= \frac{2(2 \pm \sqrt{10})}{2 \cdot 2} = \frac{2 \pm \sqrt{10}}{2}$$

The solutions are $\frac{2 + \sqrt{10}}{2}$ and $\frac{2 - \sqrt{10}}{2}$, or the solution set is $\left\{\frac{2 - \sqrt{10}}{2}, \frac{2 + \sqrt{10}}{2}\right\}$.

Work Practice 2

Helpful Hint

To simplify the expression $\frac{4 \pm 2\sqrt{10}}{4}$ in the preceding example, note that 2 is factored out of both terms of the numerator *before* simplifying.

$$\frac{4 \pm 2\sqrt{10}}{4} = \frac{2(2 \pm \sqrt{10})}{2 \cdot 2} = \frac{2 \pm \sqrt{10}}{2}$$

Answers

1. $2, -\frac{1}{3}$ **2.** $\frac{4 - \sqrt{22}}{3}, \frac{4 + \sqrt{22}}{3}$

✓**Concept Check** For the quadratic equation $x^2 = 7$, which substitution is correct?

a. $a = 1, b = 0$, and $c = -7$
b. $a = 1, b = 0$, and $c = 7$
c. $a = 0, b = 0$, and $c = 7$
d. $a = 1, b = 1$, and $c = -7$

Example 3 Solve: $\frac{1}{4}m^2 - m + \frac{1}{2} = 0$.

Solution: We could use the quadratic formula with $a = \frac{1}{4}$, $b = -1$, and $c = \frac{1}{2}$. Instead, we find a simpler, equivalent standard form equation whose coefficients are not fractions.

Multiply both sides of the equation by the LCD, 4, to clear fractions.

$$4\left(\frac{1}{4}m^2 - m + \frac{1}{2}\right) = 4 \cdot 0$$

$$m^2 - 4m + 2 = 0 \quad \text{Simplify.}$$

Substitute $a = 1, b = -4$, and $c = 2$ into the quadratic formula and simplify.

$$m = \frac{-(-4) \pm \sqrt{(-4)^2 - 4(1)(2)}}{2 \cdot 1} = \frac{4 \pm \sqrt{16 - 8}}{2}$$

$$= \frac{4 \pm \sqrt{8}}{2} = \frac{4 \pm 2\sqrt{2}}{2} = \frac{2(2 \pm \sqrt{2})}{2}$$

$$= 2 \pm \sqrt{2}$$

The solutions are $2 + \sqrt{2}$ and $2 - \sqrt{2}$.

Work Practice 3

Practice 3

Solve: $\frac{1}{8}x^2 - \frac{1}{4}x - 2 = 0$.

Example 4 Solve: $x = -3x^2 - 3$.

Solution: The equation in standard form is $3x^2 + x + 3 = 0$. Thus, let $a = 3$, $b = 1$, and $c = 3$ in the quadratic formula.

$$x = \frac{-1 \pm \sqrt{1^2 - 4(3)(3)}}{2 \cdot 3} = \frac{-1 \pm \sqrt{1 - 36}}{6} = \frac{-1 \pm \sqrt{-35}}{6} = \frac{-1 \pm i\sqrt{35}}{6}$$

The solutions are $\frac{-1 + i\sqrt{35}}{6}$ and $\frac{-1 - i\sqrt{35}}{6}$, or in the form $a + bi$, $-\frac{1}{6} + \frac{\sqrt{35}}{6}i$ and $-\frac{1}{6} - \frac{\sqrt{35}}{6}i$.

Work Practice 4

Practice 4

Solve: $x = -2x^2 - 2$.

✓**Concept Check** What is the first step in solving $-3x^2 = 5x - 4$ using the quadratic formula?

In Example 1, the equation $3x^2 + 16x + 5 = 0$ had 2 real roots, $-\frac{1}{3}$ and -5. In Example 4, the equation $3x^2 + x + 3 = 0$ (written in standard form) had no real roots. How do their related graphs compare? Recall that the x-intercepts of $f(x) = 3x^2 + 16x + 5$ occur where $f(x) = 0$ or where $3x^2 + 16x + 5 = 0$. Since

Answers

3. $1 - \sqrt{17}, 1 + \sqrt{17}$

4. $\frac{-1 - i\sqrt{15}}{4}, \frac{-1 + i\sqrt{15}}{4}$ or $-\frac{1}{4} \pm \frac{\sqrt{15}}{4}i$

✓**Concept Check Answers**

a

Write the equation in standard form.

this equation has 2 real roots, the graph has 2 x-intercepts. Similarly, since the equation $3x^2 + x + 3 = 0$ has no real roots, the graph of $f(x) = 3x^2 + x + 3$ has no x-intercepts.

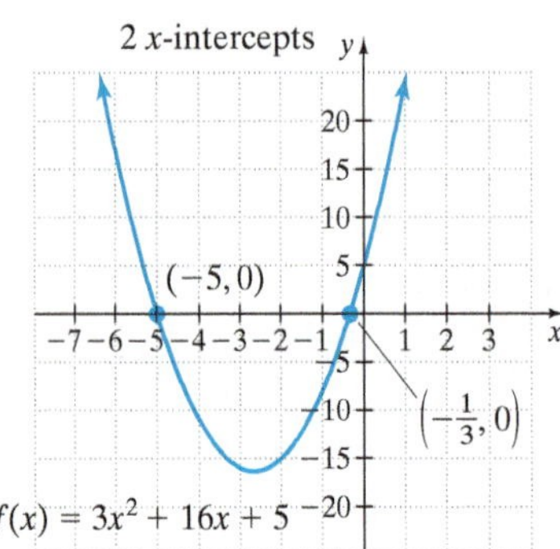

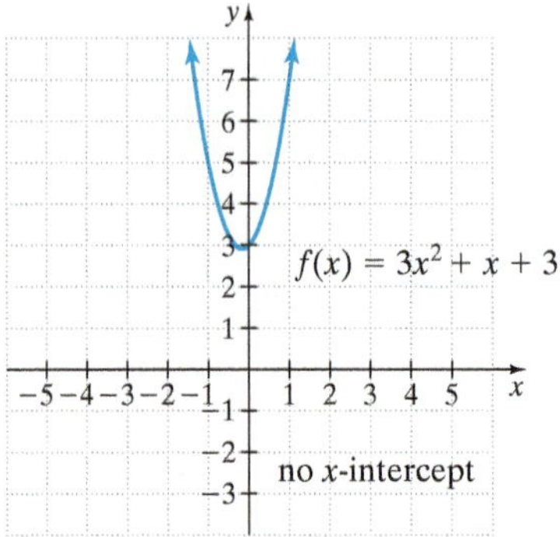

Objective 2 Using the Discriminant 

In the quadratic formula, $x = \dfrac{-b \pm \sqrt{b^2 - 4ac}}{2a}$, the radicand $b^2 - 4ac$ is called the **discriminant** because, by knowing its value, we can **discriminate** among the possible number and type of solutions of a quadratic equation. Possible values of the discriminant and their meanings are summarized next.

Discriminant

The following table corresponds the discriminant $b^2 - 4ac$ of a quadratic equation of the form $ax^2 + bx + c = 0$ with the number and type of solutions of the equation.

$b^2 - 4ac$	Number and Type of Solutions
Positive	Two real solutions
Zero	One real solution
Negative	Two complex but not real solutions

Practice 5

Use the discriminant to determine the number and type of solutions of each quadratic equation.

a. $x^2 - 6x + 9 = 0$
b. $x^2 - 3x - 1 = 0$
c. $7x^2 + 11 = 0$

Example 5 Use the discriminant to determine the number and type of solutions of each quadratic equation.

a. $x^2 + 2x + 1 = 0$ **b.** $3x^2 + 2 = 0$ **c.** $2x^2 - 7x - 4 = 0$

Solution:

a. In $x^2 + 2x + 1 = 0$, $a = 1$, $b = 2$, and $c = 1$. Thus,

$$b^2 - 4ac = 2^2 - 4(1)(1) = 0$$

Since $b^2 - 4ac = 0$, this quadratic equation has one real solution.

b. In this equation, $a = 3$, $b = 0$, $c = 2$. Then $b^2 - 4ac = 0 - 4(3)(2) = -24$. Since $b^2 - 4ac$ is negative, the quadratic equation has two complex but not real solutions.

c. In this equation, $a = 2$, $b = -7$, and $c = -4$. Then

$$b^2 - 4ac = (-7)^2 - 4(2)(-4) = 81$$

Since $b^2 - 4ac$ is positive, the quadratic equation has two real solutions.

Work Practice 5

Answers

5. a. one real solution **b.** two real solutions **c.** two complex but not real solutions

The discriminant helps us determine the number and type of solutions of a quadratic equation, $ax^2 + bx + c = 0$. Recall that the solutions of this equation are the same as the x-intercepts of its related graph $f(x) = ax^2 + bx + c$. This means that the discriminant of $ax^2 + bx + c = 0$ also tells us the number of x-intercepts for the graph of $f(x) = ax^2 + bx + c$ or, equivalently, $y = ax^2 + bx + c$.

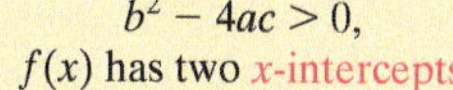

Graph of $f(x) = ax^2 + bx + c$ or $y = ax^2 + bx + c$

$b^2 - 4ac > 0$, $f(x)$ has two x-intercepts

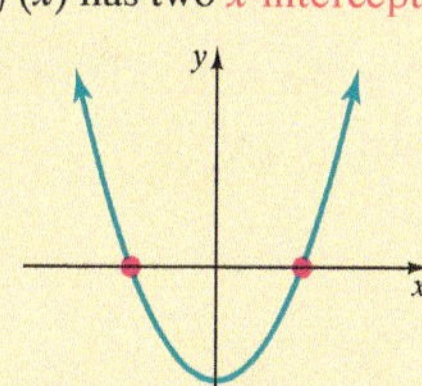

$b^2 - 4ac = 0$, $f(x)$ has one x-intercept

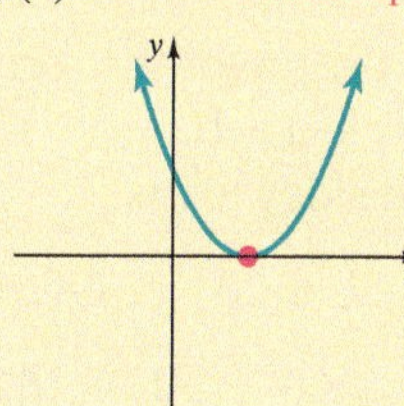

$b^2 - 4ac < 0$, $f(x)$ has no x-intercepts

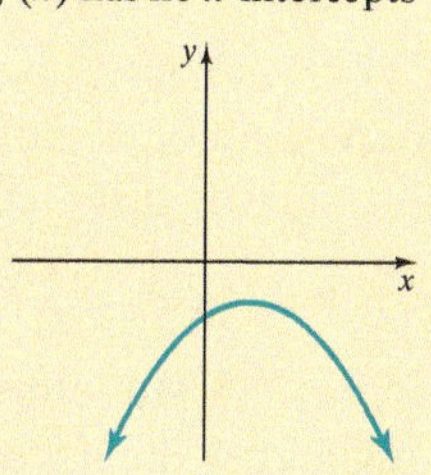

Objective 3 Solving Problems Modeled by Quadratic Equations

The quadratic formula is useful in solving problems that are modeled by quadratic equations.

Example 6 Calculating Distance Saved

At a local university, students often leave the sidewalk and cut across the lawn to save walking distance. Given the diagram below of a favorite place to cut across the lawn, approximate how many feet of walking distance a student saves by cutting across the lawn instead of walking on the sidewalk.

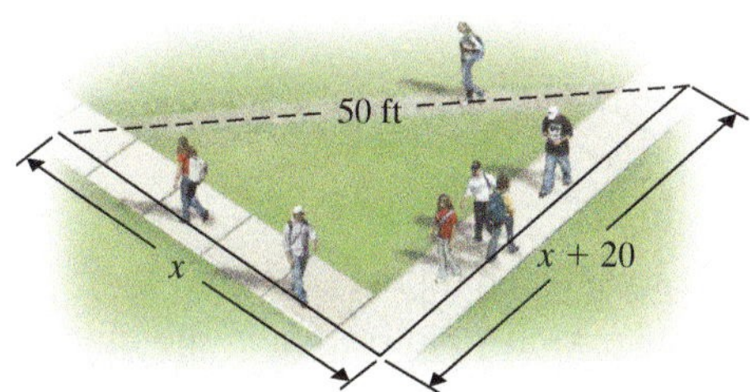

Solution:

1. UNDERSTAND. Read and reread the problem. In the diagram, notice that a triangle is formed. Since the corner of the block forms a right angle, we use the Pythagorean theorem for right triangles. You may want to review this theorem.
2. TRANSLATE. By the Pythagorean theorem, we have

   In words: $(\text{leg})^2 + (\text{leg})^2 = (\text{hypotenuse})^2$

   Translate: $x^2 + (x + 20)^2 = 50^2$
3. SOLVE. Use the quadratic formula to solve.

$$x^2 + x^2 + 40x + 400 = 2500 \quad \text{Square } (x + 20) \text{ and } 50.$$
$$2x^2 + 40x - 2100 = 0 \quad \text{Set the equation equal to 0.}$$
$$x^2 + 20x - 1050 = 0 \quad \text{Divide by 2.}$$

(*Continued on next page*)

Practice 6

Given the diagram, find how many feet of walking distance a person can save by cutting across the lawn instead of walking on the sidewalk.

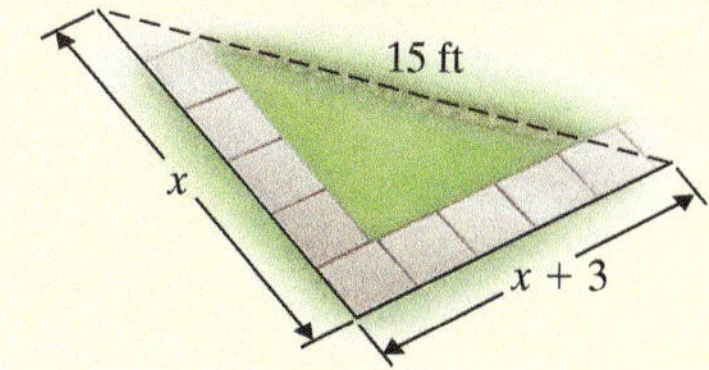

Answer

6. 6 ft

Here, $a = 1, b = 20, c = -1050$. By the quadratic formula,

$$x = \frac{-20 \pm \sqrt{20^2 - 4(1)(-1050)}}{2 \cdot 1}$$

$$= \frac{-20 \pm \sqrt{400 + 4200}}{2} = \frac{-20 \pm \sqrt{4600}}{2}$$

$$= \frac{-20 \pm \sqrt{100 \cdot 46}}{2} = \frac{-20 \pm 10\sqrt{46}}{2}$$

$$= -10 \pm 5\sqrt{46} \quad \text{Simplify.}$$

4. INTERPRET

Check: Your calculations in the quadratic formula. The length of a side of a triangle can't be negative, so we reject $-10 - 5\sqrt{46}$. Since $-10 + 5\sqrt{46} \approx 24$ feet, the walking distance along the sidewalk is

$$x + (x + 20) \approx 24 + (24 + 20) = 68 \text{ feet}$$

State: A student saves about $68 - 50$ or 18 feet of walking distance by cutting across the lawn.

Work Practice 6

Practice 7

A toy rocket is shot upward from the top of a building, 45 feet high, with an initial velocity of 20 feet per second. The height h in feet of the rocket after t seconds is

$$h = -16t^2 + 20t + 45$$

How long after the rocket is launched will it strike the ground? Round to the nearest tenth of a second.

Example 7 Calculating Landing Time

An object is thrown upward from the top of a 200-foot cliff with a velocity of 12 feet per second. The height h in feet of the object after t seconds is

$$h = -16t^2 + 12t + 200$$

How long after the object is thrown will it strike the ground? Round to the nearest tenth of a second.

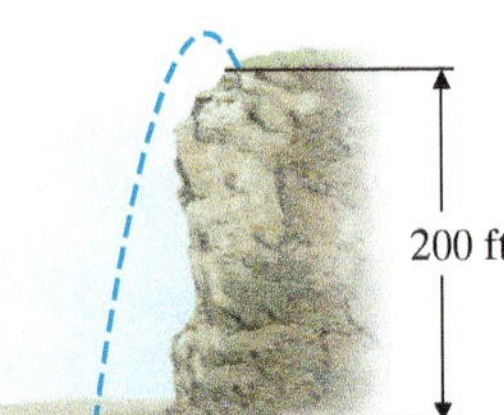

Solution:

1. UNDERSTAND. Read and reread the problem.

2. TRANSLATE. Since we want to know when the object strikes the ground, we want to know when the height $h = 0$, or

$$0 = -16t^2 + 12t + 200$$

3. SOLVE. First we divide both sides of the equation by -4.

$$0 = 4t^2 - 3t - 50 \quad \text{Divide both sides by } -4.$$

Here, $a = 4$, $b = -3$, and $c = -50$. By the quadratic formula,

$$t = \frac{-(-3) \pm \sqrt{(-3)^2 - 4(4)(-50)}}{2 \cdot 4}$$

$$= \frac{3 \pm \sqrt{9 + 800}}{8}$$

$$= \frac{3 \pm \sqrt{809}}{8}$$

Answer

7. 2.4 sec

4. INTERPRET.

Check: We check our calculations from the quadratic formula. Since the time won't be negative, we reject the proposed solution

$$\frac{3 - \sqrt{809}}{8}$$

State: The time it takes for the object to strike the ground is exactly

$$\frac{3 + \sqrt{809}}{8} \text{ seconds} \approx 3.9 \text{ seconds}$$

Work Practice 7

Vocabulary, Readiness & Video Check

Fill in each blank.

1. The quadratic formula is ____________.

2. For $2x^2 + x + 1 = 0$, if $a = 2$, then $b =$ ________ and $c =$ ________.

3. For $5x^2 - 5x - 7 = 0$, if $a = 5$, then $b =$ ________ and $c =$ ________.

4. For $7x^2 - 4 = 0$, if $a = 7$, then $b =$ ________ and $c =$ ________.

5. For $x^2 + 9 = 0$, if $c = 9$, then $a =$ ________ and $b =$ ________.

6. The correct simplified form of $\frac{5 \pm 10\sqrt{2}}{5}$ is ________.

a. $1 \pm 10\sqrt{2}$ **b.** $2\sqrt{2}$ **c.** $1 \pm 2\sqrt{2}$ **d.** $\pm 5\sqrt{2}$

Martin-Gay Interactive Videos

See Video 18.2

Watch the section lecture video and answer the following questions.

Objective 1 **7.** Based on Examples 1–3, answer the following.

a. Must a quadratic equation be written in standard form in order to use the quadratic formula? Why or why not?

b. Must fractions be cleared from an equation before using the quadratic formula? Why or why not?

Objective 2 **8.** Based on Example 4 and the lecture before, complete the following statements. The discriminant is the ________ in the quadratic formula and can be used to find the number and type of solutions of a quadratic equation without ________ the equation. To use the discriminant, the quadratic equation needs to be written in ________ form.

Objective 3 **9.** In Example 5, the value of x is found, which is then used to find the dimensions of the triangle. Yet all this work still does solve the problem. Explain.

18.2 Exercise Set MyMathLab®

Use the quadratic formula to solve each equation. These equations have real number solutions only. See Examples 1 through 3.

1. $m^2 + 5m - 6 = 0$
2. $p^2 + 11p - 12 = 0$
3. $2y = 5y^2 - 3$
4. $5x^2 - 3 = 14x$
5. $x^2 - 6x + 9 = 0$
6. $y^2 + 10y + 25 = 0$
7. $x^2 + 7x + 4 = 0$
8. $y^2 + 5y + 3 = 0$
9. $8m^2 - 2m = 7$
10. $11n^2 - 9n = 1$
11. $3m^2 - 7m = 3$
12. $x^2 - 13 = 5x$
13. $\frac{1}{2}x^2 - x - 1 = 0$
14. $\frac{1}{6}x^2 + x + \frac{1}{3} = 0$
15. $\frac{2}{5}y^2 + \frac{1}{5}y = \frac{3}{5}$
16. $\frac{1}{8}x^2 + x = \frac{5}{2}$
17. $\frac{1}{3}y^2 = y + \frac{1}{6}$
18. $\frac{1}{2}y^2 = y + \frac{1}{2}$
19. $x^2 + 5x = -2$
20. $y^2 - 8 = 4y$
21. $(m + 2)(2m - 6) = 5(m - 1) - 12$
22. $7p(p - 2) + 2(p + 4) = 3$

MIXED PRACTICE

Use the quadratic formula to solve each equation. These equations have real solutions and complex but not real solutions. See Examples 1 through 4.

23. $x^2 + 6x + 13 = 0$
24. $x^2 + 2x + 2 = 0$
25. $(x + 5)(x - 1) = 2$
26. $x(x + 6) = 2$
27. $6 = -4x^2 + 3x$
28. $2 = -9x^2 - x$
29. $\frac{x^2}{3} - x = \frac{5}{3}$
30. $\frac{x^2}{2} - 3 = -\frac{9}{2}x$
31. $10y^2 + 10y + 3 = 0$
32. $3y^2 + 6y + 5 = 0$
33. $x(6x + 2) = 3$
34. $x(7x + 1) = 2$
35. $\frac{2}{5}y^2 + \frac{1}{5}y + \frac{3}{5} = 0$
36. $\frac{1}{8}x^2 + x + \frac{5}{2} = 0$
37. $\frac{1}{2}y^2 = y - \frac{1}{2}$
38. $\frac{2}{3}x^2 - \frac{20}{3}x = -\frac{100}{6}$
39. $(n - 2)^2 = 2n$
40. $\left(p - \frac{1}{2}\right)^2 = \frac{p}{2}$

Use the discriminant to determine the number and type of solutions of each equation. See Example 5.

41. $x^2 - 5 = 0$
42. $x^2 - 7 = 0$
43. $4x^2 + 12x = -9$
44. $9x^2 + 1 = 6x$
45. $3x = -2x^2 + 7$
46. $3x^2 = 5 - 7x$
47. $6 = 4x - 5x^2$
48. $8x = 3 - 9x^2$
49. $9x - 2x^2 + 5 = 0$
50. $5 - 4x + 12x^2 = 0$

Solve. See Examples 7 and 8.

51. Nancy, Thelma, and John Varner live on a corner lot. Often, neighborhood children cut across their lot to save walking distance. Given the diagram below, approximate to the nearest foot how many feet of walking distance is saved by cutting across their property instead of walking around the lot.

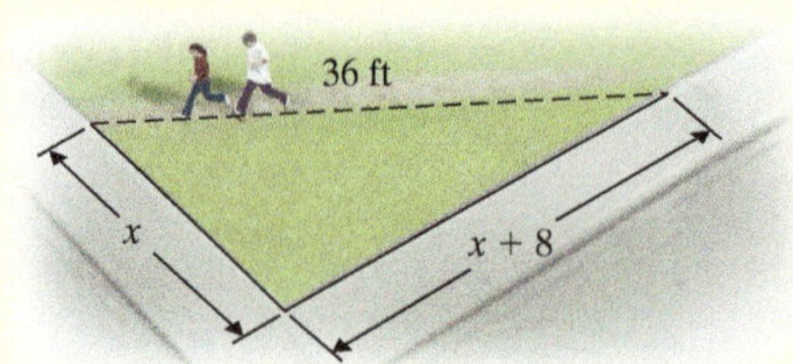

△ **52.** Given the diagram below, approximate to the nearest foot how many feet of walking distance a person saves by cutting across the lawn instead of walking on the sidewalk.

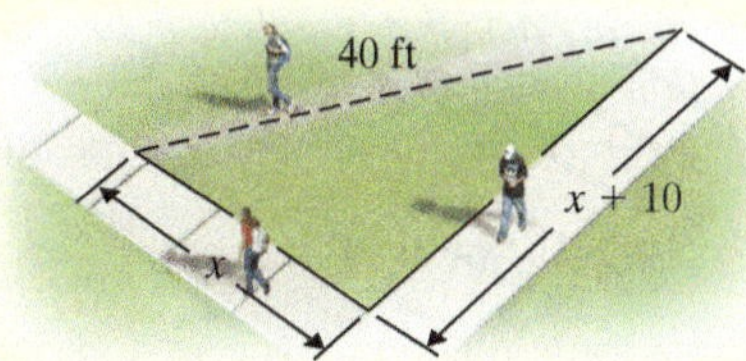

△ **53.** The hypotenuse of an isosceles right triangle is 2 centimeters longer than either of its legs. Find the exact length of each side. (*Hint:* An isosceles right triangle is a right triangle whose legs are the same length.)

△ **54.** The hypotenuse of an isosceles right triangle is one meter longer than either of its legs. Find the length of each side.

△ **55.** Bailey's rectangular dog pen for his Irish setter must have an area of 400 square feet. Also, the length must be 10 feet longer than the width. Find the dimensions of the pen.

△ **56.** An entry in the Peach Festival Poster Contest must be rectangular and have an area of 1200 square inches. Furthermore, its length must be 20 inches longer than its width. Find the dimensions each entry must have.

△ **57.** A holding pen for cattle must be square and have a diagonal length of 100 meters.

a. Find the length of a side of the pen.

b. Find the area of the pen.

△ **58.** A rectangle is three times longer than it is wide. It has a diagonal of length 50 centimeters.

a. Find the dimensions of the rectangle.

b. Find the perimeter of the rectangle.

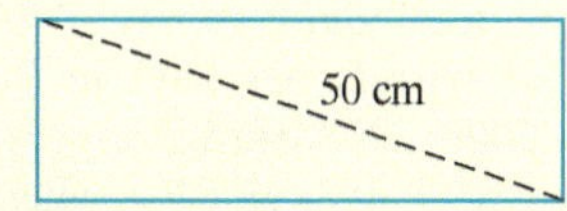

△ **59.** The heaviest reported door in the world is the 708.6 ton radiation shield door in the National Institute for Fusion Science at Toki, Japan. If the height of the door is 1.1 feet longer than its width, and its front area (neglecting depth) is 1439.9 square feet, find its width and height [Interesting note: The door is 6.6 feet thick.] (*Source: Guinness World Records*)

△ **60.** Christi and Robbie Wegmann are constructing a rectangular stained glass window whose length is 7.3 inches longer than its width. If the area of the window is 569.9 square inches, find its width and length.

△ **61.** The base of a triangle is four more than twice its height. If the area of the triangle is 42 square centimeters, find its base and height.

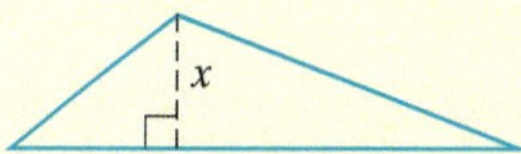

62. If a point B divides a line segment such that the smaller portion is to the larger portion as the larger is to the whole, the whole is the length of the *golden ratio.*

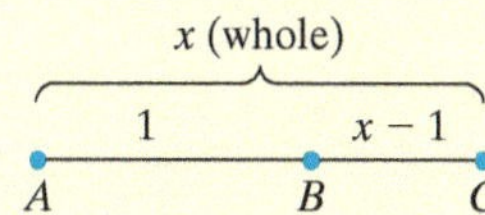

The golden ratio was thought by the Greeks to be the most pleasing to the eye, and many of their buildings contained numerous examples of the golden ratio. The value of the golden ratio is the positive solution of

$$\begin{matrix}\text{(smaller)} \\ \text{(larger)}\end{matrix}\ \frac{x-1}{1} = \frac{1}{x}\ \begin{matrix}\text{(larger)} \\ \text{(whole)}\end{matrix}$$

Find this value.

The Wollomombi Falls in Australia have a height of 1100 feet. A pebble is thrown upward from the top of the falls with an initial velocity of 20 feet per second. The height of the pebble h after t seconds is given by the equation $h = -16t^2 + 20t + 1100$. Use this equation for Exercises 63 and 64.

63. How long after the pebble is thrown will it hit the ground? Round to the nearest tenth of a second.

64. How long after the pebble is thrown will it be 550 feet from the ground? Round to the nearest tenth of a second.

A ball is thrown downward from the top of a 180-foot building with an initial velocity of 20 feet per second. The height of the ball h after t seconds is given by the equation $h = -16t^2 - 20t + 180$. Use this equation to answer Exercises 65 and 66.

65. How long after the ball is thrown will it strike the ground? Round the result to the nearest tenth of a second.

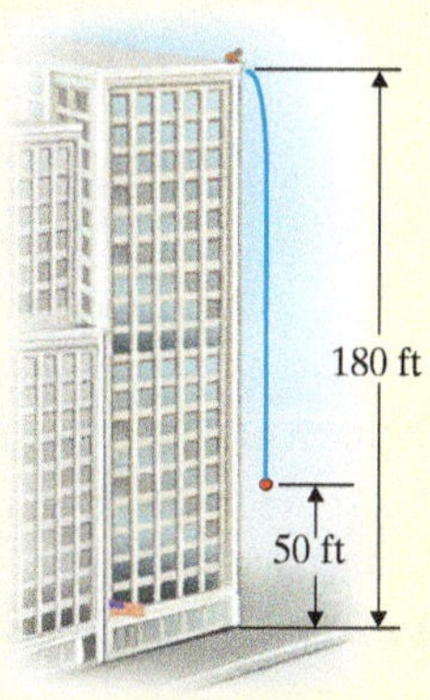

66. How long after the ball is thrown will it be 50 feet from the ground? Round the result to the nearest tenth of a second.

REVIEW AND PREVIEW

Solve each equation. See Sections 12.5 and 17.6.

67. $\sqrt{5x-2}=3$

68. $\sqrt{y+2}+7=12$

69. $\frac{1}{x}+\frac{2}{5}=\frac{7}{x}$

70. $\frac{10}{z}=\frac{5}{z}-\frac{1}{3}$

Factor. See Sections 13.3 through 13.5.

71. x^4+x^2-20

72. $2y^4+11y^2-6$

73. z^4-13z^2+36

74. x^4-1

CONCEPT EXTENSIONS

For each quadratic equation, choose the correct substitution for a, b, and c in the standard form $ax^2+bx+c=0$.

75. $x^2=-10$

a. $a=1, b=0, c=-10$
b. $a=1, b=0, c=10$
c. $a=0, b=1, c=-10$
d. $a=1, b=1, c=10$

76. $x^2+5=-x$

a. $a=1, b=5, c=-1$
b. $a=1, b=-1, c=5$
c. $a=1, b=5, c=1$
d. $a=1, b=1, c=5$

77. Solve Exercise 1 by factoring. Explain the result.

78. Solve Exercise 2 by factoring. Explain the result.

Use the quadratic formula and a calculator to approximate each solution to the nearest tenth.

79. $2x^2-6x+3=0$

80. $3.6x^2+1.8x-4.3=0$

The accompanying graph shows the daily low temperatures for one week in New Orleans, Louisiana.

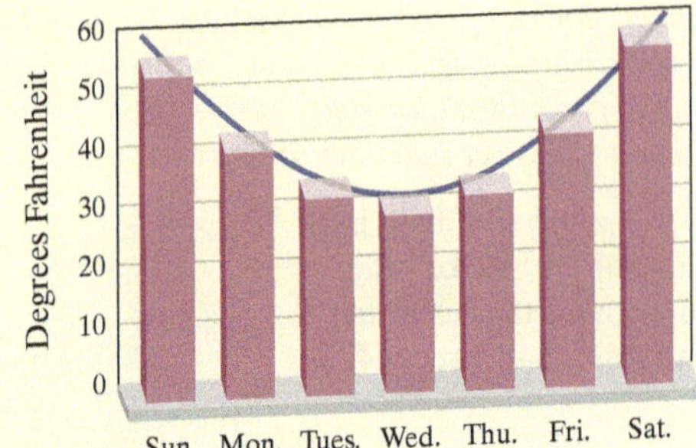

81. Between which days of the week was there the greatest decrease in the low temperature?

82. Between which days of the week was there the greatest increase in the low temperature?

83. Which day of the week had the lowest low temperature?

84. Use the graph to estimate the low temperature on Thursday.

Notice that the shape of the temperature graph is similar to the curve drawn. In fact, this graph can be modeled by the quadratic function $f(x)=3x^2-18x+56$, where $f(x)$ is the temperature in degrees Fahrenheit and x is the number of days from Sunday. (This graph is shown in blue.) Use this function to answer Exercises 85 and 86.

85. Use the quadratic function given to approximate the temperature on Thursday. Does your answer agree with the graph?

86. Use the function given and the quadratic formula to find when the temperature was 35°F. [*Hint:* Let $f(x)=35$ and solve for x.] Round your answer to one decimal place and interpret your result. Does your answer agree with the graph?

87. The number of college students in the United States can be modeled by the quadratic function $f(x)=22x^2+274x+15{,}628$, where $f(x)$ is the number of college students in thousands of students, and x is the number of years after 2000. (*Source:* Based on data from the U.S. Department of Education)

a. Find the number of college students in the United States in 2010.

b. If the trend described by this model continues, find the year after 2000 in which the population of American college students reaches 24,500 students.

88. The projected number of Wi-Fi-enabled cell phones in the United States can be modeled by the quadratic function $c(x)=-0.4x^2+21x+35$, where $c(x)$ is the projected number of Wi-Fi-enabled cell phones in millions and x is the number of years after 2009. Round to the nearest million. (*Source:* Techcrunchies.com)

a. Find the number of Wi-Fi-enabled cell phones in the United States in 2010.

b. Find the estimated number of Wi-Fi-enabled cell phones in the United States in 2012.

c. If the trend described by this model continues, find the year in which the projected number of Wi-Fi-enabled cell phones in the United States reaches 150 million.

89. The average total daily supply y of motor gasoline (in thousands of barrels per day) in the United States for the period 2000–2008 can be approximated by the equation $y=-10x^2+193x+8464$, where x is the number of years after 2000. (*Source:* Based on data from the Energy Information Administration)

a. Find the average total daily supply of motor gasoline in 2004.

b. According to this model, in what year, from 2000 to 2008, was the average total daily supply of gasoline 9325 thousand barrels per day?

c. According to this model, in what year, from 2009 on, will the average total supply of gasoline be 9325 thousand barrels per day?

90. The relationship between body weight and the Recommended Dietary Allowance (RDA) for vitamin A in children up to age 10 is modeled by the quadratic equation $y=0.149x^2-4.475x+406.478$, where y is the RDA

for vitamin A in micrograms for a child whose weight is x pounds. (*Source:* Based on data from the Food and Nutrition Board, National Academy of Sciences–Institute of Medicine, 1989)

a. Determine the vitamin A requirements of a child who weighs 35 pounds.

b. What is the weight of a child whose RDA of vitamin A is 600 micrograms? Round your answer to the nearest pound.

The solutions of the quadratic equation $ax^2 + bx + c = 0$ *are* $\dfrac{-b + \sqrt{b^2 - 4ac}}{2a}$ *and* $\dfrac{-b - \sqrt{b^2 - 4ac}}{2a}$.

91. Show that the sum of these solutions is $\dfrac{-b}{a}$.

92. Show that the product of these solutions is $\dfrac{c}{a}$.

Use the quadratic formula to solve each quadratic equation.

93. $3x^2 - \sqrt{12}x + 1 = 0$

($Hint$: $a = 3, b = -\sqrt{12}, c = 1$)

94. $5x^2 + \sqrt{20}x + 1 = 0$

95. $x^2 + \sqrt{2}x + 1 = 0$

96. $x^2 - \sqrt{2}x + 1 = 0$

97. $2x^2 - \sqrt{3}x - 1 = 0$

98. $7x^2 + \sqrt{7}x - 2 = 0$

99. Use a graphing calculator to solve Exercises 63 and 65.

100. Use a graphing calculator to solve Exercises 64 and 66.

Recall that the discriminant also tells us the number of x-intercepts of the related function.

101. Check the results of Exercise 49 by graphing $y = 9x - 2x^2 + 5$.

102. Check the results of Exercise 50 by graphing $y = 5 - 4x + 12x^2$.

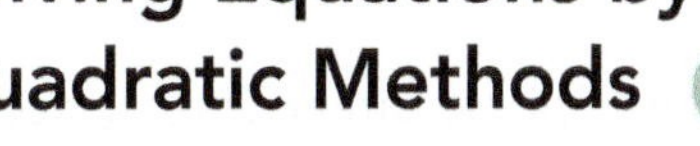

18.3 Solving Equations by Using Quadratic Methods

Objective 1 Solving Equations That Are Quadratic in Form

In this section, we discuss various types of equations that can be solved in part by using the methods for solving quadratic equations.

Once each equation is simplified, you may want to use these steps when deciding which method to use to solve the quadratic equation.

Objectives

1. Solve Various Equations That Are Quadratic in Form.
2. Solve Problems That Lead to Quadratic Equations.

Solving a Quadratic Equation

Step 1: If the equation is in the form $(ax + b)^2 = c$, use the square root property and solve. If not, go to Step 2.

Step 2: Write the equation in standard form: $ax^2 + bx + c = 0$.

Step 3: Try to solve the equation by the factoring method. If not possible, go to Step 4.

Step 4: Solve the equation by the quadratic formula.

The first example is a radical equation that becomes a quadratic equation once we square both sides.

Practice 1

Solve: $x - \sqrt{x+1} - 5 = 0$.

Example 1 Solve: $x - \sqrt{x} - 6 = 0$.

Solution: Recall that to solve a radical equation, first get the radical alone on one side of the equation. Then square both sides.

$$\begin{aligned} x - 6 &= \sqrt{x} && \text{Add } \sqrt{x} \text{ to both sides.} \\ (x-6)^2 &= (\sqrt{x})^2 && \text{Square both sides.} \\ x^2 - 12x + 36 &= x \\ x^2 - 13x + 36 &= 0 && \text{Set the equation equal to 0.} \\ (x-9)(x-4) &= 0 \\ x - 9 = 0 \quad \text{or} \quad x - 4 &= 0 \\ x = 9 \qquad\qquad x &= 4 \end{aligned}$$

Check: Let $x = 9$. Let $x = 4$.

$$\begin{aligned} x - \sqrt{x} - 6 &= 0 & x - \sqrt{x} - 6 &= 0 \\ 9 - \sqrt{9} - 6 &\stackrel{?}{=} 0 & 4 - \sqrt{4} - 6 &\stackrel{?}{=} 0 \\ 9 - 3 - 6 &\stackrel{?}{=} 0 & 4 - 2 - 6 &\stackrel{?}{=} 0 \\ 0 &= 0 \quad \text{True} & -4 &= 0 \quad \text{False} \end{aligned}$$

The solution is 9 or the solution set is {9}.

Work Practice 1

Practice 2

Solve:

$$\frac{5x}{x+1} - \frac{x+4}{x} = \frac{3}{x(x+1)}.$$

Example 2 Solve: $\dfrac{3x}{x-2} - \dfrac{x+1}{x} = \dfrac{6}{x(x-2)}$.

Solution: In this equation, x cannot be either 2 or 0 because these values cause denominators to equal zero. To solve for x, we first multiply both sides of the equation by $x(x-2)$ to clear the fractions. By the distributive property, this means that we multiply each term by $x(x-2)$.

$$x(x-2)\left(\frac{3x}{x-2}\right) - x(x-2)\left(\frac{x+1}{x}\right) = x(x-2)\left[\frac{6}{x(x-2)}\right]$$

$$\begin{aligned} 3x^2 - (x-2)(x+1) &= 6 && \text{Simplify.} \\ 3x^2 - (x^2 - x - 2) &= 6 && \text{Multiply.} \\ 3x^2 - x^2 + x + 2 &= 6 \\ 2x^2 + x - 4 &= 0 && \text{Simplify.} \end{aligned}$$

This equation cannot be factored using integers, so we solve by the quadratic formula.

$$\begin{aligned} x &= \frac{-1 \pm \sqrt{1^2 - 4(2)(-4)}}{2 \cdot 2} && \text{Use } a = 2, b = 1, \text{ and } c = -4 \text{ in the quadratic formula.} \\ &= \frac{-1 \pm \sqrt{1 + 32}}{4} && \text{Simplify.} \\ &= \frac{-1 \pm \sqrt{33}}{4} \end{aligned}$$

Neither proposed solution will make the denominators 0.

The solutions are $\dfrac{-1+\sqrt{33}}{4}$ and $\dfrac{-1-\sqrt{33}}{4}$ or the solution set is $\left\{\dfrac{-1+\sqrt{33}}{4}, \dfrac{-1-\sqrt{33}}{4}\right\}$.

Work Practice 2

Answers

1. 8 **2.** $\dfrac{5-\sqrt{137}}{8}, \dfrac{5+\sqrt{137}}{8}$

Example 3 Solve: $p^4 - 3p^2 - 4 = 0$.

Solution: First we factor the trinomial.

$$p^4 - 3p^2 - 4 = 0$$

$$(p^2 - 4)(p^2 + 1) = 0 \quad \text{Factor.}$$

$$(p - 2)(p + 2)(p^2 + 1) = 0 \quad \text{Factor further.}$$

$$p - 2 = 0 \quad \text{or} \quad p + 2 = 0 \quad \text{or} \quad p^2 + 1 = 0 \quad \text{Set each factor equal to 0 and solve.}$$

$$p = 2 \qquad p = -2 \qquad p^2 = -1$$

$$p = \pm\sqrt{-1} = \pm i$$

The solutions are 2, -2, i, and $-i$.

Work Practice 3

Practice 3

Solve: $p^4 - 7p^2 - 144 = 0$.

Helpful Hint

Example 3 can be solved using substitution also. Think of $p^4 - 3p^2 - 4 = 0$ as

$$(p^2)^2 - 3p^2 - 4 = 0 \quad \text{Then let } x = p^2 \text{ and solve and substitute back. The solutions will be the same.}$$

$$x^2 - 3x - 4 = 0$$

✓Concept Check

a. True or false? The maximum number of solutions that a quadratic equation can have is 2.

b. True or false? The maximum number of solutions that an equation in quadratic form can have is 2.

Example 4 Solve: $(x - 3)^2 - 3(x - 3) - 4 = 0$.

Solution: Notice that the quantity $(x - 3)$ is repeated in this equation. Sometimes it is helpful to substitute a variable (in this case other than x) for the repeated quantity. We will let $y = x - 3$. Then

$$(x - 3)^2 - 3(x - 3) - 4 = 0$$

becomes

$$y^2 - 3y - 4 = 0 \quad \text{Let } x - 3 = y.$$

$$(y - 4)(y + 1) = 0 \quad \text{Factor.}$$

To solve, we use the zero factor property.

$$y - 4 = 0 \quad \text{or} \quad y + 1 = 0 \quad \text{Set each factor equal to 0.}$$

$$y = 4 \qquad y = -1 \quad \text{Solve.}$$

To find values of x, we substitute back. That is, we substitute $x - 3$ for y.

$$x - 3 = 4 \quad \text{or} \quad x - 3 = -1$$

$$x = 7 \qquad x = 2$$

Both 2 and 7 check. The solutions are 2 and 7.

Work Practice 4

Practice 4

Solve:
$(x + 2)^2 - 2(x + 2) - 3 = 0$.

Helpful Hint When using substitution, don't forget to substitute back to the original variable.

Answers

3. $4, -4, 3i, -3i$ **4.** $1, -3$

✓Concept Check Answers

a. true **b.** false

Practice 5

Solve: $x^{2/3} - 5x^{1/3} + 4 = 0$.

Example 5

Solve: $x^{2/3} - 5x^{1/3} + 6 = 0$.

Solution: The key to solving this equation is recognizing that $x^{2/3} = (x^{1/3})^2$. We replace $x^{1/3}$ with m so that

$$(x^{1/3})^2 - 5x^{1/3} + 6 = 0$$

becomes

$$m^2 - 5m + 6 = 0$$

Now we solve by factoring.

$$m^2 - 5m + 6 = 0$$

$$(m - 3)(m - 2) = 0 \quad \text{Factor.}$$

$$m - 3 = 0 \quad \text{or} \quad m - 2 = 0 \quad \text{Set each factor equal to 0.}$$

$$m = 3 \qquad m = 2$$

Since $m = x^{1/3}$, we have

$$x^{1/3} = 3 \quad \text{or} \quad x^{1/3} = 2$$

$$x = 3^3 = 27 \quad \text{or} \quad x = 2^3 = 8$$

Both 8 and 27 check. The solutions are 8 and 27.

Work Practice 5

Objective 2 Solving Problems That Lead to Quadratic Equations

The next example is a work problem. This problem is modeled by a rational equation that simplifies to a quadratic equation.

Practice 6

Together, Katy and Steve can groom all the dogs at the Barkin' Doggie Day Care in 4 hours. Alone, Katy can groom the dogs 1 hour faster than Steve can groom the dogs alone. Find the time in which each of them can groom the dogs alone.

Example 6 Finding Work Time

Together, an experienced word processor and an apprentice word processor can create a word document in 6 hours. Alone, the experienced word processor can create the document 2 hours faster than the apprentice word processor can. Find the time in which each person can create the word document alone.

Solution:

1. UNDERSTAND. Read and reread the problem. The key idea here is the relationship between the *time* (hours) it takes to complete the job and the *part of the job* completed in one unit of time (hour). For example, because they can complete the job together in 6 hours, the *part of the job* they can complete in 1 hour is $\frac{1}{6}$.

Let

x = the *time* in hours it takes the apprentice word processor to complete the job alone

$x - 2$ = the *time* in hours it takes the experienced word processor to complete the job alone

We can summarize in a chart the information discussed.

	Total Hours to Complete Job	Part of Job Completed in 1 Hour
Apprentice Word Processor	x	$\frac{1}{x}$
Experienced Word Processor	$x - 2$	$\frac{1}{x-2}$
Together	6	$\frac{1}{6}$

Answers

5. 1, 64 **6.** Katy: $\frac{7 + \sqrt{65}}{2} \approx 7.5$ hours; Steve: $\frac{9 + \sqrt{65}}{2} \approx 8.5$ hours

2. TRANSLATE.

In words:	part of job completed by apprentice word processor in 1 hour	added to	part of job completed by experienced word processor in 1 hour	is equal to	part of job completed together in 1 hour
	↓	↓	↓	↓	↓
Translate:	$\frac{1}{x}$	$+$	$\frac{1}{x-2}$	$=$	$\frac{1}{6}$

3. SOLVE.

$$\frac{1}{x} + \frac{1}{x-2} = \frac{1}{6}$$

$$6x(x-2)\left(\frac{1}{x} + \frac{1}{x-2}\right) = 6x(x-2)\cdot\frac{1}{6}$$ Multiply both sides by the LCD, $6x(x-2)$.

$$6x(x-2)\cdot\frac{1}{x} + 6x(x-2)\cdot\frac{1}{x-2} = 6x(x-2)\cdot\frac{1}{6}$$ Use the distributive property.

$$6(x-2) + 6x = x(x-2)$$
$$6x - 12 + 6x = x^2 - 2x$$
$$0 = x^2 - 14x + 12$$

Now we can substitute $a = 1$, $b = -14$, and $c = 12$ into the quadratic formula and simplify.

$$x = \frac{-(-14) \pm \sqrt{(-14)^2 - 4(1)(12)}}{2\cdot 1} = \frac{14 \pm \sqrt{148}^*}{2}$$

Using a calculator or a square root table, we see that $\sqrt{148} \approx 12.2$ rounded to one decimal place. Thus,

$$x \approx \frac{14 \pm 12.2}{2}$$

$$x \approx \frac{14 + 12.2}{2} = 13.1 \quad \text{or} \quad x \approx \frac{14 - 12.2}{2} = 0.9$$

4. INTERPRET.

Check: If the apprentice word processor completes the job alone in 0.9 hours, the experienced word processor completes the job alone in $x - 2 = 0.9 - 2 = -1.1$ hours. Since this is not possible, we reject the solution of 0.9. The approximate solution thus is 13.1 hours.

State: The apprentice word processor can complete the job alone in approximately 13.1 hours, and the experienced word processor can complete the job alone in approximately

$$x - 2 = 13.1 - 2 = 11.1 \text{ hours}$$

Work Practice 6

Example 7 Finding Driving Speeds

Beach and Fargo are about 400 miles apart. A salesperson travels from Fargo to Beach one day at a certain speed. She returns to Fargo the next day and drives 10 mph faster.

(Continued on next page)

(*This expression can be simplified further, but this will suffice because we are approximating.)

Practice 7

The 36-km S-shaped Hangzhou Bay Bridge is the longest cross-sea bridge in the world, linking Ningbo and Shanghai, China. A merchant drives over the bridge one morning from Ningbo to Shanghai in very heavy traffic and returns home that night driving 50 km per hour faster. The total travel time was 1.3 hours. Find the speed to Shanghai and the return speed to Ningbo.

Answer

7. to Shanghai: 40 km/hr; to Ningbo: 90 km/hr

Her total travel time was $14\frac{2}{3}$ hours. Find her speed to Beach and the return speed to Fargo.

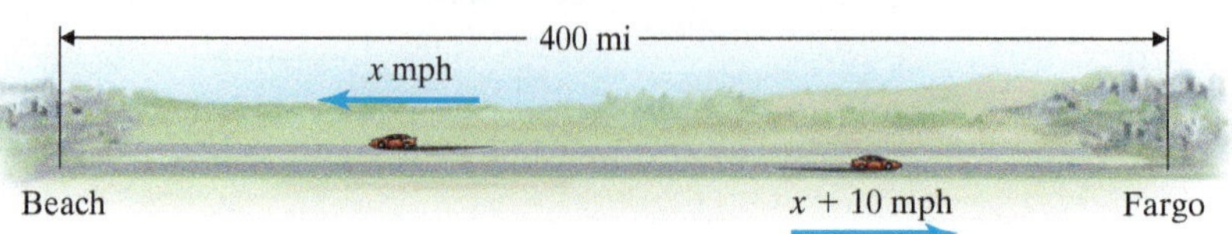

Solution:

1. UNDERSTAND. Read and reread the problem. Let

$x =$ the speed to Beach, so

$x + 10 =$ the return speed to Fargo

Then organize the given information in a table.

	distance =	rate ·	time
To Beach	400	x	$\frac{400}{x}$ ← distance ← rate
Return to Fargo	400	$x + 10$	$\frac{400}{x+10}$ ← distance ← rate

Helpful Hint Since $d = rt$, $t = \frac{d}{r}$. The time column was completed using $\frac{d}{r}$.

2. TRANSLATE.

In words: time to Beach + return time to Fargo = $14\frac{2}{3}$ hours

Translate: $\frac{400}{x} + \frac{400}{x+10} = \frac{44}{3}$

3. SOLVE.

$$\frac{400}{x} + \frac{400}{x+10} = \frac{44}{3}$$

$$\frac{100}{x} + \frac{100}{x+10} = \frac{11}{3}$$ Divide both sides by 4.

$$3x(x+10)\left(\frac{100}{x} + \frac{100}{x+10}\right) = 3x(x+10)\cdot\frac{11}{3}$$ Multiply both sides by the LCD, $3x(x + 10)$.

$$3x(x+10)\cdot\frac{100}{x} + 3x(x+10)\cdot\frac{100}{x+10} = 3x(x+10)\cdot\frac{11}{3}$$ Use the distributive property.

$$3(x+10)\cdot 100 + 3x\cdot 100 = x(x+10)\cdot 11$$

$$300x + 3000 + 300x = 11x^2 + 110x$$

$$0 = 11x^2 - 490x - 3000$$ Set equation equal to 0.

$$0 = (11x + 60)(x - 50)$$ Factor.

$11x + 60 = 0$ or $x - 50 = 0$ Set each factor equal to 0.

$x = -\frac{60}{11} = -5\frac{5}{11}$ or $x = 50$

4. INTERPRET.

Check: The speed is not negative, so it's not $-5\frac{5}{11}$. The number 50 does check.

State: The speed to Beach was 50 mph, and her return speed to Fargo was 60 mph.

Work Practice 7

Vocabulary, Readiness & Video Check

See Video 18.3

Watch the section lecture video and answer the following questions.

Objective 1 **1.** From Examples 1 and 2, what's the main thing to remember when using a substitution in order to solve an equation by quadratic methods?

Objective 2 **2.** In Example 4, the translated equation is actually a rational equation. Explain how we end up solving it using quadratic methods.

18.3 Exercise Set MyMathLab®

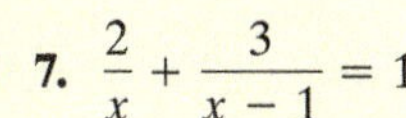

Solve. See Example 1.

1. $2x = \sqrt{10 + 3x}$

2. $3x = \sqrt{8x + 1}$

3. $x - 2\sqrt{x} = 8$

4. $x - \sqrt{2x} = 4$

5. $\sqrt{9x} = x + 2$

6. $\sqrt{16x} = x + 3$

Solve. See Example 2.

7. $\frac{2}{x} + \frac{3}{x - 1} = 1$

8. $\frac{6}{x^2} = \frac{3}{x + 1}$

9. $\frac{3}{x} + \frac{4}{x + 2} = 2$

10. $\frac{5}{x - 2} + \frac{4}{x + 2} = 1$

11. $\frac{7}{x^2 - 5x + 6} = \frac{2x}{x - 3} - \frac{x}{x - 2}$

12. $\frac{11}{2x^2 + x - 15} = \frac{5}{2x - 5} - \frac{x}{x + 3}$

Solve. See Example 3.

13. $p^4 - 16 = 0$

14. $x^4 + 2x^2 - 3 = 0$

15. $4x^4 + 11x^2 = 3$

16. $z^4 = 81$

17. $z^4 - 13z^2 + 36 = 0$

18. $9x^4 + 5x^2 - 4 = 0$

Solve. See Examples 4 and 5.

19. $x^{2/3} - 3x^{1/3} - 10 = 0$

20. $x^{2/3} + 2x^{1/3} + 1 = 0$

21. $(5n + 1)^2 + 2(5n + 1) - 3 = 0$

22. $(m - 6)^2 + 5(m - 6) + 4 = 0$

23. $2x^{2/3} - 5x^{1/3} = 3$

24. $3x^{2/3} + 11x^{1/3} = 4$

25. $1 + \frac{2}{3t - 2} = \frac{8}{(3t - 2)^2}$

26. $2 - \frac{7}{x + 6} = \frac{15}{(x + 6)^2}$

27. $20x^{2/3} - 6x^{1/3} - 2 = 0$

28. $4x^{2/3} + 16x^{1/3} = -15$

MIXED PRACTICE

Solve. See Examples 1 through 5.

29. $a^4 - 5a^2 + 6 = 0$

30. $x^4 - 12x^2 + 11 = 0$

31. $\frac{2x}{x - 2} + \frac{x}{x + 3} = -\frac{5}{x + 3}$

32. $\frac{5}{x - 3} + \frac{x}{x + 3} = \frac{19}{x^2 - 9}$

33. $(p + 2)^2 = 9(p + 2) - 20$

34. $2(4m - 3)^2 - 9(4m - 3) = 5$

35. $2x = \sqrt{11x + 3}$

36. $4x = \sqrt{2x + 3}$

37. $x^{2/3} - 8x^{1/3} + 15 = 0$

38. $x^{2/3} - 2x^{1/3} - 8 = 0$

39. $y^3 + 9y - y^2 - 9 = 0$

40. $x^3 + x - 3x^2 - 3 = 0$

41. $2x^{2/3} + 3x^{1/3} - 2 = 0$

42. $6x^{2/3} - 25x^{1/3} - 25 = 0$

43. $x^{-2} - x^{-1} - 6 = 0$

44. $y^{-2} - 8y^{-1} + 7 = 0$

45. $x - \sqrt{x} = 2$

46. $x - \sqrt{3x} = 6$

47. $\frac{x}{x - 1} + \frac{1}{x + 1} = \frac{2}{x^2 - 1}$

48. $\frac{x}{x-5} + \frac{5}{x+5} = -\frac{1}{x^2-25}$

49. $p^4 - p^2 - 20 = 0$

50. $x^4 - 10x^2 + 9 = 0$

51. $(x + 3)(x^2 - 3x + 9) = 0$

52. $(x - 6)(x^2 + 6x + 36) = 0$

53. $1 = \frac{4}{x-7} + \frac{5}{(x-7)^2}$

54. $3 + \frac{1}{2p+4} = \frac{10}{(2p+4)^2}$

55. $27y^4 + 15y^2 = 2$

56. $8z^4 + 14z^2 = -5$

57. $x - \sqrt{19 - 2x} - 2 = 0$

58. $x - \sqrt{17 - 4x} - 3 = 0$

Solve. For Exercises 59 and 60, the solutions have been started for you. See Examples 6 and 7.

59. Roma Sherry drove 330 miles from her hometown to Tucson. During her return trip, she was able to increase her speed by 11 miles per hour. If her return trip took 1 hour less time, find her original speed and her speed returning home.

Start the solution:

1. UNDERSTAND the problem. Reread it as many times as needed. Let

x = original speed

$x + 11$ = return-trip speed

Organize the information in a table.

	distance =	*rate* ·	*time*	
To Tucson	330	x	$\frac{330}{x}$	← distance ← rate
Return trip	330	___	$\frac{330}{\_\_\_}$	← distance ← rate

2. TRANSLATE into an equation. (Fill in the blanks below.)

Time to Tucson	equals	Return trip time	plus	1 hour
↓	↓	↓	↓	↓
___	=	___	+	1

Finish with:

3. SOLVE and 4. INTERPRET

60. A salesperson drove to Portland, a distance of 300 miles. During the last 80 miles of his trip, heavy rainfall forced him to decrease his speed by 15 miles per hour. If his total driving time was 6 hours, find his original speed and his speed during the rainfall.

Start the solution:

1. UNDERSTAND the problem. Reread it as many times as needed. Let

x = original speed

$x - 15$ = rainfall speed

Organize the information in a table.

	distance =	*rate* ·	*time*	
First part of trip	300 − 80, or 220	x	$\frac{220}{x}$	← distance ← rate
Heavy rainfall part of trip	80	$x - 15$	$\frac{80}{\_\_\_}$	← distance ← rate

2. TRANSLATE into an equation. (Fill in the blanks below.)

Time during first part of trip	plus	Time during heavy rainfall	equals	6 hr
↓	↓	↓	↓	↓
___	+	___	=	6

Finish with:

3. SOLVE and 4. INTERPRET

61. A jogger ran 3 miles, decreased her speed by 1 mile per hour, and then ran another 4 miles. If her total time jogging was $1\frac{3}{5}$ hours, find her speed for each part of her run.

62. Mark Keaton's workout consists of jogging for 3 miles and then riding his bike for 5 miles at a speed 4 miles per hour faster than he jogs. If his total workout time is 1 hour, find his jogging speed and his biking speed.

63. A Chinese restaurant in Mandeville, Louisiana, has a large goldfish pond around the restaurant. Suppose that an inlet pipe and a hose together can fill the pond in 8 hours. The inlet pipe alone can complete the job in one hour less time than the hose alone. Find the time that the hose can complete the job alone and the time that the inlet pipe can complete the job alone. Round each to the nearest tenth of an hour.

64. A water tank on a farm in Flatonia, Texas, can be filled with a large inlet pipe and a small inlet pipe in 3 hours. The large inlet pipe alone can fill the tank in 2 hours less time than the small inlet pipe alone. Find the time to the nearest tenth of an hour each pipe can fill the tank alone.

65. Bill Shaughnessy and his son Billy can clean the house together in 4 hours. When the son works alone, it takes him an hour longer to clean than it takes his dad alone. Find how long to the nearest tenth of an hour it takes the son to clean alone.

66. Together, Noodles and Freckles eat a 50-pound bag of dog food in 30 days. Noodles by himself eats a 50-pound bag in 2 weeks less time than Freckles does by himself. How many days to the nearest whole day would a 50-pound bag of dog food last Freckles?

67. The product of a number and 4 less than the number is 96. Find the number.

68. A whole number increased by its square is two more than twice itself. Find the number.

△ **69.** Suppose that an open box is to be made from a square sheet of cardboard by cutting out squares from each corner as shown and then folding along the dotted lines. If the box is to have a volume of 300 cubic inches, find the original dimensions of the sheet of cardboard.

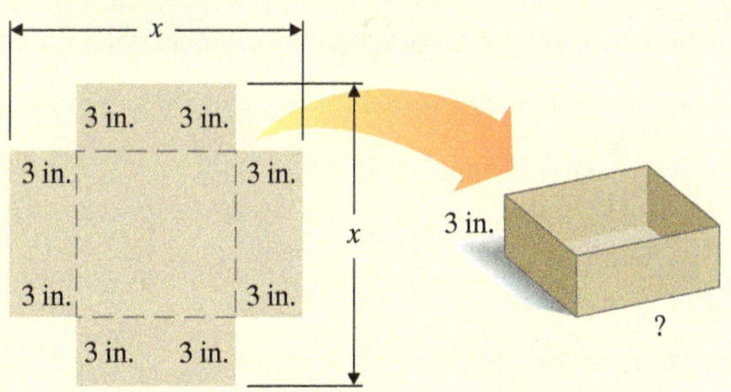

a. The ? in the drawing above will be the length (and the width) of the box as shown. Represent this length in terms of x.

b. Use the formula for volume of a box, $V = l \cdot w \cdot h$, to write an equation in x.

c. Solve the equation for x and give the dimensions of the sheet of cardboard. Check your solution.

△ **70.** Suppose that an open box is to be made from a square sheet of cardboard by cutting out squares from each corner as shown and then folding along the dotted lines. If the box is to have a volume of 128 cubic inches, find the original dimensions of the sheet of cardboard.

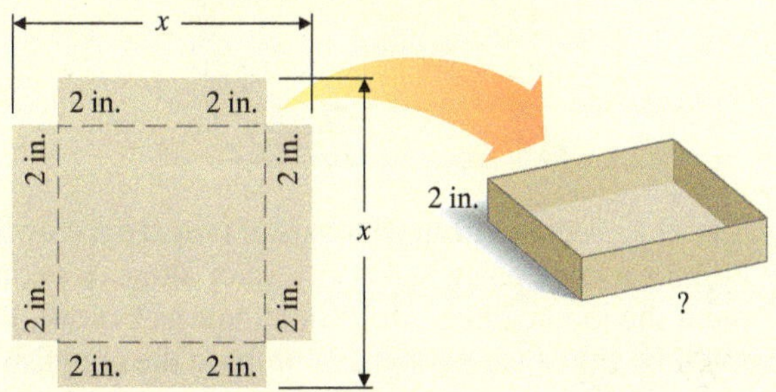

a. The ? in the drawing above will be the length (and the width) of the box as shown. Represent this length in terms of x.

b. Use the formula for volume of a box, $V = l \cdot w \cdot h$, to write an equation in x.

c. Solve the equation for x and give the dimensions of the sheet of cardboard. Check your solution.

△ **71.** A sprinkler that sprays water in a circular pattern is to be used to water a square garden. If the area of the garden is 920 square feet, find the smallest whole number *radius* that the sprinkler can be adjusted to so that the entire garden is watered.

△ **72.** Suppose that a square field has an area of 6270 square feet. See Exercise 71 and find a new sprinkler radius.

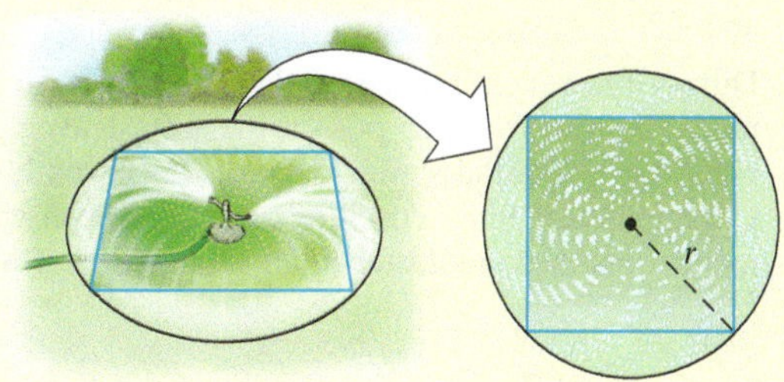

REVIEW AND PREVIEW

Solve each inequality. See Section 9.6.

73. $\frac{5x}{3} + 2 \le 7$

74. $\frac{2x}{3} + \frac{1}{6} \ge 2$

75. $\frac{y - 1}{15} > -\frac{2}{5}$

76. $\frac{z - 2}{12} < \frac{1}{4}$

Find the domain and range of each graphed relation. Decide which relations are also functions. See Section 10.6.

77. 

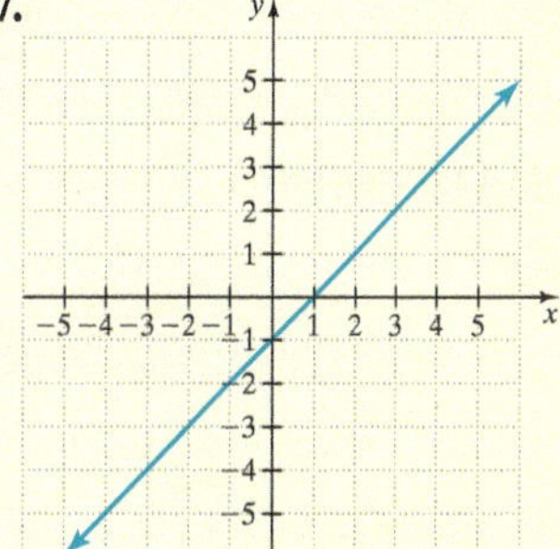

78. 

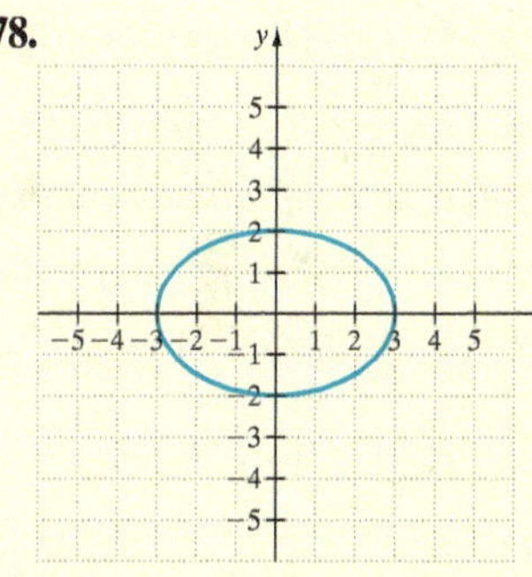

79. **80.**

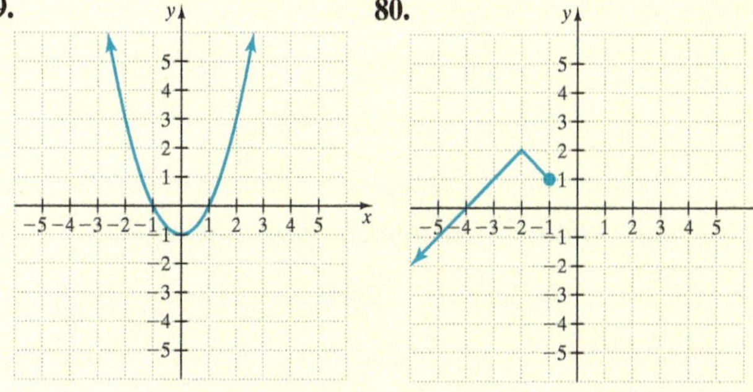

CONCEPT EXTENSIONS

Solve.

81. $5y^3 + 45y - 5y^2 - 45 = 0$

82. $10x^3 + 10x - 30x^2 - 30 = 0$

83. $3x^{-2} - 3x^{-1} - 18 = 0$

84. $2y^{-2} - 16y^{-1} + 14 = 0$

85. $2x^3 = -54$

86. $y^3 - 216 = 0$

87. Write a polynomial equation that has three solutions: 2, 5, and -7.

88. Write a polynomial equation that has three solutions: 0, $2i$, and $-2i$.

89. During the seventh stage of the 2010 Paris–Nice bicycle race, Thomas Voeckler posted the fastest average speed, but Alberto Contador won the race. The seventh stage was 119 kilometers long. Voeckler's average speed was 0.0034 meters per second faster than Contador's. Traveling at these average speeds, Contador took 3 seconds longer than Voeckler to complete the race stage. (*Source:* Based on data from cyclingnews.com)

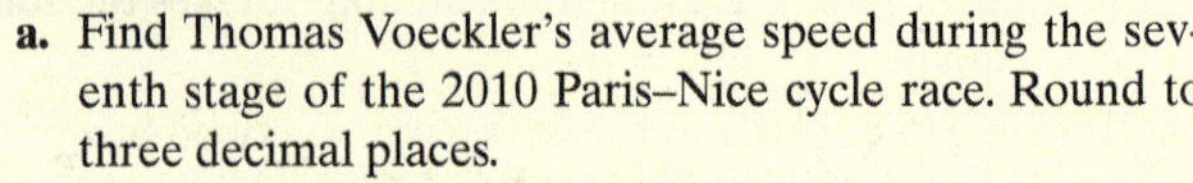

a. Find Thomas Voeckler's average speed during the seventh stage of the 2010 Paris–Nice cycle race. Round to three decimal places.

b. Find Alberto Contador's average speed during the seventh stage of the 2010 Paris–Nice cycle race. Round to three decimal places.

c. Convert Voeckler's average speed to miles per hour. Round to three decimal places.

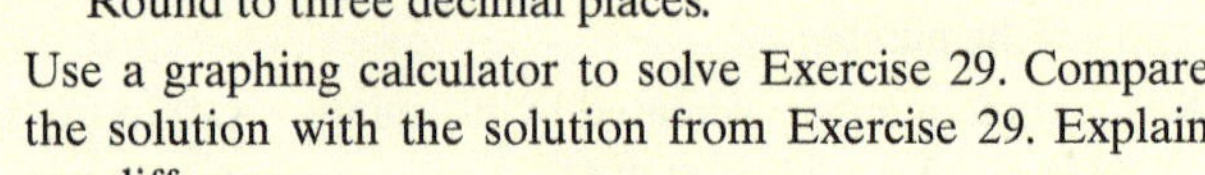

90. Use a graphing calculator to solve Exercise 29. Compare the solution with the solution from Exercise 29. Explain any differences.

Integrated Review Sections 18.1–18.3

Summary on Solving Quadratic Equations

Use the square root property to solve each equation.

1. $x^2 - 10 = 0$

2. $x^2 - 14 = 0$

3. $(x - 1)^2 = 8$

4. $(x + 5)^2 = 12$

Solve each equation by completing the square.

5. $x^2 + 2x - 12 = 0$

6. $x^2 - 12x + 11 = 0$

7. $3x^2 + 3x = 5$

8. $16y^2 + 16y = 1$

Use the quadratic formula to solve each equation

9. $2x^2 - 4x + 1 = 0$

10. $\frac{1}{2}x^2 + 3x + 2 = 0$

11. $x^2 + 4x = -7$

12. $x^2 + x = -3$

Solve each equation. Use a method of your choice.

13. $x^2 + 3x + 6 = 0$

14. $2x^2 + 18 = 0$

15. $x^2 + 17x = 0$

16. $4x^2 - 2x - 3 = 0$

17. $(x - 2)^2 = 27$

18. $\frac{1}{2}x^2 - 2x + \frac{1}{2} = 0$

19. $3x^2 + 2x = 8$

20. $2x^2 = -5x - 1$

21. $x(x - 2) = 5$

22. $x^2 - 31 = 0$

23. $5x^2 - 55 = 0$

24. $5x^2 + 55 = 0$

25. $x(x + 5) = 66$

26. $5x^2 + 6x - 2 = 0$

27. $2x^2 + 3x = 1$

28. $x - \sqrt{13 - 3x} - 3 = 0$

29. $\frac{5x}{x - 2} - \frac{x + 1}{x} = \frac{3}{x(x - 2)}$

△ **30.** The diagonal of a square room measures 20 feet. Find the exact length of a side of the room. Then approximate the length to the nearest tenth of a foot.

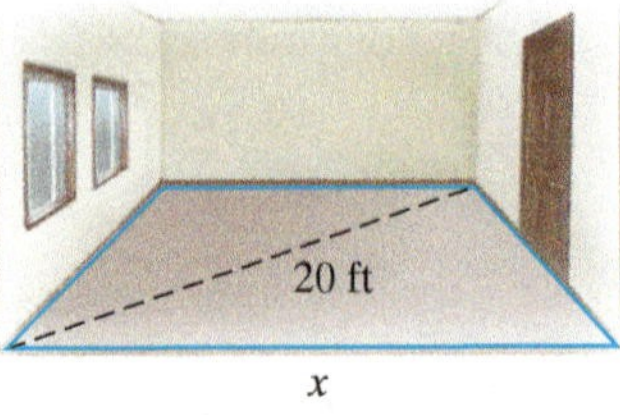

31. Together, Jack and Lucy Hoag can prepare a crawfish boil for a large party in 4 hours. Lucy alone can complete the job in 2 hours less time than Jack alone. Find the time that each person can prepare the crawfish boil alone. Round each time to the nearest tenth of an hour.

32. Diane Gray exercises at Total Body Gym. On the treadmill, she runs 5 miles, increases her speed by 1 mile per hour, and runs an additional 2 miles. If her total time on the treadmill is $1\frac{1}{3}$ hours, find her speed during each part of her run.

18.4 Nonlinear Inequalities in One Variable 

Objective 1 Solving Polynomial Inequalities

Objectives

1 Solve Polynomial Inequalities of Degree 2 or Greater.

2 Solve Inequalities That Contain Rational Expressions with Variables in the Denominator.

Just as we can solve linear inequalities in one variable, so can we also solve quadratic inequalities in one variable. A **quadratic inequality** is an inequality that can be written so that one side is a quadratic expression and the other side is 0. Here are examples of quadratic inequalities in one variable. Each is written in **standard form.**

$$x^2 - 10x + 7 \le 0 \qquad 3x^2 + 2x - 6 > 0$$

$$2x^2 + 9x - 2 < 0 \qquad x^2 - 3x + 11 \ge 0$$

A solution of a quadratic inequality in one variable is a value of the variable that makes the inequality a true statement.

The value of an expression such as $x^2 - 3x - 10$ will sometimes be positive, sometimes negative, and sometimes 0, depending on the value substituted for x. To solve the inequality $x^2 - 3x - 10 < 0$, we are looking for all values of x that make the expression $x^2 - 3x - 10$ **less than 0,** or **negative.** To understand how we find these values, we'll study the graph of the quadratic function $y = x^2 - 3x - 10$.

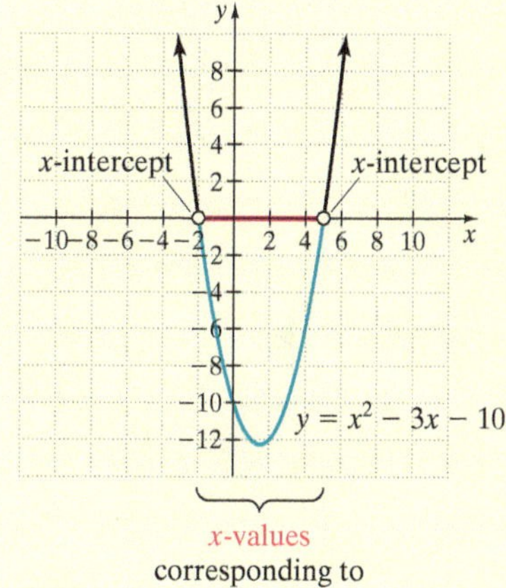

Notice that the x-values for which y is positive are separated from the x-values for which y is negative by the x-intercepts. (Recall that the x-intercepts correspond to values of x for which $y = 0$.) Thus, the solution set of $x^2 - 3x - 10 < 0$ consists of all real numbers from -2 to 5 or, in interval notation, $(-2, 5)$.

It is not necessary to graph $y = x^2 - 3x - 10$ to solve the related inequality $x^2 - 3x - 10 < 0$. Instead, we can draw a number line representing the x-axis and keep the following in mind: *A region on the number line for which the value of* $x^2 - 3x - 10$ *is positive is separated from a region on the number line for which the value of* $x^2 - 3x - 10$ *is negative by a value for which the expression is* 0.

Let's find these values for which the expression is 0 by solving the related equation:

$$x^2 - 3x - 10 = 0$$

$$(x - 5)(x + 2) = 0 \qquad \text{Factor.}$$

$$x - 5 = 0 \quad \text{or} \quad x + 2 = 0 \qquad \text{Set each factor equal to 0.}$$

$$x = 5 \qquad x = -2 \qquad \text{Solve.}$$

These two numbers, -2 and 5, divide the number line into three regions. We will call the regions A, B, and C. These regions are important because, if the value of $x^2 - 3x - 10$ is negative when a number from a region is substituted for x, then $x^2 - 3x - 10$ is negative when any number in that region is substituted for x. The same is true if the value of $x^2 - 3x - 10$ is positive for a particular value of x in a region.

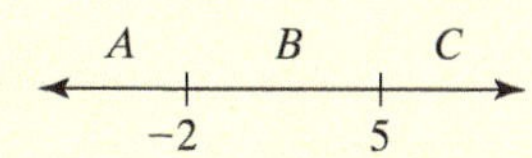

To see whether the inequality $x^2 - 3x - 10 < 0$ is true or false in each region, we choose a test point from each region and substitute its value for x in the inequality $x^2 - 3x - 10 < 0$. If the resulting inequality is true, the region containing the test point is a solution region.

Region	Test Point Value	$(x - 5)(x + 2) < 0$	Result
A	-3	$(-8)(-1) < 0$	False
B	0	$(-5)(2) < 0$	True
C	6	$(1)(8) < 0$	False

The values in region B satisfy the inequality. The numbers -2 and 5 are not included in the solution set since the inequality symbol is $<$. The solution set is $(-2, 5)$, and its graph is shown.

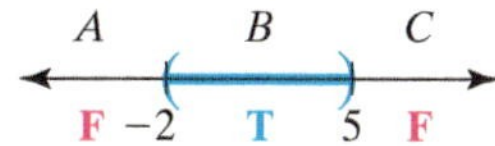

Practice 1

Solve: $(x-4)(x+3)>0$.

Example 1 Solve: $(x+3)(x-3)>0$.

Solution: First we solve the related equation, $(x+3)(x-3)=0$.

$$(x+3)(x-3)=0$$
$$x+3=0 \quad \text{or} \quad x-3=0$$
$$x=-3 \qquad\qquad x=3$$

The two numbers -3 and 3 separate the number line into three regions, A, B, and C.

A B C
−3 3

Now we substitute the value of a test point from each region. If the test value satisfies the inequality, every value in the region containing the test value is a solution.

Region	Test Point Value	$(x+3)(x-3)>0$	Result
A	-4	$(-1)(-7)>0$	True
B	0	$(3)(-3)>0$	False
C	4	$(7)(1)>0$	True

The points in regions A and C satisfy the inequality. The numbers -3 and 3 are not included in the solution since the inequality symbol is $>$. The solution set is $(-\infty,-3)\cup(3,\infty)$, and its graph is shown.

A B C
T −3 F 3 T

Work Practice 1

The following steps may be used to solve a polynomial inequality.

Solving a Polynomial Inequality

Step 1: Write the inequality in standard form and then solve the related equation.

Step 2: Separate the number line into regions with the solutions from Step 1.

Step 3: For each region, choose a test point and determine whether its value satisfies the *original inequality.*

Step 4: The solution set includes the regions whose test point value is a solution. If the inequality symbol is $\le$ or $\ge$, the values from Step 1 are solutions; if $<$ or $>$, they are not.

✓**Concept Check** When choosing a test point in Step 3, why would the solutions from Step 1 not make good choices for test points?

Practice 2

Solve: $x^2-8x\le 0$.

Example 2 Solve: $x^2-4x\le 0$.

Solution: First we solve the related equation, $x^2-4x=0$.

$$x^2-4x=0$$
$$x(x-4)=0$$
$$x=0 \quad \text{or} \quad x=4$$

Answers

1. $(-\infty,-3)\cup(4,\infty)$ **2.** $[0,8]$

✓**Concept Check Answer**

The solutions found in Step 1 have a value of 0 in the original inequality.

The numbers 0 and 4 separate the number line into three regions, A, B, and C.

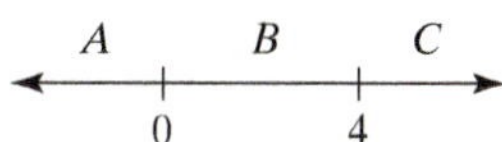

We check a test value in each region in the original inequality. Values in region B satisfy the inequality. The numbers 0 and 4 are included in the solution since the inequality symbol is $\le$. The solution set is $[0, 4]$, and its graph is shown.

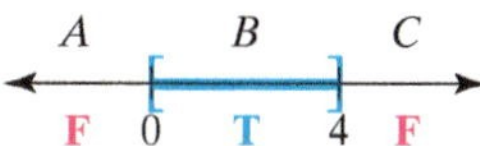

Work Practice 2

Example 3 Solve: $(x + 2)(x - 1)(x - 5) \le 0$.

Practice 3

Solve: $(x + 3)(x - 2)(x + 1) \le 0$.

Solution: First we solve $(x + 2)(x - 1)(x - 5) = 0$. By inspection, we see that the solutions are -2, 1, and 5. They separate the number line into four regions, A, B, C, and D. Next we check test points from each region.

Region	Test Point Value	$(x + 2)(x - 1)(x - 5) \le 0$	Result
A	-3	$(-1)(-4)(-8) \le 0$	True
B	0	$(2)(-1)(-5) \le 0$	False
C	2	$(4)(1)(-3) \le 0$	True
D	6	$(8)(5)(1) \le 0$	False

The solution set is $(-\infty, -2] \cup [1, 5]$, and its graph is shown. We include the numbers -2, 1, and 5 because the inequality symbol is $\le$.

Work Practice 3

Objective 2 Solving Rational Inequalities

Inequalities containing rational expressions with variables in the denominator are solved by using a similar procedure.

Example 4 Solve: $\dfrac{x + 2}{x - 3} \le 0$.

Practice 4

Solve: $\dfrac{x - 5}{x + 4} \le 0$.

Solution: First we find all values that make the denominator equal to 0. To do this, we solve $x - 3 = 0$ and find that $x = 3$.

Next, we solve the related equation $\dfrac{x + 2}{x - 3} = 0$.

$$\frac{x + 2}{x - 3} = 0$$

$$x + 2 = 0 \quad \text{Multiply both sides by the LCD, } x - 3.$$

$$x = -2$$

Now we place these numbers on a number line and proceed as before, checking test point values in the original inequality.

(*Continued on next page*)

Answers

3. $(-\infty, -3] \cup [-1, 2]$ **4.** $(-4, 5]$

Choose −3 from region *A*.

$$\frac{x+2}{x-3} \le 0$$

$$\frac{-3+2}{-3-3} \le 0$$

$$\frac{-1}{-6} \le 0$$

$$\frac{1}{6} \le 0 \quad \text{False}$$

Choose 0 from region *B*.

$$\frac{x+2}{x-3} \le 0$$

$$\frac{0+2}{0-3} \le 0$$

$$-\frac{2}{3} \le 0 \quad \text{True}$$

Choose 4 from region *C*.

$$\frac{x+2}{x-3} \le 0$$

$$\frac{4+2}{4-3} \le 0$$

$$6 \le 0 \quad \text{False}$$

The solution set is $[-2, 3)$. This interval includes -2 because -2 satisfies the original inequality. This interval does not include 3 because 3 would make the denominator 0.

A B C
F −2 T 3 F

Work Practice 4

The following steps may be used to solve a rational inequality with variables in the denominator.

Solving a Rational Inequality

Step 1: Solve for values that make all denominators 0.

Step 2: Solve the related equation.

Step 3: Separate the number line into regions with the solutions from Steps 1 and 2.

Step 4: For each region, choose a test point and determine whether its value satisfies the *original inequality.*

Step 5: The solution set includes the regions whose test point value is a solution. Check whether to include values from Step 2. Be sure *not* to include values that make any denominator 0.

Practice 5

Solve: $\frac{7}{x+3} < 5.$

Example 5 Solve: $\frac{5}{x+1} < -2.$

Solution: First we find values for x that make the denominator equal to 0.

$$x + 1 = 0$$
$$x = -1$$

Next we solve $\frac{5}{x+1} = -2.$

$$(x+1)\cdot\frac{5}{x+1} = (x+1)\cdot -2 \quad \text{Multiply both sides by the LCD, } x+1.$$

$$5 = -2x - 2 \quad \text{Simplify.}$$

$$7 = -2x$$

$$-\frac{7}{2} = x$$

A B C
$-\frac{7}{2}$ -1

Answer

5. $(-\infty, -3) \cup \left(-\frac{8}{5}, \infty\right)$

We use these two solutions to divide a number line into three regions and choose test points. Only a test point value from region B satisfies the *original inequality.* The solution set is $\left(-\frac{7}{2}, -1\right)$, and its graph is shown.

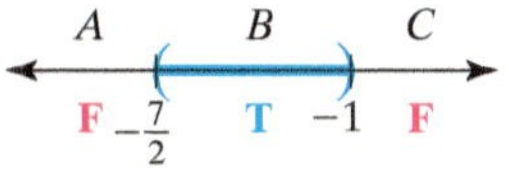

Work Practice 5

Vocabulary, Readiness & Video Check

Write the graphed solution set in interval notation.

1. 

2. 

3. 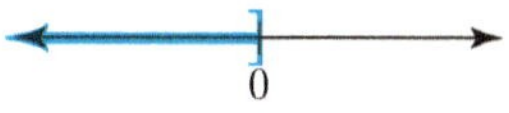

4. 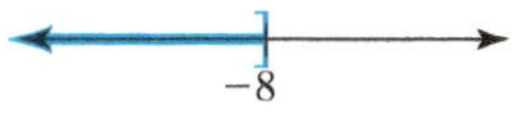

5. 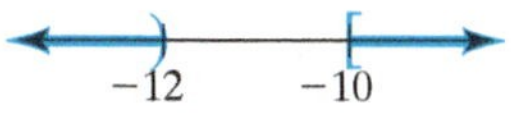

6. 

Martin-Gay Interactive Videos

See Video 18.4

Watch the section lecture video and answer the following questions.

Objective 1 7. From Examples 1–3, how does solving a related equation help you solve a polynomial inequality? Are the solutions to the related equation ever solutions to the inequality?

Objective 2 8. In Example 4, one of the values that separates the number line into regions is 4. The inequality is ≥, so why isn't 4 included in the solution set?

18.4 Exercise Set MyMathLab®

Solve each polynomial inequality. Write the solution set in interval notation. See Examples 1 through 3.

1. $(x + 1)(x + 5) > 0$
2. $(x + 1)(x + 5) \leq 0$
3. $(x - 3)(x + 4) \leq 0$
4. $(x + 4)(x - 1) > 0$
5. $x^2 - 7x + 10 \leq 0$
6. $x^2 + 8x + 15 \geq 0$
7. $3x^2 + 16x < -5$
8. $2x^2 - 5x < 7$
9. $(x - 6)(x - 4)(x - 2) > 0$
10. $(x - 6)(x - 4)(x - 2) \leq 0$
11. $x(x - 1)(x + 4) \leq 0$
12. $x(x - 6)(x + 2) > 0$
13. $(x^2 - 9)(x^2 - 4) > 0$
14. $(x^2 - 16)(x^2 - 1) \leq 0$

Solve each inequality. Write the solution set in interval notation. See Example 4.

15. $\dfrac{x + 7}{x - 2} < 0$
16. $\dfrac{x - 5}{x - 6} > 0$
17. $\dfrac{5}{x + 1} > 0$
18. $\dfrac{3}{y - 5} < 0$
19. $\dfrac{x + 1}{x - 4} \geq 0$
20. $\dfrac{x + 1}{x - 4} \leq 0$

Solve each inequality. Write the solution set in interval notation. See Example 5.

21. $\frac{3}{x-2} < 4$

22. $\frac{-2}{y+3} > 2$

23. $\frac{x^2+6}{5x} \ge 1$

24. $\frac{y^2+15}{8y} \le 1$

25. $\frac{x+2}{x-3} < 1$

26. $\frac{x-1}{x+4} > 2$

MIXED PRACTICE

Solve each inequality. Write the solution set in interval notation.

27. $(2x-3)(4x+5) \le 0$

28. $(6x+7)(7x-12) > 0$

29. $x^2 > x$

30. $x^2 < 25$

31. $(2x-8)(x+4)(x-6) \le 0$

32. $(3x-12)(x+5)(2x-3) \ge 0$

33. $6x^2 - 5x \ge 6$

34. $12x^2 + 11x \le 15$

35. $4x^3 + 16x^2 - 9x - 36 > 0$

36. $x^3 + 2x^2 - 4x - 8 < 0$

37. $x^4 - 26x^2 + 25 \ge 0$

38. $16x^4 - 40x^2 + 9 \le 0$

39. $(2x-7)(3x+5) > 0$

40. $(4x-9)(2x+5) < 0$

41. $\frac{x}{x-10} < 0$

42. $\frac{x+10}{x-10} > 0$

43. $\frac{x-5}{x+4} \ge 0$

44. $\frac{x-3}{x+2} \le 0$

45. $\frac{x(x+6)}{(x-7)(x+1)} \ge 0$

46. $\frac{(x-2)(x+2)}{(x+1)(x-4)} \le 0$

47. $\frac{-1}{x-1} > -1$

48. $\frac{4}{y+2} < -2$

49. $\frac{x}{x+4} \le 2$

50. $\frac{4x}{x-3} \ge 5$

51. $\frac{z}{z-5} \ge 2z$

52. $\frac{p}{p+4} \le 3p$

53. $\frac{(x+1)^2}{5x} > 0$

54. $\frac{(2x-3)^2}{x} < 0$

REVIEW AND PREVIEW

Recall that the graph of $f(x) + K$ is the same as the graph of $f(x)$ shifted K units upward if $K > 0$ and $|K|$ units downward if $K < 0$. Use the graph of $f(x) = |x|$ below to sketch the graph of each function. See Section 15.3.

55. $g(x) = |x| + 2$

56. $H(x) = |x| - 2$

57. $f(x) = |x| - 1$

58. $h(x) = |x| + 5$

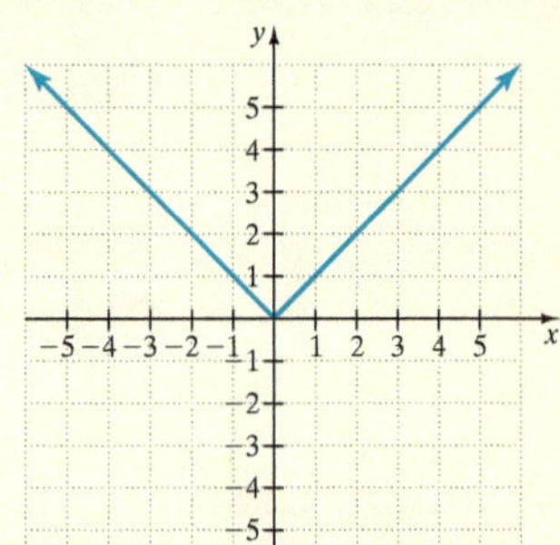

Use the graph of $f(x) = x^2$ below to sketch the graph of each function.

59. $F(x) = x^2 - 3$

60. $h(x) = x^2 - 4$

61. $H(x) = x^2 + 1$

62. $g(x) = x^2 + 3$

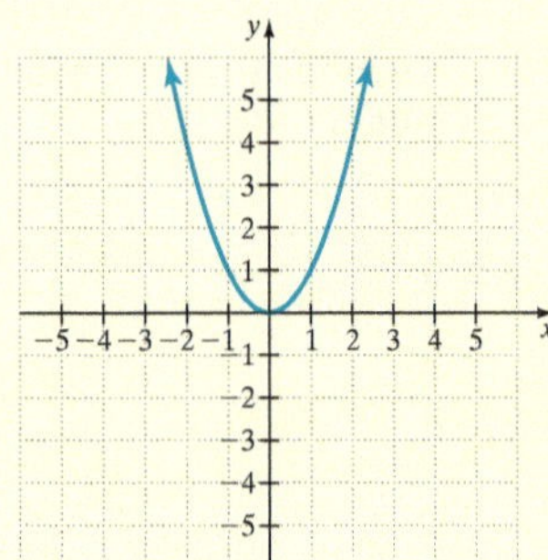

CONCEPT EXTENSIONS

63. Explain why $\frac{x+2}{x-3} > 0$ and $(x+2)(x-3) > 0$ have the same solutions.

64. Explain why $\frac{x+2}{x-3} \ge 0$ and $(x+2)(x-3) \ge 0$ do not have the same solutions.

Find all numbers that satisfy each of the following.

65. A number minus its reciprocal is less than zero. Find the numbers.

66. Twice a number added to its reciprocal is nonnegative. Find the numbers.

67. The total profit function $P(x)$ for a company producing x thousand units is given by

$$P(x) = -2x^2 + 26x - 44$$

Find the values of x for which the company makes a profit. [*Hint:* The company makes a profit when $P(x) > 0$.]

68. A projectile is fired straight up from the ground with an initial velocity of 80 feet per second. Its height $s(t)$ in feet at any time t is given by the function

$$s(t) = -16t^2 + 80t$$

Find the interval of time for which the height of the projectile is greater than 96 feet.

Use a graphing calculator to check each exercise.

69. Exercise 37 **70.** Exercise 38

71. Exercise 39 **72.** Exercise 40

18.5 Quadratic Functions and Their Graphs

Objective 1 Graphing $f(x) = x^2 + k$

Objectives

1 Graph Quadratic Functions of the Form $f(x) = x^2 + k$.

2 Graph Quadratic Functions of the Form $f(x) = (x - h)^2$.

3 Graph Quadratic Functions of the Form $f(x) = (x - h)^2 + k$.

4 Graph Quadratic Functions of the Form $f(x) = ax^2$.

5 Graph Quadratic Functions of the Form $f(x) = a(x - h)^2 + k$.

We first graphed the quadratic equation $y = x^2$ in Section 15.2. In Sections 15.2 and 15.3, we learned that this graph defines a function, and we wrote $y = x^2$ as $f(x) = x^2$. In these sections, we discovered that the graph of a quadratic function is a parabola opening upward or downward. In this section, we continue our study of quadratic functions and their graphs. (Much of the contents of this section is a review of shifting and reflecting techniques from Section 15.3, but specific to quadratic functions.)

First, let's recall the definition of a quadratic function.

Quadratic Function

A quadratic function is a function that can be written in the form $f(x) = ax^2 + bx + c$, where a, b, and c are real numbers and $a \neq 0$.

Notice that equations of the form $y = ax^2 + bx + c$, where $a \neq 0$, define quadratic functions, since y is a function of x or $y = f(x)$.

Recall that if $a > 0$, the parabola opens upward and if $a < 0$, the parabola opens downward. Also, the vertex of a parabola is the lowest point if the parabola opens upward and the highest point if the parabola opens downward. The axis of symmetry is the vertical line that passes through the vertex.

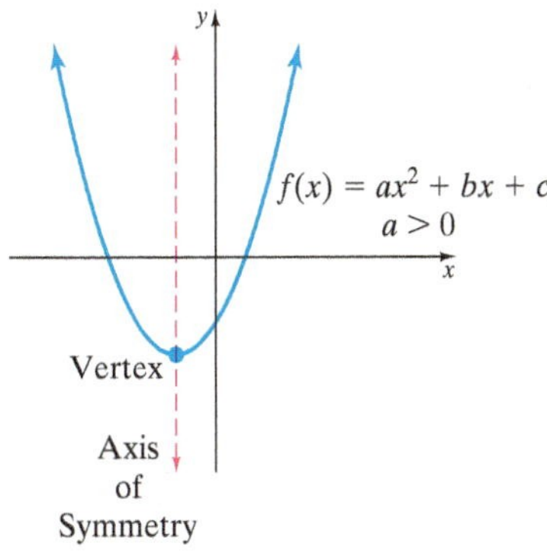

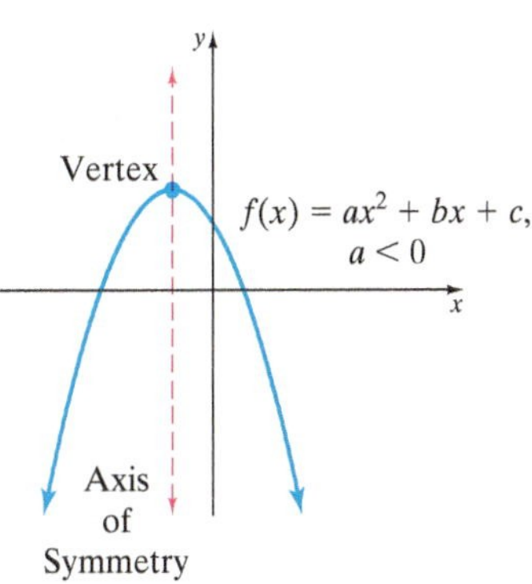

Practice 1

Graph $f(x) = x^2$ and $g(x) = x^2 - 4$ on the same set of axes.

Example 1

Graph $f(x) = x^2$ and $g(x) = x^2 + 6$ on the same set of axes.

Solution: First we construct a table of values for $f(x)$ and plot the points. Notice that for each x-value, the corresponding value of $g(x)$ must be 6 more than the corresponding value of $f(x)$ since $f(x) = x^2$ and $g(x) = x^2 + 6$. In other words, the graph of $g(x) = x^2 + 6$ is the same as the graph of $f(x) = x^2$ shifted upward 6 units. The axis of symmetry for both graphs is the y-axis.

x	$f(x) = x^2$	$g(x) = x^2 + 6$
−2	4	10
−1	1	7
0	0	6
1	1	7
2	4	10

Each y-value is increased by 6.

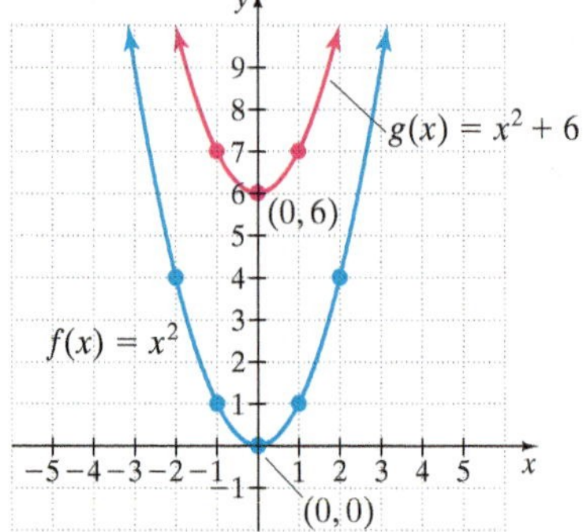

Work Practice 1

In general, we have the following properties.

Graphing the Parabola Defined by $f(x) = x^2 + k$

If k is positive, the graph of $f(x) = x^2 + k$ is the graph of $y = x^2$ shifted upward k units.
If k is negative, the graph of $f(x) = x^2 + k$ is the graph of $y = x^2$ shifted downward $|k|$ units.
The vertex is $(0, k)$, and the axis of symmetry is the y-axis.

Practice 2

Graph each function.

a. $f(x) = x^2 - 5$
b. $g(x) = x^2 + 3$

Example 2

Graph each function.

a. $F(x) = x^2 + 2$ **b.** $g(x) = x^2 - 3$

Solution:

a. $F(x) = x^2 + 2$

The graph of $F(x) = x^2 + 2$ is obtained by shifting the graph of $y = x^2$ upward 2 units.

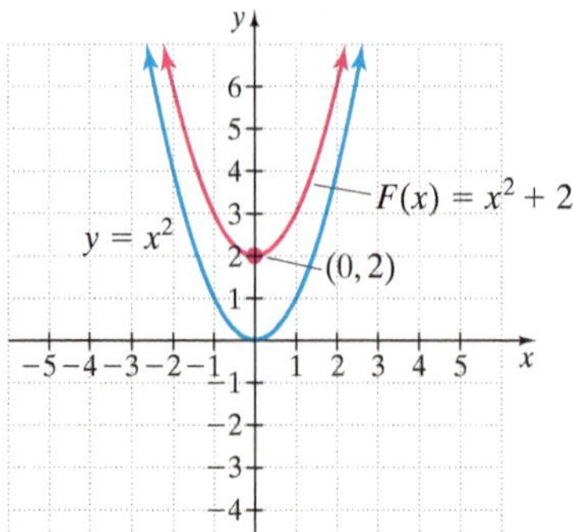

Answers

1.

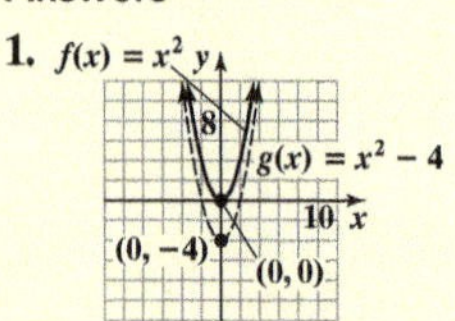

2. a.

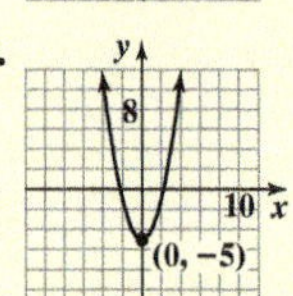

b.

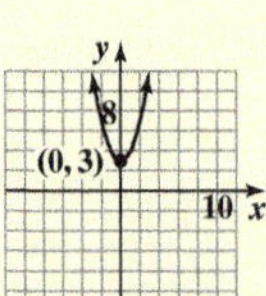

b. $g(x) = x^2 - 3$

The graph of $g(x) = x^2 - 3$ is obtained by shifting the graph of $y = x^2$ downward 3 units.

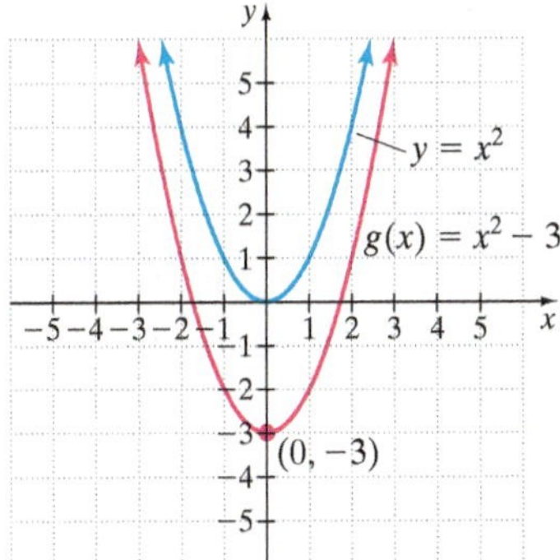

Work Practice 2

Objective 2 Graphing $f(x) = (x - h)^2$

Now we will graph functions of the form $f(x) = (x - h)^2$.

Example 3 Graph $f(x) = x^2$ and $g(x) = (x - 2)^2$ on the same set of axes.

Solution: By plotting points, we see that for each x-value, the corresponding value of $g(x)$ is the same as the value of $f(x)$ when the x-value is increased by 2. Thus, the graph of $g(x) = (x - 2)^2$ is the graph of $f(x) = x^2$ shifted to the right 2 units. The axis of symmetry for the graph of $g(x) = (x - 2)^2$ is also shifted 2 units to the right and is the line $x = 2$.

x	$f(x) = x^2$	x	$g(x) = (x - 2)^2$
−2	4	0	4
−1	1	1	1
0	0	2	0
1	1	3	1
2	4	4	4

Each x-value increased by 2 corresponds to same y-value.

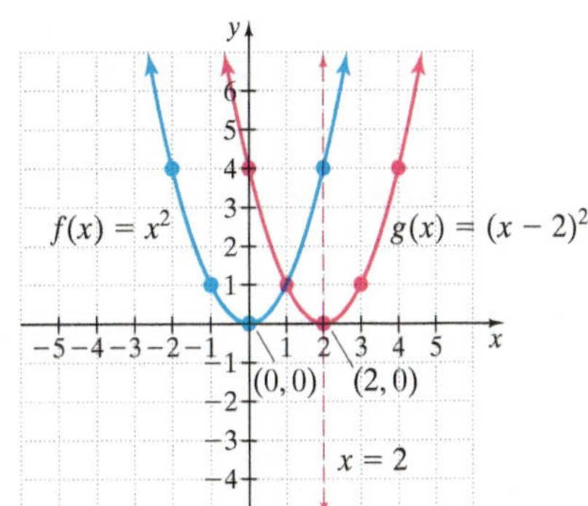

Work Practice 3

Practice 3

Graph $f(x) = x^2$ and $g(x) = (x + 6)^2$ on the same set of axes.

In general, we have the following properties.

Graphing the Parabola Defined by $f(x) = (x - h)^2$

If h is positive, the graph of $f(x) = (x - h)^2$ is the graph of $y = x^2$ shifted to the right h units.

If h is negative, the graph of $f(x) = (x - h)^2$ is the graph of $y = x^2$ shifted to the left $|h|$ units.

The vertex is $(h, 0)$, and the axis of symmetry is the vertical line $x = h$.

Answer

3.

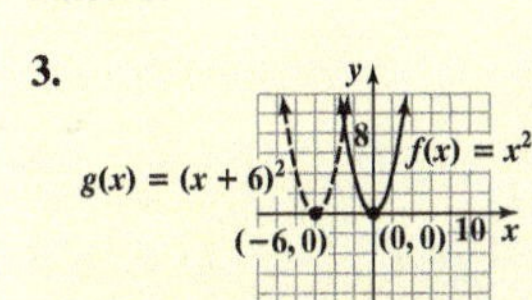

Practice 4

Graph each function.

a. $G(x) = (x + 4)^2$

b. $H(x) = (x - 7)^2$

Example 4

Graph each function.

a. $G(x) = (x - 3)^2$

b. $F(x) = (x + 1)^2$

Solution:

a. The graph of $G(x) = (x - 3)^2$ is obtained by shifting the graph of $y = x^2$ to the right 3 units. The graph of $G(x)$ is below on the left.

b. The equation $F(x) = (x + 1)^2$ can be written as $F(x) = [x - (-1)]^2$. The graph of $F(x) = [x - (-1)]^2$ is obtained by shifting the graph of $y = x^2$ to the left 1 unit. The graph of $F(x)$ is below on the right.

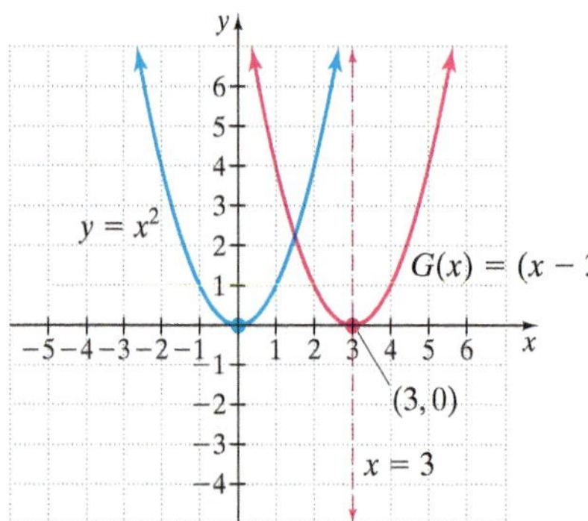

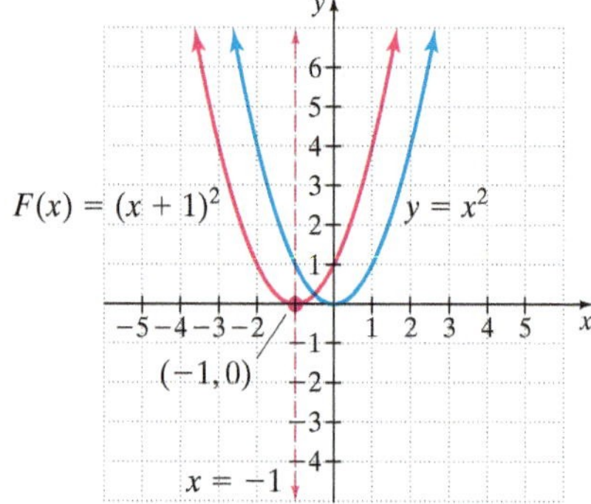

Work Practice 4

Objective 3 Graphing $f(x) = (x - h)^2 + k$

As we will see in graphing functions of the form $f(x) = (x - h)^2 + k$, it is possible to combine vertical and horizontal shifts.

> **Graphing the Parabola Defined by $f(x) = (x - h)^2 + k$**
>
> The parabola has the same shape as $y = x^2$.
> The vertex is (h, k), and the axis of symmetry is the vertical line $x = h$.

Practice 5

Graph $f(x) = (x + 2)^2 + 2$.

Example 5

Graph $F(x) = (x - 3)^2 + 1$.

Solution: The graph of $F(x) = (x - 3)^2 + 1$ is the graph of $y = x^2$ shifted 3 units to the right and 1 unit up. The vertex is then $(3, 1)$, and the axis of symmetry is $x = 3$. A few ordered pair solutions are plotted to aid in graphing.

x	$F(x) = (x - 3)^2 + 1$
1	5
2	2
4	2
5	5

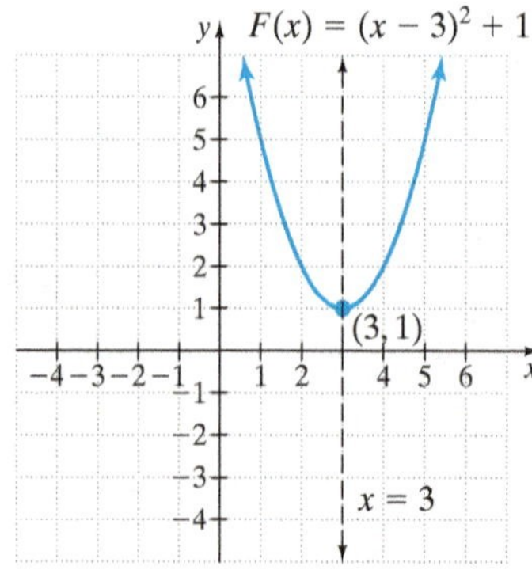

Work Practice 5

Answers

4. a.

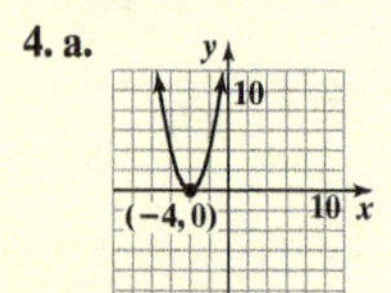

b.

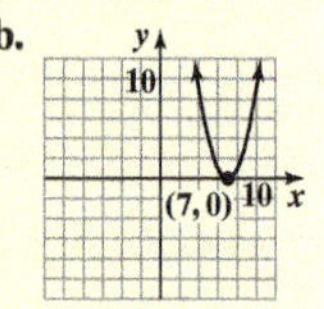

5.

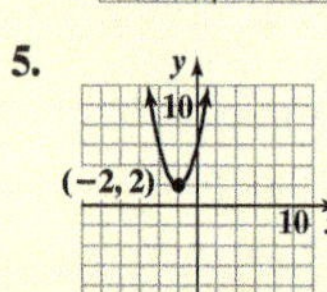

Objective 4 Graphing $f(x) = ax^2$

Next, we discover the change in the shape of a graph when the coefficient of x^2 is not 1.

Example 6 Graph $f(x) = x^2, g(x) = 3x^2$, and $h(x) = \frac{1}{2}x^2$ on the same set of axes.

Solution: Comparing the tables of values, we see that for each x-value, the corresponding value of $g(x)$ is triple the corresponding value of $f(x)$. Similarly, the value of $h(x)$ is half the value of $f(x)$.

x	$f(x) = x^2$
−2	4
−1	1
0	0
1	1
2	4

x	$g(x) = 3x^2$
−2	12
−1	3
0	0
1	3
2	12

x	$h(x) = \frac{1}{2}x^2$
−2	2
−1	$\frac{1}{2}$
0	0
1	$\frac{1}{2}$
2	2

The result is that the graph of $g(x) = 3x^2$ is narrower than the graph of $f(x) = x^2$, and the graph of $h(x) = \frac{1}{2}x^2$ is wider. The vertex for each graph is (0, 0), and the axis of symmetry is the y-axis.

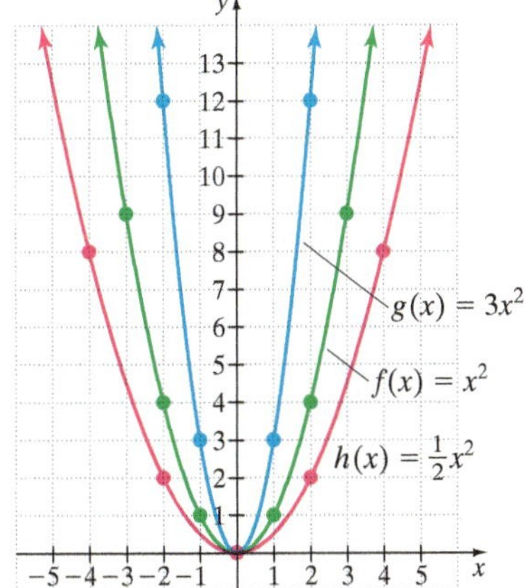

Work Practice 6

Practice 6

Graph $f(x) = x^2, g(x) = 4x^2$, and $h(x) = \frac{1}{4}x^2$ on the same set of axes.

Graphing the Parabola Defined by $f(x) = ax^2$

If a is positive, the parabola opens upward, and if a is negative, the parabola opens downward.
If $|a| > 1$, the graph of the parabola is narrower than the graph of $y = x^2$.
If $|a| < 1$, the graph of the parabola is wider than the graph of $y = x^2$.

Answer

6. 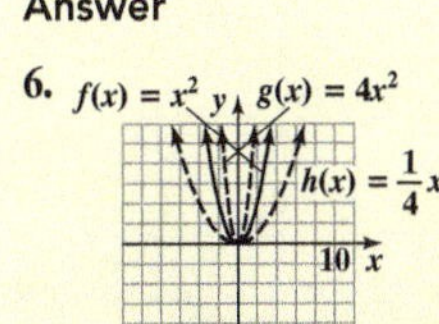

Practice 7

Graph $f(x) = -\frac{1}{2}x^2$.

Example 7 Graph $f(x) = -2x^2$.

Solution: Because $a = -2$, a negative value, this parabola opens downward. Since $|-2| = 2$ and $2 > 1$, the parabola is narrower than the graph of $y = x^2$. The vertex is $(0, 0)$, and the axis of symmetry is the y-axis. We verify this by plotting a few points.

x	$f(x) = -2x^2$
−2	−8
−1	−2
0	0
1	−2
2	−8

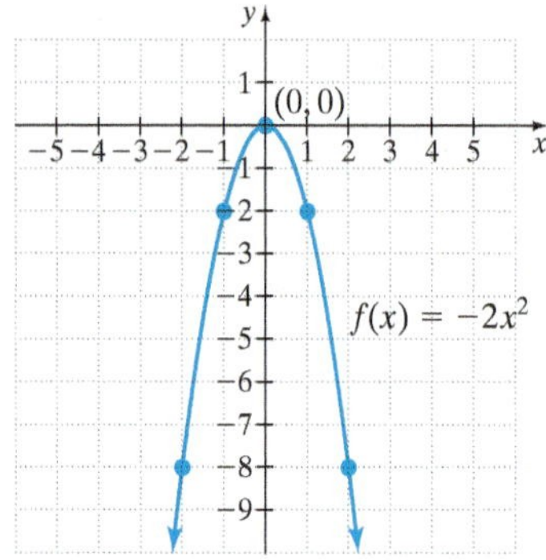

Work Practice 7

Objective 5 Graphing $f(x) = a(x - h)^2 + k$

Now we will see the shape of the graph of a quadratic function of the form $f(x) = a(x - h)^2 + k$.

Practice 8

Graph $h(x) = \frac{1}{3}(x - 4)^2 - 3$.

Example 8 Graph $g(x) = \frac{1}{2}(x + 2)^2 + 5$. Find the vertex and the axis of symmetry.

Solution: The function $g(x) = \frac{1}{2}(x + 2)^2 + 5$ may be written as $g(x) = \frac{1}{2}[x - (-2)]^2 + 5$. Thus, this graph is the same as the graph of $y = x^2$ shifted 2 units to the left and 5 units up, and it is wider because a is $\frac{1}{2}$. The vertex is $(-2, 5)$, and the axis of symmetry is $x = -2$. We plot a few points to verify.

x	$g(x) = \frac{1}{2}(x + 2)^2 + 5$
−4	7
−3	$5\frac{1}{2}$
−2	5
−1	$5\frac{1}{2}$
0	7

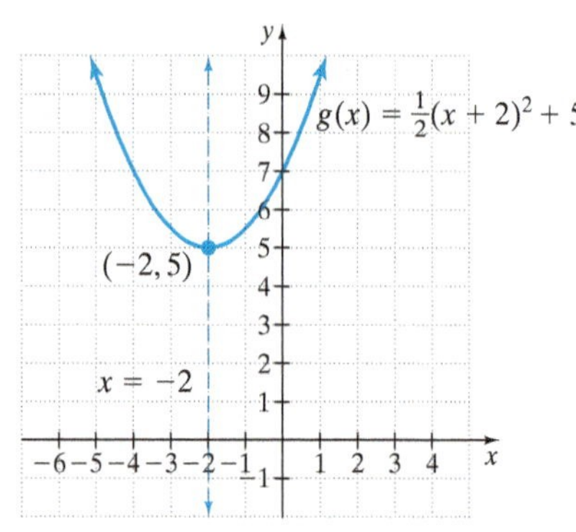

Work Practice 8

Answers

7. 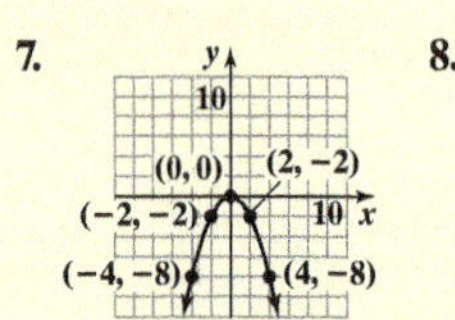

8. 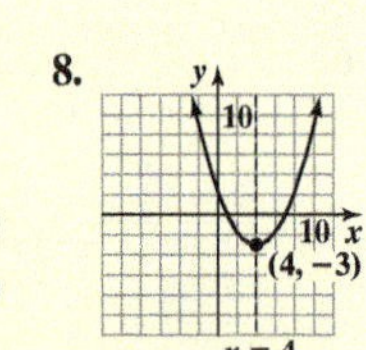

In general, the following holds

Graph of a Quadratic Function

The graph of a quadratic function written in the form $f(x) = a(x - h)^2 + k$ is a parabola with vertex (h, k).

If $a > 0$, the parabola opens upward.
If $a < 0$, the parabola opens downward.

The axis of symmetry is the line whose equation is $x = h$.

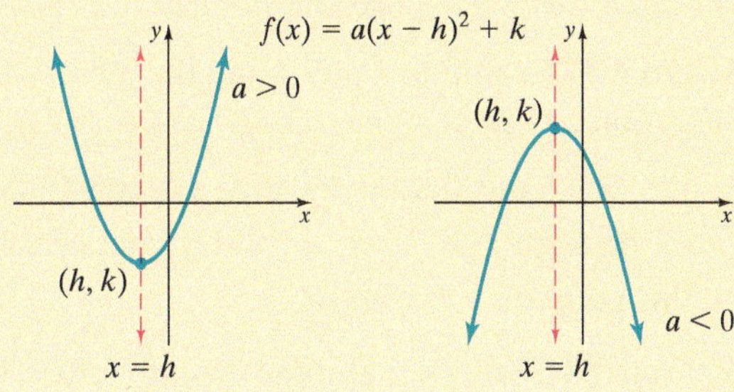

✓**Concept Check** Which description of the graph of $f(x) = -0.35(x + 3)^2 - 4$ is correct?

a. The graph opens downward and has its vertex at $(-3, 4)$.
b. The graph opens upward and has its vertex at $(-3, 4)$.
c. The graph opens downward and has its vertex at $(-3, -4)$.
d. The graph is narrower than the graph of $y = x^2$.

✓**Concept Check Answer**
c

Graphing Calculator Explorations

Use a graphing calculator to graph the first function of each pair that follows. Then use its graph to predict the graph of the second function. Check your prediction by graphing both functions on the same set of axes.

1. $F(x) = \sqrt{x}; G(x) = \sqrt{x} + 1$
2. $g(x) = x^3; H(x) = x^3 - 2$
3. $H(x) = |x|; f(x) = |x - 5|$
4. $h(x) = x^3 + 2; g(x) = (x - 3)^3 + 2$
5. $f(x) = |x + 4|; F(x) = |x + 4| + 3$
6. $G(x) = \sqrt{x} - 2; g(x) = \sqrt{x - 4} - 2$

Vocabulary, Readiness & Video Check

Use the choices below to fill in each blank. Some choices will be used more than once.

upward highest parabola downward lowest quadratic

1. A(n) __________ function is one that can be written in the form $f(x) = ax^2 + bx + c, a \neq 0$.
2. The graph of a quadratic function is a(n) __________ opening __________ or __________.
3. If $a > 0$, the graph of the quadratic function opens __________.
4. If $a < 0$, the graph of the quadratic function opens __________.

5. The vertex of a parabola is the __________ point if $a > 0$.

6. The vertex of a parabola is the __________ point if $a < 0$.

State the vertex of the graph of each quadratic function.

7. $f(x) = x^2$ **8.** $f(x) = -5x^2$ **9.** $g(x) = (x - 2)^2$ **10.** $g(x) = (x + 5)^2$

11. $f(x) = 2x^2 + 3$ **12.** $h(x) = x^2 - 1$ **13.** $g(x) = (x + 1)^2 + 5$ **14.** $h(x) = (x - 10)^2 - 7$

Martin-Gay Interactive Videos *Watch the section lecture video and answer the following questions.*

See Video 18.5

Objective 1 **15.** From Examples 1 and 2 and the lecture before, how do graphs of the form $f(x) = x^2 + k$ differ from the graph of $y = x^2$? Consider the location of the vertex $(0, k)$ on graphs of the form $f(x) = x^2 + k$—by what other name do we call this point on a graph?

Objective 2 **16.** From Example 3 and the lecture before, how do graphs of the form $f(x) = (x - h)^2$ differ from the graph of $y = x^2$? Consider the location of the vertex $(h, 0)$ on graphs of the form $f(x) = (x - h)^2$—by what other name do we call this point on a graph?

Objective 3 **17.** From Example 4 and the lecture before, what general information does the equation $f(x) = (x - h)^2 + k$ tell us about its graph?

Objective 4 **18.** From the lecture before Example 5, besides the direction a parabola opens, what other graphing information can the value of a tell us?

Objective 5 **19.** In Examples 6 and 7, what four properties of the graph did we learn from the equation that helped us locate and draw the general shape of the parabola?

18.5 Exercise Set MyMathLab®

MIXED PRACTICE

Sketch the graph of each quadratic function. Label the vertex and sketch and label the axis of symmetry. See Examples 1 through 5.

1. $f(x) = x^2 - 1$ **2.** $g(x) = x^2 + 3$

3. $h(x) = x^2 + 5$ **4.** $h(x) = x^2 - 4$

5. $g(x) = x^2 + 7$ **6.** $f(x) = x^2 - 2$

7. $f(x) = (x - 5)^2$ **8.** $g(x) = (x + 5)^2$

9. $h(x) = (x + 2)^2$ **10.** $H(x) = (x - 1)^2$

11. $G(x) = (x + 3)^2$ **12.** $f(x) = (x - 6)^2$

13. $f(x) = (x - 2)^2 + 5$ **14.** $g(x) = (x - 6)^2 + 1$

15. $h(x) = (x + 1)^2 + 4$ **16.** $G(x) = (x + 3)^2 + 3$

17. $g(x) = (x + 2)^2 - 5$ **18.** $h(x) = (x + 4)^2 - 6$

Sketch the graph of each quadratic function. Label the vertex, and sketch and label the axis of symmetry. See Examples 6 and 7.

19. $H(x) = 2x^2$ **20.** $f(x) = 5x^2$ **21.** $h(x) = \frac{1}{3}x^2$

22. $f(x) = -\frac{1}{4}x^2$ **23.** $g(x) = -x^2$ **24.** $g(x) = -3x^2$

Sketch the graph of each quadratic function. Label the vertex and sketch and label the axis of symmetry. See Example 8.

25. $f(x) = 2(x - 1)^2 + 3$ **26.** $g(x) = 4(x - 4)^2 + 2$

27. $h(x) = -3(x + 3)^2 + 1$ **28.** $f(x) = -(x - 2)^2 - 6$

29. $H(x) = \frac{1}{2}(x - 6)^2 - 3$ **30.** $G(x) = \frac{1}{5}(x + 4)^2 + 3$

MIXED PRACTICE

Sketch the graph of each quadratic function. Label the vertex and sketch and label the axis of symmetry.

31. $f(x) = -(x-2)^2$
32. $g(x) = -(x+6)^2$
33. $F(x) = -x^2 + 4$
34. $H(x) = -x^2 + 10$
35. $F(x) = 2x^2 - 5$
36. $g(x) = \frac{1}{2}x^2 - 2$
37. $h(x) = (x-6)^2 + 4$
38. $f(x) = (x-5)^2 + 2$
39. $F(x) = \left(x + \frac{1}{2}\right)^2 - 2$
40. $H(x) = \left(x + \frac{1}{2}\right)^2 - 3$
41. $F(x) = \frac{3}{2}(x+7)^2 + 1$
42. $g(x) = -\frac{3}{2}(x-1)^2 - 5$
43. $f(x) = \frac{1}{4}x^2 - 9$
44. $H(x) = \frac{3}{4}x^2 - 2$
45. $G(x) = 5\left(x + \frac{1}{2}\right)^2$
46. $F(x) = 3\left(x - \frac{3}{2}\right)^2$
47. $h(x) = -(x-1)^2 - 1$
48. $f(x) = -3(x+2)^2 + 2$
49. $g(x) = \sqrt{3}(x+5)^2 + \frac{3}{4}$
50. $G(x) = \sqrt{5}(x-7)^2 - \frac{1}{2}$
51. $h(x) = 10(x+4)^2 - 6$
52. $h(x) = 8(x+1)^2 + 9$
53. $f(x) = -2(x-4)^2 + 5$
54. $G(x) = -4(x+9)^2 - 1$

REVIEW AND PREVIEW

Add the proper constant to each binomial so that the resulting trinomial is a perfect square trinomial. See Section 18.1.

55. $x^2 + 8x$
56. $y^2 + 4y$
57. $z^2 - 16z$
58. $x^2 - 10x$
59. $y^2 + y$
60. $z^2 - 3z$

Solve by completing the square. See Section 18.1.

61. $x^2 + 4x = 12$
62. $y^2 + 6y = -5$
63. $z^2 + 10z - 1 = 0$
64. $x^2 + 14x + 20 = 0$
65. $z^2 - 8z = 2$
66. $y^2 - 10y = 3$

CONCEPT EXTENSIONS

Solve. See the Concept Check in this section.

67. Which description of $f(x) = -213(x-0.1)^2 + 3.6$ is correct?

Graph Opens	Vertex
a. upward	$(0.1, 3.6)$
b. upward	$(-213, 3.6)$
c. downward	$(0.1, 3.6)$
d. downward	$(-0.1, 3.6)$

68. Which description of $f(x) = 5\left(x + \frac{1}{2}\right)^2 + \frac{1}{2}$ is correct?

Graph Opens	Vertex
a. upward	$\left(\frac{1}{2}, \frac{1}{2}\right)$
b. upward	$\left(-\frac{1}{2}, \frac{1}{2}\right)$
c. downward	$\left(\frac{1}{2}, -\frac{1}{2}\right)$
d. downward	$\left(-\frac{1}{2}, -\frac{1}{2}\right)$

Write the equation of the parabola that has the same shape as $f(x) = 5x^2$ but with the following vertex.

69. $(2, 3)$
70. $(1, 6)$
71. $(-3, 6)$
72. $(4, -1)$

The shifting properties covered in this section apply to the graphs of all functions. Given the graph of $y = f(x)$ below, sketch the graph of each of the following.

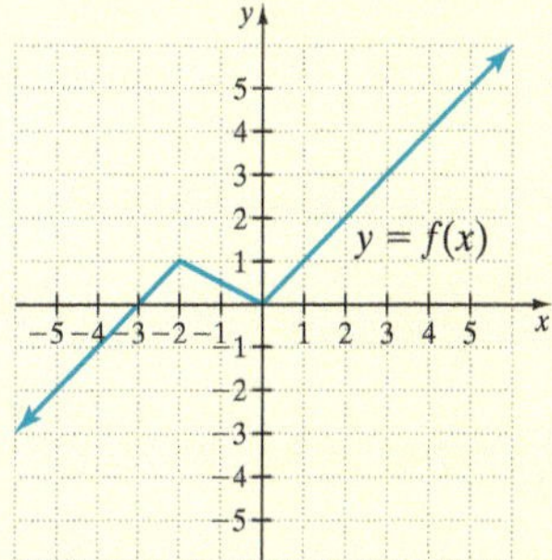

73. $y = f(x) + 1$
74. $y = f(x) - 2$
75. $y = f(x-3)$
76. $y = f(x+3)$
77. $y = f(x+2) + 2$
78. $y = f(x-1) + 1$
79. The quadratic function $f(x) = 2158x^2 - 10{,}339x + 6731$ approximates the number of text messages sent in the United States each month between 2000 and 2008, where x is the number of years past 2000 and $f(x)$ is the number of text messages sent in the U.S. each month in millions. (*Source:* cellsigns)
    - **a.** Use this function to find the number of text messages sent in the U.S. each month in 2010.
    - **b.** Use this function to estimate the number of text messages sent in the U.S. each month in 2014.
80. Use the function in Exercise 79.
    - **a.** Use this function to predict the number of text messages sent in the U.S. each month in 2018.
    - **b.** Look up the current number of cell phone subscribers in the U.S.
    - **c.** Based on your answers for parts **a.** and **b.**, discuss some possible limitations of using this quadratic function to predict data.

18.6 Further Graphing of Quadratic Functions

Objectives

1 Write Quadratic Functions in the Form $y = a(x - h)^2 + k$.

2 Derive a Formula for Finding the Vertex of a Parabola.

3 Find the Minimum or Maximum Value of a Quadratic Function.

Objective 1 Writing Quadratic Functions in the Form $y = a(x - h)^2 + k$

We know that the graph of a quadratic function is a parabola. If a quadratic function is written in the form

$$f(x) = a(x - h)^2 + k$$

we can easily find the vertex (h, k) and graph the parabola. To write a quadratic function in this form, complete the square. (See Section 18.1 for a review of completing the square.)

Example 1 Graph $f(x) = x^2 - 4x - 12$. Find the vertex and any intercepts.

Solution: The graph of this quadratic function is a parabola. To find the vertex of the parabola, we will write the function in the form $y = (x - h)^2 + k$. To do this, we complete the square on the binomial $x^2 - 4x$. To simplify our work, we let $f(x) = y$.

$y = x^2 - 4x - 12$ Let $f(x) = y$.

$y + 12 = x^2 - 4x$ Add 12 to both sides to get the x-variable terms alone.

Now we add the square of half of -4 to both sides.

$$\frac{1}{2}(-4) = -2 \quad \text{and} \quad (-2)^2 = 4$$

$y + 12 + 4 = x^2 - 4x + 4$ Add 4 to both sides.

$y + 16 = (x - 2)^2$ Factor the trinomial.

$y = (x - 2)^2 - 16$ Subtract 16 from both sides.

$f(x) = (x - 2)^2 - 16$ Replace y with $f(x)$.

From this equation, we can see that the vertex of the parabola is $(2, -16)$, a point in quadrant IV, and the axis of symmetry is the line $x = 2$.

Notice that $a = 1$. Since $a > 0$, the parabola opens upward. This parabola opening upward with vertex $(2, -16)$ will have two x-intercepts and one y-intercept. (See the Helpful Hint after this example.)

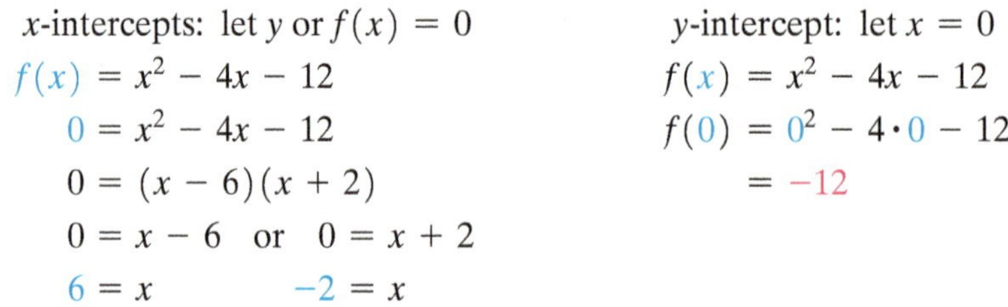

x-intercepts: let y or $f(x) = 0$

$f(x) = x^2 - 4x - 12$

$0 = x^2 - 4x - 12$

$0 = (x - 6)(x + 2)$

$0 = x - 6$ or $0 = x + 2$

$6 = x$ $\quad -2 = x$

y-intercept: let $x = 0$

$f(x) = x^2 - 4x - 12$

$f(0) = 0^2 - 4 \cdot 0 - 12$

$= -12$

The two x-intercepts are $(6, 0)$ and $(-2, 0)$. The y-intercept is $(0, -12)$. The sketch of $f(x) = x^2 - 4x - 12$ is shown.

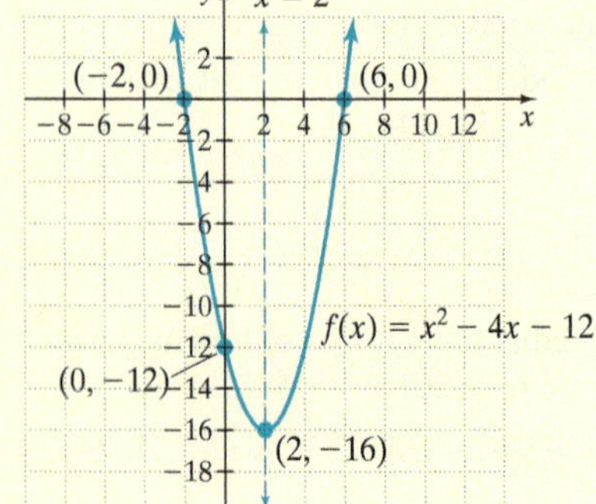

Notice that the axis of symmetry is always halfway between the x-intercepts. For this example, halfway between -2 and 6 is $\frac{-2 + 6}{2} = 2$, and the axis of symmetry is $x = 2$.

Work Practice 1

Practice 1

Graph $g(x) = x^2 - 2x - 3$. Find the vertex and any intercepts.

Answer

1.

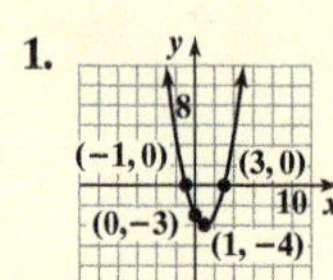

Helpful Hint

Parabola Opens Upward
Vertex in I or II: no x-intercept
Vertex in III or IV: 2 x-intercepts

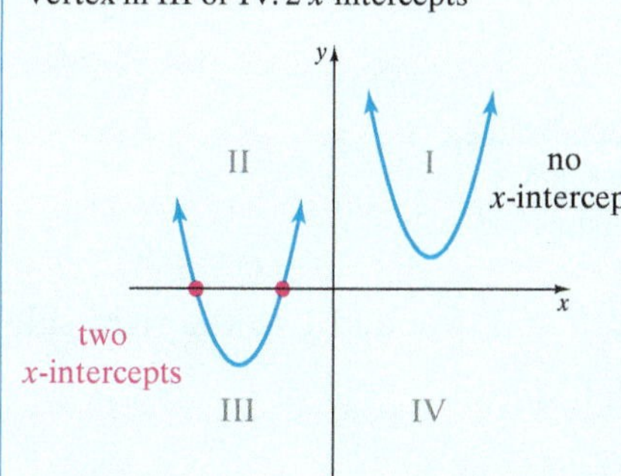

Parabola Opens Downward
Vertex in I or II: 2 x-intercepts
Vertex in III or IV: no x-intercept.

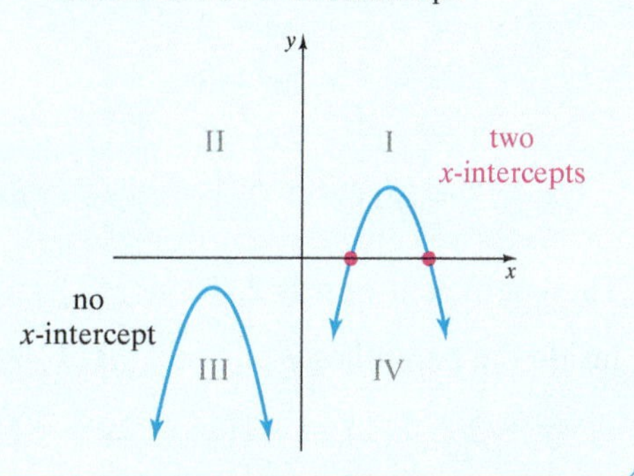

Example 2 Graph $f(x) = 3x^2 + 3x + 1$. Find the vertex and any intercepts.

Solution: Replace $f(x)$ with y and complete the square on x to write the equation in the form $y = a(x - h)^2 + k$.

$$y = 3x^2 + 3x + 1 \quad \text{Replace } f(x) \text{ with } y.$$
$$y - 1 = 3x^2 + 3x \quad \text{Isolate } x\text{-variable terms.}$$

Factor 3 from the terms $3x^2 + 3x$ so that the coefficient of x^2 is 1.

$$y - 1 = 3(x^2 + x) \quad \text{Factor out 3.}$$

The coefficient of x in the parentheses above is 1. Then $\frac{1}{2}(1) = \frac{1}{2}$ and $\left(\frac{1}{2}\right)^2 = \frac{1}{4}$. Since we are adding $\frac{1}{4}$ inside the parentheses, we are really adding $3\left(\frac{1}{4}\right)$, so we *must* add $3\left(\frac{1}{4}\right)$ to the left side.

$$y - 1 + 3\left(\frac{1}{4}\right) = 3\left(x^2 + x + \frac{1}{4}\right)$$
$$y - \frac{1}{4} = 3\left(x + \frac{1}{2}\right)^2 \quad \text{Simplify the left side and factor the right side.}$$
$$y = 3\left(x + \frac{1}{2}\right)^2 + \frac{1}{4} \quad \text{Add } \frac{1}{4} \text{ to both sides.}$$
$$f(x) = 3\left(x + \frac{1}{2}\right)^2 + \frac{1}{4} \quad \text{Replace } y \text{ with } f(x).$$

Then $a = 3$, $h = -\frac{1}{2}$, and $k = \frac{1}{4}$. This means that the parabola opens upward with vertex $\left(-\frac{1}{2}, \frac{1}{4}\right)$ and that the axis of symmetry is the line $x = -\frac{1}{2}$.

To find the y-intercept, let $x = 0$. Then

$$f(0) = 3(0)^2 + 3(0) + 1 = 1$$

Thus, the y-intercept is $(0, 1)$.

This parabola has no x-intercepts since the vertex is in the second quadrant and the parabola opens upward. Use the vertex, axis of symmetry, and y-intercept to sketch the parabola.

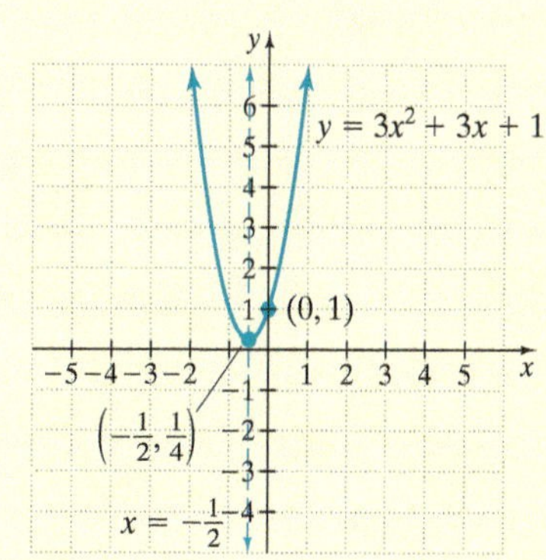

Work Practice 2

Practice 2

Graph $g(x) = 4x^2 + 4x + 3$. Find the vertex and any intercepts.

Answer

2. $\left(-\frac{1}{2}, 2\right)$

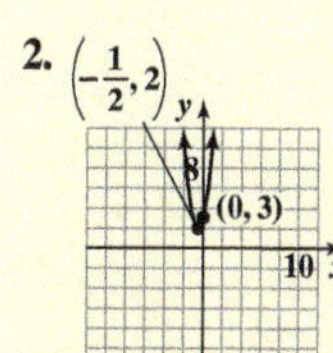

Practice 3

Graph $g(x) = -x^2 + 5x + 6$. Find the vertex and any intercepts.

Example 3 Graph $f(x) = -x^2 - 2x + 3$. Find the vertex and any intercepts.

Solution: We write $f(x)$ in the form $a(x - h)^2 + k$ by completing the square. First we replace $f(x)$ with y.

$$f(x) = -x^2 - 2x + 3$$
$$y = -x^2 - 2x + 3$$
$$y - 3 = -x^2 - 2x \quad \text{Subtract 3 from both sides to get the } x\text{-variable terms alone.}$$
$$y - 3 = -1(x^2 + 2x) \quad \text{Factor } -1 \text{ from the terms } -x^2 - 2x.$$

The coefficient of x is 2. Then $\frac{1}{2}(2) = 1$ and $1^2 = 1$. We add 1 to the right side inside the parentheses and add $-1(1)$ to the left side.

$$y - 3 - 1(1) = -1(x^2 + 2x + 1)$$
$$y - 4 = -1(x + 1)^2 \quad \text{Simplify the left side and factor the right side.}$$
$$y = -1(x + 1)^2 + 4 \quad \text{Add 4 to both sides.}$$
$$f(x) = -1(x + 1)^2 + 4 \quad \text{Replace } y \text{ with } f(x).$$

Helpful Hint This can be written as $f(x) = -1[x - (-1)]^2 + 4$. Notice that the vertex is $(-1, 4)$.

Since $a = -1$, the parabola opens downward with vertex $(-1, 4)$ and axis of symmetry $x = -1$.

To find the y-intercept, we let $x = 0$ and solve for y. Then

$$f(0) = -0^2 - 2(0) + 3 = 3$$

Thus, $(0, 3)$ is the y-intercept.

To find the x-intercepts, we let y or $f(x) = 0$ and solve for x.

$$f(x) = -x^2 - 2x + 3$$
$$0 = -x^2 - 2x + 3 \quad \text{Let } f(x) = 0.$$

Now we divide both sides by -1 so that the coefficient of x^2 is 1.

$$\frac{0}{-1} = \frac{-x^2}{-1} - \frac{2x}{-1} + \frac{3}{-1} \quad \text{Divide both sides by } -1.$$
$$0 = x^2 + 2x - 3 \quad \text{Simplify.}$$
$$0 = (x + 3)(x - 1) \quad \text{Factor.}$$
$$x + 3 = 0 \quad \text{or} \quad x - 1 = 0 \quad \text{Set each factor equal to 0.}$$
$$x = -3 \qquad x = 1 \quad \text{Solve.}$$

The x-intercepts are $(-3, 0)$ and $(1, 0)$. Use these points to sketch the parabola.

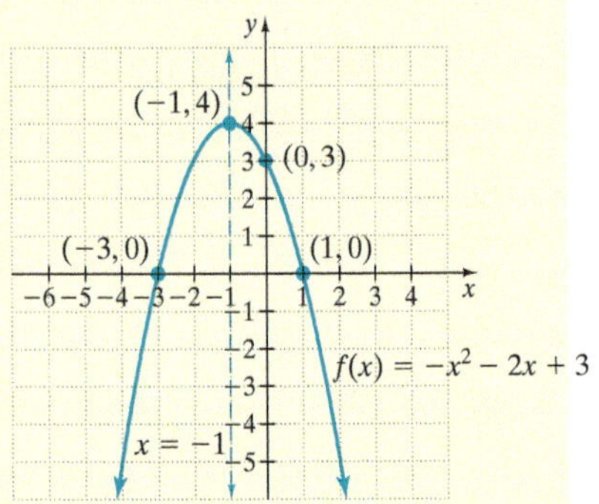

Work Practice 3

Objective 2 Deriving a Formula for Finding the Vertex

There is also a formula that may be used to find the vertex of a parabola. Now that we have practiced completing the square, we will show that the x-coordinate of the vertex of the graph of $f(x)$ or $y = ax^2 + bx + c$ can be found by the formula $x = \frac{-b}{2a}$. To do so, we complete the square on x and write the equation in the form $y = a(x - h)^2 + k$.

First, isolate the x-variable terms by subtracting c from both sides.

$$y = ax^2 + bx + c$$
$$y - c = ax^2 + bx$$

Answer

3. (2.5, 12.25), (0, 6), (6, 0), (−1, 0)

Next, factor a from the terms $ax^2 + bx$.

$$y - c = a\left(x^2 + \frac{b}{a}x\right)$$

Next, add the square of half of $\frac{b}{a}$, or $\left(\frac{b}{2a}\right)^2 = \frac{b^2}{4a^2}$, to the right side inside the parentheses. Because of the factor a, what we really added was $a\left(\frac{b^2}{4a^2}\right)$, and this must be added to the left side.

$$y - c + a\left(\frac{b^2}{4a^2}\right) = a\left(x^2 + \frac{b}{a}x + \frac{b^2}{4a^2}\right)$$

$$y - c + \frac{b^2}{4a} = a\left(x + \frac{b}{2a}\right)^2 \quad \text{Simplify the left side and factor the right side.}$$

$$y = a\left(x + \frac{b}{2a}\right)^2 + c - \frac{b^2}{4a} \quad \text{Add } c \text{ to both sides and subtract } \frac{b^2}{4a} \text{ from both sides.}$$

Compare this form with $f(x)$ or $y = a(x - h)^2 + k$ and see that h is $\frac{-b}{2a}$, which means that the x-coordinate of the vertex of the graph of $f(x) = ax^2 + bx + c$ is $\frac{-b}{2a}$.

Vertex Formula

The graph of $f(x) = ax^2 + bx + c$, when $a \neq 0$, is a parabola with vertex

$$\left(\frac{-b}{2a}, f\left(\frac{-b}{2a}\right)\right)$$

Let's use this formula to find the vertex of the parabola we graphed in Example 1.

Example 4 Find the vertex of the graph of $f(x) = x^2 - 4x - 12$.

Solution: In the quadratic function $f(x) = x^2 - 4x - 12$, notice that $a = 1$, $b = -4$, and $c = -12$. Then

$$\frac{-b}{2a} = \frac{-(-4)}{2(1)} = 2$$

The x-value of the vertex is 2. To find the corresponding $f(x)$ or y-value, find $f(2)$. Then

$$f(2) = 2^2 - 4(2) - 12 = 4 - 8 - 12 = -16$$

The vertex is $(2, -16)$. These results agree with our findings in Example 1.

Work Practice 4

Practice 4

Find the vertex of the graph of $g(x) = x^2 - 2x - 3$.

Objective 3 Finding Minimum and Maximum Values 

The vertex of a parabola gives us some important information about its corresponding quadratic function. The quadratic function whose graph is a parabola that opens upward has a minimum value, and the quadratic function whose graph is a parabola that opens downward has a maximum value. The $f(x)$ or y-value of the vertex is the minimum or maximum value of the function.

Answer

4. $(1, -4)$

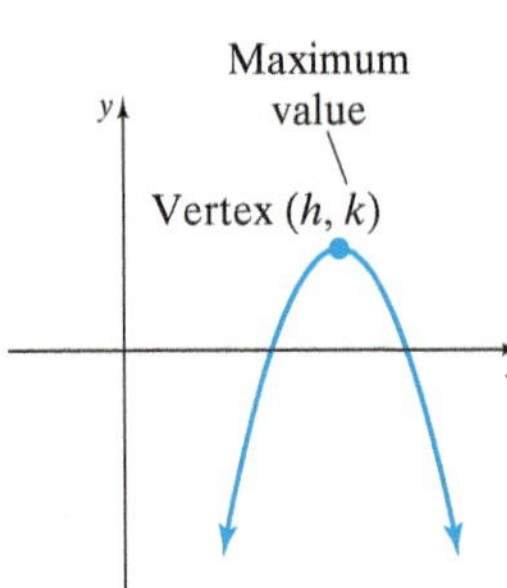

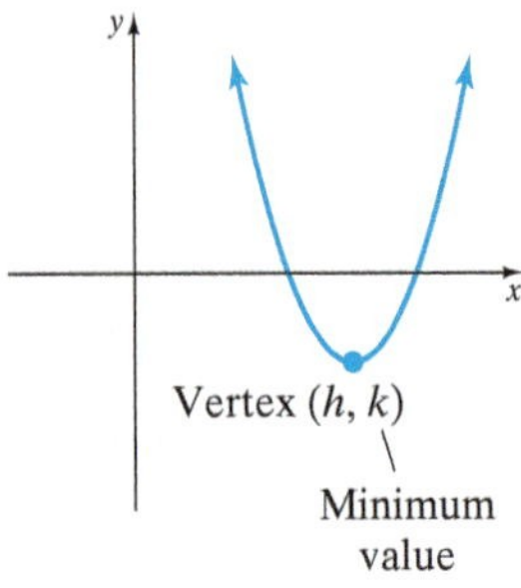

✓**Concept Check** Without making any calculations, tell whether the graph of $f(x) = 7 - x - 0.3x^2$ has a maximum value or a minimum value. Explain your reasoning.

Example 5 Finding Maximum Height

A rock is thrown upward from the ground. Its height in feet above ground after t seconds is given by the function $f(t) = -16t^2 + 20t$. Find the maximum height of the rock and the number of seconds it took for the rock to reach its maximum height.

Solution:

1. UNDERSTAND. The maximum height of the rock is the largest value of $f(t)$. Since the function $f(t) = -16t^2 + 20t$ is a quadratic function, its graph is a parabola. It opens downward since $-16 < 0$. Thus, the maximum value of $f(t)$ is the $f(t)$ or y-value of the vertex of the graph.
2. TRANSLATE. To find the vertex (h, k), notice that for $f(t) = -16t^2 + 20t$, $a = -16$, $b = 20$, and $c = 0$. We will use these values and the vertex formula

$$\left(\frac{-b}{2a}, f\left(\frac{-b}{2a}\right)\right)$$

3. SOLVE.

$$h = \frac{-b}{2a} = \frac{-20}{-32} = \frac{5}{8}$$

$$f\left(\frac{5}{8}\right) = -16\left(\frac{5}{8}\right)^2 + 20\left(\frac{5}{8}\right)$$
$$= -16\left(\frac{25}{64}\right) + \frac{25}{2}$$
$$= -\frac{25}{4} + \frac{50}{4} = \frac{25}{4}$$

4. INTERPRET. The graph of $f(t)$ is a parabola opening downward with vertex $\left(\frac{5}{8}, \frac{25}{4}\right)$. This means that the rock's maximum height is $\frac{25}{4}$ feet, or $6\frac{1}{4}$ feet, which was reached in $\frac{5}{8}$ second.

Work Practice 5

Practice 5

A ball is tossed upward from the ground. Its height in feet above ground after t seconds is given by the function $h(t) = -16t^2 + 24t$. Find the maximum height of the ball and the number of seconds it took for the ball to reach the maximum height.

Answer

5. Maximum height 9 feet in $\frac{3}{4}$ second

✓**Concept Check Answer**

$f(x)$ has a maximum value since it opens downward.

Vocabulary, Readiness & Video Check

Fill in each blank.

1. If a quadratic function is in the form $f(x) = a(x - h)^2 + k$, the vertex of its graph is ______.
2. The graph of $f(x) = ax^2 + bx + c$, $a \neq 0$, is a parabola whose vertex has x-value ______.

Martin-Gay Interactive Videos *Watch the section lecture video and answer the following questions.*

See Video 18.6

Objective 1 3. From Example 1, how does writing a quadratic function in the form $f(x) = a(x - h)^2 + k$ help us graph the function? What procedure can we use to write a quadratic function in this form?

Objective 2 4. From Example 2, how can locating the vertex and knowing whether the parabola opens upward or downward potentially help save unnecessary work? Explain.

Objective 3 5. From Example 4, when an application involving a quadratic function asks for the maximum or minimum, what part of a parabola should we find?

18.6 Exercise Set MyMathLab®

Fill in each blank.

	Parabola Opens	Vertex Location	Number of x-intercept(s)	Number of y-intercept(s)
1.	up	Q I		
2.	up	Q III		
3.	down	Q II		
4.	down	Q IV		
5.	up	x-axis		
6.	down	x-axis		
7.		Q III	0	
8.		Q I	2	
9.		Q IV	2	
10.		Q II	0	

Find the vertex of the graph of each quadratic function. See Examples 1 through 4.

11. $f(x) = x^2 + 8x + 7$

12. $f(x) = x^2 + 6x + 5$

13. $f(x) = -x^2 + 10x + 5$

14. $f(x) = -x^2 - 8x + 2$

15. $f(x) = 5x^2 - 10x + 3$

16. $f(x) = -3x^2 + 6x + 4$

17. $f(x) = -x^2 + x + 1$

18. $f(x) = x^2 - 9x + 8$

Match each function with its graph. See Examples 1 through 4.

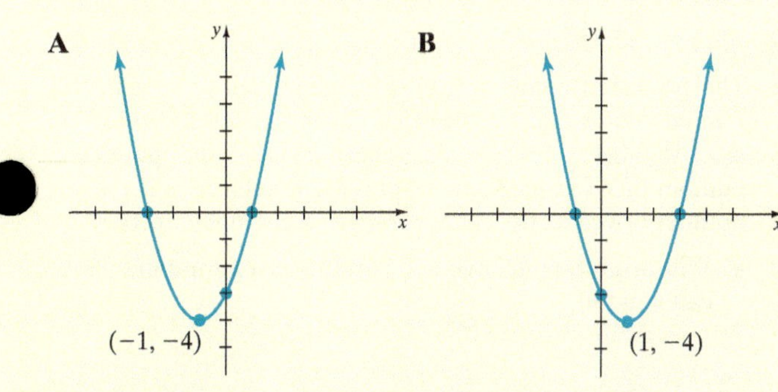

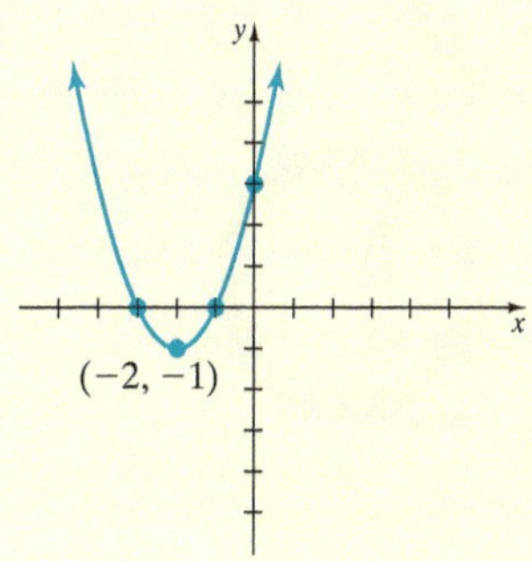

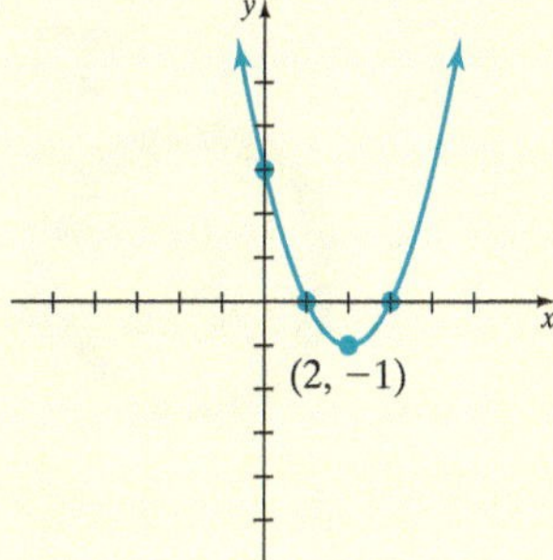

19. $f(x) = x^2 - 4x + 3$

20. $f(x) = x^2 + 2x - 3$

21. $f(x) = x^2 - 2x - 3$

22. $f(x) = x^2 + 4x + 3$

MIXED PRACTICE

Find the vertex of the graph of each quadratic function. Determine whether the graph opens upward or downward, find any intercepts, and sketch the graph. See Examples 1 through 4.

23. $f(x) = x^2 + 4x - 5$

24. $f(x) = x^2 + 2x - 3$

25. $f(x) = -x^2 + 2x - 1$

26. $f(x) = -x^2 + 4x - 4$

27. $f(x) = x^2 - 4$

28. $f(x) = x^2 - 1$

29. $f(x) = 4x^2 + 4x - 3$

30. $f(x) = 2x^2 - x - 3$

31. $f(x) = \frac{1}{2}x^2 + 4x + \frac{15}{2}$
32. $f(x) = \frac{1}{5}x^2 + 2x + \frac{9}{5}$
33. $f(x) = x^2 - 6x + 5$
34. $f(x) = x^2 - 4x + 3$
35. $f(x) = x^2 - 4x + 5$
36. $f(x) = x^2 - 6x + 11$
37. $f(x) = 2x^2 + 4x + 5$
38. $f(x) = 3x^2 + 12x + 16$
39. $f(x) = -2x^2 + 12x$
40. $f(x) = -4x^2 + 8x$
41. $f(x) = x^2 + 1$
42. $f(x) = x^2 + 4$
43. $f(x) = x^2 - 2x - 15$
44. $f(x) = x^2 - x - 12$
45. $f(x) = -5x^2 + 5x$
46. $f(x) = 3x^2 - 12x$
47. $f(x) = -x^2 + 2x - 12$
48. $f(x) = -x^2 + 8x - 17$
49. $f(x) = 3x^2 - 12x + 15$
50. $f(x) = 2x^2 - 8x + 11$
51. $f(x) = x^2 + x - 6$
52. $f(x) = x^2 + 3x - 18$
53. $f(x) = -2x^2 - 3x + 35$
54. $f(x) = 3x^2 - 13x - 10$

Solve. See Example 5.

55. If a projectile is fired straight upward from the ground with an initial speed of 96 feet per second, then its height h in feet after t seconds is given by the equation

$$h(t) = -16t^2 + 96t$$

Find the maximum height of the projectile.

56. If Rheam Gaspar throws a ball upward with an initial speed of 32 feet per second, then its height h in feet after t seconds is given by the equation

$$h(t) = -16t^2 + 32t$$

Find the maximum height of the ball.

57. The cost C in dollars of manufacturing x bicycles at Holladay's Production Plant is given by the function

$$C(x) = 2x^2 - 800x + 92{,}000$$

a. Find the number of bicycles that must be manufactured to minimize the cost.

b. Find the minimum cost.

58. The Utah Ski Club sells calendars to raise money. The profit P, in cents, from selling x calendars is given by the equation $P(x) = 360x - x^2$.

a. Find how many calendars must be sold to maximize profit.

b. Find the maximum profit.

59. Find two numbers whose sum is 60 and whose product is as large as possible. [*Hint:* Let x and $60 - x$ be the two positive numbers. Their product can be described by the function $f(x) = x(60 - x)$.]

60. Find two numbers whose sum is 11 and whose product is as large as possible. (Use the hint for Exercise 59.)

61. Find two numbers whose difference is 10 and whose product is as small as possible. (Use the hint for Exercise 59.)

62. Find two numbers whose difference is 8 and whose product is as small as possible. (Use the hint for Exercise 59.)

△ 63. The length and width of a rectangle must have a sum of 40. Find the dimensions of the rectangle that will have the maximum area. (Use the hint for Exercise 59.)

△ 64. The length and width of a rectangle must have a sum of 50. Find the dimensions of the rectangle that will have maximum area. (Use the hint for Exercise 59.)

REVIEW AND PREVIEW

Sketch the graph of each function. See Section 18.5.

65. $f(x) = x^2 + 2$
66. $f(x) = (x - 3)^2$
67. $g(x) = x + 2$
68. $h(x) = x - 3$
69. $f(x) = (x + 5)^2 + 2$
70. $f(x) = 2(x - 3)^2 + 2$
71. $f(x) = 3(x - 4)^2 + 1$
72. $f(x) = (x + 1)^2 + 4$
73. $f(x) = -(x - 4)^2 + \frac{3}{2}$
74. $f(x) = -2(x + 7)^2 + \frac{1}{2}$

CONCEPT EXTENSIONS

Without calculating, tell whether each graph has a minimum value or a maximum value. See the Concept Check in this section.

75. $f(x) = 2x^2 - 5$
76. $g(x) = -7x^2 + x + 1$
77. $f(x) = 3 - \frac{1}{2}x^2$
78. $G(x) = 3 - \frac{1}{2}x + 0.8x^2$

Find the vertex of the graph of each quadratic function. Determine whether the graph opens upward or downward, find the y-intercept, approximate the x-intercepts to one decimal place, and sketch the graph.

79. $f(x) = x^2 + 10x + 15$
80. $f(x) = x^2 - 6x + 4$
81. $f(x) = 3x^2 - 6x + 7$
82. $f(x) = 2x^2 + 4x - 1$

Find the maximum or minimum value of each function. Approximate to two decimal places.

83. $f(x) = 2.3x^2 - 6.1x + 3.2$
84. $f(x) = 7.6x^2 + 9.8x - 2.1$
85. $f(x) = -1.9x^2 + 5.6x - 2.7$
86. $f(x) = -5.2x^2 - 3.8x + 5.1$

87. The projected number of Wi-Fi-enabled cell phones in the United States can be modeled by the quadratic function $c(x) = -0.4x^2 + 21x + 35$, where $c(x)$ is the projected number of Wi-Fi-enabled cell phones in millions and x is the number of years after 2009. (*Source:* Techcrunchies.com)

a. Will this function have a maximum or a minimum? How can you tell?

b. According to this model, in what year will the number of Wi-Fi-enabled cell phones in the United States be at its maximum or minimum?

c. What is the maximum/minimum number of Wi-Fi-enabled cell phones predicted? Round to the nearest whole million.

88. Methane, a gas produced by landfills, natural gas systems, and coal mining, contributes to the greenhouse effect and global warming. Projected methane emissions in the United States can be modeled by the quadratic function

$$f(x) = -0.072x^2 + 1.93x + 173.9$$

where $f(x)$ is the amount of methane produced in million metric tons and x is the number of years after 2000. (*Source:* Based on data from the U.S. Environmental Protection Agency, 2000–2020)

a. According to this model, what will U.S. emissions of methane be in 2018? (Round to 2 decimal places.)

b. Will this function have a maximum or a minimum? How can you tell?

c. In what year will methane emissions in the United States be at their maximum/minimum? Round to the nearest whole year.

d. What is the level of methane emissions for that year? (Use your rounded answer from part (c).) (Round this answer to 2 decimal places.)

Use a graphing calculator to check each exercise.

89. Exercise 37

90. Exercise 38

91. Exercise 47

92. Exercise 48

Chapter 18 Vocabulary Check

Fill in each blank with one of the words or phrases listed below.

quadratic formula	quadratic	discriminant	$\pm\sqrt{b}$	$(h, 0)$
completing the square	quadratic inequality	(h, k)	$(0, k)$	$\frac{-b}{2a}$

1. The ______________ helps us find the number and type of solutions of a quadratic equation.

2. If $a^2 = b$, then $a =$ ______________.

3. The graph of $f(x) = ax^2 + bx + c$, where a is not 0, is a parabola whose vertex has x-value ______________.

4. A ______________ is an inequality that can be written so that one side is a quadratic expression and the other side is 0.

5. The process of writing a quadratic equation so that one side is a perfect square trinomial is called ______________.

6. The graph of $f(x) = x^2 + k$ has vertex ______________.

7. The graph of $f(x) = (x - h)^2$ has vertex ______________.

8. The graph of $f(x) = (x - h)^2 + k$ has vertex ______________.

9. The formula $x = \frac{-b \pm \sqrt{b^2 - 4ac}}{2a}$ is called the ______________.

10. A ______________ equation can be written in the form $ax^2 + bx + c = 0$, where a, b, and c are real numbers and a is not 0.

Helpful Hint

Are you preparing for your test? Don't forget to take the Chapter 18 Test on page 1262. Then check your answers at the back of the text and use the Chapter Test Prep Videos to see the fully worked-out solutions to any of the exercises you want to review.

18 Chapter Highlights

Definitions and Concepts	Examples
Section 18.1 Solving Quadratic Equations by Completing the Square	
Square Root Property If b is a real number and if $a^2 = b$, then $a = \pm\sqrt{b}$.	Solve: $(x+3)^2 = 14$. $x + 3 = \pm\sqrt{14}$ $x = -3 \pm \sqrt{14}$
To solve a quadratic equation in *x* by completing the square	Solve: $3x^2 - 12x - 18 = 0$.
Step 1: If the coefficient of x^2 is not 1, divide both sides of the equation by the coefficient of x^2.	**1.** $x^2 - 4x - 6 = 0$
Step 2: Isolate the variable terms.	**2.** $x^2 - 4x = 6$
Step 3: Complete the square by adding the square of half of the coefficient of x to both sides.	**3.** $\frac{1}{2}(-4) = -2$ and $(-2)^2 = 4$ $x^2 - 4x + 4 = 6 + 4$
Step 4: Write the resulting trinomial as the square of a binomial.	**4.** $(x-2)^2 = 10$
Step 5: Apply the square root property and solve for x.	**5.** $x - 2 = \pm\sqrt{10}$ $x = 2 \pm \sqrt{10}$
Section 18.2 Solving Quadratic Equations by the Quadratic Formula	
A quadratic equation written in the form $ax^2 + bx + c = 0$ has solutions $x = \frac{-b \pm \sqrt{b^2 - 4ac}}{2a}$	Solve: $x^2 - x - 3 = 0$. $a = 1, b = -1, c = -3$ $x = \frac{-(-1) \pm \sqrt{(-1)^2 - 4(1)(-3)}}{2 \cdot 1}$ $x = \frac{1 \pm \sqrt{13}}{2}$
Section 18.3 Solving Equations by Using Quadratic Methods	
Substitution is often helpful in solving an equation that contains a repeated variable expression.	Solve: $(2x+1)^2 - 5(2x+1) + 6 = 0$. Let $m = 2x + 1$. Then $m^2 - 5m + 6 = 0$ Let $m = 2x + 1$. $(m-3)(m-2) = 0$ $m = 3$ or $m = 2$ $2x + 1 = 3$ or $2x + 1 = 2$ Substitute back. $x = 1$ or $x = \frac{1}{2}$

Definitions and Concepts	Examples

Section 18.4 Nonlinear Inequalities in One Variable

To solve a polynomial inequality

Step 1: Write the inequality in standard form.

Step 2: Solve the related equation.

Step 3: Use solutions from Step 2 to separate the number line into regions.

Step 4: Use test points to determine whether values in each region satisfy the original inequality.

Step 5: Write the solution set as the union of regions whose test point value is a solution.

Solve: $x^2 \geq 6x$.

1. $x^2 - 6x \geq 0$

2. $x^2 - 6x = 0$

$x(x - 6) = 0$

$x = 0 \quad \text{or} \quad x = 6$

3.

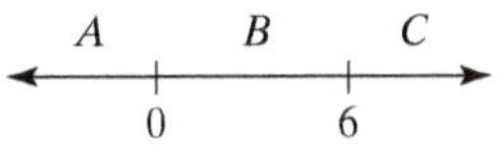

4.

Region	Test Point Value	$x^2 \geq 6x$	Result
A	−2	$(-2)^2 \geq 6(-2)$	True
B	1	$1^2 \geq 6(1)$	False
C	7	$7^2 \geq 6(7)$	True

5.

The solution set is $(-\infty, 0] \cup [6, \infty)$.

To solve a rational inequality

Step 1: Solve for values that make all denominators 0.

Step 2: Solve the related equation.

Step 3: Use solutions from Steps 1 and 2 to separate the number line into regions.

Step 4: Use test points to determine whether values in each region satisfy the original inequality.

Step 5: Write the solution set as the union of regions whose test point value is a solution.

Solve: $\dfrac{6}{x - 1} < -2$.

1. $x - 1 = 0$ Set denominator equal to 0.

$x = 1$

2. $\dfrac{6}{x - 1} = -2$

$6 = -2(x - 1)$ Multiply by $(x - 1)$.

$6 = -2x + 2$

$4 = -2x$

$-2 = x$

3.

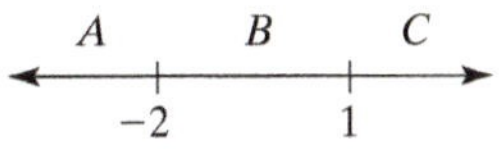

4. Only a test value from region *B* satisfies the original inequality.

5.

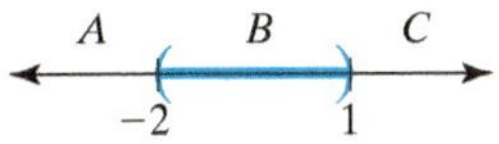

The solution set is $(-2, 1)$.

Definitions and Concepts	Examples

Section 18.5 Quadratic Functions and Their Graphs

Graph of a Quadratic Function

The graph of a quadratic function written in the form $f(x) = a(x - h)^2 + k$ is a parabola with vertex (h, k). If $a > 0$, the parabola opens upward; if $a < 0$, the parabola opens downward. The axis of symmetry is the line whose equation is $x = h$.

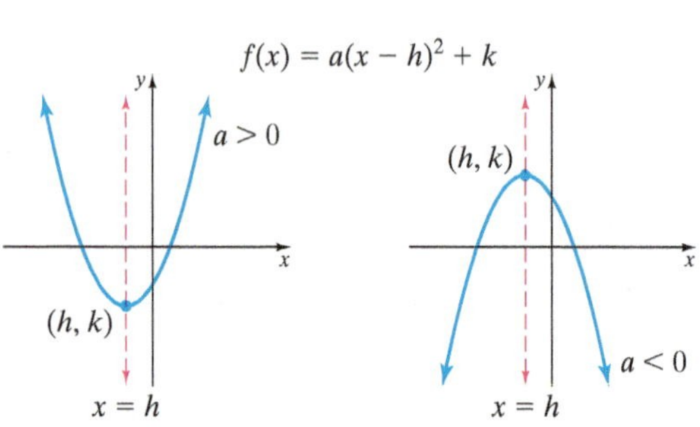

Graph $g(x) = 3(x - 1)^2 + 4$.

The graph is a parabola with vertex $(1, 4)$ and axis of symmetry $x = 1$. Since $a = 3$ is positive, the graph opens upward.

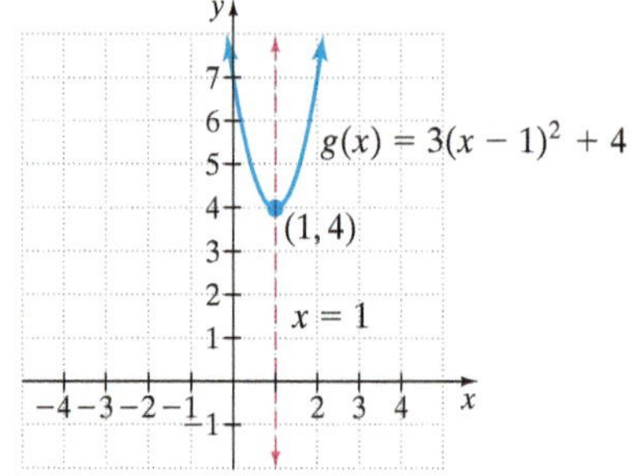

Section 18.6 Further Graphing of Quadratic Functions

The graph of $f(x) = ax^2 + bx + c$, where $a \neq 0$, is a parabola with vertex

$$\left(\frac{-b}{2a}, f\left(\frac{-b}{2a}\right)\right)$$

Graph $f(x) = x^2 - 2x - 8$. Find the vertex and x- and y-intercepts.

$$\frac{-b}{2a} = \frac{-(-2)}{2 \cdot 1} = 1$$

$$f(1) = 1^2 - 2(1) - 8 = -9$$

The vertex is $(1, -9)$.

$$0 = x^2 - 2x - 8$$
$$0 = (x - 4)(x + 2)$$
$$x = 4 \quad \text{or} \quad x = -2$$

The x-intercepts are $(4, 0)$ and $(-2, 0)$.

$$f(0) = 0^2 - 2 \cdot 0 - 8 = -8$$

The y-intercept is $(0, -8)$.

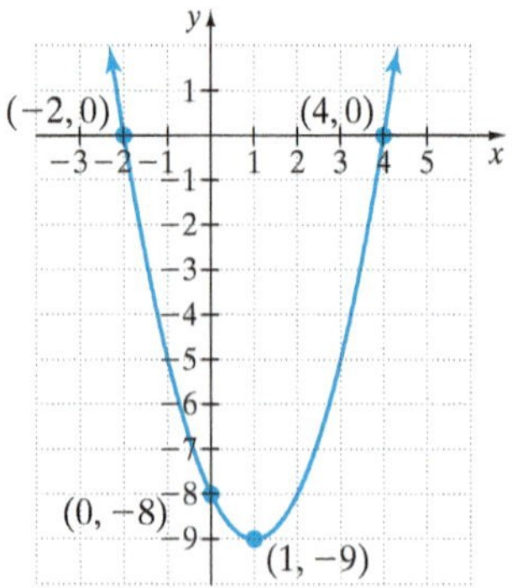

Chapter 18 Review

(18.1) Solve by factoring.

1. $x^2 - 15x + 14 = 0$
2. $7a^2 = 29a + 30$

Solve by using the square root property.

3. $4m^2 = 196$
4. $(5x - 2)^2 = 2$

Solve by completing the square.

5. $z^2 + 3z + 1 = 0$
6. $(2x + 1)^2 = x$
7. If P dollars are originally invested, the formula $A = P(1 + r)^2$ gives the amount A in an account paying interest rate r compounded annually after 2 years. Find the interest rate r such that \$2500 increases to \$2717 in 2 years. Round the result to the nearest hundredth of a percent.
8. △ Two ships leave a port at the same time and travel at the same speed. One ship is traveling due north and the other due east. In a few hours, the ships are 150 miles apart. How many miles has each ship traveled? Give an exact answer and a one-decimal-place approximation.

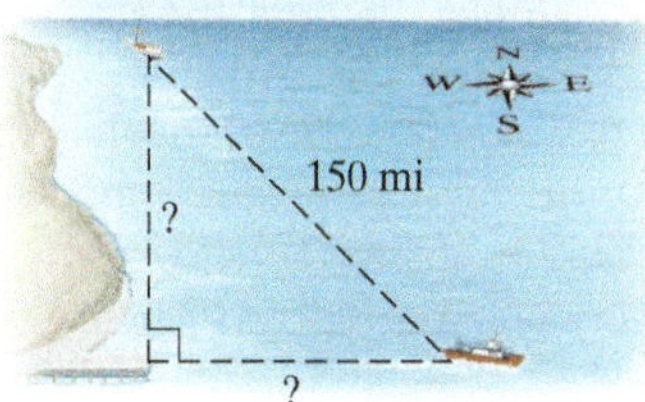

(18.2) If the discriminant of a quadratic equation has the given value, determine the number and type of solutions of the equation.

9. -8
10. 48
11. 100
12. 0

Solve by using the quadratic formula.

13. $x^2 - 16x + 64 = 0$
14. $x^2 + 5x = 0$
15. $2x^2 + 3x = 5$
16. $9x^2 + 4 = 2x$
17. $6x^2 + 7 = 5x$
18. $(2x - 3)^2 = x$
19. Cadets graduating from military school usually toss their hats high into the air at the end of the ceremony. One cadet threw his hat so that its distance $d(t)$ in feet above the ground t seconds after it was thrown was $d(t) = -16t^2 + 30t + 6$.

   a. Find the distance above the ground of the hat 1 second after it was thrown.

   b. Find the time it takes the hat to hit the ground. Give an exact time and a one-decimal-place approximation.

20. △ The hypotenuse of an isosceles right triangle is 6 centimeters longer than either of the legs. Find the length of the legs.

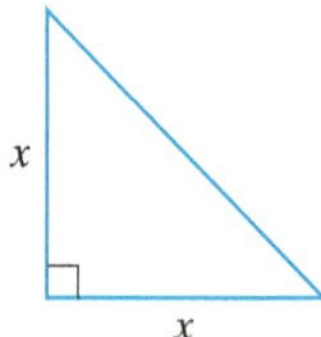

(18.3) Solve each equation for the variable.

21. $x^3 = 27$
22. $y^3 = -64$
23. $\frac{5}{x} + \frac{6}{x - 2} = 3$
24. $x^4 - 21x^2 - 100 = 0$
25. $x^{2/3} - 6x^{1/3} + 5 = 0$
26. $5(x + 3)^2 - 19(x + 3) = 4$
27. $a^6 - a^2 = a^4 - 1$
28. $y^{-2} + y^{-1} = 20$
29. Two postal workers, Jerome Grant and Tim Bozik, can sort a stack of mail in 5 hours. Working alone, Tim can sort the mail in 1 hour less time than Jerome can. Find the time that each postal worker can sort the mail alone. Round the result to one decimal place.
30. A negative number decreased by its reciprocal is $-\frac{24}{5}$. Find the number.

(18.4) Solve each inequality for x. Write each solution set in interval notation.

31. $2x^2 - 50 \le 0$
32. $\frac{1}{4}x^2 < \frac{1}{16}$
33. $(x^2 - 4)(x^2 - 25) \le 0$
34. $(x^2 - 16)(x^2 - 1) > 0$
35. $\frac{x - 5}{x - 6} < 0$
36. $\frac{(4x + 3)(x - 5)}{x(x + 6)} > 0$
37. $(x + 5)(x - 6)(x + 2) \le 0$
38. $x^3 + 3x^2 - 25x - 75 > 0$
39. $\frac{x^2 + 4}{3x} \le 1$
40. $\frac{3}{x - 2} > 2$

(18.5) Sketch the graph of each function. Label the vertex and the axis of symmetry.

41. $f(x) = x^2 - 4$
42. $g(x) = x^2 + 7$
43. $H(x) = 2x^2$
44. $h(x) = -\frac{1}{3}x^2$
45. $F(x) = (x - 1)^2$
46. $G(x) = (x + 5)^2$
47. $f(x) = (x - 4)^2 - 2$
48. $f(x) = -3(x - 1)^2 + 1$

(18.6) Sketch the graph of each function. Find the vertex and the intercepts.

49. $f(x) = x^2 + 10x + 25$
50. $f(x) = -x^2 + 6x - 9$
51. $f(x) = 4x^2 - 1$
52. $f(x) = -5x^2 + 5$
53. Find the vertex of the graph of $f(x) = -3x^2 - 5x + 4$. Determine whether the graph opens upward or downward, find the y-intercept, approximate the x-intercepts to one decimal place, and sketch the graph.
54. The function $h(t) = -16t^2 + 120t + 300$ gives the height in feet of a projectile fired from the top of a building after t seconds.
    a. When will the object reach a height of 350 feet? Round your answer to one decimal place.
    b. Explain why part (a) has two answers.
55. Find two numbers whose product is as large as possible, given that their sum is 420.
56. Write an equation of a quadratic function whose graph is a parabola that has vertex $(-3, 7)$. Let the value of a be $-\frac{7}{9}$.

MIXED REVIEW

Solve each equation or inequality.

57. $x^2 - x - 30 = 0$
58. $10x^2 = 3x + 4$
59. $9y^2 = 36$
60. $(9n + 1)^2 = 9$
61. $x^2 + x + 7 = 0$
62. $(3x - 4)^2 = 10x$
63. $x^2 + 11 = 0$
64. $x^2 + 7 = 0$
65. $(5a - 2)^2 - a = 0$
66. $\frac{7}{8} = \frac{8}{x^2}$
67. $x^{2/3} - 6x^{1/3} = -8$
68. $(2x - 3)(4x + 5) \geq 0$
69. $\frac{x(x + 5)}{4x - 3} \geq 0$
70. $\frac{3}{x - 2} > 2$
71. The busiest airport in the world is the Hartsfield International Airport in Atlanta, Georgia. The total amount of passenger traffic through Atlanta during the period 2000 through 2010 can be modeled by the equation $y = -32x^2 + 1733x + 76{,}362$, where y is the number of passengers enplaned and deplaned in thousands, and x is the number of years after 2000. (*Source*: Based on data from Airports Council International)
    a. Estimate the passenger traffic at Atlanta's Hartsfield International Airport in 2015.
    b. According to this model, in what year will the passenger traffic at Atlanta's Hartsfield International Airport first reach 99,000 thousand passengers?

Chapter 18 Test

Step-by-step test solutions are found on the Chapter Test Prep Videos. Where available: MyMathLab® or YouTube

Solve each equation.

1. $5x^2 - 2x = 7$
2. $(x + 1)^2 = 10$
3. $m^2 - m + 8 = 0$
4. $u^2 - 6u + 2 = 0$
5. $7x^2 + 8x + 1 = 0$
6. $y^2 - 3y = 5$
7. $\frac{4}{x + 2} + \frac{2x}{x - 2} = \frac{6}{x^2 - 4}$
8. $x^5 + 3x^4 = x + 3$
9. $x^6 + 1 = x^4 + x^2$
10. $(x + 1)^2 - 15(x + 1) + 56 = 0$

Solve by completing the square.

11. $x^2 - 6x = -2$
12. $2a^2 + 5 = 4a$

Solve each inequality for x. Write the solution set in interval notation.

13. $2x^2 - 7x > 15$

14. $(x^2 - 16)(x^2 - 25) \geq 0$

15. $\dfrac{5}{x+3} < 1$

16. $\dfrac{7x - 14}{x^2 - 9} \leq 0$

Graph each function. Label the vertex.

17. $f(x) = 3x^2$

18. $G(x) = -2(x - 1)^2 + 5$

Graph each function. Find and label the vertex, y-intercept, and x-intercepts (if any).

19. $h(x) = x^2 - 4x + 4$

20. $F(x) = 2x^2 - 8x + 9$

21. Dave and Sandy Hartranft can paint a room together in 4 hours. Working alone, Dave can paint the room in 2 hours less time than Sandy can. Find how long it takes Sandy to paint the room alone.

22. A stone is thrown upward from a bridge. The stone's height in feet, $s(t)$, above the water t seconds after the stone is thrown is a function given by the equation $s(t) = -16t^2 + 32t + 256$.

a. Find the maximum height of the stone.

b. Find the time it takes the stone to hit the water. Round the answer to two decimal places.

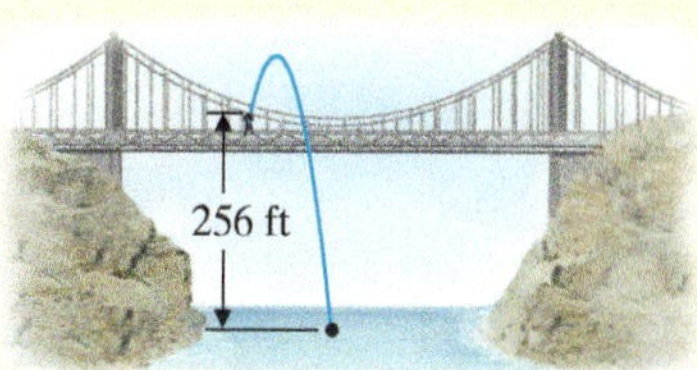

23. Given the diagram shown, approximate to the nearest tenth of a foot how many feet of walking distance a person saves by cutting across the lawn instead of walking on the sidewalk.

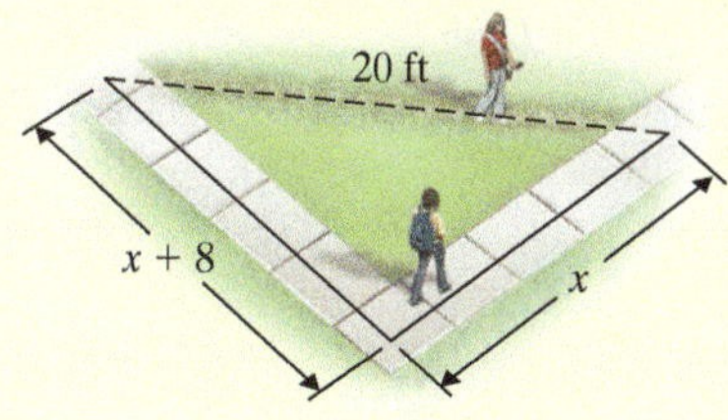

Chapters 1–18 Cumulative Review

1. Evaluate $2x + y$ for $x = 3$ and $y = -5$.

2. Solve: $|3x - 2| = -5$

3. Simplify each expression by combining like terms.

a. $4x + 6x$

b. $y - 5y$

c. $3x^2 + 5x^2 - 2$

4. Use the addition method to solve the system.
$$\begin{cases} -6x + y = 5 \\ 4x - 2y = 6 \end{cases}$$

5. Solve the following system of equations by graphing.
$$\begin{cases} 2x + y = 7 \\ 2y = -4x \end{cases}$$

6. Simplify. Use positive exponents to write each answer.

a. $(a^{-2}bc^3)^{-3}$ **b.** $\left(\dfrac{a^{-4}b^2}{c^3}\right)^{-2}$

c. $\left(\dfrac{3a^8b^2}{12a^5b^5}\right)^{-2}$

7. Solve the system: $\begin{cases} 7x - 3y = -14 \\ -3x + y = 6 \end{cases}$

8. Multiply.

a. $(4a - 3)(7a - 2)$

b. $(2a + b)(3a - 5b)$

9. Simplify each quotient.

a. $\dfrac{x^5}{x^2}$ **b.** $\dfrac{4^7}{4^3}$

c. $\dfrac{(-3)^5}{(-3)^2}$ **d.** $\dfrac{s^2}{t^3}$

e. $\dfrac{2x^5y^2}{xy}$

10. Factor.

a. $9x^3 + 27x^2 - 15x$

b. $2x(3y - 2) - 5(3y - 2)$

c. $2xy + 6x - y - 3$

11. If $P(x) = 2x^3 - 4x^2 + 5$,

a. Find $P(2)$ by substitution.

b. Use synthetic division to find the remainder when $P(x)$ is divided by $x - 2$.

12. Factor: $x^2 - 2x - 48$

13. Solve: $(5x - 1)(2x^2 + 15x + 18) = 0$

14. Factor: $2ax^2 - 12axy + 18ay^2$

15. Simplify the rational expression.
$$\frac{2x^2}{10x^3 - 2x^2}$$

16. Solve: $2(a^2 + 2) - 8 = -2a(a - 2) - 5$

17. Simplify: $\dfrac{x^{-1} + 2xy^{-1}}{x^{-2} - x^{-2}y^{-1}}$

18. Find the vertex and any intercepts of $f(x) = x^2 + x - 12$.

19. Factor: $4m^4 - 4m^2 + 1$

20. Simplify: $\dfrac{x^2 - 4x + 4}{2 - x}$

21. The square of a number plus three times the number is 70. Find the number.

22. Subtract: $\dfrac{a + 1}{a^2 - 6a + 8} - \dfrac{3}{16 - a^2}$

23. Use the product rule to simplify.

a. $\sqrt{25x^3}$ **b.** $\sqrt[3]{54x^6y^8}$

c. $\sqrt[4]{81z^{11}}$

24. Simplify: $\dfrac{(2a)^{-1} + b^{-1}}{a^{-1} + (2b)^{-1}}$

25. Rationalize the denominator of each expression.

a. $\dfrac{2}{\sqrt{5}}$ **b.** $\dfrac{2\sqrt{16}}{\sqrt{9x}}$

c. $\sqrt[3]{\dfrac{1}{2}}$

26. Divide $x^3 - 3x^2 - 10x + 24$ by $x + 3$.

27. Solve: $\sqrt{2x + 5} + \sqrt{2x} = 3$

28. If $P(x) = 4x^3 - 2x^2 + 3$,

a. Find $P(-2)$ by substitution.

b. Use synthetic division to find the remainder when $P(x)$ is divided by $x + 2$.

29. Solve: $\dfrac{x}{2} + \dfrac{8}{3} = \dfrac{1}{6}$

30. Solve: $\dfrac{x + 3}{x^2 + 5x + 6} = \dfrac{3}{2x + 4} - \dfrac{1}{x + 3}$

31. The quotient of a number and 6, minus $\frac{5}{3}$, is the quotient of the number and 2. Find the number.

32. Mr. Briley can roof his house in 24 hours. His son can roof the same house in 40 hours. If they work together, how long will it take to roof the house?

33. Suppose that y varies directly as x. If y is 5 when x is 30, find the constant of variation and the direct variation equation.

34. Suppose that y varies inversely as x. If y is 8 when x is 24, find the constant of variation and the inverse variation equation.

35. Simplify. Assume that the variables represent any real number.

a. $\sqrt{(-3)^2}$ **b.** $\sqrt{x^2}$

c. $\sqrt[4]{(x - 2)^4}$ **d.** $\sqrt[3]{(-5)^3}$

e. $\sqrt[5]{(2x - 7)^5}$ **f.** $\sqrt{25x^2}$

g. $\sqrt{x^2 + 2x + 1}$

36. Simplify. Assume that the variables represent any real number.

a. $\sqrt{(-2)^2}$ **b.** $\sqrt{y^2}$

c. $\sqrt[4]{(a - 3)^4}$ **d.** $\sqrt[3]{(-6)^3}$

e. $\sqrt[5]{(3x - 1)^5}$

37. Use rational exponents to simplify. Assume that variables represent positive numbers.

a. $\sqrt[8]{x^4}$ **b.** $\sqrt[6]{25}$

c. $\sqrt[4]{r^2s^6}$

38. Use rational exponents to simplify. Assume that variables represent positive numbers.

a. $\sqrt[4]{5^2}$ **b.** $\sqrt[12]{x^3}$

c. $\sqrt[6]{x^2y^4}$

39. Divide. Write in the form $a + bi$.

a. $\dfrac{2 + i}{1 - i}$ **b.** $\dfrac{7}{3i}$

40. Write each product in the form of $a + bi$.

a. $3i(5 - 2i)$ **b.** $(6 - 5i)^2$

c. $(\sqrt{3} + 2i)(\sqrt{3} - 2i)$

41. Use the square root property to solve $(x + 1)^2 = 12$.

42. Use the square root property to solve $(y - 1)^2 = 24$.

43. Solve: $x - \sqrt{x} - 6 = 0$

44. Use the quadratic formula to solve $m^2 = 4m + 8$.

Exponential and Logarithmic Functions

19

Sections

Check Your Progress

In this chapter, we discuss two closely related functions: exponential and logarithmic functions. These functions are vital to applications in economics, finance, engineering, the sciences, education, and other fields. Models of tumor growth and learning curves are two examples of the uses of exponential and logarithmic functions.

A compact fluorescent lamp (or light) (CFL) is a type of fluorescent light that is quickly gaining popularity for many reasons. Compared to an incandescent bulb, CFLs use less power and last between 8 and 15 times as long. Although a CFL has a higher price, the savings per bulb are substantial (possibly $30 per life of bulb). Many CFLs are now manufactured to replace an incandescent bulb and can fit into existing fixtures. It should be noted that since CFLs are a type of fluorescent light, they do contain a small amount of mercury.

Although we have no direct applications in this chapter, it should be noted that the light output of a CFL decays exponentially. By the end of their lives, they produce 70–80% of their original output, with the fastest losses occurring soon after the light is first used. Also, it should be noted that the response of the human eye to light is logarithmic.

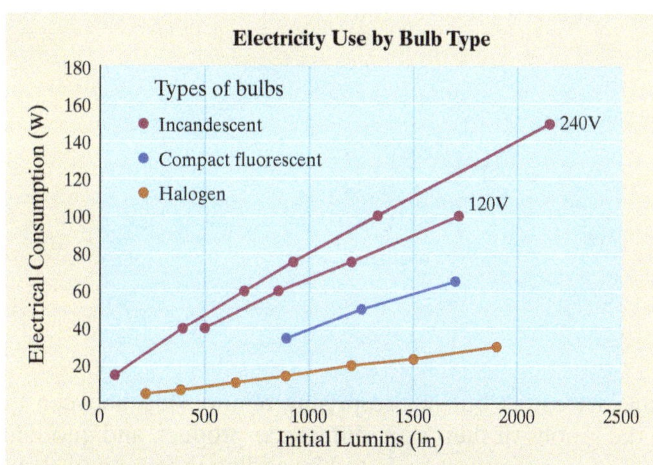

(*Note:* Lower points correspond to lower energy use.)

19.1 The Algebra of Functions; Composite Functions

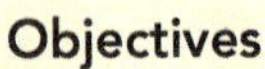

Objectives

1 Add, Subtract, Multiply, and Divide Functions.

2 Construct Composite Functions.

Objective 1 Adding, Subtracting, Multiplying, and Dividing Functions

As we have seen in earlier chapters, it is possible to add, subtract, multiply, and divide functions. Although we have not stated it as such, the sums, differences, products, and quotients of functions are themselves functions. For example, if $f(x) = 3x$ and $g(x) = x + 1$, their product, $f(x) \cdot g(x) = 3x(x + 1) = 3x^2 + 3x$, is a new function. We can use the notation $(f \cdot g)(x)$ to denote this new function. Finding the sum, difference, product, and quotient of functions to generate new functions is called the **algebra of functions.**

Algebra of Functions

Let f and g be functions. New functions from f and g are defined as follows.

Sum	$(f + g)(x) = f(x) + g(x)$
Difference	$(f - g)(x) = f(x) - g(x)$
Product	$(f \cdot g)(x) = f(x) \cdot g(x)$
Quotient	$\left(\dfrac{f}{g}\right)(x) = \dfrac{f(x)}{g(x)},\quad g(x) \neq 0$

Practice 1

If $f(x) = x + 2$ and $g(x) = 3x + 5$, find

a. $(f + g)(x)$ **b.** $(f - g)(x)$

c. $(f \cdot g)(x)$ **d.** $\left(\dfrac{f}{g}\right)(x)$

Example 1 If $f(x) = x - 1$ and $g(x) = 2x - 3$, find

a. $(f + g)(x)$ **b.** $(f - g)(x)$ **c.** $(f \cdot g)(x)$ **d.** $\left(\dfrac{f}{g}\right)(x)$

Solution: Use the algebra of functions and replace $f(x)$ by $x - 1$ and $g(x)$ by $2x - 3$. Then we simplify.

a.
$$\begin{aligned}(f + g)(x) &= f(x) + g(x)\\ &= (x - 1) + (2x - 3) = 3x - 4\end{aligned}$$

b.
$$\begin{aligned}(f - g)(x) &= f(x) - g(x)\\ &= (x - 1) - (2x - 3)\\ &= x - 1 - 2x + 3\\ &= -x + 2\end{aligned}$$

c.
$$\begin{aligned}(f \cdot g)(x) &= f(x) \cdot g(x)\\ &= (x - 1)(2x - 3)\\ &= 2x^2 - 5x + 3\end{aligned}$$

d.
$$\left(\frac{f}{g}\right)(x) = \frac{f(x)}{g(x)} = \frac{x - 1}{2x - 3}, \text{ where } x \neq \frac{3}{2}$$

Work Practice 1

There is an interesting but not surprising relationship between the graphs of functions and the graphs of their sum, difference, product, and quotient. For example, the graph of $(f + g)(x)$ can be found by adding the graph of $f(x)$ to the graph of $g(x)$. We add two graphs by adding y-values of corresponding x-values.

Answers

1. a. $4x + 7$ **b.** $-2x - 3$
c. $3x^2 + 11x + 10$
d. $\dfrac{x + 2}{3x + 5}$, where $x \neq -\dfrac{5}{3}$

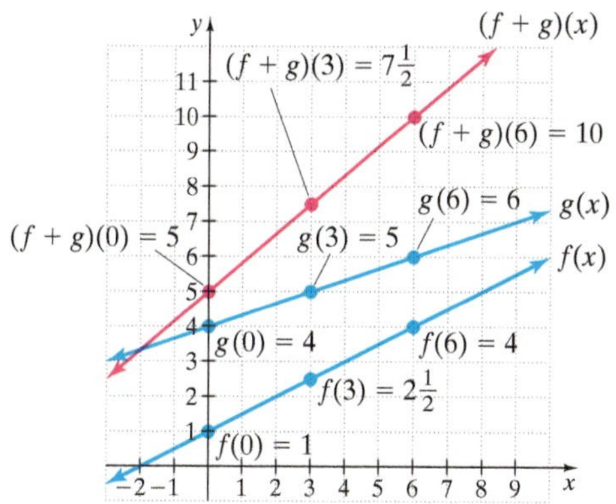

Objective 2 Constructing Composite Functions 

Another way to combine functions is called **function composition.** To understand this new way of combining functions, study the diagrams below. The left diagram shows an illustration by tables, and the right diagram is the same illustration but by thermometers. In both illustrations, we show degrees Celsius $f(x)$ as a function of degrees Fahrenheit x, and then Kelvins $g(x)$ as a function of degrees Celsius x. (The Kelvin scale is a temperature scale devised by Lord Kelvin in 1848.) The first function we will call f, and the second function we will call g.

degrees Fahrenheit | degrees Celsius | degrees Celsius | Kelvins

212 → 100; 68 → 20; 32 → 0; −13 → −25 (x → $f(x)$, f)

100 → 373.15; 20 → 293.15; 0 → 273.15; −25 → 248.15 (x → $g(x)$, g)

Table Illustration

x = Degrees Fahrenheit (Input)	−13	32	68	212
$f(x)$ = Degrees Celsius (Output)	−25	0	20	100

x = Degrees Celsius (Input)	−25	0	20	100
$g(x)$ = Kelvins (Output)	248.15	273.15	293.15	373.15

Suppose that we want a function that shows a direct conversion from degrees Fahrenheit to Kelvins. In other words, suppose that a function is needed that shows Kelvins as a function of degrees Fahrenheit. This can easily be done because the output of the first function $f(x)$ is the same as the input of the second function. If we use $g(f(x))$ to represent this, then we get the diagrams below.

x = Degrees Fahrenheit (Input)	−13	32	68	212
$g(f(x))$ = Kelvins (Output)	248.15	273.15	293.15	373.15

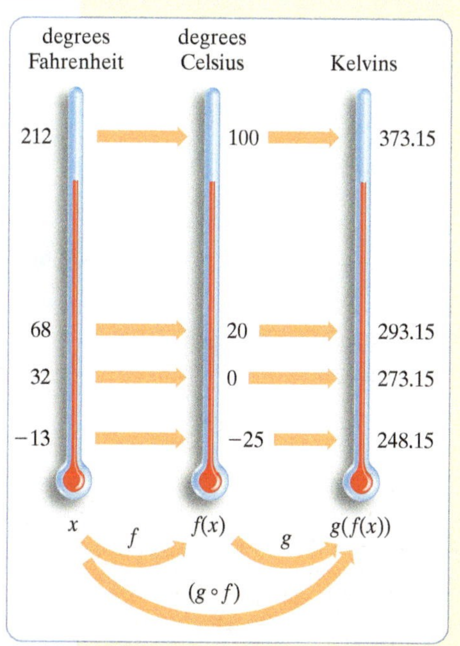

For example $g(f(-13)) = 248.15$, and so on.

Since the output of the first function is used as the input of the second function, we write the new function as $g(f(x))$. The new function is formed from the composition of the other two functions. The mathematical symbol for this composition is $(g \circ f)(x)$. Thus, $(g \circ f)(x) = g(f(x))$.

It is possible to find an equation for the composition of the two functions f and g. In other words, we can find a function that converts degrees Fahrenheit directly to Kelvins. The function $f(x) = \frac{5}{9}(x - 32)$ converts degrees Fahrenheit to degrees Celsius, and the function $g(x) = x + 273.15$ converts degrees Celsius to Kelvins. Thus,

$$(g \circ f)(x) = g(f(x)) = g\left(\frac{5}{9}(x - 32)\right) = \frac{5}{9}(x - 32) + 273.15$$

In general, the notation $g(f(x))$ means "g composed with f" and can be written as $(g \circ f)(x)$. Also, $f(g(x))$, or $(f \circ g)(x)$, means "f composed with g."

Composition of Functions

The composition of functions f and g is

$$(f \circ g)(x) = f(g(x))$$

Helpful Hint

$(f \circ g)(x)$ does not mean the same as $(f \cdot g)(x)$.

$$(f \circ g)(x) = f(g(x)) \text{ while } (f \cdot g)(x) = f(x) \cdot g(x)$$

↑ Composition of functions ↑ Multiplication of functions

Practice 2

If $f(x) = x^2 + 1$ and $g(x) = 3x - 5$, find

a. $(f \circ g)(4)$ $(g \circ f)(4)$ **b.** $(f \circ g)(x)$ $(g \circ f)(x)$

Helpful Hint

In Examples 2 and 3, notice that $(g \circ f)(x) \neq (f \circ g)(x)$. In general, $(g \circ f)(x)$ *may* or *may not* equal $(f \circ g)(x)$.

Example 2 If $f(x) = x^2$ and $g(x) = x + 3$, find each composition.

a. $(f \circ g)(2)$ and $(g \circ f)(2)$ **b.** $(f \circ g)(x)$ and $(g \circ f)(x)$

Solution:

a. $(f \circ g)(2) = f(g(2))$

$= f(5)$ Replace $g(2)$ with 5. [Since $g(x) = x + 3$, then $g(2) = 2 + 3 = 5$.]

$= 5^2 = 25$

$(g \circ f)(2) = g(f(2))$

$= g(4)$ Since $f(x) = x^2$, then $f(2) = 2^2 = 4$.

$= 4 + 3 = 7$

b. $(f \circ g)(x) = f(g(x))$

$= f(x + 3)$ Replace $g(x)$ with $x + 3$.

$= (x + 3)^2$ $f(x + 3) = (x + 3)^2$

$= x^2 + 6x + 9$ Square $(x + 3)$.

$(g \circ f)(x) = g(f(x))$

$= g(x^2)$ Replace $f(x)$ with x^2.

$= x^2 + 3$ $g(x^2) = x^2 + 3$

Work Practice 2

Practice 3

If $f(x) = x^2 + 5$ and $g(x) = x + 3$, find each composition.

a. $(f \circ g)(x)$ **b.** $(g \circ f)(x)$

Example 3 If $f(x) = |x|$ and $g(x) = x - 2$, find each composition.

a. $(f \circ g)(x)$ **b.** $(g \circ f)(x)$

Solution:

a. $(f \circ g)(x) = f(g(x)) = f(x - 2) = |x - 2|$

b. $(g \circ f)(x) = g(f(x)) = g(|x|) = |x| - 2$

Work Practice 3

Answers

2. a. 50; 46 **b.** $9x^2 - 30x + 26$; $3x^2 - 2$

3. a. $x^2 + 6x + 14$ **b.** $x^2 + 8$

Example 4 If $f(x) = 5x$, $g(x) = x - 2$, and $h(x) = \sqrt{x}$, write each function as a composition using two of the given functions.

a. $F(x) = \sqrt{x - 2}$ **b.** $G(x) = 5x - 2$

Solution:

a. Notice the order in which the function F operates on an input value x. First, 2 is subtracted from x. This is the function $g(x) = x - 2$. Then the square root *of that result* is taken. The square root function is $h(x) = \sqrt{x}$. This means that $F = h \circ g$. To check, we find $h \circ g$.

$$F(x) = (h \circ g)(x) = h(g(x)) = h(x - 2) = \sqrt{x - 2}$$

b. Notice the order in which the function G operates on an input value x. First, x is multiplied by 5, and then 2 is subtracted from the result. This means that $G = g \circ f$. To check, we find $g \circ f$.

$$G(x) = (g \circ f)(x) = g(f(x)) = g(5x) = 5x - 2$$

Work Practice 4

Practice 4

If $f(x) = 3x$, $g(x) = x - 4$, and $h(x) = |x|$, write each function as a composition using two of the given functions.

a. $F(x) = |x - 4|$

b. $G(x) = 3x - 4$

Answers

4. a. $(h \circ g)(x)$ **b.** $(g \circ f)(x)$

Graphing Calculator Explorations

If $f(x) = \frac{1}{2}x + 2$ and $g(x) = \frac{1}{3}x^2 + 4$, then

$$\begin{aligned}(f + g)(x) &= f(x) + g(x)\\ &= \left(\frac{1}{2}x + 2\right) + \left(\frac{1}{3}x^2 + 4\right)\\ &= \frac{1}{3}x^2 + \frac{1}{2}x + 6.\end{aligned}$$

To visualize this addition of functions with a graphing calculator, graph

$$Y_1 = \frac{1}{2}x + 2, \quad Y_2 = \frac{1}{3}x^2 + 4, \quad Y_3 = \frac{1}{3}x^2 + \frac{1}{2}x + 6$$

Use a TABLE feature to verify that for a given x-value, $Y_1 + Y_2 = Y_3$. For example, verify that when $x = 0$, $Y_1 = 2$, $Y_2 = 4$, and $Y_3 = 2 + 4 = 6$.

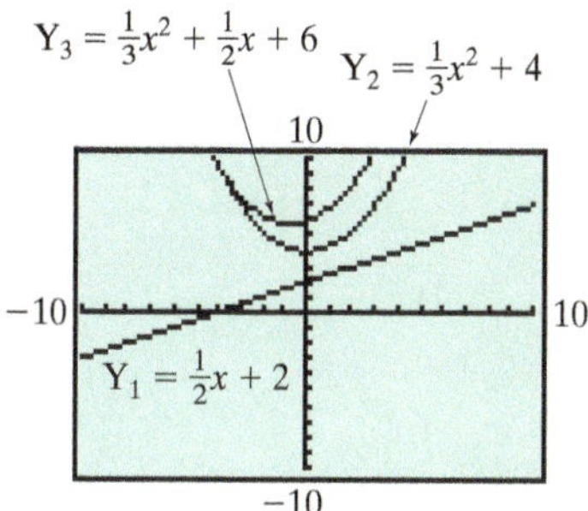

Vocabulary, Readiness & Video Check

Match each function with its definition.

1. $(f \circ g)(x)$	**4.** $(g \circ f)(x)$	**A.** $g(f(x))$	**D.** $\frac{f(x)}{g(x)}, g(x) \neq 0$
2. $(f \cdot g)(x)$	**5.** $\left(\frac{f}{g}\right)(x)$	**B.** $f(x) + g(x)$	**E.** $f(x) \cdot g(x)$
3. $(f - g)(x)$	**6.** $(f + g)(x)$	**C.** $f(g(x))$	**F.** $f(x) - g(x)$

See Video 19.1

Watch the section lecture video and answer the following questions.

Objective 1 7. From Example 1 and the lecture before, we know that $(f+g)(x) = f(x) + g(x)$. Use this fact to explain two ways you can find $(f+g)(2)$.

Objective 2 8. From Example 3, given two functions $f(x)$ and $g(x)$, can $f(g(x))$ ever equal $g(f(x))$?

19.1 Exercise Set MyMathLab®

For the functions f and g, find **a.** $(f+g)(x)$, **b.** $(f-g)(x)$, **c.** $(f\cdot g)(x)$, *and* **d.** $\left(\frac{f}{g}\right)(x)$. *See Example 1.*

1. $f(x) = x - 7, g(x) = 2x + 1$
2. $f(x) = x + 4, g(x) = 5x - 2$
3. $f(x) = x^2 + 1, g(x) = 5x$
4. $f(x) = x^2 - 2, g(x) = 3x$
5. $f(x) = \sqrt[3]{x}, g(x) = x + 5$
6. $f(x) = \sqrt[3]{x}, g(x) = x - 3$
7. $f(x) = -3x, g(x) = 5x^2$
8. $f(x) = 4x^3, g(x) = -6x$

If $f(x) = x^2 - 6x + 2$, $g(x) = -2x$, *and* $h(x) = \sqrt{x}$, *find each composition. See Example 2.*

9. $(f\circ g)(2)$
10. $(h\circ f)(-2)$
11. $(g\circ f)(-1)$
12. $(f\circ h)(1)$
13. $(g\circ h)(0)$
14. $(h\circ g)(0)$

Find $(f\circ g)(x)$ *and* $(g\circ f)(x)$. *See Examples 2 and 3.*

15. $f(x) = x^2 + 1, g(x) = 5x$
16. $f(x) = x - 3, g(x) = x^2$
17. $f(x) = 2x - 3, g(x) = x + 7$
18. $f(x) = x + 10, g(x) = 3x + 1$
19. $f(x) = x^3 + x - 2, g(x) = -2x$
20. $f(x) = -4x, g(x) = x^3 + x^2 - 6$
21. $f(x) = |x|; g(x) = 10x - 3$
22. $f(x) = |x|; g(x) = 14x - 8$
23. $f(x) = \sqrt{x}, g(x) = -5x + 2$
24. $f(x) = 7x - 1, g(x) = \sqrt[3]{x}$

If $f(x) = 3x$, $g(x) = \sqrt{x}$, *and* $h(x) = x^2 + 2$, *write each function as a composition using two of the given functions. See Example 4.*

25. $H(x) = \sqrt{x^2 + 2}$
26. $G(x) = \sqrt{3x}$
27. $F(x) = 9x^2 + 2$
28. $H(x) = 3x^2 + 6$
29. $G(x) = 3\sqrt{x}$
30. $F(x) = x + 2$

Find f(x) and g(x) so that the given function $h(x) = (f\circ g)(x)$.

31. $h(x) = (x + 2)^2$
32. $h(x) = |x - 1|$
33. $h(x) = \sqrt{x + 5} + 2$
34. $h(x) = (3x + 4)^2 + 3$
35. $h(x) = \dfrac{1}{2x - 3}$
36. $h(x) = \dfrac{1}{x + 10}$

REVIEW AND PREVIEW

Solve each equation for y. See Section 9.5.

37. $x = y + 2$

38. $x = y - 5$

39. $x = 3y$

40. $x = -6y$

41. $x = -2y - 7$

42. $x = 4y + 7$

CONCEPT EXTENSIONS

Given that $f(-1) = 4$ $\quad g(-1) = -4$
$f(0) = 5$ $\quad g(0) = -3$
$f(2) = 7$ $\quad g(2) = -1$
$f(7) = 1$ $\quad g(7) = 4$

find each function value.

43. $(f + g)(2)$

44. $(f - g)(7)$

45. $(f \circ g)(2)$

46. $(g \circ f)(2)$

47. $(f \cdot g)(7)$

48. $(f \cdot g)(0)$

49. $\left(\frac{f}{g}\right)(-1)$

50. $\left(\frac{g}{f}\right)(-1)$

51. If you are given $f(x)$ and $g(x)$, explain in your own words how to find $(f \circ g)(x)$ and then how to find $(g \circ f)(x)$.

52. Given $f(x)$ and $g(x)$, describe in your own words the difference between $(f \circ g)(x)$ and $(f \cdot g)(x)$.

Solve.

53. Businesspeople are concerned with cost functions, revenue functions, and profit functions. Recall that the profit $P(x)$ obtained from x units of a product is equal to the revenue $R(x)$ from selling the x units minus the cost $C(x)$ of manufacturing the x units. Write an equation expressing this relationship among $C(x)$, $R(x)$, and $P(x)$.

54. Suppose the revenue $R(x)$ for x units of a product can be described by $R(x) = 25x$, and the cost $C(x)$ can be described by $C(x) = 50 + x^2 + 4x$. Find the profit $P(x)$ for x units. (See Exercise 53.)

19.2 Inverse Functions 

Objectives

1 Determine Whether a Function Is a One-to-One Function.

2 Use the Horizontal Line Test to Decide Whether a Function Is a One-to-One Function.

3 Find the Inverse of a Function.

4 Find the Equation of the Inverse of a Function.

5 Graph Functions and Their Inverses.

6 Determine Whether Two Functions Are Inverses of Each Other.

Objective 1 Determining Whether a Function Is One-to-One

In the next three sections, we begin a study of two new functions: exponential and logarithmic functions. As we learn more about these functions, we will discover that they share a special relation to each other: They are inverses of each other.

Before we study these functions, we need to learn about inverses. We begin by defining one-to-one functions.

Study the following table.

Degrees Fahrenheit (Input)	−31	−13	32	68	149	212
Degrees Celsius (Output)	−35	−25	0	20	65	100

Recall that since each degrees Fahrenheit (input) corresponds to exactly one degrees Celsius (output), this pairing of inputs and outputs does describe a function. Also notice that each output corresponds to exactly one input. This type of function is given a special name—a one-to-one function.

Does the set $f = \{(0, 1), (2, 2), (-3, 5), (7, 6)\}$ describe a one-to-one function? It is a function since each x-value corresponds to a unique y-value. For this particular function f, each y-value also corresponds to a unique x-value. Thus, this function is also a **one-to-one function.**

One-to-One Function

For a **one-to-one function,** each x-value (input) corresponds to only one y-value (output), and each y-value (output) corresponds to only one x-value (input).

Practice 1

Determine whether each function described is one-to-one.

a. $f = \{(4, -3), (3, -4), (2, 7), (5, 0)\}$

b. $g = \{(8, 4), (-2, 0), (6, 4), (2, 6)\}$

c. $h = \{(2, 4), (1, 3), (4, 6), (-2, 4)\}$

d.

Year	Federal Minimum Wage
1950	\$0.75
1963	\$1.25
1968	\$1.60
1975	\$2.10
1997	\$5.15
2008	\$6.55

e.

f.

State → Mean SAT Math Scores

North Dakota (highest) → 617
Connecticut → 516
Indiana → 509
Maryland → 509
Texas → 506
Washington, D.C. (lowest) → 472

Answers

1. a. one-to-one **b.** not one-to-one **c.** not one-to-one **d.** one-to-one **e.** not one-to-one **f.** not one-to-one

Example 1 Determine whether each function described is one-to-one.

a. $f = \{(6, 2), (5, 4), (-1, 0), (7, 3)\}$

b. $g = \{(3, 9), (-4, 2), (-3, 9), (0, 0)\}$

c. $h = \{(1, 1), (2, 2), (10, 10), (-5, -5)\}$

d.

Mineral (Input)	Talc	Gypsum	Diamond	Topaz	Stibnite
Hardness on the Mohs Scale (Output)	1	2	10	8	2

e.

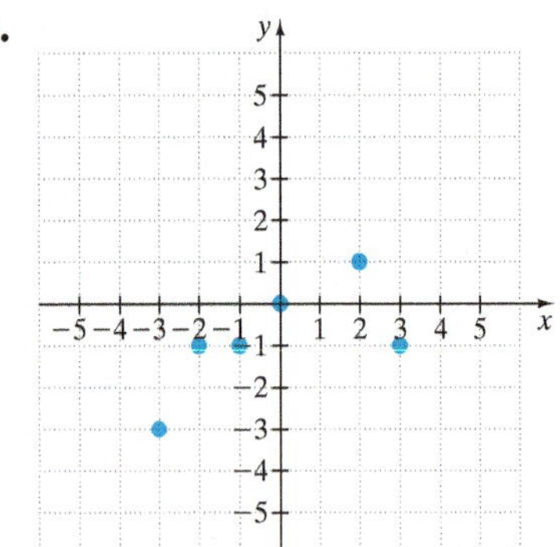

f.

Cities	Percent of Cell Phone Subscribers Who Text
El Paso (highest)	57
Buffalo	47
Austin	49
Tulsa	49
Charleston	36
Detroit	45

Solution:

a. f is one-to-one since each y-value corresponds to only one x-value.

b. g is not one-to-one because the y-value 9 in $(3, 9)$ and $(-3, 9)$ corresponds to different x-values.

c. h is a one-to-one function since each y-value corresponds to only one x-value.

d. This table does not describe a one-to-one function since the output 2 corresponds to two inputs, gypsum and stibnite.

e. This graph does not describe a one-to-one function since the y-value -1 corresponds to three x-values, -2, -1, and 3. (See the graph below.)

f. The mapping is not one-to-one since 49% corresponds to Austin and Tulsa.

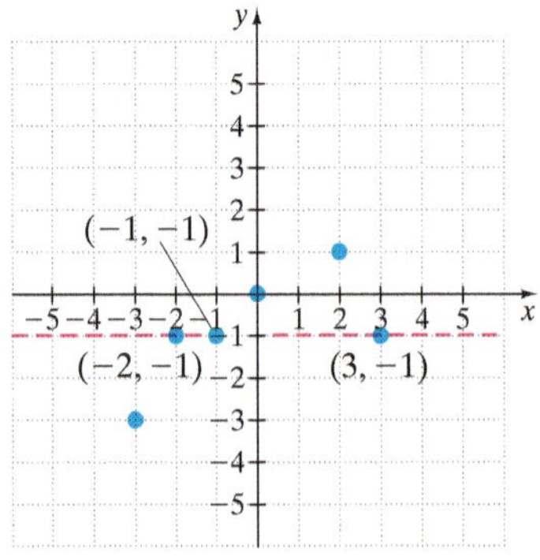

Work Practice 1

Objective 2 Using the Horizontal Line Test

Recall that we recognize the graph of a function when it passes the vertical line test. Since every x-value of the function corresponds to exactly one y-value, each vertical line intersects the function's graph at most once. The graph shown next, for instance, is the graph of a function.

Is this function a *one-to-one* function? The answer is no. To see why not, notice that the y-value of the ordered pair $(-3, 3)$, for example, is the same as the y-value of the ordered pair $(3, 3)$. In other words, the y-value 3 corresponds to two x-values, -3 and 3. This function is therefore not one-to-one.

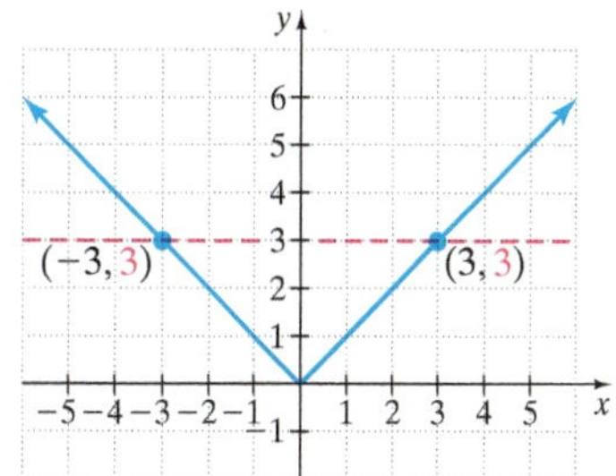

Not a one-to-one function.

To test whether a graph is the graph of a one-to-one function, apply the vertical line test to see if it is a function and then apply the similar **horizontal line test** to see if it is a one-to-one function.

Horizontal Line Test

If every horizontal line intersects the graph of a function at most once, then the function is a one-to-one function.

Example 2 Determine whether each graph is the graph of a one-to-one function.

a.

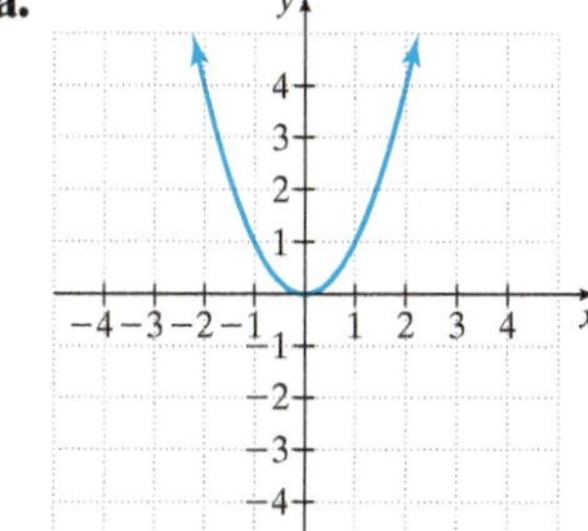

b.

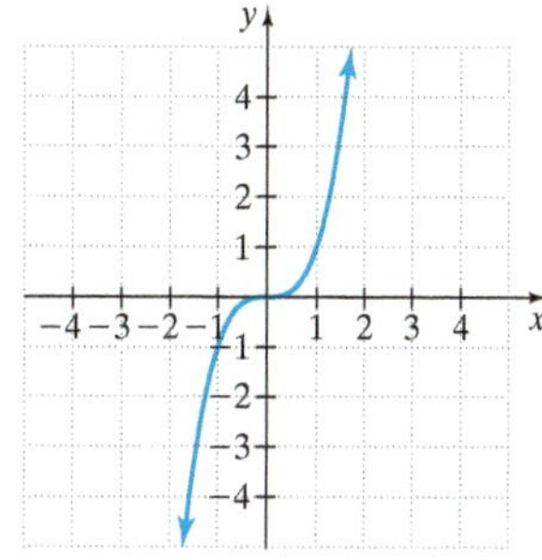

c.

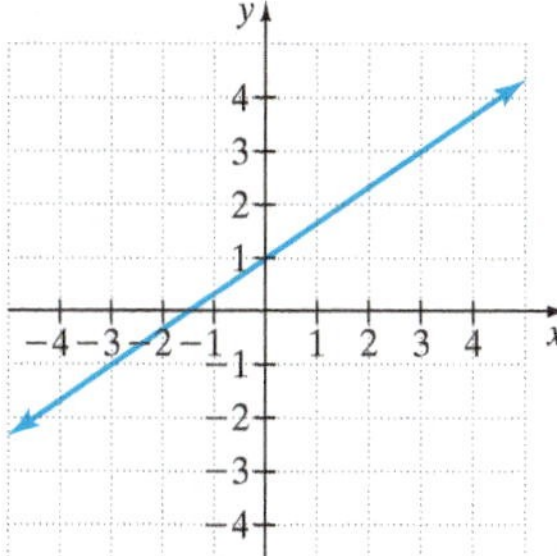

d.

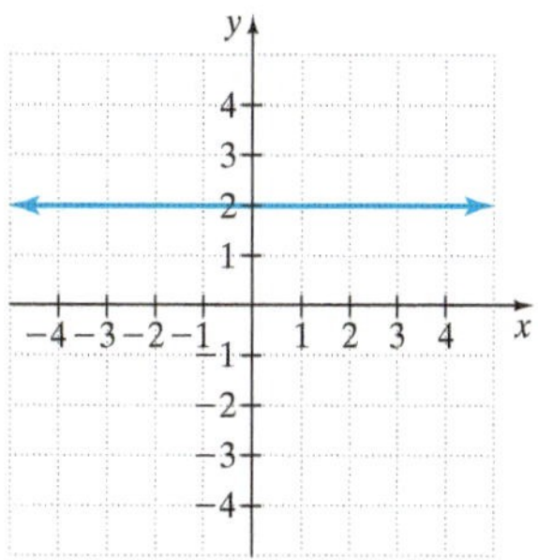

e.

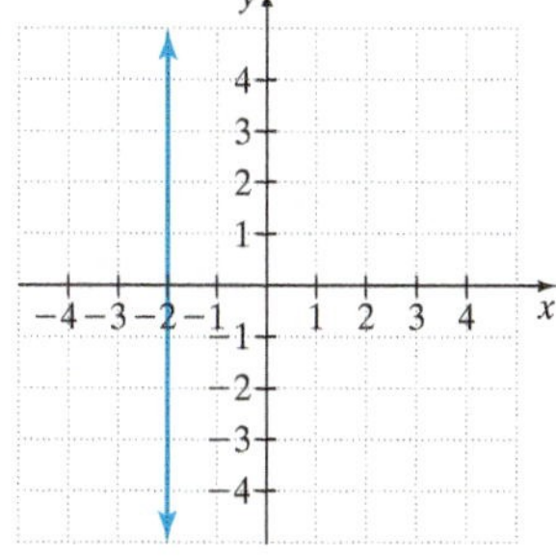

Solution: Graphs **a, b, c,** and **d** all pass the vertical line test, so only these graphs are graphs of functions. But, of these, only **b** and **c** pass the horizontal line test, so only **b** and **c** are graphs of one-to-one functions.

Work Practice 2

Practice 2

Determine whether each graph is the graph of a one-to-one function.

a.

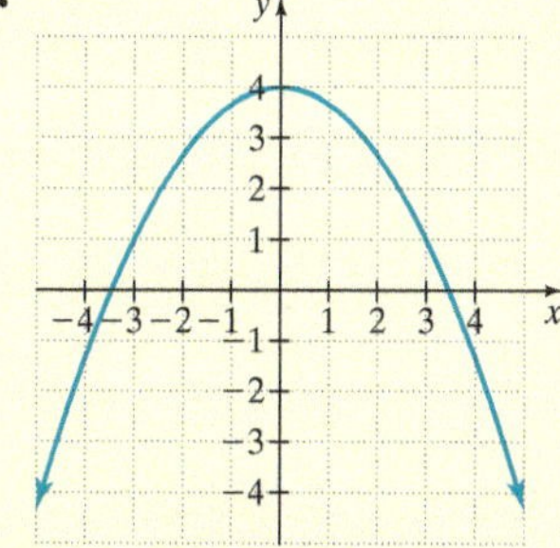

b.

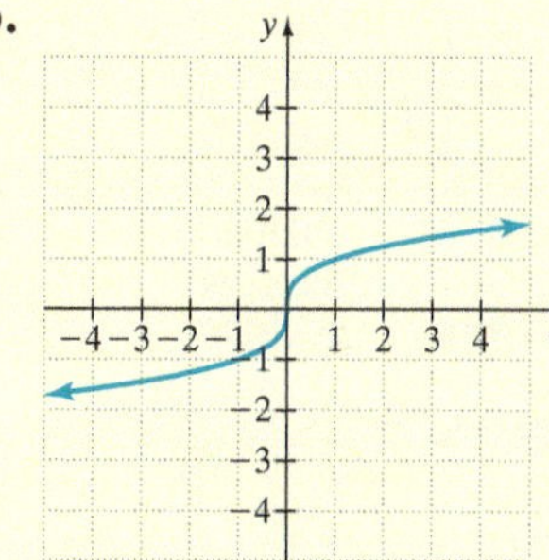

c.

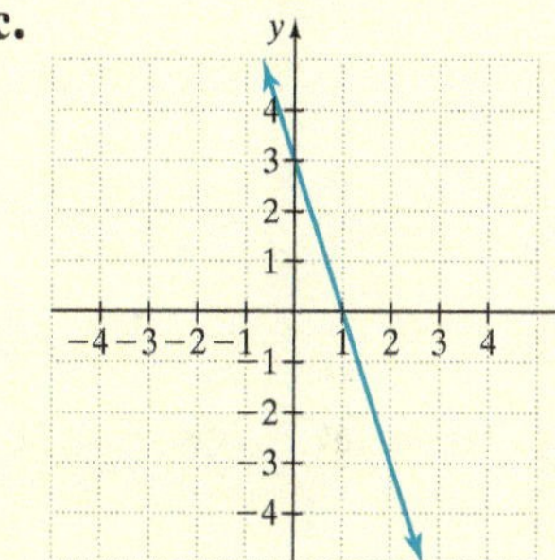

d.

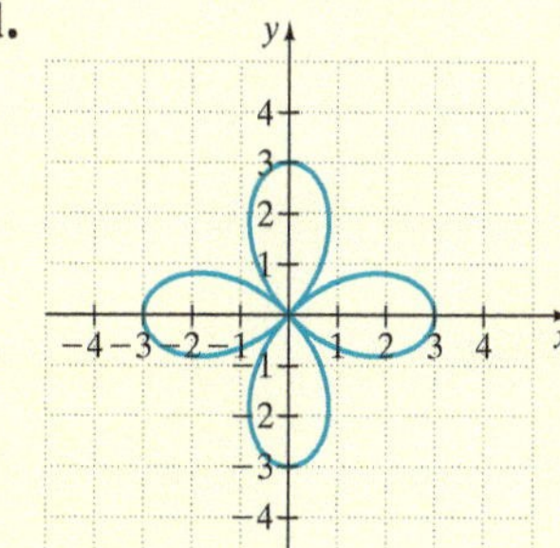

e.

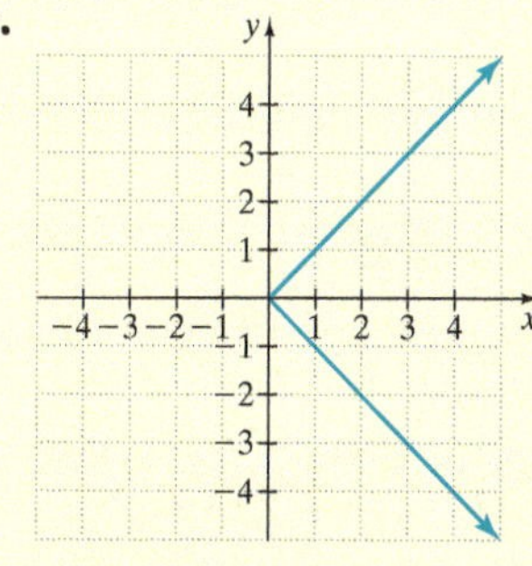

Answers

2. a. no, not one-to-one **b.** yes
c. yes **d.** no, not a function
e. no, not a function

Helpful Hint

All linear equations are one-to-one functions except those whose graphs are horizontal or vertical lines. A vertical line does not pass the vertical line test and hence is not the graph of a function. A horizontal line is the graph of a function but does not pass the horizontal line test and hence is not the graph of a one-to-one function.

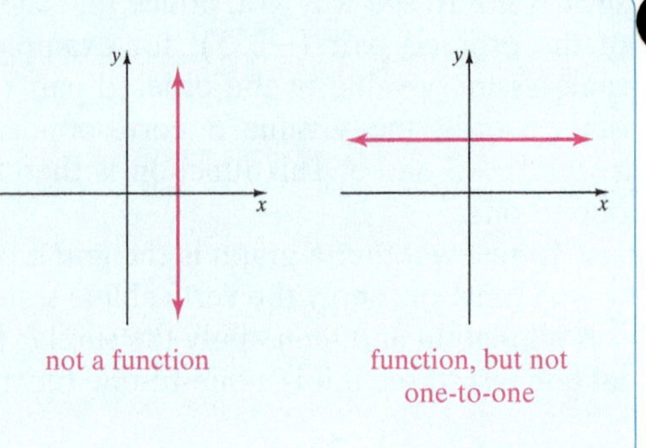

Objective 3 Finding the Inverse of a Function 

One-to-one functions are special in that their graphs pass both the vertical and horizontal line tests. They are special, too, in another sense: For each one-to-one function, we can find its **inverse function** by switching the coordinates of the ordered pairs of the function, or the inputs and the outputs. For example, the inverse of the one-to-one function

Degrees Fahrenheit (Input)	−31	−13	32	68	149.	212
Degrees Celsius (Output)	−35	−25	0	20	65	100

is the function

Degrees Celsius (Input)	−35	−25	0	20	65	100
Degrees Fahrenheit (Output)	−31	−13	32	68	149	212

f

Input: degrees Fahrenheit — Output: degrees Celsius

212 → 100
68 → 20
32 → 0
−13 → −25

f^{-1}

Notice that the ordered pair $(-31, -35)$ of the function, for example, becomes the ordered pair $(-35, -31)$ of its inverse.

Also, the inverse of the one-to-one function $f = \{(2, -3), (5, 10), (9, 1)\}$ is $\{(-3, 2), (10, 5), (1, 9)\}$. For a function f, we use the notation f^{-1}, read "f inverse," to denote its inverse function. Notice that since the coordinates of each ordered pair have been switched, the domain (set of inputs) of f is the range (set of outputs) of f^{-1}, and the range of f is the domain of f^{-1}.

Inverse Function

The inverse of a one-to-one function f is the one-to-one function f^{-1} that consists of the set of all ordered pairs (y, x) where (x, y) belongs to f.

Helpful Hint

If a function is not one-to-one, it does not have an inverse function.

Practice 3

Find the inverse of the one-to-one function.

$f = \{(3, 4), (-2, 0), (2, 8), (6, 6)\}$

Example 3 Find the inverse of the one-to-one function.

$$f = \{(0, 1), (-2, 7), (3, -6), (4, 4)\}$$

Solution: $f^{-1} = \{(1, 0), (7, -2), (-6, 3), (4, 4)\}$ Switch coordinates of each ordered pair.

Work Practice 3

Answer

3. $f^{-1} = \{(4, 3), (0, -2), (8, 2), (6, 6)\}$

Helpful Hint

The symbol f^{-1} is the single symbol that denotes the inverse of the function f.

It is read as "f inverse." This symbol *does not mean* $\frac{1}{f}$.

✓**Concept Check** Suppose that f is a one-to-one function and that $f(1) = 5$.

a. Write the corresponding ordered pair.

b. Write one point that we know must belong to the inverse function f^{-1}.

Objective 4 Finding the Equation of the Inverse of a Function

If a one-to-one function f is defined as a set of ordered pairs, we can find f^{-1} by interchanging the x- and y-coordinates of the ordered pairs. If a one-to-one function f is given in the form of an equation, we can find f^{-1} by using a similar procedure.

Finding the Inverse of a One-to-One Function $f(x)$

Step 1: Replace $f(x)$ with y.

Step 2: Interchange x and y.

Step 3: Solve the equation for y.

Step 4: Replace y with the notation $f^{-1}(x)$.

Example 4 Find an equation of the inverse of $f(x) = x + 3$.

Solution: $f(x) = x + 3$

Step 1: $y = x + 3$ — Replace $f(x)$ with y.

Step 2: $x = y + 3$ — Interchange x and y.

Step 3: $x - 3 = y$ — Solve for y.

Step 4: $f^{-1}(x) = x - 3$ — Replace y with $f^{-1}(x)$.

The inverse of $f(x) = x + 3$ is $f^{-1}(x) = x - 3$. Notice that, for example,

$f(1) = 1 + 3 = 4$ and $f^{-1}(4) = 4 - 3 = 1$

Ordered pair: $(1, 4)$ Ordered pair: $(4, 1)$

The coordinates are switched, as expected.

Work Practice 4

Practice 4

Find the equation of the inverse of $f(x) = 6 - x$.

Answer

4. $f^{-1}(x) = 6 - x$

✓**Concept Check Answer**

a. $(1, 5)$ **b.** $(5, 1)$

Practice 5

Find the equation of the inverse of $f(x) = 5x + 2$. Graph f and f^{-1} on the same set of axes.

Example 5 Find the equation of the inverse of $f(x) = 3x - 5$. Graph f and f^{-1} on the same set of axes.

Solution: $f(x) = 3x - 5$

Step 1: $y = 3x - 5$ Replace $f(x)$ with y.

Step 2: $x = 3y - 5$ Interchange x and y.

Step 3: $3y = x + 5$ Solve for y.

$$y = \frac{x+5}{3}$$

Step 4: $f^{-1}(x) = \frac{x+5}{3}$ Replace y with $f^{-1}(x)$.

Now we graph $f(x)$ and $f^{-1}(x)$ on the same set of axes. Both $f(x) = 3x - 5$ and $f^{-1}(x) = \frac{x+5}{3}$ are linear functions, so each graph is a line.

$f(x) = 3x - 5$	
x	$y = f(x)$
1	−2
0	−5
$\frac{5}{3}$	0

$f^{-1}(x) = \frac{x+5}{3}$	
x	$y = f^{-1}(x)$
−2	1
−5	0
0	$\frac{5}{3}$

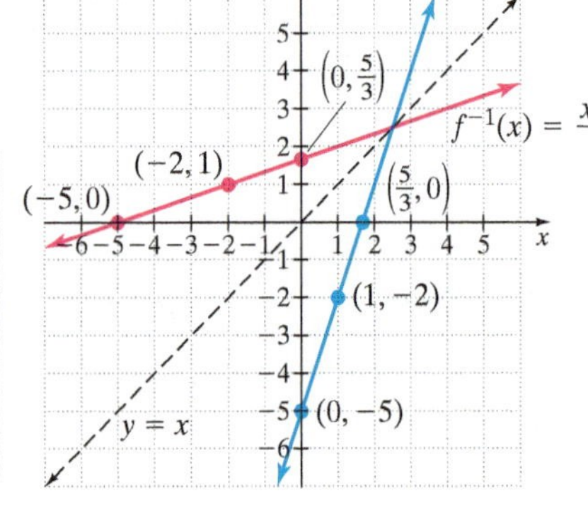

Work Practice 5

Objective 5 Graphing Inverse Functions 

Notice that the graphs of f and f^{-1} in Example 5 are mirror images of each other, and the "mirror" is the dashed line $y = x$. This is true for every function and its inverse. For this reason, we say that *the graphs of f and f^{-1} are symmetric about the line $y = x$.*

To see why this happens, study the graph of a few ordered pairs and their switched coordinates in the diagram below.

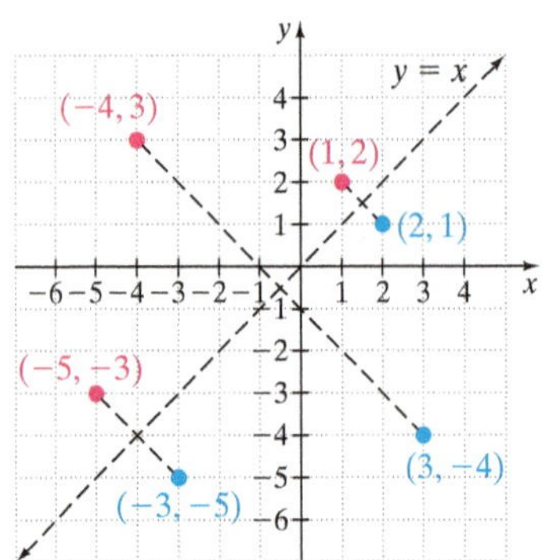

Answer

5. $f^{-1}(x) = \frac{x-2}{5}$

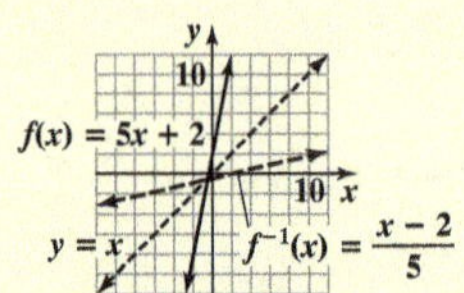

Example 6 Graph the inverse of each function.

Solution: The function is graphed in blue and the inverse is graphed in red.

a.

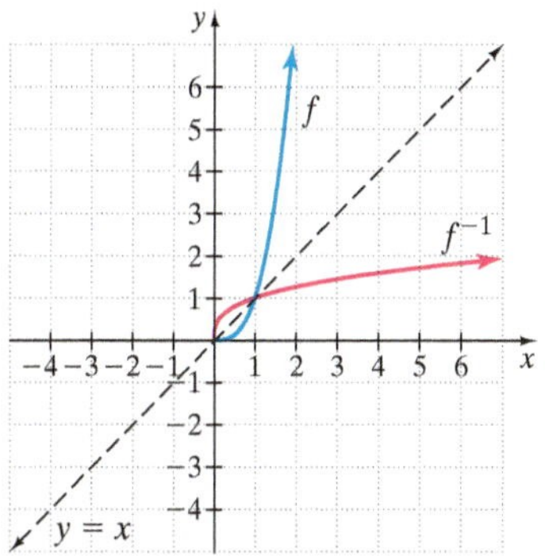

b.

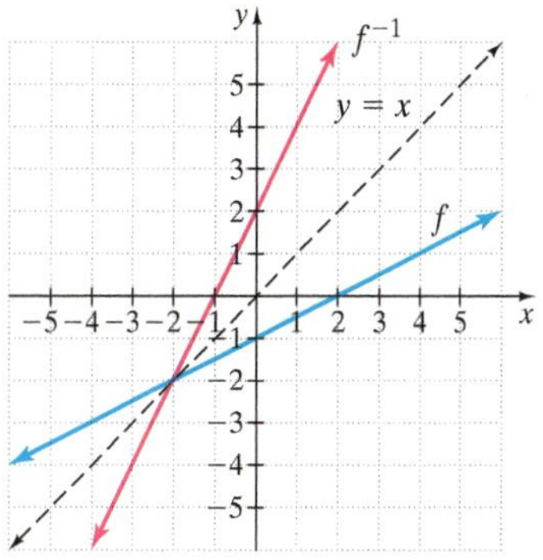

Work Practice 6

Practice 6

Graph the inverse of each function.

a.

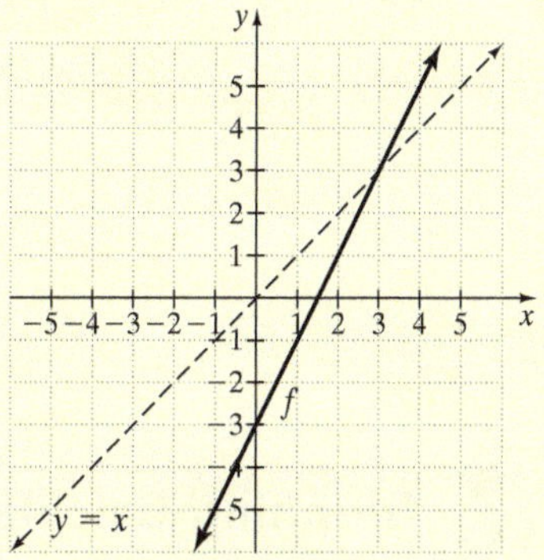

b.

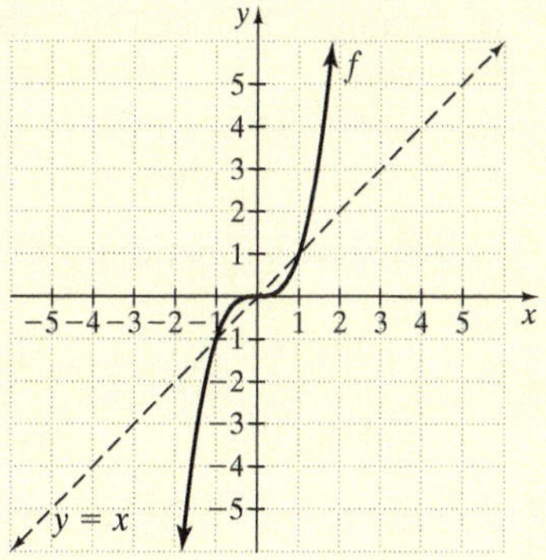

Objective 6 Determining Whether Functions Are Inverses of Each Other

Notice in the table of values in Example 5 that $f(0) = -5$ and $f^{-1}(-5) = 0$, as expected. Also, for example, $f(1) = -2$ and $f^{-1}(-2) = 1$. In words, we say that for some input x, the function f^{-1} takes the output of x, called $f(x)$, back to x.

$$x \rightarrow f(x) \quad \text{and} \quad f^{-1}(f(x)) \rightarrow x$$

$$f(0) = -5 \quad \text{and} \quad f^{-1}(-5) = 0$$

$$f(1) = -2 \quad \text{and} \quad f^{-1}(-2) = 1$$

In general,

> If f is a one-to-one function, then the inverse of f is the function f^{-1} such that
>
> $$(f^{-1} \circ f)(x) = x \quad \text{and} \quad (f \circ f^{-1})(x) = x$$

Practice 7

Show that if $f(x) = 4x - 1$, then $f^{-1}(x) = \dfrac{x+1}{4}$.

Example 7 Show that if $f(x) = 3x + 2$, then $f^{-1}(x) = \dfrac{x-2}{3}$.

Solution: See that $(f^{-1} \circ f)(x) = x$ and $(f \circ f^{-1})(x) = x$.

$$\begin{aligned}(f^{-1} \circ f)(x) &= f^{-1}(f(x)) \\ &= f^{-1}(3x + 2) \quad \text{Replace } f(x) \text{ with } 3x + 2. \\ &= \frac{3x + 2 - 2}{3} \\ &= \frac{3x}{3} \\ &= x\end{aligned}$$

(*Continued on next page*)

Answer

6. a.

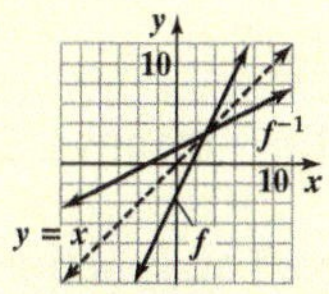

b.

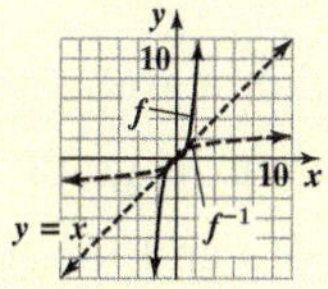

7. $f(f^{-1}(x)) = 4\left(\dfrac{x+1}{4}\right) - 1$

$= x + 1 - 1 = x$

$f^{-1}(f(x)) = \dfrac{(4x-1)+1}{4}$

$= \dfrac{4x}{4} = x$

$$(f \circ f^{-1})(x) = f(f^{-1}(x))$$
$$= f\left(\frac{x-2}{3}\right) \quad \text{Replace } f^{-1}(x) \text{ with } \frac{x-2}{3}.$$
$$= 3\left(\frac{x-2}{3}\right) + 2$$
$$= x - 2 + 2$$
$$= x$$

Work Practice 7

Graphing Calculator Explorations

A graphing calculator can be used to visualize the results of Example 7. Recall that the graph of a function f and its inverse f^{-1} are mirror images of each other across the line $y = x$. To see this for the function from Example 7, use a square window and graph

the given function: $Y_1 = 3x + 2$

its inverse: $Y_2 = \frac{x-2}{3}$

and the line: $Y_3 = x$

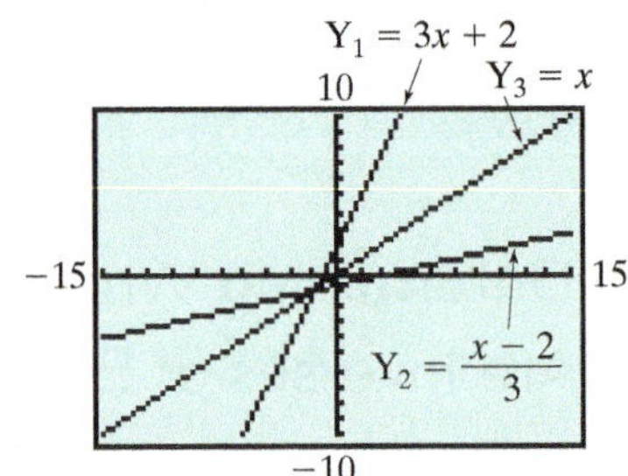

See Exercises 67–70 in Exercise Set 19.2.

Vocabulary, Readiness & Video Check

Use the choices below to fill in each blank. Some choices will not be used, and some will be used more than once.

vertical	(3, 7)	(11, 2)	$y = x$	x	true
horizontal	(7, 3)	(2, 11)	$\frac{1}{f}$	the inverse of f	false

1. If $f(2) = 11$, the corresponding ordered pair is ____________.

2. If (7, 3) is an ordered pair solution of $f(x)$, and $f(x)$ has an inverse, then an ordered pair solution of $f^{-1}(x)$ is ____________.

3. The symbol f^{-1} means ____________.

4. True or false: The function notation $f^{-1}(x)$ means $\frac{1}{f(x)}$. ____________

5. To tell whether a graph is the graph of a function, use the ____________ line test.

6. To tell whether the graph of a function is also a one-to-one function, use the ____________ line test.

7. The graphs of f and f^{-1} are symmetric about the line ____________.

8. Two functions are inverses of each other if $(f \circ f^{-1})(x) =$ ____________ and $(f^{-1} \circ f)(x) =$ ____________.

Martin-Gay Interactive Videos

See Video 19.2

Watch the section lecture video and answer the following questions.

Objective 1 **9.** From Example 1 and the definition before, what makes a one-to-one function different from other types of functions?

Objective 2 **10.** From Examples 2 and 3, if a graph passes the horizontal line test, but not the vertical line test, is it a one-to-one function? Explain.

Objective 3 **11.** From Example 4 and the lecture before, if you find the inverse of a one-to-one function, is this inverse function also a one-to-one function? How do you know?

Objective 4 **12.** From Examples 5 and 6, explain why the interchanging of x and y when finding an inverse equation makes sense given the definition of an inverse function.

Objective 5 **13.** From Example 7, if you have the equation or graph of a one-to-one function, how can you graph its inverse without finding the inverse's equation?

Objective 6 **14.** Based on Example 8 and the lecture before, what's wrong with the following statement? "If f is a one-to-one function, you can prove that f and f^{-1} are inverses of each other by showing that $f(f^{-1}(x)) = f^{-1}(f(x))$."

19.2 Exercise Set MyMathLab®

Determine whether each function is a one-to-one function. If it is one-to-one, list the inverse function by switching coordinates, or inputs and outputs. See Examples 1 and 3.

1. $f = \{(-1, -1), (1, 1), (0, 2), (2, 0)\}$

2. $g = \{(8, 6), (9, 6), (3, 4), (-4, 4)\}$

3. $h = \{(10, 10)\}$

4. $r = \{(1, 2), (3, 4), (5, 6), (6, 7)\}$

5. $f = \{(11, 12), (4, 3), (3, 4), (6, 6)\}$

6. $g = \{(0, 3), (3, 7), (6, 7), (-2, -2)\}$

7.

Month of 2009 (Input)	July	August	September	October	November	December
Unemployment Rate in Percent (Output)	9.4	9.7	9.8	10.1	10.0	10.0

(*Source:* U.S. Bureau of Labor Statistics)

8.

State (Input)	Texas	Massachusetts	Nevada	Idaho	Wisconsin
Number of Two-Year Colleges (Output)	70	22	3	3	31

(*Source:* University of Texas at Austin)

9.

State (Input)	California	Alaska	Indiana	Louisiana	New Mexico	Ohio
Rank in Population (Output)	1	47	16	25	36	7

(*Source:* U.S. Bureau of the Census)

△ **10.**

Shape (Input)	Triangle	Pentagon	Quadrilateral	Hexagon	Decagon
Number of Sides (Output)	3	5	4	6	10

Given the one-to-one function $f(x) = x^3 + 2$, find the following. [Hint: You do not need to find the equation for $f^{-1}(x)$.]

11. **a.** $f(1)$ **b.** $f^{-1}(3)$

12. **a.** $f(0)$ **b.** $f^{-1}(2)$

13. **a.** $f(-1)$ **b.** $f^{-1}(1)$

14. **a.** $f(-2)$ **b.** $f^{-1}(-6)$

Determine whether the graph of each function is the graph of a one-to-one function. See Example 2.

15.

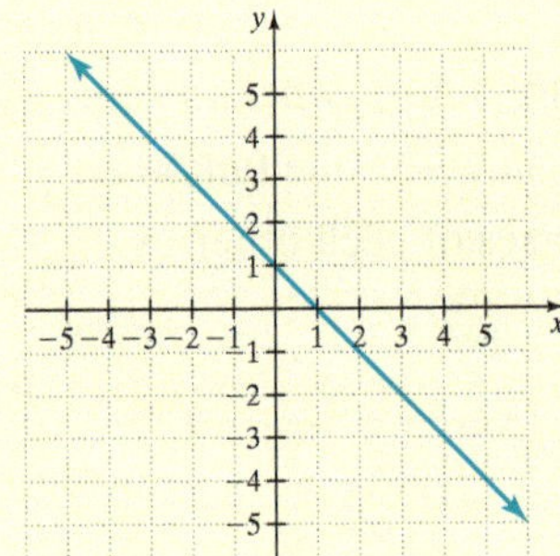

16.

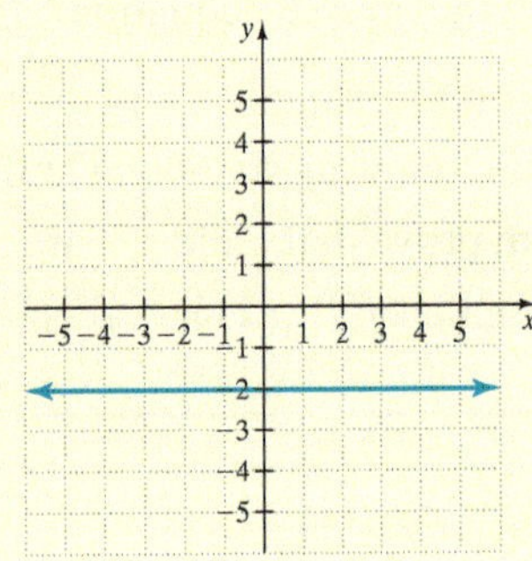

17.

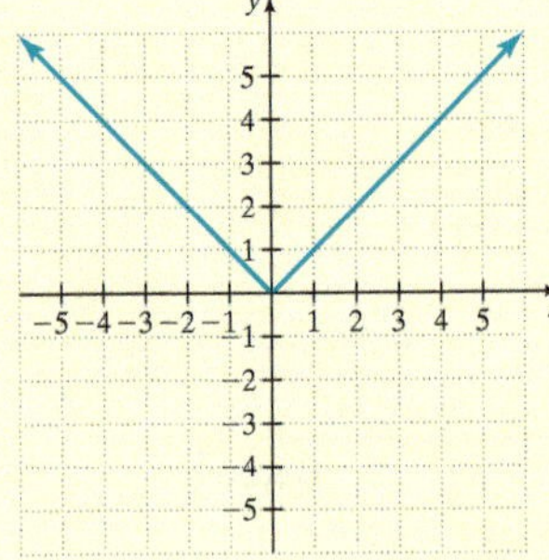

18.

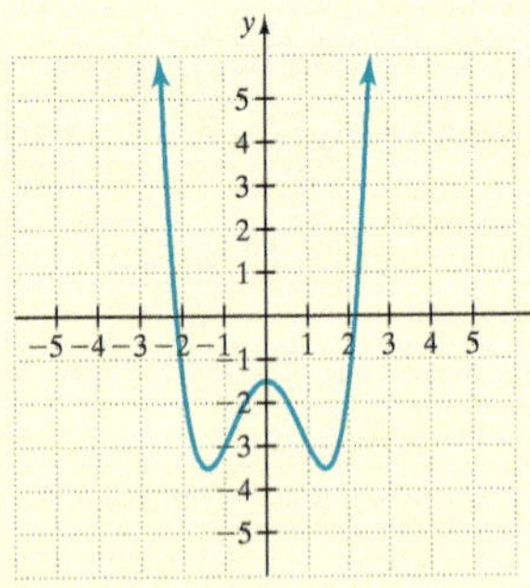

19.

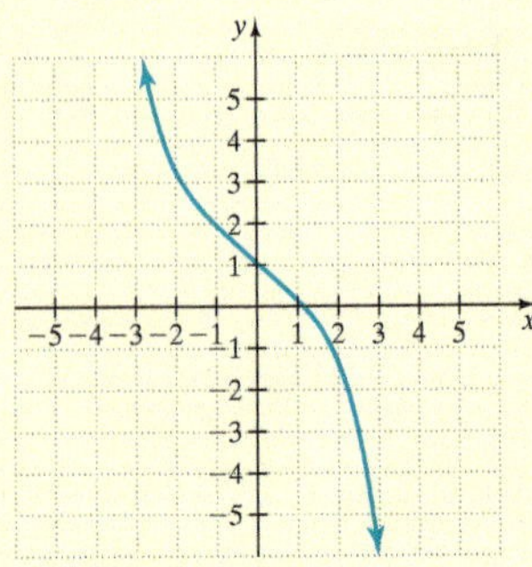

20.

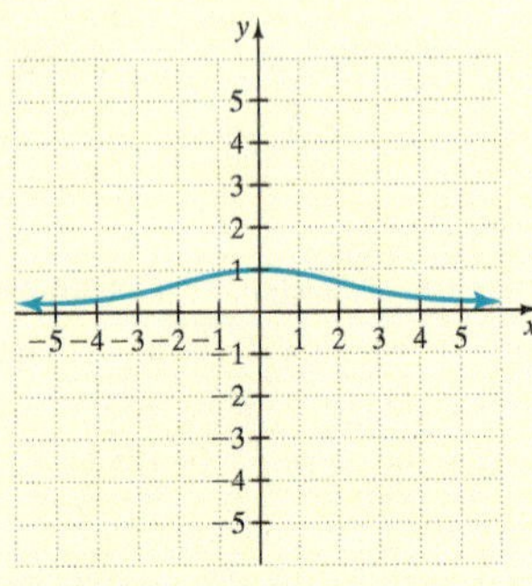

21.

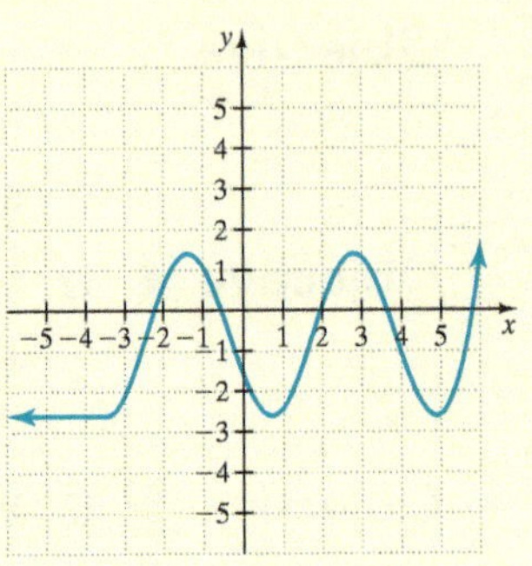

22.

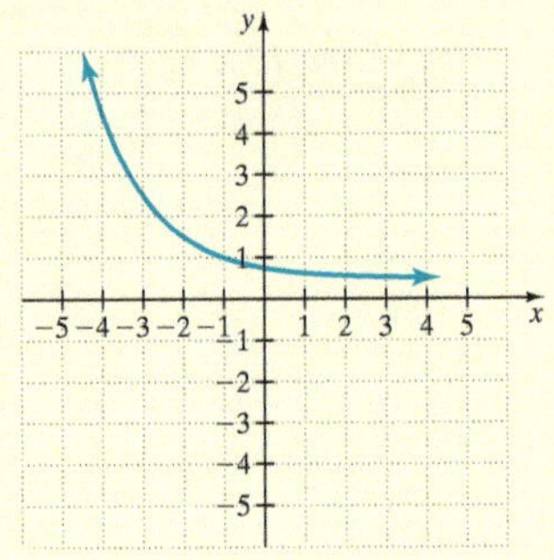

MIXED PRACTICE

Each of the following functions is one-to-one. Find the inverse of each function and graph the function and its inverse on the same set of axes. See Examples 4 and 5.

23. $f(x) = x + 4$

24. $f(x) = x - 5$

25. $f(x) = 2x - 3$

26. $f(x) = 4x + 9$

27. $f(x) = \frac{1}{2}x - 1$

28. $f(x) = -\frac{1}{2}x + 2$

29. $f(x) = x^3$

30. $f(x) = x^3 - 1$

Find the inverse of each one-to-one function. See Examples 4 and 5.

31. $f(x) = 5x + 2$

32. $f(x) = 6x - 1$

33. $f(x) = \frac{x - 2}{5}$

34. $f(x) = \frac{x - 3}{2}$

35. $f(x) = \sqrt[3]{x}$

36. $f(x) = \sqrt[3]{x + 1}$

37. $f(x) = \frac{5}{3x + 1}$

38. $f(x) = \frac{7}{2x + 4}$

39. $f(x) = (x + 2)^3$

40. $f(x) = (x - 5)^3$

Graph the inverse of each function on the same set of axes. See Example 6.

41.

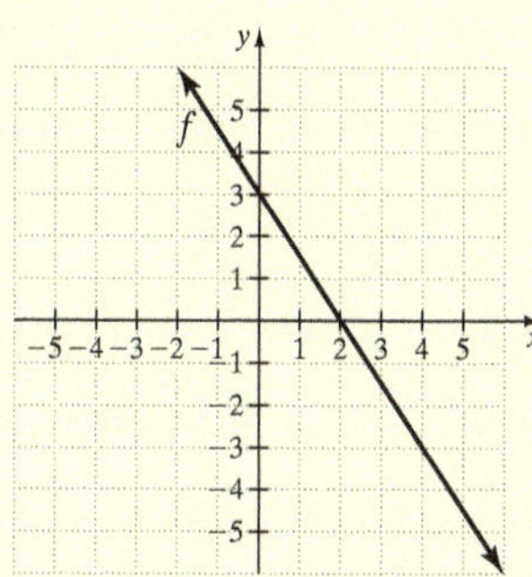

42.

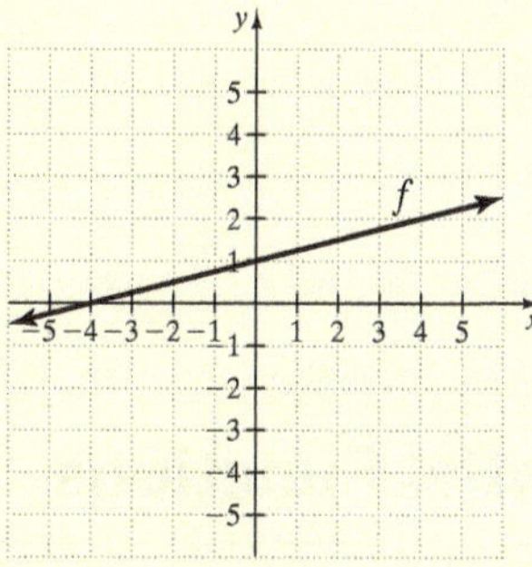

43.

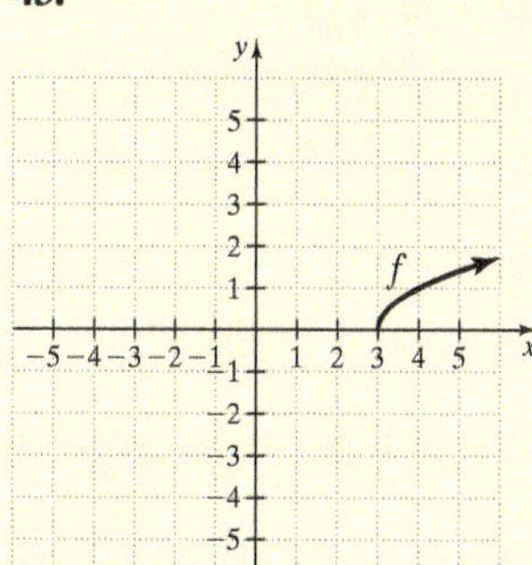

44.

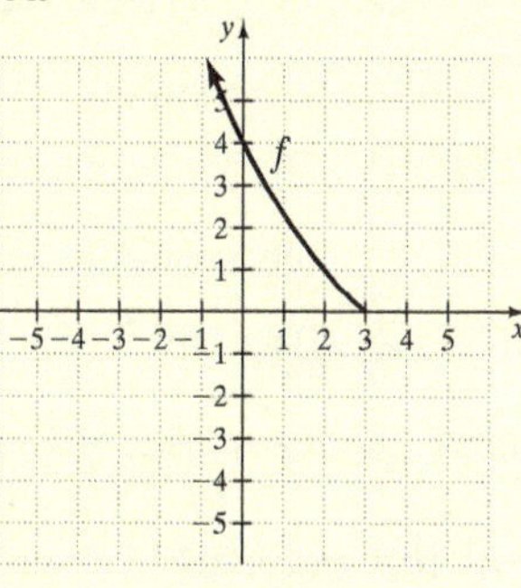

45.

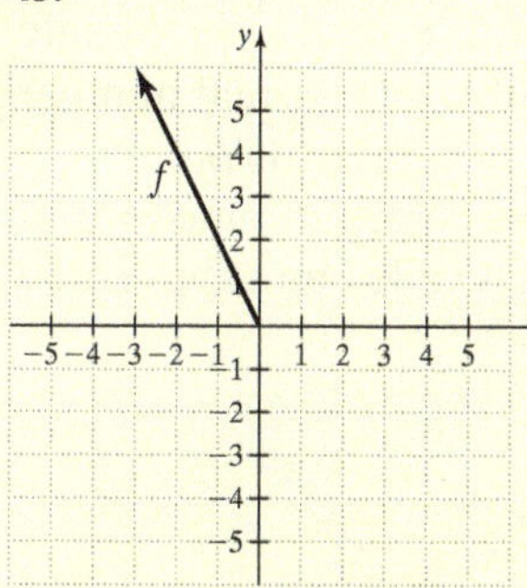

46.

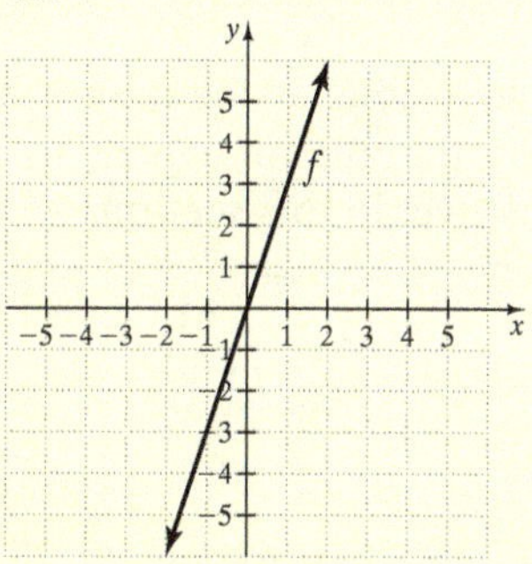

Solve. See Example 7.

47. If $f(x) = 2x + 1$, show that $f^{-1}(x) = \dfrac{x - 1}{2}$.

48. If $f(x) = 3x - 10$, show that $f^{-1}(x) = \dfrac{x + 10}{3}$.

49. If $f(x) = x^3 + 6$, show that $f^{-1}(x) = \sqrt[3]{x - 6}$.

50. If $f(x) = x^3 - 5$, show that $f^{-1}(x) = \sqrt[3]{x + 5}$.

REVIEW AND PREVIEW

Evaluate each of the following. See Section 17.2.

51. $25^{1/2}$

52. $49^{1/2}$

53. $16^{3/4}$

54. $27^{2/3}$

55. $9^{-3/2}$

56. $81^{-3/4}$

If $f(x) = 3^x$, find the following. In Exercises 59 and 60, give an exact answer and a two-decimal-place approximation. See Sections 12.1, 15.2, and 17.2.

57. $f(2)$

58. $f(0)$

59. $f\left(\frac{1}{2}\right)$

60. $f\left(\frac{2}{3}\right)$

CONCEPT EXTENSIONS

Solve. See the Concept Check in this section.

61. Suppose that f is a one-to-one function and that $f(2) = 9$.

a. Write the corresponding ordered pair.

b. Name one ordered pair that we know is a solution of the inverse of f, or f^{-1}.

62. Suppose that F is a one-to-one function and that $F\left(\frac{1}{2}\right) = -0.7$.

a. Write the corresponding ordered pair.

b. Name one ordered pair that we know is a solution of the inverse of F, or F^{-1}.

For Exercises 63 and 64,

a. *Write the ordered pairs for $f(x)$ whose points are highlighted. (Include the points whose coordinates are given.)*

b. *Write the corresponding ordered pairs for the inverse of f, f^{-1}.*

c. *Graph the ordered pairs for f^{-1} found in part* **b.**

d. *Graph $f^{-1}(x)$ by drawing a smooth curve through the plotted points.*

63.

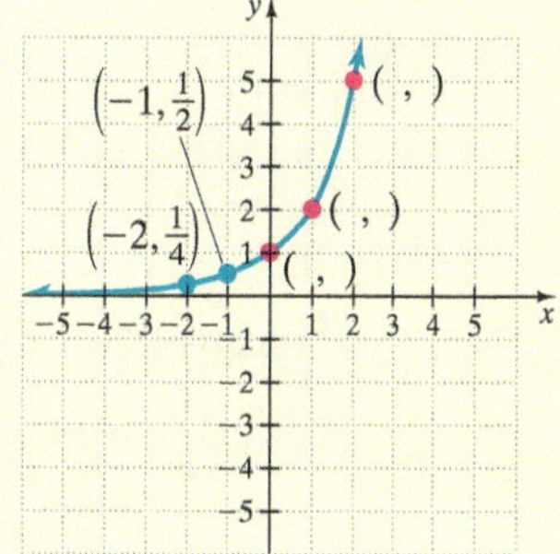

64.

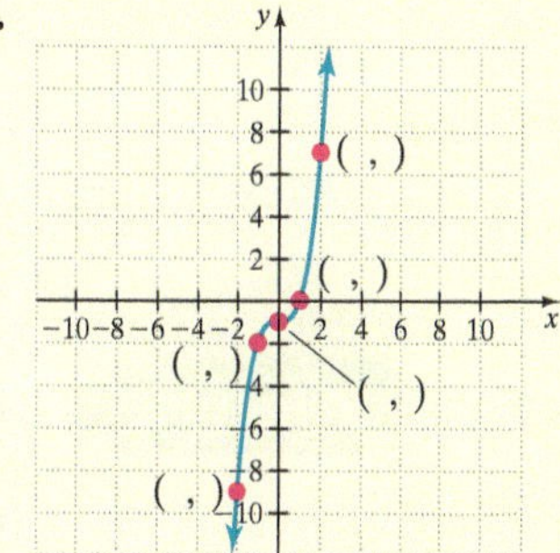

65. If you are given the graph of a function, describe how you can tell from the graph whether the function has an inverse.

66. Describe the appearance of the graphs of a function and its inverse.

Find the inverse of each given one-to-one function. Then use a graphing calculator to graph the function and its inverse on a square window.

67. $f(x) = 3x + 1$

68. $f(x) = -2x - 6$

69. $f(x) = \sqrt[3]{x+1}$

70. $f(x) = x^3 - 3$

19.3 Exponential Functions

Objectives

1 Graph Exponential Functions.

2 Solve Equations of the Form $b^x = b^y$.

3 Solve Problems Modeled by Exponential Equations.

Objective 1 Graphing Exponential Functions 

In earlier chapters, we gave meaning to exponential expressions such as 2^x, where x is a rational number. For example,

$$2^3 = 2 \cdot 2 \cdot 2 \qquad \text{Three factors; each factor is 2}$$

$$2^{3/2} = (2^{1/2})^3 = \sqrt{2} \cdot \sqrt{2} \cdot \sqrt{2} \qquad \text{Three factors; each factor is } \sqrt{2}$$

When x is an irrational number (for example, $\sqrt{3}$), what meaning can we give to $2^{\sqrt{3}}$?

It is beyond the scope of this book to give precise meaning to 2^x if x is irrational. We can confirm your intuition and say that $2^{\sqrt{3}}$ is a real number and, since $1 < \sqrt{3} < 2$, $2^1 < 2^{\sqrt{3}} < 2^2$. We can also use a calculator and approximate $2^{\sqrt{3}}$: $2^{\sqrt{3}} \approx 3.321997$. In fact, as long as the base b is positive, b^x is a real number for all real numbers x. Finally, the rules of exponents apply whether x is rational or irrational as long as b is positive. In this section, we are interested in functions of the form $f(x) = b^x$, where $b > 0$. A function of this form is called an **exponential function.**

Exponential Function

A function of the form

$$f(x) = b^x$$

is called an **exponential function** if $b > 0$, b is not 1, and x is a real number.

Next, we practice graphing exponential functions.

Example 1 Graph the exponential functions defined by $f(x) = 2^x$ and $g(x) = 3^x$ on the same set of axes.

Solution: Graph each function by plotting points. Set up a table of values for each of the two functions.

Practice 1

Graph the exponential functions defined by $f(x) = 2^x$ and $g(x) = 7^x$ on the same set of axes.

Answer

1.

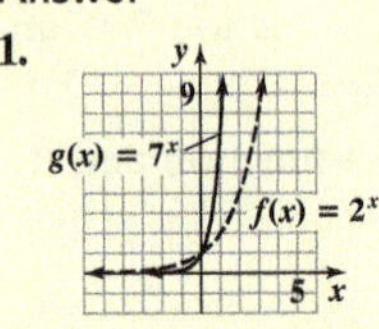

If each set of points is plotted and connected with a smooth curve, the following graphs result.

$f(x) = 2^x$	x	0	1	2	3	-1	-2
	$f(x)$	1	2	4	8	$\frac{1}{2}$	$\frac{1}{4}$

$g(x) = 3^x$	x	0	1	2	3	-1	-2
	$g(x)$	1	3	9	27	$\frac{1}{3}$	$\frac{1}{9}$

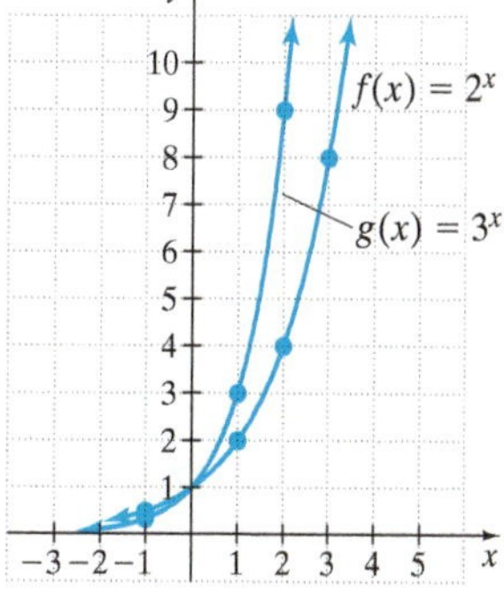

Work Practice 1

A number of things should be noted about the two graphs of exponential functions in Example 1. First, the graphs show that $f(x) = 2^x$ and $g(x) = 3^x$ are one-to-one functions since each graph passes the vertical and horizontal line tests. The y-intercept of each graph is $(0, 1)$, but neither graph has an x-intercept. From the graph, we can see that the domain of each function is all real numbers and that the range is $(0, \infty)$. We can also see that as x-values are increasing, y-values are increasing as well.

Example 2 Graph the exponential functions $y = \left(\frac{1}{2}\right)^x$ and $y = \left(\frac{1}{3}\right)^x$ on the same set of axes.

Solution: As before, plot points and connect them with a smooth curve.

$y = \left(\frac{1}{2}\right)^x$	x	0	1	2	3	-1	-2
	y	1	$\frac{1}{2}$	$\frac{1}{4}$	$\frac{1}{8}$	2	4

$y = \left(\frac{1}{3}\right)^x$	x	0	1	2	3	-1	-2
	y	1	$\frac{1}{3}$	$\frac{1}{9}$	$\frac{1}{27}$	3	9

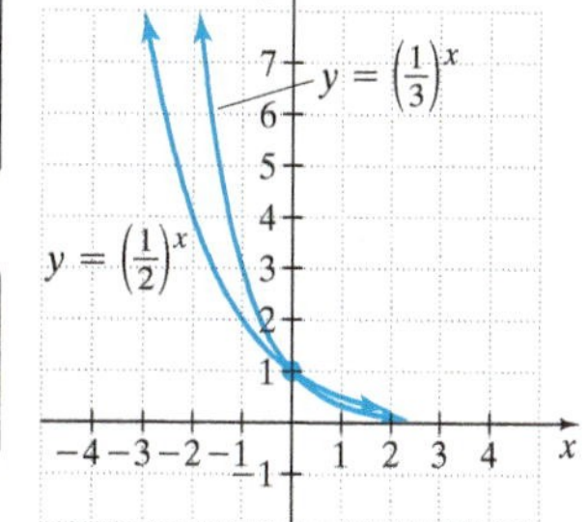

Work Practice 2

Each function in Example 2 again is a one-to-one function. The y-intercept of both is $(0, 1)$. The domain is the set of all real numbers, and the range is $(0, \infty)$.

Notice the difference between the graphs of Example 1 and the graphs of Example 2. An exponential function is always increasing if the base is greater than 1.

When the base is between 0 and 1, the graph is always decreasing. The figures on the next page summarize these characteristics of exponential functions.

Practice 2

Graph the exponential functions $f(x) = \left(\frac{1}{3}\right)^x$ and $g(x) = \left(\frac{1}{5}\right)^x$ on the same set of axes.

Answer

2. $f(x) = \left(\frac{1}{3}\right)^x$, $g(x) = \left(\frac{1}{5}\right)^x$

$f(x) = b^x, \quad b > 0, \quad b \neq 1$

- one-to-one function
- y-intercept $(0, 1)$
- no x-intercept
- domain: $(-\infty, \infty)$
- range: $(0, \infty)$

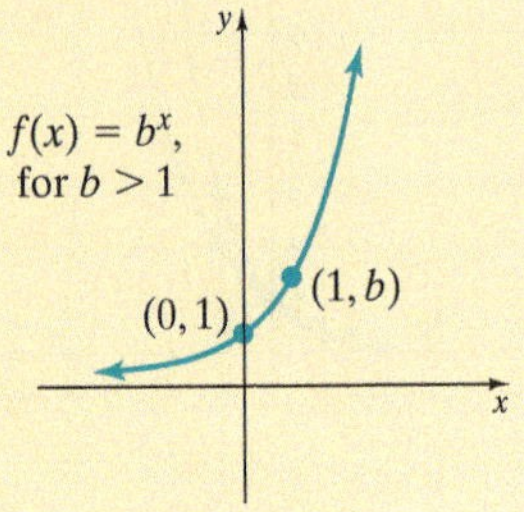

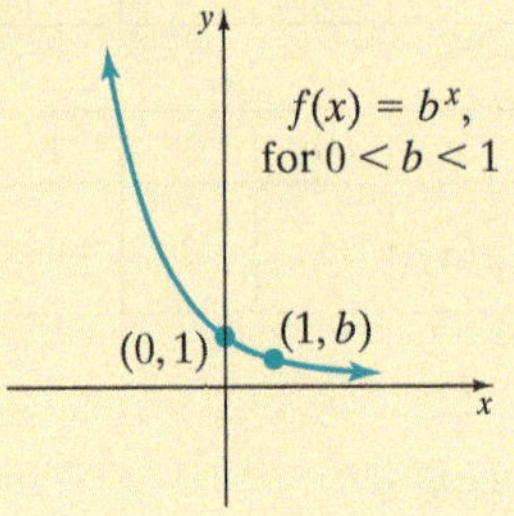

Practice 3

Graph the exponential function $f(x) = 2^{x-3}$.

Example 3

Graph the exponential function $f(x) = 3^{x+2}$.

Solution: As before, we find and plot a few ordered pair solutions. Then we connect the points with a smooth curve.

$f(x) = 3^{x+2}$	
x	$f(x)$
0	9
−1	3
−2	1
−3	$\frac{1}{3}$
−4	$\frac{1}{9}$

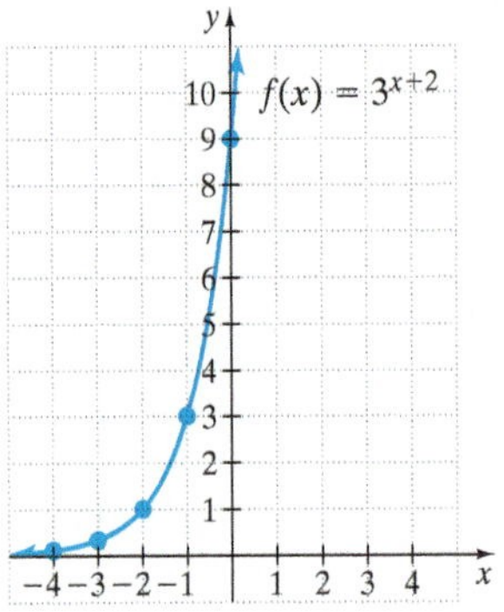

Work Practice 3

✓**Concept Check** Which functions are exponential functions?

a. $f(x) = x^3$ **b.** $g(x) = \left(\frac{2}{3}\right)^x$ **c.** $h(x) = 5^{x-2}$ **d.** $w(x) = (2x)^2$

Objective 2 Solving Equations of the Form $b^x = b^y$

We have seen that an exponential function $y = b^x$ is a one-to-one function. Another way of stating this fact is a property that we can use to solve exponential equations.

Uniqueness of b^x

Let $b > 0$ and $b \neq 1$. Then $b^x = b^y$ is equivalent to $x = y$.

Thus, one way to solve an exponential equation depends on whether it's possible to write each side of the equation with the same base; that is, $b^x = b^y$. We solve by this method first.

Answer

3.

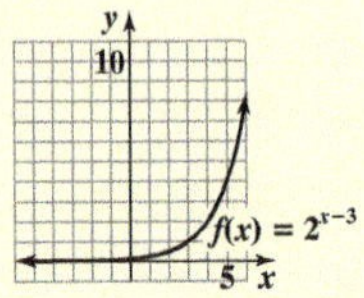

✓**Concept Check Answer**

b, c

Example 4 Solve each equation for x.

a. $2^x = 16$ **b.** $9^x = 27$ **c.** $4^{x+3} = 8^x$

Solution:

a. We write 16 as a power of 2 and then use the uniqueness of b^x to solve.

$$2^x = 16$$
$$2^x = 2^4$$

Since the bases are the same and are nonnegative, by the uniqueness of b^x, we then have that the exponents are equal. Thus,

$$x = 4$$

The solution is 4, or the solution set is $\{4\}$.

b. Notice that both 9 and 27 are powers of 3.

$$9^x = 27$$
$$(3^2)^x = 3^3 \quad \text{Write 9 and 27 as powers of 3.}$$
$$3^{2x} = 3^3$$
$$2x = 3 \quad \text{Apply the uniqueness of } b^x.$$
$$x = \frac{3}{2} \quad \text{Divide by 2.}$$

To check, replace x with $\frac{3}{2}$ in the original expression, $9^x = 27$. The solution is $\frac{3}{2}$.

c. Write both 4 and 8 as powers of 2.

$$4^{x+3} = 8^x$$
$$(2^2)^{x+3} = (2^3)^x$$
$$2^{2x+6} = 2^{3x}$$
$$2x + 6 = 3x \quad \text{Apply the uniqueness of } b^x.$$
$$6 = x \quad \text{Subtract } 2x \text{ from both sides.}$$

The solution is 6.

Work Practice 4

Practice 4

Solve each equation for x.

a. $3^x = 9$

b. $8^x = 16$

c. $125^x = 25^{x-2}$

There is one major problem with the preceding technique. Often the two sides of an equation cannot easily be written as powers of a common base. We explore how to solve an equation such as $4 = 3^x$ with the help of **logarithms** later.

Objective 3 Solving Problems Modeled by Exponential Equations

The bar graph on the next page shows the increase in the number of cellular phone users. Notice that the graph of the exponential function $y = 136.76(1.107)^x$ approximates the heights of the bars. This is just one example of how the world abounds with patterns that can be modeled by exponential functions. To make these applications realistic, we use numbers that warrant a calculator.

Answers

4. a. 2 **b.** $\frac{4}{3}$ **c.** -4

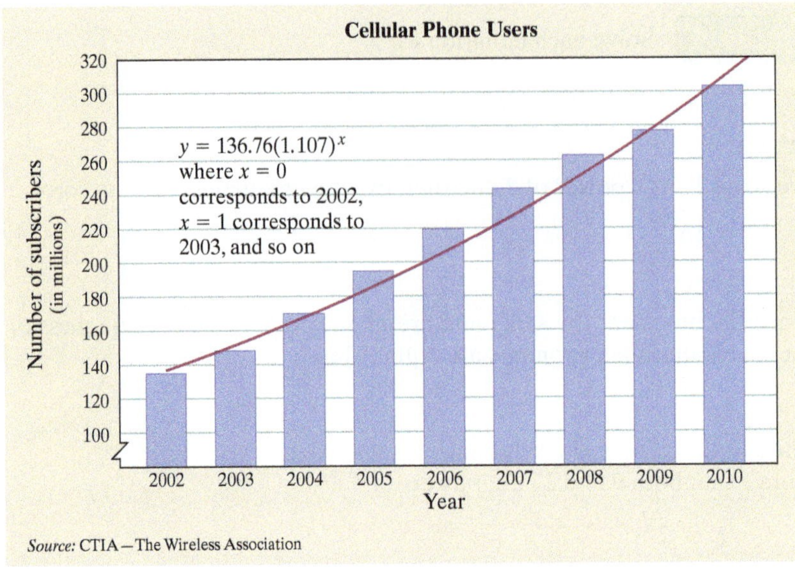

Another application of an exponential function has to do with interest rates on loans.

The exponential function defined by $A = P\left(1 + \frac{r}{n}\right)^{nt}$ models the dollars A accrued (or owed) after P dollars are invested (or loaned) at an annual rate of interest r compounded n times each year for t years. This function is known as the compound interest formula.

Practice 5

Find the amount owed at the end of 4 years if \$3000 is loaned at a rate of 7% compounded semiannually (twice a year).

Example 5 Using the Compound Interest Formula

Find the amount owed at the end of 5 years if \$1600 is loaned at a rate of 9% compounded monthly.

Solution: We use the formula $A = P\left(1 + \frac{r}{n}\right)^{nt}$, with the following values.

$P = \$1600$ (the amount of the loan)

$r = 9\% = 0.09$ (the annual rate of interest)

$n = 12$ (the number of times interest is compounded each year)

$t = 5$ (the duration of the loan, in years)

$$A = P\left(1 + \frac{r}{n}\right)^{nt} \quad \text{Compound interest formula}$$

$$= 1600\left(1 + \frac{0.09}{12}\right)^{12(5)} \quad \text{Substitute known values.}$$

$$= 1600(1.0075)^{60}$$

To approximate A, use the $\boxed{y^x}$ or $\boxed{\wedge}$ key on your calculator.

2505.0896

Thus, the amount A owed is approximately \$2505.09.

Work Practice 5

Answer

5. \$3950.43

Example 6 Estimating Percent of Radioactive Material

As a result of a nuclear accident, radioactive debris was carried through the atmosphere. One immediate concern was the impact that the debris had on the milk supply. The percent y of radioactive material in raw milk after t days is estimated by $y = 100(2.7)^{-0.1t}$. Estimate the expected percent of radioactive material in the milk after 30 days.

Solution: Replace t with 30 in the given equation.

$$y = 100(2.7)^{-0.1t}$$
$$= 100(2.7)^{-0.1(30)} \quad \text{Let } t = 30.$$
$$= 100(2.7)^{-3}$$

To approximate the percent y, the following keystrokes may be used on a scientific calculator.

[2.7] [y^x] [3] [+/−] [=] [×] [100] [=]

The display should read

[5.0805263]

Thus, approximately 5% of the radioactive material still remained in the milk supply after 30 days.

Work Practice 6

Practice 6

The percent p of light that passes through n successive sheets of a particular glass is given approximately by the function $p(n) = 100(2.7)^{-0.05n}$. Estimate the expected percent of light that will pass through the following numbers of sheets of glass. Round each to the nearest hundredth of a percent.

a. 2 sheets of glass

b. 10 sheets of glass

Answers

6. a. 90.54% **b.** 60.86%

Graphing Calculator Explorations

We can use a graphing calculator and its TRACE feature to solve Example 6 graphically.

To estimate the expected percent of radioactive material in the milk after 30 days, enter $Y_1 = 100(2.7)^{-0.1x}$. (The variable t in Example 6 is changed to x here to accommodate our work better on the graphing calculator.) The graph does not appear on a standard viewing window, so we need to determine an appropriate viewing window. Because it doesn't make sense to look at radioactivity *before* the nuclear accident, we use Xmin = 0. We are interested in finding the percent of radioactive material in the milk when $x = 30$, so we choose Xmax = 35 to leave enough space to see the graph at $x = 30$. Because the values of y are percents, it seems appropriate that $0 \le y \le 100$. (We also use Xscl = 1 and Yscl = 10.) Now we graph the function.

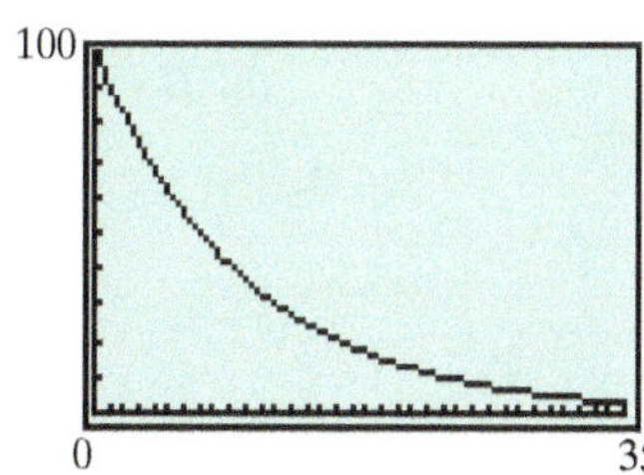

We can use the TRACE feature to obtain an approximation of the expected percent of radioactive material in the milk when $x = 30$. (A TABLE feature may also be used to approximate the percent.) To obtain a better approximation, let's use the ZOOM feature several times to zoom in near $x = 30$.

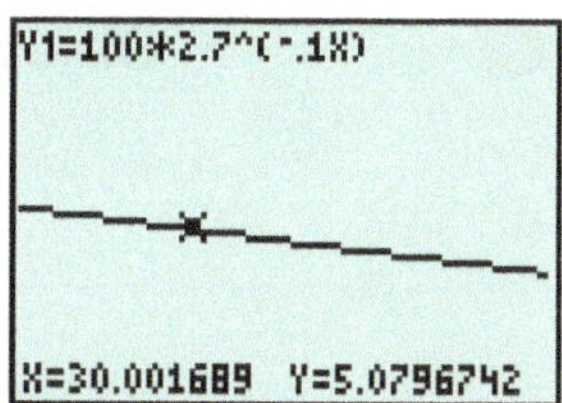

The percent of radioactive material in the milk 30 days after the nuclear accident was 5.08%, accurate to two decimal places.

Use a graphing calculator to find each percent. Approximate your solutions so that they are accurate to two decimal places.

1. Estimate the expected percent of radioactive material in the milk 2 days after the nuclear accident.

2. Estimate the expected percent of radioactive material in the milk 10 days after the nuclear accident.

3. Estimate the expected percent of radioactive material in the milk 15 days after the nuclear accident.

4. Estimate the expected percent of radioactive material in the milk 25 days after the nuclear accident.

Vocabulary, Readiness & Video Check

Use the choices to fill in each blank.

1. A function such as $f(x) = 2^x$ is a(n) __________ function.

A. linear **B.** quadratic **C.** exponential

2. If $7^x = 7^y$, then __________.

A. $x = 7^y$ **B.** $x = y$ **C.** $y = 7^x$ **D.** $7 = 7^y$

Answer the questions about the graph of $y = 2^x$, shown to the right.

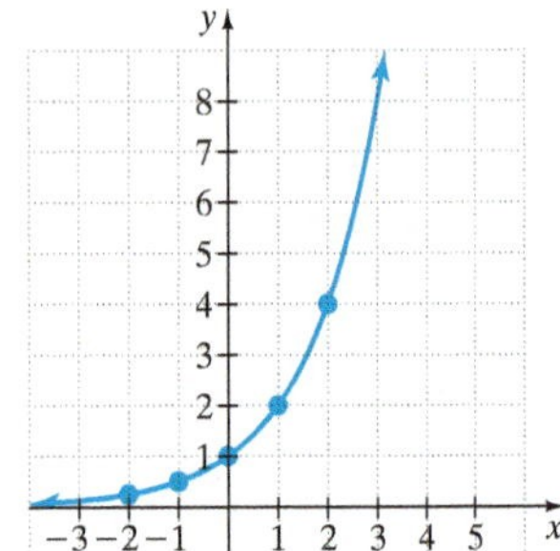

3. Is this a function? ________

4. Is this a one-to-one function? ________

5. Is there an x-intercept? ________ If so, name the coordinates. ________

6. Is there a y-intercept? ________ If so, name the coordinates. ________

7. The domain of this function, in interval notation, is ________.

8. The range of this function, in interval notation, is ________.

Martin-Gay Interactive Videos

See Video 19.3

Watch the section lecture video and answer the following questions.

Objective 1 **9.** From the lecture before Example 1, what's the main difference between a polynomial function and an exponential function?

Objective 2 **10.** From Examples 2 and 3, you can apply the uniqueness of b^x to solve an exponential equation only if you're able to do what?

Objective 3 **11.** For Example 4, write the equation and find how much uranium will remain after 101 days. Round your answer to the nearest tenth.

19.3 Exercise Set MyMathLab®

Graph each exponential function. See Examples 1 through 3.

1. $y = 5^x$ **2.** $y = 4^x$

3. $y = 2^x + 1$ **4.** $y = 3^x - 1$

5. $y = \left(\frac{1}{4}\right)^x$ **6.** $y = \left(\frac{1}{5}\right)^x$

7. $y = \left(\frac{1}{2}\right)^x - 2$ **8.** $y = \left(\frac{1}{3}\right)^x + 2$

9. $y = -2^x$ **10.** $y = -3^x$

11. $y = -\left(\frac{1}{4}\right)^x$ **12.** $y = -\left(\frac{1}{5}\right)^x$

13. $f(x) = 2^{x+1}$ **14.** $f(x) = 3^{x-1}$

15. $f(x) = 4^{x-2}$ **16.** $f(x) = 2^{x+3}$

Match each exponential equation with its graph below. See Examples 1 through 3.

17. $f(x) = \left(\frac{1}{2}\right)^x$ **18.** $f(x) = \left(\frac{1}{4}\right)^x$

19. $f(x) = 2^x$ **20.** $f(x) = 3^x$

A.

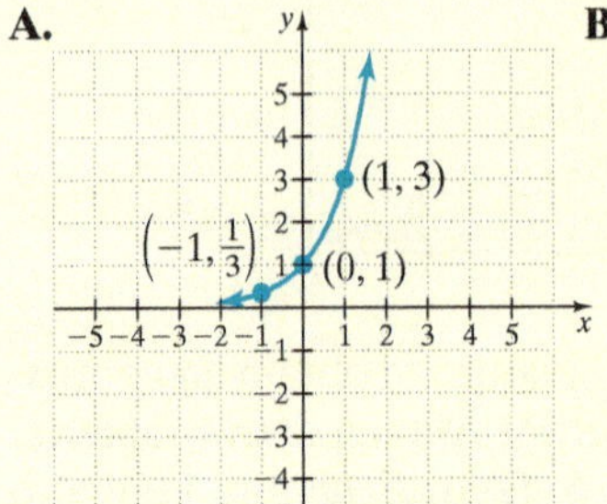

B.

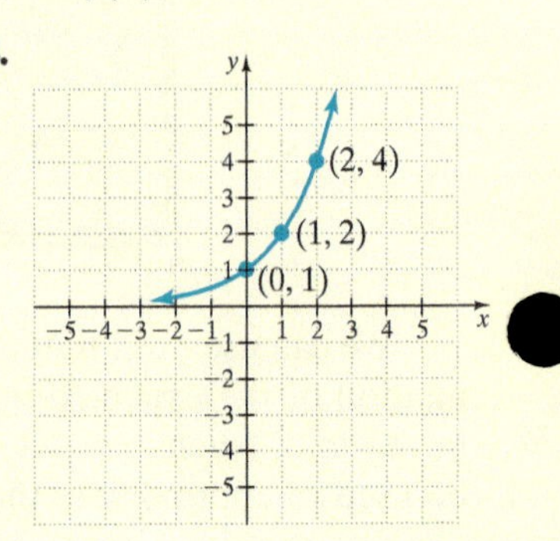

C.

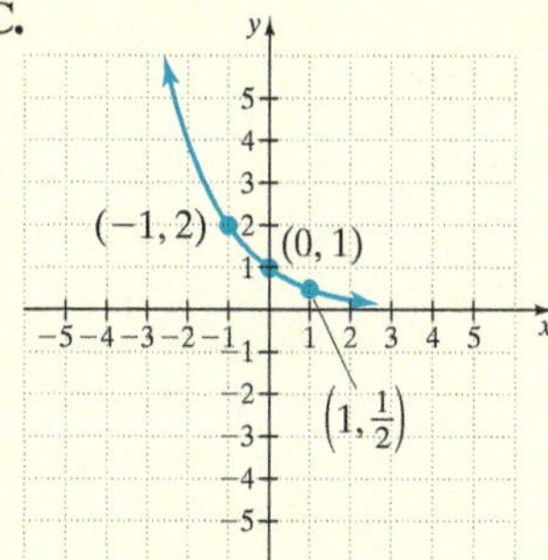

D.

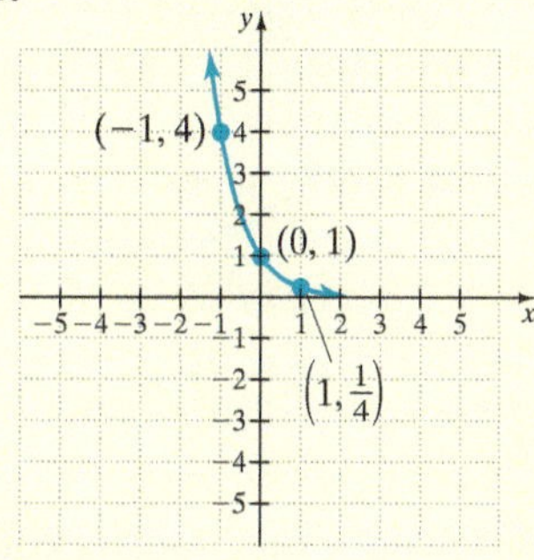

Solve each equation for x. See Example 4.

21. $3^x = 27$

22. $6^x = 36$

23. $16^x = 8$

24. $64^x = 16$

25. $32^{2x-3} = 2$

26. $9^{2x+1} = 81$

27. $\frac{1}{4} = 2^{3x}$

28. $\frac{1}{27} = 3^{2x}$

29. $5^x = 625$

30. $2^x = 64$

31. $4^x = 8$

32. $32^x = 4$

33. $27^{x+1} = 9$

34. $125^{x-2} = 25$

35. $81^{x-1} = 27^{2x}$

36. $4^{3x-7} = 32^{2x}$

Solve. Unless otherwise indicated, round results to one decimal place. See Example 6.

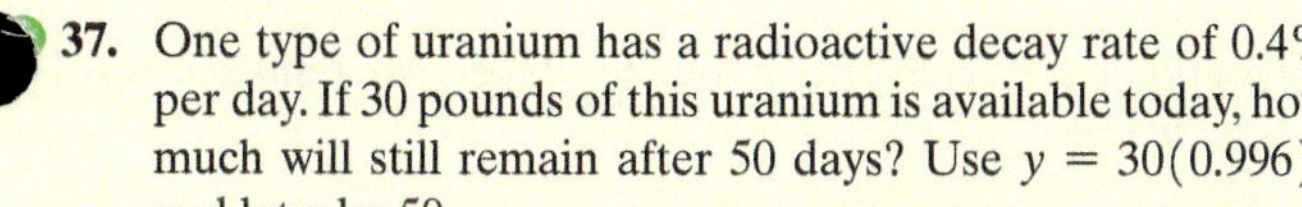

37. One type of uranium has a radioactive decay rate of 0.4% per day. If 30 pounds of this uranium is available today, how much will still remain after 50 days? Use $y = 30(0.996)^x$ and let x be 50.

38. The nuclear waste from an atomic energy plant decays at a rate of 3% each century. If 150 pounds of nuclear waste is disposed of, how much of it will still remain after 10 centuries? Use $y = 150(0.97)^x$, and let x be 10.

39. Cheese production in the United States is currently growing at a rate of 3% per year. The equation $y = 8.6(1.03)^x$ models the cheese production in the United States from 2003 to 2009. In this equation, y is the amount of cheese produced, in billions of pounds, and x represents the number of years after 2003. Round answers to the nearest tenth of a billion. (*Source:* National Agricultural Statistics Service)

a. Estimate the total cheese production in the United States in 2007.

b. Assuming this equation continues to be valid, use the equation to estimate the total amount of cheese produced in the United States in 2015.

40. Retail revenue from shopping on the Internet is currently growing at rate of 26% per year. In 2003, a total of $39 billion in revenue was collected through Internet retail sales. Answer the following questions using $y = 39(1.26)^t$, where y is Internet revenues in billions of dollars and t is the number of years after 2003. Round answers to the nearest tenth of a billion dollars. (*Source:* U.S. Bureau of the Census)

a. According to the model, what level of retail revenues from Internet shopping was expected in 2005?

b. If the given model continues to be valid, estimate the level of Internet shopping revenues in 2015.

41. The equation $y = 140{,}242(1.083)^x$ models the number of American college students who studied abroad each year from 2000 through 2009. In the equation, y is the number of American students studying abroad, and x represents the number of years after 2000. Round answers to the nearest whole. (*Source:* Based on data from Institute of International Education, Open Doors)

a. Estimate the number of American students studying abroad in 2004.

b. Assuming this equation continues to be valid, use this equation to estimate the number of American students studying abroad in 2015.

42. Carbon dioxide (CO_2) is a greenhouse gas that contributes to global warming. Partially due to the combustion of fossil fuel, the amount of CO_2 in Earth's atmosphere has been increasing by 0.5% annually over the past century. In 2000, the concentration of CO_2 in the atmosphere was 369.4 parts per million by volume. To make the following estimates, use $y = 369.4(1.005)^t$, where y is the concentration of CO_2 in parts per million by volume and t is the number of years after 2000. (*Sources:* Based on data from the United Nations Environment Programme and the Carbon Dioxide Information Analysis Center)

a. Estimate the concentration of CO_2 in the atmosphere in the year 2015.

b. Estimate the concentration of CO_2 in the atmosphere in the year 2030.

The equation $y = 136.76(1.107)^x$ gives the number of cellular phone users y (in millions) in the United States for the years 2002 through 2010. In this equation, $x = 0$ corresponds to 2002, $x = 1$ corresponds to 2003, and so on. Use this model to solve Exercises 43 and 44. Round answers to the nearest tenth of a million.

43. Estimate the number of cell phone users in the year 2012.

44. Estimate the number of cell phone users in 2014.

45. An unusually wet spring has caused the size of the Cape Cod mosquito population to increase by 8% each day. If an estimated 200,000 mosquitoes are on Cape Cod on May 12, find how many mosquitoes will inhabit the Cape on May 25. Use $y = 200{,}000(1.08)^x$, where x is number of days since May 12. Round to the nearest thousand.

46. The atmospheric pressure p, in pascals, on a weather balloon decreases with increasing height. This pressure, measured in millimeters of mercury, is related to the number of kilometers h above sea level by the function $p(h) = 760(2.7)^{-0.145h}$. Round to the nearest tenth of a pascal.

a. Find the atmospheric pressure at a height of 1 kilometer.

b. Find the atmospheric pressure at a height of 10 kilometers.

Solve. Use $A = P\left(1 + \frac{r}{n}\right)^{nt}$. Round answers to two decimal places. See Example 5.

47. Find the amount Erica owes at the end of 3 years if $6000 is loaned to her at a rate of 8% compounded monthly.

48. Find the amount owed at the end of 5 years if \$3000 is loaned at a rate of 10% compounded quarterly.

49. Find the total amount Janina has in a college savings account if \$2000 was invested and earned 6% compounded semiannually for 12 years.

50. Find the amount accrued if \$500 is invested and earns 7% compounded monthly for 4 years.

REVIEW AND PREVIEW

Solve each equation. See Sections 9.3 and 13.6.

51. $5x - 2 = 18$

52. $3x - 7 = 11$

53. $3x - 4 = 3(x + 1)$

54. $2 - 6x = 6(1 - x)$

55. $x^2 + 6 = 5x$

56. $18 = 11x - x^2$

By inspection, find the value for x that makes each statement true. See Section 12.1.

57. $2^x = 8$

58. $3^x = 9$

59. $5^x = \frac{1}{5}$

60. $4^x = 1$

CONCEPT EXTENSIONS

Is the given function an exponential function? See the Concept Check in this section.

61. $f(x) = 1.5x^2$

62. $g(x) = 3^x$

63. $h(x) = \left(\frac{1}{2}x\right)^2$

64. $F(x) = 0.4^{x+1}$

Match each exponential function with its graph.

65. $f(x) = 2^{-x}$

66. $f(x) = \left(\frac{1}{2}\right)^{-x}$

67. $f(x) = 4^{-x}$

68. $f(x) = \left(\frac{1}{3}\right)^{-x}$

A.

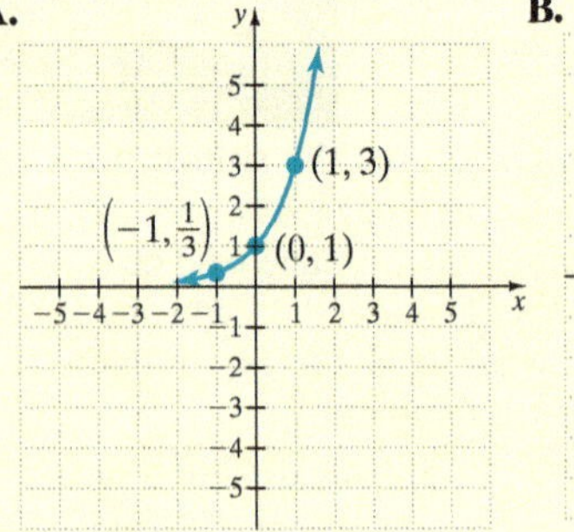

B.

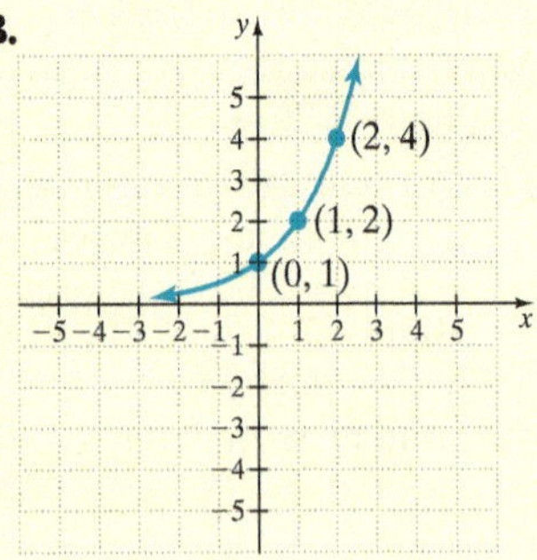

C. **D.**

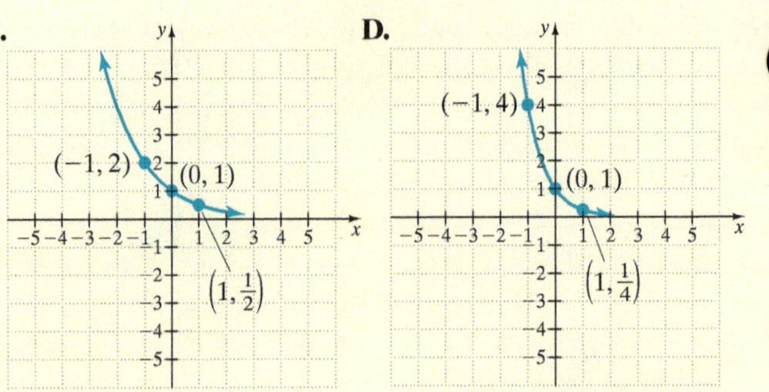

69. Explain why the graph of an exponential function $y = b^x$ contains the point $(1, b)$.

70. Explain why an exponential function $y = b^x$ has a y-intercept of $(0, 1)$.

Graph.

71. $y = |3^x|$

72. $y = \left|\left(\frac{1}{3}\right)^x\right|$

73. $y = 3^{|x|}$

74. $y = \left(\frac{1}{3}\right)^{|x|}$

75. Graph $y = 2^x$ and $y = \left(\frac{1}{2}\right)^{-x}$ on the same set of axes. Describe what you see and why.

76. Graph $y = 2^x$ and $x = 2^y$ on the same set of axes. Describe what you see.

Use a graphing calculator to solve. Estimate your results to two decimal places.

77. Verify the results of Exercise 37.

78. Verify the results of Exercise 38.

79. From Exercise 37, estimate the number of pounds of uranium that will be available after 100 days.

80. From Exercise 37, estimate the number of pounds of uranium that will be available after 120 days.

19.4 Exponential Growth and Decay Functions

Objectives

1 Model Exponential Growth.

2 Model Exponential Decay.

Now that we can graph exponential functions, let's learn about exponential growth and exponential decay.

A quantity that grows or decays by the same percent at regular time periods is said to have **exponential growth** or **exponential decay.** There are many real-life examples of exponential growth and decay, such as population, bacteria, viruses, and radioactive substances, just to name a few.

Recall the graphs of exponential functions.

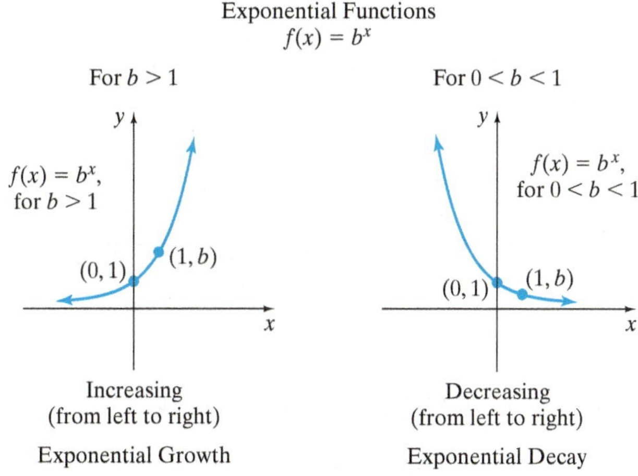

Objective 1 Modeling Exponential Growth

We begin with exponential growth, as described below.

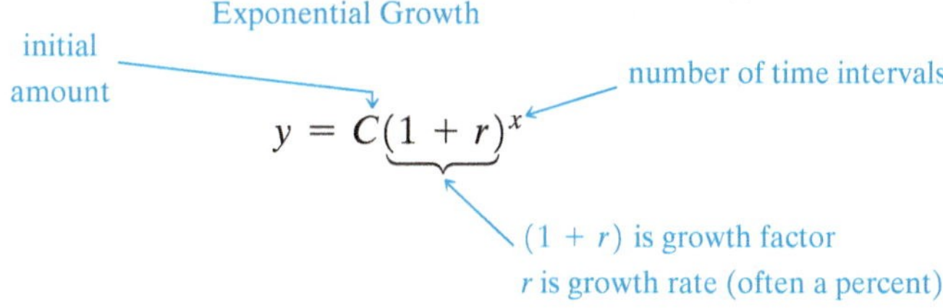

Example 1 In 1995, let's suppose a town named Jackson had a population of 15,500 and was consistently increasing by 10% per year. If this yearly increase continued, estimate the city's population in 2015. (Round to the nearest whole.)

Solution: Let's begin to understand by calculating the city's population each year:

Time Interval	$x = 1$	$x = 2$	3	4	5	and so on ...
Year	1996	1997	1998	1999	2000	
Population	17,050	18,755	20,631	22,694	24,963	

$15{,}500 + 0.10(15{,}500)$ $17{,}050 + 0.10(17{,}050)$

This is an example of exponential growth, so let's use our formula with

$C = 15{,}500; r = 0.10; x = 2015 - 1995 = 20$

(*Continued on next page*)

Practice 1

In 2000, the town of Jackson (from Example 1) had a population of 25,000 and started consistently increasing by 12% per year. If this yearly increase continued, estimate the city's population in 2015. Round to the nearest whole.

Answer

1. 136,839

Then,

$$\begin{aligned} y &= C(1 + r)^x \\ &= 15{,}500(1 + 0.10)^{20} \\ &= 15{,}500(1.1)^{20} \\ &\approx 104{,}276 \end{aligned}$$

In 2015, we estimate the population of Jackson to be 104,276.

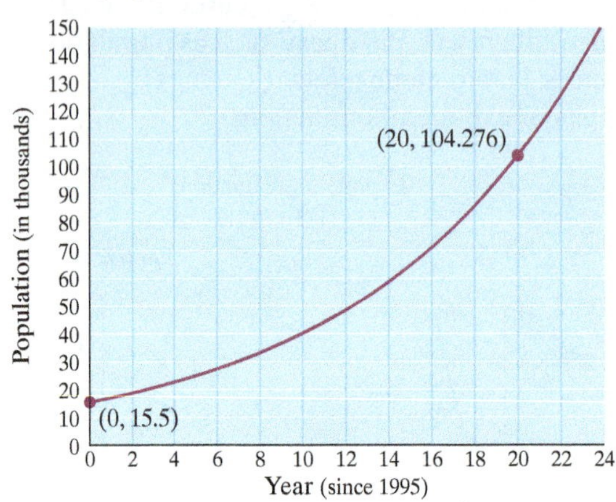

Work Practice 1

Note: The exponential growth formula, $y = C(1 + r)^x$, should remind you of the compound interest formula from the previous section, $A = P(1 + \frac{r}{n})^{nt}$. In fact, if the number of compoundings per year, n, is 1, the interest formula becomes $A = P(1 + r)^t$, which is the exponential growth formula written with different variables.

Objective 2 Modeling Exponential Decay 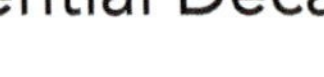

Now let's study exponential decay.

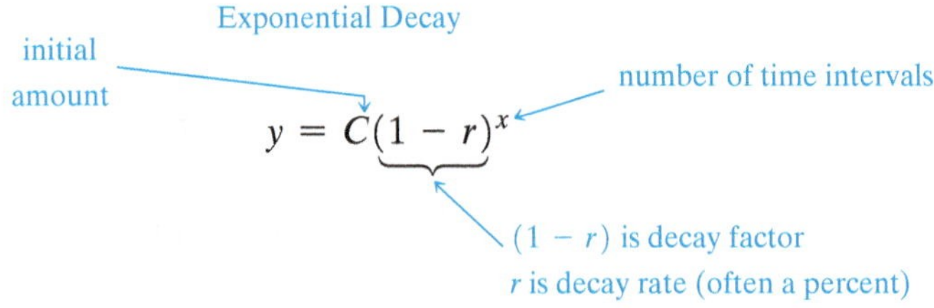

Practice 2

A tournament with 800 persons is played so that after each round, the number of players decreases by 30%. Find the number of players after round 9. Round your answer to the nearest whole.

Example 2 A large golf country club holds a singles tournament each year. At the start of the tournament for a particular year, there are 512 players. After each round, half the players are eliminated. How many players remain after 6 rounds?

Solution: This is an example of exponential decay.

Let's begin to understand by calculating the number of players after a few rounds.

Round (same as interval)	1	2	3	4	and so on ...
Players (at end of round)	256	128	64	32	

$512 - 0.50(512)$ (for round 1) $256 - 0.50(256)$ (for round 2)

Here, $C = 512$; $r = \frac{1}{2}$ or $50\% = 0.50$; $x = 6$

Answer

2. 32

Thus,

$$y = 512(1 - 0.50)^6$$
$$= 512(0.50)^6$$
$$= 8$$

After 6 rounds, there are 8 players remaining.

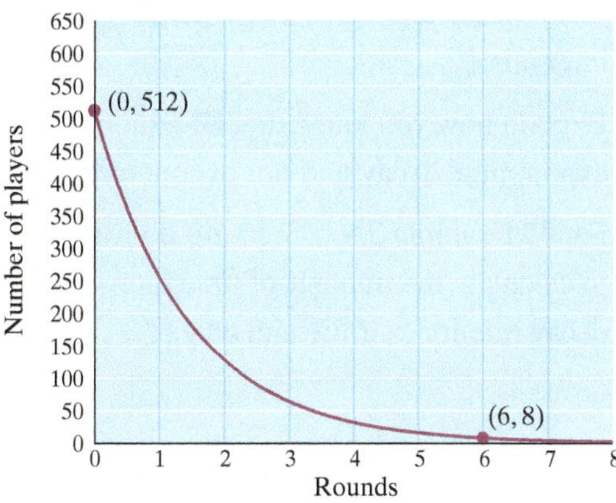

Work Practice 2

The **half-life** of a substance is the amount of time it takes for half of the substance to decay.

Example 3 A form of DDT pesticide (banned in 1972) has a half-life of approximately 15 years. If a storage unit had 400 pounds of DDT, find how much DDT is remaining after 72 years. Round to the nearest tenth of a pound.

Solution: Here, we need to be careful because each time interval is 15 years, the half-life.

Time Interval	1	2	3	4	5	and so on ...
Years Passed	15	$2 \cdot 15 = 30$	45	60	75	
Pounds of DDT	200	100	50	25	12.5	

From the table, we see that after 72 years, between 4 and 5 intervals, there should be between 12.5 and 25 pounds of DDT remaining.

Let's calculate x, the number of time intervals.

$$x = \frac{72 \text{ (years)}}{15 \text{ (half-life)}} = 4.8$$

Now, using our exponential decay formula and the definition of half-life, for each time interval x, the decay rate r is $\frac{1}{2}$ or 50% or 0.50.

$$y = 400(1 - 0.50)^{4.8}$$ — time intervals for 72 years (400: original amount; 0.50: decay rate)

$$y = 400(0.50)^{4.8}$$
$$y \approx 14.4$$

In 72 years, 14.4 pounds of DDT remain.

Work Practice 3

Practice 3

Use the information from Example 3 and calculate how much of a 500-gram sample of DDT will remain after 51 years. Round to the nearest tenth of a gram.

Answer

3. 47.4 g

Vocabulary, Readiness & Video Check

Martin-Gay Interactive Videos

See Video 19.4

Watch the section lecture video and answer the following questions.

Objective 1 1. Example 1 reviews exponential growth. Explain how you find the growth rate and the correct number of time intervals.

Objective 2 2. Explain how you know that Example 2 has to do with exponential decay, and not exponential growth.

Objective 3 3. For Example 3, which has to do with half-life, explain how to calculate the number of time intervals. Also, what is the decay rate for half-life and why?

19.4 Exercise Set MyMathLab®

Practice using the exponential growth formula by completing the table below. Round final amounts to the nearest whole. See Example 1.

	Original Amount	Growth Rate per Year	Number of Years, x	Final Amount after x Years of Growth
1.	305	5%	8	
2.	402	7%	5	
3.	2000	11%	41	
4.	1000	47%	19	
5.	17	29%	28	
6.	29	61%	12	

Practice using the exponential decay formula by completing the table below. Round final amounts to the nearest whole. See Example 2.

	Original Amount	Decay Rate per Year	Number of Years, x	Final Amount after x Years of Decay
7.	305	5%	8	
8.	402	7%	5	
9.	10,000	12%	15	
10.	15,000	16%	11	
11.	207,000	32%	25	
12.	325,000	29%	31	

MIXED PRACTICE

Solve. Unless noted otherwise, round answers to the nearest whole. See Examples 1 and 2.

13. Suppose a city with population 500,000 has been growing at a rate of 3% per year. If this rate continues, find the population of this city in 12 years.

14. Suppose a city with population 320,000 has been growing at a rate of 4% per year. If this rate continues, find the population of this city in 20 years.

15. The number of employees at a certain company has been decreasing each year by 5%. If the company currently has 640 employees and this rate continues, find the number of employees in 10 years.

16. The number of students attending summer school at a local community college has been decreasing each year by 7%. If 984 students currently attend summer school and this rate continues, find the number of students attending summer school in 5 years.

17. National Park Service personnel are trying to increase the size of the bison population of Theodore Roosevelt National Park. If 260 bison currently live in the park, and if the population's rate of growth is 2.5% annually, find how many bison there should be in 10 years.

18. The size of the rat population of a wharf area grows at a rate of 8% monthly. If there are 200 rats in January, find how many rats should be expected by next January.

19. A rare isotope of a nuclear material is very unstable, decaying at a rate of 15% each second. Find how much isotope remains 10 seconds after 5 grams of the isotope is created.

20. An accidental spill of 75 grams of radioactive material in a local stream has led to the presence of radioactive debris decaying at a rate of 4% each day. Find how much debris still remains after 14 days.

Practice using the exponential decay formula with half-lives by completing the table below. The first row has been completed for you. See Example 3.

	Original Amount	Half-Life (in years)	Number of Years	Time Intervals, $x = \left(\frac{\text{Years}}{\text{Half-Life}}\right)$ Rounded to Tenths if Needed	Final Amount after x Time Intervals (rounded to tenths)	Is Your Final Amount Reasonable?
	60	8	10	$\frac{10}{8} = 1.25$	25.2	yes
21.	**a.** 40	7	14			
	b. 40	7	11			
22.	**a.** 200	12	36			
	b. 200	12	40			
23.	21	152	500			
24.	35	119	500			

Solve. Round answers to the nearest tenth.

25. A form of nickel has a half-life of 96 years. How much of a 30-gram sample is left after 250 years?

26. A form of uranium has a half-life of 72 years. How much of a 100-gram sample is left after 500 years?

REVIEW AND PREVIEW

By inspection, find the value for x that makes each statement true. See Sections 12.1 and 19.3.

27. $2^x = 8$ **28.** $3^x = 9$ **29.** $5^x = \frac{1}{5}$ **30.** $4^x = 1$

CONCEPT EXTENSIONS

31. An item is on sale for 40% off its original price. If it is then marked down an additional 60%, does this mean the item is free? Discuss why or why not.

32. Uranium U-232 has a half-life of 72 years. What eventually happens to a 10-gram sample? Does it ever completely decay and disappear? Discuss why or why not.

19.5 Logarithmic Functions

Objective 1 Using Logarithmic Notation

Objectives

1. Write Exponential Equations with Logarithmic Notation and Write Logarithmic Equations with Exponential Notation.
2. Solve Logarithmic Equations by Using Exponential Notation.
3. Identify and Graph Logarithmic Functions.

Since the exponential function $f(x) = 2^x$ is a one-to-one function, it has an inverse.

We can create a table of values for f^{-1} by switching the coordinates in the accompanying table of values for $f(x) = 2^x$.

x	$y = f(x)$
-3	$\frac{1}{8}$
-2	$\frac{1}{4}$
-1	$\frac{1}{2}$
0	1
1	2
2	4
3	8

x	$y = f^{-1}(x)$
$\frac{1}{8}$	-3
$\frac{1}{4}$	-2
$\frac{1}{2}$	-1
1	0
2	1
4	2
8	3

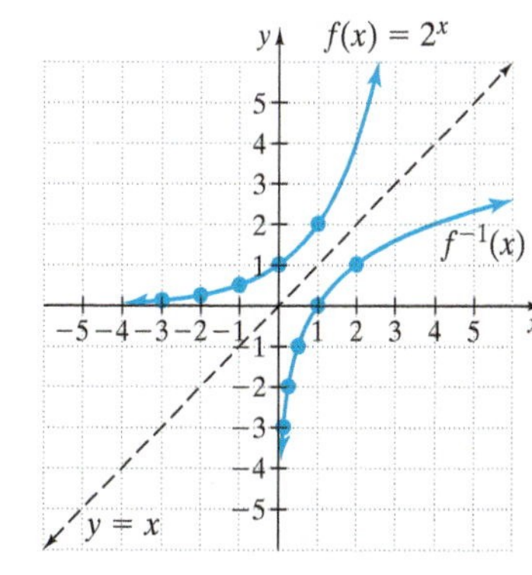

The graphs of $f(x)$ and its inverse are shown on the previous page. Notice that the graphs of f and f^{-1} are symmetric about the line $y = x$, as expected.

Now we would like to be able to write an equation for f^{-1}. To do so, we follow the steps for finding an inverse.

$$f(x) = 2^x$$

Step 1: Replace $f(x)$ by y. $\quad y = 2^x$

Step 2: Interchange x and y. $\quad x = 2^y$

Step 3: Solve for y.

At this point, we are stuck. To solve this equation for y, a new notation, the **logarithmic notation,** is needed.

The symbol $\log_b x$ means "the power to which b is raised to produce a result of x." In other words,

$$\log_b x = y \quad \text{means} \quad b^y = x$$

We say that $\log_b x$ is "the logarithm of x to the base b" or "the log of x to the base b."

Logarithmic Definition

If $b > 0$ and $b \neq 1$, then

$$y = \log_b x \text{ means } x = b^y$$

for every $x > 0$ and every real number y.

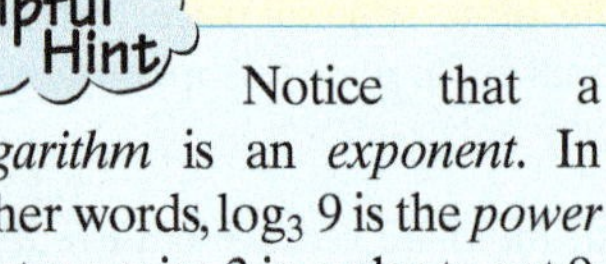

Notice that a *logarithm* is an *exponent.* In other words, $\log_3 9$ is the *power* that we raise 3 in order to get 9.

Before returning to the function $x = 2^y$ and solving it for y in terms of x, let's practice using the new notation $\log_b x$.

It is important to be able to write exponential equations from logarithmic notation and vice versa. The following table shows examples of both forms.

Logarithmic Equation	Corresponding Exponential Equation
$\log_3 9 = 2$	$3^2 = 9$
$\log_6 1 = 0$	$6^0 = 1$
$\log_2 8 = 3$	$2^3 = 8$
$\log_4 \frac{1}{16} = -2$	$4^{-2} = \frac{1}{16}$
$\log_8 2 = \frac{1}{3}$	$8^{1/3} = 2$

Example 1 Write each as an exponential equation.

a. $\log_5 25 = 2$ **b.** $\log_6 \frac{1}{6} = -1$ **c.** $\log_2 \sqrt{2} = \frac{1}{2}$ **d.** $\log_7 x = 5$

Solution:

a. $\log_5 25 = 2$ means $5^2 = 25$

b. $\log_6 \frac{1}{6} = -1$ means $6^{-1} = \frac{1}{6}$

c. $\log_2 \sqrt{2} = \frac{1}{2}$ means $2^{1/2} = \sqrt{2}$

d. $\log_7 x = 5$ means $7^5 = x$

Work Practice 1

Practice 1

Write each as an exponential equation.

a. $\log_3 81 = 4$ **b.** $\log_5 \frac{1}{5} = -1$

c. $\log_7 \sqrt{7} = \frac{1}{2}$ **d.** $\log_{13} y = 4$

Answers

1. a. $3^4 = 81$ **b.** $5^{-1} = \frac{1}{5}$ **c.** $7^{1/2} = \sqrt{7}$ **d.** $13^4 = y$

Example 2 Write each as a logarithmic equation.

a. $9^3 = 729$ **b.** $6^{-2} = \frac{1}{36}$ **c.** $5^{1/3} = \sqrt[3]{5}$ **d.** $\pi^4 = x$

Solution:

a. $9^3 = 729$ means $\log_9 729 = 3$

b. $6^{-2} = \frac{1}{36}$ means $\log_6 \frac{1}{36} = -2$

c. $5^{1/3} = \sqrt[3]{5}$ means $\log_5 \sqrt[3]{5} = \frac{1}{3}$

d. $\pi^4 = x$ means $\log_\pi x = 4$

Work Practice 2

Practice 2

Write each as a logarithmic equation.

a. $4^3 = 64$ **b.** $6^{1/3} = \sqrt[3]{6}$

c. $5^{-3} = \frac{1}{125}$ **d.** $\pi^7 = z$

Example 3 Find the value of each logarithmic expression.

a. $\log_4 16$ **b.** $\log_{10} \frac{1}{10}$ **c.** $\log_9 3$

Solution:

a. $\log_4 16 = 2$ because $4^2 = 16$

b. $\log_{10} \frac{1}{10} = -1$ because $10^{-1} = \frac{1}{10}$

c. $\log_9 3 = \frac{1}{2}$ because $9^{1/2} = \sqrt{9} = 3$

Work Practice 3

Practice 3

Find the value of each logarithmic expression.

a. $\log_3 9$ **b.** $\log_2 \frac{1}{8}$

c. $\log_{49} 7$

Helpful Hint

Another method for evaluating logarithms such as those in Example 3 is to set the expression equal to x and then write them in exponential form to find x. For example:

a. $\log_4 16 = x$ means $4^x = 16$. Since $4^2 = 16$, $x = 2$ or $\log_4 16 = 2$.

b. $\log_{10} \frac{1}{10} = x$ means $10^x = \frac{1}{10}$. Since $10^{-1} = \frac{1}{10}$, $x = -1$ or $\log_{10} \frac{1}{10} = -1$.

c. $\log_9 3 = x$ means $9^x = 3$. Since $9^{1/2} = 3$, $x = \frac{1}{2}$ or $\log_9 3 = \frac{1}{2}$.

Objective 2 Solving Logarithmic Equations 

The ability to interchange the logarithmic and exponential forms of a statement is often the key to solving logarithmic equations.

Answers

2. a. $\log_4 64 = 3$ **b.** $\log_6 \sqrt[3]{6} = \frac{1}{3}$ **c.** $\log_5 \frac{1}{125} = -3$ **d.** $\log_\pi z = 7$

3. a. 2 **b.** -3 **c.** $\frac{1}{2}$

Practice 4

Solve each equation for x.

a. $\log_5 \frac{1}{25} = x$

b. $\log_x 8 = 3$

c. $\log_6 x = 2$

d. $\log_{13} 1 = x$

e. $\log_h 1 = x$

Example 4

Solve each equation for x.

a. $\log_4 \frac{1}{4} = x$ **b.** $\log_5 x = 3$ **c.** $\log_x 25 = 2$ **d.** $\log_3 1 = x$ **e.** $\log_b 1 = x$

Solution:

a. $\log_4 \frac{1}{4} = x$ means $4^x = \frac{1}{4}$. Solve $4^x = \frac{1}{4}$ for x.

$$4^x = \frac{1}{4}$$
$$4^x = 4^{-1}$$

Since the bases are the same, by the uniqueness of b^x, we have that

$$x = -1$$

The solution is -1 or the solution set is $\{-1\}$. To check, see that $\log_4 \frac{1}{4} = -1$, since $4^{-1} = \frac{1}{4}$.

b. $\log_5 x = 3$

$5^3 = x$ Write as an exponential equation.

$125 = x$

The solution is 125.

c. $\log_x 25 = 2$

$x^2 = 25$ Write as an exponential equation. Here $x > 0, x \neq 1$.

$x = 5$

Even though $(-5)^2 = 25$, the base b of a logarithm must be positive. The solution is 5.

d. $\log_3 1 = x$

$3^x = 1$ Write as an exponential equation.

$3^x = 3^0$ Write 1 as 3^0.

$x = 0$ Use the uniqueness of b^x.

The solution is 0.

e. $\log_b 1 = x$

$b^x = 1$ Write as an exponential equation. Here, $b > 0$ and $b \neq 1$.

$b^x = b^0$ Write 1 as b^0.

$x = 0$ Apply the uniqueness of b^x.

The solution is 0.

Work Practice 4

In Example 4e, we proved an important property of logarithms. That is, $\log_b 1$ is always 0. This property as well as two important others are given next.

Properties of Logarithms

If b is a real number, $b > 0$, and $b \neq 1$, then

1. $\log_b 1 = 0$
2. $\log_b b^x = x$
3. $b^{\log_b x} = x$

Answers

4. a. -2 b. 2 c. 36 d. 0 e. 0

To see that **2.** $\log_b b^x = x$, change the logarithmic form to exponential form. Then, $\log_b b^x = x$ means $b^x = b^x$. In exponential form, the statement is true, so in logarithmic form, the statement is also true.

To understand **3.** $b^{\log_b x} = x$, write this exponential equation as an equivalent logarithm.

Example 5 Simplify.

a. $\log_3 3^2$ **b.** $\log_7 7^{-1}$ **c.** $5^{\log_5 3}$ **d.** $2^{\log_2 6}$

Solution:

a. From Property 2, $\log_3 3^2 = 2$.

b. From Property 2, $\log_7 7^{-1} = -1$.

c. From Property 3, $5^{\log_5 3} = 3$.

d. From Property 3, $2^{\log_2 6} = 6$.

Work Practice 5

Practice 5

Simplify.

a. $\log_5 5^4$ **b.** $\log_9 9^{-2}$

c. $6^{\log_6 5}$ **d.** $7^{\log_7 4}$

Objective 3 Graphing Logarithmic Functions

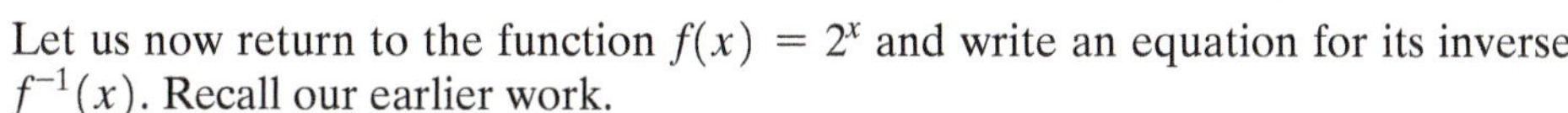

Let us now return to the function $f(x) = 2^x$ and write an equation for its inverse, $f^{-1}(x)$. Recall our earlier work.

$$f(x) = 2^x$$

Step 1: Replace $f(x)$ by y. $\quad y = 2^x$

Step 2: Interchange x and y. $\quad x = 2^y$

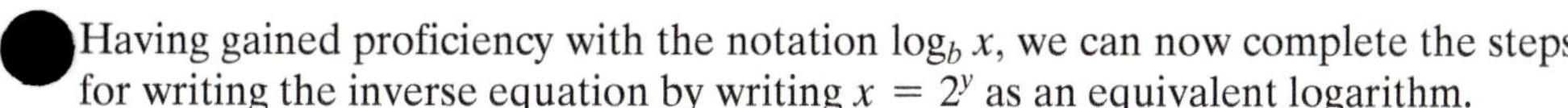

Having gained proficiency with the notation $\log_b x$, we can now complete the steps for writing the inverse equation by writing $x = 2^y$ as an equivalent logarithm.

Step 3: Solve for y. $\quad y = \log_2 x$

Step 4: Replace y with $f^{-1}(x)$. $\quad f^{-1}(x) = \log_2 x$

Thus, $f^{-1}(x) = \log_2 x$ defines a function that is the inverse function of the function $f(x) = 2^x$. The function $f^{-1}(x)$ or $y = \log_2 x$ is called a **logarithmic function.**

Logarithmic Function

If x is a positive real number, b is a constant positive real number, and b is not 1, then a **logarithmic function** is a function that can be defined by

$$f(x) = \log_b x$$

The domain of f is the set of positive real numbers, and the range of f is the set of real numbers.

✓ **Concept Check** Let $f(x) = \log_3 x$ and $g(x) = 3^x$. These two functions are inverses of each other. Since (2, 9) is an ordered pair solution of $g(x)$ or $g(2) = 9$, what ordered pair do we know to be a solution of $f(x)$? Also, find $f(9)$. Explain why.

We can explore logarithmic functions by graphing them.

Example 6 Graph the logarithmic function $y = \log_2 x$.

Solution: First we write the equation with exponential notation as $2^y = x$. Then we find some ordered pair solutions that satisfy this equation. Finally, we plot the points and connect them with a smooth curve. The domain of this function is $(0, \infty)$, and the range is all real numbers.

(*Continued on next page*)

Practice 6

Graph the logarithmic function $y = \log_9 x$.

Answers

5. a. 4 **b.** −2 **c.** 5 **d.** 4

6.

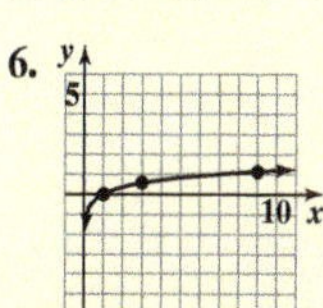

✓ **Concept Check Answer**

(9, 2); $f(9) = 2$; answers may vary

Since $x = 2^y$ is solved for x, we choose y-values and compute corresponding x-values.

If $y = 0, x = 2^0 = 1$

If $y = 1, x = 2^1 = 2$

If $y = 2, x = 2^2 = 4$

If $y = -1, x = 2^{-1} = \frac{1}{2}$

$x = 2^y$	y
1	0
2	1
4	2
$\frac{1}{2}$	−1

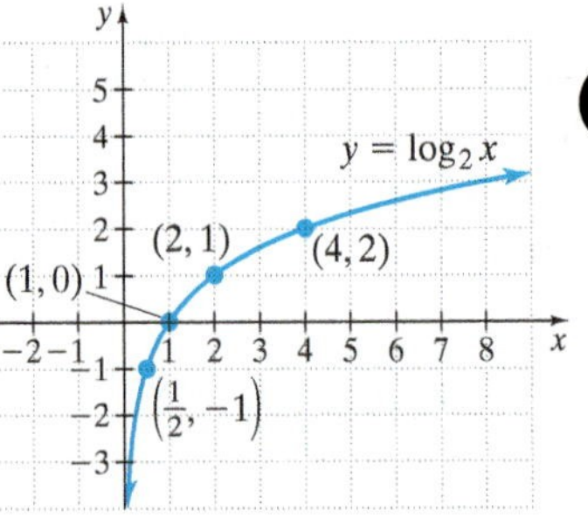

Notice that the x-intercept is $(1, 0)$ and there is no y-intercept.

Work Practice 6

Practice 7

Graph the logarithmic function $y = \log_{1/4} x$.

Example 7 Graph the logarithmic function $f(x) = \log_{1/3} x$.

Solution: Replace $f(x)$ with y and write the result with exponential notation.

$f(x) = \log_{1/3} x$

$y = \log_{1/3} x$ Replace $f(x)$ with y.

$\left(\frac{1}{3}\right)^y = x$ Write in exponential form.

Now we can find ordered pair solutions that satisfy $\left(\frac{1}{3}\right)^y = x$, plot these points, and connect them with a smooth curve.

If $y = 0, x = \left(\frac{1}{3}\right)^0 = 1$

If $y = 1, x = \left(\frac{1}{3}\right)^1 = \frac{1}{3}$

If $y = -1, x = \left(\frac{1}{3}\right)^{-1} = 3$

If $y = -2, x = \left(\frac{1}{3}\right)^{-2} = 9$

$x = \left(\frac{1}{3}\right)^y$	y
1	0
$\frac{1}{3}$	1
3	−1
9	−2

The domain of this function is $(0, \infty)$, and the range is the set of all real numbers. The x-intercept is $(1, 0)$ and there is no y-intercept.

Work Practice 7

The following figures summarize characteristics of logarithmic functions.

$f(x) = \log_b x, b > 0, b \neq 1$

- one-to-one function
- x-intercept $(1, 0)$
- no y-intercept
- domain: $(0, \infty)$
- range: $(-\infty, \infty)$

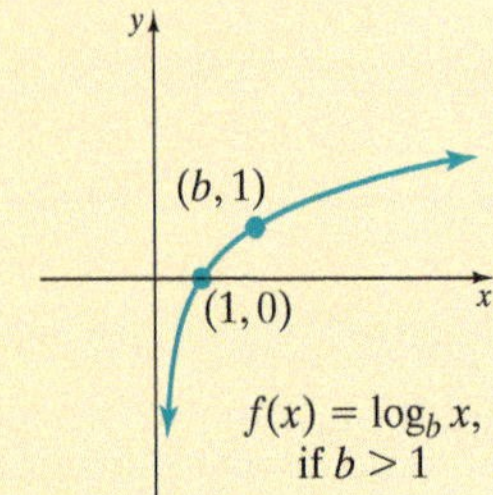

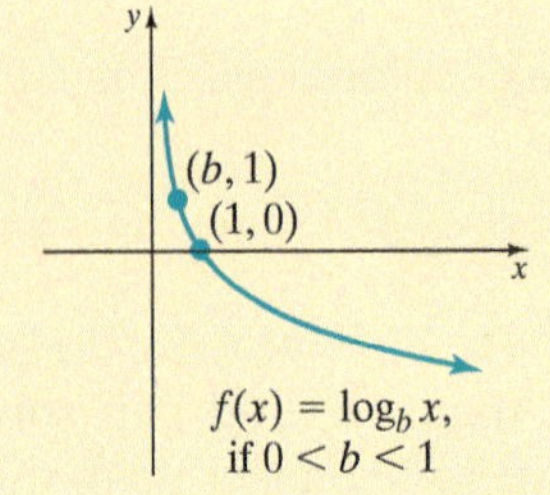

Answer

7.

Vocabulary, Readiness & Video Check

Use the choices to fill in each blank.

1. A function such as $y = \log_2 x$ is a(n) __________ function.

   A. linear B. logarithmic C. quadratic D. exponential

2. If $y = \log_2 x$, then __________.

   A. $x = y$ B. $2^x = y$ C. $2^y = x$ D. $2y = x$

Answer the questions about the graph of $y = \log_2 x$, shown below.

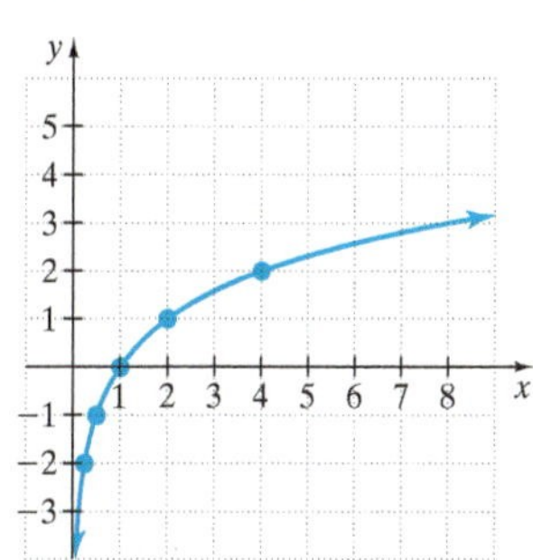

3. Is this a one-to-one function? __________
4. Is there an x-intercept? __________ If so, name the coordinates. __________
5. Is there a y-intercept? __________ If so, name the coordinates. __________
6. The domain of this function, in interval notation, is __________.
7. The range of this function, in interval notation, is __________.

Martin-Gay Interactive Videos *Watch the section lecture video and answer the following questions.*

See Video 19.5

Objective 1 8. Notice from the definition of a logarithm and from Examples 1–4 that a logarithmic statement equals the power in the exponent statement, such as $b^y = x$. What conclusion can you make about logarithms and exponents?

Objective 2 9. From Examples 8 and 9, how do you solve a logarithmic equation?

Objective 3 10. In Example 12, why is it easier to choose values for y when finding ordered pairs for the graph?

19.5 Exercise Set MyMathLab®

Write each as an exponential equation. See Example 1.

1. $\log_6 36 = 2$
2. $\log_2 32 = 5$
3. $\log_3 \frac{1}{27} = -3$
4. $\log_5 \frac{1}{25} = -2$
5. $\log_{10} 1000 = 3$
6. $\log_{10} 10 = 1$
7. $\log_9 x = 4$
8. $\log_8 y = 7$
9. $\log_\pi \frac{1}{\pi^2} = -2$
10. $\log_e \frac{1}{e} = -1$
11. $\log_7 \sqrt{7} = \frac{1}{2}$
12. $\log_{11} \sqrt[4]{11} = \frac{1}{4}$
13. $\log_{0.7} 0.343 = 3$
14. $\log_{1.2} 1.44 = 2$
15. $\log_3 \frac{1}{81} = -4$
16. $\log_{1/4} 16 = -2$

Write each as a logarithmic equation. See Example 2.

17. $2^4 = 16$
18. $5^3 = 125$
19. $10^2 = 100$
20. $10^4 = 10{,}000$
21. $\pi^3 = x$
22. $\pi^5 = y$
23. $10^{-1} = \frac{1}{10}$
24. $10^{-2} = \frac{1}{100}$

25. $4^{-2} = \frac{1}{16}$
26. $3^{-4} = \frac{1}{81}$
27. $5^{1/2} = \sqrt{5}$
28. $4^{1/3} = \sqrt[3]{4}$

Find the value of each logarithmic expression. See Examples 3 and 5.

29. $\log_2 8$
30. $\log_3 9$
31. $\log_3 \frac{1}{9}$
32. $\log_2 \frac{1}{32}$
33. $\log_{25} 5$
34. $\log_8 \frac{1}{2}$
35. $\log_{1/2} 2$
36. $\log_{2/3} \frac{4}{9}$
37. $\log_6 1$
38. $\log_9 9$
39. $\log_{10} 100$
40. $\log_{10} \frac{1}{10}$
41. $\log_3 81$
42. $\log_2 16$
43. $\log_4 \frac{1}{64}$
44. $\log_3 \frac{1}{9}$

Solve. See Example 4.

45. $\log_3 9 = x$
46. $\log_2 8 = x$
47. $\log_3 x = 4$
48. $\log_2 x = 3$
49. $\log_x 49 = 2$
50. $\log_x 8 = 3$
51. $\log_2 \frac{1}{8} = x$
52. $\log_3 \frac{1}{81} = x$
53. $\log_3 \frac{1}{27} = x$
54. $\log_5 \frac{1}{125} = x$
55. $\log_8 x = \frac{1}{3}$
56. $\log_9 x = \frac{1}{2}$
57. $\log_4 16 = x$
58. $\log_2 16 = x$
59. $\log_{3/4} x = 3$
60. $\log_{2/3} x = 2$
61. $\log_x 100 = 2$
62. $\log_x 27 = 3$
63. $\log_2 2^4 = x$
64. $\log_6 6^{-2} = x$
65. $3^{\log_3 5} = x$
66. $5^{\log_5 7} = x$
67. $\log_x \frac{1}{7} = \frac{1}{2}$
68. $\log_x 2 = -\frac{1}{3}$

Simplify. See Example 5.

69. $\log_5 5^3$
70. $\log_6 6^2$
71. $2^{\log_2 3}$
72. $7^{\log_7 4}$
73. $\log_9 9$
74. $\log_2 2$
75. $\log_8 (8)^{-1}$
76. $\log_{11} (11)^{-1}$

Graph each logarithmic function. Label any intercepts. See Examples 6 and 7.

77. $y = \log_3 x$
78. $y = \log_8 x$
79. $f(x) = \log_{1/4} x$
80. $f(x) = \log_{1/2} x$
81. $f(x) = \log_5 x$
82. $f(x) = \log_6 x$
83. $f(x) = \log_{1/6} x$
84. $f(x) = \log_{1/5} x$

REVIEW AND PREVIEW

Simplify each rational expression. See Section 14.1.

85. $\frac{x + 3}{3 + x}$
86. $\frac{x - 5}{5 - x}$
87. $\frac{x^2 - 8x + 16}{2x - 8}$
88. $\frac{x^2 - 3x - 10}{2 + x}$

Add or subtract as indicated. See Sections 14.3 and 14.4.

89. $\frac{2}{x} + \frac{3}{x^2}$
90. $\frac{5}{y + 1} - \frac{4}{y - 1}$
91. $\frac{3x}{x + 3} + \frac{9}{x + 3}$
92. $\frac{m^2}{m + 1} - \frac{1}{m + 1}$

CONCEPT EXTENSIONS

Solve. See the Concept Check in this section.

93. Let $f(x) = \log_5 x$. Then $g(x) = 5^x$ is the inverse of $f(x)$. The ordered pair $(2, 25)$ is a solution of the function $g(x)$.
   a. Write this solution using function notation.
   b. Write an ordered pair that we know to be a solution of $f(x)$.
   c. Use the answer to part b and write the solution using function notation.

94. Let $f(x) = \log_{0.3} x$. Then $g(x) = 0.3^x$ is the inverse of $f(x)$. The ordered pair $(3, 0.027)$ is a solution of the function $g(x)$.
   a. Write this solution using function notation.
   b. Write an ordered pair that we know to be a solution of $f(x)$.
   c. Use the answer to part b and write the solution using function notation.

95. Explain why negative numbers are not included as logarithmic bases.

96. Explain why 1 is not included as a logarithmic base.

Solve by first writing as an exponent.

97. $\log_7 (5x - 2) = 1$
98. $\log_3 (2x + 4) = 2$
99. Simplify: $\log_3(\log_5 125)$
100. Simplify: $\log_7(\log_4(\log_2 16))$

Graph each function and its inverse function on the same set of axes. Label any intercepts.

101. $y = 4^x; y = \log_4 x$
102. $y = 3^x; y = \log_3 x$
103. $y = \left(\frac{1}{3}\right)^x; y = \log_{1/3} x$
104. $y = \left(\frac{1}{2}\right)^x; y = \log_{1/2} x$

105. Explain why the graph of the function $y = \log_b x$ contains the point $(1, 0)$ no matter what b is.

106. $\log_3 10$ is between which two integers? Explain your answer.

107. The formula $\log_{10}(1 - k) = \frac{-0.3}{H}$ models the relationship between the half-life H of a radioactive material and its rate of decay k. Find the rate of decay of the iodine isotope I-131 if its half-life is 8 days. Round to four decimal places.

108. The formula $\text{pH} = -\log_{10}(\text{H}^+)$ provides the pH for a liquid, where H^+ stands for the concentration of hydronium ions. Find the pH of lemonade, whose concentration of hydronium ions is 0.0050 moles/liter. Round to the nearest tenth.

19.6 Properties of Logarithms 

In the previous section, we explored some basic properties of logarithms. We now introduce and explore additional properties. Because a logarithm is an exponent, logarithmic properties are just restatements of exponential properties.

Objectives

1. Use the Product Property of Logarithms.
2. Use the Quotient Property of Logarithms.
3. Use the Power Property of Logarithms.
4. Use the Properties of Logarithms Together.

Objective 1 Using the Product Property

The first of these properties is called the **product property of logarithms** because it deals with the logarithm of a product.

Product Property of Logarithms

If x, y, and b are positive real numbers and $b \neq 1$, then

$$\log_b xy = \log_b x + \log_b y$$

To prove this, let $\log_b x = M$ and $\log_b y = N$. Now write each logarithm with exponential notation.

$\log_b x = M$ is equivalent to $b^M = x$

$\log_b y = N$ is equivalent to $b^N = y$

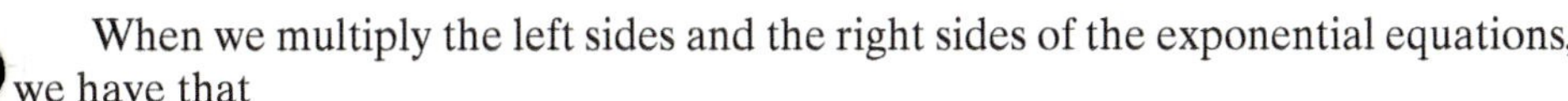

When we multiply the left sides and the right sides of the exponential equations, we have that

$$xy = (b^M)(b^N) = b^{M+N}$$

If we write the equation $xy = b^{M+N}$ in equivalent logarithmic form, we have

$$\log_b xy = M + N$$

But since $M = \log_b x$ and $N = \log_b y$, we can write

$$\log_b xy = \log_b x + \log_b y \quad \text{Let } M = \log_b x \text{ and } N = \log_b y.$$

In other words, the logarithm of a product is the sum of the logarithms of the factors. This property is sometimes used to simplify logarithmic expressions.

In the examples that follow, assume that variables represent positive numbers.

Example 1 Write each sum as a single logarithm.

a. $\log_{11} 10 + \log_{11} 3$ **b.** $\log_3 \frac{1}{2} + \log_3 12$ **c.** $\log_2(x + 2) + \log_2 x$

Solution: In each case, both terms have a common logarithmic base.

a. $\log_{11} 10 + \log_{11} 3 = \log_{11}(10 \cdot 3)$ Apply the product property.

$= \log_{11} 30$

b. $\log_3 \frac{1}{2} + \log_3 12 = \log_3\left(\frac{1}{2} \cdot 12\right) = \log_3 6$

c. $\log_2(x + 2) + \log_2 x = \log_2[(x + 2) \cdot x] = \log_2(x^2 + 2x)$

Work Practice 1

Practice 1

Write each sum as a single logarithm.

a. $\log_8 5 + \log_8 3$

b. $\log_2 \frac{1}{3} + \log_2 18$

c. $\log_5(x - 1) + \log_5(x + 1)$

Helpful Hint

Check your logarithm properties. Make sure you understand that $\log_2(x + 2)$ *is not* $\log_2 x + \log_2 2$.

Answers

1. **a.** $\log_8 15$ **b.** $\log_2 6$ **c.** $\log_5(x^2 - 1)$

Objective 2 Using the Quotient Property 

The second property is the **quotient property of logarithms.**

Quotient Property of Logarithms

If x, y, and b are positive real numbers and $b \neq 1$, then

$$\log_b \frac{x}{y} = \log_b x - \log_b y$$

The proof of the quotient property of logarithms is similar to the proof of the product property. Notice that the quotient property says that the logarithm of a quotient is the difference of the logarithms of the dividend and divisor.

✓Concept Check Which of the following is the correct way to rewrite $\log_5 \frac{7}{2}$?

a. $\log_5 7 - \log_5 2$ **b.** $\log_5(7 - 2)$ **c.** $\frac{\log_5 7}{\log_5 2}$ **d.** $\log_5 14$

Practice 2

Write each difference as a single logarithm.

a. $\log_5 18 - \log_5 6$

b. $\log_6 x - \log_6 3$

c. $\log_4(x^2 + 1) - \log_4(x^2 + 3)$

Example 2

Write each difference as a single logarithm.

a. $\log_{10} 27 - \log_{10} 3$ **b.** $\log_5 8 - \log_5 x$ **c.** $\log_3(x^2 + 5) - \log_3(x^2 + 1)$

Solution: In each case, both terms have a common logarithmic base.

a. $\log_{10} 27 - \log_{10} 3 = \log_{10} \frac{27}{3} = \log_{10} 9$

b. $\log_5 8 - \log_5 x = \log_5 \frac{8}{x}$

c. $\log_3(x^2 + 5) - \log_3(x^2 + 1) = \log_3 \frac{x^2 + 5}{x^2 + 1}$ Apply the quotient property.

Work Practice 2

Objective 3 Using the Power Property

The third and final property we introduce is the **power property of logarithms.**

Power Property of Logarithms

If x and b are positive real numbers, $b \neq 1$, and r is a real number, then

$$\log_b x^r = r \log_b x$$

Practice 3

Use the power property to rewrite each expression.

a. $\log_7 x^8$ **b.** $\log_5 \sqrt[4]{7}$

Example 3

Use the power property to rewrite each expression.

a. $\log_5 x^3$ **b.** $\log_4 \sqrt{2}$

Solution:

a. $\log_5 x^3 = 3 \log_5 x$

b. $\log_4 \sqrt{2} = \log_4 2^{1/2} = \frac{1}{2} \log_4 2$

Work Practice 3

Answers

2. a. $\log_5 3$ **b.** $\log_6 \frac{x}{3}$ **c.** $\log_4 \frac{x^2 + 1}{x^2 + 3}$

3. a. $8 \log_7 x$ **b.** $\frac{1}{4} \log_5 7$

✓Concept Check Answer

a

Objective 4 Using the Properties Together 

Many times, we must use more than one property of logarithms to simplify a logarithmic expression.

Example 4 Write as a single logarithm.

a. $2\log_5 3 + 3\log_5 2$ **b.** $3\log_9 x - \log_9(x+1)$ **c.** $\log_4 25 + \log_4 3 - \log_4 5$

Solution: In each case, all terms have a common logarithmic base.

a. $2\log_5 3 + 3\log_5 2 = \log_5 3^2 + \log_5 2^3$ Apply the power property.

$= \log_5 9 + \log_5 8$

$= \log_5(9\cdot 8)$

$= \log_5 72$ Apply the product property.

b. $3\log_9 x - \log_9(x+1) = \log_9 x^3 - \log_9(x+1)$ Apply the power property.

$= \log_9 \dfrac{x^3}{x+1}$ Apply the quotient property.

c. Use both the product and quotient properties.

$\log_4 25 + \log_4 3 - \log_4 5 = \log_4(25\cdot 3) - \log_4 5$ Apply the product property.

$= \log_4 75 - \log_4 5$ Simplify.

$= \log_4 \dfrac{75}{5}$ Apply the quotient property.

$= \log_4 15$ Simplify.

Work Practice 4

Practice 4

Write as a single logarithm.

a. $2\log_5 4 + 5\log_5 2$

b. $2\log_8 x - \log_8(x+3)$

c. $\log_7 12 + \log_7 5 - \log_7 4$

Example 5 Write each expression as sums or differences of multiples of logarithms.

a. $\log_3 \dfrac{5\cdot 7}{4}$ **b.** $\log_2 \dfrac{x^5}{y^2}$

Solution:

a. $\log_3 \dfrac{5\cdot 7}{4} = \log_3(5\cdot 7) - \log_3 4$ Apply the quotient property.

$= \log_3 5 + \log_3 7 - \log_3 4$ Apply the product property.

b. $\log_2 \dfrac{x^5}{y^2} = \log_2(x^5) - \log_2(y^2)$ Apply the quotient property.

$= 5\log_2 x - 2\log_2 y$ Apply the power property.

Work Practice 5

Practice 5

Write each expression as sums or differences of multiples of logarithms.

a. $\log_5 \dfrac{4\cdot 3}{7}$ **b.** $\log_4 \dfrac{a^2}{b^5}$

Helpful Hint

Notice that we are not able to further simplify a logarithmic expression such as $\log_5(2x-1)$. None of the basic properties gives a way to write the logarithm of a difference (or sum) in some equivalent form.

✓**Concept Check** What is wrong with the following?

$\log_{10}(x^2+5) = \log_{10} x^2 + \log_{10} 5$
$= 2\log_{10} x + \log_{10} 5$

Use a numerical example to demonstrate that the result is incorrect.

Answers

4. a. $\log_5 512$ **b.** $\log_8 \dfrac{x^2}{x+3}$ **c.** $\log_7 15$

5. a. $\log_5 4 + \log_5 3 - \log_5 7$

b. $2\log_4 a - 5\log_4 b$

✓**Concept Check Answer**

The properties do not give any way to simplify the logarithm of a sum; answers may vary.

Practice 6

If $\log_b 5 = 0.83$ and $\log_b 3 = 0.56$, use the properties of logarithms to evaluate.

a. $\log_b 15$ **b.** $\log_b 25$
c. $\log_b \sqrt{3}$

Example 6 If $\log_b 2 = 0.43$ and $\log_b 3 = 0.68$, use the properties of logarithms to evaluate.

a. $\log_b 6$ **b.** $\log_b 9$ **c.** $\log_b \sqrt{2}$

Solution:

a.
$$\begin{aligned} \log_b 6 &= \log_b(2 \cdot 3) && \text{Write 6 as } 2 \cdot 3. \\ &= \log_b 2 + \log_b 3 && \text{Apply the product property.} \\ &= 0.43 + 0.68 && \text{Substitute given values.} \\ &= 1.11 && \text{Simplify.} \end{aligned}$$

b.
$$\begin{aligned} \log_b 9 &= \log_b 3^2 && \text{Write 9 as } 3^2. \\ &= 2 \log_b 3 \\ &= 2(0.68) && \text{Substitute 0.68 for } \log_b 3. \\ &= 1.36 && \text{Simplify.} \end{aligned}$$

c. First, recall that $\sqrt{2} = 2^{1/2}$. Then

$$\begin{aligned} \log_b \sqrt{2} &= \log_b 2^{1/2} && \text{Write } \sqrt{2} \text{ as } 2^{1/2}. \\ &= \frac{1}{2}\log_b 2 && \text{Apply the power property.} \\ &= \frac{1}{2}(0.43) && \text{Substitute the given value.} \\ &= 0.215 && \text{Simplify.} \end{aligned}$$

Work Practice 6

A summary of the basic properties of logarithms that we have developed so far is given next.

Properties of Logarithms

If x, y, and b are positive real numbers, $b \neq 1$, and r is a real number, then

1. $\log_b 1 = 0$

2. $\log_b b^x = x$

3. $b^{\log_b x} = x$

4. $\log_b xy = \log_b x + \log_b y$ Product property.

5. $\log_b \frac{x}{y} = \log_b x - \log_b y$ Quotient property.

6. $\log_b x^r = r \log_b x$ Power property.

Answers

6. a. 1.39 **b.** 1.66 **c.** 0.28

Vocabulary, Readiness & Video Check

Select the correct choice.

1. $\log_b 12 + \log_b 3 = \log_b$ ____
a. 36 **b.** 15 **c.** 4 **d.** 9

2. $\log_b 12 - \log_b 3 = \log_b$ ____
a. 36 **b.** 15 **c.** 4 **d.** 9

3. $7 \log_b 2 =$ ____
a. $\log_b 14$ **b.** $\log_b 2^7$ **c.** $\log_b 7^2$ **d.** $(\log_b 2)^7$

4. $\log_b 1 =$ ____
a. b **b.** 1 **c.** 0 **d.** no answer

5. $b^{\log_b x} =$ ____
a. x **b.** b **c.** 1 **d.** 0

6. $\log_5 5^2 =$ ____
a. 25 **b.** 2 **c.** 5^{5^2} **d.** 32

Martin-Gay Interactive Videos *Watch the section lecture video and answer the following questions.*

See Video 19.6

Objective 1 7. Can the product property of logarithms be used again on the bottom line of Example 2 to write $\log_{10}(10x^2 + 20)$ as a sum of logarithms, $\log_{10} 10x^2 + \log_{10} 20$? Explain.

Objective 2 8. From Example 3 and the lecture before, what must be true about bases before you can apply the quotient property of logarithms?

Objective 3 9. Based on Example 5, explain why $\log_2 \frac{1}{x} = -\log_2 x$.

Objective 4 10. From the lecture before Example 6, where do the logarithmic properties come from?

19.6 Exercise Set MyMathLab®

Write each sum as a single logarithm. Assume that variables represent positive numbers. See Example 1.

1. $\log_5 2 + \log_5 7$
2. $\log_3 8 + \log_3 4$
3. $\log_4 9 + \log_4 x$
4. $\log_2 x + \log_2 y$
5. $\log_6 x + \log_6 (x + 1)$
6. $\log_5 y^3 + \log_5 (y - 7)$
7. $\log_{10} 5 + \log_{10} 2 + \log_{10}(x^2 + 2)$
8. $\log_6 3 + \log_6 (x + 4) + \log_6 5$

Write each difference as a single logarithm. Assume that variables represent positive numbers. See Example 2.

9. $\log_5 12 - \log_5 4$
10. $\log_7 20 - \log_7 4$
11. $\log_3 8 - \log_3 2$
12. $\log_5 12 - \log_5 3$
13. $\log_2 x - \log_2 y$
14. $\log_3 12 - \log_3 z$
15. $\log_2(x^2 + 6) - \log_2(x^2 + 1)$
16. $\log_7(x + 9) - \log_7(x^2 + 10)$

Use the power property to rewrite each expression. See Example 3.

17. $\log_3 x^2$
18. $\log_2 x^5$
19. $\log_4 5^{-1}$
20. $\log_6 7^{-2}$
21. $\log_5 \sqrt{y}$
22. $\log_5 \sqrt[3]{x}$

MIXED PRACTICE

Write each as a single logarithm. Assume that variables represent positive numbers. See Example 4.

23. $\log_2 5 + \log_2 x^3$
24. $\log_5 2 + \log_5 y^2$
25. $3\log_4 2 + \log_4 6$
26. $2\log_3 5 + \log_3 2$
27. $3\log_5 x + 6\log_5 z$
28. $2\log_7 y + 6\log_7 z$
29. $\log_4 2 + \log_4 10 - \log_4 5$
30. $\log_6 18 + \log_6 2 - \log_6 9$
31. $\log_7 6 + \log_7 3 - \log_7 4$
32. $\log_8 5 + \log_8 15 - \log_8 20$
33. $\log_{10} x - \log_{10}(x + 1) + \log_{10}(x^2 - 2)$
34. $\log_9(4x) - \log_9(x - 3) + \log_9(x^3 + 1)$
35. $3\log_2 x + \frac{1}{2}\log_2 x - 2\log_2(x + 1)$
36. $2\log_5 x + \frac{1}{3}\log_5 x - 3\log_5(x + 5)$
37. $2\log_8 x - \frac{2}{3}\log_8 x + 4\log_8 x$
38. $5\log_6 x - \frac{3}{4}\log_6 x + 3\log_6 x$

MIXED PRACTICE

Write each expression as a sum or difference of multiples of logarithms. Assume that variables represent positive numbers. See Example 5.

39. $\log_3 \frac{4y}{5}$
40. $\log_7 \frac{5x}{4}$
41. $\log_4 \frac{5}{9z}$
42. $\log_9 \frac{7}{8y}$
43. $\log_2 \frac{x^3}{y}$
44. $\log_5 \frac{x}{y^4}$
45. $\log_b \sqrt{7x}$
46. $\log_b \sqrt{\frac{3}{y}}$
47. $\log_6 x^4 y^5$
48. $\log_2 y^3 z$
49. $\log_5 x^3(x + 1)$
50. $\log_3 x^2(x - 9)$
51. $\log_6 \frac{x^2}{x + 3}$
52. $\log_3 \frac{(x + 5)^2}{x}$

If $\log_b 3 = 0.5$ and $\log_b 5 = 0.7$, evaluate each expression. See Example 6.

53. $\log_b 15$
54. $\log_b 25$
55. $\log_b \frac{5}{3}$
56. $\log_b \frac{3}{5}$
57. $\log_b \sqrt{5}$
58. $\log_b \sqrt[4]{3}$

If $\log_b 2 = 0.43$ and $\log_b 3 = 0.68$, evaluate each expression. See Example 6.

59. $\log_b 8$

60. $\log_b 81$

61. $\log_b \frac{3}{9}$

62. $\log_b \frac{4}{32}$

63. $\log_b \sqrt{\frac{2}{3}}$

64. $\log_b \sqrt{\frac{3}{2}}$

REVIEW AND PREVIEW

Graph both functions on the same set of axes. See Sections 19.3 and 19.5.

65. $y = 10^x$

66. $y = \log_{10} x$.

Evaluate each expression. See Section 19.5.

67. $\log_{10} 100$

68. $\log_{10}\frac{1}{10}$

69. $\log_7 7^2$

70. $\log_7 \sqrt{7}$

CONCEPT EXTENSIONS

Solve. See the Concept Checks in this section.

71. Which of the following is the correct way to rewrite $\log_3 \frac{14}{11}$?

a. $\frac{\log_3 14}{\log_3 11}$

b. $\log_3 14 - \log_3 11$

c. $\log_3(14 - 11)$

d. $\log_3 154$

72. Which of the following is the correct way to rewrite $\log_9 \frac{21}{3}$?

a. $\log_9 7$

b. $\log_9(21 - 3)$

c. $\frac{\log_9 21}{\log_9 3}$

d. $\log_9 21 - \log_9 3$

Answer the following true or false. Study your logarithm properties carefully before answering.

73. $\log_2 x^3 = 3\log_2 x$

74. $\log_3(x + y) = \log_3 x + \log_3 y$

75. $\frac{\log_7 10}{\log_7 5} = \log_7 2$

76. $\log_7 \frac{14}{8} = \log_7 14 - \log_7 8$

77. $\frac{\log_7 x}{\log_7 y} = (\log_7 x) - (\log_7 y)$

78. $(\log_3 6) \cdot (\log_3 4) = \log_3 24$

79. It is true that $\log_b 8 = \log_b(8 \cdot 1) = \log_b 8 + \log_b 1$. Explain how $\log_b 8$ can equal $\log_b 8 + \log_b 1$.

80. It is true that $\log_b 7 = \log_b \frac{7}{1} = \log_b 7 - \log_b 1$. Explain how $\log_b 7$ can equal $\log_b 7 - \log_b 1$.

Integrated Review Sections 19.1–19.6

Functions and Properties of Logarithms

If $f(x) = x - 6$ and $g(x) = x^2 + 1$, find each value.

1. $(f + g)(x)$

2. $(f - g)(x)$

3. $(f \cdot g)(x)$

4. $\left(\frac{f}{g}\right)(x)$

If $f(x) = \sqrt{x}$ and $g(x) = 3x - 1$, find each function.

5. $(f \circ g)(x)$

6. $(g \circ f)(x)$

Determine whether each is a one-to-one function. If it is, find its inverse.

7. $f = \{(-2,6), (4,8), (2,-6), (3,3)\}$

8. $g = \{(4,2), (-1,3), (5,3), (7,1)\}$

Determine from the graph whether each function is one-to-one.

9.

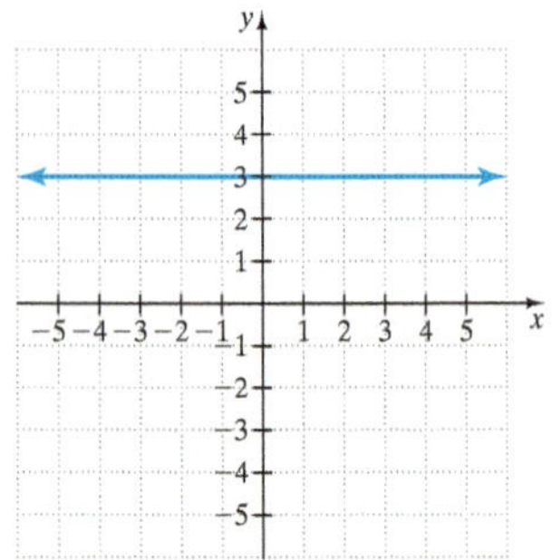

10.

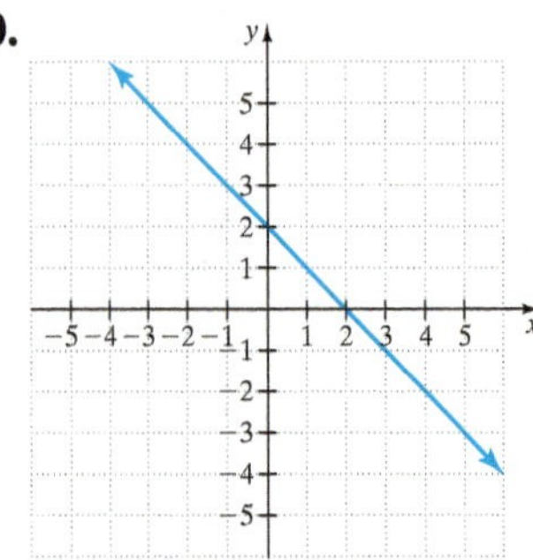

11.

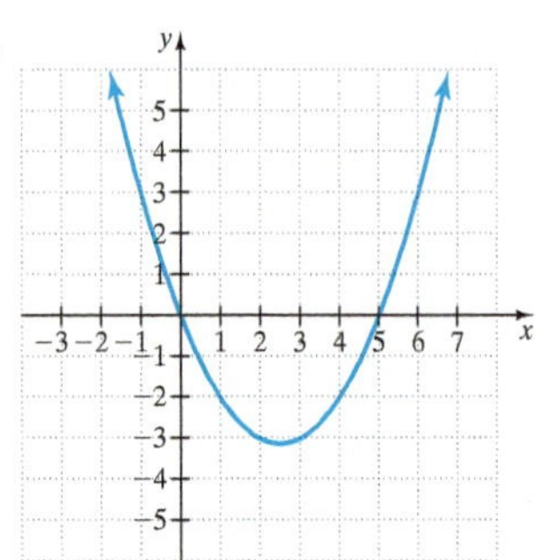

Each function listed is one-to-one. Find the inverse of each function.

12. $f(x) = 3x$ **13.** $f(x) = x + 4$ **14.** $f(x) = 5x - 1$ **15.** $f(x) = 3x + 2$

Graph each function.

16. $y = \left(\frac{1}{2}\right)^x$ **17.** $y = 2^x + 1$ **18.** $y = \log_3 x$ **19.** $y = \log_{1/3} x$

Solve.

20. $2^x = 8$ **21.** $9 = 3^{x-5}$ **22.** $4^{x-1} = 8^{x+2}$ **23.** $25^x = 125^{x-1}$

24. $\log_4 16 = x$ **25.** $\log_{49} 7 = x$ **26.** $\log_2 x = 5$ **27.** $\log_x 64 = 3$

28. $\log_x \frac{1}{125} = -3$ **29.** $\log_3 x = -2$

Write each as a single logarithm.

30. $5 \log_2 x$ **31.** $x \log_2 5$ **32.** $3 \log_5 x - 5 \log_5 y$ **33.** $9 \log_5 x + 3 \log_5 y$

34. $\log_2 x + \log_2(x - 3) - \log_2(x^2 + 4)$ **35.** $\log_3 y - \log_3(y + 2) + \log_3(y^3 + 11)$

Write each expression as sums or differences of multiples of logarithms.

36. $\log_7 \frac{9x^2}{y}$ **37.** $\log_6 \frac{5y}{z^2}$

38. An unusually wet spring has caused the size of the mosquito population in a community to increase by 6% each day. If an estimated 100,000 mosquitoes are in the community on April 1, find how many mosquitoes will inhabit the community on April 17. Round to the nearest thousand.

19.7 Common Logarithms, Natural Logarithms, and Change of Base

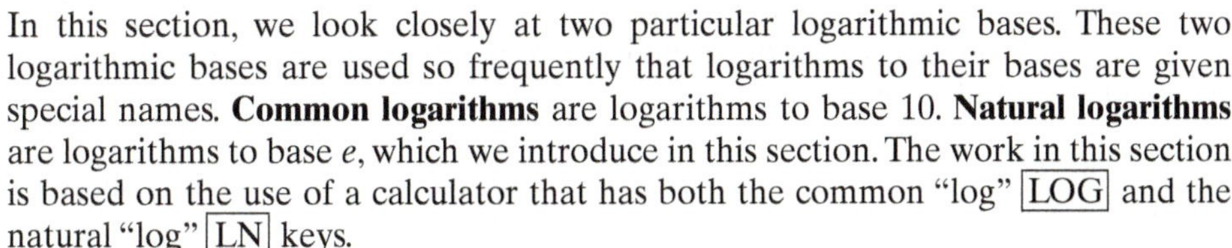

In this section, we look closely at two particular logarithmic bases. These two logarithmic bases are used so frequently that logarithms to their bases are given special names. **Common logarithms** are logarithms to base 10. **Natural logarithms** are logarithms to base e, which we introduce in this section. The work in this section is based on the use of a calculator that has both the common "log" LOG and the natural "log" LN keys.

Objectives

1. Identify Common Logarithms and Approximate Them by Calculator.
2. Evaluate Common Logarithms of Powers of 10.
3. Identify Natural Logarithms and Approximate Them by Calculator.
4. Evaluate Natural Logarithms of Powers of e.
5. Use the Change of Base Formula.

Objective 1 Approximating Common Logarithms

Logarithms to base 10, common logarithms, are used frequently because our number system is a base 10 decimal system. The notation $\log x$ means the same as $\log_{10} x$.

Common Logarithms

$\log x$ means $\log_{10} x$

Practice 1

Use a calculator to approximate log 15 to four decimal places.

Example 1 Use a calculator to approximate log 7 to four decimal places.

Solution: Press the following sequence of keys.

[7] [LOG] or [LOG] [7] [ENTER]

To four decimal places,

$\log 7 \approx 0.8451$

Work Practice 1

Objective 2 Evaluating Common Logarithms of Powers of 10

To evaluate the common log of a power of 10, a calculator is not needed. According to the property of logarithms,

$\log_b b^x = x$

It follows that if b is replaced with 10, we have

$\log 10^x = x$

Helpful Hint

Remember that the understood base here is 10.

Practice 2

Find the exact value of each logarithm.

a. $\log\frac{1}{100}$ **b.** $\log 100{,}000$

c. $\log\sqrt[5]{10}$ **d.** $\log 0.001$

Example 2 Find the exact value of each logarithm.

a. $\log 10$ **b.** $\log 1000$ **c.** $\log\frac{1}{10}$ **d.** $\log\sqrt{10}$

Solution:

a. $\log 10 = \log 10^1 = 1$

b. $\log 1000 = \log 10^3 = 3$

c. $\log\frac{1}{10} = \log 10^{-1} = -1$

d. $\log\sqrt{10} = \log 10^{1/2} = \frac{1}{2}$

Work Practice 2

As we will soon see, equations containing common logarithms are useful models of many natural phenomena.

Practice 3

Solve $\log x = 3.4$ for x. Give an exact solution, and then approximate the solution to four decimal places.

The understood base is 10.

Example 3 Solve $\log x = 1.2$ for x. Give an exact solution and then approximate the solution to four decimal places.

Solution: Remember that the base of a common logarithm is understood to be 10.

$\log x = 1.2$

$10^{1.2} = x$ Write with exponential notation.

The exact solution is $10^{1.2}$. To four decimal places, $x \approx 15.8489$.

Work Practice 3

The Richter scale measures the intensity, or magnitude, of an earthquake. The formula for the magnitude R of an earthquake is $R = \log\left(\frac{a}{T}\right) + B$, where a is the amplitude in micrometers of the vertical motion of the ground at the recording station, T is the number of seconds between successive seismic waves, and B is an adjustment factor that takes into account the weakening of the seismic wave as the distance increases from the epicenter of the earthquake.

Answers

1. 1.1761 **2. a.** -2 **b.** 5 **c.** $\frac{1}{5}$ **d.** -3 **3.** $10^{3.4} \approx 2511.8864$

Example 4 Finding the Magnitude of an Earthquake

Find an earthquake's magnitude on the Richter scale if a recording station measures an amplitude of 300 micrometers and 2.5 seconds between waves. Assume that B is 4.2. Approximate the solution to the nearest tenth.

Solution: Substitute the known values into the formula for earthquake intensity.

$$R = \log\left(\frac{a}{T}\right) + B \quad \text{Richter scale formula}$$

$$= \log\left(\frac{300}{2.5}\right) + 4.2 \quad \text{Let } a = 300,\ T = 2.5, \text{ and } B = 4.2.$$

$$= \log(120) + 4.2$$

$$\approx 2.1 + 4.2 \quad \text{Approximate log 120 by 2.1.}$$

$$= 6.3$$

This earthquake had a magnitude of 6.3 on the Richter scale.

Work Practice 4

Practice 4

Find an earthquake's magnitude on the Richter scale if a recording station measures an amplitude of 450 micrometers and 4.2 seconds between waves with $B = 3.6$. Approximate the solution to the nearest tenth.

Objective 3 Approximating Natural Logarithms

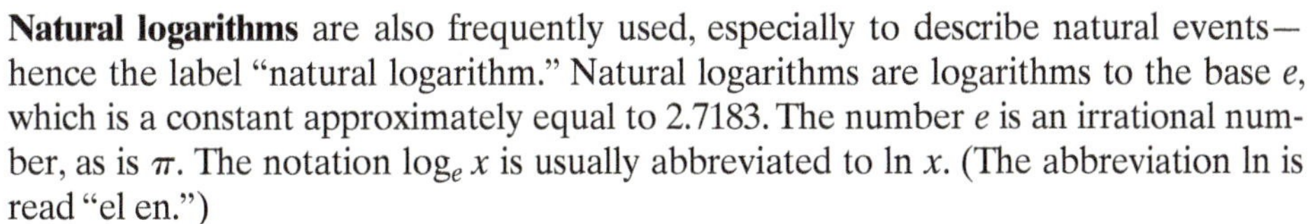

Natural logarithms are also frequently used, especially to describe natural events—hence the label "natural logarithm." Natural logarithms are logarithms to the base e, which is a constant approximately equal to 2.7183. The number e is an irrational number, as is π. The notation $\log_e x$ is usually abbreviated to $\ln x$. (The abbreviation ln is read "el en.")

Natural Logarithms

$\ln x$ means $\log_e x$

The graph of $y = \ln x$ is shown to the right.

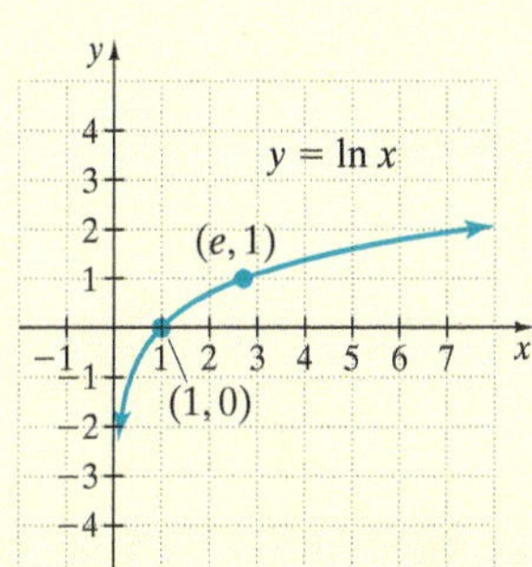

Example 5

Use a calculator to approximate ln 8 to four decimal places.

Solution: Press the following sequence of keys.

[8] [LN] or [LN] [8] [ENTER]

To four decimal places,

$\ln 8 \approx 2.0794$

Work Practice 5

Practice 5

Use a calculator to approximate ln 13 to four decimal places.

Objective 4 Evaluating Natural Logarithms of Powers of e

As a result of the property $\log_b b^x = x$, we know that $\log_e e^x = x$, or $\mathbf{\ln e^x = x}$.

Since $\ln e^x = x$, $\ln e^5 = 5$, $\ln e^{22} = 22$, and so on. Also,

$\ln e^1 = 1$ or simply $\ln e = 1$.

That is why the graph of $y = \ln x$ shown above in the margin passes through $(e, 1)$. If $x = e$, then $y = \ln e = 1$, thus the ordered pair is $(e, 1)$.

Answers

4. 5.6 **5.** 2.5649

Practice 6

Find the exact value of each natural logarithm.

a. $\ln e^4$ **b.** $\ln\sqrt[3]{e}$

Example 6

Find the exact value of each natural logarithm.

a. $\ln e^3$ **b.** $\ln\sqrt[5]{e}$

Solution:

a. $\ln e^3 = 3$ **b.** $\ln\sqrt[5]{e} = \ln e^{1/5} = \frac{1}{5}$

Work Practice 6

Practice 7

Solve $\ln 5x = 8$. Give an exact solution and then approximate the solution to four decimal places.

The understood base is e.

Example 7

Solve $\ln 3x = 5$. Give an exact solution and then approximate the solution to four decimal places.

Solution: Remember that the base of a natural logarithm is understood to be e.

$\ln 3x = 5$

$e^5 = 3x$ Write with exponential notation.

$\frac{e^5}{3} = x$ Solve for x.

The exact solution is $\frac{e^5}{3}$. To four decimal places,

$x \approx 49.4711.$

Work Practice 7

Recall from Section 19.3 the formula $A = P\left(1 + \frac{r}{n}\right)^{nt}$ for compound interest, where n represents the number of compoundings per year. When interest is compounded continuously, the formula $A = Pe^{rt}$ is used, where r is the annual interest rate, and interest is compounded continuously for t years.

Practice 8

Find the amount owed at the end of 4 years if \$2400 is borrowed at a rate of 6% compounded continuously.

Example 8 Finding Final Loan Payment

Find the amount owed at the end of 5 years if \$1600 is loaned at a rate of 9% compounded continuously.

Solution: Use the formula $A = Pe^{rt}$, where

$P = \$1600$ (the amount of the loan)

$r = 9\% = 0.09$ (the rate of interest)

$t = 5$ (the 5-year duration of the loan)

$A = Pe^{rt}$

$= 1600e^{0.09(5)}$ Substitute in known values.

$= 1600e^{0.45}$

Now we can use a calculator to approximate the solution.

$A \approx 2509.30$

The total amount of money owed is \$2509.30.

Work Practice 8

Answers

6. a. 4 **b.** $\frac{1}{3}$ **7.** $\frac{e^8}{5} \approx 596.1916$

8. \$3051.00

Objective 5 Using the Change of Base Formula 

Calculators are handy tools for approximating natural and common logarithms. Unfortunately, some calculators cannot be used to approximate logarithms to bases other than e or 10—at least not directly. In such cases, we use the change of base formula.

Change of Base

If a, b, and c are positive real numbers and neither b nor c is 1, then

$$\log_b a = \frac{\log_c a}{\log_c b}$$

Example 9 Approximate $\log_5 3$ to four decimal places.

Solution: Use the change of base property to write $\log_5 3$ as a quotient of logarithms to base 10.

$$\log_5 3 = \frac{\log 3}{\log 5}$$ Use the change of base property. In the change of base property, we let $a = 3$, $b = 5$, and $c = 10$.

$$\approx \frac{0.4771213}{0.69897}$$ Approximate logarithms by calculator.

$$\approx 0.6826063$$ Simplify by calculator.

To four decimal places, $\log_5 3 \approx 0.6826$.

Work Practice 9

Practice 9

Approximate $\log_8 5$ to four decimal places.

Answer

9. 0.7740

✓Concept Check If a graphing calculator cannot directly evaluate logarithms to base 5, describe how you could use the graphing calculator to graph the function $f(x) = \log_5 x$.

✓Concept Check Answer

$$f(x) = \frac{\log x}{\log 5}$$

Vocabulary, Readiness & Video Check

Use the choices to fill in each blank.

1. The base of log 7 is ________.

a. e **b.** 7 **c.** 10 **d.** no answer

2. The base of ln 7 is ________.

a. e **b.** 7 **c.** 10 **d.** no answer

3. $\log_{10} 10^7 =$ ________.

a. e **b.** 7 **c.** 10 **d.** no answer

4. $\log_7 1 =$ ________.

a. e **b.** 7 **c.** 10 **d.** 0

5. $\log_e e^5 =$ ________.

a. e **b.** 5 **c.** 0 **d.** 1

6. Study exercise 5 to the left. Then answer: $\ln e^5 =$ ________.

a. e **b.** 5 **c.** 0 **d.** 1

7. $\log_2 7 =$ ________ (There may be more than one answer.)

a. $\frac{\log 7}{\log 2}$ **b.** $\frac{\ln 7}{\ln 2}$ **c.** $\frac{\log 2}{\log 7}$ **d.** $\log \frac{7}{2}$

Martin-Gay Interactive Videos *Watch the section lecture video and answer the following questions.*

See Video 19.7

Objective 1 **8.** From Example 1 and the lecture before, what is the understood base of a common logarithm?

Objective 2 **9.** From Example 2, why can you find exact values of common logarithms of powers of 10?

Objective 3 **10.** From Example 4 and the lecture before, what is the understood base of a natural logarithm?

Objective 4 **11.** In Examples 5 and 6, consider how the expression is rewritten and the resulting answer. What logarithm property is actually used here?

Objective 5 **12.** From Example 8, what two equivalent fractions will give you the exact value of $\log_6 4$?

19.7 Exercise Set MyMathLab®

MIXED PRACTICE

Use a calculator to approximate each logarithm to four decimal places. See Examples 1 and 5.

1. $\log 8$
2. $\log 6$
3. $\log 2.31$
4. $\log 4.86$
5. $\ln 2$
6. $\ln 3$
7. $\ln 0.0716$
8. $\ln 0.0032$
9. $\log 12.6$
10. $\log 25.9$
11. $\ln 5$
12. $\ln 7$
13. $\log 41.5$
14. $\ln 41.5$

MIXED PRACTICE

Find the exact value. See Examples 2 and 6.

15. $\log 100$
16. $\log 10{,}000$
17. $\log \frac{1}{1000}$
18. $\log \frac{1}{100}$
19. $\ln e^2$
20. $\ln e^4$
21. $\ln \sqrt[4]{e}$
22. $\ln \sqrt[5]{e}$
23. $\log 10^3$
24. $\log 10^7$
25. $\ln e^{-7}$
26. $\ln e^{-5}$
27. $\log 0.0001$
28. $\log 0.001$
29. $\ln \sqrt{e}$
30. $\log \sqrt{10}$

Solve each equation for x. Give an exact solution and a four-decimal-place approximation. See Examples 3 and 7.

31. $\ln 2x = 7$
32. $\ln 5x = 9$
33. $\log x = 1.3$
34. $\log x = 2.1$
35. $\log 2x = 1.1$
36. $\log 3x = 1.3$
37. $\ln x = 1.4$
38. $\ln x = 2.1$
39. $\ln(3x - 4) = 2.3$
40. $\ln(2x + 5) = 3.4$
41. $\log x = 2.3$
42. $\log x = 3.1$
43. $\ln x = -2.3$
44. $\ln x = -3.7$
45. $\log(2x + 1) = -0.5$
46. $\log(3x - 2) = -0.8$
47. $\ln 4x = 0.18$
48. $\ln 3x = 0.76$

Approximate each logarithm to four decimal places. See Example 9.

49. $\log_2 3$
50. $\log_3 2$
51. $\log_{1/2} 5$
52. $\log_{1/3} 2$
53. $\log_4 9$
54. $\log_9 4$
55. $\log_3 \frac{1}{6}$
56. $\log_6 \frac{2}{3}$
57. $\log_8 6$
58. $\log_6 8$

Use the formula $R = \log\left(\frac{a}{T}\right) + B$ to find the intensity R on the Richter scale of the earthquakes that fit the descriptions given. Round answers to one decimal place. See Example 4.

59. Amplitude a is 200 micrometers, time T between waves is 1.6 seconds, and B is 2.1.

60. Amplitude a is 150 micrometers, time T between waves is 3.6 seconds, and B is 1.9.

61. Amplitude a is 400 micrometers, time T between waves is 2.6 seconds, and B is 3.1.

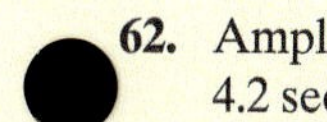

62. Amplitude a is 450 micrometers, time T between waves is 4.2 seconds, and B is 2.7.

Use the formula $A = Pe^{rt}$ to solve. See Example 8.

63. Find how much money Dana Jones has after 12 years if \$1400 is invested at 8% interest compounded continuously.

64. Determine the amount in an account in which \$3500 earns 6% interest compounded continuously for 1 year.

65. Find the amount of money Barbara Mack owes at the end of 4 years if 6% interest is compounded continuously on her \$2000 debt.

66. Find the amount of money for which a \$2500 certificate of deposit is redeemable if it has been paying 10% interest compounded continuously for 3 years.

REVIEW AND PREVIEW

Solve each equation for x. See Sections 9.3 and 13.6.

67. $6x - 3(2 - 5x) = 6$

68. $2x + 3 = 5 - 2(3x - 1)$

69. $2x + 3y = 6x$

70. $4x - 8y = 10x$

71. $x^2 + 7x = -6$

72. $x^2 + 4x = 12$

Solve each system of equations. See Section 11.2.

73. $\begin{cases} x + 2y = -4 \\ 3x - y = 9 \end{cases}$

74. $\begin{cases} 5x + y = 5 \\ -3x - 2y = -10 \end{cases}$

CONCEPT EXTENSIONS

75. Use a calculator to try to approximate log 0. Describe what happens and explain why.

76. Use a calculator to try to approximate ln 0. Describe what happens and explain why.

77. Without using a calculator, explain which of log 50 or ln 50 must be larger and why.

78. Without using a calculator, explain which of $\log 50^{-1}$ or $\ln 50^{-1}$ must be larger and why.

Graph each function by finding ordered pair solutions, plotting the solutions, and then drawing a smooth curve through the plotted points.

79. $f(x) = e^x$

80. $f(x) = e^{2x}$

81. $f(x) = e^{-3x}$

82. $f(x) = e^{-x}$

83. $f(x) = e^x + 2$

84. $f(x) = e^x - 3$

85. $f(x) = e^{x-1}$

86. $f(x) = e^{x+4}$

87. $f(x) = 3e^x$

88. $f(x) = -2e^x$

89. $f(x) = \ln x$

90. $f(x) = \log x$

91. $f(x) = -2 \log x$

92. $f(x) = 3 \ln x$

93. $f(x) = \log (x + 2)$

94. $f(x) = \log (x - 2)$

95. $f(x) = \ln x - 3$

96. $f(x) = \ln x + 3$

97. Graph $f(x) = e^x$ (Exercise 79), $f(x) = e^x + 2$ (Exercise 83), and $f(x) = e^x - 3$ (Exercise 84) on the same screen. Discuss any trends shown on the graphs.

98. Graph $f(x) = \ln x$ (Exercise 89), $f(x) = \ln x - 3$ (Exercise 95), and $f(x) = \ln x + 3$ (Exercise 96) on the same screen. Discuss any trends shown on the graphs.

19.8 Exponential and Logarithmic Equations and Problem Solving

Objectives

1 Solve Exponential Equations.

2 Solve Logarithmic Equations.

3 Solve Problems That Can Be Modeled by Exponential and Logarithmic Equations.

Objective 1 Solving Exponential Equations

In Section 19.3 we solved exponential equations such as $2^x = 16$ by writing 16 as a power of 2 and applying the uniqueness of b^x.

$2^x = 16$

$2^x = 2^4$ Write 16 as 2^4.

$x = 4$ Use the uniqueness of b^x.

Solving the equation in this manner is possible since 16 is a power of 2. If solving an equation such as $2^x =$ *a number,* where the number is not a power of 2, we use logarithms. For example, to solve an equation such as $3^x = 7$, we use the fact that $f(x) = \log_b x$ is a one-to-one function. Another way of stating this fact is as a property of equality.

Logarithm Property of Equality

Let a, b, and c be real numbers such that $\log_b a$ and $\log_b c$ are real numbers and b is not 1. Then

$\log_b a = \log_b c$ is equivalent to $a = c$

Practice 1

Solve: $5^x = 9$.

Example 1 Solve: $3^x = 7$.

Solution: To solve, we use the logarithm property of equality and take the logarithm of both sides. For this example, we use the common logarithm.

$$3^x = 7$$
$$\log 3^x = \log 7 \quad \text{Take the common logarithm of both sides.}$$
$$x \log 3 = \log 7 \quad \text{Apply the power property of logarithms.}$$
$$x = \frac{\log 7}{\log 3} \quad \text{Divide both sides by log 3.}$$

The exact solution is $\frac{\log 7}{\log 3}$. If a decimal approximation is preferred,

$$\frac{\log 7}{\log 3} \approx \frac{0.845098}{0.4771213} \approx 1.7712 \text{ to four decimal places.}$$

The solution is $\frac{\log 7}{\log 3}$, or *approximately* 1.7712.

Work Practice 1

Objective 2 Solving Logarithmic Equations

By applying the appropriate properties of logarithms, we can solve a broad variety of logarithmic equations.

Practice 2

Solve: $\log_2(x - 1) = 5$.

Example 2 Solve: $\log_4(x - 2) = 2$.

Solution: Notice that $x - 2$ must be positive, so x must be greater than 2. With this in mind, we first write the equation with exponential notation.

$$\log_4(x - 2) = 2$$
$$4^2 = x - 2$$
$$16 = x - 2$$
$$18 = x \quad \text{Add 2 to both sides.}$$

Check: To check, we replace x with 18 in the original equation.

$$\log_4(x - 2) = 2$$
$$\log_4(18 - 2) \stackrel{?}{=} 2 \quad \text{Let } x = 18.$$
$$\log_4 16 \stackrel{?}{=} 2$$
$$4^2 = 16 \quad \text{True}$$

The solution is 18.

Work Practice 2

Practice 3

Solve:
$\log_5 x + \log_5(x + 4) = 1$.

Example 3 Solve: $\log_2 x + \log_2(x - 1) = 1$.

Solution: Notice that $x - 1$ must be positive, so x must be greater than 1. We use the product property on the left side of the equation.

$$\log_2 x + \log_2(x - 1) = 1$$
$$\log_2 x(x - 1) = 1 \quad \text{Apply the product property.}$$
$$\log_2(x^2 - x) = 1$$

Answers

1. $\frac{\log 9}{\log 5} \approx 1.3652$ 2. 33 3. 1

Next we write the equation with exponential notation and solve for x.

$$2^1 = x^2 - x$$
$$0 = x^2 - x - 2 \quad \text{Subtract 2 from both sides.}$$
$$0 = (x - 2)(x + 1) \quad \text{Factor.}$$
$$0 = x - 2 \quad \text{or} \quad 0 = x + 1 \quad \text{Set each factor equal to 0.}$$
$$2 = x \qquad -1 = x$$

Recall that -1 cannot be a solution because x must be greater than 1. If we forgot this, we would still reject -1 after checking. To see this, we replace x with -1 in the original equation.

$$\log_2 x + \log_2(x - 1) = 1$$
$$\log_2(-1) + \log_2(-1 - 1) \stackrel{?}{=} 1 \quad \text{Let } x = -1.$$

Because the logarithm of a negative number is undefined, -1 is rejected. Check to see that the solution is 2.

Work Practice 3

Example 4 Solve: $\log(x + 2) - \log x = 2$.

We use the quotient property of logarithms on the left side of the equation.

Solution: $\log(x + 2) - \log x = 2$

$$\log\frac{x + 2}{x} = 2 \quad \text{Apply the quotient property.}$$
$$10^2 = \frac{x + 2}{x} \quad \text{Write using exponential notation.}$$
$$100 = \frac{x + 2}{x} \quad \text{Simplify.}$$
$$100x = x + 2 \quad \text{Multiply both sides by } x.$$
$$99x = 2 \quad \text{Subtract } x \text{ from both sides.}$$
$$x = \frac{2}{99} \quad \text{Divide both sides by 99.}$$

Verify that the solution is $\frac{2}{99}$.

Work Practice 4

Objective 3 Solving Problems Modeled by Exponential and Logarithmic Equations

Logarithmic and exponential functions are used in a variety of scientific, technical, and business settings. A few examples follow.

Example 5 Estimating Population Size

The population size y of a community of lemmings varies according to the relationship $y = y_0e^{0.15t}$. In this formula, t is time in months, and y_0 is the initial population at time 0. Estimate the population after 6 months if there were originally 5000 lemmings.

Solution: We substitute 5000 for y_0 and 6 for t.

$$y = y_0e^{0.15t}$$
$$= 5000e^{0.15(6)} \quad \text{Let } t = 6 \text{ and } y_0 = 5000.$$
$$= 5000e^{0.9} \quad \text{Multiply.}$$

(Continued on next page)

Practice 4

Solve: $\log(x + 3) - \log x = 1$.

Practice 5

The population size y of a group of rabbits varies according to the relationship $y = y_0e^{0.916t}$. In this formula, t is time in years and y_0 is the initial population at time $t = 0$. Estimate the population in three years if there were originally 60 rabbits.

Answers

4. $\frac{1}{3}$ 5. 937 rabbits

Using a calculator, we find that $y \approx 12{,}298.016$. In 6 months, the population will be approximately 12,300 lemmings.

Work Practice 5

Practice 6

How long does it take for an investment of \$3000 to double if it is invested at 7% interest compounded monthly? Round to the nearest year.

Answer

6. 10 yr

Example 6 Doubling an Investment

How long does it take an investment of \$2000 to double if it is invested at 5% interest compounded quarterly? The necessary formula is $A = P\left(1 + \frac{r}{n}\right)^{nt}$, where A is the accrued (or owed) amount, P is the principal invested, r is the annual rate of interest, n is the number of compounding periods per year, and t is the number of years.

Solution: We are given that $P = \$2000$ and $r = 5\% = 0.05$. Compounding quarterly means 4 times a year, so $n = 4$. The investment is to double, so A must be \$4000. Substitute these values and solve for t.

$$A = P\left(1 + \frac{r}{n}\right)^{nt}$$

$$4000 = 2000\left(1 + \frac{0.05}{4}\right)^{4t} \quad \text{Substitute in known values.}$$

$$4000 = 2000(1.0125)^{4t} \quad \text{Simplify } 1 + \frac{0.05}{4}.$$

$$2 = (1.0125)^{4t} \quad \text{Divide both sides by 2000.}$$

$$\log 2 = \log 1.0125^{4t} \quad \text{Take the logarithm of both sides.}$$

$$\log 2 = 4t(\log 1.0125) \quad \text{Apply the power property.}$$

$$\frac{\log 2}{4 \log 1.0125} = t \quad \text{Divide both sides by } 4 \log 1.0125.$$

$$13.949408 \approx t \quad \text{Approximate by calculator.}$$

Thus, it takes nearly 14 years for the money to double in value.

Work Practice 6

Graphing Calculator Explorations

Use a graphing calculator to find how long it takes an investment of \$1500 to triple if it is invested at 8% interest compounded monthly.

First, let $P = \$1500$, $r = 0.08$, and $n = 12$ (for 12 months) in the formula

$$A = P\left(1 + \frac{r}{n}\right)^{nt}$$

Notice that when the investment has tripled, the accrued amount A is \$4500. Thus,

$$4500 = 1500\left(1 + \frac{0.08}{12}\right)^{12t}$$

Determine an appropriate viewing window and enter and graph the equations

$$Y_1 = 1500\left(1 + \frac{0.08}{12}\right)^{12x}$$

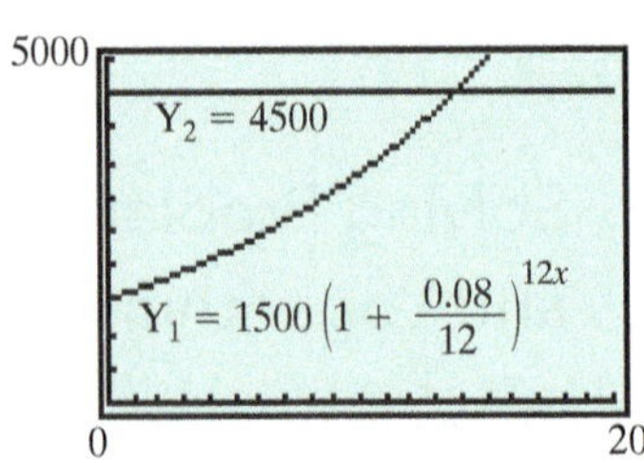

and

$$Y_2 = 4500$$

The point of intersection of the two curves is the solution. The x-coordinate tells how long it takes for the investment to triple.

Use a TRACE feature or an INTERSECT feature to approximate the coordinates of the point of intersection of the two curves. It takes approximately 13.78 years, or 13 years and 9 months, for the investment to triple in value to \$4500.

Use this graphical solution method to solve each problem. Round each answer to the nearest hundredth.

1. Find how long it takes an investment of \$5000 to grow to \$6000 if it is invested at 5% interest compounded quarterly.
2. Find how long it takes an investment of \$1000 to double if it is invested at 4.5% interest compounded daily. (Use 365 days in a year.)
3. Find how long it takes an investment of \$10,000 to quadruple if it is invested at 6% interest compounded monthly.
4. Find how long it takes \$500 to grow to \$800 if it is invested at 4% interest compounded semiannually.

Vocabulary, Readiness & Video Check

Martin-Gay Interactive Videos

See Video 19.8 

Watch the section lecture video and answer the following questions.

Objective 1 1. From the lecture before Example 1, explain why $\ln(4x - 2) = \ln 3$ is equivalent to $4x - 2 = 3$.

Objective 2 2. Why is the possible solution of -8 rejected in Example 3?

Objective 3 3. For Example 4, write the equation and find the number of years it takes \$1000 to double at 7% interest compounded monthly. Explain the similarity to the answer to Example 4. Round your answer to the nearest tenth.

19.8 Exercise Set MyMathLab®

Solve each equation. Give an exact solution and approximate the solution to four decimal places. See Example 1.

1. $3^x = 6$
2. $4^x = 7$
3. $3^{2x} = 3.8$
4. $5^{3x} = 5.6$
5. $2^{x-3} = 5$
6. $8^{x-2} = 12$
7. $9^x = 5$
8. $3^x = 11$
9. $4^{x+7} = 3$
10. $6^{x+3} = 2$

MIXED PRACTICE

Solve each equation. See Examples 1 through 4.

11. $\log_2(x + 5) = 4$
12. $\log_2(x - 5) = 3$
13. $\log_4 2 + \log_4 x = 0$
14. $\log_3 5 + \log_3 x = 1$
15. $\log_2 6 - \log_2 x = 3$
16. $\log_4 10 - \log_4 x = 2$
17. $\log_6(x^2 - x) = 1$
18. $\log_2(x^2 + x) = 1$
19. $\log_4 x + \log_4(x + 6) = 2$
20. $\log_3 x + \log_3(x + 6) = 3$
21. $\log_5(x + 3) - \log_5 x = 2$
22. $\log_6(x + 2) - \log_6 x = 2$
23. $7^{3x-4} = 11$
24. $5^{2x-6} = 12$
25. $\log_4(x^2 - 3x) = 1$
26. $\log_8(x^2 - 2x) = 1$

27. $e^{6x} = 5$

28. $e^{2x} = 8$

29. $\log_3 x^2 = 4$

30. $\log_2 x^2 = 6$

31. $\ln 5 + \ln x = 0$

32. $\ln 3 + \ln(x - 1) = 0$

33. $3 \log x - \log x^2 = 2$

34. $2 \log x - \log x = 3$

35. $\log_4 x - \log_4(2x - 3) = 3$

36. $\log_2 x - \log_2(3x + 5) = 4$

37. $\log_2 x + \log_2(3x + 1) = 1$

38. $\log_3 x + \log_3(x - 8) = 2$

39. $\log_2 x + \log_2(x + 5) = 1$

40. $\log_4 x + \log_4(x + 7) = 1$

Solve. See Example 5.

41. The size of the wolf population at Isle Royale National Park increases according to the formula $y = y_0 e^{0.043t}$. In this formula, t is time in years and y_0 is the initial population at time 0. If the size of the current population is 83 wolves, find how many there should be in 5 years. Round to the nearest whole number.

42. The number of victims of a flu epidemic is increasing according to the formula $y = y_0 e^{0.075t}$. In this formula, t is time in weeks and y_0 is the given population at time 0. If 20,000 people are currently infected, how many might be infected in 3 weeks? Round to the nearest whole number.

43. The population of the Cook Islands is decreasing according to the formula $y = y_0 e^{-0.0277t}$. In this formula, t is the time in years and y_0 is the initial population at time 0. If the size of the population in 2010 was 11,488, use the formula to predict the population of the Cook Islands in the year 2025. Round to the nearest whole number. (*Source: The World Almanac*)

44. The population of Saint Barthelemy is decreasing according to the formula $y = y_0 e^{-0.0034t}$. In this formula, t is the time in years and y_0 is the initial population at time 0. If the size of the population in 2010 was 6852, use the formula to predict the population of Saint Barthelemy in the year 2025. Round to the nearest whole number. (*Source: The World Almanac*)

Use the formula $A = P\left(1 + \frac{r}{n}\right)^{nt}$ *to solve these compound interest problems. Round to the nearest tenth. See Example 6.*

45. Find how long it takes \$600 to double if it is invested at 7% interest compounded monthly.

46. Find how long it takes \$600 to double if it is invested at 12% interest compounded monthly.

47. Find how long it takes a \$1200 investment to earn \$200 interest if it is invested at 9% interest compounded quarterly.

48. Find how long it takes a \$1500 investment to earn \$200 interest if it is invested at 10% compounded semiannually.

49. Find how long it takes \$1000 to double if it is invested at 8% interest compounded semiannually.

50. Find how long it takes \$1000 to double if it is invested at 8% interest compounded monthly.

The formula $w = 0.00185h^{2.67}$ *is used to estimate the normal weight w of a boy h inches tall. Use this formula to solve the height-weight problems. Round to the nearest tenth.*

51. Find the expected weight of a boy who is 35 inches tall.

52. Find the expected weight of a boy who is 43 inches tall.

53. Find the expected height of a boy who weighs 85 pounds.

54. Find the expected height of a boy who weighs 140 pounds.

The formula $P = 14.7e^{-0.21x}$ *gives the average atmospheric pressure P, in pounds per square inch, at an altitude x, in miles above sea level. Use this formula to solve these pressure problems. Round answers to the nearest tenth.*

55. Find the average atmospheric pressure of Denver, which is 1 mile above sea level.

56. Find the average atmospheric pressure of Pikes Peak, which is 2.7 miles above sea level.

57. Find the elevation of a Delta jet if the atmospheric pressure outside the jet is 7.5 lb/sq in.

58. Find the elevation of a remote Himalayan peak if the atmospheric pressure atop the peak is 6.5 lb/sq in.

Psychologists call the graph of the formula $t = \frac{1}{c}\ln\left(\frac{A}{A - N}\right)$ *the learning curve, since the formula relates time t passed, in weeks, to a measure N of learning achieved, to a measure A of maximum learning possible, and to a measure c of an individual's learning style. Round to the nearest week.*

59. Norman is learning to type. If he wants to type at a rate of 50 words per minute (N is 50) and his expected maximum rate is 75 words per minute (A is 75), find how many weeks it should take him to achieve his goal. Assume that c is 0.09.

60. An experiment with teaching chimpanzees sign language shows that a typical chimp can master a maximum of 65 signs. Find how many weeks it should take a chimpanzee to master 30 signs if c is 0.03.

61. Janine is working on her dictation skills. She wants to take dictation at a rate of 150 words per minute and believes that the maximum rate she can hope for is 210 words per minute. Find how many weeks it should take her to achieve the 150-words-per-minute level if c is 0.07.

62. A psychologist is measuring human capability to memorize nonsense syllables. Find how many weeks it should take a subject to learn 15 nonsense syllables if the maximum possible to learn is 24 syllables and c is 0.17.

REVIEW AND PREVIEW

If $x = -2$, $y = 0$, and $z = 3$, find the value of each expression. See Sections 4.3 and 12.3.

63. $\dfrac{x^2 - y + 2z}{3x}$

64. $\dfrac{x^3 - 2y + z}{2z}$

65. $\dfrac{3z - 4x + y}{x + 2z}$

66. $\dfrac{4y - 3x + z}{2x + y}$

Find the inverse function of each one-to-one function. See Section 19.2.

67. $f(x) = 5x + 2$

68. $f(x) = \dfrac{x - 3}{4}$

CONCEPT EXTENSIONS

The formula $y = y_0e^{kt}$ gives the population size y of a population that experiences an annual rate of population growth k (given as a decimal). In this formula, t is time in years and y_0 is the initial population at time 0. Use this formula to solve Exercises 69 and 70.

69. In 2010, the population of Michigan was approximately 9,939,000 and decreasing according to the formula $y = y_0e^{-0.003t}$. Assume that the population continues to decrease according to the given formula and predict how many years after which the population of Michigan will be 9,500,000. (*Hint:* Let $y_0 = 9{,}939{,}000$ and $y = 9{,}500{,}000$; solve for t.) (*Source:* U.S. Bureau of the Census)

70. In 2010, the population of Illinois was approximately 12,830,000 and increasing according to the formula $y = y_0e^{0.005t}$. Assume that the population continues to increase according to the given formula and predict how many years after which the population of Illinois will be 13,500,000. (See the Hint for Exercise 69.) (*Source:* U.S. Bureau of the Census)

71. When solving a logarithmic equation, explain why you must check possible solutions in the original equation.

72. Solve $5^x = 9$ by taking the common logarithm of both sides of the equation. Next, solve this equation by taking the natural logarithm of both sides. Compare your solutions. Are they the same? Why or why not?

Use a graphing calculator to solve each equation. For example, to solve Exercise 73, let $Y_1 = e^{0.3x}$ and $Y_2 = 8$ and graph the equations. The x-value of the point of intersection is the solution. Round all solutions to two decimal places.

73. $e^{0.3x} = 8$

74. $10^{0.5x} = 7$

75. $2\log(-5.6x + 1.3) + x + 1 = 0$

76. $\ln(1.3x - 2.1) + 3.5x - 5 = 0$

77. Check Exercise 23.

78. Check Exercise 24.

79. Check Exercise 31.

80. Check Exercise 32.

Chapter 19 Vocabulary Check

Fill in each blank with one of the words or phrases listed below. Some words or phrases may be used more than once.

inverse	common	composition	symmetric	exponential
vertical	logarithmic	natural	half-life	horizontal

1. For a one-to-one function, we can find its ________ function by switching the coordinates of the ordered pairs of the function.

2. The ________ of functions f and g is $(f \circ g)(x) = f(g(x))$.

3. A function of the form $f(x) = b^x$ is called a(n) ________ function if $b > 0$, b is not 1, and x is a real number.

4. The graphs of f and f^{-1} are ________ about the line $y = x$.

5. ________ logarithms are logarithms to base e.

6. ________ logarithms are logarithms to base 10.

7. To see whether a graph is the graph of a one-to-one function, apply the ________ line test to see whether it is a function and then apply the ________ line test to see whether it is a one-to-one function.

8. A(n) ________ function is a function that can be defined by $f(x) = \log_b x$, where x is a positive real number, b is a constant positive real number, and b is not 1.

9. ________ is the amount of time it takes for half of the amount of a substance to decay.

10. A quantity that grows or decays by the same percent at regular time periods is said to have ________ growth or decay.

Helpful Hint Are you preparing for your test? Don't forget to take the Chapter 19 Test on page 1328. Then check your answers at the back of the text and use the Chapter Test Prep Videos to see the fully worked-out solutions to any of the exercises you want to review.

19 Chapter Highlights

Definitions and Concepts	Examples
Section 19.1 The Algebra of Functions; Composite Functions	
Algebra of Functions Sum $(f+g)(x) = f(x) + g(x)$ Difference $(f-g)(x) = f(x) - g(x)$ Product $(f \cdot g)(x) = f(x) \cdot g(x)$ Quotient $\left(\frac{f}{g}\right)(x) = \frac{f(x)}{g(x)}, g(x) \neq 0$	If $f(x) = 7x$ and $g(x) = x^2 + 1$, $(f+g)(x) = f(x) + g(x) = 7x + x^2 + 1$ $(f-g)(x) = f(x) - g(x) = 7x - (x^2+1)$ $= 7x - x^2 - 1$ $(f \cdot g)(x) = f(x) \cdot g(x) = 7x(x^2+1)$ $= 7x^3 + 7x$ $\left(\frac{f}{g}\right)(x) = \frac{f(x)}{g(x)} = \frac{7x}{x^2+1}$
Composite Functions The notation $(f \circ g)(x)$ means "f composed with g." $(f \circ g)(x) = f(g(x))$ $(g \circ f)(x) = g(f(x))$	If $f(x) = x^2 + 1$ and $g(x) = x - 5$, find $(f \circ g)(x)$. $(f \circ g)(x) = f(g(x))$ $= f(x-5)$ $= (x-5)^2 + 1$ $= x^2 - 10x + 26$
Section 19.2 Inverse Functions	
If f is a function, then f is a **one-to-one function** only if each y-value (output) corresponds to only one x-value (input). **Horizontal Line Test** If every horizontal line intersects the graph of a function at most once, then the function is a one-to-one function.	Determine whether each graph is a one-to-one function. **A** **B** **C** Graphs **A** and **C** pass the vertical line test, so only these are graphs of functions. Of graphs **A** and **C**, only graph **A** passes the horizontal line test, so only graph **A** is the graph of a one-to-one function.
The **inverse** of a one-to-one function f is the one-to-one function f^{-1} that is the set of all ordered pairs (b, a) such that (a, b) belongs to f. **To Find the Inverse of a One-to-One Function $f(x)$** **Step 1:** Replace $f(x)$ with y. **Step 2:** Interchange x and y. **Step 3:** Solve for y. **Step 4:** Replace y with $f^{-1}(x)$.	Find the inverse of $f(x) = 2x + 7$. $y = 2x + 7$ Replace $f(x)$ with y. $x = 2y + 7$ Interchange x and y. $2y = x - 7$ Solve for y. $y = \frac{x-7}{2}$ $f^{-1}(x) = \frac{x-7}{2}$ Replace y with $f^{-1}(x)$. The inverse of $f(x) = 2x + 7$ is $f^{-1}(x) = \frac{x-7}{2}$.

Definitions and Concepts	Examples

Section 19.3 Exponential Functions

A function of the form $f(x) = b^x$ is an **exponential function,** where $b > 0$, $b \neq 1$, and x is a real number.

Graph the exponential function $y = 4^x$.

x	y
-2	$\frac{1}{16}$
-1	$\frac{1}{4}$
0	1
1	4
2	16

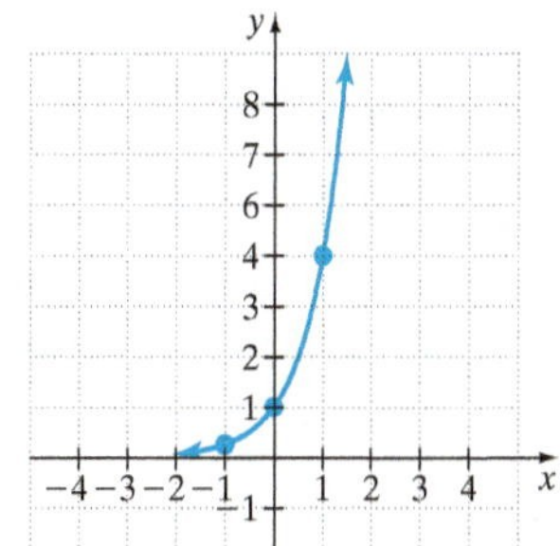

Uniqueness of b^x

If $b > 0$ and $b \neq 1$, then $b^x = b^y$ is equivalent to $x = y$.

Solve $2^{x+5} = 8$.

$2^{x+5} = 2^3$ Write 8 as 2^3.

$x + 5 = 3$ Use the uniqueness of b^x.

$x = -2$ Subtract 5 from both sides.

Section 19.4 Exponential Growth and Decay Functions

A quantity that grows or decays by the same percent at regular time periods is said to have **exponential growth** or **exponential decay.**

Exponential Growth

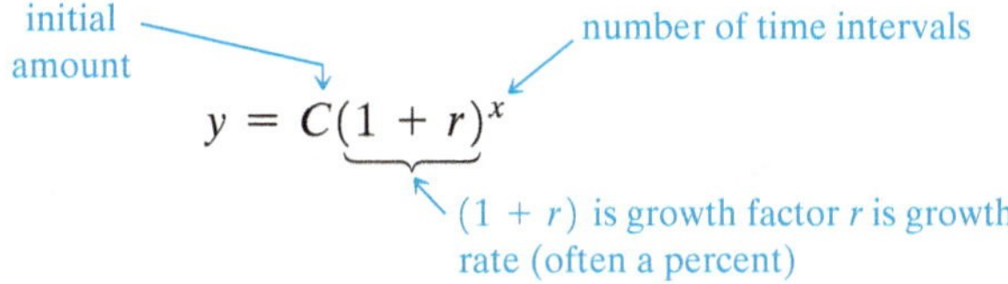

$(1 + r)$ is growth factor r is growth rate (often a percent)

A city has a current population of 37,000 that has been increasing at a rate of 3% per year. At this rate, find the city's population in 20 years.

$$y = C(1 + r)^x$$
$$y = 37{,}000(1 + 0.03)^{20}$$
$$y \approx 66{,}826.12$$

In 20 years, the predicted population of the city is 66,826.

Exponential Decay

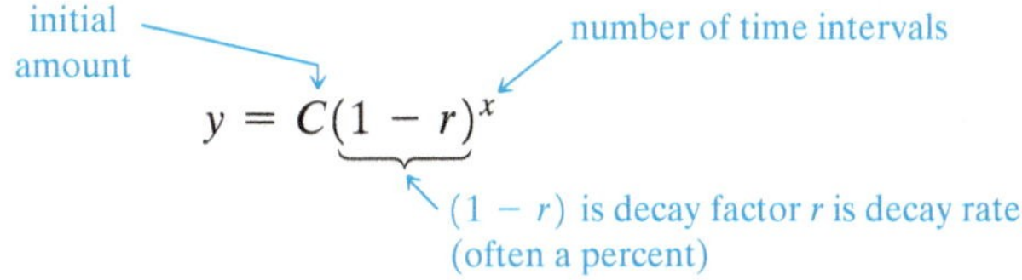

$(1 - r)$ is decay factor r is decay rate (often a percent)

A city has a current population of 37,000 that has been decreasing at a rate of 3% per year. At this rate, find the city's population in 20 years.

$$y = C(1 - r)^x$$
$$y = 37{,}000(1 - 0.03)^{20}$$
$$y \approx 20{,}120.39$$

In 20 years, the predicted population of the city is 20,120.

Section 19.5 Logarithmic Functions

Logarithmic Definition

If $b > 0$ and $b \neq 1$, then

$$y = \log_b x \quad \text{means} \quad x = b^y$$

for any positive number x and real number y.

Logarithmic Form	Corresponding Exponential Statement
$\log_5 25 = 2$	$5^2 = 25$
$\log_9 3 = \frac{1}{2}$	$9^{1/2} = 3$

Properties of Logarithms

If b is a real number, $b > 0$, and $b \neq 1$, then

$$\log_b 1 = 0, \quad \log_b b^x = x, \quad b^{\log_b x} = x$$

$$\log_5 1 = 0, \quad \log_7 7^2 = 2, \quad 3^{\log_3 6} = 6$$

(continued)

Definitions and Concepts	Examples

Section 19.5 Logarithmic Functions (*continued*)

Logarithmic Function

If $b > 0$ and $b \neq 1$, then a **logarithmic function** is a function that can be defined as

$$f(x) = \log_b x$$

The domain of f is the set of positive real numbers, and the range of f is the set of real numbers.

Graph $y = \log_3 x$.

Write $y = \log_3 x$ as $3^y = x$. Plot the ordered pair solutions listed in the table and connect them with a smooth curve.

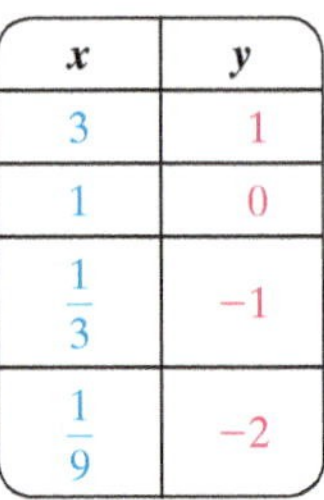

x	y
3	1
1	0
$\frac{1}{3}$	−1
$\frac{1}{9}$	−2

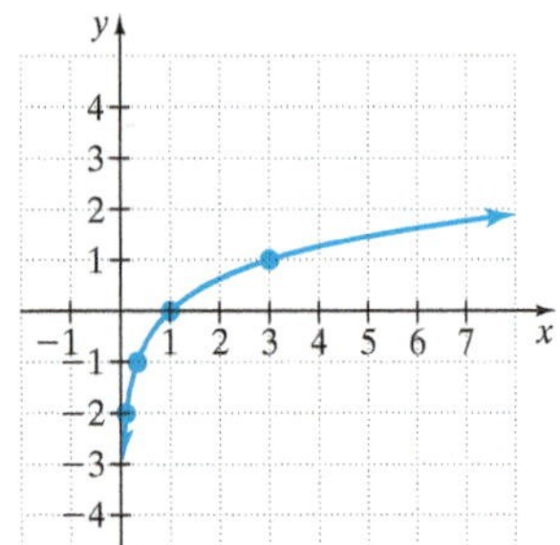

Section 19.6 Properties of Logarithms

Let x, y, and b be positive numbers and $b \neq 1$.

Product Property

$$\log_b xy = \log_b x + \log_b y$$

Quotient Property

$$\log_b \frac{x}{y} = \log_b x - \log_b y$$

Power Property

$$\log_b x^r = r \log_b x$$

Write as a single logarithm.

$$2\log_5 6 + \log_5 x - \log_5(y+2)$$

$= \log_5 6^2 + \log_5 x - \log_5(y+2)$ Power property

$= \log_5 36 \cdot x - \log_5(y+2)$ Product property

$= \log_5 \frac{36x}{y+2}$ Quotient property

Section 19.7 Common Logarithms, Natural Logarithms, and Change of Base

Common Logarithms

$\log x$ means $\log_{10} x$

Natural Logarithms

$\ln x$ means $\log_e x$

Continuously Compounded Interest Formula

$$A = Pe^{rt}$$

where r is the annual interest rate for P dollars invested for t years.

$\log 5 = \log_{10} 5 \approx 0.69897$

$\ln 7 = \log_e 7 \approx 1.94591$

Find the amount in an account at the end of 3 years if \$1000 is invested at an interest rate of 4% compounded continuously.

Here, $t = 3$ years, $P = \$1000$, and $r = 0.04$.

$$A = Pe^{rt}$$
$$= 1000e^{0.04(3)}$$
$$\approx \$1127.50$$

Section 19.8 Exponential and Logarithmic Equations and Problem Solving

Logarithm Property of Equality

Let $\log_b a$ and $\log_b c$ be real numbers and $b \neq 1$. Then

$\log_b a = \log_b c$ is equivalent to $a = c$

Solve $2^x = 5$.

$\log 2^x = \log 5$ Logarithm property of equality

$x \log 2 = \log 5$ Power property

$x = \frac{\log 5}{\log 2}$ Divide both sides by log 2.

$x \approx 2.3219$ Use a calculator.

Chapter 19 Review

(19.1) If $f(x) = x - 5$ and $g(x) = 2x + 1$, find

1. $(f + g)(x)$

2. $(f - g)(x)$

3. $(f \cdot g)(x)$

4. $\left(\frac{g}{f}\right)(x)$

If $f(x) = x^2 - 2$, $g(x) = x + 1$, and $h(x) = x^3 - x^2$, find each composition.

5. $(f \circ g)(x)$

6. $(g \circ f)(x)$

7. $(h \circ g)(2)$

8. $(f \circ f)(x)$

9. $(f \circ g)(-1)$

10. $(h \circ h)(2)$

(19.2) Determine whether each function is a one-to-one function. If it is one-to-one, list the elements of its inverse.

11. $h = \{(-9, 14), (6, 8), (-11, 12), (15, 15)\}$

12. $f = \{(-5, 5), (0, 4), (13, 5), (11, -6)\}$

13.

U.S. Region (Input)	Northeast	Midwest	South	West
Rank in Housing Starts for 2009 (Output)	4	3	1	2

△ **14.**

Shape (Input)	Square	Triangle	Parallelogram	Rectangle
Number of Sides (Output)	4	3	4	4

Given that $f(x) = \sqrt{x + 2}$ is a one-to-one function, find the following.

15. a. $f(7)$

b. $f^{-1}(3)$

16. a. $f(-1)$

b. $f^{-1}(1)$

Determine whether each function is a one-to-one function.

17.

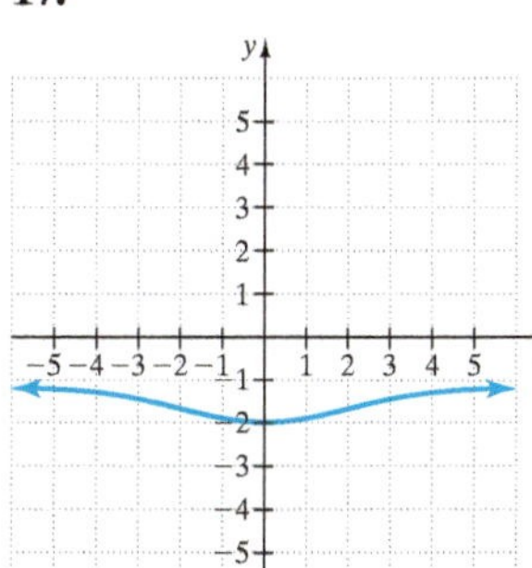

18.

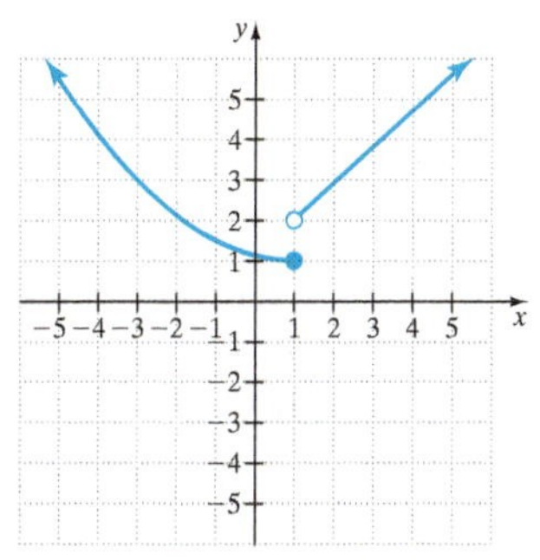

19.

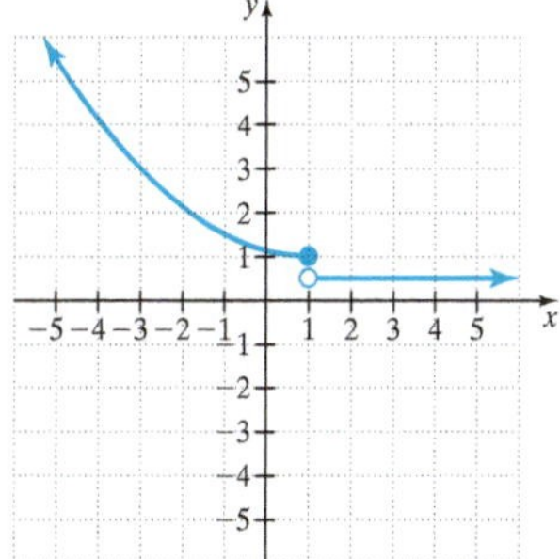

20.

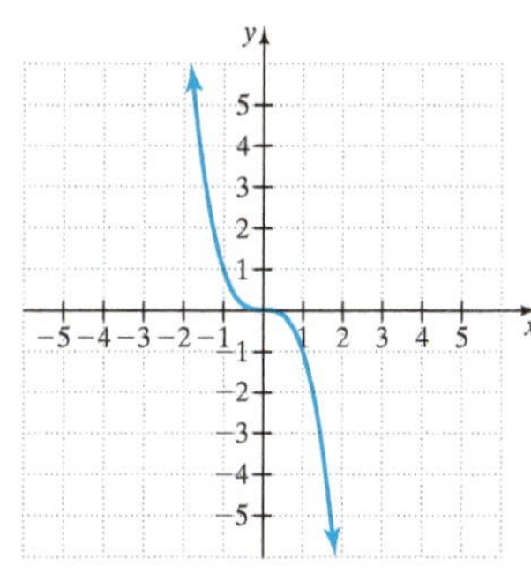

Find an equation defining the inverse function of the given one-to-one function.

21. $f(x) = x - 9$

22. $f(x) = x + 8$

23. $f(x) = 6x + 11$

24. $f(x) = 12x - 1$

25. $f(x) = x^3 - 5$

26. $f(x) = \sqrt[3]{x + 2}$

27. $g(x) = \frac{12x - 7}{6}$

28. $r(x) = \frac{13x - 5}{2}$

Graph each one-to-one function and its inverse on the same set of axes.

29. $f(x) = -2x + 3$

30. $f(x) = 5x - 5$

(19.3) Solve each equation for x.

31. $4^x = 64$

32. $3^x = \frac{1}{9}$

33. $2^{3x} = \frac{1}{16}$

34. $5^{2x} = 125$

35. $9^{x+1} = 243$

36. $8^{3x-2} = 4$

Graph each exponential function.

37. $y = 3^x$

38. $y = \left(\frac{1}{3}\right)^x$

39. $y = 2^{x-4}$

40. $y = 2^x + 4$

Use the formula $A = P\left(1 + \frac{r}{n}\right)^{nt}$ to solve the interest problems. In this formula,

A = amount accrued (or owed)

P = principal invested (or loaned)

r = rate of interest

n = number of compounding periods per year

t = time in years

41. Find the amount accrued if \$1600 is invested at 9% interest compounded semiannually for 7 years.

42. A total of \$800 is invested in a 7% certificate of deposit for which interest is compounded quarterly. Find the value that this certificate will have at the end of 5 years.

(19.4) *Solve. Round each answer to the nearest whole.*

43. The city of Henderson, Nevada, has been growing at a rate of 4.4% per year since the year 2000. If the population of Henderson was 79,087 in 2000 and this rate continues, predict the city's population in 2020.

44. The city of Raleigh, North Carolina, has been growing at a rate of 4.2% per year since the year 2000. If the population of Raleigh was 287,370 in 2000 and this rate continues, predict the city's population in 2018.

45. A summer camp tournament starts with 1024 players. After each round, half the players are eliminated. How many players remain after 7 rounds?

46. The bear population in a certain national park is decreasing by 11% each year. If this rate continues, and there is currently an estimated bear population of 1280, find the bear population in 6 years.

(19.5) *Write each equation with logarithmic notation.*

47. $49 = 7^2$

48. $2^{-4} = \frac{1}{16}$

Write each logarithmic equation with exponential notation.

49. $\log_{1/2} 16 = -4$

50. $\log_{0.4} 0.064 = 3$

Solve for x.

51. $\log_4 x = -3$

52. $\log_3 x = 2$

53. $\log_3 1 = x$

54. $\log_4 64 = x$

55. $\log_4 4^5 = x$

56. $\log_7 7^{-2} = x$

57. $5^{\log_5 4} = x$

58. $2^{\log_2 9} = x$

59. $\log_2(3x - 1) = 4$

60. $\log_3(2x + 5) = 2$

61. $\log_4(x^2 - 3x) = 1$

62. $\log_8(x^2 + 7x) = 1$

Graph each pair of equations on the same coordinate system.

63. $y = 2^x$ and $y = \log_2 x$

64. $y = \left(\frac{1}{2}\right)^x$ and $y = \log_{1/2} x$

(19.6) *Write each of the following as single logarithms.*

65. $\log_3 8 + \log_3 4$

66. $\log_2 6 + \log_2 3$

67. $\log_7 15 - \log_7 20$

68. $\log 18 - \log 12$

69. $\log_{11} 8 + \log_{11} 3 - \log_{11} 6$

70. $\log_5 14 + \log_5 3 - \log_5 21$

71. $2 \log_5 x - 2 \log_5(x + 1) + \log_5 x$

72. $4 \log_3 x - \log_3 x + \log_3(x + 2)$

Use properties of logarithms to write each expression as a sum or difference of multiples of logarithms.

73. $\log_3 \frac{x^3}{x + 2}$

74. $\log_4 \frac{x + 5}{x^2}$

75. $\log_2 \frac{3x^2y}{z}$

76. $\log_7 \frac{yz^3}{x}$

If $\log_b 2 = 0.36$ *and* $\log_b 5 = 0.83$, *find the following.*

77. $\log_b 50$

78. $\log_b \frac{4}{5}$

(19.7) *Use a calculator to approximate the logarithm to four decimal places.*

79. $\log 3.6$

80. $\log 0.15$

81. $\ln 1.25$

82. $\ln 4.63$

Find the exact value.

83. $\log 1000$

84. $\log \frac{1}{10}$

85. $\ln \frac{1}{e}$

86. $\ln e^4$

Solve each equation for x.

87. $\ln(2x) = 2$

88. $\ln(3x) = 1.6$

89. $\ln(2x - 3) = -1$

90. $\ln(3x + 1) = 2$

Use the formula $\ln \frac{I}{I_0} = -kx$ *to solve the radiation problems in Exercises 91 and 92. In this formula,*

x = depth in millimeters
I = intensity of radiation
I_0 = initial intensity
k = a constant measure dependent on the material

Round answers to two decimal places.

91. Find the depth at which the intensity of the radiation passing through a lead shield is reduced to 3% of the original intensity if the value of k is 2.1.

92. If k is 3.2, find the depth at which 2% of the original radiation will penetrate.

Approximate the logarithm to four decimal places.

93. $\log_5 1.6$

94. $\log_3 4$

Use the formula $A = Pe^{rt}$ *to solve the interest problems in which interest is compounded continuously. In this formula,*

A = amount accrued (or owed)
P = principal invested (or loaned)
r = rate of interest
t = time in years

95. Bank of New York offers a 5-year, 3% continuously compounded investment option. Find the amount accrued if \$1450 is invested.

96. Find the amount to which a \$940 investment grows if it is invested at 4% compounded continuously for 3 years.

(19.8) Solve each exponential equation for x. Give an exact solution and approximate the solution to four decimal places.

97. $3^{2x} = 7$

98. $6^{3x} = 5$

99. $3^{2x+1} = 6$

100. $4^{3x+2} = 9$

101. $5^{3x-5} = 4$

102. $8^{4x-2} = 3$

103. $5^{x-1} = \frac{1}{2}$

104. $4^{x+5} = \frac{2}{3}$

Solve the equation for x.

105. $\log_5 2 + \log_5 x = 2$

106. $\log_3 x + \log_3 10 = 2$

107. $\log(5x) - \log(x + 1) = 4$

108. $-\log_6(4x + 7) + \log_6 x = 1$

109. $\log_2 x + \log_2 2x - 3 = 1$

110. $\log_3(x^2 - 8x) = 2$

Use the formula $y = y_0e^{kt}$ to solve the population growth problems. In this formula,

y = size of population

y_0 = initial count of population

k = rate of growth written as a decimal

t = time

Round each answer to the nearest tenth.

111. In 1987, the population of California condors was only 27 birds. They were all brought in from the wild and an intensive breeding program was instituted. If we assume a yearly growth rate of 11.4%, how long did it take the condor population to reach 347 California condors? (*Source:* California Department of Fish and Game)

112. France is experiencing an annual growth rate of 0.4%. In 2010, the population of France was approximately 65,822,000. How long will it take for the population to reach 70,000,000? Round to the nearest tenth. (*Source:* Population Reference Bureau)

113. In 2010, the population of Australia was approximately 22,600,000. How long will it take Australia to double its population if its growth rate is 0.7% annually? Round to the nearest tenth. (*Source:* Population Reference Bureau)

114. Israel's population is increasing in size at a rate of 1.6% per year. How long will it take for its population of 7,746,400 to double in size? Round to the nearest tenth. (*Source:* Population Reference Bureau)

Use the compound interest equation $A = P\left(1 + \frac{r}{n}\right)^{nt}$ to solve the following. (See the directions for Exercises 41 and 42 for an explanation of this formula.) Round answers to the nearest tenth.

115. Find how long it will take a $5000 investment to grow to $10,000 if it is invested at 8% interest compounded quarterly.

116. An investment of $6000 grew to $10,000 while the money was invested at 6% interest compounded monthly. Find how long it was invested.

Use a graphing calculator to solve each equation. Round all solutions to two decimal places.

117. $e^x = 2$

118. $10^{0.3x} = 7$

MIXED REVIEW

Solve each equation.

119. $3^x = \frac{1}{81}$

120. $7^{4x} = 49$

121. $8^{3x-2} = 32$

122. $9^{x-2} = 27$

123. $\log_4 4 = x$

124. $\log_3 x = 4$

125. $\log_5(x^2 - 4x) = 1$

126. $\log_4(3x - 1) = 2$

127. $\ln x = -3.2$

128. $\log_5 x + \log_5 10 = 2$

129. $\ln x - \ln 2 = 1$

130. $\log_6 x - \log_6(4x + 7) = 1$

Chapter 19 Test

Step-by-step test solutions are found on the Chapter Test Prep Videos. Where available: MyMathLab® or YouTube

If $f(x) = x$ *and* $g(x) = 2x - 3$, *find the following.*

1. $(f \cdot g)(x)$

2. $(f - g)(x)$

If $f(x) = x$, $g(x) = x - 7$, *and* $h(x) = x^2 - 6x + 5$, *find the following.*

3. $(f \circ h)(0)$

4. $(g \circ f)(x)$

5. $(g \circ h)(x)$

On the same set of axes, graph the given one-to-one function and its inverse.

6. $f(x) = 7x - 14$

Determine whether the given graph is the graph of a one-to-one function.

7. **8.**

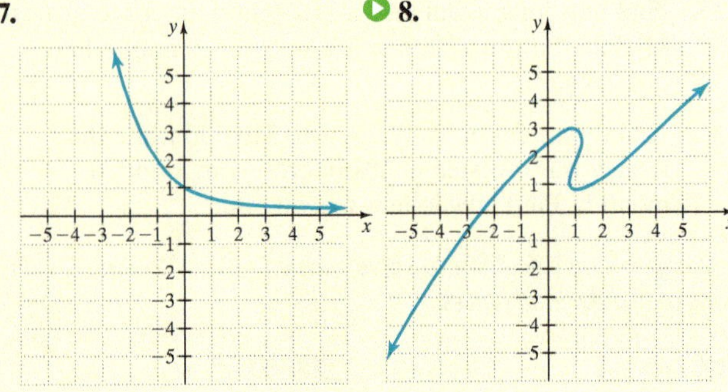

Determine whether each function is one-to-one. If it is one-to-one, find an equation or a set of ordered pairs that defines the inverse function of the given function.

9. $f(x) = 6 - 2x$

10. $f = \{(0, 0), (2, 3), (-1, 5)\}$

11.

Word (Input)	Dog	Cat	House	Desk	Circle
First Letter of Word (Output)	d	c	h	d	c

Use the properties of logarithms to write each expression as a single logarithm.

12. $\log_3 6 + \log_3 4$

13. $\log_5 x + 3\log_5 x - \log_5(x + 1)$

14. Write the expression $\log_6 \frac{2x}{y^3}$ as the sum or difference of multiples of logarithms.

15. If $\log_b 3 = 0.79$ and $\log_b 5 = 1.16$, find the value of $\log_b \frac{3}{25}$.

16. Approximate $\log_7 8$ to four decimal places.

17. Solve $8^{x-1} = \frac{1}{64}$ for x. Give an exact solution.

18. Solve $3^{2x+5} = 4$ for x. Give an exact solution and approximate the solution to four decimal places.

Solve each logarithmic equation for x. Give an exact solution.

19. $\log_3 x = -2$

20. $\ln \sqrt{e} = x$

21. $\log_8(3x - 2) = 2$

22. $\log_5 x + \log_5 3 = 2$

23. $\log_4(x + 1) - \log_4(x - 2) = 3$

24. Solve $\ln(3x + 7) = 1.31$ accurate to four decimal places.

25. Graph $y = \left(\frac{1}{2}\right)^x + 1$.

26. Graph the functions $y = 3^x$ and $y = \log_3 x$ on the same coordinate system.

Use the formula $A = P\left(1 + \frac{r}{n}\right)^{nt}$ *to solve Exercises 27–29.*

27. Find the amount in an account if \$4000 is invested for 3 years at 9% interest compounded monthly.

28. Find how long it will take \$2000 to grow to \$3000 if the money is invested at 7% interest compounded semiannually. Round to the nearest whole.

29. Suppose you have \$3000 to invest. Which investment, rounded to the nearest dollar, yields the greater return over 10 years: 6.5% compounded semiannually or 6% compounded monthly? How much more is yielded by the better investment?

Solve. Round answers to the nearest whole.

30. Suppose a city with population of 150,000 has been decreasing at a rate of 2% per year. If this rate continues, predict the population of the city in 20 years.

31. The prairie dog population of the Grand Forks area now stands at 57,000 animals. If the population is growing at a rate of 2.6% annually, how many prairie dogs will there be in that area 5 years from now?

32. In an attempt to save an endangered species of wood duck, naturalists would like to increase the wood duck population from 400 to 1000 ducks. If the annual population growth rate is 6.2%, how long will it take the naturalists to reach their goal? Round to the nearest whole year.

The reliability of a new model of CD player can be described by the exponential function $R(t) = 2.7^{-(1/3)t}$, where the reliability R is the probability (as a decimal) that the CD player is still working t years after it is manufactured. Round answers to the nearest hundredth. Then write your answers as percents.

33. What is the probability that the CD player will still work half a year after it is manufactured?

34. What is the probability that the CD player will still work 2 years after it is manufactured?

Chapters 1–19 Cumulative Review

1. Divide and simplify: $\frac{5}{16} \div \frac{3}{4}$

2. Solve: $\frac{1}{3}(x - 2) = \frac{1}{4}(x + 1)$

3. Graph: $f(x) = x^2$

4. Find an equation of the line through $(-2, 6)$ and perpendicular to $f(x) = -3x + 4$. Write the equation using function notation.

5. Solve the system.

$$\begin{cases} x - 5y - 2z = 6 \\ -2x + 10y + 4z = -12 \\ \frac{1}{2}x - \frac{5}{2}y - z = 3 \end{cases}$$

6. Line l and line m are parallel lines cut by transversal t. Find the values of x and y.

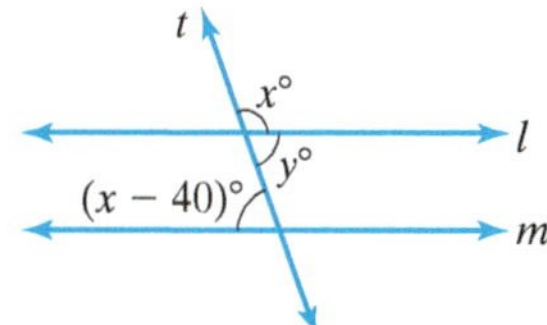

7. Simplify: $3(4 - 7) + (-2) - 5$

8. Use the power rules to simplify the following. Use positive exponents to write all results.

a. $(4a^3)^2$

b. $\left(-\frac{2}{3}\right)^3$

c. $\left(\frac{4a^5}{b^3}\right)^3$

d. $\left(\frac{3^{-2}}{x}\right)^{-3}$

e. $(a^{-2}b^3c^{-4})^{-2}$

9. For the ICL Production Company, the rational function $C(x) = \frac{2.6x + 10{,}000}{x}$ describes the company's cost per disc of pressing x compact discs. Find the cost per disc for pressing:

a. 100 compact discs

b. 1000 compact discs

10. Multiply.

a. $(3x - 1)^2$

b. $\left(\frac{1}{2}x + 3\right)\left(\frac{1}{2}x - 3\right)$

c. $(2x - 5)(6x + 7)$

11. Solve: $2 - 6 = -5(x + 4) - 39$

12. Perform the indicated operations and simplify if possible.

$$\frac{5}{x - 2} + \frac{3}{x^2 + 4x + 4} - \frac{6}{x + 2}$$

13. Divide: $\frac{8x^2y^2 - 16xy + 2x}{4xy}$

14. Simplify each complex fraction.

a. $\frac{\frac{a}{5}}{\frac{a - 1}{10}}$

b. $\frac{\frac{3}{2 + a} + \frac{6}{2 - a}}{\frac{5}{a + 2} - \frac{1}{a - 2}}$

c. $\frac{x^{-1} + y^{-1}}{xy}$

15. Factor: $3m^2 - 24m - 60$

16. Factor: $5x^2 - 85x + 350$

17. Subtract: $\frac{3x^2 + 2x}{x - 1} - \frac{10x - 5}{x - 1}$

18. Use synthetic division to divide: $(8x^2 - 12x - 7) \div (x - 2)$

19. Simplify the following expressions.
 a. $\sqrt[4]{81}$ b. $\sqrt[5]{-243}$
 c. $-\sqrt{25}$ d. $\sqrt[4]{-81}$
 e. $\sqrt[3]{64x^3}$

20. Solve: $\dfrac{1}{a+5} = \dfrac{1}{3a+6} - \dfrac{a+2}{a^2+7a+10}$

21. Use rational exponents to write as a single radical.
 a. $\sqrt{x}\cdot\sqrt[4]{x}$ b. $\dfrac{\sqrt{x}}{\sqrt[3]{x}}$
 c. $\sqrt[3]{3}\cdot\sqrt{2}$

22. Suppose that y varies directly as x. If $y = \dfrac{1}{2}$ when $x = 12$, find the constant of variation and the direct variation equation.

23. Multiply.
 a. $\sqrt{3}(5+\sqrt{30})$
 b. $(\sqrt{5}-\sqrt{6})(\sqrt{7}+1)$
 c. $(7\sqrt{x}+5)(3\sqrt{x}-\sqrt{5})$
 d. $(4\sqrt{3}-1)^2$
 e. $(\sqrt{2x}-5)(\sqrt{2x}+5)$
 f. $(\sqrt{x-3}+5)^2$

24. Find each root. Assume that all variables represent non-negative real numbers.
 a. $\sqrt{9}$ b. $\sqrt[3]{-27}$ c. $\sqrt{\dfrac{9}{64}}$
 d. $\sqrt[4]{x^{12}}$ e. $\sqrt[3]{-125y^6}$

25. Rationalize the denominator of $\dfrac{\sqrt[4]{x}}{\sqrt[4]{81y^5}}$.

26. Multiply.
 a. $a^{1/4}(a^{3/4}-a^8)$
 b. $(x^{1/2}-3)(x^{1/2}+5)$

27. Solve: $\sqrt{4-x} = x-2$

28. Use the quotient rule to divide and simplify if possible.
 a. $\dfrac{\sqrt{54}}{\sqrt{6}}$ b. $\dfrac{\sqrt{108a^2}}{3\sqrt{3}}$
 c. $\dfrac{3\sqrt[3]{81a^5b^{10}}}{\sqrt[3]{3b^4}}$

29. Solve $3x^2 - 9x + 8 = 0$ by completing the square.

30. Add or subtract as indicated.
 a. $\dfrac{\sqrt{20}}{3}+\dfrac{\sqrt{5}}{4}$ b. $\sqrt[3]{\dfrac{24x}{27}}-\dfrac{\sqrt[3]{3x}}{2}$

31. Solve: $\dfrac{3x}{x-2}-\dfrac{x+1}{x}=\dfrac{6}{x(x-2)}$

32. Rationalize the denominator. $\sqrt[3]{\dfrac{27}{m^4n^8}}$

33. Solve: $x^2 - 4x \le 0$

34. Find the length of the unknown side of the triangle.

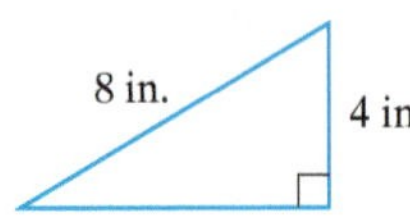

35. Graph: $F(x) = (x-3)^2 + 1$

36. Find the following powers of i.
 a. i^8 b. i^{21}
 c. i^{42} d. i^{-13}

37. Solve: $17 - 7x + 3 = -3x + 21 - 3x$

38. Solve $4x^2 + 8x - 1 = 0$ by completing the square.

39. Find an equation of the inverse of $f(x) = x + 3$.

40. Solve by using the quadratic formula.

$$\left(x-\frac{1}{2}\right)^2=\frac{x}{2}$$

41. Find the value of each logarithmic expression.
 a. $\log_4 16$ b. $\log_{10}\dfrac{1}{10}$ c. $\log_9 3$

42. Graph: $f(x) = -(x+1)^2 + 1$

20

Conic Sections

The original Ferris wheel was named after its designer, George Washington Gale Ferris, Jr., a trained engineer who produced the first Ferris wheel for the 1893 World's Columbian Exposition in Chicago. This very first wheel was 264 feet high and was the Columbian Exposition's most noticeable attraction. Since then, Ferris wheels have gotten ever taller, have been built with ever greater capacities, and have changed their designations from Ferris wheels to giant observation wheels because of their closed capsules. In Exercise 92 of Section 20.1, you will explore the dimensions of the Singapore Flyer, a record-breaking giant observation wheel.

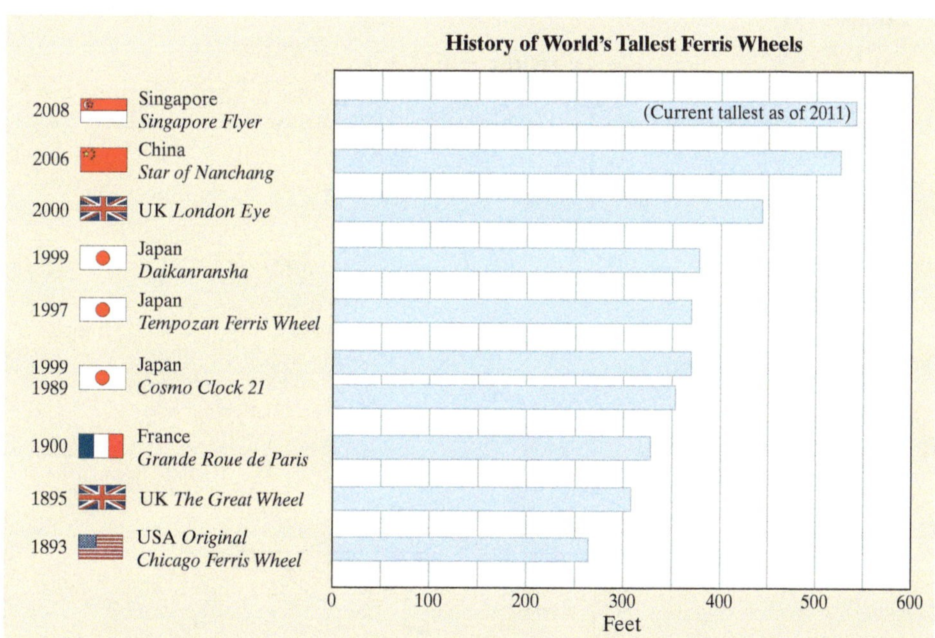

Sections

Check Your Progress

In Chapter 18, we analyzed some of the important connections between a parabola and its equation. Parabolas are interesting in their own right but are more interesting still because they are part of a collection of curves known as conic sections. This chapter is devoted to quadratic equations in two variables and their conic section graphs: the parabola, circle, ellipse, and hyperbola.

20.1 The Parabola and the Circle 

Objectives

1. Graph Parabolas of the Form $x = a(y - k)^2 + h$ and $y = a(x - h)^2 + k$.
2. Graph Circles of the Form $(x - h)^2 + (y - k)^2 = r^2$.
3. Find the Center and the Radius of a Circle, Given Its Equation.
4. Write an Equation of a Circle, Given Its Center and Radius.

Conic sections are named so because each conic section is the intersection of a right circular cone and a plane. The circle, parabola, ellipse, and hyperbola are the conic sections.

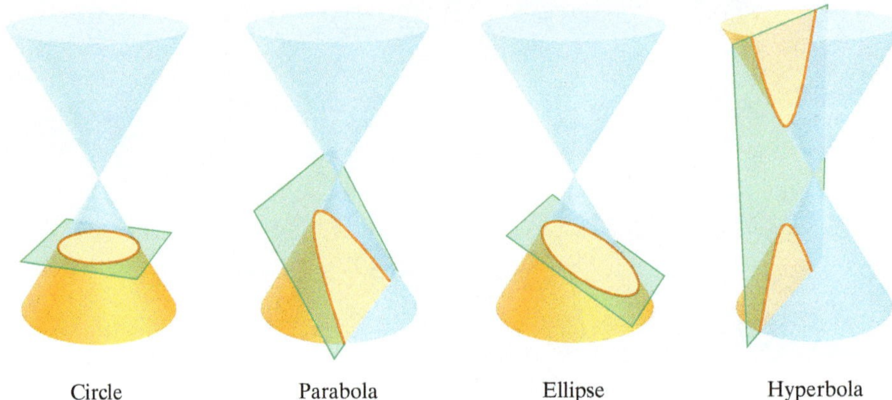

Circle Parabola Ellipse Hyperbola

Objective 1 Graphing Parabolas

Thus far, we have seen that $f(x)$ or $y = a(x - h)^2 + k$ is the equation of a parabola that opens upward if $a > 0$ or downward if $a < 0$. Parabolas can also open left or right or even on a slant. Equations of these parabolas are not functions of x, of course, since a parabola opening any way other than upward or downward fails the vertical line test. In this section, we introduce parabolas that open to the left and to the right. Parabolas opening on a slant will not be developed in this book.

Just as $y = a(x - h)^2 + k$ is the equation of a parabola that opens upward or downward, $x = a(y - k)^2 + h$ is the equation of a parabola that opens to the right or to the left. The parabola opens to the right if $a > 0$ and to the left if $a < 0$. The parabola has vertex (h, k), and its axis of symmetry is the line $y = k$.

Parabolas

$$y = a(x - h)^2 + k$$

$a > 0$ (h, k) $x = h$

(h, k) $a < 0$ $x = h$

$$x = a(y - k)^2 + h$$

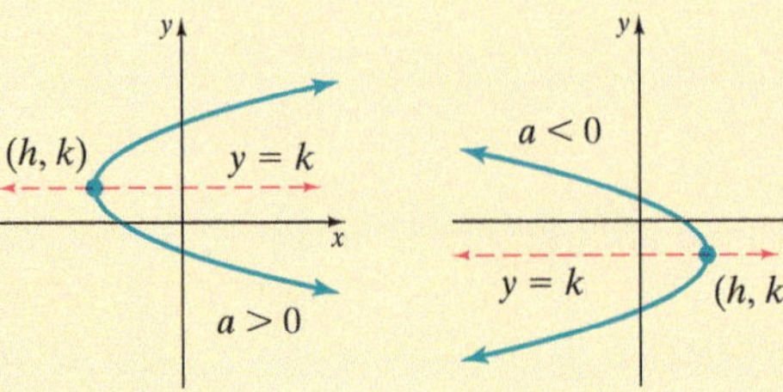

The equations $y = a(x - h)^2 + k$ and $x = a(y - k)^2 + h$ are called **standard forms.**

✓Concept Check Does the graph of the parabola given by the equation $x = -3y^2$ open to the left, to the right, upward, or downward?

Example 1

Graph the parabola $x = 2y^2$.

Solution: Written in standard form, the equation $x = 2y^2$ is $x = 2(y - 0)^2 + 0$ with $a = 2$, $k = 0$, and $h = 0$. Its graph is a parabola with vertex $(0, 0)$, and its axis of symmetry is the line $y = 0$. Since $a > 0$, this parabola opens to the right. The table shows a few more ordered pair solutions of $x = 2y^2$. Its graph is also shown.

x	y
8	−2
2	−1
0	0
2	1
8	2

Work Practice 1

Practice 1

Graph the parabola $x = \frac{1}{2}y^2$.

Example 2

Graph the parabola $x = -3(y - 1)^2 + 2$.

Solution: The equation $x = -3(y - 1)^2 + 2$ is in the form $x = a(y - k)^2 + h$ with $a = -3$, $k = 1$, and $h = 2$. Since $a < 0$, the parabola opens to the left. The vertex (h, k) is $(2, 1)$, and the axis of symmetry is the line $y = 1$. When $y = 0$, $x = -1$, so the x-intercept is $(-1, 0)$. Again, we obtain a few ordered pair solutions and then graph the parabola.

x	y
2	1
−1	0
−1	2
−10	3
−10	−1

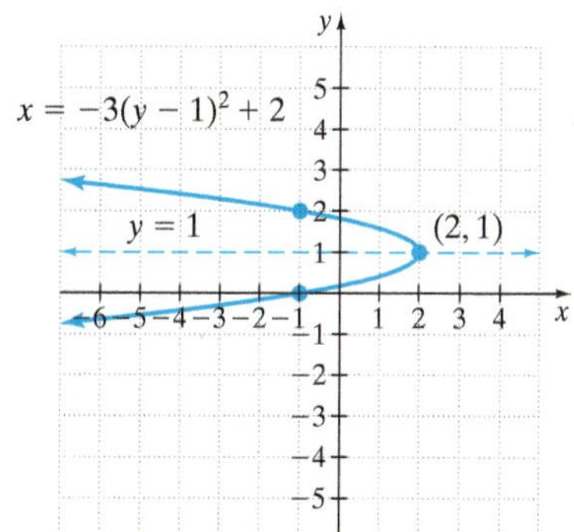

Work Practice 2

Practice 2

Graph the parabola $x = -2(y + 4)^2 - 1$.

Practice 3

Graph $y = -x^2 + 4x + 6$.

Example 3

Graph $y = -x^2 - 2x + 15$.

Solution: Complete the square on x to write the equation in standard form.

$y - 15 = -x^2 - 2x$ Subtract 15 from both sides.

$y - 15 = -1(x^2 + 2x)$ Factor −1 from the terms $-x^2 - 2x$.

(*Continued on next page*)

Answers

1. 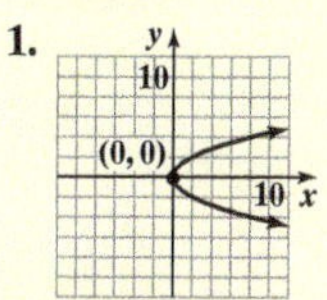

2. 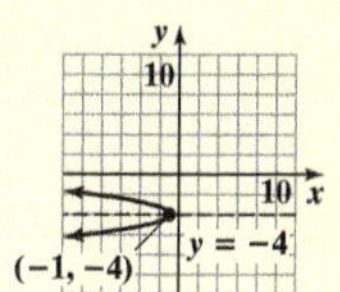

3. 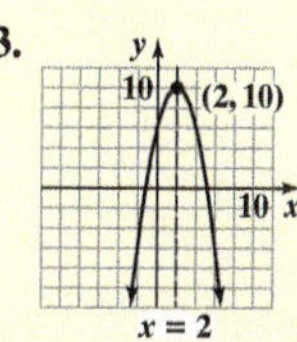

✓Concept Check Answer

to the left

The coefficient of x is 2. Find the square of half of 2.

$$\frac{1}{2}(2) = 1 \quad \text{and} \quad 1^2 = 1$$

$$y - 15 - 1(1) = -1(x^2 + 2x + 1) \quad \text{Add } -1(1) \text{ to both sides.}$$

$$y - 16 = -1(x + 1)^2 \quad \text{Simplify the left side and factor the right side.}$$

$$y = -(x + 1)^2 + 16 \quad \text{Add 16 to both sides.}$$

The equation is now in standard form $y = a(x - h)^2 + k$ with $a = -1, h = -1$, and $k = 16$.

The vertex is then (h, k), or $(-1, 16)$.

A second method for finding the vertex is by using the formula $\frac{-b}{2a}$.

$$x = \frac{-(-2)}{2(-1)} = \frac{2}{-2} = -1$$

$$y = -(-1)^2 - 2(-1) + 15 = -1 + 2 + 15 = 16$$

Again, we see that the vertex is $(-1, 16)$, and the axis of symmetry is the vertical line $x = -1$. The y-intercept is $(0, 15)$. Now we can use a few more ordered pair solutions to graph the parabola.

x	y
−1	16
0	15
−2	15
1	12
−3	12
3	0
−5	0

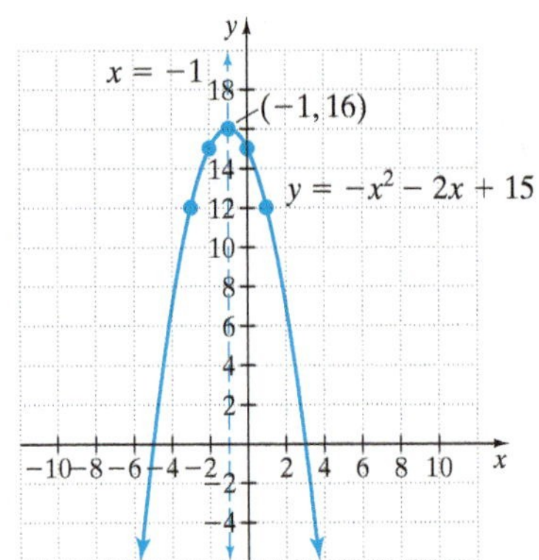

Work Practice 3

Practice 4

Graph $x = 3y^2 + 6y + 4$.

Example 4 Graph $x = 2y^2 + 4y + 5$.

Solution: Notice that this equation is quadratic in y, so its graph is a parabola that opens to the left or the right. We can complete the square on y, or we can use the formula $\frac{-b}{2a}$ to find the vertex.

Since the equation is quadratic in y, the formula gives us the y-value of the vertex.

$$y = \frac{-4}{2 \cdot 2} = \frac{-4}{4} = -1$$

$$x = 2(-1)^2 + 4(-1) + 5 = 2 \cdot 1 - 4 + 5 = 3$$

Answer

4. 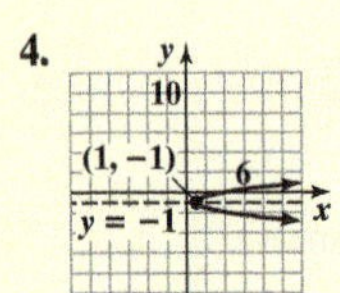

The vertex is $(3, -1)$, and the axis of symmetry is the line $y = -1$. The parabola opens to the right since $a > 0$. The x-intercept is $(5, 0)$.

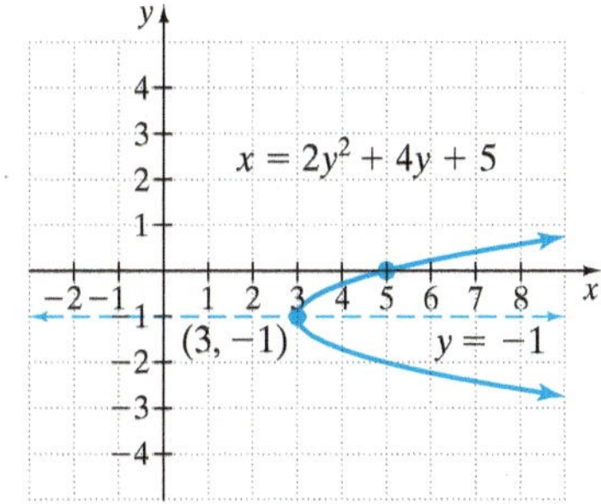

Work Practice 4

Objective 2 Graphing Circles

Another conic section is the **circle.** A circle is the set of all points in a plane that are the same distance from a fixed point called the **center.** The distance is called the **radius** of the circle. To find a standard equation for a circle, let (h, k) represent the center of the circle and let (x, y) represent any point on the circle. The distance between (h, k) and (x, y) is defined to be the circle's radius, r units. We can find this distance r by using the distance formula.

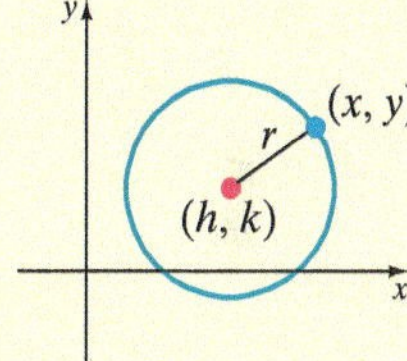

$$r = \sqrt{(x - h)^2 + (y - k)^2}$$

$$r^2 = (x - h)^2 + (y - k)^2 \quad \text{Square both sides.}$$

Circle

The graph of $(x - h)^2 + (y - k)^2 = r^2$ is a circle with center (h, k) and radius r.

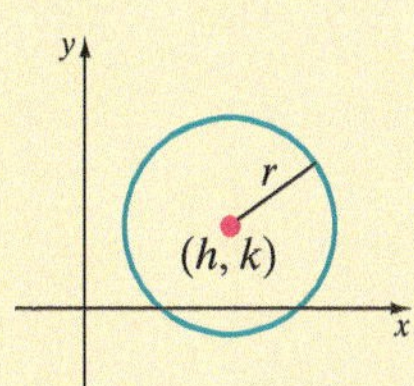

The equation $(x - h)^2 + (y - k)^2 = r^2$ is called **standard form.**

If an equation can be written in the standard form

$$(x - h)^2 + (y - k)^2 = r^2$$

then its graph is a circle, which we can draw by graphing the center (h, k) and using the radius r.

Helpful Hint

Notice that the radius is the *distance* from the center of the circle to any point of the circle. Also notice that the *midpoint* of a diameter of a circle is the center of the circle.

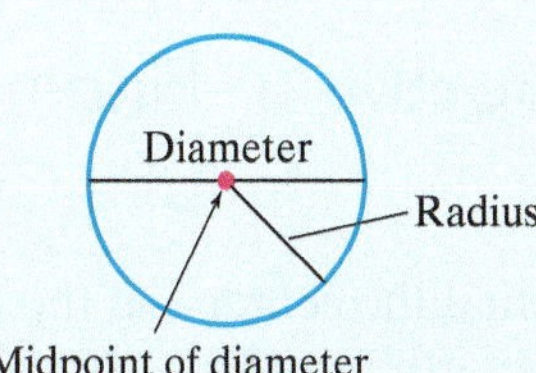

Practice 5

Graph $x^2 + y^2 = 25$.

Example 5 Graph $x^2 + y^2 = 4$.

Solution: The equation can be written in standard form as

$$(x - 0)^2 + (y - 0)^2 = 2^2$$

The center of the circle is $(0, 0)$, and the radius is 2. Its graph is shown.

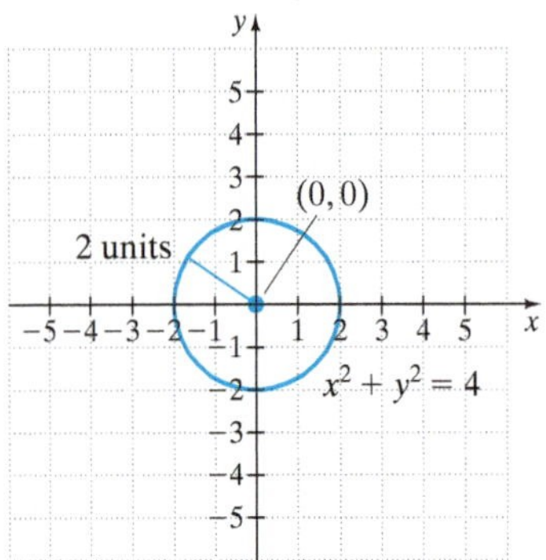

Work Practice 5

Helpful Hint

Notice the difference between the equation of a circle and the equation of a parabola. The equation of a circle contains both x^2 and y^2 terms on the same side of the equation with equal coefficients. The equation of a parabola has either an x^2 term or a y^2 term but not both.

Practice 6

Graph
$(x - 3)^2 + (y + 2)^2 = 4$.

Example 6 Graph $(x + 1)^2 + y^2 = 8$.

Solution: The equation can be written as $(x + 1)^2 + (y - 0)^2 = 8$ with $h = -1, k = 0$, and $r = \sqrt{8}$. The center is $(-1, 0)$, and the radius is $\sqrt{8} = 2\sqrt{2} \approx 2.8$.

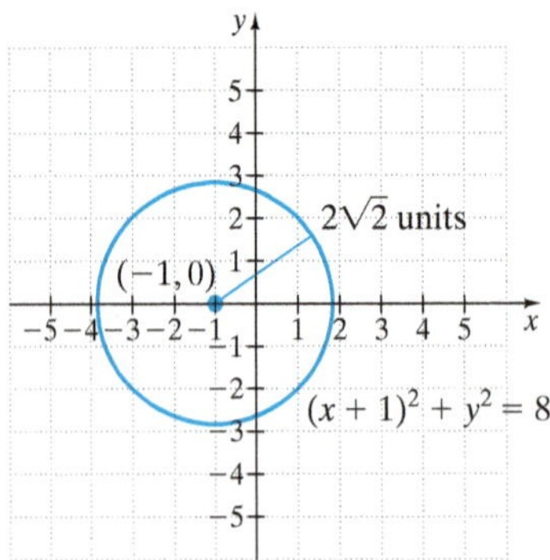

Work Practice 6

✓**Concept Check** In the graph of the equation $(x - 3)^2 + (y - 2)^2 = 5$, what is the distance between the center of the circle and any point on the circle?

Objective 3 Finding the Center and the Radius of a Circle

To find the center and the radius of a circle from its equation, write the equation in standard form. To write the equation of a circle in standard form, we complete the square on both x and y.

Answers

5. 6.

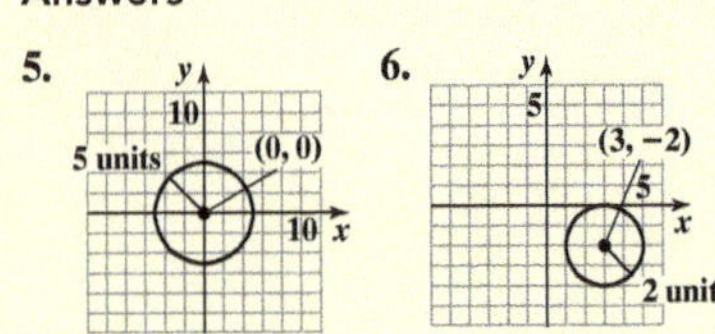

✓**Concept Check Answer**

$\sqrt{5}$ units

Example 7 Graph $x^2 + y^2 + 4x - 8y = 16$.

Solution: Since this equation contains x^2 and y^2 terms on the same side of the equation with equal coefficients, its graph is a circle. To write the equation in standard form, group the terms involving x and the terms involving y and then complete the square on each variable.

$$(x^2 + 4x) + (y^2 - 8y) = 16$$

Thus, $\frac{1}{2}(4) = 2$ and $2^2 = 4$. Also, $\frac{1}{2}(-8) = -4$ and $(-4)^2 = 16$. Add 4 and then 16 to both sides.

$$(x^2 + 4x + 4) + (y^2 - 8y + 16) = 16 + 4 + 16$$

$$(x + 2)^2 + (y - 4)^2 = 36 \quad \text{Factor.}$$

This circle has center $(-2, 4)$ and radius 6, as shown.

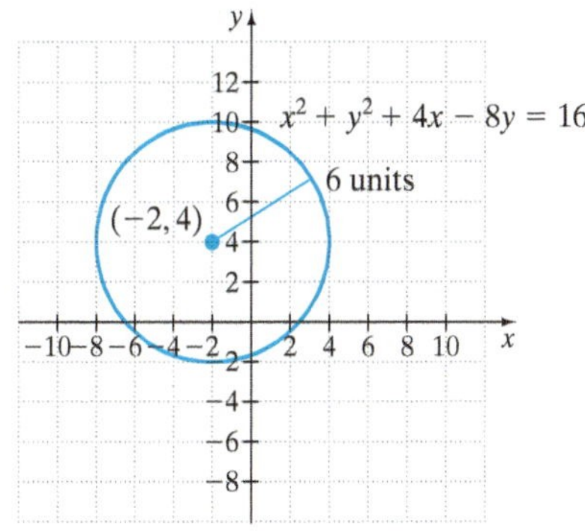

Work Practice 7

Practice 7

Graph $x^2 + y^2 + 6x - 2y = 6$.

Objective 4 Writing Equations of Circles

Since a circle is determined entirely by its center and radius, this information is all we need to write an equation of a circle.

Example 8 Find an equation of the circle with center $(-7, 3)$ and radius 10.

Solution: Using the given values $h = -7$, $k = 3$, and $r = 10$, we write the equation

$$(x - h)^2 + (y - k)^2 = r^2$$

or

$$[x - (-7)]^2 + (y - 3)^2 = 10^2 \quad \text{Substitute the given values.}$$

or

$$(x + 7)^2 + (y - 3)^2 = 100$$

Work Practice 8

Practice 8

Find an equation of the circle with center $(-2, -5)$ and radius 9.

Answers

7.

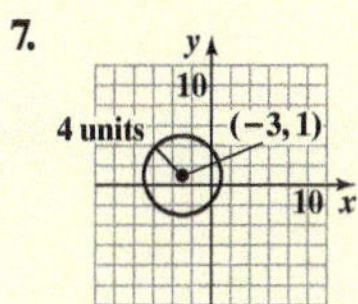

8. $(x + 2)^2 + (y + 5)^2 = 81$

Graphing Calculator Explorations

To graph an equation such as $x^2 + y^2 = 25$ with a graphing calculator, we first solve the equation for y.

$$x^2 + y^2 = 25$$
$$y^2 = 25 - x^2$$
$$y = \pm\sqrt{25 - x^2}$$

The graph of $y = \sqrt{25 - x^2}$ will be the top half of the circle, and the graph of $y = -\sqrt{25 - x^2}$ will be the bottom half of the circle.

To graph, press $\boxed{Y =}$ and enter $Y_1 = \sqrt{25 - x^2}$ and $Y_2 = -\sqrt{25 - x^2}$. Insert parentheses around $25 - x^2$ so that $\sqrt{25 - x^2}$ and not $\sqrt{25} - x^2$ is graphed.

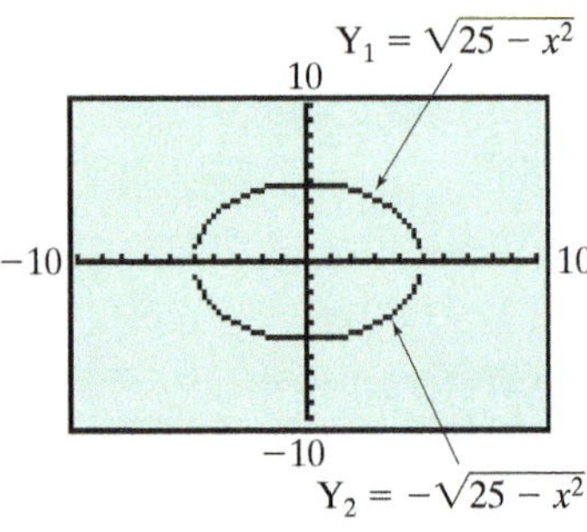

The graph above does not appear to be a circle because we are currently using a standard window and the screen is rectangular. This causes the tick marks on the x-axis to be farther apart than the tick marks on the y-axis and, thus, creates the distorted circle. If we want the graph to appear circular, we must define a square window by using a feature of the graphing calculator or by redefining the window to show the x-axis from −15 to 15 and the y-axis from −10 to 10. Using a square window, the graph appears as shown below.

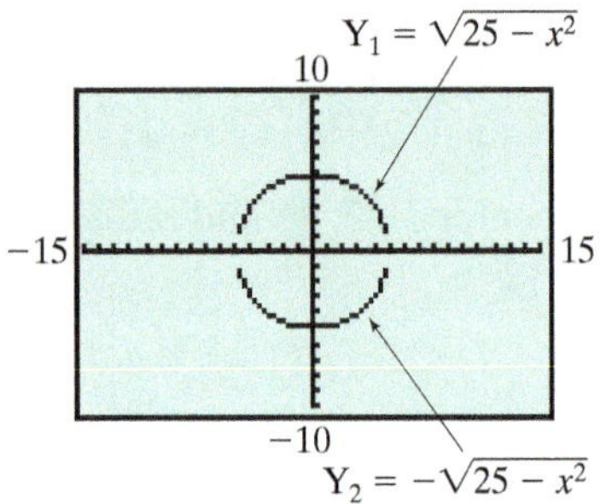

Use a graphing calculator to graph each circle.

1. $x^2 + y^2 = 55$

2. $x^2 + y^2 = 20$

3. $5x^2 + 5y^2 = 50$

4. $6x^2 + 6y^2 = 105$

5. $2x^2 + 2y^2 - 34 = 0$

6. $4x^2 + 4y^2 - 48 = 0$

7. $7x^2 + 7y^2 - 89 = 0$

8. $3x^2 + 3y^2 - 35 = 0$

Vocabulary, Readiness & Video Check

Use the choices below to fill in each blank. Some choices may be used more than once.

radius	center	vertex
diameter	circle	conic sections

1. The circle, parabola, ellipse, and hyperbola are called the __________.

2. For a parabola that opens upward, the lowest point is the __________.

3. A __________ is the set of all points in a plane that are the same distance from a fixed point. The fixed point is called the __________.

4. The midpoint of a diameter of a circle is the __________.

5. The distance from the center of a circle to any point of the circle is called the __________.

6. Twice a circle's radius is its __________.

Martin-Gay Interactive Videos

See Video 20.1

Watch the section lecture video and answer the following questions.

Objective 1 7. Based on Example 1 and the lecture before, would you say that parabolas of the form $x = a(y - k)^2 + h$ are functions? Why or why not?

Objective 2 8. Based on the lecture before Example 2, what would be the standard form of a circle with its center at the origin? Simplify your answer.

Objective 3 9. From Example 3, if you know the center and radius of a circle, how can you write that circle's equation?

Objective 4 10. From Example 4, why do we need to complete the square twice when writing this equation of a circle in standard form?

20.1 Exercise Set MyMathLab®

The graph of each equation is a parabola. Determine whether the parabola opens upward, downward, to the left, or to the right. Do not graph. See Examples 1 through 4.

1. $y = x^2 - 7x + 5$
2. $y = -x^2 + 16$
3. $x = -y^2 - y + 2$
4. $x = 3y^2 + 2y - 5$
5. $y = -x^2 + 2x + 1$
6. $x = -y^2 + 2y - 6$

The graph of each equation is a parabola. Find the vertex of the parabola and then graph it. See Examples 1 through 4.

7. $x = 3y^2$
8. $x = 5y^2$
9. $x = -2y^2$
10. $x = -4y^2$
11. $y = -4x^2$
12. $y = -2x^2$
13. $x = (y - 2)^2 + 3$
14. $x = (y - 4)^2 - 1$
15. $y = -3(x - 1)^2 + 5$
16. $y = -4(x - 2)^2 + 2$
17. $x = y^2 + 6y + 8$
18. $x = y^2 - 6y + 6$
19. $y = x^2 + 10x + 20$
20. $y = x^2 + 4x - 5$
21. $x = -2y^2 + 4y + 6$
22. $x = 3y^2 + 6y + 7$

The graph of each equation is a circle. Find the center and the radius and then graph the circle. See Examples 5 through 7.

23. $x^2 + y^2 = 9$
24. $x^2 + y^2 = 25$
25. $x^2 + (y - 2)^2 = 1$
26. $(x - 3)^2 + y^2 = 9$
27. $(x - 5)^2 + (y + 2)^2 = 1$
28. $(x + 3)^2 + (y + 3)^2 = 4$
29. $x^2 + y^2 + 6y = 0$
30. $x^2 + 10x + y^2 = 0$
31. $x^2 + y^2 + 2x - 4y = 4$
32. $x^2 + y^2 + 6x - 4y = 3$
33. $(x + 2)^2 + (y - 3)^2 = 7$
34. $(x + 1)^2 + (y - 2)^2 = 5$
35. $x^2 + y^2 - 4x - 8y - 2 = 0$
36. $x^2 + y^2 - 2x - 6y - 5 = 0$

Hint: For Exercises 37 through 42, first divide the equation through by the coefficient of x^2 (or y^2).

37. $3x^2 + 3y^2 = 75$
38. $2x^2 + 2y^2 = 18$
39. $6(x - 4)^2 + 6(y - 1)^2 = 24$
40. $7(x - 1)^2 + 7(y - 3)^2 = 63$
41. $4(x + 1)^2 + 4(y - 3)^2 = 12$
42. $5(x - 2)^2 + 5(y + 1) = 50$

Write an equation of the circle with the given center and radius. See Example 8.

43. $(2, 3); 6$
44. $(-7, 6); 2$
45. $(0, 0); \sqrt{3}$
46. $(0, -6); \sqrt{2}$
47. $(-5, 4); 3\sqrt{5}$
48. the origin; $4\sqrt{7}$

MIXED PRACTICE

Sketch the graph of each equation. If the graph is a parabola, find its vertex. If the graph is a circle, find its center and radius.

49. $x = y^2 - 3$
50. $x = y^2 + 2$
51. $y = (x - 2)^2 - 2$
52. $y = (x + 3)^2 + 3$
53. $x^2 + y^2 = 1$
54. $x^2 + y^2 = 49$
55. $x = (y + 3)^2 - 1$
56. $x = (y - 1)^2 + 4$
57. $(x - 2)^2 + (y - 2)^2 = 16$
58. $(x + 3)^2 + (y - 1)^2 = 9$
59. $x = -(y - 1)^2$
60. $x = -2(y + 5)^2$
61. $(x - 4)^2 + y^2 = 7$
62. $x^2 + (y + 5)^2 = 5$
63. $y = 5(x + 5)^2 + 3$
64. $y = 3(x - 4)^2 + 2$

65. $\frac{x^2}{8} + \frac{y^2}{8} = 2$
66. $2x^2 + 2y^2 = \frac{1}{2}$
67. $y = x^2 + 7x + 6$
68. $y = x^2 - 2x - 15$
69. $x^2 + y^2 + 2x + 12y - 12 = 0$
70. $x^2 + y^2 + 6x + 10y - 2 = 0$
71. $x = y^2 + 8y - 4$
72. $x = y^2 + 6y + 2$
73. $x^2 - 10y + y^2 + 4 = 0$
74. $x^2 + y^2 - 8y + 5 = 0$
75. $x = -3y^2 + 30y$
76. $x = -2y^2 - 4y$
77. $5x^2 + 5y^2 = 25$
78. $\frac{x^2}{3} + \frac{y^2}{3} = 2$
79. $y = 5x^2 - 20x + 16$
80. $y = 4x^2 - 40x + 105$

REVIEW AND PREVIEW

Graph each equation. See Sections 10.2 and 10.3.

81. $y = 2x + 5$
82. $y = -3x + 3$
83. $y = 3$
84. $x = -2$

Rationalize each denominator and simplify if possible. See Section 17.5.

85. $\frac{1}{\sqrt{3}}$
86. $\frac{\sqrt{5}}{\sqrt{8}}$
87. $\frac{4\sqrt{7}}{\sqrt{6}}$
88. $\frac{10}{\sqrt{5}}$

CONCEPT EXTENSIONS

For Exercises 89 and 90, explain the error in each statement.

89. The graph of $x = 5(y + 5)^2 + 1$ is a parabola with vertex $(-5, 1)$ and opening to the right.

90. The graph of $x^2 + (y + 3)^2 = 10$ is a circle with center $(0, -3)$ and radius 5.

91. ***The Sarsen Circle*** The first image that comes to mind when one thinks of Stonehenge is the very large sandstone blocks with sandstone lintels across the top. The Sarsen Circle of Stonehenge is the outer circle of the sandstone blocks, each of which weighs up to 50 tons. There were originally 30 of these monolithic blocks, but only 17 remain upright to this day. The "altar stone" lies at the center of this circle, which has a diameter of 33 meters. (Refer to the sketch at the top of the right column.)

   a. What is the radius of the Sarsen Circle?
   b. What is the circumference of the Sarsen Circle? Round your result to 2 decimal places.
   c. Since there were originally 30 Sarsen stones located on the circumference, how far apart would the centers of the stones have been? Round to the nearest tenth of a meter.
   d. Using the axes in the drawing, what are the coordinates of the center of the circle?
   e. Use parts (a) and (d) to write the equation of the Sarsen Circle.

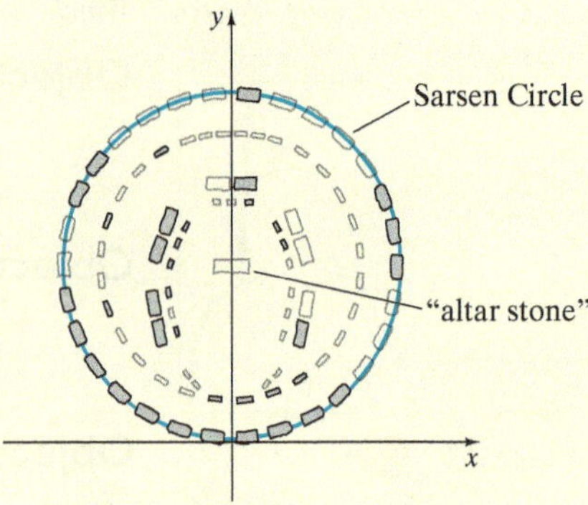

92. One of the largest observation wheels in the world is the Singapore Flyer. From the Flyer, you can see up to 45 km away. Each of the 28 enclosed capsules holds 28 passengers and completes a full rotation every 32 minutes. Its diameter is 150 meters, and the height of this giant wheel is 165 meters. (*Source:* singaporeflyer.com)

   a. What is the radius of the Singapore Flyer?
   b. How close is the wheel to the ground?
   c. How high is the center of the wheel from the ground?
   d. Using the axes in the drawing, what are the coordinates of the center of the wheel?
   e. Use parts (a) and (d) to write an equation of the Singapore Flyer.

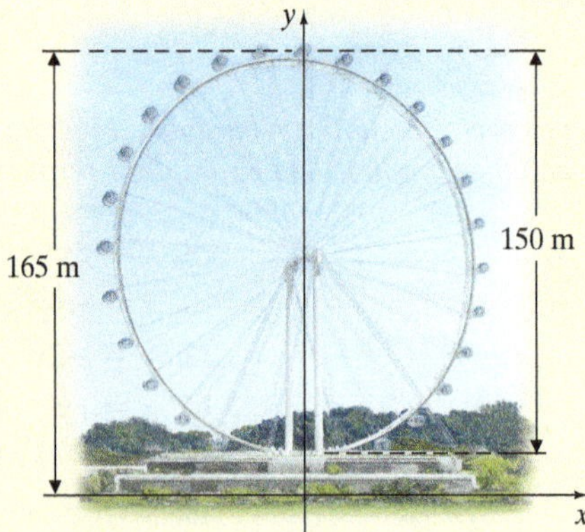

93. In 1893, Pittsburgh bridge builder George Ferris designed and built a gigantic revolving steel wheel whose height was 264 feet and diameter was 250 feet. This Ferris wheel opened at the 1893 exposition in Chicago. It had 36 wooden cars, each capable of holding 60 passengers. (*Source: The Handy Science Answer Book*)

   a. What was the radius of this Ferris wheel?
   b. How close was the wheel to the ground?
   c. How high was the center of the wheel from the ground?

d. Using the axes in the drawing, what are the coordinates of the center of the wheel?

e. Use parts (a) and (d) to write an equation of the wheel.

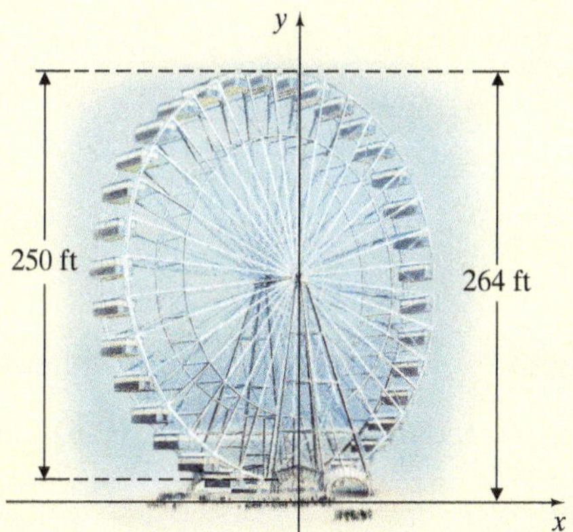

94. The world's largest-diameter Ferris wheel currently operating is the Cosmo Clock 21 at Yokohama City, Japan. It has a 60-armed wheel, its diameter is 100 meters, and it has a height of 105 meters. (*Source: The Handy Science Answer Book*)

a. What is the radius of this Ferris wheel?

b. How close is the wheel to the ground?

c. How high is the center of the wheel from the ground?

d. Using the axes in the drawing, what are the coordinates of the center of the wheel?

e. Use parts (a) and (d) to write an equation of the wheel.

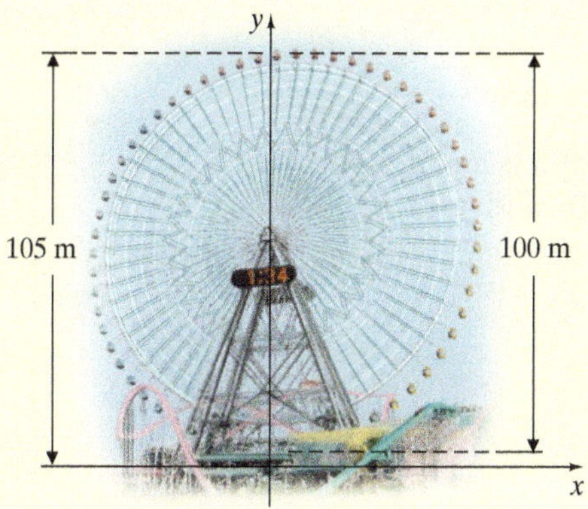

95. If you are given a list of equations of circles and parabolas and none are in standard form, explain how you would determine which is an equation of a circle and which is an equation of a parabola. Explain also how you would distinguish the upward or downward parabolas from the left-opening or right-opening parabolas.

96. Determine whether the triangle with vertices $(2, 6)$, $(0, -2)$, and $(5, 1)$ is an isosceles triangle.

Solve.

97. Two surveyors need to find the distance across a lake. They place a reference pole at point A in the diagram. Point B is 3 meters east and 1 meter north of the reference point A. Point C is 19 meters east and 13 meters north of point A. Find the distance across the lake, from B to C.

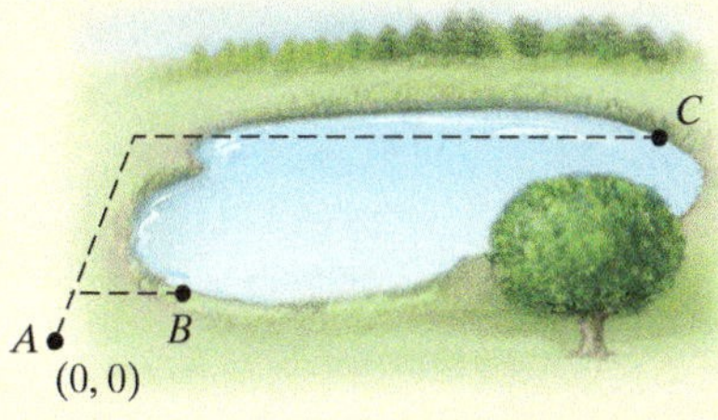

98. A bridge constructed over a bayou has a supporting arch in the shape of a parabola. Find an equation of the parabolic arch if the length of the road over the arch is 100 meters and the maximum height of the arch is 40 meters.

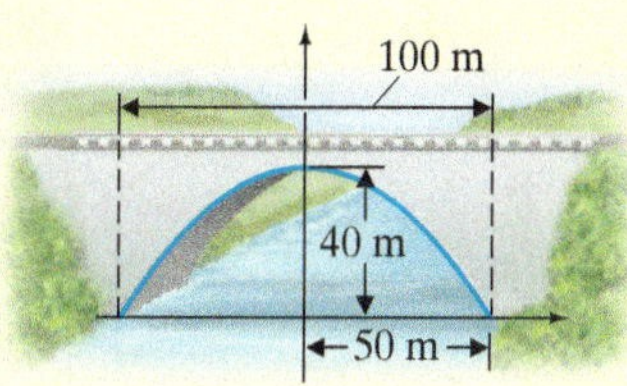

Use a graphing calculator to verify each exercise. Use a square viewing window.

99. Exercise 77 **100.** Exercise 78

101. Exercise 79 **102.** Exercise 80

20.2 The Ellipse and the Hyperbola

Objective 1 Graphing Ellipses

Objectives

1 Define and Graph an Ellipse.

2 Define and Graph a Hyperbola.

An **ellipse** can be thought of as the set of points in a plane such that the sum of the distances of those points from two fixed points is constant. Each of the two fixed points is called a **focus.** (The plural of focus is **foci.**) The point midway between the foci is called the **center.**

An ellipse may be drawn by hand by using two thumbtacks, a piece of string, and a pencil. Secure the two thumbtacks in a piece of cardboard, for example, and tie each end of the string to a tack. Use your pencil to pull the string tight and draw the ellipse, as shown on the next page. The two thumbtacks are the foci of the drawn ellipse.

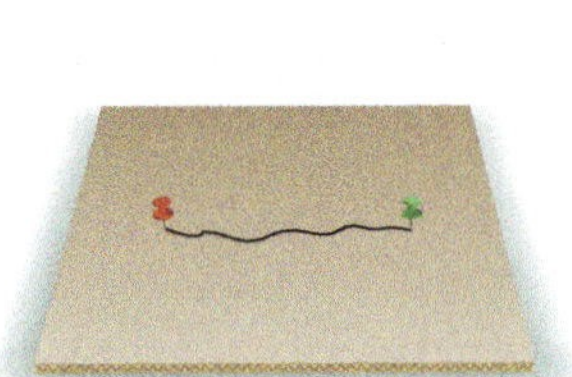

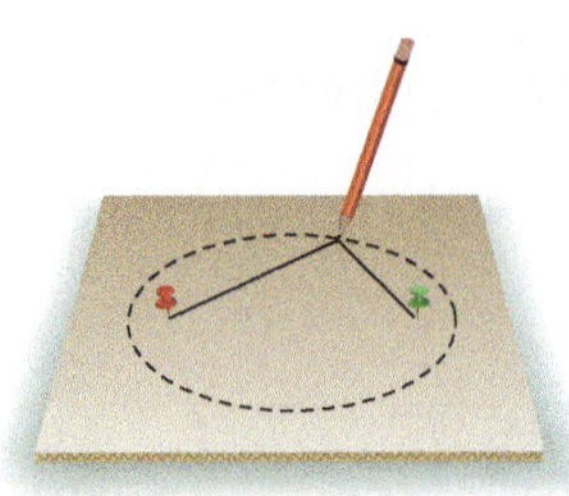

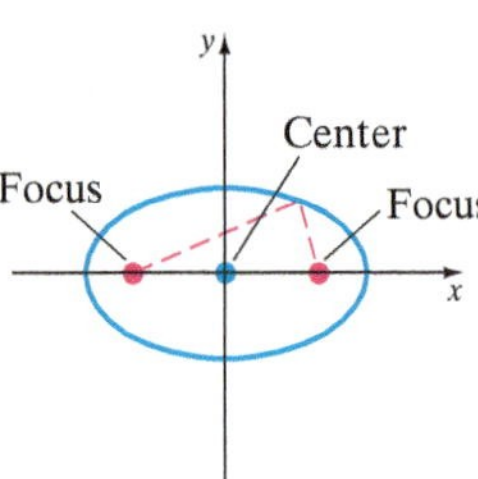

Ellipse with Center (0, 0)

The graph of an equation of the form $\frac{x^2}{a^2} + \frac{y^2}{b^2} = 1$ is an ellipse with center $(0, 0)$. The x-intercepts are $(a, 0)$ and $(-a, 0)$, and the y-intercepts are $(0, b)$, and $(0, -b)$.

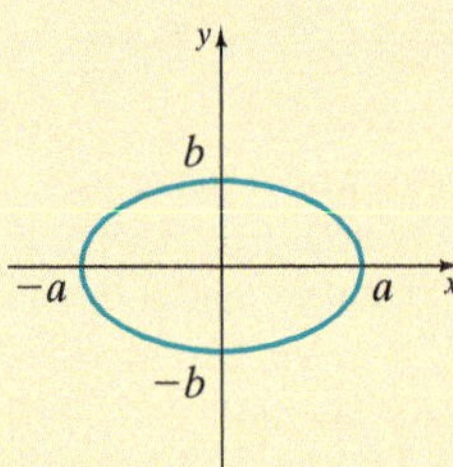

The **standard form** of an ellipse with center $(0, 0)$ is $\frac{x^2}{a^2} + \frac{y^2}{b^2} = 1$.

Practice 1

Graph $\frac{x^2}{25} + \frac{y^2}{4} = 1$.

Example 1 Graph $\frac{x^2}{9} + \frac{y^2}{16} = 1$.

Solution: The equation is of the form $\frac{x^2}{a^2} + \frac{y^2}{b^2} = 1$, with $a = 3$ and $b = 4$, so its graph is an ellipse with center $(0, 0)$, x-intercepts $(3, 0)$ and $(-3, 0)$, and y-intercepts $(0, 4)$ and $(0, -4)$.

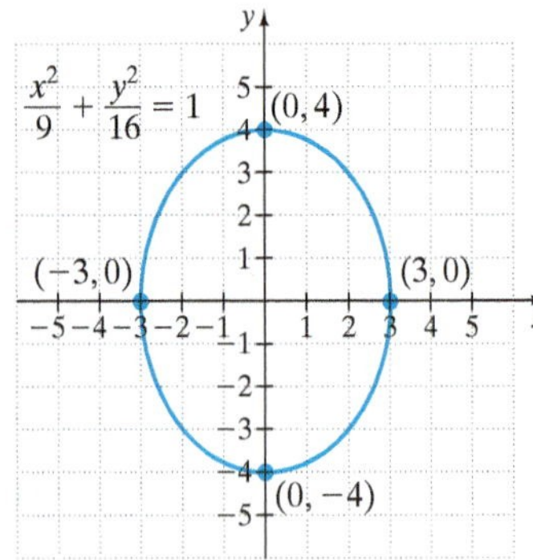

Work Practice 1

Practice 2

Graph $9x^2 + 4y^2 = 36$.

Example 2 Graph $4x^2 + 16y^2 = 64$.

Solution: Although this equation contains a sum of squared terms in x and y on the same side of the equation, this is not the equation of a circle since the coefficients of x^2 and y^2 are not the same. The graph of this equation is an ellipse. Since the standard form of the equation of an ellipse has 1 on one side, divide both sides of this equation by 64.

$$4x^2 + 16y^2 = 64$$

$$\frac{4x^2}{64} + \frac{16y^2}{64} = \frac{64}{64} \quad \text{Divide both sides by 64.}$$

$$\frac{x^2}{16} + \frac{y^2}{4} = 1 \quad \text{Simplify.}$$

Answers

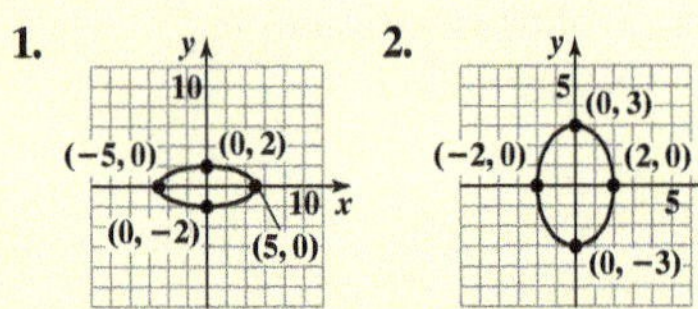

We now recognize the equation of an ellipse with $a = 4$ and $b = 2$. This ellipse has center $(0, 0)$, x-intercepts $(4, 0)$ and $(-4, 0)$, and y-intercepts $(0, 2)$ and $(0, -2)$.

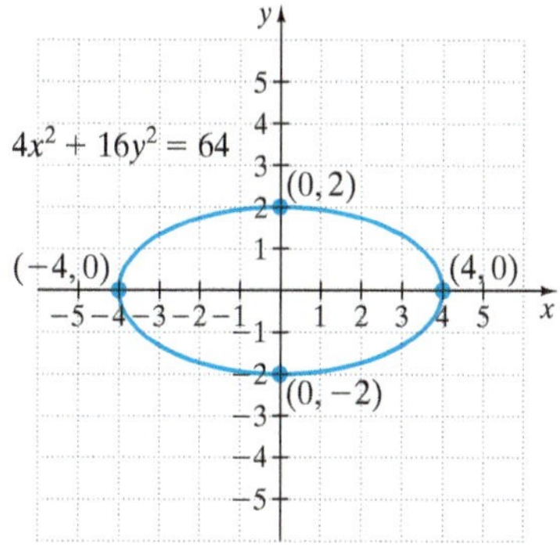

Work Practice 2

The center of an ellipse is not always $(0, 0)$, as shown in the next example.

Ellipse with Center (h, k)

The standard form of the equation of an ellipse with center (h, k) is

$$\frac{(x-h)^2}{a^2} + \frac{(y-k)^2}{b^2} = 1$$

Example 3 Graph $\frac{(x+3)^2}{25} + \frac{(y-2)^2}{36} = 1$.

Solution: The center of this ellipse is found in a way that is similar to finding the center of a circle. This ellipse has center $(-3, 2)$. Notice that $a = 5$ and $b = 6$. To find four points on the graph of the ellipse, first graph the center, $(-3, 2)$. Since $a = 5$, count 5 units right and then 5 units left of the point with coordinates $(-3, 2)$. Next, since $b = 6$, start at $(-3, 2)$ and count 6 units up and then 6 units down to find two more points on the ellipse.

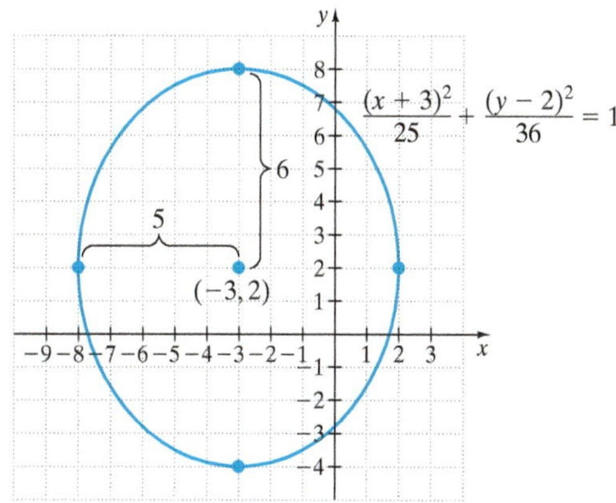

Work Practice 3

Practice 3

Graph $\frac{(x-4)^2}{49} + \frac{(y+1)^2}{81} = 1$.

✓**Concept Check** In the graph of the equation $\frac{x^2}{64} + \frac{y^2}{36} = 1$, which distance is longer: the distance between the x-intercepts or the distance between the y-intercepts? How much longer? Explain.

Answer

3.

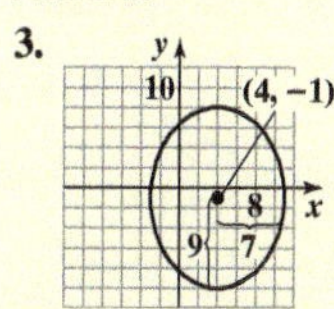

✓**Concept Check Answer**

x-intercepts, by 4 units

Objective 2 Graphing Hyperbolas 

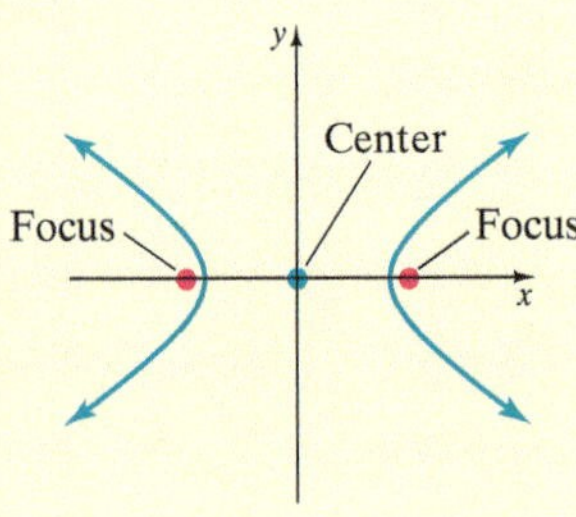

The final conic section is the **hyperbola.** A hyperbola is the set of points in a plane such that the absolute value of the difference of the distances from two fixed points is constant. Each of the two fixed points is called a **focus.** The point midway between the foci is called the **center.**

Using the distance formula, we can show that the graph of $\frac{x^2}{a^2} - \frac{y^2}{b^2} = 1$ is a hyperbola with center $(0, 0)$ and x-intercepts $(a, 0)$ and $(-a, 0)$. Also, the graph of $\frac{y^2}{b^2} - \frac{x^2}{a^2} = 1$ is a hyperbola with center $(0, 0)$ and y-intercepts $(0, b)$ and $(0, -b)$.

Hyperbola with Center (0, 0)

The graph of an equation of the form $\frac{x^2}{a^2} - \frac{y^2}{b^2} = 1$ is a hyperbola with center $(0, 0)$ and x-intercepts $(a, 0)$ and $(-a, 0)$.

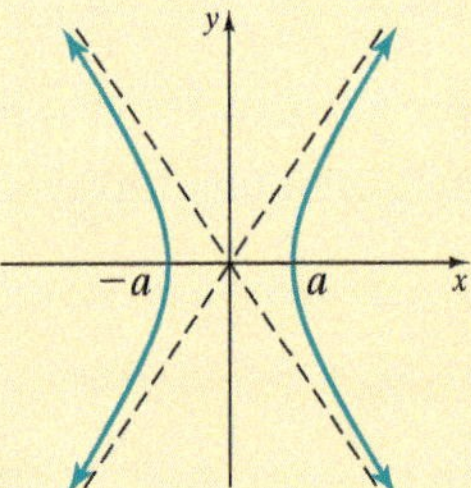

The graph of an equation of the form $\frac{y^2}{b^2} - \frac{x^2}{a^2} = 1$ is a hyperbola with center $(0, 0)$ and y-intercepts $(0, b)$ and $(0, -b)$.

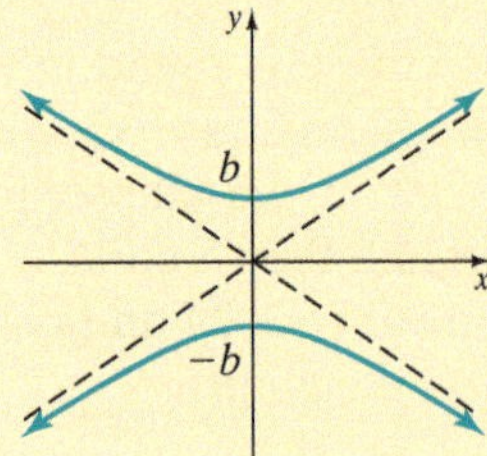

The equations $\frac{x^2}{a^2} - \frac{y^2}{b^2} = 1$ and $\frac{y^2}{b^2} - \frac{x^2}{a^2} = 1$ are the **standard forms** for the equation of a hyperbola.

Helpful Hint

Notice the difference between the equation of an ellipse and a hyperbola. The equation of the ellipse contains x^2 and y^2 terms on the same side of the equation with same-sign coefficients. For a hyperbola, the coefficients on the same side of the equation have different signs.

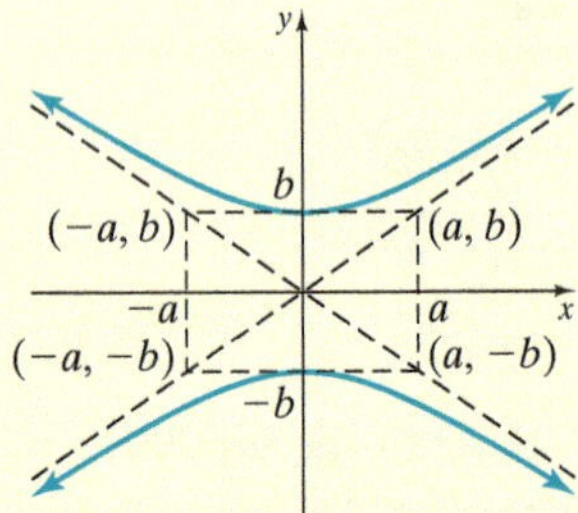

Graphing a hyperbola such as $\frac{y^2}{b^2} - \frac{x^2}{a^2} = 1$ is made easier by recognizing one of its important characteristics. Examining the figure to the left, notice how the sides of the branches of the hyperbola extend indefinitely and seem to approach the dashed lines in the figure. These dashed lines are called the **asymptotes** of the hyperbola.

To sketch these lines, or asymptotes, draw a rectangle with vertices (a, b), $(-a, b)$, $(a, -b)$, and $(-a, -b)$. The asymptotes of the hyperbola are the extended diagonals of this rectangle.

Example 4 Graph $\frac{x^2}{16} - \frac{y^2}{25} = 1$.

Solution: This equation has the form $\frac{x^2}{a^2} - \frac{y^2}{b^2} = 1$, with $a = 4$ and $b = 5$. Thus, its graph is a hyperbola that opens to the left and right. It has center $(0, 0)$ and x-intercepts $(4, 0)$ and $(-4, 0)$. To aid in graphing the hyperbola, we first sketch its asymptotes. The extended diagonals of the rectangle with corners $(4, 5)$, $(4, -5)$, $(-4, 5)$, and $(-4, -5)$ are the asymptotes of the hyperbola. Then we use the asymptotes to aid in sketching the hyperbola.

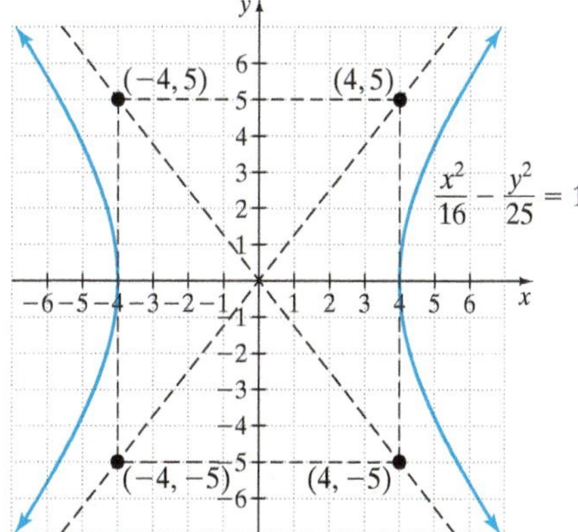

Work Practice 4

Practice 4

Graph $\frac{x^2}{9} - \frac{y^2}{16} = 1$.

Example 5 Graph $4y^2 - 9x^2 = 36$.

Solution: Since this is a difference of squared terms in x and y on the same side of the equation, its graph is a hyperbola as opposed to an ellipse or a circle. The standard form of the equation of a hyperbola has a 1 on one side, so divide both sides of the equation by 36.

$$4y^2 - 9x^2 = 36$$

$$\frac{4y^2}{36} - \frac{9x^2}{36} = \frac{36}{36} \quad \text{Divide both sides by 36.}$$

$$\frac{y^2}{9} - \frac{x^2}{4} = 1 \quad \text{Simplify.}$$

The equation is of the form $\frac{y^2}{b^2} - \frac{x^2}{a^2} = 1$, with $a = 2$ and $b = 3$, so the hyperbola is centered at $(0, 0)$ with y-intercepts $(0, 3)$ and $(0, -3)$. The sketch of the hyperbola is shown.

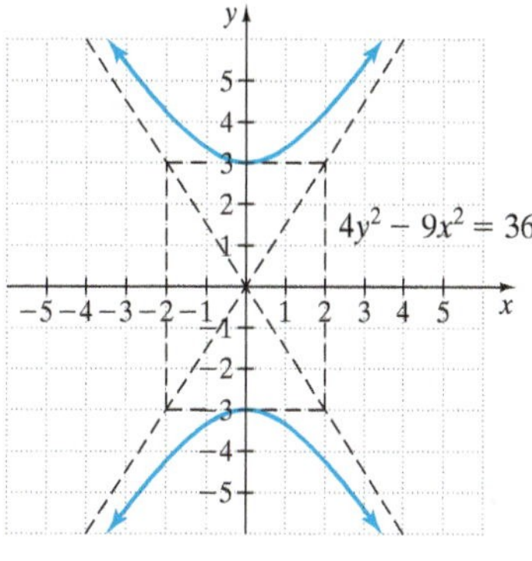

Work Practice 5

Practice 5

Graph $9y^2 - 25x^2 = 225$.

Although this is beyond the scope of this text, the standard forms of the equations of hyperbolas with center (h, k) are given next. The Concept Extensions section in Exercise Set 20.2 contains some hyperbolas of this form.

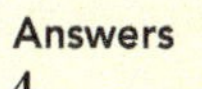

Answers

4.

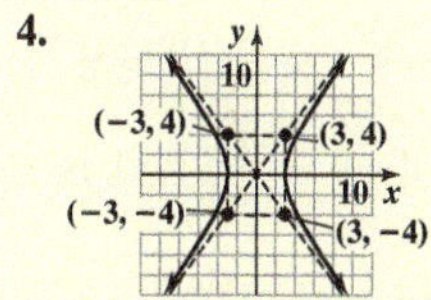

5.

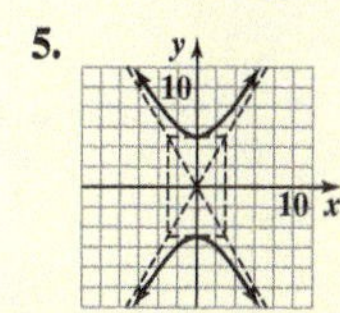

Hyperbola with Center (h, k)

Standard forms of the equations of hyperbolas with center (h, k) are:

$$\frac{(x-h)^2}{a^2} - \frac{(y-k)^2}{b^2} = 1 \qquad \frac{(y-k)^2}{b^2} - \frac{(x-h)^2}{a^2} = 1$$

Graphing Calculator Explorations

To graph an ellipse by using a graphing calculator, use the same procedure as for graphing a circle. For example, to graph $x^2 + 3y^2 = 22$, first solve for y.

$$3y^2 = 22 - x^2$$
$$y^2 = \frac{22 - x^2}{3}$$
$$y = \pm\sqrt{\frac{22 - x^2}{3}}$$

Next, press the $\boxed{Y =}$ key and enter $Y_1 = \sqrt{\frac{22 - x^2}{3}}$ and $Y_2 = -\sqrt{\frac{22 - x^2}{3}}$. (Insert two sets of parentheses in the radicand as $\sqrt{((22 - x^2)/3)}$ so that the desired graph is obtained.) The graph appears as shown.

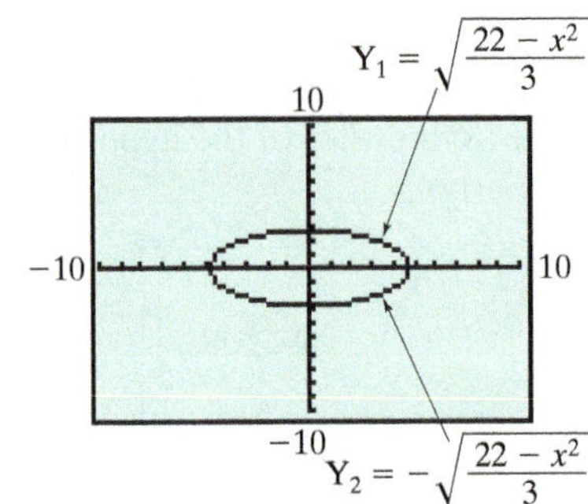

Use a graphing calculator to graph each ellipse.

1. $10x^2 + y^2 = 32$
2. $x^2 + 6y^2 = 35$
3. $20x^2 + 5y^2 = 100$
4. $4y^2 + 12x^2 = 48$
5. $7.3x^2 + 15.5y^2 = 95.2$
6. $18.8x^2 + 36.1y^2 = 205.8$

Vocabulary, Readiness & Video Check

Use the choices below to fill in each blank. Some choices will be used more than once and some not at all.

ellipse	$(0, 0)$	x	$(a, 0)$ and $(-a, 0)$	$(0, a)$ and $(0, -a)$	focus
hyperbola	center	y	$(b, 0)$ and $(-b, 0)$	$(0, b)$ and $(0, -b)$	

1. A(n) ______________ is the set of points in a plane such that the absolute value of the differences of their distances from two fixed points is constant.
2. A(n) ______________ is the set of points in a plane such that the sum of their distances from two fixed points is constant.

For Exercises 1 and 2 above,

3. The two fixed points are each called a(n) ______________.
4. The point midway between the foci is called the ______________.
5. The graph of $\frac{x^2}{a^2} - \frac{y^2}{b^2} = 1$ is a(n) ______________ with center ______________ and ______________-intercepts of ______________.
6. The graph of $\frac{x^2}{a^2} + \frac{y^2}{b^2} = 1$ is a(n) ______________ with center ______________ and x-intercepts of ______________.

Martin-Gay Interactive Videos *Watch the section lecture video and answer the following questions.*

See Video 20.2

Objective 1 **7.** From Example 1, what information do the values of a and b give us about the graph of an ellipse? Answer this same question for Example 2.

Objective 2 **8.** From Example 3, we know the points (a, b), $(a, -b)$, $(-a, b)$, and $(-a, -b)$ are not part of the graph. Explain the role of these points.

20.2 Exercise Set MyMathLab®

Identify the graph of each equation as an ellipse or a hyperbola. Do not graph. See Examples 1 through 5.

1. $\frac{x^2}{16} + \frac{y^2}{4} = 1$ **2.** $\frac{x^2}{16} - \frac{y^2}{4} = 1$

3. $x^2 - 5y^2 = 3$ **4.** $-x^2 + 5y^2 = 3$

5. $-\frac{y^2}{25} + \frac{x^2}{36} = 1$ **6.** $\frac{y^2}{25} + \frac{x^2}{36} = 1$

Sketch the graph of each equation. See Examples 1 and 2.

7. $\frac{x^2}{4} + \frac{y^2}{25} = 1$ **8.** $\frac{x^2}{16} + \frac{y^2}{9} = 1$

9. $\frac{x^2}{9} + y^2 = 1$ **10.** $x^2 + \frac{y^2}{4} = 1$

11. $9x^2 + y^2 = 36$ **12.** $x^2 + 4y^2 = 16$

13. $4x^2 + 25y^2 = 100$ **14.** $36x^2 + y^2 = 36$

Sketch the graph of each equation. See Example 3.

15. $\frac{(x + 1)^2}{36} + \frac{(y - 2)^2}{49} = 1$

16. $\frac{(x - 3)^2}{9} + \frac{(y + 3)^2}{16} = 1$

17. $\frac{(x - 1)^2}{4} + \frac{(y - 1)^2}{25} = 1$

18. $\frac{(x + 3)^2}{16} + \frac{(y + 2)^2}{4} = 1$

Sketch the graph of each equation. See Examples 4 and 5.

19. $\frac{x^2}{4} - \frac{y^2}{9} = 1$ **20.** $\frac{x^2}{36} - \frac{y^2}{36} = 1$

21. $\frac{y^2}{25} - \frac{x^2}{16} = 1$ **22.** $\frac{y^2}{25} - \frac{x^2}{49} = 1$

23. $x^2 - 4y^2 = 16$ **24.** $4x^2 - y^2 = 36$

25. $16y^2 - x^2 = 16$ **26.** $4y^2 - 25x^2 = 100$

MIXED PRACTICE

Graph each equation. See Examples 1 through 5.

27. $\frac{y^2}{36} = 1 - x^2$ **28.** $\frac{x^2}{36} = 1 - y^2$

29. $4(x - 1)^2 + 9(y + 2)^2 = 36$

30. $25(x + 3)^2 + 4(y - 3)^2 = 100$

31. $8x^2 + 2y^2 = 32$ **32.** $3x^2 + 12y^2 = 48$

33. $25x^2 - y^2 = 25$ **34.** $x^2 - 9y^2 = 9$

MIXED PRACTICE—SECTIONS 20.1, 20.2

Identify whether each equation, when graphed, will be a parabola, circle, ellipse, or hyperbola. Sketch the graph of each equation.

If a parabola, label the vertex.
If a circle, label the center and note the radius.
If an ellipse, label the center.
If a hyperbola, label the x- or y-intercepts.

35. $(x - 7)^2 + (y - 2)^2 = 4$ **36.** $y = x^2 + 4$

37. $y = x^2 + 12x + 36$ **38.** $\frac{x^2}{4} + \frac{y^2}{9} = 1$

39. $\frac{y^2}{9} - \frac{x^2}{9} = 1$ **40.** $\frac{x^2}{16} - \frac{y^2}{4} = 1$

41. $\frac{x^2}{16} + \frac{y^2}{4} = 1$ **42.** $x^2 + y^2 = 16$

43. $x = y^2 + 4y - 1$ **44.** $x = -y^2 + 6y$

45. $9x^2 - 4y^2 = 36$ **46.** $9x^2 + 4y^2 = 36$

47. $\frac{(x - 1)^2}{49} + \frac{(y + 2)^2}{25} = 1$ **48.** $y^2 = x^2 + 16$

49. $\left(x + \frac{1}{2}\right)^2 + \left(y - \frac{1}{2}\right)^2 = 1$ **50.** $y = -2x^2 + 4x - 3$

REVIEW AND PREVIEW

Perform the indicated operations. See Sections 12.1 and 12.2.

51. $(2x^3)(-4x^2)$

52. $2x^3 - 4x^3$

53. $-5x^2 + x^2$

54. $(-5x^2)(x^2)$

CONCEPT EXTENSIONS

The graph of each equation is an ellipse. Determine which distance is longer, the distance between the x-intercepts or the distance between the y-intercepts. How much longer? See the Concept Check in this section.

55. $\frac{x^2}{16} + \frac{y^2}{25} = 1$

56. $\frac{x^2}{100} + \frac{y^2}{49} = 1$

57. $4x^2 + y^2 = 16$

58. $x^2 + 4y^2 = 36$

59. If you are given a list of equations of circles, parabolas, ellipses, and hyperbolas, explain how you could distinguish the different conic sections by looking at their equations.

60. We know that $x^2 + y^2 = 25$ is the equation of a circle. Rewrite the equation so that the right side is equal to 1. Which type of conic section does this equation form resemble? In fact, the circle is a special case of this type of conic section. Describe the conditions under which this type of conic section is a circle.

The orbits of stars, planets, comets, asteroids, and satellites all have the shape of one of the conic sections. Astronomers use a measure called eccentricity to describe the shape and elongation of an orbital path. For the circle and ellipse, eccentricity e is calculated with the formula $e = \frac{c}{d}$, where $c^2 = |a^2 - b^2|$ and d is the larger value of a or b. For a hyperbola, eccentricity e is calculated with the formula $e = \frac{c}{d}$, where $c^2 = a^2 + b^2$ and the value of d is equal to a if the hyperbola has x-intercepts or equal to b if the hyperbola has y-intercepts. Use equations A–H to answer Exercises 61 through 70.

A. $\frac{x^2}{36} - \frac{y^2}{13} = 1$ **B.** $\frac{x^2}{4} + \frac{y^2}{4} = 1$ **C.** $\frac{x^2}{25} + \frac{y^2}{16} = 1$

D. $\frac{y^2}{25} - \frac{x^2}{39} = 1$ **E.** $\frac{x^2}{17} + \frac{y^2}{81} = 1$ **F.** $\frac{x^2}{36} + \frac{y^2}{36} = 1$

G. $\frac{x^2}{16} - \frac{y^2}{65} = 1$ **H.** $\frac{x^2}{144} + \frac{y^2}{140} = 1$

61. Identify the type of conic section represented by each of the equations A–H.

62. For each of the equations A–H, identify the values of a^2 and b^2.

63. For each of the equations A–H, calculate the values of c^2 and c.

64. For each of the equations A–H, find the value of d.

65. For each of the equations A–H, calculate the eccentricity e.

66. What do you notice about the values of e for the equations you identified as ellipses?

67. What do you notice about the values of e for the equations you identified as circles?

68. What do you notice about the values of e for the equations you identified as hyperbolas?

69. The eccentricity of a parabola is exactly 1. Use this information and the observations you made in Exercises 66, 67, and 68 to describe a way that could be used to identify the type of conic section based on its eccentricity value.

70. Graph each of the conic sections given in equations A–H. What do you notice about the shape of the ellipses for increasing values of eccentricity? Which is the most elliptical? Which is the least elliptical, that is, the most circular?

71. A planet's orbit about the sun can be described as an ellipse. Consider the sun as the origin of a rectangular coordinate system. Suppose that the x-intercepts of the elliptical path of the planet are $\pm 130{,}000{,}000$ and that the y-intercepts are $\pm 125{,}000{,}000$. Write the equation of the elliptical path of the planet.

72. Comets orbit the sun in elongated ellipses. Consider the sun as the origin of a rectangular coordinate system. Suppose that the equation of the path of the comet is

$$\frac{(x - 1{,}782{,}000{,}000)^2}{3.42 \times 10^{23}} + \frac{(y - 356{,}400{,}000)^2}{1.368 \times 10^{22}} = 1$$

Find the center of the path of the comet.

73. Use a graphing calculator to verify Exercise 46.

74. Use a graphing calculator to verify Exercise 12.

For Exercises 75 through 80, see the example below.

Example

Sketch the graph of $\frac{(x-2)^2}{25} - \frac{(y-1)^2}{9} = 1$.

Solution

This hyperbola has center $(2, 1)$. Notice that $a = 5$ and $b = 3$.

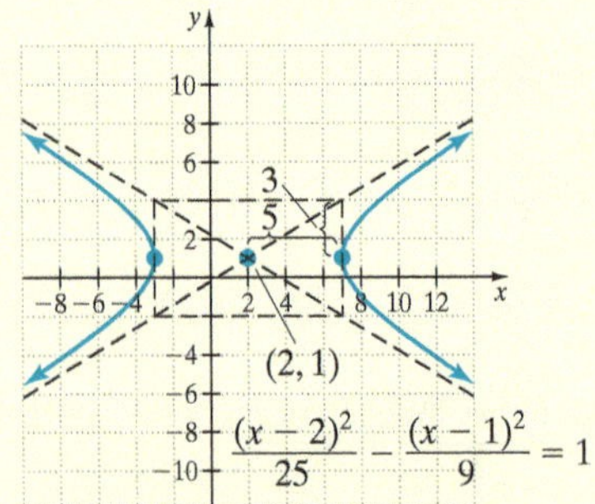

Sketch the graph of each equation.

75. $\frac{(x-1)^2}{4} - \frac{(y+1)^2}{25} = 1$

76. $\frac{(x+2)^2}{9} - \frac{(y-1)^2}{4} = 1$

77. $\frac{y^2}{16} - \frac{(x+3)^2}{9} = 1$

78. $\frac{(y+4)^2}{4} - \frac{x^2}{25} = 1$

79. $\frac{(x+5)^2}{16} - \frac{(y+2)^2}{25} = 1$

80. 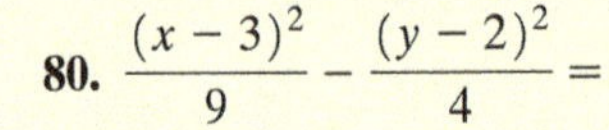 $\frac{(x-3)^2}{9} - \frac{(y-2)^2}{4} = 1$

Sections 20.1–20.2 Integrated Review

Graphing Conic Sections

Following is a summary of conic sections.

Conic Sections

	Standard Form	Graph
Parabola	$y = a(x - h)^2 + k$	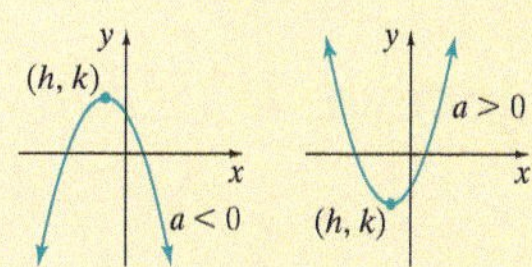
Parabola	$x = a(y - k)^2 + h$	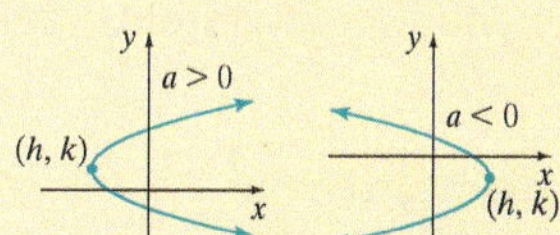
Circle	$(x - h)^2 + (y - k)^2 = r^2$	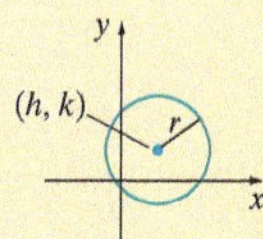
Ellipse center (0, 0)	$\frac{x^2}{a^2} + \frac{y^2}{b^2} = 1$	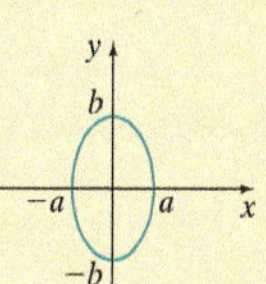
Hyperbola center (0, 0)	$\frac{x^2}{a^2} - \frac{y^2}{b^2} = 1$	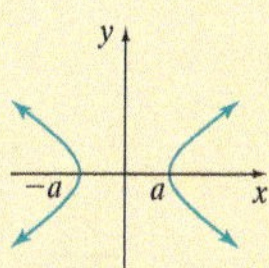
Hyperbola center (0, 0)	$\frac{y^2}{b^2} - \frac{x^2}{a^2} = 1$	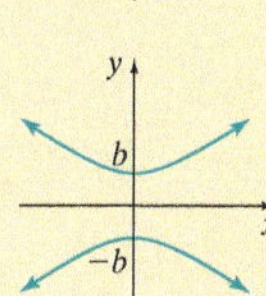

Identify whether each equation, when graphed, will be a parabola, circle, ellipse, or hyperbola. Then graph each equation.

1. $(x - 7)^2 + (y - 2)^2 = 4$

2. $y = x^2 + 4$

3. $y = x^2 + 12x + 36$

4. $\frac{x^2}{4} + \frac{y^2}{9} = 1$

5. $\frac{y^2}{9} - \frac{x^2}{9} = 1$

6. $\frac{x^2}{16} - \frac{y^2}{4} = 1$

7. $\frac{x^2}{16} + \frac{y^2}{4} = 1$

8. $x^2 + y^2 = 16$

9. $x = y^2 + 4y - 1$

10. $x = -y^2 + 6y$

11. $9x^2 - 4y^2 = 36$

12. $9x^2 + 4y^2 = 36$

13. $\frac{(x - 1)^2}{49} + \frac{(y + 2)^2}{25} = 1$

14. $y^2 = x^2 + 16$

15. $\left(x + \frac{1}{2}\right)^2 + \left(y - \frac{1}{2}\right)^2 = 1$

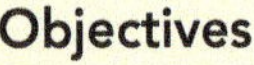 20.3 Solving Nonlinear Systems of Equations

Objectives

1 Solve a Nonlinear System by Substitution.

2 Solve a Nonlinear System by Elimination.

In Chapter 11, we used graphing, substitution, and elimination methods to find solutions of systems of linear equations in two variables. We now apply these same methods to nonlinear systems of equations in two variables. A **nonlinear system of equations** is a system of equations at least one of which is not linear. Since we will be graphing the equations in each system, we are interested in real number solutions only.

Objective 1 Solving Nonlinear Systems by Substitution

First, nonlinear systems are solved by the substitution method.

Practice 1

Solve the system

$$\begin{cases} x^2 - 4y = 4 \\ x + y = -1 \end{cases}.$$

Example 1 Solve the system

$$\begin{cases} x^2 - 3y = 1 \\ x - y = 1 \end{cases}$$

Solution: We can solve this system by substitution if we solve one equation for one of the variables. Solving the first equation for x is not the best choice since doing so introduces a radical. Also, solving for y in the first equation introduces a fraction. We solve the second equation for y.

$x - y = 1$ Second equation

$x - 1 = y$ Solve for y.

Replace y with $x - 1$ in the first equation, and then solve for x.

$$\begin{aligned} x^2 - 3y &= 1 \quad \text{First equation} \\ x^2 - 3(x - 1) &= 1 \quad \text{Replace } y \text{ with } x - 1. \\ x^2 - 3x + 3 &= 1 \\ x^2 - 3x + 2 &= 0 \\ (x - 2)(x - 1) &= 0 \\ x = 2 \quad \text{or} \quad x &= 1 \end{aligned}$$

Let $x = 2$ and then let $x = 1$ in the equation $y = x - 1$ to find corresponding y-values.

Let $x = 2$.	Let $x = 1$.
$y = x - 1$	$y = x - 1$
$y = 2 - 1 = 1$	$y = 1 - 1 = 0$

The solutions are (2, 1) and (1, 0), or the solution set is $\{(2, 1), (1, 0)\}$. Check both solutions in both equations. Both solutions satisfy both equations, so both are solutions of the system. The graph of each equation in the system is shown next. Intersections of the graphs are at (2, 1) and (1, 0).

Answer

1. $(-4, 3), (0, -1)$

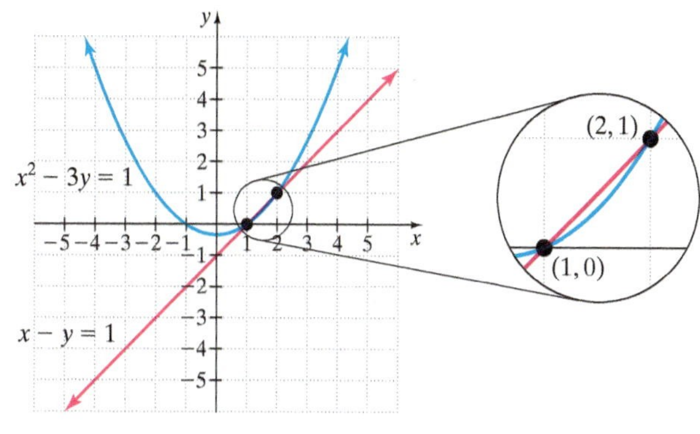

Work Practice 1

Example 2 Solve the system

$$\begin{cases} y = \sqrt{x} \\ x^2 + y^2 = 6 \end{cases}$$

Solution: This system is ideal for substitution since y is expressed in terms of x in the first equation. Notice that if $y = \sqrt{x}$, then both x and y must be nonnegative if they are real numbers. Substitute $\sqrt{x}$ for y in the second equation, and solve for x.

$$x^2 + y^2 = 6$$
$$x^2 + (\sqrt{x})^2 = 6 \quad \text{Let } y = \sqrt{x}.$$
$$x^2 + x = 6$$
$$x^2 + x - 6 = 0$$
$$(x + 3)(x - 2) = 0$$
$$x = -3 \quad \text{or} \quad x = 2$$

The solution -3 is discarded because we have noted that x must be nonnegative. To see this, let $x = -3$ in the first equation. Then let $x = 2$ in the first equation to find a corresponding y-value.

Let $x = -3$.
$$y = \sqrt{x}$$
$$y = \sqrt{-3} \quad \text{Not a real number}$$

Let $x = 2$.
$$y = \sqrt{x}$$
$$y = \sqrt{2}$$

Since we are interested only in real number solutions, the only solution is $(2, \sqrt{2})$. Check to see that this solution satisfies both equations. The graph of each equation in the system is shown to the right.

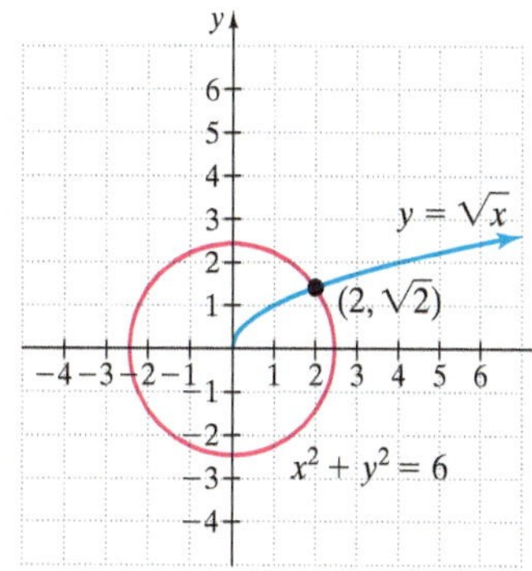

Work Practice 2

Practice 2

Solve the system
$$\begin{cases} y = -\sqrt{x} \\ x^2 + y^2 = 20 \end{cases}.$$

Answer

2. $(4, -2)$

Practice 3

Solve the system $\begin{cases} x^2 + y^2 = 9 \\ x - y = 5 \end{cases}$.

Example 3 Solve the system

$$\begin{cases} x^2 + y^2 = 4 \\ x + y = 3 \end{cases}$$

Solution: We use the substitution method and solve the second equation for x.

$$x + y = 3 \qquad \text{Second equation}$$
$$x = 3 - y$$

Now we let $x = 3 - y$ in the first equation.

$$x^2 + y^2 = 4 \qquad \text{First equation}$$
$$(3 - y)^2 + y^2 = 4 \qquad \text{Let } x = 3 - y.$$
$$9 - 6y + y^2 + y^2 = 4$$
$$2y^2 - 6y + 5 = 0$$

By the quadratic formula, where $a = 2$, $b = -6$, and $c = 5$, we have

$$y = \frac{6 \pm \sqrt{(-6)^2 - 4 \cdot 2 \cdot 5}}{2 \cdot 2} = \frac{6 \pm \sqrt{-4}}{4}$$

Since $\sqrt{-4}$ is not a real number, there is no real solution, or $\varnothing$. Graphically, the circle and the line do not intersect, as shown below.

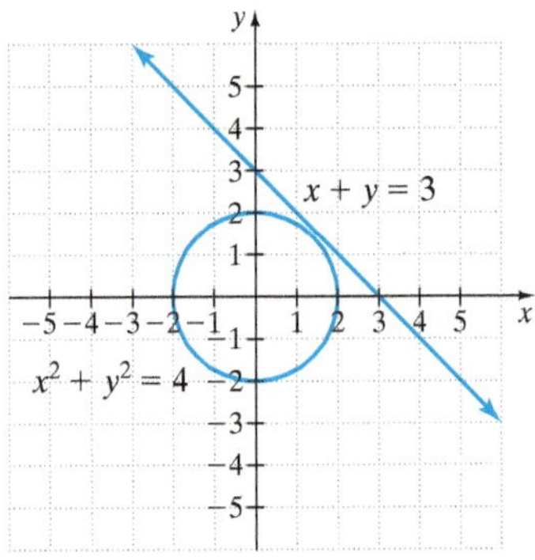

Work Practice 3

✓**Concept Check** Without solving, how can you tell that $x^2 + y^2 = 9$ and $x^2 + y^2 = 16$ do not have any points of intersection?

Objective 2 Solving Nonlinear Systems by Elimination 

Some nonlinear systems may be solved by the elimination method.

Practice 4

Solve the system
$\begin{cases} x^2 + 4y^2 = 16 \\ x^2 - y^2 = 1 \end{cases}$.

Example 4 Solve the system

$$\begin{cases} x^2 + 2y^2 = 10 \\ x^2 - y^2 = 1 \end{cases}$$

Answers

3. $\varnothing$ **4.** $(2, \sqrt{3}), (2, -\sqrt{3}), (-2, \sqrt{3}), (-2, -\sqrt{3})$

✓**Concept Check Answer**

$x^2 + y^2 = 9$ is a circle inside the circle $x^2 + y^2 = 16$; therefore, they do not have any points of intersection.

Solution: We will use the elimination, or addition, method to solve this system. To eliminate x^2 when we add the two equations, multiply both sides of the second equation by -1. Then

$$\begin{cases} x^2 + 2y^2 = 10 \\ (-1)(x^2 - y^2) = -1 \cdot 1 \end{cases} \quad \text{is equivalent to} \quad \begin{cases} x^2 + 2y^2 = 10 \\ -x^2 + y^2 = -1 \end{cases} \quad \text{Add.}$$

$$3y^2 = 9$$

$$y^2 = 3 \quad \text{Divide both sides by 3.}$$

$$y = \pm\sqrt{3}$$

To find the corresponding x-values, we let $y = \sqrt{3}$ and $y = -\sqrt{3}$ in either original equation. We choose the second equation.

Let $y = \sqrt{3}$.

$$x^2 - y^2 = 1$$
$$x^2 - (\sqrt{3})^2 = 1$$
$$x^2 - 3 = 1$$
$$x^2 = 4$$
$$x = \pm\sqrt{4} = \pm 2$$

Let $y = -\sqrt{3}$.

$$x^2 - y^2 = 1$$
$$x^2 - (-\sqrt{3})^2 = 1$$
$$x^2 - 3 = 1$$
$$x^2 = 4$$
$$x = \pm\sqrt{4} = \pm 2$$

The solutions are $(2, \sqrt{3})$, $(-2, \sqrt{3})$, $(2, -\sqrt{3})$, and $(-2, -\sqrt{3})$. Check all four ordered pairs in both equations of the system. The graph of each equation in this system is shown.

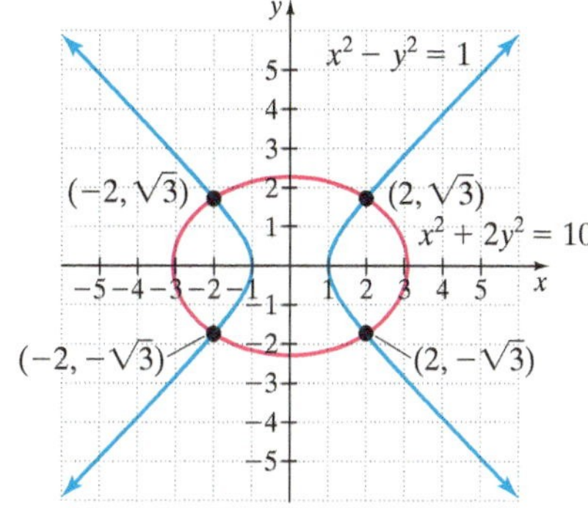

Work Practice 4

Vocabulary, Readiness & Video Check

Martin-Gay Interactive Videos *Watch the section lecture video and answer the following questions.*

See Video 20.3

Objective 1 1. In Example 1, why do we choose not to solve either equation for y?

Objective 2 2. In Example 2, what important reminder is made as the second equation is multiplied by a number to get opposite coefficients of x?

20.3 Exercise Set MyMathLab®

MIXED PRACTICE

Solve each nonlinear system of equations for real solutions. See Examples 1 through 4.

1. $\begin{cases} x^2 + y^2 = 25 \\ 4x + 3y = 0 \end{cases}$

2. $\begin{cases} x^2 + y^2 = 25 \\ 3x + 4y = 0 \end{cases}$

3. $\begin{cases} x^2 + 4y^2 = 10 \\ y = x \end{cases}$

4. $\begin{cases} 4x^2 + y^2 = 10 \\ y = x \end{cases}$

5. $\begin{cases} y^2 = 4 - x \\ x - 2y = 4 \end{cases}$

6. $\begin{cases} x^2 + y^2 = 4 \\ x + y = -2 \end{cases}$

7. $\begin{cases} x^2 + y^2 = 9 \\ 16x^2 - 4y^2 = 64 \end{cases}$

8. $\begin{cases} 4x^2 + 3y^2 = 35 \\ 5x^2 + 2y^2 = 42 \end{cases}$

9. $\begin{cases} x^2 + 2y^2 = 2 \\ x - y = 2 \end{cases}$

10. $\begin{cases} x^2 + 2y^2 = 2 \\ x^2 - 2y^2 = 6 \end{cases}$

11. $\begin{cases} y = x^2 - 3 \\ 4x - y = 6 \end{cases}$

12. $\begin{cases} y = x + 1 \\ x^2 - y^2 = 1 \end{cases}$

13. $\begin{cases} y = x^2 \\ 3x + y = 10 \end{cases}$

14. $\begin{cases} 6x - y = 5 \\ xy = 1 \end{cases}$

15. $\begin{cases} y = 2x^2 + 1 \\ x + y = -1 \end{cases}$

16. $\begin{cases} x^2 + y^2 = 9 \\ x + y = 5 \end{cases}$

17. $\begin{cases} y = x^2 - 4 \\ y = x^2 - 4x \end{cases}$

18. $\begin{cases} x = y^2 - 3 \\ x = y^2 - 3y \end{cases}$

19. $\begin{cases} 2x^2 + 3y^2 = 14 \\ -x^2 + y^2 = 3 \end{cases}$

20. $\begin{cases} 4x^2 - 2y^2 = 2 \\ -x^2 + y^2 = 2 \end{cases}$

21. $\begin{cases} x^2 + y^2 = 1 \\ x^2 + (y + 3)^2 = 4 \end{cases}$

22. $\begin{cases} x^2 + 2y^2 = 4 \\ x^2 - y^2 = 4 \end{cases}$

23. $\begin{cases} y = x^2 + 2 \\ y = -x^2 + 4 \end{cases}$

24. $\begin{cases} x = -y^2 - 3 \\ x = y^2 - 5 \end{cases}$

25. $\begin{cases} 3x^2 + y^2 = 9 \\ 3x^2 - y^2 = 9 \end{cases}$

26. $\begin{cases} x^2 + y^2 = 25 \\ x = y^2 - 5 \end{cases}$

27. $\begin{cases} x^2 + 3y^2 = 6 \\ x^2 - 3y^2 = 10 \end{cases}$

28. $\begin{cases} x^2 + y^2 = 1 \\ y = x^2 - 9 \end{cases}$

29. $\begin{cases} x^2 + y^2 = 36 \\ y = \frac{1}{6}x^2 - 6 \end{cases}$

30. $\begin{cases} x^2 + y^2 = 16 \\ y = -\frac{1}{4}x^2 + 4 \end{cases}$

31. $\begin{cases} y = \sqrt{x} \\ x^2 + y^2 = 12 \end{cases}$

32. $\begin{cases} y = \sqrt{x} \\ x^2 + y^2 = 20 \end{cases}$

REVIEW AND PREVIEW

Graph each inequality in two variables. See Section 16.4.

33. $x > -3$

34. $y \le 1$

35. $y < 2x - 1$

36. $3x - y \le 4$

Find the perimeter of each geometric figure. See Section 12.2.

△ 37. x inches; (2x − 5) inches; (5x − 20) inches

38.

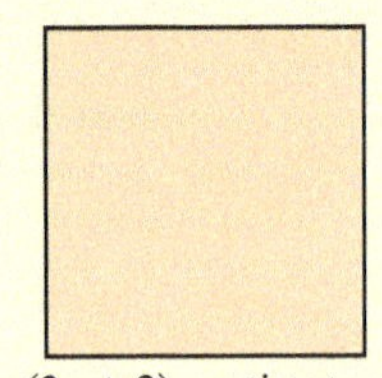

(3x + 2) centimeters

△ 39. $(x^2 + 3x + 1)$ meters

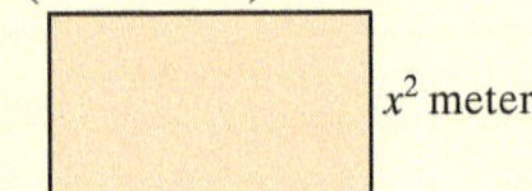

x^2 meters

△ 40.

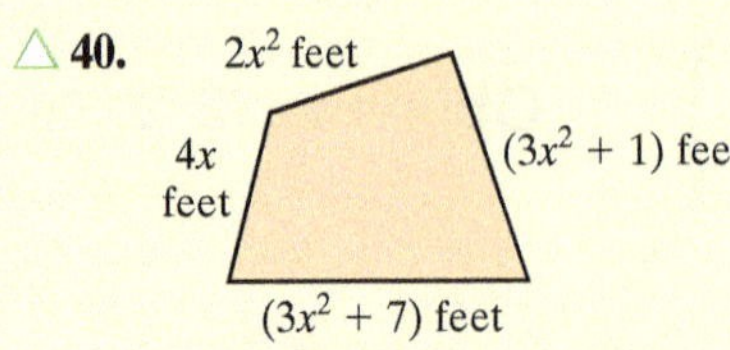

CONCEPT EXTENSIONS

For the exercises below, see the Concept Check in this section.

41. Without graphing, how can you tell that the graphs of $x^2 + y^2 = 1$ and $x^2 + y^2 = 4$ do not have any points of intersection?

42. Without solving, how can you tell that the graphs of $y = 2x + 3$ and $y = 2x + 7$ do not have any points of intersection?

43. How many real solutions are possible for a system of equations whose graphs are a circle and a parabola? Draw diagrams to illustrate each possibility.

44. How many real solutions are possible for a system of equations whose graphs are an ellipse and a line? Draw diagrams to illustrate each possibility.

Solve.

45. The sum of the squares of two numbers is 130. The difference of the squares of the two numbers is 32. Find the two numbers.

46. The sum of the squares of two numbers is 20. Their product is 8. Find the two numbers.

△ 47. During the development stage of a new rectangular keypad for a security system, it was decided that the area of the rectangle should be 285 square centimeters and the perimeter should be 68 centimeters. Find the dimensions of the keypad.

△ 48. A rectangular holding pen for cattle is to be designed so that its perimeter is 92 feet and its area is 525 feet. Find the dimensions of the holding pen.

Recall that in business, a demand function expresses the quantity of a commodity demanded as a function of the commodity's unit price. A supply function expresses the quantity of a commodity supplied as a function of the commodity's unit price. When the quantity produced and supplied is equal to the quantity demanded, then we have what is called ***market equilibrium.***

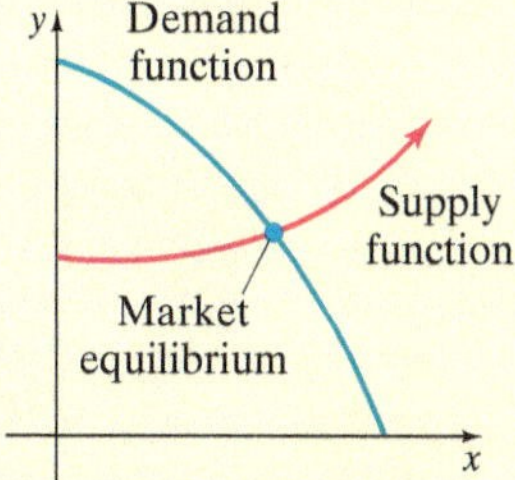

49. The demand function for a certain compact disc is given by the function

$$p = -0.01x^2 - 0.2x + 9$$

and the corresponding supply function is given by

$$p = 0.01x^2 - 0.1x + 3$$

where p is in dollars and x is in thousands of units. Find the equilibrium quantity and the corresponding price by solving the system consisting of the two given equations.

50. The demand function for a certain style of picture frame is given by the function

$$p = -2x^2 + 90$$

and the corresponding supply function is given by

$$p = 9x + 34$$

where p is in dollars and x is in thousands of units. Find the equilibrium quantity and the corresponding price by solving the system consisting of the two given equations.

 *Use a graphing calculator to verify the results of each exercise.*

51. Exercise 3
52. Exercise 4
53. Exercise 23
54. Exercise 24

20.4 Nonlinear Inequalities and Systems of Inequalities

Objective 1 Graphing Nonlinear Inequalities

We can graph a nonlinear inequality in two variables such as $\frac{x^2}{9} + \frac{y^2}{16} \le 1$ in a way similar to the way we graphed a linear inequality in two variables in Section 16.4. First, graph the related equation $\frac{x^2}{9} + \frac{y^2}{16} = 1$. The graph of the equation is our boundary. Then, using test points, we determine and shade the region whose points satisfy the inequality.

Example 1 Graph $\frac{x^2}{9} + \frac{y^2}{16} \le 1$.

Solution: First, graph the equation $\frac{x^2}{9} + \frac{y^2}{16} = 1$. Sketch a solid curve since the graph of $\frac{x^2}{9} + \frac{y^2}{16} \le 1$ includes the graph of $\frac{x^2}{9} + \frac{y^2}{16} = 1$. The graph is an

(Continued on next page)

Objectives

1 Graph a Nonlinear Inequality.

2 Graph a System of Nonlinear Inequalities.

Practice 1

Graph $\frac{x^2}{36} + \frac{y^2}{16} \ge 1$.

Answer

1.

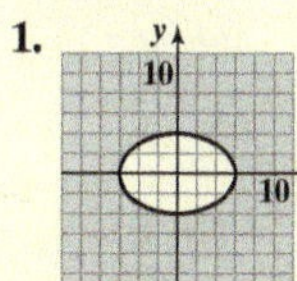

ellipse, and it divides the plane into two regions, the "inside" and the "outside" of the ellipse. To determine which region contains the solutions, select a test point in either region and determine whether the coordinates of the point satisfy the inequality. We choose (0, 0) as the test point.

$$\frac{x^2}{9} + \frac{y^2}{16} \le 1$$

$$\frac{0^2}{9} + \frac{0^2}{16} \le 1 \quad \text{Let } x = 0 \text{ and } y = 0.$$

$$0 \le 1 \quad \text{True}$$

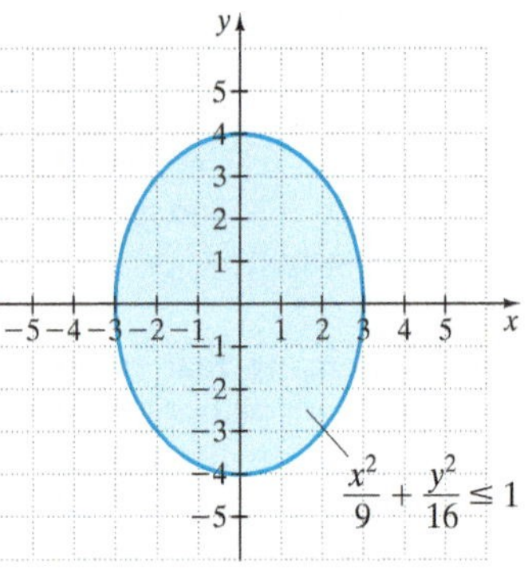

Since this statement is true, the solution set is the region containing (0, 0). The graph of the solution set includes the points on and inside the ellipse, as shaded in the figure.

Work Practice 1

Practice 2

Graph $16y^2 > 9x^2 + 144$.

Example 2 Graph $4y^2 > x^2 + 16$.

Solution: The related equation is $4y^2 = x^2 + 16$. Subtract x^2 from both sides and divide both sides by 16, and we have $\frac{y^2}{4} - \frac{x^2}{16} = 1$, which is a hyperbola. Graph the hyperbola as a dashed curve since the graph of $4y^2 > x^2 + 16$ does *not* include the graph of $4y^2 = x^2 + 16$. The hyperbola divides the plane into three regions. Select a test point in each region—not on a boundary line—to determine whether that region contains solutions of the inequality.

Test Region *A* with (0, 4)	**Test Region *B* with (0, 0)**	**Test Region *C* with (0, −4)**
$4y^2 > x^2 + 16$	$4y^2 > x^2 + 16$	$4y^2 > x^2 + 16$
$4(4)^2 > 0^2 + 16$	$4(0)^2 > 0^2 + 16$	$4(-4)^2 > 0^2 + 16$
$64 > 16$ True	$0 > 16$ False	$64 > 16$ True

The graph of the solution set includes the shaded regions *A* and *C* only, not the boundary.

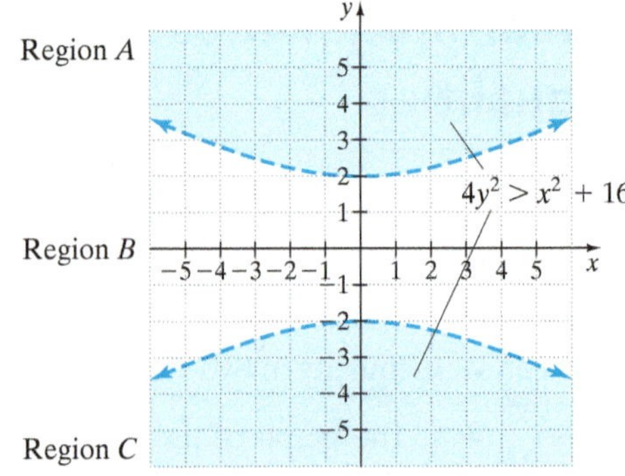

Work Practice 2

Answer

2.

Objective 2 Graphing Systems of Nonlinear Inequalities

In Section 16.4 we graphed systems of linear inequalities. Recall that the graph of a system of inequalities is the intersection of the graphs of the inequalities.

Example 3 Graph the system

$$\begin{cases} x \le 1 - 2y \\ y \le x^2 \end{cases}$$

Solution: We graph each inequality on the same set of axes. The intersection is shown in the third graph below. It is the darkest shaded (appears purple) region along with its boundary lines. The coordinates of the points of intersection can be found by solving the related system.

$$\begin{cases} x = 1 - 2y \\ y = x^2 \end{cases}$$

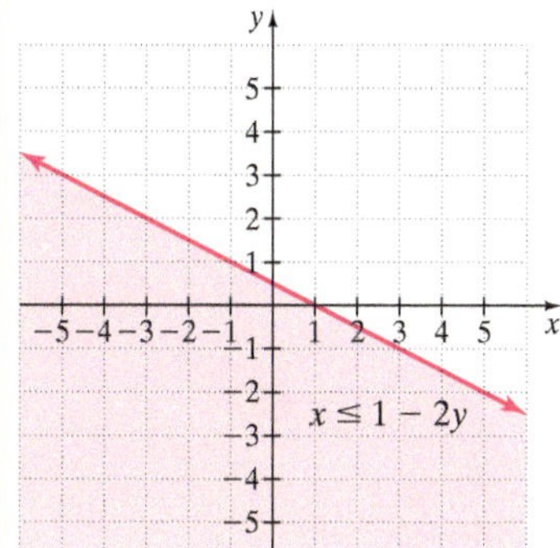

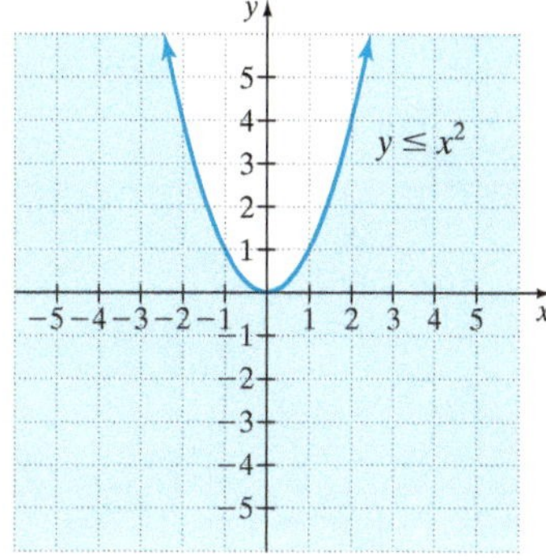

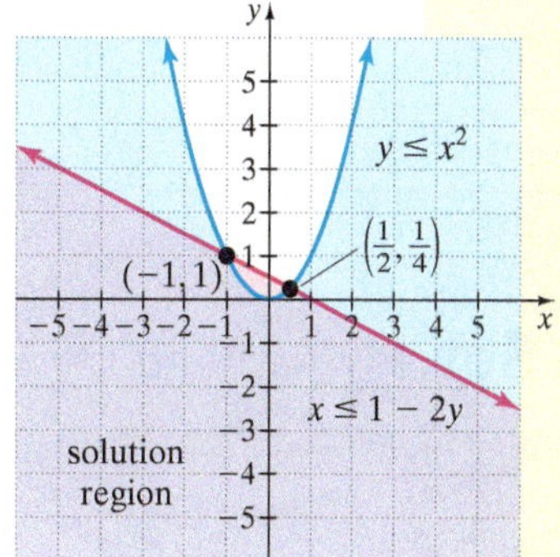

Work Practice 3

Practice 3

Graph the system
$$\begin{cases} y \ge x^2 \\ y \le -3x + 2 \end{cases}$$

Example 4 Graph the system

$$\begin{cases} x^2 + y^2 < 25 \\ \dfrac{x^2}{9} - \dfrac{y^2}{25} < 1 \\ y < x + 3 \end{cases}$$

Solution: We graph each inequality. The graph of $x^2 + y^2 < 25$ contains points "inside" the circle that has center $(0, 0)$ and radius 5. The graph of $\frac{x^2}{9} - \frac{y^2}{25} < 1$ is the region between the two branches of the hyperbola with x-intercepts -3 and 3 and center $(0, 0)$. The graph of $y < x + 3$ is the region "below" the line with slope 1 and y-intercept $(0, 3)$. The graph of the solution set of the system is the intersection of all the graphs, the darkest shaded region shown. The boundary of this region is not part of the solution.

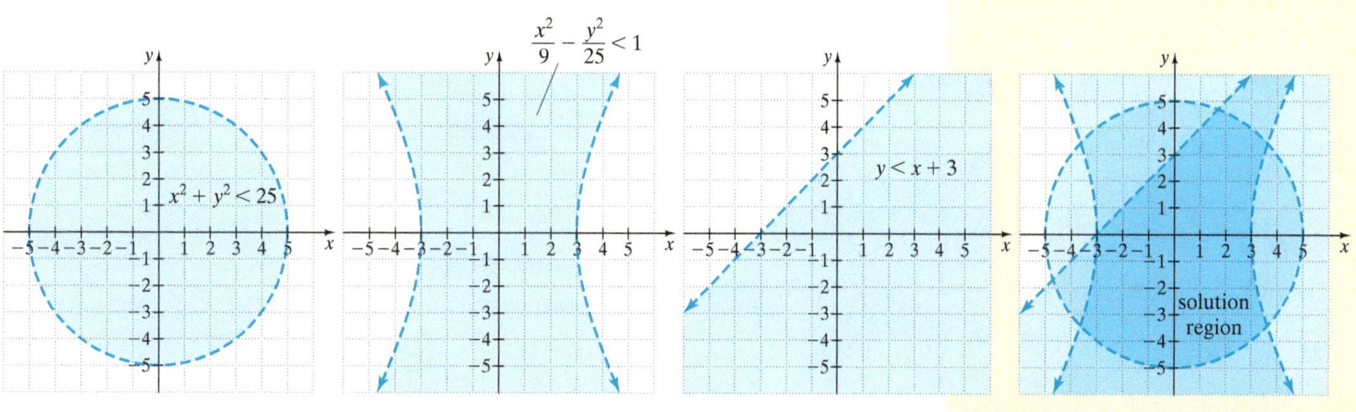

Work Practice 4

Practice 4

Graph the system
$$\begin{cases} x^2 + y^2 < 16 \\ \dfrac{x^2}{4} - \dfrac{y^2}{9} < 1. \\ y < x + 3 \end{cases}$$

Answers

3. 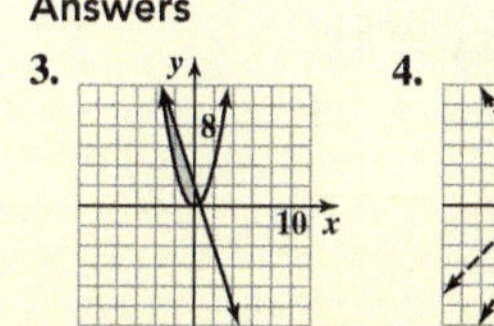

4. 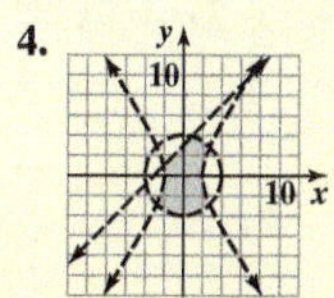

Vocabulary, Readiness & Video Check

Martin-Gay Interactive Videos

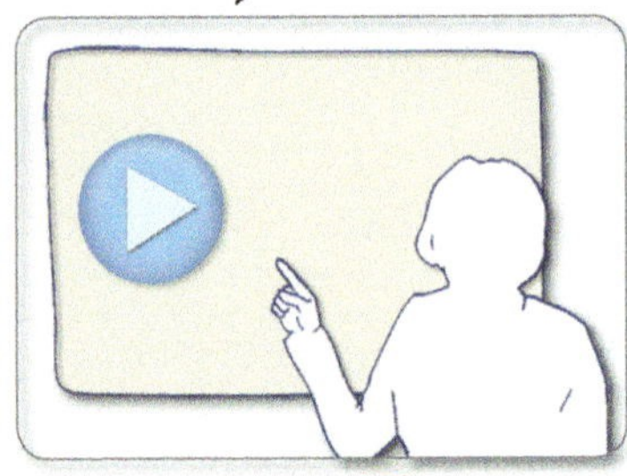

See Video 20.4 

Watch the section lecture video and answer the following questions.

Objective 1 **1.** From Example 1, explain the similarities between graphing linear inequalities and graphing nonlinear inequalities.

Objective 2 **2.** From Example 2, describe one possible illustration of graphs of two circle inequalities in which the system has no solution—that is, the graph of the inequalities in the system do not overlap.

20.4 Exercise Set MyMathLab®

Graph each inequality. See Examples 1 and 2.

1. $y < x^2$

2. $y < -x^2$

3. $x^2 + y^2 \geq 16$

4. $x^2 + y^2 < 36$

5. $\frac{x^2}{4} - y^2 < 1$

6. $x^2 - \frac{y^2}{9} \geq 1$

7. $y > (x - 1)^2 - 3$

8. $y > (x + 3)^2 + 2$

9. $x^2 + y^2 \leq 9$

10. $x^2 + y^2 > 4$

11. $y > -x^2 + 5$

12. $y < -x^2 + 5$

13. $\frac{x^2}{4} + \frac{y^2}{9} \leq 1$

14. $\frac{x^2}{25} + \frac{y^2}{4} \geq 1$

15. $\frac{y^2}{4} - x^2 \leq 1$

16. $\frac{y^2}{16} - \frac{x^2}{9} > 1$

17. $y < (x - 2)^2 + 1$

18. $y > (x - 2)^2 + 1$

19. $y \leq x^2 + x - 2$

20. $y > x^2 + x - 2$

Graph each system. See Examples 3 and 4.

21. $\begin{cases} 4x + 3y \geq 12 \\ x^2 + y^2 < 16 \end{cases}$

22. $\begin{cases} 3x - 4y \leq 12 \\ x^2 + y^2 < 16 \end{cases}$

23. $\begin{cases} x^2 + y^2 \leq 9 \\ x^2 + y^2 \geq 1 \end{cases}$

24. $\begin{cases} x^2 + y^2 \geq 9 \\ x^2 + y^2 \geq 16 \end{cases}$

25. $\begin{cases} y > x^2 \\ y \geq 2x + 1 \end{cases}$

26. $\begin{cases} y \leq -x^2 + 3 \\ y \leq 2x - 1 \end{cases}$

27. $\begin{cases} x^2 + y^2 > 9 \\ y > x^2 \end{cases}$

28. $\begin{cases} x^2 + y^2 \leq 9 \\ y < x^2 \end{cases}$

29. $\begin{cases} \frac{x^2}{4} + \frac{y^2}{9} \geq 1 \\ x^2 + y^2 \geq 4 \end{cases}$

30. $\begin{cases} x^2 + (y - 2)^2 \geq 9 \\ \frac{x^2}{4} + \frac{y^2}{25} < 1 \end{cases}$

31. $\begin{cases} x^2 - y^2 \geq 1 \\ y \geq 0 \end{cases}$

32. $\begin{cases} x^2 - y^2 \geq 1 \\ x \geq 0 \end{cases}$

33. $\begin{cases} x + y \geq 1 \\ 2x + 3y < 1 \\ x > -3 \end{cases}$

34. $\begin{cases} x - y < -1 \\ 4x - 3y > 0 \\ y > 0 \end{cases}$

35. $\begin{cases} x^2 - y^2 < 1 \\ \frac{x^2}{16} + y^2 \leq 1 \\ x \geq -2 \end{cases}$

36. $\begin{cases} x^2 - y^2 \geq 1 \\ \frac{x^2}{16} + \frac{y^2}{4} \leq 1 \\ y \geq 1 \end{cases}$

REVIEW AND PREVIEW

Determine whether each graph is the graph of a function. See Section 10.6.

37.

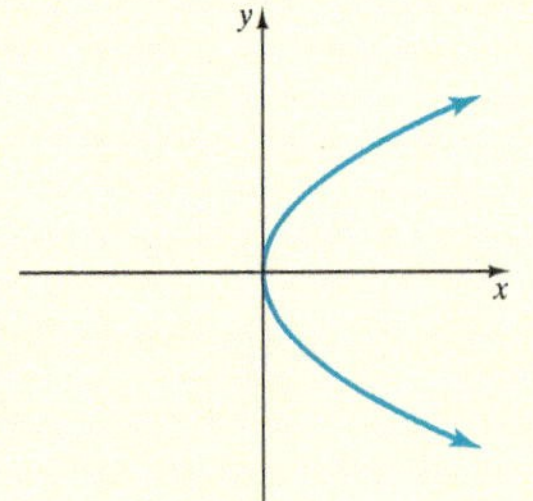

38.

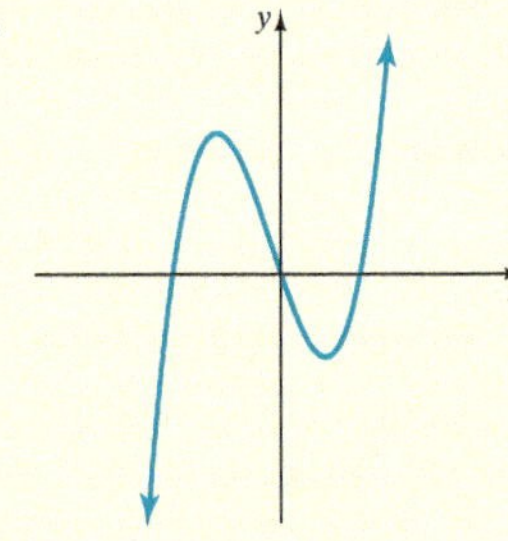

39.

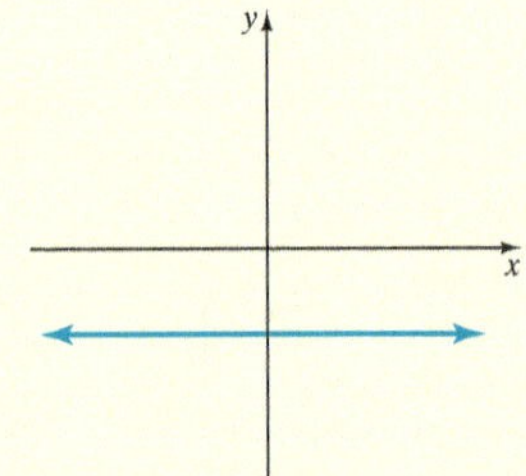

40.

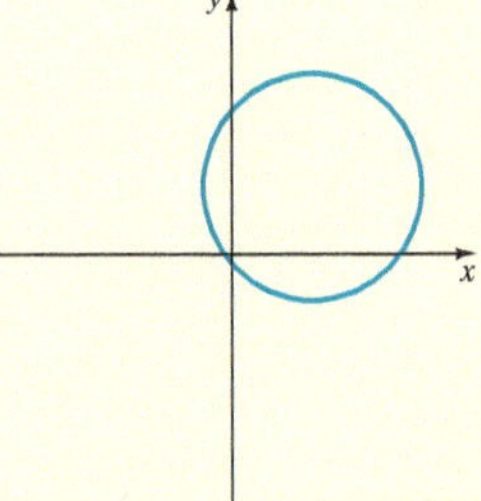

Find each function value if $f(x) = 3x^2 - 2$. See Section 10.6.

41. $f(-1)$

42. $f(-3)$

43. $f(a)$

44. $f(b)$

CONCEPT EXTENSIONS

45. Discuss how graphing a linear inequality such as $x + y < 9$ is similar to graphing a nonlinear inequality such as $x^2 + y^2 < 9$.

46. Discuss how graphing a linear inequality such as $x + y < 9$ is different from graphing a nonlinear inequality such as $x^2 + y^2 < 9$.

47. Graph the system $\begin{cases} y \le x^2 \\ y \ge x + 2 \\ x \ge 0 \\ y \ge 0 \end{cases}$.

48. Graph the system: $\begin{cases} x \ge 0 \\ y \ge 0 \\ y \ge x^2 + 1 \\ y \le 4 - x \end{cases}$

Chapter 20 Vocabulary Check

Fill in each blank with one of the words or phrases listed below.

circle	ellipse	hyperbola
conic sections	vertex	diameter
center	radius	nonlinear system of equations

1. A(n) ______________ is the set of all points in a plane that are the same distance from a fixed point, called the ______________.
2. A(n) ______________ is a system of equations at least one of which is not linear.
3. A(n) ______________ is the set of points in a plane such that the sum of the distances of those points from two fixed points is a constant.
4. In a circle, the distance from the center to a point of the circle is called its ______________.
5. A(n) ______________ is the set of points in a plane such that the absolute value of the difference of the distance from two fixed points is constant.
6. The circle, parabola, ellipse, and hyperbola are called the ______________.
7. For a parabola that opens upward, the lowest point is the ______________.
8. Twice a circle's radius is its ______________.

Helpful Hint Are you preparing for your test? Don't forget to take the Chapter 20 Test on page 1363. Then check your answers at the back of the text and use the Chapter Test Prep Videos to see the fully worked-out solutions to any of the exercises you want to review.

20 Chapter Highlights

Definitions and Concepts	Examples
Section 20.1 The Parabola and the Circle	
Parabolas $y = a(x - h)^2 + k$ 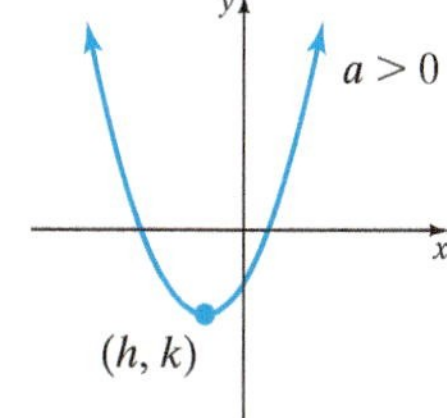 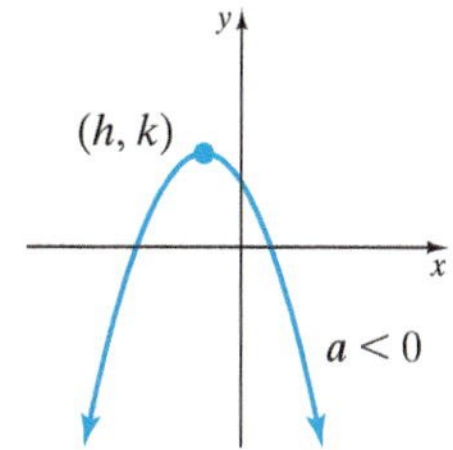	Graph $x = 3y^2 - 12y + 13$. $x - 13 = 3y^2 - 12y$ $x - 13 + 3(4) = 3(y^2 - 4y + 4)$ Add 3(4) to both sides. $x = 3(y - 2)^2 + 1$ Since $a = 3$, this parabola opens to the right with vertex $(1, 2)$. Its axis of symmetry is $y = 2$. The x-intercept is $(13, 0)$.

(continued)

Definitions and Concepts	Examples

Section 20.1 The Parabola and the Circle (*continued*)

$x = a(y - k)^2 + h$

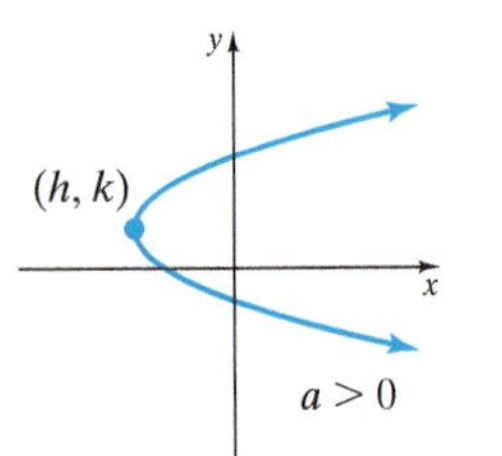

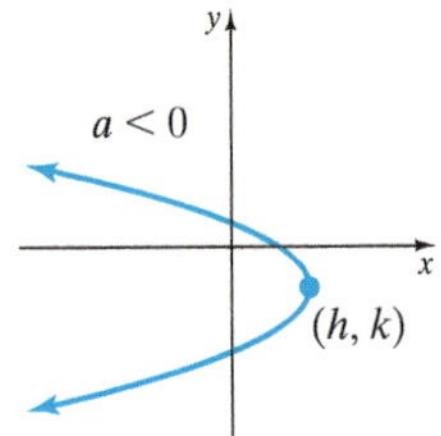

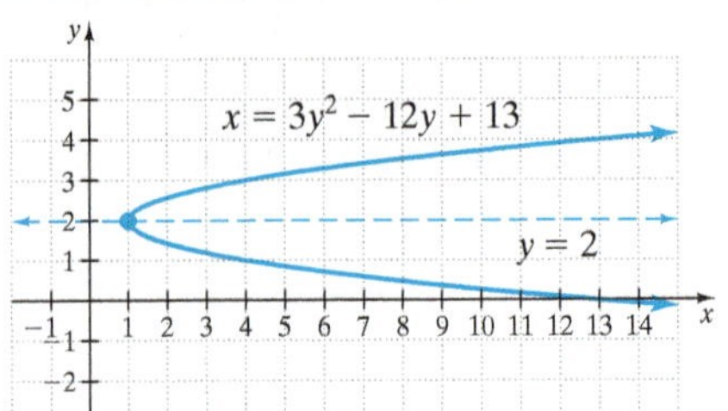

Circle

The graph of $(x - h)^2 + (y - k)^2 = r^2$ is a circle with center (h, k) and radius r.

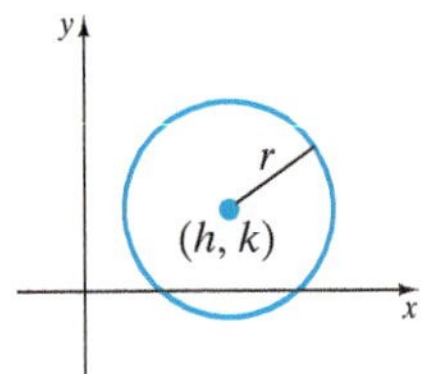

Graph $x^2 + (y + 3)^2 = 5$.

This equation can be written as

$$(x - 0)^2 + (y + 3)^2 = 5 \text{ with } h = 0,$$
$$k = -3, \text{ and } r = \sqrt{5}.$$

The center of this circle is $(0, -3)$, and the radius is $\sqrt{5}$.

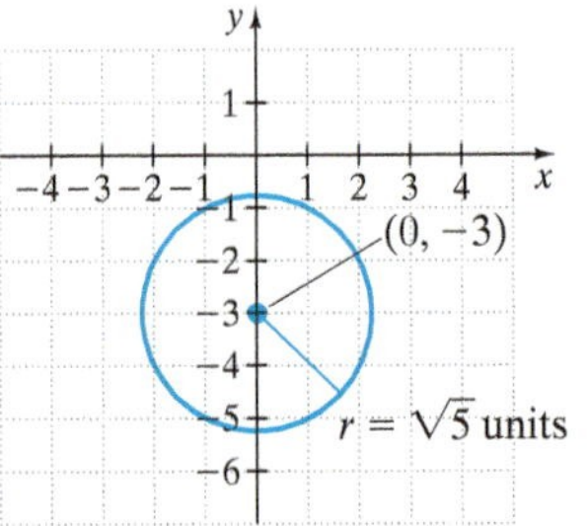

Section 20.2 The Ellipse and the Hyperbola

Ellipse with Center (0, 0)

The graph of an equation of the form $\frac{x^2}{a^2} + \frac{y^2}{b^2} = 1$ is an ellipse with center $(0, 0)$. The x-intercepts are $(a, 0)$ and $(-a, 0)$, and the y-intercepts are $(0, b)$ and $(0, -b)$.

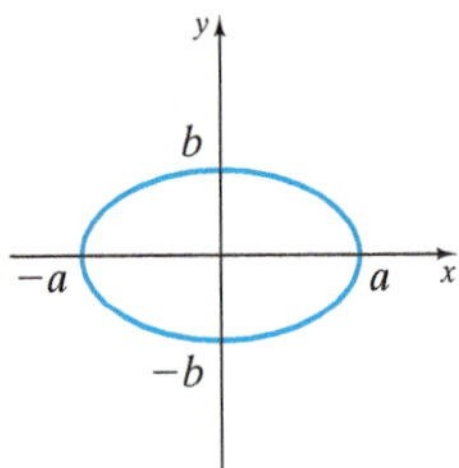

Graph $4x^2 + 9y^2 = 36$.

$$\frac{x^2}{9} + \frac{y^2}{4} = 1 \quad \text{Divide by 36.}$$

$$\frac{x^2}{3^2} + \frac{y^2}{2^2} = 1$$

The ellipse has center $(0, 0)$, x-intercepts $(3, 0)$ and $(-3, 0)$, and y-intercepts $(0, 2)$ and $(0, -2)$.

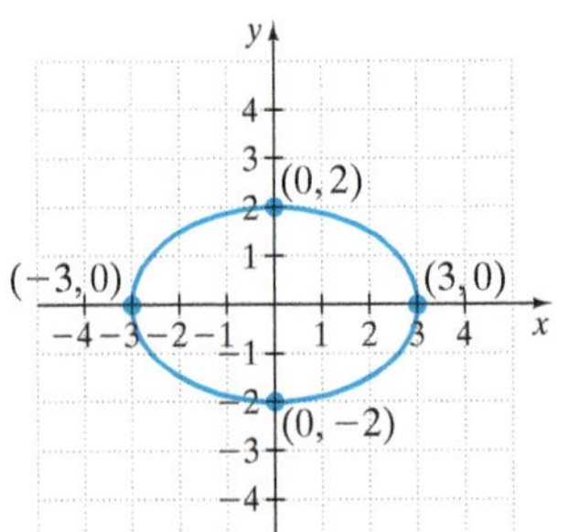

Hyperbola with Center (0, 0)

The graph of an equation of the form $\frac{x^2}{a^2} - \frac{y^2}{b^2} = 1$ is a hyperbola with center $(0, 0)$ and x-intercepts $(a, 0)$ and $(-a, 0)$.

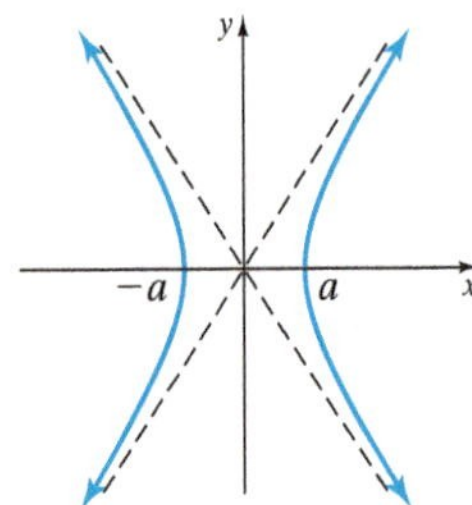

Definitions and Concepts	Examples

Section 20.2 The Ellipse and the Hyperbola (*continued*)

The graph of an equation of the form $\frac{y^2}{b^2} - \frac{x^2}{a^2} = 1$ is a hyperbola with center $(0, 0)$ and y-intercepts $(0, b)$ and $(0, -b)$.

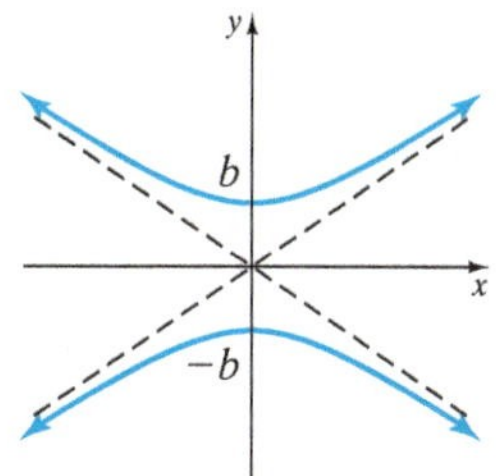

Graph $\frac{x^2}{9} - \frac{y^2}{4} = 1$. Here $a = 3$ and $b = 2$.

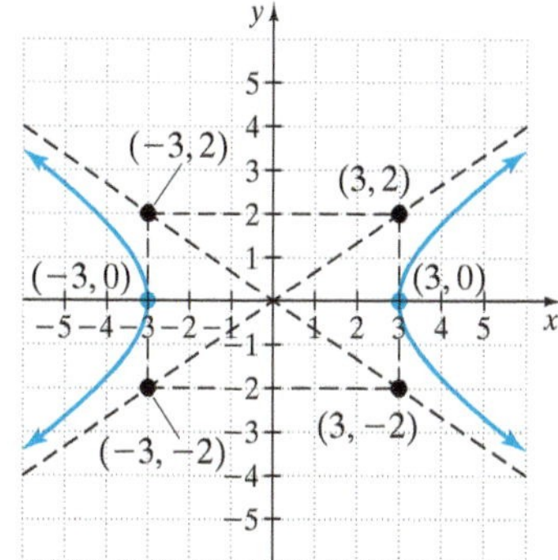

Section 20.3 Solving Nonlinear Systems of Equations

A **nonlinear system of equations** is a system of equations at least one of which is not linear. Both the substitution method and the elimination method may be used to solve a nonlinear system of equations.

Solve the nonlinear system $\begin{cases} y = x + 2 \\ 2x^2 + y^2 = 3 \end{cases}$.

Substitute $x + 2$ for y in the second equation.

$$2x^2 + y^2 = 3$$
$$2x^2 + (x + 2)^2 = 3$$
$$2x^2 + x^2 + 4x + 4 = 3$$
$$3x^2 + 4x + 1 = 0$$
$$(3x + 1)(x + 1) = 0$$
$$x = -\frac{1}{3}, x = -1$$

If $x = -\frac{1}{3}$, $y = x + 2 = -\frac{1}{3} + 2 = \frac{5}{3}$.

If $x = -1$, $y = x + 2 = -1 + 2 = 1$.

The solutions are $\left(-\frac{1}{3}, \frac{5}{3}\right)$ and $(-1, 1)$.

Section 20.4 Nonlinear Inequalities and Systems of Inequalities

The graph of a system of inequalities is the intersection of the graphs of the inequalities.

Graph the system $\begin{cases} x \geq y^2 \\ x + y \leq 4 \end{cases}$.

The graph of the system is the purple shaded region along with its boundary lines.

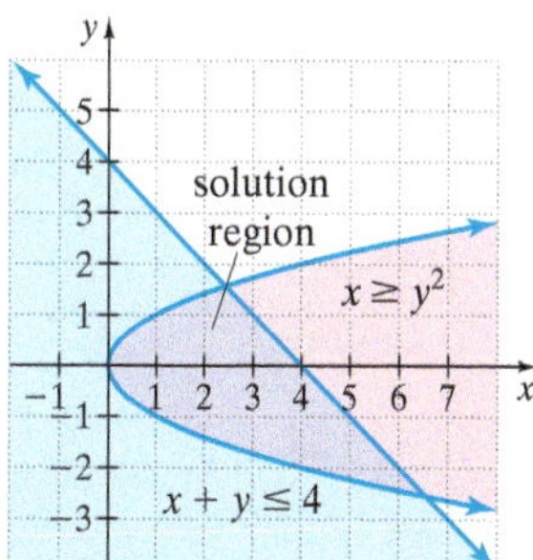

Chapter 20 Review

(20.1) *Write an equation of the circle with the given center and radius.*

1. center $(-4, 4)$, radius 3
2. center $(5, 0)$, radius 5
3. center $(-7, -9)$, radius $\sqrt{11}$
4. center $(0, 0)$, radius $\frac{7}{2}$

Sketch the graph of the equation. If the graph is a circle, find its center. If the graph is a parabola, find its vertex.

5. $x^2 + y^2 = 7$
6. $x = 2(y - 5)^2 + 4$
7. $x = -(y + 2)^2 + 3$
8. $(x - 1)^2 + (y - 2)^2 = 4$
9. $y = -x^2 + 4x + 10$
10. $x = -y^2 - 4y + 6$
11. $x = \frac{1}{2}y^2 + 2y + 1$
12. $y = -3x^2 + \frac{1}{2}x + 4$
13. $x^2 + y^2 + 2x + y = \frac{3}{4}$
14. $x^2 + y^2 - 3y = \frac{7}{4}$
15. $4x^2 + 4y^2 + 16x + 8y = 1$
16. $3x^2 + 3y^2 + 18x - 12y = -12$

(20.1, 20.2) *Graph each equation.*

17. $x^2 - \frac{y^2}{4} = 1$
18. $x^2 + \frac{y^2}{4} = 1$
19. $4y^2 + 9x^2 = 36$
20. $-5x^2 + 25y^2 = 125$
21. $x^2 - y^2 = 1$
22. $\frac{(x+3)^2}{9} + \frac{(y-4)^2}{25} = 1$
23. $y = x^2 + 9$
24. $36y^2 - 49x^2 = 1764$
25. $x = 4y^2 - 16$
26. $y = x^2 + 4x + 6$
27. $y^2 + 2(x - 1)^2 = 8$
28. $x - 4y = y^2$
29. $x^2 - 4 = y^2$
30. $x^2 = 4 - y^2$
31. $36y^2 = 576 + 16x^2$
32. $3(x - 7)^2 + 3(y + 4)^2 = 1$

(20.3) *Solve each system of equations.*

33. $\begin{cases} y = 2x - 4 \\ y^2 = 4x \end{cases}$
34. $\begin{cases} x^2 + y^2 = 4 \\ x - y = 4 \end{cases}$
35. $\begin{cases} y = x + 2 \\ y = x^2 \end{cases}$
36. $\begin{cases} 4x - y^2 = 0 \\ 2x^2 + y^2 = 16 \end{cases}$
37. $\begin{cases} x^2 + 4y^2 = 16 \\ x^2 + y^2 = 4 \end{cases}$
38. $\begin{cases} x^2 + 2y = 9 \\ 5x - 2y = 5 \end{cases}$
39. $\begin{cases} y = 3x^2 + 5x - 4 \\ y = 3x^2 - x + 2 \end{cases}$
40. $\begin{cases} x^2 - 3y^2 = 1 \\ 4x^2 + 5y^2 = 21 \end{cases}$

△ 41. Find the length and the width of a room whose area is 150 square feet and whose perimeter is 50 feet.

42. What is the greatest number of real number solutions possible for a system of two equations whose graphs are an ellipse and a hyperbola?

(20.4) *Graph each inequality or system of inequalities.*

43. $y \le -x^2 + 3$
44. $x < y^2 - 1$
45. $x^2 + y^2 < 9$
46. $\frac{x^2}{4} + \frac{y^2}{9} \ge 1$
47. $\begin{cases} 3x + 4y \le 12 \\ x - 2y > 6 \end{cases}$
48. $\begin{cases} x^2 + y^2 \le 16 \\ x^2 + y^2 \ge 4 \end{cases}$
49. $\begin{cases} x^2 + y^2 < 4 \\ x^2 - y^2 \le 1 \end{cases}$
50. $\begin{cases} x^2 + y^2 < 4 \\ y \ge x^2 - 1 \\ x \ge 0 \end{cases}$

MIXED REVIEW

51. Write an equation of the circle with center $(-7, 8)$ and radius 5.

Graph each equation.

52. $y = x^2 + 6x + 9$
53. $x = y^2 + 6y + 9$
54. $\frac{y^2}{4} - \frac{x^2}{16} = 1$
55. $\frac{y^2}{4} + \frac{x^2}{16} = 1$
56. $\frac{(x-2)^2}{4} + (y-1)^2 = 1$
57. $y^2 = x^2 + 6$
58. $y^2 + (x - 2)^2 = 10$
59. $3x^2 + 6x + 3y^2 = 9$
60. $x^2 + y^2 - 8y = 0$
61. $6(x - 2)^2 + 9(y + 5)^2 = 36$
62. $\frac{x^2}{16} - \frac{y^2}{25} = 1$

Solve each system of equations.

63. $\begin{cases} y = x^2 - 5x + 1 \\ y = -x + 6 \end{cases}$
64. $\begin{cases} x^2 + y^2 = 10 \\ 9x^2 + y^2 = 18 \end{cases}$

Graph each inequality or system of inequalities.

65. $x^2 - y^2 < 1$
66. $\begin{cases} y > x^2 \\ x + y \ge 3 \end{cases}$

Chapter 20 Test

Step-by-step test solutions are found on the Chapter Test Prep Videos. Where available: 

Sketch the graph of each equation.

1. $x^2 + y^2 = 36$
2. $x^2 - y^2 = 36$
3. $16x^2 + 9y^2 = 144$
4. $y = x^2 - 8x + 16$
5. $x^2 + y^2 + 6x = 16$
6. $x = y^2 + 8y - 3$
7. $\dfrac{(x-4)^2}{16} + \dfrac{(y-3)^2}{9} = 1$
8. $y^2 - x^2 = 1$

Solve each system.

9. $\begin{cases} x^2 + y^2 = 169 \\ 5x + 12y = 0 \end{cases}$
10. $\begin{cases} x^2 + y^2 = 26 \\ x^2 - 2y^2 = 23 \end{cases}$
11. $\begin{cases} y = x^2 - 5x + 6 \\ y = 2x \end{cases}$
12. $\begin{cases} x^2 + 4y^2 = 5 \\ y = x \end{cases}$

Graph each system.

13. $\begin{cases} 2x + 5y \geq 10 \\ y \geq x^2 + 1 \end{cases}$
14. $\begin{cases} \dfrac{x^2}{4} + y^2 \leq 1 \\ x + y > 1 \end{cases}$
15. $\begin{cases} x^2 + y^2 > 1 \\ \dfrac{x^2}{4} - y^2 \geq 1 \end{cases}$
16. $\begin{cases} x^2 + y^2 \geq 4 \\ x^2 + y^2 < 16 \\ y \geq 0 \end{cases}$

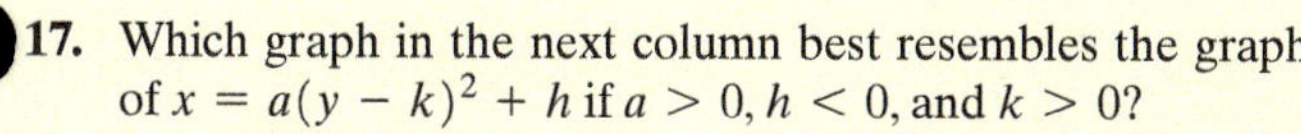

17. Which graph in the next column best resembles the graph of $x = a(y-k)^2 + h$ if $a > 0$, $h < 0$, and $k > 0$?

A.

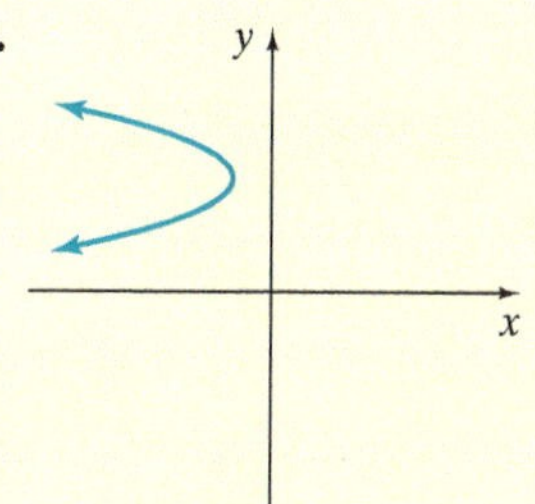

B.

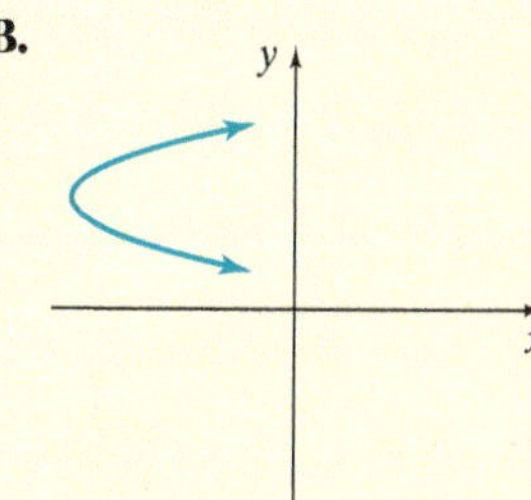

C.

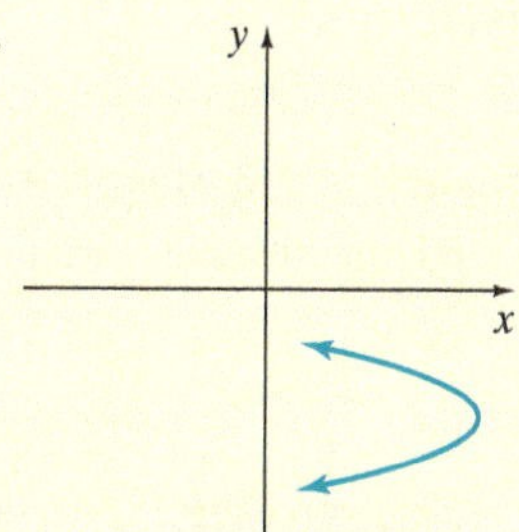

D.

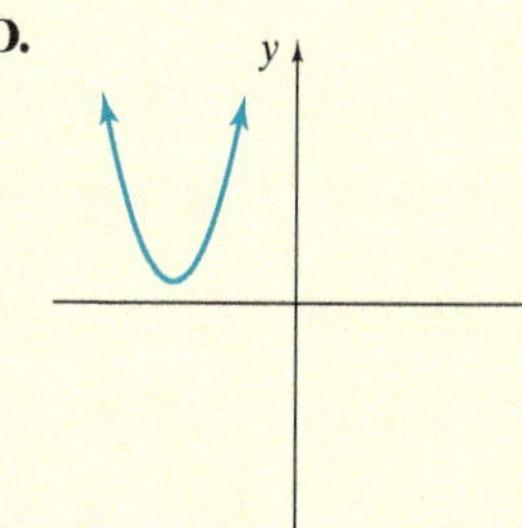

18. A bridge has an arch in the shape of half an ellipse. If the equation of the ellipse, measured in feet, is $100x^2 + 225y^2 = 22{,}500$, find the height of the arch from the road and the width of the arch.

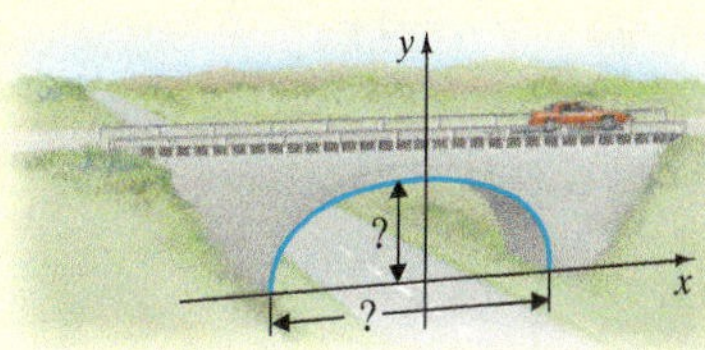

Chapters 1–20 Cumulative Review

1. Solve $2x \geq 0$ *and* $4x - 1 \leq -9$.
2. Solve $3x + 4 > 1$ *and* $2x - 5 \leq 9$. Write the solution in interval notation.
3. Solve $5x - 3 \leq 10$ *or* $x + 1 \geq 5$.
4. Find the slope of the line that goes through $(3, 2)$ and $(1, -4)$.

Solve.

5. $|5w + 3| = 7$
6. Two planes leave Greensboro, one traveling north and the other south. After 2 hours they are 650 miles apart. If one plane is flying 25 mph faster than the other, what is the speed of each?
7. $\left|\dfrac{x}{2} - 1\right| = 11$
8. Use the quotient rule to simplify.
   a. $\dfrac{4^8}{4^3}$
   b. $\dfrac{y^{11}}{y^5}$
   c. $\dfrac{32x^7}{4x^6}$
   d. $\dfrac{18a^{12}b^6}{12a^8b^6}$
9. Solve $|3x + 2| = |5x - 8|$.
10. Factor.
    a. $3y^2 + 14y + 15$
    b. $20a^5 + 54a^4 + 10a^3$
11. Solve for m: $|m - 6| < 2$.
12. Perform the indicated operation and simplify if possible.
    $\dfrac{2}{3a - 15} - \dfrac{a}{25 - a^2}$

13. Simplify $\dfrac{x^{-1} + 2xy^{-1}}{x^{-2} - x^{-2}y^{-1}}$.

14. Simplify each complex fraction.

a. $(a^{-1} - b^{-1})^{-1}$ **b.** $\dfrac{2 - \dfrac{1}{x}}{4x - \dfrac{1}{x}}$

15. Solve $|2x + 9| + 5 > 3$.

16. Solve $\dfrac{2}{x + 3} = \dfrac{1}{x^2 - 9} - \dfrac{1}{x - 3}$.

17. Use the remainder theorem and synthetic division to find $P(4)$ if $P(x) = 4x^6 - 25x^5 + 35x^4 + 17x^2$.

18. Suppose that y varies inversely as x. If $y = 3$ when $x = \dfrac{2}{3}$, find the constant of variation and the inverse variation equation.

19. Find the cube roots.

a. $\sqrt[3]{1}$ **b.** $\sqrt[3]{-64}$ **c.** $\sqrt[3]{\dfrac{8}{125}}$

d. $\sqrt[3]{x^6}$ **e.** $\sqrt[3]{-27x^9}$

20. Multiply and simplify if possible.

a. $\sqrt{5}(2 + \sqrt{15})$

b. $(\sqrt{3} - \sqrt{5})(\sqrt{7} - 1)$

c. $(2\sqrt{5} - 1)^2$

d. $(3\sqrt{2} + 5)(3\sqrt{2} - 5)$

21. Multiply.

a. $z^{2/3}(z^{1/3} - z^5)$

b. $(x^{1/3} - 5)(x^{1/3} + 2)$

22. Rationalize the denominator $\dfrac{-2}{\sqrt{3} + 3}$.

23. Use the quotient rule to divide, and simplify if possible.

a. $\dfrac{\sqrt{20}}{\sqrt{5}}$ **b.** $\dfrac{\sqrt{50x}}{2\sqrt{2}}$

c. $\dfrac{7\sqrt[3]{48x^4y^8}}{\sqrt[3]{6y^2}}$ **d.** $\dfrac{2\sqrt[4]{32a^8b^6}}{\sqrt[4]{a^{-1}b^2}}$

24. Solve $\sqrt{2x - 3} = x - 3$.

25. Add or subtract as indicated.

a. $\dfrac{\sqrt{45}}{4} - \dfrac{\sqrt{5}}{3}$ **b.** $\sqrt[3]{\dfrac{7x}{8}} + 2\sqrt[3]{7x}$

26. Use the discriminant to determine the number and type of solutions for $9x^2 - 6x = -4$.

27. Rationalize the denominator of $\sqrt{\dfrac{7x}{3y}}$.

28. Solve $\dfrac{4}{x - 2} - \dfrac{x}{x + 2} = \dfrac{16}{x^2 - 4}$.

29. Solve $\sqrt{2x - 3} = 9$.

30. Solve $6x^2 + 11x \le 10$.

31. Find the following powers of i.

a. i^7 **b.** i^{20}

c. i^{46} **d.** i^{-12}

32. Graph $f(x) = (x + 2)^2 - 1$.

33. Solve $p^2 + 2p = 4$ by completing the square.

34. Find the maximum value of $f(x) = -x^2 - 6x + 4$.

35. Solve: $\dfrac{1}{4}m^2 - m + \dfrac{1}{2} = 0$.

36. Find the inverse of $f(x) = \dfrac{x + 1}{2}$.

37. Solve: $p^4 - 3p^2 - 4 = 0$.

38. Use the quotient rule to simplify.

a. $\dfrac{\sqrt{32}}{\sqrt{4}}$ **b.** $\dfrac{\sqrt[3]{240y^2}}{5\sqrt[3]{3y^{-4}}}$

c. $\dfrac{\sqrt[5]{64x^9y^2}}{\sqrt[5]{2x^2y^{-8}}}$

39. Solve $\dfrac{x + 2}{x - 3} \le 0$.

40. Graph $4x^2 + 9y^2 = 36$.

41. Graph $g(x) = \dfrac{1}{2}(x + 2)^2 + 5$. Find the vertex and the axis of symmetry.

42. Solve each equation for x.

a. $64^x = 4$ **b.** $125^{x-3} = 25$

c. $\dfrac{1}{81} = 3^{2x}$

43. Find the vertex of the graph of $f(x) = x^2 - 4x - 12$.

44. Graph the system: $\begin{cases} x + 2y < 8 \\ y \ge x^2 \end{cases}$

45. Find the distance between $(2, -5)$ and $(1, -4)$. Give an exact distance and a three-decimal-place approximation.

46. Solve the system $\begin{cases} x^2 + y^2 = 36 \\ y = x + 6 \end{cases}$

Sequences, Series, and the Binomial Theorem

21

Sections

Check Your Progress

Having explored in some depth the concept of function, we turn now in this final chapter to *sequences.* In one sense, a sequence is simply an ordered list of numbers. In another sense, a sequence is itself a function. Phenomena modeled by such functions are everywhere around us. The starting place for all mathematics is the sequence of natural numbers: 1, 2, 3, 4, and so on.

Sequences lead us to *series,* which are a sum of ordered numbers. Through series, we gain new insight, for example about the expansion of a binomial $(a + b)^n$, the concluding topic of this book.

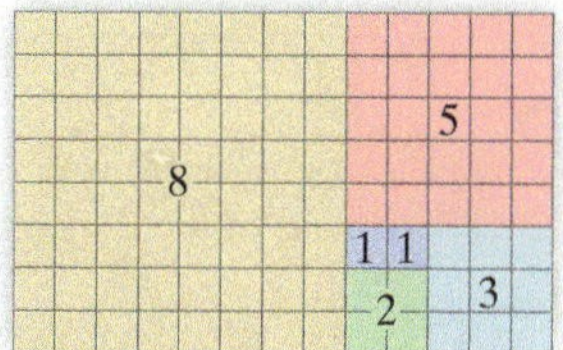

A tiling with squares whose sides are successive Fibonacci numbers in length

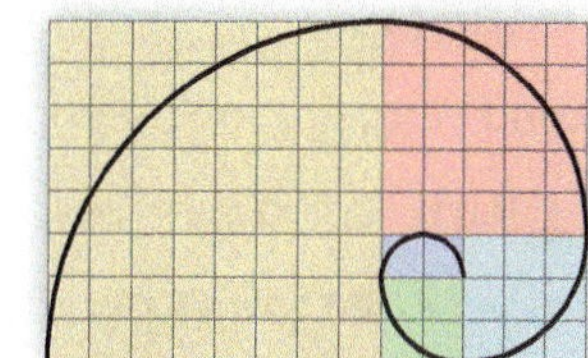
A Fibonacci spiral, created by drawing arcs connecting the opposite corners of squares in the Fibonacci tiling

The Fibonacci sequence is a special sequence in which the first two terms are 1 and each term thereafter is the sum of the two previous terms:

1, 1, 2, 3, 5, 8, 13, 21, . . .

The Fibonacci numbers are named after Leonardo of Pisa, known as Fibonacci, although there is some evidence that these numbers had been described earlier in India.

There are numerous interesting facts about this sequence, and some are shown on the diagrams on this page. In Section 21.1, Exercise 46, you will have the opportunity to check a formula for this sequence.

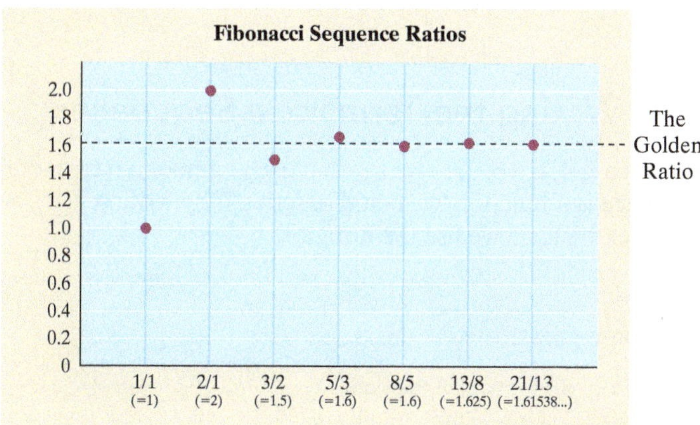

The ratio of successive numbers in the Fibonacci sequence approaches a number called the golden ratio or golden number, which is approximately 1.618034.

21.1 Sequences 

Objectives

1 Write the Terms of a Sequence Given Its General Term.

2 Find the General Term of a Sequence.

3 Solve Applications That Involve Sequences.

Suppose that a town's present population of 100,000 is growing by 5% each year. After the first year, the town's population will be

$$100{,}000 + 0.05(100{,}000) = 105{,}000$$

After the second year, the town's population will be

$$105{,}000 + 0.05(105{,}000) = 110{,}250$$

After the third year, the town's population will be

$$110{,}250 + 0.05(110{,}250) \approx 115{,}763$$

If we continue to calculate, the town's yearly population can be written as the **infinite sequence** of numbers

105,000, 110,250, 115,763, . . .

If we decide to stop calculating after a certain year (say, the fourth year), we obtain the **finite sequence**

105,000, 110,250, 115,763, 121,551

Sequences

An infinite sequence is a function whose domain is the set of natural numbers $\{1, 2, 3, 4, \ldots\}$.

A finite sequence is a function whose domain is the set of natural numbers $\{1, 2, 3, 4, \ldots, n\}$, where n is some natural number.

Objective 1 Writing the Terms of a Sequence 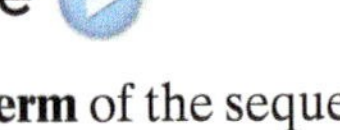

Given the sequence 2, 4, 8, 16, . . . , we say that each number is a **term** of the sequence. Because a sequence is a function, we could describe it by writing $f(n) = 2^n$, where n is a natural number. Instead, we use the notation

$$a_n = 2^n$$

Some function values are

$a_1 = 2^1 = 2$ First term of the sequence
$a_2 = 2^2 = 4$ Second term
$a_3 = 2^3 = 8$ Third term
$a_4 = 2^4 = 16$ Fourth term
$a_{10} = 2^{10} = 1024$ Tenth term

The nth term of the sequence a_n is called the **general term.**

Helpful Hint

If it helps, think of a sequence as simply a list of values in which a position is assigned. For the sequence directly above,

Value:	2,	4,	8,	16, . . . ,	1024
	↑	↑	↑	↑	↑
Position	1st	2nd	3rd	4th	10th

Example 1 Write the first five terms of the sequence whose general term is given by

$$a_n = n^2 - 1$$

Solution: Evaluate a_n, where n is 1, 2, 3, 4, and 5.

$a_n = n^2 - 1$

$a_1 = 1^2 - 1 = 0$ Replace n with 1.

$a_2 = 2^2 - 1 = 3$ Replace n with 2.

$a_3 = 3^2 - 1 = 8$ Replace n with 3.

$a_4 = 4^2 - 1 = 15$ Replace n with 4.

$a_5 = 5^2 - 1 = 24$ Replace n with 5.

Thus, the first five terms of the sequence $a_n = n^2 - 1$ are 0, 3, 8, 15, and 24.

Work Practice 1

Practice 1

Write the first five terms of the sequence whose general term is given by $a_n = 5 + n^2$.

Example 2 If the general term of a sequence is given by $a_n = \frac{(-1)^n}{3n}$, find

a. the first term of the sequence **b.** a_8

c. the one-hundredth term of the sequence **d.** a_{15}

Solution:

a. $a_1 = \frac{(-1)^1}{3(1)} = -\frac{1}{3}$ Replace n with 1.

b. $a_8 = \frac{(-1)^8}{3(8)} = \frac{1}{24}$ Replace n with 8.

c. $a_{100} = \frac{(-1)^{100}}{3(100)} = \frac{1}{300}$ Replace n with 100.

d. $a_{15} = \frac{(-1)^{15}}{3(15)} = -\frac{1}{45}$ Replace n with 15.

Work Practice 2

Practice 2

If the general term of a sequence is given by $a_n = \frac{(-1)^n}{5n}$, find

a. the first term of the sequence

b. a_4

c. The thirtieth term of the sequence

d. a_{19}

Objective 2 Finding the General Term of a Sequence 

Suppose we know the first few terms of a sequence and want to find a general term that fits the pattern of the first few terms.

Example 3 Find a general term a_n of the sequence whose first few terms are given.

a. 1, 4, 9, 16, ... **b.** $\frac{1}{1}, \frac{1}{2}, \frac{1}{3}, \frac{1}{4}, \frac{1}{5}, \ldots$

c. −3, −6, −9, −12, ... **d.** $\frac{1}{2}, \frac{1}{4}, \frac{1}{8}, \frac{1}{16}, \ldots$

Solution:

a. These numbers are the squares of the first four natural numbers, so a general term might be $a_n = n^2$.

b. These numbers are the reciprocals of the first five natural numbers, so a general term might be $a_n = \frac{1}{n}$.

(Continued on next page)

Practice 3

Find the general term a_n of the sequence whose first few terms are given.

a. 1, 3, 5, 7, ...

b. 3, 9, 27, 81, ...

c. $\frac{1}{2}, \frac{2}{3}, \frac{3}{4}, \frac{4}{5}, \ldots$

d. $-\frac{1}{2}, -\frac{1}{3}, -\frac{1}{4}, -\frac{1}{5}, \ldots$

Answers

1. 6, 9, 14, 21, 30 **2. a.** $-\frac{1}{5}$ **b.** $\frac{1}{20}$ **c.** $\frac{1}{150}$ **d.** $-\frac{1}{95}$ **3. a.** $a_n = 2n - 1$ **b.** $a_n = 3^n$ **c.** $a_n = \frac{n}{n+1}$ **d.** $a_n = -\frac{1}{n+1}$

c. These numbers are the product of -3 and the first four natural numbers, so a general term might be $a_n = -3n$.

d. Notice that the denominators double each time.

$$\frac{1}{2}, \frac{1}{2\cdot 2}, \frac{1}{2(2\cdot 2)}, \frac{1}{2(2\cdot 2\cdot 2)}$$

or

$$\frac{1}{2^1}, \frac{1}{2^2}, \frac{1}{2^3}, \frac{1}{2^4}$$

We might then suppose that the general term is $a_n = \frac{1}{2^n}$.

Work Practice 3

Objective 3 Solving Applications Modeled by Sequences

Sequences model many phenomena of the physical world, as illustrated by the following example.

Practice 4

The value v, in dollars, of an office copier depreciates according to the sequence $v_n = 3950(0.8)^n$, where n is the time in years. Find the value of the copier after three years.

Answer

4. \$2022.40

Example 4 Finding a Puppy's Weight Gain

The amount of weight, in pounds, a puppy gains in each month of its first year is modeled by a sequence whose general term is $a_n = n + 4$, where n is the number of the month. Write the first five terms of the sequence and find how much weight the puppy should gain in its fifth month.

Solution: Evaluate $a_n = n + 4$ when n is 1, 2, 3, 4, and 5.

$$a_1 = 1 + 4 = 5$$
$$a_2 = 2 + 4 = 6$$
$$a_3 = 3 + 4 = 7$$
$$a_4 = 4 + 4 = 8$$
$$a_5 = 5 + 4 = 9$$

The puppy should gain 9 pounds in its fifth month.

Work Practice 4

Vocabulary, Readiness & Video Check

Use the choices below to fill in each blank.

infinite finite general

1. The nth term of the sequence a_n is called the ________ term.

2. A(n) ________ sequence is a function whose domain is $\{1, 2, 3, 4, \ldots, n\}$ where n is some natural number.

3. A(n) ________ sequence is a function whose domain is $\{1, 2, 3, 4, \ldots\}$.

Write the first term of each sequence.

4. $a_n = 7^n$; $a_1 =$ ________.

5. $a_n = \frac{(-1)^n}{n}$; $a_1 =$ ________.

6. $a_n = (-1)^n \cdot n^4$; $a_1 =$ ________.

Martin-Gay Interactive Videos *Watch the section lecture video and answer the following questions.*

See Video 21.1

Objective 1 7. Based on the lecture before Example 1, complete the following statements. A sequence is a ________ whose ________ is the set of natural numbers. We use ________ to mean the general term of a sequence.

Objective 2 8. In Example 3, why can't the general term be $a_n = (-2)^n$?

Objective 3 9. For Example 4, write the equation for the specific term and find the allowance amount for day 9 of the vacation.

21.1 Exercise Set MyMathLab®

Write the first five terms of each sequence, whose general term is given. See Example 1.

1. $a_n = n + 4$
2. $a_n = 5 - n$
3. $a_n = (-1)^n$
4. $a_n = (-2)^n$
5. $a_n = \frac{1}{n+3}$
6. $a_n = \frac{1}{7-n}$
7. $a_n = 2n$
8. $a_n = -6n$
9. $a_n = -n^2$
10. $a_n = n^2 + 2$
11. $a_n = 2^n$
12. $a_n = 3^{n-2}$
13. $a_n = 2n + 5$
14. $a_n = 1 - 3n$
15. $a_n = (-1)^n n^2$
16. $a_n = (-1)^{n+1}(n - 1)$

Find the indicated term for each sequence, whose general term is given. See Example 2.

17. $a_n = 3n^2; a_5$
18. $a_n = -n^2; a_{15}$
19. $a_n = 6n - 2; a_{20}$
20. $a_n = 100 - 7n; a_{50}$
21. $a_n = \frac{n+3}{n}; a_{15}$
22. $a_n = \frac{n}{n+4}; a_{24}$
23. $a_n = (-3)^n; a_6$
24. $a_n = 5^{n+1}; a_3$
25. $a_n = \frac{n-2}{n+1}; a_6$
26. $a_n = \frac{n+3}{n+4}; a_8$
27. $a_n = \frac{(-1)^n}{n}; a_8$
28. $a_n = \frac{(-1)^n}{2n}; a_{100}$
29. $a_n = -n^2 + 5; a_{10}$
30. $a_n = 8 - n^2; a_{20}$
31. $a_n = \frac{(-1)^n}{n+6}; a_{19}$
32. $a_n = \frac{n-4}{(-2)^n}; a_6$

Find a general term a_n for each sequence, whose first four terms are given. See Example 3.

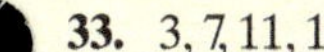

33. 3, 7, 11, 15
34. 2, 7, 12, 17
35. −2, −4, −8, −16
36. −4, 16, −64, 256
37. $\frac{1}{3}, \frac{1}{9}, \frac{1}{27}, \frac{1}{81}$
38. $\frac{2}{5}, \frac{2}{25}, \frac{2}{125}, \frac{2}{625}$

Solve. See Example 4.

39. The distance, in feet, that a Thermos dropped from a cliff falls in each consecutive second is modeled by a sequence whose general term is $a_n = 32n - 16$, where n is the number of seconds. Find the distance the Thermos falls in the second, third, and fourth seconds.

40. The population size of a culture of bacteria triples every hour such that its size is modeled by the sequence $a_n = 50(3)^{n-1}$, where n is the number of the hour just beginning. Find the size of the culture at the beginning of the fourth hour and the size of the culture at the beginning of the first hour.

41. Mrs. Laser agrees to give her son Mark an allowance of \$0.10 on the first day of his 14-day vacation, \$0.20 on the second day, \$0.40 on the third day, and so on. Write an equation of a sequence whose terms correspond to Mark's allowance. Find the allowance Mark will receive on the last day of his vacation.

42. A small theater has 10 rows with 12 seats in the first row, 15 seats in the second row, 18 seats in the third row, and so on. Write an equation of a sequence whose terms correspond to the seats in each row. Find the number of seats in the eighth row.

43. The number of cases of a new infectious disease is doubling every year such that the number of cases is modeled by a sequence whose general term is $a_n = 75(2)^{n-1}$, where n is the number of the year just beginning. Find how many cases there will be at the beginning of the sixth year. Find how many cases there were at the beginning of the first year.

44. A new college had an initial enrollment of 2700 students in 2000, and each year the enrollment increases by 150 students. Find the enrollment for each of 5 years, beginning with 2000.

45. An endangered species of sparrow had an estimated population of 800 in 2000, and scientists predicted that its population would decrease by half each year. Estimate the population in 2004. Estimate the year the sparrow was extinct.

46. A **Fibonacci sequence** is a special type of sequence in which the first two terms are 1 and each term thereafter is the sum of the two previous terms: 1, 1, 2, 3, 5, 8, etc. The formula for the nth Fibonacci term is $a_n = \frac{1}{\sqrt{5}}\left[\left(\frac{1+\sqrt{5}}{2}\right)^n - \left(\frac{1-\sqrt{5}}{2}\right)^n\right]$. Verify that the first two terms of the Fibonacci sequence are each 1.

REVIEW AND PREVIEW

Sketch the graph of each quadratic function. See Section 18.5.

47. $f(x) = (x-1)^2 + 3$

48. $f(x) = (x-2)^2 + 1$

49. $f(x) = 2(x+4)^2 + 2$

50. $f(x) = 3(x-3)^2 + 4$

Find the distance between each pair of points. See Section 17.3.

51. $(-4, -1)$ and $(-7, -3)$

52. $(-2, -1)$ and $(-1, 5)$

53. $(2, -7)$ and $(-3, -3)$

54. $(10, -14)$ and $(5, -11)$

CONCEPT EXTENSIONS

Find the first five terms of each sequence. Round each term after the first to four decimal places.

55. $a_n = \frac{1}{\sqrt{n}}$

56. $\frac{\sqrt{n}}{\sqrt{n}+1}$

57. $a_n = \left(1 + \frac{1}{n}\right)^n$

58. $a_n = \left(1 + \frac{0.05}{n}\right)^n$

21.2 Arithmetic and Geometric Sequences

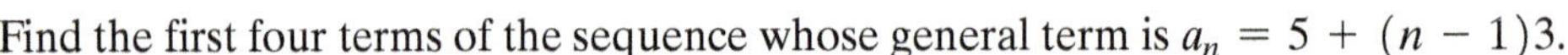

Objectives

1. Identify Arithmetic Sequences and Their Common Differences.
2. Identify Geometric Sequences and Their Common Ratios.

Objective 1 Identifying Arithmetic Sequences

Find the first four terms of the sequence whose general term is $a_n = 5 + (n-1)3$.

$$a_1 = 5 + (1-1)3 = 5 \quad \text{Replace } n \text{ with 1.}$$
$$a_2 = 5 + (2-1)3 = 8 \quad \text{Replace } n \text{ with 2.}$$
$$a_3 = 5 + (3-1)3 = 11 \quad \text{Replace } n \text{ with 3.}$$
$$a_4 = 5 + (4-1)3 = 14 \quad \text{Replace } n \text{ with 4.}$$

The first four terms are 5, 8, 11, and 14. Notice that the difference of any two successive terms is 3.

$$8 - 5 = 3$$
$$11 - 8 = 3$$
$$14 - 11 = 3$$
$$\vdots$$
$$a_n - a_{n-1} = 3$$

↑ nth term ↑ previous term

Because the difference of any two successive terms is a constant, we call the sequence an **arithmetic sequence,** or an **arithmetic progression.** The constant difference d in successive terms is called the **common difference.** In this example, d is 3.

Arithmetic Sequence and Common Difference

An **arithmetic sequence** is a sequence in which each term (after the first) differs from the preceding term by a constant amount d. The constant d is called the **common difference** of the sequence.

The sequence 2, 6, 10, 14, 18, . . . is an arithmetic sequence. Its common difference is 4. Given the first term a_1 and the common difference d of an arithmetic sequence, we can find any term of the sequence.

Example 1 Write the first five terms of the arithmetic sequence whose first term is 7 and whose common difference is 2.

Solution:

$a_1 = 7$

$a_2 = 7 + 2 = 9$

$a_3 = 9 + 2 = 11$

$a_4 = 11 + 2 = 13$

$a_5 = 13 + 2 = 15$

The first five terms are 7, 9, 11, 13, 15.

Work Practice 1

Practice 1

Write the first five terms of the arithmetic sequence whose first term is 4 and whose common difference is 5.

Notice the general pattern of the terms in Example 1.

$a_1 = 7$

$a_2 = 7 + 2 = 9$ or $a_2 = a_1 + d$

$a_3 = 9 + 2 = 11$ or $a_3 = a_2 + d = (a_1 + d) + d = a_1 + 2d$

$a_4 = 11 + 2 = 13$ or $a_4 = a_3 + d = (a_1 + 2d) + d = a_1 + 3d$

$a_5 = 13 + 2 = 15$ or $a_5 = a_4 + d = (a_1 + 3d) + d = a_1 + 4d$

(subscript − 1) is multiplier

The pattern on the right suggests that the general term a_n of an arithmetic sequence is given by

$$a_n = a_1 + (n - 1)d$$

General Term of an Arithmetic Sequence

The general term a_n of an arithmetic sequence is given by

$$a_n = a_1 + (n - 1)d$$

where a_1 is the first term and d is the common difference.

Example 2 Consider the arithmetic sequence whose first term is 3 and whose common difference is −5.

a. Write an expression for the general term a_n.
b. Find the twentieth term of this sequence.

Solution:

a. Since this is an arithmetic sequence, the general term a_n is given by $a_n = a_1 + (n - 1)d$. Here, $a_1 = 3$ and $d = -5$, so

$a_n = 3 + (n - 1)(-5)$ Let $a_1 = 3$ and $d = -5$.

$= 3 - 5n + 5$ Multiply.

$= 8 - 5n$ Simplify.

b. $a_n = 8 - 5n$

$a_{20} = 8 - 5 \cdot 20$ Let $n = 20$.

$= 8 - 100 = -92$

Work Practice 2

Practice 2

Consider the arithmetic sequence whose first term is 2 and whose common difference is −3.

a. Write an expression for the general term a_n.
b. Find the twelfth term of the sequence.

Answers

1. 4, 9, 14, 19, 24 **2. a.** $a_n = 5 - 3n$ **b.** −31

Practice 3

Find the ninth term of the arithmetic sequence whose first three terms are 3, 9, and 15.

Example 3 Find the eleventh term of the arithmetic sequence whose first three terms are 2, 9, and 16.

Solution: Since the sequence is arithmetic, the eleventh term is

$$a_{11} = a_1 + (11 - 1)d = a_1 + 10d$$

We know a_1 is the first term of the sequence, so $a_1 = 2$. Also, d is the constant difference of terms, so $d = a_2 - a_1 = 9 - 2 = 7$. Thus,

$$\begin{aligned} a_{11} &= a_1 + 10d \\ &= 2 + 10 \cdot 7 \quad \text{Let } a_1 = 2 \text{ and } d = 7. \\ &= 72 \end{aligned}$$

Work Practice 3

Practice 4

If the third term of an arithmetic sequence is 23 and the eighth term is 63, find the sixth term.

Example 4 If the third term of an arithmetic sequence is 12 and the eighth term is 27, find the fifth term.

Solution: We need to find a_1 and d to write the general term, which then enables us to find a_5, the fifth term. The given facts about terms a_3 and a_8 lead to a system of linear equations.

$$\begin{cases} a_3 = a_1 + (3 - 1)d \\ a_8 = a_1 + (8 - 1)d \end{cases} \quad \text{or} \quad \begin{cases} 12 = a_1 + 2d \\ 27 = a_1 + 7d \end{cases}$$

Next, we solve the system $\begin{cases} 12 = a_1 + 2d \\ 27 = a_1 + 7d \end{cases}$ by elimination. Multiply both sides of the second equation by -1 so that

$$\begin{cases} 12 = a_1 + 2d \\ -1(27) = -1(a_1 + 7d) \end{cases} \quad \text{simplifies to} \quad \begin{cases} 12 = a_1 + 2d \\ -27 = -a_1 - 7d \end{cases}$$

$$\begin{aligned} -15 &= -5d \quad \text{Add the equations.} \\ 3 &= d \quad \text{Divide both sides by } -5. \end{aligned}$$

To find a_1, let $d = 3$ in $12 = a_1 + 2d$. Then

$$\begin{aligned} 12 &= a_1 + 2(3) \\ 12 &= a_1 + 6 \\ 6 &= a_1 \end{aligned}$$

Thus, $a_1 = 6$ and $d = 3$, so

$$\begin{aligned} a_n &= 6 + (n - 1)(3) \\ &= 6 + 3n - 3 \\ &= 3 + 3n \end{aligned}$$

and

$$a_5 = 3 + 3 \cdot 5 = 18$$

Work Practice 4

Practice 5

A starting salary for a consulting company is $57,000 per year with guaranteed annual increases of $2200 for the next 4 years. Write the general term for the arithmetic sequence that models the potential annual salaries and find the salary for the third year.

Example 5 Finding Salary

Donna Theime has an offer for a job starting at $40,000 per year and guaranteeing her a raise of $1600 per year for the next 5 years. Write the general term for the arithmetic sequence that models Donna's potential annual salaries and find her salary for the fourth year.

Answers

3. 51 **4.** 47 **5.** $a_n = 54{,}800 + 2200n$; $61,400

Solution: The first term, a_1, is 40,000, and d is 1600. So

$$a_n = 40{,}000 + (n - 1)(1600) = 38{,}400 + 1600n$$
$$a_4 = 38{,}400 + 1600 \cdot 4 = 44{,}800$$

Her salary for the fourth year will be $44,800.

Work Practice 5

Objective 2 Identifying Geometric Sequences

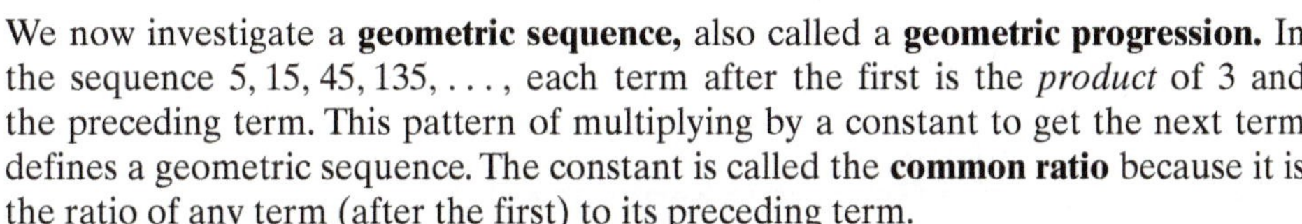

We now investigate a **geometric sequence,** also called a **geometric progression.** In the sequence 5, 15, 45, 135, . . . , each term after the first is the *product* of 3 and the preceding term. This pattern of multiplying by a constant to get the next term defines a geometric sequence. The constant is called the **common ratio** because it is the ratio of any term (after the first) to its preceding term.

$$\frac{15}{5} = 3$$
$$\frac{45}{15} = 3$$
$$\frac{135}{45} = 3$$
$$\vdots$$
$$\frac{a_n}{a_{n-1}} = 3$$

*n*th term → a_n
previous term → a_{n-1}

Geometric Sequence and Common Ratio

A **geometric sequence** is a sequence in which each term (after the first) is obtained by multiplying the preceding term by a constant r. The constant r is called the **common ratio** of the sequence.

The sequence $12, 6, 3, \frac{3}{2}, \ldots$ is geometric since each term after the first is the product of the previous term and $\frac{1}{2}$.

Example 6 Write the first five terms of a geometric sequence whose first term is 7 and whose common ratio is 2.

Solution:

$$a_1 = 7$$
$$a_2 = 7(2) = 14$$
$$a_3 = 14(2) = 28$$
$$a_4 = 28(2) = 56$$
$$a_5 = 56(2) = 112$$

The first five terms are 7, 14, 28, 56, and 112.

Work Practice 6

Practice 6

Write the first four terms of a geometric sequence whose first term is 8 and whose common ratio is −3.

Answer

6. 8, −24, 72, −216

Notice the general pattern of the terms in Example 6.

$a_1 = 7$

$a_2 = 7(2) = 14$ or $a_2 = a_1(r)$

$a_3 = 14(2) = 28$ or $a_3 = a_2(r) = (a_1 \cdot r) \cdot r = a_1 r^2$

$a_4 = 28(2) = 56$ or $a_4 = a_3(r) = (a_1 \cdot r^2) \cdot r = a_1 r^3$

$a_5 = 56(2) = 112$ or $a_5 = a_4(r) = (a_1 \cdot r^3) \cdot r = a_1 r^4$

(subscript − 1) is power

The pattern on the right above suggests that the general term of a geometric sequence is given by $a_n = a_1 r^{n-1}$.

General Term of a Geometric Sequence

The general term a_n of a geometric sequence is given by

$$a_n = a_1 r^{n-1}$$

where a_1 is the first term and r is the common ratio.

Practice 7

Find the seventh term of the geometric sequence whose first term is 64 and whose common ratio is $\frac{1}{4}$.

Example 7 Find the eighth term of the geometric sequence whose first term is 12 and whose common ratio is $\frac{1}{2}$.

Solution: Since this is a geometric sequence, the general term a_n is given by

$$a_n = a_1 r^{n-1}$$

Here $a_1 = 12$ and $r = \frac{1}{2}$, so $a_n = 12\left(\frac{1}{2}\right)^{n-1}$. Evaluate a_n for $n = 8$.

$$a_8 = 12\left(\frac{1}{2}\right)^{8-1} = 12\left(\frac{1}{2}\right)^7 = 12\left(\frac{1}{128}\right) = \frac{3}{32}$$

Work Practice 7

Practice 8

Find the seventh term of the geometric sequence whose first three terms are −3, 6, and −12.

Example 8 Find the fifth term of the geometric sequence whose first three terms are 2, −6, and 18.

Solution: Since the sequence is geometric and $a_1 = 2$, the fifth term must be $a_1 r^{5-1}$, or $2r^4$. We know that r is the common ratio of terms, so r must be $\frac{-6}{2}$, or −3. Thus,

$$a_5 = 2r^4$$

$$a_5 = 2(-3)^4 = 162$$

Work Practice 8

Practice 9

If the second term of a geometric sequence is $\frac{9}{2}$ and the third term is $\frac{27}{4}$, find the first term and the common ratio.

Example 9 If the second term of a geometric sequence is $\frac{5}{4}$ and the third term is $\frac{5}{16}$, find the first term and the common ratio.

Answers

7. $\frac{1}{64}$ **8.** −192 **9.** $a_1 = 3$; $r = \frac{3}{2}$

Solution: Notice that $\frac{5}{16} \div \frac{5}{4} = \frac{1}{4}$, so $r = \frac{1}{4}$. Then

$$a_2 = a_1\left(\frac{1}{4}\right)^{2-1}$$

$$\frac{5}{4} = a_1\left(\frac{1}{4}\right)^1, \quad \text{or} \quad a_1 = 5 \quad \text{Replace } a_2 \text{ with } \frac{5}{4}.$$

The first term is 5.

Work Practice 9

Example 10 Predicting Population of a Bacterial Culture

The population size of a bacterial culture growing under controlled conditions is doubling each day. Predict how large the culture will be at the beginning of day 7 if it measures 10 units at the beginning of day 1.

Solution: Since the culture doubles in size each day, the population sizes are modeled by a geometric sequence. Here $a_1 = 10$ and $r = 2$. Thus,

$$a_n = a_1r^{n-1} = 10(2)^{n-1} \quad \text{and} \quad a_7 = 10(2)^{7-1} = 640$$

The bacterial culture should measure 640 units at the beginning of day 7.

Work Practice 10

Practice 10

After applying a test antibiotic, the population of a bacterial culture is reduced by one-half every day. Predict how large the culture will be at the start of day 7 if it measures 4800 units at the beginning of day 1.

Answer

10. 75 units

Vocabulary, Readiness & Video Check

Use the choices below to fill in each blank. Some choices may be used more than once and some not at all.

first	arithmetic	difference
last	geometric	ratio

1. A(n) __________ sequence is one in which each term (after the first) is obtained by multiplying the preceding term by a constant r. The constant r is called the common __________.

2. A(n) __________ sequence is one in which each term (after the first) differs from the preceding term by a constant amount d. The constant d is called the common __________.

3. The general term of an arithmetic sequence is $a_n = a_1 + (n - 1)d$, where a_1 is the __________ term and d is the common __________.

4. The general term of a geometric sequence is $a_n = a_1r^{n-1}$, where a_1 is the __________ term and r is the common __________.

Martin-Gay Interactive Videos

See Video 21.2

Watch the section lecture video and answer the following questions.

Objective 1 **5.** From the lecture before Example 1, what makes a sequence an arithmetic sequence?

Objective 2 **6.** From the lecture before Example 3, what's the difference between an arithmetic and a geometric sequence?

21.2 Exercise Set MyMathLab®

Write the first five terms of the arithmetic or geometric sequence, whose first term, a_1, and common difference, d, or common ratio, r, are given. See Examples 1 and 6.

1. $a_1 = 4; d = 2$
2. $a_1 = 3; d = 10$
3. $a_1 = 6; d = -2$
4. $a_1 = -20; d = 3$
5. $a_1 = 1; r = 3$
6. $a_1 = -2; r = 2$
7. $a_1 = 48; r = \frac{1}{2}$
8. $a_1 = 1; r = \frac{1}{3}$

Find the indicated term of each sequence. See Examples 2 and 7.

9. The eighth term of the arithmetic sequence whose first term is 12 and whose common difference is 3
10. The twelfth term of the arithmetic sequence whose first term is 32 and whose common difference is -4
11. The fourth term of the geometric sequence whose first term is 7 and whose common ratio is -5
12. The fifth term of the geometric sequence whose first term is 3 and whose common ratio is 3
13. The fifteenth term of the arithmetic sequence whose first term is -4 and whose common difference is -4
14. The sixth term of the geometric sequence whose first term is 5 and whose common ratio is -4

Find the indicated term of each sequence. See Examples 3 and 8.

15. The ninth term of the arithmetic sequence 0, 12, 24, ...
16. The thirteenth term of the arithmetic sequence $-3, 0, 3, \ldots$
17. The twenty-fifth term of the arithmetic sequence 20, 18, 16, ...
18. The ninth term of the geometric sequence 5, 10, 20, ...
19. The fifth term of the geometric sequence $2, -10, 50, \ldots$
20. The sixth term of the geometric sequence $\frac{1}{2}, \frac{3}{2}, \frac{9}{2}, \ldots$

Find the indicated term of each sequence. See Examples 4 and 9.

21. The eighth term of the arithmetic sequence whose fourth term is 19 and whose fifteenth term is 52
22. If the second term of an arithmetic sequence is 6 and the tenth term is 30, find the twenty-fifth term.
23. If the second term of an arithmetic progression is -1 and the fourth term is 5, find the ninth term.
24. If the second term of a geometric progression is 15 and the third term is 3, find a_1 and r.
25. If the second term of a geometric progression is $-\frac{4}{3}$ and the third term is $\frac{8}{3}$, find a_1 and r.
26. If the third term of a geometric sequence is 4 and the fourth term is -12, find a_1 and r.
27. Explain why 14, 10, and 6 may be the first three terms of an arithmetic sequence when it appears we are subtracting instead of adding to get the next term.
28. Explain why 80, 20, and 5 may be the first three terms of a geometric sequence when it appears we are dividing instead of multiplying to get the next term.

MIXED PRACTICE

Given are the first three terms of a sequence that is either arithmetic or geometric. If the sequence is arithmetic, find a_1 and d. If the sequence is geometric, find a_1 and r.

29. 2, 4, 6
30. 8, 16, 24
31. 5, 10, 20
32. 2, 6, 18
33. $\frac{1}{2}, \frac{1}{10}, \frac{1}{50}$
34. $\frac{2}{3}, \frac{4}{3}, 2$
35. $x, 5x, 25x$
36. $y, -3y, 9y$
37. $p, p + 4, p + 8$
38. $t, t - 1, t - 2$

Find the indicated term of each sequence.

39. The twenty-first term of the arithmetic sequence whose first term is 14 and whose common difference is $\frac{1}{4}$
40. The fifth term of the geometric sequence whose first term is 8 and whose common ratio is -3
41. The fourth term of the geometric sequence whose first term is 3 and whose common ratio is $-\frac{2}{3}$
42. The fourth term of the arithmetic sequence whose first term is 9 and whose common difference is 5
43. The fifteenth term of the arithmetic sequence $\frac{3}{2}, 2, \frac{5}{2}, \ldots$
44. The eleventh term of the arithmetic sequence $2, \frac{5}{3}, \frac{4}{3}, \ldots$
45. The sixth term of the geometric sequence $24, 8, \frac{8}{3}, \ldots$
46. The eighteenth term of the arithmetic sequence $5, 2, -1, \ldots$
47. If the third term of an arithmetic sequence is 2 and the seventeenth term is -40, find the tenth term.
48. If the third term of a geometric sequence is -28 and the fourth term is -56, find a_1 and r.

Solve. See Examples 5 and 10.

49. An auditorium has 54 seats in the first row, 58 seats in the second row, 62 seats in the third row, and so on. Find the general term of this arithmetic sequence and the number of seats in the twentieth row.

50. A triangular display of cans in a grocery store has 20 cans in the first row, 17 cans in the next row, and so on, in an arithmetic sequence. Find the general term and the number of cans in the fifth row. Find how many rows there are in the display and how many cans are in the top row.

51. The initial size of a virus culture is 6 units, and it triples its size every day. Find the general term of the geometric sequence that models the culture's size.

52. A real estate investment broker predicts that a certain property will increase in value 15% each year. Thus, the yearly property values can be modeled by a geometric sequence whose common ratio r is 1.15. If the initial property value was \$500,000, write the first four terms of the sequence and predict the value at the end of the third year.

53. A rubber ball is dropped from a height of 486 feet, and it continues to bounce one-third the height from which it last fell. Write out the first five terms of this geometric sequence and find the general term. Find how many bounces it takes for the ball to rebound less than 1 foot.

54. On the first swing, the length of the arc through which a pendulum swings is 50 inches. The length of each successive swing is 80% of the preceding swing. Determine whether this sequence is arithmetic or geometric. Find the length of the fourth swing.

55. Jose takes a job that offers a monthly starting salary of \$4000 and guarantees him a monthly raise of \$125 during his first year of training. Find the general term of this arithmetic sequence and his monthly salary at the end of his training.

56. At the beginning of Claudia Schaffer's exercise program, she rides 15 minutes on the Lifecycle. Each week, she increases her riding time by 5 minutes. Write the general term of this arithmetic sequence, and find her riding time after 7 weeks. Find how many weeks it takes her to reach a riding time of 1 hour.

57. If a radioactive element has a half-life of 3 hours, then x grams of the element dwindles to $\frac{x}{2}$ grams after 3 hours. If a nuclear reactor has 400 grams of that radioactive element, find the amount of radioactive material after 12 hours.

REVIEW AND PREVIEW

Evaluate. See Sections 1.7, 2.2, 2.3, and 4.5.

58. $5(1) + 5(2) + 5(3) + 5(4)$

59. $\frac{1}{3(1)} + \frac{1}{3(2)} + \frac{1}{3(3)}$

60. $2(2 - 4) + 3(3 - 4) + 4(4 - 4)$

61. $3^0 + 3^1 + 3^2 + 3^3$

62. $\frac{1}{4(1)} + \frac{1}{4(2)} + \frac{1}{4(3)}$

63. $\frac{8-1}{8+1} + \frac{8-2}{8+2} + \frac{8-3}{8+3}$

CONCEPT EXTENSIONS

Write the first four terms of the arithmetic or geometric sequence, whose first term, a_1, and common difference, d, or common ratio, r, are given.

64. $a_1 = \$3720, d = -\$268.50$

65. $a_1 = \$11{,}782.40, r = 0.5$

66. $a_1 = 26.8, r = 2.5$

67. $a_1 = 19.652, d = -0.034$

68. Describe a situation in your life that can be modeled by a geometric sequence. Write an equation for the sequence.

69. Describe a situation in your life that can be modeled by an arithmetic sequence. Write an equation for the sequence.

21.3 Series 

Objective 1 Identifying Finite and Infinite Series and Using Summation Notation

Objectives

1. Identify Finite and Infinite Series and Use Summation Notation.
2. Find Partial Sums.

A person who conscientiously saves money by saving first \$100 and then saving \$10 more each month than he saved the preceding month is saving money according to the arithmetic sequence

$$a_n = 100 + 10(n - 1)$$

Following this sequence, he can predict how much money he should save for any particular month. But if he also wants to know how much money *in total* he has saved, say, by the fifth month, he must find the *sum* of the first five terms of the sequence

$$\underbrace{100}_{a_1} + \underbrace{100 + 10}_{a_2} + \underbrace{100 + 20}_{a_3} + \underbrace{100 + 30}_{a_4} + \underbrace{100 + 40}_{a_5}$$

A sum of the terms of a sequence is called a **series** (the plural is also "series"). As our example here suggests, series are frequently used to model financial and natural phenomena.

A series is a **finite series** if it is the sum of a finite number of terms. A series is an **infinite series** if it is the sum of all the terms of an infinite sequence. For example,

Sequence	**Series**	
$5, 9, 13$	$5 + 9 + 13$	Finite; sum of 3 terms
$5, 9, 13, \ldots$	$5 + 9 + 13 + \cdots$	Infinite
$4, -2, 1, -\frac{1}{2}, \frac{1}{4}$	$4 + (-2) + 1 + \left(-\frac{1}{2}\right) + \left(\frac{1}{4}\right)$	Finite; sum of 5 terms
$4, -2, 1, \ldots$	$4 + (-2) + 1 + \cdots$	Infinite
$3, 6, \ldots, 99$	$3 + 6 + \cdots + 99$	Finite; sum of 33 terms

A shorthand notation for denoting a series when the general term of the sequence is known is called **summation notation.** The Greek uppercase letter **sigma,** Σ, is used to mean "sum." The expression $\sum_{n=1}^{5}(3n + 1)$ is read "the sum of $3n + 1$ as n goes from 1 to 5"; this expression means the sum of the first five terms of the sequence whose general term is $a_n = 3n + 1$. Often, the variable i is used instead of n in summation notation: $\sum_{i=1}^{5}(3i + 1)$. Whether we use n, i, k, or some other variable, the variable is called the **index of summation.** The notation $i = 1$ below the symbol Σ indicates the beginning value of i, and the number 5 above the symbol Σ indicates the ending value of i. Thus, the terms of the sequence are found by successively replacing i with the natural numbers 1, 2, 3, 4, 5. To find the sum, we write out the terms and then add.

$$\begin{aligned}\sum_{i=1}^{5}(3i + 1) &= (3\cdot 1 + 1) + (3\cdot 2 + 1) + (3\cdot 3 + 1)\\ &\quad + (3\cdot 4 + 1) + (3\cdot 5 + 1)\\ &= 4 + 7 + 10 + 13 + 16 = 50\end{aligned}$$

Practice 1

Evaluate.

a. $\sum_{i=0}^{4}\frac{i - 3}{4}$ **b.** $\sum_{i=2}^{5}3^i$

Example 1 Evaluate.

a. $\sum_{i=0}^{6}\frac{i - 2}{2}$ **b.** $\sum_{i=3}^{5}2^i$

Solution:

a.
$$\begin{aligned}\sum_{i=0}^{6}\frac{i - 2}{2} &= \frac{0 - 2}{2} + \frac{1 - 2}{2} + \frac{2 - 2}{2} + \frac{3 - 2}{2} + \frac{4 - 2}{2} + \frac{5 - 2}{2} + \frac{6 - 2}{2}\\ &= (-1) + \left(-\frac{1}{2}\right) + 0 + \frac{1}{2} + 1 + \frac{3}{2} + 2\\ &= \frac{7}{2}, \text{ or } 3\frac{1}{2}\end{aligned}$$

b.
$$\begin{aligned}\sum_{i=3}^{5}2^i &= 2^3 + 2^4 + 2^5\\ &= 8 + 16 + 32\\ &= 56\end{aligned}$$

Work Practice 1

Answers

1. a. $-\frac{5}{4}$ **b.** 360

Example 2 Write each series with summation notation.

a. $3 + 6 + 9 + 12 + 15$ **b.** $\frac{1}{2} + \frac{1}{4} + \frac{1}{8} + \frac{1}{16}$

Solution:

a. Since the *difference* of each term and the preceding term is 3, the terms correspond to the first five terms of the arithmetic sequence $a_n = a_1 + (n - 1)d$ with $a_1 = 3$ and $d = 3$. So $a_n = 3 + (n - 1)3 = 3n$ when simplified. Thus, in summation notation,

$$3 + 6 + 9 + 12 + 15 = \sum_{i=1}^{5} 3i.$$

b. Since each term is the *product* of the preceding term and $\frac{1}{2}$, these terms correspond to the first four terms of the geometric sequence $a_n = a_1 r^{n-1}$. Here $a_1 = \frac{1}{2}$ and $r = \frac{1}{2}$, so $a_n = \left(\frac{1}{2}\right)\left(\frac{1}{2}\right)^{n-1} = \left(\frac{1}{2}\right)^{1+(n-1)} = \left(\frac{1}{2}\right)^n$. In summation notation,

$$\frac{1}{2} + \frac{1}{4} + \frac{1}{8} + \frac{1}{16} = \sum_{i=1}^{4} \left(\frac{1}{2}\right)^i$$

Work Practice 2

Practice 2

Write each series with summation notation.

a. $5 + 10 + 15 + 20 + 25 + 30$

b. $\frac{1}{5} + \frac{1}{25} + \frac{1}{125} + \frac{1}{625}$

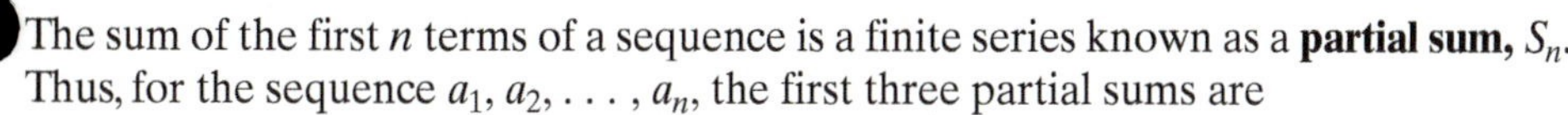

Objective 2 Finding Partial Sums

The sum of the first n terms of a sequence is a finite series known as a **partial sum,** S_n. Thus, for the sequence $a_1, a_2, \ldots, a_n$, the first three partial sums are

$$S_1 = a_1$$
$$S_2 = a_1 + a_2$$
$$S_3 = a_1 + a_2 + a_3$$

In general, S_n is the sum of the first n terms of a sequence.

$$S_n = \sum_{i=1}^{n} a_n$$

Example 3 Find the sum of the first three terms of the sequence whose general term is $a_n = \frac{n + 3}{2n}$.

Solution:

$$S_3 = \sum_{i=1}^{3} \frac{i + 3}{2i} = \frac{1 + 3}{2 \cdot 1} + \frac{2 + 3}{2 \cdot 2} + \frac{3 + 3}{2 \cdot 3}$$
$$= 2 + \frac{5}{4} + 1 = 4\frac{1}{4}$$

Work Practice 3

Practice 3

Find the sum of the first four terms of the sequence whose general term is $a_n = \frac{2 + 3n}{n^2}$.

Answers

2. a. $\sum_{i=1}^{6} 5i$ **b.** $\sum_{i=1}^{4} \left(\frac{1}{5}\right)^i$ **3.** $\frac{655}{72}$ or $9\frac{7}{72}$

The next example illustrates how these sums model real-life phenomena.

Practice 4

The number of new strawberry plants growing in a garden each year is a sequence defined by $a_n = n(2n - 1)$, where n is the number of years after planting a strawberry plant. Find the total number of strawberry plants after 5 years.

Answer

4. 95 plants

Example 4 Number of Baby Gorillas Born

The number of baby gorillas born at the San Diego Zoo is a sequence defined by $a_n = n(n - 1)$, where n is the number of years the zoo has owned gorillas. Find the *total* number of baby gorillas born in the *first 4 years*.

Solution: To solve, find the sum

$$
\begin{aligned}
S_4 &= \sum_{i=1}^{4} i(i-1) \\
&= 1(1-1) + 2(2-1) + 3(3-1) + 4(4-1) \\
&= 0 + 2 + 6 + 12 = 20
\end{aligned}
$$

Twenty gorillas were born in the first 4 years.

Work Practice 4

Vocabulary, Readiness & Video Check

Use the choices below to fill in each blank. Not all choices will be used.

index of summation	infinite	sigma	1	7
partial sum	finite	summation	5	

1. A series is a(n) ______________ series if it is the sum of all the terms of an infinite sequence.

2. A series is a(n) ______________ series if it is the sum of a finite number of terms.

3. A shorthand notation for denoting a series when the general term of the sequence is known is called ______________ notation.

4. In the notation $\sum_{i=1}^{7}(5i - 2)$, the Σ is the Greek uppercase letter ______________ and the i is called the ______________.

5. The sum of the first n terms of a sequence is a finite series known as a(n) ______________.

6. For the notation in Exercise 4 above, the beginning value of i is ______________ and the ending value of i is ______________.

Martin-Gay Interactive Videos *Watch the section lecture video and answer the following questions.*

See Video 21.3

Objective 1 **7.** From the lecture before Example 1, for the series with the summation notation $\sum_{i=2}^{10} \frac{(-1)^i}{i}$, identify/explain each piece of the notation: Σ, i, 2, 10, $\frac{(-1)^i}{i}$.

Objective 2 **8.** From Example 2 and the lecture before, if you're finding the series S_7 of a sequence, what are you actually finding?

21.3 Exercise Set MyMathLab®

Evaluate. See Example 1.

1. $\sum_{i=1}^{4}(i-3)$
2. $\sum_{i=1}^{5}(i+6)$
3. $\sum_{i=4}^{7}(2i+4)$
4. $\sum_{i=2}^{3}(5i-1)$
5. $\sum_{i=2}^{4}(i^2-3)$
6. $\sum_{i=3}^{5}i^3$
7. $\sum_{i=1}^{3}\left(\frac{1}{i+5}\right)$
8. $\sum_{i=2}^{4}\left(\frac{2}{i+3}\right)$
9. $\sum_{i=1}^{3}\frac{1}{6i}$
10. $\sum_{i=1}^{3}\frac{1}{3i}$
11. $\sum_{i=2}^{6}3i$
12. $\sum_{i=3}^{6}-4i$
13. $\sum_{i=3}^{5}i(i+2)$
14. $\sum_{i=2}^{4}i(i-3)$
15. $\sum_{i=1}^{5}2^i$
16. $\sum_{i=1}^{4}3^{i-1}$
17. $\sum_{i=1}^{4}\frac{4i}{i+3}$
18. $\sum_{i=2}^{5}\frac{6-i}{6+i}$

Write each series with summation notation. See Example 2.

19. $1+3+5+7+9$
20. $4+7+10+13$
21. $4+12+36+108$
22. $5+10+20+40+80+160$
23. $12+9+6+3+0+(-3)$
24. $5+1+(-3)+(-7)$
25. $12+4+\frac{4}{3}+\frac{4}{9}$
26. $80+20+5+\frac{5}{4}+\frac{5}{16}$
27. $1+4+9+16+25+36+49$
28. $1+(-4)+9+(-16)$

Find each partial sum. See Example 3.

29. Find the sum of the first two terms of the sequence whose general term is $a_n = (n+2)(n-5)$.
30. Find the sum of the first two terms of the sequence whose general term is $a_n = n(n-6)$.
31. Find the sum of the first six terms of the sequence whose general term is $a_n = (-1)^n$.
32. Find the sum of the first seven terms of the sequence whose general term is $a_n = (-1)^{n-1}$.
33. Find the sum of the first four terms of the sequence whose general term is $a_n = (n+3)(n+1)$.
34. Find the sum of the first five terms of the sequence whose general term is $a_n = \frac{(-1)^n}{2n}$.
35. Find the sum of the first four terms of the sequence whose general term is $a_n = -2n$.
36. Find the sum of the first five terms of the sequence whose general term is $a_n = (n-1)^2$.
37. Find the sum of the first three terms of the sequence whose general term is $a_n = -\frac{n}{3}$.
38. Find the sum of the first three terms of the sequence whose general term is $a_n = (n+4)^2$.

Solve. See Example 4.

39. A gardener is making a triangular planting with 1 tree in the first row, 2 trees in the second row, 3 trees in the third row, and so on for 10 rows. Write the sequence that describes the number of trees in each row. Find the total number of trees planted.

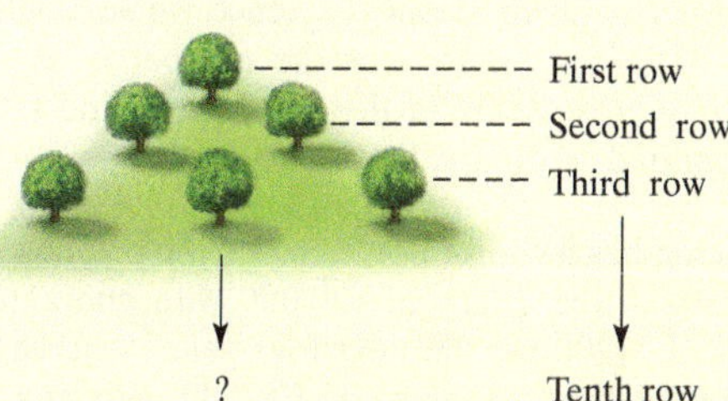

40. Some surfers at the beach form a human pyramid with 2 surfers in the top row, 3 surfers in the second row, 4 surfers in the third row, and so on. If there are 6 rows in the pyramid, write the sequence that describes the number of surfers in each row of the pyramid. Find the total number of surfers.

41. A culture of fungus starts with 6 units and grows according to the sequence defined by $a_n = 6 \cdot 2^{n-1}$, where n is the number of fungus units at the end of the day. Find the total number of fungus units there will be at the end of the fifth day.
42. A bacterial colony begins with 100 bacteria and grows according to the sequence defined by $a_n = 100 \cdot 2^{n-1}$, where n is the number of 6-hour periods. Find the total number of bacteria there will be after 24 hours.

43. The number of species born each year in a new aquarium forms a sequence whose general term is $a_n = (n + 1)(n + 3)$. Find the number of species born in the fourth year, and find the total number born in the first four years.

44. The number of otters born each year in a new aquarium forms a sequence whose general term is $a_n = (n - 1)(n + 3)$. Find the number of otters born in the third year and find the total number of otters born in the first three years.

45. The number of opossums killed each month on a new highway forms the sequence whose general term is $a_n = (n + 1)(n + 2)$, where n is the number of the month. Find the number of opossums killed in the fourth month and find the total number killed in the first four months.

46. In 2007, the population of the Northern Spotted Owl continued to decline, and the owl remained on the endangered species list as old-growth Northwest forests were logged. The size of the decrease in the population in a given year can be estimated by $200 - 6n$ pairs of birds. Find the decrease in population in 2010 if year 1 is 2007. Find the estimated total decrease in the spotted owl population for the years 2007 through 2010. (*Source:* United States Forest Service)

47. The amount of decay in pounds of a radioactive isotope each year is given by the sequence whose general term is $a_n = 100(0.5)^n$, where n is the number of the year. Find the amount of decay in the fourth year and find the total amount of decay in the first four years.

48. A person has a choice between two job offers. Job A has an annual starting salary of \$30,000 with guaranteed annual raises of \$1200 for the next four years, whereas job B has an annual starting salary of \$28,000 with guaranteed annual raises of \$2500 for the next four years. Compare the fifth partial sums for each sequence to determine which job would pay more money over the next 5 years.

49. A pendulum swings a length of 40 inches on its first swing. Each successive swing is $\frac{4}{5}$ of the preceding swing. Find the length of the fifth swing and the total length swung during the first five swings. (Round to the nearest tenth of an inch.)

50. Explain the difference between a sequence and a series.

REVIEW AND PREVIEW

Evaluate. See Section 4.6.

51. $\dfrac{5}{1 - \frac{1}{2}}$ **52.** $\dfrac{-3}{1 - \frac{1}{7}}$ **53.** $\dfrac{\frac{1}{3}}{1 - \frac{1}{10}}$

54. $\dfrac{\frac{6}{11}}{1 - \frac{1}{10}}$ **55.** $\dfrac{3(1 - 2^4)}{1 - 2}$ **56.** $\dfrac{2(1 - 5^3)}{1 - 5}$

57. $\frac{10}{2}(3 + 15)$ **58.** $\frac{12}{2}(2 + 19)$

CONCEPT EXTENSIONS

59. a. Write the sum $\sum_{i=1}^{7}(i + i^2)$ without summation notation.

b. Write the sum $\sum_{i=1}^{7} i + \sum_{i=1}^{7} i^2$ without summation notation.

c. Compare the results of parts (a) and (b).

d. Do you think the following is true or false? Explain your answer.

$$\sum_{i=1}^{n}(a_n + b_n) = \sum_{i=1}^{n} a_n + \sum_{i=1}^{n} b_n$$

60. a. Write the sum $\sum_{i=1}^{6} 5i^3$ without summation notation.

b. Write the expression $5 \cdot \sum_{i=1}^{6} i^3$ without summation notation.

c. Compare the results of parts (a) and (b).

d. Do you think the following is true or false? Explain your answer.

$$\sum_{i=1}^{n} c \cdot a_n = c \cdot \sum_{i=1}^{n} a_n, \text{ where } c \text{ is a constant}$$

Integrated Review Sections 21.1–21.3

Sequences and Series

Write the first five terms of each sequence, whose general term is given.

1. $a_n = n - 3$ **2.** $a_n = \dfrac{7}{1 + n}$

3. $a_n = 3^{n-1}$ **4.** $a_n = n^2 - 5$

Find the indicated term for each sequence.

5. $(-2)^n$; a_6 **6.** $-n^2 + 2$; a_4

7. $\dfrac{(-1)^n}{n}$; a_{40} **8.** $\dfrac{(-1)^n}{2n}$; a_{41}

Write the first five terms of the arithmetic or geometric sequence, whose first term is a_1 and whose common difference, d, or common ratio, r, is given.

9. $a_1 = 7$; $d = -3$ **10.** $a_1 = -3$; $r = 5$

11. $a_1 = 45$; $r = \dfrac{1}{3}$ **12.** $a_1 = -12$; $d = 10$

Find the indicated term of each sequence.

13. The tenth term of the arithmetic sequence whose first term is 20 and whose common difference is 9

14. The sixth term of the geometric sequence whose first term is 64 and whose common ratio is $\frac{3}{4}$

15. The seventh term of the geometric sequence $6, -12, 24, \ldots$

16. The twentieth term of the arithmetic sequence $-100, -85, -70, \ldots$

17. The fifth term of the arithmetic sequence whose fourth term is -5 and whose tenth term is -35

18. The fifth term of a geometric sequence whose fourth term is 1 and whose seventh term is $\frac{1}{8}$

Evaluate.

19. $\sum_{i=1}^{4} 5i$

20. $\sum_{i=1}^{7} (3i + 2)$

21. $\sum_{i=3}^{7} 2^{i-4}$

22. $\sum_{i=2}^{5} \frac{i}{i+1}$

Find each partial sum.

23. Find the sum of the first three terms of the sequence whose general term is $a_n = n(n - 4)$.

24. Find the sum of the first ten terms of the sequence whose general term is $a_n = (-1)^n(n + 1)$.

21.4 Partial Sums of Arithmetic and Geometric Sequences 

Objectives

1. Find the Partial Sum of an Arithmetic Sequence.
2. Find the Partial Sum of a Geometric Sequence.
3. Find the Sum of the Terms of an Infinite Geometric Sequence.

Objective 1 Finding Partial Sums of Arithmetic Sequences

Partial sums S_n are relatively easy to find when n is small—that is, when the number of terms to add is small. But when n is large, finding S_n can be tedious. For a large n, S_n is still relatively easy to find if the addends are terms of an arithmetic sequence or a geometric sequence.

For an arithmetic sequence, $a_n = a_1 + (n - 1)d$ for some first term a_1 and some common difference d. So S_n, the sum of the first n terms, is

$$S_n = a_1 + (a_1 + d) + (a_1 + 2d) + \cdots + (a_1 + (n - 1)d)$$

We might also find S_n by working backward from the nth term a_n, finding the preceding term a_{n-1}, by subtracting d each time.

$$S_n = a_n + (a_n - d) + (a_n - 2d) + \cdots + (a_n - (n - 1)d)$$

Now add the left sides of these two equations and add the right sides.

$$2S_n = (a_1 + a_n) + (a_1 + a_n) + (a_1 + a_n) + \cdots + (a_1 + a_n)$$

The d terms subtract out, leaving n sums of the first term, a_1, and last term, a_n. Thus, we write

$$2S_n = n(a_1 + a_n)$$

or

$$S_n = \frac{n}{2}(a_1 + a_n)$$

Partial Sum S_n of an Arithmetic Sequence

The partial sum S_n of the first n terms of an arithmetic sequence is given by

$$S_n = \frac{n}{2}(a_1 + a_n)$$

where a_1 is the first term of the sequence and a_n is the nth term.

Practice 1

Use the partial sum formula to find the sum of the first five terms of the arithmetic sequence 2, 9, 16, 23, 30, . . .

Example 1 Use the partial sum formula to find the sum of the first six terms of the arithmetic sequence 2, 5, 8, 11, 14, 17, ...

Solution: Use the formula for S_n of an arithmetic sequence, replacing n with 6, a_1 with 2, and a_n with 17.

$$S_n = \frac{n}{2}(a_1 + a_n)$$

$$S_6 = \frac{6}{2}(2 + 17) = 3(19) = 57$$

Work Practice 1

Practice 2

Find the sum of the first 50 positive integers.

Example 2 Find the sum of the first 30 positive integers.

Solution: Because 1, 2, 3, . . . , 30 is an arithmetic sequence, use the formula for S_n with $n = 30$, $a_1 = 1$, and $a_n = 30$. Thus,

$$S_n = \frac{n}{2}(a_1 + a_n)$$

$$S_{30} = \frac{30}{2}(1 + 30) = 15(31) = 465$$

Work Practice 2

Practice 3

An ice sculptor is creating a gigantic castle-facade ice sculpture for First Night festivities in Boston. To get the volume of ice necessary, large blocks of ice were stacked atop each other in 10 rows. The topmost row comprised 6 blocks of ice, the next row 7 blocks of ice, and so on, forming an arithmetic sequence. Find the total number of ice blocks needed if there were 15 blocks in the bottom row.

Example 3 Stacking Rolls of Carpet

Rolls of carpet are stacked in 20 rows with 3 rolls in the top row, 4 rolls in the next row, and so on, forming an arithmetic sequence. Find the total number of carpet rolls if there are 22 rolls in the bottom row.

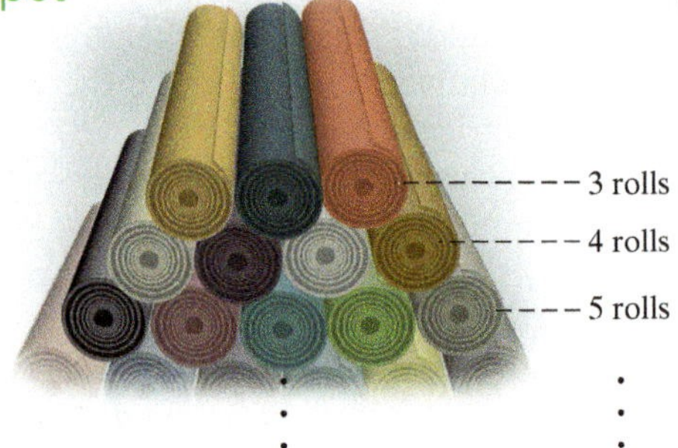

Solution: The list 3, 4, 5, . . . , 22 is the first 20 terms of an arithmetic sequence. Use the formula for S_n with $a_1 = 3$, $a_n = 22$, and $n = 20$ terms. Thus,

$$S_{20} = \frac{20}{2}(3 + 22) = 10(25) = 250$$

There are a total of 250 rolls of carpet.

Work Practice 3

Objective 2 Finding Partial Sums of Geometric Sequences

We can also derive a formula for the partial sum S_n of the first n terms of a geometric sequence. If $a_n = a_1 r^{n-1}$, then

$$S_n = \underset{\substack{\uparrow \\ \text{1st term}}}{a_1} + \underset{\substack{\uparrow \\ \text{2nd term}}}{a_1 r} + \underset{\substack{\uparrow \\ \text{3rd term}}}{a_1 r^2} + \cdots + \underset{\substack{\uparrow \\ n\text{th term}}}{a_1 r^{n-1}}$$

Answers
1. 80 **2.** 1275 **3.** 105 blocks of ice

Multiply each side of the equation by $-r$.

$$-rS_n = -a_1r - a_1r^2 - a_1r^3 - \cdots - a_1r^n$$

Add the two equations.

$$S_n - rS_n = a_1 + (a_1r - a_1r) + (a_1r^2 - a_1r^2) + (a_1r^3 - a_1r^3) + \cdots - a_1r^n$$
$$S_n - rS_n = a_1 - a_1r^n$$

Now factor each side.

$$S_n(1 - r) = a_1(1 - r^n)$$

Solve for S_n by dividing both sides by $1 - r$. Thus,

$$S_n = \frac{a_1(1 - r^n)}{1 - r}$$

as long as r is not 1.

Partial Sum S_n of a Geometric Sequence

The partial sum S_n of the first n terms of a geometric sequence is given by

$$S_n = \frac{a_1(1 - r^n)}{1 - r}$$

where a_1 is the first term of the sequence, r is the common ratio, and $r \neq 1$.

Example 4 Find the sum of the first six terms of the geometric sequence 5, 10, 20, 40, 80, 160.

Solution: Use the formula for the partial sum S_n of the terms of a geometric sequence. Here, $n = 6$, the first term $a_1 = 5$, and the common ratio $r = 2$.

$$S_n = \frac{a_1(1 - r^n)}{1 - r}$$

$$S_6 = \frac{5(1 - 2^6)}{1 - 2} = \frac{5(-63)}{-1} = 315$$

Work Practice 4

Practice 4

Find the sum of the first five terms of the geometric sequence $32, 8, 2, \frac{1}{2}, \frac{1}{8}$.

Example 5 Finding Amount of Donation

A grant from an alumnus to a university specified that the university was to receive \$800,000 during the first year and 75% of the preceding year's donation during each of the following 5 years. Find the total amount donated during the 6 years.

Solution: The donations are modeled by the first six terms of a geometric sequence. Evaluate S_n when $n = 6$, $a_1 = 800{,}000$, and $r = 0.75$.

$$S_6 = \frac{800{,}000[1 - (0.75)^6]}{1 - 0.75}$$
$$= \$2{,}630{,}468.75$$

The total amount donated during the 6 years was \$2,630,468.75.

Work Practice 5

Practice 5

A new youth center is being established in a downtown urban area. A philanthropic charity has agreed to help it get off the ground. The charity has pledged to donate \$250,000 in the first year, with 80% of the preceding year's donation for each of the following 6 years. Find the total amount donated during the 7 years.

Answers

4. $\frac{341}{8}$ or $42\frac{5}{8}$ **5.** \$987,856

Objective 3 Finding Sums of Terms of Infinite Geometric Sequences

Is it possible to find the sum of all the terms of an infinite sequence? Examine the partial sums of the geometric sequence $\frac{1}{2}, \frac{1}{4}, \frac{1}{8}, \ldots$.

$$S_1 = \frac{1}{2}$$

$$S_2 = \frac{1}{2} + \frac{1}{4} = \frac{3}{4}$$

$$S_3 = \frac{1}{2} + \frac{1}{4} + \frac{1}{8} = \frac{7}{8}$$

$$S_4 = \frac{1}{2} + \frac{1}{4} + \frac{1}{8} + \frac{1}{16} = \frac{15}{16}$$

$$S_5 = \frac{1}{2} + \frac{1}{4} + \frac{1}{8} + \frac{1}{16} + \frac{1}{32} = \frac{31}{32}$$

$$\vdots$$

$$S_{10} = \frac{1}{2} + \frac{1}{4} + \frac{1}{8} + \cdots + \frac{1}{2^{10}} = \frac{1023}{1024}$$

Even though each partial sum is larger than the preceding partial sum, we see that each partial sum is closer to 1 than the preceding partial sum. If n gets larger and larger, then S_n gets closer and closer to 1. We say that 1 is the **limit** of S_n and that 1 is the sum of the terms of this infinite sequence. In general, if $|r| < 1$, the following formula gives the sum of the terms of an infinite geometric sequence.

Sum of the Terms of an Infinite Geometric Sequence

The sum S_∞ of the terms of an infinite geometric sequence is given by

$$S_\infty = \frac{a_1}{1 - r}$$

where a_1 is the first term of the sequence, r is the common ratio, and $|r| < 1$. If $|r| \geq 1$, S_∞ does not exist.

What happens for other values of r? For example, in the following geometric sequence, $r = 3$.

6, 18, 54, 162, . . .

Here, as n increases, the sum S_n increases also. This time, though, S_n does not get closer and closer to a fixed number but instead increases without bound.

Practice 6

Find the sum of the terms of the geometric sequence $7, \frac{7}{4}, \frac{7}{16}, \frac{7}{64}, \ldots$.

Example 6 Find the sum of the terms of the geometric sequence $2, \frac{2}{3}, \frac{2}{9}, \frac{2}{27}, \ldots$.

Solution: For this geometric sequence, $r = \frac{1}{3}$. Since $|r| < 1$, we may use the formula for S_∞ of a geometric sequence with $a_1 = 2$ and $r = \frac{1}{3}$.

$$S_\infty = \frac{a_1}{1 - r} = \frac{2}{1 - \frac{1}{3}} = \frac{2}{\frac{2}{3}} = 3$$

Work Practice 6

Answer

6. $\frac{28}{3}$ or $9\frac{1}{3}$

The formula for the sum of the terms of an infinite geometric sequence can be used to write a repeating decimal as a fraction. For example,

$$0.33\overline{3} = \frac{3}{10} + \frac{3}{100} + \frac{3}{1000} + \cdots$$

This sum is the sum of the terms of an infinite geometric sequence whose first term a_1 is $\frac{3}{10}$ and whose common ratio r is $\frac{1}{10}$. Using the formula for S_∞,

$$S_\infty = \frac{a_1}{1 - r} = \frac{\frac{3}{10}}{1 - \frac{1}{10}} = \frac{1}{3}$$

So, $0.33\overline{3} = \frac{1}{3}$.

Example 7 Distance Traveled by a Pendulum

On its first pass, a pendulum swings through an arc whose length is 24 inches. On each pass thereafter, the arc length is 75% of the arc length on the preceding pass. Find the total distance the pendulum travels before it comes to rest.

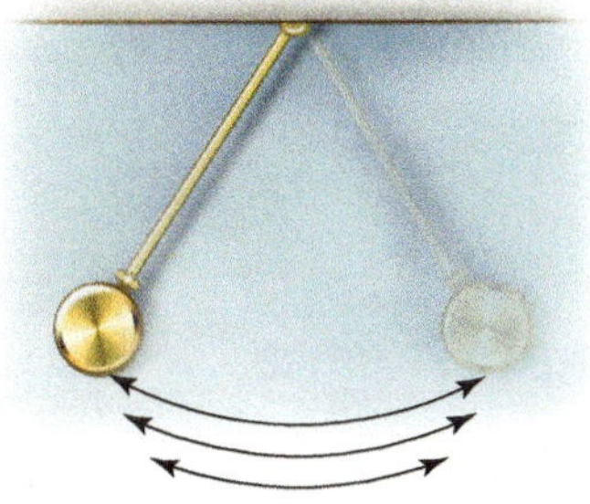

Solution: We must find the sum of the terms of an infinite geometric sequence whose first term, a_1, is 24 and whose common ratio, r, is 0.75. Since $|r| < 1$, we may use the formula for S_∞.

$$S_\infty = \frac{a_1}{1 - r} = \frac{24}{1 - 0.75} = \frac{24}{0.25} = 96$$

The pendulum travels a total distance of 96 inches before it comes to rest.

Work Practice 7

Practice 7

The manufacturers of the "perpetual bouncing ball" claim that the ball rises to 96% of its dropped height on each bounce of the ball. Find the total distance the ball travels before it comes to rest if it is dropped from a height of 36 inches.

Answer

7. 1764 in.

Vocabulary, Readiness & Video Check

Decide whether each sequence is geometric or arithmetic.

1. 5, 10, 15, 20, 25, ...; ________
2. 5, 10, 20, 40, 80, ...; ________
3. −1, 3, −9, 27, −81, ...; ________
4. −1, 1, 3, 5, 7, ...; ________
5. −7, 0, 7, 14, 21, ...; ________
6. −7, 7, −7, 7, −7, ...; ________

Martin-Gay Interactive Videos

See Video 21.4

Watch the section lecture video and answer the following questions.

Objective 1 7. From Example 1, suppose you are asked to find the sum of the first 100 terms of an arithmetic sequence in which you are given only the first few terms. You need the 100th term for the partial sum formula—how can you find this term without actually writing down the first 100 terms?

Objective 2 8. From the lecture before Example 2, we know $r \neq 1$ in the partial sum formula because it would make the denominator 0. What would a geometric sequence with $r = 1$ look like? How could the partial sum of such a sequence be found (without the given formula)?

Objective 3 9. From the lecture before Example 3, why can't you find S_∞ for the geometric sequence $-1, -3, -9, -27, \ldots$?

21.4 Exercise Set MyMathLab®

Use the partial sum formula to find the partial sum of the given arithmetic or geometric sequence. See Examples 1 and 4.

1. Find the sum of the first six terms of the arithmetic sequence $1, 3, 5, 7, \ldots$.
2. Find the sum of the first seven terms of the arithmetic sequence $-7, -11, -15, \ldots$.
3. Find the sum of the first five terms of the geometric sequence $4, 12, 36, \ldots$.
4. Find the sum of the first eight terms of the geometric sequence $-1, 2, -4, \ldots$.
5. Find the sum of the first six terms of the arithmetic sequence $3, 6, 9, \ldots$.
6. Find the sum of the first four terms of the arithmetic sequence $-4, -8, -12, \ldots$.
7. Find the sum of the first four terms of the geometric sequence $2, \frac{2}{5}, \frac{2}{25}, \ldots$.
8. Find the sum of the first five terms of the geometric sequence $\frac{1}{3}, -\frac{2}{3}, \frac{4}{3}, \ldots$.

Solve. See Example 2.

9. Find the sum of the first ten positive integers.
10. Find the sum of the first eight negative integers.
11. Find the sum of the first four positive odd integers.
12. Find the sum of the first five negative odd integers.

Find the sum of the terms of each infinite geometric sequence. See Example 6.

13. $12, 6, 3, \ldots$
14. $45, 15, 5, \ldots$
15. $\frac{1}{10}, \frac{1}{100}, \frac{1}{1000}, \ldots$
16. $\frac{3}{5}, \frac{3}{20}, \frac{3}{80}, \ldots$
17. $-10, -5, -\frac{5}{2}, \ldots$
18. $-16, -4, -1, \ldots$
19. $2, -\frac{1}{4}, \frac{1}{32}, \ldots$
20. $-3, \frac{3}{5}, -\frac{3}{25}, \ldots$
21. $\frac{2}{3}, -\frac{1}{3}, \frac{1}{6}, \ldots$
22. $6, -4, \frac{8}{3}, \ldots$

MIXED PRACTICE

Solve.

23. Find the sum of the first ten terms of the sequence $-4, 1, 6, \ldots, 41$ where 41 is the tenth term.
24. Find the sum of the first twelve terms of the sequence $-3, -13, -23, \ldots, -113$ where -113 is the twelfth term.
25. Find the sum of the first seven terms of the sequence $3, \frac{3}{2}, \frac{3}{4}, \ldots$.
26. Find the sum of the first five terms of the sequence $-2, -6, -18, \ldots$.
27. Find the sum of the first five terms of the sequence $-12, 6, -3, \ldots$.
28. Find the sum of the first four terms of the sequence $-\frac{1}{4}, -\frac{3}{4}, -\frac{9}{4}, \ldots$.
29. Find the sum of the first twenty terms of the sequence $\frac{1}{2}, \frac{1}{4}, 0, \ldots, -\frac{17}{4}$ where $-\frac{17}{4}$ is the twentieth term.
30. Find the sum of the first fifteen terms of the sequence $-5, -9, -13, \ldots, -61$ where -61 is the fifteenth term.
31. If a_1 is 8 and r is $-\frac{2}{3}$, find S_3.
32. If a_1 is 10, a_{18} is $\frac{3}{2}$, and d is $-\frac{1}{2}$, find S_{18}.

Solve. See Example 3.

33. Modern Car Company has come out with a new car model. Market analysts predict that 4000 cars will be sold in the first month and that sales will drop by 50 cars per month after that during the first year. Write out the first five terms of the sequence and find the number of sold cars predicted for the twelfth month. Find the total predicted number of sold cars for the first year.

34. A company that sends faxes charges \$3 for the first page sent and \$0.10 less than the preceding page for each additional page sent. The cost per page forms an arithmetic sequence. Write the first five terms of this sequence and use a partial sum to find the cost of sending a nine-page document.

35. Sal has two job offers: Firm A starts at \$22,000 per year and guarantees raises of \$1000 per year, whereas Firm B starts at \$20,000 and guarantees raises of \$1200 per year. Over a 10-year period, determine the more profitable offer.

36. The game of pool uses 15 balls numbered 1 to 15. In the variety called rotation, a player who sinks a ball receives as many points as the number on the ball. Use an arithmetic series to find the score of a player who sinks all 15 balls.

Solve. See Example 5.

37. A woman made \$30,000 during the first year she owned her business and made an additional 10% over the previous year in each subsequent year. Find how much she made during her fourth year of business. Find her total earnings during the first four years.

38. In free fall, a parachutist falls 16 feet during the first second, 48 feet during the second second, 80 feet during the third second, and so on. Find how far she falls during the eighth second. Find the total distance she falls during the first 8 seconds.

39. A trainee in a computer company takes 0.9 times as long to assemble each computer as he took to assemble the preceding computer. If it took him 30 minutes to assemble the first computer, find how long it takes him to assemble the fifth computer. Find the total time he takes to assemble the first five computers (round to the nearest minute).

40. On a gambling trip to Reno, Carol doubled her bet each time she lost. If her first losing bet was \$5 and she lost six consecutive bets, find how much she lost on the sixth bet. Find the total amount lost on these six bets.

Solve. See Example 7.

41. A ball is dropped from a height of 20 feet and repeatedly rebounds to a height that is $\frac{4}{5}$ of its previous height. Find the total distance the ball covers before it comes to rest.

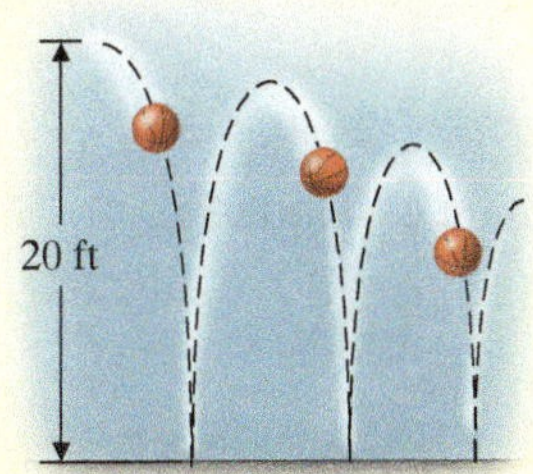

42. A rotating flywheel coming to rest makes 300 revolutions in the first minute and in each minute thereafter makes $\frac{2}{5}$ as many revolutions as in the preceding minute. Find how many revolutions the wheel makes before it comes to rest.

MIXED PRACTICE

Solve.

43. In the pool game of rotation, player A sinks balls numbered 1 to 9, and player B sinks the rest of the balls. Use an arithmetic series to find each player's score (see Exercise 36).

44. A godfather deposited \$250 in a savings account on the day his godchild was born. On each subsequent birthday, he deposited \$50 more than he had deposited the previous year. Find how much money he deposited on his godchild's twenty-first birthday. Find the total amount deposited over the 21 years.

45. During the holiday rush, a business can rent a computer system for \$200 the first day, with the rental fee decreasing \$5 for each additional day. Find the fee paid for 20 days during the holiday rush.

46. The spraying of a field with insecticide killed 6400 weevils the first day, 1600 the second day, 400 the third day, and so on. Find the total number of weevils killed during the first 5 days.

47. A college student humorously asks his parents to charge him room and board according to this geometric sequence: \$0.01 for the first day of the month, \$0.02 for the second day, \$0.04 for the third day, and so on. Find the total room and board he would pay for 30 days.

48. Following its television advertising campaign, a bank attracted 80 new customers the first day, 120 the second day, 160 the third day, and so on in an arithmetic sequence. Find how many new customers were attracted during the first 5 days following its television campaign.

REVIEW AND PREVIEW

Evaluate. See Section 1.5 and 1.6.

49. $6 \cdot 5 \cdot 4 \cdot 3 \cdot 2 \cdot 1$

50. $8 \cdot 7 \cdot 6 \cdot 5 \cdot 4 \cdot 3 \cdot 2 \cdot 1$

51. $\frac{3 \cdot 2 \cdot 1}{2 \cdot 1}$

52. $\frac{5 \cdot 4 \cdot 3 \cdot 2 \cdot 1}{3 \cdot 2 \cdot 1}$

Multiply. See Section 12.4.

53. $(x + 5)^2$

54. $(x - 2)^2$

55. $(2x - 1)^3$

56. $(3x + 2)^3$

CONCEPT EXTENSIONS

57. Write $0.88\overline{8}$ as an infinite geometric series and use the formula for S_∞ to write it as a rational number.

58. Write $0.54\overline{54}$ as an infinite geometric series and use the formula S_∞ to write it as a rational number.

59. Explain whether the sequence 5, 5, 5, . . . is arithmetic, geometric, neither, or both.

60. Describe a situation in everyday life that can be modeled by an infinite geometric series.

21.5 The Binomial Theorem 

Objectives

1. Use Pascal's Triangle to Expand Binomials.
2. Evaluate Factorials.
3. Use the Binomial Theorem to Expand Binomials.
4. Find the nth Term in the Expansion of a Binomial Raised to a Positive Power.

In this section, we learn how to **expand** binomials of the form $(a + b)^n$ easily. Expanding a binomial such as $(a + b)^n$ means to write the factored form as a sum. First, we review the patterns in the expansions of $(a + b)^n$.

$(a + b)^0 = 1$	1 term
$(a + b)^1 = a + b$	2 terms
$(a + b)^2 = a^2 + 2ab + b^2$	3 terms
$(a + b)^3 = a^3 + 3a^2b + 3ab^2 + b^3$	4 terms
$(a + b)^4 = a^4 + 4a^3b + 6a^2b^2 + 4ab^3 + b^4$	5 terms
$(a + b)^5 = a^5 + 5a^4b + 10a^3b^2 + 10a^2b^3 + 5ab^4 + b^5$	6 terms

Notice the following patterns.

1. The expansion of $(a + b)^n$ contains $n + 1$ terms. For example, for $(a + b)^3$, $n = 3$, and the expansion contains $3 + 1$ terms, or 4 terms.
2. The first term of the expansion of $(a + b)^n$ is a^n, and the last term is b^n.
3. The powers of a decrease by 1 for each term, whereas the powers of b increase by 1 for each term.
4. For each term of the expansion of $(a + b)^n$, the sum of the exponents of a and b is n. (For example, the sum of the exponents of $5a^4b$ is $4 + 1$, or 5, and the sum of the exponents of $10a^3b^2$ is $3 + 2$, or 5.)

Objective 1 Using Pascal's Triangle

There are patterns in the coefficients of the terms as well. Written in a triangular array, the coefficients are called **Pascal's triangle.**

$(a + b)^0$:						1						$n = 0$
$(a + b)^1$:					1		1					$n = 1$
$(a + b)^2$:				1		2		1				$n = 2$
$(a + b)^3$:			1		3		3		1			$n = 3$
$(a + b)^4$:		1		4		6		4		1		$n = 4$
$(a + b)^5$:	1		5		10		10		5		1	$n = 5$

Each row in Pascal's triangle begins and ends with 1. Any other number in a row is the sum of the two closest numbers above it. Using this pattern, we can write the next row, for $n = 6$, by first writing the number 1. Then we can add the consecutive numbers in the row for $n = 5$ and write each sum between and below the pair. We complete the row by writing a 1.

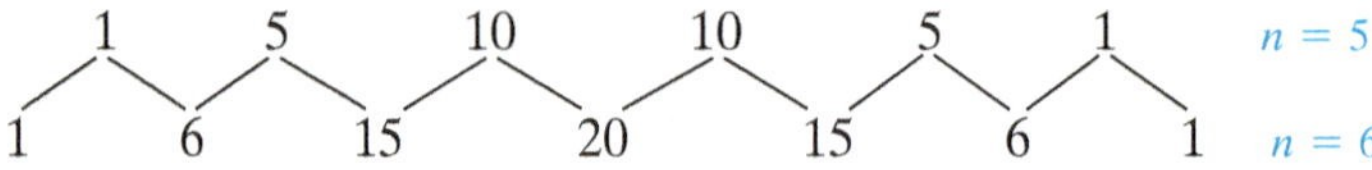

We can use Pascal's triangle and the patterns noted to expand $(a + b)^n$ without actually multiplying any terms.

Practice 1

Expand $(p + r)^7$.

Example 1 Expand $(a + b)^6$.

Solution: Using the $n = 6$ row of Pascal's triangle as the coefficients and following the patterns noted, $(a + b)^6$ can be expanded as

$$a^6 + 6a^5b + 15a^4b^2 + 20a^3b^3 + 15a^2b^4 + 6ab^5 + b^6$$

Work Practice 1

Answer

1. $p^7 + 7p^6r + 21p^5r^2 + 35p^4r^3 + 35p^3r^4 + 21p^2r^5 + 7pr^6 + r^7$

Objective 2 Evaluating Factorials 

For a large n, the use of Pascal's triangle to find coefficients for $(a + b)^n$ can be tedious. An alternative method for determining these coefficients is based on the concept of a **factorial.**

The **factorial of *n*,** written $n!$ (read "n factorial"), is the product of the first n consecutive natural numbers.

Factorial of *n*: *n*!

If n is a natural number, then $n! = n(n-1)(n-2)(n-3)\cdot\cdots\cdot 3\cdot 2\cdot 1$. The factorial of 0, written 0!, is defined to be 1.

For example, $3! = 3\cdot 2\cdot 1 = 6$, $5! = 5\cdot 4\cdot 3\cdot 2\cdot 1 = 120$, and $0! = 1$.

Example 2 Evaluate each expression.

a. $\dfrac{5!}{6!}$ **b.** $\dfrac{10!}{7!3!}$ **c.** $\dfrac{3!}{2!1!}$ **d.** $\dfrac{7!}{7!0!}$

Solution:

a. $\dfrac{5!}{6!} = \dfrac{5\cdot 4\cdot 3\cdot 2\cdot 1}{6\cdot 5\cdot 4\cdot 3\cdot 2\cdot 1} = \dfrac{1}{6}$

b. $\dfrac{10!}{7!3!} = \dfrac{10\cdot 9\cdot 8\cdot 7!}{7!\cdot 3\cdot 2\cdot 1} = \dfrac{10\cdot 9\cdot 8}{3\cdot 2\cdot 1} = 10\cdot 3\cdot 4 = 120$

c. $\dfrac{3!}{2!1!} = \dfrac{3\cdot 2\cdot 1}{2\cdot 1\cdot 1} = 3$

d. $\dfrac{7!}{7!0!} = \dfrac{7!}{7!\cdot 1} = 1$

Work Practice 2

Practice 2

Evaluate each expression.

a. $\dfrac{6!}{7!}$ **b.** $\dfrac{8!}{4!2!}$ **c.** $\dfrac{5!}{4!1!}$ **d.** $\dfrac{9!}{9!0!}$

Helpful Hint

We can use a calculator with a factorial key to evaluate a factorial. A calculator uses scientific notation for large results.

Objective 3 Using the Binomial Theorem 

It can be proved, although we won't do so here, that the coefficients of terms in the expansion of $(a + b)^n$ can be expressed in terms of factorials. Following patterns 1 through 4 given earlier and using the factorial expressions of the coefficients, we have what is known as the **binomial theorem.**

Binomial Theorem

If n is a positive integer, then

$$(a + b)^n = a^n + \frac{n}{1!}a^{n-1}b^1 + \frac{n(n-1)}{2!}a^{n-2}b^2 + \frac{n(n-1)(n-2)}{3!}a^{n-3}b^3 + \cdots + b^n$$

Answers

2. a. $\frac{1}{7}$ **b.** 840 **c.** 5 **d.** 1

We call the formula for $(a + b)^n$ given by the binomial theorem the **binomial formula.**

Practice 3

Use the binomial theorem to expand $(a + b)^9$.

Example 3 Use the binomial theorem to expand $(x + y)^{10}$.

Solution: Let $a = x$, $b = y$, and $n = 10$ in the binomial formula.

$$(x+y)^{10} = x^{10} + \frac{10}{1!}x^9y + \frac{10\cdot 9}{2!}x^8y^2 + \frac{10\cdot 9\cdot 8}{3!}x^7y^3 + \frac{10\cdot 9\cdot 8\cdot 7}{4!}x^6y^4$$
$$+ \frac{10\cdot 9\cdot 8\cdot 7\cdot 6}{5!}x^5y^5 + \frac{10\cdot 9\cdot 8\cdot 7\cdot 6\cdot 5}{6!}x^4y^6$$
$$+ \frac{10\cdot 9\cdot 8\cdot 7\cdot 6\cdot 5\cdot 4}{7!}x^3y^7$$
$$+ \frac{10\cdot 9\cdot 8\cdot 7\cdot 6\cdot 5\cdot 4\cdot 3}{8!}x^2y^8$$
$$+ \frac{10\cdot 9\cdot 8\cdot 7\cdot 6\cdot 5\cdot 4\cdot 3\cdot 2}{9!}xy^9 + y^{10}$$
$$= x^{10} + 10x^9y + 45x^8y^2 + 120x^7y^3 + 210x^6y^4 + 252x^5y^5 + 210x^4y^6$$
$$+ 120x^3y^7 + 45x^2y^8 + 10xy^9 + y^{10}$$

Work Practice 3

Practice 4

Use the binomial theorem to expand $(a + 5b)^3$.

Example 4 Use the binomial theorem to expand $(x + 2y)^5$.

Solution: Let $a = x$ and $b = 2y$ in the binomial formula.

$$(x+2y)^5 = x^5 + \frac{5}{1!}x^4(2y) + \frac{5\cdot 4}{2!}x^3(2y)^2 + \frac{5\cdot 4\cdot 3}{3!}x^2(2y)^3$$
$$+ \frac{5\cdot 4\cdot 3\cdot 2}{4!}x(2y)^4 + (2y)^5$$
$$= x^5 + 10x^4y + 40x^3y^2 + 80x^2y^3 + 80xy^4 + 32y^5$$

Work Practice 4

Practice 5

Use the binomial theorem to expand $(3x - 2y)^3$.

Example 5 Use the binomial theorem to expand $(3m - n)^4$.

Solution: Let $a = 3m$ and $b = -n$ in the binomial formula.

$$(3m-n)^4 = (3m)^4 + \frac{4}{1!}(3m)^3(-n) + \frac{4\cdot 3}{2!}(3m)^2(-n)^2$$
$$+ \frac{4\cdot 3\cdot 2}{3!}(3m)(-n)^3 + (-n)^4$$
$$= 81m^4 - 108m^3n + 54m^2n^2 - 12mn^3 + n^4$$

Work Practice 5

Objective 4 Finding the *n*th Term of a Binomial Expansion

Sometimes it is convenient to find a specific term of a binomial expansion without writing out the entire expansion. By studying the expansion of binomials, a pattern forms for each term. This pattern is most easily stated for the $(r + 1)$st term.

Answers

3. $a^9 + 9a^8b + 36a^7b^2 + 84a^6b^3 + 126a^5b^4 + 126a^4b^5 + 84a^3b^6 + 36a^2b^7 + 9ab^8 + b^9$
4. $a^3 + 15a^2b + 75ab^2 + 125b^3$
5. $27x^3 - 54x^2y + 36xy^2 - 8y^3$

(r + 1)st Term in a Binomial Expansion

The $(r + 1)$st term of the expansion of $(a + b)^n$ is $\frac{n!}{r!(n - r)!}a^{n-r}b^r$.

Example 6 Find the eighth term in the expansion of $(2x - y)^{10}$.

Solution: Use the formula with $n = 10, a = 2x, b = -y$, and $r + 1 = 8$. Notice that, since $r + 1 = 8, r = 7$.

$$\begin{aligned}\frac{n!}{r!(n - r)!}a^{n-r}b^r &= \frac{10!}{7!3!}(2x)^3(-y)^7 \\ &= 120(8x^3)(-y^7) \\ &= -960x^3y^7\end{aligned}$$

Work Practice 6

Practice 6

Find the seventh term in the expansion of $(x - 4y)^{11}$.

Answer

6. $1{,}892{,}352x^5y^6$

Vocabulary, Readiness & Video Check

Fill in each blank.

1. $0! =$ ____ **2.** $1! =$ ____ **3.** $4! =$ ____ **4.** $2! =$ ____ **5.** $3!0! =$ ____ **6.** $0!2! =$ ____

Martin-Gay Interactive Videos

See Video 21.5 

Watch the section lecture video and answer the following questions.

Objective 1 **7.** From Example 1 and the lecture before, when expanding a binomial such as $(x + y)^7$, what does Pascal's triangle tell you? What does the power on the binomial tell you?

Objective 2 **8.** From Example 2 and the lecture before, write the definition of 4! and evaluate it. What is the value of 0!?

Objective 3 **9.** From Example 4, what point is made about the terms of a binomial when applying the binomial theorem?

Objective 4 **10.** In Example 5, we are looking for the 4th term, so why do we let $r = 3$?

21.5 Exercise Set MyMathLab®

Use Pascal's triangle to expand the binomial. See Example 1.

1. $(m + n)^3$

2. $(x + y)^4$

3. $(c + d)^5$

4. $(a + b)^6$

5. $(y - x)^5$

6. $(q - r)^7$

7. Explain how to generate a row of Pascal's triangle.

8. Write the $n = 8$ row of Pascal's triangle.

Evaluate each expression. See Example 2.

9. $\frac{8!}{7!}$

10. $\frac{6!}{0!}$

11. $\frac{7!}{5!}$

12. $\frac{8!}{5!}$

13. $\frac{10!}{7!2!}$

14. $\frac{9!}{5!3!}$

15. $\frac{8!}{6!0!}$

16. $\frac{10!}{4!6!}$

MIXED PRACTICE

Use the binomial formula to expand each binomial. See Examples 3 through 5.

17. $(a + b)^7$
18. $(x + y)^8$
19. $(a + 2b)^5$
20. $(x + 3y)^6$
21. $(q + r)^9$
22. $(b + c)^6$
23. $(4a + b)^5$
24. $(3m + n)^4$
25. $(5a - 2b)^4$
26. $(m - 4)^6$
27. $(2a + 3b)^3$
28. $(4 - 3x)^5$
29. $(x + 2)^5$
30. $(3 + 2a)^4$

Find the indicated term. See Example 6.

31. The fifth term of the expansion of $(c - d)^5$
32. The fourth term of the expansion of $(x - y)^6$
33. The eighth term of the expansion of $(2c + d)^7$
34. The tenth term of the expansion of $(5x - y)^9$
35. The fourth term of the expansion of $(2r - s)^5$
36. The first term of the expansion of $(3q - 7r)^6$
37. The third term of the expansion of $(x + y)^4$
38. The fourth term of the expansion of $(a + b)^8$
39. The second term of the expansion of $(a + 3b)^{10}$
40. The third term of the expansion of $(m + 5n)^7$

REVIEW AND PREVIEW

Sketch the graph of each function. Decide whether each function is one-to-one. See Sections 10.6 and 19.2.

41. $f(x) = |x|$
42. $g(x) = 3(x - 1)^2$
43. $H(x) = 2x + 3$
44. $F(x) = -2$
45. $f(x) = x^2 + 3$
46. $h(x) = -(x + 1)^2 - 4$

CONCEPT EXTENSIONS

47. Expand the expression $(\sqrt{x} + \sqrt{3})^5$.
48. Find the term containing x^2 in the expansion of $(\sqrt{x} - \sqrt{5})^6$.

Evaluate the following.

The notation $\binom{n}{r}$ means $\frac{n!}{r!(n - r)!}$. For example,

$$\binom{5}{3} = \frac{5!}{3!(5 - 3)!} = \frac{5!}{3!2!} = \frac{5 \cdot 4 \cdot 3 \cdot 2 \cdot 1}{(3 \cdot 2 \cdot 1) \cdot (2 \cdot 1)} = 10.$$

49. $\binom{9}{5}$
50. $\binom{4}{3}$
51. $\binom{8}{2}$
52. $\binom{12}{11}$
53. Show that $\binom{n}{n} = 1$ for any whole number n.

Chapter 21 Vocabulary Check

Fill in each blank with one of the words or phrases listed below.

general term	common difference	finite sequence	common ratio	Pascal's triangle
infinite sequence	factorial of n	arithmetic sequence	geometric sequence	series

1. A(n) __________ is a function whose domain is the set of natural numbers $\{1, 2, 3, \ldots, n\}$, where n is some natural number.
2. The __________, written $n!$, is the product of the first n consecutive natural numbers.
3. A(n) __________ is a function whose domain is the set of natural numbers.
4. A(n) __________ is a sequence in which each term (after the first) is obtained by multiplying the preceding term by a constant amount r. The constant r is called the __________ of the sequence.
5. The sum of the terms of a sequence is called a(n) __________.
6. The nth term of the sequence a_n is called the __________.

7. A(n) ______________ is a sequence in which each term (after the first) differs from the preceding term by a constant amount d. The constant d is called the ______________ of the sequence.

8. A triangular array of the coefficients of the terms of the expansions of $(a + b)^n$ is called ______________.

Helpful Hint Are you preparing for your test? Don't forget to take the Chapter 21 Test on page 1399. Then check your answers at the back of the text and use the Chapter Test Prep Videos to see the fully worked-out solutions to any of the exercises you want to review.

21 Chapter Highlights

Definitions and Concepts	Examples
Section 21.1 Sequences	
An **infinite sequence** is a function whose domain is the set of natural numbers $\{1, 2, 3, 4, \ldots\}$.	**Infinite Sequence** $2, 4, 6, 8, 10, \ldots$
A **finite sequence** is a function whose domain is the set of natural numbers $\{1, 2, 3, 4, \ldots, n\}$, where n is some natural number.	**Finite Sequence** $1, -2, 3, -4, 5, -6$
The notation a_n, where n is a natural number, denotes a sequence.	Write the first four terms of the sequence whose general term is $a_n = n^2 + 1$. $a_1 = 1^2 + 1 = 2$ $a_2 = 2^2 + 1 = 5$ $a_3 = 3^2 + 1 = 10$ $a_4 = 4^2 + 1 = 17$
Section 21.2 Arithmetic and Geometric Sequences	
An **arithmetic sequence** is a sequence in which each term differs from the preceding term by a constant amount d, called the **common difference.**	**Arithmetic Sequence** $5, 8, 11, 14, 17, 20, \ldots$ Here, $a_1 = 5$ and $d = 3$.
The **general term** a_n of an arithmetic sequence is given by $a_n = a_1 + (n - 1)d$ where a_1 is the first term and d is the common difference.	The general term is $a_n = a_1 + (n - 1)d$ or $a_n = 5 + (n - 1)3$
A **geometric sequence** is a sequence in which each term is obtained by multiplying the preceding term by a constant r, called the **common ratio.**	**Geometric Sequence** $12, -6, 3, -\frac{3}{2}, \ldots$ Here, $a_1 = 12$ and $r = -\frac{1}{2}$.
The **general term** a_n of a geometric sequence is given by $a_n = a_1 r^{n-1}$ where a_1 is the first term and r is the common ratio.	The general term is $a_n = a_1 r^{n-1}$ or $a_n = 12\left(-\frac{1}{2}\right)^{n-1}$

Definitions and Concepts	Examples
Section 21.3 Series	
A sum of the terms of a sequence is called a **series.**	**Sequence** / **Series** 3, 7, 11, 15 — 3 + 7 + 11 + 15 — finite 3, 7, 11, 15, . . . — 3 + 7 + 11 + 15 + ⋯ — infinite
A shorthand notation for denoting a series is called **summation notation:** $\sum_{i=1}^{4}$ → Greek letter sigma used to mean sum; index of summation → $i=1$	$\sum_{i=1}^{4} 3^i = 3^1 + 3^2 + 3^3 + 3^4$ $= 3 + 9 + 27 + 81$ $= 120$
Section 21.4 Partial Sums of Arithmetic and Geometric Sequences	
Partial sum, S_n, of the first n terms of an arithmetic sequence: $S_n = \frac{n}{2}(a_1 + a_n)$ where a_1 is the first term and a_n is the nth term.	The sum of the first five terms of the arithmetic sequence 12, 24, 36, 48, 60, . . . is $S_5 = \frac{5}{2}(12 + 60) = 180$
Partial sum, S_n, of the first n terms of a geometric sequence: $S_n = \frac{a_1(1 - r^n)}{1 - r}$ where a_1 is the first term, r is the common ratio, and $r \neq 1$.	The sum of the first five terms of the geometric sequence 15, 30, 60, 120, 240, . . . is $S_5 = \frac{15(1 - 2^5)}{1 - 2} = 465$
Sum of the terms of an infinite geometric sequence: $S_\infty = \frac{a_1}{1 - r}$ where a_1 is the first term, r is the common ratio, and $\lvert r \rvert < 1$. (If $\lvert r \rvert \geq 1$, S_∞ does not exist.)	The sum of the terms of the infinite geometric sequence $1, \frac{1}{3}, \frac{1}{9}, \frac{1}{27}, \ldots$ is $S_\infty = \frac{1}{1 - \frac{1}{3}} = \frac{3}{2}$
Section 21.5 The Binomial Theorem	
The **factorial of *n*,** written $n!$, is the product of the first n consecutive natural numbers.	$5! = 5 \cdot 4 \cdot 3 \cdot 2 \cdot 1 = 120$
Binomial Theorem If n is a positive integer, then $(a + b)^n = a^n + \frac{n}{1!}a^{n-1}b^1 + \frac{n(n-1)}{2!}a^{n-2}b^2 + \frac{n(n-1)(n-2)}{3!}a^{n-3}b^3 + \cdots + b^n$	Expand $(3x + y)^4$. $(3x + y)^4 = (3x)^4 + \frac{4}{1!}(3x)^3(y)^1 + \frac{4 \cdot 3}{2!}(3x)^2(y)^2 + \frac{4 \cdot 3 \cdot 2}{3!}(3x)^1y^3 + y^4$ $= 81x^4 + 108x^3y + 54x^2y^2 + 12xy^3 + y^4$

Chapter 21 Review

(21.1) Find the indicated term(s) of the given sequence.

1. The first five terms of the sequence $a_n = -3n^2$
2. The first five terms of the sequence $a_n = n^2 + 2n$
3. The one-hundredth term of the sequence $a_n = \frac{(-1)^n}{100}$
4. The fiftieth term of the sequence $a_n = \frac{2n}{(-1)^n}$
5. The general term a_n of the sequence $\frac{1}{6}, \frac{1}{12}, \frac{1}{18}, \ldots$
6. The general term a_n of the sequence $-1, 4, -9, 16, \ldots$

Solve the following applications.

7. The distance in feet that an olive falling from rest in a vacuum will travel during each second is given by an arithmetic sequence whose general term is $a_n = 32n - 16$, where n is the number of the second. Find the distance the olive will fall during the fifth, sixth, and seventh seconds.
8. A culture of yeast measures 80 and doubles every day in a geometric progression, where n is the number of the day just ending. Write the measure of the yeast culture for the end of the next 5 days. Find how many days it takes the yeast culture to measure at least 10,000.
9. The Colorado Forest Service reported that western pine beetle infestation, which kills trees, affected approximately 660,000 acres of lodgepole forests in Colorado in 2006. The forest service predicted that during the next 5 years, the beetles would infest twice the number of acres per year as the year before. Write out the first 5 terms of this geometric sequence and find the number of acres of infested trees there were in 2010.
10. The first row of an amphitheater contains 50 seats, and each row thereafter contains 8 additional seats. Write the first ten terms of this arithmetic progression and find the number of seats in the tenth row.

(21.2)

11. Find the first five terms of the geometric sequence whose first term is -2 and whose common ratio is $\frac{2}{3}$.
12. Find the first five terms of the arithmetic sequence whose first term is 12 and whose common difference is -1.5.
13. Find the thirtieth term of the arithmetic sequence whose first term is -5 and whose common difference is 4.
14. Find the eleventh term of the arithmetic sequence whose first term is 2 and whose common difference is $\frac{3}{4}$.
15. Find the twentieth term of the arithmetic sequence whose first three terms are 12, 7, and 2.
16. Find the sixth term of the geometric sequence whose first three terms are 4, 6, and 9.
17. If the fourth term of an arithmetic sequence is 18 and the twentieth term is 98, find the first term and the common difference.
18. If the third term of a geometric sequence is -48 and the fourth term is 192, find the first term and the common ratio.
19. Find the general term of the sequence $\frac{3}{10}, \frac{3}{100}, \frac{3}{1000}, \ldots$
20. Find a general term that satisfies the terms shown for the sequence 50, 58, 66, ...

Determine whether each of the following sequences is arithmetic, geometric, or neither. If a sequence is arithmetic, find a_1 and d. If a sequence is geometric, find a_1 and r.

21. $\frac{8}{3}, 4, 6, \ldots$
22. $-10.5, -6.1, -1.7$
23. $7x, -14x, 28x$
24. $3x^2, 9x^4, 81x^8, \ldots$

Solve the following applications.

25. To test the bounce of a racquetball, the ball is dropped from a height of 8 feet. The ball is judged "good" if it rebounds at least 75% of its previous height with each bounce. Write out the first six terms of this geometric sequence (round to the nearest tenth). Determine if a ball that rebounds to a height of 2.5 feet after the fifth bounce is "good."
26. A display of oil cans in an auto parts store has 25 cans in the bottom row, 21 cans in the next row, and so on, in an arithmetic progression. Find the general term and the number of cans in the top row.

27. Suppose that you save \$1 the first day of a month, \$2 the second day, \$4 the third day, and continue to double your savings each day. Write the general term of this geometric sequence and find the amount you will save on the tenth day. Estimate the amount you will save on the thirtieth day of the month and check your estimate with a calculator.

28. On the first swing, the length of an arc through which a pendulum swings is 30 inches. The length of the arc for each successive swing is 70% of the preceding swing. Find the length of the arc for the fifth swing.

29. Rosa takes a job that has a monthly starting salary of \$900 and guarantees her a monthly raise of \$150 during her 6-month training period. Find the general term of this sequence and her salary at the end of her training.

30. A sheet of paper is $\frac{1}{512}$-inch thick. By folding the sheet in half, the total thickness will be $\frac{1}{256}$ inch. A second fold produces a total thickness of $\frac{1}{128}$ inch. Estimate the thickness of the stack after 15 folds and then check your estimate with a calculator.

(21.3) *Write out the terms and find the sum for each of the following.*

31. $\sum_{i=1}^{5}(2i - 1)$

32. $\sum_{i=1}^{5}i(i + 2)$

33. $\sum_{i=2}^{4}\frac{(-1)^i}{2i}$

34. $\sum_{i=3}^{5}5(-1)^{i-1}$

Write the sum with Σ notation.

35. $1 + 3 + 9 + 27 + 81 + 243$

36. $6 + 2 + (-2) + (-6) + (-10) + (-14) + (-18)$

37. $\frac{1}{4} + \frac{1}{16} + \frac{1}{64} + \frac{1}{256}$

38. $1 + \left(-\frac{3}{2}\right) + \frac{9}{4}$

Solve.

39. A yeast colony begins with 20 yeast and doubles every 8 hours. Write the sequence that describes the growth of the yeast and find the total yeast after 48 hours.

40. The number of cranes born each year in a new aviary forms a sequence whose general term is $a_n = n^2 + 2n - 1$. Find the number of cranes born in the fourth year and the total number of cranes born in the first four years.

41. Harold has a choice between two job offers. Job A has an annual starting salary of \$39,500 with guaranteed annual raises of \$2200 for the next four years, whereas job B has an annual starting salary of \$41,000 with guaranteed annual raises of \$1400 for the next four years. Compare the salaries for the fifth year under each job offer.

42. A sample of radioactive waste is decaying such that the amount decaying in kilograms during year n is $a_n = 200(0.5)^n$. Find the amount of decay in the third year and the total amount of decay in the first three years.

(21.4) *Find the partial sum of the given sequence.*

43. S_4 of the sequence $a_n = (n - 3)(n + 2)$

44. S_6 of the sequence $a_n = n^2$

45. S_5 of the sequence $a_n = -8 + (n - 1)3$

46. S_3 of the sequence $a_n = 5(4)^{n-1}$

47. The sixth partial sum of the sequence 15, 19, 23, . . .

48. The ninth partial sum of the sequence 5, −10, 20, . . .

49. The sum of the first 30 odd positive integers

50. The sum of the first 20 positive multiples of 7

51. The sum of the first 20 terms of the sequence 8, 5, 2, . . .

52. The sum of the first eight terms of the sequence $\frac{3}{4}, \frac{9}{4}, \frac{27}{4}, \ldots$

53. S_4 if $a_1 = 6$ and $r = 5$

54. S_{100} if $a_1 = -3$ and $d = -6$

Find the sum of each infinite geometric sequence.

55. $5, \frac{5}{2}, \frac{5}{4}, \ldots$

56. $18, -2, \frac{2}{9}, \ldots$

57. $-20, -4, -\frac{4}{5}, \ldots$

58. 0.2, 0.02, 0.002, . . .

Solve.

59. A frozen yogurt store owner cleared \$20,000 the first year he owned his business and made an additional 15% over the previous year in each subsequent year. Find how much he made during his fourth year of business. Find his total earnings during the first 4 years (round to the nearest dollar).

60. On his first morning in a television assembly factory, a trainee takes 0.8 times as long to assemble each television as he took to assemble the one before. If it took him 40 minutes to assemble the first television, find how long it takes him to assemble the fourth television. Find the total time he takes to assemble the first four televisions (round to the nearest minute).

61. During the harvest season, a farmer can rent a combine machine for \$100 the first day, with the rental fee decreasing \$7 for each additional day. Find how much the farmer pays for the rental on the seventh day. Find how much total rent the farmer pays for 7 days.

62. A rubber ball is dropped from a height of 15 feet and rebounds 80% of its previous height after each bounce. Find the total distance the ball travels before it comes to rest.

63. After a pond was sprayed once with insecticide, 1800 mosquitoes were killed the first day, 600 the second day, 200 the third day, and so on. Find the total number of mosquitoes killed during the first 6 days after the spraying (round to the nearest mosquito).

64. See Exercise 63. Find the day on which the insecticide is no longer effective and find the total number of mosquitoes killed (round to the nearest mosquito).

65. Use the formula S_∞ to write $0.55\overline{5}$ as a fraction.

66. A movie theater has 27 seats in the first row, 30 seats in the second row, 33 seats in the third row, and so on. Find the total number of seats in the theater if there are 20 rows.

(21.5) *Use Pascal's triangle to expand each binomial.*

67. $(x + z)^5$

68. $(y - r)^6$

69. $(2x + y)^4$

70. $(3y - z)^4$

Use the binomial formula to expand the following.

71. $(b + c)^8$

72. $(x - w)^7$

73. $(4m - n)^4$

74. $(p - 2r)^5$

Find the indicated term.

75. The fourth term of the expansion of $(a + b)^7$

76. The eleventh term of the expansion of $(y + 2z)^{10}$

MIXED REVIEW

77. Evaluate: $\sum_{i=1}^{4} i^2(i + 1)$

78. Find the fifteenth term of the arithmetic sequence whose first three terms are 14, 8, and 2.

79. Find the sum of the infinite geometric sequence 27, 9, 3, 1, ...

80. Expand: $(2x - 3)^4$

Chapter 21 Test

Step-by-step test solutions are found on the Chapter Test Prep Videos. Where available: MyMathLab® or 

Find the indicated term(s) of the given sequence.

1. The first five terms of the sequence $a_n = \frac{(-1)^n}{n + 4}$

2. The eightieth term of the sequence $a_n = 10 + 3(n - 1)$

3. The general term of the sequence $\frac{2}{5}, \frac{2}{25}, \frac{2}{125}, \ldots$

4. The general term of the sequence $-9, 18, -27, 36, \ldots$

Find the partial sum of the given sequence.

5. S_5 of the sequence $a_n = 5(2)^{n-1}$

6. S_{30} of the sequence $a_n = 18 + (n - 1)(-2)$

7. S_∞ of the sequence $a_1 = 24$ and $r = \frac{1}{6}$

8. S_∞ of the sequence $\frac{3}{2}, -\frac{3}{4}, \frac{3}{8}, \ldots$

9. $\sum_{i=1}^{4} i(i - 2)$

10. $\sum_{i=2}^{4} 5(2)^i(-1)^{i-1}$

Expand each binomial.

11. $(a - b)^6$

12. $(2x + y)^5$

Solve the following applications.

13. The population of a small town is growing yearly according to the sequence defined by $a_n = 250 + 75(n - 1)$, where n is the number of the year just beginning. Predict the population at the beginning of the tenth year. Find the town's initial population.

14. A gardener is making a triangular planting with one shrub in the first row, three shrubs in the second row, five shrubs in the third row, and so on, for eight rows. Write the finite series of this sequence and find the total number of shrubs planted.

15. A pendulum swings through an arc of length 80 centimeters on its first swing. On each successive swing, the length of the arc is $\frac{3}{4}$ the length of the arc on the preceding swing. Find the length of the arc on the fourth swing and find the total arc length for the first four swings.

16. See Exercise 15. Find the total arc length before the pendulum comes to rest.

17. A parachutist in free fall falls 16 feet during the first second, 48 feet during the second second, 80 feet during the third second, and so on. Find how far he falls during the tenth second. Find the total distance he falls during the first 10 seconds.

18. Use the formula S_∞ to write $0.42\overline{42}$ as a fraction.

Chapters 1–21 Cumulative Review

1. Evaluate.
   a. $(-3)^2$ b. -3^2
2. Simplify each expression.
   a. $3a - (4a + 3)$
   b. $(5x - 3) + (2x + 6)$
   c. $4(2x - 5) - 3(5x + 1)$
3. Simplify $2y - 6 + 4y + 8$.
4. Sara bought a digital camera for \$344.50 including tax. If the tax rate is 6%, what was the price of the camera before taxes?
5. Find an equation of the line with y-intercept $(0, -3)$ and slope of $\frac{1}{4}$.
6. Find an equation of the line through $(3, -2)$ and parallel to $3x - 2y = 6$. Write the equation using function notation.
7. Find an equation of the line through $(2, 5)$ and $(-3, 4)$. Write the equation in standard form.
8. Solve $y^3 + 5y^2 - y = 5$.
9. Use synthetic division to divide $x^4 - 2x^3 - 11x^2 + 5x + 34$ by $x + 2$.
10. Perform the indicated operation and simplify if possible. $\frac{5}{3a - 6} - \frac{a}{a - 2} + \frac{3 + 2a}{5a - 10}$
11. Simplify the following.
    a. $\sqrt{50}$ b. $\sqrt[3]{24}$
    c. $\sqrt{26}$ d. $\sqrt[4]{32}$
12. Solve $\sqrt{3x + 6} - \sqrt{7x - 6} = 0$.
13. Use the formula $A = P(1 + r)^t$ to find the interest rate r if \$2000 compounded annually grows to \$2420 in 2 years.
14. Rationalize each denominator.
    a. $\sqrt[3]{\frac{4}{3x}}$ b. $\frac{\sqrt{2} + 1}{\sqrt{2} - 1}$
15. Solve $(x - 3)^2 - 3(x - 3) - 4 = 0$.
16. Solve $\frac{10}{(2x + 4)^2} - \frac{1}{2x + 4} = 3$.
17. Solve $\frac{5}{x + 1} < -2$.
18. Graph $f(x) = (x + 2)^2 - 6$. Find the vertex and axis of symmetry.
19. A rock is thrown upward from the ground. Its height in feet above ground after t seconds is given by the function $f(t) = -16t^2 + 20t$. Find the maximum height of the rock and the number of seconds it took for the rock to reach its maximum height.
20. Find the vertex of $f(x) = x^2 + 3x - 18$.
21. If $f(x) = x^2$ and $g(x) = x + 3$, find each composition.
    a. $(f \circ g)(2)$ and $(g \circ f)(2)$
    b. $(f \circ g)(x)$ and $(g \circ f)(x)$
22. Find the inverse of $f(x) = -2x + 3$.
23. Find the inverse of the one-to-one function. $f = \{(0, 1), (-2, 7), (3, -6), (4, 4)\}$.
24. If $f(x) = x^2 - 2$ and $g(x) = x + 1$, find each composition.
    a. $(f \circ g)(2)$ and $(g \circ f)(2)$
    b. $(f \circ g)(x)$ and $(g \circ f)(x)$
25. Solve each equation for x.
    a. $2^x = 16$ b. $9^x = 27$
    c. $4^{x+3} = 8^x$
26. Solve each equation.
    a. $\log_2 32 = x$ b. $\log_4 \frac{1}{64} = x$
    c. $\log_{1/2} x = 5$
27. Simplify.
    a. $\log_3 3^2$ b. $\log_7 7^{-1}$
    c. $5^{\log_5 3}$ d. $2^{\log_2 6}$
28. Solve each equation for x.
    a. $4^x = 64$ b. $8^x = 32$
    c. $9^{x+4} = 243^x$
29. Write each sum as a single logarithm.
    a. $\log_{11} 10 + \log_{11} 3$
    b. $\log_3 \frac{1}{2} + \log_3 12$
    c. $\log_2(x + 2) + \log_2 x$
30. Find the exact value.
    a. $\log 100{,}000$ b. $\log 10^{-3}$
    c. $\ln \sqrt[5]{e}$ d. $\ln e^4$
31. Find the amount owed at the end of 5 years if \$1600 is loaned at a rate of 9% compounded continuously.
32. Write each expression as a single logarithm.
    a. $\log_6 5 + \log_6 4$
    b. $\log_8 12 - \log_8 4$
    c. $2 \log_2 x + 3 \log_2 x - 2 \log_2(x - 1)$
33. Solve: $3^x = 7$.
34. Using $A = P\left(1 + \frac{r}{n}\right)^{nt}$, find how long it takes \$5000 to double if it is invested at 2% interest compounded quarterly. Round to the nearest tenth.
35. Solve $\log_4(x - 2) = 2$.
36. Solve $\log_4 10 - \log_4 x = 2$.
37. Graph $\frac{x^2}{16} - \frac{y^2}{25} = 1$.

38. Find the distance between $(8, 5)$ and $(-2, 4)$.

39. Solve the system $\begin{cases} y = \sqrt{x} \\ x^2 + y^2 = 6 \end{cases}$.

40. Solve the system $\begin{cases} x^2 + y^2 = 36 \\ x - y = 6 \end{cases}$.

41. Graph $\frac{x^2}{9} + \frac{y^2}{16} \le 1$.

42. Graph $\begin{cases} y \ge x^2 \\ y \le 4 \end{cases}$.

43. Write the first five terms of the sequence whose general term is given by $a_n = n^2 - 1$.

44. If the general term of a sequence is $a_n = \frac{n}{n + 4}$, find a_8.

45. Find the eleventh term of the arithmetic sequence whose first three terms are 2, 9, and 16.

46. Find the sixth term of the geometric sequence 2, 10, 50, ...

47. Evaluate.

a. $\sum_{i=0}^{6} \frac{i - 2}{2}$ **b.** $\sum_{i=3}^{5} 2^i$

48. Evaluate.

a. $\sum_{i=0}^{4} i(i + 1)$ **b.** $\sum_{i=0}^{3} 2^i$

49. Find the sum of the first 30 positive integers.

50. Find the third term of the expansion of $(x - y)^6$.

Tables

A.1 Table of Geometric Figures

Plane Figures Have Length and Width but No Thickness or Depth		
Name	**Description**	**Figure**
Polygon	Union of three or more coplanar line segments that intersect with each other only at each endpoint, with each endpoint shared by two segments.	
Triangle	Polygon with three sides (sum of measures of three angles is 180°).	
Scalene Triangle	Triangle with no sides of equal length.	
Isosceles Triangle	Triangle with two sides of equal length.	
Equilateral Triangle	Triangle with all sides of equal length.	
Right Triangle	Triangle that contains a right angle.	leg, hypotenuse, leg
Quadrilateral	Polygon with four sides (sum of measures of four angles is 360°).	
Trapezoid	Quadrilateral with exactly one pair of opposite sides parallel.	base, leg, parallel sides, leg, base
Isosceles Trapezoid	Trapezoid with legs of equal length.	
Parallelogram	Quadrilateral with both pairs of opposite sides parallel.	
Rhombus	Parallelogram with all sides of equal length.	
Rectangle	Parallelogram with four right angles.	

Plane Figures Have Length and Width but No Thickness or Depth (*continued*)		
Name	Description	Figure
Square	Rectangle with all sides of equal length.	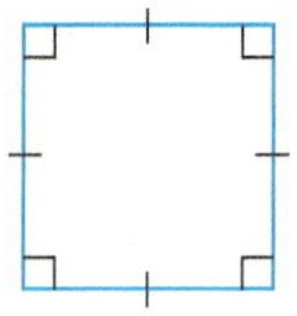
Circle	All points in a plane the same distance from a fixed point called the **center.**	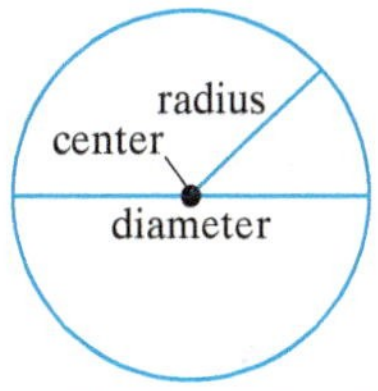 radius, center, diameter

Solid Figures Have Length, Width, and Height or Depth		
Name	Description	Figure
Rectangular Solid	A solid with six sides, all of which are rectangles.	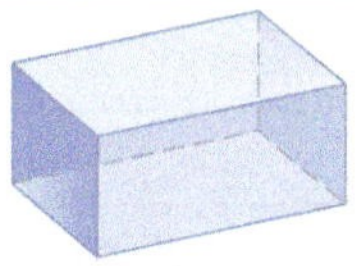
Cube	A rectangular solid whose six sides are squares.	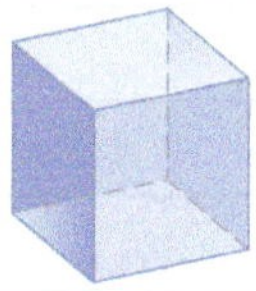
Sphere	All points the same distance from a fixed point called the **center.**	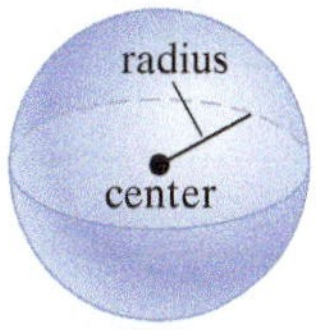 radius, center
Right Circular Cylinder	A cylinder having two circular bases that are perpendicular to its altitude.	
Right Circular Cone	A cone with a circular base that is perpendicular to its altitude.	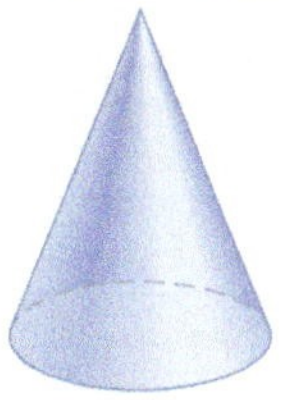

A.2 Table of Percents, Decimals, and Fraction Equivalents

Percent	Decimal	Fraction
1%	0.01	$\frac{1}{100}$
5%	0.05	$\frac{1}{20}$
10%	0.1	$\frac{1}{10}$
12.5% or $12\frac{1}{2}$%	0.125	$\frac{1}{8}$
$16.\overline{6}$% or $16\frac{2}{3}$%	$0.1\overline{6}$	$\frac{1}{6}$
20%	0.2	$\frac{1}{5}$
25%	0.25	$\frac{1}{4}$
30%	0.3	$\frac{3}{10}$
$33.\overline{3}$% or $33\frac{1}{3}$%	$0.\overline{3}$	$\frac{1}{3}$
37.5% or $37\frac{1}{2}$%	0.375	$\frac{3}{8}$
40%	0.4	$\frac{2}{5}$
50%	0.5	$\frac{1}{2}$
60%	0.6	$\frac{3}{5}$
62.5% or $62\frac{1}{2}$%	0.625	$\frac{5}{8}$
$66.\overline{6}$% or $66\frac{2}{3}$%	$0.\overline{6}$	$\frac{2}{3}$
70%	0.7	$\frac{7}{10}$
75%	0.75	$\frac{3}{4}$
80%	0.8	$\frac{4}{5}$
$83.\overline{3}$% or $83\frac{1}{3}$%	$0.8\overline{3}$	$\frac{5}{6}$
87.5% or $87\frac{1}{2}$%	0.875	$\frac{7}{8}$
90%	0.9	$\frac{9}{10}$
100%	1.0	1
110%	1.1	$1\frac{1}{10}$
125%	1.25	$1\frac{1}{4}$
$133.\overline{3}$% or $133\frac{1}{3}$%	$1.\overline{3}$	$1\frac{1}{3}$
150%	1.5	$1\frac{1}{2}$
$166.\overline{6}$% or $166\frac{2}{3}$%	$1.\overline{6}$	$1\frac{2}{3}$
175%	1.75	$1\frac{3}{4}$
200%	2.0	2

A.3 Table on Finding Common Percents of a Number

Common Percent Equivalences*	Shortcut Method for Finding Percent	Examples
$1\% = 0.01 \left(\text{or } \frac{1}{100}\right)$	To find 1% of a number, multiply by 0.01. To do so, move the decimal point 2 places to the left.	1% of 210 is 2.10 or 2.1. 1% of 1500 is 15. 1% of 8.6 is 0.086.
$10\% = 0.1 \left(\text{or } \frac{1}{10}\right)$	To find 10% of a number, multiply by 0.1, or move the decimal point of the number 1 place to the left.	10% of 140 is 14. 10% of 30 is 3. 10% of 17.6 is 1.76.
$25\% = \frac{1}{4}$	To find 25% of a number, find $\frac{1}{4}$ of the number, or divide the number by 4.	25% of 20 is $\frac{20}{4}$ or 5. 25% of 8 is 2. 25% of 10 is $\frac{10}{4}$ or $2\frac{1}{2}$.
$50\% = \frac{1}{2}$	To find 50% of a number, find $\frac{1}{2}$ of the number, or divide the number by 2.	50% of 64 is $\frac{64}{2}$ or 32. 50% of 1000 is 500. 50% of 9 is $\frac{9}{2}$ or $4\frac{1}{2}$.
$100\% = 1$	To find 100% of a number, multiply the number by 1. In other words, 100% of a number is the number.	100% of 98 is 98. 100% of 1407 is 1407. 100% of 18.4 is 18.4.
$200\% = 2$	To find 200% of a number, multiply the number by 2.	200% of 31 is $31 \cdot 2$ or 62. 200% of 750 is 1500. 200% of 6.5 is 13.

*See Appendix A.2.

A.4 Table of Squares and Square Roots

n	n^2	$\sqrt{n}$	n	n^2	$\sqrt{n}$
1	1	1.000	51	2601	7.141
2	4	1.414	52	2704	7.211
3	9	1.732	53	2809	7.280
4	16	2.000	54	2916	7.348
5	25	2.236	55	3025	7.416
6	36	2.449	56	3136	7.483
7	49	2.646	57	3249	7.550
8	64	2.828	58	3364	7.616
9	81	3.000	59	3481	7.681
10	100	3.162	60	3600	7.746
11	121	3.317	61	3721	7.810
12	144	3.464	62	3844	7.874
13	169	3.606	63	3969	7.937
14	196	3.742	64	4096	8.000
15	225	3.873	65	4225	8.062
16	256	4.000	66	4356	8.124
17	289	4.123	67	4489	8.185
18	324	4.243	68	4624	8.246
19	361	4.359	69	4761	8.307
20	400	4.472	70	4900	8.367
21	441	4.583	71	5041	8.426
22	484	4.690	72	5184	8.485
23	529	4.796	73	5329	8.544
24	576	4.899	74	5476	8.602
25	625	5.000	75	5625	8.660
26	676	5.099	76	5776	8.718
27	729	5.196	77	5929	8.775
28	784	5.292	78	6084	8.832
29	841	5.385	79	6241	8.888
30	900	5.477	80	6400	8.944
31	961	5.568	81	6561	9.000
32	1024	5.657	82	6724	9.055
33	1089	5.745	83	6889	9.110
34	1156	5.831	84	7056	9.165
35	1225	5.916	85	7225	9.220
36	1296	6.000	86	7396	9.274
37	1369	6.083	87	7569	9.327
38	1444	6.164	88	7744	9.381
39	1521	6.245	89	7921	9.434
40	1600	6.325	90	8100	9.487
41	1681	6.403	91	8281	9.539
42	1764	6.481	92	8464	9.592
43	1849	6.557	93	8649	9.644
44	1936	6.633	94	8836	9.695
45	2025	6.708	95	9025	9.747
46	2116	6.782	96	9216	9.798
47	2209	6.856	97	9409	9.849
48	2304	6.928	98	9604	9.899
49	2401	7.000	99	9801	9.950
50	2500	7.071	100	10,000	10.000

A.5 Geometric Formulas 

Rectangle

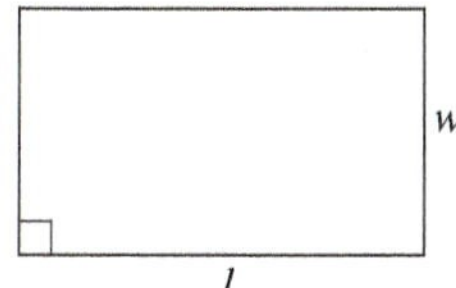

Perimeter: $P = 2l + 2w$
Area: $A = lw$

Square

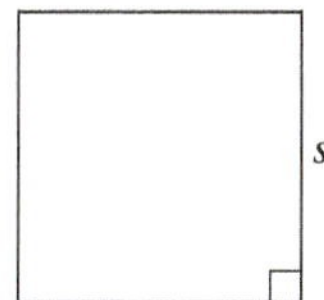

Perimeter: $P = 4s$
Area: $A = s^2$

Triangle

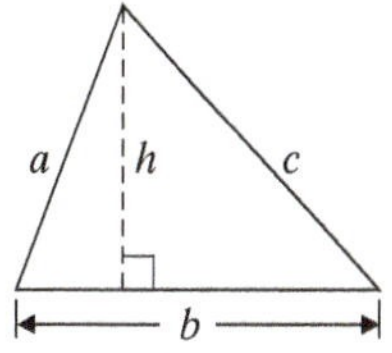

Perimeter: $P = a + b + c$
Area: $A = \frac{1}{2}bh$

Sum of Angles of Triangle

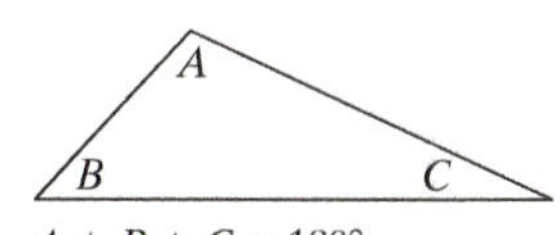

$A + B + C = 180°$
The sum of the measures of the three angles is 180°.

Pythagorean Theorem (for right triangles)

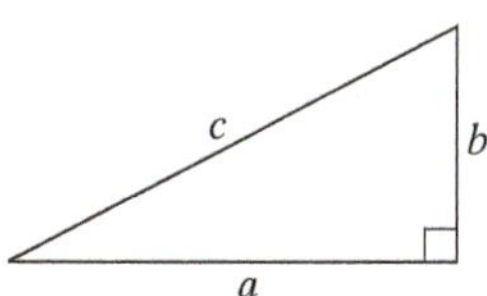

Perimeter: $P = a + b + c$
Area: $A = \frac{1}{2}ab$
One 90° (right) angle

Isosceles Triangle

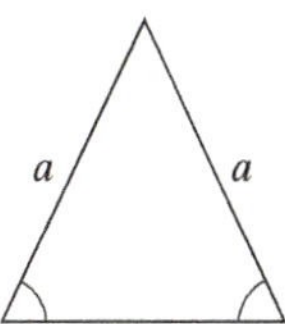

Triangle has:
two equal sides and
two equal angles.

Equilateral Triangle

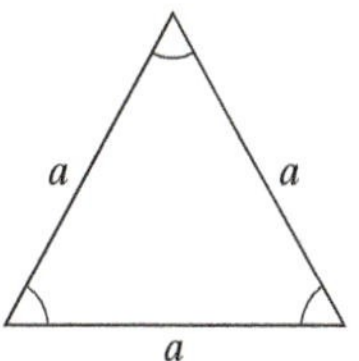

Triangle has:
three equal sides and
three equal angles.
Measure of each angle is 60°.

Trapezoid

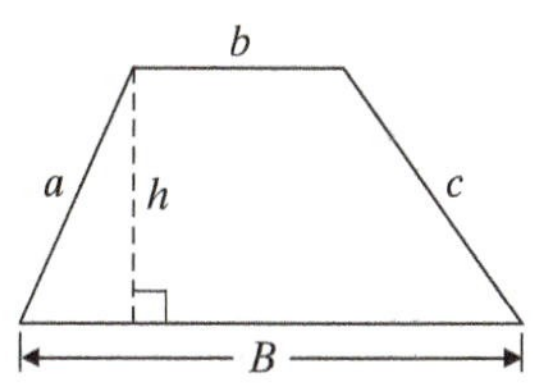

Perimeter: $P = a + b + c + B$
Area: $A = \frac{1}{2}h(B + b)$

Parallelogram

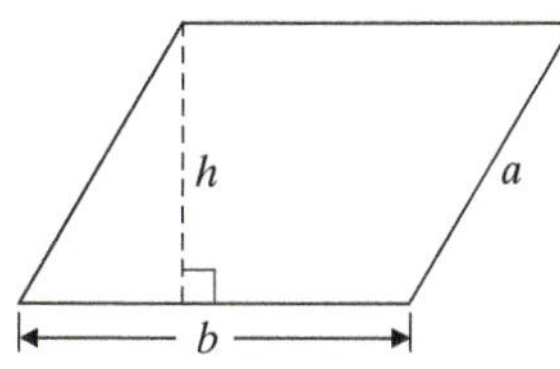

Perimeter: $P = 2a + 2b$
Area: $A = bh$

Circle

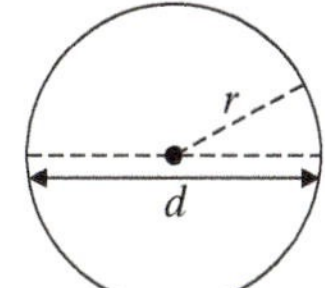

Circumference: $C = \pi d$
$C = 2\pi r$
Area: $A = \pi r^2$

Rectangular Solid

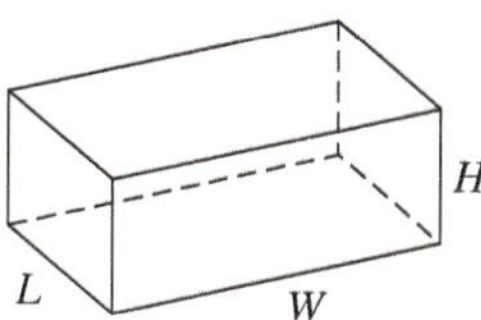

Volume: $V = LWH$
Surface Area:
$S = 2LW + 2HL + 2HW$

Cube

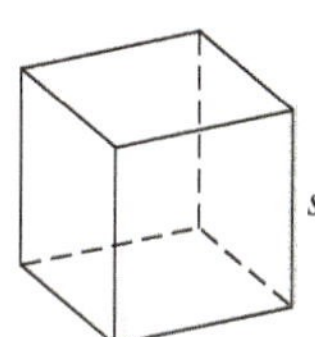

Volume: $V = s^3$
Surface Area: $S = 6s^2$

Cone

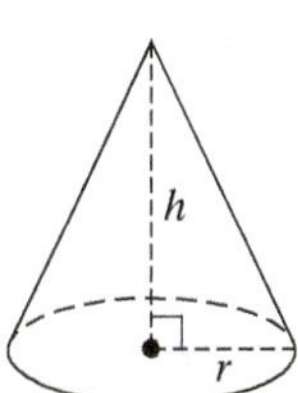

Volume: $V = \frac{1}{3}\pi r^2 h$
Lateral Surface Area:
$S = \pi r\sqrt{r^2 + h^2}$

Right Circular Cylinder

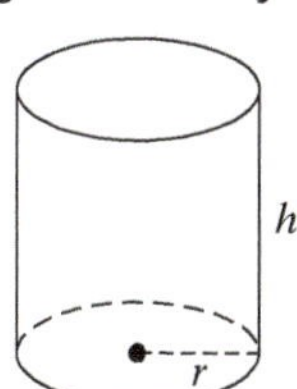

Volume: $V = \pi r^2 h$
Surface Area: $S = 2\pi r^2 + 2\pi rh$

Sphere

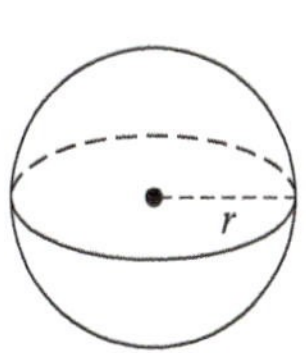

Volume: $V = \frac{4}{3}\pi r^3$
Surface Area: $S = 4\pi r^2$

Square-Based Pyramid

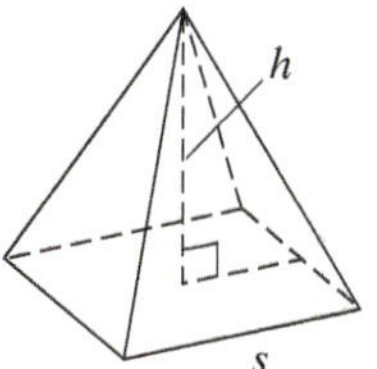

Volume: $V = \frac{1}{3}s^2 h$

Appendix

B Review of Algebra Topics

Recall that equations model many real-life problems. For example, we can use a linear equation to calculate the increase in the number (in millions) of Wi-Fi-enabled cell phones.

Wi-Fi-enabled cell phones let you carry your Internet access with you. There are already several of these smartphones available, and this technology will continue to expand. Estimated numbers of Wi-Fi-enabled cell phones in the United States for various years are shown below.

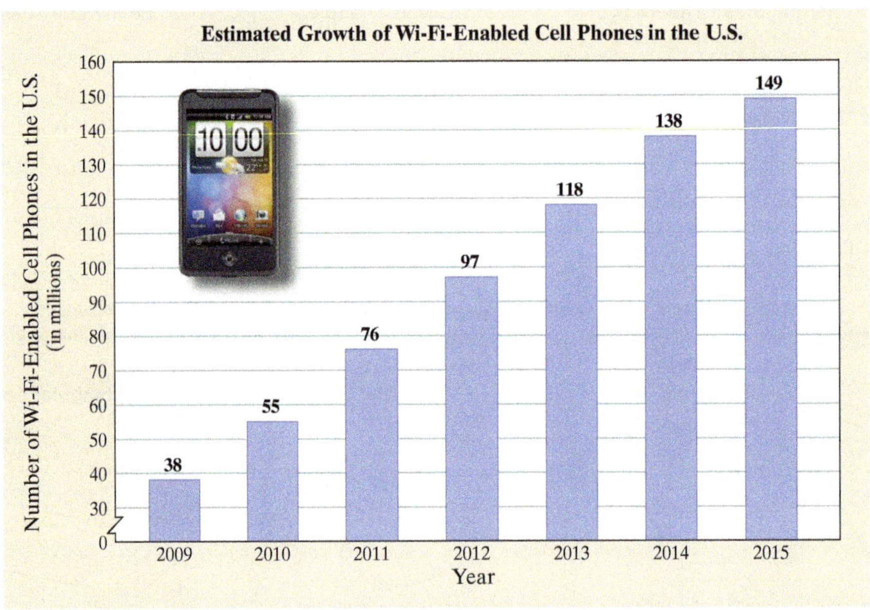

To find the estimated increase in the number of Wi-Fi-enabled cell phones in the United States from 2014 to 2015, for example, we can use the equation below.

In words:	Increase in cell phones	is	cell phones in 2015	minus	cell phones in 2014
Translate:	x	$=$	149	$-$	138

Since our variable x (increase in Wi-Fi-enabled cell phones) is by itself on one side of the equation, we can find the value of x by simplifying the right side.

$$x = 11$$

The estimated increase in the number of Wi-Fi-enabled cell phones from 2014 to 2015 is 11 million.

The **equation** $x = 149 - 138$ is a linear equation in one variable. In this section, we review solving linear equations and quadratic equations that can be solved by factoring. We will learn other methods for solving quadratic equations in Chapter 18.

B.1 Equations (Linear and Quadratic Solved by Factoring)

Objective 1 Solving Linear and Quadratic Equations 

Example 1 Solve: $2(x-3) = 5x - 9$.

Solution: First, use the distributive property.

$$2(x-3) = 5x - 9$$

$$2x - 6 = 5x - 9 \quad \text{Use the distributive property.}$$

Next, get variable terms on the same side of the equation by subtracting $5x$ from both sides.

$$2x - 6 - 5x = 5x - 9 - 5x \quad \text{Subtract } 5x \text{ from both sides.}$$

$$-3x - 6 = -9 \quad \text{Simplify.}$$

$$-3x - 6 + 6 = -9 + 6 \quad \text{Add 6 to both sides.}$$

$$-3x = -3 \quad \text{Simplify.}$$

$$\frac{-3x}{-3} = \frac{-3}{-3} \quad \text{Divide both sides by } -3.$$

$$x = 1$$

Let $x = 1$ in the original equation to see that 1 is the solution.

Work Practice 1

Don't forget, if an equation contains fractions, you may want to first clear the equation of fractions by multiplying both sides of the equation by the *least common denominator* (LCD) of all fractions in the equation.

Example 2 Solve for y: $\frac{y}{3} - \frac{y}{4} = \frac{1}{6}$.

Solution: First, clear the equation of fractions by multiplying both sides of the equation by 12, the LCD of denominators 3, 4, and 6.

$$\frac{y}{3} - \frac{y}{4} = \frac{1}{6}$$

$$12\left(\frac{y}{3} - \frac{y}{4}\right) = 12\left(\frac{1}{6}\right) \quad \text{Multiply both sides by the LCD, 12.}$$

$$12\left(\frac{y}{3}\right) - 12\left(\frac{y}{4}\right) = 2 \quad \text{Apply the distributive property.}$$

$$4y - 3y = 2 \quad \text{Simplify.}$$

$$y = 2 \quad \text{Simplify.}$$

(*Continued on next page*)

Objective

1 Solve Linear and Quadratic Equations.

Practice 1

Solve: $3(x-5) = 6x - 3$.

Practice 2

Solve for y: $\frac{y}{2} - \frac{y}{5} = \frac{1}{4}$.

Answers

1. -4 **2.** $\frac{5}{6}$

Check: To check, let $y = 2$ in the original equation.

$$\frac{y}{3} - \frac{y}{4} = \frac{1}{6} \quad \text{Original equation}$$

$$\frac{2}{3} - \frac{2}{4} \stackrel{?}{=} \frac{1}{6} \quad \text{Let } y = 2.$$

$$\frac{8}{12} - \frac{6}{12} \stackrel{?}{=} \frac{1}{6} \quad \text{Write fractions with the LCD.}$$

$$\frac{2}{12} \stackrel{?}{=} \frac{1}{6} \quad \text{Subtract.}$$

$$\frac{1}{6} = \frac{1}{6} \quad \text{Simplify.}$$

This is a true statement, so the solution is 2.

Work Practice 2

Practice 3

Solve:
$8(x^2 + 3) + 4 = -8x(x + 3) + 19$.

Example 3 Solve: $3(x^2 + 4) + 5 = -6(x^2 + 2x) + 13$.

Solution: Rewrite the equation so that one side is 0.

$$3(x^2 + 4) + 5 = -6(x^2 + 2x) + 13$$

$$3x^2 + 12 + 5 = -6x^2 - 12x + 13 \quad \text{Apply the distributive property.}$$

$$9x^2 + 12x + 4 = 0 \quad \text{Rewrite the equation so that one side is 0.}$$

$$(3x + 2)(3x + 2) = 0 \quad \text{Factor.}$$

$$3x + 2 = 0 \quad \text{or} \quad 3x + 2 = 0 \quad \text{Set each factor equal to 0.}$$

$$3x = -2 \quad \text{or} \quad 3x = -2$$

$$x = -\frac{2}{3} \quad \text{or} \quad x = -\frac{2}{3} \quad \text{Solve each equation.}$$

The solution is $-\frac{2}{3}$. Check by substituting $-\frac{2}{3}$ into the original equation.

Work Practice 3

Practice 4

Solve for x:
$x - \frac{x - 2}{12} = \frac{x + 3}{4} + \frac{1}{4}$.

Example 4 Solve for x: $\frac{x + 5}{2} + \frac{1}{2} = 2x - \frac{x - 3}{8}$.

Solution: Multiply both sides of the equation by 8, the LCD of 2 and 8.

$$8\left(\frac{x + 5}{2} + \frac{1}{2}\right) = 8\left(2x - \frac{x - 3}{8}\right) \quad \text{Multiply both sides by 8.}$$

$$8\left(\frac{x + 5}{2}\right) + 8\cdot\frac{1}{2} = 8\cdot 2x - 8\left(\frac{x - 3}{8}\right) \quad \text{Apply the distributive property.}$$

$$4(x + 5) + 4 = 16x - (x - 3) \quad \text{Simplify.}$$

$$4x + 20 + 4 = 16x - x + 3 \quad \text{Use the distributive property to remove parentheses.}$$

$$4x + 24 = 15x + 3 \quad \text{Combine like terms.}$$

$$-11x + 24 = 3 \quad \text{Subtract } 15x \text{ from both sides.}$$

$$-11x = -21 \quad \text{Subtract 24 from both sides.}$$

$$\frac{-11x}{-11} = \frac{-21}{-11} \quad \text{Divide both sides by } -11.$$

$$x = \frac{21}{11} \quad \text{Simplify.}$$

Helpful Hint When we multiply both sides of an equation by a number, the distributive property tells us that each term of the equation is multiplied by the number.

Answers

3. $-\frac{3}{4}$ 4. $\frac{5}{4}$

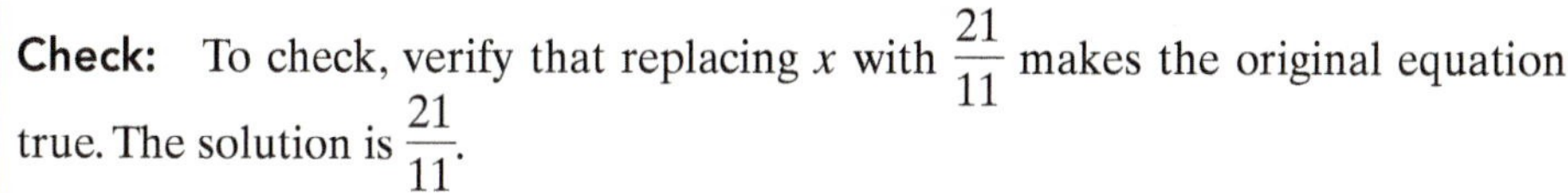

Check: To check, verify that replacing x with $\frac{21}{11}$ makes the original equation true. The solution is $\frac{21}{11}$.

Work Practice 4

Example 5 Solve: $2x^2 = \frac{17}{3}x + 1$.

Solution:

$$2x^2 = \frac{17}{3}x + 1$$

$$3(2x^2) = 3\left(\frac{17}{3}x + 1\right) \quad \text{Clear the equation of fractions.}$$

$$6x^2 = 17x + 3 \quad \text{Apply the distributive property.}$$

$$6x^2 - 17x - 3 = 0 \quad \text{Rewrite the equation in standard form.}$$

$$(6x + 1)(x - 3) = 0 \quad \text{Factor.}$$

$$6x + 1 = 0 \quad \text{or} \quad x - 3 = 0 \quad \text{Set each factor equal to zero.}$$

$$6x = -1$$

$$x = -\frac{1}{6} \quad \text{or} \quad x = 3 \quad \text{Solve each equation.}$$

The solutions are $-\frac{1}{6}$ and 3.

Work Practice 5

Practice 5

Solve: $4x^2 = \frac{15}{2}x + 1$.

Answer

5. $-\frac{1}{8}, 2$

B.1 Exercise Set MyMathLab®

Solve each equation. See Examples 1 through 5.

1. $x^2 + 11x + 24 = 0$
2. $y^2 - 10y + 24 = 0$
3. $3x - 4 - 5x = x + 4 + x$
4. $13x - 15x + 8 = 4x + 2 - 24$
5. $12x^2 + 5x - 2 = 0$
6. $3y^2 - y - 14 = 0$
7. $z^2 + 9 = 10z$
8. $n^2 + n = 72$
9. $5(y + 4) = 4(y + 5)$
10. $6(y - 4) = 3(y - 8)$
11. $0.6x - 10 = 1.4x - 14$
12. $0.3x + 2.4 = 0.1x + 4$
13. $x(5x + 2) = 3$
14. $n(2n - 3) = 2$
15. $6x - 2(x - 3) = 4(x + 1) + 4$
16. $10x - 2(x + 4) = 8(x - 2) + 6$
17. $\frac{3}{8} + \frac{b}{3} = \frac{5}{12}$
18. $\frac{a}{2} + \frac{7}{4} = 5$
19. $x^2 - 6x = x(8 + x)$
20. $n(3 + n) = n^2 + 4n$
21. $\frac{z^2}{6} - \frac{z}{2} - 3 = 0$
22. $\frac{c^2}{20} - \frac{c}{4} + \frac{1}{5} = 0$
23. $-z + 3(2 + 4z) = 6(z + 1) + 5z$
24. $4(m - 6) - m = 8(m - 3) - 5m$
25. $\frac{x^2}{2} + \frac{x}{20} = \frac{1}{10}$
26. $\frac{y^2}{30} = \frac{y}{15} + \frac{1}{2}$
27. $\frac{4t^2}{5} = \frac{t}{5} + \frac{3}{10}$

28. $\frac{5x^2}{6} - \frac{7x}{2} + \frac{2}{3} = 0$

29. $\frac{3t + 1}{8} = \frac{5 + 2t}{7} + 2$

30. $4 - \frac{2z + 7}{9} = \frac{7 - z}{12}$

31. $\frac{m - 4}{3} - \frac{3m - 1}{5} = 1$

32. $\frac{n + 1}{8} - \frac{2 - n}{3} = \frac{5}{6}$

33. $3x^2 = -x$

34. $y^2 = -5y$

35. $x(x - 3) = x^2 + 5x + 7$

36. $z^2 - 4z + 10 = z(z - 5)$

37. $3(t - 8) + 2t = 7 + t$

38. $7c - 2(3c + 1) = 5(4 - 2c)$

39. $-3(x - 4) + x = 5(3 - x)$

40. $-4(a + 1) - 3a = -7(2a - 3)$

41. $(x - 1)(x + 4) = 24$

42. $(2x - 1)(x + 2) = -3$

43. $\frac{x^2}{4} - \frac{5}{2}x + 6 = 0$

44. $\frac{x^2}{18} + \frac{x}{2} + 1 = 0$

45. $y^2 + \frac{1}{4} = -y$

46. $\frac{x^2}{10} + \frac{5}{2} = x$

47. Which solution strategies are incorrect? Why?
   a. Solve $(y - 2)(y + 2) = 4$ by setting each factor equal to 4.
   b. Solve $(x + 1)(x + 3) = 0$ by setting each factor equal to 0.
   c. Solve $z^2 + 5z + 6 = 0$ by factoring $z^2 + 5z + 6$ and setting each factor equal to 0.
   d. Solve $x^2 + 6x + 8 = 10$ by factoring $x^2 + 6x + 8$ and setting each factor equal to 0.

48. Describe two ways a linear equation differs from a quadratic equation.

Find the value of K such that the equations are equivalent.

49. $3.2x + 4 = 5.4x - 7$
$3.2x = 5.4x + K$

50. $-7.6y - 10 = -1.1y + 12$
$-7.6y = -1.1y + K$

51. $\frac{x}{6} + 4 = \frac{x}{3}$
$x + K = 2x$

52. $\frac{5x}{4} + \frac{1}{2} = \frac{x}{2}$
$5x + K = 2x$

Solve and check.

53. $2.569x = -12.48534$

54. $-9.112y = -47.537304$

55. $2.86z - 8.1258 = -3.75$

56. $1.25x - 20.175 = -8.15$

B.2 Mixture and Uniform Motion Problem Solving

Objectives

1 Solve Mixture Problems.

2 Solve Distance or Uniform Motion Problems.

This section is devoted to solving problems in the categories listed. The same problem-solving steps used in the chapters are followed in this section. They are listed below for review.

General Strategy for Problem Solving

1. UNDERSTAND the problem. During this step, become comfortable with the problem. Some ways of doing this are as follows:
   Read and reread the problem.
   Choose a variable to represent the unknown.
   Construct a drawing whenever possible.
   Propose a solution and check. Pay careful attention to how you check your proposed solution. This will help when writing an equation to model the problem.
2. TRANSLATE the problem into an equation.
3. SOLVE the equation.
4. INTERPRET the results: *Check* the proposed solution in the stated problem and *state* your conclusion.

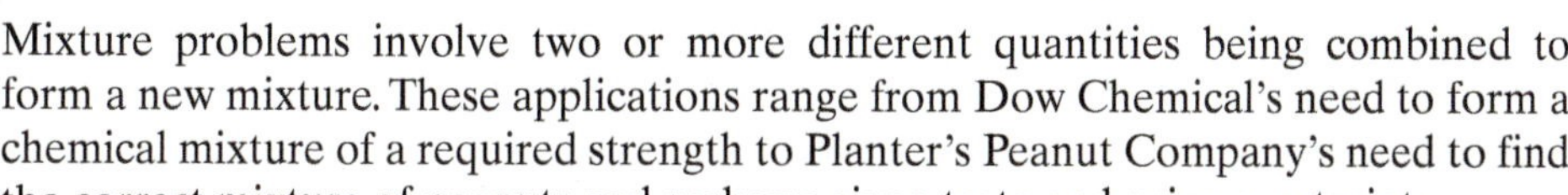

Objective 1 Solving Mixture Problems

Mixture problems involve two or more different quantities being combined to form a new mixture. These applications range from Dow Chemical's need to form a chemical mixture of a required strength to Planter's Peanut Company's need to find the correct mixture of peanuts and cashews given taste and price constraints.

Example 1 Calculating Percent for a Lab Experiment

A chemist working on his doctoral degree at Massachusetts Institute of Technology needs 12 liters of a 50% acid solution for a lab experiment. The stockroom has only 40% and 70% solutions. How much of each solution should be mixed together to form 12 liters of a 50% solution?

Solution:

1. UNDERSTAND. First, read and reread the problem a few times. Next, guess a solution. Suppose that we need 7 liters of the 40% solution. Then we need $12 - 7 = 5$ liters of the 70% solution. To see if this is indeed the answer, find the amount of pure acid in 7 liters of the 40% solution, in 5 liters of the 70% solution, and in 12 liters of a 50% solution, the required amount and strength.

number of liters	×	acid strength	=	amount of pure acid
7 liters	×	40%	=	7(0.40) or 2.8 liters
5 liters	×	70%	=	5(0.70) or 3.5 liters
12 liters	×	50%	=	12(0.50) or 6 liters

Since 2.8 liters + 3.5 liters = 6.3 liters and not 6 liters, our guess is incorrect, but we have gained some valuable insight into how to model and check this problem.
Let

x = number of liters of 40% solution; then

$12 - x$ = number of liters of 70% solution.

2. TRANSLATE. To help us translate to an equation, the following table summarizes the information given. Recall that the amount of acid in each solution is found by multiplying the acid strength of each solution by the number of liters.

	No. of Liters	·	Acid Strength	=	Amount of Acid
40% Solution	x		40%		$0.40x$
70% Solution	$12 - x$		70%		$0.70(12 - x)$
50% Solution Needed	12		50%		$0.50(12)$

The amount of acid in the final solution is the sum of the amounts of acid in the two beginning solutions.

In words:	acid in 40% solution	+	acid in 70% solution	=	acid in 50% mixture
Translate:	$0.40x$	+	$0.70(12 - x)$	=	$0.50(12)$

3. SOLVE.

$$0.40x + 0.70(12 - x) = 0.50(12)$$

$$0.4x + 8.4 - 0.7x = 6 \quad \text{Apply the distributive property.}$$

$$-0.3x + 8.4 = 6 \quad \text{Combine like terms.}$$

$$-0.3x = -2.4 \quad \text{Subtract 8.4 from both sides.}$$

$$x = 8 \quad \text{Divide both sides by } -0.3.$$

(*Continued on next page*)

Practice 1

How much 20% dye solution and 50% dye solution should be mixed to obtain 6 liters of a 40% solution?

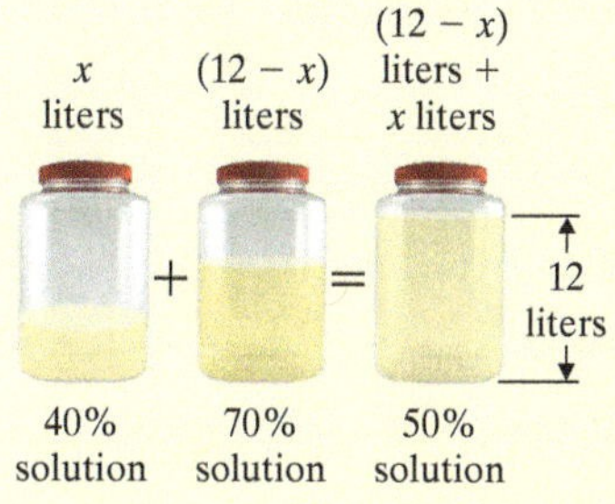

Answer

1. 2 liters of the 20% solution; 4 liters of the 50% solution

4. INTERPRET.

Check: To check, recall how we checked our guess.

State: If 8 liters of the 40% solution are mixed with $12 - 8$ or 4 liters of the 70% solution, the result is 12 liters of a 50% solution.

Work Practice 1

The next example has to do with finding an unknown number of a certain denomination of coin or bill. These problems are extremely useful in that they help you understand the difference between the number of coins or bills and the total value of the money.

For example, suppose there are seven \$5 bills. The *number* of \$5 bills is 7 and the *total value* of the money is $\$5(7) = \$35$.

Study the table below for more examples.

Denomination of Coin or Bill	Number of Coins or Bills	Value of Coins or Bills
20-dollar bills	17	$\$20(17) = \$340$
nickels	31	$\$0.05(31) = \$1.55$
quarters	x	$\$0.25(x) = \$0.25x$

Practice 2

A stack of \$5 and \$20 bills was counted by the treasurer of an organization. The total value of the money was \$1710 and there were 47 more \$5 bills than \$20 bills. Find the number of each type of bill.

Example 2 Finding Numbers of Denominations

Part of the proceeds from a local talent show was \$2420 worth of \$10 and \$20 bills. If there were 37 more \$20 bills than \$10 bills, find the number of each denomination.

Solution:

1. UNDERSTAND the problem. To do so, read and reread the problem. If you'd like, let's guess a solution. Suppose that there are 25 \$10 bills. Since there are 37 more \$20 bills, we have $25 + 37 = 62$ \$20 bills. The total amount of money is $\$10(25) + \$20(62) = \$1490$, less than the given amount of \$2420. Remember that our purpose for guessing is to help us better understand the problem.

We are looking for the number of each denomination, so we let

$x =$ number of \$10 bills

There are 37 more \$20 bills, so

$x + 37 =$ number of \$20 bills

2. TRANSLATE. To help us translate to an equation, study the table below.

Denomination	Number of Bills	Value of Bills (in dollars)
\$10 bills	x	$10x$
\$20 bills	$x + 37$	$20(x + 37)$

Since the total value of these bills is \$2420, we have

In words:	value of \$10 bills	plus	value of \$20 bills	is	2420
	↓	↓	↓	↓	↓
Translate:	$10x$	$+$	$20(x + 37)$	$=$	2420

3. SOLVE:

$$10x + 20x + 740 = 2420 \quad \text{Use the distributive property.}$$
$$30x + 740 = 2420 \quad \text{Add like terms.}$$
$$30x = 1680 \quad \text{Subtract 740 from both sides.}$$
$$\frac{30x}{30} = \frac{1680}{30} \quad \text{Divide both sides by 30.}$$
$$x = 56$$

Answer

2. 106 \$5 bills; 59 \$20 bills

4. INTERPRET the results.

Check: Since x represents the number of \$10 bills, we have 56 \$10 bills and 56 + 37, or 93, \$20 bills. The total amount of these bills is \$10(56) + \$20(93) = \$2420, the correct total.

State: There are 56 \$10 bills and 93 \$20 bills.

Work Practice 2

Objective 2 Solving Distance or Uniform Motion Problems

This example involves the distance formula $d = r \cdot t$.

Example 3 Finding Time Given Rate and Distance

Marie Antonio, a bicycling enthusiast, rode her 21-speed at an average speed of 18 miles per hour on level roads and then slowed down to an average speed of 10 mph on the hilly roads of the trip. If she covered a distance of 98 miles, how long did the entire trip take if traveling the level roads took the same time as traveling the hilly roads?

Solution:

1. UNDERSTAND the problem. To do so, read and reread the problem. The formula $d = r \cdot t$ is needed. At this time, let's guess a solution. Suppose that she spent 2 hours traveling on the level roads. This means that she also spent 2 hours traveling on the hilly roads, since the times spent were the same. What is her total distance? Her distance on the level roads is rate · time = 18(2) = 36 miles. Her distance on the hilly roads is rate · time = 10(2) = 20 miles. This gives a total distance of 36 miles + 20 miles = 56 miles, not the correct distance of 98 miles. Remember that the purpose of guessing a solution is not to guess correctly (although this may happen) but to help better understand the problem and how to model it with an equation. We are looking for the length of the entire trip, so we begin by letting

x = the time spent on level roads

Because the same amount of time was spent on hilly roads, then also

x = the time spent on hilly roads

2. TRANSLATE. To help us translate to an equation, we now summarize the information from the problem in the following table. Fill in the rates given and the variables used to represent the times, and use the formula $d = r \cdot t$ to fill in the distance column.

	Rate ·	Time =	Distance
Level	18	x	$18x$
Hilly	10	x	$10x$

We'll draw a diagram to help visualize distance.

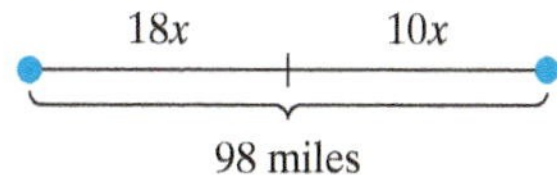

Since the entire trip covered 98 miles, we have that

In words: total distance = level distance + hilly distance

Translate: $98 = 18x + 10x$

(Continued on next page)

Practice 3

Sat Tranh took a short hike with his friends up Mt. Wachusett. They hiked uphill at a steady pace of 1.5 miles per hour, and downhill at a rate of 4 miles per hour. If the time to climb the mountain took an hour more than the time to hike down, how long did the entire hike take?

Answer

3. 2.2 hours

3. SOLVE.

$$98 = 28x \quad \text{Add like terms.}$$

$$\frac{98}{28} = \frac{28x}{28} \quad \text{Divide both sides by 28.}$$

$$3.5 = x$$

4. INTERPRET the results.

Check: Recall that x represents the time spent on the level portion of the trip and also the time spent on the hilly portion. If Marie rode for 3.5 hours at 18 mph, her distance was $18(3.5) = 63$ miles. If Marie rode for 3.5 hours at 10 mph, her distance was $10(3.5) = 35$ miles. The total distance was 63 miles + 35 miles = 98 miles, the required distance.

State: The time of the entire trip was then 3.5 hours + 3.5 hours or 7 hours.

Work Practice 3

Practice 4

The Kansas City Southern Railway has a station in Mexico City, Mexico. Suppose two trains leave Mexico City at the same time. One travels east and the other west at a speed that is 10 mph slower. In 1.5 hours, the trains are 171 miles apart. Find the speed of each train.

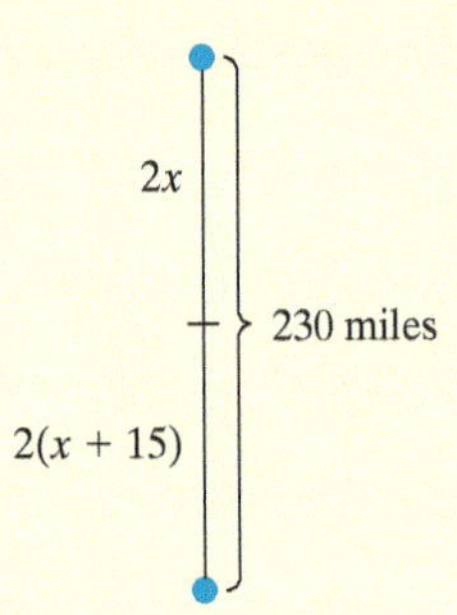

Answer

4. eastbound: 62 mph; westbound: 52 mph

Example 4 Finding Train Speeds

The Kansas City Southern Railway operates in 10 states and Mexico. Suppose two trains leave Neosho, Missouri, at the same time. One travels north and the other travels south at a speed that is 15 miles per hour faster. In 2 hours, the trains are 230 miles apart. Find the speed of each train.

Solution:

1. UNDERSTAND the problem. Read and reread the problem. Guess a solution and check. Let's let

x = speed of train traveling north

Because the train traveling south is 15 mph faster, we have

$x + 15$ = speed of train traveling south

2. TRANSLATE. Just as for Example 3, let's summarize our information in a table. Use the formula $d = r \cdot t$ to fill in the distance column.

	r	$\cdot\ t$	$= d$
North Train	x	2	$2x$
South Train	$x + 15$	2	$2(x + 15)$

Since the total distance between the trains is 230 miles, we have

In words:	north train distance	+	south train distance	=	total distance
	↓		↓		↓
Translate:	$2x$	+	$2(x + 15)$	=	230

3. SOLVE.

$$2x + 2x + 30 = 230 \quad \text{Use the distributive property.}$$

$$4x + 30 = 230 \quad \text{Combine like terms.}$$

$$4x = 200 \quad \text{Subtract 30 from both sides.}$$

$$\frac{4x}{4} = \frac{200}{4} \quad \text{Divide both sides by 4.}$$

$$x = 50 \quad \text{Simplify.}$$

4. INTERPRET the results.

Check: Recall that x is the speed of the train traveling north, or 50 mph. In 2 hours, this train travels a distance of $2(50) = 100$ miles. The speed of the train traveling south is $x + 15$ or $50 + 15 = 65$ mph. In 2 hours, this train travels $2(65) = 130$ miles. The total distance between the trains is 100 miles $+ 130$ miles $= 230$ miles, the required distance.

State: The northbound train's speed is 50 mph and the southbound train's speed is 65 mph.

Work Practice 4

B.2 Exercise Set MyMathLab®

Objective 1 *Solve. For each exercise, a table is given for you to complete and use to write an equation that models the situation. See Example 1.*

1. How much pure acid should be mixed with 2 gallons of a 40% acid solution in order to get a 70% acid solution?

	Number of Gallons ·	Acid Strength =	Amount of Acid
Pure Acid		100%	
40% Acid Solution			
70% Acid Solution Needed			

2. How many cubic centimeters (cc) of a 25% antibiotic solution should be added to 10 cubic centimeters of a 60% antibiotic solution in order to get a 30% antibiotic solution?

	Number of Cubic cm ·	Antibiotic Strength =	Amount of Antibiotic
25% Antibiotic Solution			
60% Antibiotic Solution			
30% Antibiotic Solution Needed			

3. Community Coffee Company wants to create a new flavor of Cajun coffee. How many pounds of coffee worth \$7 per pound should be added to 14 pounds of coffee worth \$4 per pound to get a mixture worth \$5 per pound?

	Number of Pounds ·	Cost per Pound =	Value
\$7 per lb Coffee			
\$4 per lb Coffee			
\$5 per lb Coffee Wanted			

4. Planter's Peanut Company wants to mix 20 pounds of peanuts worth \$3 a pound with cashews worth \$5 a pound in order to try out a mix worth \$3.50 a pound. How many pounds of cashews should be added to the peanuts?

	Number of Pounds ·	Cost per Pound =	Value
\$3 per lb Peanuts			
\$5 per lb Cashews			
\$3.50 per lb Mixture Wanted			

Complete the table. The first and sixth rows have been completed for you. See Example 2.

		Number of Coins or Bills	Value of Coins or Bills (in dollars)
	pennies	x	$0.01x$
5.	dimes	y	
6.	quarters	z	
7.	nickels	$(x + 7)$	
8.	half-dollars	$(20 - z)$	
	\$5 bills	$9x$	$5(9x)$
9.	\$20 bills	$4y$	
10.	\$100 bills	$97z$	
11.	\$50 bills	$(35 - x)$	
12.	\$10 bills	$(15 - y)$	

Solve. See Examples 1 and 2.

13. Part of the proceeds from a garage sale was \$280 worth of \$5 and \$10 bills. If there were 20 more \$5 bills than \$10 bills, find the number of each denomination.

	Number of Bills	Value of Bills
\$5 bills		
\$10 bills		
Total		

14. A bank teller is counting \$20 and \$50 bills. If there are six times as many \$20 bills as \$50 bills and the total amount of money is \$3910, find the number of each denomination.

	Number of Bills	Value of Bills
\$20 bills		
\$50 bills		
Total		

15. How much of an alloy that is 20% copper should be mixed with 200 ounces of an alloy that is 50% copper in order to get an alloy that is 30% copper?

16. How much water should be added to 30 gallons of a solution that is 70% antifreeze in order to get a mixture that is 60% antifreeze?

17. A new self-tanning lotion for everyday use is to be sold. First, a speculative lotion mixture is made by mixing 800 ounces of everyday moisturizing lotion worth \$0.30 per ounce with self-tanning lotion worth \$3 per ounce. If the speculative lotion is to cost \$1.20 per ounce, how many ounces of the self-tanning lotion should be in the mixture?

18. The owner of a local chocolate shop wants to develop a new trail mix. How many pounds of chocolate-covered peanuts worth \$5 a pound should be mixed with 10 pounds of granola bites worth \$2 a pound to get a mixture worth \$3 a pound?

Objective 2 *Solve. See Example 3.*

19. Two cars leave Richmond, Virginia, at the same time after visiting the nearby Richmond International Speedway. The cars travel in opposite directions, one traveling north at 56 mph and one traveling south at 47 mph. When will the two cars be 206 miles apart?

20. Two cars leave Las Vegas, Nevada, at the same time after visiting the Las Vegas Motor Speedway. The cars travel in opposite directions, one traveling northeast at 65 mph and one traveling southwest at 41 mph. When will the two cars be 530 miles apart?

Solve. See Example 4.

21. Suppose two trains leave Corpus Christi, Texas, at the same time, traveling in opposite directions. One train travels 10 mph faster than the other. In 2.5 hours, the trains are 205 miles apart. Find the speed of each train.

22. Suppose two trains leave Cleveland, Ohio, at the same time, traveling in opposite directions. One train travels 8 mph faster than the other. In 1.5 hours, the trains are 162 miles apart. Find the speed of each train.

Solve. See Examples 3 and 4.

23. A truck and a van leave the same location at the same time and travel in opposite directions. The truck's speed is 52 mph and the van's speed is 63 mph. When will the truck and the van be 460 miles apart?

24. Two cars leave the same location at the same time and travel in opposite directions. One car's speed is 65 mph and the other car's speed is 45 mph. When will the two cars be 330 miles apart?

25. Two cars leave Pecos, Texas, at the same time and both travel east on Interstate 20. The first car's speed is 70 mph and the second car's speed is 58 mph. When will the cars be 30 miles apart?

26. Two cars leave Savannah, Georgia, at the same time and both travel north on Interstate 95. The first car's speed is 40 mph and the second car's speed is 50 mph. When will the cars be 20 miles apart?

27. A jet plane traveling at 500 mph overtakes a propeller plane traveling at 200 mph that had a 2-hour head start. How far from the starting point are the planes?

28. How long will it take a bus traveling at 60 miles per hour to overtake a car traveling at 40 mph if the car had a 1.5-hour head start?

29. A bus traveled on a level road for 3 hours at an average speed 20 miles per hour faster than it traveled on a winding road. The time spent on the winding road was 4 hours. Find the average speed on the level road if the entire trip was 305 miles.

30. The Jones family drove to Disneyland at 50 miles per hour and returned on the same route at 40 mph. Find the distance to Disneyland if the total driving time was 7.2 hours.

31. Alan and Dave Schaferkötter leave from the same point bicycling in opposite directions, Alan bicycling at 5.5 miles per hour and Dave at 6.5 mph. How long will they be able to talk on their long-range walkie-talkies if the walkie-talkies have a 35-mile range?

32. Kathleen and Cade Williams leave simultaneously from the same point hiking in opposite directions, Kathleen walking at 4 miles per hour and Cade at 5 mph. How long can they talk on their walkie-talkies if the walkie-talkies have a 20-mile radius?

33. Two hikers are 11 miles apart and walking toward each other. They meet in 2 hours. Find the rate of each hiker if one hiker walks 1.1 mph faster than the other.

34. Nedra and Latonya Dominguez are 12 miles apart hiking toward each other. How long will it take them to meet if Nedra walks at 3 mph and Latonya walks 1 mph faster?

35. Mark Martin can row upstream at 5 mph and downstream at 11 mph. If Mark starts rowing upstream until he gets tired and then rows downstream to his starting point, how far did Mark row if the entire trip took 4 hours?

36. On a 255-mile trip, Gary Alessandrini traveled at an average speed of 70 mph, got a speeding ticket, and then traveled at 60 mph for the remainder of the trip. If the entire trip took 4.5 hours and the speeding ticket stop took 30 minutes, how long did Gary speed before getting stopped?

37. A stack of \$20, \$50, and \$100 bills was retrieved as part of an FBI investigation. There were 46 more \$50 bills than \$100 bills. Also, the number of \$20 bills was 7 times the number of \$100 bills. If the total value of the money was \$9550, find the number of each type of bill.

38. A man places his pocket change in a jar every day. The jar is full and his children have counted the change. The total value is \$44.86. Let x represent the number of quarters, and use the information below to find the number of each type of coin.

    There are:
    136 more dimes than quarters
    8 times as many nickels as quarters
    32 more than 16 times as many pennies as quarters

B.3 Graphing 

Objective 1 Plotting Ordered Pairs

Graphs are widely used today in newspapers, magazines, and all forms of newsletters. A few examples of graphs are shown here.

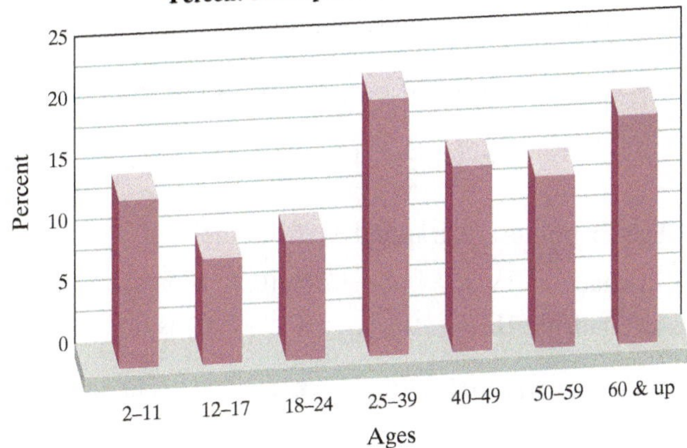

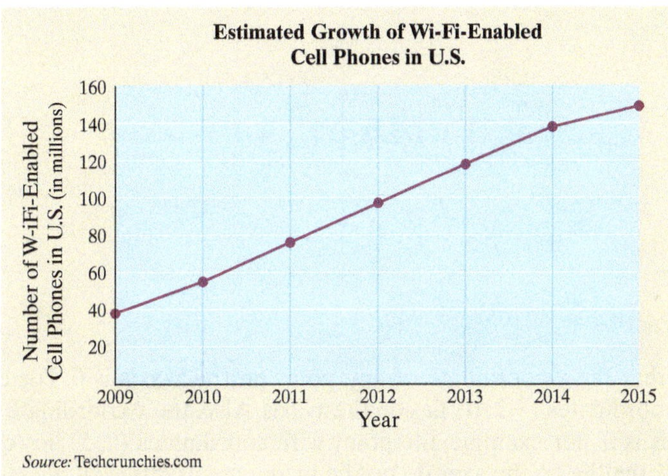

To review how to read these graphs, we review their origin—the rectangular coordinate system. One way to locate points on a plane is by using a **rectangular coordinate system,** which is also called a **Cartesian coordinate system** after its inventor, René Descartes (1596–1650). The next diagram to the left shows the rectangular coordinate system. For further review of this system, see Section 10.1.

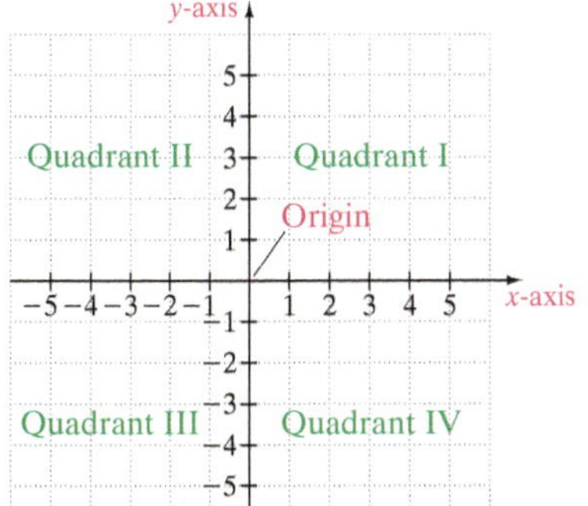

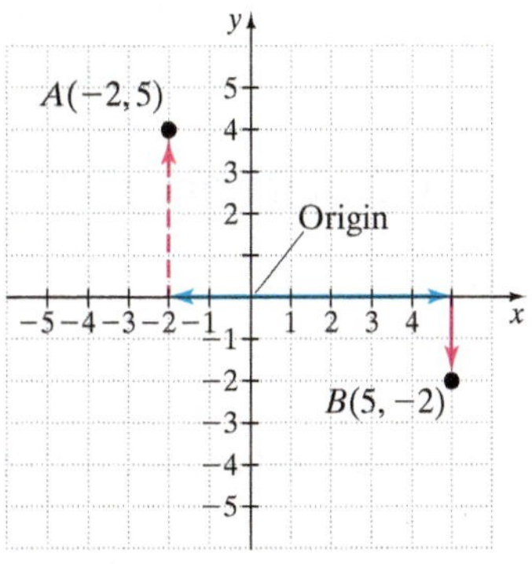

Recall that the location of point A in the figure above is described as 2 units to the left of the origin along the x-axis and 5 units upward parallel to the y-axis. Thus, we identify point A with the ordered pair $(-2, 5)$. Notice that the order of these numbers

Objectives

1 Plot Ordered Pairs. 

2 Graph Linear Equations.

is *critical.* The x-value -2 is called the ***x*-coordinate** and is associated with the x-axis. The y-value 5 is called the ***y*-coordinate** and is associated with the y-axis. Compare the location of point A with the location of point B, which corresponds to the ordered pair $(5, -2)$.

Keep in mind that **each ordered pair corresponds to exactly one point in the real plane and that each point in the plane corresponds to exactly one ordered pair.** Thus, we may refer to the ordered pair (x, y) as the point (x, y).

Practice 1

Plot each ordered pair on a Cartesian coordinate system and name the quadrant or axis in which the point is located.

a. $(3, -4)$
b. $(0, -2)$
c. $(-2, 4)$
d. $(4, 0)$
e. $\left(-1\frac{1}{2}, -2\right)$
f. $(2.5, 3.5)$

Example 1 Plot each ordered pair on a Cartesian coordinate system and name the quadrant or axis in which the point is located.

a. $(2, -1)$ **b.** $(0, 5)$ **c.** $(-3, 5)$

d. $(-2, 0)$ **e.** $\left(-\frac{1}{2}, -4\right)$ **f.** $(1.5, 1.5)$

Solution: The six points are graphed as shown below.

a. $(2, -1)$ is in quadrant IV. **b.** $(0, 5)$ is on the y-axis.

c. $(-3, 5)$ is in quadrant II. **d.** $(-2, 0)$ is on the x-axis.

e. $\left(-\frac{1}{2}, -4\right)$ is in quadrant III. **f.** $(1.5, 1.5)$ is in quadrant I.

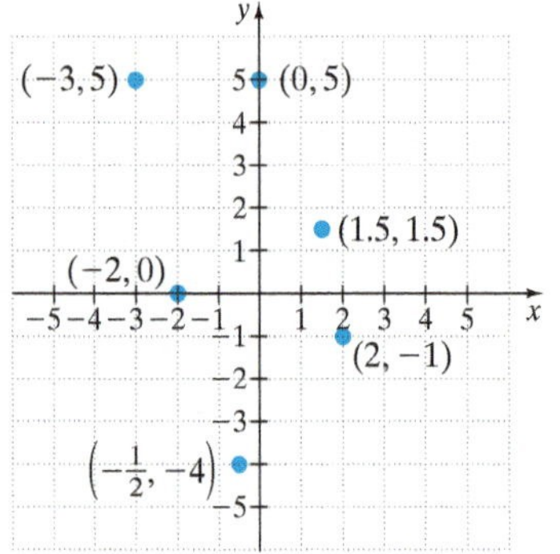

Work Practice 1

Notice that the y-coordinate of any point on the x-axis is 0. For example, the point with coordinates $(-2, 0)$ lies on the x-axis. Also, the x-coordinate of any point on the y-axis is 0. For example, the point with coordinates $(0, 5)$ lies on the y-axis. These points that lie on the axes do not lie in any quadrants.

✓**Concept Check** Which of the following correctly describes the location of the point $(3, -6)$ in a rectangular coordinate system?

a. 3 units to the left of the y-axis and 6 units above the x-axis
b. 3 units above the x-axis and 6 units to the left of the y-axis
c. 3 units to the right of the y-axis and 6 units below the x-axis
d. 3 units below the x-axis and 6 units to the right of the y-axis

Objective 2 Graphing Linear Equations 

Recall that an equation such as $3x - y = 12$ is called a linear equation in two variables, and **the graph of every linear equation in two variables is a line.**

Linear Equation in Two Variables

A linear equation in two variables is an equation that can be written in the form

$$Ax + By = C$$

where A and B are not both 0. This form is called **standard form.**

Answers

1. a. Quadrant IV **b.** y-axis
c. Quadrant II **d.** x-axis
e. Quadrant III **f.** Quadrant I

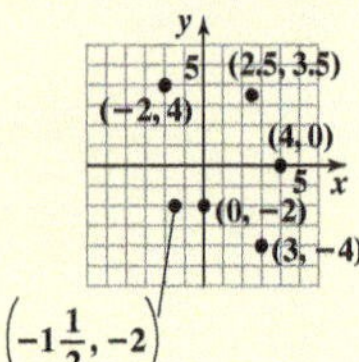

✓**Concept Check Answer**

c

Some examples of equations in standard form:

$$3x - y = 12$$
$$-2.1x + 5.6y = 0$$

Helpful Hint

Remember: A linear equation is written in standard form when all of the variable terms are on one side of the equation and the constant is on the other side.

Example 2 Graph the equation $y = -2x + 3$.

Solution: This is a linear equation. (In standard form it is $2x + y = 3$.) Find three ordered pair solutions, and plot the ordered pairs. The line through the plotted points is the graph. Since the equation is solved for y, let's choose three x-values. We'll choose 0, 2, and then -1 for x to find our three ordered pair solutions.

Let $x = 0$	Let $x = 2$	Let $x = -1$
$y = -2x + 3$	$y = -2x + 3$	$y = -2x + 3$
$y = -2 \cdot 0 + 3$	$y = -2 \cdot 2 + 3$	$y = -2(-1) + 3$
$y = 3$ Simplify.	$y = -1$ Simplify.	$y = 5$ Simplify.

The three ordered pairs $(0, 3)$, $(2, -1)$, and $(-1, 5)$ are listed in the table and the graph is shown.

x	y
0	3
2	-1
-1	5

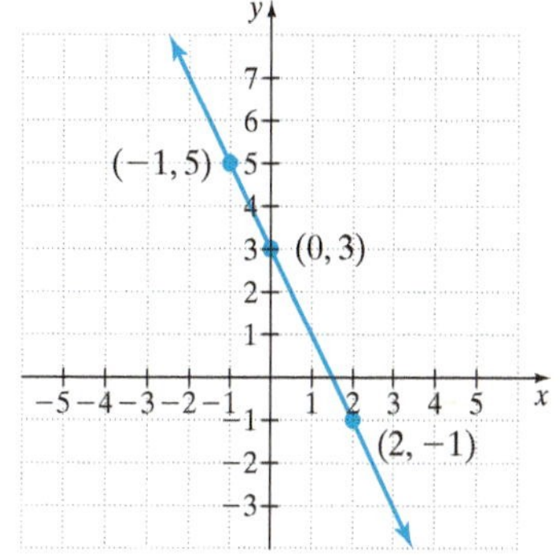

Work Practice 2

Notice that the graph crosses the y-axis at the point $(0, 3)$. This point is called the **y-intercept.** (You may sometimes see just the number 3 called the y-intercept.) This graph also crosses the x-axis at the point $\left(\frac{3}{2}, 0\right)$. This point is called the **x-intercept.** (You may also see just the number $\frac{3}{2}$ called the x-intercept.)

Since every point on the y-axis has an x-value of 0, we can find the y-intercept of a graph by letting $x = 0$ and solving for y. Also, every point on the x-axis has a y-value of 0. To find the x-intercept, we let $y = 0$ and solve for x.

Finding x- and y-Intercepts

To find an x-intercept, let $y = 0$ and solve for x.
To find a y-intercept, let $x = 0$ and solve for y.

Example 3 Graph the linear equation $y = \frac{1}{3}x$.

Solution: To graph, we find ordered pair solutions, plot the ordered pairs, and draw a line through the plotted points. We will choose x-values and substitute in the equation. To avoid fractions, we choose x-values that are multiples of 3. To find the y-intercept, we let $x = 0$.

(*Continued on next page*)

Practice 2

Graph the equation $y = -3x - 2$.

Practice 3

Graph the linear equation $y = -\frac{1}{2}x$.

Answers

2. 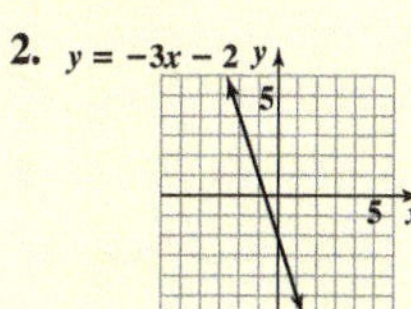

3. 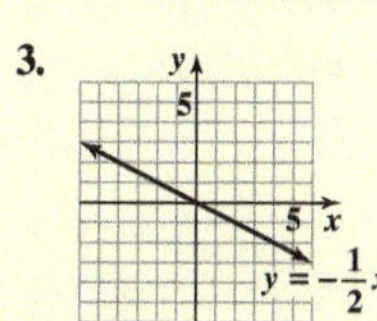

Helpful Hint Notice that by using multiples of 3 for x, we avoid fractions.

Helpful Hint Since the equation $y = \frac{1}{3}x$ is solved for y, we choose x-values for finding points. This way, we simply need to evaluate an expression to find the y-value, as shown.

$$y = \frac{1}{3}x$$

If $x = 0$, then $y = \frac{1}{3}(0)$, or 0.

If $x = 6$, then $y = \frac{1}{3}(6)$, or 2.

If $x = -3$, then $y = \frac{1}{3}(-3)$, or -1.

x	y
0	0
6	2
−3	−1

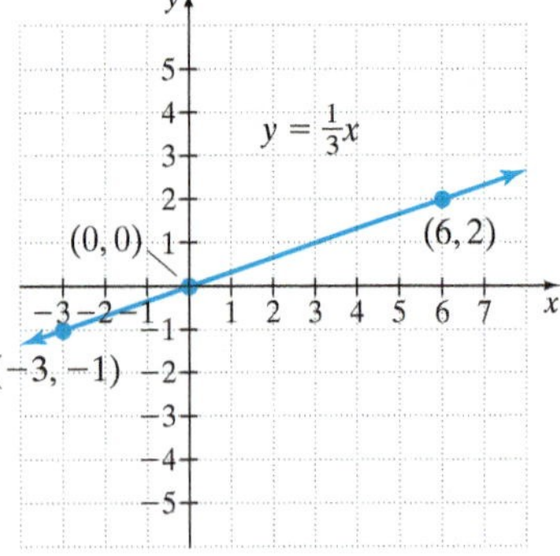

This graph crosses the x-axis at $(0, 0)$ and the y-axis at $(0, 0)$. This means that the x-intercept is $(0, 0)$ and that the y-intercept is $(0, 0)$.

Work Practice 3

B.3 Exercise Set MyMathLab®

Determine the coordinates of each point on the graph.

1. Point A
2. Point B
3. Point C
4. Point D
5. Point E
6. Point F
7. Point G
8. Point H

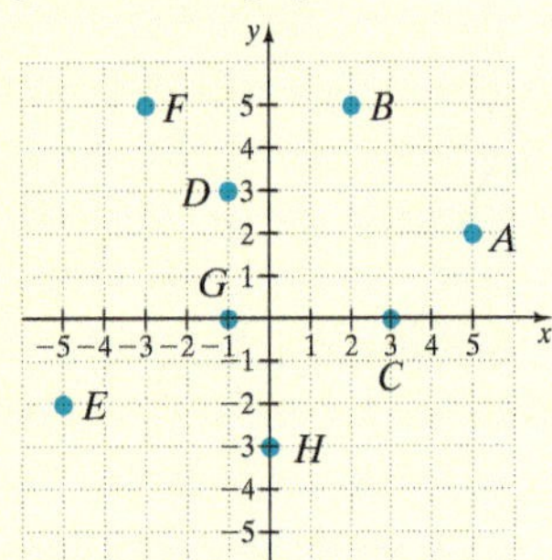

Without graphing, visualize the location of each point. Then give its location by quadrant or x- or y-axis. See Example 1.

9. $(2, 3)$
10. $(0, 5)$
11. $(-2, 7)$
12. $(-3, 0)$
13. $(-1, -4)$
14. $(4, -2)$
15. $(0, -100)$
16. $(10, 30)$
17. $(-10, -30)$
18. $(0, 0)$
19. $(-87, 0)$
20. $(-42, 17)$

Given that x is a positive number and that y is a positive number, determine the quadrant or axis in which each point lies.

21. $(x, -y)$
22. $(-x, y)$
23. $(x, 0)$
24. $(0, -y)$
25. $(-x, -y)$
26. $(0, 0)$

Graph each linear equation. See Examples 2 and 3.

27. $y = -x - 2$
28. $y = -2x + 1$
29. $3x - 4y = 8$
30. $x - 9y = 3$
31. $y = \frac{1}{3}x$
32. $y = \frac{3}{2}x$
33. $y + 4 = 0$
34. $x = -1.5$

Recall that if $f(2) = 7$, for example, this corresponds to the ordered pair $(2, 7)$ on the graph of f. Use this information and the graphs of f and g below to answer Exercises 35 through 42.

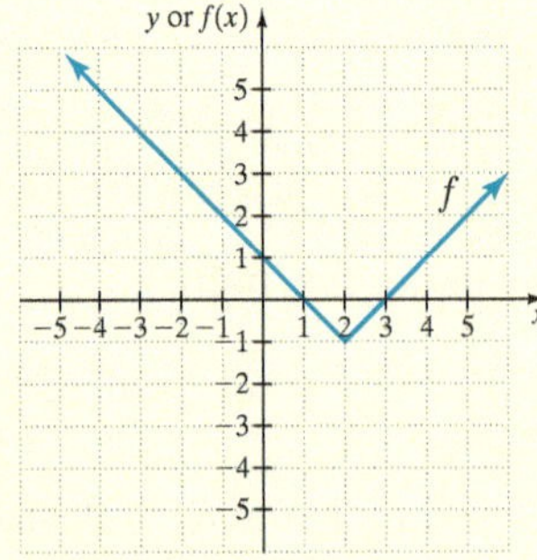

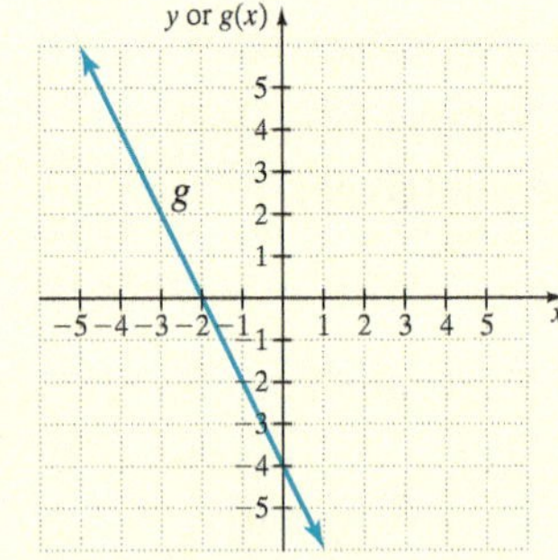

35. $f(4) =$
36. $f(0) =$
37. $g(0) =$
38. $g(-1) =$
39. Find all values for x such that $f(x) = 0$.
40. Find all values for x such that $g(x) = 0$.
41. If $(-1, -2)$ is a point on the graph of g, write this using function notation.
42. If $(-1, 2)$ is a point on the graph of f, write this using function notation.

B.4 Polynomials and Factoring

Objective 1 Operations on Polynomials

Objectives

1 Review Operations on Polynomials.

2 Review Factoring Polynomials.

B.4 Exercise Set, Part 1 MyMathLab®

Perform each indicated operation.

1. $(-y^2 + 6y - 1) + (3y^2 - 4y - 10)$
2. $(5z^4 - 6z^2 + z + 1) - (7z^4 - 2z + 1)$
3. Subtract $(x - 5)$ from $(x^2 - 6x + 2)$.
4. $(2x^2 + 6x - 5) + (5x^2 - 10x)$
5. $(5x - 3)^2$
6. $(5x^2 - 14x - 3) \div (5x + 1)$
7. $(2x^4 - 3x^2 + 5x - 2) \div (x + 2)$
8. $(4x - 1)(x^2 - 3x - 2)$

Objective 2 Factoring Strategies

The key to proficiency in factoring polynomials is to practice until you are comfortable with each technique. A strategy for factoring polynomials completely is given next.

Factoring a Polynomial

Step 1: Are there any common factors? If so, factor out the greatest common factor.

Step 2: How many terms are in the polynomial?

a. If there are *two* terms, decide if one of the following formulas may be applied:

i. Difference of two squares: $a^2 - b^2 = (a - b)(a + b)$

ii. Difference of two cubes: $a^3 - b^3 = (a - b)(a^2 + ab + b^2)$

iii. Sum of two cubes: $a^3 + b^3 = (a + b)(a^2 - ab + b^2)$

b. If there are *three* terms, try one of the following:

i. Perfect square trinomial: $a^2 + 2ab + b^2 = (a + b)^2$
$a^2 - 2ab + b^2 = (a - b)^2$

ii. If not a perfect square trinomial, factor by using the methods presented in Sections 13.2 through 13.4.

c. If there are *four* or more terms, try factoring by grouping.

Step 3: See whether any factors in the factored polynomial can be factored further.

A few examples are worked for you below.

Example 1 Factor each polynomial completely.

a. $8a^2b - 4ab$ **b.** $36x^2 - 9$ **c.** $2x^2 - 5x - 7$

d. $5p^2 + 5 + qp^2 + q$ **e.** $9x^2 + 24x + 16$ **f.** $y^2 + 25$

(*Continued on next page*)

Practice 1

Factor each polynomial completely.

a. $12x^2y - 3xy$

b. $49x^2 - 4$

c. $5x^2 + 2x - 3$

d. $3x^2 + 6 + x^3 + 2x$

e. $4x^2 + 20x + 25$

f. $b^2 + 100$

Answers

1. a. $3xy(4x - 1)$ **b.** $(7x + 2)(7x - 2)$ **c.** $(5x - 3)(x + 1)$ **d.** $(x^2 + 2)(3 + x)$ **e.** $(2x + 5)^2$ **f.** prime

Solution:

a. Step 1: The terms have a common factor of $4ab$, which we factor out.

$$8a^2b - 4ab = 4ab(2a - 1)$$

Step 2: There are two terms, but the binomial $2a - 1$ is not the difference of two squares or the sum or difference of two cubes.

Step 3: The factor $2a - 1$ cannot be factored further.

b. Step 1: Factor out a common factor of 9.

$$36x^2 - 9 = 9(4x^2 - 1)$$

Step 2: The factor $4x^2 - 1$ has two terms, and it is the difference of two squares.

$$9(4x^2 - 1) = 9(2x + 1)(2x - 1)$$

Step 3: No factor with more than one term can be factored further.

c. Step 1: The terms of $2x^2 - 5x - 7$ contain no common factor other than 1 or -1.

Step 2: There are three terms. The trinomial is not a perfect square, so we factor by methods from Section 13.3 or 13.4.

$$2x^2 - 5x - 7 = (2x - 7)(x + 1)$$

Step 3: No factor with more than one term can be factored further.

d. Step 1: There is no common factor of all terms of $5p^2 + 5 + qp^2 + q$.

Step 2: The polynomial has four terms, so try factoring by grouping.

$$\begin{aligned} 5p^2 + 5 + qp^2 + q &= (5p^2 + 5) + (qp^2 + q) \quad \text{Group the terms.} \\ &= 5(p^2 + 1) + q(p^2 + 1) \\ &= (p^2 + 1)(5 + q) \end{aligned}$$

Step 3: No factor can be factored further.

e. Step 1: The terms of $9x^2 + 24x + 16$ contain no common factor other than 1 or -1.

Step 2: The trinomial $9x^2 + 24x + 16$ is a perfect square trinomial, and $9x^2 + 24x + 16 = (3x + 4)^2$.

Step 3: No factor can be factored further.

f. Step 1: There is no common factor of $y^2 + 25$ other than 1.

Step 2: This binomial is the sum of two squares and is prime.

Step 3: The binomial $y^2 + 25$ cannot be factored further.

Work Practice 1

Example 2 Factor each polynomial completely.

a. $27a^3 - b^3$

b. $3n^2m^4 - 48m^6$

c. $2x^2 - 12x + 18 - 2z^2$

d. $8x^4y^2 + 125xy^2$

e. $(x - 5)^2 - 49y^2$

Solution:

a. This binomial is the difference of two cubes.

$$\begin{aligned} 27a^3 - b^3 &= (3a)^3 - b^3 \\ &= (3a - b)[(3a)^2 + (3a)(b) + b^2] \\ &= (3a - b)(9a^2 + 3ab + b^2) \end{aligned}$$

Practice 2

Factor each polynomial completely.

a. $64x^3 + y^3$

b. $7x^2y^2 - 63y^4$

c. $3x^2 + 12x + 12 - 3b^2$

d. $x^5y^4 + 27x^2y$

e. $(x + 7)^2 - 81y^2$

Answers

2. a. $(4x + y)(16x^2 - 4xy + y^2)$
b. $7y^2(x + 3y)(x - 3y)$
c. $3(x + 2 + b)(x + 2 - b)$
d. $x^2y(xy + 3)(x^2y^2 - 3xy + 9)$
e. $(x + 7 + 9y)(x + 7 - 9y)$

b. $3n^2m^4 - 48m^6 = 3m^4(n^2 - 16m^2)$ Factor out the GCF, $3m^4$.

$= 3m^4(n + 4m)(n - 4m)$ Factor the difference of squares.

c. $2x^2 - 12x + 18 - 2z^2 = 2(x^2 - 6x + 9 - z^2)$ The GCF is 2.

$= 2[(x^2 - 6x + 9) - z^2]$ Group the first three terms together.

$= 2[(x - 3)^2 - z^2]$ Factor the perfect square trinomial.

$= 2[(x - 3) + z][(x - 3) - z]$ Factor the difference of squares.

$= 2(x - 3 + z)(x - 3 - z)$

d. $8x^4y^2 + 125xy^2 = xy^2(8x^3 + 125)$ The GCF is xy^2.

$= xy^2[(2x)^3 + 5^3]$

$= xy^2(2x + 5)[(2x)^2 - (2x)(5) + 5^2]$ Factor the sum of cubes.

$= xy^2(2x + 5)(4x^2 - 10x + 25)$

e. This binomial is the difference of squares.

$$(x - 5)^2 - 49y^2 = (x - 5)^2 - (7y)^2$$
$$= [(x - 5) + 7y][(x - 5) - 7y]$$
$$= (x - 5 + 7y)(x - 5 - 7y)$$

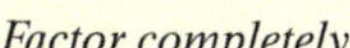
Work Practice 2

B.4 Exercise Set, Part 2 MyMathLab®

Factor completely.

9. $x^2 - 8x + 16 - y^2$
10. $12x^2 - 22x - 20$
11. $x^4 - x$
12. $(2x + 1)^2 - 3(2x + 1) + 2$
13. $14x^2y - 2xy$
14. $24ab^2 - 6ab$
15. $4x^2 - 16$
16. $9x^2 - 81$
17. $3x^2 - 8x - 11$
18. $5x^2 - 2x - 3$
19. $4x^2 + 8x - 12$
20. $6x^2 - 6x - 12$
21. $4x^2 + 36x + 81$
22. $25x^2 + 40x + 16$
23. $8x^3 + 125y^3$
24. $27x^3 - 64y^3$
25. $64x^2y^3 - 8x^2$
26. $27x^5y^4 - 216x^2y$
27. $(x + 5)^3 + y^3$
28. $(y - 1)^3 + 27x^3$
29. $(5a - 3)^2 - 6(5a - 3) + 9$
30. $(4r + 1)^2 + 8(4r + 1) + 16$
31. $7x^2 - 63x$
32. $20x^2 + 23x + 6$
33. $ab - 6a + 7b - 42$
34. $20x^2 - 220x + 600$
35. $x^4 - 1$
36. $15x^2 - 20x$
37. $10x^2 - 7x - 33$
38. $45m^3n^3 - 27m^2n^2$
39. $5a^3b^3 - 50a^3b$
40. $x^4 + x$
41. $16x^2 + 25$
42. $20x^3 + 20y^3$
43. $10x^3 - 210x^2 + 1100x$
44. $9y^2 - 42y + 49$
45. $64a^3b^4 - 27a^3b$
46. $y^4 - 16$
47. $2x^3 - 54$
48. $2sr + 10s - r - 5$
49. $3y^5 - 5y^4 + 6y - 10$

50. $64a^2 + b^2$

51. $100z^3 + 100$

52. $250x^4 - 16x$

53. $4b^2 - 36b + 81$

54. $2a^5 - a^4 + 6a - 3$

55. $(y - 6)^2 + 3(y - 6) + 2$

56. $(c + 2)^2 - 6(c + 2) + 5$

△ **57.** Express the area of the shaded region as a polynomial. Factor the polynomial completely.

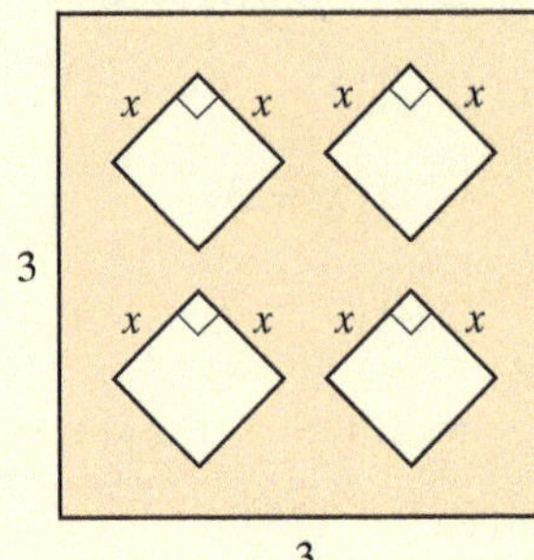

B.5 Rational Expressions

Objective

1 Perform Operations on Rational Expressions and Solve Equations Containing Rational Expressions.

Objective 1 Performing Operations on Rational Expressions and Solving Equations Containing Rational Expressions

It is very important that you understand the difference between an expression and an equation containing rational expressions. An equation contains an equal sign; an expression does not.

Expression to Be Simplified

$$\frac{x}{2} + \frac{x}{6}$$

Write both rational expressions with the LCD, 6, as the denominator.

$$\frac{x}{2} + \frac{x}{6} = \frac{x \cdot 3}{2 \cdot 3} + \frac{x}{6}$$
$$= \frac{3x}{6} + \frac{x}{6}$$
$$= \frac{4x}{6} = \frac{2x}{3}$$

Equation to Be Solved

$$\frac{x}{2} + \frac{x}{6} = \frac{2}{3}$$

Multiply both sides by the LCD, 6.

$$6\left(\frac{x}{2} + \frac{x}{6}\right) = 6\left(\frac{2}{3}\right)$$
$$3x + x = 4$$
$$4x = 4$$
$$x = 1$$

Check to see that the solution is 1.

Helpful Hint

Remember: Equations can be cleared of fractions; expressions cannot.

Example 1 Multiply. $\frac{x^3 - 1}{-3x + 3} \cdot \frac{15x^2}{x^2 + x + 1}$

Solution:

$$\frac{x^3 - 1}{-3x + 3} \cdot \frac{15x^2}{x^2 + x + 1} = \frac{(x - 1)(x^2 + x + 1)}{-3(x - 1)} \cdot \frac{15x^2}{x^2 + x + 1} \quad \text{Factor.}$$
$$= \frac{(x - 1)(x^2 + x + 1) \cdot 3 \cdot 5x^2}{-1 \cdot 3(x - 1)(x^2 + x + 1)} \quad \text{Factor.}$$
$$= \frac{5x^2}{-1} = -5x^2 \quad \text{Simplest form}$$

Work Practice 1

Practice 1

Multiply.

a. $\frac{2 + 5n}{3n} \cdot \frac{6n + 3}{5n^2 - 3n - 2}$

b. $\frac{x^3 - 8}{-6x + 12} \cdot \frac{6x^2}{x^2 + 2x + 4}$

Answers

1. a. $\frac{2n + 1}{n(n - 1)}$ **b.** $-x^2$

Example 2 Divide. $\frac{8m^2}{3m^2 - 12} \div \frac{40}{2 - m}$

Solution:

$$\frac{8m^2}{3m^2 - 12} \div \frac{40}{2 - m} = \frac{8m^2}{3m^2 - 12} \cdot \frac{2 - m}{40}$$ Multiply by the reciprocal of the divisor.

$$= \frac{8m^2(2 - m)}{3(m + 2)(m - 2) \cdot 40}$$ Factor and multiply.

$$= \frac{8\,m^2 \cdot -1(m - 2)}{3(m + 2)\,(m - 2)\,\cdot 8 \cdot 5}$$ Write $(2 - m)$ as $-1(m - 2)$.

$$= -\frac{m^2}{15(m + 2)}$$ Simplify.

Work Practice 2

Practice 2

Divide.

a. $\frac{6y^3}{3y^2 - 27} \div \frac{42}{3 - y}$

b. $\frac{10x^2 + 23x - 5}{5x^2 - 51x + 10} \div \frac{2x^2 + 9x + 10}{7x^2 - 68x - 20}$

Example 3 Perform the indicated operation.

$$\frac{3}{x + 2} + \frac{2x}{x - 2}$$

Solution: The LCD is the product of the two denominators: $(x + 2)(x - 2)$.

$$\frac{3}{x+2} + \frac{2x}{x-2} = \frac{3 \cdot (x-2)}{(x+2) \cdot (x-2)} + \frac{2x \cdot (x+2)}{(x-2) \cdot (x+2)}$$ Write equivalent rational expressions.

$$= \frac{3x - 6}{(x + 2)(x - 2)} + \frac{2x^2 + 4x}{(x + 2)(x - 2)}$$ Multiply in the numerators.

$$= \frac{3x - 6 + 2x^2 + 4x}{(x + 2)(x - 2)}$$ Add the numerators.

$$= \frac{2x^2 + 7x - 6}{(x + 2)(x - 2)}$$ Simplify the numerator.

Work Practice 3

Practice 3

Perform the indicated operation.

a. $\frac{4}{p^3q} + \frac{3}{5p^4q}$

b. $\frac{4}{y + 3} + \frac{5y}{y - 3}$

c. $\frac{3z - 18}{z - 5} - \frac{3}{5 - z}$

Example 4 Solve: $\frac{2x}{x - 3} + \frac{6 - 2x}{x^2 - 9} = \frac{x}{x + 3}$.

Solution: We factor the second denominator to find that the LCD is $(x + 3)(x - 3)$. We multiply both sides of the equation by $(x + 3)(x - 3)$. By the distributive property, this is the same as multiplying each term by $(x + 3)(x - 3)$.

$$\frac{2x}{x - 3} + \frac{6 - 2x}{x^2 - 9} = \frac{x}{x + 3}$$

$$(x + 3)(x - 3) \cdot \frac{2x}{x - 3} + (x + 3)(x - 3) \cdot \frac{6 - 2x}{(x + 3)(x - 3)} = (x + 3)(x - 3)\left(\frac{x}{x + 3}\right)$$

$$2x(x + 3) + (6 - 2x) = x(x - 3)$$ Simplify.

$$2x^2 + 6x + 6 - 2x = x^2 - 3x$$ Use the distributive property.

(*Continued on next page*)

Practice 4

Solve: $\frac{2}{x - 2} - \frac{5 + 2x}{x^2 - 4} = \frac{x}{x + 2}$.

Answers

2. a. $-\frac{y^3}{21(y + 3)}$ b. $\frac{7x + 2}{x + 2}$

3. a. $\frac{20p + 3}{5p^4q}$ b. $\frac{5y^2 + 19y - 12}{(y + 3)(y - 3)}$ c. 3

4. 1

Next we solve this quadratic equation by the factoring method. To do so, we first write the equation so that one side is 0.

$$x^2 + 7x + 6 = 0$$
$$(x + 6)(x + 1) = 0 \quad \text{Factor.}$$
$$x = -6 \quad \text{or} \quad x = -1 \quad \text{Set each factor equal to 0.}$$

Neither -6 nor -1 makes any denominator 0, so they are both solutions. The solutions are -6 and -1.

Work Practice 4

B.5 Exercise Set MyMathLab®

Perform each indicated operation and simplify, or solve the equation for the variable.

1. $\frac{x}{2} = \frac{1}{8} + \frac{x}{4}$
2. $\frac{x}{4} = \frac{3}{2} + \frac{x}{10}$
3. $\frac{1}{8} + \frac{x}{4}$
4. $\frac{3}{2} + \frac{x}{10}$
5. $\frac{4}{x+2} - \frac{2}{x-1}$
6. $\frac{5}{x-2} - \frac{10}{x+4}$
7. $\frac{4}{x+2} = \frac{2}{x-1}$
8. $\frac{5}{x-2} = \frac{10}{x+4}$
9. $\frac{2}{x^2-4} = \frac{1}{x+2} - \frac{3}{x-2}$
10. $\frac{3}{x^2-25} = \frac{1}{x+5} + \frac{2}{x-5}$
11. $\frac{5}{x^2-3x} + \frac{4}{2x-6}$
12. $\frac{5}{x^2-3x} \div \frac{4}{2x-6}$
13. $\frac{x-1}{x+1} + \frac{x+7}{x-1} = \frac{4}{x^2-1}$
14. $\left(1 - \frac{y}{x}\right) \div \left(1 - \frac{x}{y}\right)$
15. $\frac{a^2-9}{a-6} \cdot \frac{a^2-5a-6}{a^2-a-6}$
16. $\frac{2}{a-6} + \frac{3a}{a^2-5a-6} - \frac{a}{5a+5}$
17. $\frac{2x+3}{3x-2} = \frac{4x+1}{6x+1}$
18. $\frac{5x-3}{2x} = \frac{10x+3}{4x+1}$

Helpful Hint

Remember: Equations can be cleared of fractions; expressions cannot.

19. $\frac{a}{9a^2-1} + \frac{2}{6a-2}$
20. $\frac{3}{4a-8} - \frac{a+2}{a^2-2a}$
21. $-\frac{3}{x^2} - \frac{1}{x} + 2 = 0$
22. $\frac{x}{2x+6} + \frac{5}{x^2-9}$
23. $\frac{x-8}{x^2-x-2} + \frac{2}{x-2}$
24. $\frac{x-8}{x^2-x-2} + \frac{2}{x-2} = \frac{3}{x+1}$
25. $\frac{3}{a} - 5 = \frac{7}{a} - 1$
26. $\frac{7}{3z-9} + \frac{5}{z}$

Use $\frac{x}{5} - \frac{x}{4} = \frac{1}{10}$ *and* $\frac{x}{5} - \frac{x}{4} + \frac{1}{10}$ *for Exercises 27 and 28.*

27. **a.** Which one above is an expression?
 b. Describe the first step to simplify this expression.
 c. Simplify the expression.

28. **a.** Which one above is an equation?
 b. Describe the first step to solve this equation.
 c. Solve the equation.

For each exercise, choose the correct statement. Each figure represents a real number, and no denominators are 0.*

29. **a.** $\frac{\triangle + \square}{\triangle} = \square$ **b.** $\frac{\triangle + \square}{\triangle} = 1 + \frac{\square}{\triangle}$
 c. $\frac{\triangle + \square}{\triangle} = \frac{\square}{\triangle}$ **d.** $\frac{\triangle + \square}{\triangle} = 1 + \square$
 e. $\frac{\triangle + \square}{\triangle - \square} = -1$

*My thanks to Kelly Champagne for permission to use her Exercises for 29 through 33.

30. **a.** $\frac{\triangle}{\square} + \frac{\square}{\triangle} = \frac{\triangle + \square}{\square + \triangle} = 1$

b. $\frac{\triangle}{\square} + \frac{\square}{\triangle} = \frac{\triangle + \square}{\triangle\square}$

c. $\frac{\triangle}{\square} + \frac{\square}{\triangle} = \triangle\triangle + \square\square$

d. $\frac{\triangle}{\square} + \frac{\square}{\triangle} = \frac{\triangle\triangle + \square\square}{\square\triangle}$

e. $\frac{\triangle}{\square} + \frac{\square}{\triangle} = \frac{\triangle\square}{\square\triangle} = 1$

31. **a.** $\frac{\triangle}{\square} \cdot \frac{\bigcirc}{\square} = \frac{\triangle\bigcirc}{\square}$ **b.** $\frac{\triangle}{\square} \cdot \frac{\bigcirc}{\square} = \triangle\bigcirc$

c. $\frac{\triangle}{\square} \cdot \frac{\bigcirc}{\square} = \frac{\triangle + \bigcirc}{\square + \square}$ **d.** $\frac{\triangle}{\square} \cdot \frac{\bigcirc}{\square} = \frac{\triangle\bigcirc}{\square\square}$

32. **a.** $\frac{\triangle}{\square} \div \frac{\bigcirc}{\triangle} = \frac{\triangle\triangle}{\square\bigcirc}$ **b.** $\frac{\triangle}{\square} \div \frac{\bigcirc}{\triangle} = \frac{\bigcirc\square}{\triangle\triangle}$

c. $\frac{\triangle}{\square} \div \frac{\bigcirc}{\triangle} = \frac{\bigcirc}{\square}$ **d.** $\frac{\triangle}{\square} \div \frac{\bigcirc}{\triangle} = \frac{\triangle + \triangle}{\square + \bigcirc}$

33. **a.** $\frac{\frac{\triangle + \square}{\bigcirc}}{\frac{\triangle}{\bigcirc}} = \square$ **b.** $\frac{\frac{\triangle + \square}{\bigcirc}}{\frac{\triangle}{\bigcirc}} = \frac{\triangle\triangle + \triangle\square}{\bigcirc\bigcirc}$

c. $\frac{\frac{\triangle + \square}{\bigcirc}}{\frac{\triangle}{\bigcirc}} = 1 + \square$ **d.** $\frac{\frac{\triangle + \square}{\bigcirc}}{\frac{\triangle}{\bigcirc}} = \frac{\triangle + \square}{\triangle}$

Appendix C Stretching and Compressing Graphs of Absolute Value Functions

In Section 15.3, we learned to shift and reflect graphs of common functions: $f(x) = x$, $f(x) = x^2$, $f(x) = |x|$, and $f(x) = \sqrt{x}$. Since other common functions are studied throughout this text, in this appendix we concentrate on the absolute value function.

Recall that the graph of $h(x) = -|x - 1| + 2$, for example, is the same as the graph of $f(x) = |x|$ reflected about the x-axis, moved 1 unit to the right and 2 units upward. In other words,

$$h(x) = -|x - 1| + 2$$

opens downward (the $-$ sign); (1, 2) location of vertex of V shape (the 1 and the 2)

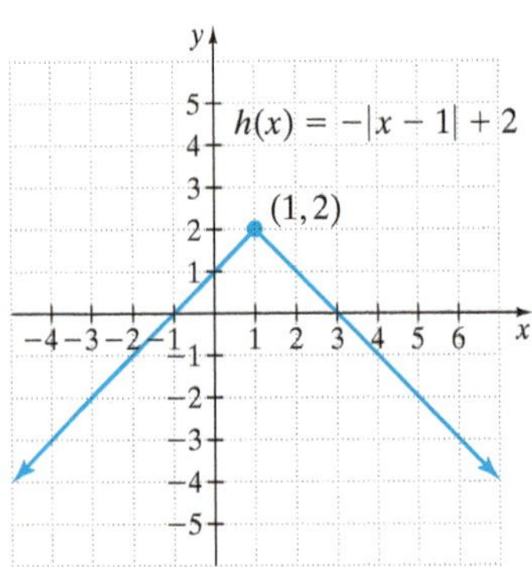

Let's now study the graphs of a few other absolute value functions.

Example 1 Graph $h(x) = 2|x|$ and $g(x) = \frac{1}{2}|x|$.

Solution: Let's find and plot ordered pair solutions for the functions.

x	$h(x)$	$g(x)$
-2	4	1
-1	2	$\frac{1}{2}$
0	0	0
1	2	$\frac{1}{2}$
2	4	2

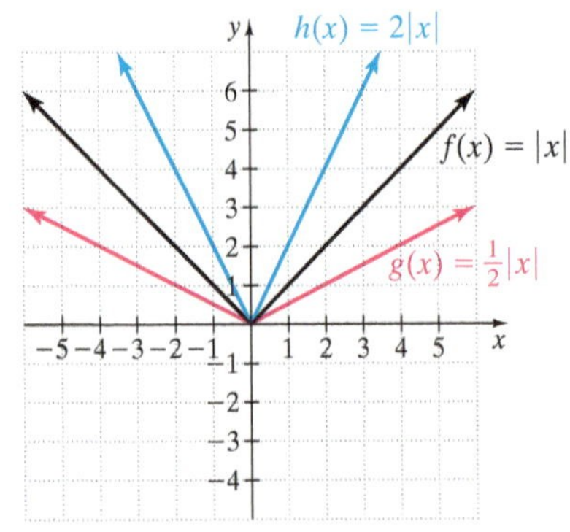

Notice that the graph of $h(x) = 2|x|$ is narrower than the graph of $f(x) = |x|$ and that the graph of $g(x) = \frac{1}{2}|x|$ is wider than the graph of $f(x) = |x|$.

In general, for the absolute value function, we have the following:

The Graph of the Absolute Value Function

The graph of $f(x) = a|x - h| + k$

- Has vertex (h, k) and is V shaped.
- Opens up if $a > 0$ and down if $a < 0$.
- If $|a| < 1$, the graph is wider than the graph of $y = |x|$.
- If $|a| > 1$, the graph is narrower than the graph of $y = |x|$.

Example 2 Graph $f(x) = -\frac{1}{3}|x + 2| + 4$.

Solution: Let's write this function in the form $f(x) = a|x - h| + k$. For our function, we have $f(x) = -\frac{1}{3}|x - (-2)| + 4$. Thus,

- vertex is $(-2, 4)$
- since $a < 0$, V shape opens down
- since $|a| = \left|-\frac{1}{3}\right| = \frac{1}{3} < 1$, the graph is wider than $y = |x|$

We will also find and plot ordered pair solutions.

If $x = -5$, $f(-5) = -\frac{1}{3}|-5 + 2| + 4$, or 3

If $x = 1$, $f(1) = -\frac{1}{3}|1 + 2| + 4$, or 3

If $x = 3$, $f(3) = -\frac{1}{3}|3 + 2| + 4$, or $\frac{7}{3}$, or $2\frac{1}{3}$

x	$f(x)$
-5	3
1	3
3	$2\frac{1}{3}$

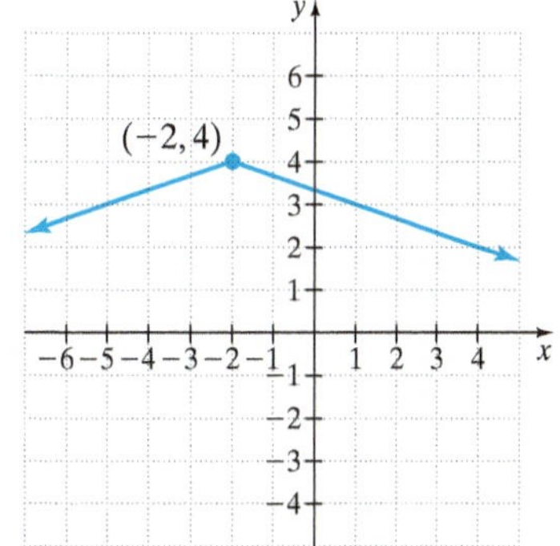

C Exercise Set MyMathLab®

Sketch the graph of each function. Label the vertex of the V shape.

1. $f(x) = 3|x|$
2. $f(x) = 5|x|$
3. $f(x) = \frac{1}{4}|x|$
4. $f(x) = \frac{1}{3}|x|$
5. $g(x) = 2|x| + 3$
6. $g(x) = 3|x| + 2$
7. $h(x) = -\frac{1}{2}|x|$
8. $h(x) = -\frac{1}{3}|x|$
9. $f(x) = 4|x - 1|$
10. $f(x) = 3|x - 2|$
11. $g(x) = -\frac{1}{3}|x| - 2$
12. $g(x) = -\frac{1}{2}|x| - 3$
13. $f(x) = -2|x - 3| + 4$
14. $f(x) = -3|x - 1| + 5$
15. $f(x) = \frac{2}{3}|x + 2| - 5$
16. $f(x) = \frac{3}{4}|x + 1| - 4$

Appendix D

An Introduction to Using a Graphing Utility

The Viewing Window and Interpreting Window Settings

In this appendix, we will use the term **graphing utility** to mean a graphing calculator or a computer software graphing package. All graphing utilities graph equations by plotting points on a screen. While plotting several points can be slow and sometimes tedious for us, a graphing utility can quickly and accurately plot hundreds of points. How does a graphing utility show plotted points? A computer or calculator screen is made up of a grid of small rectangular areas called **pixels.** If a pixel contains a point to be plotted, the pixel is turned "on"; otherwise, the pixel remains "off." The graph of an equation is then a collection of pixels turned "on." The graph of $y = 3x + 1$ from a graphing calculator is shown in Figure A-1. Notice the irregular shape of the line caused by the rectangular pixels.

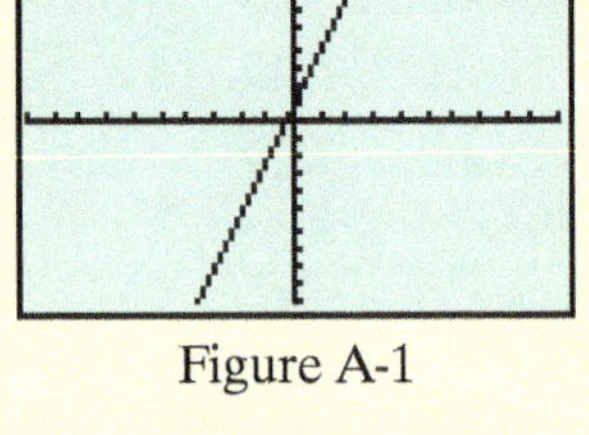

Figure A-1

The portion of the coordinate plane shown on the screen in Figure A-1 is called the **viewing window** or the **viewing rectangle.** Notice the x-axis and the y-axis on the graph. While tick marks are shown on the axes, they are not labeled. This means that from this screen alone, we do not know how many units each tick mark represents. To see what each tick mark represents and the minimum and maximum values on the axes, check the window setting of the graphing utility. It defines the viewing window. The window of the graph of $y = 3x + 1$ shown in Figure A-1 has the following settings (Figure A-2):

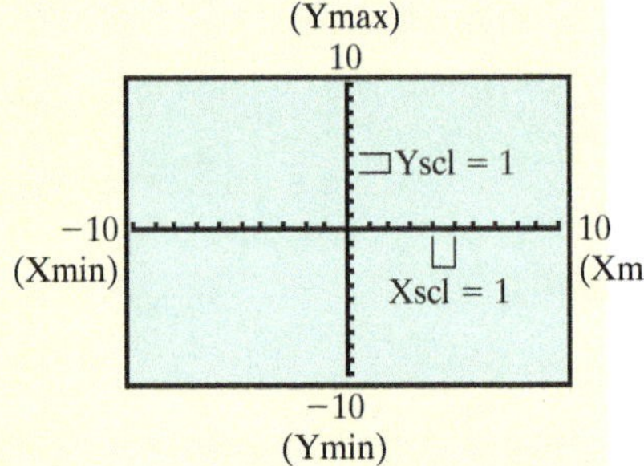

Figure A-2

Xmin = −10 The minimum x-value is −10.
Xmax = 10 The maximum x-value is 10.
Xscl = 1 The x-axis scale is 1 unit per tick mark.
Ymin = −10 The minimum y-value is −10.
Ymax = 10 The maximum y-value is 10.
Yscl = 1 The y-axis scale is 1 unit per tick mark.

By knowing the scale, we can find the minimum and the maximum values on the axes simply by counting tick marks. For example, if both the Xscl (x-axis scale) and the Yscl (y-axis scale) are 1 unit per tick mark on the graph in Figure A-3, we can count the tick marks and find that the minimum x-value is -10 and the maximum x-value is 10. Also, the minimum y-value is -10 and the maximum y-value is 10. If the Xscl changes to 2 units per tick mark (shown in Figure A-4), by counting tick marks, we see that the minimum x-value is now -20 and the maximum x-value is now 20.

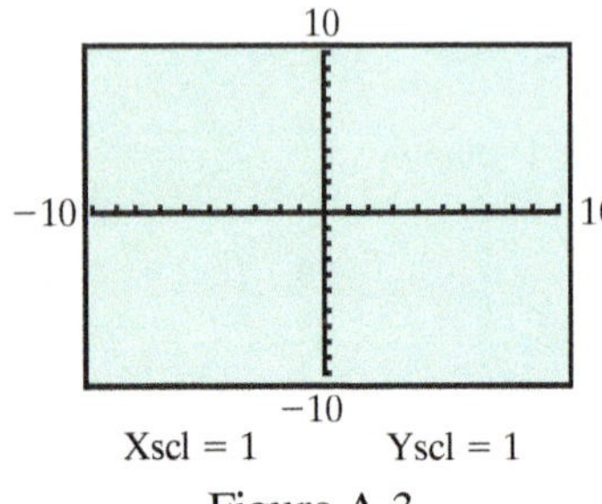

Figure A-3

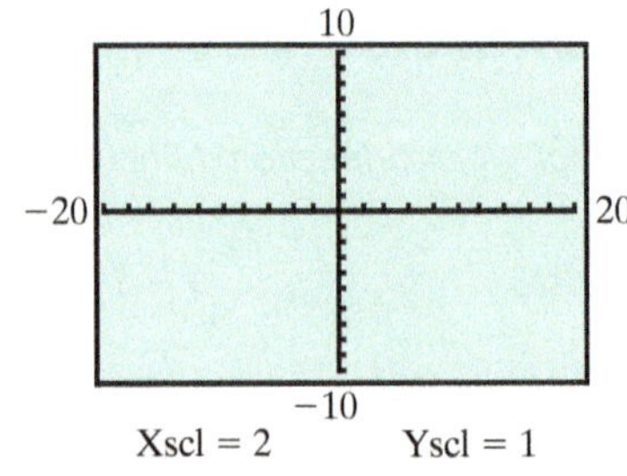

Figure A-4

It is also true that if we know the Xmin and the Xmax values, we can calculate the Xscl by the displayed axes. For example, the Xscl of the graph in Figure A-5 must be 3 units per tick mark for the maximum and minimum x-values to be as shown. Also, the Yscl of that graph must be 2 units per tick mark for the maximum and minimum y-values to be as shown.

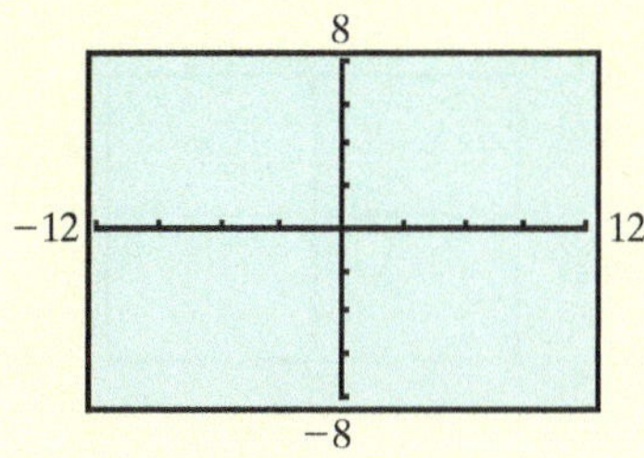

Figure A-5

We will call the viewing window in Figure A-3 a *standard* viewing window or rectangle. Although a standard viewing window is sufficient for much of this text, special care must be taken to ensure that all key features of a graph are shown. Figures A-6, A-7, and A-8 show the graph of $y = x^2 + 11x - 1$ on three different viewing windows. Note that certain viewing windows for this equation are misleading.

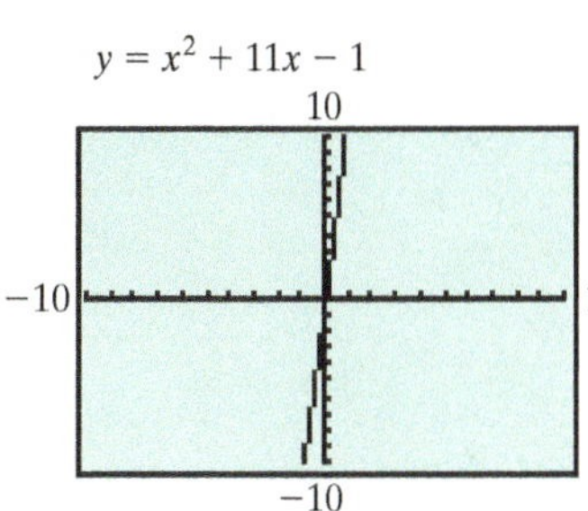

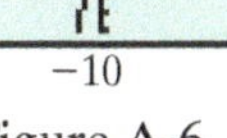

Figure A-6

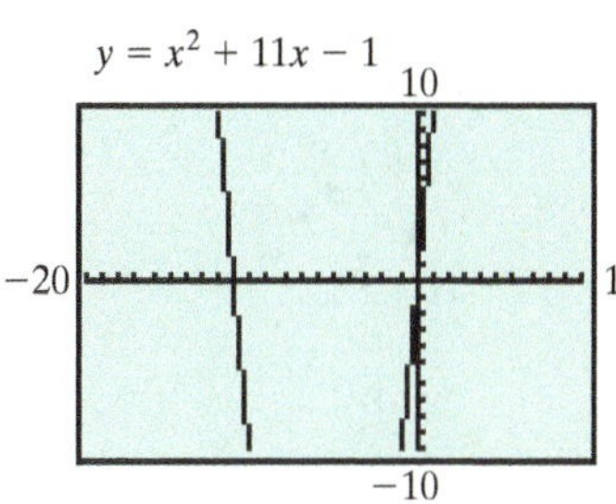

Figure A-7

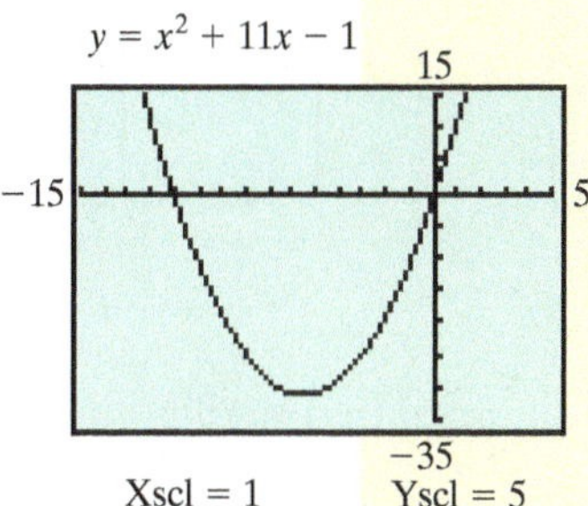

Figure A-8

How do we ensure that all distinguishing features of the graph of an equation are shown? It helps to know about the equation that is being graphed. For example, the equation $y = x^2 + 11x - 1$ is not a linear equation and its graph is not a line. This equation is a quadratic equation and, therefore, its graph is a parabola. By knowing this information, we know that the graph shown in Figure A-6, although correct, is misleading. Of the three viewing rectangles shown, the graph in Figure A-8 is best because it shows more of the distinguishing features of the parabola. Properties of equations needed for graphing will be studied in this text.

The Viewing Window and Interpreting Window Settings Exercise Set

In Exercises 1–4, determine whether all ordered pairs listed will lie within a standard viewing rectangle.

1. $(-9, 0), (5, 8), (1, -8)$
2. $(4, 7), (0, 0), (-8, 9)$
3. $(-11, 0), (2, 2), (7, -5)$
4. $(3, 5), (-3, -5), (15, 0)$

In Exercises 5–10, choose an Xmin, Xmax, Ymin, and Ymax so that all ordered pairs listed will lie within the viewing rectangle.

5. $(-90, 0), (55, 80), (0, -80)$
6. $(4, 70), (20, 20), (-18, 90)$
7. $(-11, 0), (2, 2), (7, -5)$
8. $(3, 5), (-3, -5), (15, 0)$
9. $(200, 200), (50, -50), (70, -50)$
10. $(40, 800), (-30, 500), (15, 0)$

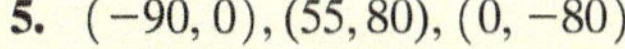

Write the window setting for each viewing window shown. Use the following format:

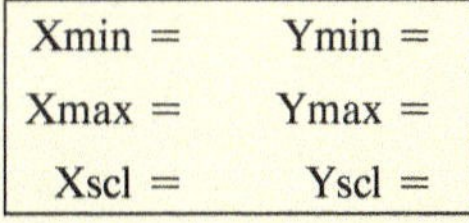

Xmin =	Ymin =
Xmax =	Ymax =
Xscl =	Yscl =

11.

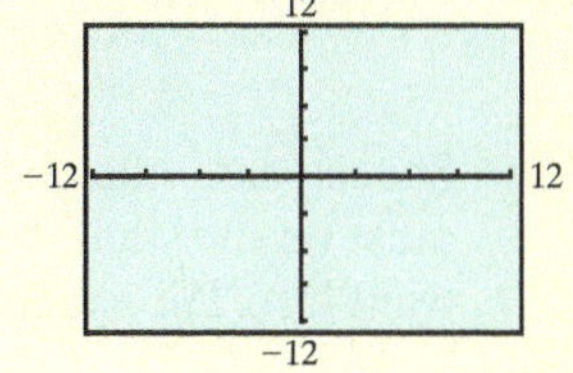

12.

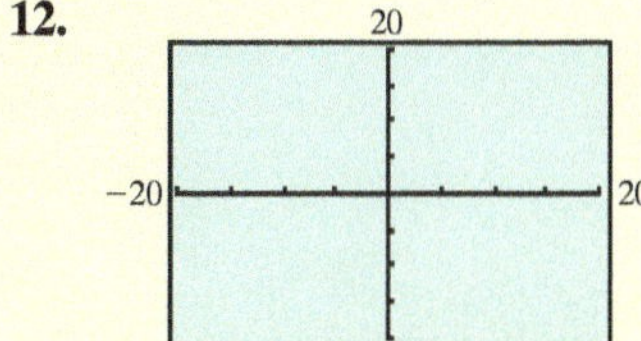

13.

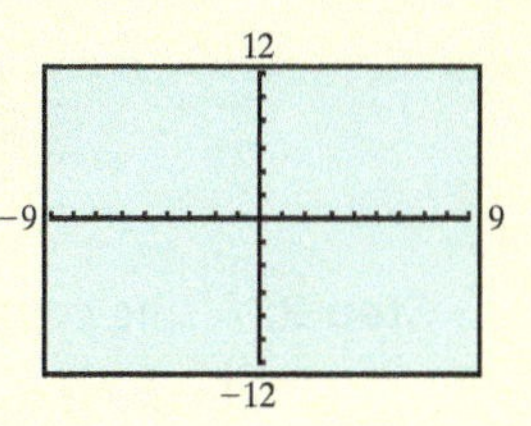

14.

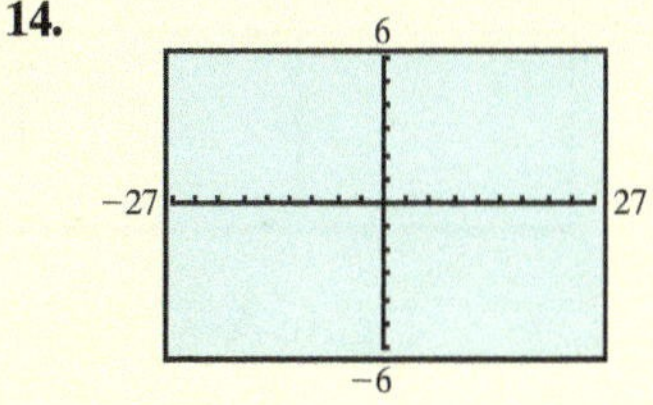

15.

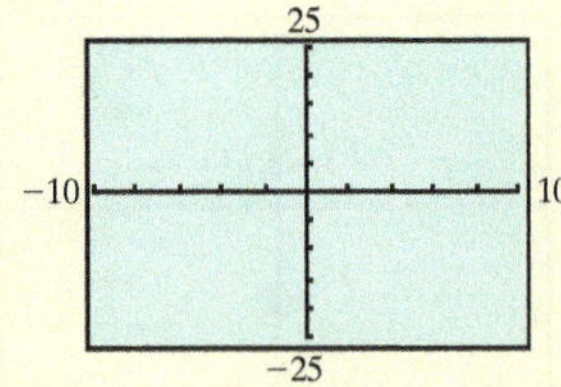

16.

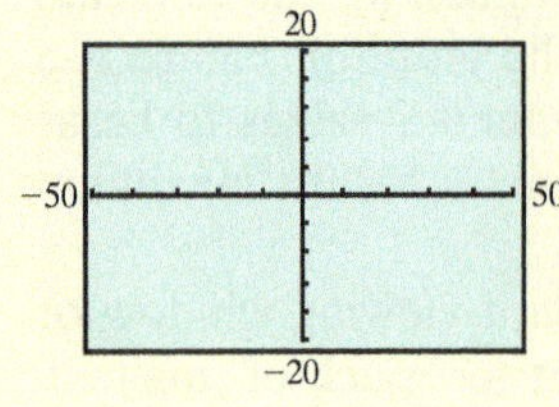

19.

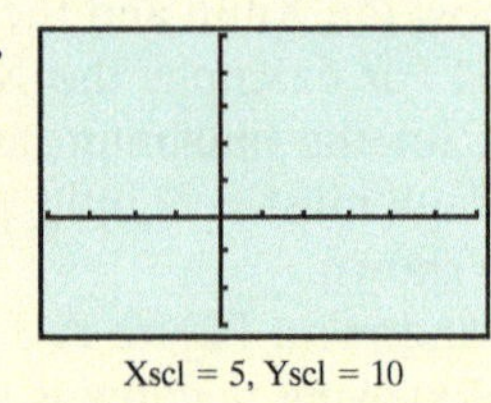

Xscl = 5, Yscl = 10

20.

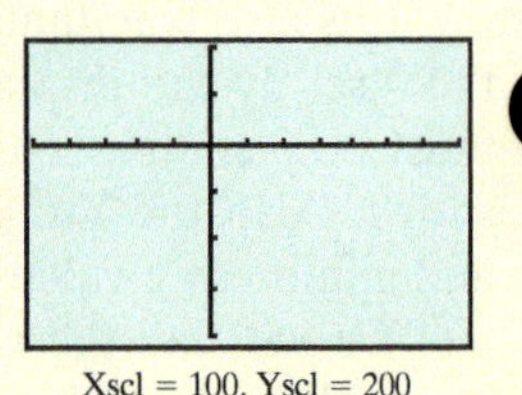

Xscl = 100, Yscl = 200

17.

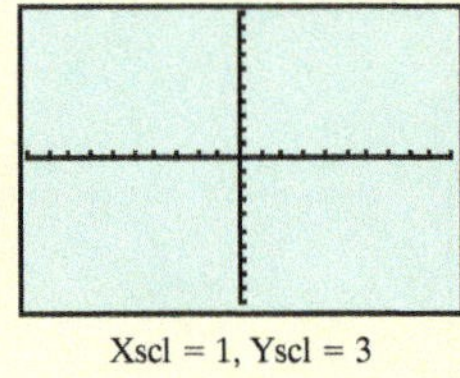

Xscl = 1, Yscl = 3

18.

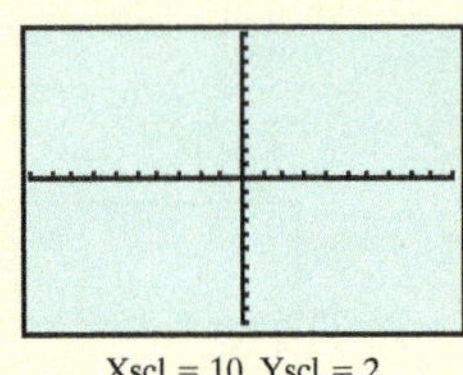

Xscl = 10, Yscl = 2

Graphing Equations and Square Viewing Window

In general, the following steps may be used to graph an equation on a standard viewing window.

> **Graphing an Equation in X and Y with a Graphing Utility on a Standard Viewing Window**
>
> **Step 1:** Solve the equation for y.
>
> **Step 2:** Using your graphing utility, enter the equation in the form Y = *expression involving x*.
>
> **Step 3:** Activate the graphing utility.

Special care must be taken when entering the *expression involving x* in Step 2. You must be sure that the graphing utility you are using interprets the expression as you want it to. For example, let's graph $3y = 4x$. To do so,

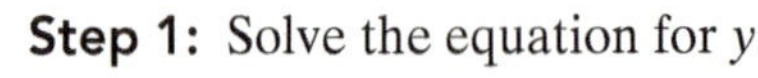

Step 1: Solve the equation for y.

$$3y = 4x$$

$$\frac{3y}{3} = \frac{4x}{3}$$

$$y = \frac{4}{3}x$$

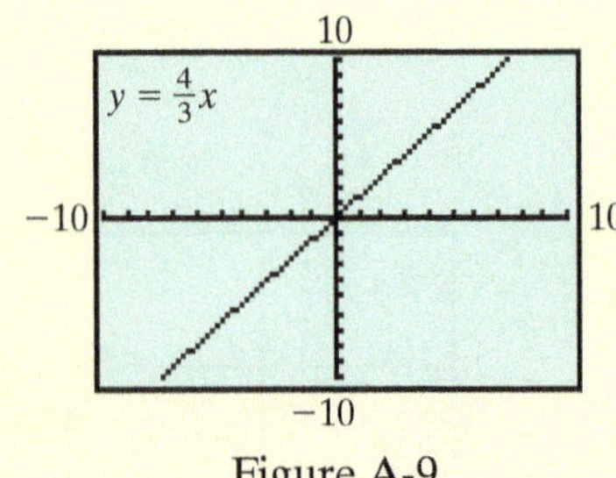

Figure A-9

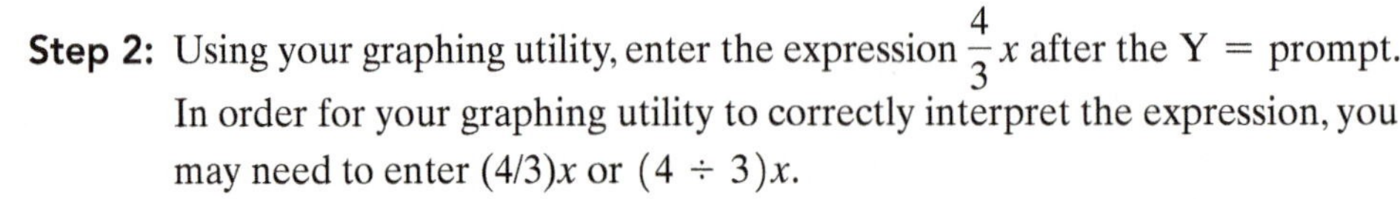

Step 2: Using your graphing utility, enter the expression $\frac{4}{3}x$ after the Y = prompt. In order for your graphing utility to correctly interpret the expression, you may need to enter $(4/3)x$ or $(4 \div 3)x$.

Step 3: Activate the graphing utility. The graph should appear as in Figure A-9.

Distinguishing features of the graph of a line include showing all the intercepts of the line. For example, the window of the graph of the line in Figure A-10 does not show both intercepts of the line, but the window of the graph of the same line in Figure A-11 does show both intercepts. Notice the notation below each graph. This is a shorthand notation of the range setting of the graph. This notation means [Xmin, Xmax] by [Ymin, Ymax].

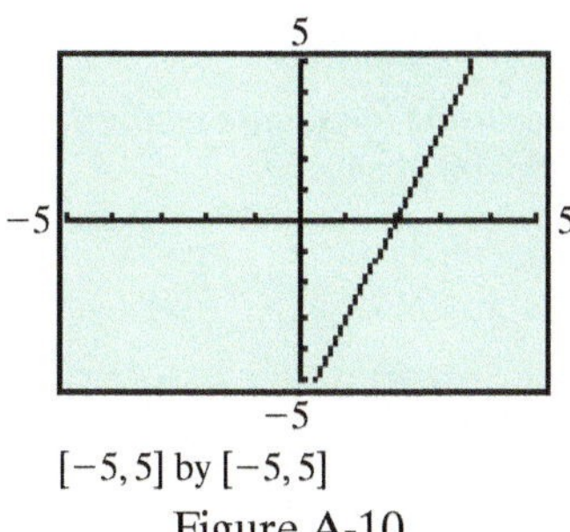

[−5, 5] by [−5, 5]
Figure A-10

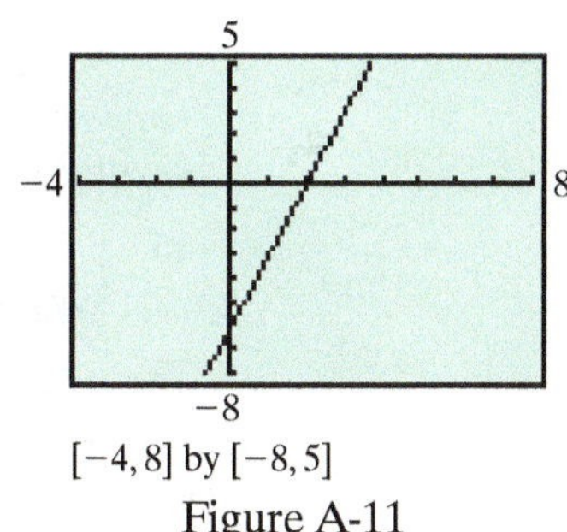

[−4, 8] by [−8, 5]
Figure A-11

On a standard viewing window, the tick marks on the y-axis are closer together than the tick marks on the x-axis. This happens because the viewing window is a rectangle, and so 10 equally spaced tick marks on the positive y-axis will be closer together than 10 equally spaced tick marks on the positive x-axis. This causes the appearance of graphs to be distorted.

For example, notice the different appearances of the same line graphed using different viewing windows. The line in Figure A-12 is distorted because the tick marks along the x-axis are farther apart than the tick marks along the y-axis. The graph of the same line in Figure A-13 is not distorted because the viewing rectangle has been selected so that there is equal spacing between tick marks on both axes.

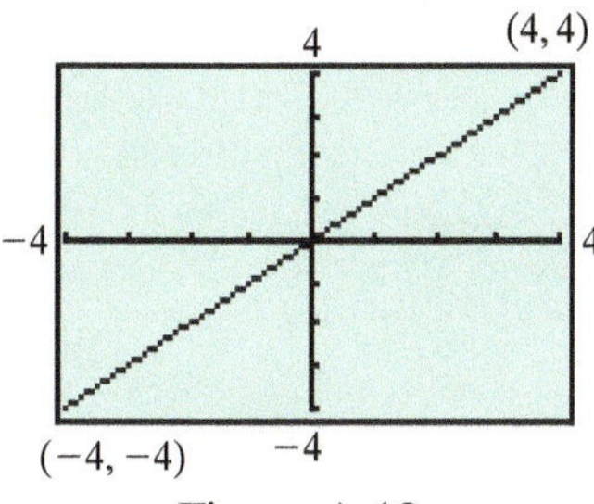

Figure A-12

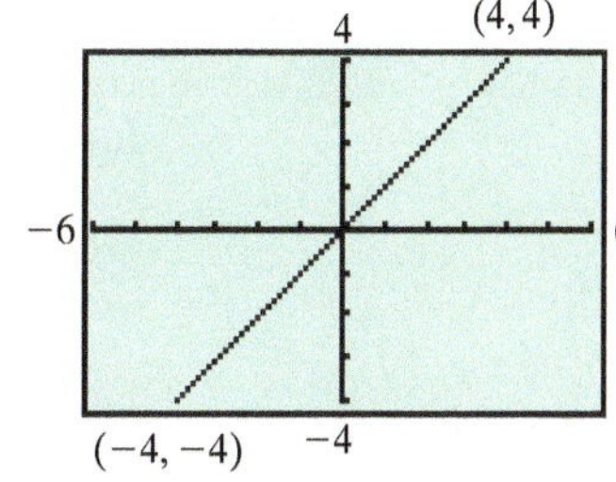

Figure A-13

We say that the line in Figure A-13 is graphed on a *square* setting. Some graphing utilities have a built-in program that, if activated, will automatically provide a square setting. A square setting is especially helpful when we are graphing circles or perpendicular lines, or when a true geometric perspective is desired. Some examples of square screens are shown in Figures A-14 and A-15.

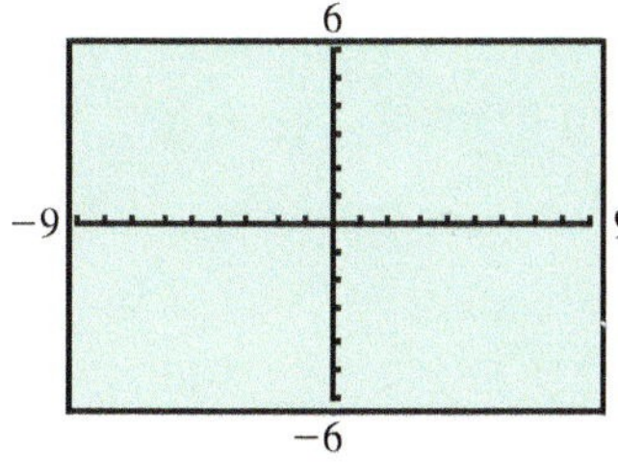

Figure A-14

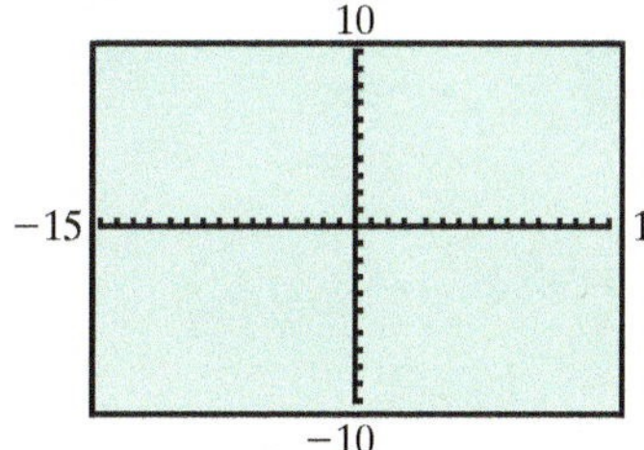

Figure A-15

Other features of a graphing utility such as Trace, Zoom, Intersect, and Table are discussed in appropriate Graphing Calculator Explorations in this text.

Graphing Equations and Square Viewing Window Exercise Set

Graph each linear equation in two variables using the two different range settings given. Determine which setting shows all intercepts of the line.

1. $y = 2x + 12$
Setting A: $[-10, 10]$ by $[-10, 10]$
Setting B: $[-10, 10]$ by $[-10, 15]$

2. $y = -3x + 25$
Setting A: $[-5, 5]$ by $[-30, 10]$
Setting B: $[-10, 10]$ by $[-10, 30]$

3. $y = -x - 41$
Setting A: $[-50, 10]$ by $[-10, 10]$
Setting B: $[-50, 10]$ by $[-50, 15]$

4. $y = 6x - 18$
Setting A: $[-10, 10]$ by $[-20, 10]$
Setting B: $[-10, 10]$ by $[-10, 10]$

5. $y = \frac{1}{2}x - 15$
Setting B: $[-10, 10]$ by $[-20, 10]$
Setting B: $[-10, 35]$ by $[-20, 15]$

6. $y = -\frac{2}{3}x - \frac{29}{3}$
Setting A: $[-10, 10]$ by $[-10, 10]$
Setting B: $[-15, 5]$ by $[-15, 5]$

The graph of each equation is a line. Use a graphing utility and a standard viewing window to graph each equation.

7. $3x = 5y$
8. $7y = -3x$
9. $9x - 5y = 30$
10. $4x + 6y = 20$
11. $y = -7$
12. $y = 2$
13. $x + 10y = -5$
14. $x - 5y = 9$

Graph the following equations using the square setting given. Some keystrokes that may be helpful are given.

15. $y = \sqrt{x}$ $[-12, 12]$ by $[-8, 8]$
Suggested keystrokes: $\sqrt{} x$

16. $y = \sqrt{2x}$ $[-12, 12]$ by $[-8, 8]$
Suggested keystrokes: $\sqrt{}(2x)$

17. $y = x^2 + 2x + 1$ $[-15, 15]$ by $[-10, 10]$
Suggested keystrokes: x^2 $+ 2x + 1$

18. $y = x^2 - 5$ $[-15, 15]$ by $[-10, 10]$
Suggested keystrokes: x^2 $- 5$

19. $y = |x|$ $[-9, 9]$ by $[-6, 6]$
Suggested keystrokes: *ABS* (x)

20. $y = |x - 2|$ $[-9, 9]$ by $[-6, 6]$
Suggested keystrokes: *ABS* $(x - 2)$

Graph each line. Use a standard viewing window; then, if necessary, change the viewing window so that all intercepts of each line show.

21. $x + 2y = 30$
22. $1.5x - 3.7y = 40.3$

Appendix E

Solving Systems of Equations by Matrices

Objectives

1 Use Matrices to Solve a System of Two Equations.

2 Use Matrices to Solve a System of Three Equations.

By now, you may have noticed that the solution of a system of equations depends on the coefficients of the equations in the system and not on the variables. In this section, we introduce solving a system of equations by a **matrix.**

Objective 1 Using Matrices to Solve a System of Two Equations

A matrix (plural: **matrices**) is a rectangular array of numbers. The following are examples of matrices.

$$\begin{bmatrix} 1 & 0 \\ 0 & 1 \end{bmatrix} \quad \begin{bmatrix} 2 & 1 & 3 & -1 \\ 0 & -1 & 4 & 5 \\ -6 & 2 & 1 & 0 \end{bmatrix} \quad \begin{bmatrix} a & b & c \\ d & e & f \end{bmatrix}$$

The numbers aligned horizontally in a matrix are in the same **row.** The numbers aligned vertically are in the same **column.**

$$\begin{array}{l} \text{row 1} \rightarrow \\ \text{row 2} \rightarrow \end{array} \begin{bmatrix} 2 & 1 & 0 \\ -1 & 6 & 2 \end{bmatrix}$$

column 1, column 2, column 3

This matrix has 2 rows and 3 columns. It is called a 2×3 (read "two by three") matrix.

To see the relationship between systems of equations and matrices, study the example below.

Helpful Hint Before writing the corresponding matrix associated with a system of equations, make sure that the equations are written in standard form.

System of Equations (in standard form)		**Corresponding Matrix**	
$2x - 3y = 6$	Equation 1	$2 \quad -3 \;\vdots\; 6$	Row 1
$x + y = 0$	Equation 2	$1 \quad 1 \;\vdots\; 0$	Row 2

$$\begin{cases} 2x - 3y = 6 \\ x + y = 0 \end{cases} \qquad \left[\begin{array}{cc|c} 2 & -3 & 6 \\ 1 & 1 & 0 \end{array}\right]$$

Notice that the rows of the matrix correspond to the equations in the system. The coefficients of each variable are placed to the left of a vertical dashed line. The constants are placed to the right. Each of the numbers in the matrix is called an **element.**

The method of solving systems by matrices is to write this matrix as an equivalent matrix from which we easily identify the solution. Two matrices are equivalent if they represent systems that have the same solution set. The following **row operations** can be performed on matrices, and the result is an equivalent matrix.

Elementary Row Operations

1. Any two rows in a matrix may be interchanged.
2. The elements of any row may be multiplied (or divided) by the same nonzero number.
3. The elements of any row may be multiplied (or divided) by a nonzero number and added to their corresponding elements in any other row.

Helpful Hint Notice that these *row* operations are the same operations that we can perform on *equations* in a system.

To solve a system of two equations in x and y by matrices, write the corresponding matrix associated with the system. Then use elementary row operations to write equivalent matrices until you have a matrix of the form

$$\left[\begin{array}{cc|c} 1 & a & b \\ 0 & 1 & c \end{array}\right],$$

where a, b, and c are constants. Why? If a matrix associated with a system of equations is in this form, we can easily solve for x and y. For example,

Matrix		**System of Equations**
$\left[\begin{array}{cc\|c} 1 & 2 & -3 \\ 0 & 1 & 5 \end{array}\right]$	corresponds to	$\begin{cases} 1x + 2y = -3 \\ 0x + 1y = 5 \end{cases}$ or $\begin{cases} x + 2y = -3 \\ y = 5 \end{cases}$

In the second equation, we have $y = 5$. Substituting this in the first equation, we have $x + 2(5) = -3$ or $x = -13$. The solution of the system is the ordered pair $(-13, 5)$.

Practice 1

Use matrices to solve the system.

$$\begin{cases} x + 4y = -2 \\ 3x - y = 7 \end{cases}$$

Example 1 Use matrices to solve the system.

$$\begin{cases} x + 3y = 5 \\ 2x - y = -4 \end{cases}$$

Solution: The corresponding matrix is $\left[\begin{array}{cc|c} 1 & 3 & 5 \\ 2 & -1 & -4 \end{array}\right]$. We use elementary row operations to write an equivalent matrix that looks like $\left[\begin{array}{cc|c} 1 & a & b \\ 0 & 1 & c \end{array}\right]$.

For the matrix given, the element in the first row, first column is already 1, as desired. Next we write an equivalent matrix with a 0 below the 1. To do this, we multiply row 1 by -2 and add to row 2. *We will change only row 2.*

$$\left[\begin{array}{cc|c} 1 & 3 & 5 \\ -2(1) + 2 & -2(3) + (-1) & -2(5) + (-4) \end{array}\right] \text{ simplifies to } \left[\begin{array}{cc|c} 1 & 3 & 5 \\ 0 & -7 & -14 \end{array}\right]$$

(In each entry of the second row: -2 times the row 1 element, plus the row 2 element.)

row 1 element ↑ row 2 element ↑ row 1 element ↑ row 2 element ↑ row 1 element ↑ row 2 element ↑

Now we change the -7 to a 1 by use of an elementary row operation. We divide row 2 by -7, and then

$$\left[\begin{array}{cc|c} 1 & 3 & 5 \\ \frac{0}{-7} & \frac{-7}{-7} & \frac{-14}{-7} \end{array}\right] \text{ simplifies to } \left[\begin{array}{cc|c} 1 & 3 & 5 \\ 0 & 1 & 2 \end{array}\right]$$

This last matrix corresponds to the system

$$\begin{cases} x + 3y = 5 \\ y = 2 \end{cases}$$

To find x, we let $y = 2$ in the first equation, $x + 3y = 5$.

$x + 3y = 5$ First equation

$x + 3(2) = 5$ Let $y = 2$.

$x = -1$

The ordered pair solution is $(-1, 2)$. Check to see that this ordered pair satisfies both equations.

Work Practice 1

Answer

1. $(2, -1)$

Example 2 Use matrices to solve the system.

$$\begin{cases} 2x - y = 3 \\ 4x - 2y = 5 \end{cases}$$

Solution: The corresponding matrix is $\left[\begin{array}{cc|c} 2 & -1 & 3 \\ 4 & -2 & 5 \end{array}\right]$. To get 1 in the row 1, column 1 position, we divide the elements of row 1 by 2.

$$\left[\begin{array}{cc|c} \frac{2}{2} & -\frac{1}{2} & \frac{3}{2} \\ 4 & -2 & 5 \end{array}\right] \text{ simplifies to } \left[\begin{array}{cc|c} 1 & -\frac{1}{2} & \frac{3}{2} \\ 4 & -2 & 5 \end{array}\right]$$

To get 0 under the 1, we multiply the elements of row 1 by -4 and add the new elements to the elements of row 2.

$$\left[\begin{array}{cc|c} 1 & -\frac{1}{2} & \frac{3}{2} \\ -4(1) + 4 & -4\left(-\frac{1}{2}\right) - 2 & -4\left(\frac{3}{2}\right) + 5 \end{array}\right] \text{ simplifies to } \left[\begin{array}{cc|c} 1 & -\frac{1}{2} & \frac{3}{2} \\ 0 & 0 & -1 \end{array}\right]$$

The corresponding system is $\begin{cases} x - \frac{1}{2}y = \frac{3}{2} \\ 0 = -1 \end{cases}$. The equation $0 = -1$ is false for all y or x values; hence, the system is inconsistent and has no solution. The solution set is { } or $\varnothing$.

Work Practice 2

Practice 2

Use matrices to solve the system.

$$\begin{cases} x - 3y = 3 \\ -2x + 6y = 4 \end{cases}$$

✓Concept Check Consider the system

$$\begin{cases} 2x - 3y = 8 \\ x + 5y = -3 \end{cases}$$

What is wrong with its corresponding matrix shown below?

$$\left[\begin{array}{cc|c} 2 & -3 & 8 \\ 0 & 5 & -3 \end{array}\right]$$

Objective 2 Using Matrices to Solve a System of Three Equations

To solve a system of three equations in three variables using matrices, we will write the corresponding matrix in the form

$$\left[\begin{array}{ccc|c} 1 & a & b & d \\ 0 & 1 & c & e \\ 0 & 0 & 1 & f \end{array}\right]$$

Answer

2. { } or $\varnothing$

✓Concept Check Answer

matrix should be $\left[\begin{array}{cc|c} 2 & -3 & 8 \\ 1 & 5 & -3 \end{array}\right]$

Practice 3

Use matrices to solve the system.

$$\begin{cases} x + 3y - z = 0 \\ 2x + y + 3z = 5 \\ -x - 2y + 4z = 7 \end{cases}$$

Example 3 Use matrices to solve the system.

$$\begin{cases} x + 2y + z = 2 \\ -2x - y + 2z = 5 \\ x + 3y - 2z = -8 \end{cases}$$

Solution: The corresponding matrix is $\left[\begin{array}{ccc|c} 1 & 2 & 1 & 2 \\ -2 & -1 & 2 & 5 \\ 1 & 3 & -2 & -8 \end{array}\right]$. Our goal is to write an equivalent matrix with 1's along the diagonal (see the numbers in red) and 0's below the 1's. The element in row 1, column 1 is already 1. Next we get 0's for each element in the rest of column 1. To do this, first we multiply the elements of row 1 by 2 and add the new elements to row 2. Also, we multiply the elements of row 1 by −1 and add the new elements to the elements of row 3. We *do not change row 1.* Then

$$\left[\begin{array}{ccc|c} 1 & 2 & 1 & 2 \\ 2(1) - 2 & 2(2) - 1 & 2(1) + 2 & 2(2) + 5 \\ -1(1) + 1 & -1(2) + 3 & -1(1) - 2 & -1(2) - 8 \end{array}\right] \text{ simplifies to } \left[\begin{array}{ccc|c} 1 & 2 & 1 & 2 \\ 0 & 3 & 4 & 9 \\ 0 & 1 & -3 & -10 \end{array}\right]$$

We continue down the diagonal and use elementary row operations to get 1 where the element 3 is now. To do this, we interchange rows 2 and 3.

$$\left[\begin{array}{ccc|c} 1 & 2 & 1 & 2 \\ 0 & 3 & 4 & 9 \\ 0 & 1 & -3 & -10 \end{array}\right] \text{ is equivalent to } \left[\begin{array}{ccc|c} 1 & 2 & 1 & 2 \\ 0 & 1 & -3 & -10 \\ 0 & 3 & 4 & 9 \end{array}\right]$$

Next we want the new row 3, column 2 element to be 0. We multiply the elements of row 2 by −3 and add the result to the elements of row 3.

$$\left[\begin{array}{ccc|c} 1 & 2 & 1 & 2 \\ 0 & 1 & -3 & -10 \\ -3(0) + 0 & -3(1) + 3 & -3(-3) + 4 & -3(-10) + 9 \end{array}\right] \text{ simplifies to}$$

$$\left[\begin{array}{ccc|c} 1 & 2 & 1 & 2 \\ 0 & 1 & -3 & -10 \\ 0 & 0 & 13 & 39 \end{array}\right]$$

Finally, we divide the elements of row 3 by 13 so that the final diagonal element is 1.

$$\left[\begin{array}{ccc|c} 1 & 2 & 1 & 2 \\ 0 & 1 & -3 & -10 \\ \frac{0}{13} & \frac{0}{13} & \frac{13}{13} & \frac{39}{13} \end{array}\right] \text{ simplifies to } \left[\begin{array}{ccc|c} 1 & 2 & 1 & 2 \\ 0 & 1 & -3 & -10 \\ 0 & 0 & 1 & 3 \end{array}\right]$$

This matrix corresponds to the system

$$\begin{cases} x + 2y + z = 2 \\ y - 3z = -10 \\ z = 3 \end{cases}$$

We identify the z-coordinate of the solution as 3. Next, we replace z with 3 in the second equation and solve for y.

Answer

3. $(-1, 1, 2)$

$$y - 3z = -10 \quad \text{Second equation}$$
$$y - 3(3) = -10 \quad \text{Let } z = 3.$$
$$y = -1$$

To find x, we let $z = 3$ and $y = -1$ in the first equation.

$$x + 2y + z = 2 \quad \text{First equation}$$
$$x + 2(-1) + 3 = 2 \quad \text{Let } z = 3 \text{ and } y = -1.$$
$$x = 1$$

The ordered triple solution is $(1, -1, 3)$. Check to see that it satisfies all three equations in the original system.

Work Practice 3

E Exercise Set MyMathLab® 

Solve each system of linear equations using matrices. See Example 1.

1. $\begin{cases} x + y = 1 \\ x - 2y = 4 \end{cases}$

2. $\begin{cases} 2x - y = 8 \\ x + 3y = 11 \end{cases}$

3. $\begin{cases} x + 3y = 2 \\ x + 2y = 0 \end{cases}$

4. $\begin{cases} 4x - y = 5 \\ 3x + 3y = 0 \end{cases}$

Solve each system of linear equations using matrices. See Example 2.

5. $\begin{cases} x - 2y = 4 \\ 2x - 4y = 4 \end{cases}$

6. $\begin{cases} -x + 3y = 6 \\ 3x - 9y = 9 \end{cases}$

7. $\begin{cases} 3x - 3y = 9 \\ 2x - 2y = 6 \end{cases}$

8. $\begin{cases} 9x - 3y = 6 \\ -18x + 6y = -12 \end{cases}$

Solve each system of linear equations using matrices. See Example 3.

9. $\begin{cases} x + y = 3 \\ 2y = 10 \\ 3x + 2y - 4z = 12 \end{cases}$

10. $\begin{cases} 5x = 5 \\ 2x + y = 4 \\ 3x + y - 5z = -15 \end{cases}$

11. $\begin{cases} 2y - z = -7 \\ x + 4y + z = -4 \\ 5x - y + 2z = 13 \end{cases}$

12. $\begin{cases} 4y + 3z = -2 \\ 5x - 4y = 1 \\ -5x + 4y + z = -3 \end{cases}$

MIXED PRACTICE

Solve each system of linear equations using matrices. See Examples 1 through 3.

13. $\begin{cases} x - 4 = 0 \\ x + y = 1 \end{cases}$

14. $\begin{cases} 3y = 6 \\ x + y = 7 \end{cases}$

15. $\begin{cases} x + y + z = 2 \\ 2x - z = 5 \\ 3y + z = 2 \end{cases}$

16. $\begin{cases} x + 2y + z = 5 \\ x - y - z = 3 \\ y + z = 2 \end{cases}$

17. $\begin{cases} 5x - 2y = 27 \\ -3x + 5y = 18 \end{cases}$

18. $\begin{cases} 4x - y = 9 \\ 2x + 3y = -27 \end{cases}$

19. $\begin{cases} 4x - 7y = 7 \\ 12x - 21y = 24 \end{cases}$

20. $\begin{cases} 2x - 5y = 12 \\ -4x + 10y = 20 \end{cases}$

21. $\begin{cases} 4x - y + 2z = 5 \\ 2y + z = 4 \\ 4x + y + 3z = 10 \end{cases}$

22. $\begin{cases} 5y - 7z = 14 \\ 2x + y + 4z = 10 \\ 2x + 6y - 3z = 30 \end{cases}$

23. $\begin{cases} 4x + y + z = 3 \\ -x + y - 2z = -11 \\ x + 2y + 2z = -1 \end{cases}$

24. $\begin{cases} x + y + z = 9 \\ 3x - y + z = -1 \\ -2x + 2y - 3z = -2 \end{cases}$

CONCEPT EXTENSIONS

Solve. See the Concept Check in this section.

25. For the system $\begin{cases} x + z = 7 \\ y + 2z = -6 \\ 3x - y = 0 \end{cases}$, which is the correct corresponding matrix?

a. $\left[\begin{array}{cc|c} 1 & 1 & 7 \\ 1 & 2 & -6 \\ 3 & -1 & 0 \end{array}\right]$

b. $\left[\begin{array}{ccc|c} 1 & 0 & 1 & 7 \\ 1 & 2 & 0 & -6 \\ 3 & -1 & 0 & 0 \end{array}\right]$

c. $\left[\begin{array}{ccc|c} 1 & 0 & 1 & 7 \\ 0 & 1 & 2 & -6 \\ 3 & -1 & 0 & 0 \end{array}\right]$

Appendix F

Solving Systems of Equations by Determinants

Objectives

1 Define and Evaluate a 2×2 Determinant.

2 Use Cramer's Rule to Solve a System of Two Linear Equations in Two Variables.

3 Define and Evaluate a 3×3 Determinant.

4 Use Cramer's Rule to Solve a System of Three Linear Equations in Three Variables.

We have solved systems of two linear equations in two variables in four different ways: graphically, by substitution, by elimination, and by matrices. Now we analyze another method, called **Cramer's rule.**

Objective 1 Evaluating 2×2 Determinants

Recall that a matrix is a rectangular array of numbers. If a matrix has the same number of rows and columns, it is called a **square matrix.** Examples of square matrices are

$$\begin{bmatrix} 1 & 6 \\ 5 & 2 \end{bmatrix} \quad \begin{bmatrix} 2 & 4 & 1 \\ 0 & 5 & 2 \\ 3 & 6 & 9 \end{bmatrix}$$

A **determinant** is a real number associated with a square matrix. The determinant of a square matrix is denoted by placing vertical bars about the array of numbers. Thus,

The determinant of the square matrix $\begin{bmatrix} 1 & 6 \\ 5 & 2 \end{bmatrix}$ is $\begin{vmatrix} 1 & 6 \\ 5 & 2 \end{vmatrix}$.

The determinant of the square matrix $\begin{bmatrix} 2 & 4 & 1 \\ 0 & 5 & 2 \\ 3 & 6 & 9 \end{bmatrix}$ is $\begin{vmatrix} 2 & 4 & 1 \\ 0 & 5 & 2 \\ 3 & 6 & 9 \end{vmatrix}$.

We define the determinant of a 2×2 matrix first. (Recall that 2×2 is read "two by two." It means that the matrix has 2 rows and 2 columns.)

Determinant of a 2×2 Matrix

$$\begin{vmatrix} a & b \\ c & d \end{vmatrix} = ad - bc$$

Example 1 Evaluate each determinant

a. $\begin{vmatrix} -1 & 2 \\ 3 & -4 \end{vmatrix}$ **b.** $\begin{vmatrix} 2 & 0 \\ 7 & -5 \end{vmatrix}$

Solution: First we identify the values of a, b, c, and d. Then we perform the evaluation.

a. Here $a = -1, b = 2, c = 3$, and $d = -4$.

$$\begin{vmatrix} -1 & 2 \\ 3 & -4 \end{vmatrix} = ad - bc = (-1)(-4) - (2)(3) = -2$$

b. In this example, $a = 2, b = 0, c = 7$, and $d = -5$.

$$\begin{vmatrix} 2 & 0 \\ 7 & -5 \end{vmatrix} = ad - bc = 2(-5) - (0)(7) = -10$$

Objective 2 Using Cramer's Rule to Solve a System of Two Linear Equations

To develop Cramer's rule, we solve the system $\begin{cases} ax + by = h \\ cx + dy = k \end{cases}$ using elimination. First, we eliminate y by multiplying both sides of the first equation by d and both sides of the second equation by $-b$ so that the coefficients of y are opposites. The result is that

$$\begin{cases} d(ax + by) = d \cdot h \\ -b(cx + dy) = -b \cdot k \end{cases} \quad \text{simplifies to} \quad \begin{cases} adx + bdy = hd \\ -bcx - bdy = -kb \end{cases}$$

We now add the two equations and solve for x.

$$\begin{aligned} adx + bdy &= hd \\ -bcx - bdy &= -kb \\ \hline adx - bcx &= hd - kb \quad \text{Add the equations.} \\ (ad - bc)x &= hd - kb \\ x &= \frac{hd - kb}{ad - bc} \quad \text{Solve for } x. \end{aligned}$$

When we replace x with $\frac{hd - kb}{ad - bc}$ in the equation $ax + by = h$ and solve for y, we find that $y = \frac{ak - ch}{ad - bc}$.

Notice that the numerator of the value of x is the determinant of

$$\begin{vmatrix} h & b \\ k & d \end{vmatrix} = hd - kb$$

Also, the numerator of the value of y is the determinant of

$$\begin{vmatrix} a & h \\ c & k \end{vmatrix} = ak - hc$$

Finally, the denominators of the values of x and y are the same and are the determinant of

$$\begin{vmatrix} a & b \\ c & d \end{vmatrix} = ad - bc$$

This means that the values of x and y can be written in determinant notation:

$$x = \frac{\begin{vmatrix} h & b \\ k & d \end{vmatrix}}{\begin{vmatrix} a & b \\ c & d \end{vmatrix}} \quad \text{and} \quad y = \frac{\begin{vmatrix} a & h \\ c & k \end{vmatrix}}{\begin{vmatrix} a & b \\ c & d \end{vmatrix}}$$

For convenience, we label the determinants D, Dx, and Dy.

x-coefficients

y-coefficients

$$\begin{vmatrix} a & b \\ c & d \end{vmatrix} = D \qquad \begin{vmatrix} h & b \\ k & d \end{vmatrix} = D_x \qquad \begin{vmatrix} a & h \\ c & k \end{vmatrix} = D_y$$

x-column replaced by constants

y-column replaced by constants

These determinant formulas for the coordinates of the solution of a system are known as **Cramer's rule.**

Cramer's Rule for Two Linear Equations in Two Variables

The solution of the system $\begin{cases} ax + by = h \\ cx + dy = k \end{cases}$ is given by

$$x = \frac{\begin{vmatrix} h & b \\ k & d \end{vmatrix}}{\begin{vmatrix} a & b \\ c & d \end{vmatrix}} = \frac{D_x}{D} \quad y = \frac{\begin{vmatrix} a & h \\ c & k \end{vmatrix}}{\begin{vmatrix} a & b \\ c & d \end{vmatrix}} = \frac{D_y}{D}$$

as long as $D = ad - bc$ is not 0.

When $D = 0$, the system is either inconsistent or the equations are dependent. When this happens, we need to use another method to see which is the case.

Example 2 Use Cramer's rule to solve the system

$$\begin{cases} 3x + 4y = -7 \\ x - 2y = -9 \end{cases}$$

Solution: First we find D, Dx, and Dy.

$$\begin{cases} \overset{a}{\overset{\downarrow}{3}}x + \overset{b}{\overset{\downarrow}{4}}y = \overset{h}{\overset{\downarrow}{-7}} \\ \underset{c}{\underset{\uparrow}{x}} - \underset{d}{\underset{\uparrow}{2}}y = \underset{k}{\underset{\uparrow}{-9}} \end{cases}$$

$$D = \begin{vmatrix} a & b \\ c & d \end{vmatrix} = \begin{vmatrix} 3 & 4 \\ 1 & -2 \end{vmatrix} = 3(-2) - 4(1) = -10$$

$$D_x = \begin{vmatrix} h & b \\ k & d \end{vmatrix} = \begin{vmatrix} -7 & 4 \\ -9 & -2 \end{vmatrix} = (-7)(-2) - 4(-9) = 50$$

$$D_y = \begin{vmatrix} a & h \\ c & k \end{vmatrix} = \begin{vmatrix} 3 & -7 \\ 1 & -9 \end{vmatrix} = 3(-9) - (-7)(1) = -20$$

Then $x = \dfrac{D_x}{D} = \dfrac{50}{-10} = -5$ and $y = \dfrac{D_y}{D} = \dfrac{-20}{-10} = 2$.

The ordered pair solution is $(-5, 2)$.
As always, check the solution in both original equations.

Example 3 Use Cramer's rule to solve the system

$$\begin{cases} 5x + y = 5 \\ -7x - 2y = -7 \end{cases}$$

Solution: First we find D, D_x, and D_y.

$$D = \begin{vmatrix} 5 & 1 \\ -7 & -2 \end{vmatrix} = 5(-2) - (-7)(1) = -3$$

$$D_x = \begin{vmatrix} 5 & 1 \\ -7 & -2 \end{vmatrix} = 5(-2) - (-7)(1) = -3$$

$$D_y = \begin{vmatrix} 5 & 5 \\ -7 & -7 \end{vmatrix} = 5(-7) - 5(-7) = 0$$

Then

$$x = \frac{D_x}{D} = \frac{-3}{-3} = 1 \quad y = \frac{D_y}{D} = \frac{0}{-3} = 0$$

The ordered pair solution is $(1, 0)$.

Objective 3 Evaluating 3 × 3 Determinants 

A 3 × 3 determinant can be used to solve a system of three equations in three variables. The determinant of a 3 × 3 matrix, however, is considerably more complex than a 2 × 2 one.

Determinant of a 3 × 3 Matrix

$$\begin{vmatrix} a_1 & b_1 & c_1 \\ a_2 & b_2 & c_2 \\ a_3 & b_3 & c_3 \end{vmatrix} = a_1 \cdot \begin{vmatrix} b_2 & c_2 \\ b_3 & c_3 \end{vmatrix} - a_2 \cdot \begin{vmatrix} b_1 & c_1 \\ b_3 & c_3 \end{vmatrix} + a_3 \cdot \begin{vmatrix} b_1 & c_1 \\ b_2 & c_2 \end{vmatrix}$$

Notice that the determinant of a 3 × 3 matrix is related to the determinants of three 2 × 2 matrices. Each determinant of these 2 × 2 matrices is called a **minor,** and every element of a 3 × 3 matrix has a minor associated with it. For example, the minor of c_2 is the determinant of the 2 × 2 matrix found by deleting the row and column containing c_2.

$$\begin{matrix} a_1 & b_1 & c_1 \\ a_2 & b_2 & c_2 \\ a_3 & b_3 & c_3 \end{matrix} \quad \text{The minor of } c_2 \text{ is} \quad \begin{vmatrix} a_1 & b_1 \\ a_3 & b_3 \end{vmatrix}$$

Also, the minor of element a_1 is the determinant of the 2 × 2 matrix that has no row or column containing a_1.

$$\begin{matrix} a_1 & b_1 & c_1 \\ a_2 & b_2 & c_2 \\ a_3 & b_3 & c_3 \end{matrix} \quad \text{The minor of } a_1 \text{ is} \quad \begin{vmatrix} b_2 & c_2 \\ b_3 & c_3 \end{vmatrix}$$

So the determinant of a 3 × 3 matrix can be written as

$$a_1 \cdot (\text{minor of } a_1) - a_2 \cdot (\text{minor of } a_2) + a_3 \cdot (\text{minor of } a_3)$$

Finding the determinant by using minors of elements in the first column is called **expanding** by the minors of the first column. *The value of a determinant can be found by expanding by the minors of any row or column.* The following **array of signs** is helpful in determining whether to add or subtract the product of an element and its minor.

$$\begin{matrix} + & - & + \\ - & + & - \\ + & - & + \end{matrix}$$

If an element is in a position marked +, we add. If marked −, we subtract.

Example 4 Evaluate by expanding by the minors of the given row or column.

$$\begin{vmatrix} 0 & 5 & 1 \\ 1 & 3 & -1 \\ -2 & 2 & 4 \end{vmatrix}$$

a. First column **b.** Second row

Solution:

a. The elements of the first column are 0, 1, and −2. The first column of the array of signs is +, −, +.

$$\begin{vmatrix} 0 & 5 & 1 \\ 1 & 3 & -1 \\ -2 & 2 & 4 \end{vmatrix} = 0 \cdot \begin{vmatrix} 3 & -1 \\ 2 & 4 \end{vmatrix} - 1 \cdot \begin{vmatrix} 5 & 1 \\ 2 & 4 \end{vmatrix} + (-2) \cdot \begin{vmatrix} 5 & 1 \\ 3 & -1 \end{vmatrix}$$

$$= 0(12 - (-2)) - 1(20 - 2) + (-2)(-5 - 3)$$

$$= 0 - 18 + 16 = -2$$

b. The elements of the second row are 1, 3, and −1. This time, the signs begin with − and again alternate.

$$\begin{vmatrix} 0 & 5 & 1 \\ 1 & 3 & -1 \\ -2 & 2 & 4 \end{vmatrix} = -1 \cdot \begin{vmatrix} 5 & 1 \\ 2 & 4 \end{vmatrix} + 3 \cdot \begin{vmatrix} 0 & 1 \\ -2 & 4 \end{vmatrix} - (-1) \cdot \begin{vmatrix} 0 & 5 \\ -2 & 2 \end{vmatrix}$$

$$= -1(20 - 2) + 3(0 - (-2)) - (-1)(0 - (-10))$$

$$= -18 + 6 + 10 = -2$$

Notice that the determinant of the 3 × 3 matrix is the same regardless of the row or column you select to expand by.

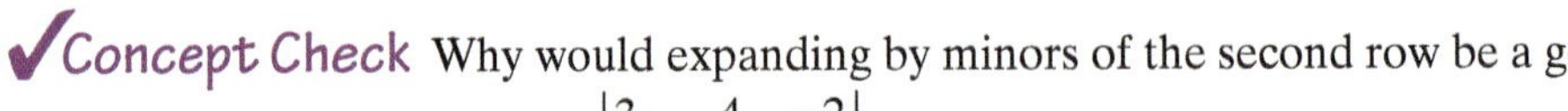

✓Concept Check Why would expanding by minors of the second row be a good choice for the determinant $\begin{vmatrix} 3 & 4 & -2 \\ 5 & 0 & 0 \\ 6 & -3 & 7 \end{vmatrix}$?

Objective 4 Using Cramer's Rule to Solve a System of Three Linear Equations

A system of three equations in three variables may be solved with Cramer's rule also. Using the elimination process to solve a system with unknown constants as coefficients leads to the following.

Cramer's Rule for Three Equations in Three Variables

The solution of the system $\begin{cases} a_1x + b_1y + c_1z = k_1 \\ a_2x + b_2y + c_2z = k_2 \\ a_3x + b_3y + c_3z = k_3 \end{cases}$ is given by

$$x = \frac{D_x}{D} \quad y = \frac{D_y}{D} \quad \text{and} \quad z = \frac{D_z}{D}$$

✓Concept Check Answer
Two elements of the second row are 0, which makes calculations easier.

where

$$D = \begin{vmatrix} a_1 & b_1 & c_1 \\ a_2 & b_2 & c_2 \\ a_3 & b_3 & c_3 \end{vmatrix} \quad D_x = \begin{vmatrix} k_1 & b_1 & c_1 \\ k_2 & b_2 & c_2 \\ k_3 & b_3 & c_3 \end{vmatrix}$$

$$D_y = \begin{vmatrix} a_1 & k_1 & c_1 \\ a_2 & k_2 & c_2 \\ a_3 & k_3 & c_3 \end{vmatrix} \quad D_z = \begin{vmatrix} a_1 & b_1 & k_1 \\ a_2 & b_2 & k_2 \\ a_3 & b_3 & k_3 \end{vmatrix}$$

as long as D is not 0.

Example 5 Use Cramer's rule to solve the system

$$\begin{cases} x - 2y + z = 4 \\ 3x + y - 2z = 3 \\ 5x + 5y + 3z = -8 \end{cases}$$

Solution: First we find D, D_x, D_y, and D_z. Beginning with D, we expand by the minors of the first column.

$$D = \begin{vmatrix} 1 & -2 & 1 \\ 3 & 1 & -2 \\ 5 & 5 & 3 \end{vmatrix} = 1 \cdot \begin{vmatrix} 1 & -2 \\ 5 & 3 \end{vmatrix} - 3 \cdot \begin{vmatrix} -2 & 1 \\ 5 & 3 \end{vmatrix} + 5 \cdot \begin{vmatrix} -2 & 1 \\ 1 & -2 \end{vmatrix}$$

$$= 1(3 - (-10)) - 3(-6 - 5) + 5(4 - 1)$$

$$= 13 + 33 + 15 = 61$$

$$D_x = \begin{vmatrix} 4 & -2 & 1 \\ 3 & 1 & -2 \\ -8 & 5 & 3 \end{vmatrix} = 4 \cdot \begin{vmatrix} 1 & -2 \\ 5 & 3 \end{vmatrix} - 3 \cdot \begin{vmatrix} -2 & 1 \\ 5 & 3 \end{vmatrix} + (-8) \cdot \begin{vmatrix} -2 & 1 \\ 1 & -2 \end{vmatrix}$$

$$= 4(3 - (-10)) - 3(-6 - 5) + (-8)(4 - 1)$$

$$= 52 + 33 - 24 = 61$$

$$D_y = \begin{vmatrix} 1 & 4 & 1 \\ 3 & 3 & -2 \\ 5 & -8 & 3 \end{vmatrix} = 1 \cdot \begin{vmatrix} 3 & -2 \\ -8 & 3 \end{vmatrix} - 3 \cdot \begin{vmatrix} 4 & 1 \\ -8 & 3 \end{vmatrix} + 5 \cdot \begin{vmatrix} 4 & 1 \\ 3 & -2 \end{vmatrix}$$

$$= 1(9 - 16) - 3(12 - (-8)) + 5(-8 - 3)$$

$$= -7 - 60 - 55 = -122$$

$$D_z = \begin{vmatrix} 1 & -2 & 4 \\ 3 & 1 & 3 \\ 5 & 5 & -8 \end{vmatrix} = 1 \cdot \begin{vmatrix} 1 & 3 \\ 5 & -8 \end{vmatrix} - 3 \cdot \begin{vmatrix} -2 & 4 \\ 5 & -8 \end{vmatrix} + 5 \cdot \begin{vmatrix} -2 & 4 \\ 1 & 3 \end{vmatrix}$$

$$= 1(-8 - 15) - 3(16 - 20) + 5(-6 - 4)$$

$$= -23 + 12 - 50 = -61$$

From these determinants, we calculate the solution:

$$x = \frac{D_x}{D} = \frac{61}{61} = 1 \quad y = \frac{D_y}{D} = \frac{-122}{61} = -2 \quad z = \frac{D_z}{D} = \frac{-61}{61} = -1$$

The ordered triple solution is $(1, -2, -1)$. Check this solution by verifying that it satisfies each equation of the system.

F Exercise Set MyMathLab® 

Evaluate. See Example 1.

1. $\begin{vmatrix} 3 & 5 \\ -1 & 7 \end{vmatrix}$

2. $\begin{vmatrix} -5 & 1 \\ 0 & -4 \end{vmatrix}$

3. $\begin{vmatrix} 9 & -2 \\ 4 & -3 \end{vmatrix}$

4. $\begin{vmatrix} 4 & 0 \\ 9 & 8 \end{vmatrix}$

5. $\begin{vmatrix} -2 & 9 \\ 4 & -18 \end{vmatrix}$

6. $\begin{vmatrix} -40 & 8 \\ 70 & -14 \end{vmatrix}$

Use Cramer's rule, if possible, to solve each system of linear equations. See Examples 2 and 3.

7. $\begin{cases} 2y - 4 = 0 \\ x + 2y = 5 \end{cases}$

8. $\begin{cases} 4x - y = 5 \\ 3x - 3 = 0 \end{cases}$

9. $\begin{cases} 3x + y = 1 \\ 2y = 2 - 6x \end{cases}$

10. $\begin{cases} y = 2x - 5 \\ 8x - 4y = 20 \end{cases}$

11. $\begin{cases} 5x - 2y = 27 \\ -3x + 5y = 18 \end{cases}$

12. $\begin{cases} 4x - y = 9 \\ 2x + 3y = -27 \end{cases}$

Evaluate. See Example 4.

13. $\begin{vmatrix} 2 & 1 & 0 \\ 0 & 5 & -3 \\ 4 & 0 & 2 \end{vmatrix}$

14. $\begin{vmatrix} -6 & 4 & 2 \\ 1 & 0 & 5 \\ 0 & 3 & 1 \end{vmatrix}$

15. $\begin{vmatrix} 4 & -6 & 0 \\ -2 & 3 & 0 \\ 4 & -6 & 1 \end{vmatrix}$

16. $\begin{vmatrix} 5 & 2 & 1 \\ 3 & -6 & 0 \\ -2 & 8 & 0 \end{vmatrix}$

17. $\begin{vmatrix} 3 & 6 & -3 \\ -1 & -2 & 3 \\ 4 & -1 & 6 \end{vmatrix}$

18. $\begin{vmatrix} 2 & -2 & 1 \\ 4 & 1 & 3 \\ 3 & 1 & 2 \end{vmatrix}$

Use Cramer's rule, if possible, to solve each system of linear equations. See Example 5.

19. $\begin{cases} 3x + z = -1 \\ -x - 3y + z = 7 \\ 3y + z = 5 \end{cases}$

20. $\begin{cases} 4y - 3z = -2 \\ 8x - 4y = 4 \\ -8x + 4y + z = -2 \end{cases}$

21. $\begin{cases} x + y + z = 8 \\ 2x - y - z = 10 \\ x - 2y + 3z = 22 \end{cases}$

22. $\begin{cases} 5x + y + 3z = 1 \\ x - y - 3z = -7 \\ -x + y = 1 \end{cases}$

MIXED PRACTICE

Evaluate.

23. $\begin{vmatrix} 10 & -1 \\ -4 & 2 \end{vmatrix}$

24. $\begin{vmatrix} -6 & 2 \\ 5 & -1 \end{vmatrix}$

25. $\begin{vmatrix} 1 & 0 & 4 \\ 1 & -1 & 2 \\ 3 & 2 & 1 \end{vmatrix}$

26. $\begin{vmatrix} 0 & 1 & 2 \\ 3 & -1 & 2 \\ 3 & 2 & -2 \end{vmatrix}$

27. $\begin{vmatrix} \frac{3}{4} & \frac{5}{2} \\ -\frac{1}{6} & \frac{7}{3} \end{vmatrix}$

28. $\begin{vmatrix} \frac{5}{7} & \frac{1}{3} \\ \frac{6}{7} & \frac{2}{3} \end{vmatrix}$

29. $\begin{vmatrix} 4 & -2 & 2 \\ 6 & -1 & 3 \\ 2 & 1 & 1 \end{vmatrix}$

30. $\begin{vmatrix} 1 & 5 & 0 \\ 7 & 9 & -4 \\ 3 & 2 & -2 \end{vmatrix}$

31. $\begin{vmatrix} -2 & 5 & 4 \\ 5 & -1 & 3 \\ 4 & 1 & 2 \end{vmatrix}$

32. $\begin{vmatrix} 5 & -2 & 4 \\ -1 & 5 & 3 \\ 1 & 4 & 2 \end{vmatrix}$

Use Cramer's rule, if possible, to solve each system of linear equations.

33. $\begin{cases} 2x - 5y = 4 \\ x + 2y = -7 \end{cases}$

34. $\begin{cases} 3x - y = 2 \\ -5x + 2y = 0 \end{cases}$

35. $\begin{cases} 4x + 2y = 5 \\ 2x + y = -1 \end{cases}$

36. $\begin{cases} 3x + 6y = 15 \\ 2x + 4y = 3 \end{cases}$

37. $\begin{cases} 2x + 2y + z = 1 \\ -x + y + 2z = 3 \\ x + 2y + 4z = 0 \end{cases}$

38. $\begin{cases} 2x - 3y + z = 5 \\ x + y + z = 0 \\ 4x + 2y + 4z = 4 \end{cases}$

39. $\begin{cases} \frac{2}{3}x - \frac{3}{4}y = -1 \\ -\frac{1}{6}x + \frac{3}{4}y = \frac{5}{2} \end{cases}$

40. $\begin{cases} \frac{1}{2}x - \frac{1}{3}y = -3 \\ \frac{1}{8}x + \frac{1}{6}y = 0 \end{cases}$

41. $\begin{cases} 0.7x - 0.2y = -1.6 \\ 0.2x - y = -1.4 \end{cases}$

42. $\begin{cases} -0.7x + 0.6y = 1.3 \\ 0.5x - 0.3y = -0.8 \end{cases}$

43. $\begin{cases} -2x + 4y - 2z = 6 \\ x - 2y + z = -3 \\ 3x - 6y + 3z = -9 \end{cases}$

44. $\begin{cases} -x - y + 3z = 2 \\ 4x + 4y - 12z = -8 \\ -3x - 3y + 9z = 6 \end{cases}$

45. $\begin{cases} x - 2y + z = -5 \\ 3y + 2z = 4 \\ 3x - y = -2 \end{cases}$

46. $\begin{cases} 4x + 5y = 10 \\ 3y + 2z = -6 \\ x + y + z = 3 \end{cases}$

CONCEPT EXTENSIONS

Find the value of x such that each is a true statement.

47. $\begin{vmatrix} 1 & x \\ 2 & 7 \end{vmatrix} = -3$

48. $\begin{vmatrix} 6 & 1 \\ -2 & x \end{vmatrix} = 26$

49. If all the elements in a single row of a square matrix are zero, to what does the determinant evaluate? Explain your answer.

50. If all the elements in a single column of a square matrix are 0, to what does the determinant evaluate? Explain your answer.

51. Suppose you are interested in finding the determinant of a 4×4 matrix. Study the pattern shown in the array of signs for a 3×3 matrix. Use the pattern to expand the array of signs for use with a 4×4 matrix.

52. Why would expanding by minors of the second row be a good choice for the determinant $\begin{vmatrix} 3 & 4 & -2 \\ 5 & 0 & 0 \\ 6 & -3 & 7 \end{vmatrix}$?

Find the value of each determinant. To evaluate a 4×4 determinant, select any row or column and expand by the minors. The array of signs for a 4×4 determinant is the same as for a 3×3 determinant except expanded.

53. $\begin{vmatrix} 5 & 0 & 0 & 0 \\ 0 & 4 & 2 & -1 \\ 1 & 3 & -2 & 0 \\ 0 & -3 & 1 & 2 \end{vmatrix}$

54. $\begin{vmatrix} 1 & 7 & 0 & -1 \\ 1 & 3 & -2 & 0 \\ 1 & 0 & -1 & 2 \\ 0 & -6 & 2 & 4 \end{vmatrix}$

55. $\begin{vmatrix} 4 & 0 & 2 & 5 \\ 0 & 3 & -1 & 1 \\ 0 & 0 & 2 & 0 \\ 0 & 0 & 0 & 1 \end{vmatrix}$

56. $\begin{vmatrix} 2 & 0 & -1 & 4 \\ 6 & 0 & 4 & 1 \\ 2 & 4 & 3 & -1 \\ 4 & 0 & 5 & -4 \end{vmatrix}$

Contents of Student Resources

Student Resources

STUDY SKILLS BUILDERS

Attitude and Study Tips

Study Skills Builder 1

Have You Decided to Complete This Course Successfully?

Ask yourself if one of your current goals is to complete this course successfully.

If it is not a goal of yours, ask yourself why. One common reason is fear of failure. Amazingly enough, fear of failure alone can be strong enough to keep many of us from doing our best in any endeavor.

Another common reason is that you simply haven't taken the time to think about or write down your goals for this course. To help accomplish this, answer the questions below.

Exercises

1. Write down your goal(s) for this course.
2. Now list steps you will take to make sure your goal(s) in Exercise **1** is/are accomplished.
3. Rate your commitment to this course with a number between 1 and 5. Use the diagram below to help.

High Commitment		Average Commitment		Not Committed at All
5	4	3	2	1

4. If you have rated your personal commitment level (from the exercise above) as a 1, 2, or 3, list the reasons why this is so. Then determine whether it is possible to increase your commitment level to a 4 or 5.

Good luck, and don't forget that a positive attitude will make a big difference.

Study Skills Builder 2

Tips for Studying for an Exam

To prepare for an exam, try the following study techniques:

- Start the study process days before your exam.
- Make sure that you are up to date on your assignments.
- If there is a topic that you are unsure of, use one of the many resources that are available to you. For example,

  See your instructor.
  View a lecture video on the topic.
  Visit a learning resource center on campus.
  Read the textbook material and examples on the topic.

- Reread your notes and carefully review the Chapter Highlights at the end of any chapter.
- Work the review exercises at the end of the chapter.
- Find a quiet place to take the Chapter Test found at the end of the chapter. Do not use any resources when taking this sample test. This way, you will have a clear indication of how prepared you are for your exam. Check your answers and use the Chapter Test Prep Videos to make sure that you correct any missed exercises.

Good luck, and keep a positive attitude.

Exercises

Let's see how you did on your last exam.

1. How many days before your last exam did you start studying for that exam?
2. Were you up to date on your assignments at that time or did you need to catch up on assignments?
3. List the most helpful text supplement (if you used one).
4. List the most helpful campus supplement (if you used one).
5. List your process for preparing for a mathematics test.
6. Was this process helpful? In other words, were you satisfied with your performance on your exam?
7. If not, what changes can you make in your process that will make it more helpful to you?

Study Skills Builder 3

What to Do the Day of an Exam

Your first exam may be soon. On the day of an exam, don't forget to try the following:

- Allow yourself plenty of time to arrive.
- Read the directions on the test carefully.
- Read each problem carefully as you take your test. Make sure that you answer the question asked.
- Watch your time and pace yourself so that you may attempt each problem on your test.
- Check your work and answers.
- ***Do not turn your test in early.*** If you have extra time, spend it double-checking your work.

Good luck!

Exercises

Answer the following questions based on your most recent mathematics exam, whenever that was.

1. How soon before class did you arrive?
2. Did you read the directions on the test carefully?
3. Did you make sure you answered the question asked for each problem on the exam?
4. Were you able to attempt each problem on your exam?
5. If your answer to Exercise **4** is no, list reasons why.
6. Did you have extra time on your exam?
7. If your answer to Exercise **6** is yes, describe how you spent that extra time.

Study Skills Builder 4

Are You Satisfied with Your Performance on a Particular Quiz or Exam?

If not, don't forget to analyze your quiz or exam and look for common errors. Were most of your errors a result of:

- *Carelessness?* Did you turn in your quiz or exam before the allotted time expired? If so, resolve to use any extra time to check your work.
- *Running out of time?* Answer the questions you are sure of first. Then attempt the questions you are unsure of, and delay checking your work until all questions have been answered.
- *Not understanding a concept?* If so, review that concept and correct your work so that you make sure you understand it before the next quiz or the final exam.
- *Test conditions?* When studying for a quiz or exam, make sure you place yourself in conditions similar to test conditions. For example, before your next quiz or exam, take a sample test without the aid of your notes or text.

(For a sample test, see your instructor or use the Chapter Test at the end of each chapter.)

Exercises

1. Have you corrected all your previous quizzes and exams?
2. List any errors you have found common to two or more of your graded papers.
3. Is one of your common errors not understanding a concept? If so, are you making sure you understand all the concepts for the next quiz or exam?
4. Is one of your common errors making careless mistakes? If so, are you now taking all the time allotted to check over your work so that you can minimize the number of careless mistakes?
5. Are you satisfied with your grades thus far on quizzes and tests?
6. If your answer to Exercise **5** is no, are there any more suggestions you can make to your instructor or yourself to help? If so, list them here and share them with your instructor.

Study Skills Builder 5

How Are You Doing?

If you haven't done so yet, take a few moments and think about how you are doing in this course. Are you working toward your goal of successfully completing this course? Is your performance on homework, quizzes, and tests satisfactory? If not, you might want to see your instructor to see if he/she has any suggestions on how you can improve your performance. Reread Section 1.1 for ideas on places to get help with your mathematics course.

Exercises

Answer the following.

1. List any textbook supplements you are using to help you through this course.
2. List any campus resources you are using to help you through this course.
3. Write a short paragraph describing how you are doing in your mathematics course.
4. If improvement is needed, list ways that you can work toward improving your situation as described in Exercise **3**.

Study Skills Builder 6

Are You Preparing for Your Final Exam?

To prepare for your final exam, try the following study techniques:

- Review the material that you will be responsible for on your exam. This includes material from your textbook, your notebook, and any handouts from your instructor.
- Review any formulas that you may need to memorize.
- Check to see if your instructor or mathematics department will be conducting a final exam review.
- Check with your instructor to see whether final exams from previous semesters/quarters are available to students for review.
- Use your previously taken exams as a practice final exam. To do so, rewrite the test questions in mixed order on blank sheets of paper. This will help you prepare for exam conditions.
- If you are unsure of a few concepts, see your instructor or visit a learning lab for assistance. Also, view the video segment of any troublesome sections.
- If you need further exercises to work, try the Cumulative Reviews at the ends of the chapters.

Once again, good luck! I hope you are enjoying this textbook and your mathematics course.

Organizing Your Work

Study Skills Builder 7

Learning New Terms

Many of the terms used in this text may be new to you. It will be helpful to make a list of new mathematical terms and symbols as you encounter them and to review them frequently. Placing these new terms (including page references) on 3×5 index cards might help you later when you're preparing for a quiz.

Exercises

1. Name one way you might place a word and its definition on a 3×5 card.
2. How do new terms stand out in this text so that they can be found?

Study Skills Builder 8

Are You Organized?

Have you ever had trouble finding a completed assignment? When it's time to study for a test, are your notes neat and organized? Have you ever had trouble reading your own mathematics handwriting? (Be honest—I have.)

When any of these things happen, it's time to get organized. Here are a few suggestions:

- Write your notes and complete your homework assignments in a notebook with pockets (spiral or ring binder).
- Take class notes in this notebook, and then follow the notes with your completed homework assignment.
- When you receive graded papers or handouts, place them in the notebook pocket so that you will not lose them.
- Mark (possibly with an exclamation point) any note(s) that seem extra important to you.
- Mark (possibly with a question mark) any notes or homework that you are having trouble with.
- See your instructor or a math tutor for help with the concepts or exercises that you are having trouble understanding.
- If you are having trouble reading your own handwriting, *slow down* and write your mathematics work clearly!

Exercises

1. Have you been completing your assignments on time?
2. Have you been correcting any exercises you may be having difficulty with?
3. If you are having trouble understanding a mathematical concept or correcting any homework exercises, have you visited your instructor, a tutor, or your campus math lab?
4. Are you taking lecture notes in your mathematics course? (By the way, these notes should include worked-out examples solved by your instructor.)
5. Is your mathematics course material (handouts, graded papers, lecture notes) organized?
6. If your answer to Exercise **5** is no, take a moment and review your course material. List at least two ways that you might better organize it.

Study Skills Builder 9

Organizing a Notebook

It's never too late to get organized. If you need ideas about organizing a notebook for your mathematics course, try some of these:

- Use a spiral or ring binder notebook with pockets and use it for mathematics only.
- Start each page by writing the book's section number you are working on at the top.
- When your instructor is lecturing, take notes. *Always* include any examples your instructor works for you.
- Place your worked-out homework exercises in your notebook immediately after the lecture notes from that section. This way, a section's worth of material is together.
- Homework exercises: Attempt and check all assigned homework.
- Place graded quizzes in the pockets of your notebook or a special section of your binder.

Exercises

Check your notebook organization by answering the following questions.

1. Do you have a spiral or ring binder notebook for your mathematics course only?
2. Have you ever had to flip through several sheets of notes and work in your mathematics notebook to determine what section's work you are in?
3. Are you now writing the textbook's section number at the top of each notebook page?
4. Have you ever lost or had trouble finding a graded quiz or test?
5. Are you now placing all your graded work in a dedicated place in your notebook?
6. Are you attempting all of your homework and placing all of your work in your notebook?
7. Are you checking and correcting your homework in your notebook? If not, why not?
8. Are you writing in your notebook the examples your instructor works for you in class?

Study Skills Builder 10

How Are Your Homework Assignments Going?

It is very important in mathematics to keep up with homework. Why? Many concepts build on each other. Often your understanding of a day's concepts depends on an understanding of the previous day's material.

Remember that completing your homework assignment involves a lot more than attempting a few of the problems assigned.

To complete a homework assignment, remember these four things:

- Attempt all of it.
- Check it.
- Correct it.
- If needed, ask questions about it.

Exercises

Take a moment and review your completed homework assignments. Answer the questions below based on this review.

1. Approximate the fraction of your homework you have attempted.
2. Approximate the fraction of your homework you have checked (if possible).
3. If you are able to check your homework, have you corrected it when errors have been found?
4. When working homework, if you do not understand a concept, what do you do?

MyMathLab and MathXL

Study Skills Builder 11

Tips for Turning in Your Homework on Time

It is very important to keep up with your mathematics homework assignments. Why? Many concepts in mathematics build upon each other.

Remember these four tips to help ensure your work is completed on time:

- Know the assignments and due dates set by your instructor.
- Do not wait until the last minute to submit your homework.
- Set a goal to submit your homework 6–8 hours before the scheduled due date in case you have unexpected technology trouble.
- Schedule enough time to complete each assignment.

Following the tips above will also help you avoid potentially losing points for late or missed assignments.

Exercises

Take a moment to consider your work on your homework assignments to date and answer the following questions:

1. What percentage of your assignments have you turned in on time?
2. Why might it be a good idea to submit your homework 6–8 hours before the scheduled deadline?
3. If you have missed submitting any homework by the due date, list some of the reasons why this occurred.
4. What steps do you plan to take in the future to ensure your homework is submitted on time?

Study Skills Builder 12

Tips for Doing Your Homework Online

Practice is one of the main keys to success in any mathematics course. Did you know that MyMathLab/MathXL provides you with **immediate feedback** for each exercise? If you are incorrect, you are given hints to work the exercise correctly. You have **unlimited practice opportunities** and can rework any exercises you have trouble with until you master them, and you can submit homework assignments unlimited times before the deadline.

Remember these success tips when doing your homework online:

- Attempt all assigned exercises.
- Write down (neatly) your step-by-step work for each exercise before entering your answer.
- Use the immediate feedback provided by the program to help you check and correct your work for each exercise.
- Rework any exercises you have trouble with until you master them.
- Work through your homework assignment as many times as necessary until you are satisfied.

Exercises

Take a moment to think about your homework assignments to date and answer the following:

1. Have you attempted all assigned exercises?
2. Of the exercises attempted, have you also written out your work before entering your answer—so that you can check it?
3. Are you familiar with how to enter answers using the MathXL player so that you avoid answer-entry-type errors?
4. List some ways the immediate feedback and practice supports have helped you with your homework. If you have not used these supports, how do you plan to use them with the success tips above on your next assignment?

Study Skills Builder 13

Organizing Your Work

Have you ever used any readily available paper (such as the back of a flyer, another course assignment, Post-its, etc.) to work out your homework exercises before entering the answer in MathXL? To save time, have you ever entered answers directly into MathXL without working the exercises on paper? When it's time to study, have you ever been unable to find your completed work or read and follow your own mathematics handwriting?

When any of these things happen, it's time to get organized. Here are some suggestions:

- Write your step-by-step work for each homework exercise (neatly) on lined, loose-leaf paper and keep this in a 3-ring binder.
- Refer to your step-by-step work when you receive feedback that your answer is incorrect in MathXL. Double-check using the steps and hints provided by the program and correct your work accordingly.
- Keep your written homework with your class notes for that section.
- Identify any exercises you are having trouble with and ask questions about them.
- Keep all graded quizzes and tests in this binder as well to study later.

If you follow the suggestions above, you and your instructor or tutor will be able to follow your steps and correct any mistakes. You will also have a written copy of your work to refer to later to ask questions and study for tests.

Exercises

1. Why is it important that you write out your step-by-step work for homework exercises and keep a hard copy of all work submitted online?
2. If you have gotten an incorrect answer, are you able to follow your steps and find your error?
3. If you were asked today to review your previous homework assignments and first test, could you find them? If not, list some ways you might better organize your work.

Study Skills Builder 14

Getting Help with Your Homework Assignments

There are many resources available to you through MathXL to help you work through any homework exercises you may have trouble with. It is important that you know what these resources are and know when and how to use them.

Let's review the features found in the homework exercises:

- **Help Me Solve This**—provides step-by-step help for the exercise you are working. You must work an additional exercise of the same type (without this help) before you can get credit for having worked it correctly.
- **View an Example**—allows you to view a correctly worked exercise similar to the one you are having trouble with. You can then go back to your original exercise and work it on your own.
- **E-Book**—allows you to read examples from your text and find similar exercises.
- **Video**—your text author, Elayn Martin-Gay, works an exercise similar to the one you need help with. **Not all exercises have an accompanying video clip.
- **Ask My Instructor**—allows you to e-mail your instructor for help with an exercise.

Exercises

1. How does the "Help Me Solve This" feature work?
2. If the "View an Example" feature is used, is it necessary to work an additional problem before continuing the assignment?
3. When might be a good time to use the "Video" feature? Do all exercises have an accompanying video clip?
4. Which of the features above have you used? List those you found the most helpful to you.
5. If you haven't used the features discussed, list those you plan to try on your next homework assignment.

Study Skills Builder 15

Tips for Preparing for an Exam

Did you know that you can rework your previous homework assignments in MyMathLab and MathXL? This is a great way to prepare for tests. To do this, open a previous homework assignment and click "similar exercise." This will generate new exercises similar to the homework you have submitted. You can then rework the exercises and assignments until you feel confident that you understand them.

To prepare for an exam, follow these tips:

- Review your written work for your previous homework assignments along with your class notes.
- Identify any exercises or topics that you have questions on or have difficulty understanding.
- Rework your previous assignments in MyMathLab and MathXL until you fully understand them and can do them without help.
- Get help for any topics you feel unsure of or for which you have questions.

Exercises

1. Are your current homework assignments up to date and is your written work for them organized in a binder or notebook? If the answer is no, it's time to get organized. For tips on this, see Study Skills Builder 13—Organizing Your Work.
2. How many days in advance of an exam do you usually start studying?
3. List some ways you think that working previous homework assignments can help you prepare for your test.
4. List two or three resources you can use to get help for any topics you are unsure of or have questions on.

Good luck!

Study Skills Builder 16

How Well Do You Know the Resources Available to You in MyMathLab?

There are many helpful resources available to you in MyMathLab. Let's take a moment to locate and explore a few of them now. Go into your MyMathLab course, and visit the Multimedia Library, Tools for Success, and E-Book.

Let's see what you found.

Exercises

1. List the resources available to you in the Multimedia Library.
2. List the resources available to you in the Tools for Success folder.
3. Where did you find the English/Spanish Audio Glossary?
4. Can you view videos from the E-Book?
5. Did you find any resources you did not know about? If so, which ones?
6. Which resources have you used most often or find most helpful?

Additional Help Inside and Outside Your Textbook

Study Skills Builder 17

How Well Do You Know Your Textbook?

The questions below will help determine whether you are familiar with your textbook. For additional information, see Section 1.1 in this text.

1. What does the ▶ icon mean?
2. What does the ✎ icon mean?
3. What does the △ icon mean?
4. Where can you find a review for each chapter? What answers to this review can be found in the back of your text?
5. Each chapter contains an overview of the chapter along with examples. What is this feature called?
6. Each chapter contains a review of vocabulary. What is this feature called?
7. There are practice exercises that are contained in this text. What are they and how can they be used?
8. This text contains a student section in the back entitled Student Resources. List the contents of this section and how they might be helpful.
9. What exercise answers are available in this text? Where are they located?

Study Skills Builder 18

Are You Familiar with Your Textbook Supplements?

Below is a review of some of the student supplements available for additional study. Check to see if you are using the ones most helpful to you.

- Chapter Test Prep Videos. These videos provide video clip solutions to the Chapter Test exercises in this text. You will find them extremely useful when studying for tests or exams.
- Interactive DVD Lecture Series. These are keyed to each section of the text. The material is presented by me, Elayn Martin-Gay, and I have placed a ▶ by the exercises in the text that I have worked on the video.
- The *Student Solutions Manual*. This contains worked-out solutions to odd-numbered exercises as well as every exercise in the Integrated Reviews, Chapter Reviews, Chapter Tests, and Cumulative Reviews.
- Pearson Tutor Center. Mathematics questions may be phoned, faxed, or e-mailed to this center.
- MyMathLab is a text-specific online course. MathXL is an online homework, tutorial, and assessment system.

Take a moment and determine whether these are available to you.

As usual, your instructor is your best source of information.

Exercises

Let's see how you are doing with textbook supplements.

1. Name one way the Lecture Videos can be helpful to you.

2. Name one way the Chapter Test Prep Videos can help you prepare for a chapter test.

3. List any textbook supplements that you have found useful.

4. Have you located and visited a learning resource lab located on your campus?

5. List the textbook supplements that are currently housed in your campus' learning resource lab.

Study Skills Builder 19

Are You Getting All the Mathematics Help That You Need?

Remember that, in addition to your instructor, there are many places to get help with your mathematics course. For example:

- This text has an accompanying video lesson for every section, and the CD in this text contains worked-out solutions to every Chapter Test exercise.
- The back of the book contains answers to odd-numbered exercises.
- A *Student Solutions Manual* is available that contains worked-out solutions to odd-numbered exercises as well as solutions to every exercise in the Integrated Reviews, Chapter Reviews, Chapter Tests, and Cumulative Reviews.
- Don't forget to check with your instructor for other local resources available to you, such as a tutor center.

Exercises

1. List items you find helpful in the text and all student supplements to this text.

2. List all the campus help that is available to you for this course.

3. List any help (besides the textbook) from Exercises **1** and **2** above that you are using.

4. List any help (besides the textbook) that you feel you should try.

5. Write a goal for yourself that includes trying everything you listed in Exercise **4** during the next week.

Bigger Picture—Study Guide Outline

OUTLINE: PART 1

Operations on Sets of Numbers and Solving Equations

I. Operations on Sets of Numbers

A. Whole Numbers

1. Add or Subtract: (Sec. 1.3)

$$\begin{array}{r} 14 \\ +39 \\ \hline 53 \end{array} \qquad \begin{array}{r} 300 \\ -\ 27 \\ \hline 273 \end{array}$$

2. Multiply or Divide: (Sec. 1.5, 1.6)

$$\begin{array}{r} 238 \\ \times\ 47 \\ \hline 1666 \\ 9520 \\ \hline 11{,}186 \end{array} \qquad \begin{array}{r} 127 \text{ R } 2 \\ 7\overline{)891} \\ -7 \\ \hline 19 \\ -14 \\ \hline 51 \\ -49 \\ \hline 2 \end{array}$$

3. Exponent: (Sec. 1.7)

$$3^4 = \underbrace{3 \cdot 3 \cdot 3 \cdot 3}_{\text{4 factors of 3}} = 81$$

4. Order of Operations: (Sec. 1.7)

$$\begin{aligned} 24 \div 3 \cdot 2 - (2 + 8) &= 24 \div 3 \cdot 2 - (10) && \text{Simplify within parentheses.} \\ &= 8 \cdot 2 - 10 && \text{Multiply or divide from left to right.} \\ &= 16 - 10 && \text{Multiply or divide from left to right.} \\ &= 6 && \text{Add or subtract from left to right.} \end{aligned}$$

5. Square Root: (Sec. 7.3)

$\sqrt{25} = 5$ because $5 \cdot 5 = 25$ and 5 is a positive number.

B. Integers

1. Add: (Sec. 2.2) $-5 + (-2) = -7$ Adding like signs: Add absolute values. Attach the common sign.

$-5 + 2 = -3$ Adding unlike signs: Subtract absolute values. Attach the sign of the number with the larger absolute value.

2. Subtract: Add the first number to the opposite of the second number. (Sec. 2.3)

$$7 - 10 = 7 + (-10) = -3$$

3. Multiply or Divide: Multiply or divide as usual. If the signs of the two numbers are the same, the answer is positive. If the signs of the two numbers are different, the answer is negative. (Sec. 2.4)

$$-5 \cdot 5 = -25, \quad \frac{-32}{-8} = 4$$

C. Fractions

1. **Simplify:** Factor the numerator and denominator. Then divide out factors of 1 by dividing out common factors in the numerator and denominator. (Sec. 4.2)

   Simplify: $\frac{20}{28} = \frac{4 \cdot 5}{4 \cdot 7} = \frac{5}{7}$

2. **Multiply:** Numerator times numerator over denominator times denominator. (Sec. 4.3)

   $\frac{5}{9} \cdot \frac{2}{7} = \frac{10}{63}$

3. **Divide:** First fraction times the reciprocal of the second fraction. (Sec. 4.3)

   $\frac{2}{11} \div \frac{3}{4} = \frac{2}{11} \cdot \frac{4}{3} = \frac{8}{33}$

4. **Add or Subtract:** Must have same denominators. If not, find the LCD, and write each fraction as an equivalent fraction with the LCD as denominator. (Sec. 4.4, 4.5)

   $\frac{2}{5} + \frac{1}{15} = \frac{2}{5} \cdot \frac{3}{3} + \frac{1}{15} = \frac{6}{15} + \frac{1}{15} = \frac{7}{15}$

D. Decimals

1. **Add or Subtract:** Line up decimal points. (Sec. 5.2)

$$\begin{array}{r} 1.27 \\ +\ 0.6 \\ \hline 1.87 \end{array}$$

2. **Multiply:** (Sec. 5.3)

$$\begin{array}{rl} 2.56 & \text{2 decimal places} \\ \times\ 3.2 & \text{1 decimal place} \\ \hline 512 & 2 + 1 = 3 \\ 7680 & \\ \hline 8.192 & \text{3 decimal places} \end{array}$$

3. **Divide:** (Sec. 5.4) $8\overline{)5.6}$ = 0.7 $\qquad 0.6\overline{)0.786}$ = 1.31

II. Solving Equations

A. Equations in General: Simplify both sides of the equation by removing parentheses and adding any like terms. Then use the addition property to write variable terms on one side and constants (or numbers) on the other side. Then use the multiplication property to solve for the variable by dividing both sides of the equation by the coefficient of the variable. (Sec. 3.3)

Solve: $2(x - 5) = 80$

$2x - 10 = 80$ Use the distributive property.

$2x - 10 + 10 = 80 + 10$ Add 10 to both sides.

$2x = 90$ Simplify.

$\frac{2x}{2} = \frac{90}{2}$ Divide both sides by 2.

$x = 45$ Simplify.

B. Proportions: Set cross products equal to each other. Then solve. (Sec. 6.1)

$\frac{14}{3} = \frac{2}{n}$ or $14 \cdot n = 3 \cdot 2$ or $14 \cdot n = 6$ or $n = \frac{6}{14} = \frac{3}{7}$

C. Percent Problems

1. **Solved by Equations:** Remember that "of" means multiplication and "is" means equals. (Sec. 6.3)

   "12% of some number is 6" translates to

   $$12\% \cdot n = 6 \text{ or } 0.12 \cdot n = 6 \text{ or } n = \frac{6}{0.12} \text{ or } n = 50$$

2. **Solved by Proportions:** Remember that percent, p, is identified by % or "percent"; base, b, usually appears after "of"; and amount, a, is the part compared to the whole. (Sec. 6.4)

   "12% of some number is 6" translates to

   $$\frac{6}{b} = \frac{12}{100} \text{ or } 6 \cdot 100 = b \cdot 12 \text{ or } \frac{600}{12} = b \text{ or } 50 = b$$

OUTLINE: PART 2

Simplifying Expressions and Solving Equations and Inequalities

I. Simplifying Expressions

A. Exponents (Sec. 12.1, 12.5)

$$x^7 \cdot x^5 = x^{12};\ (x^7)^5 = x^{35};\ \frac{x^7}{x^5} = x^2;\ x^0 = 1;\ 8^{-2} = \frac{1}{8^2} = \frac{1}{64}$$

B. Polynomials

1. **Add:** Combine like terms. (Sec. 12.2)

   $$\begin{aligned}(3y^2 + 6y + 7) + (9y^2 - 11y - 15) &= 3y^2 + 6y + 7 + 9y^2 - 11y - 15\\ &= 12y^2 - 5y - 8\end{aligned}$$

2. **Subtract:** Change the sign of the terms of the polynomial being subtracted, and then add. (Sec. 12.2)

   $$\begin{aligned}(3y^2 + 6y + 7) - (9y^2 - 11y - 15) &= 3y^2 + 6y + 7 - 9y^2 + 11y + 15\\ &= -6y^2 + 17y + 22\end{aligned}$$

3. **Multiply:** Multiply each term of one polynomial by each term of the other polynomial. (Sec. 12.3, 12.4)

   $$\begin{aligned}(x + 5)(2x^2 - 3x + 4) &= x(2x^2 - 3x + 4) + 5(2x^2 - 3x + 4)\\ &= 2x^3 - 3x^2 + 4x + 10x^2 - 15x + 20\\ &= 2x^3 + 7x^2 - 11x + 20\end{aligned}$$

4. **Divide:** (Sec. 12.6)

   a. To divide by a monomial, divide each term of the polynomial by the monomial.

   $$\frac{8x^2 + 2x - 6}{2x} = \frac{8x^2}{2x} + \frac{2x}{2x} - \frac{6}{2x} = 4x + 1 - \frac{3}{x}$$

b. To divide by a polynomial other than a monomial, use long division.

$$\begin{array}{r} x - 6 + \dfrac{40}{2x+5} \\ 2x + 5\overline{)2x^2 - 7x + 10} \\ \underline{2x^2 + 5x} \\ -12x + 10 \\ \underline{-12x - 30} \\ 40 \end{array}$$

C. Factoring Polynomials

See the Chapter 13 Integrated Review for steps.

$$\begin{aligned} 3x^4 - 78x^2 + 75 &= 3(x^4 - 26x^2 + 25) && \text{Factor out GCF—always first step.} \\ &= 3(x^2 - 25)(x^2 - 1) && \text{Factor trinomial.} \\ &= 3(x + 5)(x - 5)(x + 1)(x - 1) && \text{Factor further—each a difference of squares.} \end{aligned}$$

D. Rational Expressions

1. **Simplify:** Factor the numerator and denominator. Then divide out factors of 1 by dividing out common factors in the numerator and denominator. (Sec. 14.1)

$$\frac{x^2 - 9}{7x^2 - 21x} = \frac{(x + 3)(x - 3)}{7x(x - 3)} = \frac{x + 3}{7x}$$

2. **Multiply:** Multiply numerators, and then multiply denominators. (Sec. 14.2)

$$\frac{5z}{2z^2 - 9z - 18} \cdot \frac{22z + 33}{10z} = \frac{5 \cdot z}{(2z + 3)(z - 6)} \cdot \frac{11(2z + 3)}{2 \cdot 5 \cdot z} = \frac{11}{2(z - 6)}$$

3. **Divide:** First fraction times the reciprocal of the second fraction. (Sec. 14.2)

$$\frac{14}{x + 5} \div \frac{x + 1}{2} = \frac{14}{x + 5} \cdot \frac{2}{x + 1} = \frac{28}{(x + 5)(x + 1)}$$

4. **Add or Subtract:** Must have same denominator. If not, find the LCD and write each fraction as an equivalent fraction with the LCD as denominator. (Sec. 14.4)

$$\begin{aligned} \frac{9}{10} - \frac{x + 1}{x + 5} &= \frac{9(x + 5)}{10(x + 5)} - \frac{10(x + 1)}{10(x + 5)} \\ &= \frac{9x + 45 - 10x - 10}{10(x + 5)} = \frac{-x + 35}{10(x + 5)} \end{aligned}$$

E. Radicals

1. **Simplify Square Roots:** If possible, factor the radicand so that one factor is a perfect square. Then use the product rule and simplify. (Sec. 17.3)

$$\sqrt{75} = \sqrt{25 \cdot 3} = \sqrt{25} \cdot \sqrt{3} = 5\sqrt{3}$$

2. **Add or Subtract:** Only like radicals (same index and radicand) can be added or subtracted. (Sec. 17.4)

$$8\sqrt{10} - \sqrt{40} + \sqrt{5} = 8\sqrt{10} - 2\sqrt{10} + \sqrt{5} = 6\sqrt{10} + \sqrt{5}$$

3. **Multiply or Divide:** (Sec. 17.4) $\sqrt{a} \cdot \sqrt{b} = \sqrt{ab}$; $\dfrac{\sqrt{a}}{\sqrt{b}} = \sqrt{\dfrac{a}{b}}$.

$$\sqrt{11} \cdot \sqrt{3} = \sqrt{33}; \frac{\sqrt{140}}{\sqrt{7}} = \sqrt{\frac{140}{7}} = \sqrt{20} = \sqrt{4 \cdot 5} = 2\sqrt{5}$$

4. Rationalizing the Denominator: (Sec. 17.5)

a. If denominator is one term,

$$\frac{5}{\sqrt{11}} = \frac{5 \cdot \sqrt{11}}{\sqrt{11} \cdot \sqrt{11}} = \frac{5\sqrt{11}}{11}$$

b. If denominator is two terms, multiply by 1 in the form of $\frac{\text{conjugate of denominator}}{\text{conjugate of denominator}}$.

$$\frac{13}{3+\sqrt{2}} = \frac{13}{3+\sqrt{2}} \cdot \frac{3-\sqrt{2}}{3-\sqrt{2}} = \frac{13(3-\sqrt{2})}{9-2} = \frac{13(3-\sqrt{2})}{7}$$

II. Solving Equations

A. Linear Equations: Power on variable is 1 and there are no variables in denominator. (Sec. 9.3)

$7(x-3) = 4x + 6$	Linear equation (If fractions, multiply by LCD.)
$7x - 21 = 4x + 6$	Use the distributive property.
$3x - 21 = 6$	Subtract $4x$ from both sides.
$3x = 27$	Add 21 to both sides.
$x = 9$	Divide both sides by 3.

B. Quadratic and Higher-Degree Equations: Solve: First write the equation in standard form (one side is 0).

1. If the polynomial on one side factors, solve by factoring. (Sec. 13.6, 18.1, and 18.3)
2. If the polynomial does not factor, solve by the quadratic formula. (Sec. 18.2)

By Factoring:

$$x^2 + x = 6$$
$$x^2 + x - 6 = 0$$
$$(x-2)(x+3) = 0$$
$$x - 2 = 0 \text{ or } x + 3 = 0$$
$$x = 2 \text{ or } \quad x = -3$$

By Quadratic Formula:

$$x^2 + x = 5$$
$$x^2 + x - 5 = 0$$
$$a = 1, b = 1, c = -5$$
$$x = \frac{-1 \pm \sqrt{1^2 - 4(1)(-5)}}{2 \cdot 1}$$
$$= \frac{-1 \pm \sqrt{21}}{2}$$

C. Equations with Rational Expressions: Make sure the proposed solution does not make a denominator 0. (Sec. 14.5)

$\frac{3}{x} - \frac{1}{x-1} = \frac{4}{x-1}$	Equation with rational expressions
$x(x-1) \cdot \frac{3}{x} - x(x-1) \cdot \frac{1}{x-1} = x(x-1) \cdot \frac{4}{x-1}$	Multiply through by $x(x-1)$.
$3(x-1) - x \cdot 1 = x \cdot 4$	Simplify.
$3x - 3 - x = 4x$	Use the distributive property.
$-3 = 2x$	Simplify and move variable terms to right side.
$-\frac{3}{2} = x$	Divide both sides by 2.

D. Proportions: An equation with two ratios equal. Set cross products equal, and then solve. Make sure the proposed solution does not make the denominator 0. (Sec. 14.6)

$$\frac{5}{x} = \frac{9}{2x-3}$$

$$5(2x-3) = 9 \cdot x \quad \text{Set cross products equal.}$$

$$10x - 15 = 9x \quad \text{Multiply.}$$

$$x = 15 \quad \text{Write equation with variable terms on one side and constants on the other.}$$

E. Absolute Value Equations (Sec. 16.2)

$$|3x-1| = 8$$
$$3x - 1 = 8 \quad \text{or} \quad 3x - 1 = -8$$
$$3x = 9 \quad \text{or} \quad 3x = -7$$
$$x = 3 \quad \text{or} \quad x = -\frac{7}{3}$$

$$|x-5| = |x+1|$$
$$x - 5 = x + 1 \quad \text{or} \quad x - 5 = -(x+1)$$
$$\underbrace{-5 = 1}_{\text{No solution}} \quad \text{or} \quad x - 5 = -x - 1$$
$$\text{or} \quad 2x = 4$$
$$x = 2$$

F. Equations with Radicals (Sec. 17.6)

$$\sqrt{5x+10} - 2 = x$$
$$\sqrt{5x+10} = x + 2$$
$$(\sqrt{5x+10})^2 = (x+2)^2$$
$$5x + 10 = x^2 + 4x + 4$$
$$0 = x^2 - x - 6$$
$$0 = (x-3)(x+2)$$
$$x - 3 = 0 \quad \text{or} \quad x + 2 = 0$$
$$x = 3 \quad \text{or} \quad x = -2$$

Both solutions check.

G. Exponential Equations (Sec. 19.3, 19.8)

$$9^x = 27^{x+1}$$
$$(3^2)^x = (3^3)^{x+1}$$
$$3^{2x} = 3^{3x+3}$$
$$2x = 3x + 3$$
$$-3 = x$$

$$5^x = 7$$
$$\log 5^x = \log 7$$
$$x \log 5 = \log 7$$
$$x = \frac{\log 7}{\log 5}$$

H. Logarithmic Equations (Sec. 19.8)

$$\log 7 + \log(x+3) = \log 5$$
$$\log 7(x+3) = \log 5$$
$$7(x+3) = 5$$
$$7x + 21 = 5$$
$$7x = -16$$
$$x = -\frac{16}{7}$$

Study Guide

III. Solving Inequalities

A. Linear Inequalities (Sec. 9.6)

$$-3(x+2) \ge 6$$
$$-3x - 6 \ge 6$$
$$-3x \ge 12$$
$$\frac{-3x}{-3} \le \frac{12}{-3}$$
$$x \le -4 \quad \text{or} \quad (\infty, -4]$$

B. Compound Inequalities (Sec 16.1)

$x \le 3$ and $x < -7$ | $x \le 3$ or $x < -7$

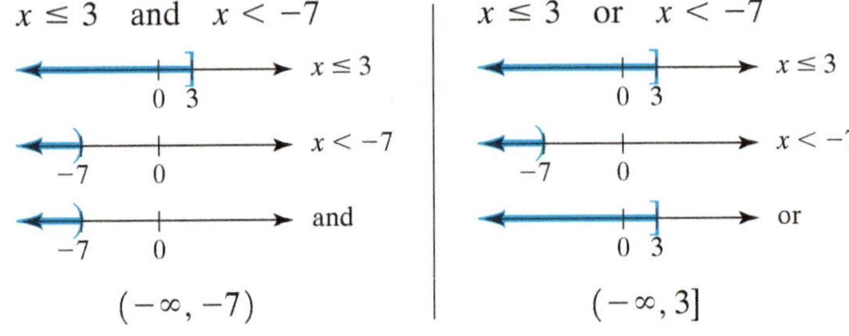

$(-\infty, -7)$ | $(-\infty, 3]$

C. Absolute Value Inequalities (Sec. 16.3)

$$|x-5| - 8 < -2$$
$$|x-5| < 6$$
$$-6 < x - 5 < 6$$
$$-1 < x < 11$$
$$(-1, 11)$$

$$|2x+1| \ge 17$$
$$2x + 1 \ge 17 \quad \text{or} \quad 2x + 1 \le -17$$
$$2x \ge 16 \quad \text{or} \quad 2x \le -18$$
$$x \ge 8 \quad \text{or} \quad x \le -9$$
$$(-\infty, -9] \cup [8, \infty)$$

D. Nonlinear Inequalities (Sec. 18.4)

$$x^2 - x < 6$$
$$x^2 - x - 6 < 0$$
$$(x-3)(x+2) < 0$$

✗ ✓ ✗
−2 3

$$(-2, 3)$$

$$\frac{x-5}{x+1} \ge 0$$

✓ ✗ ✓
−1 5

$$(-\infty, -1) \cup [5, \infty)$$

Practice Final Exam

Chapters 1–8

Simplify by performing the indicated operations.

1. $2^3 \cdot 5^2$
2. $16 + 9 \div 3 \cdot 4 - 7$
3. $18 - 24$
4. $5 \cdot (-20)$
5. $\sqrt{49}$
6. $(-5)^3 - 24 \div (-3)$
7. $0 \div 49$
8. $62 \div 0$
9. $-\dfrac{8}{15y} - \dfrac{2}{15y}$
10. $\dfrac{11}{12} - \dfrac{3}{8} + \dfrac{5}{24}$
11. $\dfrac{3a}{8} \cdot \dfrac{16}{6a^3}$
12. $-\dfrac{16}{3} \div -\dfrac{3}{12}$
13. $\dfrac{19}{-2\frac{3}{11}}$
14. $\dfrac{0.23 + 1.63}{-0.3}$
15. 10.2×4.01
16. Write 0.6% as a decimal.
17. Write 6.1 as a percent.
18. Write $\dfrac{3}{8}$ as a percent.
19. Write 0.345 as a fraction.
20. Write $-\dfrac{13}{26}$ as a decimal.
21. Round 34.8923 to the nearest tenth.

Evaluate each expression for the given replacement value(s).

22. $5(x^3 - 2)$ for $x = 2$
23. $10 - y^2$ for $y = -3$
24. $x \div y$ for $x = \dfrac{1}{2}$ and $y = 3\dfrac{7}{8}$
25. Simplify: $-(3z + 2) - 5z - 18$

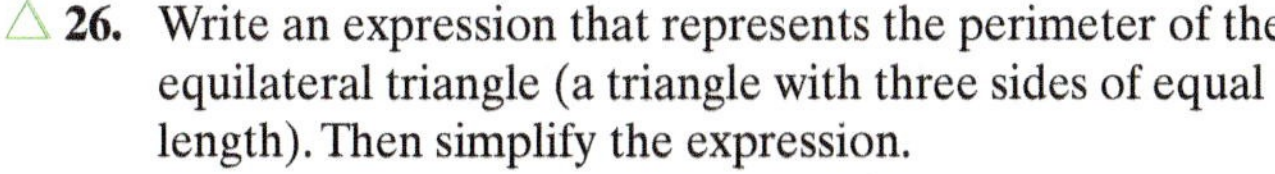

△ 26. Write an expression that represents the perimeter of the equilateral triangle (a triangle with three sides of equal length). Then simplify the expression.

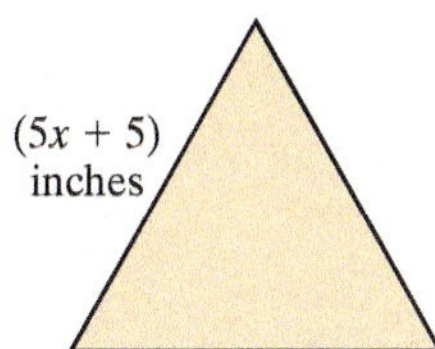

Solve each equation.

27. $\dfrac{n}{-7} = 4$
28. $-4x + 7 = 15$
29. $-4(x - 11) - 34 = 10 - 12$
30. $\dfrac{x}{5} + x = -\dfrac{24}{5}$
31. $2(x + 5.7) = 6x - 3.4$
32. $\dfrac{5}{y + 1} = \dfrac{4}{y + 2}$

△ 33. Find the perimeter and area.

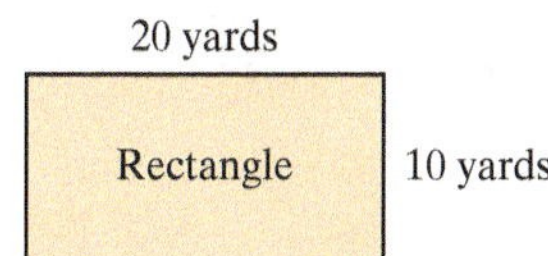

34. Find the average of −12, −13, 0, and 9.

Solve.

35. The difference of three times a number and five times the same number is 4. Find the number.
36. During a 258-mile trip, a car used $10\dfrac{3}{4}$ gallons of gas. How many miles would we expect the car to travel on 1 gallon of gas?
37. In a 10-kilometer race, there are 112 more men entered than women. Find the number of female runners if the total number of runners in the race is 600.
38. In a sample of 85 fluorescent bulbs, 3 were found to be defective. At this rate, how many defective bulbs should be found in 510 bulbs?
39. A $120 framed picture is on sale for 15% off. Find the discount and the sale price.
40. Approximate to the nearest hundredth of a centimeter the length of the missing side of a right triangle with legs of 4 centimeters each.
41. Find the complement of a 78° angle.
42. Given that $m \parallel n$, find the measures of x, y, and z.

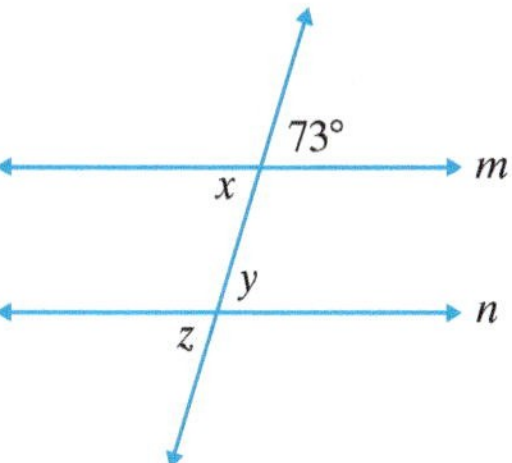

△ 43. Find the perimeter and area.

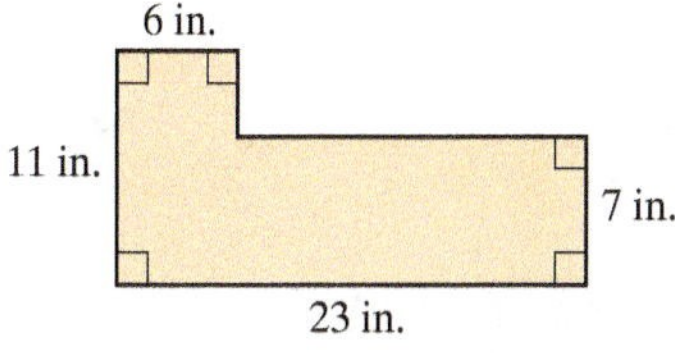

△44. Find the circumference and area. Give the exact values and then approximations using $\pi \approx 3.14$.

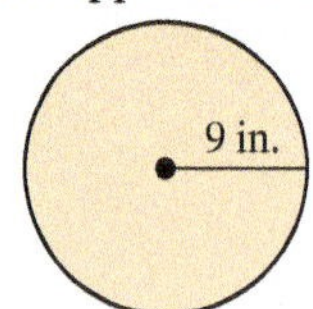

Convert.

45. $2\frac{1}{2}$ gallons to quarts

46. 2.4 kilograms to grams

Chapters 9–21

Evaluate.

47. -3^4

48. 4^{-3}

Perform the indicated operations and simplify if possible.

49. $7 + 2(5y - 3)$

50. $\begin{array}{r} 5x^3 + x^2 + 5x - 2 \\ -(8x^3 - 4x^2 + x - 7) \\ \hline \end{array}$

51. $(4x - 2)^2$

52. $(3x + 7)(x^2 + 5x + 2)$

Factor.

53. $6t^2 - t - 5$

54. $180 - 5x^2$

55. $3a^2 + 3ab - 7a - 7b$

56. $8y^3 - 64$

Simplify. Write answers with positive exponents only.

57. $\left(\dfrac{x^2y^3}{x^3y^{-4}}\right)^2$

58. $\dfrac{5 - \frac{1}{y^2}}{\frac{1}{y} + \frac{2}{y^2}}$

Perform the indicated operations and simplify if possible.

59. $\dfrac{x^2 - 9}{x^2 - 3x} \div \dfrac{xy + 5x + 3y + 15}{2x + 10}$

60. $\dfrac{5a}{a^2 - a - 6} - \dfrac{2}{a - 3}$

61. $(x^2 + 7x + 10) \div (x + 5)$
(Long division)

Solve each equation or inequality.

62. $4(n - 5) = -(4 - 2n)$

63. $(3x - 5)(x + 2) = -6$

64. $-5(x - 1) + 6 \leq -3(x + 4) + 1$

65. $y^2 - 3y = 5$

66. $\dfrac{4}{y} - \dfrac{5}{3} = -\dfrac{1}{5}$

67. $\dfrac{a}{a - 3} = \dfrac{3}{a - 3} - \dfrac{3}{2}$

68. $x = \sqrt{x - 2} + 2$

69. $|x - 5| - 4 < -2$

Graph the following.

70. $5x - 7y = 10$

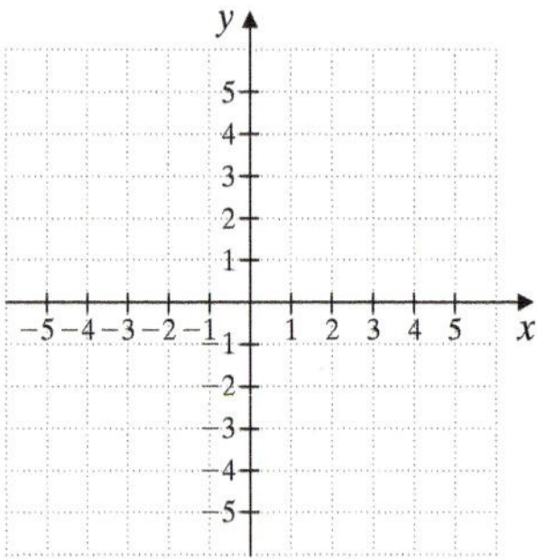

71. $y > -4x$

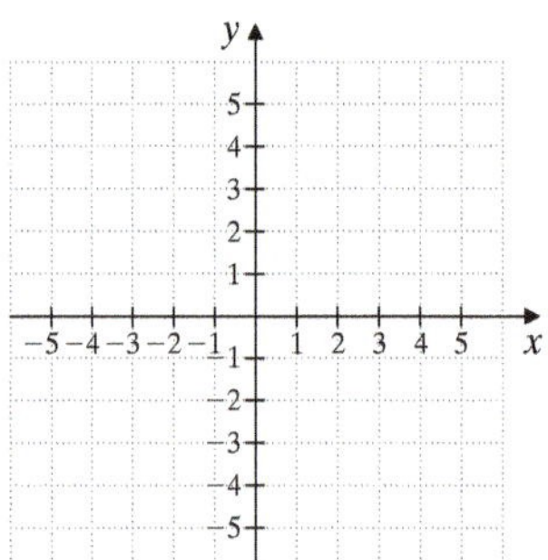

Find the slope of each line.

72. through $(6, -5)$ and $(-1, 2)$

73. $-3x + y = 5$

Write equations of the following lines. Write each equation in standard form.

74. through $(2, -5)$ and $(1, 3)$

75. slope $\frac{1}{8}$; y-intercept $(0, 12)$

Solve each system of equations.

76. $\begin{cases} 3x - 2y = -14 \\ x + 3y = -1 \end{cases}$

77. $\begin{cases} 4x - 6y = 7 \\ -2x + 3y = 0 \end{cases}$

Answer the questions about functions.

78. If $h(x) = x^3 - x$, find

a. $h(-1)$

b. $h(0)$

c. $h(4)$

79. Identify the x- and y-intercepts. Then find the domain and range of the function graphed.

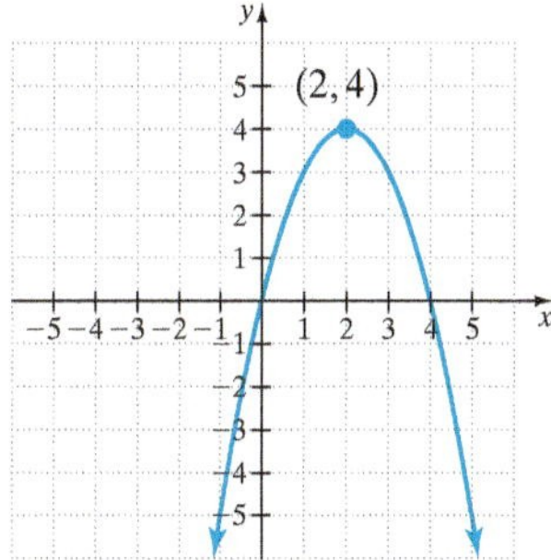

Simplify.

80. $\sqrt{216}$

81. $\left(\frac{8x^3}{27}\right)^{2/3}$

Perform the indicated operations and simplify if possible.

82. $\sqrt{125x^3} - 3\sqrt{20x^3}$

83. $(\sqrt{5} + 5)(\sqrt{5} - 5)$

84. $\sqrt{3}(\sqrt{16} - \sqrt{2})$

Rationalize each denominator.

85. $\sqrt{\frac{9}{y}}$

86. $\frac{4 - \sqrt{x}}{4 + 2\sqrt{x}}$

Graph the quadratic equation. Label the vertex and the intercepts with their coordinates.

87. $h(x) = x^2 - 4x + 4$

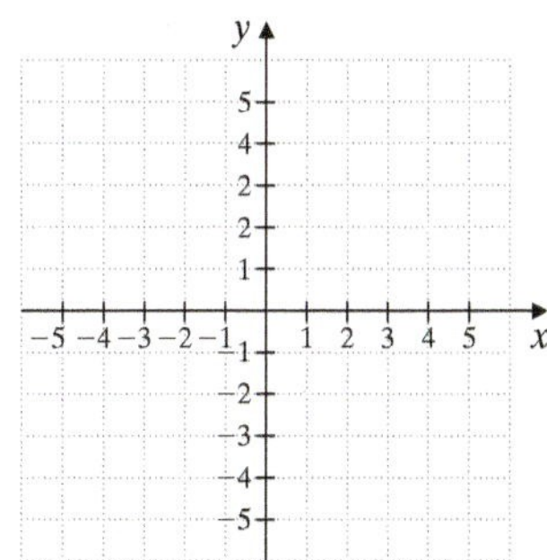

Solve each application.

88. A number plus five times its reciprocal is equal to six. Find the number.

89. A number increased by two-thirds of the number is 35. Find the number.

90. California has more public libraries than any other state. It has 387 more public libraries than Ohio. If the total number of public libraries for these states is 1827, find the number of public libraries in California and the number in Ohio. (*Source*: Institute of Museum and Library Services)

91. A pleasure boat traveling down the Red River takes the same time to go 14 miles upstream as it takes to go 16 miles downstream. If the current of the river is 2 miles per hour, find the speed of the boat in still water.

Solve.

92. Suppose that W is inversely proportional to V. If $W = 20$ when $V = 12$ find W when $V = 15$

93. Given the diagram shown, approximate to the nearest foot how many feet of walking distance a person saves by cutting across the lawn instead of walking on the sidewalk.

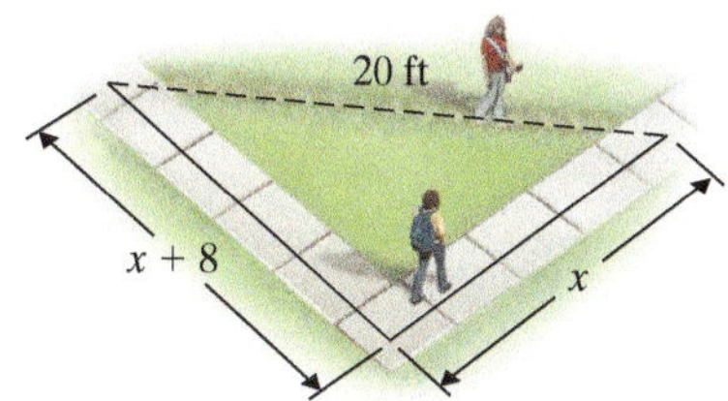

94. A stone is thrown upward from a bridge. The stone's height in feet, $s(t)$, above the water t seconds after the stone is thrown is a function given by the equation

$$s(t) = -16t^2 + 32t + 256$$

a. Find the maximum height of the stone.

b. Find the time it takes the stone to hit the water. Round the answer to two decimal places.

Perform the indicated operation and simplify. Write the result in the form $a + bi$.

95. $-\sqrt{-8}$

96. $(12 - 6i) - (12 - 3i)$

97. $(4 + 3i)^2$

98. $\frac{1 + 4i}{1 - i}$

99. If $g(x) = x - 7$ and $h(x) = x^2 - 6x + 5$, find $(g \circ h)(x)$.

100. Determine whether $f(x) = 6 - 2x$ is a one-to-one function. If it is, find its inverse.

101. Use properties of logarithms to write the expression as a single logarithm.

$$\log_5 x + 3\log_5 x - \log_5(x + 1)$$

Solve. Give exact solutions.

102. $8^{x-1} = \frac{1}{64}$

103. $3^{2x+5} = 4$ Give an exact solution and a 4-decimal-place approximation.

104. $\log_8(3x - 2) = 2$

105. $\log_4(x + 1) - \log_4(x - 2) = 3$

106. $\ln \sqrt{e} = x$

107. Graph $y = \left(\frac{1}{2}\right)^x + 1$.

108. The prairie dog population of the Grand Forks area now stands at 57,000 animals. If the population is growing at a rate of 2.6% annually, how many prairie dogs will there be in that area 5 years from now?

Sketch the graph of each equation.

109. $x^2 - y^2 = 36$

110. $16x^2 + 9y^2 = 144$

111. $x^2 + y^2 + 6x = 16$

112. Solve the system: $\begin{cases} x^2 + y^2 = 26 \\ x^2 - 2y^2 = 23 \end{cases}$

113. Find the first five terms of the sequence $a_n = \dfrac{(-1)^n}{n + 4}$.

114. Find the partial sum S_5 of the sequence $a_n = 5(2)^{n-1}$.

115. Find S_∞ of the sequence $\frac{3}{2}, -\frac{3}{4}, \frac{3}{8}, \ldots$

116. Find $\sum_{i=1}^{4} i(i - 2)$.

117. Expand: $(2x + y)^5$

Answers to Selected Exercises

Chapter 1 The Whole Numbers

Section 1.2

Vocabulary, Readiness & Video Check **1.** whole **3.** words **5.** period **7.** hundreds **9.** 80,000

Exercise Set 1.2 **1.** tens **3.** thousands **5.** hundred-thousands **7.** millions **9.** three hundred fifty-four **11.** eight thousand, two hundred seventy-nine **13.** twenty-six thousand, nine hundred ninety **15.** two million, three hundred eighty-eight thousand **17.** twenty-four million, three hundred fifty thousand, one hundred eighty-five **19.** three hundred twenty-one thousand, eight hundred **21.** two thousand, seven hundred twenty **23.** sixteen million, three hundred thousand **25.** fourteen thousand, four hundred thirty-three **27.** eighteen thousand, twenty-seven **29.** 6587 **31.** 59,800 **33.** 13,601,011 **35.** 7,000,017 **37.** 260,997 **39.** 395 **41.** 2080 **43.** 200,300,000 **45.** 565 **47.** 200 + 9 **49.** 3000 + 400 + 70 **51.** 80,000 + 700 + 70 + 4 **53.** 60,000 + 6000 + 40 + 9 **55.** 30,000,000 + 9,000,000 + 600,000 + 80,000 **57.** 1786 **59.** Mount Baker **61.** Glacier Peak **63.** German shepherd **65.** Labrador retriever; seventy-five **67.** 70 pounds **69.** 9861 **71.** no; one hundred five **73.** answers may vary **75.** 1,000,000,000,000

Section 1.3

Calculator Explorations **1.** 134 **3.** 340 **5.** 2834 **7.** 770 **9.** 109 **11.** 8978

Vocabulary, Readiness & Video Check **1.** number **3.** 0 **5.** minuend; subtrahend; difference **7.** order; commutative **9.** place; right; left **11.** triangle; 3

Exercise Set 1.3 **1.** 36 **3.** 292 **5.** 49 **7.** 5399 **9.** 209,078 **11.** 25 **13.** 212 **15.** 11,926 **17.** 16,717 **19.** 35,901 **21.** 632,389 **23.** 600 **25.** 25 **27.** 288 **29.** 168 **31.** 5723 **33.** 504 **35.** 79 **37.** 32,711 **39.** 5041 **41.** 31,213 **43.** 1034 **45.** 9 **47.** 8518 **49.** 22,876 **51.** 25 ft **53.** 24 in. **55.** 29 in. **57.** 44 m **59.** 2093 **61.** 266 **63.** 20 **65.** 544 **67.** 72 **69.** 88 **71.** 3170 thousand **73.** \$619 **75.** 264,000 sq mi **77.** 283,000 sq mi **79.** 340 ft **81.** 264 pages **83.** 31,300,000 **85.** 100 dB **87.** 58 dB **89.** 3444 **91.** 124 ft **93.** California **95.** 529 stores **97.** Pennsylvania and New York **99.** 5894 mi **101.** minuend: 48; subtrahend: 1 **103.** minuend: 70; subtrahend: 7 **105.** answers may vary **107.** correct **109.** incorrect; 530 **111.** incorrect: 685 **113.** correct **115.** $\begin{array}{r} 526\underline{9} \\ -\underline{2385} \\ 288\underline{4} \end{array}$ **117.** answers may vary **119.** no; 1089 more pages

Section 1.4

Vocabulary, Readiness & Video Check **1.** graph **3.** 70; 60 **5.** 3 is the place we're rounding to (tens), and the digit to the right of this place is 5 or greater, so we need to add 1 to the 3. **7.** Each circled digit is to the right of the place value being rounded to and is used to determine whether or not we add 1 to the digit in the place value being rounded to.

Exercise Set 1.4 **1.** 420 **3.** 640 **5.** 2800 **7.** 500 **9.** 21,000 **11.** 34,000 **13.** 328,500 **15.** 36,000 **17.** 39,990 **19.** 30,000,000 **21.** 5280; 5300; 5000 **23.** 9440; 9400; 9000 **25.** 14,880; 14,900; 15,000 **27.** 311,000 miles **29.** 60,100 days **31.** \$190,000,000,000 **33.** \$3,200,000 **35.** US: 332,000,000; India: 894,000,000 **37.** 130 **39.** 80 **41.** 5700 **43.** 300 **45.** 11,400 **47.** incorrect **49.** correct **51.** correct **53.** \$3400 **55.** 900 mi **57.** 6000 ft **59.** Springfield is larger by approximately 40,000. **61.** The increase was 55,000. **63.** 66,700,000; 67,000,000; 70,000,000 **65.** 57,800,000; 58,000,000; 60,000,000 **67.** 5723, for example **69.** **a.** 8550 **b.** 8649 **71.** answers may vary **73.** 140 m

Section 1.5

Calculator Explorations **1.** 3456 **3.** 15,322 **5.** 272,291

Vocabulary, Readiness & Video Check **1.** 0 **3.** product; factor **5.** grouping; associative **7.** length **9.** distributive **11.** Area is measured in square units, and here we have meters times meters, or square meters; the correct answer is 63 *square* meters, or the correct units are square meters.

Exercise Set 1.5 **1.** 24 **3.** 0 **5.** 0 **7.** 87 **9.** $6 \cdot 3 + 6 \cdot 8$ **11.** $4 \cdot 3 + 4 \cdot 9$ **13.** $20 \cdot 14 + 20 \cdot 6$ **15.** 512 **17.** 3678 **19.** 1662 **21.** 6444 **23.** 1157 **25.** 24,418 **27.** 24,786 **29.** 15,600 **31.** 0 **33.** 6400 **35.** 48,126 **37.** 142,506 **39.** 2,369,826 **41.** 64,790 **43.** 3,949,935 **45.** area: 63 sq m; perimeter: 32 m **47.** area: 680 sq ft; perimeter: 114 ft **49.** 240,000 **51.** 300,000 **53.** c **55.** c **57.** 880 **59.** 4200 **61.** 4480 **63.** 375 cal **65.** \$3290 **67.** **a.** 20 **b.** 100 **c.** 2000 lb **69.** 8800 sq ft **71.** 56,000 sq ft **73.** 5828 pixels **75.** 2100 characters **77.** 1280 cal **79.** \$10, \$60; \$10, \$200; \$12, \$36; \$12, \$36: total cost \$372 **81.** 1,440,000 tea bags

83. 135 **85.** 2144 **87.** 23 **89.** 15 **91.** $5 \cdot 6$ or $6 \cdot 5$ **93. a.** $5 + 5 + 5$ or $3 + 3 + 3 + 3 + 3$ **b.** answers may vary

95.
```
   203
 ×  14
   812
  2030
  2842
```
97.
```
   42
  ×93
```
99. answers may vary **101.** 506 windows

Section 1.6

Calculator Explorations **1.** 53 **3.** 62 **5.** 261 **7.** 0

Vocabulary, Readiness & Video Check **1.** quotient; dividend; divisor **3.** 1 **5.** undefined **7.** 0 **9.** $202 \cdot 102 + 15 = 20{,}619$ **11.** addition and division

Exercise Set 1.6 **1.** 6 **3.** 12 **5.** 0 **7.** 31 **9.** 1 **11.** 8 **13.** undefined **15.** 1 **17.** 0 **19.** 9 **21.** 29 **23.** 74 **25.** 338 **27.** undefined **29.** 9 **31.** 25 **33.** 68 R 3 **35.** 236 R 5 **37.** 38 R 1 **39.** 326 R 4 **41.** 13 **43.** 49 **45.** 97 R 8 **47.** 209 R 11 **49.** 506 **51.** 202 R 7 **53.** 54 **55.** 99 R 100 **57.** 202 R 15 **59.** 579 R 72 **61.** 17 **63.** 511 R 3 **65.** 2132 R 32 **67.** 6080 **69.** 23 R 2 **71.** 5 R 25 **73.** 20 R 2 **75.** 33 students **77.** 165 lb **79.** 310 yd **81.** 89 bridges **83.** 11 light poles **85.** 5 mi **87.** 1760 yd **89.** 20 **91.** 387 **93.** 79 **95.** 74° **97.** 9278 **99.** 15,288 **101.** 679 **103.** undefined **105.** 9 R 12 **107.** c **109.** b **111.** $180,845,200 **113.** increase; answers may vary **115.** no; answers may vary **117.** 12 ft **119.** answers may vary **121.** 5 R 1

Integrated Review **1.** 194 **2.** 6555 **3.** 4524 **4.** 562 **5.** 67 **6.** undefined **7.** 1 **8.** 5 **9.** 0 **10.** 0 **11.** 0 **12.** 3 **13.** 63 **14.** 9 **15.** 138 **16.** 276 **17.** 1169 **18.** 9826 **19.** 182 R 4 **20.** 79,317 **21.** 1099 R 2 **22.** 111 R 1 **23.** 663 R 24 **24.** 1076 R 60 **25.** 1037 **26.** 9899 **27.** 30,603 **28.** 47,500 **29.** 71 **30.** 558 **31.** 6 R 8 **32.** 53 **33.** 183 **34.** 231 **35.** 9740; 9700; 10,000 **36.** 1430; 1400; 1000 **37.** 20,800; 20,800; 21,000 **38.** 432,200; 432,200; 432,000 **39.** perimeter: 24 ft; area: 36 sq ft **40.** perimeter: 42 in.; area: 98 sq in. **41.** 28 mi **42.** 26 m **43.** 24 **44.** 124 **45.** Lake Pontchartrain Bridge; 2175 ft **46.** $5562

Section 1.7

Calculator Explorations **1.** 4096 **3.** 3125 **5.** 2048 **7.** 2526 **9.** 4295 **11.** 8

Vocabulary, Readiness & Video Check **1.** base; exponent **3.** addition **5.** division **7.** 1 **9.** The area of a rectangle is length · width. A square is a special rectangle where length = width. Thus, the area of a square is side · side or $(\text{side})^2$.

Exercise Set 1.7 **1.** 4^3 **3.** 7^6 **5.** 12^3 **7.** $6^2 \cdot 5^3$ **9.** $9 \cdot 8^2$ **11.** $3 \cdot 2^4$ **13.** $3 \cdot 2^4 \cdot 5^5$ **15.** 64 **17.** 125 **19.** 32 **21.** 1 **23.** 7 **25.** 128 **27.** 256 **29.** 256 **31.** 729 **33.** 144 **35.** 100 **37.** 20 **39.** 729 **41.** 192 **43.** 162 **45.** 21 **47.** 7 **49.** 5 **51.** 16 **53.** 46 **55.** 8 **57.** 64 **59.** 83 **61.** 2 **63.** 48 **65.** 4 **67.** undefined **69.** 59 **71.** 52 **73.** 44 **75.** 12 **77.** 21 **79.** 3 **81.** 43 **83.** 8 **85.** 16 **87.** area: 49 sq m; perimeter: 28 m **89.** area: 529 sq mi; perimeter: 92 mi **91.** true **93.** false **95.** $(2 + 3) \cdot 6 - 2$ **97.** $24 \div (3 \cdot 2) + 2 \cdot 5$ **99.** 1260 ft **101.** 6,384,814 **103.** answers may vary

Section 1.8

Vocabulary, Readiness & Video Check **1.** expression **3.** expression; variables **5.** equation **7.** multiplication **9.** decreased by

Exercise Set 1.8 **1.** 28; 14; 147; 3 **3.** 152; 152; 0; undefined **5.** 57; 55; 56; 56 **7.** 9 **9.** 8 **11.** 6 **13.** 5 **15.** 117 **17.** 94 **19.** 5 **21.** 34 **23.** 20 **25.** 4 **27.** 4 **29.** 0 **31.** 33 **33.** 125 **35.** 121 **37.** 100 **39.** 60 **41.** 4 **43.** 16; 64; 144; 256 **45.** yes **47.** no **49.** no **51.** yes **53.** no **55.** yes **57.** 12 **59.** 6 **61.** 4 **63.** none **65.** 11 **67.** $x + 8$ **69.** $x + 8$ **71.** $20 - x$ **73.** $512x$ **75.** $8 \div x$ or $\frac{8}{x}$ **77.** $5x + (17 + x)$ **79.** $5x$ **81.** $11 - x$ **83.** $x - 5$ **85.** $6 \div x$ or $\frac{6}{x}$ **87.** $50 - 8x$ **89.** 274,657 **91.** 777 **93.** $5x$; answers may vary **95.** As t gets larger $16t^2$ gets larger.

Chapter 1 Vocabulary Check **1.** whole numbers **2.** perimeter **3.** place value **4.** exponent **5.** area **6.** digits **7.** variable **8.** equation **9.** expression **10.** solution **11.** set **12.** sum **13.** divisor **14.** dividend **15.** quotient **16.** factor **17.** product **18.** minuend **19.** subtrahend **20.** difference **21.** addend

Chapter 1 Review **1.** tens **3.** seven thousand, six hundred forty **5.** $3000 + 100 + 50 + 8$ **7.** 81,900 **9.** 467,000,000 **11.** Oceania/Australia **13.** 67 **15.** 65 **17.** 449 **19.** 3914 **21.** 4211 **23.** 1334 **25.** 17,897 **27.** 7523 mi **29.** 216 ft **31.** 82 million or 82,000,000 **33.** May **35.** $110 **37.** 40 **39.** 880 **41.** 3800 **43.** 40,000,000 **45.** 7300 **47.** 2700 mi **49.** 2208 **51.** 2280 **53.** 560 **55.** 0 **57.** 16,994 **59.** 113,634 **61.** 411,426 **63.** 1500 **65.** $898 **67.** 91 sq mi **69.** 7 **71.** 5 R 2 **73.** undefined **75.** 33 R 2 **77.** 24 R 2 **79.** 506 R 10 **81.** 2793 R 140 **83.** 18 R 2 **85.** 27 boxes **87.** 51 **89.** 64 **91.** 405 **93.** 16 **95.** 15 **97.** 12 **99.** 42 **101.** 9 **103.** 6 **105.** 40 **107.** 5 **109.** 49 sq m **111.** 5 **113.** undefined **115.** 121 **117.** 4 **119.** $x - 5$ **121.** $10 \div x$ or $\frac{10}{x}$ **123.** yes **125.** no **127.** 11 **129.** 14 **131.** 417 **133.** 2196 **135.** 1101 **137.** 458 R 8 **139.** 70,848 **141.** 1644 **143.** 840 **145.** 12 **147.** no **149.** 53 full boxes with 18 left over

Chapter 1 Test **1.** eighty-two thousand, four hundred twenty-six **2.** 402,550 **3.** 141 **4.** 113 **5.** 14,880 **6.** 766 R 42 **7.** 200 **8.** 98 **9.** 0 **10.** undefined **11.** 33 **12.** 21 **13.** 48 **14.** 36 **15.** 5,698,000 **16.** 82 **17.** 52,000 **18.** 13,700 **19.** 1600 **20.** 92 **21.** 122 **22.** 1605 **23.** 7 R 2 **24.** \$17 **25.** \$126 **26.** 360 cal **27.** \$7905 **28.** 20 cm; 25 sq cm **29.** 60 yd; 200 sq yd **30.** 30 **31.** 1 **32. a.** $x \div 17$ or $\frac{x}{17}$ **b.** $2x - 20$ **33.** yes **34.** 10

Chapter 2 Integers and Introduction to Solving Equations

Section 2.1

Vocabulary, Readiness & Video Check **1.** integers **3.** inequality symbols **5.** is less than; is greater than **7.** absolute value **9.** number of feet a miner works underground **11.** negative **13.** opposite of

Exercise Set 2.1 **1.** −1235 **3.** +14,433 **5.** +120 **7.** −11,810 **9.** −3140 million **11.** −160, −147; Guillermo **13.** −2

15. 0 1 2 3 4 5 6 7 8 9 **17.** −5 −4 −3 −2 −1 0 1 2 3 4 **19.** 0 2 4 6 8 10 12 14 16 18

21. −8 −7 −6 −5 −4 −3 −2 −1 0 1 **23.** > **25.** < **27.** > **29.** < **31.** 5 **33.** 8 **35.** 0 **37.** 55 **39.** −5 **41.** 4 **43.** −23 **45.** 85 **47.** 7 **49.** −20 **51.** −3 **53.** 43 **55.** 15 **57.** 33 **59.** 6 **61.** −2 **63.** 32 **65.** −7 **67.** < **69.** < **71.** = **73.** < **75.** > **77.** < **79.** > **81.** < **83.** 31; −31 **85.** 28; 28 **87.** Caspian Sea **89.** Lake Superior **91.** iodine **93.** oxygen **95.** 13 **97.** 35 **99.** 360 **101.** $-|-8|, -|3|, 2^2, -(-5)$ **103.** $-|-6|, -|1|, |-1|, -(-6)$ **105.** $-10, -|-9|, -(-2), |-12|, 5^2$ **107.** a, d **109.** 8 **111.** false **113.** true **115.** false **117.** answers may vary **119.** no; answers may vary

Section 2.2

Calculator Explorations **1.** −159 **3.** 44 **5.** −894,855

Vocabulary, Readiness & Video Check **1.** 0 **3.** *a* **5.** Negative; the numbers have different signs and the sign of the sum is the same as the sign of the number with the larger absolute value, −6. **7.** The diver's current depth is 231 feet below the surface.

Exercise Set 2.2

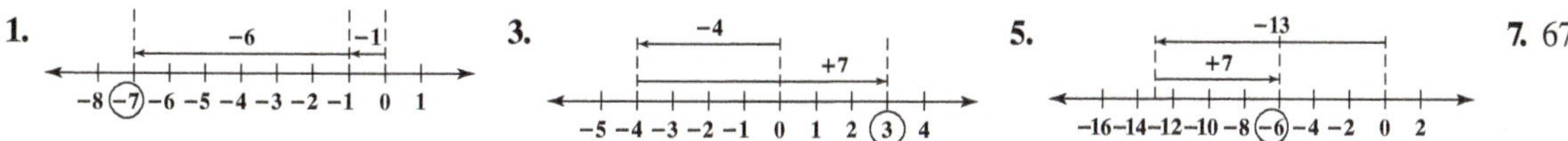

7. 67 **9.** −10 **11.** 0 **13.** 4 **15.** −6 **17.** −2 **19.** −9 **21.** −24 **23.** −840 **25.** 7 **27.** −3 **29.** −30 **31.** 40 **33.** −20 **35.** −125 **37.** −7 **39.** −246 **41.** 16 **43.** 13 **45.** −28 **47.** −11 **49.** 20 **51.** −34 **53.** −1 **55.** 0 **57.** −42 **59.** −70 **61.** 3 **63.** −21 **65.** −6 + 25; 19 **67.** −31 + (−9) + 30; −10 **69.** 0 + (−215) + (−16) = −231; 231 ft below the surface **71.** Dufner: −7; Furyk: −2 **73.** \$41,733,000,000 **75.** \$67,655,000,000 **77.** 2°C **79.** \$13,609 **81.** −2°F **83.** −7535 m **85.** 44 **87.** 141 **89.** answers may vary **91.** −3 **93.** −22 **95.** true **97.** false **99.** answers may vary

Section 2.3

Vocabulary, Readiness & Video Check **1.** b **3.** d **5.** additive inverse **7.** to follow the order of operations

Exercise Set 2.3 **1.** 0 **3.** 3 **5.** −5 **7.** 22 **9.** 3 **11.** −20 **13.** −12 **15.** −13 **17.** 508 **19.** −14 **21.** −4 **23.** −12 **25.** −25 − 17; −42 **27.** −22 − (−3); −19 **29.** 2 − (−12); 14 **31.** −56 **33.** −5 **35.** −145 **37.** −37 **39.** 3 **41.** 1 **43.** −1 **45.** −31 **47.** 44 **49.** −32 **51.** −9 **53.** 14 **55.** −11 **57.** 31 **59.** 12 **61.** 20 **63.** 12°F **65.** 4°F **67.** 263°F **69.** 14 strokes **71.** −10°C **73.** 154 ft **75.** 69 ft **77.** 652 ft **79.** 144 ft **81.** 1197°F **83.** −\$34 billion **85.** $-5 + x$ **87.** $-20 - x$ **89.** 5 **91.** 1058 **93.** answers may vary **95.** 16 **97.** −20 **99.** −4 **101.** 0 **103.** −14 **105.** false **107.** answers may vary

Section 2.4

Vocabulary, Readiness & Video Check **1.** negative **3.** positive **5.** 0 **7.** undefined **9.** multiplication **11.** The phrase "lost four yards" in the example translates to the negative number −4.

Exercise Set 2.4 **1.** 12 **3.** −36 **5.** −81 **7.** 0 **9.** 48 **11.** −12 **13.** 80 **15.** 0 **17.** −15 **19.** −9 **21.** −27 **23.** −36 **25.** −64 **27.** −8 **29.** −5 **31.** 7 **33.** 0 **35.** undefined **37.** −14 **39.** 0 **41.** −15 **43.** −63 **45.** 42 **47.** −24 **49.** 49 **51.** −5 **53.** −9 **55.** −6 **57.** 120 **59.** −1080 **61.** undefined **63.** −6 **65.** −7 **67.** 3 **69.** −1 **71.** −32 **73.** 180 **75.** 1 **77.** −30 **79.** −1104 **81.** −2870 **83.** −56 **85.** −18 **87.** 35 **89.** −1 **91.** undefined **93.** 6 **95.** 16; 4 **97.** 0; 0 **99.** −54 ÷ 9; −6 **101.** −42(−6); 252 **103.** $-71 \cdot x$ or $-71x$ **105.** $-16 - x$ **107.** $-29 + x$ **109.** $\frac{x}{-33}$ or $x \div (-33)$ **111.** $3 \cdot (-4) = -12$; a loss of 12 yd **113.** $5 \cdot (-20) = -100$; a depth of 100 feet **115.** −210°C **117.** −189°C **119.** −\$11 million per month **121. a.** −26,932 movie screens **b.** −6733 movie screens per year **123.** 109 **125.** 8 **127.** −19 **129.** −28 **131.** −8 **133.** negative **135.** $(-5)^{17}, (-2)^{17}, (-2)^{12}, (-5)^{12}$ **137.** answers may vary

Integrated Review **1.** −50; +122 or 122 **2.** 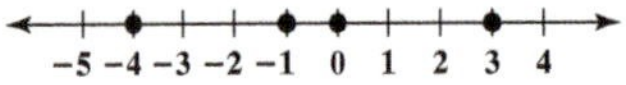 **3.** > **4.** < **5.** < **6.** > **7.** 3 **8.** 9 **9.** −4 **10.** 5 **11.** −11 **12.** 3 **13.** −64 **14.** 0 **15.** 12 **16.** −20 **17.** −48 **18.** −9 **19.** 10 **20.** −2 **21.** 106 **22.** −3 **23.** 0 **24.** 4 **25.** 42 **26.** 6 **27.** 19 **28.** −900 **29.** −12 **30.** −19 **31.** undefined **32.** 0 **33.** $-12-(-8); -4$ **34.** $-17+(-27); -44$ **35.** $-5(-25); 125$ **36.** $-100 \div (-5); 20$ **37.** $\frac{x}{-17}$ or $x \div (-17)$ **38.** $-3+x$ **39.** $x-(-18)$ **40.** $-7 \cdot x$ or $-7x$ **41.** 9 **42.** −15 **43.** 27 **44.** 33 **45.** −15 **46.** −4

Section 2.5

Calculator Explorations **1.** 48 **3.** −258

Vocabulary, Readiness & Video Check **1.** division **3.** average **5.** subtraction **7.** A fraction bar means divided by and it is a grouping symbol. **9.** Finding the average is a good application of both order of operations and adding and dividing integers.

Exercise Set 2.5 **1.** −125 **3.** −64 **5.** 32 **7.** −8 **9.** −11 **11.** −43 **13.** −8 **15.** 17 **17.** −1 **19.** 4 **21.** −3 **23.** 16 **25.** 13 **27.** −77 **29.** 80 **31.** 256 **33.** 53 **35.** 4 **37.** −64 **39.** 4 **41.** 16 **43.** −27 **45.** 34 **47.** 65 **49.** −7 **51.** 36 **53.** −117 **55.** 30 **57.** −3 **59.** −30 **61.** 1 **63.** −12 **65.** 0 **67.** −20 **69.** 9 **71.** −16 **73.** −128 **75.** 1 **77.** −50 **79.** −2 **81.** −19 **83.** 18 **85.** −1 **87.** no; answers may vary **89.** 4050 **91.** 45 **93.** 32 in. **95.** 30 ft **97.** $2 \cdot (7-5) \cdot 3$ **99.** $-6 \cdot (10-4)$ **101.** answers may vary **103.** answers may vary **105.** 20,736 **107.** 8900 **109.** 9

Section 2.6

Vocabulary, Readiness & Video Check **1.** expression **3.** equation; expression **5.** solution **7.** addition **9.** an equal sign **11.** original; true

Exercise Set 2.6 **1.** yes **3.** no **5.** yes **7.** yes **9.** 18 **11.** −12 **13.** 9 **15.** −17 **17.** 4 **19.** −4 **21.** −14 **23.** −17 **25.** 0 **27.** 1 **29.** −7 **31.** −50 **33.** −25 **35.** 36 **37.** 21 **39.** 12 **41.** −80 **43.** −2 **45.** $x-(-2)$ **47.** $-6 \cdot x$ or $-6x$ **49.** $-15+x$ **51.** $-8 \div x$ or $\frac{-8}{x}$ **53.** 41,574 **55.** −409 **57.** answers may vary **59.** answers may vary

Chapter 2 Vocabulary Check **1.** opposites **2.** absolute value **3.** integers **4.** negative **5.** positive **6.** inequality symbols **7.** solution **8.** average **9.** expression **10.** equation **11.** is less than; is greater than **12.** addition **13.** multiplication

Chapter 2 Review **1.** −1572 **3.** 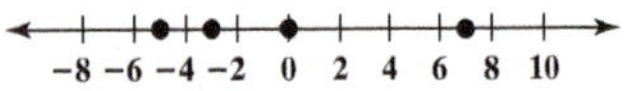 **5.** 11 **7.** −8 **9.** −16 **11.** > **13.** > **15.** 18 **17.** false **19.** true **21.** 2 **23.** −5 **25.** Elevator D **27.** 2 **29.** 4 **31.** −23 **33.** −21 **35.** 0 **37.** −151 **39.** −20°C **41.** −21 **43.** 8 **45.** −24 **47.** 20 **49.** 0 **51.** 0 **53.** −10 **55.** 692 ft **57.** −14 or 14 ft below ground **59.** true **61.** 21 **63.** −64 **65.** 25 **67.** 0 **69.** −5 **71.** 0 **73.** −20 **75.** 38 **77.** $(-5)(2) = -10$ **79.** $-1024 \div 4 = -256$ **81.** 49 **83.** 0 **85.** −16 **87.** −32 **89.** 7 **91.** 39 **93.** −2 **95.** −3 **97.** −5 **99.** −1 **101.** no **103.** −13 **105.** −3 **107.** −13 **109.** 44 **111.** −19 **113.** 6 **115.** −15 **117.** 48 **119.** 21 **121.** −27°C **123.** 13,118 ft **125.** 2 **127.** −5 **129.** −20 **131.** −21 **133.** 12 **135.** −200

Chapter 2 Test **1.** 3 **2.** −6 **3.** −100 **4.** 4 **5.** −30 **6.** 12 **7.** 65 **8.** 5 **9.** 12 **10.** −6 **11.** 50 **12.** −2 **13.** −11 **14.** −46 **15.** −117 **16.** 3456 **17.** 28 **18.** −213 **19.** −2 **20.** 2 **21.** −5 **22.** −32 **23.** −17 **24.** 1 **25.** −1 **26.** 88 ft below sea level **27.** 45 **28.** 31,642 **29.** 3820 ft below sea level **30.** −4 **31.** **a.** $17 \cdot x$ or $17x$ **b.** $20-x$ **32.** 5 **33.** −28 **34.** −20 **35.** −4

Cumulative Review Chapters 1–2 **1.** hundred-thousands; Sec. 1.2, Ex. 1 **3.** thousands; Sec. 1.2, Ex. 2 **5.** ten-millions; Sec. 1.2, Ex. 3 **7.** **a.** < **b.** > **c.** >; Sec. 2.1, Ex. 3 **9.** 39; Sec. 1.3, Ex. 3 **11.** 7321; Sec. 1.3, Ex. 6 **13.** 36,184 mi; Sec. 1.3, Ex. 11 **15.** 570; Sec. 1.4, Ex. 1 **17.** 1800; Sec. 1.4, Ex. 5 **19.** **a.** $5 \cdot 6 + 5 \cdot 5$ **b.** $20 \cdot 4 + 20 \cdot 7$ **c.** $2 \cdot 7 + 2 \cdot 9$; Sec. 1.5, Ex. 2 **21.** 78,875; Sec. 1.5, Ex. 5 **23.** **a.** 6 **b.** 8 **c.** 7; Sec. 1.6, Ex. 1 **25.** 741; Sec. 1.6, Ex. 4 **27.** 12 cards each; 10 cards left over; Sec. 1.6, Ex. 11 **29.** 81; Sec. 1.7, Ex. 5 **31.** 6; Sec. 1.7, Ex. 6 **33.** 180; Sec. 1.7, Ex. 8 **35.** 2; Sec. 1.7, Ex. 13 **37.** 14; Sec. 1.8, Ex. 1 **39.** **a.** 9 **b.** 8 **c.** 0; Sec. 2.1, Ex. 4 **41.** 23; Sec. 2.2, Ex. 7 **43.** 22; Sec. 2.3, Ex. 12 **45.** −21; Sec. 2.4, Ex. 1 **47.** 0; Sec. 2.4, Ex. 3 **49.** −16; Sec. 2.5, Ex. 8

Chapter 3 Solving Equations and Problem Solving

Section 3.1

Vocabulary, Readiness & Video Check **1.** expression; term **3.** combine like terms **5.** variable; constant **7.** associative **9.** numerical coefficient **11.** distributive property **13.** addition; multiplication; P = perimeter, A = area

Exercise Set 3.1 **1.** $8x$ **3.** $-n$ **5.** $-2c$ **7.** $-6x$ **9.** $12a-5$ **11.** $42x$ **13.** $-33y$ **15.** $72a$ **17.** $2y+6$ **19.** $3a-18$ **21.** $-12x-28$ **23.** $2x+1$ **25.** $15c+3$ **27.** $-21n+20$ **29.** $7w+15$ **31.** $11x-8$ **33.** $-2y+16$ **35.** $-2y$ **37.** $-7z$ **39.** $8d-3c$ **41.** $6y-14$ **43.** $-q$ **45.** $2x+22$ **47.** $-3x-35$ **49.** $-3z-15$ **51.** $-6x+6$ **53.** $3x-30$ **55.** $-r+8$

57. $-7n + 3$ **59.** $9z - 14$ **61.** -6 **63.** $-4x + 10$ **65.** $2x + 20$ **67.** $7a + 12$ **69.** $3y + 5$ **71.** $(14y + 22)$ m **73.** $(11a + 12)$ ft **75.** $(-25x + 55)$ in. **77.** $36y$ sq in. **79.** $(32x - 64)$ sq km **81.** $(60y + 20)$ sq mi **83.** 4700 sq ft **85.** 64 ft **87.** $(3x + 6)$ ft **89.** -3 **91.** 8 **93.** 0 **95.** incorrect; $15x - 10$ **97.** incorrect; $2xy$ **99.** correct **101.** incorrect; $4y - 12 + 11$ or $4y - 1$ **103.** distributive **105.** associative **107.** $(20x + 16)$ sq mi **109.** $4824q + 12{,}274$ **111.** answers may vary **113.** answers may vary

Section 3.2

Vocabulary, Readiness & Video Check **1.** equivalent **3.** simplifying **5.** addition **7.** Simplify the left side of the equation by combining like terms. **9.** addition property of equality.

Exercise Set 3.2 **1.** 6 **3.** 8 **5.** -4 **7.** 6 **9.** -1 **11.** 18 **13.** -8 **15.** -50 **17.** 3 **19.** -22 **21.** -6 **23.** 24 **25.** -30 **27.** 12 **29.** 4 **31.** 3 **33.** -3 **35.** 1 **37.** 5 **39.** -11 **41.** -4 **43.** -3 **45.** -1 **47.** 0 **49.** 3 **51.** -6 **53.** -35 **55.** 10 **57.** -2 **59.** 0 **61.** 28 **63.** -5 **65.** -28 **67.** -28 **69.** 5 **71.** -4 **73.** $-7 + x$ **75.** $x - 11$ **77.** $-13x$ **79.** $\frac{x}{-12}$ or $-\frac{x}{12}$ **81.** $-11x + 5$ **83.** $-10 - 7x$ **85.** $4x + 7$ **87.** $2x - 17$ **89.** $-6(x + 15)$ **91.** $\frac{45}{-5x}$ or $-\frac{45}{5x}$ **93.** $\frac{17}{x} + (-15)$ or $\frac{17}{x} - 15$ **95.** California **97.** \$88 billion **99.** answers may vary **101.** no; answers may vary **103.** answers may vary **105.** 67,896 **107.** -48 **109.** 42

Integrated Review **1.** expression **2.** equation **3.** equation **4.** expression **5.** simplify **6.** solve **7.** $8x$ **8.** $-4y$ **9.** $-2a - 2$ **10.** $5a - 26$ **11.** $-8x - 14$ **12.** $-6x + 30$ **13.** $5y - 10$ **14.** $15x - 31$ **15.** $(12x - 6)$ sq m **16.** $(2x + 9)$ ft **17.** -4 **18.** -3 **19.** -10 **20.** 6 **21.** -15 **22.** -120 **23.** -5 **24.** -13 **25.** -24 **26.** -54 **27.** 12 **28.** -42 **29.** 2 **30.** 2 **31.** -3 **32.** 5 **33.** -5 **34.** 6 **35.** $x - 10$ **36.** $-20 + x$ **37.** $10x$ **38.** $\frac{10}{x}$ **39.** $-2x + 5$ **40.** $-4(x - 1)$

Section 3.3

Calculator Explorations **1.** yes **3.** no **5.** yes

Vocabulary, Readiness & Video Check **1.** $3x - 9 + x - 16$; $5(2x + 6) - 1 = 39$ **3.** addition **5.** distributive **7.** the addition property of equality; to make sure we get an equivalent equation **9.** gives; amounts to

Exercise Set 3.3 **1.** -12 **3.** -3 **5.** 1 **7.** -45 **9.** -9 **11.** 6 **13.** 8 **15.** 5 **17.** 0 **19.** -5 **21.** -22 **23.** 6 **25.** -11 **27.** -7 **29.** -5 **31.** 270 **33.** 5 **35.** 3 **37.** 9 **39.** -6 **41.** 11 **43.** 3 **45.** 4 **47.** -4 **49.** 3 **51.** -1 **53.** -4 **55.** -5 **57.** 0 **59.** 4 **61.** 1 **63.** -30 **65.** $-42 + 16 = -26$ **67.** $-5(-29) = 145$ **69.** $3(-14 - 2) = -48$ **71.** $\frac{100}{2(50)} = 1$ **73.** 122,000,000 returns **75.** 37,000,000 returns **77.** 33 **79.** -37 **81.** b **83.** a **85.** $6x - 10 = 5x - 7$; $6x = 5x + 3$; $x = 3$ **87.** 0 **89.** -4 **91.** no; answers may vary

Section 3.4

Vocabulary, Readiness & Video Check **1.** The phrase is "a number subtracted from -20" so -20 goes first and we subtract the number from that. **3.** The original application asks for the fastest speeds of a pheasant and a falcon. The value of x is the speed in mph for a pheasant, so the falcon's speed still needs to be found.

Exercise Set 3.4 **1.** $-5 + x = -7$ **3.** $3x = 27$ **5.** $-20 - x = 104$ **7.** $2x = 108$ **9.** $5(-3 + x) = -20$ **11.** $9 + 3x = 33$; 8 **13.** $3 + 4 + x = 16$; 9 **15.** $x - 3 = \frac{10}{5}$; 5 **17.** $30 - x = 3(x + 6)$; 3 **19.** $5x - 40 = x + 8$; 12 **21.** $3(x - 5) = \frac{108}{12}$; 8 **23.** $4x = 30 - 2x$; 5 **25.** California: 55 votes; Florida: 27 votes **27.** falcon: 185 mph; pheasant: 37 mph **29.** *The New York Times*: 1439 thousand; *The Los Angeles Times*: 1055 thousand **31.** Coca-Cola: \$72 billion; Disney: \$29 billion **33.** Xbox: \$330; games \$110 **35.** 5470 miles **37.** Beaver Stadium: 107,282; Michigan Stadium: 106,201 **39.** China: 140 million; Spain: 70 million **41.** US: 8086 cars per day; Germany: 16,172 cars per day **43.** \$225 **45.** 93 **47.** USA: 310,640 thousand; China: 195,140 thousand **49.** 590 **51.** 1000 **53.** 3000 **55.** yes; answers may vary **57.** \$216,200 **59.** \$549

Chapter 3 Vocabulary Check **1.** simplified; combined. **2.** like **3.** variable **4.** algebraic expression **5.** terms **6.** numerical coefficient **7.** evaluating the expression **8.** constant **9.** equation **10.** solution **11.** distributive **12.** multiplication **13.** addition

Chapter 3 Review **1.** $10y - 15$ **3.** $-6a - 7$ **5.** $2x + 10$ **7.** $11x - 12$ **9.** $-5a + 4$ **11.** $16y - 5$ **13.** $(4x + 6)$ yd **15.** $(6x - 3)$ sq yd **17.** -2 **19.** -7 **21.** -12 **23.** -6 **25.** -25 **27.** -2 **29.** -8 **31.** 5 **33.** -63 **35.** 5 **37.** -6 **39.** $-5x$ **41.** $-5 + x$ **43.** $-5x - 50$ **45.** $\frac{70}{x + 6}$ **47.** 21 **49.** 2 **51.** 11 **53.** -15 **55.** -2 **57.** -1 **59.** 0 **61.** $20 - (-8) = 28$ **63.** $\frac{-75}{5 + 20} = -3$ **65.** $2x - 8 = 40$ **67.** $\frac{x}{2} - 12 = 10$ **69.** 5 **71.** 2386 votes **73.** $-11x$ **75.** $22x - 19$ **77.** -1 **79.** 13

81. -22 **83.** -15 **85.** -5 **87.** 2 **89.** 0 **91.** 5 **93.** Hawaii: 4371 mi; Delaware: 6302 mi

Chapter 3 Test **1.** $-5x + 5$ **2.** $-6y - 14$ **3.** $-8z - 20$ **4.** $(15x + 15)$ in. **5.** $(12x - 4)$ sq m **6.** -6 **7.** -6 **8.** 24 **9.** -2 **10.** 6 **11.** 3 **12.** -2 **13.** 0 **14.** 4 **15.** $-23 + x$ **16.** $-2 - 3x$ **17.** $2 \cdot 5 + (-15) = -5$ **18.** $3x + 6 = -30$ **19.** -2 **20.** 8 free throws **21.** 244 women

Cumulative Review Chapters 1–3 **1.** three hundred eight million, sixty-three thousand, five hundred fifty-seven; Sec. 1.2, Ex. 7 **3.** 13 in.; Sec. 1.3, Ex. 9 **5.** 726; Sec. 1.3, Ex. 8 **7.** 249,000; Sec. 1.4, Ex. 3 **9.** 200; Sec. 1.5, Ex. 3a **11.** 208; Sec. 1.6, Ex. 5 **13.** 7; Sec. 1.7, Ex. 9 **15.** 26; Sec. 1.8, Ex. 4 **17.** 20 is a solution; 26 and 40 are not solutions.; Sec. 1.8, Ex. 7 **19.** 3; Sec. 2.2, Ex. 1 **21.** -25; Sec. 2.2, Ex. 5 **23.** 23; Sec. 2.2, Ex. 7 **25.** -14; Sec. 2.3, Ex. 2 **27.** 11; Sec. 2.3, Ex. 3 **29.** -4; Sec. 2.3, Ex. 4 **31.** -2; Sec. 2.4, Ex. 10 **33.** 5; Sec. 2.4, Ex. 11 **35.** -16; Sec. 2.4, Ex. 12 **37.** 9; Sec. 2.5, Ex. 1 **39.** -9; Sec. 2.5, Ex. 2 **41.** $6y + 2$; Sec. 3.1, Ex. 2 **43.** not a solution; Sec. 2.6, Ex. 1 **45.** 3; Sec. 2.6, Ex. 7 **47.** 12; Sec. 3.2, Ex. 7 **49.** software: \$420; computer system: \$1680; Sec. 3.4, Ex. 4

Chapter 4 Fractions and Mixed Numbers

Section 4.1

Vocabulary, Readiness & Video Check **1.** fraction; denominator; numerator **3.** improper; proper; mixed number **5.** equal; improper **7.** how many equal parts to divide each whole number into **9.** addition; +

Exercise Set 4.1 **1.** numerator: 1; denominator: 2; proper **3.** numerator: 10; denominator: 3; improper **5.** numerator: 15; denominator: 15; improper **7.** $\frac{1}{3}$ **9. a.** $\frac{11}{4}$ **b.** $2\frac{3}{4}$ **11. a.** $\frac{23}{6}$ **b.** $3\frac{5}{6}$ **13.** $\frac{7}{12}$ **15.** $\frac{3}{7}$ **17.** $\frac{4}{9}$ **19. a.** $\frac{4}{3}$ **b.** $1\frac{1}{3}$ **21. a.** $\frac{11}{2}$ **b.** $5\frac{1}{2}$ **23.** $\frac{1}{6}$ **25.** $\frac{5}{8}$ **27.** **29.** **31.** **33.** $\frac{42}{131}$ of the students **35. a.** 89 **b.** $\frac{89}{131}$ of the students **37.** $\frac{7}{44}$ of the presidents **39.** $\frac{15}{28}$ of the tropical storms **41.** $\frac{11}{31}$ of the month **43.** $\frac{10}{31}$ of the class **45. a.** $\frac{33}{50}$ of the states **b.** 17 states **c.** $\frac{17}{50}$ of the states **47. a.** $\frac{21}{50}$ of the marbles **b.** 29 **c.** $\frac{29}{50}$ of the marbles

49. $\frac{1}{4}$ (0, 1) **51.** $\frac{4}{7}$ (0, 1) **53.** $\frac{8}{5}$ (0, 1, 2) **55.** $\frac{7}{3}$ (0, 1, 2, 3)

57. 1 **59.** -5 **61.** 0 **63.** 1 **65.** undefined **67.** 3 **69.** $\frac{7}{3}$ **71.** $\frac{18}{5}$ **73.** $\frac{53}{8}$ **75.** $\frac{83}{7}$ **77.** $\frac{187}{20}$ **79.** $\frac{500}{3}$ **81.** $3\frac{2}{5}$ **83.** $4\frac{5}{8}$ **85.** $3\frac{2}{15}$ **87.** 15 **89.** $1\frac{7}{175}$ **91.** $6\frac{65}{112}$ **93.** 9 **95.** 125 **97.** $\frac{-11}{2}; \frac{11}{-2}$ **99.** $\frac{13}{-15}; -\frac{13}{15}$ **101.** answers may vary **103.** **105.** 7 **107.** $\frac{640}{5067}$ of the stores **109.** $\frac{1500}{1580}$ of the affiliates

Section 4.2

Calculator Explorations **1.** $\frac{4}{7}$ **3.** $\frac{20}{27}$ **5.** $\frac{8}{15}$ **7.** $\frac{2}{9}$

Vocabulary, Readiness & Video Check **1.** prime factorization **3.** prime **5.** equivalent **7.** Check that every factor is a prime number and check that the product of the factors is the original number **9.** You can simplify the two fractions and then compare them. $\frac{3}{9}$ and $\frac{6}{18}$ both simplify to $\frac{1}{3}$ so the original fractions are equivalent.

Exercise Set 4.2 **1.** $2^2 \cdot 5$ **3.** $2^4 \cdot 3$ **5.** 3^4 **7.** $2 \cdot 3^4$ **9.** $2 \cdot 5 \cdot 11$ **11.** $5 \cdot 17$ **13.** $2^4 \cdot 3 \cdot 5$ **15.** $2^2 \cdot 3^2 \cdot 23$ **17.** $\frac{1}{4}$ **19.** $\frac{2x}{21}$ **21.** $\frac{7}{8}$ **23.** $\frac{2}{3}$ **25.** $\frac{7}{10}$ **27.** $-\frac{7}{9}$ **29.** $\frac{5x}{6}$ **31.** $\frac{27}{64}$ **33.** $\frac{5x}{8}$ **35.** $-\frac{5}{8}$ **37.** $\frac{3x^2y}{2}$ **39.** $\frac{3}{4}$ **41.** $\frac{5}{8z}$ **43.** $\frac{3}{14}$ **45.** $-\frac{11}{17y}$ **47.** $\frac{7}{8}$ **49.** $14a^2$ **51.** equivalent **53.** not equivalent **55.** equivalent **57.** equivalent **59.** not equivalent **61.** not equivalent **63.** $\frac{1}{4}$ of a shift **65.** $\frac{1}{2}$ mi **67. a.** $\frac{8}{25}$ **b.** 34 states **c.** $\frac{17}{25}$ **69.** $\frac{5}{12}$ of the wall **71. a.** 8 **b.** $\frac{4}{25}$ **73.** $\frac{13}{232}$ of U.S astronauts **75.** -3 **77.** -14 **79.** answers may vary **81.** $\frac{3}{5}$ **83.** $\frac{9}{25}$ **85.** $\frac{1}{25}$ **87.** $2^2 \cdot 3^5 \cdot 5 \cdot 7$ **89.** answers may vary **91.** no; answers may vary **93.** $\frac{3}{50}$ **95.** answers may vary **97.** $\frac{2}{25}$ **99.** answers may vary **101.** 786, 222, 900, 1470 **103.** 6; answers may vary

Section 4.3

Vocabulary, Readiness & Video Check **1.** $\frac{a \cdot c}{b \cdot d}$ **3.** $\frac{2 \cdot 2 \cdot 2}{7}; \frac{2}{7} \cdot \frac{2}{7} \cdot \frac{2}{7}$ **5.** $\frac{a \cdot d}{b \cdot c}$ **7.** We have a negative fraction times a positive fraction and a negative number times a positive number is a negative number. **9.** numerator; denominator **11.** radius $= \frac{1}{2} \cdot$ diameter

Exercise Set 4.3 **1.** $\frac{18}{77}$ **3.** $-\frac{5}{28}$ **5.** $\frac{1}{15}$ **7.** $\frac{18x}{55}$ **9.** $\frac{3a^2}{4}$ **11.** $\frac{x^2}{y}$ **13.** 0 **15.** $-\frac{17}{25}$ **17.** $\frac{1}{56}$ **19.** $\frac{1}{125}$ **21.** $\frac{4}{9}$ **23.** $-\frac{4}{27}$ **25.** $\frac{4}{5}$ **27.** $-\frac{1}{6}$ **29.** $-\frac{16}{9x}$ **31.** $\frac{121y}{60}$ **33.** $-\frac{1}{6}$ **35.** $\frac{x}{25}$ **37.** $\frac{10}{27}$ **39.** $\frac{18x^2}{35}$ **41.** $\frac{10}{9}$ **43.** $-\frac{1}{4}$ **45.** $\frac{9}{16}$ **47.** xy^2 **49.** $\frac{77}{2}$ **51.** $-\frac{36}{x}$ **53.** $\frac{3}{49}$ **55.** $-\frac{19y}{7}$ **57.** $\frac{4}{11}$ **59.** $\frac{8}{3}$ **61.** $\frac{15x}{4}$ **63.** $\frac{8}{9}$ **65.** $\frac{1}{60}$ **67.** b **69.** $\frac{2}{5}$ **71.** $-\frac{5}{3}$ **73. a.** $\frac{1}{3}$ **b.** $\frac{12}{25}$ **75. a.** $-\frac{36}{55}$ **b.** $-\frac{44}{45}$ **77.** yes **79.** no **81.** 50 **83.** 20 **85.** 112 **87.** 108 million **89.** 868 mi **91.** $\frac{3}{16}$ in. **93.** \$1838 **95.** 50 libraries **97.** $\frac{1}{14}$ sq ft **99.** 3840 mi **101.** 2400 mi **103.** 201 **105.** 196 **107.** answers may vary **109.** 5 **111.** 39,239,250 people **113.** 8505

Section 4.4

Vocabulary, Readiness & Video Check **1.** like; unlike **3.** $-\frac{a}{b}$ **5.** least common denominator (LCD) **7.** numerators; denominator **9.** $P = \frac{5}{12} + \frac{7}{12} + \frac{5}{12} + \frac{7}{12}$; 2 meters **11.** Multiplying by 1 does not change the value of the fraction.

Exercise Set 4.4 **1.** $\frac{7}{11}$ **3.** $\frac{2}{3}$ **5.** $-\frac{1}{4}$ **7.** $-\frac{1}{2}$ **9.** $\frac{2}{3x}$ **11.** $-\frac{x}{9}$ **13.** $\frac{6}{11}$ **15.** $\frac{3}{4}$ **17.** $-\frac{3}{y}$ **19.** $-\frac{19}{33}$ **21.** $-\frac{1}{3}$ **23.** $\frac{7a-3}{4}$ **25.** $\frac{9}{10}$ **27.** $-\frac{13x}{14}$ **29.** $\frac{9x+1}{15}$ **31.** $-\frac{x}{2}$ **33.** $-\frac{2}{3}$ **35.** $\frac{3x}{4}$ **37.** $\frac{5}{4}$ **39.** $\frac{2}{5}$ **41.** 1 in. **43.** 2 m **45.** $\frac{7}{10}$ mi **47.** $\frac{13}{50}$ **49.** $\frac{7}{25}$ **51.** $\frac{1}{50}$ **53.** 45 **55.** 72 **57.** 150 **59.** $24x$ **61.** 126 **63.** 168 **65.** 14 **67.** 20 **69.** 25 **71.** $56x$ **73.** $8y$ **75.** $20a$ **77.** $\frac{54}{100}; \frac{50}{100}; \frac{46}{100}; \frac{50}{100}; \frac{15}{100}; \frac{65}{100}; \frac{45}{100}; \frac{52}{100}; \frac{60}{100}; \frac{61}{100}; \frac{48}{100}; \frac{50}{100}$ **79.** drugs, health and beauty aids **81.** 9 **83.** 125 **85.** 49 **87.** 24 **89.** $\frac{2}{7} + \frac{9}{7} = \frac{11}{7}$ **91.** answers may vary **93.** 1; answers may vary **95.** $814x$ **97.** answers may vary **99.** a, b, and d

Section 4.5

Calculator Explorations **1.** $\frac{37}{80}$ **3.** $\frac{95}{72}$ **5.** $\frac{394}{323}$

Vocabulary, Readiness & Video Check **1.** equivalent; least common denominator **3.** $\frac{4}{24}; \frac{15}{24}; \frac{19}{24}$ **5.** expression; equation **7.** They are unlike terms and so cannot be combined. **9.** $\frac{3}{2} + \frac{1}{3}$; 6

Exercise Set 4.5 **1.** $\frac{5}{6}$ **3.** $\frac{1}{6}$ **5.** $-\frac{4}{33}$ **7.** $-\frac{3}{14}$ **9.** $\frac{3x}{5}$ **11.** $\frac{24-y}{12}$ **13.** $\frac{11}{36}$ **15.** $-\frac{44}{7}$ **17.** $\frac{89a}{99}$ **19.** $\frac{4y-1}{6}$ **21.** $\frac{x+6}{2x}$ **23.** $-\frac{8}{33}$ **25.** $\frac{3}{14}$ **27.** $\frac{11y-10}{35}$ **29.** $-\frac{11}{36}$ **31.** $\frac{1}{20}$ **33.** $\frac{33}{56}$ **35.** $\frac{17}{16}$ **37.** $-\frac{2}{9}$ **39.** $-\frac{11}{30}$ **41.** $\frac{15+11y}{33}$ **43.** $-\frac{53}{42}$ **45.** $\frac{7x}{8}$ **47.** $\frac{11}{18}$ **49.** $\frac{44a}{39}$ **51.** $-\frac{11}{60}$ **53.** $\frac{5y+9}{9y}$ **55.** $\frac{19}{20}$ **57.** $\frac{56}{45}$ **59.** $\frac{40+9x}{72x}$ **61.** $-\frac{5}{24}$ **63.** $\frac{37x-20}{56}$ **65.** $<$ **67.** $>$ **69.** $>$ **71.** $\frac{13}{12}$ **73.** $\frac{1}{4}$ **75.** $\frac{11}{6}$ **77.** $\frac{34}{15}$ or $2\frac{4}{15}$ cm **79.** $\frac{17}{10}$ or $1\frac{7}{10}$ m **81.** $x + \frac{1}{2}$ **83.** $-\frac{3}{8} - x$ **85.** $\frac{7}{100}$ mph **87.** $\frac{5}{8}$ in. **89.** $\frac{49}{100}$ of students **91.** $\frac{47}{32}$ in. **93.** $\frac{19}{25}$ **95.** $\frac{1}{25}$ **97.** $\frac{79}{100}$ **99.** -20 **101.** -6 **103.** $\frac{3}{5} + \frac{4}{5} = \frac{7}{5}$ **105.** $\frac{223}{540}$ **107.** $\frac{49}{44}$ **109.** answers may vary **111.** standard mail

Integrated Review **1.** $\frac{3}{7}$ **2.** $\frac{5}{4}$ or $1\frac{1}{4}$ **3.** $\frac{73}{85}$ **4.** 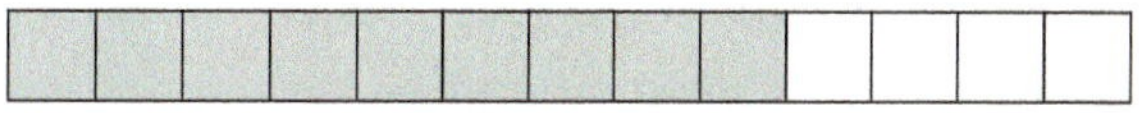 **5.** -1 **6.** 17 **7.** 0 **8.** undefined **9.** $5 \cdot 13$ **10.** $2 \cdot 5 \cdot 7$ **11.** $3^2 \cdot 5 \cdot 7$ **12.** $3^2 \cdot 7^2$ **13.** $\frac{1}{7}$ **14.** $\frac{6}{5}$ **15.** $-\frac{14}{15}$ **16.** $-\frac{9}{10}$ **17.** $\frac{2x}{5}$ **18.** $\frac{3}{8y}$ **19.** $\frac{11z^2}{14}$ **20.** $\frac{7}{11ab^2}$ **21.** not equivalent **22.** equivalent **23. a.** $\frac{1}{25}$ **b.** 48 **c.** $\frac{24}{25}$ **24. a.** $\frac{2}{5}$ **b.** 54,000 **c.** $\frac{3}{5}$ **25.** 30 **26.** 14 **27.** 90

28. $\frac{28}{36}$ **29.** $\frac{55}{75}$ **30.** $\frac{40}{48}$ **31.** $\frac{4}{5}$ **32.** $-\frac{2}{5}$ **33.** $\frac{3}{25}$ **34.** $\frac{1}{3}$ **35.** $\frac{4}{5}$ **36.** $\frac{5}{9}$ **37.** $-\frac{1}{6y}$ **38.** $\frac{3x}{2}$ **39.** $\frac{1}{18}$ **40.** $\frac{4}{21}$ **41.** $-\frac{7}{48z}$ **42.** $-\frac{9}{50}$ **43.** $\frac{37}{40}$ **44.** $\frac{11}{36}$ **45.** $\frac{11}{18}$ **46.** $\frac{25y + 12}{50}$ **47.** $\frac{2}{3} \cdot x$ or $\frac{2}{3}x$ **48.** $x \div \left(-\frac{1}{5}\right)$ **49.** $-\frac{8}{9} - x$ **50.** $\frac{6}{11} + x$ **51.** 1020 **52.** 24 lots **53.** $\frac{3}{4}$ ft

Section 4.6

Vocabulary, Readiness & Video Check **1.** complex **3.** division **5.** addition **7.** distributive property **9.** Since x is squared and the replacement value is negative, we use parentheses to make sure the whole value of x is squared. Without parentheses, the exponent would not apply to the negative sign.

Exercise Set 4.6 **1.** $\frac{1}{6}$ **3.** $\frac{7}{3}$ **5.** $\frac{x}{6}$ **7.** $\frac{23}{22}$ **9.** $\frac{2x}{13}$ **11.** $\frac{17}{60}$ **13.** $\frac{5}{8}$ **15.** $\frac{35}{9}$ **17.** $-\frac{17}{45}$ **19.** $\frac{11}{8}$ **21.** $\frac{29}{10}$ **23.** $\frac{27}{32}$ **25.** $\frac{1}{100}$ **27.** $\frac{9}{64}$ **29.** $\frac{7}{6}$ **31.** $-\frac{2}{5}$ **33.** $-\frac{2}{9}$ **35.** $\frac{11}{9}$ **37.** $\frac{5a}{2}$ **39.** $\frac{7}{2}$ **41.** $\frac{9}{20}$ **43.** $-\frac{13}{2}$ **45.** $\frac{9}{25}$ **47.** $-\frac{5}{32}$ **49.** 1 **51.** $-\frac{2}{5}$ **53.** $-\frac{11}{40}$ **55.** $\frac{x + 6}{16}$ **57.** $\frac{7}{2}$ or $3\frac{1}{2}$ **59.** $\frac{49}{6}$ or $8\frac{1}{6}$ **61.** no; answers may very **63.** $\frac{5}{8}$ **65.** $\frac{11}{56}$ **67.** halfway between a and b **69.** false **71.** true **73.** true **75.** addition: answers may vary **77.** subtraction, multiplication, addition, division **79.** division, multiplication, subtraction, addition **81.** $-\frac{77}{16}$ **83.** $-\frac{55}{16}$

Section 4.7

Calculator Explorations **1.** $\frac{280}{11}$ **3.** $\frac{3776}{35}$ **5.** $26\frac{1}{14}$ **7.** $92\frac{3}{10}$

Vocabulary, Readiness & Video Check **1.** mixed number **3.** round **5.** The denominator of the mixed number we're graphing, $-3\frac{4}{5}$, is 5. **7.** The fractional part of a mixed number should always be a proper fraction. **9.** We're adding two mixed numbers with unlike signs, so the answer has the sign of the mixed number with the larger absolute value, which in this case is negative.

Exercise Set 4.7

1.
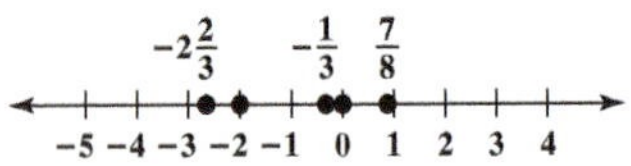

3.
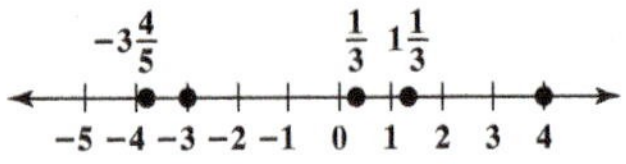

5. b **7.** a **9.** $\frac{8}{21}$ **11.** $4\frac{3}{8}$ **13.** $7\frac{7}{10}$; 8 **15.** $23\frac{31}{35}$; 24 **17.** $12\frac{1}{2}$ **19.** $5\frac{1}{2}$ **21.** $18\frac{2}{3}$ **23.** a **25.** b **27.** $6\frac{2}{3}$; 7 **29.** $13\frac{11}{14}$; 14 **31.** $17\frac{7}{25}$ **33.** $47\frac{53}{84}$ **35.** $25\frac{5}{14}$ **37.** $13\frac{13}{24}$ **39.** $2\frac{3}{5}$; 3 **41.** $7\frac{5}{14}$; 7 **43.** $\frac{24}{25}$ **45.** $3\frac{5}{9}$ **47.** $15\frac{3}{4}$ **49.** 4 **51.** $5\frac{11}{14}$ **53.** $6\frac{2}{9}$ **55.** $\frac{25}{33}$ **57.** $35\frac{13}{18}$ **59.** $2\frac{1}{2}$ **61.** $72\frac{19}{30}$ **63.** $\frac{11}{14}$ **65.** $5\frac{4}{7}$ **67.** $13\frac{16}{33}$ **69.** $-5\frac{2}{7} - x$ **71.** $1\frac{9}{10} \cdot x$ **73.** $3\frac{3}{16}$ mi **75.** $9\frac{2}{5}$ in. **77.** $7\frac{13}{20}$ in. **79.** $3\frac{1}{2}$ sq yd **81.** $\frac{15}{16}$ sq in. **83.** $21\frac{5}{24}$ m **85.** no; she will be short $\frac{1}{2}$ ft **87.** $4\frac{2}{3}$ m **89.** $9\frac{3}{4}$ min **91.** $1\frac{4}{5}$ min **93.** $-10\frac{3}{25}$ **95.** $-24\frac{7}{8}$ **97.** $-13\frac{59}{60}$ **99.** $4\frac{2}{7}$ **101.** $-1\frac{23}{24}$ **103.** $\frac{73}{1000}$ **105.** $1x$ or x **107.** $1a$ or a **109.** a, b, c **111.** Incorrect; to divide mixed numbers, first write each mixed number as an improper fraction. **113.** answers may vary **115.** answers may vary **117.** answers may vary

Section 4.8

Vocabulary, Readiness & Video Check **1.** 6 **3.** 15 **5.** addition property of equality **7.** We multiply by 12 because it is the LCD of all fractions in the equation. The equation no longer contains fractions.

Exercise Set 4.8 **1.** $-\frac{2}{3}$ **3.** $\frac{1}{13}$ **5.** $\frac{4}{5}$ **7.** $\frac{11}{12}$ **9.** $-\frac{7}{10}$ **11.** $\frac{11}{16}$ **13.** $\frac{11}{18}$ **15.** $\frac{2}{7}$ **17.** 12 **19.** −27 **21.** $\frac{27}{8}$ **23.** $\frac{1}{21}$ **25.** $\frac{2}{11}$ **27.** $-\frac{3}{10}$ **29.** 1 **31.** 10 **33.** −1 **35.** −15 **37.** $\frac{3x - 28}{21}$ **39.** $\frac{y + 10}{2}$ **41.** $\frac{7x}{15}$ **43.** $\frac{4}{3}$ **45.** 2 **47.** $\frac{21}{10}$ **49.** $-\frac{1}{14}$ **51.** −3 **53.** $-\frac{1}{9}$ **55.** 50 **57.** $-\frac{3}{7}$ **59.** 4 **61.** $\frac{3}{5}$ **63.** $-\frac{1}{24}$ **65.** $-\frac{5}{14}$ **67.** 4 **69.** −36 **71.** 57,200 **73.** 330 **75.** answers may vary **77.** $\frac{112}{11}$ **79.** area: $\frac{3}{16}$ sq in.: perimeter: 2 in.

Chapter 4 Vocabulary Check **1.** reciprocals **2.** composite number **3.** equivalent **4.** improper fraction **5.** prime number **6.** simplest form **7.** proper fraction **8.** mixed number **9.** numerator; denominator **10.** prime factorization **11.** undefined **12.** 0 **13.** like **14.** least common denominator **15.** complex fraction **16.** cross products

Chapter 4 Review **1.** $\frac{2}{6}$ **3.** $\frac{7}{3}$ or $2\frac{1}{3}$ **5.** $\frac{11}{12}$ **7.** -1 **9.** 0 **11.** number line from 0 to 1 with point at $\frac{7}{9}$ **13.** number line from 0 to 2 with point at $\frac{5}{4}$
15. $3\frac{3}{4}$ **17.** $\frac{11}{5}$ **19.** $\frac{3}{7}$ **21.** $-\frac{1}{3x}$ **23.** $\frac{29}{32c}$ **25.** $\frac{5x}{3y^2}$ **27.** $\frac{2}{3}$ of a foot **29.** no **31.** $\frac{3}{10}$ **33.** $\frac{9}{x^2}$ **35.** $-\frac{1}{27}$ **37.** -2 **39.** $\frac{27}{2}$
41. $\frac{12}{7}$ **43.** $\frac{77}{48}$ sq ft **45.** $\frac{10}{11}$ **47.** $-\frac{1}{3}$ **49.** $\frac{4y-3}{21}$ **51.** $3x$ **53.** 20 **55.** $49a$ **57.** 40 **59.** $\frac{3}{4}$ of his homework **61.** $\frac{11}{18}$
63. $-\frac{1}{12}$ **65.** $\frac{25x+2}{55}$ **67.** $\frac{7y}{36}$ **69.** $\frac{4y+45}{9y}$ **71.** $\frac{91}{150}$ **73.** $\frac{19}{9}$ m **75.** $\frac{21}{50}$ of the donors **77.** $\frac{4x}{7}$ **79.** -2 **81.** $\frac{15}{4}$ **83.** $\frac{8}{13}$
85. $\frac{29}{110}$ **87.** $20\frac{7}{24}$ **89.** $5\frac{1}{5}$; 6 **91.** 22 mi **93.** each measurement is $4\frac{1}{4}$ in. **95.** $-27\frac{5}{14}$ **97.** $1\frac{5}{27}$ **99.** $\frac{5}{6}$ **101.** -10 **103.** 15
105. 5 **107.** $\frac{1}{4}$ **109.** $\frac{1}{5}$ **111.** $\frac{11x}{12}$ **113.** $-6\frac{2}{5}$ **115.** $2\frac{19}{35}$; 2 **117.** $-\frac{1}{12}$ **119.** $-\frac{4}{9}$ **121.** $44\frac{1}{2}$ yd

Chapter 4 Test **1.** $\frac{7}{16}$ **2.** $\frac{23}{3}$ **3.** $18\frac{3}{4}$ **4.** $\frac{4}{35}$ **5.** $-\frac{3x}{5}$ **6.** not equivalent **7.** equivalent **8.** $2^2 \cdot 3 \cdot 7$ **9.** $3^2 \cdot 5 \cdot 11$ **10.** $\frac{4}{3}$
11. $-\frac{4}{3}$ **12.** $\frac{8x}{9}$ **13.** $\frac{x-21}{7x}$ **14.** y^2 **15.** $\frac{16}{45}$ **16.** $\frac{9a+4}{10}$ **17.** $-\frac{2}{3y}$ **18.** $\frac{1}{a^2}$ **19.** $\frac{3}{4}$ **20.** $14\frac{1}{40}$ **21.** $16\frac{8}{11}$ **22.** $\frac{64}{3}$ or $21\frac{1}{3}$ **23.** $22\frac{1}{2}$
24. $-\frac{5}{3}$ or $-1\frac{2}{3}$ **25.** $\frac{9}{16}$ **26.** $\frac{3}{8}$ **27.** $\frac{11}{12}$ **28.** $\frac{3}{4x}$ **29.** $\frac{76}{21}$ or $3\frac{13}{21}$ **30.** -2 **31.** -4 **32.** 1 **33.** $\frac{5}{2}$ **34.** $\frac{4}{31}$ **35.** $3\frac{3}{4}$ ft **36.** $\frac{23}{50}$
37. $\frac{13}{50}$ **38.** \$2820 **39.** perimeter: $3\frac{1}{3}$ ft; area: $\frac{2}{3}$ sq ft **40.** 24 mi

Cumulative Review Chapters 1–4 **1.** five hundred forty-six; Sec. 1.2. Ex. 5 **3.** twenty-seven thousand, thirty-four; Sec 1.2, Ex. 6 **5.** 759; Sec. 1.3, Ex. 1 **7.** 514; Sec. 1.3, Ex. 7 **9.** 278,000; Sec. 1.4, Ex. 2 **11.** 57,600 megabytes; Sec. 1.5, Ex. 7 **13.** 7089 R 5; Sec. 1.6, Ex. 7 **15.** 7^3; Sec. 1.7, Ex. 1 **17.** $3^4 \cdot 9^3$; Sec. 1.7, Ex. 4 **19.** 6; Sec. 1.8, Ex. 2 **21.** -7188; Sec. 2.1, Ex. 1 **23.** -4; Sec. 2.2, Ex. 3 **25.** 3; Sec. 2.3, Ex. 9 **27.** 25; Sec. 2.4, Ex. 8 **29.** -16; Sec. 2.5, Ex. 8 **31.** $6y+2$; Sec. 3.1, Ex. 2 **33.** -14; Sec. 3.2, Ex. 4 **35.** -1; Sec. 3.3, Ex. 2 **37.** $\frac{2}{5}$; Sec. 4.1, Ex. 3 **39. a.** $\frac{38}{9}$ **b.** $\frac{19}{11}$; Sec. 4.1, Ex. 20 **41.** $\frac{7x}{11}$; Sec. 4.2, Ex. 5 **43.** $2\frac{11}{12}$; Sec. 4.7, Ex. 2 **45.** $\frac{5}{12}$; Sec. 4.3, Ex. 11

Chapter 5 Decimals

Section 5.1

Vocabulary, Readiness & Video Check **1.** words; standard form **3.** decimals **5.** tenths; tens **7.** as "and" **9.** Reading a decimal correctly gives you the correct place value, which tells you the denominator of your equivalent fraction. **11.** When rounding, we look to the digit to the right of the place value we're rounding to. In this case we look to the hundredths-place digit, which is 7.

Exercise Set 5.1 **1.** five and sixty-two hundredths **3.** sixteen and twenty-three hundredths **5.** negative two hundred five thousandths **7.** one hundred sixty-seven and nine thousandths **9.** three thousand and four hundredths **11.** one hundred five and six tenths **13.** two and forty-three hundredths

15.

Preprinted Name
Preprinted Address
Current date
DATE
PAY TO THE ORDER OF *R.W. Financial* \$ *321.42*
Three hundred twenty-one and 42/100 DOLLARS
FOR *Signature*

17.

Preprinted Name
Preprinted Address
Current date
DATE
PAY TO THE ORDER OF *Verizon* \$ *91.68*
Ninety-one and 68/100 DOLLARS
FOR *Signature*

19. 2.8 **21.** 9.08 **23.** -705.625 **25.** 0.0046 **27.** $\frac{7}{10}$ **29.** $\frac{27}{100}$ **31.** $\frac{2}{5}$ **33.** $5\frac{2}{5}$ **35.** $-\frac{29}{500}$ **37.** $7\frac{1}{125}$ **39.** $15\frac{401}{500}$ **41.** $\frac{601}{2000}$
43. 0.8; $\frac{8}{10}$ or $\frac{4}{5}$ **45.** seventy-seven thousandths; $\frac{77}{1000}$ **47.** $<$ **49.** $<$ **51.** $<$ **53.** $=$ **55.** $<$ **57.** $>$ **59.** $<$ **61.** $>$ **63.** 0.6
65. 98,210 **67.** -0.23 **69.** 0.594 **71.** 3.1 **73.** 3.142 **75.** \$27 **77.** \$0.20 **79.** 0.7 in. **81.** 2.07 min **83.** \$68 **85.** 225 days

87. 5766 **89.** 35 **91.** b **93.** a **95.** answers may vary **97.** 7.12 **99.** $\frac{26,849,577}{100,000,000,000}$ **101.** answers may vary **103.** answers may vary **105.** 0.26499, 0.25786 **107.** 0.10299, 0.1037, 0.1038, 0.9 **109.** $3600 million

Section 5.2

Calculator Explorations **1.** 328.742 **3.** 5.2414 **5.** 865.392

Vocabulary, Readiness & Video Check **1.** last **3.** like **5.** false **7.** Lining up the decimal points also lines up place values, so we only add or subtract digits in the same place values. **9.** So the subtraction can be written vertically with decimal points lined up. **11.** perimeter

Exercise Set 5.2 **1.** 7.7 **3.** 10.35 **5.** 27.0578 **7.** −8.57 **9.** 10.33 **11.** 465.56; $\begin{array}{r}230\\+\ 230\\\hline 460\end{array}$ **13.** 115.123; $\begin{array}{r}100\\6\\+\ \ 9\\\hline 115\end{array}$ **15.** 50.409 **17.** 4.4 **19.** 15.3 **21.** 598.23 **23.** 1.83; $6 - 4 = 2$ **25.** 876.6; $\begin{array}{r}1000\\-\ 100\\\hline 900\end{array}$ **27.** 194.4 **29.** −6.32 **31.** −6.15 **33.** 3.1 **35.** 2.9988 **37.** 16.3 **39.** 3.1 **41.** −5.62 **43.** 363.36 **45.** −549.8 **47.** 861.6 **49.** 115.123 **51.** 0.088 **53.** −180.44 **55.** −1.1 **57.** 3.81 **59.** 3.39 **61.** 1.61 **63.** no **65.** yes **67.** no **69.** $6.9x + 6.9$ **71.** $3.47y - 10.97$ **73.** $7.52 **75.** −$0.42 **77.** 28.56 m **79.** 14.36 in. **81.** 195.8 mph **83.** 11.8 texts per day **85.** $2042.5 million **87.** 326.3 in. **89.** 67.44 ft **91.** 13.462 mph **93.** Switzerland **95.** 7.94 lb

97.

Country	Pounds of Chocolate per Person
Switzerland	26.24
Ireland	21.83
UK	20.94
Austria	19.40
Belgium	18.30

99. 138 **101.** $\frac{4}{9}$ **103.** incorrect; $\begin{array}{r}9.200\\8.630\\+\ +\ 4.005\\\hline 21.835\end{array}$ **105.** 6.08 in. **107.** $1.20 **109.** 1 nickel, 1 dime, and 2 pennies; 3 nickels and 2 pennies; 1 dime and 7 pennies; 2 nickels and 7 pennies **111.** answers may vary **113.** answers may vary **115.** $22.181x - 22.984$

Section 5.3

Vocabulary, Readiness & Video Check **1.** sum **3.** right; zeros **5.** circumference **7.** Whether we placed the decimal point correctly in our product **9.** $3(5.7) - (-0.2)$ **11.** This is an application problem and needs units attached. The complete answer is 24.8 grams.

Exercise Set 5.3 **1.** 1.36 **3.** 0.6 **5.** −17.595 **7.** 55.008 **9.** 28.56; $7 \times 4 = 28$ **11.** 0.1041 **13.** 8.23854; $\begin{array}{r}1\\\times\ 8\\\hline 8\end{array}$ **15.** 11.2746 **17.** 65 **19.** 0.83 **21.** −7093 **23.** 70 **25.** 0.0983 **27.** 0.02523 **29.** 0.0492 **31.** 14,790 **33.** 1.29 **35.** −9.3762 **37.** 0.5623 **39.** 36.024 **41.** 1,500,000,000 **43.** 49,800,000 **45.** −0.6 **47.** 17.3 **49.** no **51.** yes **53.** 10π cm ≈ 31.4 cm **55.** 18.2π yd ≈ 57.148 yd **57.** $715.20 **59.** 24.8 g **61.** 11.201 sq in. **63.** 250π ft ≈ 785 ft **65.** 135π m ≈ 423.9 m **67.** 64.9605 in. **69. a.** 62.8 m and 125.6 m **b.** yes **71.** $730 **73.** 786.9 Canadian dollars **75.** 1024.67 New Zealand dollars **77.** 486 **79.** −9 **81.** 3.64 **83.** 3.56 **85.** −0.1105 **87.** 3,831,600 mi **89.** answers may vary **91.** answers may vary

Section 5.4

Calculator Explorations **1.** not reasonable **3.** reasonable

Vocabulary, Readiness & Video Check **1.** quotient; divisor; dividend **3.** left; zeros **5.** a whole number **7.** We just need to know how to move the decimal point. 1000 has three zeros, so we move the decimal point in the decimal number three places to the left. **9.** We want the answer rounded to the nearest tenth, so we go to one extra place value, to the hundredths place, in order to round.

Exercise Set 5.4 **1.** 4.6 **3.** 0.094 **5.** 300 **7.** 7.3 **9.** 6.6; $6\overline{)36}$ with quotient 6 **11.** 0.413 **13.** −600 **15.** 7 **17.** 4.8 **19.** 2100 **21.** 5.8 **23.** 5.5 **25.** 9.8; $7\overline{)70}$ with quotient 10 **27.** 9.6 **29.** 45 **31.** 54.592 **33.** 0.0055 **35.** 23.87 **37.** 114.0 **39.** 0.83397 **41.** 2.687 **43.** −0.0129 **45.** 12.6 **47.** 1.31 **49.** 0.045625 **51.** 0.413 **53.** −8 **55.** −7.2 **57.** 1400 **59.** 30 **61.** −58,000 **63.** −0.69 **65.** 0.024 **67.** 65 **69.** −5.65 **71.** −7.0625

73. yes **75.** no **77.** 11 qt **79.** 5.1 m **81.** 11.4 boxes **83.** 24 tsp **85.** 8 days **87.** 146.6 mi per week **89.** 345.5 thousand books per hr **91.** $\frac{21}{50}$ **93.** $-\frac{1}{10}$ **95.** 4.26 **97.** 1.578 **99.** −26.66 **101.** 904.29 **103.** c **105.** b **107.** 85.5 **109.** 8.6 ft **111.** answers may vary **113.** 65.2 − 82.6 knots **115.** 27.3 m

Integrated Review **1.** 2.57 **2.** 4.05 **3.** 8.9 **4.** 3.5 **5.** 0.16 **6.** 0.24 **7.** 0.27 **8.** 0.52 **9.** −4.8 **10.** 6.09 **11.** 75.56 **12.** 289.12 **13.** −24.974 **14.** −43.875 **15.** −8.6 **16.** 5.4 **17.** −280 **18.** 1600 **19.** 224.938 **20.** 145.079 **21.** 0.56 **22.** −0.63 **23.** 27.6092 **24.** 145.6312 **25.** 5.4 **26.** −17.74 **27.** −414.44 **28.** −1295.03 **29.** −34 **30.** −28 **31.** 116.81 **32.** 18.79 **33.** 156.2 **34.** 1.562 **35.** 25.62 **36.** 5.62 **37.** Exact: 204.1 mi; Estimate: 200 mi **38.** $0.81 **39.** $8.8 billion or $8,800,000,000

Section 5.5

Vocabulary, Readiness & Video Check **1.** false **3.** false **5.** We place a bar over just the repeating digits and only 6 repeats in our decimal answer. **7.** The fraction bar serves as a grouping symbol **9.** $4(0.3) - (-2.4)$

Exercise Set 5.5 **1.** 0.2 **3.** 0.68 **5.** 0.75 **7.** −0.08 **9.** 2.25 **11.** $0.91\overline{6}$ **13.** 0.425 **15.** 0.45 **17.** $-0.\overline{3}$ **19.** 0.4375 **21.** $0.\overline{63}$ **23.** 5.85 **25.** 0.624 **27.** −0.33 **29.** 0.44 **31.** 0.6 **33.** 0.62 **35.** 0.86 **37.** 0.02 **39.** < **41.** = **43.** < **45.** < **47.** < **49.** > **51.** < **53.** < **55.** 0.32, 0.34, 0.35 **57.** 0.49, 0.491, 0.498 **59.** 5.23, $\frac{42}{8}$, 5.34 **61.** 0.612, $\frac{5}{8}$, 0.649 **63.** 0.59 **65.** −3 **67.** 5.29 **69.** 9.24 **71.** 0.2025 **73.** −1.29 **75.** −15.4 **77.** −3.7 **79.** 25.65 sq in. **81.** 0.248 sq yd **83.** 5.76 **85.** 5.7 **87.** 3.6 **89.** $\frac{77}{50}$ **91.** $\frac{5}{2}$ **93.** = 1 **95.** > 1 **97.** < 1 **99.** 0.057 **101.** 6300 stations **103.** answers may vary

Section 5.6

Vocabulary, Readiness & Video Check **1.** So that we are no longer working with decimals.

Exercise Set 5.6 **1.** 5.9 **3.** −0.43 **5.** 10.2 **7.** −4.5 **9.** 4 **11.** 0.45 **13.** 4.2 **15.** −4 **17.** 1.8 **19.** 10 **21.** 7.6 **23.** 60 **25.** −0.07 **27.** 20 **29.** 0.0148 **31.** −8.13 **33.** 1.5 **35.** −1 **37.** −7 **39.** 7 **41.** 53.2 **43.** $3x - 16$ **45.** $\frac{3}{5x}$ **47.** $\frac{13x}{21}$ **49.** 3.7 **51.** $6x - 0.61$ **53.** $-2y + 6.8$ **55.** 9.1 **57.** −3 **59.** $-4z + 16.67$ **61.** 15.7 **63.** 5.85 **65.** $-2.1z - 10.1$ **67.** answers may vary **69.** answers may vary **71.** 7.683 **73.** 4.683

Section 5.7

Vocabulary, Readiness & Video Check **1.** average **3.** mean (or average) **5.** grade point average **7.** Place the data numbers in numerical order (or verify that they already are)

Exercise Set 5.7 **1.** mean: 21; median: 23; no mode **3.** mean: 8.1; median: 8.2; mode: 8.2 **5.** mean: 0.5; median: 0.5; mode: 0.2 and 0.5 **7.** mean: 370.9; median: 313.5; no mode **9.** 1911.6 ft **11.** 1601 ft **13.** answers may vary **15.** 2.79 **17.** 3.64 **19.** 6.8 **21.** 6.9 **23.** 88.5 **25.** 73 **27.** 70 and 71 **29.** 9 rates **31.** $\frac{1}{3}$ **33.** $\frac{3}{5y}$ **35.** $\frac{11}{15}$ **37.** 35, 35, 37, 43 **39.** yes; answers may vary

Chapter 5 Vocabulary Check **1.** decimal **2.** numerator; denominator **3.** vertically **4.** and **5.** sum **6.** mode **7.** circumference **8.** median; mean **9.** mean **10.** standard form

Chapter 5 Review **1.** tenths **3.** negative twenty-three and forty-five hundredths **5.** one hundred nine and twenty-three hundredths **7.** 8.06 **9.** 16,025.0014 **11.** $\frac{4}{25}$ **13.** 0.00231 **15.** > **17.** < **19.** 0.6 **21.** −42.90 **23.** 887,000,000 **25.** 18.1 **27.** −7.28 **29.** 320.312 **31.** 1.7 **33.** −1324.5 **35.** 65.02 **37.** 52.6 mi **39.** 22.2 in. **41.** 72 **43.** −78.246 **45.** 14π m ≈ 43.96 m **47.** 0.0877 **49.** 70 **51.** 8.059 **53.** 0.02365 **55.** 7.3 m **57.** 0.8 **59.** $2.\overline{3}$ or 2.333 **61.** = **63.** < **65.** 0.832, 0.837, 0.839 **67.** 0.42, $\frac{3}{7}$, 0.43 **69.** −11.94 **71.** 7.26 **73.** 55 **75.** 6.9 sq ft **77.** 0.3 **79.** 8.6 **81.** 1.98 **83.** −20 **85.** mean: 17.8; median: 14; no mode **87.** mean: 24,500; median: 20,000; mode: 20,000 **89.** 3.25 **91.** two hundred and thirty-two ten-thousandths **93.** $\frac{847}{10{,}000}$ **95.** −0.07 **97.** 51.057 **99.** < **101.** 3.115 **103.** $3646.00 **105.** 5.26 **107.** −14.907 **109.** −7.245 **111.** 23.904 **113.** yes **115.** 3.6 **117.** mean: 619.17; median: 647.5; mode: 327

Chapter 5 Test **1.** forty-five and ninety-two thousandths **2.** 3000.059 **3.** 17.595 **4.** −51.20 or −51.2 **5.** −20.42 **6.** 40.902 **7.** 0.037 **8.** 34.9 **9.** 0.862 **10.** < **11.** < **12.** $\frac{69}{200}$ **13.** $-24\frac{73}{100}$ **14.** −0.5 **15.** 0.941 **16.** 1.93 **17.** −6.2 **18.** $0.5x - 13.4$ **19.** −3 **20.** 3.7 **21.** mean: 38.4; median: 42; no mode **22.** mean: 12.625; median: 12.5; mode: 12 and 16 **23.** 3.07 **24.** 4,583,000,000 **25.** 2.31 sq mi **26.** 18π mi ≈ 56.52 mi **27. a.** 9904 sq ft **b.** 198.08 oz **28.** 54 mi

Cumulative Review Chapters 1–5 **1.** seventy-two; Sec. 1.2, Ex. 4 **3.** five hundred forty-six; Sec. 1.2, Ex. 5 **5.** 759; Sec. 1.3, Ex. 1 **7.** 514; Sec. 1.3, Ex. 7 **9.** 278,000; Sec. 1.4, Ex. 2 **11.** 20,296; Sec. 1.5, Ex. 4 **13. a.** 7 **b.** 12 **c.** 1 **d.** 1 **e.** 20 **f.** 1; Sec. 1.6, Ex. 2 **15.** 7; Sec. 1.7, Ex. 9 **17.** 81; Sec. 1.7, Ex. 5 **19.** 81; Sec. 1.7, Ex. 7 **21.** 2; Sec. 1.8, Ex. 3 **23. a.** −13 **b.** 2 **c.** 0; Sec. 2.1, Ex. 5 **25.** −23; Sec. 2.2, Ex. 4 **27.** 180; Sec. 1.7, Ex. 8 **29.** −49; Sec. 2.4, Ex. 9 **31.** 25; Sec. 2.4, Ex. 8 **33.** $\frac{4}{3}$; $1\frac{1}{3}$; Sec. 4.1, Ex. 10 **35.** $\frac{11}{4}$; $2\frac{3}{4}$; Sec. 4.1, Ex. 11 **37.** $2^2 \cdot 3^2 \cdot 7$; Sec. 4.2, Ex. 3 **39.** $-\frac{36}{13}$; Sec. 4.2, Ex. 8 **41.** equivalent; Sec. 4.2, Ex. 10 **43.** $\frac{10}{33}$; Sec. 4.3, Ex. 1 **45.** $\frac{1}{8}$; Sec. 4.3, Ex. 2 **47.** −24; Sec. 3.2, Ex. 3 **49.** 829.6561; Sec. 5.2, Ex. 2 **51.** 18.408; Sec. 5.3, Ex. 1

Chapter 6 Ratio, Proportion, and Percent

Section 6.1

Vocabulary, Readiness & Video Check **1.** proportion, ratio **3.** In order to simplify the ratio, quantities must have the same units of measurement to divide out. **5.** There are also many ways to set up an incorrect proportion, so just checking your solution in your proportion isn't enough. You need to determine if your solution is reasonable from the relationships given in the stated application.

Exercise Set 6.1 **1.** $\frac{2}{15}$ **3.** $\frac{5}{6}$ **5.** $\frac{1}{10}$ **7.** $\frac{19}{18}$ **9.** $\frac{77}{100}$ **11.** $\frac{2}{7}$ **13.** $\frac{47}{25}$ **15.** $\frac{17}{40}$ **17.** $\frac{15}{1}$ **19.** $\frac{2}{3}$ **21.** $\frac{10}{21}$ **23.** 4 **25.** $\frac{50}{9}$ **27.** $\frac{21}{4}$ **29.** 7 **31.** −3 **33.** $\frac{14}{9}$ **35.** 5 **37.** −20 **39.** 123 lb **41.** 165 cal **43.** 150 cal **45. a.** 18 tsp **b.** 6 tbsp **47.** 6 people **49.** 112 ft; 11-in. difference **51.** 102.9 mg **53.** 1248 ft **55.** 1125 sales **57.** 113 million phones **59.** 2.4 c **61. a.** 0.1 gal **b.** 13 fl oz **63. a.** 2062.5 mg **b.** no **65.** $2^3 \cdot 5^2$ **67.** 2^5 **69.** answers may vary **71.** 0.8 ml **73.** 1.25 ml

Section 6.2

Vocabulary, Readiness & Video Check **1.** Percent **3.** percent **5.** 0.01 **7.** Percent means "per 100." **9.** 1

Exercise Set 6.2 **1.** 96% **3.** football; 37% **5.** 50% **7.** 0.41 **9.** 0.06 **11.** 1.00 or 1 **13.** 0.736 **15.** 0.028 **17.** 0.006 **19.** 3.00 or 3 **21.** 0.3258 **23.** $\frac{2}{25}$ **25.** $\frac{1}{25}$ **27.** $\frac{9}{200}$ **29.** $\frac{7}{4}$ or $1\frac{3}{4}$ **31.** $\frac{1}{16}$ **33.** $\frac{31}{300}$ **35.** $\frac{179}{800}$ **37.** 22% **39.** 0.6% **41.** 530% **43.** 5.6% **45.** 22.28% **47.** 300% **49.** 70% **51.** 70% **53.** 80% **55.** 68% **57.** $37\frac{1}{2}$% **59.** $33\frac{1}{3}$% **61.** 450% **63.** 190% **65.** 81.82% **67.** 26.67% **69.** 0.6; $\frac{3}{5}$; $23\frac{1}{2}$%; $\frac{47}{200}$; 80%; 0.8; $0.333\overline{3}$; $\frac{1}{3}$; 87.5%; 0.875; 0.075; $\frac{3}{40}$ **71.** 2; 2; 280%; $2\frac{4}{5}$; 7.05; $7\frac{1}{20}$; 454%; 4.54 **73.** 0.38; $\frac{19}{50}$ **75.** 0.358; $\frac{179}{500}$ **77.** 0.91; $\frac{91}{100}$ **79.** 0.005; $\frac{1}{200}$ **81.** 0.142; $\frac{71}{500}$ **83.** 78.1% **85.** 0.7% **87.** 0.40 or 0.4 **89.** $\frac{11}{36}$ **91.** $1\frac{5}{6}$ **93. a.** 52.9% **b.** 52.86% **95.** b, d **97.** 4% **99.** 75% **101.** greater **103.** 0.266; 26.6% **105.** biomedical engineers **107.** 0.39 **109.** answers may vary

Section 6.3

Vocabulary, Readiness & Video Check **1.** is **3.** amount; base; percent **5.** greater **7.** "of" translates to multiplication; "is" (or something equivalent) translates to an equal sign; "what" or "unknown" translates to our variable

Exercise Set 6.3 **1.** $18\% \cdot 81 = x$ **3.** $20\% \cdot x = 105$ **5.** $0.6 = 40\% \cdot x$ **7.** $x \cdot 80 = 3.8$ **9.** $x = 9\% \cdot 43$ **11.** $x \cdot 250 = 150$ **13.** 3.5 **15.** 28.7 **17.** 10 **19.** 600 **21.** 110% **23.** 34% **25.** 1 **27.** 645 **29.** 500 **31.** 5.16% **33.** 25.2 **35.** 35% **37.** 35 **39.** 0.624 **41.** 0.5% **43.** 145 **45.** 63% **47.** 4% **49.** 30 **51.** $3\frac{7}{11}$ **53.** $\frac{17}{12} = \frac{x}{20}$ **55.** $\frac{8}{9} = \frac{14}{x}$ **57.** c **59.** b **61.** answers may vary **63.** b **65.** c **67.** c **69.** a **71.** a **73.** answers may vary **75.** 686.625 **77.** 12,285

Section 6.4

Vocabulary, Readiness & Video Check **1.** amount; base; percent **3.** amount **5.** 45 follows the word "of" so it is the base.

Exercise Set 6.4 **1.** $\frac{a}{45} = \frac{98}{100}$ **3.** $\frac{a}{150} = \frac{4}{100}$ **5.** $\frac{14.3}{b} = \frac{26}{100}$ **7.** $\frac{84}{b} = \frac{35}{100}$ **9.** $\frac{70}{400} = \frac{p}{100}$ **11.** $\frac{8.2}{82} = \frac{p}{100}$ **13.** 26 **15.** 18.9 **17.** 600 **19.** 10 **21.** 120% **23.** 28% **25.** 37 **27.** 1.68 **29.** 1000 **31.** 210% **33.** 55.18 **35.** 45% **37.** 75 **39.** 0.864 **41.** 0.5% **43.** 140 **45.** 9.6 **47.** 113% **49.** $-\frac{7}{8}$ **51.** $3\frac{2}{15}$ **53.** 0.7 **55.** 2.19 **57.** answers may vary **59.** no; $a = 16$ **61.** yes **63.** answers may vary **65.** 12,011.2 **67.** 7270.6

Integrated Review **1.** $\frac{9}{10}$ **2.** $\frac{9}{25}$ **3.** $\frac{43}{50}$ **4.** $\frac{8}{23}$ **5.** $\frac{2}{3}$ **6.** $\frac{9}{2}$ **7.** 25 **8.** 35 **9.** 86 weeks **10.** 7 boxes **11.** 94% **12.** 17% **13.** 37.5% **14.** 350% **15.** 470% **16.** 800% **17.** 45% **18.** 106% **19.** 675% **20.** 325% **21.** 2% **22.** 6% **23.** 0.71 **24.** 0.31 **25.** 0.03 **26.** 0.04 **27.** 2.24 **28.** 7 **29.** 0.029 **30.** 0.066 **31.** 0.07; $\frac{7}{100}$ **32.** 0.05; $\frac{1}{20}$ **33.** 0.068; $\frac{17}{250}$ **34.** 0.1125; $\frac{9}{80}$ **35.** 0.74; $\frac{37}{50}$ **36.** 0.45; $\frac{9}{20}$ **37.** 0.163; $\frac{49}{300}$ **38.** 0.127; $\frac{19}{150}$ **39.** 13.5 **40.** 100 **41.** 350 **42.** 120% **43.** 28% **44.** 76 **45.** 34 **46.** 130% **47.** 46% **48.** 37.8 **49.** 150 **50.** 62

Section 6.5

Vocabulary, Readiness & Video Check **1.** The price of the home was $175,000.

Exercise Set 6.5 **1.** 1600 bolts **3.** 8.8 pounds **5.** 14% **7.** 13,600 screens **9.** 17% **11.** 496 chairs; 5704 chairs **13.** 108,680 physician assistants **15.** 1,049,841 **17.** 30% **19.** 50% **21.** 12.5% **23.** 29.2% **25.** $175,000 **27.** 31.2 hr **29.** $867.87; $20,153.87 **31.** 35 ft **33.** increase: $1328; tuition in 2013–2014: $10,007 **35.** increase: 164,082 associate degrees; 2020–2021: 1,107,082 associate degrees **37.** 30; 60% **39.** 52; 80% **41.** 2; 25% **43.** 120; 75% **45.** 44% **47.** 137.5% **49.** 69.5% **51.** 12.9% **53.** 8.3% **55.** 5.4% **57.** 141.4% **59.** 62.3% **61.** 4.56 **63.** 11.18 **65.** $\frac{1}{24}$ **67.** $\frac{28}{39}$ **69.** The increased number is double the original number. **71.** answers may vary **73.** percent increase $= \frac{30}{150} = 20\%$ **75.** False; the percents are different.

Section 6.6

Vocabulary, Readiness & Video Check **1.** sales tax **3.** commission **5.** sale price **7.** We write the commission rate as a percent.

Exercise Set 6.6 **1.** $7.50 **3.** $858.93 **5.** 7% **7. a.** $120 **b.** $130.20 **9.** $117; $1917 **11.** $485 **13.** 6% **15.** $16.10; $246.10 **17.** $53,176.04 **19.** 14% **21.** $4888.50 **23.** $185,500 **25.** $8.90; $80.10 **27.** $98.25; $98.25 **29.** $143.50; $266.50 **31.** $3255; $18,445 **33.** $45; $255 **35.** $27.45; $332.45 **37.** $3.08; $59.08 **39.** $7074 **41.** 8% **43.** 1200 **45.** 132 **47.** 16 **49.** d **51.** $4.00; $6.00; $8.00 **53.** $7.20; $10.80; $14.40 **55.** a discount of 60% is better; answers may vary **57.** $26,838.45

Section 6.7

Calculator Explorations **1.** $936.31 **3.** $9674.77 **5.** $634.49

Vocabulary, Readiness & Video Check **1.** simple **3.** Compound **5.** Total amount **7.** principal

Exercise Set 6.7 **1.** $32 **3.** $73.60 **5.** $750 **7.** $33.75 **9.** $700 **11.** $101,562.50; $264,062.50 **13.** $5562.50 **15.** $14,280 **17.** $46,815.37 **19.** $2327.14 **21.** $58,163.65 **23.** $2915.75 **25.** $2938.66 **27.** $2971.89 **29.** 32 yd **31.** 35 m **33.** $\frac{9x}{20}$ **35.** $-\frac{131}{225}$ **37.** answers may vary **39.** answers may vary

Chapter 6 Vocabulary Check **1.** of **2.** is **3.** Percent **4.** Compound interest **5.** amount; base **6.** 100% **7.** 0.01 **8.** $\frac{1}{100}$ **9.** base; amount **10.** Percent of decrease **11.** Percent of increase **12.** Sales tax **13.** Total price **14.** Commission **15.** Amount of discount **16.** Sale price **17.** Proportion **18.** ratio

Chapter 6 Review **1.** $\frac{1}{5}$ **3.** 6 **5.** 312.5 **7.** 9 **9.** 3 **11.** 675 parts **13.** 37% **15.** 0.26 **17.** 0.035 **19.** 2.75 **21.** 0.4785 **23.** 160% **25.** 7.6% **27.** 71% **29.** 600% **31.** $\frac{7}{100}$ **33.** $\frac{1}{4}$ **35.** $\frac{51}{500}$ **37.** $\frac{1}{3}$ **39.** 40% **41.** $58\frac{1}{3}\%$ **43.** 125% **45.** 6.25% **47.** 100,000 **49.** 23% **51.** 108.8 **53.** 418 **55.** 159.6 **57.** 110% **59.** 66% **61.** 20.9% **63.** $206,400 **65.** $273.75 **67.** $5000 **69.** discount: $900; sale price: $2100 **71.** $160 **73.** $30,104.61 **75.** $180.61 **77.** 0.038 **79.** 54% **81.** $\frac{47}{100}$ **83.** $37\frac{1}{2}\%$ **85.** 268.75 **87.** 708.48 **89.** 300% **91.** 560 **93.** 26% **95.** $617.70 **97.** 12.5% **99.** $11,687.50

Chapter 6 Test **1.** 0.85 **2.** 5 **3.** 0.006 **4.** 5.6% **5.** 610% **6.** 35% **7.** $1\frac{1}{5}$ **8.** $\frac{77}{200}$ **9.** $\frac{1}{500}$ **10.** 55% **11.** 37.5% **12.** 175% **13.** 20% **14.** $\frac{16}{25}$ **15.** 33.6 **16.** 1250 **17.** 75% **18.** 38.4 lb **19.** $56,750 **20.** $383.21 **21.** 5% **22.** discount: $18; sale price: $102 **23.** $395 **24.** 9% **25.** $647.50 **26.** $2005.63 **27.** $427 **28.** 4.5% **29.** $\frac{15}{2}$ **30.** −6 **31.** 18 bulbs

Cumulative Review Chapters 1–6 **1.** 20,296; Sec. 1.5, Ex. 4 **3.** −10; Sec. 2.3, Ex. 8 **5.** 1; Sec. 2.6, Ex. 2 **7.** 2; Sec. 3.3, Ex. 4 **9.** $\frac{21}{7}$; Sec. 4.4, Ex. 20 **11.** $-\frac{10}{27}$; Sec. 4.2, Ex. 6 **13.** $\frac{7}{10}$; Sec. 4.3, Ex. 13 **15.** $-\frac{1}{2}$; Sec. 4.4, Ex. 9 **17.** $\frac{1}{28}$; Sec. 4.5, Ex. 5

19. $\frac{3}{2}$; Sec. 4.6, Ex. 2 **21.** 3; Sec. 4.8, Ex. 9 **23. a.** $\frac{38}{9}$ **b.** $\frac{19}{11}$; Sec. 4.1, Ex. 20 **25.** $\frac{1}{8}$; Sec. 5.1, Ex. 9 **27.** $-105\frac{83}{1000}$; Sec. 5.1, Ex. 11 **29.** 67.69; Sec. 5.2, Ex. 6 **31.** 76.8; Sec. 5.3, Ex. 5 **33.** −76,300; Sec. 5.3, Ex. 7 **35.** 50; Sec. 5.4, Ex. 10 **37.** 80.5; Sec. 5.7, Ex. 4 **39.** $\frac{50}{63}$; Sec. 6.1, Ex. 3 **41.** 63; Sec. 6.1, Ex. 4 **43.** 17.5 mi; Sec. 6.3, Ex. 1 **45.** $\frac{19}{1000}$; Sec. 6.2, Ex. 9 **47.** $\frac{1}{3}$; Sec. 6.2, Ex. 11 **49.** 21; Sec. 6.3, Ex. 7

Chapter 7 Graphs and Triangle Applications

Section 7.1

Vocabulary, Readiness & Video Check **1.** bar **3.** line **5.** Count the number of symbols and multiply this number by how much each symbol stands for (from the key). **7.** bar graph

Exercise Set 7.1 **1.** Kansas **3.** 5.5 million or 5,500,000 acres **5.** Texas **7.** Montana (or Oklahoma) and Washington **9.** 78,000 **11.** 2006 **13.** 30,000 **15.** 86,000 wildfires/year **17.** September **19.** 77 **21.** $\frac{2}{77}$ **23.** Tokyo, Japan; about 34.7 million or 34,700,000 **25.** New York; 21.6 million or 21,600,000 **27.** approximately 2 million

29.

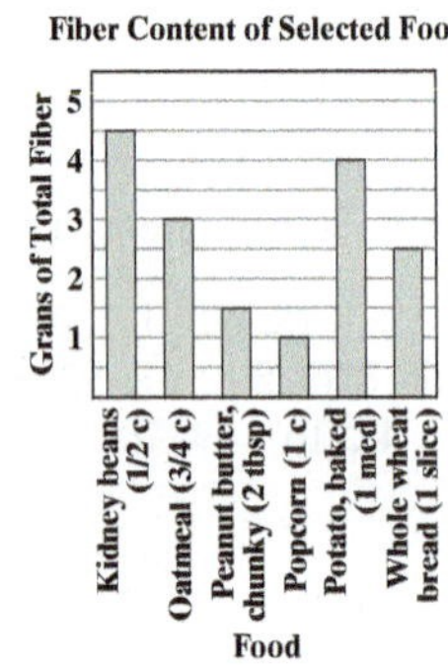

31.

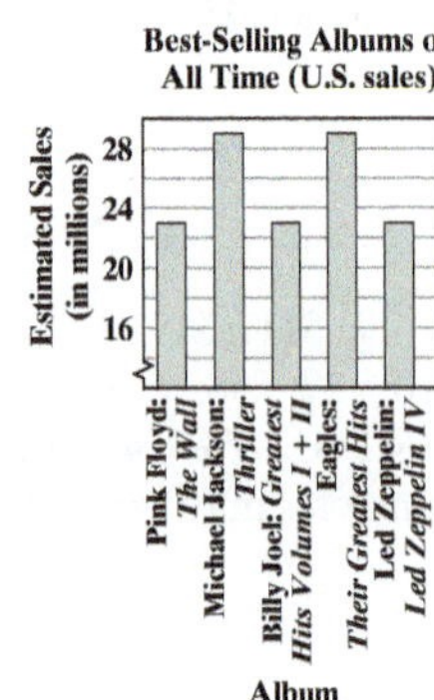

33. 15 adults **35.** 61 adults **37.** 24 adults **39.** 12 adults **41.** $\frac{9}{100}$ **43.** 45–54 **45.** 21 million householders **47.** 44 million householders **49.** 4 million householders

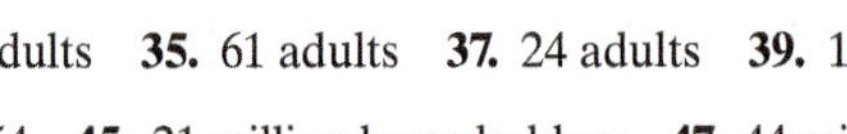

51. |; 1 **53.** 卌 |||; 8 **55.** 卌 |; 6 **57.** 卌 |; 6 **59.** ||; 2

61.

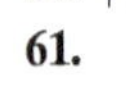

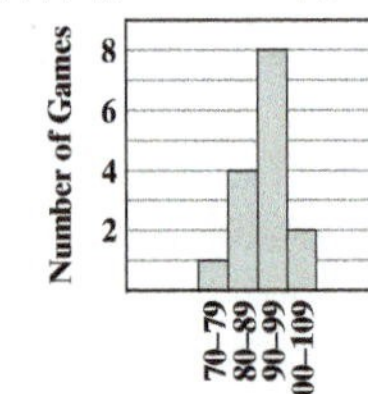

63. 8.4 goals/game **65.** 2003 **67.** increase **69.** 2001, 2007 **71.** 3.6 **73.** 6.2 **75.** 25% **77.** 34% **79.** 83°F **81.** Sunday; 68°F **83.** Tuesday; 13°F **85.** answers may vary

Section 7.2

Vocabulary, Readiness & Video Check **1.** circle **3.** 360 **5.** 100%

Exercise Set 7.2 **1.** parent or guardian's home **3.** $\frac{9}{35}$ **5.** $\frac{9}{16}$ **7.** Asia **9.** 37% **11.** 17,100,000 sq mi **13.** 2,850,000 sq mi **15.** 55% **17.** nonfiction **19.** 31,400 books **21.** 27,632 books **23.** 25,120 books

25.

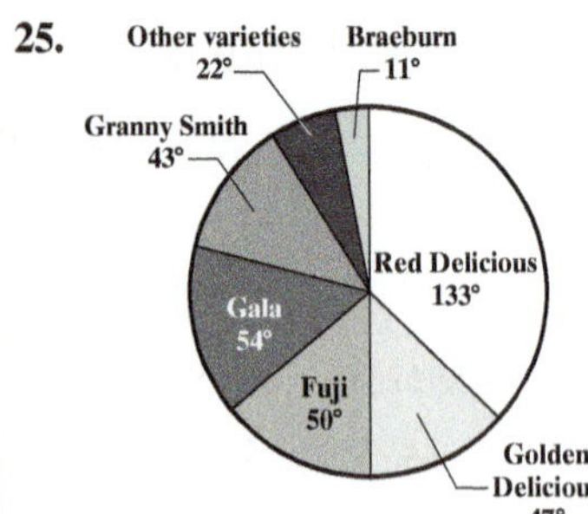

27.

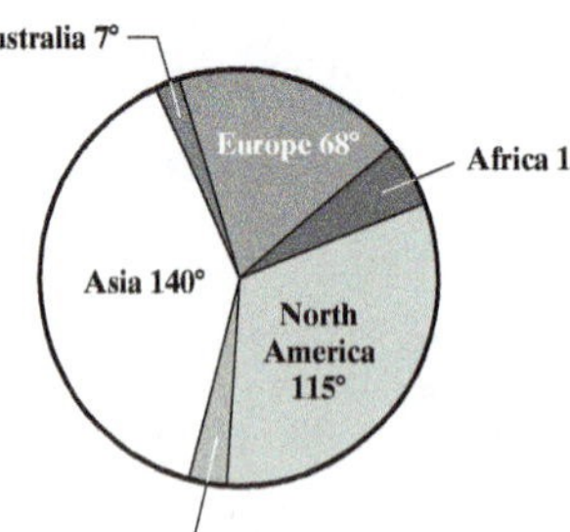

29. $2^2 \times 5$ **31.** $2^3 \times 5$ **33.** 5×17 **35.** answers may vary **37.** 129,600,002 sq km **39.** 55,542,858 sq km **41.** 672 respondents **43.** 2408 respondents **45.** $\frac{12}{31}$ **47.** no; answers may vary

Integrated Review **1.** 700,000 **2.** 725,000 **3.** registered nurses **4.** food preparation and service workers **5.** Oroville Dam; 755 ft **6.** New Bullards Bar Dam; 635 ft **7.** 15 ft **8.** 4 dams **9.** Thursday and Saturday; 100°F **10.** Monday; 82°F **11.** Sunday, Monday, and Tuesday **12.** Wednesday, Thursday, Friday, and Saturday **13.** 70 qt containers **14.** 52 qt containers **15.** 2 qt containers **16.** 6 qt containers **17.** ||; 2 **18.** |; 1 **19.** |||; 3 **20.** 卌 |; 6 **21.** 卌; 5 **22.**

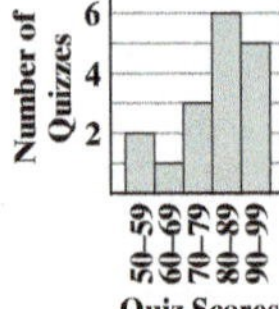

Section 7.3

Calculator Explorations **1.** 32 **3.** 3.873 **5.** 9.849

Vocabulary, Readiness & Video Check **1.** 10; −10 **3.** radical **5.** perfect squares **7.** c^2; b^2 **9.** The square roots of 49 are 7 and −7. The radical sign means the positive square root only, so $\sqrt{49} = 7$. **11.** The hypotenuse is the side across from the right angle.

Exercise Set 7.3 **1.** 2 **3.** 11 **5.** $\frac{1}{9}$ **7.** $\frac{4}{8} = \frac{1}{2}$ **9.** 1.732 **11.** 3.873 **13.** 5.568 **15.** 5.099 **17.** 6, 7 **19.** 10, 11 **21.** 16 **23.** 9.592 **25.** $\frac{7}{12}$ **27.** 8.426 **29.** 13 in. **31.** 6.633 cm **33.** 52.802 m **35.** 117 mm **37.** 5 **39.** 12 **41.** 17.205 **43.** 44.822 **45.** 42.426 **47.** 1.732 **49.** 8.5 **51.** 141.42 yd **53.** 25.0 ft **55.** 340 ft **57.** $\frac{5}{6}$ **59.** $\frac{x}{30}$ **61.** $\frac{21}{13y}$ **63.** $\frac{9x}{64}$ **65.** 6 **67.** 10 **69.** answers may vary **71.** yes **73.** $\sqrt{80} - 6 \approx 2.94$ in.

Section 7.4

Vocabulary, Readiness & Video Check **1.** false **3.** true **5.** false **7.** The ratios of corresponding sides are the same.

Exercise Set 7.4 **1.** congruent; SSS **3.** not congruent **5.** congruent; ASA **7.** congruent; SAS **9.** $\frac{2}{1}$ **11.** $\frac{3}{2}$ **13.** 4.5 **15.** 6 **17.** 5 **19.** 13.5 **21.** 17.5 **23.** 10 **25.** 28.125 **27.** 10 **29.** 520 ft **31.** 500 ft **33.** 14.4 ft **35.** 52 neon tetras **37.** 381 ft **39.** 4.01 **41.** −1.23 **43.** $3\frac{8}{9}$ in.; no **45.** 8.4 **47.** answers may vary **49.** 200 ft, 300 ft, 425 ft

Section 7.5

Vocabulary, Readiness & Video Check **1.** outcome **3.** probability **5.** 0 **7.** The number of outcomes equals the number of branches you have at the end.

Exercise Set 7.5

1. M: 1, 2, 3; A: 1, 2, 3; T: 1, 2, 3; H: 1, 2, 3 — 12 outcomes
3. Red, Blue, Yellow — 3 outcomes
5. 1: 1, 2, 3, 4; 2: 1, 2, 3, 4; 3: 1, 2, 3, 4; 4: 1, 2, 3, 4 — 16 outcomes
7. Red: 1, 2, 3, 4; Blue: 1, 2, 3, 4; Yellow: 1, 2, 3, 4 — 12 outcomes
9. H: 1, 2, 3, 4; T: 1, 2, 3, 4 — 8 outcomes

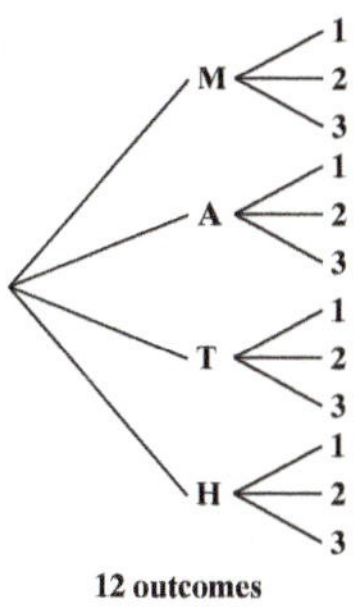

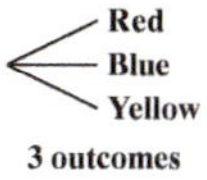

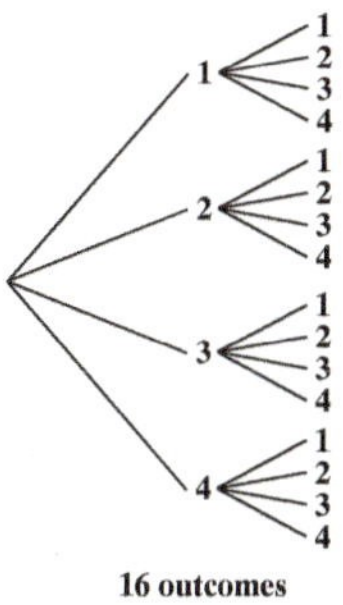

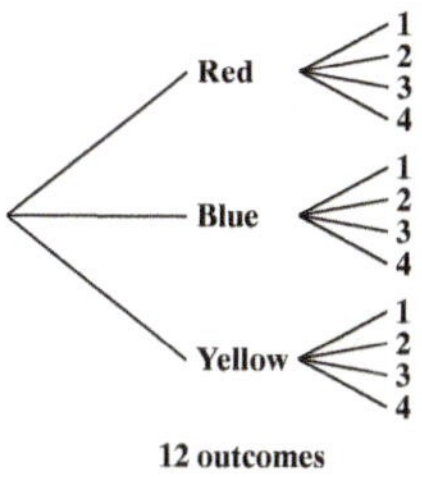

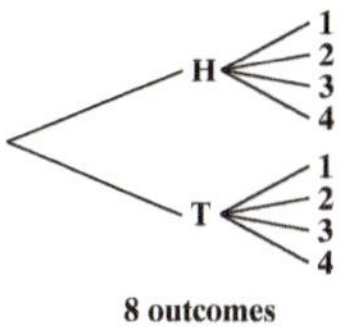

11. $\frac{1}{6}$ **13.** $\frac{1}{3}$ **15.** $\frac{1}{2}$ **17.** $\frac{2}{3}$ **19.** $\frac{1}{3}$ **21.** 1 **23.** $\frac{2}{3}$ **25.** $\frac{1}{7}$ **27.** $\frac{2}{7}$ **29.** $\frac{4}{7}$ **31.** $\frac{19}{100}$ **33.** $\frac{1}{20}$ **35.** $\frac{5}{6}$ **37.** $\frac{1}{6}$ **39.** $6\frac{2}{3}$ **41.** $\frac{1}{52}$ **43.** $\frac{1}{13}$ **45.** $\frac{1}{4}$ **47.** $\frac{1}{2}$ **49.** $\frac{5}{36}$ **51.** 0 **53.** answers may vary

Chapter 7 Vocabulary Check **1.** Congruent **2.** Similar **3.** leg **4.** leg **5.** hypotenuse **6.** right **7.** Pythagorean **8.** bar **9.** outcomes **10.** pictograph **11.** tree diagram **12.** experiment **13.** circle **14.** probability **15.** histogram; class interval; class frequency

Chapter 7 Review **1.** 2,250,000 **3.** South **5.** South, West **7.** 30% **9.** 1990, 2000, 2010, 2012 **11.** 962 (exact number) **13.** 930 **15.** 25 **17.** 1 employee **19.** 18 employees **21.** 卌; 5 **23.** ||||; 4 **25.** mortgage payment **27.** $1225 **29.** $\frac{39}{160}$ **31.** 31 **33.** 5 **35.** 8 **37.** 3.464 **39.** 0 **41.** 7.071 **43.** $\frac{2}{5}$ **45.** 13 **47.** 10.7 **49.** 28.28 cm **51.** congruent; ASA **53.** $13\frac{1}{3}$ **55.** approximately 33 ft

Answers to Selected Exercises

57. **59.**

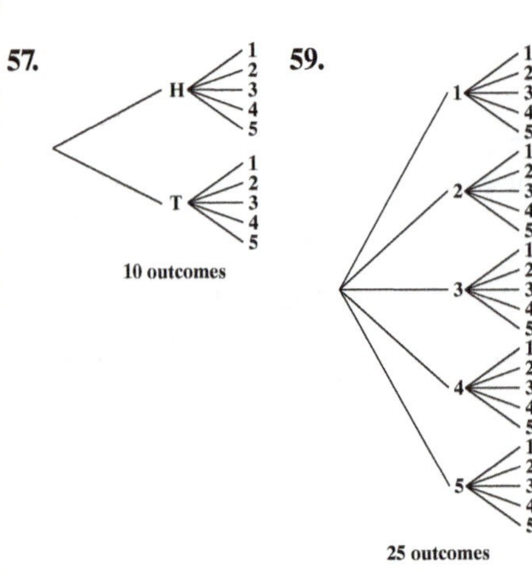

61. $\frac{1}{6}$ **63.** $\frac{1}{5}$ **65.** $\frac{3}{5}$ **67.** $\frac{1}{4}$ **69.** $\frac{1}{4}$ **71.** Insight 2WD Manual; 66 mpg **73.** Insight 2WD CVT; 55 mpg **75.** 6 **77.** 10.247 **79.** 86.6 **81.** 12

Chapter 7 Test **1.** \$225 **2.** 3rd week; \$350 **3.** \$1100 **4.** June, August, September **5.** February; 3 cm **6.** March and November **7.**

Most Common Blood Types by Percent in the Population

Percent of Population; Blood Type: O+, A+, B+, O−, A−, AB+, B−, AB−

8. 1.7% **9.** 2004, 2005, 2007 **10.** 2005–2006, 2007–2008, 2009–2010, 2011–2012 **11.** $\frac{17}{40}$ **12.** $\frac{31}{22}$ **13.** 45 million **14.** 12 million **15.** 9 students **16.** 11 students **17.** |, 1; |||, 3; ||||, 4; ~~||||~~, 5; ~~||||~~ |||, 8; ||||, 4

18.

Number of Students; Scores: 40–49, 50–59, 60–69, 70–79, 80–89, 90–99

19. 7 **20.** 12.530 **21.** $\frac{8}{10} = \frac{4}{5}$ **22.** 5.66 cm **23.** 7.5 **24.** 69 ft

25. **26.**

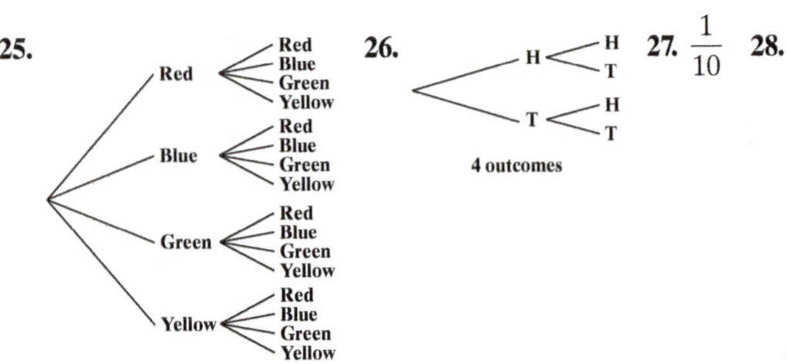

27. $\frac{1}{10}$ **28.** $\frac{1}{5}$

Cumulative Review Chapters 1–7 **1.** 47; Sec. 1.7, Ex. 12 **3.** −12; Sec. 2.3, Ex. 11 **5.** −3; Sec. 3.2, Ex. 2 **7.** 2; Sec. 4.8, Ex. 7 **9.** $7\frac{17}{24}$; Sec. 4.7, Ex. 9 **11.** $5\frac{9}{10}$; Sec. 5.1, Ex. 8 **13.** 3.432; Sec. 5.2, Ex. 5 **15.** 0.8496; Sec. 5.3, Ex. 2 **17.** −0.052; Sec. 5.4, Ex. 3 **19.** 4.09; Sec. 5.5, Ex. 13 **21.** 0.25; Sec. 5.5, Ex. 1 **23.** 0.7; Sec. 5.6, Ex. 5 **25.** 8.944; Sec. 7.3, Ex. 7b **27.** $\frac{1}{6}$; Sec. 7.3, Ex. 5 **29.** 4.90 in.; Sec. 7.3, Ex. 11 **31.** $\frac{31}{2}$; Sec. 6.1, Ex. 5 **33.** $\frac{12}{19}$; Sec. 7.4, Ex. 2 **35.** 0.046; Sec. 6.2, Ex. 4 **37.** 0.0074; Sec. 6.2, Ex. 6 **39.** 21; Sec. 6.3, Ex. 7 **41.** 52; Sec. 6.4, Ex. 9 **43.** 8.5%; Sec. 6.6, Ex. 2 **45.** \$6772.12; Sec. 6.7, Ex. 5 **47.** 57; Sec. 5.7, Ex. 3 **49.** $\frac{1}{3}$; Sec. 7.5, Ex. 4

Chapter 8 Geometry and Measurement

Section 8.1

Vocabulary, Readiness & Video Check **1.** plane **3.** Space **5.** ray **7.** straight **9.** acute **11.** Parallel; intersecting **13.** degrees **15.** vertical **17.** $\angle WUV, \angle VUW, \angle U, \angle x$ **19.** $180° - 17° = 163°$

Exercise Set 8.1 **1.** line; line CD or line l or $\overleftrightarrow{CD}$ **3.** line segment; line segment MN or $\overline{MN}$ **5.** angle; $\angle GHI$ or $\angle IHG$ or $\angle H$ **7.** ray; ray UW or $\overrightarrow{UW}$ **9.** $\angle CPR, \angle RPC$ **11.** $\angle TPM, \angle MPT$ **13.** straight **15.** right **17.** obtuse **19.** acute **21.** 67°

23. 163° **25.** 32° **27.** 30° **29.** $\angle MNP$ and $\angle RNO$; $\angle PNQ$ and $\angle QNR$ **31.** $\angle SPT$ and $\angle TPQ$; $\angle SPR$ and $\angle RPQ$; $\angle SPT$ and $\angle SPR$; $\angle TPQ$ and $\angle QPR$ **33.** 27° **35.** 132° **37.** $m\angle x = 30°$; $m\angle y = 150°$; $m\angle z = 30°$ **39.** $m\angle x = 77°$; $m\angle y = 103°$; $m\angle z = 77°$ **41.** $m\angle x = 100°$; $m\angle y = 80°$; $m\angle z = 100°$ **43.** $m\angle x = 134°$; $m\angle y = 46°$; $m\angle z = 134°$ **45.** $\angle ABC$ or $\angle CBA$ **47.** $\angle DBE$ or $\angle EBD$ **49.** 15° **51.** 50° **53.** 65° **55.** 95° **57.** $\frac{9}{8}$ or $1\frac{1}{8}$ **59.** $\frac{7}{32}$ **61.** $\frac{5}{6}$ **63.** $1\frac{1}{3}$ **65.** 360° **67.** 54.8° **69.** false; answers may vary **71.** true **73.** $m\angle a = 60°$; $m\angle b = 50°$; $m\angle c = 110°$; $m\angle d = 70°$; $m\angle e = 120°$ **75.** no; answers may vary **77.** 45°; 45°

Section 8.2

Vocabulary, Readiness & Video Check **1.** perimeter **3.** π **5.** $\frac{22}{7}$ (or 3.14); $3.14\left(\text{or } \frac{22}{7}\right)$ **7.** Opposite sides of a rectangle have the same measure, so we can just find the sum of the measures of all four sides.

Exercise Set 8.2 **1.** 64 ft **3.** 120 cm **5.** 21 in. **7.** 48 ft **9.** 42 in. **11.** 155 cm **13.** 21 ft **15.** 624 ft **17.** 346 yd **19.** 22 ft **21.** \$55 **23. a.** 8 **b.** 72 in. **25.** 28 in. **27.** \$36.12 **29.** 96 m **31.** 66 ft **33.** 74 cm **35.** 17π cm; 53.38 cm **37.** 16π mi; 50.24 mi **39.** 26π m; 81.64 m **41.** 15π ft; 47.1 ft **43.** 12,560 ft **45.** 30.7 mi **47.** 14π cm $\approx$ 43.96 cm **49.** 40 mm **51.** 84 ft **53.** 23 **55.** 1 **57.** 6 **59.** 10 **61. a.** width: 30 yd; length: 40 yd **b.** 140 yd **63.** b **65. a.** 62.8 m; 125.6 m **b.** yes **67.** answers may vary **69.** 27.4 m **71.** 75.4 m

Section 8.3

Vocabulary, Readiness & Video Check **1.** surface area **3.** Area **5.** square **7.** We don't have a formula for an L-shaped figure, so we divide it into two rectangles, use the formula to find the area of each, and then add these two areas.

Exercise Set 8.3 **1.** 7 sq m **3.** $9\frac{3}{4}$ sq yd **5.** 15 sq yd **7.** 2.25π sq in. $\approx$ 7.065 sq in. **9.** 36.75 sq ft **11.** 28 sq m **13.** 22 sq yd **15.** $36\frac{3}{4}$ sq ft **17.** $22\frac{1}{2}$ sq in. **19.** 25 sq cm **21.** 86 sq mi **23.** 24 sq cm **25.** 36π sq in. $\approx 113\frac{1}{7}$ sq in. **27.** $V = 72$ cu in.; $SA = 108$ sq in. **29.** $V = 512$ cu cm; $SA = 384$ sq cm **31.** $V = 4\pi$ cu yd $\approx 12\frac{4}{7}$ cu yd; $SA = (2\pi\sqrt{13} + 4\pi)$ sq yd $\approx$ 35.20 sq yd **33.** $V = \frac{500}{3}\pi$ cu in. $\approx 523\frac{17}{21}$ cu in.; $SA = 100\pi$ sq in. $\approx 314\frac{2}{7}$ sq in. **35.** $V = 9\pi$ cu in. $\approx 28\frac{2}{7}$ cu in. **37.** $V = 75$ cu cm **39.** $2\frac{10}{27}$ cu in. **41.** $V = 8.4$ cu ft; $SA = 26$ sq ft **43.** 113,625 sq ft **45.** 168 sq ft **47.** 960 cu cm **49.** 9200 sq ft **51.** $V = \frac{1372}{3}\pi$ cu in. or $457\frac{1}{3}\pi$ cu in.; $SA = 196\pi$ sq in. **53. a.** 381 sq ft **b.** 4 squares **55.** $V = 5.25\pi$ cu in. **57.** 4π sq ft $\approx$ 12.56 sq ft **59.** $V = 4.5\pi$ cu m; 14.13 cu m **61.** 168 sq ft **63.** $10\frac{5}{6}$ cu in. **65.** 8.8 cu in. **67.** 25 **69.** 9 **71.** 5 **73.** 20 **75.** perimeter **77.** area **79.** area **81.** perimeter **83.** 12-in. pizza **85.** 2093.33 cu m **87.** no; answers may vary **89.** 7.74 sq in. **91.** 298.5 sq m **93.** no; answers may vary

Integrated Review **1.** 153°; 63° **2.** $m\angle x = 75°$; $m\angle y = 105°$; $m\angle z = 75°$ **3.** $m\angle x = 128°$; $m\angle y = 52°$; $m\angle z = 128°$ **4.** $m\angle x = 52°$ **5.** 4.6 in. **6.** $4\frac{1}{4}$ in. **7.** 20 m; 25 sq m **8.** 12 ft; 6 sq ft **9.** 10π cm $\approx$ 31.4 cm; 25π sq cm $\approx$ 78.5 sq cm **10.** 32 mi; 44 sq mi **11.** 54 cm; 143 sq cm **12.** 62 ft; 238 sq ft **13.** $V = 64$ cu in.; $SA = 96$ sq in. **14.** $V = 30.6$ cu ft; $SA = 63$ sq ft **15.** $V = 400$ cu cm **16.** $V = 4\frac{1}{2}\pi$ cu mi $\approx 14\frac{1}{7}$ cu mi

Section 8.4

Vocabulary, Readiness & Video Check **1.** meter **3.** yard **5.** feet **7.** feet **9.** Both mean addition; $5\frac{2}{5} = 5 + \frac{2}{5}$ and 5 ft 2 in. = 5 ft + 2 in. **11.** Since the metric system is based on base 10, we just need to move the decimal point to convert from one unit to another.

Exercise Set 8.4 **1.** 5 ft **3.** 36 ft **5.** 8 mi **7.** 102 in. **9.** $3\frac{1}{3}$ yd **11.** 33,792 ft **13.** 4.5 yd **15.** 0.25 ft **17.** 13 yd 1 ft **19.** 7 ft 1 in. **21.** 1 mi 4720 ft **23.** 62 in. **25.** 26 ft **27.** 84 in. **29.** 11 ft 2 in. **31.** 22 yd 1 ft **33.** 6 ft 5 in. **35.** 7 ft 6 in. **37.** 14 ft 4 in. **39.** 83 yd 1 ft **41.** 6000 cm **43.** 4 cm **45.** 0.5 km **47.** 1.7 m **49.** 15 m **51.** 42,000 cm **53.** 7000 m **55.** 83 mm **57.** 0.201 dm **59.** 40 mm **61.** 8.94 m **63.** 2.94 m or 2940 mm **65.** 1.29 cm or 12.9 mm **67.** 12.64 km or 12,640 m **69.** 54.9 m **71.** 1.55 km **73.** $348\frac{2}{3}$; 12,552 **75.** $11\frac{2}{3}$; 420 **77.** 5000; 0.005; 500 **79.** 0.065; 65; 0.000065 **81.** 342,000; 342,000,000; 34,200,000 **83.** 10 ft 6 in. **85.** 5100 ft **87.** 5.0 times **89.** 26.7 mm **91.** 15 ft 9 in. **93.** 3.35 m **95.** $121\frac{1}{3}$ yd **97.** $\frac{21}{100}$ **99.** 0.13 **101.** 0.25 **103.** no **105.** yes **107.** no **109.** Estimate: 13 yd **111.** answers may vary; for example, $1\frac{1}{3}$ yd or 48 in. **113.** answers may vary **115.** 334.89 sq m

Section 8.5

Vocabulary, Readiness & Video Check **1.** Mass **3.** gram **5.** 2000 **7.** We can't subtract 9 oz from 4 oz, so we borrow 1 lb(= 16 oz) from 12 lb to add to the 4 oz; 12 lb 4 oz becomes 11 lb 20 oz. **9.** 18.50 dg

Exercise Set 8.5 **1.** 32 oz **3.** 10,000 lb **5.** 9 tons **7.** $3\frac{3}{4}$ lb **9.** $1\frac{3}{4}$ tons **11.** 204 oz **13.** 9800 lb **15.** 76 oz **17.** 1.5 tons **19.** $\frac{1}{20}$ lb **21.** 92 oz **23.** 161 oz **25.** 5 lb 9 oz **27.** 53 lb 10 oz **29.** 8 tons 750 lb **31.** 3 tons 175 lb **33.** 8 lb 11 oz **35.** 31 lb 2 oz **37.** 1 ton 700 lb **39.** 0.5 kg **41.** 4000 mg **43.** 25,000 g **45.** 0.048 g **47.** 0.0063 kg **49.** 15,140 mg **51.** 6250 g **53.** 350,000 cg **55.** 13.5 mg **57.** 5.815 g or 5815 mg **59.** 1850 mg or 1.85 g **61.** 1360 g or 1.36 kg **63.** 13.52 kg **65.** 2.125 kg **67.** 200,000; 3,200,000 **69.** $\frac{269}{400}$ or 0.6725; 21,520 **71.** 0.5; 0.0005; 50 **73.** 21,000; 21,000,000; 2,100,000 **75.** 8.064 kg **77.** 30 mg **79.** 5 lb 8 oz **81.** 35 lb 14 oz **83.** 6 lb 15.4 oz **85.** 144 mg **87.** 6.12 kg **89.** 130 lb **91.** 211 lb **93.** 0.16 **95.** 0.875 **97.** no **99.** yes **101.** no **103.** answers may vary; for example, 250 mg or 0.25 g **105.** true **107.** answers may vary

Section 8.6

Vocabulary, Readiness & Video Check **1.** capacity **3.** fluid ounces **5.** cups **7.** quarts **9.** We can't subtract 3 qt from 0 qt, so we borrow 1 gal(= 4 qt) from 3 gal to get 2 gal 4 qt. **11.** 0.45 dal

Exercise Set 8.6 **1.** 4 c **3.** 16 pt **5.** $3\frac{1}{2}$ gal **7.** 5 pt **9.** 8 c **11.** $3\frac{3}{4}$ qt **13.** $10\frac{1}{2}$ qt **15.** 9 c **17.** 23 qt **19.** $\frac{1}{4}$ pt **21.** 14 gal 2 qt **23.** 4 gal 3 qt 1 pt **25.** 22 pt **27.** 13 gal 2 qt **29.** 4 c 4 fl oz **31.** 1 gal 1 qt **33.** 2 gal 3 qt 1 pt **35.** 17 gal **37.** 4 gal 3 qt **39.** 5000 ml **41.** 0.00016 kl **43.** 5.6 L **45.** 320 cl **47.** 0.41 kl **49.** 0.064 L **51.** 160 L **53.** 3600 ml **55.** 19.3 L **57.** 4.5 L or 4500 ml **59.** 8410 ml or 8.41 L **61.** 16,600 ml or 16.6 L **63.** 3840 ml **65.** 162.4 L **67.** 336; 84; 168 **69.** $\frac{1}{4}$; 1; 2 **71.** 1.59 L **73.** 18.954 L **75.** 4.3 fl oz **77.** yes **79.** $0.316 **81.** $\frac{4}{5}$ **83.** $\frac{3}{5}$ **85.** $\frac{9}{10}$ **87.** no **89.** no **91.** less than; answers may vary **93.** answers may vary **95.** 128 fl oz **97.** 1.5 cc **99.** 2.7 cc **101.** 54 u or 0.54 cc **103.** 86 u or 0.86 cc

Section 8.7

Vocabulary, Readiness & Video Check **1.** 1 L $\approx$ 0.26 gal or 3.79 L $\approx$ 1 gal **3.** $F = 1.8C + 32$; 27

Exercise Set 8.7 **1.** 25.57 fl oz **3.** 218.44 cm **5.** 40 oz **7.** 57.66 mi **9.** 3.77 gal **11.** 13.5 kg **13.** 1.5; $1\frac{2}{3}$; 150; 60 **15.** 55; 5500; 180; 2160 **17.** 3.94 in. **19.** 80.5 kph **21.** 0.008 oz **23.** 229.6 ft **25.** 9.92 billion mi **27.** yes **29.** 2790 mi **31.** 90 mm **33.** 112.5 g **35.** 104 mph **37.** 26.24 ft **39.** 3 mi **41.** 8 fl oz **43.** b **45.** b **47.** c **49.** d **51.** d **53.** 25°C **55.** 40°C **57.** 122°F **59.** 239°F **61.** −6.7°C **63.** 61.2°C **65.** 197.6°F **67.** 54.3°F **69.** 56.7°C **71.** 80.6°F **73.** 21.1°C **75.** 244.4°F **77.** 7232°F **79.** 29 **81.** 36 **83.** yes **85.** no **87.** no **89.** yes **91.** 2.13 sq m **93.** 1.19 sq m **95.** 1.69 sq m **97.** 4,000,000,000°C **99.** answers may vary

Chapter 8 Vocabulary Check **1.** Weight **2.** Mass **3.** meter **4.** unit fractions **5.** gram **6.** liter **7.** line segment **8.** complementary **9.** line **10.** perimeter **11.** angle; vertex **12.** Area **13.** ray **14.** transversal **15.** straight **16.** volume **17.** vertical **18.** adjacent **19.** obtuse **20.** right **21.** acute **22.** supplementary **23.** surface area

Chapter 8 Review **1.** right **3.** acute **5.** 65° **7.** 58° **9.** 90° **11.** $\angle a$ and $\angle b$; $\angle b$ and $\angle c$; $\angle c$ and $\angle d$; $\angle d$ and $\angle a$ **13.** $m\angle x = 100°$; $m\angle y = 80°$; $m\angle z = 80°$ **15.** $m\angle x = 53°$; $m\angle y = 53°$; $m\angle z = 127°$ **17.** 69 m **19.** 36 m **21.** 32 ft **23.** 5.338 in. **25.** 240 sq ft **27.** 600 sq cm **29.** 49π sq ft $\approx$ 153.86 sq ft **31.** 119 sq in. **33.** 144 sq m **35.** 432 sq ft **37.** $V = 15\frac{5}{8}$ cu in.; $SA = 37\frac{1}{2}$ sq in. **39.** $V = 20{,}000\pi$ cu cm $\approx$ 62,800 cu cm **41.** $2\frac{2}{3}$ cu ft **43.** $7\frac{1}{2}$ cu ft **45.** 9 ft **47.** 7920 ft **49.** 17 yd 1 ft **51.** 4200 cm **53.** 0.01218 m **55.** 21 yd 1 ft **57.** 9.5 cm or 95 mm **59.** 169 yd 2 ft **61.** 617.5 km **63.** $4\frac{1}{8}$ lb **65.** 3 lb 4 oz **67.** 0.027 g **69.** 21 dag **71.** 3 lb 9 oz **73.** 21.5 mg **75.** 4 lb 4 oz **77.** 14 qt **79.** 7 pt **81.** 4 qt 1 pt **83.** 3800 ml **85.** 3060 cl **87.** 1 gal 1 qt **89.** 736 ml or 0.736 L **91.** 2 gal 3 qt **93.** 10.88 L **95.** 22.96 ft **97.** 4.55 gal **99.** 425.25 g **101.** 2.36 in. **103.** 107.6°F **105.** 5.2°C **107.** 1.7°C **109.** 108° **111.** 95° **113.** 27.3 in. **115.** 1624 sq m **117.** $346\frac{1}{2}$ cu in. **119.** 75 in. **121.** 500 cm **123.** 1.4 g **125.** 186.8°F **127.** 9117 m or 9.117 km **129.** 8 gal 1 qt

Chapter 8 Test **1.** 12° **2.** 56° **3.** 50° **4.** $m\angle x = 118°$; $m\angle y = 62°$; $m\angle z = 118°$ **5.** $m\angle x = 73°$; $m\angle y = 73°$; $m\angle z = 73°$ **6.** 6.2 m **7.** 10 in. **8.** circumference = 18π in. $\approx$ 56.52 in.; area = 81π sq in. $\approx$ 254.34 sq in. **9.** perimeter = 24.6 yd; area = 37.1 sq yd **10.** perimeter = 68 in.; area = 185 sq in. **11.** $62\frac{6}{7}$ cu in. **12.** 30 cu ft **13.** 16 in. **14.** 18 cu ft **15.** 62 ft; $115.94 **16.** 23 ft 4 in. **17.** 10 qt **18.** $1\frac{7}{8}$ lb **19.** 5600 lb **20.** $4\frac{3}{4}$ gal **21.** 0.04 g **22.** 2400 g **23.** 36 mm

24. 0.43 g **25.** 830 ml **26.** 1 gal 2 qt **27.** 3 lb 13 oz **28.** 8 ft 3 in. **29.** 2 gal 3 qt **30.** 66 mm or 6.6 cm **31.** 2.256 km or 2256 m **32.** 28.9°C **33.** 54.7°F **34.** 5.6 m **35.** 4 gal 3 qt **36.** 91.4 m **37.** 16 ft 6 in.

Cumulative Review Chapters 1–8 **1.** 5; Sec. 3.3, Ex. 1 **3. a.** $\frac{16}{625}$ **b.** $\frac{1}{16}$; Sec. 4.3, Ex. 10 **5.** $9\frac{3}{10}$; Sec. 4.7, Ex. 11 **7.** $20x - 10.9$; Sec. 5.2, Ex. 13 **9.** 28.4405; Sec. 5.5, Ex. 14 **11.** =; Sec. 5.5, Ex. 9 **13.** −1.3; Sec. 5.6, Ex. 6 **15.** 424 ft; Sec. 7.3, Ex. 12 **17. a.** $\frac{5}{7}$ **b.** $\frac{7}{24}$; Sec. 6.1, Ex. 3 **19.** no; Sec. 2.6, Ex. 1 **21.** −6; Sec. 2.6, Ex. 8 **23.** 22.4 cc; Sec. 6.1, Ex. 7 **25.** 17%; Sec. 6.2, Ex. 1 **27.** 200; Sec. 6.3, Ex. 10 **29.** 2.7; Sec. 6.4, Ex. 7 **31.** 32%; Sec. 6.5, Ex. 5 **33.** sales tax: $6.41; total price: $91.91; Sec. 6.6, Ex. 1 **35.** 57; Sec. 5.7, Ex. 3 **37.** $\frac{1}{4}$; Sec. 7.5, Ex. 5 **39.** 42°; Sec. 8.1, Ex. 4 **41.** 96 in.; Sec. 8.4, Ex. 1 **43.** 4 tons 1650 lb; Sec. 8.4, Ex. 4 **45.** 15°C; Sec. 8.7, Ex. 6

Chapter 9 Equations, Inequalities, and Problem Solving

Section 9.1

Vocabulary, Readiness & Video Check **1.** whole **3.** inequality **5.** real **7.** 0 **9.** To form a true statement: $0 < 7$ **11.** 0 belongs to the whole numbers, the integers, the rational numbers, and the real numbers; because 0 is a rational number, it cannot also be an irrational number.

Exercise Set 9.1 **1.** < **3.** > **5.** = **7.** < **9.** $32 < 212$ **11.** $30 \le 45$ **13.** true **15.** false **17.** true **19.** false **21.** $20 \le 25$ **23.** $6 > 0$ **25.** $-12 < -10$ **27.** $7 < 11$ **29.** $5 \ge 4$ **31.** $15 \ne -2$ **33.** 14,494; −282 **35.** −27,724 **37.** 475; −195

39. 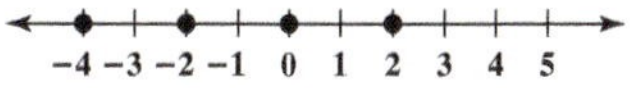

41. 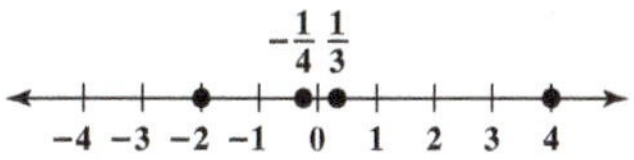

43. 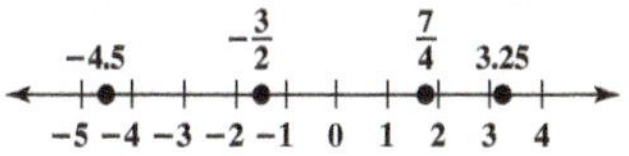

45. whole, integers, rational, real **47.** integers, rational, real **49.** natural, whole, integers, rational, real **51.** rational, real **53.** false **55.** true **57.** false **59.** false **61.** > **63.** = **65.** < **67.** < **69.** 40 million > 14 million, or 40,000,000 > 14,000,000 (exact number) **71.** 40 million pounds less, or −40 million (exact number) **73.** $-0.04 > -26.7$ **75.** sun **77.** sun **79.** answers may vary

Section 9.2

Vocabulary, Readiness & Video Check **1.** commutative property of addition **3.** distributive property **5.** associative property of addition **7.** opposites or additive inverses **9.** 2 is outside the parentheses, so the point is made that you should distribute the −9 only to the terms within the parentheses and not also to the 2.

Exercise Set 9.2 **1.** $16 + x$ **3.** $y \cdot (-4)$ **5.** yx **7.** $13 + 2x$ **9.** $x \cdot (yz)$ **11.** $(2 + a) + b$ **13.** $(4a) \cdot b$ **15.** $a + (b + c)$ **17.** $17 + b$ **19.** $24y$ **21.** y **23.** $26 + a$ **25.** $-72x$ **27.** s **29.** $-\frac{5}{2}x$ **31.** $4x + 4y$ **33.** $9x - 54$ **35.** $6x + 10$ **37.** $28x - 21$ **39.** $18 + 3x$ **41.** $-2y + 2z$ **43.** $-y - \frac{5}{3}$ **45.** $5x + 20m + 10$ **47.** $8m - 4n$ **49.** $-5x - 2$ **51.** $-r + 3 + 7p$ **53.** $3x + 4$ **55.** $-x + 3y$ **57.** $6r + 8$ **59.** $-36x - 70$ **61.** $-1.6x - 2.5$ **63.** $4(1 + y)$ **65.** $11(x + y)$ **67.** $-1(5 + x)$ **69.** $30(a + b)$ **71.** commutative property of multiplication **73.** associative property of addition **75.** commutative property of addition **77.** associative property of multiplication **79.** identity element for addition **81.** distributive property **83.** multiplicative inverse property **85.** identity element for multiplication **87.** 2 **89.** −23 **91.** −25 **93.** $-8; \frac{1}{8}$ **95.** $-x; \frac{1}{x}$ **97.** $2x; -2x$ **99.** false **101.** no **103.** yes **105.** yes **107.** yes **109. a.** commutative property of addition **b.** commutative property of addition **c.** associative property of addition **111.** answers may vary **113.** answers may vary

Section 9.3

Calculator Explorations **1.** solution **3.** not a solution **5.** solution

Vocabulary, Readiness & Video Check **1.** equation **3.** expression **5.** expression **7.** equation **9.** 3; distributive property, addition property of equality, multiplication property of equality **11.** The number of decimal places in each number helps us determine the smallest power of 10 we can multiply through by so we are no longer dealing with decimals.

Exercise Set 9.3 **1.** −6 **3.** 3 **5.** 1 **7.** $\frac{3}{2}$ **9.** 0 **11.** −1 **13.** 4 **15.** −4 **17.** −3 **19.** 2 **21.** 50 **23.** 1 **25.** $\frac{7}{3}$ **27.** 0.2 **29.** all real numbers **31.** no solution **33.** no solution **35.** all real numbers **37.** 18 **39.** $\frac{19}{9}$ **41.** $\frac{14}{3}$ **43.** 13 **45.** 4 **47.** all real numbers **49.** $-\frac{3}{5}$ **51.** −5 **53.** 10 **55.** no solution **57.** 3 **59.** −17 **61.** $(6x - 8)$ m **63.** $-8 - x$ **65.** $-3 + 2x$ **67.** $9(x + 20)$ **69. a.** all real numbers **b.** answers may vary **c.** answers may vary **71.** a **73.** b **75.** c **77.** answers may vary **79. a.** $x + x + x + 2x + 2x = 28$ **b.** $x = 4$ **c.** x cm = 4 cm; $2x$ cm = 8 cm **81.** answers may vary **83.** 15.3 **85.** −0.2

Integrated Review **1.** whole, integers, rational, real **2.** natural, whole, integers, rational, real **3.** rational, real **4.** natural, whole, integers, rational, real **5.** integers, rational, real **6.** rational, real **7.** rational, real **8.** irrational, real **9.** $7d - 11$ **10.** $9z + 48$ **11.** 16 **12.** $-12x - 15$ **13.** 5 **14.** -1 **15.** -2 **16.** -2 **17.** $-\frac{5}{6}$ **18.** $\frac{1}{6}$ **19.** 1 **20.** 6 **21.** 4 **22.** 1 **23.** $\frac{9}{5}$ **24.** $-\frac{6}{5}$ **25.** all real numbers **26.** all real numbers **27.** 0 **28.** -1.6 **29.** $\frac{4}{19}$ **30.** $-\frac{5}{19}$ **31.** $\frac{7}{2}$ **32.** $-\frac{1}{4}$

Section 9.4

Vocabulary, Readiness & Video Check **1.** $2x$; $2x - 31$ **3.** $x + 5$; $2(x + 5)$ **5.** $20 - y$; $\frac{20 - y}{3}$ or $(20 - y) \div 3$ **7.** in the statement of the application **9.** That the three angle measures are consecutive even integers and that they sum to 180°.

Exercise Set 9.4 **1.** $2x + 7 = x + 6$; -1 **3.** $3x - 6 = 2x + 8$; 14 **5.** -25 **7.** $-\frac{3}{4}$ **9.** 3 in.; 6 in.; 16 in. **11.** 1st piece: 5 in.; 2nd piece: 10 in.; 3rd piece: 25 in. **13.** Pennsylvania: 494 million pounds; New York: 720 million pounds **15.** 172 mi **17.** 25 mi **19.** 1st angle: 37.5°; 2nd angle: 37.5°; 3rd angle: 105° **21.** A: 60°; B: 120°; C: 120°; D: 60° **23.** $3x + 3$ **25.** $x + 2$; $x + 4$; $2x + 4$ **27.** $x + 1$; $x + 2$; $x + 3$; $4x + 6$ **29.** $x + 2$; $x + 4$; $2x + 6$ **31.** 234, 235 **33.** Belgium: 32; France: 33; Spain: 34 **35.** 5 ft, 12 ft **37.** Maglev: 361 mph; TGV: 357.2 mph **39.** 43°, 137° **41.** 58°, 60°, 62° **43.** 1 **45.** 280 mi **47.** Stanford; 20; Wisconsin: 14 **49.** Montana: 56 counties; California: 58 counties **51.** Neptune: 14 satellites; Uranus: 27 satellites; Saturn: 62 satellites **53.** -16 **55.** Sahara: 3,500,000 sq mi; Gobi: 500,000 sq mi **57.** Jamaica: 4; Cuba: 5; New Zealand: 6 **59.** females: 30,184; males: 28,604 **61.** 34.5°; 34.5°; 111° **63.** Hawaii **65.** Florida: \$56.2: million; California: \$50.0 million **67.** answers may vary **69.** 34 **71.** 225π **73.** 15 ft by 24 ft **75.** 5400 chirps per hour; 129,600 chirps per day; 47,304,000 chirps per year **77.** answers may vary **79.** answers may vary **81.** c

Section 9.5

Vocabulary, Readiness & Video Check **1.** relationships **3.** That the process of solving this equation for x—dividing both sides by 5, the coefficient of x—is the same process used to solve a formula for a specific variable. Treat whatever is multiplied by that specific variable as the coefficient—the coefficient is all the factors except that specific variable.

Exercise Set 9.5 **1.** $h = 3$ **3.** $h = 3$ **5.** $h = 20$ **7.** $c = 12$ **9.** $r = 2.5$ **11.** $h = \frac{f}{5g}$ **13.** $w = \frac{V}{lh}$ **15.** $y = 7 - 3x$ **17.** $R = \frac{A - P}{PT}$ **19.** $A = \frac{3V}{h}$ **21.** $a = P - b - c$ **23.** $h = \frac{S - 2\pi r^2}{2\pi r}$ **25. a.** area: 103.5 sq ft; perimeter: 41 ft **b.** baseboard: perimeter; carpet: area **27. a.** area: 480 sq in.; perimeter: 120 in. **b.** frame: perimeter; glass: area **29.** 120 ft **31.** -10°C **33.** 6.25 hr **35.** length: 78 ft; width: 52 ft **37.** 18 ft, 36 ft, 48 ft **39.** 306 mi **41.** 61.5°F **43.** 60 chirps per minute **45.** increases **47.** 96 piranhas **49.** 2 bags **51.** one 16-in. pizza **53.** 13 in. **55.** 2.25 hr **57.** 12,090 ft **59.** 50°C **61.** 10.1 min **63.** 686,664 cu in. **65.** 449 cu in. **67.** 333°F **69.** 0.32 **71.** 2.00 or 2 **73.** 17% **75.** 720% **77.** $V = G(N - R)$ **79.** multiplies the volume by 8; answers may vary **81.** $53\frac{1}{3}$ **83.** ○ = (△ − □) / □ **85.** 44.3 sec **87.** $P = 3{,}200{,}000$ **89.** $V = 113.1$

Section 9.6

Vocabulary, Readiness & Video Check **1.** expression **3.** inequality **5.** equation **7.** -5 **9.** 4.1 **11.** A parenthesis indicates $>$ or $<$; a bracket indicates $\geq$ or $\leq$. **13.** $[-2, \infty)$ **15.** is greater than; $>$

Exercise Set 9.6 **1.**  $(-\infty, -1]$ **3.** 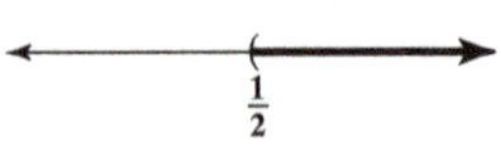 $\left(\frac{1}{2}, \infty\right)$ **5.**  $(-\infty, 4)$ **7.** 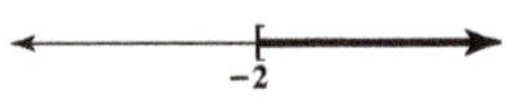 $[-2, \infty)$ **9.** $(-1, 3)$ **11.** $[0, 2)$ **13.** $[-5, \infty)$ **15.** $(-\infty, 9)$ **17.** $(-3, \infty)$ **19.** $(-\infty, 1]$  **21.** $(-\infty, -3)$  **23.** $[-2, \infty)$ **25.** $(-\infty, 0)$ **27.** $\left[-\frac{8}{3}, \infty\right)$ **29.** $(3, \infty)$ **31.** $(-15, \infty)$ **33.** $[-11, \infty)$ **35.** $\left(\frac{1}{4}, \infty\right)$ **37.** $[-12, \infty)$ **39.** $(-\infty, 0)$ **41.** $(-3, \infty)$ **43.** $\left[-\frac{2}{3}, \infty\right)$ **45.** $(-\infty, -2]$ **47.** $(-13, \infty)$ **49.** $(-\infty, -8]$ **51.** $(4, \infty)$ **53.** $\left(-\infty, \frac{5}{4}\right]$ **55.** $\left(\frac{8}{3}, \infty\right)$ **57.** $[0, \infty)$ **59.** all numbers greater than -10 **61.** 35 cm **63.** at least 193 **65.** 86 people **67.** at least 35 min **69.** 81 **71.** 1 **73.** $\frac{49}{64}$ **75.** about 34,100 **77.** 2010 and 2011 **79.** 2008 **81.** $>$ **83.** $\geq$ **85.** when multiplying or dividing by a negative number **87.** final exam score ≥ 78.5

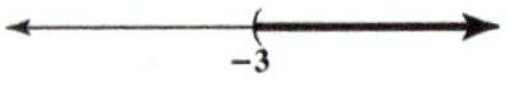

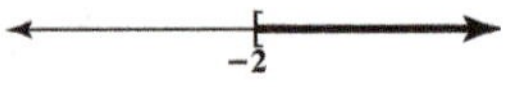

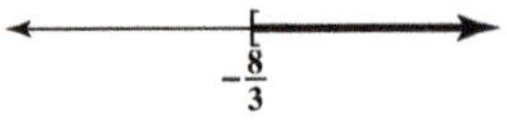

Chapter 9 Vocabulary Check **1.** linear equation in one variable **2.** equivalent inequalities **3.** formula **4.** linear inequality in one variable **5.** all real numbers **6.** no solution **7.** the same **8.** reversed **9.** opposites **10.** reciprocals

Chapter 9 Review **1.** < **3.** > **5.** < **7.** = **9.** > **11.** $4 \geq -3$ **13.** $0.03 < 0.3$ **15. a.** 1, 3 **b.** 0, 1, 3 **c.** −6, 0, 1, 3 **d.** $-6, 0, 1, 1\frac{1}{2}, 3, 9.62$ **e.** π **f.** all numbers in set **17.** Friday **19.** commutative property of addition **21.** distributive property **23.** associative property of addition **25.** distributive property **27.** multiplicative inverse property **29.** −4 **31.** 2 **33.** no solution **35.** $\frac{3}{4}$ **37.** 20 **39.** $\frac{23}{7}$ **41.** 6665.5 in. **43.** national battlefields: 11; national memorials: 29 **45.** 3 **47.** $w = 9$ **49.** $m = \frac{y - b}{x}$ **51.** $x = \frac{2y - 7}{5}$ **53.** $d = \frac{C}{\pi}$ **55.** 15 m **57.** 1 hr and 20 min **59.** −2 $(-\infty, -2]$ **61.** $(-\infty, 1]$ **63.** $(-\infty, 10]$ **65.** $(-\infty, -4)$ **67.** $(9, \infty)$ **69.** $\left(-\infty, \frac{7}{4}\right)$ **71.** at least \$2500 **73.** 4 **75.** $-\frac{3}{2}$ **77.** all real numbers **79.** −13 **81.** $h = \frac{3V}{A}$ **83.** $(-4, \infty)$ −4

Chapter 9 Test **1.** $|-7| > 5$ **2.** $9 + 5 \geq 4$ **3. a.** 1, 7 **b.** 0, 1, 7 **c.** −5, −1, 0, 1, 7 **d.** $-5, -1, 0, \frac{1}{4}, 1, 7, 11.6$ **e.** $\sqrt{7}, 3\pi$ **f.** all numbers in set **4.** associative property of addition **5.** commutative property of multiplication **6.** distributive property **7.** multiplicative inverse property **8.** 9 **9.** −3 **10.** $10y + 1$ **11.** $-2x + 10$ **12.** 8 **13.** no solution **14.** 0 **15.** 27 **16.** 3 **17.** 0.25 **18.** $\frac{25}{7}$ **19.** 21 **20.** 7 gal **21.** $x = 6$ **22.** $h = \frac{V}{\pi r^2}$ **23.** $y = \frac{3x - 10}{4}$ **24.** $(-\infty, -2)$ −2 **25.** $(-\infty, -8]$ **26.** $[11, \infty)$ **27.** $\left(\frac{2}{5}, \infty\right)$ **28.** California: 1107; Ohio: 720

Cumulative Review Chapters 1–9 **1. a.** 4 **b.** −5 **c.** −6; Sec. 2.1, Ex. 6 **3.** −19; Sec. 2.5, Ex. 10 **5.** −8; Sec. 2.2, Ex. 9 **7.** 4; Sec. 2.2, Ex. 10 **9.** −10; Sec. 2.2, Ex. 14 **11.** $-\frac{8}{33}$; Sec. 4.5, Ex. 4 **13.** $\frac{6 - x}{3}$; Sec. 4.5, Ex. 6 **15.** $-\frac{8}{9}$; Sec. 4.6, Ex. 7 **17.** $-5\frac{3}{5}$; Sec. 4.7, Ex. 21 **19.** $\frac{14}{15}$; Sec. 4.7, Ex. 22 **21.** 4.09; Sec. 5.5, Ex. 13 **23.** 7; Sec. 7.3, Ex. 1 **25.** $\frac{2}{5}$; Sec. 7.3, Ex. 6 **27.** 2; Sec. 5.6, Ex. 3 **29.** true; Sec. 9.1, Ex. 3 **31.** true; Sec. 9.1, Ex. 4 **33.** false; Sec. 9.1, Ex. 5 **35.** true; Sec. 9.1, Ex. 6 **37.** $15 - 10z$; Sec. 9.2, Ex. 8 **39.** $3x + 17$; Sec. 9.2, Ex. 12 **41.** 0; Sec. 9.3, Ex. 4 **43.** Democrats: 206; Republicans: 234; Sec. 9.4, Ex. 4 **45.** 79.2 yr; Sec. 9.5, Ex. 1 **47.** −1 $(-\infty, -1)$; Sec. 9.6, Ex. 2 **49.** $[1, \infty)$; Sec. 9.6, Ex. 9

Chapter 10 Graphing

Section 10.1

Vocabulary, Readiness & Video Check **1.** x-axis; y-axis **3.** quadrants; four **5.** one **7.** horizontal: top tourist countries; vertical: number of arrivals (in millions) to these countries **9.** Data occurring in pairs of numbers can be written as ordered pairs, called paired data, and then graphed on a coordinate system. **11.** a linear equation in one variable

Exercise Set 10.1 **1.** France **3.** France, U.S., Spain, Italy, and China **5.** 43 million **7.** 71,000 **9.** 2011; 103,000 **11.** 15.9 **13.** from 1998 to 2000 **15.** 2014 **17.**

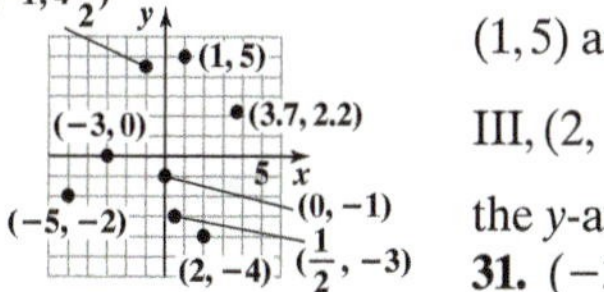

(1, 5) and (3.7, 2.2) are in quadrant I, $\left(-1, 4\frac{1}{2}\right)$ is in quadrant II, (−5, −2) is in quadrant III, (2, −4) and $\left(\frac{1}{2}, -3\right)$ are in quadrant IV, (−3, 0) lies on the x-axis, (0, −1) lies on the y-axis **19.** (0, 0) **21.** (3, 2) **23.** (−2, −2) **25.** (2, −1) **27.** (0, −3) **29.** (1, 3) **31.** (−3, −1) **33. a.** (2006, 25.5), (2007, 26.3), (2008, 27.7), (2009, 29.4), (2010, 31.8) **b.** In the year 2010, the worldwide box office was \$31.8 billion. **c.**

Worldwide Box Office

Dollars (in billions): 25, 26, 27, 28, 29, 30, 31, 32

2006 2007 2008 2009 2010

Year

d. The worldwide box office increased every year. **35. a.** (0.50, 10), (0.75, 12), (1.00, 15), (1.25, 16), (1.50, 18), (1.50, 19), (1.75, 19), (2.00, 20) **b.** When Minh studied 1.25 hours, her quiz score was 16. **c.**

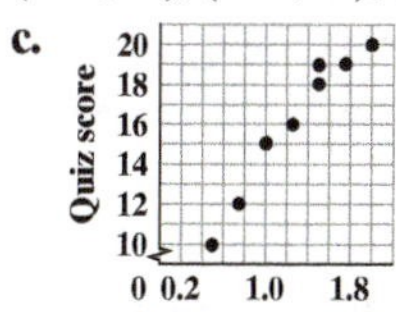

d. answers may vary **37. a.** (2313, 2), (2085, 1), (2711, 21), (2869, 39), (2920, 42), (4038, 99), (1783, 0), (2493, 9) **b.**

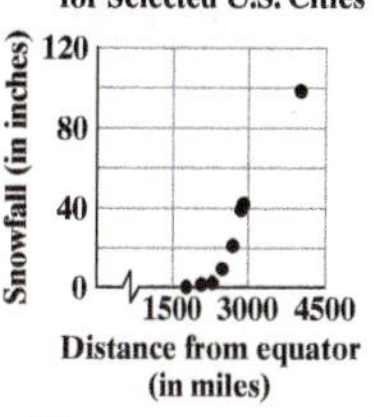

c. The farther from the equator, the more snowfall. **39.** yes; no; yes **41.** yes; yes **43.** no; yes; yes **45.** (−4, −2), (4, 0) **47.** (−8, −5), (16, 1) **49.** 0; 7; $-\frac{2}{7}$ **51.** 2; 2; 5 **53.** 0; −3; 2 **55.** 2; 6; 3 **57.** −12; 5; −6 **59.** $\frac{5}{7}$; $\frac{5}{2}$; −1

61. 0; −5; −2 **63.** 2; 1; −6 **65. a.** 13,000; 21,000; 29,000 **b.** 45 desks **67. a.** 5.52; 6.00; 6.48 **b.** year 9; 2009 **c.** 2016 **d.** In 2005, the average cinema admission price was $6.48. **69.** In 2010, there were 3755 Walmart stores in the U.S.

71. year 8: 100 stores; year 9: 105 stores; year 10: 100 stores **73.** The y-values are all 0. **75.** $y = 5 - x$ **77.** $y = -\frac{1}{2}x + \frac{5}{4}$ **79.** $y = -2x$ **81.** $y = \frac{1}{3}x - 2$ **83.** false **85.** true **87.** negative; negative **89.** positive; negative **91.** 0; 0 **93.** y **95.** no; answers may vary **97.** answers may vary **99.** (4, −7) **101.** 26 units **103. a.** (−2, 6) **b.** 28 units **c.** 45 sq units

Section 10.2

Graphing Calculator Explorations

1.

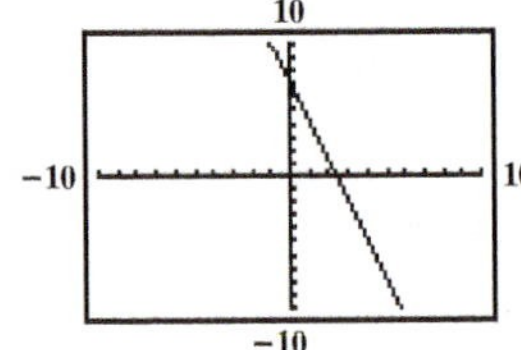

3.

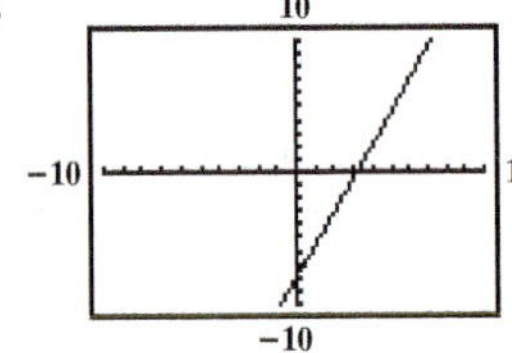

5.

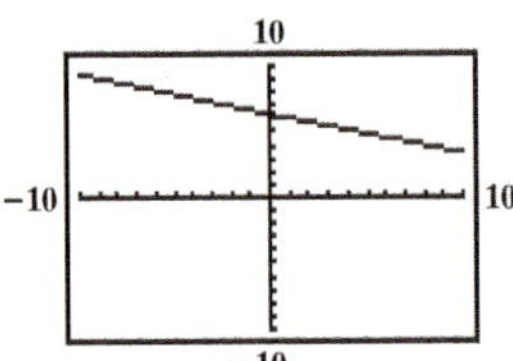

Vocabulary, Readiness & Video Check **1.** In the definition, x and y both have an understood power of 1. Example 3 shows an equation where y has a power of 2, so it is not a linear equation in two variables. **3.** An infinite number of points make up the line and each point corresponds to an ordered pair that is a solution of the linear equation in two variables.

Exercise Set 10.2 **1.** yes **3.** yes **5.** no **7.** yes **9.**

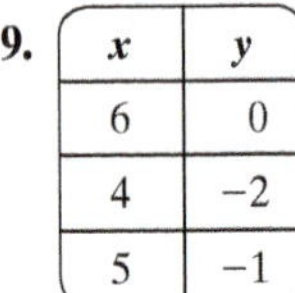

x	y
6	0
4	−2
5	−1

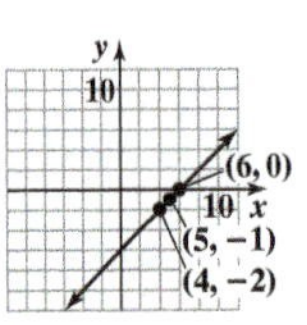

11.

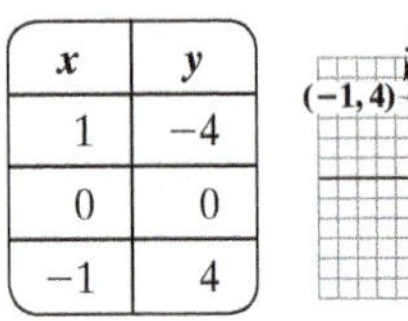

x	y
1	−4
0	0
−1	4

13.

x	y
0	0
6	2
−3	−1

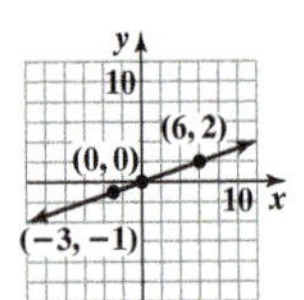

15.

x	y
0	3
1	−1
2	−5

17.

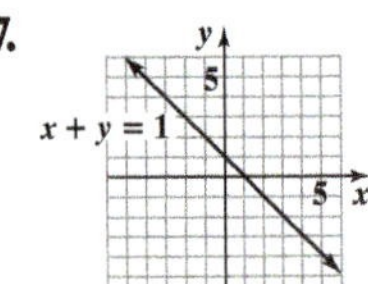

19.

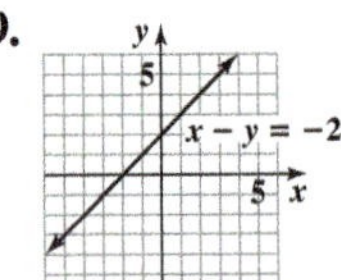

21.

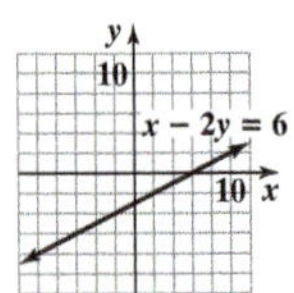

23.

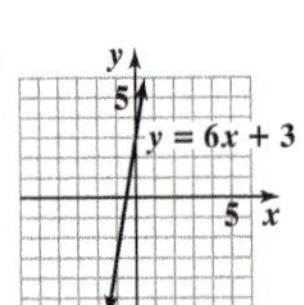

25.

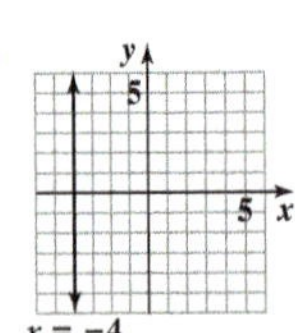

27.

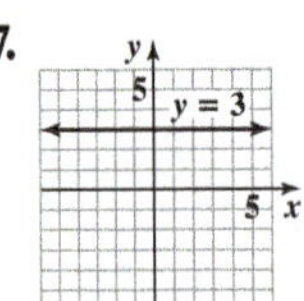

29.

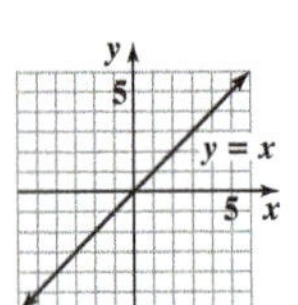

31.

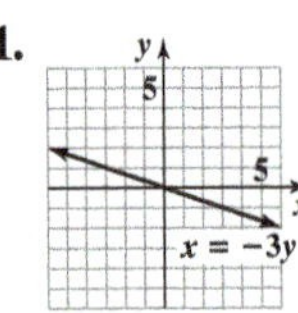

33.

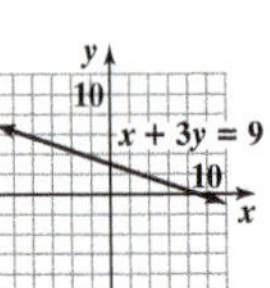

35.

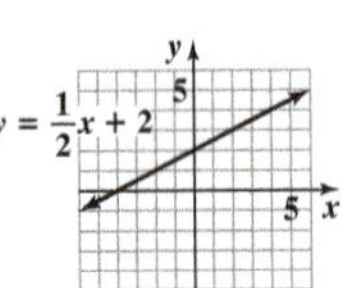

37.

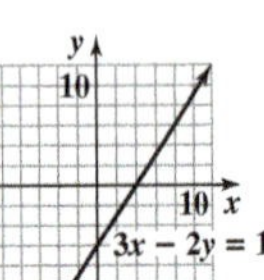

39.

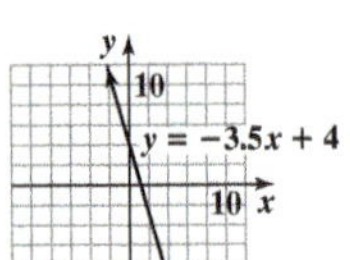

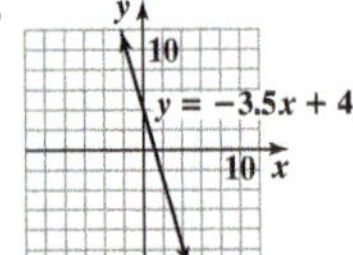

41.

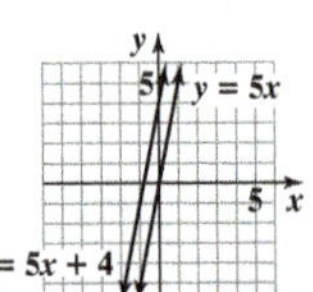

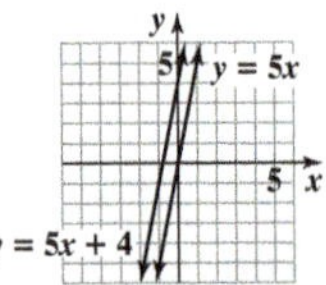

43.

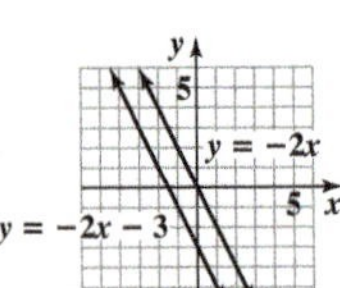

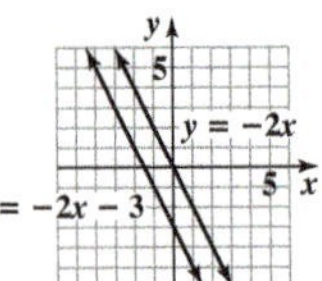

45.

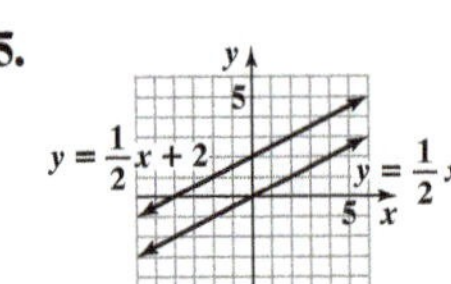

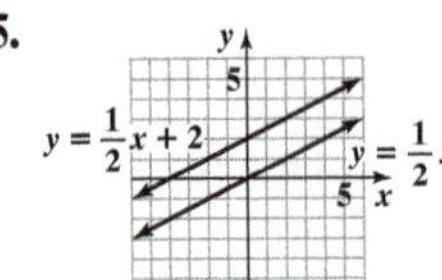

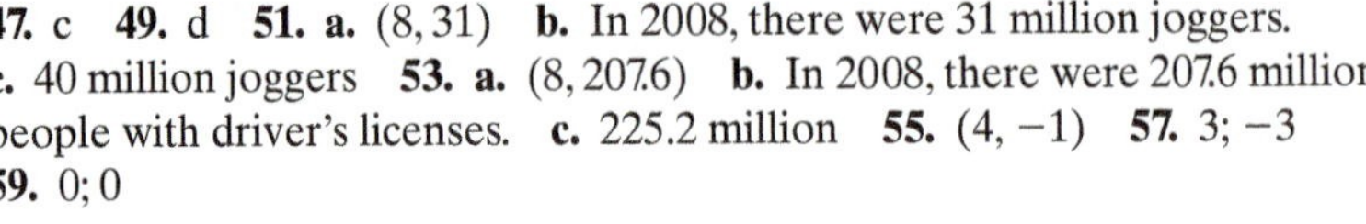

47. c **49.** d **51. a.** (8, 31) **b.** In 2008, there were 31 million joggers. **c.** 40 million joggers **53. a.** (8, 207.6) **b.** In 2008, there were 207.6 million people with driver's licenses. **c.** 225.2 million **55.** (4, −1) **57.** 3; −3 **59.** 0; 0

61. $y = x + 5$

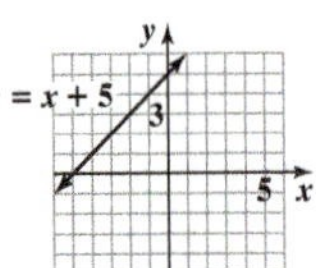

63. $2x + 3y = 6$

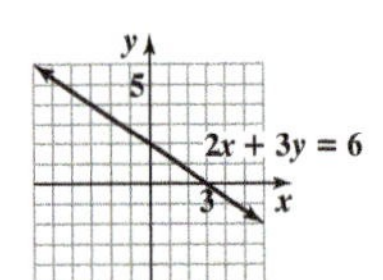

65. $x + y = 12$; $y = 9$ cm **67.** answers may vary **69.** 0; 1; 1; 4; 4

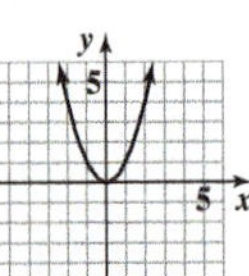

Section 10.3

Graphing Calculator Explorations

1. 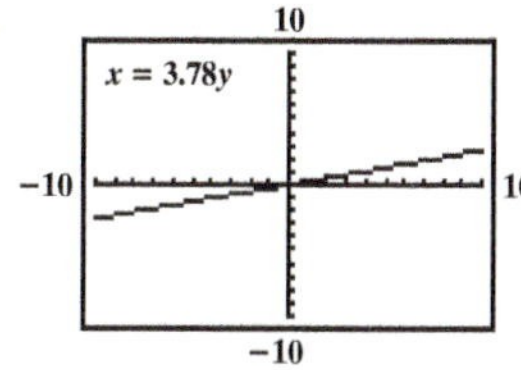

3. 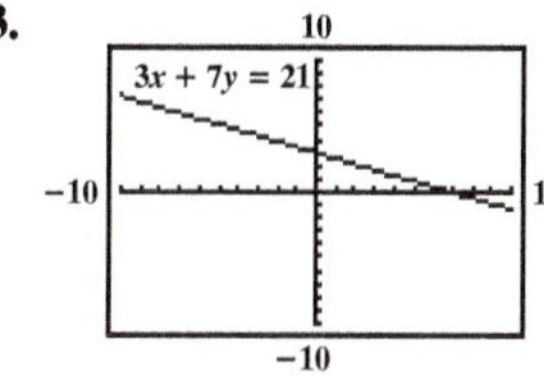

5. 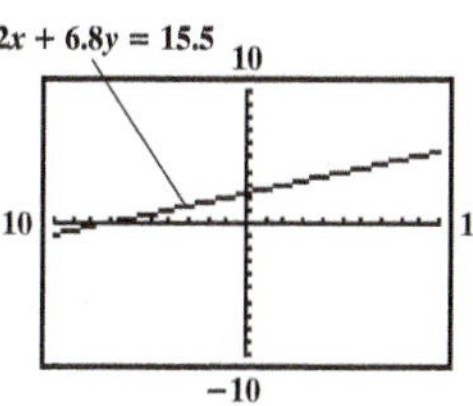

Vocabulary, Readiness & Video Check **1.** linear **3.** horizontal **5.** y-intercept **7.** $y; x$ **9.** Because x-intercepts lie on the x-axis; because y-intercepts lie on the y-axis. **11.** For a horizontal line, the coefficient of x will be 0 and the coefficient of y will be 1; for a vertical line, the coefficient of y will be 0 and the coefficient of x will be 1.

Exercise Set 10.3 **1.** $(-1, 0); (0, 1)$ **3.** $(-2, 0); (2, 0); (0, -2)$ **5.** $(-2, 0); (1, 0); (3, 0); (0, 3)$ **7.** $(-1, 0); (1, 0); (0, 1); (0, -2)$ **9.** infinite **11.** 0 **13.** 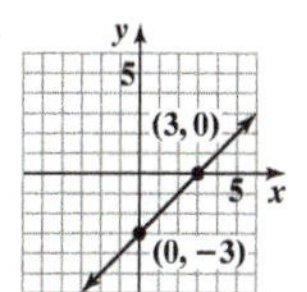

15. 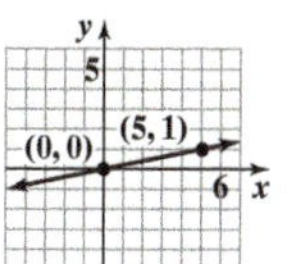

17. 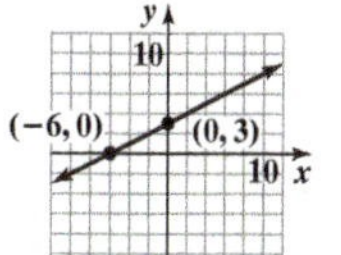

19.

21. 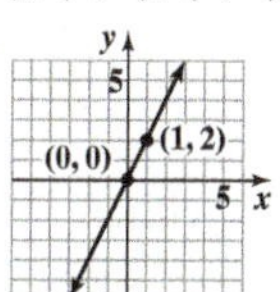

23. 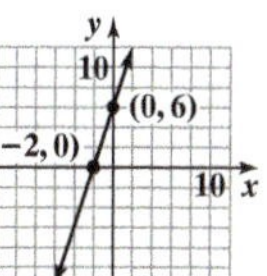

25. 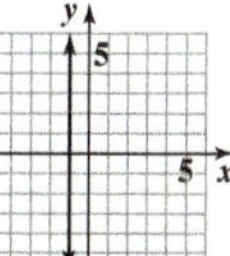

27. 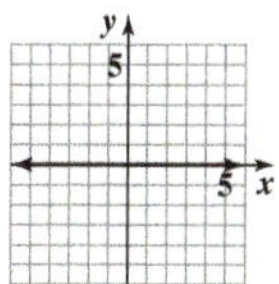

29.

31. 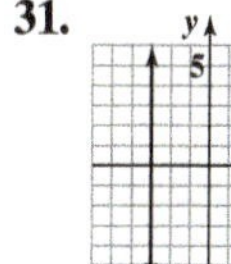

33. 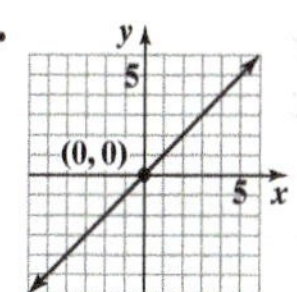

35. 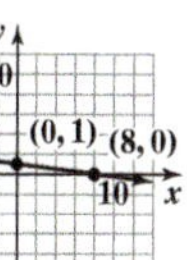

37. 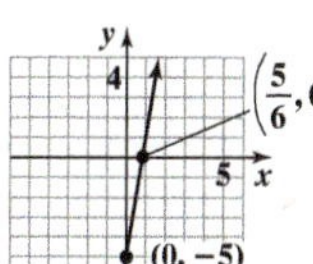

39. 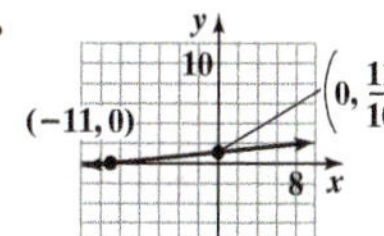

41. 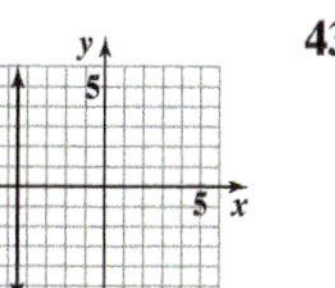

43.

45. 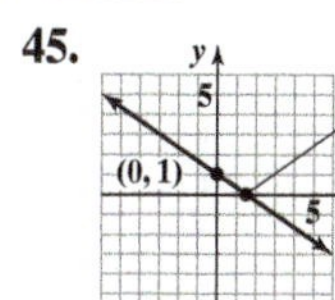

47. 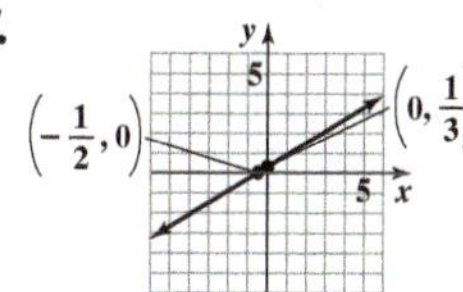

49. C **51.** E **53.** B **55.** $\frac{3}{2}$ **57.** 6 **59.** $-\frac{6}{5}$ **61.** false **63.** true **65.** $(0, 200)$; no chairs and 200 desks are manufactured. **67.** 300 chairs **69. a.** $(62, 0)$ **b.** 62 years after 2002, 0 people will attend the movies at the theater. **c.** answers may vary **71.** $x = 1$ **73.** answers may vary **75.** answers may vary

Section 10.4

Graphing Calculator Explorations

1. 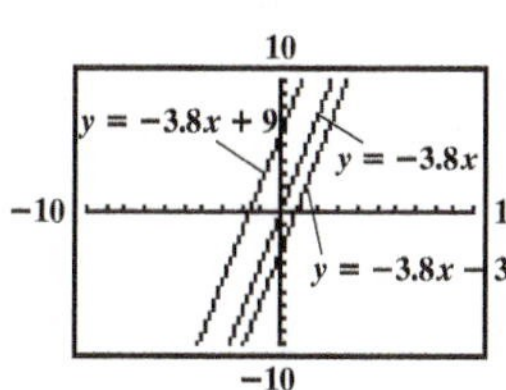

3. 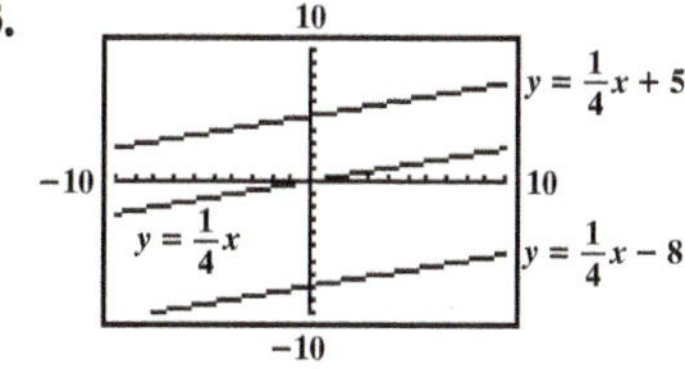

Vocabulary, Readiness & Video Check **1.** slope **3.** 0 **5.** positive **7.** $y; x$ **9.** solve the equation for y; the slope is the coefficient of x. **11.** slope-intercept form; this form makes the slope easy to see, and you need to compare slopes to determine if two lines are parallel or perpendicular.

Exercise Set 10.4 **1.** -1 **3.** undefined **5.** $-\frac{2}{3}$ **7.** 0 **9.** $m = -\frac{4}{3}$ **11.** undefined slope **13.** $m = \frac{5}{2}$ **15.** negative **17.** undefined **19.** upward **21.** horizontal **23.** line 1 **25.** line 2 **27.** D **29.** B **31.** E **33.** undefined slope **35.** $m = 0$ **37.** undefined slope **39.** $m = 0$ **41.** $m = 5$ **43.** $m = -0.3$ **45.** $m = -2$ **47.** $m = \frac{2}{3}$ **49.** undefined slope **51.** $m = \frac{1}{2}$ **53.** $m = 0$ **55.** $m = -\frac{3}{4}$ **57.** $m = 4$ **59. a.** 1 **b.** -1 **61. a.** $\frac{9}{11}$ **b.** $-\frac{11}{9}$ **63.** neither **65.** neither **67.** parallel **69.** perpendicular **71.** $\frac{3}{5}$ **73.** 12.5% **75.** 40% **77.** 37%; 35% **79.** $m = \frac{1}{1}$ or 1; Every 1 year, there are 1 million more U.S. households with televisions. **81.** $m = 0.47$; It costs \$0.47 per 1 mile to own and operate a compact car. **83.** $y = 2x - 14$

85. $y = -6x - 11$ **87.** $m = \frac{1}{2}$ **89.** answers may vary **91.** 2005 to 2006 **93.** 2000; 28.5 mi per gallon **95.** from 2008 to 2009 **97.** $x = 6$ **99. a.** (2007, 2207); (2010, 2333) **b.** 42 **c.** For the years 2007 through 2010, the number of heart transplants increased at a rate of 42 per year. **101.** The slope through $(-3, 0)$ and $(1, 1)$ is $\frac{1}{4}$. The slope through $(-3, 0)$ and $(-4, 4)$ is -4. The product of the slopes is -1, so the sides are perpendicular. **103.** -0.25 **105.** 0.875 **107.** The line becomes steeper.

Integrated Review **1.** $m = 2$ **2.** $m = 0$ **3.** $m = -\frac{2}{3}$ **4.** undefined slope **5.**

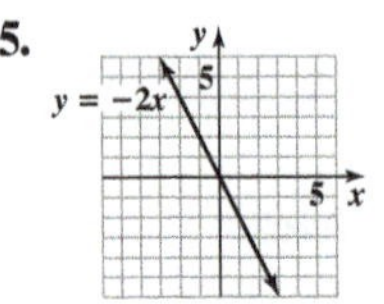

6.

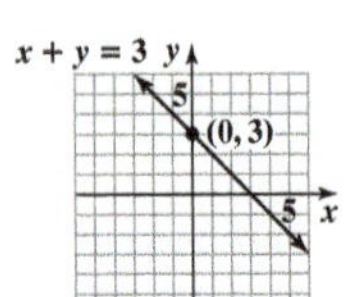

7.

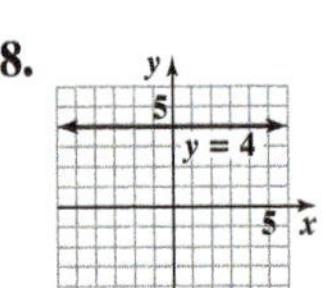

8.

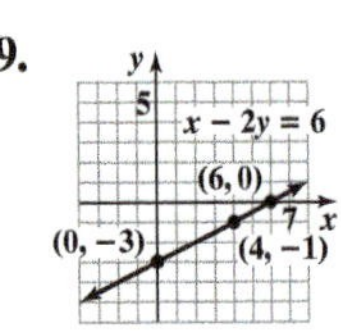

9.

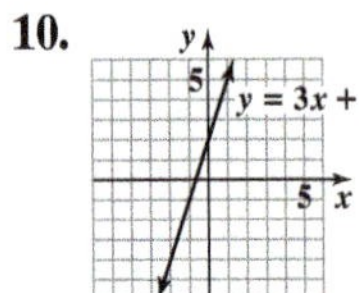

10.

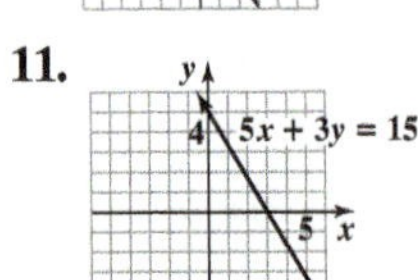

11. 5x + 3y = 15

12.

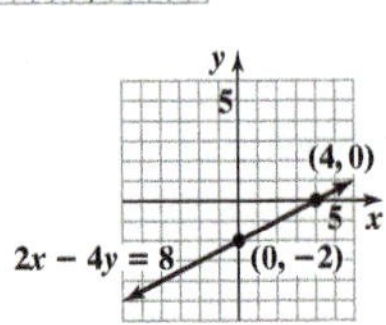

13. parallel **14.** neither **15. a.** (0, 587) **b.** In 2000, there were 587 thousand bridges on public roads. **c.** 1.7 **d.** For the years 2000 through 2009, the number of bridges on public roads increased at a rate of 1.7 thousand per year. **16. a.** (9, 26.6) **b.** In 2009, the revenue for online advertising was \$26.6 billion.

Section 10.5

Graphing Calculator Explorations

1.

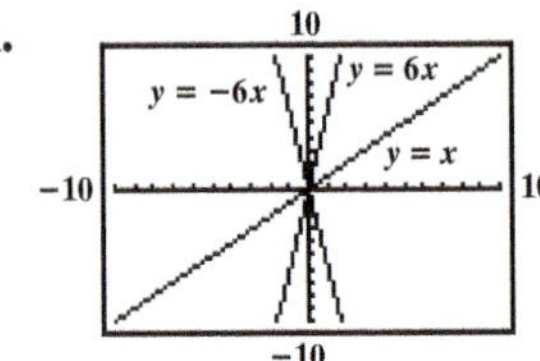

3.

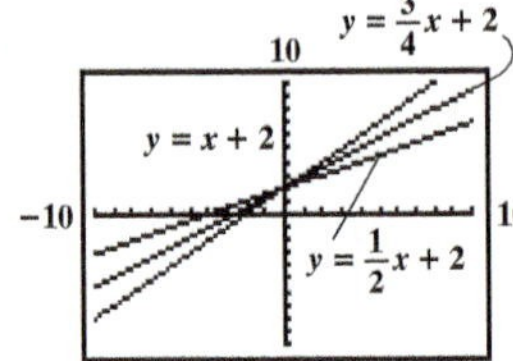

5.

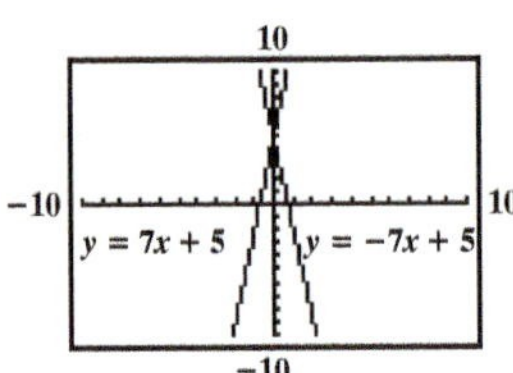

Vocabulary, Readiness & Video Check **1.** slope-intercept; m; b **3.** y-intercept; fraction **5.** Write the equation with x- and y-terms on one side of the equal sign and a constant on the other side. **7.** Example 6: $y = -3$; Example 7: $x = -2$

Exercise Set 10.5 **1.**

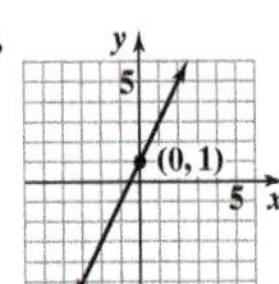

3.

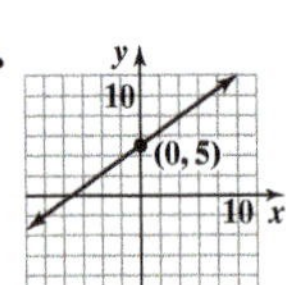

5.

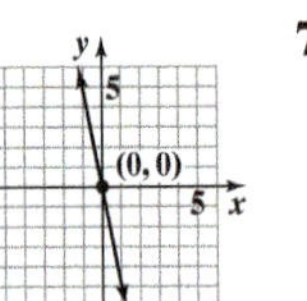

7. (0, 6)

9.

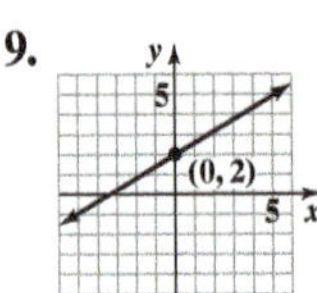

11.

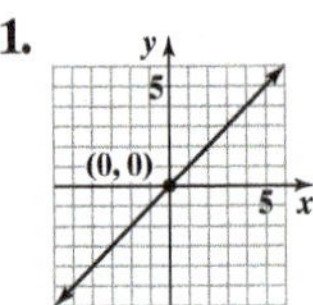

13. $y = 5x + 3$ **15.** $y = -4x - \frac{1}{6}$ **17.** $y = \frac{2}{3}x$ **19.** $y = -8$ **21.** $y = -\frac{1}{5}x + \frac{1}{9}$ **23.** $-6x + y = -10$ **25.** $8x + y = -13$ **27.** $3x - 2y = 27$ **29.** $x + 2y = -3$ **31.** $2x - y = 4$ **33.** $8x - y = -11$ **35.** $4x - 3y = -1$ **37.** $8x + 13y = 0$ **39.** $x = 0$ **41.** $y = 3$ **43.** $x = -\frac{7}{3}$ **45.** $y = 2$ **47.** $y = 5$ **49.** $x = 6$ **51.** $y = -\frac{1}{2}x + \frac{5}{3}$ **53.** $y = -x + 17$ **55.** $x = -\frac{3}{4}$ **57.** $y = x + 16$ **59.** $y = -5x + 7$ **61.** $y = 7$ **63.** $y = \frac{3}{2}x$ **65.** $y = -3$ **67.** $y = -\frac{4}{7}x - \frac{18}{7}$ **69. a.** $s = 32t$ **b.** 128 ft/sec **71. a.** $y = -33x + 356$ **b.** 92 thousand gasoline-electric hybrids **73. a.** $y = 0.6x + 85$ **b.** 93.4 persons per sq mi **75. a.** $y = -1.75x + 60$ **b.** 32 million **77. a.** $S = -1000p + 13{,}000$ **b.** 9500 Fun Noodles **79.** -1 **81.** 5 **83.** no **85.** yes **87.** point-slope **89.** slope-intercept **91.** horizontal **93.** answers may vary **95. a.** $3x - y = -5$ **b.** $x + 3y = 5$ **97. a.** $3x + 2y = -1$ **b.** $2x - 3y = 21$

Section 10.6

Vocabulary, Readiness & Video Check **1.** relation **3.** range **5.** vertical **7.** A relation is a set of ordered pairs and an equation in two variables defines a set of ordered pairs. Therefore, an equation in two variables can also define a relation.
9. A vertical line represents one x-value paired with many y-values. A function allows only an x-value paired with exactly one y-value, so if a vertical line intersects a graph more than once, there's an x-value paired with more than one y-value, and we don't have a function.

Exercise Set 10.6 **1.** $\{-7, 0, 2, 10\}$; $\{-7, 0, 4, 10\}$ **3.** $\{0, 1, 5\}$; $\{-2\}$ **5.** yes **7.** no **9.** no **11.** yes **13.** yes **15.** no **17.** yes **19.** yes **21.** yes **23.** no **25.** no **27.** 9:30 p.m. **29.** January 1 and December 1 **31.** yes; it passes the vertical line test **33.** \$4.25 per hour **35.** 2009 **37.** yes; answers may vary **39.** \$1.50 **41.** more than 1 ounce and less than or equal to 2 ounces **43.** yes; answers may vary **45.** $-9, -5, 1$ **47.** $6, 2, 11$ **49.** $-6, 0, 9$ **51.** $2, 0, 3$ **53.** $5, 0, -20$ **55.** $5, 3, 35$ **57.** $(3, 6)$ **59.** $\left(0, -\frac{1}{2}\right)$ **61.** $(-2, 9)$ **63.** $(-\infty, \infty)$ **65.** all real numbers except -5 or $(-\infty, -5) \cup (-5, \infty)$ **67.** $(-\infty, \infty)$ **69.** domain: $(-\infty, \infty)$; range: $[-4, \infty)$ **71.** domain: $(-\infty, \infty)$; range: $(-\infty, \infty)$ **73.** domain: $(-\infty, \infty)$; range: $\{2\}$ **75.** -1 **77.** -1 **79.** $-1, 5$ **81.** $(-2, 1)$ **83.** $(-3, -1)$ **85.** $f(-5) = 12$ **87.** $(3, -4)$ **89.** $f(5) = 0$ **91. a.** 166.38 cm **b.** 148.25 cm **93.** answers may vary **95.** $f(x) = x + 7$ **97. a.** 11 **b.** $2a + 7$ **99. a.** 16 **b.** $a^2 + 7$

Chapter 10 Vocabulary Check **1.** solution **2.** y-axis **3.** linear **4.** x-intercept **5.** standard **6.** y-intercept **7.** slope-intercept **8.** point-slope **9.** y **10.** x-axis **11.** x **12.** slope **13.** function **14.** domain **15.** range **16.** relation

Chapter 10 Review **1.** **3.**

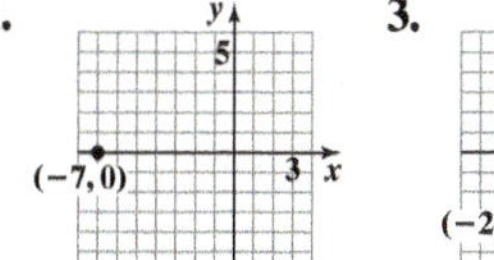

5. **7. a.** (5.00, 50), (8.50, 100), (20.00, 250), (27.00, 500) **b.**

9. no; yes **11.** yes; yes **13.** (7, 44) **15.** $(-3, 0)$; $(1, 3)$; $(9, 9)$

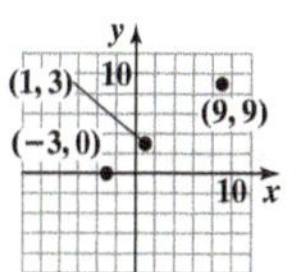

17. 2005; 2500; 7000

19. **21.**

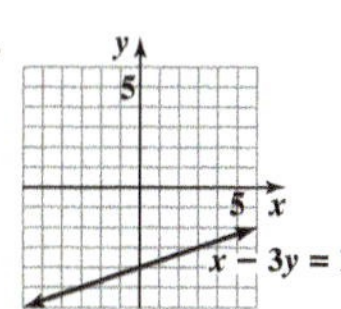

23.

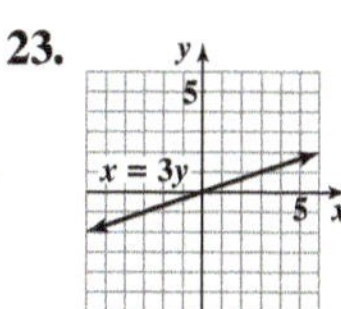

25.

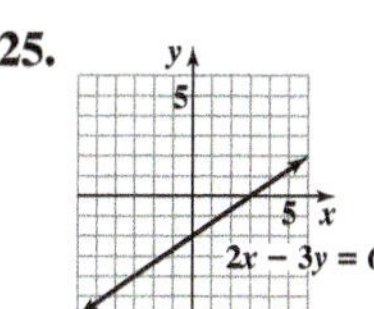

27. $(4, 0)$, $(0, -2)$

29. $(-2, 0)$, $(2, 0)$, $(0, 2)$, $(0, -2)$ **31.**

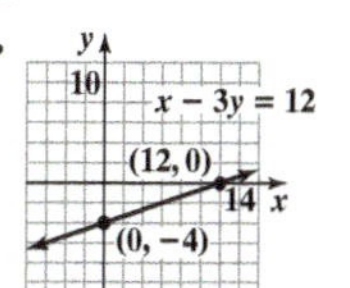

33.

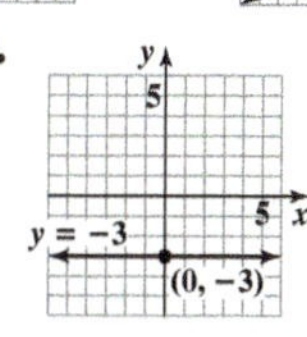

35. **37.**

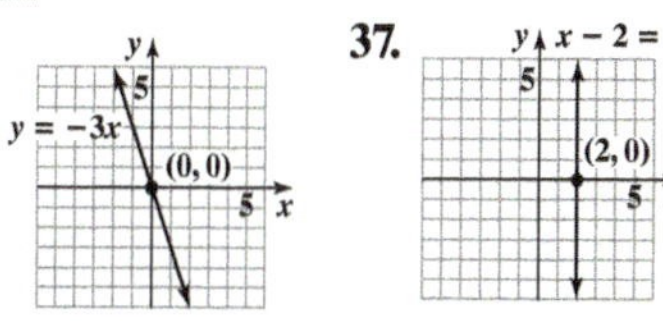

39. $m = -\frac{3}{4}$ **41.** d **43.** c **45.** $\frac{3}{4}$ **47.** 4 **49.** 3 **51.** 0 **53.** perpendicular **55.** neither **57.** $m = 44$; Every 1 year, 44 thousand (44,000) more students graduate with an associate's degree. **59.** $m = -3$; $(0, 7)$ **61.** $m = 0$; $(0, 2)$

63. **65.**

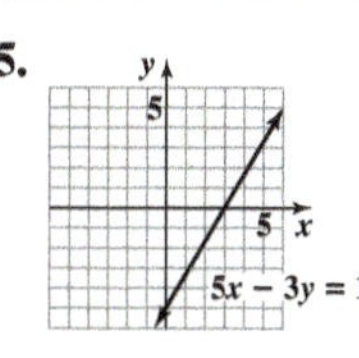

67. $y = -5x + \frac{1}{2}$ **69.** c **71.** b **73.** $(0, 1859)$ **75.** $3x + y = -5$ **77.** $y = -3$ **79.** $6x + y = 11$ **81.** $x + y = 6$ **83.** $x = 5$ **85.** $x = 6$ **87.** no **89.** yes **91.** no **93.** no **95. a.** 6 **b.** 10 **c.** 5 **97. a.** 45 **b.** -35 **c.** 0 **99.** $(-\infty, \infty)$

101. domain: $[-3, 5]$; range: $[-4, 2]$ **103.** domain: $\{3\}$; range: $(-\infty, \infty)$ **105.** 7; -1; -3 **107.** $(3, 0)$; $(0, -2)$

109. **111.**

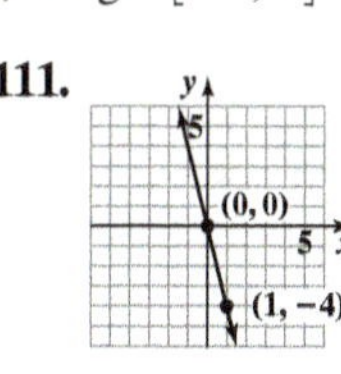

113.

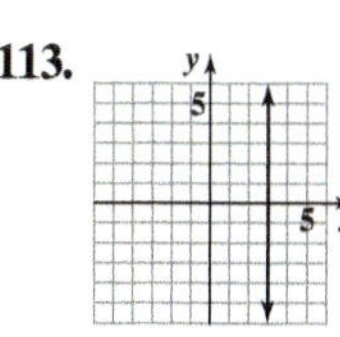

115. $m = -1$ **117.** $m = 2$ **119.** $m = \frac{2}{3}$; $(0, -5)$ **121.** $5x + y = 8$ **123.** $4x + y = -3$ **125.** $y = 238x + 2134$

Chapter 10 Test

1.

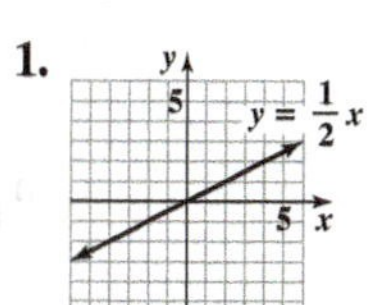

2.

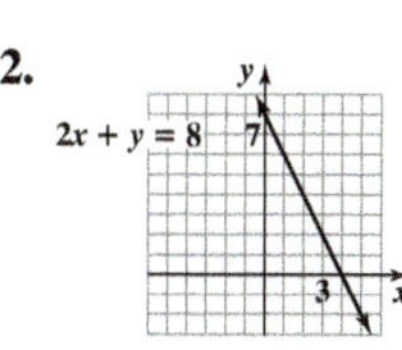

3.

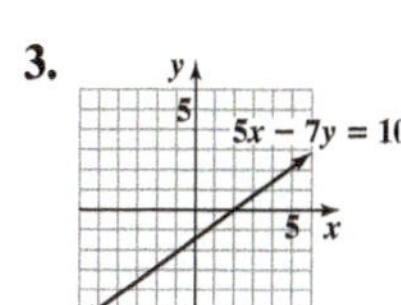

4.

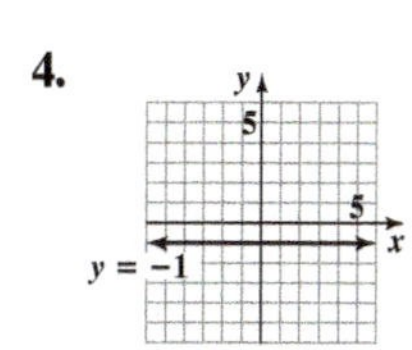

5.

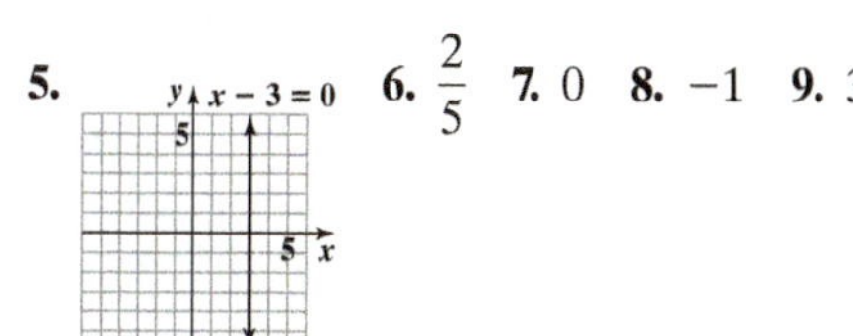

6. $\frac{2}{5}$ **7.** 0 **8.** -1 **9.** 3

10. undefined **11.** $m = \frac{7}{3}$; $\left(0, -\frac{2}{3}\right)$ **12.** neither **13.** $x + 4y = 10$ **14.** $7x + 6y = 0$ **15.** $8x + y = 11$ **16.** $x = -5$

17. $x - 8y = -96$ **18.** yes **19.** no **20. a.** 0 **b.** 0 **c.** 60 **21.** all real numbers except -1 or $(-\infty, -1) \cup (-1, \infty)$ **22. a.** x-intercepts: $(0, 0), (4, 0)$; y-intercept: $(0, 0)$ **b.** domain: $(-\infty, \infty)$; range: $(-\infty, 4]$ **23. a.** x-intercept: $(2, 0)$; y-intercept; $(0, -2)$ **b.** domain: $(-\infty, \infty)$; range: $(-\infty, \infty)$ **24.** (7, 20) **25.** 210 liters **26.** 490 liters **27.** July **28.** 63°F **29.** January, February, March, November, December **30. a.** (2003, 66.0); (2004, 65.4); (2005, 65.4); (2006, 65.6); (2007, 64.9); (2008, 63.7); (2009, 62.1)

b. 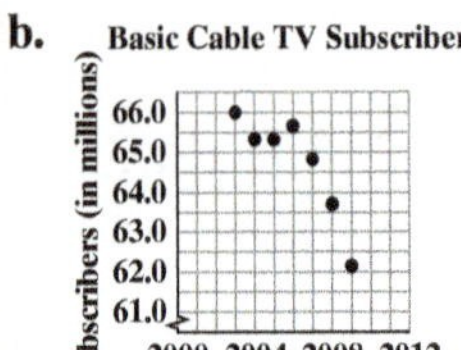

31. a. $m = -8$; Every 1 year, 8 million fewer movie tickets are sold. **b.** (0, 1380), (5, 1340) **c.** $y = -8x + 1380$ **d.** In 2015, we predict that 1300 million movie tickets will be sold in the U.S. and Canada.

Cumulative Review Chapters 1–10 **1.** $22 + x$; Sec. 9.2, Ex. 5 **3. a.** $>$ **b.** $<$ **c.** $<$; Sec. 9.1, Ex. 10 **5.** 2; Sec. 9.3, Ex. 1 **7.** no solution; Sec. 9.3, Ex. 6 **9.** 4; Sec. 9.3, Ex. 5 **11.** 603 and 605; Sec. 9.4, Ex. 7 **13.** shorter piece: 2 ft; longer piece: 8 ft; Sec. 9.4, Ex. 3 **15.** 40 ft; Sec. 9.5, Ex. 2 **17.** $\frac{y - b}{m} = x$; Sec. 9.5, Ex. 6 **19.** all numbers less than 11; Sec. 9.6, Ex. 10 **21.**

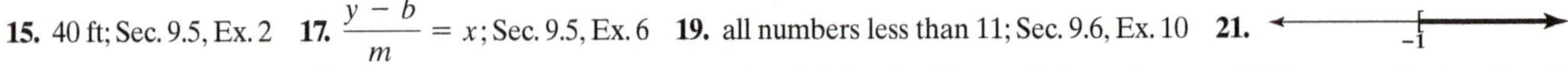

$[-1, \infty)$; Sec. 9.6, Ex. 1 **23.** $[2, \infty)$; Sec. 9.6, Ex. 5 **25. a.** solution **b.** not a solution **c.** solution; Sec. 10.1, Ex. 5 **27. a.** yes **b.** yes **c.** no **d.** yes; Sec. 10.2, Ex. 1 **29.** 0; Sec. 10.4, Ex. 6 **31.** $y = \frac{1}{4}x - 3$; Sec. 10.5, Ex. 3

Chapter 11 Solving Systems of Linear Equations

Section 11.1

Graphing Calculator Explorations **1.** (0.37, 0.23) **3.** (0.03, −1.89)

Vocabulary, Readiness & Video Check **1.** dependent **3.** consistent **5.** inconsistent **7.** The ordered pair must satisfy all equations of the system in order to be a solution of the system, so we must check that the ordered pair is a solution of both equations. **9.** Writing the equations of a system in slope-intercept form lets you see their slope and y-intercept. Different slopes mean one solution; same slope with different y-intercepts means no solution; same slope with same y-intercept means infinite number of solutions.

Exercise Set 11.1 **1.** one solution, $(-1, 3)$ **3.** infinite number of solutions **5. a.** no **b.** yes **7. a.** yes **b.** no **9. a.** yes **b.** yes **11. a.** no **b.** no

13. 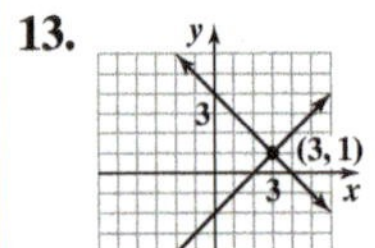

15. 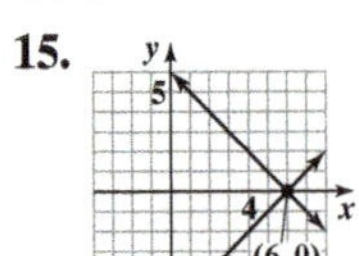

17. 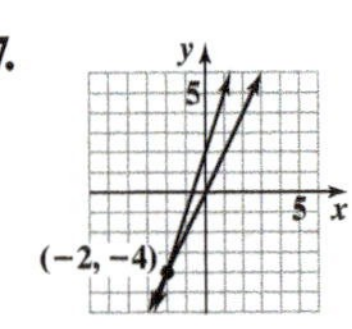

19. 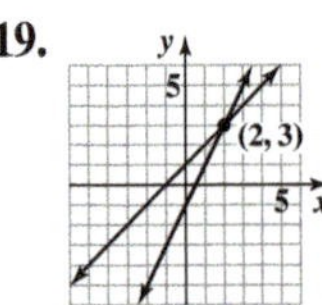

21. 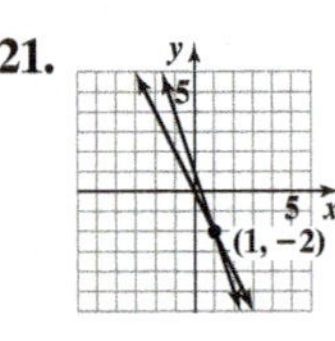

23. 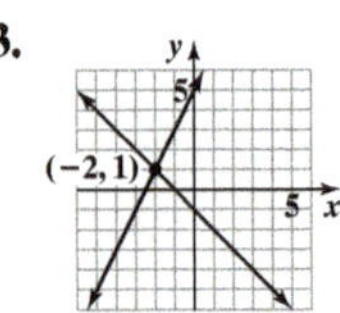

25. no solution 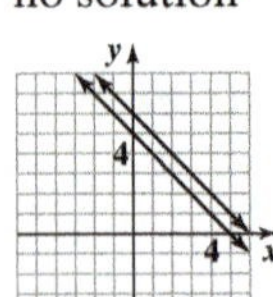

27. 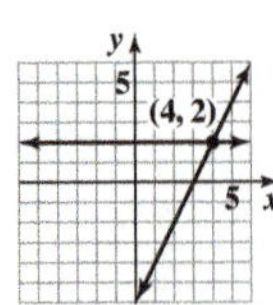

29. 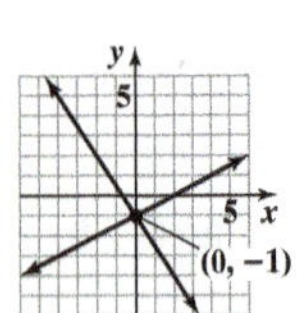

31. no solution; { } or ∅

33. infinite number of solutions; $\{(x, y) \mid y - 3x = -2\}$ or $\{(x, y) \mid 6x - 2y = 4\}$

35. (3, −1)

37. (−5, −7)

39. (5, 4)

41. infinite number of solutions; $\{(x, y) \mid 6x - y = 4\}$ or $\left\{(x, y) \middle| \frac{1}{2}y = -2 + 3x\right\}$ 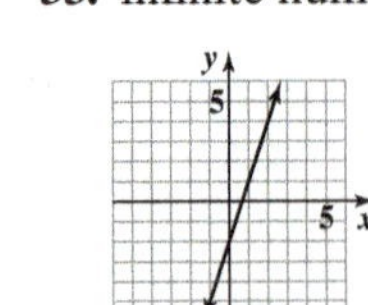

43. intersecting, one solution

45. parallel; no solution **47.** identical lines; infinite number of solutions **49.** intersecting; one solution **51.** intersecting; one solution **53.** identical lines; infinite number of solutions **55.** parallel; no solution **57.** 2 **59.** $-\frac{2}{5}$ **61.** 2

63. answers may vary; possible answer 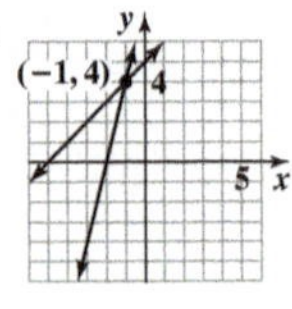

65. answers may vary; possible answer 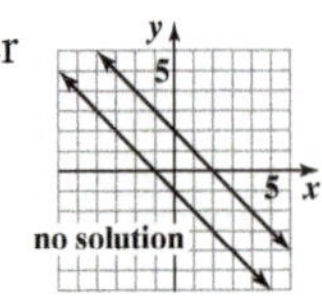

67. answers may vary **69.** 1988–1989; 2001–2002

71. 2003, 2004, 2005, 2006, 2007 **73.** answers may vary **75. a.** $(4, 9)$ **b.**

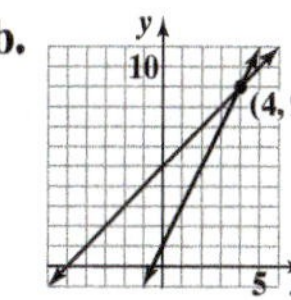

c. yes **77.** answers may vary

Section 11.2

Vocabulary, Readiness & Video Check **1.** $(1, 4)$ **3.** infinite number of solutions **5.** $(0, 0)$ **7.** You solved one equation for a variable. Now be sure to substitute this expression for the variable into the *other* equation.

Exercise Set 11.2 **1.** $(2, 1)$ **3.** $(-3, 9)$ **5.** $(2, 7)$ **7.** $\left(-\frac{1}{5}, \frac{43}{5}\right)$ **9.** $(2, -1)$ **11.** $(-2, 4)$ **13.** $(4, 2)$ **15.** $(-2, -1)$ **17.** no solution; { } or ∅ **19.** $(3, -1)$ **21.** $(3, 5)$ **23.** $\left(\frac{2}{3}, -\frac{1}{3}\right)$ **25.** $(-1, -4)$ **27.** $(-6, 2)$ **29.** $(2, 1)$ **31.** no solution; { } or ∅ **33.** infinite number of solutions; $\left\{(x, y)\middle|\frac{1}{3}x - y = 2\right\}$ or $\{(x, y)|x - 3y = 6\}$ **35.** $\left(\frac{1}{2}, 2\right)$ **37.** $(1, -3)$ **39.** $-6x - 4y = -12$ **41.** $-12x + 3y = 9$ **43.** $5n$ **45.** $-15b$ **47.** answers may vary **49.** no; answers may vary **51. c;** answers may vary **53. a.** $(13, 492)$ **b.** In $1970 + 13 = 1983$, the number of men and women receiving bachelor's degrees was the same. **c.** answers may vary

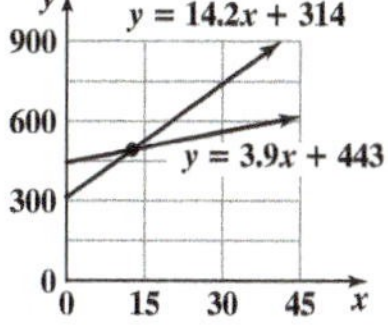

55. $(-2.6, 1.3)$ **57.** $(3.28, 2.11)$

Section 11.3

Vocabulary, Readiness & Video Check **1.** false **3.** true **5.** The multiplication property of equality; be sure to multiply *both* sides of the equation by the number chosen.

Exercise Set 11.3 **1.** $(1, 2)$ **3.** $(2, -3)$ **5.** $(-2, -5)$ **7.** $(5, -2)$ **9.** $(-7, 5)$ **11.** $(6, 0)$ **13.** no solution; { } or ∅ **15.** infinite number of solutions; $\{(x, y)|-x + 5y = -1\}$ or $\{(x, y)|3x - 15y = 3\}$ **17.** $\left(2, -\frac{1}{2}\right)$ **19.** $(-2, 0)$ **21.** $(1, -1)$ **23.** no solution; { } or ∅ **25.** $\left(\frac{12}{11}, -\frac{4}{11}\right)$ **27.** $\left(\frac{3}{2}, 3\right)$ **29.** infinite number of solutions; $\left\{(x, y)\middle|\frac{10}{3}x + 4y = -4\right\}$ or $\{(x, y)|5x + 6y = -6\}$ **31.** $(1, 6)$ **33.** $\left(-\frac{1}{2}, -2\right)$ **35.** infinite number of solutions; $\left\{(x, y)\middle|\frac{x}{3} - y = 2\right\}$ or $\left\{(x, y)\middle|-\frac{x}{2} + \frac{3y}{2} = -3\right\}$ **37.** $\left(-\frac{2}{3}, \frac{2}{5}\right)$ **39.** $(2, 4)$ **41.** $(-0.5, 2.5)$ **43.** $(2, 5)$ **45.** $(-3, 2)$ **47.** $(0, 3)$ **49.** $(5, 7)$ **51.** $\left(\frac{1}{3}, 1\right)$ **53.** infinite number of solutions; $\left\{(x, y)\middle|\frac{x + 2}{2} = \frac{y + 11}{3}\right\}$ or $\left\{(x, y)\middle|\frac{x}{2} = \left(\frac{2y + 16}{6}\right)\right\}$ **55.** $(-8.9, 10.6)$ **57.** $2x + 6 = x - 3$ **59.** $20 - 3x = 2$ **61.** $4(n + 6) = 2n$ **63.** 2; $6x - 2y = -24$ **65. b;** answers may vary **67.** answers may vary **69. a.** $b = 15$ **b.** any real number except 15 **71.** $(-4.2, 9.6)$ **73. a.** $(7, 22)$ **b.** In 2015 $(2008 + 7)$, the percent of workers age 25–34 and the percent of workers age 55 and older were the same. **c.** 22% of the workforce for each of these age groups.

Integrated Review **1.** $(2, 5)$ **2.** $(4, 2)$ **3.** $(5, -2)$ **4.** $(6, -14)$ **5.** $(-3, 2)$ **6.** $(-4, 3)$ **7.** $(0, 3)$ **8.** $(-2, 4)$ **9.** $(5, 7)$ **10.** $(-3, -23)$ **11.** $\left(\frac{1}{3}, 1\right)$ **12.** $\left(-\frac{1}{4}, 2\right)$ **13.** no solution; { } or ∅ **14.** infinite number of solutions; $\{(x, y)|-x + 2y = 3\}$ or $\{(x, y)|3x - 6y = -9\}$ **15.** $(0.5, 3.5)$ **16.** $(-0.75, 1.25)$ **17.** infinite number of solutions; $\{(x, y)|x = 3y - 7\}$ or $\{(x, y)|2x - 6y = -14\}$ **18.** no solution; { } or ∅ **19.** $(7, -3)$ **20.** $(-1, -3)$ **21.** answers may vary **22.** answers may vary

Section 11.4

Vocabulary, Readiness & Video Check **1.** a, b, d **3.** yes; answers may vary **5.** Once we have one equation in two variables, we need to get another equation in the *same* two variables, giving us a system of two equations in two variables. We solve this new system to find the value of two variables. We then substitute these values into an original equation to find the value of the third.

Exercise Set 11.4 **1.** $(-1, 5, 2)$ **3.** $(-2, 5, 1)$ **5.** $(-2, 3, -1)$ **7.** $\{(x, y, z)|x - 2y + z = -5\}$ **9.** ∅ **11.** $(0, 0, 0)$ **13.** $(-3, -35, -7)$ **15.** $(6, 22, -20)$ **17.** ∅ **19.** $(3, 2, 2)$ **21.** $\{(x, y, z)|x + 2y - 3z = 4\}$ **23.** $(-3, -4, -5)$ **25.** $\left(0, \frac{1}{2}, -4\right)$ **27.** $(12, 6, 4)$ **29.** 15 and 30 **31.** 5 **33.** $-\frac{5}{3}$ **35.** answers may vary **37.** answers may vary **39.** $(1, 1, -1)$ **41.** $(1, 1, 0, 2)$ **43.** $(1, -1, 2, 3)$ **45.** answers may vary

Section 11.5

Vocabulary, Readiness & Video Check **1.** Up to now we've been choosing one variable/unknown and translating to one equation. To solve by a system of equations, we'll choose two variables to represent two unknowns and translate to two equations. **3.** The ordered triple still needs to be interpreted in the context of the application. Each value actually represents the angle measure of a triangle, in degrees.

Exercise Set 11.5 **1.** c **3.** b **5.** a **7.** $\begin{cases} x + y = 15 \\ x - y = 7 \end{cases}$ **9.** $\begin{cases} x + y = 6500 \\ x = y + 800 \end{cases}$ **11.** 33 and 50 **13.** 10 and 8 **15.** 14 and -3 **17.** Cabrera: 126; Rodriguez: 125 **19.** child's ticket: \$18; adult's ticket: \$29 **21.** quarters: 53; nickels: 27 **23.** McDonald's: \$73.50; The Ohio Art Company: \$3.50 **25.** daily fee: \$32; mileage charge: \$0.25 per mi **27.** distance downstream = distance upstream = 18 mi; time downstream: 2 hr; time upstream: $4\frac{1}{2}$ hr; still water: 6.5 mph; current: 2.5 mph **29.** still air: 455 mph; wind: 65 mph **31.** $4\frac{1}{2}$ hr **33.** 12% solution: $7\frac{1}{2}$ liters; 4% solution: $4\frac{1}{2}$ liters **35.** \$4.95 beans: 113 lb; \$2.65 beans: 87 lb **37.** 60°, 30° **39.** 20°, 70° **41.** number sold at \$9.50: 23; number sold at \$7.50: 67 **43.** $2\frac{1}{4}$ mph and $2\frac{3}{4}$ mph **45.** 30%: 50 gal; 60%: 100 gal **47.** length: 42 in.; width: 30 in. **49. a.** 2010 **b.** answers may vary **51. a.** answers may vary **b.** 2003 **53.** $x = 75; y = 105$ **55.** 625 units **57.** 3000 units **59.** 1280 units **61. a.** $R(x) = 450x$ **b.** $C(x) = 200x + 6000$ **c.** 24 desks **63** 2 units of Mix A; 3 units of Mix B; 1 unit of Mix C **65.** 5 in.; 7 in.; 7 in.; 10 in. **67.** 18, 13, and 9 **69.** free throws: 594; 2-pt field goals: 566; 3-pt field goals: 145 **71.** $x = 60; y = 55; z = 65$ **73.** $(3, \infty)$ **75.** $\left[\frac{1}{2}, \infty\right)$ **77.** a **79.** width: 9 ft; length: 15 ft **81.** $a = 3, b = 4, c = -1$ **83.** $a = 0.5, b = 24.5, c = 849$; 2015: 1774 thousand students **85.** $(7, 215)$

Chapter 11 Vocabulary Check **1.** dependent **2.** system of linear equations **3.** consistent **4.** solution **5.** addition; substitution **6.** inconsistent **7.** independent

Chapter 11 Review **1. a.** no **b.** yes **c.** no **3. a.** no **b.** no **c.** yes
5. (3, 2) **7.**

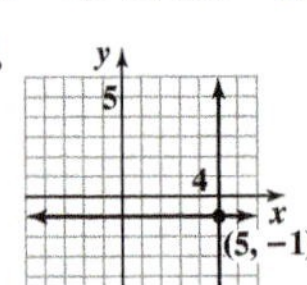

9.

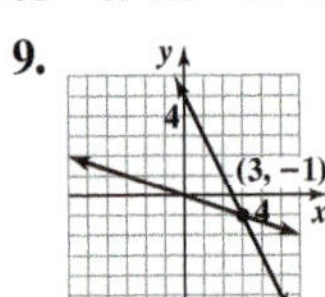

11. no solution; { } or ∅

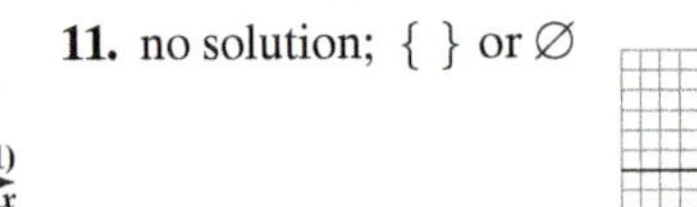

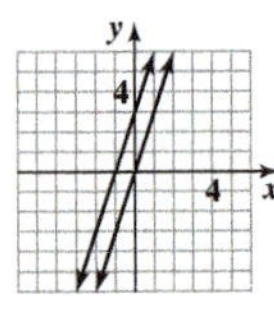

13. $(-1, 4)$ **15.** $(3, -2)$

17. infinite number of solutions; $\{(x, y) | 4y = 2x + 6\}$ or $\{(x, y) | x - 2y = -3\}$ **19.** no solution; { } or ∅ **21.** $(-6, 2)$ **23.** $(3, 7)$ **25.** infinite number of solutions; $\{(x, y) | 2x - 6y = -1\}$ or $\left\{(x, y) \middle| -x + 3y = \frac{1}{2}\right\}$ **27.** $(8, -6)$ **29.** $(2, 0, 2)$ **31.** $\left(-\frac{1}{2}, \frac{3}{4}, 1\right)$ **33.** ∅ **35.** $(1, 1, -2)$ **37.** -6 and 22 **39.** current of river: 3.2 mph; speed in still water: 21.1 mph **41.** egg: \$0.40; strip of bacon: \$0.65 **43.** 17 pennies; 20 nickels; 16 dimes **45.** two sides: 22 cm each; third side: 29 cm **47.**

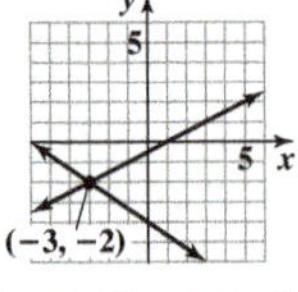

49. $(3, 2)$

51. $\left(1\frac{1}{2}, -3\right)$ **53.** infinite number of solutions; $\{(x, y) | 3x - y = 4\}$ or $\{(x, y) | 4y = 12x - 16\}$ **55.** $(-5, 2)$ **57.** $(-1, 3, 5)$ **59.** 4 and 8 **61.** 24 nickels and 41 dimes **63.** 28 units, 42 units, 56 units

Chapter 11 Test **1.** false **2.** false **3.** true **4.** false **5.** no **6.** yes **7.** $(-4, 2)$

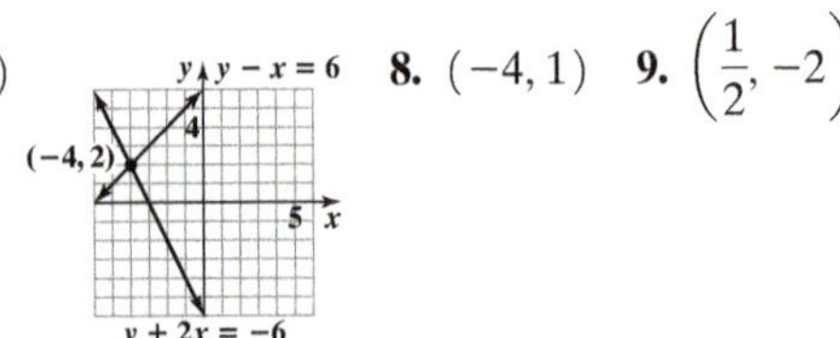

8. $(-4, 1)$ **9.** $\left(\frac{1}{2}, -2\right)$

10. $(4, -2)$ **11.** no solution; { } or ∅ **12.** $(4, -5)$ **13.** $(7, 2)$ **14.** $(5, -2)$ **15.** 78, 46 **16.** 120 cc **17.** Texas: 248 thousand; Missouri: 108 thousand **18.** 3 mph; 6 mph **19.** $(-1, -2, 4)$ **20.** ∅ **21.** 23°, 45°, 112°

Cumulative Review Chapters 1–11 **1. a.** > **b.** < **c.** <; Sec. 9.1, Ex. 10 **3. a.** distributive property **b.** associative property of addition **c.** identity element for addition **d.** commutative property of multiplication **e.** multiplicative inverse property **f.** additive inverse property **g.** commutative and associative properties of multiplication; Sec. 9.2, Exs. 15–21 **5.** $6y + 2$; Sec. 3.1, Ex. 2 **7.** -15; Sec. 3.2, Ex. 6 **9.** -24; Sec. 3.2, Ex. 3 **11.** 12; Sec. 9.3, Ex. 3 **13.** 10; Sec. 9.4, Ex. 2 **15.** $x = \frac{y - b}{m}$; Sec. 9.5, Ex. 6

17. $[2, \infty)$; Sec. 9.6, Ex. 5 **19.** 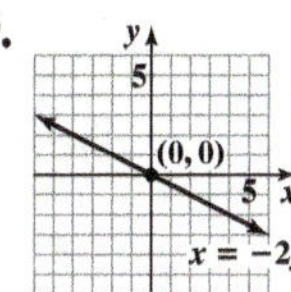

; Sec. 10.3, Ex. 7 **21.** $-\frac{8}{3}$; Sec. 10.4, Ex. 1 **23.** slope: $\frac{3}{4}$; y-intercept: $(0, 6)$; Sec. 10.4, Ex. 3 **25.** slope: $\frac{3}{4}$; y-intercept: $(0, -1)$; Sec. 10.4, Ex. 5 **27.** $y = -2x + 3$; $2x + y = 3$; Sec. 10.5, Ex. 4 **29.** $x = -1$; Sec. 10.5, Ex. 6 **31.** domain: $\{-1, 0, 3\}$; range: $\{-2, 0, 2, 3\}$; Sec. 10.6, Ex. 1 **33. a.** function **b.** not a function; Sec. 10.6, Ex. 2 **35.** one solution; Sec. 11.1, Ex. 8 **37.** $\left(6, \frac{1}{2}\right)$; Sec. 11.2, Ex. 3 **39.** $(6, 1)$; Sec. 11.3, Ex. 1 **41.** $(-4, 2, -1)$; Sec. 11.4, Ex. 1 **43.** 7 and 11; Sec. 11.5, Ex. 3

Chapter 12 Exponents and Polynomials

Section 12.1

Vocabulary, Readiness & Video Check **1.** exponent **3.** add **5.** 1 **7.** Example 4 can be written as $-4^2 = -1 \cdot 4^2$, which is similar to Example 7, $4 \cdot 3^2$, and shows why the negative sign should not be considered part of the base when there are no parentheses. **9.** Be careful not to confuse the power rule with the product rule. The power rule involves a power raised to a power (exponents are multiplied), and the product rule involves a product (exponents are added). **11.** the quotient rule

Exercise Set 12.1 **1.** exponent: 2; base: 3 **3.** exponent: 2; base: 4 **5.** exponent: 2; base: x **7.** 49 **9.** −5 **11.** −16 **13.** 16 **15.** 0.00001 **17.** $\frac{1}{81}$ **19.** 224 **21.** −250 **23.** 4 **25.** 135 **27.** 150 **29.** $\frac{32}{5}$ **31.** x^7 **33.** $(-3)^{12}$ **35.** $15y^5$ **37.** $x^{19}y^6$ **39.** $-72m^3n^8$ **41.** $-24z^{20}$ **43.** $20x^5$ sq ft **45.** x^{36} **47.** p^8q^8 **49.** $8a^{15}$ **51.** $x^{10}y^{15}$ **53.** $49a^4b^{10}c^2$ **55.** $\frac{r^9}{s^9}$ **57.** $\frac{m^5p^5}{n^5}$ **59.** $\frac{4x^2z^2}{y^{10}}$ **61.** $64z^{10}$ sq dm **63.** $27y^{12}$ cu ft **65.** x^2 **67.** −64 **69.** p^6q^5 **71.** $\frac{y^3}{2}$ **73.** 1 **75.** 1 **77.** −7 **79.** 2 **81.** −81 **83.** $\frac{1}{64}$ **85.** b^6 **87.** a^9 **89.** $-16x^7$ **91.** $a^{11}b^{20}$ **93.** $26m^9n^7$ **95.** z^{40} **97.** $64a^3b^3$ **99.** $36x^2y^2z^6$ **101.** $3x$ **103.** $81x^2y^2$ **105.** 9 **107.** $\frac{y^{15}}{8x^{12}}$ **109.** $2x^2y$ **111.** 2 **113.** $\frac{x^{18}}{4y^{22}}$ **115.** $-b^5$ **117.** $2y - 10$ **119.** $-x - 4$ **121.** $-x + 5$ **123.** c **125.** e **127.** answers may vary **129.** answers may vary **131.** 343 cu m **133.** volume **135.** answers may vary **137.** answers may vary **139.** x^{9a} **141.** a^{5b} **143.** x^{5a} **145.** \$1045.85

Section 12.2

Graphing Calculator Explorations **1.** $x^3 - 4x^2 + 7x - 8$ **3.** $-2.1x^2 - 3.2x - 1.7$ **5.** $7.69x^2 - 1.26x + 5.3$

Vocabulary, Readiness & Video Check **1.** binomial **3.** trinomial **5.** constant **7.** The degree of the polynomial is the greatest degree of any of its terms, so we need to find the degree of each term first. **9.** simplifying it

Exercise Set 12.2 **1.** 1; binomial **3.** 3; none of these **5.** 6; trinomial **7.** 8; binomial **9.** 3 **11.** 2 **13.** 57 **15.** 499 **17.** 1 **19.** $-\frac{11}{16}$ **21.** 1134 ft **23.** 1006 ft **25.** 164 thousand **27.** $23x^2$ **29.** $12x^2 - y$ **31.** $7s$ **33.** $-1.1y^2 + 4.8$ **35.** $-\frac{7}{12}x^3 + \frac{7}{5}x^2 + 6$ **37.** $5a^2 - 9ab + 16b^2$ **39.** $-3x^2 + 10$ **41.** $-x^2 + 14$ **43.** $-2x + 9$ **45.** $2x^2 + 7x - 16$ **47.** $8t^2 - 4$ **49.** $-2z^2 - 16z + 6$ **51.** $2x^3 - 2x^2 + 7x + 2$ **53.** $-y^2 - 3y - 1$ **55.** $2x^2 + 11x$ **57.** $-16x^4 + 8x + 9$ **59.** $7x^2 + 14x + 18$ **61.** $3x - 3$ **63.** $7x^2 - 4x + 2$ **65.** $62x^2 + 5$ **67.** $7x^2 - 2x + 2$ **69.** $12x + 2$ **71.** $4y^2 + 12y + 19$ **73.** $4x^2 + 7x + x^2 + 5x$; $5x^2 + 12x$ **75.** $-2a - b + 1$ **77.** $3x^2 + 5$ **79.** $6x^2 - 2xy + 19y^2$ **81.** $8r^2s + 16rs - 8 + 7r^2s^2$ **83.** $-5.42x^2 + 7.75x - 19.61$ **85.** $3.7y^4 - 0.7y^3 + 2.2y - 4$ **87.** $6x^2$ **89.** $-12x^8$ **91.** $200x^3y^2$ **93.** $18x + 44$ **95.** $(x^2 + 7x + 4)$ ft **97.** $(3y^2 + 4y + 11)$ m **99.** 7.33 million **101.** answers may vary **103.** answers may vary **105.** b **107.** e **109. a.** $4z$ **b.** $3z^2$ **c.** $-4z$ **d.** $3z^2$; answers may vary **111. a.** m^3 **b.** $3m$ **c.** $-m^3$ **d.** $-3m$; answers may vary **113.** $3x^{2a} + 2x^a + 0.7$ **115.** $4x^{2y} + 2x^y - 11$ **117.** $4x^2 - 3x + 6$ **119.** $-x^2 - 6x + 10$ **121.** $3x^2 - 12x + 13$ **123. a.** $2a - 3$ **b.** $-2x - 3$ **c.** $2x + 2h - 3$ **125. a.** $4a$ **b.** $-4x$ **c.** $4x + 4h$ **127.** 2; 2 **129.** 4; 3; 3; 4 **131.** $2x^2 + 4xy$

Section 12.3

Vocabulary, Readiness & Video Check **1.** distributive **3.** $(5y - 1)(5y - 1)$ **5.** No. The monomials are unlike terms. **7.** Yes. The parentheses have been removed for the vertical format, but every term in the first polynomial is still distributed to every term in the second polynomial.

Exercise Set 12.3 **1.** $-28n^{10}$ **3.** $-12.4x^{12}$ **5.** $-\frac{2}{15}y^3$ **7.** $-24x^8$ **9.** $6x^2 + 15x$ **11.** $-2a^2 - 8a$ **13.** $6x^3 - 9x^2 + 12x$ **15.** $-6a^4 + 4a^3 - 6a^2$ **17.** $-4x^3y + 7x^2y^2 - xy^3 - 3y^4$ **19.** $4x^4 - 3x^3 + \frac{1}{2}x^2$ **21.** $x^2 + 7x + 12$ **23.** $a^2 + 5a - 14$

25. $x^2 + \frac{1}{3}x - \frac{2}{9}$ **27.** $12x^4 + 25x^2 + 7$ **29.** $4y^2 - 16y + 16$ **31.** $12x^2 - 29x + 15$ **33.** $9x^4 + 6x^2 + 1$ **35. a.** $-4y^4$ **b.** $3y^2$ **c.** answers may vary **37.** $x^3 - 5x^2 + 13x - 14$ **39.** $x^4 + 5x^3 - 3x^2 - 11x + 20$ **41.** $10a^3 - 27a^2 + 26a - 12$ **43.** $x^3 + 6x^2 + 12x + 8$ **45.** $8y^3 - 36y^2 + 54y - 27$ **47.** $12x^2 - 64x - 11$ **49.** $10x^3 + 22x^2 - x - 1$ **51.** $2x^4 + 3x^3 - 58x^2 + 4x + 63$ **53.** $8.4y^7$ **55.** $-3x^3 - 6x^2 + 24x$ **57.** $2x^2 + 39x + 19$ **59.** $x^2 - \frac{2}{7}x - \frac{3}{49}$ **61.** $9y^2 + 30y + 25$ **63.** $a^3 - 2a^2 - 18a + 24$ **65.** $8x^3 - 60x^2 + 150x - 125$ **67.** $32x^3 + 48x^2 - 6x - 20$ **69.** $6x^4 - 8x^3 - 7x^2 + 22x - 12$ **71.** $(4x^2 - 25)$ sq yd **73.** $(6x^2 - 4x)$ sq in. **75.** $5a + 15a = 20a$; $5a - 15a = -10a$; $5a \cdot 15a = 75a^2$; $\frac{5a}{15a} = \frac{1}{3}$ **77.** $-3y^5 + 9y^4$, cannot be simplified; $-3y^5 - 9y^4$, cannot be simplified; $-3y^5 \cdot 9y^4 = -27y^9$; $\frac{-3y^5}{9y^4} = -\frac{y}{3}$ **79. a.** $6x + 12$ **b.** $9x^2 + 36x + 35$; answers may vary **81.** $13x - 7$ **83.** $30x^2 - 28x + 6$ **85.** $-7x + 5$ **87.** $x^2 + 3x$ **89.** $x^2 + 5x + 6$ **91.** $11a$ **93.** $25x^2 + 4y^2$ **95. a.** $a^2 - b^2$ **b.** $4x^2 - 9y^2$ **c.** $16x^2 - 49$ **d.** answers may vary **97.** $(x^2 + 6x + 5)$ sq units

Section 12.4

Vocabulary, Readiness & Video Check **1.** false **3.** false **5.** a binomial times a binomial **7.** Multiplying gives you four terms, and the two like terms will always subtract out.

Exercise Set 12.4 **1.** $x^2 + 7x + 12$ **3.** $x^2 + 5x - 50$ **5.** $5x^2 + 4x - 12$ **7.** $20y^2 - 125y + 30$ **9.** $6x^2 + 13x - 5$ **11.** $x^2 + \frac{1}{3}x - \frac{2}{9}$ **13.** $x^2 + 4x + 4$ **15.** $4x^2 - 4x + 1$ **17.** $9a^2 - 30a + 25$ **19.** $25x^2 + 90x + 81$ **21.** $a^2 - 49$ **23.** $9x^2 - 1$ **25.** $9x^2 - \frac{1}{4}$ **27.** $81x^2 - y^2$ **29.** $4x^2 - 0.01$ **31.** $a^2 + 9a + 20$ **33.** $a^2 + 14a + 49$ **35.** $12a^2 - a - 1$ **37.** $x^2 - 4$ **39.** $9a^2 + 6a + 1$ **41.** $4x^3 - x^2y^4 + 4xy - y^5$ **43.** $x^3 - 3x^2 - 17x + 3$ **45.** $4a^2 - 12a + 9$ **47.** $25x^2 - 36z^2$ **49.** $x^{10} - 8x^5 + 15$ **51.** $x^2 - 0.64$ **53.** $a^7 - 3a^3 + 11a^4 - 33$ **55.** $3x^2 - 12x + 12$ **57.** $6b^2 - b - 35$ **59.** $49p^2 - 64$ **61.** $\frac{1}{9}a^4 - 49$ **63.** $15x^4 - 5x^3 + 10x^2$ **65.** $4r^2 - 9s^2$ **67.** $9x^2 - 42xy + 49y^2$ **69.** $16x^2 - 25$ **71.** $64x^2 + 64x + 16$ **73.** $a^2 - \frac{1}{4}y^2$ **75.** $\frac{1}{25}x^2 - y^2$ **77.** $3a^3 + 2a^2 + 1$ **79.** $(2x + 1)(2x + 1)$ sq ft or $(4x^2 + 4x + 1)$ sq ft **81.** $\frac{5b^5}{7}$ **83.** $-2a^{10}b^5$ **85.** $\frac{2y^8}{3}$ **87.** $\frac{1}{3}$ **89.** 1 **91.** c **93.** d **95.** 2; 2 **97.** $(x^4 - 3x^2 + 1)$ sq m **99.** $(24x^2 - 32x + 8)$ sq m **101.** $(x^2 + 10x + 25)$ sq units **103.** answers may vary **105.** answers may vary **107.** answers may vary **109.** $x^2 + 2xy + y^2 - 9$ **111.** $a^2 - 6a + 9 - b^2$

Integrated Review **1.** $35x^5$ **2.** $32y^9$ **3.** -16 **4.** 16 **5.** $2x^2 - 9x - 5$ **6.** $3x^2 + 13x - 10$ **7.** $3x - 4$ **8.** $4x + 3$ **9.** $7x^6y^2$ **10.** $\frac{10b^6}{7}$ **11.** $144m^{14}n^{12}$ **12.** $64y^{27}z^{30}$ **13.** $48y^2 - 27$ **14.** $98x^2 - 2$ **15.** $x^{63}y^{45}$ **16.** $27x^{27}$ **17.** $2x^2 - 2x - 6$ **18.** $6x^2 + 13x - 11$ **19.** $2.5y^2 - 6y - 0.2$ **20.** $8.4x^2 - 6.8x - 5.7$ **21.** $x^2 + 8xy + 16y^2$ **22.** $y^2 - 18yz + 81z^2$ **23.** $2x + 8y$ **24.** $2y - 18z$ **25.** $7x^2 - 10xy + 4y^2$ **26.** $-a^2 - 3ab + 6b^2$ **27.** $x^3 + 2x^2 - 16x + 3$ **28.** $x^3 - 2x^2 - 5x - 2$ **29.** $6x^5 + 20x^3 - 21x^2 - 70$ **30.** $20x^7 + 25x^3 - 4x^4 - 5$ **31.** $2x^3 - 19x^2 + 44x - 7$ **32.** $5x^3 + 9x^2 - 17x + 3$ **33.** cannot simplify **34.** $25x^3y^3$ **35.** $125x^9$ **36.** $\frac{x^3}{y^3}$ **37.** $2x$ **38.** x^2

Section 12.5

Graphing Calculator Explorations **1.** 5.31 EE 3 **3.** 6.6 EE −9 **5.** 1.5×10^{13} **7.** 8.15×10^{19}

Vocabulary, Readiness & Video Check **1.** $\frac{1}{x^3}$ **3.** scientific notation **5.** A negative exponent has nothing to do with the sign of the simplified result. **7.** When you move the decimal point to the left, the sign of the exponent will be positive; when you move the decimal point to the right, the sign of the exponent will be negative. **9.** the quotient rule

Exercise Set 12.5 **1.** $\frac{1}{64}$ **3.** $\frac{1}{81}$ **5.** $\frac{7}{x^3}$ **7.** 32 **9.** -64 **11.** $\frac{8}{15}$ **13.** p^3 **15.** $\frac{q^4}{p^5}$ **17.** $\frac{1}{x^3}$ **19.** z^3 **21.** $\frac{4}{9}$ **23.** $-p^4$ **25.** -2 **27.** x^4 **29.** p^4 **31.** m^{11} **33.** r^6 **35.** $\frac{1}{x^{15}y^9}$ **37.** $\frac{1}{x^4}$ **39.** $\frac{1}{a^2}$ **41.** $4k^3$ **43.** $3m$ **45.** $-\frac{4a^5}{b}$ **47.** $-\frac{6x^2}{y^3}$ **49.** $\frac{a^{30}}{b^{12}}$ **51.** $\frac{1}{x^{10}y^6}$ **53.** $\frac{z^2}{4}$ **55.** $\frac{1}{32x^5}$ **57.** $\frac{49a^4}{b^6}$ **59.** $a^{24}b^8$ **61.** x^9y^{19} **63.** $-\frac{y^8}{8x^2}$ **65.** $-\frac{6x}{7y^2}$ **67.** $\frac{25b^{33}}{a^{16}}$ **69.** 7.8×10^4 **71.** 1.67×10^{-6} **73.** 6.35×10^{-3} **75.** 1.16×10^6 **77.** 2×10^9 **79.** 2.4×10^3 **81.** 0.0000000008673 **83.** 0.033 **85.** 20,320 **87.** 700,000,000 **89.** 9,460,000,000,000 **91.** 1.84×10^{11} **93.** 155,000,000,000 **95.** 35,000 **97.** 0.000036 **99.** 0.000000000000000028

101. 0.0000005 **103.** 200,000 **105.** $\frac{5x^3}{3}$ **107.** $\frac{5z^3y^2}{7}$ **109.** $5y - 6 + \frac{5}{y}$ **111.** $\frac{27}{x^6z^3}$ cu in. **113.** $9a^{13}$ **115.** -5 **117.** answers may vary **119. a.** 1.3×10^1 **b.** 4.4×10^7 **c.** 6.1×10^{-2} **121.** answers may vary **123.** a^m **125.** $27y^{6z}$ **127.** -394.5 **129.** 1.3 sec

Section 12.6

Vocabulary, Readiness & Video Check **1.** dividend, quotient, divisor **3.** a^2 **5.** y **7.** the common denominator

Exercise Set 12.6 **1.** $12x^3 + 3x$ **3.** $4x^3 - 6x^2 + x + 1$ **5.** $5p^2 + 6p$ **7.** $-\frac{3}{2x} + 3$ **9.** $-3x^2 + x - \frac{4}{x^3}$ **11.** $-1 + \frac{3}{2x} - \frac{7}{4x^4}$ **13.** $x + 1$ **15.** $2x + 3$ **17.** $2x + 1 + \frac{7}{x - 4}$ **19.** $3a^2 - 3a + 1 + \frac{2}{3a + 2}$ **21.** $4x + 3 - \frac{2}{2x + 1}$ **23.** $2x^2 + 6x - 5 - \frac{2}{x - 2}$ **25.** $x + 6$ **27.** $x^2 + 3x + 9$ **29.** $-3x + 6 - \frac{11}{x + 2}$ **31.** $2b - 1 - \frac{6}{2b - 1}$ **33.** $ab - b^2$ **35.** $4x + 9$ **37.** $x + 4xy - \frac{y}{2}$ **39.** $2b^2 + b + 2 - \frac{12}{b + 4}$ **41.** $5x - 2 + \frac{2}{x + 6}$ **43.** $x^2 - \frac{12x}{5} - 1$ **45.** $6x - 1 - \frac{1}{x + 3}$ **47.** $6x - 1$ **49.** $-x^3 + 3x^2 - \frac{4}{x}$ **51.** $x^2 + 3x + 9$ **53.** $y^2 + 5y + 10 + \frac{24}{y - 2}$ **55.** $-6x - 12 - \frac{19}{x - 2}$ **57.** $x^3 - x^2 + x$ **59.** $2a^3 + 2a$ **61.** $2x^3 + 14x^2 - 10x$ **63.** $-3x^2y^3 - 21x^3y^2 - 24xy$ **65.** $9a^2b^3c + 36ab^2c - 72ab$ **67.** $(3x^3 + x - 4)$ ft **69.** c **71.** answers may vary **73.** $(2x + 5)$ m **75.** $9x^{7a} - 6x^{5a} + 7x^{2a} - 1$

Section 12.7

Vocabulary, Readiness & Video Check **1.** The last number n is the remainder and the other numbers are the coefficients of the variables in the quotient; the degree of the quotient is one less than the degree of the dividend

Exercise Set 12.7 **1.** $x + 8$ **3.** $x - 1$ **5.** $x^2 - 5x - 23 - \frac{41}{x - 2}$ **7.** $4x + 8 + \frac{7}{x - 2}$ **9.** 3 **11.** 73 **13.** -8 **15.** $x^2 + \frac{2}{x - 3}$ **17.** $6x + 7 + \frac{1}{x + 1}$ **19.** $2x^3 - 3x^2 + x - 4$ **21.** $3x - 9 + \frac{12}{x + 3}$ **23.** $3x^2 - \frac{9}{2}x + \frac{7}{4} + \frac{47}{8(x - \frac{1}{2})}$ **25.** $3x^2 + 3x - 3$ **27.** $3x^2 + 4x - 8 + \frac{20}{x + 1}$ **29.** $x^2 + x + 1$ **31.** $2x^2 - 3 - \frac{2}{x + 6}$ **33.** 1 **35.** -133 **37.** 3 **39.** $-\frac{187}{81}$ **41.** $\frac{95}{32}$ **43.** answers may vary **45.** $-\frac{5}{6}$ **47.** 54 **49.** 8 **51.** -32 **53.** 48 **55.** 25 **57.** -2 **59.** yes **61.** no **63.** $(x^3 - 5x^2 + 2x - 1)$ cm **65.** $x^3 + \frac{5}{3}x^2 + \frac{5}{3}x + \frac{8}{3} + \frac{8}{3(x - 1)}$ **67.** $(x + 3)(x^2 + 4) = x^3 + 3x^2 + 4x + 12$ **69.** 0 **71.** $x^3 + 2x^2 + 7x + 28$

Chapter 12 Vocabulary Check **1.** term **2.** FOIL **3.** trinomial **4.** degree of a polynomial **5.** binomial **6.** coefficient **7.** degree of a term **8.** monomial **9.** polynomials **10.** distributive

Chapter 12 Review **1.** base: 7; exponent: 9 **3.** base: 5; exponent: 4 **5.** 512 **7.** -36 **9.** 1 **11.** y^9 **13.** $-6x^{11}$ **15.** x^8 **17.** $81y^{24}$ **19.** x^5 **21.** a^4b^3 **23.** $\frac{x^3y^4}{4}$ **25.** $40a^{19}$ **27.** 3 **29.** b **31.** 7 **33.** 8 **35.** 5 **37.** 5 **39.** 4000 ft; 3984 ft; 3856 ft; 3600 ft **41.** $15a^2 + 4a$ **43.** $-6a^2b - 3b^2 - q^2$ **45.** $8x^2 + 3x + 6$ **47.** $-7y^2 - 1$ **49.** $4x - 13y$ **51.** 290 **53.** $(6x^2y - 12x + 12)$ cm **55.** $8a + 28$ **57.** $-7x^3 - 35x$ **59.** $-6a^4 + 8a^2 - 2a$ **61.** $2x^2 - 12x - 14$ **63.** $x^2 - 18x + 81$ **65.** $4a^2 + 27a - 7$ **67.** $25x^2 + 20x + 4$ **69.** $x^4 + 7x^3 + 4x^2 + 23x - 35$ **71.** $x^4 + 4x^3 + 4x^2 - 16$ **73.** $x^3 + 21x^2 + 147x + 343$ **75.** $x^2 + 14x + 49$ **77.** $9x^2 - 42x + 49$ **79.** $25x^2 - 90x + 81$ **81.** $49x^2 - 16$ **83.** $4x^2 - 36$ **85.** $(9x^2 - 6x + 1)$ sq m **87.** $\frac{1}{49}$ **89.** $\frac{2}{x^4}$ **91.** 125 **93.** $\frac{17}{16}$ **95.** x^8 **97.** r **99.** c^4 **101.** $\frac{1}{x^6y^{13}}$ **103.** a^{11m} **105.** $27x^3y^{6z}$ **107.** 2.7×10^{-4} **109.** 8.08×10^7 **111.** 9.1×10^7 **113.** 867,000 **115.** 0.00086 **117.** 1,431,280,000,000,000 **119.** 0.016 **121.** $\frac{1}{7} + \frac{3}{x} + \frac{7}{x^2}$ **123.** $a + 1 + \frac{6}{a - 2}$ **125.** $a^2 + 3a + 8 + \frac{22}{a - 2}$ **127.** $2x^3 - x^2 + 2 - \frac{1}{2x - 1}$ **129.** $\left(5x - 1 + \frac{20}{x^2}\right)$ ft **131.** $3x^2 + 6x + 24 + \frac{44}{x - 2}$ **133.** $x^4 - x^3 + x^2 - x + 1 - \frac{2}{x + 1}$ **135.** $3x^3 + 13x^2 + 51x + 204 + \frac{814}{x - 4}$ **137.** 3043 **139.** $-\frac{1}{8}$ **141.** $\frac{2x^6}{3}$ **143.** $\frac{x^{16}}{16y^{12}}$ **145.** $11x - 5$ **147.** $5y^2 - 3y - 1$ **149.** $28x^3 + 12x$ **151.** $x^3 + x^2 - 18x + 18$ **153.** $25x^2 + 40x + 16$ **155.** $4a - 1 + \frac{2}{a^2} - \frac{5}{2a^3}$ **157.** $2x^2 + 7x + 5 + \frac{19}{2x - 3}$

Chapter 12 Test **1.** 32 **2.** 81 **3.** −81 **4.** $\frac{1}{64}$ **5.** $-15x^{11}$ **6.** y^5 **7.** $\frac{1}{r^5}$ **8.** $\frac{y^{14}}{x^2}$ **9.** $\frac{1}{6xy^8}$ **10.** 5.63×10^5 **11.** 8.63×10^{-5} **12.** 0.0015 **13.** 62,300 **14.** 0.036 **15. a.** 4, 3; 7, 3; 1, 4; −2, 0 **b.** 4 **16.** $-2x^2 + 12xy + 11$ **17.** $16x^3 + 7x^2 - 3x - 13$ **18.** $-3x^3 + 5x^2 + 4x + 5$ **19.** $x^3 + 8x^2 + 3x - 5$ **20.** $3x^3 + 22x^2 + 41x + 14$ **21.** $6x^4 - 9x^3 + 21x^2$ **22.** $3x^2 + 16x - 35$ **23.** $9x^2 - \frac{1}{25}$ **24.** $16x^2 - 16x + 4$ **25.** $64x^2 + 48x + 9$ **26.** $x^4 - 81b^2$ **27.** 1001 ft; 985 ft; 857 ft; 601 ft **28.** $(4x^2 - 9)$ sq in. **29.** $\frac{x}{2y} + 3 - \frac{7}{8y}$ **30.** $x + 2$ **31.** $9x^2 - 6x + 4 - \frac{16}{3x + 2}$ **32. a.** 960 ft **b.** 953.44 ft **33.** $4x^3 - 15x^2 + 45x - 136 + \frac{407}{x + 3}$ **34.** 91

Cumulative Review Chapters 1–12 **1. a.** true **b.** true **c.** false **d.** true; Sec. 9.1, Ex. 3–6 **3.** $\frac{5}{12}$; Sec. 4.3, Ex. 11 **5. a.** 81 **b.** 6 **c.** 81 **d.** 180; Sec. 1.7, Ex. 5–8 **7. a.** −8 **b.** 4 **c.** −5; Sec. 2.2, Ex. 9–11 **9.** −10; Sec. 2.3, Ex. 8 **11. a.** 3 **b.** −14 **c.** 11 **d.** −4; Sec. 2.3, Ex. 1–4 **13. a.** $(5 + 4) + 6$ **b.** $-1 \cdot (2 \cdot 5)$; **c.** $m + (n + 9)$ **d.** $x \cdot (y \cdot 12)$; Sec. 9.2, Ex. 2 **15. a.** $22 + x$ **b.** $-21x$; Sec. 9.2, Ex. 5 and 6 **17.** $5x + 20$; Sec. 3.1, Ex. 8 **19.** 1; Sec. 2.6, Ex. 2 **21.** −24; Sec. 3.2, Ex. 3 **23.** −9; Sec. 3.4, Ex. 2 **25.** 10; Sec. 9.4, Ex. 2 **27.** width: 4 ft; length: 10 ft; Sec. 9.5, Ex. 4 **29.** $\frac{5F - 160}{9} = C$; Sec. 9.5, Ex. 8 **31.** $(-4, 2]$; 

; Sec. 9.6, Ex. 3 **33. a.** (0, 12) **b.** (2, 6) **c.** (−1, 15); Sec. 10.1, Ex. 6 **35.** 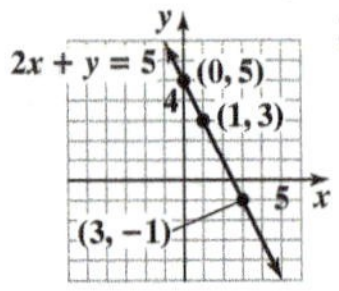

; Sec. 10.2, Ex. 2 **37.** 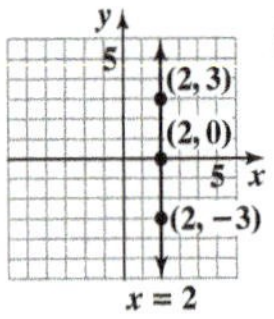

; Sec. 10.3, Ex. 9 **39.** undefined slope; Sec. 10.4, Ex. 7 **41.** $9x^2 - 6x - 1$ (Sec. 12.2, Ex. 11) **43.** $-6x^7$; Sec. 12.1, Ex. 4 **45.** $12x^3 - 12x^2 - 9x + 2$; Sec. 12.2, Ex. 10 **47.** $4x^2 - 4xy + y^2$; Sec. 12.3, Ex. 6 **49.** $3m + 1$; Sec. 12.6, Ex. 1

Chapter 13 Factoring Polynomials

Section 13.1

Vocabulary, Readiness & Video Check **1.** factors **3.** least **5.** false **7.** The GCF of a list of numbers is the largest number that is a factor of all numbers in the list. **9.** When factoring out a GCF, the number of terms in the other factor should have the same number of terms as your original polynomial.

Exercise Set 13.1 **1.** 4 **3.** 6 **5.** 1 **7.** y^2 **9.** z^7 **11.** xy^2 **13.** 7 **15.** $4y^3$ **17.** $5x^2$ **19.** $3x^3$ **21.** $9x^2y$ **23.** $10a^6b$ **25.** $3(a + 2)$ **27.** $15(2x - 1)$ **29.** $x^2(x + 5)$ **31.** $2y^3(3y + 1)$ **33.** $4(x - 2y + 1)$ **35.** $3x(2x^2 - 3x + 4)$ **37.** $a^2b^2(a^5b^4 - a + b^3 - 1)$ **39.** $4(2x^5 + 4x^4 - 5x^3 + 3)$ **41.** $\frac{1}{3}x(x^3 + 2x^2 - 4x^4 + 1)$ **43.** $(x^2 + 2)(y + 3)$ **45.** $(y + 4)(z - 3)$ **47.** $(z^2 - 6)(r + 1)$ **49.** $-2(x + 7)$ **51.** $-x^5(2 - x^2)$ **53.** $-3a^2(2a^2 - 3a + 1)$ **55.** $(x + 2)(x^2 + 5)$ **57.** $(x + 3)(5 + y)$ **59.** $(3x - 2)(2x^2 + 5)$ **61.** $(5m^2 + 6n)(m + 1)$ **63.** $(y - 4)(2 + x)$ **65.** $(2x - 1)(x^2 + 4)$ **67.** not factorable by grouping **69.** $(x - 2y)(4x - 3)$ **71.** $(5q - 4p)(q - 1)$ **73.** $x(x^2 + 1)(2x + 5)$ **75.** $2(2y - 7)(3x^2 - 1)$ **77.** $2x(16y - 9x)$ **79.** $(x + 2)(y - 3)$ **81.** $7xy(2x^2 + x - 1)$ **83.** $(4x - 1)(7x^2 + 3)$ **85.** $-8x^8y^5(5y + 2x)$ **87.** $3(2a + 3b^2)(a + b)$ **89.** $x^2 + 7x + 10$ **91.** $b^2 - 3b - 4$ **93.** 2, 6 **95.** −1, −8 **97.** −2, 5 **99.** b **101.** factored **103.** not factored **105.** answers may vary **107.** answers may vary **109. a.** 1200 million **b.** 1120 million **c.** $-20(x^2 - 15x - 6)$ **111.** $12x^3 - 2x$; $2x(6x^2 - 1)$ **113.** $(n^3 - 6)$ units **115.** $(x^n + 2)(x^n + 3)$ **117.** $(3x^n - 5)(x^n + 7)$

Section 13.2

Vocabulary, Readiness & Video Check **1.** true **3.** false **5.** +5 **7.** −3 **9.** +2 **11.** 15 is positive, so its factors would have to be either both positive or both negative. Since the factors need to sum to −8, both factors must be negative.

Exercise Set 13.2 **1.** $(x + 6)(x + 1)$ **3.** $(y - 9)(y - 1)$ **5.** $(x - 3)(x - 3)$ or $(x - 3)^2$ **7.** $(x - 6)(x + 3)$ **9.** $(x + 10)(x - 7)$ **11.** prime **13.** $(x + 5y)(x + 3y)$ **15.** $(a^2 - 5)(a^2 + 3)$ **17.** $(m + 13)(m + 1)$ **19.** $(t - 2)(t + 12)$ **21.** $(a - 2b)(a - 8b)$ **23.** $2(z + 8)(z + 2)$ **25.** $2x(x - 5)(x - 4)$ **27.** $(x - 4y)(x + y)$ **29.** $(x + 12)(x + 3)$ **31.** $(x - 2)(x + 1)$ **33.** $(r - 12)(r - 4)$ **35.** $(x + 2y)(x - y)$ **37.** $3(x + 5)(x - 2)$ **39.** $3(x - 18)(x - 2)$ **41.** $(x - 24)(x + 6)$ **43.** prime **45.** $(x - 5)(x - 3)$ **47.** $6x(x + 4)(x + 5)$ **49.** $4y(x^2 + x - 3)$ **51.** $(x - 7)(x + 3)$ **53.** $(x + 5y)(x + 2y)$ **55.** $2(t + 8)(t + 4)$ **57.** $x(x - 6)(x + 4)$ **59.** $2t^3(t - 4)(t - 3)$ **61.** $5xy(x - 8y)(x + 3y)$ **63.** $3(m - 9)(m - 6)$ **65.** $-1(x - 11)(x - 1)$ **67.** $\frac{1}{2}(y - 11)(y + 2)$ **69.** $x(xy - 4)(xy + 5)$ **71.** $2x^2 + 11x + 5$ **73.** $15y^2 - 17y + 4$ **75.** $9a^2 + 23ab - 12b^2$ **77.** $x^2 + 5x - 24$ **79.** answers may vary **81.** $2x^2 + 28x + 66$; $2(x + 3)(x + 11)$ **83.** $-16(t - 5)(t + 1)$ **85.** $\left(x + \frac{1}{4}\right)\left(x + \frac{1}{4}\right)$ or $\left(x + \frac{1}{4}\right)^2$ **87.** $(x + 1)(z - 10)(z + 7)$ **89.** $(x^n + 10)(x^n - 2)$ **91.** 5; 8; 9 **93.** 3; 4 **95.** 8; 16 **97.** 6; 26

Section 13.3

Vocabulary, Readiness & Video Check **1.** perfect square trinomial **3.** perfect square trinomial **5.** d **7.** Consider the factors of the first and last terms and the signs of the trinomial. Continue to check by multiplying until you get the middle term of the trinomial. **9.** The first and last terms are squares, a^2 and b^2, and the middle term is $2 \cdot a \cdot b$ or $-2 \cdot a \cdot b$.

Exercise Set 13.3 **1.** $x + 4$ **3.** $10x - 1$ **5.** $5x - 2$ **7.** $(2x + 3)(x + 5)$ **9.** $(y - 1)(8y - 9)$ **11.** $(2x + 1)(x - 5)$ **13.** $(4r - 1)(5r + 8)$ **15.** $(10x + 1)(x + 3)$ **17.** prime **19.** $(3x - 5y)(2x - y)$ **21.** $(3m - 5)(5m + 3)$ **23.** $x(3x + 2)(4x + 1)$ **25.** $3(7b + 5)(b - 3)$ **27.** $(3z + 4)(4z - 3)$ **29.** $2y^2(3x - 10)(x + 3)$ **31.** $(2x - 7)(2x + 3)$ **33.** $-1(x - 6)(x + 4)$ **35.** $x(4x + 3)(x - 3)$ **37.** $(4x - 9)(6x - 1)$ **39.** $(x + 11)^2$ **41.** $(x - 8)^2$ **43.** $(4a - 3)^2$ **45.** $(x^2 + 2)^2$ **47.** $2(n - 7)^2$ **49.** $(4y + 5)^2$ **51.** $(2x + 11)(x - 9)$ **53.** $(8x + 3)(3x + 4)$ **55.** $(3a + b)(a + 3b)$ **57.** $(x - 4)(x - 5)$ **59.** $(p + 6q)^2$ **61.** $(xy - 5)^2$ **63.** $b(8a - 3)(5a + 3)$ **65.** $2x(3x + 2)(5x + 3)$ **67.** $2y(3y + 5)(y - 3)$ **69.** $5x^2(2x - y)(x + 3y)$ **71.** $-1(2x - 5)(7x - 2)$ **73.** $p^2(4p - 5)^2$ **75.** $(3x - 2)(x + 1)$ **77.** $(4x + 9y)(2x - 3y)$ **79.** prime **81.** $(3x - 4y)^2$ **83.** $(6x - 7)(3x + 2)$ **85.** $(7t + 1)(t - 4)$ **87.** $(7p + 1)(7p - 2)$ **89.** $m(m + 9)^2$ **91.** prime **93.** $a(6a^2 + b^2)(a^2 + 6b^2)$ **95.** $x^2 - 4$ **97.** $a^3 + 27$ **99.** 25–34 **101.** answers may vary **103.** no **105.** answers may vary **107.** $4x^2 + 21x + 5$; $(4x + 1)(x + 5)$ **109.** $\left(2x + \frac{1}{2}\right)^2$ **111.** $(y - 1)^2(4x^2 + 10x + 25)$ **113.** 8 **115.** $a^2 + 2ab + b^2$ **117.** 2;14 **119.** 2 **121.** $-3xy^2(4x - 5)(x + 1)$ **123.** $(y - 1)^2(2x + 5)^2$ **125.** $(3x^n + 2)(x^n + 5)$ **127.** answers may vary

Section 13.4

Vocabulary, Readiness & Video Check **1.** a **3.** b **5.** This gives us a four-term polynomial, which may be factored by grouping.

Exercise Set 13.4 **1.** $(x + 3)(x + 2)$ **3.** $(y + 8)(y - 2)$ **5.** $(8x - 5)(x - 3)$ **7.** $(5x^2 - 3)(x^2 + 5)$ **9. a.** 9, 2 **b.** $9x + 2x$ **c.** $(3x + 1)(2x + 3)$ **11. a.** $-20, -3$ **b.** $-20x - 3x$ **c.** $(3x - 4)(5x - 1)$ **13.** $(3y + 2)(7y + 1)$ **15.** $(7x - 11)(x + 1)$ **17.** $(5x - 2)(2x - 1)$ **19.** $(2x - 5)(x - 1)$ **21.** $(2x + 3)^2$ **23.** $(2x + 3)(2x - 7)$ **25.** $(5x - 4)(2x - 3)$ **27.** $x(2x + 3)(x + 5)$ **29.** $2(8y - 9)(y - 1)$ **31.** $(2x - 3)(3x - 2)$ **33.** $3(3a + 2)(6a - 5)$ **35.** $a(4a + 1)(5a + 8)$ **37.** $3x(4x + 3)(x - 3)$ **39.** $y(3x + y)(x + y)$ **41.** prime **43.** $5(x + 5y)^2$ **45.** $6(a + b)(4a - 5b)$ **47.** $p^2(15p + q)(p + 2q)$ **49.** $2(9a^2 - 2)^2$ **51.** $(7 + x)(5 + x)$ or $(x + 7)(x + 5)$ **53.** $(6 - 5x)(1 - x)$ or $(5x - 6)(x - 1)$ **55.** $x^2 - 4$ **57.** $y^2 + 8y + 16$ **59.** $81z^2 - 25$ **61.** $x^3 - 27$ **63.** $10x^2 + 45x + 45$; $5(2x + 3)(x + 3)$ **65.** $(x^n + 2)(x^n + 3)$ **67.** $(3x^n - 5)(x^n + 7)$ **69.** answers may vary

Section 13.5

Graphing Calculator Explorations

	$x^2 - 2x + 1$	$x^2 - 2x - 1$	$(x - 1)^2$
$x = 5$	16	14	16
$x = -3$	16	14	16
$x = 2.7$	2.89	0.89	2.89
$x = -12.1$	171.61	169.61	171.61
$x = 0$	1	−1	1

Vocabulary, Readiness & Video Check **1.** difference of two cubes **3.** sum of two cubes **5.** $(7x)^2$ **7.** $(2y)^3$ **9.** In order to recognize the binomial as a difference of squares and also to identify the terms to use in the special factoring formula. **11.** First rewrite the original binomial with terms writtten as cubes. Answers will then vary depending on your interpretation.

Exercise Set 13.5 **1.** $(x + 2)(x - 2)$ **3.** $(9p + 1)(9p - 1)$ **5.** $(5y - 3)(5y + 3)$ **7.** $(11m + 10n)(11m - 10n)$ **9.** $(xy - 1)(xy + 1)$ **11.** $\left(x - \frac{1}{2}\right)\left(x + \frac{1}{2}\right)$ **13.** $-1(2r + 1)(2r - 1)$ **15.** prime **17.** $(-6 + x)(6 + x)$ or $-1(6 + x)(6 - x)$ **19.** $(m^2 + 1)(m + 1)(m - 1)$ **21.** $(m^2 + n^9)(m^2 - n^9)$ **23.** $(x + 5)(x^2 - 5x + 25)$ **25.** $(2a - 1)(4a^2 + 2a + 1)$ **27.** $(m + 3n)(m^2 - 3mn + 9n^2)$ **29.** $5(k + 2)(k^2 - 2k + 4)$ **31.** $(xy - 4)(x^2y^2 + 4xy + 16)$ **33.** $2(5r - 4t)(25r^2 + 20rt + 16t^2)$ **35.** $(r + 8)(r - 8)$ **37.** $(x + 13y)(x - 13y)$ **39.** $(3 - t)(9 + 3t + t^2)$ **41.** $2(3r + 2)(3r - 2)$ **43.** $x(3y + 2)(3y - 2)$ **45.** $8(m + 2)(m^2 - 2m + 4)$ **47.** $xy(y - 3z)(y + 3z)$ **49.** $4(3x - 4y)(3x + 4y)$ **51.** $9(4 - 3x)(4 + 3x)$ **53.** $(xy - z^2)(x^2y^2 + xyz^2 + z^4)$ **55.** $\left(7 - \frac{3}{5}m\right)\left(7 + \frac{3}{5}m\right)$ **57.** $(t + 7)(t^2 - 7t + 49)$ **59.** $n(n^2 + 49)$ **61.** $x^2(x^2 + 9)(x + 3)(x - 3)$ **63.** $pq(8p + 9q)(8p - 9q)$ **65.** $xy^2(27xy + 1)$ **67.** $a(5a - 4b)(25a^2 + 20ab + 16b^2)$ **69.** $16x^2(x + 2)(x - 2)$ **71.** 6 **73.** −2 **75.** 2 **77.** $(x + 2 + y)(x + 2 - y)$ **79.** $(a + 4)(a - 4)(b - 4)$ **81.** $(x + 3 + 2y)(x + 3 - 2y)$ **83.** $(x^n + 10)(x^n - 10)$ **85.** $(x + 6)$ **87.** answers may vary **89. a.** 2560 ft **b.** 1920 ft **c.** 13 sec **d.** $16(13 - t)(13 + t)$ **91. a.** 1456 ft **b.** 816 ft **c.** 10 sec **d.** $16(10 + t)(10 - t)$

Integrated Review **1.** $(x+y)^2$ **2.** $(x-y)^2$ **3.** $(a+12)(a-1)$ **4.** $(a-10)(a-1)$ **5.** $(a+2)(a-3)$ **6.** $(a-1)^2$ **7.** $(x+1)^2$ **8.** $(x+2)(x-1)$ **9.** $(x+1)(x+3)$ **10.** $(x+3)(x-2)$ **11.** $(x+3)(x+4)$ **12.** $(x+4)(x-3)$ **13.** $(x+4)(x-1)$ **14.** $(x-5)(x-2)$ **15.** $(x+5)(x-3)$ **16.** $(x+6)(x+5)$ **17.** $(x-6)(x+5)$ **18.** $(x+8)(x+3)$ **19.** $2(x+7)(x-7)$ **20.** $3(x+5)(x-5)$ **21.** $(x+3)(x+y)$ **22.** $(y-7)(3+x)$ **23.** $(x+8)(x-2)$ **24.** $(x-7)(x+4)$ **25.** $4x(x+7)(x-2)$ **26.** $6x(x-5)(x+4)$ **27.** $2(3x+4)(2x+3)$ **28.** $(2a-b)(4a+5b)$ **29.** $(2a+b)(2a-b)$ **30.** $(4-3x)(7+2x)$ **31.** $(5-2x)(4+x)$ **32.** prime **33.** prime **34.** $(3y+5)(2y-3)$ **35.** $(4x-5)(x+1)$ **36.** $y(x+y)(x-y)$ **37.** $4(t^2+9)$ **38.** $(x+1)(x+y)$ **39.** $(x+1)(a+2)$ **40.** $9x(2x^2-7x+1)$ **41.** $4a(3a^2-6a+1)$ **42.** $(x+16)(x-2)$ **43.** prime **44.** $(4a-7b)^2$ **45.** $(5p-7q)^2$ **46.** $(7x+3y)(x+3y)$ **47.** $(5-2y)(25+10y+4y^2)$ **48.** $(4x+3)(16x^2-12x+9)$ **49.** $-(x-5)(x+6)$ **50.** $-(x-2)(x-4)$ **51.** $(7-x)(2+x)$ **52.** $(3+x)(1-x)$ **53.** $3x^2y(x+6)(x-4)$ **54.** $2xy(x+5y)(x-y)$ **55.** $5xy^2(x-7y)(x-y)$ **56.** $4x^2y(x-5)(x+3)$ **57.** $3xy(4x^2+81)$ **58.** $2xy^2(3x^2+4)$ **59.** $(2+x)(2-x)$ **60.** $(3+y)(3-y)$ **61.** $(s+4)(3r-1)$ **62.** $(x-2)(x^2+3)$ **63.** $(4x-3)(x-2y)$ **64.** $(2x-y)(2x+7z)$ **65.** $6(x+2y)(x+y)$ **66.** $2(x+4y)(6x-y)$ **67.** $(x+3)(y+2)(y-2)$ **68.** $(y+3)(y-3)(x^2+3)$ **69.** $(5+x)(x+y)$ **70.** $(x-y)(7+y)$ **71.** $(7t-1)(2t-1)$ **72.** prime **73.** $(3x+5)(x-1)$ **74.** $(7x-2)(x+3)$ **75.** $(x+12y)(x-3y)$ **76.** $(3x-2y)(x+4y)$ **77.** $(1-10ab)(1+2ab)$ **78.** $(1+5ab)(1-12ab)$ **79.** $(3+x)(3-x)(1+x)(1-x)$ **80.** $(3+x)(3-x)(2+x)(2-x)$ **81.** $(x+4)(x-4)(x^2+2)$ **82.** $(x+5)(x-5)(x^2+3)$ **83.** $(x-15)(x-8)$ **84.** $(y+16)(y+6)$ **85.** $2x(3x-2)(x-4)$ **86.** $2y(3y+5)(y-3)$ **87.** $(3x-5y)(9x^2+15xy+25y^2)$ **88.** $(6y-z)(36y^2+6yz+z^2)$ **89.** $(xy+2z)(x^2y^2-2xyz+4z^2)$ **90.** $(3ab+2)(9a^2b^2-6ab+4)$ **91.** $2xy(1+6x)(1-6x)$ **92.** $2x(x+3)(x-3)$ **93.** $(x+2)(x-2)(x+6)$ **94.** $(x-2)(x+6)(x-6)$ **95.** $2a^2(3a+5)$ **96.** $2n(2n-3)$ **97.** $(a^2+2)(a+2)$ **98.** $(a-b)(1+x)$ **99.** $(x+2)(x-2)(x+7)$ **100.** $(a+3)(a-3)(a+5)$ **101.** $(x-y+z)(x-y-z)$ **102.** $(x+2y+3)(x+2y-3)$ **103.** $(9+5x+1)(9-5x-1)$ **104.** $(b+4a+c)(b-4a-c)$ **105.** answers may vary **106.** yes; $9(x^2+9y^2)$ **107.** a, c **108.** b, c

Section 13.6

Graphing Calculator Explorations **1.** $-0.9, 2.2$ **3.** no real solution **5.** $-1.8, 2.8$

Vocabulary, Readiness & Video Check **1.** quadratic **3.** 3, −5 **5.** One side of the equation must be a factored polynomial and the other side must be zero. **7.** To find the x-intercepts of any graph in two variables, we let $y = 0$. Doing this with our quadratic equation gives us an equation $= 0$, which we can try to solve by factoring.

Exercise Set 13.6 **1.** 6, 7 **3.** 2, −1 **5.** −9, −17 **7.** 0, −6 **9.** 0, 8 **11.** $-\frac{3}{2}, \frac{5}{4}$ **13.** $\frac{7}{2}, -\frac{2}{7}$ **15.** $\frac{1}{2}, -\frac{1}{3}$ **17.** −0.2, −1.5 **19.** 9, 4 **21.** −4, 2 **23.** 0, 7 **25.** 8, −4 **27.** 4, −4 **29.** −3, 12 **31.** $\frac{7}{3}, -2$ **33.** −5, 5 **35.** $-2, \frac{1}{6}$ **37.** 0, 4, 8 **39.** $\frac{3}{4}$ **41.** $-\frac{1}{2}, 0, \frac{1}{2}$ **43.** $-\frac{3}{8}, 0, \frac{1}{2}$ **45.** −3, 2 **47.** −20, 0 **49.** $\frac{17}{2}$ **51.** $-\frac{1}{2}, \frac{1}{2}$ **53.** $-\frac{3}{2}, -\frac{1}{2}, 3$ **55.** −5, 3 **57.** $-\frac{5}{6}, \frac{6}{5}$ **59.** $2, -\frac{4}{5}$ **61.** $-\frac{4}{3}, 5$ **63.** −4, 3 **65.** $\frac{8}{3}, -9, 0$ **67.** −7 **69.** $0, \frac{3}{2}$ **71.** 0, 1, −1 **73.** $-6, \frac{4}{3}$ **75.** $\frac{6}{7}, 1$ **77.** $\left(-\frac{4}{3}, 0\right), (1, 0)$ **79.** $(-2, 0), (5, 0)$ **81.** $(-6, 0), \left(\frac{1}{2}, 0\right)$ **83.** e **85.** b **87.** c **89.** $\frac{47}{45}$ **91.** $\frac{17}{60}$ **93.** $\frac{15}{8}$ **95.** $\frac{7}{10}$ **97.** didn't write equation in standard form; should be $x = 4$ or $x = -2$ **99.** answers may vary; for example $(x-6)(x+1) = 0$ **101.** answers may vary; for example, $x^2 - 12x + 35 = 0$ **103. a.** 300; 304; 276; 216; 124; 0; −156 **b.** 5 sec **c.** 304 ft **d.**

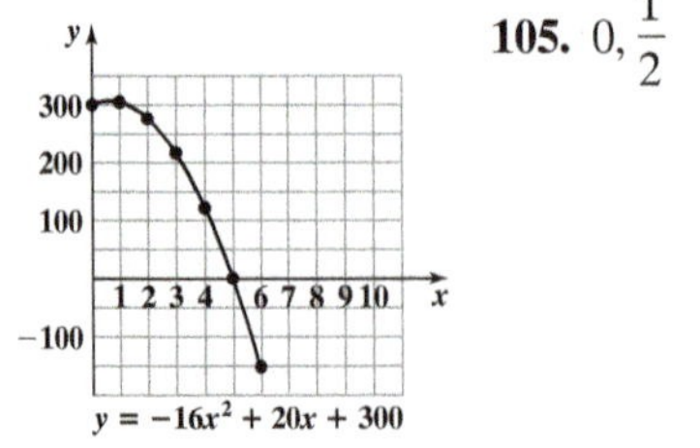

105. $0, \frac{1}{2}$ **107.** 0, −15

Section 13.7

Vocabulary, Readiness & Video Check **1.** In applications, the context of the problem needs to be considered. Each exercise resulted in both a positive and a negative solution, and a negative solution is not appropriate for any of the problems.

Exercise Set 13.7 **1.** width $= x$; length $= x + 4$ **3.** x and $x + 2$ if x is an odd integer **5.** base $= x$; height $= 4x + 1$ **7.** 11 units **9.** 15 cm, 13 cm, 70 cm, 22 cm **11.** base = 16 mi; height = 6 mi **13.** 5 sec **15.** width = 5 cm; length = 6 cm **17.** 54 diagonals **19.** 10 sides **21.** −12 or 11 **23.** 14, 15 **25.** 13 feet **27.** 5 in. **29.** 12 mm, 16 mm, 20 mm **31.** 10 km **33.** 36 ft **35.** 9.5 sec **37.** 20% **39.** length: 15 mi; width: 8 mi **41.** 105 units **43.** 2 million **45.** 1.9 million **47.** 2003 **49.** answers may vary **51.** $\frac{4}{7}$ **53.** $\frac{3}{2}$ **55.** $\frac{1}{3}$ **57.** slow boat: 8 mph; fast boat: 15 mph **59.** 13 and 7 **61.** width: 29 m; length: 35 m **63.** answers may vary

Chapter 13 Vocabulary Check **1.** quadratic equation **2.** factoring **3.** greatest common factor **4.** perfect square trinomial **5.** difference of two squares **6.** difference of two cubes **7.** sum of two cubes **8.** 0 **9.** hypotenuse **10.** leg **11.** hypotenuse

Chapter 13 Review **1.** $2x - 5$ **3.** $4x(5x + 3)$ **5.** $(2x + 3)(3x - 5)$ **7.** $(x - 1)(3x + 2)$ **9.** $(2a + b)(5a + 7b)$ **11.** $(x + 4)(x + 2)$ **13.** prime **15.** $(x + 6y)(x - 2y)$ **17.** $2(3 - x)(12 + x)$ **19.** $10a(a - 1)(a - 10)$ **21.** $-48, 2$ **23.** $(2x + 1)(x + 6)$ **25.** $(3x + 4y)(2x - y)$ **27.** $5y(2y - 3)(y + 4)$ **29.** $2(3x - 5)^2$ **31.** $(2x + 3)(2x - 3)$ **33.** prime **35.** $(2x + 3)(4x^2 - 6x + 9)$ **37.** $2(3 - xy)(9 + 3xy + x^2y^2)$ **39.** $(4x^2 + 1)(2x + 1)(2x - 1)$ **41.** $-6, 2$ **43.** $-\frac{1}{5}, -3$ **45.** $-4, 6$ **47.** $2, 8$ **49.** $-\frac{2}{7}, \frac{3}{8}$ **51.** $-\frac{2}{5}$ **53.** 3 **55.** $0, -\frac{7}{4}, 3$ **57.** c **59.** 9 units **61.** width: 20 in.; length: 25 in. **63.** 19 and 20 **65. a.** 17.5 sec and 10 sec; The rocket reaches a height of 2800 ft on its way up and on its way back down. **b.** 27.5 sec **67.** $7(x - 9)$ **69.** $\left(m + \frac{2}{5}\right)\left(m - \frac{2}{5}\right)$ **71.** $(y + 2)(x - 1)$ **73.** $3x(x - 9)(x - 1)$ **75.** $2(x + 3)(x - 3)$ **77.** $5(x + 2)^2$ **79.** $2xy(2x - 3y)$ **81.** $3(8x^2 - x - 6)$ **83.** $(x + 3)(x + 2)(x - 2)$ **85.** $5x^2 - 9x - 2$; $(5x + 1)(x - 2)$ **87.** $-\frac{7}{2}, 4$ **89.** $0, -7, -4$ **91.** $0, 16$ **93.** length: 6 in.; width: 2 in. **95.** $28x^2 - \pi x^2$; $x^2(28 - \pi)$

Chapter 13 Test **1.** $(x + 7)(x + 4)$ **2.** $(7 - m)(7 + m)$ **3.** $(y + 11)^2$ **4.** $(a + 3)(4 - y)$ **5.** prime **6.** $(y - 12)(y + 4)$ **7.** prime **8.** $3x(3x + 1)(x + 4)$ **9.** $(3a - 7)(a + b)$ **10.** $(3x - 2)(x - 1)$ **11.** $(x + 12y)(x + 2y)$ **12.** $5(6 + x)(6 - x)$ **13.** $(6t + 5)(t - 1)$ **14.** $(y + 2)(y - 2)(x - 7)$ **15.** $x(1 + x^2)(1 + x)(1 - x)$ **16.** $-xy(y^2 + x^2)$ **17.** $(4x - 1)(16x^2 + 4x + 1)$ **18.** $8(y - 2)(y^2 + 2y + 4)$ **19.** $-9, 3$ **20.** $-7, 2$ **21.** $-7, 1$ **22.** $0, \frac{3}{2}, -\frac{4}{3}$ **23.** $0, 3, -3$ **24.** $-3, 5$ **25.** $0, \frac{5}{2}$ **26.** 17 ft **27.** 8 and 9 **28.** 7 sec **29.** hypotenuse: 25 cm; legs: 15 cm, 20 cm

Cumulative Review Chapters 1–13 **1. a.** $9 \le 11$ **b.** $8 > 1$ **c.** $3 \ne 4$; Sec. 9.1, Ex. 7 **3.** $\frac{3}{5}$; Sec. 4.2, Ex. 4 **5.** $-\frac{9}{25}$; Sec. 4.6, Ex. 5 **7.** -10; Sec. 2.2, Ex. 14 **9. a.** -21 **b.** 15 **c.** 0; Sec. 2.4, Ex. 1–3 **11. a.** $10x$ **b.** $-4y$ **c.** $8x^2 - 2$; Sec. 3.1, Ex. 1 **13.** -3; Sec. 3.2, Ex. 2 **15.** -15; Sec. 3.2, Ex. 6 **17.** -4; Sec. 3.3, Ex. 3 **19.** shorter: 2 ft; longer: 8 ft; Sec. 9.4, Ex. 3 **21.** ; Sec. 10.2, Ex. 5 **23.** $m = \frac{3}{4}$; y-intercept: $(0, -1)$; Sec. 10.4, Ex. 5 **25. a.** 250 **b.** 1; Sec. 12.1, Ex. 2 **27. a.** 2 **b.** 5 **c.** 1 **d.** 6 **e.** 0; Sec. 12.2, Ex. 1 **29.** $9x^2 - 6x - 1$; Sec. 12.2, Ex. 11 **31.** $6x^2 - 11x - 10$; Sec. 12.3, Ex. 5 **33.** $9y^2 + 6y + 1$; Sec. 12.4, Ex. 4 **35. a.** $\frac{1}{9}$ **b.** $\frac{2}{x^3}$ **c.** $\frac{3}{4}$ **d.** $\frac{1}{16}$ **e.** $\frac{1}{y^4}$; Sec. 12.5, Ex. 1 **37. a.** 3.67×10^8 **b.** 3.0×10^{-6} **c.** 2.052×10^{10} **d.** 8.5×10^{-4}; Sec. 12.5, Ex. 5 **39.** $x + 4$; Sec. 12.6, Ex. 4 **41. a.** x^3 **b.** y; Sec. 13.1, Ex. 2 **43.** $(x + 3)(x + 4)$; Sec. 13.2, Ex. 1 **45.** $(4x - 1)(2x - 5)$; Sec. 13.3, Ex. 2 **47.** $(5a + 3b)(5a - 3b)$; Sec. 13.5, Ex. 2b **49.** $3, -1$; Sec. 13.6, Ex. 1

Chapter 14 Rational Expressions

Section 14.1

Graphing Calculator Explorations

1. $\{x \mid x \text{ is a real number and } x \ne -2, x \ne 2\}$ **3.** $\left\{x \middle| x \text{ is a real number and } x \ne -4, x \ne \frac{1}{2}\right\}$

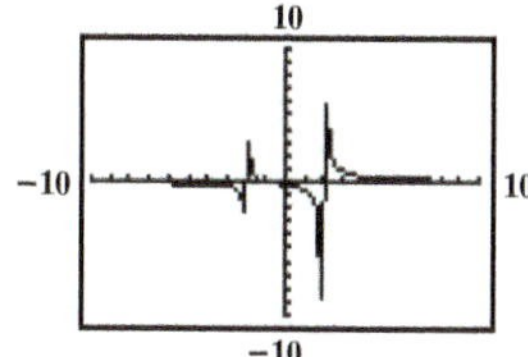

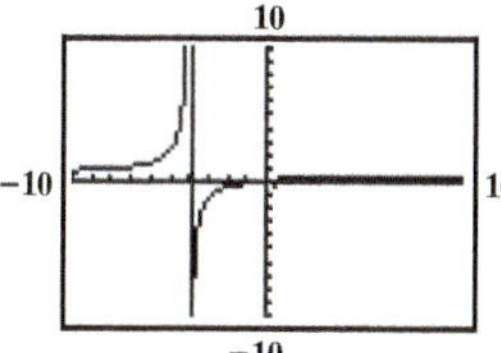

Vocabulary, Readiness & Video Check **1.** rational **3.** domain **5.** 1 **7.** $\frac{-a}{b}; \frac{a}{-b}$ **9.** Rational expressions are fractions and are therefore undefined if the denominator is zero; the domain of a rational function is all real numbers except those that make the denominator of the related rational expression equal to 0. If a denominator contains variables, set it equal to zero and solve. **11.** We insert parentheses around the numerator or denominator if it has more than one term because the negative sign needs to apply to the entire numerator or denominator.

Exercise Set 14.1 **1.** $\{x \mid x \text{ is a real number}\}$ **3.** $\{t \mid t \text{ is a real number and } t \ne 0\}$ **5.** $\{x \mid x \text{ is a real number and } x \ne 7\}$ **7.** $\left\{x \middle| x \text{ is a real number and } x \ne \frac{1}{3}\right\}$ **9.** $\{x \mid x \text{ is a real number and } x \ne -2, x \ne 0, x \ne 1\}$ **11.** $\{x \mid x \text{ is a real number and } x \ne 2, x \ne -2\}$ **13.** $\frac{-(x - 10)}{x + 8}; \frac{-x + 10}{x + 8}; \frac{x - 10}{-(x + 8)}; \frac{x - 10}{-x - 8}$ **15.** $\frac{-(5y - 3)}{y - 12}; \frac{-5y + 3}{y - 12}; \frac{5y - 3}{-(y - 12)}; \frac{5y - 3}{-y + 12}$ **17.** 1 **19.** -1

21. $\frac{1}{4(x+2)}$ **23.** -5 **25.** $\frac{7}{x}$ **27.** $\frac{1}{x-9}$ **29.** $5x+1$ **31.** $\frac{x^2}{x-2}$ **33.** $\frac{x+2}{2}$ **35.** $-(x+2)$ or $-x-2$ **37.** $\frac{11x}{6}$ **39.** $x+y$ **41.** x^2-2x+4 **43.** $-x^2-x-1$ **45.** $\frac{2y+5}{3y+4}$ **47.** $\frac{x-2}{2x^2+1}$ **49.** $\frac{1}{3x+5}$ **51.** correct **53.** correct **55.** $\frac{10}{3}, -8, -\frac{7}{3}$ **57.** $-\frac{17}{48}, \frac{2}{7}, -\frac{3}{8}$ **59. a.** \$200 million **b.** \$500 million **c.** \$300 million **d.** $\{x|x$ is a real number$\}$ **61.** 400 mg **63.** $C = 78.125$; medium **65.** 59.6% **67.** $\frac{3}{11}$ **69.** $\frac{4}{3}$ **71.** $\frac{117}{40}$ **73.** correct **75.** incorrect; $\frac{1+2}{1+3} = \frac{3}{4}$ **77.** no **79.** yes; 1 **81.** yes; -1 **83.** no; answers may vary **85.** answers may vary **87.** $0, \frac{20}{9}, \frac{60}{7}, 20, \frac{140}{3}$, 180, 380, 1980;

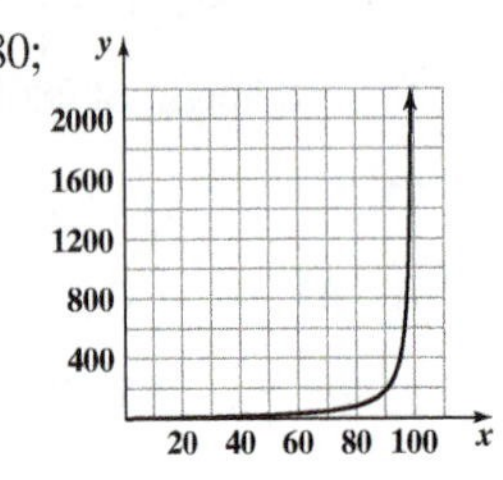

89.

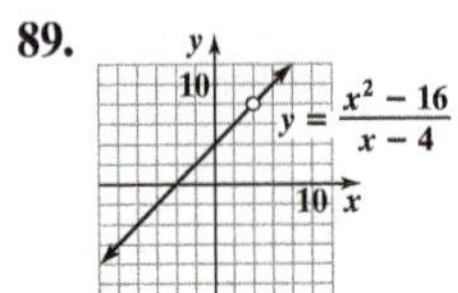

91.

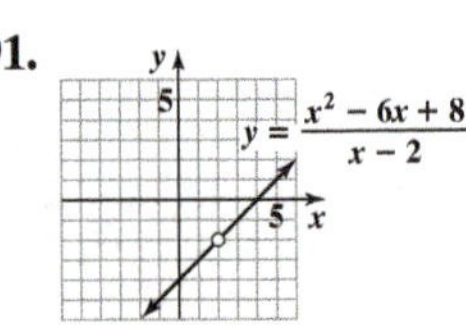

Section 14.2

Vocabulary, Readiness & Video Check **1.** reciprocals **3.** $\frac{a \cdot d}{b \cdot c}$ or $\frac{ad}{bc}$ **5.** $\frac{6}{7}$ **7.** fractions; reciprocal **9.** The units in the unit fraction consist of $\frac{\text{units converting to}}{\text{original units}}$.

Exercise Set 14.2 **1.** $\frac{21}{4y}$ **3.** x^4 **5.** $-\frac{b^2}{6}$ **7.** $\frac{x^2}{10}$ **9.** $\frac{1}{3}$ **11.** $\frac{m+n}{m-n}$ **13.** $\frac{x+5}{x}$ **15.** $\frac{(x+2)(x-3)}{(x-4)(x+4)}$ **17.** $\frac{2x^4}{3}$ **19.** $\frac{12}{y^6}$ **21.** $x(x+4)$ **23.** $\frac{3(x+1)}{x^3(x-1)}$ **25.** m^2-n^2 **27.** $-\frac{x+2}{x-3}$ **29.** $\frac{x+2}{x-3}$ **31.** $\frac{5}{6}$ **33.** $\frac{3x}{8}$ **35.** $\frac{3}{2}$ **37.** $\frac{3x+4y}{2(x+2y)}$ **39.** $\frac{2(x+2)}{x-2}$ **41.** $-\frac{y(x+2)}{4}$ **43.** $\frac{(a+5)(a+3)}{(a+2)(a+1)}$ **45.** $\frac{5}{x}$ **47.** $\frac{2(n-8)}{3n-1}$ **49.** $4x^3(x-3)$ **51.** $\frac{(a+b)^2}{a-b}$ **53.** $\frac{3x+5}{x^2+4}$ **55.** $\frac{4}{x-2}$ **57.** $\frac{a-b}{6(a^2+ab+b^2)}$ **59.** 1440 **61.** 5 **63.** 81 **65.** 73 **67.** 56.7 **69.** 1,201,500 sq ft **71.** 244.9 miles/hour **73.** 1 **75.** $-\frac{10}{9}$ **77.** $-\frac{1}{5}$ **79.**

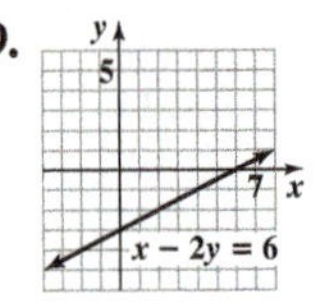

81. true **83.** false; $\frac{x^2+3x}{20}$ **85.** $\frac{2}{9(x-5)}$ sq ft **87.** $\frac{x}{2}$ **89.** $\frac{5a(2a+b)(3a-2b)}{b^2(a-b)(a+2b)}$ **91.** answers may vary

Section 14.3

Vocabulary, Readiness & Video Check **1.** $\frac{9}{11}$ **3.** $\frac{a+c}{b}$ **5.** $\frac{5-(6+x)}{x}$ **7.** We factor denominators into the smallest factors—including coefficients—so we can determine the most number of times each unique factor occurs in any one denominator for the LCD.

Exercise Set 14.3 **1.** $\frac{a+9}{13}$ **3.** $\frac{3m}{n}$ **5.** 4 **7.** $\frac{y+10}{3+y}$ **9.** $5x+3$ **11.** $\frac{4}{a+5}$ **13.** $\frac{1}{x-6}$ **15.** $\frac{5x+7}{x-3}$ **17.** $x+5$ **19.** 3 **21.** $4x^3$ **23.** $8x(x+2)$ **25.** $(x+3)(x-2)$ **27.** $3(x+6)$ **29.** $5(x-6)^2$ **31.** $6(x+1)^2$ **33.** $x-8$ or $8-x$ **35.** $(x-1)(x+4)(x+3)$ **37.** $(3x+1)(x+1)(x-1)(2x+1)$ **39.** $2x^2(x+4)(x-4)$ **41.** $\frac{6x}{4x^2}$ **43.** $\frac{24b^2}{12ab^2}$ **45.** $\frac{9y}{2y(x+3)}$ **47.** $\frac{9ab+2b}{5b(a+2)}$ **49.** $\frac{x^2+x}{x(x+4)(x+2)(x+1)}$ **51.** $\frac{18y-2}{30x^2-60}$ **53.** $2x$ **55.** $\frac{x+3}{2x-1}$ **57.** $x+1$ **59.** $\frac{1}{x^2-8}$ **61.** $\frac{6(4x+1)}{x(2x+1)}$ **63.** $\frac{29}{21}$ **65.** $-\frac{7}{12}$ **67.** $\frac{7}{30}$ **69.** d **71.** answers may vary **73.** c **75.** b **77.** $-\frac{5}{x-2}$ **79.** $\frac{7+x}{x-2}$ **81.** $\frac{20}{x-2}$ m **83.** answers may vary **85.** 95,304 Earth days **87.** answers may vary **89.** answers may vary

Section 14.4

Vocabulary, Readiness & Video Check **1.** d **3.** a **5.** The problem adds two rational expressions with denominators that are opposites of each other. Recognizing this special case can save you time and effort. If you recognize that one denominator is -1 times the other denominator, you may save time.

Exercise Set 14.4 **1.** $\frac{5}{x}$ **3.** $\frac{75a - 6b^2}{5b}$ **5.** $\frac{6x + 5}{2x^2}$ **7.** $\frac{11}{x + 1}$ **9.** $\frac{x - 6}{(x - 2)(x + 2)}$ **11.** $\frac{35x - 6}{4x(x - 2)}$ **13.** $-\frac{2}{x - 3}$ **15.** 0 **17.** $-\frac{1}{x^2 - 1}$ **19.** $\frac{5 + 2x}{x}$ **21.** $\frac{6x - 7}{x - 2}$ **23.** $-\frac{y + 4}{y + 3}$ **25.** $\frac{-5x + 14}{4x}$ or $-\frac{5x - 14}{4x}$ **27.** 2 **29.** $\frac{9x^4 - 4x^2}{21}$ **31.** $\frac{x + 2}{(x + 3)^2}$ **33.** $\frac{9b - 4}{5b(b - 1)}$ **35.** $\frac{2 + m}{m}$ **37.** $\frac{x^2 + 3x}{(x - 7)(x - 2)}$ or $\frac{x(x + 3)}{(x - 7)(x - 2)}$ **39.** $\frac{10}{1 - 2x}$ **41.** $\frac{15x - 1}{(x + 1)^2(x - 1)}$ **43.** $\frac{x^2 - 3x - 2}{(x - 1)^2(x + 1)}$ **45.** $\frac{a + 2}{2(a + 3)}$ **47.** $\frac{y(2y + 1)}{(2y + 3)^2}$ **49.** $\frac{x - 10}{2(x - 2)}$ **51.** $\frac{2x + 21}{(x + 3)^2}$ **53.** $\frac{-5x + 23}{(x - 2)(x - 3)}$ **55.** $\frac{7}{2(m - 10)}$ **57.** $\frac{2x^2 - 2x - 46}{(x + 1)(x - 6)(x - 5)}$ or $\frac{2(x^2 - x - 23)}{(x + 1)(x - 6)(x - 5)}$ **59.** $\frac{n + 4}{4n(n - 1)(n - 2)}$ **61.** 10 **63.** 2 **65.** $\frac{25a}{9(a - 2)}$ **67.** $\frac{x + 4}{(x - 2)(x - 1)}$ **69.** $x = \frac{2}{3}$ **71.** $x = -\frac{1}{2}, x = 1$ **73.** $x = -\frac{15}{2}$ **75.** $\frac{6x^2 - 5x - 3}{x(x + 1)(x - 1)}$ **77.** $\frac{4x^2 - 15x + 6}{(x - 2)^2(x + 2)(x - 3)}$ **79.** $\frac{-2x^2 + 14x + 55}{(x + 2)(x + 7)(x + 3)}$ **81.** $\frac{2x - 16}{(x + 4)(x - 4)}$ in. **83.** $\frac{P - G}{P}$ **85.** answers may vary **87.** $\left(\frac{90x - 40}{x}\right)^\circ$ **89.** answers may vary

Section 14.5

Graphing Calculator Explorations

1. **3.**

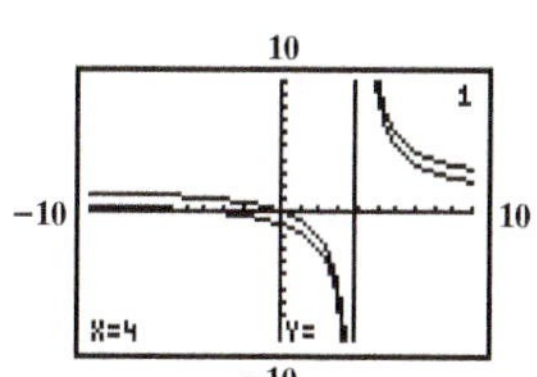

Vocabulary, Readiness & Video Check **1.** c **3.** b **5.** a **7.** These equations are solved in very different ways, so you need to determine the next correct move to make. For a linear equation, you first "move" variable terms on one side and numbers on the other; for a quadratic equation, you first set the equation equal to 0. **9.** the steps for solving an equation containing rational expressions; as if it's the only variable in the equation

Exercise Set 14.5 **1.** 30 **3.** 0 **5.** −2 **7.** −5, 2 **9.** 5 **11.** 3 **13.** 1 **15.** 5 **17.** no solution **19.** 4 **21.** −8 **23.** 6, −4 **25.** 1 **27.** 3, −4 **29.** −3 **31.** 0 **33.** −2 **35.** 8, −2 **37.** no solution **39.** 3 **41.** −11, 1 **43.** $I = \frac{E}{R}$ **45.** $B = \frac{2U - TE}{T}$ **47.** $w = \frac{Bh^2}{705}$ **49.** $G = \frac{V}{N - R}$ **51.** $r = \frac{C}{2\pi}$ **53.** $x = \frac{3y}{3 + y}$ **55.** $\frac{1}{x}$ **57.** $\frac{1}{x} + \frac{1}{2}$ **59.** $\frac{1}{3}$ **61.** (2, 0), (0, −2) **63.** (−4, 0), (−2, 0), (3, 0), (0, 4) **65.** answers may vary **67.** $\frac{5x + 9}{9x}$ **69.** no solution **71.** 100°, 80° **73.** 22.5°, 67.5° **75.** $\frac{17}{4}$

Integrated Review **1.** expression; $\frac{3 + 2x}{3x}$ **2.** expression; $\frac{18 + 5a}{6a}$ **3.** equation; 3 **4.** equation; 18 **5.** expression; $\frac{x + 1}{x(x - 1)}$ **6.** expression; $\frac{3(x + 1)}{x(x - 3)}$ **7.** equation; no solution **8.** equation; 1 **9.** expression; 10 **10.** expression; $\frac{z}{3(9z - 5)}$ **11.** expression; $\frac{5x + 7}{x - 3}$ **12.** expression; $\frac{7p + 5}{2p + 7}$ **13.** equation; 23 **14.** equation; 5 **15.** expression; $\frac{25a}{9(a - 2)}$ **16.** expression; $\frac{4x + 5}{(x + 1)(x - 1)}$ **17.** expression; $\frac{3x^2 + 5x + 3}{(3x - 1)^2}$ **18.** expression; $\frac{2x^2 - 3x - 1}{(2x - 5)^2}$ **19.** expression; $\frac{4x - 37}{5x}$ **20.** expression; $\frac{29x - 23}{3x}$ **21.** equation; $\frac{8}{5}$ **22.** equation; $-\frac{7}{3}$ **23.** answers may vary **24.** answers may vary

Section 14.6

Vocabulary, Readiness & Video Check **1.** c **3.** $\frac{1}{x}; \frac{1}{x} - 3$ **5.** $z + 5; \frac{1}{z + 5}$ **7.** $2y; \frac{11}{2y}$ **9.** There are also many ways to set up an incorrect proportion, so just checking your solution in your proportion isn't enough. You need to determine if your solution is reasonable from the relationships given in the problem. **11.** Two machines (or people) take different amounts of time to complete the task, one faster and one slower than the other. When working together, they will complete the task in less time than the faster machine, so your answer must be less than the time of the faster machine.

Exercise Set 14.6 **1.** 123 lb **3.** 165 cal **5.** $y = 21.25$ **7.** $y = 5\frac{5}{7}$ ft **9.** 2 **11.** -3 **13.** $2\frac{2}{9}$ hr **15.** $1\frac{1}{2}$ min **17.** trip to park rate: r; to park time: $\frac{12}{r}$; return trip rate: r; return time: $\frac{18}{r} = \frac{12}{r} + 1$; $r = 6$ mph **19.** 1st portion: 10 mph; cooldown: 8 mph **21.** 360 sq ft **23.** 2 **25.** \$108.00 **27.** 20 mph **29.** $y = 37\frac{1}{2}$ ft **31.** 41 mph; 51 mph **33.** 5 **35.** 217 mph **37.** 9 gal **39.** 8 mph **41.** 2.2 mph; 3.3 mph **43.** 3 hr **45.** $26\frac{2}{3}$ ft **47.** 216 nuts **49.** $666\frac{2}{3}$ mi **51.** 20 hr **53.** car: 70 mph; motorcycle: 60 mph **55.** $5\frac{1}{4}$ hr **57.** 8 **59.** first car: 64 mph; second car: 50 mph **61.** 510 mph **63.** $x = 5$ **65.** $x = 13.5$ **67.** $-\frac{4}{3}$; downward **69.** $\frac{11}{4}$; upward **71.** undefined slope; vertical **73.** 40,200 megawatts **75.** 22,512,000 people **77.** yes **79.** first pump: 28 min; second pump: 84 min **81.** none; answers may vary **83.** answers may vary **85.** $R = \frac{D}{T}$

Section 14.7

Vocabulary, Readiness & Video Check **1.** $\frac{7}{1+z}$ **3.** $\frac{1}{x^2}$ **5.** $\frac{2}{x}$ **7.** $\frac{1}{9y}$ **9.** a single fraction in the numerator and in the denominator **11.** Since a negative exponent moves its base from a numerator to a denominator of the expression only, a rational expression containing negative exponents can become a complex fraction when rewritten with positive exponents.

Exercise Set 14.7 **1.** 4 **3.** $\frac{7}{13}$ **5.** $\frac{4}{x}$ **7.** $\frac{9(x-2)}{9x^2+4}$ **9.** $2x + y$ **11.** $\frac{2(x+1)}{2x-1}$ **13.** $\frac{2x+3}{4-9x}$ **15.** $\frac{1}{x^2-2x+4}$ **17.** $\frac{x}{5(x-2)}$ **19.** $\frac{x-2}{2x-1}$ **21.** $\frac{x}{2-3x}$ **23.** $-\frac{y}{x+y}$ **25.** $-\frac{2x^3}{y(x-y)}$ **27.** $\frac{2x+1}{y}$ **29.** $\frac{x-3}{9}$ **31.** $\frac{1}{x+2}$ **33.** 2 **35.** $\frac{xy^2}{x^2+y^2}$ **37.** $\frac{2b^2+3a}{b(b-a)}$ **39.** $\frac{x}{(x+1)(x-1)}$ **41.** $\frac{1+a}{1-a}$ **43.** $\frac{x(x+6y)}{2y}$ **45.** $\frac{5a}{2(a+2)}$ **47.** $xy(5y+2x)$ **49.** $\frac{xy}{2x+5y}$ **51.** $\frac{x^2y^2}{4}$ **53.** $-9x^3y^4$ **55.** -9 **57.** a and c **59.** $\frac{770a}{770-s}$ **61.** a, b **63.** answers may vary **65.** $\frac{1+x}{2+x}$ **67.** $x(x+1)$ **69.** $\frac{x-3y}{x+3y}$ **71.** $3a^2 + 4a + 4$ **73. a.** $\frac{1}{a+h}$ **b.** $\frac{1}{a}$ **c.** $\frac{\frac{1}{a+h} - \frac{1}{a}}{h}$ **d.** $-\frac{1}{a(a+h)}$ **75. a.** $\frac{3}{a+h+1}$ **b.** $\frac{3}{a+1}$ **c.** $\frac{\frac{3}{a+h+1} - \frac{3}{a+1}}{h}$ **d.** $-\frac{3}{(a+h+1)(a+1)}$

Chapter 14 Vocabulary Check **1.** rational expression **2.** complex fraction **3.** $\frac{-a}{b}$; $\frac{a}{-b}$ **4.** denominator **5.** simplifying **6.** reciprocals **7.** least common denominator **8.** ratio **9.** proportion **10.** cross products **11.** domain

Chapter 14 Review **1.** $\{x | x \text{ is a real number}\}$ **3.** $\{x | x \text{ is a real number and } x \neq 5\}$ **5.** $\{x | x \text{ is a real number and } x \neq 0, x \neq -8\}$ **7.** -1 **9.** $\frac{1}{x-7}$ **11.** $\frac{x+a}{x-c}$ **13.** $-\frac{1}{x^2+4x+16}$ **15.** \$119 **17.** $\frac{3x^2}{y}$ **19.** $\frac{x-3}{x+2}$ **21.** $\frac{x+3}{x-4}$ **23.** $(x-6)(x-3)$ **25.** $\frac{1}{2}$ **27.** $-\frac{2(2x+3)}{y-2}$ **29.** $\frac{1}{x+2}$ **31.** $\frac{2x-10}{3x^2}$ **33.** $14x$ **35.** $\frac{10x^2y}{14x^3y}$ **37.** $\frac{x^2-3x-10}{(x+2)(x-5)(x+9)}$ **39.** $\frac{4y-30x^2}{5x^2y}$ **41.** $\frac{-2x-2}{x+3}$ **43.** $\frac{x-4}{3x}$ **45.** $\frac{x^2+2x+4}{4x}$; $\frac{x+2}{32}$ **47.** 30 **49.** no solution **51.** $\frac{9}{7}$ **53.** $b = \frac{4A}{5x^2}$ **55.** 675 parts **57.** 3 **59.** fast car speed: 30 mph; slow car speed: 20 mph **61.** $17\frac{1}{2}$ hr **63.** $x = 15$ **65.** $-\frac{7}{18y}$ **67.** $\frac{3y-1}{2y-1}$ **69.** $-\frac{x^2+9}{6x}$ **71.** $\frac{xy+1}{x}$ **73.** $\frac{1}{2x}$ **75.** $\frac{x-4}{x+4}$ **77.** $\frac{1}{x-6}$ **79.** $\frac{2}{(x+3)(x-2)}$ **81.** $\frac{1}{2}$ **83.** 1 **85.** $x = 6$ **87.** $\frac{3}{10}$ **89.** $\frac{1}{y^2-1}$

Chapter 14 Test **1.** $\{x | x \text{ is a real number}, x \neq -1, x \neq -3\}$ **2. a.** \$115 **b.** \$103 **3.** $\frac{3}{5}$ **4.** $\frac{1}{x+6}$ **5.** $\frac{1}{x^2-3x+9}$ **6.** $\frac{2m(m+2)}{m-2}$ **7.** $\frac{a+2}{a+5}$ **8.** $-\frac{1}{x+y}$ **9.** 15 **10.** $\frac{y-2}{4}$ **11.** $\frac{19x-6}{2x+5}$ **12.** $\frac{3a-4}{(a-3)(a+2)}$ **13.** $\frac{3}{x-1}$ **14.** $\frac{2(x+5)}{x(y+5)}$ **15.** $\frac{x^2+2x+35}{(x+9)(x+2)(x-5)}$ **16.** $\frac{30}{11}$ **17.** -6 **18.** no solution **19.** $-2, 5$ **20.** no solution **21.** $\frac{xz}{2y}$ **22.** $\frac{5y^2-1}{y+2}$ **23.** $b - a$ **24.** 18 bulbs **25.** 5 or 1 **26.** 30 mph **27.** $6\frac{2}{3}$ hr **28.** $x = 12$

Cumulative Review Chapters 1–14 **1. a.** $7 \cdot 6 = 42$ **b.** $2(3 + 5) = 16$ **c.** $\frac{-45}{5} = -9$; Sec. 3.3, Ex.5 **3. a.** $20 + x = 5$ **b.** $2x = -10$ **c.** $x - 11 = 168$ **d.** $3(x + 5) = -30$ **e.** $\frac{2x}{8} = 2$; Sec. 3.4, Ex.1 **5.**

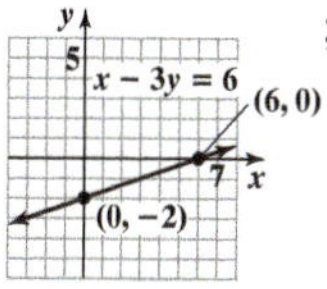

; Sec. 10.3, Ex. 6

7. a. 4^7 **b.** x^{10} **c.** y^4 **d.** y^{12} **e.** $(-5)^{15}$ **f.** a^2b^2; Sec. 12.1, Ex. 3 **9.** $12z + 16$; Sec. 12.2, Ex. 12 **11.** $27a^3 + 27a^2b + 9ab^2 + b^3$; Sec. 12.3, Ex. 8 **13. a.** $t^2 + 4t + 4$ **b.** $p^2 - 2pq + q^2$ **c.** $4x^2 + 20x + 25$ **d.** $x^4 - 14x^2y + 49y^2$; Sec. 12.4, Ex. 5 **15. a.** x^3 **b.** 81 **c.** $\frac{q^9}{p^4}$ **d.** $\frac{32}{125}$; Sec. 12.5, Ex. 2 **17.** $4x^2 - 4x + 6 + \frac{-11}{2x + 3}$; Sec. 12.6, Ex. 6 **19. a.** 4 **b.** 1 **c.** 3; Sec. 13.1, Ex. 1 **21.** $-3a(3a^4 - 6a + 1)$; Sec. 13.1, Ex. 5 **23.** $3(m + 2)(m - 10)$; Sec. 13.2, Ex. 9 **25.** $(3x + 2)(x + 3)$; Sec. 13.3, Ex. 1 **27.** $(x + 6)^2$; Sec.13.3, Ex. 8 **29.** prime polynomial; Sec. 13.5, Ex. 4b **31.** $(x + 2)(x^2 - 2x + 4)$; Sec. 13.5, Ex. 8 **33.** $(2x + 3)(x + 1)(x - 1)$; Ch. 13 Int. Rev., Ex. 2 **35.** $3(2m + n)(2m - n)$; Ch. 13 Int. Rev., Ex. 3 **37.** $-\frac{1}{2}, 4$; Sec. 13.6, Ex. 6 **39.** (1, 0), (4, 0); Sec. 13.6, Ex. 11 **41.** base: 6 m; height: 10 m; Sec. 13.7, Ex. 3 **43.** $-\frac{2(3 + x)}{x + 1}$; Sec. 14.1, Ex. 4 **45.** $\frac{2}{x(x + 1)}$; Sec. 14.2, Ex. 6 **47.** $\frac{1 + 2x}{2(2 - x)}$; Sec. 14.7, Ex. 4

Chapter 15 More on Functions and Graphs

Section 15.1

Graphing Calculator Explorations

1. $y = \frac{x}{3.5}$

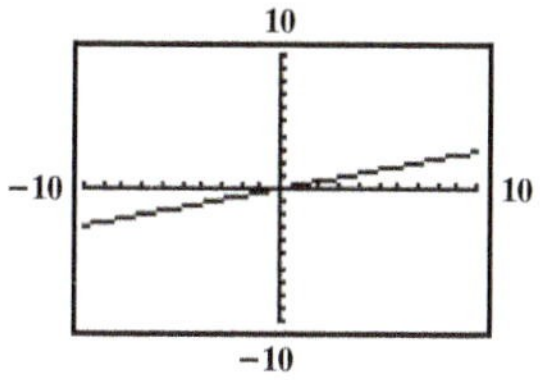

3. $y = -\frac{5.78}{2.31}x + \frac{10.98}{2.31}$

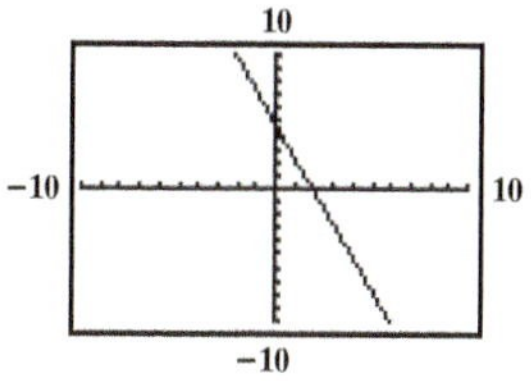

5. $y = |x| + 3.78$

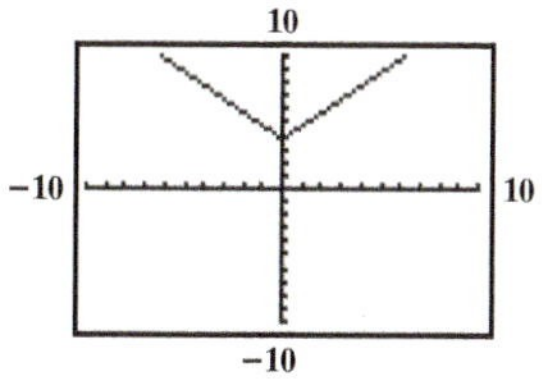

7. $y = 5.6x^2 + 7.7x + 1.5$

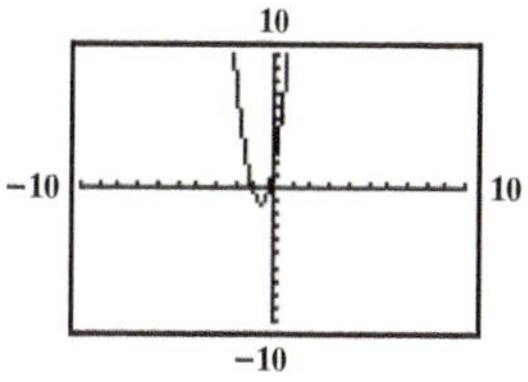

Vocabulary, Readiness & Video Check **1.** linear **3.** $m = -4$, y-intercept: (0, 12) **5.** $m = 5$, y-intercept: (0, 0) **7.** parallel **9.** neither **11.** $f(x) = mx + b$, or slope-intercept form **13.** if one of the two points given is the y-intercept

Exercise Set 15.1 **1.**

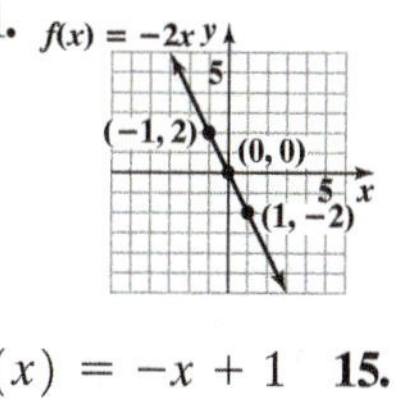

3.

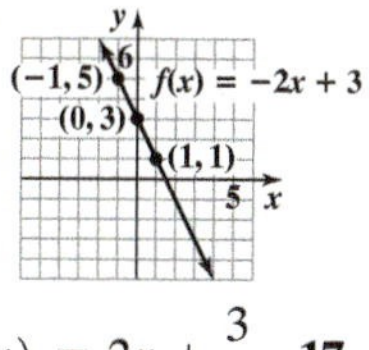

5.

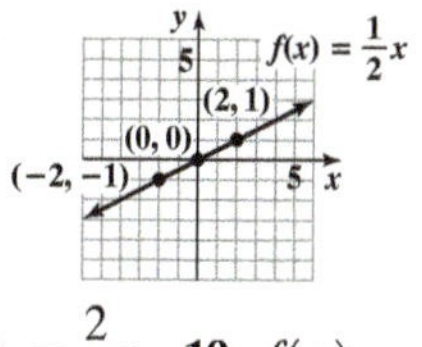

7.

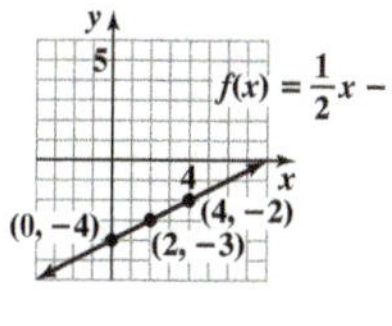

9. C **11.** D **13.** $f(x) = -x + 1$ **15.** $f(x) = 2x + \frac{3}{4}$ **17.** $f(x) = \frac{2}{7}x$ **19.** $f(x) = 3x - 1$ **21.** $f(x) = -2x - 1$ **23.** $f(x) = \frac{1}{2}x + 5$ **25.** $f(x) = -\frac{9}{10}x - \frac{27}{10}$ **27.** $f(x) = 3x - 6$ **29.** $f(x) = -2x + 1$ **31.** $f(x) = -\frac{1}{2}x - 5$ **33.** $f(x) = \frac{1}{3}x - 7$ **35.** $f(x) = -\frac{3}{8}x + \frac{5}{8}$ **37.** $f(x) = -4$ **39.** $f(x) = 5$ **41.** $f(x) = 4x - 4$ **43.** $f(x) = -3x + 1$ **45.** $f(x) = -\frac{3}{2}x - 6$ **47.** $2x - y = -7$ **49.** $f(x) = -x + 7$ **51.** $f(x) = -\frac{1}{2}x + 11$ **53.** $2x + 7y = -42$ **55.** $4x + 3y = -20$ **57.** $f(x) = -10$ **59.** $x + 2y = 2$ **61.** $f(x) = 12$ **63.** $8x - y = 47$ **65.** $x = 5$ **67.** $f(x) = -\frac{3}{8}x - \frac{29}{4}$ **69.** 28.4; In 2009, about 28.4% of students took at least one online course. **71.** 54.22; In 2016, we predict that 54.22% of students will take at least one online course. **73.** answers may vary **75.** $f(x) = -2x + 3$ **77.** $f(x) = \frac{2}{3}x + \frac{7}{3}$ **79. a.** $y = 32x$ **b.** 128 ft per sec

81. a. $y = -250x + 3500$ **b.** 1625 Frisbees **83. a.** $y = 58.1x + 2619$ **b.** 2851.4 thousand **85. a.** $y = -4820x + 297{,}000$ **b.** \$258,440 **87.** $-4x + y = 4$ **89.** $2x + y = -23$ **91.** $3x - 2y = -13$ **93.** answers may vary

Section 15.2

Graphing Calculator Explorations

1.

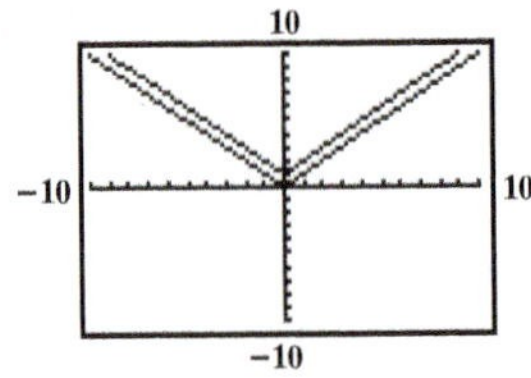

3.

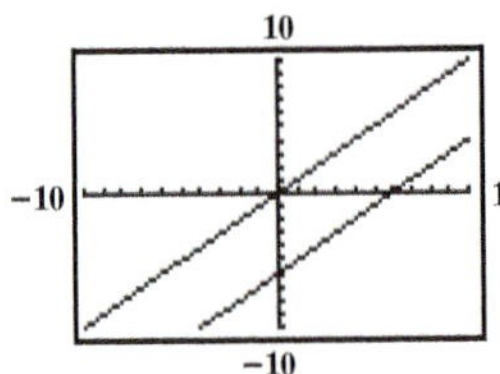

5.

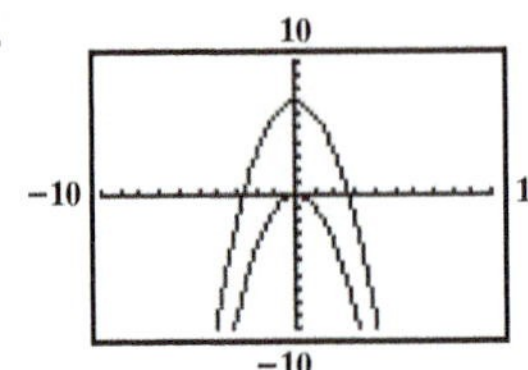

Vocabulary, Readiness & Video Check **1.** V-shaped **3.** $(-2, 1.7)$ **5.** Using function notation, the replacement value for x and the resulting $f(x)$, or y-, value corresponds to an ordered pair (x, y) solution to the function. **7.** not; shape

Exercise Set 15.2 **1.** 0 **3.** -4 **5.** 3 **7.** $(1, -10)$ **9.** $(4, 56)$ **11.** $f(-1) = -2$ **13.** $g(2) = 0$ **15.** $-4, 0$ **17.** 3 **19.** 7 **21.** $-\frac{2}{3}$ **23.** 8 **25.** 9 **27.** not a real number **29.**

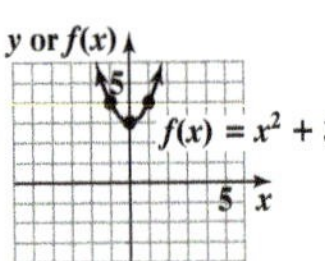

31.

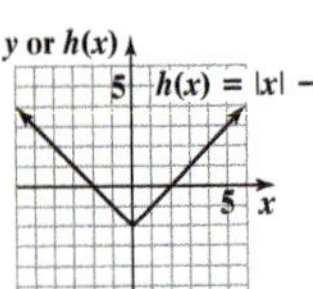

33. **35.**

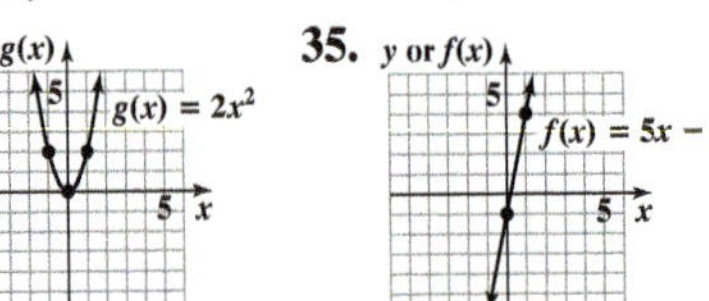

37.

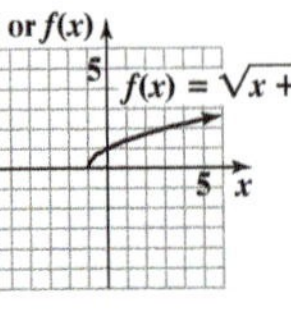

39.

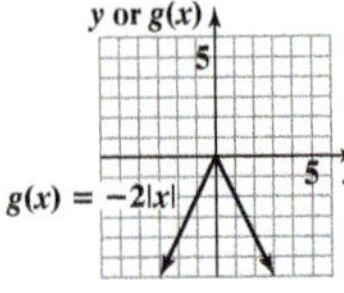

41.

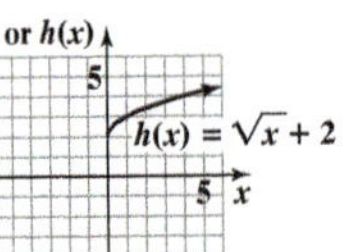

43. a. answers may vary; between 5 and 6 million **b.** 5.25 million or 5,250,000 students **45.** \$15.54 billion **47.** 25π sq cm **49.** 2744 cu in. **51.** 166.38 cm **53.** 163.2 mg **55.** infinite number **57.** -5 **59.** $-\frac{1}{10}$ **61.** b **63.** c **65.** 1997 **67.** answers may vary

69. **71.**

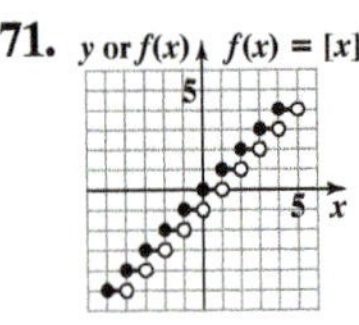

Integrated Review **1.** $m = 3$; $(0, -5)$ **2.** $m = \frac{5}{2}$; $\left(0, -\frac{7}{2}\right)$ **3.** parallel **4.** perpendicular **5.** $f(x) = -x + 7$ **6.** $f(x) = -\frac{3}{8}x - \frac{29}{4}$ **7.** $f(x) = 3x - 2$ **8.** $f(x) = -\frac{5}{4}x + 4$ **9.** $f(x) = \frac{1}{4}x - \frac{7}{2}$ **10.** $f(x) = -\frac{5}{2}x - \frac{5}{2}$

11. linear

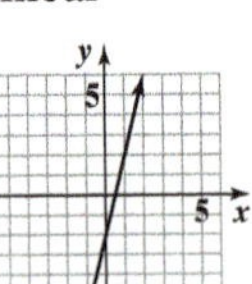

12. linear **13.** not linear

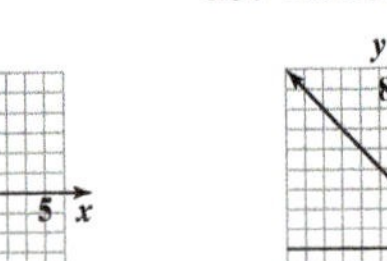

14. not linear

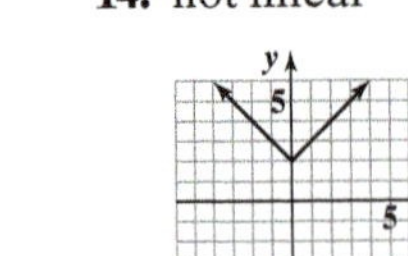

15. not linear

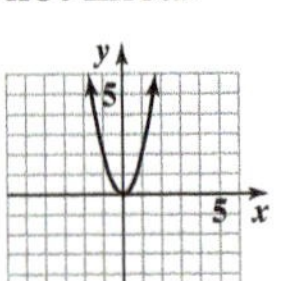

16. not linear

17. not linear

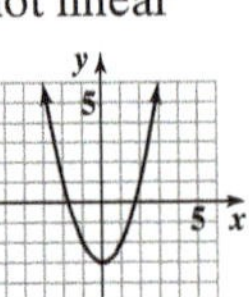

18. not linear

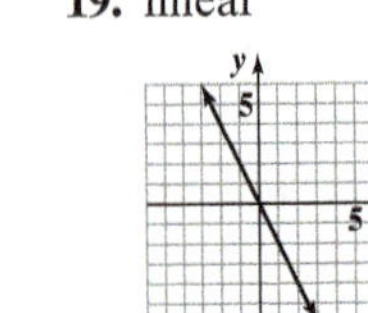

19. linear

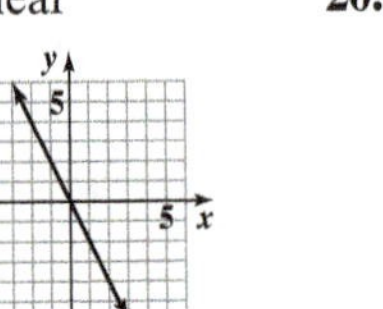

20. linear

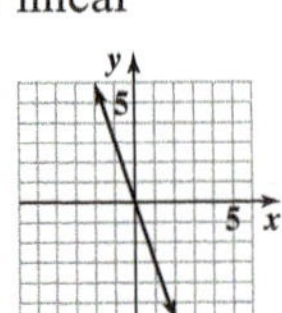

21. not linear

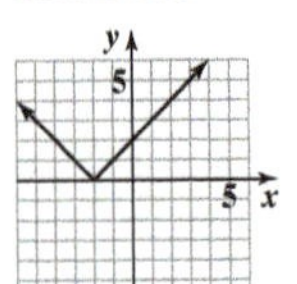

22. not linear

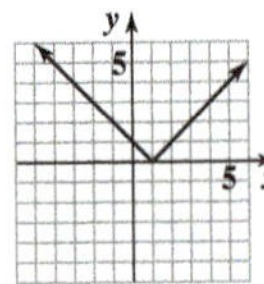

23. linear

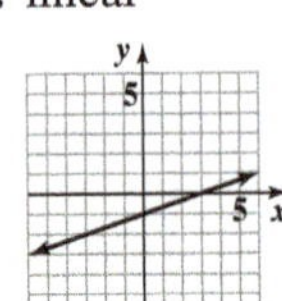

24. linear

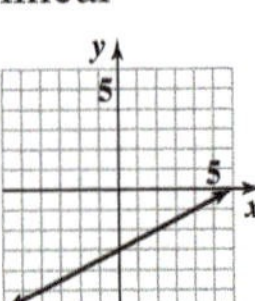

25. linear

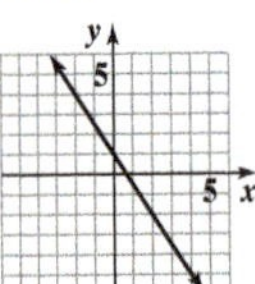

26. linear

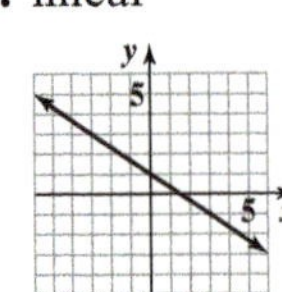

Section 15.3

Vocabulary, Readiness & Video Check **1.** C **3.** D **5.** Although $f(x) = x + 3$ isn't defined for $x = -1$, we need to clearly indicate the point where this piece of the graph ends. Therefore, we find this point and graph it as an open circle. **7.** x-axis

Exercise Set 15.3

1. 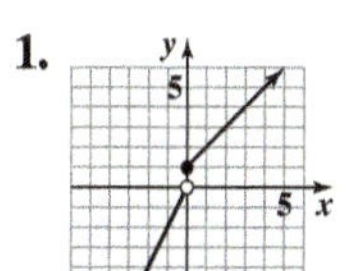 **3.** 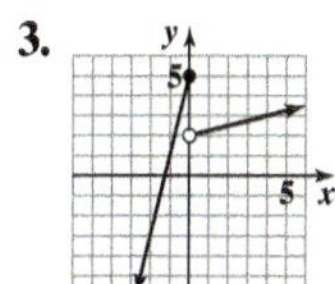 **5.** 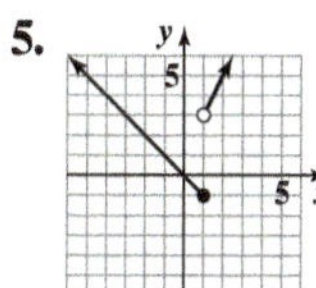 **7.** 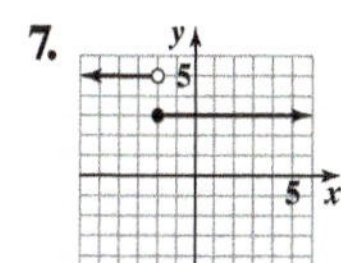 **9.** 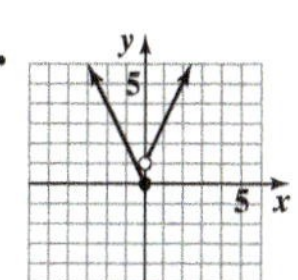 domain: $(-\infty, \infty)$; range: $[0, \infty)$

11. 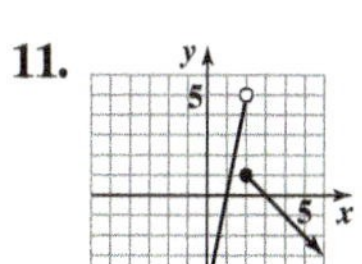 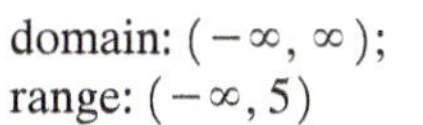 domain: $(-\infty, \infty)$; range: $(-\infty, 5)$ **13.** 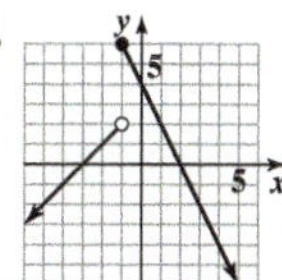 domain: $(-\infty, \infty)$; range: $(-\infty, 6]$ **15.** 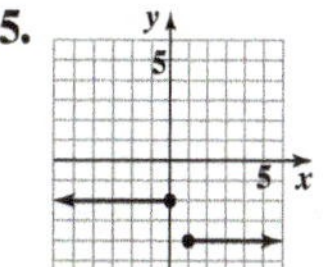 domain: $(-\infty, 0] \cup [1, \infty)$; range: $\{-4, -2\}$

17. 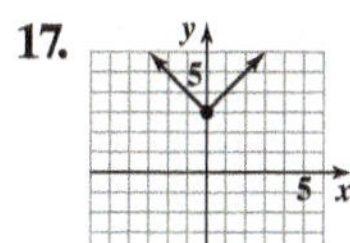 **19.** 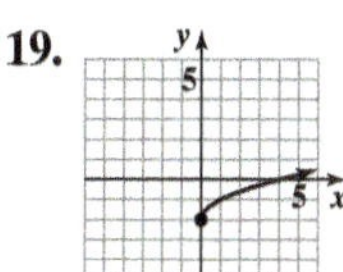 **21.** 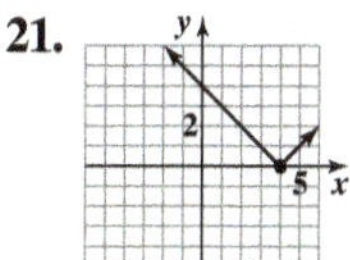 **23.** 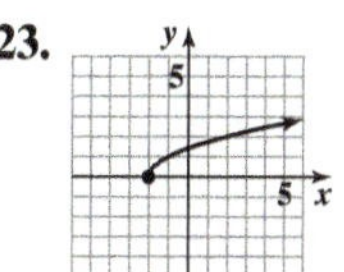 **25.** 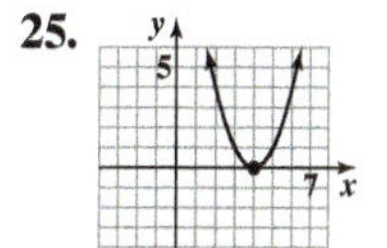 **27.** 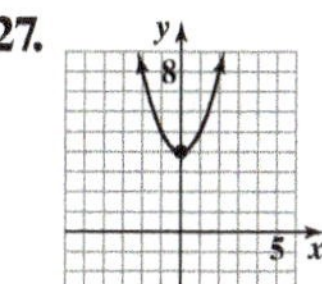

29. 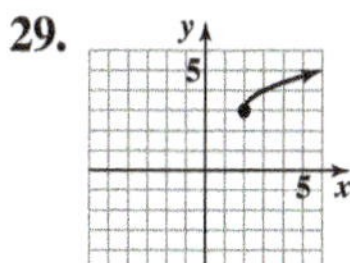 **31.** 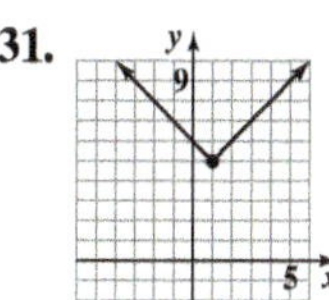 **33.** 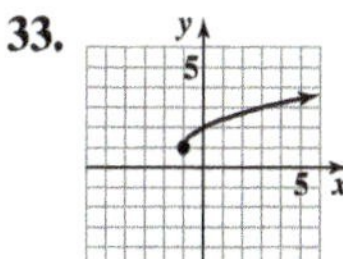 **35.** 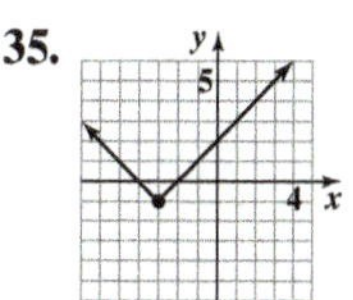 **37.** 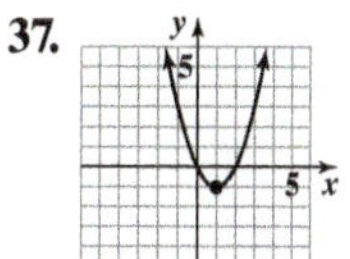 **39.** 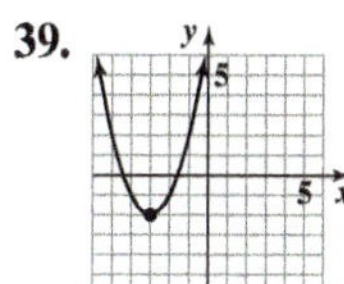

41. 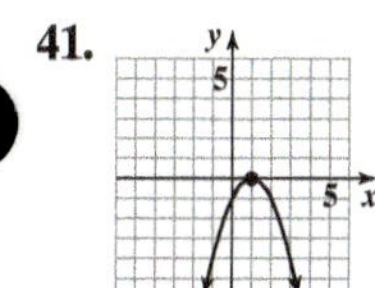 **43.** 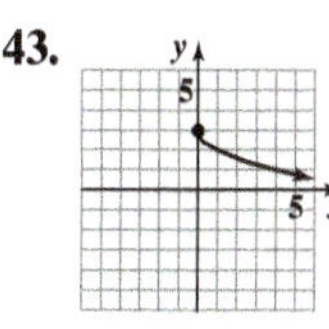 **45.** 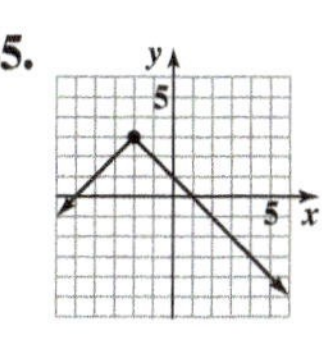 **47.** 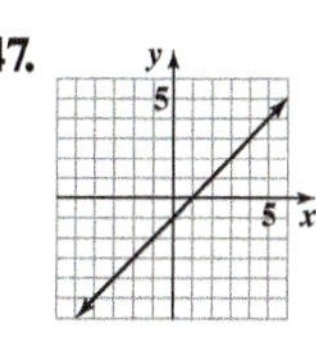 **49.** A **51.** D **53.** answers may vary

55. 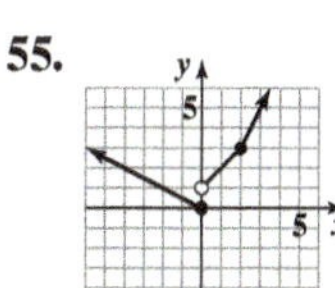 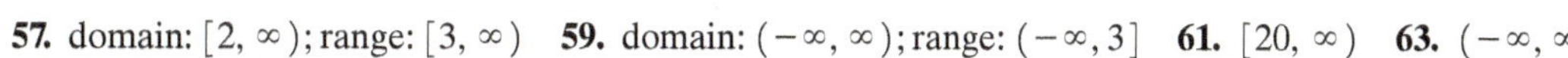 **57.** domain: $[2, \infty)$; range: $[3, \infty)$ **59.** domain: $(-\infty, \infty)$; range: $(-\infty, 3]$ **61.** $[20, \infty)$ **63.** $(-\infty, \infty)$

65. $[-103, \infty)$ **67.** 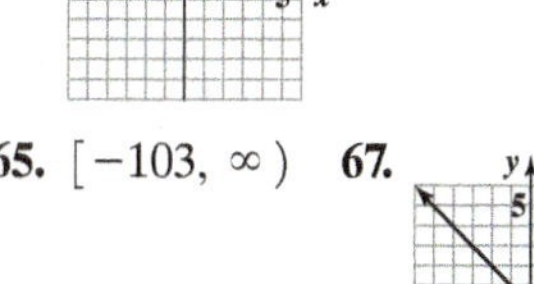 domain: $(-\infty, \infty)$; range: $[0, \infty)$ **69.** 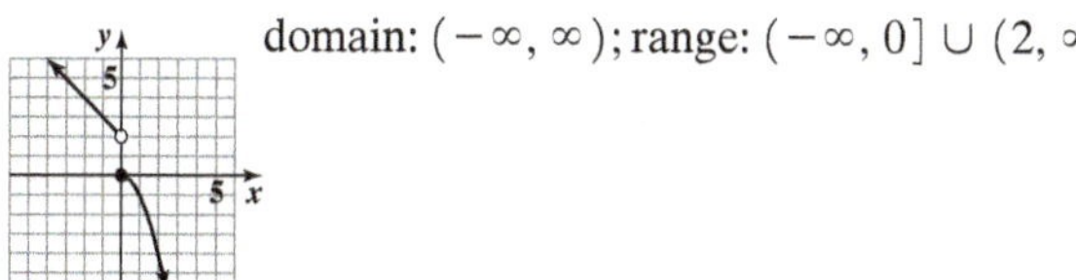 domain: $(-\infty, \infty)$; range: $(-\infty, 0] \cup (2, \infty)$

Section 15.4

Vocabulary, Readiness & Video Check **1.** direct **3.** joint **5.** inverse **7.** direct **9.** linear; slope **11.** $y = ka^2b^5$

Exercise Set 15.4 **1.** $k = \frac{1}{5}; y = \frac{1}{5}x$ **3.** $k = \frac{3}{2}; y = \frac{3}{2}x$ **5.** $k = 14; y = 14x$ **7.** $k = 0.25; y = 0.25x$ **9.** 4.05 lb **11.** 187,239 tons **13.** $k = 30; y = \frac{30}{x}$ **15.** $k = 700; y = \frac{700}{x}$ **17.** $k = 2; y = \frac{2}{x}$ **19.** $k = 0.14; y = \frac{0.14}{x}$ **21.** 54 mph **23.** 72 amps **25.** divided by 4 **27.** $x = kyz$ **29.** $r = kst^3$ **31.** $k = \frac{1}{3}; y = \frac{1}{3}x^3$ **33.** $k = 0.2; y = 0.2\sqrt{x}$ **35.** $k = 1.3; y = \frac{1.3}{x^2}$ **37.** $k = 3; y = 3xz^3$ **39.** 22.5 tons **41.** 15π cu in. **43.** 8 ft **45.** $y = kx$ **47.** $a = \frac{k}{b}$ **49.** $y = kxz$ **51.** $y = \frac{k}{x^3}$ **53.** $y = \frac{kx}{p^2}$ **55.** $C = 8\pi$ in.; $A = 16\pi$ sq in. **57.** $C = 18\pi$ cm; $A = 81\pi$ sq cm **59.** 1.2 **61.** -7 **63.** $-\frac{1}{2}$ **65.** $\frac{8}{27}$ **67.** a **69.** c

71. multiplied by 8 **73.** multiplied by 2 **75.**

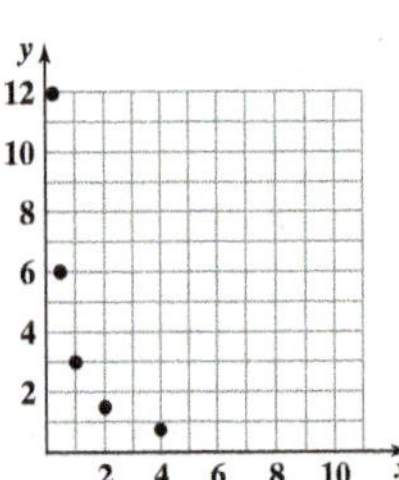

77.

Chapter 15 Vocabulary Check **1.** Parallel **2.** Slope-intercept **3.** function **4.** slope **5.** perpendicular **6.** linear function **7.** directly **8.** inversely **9.** jointly

Chapter 15 Review **1.**

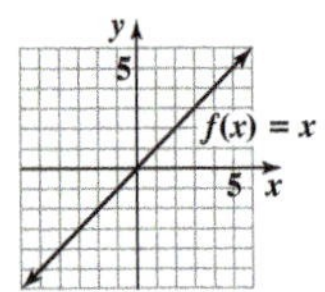

3.

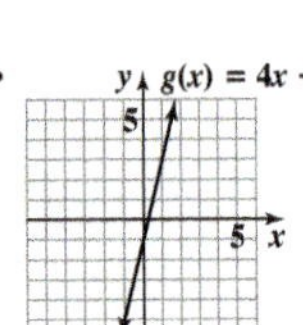

5. C **7.** B **9.** $m = \frac{2}{5}$; y-intercept $\left(0, -\frac{4}{3}\right)$ **11.** $2x - y = 12$ **13.** $11x + y = -52$ **15.** $y = -5$ **17.** $f(x) = -1$ **19.** $f(x) = -x - 2$ **21.** $f(x) = -\frac{3}{2}x - 8$ **23.** $f(x) = -\frac{3}{2}x - 1$ **25. a.** $y = 12{,}000x + 126{,}000$ **b.** \$342,000 **27.** 0 **29.** −2, 4

31. linear

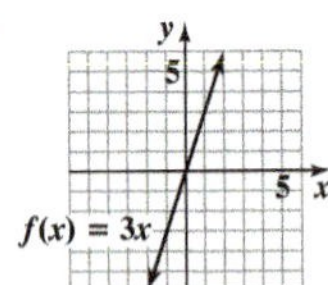

33. not linear

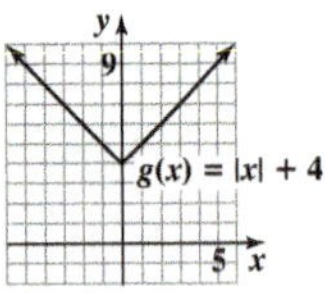

35. linear

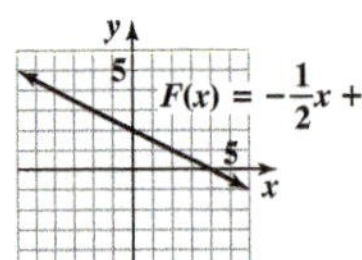

37. linear

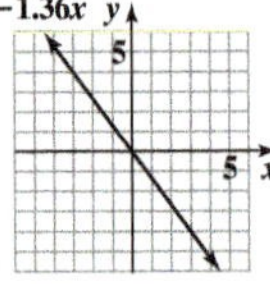

39. not linear

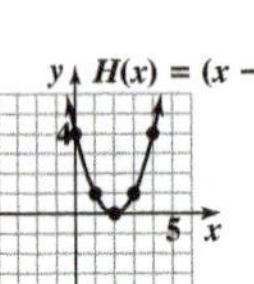

41.

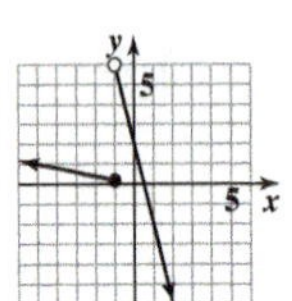

43.

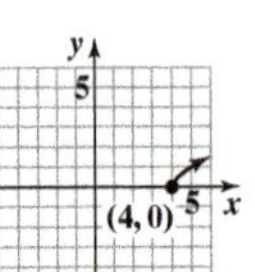

45.

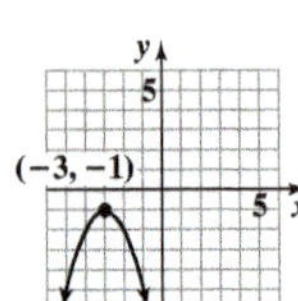

47. 9 **49.** 3.125 cu ft **51.** $f(x) = \frac{9}{2}$

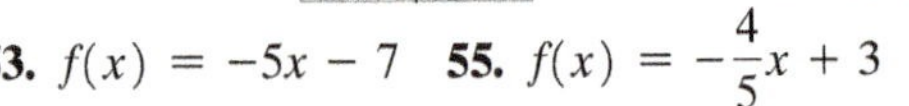

53. $f(x) = -5x - 7$ **55.** $f(x) = -\frac{4}{5}x + 3$ **57.**

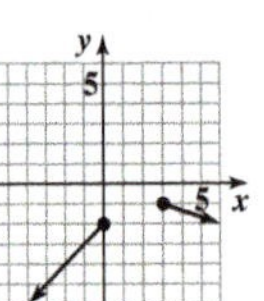

59.

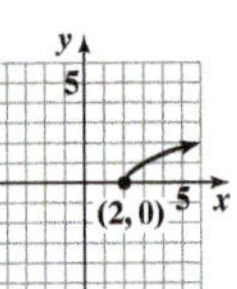

Chapter 15 Test **1.** 3 **2.** −5 **3.** 2, −2 **4.** 0 **5.**

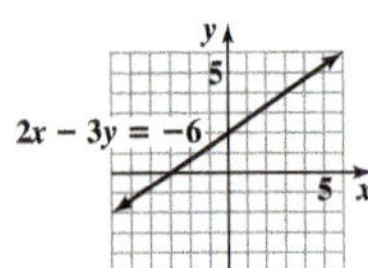

6.

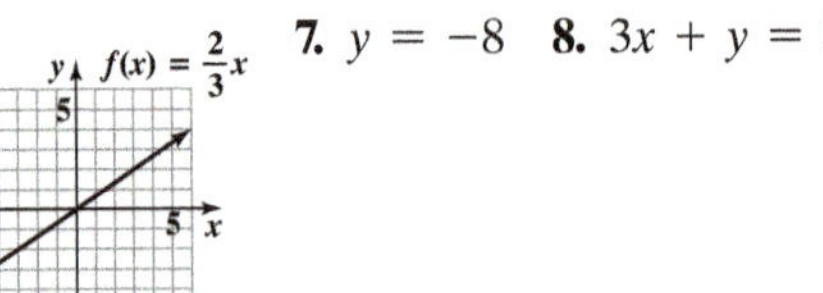

7. $y = -8$ **8.** $3x + y = 11$

9. $5x - y = 2$ **10.** $f(x) = -\frac{1}{2}x$ **11.** $f(x) = -\frac{1}{3}x + \frac{5}{3}$ **12.** $f(x) = -\frac{1}{2}x - \frac{1}{2}$ **13.** neither **14.** domain: $(-\infty, \infty)$; range: $\{5\}$; function **15.** domain: $\{-2\}$; range: $(-\infty, \infty)$; not a function **16.** domain: $(-\infty, \infty)$; range: $[0, \infty)$; function **17.** domain: $(-\infty, \infty)$; range: $(-\infty, \infty)$; function **18. a.** 81 games **b.** 79 games **c.** \$231 million **d.** 0.096; Every million dollars spent on payroll increases winnings by 0.096 game. **19.**

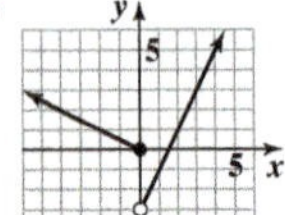

domain: $(-\infty, \infty)$; range: $(-3, \infty)$ **20.**

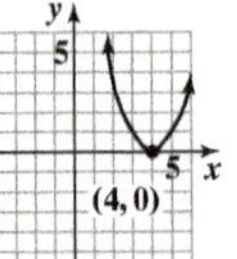

21.

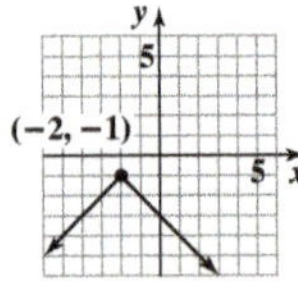

domain: $(-\infty, \infty)$; range: $(-\infty, -1]$ **22.**

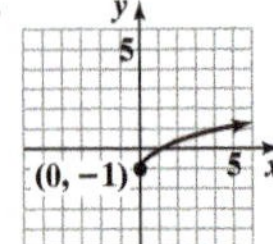

23. 16 **24.** 9 **25.** 256 ft

Cumulative Review Chapters 1–15 **1.** 47; Sec. 1.7, Ex. 12 **3.** −9; Sec. 2.5, Ex. 14 **5.** −3; Sec. 2.6, Ex. 5 **7.** $\frac{16}{3}$; Sec. 9.3, Ex. 2 **9.** $x = \frac{y - b}{m}$; Sec. 9.5, Ex. 6 **11.** $(-\infty, 4]$; Sec. 9.6, Ex. 7 **13.** 0; Sec. 10.4, Ex. 6 **15. a.** 1; (2, 1) **b.** 1; (−2, 1)

c. -3; $(0, -3)$; Sec. 10.6, Ex. 7 **17.** $(4, 2)$; Sec. 11.2, Ex. 1 **19.** $\left(-\frac{15}{7}, -\frac{5}{7}\right)$; Sec. 11.3, Ex. 6 **21.** $x + 4$; Sec. 12.6, Ex. 4 **23. a.** $6(t + 3)$ **b.** $y^5(1 - y^2)$; Sec. 13.1, Ex. 4 **25.** $(x - 2)(x + 6)$; Sec. 13.2, Ex. 3 **27.** $(2x - 3y)(5x + y)$; Sec. 13.3, Ex. 4 **29.** $(x + 2)(x^2 - 2x + 4)$; Sec. 13.5, Ex. 8 **31.** 11, -2; Sec. 13.6, Ex. 4 **33. a.** $\frac{1}{5x - 1}$ **b.** $\frac{9x + 4}{8x - 7}$; Sec. 14.1, Ex. 2 **35.** $-\frac{3(x + 1)}{5x(2x - 3)}$; Sec. 14.2, Ex. 3 **37.** $3x - 5$; Sec. 14.3, Ex. 3 **39.** $-3, -2$; Sec. 14.5, Ex. 3 **41.** $f(x) = \frac{5}{8}x - \frac{5}{2}$; Sec. 15.1, Ex. 4

Chapter 16 Inequalities and Absolute Value

Section 16.1

Vocabulary, Readiness & Video Check **1.** compound **3.** or **5.** $\cup$ **7.** and **9.** or

Exercise Set 16.1 **1.** $\{2, 3, 4, 5, 6, 7\}$ **3.** $\{4, 6\}$ **5.** $\{\ldots, -2, -1, 0, 1, \ldots\}$ **7.** $\{5, 7\}$ **9.** $\{x \mid x$ is an odd integer or $x = 2$ or $x = 4\}$ **11.** $\{2, 4\}$ **13.**  [number line: -3, 1] $(-3, 1)$ **15.** 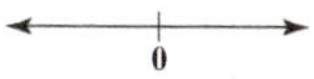 [number line: 0] $\varnothing$ **17.**  [number line: -1] $(-\infty, -1)$ **19.** $[6, \infty)$ **21.** $(-\infty, -3]$ **23.** $(4, 10)$ **25.** $(11, 17)$ **27.** $[1, 4]$ **29.** $\left[-3, \frac{3}{2}\right]$ **31.** $\left[-\frac{7}{3}, 7\right]$ **33.**  [number line: 5] $(-\infty, 5)$ **35.**  [number line: -4, 1] $(-\infty, -4] \cup [1, \infty)$ **37.** [number line: 0] $(-\infty, \infty)$ **39.** $[2, \infty)$ **41.** $(-\infty, -4) \cup (-2, \infty)$ **43.** $(-\infty, \infty)$ **45.** $\left(-\frac{1}{2}, \frac{2}{3}\right)$ **47.** $(-\infty, \infty)$ **49.** $\left[\frac{3}{2}, 6\right]$ **51.** $\left(\frac{5}{4}, \frac{11}{4}\right)$ **53.** $\varnothing$ **55.** $\left(-\infty, -\frac{56}{5}\right) \cup \left(\frac{5}{3}, \infty\right)$ **57.** $\left(-5, \frac{5}{2}\right)$ **59.** $\left(0, \frac{14}{3}\right]$ **61.** $(-\infty, -3]$ **63.** $(-\infty, 1] \cup \left(\frac{29}{7}, \infty\right)$ **65.** $\varnothing$ **67.** $\left[-\frac{1}{2}, \frac{3}{2}\right)$ **69.** $\left(-\frac{4}{3}, \frac{7}{3}\right)$ **71.** $(6, 12)$ **73.** -12 **75.** -4 **77.** $-7, 7$ **79.** 0 **81.** 2004, 2005 **83.** answers may vary **85.** $(6, \infty)$ **87.** $[3, 7]$ **89.** $(-\infty, -1)$ **91.** $-20.2° \le F \le 95°$ **93.** $67 \le$ final score ≤ 94

Section 16.2

Vocabulary, Readiness & Video Check **1.** C **3.** B **5.** D

Exercise Set 16.2 **1.** $-7, 7$ **3.** 4.2, -4.2 **5.** 7, -2 **7.** 8, 4 **9.** 5, -5 **11.** 3, -3 **13.** 0 **15.** $\varnothing$ **17.** $\frac{1}{5}$ **19.** 9, $-\frac{1}{2}$ **21.** $-\frac{5}{2}$ **23.** 4, -4 **25.** 0 **27.** $\varnothing$ **29.** 0, $\frac{14}{3}$ **31.** 2, -2 **33.** $\varnothing$ **35.** 7, -1 **37.** $\varnothing$ **39.** $\varnothing$ **41.** $-\frac{1}{8}$ **43.** $\frac{1}{2}, -\frac{5}{6}$ **45.** 2, $-\frac{12}{5}$ **47.** 3, -2 **49.** $-8, \frac{2}{3}$ **51.** $\varnothing$ **53.** 4 **55.** 13, -8 **57.** 3, -3 **59.** 8, -7 **61.** 2, 3 **63.** 2, $-\frac{10}{3}$ **65.** $\frac{3}{2}$ **67.** $\varnothing$ **69.** 31% **71.** 32.4° **73.** answers may vary **75.** answers may vary **77.** $\varnothing$ **79.** $|x| = 5$ **81.** answers may vary **83.** $|x - 1| = 5$ **85.** answers may vary **87.** $|x| = 6$ **89.** $|x - 2| = |3x - 4|$

Section 16.3

Vocabulary, Readiness & Video Check **1.** D **3.** C **5.** A **7.** The solution set involves "or" and "or" means "union."

Exercise Set 16.3 **1.**  [number line: -4, 4] $[-4, 4]$ **3.**  [number line: 1, 5] $(1, 5)$ **5.**  [number line: -5, -1] $(-5, -1)$ **7.**  [number line: -10, 3] $[-10, 3]$ **9.**  [number line: -5, 5] $[-5, 5]$ **11.** 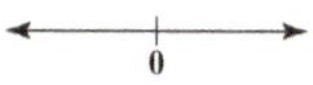 [number line: 0] $\varnothing$ **13.**  [number line: 0, 12] $[0, 12]$ **15.**  [number line: -3, 3] $(-\infty, -3) \cup (3, \infty)$ **17.**  [number line: -24, 4] $(-\infty, -24] \cup [4, \infty)$ **19.** [number line: -4, 4] $(-\infty, -4) \cup (4, \infty)$ **21.**  [number line: 0] $(-\infty, \infty)$ **23.** 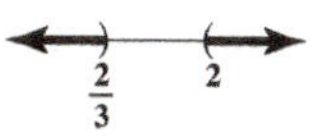 [number line: $\frac{2}{3}$, 2] $\left(-\infty, \frac{2}{3}\right) \cup (2, \infty)$ **25.**  [number line: 0] $\{0\}$ **27.** 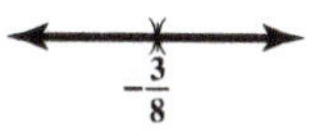 [number line: $-\frac{3}{8}$] $\left(-\infty, -\frac{3}{8}\right) \cup \left(-\frac{3}{8}, \infty\right)$ **29.**  [number line: -2, 2] $[-2, 2]$ **31.**  [number line: -1, 1] $(-\infty, -1) \cup (1, \infty)$ **33.**  [number line: -5, 11] $(-5, 11)$ **35.**  [number line: 4, 6] $(-\infty, 4) \cup (6, \infty)$ **37.** 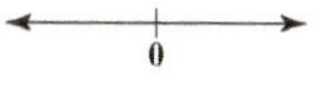 [number line: 0] $\varnothing$ **39.**  [number line: 0] $(-\infty, \infty)$ **41.**  [number line: -2, 9] $[-2, 9]$ **43.** 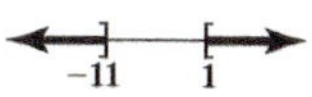 [number line: -11, 1] $(-\infty, -11] \cup [1, \infty)$ **45.** [number line: 0] $(-\infty, 0) \cup (0, \infty)$ **47.**  [number line: 0] $(-\infty, \infty)$ **49.** 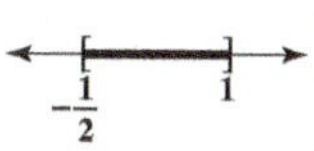 [number line: $-\frac{1}{2}$, 1] $\left[-\frac{1}{2}, 1\right]$ **51.**  [number line: -3, 0] $(-\infty, -3) \cup (0, \infty)$ **53.** 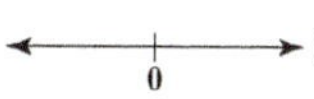 [number line: 0] $\varnothing$ **55.** 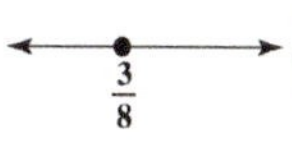 [number line: $\frac{3}{8}$] $\left\{\frac{3}{8}\right\}$ **57.** 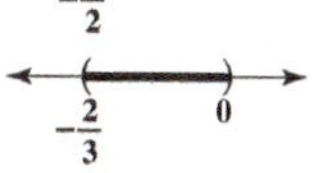 [number line: $-\frac{2}{3}$, 0] $\left(-\frac{2}{3}, 0\right)$ **59.** 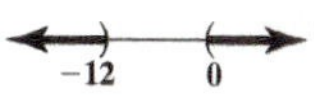 [number line: -12, 0] $(-\infty, -12) \cup (0, \infty)$ **61.**  [number line: -1, 8] $[-1, 8]$ **63.** 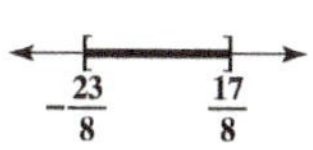 [number line: $-\frac{23}{8}$, $\frac{17}{8}$] $\left[-\frac{23}{8}, \frac{17}{8}\right]$

65. $(-2, 5)$ **67.** $5, -2$ **69.** $(-\infty, -7] \cup [17, \infty)$ **71.** $-\frac{9}{4}$ **73.** $(-2, 1)$ **75.** $2, \frac{4}{3}$ **77.** $\varnothing$ **79.** $\frac{19}{2}, -\frac{17}{2}$ **81.** $\left(-\infty, -\frac{25}{3}\right) \cup \left(\frac{35}{3}, \infty\right)$ **83.** 55 million, 97 million, 138 million **85.** -1.5 **87.** 0 **89.** $|x| < 7$ **91.** $|x| \le 5$ **93.** answers may vary **95.** $3.45 < x < 3.55$

Integrated Review **1.**  (number line: −5, 7) $(-5, 7)$ **2.**  (number line: 0) $(-\infty, \infty)$ **3.** $1, \frac{1}{2}$ **4.** $(-3, 2)$ **5.** $(-\infty, -1] \cup [1, \infty)$ **6.** $-18, -\frac{4}{3}$ **7.** $\left[-\frac{2}{3}, 4\right]$ **8.** $\varnothing$ **9.**  (number line: 1) $(-\infty, 1]$ **10.**  (number line: 0) $(-\infty, 0]$ **11.** $[0, 6]$ **12.** $-\frac{3}{2}, 1$ **13.** B **14.** E **15.** A **16.** C **17.** D

Section 16.4

Vocabulary, Readiness & Video Check **1.** linear inequality in two variables **3.** false **5.** true **7.** yes **9.** yes **11.** We find the boundary line equation by replacing the inequality symbol with =. The points on this line are solutions (line is solid) if the inequality is $\ge$ or $\le$; they are not solutions (line is dashed) if the inequality is $>$ or $<$.

Exercise Set 16.4 **1.** no; yes **3.** no; no **5.** no; yes **7.**

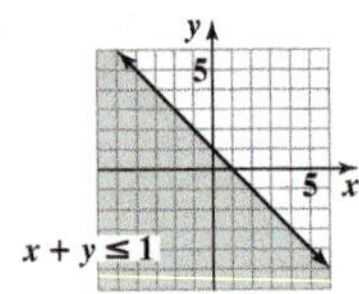

9.

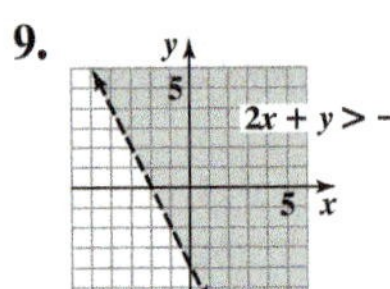

11.

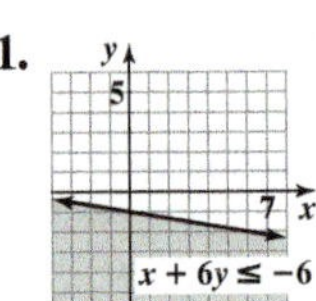

13. $2x + 5y > -10$

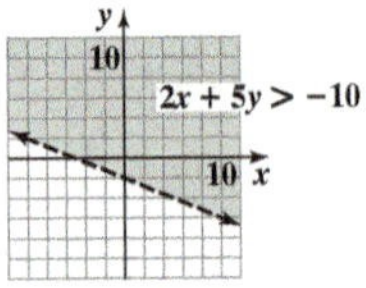

15.

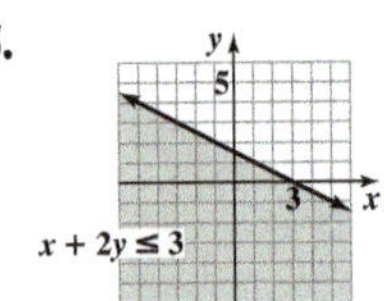

17.

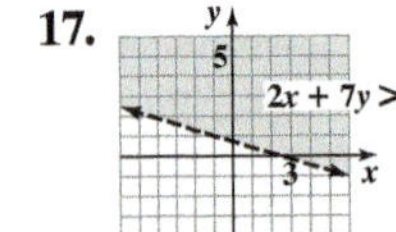

19.

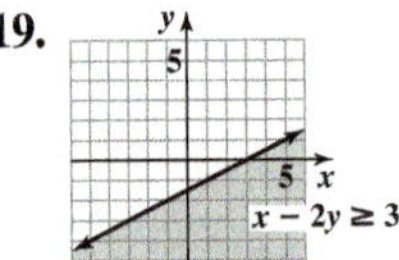

21.

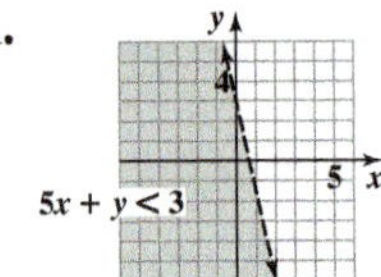

23. $4x + y < 8$

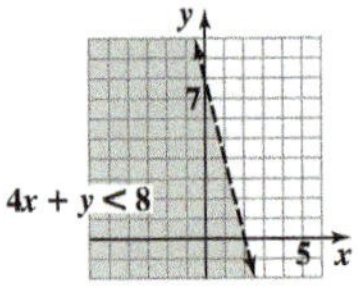

25.

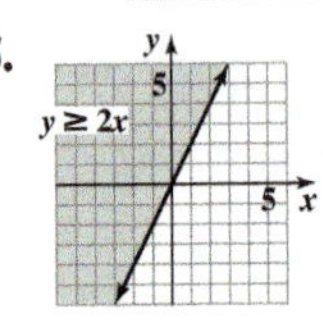

27.

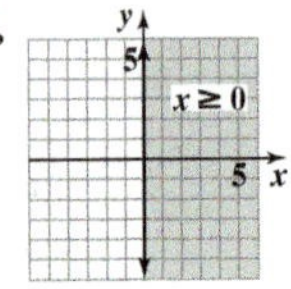

29.

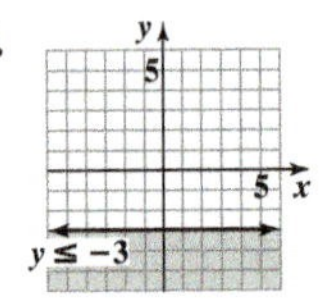

31.

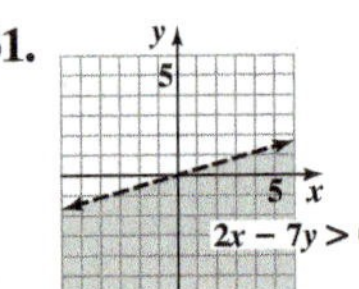

33.

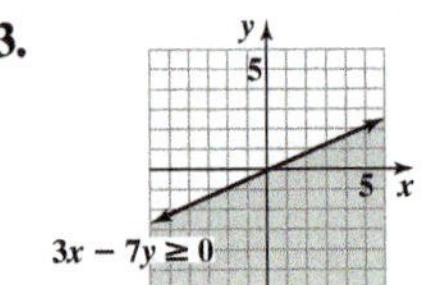

35. $x > y$

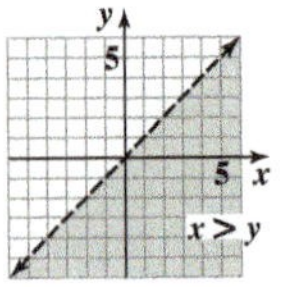

37.

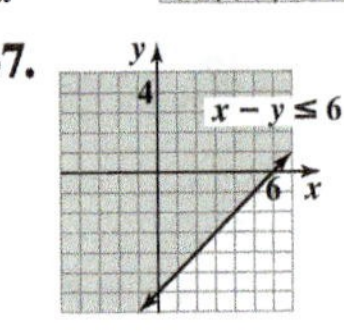

39.

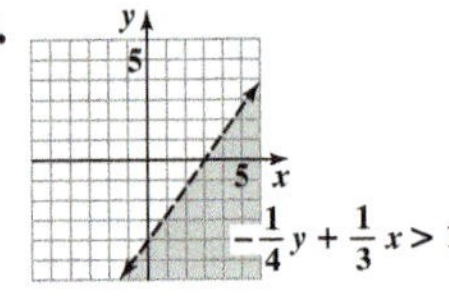

41.

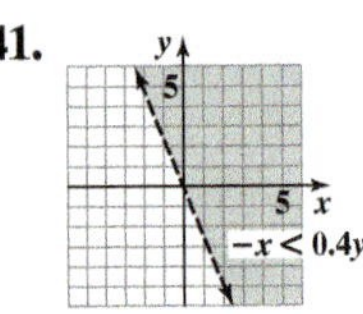

43. e **45.** c **47.** f **49.**

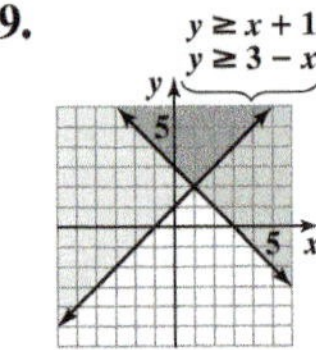

51. $\begin{cases} y < 3x - 4 \\ y \le x + 2 \end{cases}$

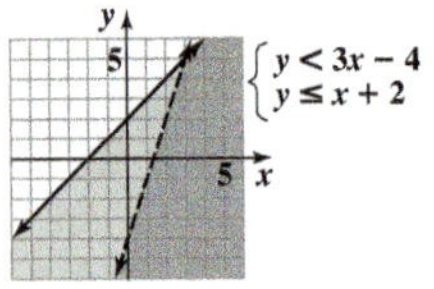

53.

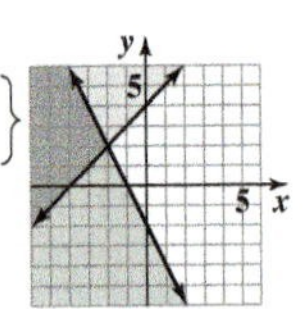

55.

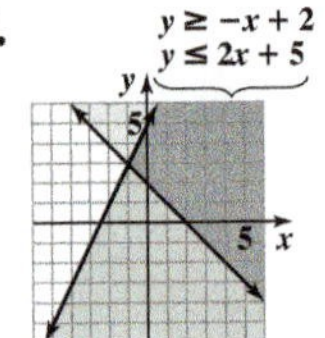

57.

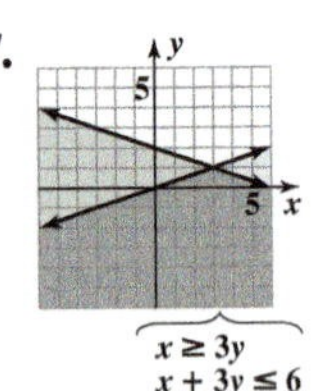

59.

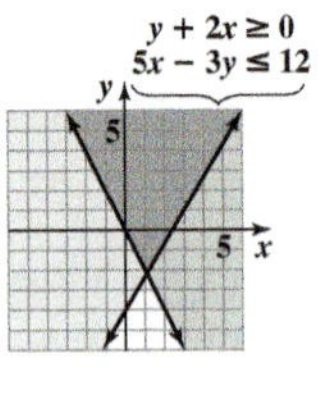

61. $3x - 4y \ge -6$, $2x + y \le 7$

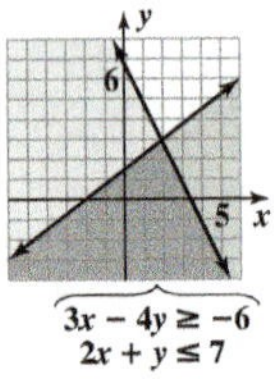

63.

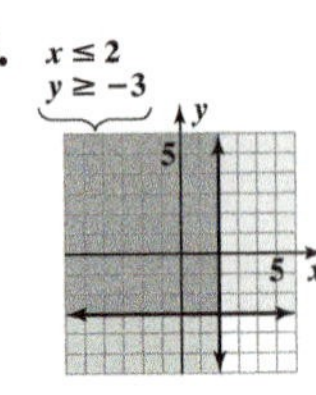

65.

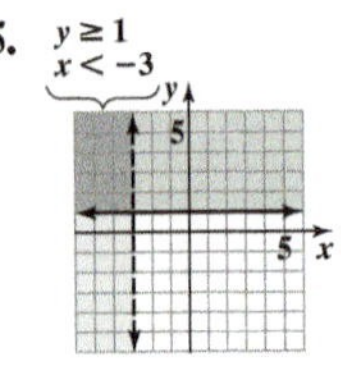

67.

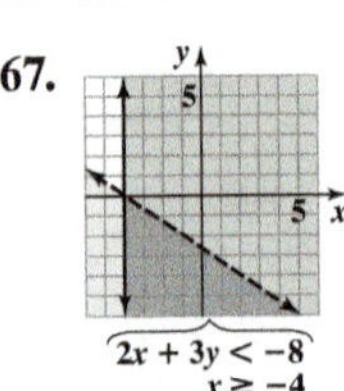

69.

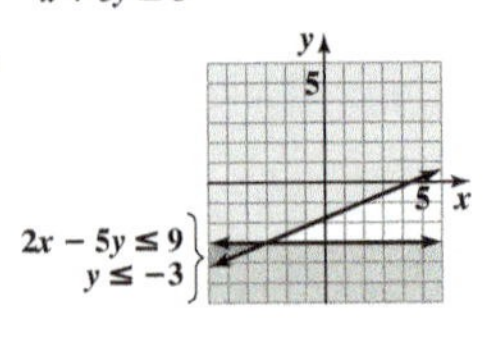

71. no solution; { } or $\varnothing$ **73.** 25 **75.** -2 **77.** yes **79.** yes **81.** $x + y \ge 13$

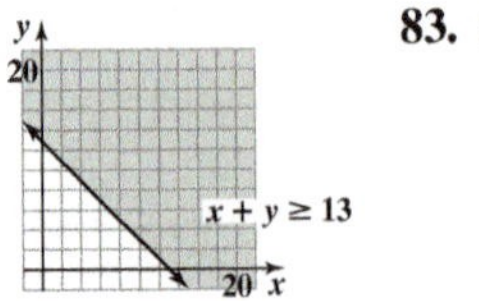

83. answers may vary

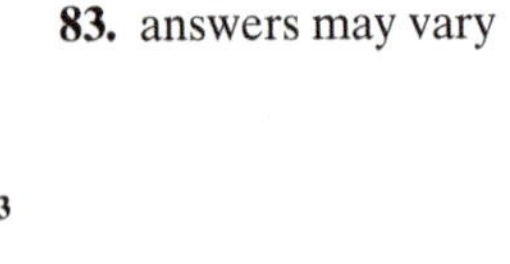

85. $2.5x + 0.25y \le 20$ (Minutes waiting; Miles traveled)

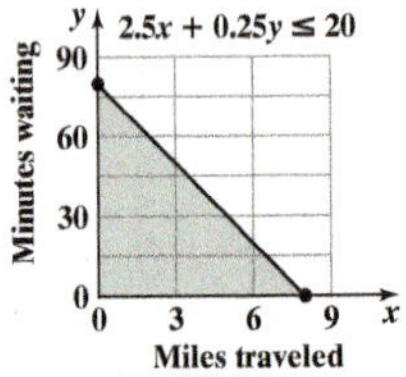

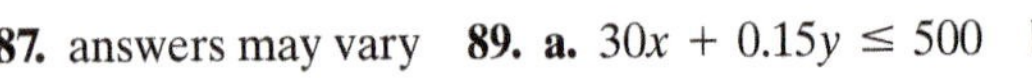

87. answers may vary **89. a.** $30x + 0.15y \le 500$ **b.**

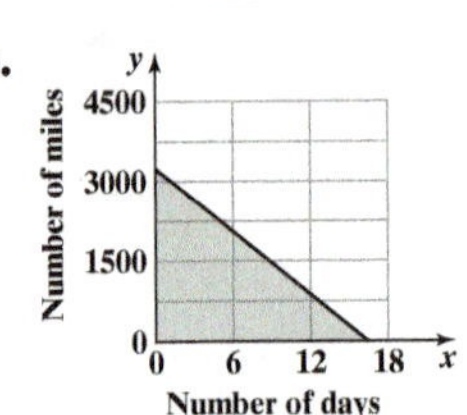

c. answers may vary

91. C **93.** D **95.** 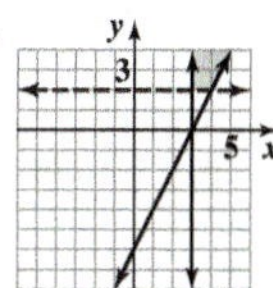 **97.** answers may vary

Chapter 16 Vocabulary Check **1.** compound inequality **2.** intersection **3.** union **4.** absolute value **5.** solution **6.** system of linear inequalities

Chapter 16 Review **1.** $\left(\frac{1}{8}, 2\right)$ **3.** $\left(\frac{7}{8}, \frac{27}{20}\right]$ **5.** $\left(\frac{11}{3}, \infty\right)$ **7.** 5, 11 **9.** $-1, \frac{11}{3}$ **11.** $-\frac{1}{6}$ **13.** ∅ **15.** $5, -\frac{1}{3}$ **17.** $7, -\frac{8}{5}$ **19.** $\left(-\frac{8}{5}, 2\right)$ **21.** $(-\infty, -3) \cup (3, \infty)$ **23.** ∅ **25.** $\left(-\infty, -\frac{22}{15}\right] \cup \left[\frac{6}{5}, \infty\right)$ **27.** $(-\infty, -27) \cup (-9, \infty)$ **29.** 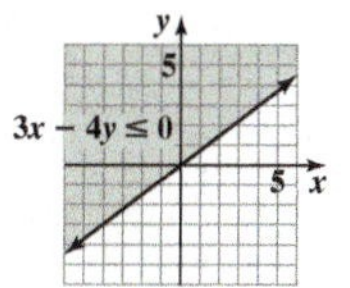

31. 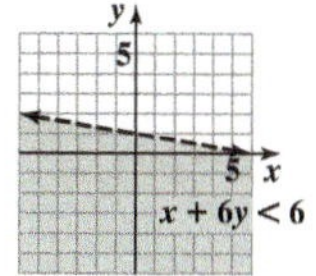

33. 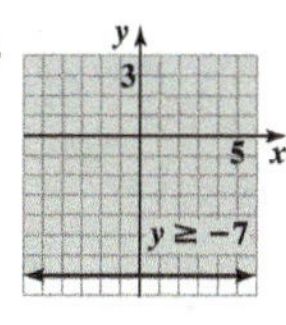

35. 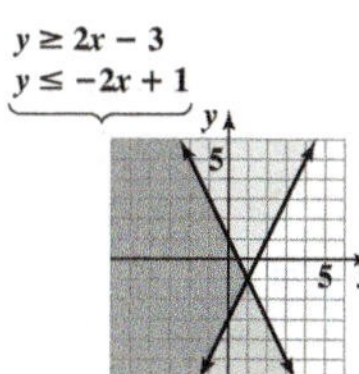

37. 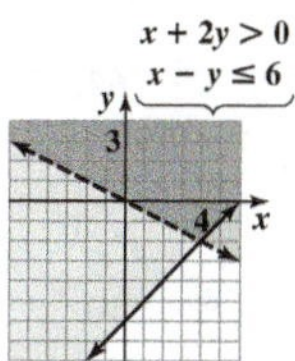

39. 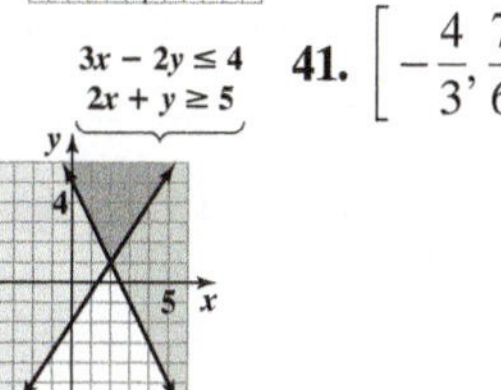

41. $\left[-\frac{4}{3}, \frac{7}{6}\right]$

43. $(5, \infty)$ **45.** ∅ **47.** $(-\infty, \infty)$ **49.** 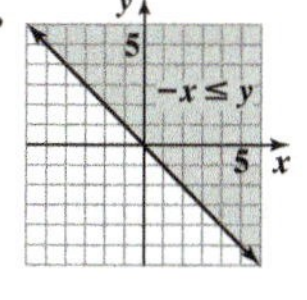

51. 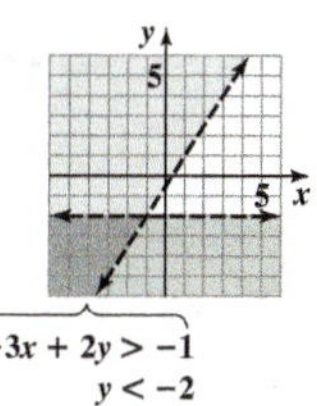

Chapter 16 Test **1.** $1, \frac{2}{3}$ **2.** ∅ **3.** $\frac{3}{2}$ **4.** $\left(\frac{3}{2}, 5\right]$ **5.** $(-\infty, -2) \cup \left(\frac{4}{3}, \infty\right)$ **6.** $(3, 7)$ **7.** $(-\infty, -5]$ **8.** $(-\infty, -2]$ **9.** $[-3, -1)$ **10.** $(-\infty, \infty)$ **11.** $3, -\frac{1}{5}$ **12.** $(-\infty, \infty)$ **13.** $\left[1, \frac{11}{2}\right)$ **14.** 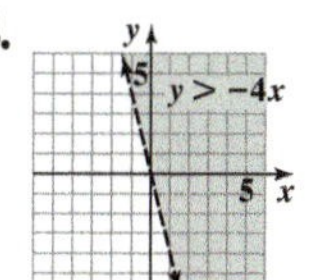

15. 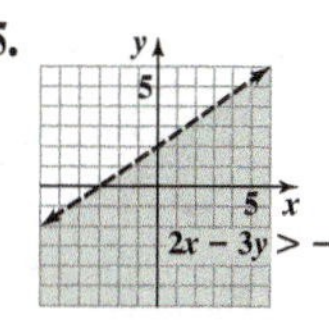

16. 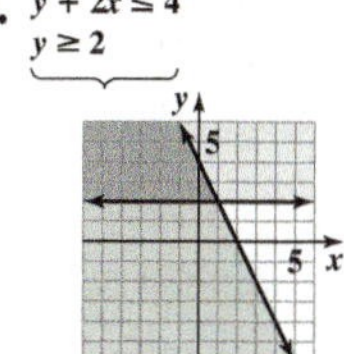

17. $2y - x \geq 1$, $x + y \geq -4$

Cumulative Review Chapters 1–16 **1. a.** $-\frac{7}{24}$ **b.** $-\frac{21}{8}$; Sec. 4.3, Ex. 16 **3.** −2; Sec. 2.5, Ex. 5 **5. a.** $10x$ **b.** $-4y$ **c.** $8x^2 - 2$; Sec. 3.1, Ex. 1 **7.** −14; Sec. 3.2, Ex. 4 **9.**

x	y
−1	−3
0	0
−3	−9

; Sec. 10.1, Ex. 7 **11. a.** x-int: $(-3, 0)$; y-int: $(0, 2)$ **b.** x-int: $(-4, 0)$, $(-1, 0)$; y-int: $(0, 1)$ **c.** x-int and y-int: $(0, 0)$ **d.** x-int: $(2, 0)$; y-int: none **e.** x-int: $(-1, 0)$, $(3, 0)$; y-int: $(0, -1)$, $(0, 2)$; Sec. 10.3, Ex. 1–5

13. parallel; Sec. 10.4, Ex. 8a **15.** $y = \frac{1}{4}x - 3$; Sec. 10.5, Ex. 3 **17.** $y = -3$; Sec. 10.5, Ex. 7 **19.** a, b, c; Sec. 10.6, Ex. 5 **21.** solution; Sec. 11.4, Ex. 1 **23.** $12x^3 - 12x^2 - 9x + 2$; Sec. 12.2, Ex. 10 **25.** $(x + 6y)(x + y)$; Sec. 13.2, Ex. 6 **27.** $\frac{3y^9}{160}$; Sec. 14.2, Ex. 4 **29.** 1; Sec. 14.3, Ex. 2 **31.** $\frac{x(3x - 1)}{(x + 1)^2(x - 1)}$; Sec. 14.4, Ex. 7 **33.** −5; Sec. 14.5, Ex. 1 **35.** no solution; { } or ∅; Sec. 11.1, Ex. 5

37. $(-2, 0)$; Sec. 11.2, Ex. 4 **39.** $\{(x, y) \mid 3x - 2y = 2\}$ or $\{(x, y) \mid -9x + 6y = -6\}$; Sec. 11.3, Ex. 4

41. $\begin{cases} -3x + 4y < 12 \\ x \geq 2 \end{cases}$; Sec. 16.4, Ex. 8 **43. a.** x^{11} **b.** $\frac{t^4}{16}$ **c.** $81y^{10}$; Sec. 12.1, Ex. 11 **45.** $-6, -\frac{3}{2}, \frac{1}{5}$; Sec. 13.6, Ex. 9 **47.** $62.45; Sec. 14.6, Ex. 1

Chapter 17 Rational Exponents, Radicals, and Complex Numbers

Section 17.1

Vocabulary, Readiness & Video Check **1.** index; radical sign; radicand **3.** is not **5.** $[0, \infty)$ **7.** $(16, 4)$ **9.** Divide the index into each exponent in the radicand. **11.** The square root of a negative number is not a real number, but the cube root of a negative number is a real number. **13.** For odd roots, there's only one root/answer whether the radicand is positive or negative, so absolute value bars aren't needed.

Exercise Set 17.1 **1.** 10 **3.** $\frac{1}{2}$ **5.** 0.01 **7.** -6 **9.** x^5 **11.** $4y^3$ **13.** 2.646 **15.** 6.164 **17.** 14.142 **19.** 4 **21.** $\frac{1}{2}$ **23.** -1 **25.** x^4 **27.** $-3x^3$ **29.** -2 **31.** not a real number **33.** -2 **35.** x^4 **37.** $2x^2$ **39.** $9x^2$ **41.** $4x^2$ **43.** 8 **45.** -8 **47.** $2|x|$ **49.** x **51.** $|x - 5|$ **53.** $|x + 2|$ **55.** -11 **57.** $2x$ **59.** y^6 **61.** $5ab^{10}$ **63.** $-3x^4y^3$ **65.** a^4b **67.** $-2x^2y$ **69.** $\frac{5}{7}$ **71.** $\frac{x^{10}}{2y}$ **73.** $-\frac{z^7}{3x}$ **75.** $\frac{x}{2}$ **77.** $\sqrt{3}$ **79.** -1 **81.** -3 **83.** $\sqrt{7}$ **85.** $[0, \infty)$;

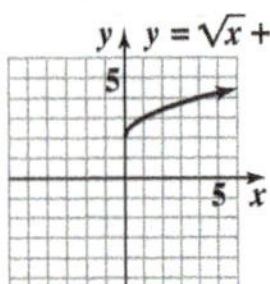

87. $[3, \infty)$; 0, 1, 2, 3

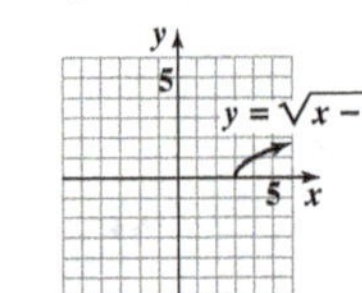

89. $(-\infty, \infty)$;

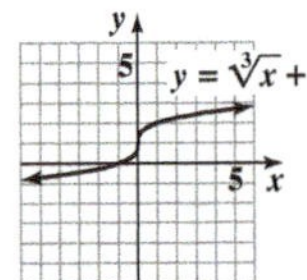

91. $(-\infty, \infty)$; 0, 1, -1, 2, -2

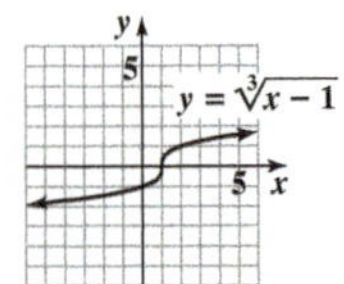

93. $-32x^{15}y^{10}$ **95.** $-60x^7y^{10}z^5$ **97.** $\frac{x^9y^5}{2}$ **99.** not a real number **101.** not a real number **103.** d **105.** d **107.** b **109.** b **111.** answers may vary **113.** 1.69 sq m **115.** 11,181 m per sec **117.** answers may vary **119.**

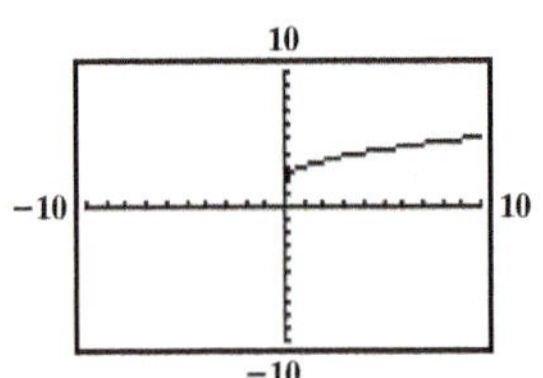

121.

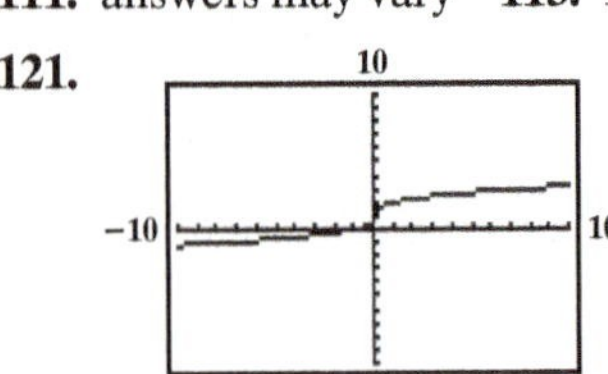

Section 17.2

Vocabulary, Readiness & Video Check **1.** true **3.** true **5.** multiply, c **7.** $-\sqrt[5]{3x}$ **9.** denominator; positive **11.** Write the radical using an equivalent fractional exponent form, simplify the fraction, and then write as a radical again.

Exercise Set 17.2 **1.** 7 **3.** 3 **5.** $\frac{1}{2}$ **7.** 13 **9.** $2\sqrt[3]{m}$ **11.** $3x^2$ **13.** -3 **15.** -2 **17.** 8 **19.** 16 **21.** not a real number **23.** $\sqrt[5]{(2x)^3}$ **25.** $\sqrt[3]{(7x + 2)^2}$ **27.** $\frac{64}{27}$ **29.** $\frac{1}{16}$ **31.** $\frac{1}{16}$ **33.** not a real number **35.** $\frac{1}{x^{1/4}}$ **37.** $a^{2/3}$ **39.** $\frac{5x^{3/4}}{7}$ **41.** $a^{7/3}$ **43.** x **45.** $3^{5/8}$ **47.** $y^{1/6}$ **49.** $8u^3$ **51.** $-b$ **53.** $\frac{1}{x^2}$ **55.** $27x^{2/3}$ **57.** $\frac{y}{z^{1/6}}$ **59.** $\frac{1}{x^{7/4}}$ **61.** $y - y^{7/6}$ **63.** $x^{5/3} - 2x^{2/3}$ **65.** $4x^{2/3} - 9$ **67.** $x^{8/3}(1 + x^{2/3})$ **69.** $x^{1/5}(x^{1/5} - 3)$ **71.** $x^{-1/3}(5 + x)$ **73.** $\sqrt{x}$ **75.** $\sqrt[3]{2}$ **77.** $2\sqrt{x}$ **79.** $\sqrt{xy}$ **81.** $\sqrt[3]{a^2b}$ **83.** $\sqrt{x + 3}$ **85.** $\sqrt[15]{y^{11}}$ **87.** $\sqrt[12]{b^5}$ **89.** $\sqrt[24]{x^{23}}$ **91.** $\sqrt{a}$ **93.** $\sqrt[6]{432}$ **95.** $\sqrt[15]{343y^5}$ **97.** $\sqrt[6]{125r^3s^2}$ **99.** $25 \cdot 3$ **101.** $16 \cdot 3$ or $4 \cdot 12$ **103.** $8 \cdot 2$ **105.** $27 \cdot 2$ **107.** A **109.** C **111.** B **113.** 1509 calories **115.** 302.0 million **117.** answers may vary **119.** $a^{1/3}$ **121.** $x^{1/5}$ **123.** 1.6818 **125.** 5.6645 **127.** $\frac{t^{1/2}}{u^{1/2}}$

Section 17.3

Vocabulary, Readiness & Video Check **1.** midpoint; point **3.** distance **5.** The indexes must be the same. **7.** The power must be 1. Any even power is a perfect square and will leave no factor in the radicand; any higher odd power can have an even power factored from it, leaving one factor remaining in the radicand. **9.** average; average

Exercise Set 17.3 **1.** $\sqrt{14}$ **3.** 2 **5.** $\sqrt[3]{36}$ **7.** $\sqrt{6x}$ **9.** $\sqrt{\frac{14}{xy}}$ **11.** $\sqrt[4]{20x^3}$ **13.** $\frac{\sqrt{6}}{7}$ **15.** $\frac{\sqrt{2}}{7}$ **17.** $\frac{\sqrt[4]{x^3}}{2}$ **19.** $\frac{\sqrt[3]{4}}{3}$ **21.** $\frac{\sqrt[4]{8}}{x^2}$ **23.** $\frac{\sqrt[3]{2x}}{3y^4\sqrt[3]{3}}$ **25.** $\frac{x\sqrt{y}}{10}$ **27.** $\frac{x\sqrt{5}}{2y}$ **29.** $-\frac{z^2\sqrt[3]{z}}{3x}$ **31.** $4\sqrt{2}$ **33.** $4\sqrt[3]{3}$ **35.** $25\sqrt{3}$ **37.** $2\sqrt{6}$ **39.** $10x^2\sqrt{x}$ **41.** $2y^2\sqrt[3]{2y}$ **43.** $a^2b\sqrt[4]{b^3}$ **45.** $y^2\sqrt{y}$ **47.** $5ab\sqrt{b}$ **49.** $-2x^2\sqrt[5]{y}$ **51.** $x^4\sqrt[3]{50x^2}$ **53.** $-4a^4b^3\sqrt{2b}$ **55.** $3x^3y^4\sqrt{xy}$ **57.** $5r^3s^4$ **59.** $2x^3y\sqrt[4]{2y}$ **61.** $\sqrt{2}$ **63.** 2 **65.** 10 **67.** x^2y **69.** $24m^2$ **71.** $\frac{15x\sqrt{2x}}{2}$ or $\frac{15x}{2}\sqrt{2x}$ **73.** $2a^2\sqrt[4]{2}$ **75.** $2xy^2\sqrt[5]{x^2}$ **77.** 5 units **79.** $\sqrt{41}$ units $\approx$ 6.403 units **81.** $\sqrt{10}$ units $\approx$ 3.162 units **83.** $\sqrt{5}$ units $\approx$ 2.236 units **85.** $\sqrt{192.58}$ units $\approx$ 13.877 units **87.** $(4,-2)$ **89.** $\left(-5, \frac{5}{2}\right)$ **91.** $(3, 0)$ **93.** $\left(-\frac{1}{2}, \frac{1}{2}\right)$ **95.** $\left(\sqrt{2}, \frac{\sqrt{5}}{2}\right)$ **97.** $(6.2, -6.65)$ **99.** $14x$ **101.** $2x^2 - 7x - 15$ **103.** y^2 **105.** $-3x - 15$ **107.** $x^2 - 8x + 16$ **109.** true **111.** false **113.** true **115.** $\frac{\sqrt[3]{64}}{\sqrt{64}} = \frac{4}{8} = \frac{1}{2}$ **117.** x^7 **119.** a^3bc^5 **121.** $z^{10}\sqrt[3]{z^2}$ **123.** $q^2r^5s\sqrt[7]{q^3r^5}$ **125.** 1.6 m **127.** **a.** 20π sq cm **b.** 211.57 sq ft

Section 17.4

Vocabulary, Readiness & Video Check **1.** Unlike **3.** Like **5.** $6\sqrt{3}$ **7.** $7\sqrt{x}$ **9.** $8\sqrt[3]{x}$ **11.** Sometimes you can't see that there are like radicals until you simplify, so you may incorrectly think you cannot add or subtract if you don't simplify first.

Exercise Set 17.4 **1.** $-2\sqrt{2}$ **3.** $10x\sqrt{2x}$ **5.** $17\sqrt{2} - 15\sqrt{5}$ **7.** $-\sqrt[3]{2x}$ **9.** $5b\sqrt{b}$ **11.** $\frac{31\sqrt{2}}{15}$ **13.** $\frac{\sqrt[3]{11}}{3}$ **15.** $\frac{5\sqrt{5x}}{9}$ **17.** $14 + \sqrt{3}$ **19.** $7 - 3y$ **21.** $6\sqrt{3} - 6\sqrt{2}$ **23.** $-23\sqrt[3]{5}$ **25.** $2a^3b\sqrt{ab}$ **27.** $20y\sqrt{2y}$ **29.** $2y\sqrt[3]{2x}$ **31.** $6\sqrt[3]{11} - 4\sqrt{11}$ **33.** $3x\sqrt[4]{x^3}$ **35.** $\frac{2\sqrt{3}}{3}$ **37.** $\frac{5x\sqrt[3]{x}}{7}$ **39.** $\frac{5\sqrt{7}}{2x}$ **41.** $\frac{\sqrt[3]{2}}{6}$ **43.** $\frac{14x\sqrt[3]{2x}}{9}$ **45.** $15\sqrt{3}$ in. **47.** $\sqrt{35} + \sqrt{21}$ **49.** $7 - 2\sqrt{10}$ **51.** $3\sqrt{x} - x\sqrt{3}$ **53.** $6x - 13\sqrt{x} - 5$ **55.** $\sqrt[3]{a^2} + \sqrt[3]{a} - 20$ **57.** $6\sqrt{2} - 12$ **59.** $2 + 2x\sqrt{3}$ **61.** $-16 - \sqrt{35}$ **63.** $x - y^2$ **65.** $3 + 2x\sqrt{3} + x^2$ **67.** $23x - 5x\sqrt{15}$ **69.** $2\sqrt[3]{2} - \sqrt[3]{4}$ **71.** $x + 1$ **73.** $x + 24 + 10\sqrt{x - 1}$ **75.** $2x + 6 - 2\sqrt{2x + 5}$ **77.** $x - 7$ **79.** $\frac{7}{x + y}$ **81.** $2a - 3$ **83.** $\frac{-2 + \sqrt{3}}{3}$ **85.** $22\sqrt{5}$ ft; 150 sq ft **87.** **a.** $2\sqrt{3}$ **b.** 3 **c.** answers may vary **89.** $2\sqrt{6} - 2\sqrt{2} - 2\sqrt{3} + 6$ **91.** answers may vary

Section 17.5

Vocabulary, Readiness & Video Check **1.** conjugate **3.** rationalizing the numerator **5.** to write an equivalent expression without a radical in the denominator **7.** No, except for the fact you're working with numerators, the process is the same.

Exercise Set 17.5 **1.** $\frac{\sqrt{14}}{7}$ **3.** $\frac{\sqrt{5}}{5}$ **5.** $\frac{2\sqrt{x}}{x}$ **7.** $\frac{4\sqrt[3]{9}}{3}$ **9.** $\frac{3\sqrt{2x}}{4x}$ **11.** $\frac{3\sqrt[3]{2x}}{2x}$ **13.** $\frac{3\sqrt{3a}}{a}$ **15.** $\frac{3\sqrt[3]{4}}{2}$ **17.** $\frac{2\sqrt{21}}{7}$ **19.** $\frac{\sqrt{10xy}}{5y}$ **21.** $\frac{\sqrt[3]{75}}{5}$ **23.** $\frac{\sqrt{6x}}{10}$ **25.** $\frac{\sqrt{3z}}{6z}$ **27.** $\frac{\sqrt[3]{6xy^2}}{3x}$ **29.** $\frac{3\sqrt[4]{2}}{2}$ **31.** $\frac{2\sqrt[4]{9x}}{3x^2}$ **33.** $\frac{5\sqrt[5]{4ab^4}}{2ab^3}$ **35.** $\sqrt{2} - x$ **37.** $5 + \sqrt{a}$ **39.** $-7\sqrt{5} - 8\sqrt{x}$ **41.** $-2(2 + \sqrt{7})$ **43.** $\frac{7(3 + \sqrt{x})}{9 - x}$ **45.** $-5 + 2\sqrt{6}$ **47.** $\frac{2a + 2\sqrt{a} + \sqrt{ab} + \sqrt{b}}{4a - b}$ **49.** $-\frac{8(1 - \sqrt{10})}{9}$ **51.** $\frac{x - \sqrt{xy}}{x - y}$ **53.** $\frac{5 + 3\sqrt{2}}{7}$ **55.** $\frac{5}{\sqrt{15}}$ **57.** $\frac{6}{\sqrt{10}}$ **59.** $\frac{2x}{7\sqrt{x}}$ **61.** $\frac{5y}{\sqrt[3]{100xy}}$ **63.** $\frac{2}{\sqrt{10}}$ **65.** $\frac{2x}{11\sqrt{2x}}$ **67.** $\frac{7}{2\sqrt[3]{49}}$ **69.** $\frac{3x^2}{10\sqrt[3]{9x}}$ **71.** $\frac{6x^2y^3}{\sqrt{6z}}$ **73.** $\frac{-7}{12 + 6\sqrt{11}}$ **75.** $\frac{3}{10 + 5\sqrt{7}}$ **77.** $\frac{x - 9}{x - 3\sqrt{x}}$ **79.** $\frac{1}{3 + 2\sqrt{2}}$ **81.** $\frac{x - 1}{x - 2\sqrt{x} + 1}$ **83.** 5 **85.** $-\frac{1}{2}, 6$ **87.** 2, 6 **89.** $r = \frac{\sqrt{A\pi}}{2\pi}$ **91.** **a.** $\frac{y\sqrt{15xy}}{6x^2}$ **b.** $\frac{y\sqrt{15xy}}{6x^2}$ **c.** answers may vary **93.** $\sqrt[3]{25}$ **95.** answers may vary **97.** answers may vary

Integrated Review **1.** 9 **2.** -2 **3.** $\frac{1}{2}$ **4.** x^3 **5.** y^3 **6.** $2y^5$ **7.** $-2y$ **8.** $3b^3$ **9.** 6 **10.** $\sqrt[4]{3y}$ **11.** $\frac{1}{16}$ **12.** $\sqrt[5]{(x + 1)^3}$ **13.** y **14.** $16x^{1/2}$ **15.** $x^{5/4}$ **16.** $4^{11/15}$ **17.** $2x^2$ **18.** $\sqrt[4]{a^3b^2}$ **19.** $\sqrt[4]{x^3}$ **20.** $\sqrt[6]{500}$ **21.** $2\sqrt{10}$ **22.** $2xy^2\sqrt[4]{x^3y^2}$ **23.** $3x\sqrt[3]{2x}$ **24.** $-2b^2\sqrt[5]{2}$ **25.** $\sqrt{5x}$ **26.** $4x$ **27.** $7y^2\sqrt{y}$ **28.** $2a^2\sqrt[4]{3}$ **29.** $2\sqrt{5} - 5\sqrt{3} + 5\sqrt{7}$ **30.** $y\sqrt[3]{2y}$ **31.** $\sqrt{15} - \sqrt{6}$ **32.** $10 + 2\sqrt{21}$ **33.** $4x^2 - 5$ **34.** $x + 2 - 2\sqrt{x + 1}$ **35.** $\frac{\sqrt{21}}{3}$ **36.** $\frac{5\sqrt[3]{4x}}{2x}$ **37.** $\frac{13 - 3\sqrt{21}}{5}$ **38.** $\frac{7}{\sqrt{21}}$ **39.** $\frac{3y}{\sqrt[3]{33y^2}}$ **40.** $\frac{x - 4}{x + 2\sqrt{x}}$

Section 17.6

Graphing Calculator Explorations **1.** 3.19 **3.** $\varnothing$ **5.** 3.23

Vocabulary, Readiness & Video Check **1.** extraneous solution **3.** $x^2 - 10x + 25$ **5.** Applying the power rule can result in an equation with more solutions than the original equation, so you need to check all proposed solutions in the original equation. **7.** Our answer is either a positive square root of a value or a negative square root of a value. We're looking for a length, which must be positive, so our answer must be the positive square root.

Exercise Set 17.6 **1.** 8 **3.** 7 **5.** $\varnothing$ **7.** 7 **9.** 6 **11.** $-\frac{9}{2}$ **13.** 29 **15.** 4 **17.** -4 **19.** $\varnothing$ **21.** 7 **23.** 9 **25.** 50 **27.** $\varnothing$ **29.** $\frac{15}{4}$ **31.** 13 **33.** 5 **35.** -12 **37.** 9 **39.** -3 **41.** 1 **43.** 1 **45.** $\frac{1}{2}$ **47.** 0, 4 **49.** $\frac{37}{4}$ **51.** $3\sqrt{5}$ ft **53.** $2\sqrt{10}$ m **55.** $2\sqrt{131}$ m $\approx$ 22.9 m **57.** $\sqrt{100.84}$ mm $\approx$ 10.0 mm **59.** 17 ft **61.** 13 ft **63.** 14,657,415 sq mi **65.** 100 ft **67.** 100 **69.** $\frac{\pi}{2}$ sec $\approx$ 1.57 sec **71.** 12.97 ft **73.** answers may vary **75.** $15\sqrt{3}$ sq mi $\approx$ 25.98 sq mi **77.** answers may vary **79.** 0.51 km **81.** function **83.** function **85.** not a function **87.** $\frac{x}{4x + 3}$ **89.** $-\frac{4z + 2}{3z}$

91.
$$\begin{aligned} \sqrt{5x - 1} + 4 &= 7 \\ \sqrt{5x - 1} &= 3 \\ (\sqrt{5x - 1})^2 &= 3^2 \\ 5x - 1 &= 9 \\ 5x &= 10 \\ x &= 2 \end{aligned}$$

93. 1 **95. a.–b.** answers may vary **97.** $-1, 0, 8, 9$ **99.** $-1, 4$

Section 17.7

Vocabulary, Readiness & Video Check **1.** complex **3.** -1 **5.** real **7.** The product rule for radicals; you need to first simplify each separate radical and have nonnegative radicands before applying the product rule. **9.** the fact that $i^2 = -1$ **11.** $i, i^2 = -1, i^3 = -i, i^4 = 1$

Exercise Set 17.7 **1.** $9i$ **3.** $i\sqrt{7}$ **5.** -4 **7.** $8i$ **9.** $2i\sqrt{6}$ **11.** $-6i$ **13.** $24i\sqrt{7}$ **15.** $-3\sqrt{6} + 0i$ **17.** $-\sqrt{14}$ **19.** $-5\sqrt{2}$ **21.** $4i$ **23.** $i\sqrt{3}$ **25.** $2\sqrt{2}$ **27.** $6 - 4i$ **29.** $-2 + 6i$ **31.** $-2 - 4i$ **33.** $-40 + 0i$ **35.** $18 + 12i$ **37.** $7 + 0i$ **39.** $12 - 16i$ **41.** $0 - 4i$ **43.** $\frac{28}{25} - \frac{21}{25}i$ **45.** $4 + i$ **47.** $\frac{17}{13} + \frac{7}{13}i$ **49.** $63 + 0i$ **51.** $2 - i$ **53.** $27 + 3i$ **55.** $-\frac{5}{2} - 2i$ **57.** $18 + 13i$ **59.** $20 + 0i$ **61.** $10 + 0i$ **63.** $2 + 0i$ **65.** $-5 + \frac{16}{3}i$ **67.** $17 + 144i$ **69.** $\frac{3}{5} - \frac{1}{5}i$ **71.** $5 - 10i$ **73.** $\frac{1}{5} - \frac{8}{5}i$ **75.** $8 - i$ **77.** $7 + 0i$ **79.** $12 - 16i$ **81.** 1 **83.** i **85.** $-i$ **87.** -1 **89.** -64 **91.** $-243i$ **93.** 40° **95.** $x^2 - 5x - 2 - \frac{6}{x - 1}$ **97.** 5 people **99.** 14 people **101.** 16.7% **103.** $-1 - i$ **105.** $0 + 0i$ **107.** $2 + 3i$ **109.** $2 + i\sqrt{2}$ **111.** $\frac{1}{2} - \frac{\sqrt{3}}{2}i$ **113.** answers may vary **115.** $6 - 3i\sqrt{3}$ **117.** yes

Chapter 17 Vocabulary Check **1.** conjugate **2.** principal square root **3.** rationalizing **4.** imaginary unit **5.** cube root **6.** index; radicand **7.** like radicals **8.** complex number **9.** distance **10.** midpoint

Chapter 17 Review **1.** 9 **3.** -2 **5.** $-\frac{1}{7}$ **7.** -6 **9.** $-a^2b^3$ **11.** $2ab^2$ **13.** $\frac{x^6}{6y}$ **15.** $|-x|$ **17.** -27 **19.** $-x$ **21.** $2|2y + z|$ **23.** y **25. a.** 3, 6 **b.** $[0, \infty)$ **c.** $f(x) = \sqrt{x} + 3$ **27.** $\frac{1}{3}$ **29.** $-\frac{1}{3}$ **31.** -27 **33.** not a real number **35.** $\frac{9}{4}$ **37.** $x^{2/3}$ **39.** $\sqrt[5]{y^4}$ **41.** $\frac{1}{\sqrt[3]{x + 2}}$ **43.** $a^{13/6}$ **45.** $\frac{1}{a^{9/2}}$ **47.** a^4b^6 **49.** $\frac{b^{5/6}}{49a^{1/4}c^{5/3}}$ **51.** 4.472 **53.** 5.191 **55.** -26.246 **57.** $\sqrt[6]{1372}$ **59.** $2\sqrt{6}$ **61.** $2x$ **63.** $2\sqrt{15}$ **65.** $3\sqrt[3]{6}$ **67.** $6x^3\sqrt{x}$ **69.** $\frac{p^8\sqrt{p}}{11}$ **71.** $\frac{y\sqrt[4]{xy^2}}{3}$ **73.** $\frac{5}{\sqrt{\pi}}$ m or $\frac{5\sqrt{\pi}}{\pi}$ m **75.** $\sqrt{197}$ units $\approx$ 14.036 units **77.** $\sqrt{73}$ units $\approx$ 8.544 units **79.** $2\sqrt{11}$ units $\approx$ 6.633 units **81.** $(-5, 5)$ **83.** $\left(-\frac{11}{2}, -2\right)$ **85.** $\left(\frac{1}{4}, -\frac{2}{7}\right)$ **87.** $-2\sqrt{5}$ **89.** $9\sqrt[3]{2}$ **91.** $\frac{15 + 2\sqrt{3}}{6}$ **93.** $17\sqrt{2} - 15\sqrt{5}$ **95.** 6 **97.** $-8\sqrt{5}$ **99.** $a - 9$ **101.** $\sqrt[3]{25x^2} - 81$ **103.** $\frac{3\sqrt{7}}{7}$ **105.** $\frac{5\sqrt[3]{2}}{2}$ **107.** $\frac{x^2y^2\sqrt[3]{15yz}}{z}$ **109.** $\frac{3\sqrt{y} + 6}{y - 4}$ **111.** $\frac{11}{3\sqrt{11}}$ **113.** $\frac{3}{7\sqrt[3]{3}}$ **115.** $\frac{xy}{\sqrt[3]{10x^2yz}}$ **117.** 32 **119.** 35 **121.** 9 **123.** $3\sqrt{2}$ cm **125.** 51.2 ft **127.** $0 + 2i\sqrt{2}$ **129.** $0 + 6i$ **131.** $15 - 4i$ **133.** -64 **135.** $-12 - 18i$ **137.** $-5 - 12i$ **139.** $\frac{3}{2} - i$ **141.** x **143.** -10 **145.** $\frac{y^5}{2x^3}$ **147.** $\frac{1}{8}$ **149.** $\frac{1}{x^{13/2}}$ **151.** $\frac{n\sqrt{3n}}{11m^5}$ **153.** $4x - 20\sqrt{x} + 25$ **155.** (4, 16) **157.** $\frac{2\sqrt{x} - 6}{x - 9}$ **159.** 5

Chapter 17 Test **1.** $6\sqrt{6}$ **2.** $-x^{16}$ **3.** $\frac{1}{5}$ **4.** 5 **5.** $\frac{4x^2}{9}$ **6.** $-a^6b^3$ **7.** $\frac{8a^{1/3}c^{2/3}}{b^{5/12}}$ **8.** $a^{7/12} - a^{7/3}$ **9.** $|4xy|$ or $4|xy|$ **10.** -27 **11.** $\frac{3\sqrt{y}}{y}$ **12.** $\frac{8 - 6\sqrt{x} + x}{8 - 2x}$ **13.** $\frac{\sqrt[3]{b^2}}{b}$ **14.** $\frac{6 - x^2}{8(\sqrt{6} - x)}$ **15.** $-x\sqrt{5x}$ **16.** $4\sqrt{3} - \sqrt{6}$ **17.** $x + 2\sqrt{x} + 1$ **18.** $\sqrt{6} - 4\sqrt{3} + \sqrt{2} - 4$ **19.** -20 **20.** 23.685 **21.** 0.019 **22.** 2, 3 **23.** $\varnothing$ **24.** 6 **25.** $0 + i\sqrt{2}$ **26.** $0 - 2i\sqrt{2}$ **27.** $0 - 3i$ **28.** $40 + 0i$ **29.** $7 + 24i$ **30.** $-\frac{3}{2} + \frac{5}{2}i$ **31.** $\frac{5\sqrt{2}}{2}$ **32.** $[-2, \infty)$; 0, 1, 2, 3;

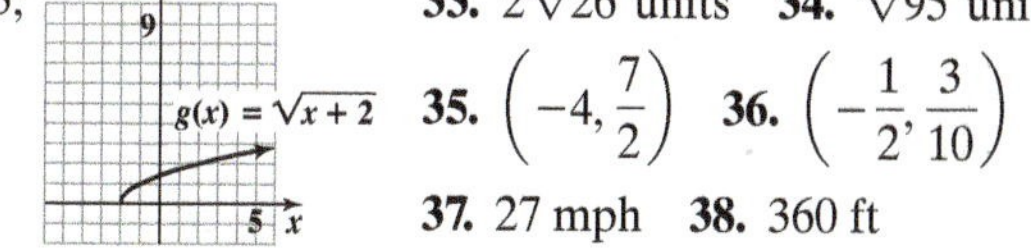

33. $2\sqrt{26}$ units **34.** $\sqrt{95}$ units **35.** $\left(-4, \frac{7}{2}\right)$ **36.** $\left(-\frac{1}{2}, \frac{3}{10}\right)$ **37.** 27 mph **38.** 360 ft

Cumulative Review Chapters 1–17 **1.** -19; Sec. 2.5, Ex. 10 **3.** 12; Sec. 9.3, Ex. 3 **5.** 2 ft, 8 ft; Sec. 9.4, Ex. 3 **7.** one; Sec. 11.1, Ex. 8 **9.** $\left(6, \frac{1}{2}\right)$; Sec. 11.2, Ex. 3 **11.** no solution; { } or $\varnothing$; Sec. 11.3, Ex. 3 **13.** 30% solution: 42 L; 80% solution: 28 L; Sec. 11.5, Ex. 6 **15. a.** -4 **b.** 11; Sec. 12.2, Ex. 4 **17.** $3m + 1$; Sec. 12.6, Ex. 1 **19.** $2x^2 + 5x + 2 + \frac{7}{x - 3}$; Sec. 12.7, Ex. 1 **21.** $(t - 8)(t - 5)$; Sec. 13.2, Ex. 8 **23. a.** $x^2 - 2x + 4$ **b.** $\frac{2}{y - 5}$; Sec. 14.1, Ex. 5 **25.** 4; Sec. 16.2, Ex. 9 **27.** $3x - 5$; Sec. 14.3, Ex. 3 **29.** $\frac{2m + 1}{m + 1}$; Sec. 14.4, Ex. 5 **31. a.** $\frac{x(x - 2)}{2(x + 2)}$ **b.** $\frac{x^2}{y^2}$; Sec. 14.7, Ex. 2 **33.** $\left[-2, \frac{8}{5}\right]$; Sec. 16.3, Ex. 3 **35.** 15 yd; Sec. 14.6, Ex. 2 **37. a.** 1 **b.** -4 **c.** $\frac{2}{5}$ **d.** x^2 **e.** $-3x^3$; Sec. 17.1, Ex. 3 **39. a.** $\frac{1}{8}$ **b.** $\frac{1}{9}$; Sec. 17.2, Ex. 3 **41.** $\frac{x - 4}{5(\sqrt{x} - 2)}$; Sec. 17.5, Ex. 7 **43.** constant of variation: 15; $u = \frac{15}{w}$; Sec. 15.4, Ex. 3

Chapter 18 Quadratic Equations and Functions

Section 18.1

Graphing Calculator Explorations

1. $-1.27, 6.27$

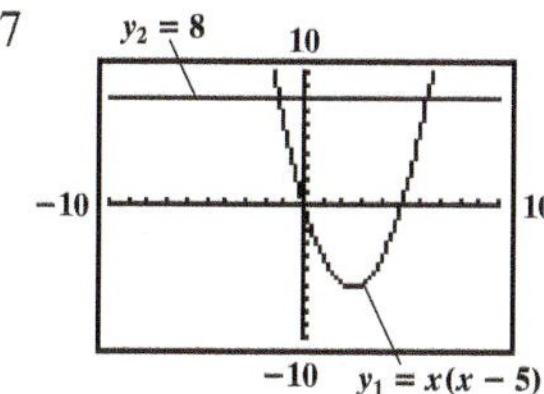

3. $-1.10, 0.90$

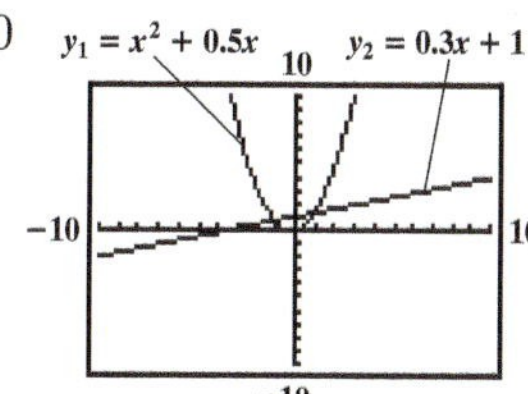

5. no real solutions

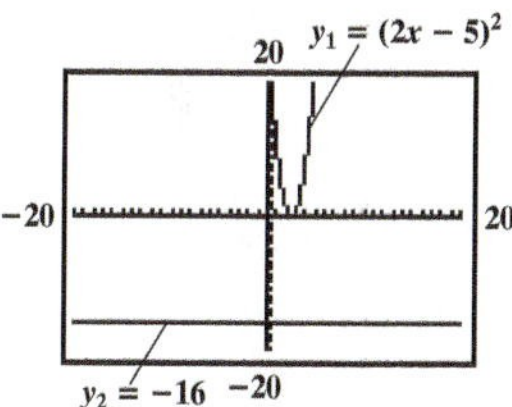

Vocabulary, Readiness & Video Check **1.** $\pm\sqrt{b}$ **3.** completing the square **5.** 9 **7.** We need a quantity squared by itself on one side of the equation. The only quantity squared is x, so we need to divide both sides by 2 before applying the square root property. **9.** We're looking for an interest rate, so a negative value does not make sense.

Exercise Set 18.1 **1.** $-4, 4$ **3.** $-\sqrt{7}, \sqrt{7}$ **5.** $-3\sqrt{2}, 3\sqrt{2}$ **7.** $-\sqrt{10}, \sqrt{10}$ **9.** $-8, -2$ **11.** $6 - 3\sqrt{2}, 6 + 3\sqrt{2}$ **13.** $\frac{3 - 2\sqrt{2}}{2}, \frac{3 + 2\sqrt{2}}{2}$ **15.** $-3i, 3i$ **17.** $-\sqrt{6}, \sqrt{6}$ **19.** $-2i\sqrt{2}, 2i\sqrt{2}$ **21.** $\frac{1 - 4i}{3}, \frac{1 + 4i}{3}$ or $\frac{1}{3} \pm \frac{4}{3}i$ **23.** $-7 - \sqrt{5}, -7 + \sqrt{5}$ **25.** $-3 - 2i\sqrt{2}, -3 + 2i\sqrt{2}$ **27.** $x^2 + 16x + 64 = (x + 8)^2$ **29.** $z^2 - 12z + 36 = (z - 6)^2$ **31.** $p^2 + 9p + \frac{81}{4} = \left(p + \frac{9}{2}\right)^2$ **33.** $x^2 + x + \frac{1}{4} = \left(x + \frac{1}{2}\right)^2$ **35.** $-5, -3$ **37.** $-3 - \sqrt{7}, -3 + \sqrt{7}$ **39.** $\frac{-1 - \sqrt{5}}{2}, \frac{-1 + \sqrt{5}}{2}$ **41.** $-1 - \sqrt{6}, -1 + \sqrt{6}$ **43.** $\frac{-1 - \sqrt{29}}{2}, \frac{-1 + \sqrt{29}}{2}$ **45.** $\frac{6 - \sqrt{30}}{3}, \frac{6 + \sqrt{30}}{3}$ **47.** $\frac{3 - \sqrt{11}}{2}, \frac{3 + \sqrt{11}}{2}$ **49.** $-4, \frac{1}{2}$ **51.** $-4 - \sqrt{15}, -4 + \sqrt{15}$ **53.** $\frac{-3 - \sqrt{21}}{3}, \frac{-3 + \sqrt{21}}{3}$ **55.** $-1, \frac{5}{2}$ **57.** $-1 - i, -1 + i$ **59.** $-3 - \sqrt{17}, -3 + \sqrt{17}$ **61.** $-2 - i\sqrt{2}, -2 + i\sqrt{2}$ **63.** $\frac{-15 - 7\sqrt{5}}{10}, \frac{-15 + 7\sqrt{5}}{10}$ **65.** $\frac{1 - i\sqrt{47}}{4}, \frac{1 + i\sqrt{47}}{4}$ or $\frac{1}{4} \pm \frac{\sqrt{47}}{4}i$ **67.** $-5 - i\sqrt{3}, -5 + i\sqrt{3}$ **69.** $-4, 1$ **71.** $\frac{2 - i\sqrt{2}}{2}, \frac{2 + i\sqrt{2}}{2}$ or $1 \pm \frac{\sqrt{2}}{2}i$ **73.** $\frac{-3 - \sqrt{69}}{6}, \frac{-3 + \sqrt{69}}{6}$ **75.** 20% **77.** 4% **79.** 9.63 sec **81.** 8.29 sec **83.** 15 ft by 15 ft **85.** $10\sqrt{2}$ cm **87.** -1 **89.** $3 + 2\sqrt{5}$ **91.** $\frac{1 - 3\sqrt{2}}{2}$ **93.** $2\sqrt{6}$ **95.** 5 **97.** complex, but not real numbers **99.** real numbers

101. complex, but not real numbers **103.** $-8x, 8x$ **105.** $-5z, 5z$ **107.** answers may vary **109.** compound; answers may vary **111.** 6 thousand scissors

Section 18.2

Vocabulary, Readiness & Video Check **1.** $x = \frac{-b \pm \sqrt{b^2 - 4ac}}{2a}$ **3.** $-5; -7$ **5.** $1; 0$ **7. a.** Yes, in order to make sure we have correct values for a, b, and c. **b.** No; clearing fractions makes the work less tedious, but it's not a necessary step. **9.** With applications, we need to make sure we answer the question(s) asked. Here we're asked how much distance is saved, so once the dimensions of the triangle are known, further calculations are needed to answer this question and solve the problem.

Exercise Set 18.2 **1.** $-6, 1$ **3.** $-\frac{3}{5}, 1$ **5.** 3 **7.** $\frac{-7-\sqrt{33}}{2}, \frac{-7+\sqrt{33}}{2}$ **9.** $\frac{1-\sqrt{57}}{8}, \frac{1+\sqrt{57}}{8}$ **11.** $\frac{7-\sqrt{85}}{6}, \frac{7+\sqrt{85}}{6}$ **13.** $1-\sqrt{3}, 1+\sqrt{3}$ **15.** $-\frac{3}{2}, 1$ **17.** $\frac{3-\sqrt{11}}{2}, \frac{3+\sqrt{11}}{2}$ **19.** $\frac{-5-\sqrt{17}}{2}, \frac{-5+\sqrt{17}}{2}$ **21.** $\frac{5}{2}, 1$ **23.** $-3-2i, -3+2i$ **25.** $-2-\sqrt{11}, -2+\sqrt{11}$ **27.** $\frac{3-i\sqrt{87}}{8}, \frac{3+i\sqrt{87}}{8}$ or $\frac{3}{8} \pm \frac{\sqrt{87}}{8}i$ **29.** $\frac{3-\sqrt{29}}{2}, \frac{3+\sqrt{29}}{2}$ **31.** $\frac{-5-i\sqrt{5}}{10}, \frac{-5+i\sqrt{5}}{10}$ or $-\frac{1}{2} \pm \frac{\sqrt{5}}{10}i$ **33.** $\frac{-1-\sqrt{19}}{6}, \frac{-1+\sqrt{19}}{6}$ **35.** $\frac{-1-i\sqrt{23}}{4}, \frac{-1+i\sqrt{23}}{4}$ or $-\frac{1}{4} \pm \frac{\sqrt{23}}{4}i$ **37.** 1 **39.** $3-\sqrt{5}, 3+\sqrt{5}$ **41.** two real solutions **43.** one real solution **45.** two real solutions **47.** two complex but not real solutions **49.** two real solutions **51.** 14 ft **53.** $(2+2\sqrt{2})$ cm, $(2+2\sqrt{2})$ cm, $(4+2\sqrt{2})$ cm **55.** width: $(-5+5\sqrt{17})$ ft; length: $(5+5\sqrt{17})$ ft **57. a.** $50\sqrt{2}$m **b.** 5000 sq m **59.** 37.4 ft by 38.5 ft **61.** base, $(2+2\sqrt{43})$ cm; height, $(-1+\sqrt{43})$ cm **63.** 8.9 sec **65.** 2.8 sec **67.** $\frac{11}{5}$ **69.** 15 **71.** $(x^2+5)(x+2)(x-2)$ **73.** $(z+3)(z-3)(z+2)(z-2)$ **75.** b **77.** answers may vary **79.** 0.6, 2.4 **81.** Sunday to Monday **83.** Wednesday **85.** 32; yes **87. a.** 20,568 students **b.** 2015 **89. a.** 9076 thousand barrels per day **b.** 2007 **c.** 2012 **91.** answers may vary **93.** $\frac{\sqrt{3}}{3}$ **95.** $\frac{-\sqrt{2}-i\sqrt{2}}{2}, \frac{-\sqrt{2}+i\sqrt{2}}{2}$ or $-\frac{\sqrt{2}}{2} \pm \frac{\sqrt{2}}{2}i$ **97.** $\frac{\sqrt{3}-\sqrt{11}}{4}, \frac{\sqrt{3}+\sqrt{11}}{4}$

99. 8.9 sec:

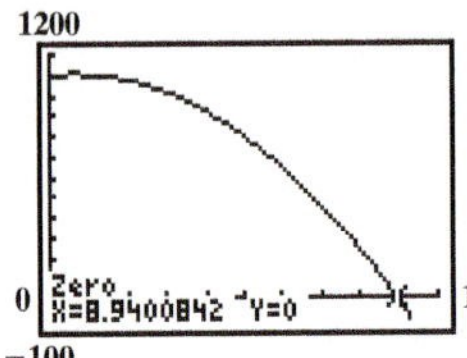

2.8 sec:

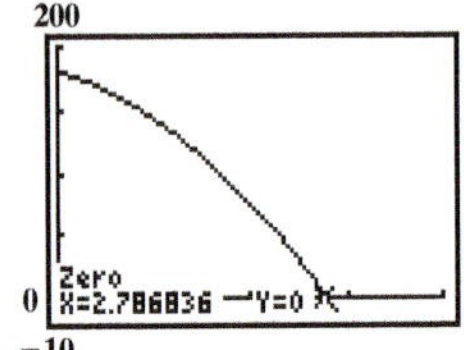

101. two real solutions

Section 18.3

Vocabulary, Readiness & Video Check **1.** The values we get for the substituted variable are *not* our final answers. Remember to always substitute back to the original variable and solve for it if necessary.

Exercise Set 18.3 **1.** 2 **3.** 16 **5.** 1, 4 **7.** $3-\sqrt{7}, 3+\sqrt{7}$ **9.** $\frac{3-\sqrt{57}}{4}, \frac{3+\sqrt{57}}{4}$ **11.** $\frac{1-\sqrt{29}}{2}, \frac{1+\sqrt{29}}{2}$ **13.** $-2, 2, -2i, 2i$ **15.** $-\frac{1}{2}, \frac{1}{2}, -i\sqrt{3}, i\sqrt{3}$ **17.** $-3, 3, -2, 2$ **19.** $125, -8$ **21.** $-\frac{4}{5}, 0$ **23.** $-\frac{1}{8}, 27$ **25.** $-\frac{2}{3}, \frac{4}{3}$ **27.** $-\frac{1}{125}, \frac{1}{8}$ **29.** $-\sqrt{2}, \sqrt{2}, -\sqrt{3}, \sqrt{3}$ **31.** $\frac{-9-\sqrt{201}}{6}, \frac{-9+\sqrt{201}}{6}$ **33.** 2, 3 **35.** 3 **37.** 27, 125 **39.** $1, -3i, 3i$ **41.** $\frac{1}{8}, -8$ **43.** $-\frac{1}{2}, \frac{1}{3}$ **45.** 4 **47.** -3 **49.** $-\sqrt{5}, \sqrt{5}, -2i, 2i$ **51.** $-3, \frac{3-3i\sqrt{3}}{2}, \frac{3+3i\sqrt{3}}{2}$ or $-3, \frac{3}{2} \pm \frac{3\sqrt{3}}{2}i$ **53.** 6, 12 **55.** $-\frac{1}{3}, \frac{1}{3}, -\frac{i\sqrt{6}}{3}, \frac{i\sqrt{6}}{3}$ **57.** 5 **59.** 55 mph: 66 mph **61.** 5 mph, then 4 mph **63.** inlet pipe: 15.5 hr; hose: 16.5 hr **65.** 8.5 hr **67.** 12 or -8 **69. a.** $(x-6)$ in. **b.** $300 = (x-6)\cdot(x-6)\cdot 3$ **c.** 16 in. by 16 in. **71.** 22 feet **73.** $(-\infty, 3]$ **75.** $(-5, \infty)$ **77.** domain: $(-\infty, \infty)$; range: $(-\infty, \infty)$; function **79.** domain: $(-\infty, \infty)$; range: $[-1, \infty)$; function **81.** $1, -3i, 3i$ **83.** $-\frac{1}{2}, \frac{1}{3}$ **85.** $-3, \frac{3-3i\sqrt{3}}{2}, \frac{3+3i\sqrt{3}}{2}$ or $-3, \frac{3}{2} \pm \frac{3\sqrt{3}}{2}i$ **87.** answers may vary **89. a.** 11.615 m/sec **b.** 11.612 m/sec **c.** 25.925 mph

Integrated Review **1.** $-\sqrt{10}, \sqrt{10}$ **2.** $-\sqrt{14}, \sqrt{14}$ **3.** $1-2\sqrt{2}, 1+2\sqrt{2}$ **4.** $-5-2\sqrt{3}, -5+2\sqrt{3}$ **5.** $-1-\sqrt{13}, -1+\sqrt{13}$ **6.** 1, 11 **7.** $\frac{-3-\sqrt{69}}{6}, \frac{-3+\sqrt{69}}{6}$ **8.** $\frac{-2-\sqrt{5}}{4}, \frac{-2+\sqrt{5}}{4}$ **9.** $\frac{2-\sqrt{2}}{2}, \frac{2+\sqrt{2}}{2}$ **10.** $-3-\sqrt{5}, -3+\sqrt{5}$ **11.** $-2-i\sqrt{3}, -2+i\sqrt{3}$ **12.** $\frac{-1-i\sqrt{11}}{2}, \frac{-1+i\sqrt{11}}{2}$ or $-\frac{1}{2} \pm \frac{\sqrt{11}}{2}i$ **13.** $\frac{-3-i\sqrt{15}}{2}, \frac{-3+i\sqrt{15}}{2}$ or $-\frac{3}{2} \pm \frac{\sqrt{15}}{2}i$ **14.** $-3i, 3i$ **15.** $-17, 0$ **16.** $\frac{1-\sqrt{13}}{4}, \frac{1+\sqrt{13}}{4}$ **17.** $2-3\sqrt{3}, 2+3\sqrt{3}$ **18.** $2-\sqrt{3}, 2+\sqrt{3}$ **19.** $-2, \frac{4}{3}$ **20.** $\frac{-5-\sqrt{17}}{4}, \frac{-5+\sqrt{17}}{4}$

21. $1 - \sqrt{6}, 1 + \sqrt{6}$ **22.** $-\sqrt{31}, \sqrt{31}$ **23.** $-\sqrt{11}, \sqrt{11}$ **24.** $-i\sqrt{11}, i\sqrt{11}$ **25.** $-11, 6$ **26.** $\frac{-3 - \sqrt{19}}{5}, \frac{-3 + \sqrt{19}}{5}$ **27.** $\frac{-3 - \sqrt{17}}{4}, \frac{-3 + \sqrt{17}}{4}$ **28.** 4 **29.** $\frac{-1 - \sqrt{17}}{8}, \frac{-1 + \sqrt{17}}{8}$ **30.** $10\sqrt{2}$ ft $\approx$ 14.1 ft **31.** Jack: 9.1 hr; Lucy: 7.1 hr **32.** 5 mph during the first part, then 6 mph

Section 18.4

Vocabulary, Readiness & Video Check **1.** $[-7, 3)$ **3.** $(-\infty, 0]$ **5.** $(-\infty, -12) \cup [-10, \infty)$ **7.** We use the solutions to the related equation to divide the number line into regions that either entirely are or entirely are not solution regions; the solutions to the related equation are solutions to the inequality only if the inequality symbol is $\leq$ or $\geq$.

Exercise Set 18.4 **1.** $(-\infty, -5) \cup (-1, \infty)$ **3.** $[-4, 3]$ **5.** $[2, 5]$ **7.** $\left(-5, -\frac{1}{3}\right)$ **9.** $(2, 4) \cup (6, \infty)$ **11.** $(-\infty, -4] \cup [0, 1]$ **13.** $(-\infty, -3) \cup (-2, 2) \cup (3, \infty)$ **15.** $(-7, 2)$ **17.** $(-1, \infty)$ **19.** $(-\infty, -1] \cup (4, \infty)$ **21.** $(-\infty, 2) \cup \left(\frac{11}{4}, \infty\right)$ **23.** $(0, 2] \cup [3, \infty)$ **25.** $(-\infty, 3)$ **27.** $\left[-\frac{5}{4}, \frac{3}{2}\right]$ **29.** $(-\infty, 0) \cup (1, \infty)$ **31.** $(-\infty, -4] \cup [4, 6]$ **33.** $\left(-\infty, -\frac{2}{3}\right] \cup \left[\frac{3}{2}, \infty\right)$ **35.** $\left(-4, -\frac{3}{2}\right) \cup \left(\frac{3}{2}, \infty\right)$ **37.** $(-\infty, -5] \cup [-1, 1] \cup [5, \infty)$ **39.** $\left(-\infty, -\frac{5}{3}\right) \cup \left(\frac{7}{2}, \infty\right)$ **41.** $(0, 10)$ **43.** $(-\infty, -4) \cup [5, \infty)$ **45.** $(-\infty, -6] \cup (-1, 0] \cup (7, \infty)$ **47.** $(-\infty, 1) \cup (2, \infty)$ **49.** $(-\infty, -8] \cup (-4, \infty)$ **51.** $(-\infty, 0] \cup \left(5, \frac{11}{2}\right]$ **53.** $(0, \infty)$

55. 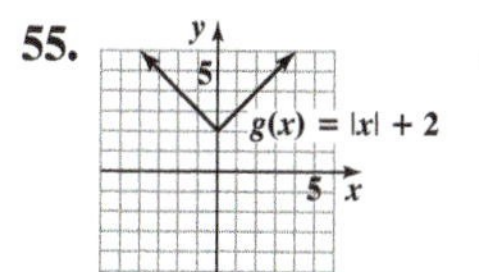

57. 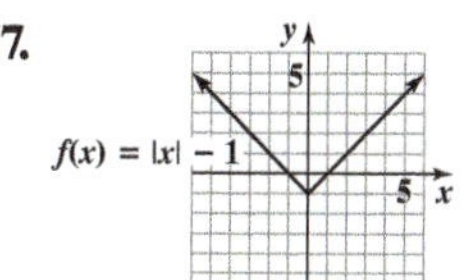

59. 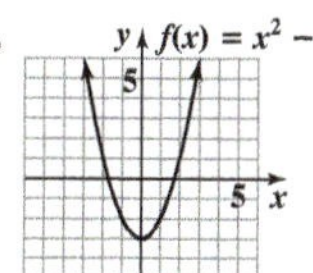

61. 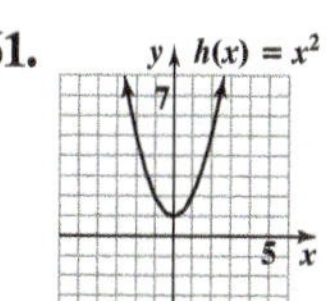

63. answers may vary **65.** $(-\infty, -1) \cup (0, 1)$, or any number less than -1 or between 0 and 1

67. x is between 2 and 11 **69.**

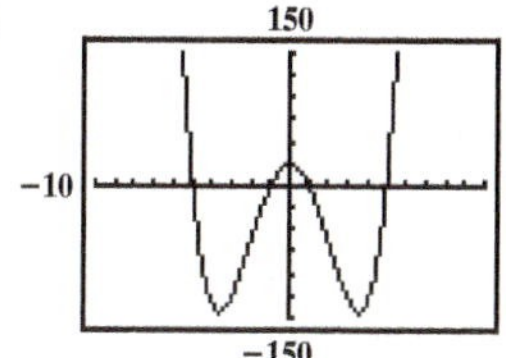

71.

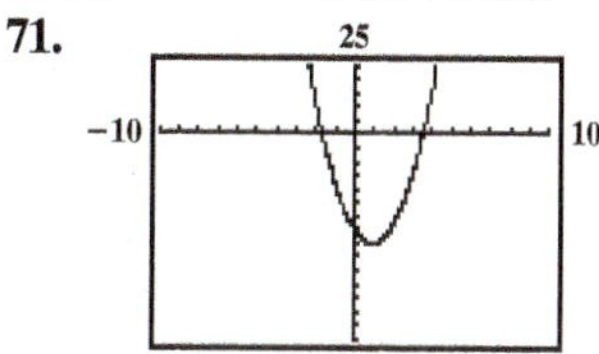

Section 18.5

Graphing Calculator Explorations

1. 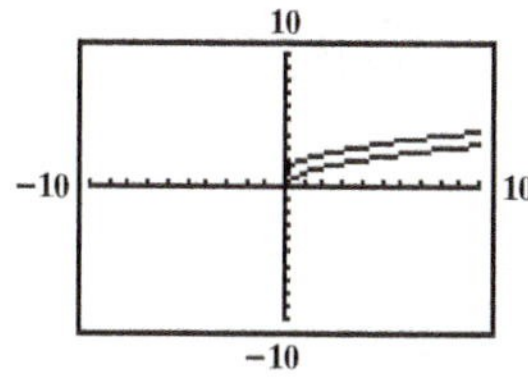

3. 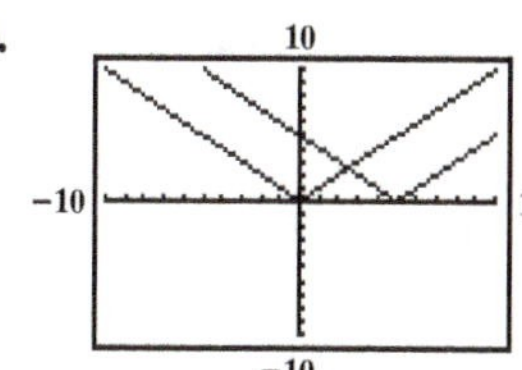

5. 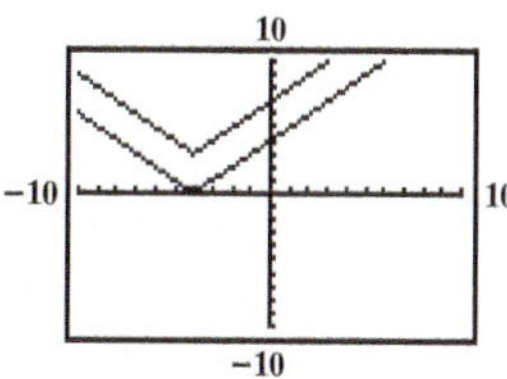

Vocabulary, Readiness & Video Check **1.** quadratic **3.** upward **5.** lowest **7.** $(0, 0)$ **9.** $(2, 0)$ **11.** $(0, 3)$ **13.** $(-1, 5)$ **15.** Graphs of the form $f(x) = x^2 + k$ shift up or down the y-axis k units from $y = x^2$; the y-intercept. **17.** The vertex, (h, k), and the axis of symmetry, $x = h$; the basic shape of $y = x^2$ does not change. **19.** the coordinates of the vertex, whether the graph opens upward or downward, whether the graph is narrower or wider than $y = x^2$, and the graph's axis of symmetry

Exercise Set 18.5 **1.** 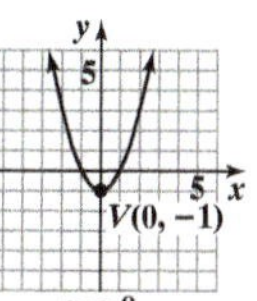

3. 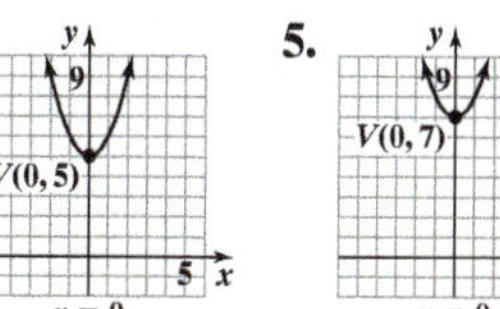

5. 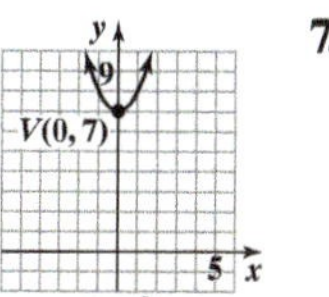

7. 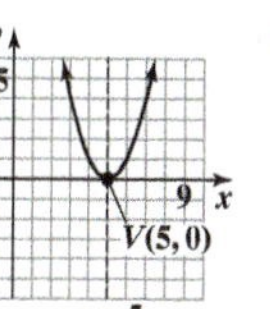

9. 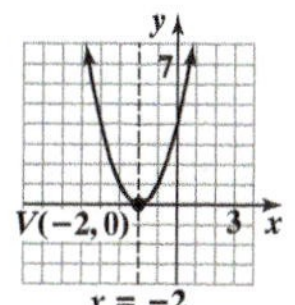

11. 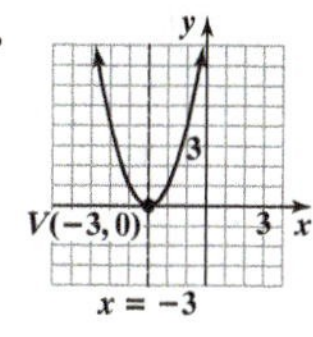

13. 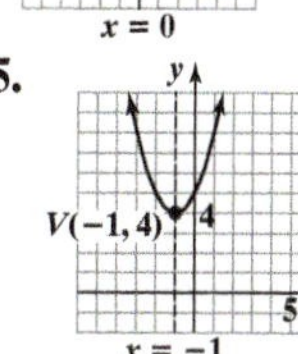

15. 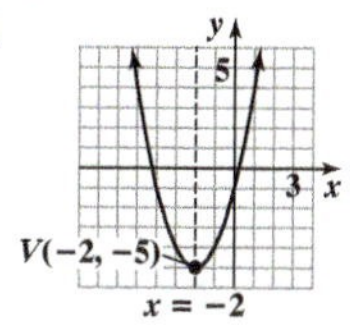

17. 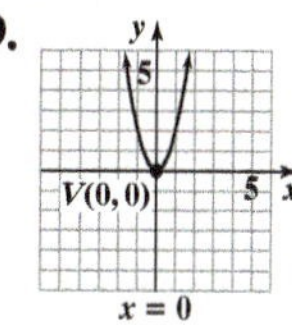

19. 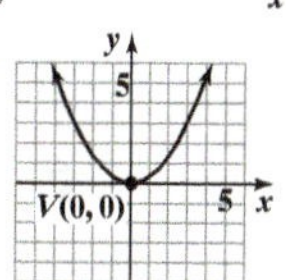

21. 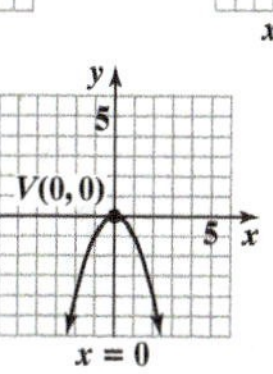

23. y 5 V(0, 0) 5 x x = 0

25. 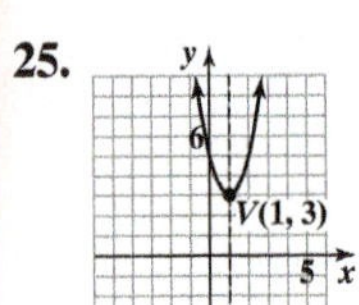

27. 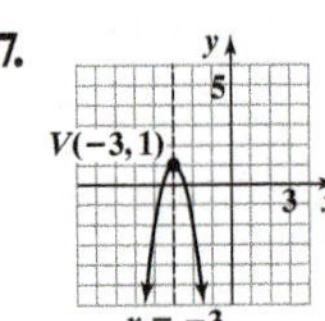

29. 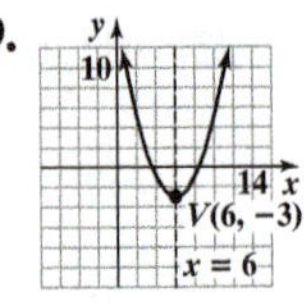

31. 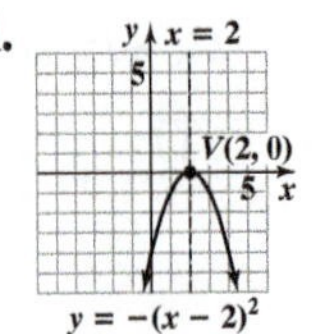

33. 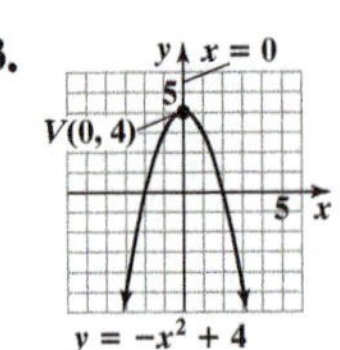

35. 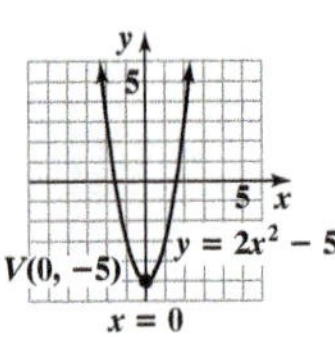

37. 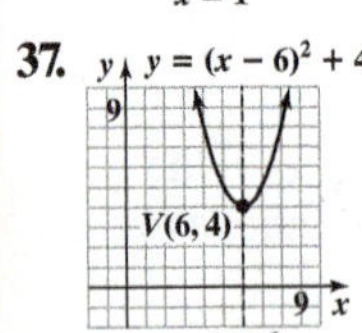

39. 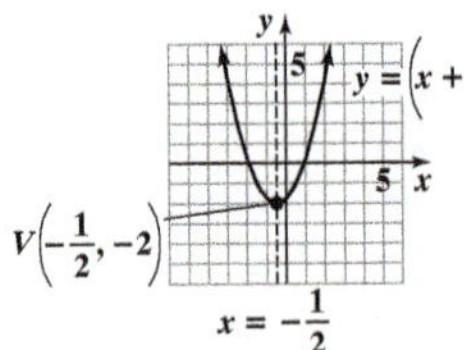

41. 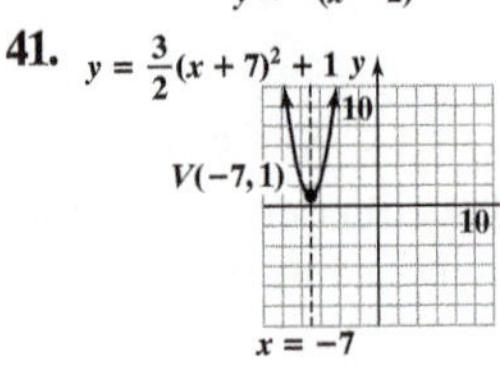

43. 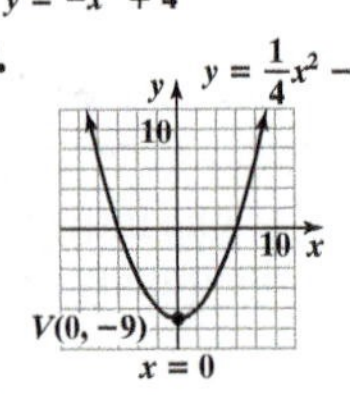

45. 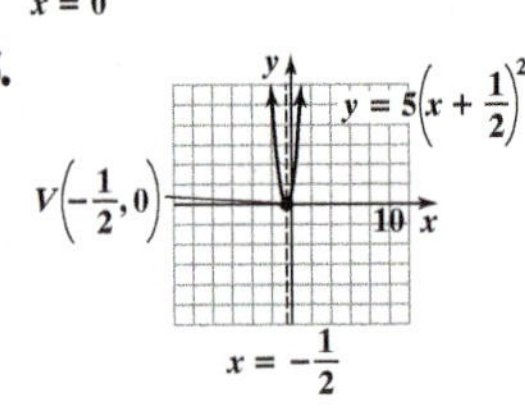

47. 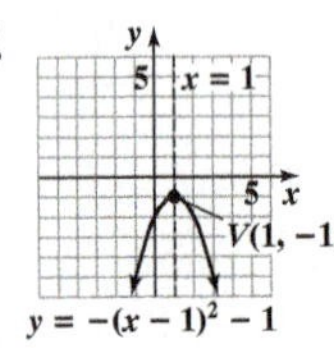

49. 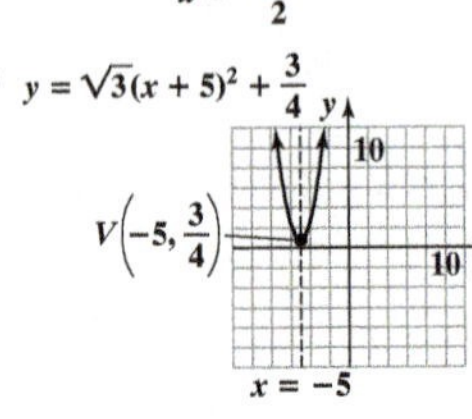

51. 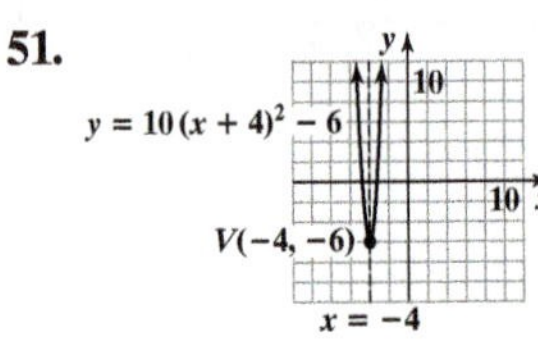

53. 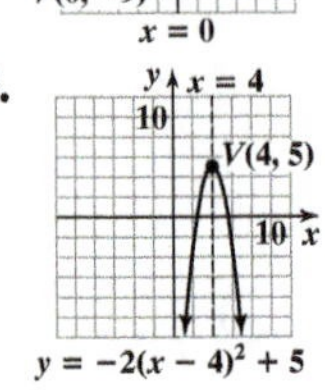

55. $x^2 + 8x + 16$ **57.** $z^2 - 16z + 64$ **59.** $y^2 + y + \frac{1}{4}$ **61.** $-6, 2$ **63.** $-5 - \sqrt{26}, -5 + \sqrt{26}$ **65.** $4 - 3\sqrt{2}, 4 + 3\sqrt{2}$

67. c **69.** $f(x) = 5(x - 2)^2 + 3$ **71.** $f(x) = 5(x + 3)^2 + 6$

73. 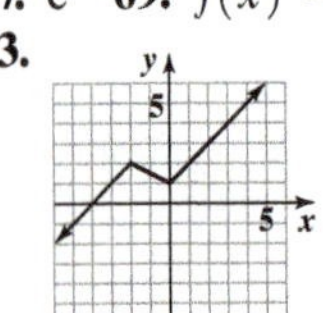

75. 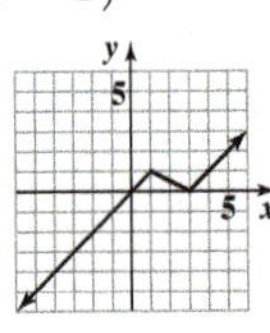

77. 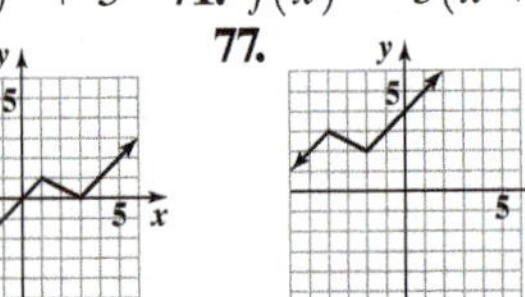

79. a. 119,141 million **b.** 284,953 million

Section 18.6

Vocabulary, Readiness & Video Check **1.** (h, k) **3.** We can immediately identify the vertex (h, k), know whether the parabola opens upward or downward, and identify its axis of symmetry; completing the square. **5.** the vertex

Exercise Set 18.6 **1.** 0; 1 **3.** 2; 1 **5.** 1; 1 **7.** down **9.** up **11.** $(-4, -9)$ **13.** $(5, 30)$ **15.** $(1, -2)$ **17.** $\left(\frac{1}{2}, \frac{5}{4}\right)$ **19.** D **21.** B

23. 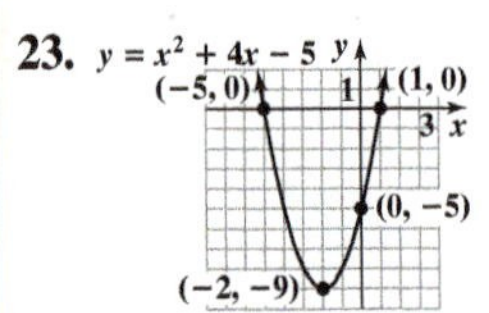

25. 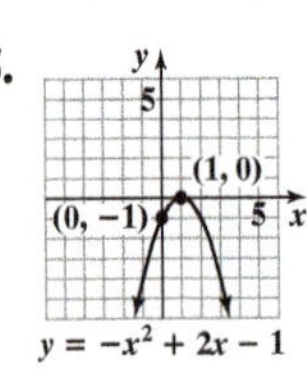

27. 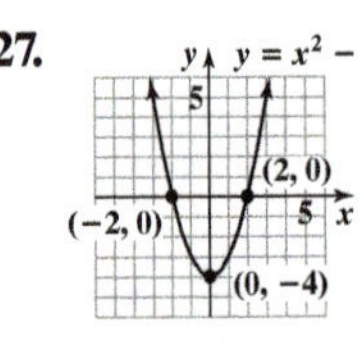

29. 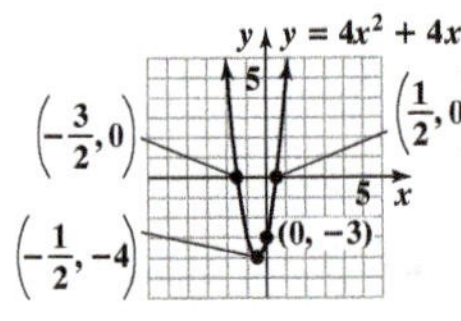

31. 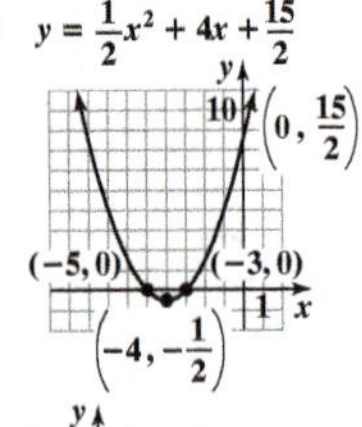

33. 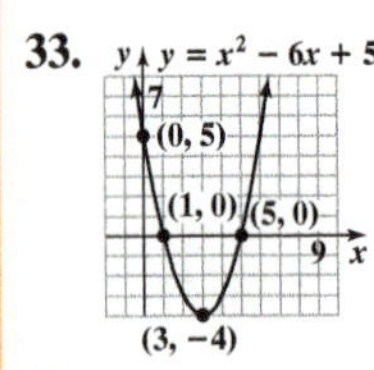

35. 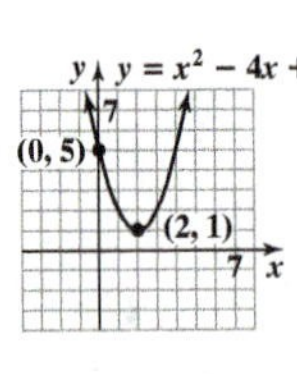

37. 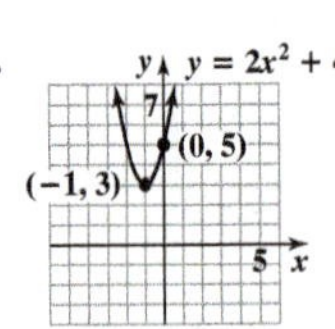

39. 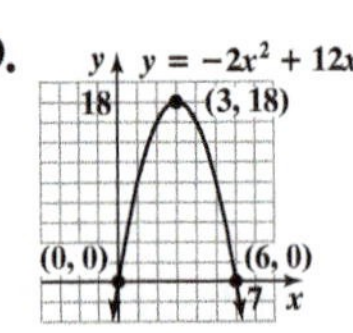

41. 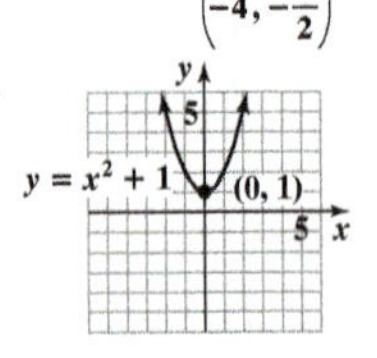

43. 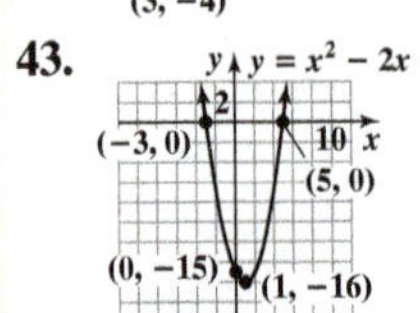

45. 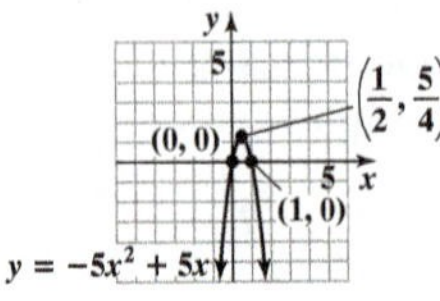

47. 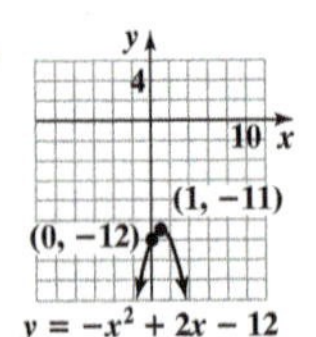

49. 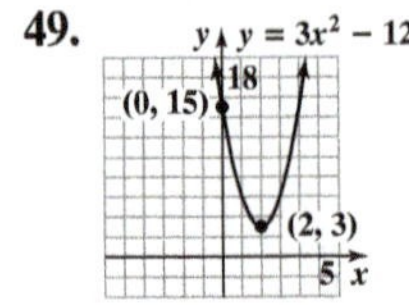

51. 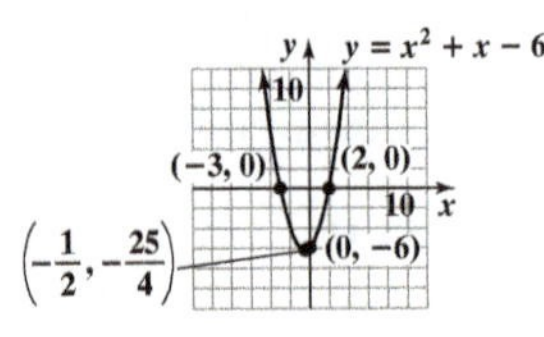

53. 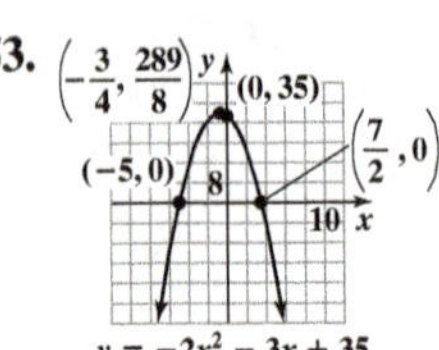

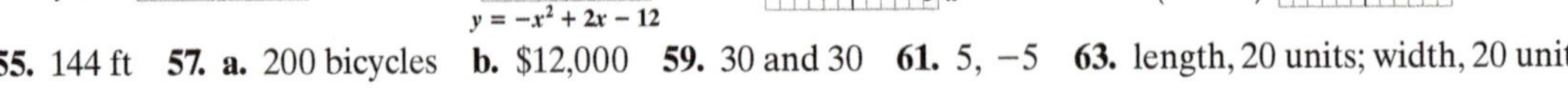
55. 144 ft **57. a.** 200 bicycles **b.** $12,000 **59.** 30 and 30 **61.** 5, −5 **63.** length, 20 units; width, 20 units

65. 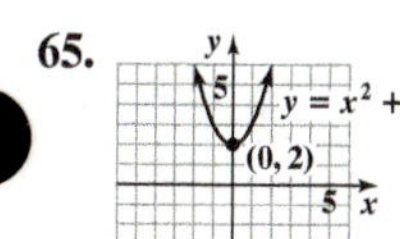

67. 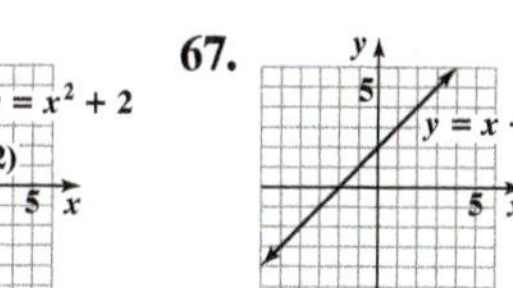

69. 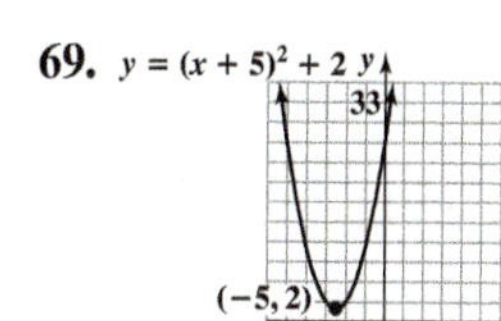

71. 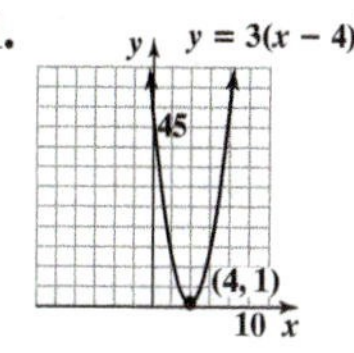

73. 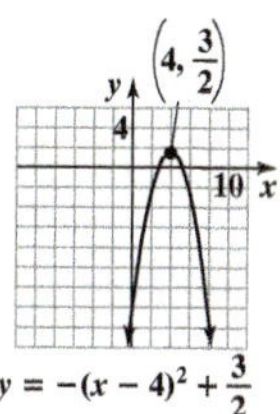

75. minimum value **77.** maximum value **79.** 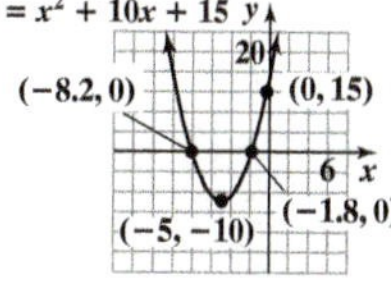

81. 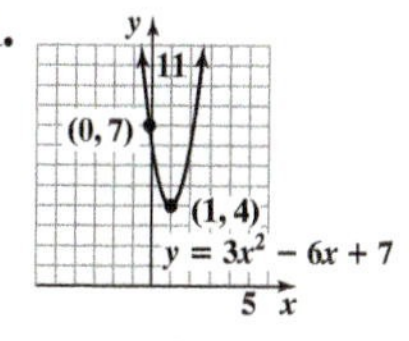

83. −0.84 **85.** 1.43

87. a. maximum; answers may vary **b.** 2035 **c.** 311 million **89.** 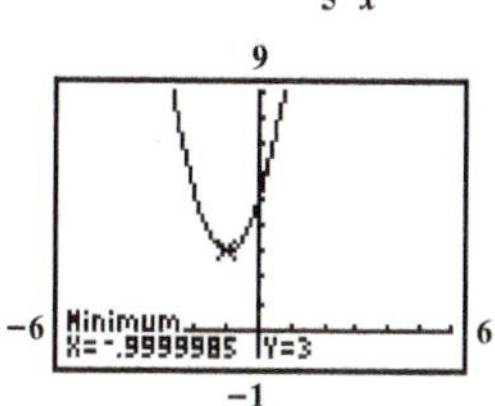

91.

Chapter 18 Vocabulary Check **1.** discriminant **2.** $\pm\sqrt{b}$ **3.** $\frac{-b}{2a}$ **4.** quadratic inequality **5.** completing the square **6.** $(0, k)$ **7.** $(h, 0)$ **8.** (h, k) **9.** quadratic formula **10.** quadratic

Chapter 18 Review **1.** 14, 1 **3.** −7, 7 **5.** $\frac{-3-\sqrt{5}}{2}, \frac{-3+\sqrt{5}}{2}$ **7.** 4.25% **9.** two complex but not real solutions **11.** two real solutions **13.** 8 **15.** $-\frac{5}{2}, 1$ **17.** $\frac{5-i\sqrt{143}}{12}, \frac{5+i\sqrt{143}}{12}$ or $\frac{5}{12} \pm \frac{\sqrt{143}}{12}i$ **19. a.** 20 ft **b.** $\frac{15+\sqrt{321}}{16}$ sec; 2.1 sec **21.** $3, \frac{-3-3i\sqrt{3}}{2}, \frac{-3+3i\sqrt{3}}{2}$ or $3, -\frac{3}{2} \pm \frac{3\sqrt{3}}{2}i$ **23.** $\frac{2}{3}, 5$ **25.** 1, 125 **27.** $-1, 1, -i, i$ **29.** Jerome: 10.5 hr; Tim: 9.5 hr **31.** $[-5, 5]$
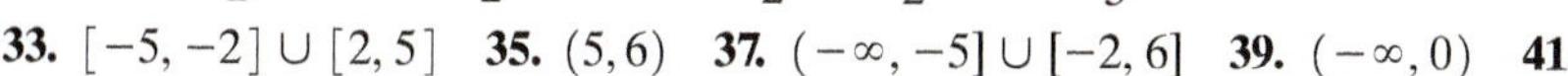
33. $[-5, -2] \cup [2, 5]$ **35.** $(5, 6)$ **37.** $(-\infty, -5] \cup [-2, 6]$ **39.** $(-\infty, 0)$ **41.** 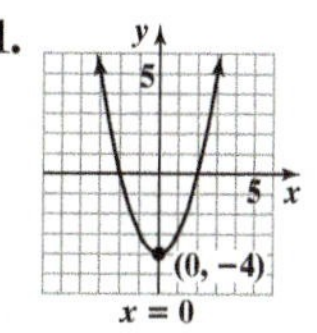

43. 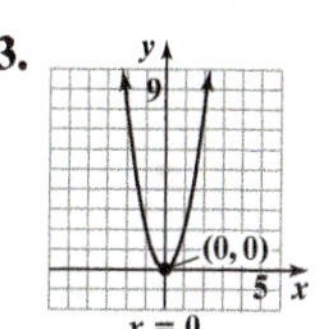

45. 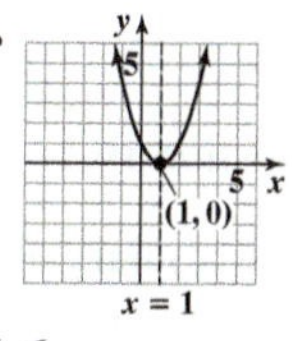

47. **49.** 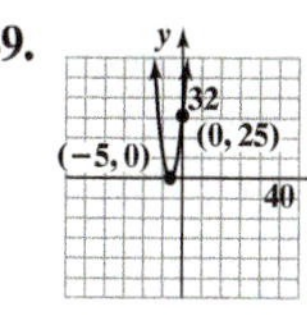

51. 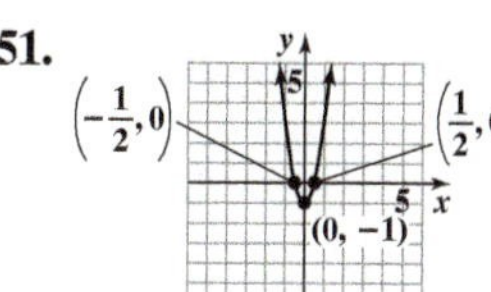

53. 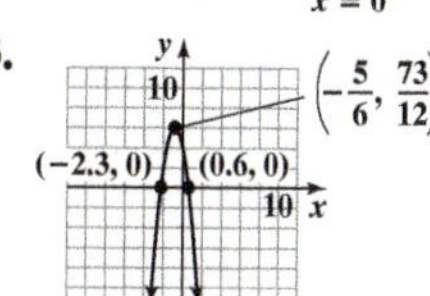

55. 210 and 210 **57.** −5, 6

59. −2, 2 **61.** $\frac{-1-3i\sqrt{3}}{2}, \frac{-1+3i\sqrt{3}}{2}$ or $-\frac{1}{2} \pm \frac{3\sqrt{3}}{2}i$ **63.** $-i\sqrt{11}, i\sqrt{11}$ **65.** $\frac{21-\sqrt{41}}{50}, \frac{21+\sqrt{41}}{50}$ **67.** 8, 64 **69.** $[-5, 0] \cup \left(\frac{3}{4}, \infty\right)$ **71. a.** 95,157 thousand passengers **b.** 2022

Chapter 18 Test **1.** $\frac{7}{5}, -1$ **2.** $-1-\sqrt{10}, -1+\sqrt{10}$ **3.** $\frac{1-i\sqrt{31}}{2}, \frac{1+i\sqrt{31}}{2}$ or $\frac{1}{2} \pm \frac{\sqrt{31}}{2}i$ **4.** $3-\sqrt{7}, 3+\sqrt{7}$ **5.** $-\frac{1}{7}, -1$ **6.** $\frac{3-\sqrt{29}}{2}, \frac{3+\sqrt{29}}{2}$ **7.** $-2-\sqrt{11}, -2+\sqrt{11}$ **8.** $-1, 1, -i, i, -3$ **9.** $-1, 1, -i, i$ **10.** 6, 7 **11.** $3-\sqrt{7}, 3+\sqrt{7}$ **12.** $\frac{2-i\sqrt{6}}{2}, \frac{2+i\sqrt{6}}{2}$ or $1 \pm \frac{\sqrt{6}}{2}i$ **13.** $\left(-\infty, -\frac{3}{2}\right) \cup (5, \infty)$ **14.** $(-\infty, -5] \cup [-4, 4] \cup [5, \infty)$ **15.** $(-\infty, -3) \cup (2, \infty)$ **16.** $(-\infty, -3) \cup [2, 3)$ **17.** 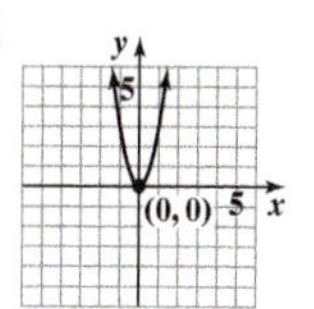

18. 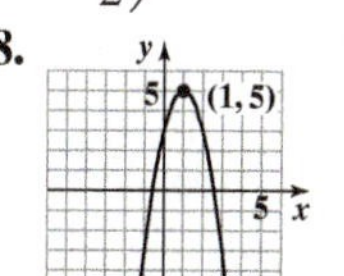

19. 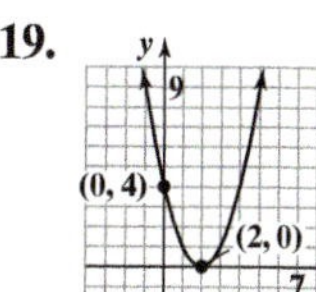

20. 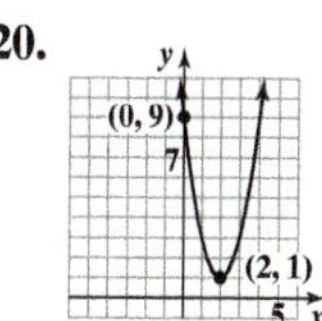

21. $(5+\sqrt{17})$ hr $\approx$ 9.12 hr

22. a. 272 ft **b.** 5.12 sec **23.** 7.2 ft

Cumulative Review Chapters 1–18 **1.** 1; Sec. 2.2, Ex. 16 **3. a.** $10x$ **b.** $-4y$ **c.** $8x^2 - 2$; Sec. 3.1, Ex. 1 **5.** no solution, { } or ∅; Sec. 11.1, Ex. 5 **7.** $(-2, 0)$; Sec. 11.2, Ex. 4 **9. a.** x^3 **b.** 256 **c.** -27 **d.** cannot be simplified **e.** $2x^4y$; Sec. 12.1, Ex. 9 **11. a.** 5 **b.** 5; Sec. 12.7, Ex. 3 **13.** $-6, -\frac{3}{2}, \frac{1}{5}$; Sec. 13.6, Ex. 9 **15.** $\frac{1}{5x - 1}$; Sec. 14.1, Ex. 2a **17.** $\frac{xy + 2x^3}{y - 1}$; Sec. 14.7, Ex. 3 **19.** $(2m^2 - 1)^2$; Sec. 13.3, Ex. 10 **21.** -10 and 7; Sec. 13.7, Ex. 2 **23. a.** $5x\sqrt{x}$ **b.** $3x^2y^2\sqrt[3]{2y^2}$ **c.** $3z^2\sqrt[4]{z^3}$; Sec. 17.3, Ex. 4 **25. a.** $\frac{2\sqrt{5}}{5}$ **b.** $\frac{8\sqrt{x}}{3x}$ **c.** $\frac{\sqrt[3]{4}}{2}$; Sec. 17.5, Ex. 1 **27.** $\frac{2}{9}$; Sec. 17.6, Ex. 5 **29.** -5; Sec. 14.5, Ex. 1 **31.** -5; Sec. 14.6, Ex. 3 **33.** $k = \frac{1}{6}$; $y = \frac{1}{6}x$; Sec. 15.4, Ex. 1 **35. a.** 3 **b.** $|x|$ **c.** $|x - 2|$ **d.** -5 **e.** $2x - 7$ **f.** $5|x|$ **g.** $|x + 1|$; Sec. 17.1, Ex. 5 **37. a.** $\sqrt{x}$ **b.** $\sqrt[3]{5}$ **c.** $\sqrt{rs^3}$; Sec. 17.2, Ex. 7 **39. a.** $\frac{1}{2} + \frac{3}{2}i$ **b.** $0 - \frac{7}{3}i$; Sec. 17.7, Ex. 5 **41.** $-1 + 2\sqrt{3}, -1 - 2\sqrt{3}$; Sec. 18.1, Ex. 3 **43.** 9; Sec. 18.3, Ex. 1

Chapter 19 Exponential and Logarithmic Functions

Section 19.1

Vocabulary, Readiness & Video Check **1.** C **3.** F **5.** D **7.** You can find $(f + g)(x)$ and then find $(f + g)(2)$ or you can find $f(2)$ and $g(2)$ and then add those results.

Exercise Set 19.1 **1. a.** $3x - 6$ **b.** $-x - 8$ **c.** $2x^2 - 13x - 7$ **d.** $\frac{x - 7}{2x + 1}$, where $x \neq -\frac{1}{2}$ **3. a.** $x^2 + 5x + 1$ **b.** $x^2 - 5x + 1$ **c.** $5x^3 + 5x$ **d.** $\frac{x^2 + 1}{5x}$, where $x \neq 0$ **5. a.** $\sqrt[3]{x} + x + 5$ **b.** $\sqrt[3]{x} - x - 5$ **c.** $x\sqrt[3]{x} + 5\sqrt[3]{x}$ **d.** $\frac{\sqrt[3]{x}}{x + 5}$, where $x \neq -5$ **7. a.** $5x^2 - 3x$ **b.** $-5x^2 - 3x$ **c.** $-15x^3$ **d.** $-\frac{3}{5x}$, where $x \neq 0$ **9.** 42 **11.** -18 **13.** 0 **15.** $(f \circ g)(x) = 25x^2 + 1$; $(g \circ f)(x) = 5x^2 + 5$ **17.** $(f \circ g)(x) = 2x + 11$; $(g \circ f)(x) = 2x + 4$ **19.** $(f \circ g)(x) = -8x^3 - 2x - 2$; $(g \circ f)(x) = -2x^3 - 2x + 4$ **21.** $(f \circ g)(x) = |10x - 3|$; $(g \circ f)(x) = 10|x| - 3$ **23.** $(f \circ g)(x) = \sqrt{-5x + 2}$; $(g \circ f)(x) = -5\sqrt{x} + 2$ **25.** $H(x) = (g \circ h)(x)$ **27.** $F(x) = (h \circ f)(x)$ **29.** $G(x) = (f \circ g)(x)$ **31.** answers may vary; for example, $g(x) = x + 2$ and $f(x) = x^2$ **33.** answers may vary; for example, $g(x) = x + 5$ and $f(x) = \sqrt{x} + 2$ **35.** answers may vary; for example, $g(x) = 2x - 3$ and $f(x) = \frac{1}{x}$ **37.** $y = x - 2$ **39.** $y = \frac{x}{3}$ **41.** $y = -\frac{x + 7}{2}$ **43.** 6 **45.** 4 **47.** 4 **49.** -1 **51.** answers may vary **53.** $P(x) = R(x) - C(x)$

Section 19.2

Vocabulary, Readiness & Video Check **1.** (2, 11) **3.** the inverse of f **5.** vertical **7.** $y = x$ **9.** Every function must have each x-value correspond to only one y-value. A one-to-one function must also have each y-value correspond to only one x-value. **11.** Yes; by the definition of an inverse function. **13.** Once you know some points of the original equation or graph, you can switch the x's and y's of these points to find points that satisfy the inverse and then graph it. You can also check that the two graphs (the original and the inverse) are symmetric about the line $y = x$.

Exercise Set 19.2 **1.** one-to-one; $f^{-1} = \{(-1, -1), (1, 1), (2, 0), (0, 2)\}$ **3.** one-to-one; $h^{-1} = \{(10, 10)\}$ **5.** one-to-one; $f^{-1} = \{(12, 11), (3, 4), (4, 3), (6, 6)\}$ **7.** not one-to-one **9.** one-to-one;

Rank in Population (Input)	1	47	16	25	36	7
State (Output)	CA	AK	IN	LA	NM	OH

11. a. 3 **b.** 1 **13. a.** 1 **b.** -1 **15.** one-to-one **17.** not one-to-one **19.** one-to-one **21.** not one-to-one **23.** $f^{-1}(x) = x - 4$ **25.** $f^{-1}(x) = \frac{x + 3}{2}$ **27.** $f^{-1}(x) = 2x + 2$ **29.** $f^{-1}(x) = \sqrt[3]{x}$ **31.** $f^{-1}(x) = \frac{x - 2}{5}$ **33.** $f^{-1}(x) = 5x + 2$

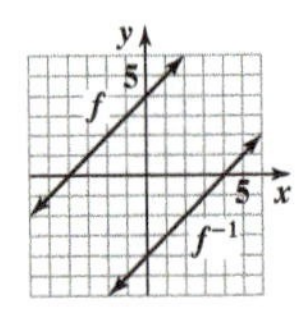

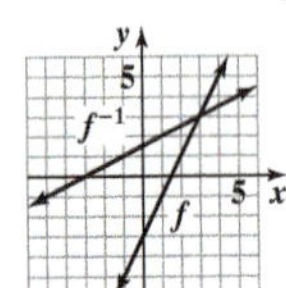

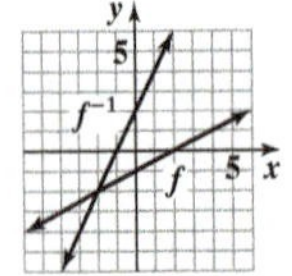

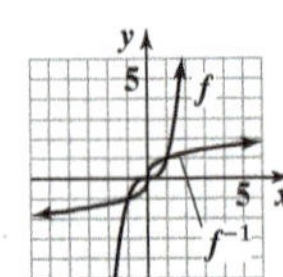

35. $f^{-1}(x) = x^3$ **37.** $f^{-1}(x) = \frac{5 - x}{3x}$ **39.** $f^{-1}(x) = \sqrt[3]{x} - 2$

41. 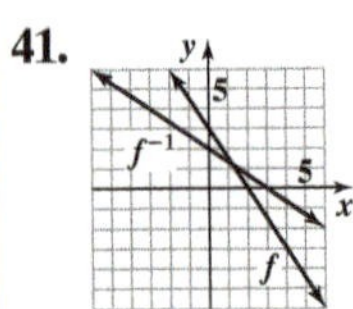 **43.** 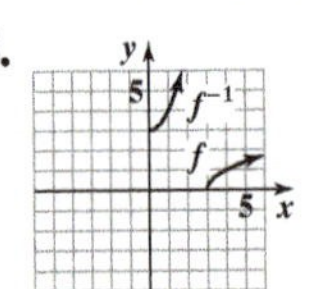 **45.** 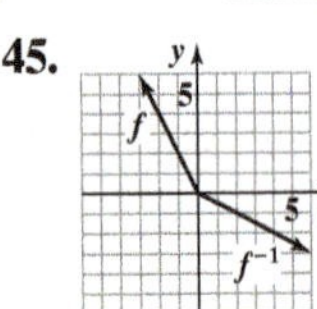

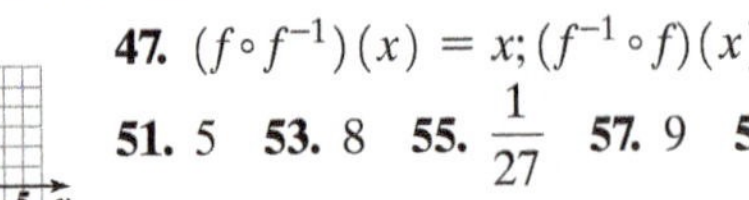

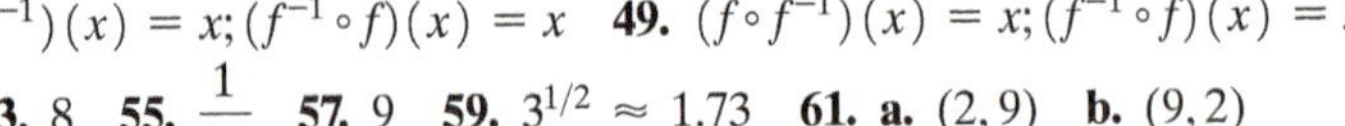

47. $(f \circ f^{-1})(x) = x$; $(f^{-1} \circ f)(x) = x$ **49.** $(f \circ f^{-1})(x) = x$; $(f^{-1} \circ f)(x) = x$ **51.** 5 **53.** 8 **55.** $\frac{1}{27}$ **57.** 9 **59.** $3^{1/2} \approx 1.73$ **61. a.** (2, 9) **b.** (9, 2)

63. a. $\left(-2, \frac{1}{4}\right), \left(-1, \frac{1}{2}\right), (0, 1), (1, 2), (2, 5)$ **b.** $\left(\frac{1}{4}, -2\right), \left(\frac{1}{2}, -1\right), (1, 0), (2, 1), (5, 2)$ **c.**

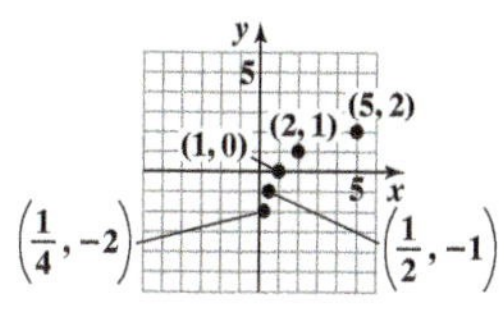

d.

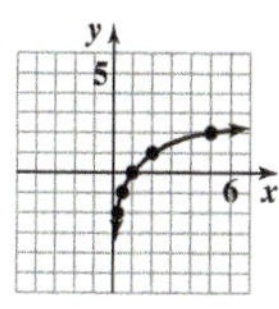

65. answers may vary **67.** $f^{-1}(x) = \frac{x-1}{3}$;

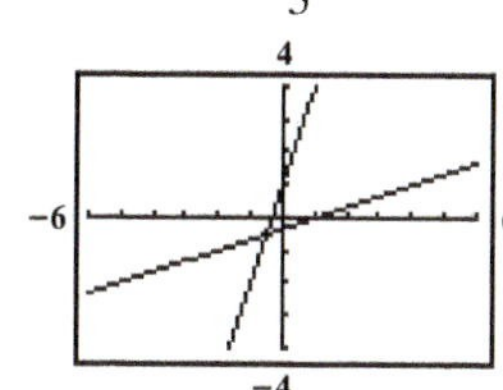

69. $f^{-1}(x) = x^3 - 1$;

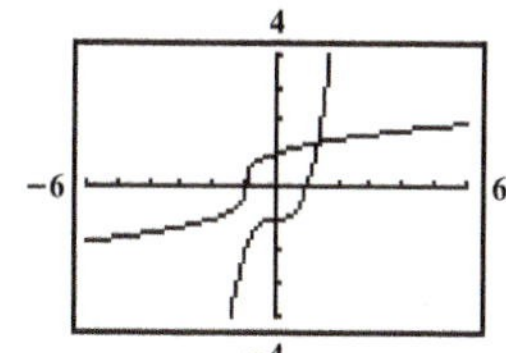

Section 19.3

Graphing Calculator Explorations

1. 81.98%;

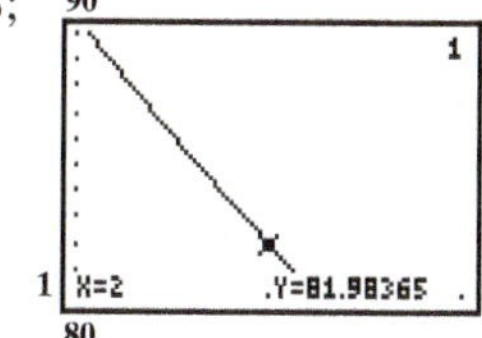

3. 22.54%

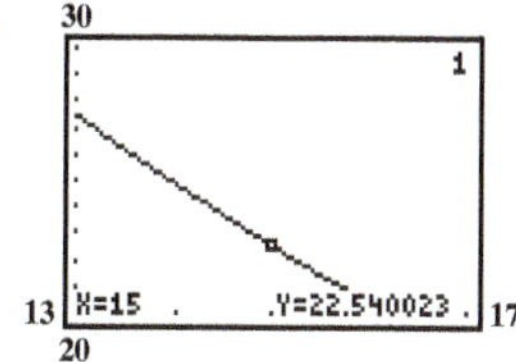

Vocabulary, Readiness & Video Check **1.** exponential **3.** yes **5.** no; none **7.** $(-\infty, \infty)$ **9.** In a polynomial function, the base is the variable and the exponent is the constant; in an exponential function, the base is the constant and the exponent is the variable. **11.** $y = 30(0.996)^{101} \approx 20.0$ lb

Exercise Set 19.3

1.

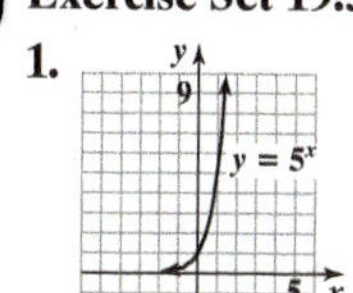

3.

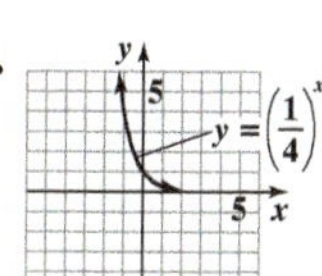

5. **7.** **9.** **11.**

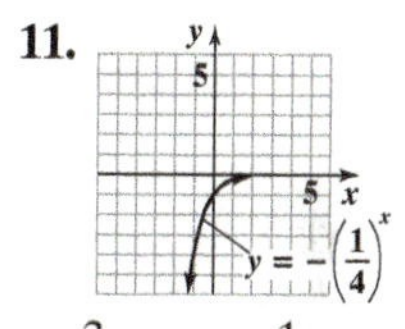

13. **15.**

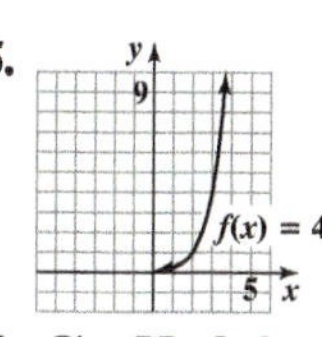

17. C **19.** B **21.** 3 **23.** $\frac{3}{4}$ **25.** $\frac{8}{5}$ **27.** $-\frac{2}{3}$ **29.** 4 **31.** $\frac{3}{2}$ **33.** $-\frac{1}{3}$ **35.** −2 **37.** 24.6 lb **39. a.** 9.7 billion lb **b.** 12.3 billion lb **41. a.** 192,927 students **b.** 463,772 students **43.** 378.0 million **45.** 544,000 mosquitoes **47.** \$7621.42 **49.** \$4065.59 **51.** 4 **53.** ∅ **55.** 2, 3 **57.** 3 **59.** −1 **61.** no **63.** no **65.** C **67.** D **69.** answers may vary

71.

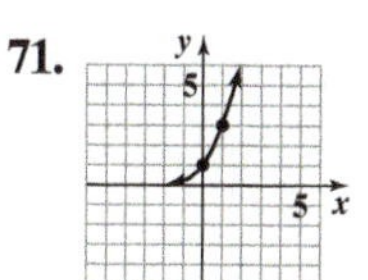

73.

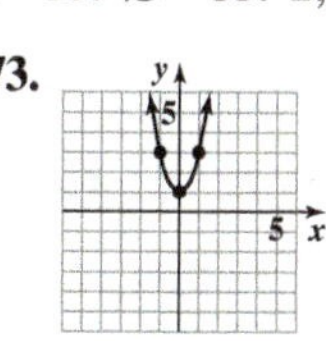

75. The graphs are the same since $\left(\frac{1}{2}\right)^{-x} = 2^x$. **77.** 24.55 lb **79.** 20.09 lb

Section 19.4

Vocabulary, Readiness & Video Check **1.** For Example 1, the growth rate is given as 5% per year. Since this is "per year," the number of time intervals is the "number of years," or 8. **3.** time intervals = years/half-life; the decay rate is 50% or $\frac{1}{2}$ because half-life is the amount of time it takes half of a substance to decay

Exercise Set 19.4 **1.** 451 **3.** 144,302 **5.** 21,231 **7.** 202 **9.** 1470 **11.** 13 **13.** 712,880 **15.** 383 **17.** 333 bison **19.** 1 g **21. a.** $\frac{14}{7} = 2$; 10; yes **b.** $\frac{11}{7} \approx 1.6$; 13.2; yes **23.** $\frac{500}{152} \approx 3.3$; 2.1; yes **25.** 4.9 g **27.** 3 **29.** −1 **31.** no; answers may vary

Section 19.5

Vocabulary, Readiness & Video Check **1.** logarithmic **3.** yes **5.** no; none **7.** $(-\infty, \infty)$ **9.** First write the equation as an equivalent exponential equation. Then solve.

Exercise Set 19.5 **1.** $6^2 = 36$ **3.** $3^{-3} = \frac{1}{27}$ **5.** $10^3 = 1000$ **7.** $9^4 = x$ **9.** $\pi^{-2} = \frac{1}{\pi^2}$ **11.** $7^{1/2} = \sqrt{7}$ **13.** $0.7^3 = 0.343$ **15.** $3^{-4} = \frac{1}{81}$ **17.** $\log_2 16 = 4$ **19.** $\log_{10} 100 = 2$ **21.** $\log_\pi x = 3$ **23.** $\log_{10}\frac{1}{10} = -1$ **25.** $\log_4\frac{1}{16} = -2$ **27.** $\log_5\sqrt{5} = \frac{1}{2}$ **29.** 3 **31.** -2 **33.** $\frac{1}{2}$ **35.** -1 **37.** 0 **39.** 2 **41.** 4 **43.** -3 **45.** 2 **47.** 81 **49.** 7 **51.** -3 **53.** -3 **55.** 2 **57.** 2 **59.** $\frac{27}{64}$ **61.** 10

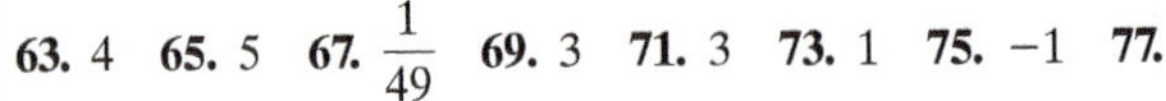

63. 4 **65.** 5 **67.** $\frac{1}{49}$ **69.** 3 **71.** 3 **73.** 1 **75.** -1 **77.**

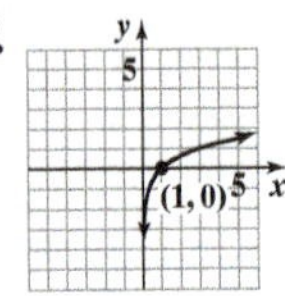

79.

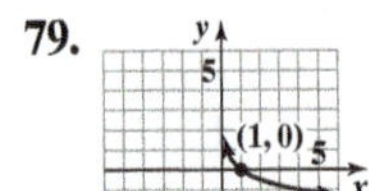

81.

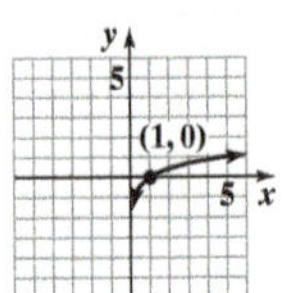

83.

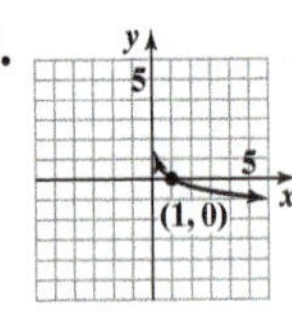

85. 1 **87.** $\frac{x-4}{2}$ **89.** $\frac{2x+3}{x^2}$ **91.** 3 **93. a.** $g(2) = 25$ **b.** $(25, 2)$ **c.** $f(25) = 2$ **95.** answers may vary **97.** $\frac{9}{5}$ **99.** 1 **101.**

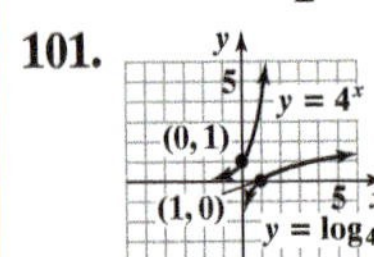

103.

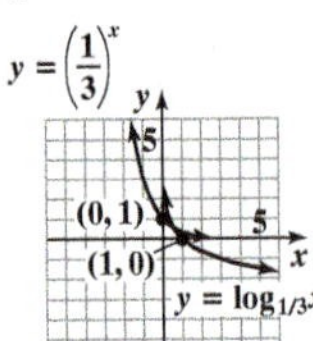

105. answers may vary **107.** 0.0827

Section 19.6

Vocabulary, Readiness & Video Check **1.** 36 **3.** $\log_b 2^7$ **5.** x **7.** No, the product property says the logarithm of a product can be written as a sum of logarithms—the expression in Example 2 is a logarithm of a sum. **9.** Since $\frac{1}{x} = x^{-1}$, this gives us $\log_2 x^{-1}$. Using the power property, we get $-1\log_2 x$ or $-\log_2 x$.

Exercise Set 19.6 **1.** $\log_5 14$ **3.** $\log_4 9x$ **5.** $\log_6(x^2 + x)$ **7.** $\log_{10}(10x^2 + 20)$ **9.** $\log_5 3$ **11.** $\log_3 4$ **13.** $\log_2\frac{x}{y}$ **15.** $\log_2\frac{x^2+6}{x^2+1}$ **17.** $2\log_3 x$ **19.** $-1\log_4 5 = -\log_4 5$ **21.** $\frac{1}{2}\log_5 y$ **23.** $\log_2 5x^3$ **25.** $\log_4 48$ **27.** $\log_5 x^3z^6$ **29.** $\log_4 4$, or 1 **31.** $\log_7\frac{9}{2}$ **33.** $\log_{10}\frac{x^3-2x}{x+1}$ **35.** $\log_2\frac{x^{7/2}}{(x+1)^2}$ **37.** $\log_8 x^{16/3}$ **39.** $\log_3 4 + \log_3 y - \log_3 5$ **41.** $\log_4 5 - \log_4 9 - \log_4 z$ **43.** $3\log_2 x - \log_2 y$ **45.** $\frac{1}{2}\log_b 7 + \frac{1}{2}\log_b x$ **47.** $4\log_6 x + 5\log_6 y$ **49.** $3\log_5 x + \log_5(x+1)$ **51.** $2\log_6 x - \log_6(x+3)$ **53.** 1.2 **55.** 0.2 **57.** 0.35 **59.** 1.29 **61.** -0.68 **63.** -0.125 **65.–66.**

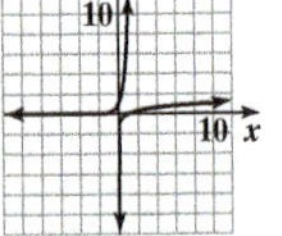

67. 2 **69.** 2 **71.** b **73.** true **75.** false **77.** false **79.** because $\log_b 1 = 0$

Integrated Review **1.** $x^2 + x - 5$ **2.** $-x^2 + x - 7$ **3.** $x^3 - 6x^2 + x - 6$ **4.** $\frac{x-6}{x^2+1}$ **5.** $\sqrt{3x-1}$ **6.** $3\sqrt{x} - 1$ **7.** one-to-one; $\{(6,-2), (8,4), (-6,2), (3,3)\}$ **8.** not one-to-one **9.** not one-to-one **10.** one-to-one **11.** not one-to-one

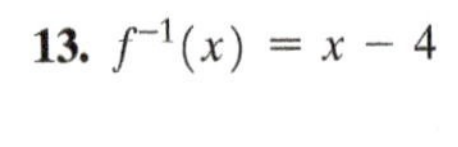

12. $f^{-1}(x) = \frac{x}{3}$ **13.** $f^{-1}(x) = x - 4$ **14.** $f^{-1}(x) = \frac{x+1}{5}$ **15.** $f^{-1}(x) = \frac{x-2}{3}$ **16.**

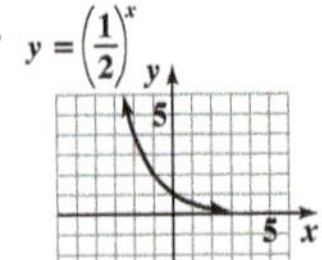

17.

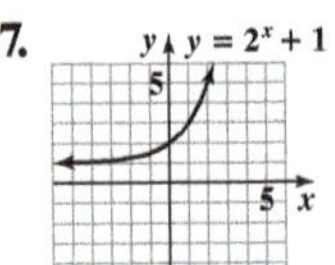

18. **19.**

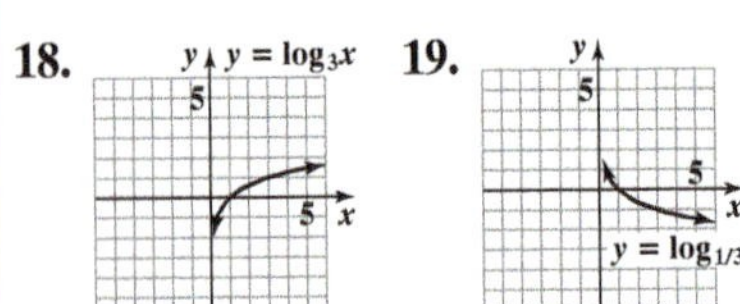

20. 3 **21.** 7 **22.** -8 **23.** 3 **24.** 2 **25.** $\frac{1}{2}$ **26.** 32 **27.** 4 **28.** 5 **29.** $\frac{1}{9}$ **30.** $\log_2 x^5$ **31.** $\log_2 5^x$ **32.** $\log_5\frac{x^3}{y^5}$ **33.** $\log_5 x^9y^3$ **34.** $\log_2\frac{x^2-3x}{x^2+4}$ **35.** $\log_3\frac{y^4+11y}{y+2}$ **36.** $\log_7 9 + 2\log_7 x - \log_7 y$ **37.** $\log_6 5 + \log_6 y - 2\log_6 z$ **38.** 254,000 mosquitoes

Section 19.7

Vocabulary, Readiness & Video Check **1.** 10 **3.** 7 **5.** 5 **7.** $\frac{\log 7}{\log 2}$ or $\frac{\ln 7}{\ln 2}$ **9.** The understood base of a common logarithm is 10. If you're finding the common logarithm of a known power of 10, then the common logarithm is the known power of 10. **11.** $\log_b b^x = x$

Exercise Set 19.7 **1.** 0.9031 **3.** 0.3636 **5.** 0.6931 **7.** −2.6367 **9.** 1.1004 **11.** 1.6094 **13.** 1.6180 **15.** 2 **17.** −3 **19.** 2 **21.** $\frac{1}{4}$ **23.** 3 **25.** −7 **27.** −4 **29.** $\frac{1}{2}$ **31.** $\frac{e^7}{2} \approx 548.3166$ **33.** $10^{1.3} \approx 19.9526$ **35.** $\frac{10^{1.1}}{2} \approx 6.2946$ **37.** $e^{1.4} \approx 4.0552$ **39.** $\frac{4 + e^{2.3}}{3} \approx 4.6581$ **41.** $10^{2.3} \approx 199.5262$ **43.** $e^{-2.3} \approx 0.1003$ **45.** $\frac{10^{-0.5} - 1}{2} \approx -0.3419$ **47.** $\frac{e^{0.18}}{4} \approx 0.2993$ **49.** 1.5850 **51.** −2.3219 **53.** 1.5850 **55.** −1.6309 **57.** 0.8617 **59.** 4.2 **61.** 5.3 **63.** \$3656.38 **65.** \$2542.50 **67.** $\frac{4}{7}$ **69.** $x = \frac{3y}{4}$ **71.** −6, −1 **73.** (2, −3) **75.** answers may vary **77.** ln 50; answers may vary **79.**

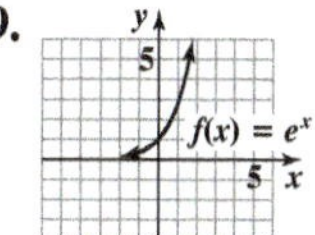

81.

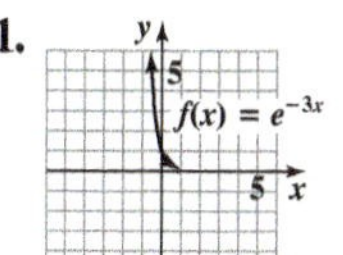

83.

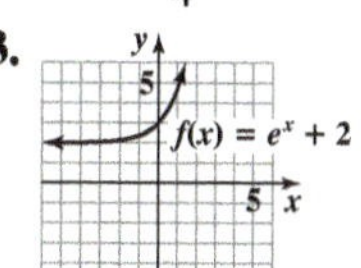

85.

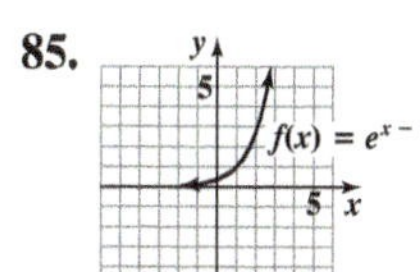

87.

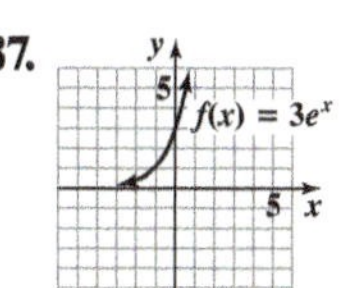

89.

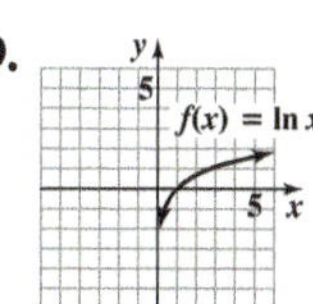

91.

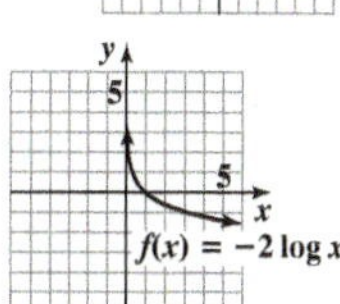

93.

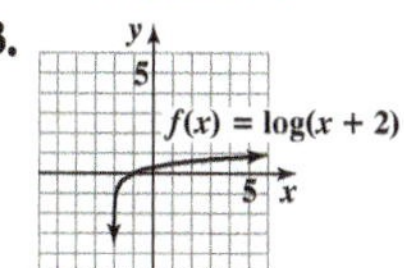

95.

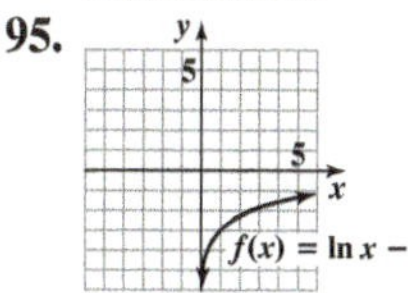

97. answers may vary;

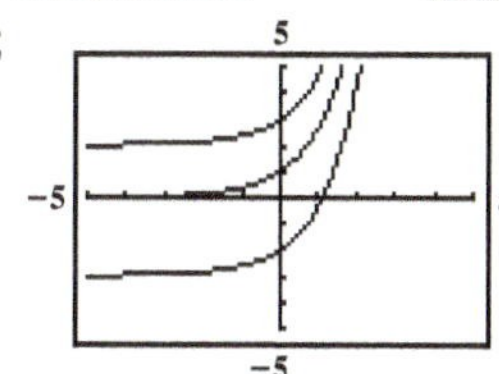

Section 19.8

Graphing Calculator Explorations **1.** 3.67 years, or 3 years and 8 months **3.** 23.16 years, or 23 years and 2 months

Vocabulary, Readiness & Video Check **1.** $\ln(4x - 2) = \ln 3$ is the same as $\log_e(4x - 2) = \log_e 3$. Therefore, from the logarithm property of equality, we know that $4x - 2 = 3$. **3.** $2000 = 1000\left(1 + \frac{0.07}{12}\right)^{12 \cdot t} \approx 9.9$ yr; as long as the interest rate and compounding are the same, it takes any amount of money the same time to double.

Exercise Set 19.8 **1.** $\frac{\log 6}{\log 3}$; 1.6309 **3.** $\frac{\log 3.8}{2 \log 3}$; 0.6076 **5.** $\frac{3 \log 2 + \log 5}{\log 2}$ or $3 + \frac{\log 5}{\log 2}$; 5.3219 **7.** $\frac{\log 5}{\log 9}$; 0.7325 **9.** $\frac{\log 3 - 7 \log 4}{\log 4}$ or $\frac{\log 3}{\log 4} - 7$; −6.2075 **11.** 11 **13.** $\frac{1}{2}$ **15.** $\frac{3}{4}$ **17.** −2, 3 **19.** 2 **21.** $\frac{1}{8}$ **23.** $\frac{4 \log 7 + \log 11}{3 \log 7}$ or $\frac{1}{3}\left(4 + \frac{\log 11}{\log 7}\right)$; 1.7441 **25.** 4, −1 **27.** $\frac{\ln 5}{6}$; 0.2682 **29.** 9, −9 **31.** $\frac{1}{5}$ **33.** 100 **35.** $\frac{192}{127}$ **37.** $\frac{2}{3}$ **39.** $\frac{-5 + \sqrt{33}}{2}$ **41.** 103 wolves **43.** 7582 inhabitants **45.** 9.9 yr **47.** 1.7 yr **49.** 8.8 yr **51.** 24.5 lb **53.** 55.7 in. **55.** 11.9 lb/sq in. **57.** 3.2 mi **59.** 12 weeks **61.** 18 weeks **63.** $-\frac{5}{3}$ **65.** $\frac{17}{4}$ **67.** $f^{-1}(x) = \frac{x - 2}{5}$ **69.** 15 yr **71.** answers may vary **73.** 6.93; **75.** −3.68; 0.19

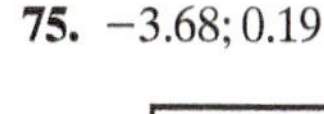

77. 1.74

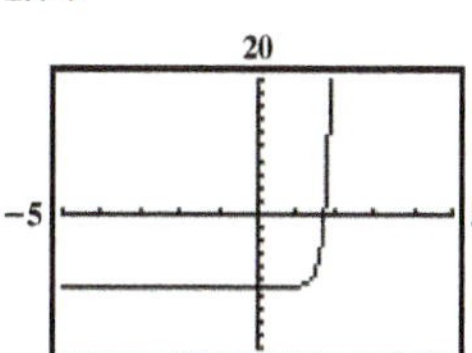

79. 0.2

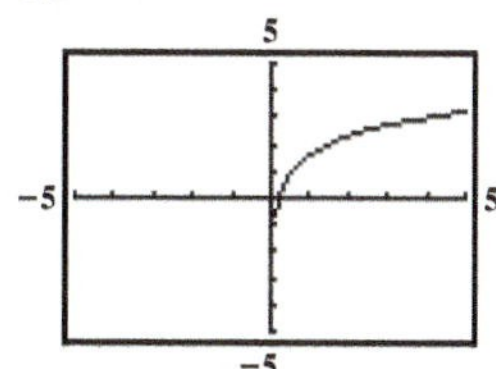

Chapter 19 Vocabulary Check **1.** inverse **2.** composition **3.** exponential **4.** symmetric **5.** Natural **6.** Common **7.** vertical; horizontal **8.** logarithmic **9.** Half-life **10.** exponential

Chapter 19 Review **1.** $3x - 4$ **3.** $2x^2 - 9x - 5$ **5.** $x^2 + 2x - 1$ **7.** 18 **9.** −2 **11.** one-to-one; $h^{-1} = \{(14, -9), (8, 6), (12, -11), (15, 15)\}$ **13.** one-to-one;

Rank in Housing Starts for 2009 (Input)	4	3	1	2
U.S. Region (Output)	Northeast	Midwest	South	West

15. a. 3 **b.** 7 **17.** not one-to-one **19.** not one-to-one **21.** $f^{-1}(x) = x + 9$ **23.** $f^{-1}(x) = \frac{x - 11}{6}$ **25.** $f^{-1}(x) = \sqrt[3]{x + 5}$

Answers to Selected Exercises

27. $g^{-1}(x) = \dfrac{6x + 7}{12}$ **29.**

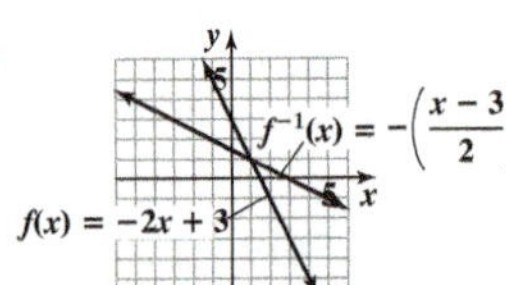

31. 3 **33.** $-\dfrac{4}{3}$ **35.** $\dfrac{3}{2}$ **37.**

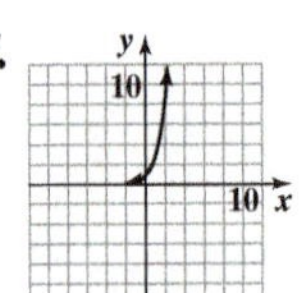

39.

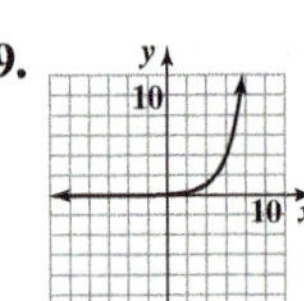

41. \$2963.11 **43.** 187,118 **45.** 8 players **47.** $\log_7 49 = 2$ **49.** $\left(\dfrac{1}{2}\right)^{-4} = 16$ **51.** $\dfrac{1}{64}$ **53.** 0 **55.** 5 **57.** 4 **59.** $\dfrac{17}{3}$ **61.** $-1, 4$

63.

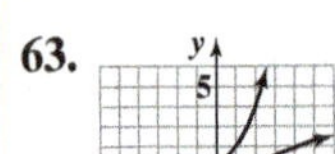

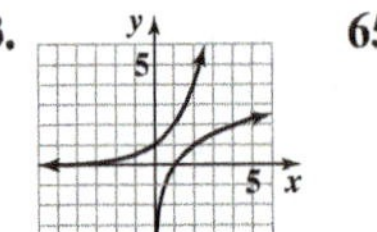

65. $\log_3 32$ **67.** $\log_7 \dfrac{3}{4}$ **69.** $\log_{11} 4$ **71.** $\log_5 \dfrac{x^3}{(x+1)^2}$ **73.** $3 \log_3 x - \log_3 (x + 2)$

75. $\log_2 3 + 2 \log_2 x + \log_2 y - \log_2 z$ **77.** 2.02 **79.** 0.5563 **81.** 0.2231 **83.** 3 **85.** -1 **87.** $\dfrac{e^2}{2}$ **89.** $\dfrac{e^{-1} + 3}{2}$ **91.** 1.67 mm

93. 0.2920 **95.** \$1684.66 **97.** $\dfrac{\log 7}{2 \log 3}$; 0.8856 **99.** $\dfrac{\log 6 - \log 3}{2 \log 3}$ or $\dfrac{1}{2}\left(\dfrac{\log 6}{\log 3} - 1\right)$; 0.3155

101. $\dfrac{\log 4 + 5 \log 5}{3 \log 5}$ or $\dfrac{1}{3}\left(\dfrac{\log 4}{\log 5} + 5\right)$; 1.9538 **103.** $\dfrac{\log \frac{1}{2} + \log 5}{\log 5}$ or $-\dfrac{\log 2}{\log 5} + 1$; 0.5693 **105.** $\dfrac{25}{2}$ **107.** $\varnothing$ **109.** $2\sqrt{2}$ **111.** 22.4 yr

113. 99.0 yr **115.** 8.8 yr **117.** 0.69;

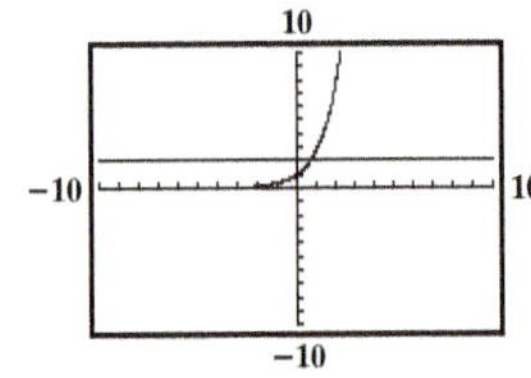

119. -4 **121.** $\dfrac{11}{9}$ **123.** 1 **125.** $-1, 5$ **127.** $e^{-3.2}$ **129.** $2e$

Chapter 19 Test **1.** $2x^2 - 3x$ **2.** $3 - x$ **3.** 5 **4.** $x - 7$ **5.** $x^2 - 6x - 2$ **6.**

7. one-to-one **8.** not one-to-one

9. one-to-one; $f^{-1}(x) = \dfrac{-x + 6}{2}$ **10.** one-to-one; $f^{-1} = \{(0, 0), (3, 2), (5, -1)\}$ **11.** not one-to-one **12.** $\log_3 24$ **13.** $\log_5 \dfrac{x^4}{x + 1}$

14. $\log_6 2 + \log_6 x - 3 \log_6 y$ **15.** -1.53 **16.** 1.0686 **17.** -1 **18.** $\dfrac{1}{2}\left(\dfrac{\log 4}{\log 3} - 5\right)$; -1.8691 **19.** $\dfrac{1}{9}$ **20.** $\dfrac{1}{2}$ **21.** 22 **22.** $\dfrac{25}{3}$ **23.** $\dfrac{43}{21}$

24. -1.0979 **25.**

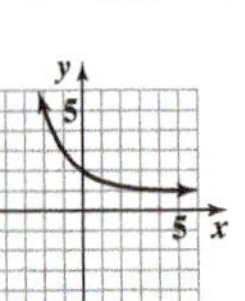

26.

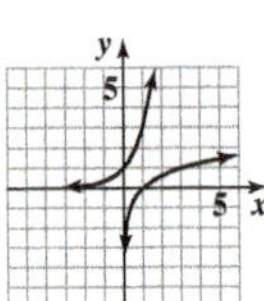

27. \$5234.58 **28.** 6 yr **29.** 6.5%; \$230 **30.** 100,141
31. 64,805 prairie dogs **32.** 15 yr **33.** 85% **34.** 52%

Cumulative Review Chapters 1–19 **1.** $\dfrac{5}{12}$; Sec. 4.3, Ex. 11 **3.**

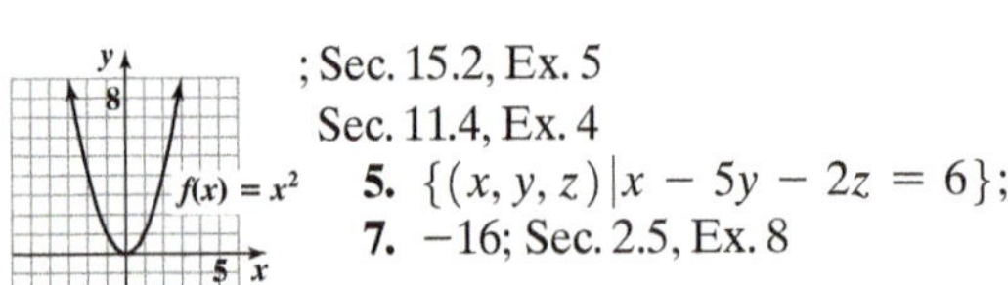

; Sec. 15.2, Ex. 5
Sec. 11.4, Ex. 4
5. $\{(x, y, z) | x - 5y - 2z = 6\}$;
7. -16; Sec. 2.5, Ex. 8

9. a. \$102.60 **b.** \$12.60; Sec. 14.1, Ex. 7 **11.** -11; Sec. 3.2, Ex. 8 **13.** $2xy - 4 + \dfrac{1}{2y}$; Sec. 12.6, Ex. 3 **15.** $3(m + 2)(m - 10)$; Sec. 13.2, Ex. 9 **17.** $3x - 5$; Sec. 14.3, Ex. 3 **19. a.** 3 **b.** -3 **c.** -5 **d.** not a real number **e.** $4x$; Sec. 17.1, Ex. 4 **21. a.** $\sqrt[4]{x^3}$ **b.** $\sqrt[6]{x}$ **c.** $\sqrt[6]{72}$; Sec. 17.2, Ex. 8 **23. a.** $5\sqrt{3} + 3\sqrt{10}$ **b.** $\sqrt{35} + \sqrt{5} - \sqrt{42} - \sqrt{6}$ **c.** $21x - 7\sqrt{5x} + 15\sqrt{x} - 5\sqrt{5}$ **d.** $49 - 8\sqrt{3}$ **e.** $2x - 25$ **f.** $x + 22 + 10\sqrt{x - 3}$; Sec. 17.4, Ex. 4 **25.** $\dfrac{\sqrt[4]{xy^3}}{3y^2}$; Sec. 17.5, Ex. 3 **27.** 3; Sec. 17.6, Ex. 4

29. $\dfrac{9 + i\sqrt{15}}{6}, \dfrac{9 - i\sqrt{15}}{6}$ or $\dfrac{3}{2} \pm \dfrac{\sqrt{15}}{6}i$; Sec. 18.1, Ex. 8 **31.** $\dfrac{-1 + \sqrt{33}}{4}, \dfrac{-1 - \sqrt{33}}{4}$; Sec. 18.3, Ex. 2 **33.** $[0, 4]$; Sec. 18.4, Ex. 2

35.

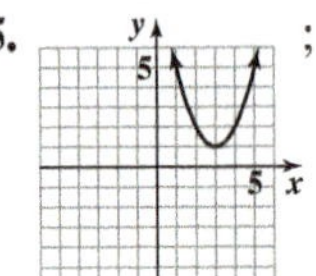

; Sec. 18.5, Ex. 5 **37.** -1; Sec. 3.3, Ex. 2 **39.** $f^{-1}(x) = x - 3$; Sec. 19.2, Ex. 4

41. a. 2 **b.** -1 **c.** $\dfrac{1}{2}$; Sec. 19.5, Ex. 3

Chapter 20 Conic Sections

Section 20.1

Graphing Calculator Explorations

1. 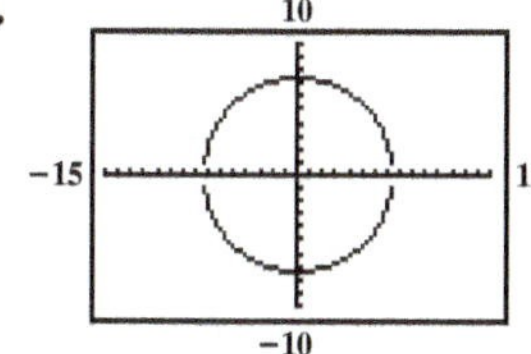
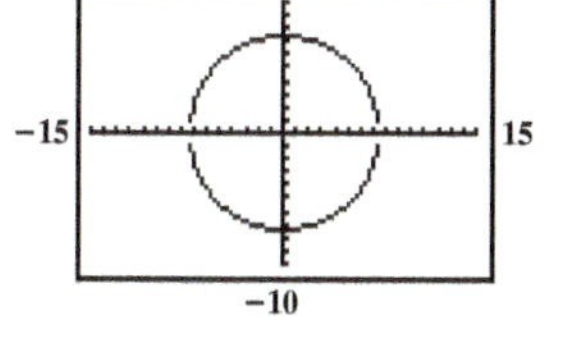

3. 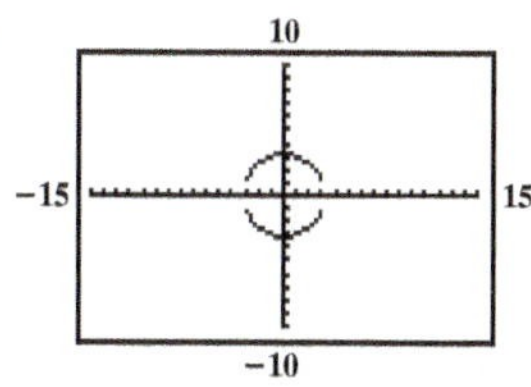
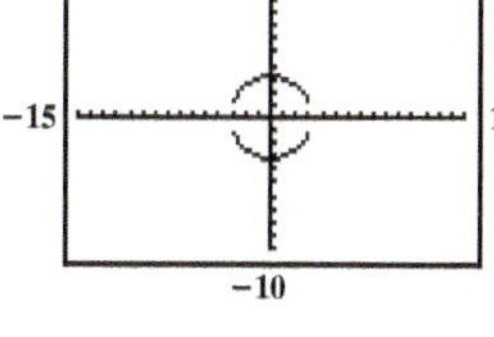

5. 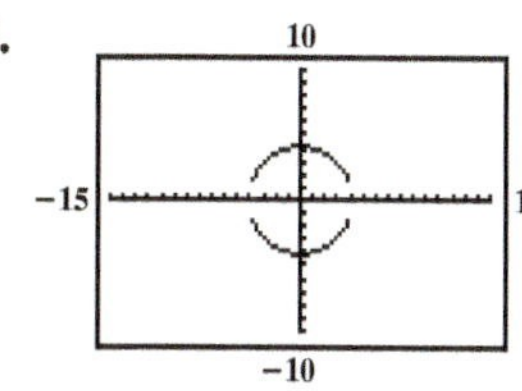
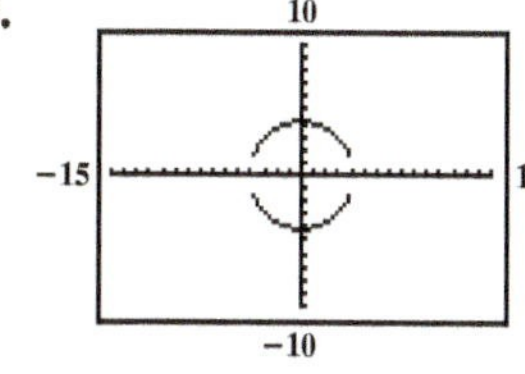

7. 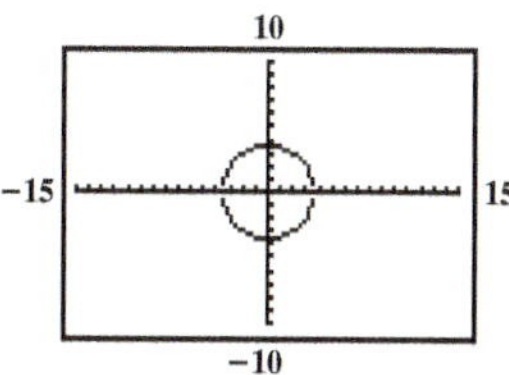
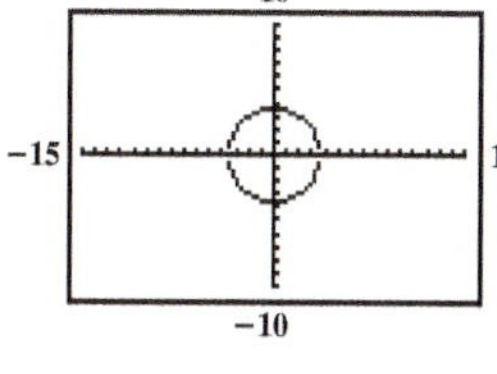

Vocabulary, Readiness & Video Check **1.** conic sections **3.** circle; center **5.** radius **7.** No, their graphs don't pass the vertical line test. **9.** The formula for the standard form of a circle identifies the center and radius, so you just need to substitute these values into this formula and simplify.

Exercise Set 20.1 **1.** upward **3.** to the left **5.** downward **7.** 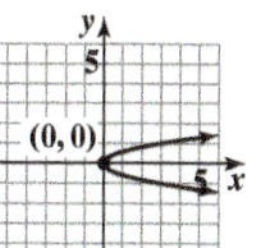

9. **11.** 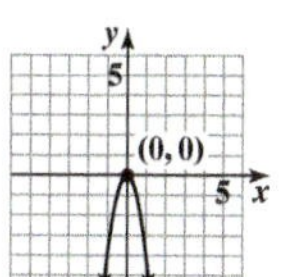

13. 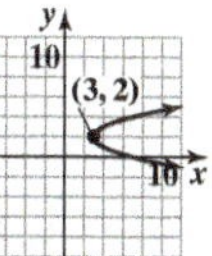

15. 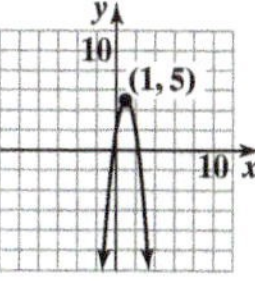

17. 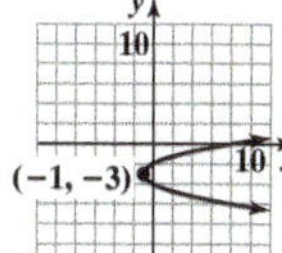

19. 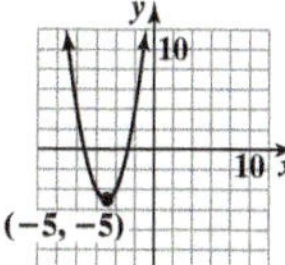

21. 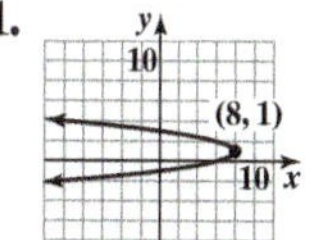

23. 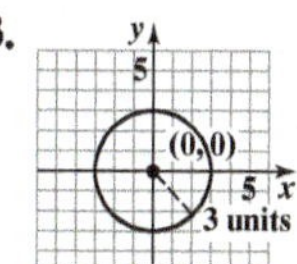

25. 

27. 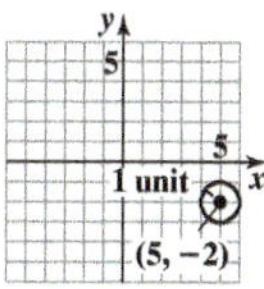

29. 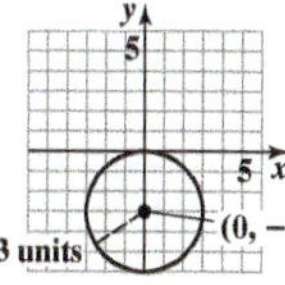

31. 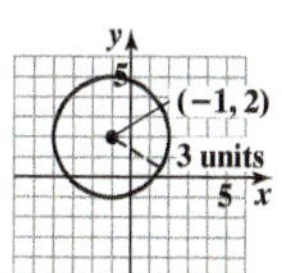

33. 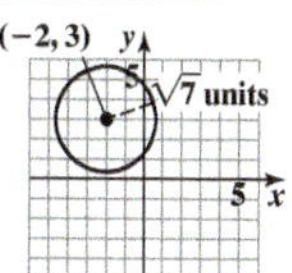

35. 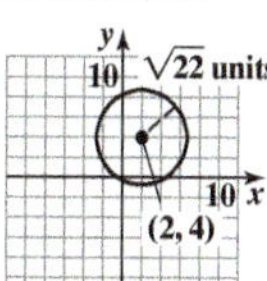

37. 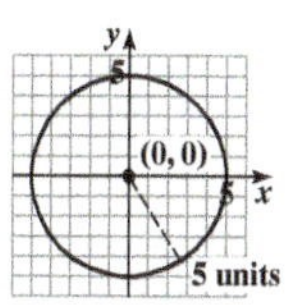

39. 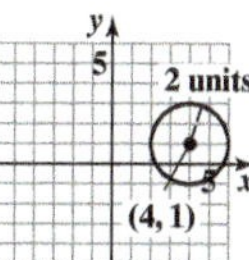

41. 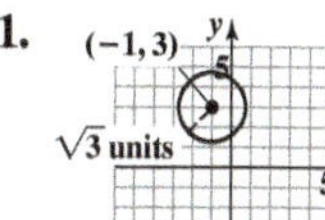

43. $(x-2)^2+(y-3)^2=36$ **45.** $x^2+y^2=3$ **47.** $(x+5)^2+(y-4)^2=45$

49. 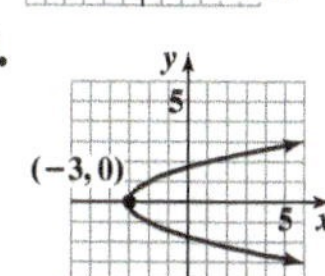

51. 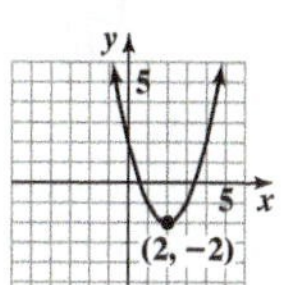

53. 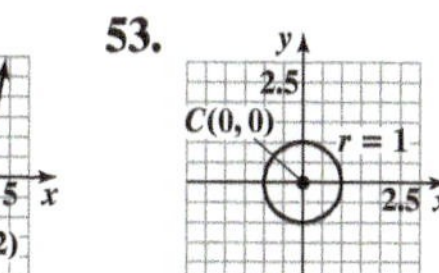

55. 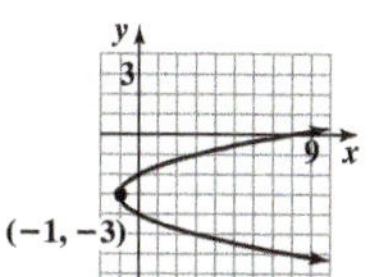

57. 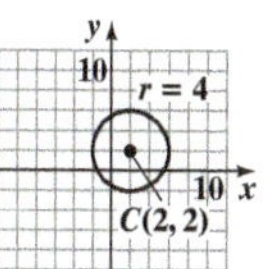

59. 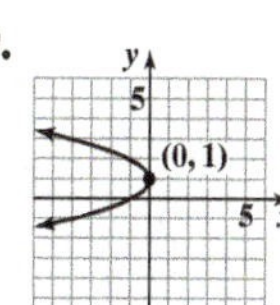

61. 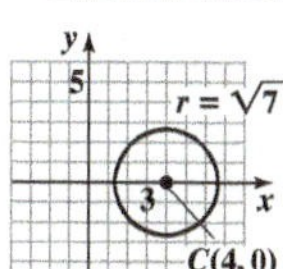

63. 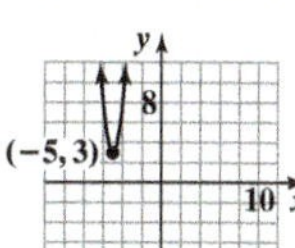

65. 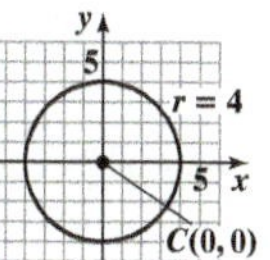

67. 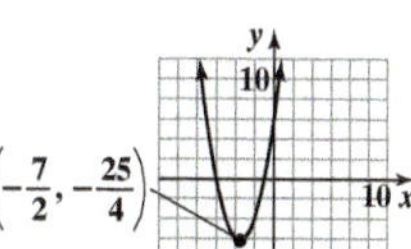

69. 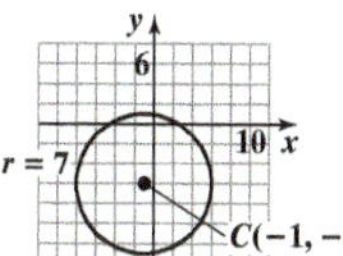

71. 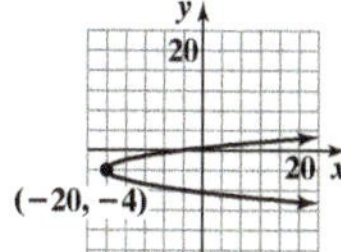

73. 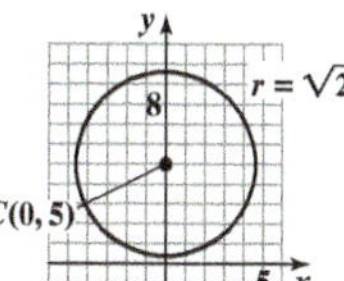
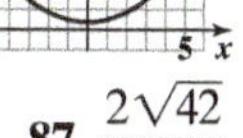

75. 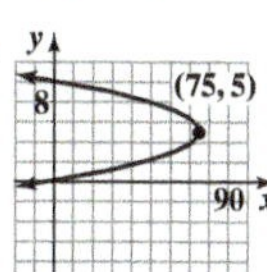

77. 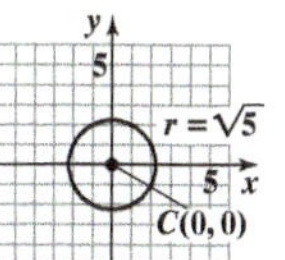

79. 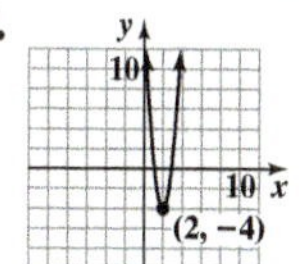

81. 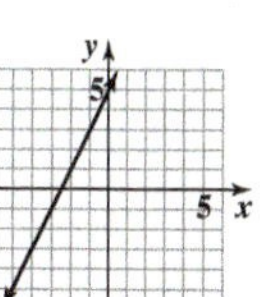
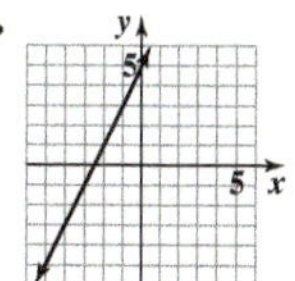
83. 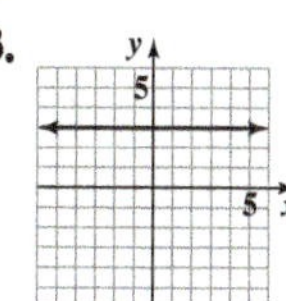
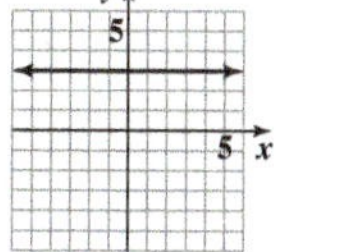

85. $\frac{\sqrt{3}}{3}$ **87.** $\frac{2\sqrt{42}}{3}$ **89.** The vertex is $(1, -5)$. **91. a.** 16.5 m **b.** 103.67 m **c.** 3.5 m **d.** $(0, 16.5)$ **e.** $x^2+(y-16.5)^2=16.5^2$ **93. a.** 125 ft **b.** 14 ft **c.** 139 ft **d.** $(0, 139)$ **e.** $x^2+(y-139)^2=125^2$ **95.** answers may vary **97.** 20 m

99. 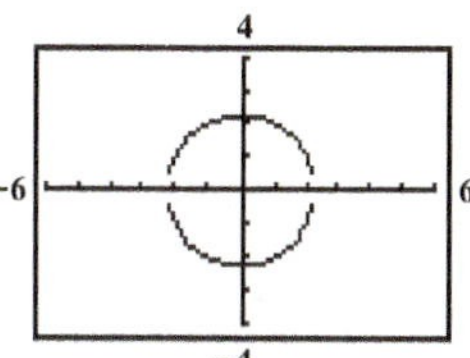

101. 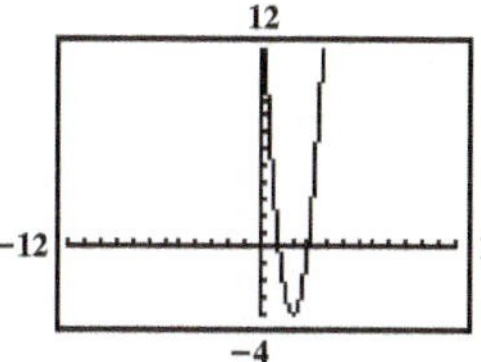

Section 20.2

Graphing Calculator Explorations

1.

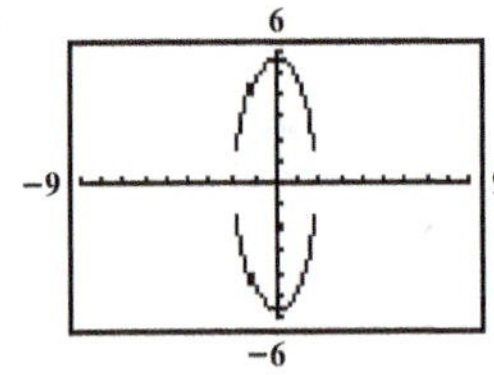

3.

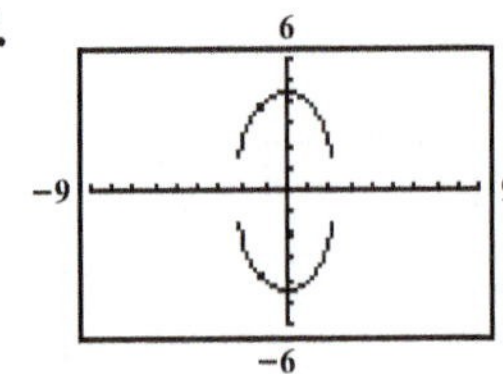

5.

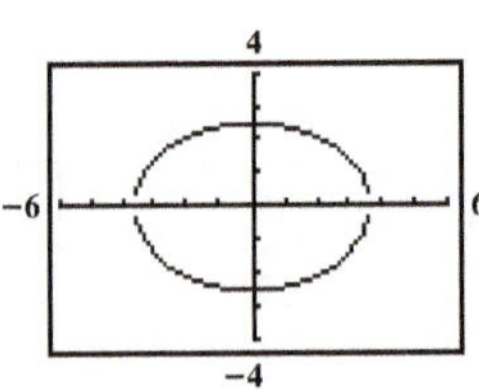

Vocabulary, Readiness & Video Check **1.** hyperbola **3.** focus **5.** hyperbola; $(0, 0)$; x; $(a, 0)$ and $(-a, 0)$ **7.** a and b give us the locations of 4 intercepts—$(a, 0)$, $(-a, 0)$, $(0, b)$, and $(0, -b)$ for $\frac{x^2}{a^2} + \frac{y^2}{b^2} = 1$ with center $(0, 0)$. For Example 2, the values of a and b also give us 4 points of the graph, just not intercepts. Here we move a distance of a units horizontally to the left and right of the center and b units above and below.

Exercise Set 20.2 **1.** ellipse **3.** hyperbola **5.** hyperbola **7.**

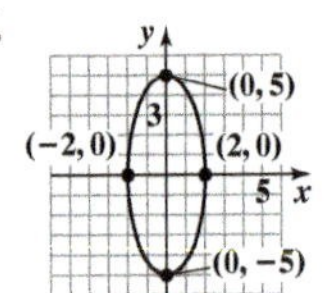

9.

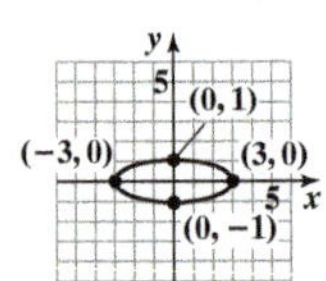

11.

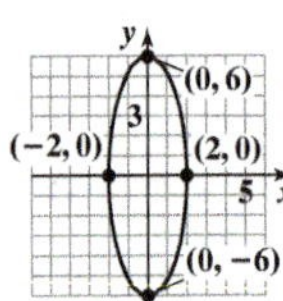

13.

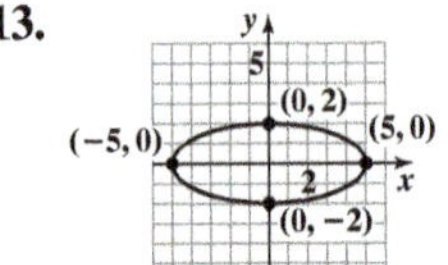

15.

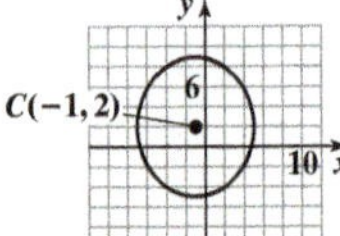

17.

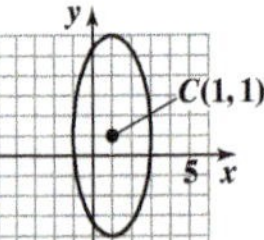

19.

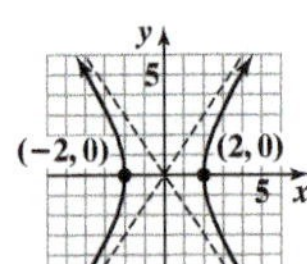

21.

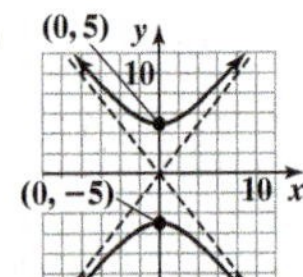

23.

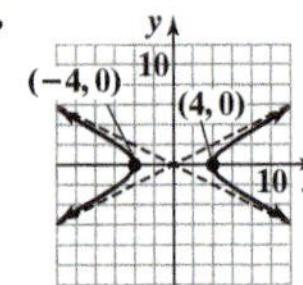

25.

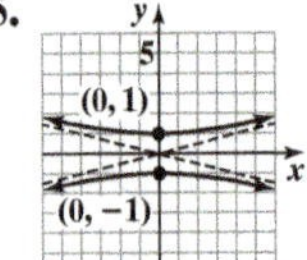

27.

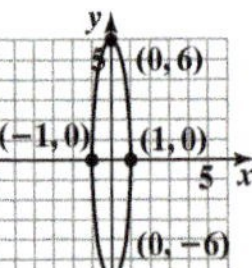

29.

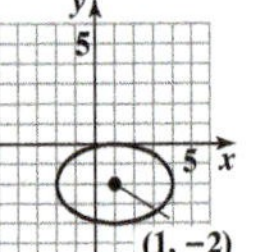

31.

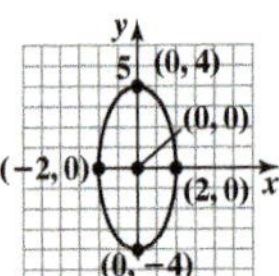

33.

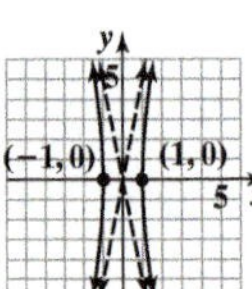

35. circle

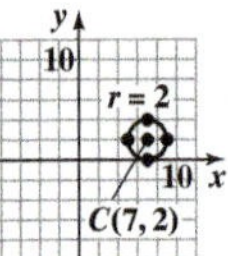

37. parabola

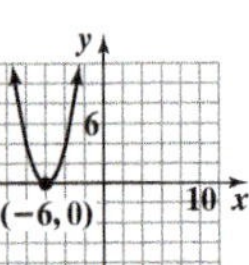

39. hyperbola

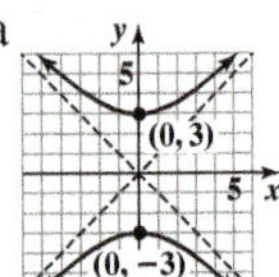

41. ellipse

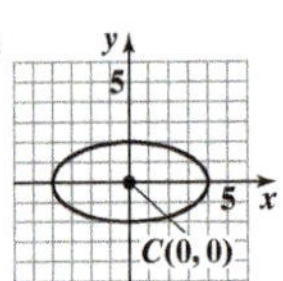

43. parabola

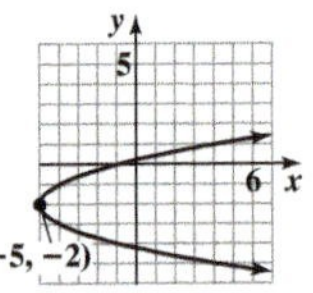

45. hyperbola

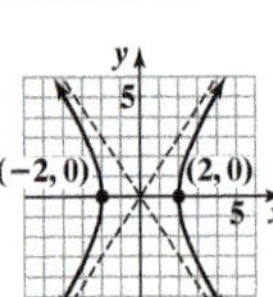

47. ellipse

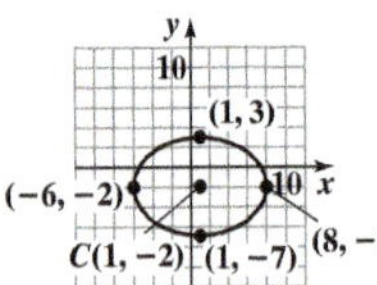

49. circle

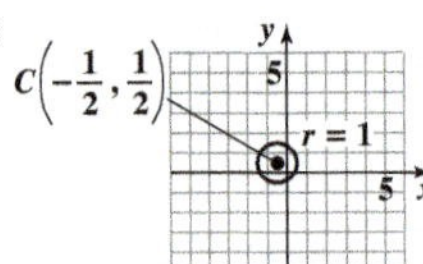

51. $-8x^5$ **53.** $-4x^2$ **55.** y-intercepts: 2 units **57.** y-intercepts: 4 units **59.** answers may vary **61.** ellipses: C, E, H; circles: B, F; hyperbolas: A, D, G

63. A: 49, 7; B: 0, 0; C: 9, 3; D: 64, 8; E: 64, 8; F: 0, 0; G: 81, 9; H: 4, 2 **65.** A: $\frac{7}{6}$; B: 0; C: $\frac{3}{5}$; D: $\frac{8}{5}$; E: $\frac{8}{9}$; F: 0; G: $\frac{9}{4}$; H: $\frac{1}{6}$ **67.** equal to zero

69. answers may vary **71.**

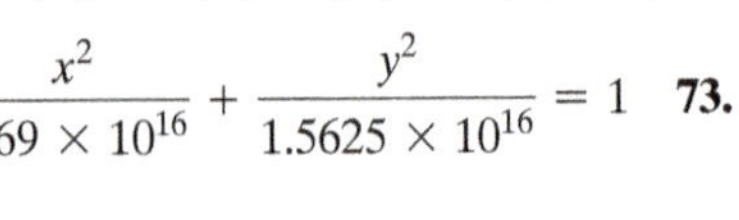

$$\frac{x^2}{1.69 \times 10^{16}} + \frac{y^2}{1.5625 \times 10^{16}} = 1$$

73.

4
−6 6
−4

75.

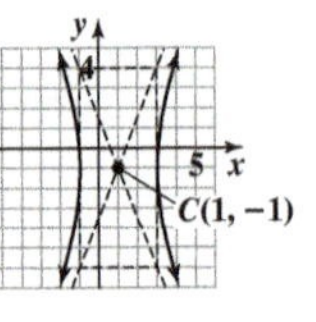

77.

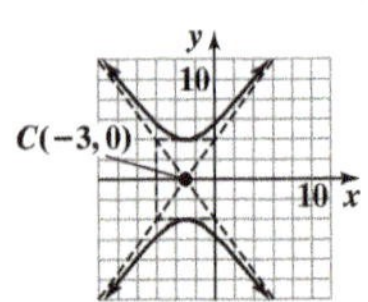

79.

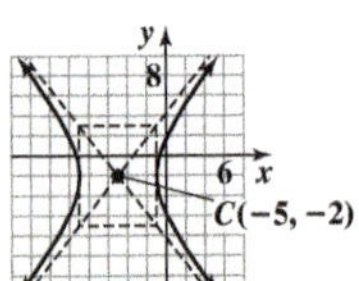

Integrated Review **1.** circle 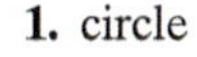 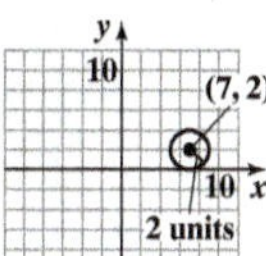

2. parabola 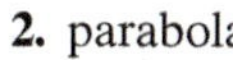 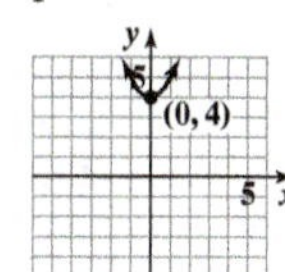

3. parabola 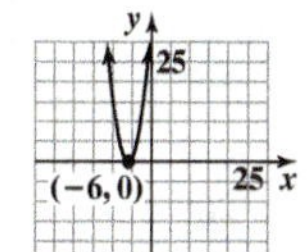

4. ellipse 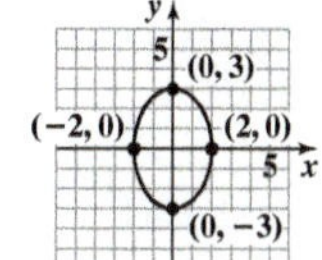

5. hyperbola 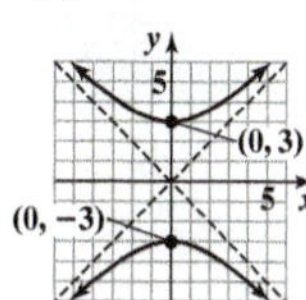

6. hyperbola 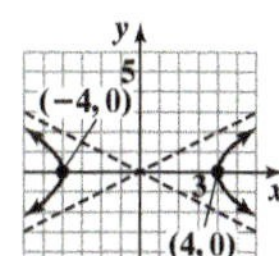

7. ellipse 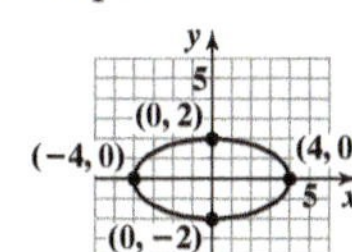

8. circle 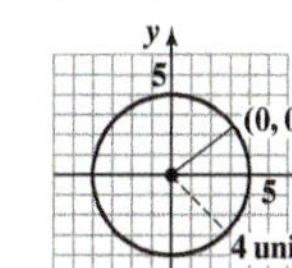

9. parabola 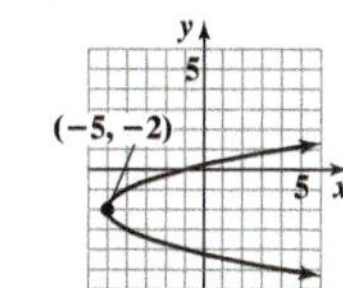

10. parabola 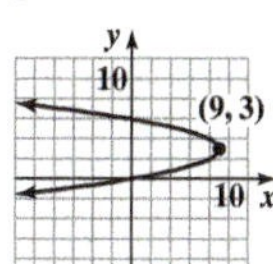

11. hyperbola 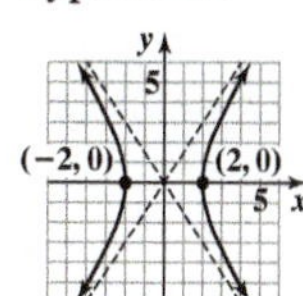

12. ellipse 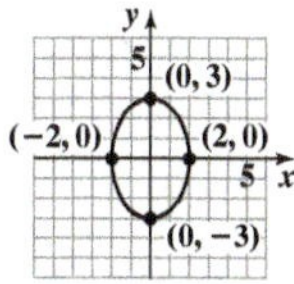

13. ellipse 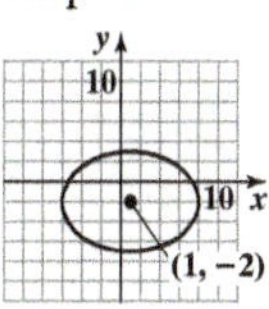

14. hyperbola 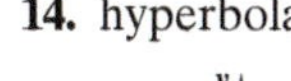 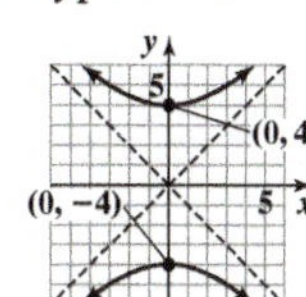

15. circle 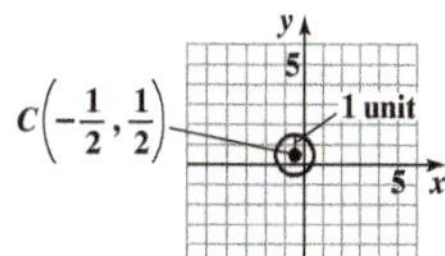

Section 20.3

Vocabulary, Readiness & Video Check **1.** Solving for y would either introduce tedious fractions (2nd equation) or a square root (1st equation) into the calculations.

Exercise Set 20.3 **1.** $(3, -4), (-3, 4)$ **3.** $(\sqrt{2}, \sqrt{2}), (-\sqrt{2}, -\sqrt{2})$ **5.** $(4, 0), (0, -2)$ **7.** $(-\sqrt{5}, -2), (-\sqrt{5}, 2), (\sqrt{5}, -2), (\sqrt{5}, 2)$ **9.** $\varnothing$ **11.** $(1, -2), (3, 6)$ **13.** $(2, 4), (-5, 25)$ **15.** $\varnothing$ **17.** $(1, -3)$ **19.** $(-1, -2), (-1, 2), (1, -2), (1, 2)$ **21.** $(0, -1)$ **23.** $(-1, 3), (1, 3)$ **25.** $(\sqrt{3}, 0), (-\sqrt{3}, 0)$ **27.** $\varnothing$ **29.** $(-6, 0), (6, 0), (0, -6)$ **31.** $(3, \sqrt{3})$

33. 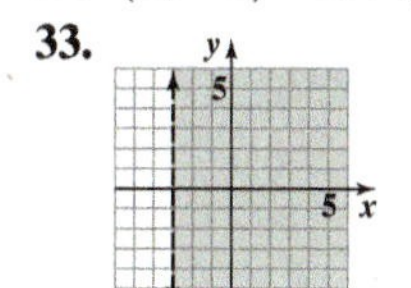 **35.** 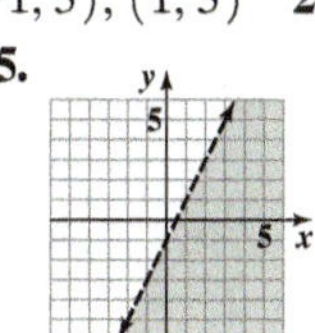 **37.** $(8x - 25)$ in. **39.** $(4x^2 + 6x + 2)$ m **41.** answers may vary **43.** 0, 1, 2, 3, or 4; answers may vary **45.** 9 and 7; 9 and −7; −9 and 7; −9 and −7 **47.** 15 cm by 19 cm **49.** 15 thousand compact discs; price: \$3.75

51. 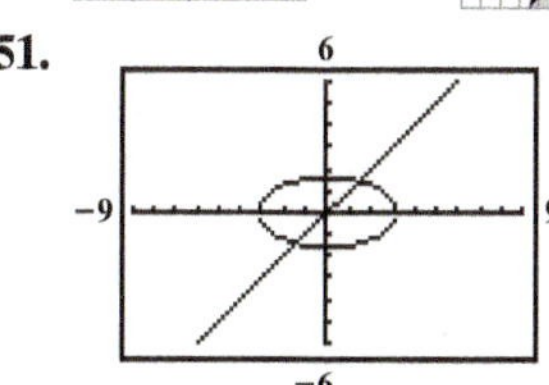

53. 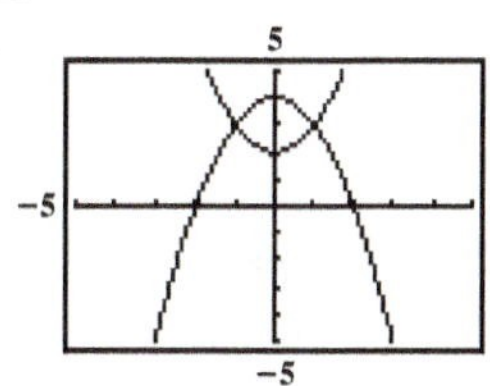

Section 20.4

Vocabulary, Readiness & Video Check **1.** For both, we graph the related equation to find the boundary and sketch it as a solid boundary for ≤ and ≥ and a dashed boundary for < or >; also, we choose a test point not on the boundary and shade that region if the test point is a solution of the original inequality or shade the other region if not.

Exercise Set 20.4 **1.** 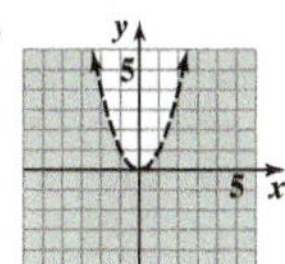 **3.** 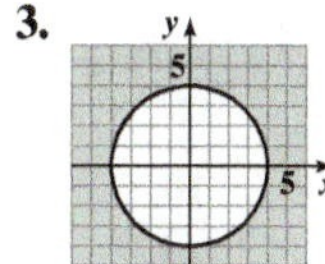 **5.** 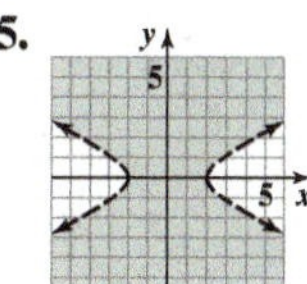 **7.** 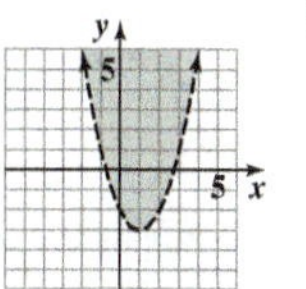 **9.** 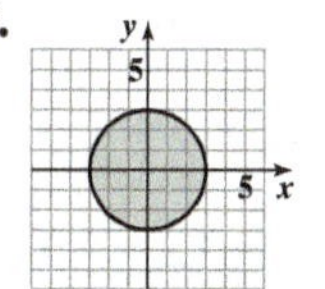 **11.** 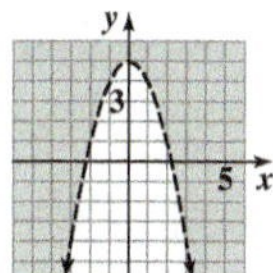

13. **15.** 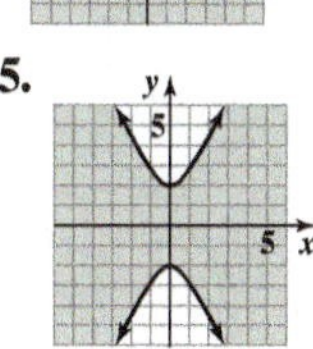 **17.** 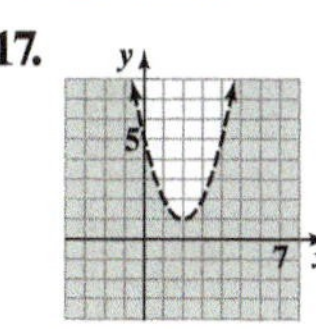 **19.** 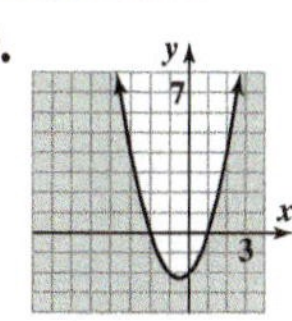 **21.** 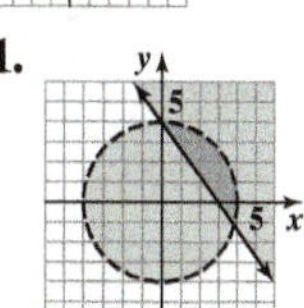 **23.** 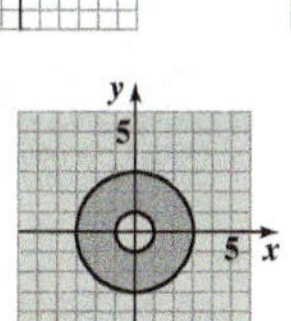

25. 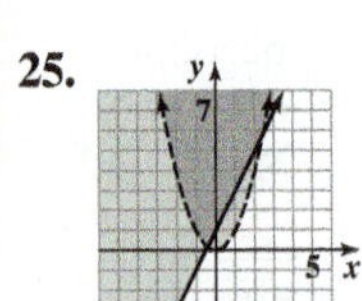 **27.** 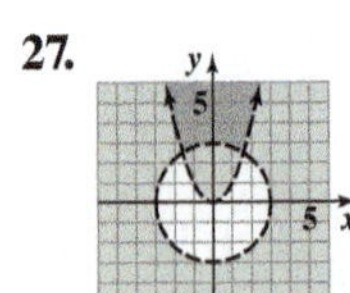 **29.** 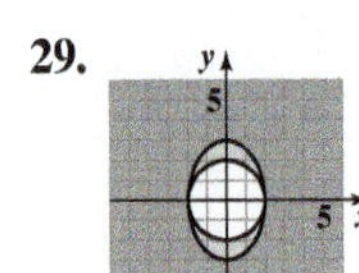 **31.** 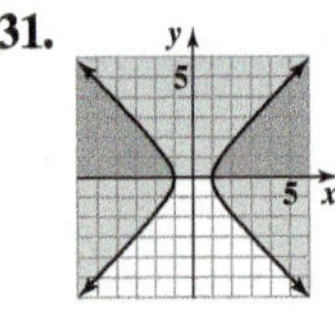 **33.** 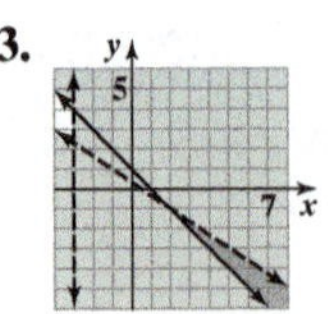 **35.** 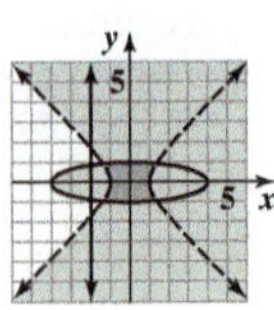

37. not a function **39.** function **41.** 1 **43.** $3a^2 - 2$ **45.** answers may vary **47.** 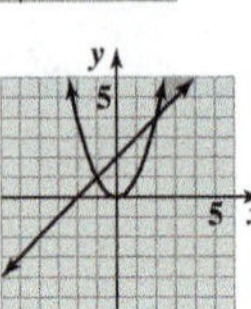

Chapter 20 Vocabulary Check **1.** circle; center **2.** nonlinear system of equations **3.** ellipse **4.** radius **5.** hyperbola **6.** conic sections **7.** vertex **8.** diameter

Chapter 20 Review **1.** $(x + 4)^2 + (y - 4)^2 = 9$ **3.** $(x + 7)^2 + (y + 9)^2 = 11$ **5.** 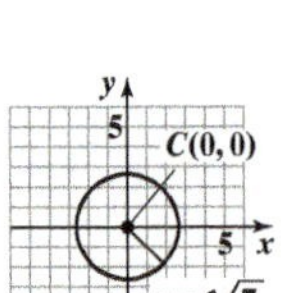

7. 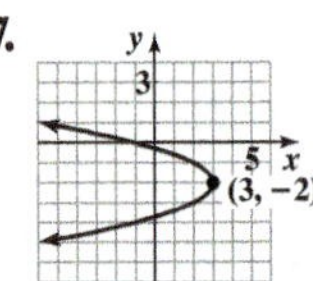

9.

11. 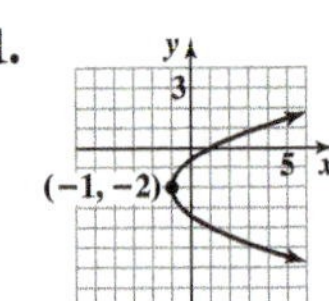

13. 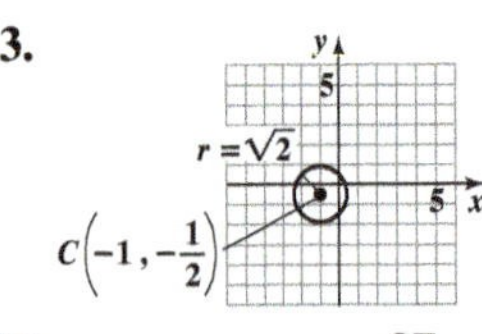

15. 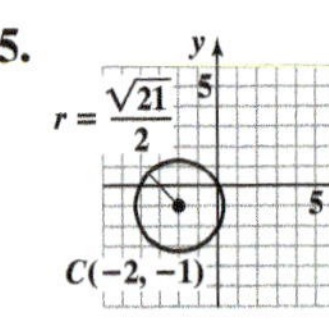

17. 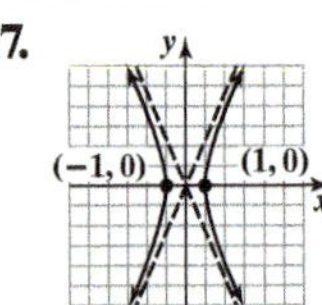

19. 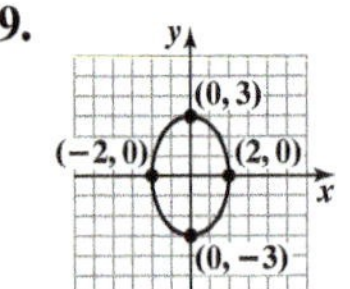

21. 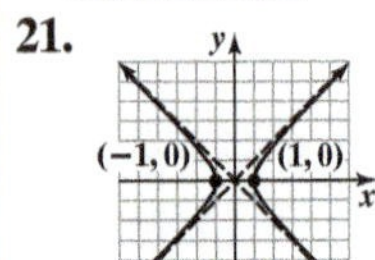

23. 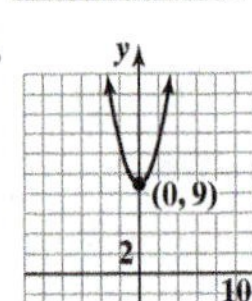

25. 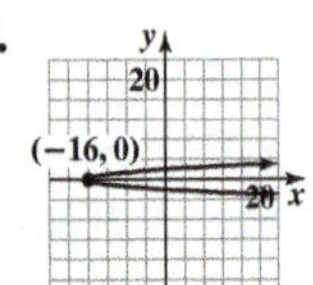

27. 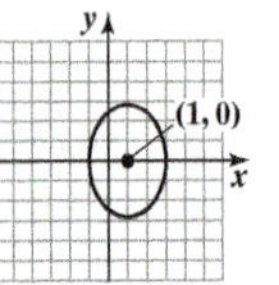

29. 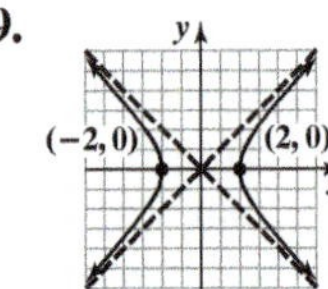

31. 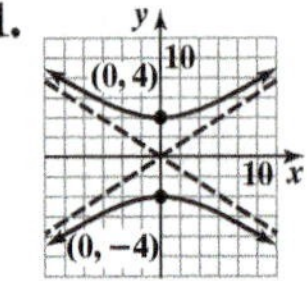

33. $(1, -2), (4, 4)$ **35.** $(-1, 1), (2, 4)$ **37.** $(0, 2), (0, -2)$ **39.** $(1, 4)$ **41.** length: 15 ft; width: 10 ft

43. 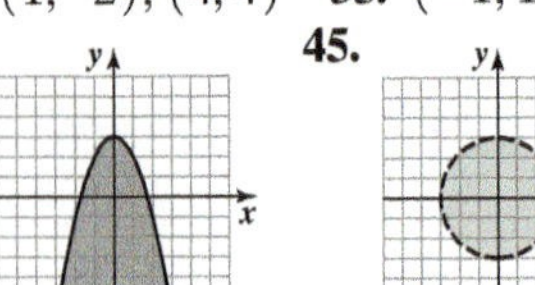 **45.** 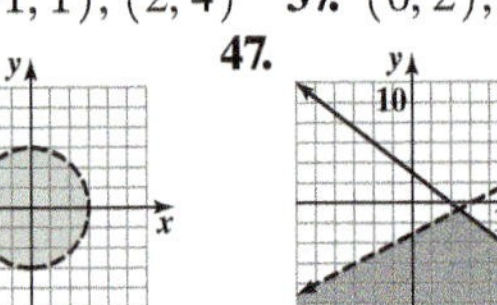 **47.** 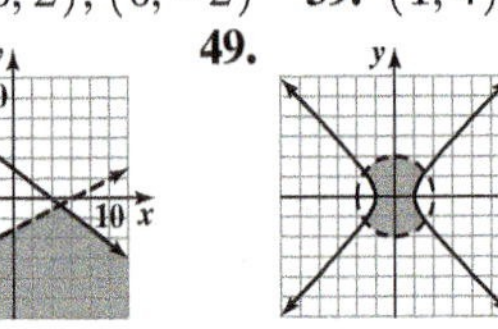 **49.** 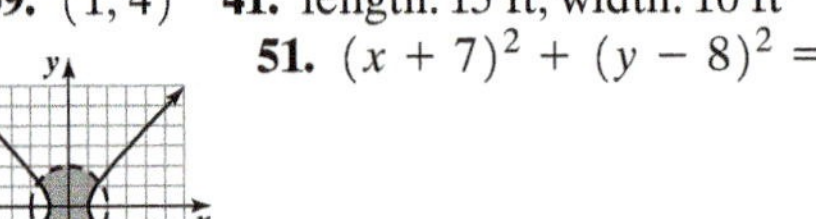 **51.** $(x + 7)^2 + (y - 8)^2 = 25$ **53.** 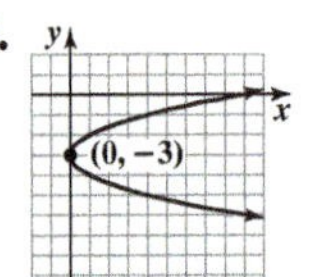

55.

57. 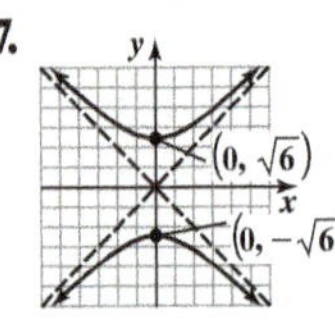

59. 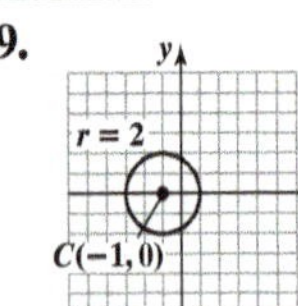

61. 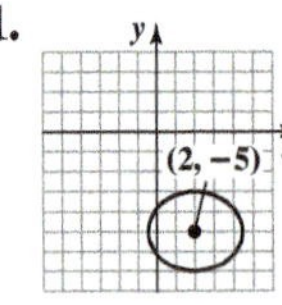

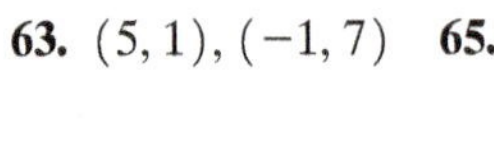
63. $(5, 1), (-1, 7)$ **65.** 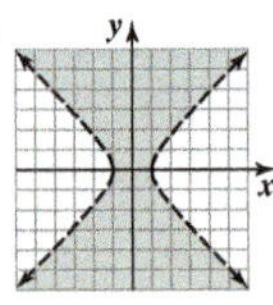

Chapter 20 Test **1.** 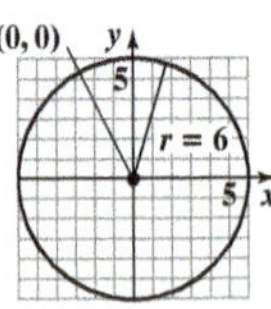

2. 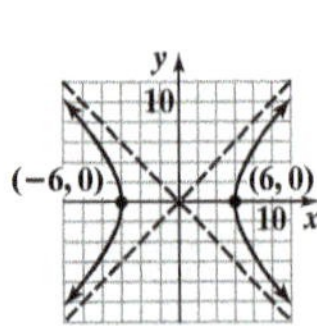

3. 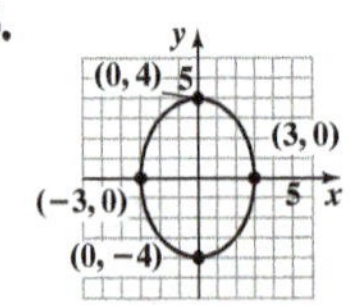

4. 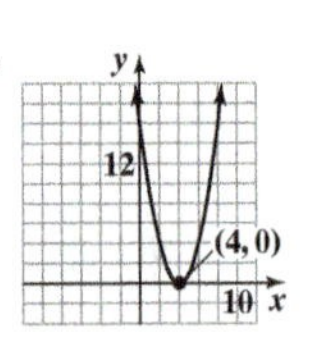

5. 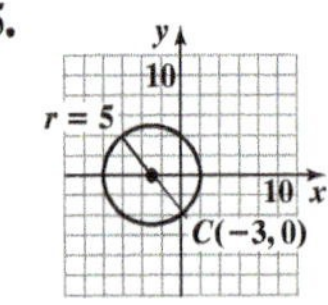

6.

7. 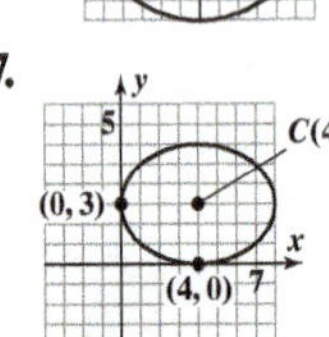

8. 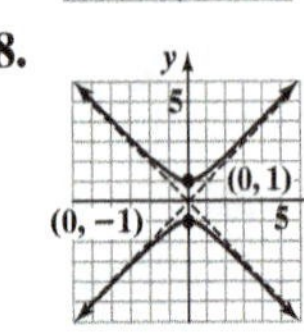

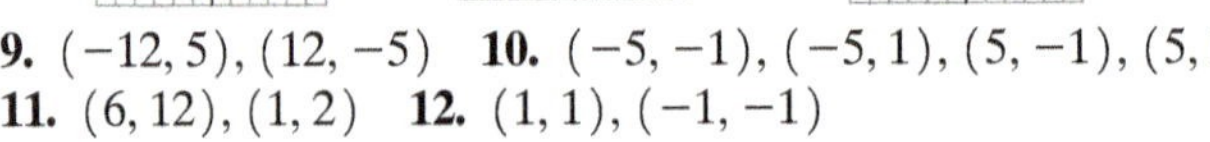
9. $(-12, 5), (12, -5)$ **10.** $(-5, -1), (-5, 1), (5, -1), (5, 1)$
11. $(6, 12), (1, 2)$ **12.** $(1, 1), (-1, -1)$

13. **14.** 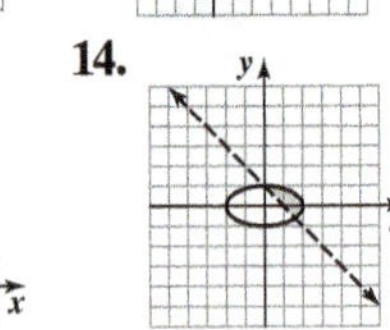 **15.** 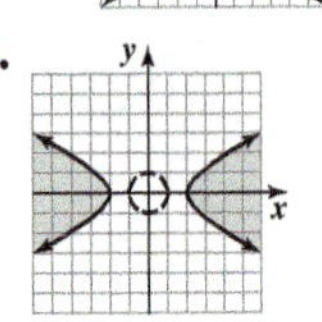 **16.** 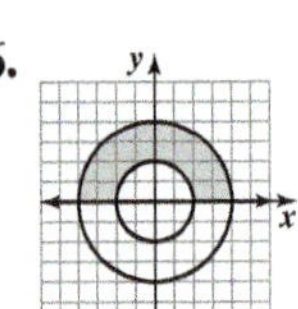 **17.** B **18.** height: 10 ft; width: 30 ft

Cumulative Review Chapters 1–20 **1.** $\varnothing$; Sec. 16.1, Ex. 3 **3.** $\left(-\infty, \frac{13}{5}\right] \cup [4, \infty)$; Sec. 16.1, Ex. 7 **5.** $-2, \frac{4}{5}$; Sec. 16.2, Ex. 2 **7.** 24, -20; Sec. 16.2, Ex. 3 **9.** $\frac{3}{4}$, 5; Sec. 16.2, Ex. 8 **11.** (4, 8); Sec. 16.3, Ex. 2 **13.** $\frac{xy + 2x^3}{y - 1}$; Sec. 14.7, Ex. 3 **15.** $(-\infty, \infty)$; Sec. 16.3, Ex. 7 **17.** 16; Sec. 12.7, Ex. 4 **19. a.** 1 **b.** -4 **c.** $\frac{2}{5}$ **d.** x^2 **e.** $-3x^3$; Sec. 17.1, Ex. 3 **21. a.** $z - z^{17/3}$ **b.** $x^{2/3} - 3x^{1/3} - 10$; Sec. 17.2, Ex. 5 **23. a.** 2 **b.** $\frac{5}{2}\sqrt{x}$ **c.** $14xy^2\sqrt[3]{x}$ **d.** $4a^2b\sqrt[4]{2a}$; Sec. 17.3, Ex. 5 **25. a.** $\frac{5\sqrt{5}}{12}$ **b.** $\frac{5\sqrt[3]{7x}}{2}$; Sec. 17.4, Ex. 3 **27.** $\frac{\sqrt{21xy}}{3y}$; Sec. 17.5, Ex. 2 **29.** 42; Sec. 17.6, Ex. 1 **31. a.** $-i$ **b.** 1 **c.** -1 **d.** 1; Sec. 17.7, Ex. 6 **33.** $-1 + \sqrt{5}, -1 - \sqrt{5}$; Sec. 18.1, Ex. 5 **35.** $2 + \sqrt{2}, 2 - \sqrt{2}$; Sec. 18.2, Ex. 3 **37.** 2, -2, i, $-i$; Sec. 18.3, Ex. 3 **39.** $[-2, 3)$; Sec. 18.4, Ex. 4
41. 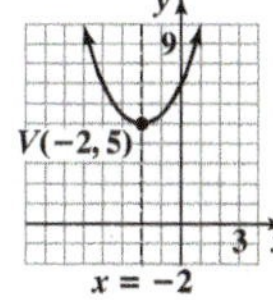

; Sec. 18.5, Ex. 8 **43.** $(2, -16)$; Sec. 18.6, Ex. 4 **45.** $\sqrt{2} \approx 1.414$; Sec. 17.3, Ex. 6

Chapter 21 Sequences, Series, and the Binomial Theorem

Section 21.1

Vocabulary, Readiness & Video Check **1.** general **3.** infinite **5.** -1 **7.** function; domain; a_n **9.** $a_9 = 0.10(2)^{9-1} = \$25.60$

Exercise Set 21.1 **1.** 5, 6, 7, 8, 9 **3.** $-1, 1, -1, 1, -1$ **5.** $\frac{1}{4}, \frac{1}{5}, \frac{1}{6}, \frac{1}{7}, \frac{1}{8}$ **7.** 2, 4, 6, 8, 10 **9.** $-1, -4, -9, -16, -25$ **11.** 2, 4, 8, 16, 32 **13.** 7, 9, 11, 13, 15 **15.** $-1, 4, -9, 16, -25$ **17.** 75 **19.** 118 **21.** $\frac{6}{5}$ **23.** 729 **25.** $\frac{4}{7}$ **27.** $\frac{1}{8}$ **29.** -95 **31.** $-\frac{1}{25}$ **33.** $a_n = 4n - 1$ **35.** $a_n = -2^n$ **37.** $a_n = \frac{1}{3^n}$ **39.** 48 ft, 80 ft, and 112 ft **41.** $a_n = 0.10(2)^{n-1}$; \$819.20 **43.** 2400 cases; 75 cases **45.** 50 sparrows in 2004: extinct in 2010 **47.** 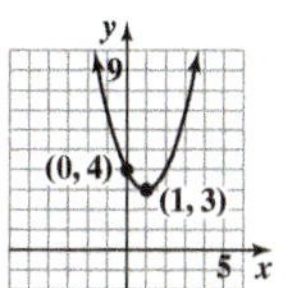

49. 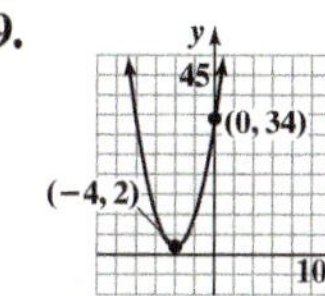

51. $\sqrt{13}$ units **53.** $\sqrt{41}$ units **55.** 1, 0.7071, 0.5774, 0.5, 0.4472 **57.** 2, 2.25, 2.3704, 2.4414, 2.4883

Section 21.2

Vocabulary, Readiness & Video Check **1.** geometric; ratio **3.** first; difference **5.** If there is a common difference between each term and its preceding term in a sequence, it's an arithmetic sequence.

Exercise Set 21.2 **1.** 4, 6, 8, 10, 12 **3.** 6, 4, 2, 0, -2 **5.** 1, 3, 9, 27, 81 **7.** 48, 24, 12, 6, 3 **9.** 33 **11.** -875 **13.** -60 **15.** 96 **17.** -28 **19.** 1250 **21.** 31 **23.** 20 **25.** $a_1 = \frac{2}{3}; r = -2$ **27.** answers may vary **29.** $a_1 = 2; d = 2$ **31.** $a_1 = 5; r = 2$ **33.** $a_1 = \frac{1}{2}; r = \frac{1}{5}$ **35.** $a_1 = x; r = 5$ **37.** $a_1 = p; d = 4$ **39.** 19 **41.** $-\frac{8}{9}$ **43.** $\frac{17}{2}$ **45.** $\frac{8}{81}$ **47.** -19 **49.** $a_n = 4n + 50$; 130 seats **51.** $a_n = 6(3)^{n-1}$ **53.** 486, 162, 54, 18, 6; $a_n = \frac{486}{3^{n-1}}$; 6 bounces **55.** $a_n = 3875 + 125n$; \$5375 **57.** 25 g **59.** $\frac{11}{18}$ **61.** 40 **63.** $\frac{907}{495}$ **65.** \$11,782.40, \$5891.20, \$2945.60, \$1472.80 **67.** 19.652, 19.618, 19.584, 19.55 **69.** answers may vary

Section 21.3

Vocabulary, Readiness & Video Check **1.** infinite **3.** summation **5.** partial sum **7.** sigma/sum, index of summation, beginning value of i, ending value of i, and general term of the sequence

Exercise Set 21.3 **1.** -2 **3.** 60 **5.** 20 **7.** $\frac{73}{168}$ **9.** $\frac{11}{36}$ **11.** 60 **13.** 74 **15.** 62 **17.** $\frac{241}{35}$ **19.** $\sum_{i=1}^{5}(2i - 1)$ **21.** $\sum_{i=1}^{4}4(3)^{i-1}$ **23.** $\sum_{i=1}^{6}(-3i + 15)$ **25.** $\sum_{i=1}^{4}\frac{4}{3^{i-2}}$ **27.** $\sum_{i=1}^{7}i^2$ **29.** -24 **31.** 0 **33.** 82 **35.** -20 **37.** -2 **39.** 1, 2, 3, …, 10; 55 trees **41.** 186 units **43.** 35 species; 82 species **45.** 30 opossums; 68 opossums **47.** 6.25 lb; 93.75 lb **49.** 16.4 in.; 134.5 in. **51.** 10 **53.** $\frac{10}{27}$ **55.** 45 **57.** 90 **59. a.** $2 + 6 + 12 + 20 + 30 + 42 + 56$ **b.** $1 + 2 + 3 + 4 + 5 + 6 + 7 + 1 + 4 + 9 + 16 + 25 + 36 + 49$ **c.** answers may vary **d.** true; answers may vary

Integrated Review **1.** $-2, -1, 0, 1, 2$ **2.** $\frac{7}{2}, \frac{7}{3}, \frac{7}{4}, \frac{7}{5}, \frac{7}{6}$ **3.** $1, 3, 9, 27, 81$ **4.** $-4, -1, 4, 11, 20$ **5.** 64 **6.** -14 **7.** $\frac{1}{40}$ **8.** $-\frac{1}{82}$ **9.** $7, 4, 1, -2, -5$ **10.** $-3, -15, -75, -375, -1875$ **11.** $45, 15, 5, \frac{5}{3}, \frac{5}{9}$ **12.** $-12, -2, 8, 18, 28$ **13.** 101 **14.** $\frac{243}{16}$ **15.** 384 **16.** 185 **17.** -10 **18.** $\frac{1}{2}$ **19.** 50 **20.** 98 **21.** $\frac{31}{2}$ **22.** $\frac{61}{20}$ **23.** -10 **24.** 5

Section 21.4

Vocabulary, Readiness & Video Check **1.** arithmetic **3.** geometric **5.** arithmetic **7.** Use the general term formula from Section 21.2 for the general term of an arithmetic sequence: $a_n = a_1 + (n-1)d$. **9.** The common ratio r is 3 for this sequence so that $|r| \geq 1$, or $|3| \geq 1$; S_∞ doesn't exist if $|r| \geq 1$.

Exercise Set 21.4 **1.** 36 **3.** 484 **5.** 63 **7.** $\frac{312}{125}$ **9.** 55 **11.** 16 **13.** 24 **15.** $\frac{1}{9}$ **17.** -20 **19.** $\frac{16}{9}$ **21.** $\frac{4}{9}$ **23.** 185 **25.** $\frac{381}{64}$ **27.** $-\frac{33}{4}$ or -8.25 **29.** $-\frac{75}{2}$ **31.** $\frac{56}{9}$ **33.** 4000, 3950, 3900, 3850, 3800; 3450 cars; 44,700 cars **35.** Firm A (Firm A, \$265,000; Firm B, \$254,000) **37.** \$39,930; \$139,230 **39.** 20 min; 123 min **41.** 180 ft **43.** Player A, 45 points; Player B, 75 points **45.** \$3050 **47.** \$10,737,418.23 **49.** 720 **51.** 3 **53.** $x^2 + 10x + 25$ **55.** $8x^3 - 12x^2 + 6x - 1$ **57.** $\frac{8}{10} + \frac{8}{100} + \frac{8}{1000} + \cdots; \frac{8}{9}$ **59.** answers may vary

Section 21.5

Vocabulary, Readiness & Video Check **1.** 1 **3.** 24 **5.** 6 **7.** Pascal's triangle you the coefficients of the terms of the expanded binomial; also, the power tells you how many terms the expansion has (1 more than the power on the binomial). **9.** The theorem is in terms of $(a+b)^n$, so if your binomial is of the form $(a-b)^n$, then remember to think of it as $(a+(-b))^n$, so your second term is $-b$.

Exercise Set 21.5 **1.** $m^3 + 3m^2n + 3mn^2 + n^3$ **3.** $c^5 + 5c^4d + 10c^3d^2 + 10c^2d^3 + 5cd^4 + d^5$ **5.** $y^5 - 5y^4x + 10y^3x^2 - 10y^2x^3 + 5yx^4 - x^5$ **7.** answers may vary **9.** 8 **11.** 42 **13.** 360 **15.** 56 **17.** $a^7 + 7a^6b + 21a^5b^2 + 35a^4b^3 + 35a^3b^4 + 21a^2b^5 + 7ab^6 + b^7$ **19.** $a^5 + 10a^4b + 40a^3b^2 + 80a^2b^3 + 80ab^4 + 32b^5$ **21.** $q^9 + 9q^8r + 36q^7r^2 + 84q^6r^3 + 126q^5r^4 + 126q^4r^5 + 84q^3r^6 + 36q^2r^7 + 9qr^8 + r^9$ **23.** $1024a^5 + 1280a^4b + 640a^3b^2 + 160a^2b^3 + 20ab^4 + b^5$ **25.** $625a^4 - 1000a^3b + 600a^2b^2 - 160ab^3 + 16b^4$ **27.** $8a^3 + 36a^2b + 54ab^2 + 27b^3$ **29.** $x^5 + 10x^4 + 40x^3 + 80x^2 + 80x + 32$ **31.** $5cd^4$ **33.** d^7 **35.** $-40r^2s^3$ **37.** $6x^2y^2$ **39.** $30a^9b$ **41.**

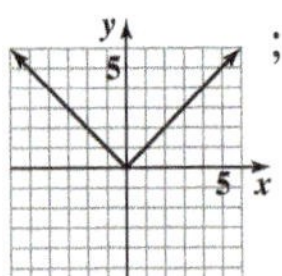

; not one-to-one **43.**

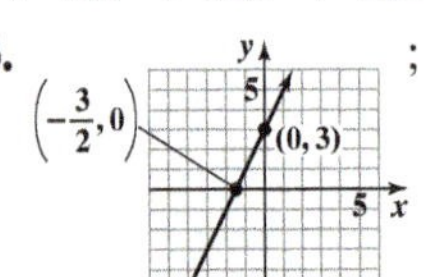

; one-to-one **45.**

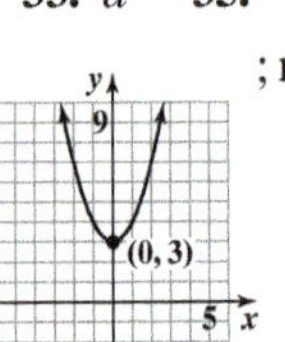

; not one-to-one

47. $x^2\sqrt{x} + 5\sqrt{3}x^2 + 30x\sqrt{x} + 30\sqrt{3}x + 45\sqrt{x} + 9\sqrt{3}$ **49.** 126 **51.** 28 **53.** answers may vary

Chapter 21 Vocabulary Check **1.** finite sequence **2.** factorial of n **3.** infinite sequence **4.** geometric sequence; common ratio **5.** series **6.** general term **7.** arithmetic sequence; common difference **8.** Pascal's triangle

Chapter 21 Review **1.** $-3, -12, -27, -48, -75$ **3.** $\frac{1}{100}$ **5.** $a_n = \frac{1}{6n}$ **7.** 144 ft, 176 ft, 208 ft **9.** 660,000; 1,320,000; 2,640,000; 5,280,000; 10,560,000; 2010: 10,560,000 infested acres **11.** $-2, -\frac{4}{3}, -\frac{8}{9}, -\frac{16}{27}, -\frac{32}{81}$ **13.** 111 **15.** -83 **17.** $a_1 = 3; d = 5$ **19.** $a_n = \frac{3}{10^n}$ **21.** $a_1 = \frac{8}{3}, r = \frac{3}{2}$ **23.** $a_1 = 7x, r = -2$ **25.** 8, 6, 4.5, 3.4, 2.5, 1.9; good **27.** $a_n = 2^{n-1}$; \$512; \$536,870,912 **29.** $a_n = 150n + 750$; \$1650/month **31.** $1 + 3 + 5 + 7 + 9 = 25$ **33.** $\frac{1}{4} - \frac{1}{6} + \frac{1}{8} = \frac{5}{24}$ **35.** $\sum_{i=1}^{6} 3^{i-1}$ **37.** $\sum_{i=1}^{4} \frac{1}{4^i}$ **39.** $a_n = 20(2)^n$; n represents the number of 8-hour periods; 1280 yeast **41.** Job A, \$48,300; Job B, \$46,600 **43.** -4 **45.** -10 **47.** 150 **49.** 900 **51.** -410 **53.** 936 **55.** 10 **57.** -25 **59.** \$30,418; \$99,868 **61.** \$58; \$553 **63.** 2696 mosquitoes **65.** $\frac{5}{9}$ **67.** $x^5 + 5x^4z + 10x^3z^2 + 10x^2z^3 + 5xz^4 + z^5$ **69.** $16x^4 + 32x^3y + 24x^2y^2 + 8xy^3 + y^4$ **71.** $b^8 + 8b^7c + 28b^6c^2 + 56b^5c^3 + 70b^4c^4 + 56b^3c^5 + 28b^2c^6 + 8bc^7 + c^8$ **73.** $256m^4 - 256m^3n + 96m^2n^2 - 16mn^3 + n^4$ **75.** $35a^4b^3$ **77.** 130 **79.** 40.5

Chapter 21 Test **1.** $-\frac{1}{5}, \frac{1}{6}, -\frac{1}{7}, \frac{1}{8}, -\frac{1}{9}$ **2.** 247 **3.** $a_n = \frac{2}{5}\left(\frac{1}{5}\right)^{n-1}$ **4.** $a_n = (-1)^n 9n$ **5.** 155 **6.** -330 **7.** $\frac{144}{5}$ **8.** 1 **9.** 10 **10.** -60 **11.** $a^6 - 6a^5b + 15a^4b^2 - 20a^3b^3 + 15a^2b^4 - 6ab^5 + b^6$ **12.** $32x^5 + 80x^4y + 80x^3y^2 + 40x^2y^3 + 10xy^4 + y^5$

13. 925 people; 250 people initially **14.** $1 + 3 + 5 + 7 + 9 + 11 + 13 + 15$; 64 shrubs **15.** 33.75 cm; 218.75 cm **16.** 320 cm **17.** 304 ft; 1600 ft **18.** $\frac{14}{33}$

Cumulative Review Chapters 1–21 **1. a.** 9 **b.** -9; Sec. 2.5, Ex. 1 and 2 **3.** $6y + 2$; Sec. 3.1, Ex. 2 **5.** $y = \frac{1}{4}x - 3$; Sec. 10.5, Ex. 3 **7.** $-x + 5y = 23$ or $x - 5y = -23$; Sec. 10.5, Ex. 5 **9.** $x^3 - 4x^2 - 3x + 11 + \frac{12}{x + 2}$; Sec. 12.7, Ex. 2 **11. a.** $5\sqrt{2}$ **b.** $2\sqrt[3]{3}$ **c.** $\sqrt{26}$ **d.** $2\sqrt[4]{2}$; Sec. 17.3, Ex. 3 **13.** 10%; Sec. 18.1, Ex. 9 **15.** 2, 7; Sec. 18.3, Ex. 4 **17.** $\left(-\frac{7}{2}, -1\right)$; Sec. 18.4, Ex. 5 **19.** $\frac{25}{4}$ ft; $\frac{5}{8}$ sec; Sec. 18.6, Ex. 5 **21. a.** 25; 7 **b.** $x^2 + 6x + 9$; $x^2 + 3$; Sec. 19.1, Ex. 2 **23.** $f^{-1} = \{(1, 0), (7, -2), (-6, 3), (4, 4)\}$; Sec. 19.2, Ex. 3 **25. a.** 4 **b.** $\frac{3}{2}$ **c.** 6; Sec. 19.3, Ex. 4 **27. a.** 2 **b.** -1 **c.** 3 **d.** 6; Sec. 19.5, Ex. 5 **29. a.** $\log_{11} 30$ **b.** $\log_3 6$ **c.** $\log_2(x^2 + 2x)$; Sec. 19.6, Ex. 1 **31.** \$2509.30; Sec. 19.7, Ex. 8 **33.** $\frac{\log 7}{\log 3} \approx 1.7712$; Sec. 19.8, Ex. 1 **35.** 18; Sec. 19.8, Ex. 2 **37.**

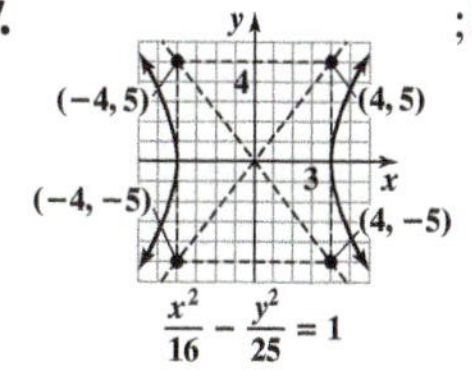

; Sec. 20.2, Ex. 4 **39.** $(2, \sqrt{2})$; Sec. 20.3, Ex. 2 **41.**

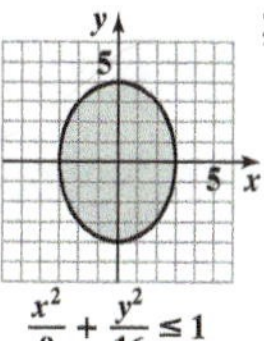

; Sec. 20.4, Ex. 1 **43.** 0, 3, 8, 15, 24; Sec. 21.1, Ex. 1 **45.** 72; Sec. 21.2, Ex. 3 **47. a.** $\frac{7}{2}$ **b.** 56; Sec. 21.3, Ex. 1 **49.** 465; Sec. 21.4, Ex. 2

Appendix B Review of Algebra Topics

Exercise Set B.1 **1.** $-3, -8$ **3.** -2 **5.** $\frac{1}{4}, -\frac{2}{3}$ **7.** 1, 9 **9.** 0 **11.** 5 **13.** $\frac{3}{5}, -1$ **15.** no solution **17.** $\frac{1}{8}$ **19.** 0 **21.** 6, -3 **23.** all real numbers **25.** $\frac{2}{5}, -\frac{1}{2}$ **27.** $\frac{3}{4}, -\frac{1}{2}$ **29.** 29 **31.** -8 **33.** $-\frac{1}{3}, 0$ **35.** $-\frac{7}{8}$ **37.** $\frac{31}{4}$ **39.** 1 **41.** $-7, 4$ **43.** 4, 6 **45.** $-\frac{1}{2}$ **47. a.** incorrect **b.** correct **c.** correct **d.** incorrect **49.** $K = -11$ **51.** $K = 24$ **53.** -4.86 **55.** 1.53

Exercise Set B.2 **1.** $1.00x + 0.40(2) = 0.70(x + 2)$; 2 gal **3.** $7x + 4(14) = 5(x + 14)$; 7 lb **5.** $0.10y$ **7.** $0.05(x + 7)$ **9.** $20(4y)$ or $80y$ **11.** $50(35 - x)$ **13.** 12 \$10 bills; 32 \$5 bills **15.** 400 oz **17.** 400 oz **19.** 2 hr **21.** 36 mph; 46 mph **23.** 4 hr **25.** $2\frac{1}{2}$ hr **27.** $666\frac{2}{3}$ mi **29.** 55 mph **31.** 2 hr 55 min **33.** 2.2 mph; 3.3 mph **35.** 27.5 mi **37.** 25 \$100 bills; 71 \$50 bills; 175 \$20 bills

Exercise Set B.3 **1.** (5, 2) **3.** (3, 0) **5.** $(-5, -2)$ **7.** $(-1, 0)$ **9.** Quadrant I **11.** Quadrant II **13.** Quadrant III **15.** y-axis **17.** Quadrant III **19.** x-axis **21.** Quadrant IV **23.** x-axis **25.** Quadrant III **27.**

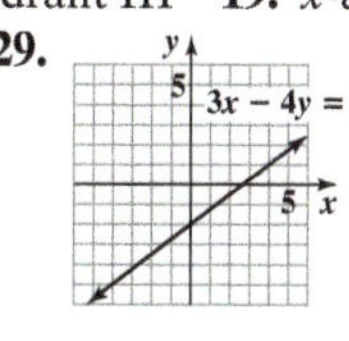

29.

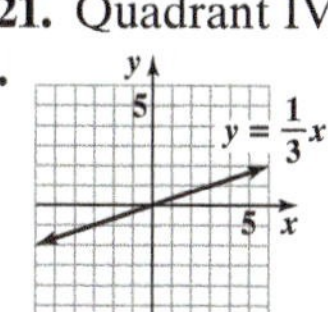

31.

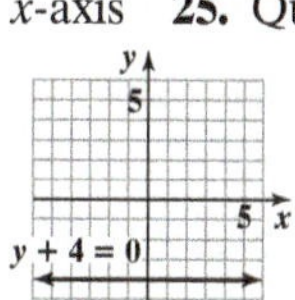

33. y + 4 = 0 **35.** 1 **37.** -4 **39.** 1, 3 **41.** $g(-1) = -2$

Exercise Set B.4 **1.** $2y^2 + 2y - 11$ **3.** $x^2 - 7x + 7$ **5.** $25x^2 - 30x + 9$ **7.** $2x^3 - 4x^2 + 5x - 5 + \frac{8}{x + 2}$ **9.** $(x - 4 + y)(x - 4 - y)$ **11.** $x(x - 1)(x^2 + x + 1)$ **13.** $2xy(7x - 1)$ **15.** $4(x + 2)(x - 2)$ **17.** $(3x - 11)(x + 1)$ **19.** $4(x + 3)(x - 1)$ **21.** $(2x + 9)^2$ **23.** $(2x + 5y)(4x^2 - 10xy + 25y^2)$ **25.** $8x^2(2y - 1)(4y^2 + 2y + 1)$ **27.** $(x + 5 + y)(x^2 + 10x + 25 - xy - 5y + y^2)$ **29.** $(5a - 6)^2$ **31.** $7x(x - 9)$ **33.** $(b - 6)(a + 7)$ **35.** $(x^2 + 1)(x + 1)(x - 1)$ **37.** $(5x - 11)(2x + 3)$ **39.** $5a^3b(b^2 - 10)$ **41.** prime **43.** $10x(x - 10)(x - 11)$ **45.** $a^3b(4b - 3)(16b^2 + 12b + 9)$ **47.** $2(x - 3)(x^2 + 3x + 9)$ **49.** $(3y - 5)(y^4 + 2)$ **51.** $100(z + 1)(z^2 - z + 1)$ **53.** $(2b - 9)^2$ **55.** $(y - 4)(y - 5)$ **57.** $A = 9 - 4x^2 = (3 + 2x)(3 - 2x)$

Exercise Set B.5 **1.** $\frac{1}{2}$ **3.** $\frac{1 + 2x}{8}$ **5.** $\frac{2(x - 4)}{(x + 2)(x - 1)}$ **7.** 4 **9.** -5 **11.** $\frac{2x + 5}{x(x - 3)}$ **13.** -2 **15.** $\frac{(a + 3)(a + 1)}{a + 2}$ **17.** $-\frac{1}{5}$ **19.** $\frac{4a + 1}{(3a + 1)(3a - 1)}$ **21.** $-1, \frac{3}{2}$ **23.** $\frac{3}{x + 1}$ **25.** -1 **27. a.** $\frac{x}{5} - \frac{x}{4} + \frac{1}{10}$ **b.** Write each rational expression term so that the denominator is the LCD, 20. **c.** $\frac{-x + 2}{20}$ **29.** b **31.** d **33.** d

Appendix C Stretching and Compressing Graphs of Absolute Value Functions

Exercise Set Appendix C

1.
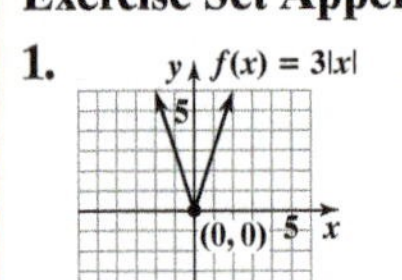

3.
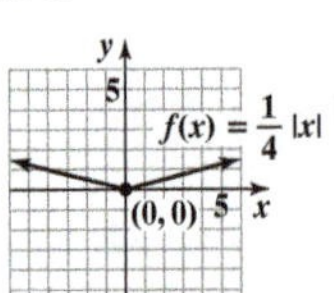

5.
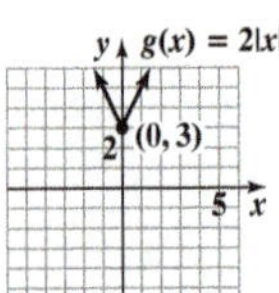

7.
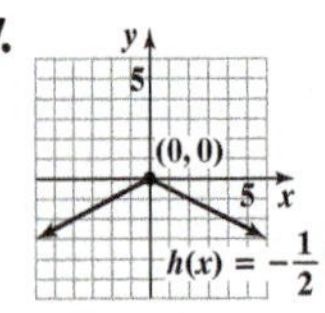

9.
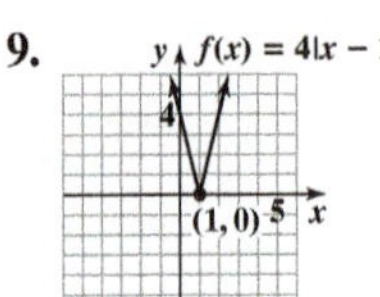

11.
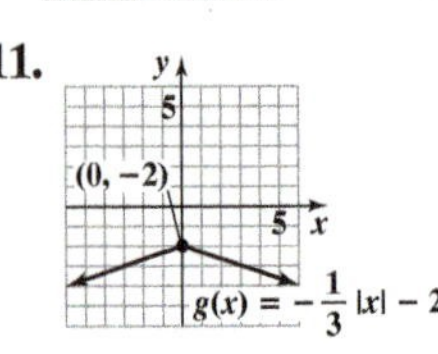

13.
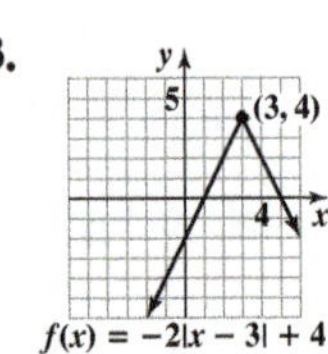

15.
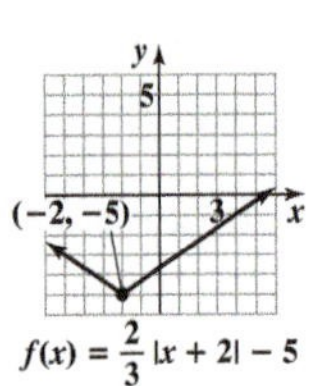

Appendix D An Introduction to Using a Graphing Utility

The Viewing Window and Interpreting Window Settings Exercise Set **1.** yes **3.** no **5.** answers may vary **7.** answers may vary **9.** answers may vary **11.** Xmin = −12 Ymin = −12 Xmax = 12 Ymax = 12 Xscl = 3 Yscl = 3 **13.** Xmin = −9 Ymin = −12 Xmax = 9 Ymax = 12 Xscl = 1 Yscl = 2 **15.** Xmin = −10 Ymin = −25 Xmax = 10 Ymax = 25 Xscl = 2 Yscl = 5 **17.** Xmin = −10 Ymin = −30 Xmax = 10 Ymax = 30 Xscl = 1 Yscl = 3 **19.** Xmin = −20 Ymin = −30 Xmax = 30 Ymax = 50 Xscl = 5 Yscl = 10

Graphing Equations and Square Viewing Window Exercise Set **1.** Setting B **3.** Setting B **5.** Setting B

7.
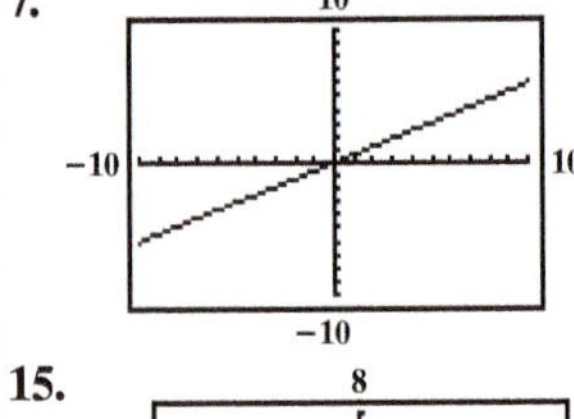

9.
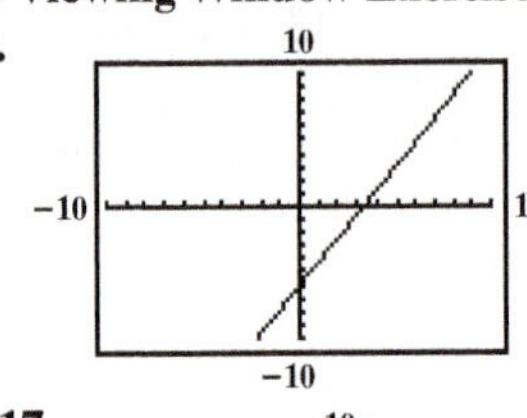

11.
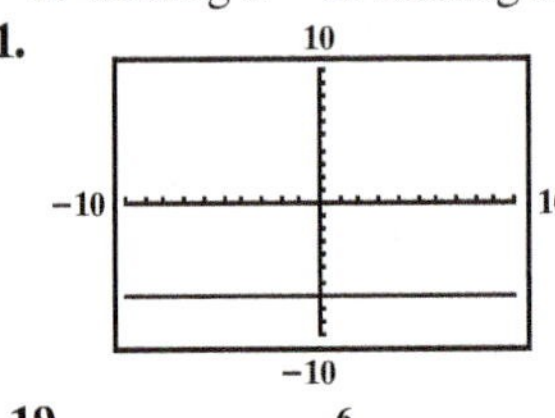

13.
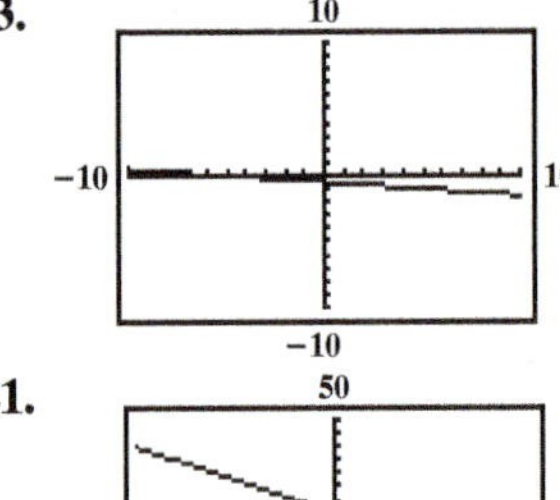

15.
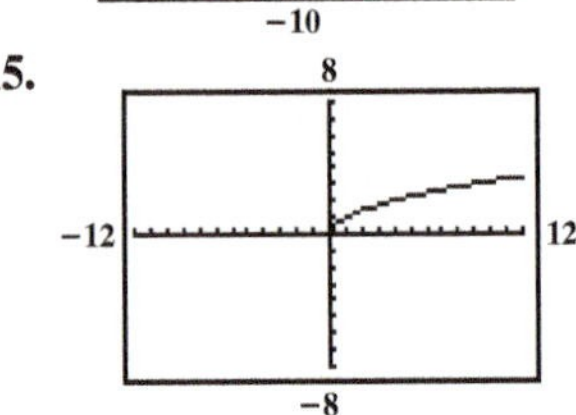

17.
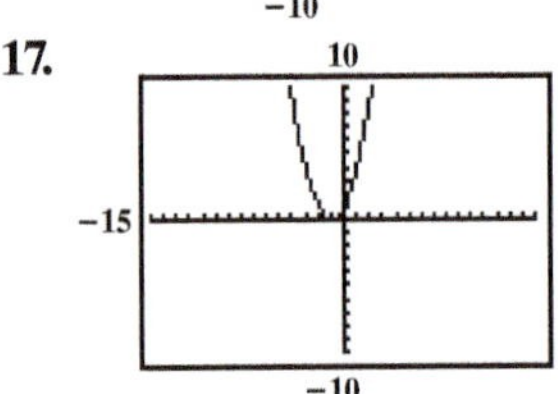

19.
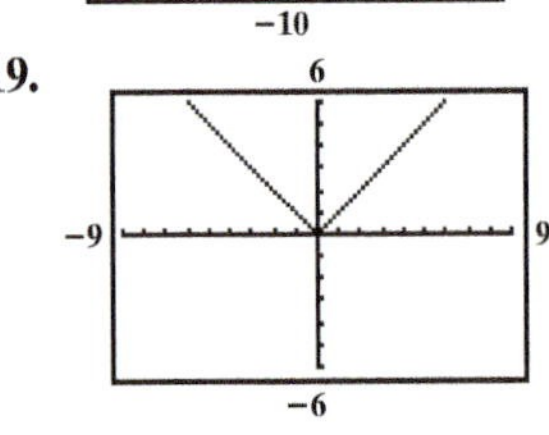

21. 50, −50, 50, −50

Appendix E Solving Systems of Equations by Matrices

1. $(2, -1)$ **3.** $(-4, 2)$ **5.** $\{\ \}$ or $\varnothing$ **7.** $\{(x, y) \mid 3x - 3y = 9\}$ **9.** $(-2, 5, -2)$ **11.** $(1, -2, 3)$ **13.** $(4, -3)$ **15.** $(2, 1, -1)$ **17.** $(9, 9)$ **19.** $\{\ \}$ or $\varnothing$ **21.** $\{\ \}$ or $\varnothing$ **23.** $(1, -4, 3)$ **25.** c

Appendix F Solving Systems of Equations by Determinants

1. 26 **3.** −19 **5.** 0 **7.** $(1, 2)$ **9.** $\{(x, y) \mid 3x + y = 1\}$ **11.** $(9, 9)$ **13.** 8 **15.** 0 **17.** 54 **19.** $(-2, 0, 5)$ **21.** $(6, -2, 4)$ **23.** 16 **25.** 15 **27.** $\frac{13}{6}$ **29.** 0 **31.** 56 **33.** $(-3, -2)$ **35.** $\{\ \}$ or $\varnothing$ **37.** $(-2, 3, -1)$ **39.** $(3, 4)$ **41.** $(-2, 1)$ **43.** $\{(x, y, z) \mid x - 2y + z = -3\}$ **45.** $(0, 2, -1)$ **47.** 5 **49.** 0; answers may vary **51.** $\begin{matrix} + & - & + & - \\ - & + & - & + \\ + & - & + & - \\ - & + & - & + \end{matrix}$ **53.** −125 **55.** 24

Practice Final Exam **1.** 200 **2.** 21 **3.** −6 **4.** −100 **5.** 7 **6.** −117 **7.** 0 **8.** undefined **9.** $-\frac{2}{3y}$ **10.** $\frac{3}{4}$ **11.** $\frac{1}{a^2}$ **12.** $\frac{64}{3}$ or $21\frac{1}{3}$ **13.** $16\frac{8}{11}$ **14.** −6.2 **15.** 40.902 **16.** 0.006 **17.** 610% **18.** 37.5% **19.** $\frac{69}{200}$ **20.** −0.5 **21.** 34.9 **22.** 30 **23.** 1 **24.** $\frac{4}{31}$ **25.** $-8z - 20$ **26.** $(15x + 15)$ in. **27.** −28 **28.** −2 **29.** 3 **30.** −4 **31.** 3.7 **32.** −6 **33.** perimeter = 60 yd; area = 200 sq yd **34.** −4 **35.** −2 **36.** 24 mi **37.** 244 women **38.** 18 bulbs **39.** discount: $18; sale price: $102 **40.** 5.66 cm

41. $12°$ **42.** $m\angle x = 73°; m\angle y = 73°; m\angle z = 73°$ **43.** perimeter = 68 in.; area = 185 sq in. **44.** circumference = 18π in. $\approx$ 56.52 in.; area = 81π sq in. $\approx$ 254.34 sq in. **45.** 10 qt **46.** 2400 g **47.** -81 **48.** $\frac{1}{64}$ **49.** $10y + 1$ **50.** $-3x^3 + 5x^2 + 4x + 5$ **51.** $16x^2 - 16x + 4$ **52.** $3x^3 + 22x^2 + 41x + 14$ **53.** $(6t + 5)(t - 1)$ **54.** $5(6 + x)(6 - x)$ **55.** $(a + b)(3a - 7)$ **56.** $y^2 + 2y + 4$ **57.** $\frac{y^{14}}{x^2}$ **58.** $\frac{5y^2 - 1}{y + 2}$ **59.** $\frac{2(x + 5)}{x(y + 5)}$ **60.** $\frac{3a - 4}{(a - 3)(a + 2)}$ **61.** $x + 2$ **62.** 8 **63.** $1, -\frac{4}{3}$ **64.** $\{x | x \geq 11\}$ **65.** $\frac{3 \pm \sqrt{29}}{2}$ **66.** $\frac{30}{11}$ **67.** no solution **68.** 2, 3 **69.** (3, 7) **70.** **71.**

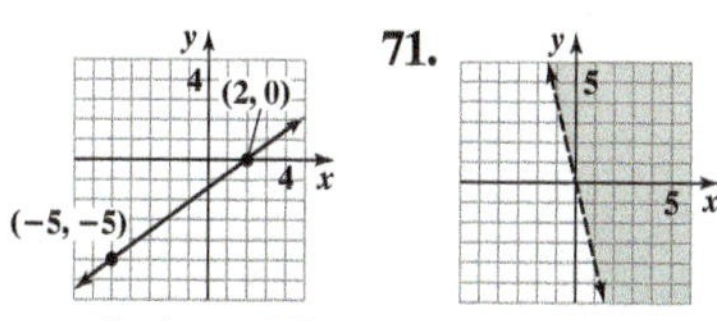

72. $m = -1$ **73.** $m = 3$ **74.** $8x + y = 11$ **75.** $x - 8y = -96$ **76.** $(-4, 1)$ **77.** no solution **78. a.** 0 **b.** 0 **c.** 60 **79.** x-intercepts: $(0, 0), (4, 0)$; y-intercept: $(0, 0)$; domain: $(-\infty, \infty)$; range: $(-\infty, 4)$ **80.** $6\sqrt{6}$ **81.** $\frac{4x^2}{9}$ **82.** $-x\sqrt{5x}$ **83.** -20 **84.** $4\sqrt{3} - \sqrt{6}$ **85.** $\frac{3\sqrt{y}}{y}$ **86.** $\frac{8 - 6\sqrt{x} + x}{8 - 2x}$ **87.**

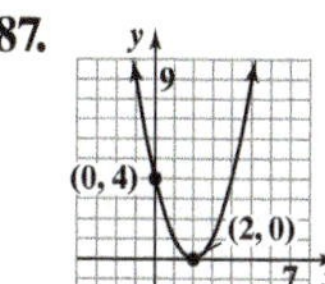

88. 5 or 1 **89.** 21 **90.** California: 1107; Ohio: 720 **91.** 30 mph **92.** 16 **93.** 7 ft **94. a.** 272 ft **b.** 5.12 sec **95.** $0 - 2i\sqrt{2}$ **96.** $0 - 3i$ **97.** $7 + 24i$ **98.** $-\frac{3}{2} + \frac{5}{2}i$ **99.** $(g \circ h)(x) = x^2 - 6x - 2$

100. one-to-one; $f^{-1}(x) = \frac{-x + 6}{2}$ **101.** $\log_5 \frac{x^4}{x + 1}$ **102.** -1 **103.** $\frac{1}{2}\left(\frac{\log 4}{\log 3} - 5\right)$; -1.8691 **104.** 22 **105.** $\frac{43}{21}$ **106.** $\frac{1}{2}$ **107.**

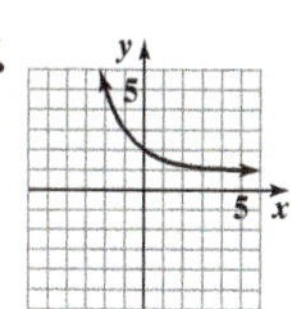

108. 64,805 prairie dogs **109.**

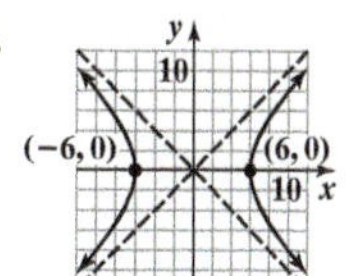

110. **111.**

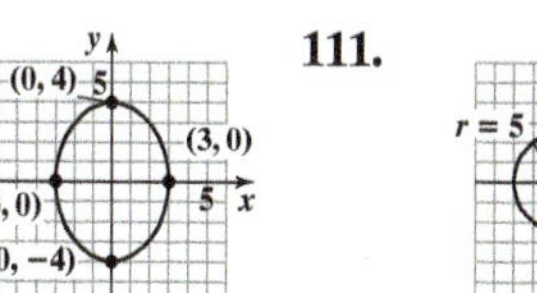

112. $(-5, -1), (-5, 1), (5, -1), (5, 1)$ **113.** $-\frac{1}{5}, \frac{1}{6}, -\frac{1}{7}, \frac{1}{8}, -\frac{1}{9}$ **114.** 155 **115.** 1 **116.** 10 **117.** $32x^5 + 80x^4y + 80x^3y^2 + 40x^2y^3 + 10xy^4 + y^5$

G

N

O

P

Q

R

S

Photo Credits

Chapter 1 **p. 6** Maciej Noskowski/Getty Images **p. 14** NASA **p. 15** Jeanne Provost/Fotolia **p. 17** Pearson Education **p. 23** Mediagram/Shutterstock **p. 24** Kikkerdirk/Fotolia **p. 35** Paylessimages/Fotolia **p. 37** Auttapon Moonsawad/Fotolia **p. 46** Greg Henry/Shutterstock **p. 50** Photographee.eu/Fotolia

Chapter 2 **p. 97** (l) Pearson Education; (r) Stephen Alvarez/Getty Images **p. 103** (l) Snehit/Fotolia; (r) Lunamarina/Fotolia **p. 111** Paul Maguire/Fotolia **p. 126** Samott/Fotolia **p. 130** James Steidl/Fotolia **p. 150** SuperStock **p. 153** (l) Chris Turner/Shutterstock; (r) Tusharkoley/Fotolia **p. 155** Itsallgood/Fotolia

Chapter 3 **p. 162** Imago stock&people/Newscom **p. 194** .shock/Fotolia **p. 196** (b) David Benton/ Shutterstock; (t) Falconhy/Fotolia **p. 198** (l) Gretchen Owen/Fotolia; (r) HelleM/Fotolia **p. 208** Greg Roden/DK images

Chapter 4 **p. 211** William87/Fotolia **p. 237** Keith Brofsky/Getty Images **p. 245** Pefkos/Fotolia **p. 264** Andres Rodriguez/Fotolia **p. 302** James Thew/Fotolia

Chapter 5 **p. 328** Dan Race/Fotolia **p. 330** AP Images **p. 350** (l) Apple Inc.; (r) Pearson Education **p. 361** Giemmephoto/Fotolia **p. 371** WavebreakMediaMicro/Fotolia **p. 377** SeanPavonePhoto/Fotolia **p. 399** Grant V. Faint/Getty Images

Chapter 6 **p. 410** Studio306fotolia/Fotolia **p. 421** Pressmaster/Fotolia **p. 426** Pearson Education **p. 450** Michaeljung/Fotolia **p. 451** Remik44992/Fotolia **p. 455** (l) Monkey Business/Fotolia; (r) Elayn Martin-Gay **p. 457** Kubais/Fotolia **p. 458** Rafa Irusta/Fotolia **p. 459** (l) Andreamuscatello/Fotolia; (r) Kletr/Fotolia **p. 461** Narvf/Fotolia **p. 465** (bl) Terex/Fotolia; (br) LuckyPhoto/Fotolia; (tl) Sashkin/Fotolia

Chapter 7 **p. 485** (t) Michael Shake/Fotolia; (b) Danr13/Fotolia

Chapter 8 **p. 559** (l) Asa Gauen/Alamy; (r) Donyanedomam/Fotolia **p. 568** Frankix/Fotolia **p. 583** (l) Jörg Hackemann/Fotolia; (r) Shuttoz/Fotolia **p. 588** Daphot75/Fotolia **p. 608** (t) Bill/Fotolia; (b) Vuktopua/Fotolia **p. 623** (l) Huaxiadragon/Fotolia; (r) Roibu/Fotolia **p. 624** Dave/Fotolia **p. 639** Worker/ Shutterstock

Chapter 9 **p. 646** (l) Calee Allen/Fotolia; (r) Steve estvanik/Fotolia; **p. 675** Matthew Cavanaugh/epa/ Newscom **p. 676** Lightwavemedia/Fotolia **p. 677** Zack Frank/Shutterstock **p. 678** Felix Mizioznikov/ Fotolia **p. 683** (l) FomaA/Fotolia; (r) Apple Inc. **p. 685** Stefano Tiraboschi/Fotolia **p. 695** Steve Geer/ Getty Images **p. 698** Colin Braley/Reuters **p. 716** W.Scott/Fotolia

Chapter 10 **p. 723** Europhotos/Shutterstock **p. 724** Stephen Coburn/Shutterstock **p. 732** Piero Cruciatti/ Dreamstime **p. 734** Grondin Franck Olivier/Dreamstime **p. 735** WilleeCole Photography/Shutterstock **p. 744** Alexander Raths/Shutterstock **p. 747** (bl) Stockbyte/Getty Images; (br) William Casey/Dreamstime; (t) Teresa Kenney/Dreamstime **p. 756** Anton Prado Photo/Shutterstock **p. 765** George Burba/Dreamstime **p. 771** (t) Gennadiy Poznyakov/Fotolia; (b) Uzi Tzur/Fotolia **p. 772** Raluca Tudor/Dreamstime **p. 776** GoodMood Photo/Fotolia **p. 780** (l) Fabinus08/Dreamstime; (r) Georgiy Pashin/Dreamstime **p. 796** Cpbackpacker/Shutterstock

Chapter 11 **p. 803** Wavebreakmedia/Shutterstock **p. 818** Comstock/Getty Images **p. 825** (t) Kzenon/ Fotolia; (b) Wavebreakmedia/Shutterstock **p. 837** Pavel Losevsky/Dreamstime **p. 845** (t) Smontgom65/ Dreamstime; (b) iStockphoto/Getty Images **p. 846** Ekaterina Semenova/Dreamstime **p. 848** Jupiterimages/ Getty Images **p. 849** (t) Zuma Press, Inc./Alamy; (b) Sport In Pictures/Alamy **p. 857** Jelle-vd-Wolf/ Shutterstock

Chapter 12 **p. 903** Astrofireball/Dreamstime **p. 904** Marek Redesiuk/Dreamstime **p. 920** Daivaluk/ Dreamstime **p. 922** Ablestock.com/Getty Images

Chapter 13 **p. 925** CandyBox Images/Shutterstock **p. 961** Anton Foltin/Dreamstime **p. 962** (t) Pavel Losevsky/Dreamstime; (b) LIU, CHIN-CHENG/Shutterstock **p. 975** Cenk Unver/Dreamstime

Chapter 14 **p. 993** (l) PCN Photography/Alamy; (r) Aspenphoto/Dreamstime **p. 1003** Wavebreakmedia/ Shutterstock **p. 1010** (l) Matthew Grant/Alamy; (r) Mediacolor's/Alamy **p. 1039** Digital Vision/Getty Images **p. 1052** (t) James Steidl/Fotolia; (b) Richard Coencas/Shutterstock **p. 1060** James Woodson/Getty Images

Chapter 15 **p. 1062** Lightpoet/Shutterstock **p. 1072** Longimanus/Shutterstock **p. 1078** Erik Lam/ Shutterstock **p. 1090** Neil Roy Johnson/Shutterstock **p. 1094** Johanna Goodyear/Shutterstock

Chapter 16 **p. 1103** Oleksiy Mark/Shutterstock

Chapter 17 **p. 1139** Shukaylova Zinaida/Shutterstock **p. 1155** Jerryb8/Dreamstime **p. 1164** Jorg Hackemann/Shutterstock **p. 1185** Marcin Pawinski/Dreamstime

Chapter 18 **p. 1204** Typhoonski/Dreamstime **p. 1211** Stockbyte/Getty Images **p. 1214** Andrew Zarivny/ Shutterstock **p. 1223** Microstock Man/Shutterstock **p. 1228** Edhar/Shutterstock **p. 1232** (t) Vibrant Image Studio/Shutterstock; (b) Phillip Minnis/Shutterstock **p. 1256** Tarasov_vl/Fotolia **p. 1257** Daniel Cviatkov Yordanov/Shutterstock **p. 1261** Keith Verkem/Dreamstime

Chapter 19 **p. 1265** Roman Milert/Dreamstime **p. 1289** Blend Images/Shutterstock **p. 1317** Franant/ Dreamstime **p. 1329** (t) Andrew Buckin/Fotolia; (b) George Doyle & Ciaran Griffin/Getty Images

Chapter 20 **p. 1331** Alex Melnick/Shutterstock